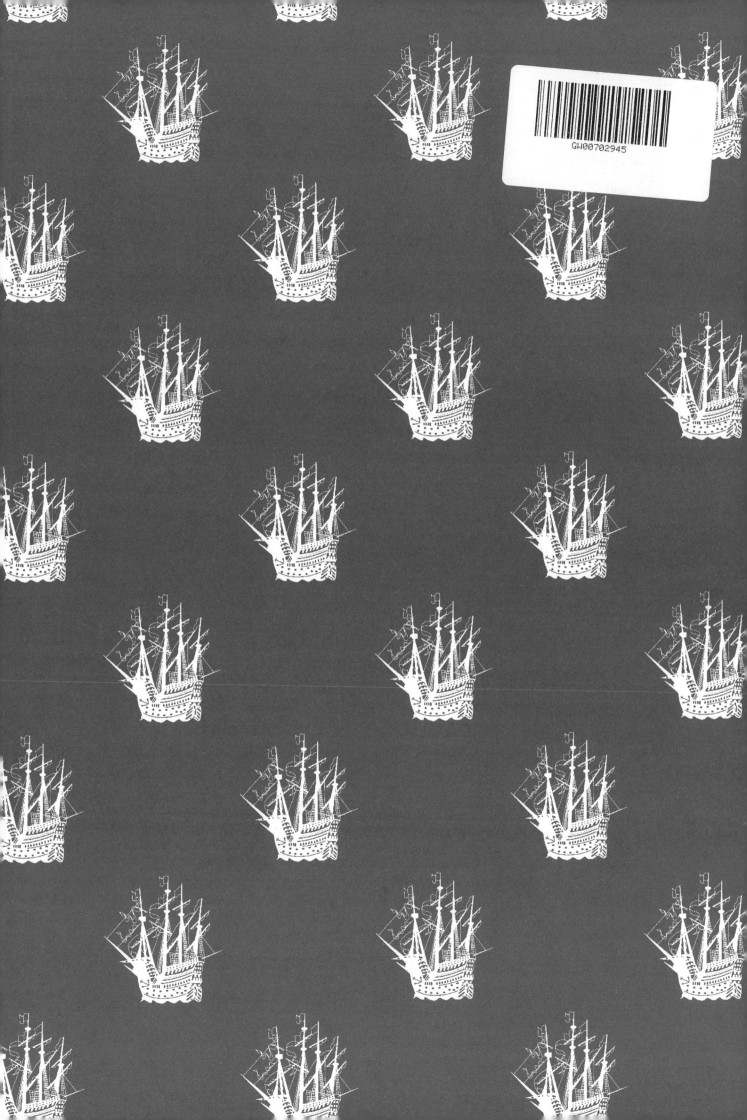

# WEAPONS OF WARRE:
The Armaments of the *Mary Rose*

edited by Alexzandra Hildred

For Michael Hildred and Margaret Rule

In memory of
Howard Blackmore
Deputy Master, the Royal Armouries

# WEAPONS OF WARRE:

The Armaments of the *Mary Rose*

edited by Alexzandra Hildred

with major contributions from

Dominic Fontana, Nicholas Hall, Robert Hardy, C.S. Knighton, David Loades,
Robert Smith, Hugh Soar, Keith Watson and Guy Wilson

and contributions from

D. Adams C. Bartlett, C. Boyton, Ruth Rynas Brown, Hector Cole, A. Crowley,
Laura Davidson, Alison Draper, Andrew Elkerton, Nick Evans, Rowena Gale, Bert Hall,
Peter Holt, Mark Hopkins, Adam Jackson, S. Jackson, the late R. King, Kirran Louise Klein,
B.W. Kooi, Douglas McElvogue, J.P. Northover, Maggie Richards, Thom Richardson,
N.A.M. Rodger, David Starley, Ann Stirland, Clare Venables, John Waller,
Alan Williams and Robin Wood

The Archaeology of the *Mary Rose*
Volume 3

2011

Published 2011 by The Mary Rose Trust Ltd
College Road, HM Naval Base, Portsmouth, England PO1 3LX

Copyright © 2011 The Mary Rose Trust Ltd
*All rights reserved*

**British Library Cataloguing in Publication Data**
A catalogue record for this book is available from the British Library

ISBN 978-0-9544029-3-8

Series Editor: Julie Gardiner
Series Editor (graphics): Peter Crossman

Designed by Julie Gardiner and Peter Crossman
Produced by Wessex Archaeology

Printed by Short Run Press Ltd, Exeter, England

**The publishers acknowledge with gratitude a grant from the
Heritage Lottery Fund towards the cost of publishing this volume**

*Cover: Device and surface decoration from bronze guns 81A3000 and 79A1232 with cast iron shot.
Photographs: Peter Crossman, Dominic Fontana and Stephen Foote. © Mary Rose Trust Ltd.
Designed by Stephen Foote with Alexzandra Hildred*

The Mary Rose Trust is a Registered Charity No. 277503

Erratum: Figures 2.73 and 2.75 on pp 90 and 92 have been transposed.
Figure 2.75 is gun 79A1279

# Contents

List of Figures . . . . . . . . . . . . . . . . . . . . . . . . . viii
List of Tables . . . . . . . . . . . . . . . . . . . . . . . . . xviii
List of Colour Plates . . . . . . . . . . . . . . . . . . . . . xxi
Contents of DVD . . . . . . . . . . . . . . . . . . . . . . . xxii
Acknowledgements . . . . . . . . . . . . . . . . . . . . . xxiii
Abstract . . . . . . . . . . . . . . . . . . . . . . . . . . . . . . xxvii
Preface . . . . . . . . . . . . . . . . . . . . . . . . . . . . . . . xxx

## PART 1: THE KING'S GREAT SHIP

### 1. The Kings 'Great Ship': the *Mary Rose* and her naval background
Overview of Henry VIII's navy,
    by David Loades . . . . . . . . . . . . . . . . . . . . 1
The *Anthony Roll*, by Alexzandra Hildred
    and C.S. Knighton . . . . . . . . . . . . . . . . . . . 2
    The manuscript, by C.S. Knighton . . . . . . . 2
    The importance and accuracy of the
        *Anthony Roll*, by Alexzandra Hildred . . . . 4
Changes in armament: an introduction,
    by Alexzandra Hildred . . . . . . . . . . . . . . . . 7
The service career of the *Mary Rose*,
    by David Loades and C.S. Knighton . . . . . . 8
The Battle of the Solent, 19 July 1545,
    by David Loades . . . . . . . . . . . . . . . . . . . . 10
The historical and archaeological importance
    of the *Mary Rose* ordnance,
    by Alexzandra Hildred . . . . . . . . . . . . . . . . 12

## PART 2. ARMING THE KING'S SHIP: THE GREAT GUNS

### 2. 'Brass' guns, by Alexzandra Hildred
Introduction . . . . . . . . . . . . . . . . . . . . . . . . . . 17
Types of gun . . . . . . . . . . . . . . . . . . . . . . . . . 22
    Gun classification . . . . . . . . . . . . . . . . . . . 24
Distribution of bronze guns . . . . . . . . . . . . . . 25
    The Main gun deck . . . . . . . . . . . . . . . . . . 26
    The Upper gun deck . . . . . . . . . . . . . . . . . 26
    The Castle deck . . . . . . . . . . . . . . . . . . . . 27
Decoration and devices, by Alexzandra Hildred
    and C.S. Knighton . . . . . . . . . . . . . . . . . . . 29
    Discussion . . . . . . . . . . . . . . . . . . . . . . . . 35
Gun mountings . . . . . . . . . . . . . . . . . . . . . . . 37
Cannon . . . . . . . . . . . . . . . . . . . . . . . . . . . . . 45
    Cannon royal 79A1276 . . . . . . . . . . . . . . . 47
Cannon 81A3003 . . . . . . . . . . . . . . . . . . . . . 51
Demi cannon . . . . . . . . . . . . . . . . . . . . . . . . 58
    Demi cannon 81A3000 . . . . . . . . . . . . . . . 59
    Demi cannon 79A1277 . . . . . . . . . . . . . . . 64
    Demi cannon 81A3002 . . . . . . . . . . . . . . . 70
Culverins . . . . . . . . . . . . . . . . . . . . . . . . . . . . 74

Culverin 81A1423 . . . . . . . . . . . . . . . . . . . . 75
Culverin 80A0976 . . . . . . . . . . . . . . . . . . . . 83
Culverin 79A1278 . . . . . . . . . . . . . . . . . . . . 87
Demi culverins . . . . . . . . . . . . . . . . . . . . . . . 89
    Demi culverin 79A1279 . . . . . . . . . . . . . . . 90
    Demi culverin 79A1232 . . . . . . . . . . . . . . . 92
Sakers . . . . . . . . . . . . . . . . . . . . . . . . . . . . . 101
Falcons . . . . . . . . . . . . . . . . . . . . . . . . . . . . 102
Isolated carriage elements . . . . . . . . . . . . . . 012
Casting and firing a bronze culverin . . . . . . 112
    Techniques of casting bronze guns . . . . . 112
    Replicating a gun . . . . . . . . . . . . . . . . . . 116
    Firing the reproduction culverin, with a
        contribution by Nicholas Hall . . . . . . 123

### 3. Wrought iron guns, by Alexzandra Hildred
Introduction . . . . . . . . . . . . . . . . . . . . . . . . 130
Identification and historical importance . . . . 131
Scope of the assemblage . . . . . . . . . . . . . . 133
Morphology . . . . . . . . . . . . . . . . . . . . . . . . 134
Port pieces . . . . . . . . . . . . . . . . . . . . . . . . . 136
    Identification and historical significance . . 136
    Scope and distribution of the assemblage . 137
    Morphology . . . . . . . . . . . . . . . . . . . . . . 140
    Catalogue . . . . . . . . . . . . . . . . . . . . . . . . 143
Slings . . . . . . . . . . . . . . . . . . . . . . . . . . . . . 199
    Identification and historical significance . . 199
    The scope and distribution of the
        assemblage . . . . . . . . . . . . . . . . . . . . . 202
    Catalogue . . . . . . . . . . . . . . . . . . . . . . . . 205
Fowlers . . . . . . . . . . . . . . . . . . . . . . . . . . . . 213
    Identification and historical significance . . 213
    Catalogue . . . . . . . . . . . . . . . . . . . . . . . . 217
Bases of Iron (swivel guns) . . . . . . . . . . . . . 221
    Identification and historical importance . . 222
    Scope and distribution of the assemblage . 227
    Method of manufacture . . . . . . . . . . . . . 236
    Evidence for loading . . . . . . . . . . . . . . . . 243
    Form 1: catalogue . . . . . . . . . . . . . . . . . . 244
    Form 2: catalogue . . . . . . . . . . . . . . . . . . 254
    Unknown types and fragments . . . . . . . . 266
Top pieces, by Ruth Rhynas Brown . . . . . . . 269
    Structure of the *Mary Rose* and
        accommodation of top pieces,
        by Alexzandra Hildred . . . . . . . . . . . . . 272
Replication of one of the *Mary Rose*
    port pieces . . . . . . . . . . . . . . . . . . . . . . . 273
    Method of manufacture . . . . . . . . . . . . . 274
    Replication of the carriage . . . . . . . . . . . 284
    Firing the port piece . . . . . . . . . . . . . . . . 285

### 4. Cast iron guns, by Alexzandra Hildred
Introduction . . . . . . . . . . . . . . . . . . . . . . . . 291
Hailshot pieces . . . . . . . . . . . . . . . . . . . . . . 292

Distribution . . . . . . . . . . . . . . . . . . . . . . .294
Morphology . . . . . . . . . . . . . . . . . . . . . .295
Radiographic study . . . . . . . . . . . . . . . . .297
The importance of hailshot pieces . . . . . .301
Analysis of a hailshot piece, by D Starley
   and Alison Draper . . . . . . . . . . . . . . . .302
Manufacture of a replica hailshot piece . .304
Firing a replica hailshot piece . . . . . . . . .305

**5. Munitions and accessories**
Shot, by Alexzandra Hildred with
   Robert Smith . . . . . . . . . . . . . . . . . . . .307
   Supply of shot . . . . . . . . . . . . . . . . . . .308
   The *Anthony Roll* . . . . . . . . . . . . . . . . .309
   Shot types, size and materials: evidence
      from other sources . . . . . . . . . . . . . .310
   Matching shot with guns . . . . . . . . . . . .311
   Shot material . . . . . . . . . . . . . . . . . . . .311
   Storage of shot . . . . . . . . . . . . . . . . . . .312
   Shot of iron . . . . . . . . . . . . . . . . . . . . .313
   Shot of stone . . . . . . . . . . . . . . . . . . . .339
   Composite shot . . . . . . . . . . . . . . . . . .348
   Shot of lead . . . . . . . . . . . . . . . . . . . . .367
   Canister shot . . . . . . . . . . . . . . . . . . . .370
Manufacture and selection of shot,
   by Alexzandra Hildred . . . . . . . . . . . . .381
   Shot production in the sixteenth century,
      with contribution by Robert Smith and
      Clare Venables . . . . . . . . . . . . . . . . .381
   Shot moulds . . . . . . . . . . . . . . . . . . . .387
   Gunner's rule, with a contribution
      by Keith Watson . . . . . . . . . . . . . . . .393
   Shot gauges . . . . . . . . . . . . . . . . . . . .401
Loading and firing, by Alexzandra Hildred . .407
   Muzzle-loading guns . . . . . . . . . . . . . . .407
   Breech-loading guns . . . . . . . . . . . . . . .407
   Positioning of equipment . . . . . . . . . . . .409
   Gun drill on the *Mary Rose*,
      by Nicholas Hall . . . . . . . . . . . . . . . .410
   Gunpowder, by Bert Hall and
      Alexzandra Hildred . . . . . . . . . . . . . .419
   Formers, with a contribution by
      Robin Wood . . . . . . . . . . . . . . . . . .435
   Powder ladles, with a contribution by
      Robin Wood . . . . . . . . . . . . . . . . . .349
   Rammers, with a contribution by
      Robin Wood . . . . . . . . . . . . . . . . . .454
   Tampions, by Alexzandra Hildred,
      Robin Wood and Rowena Gale . . . . . .462
   Wads . . . . . . . . . . . . . . . . . . . . . . . . .474
   Priming wires . . . . . . . . . . . . . . . . . . .476
   Powder dispensers . . . . . . . . . . . . . . . .479
   Powder flask . . . . . . . . . . . . . . . . . . . .484
   Linstocks . . . . . . . . . . . . . . . . . . . . . .485
   Tinder boxes . . . . . . . . . . . . . . . . . . . .499
   Gun flints . . . . . . . . . . . . . . . . . . . . . .503
   Fuses . . . . . . . . . . . . . . . . . . . . . . . . .503

Working the guns, by Alexzandra Hildred
   and Nicholas Hall . . . . . . . . . . . . . . . .504
   Commanders or mallets . . . . . . . . . . . . .507
   Sledgehammers . . . . . . . . . . . . . . . . . .507
   Chisels . . . . . . . . . . . . . . . . . . . . . . . .508
   'Crows' (crowbars) of iron . . . . . . . . . . .508
   Wooden handspikes . . . . . . . . . . . . . . .508
   Baskets . . . . . . . . . . . . . . . . . . . . . . . .508
   Leather buckets . . . . . . . . . . . . . . . . . .508
   Wooden tubs and buckets . . . . . . . . . . .510
   Funnel and sieves . . . . . . . . . . . . . . . . .511
   Shovels and brushes . . . . . . . . . . . . . . .511
   Wedges . . . . . . . . . . . . . . . . . . . . . . .512

**PART 3: HAND-HELD WEAPONS**

**6. Incendiary devices**, by Alexzandra Hildred
   Introduction . . . . . . . . . . . . . . . . . . . .519
   Historical importance . . . . . . . . . . . . . .521
   *Mary Rose* incendiary projectiles . . . . . . . . .523
   Distribution of other incendiary items . . .529
   Analysis of incendiary mixture . . . . . . . .531
   Mortar bomb, by Alexzandra Hildred and
      Andrew Elkerton, with a contribution
      by David Starley . . . . . . . . . . . . . . . .531

**7. Hand-held firearms**
Handguns . . . . . . . . . . . . . . . . . . . . . . .537
Scope and importance of the assemblage,
   by Guy Wilson . . . . . . . . . . . . . . . . .537
Inventory evidence, by Alexzandra Hildred . .543
Distribution of handguns, by
   Alexzandra Hildred . . . . . . . . . . . . . . .545
The assemblage, by Douglas McElvogue
   and Guy Wilson . . . . . . . . . . . . . . . . .545
Matching the shot to the guns,
   by Douglas McElvogue . . . . . . . . . . . .551
Associated objects, by Douglas McElvogue . .551
Replication and firing, by Alexzandra Hildred 553
Gun-shields: '*Targettes steilde with gonnes*',
   by Alexzandra Hildred . . . . . . . . . . . . .553
Scope and significance of the assemblage . . .553
Morphology . . . . . . . . . . . . . . . . . . . . .555
Distribution of the gun-shields . . . . . . . . . .560
Leather elements, with a contribution
   by Nick Evans . . . . . . . . . . . . . . . . . .561
The gun-shields . . . . . . . . . . . . . . . . . . .564
Comparative study of two gun-shields in
   the Royal Armouries Collections . . . . . .573
Suggested method of manufacture . . . . . . . .577

**8. Archery**
Scope and importance of the assemblage,
   by Alexzandra Hildred . . . . . . . . . . . . .528
Archery of Henry VIII's vessels: inventory
   evidence, by Alexzandra Hildred . . . . . .579

The inventory of 1547 . . . . . . . . . . . . . . .581
Distribution of the assemblage,
    by Alexzandra Hildred . . . . . . . . . . . . . . .583
Distribution of archers based on associated
    artefacts, by Alexzandra Hildred . . . . . . . .585
Historical importance and assessment of the
    longbow assemblage, by Robert Hardy . . .586
Longbows, by Robert Hardy . . . . . . . . . . . . .589
    Archery at sea . . . . . . . . . . . . . . . . . .593
    The longbow assemblage, by Clive Bartlett,
        Chris Boyton, Steve Jackson, Adam Jackson,
        Douglas McElvogue, Alexzandra Hildred
        and Keith Watson . . . . . . . . . . . . . . . .594
    Statistical analysis of the longbow data,
        by Keith Watson . . . . . . . . . . . . . . . . .618
    Assessment of the working capabilities of the
        longbows, by Robert Hardy, the late
        R. King, B.W. Kooi, D. Adams and
        A. Crowley . . . . . . . . . . . . . . . . . . . . .622
    Horn nock, by Alexzandra Hildred
        with the late Roy King . . . . . . . . . . . .632
    Bowstrings, by Alexzandra Hildred and
        John Waller . . . . . . . . . . . . . . . . . . . .632
Longbow chests, by Alexzandra Hildred . . . .641
Wristguards, by Hugh Soar . . . . . . . . . . . . .644
    Distribution . . . . . . . . . . . . . . . . . . . . .647
    Form . . . . . . . . . . . . . . . . . . . . . . . . . .647
    Fastening . . . . . . . . . . . . . . . . . . . . . .650
    Decoration, by Matthew Champion
        and Hugh Soar . . . . . . . . . . . . . . . . .651
    Non-leather wristguards . . . . . . . . . . . . .665
Arrows by Alexzandra Hildred with
    Keith Watson, Mark Hopkins, Adam Jackson
    and John Waller . . . . . . . . . . . . . . . . . . .665
    Introduction . . . . . . . . . . . . . . . . . . . . .665
    Distribution . . . . . . . . . . . . . . . . . . . . .666
    Form and function . . . . . . . . . . . . . . . .669
    Scope and importance of the assemblage .672
    Analysis, by Keith Watson . . . . . . . . . . . .677
    Other features of the shafts,
        by Adam Jackson . . . . . . . . . . . . . . .686
    Manufacture, by Keith Watson . . . . . . . .688
    Conclusions, by Alexzandra Hildred . . . .688
Arrow chests, by Alexzandra Hildred . . . . . .689
Arrow spacers, by Alexzandra Hildred . . . . . .693
Arrow bag, by Alexzandra Hildred
    and John Waller . . . . . . . . . . . . . . . . . .698
Handgun bolts, incendiary arrows or top
    darts?, by Alexzandra Hildred . . . . . . . . .699
Identification of archers from skeletal
    remains, by Alexzandra Hildred and
    Ann Stirland . . . . . . . . . . . . . . . . . . . . .702

**9. Hand-to-Hand Fighting**
Staff weapons, by Guy Wilson with
    Alexzandra Hildred . . . . . . . . . . . . . . . .713
Scope and importance of the assemblage . . . .713
    The *Anthony Roll* . . . . . . . . . . . . . . . .718

Bills, with a contribution by Hector Cole .719
Pikes . . . . . . . . . . . . . . . . . . . . . . . .735
Halberd . . . . . . . . . . . . . . . . . . . . . .736
Hafts, possibly for throwing darts . . . . . . .740
Edged weapons . . . . . . . . . . . . . . . . . . . . .740
    Swords and daggers, by Guy Wilson
        with Alexzandra Hildred . . . . . . . . . .740
    Sword and dagger grips, by Guy Wilson
        with Laura Davidson, Alexzandra Hildred
        and Douglas McElvogue . . . . . . . . . .765
    Daggers . . . . . . . . . . . . . . . . . . . . . . .771
    Ballock daggers, by Maggie Richards with
        Alexzandra Hildred and Guy Wilson . .776
    Scabbards, sheaths, belts and hangers,
        by Alexzandra Hildred, Laura Davidson,
        Douglas McElvogue and Guy Wilson . .793

**10. Armour and Personal Protection**
Armour for warfare, by Alexzandra Hildred
    and Alan Williams . . . . . . . . . . . . . . . .820
    Armour worn . . . . . . . . . . . . . . . . . . . .820
Protection at sea, by Alexzandra Hildred . . . .825
    Armour worn . . . . . . . . . . . . . . . . . . . .825
    Armour carried . . . . . . . . . . . . . . . . . .826
    The metallurgy of sixteenth century
        armour, by Alan Williams . . . . . . . . . .827
The *Mary Rose* armour, by
    Alexzandra Hildred . . . . . . . . . . . . . . . .828
    Possible armour and armour-related
        fragments, with contributions by
        Kirran Lousie Klein, Thom Richardson,
        Alan Williams and Guy Wilson . . . . . .829
Mail, by Thom Richardson . . . . . . . . . . . . .842
    Types and distribution of *Mary Rose* mail,
        by Alexzandra Hildred
        and Nick Evans . . . . . . . . . . . . . . . . .844
Skeletal remains suggesting the distribution of
    armour clad men, by Alexzandra Hildred
    and Ann Stirland . . . . . . . . . . . . . . . . .850
    The Upper deck . . . . . . . . . . . . . . . . .850
    The Main deck . . . . . . . . . . . . . . . . . .851
    The Orlop deck . . . . . . . . . . . . . . . . .852
    The Hold . . . . . . . . . . . . . . . . . . . . . .852

**PART 4: FIGHTING THE SHIP**

**11. Fighting the ship**
Historical evidence . . . . . . . . . . . . . . . . . .855
    Fighting tactics, by N.A.M. Rodger . . . . .855
    Documentary evidence, by
        Alexzandra Hildred with a contribution
        by Dominic Fontana . . . . . . . . . . . .860
    The theatre of war: geographical evidence
        from the Cowdray Engraving and
        GIS, by Dominic Fontana and
        Alexzandra Hildred . . . . . . . . . . . . . .871

Archaeological evidence, by
    Alexzandra Hildred ..................886
    Gunports .........................891
    Layout and operation of the gundecks ...899
Overloaded with ordnance?,
    by Alexzandra Hildred ..............917
Playing with arcs of fire: how well defended was
    the *Mary Rose*?, by Alexzandra Hildred ...919
Fighting with the ship ..................924
    The grapnel, by Douglas McElvogue ....924
    Shear hooks, by Douglas McElvogue ....925
Rules of engagement: what does the ammunition
    tell us?, by Alexzandra Hildred ........925

**12. The Beginning of the end: some reflections on the ordnance of the *Mary Rose* and avenues for future research**, by Alexzandra Hildred
Introduction ........................932

Back to the future: opening the door .......934
The 2003–5 excavations: an interim statement,
    by Alexzandra Hildred and Peter Holt ...938
Postscript: document CP 201/127,
    by C.S. Knighton and
    Alexzandra Hildred .................946

**Technical appendices**
1. *Gun Metallurgy* ......................950
    1.1 Average metal composition for the
        bronze guns .....................950
    1.2 Analysis of guns from the *Mary Rose*,
        by J.P. Northover ..................951

2. *Gunpowder (GP) analysis: analytical tables* ...965

Bibliography ..........................967
Index, by Barbara Hird .................976

# List of Figures

Fig. i    Isometric showing labelled decks, cabins
Fig. ii   The isometric showing guns and principal stowage areas
Fig. iii  Initial trench plans
Fig. iv   Trench plan from 1979
Fig. v    Flattened diagram of the ship used to show distribution plots

*Parts 1 and 2*
Fig. 1.1  The *Mary Rose* as she is depicted in the *Anthony Roll* with inventory
Fig. 1.2  The Cowdray Engraving showing the sinking of the *Mary Rose* and disposition of the fleets
Fig. 1.3  The *Beauchamp Pageant* (c. 1485) depicting hand-to-hand naval engagement
Fig. 1.4  Clinker and carvel building techniques
Fig. 1.5  Elevation of the outside of ship demonstrating the change from clinker to carvel build in the hull and a gunport lid
Fig. 1.6  *The Embarkation of Henry VIII at Dover for the Field of Cloth of Gold*
Fig. 1.7  Demi culverin 79A1232 on its carriage in the Mary Rose Museum
Fig. 2.1  Fortifications listed in the 1547 inventory
Fig. 2.2  Model of the *Mary Rose* built by Bassett-Lowke
Fig. 2.3  All guns listed in the *Anthony Roll*
Fig. 2.4  Parts of a bronze gun
Fig. 2.5  The *Mary Rose* bronze guns
Fig. 2.6  Three of four guns recovered by John Deane in 1836
Fig. 2.7  M3 culverin on carriage *in situ* seen through Upper deck half beams
Fig. 2.8  Isometric of the ship showing the actual and inferred positions of guns on the Main and Upper decks
Fig. 2.9  Plan of the Main deck with guns
Fig. 2.10 The front of the Sterncastle with the Castle deck gun showing
Fig. 2.11 Simplified deck plans with schematic gun types
Fig. 2.12 Guns with lion head dolphins
Fig. 2.13 Guns with mermen and leopard head dolphins
Fig. 2.14 Guns with rose and crown devices
Fig. 2.15 Guns with shield devices
Fig. 2.16 Gun cartouch
Fig. 2.17 Cascabel shapes
Fig. 2.18 Trunnion marks
Fig. 2.19 Faceted guns
Fig. 2.20 Gun on carriage underwater
Fig. 2.21 Guns lifted on carriages
Fig. 2.22 Sterncastle mosaic and underwater photograph of gun during excavation
Fig. 2.23 Parts of a gun carriage
Fig. 2.24 Museum gun display with reconstruction of the midships area of the Main and Upper gun decks
Fig. 2.25 Three types of carriage as recognised from the *Mary Rose* bronze guns
Fig. 2.26 Canon as depicted by Wright in 1563
Fig. 2.27 Cannon royal 79A1276; decoration and views of gun

| | | | |
|---|---|---|---|
| Fig. 2.28 | Cannon royal 79A1276; top view of gun | Fig. 2.63 | Culverin 81A1423; building the replica carriage |
| Fig. 2.29 | Cannon royal 79A1276; broken carriage underwater and interpretative sketch | Fig. 2.64 | Culverin 81A1423; carriage as lifted and loose cheek and fragments |
| Fig. 2.30 | Cannon royal 79A1276; Tudor cable in hull | Fig. 2.65 | Culverin 81A1423; divelog showing location of rear cheek |
| Fig. 2.31 | Cannon 81A3003; decoration and views of gun | Fig. 2.66 | Culverin 81A1423; reconstruction of gun on carriage |
| Fig. 2.32 | Cannon 81A3003; top view of gun | Fig. 2.67 | Culverin 80A0976; decoration and views of gun |
| Fig. 2.33 | Cannon 81A3003; underwater photograph of gun cascabel and rear of carriage | Fig. 2.68 | Culverin 80A0976; top view of gun |
| Fig. 2.34 | Cannon 81A3003; elements of the carriage | Fig. 2.69 | Culverin 80A0976; gun as lifted on deck |
| Fig. 2.35 | Cannon 81A3003; views of the gun and carriage | Fig. 2.70 | Culverin 80A0976; carriage and reconstruction of gun on carriage |
| Fig. 2.36 | Cannon 81A3003; objects found in the M6 gun bay | Fig. 2.71 | Culverin 79A1278; decoration |
| Fig. 2.37 | Cannon 81A3003; details of the carriage | Fig. 2.72 | Culverin 79A1278; views of gun |
| Fig. 2.38 | Cannon 81A3003; radiography of the carriage | Fig. 2.73 | Demi culverin 79A1279 decoration and views of gun |
| Fig. 2.39 | Demi cannon 81A3000; decoration and views of gun | Fig. 2.74 | Demi culverin 79A1279; top view of gun |
| Fig. 2.40 | Demi cannon 81A3000; top view of gun | Fig. 2.75 | Demi culverin 79A1232; decoration and views of gun |
| Fig. 2 41 | Demi cannon 81A3000; close up of muzzle damage | Fig. 2.76 | Demi culverin 79A1232; top view of gun |
| Fig. 2.42 | Demi cannon 81A3000; gun *in situ* and being raised | Fig. 2.77 | Demi culverin 79A1232; arches and lion heads |
| Fig. 2.43 | Demi cannon 81A3000; elements of the carriage | Fig. 2.78 | Demi culverin 79A1232; carriage |
| Fig. 2.44 | Demi cannon 81A3000; gun mounted on carriage and location in ship | Fig. 2.79 | Demi culverin 79A1232; tools needed for making carriage |
| Fig. 2.45 | Demi cannon 79A1277; decoration | Fig. 2.80 | Demi culverin 79A1232; making the carriage |
| Fig. 2.46 | Demi cannon 79A1277; views of gun | Fig. 2.81 | Demi culverin 79A1232; replicating the carriage |
| Fig. 2.47 | Demi cannon 79A1277; trunnion support cheek in position underwater | Fig. 2.82 | Demi culverin 79A1232; replicating the carriage |
| Fig. 2.48 | Demi cannon 79A1277; underwater sketch of broken carriage | Fig. 2.83 | Demi culverin 79A1232; mounting the gun |
| Fig. 2.49 | Demi cannon 79A1277; carriage bed reassembled | Fig. 2.84 | Demi culverin 79A1232; the original gun on reproduction carriage in the Mary Rose Museum |
| Fig. 2.50 | Demi cannon 79A1277; elements of the carriage | Fig. 2.85 | Royal Armouries saker XIX.165 |
| Fig. 2.51 | Demi cannon 79A1277; reconstruction of gun on carriage showing tight fit | Fig. 2.86 | Royal Armouries falcon XIX.245 |
| Fig. 2.52 | Demi cannon 81A3002; decoration and views of gun | Fig. 2.87 | Spare gun carriage wheels, presumably in storage, on the seabed |
| Fig. 2.53 | Demi cannon 81A3002; top view of gun | Fig. 2.88 | Gun carriage fragments 81A0124 and 81A0707 |
| Fig. 2.54 | Demi cannon 81A3002; scar on outside of gun | Fig. 2.89 | Isometric showing distribution of individual carraige elements |
| Fig. 2.55 | Demi cannon 81A3002; sketch of gun and upside down carriage as found on seabed | Fig. 2.90 | Stepped cheek fragment 81A6896 |
| Fig. 2.56 | Demi cannon 81A3002; sequence of carriage making | Fig. 2.91 | Stepped cheek piece fragment 81A0107 and axle 78A0139 |
| Fig. 2.57 | Demi cannon 81A3002; gun on display in the National Maritime Museum | Fig. 2.92 | Gun carriage 81A0486 |
| Fig. 2.58 | Culverin 81A1423; decoration | Fig. 2.93 | Reconstruction of carriage type represented by 81A0486 |
| Fig. 2.59 | Culverin 81A1423; views of gun | Fig. 2.94 | Axles |
| Fig. 2.60 | Culverin 81A1423; mermen lifting dolphins and wire in bore | Fig. 2.95 | Solid wheels |
| Fig. 2.61 | Culverin 81A1423; gun on carriage | Fig. 2.96 | Axle fragments and wheels |
| Fig. 2.62 | Culverin 81A1423; carriage | Fig. 2.97 | Axle and axle fragments |
| | | Fig. 2.98 | Making the pattern and jacket for casting a bronze gun |

Fig. 2.99 Drawing of gun in jacket and embellishments, after Biringuccio
Fig. 2.100 Overview of the foundry
Fig. 2.101 Making the wax moulds
Fig. 2.102 Making the cascabel
Fig. 2.103 Portion of steel tube used to form jacket
Fig. 2.104 Making the pattern
Fig. 2.105 Making the core
Fig. 2.106 Casting the gun
Fig. 2.107 Complete gun, as removed from mould and after finishing
Fig. 2.108 Firing trial
Fig. 2.109 Damage to carriage and target
Fig. 3.1 Anatomy of a wrought iron gun
Fig. 3.2 External appearance of wrought iron swivel gun (base) showing hoops and bands
Fig. 3.3 Construction of barrrel and chamber
Fig. 3.4 Components of wrought iron gun on carriage
Fig. 3.5 Section through a swivel gun (base) and reconstruction of use
Fig. 3.6 Port piece barrel and chamber as lifted.
Fig. 3.7 Anatomy of a port piece
Fig. 3.8 Port piece barrels and *in situ* chambers
Fig. 3.9 Distribution of port pieces
Fig. 3.10 Distributions of various possible gun positions and external starboard elevation showing Upper deck gunports
Fig. 3.11 Port piece 81A2650; reconstruction of gun on carriage
Fig. 3.12 Port piece 81A2650; details of gun
Fig. 3.13 Port piece 81A2650; details of carriage
Fig. 3.14 Port piece 81A2650; carriage
Fig. 3.15 Port piece 81A2650; montage of carriage
Fig. 3.16 Port piece 81A2650; carriage, axle and wheel
Fig. 3.17 Port piece 81A2604; reconstruction drawing of gun on carriage
Fig. 3.18 Port piece 81A2604; underwater photograph of gun as found and diver's sketch of stone shot around wheels
Fig. 3.19 Port piece 81A2604; diver's sketch of the gun position as found and gunport lid
Fig. 3.20 Port piece 81A2604; barrel and breech
Fig. 3.21 Port piece 81A2604; details of gun, radiograph of chamber, external view and cross-section
Fig. 3.22 Port piece 81A2604; carriage and detail of rope, carriage trough and elevating post
Fig. 3.23 Port piece 81A2604; axle and spoked wheels
Fig. 3.24 Port piece 81A3001; underwater photograph of gun carriage as found and reconstruction of gun on carriage without breech chamber
Fig. 3.25 Port piece 81A3001; barrel and chamber; windout radiograph showing internal staves
Fig. 3.26 Port piece 81A3001; section through the chamber; carriage and forelock
Fig. 3.27 Port piece 81A3001; wheels
Fig. 3.28 Port piece 81A3001; elevating post and axle
Fig. 3.29 Port piece 81A3001; objects associated with the gun
Fig. 3.30 Port piece 71A0169; barrel and chamber
Fig. 3.31 Port piece 71A0169; barrel construction, stave joints at breech end of barrel
Fig. 3.32 Port piece NMM KPT0017; reconstruction of gun on carriage
Fig. 3.33 Port piece NMM KPT0017; barrel, chamber, carriage, forelock and wheel
Fig. 3.34 Woolwich 1/10 port piece; gun on carriage
Fig. 3.35 Woolwich 1/10 port piece; on display in the Woolwich Rotunda
Fig. 3.36 Port piece RA XIX.1 (80A2004); barrel hoops and carriage fragment
Fig. 3.37 Port piece RA XIX.2 (80A2003); barrel
Fig. 3.38 Location of loose port piece chambers
Fig. 3.39 Port piece chambers, manufacturing sequence and named parts
Fig. 3.40 Port piece chambers
Fig. 3.41 Port piece chambers in concretion
Fig. 3.42 Port piece chamber 82A5172
Fig. 3.43 Port piece chamber 82A5176
Fig. 3.44 Port piece chambers, fragments
Fig. 3.45 Port piece chambers, fragments
Fig. 3.46 Parts of a spoked wheel
Fig. 3.47 Details of wheels
Fig. 3.48 Wheels from M11, underwater photograph and sketch
Fig. 3.49 Wheels from M11, underwater views of the group as found
Fig. 3.50 M11 spoked wheels
Fig. 3.51 M11 spoked wheels and axle
Fig. 3.52 Spoked wheel hub fragments
Fig. 3.53 Underwater views of carriage wheels in O1 and lifting the wheels
Fig. 3.54 Spoked wheels from O1
Fig. 3.55 Spoked wheel fragments from H1
Fig. 3.56 Forelocks
Fig. 3.57 Forelocks
Fig. 3.58 Possible unfinished forelocks
Fig. 3.59 Marks on forelocks
Fig. 3.60 Elevating posts
Fig. 3.61 Elevating post 80A1364
Fig. 3.62 Distribution of port piece elements
Fig. 3.63 Sling 70A0001
Fig. 3.64 Comparison of bore size relative to length of sling 81A0645 and port piece 81A2604
Fig. 3.65 Slings
Fig. 3.66 Slings: restored barrel of 79A1295
Fig. 3.67 Sling chambers
Fig. 3.68 Distribution of slings and possible sling fragments

Fig. 3.69　The Basset Lowke model of the *Mary Rose*, showing the putative positions of slings at semi-circular gunports on the open waist
Fig. 3.70　Sling 81A0645; side view and close up of barrel with shot
Fig. 3.71　Sling 81A0645; gun on carriage
Fig. 3.72　Sling 70A0001; barrel and reconstruction of gun on carriage
Fig. 3.73　Slings; bed fragment 80A1957
Fig. 3.74　Sling 82A2700; sketch from artefact card of interpretation of radiograph
Fig. 3.75　Sling 82A2700; radiograph and layered nature of one end of gun
Fig. 3.76　Sling 81A0772
Fig. 3.77　Sling chamber 80A1952
Fig. 3.78　Sling fragment 90A0099
Fig. 3.79　Sling fragment 80A2001
Fig. 3.80　Sling fragments
Fig. 3.81　Distribution of possible fowlers
Fig. 3.82　Possible fowlers
Fig. 3.83　Fowler chambers
Fig. 3.84　Fowler 78A0614
Fig. 3.85　Fowler 77A0143
Fig. 3.86　Concretion 90A0131 as recovered
Fig. 3.87　Fowler chamber 90A0127
Fig. 3.88　Fowler barrel fragment 90A0131 and wad
Fig. 3.89　Fowlers: sequence of removal of 90A0126 and 90A0125 from concretion
Fig. 3.90　Swivel gun, reconstruction in use and drawing by Wright (1563)
Fig. 3.91　Swivel guns from *Siege of Boulogne*; 'ribaldiquins'
Fig. 3.92　Anatomy of a base
Fig. 3.93　Bases; the four most complete bases
Fig. 3.94　Bases; selection of base chambers
Fig. 3.95　Bases; selection of base chambers
Fig. 3.96　Bases; distribution of bases and elements
Fig. 3.97　Internal and external elevations of the hull
Fig. 3.98　Hole in sill of U8 port and peg holes in sill
Fig. 3.99　Detail of castle deck rail with peg holes
Fig. 3.100　Bases: tiller construction and insertion
Fig. 3.101　Bases: alternative construction of chamber holder and wedge slot
Fig. 3.102　Bases: wedge slot and wedge
Fig. 3.103　Bases: suggested construction of the trunnions
Fig. 3.104　Bases; parallel sided and tapering from breech plug to mouth
Fig. 3.105　Chamber back, methods of breech plug attachment
Fig. 3.106　Bases; detail of handle attachment on 82A4076
Fig. 3.107　Bases; details of swivel yoke and peg construction
Fig. 3.108　Bases; excavation of yoke 89A0090
Fig. 3.109　Bases; proposed method of manufacture by bending a single bar
Fig. 3.110　Bases: Smith's forms 1 and 2 as represented by 82A4077 and 82A4076
Fig. 3.111　Bases; radiographs
Fig. 3.112　Base 82A4077
Fig. 3.113　Base 79A0543
Fig. 3.114　Base 79A0543; radiograph of chamber and detail of marks on back of chamber trough
Fig. 3.115　Base 79A1075
Fig. 3.116　Base 79A1075; radiograph and details of marks on back of chamber trough
Fig. 3.117　Base 82A5096; radiograph and interpretation
Fig. 3.118　Base 85A0024; radiograph, concretion as lifted
Fig. 3.119　Base 82A4076
Fig. 3.120　Base 82A4076; details
Fig. 3.121　Base 82A4076; removal of tampion and wad and radiograph of handle
Fig. 3.122　Base 82A1603
Fig. 3.123　Base 82A1603 assembly
Fig. 3.124　Base 90A0050 with 90A0027 and 90A0132, interpretation of concretion
Fig. 3.125　Concretion 90A0050 and base fragment 90A0027
Fig. 3.126　90A0050; muzzle portion 90A0132
Fig. 3.127　Base 90A0050; reconstruction including 90A0027
Fig. 3.128　Base 81A1082; radiographs
Fig. 3.129　Base 82A4080; as recovered and radiographs
Fig. 3.130　Location of base fragments of unknown type
Fig. 3.131　Base 90A0052
Fig. 3.132　Base 82A5152
Fig. 3.133　Ship in *Beauchamp Pageant* showing top darts
Fig. 3.134　The 'WA' illustration of a carrack showing a probable swivel gun in the fighting top
Fig. 3.135　The Goodwin Sands gun
Fig. 3.136　The fighting top from the *Mary Rose*
Fig. 3.137　The sinking of the ship as depicted in the Cowdray Engraving
Fig. 3.138　External elevation of starboard side showing gunport and gun below ship
Fig. 3.139　Port piece 81A3001
Fig. 3.140　Manufacture of the barrel
Fig. 3.141　Manufacture of the barrel
Fig. 3.142　Manufacture of the breech chamber
Fig. 3.143　Manufacture of the breech chamber
Fig. 3.144　Fitting the gun to the carriage
Fig. 3.145　Firing the reproduction port piece
Fig. 3.146　First firing of the port piece
Fig. 3.147　Results of the firing
Fig. 4.1　Hailshot piece 90A0028
Fig. 4.2　All four hailshot pieces
Fig. 4.3　Reconstruction of the use of a hailshot piece over rail

| | | | | |
|---|---|---|---|---|
| Fig. 4.4 | Cross-section through loaded hailshot piece, muzzle and components removed from 79A1088 | | Fig. 5.35 | Reconstruction of spiked shot and method of construction |
| Fig. 4.5 | Distribution of hailshot pieces and outboard locator for three of the four guns | | Fig. 5.36 | Radiograph of hailshot piece 79A1088 showing shot in position and recovered |
| Fig 4.6 | Anatomy of a hailshot piece | | Fig. 5.37 | Marking out lines for unfinished shot with examples of unfinished, semi-finished and finished shot |
| Fig 4.7 | Radiographs of 80A0544; 79A1088; 81A1080 | | Fig. 5.38 | Graph of all stone shot, finished and unfinished |
| Fig 4.8 | Interpretation of concretion enclosing 90A0028 | | Fig. 5.39 | Distribution of all stone shot |
| Fig 4.9 | Hailshot piece 90A0028 emerging from concretion | | Fig. 5.40 | Distribution of all unfinished shot |
| Fig 4.10 | Hailshot piece 90A0028 radiograph and metallurgical thin sections | | Fig. 5.41 | Shot 81A1387 and petrological thin sections of sample shot |
| Fig 4.11 | Replica hailshot piece compared with 80A0544 | | Fig. 5.42 | Graphs of port piece and fowler shot |
| | | | Fig. 5.43 | Graphs of all stone shot in M6 and M2 |
| | | | Fig. 5.44 | Wrapped shot 82A4199 |
| Fig 4.12 | Firing trial of the replica at Ridsdale | | Fig. 5.45 | Details of composite shot |
| Fig. 5.1 | Types of shot and moulds | | Fig. 5.46 | Composite shot |
| Fig. 5.2 | All iron and all stone shot | | Fig. 5.47 | Dice straddling castline, 81A0483 |
| Fig. 5.3 | *Anthony Roll* heading listing shot | | Fig. 5.48 | Dice shot experiment, sequence to show finished shot with triangular holes |
| Fig. 5.4 | Graph showing all shot on ships as listed in the *Anthony Roll* | | Fig. 5.49 | Interpretation of radiographs of dice shot and radiographs of 87A0070 and 81A4833 |
| Fig. 5.5 | Location of M2/M6 shot stores and area inside of U7 gunport showing shot | | Fig. 5.50 | Inset shot 82A4339 |
| Fig. 5.6 | All iron shot | | Fig. 5.51 | Analysis of composite inset shot |
| Fig. 5.7 | Iron shot in M2 and M6 | | Fig. 5.52 | Graph of composite inset shot |
| Fig. 5.8 | Round iron shot: castline finishes | | Fig. 5.53 | Distribution of composite inset shot |
| Fig. 5.9 | Round iron shot: examples of sprue | | Fig. 5.54 | Inset ball shot |
| Fig. 5.10 | Iron shot 81A3666 showing 'H' mark | | Fig. 5.55 | Graph of inset and dice and ball shot |
| Fig. 5.11 | Varieties of 'H' marks present on iron shot | | Fig. 5.56 | Analysis of 83A0239 |
| Fig. 5.12 | Varieties of 'H' marks present on iron shot | | Fig. 5.57 | Distribution of inset ball shot |
| Fig. 5.13 | Sectioned Royal Armouries shot | | Fig. 5.58 | Distribution of odd composite shot |
| Fig. 5.14 | Round iron shot: cannon and demi cannon shot | | Fig. 5.59 | Hemispherical shot |
| | | | Fig. 5.60 | Suggested reconstruction of hemispherical shot compared with illustration in Blackmore 1976 |
| Fig. 5.15 | Distribution of cannon shot | | | |
| Fig. 5.16 | Distribution of demi cannon shot | | | |
| Fig. 5.17 | Distribution of culverin shot | | Fig. 5.61 | Lead shot 81A0141 |
| Fig. 5.18 | Distribution of demi culverin shot | | Fig. 5.62 | Lead composite shot |
| Fig. 5.19 | Round iron shot: culverin and demi culverin shot | | Fig. 5.63 | Unidentified hemisphere 81A2536 |
| | | | Fig. 5.64 | Stone and lead shot |
| Fig. 5.20 | Round iron shot: saker and falcon shot | | Fig. 5.65 | Graph of lead shot |
| Fig. 5.21 | Distribution of saker shot | | Fig. 5.66 | Sprue and casting on 81A6823 |
| Fig. 5.22 | Distribution of falcon shot | | Fig. 5.67 | Lead shot: holes in 79A1181 and concentric circles on 81A5655 |
| Fig. 5.23 | Round iron shot less than 110mm and possible sling shot | | Fig. 5.68 | Distribution of lead shot |
| Fig. 5.24 | Distribution of shot less than 110mm | | Fig. 5.69 | Three forms of canister shot |
| Fig. 5.25 | Distribution of all iron shot | | Fig. 5.70 | Distribution of canister shot |
| Fig. 5.26 | All iron shot in relation to measured gun bores | | Fig. 5.71 | Cylindrical canister shot 81A1905 |
| | | | Fig. 5.72 | Canister shot 81A1905 and fragment 81A2410 |
| Fig. 5.27 | Cross bar shot 83A0289 | | | |
| Fig. 5.28 | Cross bar shot with clean hole and radiograph of 83A0289 showing structure | | Fig. 5.73 | Conical lantern shot 82A2298 |
| | | | Fig. 5.74 | Cylindrical lantern shot 81A0101 |
| Fig. 5.29 | Cross bar shot | | Fig. 5.75 | Details of 81A0101 and photograph on seabed |
| Fig. 5.30 | Distribution cross bar shot | | | |
| Fig. 5.31 | Cross bar shot 83A0342 and radiograph | | Fig. 5.76 | Cylindrical lantern shot 81A0502 and flint contents |
| Fig. 5.32 | Details of spike/shot | | | |
| Fig. 5.33 | Composite radiograph of 89A0008 | | Fig. 5.77 | Cylindrical lantern shot 81A1183 |
| Fig. 5.34 | 89A0008 during excavation | | | |

Fig. 5.78 Cylindrical lantern shot fragments 82A2696 and 81A2859
Fig. 5.79 Cylindrical lantern shot fragment 81A3061
Fig. 5.80 Cylindrical lantern shot; loose bases with markings
Fig. 5.81 Drawing of tongs with moulds
Fig. 5.82 Stone shot 72A0051 showing effects of blunt point used
Fig. 5.83 Stonemason from Sachs and Amman (1568)
Fig. 5.84 Finishing stone shot
Fig. 5.85 All shot moulds
Fig. 5.86 Distribution of gauges, moulds and gunner's rule
Fig. 5.87 Details of mould 80A1847, marking out lines
Fig. 5.88 Details of mould 72A0812
Fig. 5.89 Experimental shot cast in mould 72A0812
Fig. 5.90 The gunner's rule from the *Mary Rose* and that illustrated by Norton (1628)
Fig. 5.91 Gunner holding rule, after Wright (1563)
Fig. 5.92 Shot density vs radius
Fig. 5.93 Shot mass vs radius
Fig. 5.94 Radius of iron core vs shot radius
Fig. 5.95 Shot gauge as illustrated by Norton (1628)
Fig. 5.96 Rectangular gauges
Fig. 5.97 Stack of gauges in M5
Fig. 5.98 Paddle shaped gauges
Fig. 5.99 Gilmartin's recommended gun drill and equipment needed (1988)
Fig. 5.100 Loading the port piece breech chamber
Fig. 5.101 Port piece gun crew positions
Fig. 5.102 Loading and firing the culverin
Fig. 5.103 Loading and firing muzzle-loaders from the *Seige of Boulogne*
Fig. 5.104 Markings on casks
Fig. 5.105 Micrographs of gun powder compared with modern gunpowder
Fig. 5.106 Illustration of a former by Tartaglia
Fig. 5.107 Former 81A2278
Fig. 5.108 Distribution of small items associated with loading and firing
Fig. 5.109 Selection of ladles
Fig. 5.110 Representations of ladles by Tartaglia and Wright
Fig. 5.111 Distribution of ladles
Fig. 5.112 Distribution of rammers
Fig. 5.113 Divelog sketch of ladle and rammer for M6 cannon
Fig. 5.114 Matching ladle to guns
Fig. 5.115 Anatomy of a ladle
Fig. 5.116 Ladle heads
Fig. 5.117 Ladle heads
Fig. 5.118 Selection of rammers
Fig. 5.119 Rammer head forms
Fig. 5.120 Selection of potential sponges
Fig. 5.121 Matching rammers to guns
Fig. 5.122 Selection of rammers
Fig. 5.123 Rammer 81A2680 showing whittling by hand in manufacture
Fig. 5.124 Tampion 81A1110 and reel of tampions 81A0322
Fig. 5.125 Tampions: cross-sections of wrought iron sling 81A0645 and hailshot piece 79A1088 to show positions of the tampion
Fig. 5.126 Examples of tampion reels
Fig. 5.127 Underwater sketch of tampions in storage
Fig. 5.128 Distribution of tampions
Fig. 5.129 Tampions; radiograph of breech swivel with tampion *in situ* and tampion and wad from 82A4076
Fig. 5.130 Examples of tampions from port pieces
Fig. 5.131 Examples of tampions and wads from swivel guns
Fig. 5.132 81A2640 tampion attached to wad
Fig. 5.133 Possible tampions
Fig. 5.134 Indications of manufacture on tampion reel 83A0404
Fig. 5.135 Excavation of wad from base 82A4076
Fig. 5.136 Examples of wads
Fig. 5.137 Use of priming wires from the *Siege of Boulogne*
Fig. 5.138 The four priming wires recovered in 1979–81
Fig. 5.139 Priming wire discovered in 2004
Fig. 5.140 Powder dispensers
Fig. 5.141 Anatomy of a powder dispenser
Fig. 5.142 Powder dispensers being carried by troops in engraving by John Derricke (1581)
Fig 5.143 Powder flask 81A1517
Fig. 5.144 Possible powder horn top
Fig. 5.145 Use of linstocks from the Cowdray Engraving and *The Embarkation of Henry VIII at Dover*
Fig. 5.146 Anatomy of a linstock and head types
Fig. 5.147–51 Linstocks
Fig. 5.152 Linstock 81A1922
Fig. 5.153 Linstock fragments
Fig. 5.154 Distribution of linstocks
Fig. 5.155 Tinder boxes
Fig. 5.156 Tinder boxes
Fig. 5.157 Knife found inside tinder box 81A5967
Fig. 5.158 Tinder boxes depicted in the Cowdray Engraving
Fig. 5.159 Tools needed for working guns.
Fig. 5.160 Distribution of tools used for working guns
Fig. 5.161 Distribution of tools possibly for use in working the guns
Fig. 5.162 Mallet and possible handspike
Fig. 5.163 Use of baskets from the *Siege of Boulogne* and site book for basket in U/M11
Fig. 5.164 Leather bucket
Fig. 5.165 Wooden tub
Fig. 5.166 Possible sieve handles and reconstruction
Fig. 5.167 Wooden shovel
Fig. 5.168 Wedges, possible gun-related quoins

*Parts 3 and 4*

| | |
|---|---|
| Fig. 6.1 | Types of wildfire |
| Fig. 6.2 | Use of shot of wildfire at sea |
| Fig. 6.3 | Fire trunk/tubes serving as a light gun and guns firing trunks of wildfire |
| Fig. 6.4 | Musket firing an incendiary arrow |
| Fig. 6.5 | Arrows, trunks and garlands of wildfire |
| Fig. 6.6 | Incendiary projectile 81A2866 |
| Fig. 6.7 | Location of incendiary darts |
| Fig. 6.8 | Incendiary projectile 81A2866 as found |
| Fig. 6.9 | Flight, shaft and remains of bag containing incendiary mixture |
| Fig. 6.10 | Proposed reconstruction of the stages of manufacture and final composition of an incendiary dart |
| Fig. 6.11 | Firepots with slings and incendiary arrows |
| Fig. 6.12 | Fragments of incendiary darts |
| Fig. 6.13 | Incendiary dart at sea in fighting tops |
| Fig. 6.14 | Incendiary dart in Alnwick Castle |
| Fig. 6.15 | Reconstruction of 13in mortar shell |
| Fig. 6.16 | Reconstruction of mortar bomb |
| Fig. 6.17 | Distribution of probable mortar bomb fragments |
| Fig. 7.1. | Anatomy of a handgun |
| Fig. 7.2 | Royal Armouries matchlock RA.XII |
| Fig. 7.3. | Gun with fishtail butt of ?1588 |
| Fig. 7.4 | Matchlocks in the Royal Armouries and GARDO crest |
| Fig. 7.5 | Distribution of handguns |
| Fig. 7.6 | Harquebus stock 82A4502 |
| Fig. 7.7 | Harquebus stock 81A2679 |
| Fig. 7.8 | Gardo crest on 81A2679 |
| Fig. 7.9 | Harquebus stock fragments |
| Fig. 7.10 | Butt fragments |
| Fig. 7.11 | Stock fragment 81A3884 and partial stock reconstruction |
| Fig. 7.12 | Graeme Rimer firing a replica of the *Mary Rose* harquebus |
| Fig. 7.13 | Gun-shield 81A5771 as lifted |
| Fig. 7.14 | Gun-shield forms and anatomy of the outside |
| Fig. 7.15 | Anatomy of the inside of a gun-shield |
| Fig. 7.16 | Gun-shields in the Walters Art Gallery, external view and Art Institute of Chicago, internal view |
| Fig. 7.17 | Distribution of gun-shield elements |
| Fig. 7.18 | Distribution of numbered gun-shields in O9/10 |
| Fig. 7.19 | Radiograph of 82A1530 gun and Royal Armouries gun |
| Fig. 7.20 | Pouch-pocket 82A0577 |
| Fig. 7.21 | Grip 82A0940 and strap 82A0948 |
| Fig. 7.22 | Gun-shield 81A5771 |
| Fig. 7.23 | Pouch-pocket and spring elements associated with 81A5771 |
| Fig. 7.24 | Gun-shield 82A0992 |
| Fig. 7.25 | Details of boss on 82A0992 and 81A5771 |
| Fig. 7.26 | Gun-shield 82A1530 |
| Fig. 7.27 | Gun-shield 82A1586 |
| Fig. 7.28 | Gun-shield 82A2060 with radiograph |
| Fig. 7.29 | Gun-shield 82A2954 |
| Fig. 7.30 | Gun-shield 88A0009 |
| Fig. 7.31 | Probable gun-shield fragment 82A4465 |
| Fig. 7.32 | Royal Armouries gun-shield V34 |
| Fig. 7.33 | Royal Armouries gun-shield V38 |
| Fig. 8.1 | Distribution all archery related major groups in relation to that of Fairly Complete Skeletons |
| Fig. 8.2 | Distribution of longbows |
| Fig. 8.3 | Finger ring 81A4394 decorated with sheaf of arrows |
| Fig. 8.4 | Typical longbow and drawing showing belly and back, upper and lower limb |
| Fig. 8.5 | Distribution of heartwood/sapwood in bows |
| Fig. 8.6 | Fluting on bow |
| Fig. 8.7 | Pale bow tip with ridge at end |
| Fig. 8.8 | Sections of longbows |
| Fig. 8.9 | Graph showing range of complete lengths of bows studied |
| Fig. 8.10 | Graph showing range (of complete bows from study) found in chests 81A1862 and 81A3927 |
| Fig. 8.11 | Bows set back in the handle |
| Fig. 8.12 | Graph showing lengths of bows set back in the handle |
| Fig. 8.13 | Graph showing lengths of bows set back in the handle from chests |
| Fig. 8.14 | Reflex and deflex bows |
| Fig. 8.15 | Deflex bows |
| Fig. 8.16 | Trapeziodal bow 81A1607 |
| Fig. 8.17 | Nock slots upper tip, ridge |
| Fig. 8.18 | Upper limb with double nock and right side lower limb, single slot |
| Fig. 8.19 | Double nock on upper limb |
| Fig. 8.20 | Comparison between nock slots in the upper and lower limbs |
| Fig. 8.21 | Length range of double nocked bows |
| Fig. 8.22 | Length range of double nocked bows from chests |
| Fig. 8.23 | Bowmarks |
| Fig. 8.24 | Bowmarks |
| Fig. 8.25 | Catalogue of bowmarks |
| Fig. 8.26 | Size ranges of 102 bows |
| Fig. 8.27 | Draw-weights |
| Fig. 8.28 | Constructs for the average bow |
| Fig. 8.29 | Graph of mass versus maximum width |
| Fig. 8.30 | Comparisons of constructs |
| Fig. 8.31–4 | Selection of longbows |
| Fig. 8.35 | Growth rings on sectioned bow |
| Fig. 8.36 | Bow 81A1648 on tiller |
| Fig. 8.37 | Roy King bending fresh stave |
| Fig. 8.38 | Reproduction A1 straight and bent |
| Fig. 8.39 | Reproducion A4 on tiller |
| Fig. 8.40 | Simon Stanley shooting the reproduction |
| Fig. 8.41 | Nock; original and replica and bow tips |
| Fig. 8.42 | Nock 97A0003 and possible bowstring |

Fig. 8.43   Bowstrings; fastenings and strung bow
Fig. 8.44   Bowstrings; details of attachments
Fig. 8.45   Distribution of longbow and arrow chests
Fig. 8.46   Exploded view of longbow chest construction
Fig. 8.47   Underwater photograph of bows in chest and on deck as recovered
Fig. 8.48   Newport ship wristguard and a medieval sea battle from the *Beauchamp Pageant* with armguard on an archer's arm
Fig. 8.49   Wristguards found with the burial of the Amesbury Archer (2400–2200 cal. BC) and ivory wristguard from Southend
Fig. 8.50   Distributon of wristguards and related items
Fig. 8.51   Wristguard shapes
Fig. 8.52   Fastenings on wristguards
Fig. 8.53   Reconstuction of 81A4341 showing the position of the wristguard on the archer's forearm
Fig. 8.54   Decorated leather armguard in the British Museum bearing Plantagenet insignia
Fig. 8.55   Wristguard 82A0943
Fig. 8.56   Book covers 81A0895, 81A4131
Fig. 8.57–62 Decorated wristguards
Fig. 8.63–4 Undecorated wristguards
Fig. 8.65   Laced armguards 81A1863 and 81A4675
Fig. 8.66   Ivory wristguard 81A0815
Fig. 8.67   Signet ring decorated with letter 'K'
Fig. 8.68   Reconstruction of a spacer of arrows and illustrations of arrows
Fig. 8.69   Distribution of arrows and spacers
Fig. 8.70   Anatomy of an arrow
Fig. 8.71   Types of leaf-shaped and bodkin arrowheads
Fig. 8.72   Overlap of different arrow studies
Fig. 8.73   XRF analysis for copper and zinc of arrow shaft and arrow analysed
Fig. 8.74   Principal length measurements of the arrow shafts
Fig. 8.75   Histogram of total length of the arrow shafts
Fig. 8.76   Graphs of arrow data
Fig. 8.77   Arrow shaft profiles by wood type
Fig. 8.78   Graphs of arrow nock data
Fig. 8.79   Nock orientation relative to annual growth
Fig. 8.80   Tip cone types
Fig. 8.81   Arrow shaft profiles
Fig. 8.82   Arrow shaft data
Fig. 8.83   Arrow mass vs total length by wood type
Fig. 8.84   Location of divots in relation to annual growth
Fig. 8.85   Saw marks on face in slot
Fig. 8.86   Exploded view of arrow chest construction
Fig. 8.87   Underwater photographs of arrow chests
Fig. 8.88   Spacers 79A1204 and 82A2026
Fig. 8.89   Reconstruction drawing of spacer 79A0653 with arrows
Fig. 8.90   Extract from the Cowdray Engraving showing archers with spacer/bag
Fig. 8.91   Bag 816844 and associated arrows 81A0116
Fig. 8.92   Bag ancient quiver
Fig. 8.93   Tip and end of of possible handgun bolt 81A1093
Fig. 8.94   Group of possible handgun bolts 81A1093 as recovered
Fig. 8.95   Handgun bolts 81A1093 and unidentified possible ramrods 81A1022
Fig. 8.96   Sitebook showing contents of chest 81A1328
Fig. 8.97   Unusual small handplanes
Fig. 8.98   Lidded box 81A1049
Fig. 8.99   Scapula and example with *os acromiale*
Fig. 8.100  Divelog recording discovery of human remains FCS82, a possible archer
Fig. 8.101  Jerkin 81A0090
Fig. 8.102  Underwater photograph of 'Little John' and detail of arrows and arrow bag
Fig. 8.103  Personal possessions associated with the remains of FCS 74 and 75
Fig. 8.104  Jerkin 825026 and comb 82A5064 associated with human remains
Fig. 8.105  Divelog 82/8/58 showing 82H1000 and associated objects
Fig. 9.1    Examples of staff weapons from the Cowdray Engraving
Fig. 9.2    Anatomy of a bill, pike and halberd
Fig. 9.3    Bill from Coventry and use of bills from the *Beauchamp Pageant*
Fig. 9.4    Ceremonial halberd in the Royal Armouries
Fig. 9.5    Bill from Jamestown, USA
Fig. 9.6    Anatomy of a billhead and 'T' shaped cross-nail
Fig. 9.7    Tops of *Mary Rose* bill hafts
Fig. 9.8    Bill hafts
Fig. 9.9    Marked tops of bill hafts
Fig. 9.10   Bill haft 79A1170 showing brass nails in cross formation
Fig. 9.11   Bill in Higgins Armory Museum
Fig. 9.12   Bill 82A3590
Fig. 9.13   Bill 82A2637
Fig. 9.14   Bill haft 81A1646 with concretion and possible textile
Fig. 9.15   Distribution of staff weapons
Fig. 9.16   The 'pike garden'
Fig. 9.17   Stages in forging a bill
Fig. 9.18   Stages in forging a bill
Fig. 9.19   Anatomy of a pikehead and pike from Royal Armouries
Fig. 9.20   Remains of pike staves
Fig. 9.21   Pike stave 81A4965 showing cheek rebates
Fig. 9.22   Haft and collar, probably of a halberd
Fig. 9.23   Halberds in Royal Armouries
Fig. 9.24   Haft fragments possibly from throwing darts

Fig. 9.25   Anatomy of a sword and scabbard
Fig. 9.26   Anatomy of a dagger and sheath
Fig. 9.27   Basket hilt sword 82A3589 and details of hilt
Fig. 9.28   Radiographs of 83A0048, 82A1915 and 82A1932
Fig. 9.29   Stain of a third basket hilt sword on seabed
Fig. 9.30   Pommel and grip fragments
Fig. 9.31   Anatomy of a by-knife
Fig. 9.32   Drawing taken from effigy of Lord Parr
Fig. 9.33   Dagger from Hook Court, sword hangers and girdle end
Fig. 9.34   Distribution all sword finds
Fig. 9.35   Distribution of Fairly Complete Skeletons and with catalogue entries numbered
Fig. 9.36   Basket-hilted sword 82A3589 after removal of concretion
Fig. 9.37   Guard of basket-hilted sword 82A3589 after removal of concretion
Fig. 9.38   Metallography of the blade on basket-hilted sword 82A3589
Fig. 9.39   Similar basket-hilted swords
Fig. 9.40   Reconstructions of hilts
Fig. 9.41   Grip in concretion showing wire
Fig. 9.42   Grips 82A2054 and 82A4412
Fig. 9.43   Selection of wooden grips
Fig. 9.44   Selection of wooden grips
Fig. 9.45   Pommels and pommel fragments
Fig. 9.46   Dagger from the Thames foreshore at the Tower of London
Fig. 9.47   Pommel fragments and sheathing
Fig. 9.48   Drawing of radiograph of 82A0027 and of pommel from sword 82A3589
Fig. 9.49   Rondel dagger 82A1914 in concretion
Fig. 9.50   Rondel dagger excavated from the foreshore of the Thames at Queenhithe
Fig. 9.51   Radiograph of 82A0975/2
Fig. 9.52   Dagger 83A0022
Fig. 9.53   Construction of ballock dagger grip, lobes with pins
Fig. 9.54   Ballock dagger tang hole shapes
Fig. 9.55   Distribution of ballock daggers and associated by-knives
Fig. 9.56   Chest 81A1429 and objects
Fig. 9.57   The variety of pommel shapes found on *Mary Rose* ballock daggers
Fig. 9.58   The variey of grip sections found on *Mary Rose* ballock daggers
Fig. 9.59   Diamond-sectioned ballock dagger grips
Fig. 9.60   Octagonal-sectioned ballock dagger grips
Fig. 9.61   Hexagonal, round, ovoid and lobe-shaped ballock dagger grips
Fig. 9.62   Honed ballock dagger grips and examples with one lobe larger than the other
Fig. 9.63   Faceted ballock dagger grip, plain pommel 81A0811
Fig. 9.64   Staining under lobes for ballock dagger guard of 81A1412
Fig. 9.65   Radiographs showing pinning on 81A1845 and 82A0890
Fig. 9.66   The variety of tang buttons found on *Mary Rose* ballock daggers
Fig. 9.67   Dagger 81A0862
Fig. 9.68   Hilt 80A1350 showing the guard cut-out to take a double-edged blade and radiographs
Fig. 9.69   Diamond grip ballock daggers together with by-knives
Fig. 9.70   Octagonal, hexagonal, round and ovoid shaped ballock dagger grips with by-knives
Fig. 9.71   Scabbards, complete examples
Fig. 9.72   Decorated scabbards
Fig. 9.73   Scabbards 80A0549 and 79A1146
Fig. 9.74   Details of seam location and stitches
Fig. 9.75   Stitch types found on *Mary Rose* scabbards and sheaths
Fig. 9.76   Scabbard 83A0151
Fig. 9.77   The range of chapes found on *Mary Rose* scabbards and sheaths
Fig. 9.78   Radiographs of chapes
Fig. 9.79   Sheaths 79A1146 and 81A1567, concretion stain on tip from corroded iron chape
Fig. 9.80   Sheaths
Fig. 9.81   Radiographs of 82A1870, 82A1916, 82A1697
Fig. 9.82   Albrecht Durer's engraving *Knight, Death and Devil* and drawing from engraving by Giovanni Battista Palimba
Fig. 9.83   Sword hanger and sheath 80A1709
Fig. 9.84   Reconstruction of bifurcating belt
Fig. 9.85   Sword belt, hanger and scabbard 81A1567
Fig. 9.86   Sword belts and hangers
Fig. 9.87   Sword hangers
Fig. 10.1   Mail garments, Brigandine (Jack) and plate armour
Fig. 10.2   Comb morion; Spanish morion with pear stalk finial; sketch of helmets from Boulogne, Marquisson/Portsmouth
Fig. 10.3   Almain rivets
Fig. 10.4   Drawing of armour, from the *Siege of Boulogne*
Fig. 10.5   82A4096 breastplate
Fig. 10.6   Distribution of armour and possible fittings
Fig. 10.7   Breastplate RA 111.4572
Fig. 10.8   A sample from the breastplate
Fig. 10.9   Sample 81S0167, stack of breastplates
Fig. 10.10  Buckler 82A4212
Figs 10.11–13 Possible armour fragments
Fig. 10.14  Range of sixteenth century buckle types and those actually found
Fig. 10.15  Selection of buckles
Fig. 10.16  Types of mail and methods of manufacture
Fig. 10.17  Examples of mail fragments
Fig. 10.18  Examples of mail fragments
Fig. 10.19  Distribution of mail
Fig. 10.20  Distribution of Fairly Complete Skeletons
Fig. 10.21  Distribution of human bones

Fig. 11.1 Tactics: gaining the weather gauge
Fig. 11.2 Artist's reconstruction of demi culverin facing forwards in the Sterncastle; front of Sterncastle as it survives in museum
Fig. 11.3 Canted guns on the *Henry Grace à Dieu*, as showing in the Cowdray Engraving
Fig. 11.4 Sailing formation: line abreast
Fig. 11.5 Dardanelles gun
Fig. 11.6 Various schemes for naming of the decks
Fig. 11.7 Outboard elevation of the *Mary Rose* in 1514
Fig. 11.8 Outboard elevation of the *Mary Rose* in 1545
Fig. 11.9 Internal elevation of the *Mary Rose* showing final arrangement of decks
Fig. 11.10 The Cowdray Engraving showing the sinking of the *Mary Rose*
Fig. 11.11 Details from the Cowdray Engraving
Fig. 11.12 Digital terrain models of the Solent; aspect of the battle as in the Cowdray Engraving and with the English and French fleets
Fig. 11.13 The position of the fleets plotted on an old chart
Fig. 11.14 Positions of the opposing fleets with 250m range rings
Fig. 11.15 Fleet with ships' tracks
Fig. 11.16 Charts showing the tidal changes throughout the day
Fig. 11.17 External elevation of the ship
Fig. 11.18 Gunports
Fig. 11.19 Internal structure of the ship
Fig. 11.20 Diagonal braces
Fig. 11.21 Line of the Main deck and interpretation showing positions of rising knees and photograph of large knees on Main deck showing how they constrain gun movement
Fig. 11.22 Tiers of guns and tiller/rudder attachments from the *Anthony Roll*
Fig. 11.23 Gunport lids, external
Fig. 11.24 Gunport lids, internal
Fig. 11.25 Fittings found beside gunports
Fig. 11.26 Views of gunports and swivel holes in inner wale at Upper deck level
Fig. 11.27 Operation of gunport lids
Fig. 11.28 Location of armament stores
Fig. 11.29 Isometric with all guns found in position and spot locations for recovered handguns and hailshot pieces
Fig. 11.30 Main deck showing all guns in position
Fig. 11.31 Securing iron guns to the side of the ship
Fig. 11.32 Main deck plan with possible methods of bringing in bronze guns to reload
Fig. 11.33 Positioning of alternative types of gun out of the stern at Main deck level
Fig. 11.34 Stern elevation showing possible location of port piece or demi culverin
Fig. 11.35 Outboard elevation of Sterncastle showing gun below port and details of structure around M8 port
Fig. 11.36 Plan of the Upper deck with all postulated guns in position
Fig. 11.37 The Upper deck in the waist
Fig. 11.38 Reconstruction outboard view of upper part of ship's side in waist and outboard view showing blinds
Fig. 11.39 Possible distribution of ordnance on Castle deck in stern, bases on either side of rudder
Fig. 11.40 Possible distribution of ordnance on Castle deck in stern, demi culverins on either side of rudder
Fig. 11.41 Approximate intersection of arcs of fire of Main deck ordnance at various ranges
Fig. 11.42 Approximate intersection of arcs of fire of Upper deck ordnance at various ranges
Fig. 11.43 Protective zones afforded by proposed ordnance on the Castle deck
Fig. 11.44 Protective zones afforded by the intersections of arcs of fire on all decks
Fig. 11.45 Zones of coverage afforded by the longest range guns on all decks
Fig. 11.46 Zones of coverage afforded by weapons on a modern warship
Fig. 11.47 Detail of grapnel on *Anthony Roll* and reconstruction of grapnel
Fig. 12.1 Solent chart with Historic Wreck circle and proposed line of MOD dredge
Fig. 12.2 Data provided by remote sensing
Fig. 12.3 The ROV vehicle and sieving spoil
Fig. 12.4 The positions of the trenches into spoil and area worked north of bow
Fig. 12.5 The Site Recorder displays
Fig. 12.6 Location of ordnance recovered 2003–5
Fig. 12.7 Scabbard strap and top and ballock dagger
Fig. 12.8. Distribution of ordnance in the Forecastle area
Fig. 12.9 Tritech parametric profiler on frame
Fig. 12.10 Priming wire and dagger quillion block or part of gunner's stiletto
Fig. 12.11 Sand barge and hopper used in reburial of wreck site
Fig. 12.12 Sketch of bow showing two tiers of forward facing guns in Sterncastle and *Anthony Roll* image of *Mary Rose* for comparison
Fig. 12.13 Simple plan of *Mary Rose* with alternative gun arrangements on Castle decks
Fig. 12.14 Bowstring found in 2003 and previously recorded example
Fig. Appendix1.1 Segregation; 81A1423 and 79A1232
Fig. Appendix 1.2 81A1423 correlation of tin and antimony
Fig. Appendix 1.3 *Mary Rose* bronze ordnance segregation
Fig. Appendix 1.4 Analysis of European bronze ordnance 1500–1560
Fig. Appendix 1.5 lead isotope analysis of *Mary Rose* ordnance

# List of Tables

*Parts 1 and 2*

Table 1.1   Comparison between ordnance listed in the *Anthony Roll* and archaeological evidence from the *Mary Rose*
Table 2.1   Dimensions and weight of bronze guns
Table 2.2   *Mary Rose* bronze guns
Table 2.3   Devices and decoration on *Mary Rose* bronze guns
Table 2.4   Comparison 1524, 1540, 1546 inventories
Table 2.5   *Mary Rose* gun carriage dimensions
Table 2.6   Details of measurements for reconstructed carriage 81A3002
Table 2.7   Detailed measurements for gun carriage 81A1423
Table 2.8   Details of unassociated truck wheels
Table 2.9   Axles in store
Table 2.10 Loose axles not in store
Table 2.11 Firing schedule reproduction of culverin 81A1423
Table 3.1   Relative statistics for chamber pieces
Table 3.2   Incidence of port pieces listed on ships between 1540 and 1547
Table 3.3   Port piece basic data
Table 3.4   Port piece chambers
Table 3.5   Port piece fragments
Table 3.6   Spoked wheels
Table 3.7   Forelock characteristics
Table 3.8   Elevating posts
Table 3.9   Possible culverin equivalents to guns of the sling class
Table 3.10 Sling barrels
Table 3.11 Sling chambers
Table 3.12 Sling fragments
Table 3.13 Fowler barrels
Table 3.14 Fowler chambers
Table 3.15 Bases on vessels in 1546 and 1547
Table 3.16 Bases: summary of most complete guns
Table 3.17 Bases: summary of semi-complete examples
Table 3.18 Bases: summary of barrel fragments
Table 3.19 Bases: summary of other fragments
Table 3.20 Bases: summary of breech chambers, troughs, trunnions and wedges
Table 3.21 Barrel manufacture
Table 3.22 Manufacture of breech chamber
Table 3.23 Port piece firing sequence and results
Table 4.1   Dimensional data for hailshot pieces
Table 4.2   Incidence of hailshot pieces recorded in the Tower inventories
Table 4.3   Incidence of hailshot pieces against total iron guns on vessels in 1546
Table 4.4   Incidence of hailshot pieces listed on vessels 1546–1555
Table 5.1   Shot as listed in the *Anthony Roll*
Table 5.2   Burton list of shot about 1550
Table 5.3   Shot material from 1547 inventory
Table 5.4   Expected variability in bore and shot diameters
Table 5.5   Proposed bore sizes for slings
Table 5.6   Identification of slings in the *Mary Rose* assemblage
Table 5.7   Cross bar shot
Table 5.8   Gun bores and stone shot size
Table 5.9   Stone shot groups associated with Main deck guns
Table 5.10 Stone shot groups on the Upper deck
Table 5.11 Least weights for composite dice shot
Table 5.12 Bases recovered with shot
Table 5.13 Diameter and weight of inset ball shot
Table 5.14 Odd shot
Table 5.15 Cylindrical lantern loose ends and staves
Table 5.16 *Mary Rose* Gunner's rule specific gravity calculations (iron)
Table 5.17 *Mary Rose* rule specific gravity calculations (lead)
Table 5.18 Comparison of *Mary Rose* and Norton rules for iron and lead
Table 5.19 Listed historical weights for shot of a given diameter compared with *Mary Rose* and Norton rules
Table 5.20 Iron core radius for composite inset ball shot
Table 5.21 Shot gauges, summary data
Table 5.22 Recovered guns and shot gauge sizes
Table 5.23 Gauge sizes against iron and stone shot diameters
Table 5.24 Shot size in relation to gauges and/or formers
Table 5.25 *Anthony Roll* gunpowder summary
Table 5.26 Estimated powder expenditure for bronze guns
Table 5.27 Estimated gunpowder expenditure for all *Mary Rose* guns and shot
Table 5.28 Estimated gunpowder expenditure of all *Sun* and *Gillyflower* guns and shot
Table 5.29 Degree of fortification and gunpowder charge
Table 5.30 Weight of serpentine and corn powder for named guns
Table 5.31 Gun and projectile sizes, gunpowder charge and range
Table 5.32 Gunpowder and elements for cartridge making, *Anthony Roll* (ships)
Table 5.33 Gun type and number of formers: *Anthony Roll* (pinnaces)
Table 5.34 Components and materials of *Mary Rose* ladles
Table 5.35 Metal analysis of ladle components

Table 5.36 Estimates of ladle sizes in relation to estimates of powder weights required for different gun sizes
Table 5.37 Comparison of *Mary Rose* ladle dimensions and historic referents for gun classes
Table 5.38 Components of *Mary Rose* rammers compared with historical and actual gun sizes
Table 5.39 Comparison between historical referents for dimensions of rammers, ladles and sponges and *Mary Rose* examples
Table 5.40 Rammer diameter, gun and chamber bores and cast iron shot peaks
Table 5.41 Tampion reels from O1–2/H1
Table 5.42 Tampion reels not in storage, by sector
Table 5.43 Numbers and sizes of loose tampions
Table 5.44 Tampions in wrought iron breech-loading and cast iron muzzle-loading guns
Table 5.45 Wads
Table 5.46 Priming wires
Table 5.47 Powder dispensers
Table 5.48 Linstocks
Table 5.49 Tinder boxes
Table 5.50 Objects possibly associated with working the guns
Table 5.51 Buckets and possible tubs
Table 5.52 Wedges, possible gun-related quoins

*Parts 3 and 4*
Table 6.1 Analysis of the incendiary mix from projectile 81A2866
Table 6.2 Fragments of 'charge' associated with Tudor artefacts
Table 6.3 Mortar bomb fragments by sector
Table 7.1 Common surviving measurements of handgun stocks
Table 7.2. Size, number, and distribution of small lead round shot suitable for handguns, and estimated bore sizes
Table 7.3 Components of gun-shields recovered
Table 7.4 Principal measurements of gun-shields
Table 7.5 XRF results of metal components of gun-shields
Table 7.6 Comparative data for the Royal Armouries gun-shields
Table 8.1 Archery equipment listed in the *Anthony Roll*
Table 8.2 Distribution of longbows
Table 8.3 Contents of chest 81A5783
Table 8.4 Location of 104 complete longbows
Table 8.5 Longbow data
Table 8.6 Distribution of longbows of reflexed and deflexed shape
Table 8.7 'Slab-sided' longbows; location and length
Table 8.8 'Slab-sided' longbows; width and depth along length
Table 8.9 Round D-sectioned longbows; width and depth along length
Table 8.10 Flat D-sectioned longbows; width and depth along length
Table 8.11 D-sectioned longbows; width and depth along length
Table 8.12 Handled bow 81A3949; width and depth along length
Table 8.13 Longbows with a trapezoid section and relieved grip area
Table 8.14 Distribution of longbows with double nocks
Table 8.15 Distribution of marked longbows
Table 8.16 Timber grade and uniformity
Table 8.17 Distribution of used longbows
Table 8.18 Draw-weights of *Mary Rose* bows by location
Table 8.19 Grouped frequency table for bow weight by wood type
Table 8.20 Variables used in the preliminary analysis
Table 8.21 Mean length, mass, maximum width and maximum depth by category
Table 8.22 Testing of *Mary Rose* longbows
Table 8.23 Length and elastic modulii for selected longbows
Table 8.24 Weight and density of selected longbows
Table 8.25 Weight and density of different woods
Table 8.26 Estimated draw-weights of *Mary Rose* longbows
Table 8.27 Experimental shooting data
Table 8.28 Range and velocity tests of MRA4 Harlington
Table 8.29 Measurement of horn nocks from arrows
Table 8.30 String fit to arrow nock
Table 8.31 Contents of longbow chests
Table 8.32 Wristguards
Table 8.33 Distribution of arrows by deck
Table 8.34 Arrows in chests
Table 8.35 Arrows; wood analysis
Table 8.36 Arrows: wood analysis arrows in chests
Table 8.37 Arrows; wood identification in arrow 'bundles'
Table 8.38 Density tables of woods used in *Mary Rose* arrow assemblage
Table 8.39 Density classification
Table 8.40 Mechanical properties
Table 8.41 Specific gravity and modulus of elasticity
Table 8.42 Distribution of wood types
Table 8.43 Two-way distribution of location and length group
Table 8.44 Mean total length and 95% confidence limits by arrow chest
Table 8.45 Percentiles of binding length plus nock end length
Table 8.46 Mean and 95% confidence limits for total length and for binding length plus nock length by wood type
Table 8.47 Two-way distribution of wood type and arrow chest
Table 8.48 Distribution of binding frequency

Table 8.49 Two-way distribution of nock end diameter and tip end diameter
Table 8.50 Distribution of arrow profiles
Table 8.51 Two-way distribution of profile and arrow chest
Table 8.52 Two-way distribution of wood type and profile
Table 8.53 Arrow mass (g) by wood type
Table 8.54 Arrow mass (g) by wood type, length group and profile
Table 8.55 Contents of arrow chests
Table 8.56 Arrow spacers
Table 8.57 Contents of chest 81A1328
Table 8.58 Details of short arrows 81A1093
Table 8.59 Contents of chests possibly belonging to archers
Table 9.1 Total numbers of Morris pikes, bills and longbows listed by vessel type in the *Anthony Roll*
Table 9.2 Summary of basic measurements for bill hafts
Table 9.3 Grips and attached pommels from swords and daggers
Table 9.4 Pommel sheaths with metal analysis
Table 9.5 Ballock daggers
Table 9.6 By-Knives associated with ballock daggers
Table 9.7 Unassociated by-knives
Table 9.8 Scabbards
Table 9.9 Chapes
Table 10.1 Composition (inclusions) of breastplate 82A4096
Table 10.2 Leather straps and other items possibly for armour
Table 10.3 Tudor buckles
Table 10.4 Analysis of mail samples
Table 10.5 Type 1 (butt link) mail
Table 10.6 Type 2 and other mixed rivet/butt link mail
Table 10.7 Type 3 (rivet link) mail
Table 10.8 Mail of uncertain link form
Table 11.1 Comparison of inventory records for *Mary Rose* guns
Table 11.2 Extroplated location of guns on named ships in 1514
Table 11.3 Gun ranges from historical sources
Table 11.4 Historical gun range (average point blank in metres)
Table 11.5 Range of culverin calculated in degrees of the quadrant
Table 11.6 Gunport dimensions
Table 11.7 Gunport lid dimensions
Table 11.8 Ordnance on the Main deck
Table 11.9 Possible ordnance on the Upper deck
Table 11.10 Ordnance on the Castle deck (postulated)
Table 11.11 Arcs of fire: Main deck
Table 11.12 Arcs of fire: Upper deck
Table 11.13 Weapons and ammunition
Table 11.14 Weapon '*preference*' based on number of rounds per gun
Table 11.15 Daily expected firing capability on land
Table 11.16 Sequence based on total number shot supplied
Table 11.17 Preference list based on number guns carried
Table 11.18 Weapons and ammunition according to range
Table 11.19 Ship supported/carriage-mounted guns and shot
Table 11.20 Bronze guns/shot listed and recovered
Table 11.21 Endurance based on shot and reload rates
Table 11.22 Endurance based on volley firing (all guns loaded)
Table 12.1 Artefacts recovered from spoil mounds
Table 12.2 Artefacts recovered from the Timber Park
Table 12.3 Artefacts found on stem or western timbers
Table 12.4 Artefacts found east of stem

Table Appendix 1.1 Guns sampled for metal analysis
Table Appendix 1.2 Segregation in the bronze guns
Table Appendix 1.3 Sample analysis of the bronze guns from the *Mary Rose*
Table Appendix 1.4 Compositions of English guns of the reign of Henry VIII
Table Appendix 1.5 Analysis of European bronze ordnance 1500–1560
Table Appendix 1.6 Lead isotope analysis of bronze guns
Table Appendix 1.7 Detailed analysis of the bronze guns
Table Appendix 2.1 Gravimetric and Titrimetric determination of sulphur
Table Appendix 2.2 Determination of sulphur content (bomb calorific determinations)
Table Appendix 2.3 Emission spectophotometric analysis of powder samples
Table Appendix 2.4 Determination of metals by atomic absorbtion spectrophotometry
Table Appendix 2.5 Determination of carbon content
Table Appendix 2.6 Percentage of soluable solids in samples 81S1409 and 82S1207
Table Appendix 2.7 Elemental analysis of insoluble solids, samples 81S1409 and 82S1207
Table Appendix 2.8 Elements present in samples 81S1409 and 82S1207
Table Appendix 2.9 Ether soluble material present in samples 81S1409 and 82S1207

# List of Colour Plates

Plate 1     The *Mary Rose* as she appears in the *Anthony Roll* of 1546; painting by Geoff Hunt *The Mary Rose. Henry VIII's flagship 1545*

Plate 2     Cannon 81A3003 on carriage underwater; culverin 80A0976 on its carriage on the Upper deck in the stern; detail of Arms and Garter, bronze demi culverin 79A1232; detail of device, bronze demi cannon 81A3002

Plate 3     The bellfoundry during the casting of the reproduction culverin; firing the reproduction port piece

Plate 4     Copper alloy priming wire; gunner's rule, shot gauge and muzzle of demi culverin; scorched head of linstock; selection of objects used in firing the guns

Plate 5     Remains of wooden gun-shield; decorated copper alloy gun-shield boss; underwater shot of chest 81A1862 full of longbows; decorated leather wristguard 80A0901

Plate 6     Arrow spacer containing arrows; stain of a basket-hilted sword on the seabed; ballock dagger grip; copper alloy fittings for sword belts and scabbards

Plate 7     The only complete sword recovered; remains of the only identifiable armour breast plate; view along the surviving portion of the Main deck showing knees and gunports

Plate 8     Internal elevation of the starboard side showing decks and gunports; external view of Main deck gunport

# Contents of DVD

**Image gallery (PDF)**

*Bronze guns*
    Line drawings
    Colour photographs
*Iron guns*
    Line drawings
    Colour photographs
*Gun furniture*
    Line drawings
    Colour photographs
        Colour photographs (shot)
*Handweapons*
    Edged weapons
        Line drawings
        Colour photographs
    Handguns and gun-shields
        Line drawings
        Colour photographs
    Staff weapons
        Line drawings
        Colour photographs
*Archery*
    Line drawings
    Colour photographs
*Protective*
    Line drawings
    Colour photographs
*Ship structure*
    Line drawings
    Colour and black and white photographs
*Paintings*

**Spreadsheets, databases and recording forms**
    *Pro-forma recording sheets (PDF)*
        Arrows
        Linstocks
        Powder dispensers
        Shot
        Staff weapons
    *Longbow data sheets (PDF)*
    *Ordnance database (PDF)*
    *Site Recorder interactive database*

**Videos (three short films)**

Mary Rose *2003*
A short film about the 2003 site investigations and excavations. It outlines the Ministry of Defence's potential plans to dredge a new deeper and straighter channel into Portsmouth for the new aircraft carriers. This would have cut into the 500m zone of protection around the wreck site, and provided funding for specific work onsite. Written by Alex Hildred, narrated and produced by Stephen Foote.

*Reunion and Reburial 2003–2005*
A short film put together from 2004 site film footage and extracts from the BBC *Timewatch* programme *Secrets of the Mary Rose*. It introduces the plans for the 2005 season to recover the stem and anchor and rebury the site following the excavations of 2003–2005. Written and partly narrated by Alex Hildred, produced by Kester Keighley. Reproduced by permission.

*Making and firing* Mary Rose *guns*
This film is an excerpt of a documentary produced by Yorkshire Television and the Royal Armouries. The excerpt features the building and firing of reproductions of a wrought iron and a cast bronze gun between 1998 and 2000. Included by permission of the Board of Trustees of the Royal Armouries. Produced by Kester Keighley.

# Acknowledgements

First and foremost, those whose names appear aginst portions of the text in this volume deserve enormous thanks for their hard work and patience in seeing this publication project through to completion. In truth, however, they represent only the tip of a considerable iceberg. This volume is the result of over 25 years of data collation, analysis and interpretation. Many people have contributed to this in a variety of ways, from participating in small projects, such as shot measuring or concretion removal, through to the larger projects, such as the construction and firing of an iron port piece and a bronze culverin: all deserve equal thanks.

The information that can be gained from a study of the objects is only part of the process, without context, interpretation is incredibly difficult, therefore my first thanks are to the divers whose reports, 'divelogs', were fundamental to providing context and associations. The archaeological supervisors who kept daily journals and site books began the process of associating groups of objects and preliminary interpretation. Andrew Fielding's daily journals are a joy to read and contain his ideas and interpretation of site formation not found elsewhere. Personally, I have benefited from many discussions regarding the structure and accommodation of weapons with Andrew, although I think that he may disagree with some of my conclusions!

In 1980 it became obvious that parts of the structure of the ship would have to be dismantled and a small team was put together from those divers who had the necessary skills. The 'dismantling team' were also responsible for the recovery of most of the guns, and many of their observations are included here. Initial registration of finds and primary identification fell to the finds assistants on the boat, led by Andrew Elkerton.

Once ashore detailed recording and analysis began, and the records of the finds staff under Adele Wilkes (now Adele Berchtold) are also of great importance; individuals whose work has been used include Maggie Richards, Jenny Woodgate, John Randall Holly Arnold and Simon Ware. Ian Oxley was responsible for environmental archaeology and the processing of the many ordnance related samples. Sue Cooke, Sue Bickerton and Glenn McConnachie provided the in-house wood identification, with selected examples sent to Rowena Gale. Photographers whose pictures have been used or have contributed to understanding the collection include Christopher Dobbs, Stephen Foote, Andy Vowles, Robert Stewart, Dominic Fontana, Patrick Baker, Richard Hubbard, Kester Keighley, Hilary Painter, the late Sydney Richman and Peter Crossman.

The line drawings in this, as in the other, volumes are, unless otherwise stated, edited and compiled from originals produced over a number of years by a team of illustrators initially under the leadership of Debby Fulford (now Debby Fox). Those who have contributed greatly to this volume include Debby Fulford, Colin Bass, Arthur Evans, Roger Purkis and Clare Venables. Peter Crossman deserves special thanks for his own illustrations and for committing all of these to digital form and for rationalising the varied source. He has also created some extremely complex illustrations far beyond his remit as an archaeological illustrator. Juggling several large volumes at once with limited computer capability cannot have been easy. Andrew Elkerton as Collections Manager has been responsible for data collection since 1979, and for the creation of the computer archive. Without his knowledge of the collection and willingness to help, rationalisation of this huge resource would not have been possible.

There have been a number of small projects where staff and volunteers have come together to work on specific groups of material, or student projects have addressed specific individual or groups of objectcts. These projects have included the measuring of shot; staff weapons; edged weapons; mail; buckles; gun-shields, concretions and individual guns. So many people have helped that it is impossible to name them all. Included amongst these are; Sue Barber, Laura Davidson, Alison Draper, R. Dunham, Sue Erridge, Nick Evans, Stephen Foote, Rob George, Jo Henderson, Kirren Louise Klein, Steve Liscoe, Anne Majeswski, Peter Martin, Richard Muskett, Alex Naylor, Andrew Richards, Maggie Richards, Simon Ware, Cate Watson and the late Jean Humphrey and Sydney Richman, among others.

Other individuals who have helped include Simon Metcalf (both mail and gun-shields), Ian Eaves (armour) and John Bevan. John has helped particularly with information about the nineteenth century salvaging of the wreck but, as part of the team which found the ship under the leadership of Alex McKee, he has provided a link with this important phase of the project. Jon Adams' interpretative sketches and notes regarding the salvage cables placed under the hull were of great help.

The nomenclature for the Heraldic Devices on the guns was provided by Sir George Bellew. Robert Smith and Ruth Brown have always been willing to share their knowledge and research. Much stems from their enthusiam and expertise. Brian Awty has been a great source of information regarding the Arcana family of gun founders. Jan Piet Puype has helped sourcing parallels for some of the gun furniture recovered.

Richard Barker has provided advice regarding ship stability and deck loading over a number of years and Barbara and the late John Saunders have helped me with trying to present the firing capability afforded by her different types of weapons (Chapter 11). Ann Stirland has been working on the skeletal remains from the *Mary Rose* from the outset and her deductions regarding occupation are included within this report. Robin Wood was already studying the domestic treen when I asked whether he would look at the gun furniture, thankfully he agreed.

Analytical work has been ongoing since before the ship was raised; usually free of charge and often as student projects. This work is expensive and much would not have been undertaken without the help of particular people; Peter Jones, Brian Bourne and P. Emmott of RARDE; David Starley and Brian Guilmore from the Royal Armouries; Alan Williams, whose analysis of the *Mary Rose* sword and breastplate are included; Robert Walker and Professor Goodhew from the University of Surrey; David Peacock from the University of Southampton; and Martyn Owen from the Geology Museum; London. Of particular help was the late Ernest Pitt from Landchester Polytechnic who single-handedly undertook XRFS on the majority of metal small finds from the site. Jeol UK processed samples from several of the bronze guns. For this publication the bronze guns were resampled and interpreted by Peter Northover and these results are included as an appendix to this volume.

One of the major contributions to the understanding of early gunpowder weapons has been the wealth of form and size of wrought iron guns found on the *Mary Rose*. The chance preservation of both objects and an inventory (the *Anthony Roll*) have enabled the identification of many types of iron gun. The importance of the dated context provided by a shipwreck was demonstrated in the landmark conference held at the Tower of London in 1986, '*Guns from the Sea: Ships' Armaments in the Age of Discovery. Current Research and New Discoveries in Early Artillery*', the findings from the *Mary Rose* helped with the creation of a typology presented in Robert Smith's contribution.

Much of the interpretation of the wrought iron guns is only possible through interrogation of radiographs. The non-destructive testing facility at Portsmouth Dockyard has been fundamental to this work, undertaking literally hundreds of radiographs. Radiography of guns in the Exhibition was undertaken by Andrex UK and Quality Inspection Services Ltd. Handling these concretions is not easy, and transport often fell to the late Peter Womack.

Conservation of these objects is challenging and many items are still undergoing treatment, the results a testimony to the skills of the conservators. Many of the guns were initially worked on by a group under the direction of the late Chris O'Shea of Portsmouth City Museum Conservation Department and a team known as the 'gun crew', more recently this has devolved to the Trust under the leadership of Mark Jones.

The archery equipment from the *Mary Rose* has excited interest from the outset. As soon as the first longbow was recovered, Margaret Rule sought the advice of Robert Hardy (now a Trustee) and one of the early 'Advisory Panels' to the Mary Rose Trust was created; Robert has been an inspiration and guide ever since and has provided all of his work for inclusion within this volume. Nicholas Rodger came to the Mary Rose Trust whilst researching for '*The Safeguard of the Sea*' (Rodger 1996). At that time Andrew Fielding and I were certain that the structure of the *Mary Rose* could not accommodate guns positioned on either side of the rudder. I am now convinced that these guns must have been present. In the light of recent findings, it seems that Henry VIII's fleet was very compromised in guns which could fire ahead, again a point made by Dr Rodger. So I have much to thank him for in addition to his contribution to this volume.

Interpretation of objects and presentation of date from the recent excavations has only been achieved through the help of Martin Dean (multi-beam image), Tritech Ltd (parametric sub-bottom profiler), Justin Dix and the University of Southampton (Chirp sub-bottom profiler) with acoustic positioning supported by Sonardyne International. Interpretation of this data for presentation in this volume includes help from Kester Keighley, Stephen Foote, Colin McKewan, Peter Magowan, Nigel Boston and Peter Holt. The use of Site Recorder for the data compilation from the excavations of 2003–5 has exceeded all expectations, and some of its capabilities are illustrated in Chapter 12 and within the Site Recorder file on the DVD. The funding for this phase of the work was provided by the Ministry of Defence with special thanks due to Mike Power.

This volume would not have been possible without the advice and support of the Royal Armouries; indeed many objects would probably still be unidentified without their input. A.B.V. Norman and Howard Blackmore were supporters from the start, and Howard nurtured this link. As Deputy Master he fostered introductions to individuals who have authored many of the texts within this publication. Much of the way in which artefacts are described is fashioned on his outstanding and comprehensive publication on the Ordnance of the Tower of London. Howard also realised the importance of shipwrecks as sources for guns and sat on the Government's Advisory Committee for Historic Wreck Sites, a position which was then filled by Guy Wilson as Master of the Armouries and, between 1992 and 1996, by myself. Howard identified our Gunner's Rule (see Chapter 5) and introduced me to the Wright Manuscript in the Society of Antiquaries, a source frequently cited within this publication. Very early on he introduced me to the late Claude Blair (then Keeper of the Department of Metalwork at the Victoria and Albert Museum) who was particularly helpful regarding staff weapons, swords, armour and gun shields. Claude made me aware of the Henry VIII

Inventory; another source quoted extensively within this work. I am privileged to be a part of this inventory project, with publication of the interpretative essays imminent. Before the Mary Rose Museum was opened in 1984 we had a visit from the Royal Armouries staff when I first met many of the people who have helped with this volume and who I am honoured to consider as friends. In addition to Guy Wilson these included Graeme Rimer, Thom Richardson, Robert Smith, Ruth Brown, Nicholas Hall and John Waller. In addition to their academic input into this book, my life has certainly been richer for knowing them.

I was introduced to Professor Bert Hall through association with Bob Smith, and have benefited from many discussions regarding black powder ballistics. I met David Loades and Charles Knighton whilst working on a small essay for the publication of the *Anthony Roll*. Both agreed to submit essays. The document recently found by Charles Knighton has had a huge impact (very late on) on this volume, encapsulated in Chapter 12. They have both been incredibly helpful and continue to be so. Charles additionally underook to check a large number of references to the documentary sources cited in this voloume and I am extremely grateful that he did!

1998 saw the first of two collaborative projects between the Royal Armouries and the Mary Rose Trust. The first was to make and fire a large wrought iron port piece copied from a *Mary Rose* example and the second (in 2000/2001) was to do the same with a culverin. The results of both are fundamental to understanding the capabilities of the different guns, why they were used together and why eventually cast bronze guns became favoured over wrought iron. Different skills were required, and the knowledge that the blacksmith Chris Topp brought to the port piece project in producing a gun which could send a stone shot faster than the speed of sound is a reflection of these skills. At the inception of the project I sent a begging letter asking individuals to consider the questions they might ask of such a project, and the results were staggering. Whilst all the questions were not answered, consideration of these was built in to the project. In addition to the team comprising Guy Wilson, Robert Smith, Nicholas Hall, Graeme Rimer and John Waller, the following were particularly helpful; Douglas Armstrong, Don Hamilton, John Lightfoot, Donald Keith, Joe Simmons, Joe Guilmartin, Jeremy Green, Colin Martin and Nigel Boston. The different skills and huge amount of infrastructure required to cast bronze guns was evident when we poured the three tons of molten brass needed for the reproduction culverin at John Taylor Bellfounders in Loughborough. The skill and ingenuity in the design of the collapsible pattern by Morris Singer of Basingstoke was also evident, and necessary. Photographic logs kept by Chris Topp and Co and Morris Singer Ltd were exemplary and many of their photographs are included within this work. Rayne Foundry tested and made copies of the enigmatic hailshot pieces, also used as part of these trials.

Both large guns required carriages based on the originals. The fascinating structure of the carriages supporting bronze guns was first understood in 1984 whilst researching the original carriage for a bronze demi culverin. Colin Carpenter was instrumental in this study and has provided sound advice throughout. Nicholas Hall elegantly solved the wedging problem on the port piece carriage, and is a regular (and local) source of information and support.

All my colleagues at the Mary Rose Trust have been of great help, many have already been named above. Christopher Dobbs has been a friend and colleague since he took me on my first dive on the wreck in 1979, and has always been a huge support both personally and professionally – far more than I can even begin to acknowledge.

Peter Crossman has turned the most terrible of my sketches into presentable and understandable illustrations. Douglas McElvogue has helped with several sections of the report, including interpretation of the ship to accommodate sternchasers. John Lippiett has been encouraging and supportive, in particular in enabling the recent work on site which has had a great bearing on some conclusions presented here. Julie Gardiner, editor of this series, has shown patience and understanding and I have enjoyed working with her. I realise that her input is far greater than any contract allowed and that this has made the series a success.

The decision to add a DVD, primarily to include the archaeological drawings at a larger scale than possible on a printed page, was very last minute. The compilation of this would not have been possible without the help of Peter Crossman who rescanned the illustrations and Kester Keighley and Stephen Foote who scanned and reworked many of the photographs and videos. Peter Holt provided the Site Recorder file of the recent excavations, adding all site data we could make available at short notice. These good friends have always made time to help over many years, a belated and huge 'thank you'.

Guy Wilson has provided advice, support and friendship throughout. He has brought his scholarship to many of the sections. He has read most of the chapters and his comments and thoroughness are appreciated. The projects to reproduce and fire the guns simply would not have happened without him and the staff of the Royal Armouries. The information gained has been instrumental in understanding these guns.

Charles Knighton read the entire volume and Richard Barker much of it. I have considered all of their comments and the final text reflects this and is better for it. Charles Knighton has been particularly helpful, noting and offering alternatives for a number of inaccuracies through several drafts.

My utmost thanks are to Margaret Rule. Her inspiration, drive, skill and determination brought the

*Mary Rose* from the past into the present. The team she drew together and inspired have individually gone on to lead and inspire future generations of underwater archaeologists in a number of ways; by leading other projects, by serving on government committees which determine the future of the discipline, through holding public office in heritage agencies, and through teaching. It was a wonderful team and a fantastic adventure, it still is. On a personal level, it was Margaret who first suggested I research the guns.

**Alexzandra Hildred**
July 2010

*C.S. Knighton:* is indebted to the Most Honourable the Marquess of Salisbury for permission to publish the text of the manuscript at Hatfield House featured in Chapter 12; he thanks the Librarian at Hatfield, Mr R. Harcourt Williams, and his staff, for assistance in consulting and copying this document. He also acknowledges the privilege of working for many years in the Pepys Library at Magdalene College, Cambridge, from which his association with the *Mary Rose* developed. He is as ever grateful to Professor David Loades, his chief collaborator in the resulting ventures.

*Peter Northover:* thanks J. Spratt of the Natural History Museum, London for help with the analysis of bronze guns, and Ernst Pernicka of TU-Bergakademie Freiberg, Germany for help with the lead isotope analyses. The other analyses f were carried out in the Department of Materials, University of Oxford with the assistance of C.J. Salter. Thanks are also due to B.J. Gilmour, formerly of the Royal Amouries, for collaboration in the analysis of the samples of the guns in the Armouries collection, and to J. Riederer of the Rathgenlabor, Berlin, for unpublished data in the Berliner Datenbank.

*Thom Richardson:* thanks the late Claude Blair for reading the manuscript of his contributions and making numerous helpful suggestions, including pointing out the existence of the breastplate in Newcastle.

*Hugh Soar:* acknowledges Maggie Richards for data collection and initial study. There have been numerous correspondences over a long period regarding the wristguards. These are the names of those that have given advice as collated from the letters on file as well as those who have responded to specific questions. David Beasley, Richard Brunning, John Cherry, John Clark, Jeremy Coole, Basil Cottle, Andrew Deathe, Cannon John Halliburton, Kate Hunter, Dom Philip Jebb, Stephen Minnitt, W.F. Patterson, Mark Redknap, Veronica Soar, Paul Thompson, Abbot Watkin, and T. Woodcock

*Keith Watson:* would like to thank the staff of the Mary Rose Trust – particularly Andy Elkerton, Alexzandra Hildred, Mark Jones and Glenn McConnachie – for their support; also Chris Boyton and Hugh Soar for their expert advice so patiently and generously given.

*Guy Wilson:* would like to thank all those who have assisted in writing about the arms and armour found on the *Mary Rose*. First and foremost thanks go to Alexzandra Hildred and the team at the Mary Rose Trust who put up with persistent and often apparently inane questioning and generously showed him all he asked to see and shared so much of their knowledge. He is also grateful to the Trustees and curatorial and library staff of the Royal Armouries who allowed access and opened their records to their recently-retired colleague while they were still celebrating working life without him, and especially to Philip Abbott, Bridget Clifford, Philip Lankester and Robert Woosnam-Savage who gave up much of their time to help. In addition, despite a very heavy workload Thom Richardson spent much time advising on those parts of the armour assemblage that he was not himself writing up. Finally, he would not have got very far in trying to make sense of the fragments of sword belts without the active assistance of an old friend and colleague, John Waller, who gave generously of his compendious knowledge and made the first model to prove that our interpretation might be correct.

# Abstract

This volume presents a detailed description and assessment of the full range of armaments that were recorded on and recovered from the wreck of the *Mary Rose*. Each chapter considers the historical importance of the relevant parts of the assemblage and the relationship between the written records (mostly drawn from inventory data) and the archaeological evidence. Some categories of object are described generically with specific items chosen for fuller description and illustration to indicate the characteristics, range and peculiarities that have been observed, but many categories (especially the guns), because of their rare survival, are described and illustrated individually.

In Chapter 1, an introduction to Henry VIII's navy is provided and the value of the inventory of the King's ships, known as the *Anthony Roll*, as a historical document assessed. The service career of the ship, and her loss, are reviewed together with major changes that were made to her structure and armaments.

Above all else the *Mary Rose* was a gunship – one of the first to carry heavy guns on multiple decks. The second part of the volume therefore begins with descriptions of the guns, including experiments in replication and firing, and an assessment of missing items – some of which are known to have been salvaged in the sixteenth and nineteenth centuries (Chapters 2–4). In Chapter 2 the muzzle-loading bronze guns are described and illustrated individually, beginning with the largest. By comparison with the iron guns, each is exceptionally well preserved with fine details of mouldings, decorated lifting lugs, inscriptions and devices providing invaluable historical evidence. In some cases the gunfounder and date of casting are known, enabling a discussion of the provisioning of the ship with bronze ordnance. A mould was made from one of the culverins and a new gun cast using methods as close the original as possible. Some guns were found still mounted on, or associated with broken fragments of, their four-wheeled carriages, allowing for their reassembly and for replication; the process of which is described. Spare carriage parts, particularly wheels, can be seen to have been kept in storage. Firing trials using the replica culverin are discussed.

Chapter 3 considers the large assemblage of breech-loading wrought iron guns. The condition of many of these is poor (and some are still in concretion or undergoing conservation) but it is clear from comparison of the assemblage with historical evidence that a wide range of guns is present, including some types that have not previously been recognised. The use of radiography has provided much information on the structure and form of badly corroded objects confirming that most were constructed from a tube of iron staves bound by abutting bands and hoops. Each gun was probably provisioned with more than one breech chamber, of which several loaded examples were recovered. Some guns were carriage-mounted, the chambers and barrels resting in troughs cut from solid blocks of wood (usually elm). These carriages were generally of 'sled' construction with two wheels, the rear of the carriage being supported by a perforated wooden peg or elevating post. Lighter 'swivel guns' were designed to be elevated, loaded and fired by one individual. They were located by means of a swivel yoke and peg into a hole in a horizontal timber or the sill of a port at a convenient height above deck to be operated while standing. One of the iron port pieces and its carriage were reconstructed and firing trials conducted.

A small number of cast iron guns was recovered. These small, individually operated pieces are described in Chapter 4. The use of cast iron to manufacture a large number of identical guns is evident in the *Mary Rose* assemblage. These guns have been identified as 'hailshot pieces' because their projectiles, dice of iron fired through a rectangular bore, caused a devastating scatter at short range. They were anti-personel guns used on the upper decks of ships, as indicated by their locations in relation to the structure of the *Mary Rose*. At least one example was found fully loaded. A replica of it was made and firing trials conducted. Although numerically common in the inventories of the period, only four examples were recovered.

Chapter 5 begins with a detailed description and discussion of the variety of shot recovered from the wreck. Round iron shot is abundant as is stone shot, a considerable proportion of which is unfinished. Lead shot for handguns is present as well as a range of composite shot including iron dice with lead cores. Canister shot comprising wooden cylinders filled with shattered flint provided a lethal form of shrapnel. The relative importance of each type within the assemblage, its method of manufacture (including experimental work), size range, location on the ship and relation to the bore sizes of the guns is considered.

An enormous range of other items is associated with the operation of the guns. The assemblage includes much of the apparatus described in historical works of the time and in inventories as being required for loading and firing. It has been possible to reconstruct the operational 'toolkit', sequence of loading, gun drill and probable size of gun crews. Artefacts include ladles, rammers, tampions, priming flasks and wires, and a fine collection of elaborately carved linstocks used for holding slow matches that actually fired the guns. The process of selecting the correct shot to suit individual guns is illustrated by the presence of shot gauges and by the preservation of a gunner's rule – a device used to calculate the weight of shot of a specific material needed in preparing the appropriate weight of powder. The function and use of this extremely rare object is discussed in detail. Although gunpowder has not survived it has been possible, from historical records, to assess the likely type, composition and quantity of gunpowder needed and to speculate on the location of stores. The chapter concludes with a brief description of a variety of more general items present on the ship, examples of which would have been associated with the operation of the guns, such as buckets, , metal spikes, wooden tubs and wedges.

Historical evidence, including inventories of the time, indicate that the ship was provisioned with a large number of incendiary devices of various kinds, though few have been positively identified (Chapter 6). At least three incendiary arrows are present. These comprise a heavy wooden shaft and flight (the iron 'arrowhead' not having survived) to which was nailed a linen bag filled with an incendiary mix and covered with pitch. Fragments of mortar bombs of cast iron found with a tapering hole to take a fuse plug provide evidence of nineteenth century salvaging.

In addition to the big guns, the *Mary Rose* was also issued with a number of handguns – harquebuses – of which little survives other than fragments of the stocks and butts (Chapter 7). Nevertheless it is possible to identify a number of different known types of weapon. Few military firearms survive from the early sixteenth century, a period of transition in the development of the construction, use, and standardisation of firearms. Therefore, the remains of any standard munitions weapons are of great importance. The collection as a whole offers some intriguing and important details on the construction, form, and type of firearms being used on board a mid-sixteenth century naval vessel. The ship also carried 'gun-shields', or targets: an unusual and unwieldy firearm comprising a circular shield made from two layers of wooden laths fastened perpendicular to each other and covered with steel plates, fitted with a small breech-loading gun with a matchlock ignition system and separate chamber, firing through a hole in the centre of the shield. Although a few elaborately decorated and probably ceremonial examples are known in various armouries this is the first record of an assemblage deployed for warfare. A number of the gun-shields were found together in storage.

The archery assemblage from the *Mary Rose* is unprecedented in its size and preservation and is described in detail in Chapter 8. A total of 172 single-piece longbows have been recorded in detail and fragments of more than 3700 arrows, including large numbers of each contained in specially made wooden chests. Measurements of the longbows, including length, width and depth along the limbs, have provided information on shape and form not available elsewhere. Microscopic study of broken bows has enabled a study of the growth rings, offering some indications of the environmental conditions needed for growth. Analysis of the relative density of the bows has provided further information on the nature and quality of the wood. These studies have enabled interpretations of inherent strength/flexibility and also of the amount of physical force required to draw back the bowstring and the weight held by the archer momentarily before releasing the arrow (the draw-weight). These reflect the skill, stature and strength of the archer. Details of recorded bowmarks are presented and illustrated.

Measurement of the arrow tips has enabled suggestions on head form, and the length and diameter of the arrow nock slots and of the few horn inserts has also fuelled discussions as to possible string diameters and their potential strength (another way of suggesting the draw-weight of the bow). Analysis of the arrows has allowed identification of some of the wood and feathers used, together with the glue and binding and even some of the preparatory methods employed during manufacture.

The sheer number of items has huge implications for understanding the techniques of manufacture and the ability to produce large numbers of hand-crafted weapons and projectiles. Items missing from the assemblage include the metal heads of the arrows and the majority of bowstrings.

A fine assemblage of leather (and one ivory) archers' wristguards has been preserved. Many of these are decorated with royal emblems, religious symbols, symbols suggesting association with guilds and some that appear to be reused book covers. In addition, arrow bags and spacers and possible handgun bolts have been identified. The chapter concludes with a discussion on the posible identification of archers and their personal belongings based on pathological evidence from skeletons, objects associated with human remains, and the contents of sea-chests.

Chapter 9 considers the staff and edged weapons recovered from the wreck. In most cases these objects are repesented by their wooden (handles, shafts), copper alloy (hilts, pommels, chapes, decorative elements) and leather (straps, belts, sheaths) elements with comparatively poor survival of metal parts. The staff weapons include pikes and bills and at least one, possibly ceremonial, halberd, while the edged weapons include swords, many daggers, bi-knives and a fine if fragmentary collection of sheaths, scabbards and associated straps and fittings. The sixteenth century saw an explosion in the variety of swords and daggers being made, supporting a bewildering number of different hilt forms, and used both by civilians and the military. The chance survival of a basket hilted sword and of several other hilts provide rare dated examples of specific forms. The use of radiography has greatly enhanced understanding of many of the forms present.

Given the comparatively poor preservation of iron objects on the *Mary Rose* it is perhaps not surprising that the assemblage of possible armour is rather small. Example of sixteenth century armour survive in many collections – though these are often ceremonial items – and contemporary engravings and painting depict soldiers wearing body armour and helmets. Armour is, however, rarely mentioned in inventories and it is not know if it was issued to men on the ship. A discussion of the possible forms of armour that should have been present is followed by description and metallurgical analysis of the few surviving remains of breastplate and buckler. Possible indication of armour are provided by a range of straps, buckles and fastenings and there are a number of fragments of mail made from leaded brass.

Chapter 11 draws together all the evidence presented in the earlier chapters to present a detailed consideration of the ship as a fighting unit. The chapter is introduced with a brief discussion of known naval tactics of the time. The evidence for changes in the armaments on board the *Mary Rose* is discussed, first through analysis of the inventory evidence and then in terms of the archaeological evidence. The operation of the ship on the day of her sinking, the theatre of war and conduct of the Battle of the Solent are examined using geographical evidence provided by the Cowdray Engraving and the application of GIS, which indicate a number of inter-related factors that could have contributed to her demise.

The operation of the ship is examined through a detailed description of the location of armaments and related items on each of the decks. The probable configuration of gun deployment and possible problems encountered in their operation resulting from restrictions of space and peculiarities of hull structure are considered. The question of whether the *Mary Rose* may have been overloaded with ordnance is addressed and attempts made to define the range and arcs of fire of the guns, their relative functions (anti-ship, anti-personnel, etc) and the sustainability of engagement. The chapter concludes with a consideration of what the ammunition may tell us about the rules of engagement.

The final chapter (Chapter 12) opens with a summary of the key results of the artefact anlalyses presented in the preceding chapters and suggests important avenues for continued research. Regular monitoring of the *Mary Rose* wreck has continued and proposals to improve the deep water channel approach to Portsmouth harbour led to Ministry of Defence funding for a defined programme of site clearance and investigation in 2003–5. These works included the sampling of a percentage of the silts excavated out of the hull in 1979–82 that had been redeposited by tidal action as linear mounds on either side of the hull depression. The methods employed in this work, the recording system and interim results are presented with newly recovered objects listed and illustrated.

In a postscript, a newly discovered and transcribed document is presented which provides important corroborative information to the archaeological evidence. It shows that the King had ordered three ships, including the *Mary Rose*, to carry extra guns forward, and these were to be canted as finely as possible to match the firing capacity of the French galleys while presenting the narrowest target. It appears that anything, even major structural alteration, was being considered to achieve the desired effect. In order to place the guns slant-wise the shipwrights would have to remove knees or even masts; they told the King sensibly but bravely that if they followed his commands they would be weakening the very structure of his ships.

The volume concludes with appendices on the metallurgy of the guns and detailed tables relating to gunpowder, and is complemented by a DVD containing a selection of artefact drawings at their original scale and files containing groups of colour images. It contains some datasets; including a PDF version of the artefact database and the Site Recorder file which shows the distribution of all trenches, the ship, data and location of artefacts and structure recovered between 2003 and 2005. Several short videos have been added, including the manufacture and firing of the guns and the work undertaken onsite between 2003 and 2005.

# Preface

Sometime on the afternoon of 19 July 1545, the *Mary Rose*, the second largest of King Henry VIII's great warships, sank. As the King watched from his encampment at Southsea Common, he could not have realised what an unparalleled insight into his life and times this catastrophe would ensure. She was both a living community and a state-of-the-art fighting machine from a period of great change in both the design and structure of ships and of guns. She came to rest at an angle at the bottom of the Solent where she became partly buried in sediment. Her exposed parts gradually eroded away leaving something less than half of the entire hull to be recovered so memorably in 1982, following years of painstaking excavation. It is the purpose of this series of volumes on the archaeology of the *Mary Rose* to attempt to understand and reconstruct (on paper at least) her form, construction, layout and contents and to provide an insight into how she was used and operated.

To that end, four of the five volumes are concerned with the thematic description and interpretation of specific component parts of the vessel and her astonishing assemblage of recovered objects, in order to try and piece together many aspects of the ship as a complete, 'living' entity. The fifth volume is concerned with the conservation programme from the ship, describing and explaining the techniques used (and in some cases specifically devised) to conserve and display the hull itself and the myriad objects and many materials recovered. This volume is concerned with the armament of the ship – the very reason for her creation. From bronze guns to longbows, and including all the subsidiary equipment necessary to sustain a battle at sea, this volume aims to present all of the weapons carried and to interpret them as best as possible to try and gain an understanding of how she may have fought.

The four other volumes in the series *Archaeology of the Mary Rose* are:

*Volume 1. Sealed by Time: the loss and recovery of the* Mary Rose, by Peter Marsden (2003)
*Volume 2.* Mary Rose: *Your Noblest Shippe. Anatomy of a Tudor Warship*, edited by Peter Marsden (2008)
*Volume 4. Before the Mast: life and death aboard the Mary Rose*, edited by Julie Gardiner with Michael. J. Allen (2005)
*Volume 5. For Future Generations: conservation of a Tudor Maritime Collection*, edited by Mark Jones (2003)

Throughout this volume they are referred to by the abbreviation *AMR*.

The *Mary Rose* was essentially four-masted carrack modified to include several gundecks and high castles. She is the earliest surviving example of a ship with watertight gunports close to the waterline, a technological breakthrough that was to determine the nature of warfare at sea for the next three centuries. The only surviving illustration of her is from an inventory known as the *Anthony Roll*, of 1546, though the archaeological evidence suggests that there is some artistic licence involved in this depiction, as will be discussed in this volume. Figures i and ii indicate the main parts of the ship and the partially conjectured areas of stowage, which give some indication of the use of the vessel and are provided in order to orientate the reader.

Internally, the lowest part of the ship was the *Hold*. This held the many tons of gravel ballast that helped to stabilise the vessel. It was also a storage area. The kitchen or galley was situated in the Hold and objects associated with the cooking, serving and eating and storage of food were recovered. Both the main mast and the pump well were in the Hold.

Above the Hold was the *Orlop* deck, primarily a storage area. This included a rigging store, two cable tiers, a storage area for tampions, spare axles and wheels for gun carriages and probably the gunpowder store. In the stern there was a lantern store, and specific areas containing gun-shields and chests of longbows and arrows. Casks of food and equipment and some personal chests were located on this deck.

The *Main* deck above was the principal gundeck, supporting a large complement of heavy, carriage-mounted bronze and iron guns firing through lidded gunports. Two major shot lockers were found on the Main deck, both situated close to companionways to the deck above. Although primarily for fighting, this deck also acted as a living space. There were at least four cabins on the surviving starboard side: one for the Barber-surgeon with a second, adjacent cabin; one for the ship's carpenters; and one possibly for the ship's navigators. Similar cabins were probably situated on the, now destroyed, port side of the ship. The majority of gun furniture was recovered fron this deck.

The *Upper* deck was also a fighting area. At bow and stern the *Fore-* and *Sterncastles* extended above the Upper deck, with an open waist between. Most of the forward part of the deck was destroyed in the sinking, as was the Forecastle itself. The central, open, area was originally covered by horizontal anti-boarding netting.

On the Upper deck aft, under the Sterncastle, archery and hand-to hand fighting equipment was recovered, including chests of longbows and arrows and staff weapons such as pikes and bills, together with handguns and remains of swords and scabbards. Although this deck would have supported a number of large guns, many were absent, salvaged either in the sixteenth or nineteenth century. It is likely that the capstan and anchor handling equipment were located in the stern. The many personal and 'everyday' items recovered suggest that this deck may have provided cramped living accommodation for the officers. Very little of the deck of the Sterncastle survived but the one gun found, a bronze gun firing forward from the front of the Sterncastle, is the most ornate and controversial gun recovered.

Unfortunately none of the masts of the *Mary Rose* survives today, having probably been at least partly recovered in salvage operations during the sixteenth century. The location of the base of the main mast survives in the Hold but the position of the other three (the foremast, probably rising through the Forecastle, and the Mizzen and Bonaventure masts rising through the Sterncastle), can only be estimated from particular features of some of the surviving deck timbers. The masts may not survive but many items of standing and running rigging have done and these are enabling experts to reconstruct the rigging pattern of the ship (see *AMR* Vol. 2).

The excavation of the ship is discussed in some detail in the *Archaeology of the Mary Rose (AMR)* Vol. 1, chapter 4 (Marsden 2003) and only a summary is presented here in order to explain the nomenclature used for the various parts of the ship and the numbering system adopted for the objects recovered.

In the early years of the excavation (1972–6) the strategy was to cut exploratory trenches to answer specific questions and, until the nature and general condition of the wreck was better understood, these were confined to areas outboard. In 1977 a trench was excavated at the bow on the port side. This confirmed that the ship had not completely broken up, and established the angle of heel of 60° to starboard and the true extent of the surviving structure. The first trench excavated across the hull from port to starboard was excavated in 1978 and, although rather less than half of the hull survived, it established that the structure was coherent rather than a collection of fractured sections. From 1979 onwards the strategy was to replace individual trenches with a more open-area excavation (Fig. iii) and initially remove the modern objects and deposits. In 1980 twelve parallel trenches were established, each about 3m wide, whose sides were determined by the position of deck beams, crossing the ship from east to west, and extending far enough outboard to include the areas of scour around the hull.

Within the ship, the sides of the trenches were made to coincide with the exposed ends of the major deck beams, and as each trench was dug, the sloping timbers of each deck were found crossing it. This enabled the finds from each deck and trench to be located in a Sector. Trench 3, for example, crossed the Hold and that sector was designated H3. Where it crossed between the Orlop and Main decks, the sector was designated O3. Thereafter it crossed between the Main and Upper decks, the sector being designated M3, and so on for the Upper deck (Fig. iv).

The deposits within the hull and in the surrounding scourpits (the result of the scouring action of the water currents around the wreck) were recorded stratigraphically from top to bottom of the excavated sectors, as on a dryland excavation, with each new layer of deposit encountered given a context number (see *AMR* vol. 1, chapter 9 for an explanation and Chapter 12, below). Within the hull individual deposits were also given context numbers. Feature numbers (F1, F2, etc) were assigned later as an aid to studying particular groups of material found together. In some cases the feature was an individual item, such as a chest, basket or cask, that was to be removed intact with its contents (see below).

An area excavated under the hull across the ship at the bow was termed the *Bow Bulkhead*. As well as containing the bow anchors, this trench yielded a number of concretions of guns and shot probably derived from the Forecastle. In 2003 the stem was identified just forward of this area and, in 2004, timbers from the port side of the vessel at the bow were located. The stem was raised in 2005.

The positions of and relationships between elements of the hull and the objects found within it would be crucial to the interpretation of the remains in terms of the layout and use of the ship. A direct survey method (DSM) was adopted for the three-dimensional recording of objects, whereby a series of datum points was established and the position of key items to be surveyed were calculated by measuring the distance from them to four known datum points. This gave four results, one for each combination of three datums, and these could be averaged, and the disparity between the answers used to give an indication of the quality of the survey (Rule 1989; 1995; Adams and Rule 1990). The distances of other items from the DSM surveyed objects were measured by triangulation or straightforward tape measurement. Theoretically, therefore, the position of every object would be known in three dimensions. Unfortunately, a lack of computer facilities and funding has meant that the full DSM results have still not been processed and the finds plotted accurately.

Objects were recovered in a variety of ways and given finds numbers immediately. The divelogs were then annotated with those numbers. The divelogs list objects and frequently include sketches showing their relative positions (see, for instance, Fig. 2.48, below). Large objects, such as guns, were craned aboard by moving the dive support vessel over the item (eg, Fig. 2.42 below). These lifts required detailed planning and were scheduled to make best use of tides and had to be factored into the overall work programme. In order to minimise damage, an object was often recovered within a bespoke container with a quantity of its surrounding sediment, which might then be assigned a sample number and processed for environmental material (see

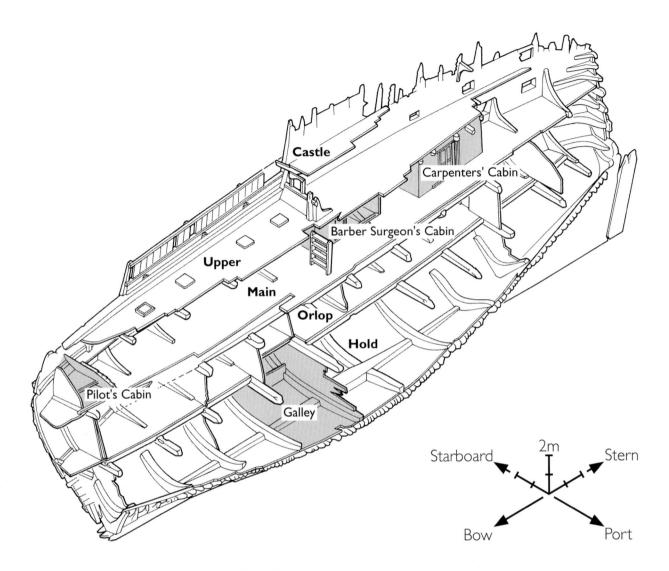

*Figure i Isometric showing the structure of the ship*

*AMR* Vol. 4, chap. 12). Many chests were also raised in this way for careful examination onshore (eg, Fig 8.47, below). Shorelogs recorded the process of excavation and preliminasry identification of objects. some sediment samples were kept for analysis. Blocks of sediment were asigned sample numbers (see below), lifted and processed ashore. Any items found therein were given appropriate artefact or bone numbers, as described below.

Large objects, especially timbers and chests, were frequently much more fragile than they appeared and could not be lifted directly into air as they were in danger of collapsing under their own weight. These were lifted by placing them in polythene lined crates lowered into the water next to them. The polythene was wrapped over and the lid secured, and the whole package was hoisted on board the recovery ship Sleipner by crane. Many chests were raised with their contents intact and these were carefully excavated onshore. Shorelogs recorded the process of excavation and preliminary identification of objects found inside.

Some sediment samples were kept for analysis.

Every object, timber, and 'sample' was given a unique, three-part alpha-numeric identifier consisting of the year of recording (which was usually the year of excavation, either from the seabed, onshore or from within the hull after it was raised), a letter to indicate the basic category of material, and a four digit sequential number (eg 81A1234; 82S2143). Numbering began again at 0001 each year. The letters used were:

A = artefact
H = human bone
S = sample
T = timber

Initially, as there were thousands of animal bone fragments and much of this material was recovered *en masse* or from within casks, groups of bones were given 'S' numbers. In should be noted that the term 'Sample' was a generic term used for non-artefact material rather

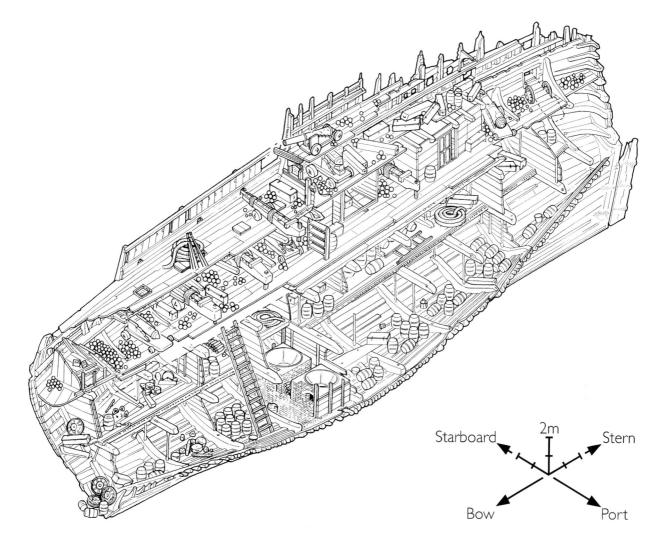

*Figure ii Isometric showing guns and principal stowage areas*

than used in the stricter sense of material obtained for specific environmental analyses familiar from dryland excavations. Human bones tended to occur in groups and so the H number could refer to a single bone or to a bone group. Detailed analysis of the human remains identified nearly 100 individuals who were represented by skulls and a reasonable portion of the overall skeleton. These were designated Fairly Complete Skeletons (FCS). Each FCS could comprise several H numbers.

**A note on spelling, scales and distribution plots**
Throughout this volume, modern spellings are used for ship names, etc, where a modern equivalent is clear and/or unambiguous, otherwise the original spelling is used. This applies also to names referred to in the various inventories that are cited throughout the text. Original spellings are used in direct quotations unless specifically indicated otherwise.

In keeping with the rest of the series, metric units and weights are used when presenting data on an object. As all the guns, shot and associated gun furniture would have been originally listed either in inches or feet, and weight given in pounds and hundredweight (cwt), in some cases imperial measurements are used to describe these. In many cases both imperial and metric units are given, dependent on whether the context is historically or archaeologically biased.

Every endeavour has been made to present illustrations of objects at sensible, recognisable scales (1:2; 1:10; 1:20 etc) and to illustrate similar objects at a consistent scale. This has not always been possible within the printed volume but the variety of scales used has been kept to an absolute minimum so that like can most easily be compared with like. In the case of guns, a scale of 4:5 has been used in many cases in order that the illustration will fit 'portrait' across the page. Deck plans and elevations are presented at 1:100 according to conventions established in *AMR* Vol. 2 for consistency with that volume. A 'flattened' grid pattern is used to show the distributions of objects where there are too

**TRENCH GRIDS 1975-8**

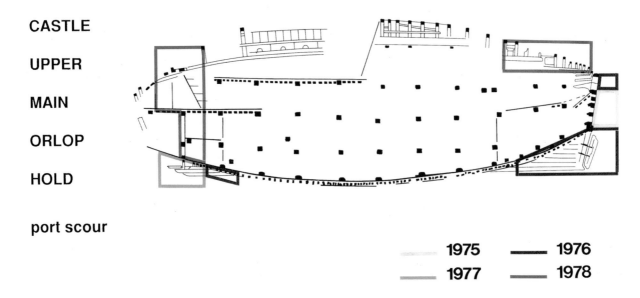

**TRENCH GRIDS 1979**

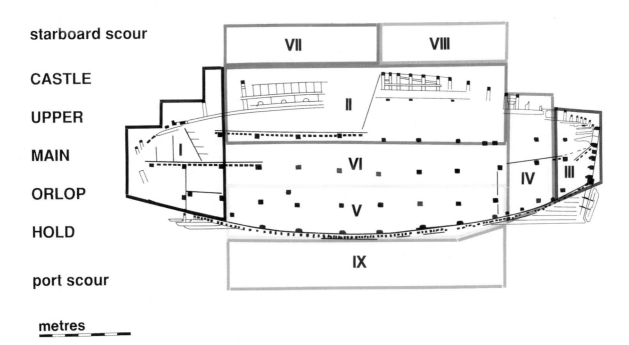

*Figure iii  Initial trench plans*

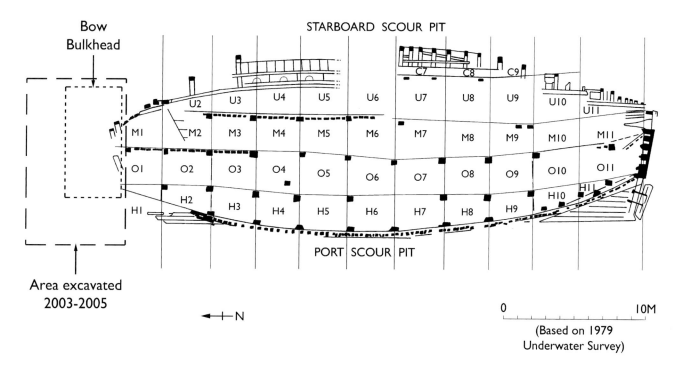

*Figure iv* Trench plan from 1979 showing areas excavated in 2003–2005

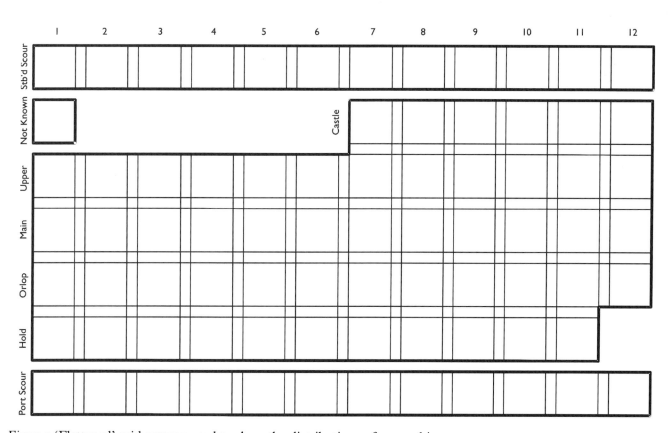

*Figure v* 'Flattened' grid pattern used to show the distributions of some objects

many, or the distribution is too complicated, to display clearly on an isometric (Fig. v).

It is recognised that illustrations of some objects (notably the guns) and the deck plans are, of necessity, rather smaller than is desirable. To compensate for this we have reproduced a selection of artefact drawings at their orginal scale and included these on the accompanying DVD. As space allows, we have also included files contaning groups of colour images and some datasets, including the Excel spreadsheets of all the ordnance and of the recent longbow study. None of these datasets has been reworked for this publication and are, therefore, not up to date with most recent reseach, which is, as ever, ongoing.

# 1. The King's 'Great Ship': the *Mary Rose* and her Naval Background

This volume, the third in the series of five recounting the archaeology of the *Mary Rose*, is devoted to consideration of the ordnance, munitions and equipment for war – the *raison d'être* for the building of the ship. As with the personal and 'everyday' objects that survived on the wreck, reported in *AMR* Vol. 4, the assemblage described here is of unprecedented size and importance to our understanding of Tudor warfare and life at sea in the sixteenth century, above and beyond the immediate, intrinsic interest of the objects themselves. Study of the armaments leads us down many paths, providing insights or prompting questions into such diverse topics as technological developments in gunfounding, changes in the manufacture and composition of gunpowder, organisation of naval supplies and battle tactics, as well as into technical details such as the operation of a gunport lid from outside of a moving ship or reloading a muzzle-loading gun with apparently insufficient room to bring it inboard.

In the chapters that follow we will begin with the description of the guns, including experiments in replication and firing (Chapters 2–4), followed by discussion of the many objects that relate to their use: the shot; the gunpowder and the items needed for loading and firing; and a discussion on gundrill (Chapter 5). This is followed by chapters dealing with other ordnance: incendiaries; handguns; and archery equipment (Chapters 6–8). The staff weapons, edged weapons and armour represent not only a precisely dated and extremely large assemblage, but one which, unusually, contains many fundamentally mundane or ordinary weapons (Chapters 9–10). The volume concludes by drawing together all the evidence to present a detailed consideration of the ship as a fighting unit (Chapter 11) and an indication of some of the major topics that still require research (Chapter 12). Each chapter considers the historical importance of the relevant parts of the assemblage and the relationship between the written records (mostly drawn from inventory data) and the archaeological evidence. Some categories of object are described generically with specific items chosen for fuller description and illustration to indicate the characteristics, range and peculiarities that have been observed, but many categories (especially the guns), because of their rare survival, are described and illustrated individually.

Like the others in this series, this volume is more than a catalogue; it represents the sum of our knowledge to date of the *Mary Rose* as a fighting vessel and a self-contained working ship and it should be read in conjunction with the other volumes. Much research continues to be done, including experimental work, computer simulation and the building and testing of objects such as guns and longbows. In the case of the latter, such work has been ongoing almost since the first longbow was recovered in 1979. Thirty years later it is still in progress, attesting to the passion and thirst for knowledge that the *Mary Rose* and her contents continue to arouse and as a testament to the overwhelming wealth of this collection.

## Overview of Henry VIII's Navy
by David Loades

Upon his accession in April 1509 Henry inherited five ships, two of which, the *Regent* and the *Sovereign*, were large warships built nearly 30 years before. His legacy included a well-developed dockyard at Portsmouth, a basic facility at Woolwich and a Clerk of the Ships, Robert Brigandine, who served from 1495 to 1523 and whose early accounts show expenditure on the *Regent* and the *Sovereign* (Oppenheim 1896a). When Henry died in 1547 he bequeathed to his successor a warfleet of over 50 vessels, 20 classed as 'Great Ships'; an Admiralty with five full-time professional officers; and three fully-developed dockyards – Woolwich, Deptford and Portsmouth (Loades 1992, Chapter 4). Naval expenditure increased accordingly. In the year which saw *Mary Rose* being fitted out (1511–12) Brigandine's total receipts amounted to around £2500. By 1547, when there was no major construction programme but with each dockyard having clerks and master craftsmen in full-time employment in addition to those who served the central Admiralty, the Treasurer's charge was above £40,000 (*LP* 1, 1393, 3608; Dasent 1890–1907, 60 x 151 [occasional warrants]; Loades 1992, 149).

The main reason for this expansion was the King's need to be prepared for war. Three times, in 1512, 1522 and 1544, he deliberately launched attacks against France and, in 1542, he provoked war with Scotland. In 1513 the Scots attacked him, and throughout the period of domestic crisis produced by his '*Great Matter*'

(1527–1540), he felt himself to be in danger from Charles V, Holy Roman Emperor and King of Spain. The Emperor, who was Catherine of Aragon's nephew, had blocked all Henry's attempts to secure an annulment of his marriage and was deeply offended by Henry's solution (Scarisbrick 1997, chaps 6 and 11). So acute was Henry's fear in 1539 that he fortified the south coast and ordered a major mobilisation both by land and sea (Colvin 1963–82, 4, 367–95). These alarms and excursions kept England in a high state of military preparedness, to which should be added Henry's own perception of himself as a warrior king, whose honour required strength at sea. It was customary for a king going to war to build and purchase ships for that purpose and Henry acquired about 25 between 1511 and 1514. It was also customary for kings to sell or dispose of these expensive machines as soon as the war was over, but Henry did not. In 1514 he mothballed a large proportion of his fleet but kept a selection of smaller vessels in commission for escort ('*wafting*') and patrol duties (Knighton and Loades 2000, 113–58). During the war he appointed a new officer, the Clerk Controller, and established a store house under its own Keeper at Erith. The duties of the Clerk of the Ships thereafter were largely confined to Portsmouth (Loades 1992, 74).

By the time of the second French war (1522–5), Henry had about 35 vessels, four shore establishments (Deptford, Woolwich, Erith and Portsmouth), and three officers. There was, however, no departmental structure and no budget. The officers accounted separately, drawing their money from the Exchequer by occasional warrant, and took their orders first from Wolsey, and then (after 1532) from Thomas Cromwell.

Naval administration was reorganised after 1545. This may have been because of the fall of Cromwell (1540), or due to the appointment of John Dudley, Viscount Lisle, as Lord Admiral (1543). Alternatively, it may have been the death in 1544 of William Gonson, long-serving and trusted Keeper of the Erith storehouse, who had eventually handled the bulk of the finances as Paymaster of the Navy. When Gonson died, a new structure was created on the lines of the recently reorganised Ordnance Office. This new administration, called the '*Council for Marine Causes*' was a department of state, responsible to the Lord Admiral. It consisted of five officers, under the presidency of the Lieutenant (or Vice-Admiral), with a clearly laid out command structure, substantial professional salaries and generous expenses (Davies 1965). This council was made responsible for maintaining and servicing the fleet, except victualling, which continued to be dealt with by private contract. It was not responsible for operations, which remained the preserve of the Lord Admiral, nor did it have a fixed budget, although the new Treasurer of the Navy handled all accounts.

By 1547, the English Crown had the most effective admiralty organisation in Europe together with a substantial navy, well equipped with up to date weapons. Warships of 1543 carried a greater number of big guns than those of 1512. English ships in the Third French War had no particular technological advantages over their opponents, but were of equal size, firepower and number, to those of France; and they were quicker and cheaper to mobilise.

Operationally, Henry VIII's navy was successful, but not overwhelmingly so. It took the offensive in 1512, and had the better of some inconclusive fighting, but in 1513 Sir Edward Howard failed in an attack on the French galleys in Britanny, and was killed in action (Spont 1897, 132–40). Throughout the Second, and most of the Third French War the fleet raided, escorted troop carriers and supported coastal sieges. There was fighting at sea, in which it acquitted itself well, but no major battle. Then in 1545, Charles V having withdrawn from the war, Francis I saw his chance to launch a major invasion of the south of England.

## The *Anthony Roll*
by Alexzandra Hildred and C.S. Knighton

The only contemporary illustration and most familiar image of the *Mary Rose* comes from the illuminated ordnance inventory of 1546, known as and hereafter referred to as the *Anthony Roll*. It shows, albeit with a certain amount of artistic licence, a purpose-built ship of war, multi-decked, fully rigged with four masts, and with guns protruding through lidded and unlidded ports (Fig. 1.1). For all its limitations, the *Anthony Roll* image has always been of fundamental importance in the interpretation of the *Mary Rose*, and new evidence suggests that it may in some respects be more accurate than we had supposed (Chapter 12). Each of the 58 illustrations in the *Roll* is accompanied by a list of '*Ordnance, Artillery* [and] *Habilliments for the War*'. A warship has to be a self-contained fighting unit in which the guns, powder, ammunition and ancillary equipment function together, with ample provision of spares. The inventory allows us to estimate the relative firing frequencies of each type of gun, and so helps us to make suggestions about the nature and endurance of any engagement (Chapter 11).

### *The manuscript*
by C.S. Knighton

The *Roll* is named after its compiler Anthony Anthony. It originally comprised three vellum rolls 0.7m wide and in total at least 15.14m long. The first Roll (with *Mary Rose* in second position) and the third now form volume 2991 in the Pepys Library at Magdalene College, Cambridge; the second roll, which remains in that format, is British Library Additional MS 22047. The illustrations and text are reassembled in their

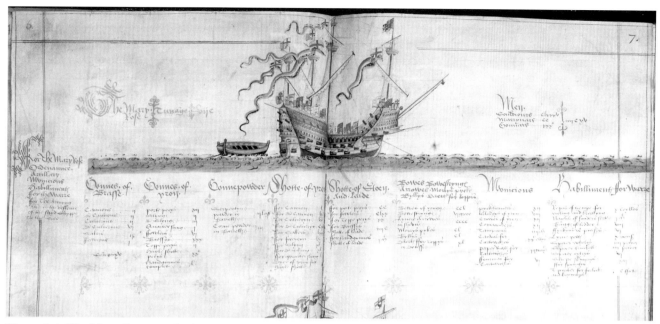

*Figure 1.1 The* Mary Rose *as she is depicted in the* Anthony Roll *with inventory (© Pepys Library, Magdalene College, Cambridge)*

original order in the Navy Records Society edition (Knighton and Loades 2000).

When he completed the manuscript Anthony was Clerk of Deliveries in the Ordnance Office. He went on to become Surveyor of the Ordnance from 1549 to his death in 1563. He also traded in beer and other commodities, as his Dutch-born father William had done. In 1530 Anthony had been appointed a groom of the Chamber and, in 1533, a gunner at the Tower. He seems to have entered the Ordnance Office as a clerk in 1537, when Henry Johnson (a business partner of his father) became Surveyor. In the same year Anthony became one of the founding masters of the Honourable Artillery Company. He clearly had administrative competence, intellectual interests and artistic skill. Among other writings he compiled a chronicle. This mentions Henry VIII and Charles V going aboard the *Mary Rose* at Dover on 30 May 1522, which Anthony perhaps witnessed (*LMR*, 46).

A theory that Anthony was the brother of the Dutch artist Cornelis Anthoniszoon can no longer be accepted, because William Anthony's will names Anthony as his only son. But it is evident that, before compiling the *Roll*, Anthony had produced several pictorial maps ('plats'), including the representation of a French raid on Brighton dated July 1545 (Payne 2000, 20-3). Anthony played a part in that summer's campaign; on 21 July, two days after the loss of the *Mary Rose*, he had an urgent message to send to Portsmouth 500 bows, 1000 sheaves of arrows, 1000 bills, 4 lasts of serpentine powder, 2 sakers, 6 falcons and falconets. Two days later came an order for all the cannon, demi cannon, culverins and demi culverins that remained in the Tower (Dasent 1890–1907, 1, 213, 215). It must have been in the following months that Anthony began his illuminated inventory; perhaps, like the Brighton *plat*, to celebrate the thwarting of the French invasion. The inclusion of the *Mary Rose* might seem tactless; in fact there was still some hope that she would be salvaged. It is impossible to say if the work was commissioned by Henry, or if Anthony submitted it on his own initiative. There is nothing quite like it; continental armouries produced illuminated inventories (Grancsay 1966, 2, 5), and it is not impossible that Anthony was familiar with the genre. But these, however elaborate, remained working catalogues, whereas Anthony's manuscript can only have been intended for display. Nevertheless Anthony's professionalism ensured that his inventories were real enough. The accuracy of his illustrations is less certain; those of the *Mary Rose* and the other Great Ships appear to be individual portraits, but this can not have been true of the entire fleet. It is also evident that Anthony's enthusiasm for his project waned as he neared the end, and the execution becomes less careful.

In 1680 the first roll was given by Charles II to Samuel Pepys, perhaps as compensation for the loss of his Admiralty secretaryship in the previous year. Since his enforced resignation Pepys had been collecting material for a history of the navy, and he knew of the '*long Parchment Roll in his Majesty's hands*' which portrayed Henry VIII's fleet (Knighton and Wilson 2001, 363). Perhaps he was invited to chose something from the Royal Collection, and had selected this. Pepys claims that the King's gift included the third roll, but it now appears that he simply stole this during the Revolution crisis of 1688 (Knighton 2004, 144). Soon afterwards the middle roll, which had been missing, surfaced at St James's Palace. Pepys could expect no favours from William III, but he was friendly with the King's Librarian, and he acquired an abstract of the middle roll and copies of some of its illustrations. The first and third rolls he then had cut and handsomely bound as a volume; each page spread thus created shows one or two ship illustrations with the corresponding inventory text below. In most cases the ships appear left and right; but unfortunately the *Mary Rose*, the *Peter Pomegranate* below, and the *Henry Grace à Dieu* before them, which were drawn in the centre of the roll, have been bisected by the gathering, and have suffered further deterioration in consequence (Knighton 1981, 37–8, 174–6).

By the terms of Pepys's will his entire library passed to his old college of Magdalene at Cambridge in 1724. Parts of Anthony's manuscript were first copied and published in 1782 (Topham 1782; Barnard 1914). The significance of the document as a whole was recognised by Sir Frederic Madden, keeper of manuscripts at the British Museum; in 1842 he spent two hours tracing a copy of the *Mary Rose*, but did not bother to reproduce the colouring, which he thought '*so much appearance of fancy on the part of the artist*' (Rogers 1980, 17, 22–4). Madden was a Portsmouth man and provided some source material for Samuel Horsey's book on the *Mary Rose*, published in the same year (Horsey 1842, 9). Madden assumed that Anthony's second Roll had been lost (else it would have come, via George III's library, into his own custody). In fact it had remained in royal hands until William IV gave it to his illegitimate daughter Lady Mary Fox, whose husband Major-General C.R. Fox became Surveyor-General of Ordnance and, therefore, Anthony's successor. In 1857 Lady Mary offered the manuscript to the British Museum; Madden immediately recognised it as Anthony's second Roll and secured it for £15.

## *The importance and accuracy of the* Anthony Roll
by Alexzandra Hildred

The value of the *Anthony Roll* lies not merely in the vessel illustrations that accompany the inventory but the fact that many of the items listed can be identified in the assemblage excavated from the *Mary Rose*. Although it is often difficult to reconcile recovered objects with unfamiliar descriptions or terms used, the inventory is the key to identifying (in many cases for the first time) types of gun that have long since fallen out of use and are unfamiliar even to ordnance experts today. In the list of ordnance provided to each ship, guns are grouped according to their material (bronze, iron, etc) and are named in descending order of size. Therefore, where we have found guns that are not easily identifiable from comparative physical remains (such as '*bases*' and '*slings*'), we can establish their identity, first, from their relative size order in the archaeological assemblage, and then by comparison with the positions of named guns in the inventory hierarchies. This has not proved infallible, however, as subsequent chapters will show, and there are some types of gun that were present on the ship in her early years that still cannot be equated with later forms with any certainty (eg '*great curtows*' and '*murderers*').

How the numbers of named guns fluctuated over time (or on the different types of vessel) and what they were replaced with (both numerically and hierarchically) is important when considering the evolution of seaborne ordnance and fighting tactics. By mixing inventory and archaeological resources, hitherto enigmatic guns can be identified and given detailed descriptions to include size, weight and form. Their potential capability can be predicted; they can be replicated, and tested. This is only possible from a study of the inter-relationship between the number and type of particular guns and the type of shot listed compared with that found in the bore of a particular type of gun, complemented with notes gathered from all sixteenth century sources. The location of these various guns about the ship, as indicated by the excavation of the *Mary Rose*, together with an analytical appraisal of the number of shot per gun, enable us to formulate an interpretation of the functional capability of the ship as a fighting unit. On a broader scale, the differences in the weapons carried on each type of vessel, together with what we can glean about the numbers and profile of the crew and the engagement sustainability, can eventually be rendered up to enable predictions to be made about the capability of the navy of Henry VIII. This is one of the challenges for the future. Our detailed analysis is presented in Chapter 11, following description of the assemblage.

In terms of basic comparison between the *Anthony Roll* and the archaeological evidence, the following can be stated.

### 1. The *Roll* lists 185 soldiers, 200 mariners and 30 gunners, 415 men in total

The total for mariners would have included the master and all the officers (including the carpenter(s)). Such figures did not normally include the captain (in this case Sir George Carew, who was also Vice-Admiral) and his personal servants, or the other senior military commanders and their servants. There may well also have been volunteer military officers aboard, serving without pay. The Barber-surgeon and chaplain (if any) would also have been 'supernumeraries'. All such totals were establishments, supplied for normal service afloat but likely to be supplemented in an emergency. The most authoritative command list for the Battle of the Solent assigns 500 men to the *Mary Rose* (Hatfield House CP 201/53), and the additional requirement would certainly have been for soldiers rather than mariners. The contemporary reports speak of between 25 and 40 survivors from a complement of about 500 (*AMR* Vol. 1, appx, 149. *LMR* 62, 64). The recovered human remains include 178 skulls, or 42% of the basic establishment given in the *Anthony Roll*, perhaps 35% of those actually aboard (see also McKee 1986).

**2. There are eight groups under the heading of 'Ordenaunce, artillary, munitions, habillimentes for the warre' beneath the illustration of the *Mary Rose* in the *Roll***

These are: guns of bronze; guns of iron; gunpowder; shot of iron; shot of stone and lead; bows, bowstrings, arrows, morris pikes, bills, darts for tops; munitions; and habiliments for war. Under each heading is a list of named items and the number of each (Fig. 1.1).

As far as the ordnance is concerned, the excavations and earlier salvage work have, not surprisingly, failed to recover the entire listed assemblage (Table 1.1). As far as the bronze (listed as '*brass*') guns are concerned there appear to be more demi cannon and culverins than there should be (Chapter 2). However, when the guns are measured, two of them are oddities; one is not quite a culverin and one is not quite a demi cannon. These are assumed to be a pair based on seabed location and design of both the guns and their carriages. The excavated bronze guns alone, and their similarity to the inventory, confirm the wreck as being that of the *Mary Rose*. There were only nine cannons in the fleet, four on the *Henry Grace à Dieu*, two on the *Mary Rose*, two on the *Jesus of Lübeck* and one on the *Galley Subtle*.

It is the list of iron guns that causes most problems, both in terms of understanding what the original guns actually looked like, how they were made and what size they were, and in interpreting the heavily corroded remains that have been recovered. The matter is further complicated by the presence of breech chambers and other items which might represent individual guns or spare elements. Examples for most of the listed types have been identified, some for the first time (Table 1.1; Chapters 3 and 4). The only type not identified is the top piece, but if these were positioned in the fighting tops they were probably thrown clear of the site during the sinking or recovered from the masts. The smaller guns are less evident than the larger ones and handguns are particularly poorly represented given the total that should have been on board (Table 1.1; Chapter 7).

The amounts of each type of gunpowder relative to the ordnance are intriguing and offer a new challenge for research into gunpowder technology, use and storage (Chapter 5). It is clear from the amounts listed that most guns must have used serpentine powder and that the small amount of corned powder listed poses a question as to what it was used for and with (see Chapter 5). Gunpowder has been found in gun bores and chambers but no store of powder casks have yet been identified, although there is some evidence for gunpowder from a pollen sample recovered from a cask found on the Main gun deck and from a sample of dark organic matter taken from around the bilge pump on the Orlop deck (*AMR* Vol. 4, 628, table 15.3 and 642).

**Table 1.1 Comparison between ordnance listed in the *Anthony Roll* and the archaeological evidence from the *Mary Rose* (including items recovered in early salvage)**

| | *Anthony Roll* | Recovered |
|---|---|---|
| *Guns of brass* | | |
| Cannon | 2 | 2 |
| Demi cannon | 2 | 2–3* |
| Culverin | 2 | 2–4* |
| Demi culverin | 6 | 2 |
| Saker | 2 | 0 |
| Falcon | 1 | 0 |
| Total bronze | 15 | 10 (67%) |
| *Guns of iron* | | |
| Port piece | 12 | 8 |
| Sling (various sizes) | 6 | 4 |
| Fowler | 6 | ?4 |
| Base | 30 | 9 (min.) |
| Top piece | 2 | 0 |
| Hailshot piece | 20 | 4 |
| Total iron | 76 | 29 (38%) |
| Total guns | 91 | 39 (43%) |
| *Handgun* | 50 | 5 (10%) |
| *Gunpowder* | | |
| Serpentine | 4,800lb/2 last | not measurable |
| Corned | 3 barrels (300–600lb) | not measurable |
| *Shot* | | |
| Iron | 580 | 1248 |
| Stone | 390 | 387 |
| Lead | 400 | 0 |
| Lead composite | 0 | 235*** |
| Other composite | 0 | 44 |
| Lead shot for handgun | 1000 | 62 |
| Canister shot | 0 | 9 (min.) |
| Total shot | 2370 | 1932 (81.5%) + 23 'odd' shot |
| *Non-gunpowder weapons* | | |
| Longbows | 250 | 179 |
| Bowstrings | 864 | ?2 |
| Arrows | 9600 | 3899 (min.)† |
| Bills | 150 | 116 (min.) |
| Morris pikes | 150 | 31 (min.) |
| Top darts | 480 | ?5 |
| Halberd | 0 | 1 |
| Total | 10,530 | 4301 (41%) 52% of all listed exc. gunpowder |

\* one gun seems too small to be a demi cannon but too large for a culverin
\*\* = 300–600lb
\*\*\* = 206 inset dice, 29 inset ball
† surviving chests could accommodate 8400

Shot is listed under the separate headings of iron or stone and lead in descending order of size for named guns (Chapter 5). This provides a vital clue in matching name to gun, suggesting size through material and position in the lists. This has been crucial in the identification of types of wrought iron gun (Chapters 3 and 5). It is clear that there was ample shot (Table 1.1) and examples are still being recovered from the wreck site during monitoring. The lead shot listed for the iron bases is actually a composite shot of cast lead containing an iron dice. The severely under-represented lead shot for handguns are very small and were probably stored close to the guns on the upper decks from where they would have been easily lost. The assemblage includes a number of composite shot that may or may not be subsumed in one of the listed categories. In addition there are fragments of canister or case shot. These objects are not listed but consisted of wooden canisters containing sharp fragments of broken stone (flint) that could be fired to produce lethal shrapnel.

The *Anthony Roll* combined with the archaeological evidence demonstrates the continued importance of longbows and arrows at sea. Every vessel carried longbows. Dedicated archers are not listed, but then, on smaller vessels, neither are soldiers. This suggests that gunners or mariners may have performed this task. Longbows and arrows were brought onboard in chests, which were found both in storage and opened and empty or partly so (below, Chapter 8 and *AMR* Vol. 4, chap. 10 and appx 1). There is no evidence for stored bowstrings, but there are plenty of casks that could have held them.

Identification of staff weapons is hampered by the lack of metal heads (lost to corrosion) and the difficulty of determining the function of some wooden hafts (Chapter 9). The locations of many of these items on the wreck suggest that they had been placed on the Upper deck ready for use or that men were standing at arms. The top darts are notably absent in the assemblage, but, like the top piece guns, if these had been in the fighting tops they could have fallen clear of the ship (Chapter 6). No halberds are listed in the *Roll* and the one example found (see Chapter 9) is very ornate so probably ceremonial rather than functional.

Munitions listed include pick-hammers, sledge-hammers, crowbars and commanders (large mallets) for working the guns; tampions to seal chambers and canvas, paper and wooden formers for making gunpowder cartridges. Wooden commanders and tampions were recovered, and one former, although the amounts do not square with the inventory (Chapter 5). Paper and canvas for cartridges has not been identified and the iron heads of the majority of tools are too vestigial to enable secure identification.

The habiliments listed include ropes for breeching guns and securing them to their carriages ('*woling*'). Ten coils are listed. The large rope coils recovered are thought to be anchor cables and, although substantial amounts of smaller ropes exist, their uses have not been identified. Bags of leather survive but not directly associated with ordnance; nails rarely survive. The six firkins with purses (barrels with leather fittings traditionally for gunpowder) have not been isolated, neither have the ten dozen lime pots. The latter may have been stored with easy access to the fighting tops and lost. The four spoked, four solid wheels and six spare axles listed exactly equal those recovered. One of the 12 sheepskins has been found. The 100ft (30m) of timber for '*forlocks and quoins*' to secure breech chambers in the iron gun carriages and to place under the bronze guns for elevation, may be represented by some of the spare timber found in the Carpenters' cabin. Armour is not listed in the *Anthony Roll* and very little has survived (Chapter 10).

**3. The illustration of the hull shows considerable detail concerning the positioning of guns, gun-ports (lidded and open), and the arrangement of the decks – most importantly of the castle decks which are all but absent from the surviving hull**

Comparing the illustration with the excavated hull is an enormous task and is dealt with in detail in Chapter 11. A few brief comments will suffice here. The ordnance is shown as being a mixture of bronze and iron and the bases or swivel guns can be identified. The Main deck is depicted with lidded ports but only five out of the seven are shown. Nine large guns (three bronze and six wrought iron) are illustrated but only seven were recovered (three bronze and four wrought iron). Two of these ports seem to be forward of the rise to the Forecastle, within the area of structure so far unexcavated. Although these guns all appear to be on a single deck, their height above the deck appears to be staggered, as though on two levels. The four at the upper level are situated at unlidded ports. The two lidded ports shown at a low level on either side of the rudder cannot be verified, but the structure here is incomplete. Two large wrought iron guns are shown at ports on the Upper deck within the Forecastle and four on the Upper deck within the Sterncastle; the recovered structure shows evidence for three of these stern ports. Interestingly the *Mary Rose* is the only vessel shown with series of guns on the Upper deck between the castles. A wrought iron sling was recovered from one of these ports. Nineteen large guns are depicted on the starboard side, just under half of the total large guns listed (39). Detailed comparisons of all the *Anthony Roll* drawings and their armaments as listed have been made (details in archive). This suggests that, at least for the first few vessels, Anthony is trying to portray accurately the number of guns listed, and each of the great ships is differently armed. It appears that he paid close attention to detail in the first few ships. Thereafter there seems to be more of a mechanistic approach and the number of guns shown does not reflect the individual inventory entries, even taking into account that only half should be visible.

As this brief description shows, most of the listed equipment for the war is represented in the recovered assemblage: indeed, many items could only be identified with reference to the *Anthony Roll*. The numbers do not always tally but some are remarkably accurate even though the hull and the assemblage are incomplete. The importance of this inventory is unquestionable and it is a remarkable turn of fate that we are able to compare it with the remains of one of the vessels it illustrates.

## Changes in Armament: an Introduction
by Alexzandra Hildred

There are two avenues of research regarding the changing armaments of the *Mary Rose* over her life: a study of the three inventories of her guns (1514, 1540, 1546); an investigation of the structural alterations visible within the ship, and attempting to date and relate these to different patterns of arming the vessel. Information from these sources can then be expanded by incorporation of informative descriptions found in the detailed inventory prepared following the death of Henry VIII in 1547, by which time the *Mary Rose* was no more (Starkey 1998). Only three vessels are listed in the three inventories compiled during Henry's reign, the *Henry Grace à Dieu* (listed as not furnished in 1540), the *Mary Rose* and the *Peter Pomegranate*.

Analysis of the changes in armament is important not only for understanding the operational history of the ship herself but also for appreciating developments in Tudor naval firepower, the wider disposition of shipbourne ordnance and the changes in fighting tactics, and the sustainability of engagement that are implied and reflected. These issues are discussed in detail in Chapter 11 and here we present just an outline of the changes indicated by the historical record.

The ship that sank is not the same as the ship 'as built' and considerable changes in the types of ordnance carried are apparent between the inventories of 1514 (PRO E 36/13 [*LP* 1, 3137(6); Knighton and Loades 2000, 113–58]), 1540 (PRO E 101/60/3 [*LP* 15, 196; Anderson 1920]) and 1546 (the *Anthony Roll*). This is explored more fully in Chapter 11 (see Tables 11.1 and 11.2). This re-arming demanded significant alteration to the structure of the vessel, including the addition and alteration of gunports and strengthening of the gun decks. Details of these alterations are provided in *AMR* Vol. 2.

The 1514 inventory was undertaken after the end of the first war with France. It lists thirteen decommissioned ships, and is mostly concerned with their rigging and armament. Unusually it identifies the location of guns on six vessels; these include the *Peter Pomegranate* but unfortunately not the *Mary Rose*.

Nevertheless the numbers and types of guns, and their position in her near-sister ship, suggest that the structure of the *Mary Rose* was markedly different in her early years to what it was when she sank.

The gun names and their sizes in 1514 differ considerably from both the later inventories, although the mixture of wrought iron and bronze guns is present at this early stage. The 78 guns listed for the *Mary Rose* in 1514 include serpentines, stone guns, murderers, curtows, falcons and falconets. Some of the terms appear to be generic and do not necessarily conform easily to later traditions.

Although the differing terminology used for locations within each vessel is confusing, interpretation is possible and suggests that the majority of guns were lighter weight anti-personnel wrought iron guns, which could be accommodated high up in the ship without compromising stability. The consistent lack of significant numbers of large guns indicates that there was no need for a continuous gun deck requiring lidded ports, the Main deck as found was not required to furnish the guns listed in 1514.

There are few illustrations of English vessels during this period, and those which we can date are often of the 1480s. The *Beauchamp Pageant* illustrations (*c.* 1485) suggest that the majority of any engagement was hand-to-hand, with only small guns visible in the waist of ships (BL Cotton MS Julius E.IV, art. 6, f. 18v). Contemporary illustrations of ships include one of an attack on Brighton (suggested as depicting a French raid of 1514, as mentioned above), drawn by Anthony Anthony in 1545 (BL Cotton MS Augustus I.i.18; Payne 2000, 20–2) and a *plat* depicting a fortification for the '*Wiche*', possibly Poole, also attributed to Anthony Anthony (BL Cotton MS Augustus I.i.29). The vessels depicted show far fewer gunports in every case than those shown either in the *Anthony Roll* or, indeed, within the hull of the *Mary Rose*, and lids are not evident. Also now dated to around 1545 is the painting of the *Embarkation of Henry VIII from Dover for the Field of Cloth of Gold* (Royal Collection RCIN 405793; the event itself of 1520). The vessels here (possibly including the *Mary Rose*) do not show a continuous row of lidded ports at main deck level (see Fig. 1.6).

In 1514 there were 78 guns (excluding handguns). Sixty-five of these were wrought iron, all but five probably less than a ton in weight. The inventory of 1540 shows a radical change, in particular the addition of nine large wrought iron port pieces (Chapter 3), all over a ton in weight. By 1540 it is estimated that the *Mary Rose* carried 15 guns over a ton in weight and four over two tons. This represents a significant increase in the sizes of iron guns carried and there is a trend across the fleet for larger carriage mounted guns in both bronze and iron. The number of bronze guns does not increase dramatically, but the types listed suggest a rationalisation into a size sequence, the basics of which were to remain essentially fixed for the remainder of the age of sail. These have comfortably recognisable names including cannon, demi cannon, culverin and sakers. For the first time we are witnessing the introduction of a number of longer range guns firing iron shot, the culverins, and the introduction of corn powder – a stronger type of gunpowder (see Chapter 5).

The addition of the nine port pieces by 1540 could not have been achieved without the cutting of a row of ports on each side of the ship to accommodate them. They demonstrate an attempt at standardisation of a single type of large gun. As battering pieces, these are best suited to horizontal firing. It is suggested that the main gun deck, as found, was built in part to accommodate these pieces. This, in turn, could not be achieved before the acceptance of the gunport lid as an integral part of shipbuilding. Although attributed to the Breton shipbuilder Descharges, sometime around 1501, this integration along both sides of the ship does not seem to have been necessary before the guns in the 1540 inventory were assigned to the *Mary Rose*. Although the *Anthony Roll* does not include lidded ports on either side of the rudder on all ships (these feature on eight out of 20) it is suggested that lidded ports were first used in this position, in what became known as the gunroom (Rodger 1997, 206). The accommodation of ports in this position is explored in Chapter 11.

The 1546 inventory shows little change for nine out of ten ships that appear both here and in 1540. The emphasis is on an upgrade in the size of the weapons. The number of guns still recorded for the *Mary Rose* is actually reduced from 96 to 91 but the emphasis changes, with heavy medium range guns placed amidships and long range culverins and demi culverins placed in the bow and stern. These changes result in a fighting platform accommodating guns offering of long, medium and short range cover with a good broadside defence. Such changes reflect not only developments in the manufacture of guns and materials employed but in the tactics used against enemy forces at sea (see Chapter 11). The relatively cheap cost of supplying iron guns rather than bronze ones is, of course, also likely to have been a major factor in the provision of armaments.

## The Service Career of the *Mary Rose*
by David Loades and C.S. Knighton

The working life of the *Mary Rose* is very patchily documented. We do not even know exactly when she was built, or where, or by whom (*AMR* Vol. 1, appx, 1–5; *LMR*, 1–4). If a reference to four new ships being fitted out at Southampton in December 1509 includes the *Mary Rose*, then she had probably been laid down before April of that year, when Henry VIII had come to the throne (*LP* 1, 287; *AMR* Vol. 1, appx, 1). It seems more likely that the two new ships for which a warrant was issued in January 1510 were the *Mary Rose* and the *Peter Pomegranate*. The tonnages given do not quite fit, but such figures were notoriously approximate, and large ships were not being laid down every day, with only five Great Ships (400 tons or more) launched between 1510 and 1515 (PRO E 404/87/2, no. 121; *LMR*, 1; Rodger 1997, 476–7). This theory would be consistent with the first specific reference to the ships, being conveyed from Portsmouth to the Thames at the end of July 1511. If correct, this suggests that they were built at Portsmouth, under the direction of Robert Brigandine, Clerk of the Ships. 'Conveyed' may well mean towed rather than sailed, because a number of accounts for fitting out the *Mary Rose* survive from September through to December, including decking, rigging, and the shipping and stocking of ordnance. The accounts conclude with payment for banners and streamers, and with the name of the master, Thomas Spert (*LP* 1, 1393(ii), 3608; *AMR* Vol. 1, appx, 11–20; *LMR*, 4). On 8 April 1512 the newly commissioned warship was one of those named in the indenture by which Sir Edward Howard took over the command of the King's warfleet (*LP* 1, 1132; *AMR* Vol. 1, appx, 24).

Howard's accounts from April to July 1512 show that the *Mary Rose* was in service with the fleet which blockaded Brest, and for part of that campaign she served as the Admiral's flagship (*LP* 1, 1453(5). *AMR* Vol. 1, appx. 27; *LMR*, 5). In October she was one of the ships which returned directly to the Thames from Southampton (*LP* 1, 1413; *AMR* Vol. 1, appx, 40). During the winter she was laid up, probably at Deptford. John Hopton's maintenance account for the *Mary Rose* and other ships runs from 28 October 1512 to 11 February 1513 (*LP* 1, 3318; *AMR* Vol. 1, appx, 42; *LMR*, 9). In February she was listed for mobilisation, and was back in service by the 1 March, when she was included in the '*charges of the army by sea*' (*LP* 1, 1661(5)). A bill outlines her summer itinerary as follows: 18 March at Woolwich; 25 March at Sandwich; 10–23 May at Dartmouth and Plymouth; 4 June at Southampton; 8 June at Portsmouth; and 22 June–15 July, back at Sandwich (*LP* 1, 2305(ii), p. 1033). We also know that on 5 April she was in Plymouth Road as Howard's flagship and, by 12 April was in action off Brest (*LP* 1, 1771; *LMR*, 15). On 25 April Sir Edward Howard was killed in a rash attack on the French galleys and, within a few days, the whole English fleet had retreated, pleading lack of victuals. When Thomas, Lord Howard, took over from his younger brother as Lord Admiral, he retained the *Mary Rose* as his flagship, but he did not act as her captain. On 9 May Edward Bray was appointed to that post (*LP* 1, 1825, 1844, 1851–2, 2305; *AMR* Vol. 1, appx, 62–7; Spont 1897, 133; *LMR*, 17–20).

Later in 1513 under this command the *Mary Rose* took part in the Flodden campaign, leading a fleet of fifteen ships to Newcastle. There Howard disembarked, taking his forces (including Bray and 201 soldiers and mariners out of the *Mary Rose*) to join the main army at

Alnwick on 4 September. At the battle five days later the Scots were routed and King James IV was killed. The men from the fleet served ashore for sixteen days before returning to the ships at Newcastle (*LP* 1, 2651–2; Hall 1548, 557; Mackie 1951, 62, 78–9, 85; *LMR*, 37). In reward for his service in the victory Howard was created Earl of Surrey in February 1514. Meanwhile the *Mary Rose* had been laid up; from 28 October Hopton listed her as one of many ships on a care and maintenance basis (*LP* 1, 3318(i, iv); *AMR* Vol. 1, appx, 90).

Thereafter detailed documentation becomes sporadic. John Brown took over as master by 2 March 1514 and the *Mary Rose* was victualled for six weeks on 23 April (*LP* 1, 2764, 3148; *AMR* Vol. 1, appx, 97, 100; *LMR*, 37). On 2 May a pilot submitted his account for navigating her over the Naze shallows off Harwich (*LP* 1, 2865; *AMR* Vol. 1, appx, 101; *LMR*, 38). She was again at sea between 23 May and 19 June, commanded by Sir Henry Sherborne (*LP* 1, 2938; *AMR* Vol. 1, appx, 102). Lord Admiral Surrey went aboard at Dover on 26 or 27 May, and though he crossed the Channel in another vessel, the *Mary Rose* was his flagship when he raided the French coast on 13 June (*LP* 1, 2946, 2959, 3000–1; *AMR* Vol. 1, appx, 103–5; *LMR*, 39–42). The war with France was petering out and no further engagements are recorded.

On 27 July the *Mary Rose* was inventoried along with other ships (*LP* 1, 3137(6); Knighton and Loades 2000, 139–42), and an indenture for receipt of her rigging was made on 9 August (*LP* 1, 3137(11); *LMR*, 43). She was laid up, probably at Deptford, where in June 1517 a '*good and able pond*' was excavated to enable her (and other ships) to be kept in safety (*LMR*, 44). In August and November 1518 there are accounts for repairing and recaulking the *Mary Rose*, so she was not neglected, but may not have been active (*LP* 2, p. 1479 [Aug]; no. 4606, p. 1407 [Nov]; *AMR* Vol. 1, appx, 111–12). In June 1520 she was one of the Great Ships which escorted Henry to his showpiece meeting with Francis I at the Field of Cloth of Gold (*LP* 3, 704, p. 239; *AMR* Vol. 1, appx, 113); but between October 1517 and December 1521 she seems, for the most part, to have been in the care of five shipkeepers (*LP* 3, 1911). In October 1520 she was 'pumped', presumably a routine job for a ship lying idle at anchor (*LP* 3, 1009; *AMR* Vol. 1, appx, 114; *LMR*, 45).

The return of war in 1522 brought the *Mary Rose* back into commission. She was at Dover on 26 May when Henry and the Emperor went aboard, as reported by Anthony Anthony (*LMR*, 46). On 4 June she was the flagship of the fleet in the Downs, from whence Sir William Fitzwilliam, the Vice-Admiral, wrote to the King extolling (once again) her excellent sailing qualities (*LP* 3, 2302; *AMR* Vol. 1, appx, 115; *LMR*, 47). Her main duty was to escort the Army Royal to France, where it captured Morlaix on 3 July; and from whence Surrey wrote to Wolsey from aboard the *Mary Rose* (*LP* 3, 2362; *AMR* Vol. 1, appx, 121; *LMR*, 53). By 4 August Fitzwilliam was debating with Wolsey whether to lay up the *Mary Rose* or not (*LP* 3, 2419; *AMR* Vol. 1, appx, 122; *LMR*, 54). On returning to Portsmouth he discharged Sir William Sydney (16 August), who had been aboard *Mary Rose* as fleet treasurer (*LP* 3, 2442; BL Royal MSS 14.B.XXV / 7.F.XIV, art. 23 [*LP* 1, 1st edn, 3980 (misdated)], and Royal App. 89, art. 1(b); *AMR* Vol. 1, appx, 123). The ship must indeed have been laid up at this point, because there are no further references to her at sea during the war, which lasted till 1525.

At some point she was in dry dock at Deptford (*LP* 4, 1714(3); *AMR* Vol. 1, appx, 130; *LMR*, 55), but she is documented at Portsmouth from November 1523 to January 1530. Throughout this period Thomas Jermyn, Clerk of the Ships, accounted for shipkeepers aboard her. The numbers fluctuated from six to 21 at different times for no apparent reason, and for some periods Jermyn described himself as master. A new dock was excavated for her between February and June 1527, when she was thoroughly caulked and repaired (*LP* 4, 6138; *AMR* Vol. 1, appx, 135–7; *LMR*, 56).

After 1531, the documentary evidence is sporadic. Some time between then and the end of 1536 the *Mary Rose* and other ships were '*new made*'; Thomas Cromwell obliquely claimed credit for this as being among works undertaken since he had come into the King's service (*LP* 10, 1231; *AMR* Vol. 1, appx, 138). The archaeological evidence attests to radical alterations, but Cromwell's statement provides the only documentary clue to their date. Subsequently William Gonson accounted for the *Mary Rose* and several other ships laid up in the Thames (PRO E 101/612/59).

In the spring and again in the autumn of 1537 the King sent four armed ships to sea under Sir John Dudley as Vice-Admiral (*LP* 12/1, 601–4; 12/2, 236, 393). The *Mary Rose* was not in either squadron, but her future and final captain, Sir George Carew, was Dudley's second-in-command on the second voyage, and must already have been in some peril because his ship (the *Jennet*, *Minion* or *Sweepstake*) was taking in so much water (*LP* 12/2, 535).

In 1539 the *Mary Rose* was mobilised, along with the rest of the fleet, when a Franco-Imperial invasion was feared, and there are a number of references to her, but no action (*LP* 14/1, 143; *AMR* Vol. 1, app, 139). We have an ordnance list from February 1540 (*LP* 15, 196; *AMR* Vol. 1, appx, 141 [misdated]), and a mention of her further mobilisation in July 1543, at the outbreak of the third French war (*LP* 18/1, 973; *AMR* Vol. 1, appx, 142). After that, she disappears again, in spite of the activity which was undoubtedly going on in 1543 and 1544. She makes her final entrance on 23 June 1545, when Dudley (now Viscount Lisle) found her in the Downs as he was preparing a pre-emptive strike against the threatened French invasion (*LP* 20/1, 1023). She

*Figure 1.2 The Cowdray Engraving showing the sinking of the* Mary Rose *and disposition of the fleets (© Society of*
MR = Mary Rose; HGD = Henry Grace à Dieu; EF = English fleet; TP = The Platform; ST = Square Tower;

features in a sequence of command lists drawn up in June/July 1545, the last of which shows that Sir George Carew had been appointed captain (*LP Addenda*, 1697–8; *AMR* Vol. 1, appx, 144–5; Hatfield House, Cecil Papers 201/51). We also now know of plans to modify her further before the confrontation in the Solent which was to be her last action (see Chapter 12).

## The Battle of the Solent, 19 July 1545
by David Loades

The confrontation at Spithead occurred during Henry's third and final war with France. Throughout the 1530s Henry's relations with the Emperor Charles V had been frosty, and he had been driven into a marriage of convenience with the French King, Francis I, in order to avoid isolation. In 1538–9 isolation had been forced upon him by a brief period of Franco-Imperial amity, and he had been compelled to seek allies among the German princes. The Cleves marriage of January 1540 had been one result (Warnicke 2000, chap 4). Henry did not relish this line of policy any more than he relished his fourth wife, and deteriorating relations between Charles V and Francis enabled him to abandon both in the early part of 1540 (*ibid.*, chaps 7–8). Charles was anxious to prevent Henry from dabbling any further in German politics, and Henry preferred an Imperial to a French alliance. Negotiations commenced in 1542, and an alliance was concluded in February 1543, committing England to a supporting role in the next round of the Habsburg/Valois wars (*LP* 18, 144). It is often said that Henry fought this war in order to recover his self-confidence after the humiliation of his fifth marriage (to Catherine Howard), but the timing of events does not substantiate this. There was always an element of personal 'machismo' about Henry's war-making, but this conflict, like that of 1522–5, was mainly an opportunistic attempt to take advantage of a continental conflict in order to strengthen the English bridgehead in France.

In fact there was little real friendship in the Anglo-Imperial alliance; each party was using the other for its own purposes. Entanglements in Scotland forced Henry to delay his French campaign until 1544, and then he concentrated upon the siege of Boulogne, virtually ignoring the Emperor's strategy. Boulogne fell in September, and simultaneously Charles made a separate peace. This left Henry in a vulnerable position, fighting on his own to retain his conquest. It was too late in the year 1544 for Francis to launch a major campaign against Bolougne, now garrisoned and supplied by Henry. When the time came, the French had a double strategy; a siege of Boulogne from the land, and seizure of some English towns with which to bargain. To succeed, both required command of the sea.

French sea power had been considerable since the acquisition of Brittany in the fifteenth century but naval infrastructure was weak and mobilisation was slow and cumbersome. Henry had a warfleet of 50 vessels which could be efficiently deployed but these were not sufficient to meet the growing French threat and they

*Antiquaries of London). Key (left to right): FF = French fleet; FG = French galleys; SC = Southsea Castle; B = baskets; RT = Round Tower*

could not be concentrated because of the need to protect other sea lanes. Peace with the Emperor meant that Francis could use some of his Mediterranean fleet for the northern campaign. These were galleys, armed with three forward-firing heavy guns. In coastal waters and calm weather they were formidable, and the English had no real answer to them, but they were extremely vulnerable to rough seas and high winds. By June 1545 Francis had brought 25 of these galleys round to the Seine estuary – the rendezvous for his Channel fleet. The English were at sea first and by early May their privateers had inflicted heavy damage on French traders, capturing or destroying nearly 50 in the space of a few days (*LP* 20/1, 689). At the end of June, Lord Lisle, the Lord Admiral, led out a war fleet numbering 30–35 vessels to disrupt French preparations. Lisle held the initiative, but his attack on the Seine on 3 July was unsuccessful (La Roncière 1899–1932, 3, 417–19). Several factors contributed to this: the effectiveness of the French galleys, sickness on board the English ships and the weather, which worsened in the few days Lisle had to wait before he received the King's authorisation for his plan (*LP* 20/1, 987, 1023). The uncertainty of communications at sea was always a serious handicap.

To some extent fate achieved what the English had failed to do. Within days of celebrating the repulse of their enemies, the French lost the flagship of Claude d'Annebault, the Admiral of France, the 1200 ton *Philip*, which caught fire and was totally destroyed. The ship to which d'Annebault transferred his flag then ran aground. But undeterred by these omens, his armada set sail and came within sight of the English coast on Saturday 18th (La Roncière 1899–1932, 417–19). Estimates of his strength vary between 123 and 235 ships, the latter (and more reliable) total comprising 150 warships, 60 transports and the 25 galleys, together carrying 30,000 men. D'Annebault had around 30 warships of the largest kind, at least fifteen of them being four-masted like the *Mary Rose*. (*AMR* Vol. 1, appx, 151; *LP* 20/1, 1007; Du Bellay 1569, 4, 285–6; Clowes 1897–1903, 1, 462; La Roncière 1899–1932, 3, 415; McKee 1973, 52–3).

Lisle, who had retreated to Portsmouth after leaving the Seine, now had, by the fullest account, 68 fighting ships, 23 of them rated at 300 tons or above, so about the same strength in large warships as the French; the smaller armed vessels included a number of privateers volunteered or commandeered for the occasion (Hatfield House, Cecil Papers 201/51–3). The French priority was an amphibious landing rather than destruction of the English fleet and the objective was Portsmouth. On the 18 July d'Annebault made a diversionary raid on the Sussex coast and the following day entered the Solent. What followed was largely determined by the vagaries of the weather. The French warships were unable to enter Portsmouth harbour, partly because of the extensive fortifications, but more because of the presence of the English fleet (Du Bellay 1569; *AMR* Vol. 1, appx, 150). At the same time Lisle was unable to move out to the attack because of the absence of wind. D'Annebault landed about 5000 of his soldiers on the Isle of Wight and drew up his fleet in three squadrons in the hope of battle. Meanwhile he sent his galleys (which were independent of the wind) to skirmish ahead and they began to fire their heavy guns at the becalmed English ships from a safe distance (Fig 1.2). In spite of later French claims to the contrary, as far as we know, they inflicted no damage.

Late in the afternoon the wind began to stir. Lisle moved his Great Ships out to the attack and the galleys retreated in some confusion. It was at this point that disaster befell the *Mary Rose*. Her rebuild may have left her somewhat top heavy or cumbersome, and we hear no more of her excellent sailing qualities after 1536. A recently discovered document suggests the potential for quite radical alterations to her structure possibly as late as 1545 (see Chapter 12). Accounts cite too much ordnance (unbreeched), over-manning and insubordination and gunports positioned too low and left open. The most plausible contemporary account states that she fired one broadside and, in coming about to fire, the other was caught by a sudden gust of wind and she heeled, enabling water to enter through her open gunports (see Chapter 11 for a more detailed consideration). Once the water had begun to enter, the end was swift and catastrophic (*LP* 20/1, 1263; *AMR* Vol. 1, appx, 149; *LMR*, 62). Only a handful of the seamen and soldiers survived and the King was a helpless spectator of the whole tragedy. Having wrought this destruction, the wind then failed again, and it was only on the ebb tide that Lisle was able to resume his advance. This he did with great skill, reducing the galleys to confusion and frustrating d'Annebault's attempts to get his warships into action. The resulting stalemate persisted for two days. Lisle's purpose being defensive, this suited him very well, and d'Annebault was unable either to bring him to battle or to force his way into Portsmouth. Even his landings on the Isle of Wight were proving fruitless and expensive because Henry had reinforced the garrison. Both admirals explored possibilities of breaking the deadlock but the English were frustrated by the weather and the French by lack of local knowledge (Dasent 1890–1907, 1, 212–13; *LP* 20/1, 1235, 1237, 1255). With his supplies running low and disease beginning to infest his overcrowded ships, d'Annebault withdrew his forces from the Isle of Wight on the 23 July and retreated the following day.

This was not the end of the campaign but the sequel was unexciting. The French fleet proceeded to support the siege of Boulogne by landing 7000 soldiers (Loades 1992, 134–5). D'Annebault remained at sea for rather more than two weeks. His galleys carried out a few small-scale raids but when Lisle, with his fleet augmented to over 100 ships, came in search of him in the first week of August, the result was inconclusive manoeuvring rather than battle. The two fleets encountered off Beachy Head on 10 August; with some exchange of fire, and we know Lisle had developed an innovatory set of battle orders for what he hoped would be a decisive engagement (*LP* 20/2; Corbett 1905, 20–4; Knighton and Loades 2000, 159). However, the wind again frustrated him, and the French withdrew under cover of darkness. 'Plague' (possibly food poisoning) was raging through his fleet and d'Annebault had no choice but to retreat to the Normandy coast and demobilise. Lisle, afflicted by a similar scourge, returned to Portsmouth (*LP* 20/2, 3, 5, 13; see *AMR* Vol. 4, 174, 608, 610). Although they had been frustrated of their battle, and lost a capital ship, the victory lay with the English. Without effective command of the sea, the siege of Boulogne failed, and the town remained in English hands when peace was concluded in 1546 (*LP* 21/1, 1014).

## The Historical and Archaeological Importance of the *Mary Rose* Ordnance
by Alexzandra Hildred

When the *Mary Rose* sank at Spithead on 19 July 1545, the science of gunfounding had been evident in the west for over 200 years. Guns are recorded as being for use on (rather than merely transported by) English Royal Ships as early as 1337–8, when the *All Hallows Cog* carried a '*certain iron instrument for firing quarrels* [crossbow-bolts] *and lead pellets, with powder, for the*

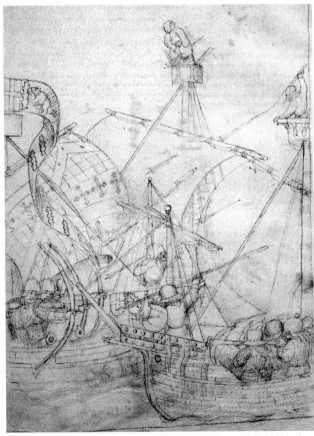

*Figure 1.3 The* Beauchamp Pageant *(c. 1485) depicting hand-to-hand naval engagement (BL Cotton MS Julius E.IV, art. 6)*

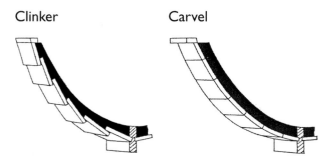

*Fig 1.4 Clinker and carvel building techniques*

*defence of the ship*' (Friel 1995, 152). Ships using guns in battle are recorded during the battle of Sluys in 1340 (De Vries 1998, 390) and in 1409–11 the *Marie of Weymouth* carried one gun of bronze and two of iron, all breech-loading (Hutchinson 1994, 156).

By 1485 the *Grace Dieu* carried 21 guns and the *Mary of the Tower* had 48, all chamber pieces. Just over ten years later, the *Sovereign* carried 141 guns and the *Regent* 181. Of the 141 guns carried by the *Sovereign*, 121 were '*serpentines*' – small breech-loaders functioning as anti-personnel weapons – the 20 remaining were '*stone guns*' carried in the waist (Friel 1995, 154; Fig. 1.3). Specific mention of serpentines for shipboard use can be found in a report dated 1513, where they are listed as weighing 261¼lb (*c.* 118.5kg) with breech chambers weighing 41lb (18.6kg) each (BL Stowe MS 146, f. 41). Although these guns varied in size and weight, their sheer numbers and locations about the ship suggest that they must have been small. During this period warfare at sea still depended on tactics of grappling, boarding and hand-to-hand combat (see Chapter 11). There is clear documentary evidence, however, that the fifteenth century saw a dramatic rise in the use of gunpowder weapons at sea (Friel 1995, 152, 153; De Vries 1998, 390).

The evolution of a vessel capable of carrying enough large ordnance to penetrate the hulls and sink other ships, in any quantity, is not in evidence until the late fifteenth or early sixteenth century. It evolved as a result of the refinement of the techniques of gun and projectile manufacture, of gunpowder production and refinement/processing and, most importantly, as a result of the changing techniques of ship building, enabling lidded gunports to be placed close to the waterline. This increased the number of large guns which could be carried and eventually led to what we know of as a 'broadside', the simultaneous firing of all guns along one side of a vessel. These lines of evolution influenced one another and eventually mingled, possibly driven by political and economic pressures. The *Mary Rose*, built between 1510 and 1511, was part of a programme of warship building devised by Henry VIII. She was one of only 20 vessels classified as Great Ships and, by 1545 she was the second largest and most heavily armed vessel in the fleet of 58, carrying 91 guns, 39 of which were large enough to be mounted only on carriages. These were the King's Ships, owned by the monarch. In times of war, the fleet was augmented with hired merchantmen. In addition to creating, arming and maintaining a fleet, Henry VIII fortified the coast of England in a manner not seen since Roman times. Armaments were required in all these establishments (see Fig. 2.1). Undoubtedly the amount of guns produced in this period was 'revolutionary'.

The *Mary Rose* was a 35 year old veteran when she sank defending Portsmouth Harbour against an invasion fleet of 150–200 vessels and 30,000–50,000 men (see above). This upper figure is twice that listed in the July 1588 muster for the Spanish Armada (Rodríguez-Salgado 1988, 124). We have no archaeological evidence that she was sunk by enemy fire, although she appears to have been within gun range of

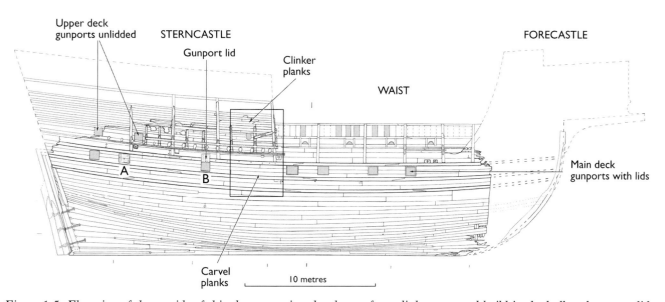

*Figure 1.5 Elevation of the outside of ship demonstrating the change from clinker to carvel build in the hull and gunport lids*

*Figure 1.6* The Embarkation of Henry VIII at Dover for the Field of Cloth of Gold *(The Royal Collection © 2003; Her Majesty Queen Elizabeth II)*

the French galleys, as we shall discuss in detail in Chapter 11. Her downfall seems to have been a combination of human error, probably born, at least in part, of inexperience in handling such a heavily armed vessel, limited manoeuvrability and weather conditions. The size and number of guns, and thus the firepower of any ship, was limited by the weight that the vessel could carry without affecting the stability of the ship. Parker (1996) in discussing the weight of ordnance relative to tonnage, argues that the true warship was a later phenomenon. Applying the same criteria, the *Mary Rose* comes very close to a figure of 4.5%, however the predicted weight of her armaments is insufficient to have led to any serious instability (see Chapter 11). Until guns could be positioned low down in the hull, the number of substantial guns which could be carried was limited. The transition between the traditional shipbuilding techniques using short overlapping planks to produce a 'clinker built' hull, and the longer flush jointed planks of the 'carvel built' hull appears to have occurred within English vessels towards the end of the fifteenth century with the building of the *Sovereign* (Fig. 1.4). This method of hull manufacture facilitated the cutting of ports. Fitting these with lids meant that these could be positioned lower in the hull, which, in turn led to an increase in the number of large guns that any ship could carry safely. Although lidded (clinker built) ports can be built into a clinker hull, there is no evidence that this was adopted. Before this larger guns could only have been accommodated at unlidded ports well above the waterline and this is still the case in the Upper deck ports of the *Mary Rose* (Fig. 1.5).

Gunport lids (distinct from cargo ports, sealed at sea), at either end of what appears to be a continuous deck below the weather deck, are depicted in the painting of Henry VIII's *Embarkation at Dover* for his meeting with Francis I in 1520 (Fig. 1.6). This has been used as an important date for the depiction of the fully-fledged lidded gunport, though the painting itself is believed to date from the 1540s (Lloyd and Thurley 1990, 56 and title page). However, there is little corroborative illustrative evidence for gunports all along a lower deck before 1546. The illustrations in the *Anthony Roll* show only the two largest vessels, the *Henry Grace à Dieu* and the *Mary Rose*, with lidded gunports on their broadsides, though more, but by no means all, show lidded ports in the stern. Illustrations of vessels at sea before the mid-sixteenth century rarely show more than a few broadside gunports (for example BL Cotton MS Augustus I.ii.64). Few ships on the Cowdray engraving, which shows the sinking of the *Mary Rose* (see Fig. 1.2) have more than three gunports on their sides.

The invention of gunport lids some half century earlier was a major advance in shipbuilding technology (see p. 8, above referring to Descharges). In this respect the evidence of the *Anthony Roll* is limited and rather negative, showing only that in 1546 lids were still far from being standard fittings.

*Mary Rose*, therefore, sits historically at a very important point in the development of both shipbuilding and in the armament of naval vessels, she herself being reconfigured at least once to accommodate a greater quantity and weight of ordnance.

*Figure 1.7 Demi culverin 79A1232 on its carriage in the Mary Rose Museum*

Archaeologically, in terms of that ordnance, her importance is equally without question. Although a great many armouries exist, with seemingly infinite numbers of guns, there are many guns listed in inventories of which no recognised examples survive. The number of surviving examples of sixteenth century guns is relatively small. Both bronze and iron were valuable commodities and, in many cases, guns were recycled for their metal. The number of original gun carriages is extremely limited and, where they do exist, many are from shipwrecks. Thus sea-recovered ordnance is virtually the only way of adding to the present information about large pieces and the carriages that supported them. Main deck guns from the starboard side of the *Mary Rose* were found *in situ* on their elm carriages, providing vital information on their mounting and operation (Fig. 1.7).

The guns carried include wrought iron pieces, some of which would sit unquestioned beside fifteenth century serpentines. The smooth bore, cast bronze muzzle-loaders do not differ greatly from those used in the seventeenth century. This mixed array in terms of size, form and materials reflects the different technology and skills used to manufacture them. Their presence together on a capital ship of mid-sixteenth century date challenges the simplistic view of a linear evolution of artillery based on the displacement of wrought iron by cast bronze and, eventually, cast iron (see Chapters 2–4). Historical information gathered from a study of inventories for vessels and fortifications during the period can be used to view the armaments within the context of sixteenth century warfare as a whole, and to look at the relative importance of specific guns and aspects of change or stasis in armaments. It is argued that the major developments which revolutionised warfare at sea were born neither in the forge nor the foundry, but in the shipyards. The revolution was not one of superseding 'obsolete' weapons with ones that were more reliable, could shoot further or be reloaded faster, but simply one of scale: the ability to carry more big guns. It was this which was eventually to herald a complete revision in the tactics of warfare at sea. What is intriguing is that many of the guns chosen were not of the newest technology available but were the stalwart types made in wrought iron in the traditional manner.

The importance of the *Mary Rose* cannot be underestimated. We have her. She is exactly dated. Her life spans a 35 year period of immense change, reflected in alterations to the hull and the armaments. Ongoing study of the structural alterations and additions, backed by a dendrochronological study of the timber elements making up the hull, should allow us to follow the evolution of structure and ordnance which could be accommodated at any time. A collection which includes weapons, specific ammunition and associated gun furniture is rare enough, especially on such a large scale. However, its association with extensive human remains, portable arms such as staff weapons, swords, longbows and arrows, personal items and everyday objects relating to the working of a Great Ship, is unparalleled.

# PART 2. ARMING THE KING'S SHIP: THE GREAT GUNS

# 2. 'Brass' Guns

## Alexzandra Hildred

## Introduction

By the time Henry VIII ascended the throne the use of guns at sea was routine. The majority of these guns were 'serpentines', anti-personnel weapons weighing about 260lb (*c.* 118kg) with breeches of about 40lb (18.1kg) (Blackmore 1976, 242). These were augmented by a small number of large bronze guns, '*curtows*' and '*bombards*' and possibly intermediately sized wrought iron guns called '*stone guns*' (Chapter 1).

Immediately Henry began a period of shipbuilding, refurbishment and acquisition (Rodger 1997, 204). This was repeated several times during his reign; not surprisingly these episodes correspond to direct or perceived external threats. Between 1535 and 1544 eight vessels were built, and seven were rebuilt, including the *Mary Rose*. Four were purchased. On land the financial input into defence was similar. Following the truce between France and the Holy Roman Empire in 1538, possibly heralding an invasion force to reclaim Papal authority in England, Henry appointed Commissioners to '*search and defend*' the coastline. His ambition was to form a chain of batteries and forts around the coast to protect against an invasion. This resulted in the thirteen Henrician castles and seven Henrician blockhouses stretching from Tilbury near London to Pendennis in Cornwall. These supplemented existing fortifications built by Henry VII and such defensive positions were extended as far north as Berwick on Tweed and as far west as Milford Haven in Pembrokeshire (Fig. 2.1). Guns were required for all these establishments in addition to those required for the '*King's Ships*'. The list is impressive, and the weapons and ammunition within these fortifications are described in the 1547 inventory of Henry VIII's possessions (Kenyon 1982). The production of wrought iron guns, and initially the purchase of (and later the home production of) cast bronze guns, increased accordingly. This need for armaments may well have stimulated the quest for cheaper raw materials and larger production centres, reviving experimentation with casting iron ordnance which had begun in the Sussex Weald in 1496 (Teesdale 1991, 11). During the reign of Henry VIII and the life of the *Mary Rose*, most '*great guns*' were of wrought iron or cast bronze.

In addition to an increased fleet requiring more guns, the ability to carry a larger battery by placing ordnance lower down in the ship (enabled by the incorporation of lidded gunports) increased the demand for large guns (Fig. 2.2). Henry's personal interest in guns was well known – in fact it was to him that Niccolo Tartaglia dedicated his *Nova Scientia* treatise on gunnery in 1537 (Lucar 1588). Self sufficiency and, therefore, independence from Europe with regards to casting bronze guns, was one of Henry's achievements. This culminated in the domestic manufacturing of cast iron guns, a cheaper alternative which did not require the importation of any raw materials.

Bronze is an alloy of copper and tin with various other elements either intentionally added, or as impurities extant within the other elements. Analysis of the *Mary Rose* 'brass' guns show that the average range in copper content is 84.71–94.46%, alloyed primarily with tin, averaging 3.63–6.77%, (see Appendix 1.1) – today, we would term them bronze guns. Guns of '*cuprum*' are listed in the Tower of London as early as 1353 (Blackmore 1976, 252). No indication of size is given, but it is assumed that these were small. In 1365 two '*magni gunnes de cupro*' are recorded (*ibid.*, 253), and by 1405 there were 23 '*canones de cupro et ferro*' (*ibid.*, 255). The types (and sizes) increase rapidly alongside the wrought iron guns. In 1514 (*ibid.*, 259) the foreign gunfounders Hans Poppenruyter and Simon Giles of Mechelen are named. '*Hunsdyche*' [Houndsditch] is also mentioned; this later became a famous London foundry (Smith 1997, 108). It is suggested that the bulk of these early bronze guns were cast abroad and imported, in particular from Flanders (Cipolla 1965, 30–4; Teesdale, 1991, 12). Other major European centres were Caen, Lyons, Ghent, Mechelen, Bruges, Paris, Dijon and Venice.

In 1523 (*LP* 3, 3034) a new material for cannon in France, not so hard, more yielding and not so fragile as 'Italian' metal, is recorded. Less material was required and it was cheaper to transport. A 60-pounder which used to weigh 10,000lb (*c.* 4536kg) would be reduced to 6000lb (*c.* 272kg). Length was also reduced: a cannon to 10ft (*c.* 3m) from 12ft (*c.* 3.7m) and culverins and bastard culverins to 8ft (*c.* 2.4m) from 14ft (*c.* 4.3m). This French metal also included a reduction in tin from 10% to 6%. The report is unsigned but Awty (1988, 73) associated it with Francesco Arcano, then en route to England. If one looks at the *Mary Rose*

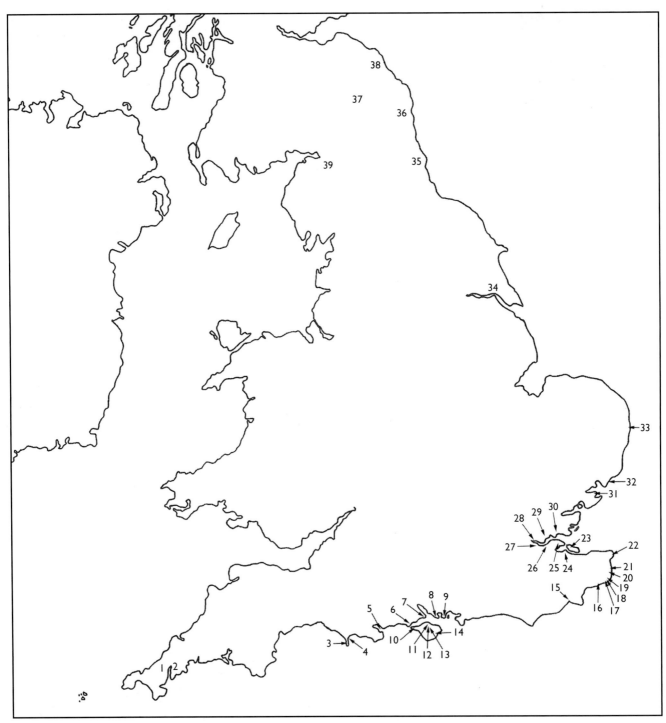

*Figure 2.1* Fortifications listed in the 1547 inventory. Key sites listed below
1. Pendennis Castle, Cornwall; 2. St. Mawes Castle, Cornwall; 3. Portland Castle, Dorset; 4. Sandsfoot Fortress, Dorset; 5. Brownsea Castle and Fortress, Dorset; 6. Hurst Castle, Hampshire; 7. Calshot Castle, Hampshire; 8. Haselworth Castle, Hampshire; 9. Portsmouth Town, fortified walls and Southsea Castle, Hampshire; 10. Sharpnode Blockhouse, Isle of Wight; 11. Yarmouth Castle, Isle of Wight; 12. Carisbrooke Castle, Isle of Wight; 13. Cowes Castle, Isle of Wight; 14. Sandown Castle, Isle of Wight; 15. Camber Castle, Sussex; 16. Sandgate Castle, Kent; 17. Archcliff Bulwark, Kent; 18. Dover Bulwark, Kent; 19. Dover Castle, Kent; 20. Walmer Castle, Kent; 21. Deal Castle, Kent; 22. Sandown Castle, Kent; 23. Sheerness Blockhouse, Kent; 24. Queenborough Castle, Kent; 25. Higham Bulwark, Kent; 26. Milton Bulwark, Kent; 27. Gravesend Bulwark, Kent; 28. The Tower of London; 29. West Tilbury Bulwark, Essex; 30. East Tilbury Bulwark, Essex; 31. Harwich Blockhouse, Essex; 32. Landghard, Suffolk; 33. Lowestoft, Suffolk; 34. Kingston upon Hull Castle, Yorkshire; 35. Newcastle upon Tyne; 36. Alnwick Castle, Northumberland; 37. Wark Castle, Northumberland; 38. Berwick upon Tweed Castle; 39. Carlisle Castle, Cumberland

*Figure 2.2* Model of the Mary Rose *built by Bassett-Lowke based on earlier interpretation of existing structure*

metallography results discussed later in this chapter (and given in Appendix 1.1) the highest single result for tin on any of the guns is 7.52%. This is from the cascabel of a gun cast in 1542 by Peter Baude (81A1423). The highest average tin composition (6.81%) is found in demi cannon 81A3002 cast by Francesco Arcano in 1535 and the lowest (3.63%) in the Owen double cannon (79A1276) cast in 1535, possibly in France. In some cases, however, the tin content is much lower (3.02%; cascabel, Owen double cannon, 79A1276). It is interesting that a number of the guns recovered from the *Mary Rose* are a great deal lighter than the annotated weights for the size of gun as given by Bourne (1587) or Norton (1628) (Table 2.1).

Bronze guns make up a very small percentage of the number guns from the *Mary Rose* and of the fleet in general. This was true also on land (Starkey 1998). The main constituent, copper, was imported initially from Hungary, the Tyrol, Saxony and Bohemia and the tin, if not English, from Spain and Germany. Trade in copper or in completed bronze guns centred around Nuremberg, Lyon, Bolzano (Italy) and Antwerp. During the life of the *Mary Rose* England began to produce enough cast guns for her home market and in the reign of Elizabeth I, enough for export. As the demand for bronze ordnance increased through the fifteenth and sixteenth centuries so did the cost. It is argued that this was one of the main reasons for the

**Table 2.1 Dimensions and weight of bronze guns**

| Named gun | Source | Bore (inches) | Length | No. of calibres | Weight (lb) |
|---|---|---|---|---|---|
| Cannon royal/8 | Norton | 8 | 12fft | 15–16 | 8000 |
|  | Bourne | 8 | 11–12ft | – | 7000–8000 |
| Cannon royal | 79A1276 | 8.5 | 9ft 9in | 12 | 4796 |
| Cannon | 81A3003 | 7.9 (8) | 9ft 9in | 13–14 | 5418 |
| Cannon of 7 | Norton | 7 | 11ft | – | 7000 |
| Demi cannon | Norton | 6.5 | 10–11ft | 20–22 | 6000 |
|  | Bourne | 6.5 | 10–12ft | 20–22 | 5000–5500 |
|  | 81A3000 | 6.5 | 9ft 9in (inc) | 15–16 (inc) | 5529 |
|  | 79A1277 | 6.3 | 12ft 3in | 20–21 | 6028 |
|  | 81A3002 | 5.5 | 8ft 8in | 17–18 | 3065 |
| Whole culverin | Norton | 5–5.5 | 12ft | 28–60 | 2200–2500 |
|  | Bourne | 5–5.5 | 12–13ft | – | 4300–4600 |
|  | 81A1423 | 5.5 | 12ft | 23–24 | 4556 |
|  | 80A0976 | 5.4 | 9ft 8in | 19–20 | 2977 |
|  | 79A1278 | 5.2 | 11ft 9in | 25–26 | 4829 |
| Demi culverin | Norton | 4.25–4.75 | 11ft | 28–60 | 2200–2500 |
|  | Bourne | 4.25–4.75 | 9–12ft | – | 1800–3200 |
|  | 79A1279 | 4.55 | 9ft 4in | 22–23 | 2299 |
|  | 79A1232 | 4.4 | 10ft 6in | 25–26 | 2845 |
| Saker | Norton | 3.5–4 | 9ft | – | 1400–1600 |
|  | Bourne | 3.5–3.4 | 8–10ft | – | 1400–1800 |
| Falcon | Norton | 2.75 | 7ft | – | 700 |
|  | Bourne | 2.75 | 7ft | – | 550–750 |

This table shows the prerequisites for the different names of guns listed by William Bourne in 1587 and Robert Norton in 1628. Length and weight of *Mary Rose* guns do not always conform to the prescribed Tudor doctrines of guns of the types listed

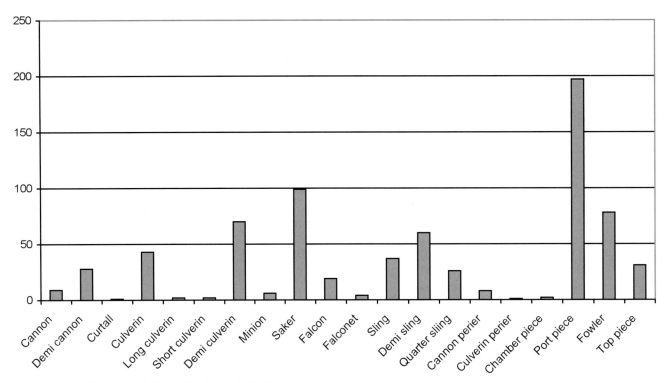

*Figure 2.3 All guns listed in the* Anthony Roll *(courtesy of Robert Smith)*

development of casting in iron and, by the middle of the sixteenth century, the nearly bankrupt Henry VIII was forced to use domestic raw materials (Cipolla 1965, 30).

Bronze is easier to cast than iron and requires a much lower temperature to achieve a molten state. There were no problems with rusting barrels, as was the case with iron guns. Teesdale argues that cast iron is inherently more brittle than bronze and iron guns were, size for size, heavier (1991, 19). However, the availability and relative cheapness of English iron, together with the acquired casting skills in bronze, opened up new potentials for casting in iron, and the Wealden cast iron gun industry was born.

In spite of the existence of a healthy domestic industry casting in both bronze and iron, the wrought iron gun was still numerically more significant on ships throughout the reign of Henry VIII. The sectional totals for the *Anthony Roll* (1546) list 310 cast bronze guns, 26 cast iron and 1268 wrought iron, excluding hailshot and handguns (Knighton and Loades 2000). One questions whether this is by desire, a reflection of the differential costs, or merely to use up existing stocks of wrought iron (Fig. 2.3). The ordnance carried onboard the *Mary Rose* mirrors this diverse technology. The fine cast bronze muzzle-loading guns were stationed in gun bays on the main gun deck beside the older wrought iron, breech-loading guns (see Figs 2.2, 2.24). She carried a total of 39 carriage-mounted guns distributed over three decks, but only fifteen of these were bronze (Figs 2.4–6).

A number of the bronze guns still bear the names of the continental gunfounders 'head hunted' by the young King, notably the Italians Francesco Arcano [Arcanus] and his sons Archangelo and Rafaelo (working in the Italian foundry in Salisbury Place, London) and the Frenchman Peter Baude, renowned for his role in establishing the cast iron gun industry in Sussex (Holinshed 1577, 832). Also named are the brothers John and Robert Owen. John is credited as being the first Englishman to cast culverins and cannon in bronze (Stow 1598, 288). The Owens and Peter Baude worked in the royal foundry in Houndsditch. It would appear that the Owens were junior to Peter Baude, who was paid four times their daily salary until 1537, at which point their salary becomes equal to half of his (Awty 1988, 62–4). Three of the *Mary Rose* guns are unnamed but, of these, one is initialled PB (81A3003), suggesting Peter Baude (and hereafter referred to as such). The other two (79A1279, 80A0976; Figs 2.5–6) may be Francesco Arcano's work. They also resemble his other guns, dodecagonal in section and without lifting lugs. These are similar to a saker in the Museum of Artillery Woolwich (Class II.2) dated 1535 and cast by Francesco, of which the intervening history is unknown.

## Types of Gun

There are several contemporary works describing the sixteenth century muzzle-loader. In the first place is

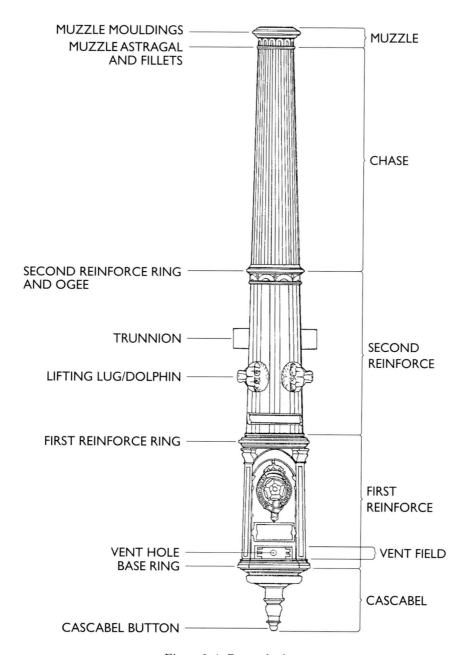

*Figure 2.4 Parts of a bronze gun*

Vanoccio Biringuccio's *De Pirotechnia*, printed in 1540. An unpublished manuscript compiled by Richard Wright in 1563 (Society of Antiquaries MS 94) gives the parameters for guns of specific types. The earliest English publication to describe particular weapons is William Bourne's '*The Art of Shooting in Great Ordnance*' written 1572/3 (BL Sloane MS 3651) and printed in 1587. The most substantial treatise is '*The Great Gunner*' by Robert Norton (1628), which includes descriptions of guns ancient and modern, with the mathematics and physics necessary to understand sixteenth century ballistics. There are also shorter written accounts such as that of John and Christopher Lad (1586: Bodleian Library, Oxford, MS Rawlinson A. 192. art. 4 / ff. 22–36; printed in Caruana 1992), and the anonymous '*Secrets of Gunmen*' (temp. James I: Bodleian MA Ashmole 343.VIII / ff. 128–139v; transcribed in Walton 1999, 389–428).

These works, referred to frequently below, are supplemented by information from other sources. Inventories (some including hand weapons and shot) survive with increasing frequency. These were often made at significant moments such as the start of a reign or the end of a war, or by commissions set up for the purpose. The key documents are the inventories of 1524 (Knighton and Loades 2000, 113–58), 1540 (*LP* 15, 196), 1546 (the *Anthony Roll*), 1547 (Starkey 1998) and 1444 (Longleat Miscellaneous MS V). Another important list of ordnance was compiled c. 1550 by James Burton (now East Sussex Record Office, SAS-CP/5/184; printed Awty 1989, 143). Among many later items in the Public Records are inventories of 1585 and

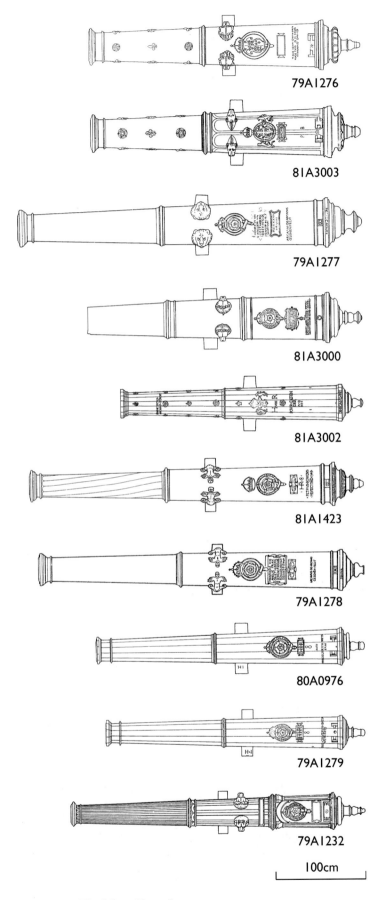

*Figure 2.5* The Mary Rose *bronze guns*

1586 with details of a lost '*Proportion*' of naval ordnance made by Sir William Winter in 1569 (Corbett 1898b, 300–11). Lists and tables from the early seventeenth century are also useful, including Bodleian MZ Ashmole 824.XXX / ff. 217v– 240v, dated 1614 and PRO SP 16/13, nos 56, 58, from Charles I's reign.

Norton further details the '*ancient forms*' and the '*foreign types*' and their influences on contemporary English ordnance, but he keeps these discrete from the English. He divides the latter into four types based on their bore diameter, the length of their chases, the '*fortification*' of their metal (wall thickness) and their use. Length is often expressed in calibres; the number of times the bore fits into the length. The largest guns are the *cannon*. The main difference between these and the other groups is the ratio of the bore (6–8in [152–203mm] in diameter) to the length from the muzzle to the base ring (the *calibre*, here ranging between 15 and 22). Their wall thickness never exceeds the diameter of the bore at the breech. The use of cannon was primarily to batter walls, curtains, bulwarks and defences, firing cast iron shot. Norton subdivides these into the *double cannon/cannon*, *royal/cannon of eight*, *whole cannon/cannon of seven* and *demi cannon*. He includes *minions* and *drakes* within the class.

Norton's second type are the *culverins*. These range in bore from 1½ –5½in (38–140mm) with a length of 28–60 calibres. Wall thickness is never less than one diameter of the bore at the breech. The subdivisions include *double culverin*, *whole culverin*, *demi culverin*, *saker*, *falcon*, *falconet*, *robinet* and *base*. These were designed to fire cast iron shot.

His third type include the *perriers*, or guns designed to fire stone shot, '*murthering*' shot (anything that will scatter) or fireballs. He includes the *cannon perrier*, the *port piece* (by this stage cast in bronze, but still chambered), the *stock-fowler*, the *sling* and the *bombard*. These have a length of eight calibres or more. The fourth type are the *mortar pieces*, *murderers* and *pettards*. These have a length of less than six calibres.

The *Mary Rose* carries examples of all four types, although not all in bronze.

## Gun classification

The sixteenth century classification system of smooth bore cast ordnance is by no means simple. Additional 'names', or 'types' can be found by scrutiny of not only length in calibres, but also by thickness of metal at predetermined

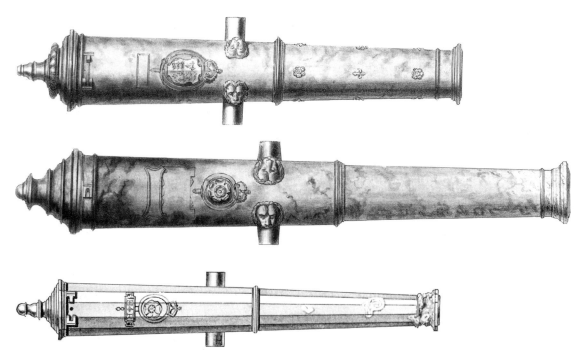

*Figure 2.6 Three of four guns recovered by John Deane in 1836 (© Portsmouth City Museum)*

locations. Thus, within even the simplest groups, the cannon and the culverins and their smaller versions, the demi cannon and demi culverins, there are numerous possibilities for alternative 'names', especially if one considers continental influence.

Names based on length include the following:

*Legitimate pieces*: The named gun adheres to the prescribed number of calibres.

*Bastard pieces*: The named gun has a larger bore and is shorter than prescribed number of calibres.

*Extraordinary pieces*: The named gun has a smaller bore and exceeds the prescribed number of calibres.

These names can be added to the existing name, either as a prefix or a suffix – *culverin bastard* or *bastard culverin*. Some of these also have 'nicknames', for example a bastard demi culverin is also termed an 'aspick' or 'aspyke'.

Names based on '*fortification*' or thickness of metal include the following:

*Common pieces*: The named gun has the same thickness of metal at the breech as the diameter of the mouth of the gun, or the bore.

*Lessened pieces*: The named gun has a thickness of metal at the breech which is less than the bore.

*Reinforced pieces*: The named gun has a thickness of metal at the breech which is greater than the bore.

It is perfectly possible for a bastard demi culverin or aspick to be common, reinforced or lessened. The reinforced pieces generally use a larger powder charge and greater range is listed (Norton 1628, 44).

## Distribution of Bronze Guns

The Anthony Roll lists fifteen guns of '*brass*' for the *Mary Rose*; two cannon, two demi cannon, two culverins, six demi culverins, two sakers and one falcon.

*Figure 2.7 M3 culverin on carriage* in situ *seen through Upper deck half beams*

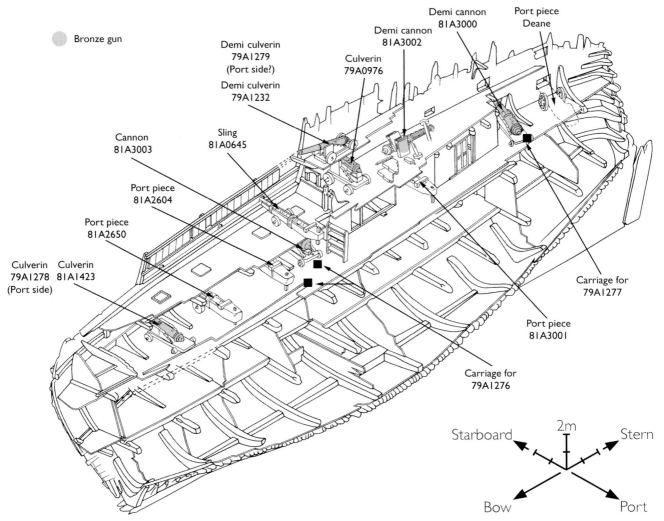

*Figure 2.8 Isometric diagram of the ship showing the actual and inferred positions of the guns on the Main, Upper and Castle decks*

Ten are known to have been recovered (Fig. 2.5) four by the nineteenth century salvaging (Fig. 2.6), and six during the recent excavations. Five of these were still on elm carriages with their muzzles protruding through open lidded gunports (Fig. 2.7), some of which were found hinged back against the starboard side of the ship, reflecting their open position at the time of sinking. The sixth had overturned and its carriage was found above it (Figs 2.8, 2.55). The positions of *in situ* ordnance and collapsed carriages are shown in Figure 2.8. as are the possible locations for four guns (79A1276, 79A1277, 79A1278 and 79A1279) recovered by John Deane. Locations postulated for the last two are based on size and style of gun; there is no archaeological evidence to back this up. Table 2.2 presents a summary of the bronze guns.

ordnance: a culverin at the bow (81A1423), a cannon amidships (81A3003) and a demi cannon in the stern (81A3000) (Fig. 2.9). These were flanked by the largest of the breech-loading wrought iron guns, the port pieces (Chapter 3) whose weight, at approximately one tonne, is nevertheless significantly less than the bronze guns. The bow and stern guns were both loaded with cast iron shot. The location of two virtually complete gun carriages in the upper silts amidships and in the stern at a higher level than the *in situ* bronze guns suggests a bilateral symmetry of the placement of the larger bronze ordnance. These have been studied and paired with the guns raised in 1836 and 1840 (79A1276, 79A1277). The possible pair to the bow culverin is culverin 79A1278, lifted in the nineteenth century (see Fig. 2.6 and below).

### The Main gun deck

Three of the seven gunports situated on the Main deck accommodated the starboard complement of bronze

### The Upper gun deck

The only *in situ* bronze gun was a culverin (80A0976) situated at an unlidded gunport just inside the

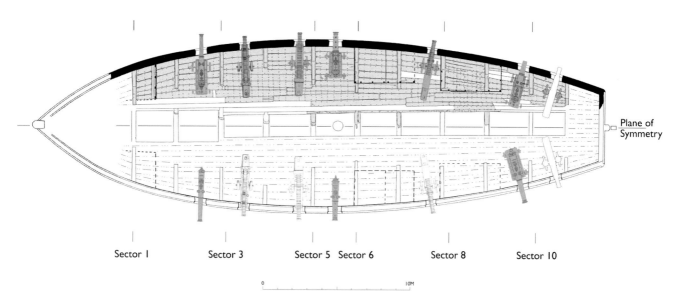

*Figure 2.9 Plan of the Main deck with guns*

Sterncastle structure (Figs 2.5, 2.8, 2.10). A second, very similar gun, although designated as a demi cannon (81A3002) was found lying north–south on the Upper deck just to the stern of this gun (Fig. 2.5, 2.8). This may be its port side pair. Both are unusually short (for either type of gun) possibly a prerequisite for this position.

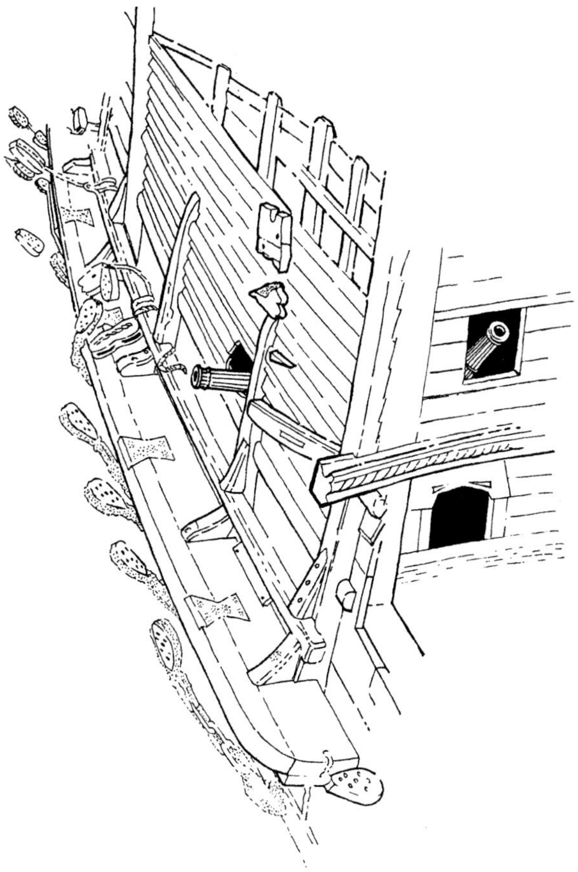

## The Castle deck

Only one *in situ* gun was found on the Castle deck, a demi culverin (79A1232). This faced forwards towards the Forecastle, possibly providing a valuable tactical role in covering the starboard bow of the ship (Figs 2.5, 2.8, 2.10). The only other demi culverin was lifted by Deane and Edwards (79A1279) and may be its pair, or it may belong to another pair of demi culverins situated elsewhere in the Sterncastle. Figure 2.11 shows a simplistic plan of the decks with known and suggested guns. Isolated carriage elements for bronze guns found in both the Fore and Sterncastles, along with loading implements and cast iron and composite shot of appropriate size, suggest the presence of the listed falcons and sakers. The partial remains of a trailing carriage (81A0486) found in the upper silts in the bow probably derived from the Forecastle (see below).

Although sharing similar attributes, each gun is unique as the method of manufacture requires an individual pattern of the final gun to be created (as discussed further below). These moulds are destroyed during the casting process.

Features (see below) seem to have been shared by the different gunfounders (Table 2.3; Fig. 2.5). The lifting lugs, known as '*dolphins*', in the form of lions' heads and winged mermen are used by Archangelo Arcano, the Owens and Peter Baude (Figs 2.12–13). Devices (as expected) are often similar, as is the placing

*Figure 2.10 The front of the Sterncastle with the forward firing castle deck gun (79A1232) and Upper deck gun 80A0976 showing*

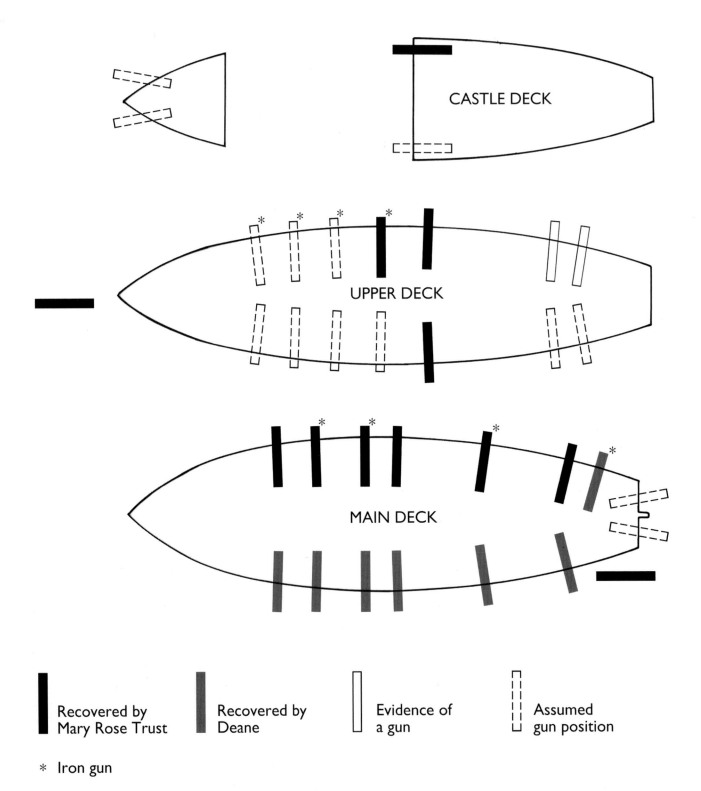

*Figure 2.11 Simplified deck plans with schematic gun types*

of relief decoration on the chase (Figs 2.14–16). Cascabel form (Figs 2.5, 2.17) and muzzle design also seem to vary among products of one foundry, yet share similarities with other workshops. The dates on the *Mary Rose* show a cluster around 1535, the time when she was altered, and, more surprisingly, 1542/3. There are four guns on the Main deck relating to this later period, a pair of culverins in the bow and a pair of demi cannon in the stern. As there is little evidence for the ship being in active service over this period she may not have had battle experience with her final array. Two guns are undated. The proposed pairs across the decks share a common or close date.

Table 2.2 *Mary Rose* bronze guns

| Type | No./Sector | Date | Founder | Length to base ring | Total length | Weight | Bore | Shot wt or diam./no. calibres |
|---|---|---|---|---|---|---|---|---|
| Cannon royal | 79A1276 M6 P** | 1535 | John & Robert Owen | 8ft 6in 2582mm | 9ft 9in 2977mm | 4796lb 2180kg | 8.5in 215mm | 68lb 12 |
| Cannon | 81A3003 M6 S | – | P.B. (?Peter Baude) | 8ft 9in 2656mm | 9ft 9in 2960mm | 5418lb 2463kg | 7.9in 200mm | none 13–14 |
| Demi cannon | 81A3000 M10 S | 1542 | John & Robert Owen | 8ft 7in 2622mm | 9ft 9in 2960mm | 5529lb 2513kg | 6.5in 165mm | 160mm 15–16 |
| Demi cannon | 79A1277 | 1542 | Arcangelo Arcano | 11ft 3351mm | 12ft 3in 3730mm | 6028lb 2740kg | 6.3in 160mm | 19lb 20–21 |
| Demi cannon | 81A3002 U7 P | 1535 | Francesco Arcano | 7ft 11in 2415mm | 8ft 8in 2635mm | 3065lb 1393kg | 5.5in 140mm | Broken† 17–18 |
| Culverin | 80A0976 U7 S | 1535 | Unnamed | 8ft 11in 2730mm | 9ft 8in 2955mm | 2977lb 1353kg | 5.4in 138mm | 128mm 19–20 |
| Culverin | 81A1423 M3 S | 1543 | Peter Baude | 11ft 3335mm | 12ft 3655mm | 4556lb 2071kg | 5.5in 140mm | Lost 23–24 |
| Culverin | 79A1278 M3 P** | 1542 | Arcangelo Arcano | 11ft 3335mm | 11ft 9in 3581mm | 4829lb* 1043kg | 5.2in 132mm | broken 25–26 |
| Demi culverin | 79A1279 ?C1 P** | – | Unnamed | 8ft 7in 2615mm | 9ft 4in 2865mm | 2299lb* 1043kg | 4.5in 115mm | 12.73lb 22–23 |
| Demi culverin | 79A1232 C1 S | 1537 | John & Robert Owen | 9ft 5in 2876mm | 10ft 6in 3210mm | 2845lb 1293kg | 4.4in 112mm | ?lost 25–26 |

All trunnions are below the centre line of the gun. Where recovered, all shot is cast iron. All weights are current weights unless marked * = taken from weight on the gun
** Recovered by the Deanes and Edwards 1836–1840
† Record destroyed: conservator reports object crumbled during extraction
P = port side gun; S = starboard side gun

## Decoration and Devices
by C.S. Knighton and Alexzandra Hildred

Many of the guns carry surface decoration and inscriptions. All are embellished with Royal Emblems, of which the most common (found on seven guns) is the Tudor Rose encircled by the Garter and surmounted by the Crown (Table 2.3; Fig. 2.14). In six cases the Crown lies over the Garter; in one case (81A1423) it sits clear above it. All but two of the Garters carry a version of the motto in Old French '*Hony soit qui mal y pense*' ('*Evil to him who evil thinks*'); those without the words are both by Archangelo Arcano (79A1277, 79A1278) and are dated 1542. Because the year began on Lady Day [25 March], the contemporary '1542' means 25 March 1542 – 24 March 1543 by modern reckoning, but, for convenience, guns are referred to here by the year dates inscribed.

The English heraldic rose is the hedgerow dog rose with five displayed petals. This device originated with the gold rose favoured by Edward I (reigned 1272–1307). During the dynastic wars of the fifteenth century a white rose was used by the House of York and a red rose was subsequently assigned to the rival House of Lancaster, so the conflict came to be called the Wars of the Roses. Henry VII, who descended from the Lancastrians and married the Yorkist heiress, devised the red and white Tudor Rose, with a double row of petals, as a symbol of dynastic and national reunion.

The Royal Crown as seen on the *Mary Rose* guns was first worn by Henry V (reigned 1413–22). It consists of a circlet of four crosses *formy* (ie, with each arm of the cross splayed outwards to the ends) alternating with four *fleurs de lys*. Springing from the circlet are four quarter-arches, surmounted by an orb and cross.

The Order of the Garter was founded by Edward III in 1348 and remains England's highest Order of Chivalry. The actual garter is still worn on ceremonial occasions and, since the fifteenth century, the Sovereign and the Knights Companion have encircled their arms with a representation of it.

A variant armorial on the guns is the Shield, either displayed alone, encircled by the Garter, or flanked by Royal Supporters. A demi cannon (81A3002) and two cannon (79A1276 and 81A3003; Table 2.3, Fig. 2.5) are embellished in this way; the first two are dated 1535, the other is undated.

The Shield is the same in all three cases, displaying the quartering of three lions *passant guardant* for England and three *fleurs de lys* for France. Three lions (sometimes understood as leopards) *or on a field gules*

79A1276

79A1277

79A1232

81A3000

*Figure 2.12 Guns with lion head dolphins*

(gold on a red background) were adopted by Richard I (reigned 1189–99). The *fleur de lys* had been the emblem of the French monarchs since Louis VII (reigned 1137–80) and were originally spread in profusion over the Shield (France ancient). Edward III added this device to the English Royal Arms *c.* 1340 symbolising his claim to the French throne which provoked the Hundred Years' War. He gave the French arms pride of place in the 1st and 4th quarters (top left, bottom right) because France was considered the senior kingdom. In the early fifteenth century the French reduced the number of *fleur de lys* to three (France modern), and Henry IV of England (reigned 1399-1413) followed suit. This form of the English Royal Arms was retained until 1603 and was therefore borne by all the Tudor monarchs.

Two Shields have Supporters (81A3002, 81A3003). Such augmentation of coats of arms, by placing creatures on either side, dates from the fifteenth century. The Supporters on the Royal Arms were not

      81A1423                            81A3003

*Figure 2.13 Guns with mermen and leopard head dolphins*

yet fixed; Henry VII preferred a dragon (proclaiming the Tudor descent from Welsh royalty) and a greyhound (a symbol of loyalty adopted by Edward III). Henry VIII used various Supporters, most commonly a lion and a dragon. The undated cannon (81A3003) bears a lion *rampant guardant* (erect and looking towards the observer) and a dragon. The demi culverin dated 1535 (81A3002) bears a greyhound and what appears to be a griffin. The griffin was a mythical beast with the head, ears, foreparts and wings of an eagle, and the hindquarters of a lion. This representation appears to have the foreparts of and a head complete with scales. The wings are quite large, more suggestive of a dragon.

All these guns carry inscriptions, sometimes within a decorative panel, cartouche or '*tabella*'. The most elaborate is on the demi cannon (81A3000), itself not otherwise much embellished (Fig. 2.16). The most ornate recovered gun (79A1232) has a similar but simpler cartouche. Both are Owen guns.

The smaller gun, dated 1537, is inscribed with the Latin version of the Royal style then current, which in translation reads: '*Henry the Eighth, by the Grace of God,*

Table 2.3 Devices and decoration on *Mary Rose* bronze guns

| Gun | Type | Surface decoration | Lifting lugs | Trunnion marks | Device | Founder | Date | Inscription |
| --- | --- | --- | --- | --- | --- | --- | --- | --- |
| 79A1276 | Cannon | single reinforce alternating *fleurs*/roses | lion heads | HI face r. trunnion | S, C, G | Owens | 1535 | *Henricus* etc in tabella |
| 81A3003 | Cannon | single reinforce alternating *fleurs*/roses | leopard heads | no | S, C, no G, lion/dragon | PB (?Baude) | No | *Henricus* etc in tabella |
| 81A3000 | Demi cannon | double reinforce plain | lion heads | no | R, C, insc. G | Owens | 1542 | *Henricus* etc in tabella |
| 79A1277 | Demi cannon | single reinforce plain | lion heads | no | G, C, G | A. Arcano | 1542 | *Henricus* etc in tabella |
| 81A3002 | Demi cannon | single reinforce alternating *fleurs*/roses faceted | no | HI face r. trunnion | S, C, ?griffin, greyhound | F. Arcano | 1535 | HVIII on gun, inscription on chase by muzzle |
| 80A0976 | Culverin | single reinforce faceted | no | HI top l. trunnion | R, C, insc. G | unknown (?FA?) | 1535 | HR in tabella, 8 below |
| 81A1423 | Culverin | single reinforce faceted & twisted | winged mermen | no | R, C above insc. G | Peter Baude | 1543 | *Vivat Rex* in tabella, HR 8 below |
| 79A1278 | Culverin | single reinforce plain | winged mermen | no | R, C, G | A. Arcano | 1542 | *Henricus* etc on gun, tabella |
| 79A1232 | Demi culverin | double reinforce fluted with columns | lion heads | no | R, C, insc. G | Owens | 1537 | *Henricus* etc in tabella |
| 79A1279 | Demi culverin | single reinforce faceted | no | HI top l. trunnion | R, C, insc. Garter | unknown (?FA?) | ? | HR in tabella, 8 below |

Devices:  C = Crown; (insc.) G = (inscribed) Garter; R = Rose; S = Shield

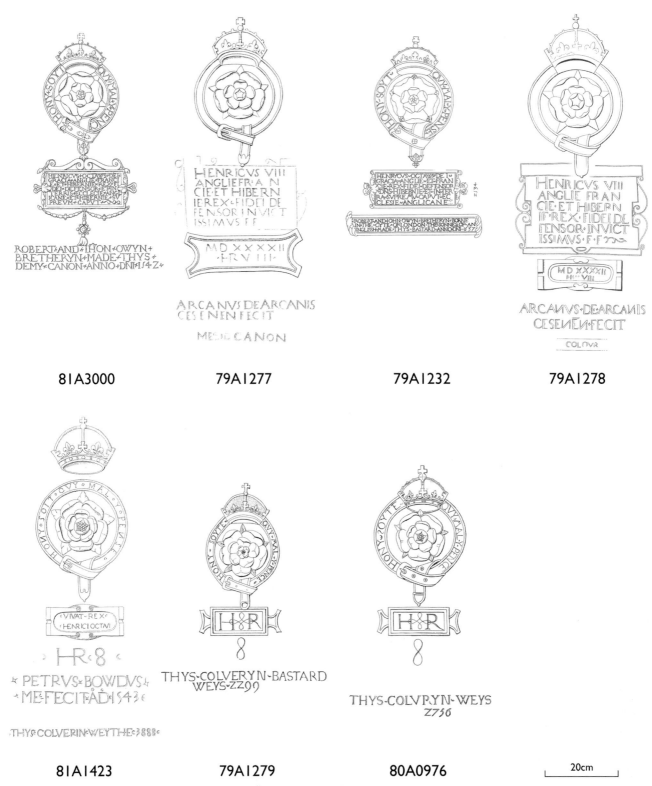

Figure 2.14 Guns with Rose and Crown devices

*King of England and France, Defender of the Faith, Lord of Ireland, and in Earth of the English Church Supreme Head*' (see below for Latin text). The claim of the English kings to be Kings of France dates from Edward III (see above) and was retained as a formality until 1801. The title '*Defender of the Faith*' was conferred on Henry VIII by Pope Leo X in 1521 as a reward for writing *Assertio Septem Sacramentorum*, a defence of Catholic belief against the Protestant reformer Martin Luther. The King's headship of the English church was established by the *Act of Supremacy* (1534), which ended English ties with the Papacy. The title '*Lord of Ireland*' was another papal honour, first given to John (reigned 1199–1216).

*Figure 2.15 Guns with Shield devices*

In January 1542 Henry VIII promoted himself King of Ireland and head of the Irish church, and the Royal style was changed to: '*Henry the Eighth, by the Grace of God, King of England, France and Ireland, Defender of the Faith, and of the Church of England and also of Ireland in Earth the Supreme Head*'. This form had been current for only a matter of months when the 81A3000 gun was cast; the Latin version inscribed on the cartouche is slightly defective (see below), perhaps because the wording was a novelty.

*Figure 2.16 Gun cartouch*

Four *Mary Rose* guns bear on one trunnion the letters 'HI', thought to stand for HENRICVS INVICTISSIMVS ('*Henry most invincible*'). This is located on the top of the left trunnion on two guns of similar form by unknown founders (79A1279, 80A0976) and on the face of the right trunnion on guns cast respectively by the Owens (79A1276) and Francesco Arcano (81A3002; Fig. 2.18).

Further details are inscribed in various permutations on the assemblage, such as the date of a gun's manufacture, the name(s) of the gunfounders, the gun type and the weight in pounds. Other marks observed include letters engraved on the tops or faces of trunnions. In the seventeenth century this usually denoted the gunfounder.

Three guns have surface decoration in the form of alternating *fleurs de lys* and roses on the chase. Two are similar, cannon 79A1276 and 81A3003. The third is a demi cannon (81A3002); in this case the decoration also partly covers the reinforce. This is the only recovered gun to have an inscription on the chase bearing the name of the gunfounder: FRANCISVS ARCANS ITALVS ('*Francisco Arcano the Italian*'). These inscriptions are made by applying relief decoration to the wooden former, probably using the wax technique now lost (see later in this Chapter).

|  |  | BAUDE GUNS | ARCANO GUNS |
|---|---|---|---|
| DEMI CULVERIN | 79A1232<br>Date 1537<br> 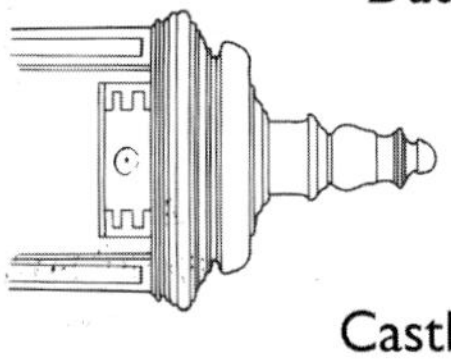<br>Castle deck<br>Firing forward starboard | | 79A1279<br>Undated<br> 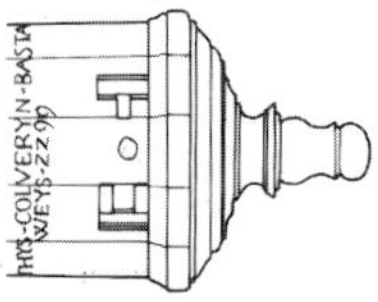<br>Location Unknown |
| CULVERIN | | | 80A0976<br>Date 1535<br> 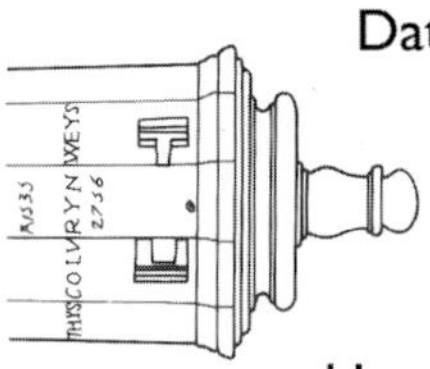<br>Upper deck<br>Stern starboard |
| CULVERIN | | 81A1423<br>Date 1543<br> 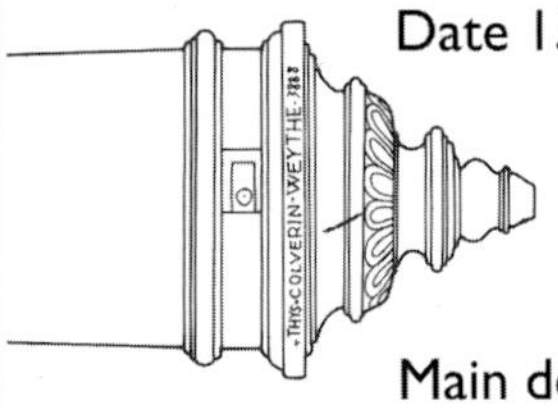<br>Main deck<br>Bow | 79A1278<br>Date 1542<br> 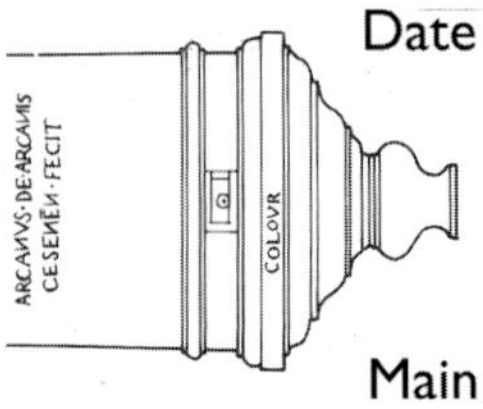<br>Main deck<br>Bow port |
| DEMI CANNON | | | 81A3002<br>Date 1535<br> 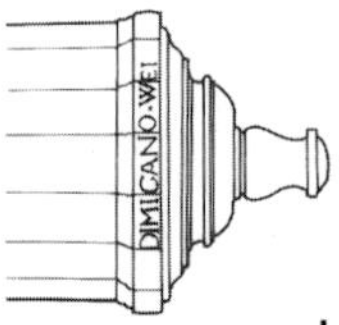<br>Upper deck<br>Stern |
| DEMI CANNON | 81A3000<br>Date 1542<br> 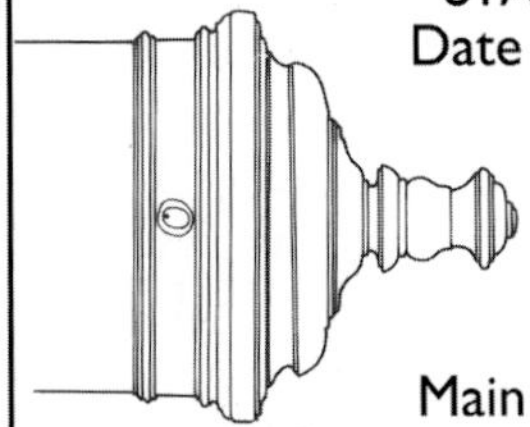<br>Main deck<br>Stern starboard | | 79A1277<br>Date 1542<br> 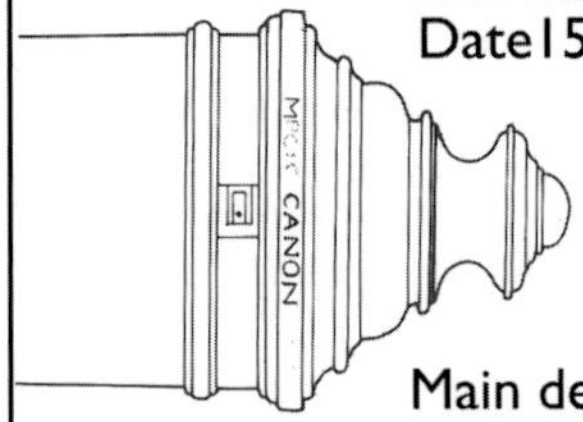<br>Main deck<br>Stern port |
| CANNON | 79A1276<br>Date 1535<br> 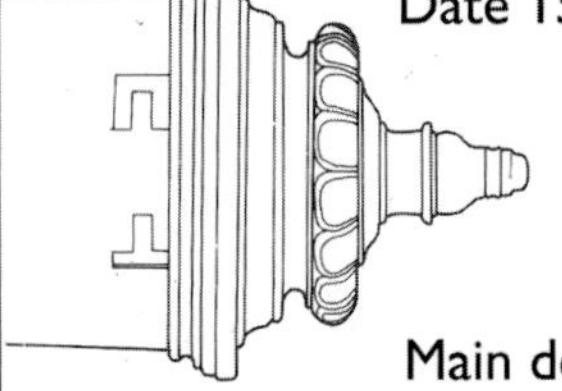<br>Main deck<br>Amidships port | 81A3003<br>Undated<br> 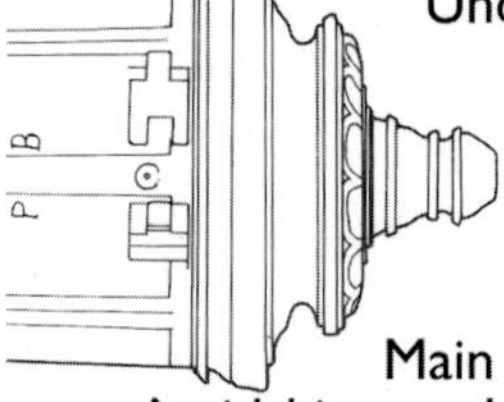<br>Main deck<br>Amidships starboard | |

*Figure 2.17 Cascabel shapes*

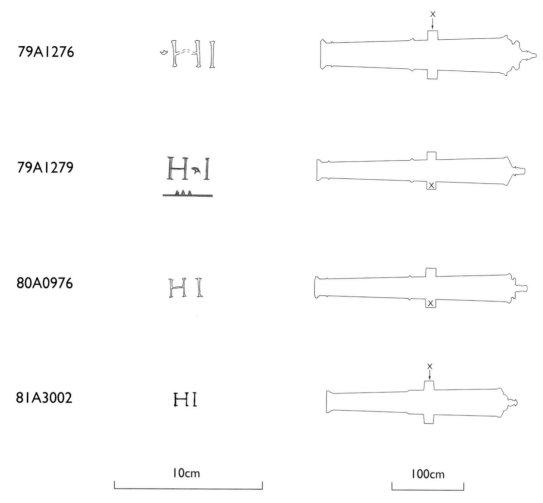

*Figure 2.18 Trunnion marks*

Four guns are faceted. Two are dodecagonal over the entire length (79A1279, 80A0976) and one is sixteen-sided (81A3002) (Fig. 2.19). The fourth has a faceted and spiralled chase (81A1423) (Fig. 2.59). Decoration in the form of elongate arches can be found on the reinforce of 81A3003, with columnar and arched grooves covering the demi culverin 79A1232. Three guns carry arched decorations on their cascabels (79A1279, 81A1423, 81A3003; Fig. 2.17).

## Discussion

Observing the distribution of guns in both functional and design/decorative terms, there is some similarity between those attributed to a similar position on the port and starboard side of the ship, despite the fact that these were manufactured by different gunfounders. This may be purely accidental, but our current knowledge about the requirements of knowing the structural constraints of specific areas before building carriages, suggests that there were a number of prerequisites, and design/decoration may only be one more of these.

The classes of guns were beginning to be standardised and it appears that the named classification in the 1546 inventory may be as a result of the dimensions of the gun rather than the type of gun achieved by the gunfounder. An example of this is the gun inscribed

**Table 2.4 Comparison 1514, 1540, 1546 inventories**

| 1514 | | 1540 | | 1546 | |
|---|---|---|---|---|---|
| Great curtow | 5 | demi cannon | 4 | cannon | 2 (2) |
| Murderer | 2 | culverin | 2 | demi cannon | 2 (3)* |
| Falcon | 2 | demi culverin | 2 | culverin | 2 (3 or 4*) |
| Falconet | 3 | saker | 5 | demi culverin | 6 (2) |
| Little brass gun | 1 | falcon | 2 | saker | 2 (0) |
| | | | | falcon | 1 (0) |
| Total | 15 | | 15 | | 15 |

Figures in brackets are guns recovered
* gun too small for demi cannon, too big for culverin

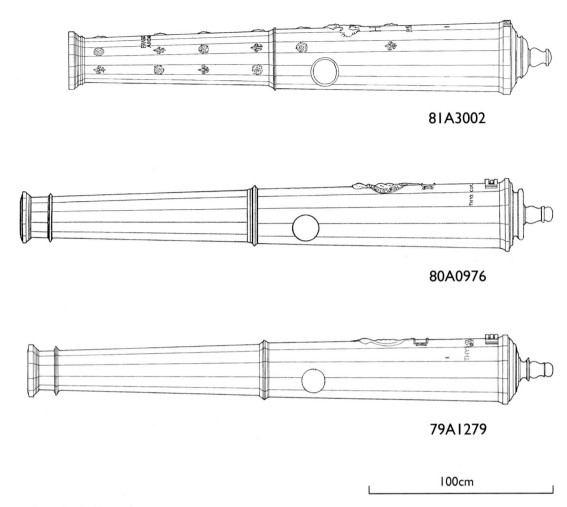

*Figure 2.19 Faceted guns*

with the words '*demi cannon*' (81A3002) which has a smaller bore and a shorter length than most guns within the demi cannon group. Although short, had this gun been unnamed, it would have been classified as a short culverin alongside culverin 80A0976, which it may have been paired with on the Upper deck in the stern (Figs 2.8, 2.19). The excavated assemblage would then correspond in both the cannon and demi cannon classes with the *Anthony Roll*. Interestingly, this is an early gun, cast by Francesco Arcano, with a relatively high tin content. There are two unnamed guns which are very similar in form and decoration (79A1279, 80A0976). They are both dodecagonal and both lack dolphins, a feature observed on all surviving guns cast by Francesco (Awty 1988, 63; Fig. 2.19).

The bronze guns carried on board include two of only nine cannon listed within the entire fleet, and differences in the armaments in 1540 and 1546 (Chapter 1) demonstrate an increase in the size of bronze guns (Table 2.4). It is unlikely that she would be re-armed with such an investment in ordnance should her seaworthiness have been questionable. The cannon were mounted amidships – one to starboard and the other to port – and are outwardly similar in appearance (Fig. 2.5). One is dated 1535, although cannon are not listed as being on the *Mary Rose* until the 1546 inventory. This suggests either that the gun had come from elsewhere and was not placed on the ship until 1542/3, or that the 1540 inventory is incorrect. The use of guns from other ships is not unexpected, nor is the obvious re-use of carriages (see 79A1277). When a ship was laid up it was customary to hand over the artillery to the ordnance office. The length of time for which a carriage was in use is unknown, but inventories often list the following categories: guns on carriages newly stocked, guns on old carriages, worn, etc, and guns without carriages (Starkey 1998, 102–44).

The total of bronze guns recovered falls five short of the *Anthony Roll*. The recovered assemblage is also different, with three demi cannon and three culverins rather than the two listed. One of the demi cannon (81A3002), however, has dimensions and weight which fit better with a culverin, albeit a short one. This allows for two 'normal' culverins and two 'short culverins'. This would leave two demi culverins and the sakers and falcons unaccounted for. All of these would be on the Upper or Castle decks and may have been salvaged by the Tudors. This makes more sense if the Deane demi culverin (79A1279) was a starboard gun rather than the pair to the *in situ* demi culverin (79A1232). Port side

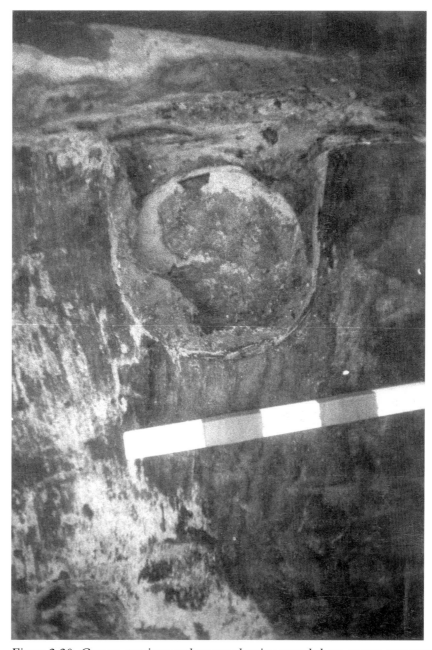

*Figure 2.20 Gun on carriage underwater showing corroded cap square over trunnions*

ordnance (if attached by recoil ropes) would be generally more accessible to contemporary salvors. The suggested pairing of the demi culverins (79A1232/79A1279) and, indeed, the two Upper deck guns (80A0976/81A3002) is hard to prove. Although the weights do not vary greatly between the proposed pairs, in each case the length is different: around 300mm in both instances (Table 2.2).

## Gun Mountings

Six of the ten bronze guns recovered were directly associated with their carriages (Fig. 2.20). The *in situ* guns were recovered, where possible, on their carriages so that the fittings securing the gun to the carriage could be studied in detail, and facsimiles made. Each gun lift was unique as different methods had to be employed to remove the gun and carriage from its particular confines within the ship. Three were raised on their carriages (81A1423, 81A3000, 81A3003) (Fig. 2.21) and two were removed from their carriages underwater (79A1232, 80A0976). The sixth gun (81A3002) was found upside down beneath its broken carriage on the Upper deck in the stern (Figs 2.8, 2.22).

Inventories list some of the components that have been identified on the *Mary Rose* gun carriages. The 1547 inventory (Starkey 1998) lists items held within the fortifications and ships at the end of Henry's reign and specifically lists carriages for different types of guns.

*Figure 2.21 Guns lifted on carriages: a) culverin 81A1423 (M3); b) demi cannon 81A3000 (M10); c) Charles Pochin washing cannon 81A3003 (M6) after lifting*

*Figure 2.22 a) Sterncastle mosaic; b) underwater photograph of gun during excavation (courtesy Patrick Baker, Western Australian Museum)*

It also lists '*Forecarriages*', which may be merely the bed and trunnion support cheeks, '*Forecarriage bolts of iron*', '*washers for wheels*', '*iron wedges*', '*elm planks for the stocking of brass ordnance*' and wheels listed for particular types of gun. Other entries list '*capsquiers*', [capsquares] '*capsquier pynnes*', '*spikes of iron*' and '*staples*'. Contemporary terms have been applied where possible; where not, descriptive terms have been invented or borrowed.

The carriages recovered share the following characteristics (Fig. 2.23). They comprise a bed, two trunnion support cheeks and two rear stepped cheeks, all of elm. Each have a foreaxle tree and a rearaxle tree of ash and four trucks of elm. The carriage is held together by a series of tenons with additional wrought iron bolts and wedges. Certain elements are additionally secured with oak pegs. Later carriages had sides made of a single cheek. Accurately dated carriages are rare and shipwrecks provide one of the best sources. The two most closely dated parallels are from the late sixteenth/early seventeenth century wreck off Alderney and the *Scheurrak S01* wreck in the Netherlands, dated 1590 (Smith 2001, 25–38). The *Mary Rose* is important for this period as it has examples of a number of types of guns with their carriages.

The beds are either made from a single plank or two planks held longitudinally by draw-tongued joints formed by loose wooden tenons mortised into the side of each plank and subsequently pegged. This technique would not require a massive bed plank, and thus a smaller, younger tree would suffice.

Biringuccio (1540, 314) in his section on '*Finishing and Mounting Guns*' writes regarding carriages for guns on board ships:

> '*There are many who customarily make these beds without any care, like two flat boards, and indeed, this form is made for the very heavy guns, or for those that are to serve to arm warships, or perhaps because of inability to find wide enough boards, since it does not matter how those are made that are put in ships and places where they are to stand still.*
>
> *Now, these boards are joined with three strong transverse pieces mortised through the thickness of the wood with two tenons at each end. They are secured from the top with a peg of holm oak or oak which passes completely through all mortises and reaches to the bottom.*

He may be describing an Italian tradition and it is interesting that the guns (where the founder is named) with jointed beds were cast by the (Italian) Arcana family (79A1277, 81A3002), although the '*stocking*' of carriages was undoubtedly separate from the manufacturing of the gun.

Slots (either single or two parallel ones on each side of the bed) are cut in the beds to accept single or double tenons on the trunnion support cheeks; these are secured with oak pegs (Fig. 2.23). In all but two cases the trunnion support cheek has a double tenon. The two exceptions are the carriage for the bastard culverin (79A1232, the most gracile carriage recovered supporting the lightest gun) and a single cheek (81A0707) found in the starboard Sterncastle scourpit. This is also slight and may be for a similar size of gun. Shallow rebates in the bed in front and behind these slots accept angled brackets bolted to the bed. These brackets are also bolted to the front and rear faces of the trunnion support cheeks which are relieved to accept them.

The trunnion support cheeks and the bed form the support for the gun. The majority of the weight is directly onto the trunnion rests and through the trunnion support cheeks to the bed. Directly beneath the trunnion rests (either on or under the bed) is the axle for the fore trucks. This is made of ash and acts to spread the load outwards to the trucks. The choice of elm for the majority of the carriage components may be based on its toughness and ability to withstand shock-loading. The components are cut so that the grain line is parallel to the lines of stress. The trunnion support cheeks are cut with the grain aligned vertically and the rear stepped cheeks are manufactured with the grain aligned horizontally, to transmit an increased load from the recoil backwards along the length of the bed plank. The separate side-cheeks are interesting as when the gun is fired the whole assembly tries to 'jump' off the fore trucks. The divisions might be to counteract the splitting which might occur in a single cheek. The adoption of single cheeks might have arisen as a result of a refinement in restraining recoil.

The tenons on the trunnion support cheeks are pegged with oak pegs which originate in the sides of the carriage bed. The holes for these are offset a quarter inch (6mm) upwards in the tenon relative to the holes in the side of the bed. Once in place, these would require drilling out before the trunnion support cheeks could be removed. It is not necessary to remove these cheeks to mount or dismount the gun.

An area must be relieved at the front of the rear stepped cheek from the underside to accommodate an iron wedge that holds the rear bracket. This is, in turn, rebated into the rear of the trunnion support cheek and the bed. In most cases the stepped rear cheek has a tenon on its front face, offset to the outside of the carriage. This tenon enters into a rebate cut into the rear face of the trunnion support cheek. The offsetting of this tenon means that the carriage has a definite left and right hand side and so it is impossible to assemble it incorrectly. Several carriages have a tenon on the rear face of the trunnion support cheek with a corresponding mortise. Wrought iron transom bolts held with iron wedges are placed between the trunnion support cheeks to secure them.

A three-sided cap square made of wrought iron is fashioned with a portion removed to enable it to be revolved about the tenon which holds the trunnion support and rear stepped cheeks together. The cap square overlaps the brackets on the front and rear faces of the trunnion support cheeks and the entire assemblage is secured by a long iron bolt originating on the front face of the trunnion support cheek. This goes through the fore bracket, cap square (front), trunnion support cheek, rear bracket and into the stepped cheek. The stepped cheeks are then wedged into position.

Long, shallow rebates are cut in the bed towards the rear in order to accept a shallow tenon located on the underside of the stepped rear cheek. An area inside the stepped rear cheek is hollowed out to house either a staple or a pronged plate which is hammered into the bed. This is wedged together with wood or wrought iron wedges which penetrate the slots cut through the rear stepped cheek.

The axle trees are usually positioned in a groove cut in the underside of the bed at the front and rear and are held with slotted bolts and wedges. The axles are of ash with the grain aligned perpendicular to the length of the bed, thereby transmitting the weight and any shock out towards the trucks. The trucks, which are convex on their inner faces to reduce rubbing and sticking, are

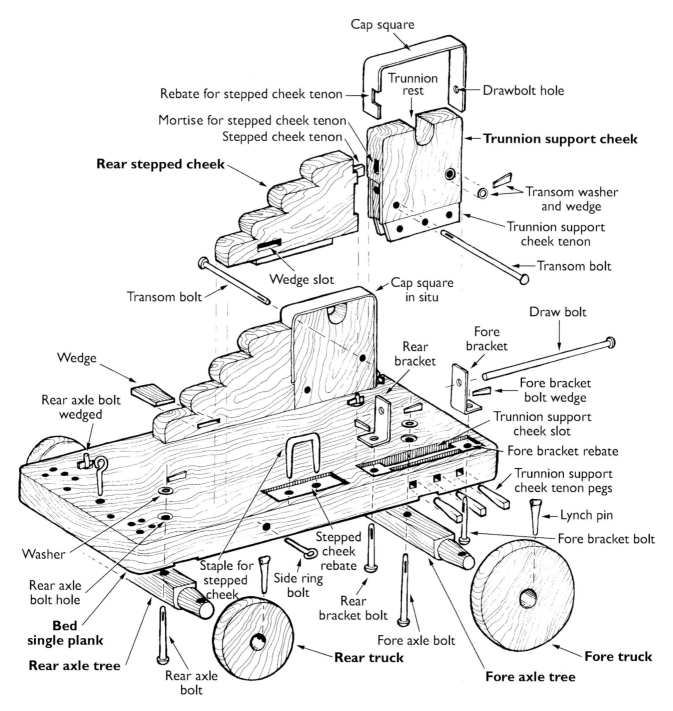

*Figure 2.23 Parts of a gun carriage*

positioned on the chamfered ends of the axle trees and held with a lynch pin of oak or wrought iron. It is necessary to remove only the cap square and one rear stepped cheek to dismount the gun. Practical experiment has shown that it is possible to assemble the carriage completely and mount the barrel (using a chain hoist) within 30 minutes.

Carriages with single bed planks have the foreaxle bolts inside the trunnion support cheeks, whilst those manufactured with jointed beds have their axle bolts outside the trunnion support cheeks. The axle bolts for the rear axle tree always occur behind the rear stepped cheek and never directly inside or outside the cheeks.

The bed thickness also varies with the weight of the gun: from 6.7in (170mm) for a cannon to 4.7in (120mm) for a culverin. The length of gun determines the length of the bed plank and the maximum width of the gun, naturally, dictates the width of the bed plank. This varies between 25in (635mm) and 34in (864mm) for single planked beds to 38in (965mm) for the largest jointed bed (Table 2.5).

The presence or absence of a jointed bed does not seem to correspond with deck, size of gun or date of

Table 2.5 *Mary Rose* gun carriage dimensions (in mm)

| Gun Type | 79A1276 Cannon | 81A3003 Cannon | 81A3000 Demi cannon | 79A1277 Demi cannon | 81A3002 Demi cannon | 81A1423 Culverin | 80A0976 Culverin | 79A1232 Demi culverin | 81A0707 ? | 81A0124 ? | 81A0107 ? |
|---|---|---|---|---|---|---|---|---|---|---|---|
| Bed L. | 2000 | 1760 | 2160 | 2060 | 2100 | 2310 | 1840 | 2050 | – | – | – |
| Type | single | single | single | jointed | jointed | single | jointed | single | – | – | – |
| Bed Th. | 100–150 | 170 | 120 | 152 | 127–140 | 127 | 127 | 152 | – | 100 | – |
| Bed W. | 870–940 | 860–900 | 730 | 960 | 387–450 | 812 | 850 | 635 | – | – | – |
| TSC L. | – | 565 | 450–460 | 450 | 420 | 559 | – | 250–270 | – | – | – |
| TSC H. | 410 | 410 | 400–420 | 400–410 | 795 | 380 | 762 | 530 | 490 | – | – |
| TSC Th. | 150 | 152–160 | 125 | 140 | 152–159 | 152 | 152 | 102 | 90 | – | – |
| TSC W. | 560 | 570 | 600 | 550 | 495–546 | 550 | 550 | 500 | – | – | – |
| Tennon type | double | double | double | double | double | double | double | single | single | – | – |
| T. rest H. | 150 | 150 | 150 | 150 | 145 | obscured | 153 | 110 | 130 | – | – |
| T. rest W. | 160 | 150 | 150 | 150 | 150 | 140 | 153 | 120 | 135 | – | – |
| Clad type | Y open | Y open | ? | ?enc. | enc. | Y enc. | ?enc. | no | no | – | – |
| RSC L. | 850–910 | 870 | 750 | 780 | 780–800 | 806 | 790 | 820 | – | – | 659 |
| RSC Th. | 120 | 135 | 120 | 125 | 152+ | 146 | 127 | 99 | – | – | 100 |
| RSC H.* | 380–390 | 420 | 400–420 | 400–430 | ? 533 | 375 | 455 | 465 | – | – | 440 |
| Foreaxle L. | – | holes 270 apart | holes 350 apart | – | inc. | – | 1708 holes 700 apart | 1270 holes 180 apart | – | – | – |
| Foreaxle W. | – | 125 | 120 | – | 100 | 120 110 | 140 | 140 | – | – | – |
| Rearaxle L. | – | 1400 | holes 450 apart inc. | – | 1145 inc. | 1450 holes 580 apart | 1410 holes 580 apart | 965 holes 340 apart | – | – | – |
| Rearaxle W. | – | 140 | – | – | 100 | 140 | 130 | 125 | – | – | – |
| RFT D. | – | 440 | 470 | **250 | 350 | 431 | 580 | 600 | – | – | – |
| RFT. Th. | 150 | 110 | 120 | 170 | 150 | 108 | 150 | 127 | – | – | – |
| LFT D. | – | 440 | 445 | **420 | 300 | 430 | 603 | 600 | – | – | – |
| LFT. Th. | – | 120 | 120 | 170 | 150 | 107 | 150 | 100–120 | – | – | – |
| RRT D. | 300 | 410 | 285 | **262 | 350 | 330 | 496 | 510 | – | – | – |
| RRT. Th. | 150 | 140 | 130 | 170 | 110 | 120 | 100 | 100 | – | – | – |
| LRT D. | 280 | 410 | 275 | **250 | 250 | 330 | 496 | 510 | – | – | – |
| LRT. Th | 130 | 150 | 140 | 120 | 120 | 120 | 100 | 100 | – | – | – |

Key:
L. = length; D. = diameter; Th. = thickness H. = height
TSC = trunnion support cheek; T.rest = trunnion rest; RSC = rear stepped cheek; R/LFT = right/left fore truck;
R/LRT = right/left rear truck; Y = trunnion rest has iron cladding; enc. = enclosed; inc. = incomplete
Trunnion support cheek tenons are 95–120mm deep
* The height of the rear stepped cheek does not include the tenon
** Based on divelogs, sensibly 81A1277 would pair with 81A3000 and some of the trucks. Eight were found close together

casting. Two of the three guns attributed to the Arcana family have jointed beds (79A1277, 81A3002); that for the third (79A1278, Tower of London XIX. 167) was not found. The only other gun carriage with a jointed bed is the culverin (80A0976) bearing the same date as the Arcana demi cannon which may have been its pair on the Upper deck (81A3002). These two beds are almost identical, the only discernible difference is that the bed for the culverin is longer than that for the demi cannon. Both these guns and a third, undated, unnamed bastard culverin, are either twelve- or sixteen-sided and do not have dolphins. These three are the only

*Figure 2.24 Museum display showing bronze and iron guns on a reconstruction of the midships area of the Main and Upper gun decks*

guns of this assemblage with these features and all three have '*H I*' incised on one trunnion. It is therefore possible that the jointed carriages originate from one workshop, possibly used by the Arcana family, reflecting a bias towards a specific design (or the availability of raw material).

The weight of the gun seems to influence the thickness of the trunnion support cheeks which vary between 4.9in (125mm) and 6.7in (170mm) for the larger guns and 3.7in (94mm) for the bastard demi culverin (79A1232). The rear stepped cheeks do not bear the weight of the gun and are often thinner than the trunnion support cheeks. In this case they are offset towards the outside of the carriages in order to present an external face flush with the trunnion support cheek. In some instances the trunnion recesses penetrate the entire width of the trunnion support cheeks while, in others, a portion is left towards the outside face of the cheek.

Once the parameters of width, depth and thickness of bed plank and the thickness of the trunnion support cheeks have been met, the slots for the trunnion support cheek tenons must be marked out and cut into the bed. These vary depending on gun width across the trunnions and are between 17.7in (450mm) for the demi cannon and 9.8in (250mm) for the bastard culverin. Although it is necessary to know the gun type at this stage in the carriage construction, it is impossible to state whether or not the exact dimensions of a particular gun were known before the carriage components were chosen and the trunnion support cheek tenon slots cut. A general clustering of measurements for similar types of gun, combined with evidence of the adaptation of certain carriage components for certain guns, suggests that exact measurements were not known. In all cases the adaptation is slight, and usually to accept exaggerated decorative features on the gun (79A1232, 81A3003).

Payments dated 1511 were made to Cornelis Johnson towards the '*new stocking and repairing of divers pieces of ordnance*' of four of the King's ships then in the Thames, including the *Mary Rose* (*LMR*, 4). This actually suggests that some of the carriages were tailored to guns already on board ships.

The trunnion rests are exactly fashioned to the trunnions; these could be finely fitted to the gun during the mounting. This 'fine tuning' is borne out by the relieving of some of the carriages. Several are lined with an iron 'pillow'. This might be to take a gun with smaller trunnions within an existing carriage. There is no indication that these are to take any trunnion taper.

The height of the trunnion support cheek appears to vary with the location of the gun within the ship and its function. The three *in situ* starboard Main deck bronze gun carriages and the two collapsed supposedly Main deck port side carriages are built with their trunnion support cheeks and rear stepped cheeks of the same height. This places a constraint on the maximum attainable elevation of the gun. The size of the main deck gunports and the height of the lower sill from the deck also dictate the minimum and maximum height of the trunnion recesses and in turn the height of the cheek. Elevation of these guns above the horizontal is very slight (Fig. 2.24).

The carriages from the Upper deck have very high trunnion support cheeks and also very large fore trucks. These features may be purely to gain access to the Upper deck gunports, or for manoeuvrability, or both.

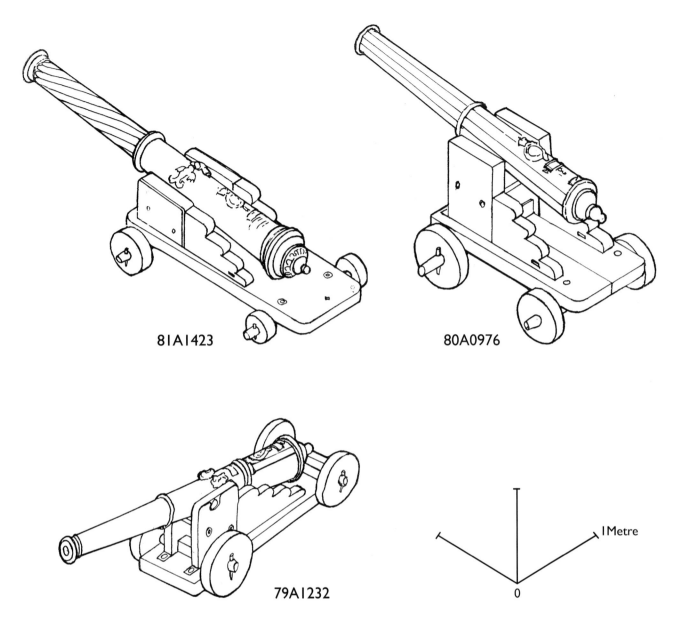

*Figure 2.25 Three types of carriage as recognised from the* Mary Rose *bronze guns: 81A1423 (Main deck); 80A0976 (Upper deck); 79A1232 (Castle deck)*

The Upper deck gunports known to house bronze guns are situated much higher above the deck than those on the Main deck. This dictates an increased angle of elevation, necessitated by the external features of the ship at this station, just aft of the start of the chain plate and rigging rail. The function of the gun may also dictate the need to elevate it. Constraints on maximum elevation of guns indicates that this is a combination of both carriage and height of the upper sill above the deck.

The thickness and diameter of the trucks differ with the position within the ship and weight of the gun. Those on the Upper and Castle decks are thinner but generally have a larger diameter than those which support the Main deck gun carriages which have smaller trucks that absorb more energy.

The ultimate location of a specific type of gun also appears to have been understood by the carpenters. It is this factor which may account for carriage differences. The variants include robust carriages with their trunnion support cheeks and rear stepped cheeks of approximately equal height and between 14.9in (380mm) and 17.3in (440mm) above the bed. These are restricted within our assemblage to the five carriages for Main deck bronze guns. In type, the guns are culverins, demi cannon and cannon (Fig. 2.25).

The second variant comprises the robust carriages where the trunnion support cheek is much higher than the rear stepped cheek. There are two carriages of this type, both for Upper deck guns, one a culverin and the other a small demi cannon (80A0976, 81A3002). The exaggerated height of the trunnion support cheek is necessitated by the height of the lower sill above the Upper deck. This is 40in (1030mm) as compared with heights of between 16.2in (410mm) and 25in (640mm) for the lower gunport sills above the Main deck. In both instances the bed is jointed and has two ancillary wooden transoms in addition to the transom bolts to reinforce the cheeks. These fit between the two trunnion support cheeks and are possibly as a result of the height of these cheeks. Both these carriages are remarkably similar with the exception of bed length and trunnion rest width. The variant (found with a demi culverin) is much lighter in form (79A1232). There is also one similar trunnion support cheek found outboard (80A0707).

All but one of the bronze gun carriages recovered share these features and methods of construction. There are, however, differences in the carriages. These include the thickness of the wood used for the bed, the thickness and height of the trunnion support cheeks and the rear stepped cheeks, the diameter and thickness of the trucks and the form of attachment used for the rear stepped cheek.

### A note on the gun descriptions and measurements

In the descriptions that follow, in accordance with the system used for the Royal Armouries guns (Blackmore 1976), two lengths are provided for each gun; the length from the mouth to the base ring; the total length, including the cascabel and button (see Fig. 2.4, above, for parts of the gun). All measurements are quoted in both imperial and metric. Original measurements were taken in millimetres and are here converted back to imperial and rounded to the nearest inch for measurements over 1ft and to a quarter inch for those of less than 1ft. Guns were weighed using a Dynafor Load Link load indicator and the inscribed weights on the guns are assumed to be in English pounds. Original weights were taken in kilograms and are here also converted back to imperial and rounded to the nearest pound, for weights over 100lb and to the nearest ounce for those of less than 100lb. Details of the measurements of major elements of the gun carriages are given in Table 2.5 and are, for brevity, omitted from the text (details in archive). The position of the trunnions is measured along the length of the gun taken from the back of the base ring to the centre of the trunnions. The number of calibres of any gun is calculated as the number of times that the bore can be divided into the length of the tube; given the incomplete nature of some of the *Mary Rose* guns this can only be an estimate in those cases.

## Cannon

Although the name '*cannon*' is often considered synonymous with any large muzzle-loading gun, by the middle of the sixteenth century the name was specific to one of a set of guns designed to throw a heavy shot, primarily for use in battery against walls. Although sporadic references occur as early as 1399 (Blackmore 1976, 255) the class is not routinely referred to until the 1530s. Blackmore suggests that the first real inventory notation is in 1540 (*ibid.*, 220). Early descriptive English references include the Richard Wright manuscript of 1563 (Society of Antiquaries MS 94) where line drawings of fat short guns mounted on two wheeled carriages accompany the text (Fig. 2.26). The guns in this section include double cannon, demi cannon, curtalls, bombards and basilisks. Biringuccio (1540, 255) states that the last three were, by this period, replaced with cannon shooting iron rather than stone shot. He describes three kinds: double cannon, cannon and demi cannon. All are in one piece, the thickness of bronze at the breech is three-quarters of the diameter of the ball and, at the mouth, it is one-third of the diameter. Length is described as 22 calibres, and the weight up to 8000–9000lb (*c.* 3629–4082kg).

Norton, published some 70 years later, describes the *fortification* of the ordinary examples from each subgroup as being seven-eighths of the diameter of their shot at the breech, seven-eighths at the trunnions, five-eighths at the Cornish (muzzle-moulding) and three-eighths at the mouth. He states that some founders gave demi cannon the full thickness of one whole diameter at the chamber, with 143–220lb (*c.* 65–100kg) of metal per pound of shot.

The cannon have the largest bore of all of the cast bronze guns within the fleet. Attributed names include cannon royal, double cannon, cannon of eight, whole cannon, cannon of seven, cannon serpentine. Inventory entries also list these by country of origin, '*Venetian*', and '*French*' being the most common. The cannon listed within the Tower of London (hereafter: the Tower) from

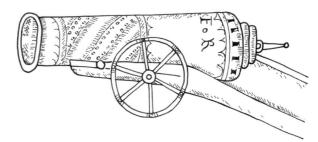

*Figure 2.26 Cannon as depicted by Wright in 1563*

1540 onwards give an indication of size and weight of guns and shot over time (Blackmore 1976, 220)

The 1540 inventory of ten vessels of varying types lists nine cannon. Five are assigned to the *Great Galley*, two to the *Peter*, and two to the *Small* (or *Less*) *Galley*. The inventory is difficult to interpret, and it appears that demi culverin are listed twice. Cannons occur, most unusually, after the first listing of demi culverin. Although featured, the *Mary Rose* does not seem to have been furnished with cannon at this time. The *Anthony Roll* of 1546 also lists nine cannon, all but one on the Great Ships: four on the *Henry Grace à Dieu* (not inventoried in 1540), two on the *Mary Rose*, two on the *Jesus of Lübeck*, the other on the *Galley Subtle*.

The largest are the double cannon or cannon royal. These have a bore of over 8in (203mm) and are listed as being of 15 calibres or 12ft (3.6m) in length and weighing 8000lb (3629kg) firing a cast iron shot of 63lb (28.6kg). The whole cannon are described as having bores of 7in, a length of 11ft (3.4m) or 18 calibres, and weighing 7000lb (3152kg). Bourne (1587) lists 11–12ft and 7000–8000lb, with calibres of 15–16. The cannon from the *Mary Rose* are all shorter and lighter than the sources suggest (Table 2.1), although the bore falls within the prescribed limits. Length in calibres is 12–14. Fortification is described as never exceeding one diameter of the bore at the breech.

The two *Mary Rose* cannon are of identical length (9ft 9in; 2.97m) and weigh 4796lb (2175kg), gun 79A1276, and 5418lb (2458kg), gun 81A3003. An interesting statement can be found in Bourne (1587, 13) in a discussion regarding the correct amount of powder for specific types of gun. Both have less metal at their breeches than the diameter of their bores, consistent with the recorded parameters:

> '*And furthermore, upon good considerations, for diuers causes, and especially for the Queenes Nauie, they have deuised to make their Ordnaunce shorter than the accustomed manner, and so by that meanes they are lighter than the peeces before time made, and yet as seruiceable as the longer in some points, shooting that weight in pouder, and the shotte that the heauier doth, in all poyntes as the other: for that mettall that is taken from the length of the peece, hurteth not the fortifiyng of the peece. And as for the making of the Cartredges for any peece, it is easie ynough to be done: for the compasse of the shotte, and the length of the Ladel, shall rule that matter well ynough.*'

Although these are extremely early cannon, including one of the earliest cast by English founders (79A1276), they are already shorter and lighter than many listed. Whether this was a prerequisite, perhaps designed especially for naval use, or whether this reflects caution in the casting of such large pieces, cannot yet be determined.

*Cannon royal 79A1276*

*Made*: Robert & John Owen in 1535
*Recovered*: John Deane & William Edwards, 1836
*Assumed location*: Main deck, amidships
*Original location*: Port side gun

Loaned by the Royal Artillery Historical Trust
(II/8)

*Statistics*
Length: 8ft 6in (2582mm); 9ft 9in (2972mm);
weight: 4806lb (2180kg); internal bore: 8½in (216mm); 12 calibres

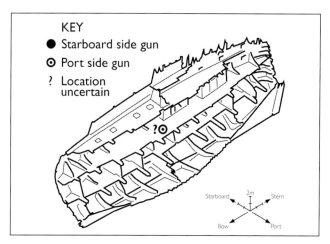

The cascabel bears a scalloped decoration and the button is elongate, a feature characteristic of Owen guns (Fig. 2.27–8). The vent cover is absent, but the attachment indicates that it opened to the right. The gun bears a strong resemblance to the *in situ* cannon (81A3003).

The reinforce bears the Royal Arms in the form of a scrolled shield displaying quarterly the arms of France Modern and England (Fig. 2.27). This is surrounded by the Garter and ensigned with a Royal Crown. The Garter bears the inscription: HONY SOYT QVY MAL Y PENCE, with the letter S reversed. Below this is a raised cartouche with an inscription now only partially legible: HENRIC[VS] OCTAVVS DEI GR[ACIA] R[EX] ANGL[IE] ET FRANC[IE] FIDEI] DEFEN[SOR] DNS [*abbreviating* DOMINVS] HIBERNIE ('*Henry the Eighth, by the Grace of God, King of England and France, Defender of the Faith, Lord of Ireland*'). Towards the breech is incised on the gun: [ROBER]T and IOHN OWYN BROTH[ERS or (E)REN] MADE THYS RO[-]YL CANON [-]TG

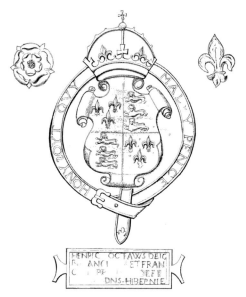

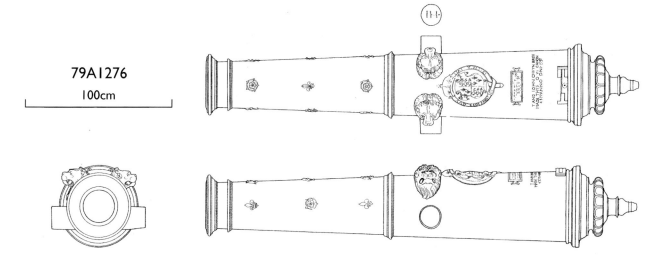

*Figure 2.27 Cannon royal 79A1276: above) detail of the Arms and Garter, tabella and inscription; below) top, side and of muzzle views*

*Figure 2.28 Cannon royal 79A1276: top view of gun*

[?remains of WTG abbreviating WEYGHTING] 4337 ANNO DNI [abbreviating DOMINI] ('*In the year of The Lord*') 1535. Both inscriptions are much defaced, and the reconstruction rests partly on notes taken by the John Deane in 1836 (see Bevan 1996). The weight is different to that inscribed on the gun (4377) by 419lb (190kg). The inscribed weight is, therefore, less than the modern weight.

The dolphins are in the form of lion heads positioned towards the front of the trunnions. The trunnion centres are at two-fifths of the length of the gun (Fig. 2.27). The face of the right trunnion is incised with the initials 'H I'. The horizontal diameter of the left trunnion is 140mm and vertical diameter 145mm: those of the right trunnion 135mm and 141mm. Forward of the trunnions are a series of mouldings, the front of which forms the midpoint between the muzzle and the vent astragal. Their centres are positioned at four-sevenths of the total length of the gun, similar to cannon 81A3003. The chase is decorated with Tudor Roses and *fleurs de lys* in relief. There is no muzzle swell, but the muzzle terminates in a prominent cornice.

Examination of the bore using a fibre optic endoscope revealed the remains of iron wire used to wrap the core during the casting of the gun. The bore narrows towards the breech. There are no obvious scars left by the corrosion of core supports. The thickness of the metal is approximately three-quarters of the bore at the breech and one-quarter at the mouth. The overall average composition of the metal for this gun is given in Appendix 2.1.

The description of the guns raised by Deane at Spithead on 15 August 1836 includes the following entry for this gun, accompanied by a pencil sketch: '*A very handsome brass gun carrying a 68lb shot. The shot and some remains of the powder and wadding drawn from it when it landed at the Gun Wharf Portsmouth.*' It was raised with the bastard demi culverin, 79A1279, together with an iron gun.

This gun is one of the earliest complete cannon, and may be the one cast at Calais by the Owens which is referred to in a letter of 29 April 1536 to Lord Lisle, Lord Deputy there, from Sir Christopher Morres, Master of the Ordnance:

'*My lord, as ye love the ij young men and would do for them, Robert and John Owen, see that the said piece may be well laden; for I have caused the said Robert and John to make xij new pieces for the King's person; and when his Grace hath seen them and the double cannon together it shall be for their preferment, I doubt not.*' (LP 10, 756; Byrne 1981, 3, 346)

It was not until the following February that the twelve new pieces were ready to be shown to the King in St James's Park. The double cannon was still lacking its axle, and it was only at the insistence of Anthony Anthony (compiler of the *Roll*, as yet a relatively junior official in the Ordnance Office) that it also featured in the display. When someone asked what this great piece was, the King, better informed in these matters, '*answered and said it was a double cannon which was cast at Calais, liking it [marvellous] well*' (LP 12/2, appx, 7; Byrne 1981, 4, 266). The terminology used suggests that this may be have been an unusual piece and the date of 1535 on the gun fits well with the description.

**The carriage**

The carriage consisted of a single bed plank. The attachment of the trunnion support cheeks to the bed was via a double tenon into slots into the bed and pegged. An inscription in the bed resembles an 'O' and '6', and is in a similar position to marks found on the carriage of the *in situ* cannon amidships. The left trunnion support cheek has the central portion missing, and none of the tenon on the underside of the cheek is present. Concretion adjacent to the broken areas suggests that it had formed before the cheek had been broken, supporting the hypothesis of later interference (see below). The tenon on the right trunnion support cheek was present, but broken around the treenail positions, suggesting that force pulled the cheek out, breaking the treenails and the tenon. Corrosion products within the bolt holes shows that the iron had decayed before the carriage was damaged.

The rear stepped cheeks (for measurements see Table 2.5) are offset on the bed to match the trunnion support cheeks and are held via a two pronged plate or

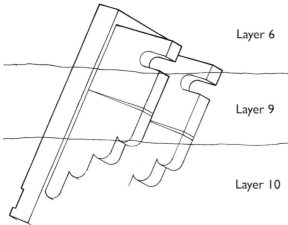

*Figure 2.29 Cannon royal 79A1276: above) the broken carriage underwater; below) interpretative sketch showing layers removed*

staple. The tenon between the side cheeks is on the front face of the stepped cheek.

## Salvage revealed by stratigraphy

Excavation in the midships area during the first week of the 1980 diving season yielded a group of associated timbers on the Main deck at a depth of 250mm below seabed level within a scoured horizon (Fig. 2.29). As the excavation progressed the timbers were identified as a semi-intact gun carriage for a bronze gun, lying east–west. The trunnion support cheeks faced the port side, and were at the same depth below the modern seabed level as the eroded Main deck beams. The south corner of the carriage bed had eroded away, causing part of the left trunnion support cheek to fall in towards the centre of the carriage. Only the rear of the carriage was embedded in the Tudor silts, the remainder of the carriage being within its own scour layer (Fig. 2.29). The position, size and style of the carriage suggests that it originally served this cannon and paired the *in situ* cannon amidships. The carriage does not fit any other known gun from the site.

The episodes of interference by salvors have left scars in the form of the cutting of the layers of sediments built up over the site. In some cases objects have fallen into these layers, providing another method of assigning a date to the salvage attempts. The depth of sediment below the carriage and its position relative to the starboard side of the ship suggest that it remained in position, possibly held by recoil ropes, for some time after the ship sank. Later disturbance resulted in the displacement of two external gunport lids (assumed M5/6 port) inboard to a depth of over 500mm below the top of the deck beams, upper edges uppermost, with no associated structure.

The carriage was clearly not in its original position. Not only had it fallen from the port side but, if we assume bilateral symmetry in the positioning of bronze guns, it was further north than it would have been originally. Parts of one of the wheels and axle were found further south, also suggesting a more southerly original location. As the lower portion (back end) was

*Figure 2.30 Cannon royal 79A1276: above) Tudor salvage cable in hull hole; below) concreted hull bolt*

within the Tudor layer (10), the mid-portion within a scoured layer (9), and the upper portion within the hard grey shelly layer (6), it is probable that the carriage was intitially displaced during the Tudor period, otherwise it would all be in a later horizon (see *AMR* Vol. 1, chap. 9 for a brief description of the stratigraphy of the wreck site). Further indications are the dislocation and breaking of an Upper deck beam (U55), wrenched towards the east, and fact that the mainmast had been wrenched free from its position just further south. Lack of disturbance to the sediments at the mast step level (in the Hold) suggests that this removal of the lower end of the main mast may have occurred during the early salvage attempts, possibly during August 1545 and might have been a contributory factor in the decision to abandon salvage.

As the gun and carriage were not found right across the ship on the starboard side (as was the Upper deck gun and carriage 81A3002) the gun cannot have fallen completely away from the port side at the time of sinking. The similar depth of the gunport lids also indicates a time lag between sinking and deposition. We know that salvage work was carried on for weeks after the ship sank. Excavation under the Sterncastle in 1982 revealed two large cables running straight down under the ship within the first layer of sandy sediments deposited after the ship sank outboard of sector 8 (see Chapter 11, Fig. 11.35 1–5 and *AMR* Vol. 1, fig. 13.2). Their orientation and length show that the Tudor salvors had managed to place at least two cables under the hull. Remains of a Tudor salvage cable were found in a hole in the hull in the stern and show the concreted remains of one of the heads of a hull bolt, presumably snagging the cable (Fig. 2.30).

Later interference was noted to both north and south. The next deck plank to the north had been pulled eastwards and the cutting of the Tudor layer 10 and several localised erosion horizons by a shelly, organic enriched silty clay (layer 100) and modern anchorage debris may be attributable to anchor drag. Mid-nineteenth century objects were found between the cheeks of the carriage, probably resulting from the nineteenth century salvage operations. It is clear from a study of the carriage that considerable force during salvage resulted in the trunnion support cheeks breaking away from the bed, leaving the tenons broken at the peg. The left (southern) stepped cheek, the rear trucks, rear axle tree and southern rear corner of the bed were damaged and the left stepped cheek bears the scar of a rope mark, possibly a breeching rope or salvage attempts on wet timber. Both trunnion support cheeks had split downwards from the centre of the trunnion rest to the bed following the grain. As this is the only carriage displaying this feature, it is unlikely that this is an inherent failing of the carriage, but more likely the result of direct force, possibly levering the gun off the carriage.

## Cannon 81A3003

> *Made*: Initialled 'PB', undated
> *Recovered*: December 1981
> *Location*: Main deck amidships
> *Original location*: in situ
>
> *Statistics*
> Length: 8ft 9in (2656mm); 9ft 9in (2960mm); weight: 5430lb (2463kg); internal bore: 8in (201mm); 13–14 calibres

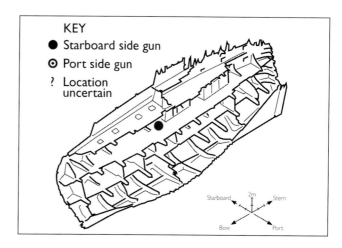

The cascabel bears a scalloped decoration (Figs 2.31–2). The bronze vent cover, which opens to the right, is in place, held by an iron pin. The figure 4873 is incised to the left of the vent cover, orientated longitudinally. This gun is 545lb (247kg) heavier than the inscribed weight.

The initials 'P B', incised into the gun towards the muzzle, suggest the gunfounder Peter Baude. The top of the reinforce is decorated with elongate arches in low relief terminating with pronounced mouldings. The reinforce bears a device intended to be the Royal Arms as on 79A1276 (Fig. 2.31). Between lies a Cross, presumably of St George. The shield is surrounded by a circular motif. The device is supported by a Lion and a Dragon, two of the Supporters of King Henry VIII. The whole device is in low relief and ensigned with a Royal Crown. Below the device is a decorative panel within which is inscribed, 'H REX VIII', a variation of the Royal Cipher.

The dolphins, in the form of two leopard heads, are positioned above and towards the front of the trunnions which taper slightly (Fig. 2.31). The trunnion centres are situated at two-fifths of the length of the gun. The horizontal diameter of the left trunnion is 136mm and the vertical 140mm. The diameter of the right trunnion is 140mm.

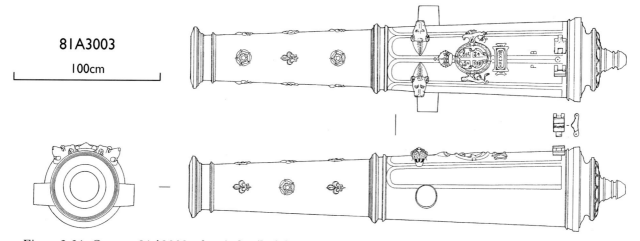

*Figure 2.31 Cannon 81A3003: above) detail of the arms, tabella and inscription; below) top, side and muzzle views*

*Figure 2.32 Cannon 81A3003: top view of gun*

Forward of the trunnions is a series of mouldings, the rearmost of which is situated halfway between the muzzle and the rear of the base ring. The centremost of these mouldings is situated at four-sevenths of the length of the entire gun. All the proportional measurements are similar to those of cannon 79A1276. The thickness of the metal is approximately three-quarters of the bore at the breech and a half at the mouth. The chase is decorated with alternating *fleurs des lys* and Tudor Roses in low relief. The muzzle does not swell, but terminates in a prominent muzzle cornice.

Examination of the bore of the gun with a fibre optic endoscope revealed remains of the iron wire used to wrap the core of the gun during the casting. These were approximately 100mm apart. There are no externally visible scars attributable to corrosion of the core supports. There is no record of shot for this gun, but the conservator recalls that the gun was loaded with cast iron shot which disintegrated upon extraction (O'Shea, pers. comm, 1986). The overall average composition of the metal for this gun is given in Appendix 2.1.

The gun was positioned on its carriage with a quoin in place beneath the cascabel (Fig. 2.33). Its muzzle protruded through an open lidded gunport.

**The carriage**
The carriage represents a robust Main deck carriage made with a single bed plank (Fig. 2.34). The bed plank has the numbers '14' and '09' on the upper side of the bed positioned towards the front of the carriage and the bed has a split towards the rear of the gun. All four corners have been rounded. A central circular groove exists on the topside of the bed, probably the remains of a central ring.

The fore bracket retaining plate was in situ, rebated beneath the bed. This has a length of 28in (710mm) and a width of 3½in (90mm). Two transom bolts originally held the trunnion support cheeks together across the bed, these were positioned 2½in (64mm) above the bed, one beneath the trunnion rest and the second 2¾in (70mm) in from the rear of the trunnion support cheek. There is a concretion filled hole in the side of the trunnion support cheek, below and just behind the trunnion rest. This has been interpreted as part of the breeching system. The fore and rear brackets have a width of 3¼–3½in (82–90mm) and are rebated in to the front and rear faces of the trunnion support cheek. The cap square which overlies these is wider.

The outer edges of the rear stepped cheeks are chamfered. These are offset towards the outside of the carriage to present a flush external face. The cheeks display a shallow tenon on their underneath surface, rebating into the bed.

The ash foreaxle is rebated beneath the bed. The bolt holes are 2¾ (70mm) in diameter. The positions of the axle bolts are inside the trunnion support cheeks, as with all the single bed carriages. The distance between the bolt hole centres is *c.* 12in (300mm) The holes are 1in (25mm) in diameter, and the heads 2¾in (70mm). The breech-loadings are of elm and domed on their inner surface. The bolt holes are 1¼in (31mm) in

*Figure 2.33 Cannon 81A3003: underwater photograph of gun cascabel and rear of carriage with quoin in situ*

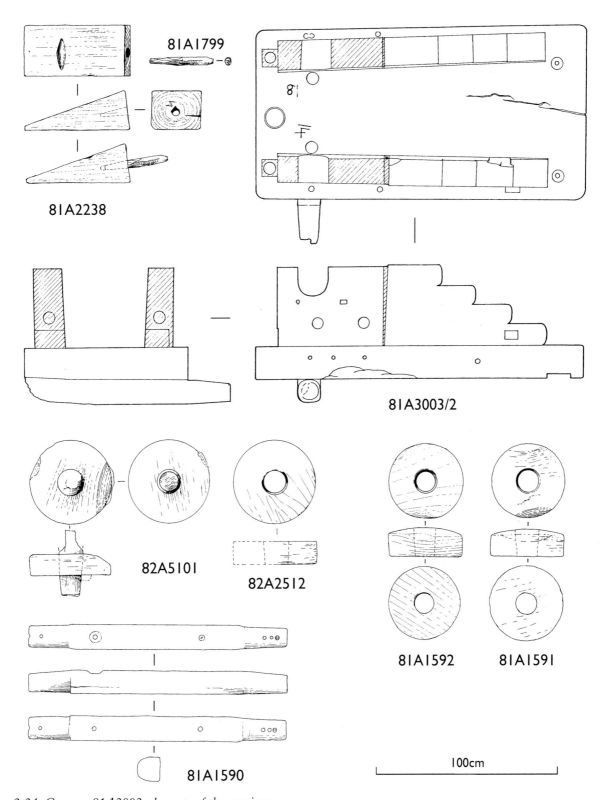

*Figure 2.34 Cannon 81A3003: elements of the carriage*

diameter and are situated 1¼in (31mm) behind the end of the stepped cheeks. The left bolt hole is directly behind the cheek, but the right is offset towards the middle of the bed. The distance centre to centre is 22¾in (570mm). The axle rebates are ¾in (19mm) deep.

The double slots for the trunnion support cheeks are 18½in (470mm) long and have a width varying between 1¼in and 2in (31–55mm). The trunnion support cheek rear bracket bolt washer/head recesses are 2¾in (70mm) in diameter and are positioned 4in (100mm) behind the slots for the trunnion support cheeks. The

*Figure 2.35 Cannon 81A3003: a) front view as lifted; b) rear of carriage as found and; c) as displayed in museum*

rear bracket would have to have a length of 6½–6¾in (165–170mm). The rectangular holes for the stepped cheek attachment are 1 x ½in (25 x 12mm).

The gun was mounted in the horizontal position, held in by a wedge or quoin, and by the fact that the cascabel cornice was rebated into the rear stepped cheeks (Fig. 2.35). This feature is noted by Biringuccio (1540, 314) in connexion with land carriages.

> *'These crosspieces [forming the carriage] should be just long enough to enable the width of the gun to enter. If there are any cornices, their shape is cut into the boards of the bed and they are fitted in, otherwise the piece would have occasion to jump about in too much space and would shoot obliquely if the gunner did not take great care.'*

The height of the carriage at the top of the trunnion rests is 1135mm above the deck with the gun mounted and in a horizontal attitude. In this position, with the front of the carriage against the starboard side of the ship, the lower sill of the port is 10mm above the bed and the upper sill only 65mm above the top of the gun. This position falls between the mouldings on the reinforce. To clear the mouldings, the bed would lie 90mm away from the gunport. In this position it is just possible to depress the gun without touching the upper sill of the port. The closer the front of the bed gets to the port, the harder it is to elevate the gun. The position of

*Figure 2.36 Cannon 81A3003: objects found in the M6 gun bay*

the impression on the wedge suggests that this functioned to retain the gun in the horizontal position. Sighting the gun, in any position, is difficult (Fig. 2.35).

The constraints of the carriage and the upper sill dictate that the gun could not function in the Main deck M3 and M4 gunport stations, but could in M5 and M10. Position M8 is just possible. Both gun positions M5 and M8 are constrained by the presence of hatches into the Orlop deck; breech-loading wrought iron guns were found serving these positions. The integration of gun type and carriage form may be in part determined by the structural layout of the ship, and guns could not always be accommodated at other stations. These constraints (particular guns in specific locations) must imply the need for a predetermined engagement strategy.

**Associations**

The area surrounding the gun served as the midships storage area for shot and equipment serving many guns. The assemblage included five powder ladles, one of the correct size to serve this gun, and six rammers, only one fitting this gun (Fig. 2.36). To the north lay a set of wooden gauges which may have been used for rapid selection of correctly sized shot for specific guns. Over 250 iron and 120 stone shot were removed from this area, contained by a partition wall to the south and the large wooden ships knees rising off the Main deck.

**Lifting**

The gun was lifted on 11 December 1981, the last bronze gun raised, and the most problematic. The wedge, rearaxle tree and trucks were removed during the excavation. The foreaxle tree and trunnion support cheeks were concreted to the junction between the starboard side and the deck. This had to be separated with explosive charges. During the sinking the rear of the carriage had lifted away from the Main deck towards the underneath of the Upper deck. This pushed the muzzle further out through the port, causing the dolphins to lodge securely against the outside of the upper lintel of the gunport. This necessitated levering the gun below the port and across the main deck towards the port side before a direct lift could be achieved.

The carriage and the gun were lifted as one unit (Fig. 2.21b), in order to study the exact nature of the carriage fittings (Fig. 2.37). Examination of the gun carriages for guns 79A1232 (demi culverin bastard) 80A976 (culverin) and 81A1423 (culverin) had

*Figure 2.37 Cannon 81A3003: a) concretion over trunnions; b) view of left trunnion support cheek from the front showing overlap of cap square and fore bracket; c) side view showing rope and shot; d) front view showing transom bolt; e) shell band on chase*

suggested the presence of a number of internal wrought iron fittings either by extant concretion, rebates, vestigial stains surrounding rebates or slots, and by the presence of bolt holes within the wood. In this instance we thought that many of the iron fittings might remain in place.

Mechanical excavation of the concretion which bound the gun to the carriage revealed unequivocally the nature of the cap square holding the gun to the carriage over the trunnion rests (Fig. 2.37a). This originated on the front face of the trunnion support cheeks (Fig. 2.37b) over the fore bracket which secured them to the bed, and was held with a large mushroom-headed bolt.

Concretion on the forward left trunnion support cheek trapped vestiges of rope and stone shot (Fig. 2.37b). A fragment of coarse canvas was excavated from within the concretion that may be the remains of a cartridge (Chapter 5). The rope was not fastened to anything, although a double loop ran along the side of

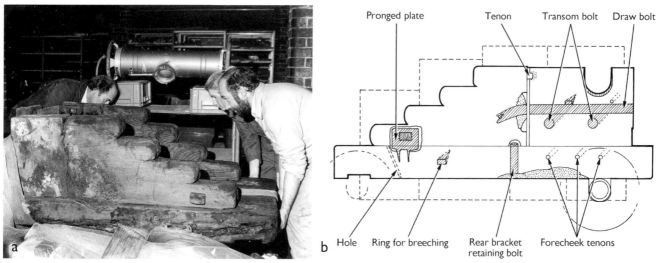

*Figure 2.38 Cannon 81A3003: a) gun carriage about to be radiographed; b) interpretation of internal fittings as revealed by radiography*

the left trunnion support cheek and hung down its front face in a ragged end. It possibly secured the gun via ring bolts found in the ship just below the gunport (Fig. 2.37c).

The upward pressure of the gun against the cap square was visible within the concretion which formed a convex bulge over the trunnion. Evidence for a ¾in (20mm) thick iron 'pillow' or sleeve was visible on the trunnion rests. The trunnions penetrated the entire width of the support cheek over which the cap square was continuous and not hinged. The concretion was only removed enough to release the gun which also resulted in the release of the two transom bolts that went across the bed between the trunnion support cheeks (Fig. 2.37d).

A prominent 300mm thick band of shells around the barrel of the gun between the reinforce mouldings and the lifting dolphins represents a stationary seabed horizon marking the interface between the ship and the seabed (Fig. 2.37e), since the lifting dolphins had been pushed through the gunport during the sinking, and the muzzle pushed into the then seabed. Oyster spat colonised the area until smothered by silts. The sediment inside the muzzle of the gun was firmly packed, presumably having been pushed inside the barrel in a coring action at the sinking. There was neither shot nor gunpowder within the barrel. Either the gun was not loaded at the time the ship sank, or the shot fell out, either as the ship heeled or on contact with the seabed.

**A Tudor gun carriage studied by radiography**
The carriage was radiographed (details in archive) in order to view the nature of the internal fittings (Fig. 2.38a).

Eight shots were taken to form a mosaic of one side of the carriage. Four were horizontal, with the aim of identifying the drawbolt. Four were at an angle from above of 40° from vertical, at the junction between the cheek and bed in order to analyse what form of pinning kept the rear stepped cheek in place and what sort of bolt was holding the rear bracket on the trunnion support cheek in position.

The major wrought iron fittings of the draw bolt, bolts, spike and internal staple were clearly visible and could be accurately measured. These did not show extensive corrosion (Fig. 2.38b). The draw bolt clearly originated at the front of the trunnion support cheek and penetrated the stepped cheek to a depth of 3in (76mm). The large concreted holes securing the transom bolts were positioned below this and a spike and bolt were revealed above the draw bolt, possibly for gun tackle. The concreted remains of the rear bracket and cap square were identified between the two cheeks. The rear bracket was bolted through the bed. This bolt, and possibly a wedge (forelock) were revealed.

The plate holding the stepped cheek in position was visible and in good condition. It was similar in form to that manufactured for the replica carriage made in 1984 for 79A1232, but showed only two iron prongs protruding into the bed, as observed in 81A1423 and 80A0976. Radiography indicated a pronged plate with a wedge. An empty hole was observed just behind the plate, as with the other carriages studied. This has been interpreted as a '*drift*', an empty hole used to accept an iron rod to dislodge the stepped cheek.

The radiography of this carriage confirmed a number of our assumptions and certain features already incorporated into reconstructions. The only elements not visually confirmed were the washers and wedges. Also revealed was the extent of internal wrought iron fittings and the nature of the migration of the corroding iron into the wood, which, although affording localised preservation of the wood, also forms a barrier to the penetration of chemicals used during conservation. These vestigial remains provide archaeological evidence of constructional technique, but create the dilemma of having to choose between preserving the evidence or the timber.

## Demi cannon

Norton (1628) lists these as having a bore of 6½in (165mm) and a length of 10–11ft (3–3.4m) weighing 6000lb (2721.6kg) with a shot weight of 30lb (13.6kg). Bourne (1587) lists 10–12ft; 3–3.7m), and 5000–5500lb (2268–2495kg). Length in calibres is 20–22. As with the double cannon, these are listed as having less than one diameter of the bore in metal at the breech. 79A1277 just exceeds this. Norton (1628, 54) states:

> 'But of late some Founders haue giuen vnto the Demy Canons the full thicknes of one whole dyametre in Mettall at their Chambers; allowing for euery one pound weight of their shot some 220 or 143 pound of Mettall (for the biggest of this kinde) and more, others in proportion for the least.'

The 1540 inventory lists nine of these guns, two on the *Minion*, two on the *Peter*, four on the *Mary Rose* and one on the *Trinity Harry*. The 1546 inventory lists 27, mostly on the ships. The greatest number is on the *Great Bark* (five) followed by the *Pauncy* (four), the *Harry Grace à Dieu* (three), with the *Mary Rose*, *Peter*, *Matthew*, *Small Bark* and *Sweepstake* each having two. The galley *Grand Mistress* has two, and the *Hart*, *Antelope* and *Swallow* have one each.

Two of the *Mary Rose* guns inscribed with the name '*demi cannon*' fit in with the historical parameters regarding bore size, approximately 6½in (*c.* 150mm). One was recovered from the Main deck in the stern (81A3000, M10) the second, recovered in the nineteenth century, may have paired this on the port side. 81A3000 is lacking the muzzle cornice and is only 9ft 9in (2960mm) in length and is therefore almost exactly 6000lb (2722kg). If complete, it would probably be longer than the 10ft minimum listed. Its probable pair (79A1277; Rotunda II/6) is of the appropriate weight for a 10ft demi cannon, but is actually 12ft 3in (3730mm) in length, 20–21 calibres. Although cast by different founders, both are dated 1542. These are long guns, but not dissimilar in length to the M3 culverin (81A1423). They are located in a similar point where the ship begins to narrow, but at the stern.

The third gun listed as a demi cannon (81A3002) does not fit with any of the historical parameters. It has a bore of 5½in (140mm) and a length of 8ft 8in (2635mm) and is only half the recommended weight. It probably pairs a culverin on the Upper deck (80A0976) and is comparable to this gun, although it is one inch shorter.

*Demi cannon 81A3000*

*Made*: John and Robert Owen in 1542
*Recovered*: November 1981
*Location*: Main deck in the stern
*Original location*: *in situ*

*Statistics*
Length: 8ft 8in (2622mm); 9ft 9in (2960mm);
weight: 5529lb (2508kg); internal bore: 6½in
(165mm); 15–16 calibres

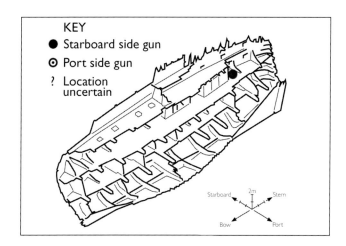

The cascabel is formed of a number of complex mouldings with an elongate button. An illegible number is incised on the base ring. There is a defined vent field with astragal and fillets (Fig. 2.39–40). The first reinforce bears in relief a version of the Royal device: a Tudor Rose within the Garter. The Garter bears the motto: HONY SOYT QVY MALY PENCE. Above the Rose and surmounting the Garter is a version of the Royal Crown. Below there is an ornately decorated cartouche with the inscription: HENRICVS OCTAVVS DEI GRACIA ANGLIE FRANCIE ET HIBERNIE REX FIDEI DEFENSOR ET IN TERRIS [*error for* TERRA] ECCLESIE ANGLICANE ET HIBERNIE [*error for* HIBERNICE] SVPREMVM CAPVT. (properly '*Henry the Eighth, by the Grace of God, King of England, France and Ireland, Defender of the Faith, of the Church of England and also of Ireland in Earth the Supreme Head*'.)

The mistakes may have been made because this version of the Royal style had only recently been introduced when the gun was made, following Henry

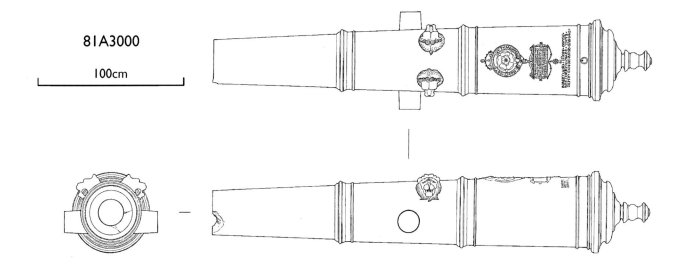

*Figure 2.39 Demi cannon 81A3000: above) detail of the arms and garter, tabella and inscription; below) top, side and muzzle views*

Figure 2.40 Demi cannon 81A3000: top view of gun

VIII's assumption of the title King of Ireland in January 1542 (see above).

Below the cartouche is inscribed directly on to the gun: ROBERT AND IHON OWYN BRETHERYN MADE THYS DEMY CANON ANNO DŇI [*abbreviating* DOMINI] 1542.

The first reinforce terminates with a series of mouldings. The second has dolphins in the form of two lion heads positioned just behind the trunnions which taper slightly. The horizontal diameter of the left trunnion is 144mm and the vertical diameter 140mm. The diameter of the right trunnion is 140mm. If the central point of the trunnions as measured from the rear of the base ring is accepted as two-fifths the length of the gun, the projected gun length would be 3612mm. Further proportional studies show very strong similarities to the port side pair to this gun (79A1277/ Royal Artillery Institute (RAI) II/6). The thickness of the metal is approximately three-quarters of the bore at the breech and a third at the mouth. The second reinforce also terminates in a series of mouldings. The chase is undecorated.

The gun has suffered substantial damage to its chase, which is substantially shorter than expected. It lacks any muzzle mouldings (Fig.2.39, lower). The damage looks like a casting flaw. The first replica culverin cast (see below) was honeycombed and flawed with cracks that increased towards the muzzle, which had naturally nearly separated from the gun. This problem is discussed by Norton (1628, 68):

> '*Some and a great many Peeces are come forth of the Furnace spoongy, or full of hony-combes and flawes, by reason that the mettall runneth not fine, or that the moulds are not throughly dryed, or well nealed: whereby eyther the Gunner that serueth with them is much endangered, they being as bad or worse to serue with, as those that are too weake and poore in mettall: for if they be loaded with so much powder as is ordinary for those sorts of Peeces … they will either breake, split, or blowingly spring their mettalls.*'

As the muzzle looks as though it has been damaged by a glancing blow from a shot, the original gun was examined to rule out any possible battle damage. Both Peter Jones and Peter Northover agree that damage is the result of a casting flaw resulting in a poor muzzle moulding which was subsequently cut off so that the gun could still be used.

Examination of the bore with a fibre optic endoscope again revealed remains of the wire used to wrap the core during casting (Fig. 2.41). These were situated 80mm apart. Two longitudinal slots are visible on the reinforce on either side of the top of the cartouche. These are corrosion scars left by the iron core supports. The overall average composition of the metal for this gun is given in Appendix 2.1.

The gun contained a cast iron shot (82A4124) with a diameter of 160mm. This was discovered 1220mm from the mouth. Remains of gunpowder were recorded (Chapter 5).

## The carriage – a study in adaptation

The gun was *in situ*, with its muzzle facing out through a lidded port situated close to a rising knee at the south. The working space was limited by the enlargement of the Carpenters' cabin to the north (Fig. 2.42a and 2.44, lower). The gun was lifted on its elm carriage on 10 November 1981 (see Fig. 2.21c). Both axle trees, left stepped cheek and all four trucks had already been recovered, the rear trucks having suffered damage

Figure 2.41 Demi cannon 81A3000: close up of muzzle damage and wire in bore

support and rear stepped cheek. The remains of the cap square still covered the right trunnion support cheek and the rear cheek had broken cleanly away from the trunnion support cheek revealing the mortise for the trunnion support cheek tenon.

The bed is made of a single elm plank and is wider at the rear, with the distance between the cheeks increasing to cope with the larger diameter of the gun towards the reinforce (Fig. 2.43). The bed had been adapted to suit this gun, or to suit the gun after foreshortening. The bed ends with two axle bolt holes that have been cut with a partial rebate for the axle beneath the bed, also cut. The position of the *in situ* axle, 8in (203mm) towards the front of the bed, had been a secondary location, no effort had been made to rebate the axle into the underside of the bed. Perhaps the damage to the gun muzzle may have prompted the re-use of an extant carriage rather than a new build.

In construction the carriage is similar to other Main deck carriages (Fig. 2.44a). The trunnion support cheeks and rear stepped cheeks are of near equal height above the bed. The trunnion support cheek is held via a double tenon penetrating slots in the bed, and pegged through the side with three oak pegs. Concretion beneath the bed obscures the origin of the fore bracket and its retaining plate, but one bolt hole is visible. Both rear bracket bolt holes are visible. The rear cheek is held in a shallow rebate by means of a staple or two pronged plate. An ancillary hole behind the staple is visible in both rebates, rather than in one only.

The alignment of concretion stains on top of the bed suggest that the axle bolts were wedged over the top. Rectangular holes in the centre of the bed at front and rear have been interpreted as holes for ring bolts. A small staple hole is visible on the side of the bed just in front of the rear axle bolt hole, this may function as a rope guide.

Radiography using a Cobalt 60 source was undertaken to answer specific questions. The use of washers was confirmed for both transom bolts, in addition to the recognition of a ring bolt positioned on the top of the bed at the front of the carriage. This had an internal diameter of 3½in (89mm) formed of ¾in (19mm) iron. Radiography also revealed a two pronged staple to secure the stepped cheek, identical to that on cannon 81A3003 (Fig. 2.38).

Divers' observations noted that the right fore truck was much smaller than the left, by 3½in (88mm). This was too great to be attributed to the camber of the deck and so the immediate structure around the gunport was examined. It was found that the knee to the south of the port had been cut away, and would easily accept the smaller right wheel. Thus gun, carriage and possibly the ship have all been adapted in order to enable this gun to function in a specific position. As the gun was cast in 1542, this was one of the later additions to the armament of the ship.

*Figure 2.42 Demi cannon 81A3000: a) gun in situ looking north; b) the gun being raised*

during the salvaging of guns in the mid-nineteenth century. This also dislodged the left stepped cheek. The carriage bed had split between the right trunnion

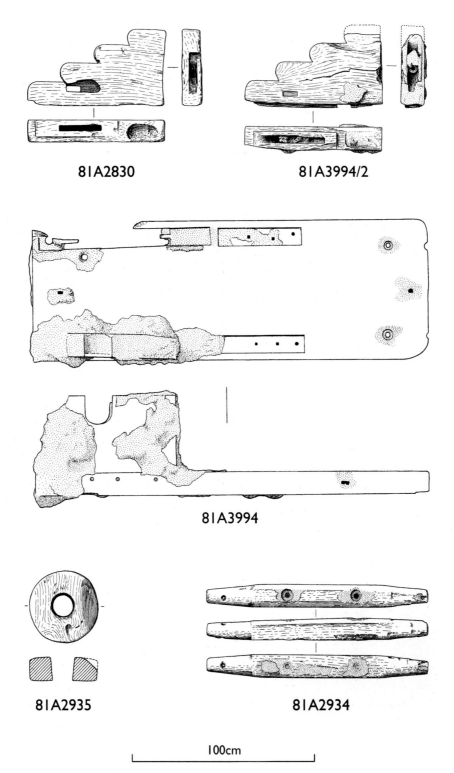

*Figure 2.43 Demi cannon 81A3000: elements of the carriage*

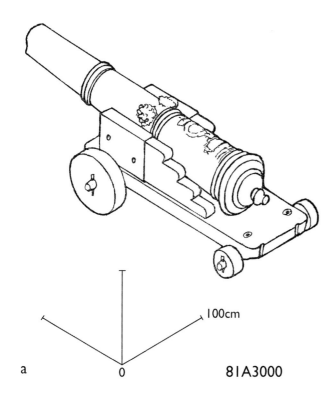

*Figure 2.44 Demi cannon 81A3000: a) gun mounted on carriage; b) location in ship*

*Demi cannon 79A1277*

*Made*: Archangelo Arcano of Cesena, 1542
*Recovered*: John Deane & William Edwards, 1836
*Assumed location*: Main deck, stern
*Original position*: Port side gun
Loaned by the Royal Artillery Historical Trust, Class II/6

*Statistics*
Length: 11ft (3351mm); 12ft 3in (3734mm); weight: 6027lb (2734kg); internal bore: 6¼in (160mm); 20–21 calibres

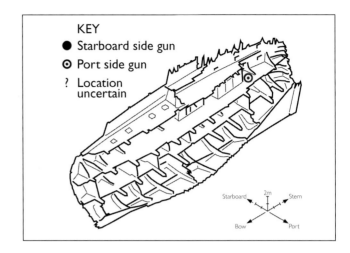

The cascabel is without decoration but the basal ring bears an inscription, partly defaced: MEDIO or MESIO CANON (Figs 2.45–6). There is a vent field and the vent arrangement is similar to that found on 79A1278 (X1X.167) and 81A1423. The reinforce bears an incomplete version of the Royal device. There is no motto visible within the Garter. Below this is a Latin inscription incised into the gun: HENRICVS VIII ANGLIE FRANCIE ET HIBERNIE REX FIDEI DEFENSOR INVICTISSIMVS FF [*abbreviating FIERI FECIT*] ('*Henry the Eighth, King of England, France and Ireland, Defender of the Faith, the Most Invincible, caused to be made*']. A raised plaque below bears the date 1542 and the Royal Cipher: MDXXXXII HR VIII. The mark of the gunfounder is incised towards the breech: ARCANVS DE ARCANIS CESENEN FECIT ('*Arcano of the Arcana of Cesena made*').

The dolphins are in the form of lion heads and are positioned over the trunnions. The centre of the trunnions are situated at two-fifths of the length of the gun taken from the rear of the base ring to the muzzle. The horizontal diameter of the left trunnion is 130mm and vertical diameter 126mm, those of the right trunnion are 122mm and 128mm. Towards the muzzle is a series of mouldings, the rear of which form the mid-point between the muzzle and the vent field astragal and fillets. The centre of these is at four-sevenths of the total length of the gun, similar to the projected proportions for 81A3000, the foreshortened starboard pair to this gun. The thickness of metal is approximately the same at the breech of the gun as the bore and just under a third of the bore at the mouth. There is no muzzle swell, but the muzzle terminates in a prominent cornice.

Examination of the bore using a fibre optic endoscope revealed the remains of wire which had been used to wrap the core of the gun during the casting. There are no scars which can be identified corroded core supports. The Deane records state that the gun was loaded with a 19lb (8.6kg) cast iron shot. The overall average composition of the metal for this gun is given in Appendix 2.1.

*Figure 2.45 Demi cannon 79A1277; armorial device and inscription*

This was the first gun found by John Deane in 1836, while surveying an area reported as routinely snagging fishermen's nets. The finding of this gun caused a debate regarding ownership between the finders (the fishermen), the potential owners (the Crown) and the salvors (Deane and Edwards); these are fully listed as are transcriptions of many of the Deane letters by Bevan

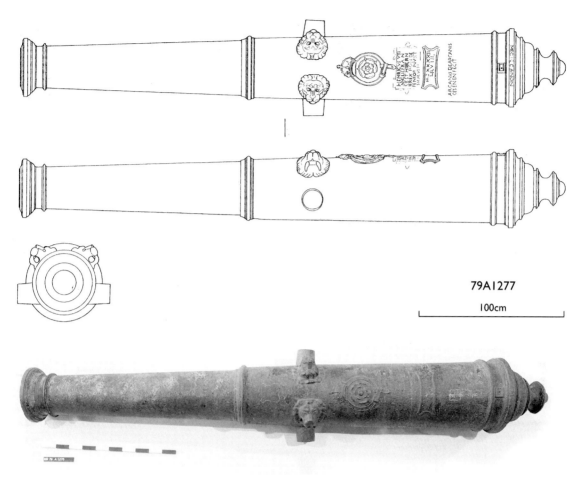

*Figure 2.46 Demi cannon 79A1277: above) top, side and muzzle views; below) top view*

(1996, 109; see also *AMR* Vol. 1, chap. 2). It was the finding of this gun which initiated further survey of this area and the eventually led to the identification of the *Mary Rose*. The gun was originally valued by the Surveyor General of the Board of Ordnance on 17 August 1836 under the title of '*old metal*' at £220 19s 0d.

**A gun carriage in pieces**

Elements forming a complete carriage were found in the sediments immediately above the gun itself (Fig. 2.47). Areas of disturbance in the stern sectors 10 and 11 were noted in 1979. This included localised damage to the deck beams supporting the Main deck, resulting in deck collapse here. At least three deck beams had been displaced westwards, towards the Orlop deck. As yet the date of this collapse has not been determined. This damage was also recorded on the Upper deck which was torn away in sector 11. The area of disturbance to the hard grey shelly layer was substantial, covering the sectors M10 and 11. Two years later, when the M11 starboard gunport was excavated, it was noted that the gun and carriage had been wrenched from the port leaving a broken axle and pair of wheels. Clearly some episodes of sedimentary disturbance can be attributed to nineteenth century interference.

Although recognised as a gun carriage, it was not realised that this was for a port side gun until excavation revealed the true orientation of the carriage and cheeks in 1981. The trunnion support cheeks faced the port side, and the carriage had fallen backwards down towards the starboard side some time after the ship sank. The upper portions of the carriage (the trunnion support cheeks and front of the bed) were at a higher level and within the hard grey shelly post-Tudor layer.

*Figure 2.47 Demi cannon 79A1277: trunnion support cheek in position underwater*

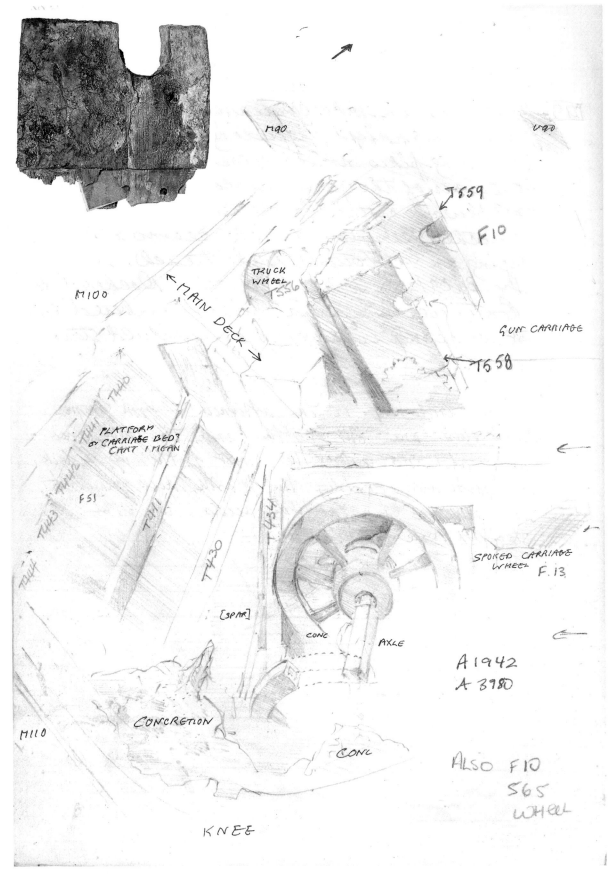

*Figure 2.48 Demi cannon 79A1277: underwater sketch by Jon Adams of gun carriage in situ, with inset showing damage to trunnion support cheek*

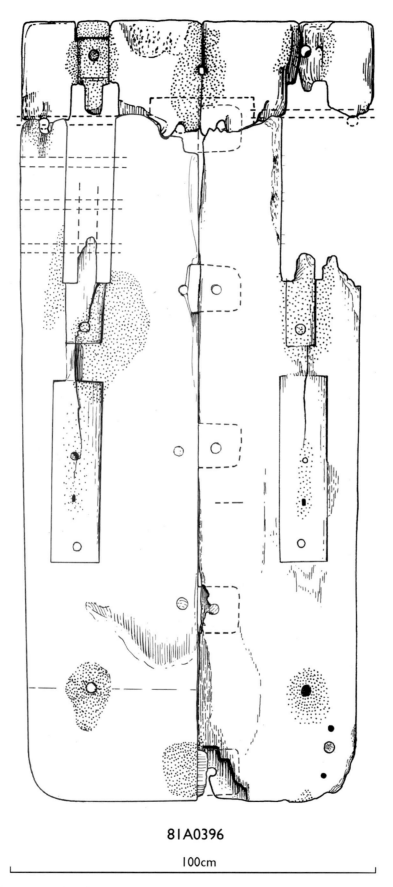

81A0396

100cm

*Figure 2.49 Demi cannon 79A1277: parts of the carriage bed reassembled*

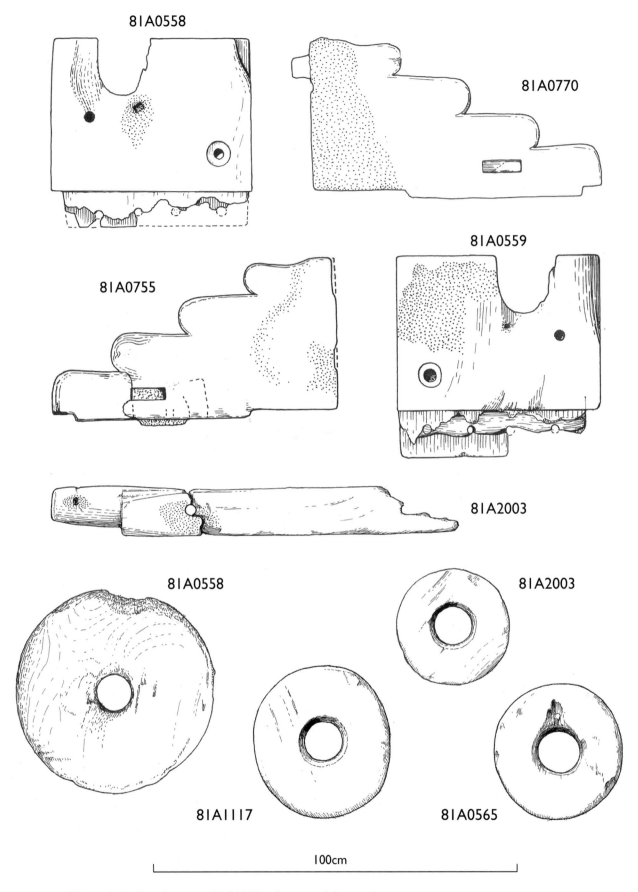

*Figure 2.50 Demi cannon 79A1277: elements of the carriage*

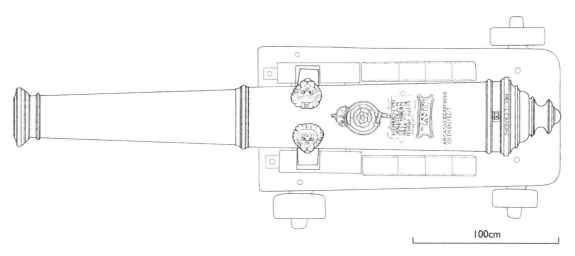

*Figure 2.51 Demi cannon 79A1277: reconstruction of gun on carriage showing close fit*

The rear of the bed was at a similar level to the rear of the *in situ* starboard demi cannon, 81A3000 (Fig. 2.48). Concretion had formed from the corrosion of the rear bracket before the gun was pulled off the carriage, indicating that the gun was on the carriage for some time after the ship sank, and the damage to the trunnion support cheeks and bed are consistent with wrenching the gun away.

The front of the bed had suffered damage and was broken into five portions. These were in the area of the carriage, but were not recognised at the time as being part of it (Fig. 2.49). These were put together some six years after excavation by studying all the elm timber of anticipated proportions from that area. The bed is still incomplete but enough exists (Fig. 2.50) to demonstrate that it was a jointed bed with the trunnion support cheeks with double tenons and pegged with four oak pegs (rather than three as in other cases). The foreaxle bolts were situated outside the trunnion support cheeks. In offering up the trunnion support cheeks to the slots in the bed it was possible to judge their distance apart. This, combined with the size of the trunnion recesses and the height above the bed of the trunnion support cheek, enables us to place the gun on this carriage with reasonable certainty (Fig. 2.51). A study of the bed elements suggests that this was broken at the first loose tenon at the position of the fore bracket bolt holes, also suggestive of directional force being applied.

A total of 15 truck wheels were excavated from the main and upper sectors in sectors 10 and 11 during 1981 and 1982. Eight of these would have served the two carriages for bronze guns. The *in situ* carriage (81A3000) still had its foreaxle and trucks. The rearaxle was at a similar level to the axles and four trucks which would have served the broken carriage. Providing the axle is complete enough to include axle tree bolt holes, assignment to a particular carriage is usually possible. In the instance of the collapsed carriage, both axles were broken. Assignment of wheels by location suggests that one of the front trucks was 21¼in (540mm) in diameter with a thickness of 6¾in (170mm), and that the other was 13¾in (350mm) with a thickness of 6¾in (170mm). The difference in size reflects that of the *in situ* gun, suggesting that this gun also may have been run up a knee or otherwise chocked up to increase the sternwards angle.

The gun is a tight fit for the bed, both in length and in width. The rear stepped cheeks show relieving to accept the widening of the gun towards the cascabel. The front transom bolt hole is very high and virtually obstructs the gun barrel. The barrel just clears this and the fact that the hole did not show the usual circular rebate for the washer or concretion staining from a wedge indicates that this may not have been used.

The cascabel of the gun extends nearly to the back of the carriage making it appear very short for this gun. The other guns recovered in the nineteenth century (79A1276, 1278, 1279) do not fit this carriage and, although it is not impossible that Tudor salvors reached the Main deck port side stern guns, it appears unlikely that they would recover the gun without the carriage. Fresh wrought iron cap squares and brackets would be harder to break without damaging the carriage and this model would have to assume either substantial damage to the port side at the stern, or suggest that the Tudors had access to the inside of the ship. Even if the latter is the case, it is unlikely that they would be able to break the gun off the carriage in the confined space between decks.

*Demi cannon 81A3002*

*Made*: Francesco Arcano in 1535
*Recovered*: 1981
*Location*: Lying north–south along the Upper deck
*Assumed location*: ?Upper deck
*Original location*: ?Port side

*Statistics*
Length: 7ft 11in (2415mm); 8ft 8in (2642mm); weight: 3065lb (1390kg); internal bore: 5½in (140mm); 17–18 calibres

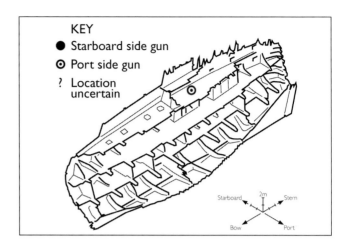

The gun is 16-sided along its entire length (Figs 2.52–3 demican 14). The base ring bears the inscription: DIMICANO WEI but the weight is obscured. The vent shows no sign of having had a cover. The reinforce bears the Royal Arms but without the Garter. The supporters are a Dragon and a Greyhound (possibly chained). The Shield is scrolled and above it is a Royal Crown. Below the Royal Arms is a version of the Royal Cipher H VIII R and below this: POVR DEFENDER AD 1535. There are two small slots towards the breech resulting from the deterioration of the iron core support.

'H I' is inscribed on the right trunnion. There are no dolphins. The trunnions are situated at two-fifths of the length of the gun. The horizontal diameter of the left trunnion is 126mm and vertical diameter 128mm; of the right trunnion 123mm and 128mm. Forward of the trunnions is a series of mouldings. The centre of these is at four-sevenths of the length of the gun taken from the end of the cascabel button. The proportions of certain features are similar to those of 79A1278/Royal Armouries XIX. 167. The thickness of the metal is approximately three-quarters of the bore at the breech and one-third at the mouth. The chase is decorated with alternating *fleurs des lys* and Tudor Roses.

The name of the gunfounder, FRANCISCVS ARCANS ITALVS, the eldest member of the family, is incised on the chase towards the muzzle. The muzzle terminates in a prominent cornice.

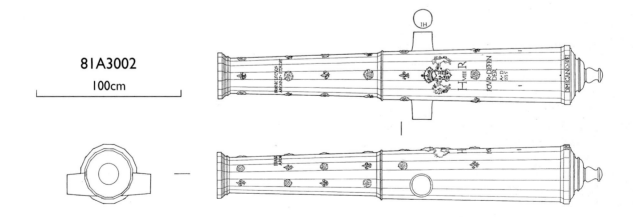

*Figure 2.52 Demi cannon 81A3002: above) armorial device on chase; below) top, side and muzzle views*

*Figure 2.53 Demi cannon 81A3002: top view of gun*

There is no record of this gun having been loaded but the conservator recalls that it was and that the cast iron shot crumbled during extraction (O'Shea, pers. comm., 1986). Examination of the bore of the gun using a fibre optic endoscope revealed traces of the iron wire used to wrap the core during the casting of the gun approximately 20–30mm apart. A scar on the outer surface of the gun has been partially filled (Fig. 2.54) but no attempts have been made to conceal a substantial scar on the inside of the barrel. The overall average composition of the metal for this gun is given in Appendix 2.1.

This gun was found lying longitudinally along the Upper deck having fallen off its carriage and towards the starboard side of the ship (see Fig. 2.22). The carriage was upside down above the gun (Fig. 2.55). A substantial portion of the bed for this carriage was found in O7, vertically below the proposed port side position, but two decks lower. This suggests hatches, companionways, or interference. There was no evidence for the latter. The carriage is similar to that of 80A0976, as is the overall design of the gun.

**Making the carriage**

The carriage elements together with the complete example from 80A0976 enabled reconstruction of this carriage for display purposes. Planks of 5–5¼in (127–140mm) thick elm with a length of 82½in (2100mm) and width of 15¼in (left) and 17¾in (right) (387–450mm) were chosen and fitted with three sets of loose tenons, each held with three straight oak pegs. One side of the bed was fashioned to be wider than the other to accommodate a centrally placed ring bolt avoiding the joint. Details of the measurements for the component parts for the original are given in Table 2.5 and for the replica in Table 2.6.

The fore and rear bracket rebates and slots were then cut into the bed to receive and secure the cheeks. The fore brackets were manufactured from wrought iron. Both brackets have a 1½in (38mm) hole for positioning of a draw bolt towards the top of the brackets. Double slots were fashioned behind the fore bracket rebate, the latter of which continues 1in

*Figure 2.54 Demi cannon 81A3002: scar on outside of gun*

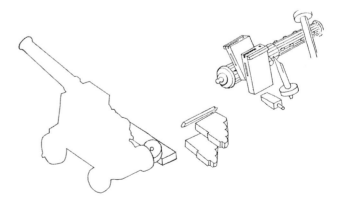

*Figure 2.55 Demi cannon 81A3002: sketch of* in situ *gun (80A0976) and upside down gun and carriage (81A3002) as found on seabed*

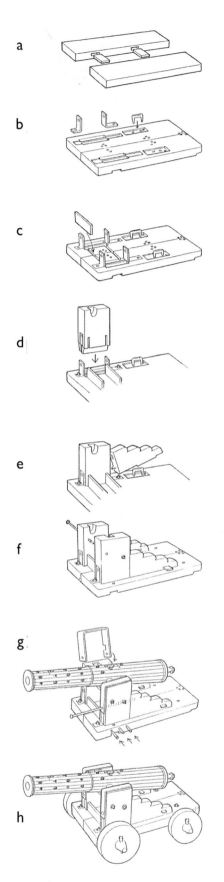

Table 2.6 **Detailed measurements for reconstructed carriage 81A3002**

| Item | Inches | Millimetres |
|---|---|---|
| Fore bracket rebate | | |
| depth | ¾ | 19 |
| width | 3¾ | 95 |
| Rear bracket rebate | | |
| depth | ¾ | 19 |
| width | 3¾ | 95 |
| Fore bracket | | |
| thickness | ½ | 13 |
| width | 3½ | 89 |
| height (l/r) | 12¾/11¾ | 324/298 |
| length (l/r) | 4½/3¼ | 114/83 |
| Bed bolt hole (fore) | | |
| diam (l/r) | 1/1¼ | 25/32 |
| Bed bolt hole (rear) | | |
| diameter | 1 | 25 |
| Rear bracket | | |
| thickness | ¾ | 19 |
| width | 3¾ | 95 |
| length | 13½ | 343 |
| Slots behind fore bracket | | |
| length | 19 | 482 |
| width | 1½ | 38 |
| Rear cheek rebate | | |
| length | 15 | 381 |
| width | 6 | 152 |
| depth | 1 | 25 |
| foreaxle rebate | | |
| depth | ¾ | 19 |
| width | 5½ | 140 |
| Rearaxle slot | | |
| width | 5¼ | 133 |

*Figure 2.56 Demi cannon 81A3002: sequence of carriage making (see text)*

(25mm) behind the slots. The rear cheek rebate (Fig. 2.56b) displays two rectangular holes for a staple situated 3½in (89mm) apart, and a clean ¾in (19mm) hole towards the rear of the rebate. The rebate was made with a bevelled edge at the rear to ease location of the rear cheek. The foreaxle rebate and rearaxle slot were then cut in the underside of the bed.

Rebates were cut into the bed between the trunnion support cheek slots to accept the elm transoms of height 6in (152mm) and width ¾in (19mm; Fig. 2.56c). The inside of the trunnion support cheek was also relieved to accept the transoms. At this stage the underside of the bed was fitted with a rebate housing a wrought iron plate which fits between the fore bracket bolt holes. The plate has a width of 4in (102 mm) and a thickness of ¾in (19mm). The transoms were then located within the bed grooves.

The trunnion support cheeks were positioned next. Each has a double tenon, pre-drilled for the oak pegs (Fig. 2.56d). The mortise in the front face of the rear stepped cheek was located over the tenon on the trunnion support cheek and the rear stepped cheek entered over the staple, a shallow tenon locating into the shallow mortise in the bed (Fig. 2.56e). The tenon was offset towards the outside face of the cheek and positioned higher on one cheek than the other. In this way it is impossible to mix up the carriage components. The two transom bolts of 1¼in (32mm) were then entered between the front cheeks. These were made with large domed heads then slotted and wedged against washers (Fig. 2.56f).

The gun was then placed into the trunnion support cheeks and the cap square, of 4½in (114mm) wide, ¾in (19mm) thick wrought iron, was pivoted around the tenon projecting between the front and rear cheeks. The cap square was cut to accept the tenon.

The weight of the gun on the front cheeks has the effect of locating the tenons deeply into the bed, enabling the oak pegs to be entered from the side of the bed. Lastly the long draw bolt was entered from the front of the carriage, going through both cheek assemblies (Fig. 2.56g). These are of 1–1½in in diameter (25–38mm), and oak wedges are entered in the side of the rear cheeks, penetrating the staple. An alternative would be to place the gun on the fore carriage first, and place the rear stepped cheeks later. The possible presence of a wedge between the trunnion support and rear stepped cheeks wedging the draw bolt makes this a viable alternative. The concretion on the

*Figure 2.57 Demi cannon 81A3002: the gun on display in the National Maritime Museum (© National Maritime Museum, B2244-B)*

original was too vestigial to be conclusive. The particular features of high trunnion support cheeks suggest that it required a similar height above the deck and therefore an Upper deck position is suggested (Fig. 2.57).

## Culverins

Norton (1628, 52) includes within this group the double, whole and demi culverin, the saker, falcon, falconet, robinet and base. All share the feature of being long relative to their bores. They are all classified as having no less than one diameter of the bore in metal thickness at the breech, with the smallest examples being thicker at the breech and of a greater number of calibres in length.

Their main function is to penetrate at an increased distance to other guns to '*shoot further & pierce deeper then those of the first kind*' [cannon], and '*to pierce & cut out in batteries what the Cannons haue shaken and loosed*' (Norton 1628, 55, 56). This is the primary reason for the increase in metal at the breech, and their length. Norton states that earlier foreign culverins were between 24 and 32 calibres in length with one diameter of thickness of metal at the breech and having not more than 150lb (68kg) of metal for each pound in weight of shot. He states that current English culverins were as much as 30–32 calibres in length.

Whole culverins are listed as being 5–5½in (127–140mm) in bore, of 12ft (3658mm) length and weighing between 4300 and 4600lb (1950–2087kg), with a shot weight of 15–25lb (6.8–11.3kg). Bourne lists 12–13ft and 4300–4800lb. They are listed in the Tower from 1514. These guns are mostly of bronze, until the middle of the century when cast iron culverins begin to appear in the Tower lists (Blackmore 1976, 224). The 1514 inventory of thirteen vessels (Knighton and Loades 2000, 113–58) lists seven culverins (no demi culverins) on four vessels: two on the *Henry Grace à Dieu*, three on the *Trinity Sovereign* and one each on the *Gabriel Royal* and *Kateryn Forteleza*. The situation by 1540 (PRO E 101/60/3) had not changed greatly, the ten listed vessels having only three culverins; two on the *Mary Rose* and one on the *Peter* (the *Henry Grace à Dieu* is listed as '*not furnished*')

The number of culverins had increased by the time Anthony Anthony undertook his inventory of 1546, but the number of vessels is also greater, totalling 58. There are 44 culverins listed. The 20 ships account for 22, on eleven of the vessels. The *Harry Grace à Dieu* has four and a further eight ships (including the *Mary Rose*) are listed with two each. Two have only a single culverin. The fifteen '*galliasses*' have 22 over nine of the vessels. The *Grand Mistress* is listed with four as is the *Anne Gallant*, interestingly these are listed as two long and two short culverins. This may be what we are finding within the *Mary Rose* assemblage. The 23 vessels classified as '*pinnaces*' and '*rowbarges*' do not carry culverins.

There are three guns inscribed with the word '*culverin*' that can be described as whole culverins, although one stands out as being different. The two which best suit the parameters are 81A1423 and 79A1278. Both appear to have the same thickness of metal at the breech as their bores and are suggested to be a pair, port and starboard, across the Main deck in the bow. They are both over 23 calibres and are within 100mm of each other in length. Both were cast in the 1540s. The third is an Upper deck gun (80A0976). This is extremely short (9ft 8in; 2946mm), and is less fortified. It is interesting that this, and the 'odd demi cannon' (81A3002) are a possible pair, also only fortified with three-quarter thickness of bore at the breech. Perhaps the reason lies in their location, on the Upper deck just aft of amidships amidst the rigging. Like the cannon below and just forward of them (79A1276, 81A3003) they may be short to suit their position. Bourne (1587, 59) reinforces this suggestion:

> '*Fyrste, for the ease of the Shyppe, for theyr shortenesse they are lyghter: and also, if that the Shyppes shoulde heelde wyth the bearyng of a Sayle, that you muste shutte the portes, especially if that the Ordnaunce bee upon the lower Orloppe, and then the shorter peece is the easyer to bee taken in, both for the shortenesse and the weyght also. In lyke manner, the shorter that the peece lyeth oute of the shyppes side, the lesse it shall annoy them in the tacklyng of the Shyppes Sayles, for if that the peece doe lye verye farre oute of the Shyppes syde, then the Sheetes and Tackes, or the Bolynes [bowlines] wyll always bee foule of the Ordnaunce, whereby it maye muche annoy them in foule weather, and so foorth.*'

## Culverin 81A1423

*Made*: Peter Baude, 1543
*Recovered*: 1981
*Location*: Main deck in the bow
*Original location*: *in situ*

*Statistics*
Length: 11ft (3355mm); 12ft (3655mm)
weight: 4556lb (2067kg); internal bore: 5½in (140mm); 23–24 calibres

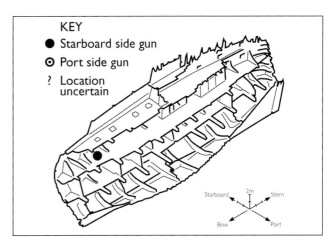

The cascabel bears a scalloped decoration and the basal ring is inscribed: THYS COLVERIN WEYTHE 3888. The gun was weighed using an S beam Load Cell giving a difference in weight to that inscribed on the gun of 668lb (304kg). The inscribed weight is therefore less than the true weight.

There is a defined vent field and reinforce moulding. The vent cover is absent (Figs 2.58–9). The reinforce bears a Tudor Rose ensigned with the Royal Crown and surrounded by the Garter which bears the inscription: HONY SOIT QVY MAL Y PENSE. The letter 'N' is reversed in both instances. The Royal Crown does not touch the Garter at any point. The device, which is a variation of one of King Henry VIII's badges, is in low relief. Below the Garter is a plaque in relief bearing the inscription: VIVAT REX HENRICI OCTAVI [*error for* HENRICVS OCTAVS] (properly '*May King Henry the Eighth live*'). Beneath this is inscribed: HR8 (the letters ligatured) and below is the further inscription: PETRVS BOWDVS ME FECIT AD 1543 ('*Peter Baude made me in the year of the Lord 1543*').

The dolphins are in the form of winged mermen and are positioned above the trunnions (Figs 2.59–60). The trunnion centres are situated at two-fifths of the length of the gun. The horizontal diameter of the left trunnion is 117mm and the vertical 122mm. The horizontal diameter of the right trunnion is 118mm and the vertical 122mm. Forward of these, a series of mouldings define the fluted and spiralled form of the chase. The rearmost moulding is situated at four-sevenths of the total length of the gun. The thickness of metal is approximately the same as the bore at the breech and between one-third and a half at the mouth. The muzzle does not swell but terminates in a prominent muzzle cornice.

Excavation of the bore of this gun yielded firstly a wad, 1000mm inside the gun, then a cast iron shot stamped with the letter 'H', and finally gunpowder. The shot has since been lost. Examination of the bore with a fibre optic endoscope revealed the remains of the iron wire used to wrap the core during the casting of the gun 20–30mm apart (Fig. 2.60c). Longitudinal slots can be seen on both sides of the maker's name. There may be a

*Fig. 2.58 Culverin 81A1423: device and inscription*

third slot just behind the left trunnion. These are the result of corrosion of the iron structure used to support the core during the casting process. The overall average composition of the metal for this gun is given in Appendix 2.1.

The area immediately to the north of the gun served as one of the major shot storage areas, affording easy access to the Upper and Castle decks via a permanent companionway. The gun sat on a trucked elm carriage with its muzzle protruding through an open lidded gunport. It was recovered on its carriage and the gun

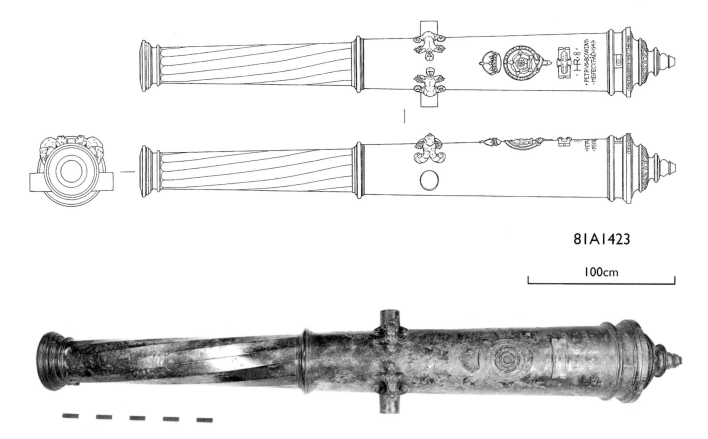

*Figure 2.59 Culverin 81A1423: above) top, side and muzzl view; below) top view of gun*

was removed by careful excavation of the corroding iron forming the cap square which held the gun on to the carriage (Figs 2.61–2). A replica of this gun was made using sixteenth century casting methods and a similar alloy. The gun was mounted and firing trials were undertaken (see below).

## The carriage: a study by excavation and replication

An exact facsimile was made to understand the constructional features of this carriage (Fig. 2.62). Traditional tools available to the Tudor carpenter were used, but not exclusively. Imperial units of measurement were used. Detailed measurements for various components are given in Table 2.7 (see also Table 2.5).

The original bed (and the facsimile) was made from a single elm plank (Figs 2.63a and 2.64). The original plank contained the first tree ring which was centrally positioned and formed a line of weakness which caused the bed to split. Although the plank is the widest of the tree and was utilised to make the bed, the outer edge of the heartwood was visible on the right side. The rear corners were rounded and the front corners chamfered to 1½in (38mm) wide at an angle of 45°.

The trunnion support cheeks were made of 6in (152mm) elm with trunnion recesses which penetrated the cheek to a width of 4½in (114mm), leaving a 1½in (38mm) portion of cheek towards the outer face. The trunnion rests were located 5in (127mm) in from the front face of the cheek. The distance from the rear of the trunnion rest to the back face of the cheek was 11½in (292mm). Concretion at the bottom of the recess suggested a lining or 'pillow' of iron.

The front and rear faces of the trunnion support cheek had rebates within the cheek for the fore and rear brackets (Fig. 2.63d). The brackets were 'L' shaped and were rebated into the bed as well as into the front and rear faces of the trunnion support cheeks. These were set into the cheek so that they could not be seen from either the inside or the outside of the carriage. The fore brackets were secured through the bed by means of a wrought iron plate extending the width of the bed, which was rebated into its underside for strength, held with bolts secured with wedges on the top of the bed. The brackets extended up the trunnion support cheek for 8½in (220mm). The rear bracket attachment was secured with the wedges above the bed rebated into the underneath of the rear stepped cheek; the bolt heads rebated directly into the wood underneath the bed. However, when the components of the original carriage became available for detailed analysis, it was realised that the space created under the stepped cheek for attachment of the rear bracket suggested that the heads were positioned on the top of the bed and the wedges below.

The trunnion support cheek extended 4in (100mm) into the bed via a double tenon, each 1¾in (45mm)

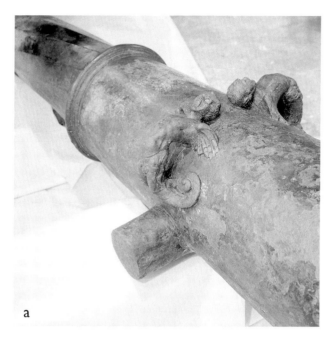

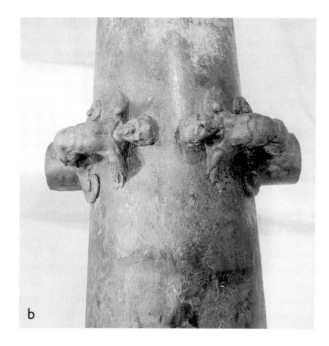

*Figure 2.60 Culverin 81A1423: a–b) mermen lifting dolphins; c) remains of iron wire in bore*

oak pegs. The holes in the tenons were offset upwards relative to those in the bed by ¼in (6mm), which had the effect of drawing the cheek deeply into the mortise.

The trunnion support cheeks were further strengthened laterally by the positioning of two wrought iron transom bolts, also held by wedges and collars. A bolt hole of 1in (25mm) diameter originated on the front face of the trunnion support cheek and continued through this for a distance of 13¾in (350mm) into the rear stepped cheek. This bolt was the last element to be fitted after the barrel and cap square were positioned.

The right rear stepped cheek was *in situ* when the gun was recovered. The tenon, which originated 12in (305mm) from the front face of the cheek, projected 1¼in (31mm) into the bed. In this area an ovoid depression had been cut to accept the wedge or head of the rear forecheek retaining bolt. Towards the rear of the cheek the tenon was shouldered in ¼–½in (6–13mm) spits. The underneath of the tenon had a 9in (229mm)

thick and chamfered from the back of the cheek towards the front. The tenons protruded centrally from the base of the cheek and were shouldered so that the cheek sat on the bed.

Three 1in (25mm) square-headed tapered oak pegs were entered from the side of the bed through the trunnion support cheek tenons. These were placed at 10in (254mm), 15in (381mm) and 21in (533mm) from the front of the bed. To secure the tenons, these would require a length of between 7¼in (184mm) and 11in (279mm). In the instances of separated carriages the trunnion support cheek has either torn at the pegs or broken away taking part of the bed with it, showing the inherent weakness at this point and the strength of the

*Figure 2.61 Culverin 81A1423: gun on carriage as raised*

**Table 2.7 Detailed measurements for gun carriage 81A1423**

| Item | Inches | Millimetres |
|---|---|---|
| Fore bracket rebate (L-shaped) | | |
| depth | ⅜ | 10 |
| width | 4¼ | 108 |
| length | 4¼ | 108 |
| Rear bracket rebate (L-shaped) | | |
| depth | ⅜ | 10 |
| width | 4¼ | 108 |
| length | 7¼ | 197 |
| Trunnion support cheek | | |
| from side of bed | 2 | 51 |
| from front | 4¾ | 120 |
| between extremities | 24¼ | 616 |
| Trunnion support cheek transom bolts | | |
| diameter | 1 | 25 |
| from front face | 5½ | 140 |
| up from bed | 5 | 127 |
| from rear face | 3½ | 89 |
| up from bed | 3½ | 89 |
| Rear stepped cheek | | |
| steps, up from front | 4½, 7, 10¾, 14¾ | 114, 178, 273, 375 |
| length (from bottom step) | 31¾, 26½, 17½, 8 | 806, 673, 445, 203 |
| Rear cheek rebate | | |
| length | 18¼ | 464 |
| width | 5¼–6 | 133–152 |
| depth | 1 | 25 |
| Foreaxle tree rebate | | |
| from front of bed | 10¼ | 260 |
| depth | 10 | 254 |
| width | 4¾ | 120 |
| Rearaxle tree rebate | | |
| from rear of bed | 13¼ | 337 |
| width | 5 | 127 |
| depth | 1 | 25 |
| Front wheels | | |
| diameter | 17 | 432 |
| thickness | 4¼ | 108 |
| Rear wheels | | |
| diameter | 13 | 330 |
| thickness | 4½ | 108 |

*Figure 2.62 Culverin 81A1423 carriage: a) top view; b) side view; c) view front*

slot cut upwards into the cheek to allow its placement over the staple; this was held with a wedge inserted from the outside of the cheek (Fig. 2.63c).

The cheek had four steps that were rounded and chamfered towards the outer face. The grain was aligned lengthways along the step and they were cut out step to step from one plank of wood. The rear stepped cheek had a tenon offset towards its outer face which sat in a mortise in the rear face of the trunnion support cheek. This tenon was shouldered and the wrought iron cap square was relieved to pivot around the tenon. The tenon was 2¼in (57mm) long and 2¾in (70mm) wide.

Rebates for the rear stepped cheeks had two small holes of ¾in by ½in (13 x 19mm), 3¾in (95mm) apart. This suggested the presence of a staple, or 'dog', inserted into the bed. The holes within the bed

*Figure 2.63 Culverin 81A1423: building replica carriage: a) bed showing mortices for fore cheeks; b) positioning the cheeks; c) placing wedge into one stepped cheek; d) gun fitted on carriage*

suggested that the staple was shouldered and tapered as it entered. Ash wedges positioned through slots in the sides of the rear stepped cheeks held them in position through the staple. The right rebate had a clean circular hole ¾in (19mm) in diameter. This may have been to facilitate removal of the cheek.

Concretion adhering to the outside face of the right trunnion support cheek contained the remains of rope, probably an attachment for breeching the gun. A concretion filled hole placed centrally at the back of the bed was probably for a ring bolt (see Fig. 2.63a and c)

The open-ended axle rebates were positioned under the bed. The axles were fastened beneath the bed by two wrought iron slotted bolts of 1in (25mm) diameter, held tightly with wedges. Elongated concretion stains surrounding a circular hole on the top of the bed suggested the presence of circular washers with wedges or keys. The mushroom heads of the bolts would have

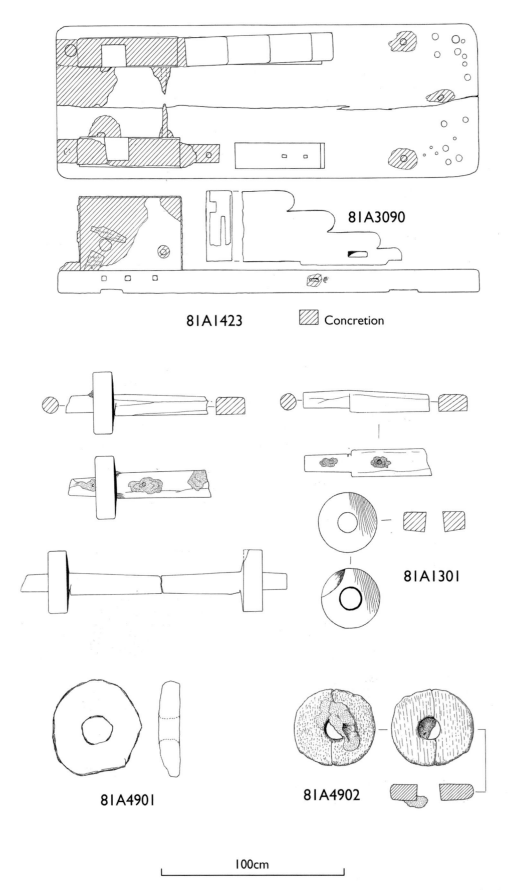

*Figure 2.64 Culverin 81A1423: top) carriage as lifted; middle) loose cheek; bottom) axle and wheels*

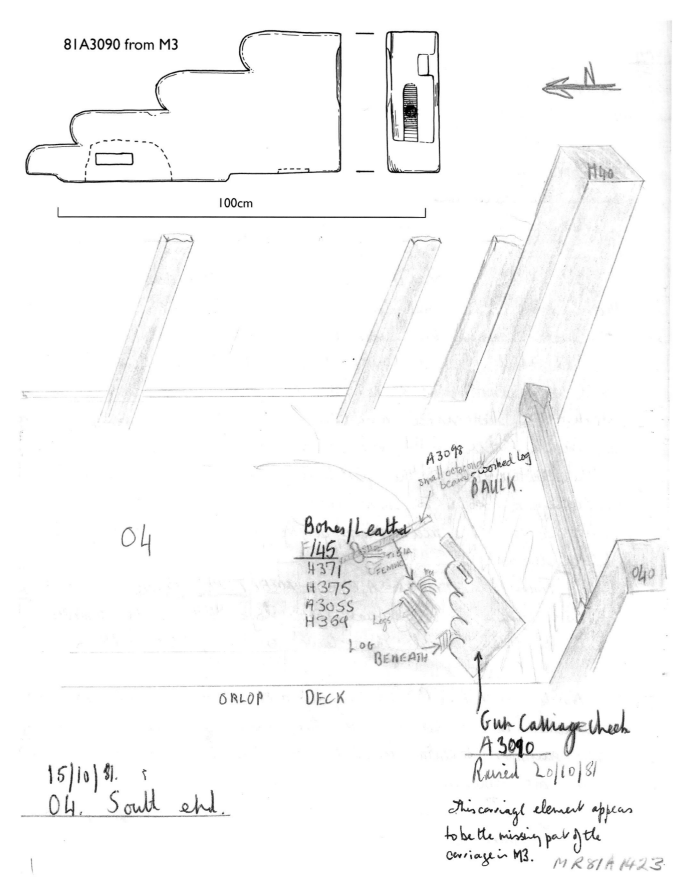

*Figure 2.65 Culverin 81A1423: divelog showing location of rear cheek in O4*

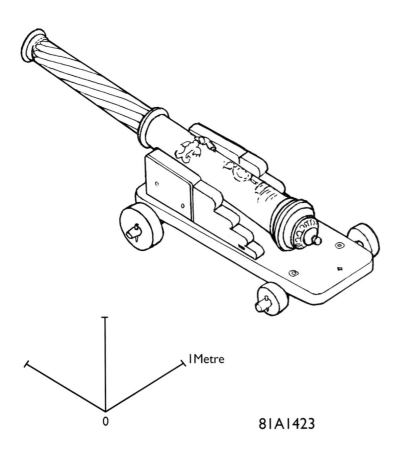

*Figure 2.66 Culverin 81A1423: reconstruction of gun on carriage*

been secured under the axles and the wedges on top, facilitating easy removal of the axles.

The rear axle tree was made of ash, with the grain aligned along its length. The rear wheels were of elm. The front wheels were larger and slightly thinner than the back ones (Table 2.7; Fig. 2.63d). It is estimated that if the gun were to rest on the bed, its muzzle would chafe against the upper sill of the gunport. There is no evidence of this from the gunport.

Analysis of the area around this gunport revealed two areas of concretion staining which may have been ring bolts. That to the left suggests a 1½in ring (38mm), with a 2in (51mm) shank level with the lower sill, 140mm away from the port. On the right side the base of the ring is 50mm above the lower sill and 103mm to the right. The port itself is not placed centrally between the supporting knees, but is 1800mm from the left knee and 730mm away from the right (see Chapter 11, Figs 11.20–22).

The left rear stepped cheek was not *in situ*. Concretion within the staple hole suggested that it had been in position until that had formed. A single left rear stepped cheek was found nearby on the deck below, 81A3090 (Figs 2.64–5). This location was not a storage area and the removal of the cheek without removal of the draw bolt would have been impossible. The concreted remains of both cap squares were evident over the trunnion support cheeks. The loss of the rear cheek may be a result of the salvaging of guns, either in the years after the sinking or of the nineteenth century salvaging. The loose cheek is without the tenon which held it to the trunnion support cheek and also shows evidence of rope marks on the third of four steps. It fits the carriage but is only 5¼in (133mm) wide, ½in (13 mm) thinner than the *in situ* right rear cheek (Figs 2.64–6).

Several features on the carriage were radiographed for further study. One of these was a concreted lump on the rear face of the left trunnion support cheek in the large bolt hole going through its length. This suggested that the bolt was wedged but did not enter the rear stepped cheek. The rear cheek would therefore have to be removed to gain access to the cap square in order to separate the gun from the carriage. This could be a contributory factor as to why the left rear cheek was no longer in place.

## Culverin 80A0976

*Made*: Gunfounder unknown, dated 1535
*Recovered*: 1980
*Location*: Upper deck in the stern
*Original location*: in situ

*Statistics*
Length: 8ft 11in (2730mm); 9ft 8in (2955mm);
weight: 2977lb (1350kg); internal bore: 5½in
(138mm); 19–20 calibres

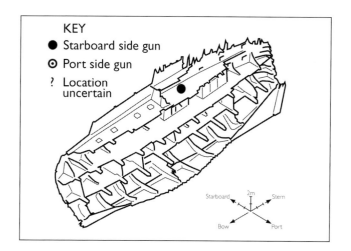

The gun is dodecagonal along its entire length and is very similar in design and decoration to the bastard demi culverin recovered in 1836 (79A1279, RAI II/7). The vent cover is not present but the hinge arrangements on the gun suggest that it would have opened to the right (Figs 2.67–8).

The date is incised towards the breech in the form: A [*abbreviating* ANNO] 1535 ('*In the year 1535*'). Beside is the inscription: THYS COLVERYN WEYS 2756. The weight is 221lb (100kg) heavier than that inscribed on the gun. A scar left by the corrosion of the iron supports for the core is visible to the right of the date.

The reinforce bears a Tudor Rose ensigned with the Royal Crown and surrounded by the Garter which bears the inscription: HONY SOYTE QVY MAL Y PENCE. The 'S' is reversed. The Crown is superimposed over the top part of the Garter and the whole device is in low relief. Below is a decorative panel in which there is a version of the Royal Cipher, an 'H' and an 'R' joined by a looped cord. Below this an incised '8'. The gun does not possess any dolphins. The trunnions (diameter 142mm) are below the centre line and the left trunnion bears the inscription 'H I' on its

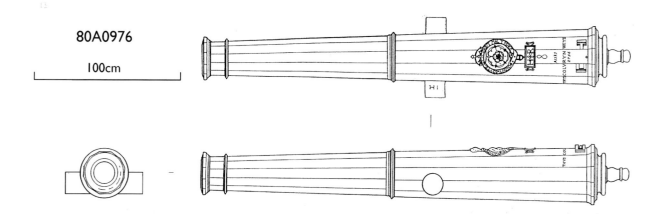

*Figure 2.67 Culverin 80A0976: above) detail of the arms, tabella and inscription; below) top, side and muzzle views*

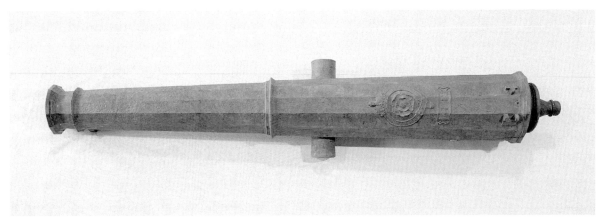

*Figure 2.68 Culverin 80A0976: top view of gun*

upper surface. The trunnion centres are positioned just over two-fifths of the length of the gun).

Forward of the trunnions is a series of mouldings. The rear of these is halfway between the muzzle and the front of the base ring. This corresponds to four-sevenths of the distance from the end of the gun to the muzzle end of the mouldings. The thickness of the metal is approximately three-quarters of the bore at the breech and one-third at the muzzle. The muzzle does not swell but terminates in a prominent muzzle cornice. A cast iron shot (80A1941), 128mm in diameter, was recovered from inside the barrel. The overall average composition of the metal for this gun is given in Appendix 2.1.

The culverin sat on an elm carriage with its muzzle protruding through an unlidded gunport cut into the Sterncastle structure (see Figs 2.8, 2.10; Chapter 11, Fig. 11.20). The port appears to be an enlarged version of a swivel port, suggesting that it belonged to a rebuilding phase. Survey of the immediate area revealed a storage ledge which contained eleven cast iron shot ranging in diameter from $4^{15}/_{16}$in (126mm) and $5^{1}/_{8}$in (130mm) pre-selected for this gun (see Fig. 5.5, lower). The weight varied from 4.4lb (2kg) to 11lb (5kg), depending on the degree of corrosion. Contemporary weights (Bourne 1587) for shot of these sizes ($4^{3}/_{4}$–$5^{1}/_{2}$in) are 14lb 14oz–23lb 2oz (6.75–10.50kg)

The gun had been pushed upwards off the Upper deck towards the underneath of the Castle deck. The hanging knees supporting the latter framed each side of the barrel, placing constraints on the angle and method of extracting the gun. It was removed from its carriage underwater, as were the axle trees, trucks and right stepped cheek (Fig. 2.69).

**The carriage**
The bed was jointed lengthways, using elm planks. The left side of the bed was wider than the right and this enabled the placement of a ring bolt in the centre of the bed without it being on a joint. The planks were held together by two loose tenons 5in (127mm) in length. These were pegged into mortises cut centrally in the side of the bed at the front and the rear of the carriage. Each draw-tongued joint was pegged with three oak pegs of 1in (25mm) diameter (Fig. 2.70).

The trunnion support cheeks were held in the bed via two double tenons which penetrated the width of the bed. The tenons were angled backwards for ease of location. It was possible to measure the difference in height of the peg holes to retain this tenon; those on the

*Figure 2.69 Culverin 80A0976: gun as lifted on deck being examined by Alex Hildred (left) and Nick Rule (right)*

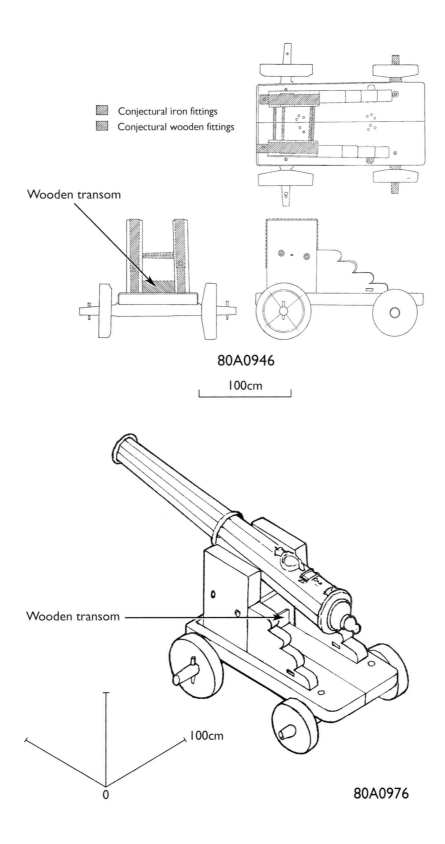

*Figure 2.70 Culverin 80A0976: above) the carriage; below) reconstruction of gun on carriage*

cheek were offset ⅛in (3mm) higher than those in the side of the bed. The oak pegs were of 1in diameter (25mm) and 14½in (368mm) length. The trunnion support cheeks were held with fore and rear brackets rebated into the cheeks and the bed. The rebates, dictating the thickness and width of the brackets, were 3½in (89mm) wide and ½in (13mm) deep. These were reinforced across the front underneath of the bed with a wrought iron plate 5in (127mm) wide. These elements were held with a 1in (25mm) bolt, wedged above the bed. The brackets were not rebated centrally into the cheek, but positioned 1½in (38mm) from the outside of the cheek and only 1in (25mm) to the inside face. The draw bolt, which ties the trunnion support cheeks and rear stepped cheeks, was 1½in (38mm) in diameter.

The trunnion rests did not penetrate the width of the cheek, leaving a 1in (25mm) haunch to the outside. Concreted remains indicated a cap square of 4in (102mm) width placed centrally over the cheek and rebated into the front face of the rear stepped cheek. The tenon which held the front and rear cheeks together was positioned 2in (51mm) below the top of the trunnion support cheek and was straight, not shouldered as in the case of the castle deck gun carriage (79A1232). It was 3in (76mm) long and 1in (25mm) wide. Two wrought iron bolts were positioned transversely through the front cheeks and carried washers and wedges. Two wooden transoms, 1in (25mm) wide and 5in (127mm) high, were rebated into the bed and the sides of the trunnion support cheeks. These were only noted in the jointed beds where the trunnion support cheeks were higher than the stepped cheeks. The stepped rear cheeks were 1in (25mm) narrower than the front cheeks. These were offset relative to the front cheeks to present a flush external face and also to allow for the widening of the gun towards the cascabel. This is reflected in the positioning of rebates in the bed. A study of the underneath of these cheeks suggested a two-pronged staple for attachment of the rear cheek.

The axle trees were slung beneath the bed, square in section where they contacted it, and shouldered and tapered to form 'arms' which accepted the trucks. These were wedged above the bed and outside the trunnion support cheeks, a feature exclusive to the jointed bed carriages. The foreaxle holes were 2in (51mm) in diameter. The thickness of the fore trucks sloped from 6¼in (158mm) in the centre to 4¼in (108mm) on the outside. These were held by wrought iron lynch pins, convex on their inner faces for ease of movement. The fore trucks on all Upper and Castle deck carriages were much larger than the rear trucks, allowing manoeuvrability and accessibility to the higher ports (Fig. 2.70).

The similarity of this carriage both in design and size of individual elements, to that for the demi cannon 81A3002 is striking. It suggests the use of the same carpentry workshop and similar constraints dictated by location within the ship and the height of the sills of the Upper deck gunports. The dates on the guns are also identical. The form of the gun, date and absence of dolphins suggest that this may also have been made by Francesco Arcano.

*Culverin 79A1278*

> *Made*: Archangelo Arcano of Cesena, 1542
> *Recovered*: John Deane & William Edwards, 1840
> *Assumed location*: Main deck in the bow
> *Original location*: Port side gun
> Currently on display in the White Tower, Tower of London. Armouries XIX.167
>
> *Statistics*
> Length: 11ft (3353mm); 11ft 9in (3581mm);
> weight: 4829lb (21904kg; as listed in Blackmore 1976); internal bore: 5¼in (132mm); 25–26 calibres

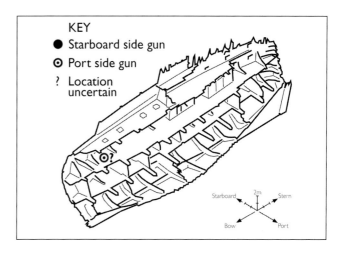

The cascabel is undecorated but a partially defaced inscription on the base ring reads: COLOVR[-] [*error for* COLVOR]. There is a vent field defined by astragal and fillets (Figs 2.71–2). This is similar to 81A1423, possibly the starboard pair to this gun. An inscription above the vent reads: ARCANVS DE ARCANIS CESENEN FECIT ('*Arcano of the Arcana of Cesenea made*'). The reinforce bears a Tudor Rose crowned and encircled with a Garter. There is no inscription within the Garter. A cartouche below bears the inscription: HENRICVS VIII ANGLIE FRANCIE ET HIBERNIE REX FIDEI DEFENSOR INVICTISSIMVS FF [*abbreviating* FIERI FECIT] ('*Henry the Eighth, King of England, France and Ireland, Defender of the Faith, caused to be made*'). A raised plaque below bears the date 1542 and the Royal Cipher: MDXXXXII HR VIII.

The dolphins, in the form of winged mermen, are situated directly over the trunnions. The trunnion centres are positioned at two-fifths of the length of the gun. Forward of the trunnions is a series of mouldings, the rear of which marks the mid-point between the muzzle and the vent field astragal and fillets. The forward point of these mouldings lies at four-sevenths of the total length of the gun. The thickness of metal is approximately equal to the bore at the breech and half of the bore at the mouth. The proportions of this gun are similar to demi cannon 81A3002. There is no muzzle swell, but the muzzle terminates in a prominent cornice. The overall average composition of the metal for this gun is given in Appendix 2.1.

This gun was raised on 27 August 1840 and was described in a letter of 5 September 1840 from John Deane and William Edwards to the Board of Ordnance as being:

> '*a brass gun in a high state of preservation, having the same inscription and very similar to the one we recovered and reported to your Honble Board 18th June 1836, but in the usual place of Dolphins or*

*Figure 2.71 Culverin 79A1278: armorial device and inscription*

*Lions' Heads, there are two Cherubins lying on their backs with extended wings, their lower extremities being divided represent each the curved tails of two serpents or fish*'.

It is recorded as having contained a cast iron shot, two wads and about two quarts of gunpowder (Bevan

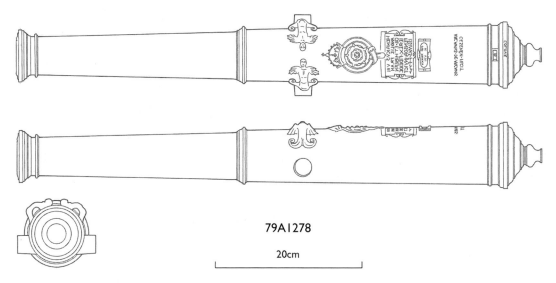

*Figure 2.72 Culverin 79A1278: top, side and muzzle view*

1996, 168, 169). If this is the pair to the M3 culverin, it indicates that the divers were working in the bow. It is also recorded that they recovered four large iron guns at the same time.

The gun was purchased by the Board of Ordnance in 1840 for £216 6s 3d and transferred in 1930 from the Rotunda Museum, Woolwich, to the Tower of London (Blackmore 1976, 60).

## Demi culverins

Demi culverins are listed as having bores of 4¼–4¾in (108–121mm), being 11ft (3.35m) in length and weighing 2200–2500lb (998–1134kg; Norton 1628) with a shot weight of 9–12lb (4.1–5.4kg). Bourne (1587) lists 9–12ft (2.7–3.7m) with a broader weight band of 1800–3200lb (816–1451kg). As with all the culverin class, these guns are expected to have a wall thickness at the breech of not less than their bore. They are listed as being in the Tower from 1514 until 1726. They seem to vary considerably in size and weight, the average weight in 1568–83 being 2200lb (998kg) (Blackmore 1976, 264–8).

The 1514 inventory of ships (Knighton and Loades 2000) does not list demi culverins and there are only two listed for the Tower, both of Humphrey Walker's making (Blackmore 1976, 259). There are none listed for the Tower in 1523. In 1540 there are two demi culverins '*mountid for shippis*' upon the Tower wharf (*ibid.*, 261). Eight are listed over five of the ten vessels within the fleet in 1540. The *Mary Rose*, *Primrose* and *Great Galley* each support two, with the *Minion* and *Small Galley* having one each. By 1546 the situation has changed dramatically, with 49 in bronze listed over 58 vessels. The 20 '*ships*' carry the majority, accounting for 31, distributed over twelve ships. The greatest number are attributed to the *Mary Rose*, supporting six, with the *Matthew* listed as having five, the *Peter* having four, and *Great Bark* with three and the *Henry Grace à Dieu* only two. Eleven bronze demi culverins are listed on the fifteen '*galliasses*', distributed over seven vessels. Four vessels carry two each and three carry one. There are seven on the smaller vessels (pinnaces and rowbarges). At this time we begin to see the rise in cast iron, with 18 cast iron demi culverins listed for the '*galliasses*'.

Only one of the six demi culverins listed for the *Mary Rose* was found in position. This was recovered from the starboard side of the ship on the Castle deck at the front of the Sterncastle. In this position, at the widest part of the ship, it fired forward, just clearing the bow (Fig. 2.10). A second demi culverin was recovered by John Deane and may be the port side pair to this gun. The four further listed demi culverins could also have been located in four vacant gunports at Upper deck level, on a second Castle deck in the stern, as sternchasers, or in the Forecastle (see Chapters 11 and 12). Any of these locations would have been relatively easily accessible to Tudor salvors. Remains of elements from at least two gun carriages in the starboard Scourpit suggests a pair to the Sterncastle bastard demi culverin, and the possibility of a further bronze gun positioned just south of the demi culverin, possibly facing out of the side of the ship.

Both guns are incised '*culverin bastard*'. Their bores, sizes and weights suggest that they are demi culverins. Both are correctly fortified, meeting the criteria listed by Bourne (1587) for demi culverins, so one wonders why they are incised '*culverin bastard*'. This suggests a rather looser terminology at this period than when Norton was writing in 1628. They are neither large in bore nor short in length for demi culverins. They are too long for bastard sakers. They may be a pair, one recovered in 1836 and the other in 1979; but there is no unequivocal archaeological evidence to support this.

Unlike many of the guns lifted in the nineteenth century, there are no comprehensive carriage elements to suggest a location for 79A1279. The two are dissimilar in appearance; the gun recovered in 1836 is 1ft (300mm) shorter than the *in situ* example. The work in 1836 recovered guns from the midships area and the stern. It is possible that 79A1279 (the 1836 gun) was actually situated further back in the stern, and that the pair to 79A1232 (at the front of the Sterncastle) is still missing. If this is the case, the pair to 79A1232 may still remain on the seabed, having fallen away from the ship, unless recovered by Tudor salvage. The gun recovered in the nineteenth century could have furnished one of the ports situated further back in the Sterncastle, with its pair also missing/recovered.

## Demi culverin 79A1279

*Made*: unknown founder, unknown date
*Recovered*: William Edwards, 1836
*Assumed location*: ?Castle deck in the stern
*Original location*: uncertain whether starboard or port
Royal Engineers Museum, Chatham

*Statistics*
Length: 8ft 7in (2615mm); 9ft 5in (2865mm); weight: 2299lb (1043kg); internal bore: 4½in (115mm); 22–23 calibres

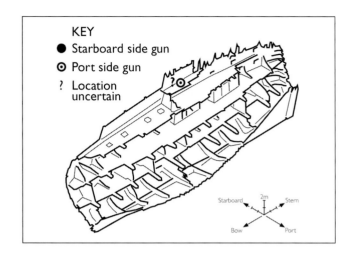

The gun is dodecagonal along its entire length (Figs 2.73–4), and bears a remarkable resemblance to a culverin recovered from the Upper deck in the stern (80A0976) (Fig. 2.83). The vent is extremely large compared with other guns, and may have been enlarged in use, then redrilled to take an iron bush. There is no vent cover, although fixings indicate that it opened to the right.

The reinforce bears a Tudor Rose ensigned with the Royal Crown (Fig. 2.73). This is surrounded by the Garter bearing the inscription: HONY SOYTE QVY MAL Y PENCE. The lower end of the Garter emblem just touches a decorative panel displaying a version of the Royal Cipher, and 'H' and 'R' joined by a looped cord. Beneath is an incised '8'. Towards the breech is inscribed: THYS COLVERYN BASTARD WEYS 2299. Forward and on either side of the inscription two scars are visible. These are the result of the corrosion of the iron core support used during the casting of the gun. There are no dolphins. The initials 'H I' are incised on the top of the left trunnion.

The trunnion centres are situated at 2/5 of the length of the gun. The horizontal diameter of the trunnions is 123mm and the vertical diameter 127mm for the left and 123 for the right. Forward of the trunnions is a series of mouldings, the front of which forms the mid-point between the muzzle and the front of the base ring. The thickness of the metal is approximately equal to the bore at the breech and about half of the bore at the mouth. There is no muzzle swell, but the muzzle terminates in a prominent cornice.

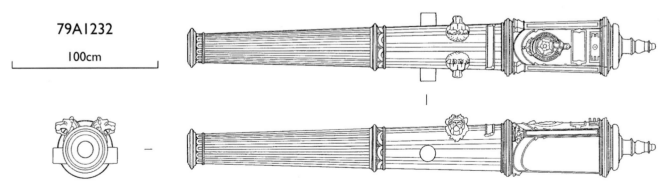

*Figure 2.73 Demi culverin 79A1279: above) detail of the device and inscription; below) top, side and muzzle views*

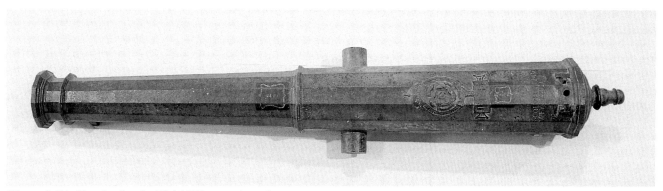

*Figure 2.74 Demi culverin 79A1279: top view of gun*

Examination of the bore with a fibre optic endoscope revealed the remains of iron wire used to reinforce the core during casting. This is approximately 1mm in diameter and 25mm apart. Corrosion of the core supports can be seen as vertical slots on the reinforce. There is no metal analysis for this gun.

The gun was landed on 15 August 1836 together with the cannon 79A1276. The possible location (pairing 79A1232) just aft of amidships means that these two guns would not have been too great a distance apart. Several components of gun carriages found in the starboard scourpit underneath the Sterncastle indicate that there were more bronze guns in the stern. One of these, a trunnion support cheek (81A0707), has a trunnion rest which could accommodate this gun. It was found outside the ship at the start of the Sterncastle. This area requires further excavation.

## Demi culverin 79A1232

Made: Robert & John Owen, 1537
Recovered: 1979
Location: Castle deck in the stern
Original location: in situ

Statistics
Length: 9ft 5in (2876mm); 10ft 6in (3210mm);
weight: 2851lb (1293kg); internal bore: 4½in (112mm); 25–26 calibres

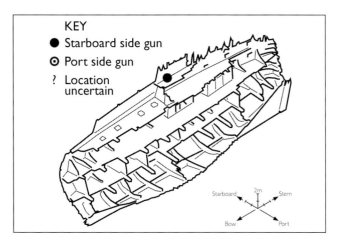

This was the first bronze gun to be found in recent times. It is the most highly decorated gun recovered from the *Mary Rose* (Figs 2.75–6). The cascabel button is elongate and the cascabel cornice is elaborate. There is a vent patch, without cover. The figure '2534' is incised on the breech. The weight is again different from that inscribed on the gun and is heavier by 311lb (141kg).

The first reinforce bears a version of the Royal device consisting of a Tudor Rose crowned and within the Garter). The Garter bears the motto: HONY SOYT QVY MALY PENSE. Below this is a raised cartouche bearing the inscription: HENRICVS OCTAVVS DEI GRACIA ANGLIE ET FRANCIE REX FIDEI DEFENSOR DŇS [*abbreviating* DOMINUS] HIBERNIE ET IN TERRA SVPREM̄V [*abbreviating* SVPREMVM] CAPVT ECCLESIE ANGLICANE ('*Henry the Eighth, by the Grace of God, King of England and France, Defender of the Faith, Lord of Ireland, and in Earth of the English Church Supreme Head*').

The reinforce is decorated with classical Roman arches springing from columns and abutted by pilasters supporting an entablature above (Fig. 2.76). The barrel is almost identical to a breech-loading handgun described as '*Henry Eights Carbine*', also dated 1537 (RA XII.1; Fig. 7.2). Longitudinal scars left by the corrosion of the iron core supports are visible forward and on either side of the cartouche. The second reinforce is faceted to form columns surmounted by arches which form part of the decoration within the second reinforce moulding. A plaque bears the

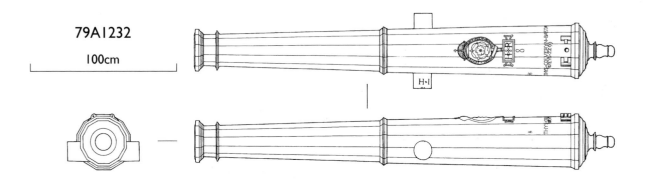

*Figure 2.75 Demi culverin 79A1232: above) detail of the arms, tabella and inscription; below) top, side and muzzle views*

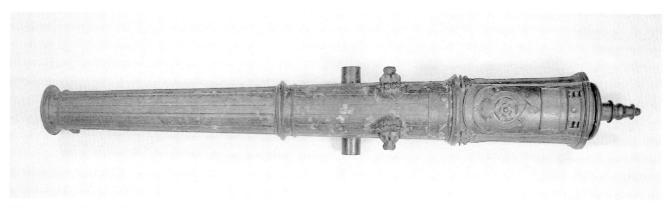

*Figure 2.76 Demi culverin 79A1232: top view of gun*

inscription: ROBERT AND JOHN OWYN BRETHERYN BORNE IN THE CYTE OF LONDON THE SONNES OF AN INGLISSH MADE THYS BASTARD ANNO DŇI [*abbreviating* DOMINI] 1537.

The dolphins are in the form of two lion heads (Fig.2.77). These are positioned in front of the trunnions. The trunnion centres are at two-fifths the length of the gun. The diameters are 103mm. The complex moulding forward of the trunnions begins at four-sevenths of the total length of the gun. This also marks the mid-point between the muzzle and the beginning of the vent patch. The proportions of features on this gun are similar to those of 81A1423. The thickness of metal is equal to the bore at the breech and about 1/3 of the bore at the mouth. The chase is multifaceted. There is no muzzle swell, but there is an elaborate muzzle cornice and a band decorated with arches in relief. Examination of the bore using a fibre optic endoscope revealed the remains of wire used to wrap the core during the casting. These are approximately 10–20mm apart. The overall average composition for the metal of this gun is given in Appendix 2.1.

The gun, still on its carriage, lay on its side, having fallen towards the starboard side as the ship sank. It was positioned on the Castle deck, at the widest part of the ship, pointing forwards. When excavated, its muzzle still protruded through a port built in the front of the Sterncastle structure. Although no shot is recorded as having been excavated from the gun, there was concretion in the bore which required mechanical excavation. It is assumed that this represented the remains of the shot (Rule, pers. comm., 1984).

This gun, a long range weapon, would have functioned to ward off an oncoming enemy vessel, firing forward at a slight angle to the centre line of the ship, a position suggested by Philip of Cleves in his naval manual '*Instruction de toutes manières de guerroyer*' (Rodger 1997, 209).

*Figure 2.77 Demi culverin 79A1232: a) arches; b) lion heads*

*Figure 2.78 Demi culverin 79A1232: the carriage*

**Replication of the gun carriage**
The carriage for this gun has unique features, in particular in the placement of the axletrees above the bed to keep the carriage as low above the deck as possible. The exaggerated height of the trunnion support cheeks allows the elevation necessary for the gun to fire over the anti-boarding netting strung across the weather deck. In order to fully understand the anatomy of the carriage, a study was undertaken of the original, combined with an experiment to reproduce a carriage using tools available to the Tudor carpenter and based on a study of tool marks on the rebates and bolt holes (Fig. 2.79). Examples of most of the tools used were found on board the *Mary Rose* (and see *AMR* Vol. 4, Chap. 8). This process was documented by stills and video photography, held in archive. The facsimile carriage is currently on display, supporting the original gun, in the Mary Rose Museum (Fig. 2.84).

As with the original, the material chosen for the bed, cheeks and trucks was elm. This was felled and allowed to air dry for six years after which it still retained a moisture content of 20%. The bed plank was cut from close to the outside of the tree as the sapwood and remains of the bark were found on the underside of the original bed plank (Fig. 2.80a). The reproduction used a similar portion of timber.

The trunnion support cheeks on the original were cut from one plank with the grain running parallel to that of the tree, possibly cut from an adjacent plank to the bed (Fig. 2.80b). The stepped rear cheeks had been cut from one plank, again taken from close to the bed plank (Fig. 2.80c), back to back (step to step). The trucks had been cut yet again from another plank. The section of timber required necessitated a tree of at least 2ft 6in (762mm) diameter, indicating a tree about 200 years old.

Not only did the grain exhibited by the carriage elements seem appropriate for the orientation of each element with respect to strength, it also demonstrated the most conservative use of timber. Had the stepped cheeks been aligned with the grain at right-angles to the bed, a tree of much wider circumference, and therefore much older, would have been required. The situation is similar with the other elements.

The choice of elm may have been for the purposes of toughness rather than ease of working. English elm is known for not splitting and is possibly the only English wood which could successfully be used to make a solid wheel with a circular hole taken out of the centre. The method of cutting out, which results in a close, multi-grained wheel, utilises this inherent strength (Fig. 2.80d). The wheel would have been sawn out of the board with a framed handsaw and the material cut away with an adze to form the domed inner surface, then finished with a hand smoothing plane with a slightly hollow sole. Elm does not yield easily to the adze so this was a strenuous operation.

The choice of ash for the axle was probably due to the fact that ash is extremely strong along its length. The foreaxle carries the majority of the weight of the gun and a direct downward force after firing which results in a reciprocal upward thrust. The axle was bevelled on its lower surface (see Fig. 2.82e). This helped to hold the front cheek tenons within their mortises, together with the draw-boring of the peg holes. It seems that this was preconceived. The axle hole had not been taken out to the full square, leaving more timber on the front cheek, thereby reinforcing the section which provides the tenon which runs the length of the front cheek.

A frame handsaw was used for the initial cutting out of all elements. An adze was used to make primary shapes and remove the bulk of excess wood with a draw-knife or draw-shave to bevel the sides of timbers, and to remove large quantities of timber from corners, edges and all surfaces. A long-based jack plane was used to obtain straight edges and a smooth surface.

Rebates and mortises were formed using a brace and bit with a bull-nosed auger. Unlike modern augers, where the ears are situated in advance of the cutting head, the bull-nosed auger has them behind the cutting head forming a curved, rounded bottom to any hole. An '*Old Woman's Tooth*' (a modern hand router), with an adjustable blade pre-set to a fixed depth protruding from a wooden base, was used to create a channel and then moved to obtain a flat base. A series of wheelwright's tools, included a scribing gauge for the centre of the wheels, was used to release material caught in the corners of mortises or inside holes. Mortising and paring chisels, and a cornering chisel or '*bruz*', were used with a series of sized mauls and the shipwright's shave. These were particularly useful for cleaning out

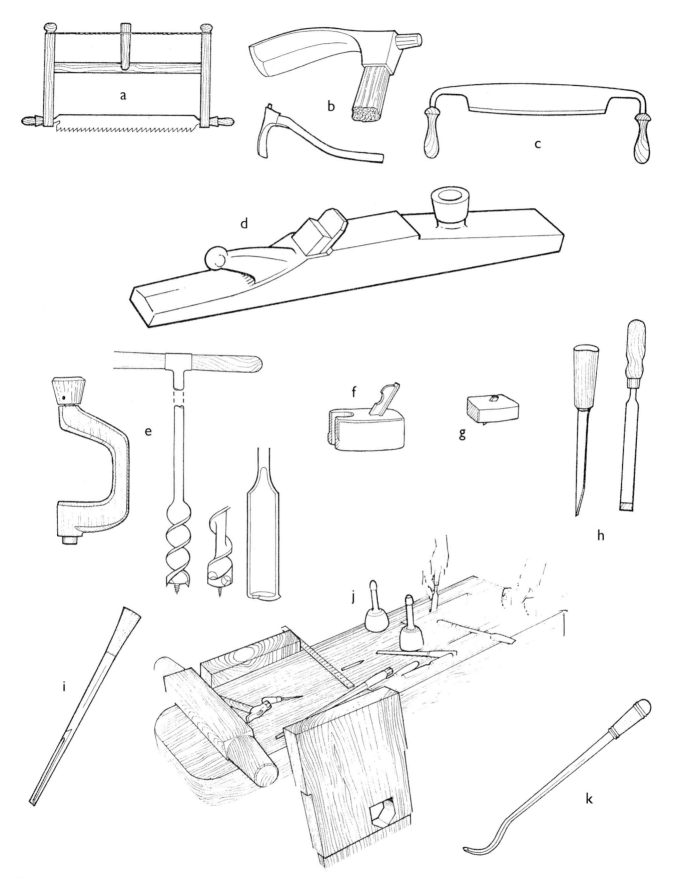

*Figure 2.79 Demi culverin 79A1232: tools needed for making carriage: a) handsaw; b) adze; c) draw-knife; d) long based jack plane; e) brace and bit with bull-nosed auger; f) 'old woman's tooth'; h) scribing gauge, mortise chisel; i) cornering chisel; j) assembling the bed, note also the mauls; k) swan neck*

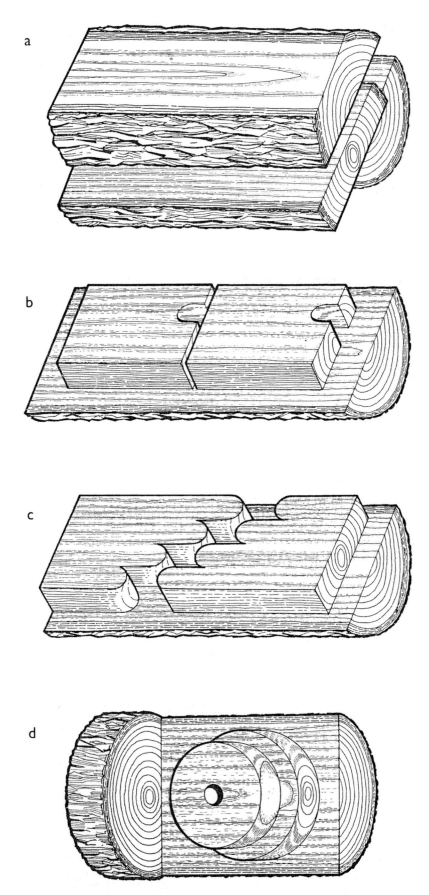

*Figure 2.80 Demi culverin 79A1232: method of cutting out portions of carriage: a) bed plank; b) trunnion support cheeks; c) stepped cheeks; d) wheels*

*Figure 2.81 Demi culverin 79A1232: replicating the carriage: above) bed with fore brackets in position; below) the rebate on the underside of the bed*

bolts. Any slots were hand-punched. The type of forelock bolt used was commonplace; similar bolts have been recovered from other sixteenth century gun sites (Arnold and Weddle 1978). The domed heads were formed by heating the bar to red hot and placing it upon an anvil and striking the top with a ball pane hammer to dress it. The slot was punched out by driving an iron tool through both sides whilst the bolt was still hot. This is the simplest form of bolt and also enables dismantling of the carriage.

Whilst fitting the metal plate rebated into the bed beneath the front cheeks, the importance of this feature was realised. It not only retains the forecheeks in position and strengthens the joint, but the entire plate would have to be pulled through the bed to tear it out of position (Fig. 2.81). The rear stepped cheek was held in position by being inserted over a 'pronged plate' (Fig. 2.82b) which retained a slot to hold a wooden wedge, like a loose tenon. The 'plate' was a design based on the nature of the holes left within the rear cheek rebate, and the housing and concretion within the rear stepped cheek. It was only by radiography of other carriages at a later date, that confirmation of the validity of the design was forthcoming. The tenon is a turned billet formed from four pieces of 1in (25mm) bar fire welded on either side and dressed into shape. The four prongs were ragged, or barbed, to insure a tight fit with the bed. The wedge, in this case of iron, helped to pull the rear cheek into the mortise.

Polythene tracings were made of all original elements detailing the features visible within the wood. The bed plank was prepared, initially by using a hand saw. It was estimated that this would have taken two people one morning of strenuous work to cut out the bed plank entirely by hand saw. The elements on the tracings were chalked onto the base and check measurements taken. These conflicted with the measurement taken of the distance between the trunnion extremities on the gun itself, and it was necessary to study the original carriage. It was found not only that the slot for the fore cheek tenon was aligned to flare outwards towards the rear of the carriage, but a detailed study of the fore cheek tenon suggested that the shoulder which made contact with the bed was chamfered, which resulted in the cheeks themselves being raked outwards. This detail could not have been reflected in the bed tracings alone.

Once the rebates for the brackets and tenons had been cut into the bed, the iron fore and rear brackets holding the trunnion support cheek were set in their rebates and wedged (Fig. 2.82a, b). To do this the bolts were knocked upwards from under the base. The wedges had their edges turned to prevent loosening. At this stage the rear cheeks were set, firstly by locating the tenon into the rear face of the trunnion support cheek, and then by pivoting the cheek down over the pronged plate and wedging. This also had the effect of drawing the cheek into the mortise and securing it to the base.

the flat surfaces between shoulders. A swan neck was used to clear out the bottom of mortises.

Wrought iron was used for the original carriage and the reproduction (Fig. 2.81). No machine tools were used; the holes were punched when the iron was red hot and the ends cut with a cold chisel and dressed on the edge. The cap squares and the fore and rear brackets were all made from 3¼in (83mm) wide and ½in (13mm) thick iron held with 1in (25mm) diameter

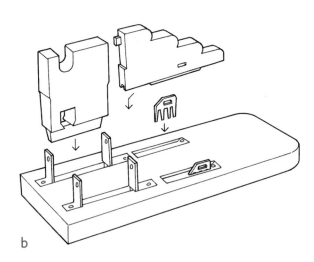

*Figure 2.82 Demi culverin 79A1232: replicating the carriage: a) fore bracket; b) assembly sketch; c) tenon on stepped cheek; d) inside view of the bed with stepped cheek; e) placing the ash axle*

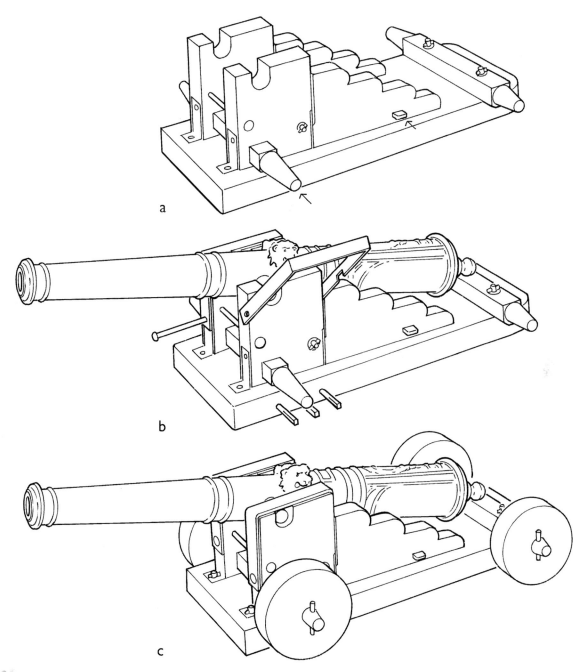

*Figure 2.83 Demi culverin 79A1232: mounting the gun*

It was demonstrated that the rear cheeks could be removed with the gun in position, so that the rear cheeks could be replaced. The tenons were copied from the original, offset towards the outside face of the cheek, so that the carriage was handed (Fig. 2.82c). The height of the bolt holes for the draw bolt also differed between left and right, reinforcing this specification. The tenon was manufactured with a barefaced shoulder to the outer face of the cheek and the capsquare cut away to accept both the tenon and shoulder (Fig. 2.82d).

The ash axle was placed through the fore carriage before the gun was mounted (Fig. 2.82e). Rectangular iron pieces were manufactured to accept the axle wedges and were wedged above the bed and turned to prevent dislocation. The axle was placed as a square timber over the bed, then fashioned to accept the trucks. These 'arms' extended out from the bed for 12¾in (324mm) (Fig. 2.83a).

Oak pegs were entered from the side of the bed through the tenons rebated into the bed to secure and retain the front cheek (Fig. 2.83b). These were originally made from a square timber and tapered over most of their length. The final few inches were left square, making a very solid joint. Alterations to the rear cheek, in the way of relieving a portion of elm to accommodate decoration on the gun, was undertaken at the time of mounting, copying the original.

*Figure 2.84 Demi culverin 79A1232: the original gun on a reproduction carriage in the Mary Rose Museum*

The cap square was offered towards the barrel from an angle of 45° and pivoted downwards, rotating around the mortise and tenon. This was eased onto the fore and rear brackets and locked together with the draw bolt. At this stage the transom bolts were placed laterally through the trunnion support cheeks and wedged, preventing splaying of these cheeks under load (Fig. 2.83b). These were of 1in (25mm) diameter and 20in (420mm) long. Washers were produced using information from vestigial remains of iron concretion, which suggested domed heads and/or washers on all outer faces. The staining from a concreted iron wedge against wood would not have appeared circular. The gun on its replica carriage is shown in Figure 2.84.

# Sakers

Sakers are described by Bourne (1587) and Norton (1628) as having a bore of 3½–4in (89–102mm), a length of 9ft (2.7m) and weighing 1400–1800lb (635–817kg; Fig. 2.85). In keeping with all guns of the culverin class, they are fortified with no less than their bore size in metal at the breech. The shot is listed as weighing 5–5½lb (2.3–2.5kg) of cast iron. Shot is sometimes listed as being composite lead and iron.

Sakers are not listed in inventories for the Tower until 1540, but rapidly become the most prevalent bronze gun listed. There are 31 listed over ten vessels in 1540, with the *Mary Rose*, *Sweepstake* and *Peter* having five each, the *Minion* four, *Great Galley* and *Primrose* three each, the *Small Galley* and *Trinity Harry* each with two and the *Jennet* and *Lion* each having one. By 1546 there are 104 distributed over the 58 vessels; fourteen of these are iron. All of the ships except the *Hoy Bark* carry sakers, totalling 50, of which nine are iron. The majority carry two, with the *Minion* carrying six, the *Small Bark* five, the *Peter*, *Pauncy*, *Trinity Harry* and *Henry Grace à Dieu* carry four. The *Mary Rose* is listed as carrying two. There are 33 listed for the '*galliasses*', four of which are iron. All but two vessels (*Grand Mistress* and *Tiger*) carry sakers. Most support two, with a maximum of four on the *Salamander* and *Swallow*. Twenty-one are attributed to the pinnaces and rowbarges, one of which is iron. These usually occur singly, but the *Falcon* is listed with four, and the *Saker* and *Phoenix* with two each. Six vessels do not carry sakers, the *Roo*, *Merlin*, *Brigandine*, *Trego Ronnyger*, *Rose in the Sun* and *Falcon in the Fetterlock*.

No sakers have been identified from the *Mary Rose*. The portion of a trailing carriage (81A0486, see below) could have supported a saker or a falcon. Trailing carriages (or fragments) have been recovered with sakers on the (later) sixteenth century shipwreck sites of Yarmouth Roads (Isle of Wight) and Teignmouth, Devon.

There are 80 iron shot listed for sakers in the *Anthony Roll*. The shot band is taken as 3–3¾in (79–96mm). There are 558 cast iron shot within this band, with the majority at 84–6mm (346 shot, 77%). Most were stored in M2 or M6, with more in M2 and there were isolated examples (see Chapter 5, Fig. 5.21). There are 12 in the starboard scour in the bow, and three in O2, possibly fallen from the port side. There are six in M12 and one cross-bar shot of the appropriate size between M10 and M12. This suggests the possibility of sakers in the stern, possibly as stern chasers and also in the Forecastle. The fact that the

*Figure 2.85 Royal Armouries saker XIX.165. Reproduced by permission*

wrought iron demi slings will have used shot of this size means that the shot is not exclusive to sakers and could also argue for the presence of slings (Chapters 3 and 5). A recently found document (see Chapter 12) suggests that a pair of these was positioned on a second castle deck firing forwards above the demi culverins.

## Falcons

Falcons are listed in inventories for the Tower of London from 1495. By the mid-sixteenth century (Bourne 1587; Norton 1628) they are listed as being 2¾in (70mm) in bore, 7ft (2.1m) in length and 550–750lb (250–340kg; Fig. 2.86). Shot is listed as being 1¾–2⅛lb (0.8–0.9kg), but composite shot of iron dice and lead is also listed. Norton (1628) attributes these with 36–40 calibres in length, reinforced with 1¼–1⅓ of their bore size in metal at the breech. As to their use, Norton states (1628, 56):

> 'Culverings *and* Demy Culverings *serue to pierce & cut out in batteries what the* Cannons *haue shaken and loosed: the Sakers and Falcons serue for Flankers, the other smaller sorts of this kind, for Field Peeces for the assaults, and to shoote at Troopes or Companies of men that are neere together. Al these shoote iron shot, but many shoote stone shot where the marks are but tender, and so they will saue much in Amonition, and yet performe as good seruice as with the iron shot they can doe.'*

Falcons are listed for vessels in the 1514 inventory. There are seventeen distributed over thirteen vessels, with the largest number being on the *Henry Grace à Dieu* (six). The *Trinity Sovereign* is listed with three, and the rest are mostly in pairs. In 1540 there are seventeen over the ten vessels in pairs or as singletons. Only one vessel, the *Peter*, has four. The *Mary Rose* is listed with two. In 1546 there are 22 listed over the 58 vessels. Fourteen are assigned to eight ships, the largest number being four on the *Pauncy*. The *Mary Rose* is listed with one. There are three on the '*galiasses*': two on the *Greyhound* and one on the *Lion*. Only three pinnaces carry falcons, one each on the *Marlion*, *Less Pinnace* and *Brigandine*. The only rowbarge carrying falcons is the *Falcon in the Fetterlock*, with two.

No falcons have been identified from the *Mary Rose*. There are 60 iron shot listed in the *Anthony Roll* for the single falcon. The shot band width is up to 2½in (47–65mm). There are 128 shot within this band, with a peak of 36 at 63mm and 24 at 64mm (see Chapter 5, Figs 5.22–3). Most were in store in M2 and M6, the only loose single examples being in M12, U7/8, port scour 2, H/O1 and M1. One of the groupings of composite shot (of dice and lead) includes nine of this size. There are six individual shot not in M2 or M6 (in C1, M7).

## Isolated Carriage Elements

A considerable number of gun carriage elements were excavated from the site. As it is not always clear whether individual elements belong with bronze or iron guns, they are discussed together here including arguments as to which type of gun they belong with. Excluding sleds, cheek pieces and beds, the total includes 56 truck wheels, 14 spoked wheels and 41 axles. As some elements are broken the actual number may be slightly less. Using the *Anthony Roll* and the types of wheels found with particular guns on the site, a good estimate would be that six of the twelve port pieces supported spoked wheels, totalling twelve around the ship, excluding the spare four pairs (eight) listed in the *Roll* and found in the bow. There is no evidence for more than three pairs around the ship, those found in gunbays M5, M8 and M11. As the Upper deck stern ports are so close to the deck it is unlikely that these could take guns with spoked wheels. The total estimates for axles excluding spares is 51, of which 32 have been found (this assumes that the two sakers and falcon listed were supported on carriages with a single axle and pair of wheels). The total number of trucks should be 90 (excluding those in store) reduced to 84 if the three small bronze guns were on spoked wheels (for which there is no evidence, see above); 49 were found. All of the elements listed as spares were recovered from H1 and O1. Four 'pairs' of spoked wheels, nine trucks and

*Figure 2.86 Royal Armouries falcon XIX.245. Reproduced by permission*

*Figure 2.87 Spare gun carriage wheels, presumably in storage, on the seabed in the bow area of the ship*

nine axles were found there (Fig. 2.87). The distribution of major elements is shown in Figure 2.89.

Diagnostic elements which can be used to suggest the presence of bronze guns in particular places include a portion of a bed (81A0124; Fig. 2.88) found on the Upper deck in the stern, just in front of the U11 gunport. Although only a fragment, this is from a 4in (102mm) thick bed with a double mortise to accept a double tenon on the trunnion support cheek and is the right front portion with the rebate for the cap square on the front right trunnion support cheek and a rebate for a strap across the underside of the bed. Although very thin for a bed (Table 2.5), suggesting a smaller bronze gun such as a demi culverin, saker or falcon, it has a double rebate which is not seen in the carriage for the Castle deck demi culverin, 79A1232. There is not enough to state whether it is from a jointed bed or not, although the inner edge is very square and the break could be in line with the join lengthways. Widthways the break could be in line with a rebate to take a wooden transom which could provide a point of weakness in thin beds (see 80A0976, carriage).

A right trunnion support cheek (81A0707), with trunnions penetrating the cheek, was found just outboard of the beginning of the Sterncastle (Fig. 2.88) almost directly above the U8 gunport. In this position it could derive from either a port side equivalent of the *in situ* Castle deck demi culverin (79A1232) or from the pair of Castle deck bronze guns suggested as having been positioned just to its south. This would be a difficult location for a gun, because of the alignment of 79A1232. The piece is 3½in thick (90mm) and has holes for two transom bolts, one higher than the other. The thickness of the cheek and the position of the bolt holes suggest a Castle deck carriage, as does the size. The tenon is single and the only carriage identified to date with this feature is for 79A1232. The arrangement for the cap square is very similar to 79A1232 and the fore and rear brackets take up the entire width of the cheek. The height of the cheek above the bed is less than for the demi culverin (18½in/470mm), and the axle position is not above the bed, differing from 79A1232. The transom bolt locations, however, dictate a gun elevated to an angle in order to clear the forward (higher) bolt. There is concretion around the transom bolt holes and the cheek has been torn at the tenon. There was a portion of the bed attached when the object was excavated, with considerable concretion. It must be concluded that this was the result of nineteenth century rather than Tudor, activity. The trunnion rest, with a diameter of 5in (130mm) would fit 79A1279, raised in the nineteenth century. This cannot fit the bed fragment 81A0124 described above as the bed is rebated for a double tennon.

A right stepped cheek was also found in the starboard Sterncastle scourpit (Fig. 2.90), SS07, 81A6896. It is incomplete, with only two steps present. It has a height of 380mm and length of 350mm and thickness of 90mm. It does not seem to fit 81A0707, and it seems impossible to pair these. The tenon position is dissimilar as well as the cheek heights. This may be resolved by actually offering up the two pieces once conservation is complete. Concretion suggesting that the cap square was retained under water is also evident. This example is slightly different from other stepped cheeks; the distance along to the first step is shorter than other carriages studied although it is most similar to 79A1079, from the C1 demi culverin. It is therefore possible that this was from a carriage for another gun station within the castles. These carriage components indicate the presence of either two or most probably three further bronze guns in the Sterncastle. Three would furnish a carriage for the demi culverin recovered in the nineteenth century and two more guns. These three guns would serve as the port side pair to 79A1232 and the two gunports further south on the Castle deck, port and starboard respectively. This would leave a further three bronze guns still unaccounted for, two sakers and a falcon.

Remains attributable to three or four carriages were found in the bow in an area containing collapsed

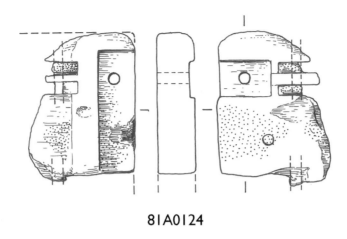

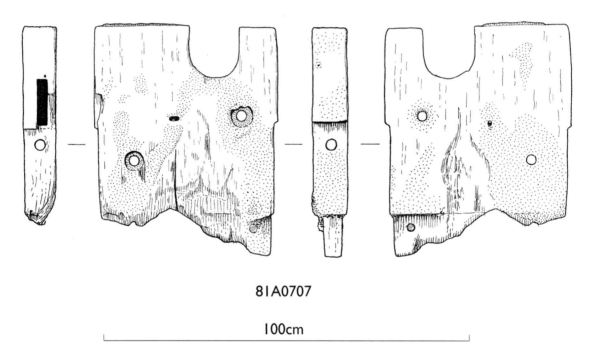

*Figure 2.88 Gun carriage fragments 81A0124 and 81A0707*

material from the Forecastle. 81A0107 is a badly eroded 4in (102mm) thick right rear stepped cheek found in O1 (Fig. 2.91), with a height of 440mm and length of 550mm. It is most similar in size and from the 81A6861 (SS7/8 and the Castle deck demi culverin 79A1232). 78A0139 is a small axle and wheel (details below) found in U1 (Fig. 2.91), a further possible pair of wheels found in U2 (79A0190, 79A0375) and, perhaps most importantly, the spade (81A0486) and a possible cheek fragment (81A0451) of a trailing carriage (Fig. 2.92). It has a length of 678mm and overall width of 480mm. The trail has an internal width of 300mm (narrowing to 250mm at the back) and retains one corroded transom bolt. The side pieces are 100mm at the back of the carriage but 50mm just in front of the transom. The transom block has a length of 220mm, width of 250mm and height of 300mm. There is concretion and bolt holes suggestive of iron straps over all the upward edges. The centre of the transom bolt is 150mm above the base, suggested diameter 30mm. The type of carriage is shown in Figure 2.93. The trail is similar to that found with a minion on the sixteenth century Yarmouth Roads wreck off the Isle of Wight (Tomalin *et al*. 1988, 75–86).

These pieces suggest that up to four guns were situated in the Forecastle, at least two of which would be bronze. Both the stepped cheek and trail retained concretion suggestive of use rather than having been in storage, although both were found in O1 (where wheels were being stored). Overall, these isolated elements

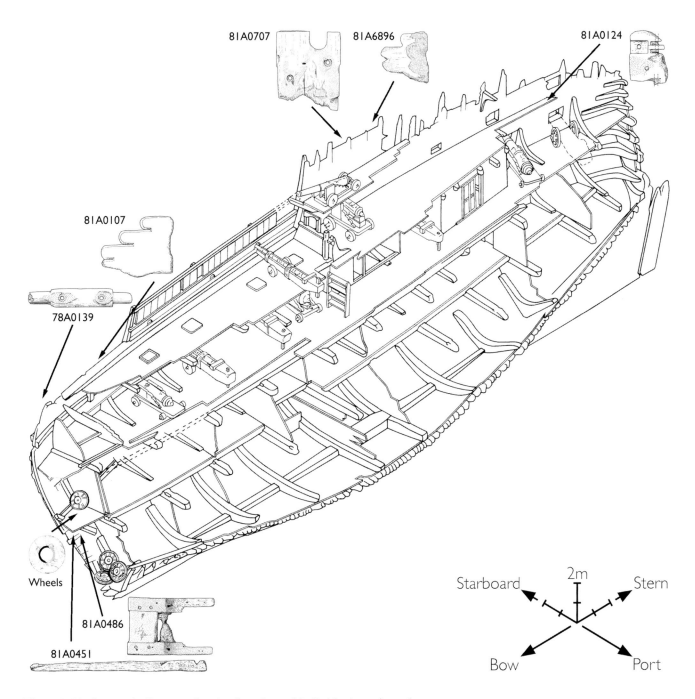

*Figure 2.89 Isometric diagram showing locations of individual carriage elements*

argue for a further four (possibly five) bronze guns, bringing the total to fourteen or fifteen, matching the *Anthony Roll* inventory.

The truck wheels and axles could be used on carriages for both bronze and iron guns, so their distribution cannot be used with any certainty to suggest locations for specific types of guns. However, some diagnostic features can be used to suggest gun types based on the length of the axle arms (much longer for guns with spoked wheels) and the width of the portion squared off to rest under the carriage, which is wider for the majority of bronze guns than for iron ones. The distance between the centres of the axle bolt holes can also be used (less with wrought iron guns) although caution must be applied; the unusual C1 demi culverin (79A1232, Fig. 2.94) with the axle above the bed has axle bolts which fit between the trunnion support cheeks and are closer together than those from the larger bronze guns (for example, culverin 80A0976, Fig. 2.94). This could be mistaken for an axle for a port piece or sling carriage, with narrower beds. Generally the space between bolt holes is more on the rearaxles of carriages for bronze guns, often located immediately behind the stepped cheek, whereas the foreaxle bolts occur inside the trunnion support cheeks. However, this is not the case with the jointed beds where the axle bolts

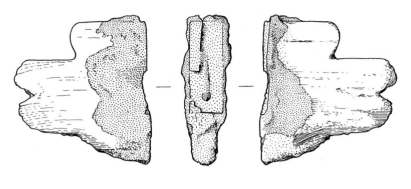

81A6896

100cm

*Figure 2.90 Stepped cheek fragment 81A6896*

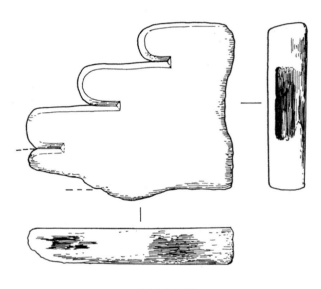

81A0107

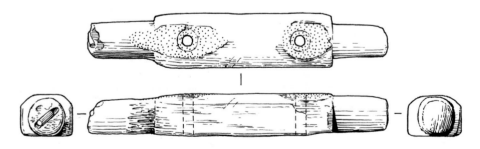

78A0139/2

100cm

*Figure 2.91 Stepped cheek piece fragment 81A0107 and axle 78A0139*

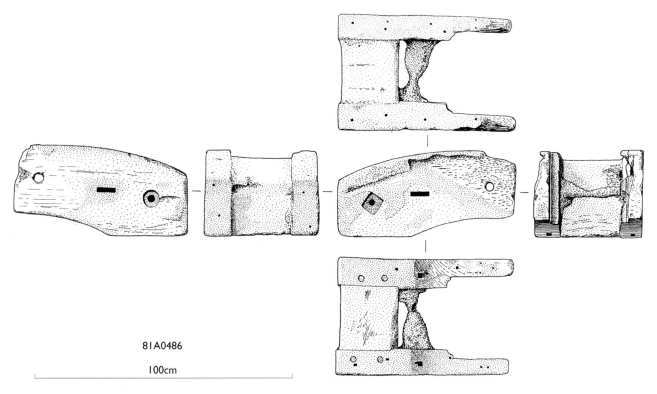

*Figure 2.92 Gun carriage 81A0486*

occur outside the trunnion support cheeks. Generally speaking the axles supporting carriages for bronze guns and the wrought iron slings appear to be more definitely worked and are sometimes hexagonal across the bed area with a very defined edge to mark the start of the arms (for example the axle for 79A1232). A contrast would be the axle for a port piece supporting spoked wheels. There the axle is nearly flat-sided on the lower edge with a slight relief to take the bed and taper to the axle arms (M5 port piece, 81A2604) (Fig. 2.94). Clearly, therefore, single items are often difficult to differentiate. We also have no identified carriage elements for wrought iron fowlers for comparison.

Excluding the eight spoked and eight solid wheels in store and those associated with *in situ* guns, fourteen solid wheels were distributed around the site (Figs

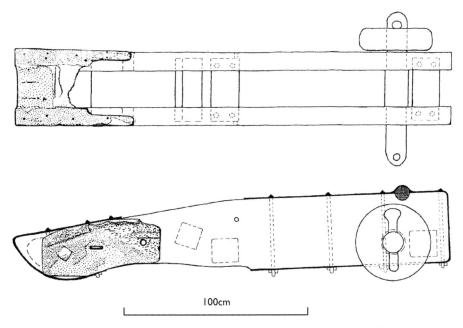

*Figure 2.93 Reconstruction of carriage type represented by 81A0486, based on carriage elements recovered from Yarmouth Roads (reconstruction by Nicholas Hall)*

## Table 2.8 Details of unassociated truck wheels

| Sector | Object | Diam. (mm) | Thickness (mm) | Axle hole diam. (mm) | Description/associations |
|---|---|---|---|---|---|
| *Trucks in store* | | | | | |
| H1 | 81A0157 | 530 | 160–120 | 130 | well formed, domed, eroded face |
| H1 | 81A0522 | 410 | 80–120 | 120–110 | well formed, domed |
| H1 | 81A1035 | 250 | 90–100 | 100 | very eroded |
| O1 | 79A1230 | 350 | 100 | 100 | very eroded, parallel faces |
| O1 | 81A0316 | 665 | 150 | 150–100 | new, parallel faces, tapering axle hole |
| O1 | 81A0392 | 460 | 115 | 115–110 | new, parallel faces, tapering axle hole |
| O1 | 81A0492 | 410 | 80 | 120–100 | parallel faces, tapering axle hole |
| O1 | 81A0476 | 410 | 110 | 110–90 | parallel faces, tapering axle hole |
| *Loose trucks* | | | | | |
| BB★ | 81A4901 | 500 | 120–100 | 150–100 | eroded, domed; looks similar to demi culverin 79A1232 |
| U1 | 78A0139 | 550 | 100–90 | 150 | eroded, slightly domed, found with axle |
| U2 | 79A0190 | 550 | 150–120 | 110–90 | domed, similar to demi culverin 79A1232 |
| U2 | 79A0375 | 500 | 150 | 100–90 | domed, similar to demi culverin 79A1232 |
| U6 | 81A0332 | 520 | 110 | 125–110 | eroded, ?parallel faces, similar to 79A1232 |
| U7 | 79A0132/2 | 350 | 150 | 200–100 | parallel, thick; foound with axle; similar to U6 sling 81A0645 |
| O8 | 82A0981 | 580 | 130–120 | 120–100 | clean, slightly domed |
| M11 | 81A1875 | 355 | 140 | 130–110 | parallel, similar to 81A3045 |
| U11 | 81A3045 | 355 | 150 | obscured | parallel, similar to 81A1875; found with axle |
| U11 | 78A0221 | 424 | ? | ? | listed as truck fragment |
| U11 | 81A1906 | 600 | 110 | 120–105 | eroded, ?parallel, similar to 79A1232 |
| SS11 | 82A2339 | 350 | 70 | eroded | eroded and gribbled |
| Unknown | 81A4902 | 430 | 90–70 | 135–120 | two halves |

BB★ = Bow bulkhead excavation. Trench under the ship at the bow

2.95–6; Table 2.8). There are no spoked wheels except those either within gunbays or in store (see Chapter 3, Table 3.5 and Figs 3.50–4). There are observable differences within the assemblage of trucks. Broadly speaking, trucks can be described as being parallel-sided or convex on their inner faces. The smaller examples are often parallel, but are generally thicker overall than the convex varieties. This suggests that the larger wheels were made thinner where possible to reduce weight but are thickened to provide adequate strength where they are entered over the axle. Most of the wheels have holes which taper in size (larger on the inside) accommodate the tapering axles, although this is not always easily recognisable and it less obvious on smaller examples (U6 sling, 81A0645).

There are only two instances where multiple truck wheels are found in one area, U2 and U11. The wheels found in U2 (79A0190, 79A0375) are both like the large, relatively thin wheels found with the demi culverin 79A1232 (see Figs 2.78, 2.82). The diameters are slightly different (550mm and 500mm) but the thickness and axle hole sizes are similar. The three found in U11 (81A3045, 78A0221, 81A1906) are all of slightly different sizes, although the single wheel found in SS11 (82A2339) is of a similar diameter to 81A0345, found with an axle, as is the single wheel found in M11 (81A1875). Unfortunately the SS11 wheel has been badly damaged by boring organisms (gribble), and the thickness cannot be compared. The size of wheel and axle bolt hole spacings is similar to the axle and wheels (81A0657) supporting the *in situ* wrought iron sling from U6 (81A0645, Chapter 3). The U7 wheel and axle (79A0132/2)) also resembles that found with the U6 gun (see Figs 2.94–7 and Chapter 3, Fig.3.71) and might represent a port side sling in a similar position to the U6 starboard example, and the U11 wheel and axle suggest the presence of a sling or gun with a similar carriage in this area. One of the wheels from U11 is large (81A01906) and resembles the front wheels of demi culverin 79A1232. A demi culverin (or pair) firing from either side of the rudder would be a sensible configuration, possibly on the Upper deck (Chapter 11).

Table 2.9 Axles in store

| Sector | Object | Length (mm) est* | Width (mm) | Depth (mm) | Bed width (mm) | Arm length/ diam at face (mm) | Description |
|---|---|---|---|---|---|---|---|
| H1 | 81A0425 | 1900 | 152 | 149 | 800 | 600/100 | v. rough, no holes; suggestion of bed area squared; ?for bronze gun with solid wheels, though squared area could be made smaller |
| H1 | 81A0489 | 1500 | 130 | 130 | 600 | 470/400/120 | bed area well cut, hexagonal, no holes; ?solid not spoked wheels; either narrow bronze or wrought iron gun |
| O1 | 81A0484 | 900 *1000 | 125 | 125 | 520 | 240/100 | semi-finished, bed area squared, arms roughly tapered with 1 lynch pin hole; size suggests wrought iron gun with solid wheels |
| O1 | 81A0485 | 1850 | 165 | 155 | 900 | 550/400/100 | roughout, no holes, area squared for bed, slight taper to arms; ?bronze gun |
| O1 | 81A0494 | 1900 | 130 | 130 | 900 | 500/100 | roughout, no holes, area slightly squared for bed; ?bronze gun |
| O1 | 81A0495 | 1300 *1400 | 150 | 130 | 580 | 450/80 | 1 bolt hole on arm 250mm from bed suggests thick wheel; would accommodate spoked wheels |
| O1 | 81A0496 | 1340 *1500 | 110 | 110 | 600 | 450/80 | 1 bolt hole on arm 250mm from bed suggests thick wheel; would accommodate spoked wheels |
| O1 | 81A0542 | 125 | 125 | ? | ? | ? | v small fragment |
| O1 | 81A0623 | 180 | ? | ? | ? | 90–72 face | well finished arm end, 1 bolt hole 70mm from face fragment |

* If enough of one half of axle exists, total length is estimated

Excluding the nine stored axles in O1 and H1, a further eight were found around the site (Figs 2.96–7; Table 2.9). Three were found with one truck still attached (78A0139, 79A0132, 81A03045), all on the Upper deck (Fig. 2.96); 78A0139 was found in U1. It has a strong chamfer and defined hexagonal shape. The arms appear to be broken at the lynch pin holes and has a current length of 800mm with arms of 180mm and 150mm and diameter at the end of the arms of approximately 100mm. The area squared-off to fit the bed is 500mm in length, similar to the U6 sling. The axle bolt holes centre to centre are 250mm, again similar to U6. The width of the axle is 140mm and its height is 100mm. The chamfered (and axle forelocks) lie facing the deck – the flat wide portion of the axle touches the lower face of the bed. The associated wheel is quite large and thin (550 x 100mm), compared with the U6 sling of 490 x 120mm.

79A0132 was found in U7 and is a fragment only 200mm in length. The truck was small (350 x 150mm), but appears parallel sided like U6. 81A3045 was found in U11, close to two other trucks, two axles and a portion of a bed for a bronze gun (81A0124). This is well formed, hexagonal in section and similar in many respects to 78A0139 but is also like the foreaxle of the *in situ* demi culverin, 79A1232. Concretion suggests that the axle was attached beneath the bed and the chamfer faced the deck. The axle is complete, with a total length of 1200mm and bed portion of 600mm with arms of 250mm in length. The axle face is 90mm in diameter increasing to 100mm at the shoulder. The axle width is 150mm and height 120mm. The bolt holes are 250mm apart. The demi culverin (79A1232) front axle had a bed portion of 650mm, length 1100mm, and the foreaxle bolt holes were inside the trunnion support cheeks and 200mm apart. The wheel for the U11 axle is much smaller (350mm diameter rather than 600mm) and is thicker. However if this faced out of the stern, we have no idea of gunport height. The U10 and U11 ports are extremely close to the deck and any gun would have to be on very small wheels. Typologically it is impossible to tell what sort of gun this supported. The other wheels are 600mm and 425mm and the axles are relatively small fragments and non-distinctive (see Tables 2.9–10). Locations for the isolated axles include U4, bow bulkhead, M11, H2 and H4. The stored axles are obviously unfinished, some drilled for a single lynch pin on one arm. Some suggestions regarding type of gun are

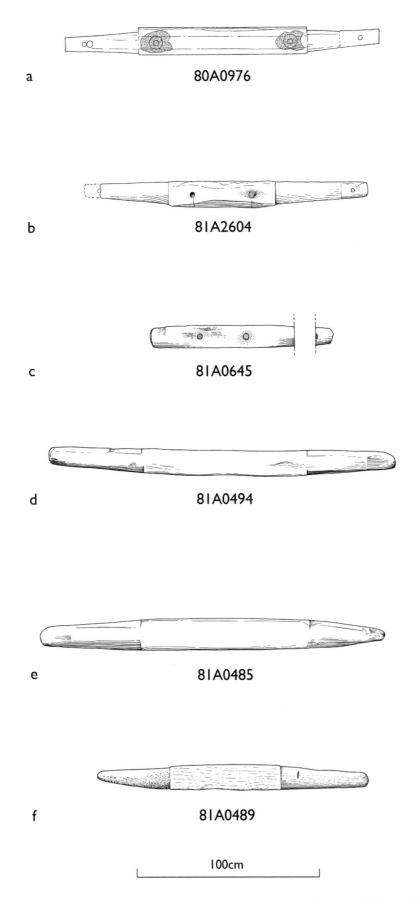

*Figure 2.94  Axles: a) culverin 80A0976; b) port piece 81A2604; c) sling 81A0645; d) roughout 81A0494; e) roughout 81A0485; f) roughout 81A0489*

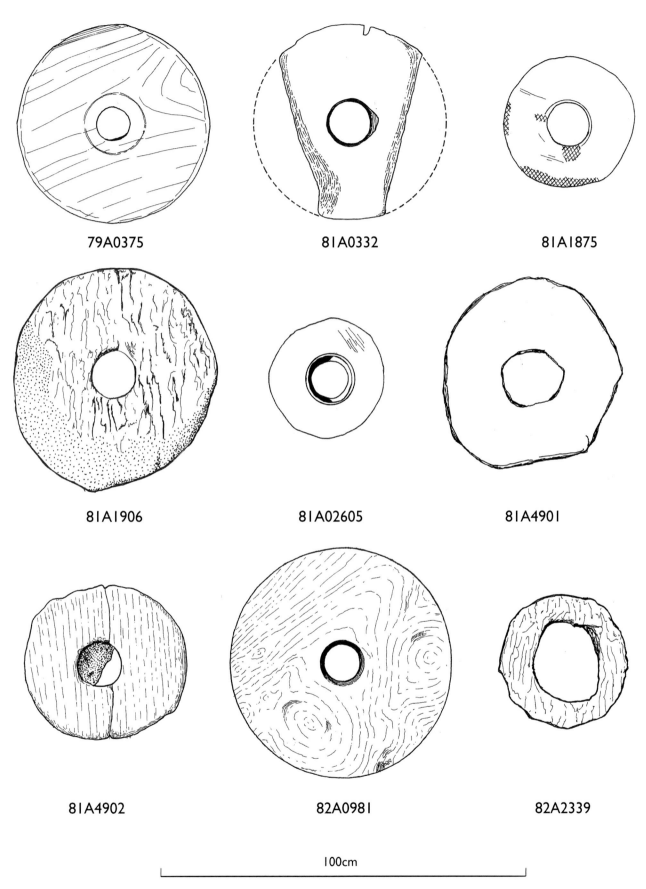

*Figure 2.95 Solid wheels*

Table 2.10 Loose axles not in store

| Sector | Object | Length (mm) | Width (mm) | Depth (mm) | Bed width (mm) | Axle bolt hole centre-to-centre | Description |
|---|---|---|---|---|---|---|---|
| U1 | 78A0139 | 800 | 140 | 100 | 500 | 250 | well finished, v square to bed, hexagonal; arm l. 1800mm, max. diam. at end 100mm for trucks, looks similar to U6 81A0645; sling but poss. narrow bronze gun; found with truck diam. 550mm |
| U7 | 79A0132 | 200 | 90 | 70 | ? | ? | frag. arm with face & 2 axle lynch pin holes close to end (outer 25mm, inner 20mm); 2 collapsed pegs *in situ* in small wheel |
| H2 | 81A1103 | 1150 | 120 | 120 | 680 | 250 | well finsihed, used axle; arm l. 200–220mm, for trucks; similar to C1 foreaxle so for ?bronze gun |
| H4 | 82A0083 | 290 | n/a | n/a | n/a | n/a | frag. not drawn & not accessible; X mark on side |
| U4 | 79A0491 | 90 | n/a | n/a | n/a | n/a | arm frag. with lynch pin hole 35–40mm from end; diam. at face 90mm; v. similar to 81A6370 |
| Unknown | 81A6370 | 110 | n/a | n/a | n/a | n/a | arm frag. with portion wooden lynch pin *in situ*, hole 25mm; diam. at end 80mm; similar to 79A0491 |
| M11 | 81A0894 | 320 | 90 | 110 | n/a | n/a | frag. squared area near bed with axle bolt hole & part arm; diam. axle at bed 90mm, at broken end 70mm, arm l. 80mm; |
| U11 | 78A0229 | 250 | n/a | n/a | n/a | n/a | 2 frags arm with 2 lynch pin holes |
| U11 | 81A0127 | 250 | n/a | n/a | n/a | n/a | well finished, 2 lynch pin holes diam. 110–70mm (face); poss. sheared at inner face of wheel suggesting solid wheel W. 110mm |
| U11 | 81A3045 | 1200 | n/a | n/a | 600 | 250 | 2 portions comprising complete axle, 1 retaining wheel diam. 350mm; similar to 81A0645 U6, also to foreaxle demi cannon 79A1232 C1 |
| BB | 90A0128 | 490 | ? | ? | ? | ? | not drawn not available, frag., retains bolt holes & bolt frags; found in large concretion (0A0050) contaninig portions of swivel gun, wrought iron fowler & chambers |

possible based on arm length and the area worked square to support the bed (see Tables 2.9–10; Figs 2.94 and 2.97).

## Casting and Firing a Bronze Culverin

### Techniques of casting bronze guns

The processes involved in the manufacture of cast ordnance in the sixteenth century required a number of stages (see Kennard 1986 for a modern interpretation, also Murphy 2001, 73–97). Initially a wooden pattern (or wood covered with clay) was produced. Whilst still malleable, this could be worked with a strickle board to resemble the finished gun. It would be longer than the gun, and would include a feeding head or large reservoir, which was filled during the metal pour and would pressurise the molten liquid. As the metal begins to cool and solidify, the pouring head serves to keep the metal insulated and fluid and able to provide the metal to fill the voids that would otherwise form. Drying of the pattern could be obtained by rotating it over a fire with the breech and feeding head resting on trestles (Fig. 2.98). The trunnions, dolphins and any relief decoration or other embellishments, together with the device, would be pinned to the surface of the pattern which would finally be coated in layers of wax (Fig. 2.99). In order to extract the pattern from the external jacket, and later the cast gun from the mould, both the pattern of wood or wood/clay and the external jacket would be destroyed. Each gun cast in this method was

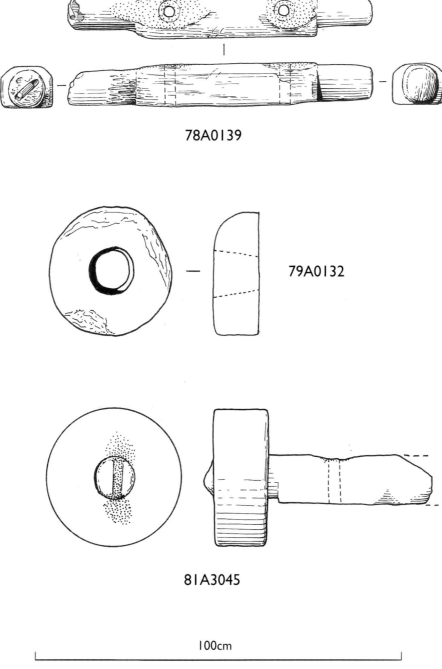

*Figure 2.96 Axle fragements and wheels*

therefore unique, as the moulds could not be re-used. Certain elements, such as the patterns for the embellishments, were re-usable.

An external jacket made of clay and externally reinforced with iron was built over the wooden pattern (Fig. 2.98d). The pattern was then removed by exerting force on its centre, the wooden pole. This would cause the built-up layers of clay to break away at the junction with the pattern (often aided by the application of heat) and collapse into the jacket, leaving a hollow mould or negative impression of the barrel. The inside of the jacket would be cleaned, the trunnions (if wood) removed, and any holes patched and the inside of the mould baked. A damp mould was very dangerous and often ruined the gun (Norton 1628, 67). The breech was made separately. The breech pattern was also made in wood and could be coated with clay and any decoration moulded within the clay. This fitted to the pattern forming the barrel. The jacket was also made in two portions, the breech and the rest of the gun, clamped together.

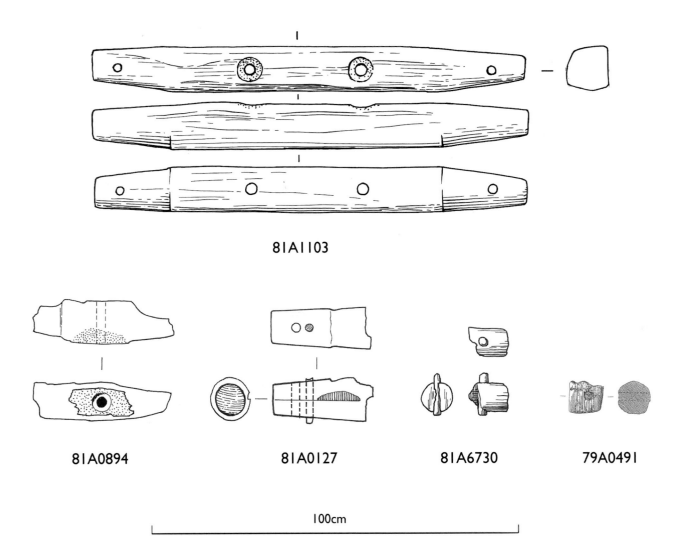

*Figure 2.97 Axle and axle fragments*

The space created by the extraction of the pattern was filled with an iron rod aligned on the axis of the proposed bore. This was covered with clay of appropriate thickness to achieve the desired bore diameter. Biringuccio (1540, 241) notes that '*some masters ... reinforce their cores by wrapping them with an iron wire*' and vestigial remains of this wire has been observed within the bores of *Mary Rose* guns. The support for the core often consisted of a central ring with four arms radiating from it, usually made of iron (Fig. 2.99). These are termed the '*crown*' (more specifically for supporting the breech end centrally) '*chaplet*', '*cruzeta*' or '*spider*'. Their ends can often be seen as iron-stained slots on guns recovered from the sea where the iron has corroded away. The number depended on the size of the gun. These supports would remain in position as the bronze was poured into the space between the jacket and core.

The mould assembly was buried vertically in the ground in a pre-moistened hole with the breech portion lowermost. This method resulted in a cast gun without a longitudinal cast-line; and with a central bore.

Multiple small guns were often created during one long casting, with guns buried at progressively deeper horizons radiating away from the furnace. The largest guns would be closest to the furnace, just below the taphole. A series of clay conduits, which could be opened or closed as necessary, conducted the molten metal into the feeding heads of the moulds. Tin was often added separately towards the end of the casting to keep the feeding head molten. The filled mould was allowed to cool before being dug up, broken to extract the gun, and the core removed from the centre.

Finishing the gun involved sawing off the feeding head; washing and brushing the external surface; the striking of the surface of the bronze with a hammer to smooth it; the searching and cleaning of the inside of the gun for flaws; and the removal of any remains of the core with a reaming tool. Three shots with common powder were fired in order to '*proof*' the gun – the last proof with a quantity of powder equal to the weight of the ball. A delightful contemporary description is given of the casting of a gun for Henry VIII in the Spanish town of Fuenterrabia in 1517–18 (*LP* 2, 4108). Pre-

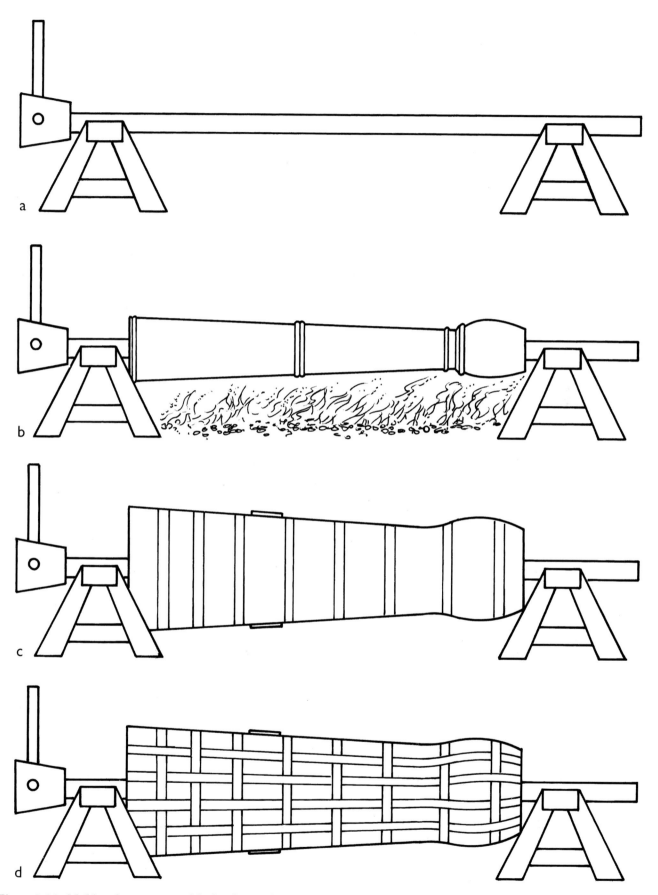

*Figure 2.98 Making the pattern and jacket for casting a bronze gun: a) mandrel; b) making the wood pattern; c) the external clay jacket; d) completed, reinforced jacket*

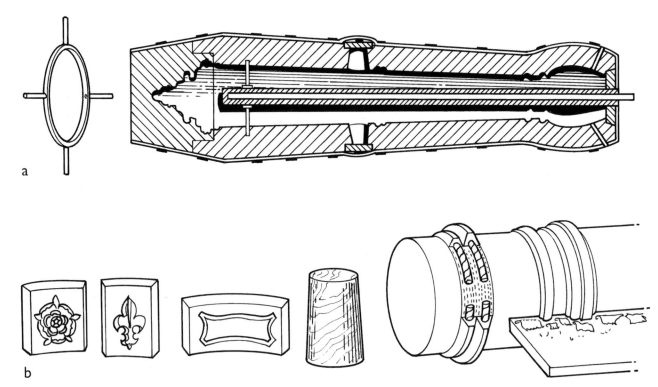

*Figure 2.99 Drawings after Biringuccio (1540): a) Crown and gun in jacket; b) embellishment*

dating both Biringuccio and Norton, it is in the form of payments by the English merchant Thomas Batcock to various individuals during the casting. It is very detailed and verifies the re-use of old guns, of certain parts of the mould armature, of the vertical casting in a pit, of the fitting of surface decoration in wax with nails, of the making of the crown support and the core (termed the '*heart*'), of the reaming of the bore and of the proofing of the gun. Six cast iron shot were made for it, and three were fired during proof. The balls were listed as weighing 13½lb (6.1kg), and the proof charges were 12, 13½ and 14lb (5.4, 6.1, 6.4kg) of '*heavy*' powder.

## Replicating a gun

In 1998 a project was created to manufacture and fire a cast bronze smooth bore muzzle-loading gun, reproduced from an original recovered from the *Mary Rose*, as part of a documentary series on the Royal Armouries filmed by Yorkshire Television. The plan from the outset was to manufacture the gun using as close as possible sixteenth century techniques of gun casting. The major source material for this was the *Pirotechnia* of Vannoccio Biringuccio, written in Venice in 1540. All stages of the project were to be filmed.

In addition to listing the specifications for the replication of the gun (size, weight, decoration), the brief included a caveat covering the failure of the gun at any part of the casting or firing and also listed a number of prerequisites for the manufacturing methods to be employed. Some of these were subsequently amended. The brief also included the provision of a photographic journal.

In choosing which of the ten surviving guns to replicate the following criteria were applied:

- The gun should be from the Main gundeck, one that may have been situated beside the wrought iron port piece that had already been manufactured and fired (see Chapter 3).
- The gun should have been recovered with its carriage so that this could also be replicated.
- The gun should be as representative of the bronze gun assemblage at this time as the port piece was of the wrought iron gun assemblage.
- The size of the gun and the intricacy of the mouldings should test the skill of the modern bronze founder.

If the gun and carriage were to be from the Main deck, this immediately discounted the most numerically significant bronze guns in use in 1546 (see above) – the demi culverins (50) and the sakers (85) and also the culverin recovered in 1840 (79A1278). Of the two cannon, two demi cannon and one culverin that qualified, the culverin was the most commonly listed type (40), although still far less numerically significant than the port pieces (218 listed). The Main deck culverin with its carriage was an extremely long gun (81A1423) cast by Peter Baude in 1543. It had a fluted and spiralled chase with winged mermen as lifting

*Figure 2.100 Overview foundry: a) the hole; b) using tiller furnace to pour the bronze*

dolphins – a gunfounder's nightmare (see Figs 2.59–61).

The original gun is 12ft (3655mm) in length with a bore of 5½in (140mm) and weighs more than the 3888lb incised on the base ring at 2071kg.

### Replicating the material

Historically '*gunmetal*' was an alloy of copper and tin and, whilst impurities did occur in the guns, generally the alloy contained 90% or more copper. The proportions of the main metals, copper and tin, vary from author to author and we wanted to replicate the original alloy. Eleven superficial samples of sound metal were taken from varying positions along the length of the gun including the dolphins and trunnions (see Appendix 2.1). The final composition chosen as being representative of the outer surface original gun was 6% tin, 0.7% antimony, 0.3% arsenic, 0.2% lead, 0.1% nickel, 0.1% silver, 92.6% copper. The impurities were not considered large enough to have an effect on the casting, but the total of 1.2% antimony+arsenic+lead was considered enough to reduce the viscosity of the melt. Arsenic is known to be a good de-oxidant. Ingots were prepared to this mixture. It is acknowledged that there will be a difference between the inside and the outside of the gun, but we decided against drilling through the wall of the gun.

### The brief

Specifications as to the methods used were drawn up as far as possible replicating the materials and processes listed above. Our first criterion was that we wanted to

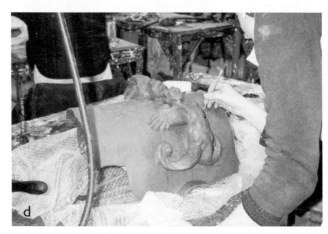

*Figure 2.101 Making the wax moulds: a) silicon rubber mould; b) external polyester being applied over; c) inside view of mould having been removed; d) wax model produced for the lifting dolphins from the silicon rubber mould*

ensure that if any modern materials or techniques were employed they should not impinge on how the gun functioned. Our second criterion was that each of the major phases of production should demonstrate the use of original techniques where possible and in a way which would enable filming. These phases included:

- Manufacturing a pattern of the finished gun.
- Preparation of an external jacket to surround the pattern.
- Extraction of pattern.
- Manufacture of a core to provide the bore of the gun.
- Insertion and stablilisation of core.
- Melting and pouring of metal.

The following criteria were considered fundamental to the project:

- The pattern had to be made in wood.
- The lifting dolphins, trunnions, cornices, reinforcing rings and surface ornamentation had to be cast with the gun.
- The gun had to be cast in a pit and backfilled with earth/sand.
- There had to be a feeding head.
- The gun had to be cast with a rod forming the bore, held in position with iron spiders.
- The full amount of metal had to be molten before pouring.
- The mixture had to replicate the original metal of the gun.
- Cooling had to be by natural agency.

The original brief included the following: a test of tensile strength, either by examination through non-destructive testing, or pressure tests on the gun before firing; analysis of samples of the metal following casting along the length of the gun to compare with the original; and the weighing of the gun after casting. A tolerance of 1mm between the new casting and the original after reaming of the bore was stipulated.

The brief was prepared and went to tender. The preferred methods statement was received from Morris Singer of Basingstoke. The wooden pattern was to be manufactured by G.F. James in Southampton, with casting by Taylors (Bellfounders) of Northampton, one of the few establishments that have a 3 ton tilter furnace capable of casting the gun in one pour and who were willing to excavate a 4.3m deep casting pit and had the height of ceiling which could enable lifting to a height of 9m (Fig. 2.100). The moulding box to form the outer jacket was made in mild steel sections by A & P Southampton. An insurance policy was taken out to cover failure of casting and on firing.

The proposal from Morris Singer was elegant in that it included the production of a collapsing wood pattern set up with lost wax detail and placed in a specially prepared circular steel-section moulding box (instead of the jacket used in earlier times). A suitable moulding compound (Mansfield red moulding sand) would be used to fill the space between the moulding box and the wooden pattern, the box and compound becoming the mould into which the molten bronze would be poured. The core would be made similarly, but the sand laid upon an iron bar. Although the sand mould would be lost during the casting, the pattern and core boxes could be re-used and the sand retrieved.

**Making the pattern**
Dimensions including internal bore, width across the trunnions and diameters along the length of the gun were taken. As the gun was to be a copy of an original piece rather than a new gun, it could not merely be made in wood and clay. The fluting and any of the details of the wooden pattern would be collapsible and the dolphins, trunnions and devices would be made using the lost wax process. This was achieved by making a ceramic shell of these features which were placed in sand cores (Fig. 2.101).

In order to get precision details for these specific areas, wax moulds were produced from the silicon rubber moulds. To ensure that the ornamentation and positioning was exact, a silicon rubber negative mould was taken of parts of the existing gun (Fig. 2.101). The upper face of the gun below the chase was coated with trichlorethylene mould release agent mixed with wax and covered with silicon rubber from the cascabel to the mouldings forward of the trunions. Modelling clay was used to form walls to prevent any running of the silicon. A polyester mother mould was then applied over the rubber. Finally a layer of fibreglass backing mixed with a hardening agent (5%) was placed in bandage strips as reinforcement. The chase was coated in fibreglass only as there were no particularly difficult features to duplicate. In Southampton a complete positive fibreglass mould was created from the negative rubber mould in order to manufacture the wooden pattern.

The wooden patterns were laminated up from timber, in sections, turned with the lathe where necessary, keyed together and turned to match. The cascabel was made in eight wooden sections keyed together and turned (Fig. 2.102a). Decoration was recarved on it and a ceramic ?? made (Fig. 2.102b–c).

**Making the 'jacket' or moulding box**
As we did not want to destroy the wooden pattern, a great deal of thought went into how elements could be made that could be extracted from the 'jacket'. Modern moulding sand was used over the wooden pattern to form the jacket in order to achieve a good surface finish. Historically this would often have been made of steel or of wooden slats bound together with steel hoops. Limitations of time and expertise led to the use of a 12ft (3.7m) long mild steel tube with an internal diameter of 3ft (0.9m) for the moulding box made of four

*Figure 2.102 Making the cascabel: a) turning the cascabel; b) silicon mould made from wooden cascabel; c) investing the wax pattern with ceramic material*

## Moulding the gun

As delivered to Taylors, the elements included the pattern equipment, metal and moulding boxes. The hole required was 4.3m deep and lined with steel shuttering to enable the moulding box to be at a low enough level to accept the pour from the tilter furnace. The lower sections were rammed in Mansfield red sand around the pattern, which had been dusted with parting powder. The two lower sections were made up on the floor, then lowered into the pit. Gradually, from the cascabel upwards, the sections were assembled (Fig. 2.104a, b).

Once the sand moulds were prepared, the timber sections were removed and the moulds dried. The lost wax items would melt away and be absorbed into the ceramic mould material. Other than the steel tube forming the moulding box. This method allowed for the incorporation of vents (Fig. 2.104c).

The pattern was then stripped out by collapsing the annular sections. The longer wooden sections (the chase) had to be removed horizontally with the tube removed from the pit and held horizontally on the crane. The moulds were then placed in the oven overnight to dry the sand ready for casting. Once the

*Figure 2.103 Portion of steel tube used to form the jacket*

interlocking sections (Fig. 2.103). The sections were: breech to base ring (500mm); reinforce to just forward of the dolphins (1500mm); the chase (1850mm); the muzzle (600mm). This was held together by 25mm thick flanges each held with eight 20mm bolts. This method actually necessitated a 'double cast', putting the tube with wood/ceramic pattern into the casting pit and ramming up the sand, withdrawing it from the pit, collapsing the internal wooden skeleton, and replacing the tube containing the sand mould for casting.

*Figure 2.104 a–b) Building up the pattern in the casting hole; c) exothermic vent arrangement; d) the core box in the hole backfilled with earth*

sand mould was made, each section had to be painted with spirit and the surface burnt, a process termed 'burning off'. During the sand cast the tube was lowered into the casting pit which had been lined with shuttering. For the final pour of molten metal, the shuttering was removed and the pit backfilled with earth (Fig. 2.104d).

## Making the core

One of the modern allowances was the use of an air set core to save time. To make the sand core a wooden box was prepared in two halves to the finished diameter, just over 5in (127mm). The central portion of the sand core consisted of a square section bar 2 x 2in (51 x 51mm) of mild steel armature with cast iron flights drilled as necessary with spigots along its length and a spider at the top end of the core to keep it upright and central. This method enabled the incorporation of channels and vents.

The core box was partially filled with sand, the bar put in place, and the rest of the box filled with sand. Once set, the wooden box was removed and the core painted with spirit and burned (Fig. 2.105). Traditionally the cores would have been strickled in loam with straw or sisal rope wrapped around the steel armature.

## Casting the gun

The date chosen for the casting was 19 July 1999, 454 years to the day after the *Mary Rose* sank. For casting, the sections of the tube containing the sand mould were placed in the pit. The crown to hold the core centrally was put in position locating into the cascabel mould

*Figure 2.105 Making the core: a) core box; b) steel crosstree*

hot liquid metal channeled along gutters into the top of a feeding head for each gun, with a number of guns being cast on a decreasing slope from longest to shortest simultaneously with thinner channels being used either to cut off or to feed particular guns.

The main bore core was supported at the top end. The core at the cascabel end was centred within the mould using a bronze and mild steel spider ring. The positioning of the shell and the cores and their method of assembly were fundamental to the process as they had to be assembled correctly and secured back into position when the bore core was lowered into it. The bore tube was lowered into the mould to locate with the lower spider and then two halves of the runner box positioned to locate the top end of the mould. Holding the core straight is of great importance to make sure that it is not displaced with the flood of molten metal and to prevent it floating upwards when the inherent buoyancy of the melt comes into play. To prevent movement, a mild steel crosstree was bolted both up and down at the top of the armature (Fig. 2.105b). Traditionally the mould included a feeding head (or gun-head) which provided a reservoir of molten metal to make up for shrinkage as the metal in the gun cooled and solidified. Tin was added to this to ensure that it remained liquid throughout. In addition to a feeding head, pencil (thin) gates of 1½in (36mm) diameter were incorporated and an exothermic feeder was placed above these to ensure that the tin bronze was sufficient fed to the gun. The hole was backfilled and banked up to the top of the moulding box.

The furnace was lit at 8 am, ready for the pour at 1 pm, as normal for casting bells. At 1pm the pour was achieved by decanting from the tilt furnace into a ladle and then into the pit (Fig. 2.106b). The metal solidified before the exothermic feeders were fully deployed and the casting was incomplete at the muzzle (Fig. 2.106c). The temperature of the melt was insufficient to last the pour. A week later, the casting had cooled and the mould was lifted out by crane, inverted and shaken. The cascabel was a beautiful copper colour and all the detail was perfect. Longitudinal cracks were evident from the position of the dolphins forward. The muzzle was severely cracked and the cornice nearly separated, almost as though it had been purposefully cut. This may be what had occurred in 81A3000, the bronze demi cannon with a cut off muzzle.

Fresh ceramic shells were made and the mould was rammed up again. A hollow 50mm square mild steel tube was used as the core arbor. The flawed gun was melted down and new ingots produced. The casting was poured at 1150°, this time, and completed successfully (Fig. 2.107).

After the initial fettling a problem arose in the removal of the core, that had become extremely hard and compressed, and the armature. This was considered to be the result of a combination of sand burn and metal shrinkage. Various methods were used for

within the tube (Fig. 2.106a). The rest of the tube was assembled in sections, and finally the core placed into the tube.

Considerable discussion took place on how the metal would have been poured into the moulds. Documentation suggests that the furnace would have been placed at a higher level than the moulds and the

*Figure 2.106 Casting the gun: a) crown; b) tilting furnace; c) incomplete gun*

leaching out the core and burning out the armature employing mechanical tools and a range of power tools, taking several days. The major difference between the replica and the original is undoubtedly the smoothness and size of bore: the bored out reproduction being ¼in (6mm) larger than the original. Although accepted at the time, this later caused problems with wall thickness. With hindsight, we should have recast the gun to the correct bore.

**The carriage**

The carriage was made to fit the specifications of the original but using parts of a carriage copied from the original for demi cannon 81A3002 that had been made to display the gun in the National Maritime Museum '*Armada*' exhibition in 1988 (see Figs 2.56–7, details above). It had been made in elm with ash axles. The main differences between the original carriage for the culverin and the one used was the height of the trunnion support cheeks. The demi cannon was an Upper deck gun, designed to be higher off the ground to meet the higher gunports on this deck. New trunnion support cheeks were made to the correct specifications. The bed for the demi cannon, although the same thickness as that for the culverin, had been made in two halves with loose pegged tenons holding it together longitudinally. As the trunnion support cheeks were higher, in addition to the cross transom bolts on all carriages, the Upper deck carriages were further reinforced with wooden transoms rebated into the sides of the trunnion support cheeks and across the bed. These were not used in the reproduction for the culverin, but did leave the rebate in the bed which may have constituted a line of weakness. The wheels were small solid elm trucks. The carriage was adapted and the new elements manufactured by Jim Sadler at Fort Nelson. The carriage was held together with wrought iron brackets and bolts with washers and wedges. The height of the trunnion rests above the bed do not enable the gun to be elevated more than 8°.

*Firing the reproduction culverin*

In a joint project between the Mary Rose Trust and the Royal Armouries in 1997–8 data were collected on the ballistic performance of a copy of a *Mary Rose* breech-loading wrought iron gun in order to compare it with

*Figure 2.107 Complete gun: a) as removed from mould; b) after finishing*

that of a smooth bore muzzle-loader (see Chapter 3 and Hall 1998, 57–67). In addition, the reproduction of bronze culverin 81A1423 was also test fired. One of the stipulations within the contract issued to the founder was that: '*The finished product must be fit for its purpose, to fire a cast iron round shot with up to half its weight in powder*'. The weight of the shot corresponded to an iron shot up to ¼in smaller than the bore, in this instance 5 9/16in (140mm), about 23lb (10.4kg). The firing trials took place at the Ministry of Defence Ranges, Shoeburyness Churchend Battery on 18–19 September 2001.

There is little in the way of instruction on fundamentals for the firing of smooth bore artillery in this period. Although windage (space allowed between the shot and the wall of the gun) and the ratio of powder to shot in weight are discussed, it appears that there was no one clear 'recipe'. The recovery of shot from within the bores of *Mary Rose* guns demonstrated varying windages, from ¼in to 1in (6–25mm). In addition to testing the casting of the reproduction gun, in effect 'proofing' it, the firings were carried out to examine specific issues:

- The effect of varying the windage on muzzle velocity and range.
- The effect of altering the ratio of powder weight to shot weight.

We also wanted to know what a fair 'point blank' range was – how far the shot would go with the gun laid level, fired from its service carriage. This would give an idea of the practical fighting range of the gun. In recording the velocity with which the shot left the barrel it is possible to predict extreme range. In recording the terminal velocity the destructive power of the gun can be estimated. We wanted to find out what the 'effective fighting range' might be, that is, the range whereby one might be expected to strike a ship and damage it. An oak reproduction of the *Mary Rose* hull at Main deck level was used as a target to observe impact damage with the target placed at 45m and 90m. As the height of the upper sill of the gunport for this gun is only 1060mm above the deck, elevation much beyond a few degrees above horizontal was impossible. Elevations were tested at 0°, 5° and 7°.

A firing schedule was prepared in advance around these questions. Iron shot was cast to be ¼in, ½in and ¾in smaller than the bore. Powder charges were pre-prepared to set weights which corresponded proportionally to ⅕, ¼ and ⅓ the weight of the shot. All the shot and powder charges were clearly marked for ease of use and recording and retrieval. This was necessarily limited by finances to a maximum of two days and ten rounds.

The schedule was designed to build up the powder charges gradually for shot of different sizes, so that each size of shot would be fired using each weight of powder to shot. It was hoped that the trials would establish what should be a typical muzzle velocity (MV). The weight of the shot and its known MV provide a good mathematical way of assessing its effectiveness.

A high speed Phantom camera housed in a splinter-proof bunker placed so as to cover both shot exit and target strike filmed the firings at 5000 frames per second. A TermA Electronic AS Doppler Radar 5000 Tracking System was used to obtain muzzle velocity

*Figure 2.108 Firing trials: a) gun on carriage; b) culverin, splinterproof, 5000 fp sec high speed film capture; c) worm, sponge, ram, bucket; d) ramming the shot while serving the vent; e) gun in front of target; f) gun firing (distance shot)*

and strike velocity. The firing sequence and resulting damage to the target are shown in Figures 2.108–9.

## The shot

The method of manufacture of an original shot (81A1012, of 5¼in (133mm), present weight 14.99lb, 6.8kg) was studied at Rayne Foundry. Metallography clearly displayed the presence of chiller crystals. This suggested that the molten iron had been poured into a cool metal mould, probably of brass, as listed in the 1547 inventory (Starkey 1998; Chapter 5). It was a high phosphoric cast iron, in the region of 1–1½%. The shot for this gun was ¼in (6mm) smaller than the bore. The costs for manufacturing of a brass mould for each size of shot was £400 against routine sand castings of £20 each. Therefore six shot were cast in sand, three of 5in (127mm) diameter and three of 5¼in (133mm) diameter. Three traditionally cast shot from Fort Nelson were supplied for the 5½in (140mm) shot, the closest fit to the bore.

## Tools

For definition and description of the accessories used in the loading and firing of the *Mary Rose* guns, see Chapter 5.

### Worm

This is used to clean the bore of any debris remaining from a previous firing. There was no evidence for worms from the *Mary Rose*, so one from the Royal Armouries

*Figure 109 Damage: a) target after shots 3, 4, 5; b) inside target following 3, 4, 5; c) target after shot 7, outer plank out; d) inside damage; e) damage to bed, split*

at Fort Nelson of a length long enough to reach the bottom of the barrel was used.

*Sponge*

Again there was no evidence for sponges, although twelve sheep skins for sponges are listed in the *Anthony Roll* inventory. It may be that these were merely wrapped around rammers or plain poles. One was manufactured to be a tight fit to the bore, the pole had a diameter of 1½in (38mm) and a length of 3.69m. The length of the sheepskin sponge head was 290mm.

*Rammer*

A rammer of the correct size for the gun was found close to it (80A1332). A rammer was manufactured to suit the finished gun, length 3.62m, head diameter 5¼in (133mm). The pole diameter was 1½in (38mm). This proved heavy to use.

**Powder**

The type of powder used was Orica NPCG black powder, coarse grained, contained in cloth bags. We reduced the ratios of powder to shot, which according to historical sources could be up to the full weight of the shot, as we were aware that modern powder was probably more effective than traditional black powder. Because of the difficulties in loose loading of gunpowder it was decided to prepare charges in silk bags. '*Formers for cartridges*' and '*Paper Royal*' were listed on the *Anthony Roll* inventory for all the ships, so cartridges were used. For ease of operation sawdust packing was placed in the cartridge on top of the powder, approximately half each by volume, simulating the sixteenth century use of packing such as straw or oakum, Normally this would be placed in loose. We had some evidence for tampions in guns that were presumably used to keep the bore dry. The top wad we used was hay rather than a traditional wooden tampion.

### Breeching the gun

A 2in (51mm) diameter breeching rope was provided to control recoil. Sisal was used instead of hemp as it was immediately available. Two eye splices were formed in each end and a short length spliced in halfway along the rope so as to form an eye that would pass over the cascabel. The breeching rope was secured to two stakes positioned 2m from the centreline of the cannon, allowing some recoil. The posts were those used with a naval field gun.

### The gundrill

As with the port piece (Chapters 3 and 5), no original record of gundrill survives. The drill therefore followed Royal Armouries usual safe practice when firing muzzle-loading guns. Each sequence began with worming followed by sponging, loading using cloth cartridge, wadding, projectile and final wad, the vent being served (tightly covered) throughout. The cartridge was pricked and the priming was completed using fine powder from a horn flask. Ignition was by electric igniter taped into place. As soon as the firing was complete, the barrel was cleaned with the worm and sponged with water.

A gun drill was prepared using a seventeenth century gunnery manual (Eldred 1646). It was abbreviated and used to ensure that adequate commonsense precautions were maintained. Drill is discussed in full in Chapter 5.

### Elevation

Most firings were undertaken at 0° elevation. The trials where the gun was elevated were assessed by a device supplied by the range. The gun was easily elevated with two persons using handspikes. The quoin was based on the only extant *Mary Rose* example, one found with cannon 81A3003. Aiming at the target for line was achieved by simple sighting along the gun, from the cartouche on the base ring through the central dip of the lifting dolphins and to the muzzle centre. In the vertical plane, all aiming was at the centre of the target, most hit were slightly higher than aimed. Elevation was set using a standard field artillery clinometer placed on a straight length of timber laid in the bore and projecting from the muzzle.

### Target details

The target replicated the three layers of structure of the hull of the *Mary Rose* at Main deck level. This was made of unseasoned oak, held with oak pins. The hull would have become seasoned with time and was also reinforced with iron nails. The structure included an outer layer of butt-jointed planks, laid over frames. On the inside there were horizontal stringers and frames.

### Ballistic results and conclusions
by Nicholas Hall

The record of rounds fired is given in Table 2.11. The best performance was extremely impressive, both visually and for the results achieved. The maximum MV was almost 502.5 m/s (1650 ft/s). Another shot achieved a probable range of about 1500m. Visibility was very poor and conditions also affected the instrumentation. The 'extreme range' shot was fired at only 7° of elevation but with a slightly lower charge and lower muzzle velocity (Qinetic 2001, 4). The trials seemed to confirm that the closer fitting the ball, the better the performance.

The use of wooden tampions between the charge and shot (as is remembered but not recorded for several *Mary Rose* guns) was not employed. Had they been made snugly there were concerns about them jamming and if too loose we assumed they would have little effect.

Given that the strength of the gun was unknown, trials began using small charges. The greatest charge was one-third of the weight of a 5½in (140mm) shot, 7.7lb (3.4kg) of powder. This gave the highest muzzle velocity. In black powder ballistics it is usually accepted that increasing the charge above one-third charge weight is counter-productive. It appeared clear that one-third generated more effective power than one-quarter. Muzzle energy (ME) is a useful measure of the gun's power and its likely effect on any target. Muzzle energy is that contained in the projectile as it leaves the muzzle. It will then have received all possible force from the burning of the powder charge. In order to compare the port piece with the culverin in terms of their weight to performance, we can calculate the ME developed per weight of gun. As the complete port piece weighed 1.1 tonnes its ME per tonne of gun, on the basis of its best trial performance, is 352.7kJ.

Before the trials, it was assumed that the culverin should manage the same MV as the port piece's best. But with a heavier projectile than that of the port piece, clearly it would then achieve a higher ME. The culverin is much heavier, so practical experiment was necessary to discover, among other things, whether there was real difference in power to weight ratio between the two guns. The high figure achieved for the culverin's maximum MV came as something of a surprise, especially since the cast iron ball is denser than one of stone and therefore needs more force from the powder to accelerate it. The culverin was bored out (see above) so that the largest projectiles we fired were in fact of 23lb (10.4kg) rather than the 18–20lb (8.2–9.1kg) which would probably have been fired originally. Using the formula for kinetic energy for ME the culverin's ME firing 23lb solid shot is 1313kJ. The gun, complete with carriage, weighs 2.36 tonnes, giving a ME per tonne of

Table 2.11 Firing schedule for reproduction culverin 81A1423

| Trial no. | Date/ time | Shot diam. (in/mm) | Shot wt (lb/kg) | Powder charge (nos/kg) | Ratio of shot wt | Elevation | Target type/ range (m) | Muzzle velocity (ft/m/sec) | Velocity at target (ft/m/sec) | Notes |
|---|---|---|---|---|---|---|---|---|---|---|
| 1 | 18/09/01 10.50 | 5 127 | 16.2 7.35 | 2.75 1.25 | 1:6 | point blank | distance | 472 144 | – | bal bounced, recoil 140mm[1] |
| 2 | 18/09/01 11.25 | 5 127 | 16.3 7.4 | 3.3 1.5 | 1:5 | point blank | distance | 533 162.5 | – | ball bounced[2] |
| 3 | 18/09/01 12.13 | 5.25 133 | 19 8.65 | 3.8 1.73 | 1:5 | aimed at target | ship side 45m | 515 157 | 508 155 | clean hole[3] |
| 7 | 19/09/01 09.22 | 5.25 133 | 19 8.65 | 4.75 2.16 | 1:4 | aimed at target | ship side 90m | 428 130.5 | 420 128 | clean hole[4] |
| 8 | 19/09/01 10.01 | 5.25 133 | 19 8.65 | 4.75 2.16 | 1:4 | 5° | – | – | – | 1st bounce at 900m[5] |
| 4 | 18/09/01 13.50 | 5.5 140 | 22.9 10.4 | 4.6 2.08 | 1:5 | aimed at target | ship side 45m | – | – | clean hole[6] |
| 5 | 18//09/01 14.20 | 5.5 140 | 22.9 10.4 | 4.75 2.16 | 1:4.8 | aimed at target | ship side 45m | 653 199 | 646 197 | clean hole[7] |
| 9 | 19/09/01 11.59 | 5.5 140 | 22.9 10.4 | 5.7 2.6 | 1:4 | 7° | – | 1492 454.9 | – | 1st bounce 1400m[8] |
| 6 | 18/09/01 15.55 | 5.5 140 | 22.9 10.4 | 7.7 3.5 | 1:3 | 7° | distance | 1648 502.5 | – | 1st bounce 1200m[9] |

*Notes:*

1. The first bounce of the ball was estimated at 150m; there were at least 3 further bounces. The carriage moved back 140 mm.
2. 100 mm movement of carriage.
3. Ist shot at target. It hit the target on the 3rd plank down making a very clean hole. There was little splintering on the outside but on the inside there was a shaving off the stringer at the top & a rib to its left. The ball actually only went through one layer, glancing off the stringer above & the rib (left on inside). There were some splinters on the inside, maximum spread 5m & 48 paces where small splinters were found.
4. Hit target at junction between 1st & 2nd plank down, just to the right of the (much larger) impact from a stone shot fired previously from a wrought iron gun. This too was a clean hole, but some splintering outside. Just missed the frame behind, still through only one layer. Hardly any damage (Fig 2.109a).
5. Carriage moved 400mm back. No damage to the carriage. The shot hit the target in the middle of the 2nd plank from the top, springing the stringer on the inside off its trenails. The outer plank was split to the edge of the target from the impact hole. We were 10in (254mm) off the aimed position on the target (to the right) (Fig. 2.109a).
6. Estimated 12 bounces of shot, 1st at about 1200m. The gun recoiled violently, pulling the stakes out of the ground & causing the gun to move 800mm backwards. The barrel also jumped vertically. There was a delay (6 seconds) between the initial explosion, which caused a huge plume, & the shot leaving the bore.
7. The target was moved to 90m. The gun was aimed on the target, using line of sight along the top of the gun. It was clear from trying to sight the gun that the touch hole was not in the centre of the top of the barrel. The elevation was 0.5°. The ball did not seem to bounce. The gun moved 200mm. The shot hit the 2nd plank from the top, left of earlier hits (Fig. 2.109a). The plank fell off the front of the target, split (Fig. 2.109c). On the inside the 3rd frame from the left was completely destroyed at the top. It was pushed to the right, shearing off a portion of the back (Fig. 2.109b–c). Most parts of the target were within 15m of the back, but several larger fragments were found up to 30m behind the target with a spread of 20m of the left of the centre of the target & 30m to the right. The target was not used after this shot.
8. This was at 5°. At this point the top of the muzzle was 1150mm off the ground. At 900mm back from the muzzle the height of the top of the gun off the ground was 1000mm. The 1st bounce was estimated at 900–1000m, after which the shot ranged to the left. It was estimated that the shot went beyond the Middle Road, 1.5km.
9. Elevation was 7°48. At this elevation the cascabel was resting on the bed, this is therefore the maximum elevation possible. It took 0.65 sec to ignite the main charge, by 0.75 sec it was all burnt. After 3 seconds the shot was still traveling at 248m/s. The cannon jumped to the right, & it moved 600mm back. A large transverse crack in the bed was noted just behind the 3rd tenon, 530mm in from the front of the bed & extending to 710mm (Fig. 2.109e).

556.4kJ. This is a considerably higher figure than for the port piece (Hall 2001, 106–16).

It might be argued though, that we had up-rated our culverin by increasing the bore size during the extraction of the core without suffering a weight penalty. But when calculating ME using a 20lb (9.1kg) shot it still achieves 1146.3kJ per tonne of gun. Even if the actual results in the sixteenth century differed somewhat as a result of such factors as the use of different powder, wadding and wooden plugs or tampions, the magnitude of the difference seems sufficiently great to give a real idea of the apparent superiority in performance between the two guns.

Nevertheless, the ballistic superiority of the bronze gun was not achieved without cost. Although effective for its weight, the culverin was, after all, a fairly hefty piece of equipment and must have presented some difficulties in the confined spaces on board ship. The ballistic results, as demonstrated by our trials with two types of gun from the *Mary Rose*, offer a clear indication of why the large muzzle-loading single piece cast gun dominated naval gunnery from the late sixteenth to the mid nineteenth century. Our culverin, though placing great demands on modern day foundry processes and firing heavier projectiles than intended, not only withstood the trials but shot well and gave no sign of strain. All the equipment performed well except the carriage, which had been built originally for static display and suffered cracks across the bed (Fig. 2.109e). This was not entirely unexpected. This early design of truck carriage places a heavy load on firing at the base of the trunnion support cheeks. The separate stepped rear cheek, while providing places to lever the barrel using handspikes, provides little support to the trunnion support cheeks. It seems very likely that this weakness eventually led to the provision of the single piece cheek to support the trunnions and for the use of the handspike. This is seen within 50 years of the *Mary Rose*'s demise in the Alderney wreck gun carriage (Monaghan and Bound 2002). This familiar pattern remained in use as long as did muzzle-loading guns; it is perhaps surprising that the massive bed on which both types of cheeks were planted took so long to disappear (Lavery 1987, 127).

# 3. Wrought Iron Guns
## Alexzandra Hildred

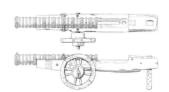

## Introduction

The ordnance from the *Mary Rose*, as listed in the *Anthony Roll* inventory of 1546, included 91 guns and 50 handguns. Of the 91 guns, all but the fifteen listed as '*gonnes of brasse*', described in Chapter 2, were made in iron, principally for economic reasons (Teesdale 1991, 9). Iron was locally available whereas both the copper and the technology required to make bronze guns were not. The difference in cost of the bronze over the iron was based on the fluctuating price of copper (see Chapter 2). Therefore, the majority of the guns at this period were made of wrought iron, manufactured by heating and hammering of iron billets, although the casting of carriage mounted iron guns was clearly under way in small numbers.

Wrought iron contains about 0.1% carbon. It also includes slag; non-metallic impurities removed from the ore in molten form within blast furnaces or bloomeries. The slag helps prevent corrosion and also allows it to be hammer-welded. The best wrought iron is obtained through the removal of slag via a repeated process of heating, hammering and rolling. Often new billets were fire-welded to begin that process. It can be both ductile and strong and can be easily forged, rolled and shaped, hot or cold. Impurities can include carbon, phosphorous, nickel and manganese. At a sustained welding temperature this carbon would be uniformly distributed.

Wrought iron is welded by hammering using a rapid succession of blows. Temperature is important, 1300–1400°C is best. If too cold, the metal may appear to weld but the joint will break under stress; if it is too hot the metal will burn and not weld properly. The experienced smith could judge how and when to work the iron based on texture, colour and feel. In the Middle Ages the welding of large pieces required several assistants, known as strikers, to use sledge-hammers under direction of the smith. Commonly three strikers were used, but there are records of 20–30 (Smith and Brown 1989, ix). To achieve a good hammer-weld the two pieces must be hammered into one another, so the direction of strike is critical. Within the recovered guns the metal is indistinguishable across the joint, with no

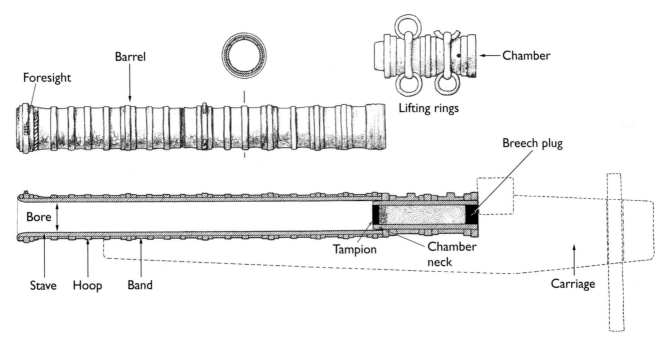

*Figure 3.1 Anatomy of a wrought iron gun*

patent slag lines, demonstrating the skill of the blacksmith.

The metallographic study of a number of early wrought iron guns (Starley 1989, 90) has shown that irons with different characteristics have been observed within specific parts of wrought iron guns. Through experience, better (stronger) pieces of wrought iron (higher carbon, finer grained, lower phosphorus) were found in portions of the gun exposed to greater pressure. Certainly with the *Mary Rose* examples, portions of the iron were so poor they could not have been used in anything but a very insignificant portion of a much larger object (C. Topp, pers. comm.).

Although wrought iron can be easily shaped, worked and joined, it cannot usefully be heated to the molten state and poured into a mould like bronze or cast iron. Elements were shaped before being assembled and fitted together. This influenced the way in which large artillery was made – essentially from long staves of iron secured with hoops and bands. This determined the form of wrought iron guns in general and had important implications on the way they were used.

Most English large wrought iron guns were breech-loaders. The difficulties in obtaining a seal between the plug and the tube to form a closed-ended stave-built gun of sufficient length may be why there are so few examples of wrought iron muzzle-loaders. An exception is seen in the bombards, where, in many cases, a separate chamber is physically joined to the tube to make a muzzle-loader. The reload speed afforded by having a separate powder chamber, together with the confined space onboard a ship, may also be one reason why breech-loaders are so numerous.

The guns are formed of two distinct elements, the tube forming the barrel (or '*hall*') and a shorter element (the breech chamber) with a plug which was located into the back of the barrel to seal it. This held the powder, and is referred to hereafter as '*chamber*' (Fig. 3.1). As this was built separately from the barrel, it could be thicker-walled, or '*reinforced*', to take the explosion of the powder without adding weight over the entire length of the barrel. Even the smaller anti-personnel wrought iron guns were breech-loaders, although these had drop-in chambers with handles (see bases, below). Traditionally, each barrel had two chambers, with the smaller classes of gun sometimes having three. If stored ready-filled these enabled rapid reloading. Contemporary inventories provide valuable information as to the relative importance of these guns, merely by their number expressed against the percentage of guns carried.

## Identification and Historical Importance

The iron guns listed for the *Mary Rose* in the *Anthony Roll* include '*port pieces*', '*slings*', '*fowlers*', '*bases*', '*hailshot pieces*' and '*top pieces*'. There are very few descriptions of these guns and their identification is based largely on a mix and match between the recovered artefacts and the few historical documents which list them (usually inventories or accounts).

These are usually listed in the same order for each ship, which, if one follows the listing for the cast bronze, and knowing how the names relate to the guns themselves, suggests a hierarchy from largest to smallest. Position within an inventory together with a description of shot type listed for named guns enables us to match names to the recovered guns from the *Mary Rose*, sometimes for the first time. The *Mary Rose* assemblage enables us to put size, weight and form to some of these guns that, until now, were no more than names, in many cases dismissed as being small, ineffective and unimportant.

The *Anthony Roll* gives the following iron guns for the *Mary Rose*:

Port pieces 12
Slings 2
Demi slings 3
Quarter sling 1
Fowlers 6
Bases 30
Top pieces 2
Hailshot pieces 20

More expansive descriptions occasionally annotate entries for guns in shore establishments relating to mounting or loading, such as the 1547 inventory (Starkey 1998). This document demonstrates the continuing importance of wrought iron guns within the fleet, with over 1200 compared with around 250 of bronze. Even if only the carriage-mounted wrought iron guns are considered, these still number nearly 500. Of these, the port pieces dominate (211) followed by the sling class (154) and fowlers (78). At this date only 67 cast iron guns are listed for the fleet. The inventories for ships are vital in understanding the usage of what may, at that time, have included some of the most up to date weapons and their study can reflect how long it took for a new technique or type of weapon to come into use. Guns were readily taken on and off ships and, in fact, ships were often fitted with new guns. The ships formed the first line of defence and periodically the first line of attack in Henry's military force and reflect very clearly the state of thinking about particular guns and their perceived strategic importance.

The importance of the excavated assemblage of wrought iron ordnance from the *Mary Rose* cannot be understated. Indeed, the finding of the first wrought iron guns by John Deane in 1836 caused an uproar (McKee 1973, 100). Two were raised alongside four cast bronze guns. Both were loaded with stone shot, but they were clearly built up of staves and hoops of wrought iron and were breech-loaders. The scholars of the day (Archibald 1840) were convinced that these had

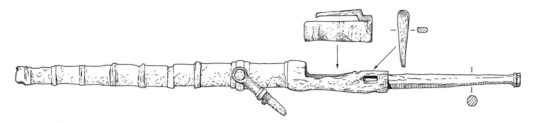

*Figure 3.2 External appearance of wrought iron swivel gun (base) showing hoops and bands*

no place alongside the fine examples of cast bronze ordnance:

> 'The one of them is of great length, formed of bars and hoops of iron, and is firmly imbedded in a large and heavy piece of timber. It must at all times have been an unwieldy and inefficient machine, and I cannot imagine that it could have co-existed, for purposes of active service on shipboard, with those highly finished pieces (of bronze) just mentioned. The gunpowder which would be suitable for the one would blow the other to pieces, and the gunners accustomed to the former would hardly be persuaded to run the risk of discharging the latter. It seems to me, therefore, that these rude pieces of olden time, if indeed they were ever on board the Mary Rose, must have been used for ballast or some other illegitimate purpose.'

Archibald was clearly wrong, as were many scholars, in the belief that the guns were inefficient, ineffective and dangerous. Wrought iron guns were used throughout the sixteenth century and numerically outnumbered both bronze and cast iron by significant amounts until at least 1555.

Having the Main deck armament of one side of the *Mary Rose* virtually intact (one gun is missing, but the carriage is present) shows us the heaviest armament and enables us to identify particular guns. The largest wrought iron guns are without doubt the carriage-mounted '*port pieces*', twelve of which are listed together with their stone shot. These were found stationed at gunports beside the largest bronze guns (see Fig. 2.24). The '*slings*' are easily identifiable as these are the only wrought iron guns listed with iron shot and can, therefore, be classified by the shot found in the bores of recovered guns. Four have been identified, all long in relation to their bores and all with the large powder chambers required to propel iron shot. Inventory research also reveals that many of these slings are '*stocked*', that is, mounted on some form of carriage. The six '*fowlers*' are the most enigmatic and have been identified as being smaller and having thinner walls than the port pieces. They were also traditionally carriage-mounted, firing shot of stone. The '*bases*' are clearly the smallest guns, as indicated by their position in the inventory and their large number, 30, listed together with lead shot. These are the swivel guns, and were mounted directly into a hole cut in a wale or gunport sill by means of an iron peg attached to a yoke over the trunnions. An illustration of 1563 by Richard Wright (Society of Antiquaries MS 94, 7) shows a gun listed as a base that is very like the *Mary Rose* examples and similar to Smith's type SW5 or SW6 (1988, 10) (Fig. 3.2). The '*top pieces*' were probably of a similar size and (presumably) design, although none has been identified. The last named type of gun is the '*hailshot piece*', 20 of which are listed. These were hitherto unknown and those from the *Mary Rose* remain the only known extant examples, identified largely by their rectangular bore and shot of iron dice, literally scattering hailshot. The four so far recovered are all of cast iron, although many inventories list them as '*forged*' and some as '*chambered*'.

Although there was clearly some difference between the guns of 1545 and those discussed by Norton in 1628, the latter is one of the few sources which describes some of the variants found on the *Mary Rose*. Chapter xiii describes '*Cannon Periors and Perieraes, the third kind and their sorts*', basically guns which fired stone shot. The items relevant to the *Mary Rose* assemblage are of the '*third sort of Ordnance of this third kind ... the Port Peeces, and Stocke Fowlers*' and the '*fourth sort of this 3 kind ... the Slings and Portingale Bases*' (ibid., 57, 58). Although, by this time, the port pieces and stock fowlers were made of cast bronze rather than the wrought iron of 70 years earlier, and the sling had become altered to a swivel gun, they were still breech-loaders. The largest stone throwers were:

> '*open at both ends, inuented to be loaded with Chambers at the Breech end, fitted close thereinto with a shouldring* [shoulder], *euen as the wooden Trees for water pypes haue tapred ends to let them close one into another; The shot and wadde being first put into the Chase, then is the Chamber to bee firmly wedged into the Tayle of the Chase and Carriage. Now in stead of round Trunnions, there are 4 square tenants* [tenons] *cast ioyning with the side of the Chase of the Peece, on eyther side two, which being let into the Block or Carriage, holdeth the whole Chase fast therein, leauing the Cornish* [muzzle ring] *lying vpon the ledge of the Ships Port, or vpon the Vawmure* [bulwark] *in a Fort, and tryced* [tricked = fitted] *vp with a rope fastned about the muzzle; The Tayle of the*

*Carriage is to rest, and to be shored vp with an vpright post or foote, full of holes to slide vp and downe in a square Mortice fitted thereunto, hauing a shiuer at the lower end thereof, with two Trestle legges morteized before vnder the blocke of the Carriage, the foote with holes hath a pinne to stay the Peece vpon any Mounture assigned'* (*ibid.*, 58).

The descriptions are very similar to the larger *Mary Rose* wrought iron guns. The breech chamber to '*hall*' male/female joint is present, the large block carriage and the wedging of the chamber to the gun and the back of the carriage. The description of the carriages fits those recovered for port pieces and slings, with an upright foot pierced with a pin as an elevating mechanism. The muzzle overshoots the bed, and possibly rests on the gunport sill. The basic differences include the use of wheels and a single axle on the *Mary Rose* examples, and the fact that the guns are not held into the carriage by tenons but by the relieving of the carriage to fit the upstanding hoops which are part of the manufacturing process of the wrought iron guns. The function is the same, it keeps the main part of the gun tight in the bed. The vertical movement upon firing is restrained in the case of the *Mary Rose* guns by ropes, possibly the '*woling*' listed in the *Anthony Roll*.

The slings and '*Portingale*' (Portuguese) bases are also described by Norton. These:

'*haue Chambers fitted into their Breeches as the Stocke-Fowlers haue, but that the Tayles that stayes their Chambers to wedge them fast (as in one continued Peece of yron whereof they are vsually welded and wrought) vnto the Tayle whereof there is a long sterne handle of iron to direct them to respect the assigned marke: They stand upon a forked Prop or Pintle vpon the ends of which the Trunnions resteth, they are loaded with their Chambers as the Stocke-Fowlers are: these shoote eyther Base and Burre, Musket or any other kind of Murthering [murdering] Shot, being put vp in bagges or Lanthornes fitted to their Calibres. And being discharged, their Chambers are to bee taken out and fil'd again, and others to be put in ready charged in the place thereof. These Peeces are vsually loaded with ¼ or ⅓ of the weight of their shot in corne powder … the lengths of the Portingale base is about 30 times her Calibre; the Sling about 12 times, the Murtherers, Port Peeces, and Fowlers 8 at the most besides their Chambers, their Chambers about 3 times their Calibre in length, and weigh the 6 or 8 part of the whole Chase*' (*ibid.*).

Norton provides a table providing an indication of the relative size of the chambers, shot and powder (Table 3.1)

Table 3.1 Relative statistics for chamber pieces (from Norton 1628, in Imperial measurements)

| Chamber | Length (in) | Height (in) | Powder weight (lb) | Shot size (in)/ type | Weight shot (lb) |
|---|---|---|---|---|---|
| Chamber | 12½ | 4½ | 7 | 6½ stone | 13 |
| Chamber | 22 | 4½ | 7 | 6 stone | 10 |
| Chamber | 24 | 4½ | 9 | 7 stone | 17 |
| Chamber | 17 | 3¼ | 5 | 5 stone | 9 |
| Sling chamber | 22 | 2½ | 3½ | 2¼ iron | 2½ iron |
| Port chamber | 16 | 3½ | 3¼ | 5½ stone | 9 |
| Base chamber | 9¼ | 1½ | ¼ | 1½ iron | 6oz iron |

## Scope of the Assemblage

There are 103 recognised items which are parts of carriage-mounted guns made of wrought iron. This includes 38 breech chambers for gunpowder, 17 '*tubes*' or '*halls*' (the main part of the gun barrel) and 48 fragments including portions of guns such as the staves, bands or hoops, remnants of lifting lugs and lifting rings and several lumps found in association with other wrought iron ordnance. These vary in size and condition. Among the smallest guns, the bases, there are a further fifteen recognised barrel elements and 25 powder chambers.

The wrought iron guns are the most challenging section of the ordnance assemblage and possibly the most rewarding. Naming the guns is difficult as so little is recorded and, being hand-built, there is considerable variation. Relatively little is known about the structure of these guns and radiographic study and sectioning reveals many contradictions with the written sources. The 'patchwork' method of manufacture means that these objects break into their component elements and differential corrosion to parts of the same piece can have a marked effect on their visual appearance.

Identification, even as to whether a fragment is a portion of a barrel or a chamber, is difficult. A chamber without the breech plug can easily be misidentified as a smaller barrel, creating a new size/type of gun. Corrosion means that identification of basic structure is difficult, some concretions are opened only to reveal dark black liquid although, more often, the woody texture of the iron is visible. It is therefore extremely difficult to 'fit' broken elements together. Observing the direction of the corrosion provides a clue, staves are laid longitudinally and hoops and bands circumferentially, so the way the woody texture of the corroding iron lies is a guide to the structural elements.

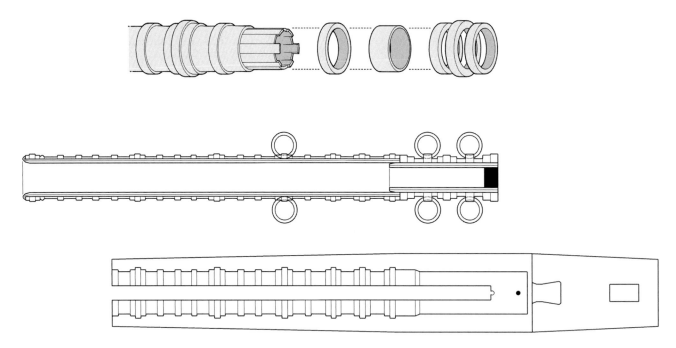

*Figure 3.3 Construction of barrrel and chamber*

The collection includes iron guns and fragments raised by John Deane and William Edwards in 1836–40 and subsequently acquired by various military institutions and museums (currently eighteen items and eleven fragments). It is not always clear whether a gun attributed to the *Mary Rose* actually is from the ship. In a few instances it is possible to suggest original locations for the larger elements recovered, such as three of the four raised with carriage portions, but the rest cannot currently be located with any confidence. Of those raised by the Mary Rose Trust, eighteen are without locations, having been raised as 'amorphous concretions' or removed from the site (albeit documented and tagged) and left in a collection area. Thirteen elements (including some which had been lifted clear of the site and held for collection) have been recovered during site monitoring since the ship was raised in 1982, including one chamber raised in 2003 and another from the recovery of the stem in 2005. There are possibly another ten (not included in the 103) in the Royal Artillery Institution, Woolwich collection that are without provenance; morphologically many of these would fit within the assemblage. Some are undoubtedly swivel fragments. John Deane refers to swivel guns, some are visible in early photographs (see below) but these have not been traced.

Another problem is accessibility. Some barrels weigh over 400kg and chambers over 150kg. No matter how accurate illustrations and drawings are, putting bits together to see if they fit is the only way to be certain about associating individual elements. This has not been possible. The treatment of the iron has also had a role to play in identification and access. The early material was treated with hydrogen reduction at high temperatures which causes the surface of the metal to become scorched and cracked. It was then coated with thick layers of microcrystalline wax, completely obscuring surface detail. A switch to long term soaking means that approximately one-third of the assemblage is still in sodium hydroxide. The assemblage is spread between dry storage areas, the museum and wet tanks, all geographically separated (see *AMR* Vol. 5, chap. 5). Only some of the nineteenth century material held in other collections has passed through our hands.

Having such a large collection deriving from a site over a short period (1979–82) meant that an assembly-line approach to recording (registration, photography, radiography, de-concretion and stabilisation) had to take place and much detailed study remains to be undertaken. As a result of these constraints not all of the elements were illustrated.

## Morphology

The port pieces, slings and possibly the fowlers were made in a similar manner, well described by Smith (1988, 11–12). A tube of wrought iron is formed of staves laid around a mandrel to butt against one another. In some cases a double layer of staves is observable. The width of the staves depends on the size of bore required. A layer of thin, wide bands and thicker, narrower hoops is made over the inner tube. Hoops can occur singly, or in sets when the central hoop is thicker than the others so it appears raised (Fig. 3.3) This process naturally forms a gun with an uneven surface which is exactly fitted to a wooden carriage, relieved to accept the hoops and bands on the barrel

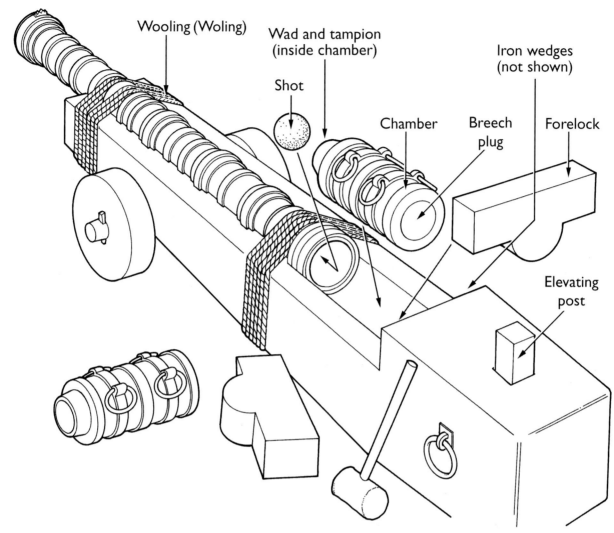

*Figure 3.4 Components of wrought iron gun on carriage*

(Fig. 3.4). This prevents longitudinal movement of the gun following firing and ropes bind the barrel to the carriage preventing upward movement on firing. One or more pairs of rings are situated over the length of the barrel, locating on to one of the hoops. A powder chamber is manufactured in a similar way (occasionally with double layers of staves) with an iron plug sealing one end. The staves project further than the outer layer of hoops and bands forming a neck which locates into the back of the barrel. Together the barrel and chamber are wedged to fit the bed with a '*forelock*' (wedge) of elm and one or two iron wedges (Fig. 3.4).

The carriage is a baulk of elm (often termed a '*stock*') hollowed out and relieved to fit the profile of the barrel with the end supporting the breech chamber merely hollowed out to enable the sliding of the breech chamber into position within the barrel. This also facilitates the acceptance of the two chambers routinely supplied with each gun, which, by nature of the method of manufacture, cannot be identical. The rear of the carriage steps upwards to form a solid block protected by an iron plate or wedge (Fig. 3.4). As the location of the chamber into the rear of the barrel necessitated a larger space than is finally utilised, an elm forelock with a projection on its lower surface fits into the hollowed-out semi-circle of the bed. Additional iron wedges can also be used to secure the fit of the gun to the bed. A rectangular hole through the rear of the bed accepts a wooden post (foot/tiller) pierced with holes and pegged to the required elevation. The carriage is supported on a single axle of ash on either two solid elm trucks (small wheels) or spoked wheels with elm fellies and hub and oak spokes. None shows any trace of iron tyres over the fellies.

The swivel guns or '*bases*' from the *Mary Rose* are not mounted on beds. They are all made with a central tube and most appear to have a second layer of hoops and bands (Figs 3.2 and 3.5). A trough is formed at the back end of the barrel into which a small powder chamber is located. The rear end of the trough is plugged with an iron plate into which an iron bar (tiller) is located to project behind the chamber trough to enable training of the piece by both vertical and horizontal movement. Trunnions are positioned on the

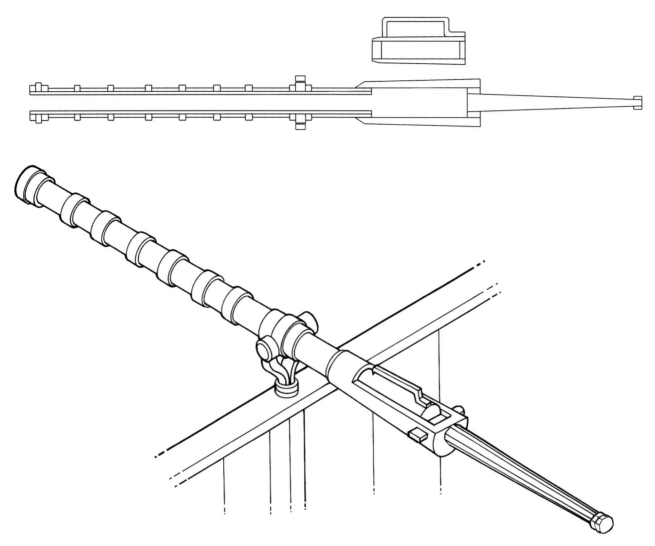

*Figure 3.5 Section through a swivel gun (base) and reconstruction of use*

barrel. These are fitted into a yoke that projects downwards to form a peg locating into a hole within the ship's structure. It is assumed that the top pieces are similar.

## Port Pieces

### Identification and historical significance

The *Anthony Roll* lists twelve port pieces for the *Mary Rose*, with 200 stone shot. The identification of port pieces (Hildred and Rule 1984) is based on a few existing historical descriptions but most reliably on their position within sixteenth century inventories, the first iron guns listed, suggesting that they are the largest of the wrought iron ordnance carried (Figs 3.6–7). Howard Blackmore provides a good description (1976, 239) based on a study of historical sources and descriptions within inventories for guns in the Tower of London. Port pieces are listed for the Tower between 1540 and 1677 and Blackmore's definition is: '*A breech loading gun, firing stone or 'muthering' shot*'

Entries include guns with a bore size of up to 10in (254mm). Barrel weights for port pieces issued from the Tower between 1559 and 1568 averaged 6½ hundredweight (*c*. 330kg), with chambers weighing about 300lb (136kg). Early inventories list these as wrought iron, but after 1589 they include weights of both chambers and halls cast in bronze or iron. As solid castings of this size were already in circulation, this indicates that an important feature of these guns was that they were breech-loading.

Norton (1628, 58), although probably describing cast bronze port pieces, provides some detail: the barrel has a length in calibres (see p. 132 above for definition) of eight, the chamber about three and the chambers take one-third to one-quarter of the shot weight in corn powder. He lists chambers of 12in, 17in, 22in and 24in (305mm, 432mm, 559mm, 610mm) length with shot size ranging 5–7in (9–17lb) (127–229mm/4.1–7.7kg), but gives a powder weight of half the shot weight. In a separate list, under the title '*port ch*', he gives a length of 16in (406mm) a height at the mouth of 3½in (89mm) firing a 5¼in (133mm) stone shot of 9lb (4kg) taking a lower powder charge of 3¼lb (8.2kg).

Their relative importance on the *Mary Rose* can be suggested by the number listed; there are only fifteen cast bronze guns listed, ranging from cannon to falcons, but the total for port pieces alone is twelve, only outnumbered by small bases and cast iron hailshot pieces.

The importance of the port piece as part of any ship's armament over time can be judged by assessing their numbers against the total number of carriage-mounted iron guns on the ships in 1540 (PRO E 101/60/3 ff. 4–5), 1546 and 1547 (Table 3.2). In 1546, 144–160 port pieces are listed on the 20 ships (the counted number differs from the total listed in the document) in many cases making up over half the carriage mounted iron guns – 218 over the 58 vessels. Generally, the smaller the vessel, the fewer port pieces it carried. Few pinnaces and none of the rowbarges carried them. These guns, must, therefore be considered as an integral part of the main armament of the largest fighting vessels, alongside the newer cast bronze guns. Therefore, they must have functioned as anti-ship guns. In the case of the *Mary Rose*, port pieces are the only wrought iron guns found stationed at gun bays fitted with lidded ports and it is interesting that the term '*port piece*' does not seem to exist in the fifteenth century, before the adoption of the lidded gunport.

*Figure 3.6 Port piece barrel and chamber as lifted*

By 1547 (Starkey 1998) all but one ship carried substantial numbers, with sixteen on the *Matthew* (*ibid.*, 7278) (Table 3.2). Fifteen '*galliasses*' carried port pieces, from thirteen on the *Swallow* (*ibid.*, 7490) to one on the *Greyhound* (*ibid.*, 7783). Port pieces are dominant among the wrought iron carriage mounted guns within the fleet, with 211 (40%) listed against 154 slings and 78 fowlers.

## Scope and distribution of the assemblage

Substantial remains of up to eight of the twelve port pieces listed have been recovered (Figs 3.7–8), six on

Table 3.2 Incidence of port pieces listed on ships between 1540 and 1547

| Vessel | 1540 port pieces | 1540 total iron | 1546 port pieces | 1546 total iron | 1547 port pieces | 1547 total iron |
|---|---|---|---|---|---|---|
| *Henry Grace à Dieu* | not furnished | | 14 | 28 | 15 | 20 |
| *Mary Rose* | 9 | 20 | 12 | 24 | not listed | |
| *Peter* | 10 | 19 | 16 | 34 | 9 | 9 |
| *Matthew* | not listed | | 16 | 20 | 16 | 18 |
| *Great Bark* | not listed | | 10 | 20 | 15 | 24 |
| *Jesus of Lübeck* | not listed | | 4 | 18 | 10 | 26 |
| *Pauncy* | not listed | | 12 | 25 | 11 | 22 |
| *Morian (of Danzig)* | not listed | | 4 | 8 | 7 | 25 |
| *Struse (Spruce of Danzig)* | not listed | | 6 | 12 | 4 | 21 |
| *Christopher of Danzig\** | not listed | | not listed | | 8 | 21 |
| *Mary (of) Hamburg* | not listed | | 6 | 10 | 9 | 25 |
| *Christopher of Bremen\** | not listed | | 2 | 4 | not listed | |
| *Trinity Harry* | ?11 | 19 | 10 | 19 | 9 | 23 |
| *Small Bark* | not listed | | 12 | 17 | 14 | 21 |
| *Sweepstake* | 9 | 15 | blank | 8 | 13 | 22 |
| *Minion* | 9 | 17 | 12 | 19 | 13 | 17 |
| *Lartique* | not listed | | 0 | 3 | not listed | |
| *Mary Thomas* | not listed | | 2 | 7 | not listed | |
| *Hoy Bark* | not listed | | 2 | 6 | 0 | 0 |
| *George* | not listed | | 2 | 3 | 1 | 6 |
| *Mary James* | not listed | | 2 | 7 | not listed | |

\*possibly the same vessel. Listed by Rodger (1997, 478) as a hulk bought in Bremen in 1545

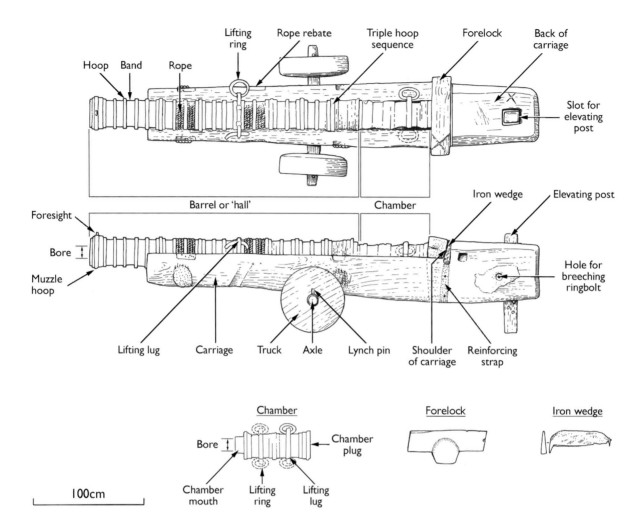

*Figure 3.7 Anatomy of a port piece*

their elm carriages. Four were recovered in 1970–82, the remainder during nineteenth century salvage.

Three on carriages were found at gun stations on the starboard side of the Main deck, with the axle and wheels of a fourth denoting the salvaging activities in 1836 and 1840 (Fig. 3.9). Assuming bilateral symmetry in the positioning of ordnance, the Main deck would have accommodated at least eight of the twelve (Fig. 3.10a). Two were side by side (M4, M5) flanked by a culverin in M3 and cannon in M6; all are of similar bore, suggests that these four guns could have been used together for 'battering' (Fig. 3.10b). The next gun bay (M8) also housed a port piece, separated from a demi cannon in M10 by the Carpenters' cabin. The final broadside gun was a port piece in M11. The lack of structure at Main deck height at the transom does not preclude the possibility that two more port pieces were situated here, either side of the rudder (Fig. 3.10c). Fifteen of the 20 ships in the *Anthony Roll* clearly depict guns on either side of the rudder low down in the hull, eight (including *Mary Rose*) are shown with lidded gun ports. Nine show 'banded', rather than smooth, muzzles suggesting wrought iron guns. It is unfortunate that, as yet, there is no archaeological evidence for gunport lids, or indeed large guns, in this area, so we can only postulate on the basis of gun size, function and the wider distribution of gun elements.

The association, however tentative, of the words '*port piece*' with lidded gunports favours placing all twelve guns on the Main deck. This is potentially possible if two are placed in the bow, between the stem and the culverin in M3 and two in the stern facing aft. This distribution is shown in Figure 3.10d. The illustration of the *Henry Grace à Dieu* in the *Anthony Roll* shows at least one, possibly two, guns on the lowermost gun deck forward of the rise up to the Forecastle (Knighton and Loades 2000, 40). The *Mary Rose* may have one, and is shown with nine guns below what appears to be the Upper deck level, although only five are lidded and four are shown at a 'higher' level. Arguments against this include the fineness of the hull towards the stem, suggesting little space forward of the cabin in M2. A port piece chamber was found associated with a bow anchor just east of the stem post and recovered in 2005.

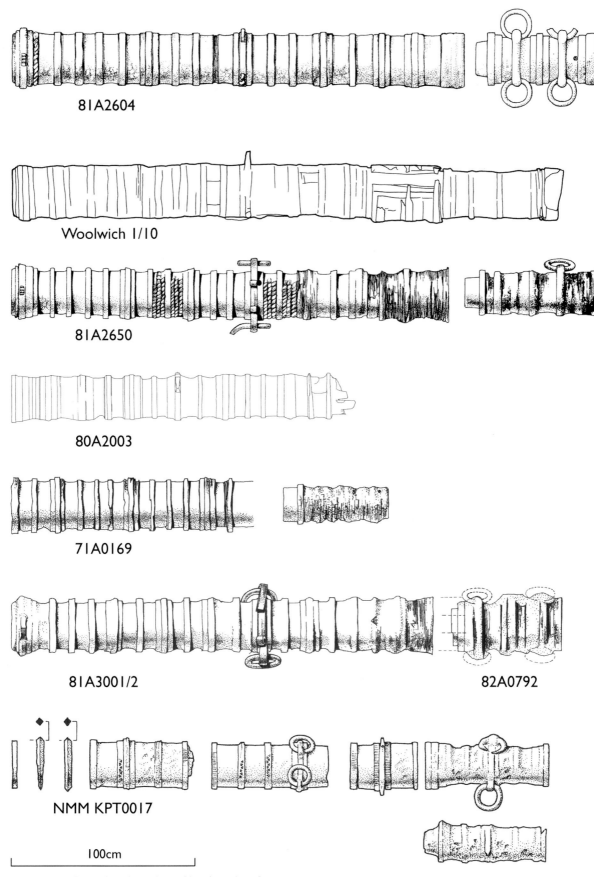

*Figure 3.8 Port piece barrels and* in situ *chambers*

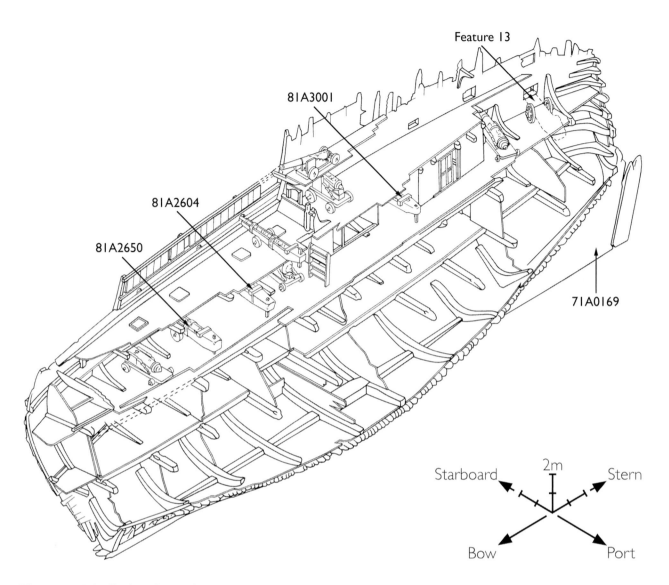

*Figure 3.9 Distribution of port pieces*

Although this could be derived from either the Main or Upper decks, finding it here suggests the positioning of large guns on the starboard side forward of Sector 3. One of two extremely large breech chambers was found outboard of M2 (85A0028) but, as this was found already stropped, we cannot be certain that this was its original position. It may have come from just under the bow, together with concreted anchors and other guns.

Other potential locations for port pieces include the four 'empty' gunports on the Upper deck in the stern: U10 and U11 port and starboard. The U10 starboard port had remains of concreted wrought iron and other elements suggestive of a wrought iron gun, but of a size certainly smaller than the Main deck port pieces. There is also the requirement to position another five bronze guns within the ship, and two of these could be on the Upper deck in the stern. The *Anthony Roll* illustration shows a fourth gunport at Upper deck level towards the stern. This portion of the ship has not survived.

## Morphology

All the guns are made by laying a number of staves parallel to one another to form an inner tube. This is reinforced by a second layer of bands and abutting hoops (or hoop sequences) laid directly over the staves. Hoop sequences are generally of three hoops, with the central one thicker than the flankers. All the hoops are higher than the bands. There is no infilling of the seams nor any sign of attempts to weld staves, except in the area of the neck of breech chambers. There is no indication of a single sheet of metal being used to produce the inner tube and, in most cases, there seems to be one stave that is substantially smaller or larger than the rest.

141

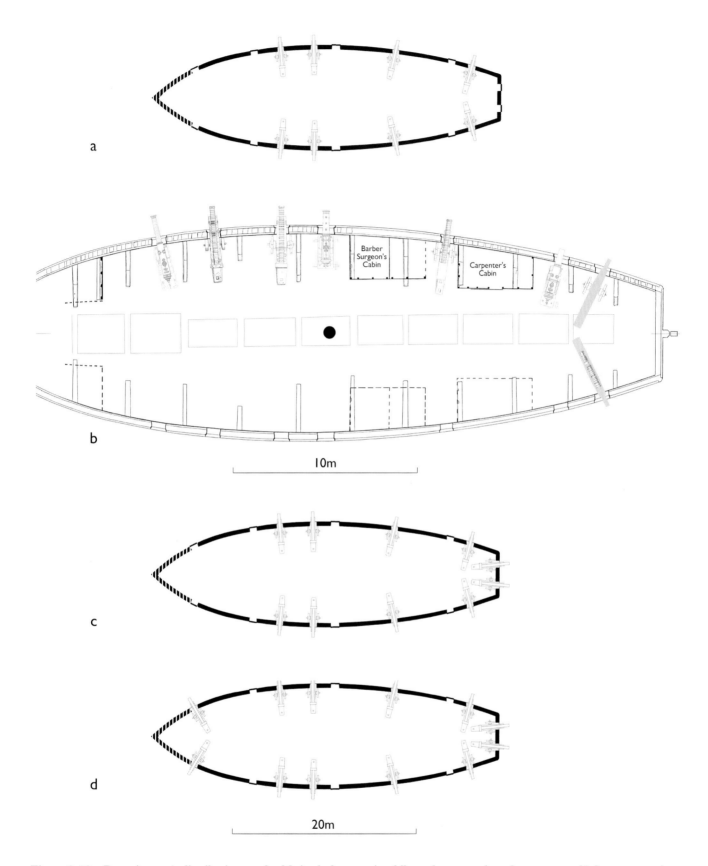

*Figure 3.10 Port pieces: a) distribution on the Main deck assuming bilateral symetry from known guns; b) best suggested positions of all guns where possible on Main deck; c) suggested gun positions with sternchaser port pieces either side of rudder; d) gun positions on the Main deck as indicated in the* Anthony Roll

Table 3.3 Port piece basic data (metric)

| Object | Sector/ date re-covered | Barrel length | Breech length (exc. neck) | Int. diam. barrel at breech | Ext. diam. barrel at breech | No. calibres | Forelock width | Length bed | Length of bed behind shoulder | Wheel type | Shot type/size |
|---|---|---|---|---|---|---|---|---|---|---|---|
| 81A2650 | M4S | 2385 | 640 | 179 | 290 | 16.89 | 140 | 3450 | 730 | solid | none |
| 81A2604 | M5S | 2420 | 590 | 195 | 280 | 15.43 | 240 | 3595 | 905 | spoked | none |
| 81A3001 | M8S | 2300 | 570 | 180 | 305 | 15.94 | 210* | 3350 | 800 | spoked | stone, 172 |
| 71A0169 | stern | 1300 inc. | 500 | 190 | 290 | inc. | n/a | n/a | n/a | n/a | stone, 184 |
| NMM KPT 0017 | 1840 | 1500 inc. | 600 | min. 160 max. 180 | ?270 | n/a | ?190 | 3000 | 700 | ?solid | stone, 155 |
| RAI 1836** | ?M5P | 2180/ 2921† | ?650 | ?200 | 280 | 14.60 | ?180–200 | 3400 | 780 | ?solid | stone |
| RA XIX.1 80A2004 | 1840 ? | c. 1982 hoops only | breech see 80A2002 | 150 | 230 | n/a | n/a | 1970 inc. | inc. | ? | Deane states stone |
| RA XIX.2 80A2003 | 1840 ? | 1880 inc. | see 80A2000 | 152–160 | 250 | n/a | n/a | n/a | n/a | n/a | stone |

\*    Two forelocks, 81A3188, 81A3129

\*\*   The gun is in a case & cannot be measured. The measurements are from old drawings. The breech chamber wedge is shown incorrectly inserted, the measurement is the depth excluding the tang. Often depth & width are similar. It is not known whether the bore measurement is from the mouth or the breech. If it is from the mouth, the breech could be up to 10mm smaller, the smallest measurement must be taken as the bore. The gun was the subject of a detailed photomosaic survey by the City University. We have the photographs, but not the interpretation

†    Total length given in 1836, 9ft 7in, 2921mm

In 1841 John Deane states in a letter to John Powell that he recovered three iron guns on beds. One can assume that these include the gun RAI (Royal Artillery Institution), the National Maritime Museum & Royal Armouries XIX.I. He also states that all contained powder and stone shot (letter dated April 1841, 72 North Street, Gosport, pers. comm. Martin Dean 1984, reprinted in Bevan 1996, 186/7)

All port pieces are large, with bore sizes 150–200mm (6–8in) and barrel lengths of 12½–13 calibres (Table 3.3 and Fig. 3.8). Including the chambers, the lengths of complete guns range from 2870mm to 3025mm (15½–17 calibres) more than the 11 calibres quoted by Norton (1628). The ratio of chamber to hall is between 2¼ (71A0169) and nearly 4 (81A3001) and the chamber lengths are 3–3½ times the internal bore of the barrel (except 71A0169 which is 2.63 times the bore). Norton states that the chambers should be about three times the bore (1628, 58). It is extremely difficult to get absolute measurements because of the varying degrees of corrosion, effects of conservation and the method of manufacture and nature of the material itself. In all cases, where measurements of the internal diameter of the barrel were taken at the mouth and at the breech, the breech was the smaller by up to 10mm. The best indication of the correct bore measurement is the diameter of the stone shot, requiring at least 6mm (¼in) windage for safe passage.

The total length of the complete guns on their carriages ranges from 4280mm (81A2604; M5, 15.43 calibres), to 3800mm (81A3001; M8, 15.94 calibres). The difference in length may reflect position within the ship – the longest gun being close to midships and the shortest further aft where the ship begins to narrow towards the stern (Fig. 3.9).

Not all the guns have been weighed. The barrel of 81A2604 weighs 421kg, and the reproduction of 81A3001 on its carriage weighs 1100kg. This is the shortest in situ gun and it also has a large bore; so this can be considered a low weight. However, by comparison, the maximum estimated weight for the heaviest bronze gun on its carriage is 3000kg; the lightest falcon is estimated at c. 500kg. The reproduction of culverin (81A1423), on its carriage, weighs 2360kg.

The carriages are all still soaking and are not available either for wood analysis or for detailed measurement so details are taken from photographs and illustrations. A summary is provided in Table 3.3, where some dimensions for iron components have been taken from radiographs.

*81A2650 Barrel with chamber on carriage*

*Location*: M4. On carriage with forelock, 2 iron wedges, axle (81A1859) and truck wheels (81A1858, 81A2830)

*Recovered*: 1981. *In situ*, muzzle protruding through lidded gunport

*Statistics*
Length: barrel: 2385mm; breech: 640mm; bore: 179mm; 16.89 calibres

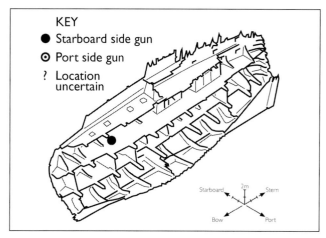

Although the chamber was *in situ* and contained a tampion and powder, no shot is associated with the gun (Figs 3.11–12). It is assumed that the gun was originally loaded and the shot lost, possibly rolling out of the barrel as the ship heeled. A second breech chamber was found concreted to the south wheel (83A0513). Two additional forelocks were found, a small one to the north (80A1723) found at a higher level, and one in M4/5 (81A1229). The elevating post was not *in situ*, but two were found nearby (80A1466 and 80A1467). Either would fit into the hole. For ease of reloading, the breech would have to be depressed and ideally the post removed, therefore the gun may have been recently reloaded.

The barrel is complete and has a single set of lifting rings. The centre of this assembly is 1035mm from the breech and 1340mm from the bore. The lugs are set into a triple hoop and hold a 25mm ring with an internal diameter of 150mm on each side. One ring is complete and the other fragmentary (Fig. 3.12a). The barrel consists of six staves forming the inner tube covered with 17 bands, *c.* 60mm in width, separated by single hoops or triple hoop sequences. The hoops are 25–30mm in width and stand above the bands by 15mm. The breech junction is strengthened with three adjacent hoops descending in height and the muzzle with three ascending hoops and what appears to be a ring, over which the staves are bent backwards. A foresight (a block with simple linear ridges) is positioned on the last hoop (Fig. 3.12d). The gun extends beyond the end of the carriage by 550mm.

The breech chamber was held with the elm forelock and has a total length of 725mm, including the neck. Originally it would have had two sets of lifting rings, towards either end. One ring survives from the rear set, with external diameter of 150mm and internal diameter of 100mm. The 85mm neck tapers 10mm over its length. It is not possible to see the stave joins at the neck where the thickness of metal is 25mm and the gap between the chamber and barrel is 12mm. The first hoop (which abuts the barrel) is 245mm in diameter and 45mm thick. The outer surface is made up of alternating bands (60–70mm wide) and hoops (width 40mm if single, and 30mm, 40mm, 30mm for sets). The back of the chamber has a diameter of 290mm and plug diameter of 100mm and ends with three hoops. The *in situ* forelock has a width of 140mm. Between the forelock and the back of the carriage were found the remains of what was described as an '*iron plate*' concreted to the carriage. This was 540m in width, 132mm in height and 53mm thick (Fig. 3.13). It is uncertain as to whether this was a loose wedge, or something fitted to the carriage. Folding iron wedges

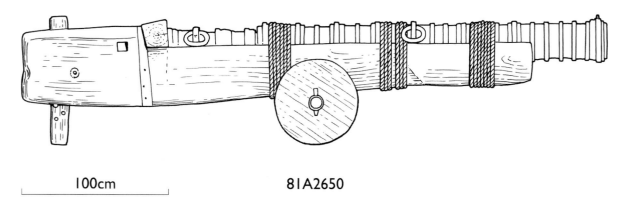

*Figure 3.11 Reconstruction drawing of gun 81A2650 on carriage*

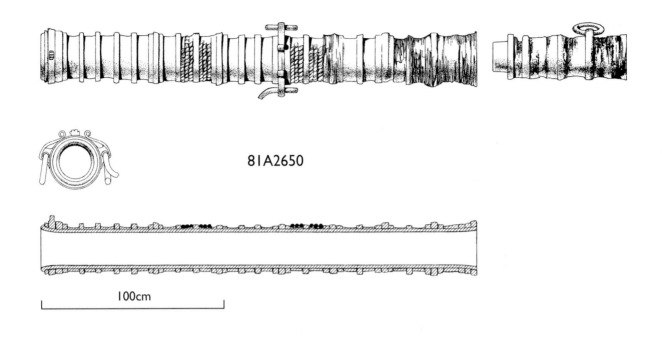

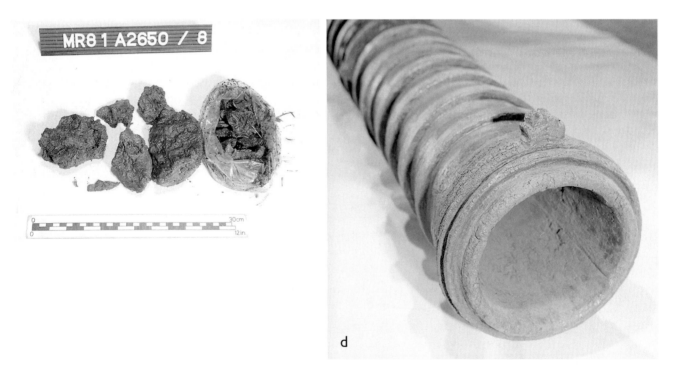

*Figure 3.12 Port piece 81A2650: a–b) external views and cross-section; c) wadding and concretion found inside the chamber; d) detail of foresight*

*Figure 3.13 Port piece 81A2650: a) forelock; b) iron wedge between forelock and back of be; c) back of carriage; d) iron wedge*

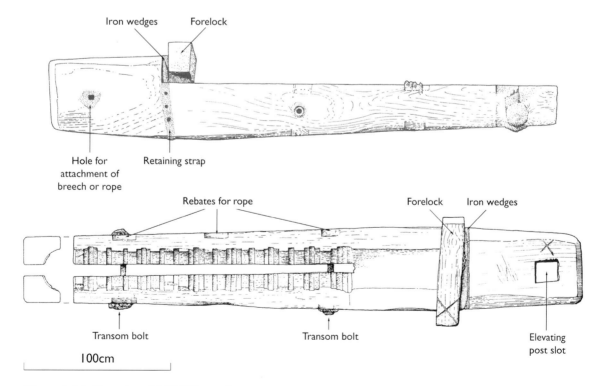

*Figure 3.14 Port piece 81A2650: carriage*

*Figure 3.15  Port piece 81A2650: montage of the carriage*

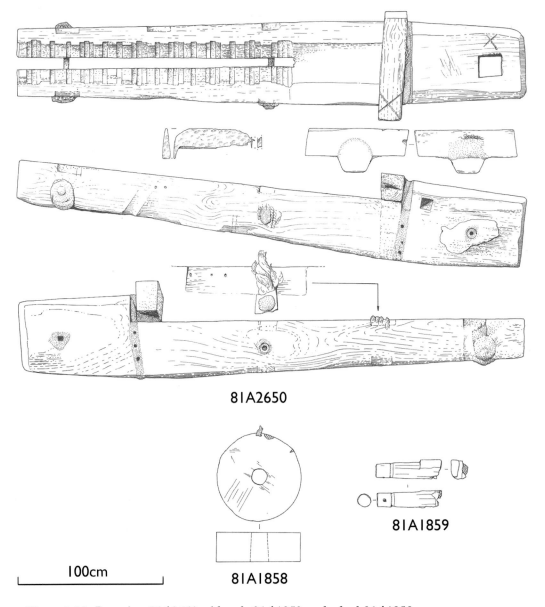

*Figure 3.16  Port piece 81A2650 with axle 81A1859 and wheel 81A1858*

were used to secure the forelock during the experimental firing of a similar gun, and John Deane describes an iron wedge and '*chock*' (the forelock) found on one of the guns recovered in 1840 (NMM KPT 0017, illustrated in Bevan 1996, pl. 24).

The gun is held on the carriage with three sets of rope binding ('*woling*' or '*wooling*') which sit in rebates 100mm, 156mm and 160mm wide cut in both top and bottom of the carriage, which holds about five turns of rope (Fig. 3.14). The presence of iron staining and three 25mm diameter bolt holes suggests a 70mm wide retaining strap at the junction with the end of the trough and the bed. This is a weak point in the carriage and may be to retain an iron plate (see above) to protect its back. The slot, which nearly divides the carriage into two, starts at the front (65mm wide) and ends 800mm from the shoulder (90mm wide), at about the junction with the breech chamber. Two horizontal, rebated, 30mm diameter holes for transom bolts are evident; rebates suggest a head of 60mm on one side and washer with wedge at the other. Concreted bolt holes behind the shoulder may have held 30mm through bolts. On the right side of the carriage this hole ends in a 25mm square hole and on the left a 60mm washer. These may have held rings for breeching ropes (Figs 3.14–5).

The 3450mm long carriage, which is cut to accept the hoops and bands on the gun, is 450 x 450mm at the back, widening to become 500mm square at the shoulder where it forms the backstop to the breech chamber. The carriage and forelock are both marked with a cross or letter 'X' (Fig. 3.14). The rectangular hole (150 x 140mm) for the elevating post is 165mm from the back of the carriage. The single ash axle (81A1859) is incomplete, having broken at the junction with the carriage. Only one arm is present. It is 440mm long and 90mm wide with a diameter at the face of 85mm. There is a single pin hole at the end of the arm to secure the wheels but no record of axle bolt holes. The north wheel (81A1858), which is domed on its inner face, is 475mm in diameter and 100mm thick with a 105mm diameter hole (Fig. 3.16). The south wheel (81A2030) is in two fragments (250/100mm long, 90/60mm wide and 25/15mm thick. Two iron staves were found outside the gunport (MRT/81/DM/350, 408, 415, 417, 418). These may have been supports for the gunport lid.

## 81A2604 Barrel with chamber on carriage

*Location*: M5, on carriage with forelock, elevating post (80A1364), axle (81A2571) & spoked wheels (81A2569, 81A2570)
*Recovered*: 1981 (elevating post 1980). *In situ* muzzle protruding through lidded gunport

*Statistics*
Length: barrel: 2420mm; breech: 590mm;
bore: 195mm; 15.43 calibres;
Weight (barrel): 421kg

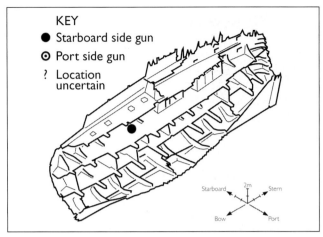

The chamber (Fig. 3.17) was *in situ* and sealed with a wooden tampion. Although no shot is recorded, it is assumed that the gun was originally loaded, though both stone and iron shot had fallen beneath the axle, held by the large spoked wheels (Fig. 3.18). The gunport lid was *in situ* and was hinged open (Fig. 3.19).

The barrel is complete and has a single set of lifting lugs (Figs 3.17 and 3.20). The centre of this assembly is 1210mm from the breech and 1250mm from the bore. The lugs are set into a triple hoop which would have held a lifting ring on each side of the gun; these have broken off.

Radiography (details in archive), just behind the muzzle hoop, showed the barrel to be composed of seven abutting staves (Fig. 3.21), measuring 120mm (2), 130mm (3) and 155mm (2), with a thickness of 15mm. The estimated thickness of the bands is also 15mm. A two-layered structure is apparent (Fig. 3.21d): an inner tube formed of staves and a single outer layer of bands and hoops or hoop sequences with a (visible, but as yet not understood) 5mm gap between the layers. The staves flare towards the mouth where the internal diameter is 205mm.

The outer layer comprises sixteen *c*. 70mm wide bands separated either by single hoops (10) or triple sets (6) *c*. 25mm in width and standing 15mm above the bands. A sequence of three hoops of equal thickness descends from the breech. The muzzle sequence includes four hoops, the first twisted to resemble rope, then a triple set with the central hoop upstanding, bearing a foresight, and finally a ring over which the staves have been bent.

The breech chamber (bore 115mm, total length including neck 650mm) was *in situ* with a tampion (see Table 3.4) and contents (82S1208). Wadding (82S1106) was recorded as coming from the barrel. Radiography clearly shows staves covered by hoops and bands (Fig. 3.21a). The external diameter at the mouth (on the first hoop) is 280mm, that of the neck at the junction with the first hoop 192mm and at the front of the neck 172mm. The thickness of staves at this position is 35mm. The number of staves cannot be ascertained, but at least three clear joins are visible. The three breech

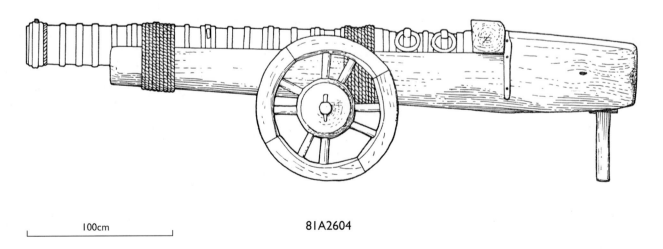

*Figure 3.17 Port piece 81A2604: reconstruction drawing of gun on carriage*

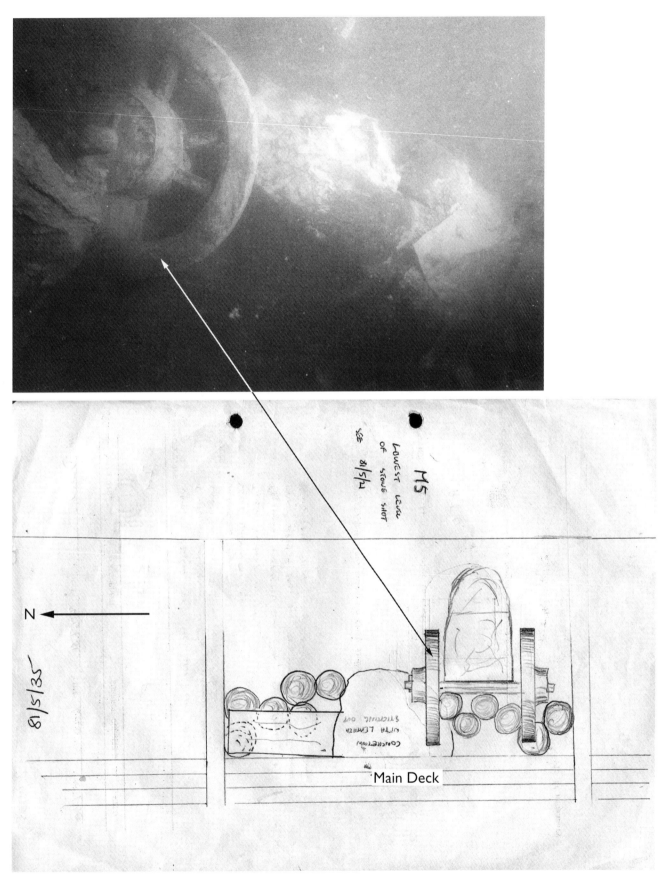

*Figure 3.18 Port piece 81A2604: a) underwater photograph of gun as found; b) diver's sketch of stone shot around wheels (divelog 81/5/35)*

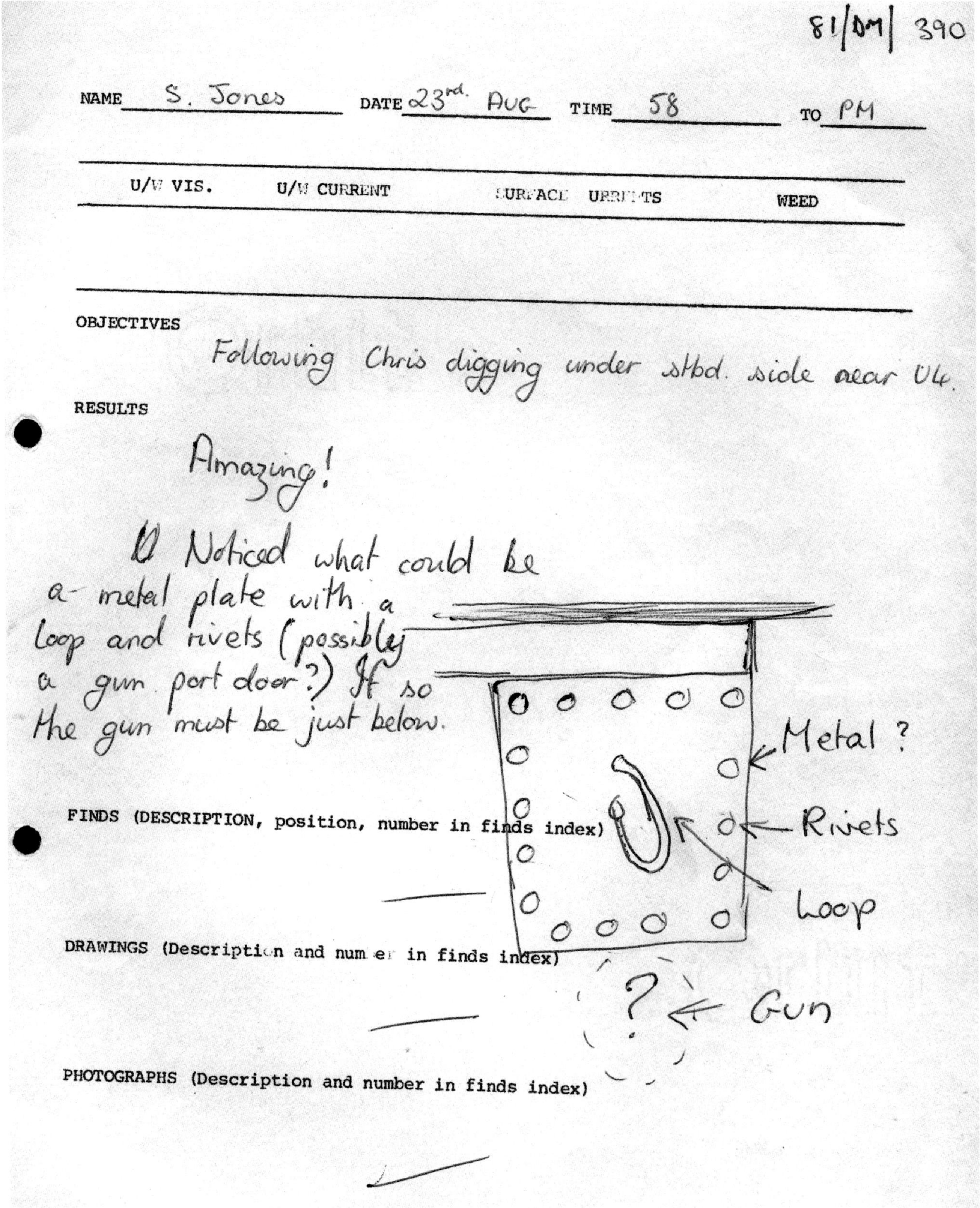

81/DM/ 390

NAME S. Jones DATE 23rd. AUG TIME 58 TO PM

U/W VIS. U/W CURRENT SURFACE CURRENTS WEED

OBJECTIVES

Following Chris digging under stbd. side near U4.

RESULTS

Amazing!

(1) Noticed what could be a metal plate with a loop and rivets (possibly a gun port door?) If so the gun must be just below.

FINDS (DESCRIPTION, position, number in finds index)

DRAWINGS (Description and number in finds index)

PHOTOGRAPHS (Description and number in finds index)



*Figure 3.19 Port piece 81A2604: diver's sketch of the gun position as found and gunport lid (divelog 81/5/30)*

hoops on the barrel locate with the two neck hoops on the chamber, providing a reinforcement of five hoops at the junction. The sequence of hoops and bands of the outer layer can be seen on Figure 3.21b. The hoops are mostly 35mm wide, and 50mm with lifting rings; the bands are 60–90mm. The two pairs of 180mm lifting

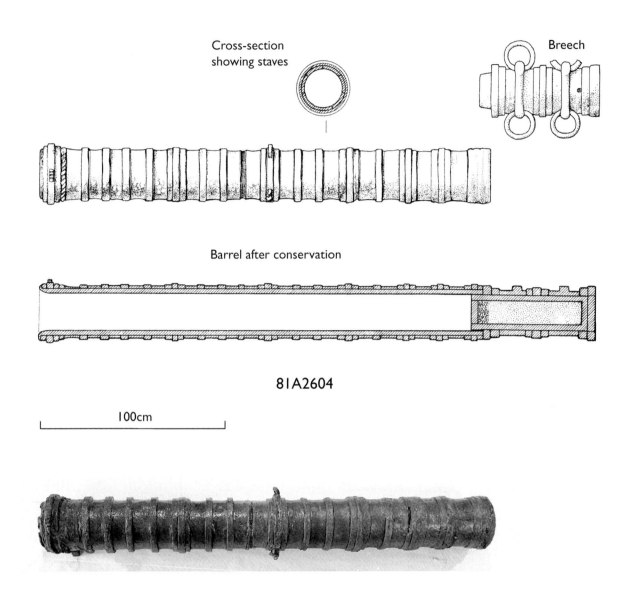

*Figure 3.20 Port piece 81A2604: barrel and breech*

rings are set in single hoops and are made from 30mm round section iron. The diameter at the back is 280mm with a plug diameter of 80mm.

The gun was held on the carriage with two sets of rope bindings, secured in rebates (200mm and 80mm length) cut in the carriage top 950mm and 2380mm from the shoulder. These accommodated about nine turns of rope each (Fig. 3.22a–b). Rope marks were also noted over the back of the carriage on the north side. Evidence for a horizontal bolt (two small holes on the right side and a rectangular slot on the left) is visible behind the shoulder. This may have held rings for securing the gun during firing. The elm forelock is marked with / \ .

The carriage (length 3595mm) is 400mm high and 500mm wide at the back, enlarging to 450 x 600mm towards the shoulder where it forms the backstop to the breech chamber (Fig. 3.22b). The rectangular hole for the elevating post is 220mm from the back of the carriage and there are two small square holes of indeterminate depth behind this on either side. There is a slot in the bed from the front of the carriage to about the junction with the breech chamber and evidence suggesting two horizontal transom bolts. The carriage is cut out to accommodate the hoops and bands on the gun (Fig. 3.22c). The gun extends over the end of the carriage by 610mm.

There is again evidence for a 70mm wide vertical strap at the end of the chamber trough (Fig. 3.16). The incomplete elevating post (80A1364, 980 x 110mm; Fig. 3.22d) was *in situ*. It has 10 holes for altering the height of the carriage and is marked with a 'W'. The single ash axle (81A2571) is 1420mm long (estimated 50mm missing). The spoked wheels are 950mm and 970mm in diameter respectively, with treads of 110mm (Fig. 3.23). 81A2570 has two '+' marks on the hub, one larger than the other. A wooden lynch pin for the north wheel is recorded.

A forelock (81A1261), which may be a spare, lay close by and a number of wooden shot gauges (110–176mm) were found (most together) beside the gun (81A1285–8; see Chapter 5). There was a spare

50cm

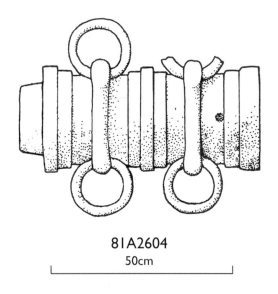

81A2604
50cm

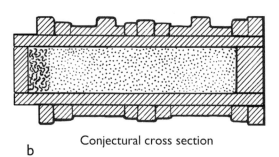

Conjectural cross section

*Figure 3.21 Port piece 81A2604: a) radiograph of chamber; b) chamber external view and cross section; c) photo showing stave junctions of inner layer at back of the gun; d) detail of construction at mouth (2 layers) and the remains of the foresight*

Table 3.4 Port piece chambers (metric)

| Object | Sector | Length | Neck length | Neck thickness | Bore mouth | Ext. diam. neck | Ext. diam. neck hoop | Sets lift rings | Ext. diam. lift ring | Ext. diam. back | Comment |
|---|---|---|---|---|---|---|---|---|---|---|---|
| 81A1084 | ?M1 | 1265 | 65 p | 32 | 88 | 140 | 230 | 3 | 305 | 280 | ?3 layers; staves, bands (mid-layer), hoops, bands (radiograph) |
| 85A0028 | ?M2 | 1250 | 50 p | 25–38 | 95 | 170 | 240 | 3 | 255 | 290 | full excav. report; radiography suggests outer hoops abutting, no obvious bands |
| 81A2650 | M4 in situ | 725 | 85 t | 25 | 85 | 165 | 245 | 2 | 280 | 290 | breech plug 55mm thick; gap at face 40mm, at neck 20mm |
| 81A2604 | M5 in situ | 650 | 60 t | 35 | 115 | 172–192 | 280 | 2 | 290 | 280 | breech plug 70mm thick; 3 rings on barrel meet 2 on breech |
| 71A0169 | stern | 570* | 70 p | 25 | 130 | 180 | 210 | ?* | – | 200–220 | 15mm diam. touch hole |
| 82A0792 | M8 | 605* | 35* | 37 | 89 | 155 | 260 | 2 | 330 | 268 | sectioned; replica made for firing trials |
| 82A4081 | U11 | 700 | inc. | inc. | 120 | 150 | 250 | 1 | 300 | 280 | hoops & bands; central triple hoops with lift rings |
| 81A4890 | ? | 360* | 730* | 7 staves | 85 | 140 | 200 | ? | – | 205 | |
| 85A0027 | SS2 | 595 | 68 | 25 | 72 | 120 | 230 | ? | – | 220 | hoops only; frags 1 ring in concretion; ?bowed ends |
| 83A0513 | M4ex c | 720 | 50 | 34 | 85–90 | – | – | ?2 | – | 310 | stave & hoop/band (radiograph) |
| 82A5097 | ? | 770* | * | * | 100 | 170 | – | 2 | – | 250 | stave, band & hoop, unclear; tampion in situ; ?2 sets rings; ?like 82A5176 |
| 82A5172 | ? | 635 | 35 p | 35 | 100 | 165 | 230 | ?2 | 250 | 260 | excavated; 8mm touch hole 120mm from back; ?double staves |
| 82A5176 | ? | 720 | 100 p | 35 | c. 100 | 170 | 250 | 2 | 250 | 250 | similar to 80A2005 & E31/104; ?abutting hoops without bands |
| 90A0001 | ?c | 805 | 90 | – | 100 | 145 | 235 | 2 | 225 | 225+ | 10mm touch hole 90mm from back |
| 03A0070 | ? | 620 | – | – | – | – | – | 2 | – | 200 | inaccessible, not drawn or radiographed |
| 05A0057 | ?c | 650 | – | – | – | – | – | – | – | 265 max. | inaccessible; 1 set lifting rings visible |
| NMM1 KPT 0017 | ? | 680 | 80 in gun | 40 | 110 | 150 | 190 | 1 | – | 230 | neck inside portion, broken gun |
| NMM2 | ? | 720 | 80 t | 20 | 80 | 122 | – | ?1 | – | 160–180 | eroded; missing end hoops at back |
| RAI | ? | c. 650 | – | – | – | – | 200 | – | – | 245 | eroded, in gun; looks similar to 71A0169 |
| 80A2000/] RAXIX.2 | ? | 670 | – | – | 110 | – | – | unclear | – | 260 | hoops & bands, similar to Woolwich |
| 80A2001/RA XIX.163 | ? | 470 | 50 | – | 140 | – | – | – | – | – | small, hoops & bands, no obvious lifting rings |
| 80A2002/ RA XIX.1 | ? | 540 | – | – | 70 | – | – | – | – | 190 | seems small, poss. not port piece |
| 80A2005/ RA XIX | | 700 | – | 25 | – | – | – | unclear | – | 254 | similar to E31/104 & 82A5176 |
| E31/104/ Powell Cotton | ? | 630 | 70 p ?lug | 80 | 80 | 170 | 230 | 2 | – | 210 | staves, 2 visible on radiograph, c. 14mm thick; hoops over; 10mm touch hole 65mm from back, similar to 80A2005 & 82A5176 |
| Ironbridge Gorge Museum | ? | 510 | – | – | – | – | – | 1 | – | 225 | similar to NMM |

\* = incomplete or badly eroded
c = in concretion, information from radiograph (Cobalt 60 source with source to film distance of c. 1 m for 2–28 hours)
p = parallel; t = tapered

154

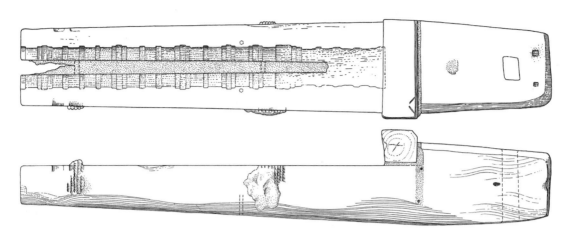

*Figure 3.22 Port piece 81A2604: a) carriage; b) detail of rope woling; c) detail of carriage trough; this is typical of the port piece carriages, the forelock is* in situ; *d) elevating post*

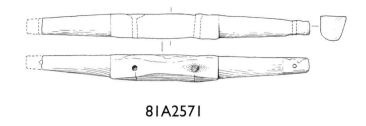
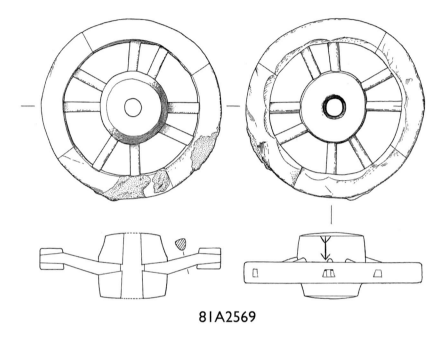
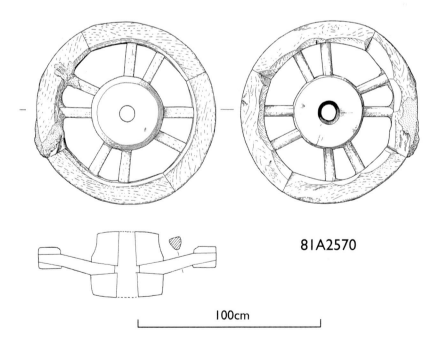

*Figure 3.23 Port piece 81A2604: axle and spoked wheels*

breech chamber concreted to the deck and the south wheel. Explosives were used to remove the concretion. An eyebolt going through the hull beside the gunport had to be cut to free the north wheel.

*81A3001 Barrel and carriage, probable chamber*

Location: M8. Two forelocks in sector (81A3188 & 3129), single chamber (82A0792), elevating post (812088), axle (81A0891, 81A3765, 81A3998) and spoked wheels (81A5688, 5689). Stone shot in barrel (82A4213)

Recovered: 1981–1982, carriage *in situ* going through gunport, barrel found beneath hull after recovery

Statistics
Length: barrel: 2300mm; breech: 570mm; bore: 180mm; 15.94 calibres

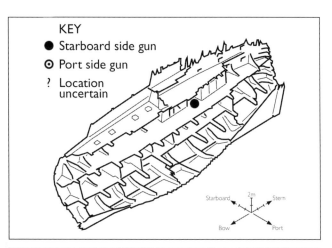

The assemblage (Fig. 3.24, below) as found included an empty carriage (Fig. 3.24, above) with axle and spoked wheels. The nature of the wood suggests that not all the ironwork was *in situ* during burial and that the gun fell off the carriage and out of the gunport at an early stage with the forelock and chamber not *in situ*. The breech chamber (82A0792) and forelocks (81A3188 and 81A3129) and elevating post (81A2088) were not in position, although they were found within the sector, together with an assemblage of stone shot (see below). The presence of shot within the barrel (172mm diam.) suggests that the gun was in the process of being loaded as the vessel sank. The environment beneath the ship

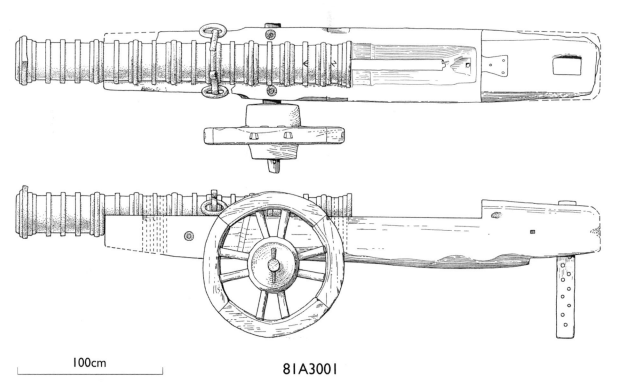

*Figure 3.24 Port piece 81A3001: above) underwater photograph of gun carriage as found; below) reconstruction drawing of gun on carriage without breech chamber, elevation*

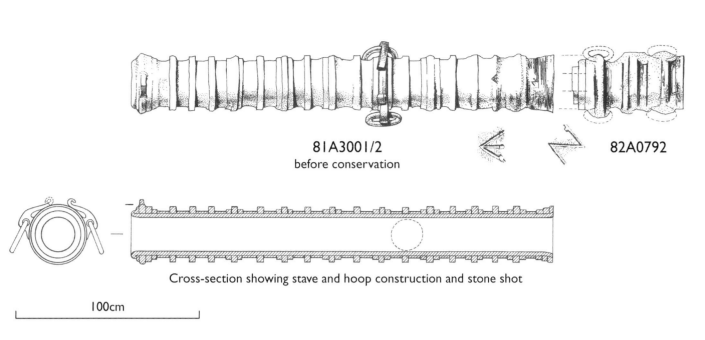

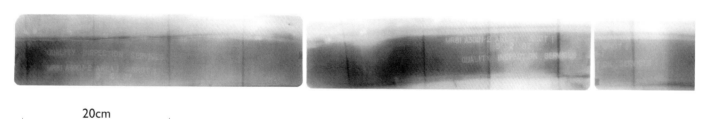

*Figure 3.25 Port piece 81A3001: barrel and chamber and windout radiograph showing nine internal staves*

appears to have protected the iron, and this gun is in the best condition of any recovered. The gunport lid was *in situ*, under the ship and hinged open (81T14453).

A chest (81A2573) had smashed down over the south wheel during the sinking and the rim was found within it (divelog 81/8/161). There is damage to the front of the carriage and the front left portion forward of the axle bolt hole was brought up separately. A portion of the front from both sides is broken, the length is probably not complete. What was described as a '*Deane bomb*' (81A2863: see Chapter 6) was found behind the carriage. There was a considerable amount of rope in the vicinity; some appeared to go around the north wheel.

Details of manufacture, such as dressing of staves and bevelling of the hoops, are visible. For this reason, this was the gun chosen to replicate (see below). The barrel has a single set of lifting lugs, the centre of which is 940mm from the breech and 1360mm from the bore (Fig. 3.25). These are formed by part of one of the

158

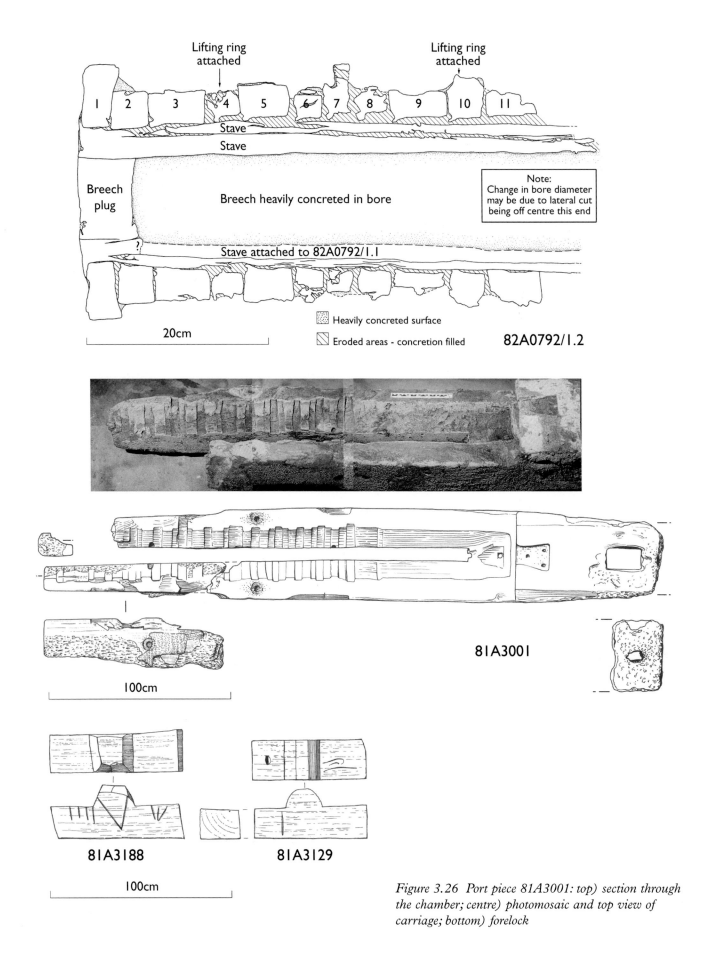

*Figure 3.26 Port piece 81A3001: top) section through the chamber; centre) photomosaic and top view of carriage; bottom) forelock*

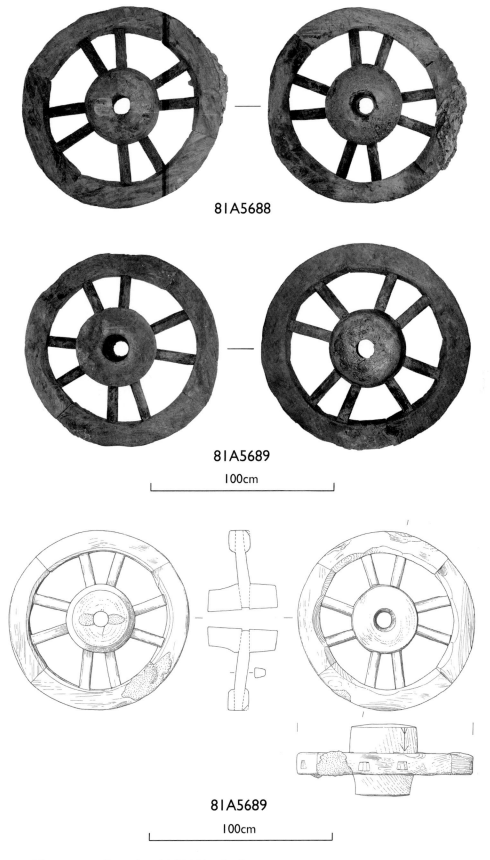

*Figure 3.27 Port piece 81A3001: wheels*

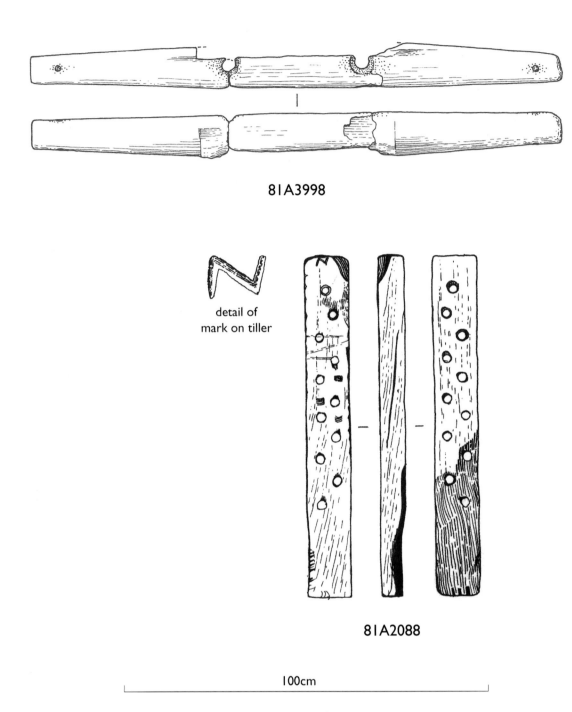

*Figure 3.28 Port piece 81A3001: elevating post and axle*

hoops and hold a 25mm ring with an internal diameter of 250mm on each side. The barrel is marked at the hoop sequence where the chamber locates with a 'Z' or 'N'; this mark is also on the tiller and on one of the forelocks.

Radiography, using a wind-out source placed inside the barrel provided an all round image of the staves. Taken towards the breech forward of the junction with the chamber, it showed that the bands and hoops abutted in a single outer layer, with the inner tube formed of nine staves 90mm (1), 95mm (6), and 105mm (2) wide (Fig. 3.25, bottom). The area into which the chamber locates shows a dense area of metal followed by a 60mm band, a triple hoop sequence (40mm each), followed by a 50mm band and a dense metallic mass forming the beginning of another triple hoop sequence. The thickness of staves at the breech is 16mm ($^5/_8$in) with a total wall thickness (including the hoops) of 61mm. The barrel flares to 190mm at the extreme back and its internal diameter at the mouth is 185mm, the mouth itself flaring to 205mm. The staves are overlain with alternating low bands (17) and high hoops or sets of triple hoops (total 38 individual hoops); all appear to be chamfered with the highest point at the centre. Each one of the three in sets has chamfers on both sides so that the height above the bands is

605mm (originally *c*. 645mm to fit carriage) and supports two sets of lifting ring set into lugs on single hoops (Fig. 3.25). The external diameter of the chamber on the neck at the shoulder ring is 25mm less than the barrel at this location. The trough recess on the carriage at the back of the chamber is 275mm wide and the external diameter of the breech 268mm. As the radiograph was inconclusive regarding structure, the chamber was sectioned (Fig. 3.26, top). This revealed two inner layers: four equal staves forming the innermost layer with three outer staves, of which one formed a trough corresponding to half the circumference and the other two equal quarters. All joins are therefore staggered between the layers. The neck tapers 30mm along its length. The innermost staves tapered from 24mm at the breech plug to 23mm at the mouth and were 75mm wide. The middle layer was consistent at 13mm over its length giving a wall thickness, excluding the outer layer, of 36 mm.

The breech plug is 70mm thick and the internal diameter is 89mm at both ends. The external sequence from the back to the front consists of the end ring (Fig. 3.26 (top), 1), a hoop (2), band (3), hoop with lifting rings (4), band (5), triple hoop (6–8), band (9), hoop with lifting rings (10), band (11), hoop – broken band (not visible on Fig. 3.26). The hoops are 38mm wide, 41mm (single) and 50mm (centre of triple sequence) thick, the bands 65mm wide and 38mm thick. Hoops at the breech and mouth are thicker up to 65mm. The lifting rings have an external diameter of 170mm and ring diameter of 25mm. These are equally positioned on either side of the triple hoop set and are well balanced to lift the chamber.

The carriage length is 3350mm. It is 280mm high at the back and 390mm wide, but this is slightly eroded (Fig. 3.26, centre). At the junction with the shoulder, the height and width are 450mm. A rectangular hole (150 x 200mm) for the elevating post is positioned 155mm from the back of the carriage which is eroded at this point. The chamber trough is 850mm long and the rest of the carriage is relieved to accept the hoops and bands on the gun, which extends beyond the carriage by 900mm (but a portion of the carriage front is missing). The carriage reduces in width by 45mm just in front of the axle bolt holes. Two forelocks were recovered (81A3188 and 81A3129, both of similar widths, 200mm; Fig. 3.26, bottom).

The remains of an eroded rebate 1750mm from the back of the carriage, just forward of the breech junction, may have taken the binding ropes. There is no evidence for a retaining strap at the shoulder, but there is a 'fishtail' plate held by three nails just behind it on top of the carriage. This may have held an iron plate over the back of the bed (see *reconstructing a port piece*, below), or may simply have taken a ring, as is seen in similar guns recovered from the Anholt wreck in Denmark (presumed fifteenth century, guns lifted 1847) now in the Royal Danish Arsenal. This feature is also present

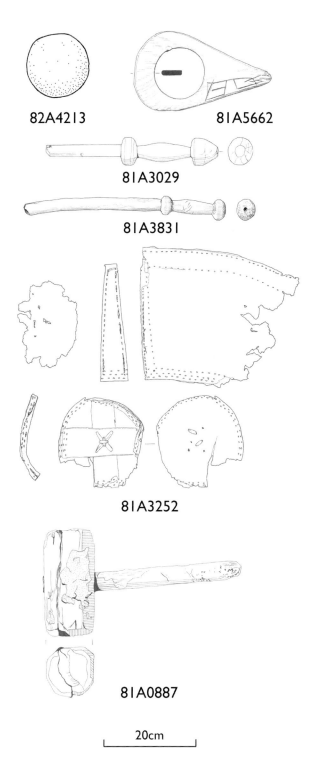

*Figure 3.29 Port piece 81A3001: objects associated with the gun: 82A4123 stone shot; 81A5662 shot gauge; 81A3029, 3831 linstocks; 81A3252 leather bucket; 81A0887 mallet*

asymmetrical, and the bands themselves vary in width (55–73mm) and are 13mm thick. Hoop width is more consistent, *c*. 28mm, with single hoops 32mm thick and flankers 25mm.

The chamber was found on the deck concreted to the north wheel. It is incomplete and has a length of

on the Woolwich and National Maritime Museum port pieces (see below). There is a slot from the front of the carriage to within 280mm of the shoulder (Fig. 3.26, bottom) and two rebated, 30mm diameter horizontal transom bolt holes 1000mm and 2000 mm from the shoulder. The 30mm diameter rebated axle bolt holes are 2250mm from the back of the carriage on either side of the gun trough. Evidence of possible ring bolts are visible behind the shoulder, situated half way up the carriage and 450mm from the back possibly for holding breeching ropes. Rope was noted in the divelogs (81/8/148) and also around one of the wheels (81/8/179).

The spoked wheels (81A5688 north, 81A5689 south) have diameters of 1100mm and 950mm and treads of 120mm and 100mm respectively (Fig. 3.27). These are held with iron lynch pins. Both wheels have broad arrows spreading outwards from the hub. The single ash axle is fragmentary (81A0891, A3765, A3998; Fig. 3.28). The thickness of the axle in the centre is 405mm and the length across the underside (where the axle touched the carriage) 435mm. The elevating post (81A2088) is also incomplete, with a surviving length of 890mm. It is 110mm thick and 75mm wide with eleven staggered holes. It is marked with a 'Z' or 'N'.

A considerable amount of gun furniture was found in the proximity of this gun, including fifteen stone shot. Although of varying sizes, the majority were finished (see Chapter 5). The *in situ* shot was 172mm. A linstock (81A3831) was concreted to the north wheel and a second was found beside the gun (81A3029) as was a shot gauge (81A5662; 100mm hole), leather bucket (81A3252) and a mallet (81A0887); possibly for securing the forelock (Fig. 3.29). Another shot gauge, with a hole of 175mm, was found nearby in M9 (82A4213).

## 71A0169 Incomplete barrel and chamber

*Location*: ?M11 port or starboard. Chamber containing tampion and powder. Stone shot in barrel.
*Recovered*: 1971, outboard stern, not certain

*Statistics*
Length: barrel: 1300mm; breech: 500mm; bore: 190mm

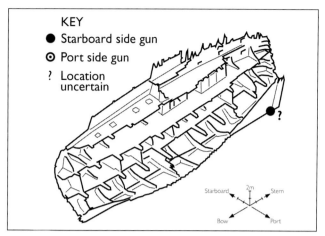

The location of this is uncertain and it could have been from either M11 port or M11 starboard, from the Upper deck in the stern, or the Main deck on either side of the rudder. The barrel is incomplete and the bore is one of the largest. The recovered gun muzzle has broken off. McKee describes it as having an 8in bore containing a 6½in shot with a 22½in powder chamber (1982, 87, 88). There is no mention of any carriage fragments (Fig. 3.30).

The barrel has an external diameter of 290mm and internal diameter of 190mm. The inner tube is made up of seven staves, curved to form the tube (Fig. 3.31). Radiography clearly demonstrated gaps between the staves with crisp edges. These are visible at the back of the barrel. Estimated stave widths are 100mm (6) and 90–100mm (1) and they are *c.* 12mm thick. The eleven bands (*c.* 70mm width) and eighteen hoops (10–30mm width). These formed the outer layer laid directly over the staves. Hoop thickness is estimated at 25mm. The sequencing of single hoops, bands and hoop sets is clearly seen in Figure 3.30. The muzzle is incomplete with a substantial portion missing as are the lifting rings and their hoop. Shot of 184mm was recovered from inside the bore.

The chamber was sealed with a wooden tampion of 100mm diameter and 50mm thickness and contained

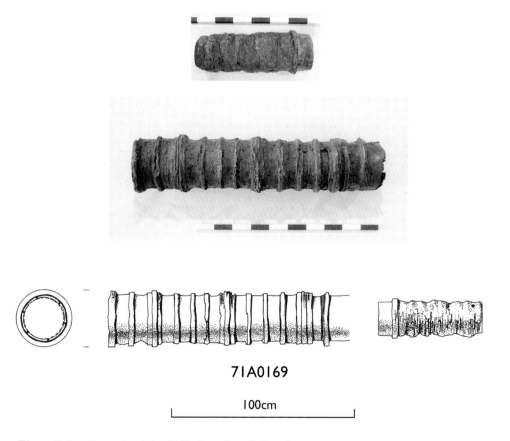

*Figure 3.30 Port piece 71A0169: barrel and chamber*

*Figure 3.31 Port piece 71A0169: barrel construction, staves at breech end of barrel*

powder (71A0169/5, not analysed). The chamber neck appears to be parallel-sided, as on the National Maritime Museum example, but is 5mm larger at the shoulder than the mouth. The touch hole is 15mm in diameter and 50mm from the end of the chamber, the breech plug must therefore be less than this. The neck of the chamber is 40mm less than the barrel and the staves are not visible; these are probably welded. The chamber to barrel fit is extremely loose. The chamber appears very even, as though made of thirteen bands only, with no hoops. The only defined hoop is at the end of the neck and there is no evidence for lifting rings. The staves at the back of the chamber are visible. These are 25mm wide.

*National Maritime Museum KPT0017*
*Incomplete barrel with breech and carriage*

> *Location*: not certain ?U10/11. Gun with breech, wooden forelock and iron wedge *in situ* on carriage. Stone shot concreted within barrel
> *Recovered*: John Deane, 1840
> *Current location*: Mary Rose Trust
>
> *Statistics*
> Length:  barrel: 1500mm; breech: 600mm; bore: 160mm

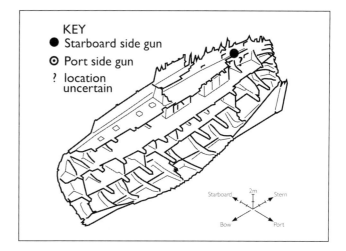

The barrel is incomplete and has broken into a number of pieces. There is an additional chamber and eroded truck wheel. This is the gun shown in a watercolour by S. Finney commissioned by the Royal Institution of Cornwall, which is captioned:

> '*A drawing of a Gun formed of Iron Bars, hooped round, with a moveable Chamber, let into a solid block of Elm, which was recovered by Mr John Deane in the Summer of 1840 from the wreck of the* Mary Rose *which sank at Spithead in the Year 1545*'. (Bevan 1996, 146, 167, pl. 24)

It appears that this is one of the guns referred to in the *Portsmouth Independent* (September 1840) as one of two on elm beds. These were brought up over the same week as the culverin (79A1278) which may be the port side gun from M3. Over the same period an anchor, pump and stump of the main mast were brought up from the midships area (Bevan 1996, 169). The shortness of the bed (3000mm) favours a position either at the extreme bow or stern of the vessel, where the ship narrows. The *Independent* further states:

> '*It may be interesting to the naval architect of the present day to learn that so large a ship as the* Mary Rose *was clinch built, as this mode of building is now entirely confined to boats and small craft*'.

This suggests that work was around the Upper deck, possibly U10/11 (starboard) where clinker planking can clearly be seen. An eroded solid wheel is currently with the gun, but this and the second chamber (without lifting rings), are not shown in the illustration. A photograph taken of this gun whilst on display in the Museum of the United Service Institution (Liscoe, pers. comm., 1986) shows the gun as it was at that time, on its bed, with a wheel positioned close to the front of the bed on the left hand side. It appears to be in parts, possibly assembled for display. A portion of the gun with lifting rings is positioned over the end of the carriage and looks out of place. The separate breech lies over the back of the gun and the breech chamber wedge is *in situ*. A longbow and swivel gun are also shown.

The gun, reconstructed in Figure 3.32, is incomplete and fragmented (Fig. 3.33). The internal

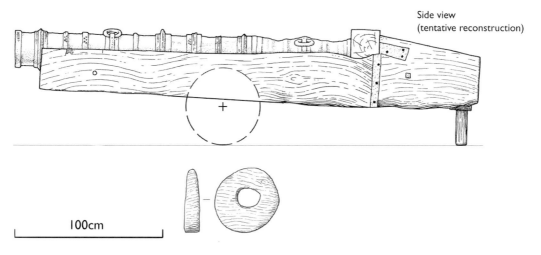

*Figure 3.32 Port piece NMM KPT0017: reconstruction drawing of gun on carriage*

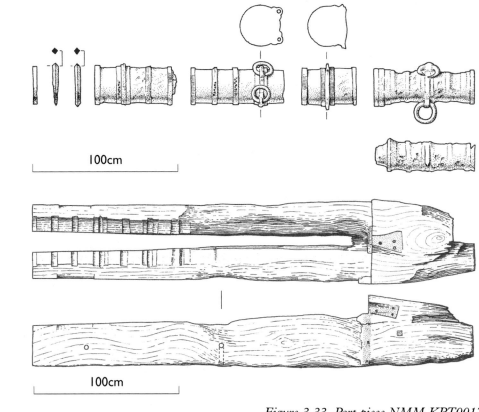

*Figure 3.33 Port piece NMM KPT0017: above) barrel, chamber and carriage; below) forelock and wheel*

diameter varies between 160mm and 180mm. The fragments consist of three barrel portions (560mm; 620mm; 400mm). The longest section has a pair of lifting rings set into a hoop. Nine of the 21 bands are present. There are nine visible staves forming the inner tube, each *c.* 80mm wide and 15mm thick. The bands are 20–25mm thick and of varying widths (*c.* 112mm). The hoops are all 25mm wide and 38mm thick. There is a series of triangular and linear impressions on eight of the hoops.

The breech chamber (length 680mm, external diameter at back of 230mm) with lifting rings has broken away from the gun, and has two of the barrel hoops attached to it. This comprises four bands separated by five hoops with a pair of 100mm diameter lifting rings of 30mm diameter section on the third hoop from the back. The neck has a length of 80mm and internal diameter of 110mm. The spare chamber is of a similar length and construction, although less defined. The neck ring on the chamber is pronounced and there is no evidence for lifting rings. The chamber is smaller at the back, with an external diameter of 160–180mm.

The remains of an eroded rebate around the top of the carriage 180mm from the front may have taken the wooling. An iron stained band, 50mm wide, is evidence for a retaining strap at the shoulder. Three 18mm holes equally spaced along the length of the strap may have held retaining spikes. On the top of the carriage, just behind the shoulder, is a 'fishtail' plate held by three nails (length 200mm, width 80mm). This may have retained an iron plate over the back of the bed (see *reconstructing a port piece*), or taken a ring. The slot, which nearly divides the bed, runs from the front of the carriage to within 150mm of the shoulder, with a tapering width of 130–70mm. Two 30mm diameter, holes for horizontal transom bolts are visible at 1000mm and 1900mm from the shoulder, the first of which has a vertical slot beneath, possibly the end of the axle bolt hole which may not penetrate the full height of the carriage. There is a 30mm square rebate 450mm from the back of the carriage on each side face, probably for securing rings.

The maximum height of the carriage at the shoulder is 450mm and the width 560mm, at the back about 280 x 360mm but it is incomplete. It has a length of 3000mm. The rectangular hole for the elevating post is positioned 155mm from the back of the carriage (this area is eroded). The chamber trough appears to be 720–1200mm in length, rebates are obvious from 1200mm, but the extra length of the trough could be surface erosion. The one truck shown with this gun is 400mm in diameter and 100mm thick (Fig. 3.33). The early illustration (Bevan 1996, pl. 24) clearly shows a small breech in position together with a forelock. The shortness of the carriage may be as a response to a restriction of space within a specific gun bay. Possible locations include the Main deck in the stern facing aft; a port side stern gun; in M1 (canted forward); or either of the U10, U11 gunports in the stern. Erosion of the back of the carriage suggests a starboard location for the gun.

*Royal Artillery Institution Woolwich Class 1/10*
*Barrel with chamber and forelock on carriage*

*Location*: unknown. Possibly M5 port side
*Recovered*: Deanes, 1836

*Statistics*
Length:  barrel: 2180mm; breech: 650mm;
 bore: 200mm; 14.60 calibres

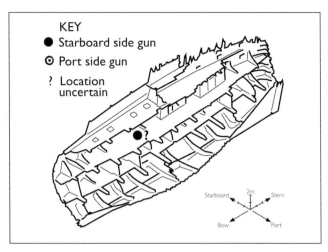

This was one of the first guns recovered from the *Mary Rose*. A pencil sketch by R.A.C. Ubsdell is accompanied by this description:

> '*A piece of ancient ordnance made of 11 iron bars measuring in circumference 2 ft 10 in secured together by stout iron hoops. Length 9 ft 7 in and with the carriage 13 ft and must have taken a shot equal to 42 lb of the present day. The iron is very much corroded. The carriage is rough hewn elm tree in which the gun is fixed in a manner that a musket is secured to the stock and is in the highest state of preservation, there is a groove under the carriage which makes it probably that it moved on a kind of slide or pivot. Part of another gun of the same sort which broke in their endeavours to raise it from the bottom when a STONE shot fell from it and it is preserved by Mr Deane*'.

Several contemporary representations exist of this gun, all differ slightly in detail. A poster titled '*Representation of guns taken up at Spithead by Messrs John Deane and William Edwards ... landed in H M Gunwarf, Portsmouth, 15th August 1836*' states the following below the gun: '*Iron gun manufactured with wrought iron bars with 33 iron hoops found in a solid bed of elm, length 9 foot 6 inches.*'

The best of these two representations (Rule 1982, 43; Bevan 1996, pl. 19) is within the Portsmouth City Museums and Record Services collection. It shows the metal fishtail plate on top of the carriage behind the forlock and the wrought iron strengthening band aligned vertically behind it. Most importantly, the only illustration of a ring is clearly depicted on the right hand side of the back of the carriage (Bevan 1996, pl. 19). The plan view shows this on the right, but not on the left. In pencil below are the words: '*Mary Rose, sunk Spithead in the reign of Henry VIII, 291 years underwater, now in repository Woolwich*'.

More detailed accounts of the raising of this gun can be found in Bevan (*op. cit.*, 110). It appears that, in the absence of John Deane, William Edwards had dived the *Mary Rose*, recovering a bronze gun on 8 August 1836, and an iron gun on the night of the 9th to avoid sharing any reward with the Gosport team who had directed Deane and himself to the site. The iron gun is described

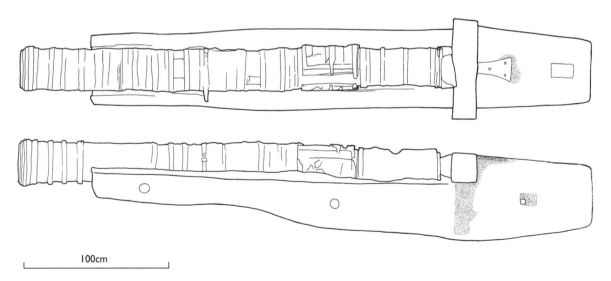

*Figure 3.34 Woolwich 1/10 port piece: gun on carriage*

*Figure 3.35 Woolwich 1/10 port piece on display in the former Woolwich Rotunda*

in the *Hampshire Telegraph*, 15 August 1836 (Bevan 1996, 111):

> '...*The two latter are objects of great curiosity, and must be of great age; the entire one is 14 feet long: they are constructed of thin iron bars; both were loaded with a stone shot, of about the size of a 32-pounder; the entire one rests on a wooden stock, the half of its circumference being embedded therein its whole length, and it had evidently moved on a slide*'.

The report further suggests that this gun was found in a similar location to the two bronze guns also recovered, a cannon (79A1276, probably M6 port side) and a demi culverin (79A1278, possibly C1 port). All were: '*discovered on the same spot, resting on some wreck, which was so completely buried in the sand that the diver could find nothing to which he could afix a rope*'. Certainly Edwards was working in the midships area when raising the two bronze guns. This might suggest that the iron gun was the closest one to these, which would be M5 or M8 port. The length of the bed favours M5 port, pairing 81A2604.

The gun (Fig. 3.34) was, until recently, displayed in the Woolwich Rotunda (Fig. 3.35). Chamber, barrel and forelock are *in situ*, but the gun is badly corroded and sections of the staves are visible. The external diameter at the mouth is 205mm and at the breech, 240mm. Clearly the gun is of similar construction to the other port pieces and, according to Ubsdell, there are eleven '*thin*' staves overlain with 33 '*hoops*' (it is not clear whether or not these are both hoops and bands on both the barrel and the breech). These are arrayed in sets of triple and single hoops flanked by bands but the exact orientation has not been recorded and estimates of width are. c. 90mm (bands) and c. 30mm (hoops). As with the M5 and M4 guns (81A2604, 81A2650) there appear to be four single hoops after the last set of three towards the muzzle (Fig. 3.34). Most of these overshoot the carriage. The gun may have had a single set of lifting lugs. The remains of the right lug is present 980mm from the breech and 1189mm from the mouth. The watercolour shows that the staves were visible in 1836 and that the lifting rings on both the chamber and the barrel were missing. The nature of the corrosion suggests that the barrel was uppermost (and the carriage protected). There are no wheels shown in any of the illustrations. The chamber looks very plain, alternating bands and hoops, but its size matches that of the gun, rather than appearing smaller as with the National Maritime Museum example and 79A0169. The chamber has a maximum external diameter at the back of 205mm.

The 3400mm long carriage is 400mm square at the back, expanding to 560 x 550mm at the junction with the shoulder. There is a slot from the front of the carriage, as with the other port pieces, but its length is undetermined. Two horizontal transom bolts are shown in Ubsdell's illustration, but holes for three (50mm diameter) are shown on the archaeological drawings, 980mm, 1850mm and 2220mm from the shoulder. There is no sign of axle attachments. A single ring is shown in the early watercolours on the right side of the carriage at the back, its position indicated by a square, rebated 25mm hole on the drawings (Fig. 3.34). This may have held breeching ropes. The rectangular hole for the elevating post is 140mm from the back of the carriage. The chamber trough is about 850mm long and the rest of the carriage is relieved to accept the hoops and bands on the gun, which extends 450mm beyond the carriage. The carriage has been reduced in width forward of the lifting ring lug of the axle bolt holes by about 50mm. There is no evidence for axle, wheels or elevating post. Evidence of an eroded rebate around the top of the carriage (1750mm from the back) just forward of the breech junction may have taken the wooling. A 120mm wide corroded reinforcing band around the back of the carriage is evident.

This gun was the subject of a photogrammetric study undertaken by the City University (London). The detailed plates show a broken shot lying on the back of the carriage and two breech chambers, one eroded and without lifting rings, and the second seemingly perfect with a single set of lifting rings extant and mounted centrally on a small chamber with long parallel neck, not unlike 90A0127 in appearance. Although of a type found on the *Mary Rose*, it is not known whether this is indeed from the ship. Much material was accessed by Woolwich and, indeed, studied at the Mary Rose Trust, some of this may be derived from the wreck (Woolwich Catalogue Miscellaneous, Class 1: 56; Kaestlin 1963, 5).

Metallurgical examination of an iron hoop from the extreme breech end of the gun (B. Mellors, Materials Quality Test Centre, Royal Arsenal West, Woolwich) indicated alpha iron together with Iron Carbide ($Fe_3C$) and areas of black oxide inclusions. In section, the iron base material was found to have a carbon content of 0.1–0.2% and displayed a fine grain. This material was clean and relatively free from non-metallics, although there were bands with significant slag content (details in archive). A Vickers hardness test indicated that the maximum or ultimate tensile strength of the iron would be 20–25 tons per square inch, equivalent of a mild steel. Chemical analysis of the transverse section (Bragg Laboratory, Sheffield) confirmed the presence of *c.* 0.1% carbon (with sulphur 0.015%; silicon <0.1%; manganese <0.05%; chromium <0.1% and various trace elements). The results of the chemical analysis show that the iron was remarkably pure, a small amount of carbon was virtually the only alloying element present. The low silicon and manganese content would have made forge welding a very skilled operation.

Taken together, these analyses suggest that the hoop was hammer forged to shape from a bar with the ends scarfed to facilitate joining by forge welding. The mixed grain size shown in the microsection indicated that, after forge welding, the hoop was not reheated above *c.* 830–860°C. The degree of corrosion varied markedly: the lower portion, shielded by contact with the wooden bed and breech chamber, had suffered relatively little metal loss whereas the upper part had almost completely dissolved following exposure to variable tidal flow and scouring. The shielded regions would have been protected by the relative stability of the stagnant corrosion layer saturated with iron salts. It is estimated that the original hoop dimensions were 210mm internal diameter, 292mm external diameter and 32mm thick.

## Royal Armouries RA XIX.1 (80A2004) Hoops, stave and bed fragments (80A1954)

*Location*: unknown
*Recovered*: 1840, purchased by the Board of Ordnance.
On display in Fort Nelson Museum of Artillery, Fareham

*Statistics*
Estimated length: 1900mm

This consists of 45 hoops and several attached stave fragments (Fig. 3.36) and a portion of carriage (80A1954). There is currently not enough structure to indicate bore. When recovered in 1840 it was on its bed with chamber *in situ*. The length was given as 1892mm with a bore of 152mm. It is listed in Blackmore (1976, 55) as having been recovered in 1840 and purchased by the Board of Ordnance in the same year together with .167 (culverin 79A1278), possibly the port side pair to the M3 starboard culverin. Also purchased was an additional chamber. The entry reads:

'*Of built- up construction, being formed of flat iron bars with iron hoops passed over them and shrunk on. The iron bars forming the bore have disappeared for over half the length, leaving the hoops unsupported. The breech and an unknown length from the muzzle end are missing. One hoop with eyes for two rings remains, one being preserved.*' (Blackmore 1976, 55)

In 1911 the gun was shown together with a breech, a central portion intact, and unsupported rings. The chamber may be incorrectly assigned to XIX.163, loaned to Portsmouth City Museums and Art Gallery in 1978. This item has been recorded as 80A2001 and is described as a separate chamber below.

This is possibly one of the number of items salvaged in the autumn of 1840, following the placing of explosive charges on the site. It may be one of the:

'*four iron 32 prs from 6 to 8 ft long, constructed of wrought iron bars and hoops, containing powder and stone shot, these are very antique and have a hole* [the Bore] *completely through, although they have no appearance of being broken, but a perfect muzzle at each end*'. (Bevan 1996, 168)

Two of these are described as being on elm beds (*Portsmouth Independent*, 10 October 1840 (*ibid.*, 169 – see above).

The existing carriage (Fig. 3.36, below; length 1970mm) consists of the rear portion hollowed out to accept the breech chamber. Although eroded, there is detail to suggest that it had a retaining strap behind the forelock. The shoulder is eroded and there is no clear indication of a fishtail plate. The rear of the bed is 320 mm in height with a width of 370mm. The back is missing centrally behind the slot for the elevating post, 80mm from the back of the carriage. The chamber recess is 270mm in width but of undetermined length. The (incomplete) height of the carriage at the shoulder is 400mm and the width 570mm. There is a 30mm square hole centrally placed along the height of the carriage 400mm behind the back, possibly for a ring bolt. A 25mm hole for a transverse transom bolt is positioned 1380mm from the back of the carriage. The slot in the bottom of the bed stops 200mm in front of the shoulder and has a width of 30mm. Just before the break it narrows with an inward sweep of about 50mm as does 81A3001 and the Woolwich carriage (Class 1:10). Blackmore (1976) described the carriage thus:

'*The bed, possibly that belonging to the gun, is formed from a heavy piece of elm of square section of which only the rear portion has been preserved and this has split longitudinally. The upper part is hollowed out to take the barrel and its chamber and has a vertical drainage slot. An opening at the rear of the bed was possibly originally a slot for a wooden upright to prevent recoil. An unknown length, probably not less than five feet, is missing from the front and some six inches from the rear. There are three rectangular holes on either side which probably held ringbolts*'

*Figure 3.36 Port piece RA XIX.1 (80A2004): above) barrel hoops; below) carriage fragment*

*Royal Armouries XIX. 2 (80A2003) Incomplete barrel, stone shot* in situ

> *Location*: unknown
> *Recovered*: 1840, purchased by the Board of Ordnance.
> Currently at Fort Nelson Museum of Artillery, Fareham
>
> *Statistics*
> Length: 1880mm; bore: 152–160mm

The breech end is intact and the gun is broken towards the muzzle to show a single stave protruding by 110mm (Fig. 3.37). The total wall thickness (staves and reinforcing hoops) is 50mm at the breech. At the broken end it is possible to count seven inner staves. The remains of a hoop carrying the lifting lugs is situated 1350mm from the breech. The outer surface consists of twelve bands of varying widths (60–80mm) and seven sets of triple hoop sequences with only six individual hoops. Five hoop sets have a central hoop of 30mm width flanked by 35mm hoops. The single hoops are 25–35mm wide. From the breech there is a descending set of hoops from the junction with the chamber (30mm, 40mm, 35mm wide). There is a stone shot in the bore held within concretion. Two chambers are listed as having been acquired at the same time as this gun, one with it (part of XIX.2, length 660mm), and XIX.5 (623mm). The former may be incorrectly assigned to XIX.163, and has been given number 80A2000. It is described below under separate chambers.

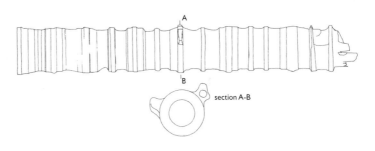

*Figure 3.37 Port piece RA XIX.2 (80A2003): barrel*

## Chambers

A total of 38 breech chambers for carriage mounted guns can be assigned to the *Mary Rose* collection, 25 raised since the 1970s and thirteen from the nineteenth century salvage. Not all of these early chambers have proof of provenance, but visually many are similar enough to be included. We do not have measurements for all of these. The total collection represents nearly 80% of the expected chambers, 48 for the 24 carriage mounted guns listed (Figs 3.39–40). Their distribution is shown in Figure 3.38.

There are 25 larger breeches (Table 3.4) and, if these are all for port pieces, we have more than two chambers for each of the twelve guns listed. Sixteen of these were raised by the Mary Rose Trust and nine by Deane *et al.* The apparently inexplicable differences in chamber size (and therefore weight) may relate to type of shot (lantern or solid stone) and, therefore, the volume of powder required, or possibly the length of the barrel. The large size of chamber found with the *in situ* sling (81A0645) means that size alone cannot be used to differentiate chambers for different types of gun. It is extremely difficult with a corroded and possibly incomplete chamber to ascertain the tube length, correct diameter and, hence, volume. It is not known how much powder was used, but in the seventeenth century Norton (*ibid.*, 1628, 58) advocated up to half the weight of shot in powder. Earlier practices (designed for serpentine powder only) suggest only partly filling the chamber (Kramer 2001, 22). If this was still practised, using chamber volume to suggest powder charge would be invalid. Recovered port piece bores are in the range 150–200mm (6–8in). Corresponding shot weights should therefore range from 4.38kg (9.7lb) at 144mm to 9.9kg (21.83lb) at 193mm. Those guns available for study together with their own chambers include the guns at Woolwich and the National

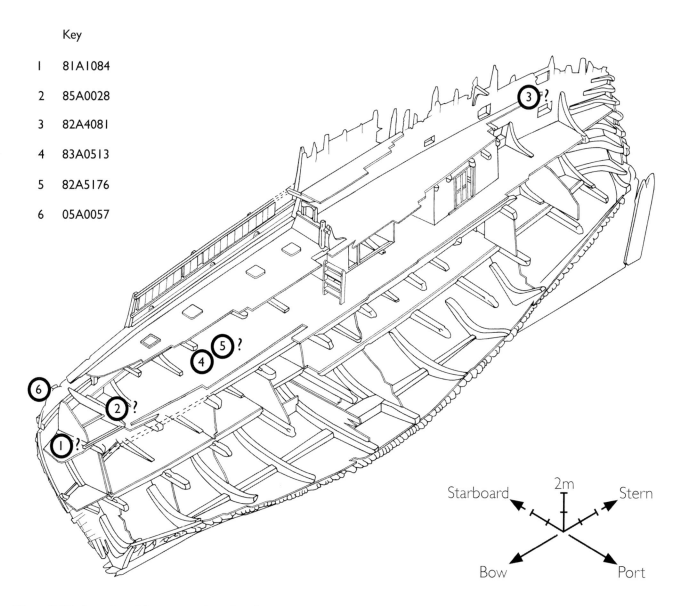

Key

1　81A1084
2　85A0028
3　82A4081
4　83A0513
5　82A5176
6　05A0057

*Figure 3.38 Location of loose port piece chambers*

Maritime Museum and *Mary Rose* guns 81A2650, 81A2604 and 71A0169. Many chambers are still soaking so have not been studied in detail.

There are 22 chambers not directly attributable to particular guns. Although divelogs suggest that there were spare chambers beside both the M4 and M5 port pieces, only one is recorded (83A0513, from M4). There are four chambers from unknown locations (81A4890, 82A5097, 82A5172, 82A5176), recovered during the 'clean up' of the timber storage areas following the 1981 and 1982 seasons. Two more were recovered from the Starboard scour (SS2) during post-excavation monitoring (85A0027, 85A0028), and a further chamber from cleaning up a seabed storage area in 2003 (03A0070) and one from the starboard side of the stem at the extreme bow in 2005 (05A0057). Chambers were taken from beside guns and put into storage areas on the seabed but, without any extant tags, identification is problematic. Another chamber was 'found' in the Ship Hall in 1990 (90A0001), possibly recovered from inside the ship during cleaning. Two chambers, possibly from concretions found in M1 and M2 respectively (Liscoe, divelog 81/2/22) are too large for any of the carriages recovered (81A1084, 85A0028, Figs 3.40 and 3.41).

Matching chambers to barrels based on size and form is extremely difficult as the fit can be altered in length with the use of differently sized wooden forelocks. The two chambers in M4, for instance, were 640mm (*in situ*) and 670mm (spare, 83A0513, still in concretion, so possibly smaller) with the *in situ* forelock having a width of 140mm and the two possible spares being 170mm and 210mm (80A1723, 81A1229). The space required to accommodate these is 780mm (*in situ*) and 840–880mm for the spare chamber and each forelock and the space available is about 880mm maximum. The *in situ* forelock was secured with one or two iron wedges. Crucial dimensions include the chamber neck to rear of barrel and obviously the fit of the barrel/chamber/forelock and wedges to the carriage. The neck to barrel fit is never tight, so trying to fit chamber to hall is not proof of association, although it can exclude certain matches. It seems also to be the case that some chambers were indeed much smaller in diameter than the gun barrel, with perhaps only some of the hoops originally conforming to the size of the barrel. The corrosion of both chamber and barrel make this even more difficult. This size variation can be seen in some of the illustrations of the nineteenth century recoveries, such as the National Maritime Museum gun (Bevan 1996, pl. 24). This is not always the case, for example the Woolwich gun clearly shows that the chamber and barrel are of a similar size (*ibid.*, pl. 19). Divelogs show where concreted items were, but these can rarely be used to identify a particular item. Many concretions were taken out of position and placed in a centralised storage area on the seabed without numbers.

The distribution of these chambers together with other port piece fragments can be seen in Figure 3.62, below.

**Construction**

The chambers all seem to be made of an inner tube of staves in the same manner as the barrels, although some staves are slightly curved (Fig. 3.39). This tube is usually covered by wide bands alternating with narrower hoops or by a layer of abutting bands. The hoops are generally higher than the bands, but not by a great deal. Although there are instances of triple hoop sequences, this feature appears less in chamber construction than in barrels. A defining constructional feature appears to be the number of pairs of lifting rings which varies between one and three (Fig. 3.39e).

The neck of the chamber is reinforced either by welding of the inner staves or, possibly, by the addition of a collar at this position. Some chambers have two layers of staves, for instance, 82A0792 where outer staves were staggered to cover the joins of the inner ones (as shown in Fig. 3.39d). This double layer exhibits itself in the form of thick staves, the corrosion of the neck suggesting several layers, or by a radiograph which has too many 'blurred' lines which cannot be understood, or by a small bore in a large iron object.

Where triple hoop sequences occur they seem to be in the centre of chambers with two sets of lifting rings (81A2650, 81A2604, 82A0792, possibly 81A4890). Where the lifting rings occur towards the ends of the chambers, they are on single hoops. In two instances, 82A4081 and the National Maritime Museum *in situ* chamber, the triple hoop sequence is in the middle supporting only one set of lifting rings. Sets of hoops are generally found at each end, stepping up towards the neck and the back. This can convey a 'bowed' appearance to some chambers (82A5176). In several instances the chamber looks as though there is though the outer layer is formed of hoops only, butting in a single layer (82A5176, 80A2005 RA, E31/104).

The lifting lugs seem always to occur in opposed pairs rather than as single central lugs. The touch hole (where visible) is positioned on the top of the cylinder, the handles to each side. The rings are set into lugs formed within a hoop (see *manufacture of port piece* below). These occur as single sets (82A4081, 83A0513, National Maritime Museum, Ironbridge Gorge Museum Trust), in pairs (81A2650, 81A2604, 82A0792, 82A5176, 80A2005, E31/104, 90A0001), or in sets of three (81A1084, 85A0028). There are seven examples where it is difficult to see any clear indication of original detail (Woolwich, 71A0169, 81A4890, 81A5097, 82A5172, 85A0027, 80A2000, 80A2002; see Fig. 3.40).

Dimensions of the chambers are summarised in Table 3.4, above. Other than the two extremely long examples (81A1084, 85A0028, both over 1250mm), those found with guns are 570mm (71A0169) to

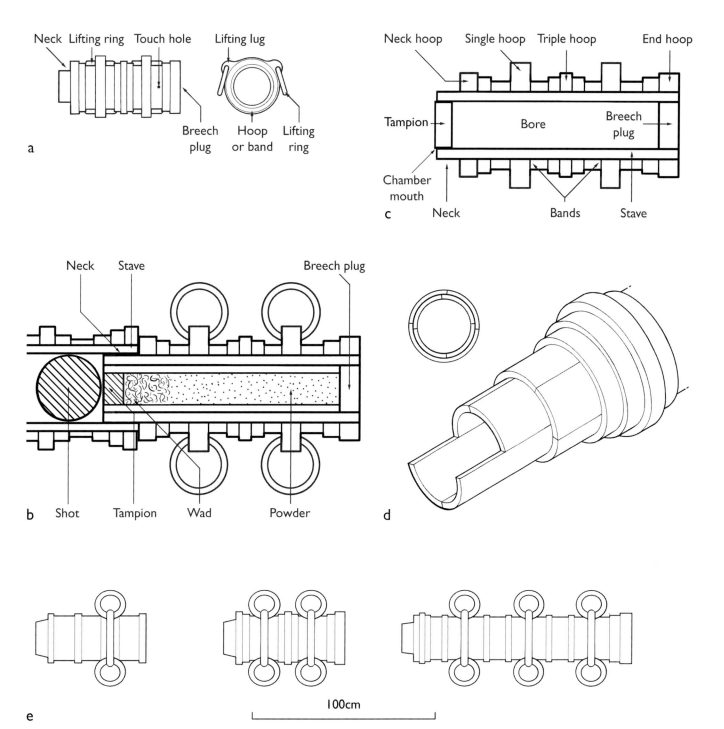

*Figure 3.39 Port piece chambers, manufacturing sequence and named parts: a) external features; b) internal features and sequence of charge; c) section; d) construction sequence; e) lifing rings on varying sizes of chamber*

725mm (81A2650) long with bores of 85–115mm (81A2650, 81A2604). The greatest diameters occur at the neck hoop (chamber/hall junction) over the lifting rings and the end hoop at the back. The internal diameter at the mouth also varies (70–115mm). Interestingly, the very long chambers have a small bore, 88mm and 95mm. This can only be achieved without changing the external profile by having a chamber of greater length but with a smaller bore and thicker walls. This prompts the question as to whether these should be assigned to the sling class of gun, reinforced enough (and long enough) to take the greater pressure of an increased powder charge for longer range. Alternatively it is possible that these were used with a very short barrel with a large bore, to some extent resembling the profile of a bombard. The external diameter of the

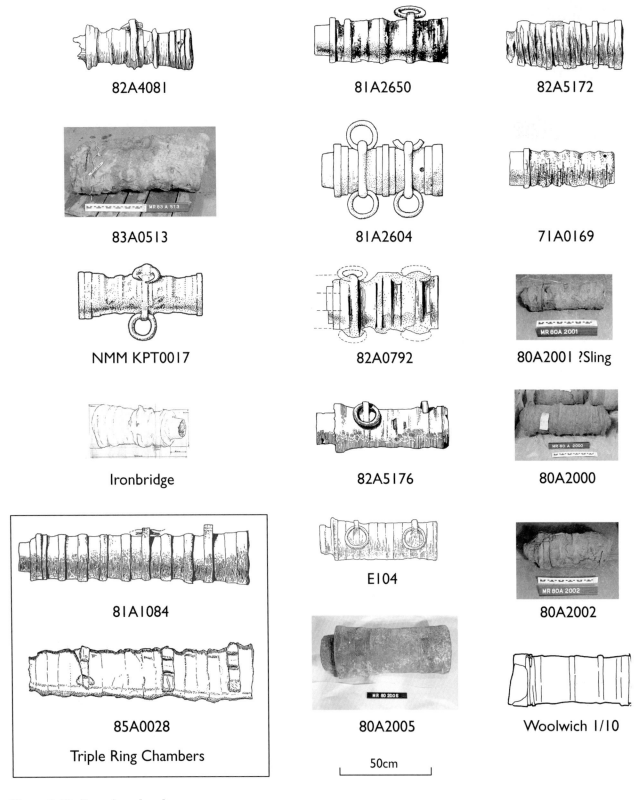

*Figure 3.40 Port piece chambers*

largest chambers is comparable with other port piece chambers. The troughs in the carriage beds have to accommodate these diameters so the external diameter remains fairly constant. Differing lengths of chamber can be accommodated by wedging, including the use of wider forelocks and the orientation and number of metal wedges. Chamber size may also be related to whether the shot is solid round stone, or lantern.

*Figure 3.41 Port piece chambers in concretion: a) 81A1084; b) 85A0028*

Although lantern shot is heavier, it is assumed that this was used as anti-personnel shot at closer range than the lighter round stone shot (see firing a port piece, below).

The smaller chambers within the assemblage are only about twice the length of the external diameter at the back, whereas the larger ones are three or more times this diameter. Volume is difficult to assess with any certainty, but using the formula $\pi r^2 l$ provides space equivalent to 17lb powder for the longest chamber 81A1084, 7lb 4oz for 81A2650 (M4), 11lb 6oz for 81A2604 (M5) and 6lb 4oz for 82A0792 (M8). (Length is obtained is by subtracting 140mm from the total chamber length, an average of 50mm for the neck and 90 mm for the breech plug). This also assumes a parallel internal bore and is inherently unreliable. The largest stone shot quoted by Norton is 7in, weighing 17lb (over 178mm/7.7kg). Our stone shot of this diameter range in weight from 16–20lb. The chamber for this is listed at 24in (609mm) and powder charge 9lb (4.1kg) (Norton 1628, 58). The closest complete chamber to this is 81A2604 at 650mm (25½in) with an estimated complete volume of this is 11.6lb (5.2kg). Shot for this gun is estimated at between 185mm and 190mm in diameter (7¼–7½). Recovered shot of this size ranges in weight between about 17lb and 23lb (7.7–10.5kg).

**81A1084 Large chamber**
*Location*: unknown
*Recovered*: uncertain; possibly removed from M1 to collection area NE of bow in July 1981

*Statistics*
Length: 1200mm; bore: 88mm

This chamber is similar to 85A0028 (Figs 3.40, 3.41a); these being the two largest known breech chambers in England. It has three muzzle hoops with alternating 70mm wide bands and low, 40–50mm wide, hoops. The diameter over the bands is 225mm and over the hoops 250mm. Two sets of lifting lugs are still present, 630mm and 975mm from the shoulder. The design suggests three equally spaced lugs, as on 85A0028. The radiograph shows eight stave lines, possibly in two layers. At the neck corrosion also suggests two layers, inner possibly 12mm thick and outer possibly 17mm, full thickness 32mm. The inner layer may consist of three equally sized staves. It is too long for any carriage recovered though the external neck diameter is similar to other port piece chambers.

**85A0028 Large chamber**
*Location*: unknown. Tampion (85A0031), wad (85A0032), powder (89S0024–25), 'rope cushion' (85A0030) from back
*Recovered*: possibly moved from M2 to Starboard scour east of SS2/3 in April 1981 (81/2/22). Found during site survey in 1985

*Statistics*
Length: 1200mm; bore: 95mm

This was found in concretion (Figs 3.40, 3.41b). The radiograph suggests a large chamber with staves forming the neck, possibly in two layers. No stave junctions can be seen beneath the 64mm wide hoops which seem to abut each other without any definite bands. The neck hoop is 38mm wide as is the last but one. The wall thickness at the neck is 25–38mm. Excavation revealed a compact, hard outer layer 50–100mm thick which peeled away easily. Beneath this was a black sulphurous layer and just beneath this a hard dry layer covering the metal. Magnetic stimuli varied between non-existent and excellent. The surface was corroded in parts, often washing away during cleaning. A 10% solution of sodium hydroxide was used to curtail rusting during the excavation. Negative casts were revealed of lug and ring sets at 200mm, 600mm, 905mm and 1005mm from the shoulder. When excavated from concretion three sets of lifting lugs with the vestigial remains of the rings were present. The chamber had a 25mm thick, 80mm diameter, tampion in situ (85A0031) domed to the inside of the chamber. Its inner face was covered in a black powdery substance and wadding (85A0032). The contents were sampled (89S0024–9). The back of the chamber had a coiled rope 'cushion' (85A0030) conforming to its diameter beneath what appeared to be an extra metal plate. The loading sequence indicated was powder, wad and tampion.

**82A4081 Chamber**
*Location*: unknown, possibly U11
*Recovered*: during site clearance following the raising of the hull

*Statistics*
Length: 700mm; bore: 120mm

The outer layer is formed of alternating bands and hoops, with one central triple set of hoops supporting a pair of lifting rings, of which one survives. Towards each end there is a sequence of hoops increasing in size. The sequence can be seen in Figure 3.40. The bands are 60–70mm wide and the hoops 25–35 mm. The hoops are raised above the staves by 40mm and the bands by 15mm. The radiograph suggests a stave thickness of 20mm. There appear to be at least three longitudinal joins. This is similar in form to the National Maritime and Ironbridge chambers.

**81A4890 Chamber**
*Location*: unknown
*Recovered*: excavated 1981

*Statistics*
Length: 360mm; bore: 85mm

This small chamber (not illustrated) is possibly incomplete and its surface is eroded. The radiograph shows a broken neck and staves but it is unclear as to how many layers are present. The corrosion suggests two layers with three inner staves, 7mm thick, and an outer layer of hoops and bands. The only really visible elements are the three centrally placed hoops, with the middle one upstanding. Towards the back there is a single band 70mm wide with two or three hoops forming the end. The hoops are *c*. 30mm wide. There appears to be a band forward of the hoop set, but this is not clear. It looks similar to 82A0792 but there are no obvious lifting lugs on this chamber.

**85A0027 Chamber**
*Location*: unknown
*Recovered*: 'outboard starboard side' after raising of the hull, possibly in a secondary position

*Statistics*
Length: 800mm (still in concretion); bore: 72mm; external diameter neck: 120mm

The outer surface is formed of fourteen abutting 50mm bands with no upstanding hoops (not illustrated). The chamber appears to bow at each end. Outwardly it is similar to 82A5176, 90A001, E31/104 and 82A5097. The fragmentary remains of a single lifting ring were found during excavation. The inside of the bore was hollow to a depth of 380mm, suggesting that it was not loaded. A small piece of rope was concreted to the centre of the chamber (89S0023). Towards the rear of the chamber a 25mm void was visible leading into it, possibly a large touch hole. This was filled with a metallic grey liquid. Two distinct layers of metal were visible below the hoops, the inner staves of 25mm and a second 'band' type of layer of 50mm.

**83A0513 Chamber**
*Location*: M4, possibly spare chamber for 81A2650
*Recovered*: excavated

*Statistics*
Length: 720mm (still in concretion); bore: 85–90mm

When recovered there was Tudor rope attached around a lifting ring. The radiograph suggests double staves, with eight horizontal lines and a breech plug 60mm thick. The wall thickness at the mouth is 34mm. It appears to be hooped and banded, with the bands 75mm wide and the hoops *c*. 40mm. The breech plug is present and is *c*. 60 mm thick. There are possible indications of two sets of lifting lugs, 150mm and 400mm from the shoulder (Fig. 3.40).

**82A5097 Incomplete chamber**
*Location*: unknown. Tampion *in situ*
*Recovered*: excavated

*Statistics*
Length: 770mm (from radiograph, still in concretion); bore 100mm

This incomplete chamber is corroded and broken at the neck. The outside shows wide bands abutting at one end of the breech, but these are missing towards the mouth. The staves are not clearly visible on the radiograph. There may have been a set of lifting lugs towards each end. The maximum diameter is 350mm. This form of abutting bands is seen on the Powell Cotton chambers, 82A5176, 90A0001 and 80A2005 (not illustrated).

## 82A5172 Incomplete chamber

*Location*: unknown. Wad, 82S1209/1. Contents sampled, 82S1209/1–9
*Recovered*: excavated

*Statistics*
Length: 635mm; bore: 100mm

This chamber is broken at the neck. The radiograph suggests two layers of staves. The hoops and bands seem to be balanced; with two hoops at either end then alternating hoops and bands with a triple hoop sequence in the centre supporting the lifting rings (Figs 3.40, 3.42). The bands are 50mm wide and the hoops 30–35mm, with the exception of the end hoop, which is 40mm. It is otherwise similar in form to 82A2650. The breech plug shows clearly as 75mm in thickness.

Excavation revealed a loose rope wad (82A1209/1), 80mm inside the mouth but there was no tampion. Following the wad, wet granular material (82S1209/2) and very liquid, extremely black contents (82S1209/5) were recovered from the centre of the chamber, while the sides produced a dry flaky material (82S1209/3) within which samples of what looked like thread were noted. Twig-like material was also adhering to the sides (82S1209/7; Hildred 9/12/82). It seems unlikely that the contents would be this well preserved if a tampion had not originally been present in the chamber.

*Figure 3.42 Port piece chamber 82A5172*

## 82A5176 Chamber

*Location*: unknown, possibly M4/5
*Recovered*: excavated

*Statistics*
Length: 720mm; bore: 100mm

The chamber is very similar in size and form to the *in situ* M4 chamber (81A2650) and it is not impossible that the two spare breeches situated between the M4 and M5 guns (83A0513 and 82A5176) could be used with either gun by varying the width of the forelock and iron wedges to ensure a good fit with the carriage. The radiograph suggests two layers of staves, the inner 15mm thick and the outer 15–20mm. At the neck, one 80mm wide stave can be seen. There are two sets of lifting lugs, each bearing one of two original rings (Figs 3.40, 3.43) These are positioned towards each end of the chamber (170mm from neck junction and 130mm from the back) resulting in a 'bowed' shape. The surface is very smooth, as though formed of bands only (40–50mm wide) and is very similar to E31/104 but with a parallel neck. Leather or textile was recovered beside the chamber mouth.

*Figure 3.43 Port piece chamber 82A5176*

## 90A0001 Chamber

*Location*: unknown
*Recovered*: found loose on barge in Ship Hall

*Statistics*
Length: 800mm (in concretion); bore: 100mm

This chamber (not illustrated) is broken at the neck with two stave lines visible on the radiograph. The stave thickness at the neck is estimated at about 12mm and the mouth may be sealed with a tampion. The outer layer appears to consist of equally sized narrow bands with an estimated thickness of 38mm. Thirteen can be counted, all abutting and *c.* 70mm wide. Two sets of lifting lugs can be seen, 270mm and 570mm from the mouth. The 10mm diameter touch hole shows clearly, entered through a band 90mm from the end, this has a diameter of 10mm. It is similar to E31/104, 80A2005 and 82A5176, but the neck taper is more obvious.

**03A0070 Chamber**
*Location*: unknown
*Recovered*: Breech recovered from clearing up of collection area on seabed in 2003 (Timber Park)

*Statistics*
Length: 620mm; maximum diameter: 200mm. No detailed recording.

**05A0057 Chamber**
*Location*: Forecastle. East of top of stem, close to anchor
*Recovered*: excavated east of stem beside large bow anchor

*Statistics*
Length: 650mm; maximum diameter: 265mm.

Extremely well concreted. One set of centrally placed lifting lugs with rings, external diameter 160mm.

**National Maritime Museum second chamber (NMM2) (incomplete)**
*Location*: unknown
*Recovered*: nineteenth century

*Statistics*
Length: 720mm; bore: 80mm

This is corroded and the surface detail is obscured. It appears to be made of 32mm wide hoops and 70mm wide bands. It has two descending hoops at the apparently tapering neck, followed by a band, hoop, band, triple hoop sequence, band, hoop, band and two or three end hoops. It seems to be similar in form to the chamber currently displayed with this gun (see KPT0017 above) possibly supporting a set of lifting rings on the central of the triple hoop assembly, similar to 82A5097 (not illustrated).

**Royal Armouries Chambers**
In addition to the two guns described above, possibly four separate chambers from the *Mary Rose* are recorded within the collection at the Royal Armouries. Blackmore (1976) states that the chamber belonging with XIX.1 (see above) was missing at that time. However a loan form issued from the Tower to Portsmouth City Museums (262, dated 1978), transferred the two guns (XIX.1 and XIX.2) without chambers. A separate loan (163) included three chambers (80A2000–2) amongst twelve 'iron fragments', though only the first seems to have actually been recognised (or described as) a chamber. Blackmore (1976, 56), however, suggests that these fragments should only have contained two portions of gun rather than any chambers. It is suggested here that loan 163 included the two chambers missing from the guns listed under 262. XIX. 5 was loaned as a chamber in its own right. As stated above, two guns illustrated in 1911 (Liscoe, pers. comm.) together with chambers, are identifiable as XIX.2 and XIX.1 so, the third should be the extra chamber listed with XIX.1.

*80A2000 Royal Armouries XIX.163. Re-identified as chamber for XIX.2*
*Location*: unknown
*Recovered*: 1840. Given by John Deane to the South Eastern Railway Company who presented it together with XIX. 5 (80A2005) to the Armouries in 1851

*Statistics*
Length: 660–70mm; bore: 110mm; external diameter: 260mm

If it is assumed that the photographs taken in 1911 had the correct chambers associated with their guns, 80A2000 (Fig. 3.40) would be the chamber for XIX.2 (80A2003). Blackmore (1976, 56) cites a length of 660mm. Hooped and banded, it resembles the chamber within the port piece owned by Woolwich. It is corroded and there are no obvious attachments for lifting rings.

*80A2001 Royal Armouries XIX. 163. Re-identified as chamber for XIX.1*
*Location*: unknown
*Recovered*: 1840 together with 80A2002, 80A2004 & culverin 79A1278

*Statistics*
Length: 470mm; bore: 50mm

This looks like the chamber photographed with XIX.1 (80A2004) in 1911 (Liscoe, pers. comm.). It is very small for a port piece, although in form it is not dissimilar to the National Maritime breeches with obvious hoops and bands (Fig. 3.40).

*80A2002 RA XIX. 163. Re-identified as spare chamber for XIX. 1*
*Location*: unknown
*Recovered*: 1840 together with 80A2001, 80A2004 & culverin 79A1278

*Statistics*
Length: 540mm; bore: 72mm

The associated port piece is small, with a listed bore of 152 mm. It looks as though the end ring is missing, otherwise this appears to be hooped and banded (Fig. 3.40).

*80A2005 RA XIX. 5 Chamber*
*Location*: unknown
Recovered: 1840. Given by John Deane to the South Eastern Railway Company who presented it together with XIX.1 to the Armouries in 1851. Listed as attached to bed fragment in 1916

*Statistics*
Length: 700mm; external diameter: 254mm; bore listed: 127mm (Blackmore 1976, 56)

This is a large solid chamber, similar to 82A5176, 90A0001 and E31/104. The neck hardly tapers and it has a bowed appearance; the hoops do not seem to come above the level of the bands (Fig. 3.40). There are no obvious signs of lifting rings.

**Powell Cotton Chamber E31/104**
*Location*: unknown
*Recovered*: probably raised in 1840. Assumed to have been bought from John Deane following a letter to John Powell dated 1841 offering items for sale from the *Mary Rose*.

*Statistics*
Length: 630mm; bore: 80–105mm

This piece was loaned to the Mary Rose Trust for radiography and illustration. It is a fine example of a breech, similar to 82A5176, 80A2005 and 90A0001. The radiographic images of 82A5176 and this breech suggest similarities in length, width, bore and method of manufacture. It looks as though 5–7 staves are held together with fourteen abutting 50mm wide bands with thicker bands at each end giving a bowed appearance (Fig. 3.40). Two sets of lifting rings are present *c*. 20mm and 500mm from the mouth. There is a 10mm diameter touch hole 65mm from the back.

**Ironbridge Gorge Museum Trust Chamber**
*Location*: unknown
*Recovered*: probably raised in 1840

*Statistics*
Length: 510mm; bore: 80mm

This is a breech chamber supporting a centrally placed single set of lifting lugs, similar to the National Maritime Museum example (Fig. 3.40). The caption in the museum states: '*A wrought iron breech from a cannon on the* Mary Rose. *This was probably fitted to a cast iron barrel and was salvaged by the Mary Rose Project and donated to the Ironbridge Gorge Museum Trust.*'

The Mary Rose Trust has no record of either the recovery nor of the donation.

*Fragments*

There are a number of broken elements which may be fragments of port pieces, those examined are listed in Table 3.5. Many are still wet and inaccessible, some are

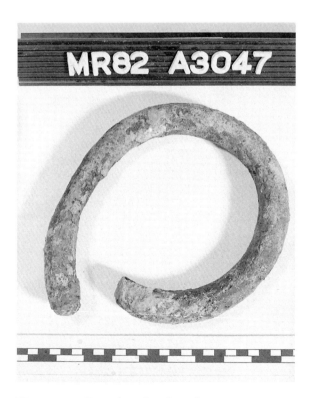

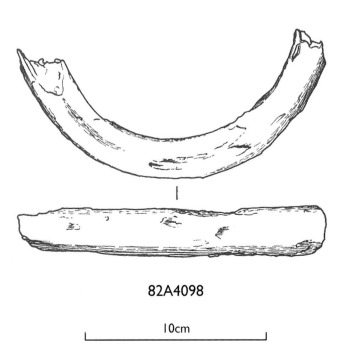

*Figure 3.44 Port piece chambers: fragments*

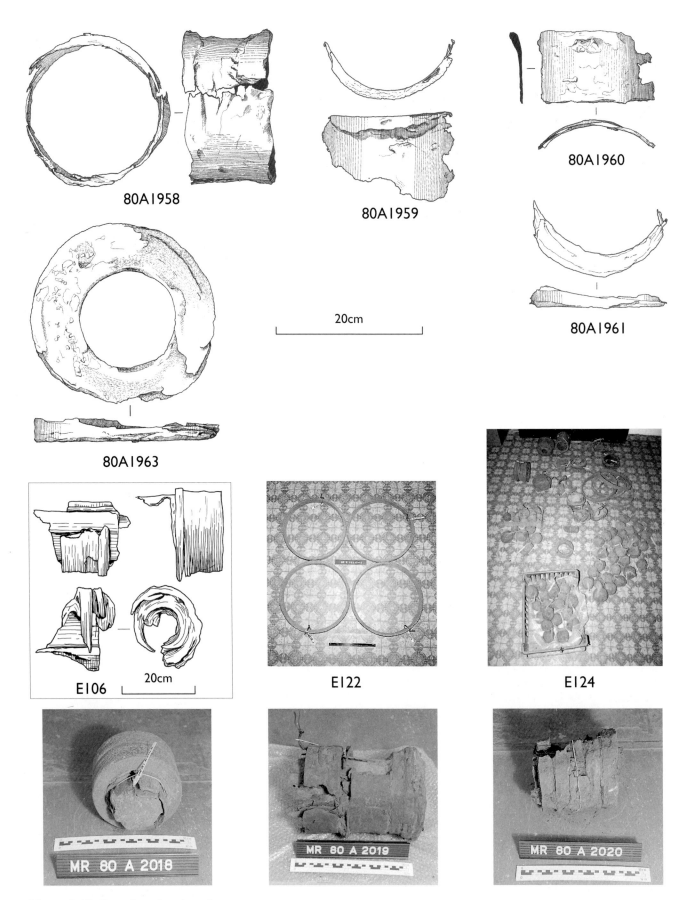

*Figure 3.45 Port piece chambers: fragments*

**Table 3.5 Port piece fragments (metric)**

| Object | Sector | Description | Measurements (mm) | Notes |
|---|---|---|---|---|
| 83A0277 | M2 | breech plug | D. 250 | in concretion; gives credence to large breeches poss. from this area |
| 81A2032 | M4 | lifting ring | D. 155; D. ring 25 | associated chamber concreted to S wheel |
| 82A3047 | SS4 | lifting ring | L. 630; D. 28 | 'curved rod' |
| 81A1267 | M5 | lifting ring | D. 145, D. ring 25 | feature 55 |
| 80A0303 | O8 | band & stave frag. | int. D. 160–180; band Th. 12; stave W. 95, L. 140 | attached together, total W. 105, L. 140 |
| 83A0555 | M12 | 2 frags | L. 140, W. 250, Th. 110; L. 200, W. 100 | 2 lumps, nor drawn or photographed; still soaking |
| 82A4098 | ? | lifting ring | D. 160, Th. 25 | raised in cube 26/10/82 |
| 83A0464 | ? | stave frags | none; weight 2kg | truck body 82, soaking |
| 83A0465 | ? | stave frags | none; weight 4kg | truck body 82, soaking |
| 83A0466 | ? | stave frags | none; weight 7.5kg | truck body 82, soaking |
| 80A1953 ?RA Deane | ? | gun portion | L. 700, int. D. 180 | band & hoop, found in stores; looks like 80A2003, RA XIX.2 |
| 80A1958 RA Deane | ? | band, worn, thin, split | D. 190, int. D. 180, L. 120, Th. 10–13,; as XIX 1/22 | XIX.776 loaned with XIX.163, not listed separately; loan 262, 25/10/78 |
| 80A1959 RA Deane | ? | band frag. | D. 185, L. 115, Th. 13; as XIX1/2 | XIX.777 loaned with XIX.163 |
| 80A1960 RA Deane* | ? | band frag. | D. 170, L. 95, Th. 10–13; as XIX.1 | XIX.778 loaned with XIX.163 |
| 80A1961 RA Deane | ? | hoop frag. | D. 190, L. 30, Th. 25; as XIX.1 | XIX.779 loaned with XIX.163 |
| 80A1962 RA Deane | ? | hoop | D. 220, L. 60, Th. 25; as XIX.1 | XIX.780 loaned with XIX.163 |
| 80A1963 RA Deane* | ? | hoop | D. 255, int. D. 135, L. 60–70, Th. 30; as XIX.1 | XIX.781 loaned with XIX.163 |
| 80A2018 RA Deane* | ? | gun frag., staves, hoops & bands | D. 200, int. D. 130, L. 175; staves 9–11 ?double layer, stave W. 55–60, Th. 12; band W. 120; as XIX.1 | XIX.783 loaned with XIX.163 |
| 80A2019 RA Deane* | ? | muzzle frag., staves alt. hoops 2 & bands 3 | L. 350, D. 250, int. D. 115 | XIX.784 loaned with XIX.163; XIX–1 painted red (80A2004); fits dimensions of rings without satves; as E106/2 |
| 80A2020 RA Deane | ? | barrel frag., staves, hoop 4, band 3 | L. 230, D. 170, int. D. 110 (corroded), as 80A2018–9 | XIX.785 loaned with XIX.163 |
| 80A2021* | ? | gun frag. | L. 590, D. 190 | no information |
| E106 1–3* RAI Deane | ? | barrel frags 1)? frag.; 2) band with staves; 3) band with staves | 1) D. 220, 210, 240 | 1 & 2 look like MR guns; 3 poss. portions of 2; E106/2 fits RA gun 80A2003 & RA 80A2019; 3 larger than 2; 106/3 looks Mr but larger |
| E122 1–4 RAI Deane | ? | 4 large hoops | all D. *c.* 180, int. D. 150, L. 40, Th. 20 with scarf joints. Pt 4 L. 45 | poss. part of XIX.1 80A2004 |
| E124 1–5 RAI Deane | ? | 5 hoops; 1–2 look like port piece hoops | | 2 frags not part of same group, 4/5 poss. part of XIX.1 |
| E125 1–3 RAI Deane | ? | 1&2 hoops, 3 unknown | D. *c.* 250 | poss. part of XIX.1 |

still in concretion and corroded making diagnosis difficult. An internal hoop/band diameter of 180mm as a minimum has been taken as diagnostic, but it is possible that the smallest may represent slings or fowlers.

The number of fragments which may represent port pieces totals 31–3 (two are smaller, but they are clearly associated with other port piece fragments), of which ten were raised by the Mary Rose Trust, eighteen belong to the Royal Armouries and are provenanced as *Mary*

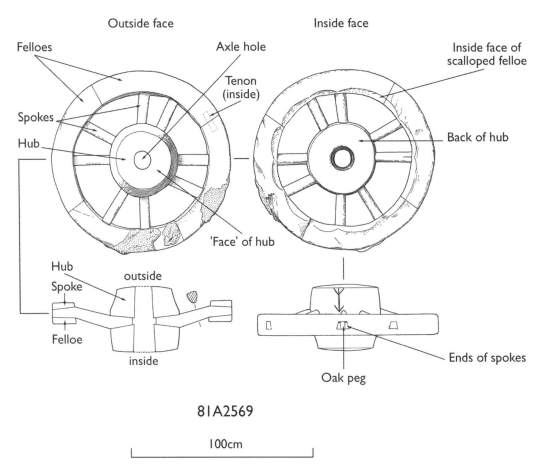

*Figure 3.46 Parts of a spoked wheel*

## Table 3.6 Spoked wheels (metric)

| Object | Sector | Hub diam. | Hub depth | Rim diam. | Rim width | Rim depth | No. spokes | Spoke length | Spoke diam. | Marks | Notes |
|---|---|---|---|---|---|---|---|---|---|---|---|
| 81A2569 | M5S | 400/300 | 350 | 1000 | 100 | 125 | 8 | 200 | 50 | yes | |
| 81A2570 | M5N | 350/320 | 350 | 990 | 122 | 148 | 8 | 185 | 60 | – | |
| 81A5688 | M8 | 380/300 | 380 | 1100 | 120 | 120 | 8 | 200 | 80 | – | iron pin |
| 81A5689 | M8 | 370/300 | 370 | 950 | 100 | 100 | 8 | 190 | 50 | yes | iron pin |
| 81A1942 | M11N | 350/300 | 350 | 1000 | 120 | 140 | 8 | 220 | 50 | – | |
| 81A3980 | M11S | 330/280 | 350 | 990 | 100 | 120 | 8 | 220 | 50 | – | inc. |
| 81A0131 | H1 | – | – | – | – | – | – | – | – | – | hub frag. |
| 81A0169 | H1 | 380/200 | 300 | ?900 | – | 100 | 7/8 | 200 | 50 | – | frags |
| 81A0170 | H1 | c. 280 | – | – | – | – | 3/8 | – | – | – | |
| 79A1277 | O1 | 350/300 | 330 | 920 | 100 | 150 | 8 | 180 | 50 | – | |
| 81A0430 | O1 | 400/320 | 380 | 1000 | 100 | 110 | 10 | 200 | 50 | – | |
| 81A0490 | O1 | 380/310 | 310 | 950 | 110 | 110 | 8 | 200 | 50 | yes | |
| 81A0429 | O1 | 350/300 | 350 | 995 | 125 | 145 | 8 | 180 | 50 | – | |
| 81A0493 | O1 | 350/300 | 360 | 950 | 110 | 120 | 8 | 180 | 50 | yes | |
| 81A0279 | ? | – | – | – | – | – | – | – | – | – | frags |
| 82A5039 | ? | – | – | – | – | – | – | – | – | – | frag. |
| 04A0072 | bow | 270+ | – | – | – | – | 4 holes | – | 45 | – | frag. |

S = south; N = north

*Figure 3.47 Spoked wheels: a) auger marks on inside of hub showing how spoke holes were augered out using a bull-nosed auger; b) wheel back showing wedge; c) peg on felloe; d) arrow mark on hub*

*Rose*, and the rest are parts of four accession numbers from the Royal Artillery Institution. The RAI group included 14 accession numbers, one of which has over 58 cast iron shot. A number of these may well be from the *Mary Rose*, and officially recognising this might enable us to identify more items resulting from the Deane excavations, in particular the swivel fragments. Those illustrated are shown in Figures 3.44–5.

The distribution (where known) is as expected, with many items from sectors containing iron guns (Fig. 3.62). The band in O8 (81A0303) may have derived from the broken *in situ* M8 starboard gun or from its port side equivalent. The lifting rings (easily separated and often missing) show distributions in the Starboard scourpit (82A3047, Fig. 3.44) and close to iron guns (81A2032; M4; 81A1267, M5), one is from an

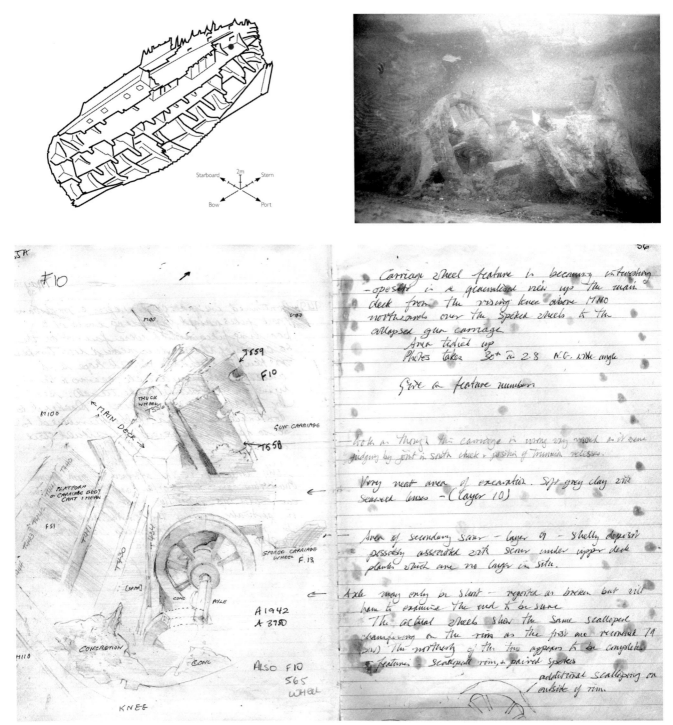

*Figure 3.48 Wheels from M11: underwater photograph and sketch (site book 1981 mu9-11, p. 55, Adams. J.)*

unknown location (82A4098, Fig. 3.44). The single breech plug found in M2 is interesting (83A0277) as this is where one of the extremely large breeches may be derived from and the location forward of M3 is another possible position for additional port pieces.

There are four barrel fragments, two of which (80A1953 and 80A2020) might be part of 80A2003 (XIX.2) (Fig. 3.45). The four band fragments (80A0303) in O8 and three in the Royal Armouries (80A1958–1960) may all have derived from the broken gun (now mostly rings) 80A2004/XIX. 1. There are six instances of hoop fragments, three from the RAI (E122, 124, 125), comprising multiple rings, and 80A1961–3, in the Royal Armouries. All are similar to the collapsed gun 80A2004/XIX.I and may be part of it (Fig. 3.36). A bed fragment, 80A1955, may relate to a port piece. This is part of a loan to Southsea Castle in 1979 and fits 80A1953. It is therefore possibly associated with Tower of London guns 80A2003/2004.

## Carriage fragments

Portions of carriages for wrought iron guns can also be used to suggest locations for port pieces. As some elements are common to all sizes of wrought iron guns, there is a potential for misidentification.

**Spoked wheels**

These are relatively easy to identify and were found in situ on two starboard side Main deck port pieces in M5 and M8. Evidence for thirteen additional spoked wheels have been identified (Fig. 3.62). Common measurements are summarised in Table 3.6.

The wheels are composed of two different woods. The nave (hub) and rim components ('*fellies*') are turned from elm (Fig. 3.46). The spokes are made of oak. This construction is often found in cart wheels.

The hub is shaped and shouldered towards the outside of the '*face*' of the wheel (Fig. 3.46). A hole is bored in the centre to accept the axle which is larger towards the inside or '*back*' of the wheel to accommodate the taper of the axle arm/stub under the bed. Holes are bored through the hub to accept the spokes. Often the orientation of the spoke is marked out by incising three thin lines on the hub (sometimes only top and bottom are marked, or only centre). The spoke holes are formed by overlapping two or three auger holes which can be seen on the inside of the axle hole (Fig. 3.47a). The spokes are tapered to form tenons as they enter these holes. The spokes are flat towards the face of the wheel and chamfered and rounded towards the back. The end of each is shouldered as it enters the felloe, relieved towards the front.

In order to secure this joint an oak wedge is entered through the felloe to expand the spoke (Fig. 3.47b). The

*Figure 3.49 Wheels from M11: two underwater views of the group as found: above) looking south; below) looking north*

ring of fellies is held by loose internal pegs placed in pre-drilled holes at the end of each (Fig. 3.47c). These are entered loosely onto the spokes and then knocked up towards the shoulder. The excess spoke is then cut and wedged.

*Figure 3.50 M11 spoked wheels*

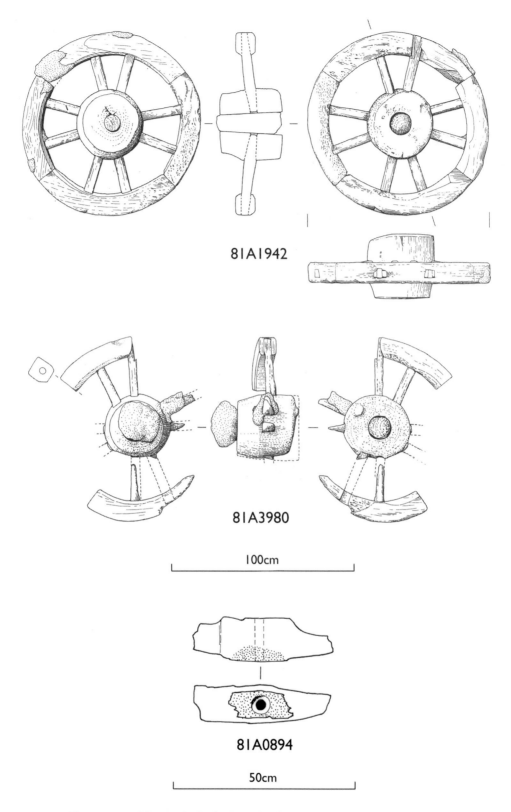

*Figure 3.51 M11 spoked wheels and axle*

In some instances the weight of the wheel has been lessened by either chamfering or removing scallop-shaped wooden pieces from the inner face of the fellies. In some instances one side of the wheel has been chamfered and the other scalloped. This is not always easy to determine, and the scalloping is not always regular, although some examples show two scallops between pairs of spokes, and one between the two spokes forming a pair (Figs 3.46, 3.48).

Marks were found on the hubs of four spoked wheels, two from *in situ* carriages (81A2569, M5; 81A5689, M8) and two from wheels in storage (81A0490, A0493). These are always positioned on the inside of the hub, either radiating from the spoke origin towards the inner face, or from the inner face towards the spoke origin (Fig. 3.47d). Other marks include incised broad arrows, one is a simple inverted 'V' at the end of a line (81A0490), others have a secondary inverted 'V' at the opposite end to form the '*flights*'. These invariably take up the entire width of the inner hub (*c.* 120–150mm). They are likely to represent the broad arrow used to designate Crown ownership.

*81A0894, 81A3980, 81A1942 Spoked wheels in M11 (Feature 13)*
*81A0894 (axle fragment)*
*81A3980 (south wheel). Diameter: 990mm; rim: 100 x 110mm; hub diameter: 330/280mm*
*81A1942 (north wheel). Diameter: 980–1000mm; rim: 100 x 100 mm; hub diameter: 300mm*

This collection of gun carriage fragments was given a feature number (F13; Fig. 3.48). The axle (81A0894) was associated with concretion (81A2091–2) and two spoked wheels. When excavation progressed around the wheels, it was realised that they were positioned directly in front of a starboard side Main deck gunport (M11) and that this was all that remained of a port piece which was stationed here. Beneath the wheels, adjacent to the Main deck, was the partition 'platform' listed as F51 (81T430, 440–1), possibly a hatch cover (Fig. 3.48; see also *AMR* vol. 2, 162). This area contained a number of other items relating to gunnery, including finished stone shot of 175mm and 176mm (81A0822, 81A0830) and an iron shot of 152mm (81A3819). The stone shot suggests an iron gun with a bore of 182mm, smaller than the port piece found in the stern (71A0169; bore 190mm, shot 184mm). Other gun furniture included a heavily concreted powder ladle (81A2217), two rammers (81A823, 81A0829, 81A1130), a shovel (81A2194), a mallet and lantern shot (divelog 81/11/275). Parts of a concreted gun were also recovered (81A2108). The area showed sedimentary disturbance suggested to be anchor drag. The presence of the carriage for a demi cannon recovered in the nineteenth century at the same level suggests that the port piece was recovered at this time. The axle and wheels are in a similar orientation as the intact demi cannon (81A3000) further north (Fig. 3.49). This position would favour a 'short' port piece.

The axle (81A0894) survives to a length of 340mm and has an *in situ* wooden pin of 60 x 30mm. The north wheel, 81A1942, has eight spokes and is missing part of one felloe (81A2681) (Figs 3.50; 3.51). Both edges of the rim are scalloped. The southern wheel (81A3980) is similar but more fragmentary. Parts of two fellies and five spokes remain (Figs 3.50b, 3.51). This also shows evidence of scalloping.

*82A5039, 81A0279, 04A0072 hub fragments*
*No location*
The first two of these are the only indications of the presence of other guns with spoked wheels on the ship (Fig. 3.52), though such fragmentary pieces are difficult

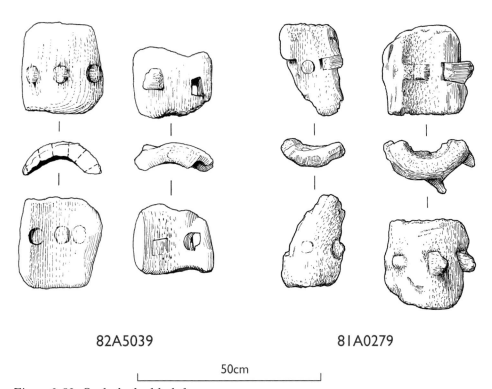

82A5039         81A0279

50cm

*Figure 3.52 Spoked wheel hub fragments*

*Figure 3.53 Spoked wheels: a and b) underwater views of carriage wheels in O1; c and d) lifting wheels*

to recognise and others may have gone unnoticed so far. The anticipated three port side iron guns would have needed six spoked wheels. A further possible hub fragment, badly eroded, was recovered from the bow area in 2004 (04A0072; not illustrated).

*Stored wheels 79A1227, 81A0429, 81A0430, 81A0490, 81A0493 in O1; 81A0131, 81A0169, 81A0170 in H1*

Five wheels of very similar size were found in storage in O1 (Figs 3.53–4). Many of these are pristine, and none displays any concretion suggestive of having had lynch pins attaching the wheel to the axle. None displays any wear over the axle hole or on the outside face of the hub, nor any evidence for iron tyres. It is suggested that these were being carried as spares for use on board. Seven are composed of four fellies and eight spokes, 81A0430 has five fellies and ten spokes but there is no reason why this should not have been paired with an eight-spoked wheel.

Three incomplete spoked wheels were recovered from H1, possibly derived from the gun carriage store in

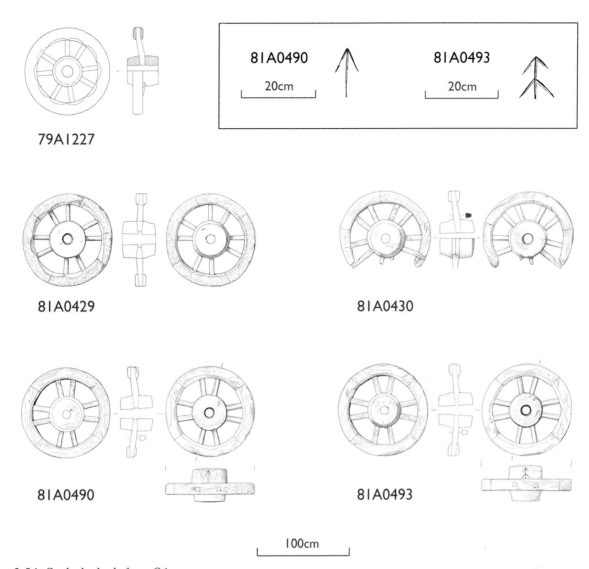

*Figure 3.54  Spoked wheels from O1*

*Figure 3.55  Spoked wheel fragments from H1*

Table 3.7 Forelock characteristics (metric)

| Object | Sector | Gun | Tang width | Length | Width | Height | Markings |
|---|---|---|---|---|---|---|---|
| 79A0191 | U2 | n/a | 250 | 550 | 275 | 85 (inc.) | |
| 81A0645 | U6 | 81A0645 (sling) | 200 | 680 | 110 | 250 | |
| 81A1561 | U10 | n/a | 200 | 520 | 190 | 240 | |
| 81A1520 | U10 | n/a | 200 | 515 | 170 | 170 | |
| 80A1875 | U10/11 | n/a | 250 | 585 | 260 | 245 | |
| 80A1723 | M4 | ? | 150 | 520 | 170 | 230 | |
| 80A1786 | O5 | ? | n/a | 840 | 160 | 150 | |
| 81A1229 | M4/5 | ?81A2650 | 250 | 710 | 210 | 240 | 'H' |
| 81A2604 | M5 | 81A2604 | 250 | 600 | 250 | 250 | Λ |
| 81A1261 | M5 | ?81A2604 | – | 545 | 150 | 185 | |
| 81A3188 | M8 | 81A3001 | 200 | 700 | 200 | 390 | IIIvIII |
| 81A3129 | M8 | 81A3001 | 200 | 620 | 200 | 180 | X |
| 81A2271 | M10 | ?M11 wheels | 220 | 560 | 270 | 200 | N/A |
| 81A2650 | M4 | 81A2650 | 250 | 660 | 140 | 170 | 'X' TOP |
| 81A0607 | O1 | blank | n/a | 750 | 235 | 235 | |
| 81A5950 | M9 | blank | n/a | 405 | – | – | |
| 81A1625 | O3 | blank | n/a | 465 | 155 | 132 | |
| 79T0114 | stern | blank | – | – | – | – | |

O1 (port). All were badly degraded. Only for 81A0169 (Fig. 3.55) could an estimate of diameter be made as there was one *in situ* spoke.

**Forelocks**

Other indicators of carriage mounted wrought iron guns are the wedges used to hold the breech chamber tight against the back of the carriage, with their grain aligned across the carriage (Fig. 3.56a). These are stout oblong blocks with a half rounded to rectangular tang which locates into the chamber groove in the carriage (Fig. 3.56b). The degree of roundness to the tang differs, between round to chamfered to form three or even five faces. To date, all identified examples are elm (eg, 81A1229, 81A1261). The *Anthony Roll* mentions '*Tymber for forelockes and koynnys*' for all the vessels with carriage mounted guns. The *Mary Rose* was listed with 100ft of this.

A total of fourteen forelocks (three *in situ*) and four potential blanks have been recovered (Fig. 3.57 and Table 3.7). One is from the sling (81A0645) situated on the Upper deck in sector 6, and the other two are from M4 (81A2650) and M5 (81A2604) respectively. The Woolwich and Greenwich port pieces were illustrated showing these *in situ*.

These items are easily identifiable and are durable. Their distribution may help to suggest positions for missing guns (see Fig. 3.62). Matching them with the type of iron gun is not easy, there are a variety of tang shapes, and size does not appear critical, Table 3.7. Some forelocks sit exactly over the carriage, only just wider than the bed (81A2604). Others are much wider (81A0645). The bed slot for the sling in U6 is 200mm, and for the port pieces is 225–300mm, although the forelocks do not have to be exactly the right size, a good fit seems to have been important.

There is no central storage area for the forelocks (see Fig. 3.62) and most reflect gun stations or at least fighting areas. There finding at least one spare forelock and often a spare chamber in sectors with *in situ* guns suggests that these either stayed in these positions, or were there specifically ready for battle. The portage of

*Figure 3.56 Forelocks: a) from 81A2271 showing tang; b) in position on gun carriage*

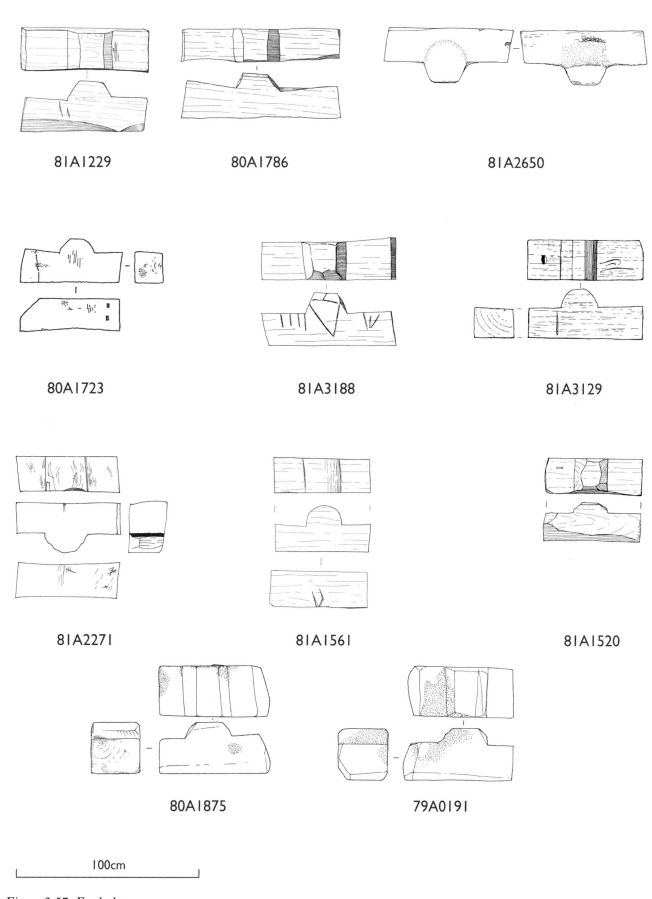

Figure 3.57 Forelocks

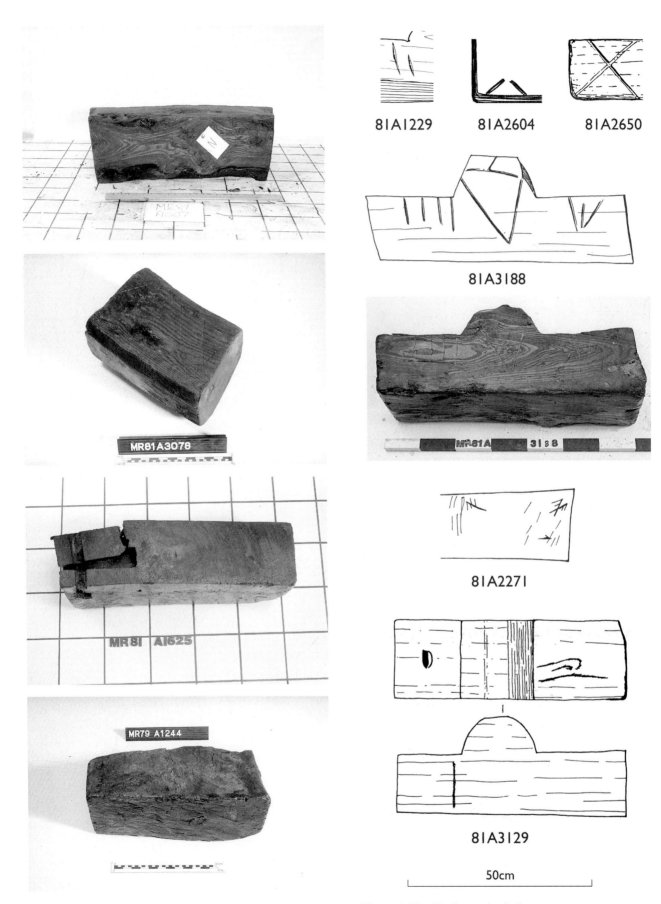

Figure 3.58  Possible unfinished forelocks

Figure 3.59  Marks on forelocks

spare wood suggests that these were vital but may have been subject to breakage. The *in situ* forelock from 81A2650 was examined in detail. The barrel and breech chamber had previously been removed from the carriage. The forelock was held fast within concretion which had formed over its top, the back of the carriage, down each side and beneath the slight overhang of the forelock each side of the carriage. It was removed using a fine chisel and hammer. It became evident that the concretion formed remains of at least one iron tapered wedge positioned between the forelock and the step at the back of the carriage. Two of these could be used to fine tune the wedging and ensure the perfect fit vital to prevent the forelock (and indeed the chamber) from flying out when the gun was fired. Additional concretion rested on a 60mm wide lip at the back of the carriage. This could have covered the back of the carriage, with additional loose iron wedges entered between this and the forelock. Rope was found over the top of the iron wedge and also around the base of the tang where it located into the chamber groove. Both forelock and gun carried the incised letter 'X'. The existence of these iron wedges is noted by John Deane in a letter to Mr Powell Cotton and is illustrated in the Royal Institution of Cornwall painting of the National Maritime Museum port piece (Bevan 1996, pl. 24). Iron wedges were found to be vital in firing the reproduction (see below).

A number of blocks have also been identified as potential blanks for forelocks (Fig. 3.58; Table 3.7).

The fourteen finished forelocks include five from the Upper, eight from the Main and one from the Orlop deck. The Upper deck examples all have a similar length (Table 3.7) but the two in U10 (81A1561 and 81A1520) are smaller in breadth and tang width than the others. This may be indicative of port pieces in the bow and U11, but smaller guns (slings or fowlers) in U10.

All of the forelocks on the Main deck were found in areas with iron guns (two *in situ* in the M4 and M5 port pieces). 80A1723 (M4) was found close to the north side of the gun beside an Upper deck hanging knee at a higher level than the gun. In this location it may have derived either from the Upper deck or the port side. The spare for the M4 gun is more likely to be 81A1229, found between the iron guns in M4 and M5. 81A1261 was just west of the gun carriage. The two loose forelocks from M8 (81A3129 and 81A3188) were both between the gun axle and wheels and the starboard side but are of slightly different sizes. The groove width on the carriage is 250mm, which would require considerable wedging of both tangs (see Table 3.7). However, the tang depth for both is similar (100mm), again fitting the groove with the aid of wedges (130mm). The forelock from M10 (81A2271) may have derived from a possible iron gun in U10, the spoked wheels and axle found in M11, or its port side equivalent. As the size of this is similar to that of the forelock of the M8 port piece, it is more likely to have serviced a port piece than a sling.

The solitary forelock from the Orlop deck was not in store (80A1786, O5) but in the upper levels above collapsed partitioning (F111) which sealed off all of O5. This should not have collapsed until some time after the ship sank, as there was a very large build up of sand beneath it. This is most likely to have derived from either the port side, or M5 starboard, probably the former.

The unfinished blocks include one each from O3 (81A1625), from the stern close to the bulwark rail (79T0114), O1 in the area of ordnance stores (81A0607) and O9 (81A3078), where a number of woodworking tools were recovered (Fig. 3.58). Six forelocks are marked (Fig. 3.59), most obviously with an 'X' incised on the top of the *in situ* forelock for 81A2650 which is also found on the carriage just behind the wedge as well as on the tang of 81A3129. 81A2604 (M5) has an inverted 'V' and 81A1229 (M4/5) has two parallel lines scored over one face. 81A3188 and 81A2271 also carry marks (Fig. 3.59; Table 3.7).

**Elevating posts**

There are nine elevating posts (Table 3.8). These are rectangular wooden posts pierced with transverse holes Figs 3.60–1). The holes are staggered across the face to

Table 3.8 Elevating posts (metric)

| Object | Sector | Gun assoc. | Length | Width | Thickness | No. holes | Marks |
|---|---|---|---|---|---|---|---|
| 72A0194 | ? | n/a | 394 | 85 | 60 | 3 | |
| 79A0186 | 79/S8 | 81A0645 (sling) | 680 | 110 | 80 | 8 | |
| 80A1364 (oak) | M5 | 81A2604 (port piece) | 980 | 110 | 90 | 10 | 'W' |
| 80A1466 (ash) | M4 | ?81A2650 (port piece) | 446 | 127 | 71 | 4* | |
| 80A1467 (oak) | M4 | ?81A2650 (port piece) | 855 | 120 | 75 | 11* | |
| 81A0617 | U10 | n/a | 485 | 90 | 68 | 6 | |
| 81A2088 | M8 | 81A3001 (port piece) | 890 | 110 | 75 | 11 | |
| 81A2848 | O6 | n/a | 435 | 105 | ? | 2 | |
| 81A3868 | H4 | n/a | 635 | 132 | 70 | 7 | |

* includes 1 in concretion

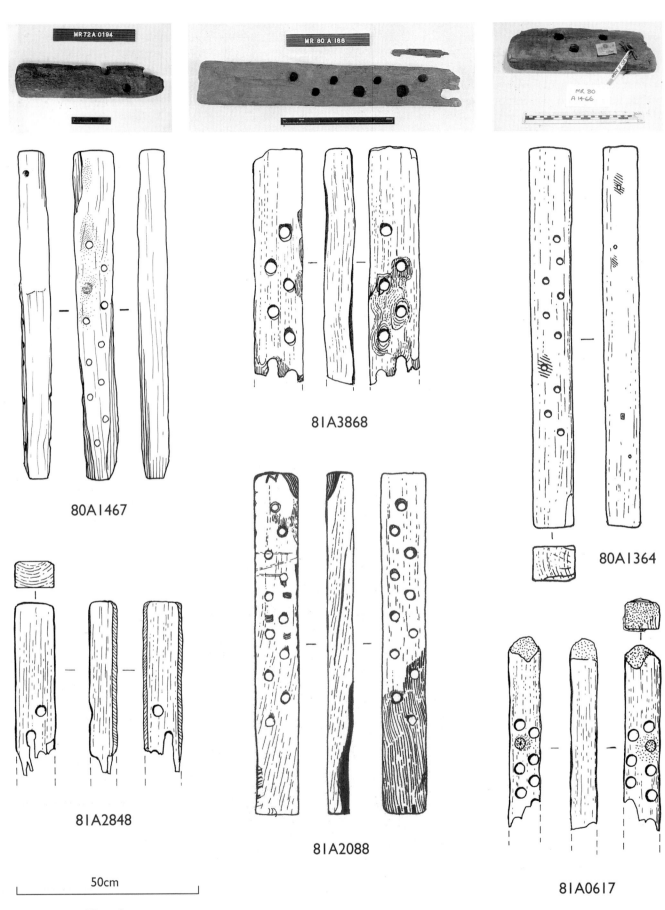

Figure 3.60 Elevating posts

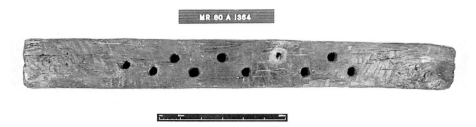

*Figure 3.61 Elevating post 80A1364*

provide maximum positions for pinning without creating a single line of weakness. These posts slot into holes cut through the block portion of the carriage (see above). The slots vary in length (130–200mm) and width (80–150mm). The one identified sling (81A0645) has a hole of 110 x 70mm. Most have their longest edge aligned along the length of the carriage, with the exception of 81A2604, which is aligned across the carriage. The 25–34mm diameter pegs are entered under the bed at right-angles to the carriage.

Norton (1628, 58 in a passage more fully quoted above, pp. 132–3) describes them thus:

'The Tayle of the Carrriage is to rest, and to be shored vp with an vpright post or foote, full of holes to slide vp and downe in a square Mortice fitted ther unto'.

Although the *Mary Rose* examples show no evidence of sheaves at the base of the post and the carriages are mounted on an axle and pair of wheels, the pinned post itself survives. Only one was found *in situ* (80A1364 with 81A2604, the port piece in M5). Two similar sized posts found at the same time in M4 (80A1466, 80A1467) may relate to the gun there although, in each

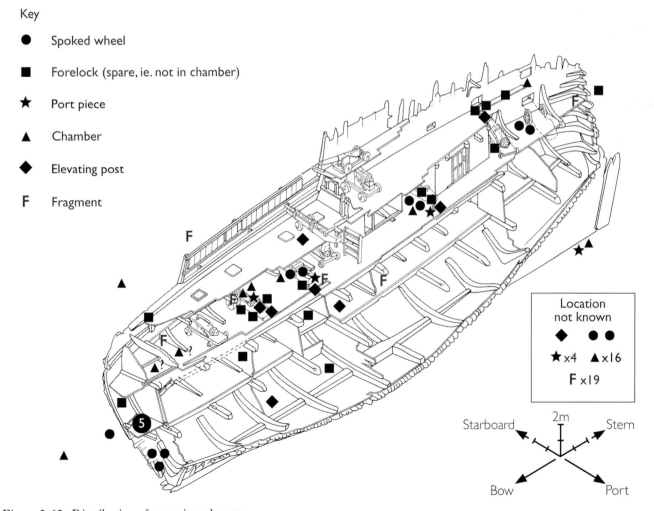

*Figure 3.62 Distribution of port piece elements*

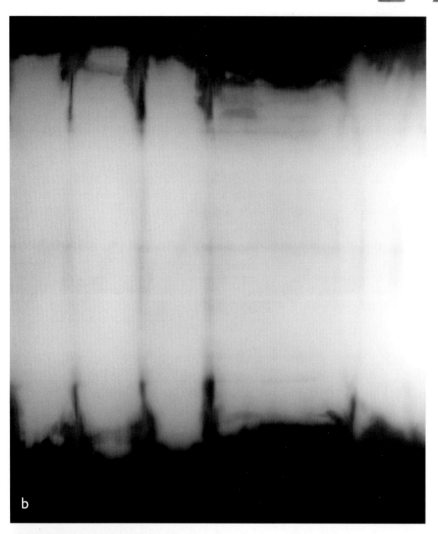

*Figure 3.63 Sling 70A0001: a) photograph of gun; b) radiograph ; c) underwater photograph, gun protruding through port*

instance, one of the holes is filled with concretion. 81A2088 was found beside the carriage 81A3001 (M8). The longest was from the *in situ* example, incised with a 'W'.

The distribution of these objects is shown in Figure 3.62. 79A0186 (Starboard scour) is recorded as being associated with the *in situ* sling (81A0645) while a loose example from U10 (81A0617) is smaller than those found in port pieces. It was associated with broken remains of a gun and carriage concreted against the starboard frames. There is one from O6 (81A2848) and one from H4 (81A386) and one unknown (72A0194).

## Slings

### Identification and historical significance

The *Anthony Roll* listing for the *Mary Rose* includes two '*slings*', three '*demi slings*' and one '*quarter sling*' together with a specified number of iron shot for each named type. These are the only iron guns listed with iron shot: slings with 20 iron shot each, demi slings with 13 and quarter slings with 50. The *Anthony Roll* records 29 slings, restricted to ships and '*galliases*'; 68 demi slings, mostly on ships and '*galliases*'; and 21 quarter slings divided equally between ships, '*galliases*' and the pinnaces and rowbarges. The shot per gun ratio for the slings is between 15 (the *Anne Gallant*) and 30 (the *Hart* and the *Antelope*). The demi slings have between 13 (the *Mary Rose*) and 40 (the *Falcon*) shot per gun and the quarter slings 13–50 (*Lion*, *Mary Rose*).

The first of these (70A0001) was identified by Alex McKee. It contained a cast iron shot (Fig. 3.63a). It had an internal bore of 95mm (3¾in) and was without chamber or bed. The other unusual feature was the length of the gun relative to the bore. Although not quite complete, the recorded length is 2295mm, giving a length in calibres (excluding the breech) of 24.16. This gun was radiographed and interpreted as being made of a single rolled sheet. It has since been re-radiographed and shows distinct lines where staves meet (Fig. 3.63b). In 1981, a second iron gun (81A0645) was recovered with an iron shot, this time on a carriage with small solid trucks, similar to those supporting the larger, wrought iron port pieces. This was *in situ*, on the Upper deck in the waist with its muzzle going out through a semi-circular hole within the blinds (Fig. 3.63c). This has a bore of 90mm (3½in) and incomplete length of 2240mm, giving a length in calibres of just under 25. A reference to the *Mary Rose* having two slings, one on either side of the foremast, has recently been found in a document at Hatfield House (see Chapter 12). In the light of this, the location of the McKee gun (north of the site) makes some sense.

These two long guns with cast iron shot represent the most complete candidates for the sling group. 81A0645 has an intact breech chamber that is both long and large for the barrel, with a length of over 800mm (including neck) and external diameter of 205mm. It is thick walled. The added length means that the volume remains large, possibly suggesting an increased powder charge to propel an iron rather than a stone shot.

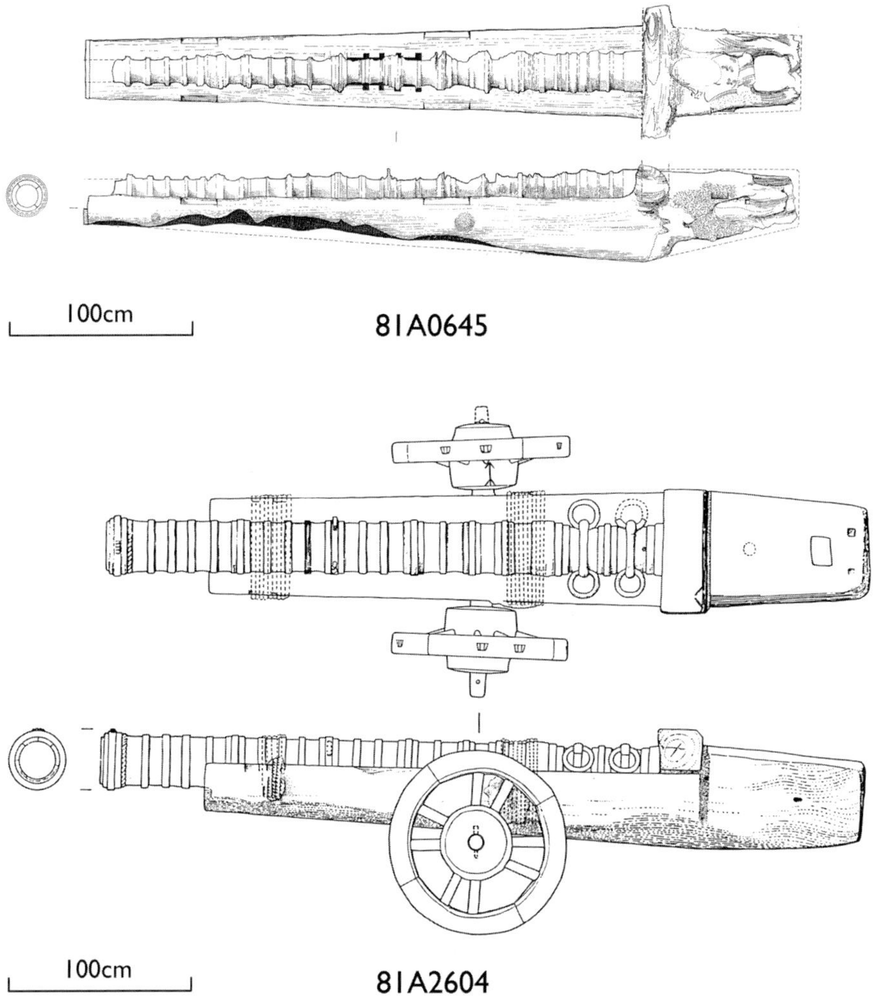

*Figure 3.64 Comparison of bore size relative to length of sling 81A0645 and port piece 81A2604*

Trying to ascertain which of the listed slings these represent is very difficult. There are no detailed English descriptions of these guns until 1628, when Norton (p. 58) describes them as swivel guns almost identical to the '*bases*' found on the *Mary Rose*, adding that: '*the lengths of the Portingale base is about 30 times her Calibre; the Sling about 12 times, the Murtherers, Port Peeces, and Fowlers 8 at the most besides their Chambers.*' Blackmore's definition (1976, 244) is: '*Originally, one of the guns of the 'serpent' type, long in relation to its calibre, later a swivel breechloader*'. He notes that seventeenth century descriptions conflict with the few early documentary references to slings, such as one for '*one hole pece of iron calld a serpentyne alias aslang*' received into the Tower in 1508 (*ibid.*, 242).

This indicates a direct connection with the long-thin '*serpentines*' of the fifteenth and early sixteenth century, often mounted on stock-like carriages with a swivel yoke (see the Cattewater guns, Redknap 1997, 76).

Slings, or '*slanges*' are listed as '*Double*', '*Whole*' (used often in place of '*ordinary*' and sometimes suggesting a muzzle-loader rather than a breech-loader), '*demi*', '*quarter*' and merely '*sling*'. Blackmore suggests that this group might represent the wrought iron equivalent of the bronze culverin class, an idea endorsed by both Robert Smith and Ruth Brown (pers. comm., Dec. 2001). Although Norton describes the sling as a swivel gun, his '*Table concerning Chamber Peeces*' (1628, 58) suggests an altogether larger gun.

He lists four un-named chambers with lengths of 12in, 17in, 22in and 24in (305mm, 432mm, 559mm and 610mm) and 3–4in (76–102mm) in height (bore at the mouth), with associated stone shot of 5–7in (127–178mm). He includes three named chambers listed separately, a sling chamber of 22in (559mm), a port chamber of 16in (406mm) and a base chamber of 9in (229mm). The sling chamber has a height of 2½in (63mm), holding 3½lb (1.6kg) of powder and taking a 2¼in (63mm) shot of iron 2⅓lb (1.1kg) in weight. This suggests a powder charge of 1½ times the shot weight! This large powder charge relative to the shot also suggests a similarity with the culverin class – large powder charge, long barrel, cast iron shot. The newly found Hatfield document (see Chapter 12) appears to confirm this.

The inventories that include items in the Tower of London (Blackmore 1976, 251–406) show slings in 1523, 1540, 1547, 1568, 1595, 1675 and 1676. At first most are of wrought iron but eventually these are replaced with cast bronze examples, with only a few entries suggesting cast iron. As guns of this size were already being cast in one piece in bronze, the decision to cast slings (and also the larger port pieces) in bronze may have been as a response to the continued importance of the breech-loader.

The inventory of thirteen vessels in 1514 (Knighton and Loades 2000, 113–58) includes slings or demi slings for twelve of the thirteen listed vessels, mostly with two chambers to one gun, so these are certainly around early in the century. In 1514 the *Henry Grace à Dieu* has '*A slyng of yron apon trotill wheles*'. The *Trinity Sovereign* has slings immediately following culverins (suggesting a gun smaller than a culverin) with the description '*slynges of yron stocked*' (*ibid.*, 119, 122). One entry, for the *John Baptist* (*ibid.*, 138), reads '*Chambers to the same with miches boltes and forelockes*', suggesting altogether a more '*serpentine*' type of gun on a swivel mechanism. One entry reads '*Half slynges of brasse ij*' for the *Great Elizabeth* (*ibid.*, 149). She is also listed with iron slings. The storehouse at Erith provides some indication of barrel length, '*Slynges of yron ryngyd containing xx fote besides the chambers*' (*ibid.*, 158). Clearly these guns were made in a number of sizes, some were of great length, most had two chambers, some were on wheeled carriages and others on swivel mountings, most fired iron shot.

The 1547 inventory (Starkey 1998) provides some additional information. All the types of sling are represented in both the shore fortifications and on vessels. The vessel inventories are simple, listing merely the number of each type under the broad heading '*guns of iron*' and the shot, but without stating the material. Position within the inventory, usually providing an indication of size, generally shows the slings after the port pieces, and demi slings after the fowlers. As both port pieces and fowlers seem to have relatively large variation in bore size, this is not particularly helpful. It is within the ships and galleys that most of the larger slings can be found, with a total of seventeen over the 53 vessels, ten of which are on the *Jesus* (*ibid.*, 7384). There are 69 demi slings distributed over nine ships, eight galleys and one rowbarge. The quarter slings are distributed throughout all the vessels. The small rowbarges in general carry few slings, an exception being the *Cloud in the Sun* which carries one quarter sling (*ibid.*, 8090). An attempt was made to assess the vessel inventories in terms of which other guns were missing when slings were included – this was inconclusive.

Entries for ships do not always include shot for slings. In a number of instances demi slings and quarter slings are listed but without shot. However the same vessel is listed with shot for bronze guns including falcons, yet that particular size of bronze gun is not present. The *Anne Gallant* is listed with culverins, demi culverins and a saker and sling. Shot is listed for all the bronze guns but not the sling. The entry for the *Unicorn* lists (7858) demi slings and for the shot (7870) '*falcon shot and sling shot*' suggesting some equivalence.

The sling group is represented in about two-thirds of the land fortifications. The Tower has five slings, ten demi slings and fifteen quarter slings (*ibid.*, 3715–17). No shot is listed for these, but iron shot is listed for culverins, demi culverins, sakers, minions, falcons and falconets (*ibid.*, 3728–34). Although shot for the sling group is listed within other entries, in many cases if all

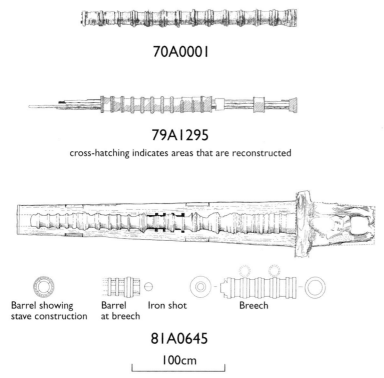

*Figure 3.65 Slings*

the bronze guns are listed with their shot, there is no differentiation in the sling shot. What we may be observing is the direct compatibility of the sling group with the culverin group of guns, hence obviating the necessity of listing shot specifically for slings. The inventory was scrutinised with this in mind. Sandown Castle is listed with slings (right after demi culverins) and '*half slinges*'. Bronze guns listed include demi culverins, sakers and falconets. Shot is listed for these guns and for falcons. No sling shot is listed. Falcons shot is listed, yet no gun. There are a number of replications of this. Walmer is listed with slings (*ibid.*, 4114) with no shot entry, yet amongst the iron shot is that for demi culverins but these guns are absent.

The assemblage for Camber Castle includes demi cannon, culverins, demi culverins, sakers and falcons. The numbers of shot listed range between 21 each (demi cannon) to 45 (demi culverin); the one saker, however, is listed with 220 shot. Four slings are listed, but no shot. If the saker shot could be used with both the saker and the four slings this would give 44 shot per gun. Providing bore diameter for a listed gun at a moment in time is always difficult. However, some of the inventory entries do provide information. The shot entry for God's House in Portsmouth lists saker shot of 3¼–3½in (83–89mm) and falcon shot of 2½in (64mm). In weight, this would correspond to 4lb 12oz–6lb 1oz and 2lb 2oz (*c*. 2.2–2.8kg, *c*. 1kg) using Bourne's tables of 1587 (1587, 14). One is more or less half the weight of the other and the tolerance is within the variance observed within the culverin/demi culverin ratio of half weights.

The shot listed for slings is usually cast iron. However, some entries include shot of dice and lead (Starkey 1998, 4538, 4539), shot of lead (*ibid.*, 4564, 6580) and stone cast about with lead (*ibid.*, 4634, for quarter slings). Similar types of shot are found within the culverin shot listings. There are a few entries for stone shot for slings (*ibid.*, 4664) but these are rare.

The inventory suggests that most slings are chamber pieces, usually with two chambers per hall. There are entries for broken chambers, for example 3967 '*Doble slinges of yrone with two broken chambers*'. The number of broken sling chambers is higher than any other group of guns, possibly a result of a large powder charge.

There are enough entries mentioning '*stocks*' to suggest that at least some of the slings were mounted. Where wheels are noted these are frequently these are frequently '*Truckells*' (eg, *ibid.*, 4578); however at Portland there were two '*Slynges of yrone with oone chamber a peice stocked and mounted with vnshod wheles*' (*ibid.*, 4719), and another sling '*lacking chambers mounted and stocked with wheles vnshodde*' (*ibid.*, 4721). The next entry describes three '*Sling chambers of yrone stocked and mounted with new Axeltries and troulces and for euery chambre a ladell and a Rammer*' (*ibid.*, 4722). Berwick has a demi sling with a ladle and sponge (*ibid.*, 6560, 6579).

There are several entries which provide tangible evidence to support the hypothesis that the sling group was the wrought iron counterpart of the culverin class of guns. Entries for Blackness clearly state '*Demy Culveryne sling of yrone*' and '*Demi Culveryne shott*' (*ibid.*, 5752, 5765). At this period a demi culverin would have a bore size of 4¼–4¾in (108–121mm), larger than the largest *Mary Rose* sling candidates.

**Table 3.9 Possible culverin equivalents to guns of the sling class**

| Type | Bore (in/mm) | Weight shot (lb/kg) | Shot size (mm) | Suggestions: slings | |
|---|---|---|---|---|---|
| Culverin | 5½/140 | 18/8.2 | 134 | – | double sling |
| Demi culverin | 4½/114 | 9/4.1 | 108 | double sling | sling* |
| Saker | 3½/89 | 5¼/2.4 | 83 | sling | demi sling* |
| Falcon | 2¾/70 | 3/1.4 | 64 | demi sling | quarter sling* |
| Falconet | 2¼/57 | 1½/0.7 | 51 | quarter sling | – |

* Preferred choice for identification

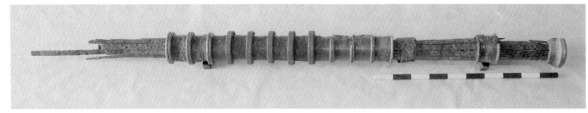

*Figure 3.66 Slings: restored barrel of 79A1295*

Although contemporary sources, like Norton, list a range of shot weights and bores for each type of gun, the standard English government weight for each is given in Table 3.9 together with two suggestions for their wrought iron sling counterpart. This gives two options; the full culverin equating to the double sling with the sling, demi sling and quarter sling equating to the demi culverin, saker and falcon; or the demi culverin equating to the double sling, the sling to the saker, demi sling to the falcon and quarter sling to the falconet. The first is reinforced by an analysis of the section totals for the *Anthony Roll* (Knighton and Loades 2000, 61). The descending order for iron guns is port piece, sling, demi culverin, saker, demi sling, quarter sling, fowler. Analysis of the shot recovered against shot listed suggests that there is nearly enough shot for either option.

Detailed study of the sizes of shot recovered (see Chapter 5) suggests the presence of excess shot of the following sizes; 95mm, 82mm and 60mm, with weights of about 3.4kg, 2.2kg, and 0.7kg (7.5lb, 4¾lb and 1½lb). Projected bore sizes would be about 95–103 mm for a sling, 83–88mm for a demi sling and 60–65mm for a quarter sling. We still cannot be definitive, and the bore range for any variation may be greater than we would expect.

## The scope and distribution of the assemblage

There are four guns which may represent a selection of the six slings listed: 70A0001, 81A0645, 78A1295 and 82A2700 (Table 3.10). Two were recovered with chambers and there are an additional four to six breech chambers, and six fragments which may be barrels or chambers.

The obvious difference between these and port pieces is the smaller bore relative to length (Fig. 3.64). The barrels of the complete port pieces do not exceed 2420mm in length (81A2604) and the greatest length measured in calibres is 16.89 (81A2650). The greatest length of sling barrel (79A1295) is 3000mm (over 35 calibres) see Table 3.10. The barrels are built of staves, with hoops and bands over these in a single layer.

Several examples (70A0001, 81A0645 and possibly 82A2700) show a reinforced area of a number of hoops forming a collar at the junction with the breech face (Fig. 3.65). Thereafter, each band is flanked by either a single hoop, or multiples of two or three hoops. The outer diameter of the gun is exceptionally large compared with the bore, giving a thick-walled tube. In both instances the external diameter of the hoops exceeds 200mm. The length of the powder chamber for 81A0645, at 755mm to the neck, is longer than many port piece chambers. The weight of the shot for its size is greater than stone and the barrel length and quoted powder charge suggest that range was a consideration.

79A1295 was accessed into the collection and is presumed to be from the Deane excavations (Fig. 3.66). It has a length of 3000mm and a bore of 85mm. Extensive corrosion has resulted in the staves being exposed for nearly half the length of the gun showing it to be made of many small staves. These appear flat rather than chamfered at the edges, unlike the profile of the port piece staves. One looks to be complete, giving the most complete length of barrel recovered. The gun has been restored with bands separated by single large hoops, although some of the original scarring suggests multiple smaller hoops (two or three) between the bands. There is no chamber associated with this gun.

The fourth gun (82A2700) is still in wet storage and is not accessible. The radiograph shows an incomplete, extremely degraded gun with a possible bore of 65–70mm and length of 1270mm. Excavation revealed a tampion and possible wad, so this may represent the remains of a corroded gun with chamber *in situ*. It has been given the location of O3, although there is some dispute and it may have derived from the Upper deck. The hoops give a maximum outer diameter of 250mm.

Chambers recovered include two breeches recovered in the nineteenth century, 80A2001 (RA XIX.163) and

**Table 3.10 Sling barrels (metric)**

| Object | Sector | Length | Bore | No. calibres | Max. ext. diam. | Width/thickness bands | Width/thickness hoops |
|---|---|---|---|---|---|---|---|
| 81A0645 | U6 | 2240 | 90–115 | 24.88 | 190–220 | 70–75/20 | 25–30/? |
| 70A0001 | ?bow | 2295 | 95 | 24.15 | 200 | 55–80/? | 25–30/? |
| 79A1295 | ? | 3000 | 85 | 35.29 | 175–255 | 85–100/8 | 30–35/20 |
| 82A2700 | ?O3 | 1270 | 65 | 19.50 | 200 | 80–130/? | 20–30/? |

80A1952 (E121; Fig. 3.67), one recovered from U6 (81A0772) and one from SS10 (90A0099; Fig. 3.67 and Table 3.11). A concretion identified as a chamber during excavation (Ewens, divelog 81/DM/147) is unavailable for study but is included based on location and size (81A0783). A chamber shown in a photograph beside the Woolwich port piece may be a sling chamber, but this is not available for study at present. There are further fragments which may represent sling fragments; three from the nineteenth century salvage excavations (78A0611–3). There is one fragment from U6 (81A0708) and two from C8 (81A0803, 81A0765) (Fig. 3.80 and Table 3.12).

The lack of chambers for these guns is confusing as they are thick walled and relatively durable. If two chambers were carried per gun, there should be twelve. A similar type of gun found on sixteenth century New World sites has been recovered with four different types of chamber (Olds 1976, 70). Some of these have upstanding hoops and others do not. Although some of the external features may be partially a result of differential corrosion, parallels can be seen with the *Mary Rose* examples. The Gulf of Mexico bombardeta chambers include those with a single sheet inner core, a mid layer of bands and a final layer of spaced hoops. These vary in length from 555mm to 845 mm. The second type have a rolled inner tube and second layer of thick but narrow hoops/bands without the final layer of spaced hoops; this is similar to the abutting hoops seen in some *Mary Rose* examples. The third type is similar, but with two hoops, one at each end to accept lifting rings. The fourth is much smaller and thought to be a remodelled *verso* (base) chamber. The condition of these seems much better than many of the *Mary Rose* examples and although attempts were made to type the chambers, there are too many variables which do not seem to match. Chambers from the *Mary Rose* seem to include examples with visible inner staves and without staves, chambers with double inner tubes and chambers with single inner tubes. Some have abutting wide hoops and others both hoops and bands in different arrangements. Detailed radiography from a number of angles with specific questions may be required together with the sectioning of some examples.

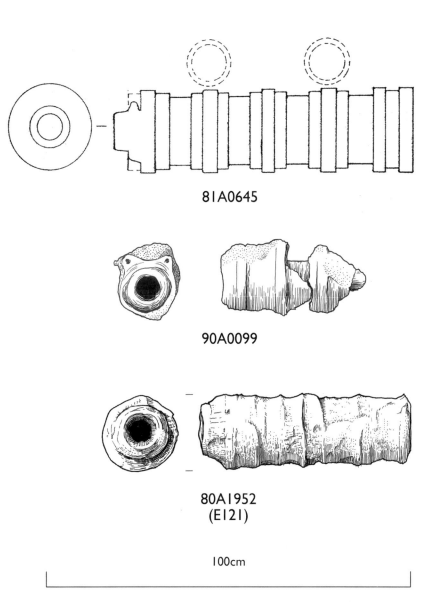

*Figure 3.67 Sling chambers*

The locations for many of the slings are unknown. The only complete example was found in the waist on the Upper deck. The remainder are from the stern on the Castle deck or under the Sterncastle (Fig. 3.68), apart from 82A2700, possibly from O3. If this location is correct, this gun may have derived from U3 port side

**Table 3.11 Sling chambers (metric)**

| Object | Sector | Length | Bore | Max. ext. diam. | Width bands | Width hoops |
|---|---|---|---|---|---|---|
| 81A0645 | U6 | 755 | 60 | 200 | – | – |
| 81A0772 | U6 | 530 | 100 | 225 | 80 | 30 |
| 80A1952 RAI | ?/?Deane | 600 | 80 | 475 | 140 | – |
| 90A0099 | SS10 | 400 | 50 | 200 | 75–80 | 30 |
| 80A2001* | ?/Deane | 470 | 50 | 140 | – | – |

\* poss. port piece, see also Table 3.4; SS = Starboard scour

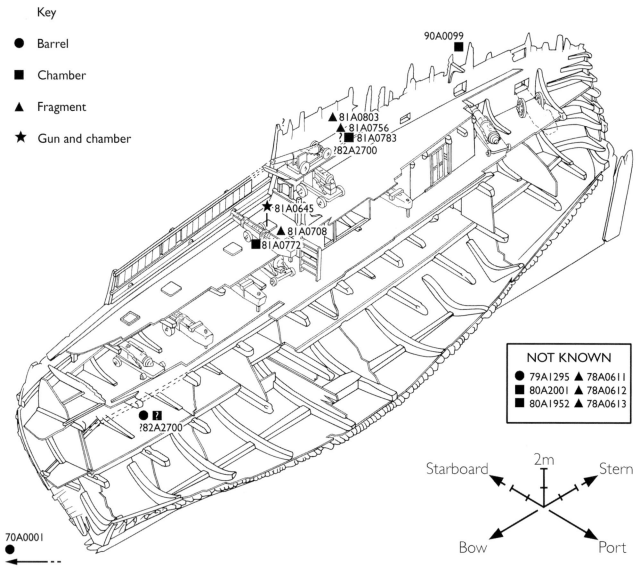

*Figure 3.68 Distribution of slings and possible sling fragments*

firing from a semi-circular gunport in the open waist as suggested in a model of the ship which has three gunports on the starboard side (Fig. 3.69). Fragments of barrel and chamber were also found in C8 and SS10, however, suggesting that not all these guns were in the waist. The Hatfield paper suggests that one may have been positioned on either side of the Forecastle (see Chapter 12).

*Figure 3.69 The Bassett Lowke model of the* Mary Rose, *showing the putative positions of slings at semi-circular gunports on the open waist*

## Barrels/barrel and chamber fragments

### 81A0645. Barrel with chamber (incomplete)

*Location*: Upper deck, U6
*Recovered*: excavated; on bed with single axle and trucks, elevating post; chamber, shot (82A4215), tampion (82A4222), wad, powder (82S1152)

*Statistics*
Length: barrel: 2240mm, breech: 755mm; bore: 90mm (barrel & breech) 34 incomplete; length in calibres: 24.88. Weight: 1200kg

The gun was first seen in 1979. It was sitting on its carriage (81A0640) on an axle (81A0640/679) with one wheel attached (81A0679). The gun barrel had become concreted to a large anchor. The barrel is broken at the muzzle. The breech is degraded in parts but has a slightly tapering neck of 80mm. The gun ends 80mm before the end of the bed (Fig. 3.70). The internal diameter at the breech is 90mm, at the bore this is 115mm.

The external diameter of the inner tube is 130mm, and external diameter on the bands is 170mm and on the hoops 220mm. It is possible to count 19 bands (width 60–75mm) and 16 single hoops (width 25–35mm) and three sets of double hoops, the central set sometimes raised (Fig. 3.70). The total wall thickness over the bands is 40mm. Radiography showed that the inner tube was formed of four large staves, three of 125mm, and one 205mm all about 20mm thick. The bands and hoops form a single layer directly over these, with spaces between all elements. The area of the breech junction is heavily reinforced, with possibly seven hoops on the barrel locating with three on the chamber. This appears to be solid rather than hooped and banded.

The chamber has an external diameter at the neck of 85mm which tapers to 110mm at the shoulder (Fig. 3.71). The external diameter over the hoops is 210mm and over the outer tube 180mm. The bore is 60mm. Radiography suggests an inner layer of three staves (two joins are visible), *c.* 15mm thick covered by two semi-circular staves forming an outer tube, also 15mm thick. The joins of the staves forming the inner tube are offset from those forming the outer tube. The hoops (40mm wide, *c.* 40mm high) appear to sit directly over this and abut each other with no bands, though the corrosion is too severe to be certain. The breech plug (diam. 50mm) is clearly visible suggesting that it is set in with the staves curling over to hold it in place.

The back of the carriage bed is extremely eroded, reflecting the time it was exposed during the excavation. In width this varies from 110mm at the front to 480mm at the forelock position and 400m at the back. The chamber and barrel were held by a wooden forelock (Fig. 3.71). The carriage (length 3840mm) is single piece of elm grooved out to take the barrel and stepped up to provide a solid block behind the chamber. Two 250 x 200 x 35mm rebates for securing ropes ('*woling*') are present with a 180 x 20mm rebate underneath,

*Figure 3.70 Sling 81A0645: above) close up of barrel; below) shot*

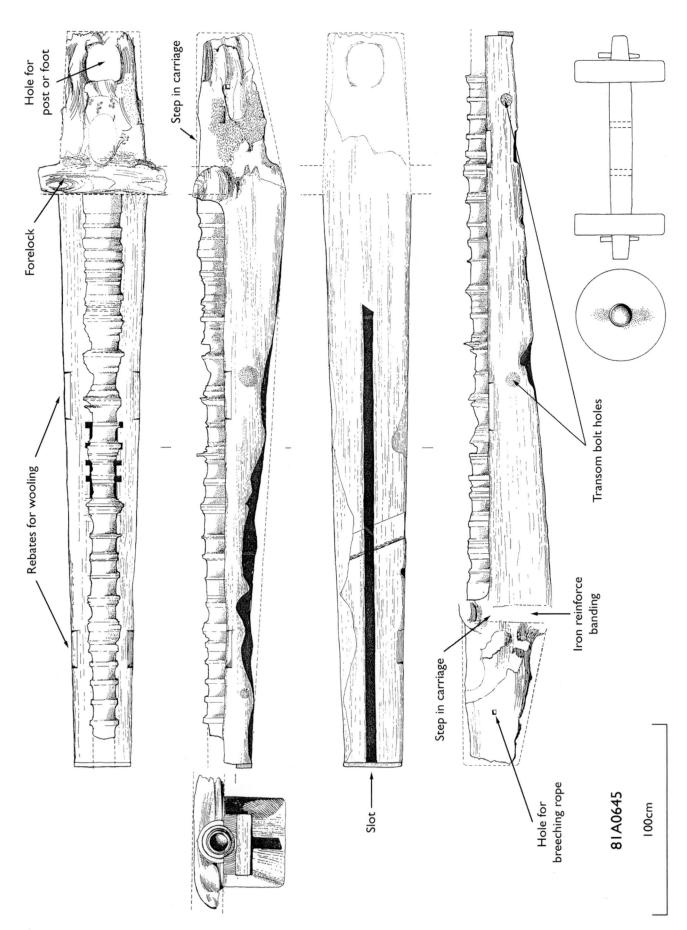

*Figure 3.71 Sling 81A0645: gun on carriage*

possibly for the single axle. There is no archive illustration showing axle bolt holes but they must be present. Also underneath is a 60mm wide slot extending 2400mm towards the back. The sides are held together by two transverse iron bolts, with iron staining displaying a head diameter of 65mm. There is another, 25mm square, bolt hole in the side of the bed towards the back, possibly for a breeching rope. The thickest part of the bed is behind the breech and there is a change in angle at the front of the forelock position. Iron reinforcing straps (100mm wide) are situated at the forelock position. The forelock itself extends over the side of the bed by approximately 100mm each side. An angled rebate, 130mm wide and 10mm deep, has been cut in the underside of the bed 1650mm from the shoulder. The elevating post (79A0184) was found close by. It has eight 25mm holes set at irregular distances and staggered across the face (not illustrated).

An iron shot (diam. 84mm) was recovered from the junction with the breech (82A4215; Fig. 3.70). Excavation of the chamber revealed fibres (82S1152/1) a tampion (82A4222) and a straw wad (82S115/4). As excavation progressed further down the breech the contents became liquid with fragments of concretion within; presumably the remains of gunpowder (82S1152, 5–8).

**70A0001 Incomplete barrel**
*Location*: unknown
*Recovered*: 100m north of the bow; no bed; shot and wadding *in situ* in barrel

*Statistics*
Length: 2295mm; bore: 95mm; 24.15 calibres

This gun was one of the contributory factors to finding the wreck site as it was the first object found during Alexander McKee's searches (McKee 1982, 86). As it was found away from the ship and exhibited a secondary layer of corrosion, it is suggested that this might have been the gun lifted and dropped during the nineteenth century salvaging (Bevan 1996, 161–91). It is complete at breech and broken towards the bore (Fig. 3.72). The outer layer is primarily formed of bands (13) separated by 12 triple hoop assemblages, with the exception of the broken endwhich has three single hoops. This latter feature is common towards the ends of all the bedded guns.

Radiography revealed three joins suggesting three staves with widths of 195mm (x2) and 175mm. They are not overlapped or hammered and the junctions show up as clean lines (see Fig. 3.63b). The bands (55–80mm wide) and hoops (25–30mm wide) are placed directly onto the staves with gaps between the bands. The gun was recovered with an iron shot (89mm diam.) and flax and wadding.

**79A1295 Incomplete barrel**
*Location*: unknown
*Recovered*: assumed by nineteenth century salvaging; found with bed fragments (80A1957)

*Statistics*
Length: 3000mm; bore: 85mm; 35.29 calibres

This was accessed to the Royal Navy Trophy Centre and loaned to the Mary Rose Trust through the Royal Naval Museum. It is badly corroded and heavily reconstructed in fibreglass. The barrel comprises eight staves each 30mm wide and 5mm thick covered by an outer layer of

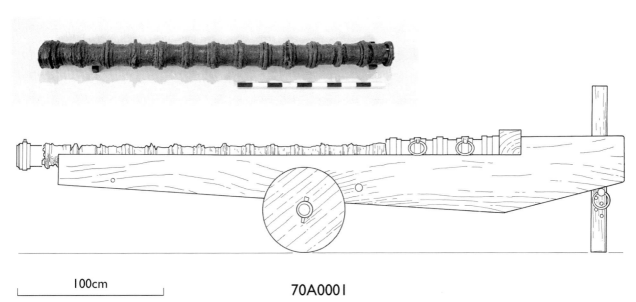

*Figure 3.72 Sling 70A0001: barrel and reconstruction drawing of gun on carriage*

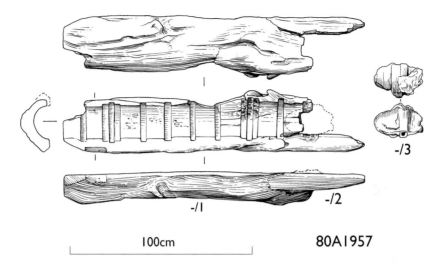

*Figure 3.73 Slings: bed fragment 80A1957*

alternating bands (85–100mm wide, 25mm thick) and hoops (30–35mm wide, 38mm thick) (Fig. 3.66).

The bed fragment (80A1957; Fig. 3.73) came from the same loan source and is considered together with the barrel. It is extremely fragmentary and consists only of a 1350mm long section of the left side with two smaller fragments. The maximum width is less than 300mm and height less than 200mm. It appears to be a finished end, with a rope rebate. Confusingly, there is no evidence for the slot in the centre of the bed, but it may be that this is where the break occurs and it has eroded to an uneven edge. There are recesses for reinforcing hoops both in triple sets (2) with the central hoop upstanding, and single hoops (4). However, this sequencing cannot be found on 79A1295 (although single hoops between triple sets can be seen on 81A0645).

The measured hoop grooves are 50mm in width, too large for the gun. It looks remarkably like port piece carriage 80A1955 and may be a fragment of a larger carriage (possibly 80A1954).

### 82A2700 Barrel fragment (possibly including chamber)

*Location*: unknown

*Recovered*: uncertain ?O3 or Sterncastle with associated wood fragments

*Statistics*
   Length: 1270mm; bore: ?60mm; 19.5 calibres

This is an extremely degraded fragment of gun barrel (Figs 3.74–5), with wooden bed fragments associated. There are hardly any hoops present but these show up as 'woody' images on the radiographs (Fig. 3.75a). One of the broken ends suggests that the inner tube is probably stave built, with a stave thickness of 20–25mm. The radiograph shows some bands and hoop sequences, apparently nine bands (width 80–130mm) separated by eight sets of triple hoops (width each hoop 20–30mm). Concretion removal suggested two layers and a ring at one end, possibly the reinforcing of the breech end as shown in 81A0645 and 70A0001. The

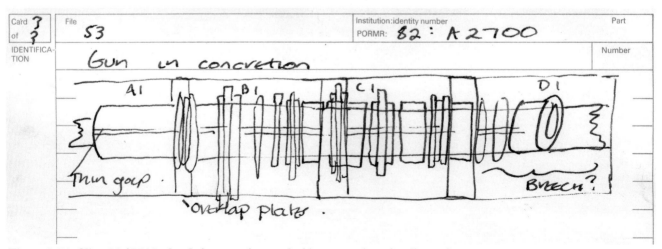

*Figure 3.74 Sling 82A2700: sketch from artefact card of interpretation of radiograph*

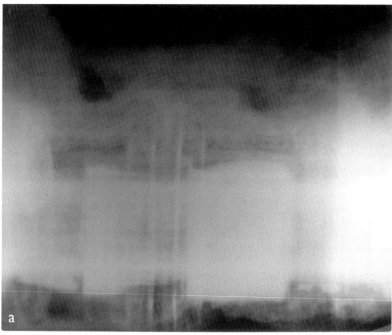
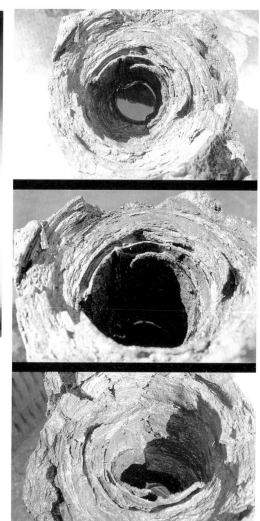

*Figure 3.75 Sling 82A2700: a) radiograph; b) layered nature of one end of the gun*

inner layer displayed a longitudinal grain to the corroded iron, suggesting inner staves 8mm thick. Around this, a second layer with circumferential grain was visible, 5mm in thickness, over which was a hoop, 45–50mm in height. The remains of a 25mm ring were found at one end of the concretion. It was thought that this may be the remains of a breech chamber, but this was not visible on the radiograph.

When recovered the gun was encased in concretion that had two distinct angular faces to it as though it had formed between the deck and side of the ship. Wood grain was impressed into the surface. A bill haft (82A5994) had become concreted onto the top of the gun. Most of the bills were on the Upper deck in the stern, possibly suggesting the original location of the piece. Severe blackening to the head end of the haft suggests that the head may have been present when it fell against the gun. All the concretion recovered from this gun was sieved after four brass aiglets (clothing fastenings) were found. These proved to be associated with remains of folded material with a visible weave. An arrow and cask stave were also within the concretion, over a hoop which had a thin layer of concretion, suggesting that it had begun to corrode before these objects became attached. Examples of all these items occur in U8/9/U10. Concreted iron gun fragments were found in U10 so this location should be considered.

A break occurs at the junction between two hoops towards one end revealing an 8mm twisted rope passing vertically down into the bore of the gun. The gun was corroding badly and removal of the concretion physically destroyed hoops and bands, the hoops being preferentially corroded rather than the bands, probably because they stood above the level of the rest of the barrel. The corrosion was not uniform over the length of the gun, with some areas markedly worse than others. The opportunity was taken to section the gun along an existing fracture line, revealing the internal structure and the junction between the concretion and the surface of the gun. The inner tube consisted of either large staves or semicircular troughs of iron, 12mm thick. Where a band covered this it was extremely dense, but only 3mm thick. It appeared that a second tube had been formed of abutting bands. Between this and the sets of hoops and bands was a 20mm gap. It looked as though the banded layer had been welded, with metal entering between the seams, though this may merely indicate portions of metal from the placing of the heated elements forming the final layer over the second falling into the joins. There was clearly one inner layer with longitudinal grain and two outer layers with circumferential grain (Fig. 3.75b).

It is not known whether this area represents a barrel or breech chamber, or the reinforced part of the breech

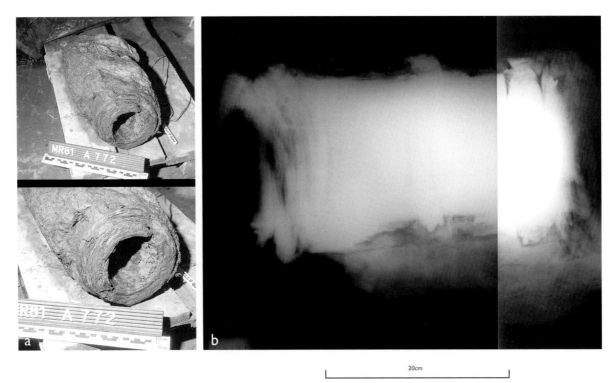

*Figure 3.76 Sling 81A0772: a) photographs; b) radiograph*

end of the barrel, but the small diameter suggests a chamber with a neck of 80–100mm diameter and thickness 35mm. A portion of fibrous textile, possibly wool, was found entering the gun vertically at a distance of 130mm from the break and could represent a powder bag. Oily, fine-grained matter was recovered from inside the bore of the gun. The remains of a negative cast of a lifting ring were visible within the outer concretion.

Investigation of the other end of the object suggested the same number of layers, but the middle layer was thought to consist of a second series of semicircular longitudinal elements rather than bands. Here the inner tube was 7mm thick and the middle section 8mm with an outer layer of hoops and bands. Twisted rope and a tampion impregnated with iron were found inside.

*Breech chamber fragments*

### 81A0772 Possible breech chamber
*Location*: ?U6
*Recovered*: excavated U6

*Statistics*
Length: 530mm; bore: 100mm

This fragment (still in wet storage) is represented by a portion of barrel or breech with lifting ring in situ (Fig. 3.76). It was classified as a breech chamber, although the radiograph does not show a solid end but does reveal what seems to be the neck extension to a chamber (unless it is merely erosion). The radiograph shows a multi-staved inner tube (at least four are visible; Fig. 3.76b). The portion consists of three hoops (30mm wide) and a band (80mm wide), a single hoop with lifting lug and then a band. The lifting/securing ring has an external diameter of 130mm and the ring diameter is 20mm. This could represent a chamber with lifting ring, or rings on the gun, though the bore size (100mm) is more in keeping with a gun rather than a breech, and is 40mm larger than the chamber for 81A0645.

### 81A0783 Possible breech chamber
*Location*: C1/2
*Recovered*: ??

*Statistics*
Length: 550mm; maximum external diameter: 195mm

Described by the excavator as a breech chamber and associated with other fragments of iron gun (81A0803). Both the radiograph and the object (not illustrated) are currently inaccessible, so identification must remain tentative.

### 80A1952/E121 Breech chamber
*Location*: unknown
*Recovered*: unknown position, nineteenth century salvaging

*Statistics*
Length: 600mm; bore: 80mm

This fragment (Fig. 3.77) is blocked at one end and appears to have a touch hole and a hole drilled through

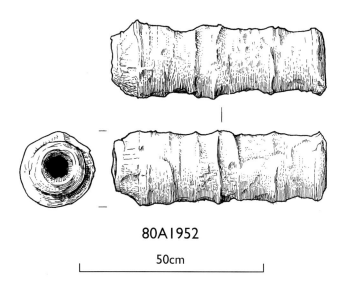

*Figure 3.77 Sling chamber 80A1952*

the blocked end (possibly for metal analysis). The external diameter at the neck is 130mm and at the shoulder 200mm, providing a wall thickness of 60mm at this point. The neck extension is 70mm with a thickness at the mouth of 25mm. Radiography suggests two stave junctions measuring 75mm and that the outer surface is covered with only a single layer of bands each 140mm wide and up to 35mm thick.

**90A0099 Chamber fragment**
*Location*: unknown
*Recovered*: on seabed SS10

*Statistics*
Length 400; bore 50mm

This chamber portion (still in wet storage) was found loose in the Starboard scourpit following recovery of the ship (Fig. 3.78). Wall thickness at the mouth is 50mm, indicating a strongly reinforced chamber. It is badly corroded but appears to be alternating hoops (30mm wide, c. 35mm thick) and bands (75–80mm wide). The object is still partially covered in concretion, with corrosion extending down to the 100mm diameter inner tube in one place. Two lifting lugs are situated on the central hoops. The chamber is incomplete and there are few diagnostic features present.

**80A2001 Breech chamber**
*Location*: unknown
*Recovered*: unknown position, nineteenth century salvaging

*Statistics*
Length: 470mm; bore: 50mm

This chamber has not been radiographed or drawn and was part of a loan from the Royal Armouries (248) grouped with a number of items listed as XIX.163. Photographic recording suggests a chamber made of alternating bands and hoops, but few in number (Fig. 3.79). Possibly the hoops are only at the mouth and back and at two places on the chamber for lifting rings. This is rather small for a port piece chamber, hence its classification within the sling group.

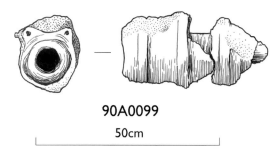

*Figure 3.78 Sling fragment 90A0099*

*Figure 3.79 Sling fragment 80A2001*

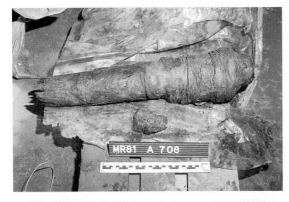

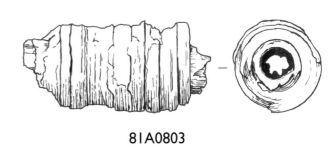
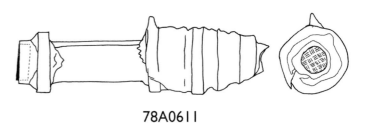
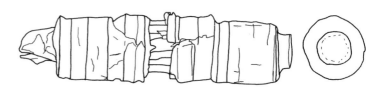

*Figure 3.80 Sling fragments 78A0611, 78A0612, 78A0613, 81A0708, 81A0756 and 81A0803*

**81A0708 Barrel or chamber fragment**
*Location*: U6
*Recovered*: excavated; U6, within concretion between the Upper deck and the gunwale

*Statistics*
Length: 680mm; bore: 70mm

This fragment (still in wet storage) resembles the U6 gun (81A0645). A comparison of the radiographs suggests that it may be smaller. The radiograph shows a tube with an external diameter of 110mm with abutting hoops to one end, bare tube to the other. Three staves are visible and possibly two layers. The hoops vary in width from 30mm to 55mm. The wall thickness over the staves is 20mm and over the hoops 65mm. This also has abutting hoops rather than the alternating hoops and bands of 81A0645. This could be an incomplete second breech chamber for that gun (Fig 3.80).

### 81A0756 Possible barrel or chamber fragment
*Location*: Sterncastle?
*Recovered*: excavated: SC8

*Statistics*
Length: 440mm; bore: 100mm

The radiograph shows three clear lines, suggesting a stave built inner tube reinforced at one end with three hoops (30mm wide) directly over the staves. The outside of the tube is 130mm in diameter, suggesting a wall thickness of 30mm and internal diameter of 100mm. There is a slight flare at the end showing the hoops. The radiograph appears to show a solid end, but there is evidence for mud and concretion within the bore (Fig. 3.80).

### 81A0803 Probable chamber
*Location*: Sterncastle
*Recovered*: SC8

*Statistics*
Length: 500mm; bore: 100mm

This very corroded tube fragment looks more like a chamber than a barrel and appears to comprise two layers of staves with a maximum wall thickness of 22mm. Nine hoops are visible, apparently sitting directly on the staves without any bands. They are 38mm thick and 30–50mm wide. One end of the object appears tapered as though forming a breech neck, but this could just be corroded staves. The junctions of the staves appear staggered (as in Fig. 3.39d), and there seem to be three, 22mm thick and up to 40mm wide. The internal diameter at the neck is 80mm (Fig. 3.80).

### 78A0611, 78A0612 and 78A0613 Barrel or chamber fragments
*Location*: unknown
*Recovered*: unknown, nineteenth century

*Statistics*
Length: 725mm, 725mm, 960mm; bore: 60/75mm, 75/80mm, 60mm

Three further barrel/breech fragments were recovered by John Deane. 78A0613 is most likely part of a barrel (Fig. 3.80). 78A0611 comprises large staves (22mm thick) possibly two quarters and a half to form the inner tube. There are no visible bands and the hoops (35mm wide and 35mm thick) appear to lie directly over this. The radiograph looks similar to 81A0645. 78A0612 appears similar to 78A0611, but with hoop widths of 30-40mm and a total estimated wall thickness of 50mm. 78A0613 appears to have two layers of staves (12mm thick) and reinforcing hoops (40mm wide, 40mm thick) but no bands. The eroded remains for two pairs of lifting rings were visible in 1981 (see Tables 3.11–12).

**Table 3.12 Sling fragments (metric)**

| Object | Sector | Length | Bore | Ext. max. diam. |
|---|---|---|---|---|
| 78A0511 | ?/Deane | 725 | 75/80 | 185 |
| 78A0612 | ?/Deane | 725 | 75/80 | 180 |
| 78A0613 | ?/Deane | 960 | 70 | 165 |
| 81A0803 | C8 | 500 | 100 | 240 |
| 81A0708 | U6 | 680 | 70 | 180 |
| 81A0756 | C8 | 440 | 100 | 210 |

## Fowlers

### Identification and historical importance

The *Anthony Roll* lists the *Mary Rose* as having six iron guns called '*fowlers*'; the same number are recorded in the 1540 inventory (PRO E 101/60/3). Blackmore (1976, 230) has traced fowlers listed as being in the Tower of London from 1540 until 1635. It seems clear that, together with port pieces, these are some of the 'newer' wrought iron guns resulting from attempts at rationalisation of wrought iron ordnance, completed by 1540, which were eventually superseded by bronze. Bronze fowlers are recorded in the Tower in 1589, however the 1547 inventory (Starkey 1998) already records five fowlers of bronze, one of which is a muzzle-loader. During the seventeenth century the bronze examples are listed with weights, the barrel weighing 5 hundredweight (254kg) and the chamber one hundredweight (50.8kg). Clearly these are relatively small. There is no good sixteenth century description of a fowler. The best is that of Robert Norton (1628, 57–8), fully transcribed above (see pp. 132–3) by which stage these were cast in bronze. The description suggests a breech-loader set into a carriage with an elevating post and held with a wedge. The major difference between these and port pieces appears to be size.

There is still a lack of information regarding their size during the sixteenth century. It is possible that these are a derivation of '*veugler*', much larger wrought iron guns common in the fifteenth century (Blackmore 1976, 230), although the description better suits port pieces. The weight listed for these suggests 2141 lb (*c.* 971kg) for one barrel and two chambers.

*'4 iron cannons called veuglers, of which each is 8 feet long, with 8 chambers for them, of which cannon 2 can throw a stone shot 9 inches in diameter, weighing in all 8,566 lb.'*

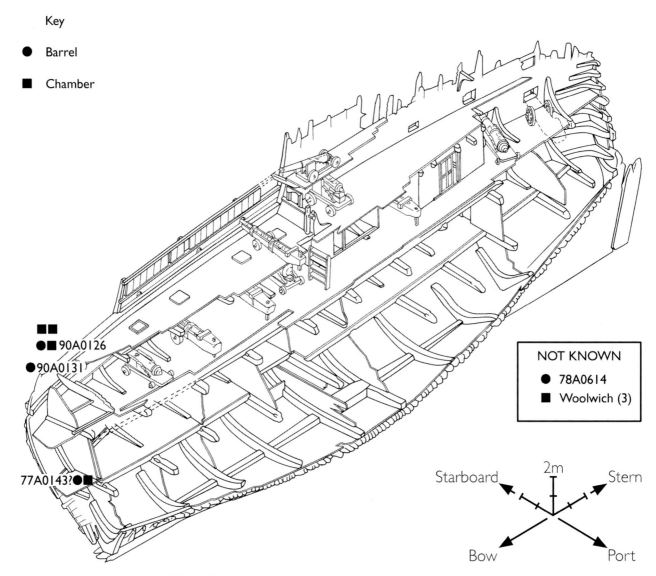

*Figure 3.81 Distribution of possible fowlers*

In 1546 fowlers are present on twelve of the 20 ships, in numbers varying from nine on the *Pauncy* to two on the *Hoy Bark*. The total listed in the *Anthony Roll* for the 20 ships is 54 (Knighton and Loades 2000, 61). Their relative importance on ships can be suggested by the numbers of named guns carried. The bases are first (small swivel guns), with 491, followed by the hailshot pieces 272. Of the carriage mounted guns, port pieces are first (144), followed by the demi slings (57), the fowlers (54), top pieces (22) and slings (20).

The position of fowlers in the inventories varies. In the summary at the end of the *Anthony Roll* (*ibid.*, 61) demi slings and quarter slings are listed after sakers but before fowlers. The *Hoy Bark*, however, has port pieces but the only slings listed are quarter slings. The fowlers are listed before these. There are a number of entries where fowlers occur after demi slings but in these cases there are no quarter slings. All we can be certain of is that they are smaller than port pieces but larger than quarter slings.

Seven of the fifteen '*galliasses*' carry seventeen fowlers between them, with six on the *New Bark* and 2–3 on each of the others. These seem to occur after demi slings, unless there are quarter slings present (the *Lion*). In this instance fowlers are listed after the quarter slings. Of the ten pinnaces only two carry fowlers, the *Saker* (2) and the *Less Pinnace* (2). The latter shows an interesting listing: saker, falconet, fowler, base. Whereas we have no confirmed indication of the size of the sling group, the parameters of a falconet are well defined. Bourne (1587, 71) lists these as having a bore of 2¼in (57mm), shot of 2in (51mm). No fowlers are recorded on the rowbarges.

Fowler shot is listed under the heading of '*shot of stone and lead*', usually 20 rounds per gun though there are exceptions: the *Mary Rose* (28), the *Henry Grace à Dieu* (12), the *Peter* (15) *Pauncy* (11), the *Grand Mistress* (30) and the *Less Pinnace* (40). The latter had no other wrought iron guns except bases. The 1547 inventory (Starkey 1998) usually refers simply to stone shot, but is sometimes more specific, as in '*Shot of stone for fowlers*

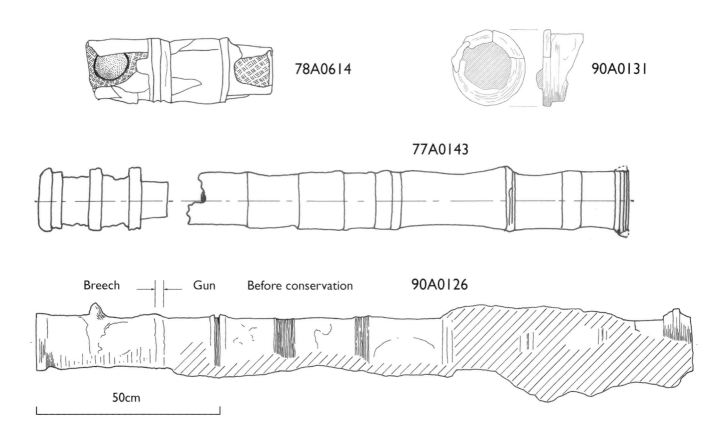

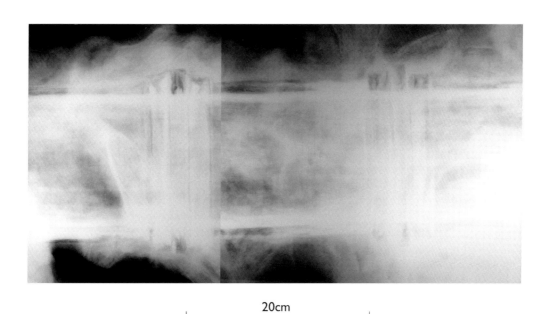

*Figure 3.82 Possible fowlers, including radiograph of 90A0126*

*Serpentynes and other'* (*ibid.*, 5548). There are two references to fowler shot of iron, and in another place '*Fowler shotte of Iron*' has been corrected to '*stone*' (3941). Several entries suggest composites, such as '*Fowlers shott of leade and stone*' (5352). Of the 51 vessels with detailed inventories in 1547, 23 have fowlers (Starkey 1998, 144–57). The fleet as a whole carried 79. The bulk of these are on the ships or galleys, although the *Peter* (ship) does not have any. The largest number are carried on the *Trinity Harry* (9), followed by the *Pauncy* (8). The *Jesus of Lübeck* and the *Sweepstake* have six each, and the others have four or fewer. These are

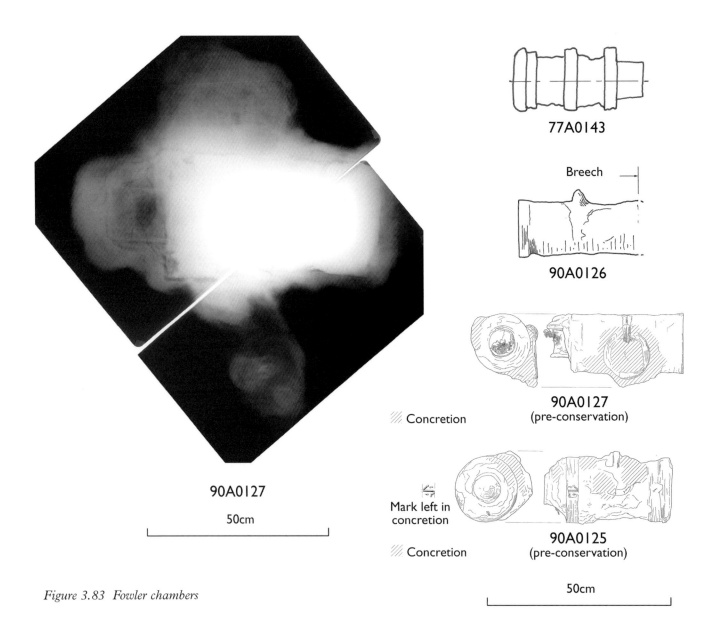

Figure 3.83 Fowler chambers

always listed under guns of iron, and where there is any detail, they are listed with two chambers each. As in 1546 the position within the listed guns varies from ship to ship but always following port pieces and usually before demi slings. The shot is undifferentiated, and one assumes that the fowlers are catered for by '*shot of stone of all sorts*'.

About 40 of the 76 fortifications support a total of 284 fowlers. The largest number are in Calais (67), Guisnes (60), Boulogne (24), Hammes (14) and Ruysbank (10). It is from this inventory that the greatest detail regarding their form is to be found. Most are iron, however there are five of bronze. Three of these are in Calais (Starkey 1998, 4808 (two guns) and 4819, the latter '*being a whole peice*', (ie, either complete or a muzzle-loader). Hammes has a further two '*not good*' (*ibid.*, 5384). It is clear that most are breech-loaders, often with two chambers per barrel. These are variously described: '*Fowlers of Irone mounted vppon high wheles withe v$^e$ chambers*' (3926), '*ij stocked and the iij$^{de}$ vnstocked with iiij$^{or}$ Chambers*' (4037), '*their stockes broken*' (4478), '*mounted vppon their cariage of Trockells with ij chambers and ij forlockes of yrone*' (6260), '*mounted vppon Truckells*' (6701; similarly 6726, 6864). An entry for the forge at Berwick lists four '*Fowlers of yrone dampned*' (6458). Other than position relative to other guns of better known dimension, there is little indication of size. Landguard Point distinguishes '*Fowlers gret with ij Chambers*' and '*Small fowlers with ij Chambers*' (7039-40). As on the ships, they invariably occur after port pieces and usually after demi slings and before quarter slings. In two cases they occur after falcons (3899, 5307) which have a bore of 2¾in (70mm; Bourne 1587, 70).

It is probable that these guns were carriage-mounted. Norton suggests that, by 1628, their breech

**Table 3.13 Fowler barrels (metric)**

| Object | Sector | Length | Bore | Max. diam. |
|---|---|---|---|---|
| 77A0143 | above O1/2 | 1180 | 110 | 150 |
| 78A0614 | ?/Deane | 520 | 110–120 | 160 |
| 90A0126 | beneath bow SS | 1426 | 110 | 190 |
| 90A0131 | beneath bow SS | 103 | 120 | 195 |

chambers were three times their bore and the barrel eight times the bore (Norton 1628, 57–8). Their size, obviously smaller than port pieces and larger than bases, suggests that they would have been situated on the Upper and Castle decks. It is precisely these locations on the *Mary Rose* which were worked both by Tudor salvors and also during the nineteenth century. The best chance of finding a representative is therefore in the Starboard scourpit or within the assemblage of wrought iron recovered by John Deane. All the examples that have been identified are from the bow bulkhead and Starboard scour in the bow (Fig. 3.81). As the size is uncertain, the distribution of stone shot might provide a potential location and there are two small stone shot (75mm), both from SS10. There are ten stone shot of 95–102mm. Where known, their locations are in the Starboard scourpit in the bow (SS01) or the stern (SS10–11). Three were recovered under the bow bulkhead and several more from the 2004 bow excavation. The large number of unfinished stone shot make matching shot to guns difficult and it is assumed that no great store of finished stone shot was carried (see Chapter 5).

There are four gun fragments which may represent fowlers: 78A0614, 77A0143, 90A0126 and 90A0131; the last two may be part of the same object (Fig. 3.82; Table 3.13). The barrels appear to be thin walled with wide bands and thin hoops; all are broken into small sections. There are four chambers, one with 77A0143 and three found in the same concretion as 90A0126, one still concreted to it. All are small, less than 350mm in length with a maximum diameter of 185mm. The outer surface is not reinforced with many hoops, possibly only three; one in the centre carrying lifting rings and one at each end, in addition to a final breech hoop (Fig. 3.83, Table 3.14). It is impossible to tell how the inner tube was constructed. A chamber of this form is shown alongside the Woolwich port piece in an early photograph, which may be a fifth example of this type.

The tentative identification of fowlers is based on the finding of a small number of similar barrel fragments and breech chambers, one of which was associated with stone shot (78A0164). Although the bore sizes (115–38mm) are larger than the thickness of the barrel walls, the size and form of the chambers and the arrangements of hoops and bands on both chamber and barrel are different to both port pieces and slings.

*Barrel and chamber fragments*

### 78A0614 Barrel fragment
*Location*: unknown
*Recovered*: unknown, nineteenth century

*Statistics*
Length: 520mm; bore: 115mm max

This was given to the Mary Rose 1967 Committee by the Dockyard and is presumed to have come from the Deane excavations. It is broken at one end and has a 3mm layer of concretion sealing the other that may be hardened silt (Fig. 3.84). It is just possible that this represents the back end of a breech chamber, although the opposite end to this is blocked with a stone shot of about 90mm. The wall thickness throughout is less than 10mm at the most and the entire object is extremely badly degraded. Over the length there are scars showing two 25mm wide hoops 140mm apart. No stave junctions are visible on the radiograph but physical examination of the piece suggests 5–7 staves with a maximum measurable thickness of 8mm. The hoops do not protrude above the level of the tube and there is no indication of any bands. A denser area, 110mm behind

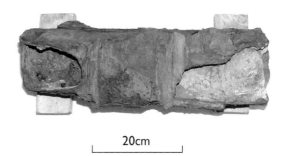

*Figure 3.84 Fowler 78A0614*

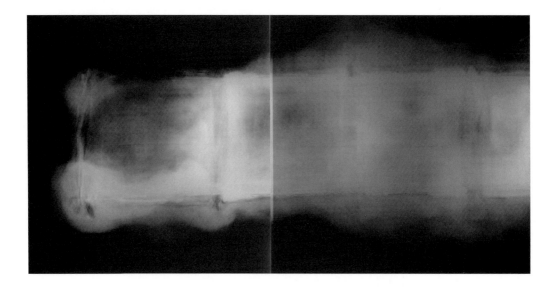

*Figure 3.85 Fowler 77A0143: above) barrel, radiograph; below) chamber*

the shot, may just represent a junction between chamber and barrel (the most reinforced part of the gun), with the shot in the barrel. The void in this area decreases to 55mm before widening again towards the back where the metal is much more degraded. This would give a chamber length of just over 300mm. This is the only evidence for a gun smaller than a port piece with a stone shot.

**77A0143 Barrel fragment and chamber**
*Location*: unknown
*Recovered*: 1977, uncertain bow above O1–2

*Statistics*
Length: 1180mm; bore 110mm; chamber: length: 270mm; bore 72mm

The gun is associated with a small chamber but no shot is recorded. It bears a broad arrow mark. The radiograph shows two clear stave junctions, with a width of 50mm, so it is clearly a stave built inner tube but there are no signs of any staves at either end (Fig. 3.85). The inner tube has a thickness of only 5mm and the maximum thickness of metal at the breech is only 30mm. The outside is covered with extremely thin alternating hoops (possibly several together) and bands, but many of the hoops are missing. Towards the breech junction there is a series of four hoops together.

The chamber is complete (Fig. 3.85). Two layers are visible, the central tube (structure unclear) within hoops and a band. The neck is 70mm in length with an external diameter at the mouth of 90mm tapering to 106mm at the shoulder. The thickness of metal at the chamber neck is 12.5mm. The plug is visible in the back and has a diameter of 75mm. There is one hoop (35mm wide) at the back of the neck to match up with the reinforced hoops on the gun, a centrally placed, 5mm wide, second hoop, and two more at the rear. These project about 37mm above the tube and 28mm above the band, which are *c*. 9mm thick and 88–85mm wide. There is one visible weld on the end hoop.

**90A0125 Chamber, 90A0126 Barrel and chamber, 90A0127 Chamber, 90A0131 Barrel fragment, all in 90A0050 concretion conglomerate**
*Location*: unknown
*Recovered*: unknown, bow

*Statistics*
90A0125 length: 370mm; bore: 75–80mm
90A0126 length: 1796mm; bore: 110mm
90A0127 length: 370mm; bore: 75–80mm
90A0131 length: 103mm; bore: 112–28mm

An extremely large concretion (2080 x 2060mm) was excavated and lifted from the eroding face of the hull depression in 1990, outboard of sector 2. Documentation suggests that it might represent a concretion previously removed from beneath the bow. The concretion (Fig. 3.86) contained swivel gun portions and a hailshot piece (90A0028) as well as part of a gun barrel with a chamber *in situ* (90A0126), two further, identical chambers (90A0125 and 90A0127) and a further fragment of gun barrel (90A0131), all of which seem most likely to represent parts of fowlers.

The radiograph of 90A0126, a barrel with chamber concreted into it, shows a gun similar to 77A014. It is

*Figure 3.86 Concretion 90A0131 as recovered*

hooped and banded, with sets of triple hoops (20mm wide each) separated by bands and single hoops (Fig. 3.82). Estimated wall thickness over the bands is 15mm and over the hoops 25mm. Four stave junctions are visible; the staves are narrower than 77A0143. The chamber has a length of 370mm with one central lifting lug. The external diameter of the neck is 80mm. There is no sign of any stave junctions, the detail is insufficient under the hoops and bands, and the neck appears solid. 90A0125 and 90A0127 are similar chambers (Figs 3.87–8; Table 3.14). Both are 370mm in length with external diameters of 165–80mm. They each have a prominent neck hoop, a wide band, a prominent central hoop to support a single lifting ring, another wide band and end with two adjacent hoops. It is impossible to see constructional details. Both contained tampions (diam. 70–75mm, 25mm thick) with small wad (30 x 20mm) behind. 90A0125 has a bore of 75mm, neck of 70mm length and a 45mm thick 20mm wide neck ring. The bands are 80mm wide. The touch hole is visible with a mark beside it. 90A0127 has a neck of 80mm in length, a bore of 70mm and a 40mm thick neck hoop. The inner tube thickness at the neck is only 5mm. The bands are 100mm and central hoop 25–30mm wide with a lifting lug and ring *in situ* (external diam. 120mm, bar thickness 25mm).

These chambers are similar in form to type 4 sixteenth century chambers from the Gulf of Mexico (Olds 1976, 70–1), although the *Mary Rose* examples are larger. The inner tubes from the Gulf of Mexico examples were made of a single rolled sheet, lap welded along its length.

Table 3.14 Fowler chambers (metric)

| Object | Sector | Length | Int. diam. | Max. diam. | Width hoops |
|---|---|---|---|---|---|
| 77A0143 | above O1–2 | 370 | 72–75 | 165 | 32 |
| 90A0126 | beneath bow SS | 370 | 75–80 | 185–180 | 20 |
| 90A0125 | beneath bow SS | 370 | 75–80 | 165–180 | 25–30 |
| 90A0127 | beneath bow SS | 370 | 75–80 | 160+ | 25–30 |

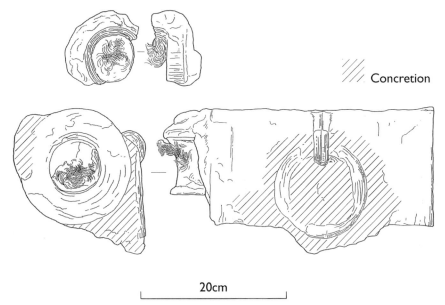

*Figure 3.87 Fowler chamber 90A0127*

*Figure 3.88 Fowler barrel fragment 90A0131 and wad*

*Figure 3.89 Fowlers: sequence of removal of: left) 90A0126; right) 90A0125 from concretion*

90A0131 is a fragment of thin walled barrel (c. 33mm) comprising a band and hoop (Fig. 3.88). The radiograph suggests a void of 112mm. The orientation of the radiograph is directly over a hoop and shows its orientation around the gun rather than along it. It is very fragile and there is not much iron present. It may well represent a portion of 90A0126 rather than another gun. As both are currently wet and inaccessible it is impossible to test this at present. If there are two guns, more barrel fragments might have been expected, although there are three breeches within the concretion.

## Bases of Iron (Swivel Guns)

The 30 bases listed for the *Mary Rose* in the *Anthony Roll* have been identified as the swivel guns recovered. These are designed to be elevated, traversed, loaded and fired by one individual. They all share the common characteristic of being located by means of a swivel yoke and peg (Smith 1988, 7) into a hole within a longitudinal member or the sill of a port at a convenient height above deck to be operated whilst standing (Fig. 3.90). All extant examples, and those described in

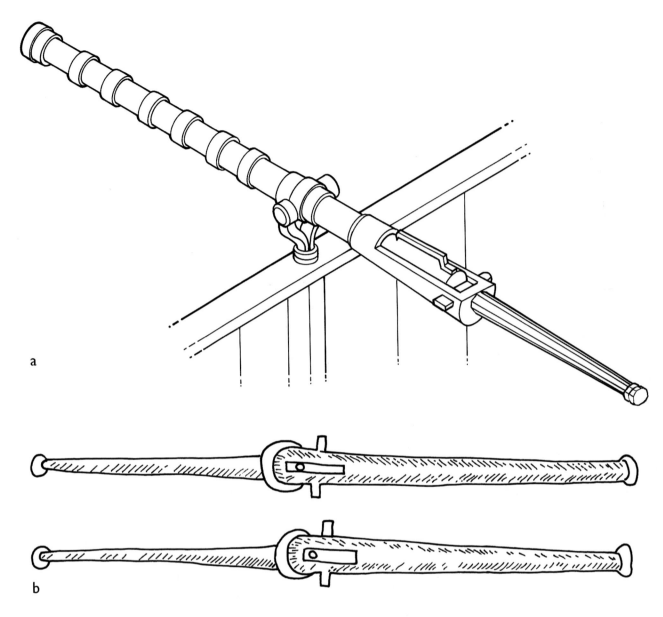

*Figure 3.90 Swivel gun: a) reconstruction in use; b) drawing by Wright (1563)*

sixteenth century inventories, have removable chambers. They can be either of wrought, forged or cast iron, cast bronze, or composite. They are primarily short-range weapons, designed to carry anti-personnel shot. They have a long history of sea usage dating back at least to the first quarter of the fifteenth century. An early illustration can be found in Wright, 1563 (Fig. 3.90b). Those recovered from the *Mary Rose* are similar to type SW5 or SW6 identified by Smith (1988, 10). They are described as swivel guns by Norton (1628, 58). He lists the length as 30 calibres and the chamber as 3. The most complete swivel gun from the *Mary Rose* has a calibre of 25.83, with a chamber of 6 calibres (82A4076). The barrel is incomplete in length, but it is impossible to say by how much. The most complete length of barrel (82A1603), which appears to be complete except for muzzle moulding and reinforce/chamber trough and tiller, has a calibre of 35.25.

## Identification and historical importance

As discussed in the introduction to this chapter, the allocation of the term '*base*' to the *Mary Rose* swivel guns has been made partly on the basis of relative size with reference to the inventories of the period and partly on the number recovered. Analysis of shot listed for bases within more detailed inventories reinforces this identification. The 1547 inventory (Starkey 1998, 102–57) identifies 51 vessels furnished with bases. Thirty-two list '*base shot of dice and lead*', thirteen merely '*base shot*' or '*shot for bases*', four vessels do not mention base shot, and one lists '*base shot of dice*'. Analysis of shot from the *Mary Rose* swivel guns, and the majority of shot of this size, are all cast lead with an internal iron dice (Walker *et al.* 1989; see Chapter 5).

Similar swivel guns have been recovered from other sixteenth century wrecks, the most notable being the Padre Island wrecks of 1554 (Arnold and Weddle 1978), thought to represent the remains of the ships *San Estaban*, *Espiritu Santo* and *Santa Maria de Ycar*. The *San Estaban* assemblage included only one swivel gun but had seventeen chambers with 38 shot of lead covered iron. The *Espiritu Santo* had five swivel guns, 29 iron chambers and one bronze, also '*lead balls with iron cores*'. At the Molasses Reef, Highborn Cay, and Bahia Mujeres wreck sites (Olds 1976) guns identified as '*versos*' were associated with wedges (forelocks) and swivel mountings. The presence of three different sizes of breech chamber suggests three different sizes of verso. Composite shot of the correct size was recovered from Molasses Reef, of diameter 10–40mm (Simmons 1988). Other examples from sixteenth century UK sites, many of them designated under the 1973 *Protection of Wrecks Act* include the Church Rocks site, the *St Anthony* and a wreck in Rill Cove. Finds from a site in Brighton Marina also include sixteenth century ordnance and composite shot of dice and lead for bases.

Between 1563 and 1589 the average weight of those issued from the Tower of London is listed as 168lb (76.2kg), with chambers of 8lb (3.6kg). Blackmore (1976, 218) describes the various forms of base as:

> '*smallest of the 'standard' list of cannon of the late 16th and 17th centuries*', adding '*the term was also used for a small-bore breechloader and, earlier, for some apparently larger guns. English writers of the 17th century also describe a Portuguese or 'Portingale' base which was clearly akin to the sling, a swivel breechloader with an iron tail or handle*'.

John Lad, writing around 1586, describes a base chamber as 9½in (241mm) long 1½in (38mm) high and containing ¼lb (113g) of powder (Caruana 1992, 25). This is an average size for the chambers from the *Mary Rose* (intense corrosion prevents the use of weight as a diagnostic feature).

An analysis of guns listed within the Tower of London suggests that the earliest examples of bases were of iron (Blackmore 1976, 218). In 1540 they are listed with '*flappis*' as well as '*great*' and '*double*' bases. An enigmatic swivel gun recovered from the sea in Dover and currently on display in Southsea Castle, Portsmouth, has a hinged chamber cover and may well be an example of a '*base with flappis*'. There is also one in the National Museum in Dublin (St John Henessy 1991, 1–3). These '*bases with flaps*' are probably the '*lidded*' guns described in recent articles (Smith 1988, 11, type SW3; Simmons 1989, 63–9).

*Figure 3.91 Swivel guns from the* Siege of Boulogne: '*ribaldiquins*'

More precise dimensions can be found in Norton (1628, 53), along with other data given for the base: bore 1¼in, weight of shot ½lb, length of piece 3½ft, weight of piece 200lb (misprinted 201). He describes these as within the seventh form of early ordnance:

> '*Nothing vnlike our Portingale-base, which with her Chamber, Tayle, and Hand Stearne, to guide and direct it vnto the assigned marke ... especially in small vessels at Sea*'. (ibid., 41)

However, that description of bases and slings in 1628 cannot necessarily be used to identify names of guns 70 years earlier and it is likely that the gun termed a 'sling' in the seventeenth century may have been different in the sixteenth century. By 1646 Eldred suggests that bases are rarely used (1646, 18).

Bases are listed in most of the inventories of armaments for ships in the sixteenth century, as well as within the Tower of London in 1540, 1547, 1559, 1589, 1595, 1617, 1620, 1635, 1675 and 1677 (Blackmore 1976, 260–32). In the *Anthony Roll* all 58 ships carried bases, ranging from 66 (on the 600 ton *Peter*) to five on the smaller vessels. The inventories occasionally differentiate between '*double*' and '*single*' bases, although actual size is not mentioned. Bases consistently outnumber any other single type of gun carried (iron or bronze) on every ship in 1546 (Knighton and Loades

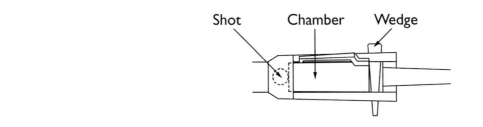

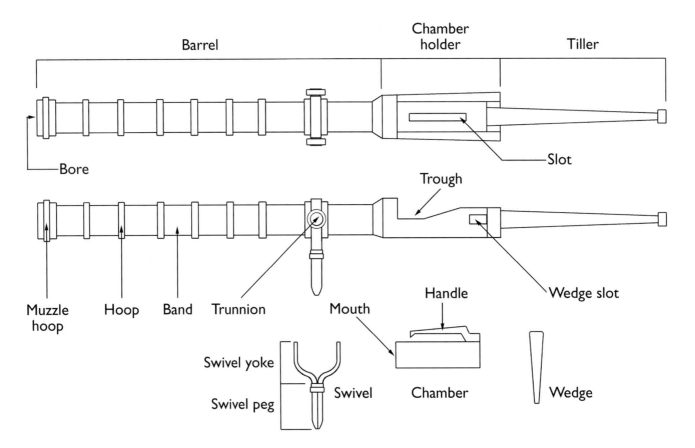

*Figure 3.92 Anatomy of a base*

2000). Out of the 1814 iron guns listed in the *Anthony Roll*, 892 (63%) are bases. Table 3.15 illustrates that bases form from 25% to more than 75% of the iron gun assemblage in 1546, and up to 100% on the smallest vessels in 1547 (Starkey 1998, 146–57). Iron gun totals include hailshot pieces and top pieces, but exclude handguns. Their relative importance at sea, therefore, is obvious.

On land, inventories of garrisons, castles, blockhouses, forts and bulwarks (*ibid.*, 102–44) indicate that the pieces of ordnance most frequently mounted within the majority of 77 sites listed were the iron slings (in 52 fortifications), bases (53) and fowlers (40) (Kenyon 1982). The 1547 inventory is much more descriptive than that of 1546, with qualifications including differentiation between '*Double*' and '*Single*' bases and occasionally listing damaged items, for instance, '*oone broken*'. Where they occur together, double appears first, suggesting a larger gun. The occasional occurrence of '*sling*' before '*base*', as '*Slyng basys of yron*' (Starkey 1998, 4010) is disconcerting given the similarity between '*bases*' and '*slings*'; but this can be dismissed as scribal error for '*single*' (the instance cited follows '*Doble basys of yron*').

The inventory can be used to formulate a series of names which may relate to different types of base. These include; '*Doble*', '*Demy*', '*Demy Carte*' and '*Croke*' bases. The first two probably relate to size, the same pre-fixes are used with smooth bore cast muzzle-loading guns of the period, probably reflecting half-weights of shot. '*Croke*' most probably indicates a base mounted on a supporting staff, or hook, as many handguns are listed with these. Other entries include one hundred and fourteen '*Prevy*' bases listed only in Boulogne (*ibid.*, 6186). Double bases and '*Shrimp*' bases suggest guns with large and small bores respectively and '*Wagon*' and '*Cart*' bases suggest those shown in pairs on carts with spoked wheels. These are visible on the engraving

Table 3.15 Bases on vessels in 1546 and 1547

| Vessel | 1546 No. bases | 1546 Total iron | % | 1547 No. bases | 1547 Total iron | % |
|---|---|---|---|---|---|---|
| Henry Grace à Dieu | 60 | 130 | 46.15 | 43 | 63 | 68 |
| Mary Rose | 30 | 76 | 39.47 | | | |
| Peter | 66 | 112 | 58.92 | 37 | 50 | 66 |
| Matthew | 48 | 72 | 66.66 | 54 | 66 | 81 |
| Great Bark | 30 | 71 | 42.25 | 29 | 97 | 29 |
| Jesus of Lübeck | 20 | 52 | 38.46 | 15 | 46 | 33 |
| Pauncy | 20 | 70 | 28.57 | 31 | 80 | 39 |
| Morian (of Danzig) | 12 | 27 | 44.44 | 12 | 40 | 30 |
| Struse (of Danzig) | 12 | 37 | 32.43 | 7 | 33 | 21 |
| Mary (of) Hamburg | 12 | 35 | 34.28 | 12 | 36 | 33 |
| Christopher of Bremen | 12 | 29 | 41.37 | 25 | 40 | 63 |
| Trinity Harry | 12 | 45 | 26.66 | 12 | 51 | 24 |
| Small Bark | 46 | 76 | 60.52 | | | |
| Sweepstake | 29 | 50 | 58 | 33 | 66 | 50 |
| Minion | 33 | 65 | 50.76 | 33 | 53 | 63 |
| Lartique | 8 | 20 | 40 | | | |
| Mary Thomas | 10 | 23 | 43.47 | | | |
| Hoy Bark | 6 | 18 | 33.33 | 5 | 5 | 100 |
| George | 8 | 15 | 53.33 | 10 | 16 | 62 |
| Mary James | 10 | 22 | 45.45 | | | |
| Grand Mistress | 12 | 31 | 38.70 | 6 | 14 | 43 |
| Anne Gallante | 12 | 26 | 46.15 | 10 | 14 | 86 |
| Hart | 12 | 26 | 46.15 | 8 | 33 | 24 |
| Antelope | 12 | 32 | 37.50 | 11 | 28 | 39 |
| Tiger | 12 | 26 | 46.15 | 12 | 26 | 46 |
| Bull | 12 | 26 | 46.15 | 11 | 33 | 33 |
| Salamander | 17 | 43 | 39.53 | 17 | 32 | 53 |
| Unicorn | 12 | 29 | 41.37 | 12 | 23 | 52 |
| Swallow | 20 | 43 | 46.51 | | | |
| Galley Subtle | 14 | 28 | 50 | 8 | 8 | 100 |
| New Bark | 20 | 43 | 46.51 | | | |
| Greyhound | 12 | 22 | 54.54 | 12 | 21 | 57 |
| Jennet | 10 | 20 | 50 | 14 | 24 | 58 |
| Lion | 18 | 36 | 50 | 17 | 37 | 46 |
| Dragon | 20 | 34 | 58.82 | 19 | 31 | 61 |
| Falcon | 20 | 26 | 76.92 | 20 | 22 | 85 |
| Saker | 12 | 16 | 75 | 12 | 14 | 86 |
| Hind | 14 | 21 | 66.66 | 13 | 20 | 65 |
| Roo Barge | 12 | 18 | 66.66 | | | |
| Phoenix | 10 | 18 | 55.55 | 18 | 20 | 90 |
| Merlin | 8 | 12 | 66.66 | | | |
| Less Pinnace | 6 | 13 | 46.15 | 10 | 13 | 63 |
| Brigandine | 10 | 12 | 83.33 | 5 | 13 | 38 |
| Hare | 12 | 16 | 75 | 10 | 10 | 100 |
| Trego Renniger | 12 | 16 | 75 | | | |
| Double Rose | 6 | 10 | 60 | 6 | 6 | 100 |
| Flower de Luce | 7 | 11 | 63.63 | 7 | 7 | 100 |
| Portcullis | 6 | 10 | 60 | 6 | 6 | 100 |
| Harp | 6 | 9 | 66.66 | 6 | 6 | 100 |
| Cloud in the Sun | 6 | 10 | 60 | 6 | 7 | 83 |
| Rose in the Sun | 6 | 10 | 60 | 7 | 7 | 100 |
| Hawthorn | 6 | 9 | 66.66 | | | |
| Ostrich Feathers | 6 | 9 | 66.66 | 6 | 6 | 100 |
| Falcon in the Feterlock | 8 | 12 | 66.66 | 7 | 7 | 100 |
| Maiden Head | 6 | 9 | 66.66 | 6 | 6 | 100 |
| Rose Slip | 6 | 9 | 66.66 | 6 | 6 | 100 |
| Gillyflower | 5 | 8 | 62.5 | | | |
| Sun | 7 | 10 | 70 | 6 | 6 | 100 |

depicting the siege of Boulogne (Figs 3.91 and 3.92). *'Red'* bases are also listed (Starkey 1998, 5471) and *'Portingle'* (ibid., eg, 5388) or the more usual *'Portingale'* in others. The first may indicate colour (suggesting a copper alloy), and the second, imports from Portugal, or of Portuguese style. The identification of forelock as the wedge between the chamber and the back of the chamber holder is suggested some entries for Carlisle, where the citadel had three *'Doble Basis of yrone with ij chambers a pece and no forlockes'* and eight *'Single Basis of yrone with ij chambers a pece and viij forlockes of yron'* (ibid., 6263–4), while in the castle were four *'Doble bases*

226

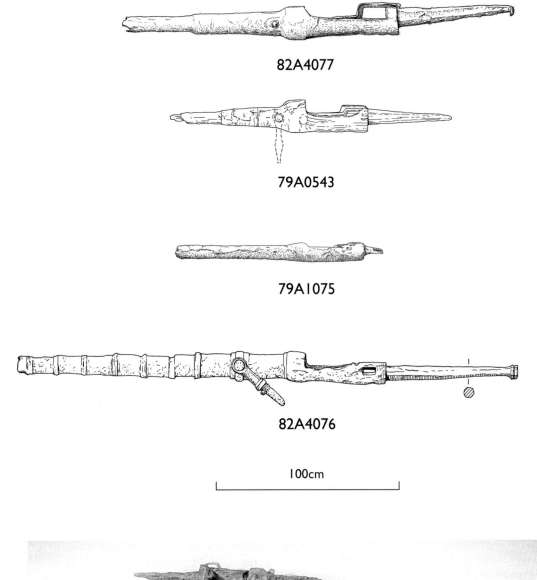

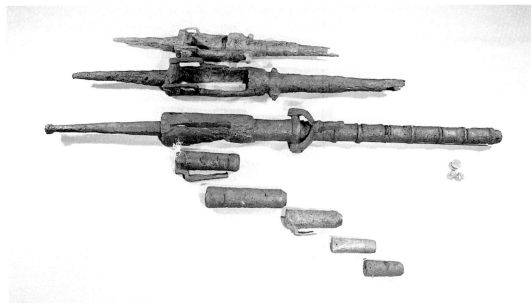

*Figure 3.93 Bases: the four most complete bases*

Table 3.16 Bases: summary of most complete guns (metric)

| Object | Sector | Form | Total length | Int. bore | Ext. bore | No. calibres | Tiller length | Chamber holder length | Barrel length | Chamber length | Chamber width | Shot diam. |
|---|---|---|---|---|---|---|---|---|---|---|---|---|
| 82A4077 | SS8 | 1 | 2125 | 50–60 | 80 | 15–18 | 625 | 595 | 905 | 450 | 140–210 | nr |
| 79A0543 | M11 | 1 | 1550 | 45–50 | 55 | ? | 450 | 405 | 695 | 315 | ? | 38 |
| 79A1075 | M1 | 1 | 1135 | 37–40 | 70 | ? | 100 | 335 | 710 | 235 | 130 | 30 |
| 82A4076 | SS9 | 2 | 2745 | 65 | 85 | 25 | 735 | 410 | 1500 | 365 | 120 | 48 |

nr = not recorded

*with vj chambers of yrone and forlokes for them*' (*ibid.*, 6234).

There are a number of entries suggesting that some bases were mounted, as Pendennis: '*Doble basis of yrone with ij chambers a pece well mounted*' (*ibid.*, 4659). Nottingham Castle had '*Basis of yron with the cariagis whereof the forestockes are naught*' (*ibid.*, 6191); much of the equipment kept here was noted as old and useless. Wark Castle on the Tweed has two '*Doble bases mounted vppon trendells with ij Chambers a pece*' and two '*Bases mounted vppon their cariages with iij Chambers*' (*ibid.*, 6320–1). Bases are also described as mounted on '*Truckells*' (eg, *ibid.*, 6678).

References to relevant shot includes iron, lead (for both single and double bases); dice and lead; stone cast about with lead; lead diced with stone and iron; and lead with dice of iron of sizes 3in, 2¾in, 2½in, 1¾in, 1½in, 1in (76–25mm).

To date, all the swivel guns recovered from the *Mary Rose* are of iron, probably wrought, and share the following characteristics: a barrel, a trough-shaped cradle (chamber holder) formed to hold a separate powder chamber, a long tiller to traverse and elevate the weapon (Fig. 3.93), and a swivel mechanism comprising a yoke and peg ('*miche*') (Fig. 3.92). The chamber is secured by a wrought iron forelock aligned transversely. This locates through slots in the sides of the chamber holder. The gun is located onto the swivel mechanism by trunnions positioned forward of the chamber holder.

The swivel yoke appears to be formed of two iron bars bent at their top to form circles to accept the trunnions, with the bars fire-welded together with an encircling band beneath the yoke. Once crimped together, the ends of the yoke extend to form a peg inserted into a hole within a gunport sill or longitudinal rail.

None of the *Mary Rose* bases shows any evidence of being carriage-mounted. It is tempting to see carriage-mounted swivel guns, where the swivel yoke is mounted onto the carriage rather than the barrel, as being earlier types of swivel guns, possibly the serpentines which are so numerous on vessels of the late fifteenth and early sixteenth centuries. Examples of this include guns found on the Cattewater wreck (assumed to be of the early 1500s) and a gun recovered from the Goodwin Sands (Redknap 1997, 77, 78 and Fig. 3.135).

## Scope and distribution of the assemblage

This is an extremely rich assemblage, with fifteen elements of barrels (Tables 3.16–19) and 25 breech chambers (Table 3.20). This excludes the swivel gun raised from the bow in 2004 (not yet available for study). As described above, condition is very variable, often hindering identification and the reuniting of components. The small size of many of these pieces means that the iron is in an even poorer state than that of the larger guns. The number of guns represented is

Table 3.17 Bases: summary of semi-complete examples (metric)

| Object | Sector | Form | Total length | Chamber holder length | Int. bore | Ext. diam. barrel | Chamber length | Tiller length | Max. tiller width | Shot diam. |
|---|---|---|---|---|---|---|---|---|---|---|
| 82A5096 | ? | 1 | 810 | 225 | 50–60 | ? | 255 | ? | ? | nv |
| 85A0024 | SS2/3 | ?1 | 1170 | ? | *c.* 35 | ? | ? | ? | ? | *c.* 33 |
| 82A4080 | SS8/9 | ?2 | 840 | >310 | *c.* 90 | ? | 310 | ? | ? | nv |
| 90A0050 | SS2/3 | ? | 665 | 315 | ? | ? | 250 | 300 | 70 | nv |
| 90A0027* | SS2/3 | 2 | 960 | n/a | *c.* 50 | 100 | n/a | ? | ? | *c.* 45 |
| 90A0132* | SS2/3 | 2 | 606 | n/a | *c.* 50 | 60 | n/a | ? | ? | ? |
| 04A0097 | ?Forecastle | ? | 1100 | ? | 45 | 70 | ? | ? | ? | <28 |
| 90A0052 | SS2/3 | ? | 850 | nr | nr | ? | 250 | 410 | 80 | |

nr = not recorded; nv = not visible on radiographs; * = possibly part of 90A0050

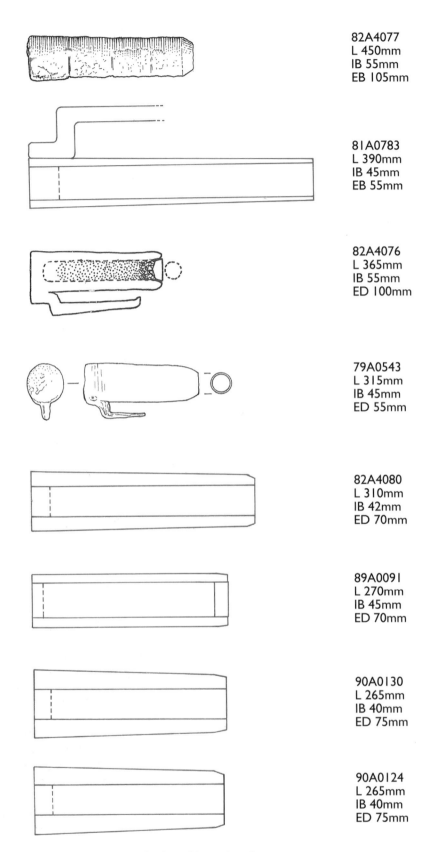

*Figure 3.94 Bases: selection of base chambers*

based on the presence of the chamber/barrel junction. Breakages seem to occur forward of the chamber junction, either just including or excluding the trunnions and peg, and towards the back of the chamber holder. A minimum of nine swivel guns (all with *in situ* chambers) have been identified using these criteria.

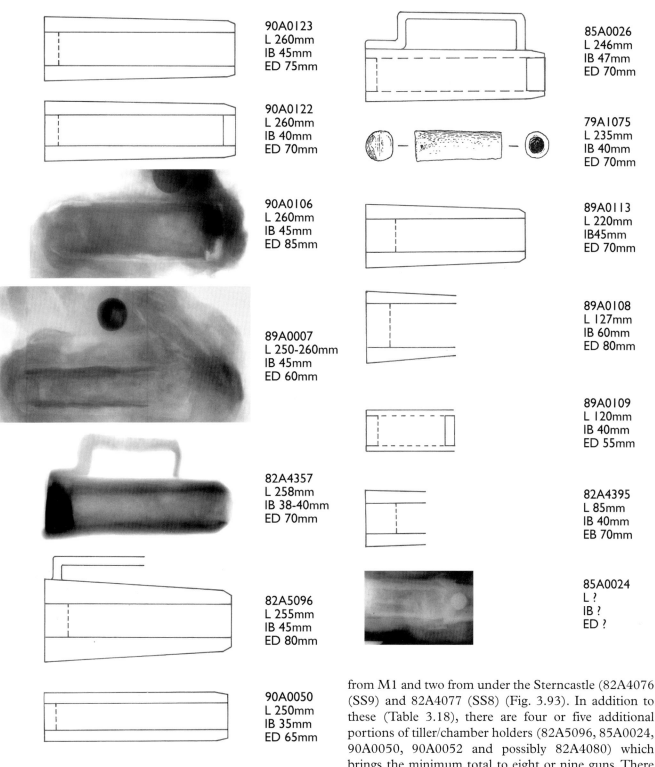

*Figure 3.95  Bases: selection of base chambers*

As can be seen from Table 3.16, four are relatively complete examples: 79A0543 from M11, 79A1075 from M1 and two from under the Sterncastle (82A4076 (SS9) and 82A4077 (SS8) (Fig. 3.93). In addition to these (Table 3.18), there are four or five additional portions of tiller/chamber holders (82A5096, 85A0024, 90A0050, 90A0052 and possibly 82A4080) which brings the minimum total to eight or nine guns. There is one more chamber holder/barrel portion (90A0027), recovered with other ordnance from a secondary location outboard of sector 2. This probably derived from the Forecastle.

The longest surviving barrel is relatively complete but is broken just forward of the trunnions (82A1603, SS9) and indicates that these guns had barrels probably in excess of 1400mm (around 4½ft). Its location argues for another base in the stern, but we cannot be certain that it is not part of 82A4080 (found close by), so is not

### Table 3.18 Bases: summary of barrel fragments (metric)

| Object | Sector | Form | Length | Int. bore | No. calibres | Ext. diam. |
|---|---|---|---|---|---|---|
| 81A5152 | SS8/9 | ?1 | 430 | 50 | ? | 75, 89 on hoop |
| 82A1603 | SS8/9 | 2 | 1460 | 40 | 35 | 120 on hoop, 72 on barrel only |
| 81A1082 | SS8/9 | ?2 | 1235 | 60–70 | ? | 130 over hoops |

counted as a separate unit. Three further barrel elements exist: 90A0132 (possibly part of 90A0050 and 90A0027), 81A5152 (muzzle fragment), and 81A1082 (Table 3.18). Two elements within this assemblage, both from the Starboard scourpit (SS8/9) appear to be stave-built with hoops and bands in a second layer over them and their identification as bases is uncertain. One has a bore of 60–70mm (81A1082) and the other of 90mm (82A4080). There is a record of a radiograph (now lost) of 82A4080 suggesting a chamber.

A further eight elements are mostly in poor condition. Five are yoke portions (71A0009, 89A0005, 89A0051, 89A0090, 81A2408) and three unidentified, possible swivel fragments (81A1852, 82A5095, 90A0063) (Table 3.19). Several items in the Woolwich Rotunda may be from the *Mary Rose*. These include three yoke fragments (MRE 123/1, 123/2, 112), including one which is slightly different from the other *Mary Rose* examples (MRE 123/2), and the back portion of a chamber holder (MRE 111). Swivel elements were listed within the items salvaged in 1836–1840 and one is shown in a photograph of a collection including *Mary Rose* items (Liscoe, pers. comm.).

Six of the guns were found with shot (79A0543, 79A1075, 82A4076, 85A0024, 90A0027, 82A1603). Windage is hard to gauge, the internal dice corrodes outwards causing expansion of the shot and the bore corrodes inwards (see Chapter 5). Tolerance varies between 5mm (90A0027) and 12mm (82A4076). Three chambers were found with shot just above their handles, encased within the concretion surrounding the chamber (85A0026, 89A0007, 90A0106). These range between 45mm and 52mm, the largest documented shot directly associated with a swivel gun.

Some guns and elements were recovered during post-excavation site monitoring, exposed as the sides of the hull depression eroded outwards. It is likely that additional elements lie within the scourpits and within the unexplored area north of the bow where further collapsed remains of Forecastle structure may lie.

Of the 25 breech chambers, nine were found with guns or within chamber holders and sixteen were recovered separately. Many are similar in shape, most flare outwards towards the back and taper inwards at the mouth. It is likely that all originally had handles, but not all have survived. A selection is shown in Figures 3.94–5 and measurements are given in Table 3.20. Inventories invariably list two chambers for each gun. Twelve base chambers were recovered during post-excavation monitoring of the hull depression, seven from within one large conglomeration of concretions just off the starboard bow area (90A0050). This conglomeration may represent guns originally located within the Forecastle.

In the Mary Rose Trust records, many swivel guns are merely listed as '*truck body lift*' with a date. These containers were positioned around the site as short-term storage areas and objects placed within them should have been recorded as such. This was not always the case, especially towards the end of the excavation

### Table 3.19 Bases: summary of other fragments

| Object | Sector | Description | Measurements (mm) |
|---|---|---|---|
| 71A0009 | ? | swivel mount; no details; lost | |
| 81A1852 | O3 | possible swivel in concretion; no info. | |
| 81A2408 | H2 | yoke frag.; 1 eye & part peg | diam. over eye 70, stirrup L.130, W. 125 |
| 82A5095 | ? | 3 frags; ?barrel frag. with trunnion type 1 | L. 480, int. bore 40–50 |
| 89A0005 | ? | poss. barrel frag.; yoke inc. peg | L. 615 |
| 89A0051 | SS5/6 | yoke frag.; no radiograph, photo or drawing; wet | L. 290, W. 215 |
| 89A0090 | SS9 | peg & part yoke with trunnion frag. *in situ*; assoc. 89A0137; good constructional detail | Ht 175, collar W. 85, collar Th. 25, trunnion diam. 55, Th. 34 |
| MRE 123/1 | RAI | yoke only; 1 side | bar Th. 20, W. 28, stirrup W. 45 |
| MRE 123/2 | RAI | incomplete yoke & part saddle of stirrup; looks different from other MRT yoes | bar W. 30, Ht 205 |
| MRE 111 | RAI | part chamber trough & tiller; looks more cast than wrought, but cold be previous conservation | side Th. 20, back 40, max. W. 195, Ht. at back 60, W. across back 130, tiller W. 40 |
| MRE 112 | RAI | yoke; part peg with collar | bar 25, Ht 270 |

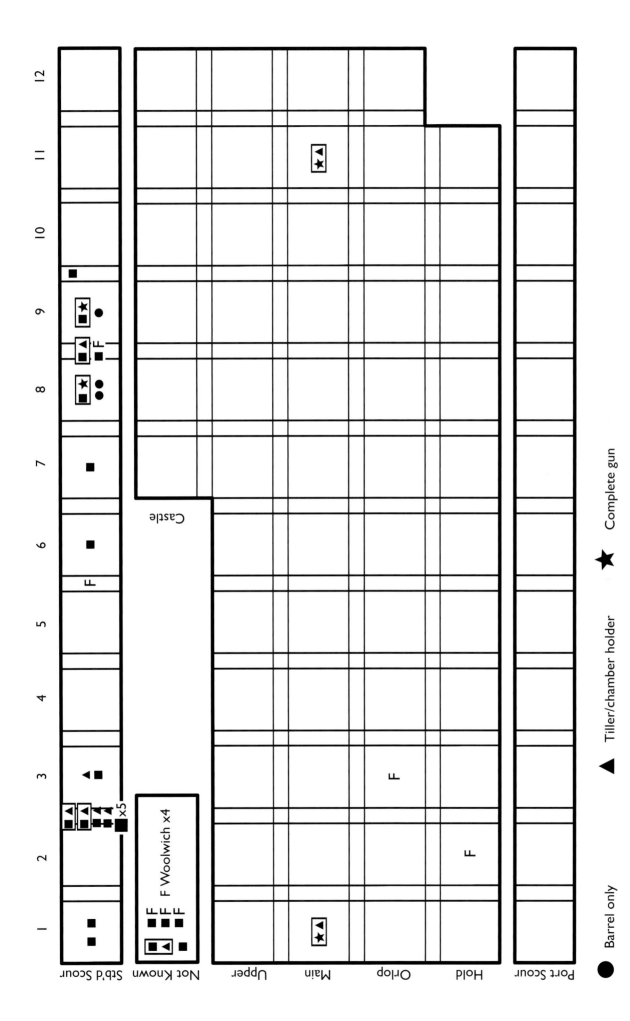

*Figure 3.96 Bases: distribution of bases and elements*

Table 3.20 Bases: summary of breech chambers, troughs, trunnions and wedges in order of chamber length (metric)

| Object | Sector | Description | Vol./wt | Breech chamber L. | Int. bore | Ext. diam. mouth | Ext. diam. back | Handle L. | Chamber trough L. | W. | Slot L. | Slot W. | Trunnions L. | W./diam. | W. across | Wedge L. | Th. | Contents |
|---|---|---|---|---|---|---|---|---|---|---|---|---|---|---|---|---|---|---|
| 82A4077 | SS8 | flares to back, mouth chamfers inwards; in gun | 7250.2 21.5 | 450 | 55 | 105 | 110 | NP | 450 | 140 | 430 | 65 | 60 | 35 | 210 | 120 | 50 | tampion, powder 83S0286, 386 |
| 81A0783 | SS7 | flares to back; loose | NR | 390 | 45 | 55 | 65 | inc. | | | | | | | | | | NR |
| 82A4076 | SS9 | flares to back & slightly at mouth; in gun | 57,027.3 | 365 | 55 | 100 | 110 | c.300 | | | 320 | 10–30 | | 60 | 275 | 200 | 20–60 | tampion 82A5179, samples 82S1228/9 |
| 79A0543 | M11 | flares to back, mouth chamfers inwards; in gun | 24,743.25 11.5 | 315 | 45 | 55 | 100–110 | c.165 | 300 | 130 | 255 | 30–50 | | 40 | 235 | 50–55 | | tampion/11, wad/10, powder/9 |
| 82A4080 | SS8/9 | flares to back inwards; in conc. trough | 40,908.8 | 310 | 42 | 60–70 | 88 | NP | | | | | | | 210 | 280 | 15–50 | powder & rope 83S0384 |
| 89A0091 | SS9/10 | poor condition; collapsed* | 34,640.6 | 270 | 45 | 70 | c.90 | NR | | | | | | | | | | tampion (45mm), also poss. paper & fabric |
| 90A0130 | SS2/3 | flares to back, mouth chamfers inwards; in conc, above 90A0050 | 41,945.7 | 265 | 40 | 75 | 90 | NV | | | | | | | | | | shot visible above, no tampion |
| 90A0124 | SS2/3 | flares to back, mouth chamfers inwards; v. degraded† | NR | c.265 | 40 | 75 | NR | NV | | | | | | | | | | in concretion |
| 90A0123 | SS2/3 | flares to back, mouth chamfers inwards** | 6.2 | 260 | 45 | 75 | 85 | NV | | | | | | | | | | NR |
| 90A0122 | SS2/3 | flares to back, mouth chamfers inwards** | 5.2 | 260 | 40 | 70 | 80 | degraded | | | | | | | | | | tampion in situ |
| 90A0106 | SS2/3 | flares to back, mouth chamfers inwards** | 36,133.0 | 260 | 45 | 75 | 85 | 175, inc. | | | | | | | | | | tampion in situ 50(mm) & shot above (52mm) |
| 89A0007 | ? | flares to back, mouth chamfers inwards | 39,227.8 | 250–260 | 45 | c.60 | c.90 | 250 | | | | | | | | | | shot 89A0038 (45mm) above handle, tampion (25mm) |
| 82A4357 | ? | flares to back, mouth chamfers inwards; in conc., poss. fuse on upper surface 83S0384 (50g) | 25,764.4 | 258 | 40 | c.70 | c.80 | c.240 | | | | | | | | | | tampion (35mm) |
| 82A5096 | ? | flares to back, mouth chamfers inwards | 37,414.0 | 255 | 40 | c.80 | c.85 | 125 | 226 | 110 | | | | | | | | NV |
| 90A0050 | BB | flares to back, mouth chamfers inwards | 28,278.0 | 250 | 35 | 60 | c.72 | NV | | | | | | | 190 | 210 | 35 | NV |
| 90A0052 | BB | flares to back, mouth chamfers inwards | NR | 250 | 28 | NR | NR | NV | | | | | | | | 250 | 60 | tampion in situ |
| 85A0026 | SS6/8 | flares to back, mouth chamfers inwards | NR | 246 | 47 | 67 | 72 | 220 | | | | | | | | | | tampion in situ |

Table 3.20 continued

| Object | Sector | Description | Vol./wt | L. | Breech chamber Int. bore | Ext. diam. mouth | Ext. diam. back | Handle L. | Chamber trough L. | W. | Slot L. | Slot W. | Trunnions L. | W./diam. across | W. | Wedge L. | Th. | Contents |
|---|---|---|---|---|---|---|---|---|---|---|---|---|---|---|---|---|---|---|
| 79A1075 | M1 | flares to back, straight mouth | 25,764.4 | 235 | 40 | 60 | 70 | NP | 360 | 70–110 | | | | | | 25–40 | | tampion, wad |
| 89A0113 | SS8/9 | flares to back, mouth chamfers inwards; degraded | 27,146.88 | 220 | 45 | 70 | 85 | NV | | | | | | | | | | NV |
| 89A0108 | SS1 | inc., flares to back, uncertain mouth form; fibre in touch hole | NR inc. | 127 inc. | 60 | 80 | 90 | degraded | | | | | | | | | | tampion 89A0142 (diam. 60mm), powder 89S0038 |
| 89A0109 | SS1 | slight flaring to back, uncertain mouth form; v. degraded*** | NR inc. | 120 inc. | 40 | 55 | 58 | NV | | | | | | | | | | tampion 89A0140 *in situ*, powder 89S0139 |
| 82A4395 | ? | inc., back & part of chamber | NR inc. | 85 inc. | 40 | 70 | 75 | inc. | | | | | | | | | | inc. |
| 85A0024 | SS2/3 | NR, in conc.mouth chamfers inwards | | | | | | present, NR | | | | | | | | | | |
| 82A2589 | BB | frags, NR | | | | | | | | | | | | | | | | |
| 82A5134 | ? | frags, NR | | | | | | | | | | | | | | | | |

Volume: the formula used to calculate the volume of the breech chambers was Pi x diameter x length of bore. This was not always possible to obtain so, in some instances, the external length has been taken with a factor removed for the breech plug. It was not possible to calculate volume is all instances.

BB = Bow bulkhead excavation: trench under the ship at the bow

* = Aiglet 89A0123 & 2 shot 89A0121–2 also in concretion
** = In concretion with 90A0050
*** = Aiglet 89A0138 also in concretion

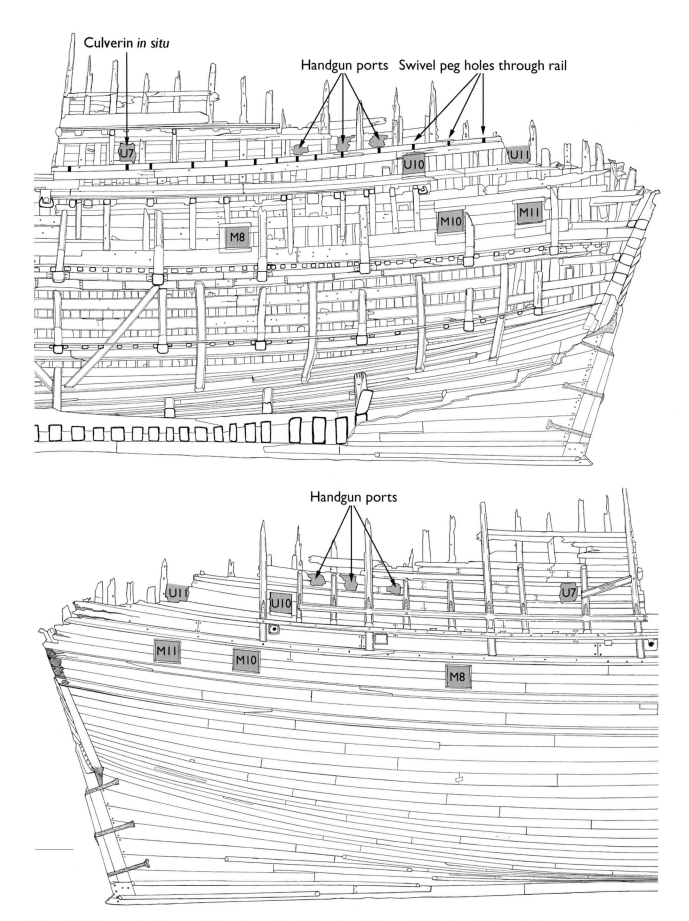

*Figure 3.97 Internal and external elevations of the hull to show holes and blocked swivel ports in the stern*

*Figure 3.98 a) Photo showing hole in sill of U8 port; b) peg holes in sill*

These locations reflect gun distribution. Of the sixteen isolated chambers (ie, not within chamber holders) five are from unknown locations (82A4357, 82A4395, 82A5096, 82A5134, 89A0007) and five were picked up during post-excavation searches along the starboard side of the hull depression; two from the bow (89A0108–9) and three in the stern (85A0027, 89A0091, 89A0113). Seven were associated with large concretions in the Starboard Scourpit in the bow (90A0050, 90A0052, 90A0106, 90A0122–4, 90A0130). One was found under the bow (82A5134) and one from around the beginning of the Sterncastle (81A0783). This and 85A0026, also found outboard in the Starboard Scourpit (SS6–8), enlarge the distribution to include the forepart of the Sterncastle. The assemblage represents 42% of the expected total of 60, two for each base listed.

The structure of the ship shows that the positioning of swivel guns altered during the life of the *Mary Rose* (Chapters 1 and 11). The Upper deck retains a sill that contains fourteen swivel peg holes. Ten holes had been rendered useless by the addition of continuous clinker planking which blocked in the ports, leaving only three ports open at Upper deck level (Fig. 3.97). One hole is within the sill of a larger port that housed a culverin (80A0976; see Chapter 2; Fig. 3.98a). It is unlikely that any of these ports supported bases when the ship sank (Chapters 7 and 11). Although the height is sufficient (940–1020mm) there is no evidence inboard for any swivel fragments and manoeuvrability is limited. The ports are too small for the guns to have fallen through these ports as the ship sank. There is no concretion within the peg holes nor has any iron migrated into the wood (Fig. 3.98b). Handguns, in contrast, were found inboard in sectors U9–11, suggesting the use of handguns at these ports in 1545.

Evidence for the positioning of swivel guns on the Castle deck at the time of sinking is revealed by two

when rapid clearance was needed prior to the lift. Attempts have been made to backtrack through records to find items put in particular truck bodies and assign these to potential locations.

The known elements cluster in the bow and stern, exactly where one would expect these to be used within the castles, and having fallen outboard into the Starboard Scourpit (Fig. 3.96). The two complete examples (79A0543 and 79A1075) at either end of the Main deck both appeared to have fallen from above, possibly from the Upper decks, or perhaps even the fighting tops (see discussion of top pieces below). The example from the stern provides the only tangible evidence that there may have been guns in the transom firing aft. There are guns in this position in nearly every illustration of 20 ships in the *Anthony Roll*. There are three unusual findspots, H2, O3 and O6 – all are fragments which are not currently available for study (81A2408, 81A1852, 89A0051).

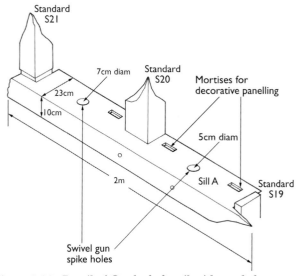

*Figure 3.99 Detail of Castle deck rail with peg holes*

sections of timber stretching between the starboard frames S24 and S19 in the Sterncastle. The forward section is 1550mm in length, 200mm wide, 100mm thick and has an eroded hole of *c.* 90mm diameter aligned vertically and positioned between S23 and S24. This sill is identical to that on the Upper deck and it is suggested that it supported one swivel gun. The second portion extends from S19–S21 and is rebated around these standards. This is 2000mm long, 230mm wide, 100mm thick and is broken at both ends. There are holes for the pegs of swivel guns between S19 and S20 and S20–S21 (Fig. 3.99). The forward one is eroded and has a diameter of 70mm; the aft has a diameter of 50mm and retains a portion of swivel peg.

Both sills supported decorative panelling that appears to have been present in two levels at the top of the Sterncastle, one at waist level above the Castle deck and one at about head height. This panelling (much of which was recovered in store on the Orlop deck) would have to be removed before combat to allow the swivel guns full traverse and elevation (see *AMR* Vol. 2 for full description). The height would be sufficient, about 1000mm above the deck. The use of bases with respect to structure is discussed more fully in Chapter 11.

## Method of manufacture

The bases from the *Mary Rose* outwardly show no clear indication of method of manufacture although, when viewed from certain angles, several tubes (82A1603) appear angular as though they are stave built. In all examples, where identification of other swivel parts confirm the gun is a base, no internal staves have been noted by conventional radiography of definite swivel guns (except 80A0132; details for each gun are in archive). It is possible that the inner tubes are made from a wrapped sheet, although coiling has been suggested, particularly for breech chambers (Carpenter, pers. comm. 1998). Magnetic particle inspection, which might suggest welds, has not been attempted. All of the accessible dry assemblage has been treated with hydrogen reduction with no prior metallographic study. Radiography of two incomplete barrels (81A1082, 82A4080, both from SS8/9) showed that these are stave-built. These have the largest bore size (70–90mm) whereas the rest fall between 37mm and 60mm. They cannot be positively identified as bases, although archive notes suggest that 82A4080 was recovered with a swivel chamber; this was not visible in the radiograph with the barrel. Composite lead shot with an inset iron dice, conventional shot for bases in this period, were found up to a size of 67.5mm, and composite inset ball shot up to 78mm (Chapter 5).

The tiller, used to elevate and traverse the gun, does not seem to be integral with the gun. How this is attached to the back of the chamber holder is uncertain,

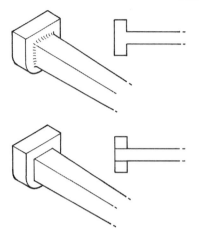

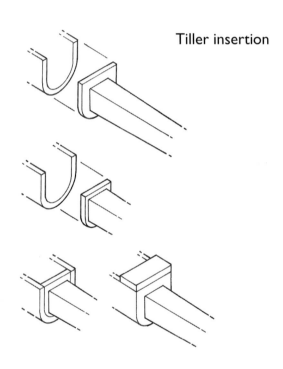

*Figure 3.100 Bases: tiller construction and insertion*

in spite of specific radiographic and visual study (Fig. 3.100).

Chamber holders appear to be made in a number of ways (Fig. 3.101). One form seems to consist of a sheet of metal (cut to have a slot), or two partly rounded staves partly welded along the centre of the base (Fig. 3.101a) creating a slot, possibly for water drainage. A second form is possibly made from a 'T' shaped sheet with a slot cut within it, or two partially welded centrally to leave a slot. Another form includes two 'L' shaped pieces welded partially along the bottom, the rest left open to form the slot. Both are wrapped over the barrel and the ends of the 'T' meet over the top of the junction with the barrel (Fig. 3.101b). There are differences in

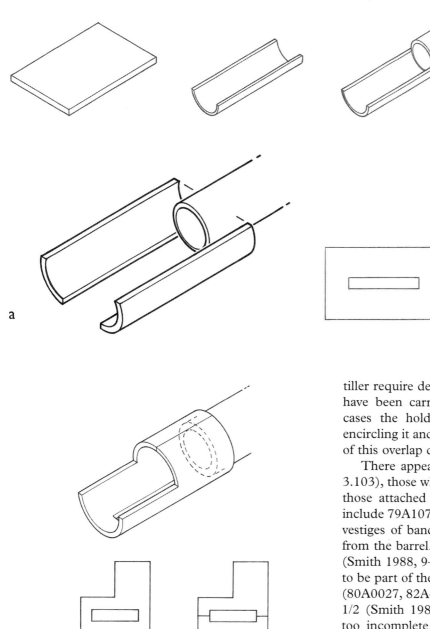

*Figure 3.101 Bases: alternative construction of chamber holder and wedge slot*

the wedge slot. This can be cut into the sides (Fig. 3.102a) or formed by welding a strip (Fig. 3.102b). In the latter examples the trough side is straight, where the slot is pierced the trough side is lower towards the muzzle, widening towards the back to accommodate the slot. The shape can vary, being either parallel sided or widening to the back.

In the *Mary Rose* examples the back of the chamber holder appears to be a separate element, or part of the tiller, but may have strengthening fillets around the sides. The nature of the junction with the barrel and the tiller require detailed study in all cases. Although some have been carried out, most are inconclusive. In all cases the holder lies over a portion of the barrel, encircling it and strengthening this junction. The length of this overlap differs (see Fig 3.100).

There appear to be two varieties of trunnion (Fig. 3.103), those which appear to be part of the barrel, and those attached to a separate hoop. Integral examples include 79A1075, 79A0543 and 82A4077. There are no vestiges of bands or hoops, and the trunnions project from the barrel. This is similar to Smith's type SW 4/5 (Smith 1988, 9–10). In two instances trunnions appear to be part of the central hoop of a triple hoop assembly (80A0027, 82A4076). These are similar to Smith's type 1/2 (Smith 1988, 7–8). Diameters vary, but most are too incomplete for comparison of either diameter or length. The position of the trunnions relative to the chamber/barrel junction varies.

The breech chambers recovered vary from complete examples to fragments of back plug only. The method of construction is still uncertain despite a number of attempts at radiography. They do not appear to be stave built. Suggestions include bored from solid, thick collars laid next to each other and welded together, a coiled band (welded), stave built or a wrapped sheet. The woody nature of corroded wrought iron enables a rudimentary suggestion of whether a fragment is longitudinal or circumferential, examples displaying both exist. A number of shapes are suggested. These include parallel-sided (Fig. 3.104a) or tapering inwards towards the mouth (Fig. 3.104b). The shape at the mouth differs: straight cut; shouldered and chamfered inwards; flaring outwards at the mouth. The chambers very in length between 220mm (85A0026) with a

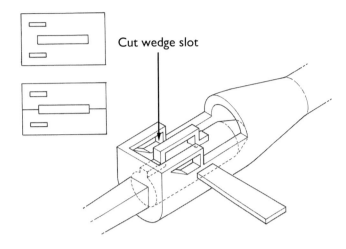

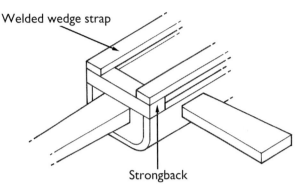

*Figure 3.102 Bases: wedge slot and wedge: a) cut wedge slot, b) welded wedge slot*

diameter at the back of 65mm and bore of 38mm, and 450mm (82A4077) with a diameter at the back of 110mm and bore of 55mm (Table 3.20). The smallest in situ chamber is from 79A1075. This has a length of 235mm and external diameter at the back of 70mm and bore of 40mm. Touch holes vary in diameter between 5mm and 10mm.

Where a study of the chamber back is possible, this seems to consist of a tapering plug, larger towards the outside of the chamber (Fig. 3.105). Diameters vary between 35mm and 20mm. The side-walls of the chamber are always thinner than the back, thickening from *c.* 8mm at the mouth to *c.* 15mm at the back. In some cases there is the suggestion of an extra fillet welded across the back, possibly to add strength. It is also possible that the back may form part of the handle, but this is complicated. The typical form of handle can be seen on Figure 3.93; the most complete example is a welded strip *c.* 30mm in width (82A4076; Fig. 3.106). There is no indication that it is turned again and welded as in the back end, although others clearly do. Where chambers have been found *in situ*, the position of the handle is rotated to the right, to lie flat against the side of the chamber holder, bringing the touch hole into the mid point of the top. Most of the chambers show radiographic evidence of handles and it is difficult to see how these would be loaded without them. The exact length and nature of the weld may be different and, indeed, this may be why some are better preserved than are others.

A number of fragments were found (Fig. 3.107) which suggest how the swivel yoke was made. It appears to consist of two separate iron bars which are turned back against themselves over the trunnions, and

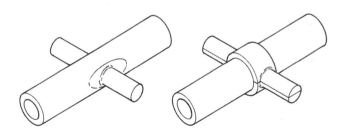

*Figure 3.103 Bases: suggested construction of the trunnions*

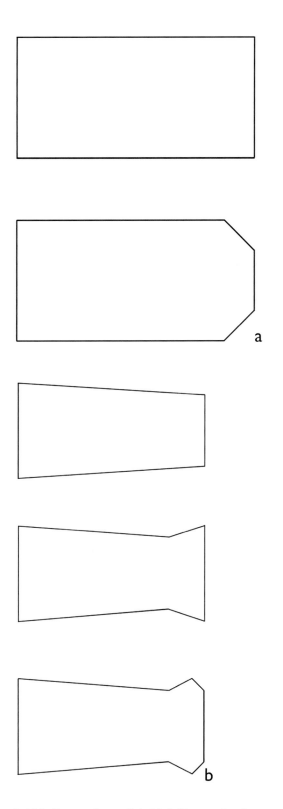

*Figure 3.104 Bases: a) parallel sided; b) tapering from breech plug to mouth*

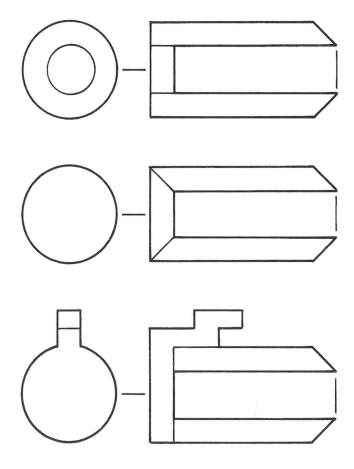

*Figure 3.105 Bases: chamber back, methods of breech plug attachment*

clenched together with a ring beneath the gun and fashioned beneath the ring to form one peg. This construction is suggested through a study of the alignment of the corroded iron elements during excavation from concretion (Fig. 3.108). A method of construction is suggested in Figure 3.109.

Detailed analysis has been undertaken into the methods of manufacture of swivel guns by Simmons (1988) and Armstrong (1997). The former investigation was of wrought iron guns raised from the sixteenth century Molasses Reef wreck. Both guns and breech chambers were sectioned and studied using a magnafluxing technique to identify the individual billets of iron within the larger objects by identifying the 'joins' between them. This technique showed the manufacturing sequence of a verso (see above) which outwardly resembled some of the *Mary Rose* examples. A slightly different type of gun (probably a '*murderer*') from the Tumbaga wreck near the Bahamas, was disassembled for conservation purposes and analysed by Armstrong. This again showed a structure outwardly similar to other of the *Mary Rose* swivels. When the complete *Mary Rose* collection is accessible it is suggested that some of these techniques be employed.

There is a wide variation within the individual elements forming both guns and chambers in terms of techniques of manufacture and in design. Length, bore size and wall thickness are the most crucial factors

*Figure 3.106 Bases: detail of handle attachment on 82A4076*

which affect performance; unfortunately the incompleteness of most of the assemblage makes comparisons of all the elements impossible.

Although a number of attempts have been made to answer questions regarding the method of construction, many remain that cannot be answered without the detailed analysis that has not yet been possible. The most obvious is one of the most difficult to answer given the variation in condition of the surviving objects, namely, is there any observable variation in size and can '*double bases*' and '*single bases*' be distinguished by observed differences in gun length, bore or ball weight? Although variations exist in bore size between 30mm and 60mm (possibly 90mm if 82A4080 is a swivel gun), there is not enough information to classify the guns further.

Questions that need to be addressed, or further addressed include the following:

### Barrel
- Construction: are the barrels formed of staves, or of one rolled sheet (if so, how and how thick?), or are both methods used? Are they made of numerous small pieces hammered together? If different methods of construction are employed, what other differences in construction exist and can these be termed different 'types'?
- Can observed differences in barrel length be related to bore? Are the inner tubes parallel sided or do they taper, are they larger at the junction with the chamber trough or at the mouth?
- Are barrels single or multi-layered? The assemblage appears to consist of guns with external hoops and bands on their barrels, as well as those without. Is one merely a more corroded form of the other? If there are indeed two different forms of outer barrel, is this reflected in differences of inner tube construction, or any other features which may be exclusive to the different forms of barrel construction? There is only one complete 'end' which may be for a swivel gun (82A5152) apparently with three hoops shrunk over the tube.

### Tiller
- No guns with integral tillers have been identified and the tiller attachment method to the back of the chamber holder is not known. Does the tiller appear to be part of the backplate? Does it enter the back of the trough or butt against it? Is it placed through a hole cut in the back and then hammered or wedged? Are there are any additional fillets to strengthen this

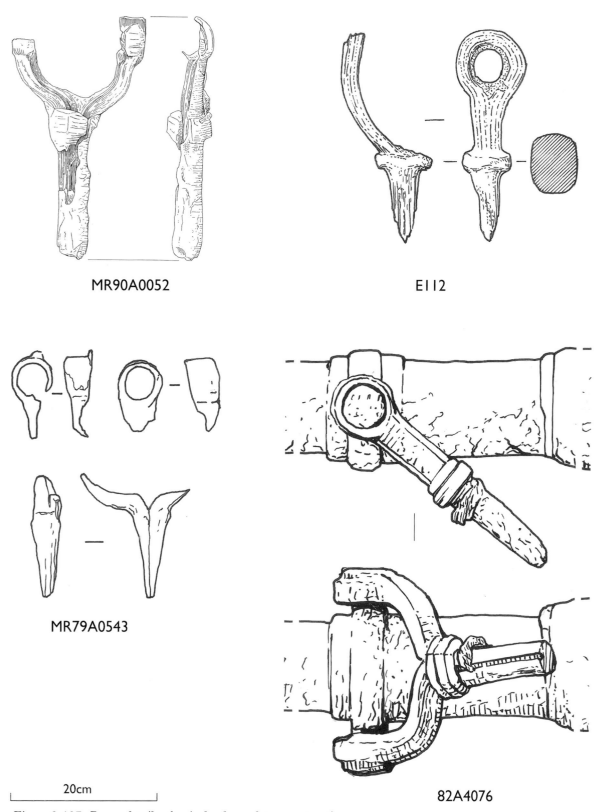

*Figure 3.107 Bases: details of swivel yoke and peg construction*

junction? Several tiller shapes have been noted; angular, tapering and hexagonal; ends are either angular (79A0543), double hooped (82A4076) or display a downward turn (82A4077). The two complete examples appear to have tillers of different length. On 82A4077 the tiller length almost equals the distance between the end of the tiller and the trunnions whereas, on 82A4076, the length is 120mm short of this (Fig. 3.93).

89A0090

*Figure 3.108 Bases: excavation of yoke 89A0090*

**Chamber holder**
- How is the chamber holder attached to the barrel and is it formed of one or more components? Is there evidence to suggest that it locates over lugs on the barrel as in Smith's 'commonest type', SW1 (Smith 1988, 8)?
- Is it formed of a sheet of metal folded over the rear end of the barrel? Is the open end the top, enlarged to accept the breech chamber, or the underside, where the slot is located? Is there any evidence for welding which might answer these questions? Another alternative might be large staves, each forming the side of the chamber holder. These would then either have the wedge slots cut into each side or have flat bars welded on to each side (Fig. 3.101).
- Is the slot in the base of the chamber trough cut out of a sheet, or is it part of a join in the chamber trough?
- Are some of the wedge slots cut into a sheet forming the chamber trough, and others made of welding a bar on to the top of each side of the chamber holder, raised at one end to form a wedge slot? If two methods of manufacture exist, what other differences exist within the guns and are these consistent (Fig 3.102)?
- How is the back of the chamber holder attached? Is it a separate semicircular piece of flat iron which is welded to the open back of the trough?

**Trunnions**
- Are the trunnions attached directly on to the barrel, and if so, how? Are the trunnions located on a hoop, and if so, how is this made? The most straightforward method would be to manufacture half the

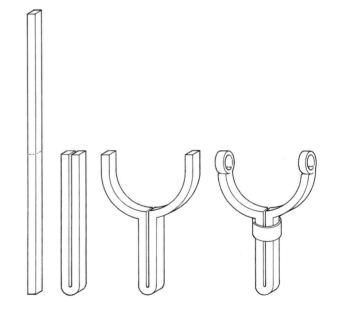

*Figure 3.109 Bases: proposed method of yoke manufacture by bending a single bar*

hoop with half of each trunnion and join them by a horizontal weld across the trunnions (Fig. 3.103).

## Chambers

- It is uncertain as to whether these are bored from solid or staved and hooped and banded as with the chambers for larger wrought iron guns. Alternatively they could be made of a single bar, coiled and then hammer welded, or of a number of bands or bands and hoops merely welded together. It is possible that they were made from a solid cylinder and bored out, or from a sheet wrapped and hammer-welded.
- Other questions include the method of making the touch hole and how to insure that the back of the chamber is secure. Is it a plug which is merely entered into the back of the rolled or stave-built chamber and then heated and hammered, or is there a more elaborate locking mechanism between the back of the chamber and the cylinder?
- Do all chambers have handles, and are they always only attached at one end (hammer-welded to the back of the chamber), or did they originally have a handle welded at both ends of the chamber? Are chambers without handles merely more eroded, handled, examples?
- Does the tapering of the chamber neck in some examples correspond with any other traits observed within the assemblage?

### Evidence for loading

Of the 25 chambers, twelve were sealed with tampions suggesting that they were loaded. 82A4077 is recorded with a sample of powder, but no shot or tampion. Four are too incomplete to tell, three are not recorded as having any contents and three more may be loaded, but tampions are not visible on the radiographs. One chamber within concretion (90A0130) clearly shows a shot just above the chamber but the chamber does not appear to have a tampion sealing the mouth. The tampions are relatively thick for their diameter (up to 35mm) and often have markedly sloping sides with the domed inner face up to 16mm less then the outer. Radiographs of three further concretions show an almost identical position of an inset shot just above the chamber (85A0026, 89A0007 and 90A0106).

On balance most appear to have been loaded. In two instances fabric or paper was noted during the excavation of the chambers (89A0113 and 89A0091). Four other chambers had fibres near to or protruding from the touch hole (89A0108, 85A0106, 82A4357, 82A4076). One *in situ* shot displayed the lifting flap with fibre protruding on it that is a feature of some of the inset shot (see Chapter 5). This could be the remains of wadding concreted to the corroding internal dice. In two instances copper alloy aiglets (lace-ends) were found extremely close to the touch holes on chambers (89A0091 and 89A0109). If not clothing remains, it is just possible that these were used as small funnels (or liners for funnels) through which priming powder was poured into the chamber. These chambers were extremely degraded, possibly exaggerated by the presence of copper alloy.

It is difficult to match loose chambers with guns. The condition of the iron makes detailed measuring difficult, and the tolerance is unknown, especially given the ability to alter the width of the wedge according to how far it is pushed through the chamber holder. Those found together within large concretions are the only easily associated extra chambers.

The large conglomeration recovered from the area of the bow partition excavation revealed the remains of two separate bases (90A0050, 90A0027) with *in situ* chambers and three more were found within the concretion. Three chambers were almost identical, 260mm in length and *c*. 80mm in diameter (90A0106, 90A0122, 90A0123.) A fourth was slightly larger, 265mm (90A0130). The chambers within the guns were *c*. 250mm (90A0050, 90A0052), the third *in situ* chamber is not recorded (84A0024). It may be that a 10mm tolerance was acceptable.

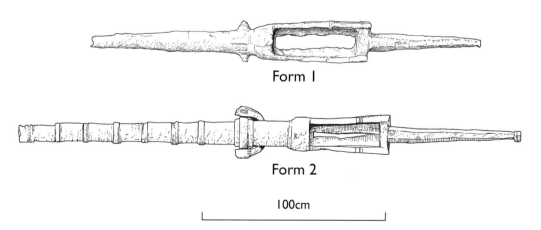

*Figure 3.110 Bases: Smith's forms 1 and 2 as represented by: above) 82A4077; below) 82A4076*

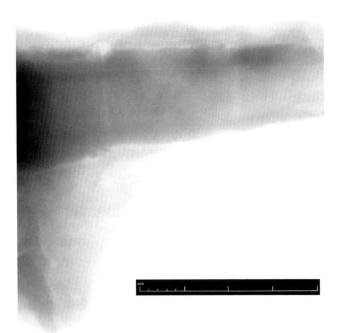

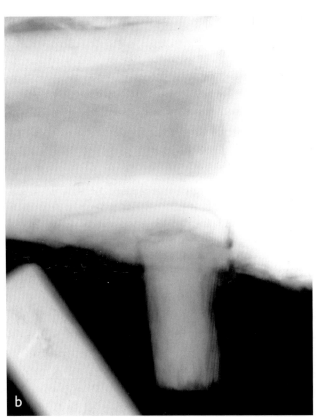

*Figure 3.111 Bases: a) radiograph of 79A0543 showing second layer of hoops and bands; b) radiograph of plate holding trunnions for 82A4077*

Smith proposed a typology of swivel guns based on form (Smith 1988, 8–11). In the absence of detailed structural analysis, matching the *Mary Rose* examples to these types is difficult. Most complete guns recovered from the *Mary Rose* seem to fall broadly within two distinct forms, based on whether the trunnions are welded directly onto the barrel (form 1, similar to Smith's 4–5), or form part of a hoop (form 2, similar to Smith's 1–2) (Fig. 3.110). Those of type 2 retain a second, outer layer of alternating hoops and bands, however radiography suggests that some of the form 1 examples may also have had a similar layer (82A4077 and 79A0543; Fig. 3.111). It is possible that presence or absence of certain features may be due to corrosion rather than inherent constructional differences.

## Form 1

Three complete guns (79A0543, 79A1075, 82A4077) are of form 1, with 82A5096 and 85A0024 as other possible examples. Measurements are given in Tables 3.16–20. These appear to have a chunky, short, tapering tiller, often broken, the end of which varies from a simple point (79A0543) to one which bends downwards slightly and is hexagonal (82A4077). The chamber holders of 82A4077 and 79A0543 have two welded straps holding the wedge. The trunnions are 140–80mm forward of the beginning of the chamber holder over the top of the gun and appear integral, although radiography seems to suggest they are on an inset plate (Fig. 3.111). This feature was observed in some of the Padre Island examples (Olds 1976, 75–84, eg, verso 328, 330). The barrel appears to be formed of a sheet, although this cannot be confirmed by visual or radiographic study.

The Padre Island swivels were made of five cylindrical sub-assemblies, each of which had been made of layered patches, lap-welded to form a planar sheet then wrapped around a parallel mandrel and held with one single longitudinal lap weld. This may be what we have and magnetic particle inspection may well show up the individual patches on the wet examples. The barrel tapers externally quite dramatically and there is no external sign of hoops or bands. 79A1075, is different (Fig. 3.115–6). Although incomplete, the wedge slots seem to be cut into the chamber holder rather than made by welding straps and there is no slot in the base of the trough. Even the most complete examples appear shorter and cruder than form 2 bases; although this could be a result of decay. The two found on the Main deck (79A0543, 79A1075) are smaller (therefore lighter) than 82A4077.

## 82A4077 Complete gun with chamber *in situ* and wedge

*Location*: unknown
*Recovered*: in concretion in SS8

*Statistics*
Length: 2125mm; bore 50mm; 15–18 calibres
Chamber weight: 21.5kg

This gun (Fig. 3.112) was excavated with its chamber *in situ* with a wedge but no shot or tampion, although powder was found within the chamber (83S0286, 0386). The tiller is crudely hexagonal with an end pointing down with a width of 80mm where it is attached to the back. The wedge slot is formed by welding two 25mm square bands, 345mm in length, to the top of the back of the chamber holder and to the side of the chamber. The wedge is pierced with a 15mm hole at its wider end, presumably to attach it to the gun. The breech chamber flares towards the back tapering inwards at the mouth. The chamber holder has a step at the back (against which the wedge lies) 65mm in width and a slot in the base (430 x 65mm). The 'grain' of the wrought iron at the back of the chamber holder goes across the trough; the sides and the tiller lie parallel with the length. Detailed radiography of this area shows that the back is separate from the trough, but could be a widening of the tiller. Above this a fillet of metal (or strongback) has been welded, hence the appearance of a cross-grained structure between tiller and trough back. The straps for the wedge slots are welded on to this. There is a suggestion of a rivet, used to help locate the trough over the barrel (or barrel into the trough) before it is placed in the furnace. A small folded sheet of lead was over the back of the chamber, possibly functioning as an apron (width 120mm).

Radiography does not demonstrate how the inner tube is formed, although it does suggest a corroded outer layer of hoops (20mm wide) and bands (100mm wide) with a maximum height above the tube of no more than 20mm. Radiography suggests that the trunnion is inset into a plate which in turn is welded onto or into the barrel (Figs 3.103 and 3.111). This may be similar to the 'jump welded' examples from the Mollasses Reef wreck (Simmons 1988). Externally this area is possibly covered by a reinforcing band cut out to take the trunnions, suggested by a very prominent line originating at the front of the chamber holder and ending just in front of the trunnions and by the increase in external barrel diameter here. The trough itself may be fire welded to this reinforcing band. Examination of the inside between the junction of the trough and the barrel reveals a small scar on each side, as though whatever was finally the sides of the trough was at one time held through the band with a rivet, perhaps to hold

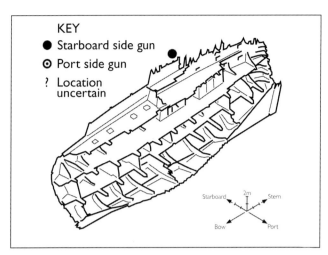

it together prior to placing it in a furnace and fire welding the two together.

The chamber holder was formed separately from the barrel (Fig. 3.112) which locates into the forward end of the trough, the end of the barrel flaring outwards forming a lip which is clearly visible inside the trough. It appears that, at this junction, the gun may have been formed of three layers, the barrel, the chamber holder, and possibly an external reinforcing band. There is also the suggestion of a band at the front of the trough, as the 'grain' to the wrought iron is along the length of the barrel on the sides of the trough but around the barrel at this point.

The chamber holder appears to be made of two half rounded plates, each end of which terminated in an 'L' (Fig. 3.101b). The ends of the 'L' are apparently welded to form a trough with the observed cut-out in the base. The edge which would ultimately be wrapped to form the top of the chamber holder at the front may also be 'L' shaped, overlapped and welded, giving the 'thickening' often visible at this point. If open at the back, the barrel can be entered into the heated trough which would then cool over the junction with the barrel. The barrel, in essence, becomes an 'inner tube' just at this point, where the shot sits, and just in front of the junction with the chamber. The tiller could be attached in a similar way, pushed into the heated trough, which would then cool over it. Externally the barrel appears as though it is wrapped with an extremely long band from the junction with the chamber holder for 500mm. It is assumed that this 'reinforced band' was in place before insertion into the chamber holder. The external diameter of the barrel is 85mm at the end of this 'reinforce', 105mm over the band, 140mm at the junction with the holder and 160mm where the trough begins. This method of manufacture would account for the changing external profile of the gun and is in keeping with techniques used.

The straps which form the wedge slot are clearly separate elements welded to the holder over what appears to be a bridging piece across the back of the chamber holder over the tiller (see Fig. 3.102b). The tiller appears to actually form the back of the chamber

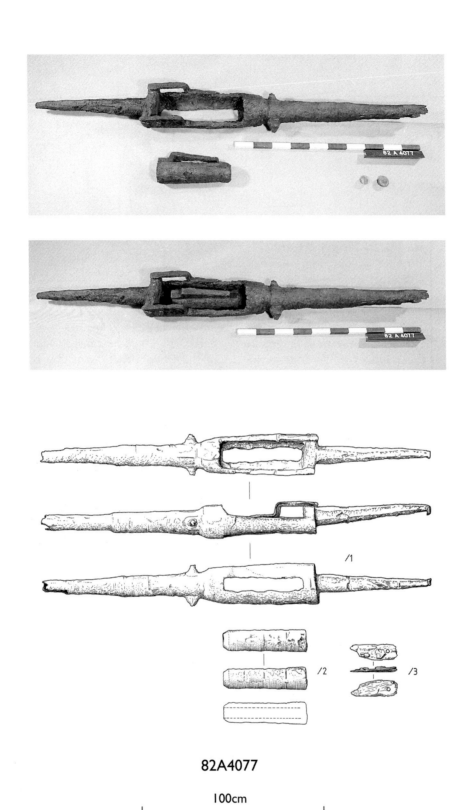

Figure 3.112 Base 82A4077

holder which has been fined down along the length of the tiller. The presence of the bridging piece is suggested by opposing 'grain' here. The trough may be one rectangular sheet, with portions cut out to form a 'T' shape, the top of which forms the front of the chamber trough at the top. Physical examination suggests that the trough is not joined longitudinally underneath, and that the slot is cut out of the metal, rather than merely an enlarged seam.

## 79A0543 Fairly complete gun with chamber

*Location:* unknown, stern
*Recovered:* M11, in concretion diagonally between deck and starboard side

*Statistics*
Length: 1550mm; bore: 45mm
Chamber weight: 11.5kg

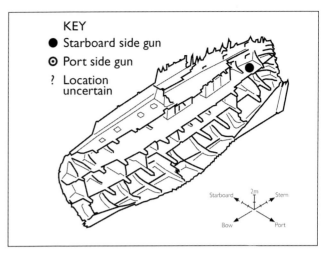

This gun lay within a hard shell layer overlying soft grey mud that formed a lens between the starboard frames and the transom. The area had been disturbed and may coincide with the removal of the M11 port piece in the nineteenth century. As it remained a feature on the seabed for a long time and was one of the first guns seen by the divers of the Mary Rose Trust, it was given the nickname '*Stiff Sid*' and became a reference point in the stern. Composite shot was also recovered from this area (79A1185, 79A1186) and a powder ladle (79A1152). A small wicker basket found between the gun and the transoms contained a ram (79A1208) and lead shot (79A1214), which was assumed to be associated with the gun. Rams are not usually used with breech-loaders and neither are powder ladles if cartridges are used, but the small size of these is interesting and they may have been for the missing falcon. Artefacts found beneath the basket included another powder ladle (79A1211) and two small solid lead shot (79A1207).

The gun was loaded with inset dice shot, 38mm in diameter, straddling the join between chamber and barrel. The chamber contained a tampion (38–43mm diam., 20mm thick), a '*fibrous wad*' and powder.

Although the tiller is short (Table 3.16) it tapers to a point and may be complete (Fig. 3.113). The width of the tiller as it joins the back is 85mm. It appears more rounded than hexagonal, possibly a result of corrosion. The direction of the iron is longitudinal on the tiller and sides of the chamber holder, but across the back. The method of manufacture is as suggested for 82A4077. This barrel also may carry a wide reinforcing band overlying part of the barrel and in turn, overlain by the chamber holder. Trunnions appear integral, but the radiograph suggests that they are on small pads which are again inset/welded to the barrel (Fig. 3.114a). Although not complete, part of the yoke and peg are present. The distance across the swivel yoke is 225mm externally and 150mm internally. It appears to be made of a 25mm bar curved over on itself at the top to encircle the trunnions, welded to form the peg and clenched. The chamber seems to have been slipped over the barrel from the rear, butting against the trunnion band (Fig. 3.113).

The barrel is incomplete and in poor condition, with a thickness of 5–8mm. No lines suggestive of a single seam, overlap or staves can be seen. The radiograph shows that the inner tube was covered with hoops and bands for at least 250mm forward of the trunnions; this feature is not visible on the barrel after conservation. Radiography proves that the hoops were placed longitudinally around the tube, as the slag inclusions show corrosion lines at right-angles to the length of the gun. Parts of three external bands are visible, 100–120mm in width separated by single 25mm wide hoops. The width across the bands tapers towards the bore from 130mm to 100mm. On the bare barrel the diameter is 66mm with the inner tube having walls of 8mm thick, suggesting a band thickness of 22mm. The hoops are 18mm and rise above the bands, but by less than 5mm. There is no indication of further bands and hoops and the walls at the last band have a thickness of only 8mm. At the mouth this is reduced to 5mm.

The chamber holder has a slot in the base (length 225mm, width 30–50mm). There is a 50mm step at the back of the chamber trough to accept the wedge. The wedge straps clearly continue to the very back of the trough, welded over the top of the block forming the backplate. Corrosion here is perpendicular to that of the wedge straps. The back of the trough is 68mm wide and appears to extend directly into the tiller. The corrosion lines suggest otherwise and the tiller may enter through the trough back, detailed radiography may corroborate this. A tapered wedge was found *in situ* behind the chamber.

The chamber tapers from the bore to the back (110–55mm) with a bore of 45mm, suggesting a maximum wall thickness of 32mm. The remains of part of a handle welded to the back and to the right of the trough when the touch hole is uppermost (see Fig. 3.106). The handle is welded at the back and may be incomplete at the front. The touch hole has a diameter of 5mm but lies within a circular depression 20mm in diameter. The radiograph shows the chamfered neck of the chamber nearly butting against what appears to be

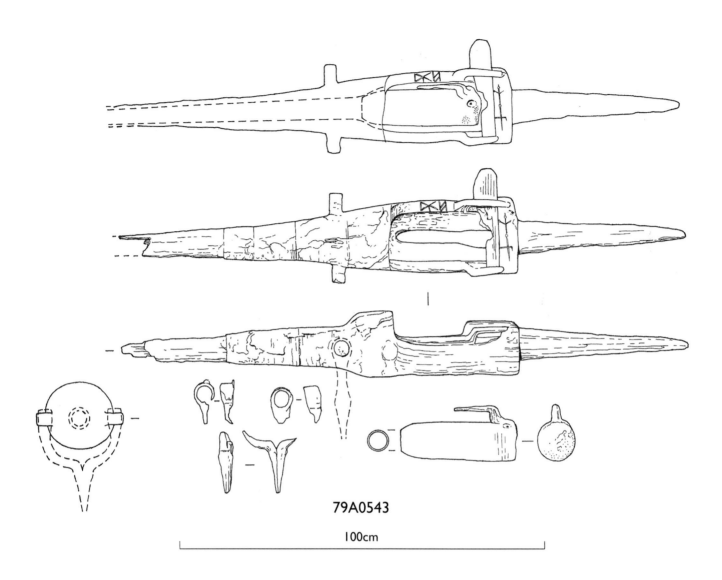

*Figure 3.113 Base 79A0543*

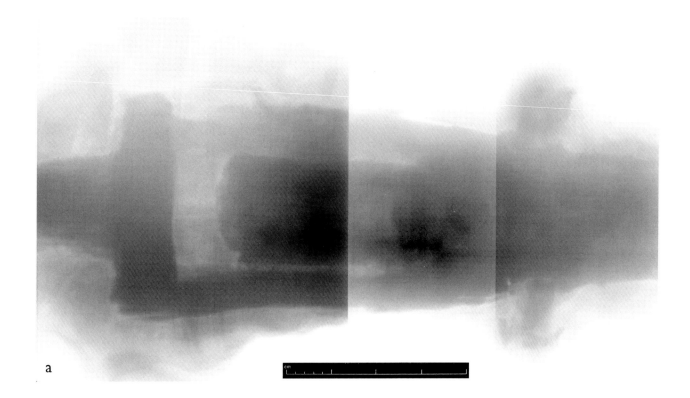

a chamfered end of the barrel, each element chamfered from the back of the gun towards the front so the join is parallel and diagonal (Fig. 3.114a). Marks across the back and on the right-hand side of the trough are simple geometric shapes, possibly incorporating a broad arrow (Fig. 3.114b).

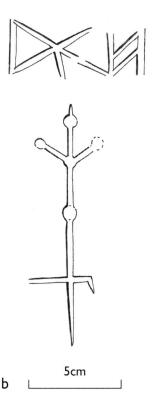

*Figure 3.114 Base 79A0543: a) radiograph of chamber of 79A0543; b) detail of marks on back of chamber trough*

**79A1075 Most of barrel and chamber holder with tiller stub and chamber**

*Location*: unknown, bow
*Recovered*: M1

*Statistics*
Length: 1135mm; bore: 37mm
Weight barrel: 10.3kg; chamber: 5.2kg

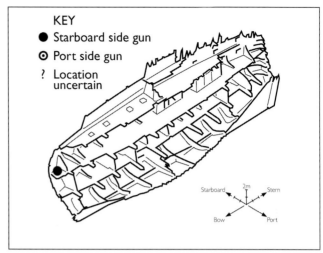

This gun contained inset dice shot of 34mm diameter and a fragmentary tampion *c.* 32mm in diameter. There is no mention of wad or powder, though the recovery of a tampion suggests that there should have been powder present.

The general condition is poor (Fig. 3.115). The tiller is present only as a 100mm stub with a width at the gun of 60–65mm. The barrel is incomplete with few visible external features. The chamber holder does not have a slot in the base and appears to have a band covering the junction with the barrel. This does not extend as far forwards as the trunnions. The right trunnion is present. Radiographic interpretation suggests this is not integral with the barrel, so it has either separated, or broken along a weld (a weld line is clearly visible), or belongs to

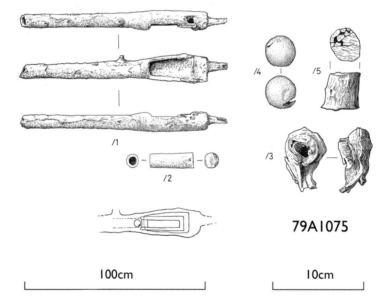

*Figure 3.115 Base 79A1075*

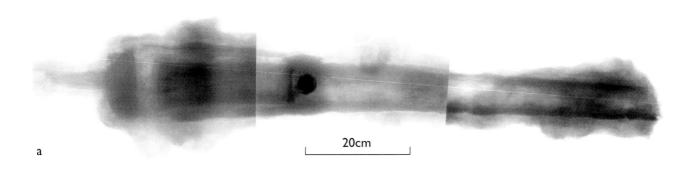

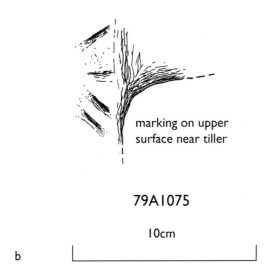

*Figure 3.116 Base 79A1075: a) radiograph; b) details of marks on back of chamber trough*

a second or third layer that is no longer visible. There is no extant wedge and the wedge slot, *c.* 40mm wide and 25mm deep, is corroded. It appears to be cut through, like the form 2 swivels, rather than formed from a strap. The back of the trough is 80mm thick and has four incised lines in a radial pattern on the top surface (Fig. 3.116b). A fragment of the yoke was found, bent over the right trunnion.

The radiograph (Fig. 3.116a) suggests a wall thickness at the chamber holder of 30mm and of the barrel of 15mm. The maximum wall thickness (18mm) is just behind the trunnions, with an internal tube diameter of 45mm at the muzzle. The 45mm thick plug at the rear of the chamber holder is located from the rear, and the tiller is apparently rebated into this for most (but not all) of the thickness of the plug. A very degraded wedge, which no longer survives, is visible wedging the chamber.

The chamber tapers slightly with no chamfer at the mouth. The plug at the rear may be entered at an angle, or even with a lap joint. There is no obvious handle scar. The touch hole has a diameter of 5mm, and the plug at the back is 45mm thick. The maximum wall thickness is 10-15mm. The area immediately in front of the chamber contains the shot, which is lifting where the iron dice inside is corroding. This area appears reinforced with a band for a distance of about 70mm.

The condition of this gun is so poor that it is impossible to be certain from the radiograph whether there is more than one layer. It is tempting to see band and hoop sequences in the external shape of the concretion with one wide band (*c.* 70mm) behind the trunnions with abutting narrower bands of 40mm in front of them. The lack of a trough slot, uncertain wedging method and uncertainty of an external layer suggest that this may be a variant of form 1. It has several features similar to Smith's SW4 swivels (Smith 1988, 9).

## 82A5096 Part tiller, holder, chamber, ?trunnions and barrel frament

*Location*: unknown
*Recovered*: unknown

*Statistics*
Length: 810mm; bore: 50mm

Inaccessible in wet storage. Radiography seems to show the end of the tiller entering and going right through the back of the chamber trough (Fig. 3.117). The tiller has a width of 35mm and the back of the trough a thickness of 50mm and wall thickness of 28mm. There is a 40mm gap between this and the chamber, which tapers inwards. The chamber rests against the barrel which itself chamfers to accommodate it, with the two edges sloping towards the front of the gun touching. The barrel is 15mm thick at this point. Two circular elements on either side of the gun just forward of where the barrel meets the chamber may be a part of the swivel yoke. The chamber is flared towards the back and tapers in at the mouth. The plug at the back is 26mm thick and the wall thickness of the chamber is 12–18mm. Remains of a handle can be seen, length 126mm, width 15mm.

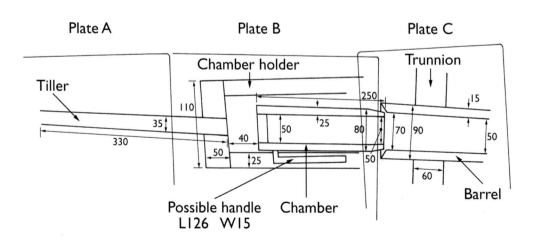

*Figure 3.117 Base 82A5096, radiograph and interpretative drawing*

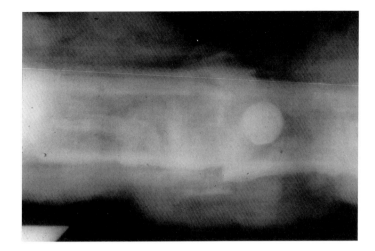

a

b

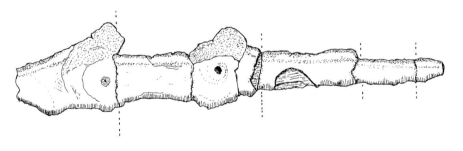

Denotes sections created as a result of multiple damage

85A0024

|—————————————| 100cm

c

*Figure 3.118 Base 85A0024: a) radiograph; b) concretion as lifted; c) interpretive drawing*

**85A0024 Partial tiller, yoke, chamber holder, shot and chamber**

*Location*: unknown, bow
*Recovered*: SS2/3; surface find during post-lift site monitoring

*Statistics*
Length: 1170mm; bore: *c.* 35mm

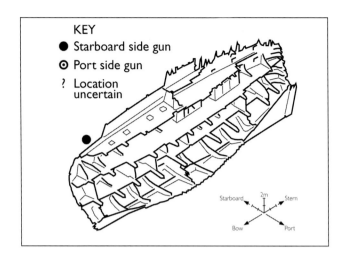

Possible trunnions and barrel pieces were also present with this gun fragment, which appears to be a form 1 base, from the tiller through to the junction with the barrel (Fig. 3.118). Radiography suggests a chamber with an *in situ* handle and a lead shot with internal iron dice concreted in position. The item is still in wet storage and there are no measurements.

*Form 2*

Three form 2 bases have been identified (82A4076, 82A1603, 90A0027/132) and possibly 90A0050. They appear to be generally in better condition than the form 1 bases. The radiographs clearly show that the inner tube is covered with alternating hoops and bands; sometimes single hoops, sometimes multiples. The trunnions are set within a triple hoop sequence, possibly on the larger central hoop (see Fig. 3.93; Figs 3.119–20). This is set further forward of the chamber mouth than the examples of form 1. The gun from the Bahamas (Armstrong 1997, 27–49) showed that this similar (although much shorter) swivel gun had half trunnions at each end of half hoops, with two half hoops welded along the joins of the trunnions (*ibid.*, 36). This is suggested for the form 2 examples. The chamber holder appears to be a separate element, slipped over the barrel behind the trunnions and held with a collar which must have been put in place before the barrel bands and trunnions, as shown in Figure 3.101a. The chamber trough is pierced with two rectangular slots to take the wedge. The tiller is often hexagonal and does not appear to enter through the back of the trough, but may be part of it. The whole trough back, including the tiller, is held by wrapping the trough over the back and welding the two pieces together. Outwardly, this appears to be the more elegant form. In only two instances can staves be seen (82A4080 and 81A1082). These both have large bores (60–90mm) and may be incorrectly placed within this category, or these may be stave built swivel barrels.

## 82A4076 Complete with wedge, chamber, tampion, shot, and yoke with peg

*Location*: unknown
*Recovered*: SS9

*Statistics*
Length: 2745mm; bore: 65mm; 25 calibres

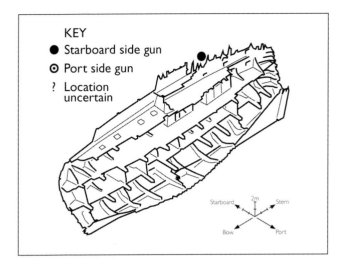

This is the most complete of the bases. The hexagonal tiller and finial and the clenched double hoop over the stirrup peg do not appear on the other guns (Fig. 3.119). There is geometric decoration on the centre top of the chamber holder and the number '4' on the chamber handle. Radiography shows that the trunnions are clearly situated on the central hoop of a triple hoop complex (Fig. 3.120a) and there is an outer layer of alternating bands (8) and hoops (7) for the entire length of the barrel forward of the trunnions (Fig. 3.120b). The point of balance was tested and is just forward of the end of the breech trough, 200mm behind the trunnions. This means that, even without the breech chamber in position, the gun would rest tiller down (which may be beneficial for loading), unless there is around 500mm missing from the muzzle.

The tiller tapers to a double hoop of 55 x 70mm. The chamber trough appears to have been made by two 25mm thick 'L' shaped staves partly welded to form the base and tapered slot (320 x 30–10mm), and welded centrally over the top, possibly overlapped (Fig. 3.92a and 3.120, c, d). There is no clear evidence for rivets placed at the ends of each 'L' to help retain the trough to the barrel. A reinforcing band towards the front covers the end of the trough staves. Side slots (200mm x 60mm) have been cut into the pieces to form the hole for the tapering wedge (Fig. 3.120, c, d). The yoke length is 370mm and internal distance between the straps over the trunnions is 180mm with a maximum width across the trunnions of 275mm. The peg is 190mm long (Fig. 3.120e).

Radiography of the trunnion face (diam. 60mm) suggests two half hoops were manufactured, each bearing half a trunnion as demonstrated by Armstrong (1997, 36; see Fig. 3.93). The centre of the trunnions are positioned 1215mm from the bore and 1510mm from the end of the tiller. The yoke appears to be made of two single strips, each folded together to form an eye which sits around the trunnion. The ends are welded to each other and the two ends further held with hoops. Radiography revealed an inner tube of internal diameter 70mm and wall thickness 14mm. Over this is a sequence of bands (thickness 12mm, length 140mm) and hoops, either singly (35mm wide), or in triple sequence with the central hoop (45mm wide) upstanding (Fig. 3.120, a, e). The hoops rise above the inner tube by a maximum of 24mm. Specifically targeted positions were designated for detailed radiography in order to examine whether the bore was stave built or not, the nature of the trunnion attachment, how the tiller enters the trough at the back, and the nature of the back of the breech chamber. A particular area of structural uncertainty is the blocking of the back of the chamber holder and the insertion of the tiller. The side elevation of all the swivels suggests that the trough is complete and plugged from the rear. As few radiographs are taken side on, this is difficult to corroborate but the corrosion along lines of slag within the manufacturing of the metal tends to support this. Whether the tiller and rear 'plug' are one unit, or the plug is placed and the tiller cut into it is unknown. The radiographs and corrosion planes suggest the latter. Although one would expect a simple weld here would be a point of weakness, most of the guns seem to have broken either side of the trunnions or halfway down the barrel rather than at the tiller junction. The back of the chamber trough has a fillet or bridging piece welded over the area of tiller, creating a cross-grained effect between the tiller and the backplate and no details could be observed. It appears that the tiller itself is drawn out at this point to fill the back of the trough, sitting in it.

The inner tube flares outwards as it locates into the chamber trough (Fig. 3.120f). The overlying band ends at the junction with the trough and there appear to be two hoops, presumably put on after the barrel was located into the trough unless these merely acted as a backstop when the trough was heated and shrunk onto the barrel. This could not be accessed if the trunnions were in position, although the other hoops and bands could be as they are much smaller.

Radiography suggests that the bore is not stave built (Fig. 3.120b). The hoops and bands were clearly visible along the entire length as a single second layer. Forward of the trunnions the bands are of differing lengths (165/105/125/120/135/130/120mm). The hoops vary

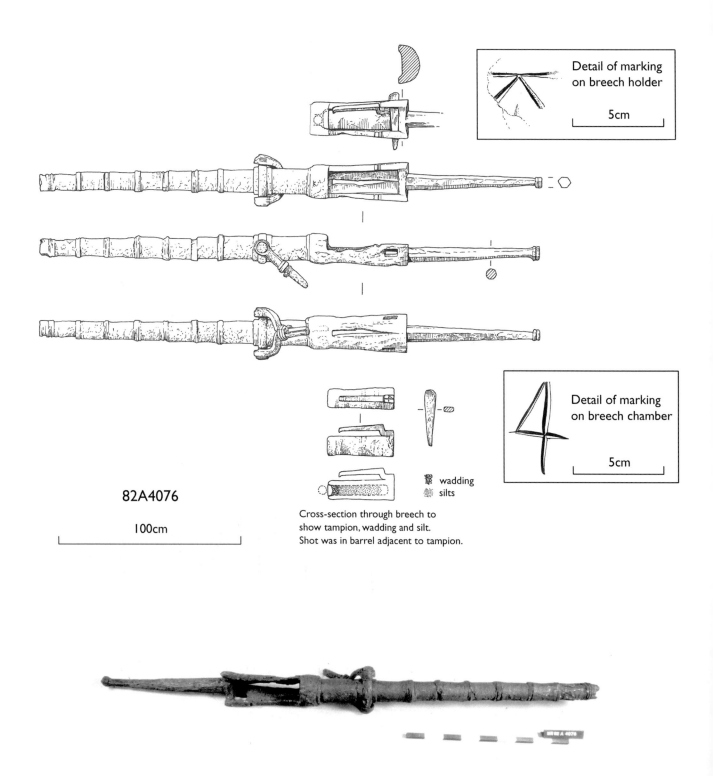

*Figure 3.119 Base 82A4076*

between 40mm and 20mm. There are three or four at the very end (15/10/10/15mm) possibly forming the muzzle. None of the 15mm thick muzzle hoops shows any indication of being much higher than the bands. This is not unexpected, as these do not need to act as anchor points within a wooden carriage as they do with the larger wrought iron guns. The chamber was *in situ* (Figs 3.106 and 3.120d).

The barrel forward of the chamber contained fibre (82S1229, possibly a wad), followed by an inset iron

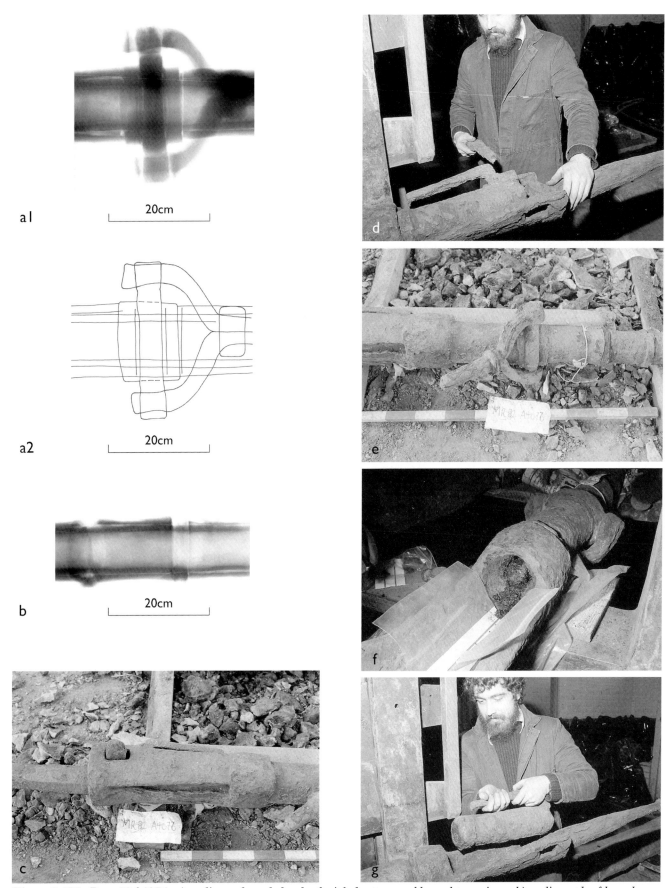

*Figure 3.120 Base 82A4076: a) radiograph and sketch of triple hoop assembly and trunnions; b) radiograph of barrel showing bands and hoops; c) side view showing chamber trough and wedge slot cut out; d) view with chamber in situ showing wedge insertion; e) swivel yoke and peg; f) Chamber trough and barrel junction; g) insertion of chamber*

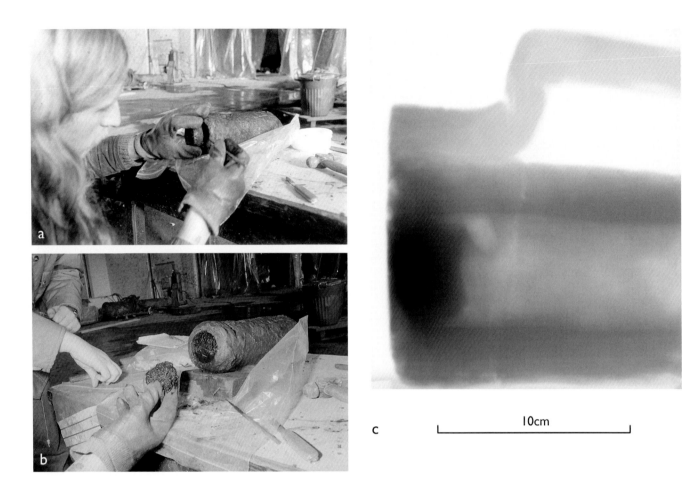

*Figure 3.121 Base 82A4076: a) removal of tampion 82A5179; b) removal of wad 82S1228; c) radiograph of handle*

dice shot (82A5180; diameter 48mm) and additional wadding interspersed with fine sediment. The breech chamber was sealed with a 20–24mm thick tampion (82A5179; Fig 3.121a) with the widest face (60mm) outermost and the domed face (45mm) innermost. Just behind this was a 50g wad (82S1228; Fig. 3.121b), followed by a thick granular deposit and lastly 0.25l of black liquid containing fibre. The wall thickness varies from 22.5mm at the front to 27.5mm by the touch hole (10mm diam.). The plug at the rear is clearly visible on the radiograph (Fig. 3.121c). The handle is complete but terminates without bending back to join the chamber at the front. It is welded to the cylinder for 80mm, has a width of 25mm and rests towards the right side when the touch hole is uppermost. Radiography suggests a sheet rather than stave or band construction, but no welds are visible.

## 82A1603 Barrel

*Location*: unknown
*Recovered*: SS8/9

*Statistics*
Length: 1460mm; bore: 40mm; 35 calibres

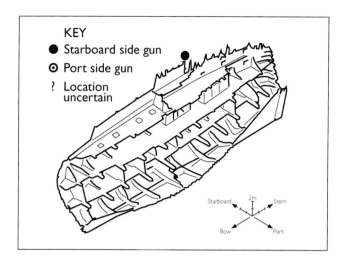

This object seems to be part of a swivel gun from the junction with the chamber holder forwards (Fig. 3.122). Radiography confirms that it is hooped and banded towards the breech. Five bands are visible, all *c.* 67mm wide with four badly degraded hoops, 13–15mm wide, that stand above the bands by only 8mm. The third hoop from the breech may have held the trunnions as this is the most pronounced and looks like a triple hoop assembly with a thicker central band (Fig. 3.123a). Another triple hoop assembly is suggested, a further band and triple hoop assembly, and then alternating bands and hoops along the length of the barrel that exist only as 'scars'. The wall thickness at the breech is about

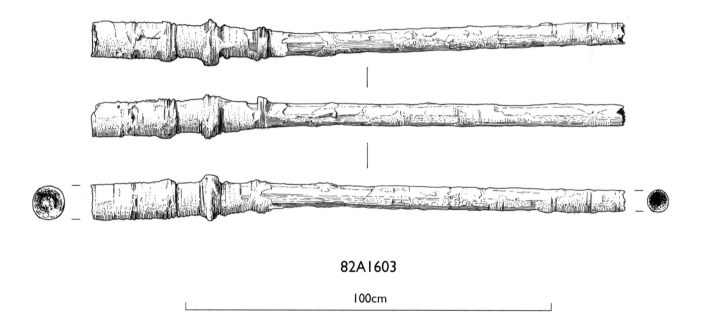

*Figure 3.122 Base 82A1603*

*Figure 3.123 Base 82A1603: a) triple hoop assembly; b) staves on inner tube*

16mm. A lead shot with an inset iron dice (82A4339), diameter 29mm and weight 128.3g, was excavated from the back of the gun. Outwardly the barrel appears to be made of eight staves of approximately 25mm width with a thickness of 5mm at the bore (Fig. 3.123b). These staves do not show up on the radiographs, which traditionally use a single directional source.

As the radiograph was inconclusive and we were uncertain as to whether the central tube was cast or wrought, attempts were made to measure the velocity of sound through the material and to compare the results with standards but the results were inconclusive and could not be compared with modern samples (details in archive). This may be the barrel to 82A5096, which is currently in wet storage and inaccessible (Fig. 3.117).

## 90A0050 Part tiller, chamber holder, chamber and barrel frag; 90A0027 Barrel fragment with swivel yoke and peg; 90A0132 Muzzle portion

*Location*: unknown, bow bulkhead?
*Recovered*: SS2/3 in concretion

*Statistics*
90A0050: length: 665mm
90A0027: length: 960mm; bore: *c.* 50mm
90A0132: length: 606mm; bore: *c.* 50mm

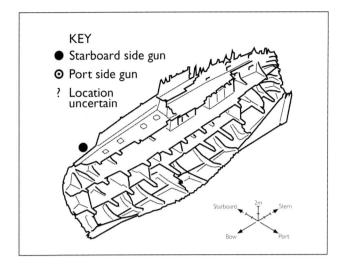

A chamber holder with *in situ* chamber and wedge and tiller was excavated from within a large concretion (90A0050; Fig. 3.124). 90A0027 and 90A0132 (Figs 3.125–6) may be part of the same gun, though this is not certain. The concretion also included three separate breech chambers (90A0122–4). The first two could have been used with 90A0050 (Fig. 3.127). The original location for this gun is assumed to be the bow bulkhead. The piece remains in wet storage and measurements were taken from the radiograph (Tables 3.17 and 3.20). The tiller has a length of 300mm and width at the back of 70mm. The chamber tapers inwards to the mouth with a wall thickness of 15mm, bore of 35mm and 20mm thick breech plug. The yoke is visible and is made of 25mm bar. The distance between the trunnions is 190mm and the eyes formed to go over the trunnions are 100mm in diameter.

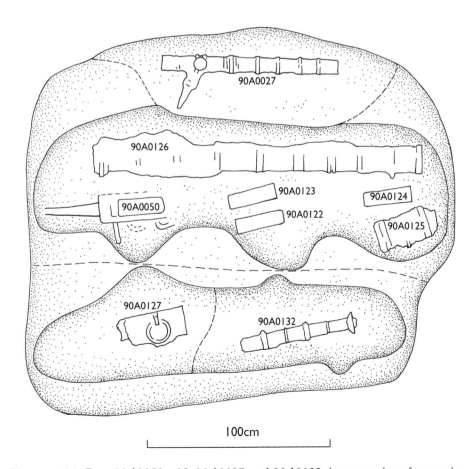

*Figure 3.124 Base 90A0050 with 90A0027 and 90A0132: interpretation of concretion*

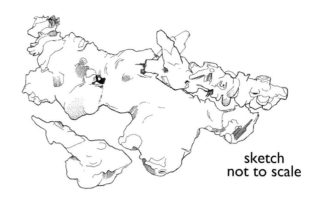

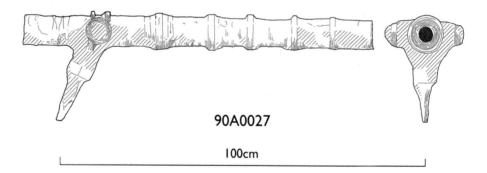

*Figure 3.125 Concretion 90A0050 and base fragment 90A0027*

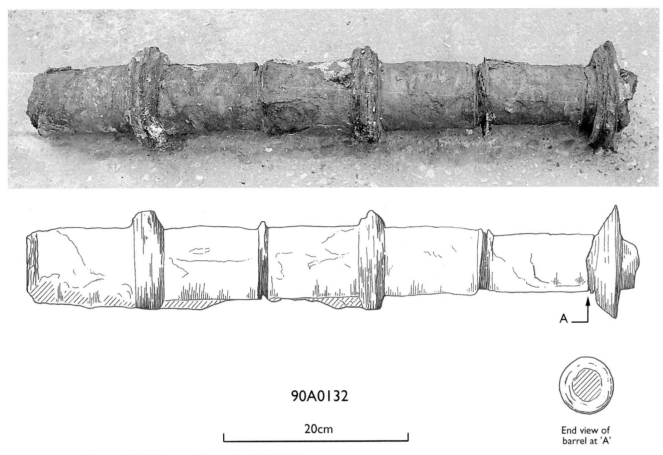

*Figure 3.126 Base 90A0050: muzzle portion 90A0132*

Both barrel fragments 90A0027 and 90A0132 have a double layer construction. There is no indication of the method of manufacture of the inner barrels but the outer surfaces are covered with alternating bands (90–140mm, 8/10mm thick) and hoops (10–20mm wide, 20mm thick). 90A0027 includes trunnions (diameter 55mm) set within a triple hoop complex with a second set towards the front of the gun and then alternating hoops and bands (Fig. 3.125). The remains of the peg are visible on the radiograph (250mm long, 30mm diameter). There is a 45mm shot 200mm behind the peg. The wall thickness varies between 10mm (behind the shot) to 35mm over a hoop. 90A0132 is a barrel portion (Fig. 3.126) with bands of 80–100mm wide and hoops 10–20mm wide, some of which may be double. It is badly degraded with an inner tube wall thickness of 5mm.

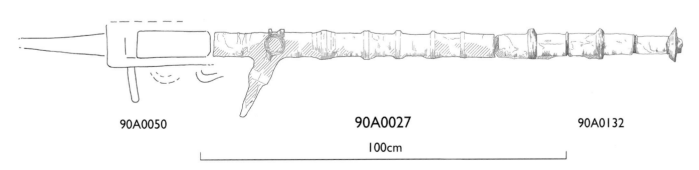

*Figure 3.127 Base 90A0050: reconstruction including 90A0027*

## 81A1082 Barrel

*Location*: unknown
*Recovered*: SS8/9 (lifted as part of 90A0025)

*Statistics*
Length: 1235mm; bore: 60–70mm

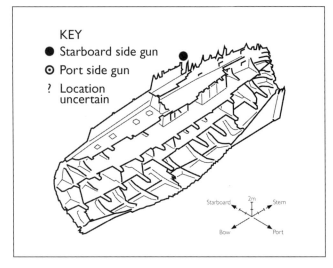

This possible swivel barrel, in extremely poor condition and currently in wet storage, was found in SS8/9, and is thought to have originated from beneath the Sterncastle (Fig. 3.128). Radiography suggests that it is stave-built with second layer of alternating bands of varying width 80mm (1), 85mm (5), 90mm (1) and 125mm at the mouth (1) and single hoops (20–25mm wide), the hoops raised at least 10mm above the bands. It is not possible, therefore, to securely identify this piece as a swivel gun. The fragment may include the mouth, which consists of 20mm hoops, one of which is angled slightly to suggest the bore. No swivel elements were found. It is impossible to count the number of staves, but two parallel lines are visible, possibly suggesting two halves of an inner tube with a wall thickness of 25mm. Other than the staves, the radiograph appears very similar to that of 82A4076 (above). On the other hand, the apparent presence on the radiograph of corroded rings (no longer on the gun) 30mm above the inner tube and 20mm above the height of the bands, is more in keeping with a gun that locates into a wooden bed, possibly a sling. Sling 82A2700 has a bore of 70mm, but this is so corroded it is difficult to tell whether it is a barrel or breech. Others are 80mm or more.

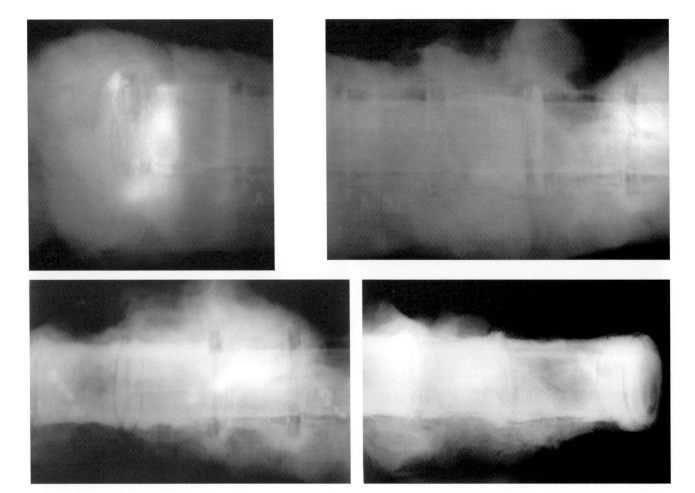

*Figure 3.128 Base 81A1082: radiograph*

## 82A4080 Barrel with chamber holder and chamber

*Location*: unknown
*Recovered*: SS8/9 in concretion

*Statistics*
Length: 840mm; bore: *c.* 90mm

This gun barrel was excavated from within a large concretion that also contained gun muzzle 82A5152. It is in wet storage but was radiographed in concretion. This shows that it is possibly (but not definitely) stave-

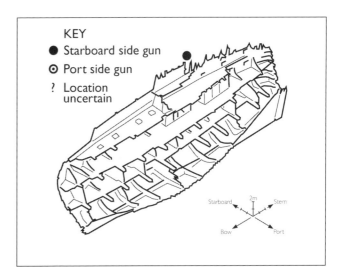

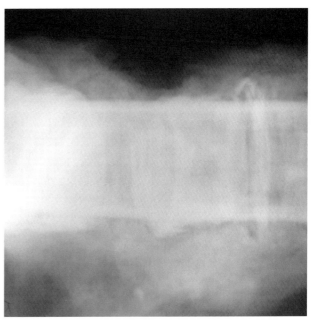

*Figure 3.129 Base 82A4080: photograph and radiograph*

Key

1  04A0097
2  90A0052
3  81A5152

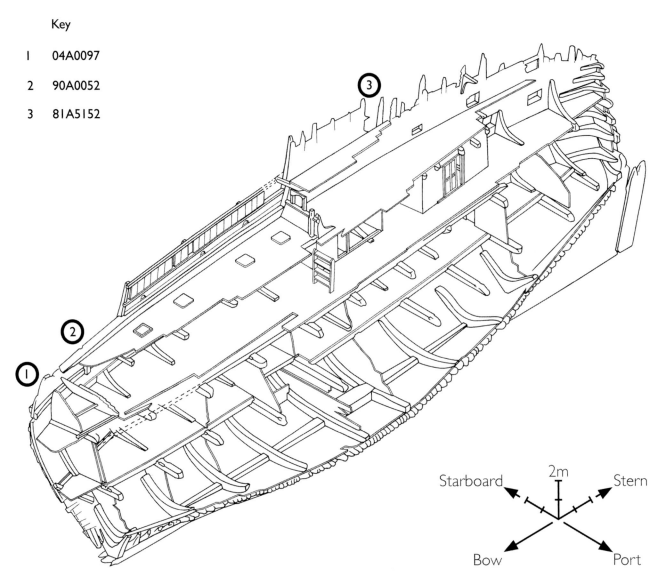

*Figure 3.130 Location of base fragments of unknown type*

built with alternating 40mm wide hoops and bands of 80–100mm width. (Fig. 3.129) Archive notes suggest a chamber trough was also present with a width at the back of 125mm, containing a chamber, mouth tapering inwards. A long tapering wedge can be seen. Thick white metal suggests a hoop assembly holding the trunnions, and there are possible yoke elements visible. There is no obvious shot. The radiograph shows the diagonal line marking the junction between chamber and inner tube and the thickness of the inner wall is 12mm at this point. There is a maximum width over the bands of 130mm and over the hoops of 160mm. The bore is recorded as 90mm in the excavation report and reinforced by the radiograph. If correct, this is a very large swivel gun compared with the rest of the assemblage, although the distance across the trunnions (210mm) is similar to 82A4076 (bore 60mm). The records indicate that black powder and rope was found 'inside', presumably within the chamber, but its excavation is not recorded (86S0030, 25g).

## Unknown types and fragments

A selection of the remaining base elements was found (Fig. 3.130); summary information on all is given in Tables 3.17–20 and details are in the archive.

**04A0097 1–3  1: base in concretion; 2: chamber holder and barrel fragment, possible chamber ; 3: ?spare chamber**

*Recovered*: Excavated: East of stem, collapsed Forecastle structure

*Statistics*

1: Length: 1100mm; maximum width: 360mm: bore: 45mm
2: Length: 970mm; width 260mm; maximum external diameter: 105mm
3: Length: 280mm; maximum width: 260mm

267

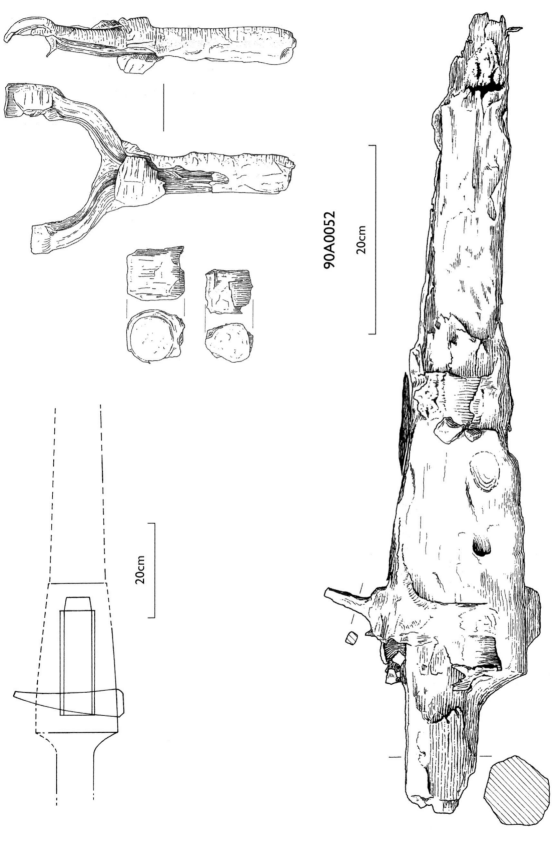

*Figure 3.131 Base 90A0052*

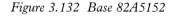

*Figure 3.132  Base 82A5152*

All fragments were heavily concreted (not illustrated). The largest retains one flat edge as though it had been resting against a deck. These fragments are obviously portions of a swivel gun including the chamber holder, but are too concreted to ascertain any details. They are not currently available for more detailed study.

**90A0052 Semi-complete tiller, holder, yoke, wedge, chamber and barrel**
*Location*: unknown, bow bulkhead?
*Recovered*: SS2/3, below concretion 90A0050

*Statistics*
Length: 850mm; chamber bore: 28mm

This semi-complete example (Fig. 3.131), was probably originally positioned beneath the bow. It was in extremely poor condition and covered in large oyster shells, breaking up during excavation. The tiller (length 410mm) is hexagonal and large; the wedge is 250mm in length and tapers from 60mm to a point. The yoke is present with a 300mm long peg. This has a hexagonal collar holding the ends together (similar to 82A4076). Twine was noted around this and behind the chamber holder at the junction with the tiller. There are no obvious hoops and bands. The swivel yoke appears to have moved relative to the gun and was loose but is well forward of the chamber trough as in the form 2 swivels. When excavated, the yoke, which is made from 25mm bar, came away with a trunnion fragment of 45mm diameter, again indicating a form 2 base. The distance between the trunnion faces is 190mm. The mouth of the chamber (length 250mm) was sealed with a tampion. There are six other fragments forming parts of the chamber trough.

**81A5152 Barrel fragment**
*Location*: unknown
*Recovered*: SS8/9 in concretion 81A4080

*Statistics*
Length: 430mm; bore: 50mm; weight: 7.5kg

This barrel fragment, including the muzzle, ends in a double hoop (Fig. 3.132). It has a wall thickness of 10mm. Radiography did not show up any detail of the internal structure and so an attempt was made to measure the velocity of sound through the material (details in archive). This gave readings close to that of modern cast iron, though not close enough for confirmation. Vague shadows on the radiograph suggest the possibility of external hoops. It remains enigmatic.

# Top Pieces
by Ruth Rhynas Brown

The '*top piece*' or '*top gun*' has a very short history in Tudor times. Their heyday corresponded with the reign of Henry VIII, and these pieces quickly passed into obscurity. Later they were confused with top darts, which, as the name implies, were javelins thrown down from the tops. Top darts are often depicted in fifteenth and sixteenth century illustrations of ships, such as in the fifteenth century manuscript *Beauchamp Pageant* in the British Library (Fig. 3.133), the tapestry of Sir John Denhym in the Cloisters Museum, New York of *c*. 1500 or the portrait of Sir John Luttrell (1550) in the Courtauld Institute Collection, London, in which the darts can be seen cascading from the broken mast.

In 1513 Cornelis Johnson, the King's gunsmith, was paid for work relating to a number of the King's ships, including for the Spanish Carrack, manufacturing or repairing '*a stone gun to the mayne mast*', with four new chambers, a '*top gonne*' for the foremast, with four new chambers and a '*new small gonne made to the myssyn mast for iiij new chambers made to the same gonne*' (BL Stowe MS 146, art. 6; *LP* 1, 1704). Both these accounts, and the inventory of the King's ships from 1514, shed light on the distribution of the guns on the ships. Of the thirteen ships named, only four are listed as specifically having some sort of top piece: the *John Baptist*, which lists two top guns, with eight chambers and '*miche, boltes and forelockes*', the *Mary Rose*, with three top guns, the *Peter Pomegranate* with one small top gun and two or four chambers and the *Christ of Greenwich* with one top gun and chamber in the highest deck. Even more illuminating is the inventory of the *Great Elizabeth* where one stone gun is listed in both the foretop and mizzen top (Knighton and Loades 2000, 138, 141, 146, 150).

No top pieces are listed in the 1540 inventory, which includes the *Mary Rose* (PRO E 101/60/3, summarised in Anderson 1920). The *Anthony Roll* (Knighton and Loades 2000) lists one top piece in each of 20 ships with five vessels, including the *Mary Rose*, having two each, a total of 31 in 50 ships in the English fleet. Most of these were the largest ships, although it included smaller vessels such as the *Christopher*, *Hoy Bark*, *George*, *Mary James*, *Tiger* and *Bull*. Each top piece has an allowance of 20 stone shot, although intriguingly the *Mary Rose* top pieces have an allowance of only 10. The listed shot is always stone, in contrast to bases or other swivel guns which had lead shot. This is highlighted in the entry for the *Trinity Harry* which specified: '*For toppe pece, shot of stoen*', and the like for the *Greyhound*.

By the time of the 1547 inventory of Henry VIII's possessions, thirteen top pieces '*Grete and smalle*' are listed in the stores of the Tower of London (Blackmore 1976, 262). Of the 52 ships listed, nine had top pieces: three on the *Minion*, two on the *Less Bark*, and single guns on six ships. However, seven guns are recorded on the *Hart*, but with only 60 stone shot (Starkey 1998, 146–57, 163). This suggests that top pieces were no

*Figure 3.133 Ship in* Beauchamp Pageant *showing top darts (BL Cotton MS Julius E.IV, art..6). Reproduced by permission*

*Figure 3.134 The 'WA' illustration of a carrack showing a probable swivel gun in the stern fighting top (© Ashmolean Museum Oxford)*

*Figure 3.135 The Goodwin Sands gun (© Mark Redknap)*

longer only used in the tops, but also as light stone throwers since the total still comes to 31. There certainly seems little continuity in their distribution; many ships previously armed with them no longer are, while others, such as the *Greyhound*, now have one.

In the 1555 inventory now at Longleat only the *Swallow* retained her single top piece. She was also armed with 100 stone shot, although it is not specified which guns these are for. A general survey of the Queen's ships in 1558 records three top pieces onboard and one in store, listed with the forged iron guns. No details are given of which ship they were actually aboard. The shot records do not list shot for top pieces, the only stone shot specifically mentioned is for port pieces. Another listing of ordnance aboard 24 ships in 1574 includes two top pieces among the forged iron ordnance (BL Add MS 48092, f. 89). By 1576 the *Swallow* has gained a second top piece (NMM CAD/C/1). These two appear to be the last still in service; they had been removed by 1585 when the *Swallow*'s guns had been replaced with port pieces and fowlers, although it is possible they have been absorbed into the list of fowlers. During this period, wrought iron guns were being removed from ships and replaced with bronze breech-loaders. By 1589 only nine forged iron fowlers and bases were left in total (Bodleian Library MS Rawlinson A.207, f. 3).

There is little other information. Most pictures of ships of the sixteenth century quite clearly display top darts – guns are a rarity. The well-known depiction of a carrack by the artist '*WA*' which dates from the late fifteenth century shows a small gun, possibly a swivel gun mounted on a shaped wooden stock, in the mizzenmast (Fig. 3.134), more like the Cattewater gun (Redknap 1997, 76) or the gun recovered from the Goodwin Sands (Fig. 3.135) than the unstocked bases so far recovered from the *Mary Rose*, however the illustration is earlier (*ibid.*, 79). A silver nef made in Nuremberg in 1503 is similar to the vessel depicted in the '*WA*' engraving and also supports a swivel gun in a heavy wooden stock in the mizzenmast (Oman 1963). A vessel shown in a window in King's College Chapel, Cambridge also displays what may be top pieces in the fighting tops. Documentary evidence also reveals guns in tops in foreign ships; for example, a near contemporary of the *Mary Rose*, the *Santa Anna*, the great carrack of the Order of Saint John, built in 1522, also carried small guns at the masthead (Muscat 2000, 19).

From the little evidence available few deductions can be made about the top piece. A fundamental question is whether it was merely a small piece placed in the mast top or a distinctive gun that would have been recognisable as a '*top piece*', wherever placed. They do not appear in every inventory, but this may be because they were thought too insignificant to count, or because of their inaccessiblility in the tops, or because they may have been included in other categories by less knowledgeable clerks. The accounts for the manu-

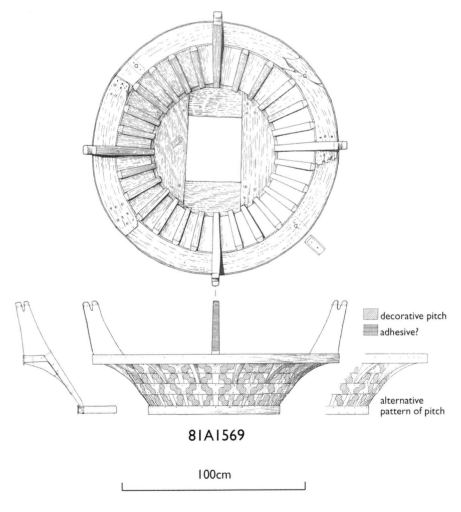

*Figure 3.136 The fighting top from the* Mary Rose

facture of top pieces and history of their appearance in inventory records indicates that they did differ in their construction from other ordnance, so that officers and clerks were able to distinguish them. As they fell out of use, they were absorbed into one of the general categories of base, sling or fowler.

It is clear, however, that the top piece was a forged iron, breech-loading gun firing stone shot. The reference to '*miches*' implies that the barrel was mounted on an iron swivel and possibly within a wooden stock. The reason for its demise was most likely the difficulty of handling it in the tight space of a fighting top and it was quickly phased out.

## The structure of the Mary Rose and the accommodation of top pieces
by Alexzandra Hildred

One fighting top was recovered from the Orlop deck (Fig. 3.136). Its position and orientation suggest that it was not *in situ* at the time of the sinking but was a spare. It is constructed as a flat-based 'basket' out of oak and pine with a central rectangular hole allowing for a mast of no more than 300mm diameter. The basket itself is only 650mm in floor diameter, flaring to 1100mm. Its construction is described in full in *AMR* Vol. 2 (chap. 14). Its size and lightweight structure indicate that it could only have held one man with difficulty, never mind a small gun. The possibility remains that it is decorative rather than functional (Endsor, *AMR* Vol. 2, chap. 14).

The *Anthony Roll* shows five tops on the *Mary Rose*, one on the foremast, two on the main, one on the mizzen and one on the bonaventure. All of the ships illustrated are portrayed with tops, the *Henry Grace à Dieu* has two to each mast, the *Peter* and *Mary Hamburg* carry two on the main and one on each of the other three (although this may be masked by a flag on the mizzen mast of the *Peter*). Others (eg, the *Christopher* and *Trinity Harry*) only carry three, on foremast, main and mizzen, and yet others (eg, the *Lartique* and *Mary Thomas*) only two, on foremast and main mast. None is shown with a gun.

None of the recovered bases was furnished with stone shot and while the lightest and shortest guns recovered are the fowlers (though these are listed as carriage mounted rather than swivel mounted), all have bores of over 100mm and are similar to the port pieces and slings.

Only two small stone shot were recovered, both from SS10 (diam. 75mm) and are the only stone shot which may be small enough to furnish the top pieces, suggesting a bore of *c*. 80mm. The items identified as incendiary darts (see Chapter 6) were found just inboard in M10 and it is not unlikely that both shot and darts were for use in bonaventure or mizzen masts. It must be concluded that, as yet, we have no guns

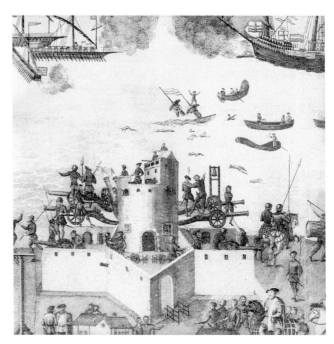

*Figure 3.137 The sinking of the ship as depicted in the Cowdray Engraving (© Society of Antiquaries of London)*

identifiable as top pieces. This is not surprising. The Cowdray engraving depicting the sinking of the *Mary Rose* shows two fighting tops clear of the water (Fig. 3.137). Any guns in these positions may either have been accessible to immediate salvors, or could have fallen some distance clear of the wreck.

## Replication of one of the Mary Rose port pieces

In January 1988 The Royal Armouries and the Mary Rose Trust decided to replicate one of the large wrought iron guns recovered from the wreck of the Mary Rose. The research design included the following elements:

- Practical research into the methods used in the construction of the original by experimentation.
- Production of a gun which could be fired.
- Testing the effectiveness of the gun against a wooden target.

The project included a detailed study of the original gun and breech chamber, replication in wrought iron, and firing trials. As we realised that the project was unique, other specialists were asked to supply questions for us to address (D. Armstrong, H. Blackmore, N. Boston, R. Brown, C. Carpenter, F. Craddock, J. Green, J. Guilmartin, D. Keith, J. Light, G. Rimer, M. Rule, J. Simmons III, J. Waller). The research team included Alex Hildred, Robert Smith, Nick Hall and Guy Wilson. The reproduction was manufactured by Chris Topp and Co., Thirsk, North Yorkshire. Sub-contractors included

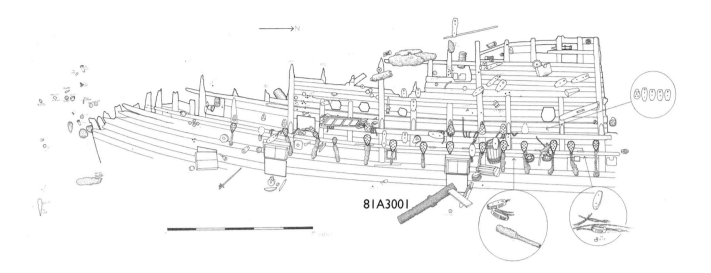

*Figure 3.138 External elevation of starboard side at the Sterncastle showing gunport and gun (81A3001) outside the ship just below the gunport*

Richard Lewis (portions of the breech), Crowder Engineers (machining and fitting of staves) and J. Pilkington (turning the wooden core). To provide the blacksmith with as much information as possible, specific measurements were taken along with photographs and descriptions. These and the 1:5 archaeological drawings were given to Chris Topp, who also made his own calculations based on study and interrogation of data. The sectioned chamber was present in the workshop for reference throughout the replication process. The gun was mounted in an elm carriage supported by a pair of spoked wheels; this was replicated by Jim Sadler at Fort Nelson, using the archaeological drawings.

The original gun chosen was (81A3001). It was found below the ship, beneath a gunport on the Main deck (M8) having separated from its carriage as the ship sank (Fig. 3.138; see above). In this burial environment corrosion of the iron was limited, and as a result the treatment of the gun to date has been restricted to long term (2 years) soaking in sodium hydroxide, mechanical cleaning and coating with waxoyl. The condition of this gun is better than all the others, with surface detail neither too badly corroded, nor obscured by cleaning/conservation. The breech chamber was not in situ, but was one of two lying beside the empty carriage in the trench. It has been mechanically cleaned and soaked in a weak solution of sodium hydroxide.

The gun is 3m in length, with an internal bore of 175mm (7¾in) (Fig. 3.139). The barrel is made up of nine internal butt jointed staves, each 16mm (⅝in) thick, reinforced by a single layer of alternating bands (17) and abutting single or triple hoops (38) (Fig. 3.140a). The staves are curved across their width and their edges were chamfered to fit more closely together.

All groups of hoops, except those directly at muzzle and at bore, are chamfered with the highest point being at the centre. Each one in a triple group has chamfers on both sides. Four hoops are distinctive – the non-visible muzzle hoop, to provide a bead over which the ends of the staves can be turned (observed on a number of guns); the sighting hoop, which has a raised sight; and a lifting ring hoop, and the last barrel hoop which is thinner than the rest and countersunk (Fig. 3.140, b–d).

The original gun was loaded with a Kentish Ragstone shot (82A4213) weighing 6.4kg and 165mm (6½in) in diameter. The powder was contained in a separate breech chamber. Radiography using a Cobalt 60 source was inconclusive regarding its construction, volume and the shape of the bore so it was sectioned longitudinally using a planer miller machine at the Fleet Maintenance and Repair workshops in Portsmouth Dockyard (Fig. 3.140e). Not only does this show that corrosion is limited to the outer few millimetres of the surfaces of the iron, but provides excellent raw material for metallographic and chemical analysis. Sectioning confirmed the presence of three layers, and questions such as the volume and shape of the bore were answered. The chamber tube was made of two layers of staves covered with a series of thirteen alternating bands and hoops. The internal layer has four equal staves 20mm (¾in) thick about 100mm (4in) wide (63mm on inside and 109mm on outside) and is encased by three staves 15mm (⅝in) thick, one covering half of the tube and the other two equal quarters. It has a tapering internal diameter of 100mm (4in) at the breech plug to 115mm (4½in) at the mouth. The neck extension into the back of the barrel is 70mm (2¾in) long. The total length of the chamber, excluding the neck, is 890mm (22½in; see Figs 3.25 and 3.26).

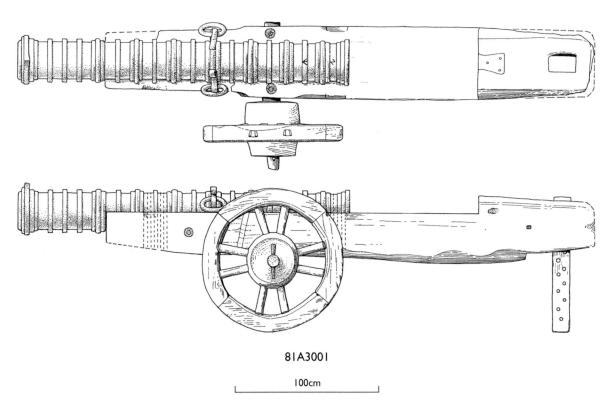

*Figure 3.139 Port piece 81A3001*

## Method of manufacture

### Material

The iron used was puddled wrought iron, primarily reclaimed from long link mooring chain dating from the early twentieth century. Originally charcoal-smelted wrought iron would have been used, but this was not cost-effective. Most of the techniques used in the manufacture were those available to the Tudor blacksmith; furnace, fire and hammering (either hot or cold). Table 3.21 shows the techniques used in each process, the time each element took, and how this may have differed from Tudor methods. The methods of heating were modern, using coal and electricity. It was not felt that the use of puddled iron affected either the working of the material nor its capabilities as a gun.

### Manufacture of the barrel

Study of the barrel confirmed a two layered construction composed of an inner tube of nine inner staves held in place by an outer layer of contiguous hoops and bands. The function of the inner tube is to provide a straight clean surface through which the shot is propelled. The gas-tightness is achieved through the method of building up the layers, combining the longitudinal strength of the staves in one direction with

Table 3.21 Manufacture of the barrel: processes, techniques and time

| Procedure | Labour (no. men) | Time per item (minutes) | Tools used | Tudor tools |
|---|---|---|---|---|
| **a. Staves (9)** | | | | |
| Upset muzzle | 1 | 20 | fire, hand, power hammer | fire, hand, power hammer |
| Curve stave | 2 | 20 | furnace, power hammer, top & bottom swages | fire, by hand |
| Turn muzzle end over | 2 | 20 | fire, by hand, forming tools | fire, by hand, forming tools |
| Straighten cold | 1–2 | 30 | hand tools, power press | hand tools |
| Mill sides | 1 | 90 | milling machine | cold chisel |
| Drill for bolt, countersink | 1 | 5 | drilling machine | hand drill |
| Cut off, turn | 1 | 10 | power saw, file | hand saw, file |
| *Total: 2385 minutes, 189 per stave* | | | | |
| **b. Hoops (35)** | | | | |
| Firge section from billets | 1 | 30 | furnace, power hammer | fire, power hammer, sledge hammer |
| Calculate length & cut | 1 | 10 | power saw | cut hot |
| Upset & scarf ends for side weld | 1 | 15 | fire, power hammer | fire, power hammer |
| Bend to near circle, overlap ends | 1 | 15 | furnace, power press | fire, by hand |
| Forge weld | 2 | 15 | fire, sledge hammer, power hammer | fire, sledge hammer, power hammer |
| Round up & size | 2 | 20 | furnace, power hammer, cone mandrel, hammers & special swages | furnace, power hammer, cone mandrel, hammers & special swages |
| Bevel (f req.) | 3 | 20 | furnace, stepped mandrel | fire, stepped mandrel |
| Flatten | 1 | 5 | power hammer, same heat as furnace | by hand |
| **c. Muzzle hoop (1)** | | | | |
| Forge section in moulding tool | 1 | 29 | furnace, power hammer, tools | fire, power hammer, tools |
| Cut to length | 1 | 5 | hot fire | hot fire |
| Scarf ends | 1 | 10 | fire, by hand | fire, by hand |
| Form into circle | 2 | 10 | fire, by hand | fire, by hand |
| Forge weld in moulding tool | 2 | 10 | fire, by hand | fire, by hand |
| Trus up circle | 2 | 10 | furnace, cone mandrel, by hand | fire, cone mandrel, by hand |
| Flatten | 2 | 10 | same heat, by hand | same heat, by hand |
| Forging the section required a lot of power | | | | |
| **d. Lifting ring hoop (1)** | | | | |
| Forge stock from billet | 2 | 10 | furnace, power hammer | fire, power hammer |
| Set in for lugs & draw down | 2 | 20 | fire, power hammer | fire, power hammer |
| Punch slots for scrolls & cut | 2 | 40 | fire, by hand | fire, by hand |
| Draw out scrolls & dress | 2 | 40 | fire, by hand | fire, by hand |
| Cut & scarf ends | 2 | 30 | fire, by hand | fire, by hand |
| Bend to rough circle | 2 | 20 | furnace, by hand | fire, by hand |
| Forge weld | 2 | 10 | fire, by hand | fire, by hand |
| Dress to true circle & size | 2 | 20 | furnace, by hand | fire, by hand |
| Punch holes | 2 | 30 | fire, by hand | fire, by hand |
| Forge 1in (6mm) rounds for rings | 1 | 20 | furnace, by hand | fire, by hand |
| Upset & scarf rings | 1 | 20 | furnace, by hand | fire |
| Bend rings to open circle | 1 | 10 | furnace, by hand | fire, by hand |
| Thread through hole, forge weld | 3 | 40 | fire by hand | fire, by hand |

**Table 3.21 continued**

| Procedure | Labour (no. men) | Time per item (minutes) | Tools used | Tudor tools |
|---|---|---|---|---|
| **e. Assembly of barrel (after manufacture of core & fabrication of stand)** | | | | |
| Assembly of staves | ? | | chain hoist | chain hoist |
| Place hoops (34) & bands (17) | 1 to size 3 to embed | 60 each 68 | furnace, cone mandrel, sledge hammer, jump ring, set hammers | fire, cone mandrel, sledge hammer, jump ring, set hammers |
| Caulking | ? | | | |
| Quenching | ? | | | |
| Core extraction | 1 | 60 | fire | fire |
| Cold working | ? | | | |
| Cutting & upsetting staves | ? | | | |
| Cold caulking | 2 | 1 day | | |

the second layer (hoops and bands) at right-angles to this. This confirms a great understanding of the materials and the stresses to which they would be put.

A wooden mandrel was made to form the bore (Fig. 3.140f). It consisted of four pieces of timber glued together to leave a rectangular hole up the centre. The outside was turned to equal the internal dimensions of the bore. A stand was fabricated from steel joist sections with a vertically projecting stub which fitted into the muzzle end of the wooden core. The breech end was relieved internally to receive a 100mm (4in) diameter iron ring. The timber core was 76mm (3in) longer than the gun at the muzzle end, and 50mm (2in) longer at the breech end. The purpose of the core was to maintain the straightness of the barrel, and to hold the barrel into the stand for the purpose of fitting the hoops and bands.

The curve required for each of the nine staves was established. Each was forged from 73mm (2⅞in) x 13mm (½in) rolled bar cut slightly longer than the required length (traditionally these would be made of smaller pieces welded end to end to form the stave). The edges were milled as dressed edges were observed on the original. These were held on the mandrel with spikes driven through the ends. The barrel was made to taper outwards from breech to bore by 13mm (½in), as observed in the original. The staves provide the longitudinal strength, conferring a resistance to bending. The ends of the staves were upset (thickened) at the muzzle 94mm (3¾in) x 13mm (½in) for about 64mm (2½in) (Fig. 3.140g). This was achieved by heating the stave at the required position and bouncing it on an anvil to make it swell.

The curve on the staves (Fig. 3.140h) was achieved by hammering using a power hammer and utilising top and bottom swages (iron blocks forming the negative of the required shape). The muzzle end was turned over (Fig. 3.140g, h). The staves were not attached to each other along their length. None of the radiographs showed any signs of attempting to weld the seams, nor the addition of lead as a filler. Both of these have been suggested as being necessary to make the inner tube gas tight. Chris Topp considered that welding these would be a near impossible task given the length of the staves, as both staves would require heating to and retaining forging temperature along their lengths.

The hoops and bands form the circumferential strength of the tube and bind the staves together. The 34 hoops were not standard sizes, with heights above the bands differing on either side, and bands differing in width. There is some consistency in single hoop width (±42mm; 1⅙in). These were forged from billet stock to approximate the original section cut to length with a power saw. The ends were upset and scarfed for a side weld. The hoop was bent into a circle and the ends overlapped and forge welded (Fig. 3.141a–b). Each was made 3mm (⅛in) below the required size on a cone mandrel (Fig. 3.141c). Any flattening was undertaken with a power hammer and bevelling by reheating in the furnace and using a stepped mandrel, bottom swage and hammer.

The muzzle hoop (Fig. 3.141d) was differently shaped to accommodate the outward turn to the ends of the staves. This was achieved using a power hammer. It was then cut and the ends scarfed. It was formed into a circle, forge welded and checked to be true on a cone mandrel. It was then flattened (flat welded lap). The sight hoop is a standard hoop but with any excess material in the area of the weld used to make the sight. This was made by hand over the anvil. The lifting hoop (Fig. 3.141e) was forged and then thickened to accommodate the holes for the lifting rings. Slots were punched for the scrolls and cut. These were drawn out, dressed, cut and the ends scarfed. The piece was bent, forge welded and dressed to a true circle and sized. Holes for the rings were punched. The rings were forged to be of 1in (25mm) diameter. These were upset and scarfed, bent and threaded through the holes in the scrolls and forge welded shut.

*Figure 3.140 Manufacture of the barrel: a) detail of hoops and surface finish on port piece 81A3001; b) sight hoop; c) lifting ring hoop; d) hoops at breech junction; e) sectioned chamber; f) staves held on mandrel; g) upset at muzzle; h) showing the curve of the staves*

*Figure 3.141 Manufacture of the barrel: a) bending a hoop; b) dressing a hoop; c) sizing cone used for hoops; d) the muzzle/sight hoop; e) lifting ring assembly; f) putting staves over core; g) measuring for muzzle hoops; h) jumping block sequence; i) burning out core*

The seventeen bands were made like the hoops, but the weld is made like a tyre, over the anvil bick. The estimate of 13mm (½in) band thickness was based on the residual left after subtracting the inner diameter from the outer diameter over the bands and the known stave thickness. This is a harder weld than the hoops. As these are small in diameter, it proved challenging to get full welding heat on the inside. The narrowest was 80mm (3⁵⁄₁₆in). These get thicker and narrower towards the breech. The originals display a dish which was probably caused by increasing the diameter over a tapered mandrel working from both ends. This was an effective method and also tested the integrity of the weld. The last hoop towards the breech had a thinner circumference and was countersunk about 6mm (¼in). As in the original, this was achieved using a grinder, but could have been effected with a cold chisel.

The core was raised to a vertical position and each stave was bolted to it by means of an 8mm countersunk bolt at the breech (upper) end, through the stave, core and 100mm (4in) iron ring (Fig. 3.141f). The muzzle end of each stave rested on an iron spacing ring of appropriate width, placed upon the stand, around the core. There is no evidence for this method being adopted, but it would have been a perfectly plausible technique.

The first hoop, the muzzle hoop, was driven on cold and hammered down into the annular trough formed by the turned over muzzle ends of the staves. This was done cold to avoid distortion of this thin ring. Once positioned, this hoop served to control the muzzle ends (lower) of the staves. The rest of the hoops and bands were heated and sized, and hammered down hot, allowing 3mm (⅛in) oversize over the staves. Each position was measured with calipers (Fig. 3.141g) and the size of the hoop adjusted hot immediately prior to fitting. By this method the accuracy of the fit of each hoop or band was ensured. The radiographs of the original were looked at to ascertain whether there was any conscious positioning of the lap welds on the hoops or bands. The welds were hard to see. It was difficult to see these on the reproduction, although easier on the bands than the hoops. Where these were visible, they were positioned under the barrel so as not to be visible.

The effect of the shrinkage of the hot rings is to compress the staves together at the ring position, but also to spring the loose staves outwards slightly further up the barrel. Therefore, in spite of the overall taper of the outer diameter of the staves, their distortion provided an obstacle to the free passage of hoops and bands at the centre of the remaining length of the barrel. This effect was most marked with the first rings, and lessened as work progressed up the barrel. This is one reason why it was felt necessary to bolt the upper ends of the staves. This was solved by making the lower rings slightly oversized. By using a heavy steel jump ring or block, with handles (Fig. 3.141h) considerable force was used to get the rings past the mid-point on the barrel. Rings were bedded on to each other by hammering while still hot, using special curved set hammers. These would also be caulked while still hot (caulking was observed on the original). After fitting each ring or hoop water was used to cool the ring and stave to prevent heating the whole barrel too much and therefore distorting the size. Each hoop or band took about an hour to size and fit using a furnace, cone mandrel and up to three men for the fitting and bedding. It was noted that there was a tendency to lose a little length from the barrel owing to the compression of the hoops as they were driven down and bedded, and that any 'out of squareness' of the rings had to be monitored and corrected. The length did not prove critical, as the projecting staves were merely cut off with enough to dress over cold. The completed barrel was chain-hoisted off the stand and the timber core burnt out (Fig. 3.141i). This was ignited with a gas burner but burnt under its own combustion in an hour.

The muzzle end was placed over the bick of an anvil whilst still supported on the hoist. The muzzle staves, already turned over, were driven down cold over the muzzle ring using a sledge hammer, fullers and set hammers. Any gaps were caulked. The muzzle was hand-finished with files to make a smooth end, two men taking a whole day. The projecting ends of the staves were cut off with an angle grinder, leaving 10mm (³⁄₈in) proud beyond the countersunk breech hoop using fullers and sledgehammers. This was ground smooth and a radius ground on the inner circumference. Cold caulking was begun on the outside face of hoop/band joints where any gaps could be seen. There is evidence on the original gun to support surface dressing. The staves were found to be very tightly held together, with no visible gaps between, forming an extremely rigid structure.

## Manufacture of the breech chamber

The breech chamber consists of three layers – two of staves and the outer of hoops and bands (see Fig. 3.140e). The inner layer comprises four staves of equal size (100mm wide, 25mm thick) tapering from plug to neck (25–20mm; 1in–¹¹⁄₁₆in). The second layer (13mm; ½in thick, non-tapered) comprises three equally sized and curved staves, one forming a semicircle and two forming one quarter each, with the joins arranged between those of the inner layer. The plug at the breech (equal to the bore width at the back) is 88mm (3½in) in diameter and at least 75mm (3in) thick. The plug tapers inwards and was loacted from the outside at the back. There is no attempt to bend the ends of the staves over the plug. The outer layer of hoops have a width and height of 38mm (1½in) and the bands have a width of 63mm (2½in) and height of 38 mm (1½in). These alternate, with the exception of one triple hoop sequence. The external sequence from the back to the front is: end ring, hoop, band, hoop, band, triple hoop, band, hoop, band, hoop, broken ?band. Although they

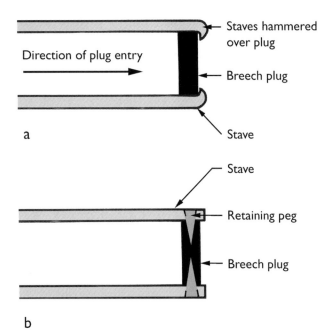

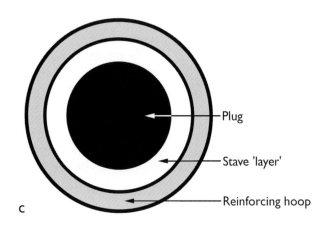

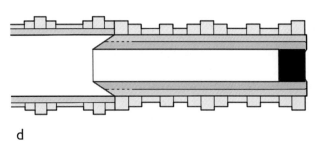

*Figure 3.142 Manufacture of the chamber: a) schematic section through back of chamber; b) possible reconstruction using retaining pegs with the breech plug; c) schematic section through plug end of breech chamber; d) schematic section through junction of barrel and chamber*

are same height above stave layer, the hoops look as though they should be higher.

The length of the breech needed to be a total of 645mm to fit in the back of the gun, allowing for a wooden forelock (200mm) and two iron wedges taking up a maximum of 50mm (2in) behind it. On other examples the neck (formed of the stave layer/s only) extends some 57–82mm (2¼–3¼in) into the back of the barrel.

Detailed 1:5 scale drawings were studied of chambers found *in situ* in port pieces on carriages (81A2604; 81A2650); two found inboard of gun 81A3001; and two large loose examples (85A0028, 81A1084) in order to elucidate the method of manufacture, and determine whether there were any consistent design criteria. Specific attributes examined were:

- Evidence of breech plug insert.
- Length of neck.
- Taper on neck.
- Presence of collar/fillets around staves at neck.
- Physical evidence for double layering, either of staves, or staves/collar/tube.
- Looseness of fit, neck to back of barrel.
- Whether any hoops from chamber locate into back of barrel, or hoops on barrel just butt against hoops on chamber.

The study indicated that the exposed breech chamber length for the reproduction should be 575mm with a neck of 70mm. It was assumed that the best method for closing the end of the chamber was the insertion of a plug to seal one end. A good method for positioning the breech plug would be to enter it from either end of the chamber and bend the staves over the back (Fig. 3.142a). However this could not be confirmed as this area on the original was obscured by one of the hoops forming the outer layer. We also wanted to look for evidence of pegs which might be driven through the plug to secure it to the chamber (Fig. 3.142b). We surmised that the evidence to support a plug might consist of visible lines forming two circles on the back of the chamber. The inner circle forming the plug and the outer forming the ring of staves (Fig. 3.142c). Unfortunately in all but one instance it was almost impossible to see any differentiation at the back of the chamber. This area was either masked by corrosion or conservation treatment, or the welding of the plug to the staves was so good as to be now invisible. In all cases it was possible to judge that a circular hoop was present as an outermost layer, but corrosion was too bad in all cases to fully understand this feature and it was impossible to determine thickness of this hoop with any certainty. In most cases the back of the breech was flat rather than domed outwards.

The exception was a broken breech chamber where some of the stave layer and the outer hoop were missing, showing the plug securely in place within the staves. The line between the plug and the staves was barely visible, there were no signs of additional pegs having been entered through the staves into the plug,

*Figure 3.143 Manufacture of the breech chamber: a) forging a breech stave; b) dishing inner stave; c) forming outer stave; d) swaging neck of chamber; e) putting on temporary rings;. f) forging breech plug; g) fitting plug; h) punching touch hole; i) putting on hoops and bands and lifting assembly*

### Table 3.22 Manufacture of breech chamber

| Procedure | Labour (no. men) | Time per item (hours) | Tools used | Tudor tools |
|---|---|---|---|---|
| Forge inner staves, curve in swages | 2 | 2 | furnace, power hammer, top & bottom swages | fire, power hammer, top & bottom swages |
| Mill inner staves to mate & fit core | 1 | 8 | milling machine | hand, chisels & files |
| Shrink temporary rings each end | 1 | 2 | fire, by hand | fire, by hand |
| Cut outer staves from rolled plate | 1 | 1 | oxy-acetylene | forged, hot cut |
| Curve outer staves hot around inner stave assembly | 3 | 2 | furnace, power press, by hand | fire, power press, by hand |
| Grind outer staves to mate | 1 | 2 | power grinder | hot cut |
| Assemble staves, clean & flux weld surfaces | 1 | 2 | hy hand | by hand |
| Shrink temporary rings on whole assembly | 2 | 2 | fire, by hand | fire, by hand |
| Cut off to length | 1 | 1 | power saw | hand saw |
| Forge weld neck end | 4 | 1 | fire, swages | fire, swages |
| Forge up plug with long end | 1 | ½ | power hammer, fire | power hammer, fire |
| Heat plug end & drift with cold plug | 3 | ½ | fire, by hand | fire, by hand |
| Forge weld plug in, heating plug & breech in separate fires | 4 | 1 | fire, power hammer, swages, by hand | fire, power hammer, swages, by hand |
| Set plug in | 3 | ½ | fire, by hand | fire, by hand |
| Punch for pegs, set pegs in | 3 | ½ | fire, by hand | fire, by hand |
| Punch touch hole | 3 | 5 mins | by hand, same heat as punching | by hand |
| Fit hoops, bands | 1 to size 3 to embed | 1 each | furnace, cone mandrel, sledge hammer, jump ring, set hammers | fire, cone mandrel, sledge hammer, jump ring, set hammers |
| Upset staves cold at plug end to secure last hoop & plug | 1 | | fire, power hammer | fire, power hammer |
| Dress off back of breech | 1 | | | |

Total for breech: 40 hours 15 minutes, three men needed for some portions
Total for project excluding core and carriage: 239.8 hours, about 7 weeks

but the corrosion was intense. The plug 'burred out' towards its outer edge just as witnessed in the chamber 82A0792. This suggested that the plug might have been entered from front to back.

The neck was another area of uncertainty. On most chambers it appears so solid that it is difficult to see the individual staves. It seems that these must have been welded. Thus two areas of good welding exist on the chamber, at the neck and of the breech plug to the staves at the back. This requires heating the chamber and plug to the same temperature to effect a good weld.

Neck lengths measured on four similar sized chambers were 60mm (2⅜in), 70mm (2¾in), and two at 80mm (3⅛in). These taper outwards towards the breech and the neck taper was 3mm (⅛in) one example and 9mm (⅜in) on three. The fit between the barrel and the chamber varies from, seemingly, exact metal to metal (the back of the neck external diameter is the same as the back of the barrel internal diameter) to a total gap of 15mm (Fig. 3.142d). There was no evidence for collars, fillets or sleeves on the port pieces studied.

On the reproduction, the neck was made parallel and then tapered. This taper would help with locating the chamber, and may be a direct result of the process of welding the staves together at the neck. A straight neck would have to be loose all around to prevent jamming, and a taper is an elegant solution. The use of additional washers was considered but these proved unnecessary.

It was not certain to begin with whether any of the hoops from the chamber located into the back of the barrel or butted against it, with only the breech neck extending into the barrel. In all studied examples, multiples of hoops on the breech abutted against multiples on the barrel to form a number of hoops at the breech/barrel junction (Fig. 3.142d). This number varied, but three on the barrel and two to three on the chamber is not uncommon.

The manufacture of the chamber required the same techniques and the same materials (Table 3.22) as the barrel, though, on this chamber, the stave joins were staggered between the layers. Once this was assembled the neck was welded and the plug inserted and welded to close the end of the breech.

The inner staves were forged over a mandrel and the curves obtained using appropriate swages (Fig. 3.143a, b). These were bent out slightly in the region of the neck end weld, to provide an outward taper. This was done so that, during the welding process (in swages under the power hammer), the edges of the inner staves would be driven together. The edges were milled to fit each other by machine. Temporary rings were shrunk on either end. The outer staves (Fig. 3.143c) were cut from rolled plate (using oxy-acetylene), then curved hot around the inner stave assembly and ground to fit. Although the semicircle outer staves were difficult to reproduce, and would have probably required a furnace to bend them, this feature made the assembly of the chamber relatively simple.

The half cylinder trough was used to support the four inner staves and the upper two quarter staves were positioned without the requirement of a core or any form of clamping other than temporary rings that were shrunk over the entire assembly (Fig. 3.143d, e). The staves were assembled and the surfaces cleaned and fluxed. A power saw was used to cut the tube to the correct length and the neck end was forge welded until the joins between the staves were no longer visible. The plug was forged up with a tong end (Fig. 3.143f). Both the plug and the chamber were heated in separate fires to forging temperature and the plug was forge welded in place (Fig. 3.143g). The plug punched for three locking pegs, put in as an extra security, and the touch hole punched (Fig. 3.143h). The hoops and bands were fitted as for the barrel (Fig. 3.143i), and the staves upset cold at the plug end to secure the last hoop and plug. The back of the breech was dressed.

In total, manufacture of the breech chamber took 40 hours 15 minutes, with three men needed for some portions. The total time taken for the project (excluding core and carriage) was 239 hours 48 minutes, in practice about seven working weeks.

**Discussion of techniques**

It was felt that original methods would have been similar, although the heating of the rings would have been more time consuming with a forge fire rather than furnace. We also used electricity or coal to heat the furnace/fire, but then our experiment was in the method of manufacture not the nature of combustion. The greatest generation of heat was required for the insertion of the breech plug, the entire chamber as well as the plug had to be up to welding temperature.

The construction of these guns demonstrates the ability to create a large and strong object from comparatively small pieces of iron (like the bricks in a house). Furthermore, it was possible to use iron of different grades selectively within the gun without compromising the integrity of the piece. One of the inner staves from the original chamber broke off and was almost impossible to work. It was of extremely low quality and could not be used for an individual small item. As an inner stave, it was fine. It is suggested that historically there was some grading of the iron. Historical records regarding supply and shipments of iron might shed light on billet size and any grading. John Light, blacksmith and research archaeologist for Parks Canada, has studied early records regarding the Basques transporting specific iron from Germany during the sixteenth century (Light, pers. comm, Jan. 1998). References were made to grading iron depending both on how many times the iron had been worked (the more the iron is worked, the better it becomes as the slag stringers are refined) and on country of origin, with trace elements even being chosen for particular purpose. Each component of our original was not studied with a view to the welding of faggots of specific sizes. Where it was possible to look at elements, these were so well welded as to obscure this information. Although it may be possible to learn about the size of faggots by employing some non-destructive examination techniques (dye penetration analysis), this was not undertaken. Chris Topp felt that size of faggot was not critical, although if given the choice, he would opt for forging a stave from a single billet.

Although the design is sophisticated, none of the techniques required was beyond the competence or resources of a Tudor blacksmith skilled in the manufacture of guns. If these guns were required in a hurry, it is possible that a relatively unskilled labour force, using low capital equipment, could make them if properly instructed. So although labour intensive, these could be made by most village blacksmiths; this is certainly not the case with cast iron guns. The blacksmiths used on this project, more used to producing ornamental ironwork, were not acquainted with the construction of guns until our first meeting in October 1997. By February 1999 a serviceable gun had been built. Although the first chamber burst, the metal was reused, and it was rebuilt. Perhaps this is one of the contributory factors to the longevity of wrought iron ordnance. It was felt that the material was extremely cleverly used, utilising the structure of the iron, longitudinally for the staves and circumferentially for the hoops and bands.

One of the questions addressed was whether there was a size limit imposed on the gun by the nature of construction. Whilst weight and length do pose an increasing problem with size, this can be overcome by employing more men, a longer core with working levels

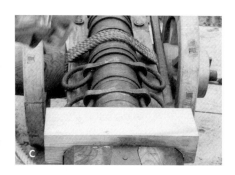

*Figure 3.144 a) detail fit gun to carriage; b) axle position, dishing wheels; c) forelock; d) placing iron wedge; e) fishtail on back of carriage*

incorporated, or by using more staves to form the tube. This may be why some of the slings studied (smaller bore, longer barrels) were made up of more staves. It was felt that real limit was the length of the powder chamber because of the difficulties of welding the breech plug. To maintain heat on a chamber the size of our largest (1200mm long) would be taxing enough and it was felt that there would be a difficulty in fitting and maintaining a breech plug on a chamber substantially longer than the longest *Mary Rose* chambers.

## Replication of the carriage

The carriage of port piece 81A3001 was replicated, including the spoked wheels (81A5688, 81A5689), axle (81A0891/3765/3998) and elevating post (81A2088) (see Figs 3.26–8). Although the wood used for this carriage has not been positively identified, the replica was fashioned of elm. The axle was made from ash as eighteen out of nineteen identified are in this wood and the elevating post from oak, which was used for seven out of nine examples identified to species.

The carriage is an important element in the working of the gun. In the case of the port piece, it is fundamental to its operation. Together with the forelock and wedges, it functions to hold the breech chamber and barrel together, forming the gun. The reproduction was based on the drawings of the original carriage and was manufactured by Jim Sadler at Fort Nelson. The timber used was a 508mm square (20 x 20in) block of elm, nearly 4m (12ft) long. It was green, and owing to time constraints, was made before the breech chamber was ready. Cutting out the barrel recess was difficult and required considerable accuracy as the series of rebates must conform to the hoops and are all there is to resist the longitudinal force that tries to drive the barrel and chamber apart upon firing (Fig. 3.144a).

The axle position was ascertained by trial and error (Fig. 3.144b). The axle was temporarily placed and two persons attempted to lift the end. The axle was moved until the easiest lift was obtained and this position was adopted. The wheels and axles were set out replicating the original angle of the hub and spokes, and slight backward angle to the ends of the axle. A large elm forelock was manufactured, based on the original. This was positioned between the chamber and the vertical face of the bed (Fig. 3.144c). Information about the operation of this is scant with the best sources being illustrations made for John Deane, showing portions of two iron wedges, and his 'sales pitch' stating that two iron wedges were driven between the wooden forelock and the back of the bed. Nicholas Hall devised the two iron 'folding wedges'. These worked against a 'fishtail' iron plate which served to protect the top of the breech step (Fig. 3.144d, e). The exact nature of these remains unknown, as the beds are still in wet storage, but for purposes of reproduction the wedge was fixed both to the bed and to a strap which extends down the sides of the bed. This is potentially an area of critical stress. The strap is shown in the Deane watercolours and evidence is visible in the archaeological drawings (see, for instance, Fig. 3.14). The free wedge was driven down between the fixed wedge and the elm forelock. The effect was to provide as near a straight push, and as much support to the chamber, as possible. Although this cannot be proved archaeologically, this is one effective method that might have been employed. The total weight of the gun on its carriage is 1100kg (October 2008).

*Firing the port piece*

Inherent within the research design was the desire to test the effectiveness of the stone firing port piece against the hull of a vessel (Hall 1998, 57–66). It was therefore decided to try and replicate a target of similar construction to the hull of the *Mary Rose*. This was based on the survey of the hull at Main deck level but, unlike the hull, the target was not reinforced with iron nails. The target was of oak (Fig. 3.145a) and consisted of an outer section of six 70mm (2¾in) thick edge-to-edge planks 280mm wide (original was 300mm). The middle section was formed of six vertical oak frames of 250mm (¾in) with 200mm (8in) gaps between the frames. The inner layer was made of three horizontal stringers 70–90mm (2¾–3¾in) thick and 340mm (original 380mm) wide with gaps of 200mm (8in). It was held together with wooden treenails.

One unsolvable problem was that, on the original vessel, ageing would probably have hardened the timber. Similarly the original iron nails would have conferred additional strength. The target was also a single element, rather than a section of a larger articulated structure.

The modern timber splintered readily and the shot penetrated the target without difficulty. The original hull would also have the additional strength of some iron nails as well as the oak treenails. Two other targets were used: a sandstone wall (shown to the right of the target in Fig. 3.145a) and, for long range, a 30m long reinforced concrete bunker filled with sand (Fig. 3.145b).

**Shot**

Most of the 378 stone shot from the *Mary Rose* have been identified as Kentish Ragstone, a glauconitic limestone from the Maidstone area of Kent (see Chapter 5). Although not petrologically identified, the shot (82A4213) recovered from inside the barrel of 81A3001 is, therefore, most likely to be Kentish Ragstone. It has a diameter of 172mm and the bore is 180mm.

For the replica firing, twelve shot were manufactured from Purbeck limestone quarried from St Aldhelm's Quarry at Worth Matravers near Swanage (Fig. 3.145c). The shot were in three sizes allowing for windage of ¼in, ½in and 1in (6mm, 13mm, 25mm). The Purbeck shot is lighter than Kentish Ragstone but has one of the closest mineralogical profiles to it. Lantern shot were also made based on drawings of *Mary Rose* examples and filled with beach stones (Fig. 3.145d). One *Mary Rose* round shot was used for comparison. The placing of a gasket between the chamber and the back of the barrel was discussed but no supporting evidence was found for this. Both twisted rope and twisted straw were tried, but these were merely blown out with no effect, so it was concluded that a metal to metal contact was sufficient. When the gun was fired the staves, which were tight against each other opened, but all the welds held on the bands and hoops: in other words, it expanded. With repeated firings captured on high speed camera, the greatest escape of gas (and smoke) was seen to be between the breech chamber and the barrel.

The original shot sequence was designed to build up the size of ball (ie, decrease windage) and the charge weight. However, the requirements of filming meant that this gradual increase was abandoned. At the same time as planning the building of the port piece, a project was devised to study and replicate true black powder using a number of original recipes. This could not be achieved in time, but is ongoing. In the absence of a true mix, the constant and repeatable nature of modern powder was considered to be the best choice. Modern gunpowder ('*black powder*') is corned, that is with the ingredients mixed wet, dried and broken into lumps which are glazed with graphite to help resist damp and slow down the rate of combustion (see Chapter 5). ICI NPCG, the slowest burning modern powder available, was chosen.

The weight of powder was a matter of concern. There are several references to loading chambers (Tartaglia 1588; Norton 1628) but, as we were using modern powder, a compromise had to be made. The

*Figure 3.145 Firing the reproduction port piece: a) gun positioned to fire at oak target; stone walls with wood target in centre; b) gun aimed at concrete bunker; c) new stone shot held in back of barrel; d) new lantern shot; e) straw wadding in position in chamber; f) tampion in position; g, h) elevating the gun to just above horizontal; i) recoil restraints*

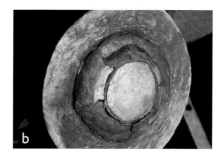

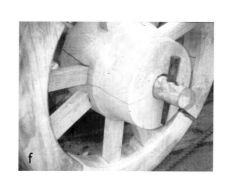

*Figure 3.146 Firing the reporduction port piece: first firing of the port piece: a) plug as made; b) plug damaged; c) neck as made; d) neck damaged; e) damage to wall; f) damage to wheel and axle*

first trials were ⅓ weight of powder to shot, not far off Norton's figure of 3¼lb (1.48kg) of powder to 9lb (4.08kg) of stone shot for a '*port*' chamber (we assumed this meant port piece chamber). There is some difficulty with these sources in working out whether serpentine or corned powder was used, and the true differences between these (see Chapter 5).

In conclusion, it was decided that the ratio of 1:3 was too great. For the second trials this was reduced to 1:4. NPCG priming powder was used throughout and all ignition was electronic. Hay or straw was used as wadding after the powder but a considerable gap was left within the chamber (as advocated by Tartaglia). The chamber was sealed with a poplar tampion (Fig. 3.145e, f).

Firing was undertaken at the British Aerospace Defence Division Ltd, Royal Ordnance Division, Bellingham, Cumbria on 26 February and 4 August 1998. Equipment supplied to record the firing was a Doppler Radar Type DR 5000 which measures muzzle velocity in metres per second and a high speed 16mm cine camera (Hadlands Type S2) capable of 5000 frames per second. For a description of the gun drill adopted, see Chapter 5.

A summary of the firing sequence and results is given in Table 3.23. The elevation of the gun was just above horizontal with a height of 1050mm to the top of the breech ring. The upper sill of the gunport in M8, where the original gun was found was only 1000mm off the deck. The peg position on the elevating post corresponded to the eighth (of eleven) peg from the deck, raising the back of the bed 500mm off the ground (Fig. 3.145g, h). Recoil was adjusted by using a sizal rope of 2in (51mm) diameter. This was taken through the rings at the back of the bed and around the centre of the back to two metal pins fixed in the ground for the short range firing. For the long range firing (130m) the ropes were attached to two ¾tonne concrete blocks (Fig. 4.145i).

### Results: 26 February 1998

The first firing session (two round stone shot) resulted in damage to the gun as well as the wall. Damage to the wall occurred to a depth of 50–70mm, the hole having a maximum height 600mm and width of 400–460mm (Fig. 3.146e, Table 3.23).

The first shot did not appear to stress the gun. This had 25mm windage and a charge weight of approx-

*Figure 3.147 Firing the reproduction port piece: a, b) damage to oak target, shot 2, exterior and interior; c, d) damage to oak target, shot 5, exterior and interior; e, f) lantern shot effect on target; g, h) sighting at bunker*

Table 3.23 Port piece firing sequence and results

| Date/time | Sequence/shot no. | Windage | Shot type/wt | Charge wt | Target type | Distance (m) | Muzzle velocity (mps) |
|---|---|---|---|---|---|---|---|
| 23/02/98 | 1 | | No shot | 1½lb/0.66kg | clay bank | c. 20 | |
| | 2 | 1in/25mm less than bore | round limestone 13lb/5.90kg | 2½lb/1.13kg | stone wall | c. 20 | 141 |
| | 3 | ¼in/12.5mm | round limestone 14¾lb/6.69kg | 4lb 11oz/2.13kg | stone wall | c. 20 | 141 |
| 04/08/98 09.33 | Round 1 Shot 2 | | round limestone 12lb/5.44kg | 3lb/1.36kg | stone wall | 17.2 | 160 |
| 10.00 | Round 2 MR shot | | round Kentish Ragstone 12lb 14oz/5.81kg | 3lb 3oz/1.45kg | oak target | 17.2 | 240 |
| 10.30 | Round 3 Shot 5 | | round limestone 13lb 11oz/6.21kg | 3lb 8oz/1.59kg | oak target | 17.2 | 298 |
| 11.00 | Round 4 Shot 13 | 1in/25mm less than bore | lantern 160 x 500mm 25lb 2oz/11.40kg | 3lb/1.36kg | oak target | 15.7 | n/a satter |
| 14.25 | Round 5 Shot 9 | | round limestone 15lb/6.80kg | 3lb 12oz/1.70kg | reinforced bunker 30m long, sand filled | 130.9m to rear of carriage | 338 |

imately one-sixth the shot weight. After the second shot (one-third weight powder to weight of shot, 7mm windage) the breech plug came out and cracked (Fig. 3.146 a, b), the inner staves were displaced, the iron wedge ejected and the elm forelock pushed up. The breech chamber was partially ejected from the bed and the staves at the chamber neck were damaged (Fig. 3.146c, d). The gun could not be fired again because of damage to the chamber and wheels (Fig. 3.146f). Obviously the chamber had been strained by the increase in ball size and powder charge (Table 3.23). The plug had set back, which contributed to the failure of the forelock by impact and the loosening of the steel wedge: a situation described by Cyprian Lucar (1588, appendix, 47): '*the sayde wedge may through the discharge of that peece flie out, and kill the Gunner.*' However, the barrel was undamaged, the bed was serviceable, the rope lashing the barrel to the bed was sound, the hoops of the chamber were undamaged and the wheels, although badly damaged, were repairable. The conclusions were that the circumferential strength – that which resists the bursting forces – imparted by the hoops and bands was adequate. Despite cracks, the strength of the bed was sufficient – those items which were damaged are precisely those for which spares are inventoried for all vessels – axles, wheels and forelock timber. The basic failure was the powder chamber, with areas of weakness being the welding of the neck and breech plug. It was also felt that the charge had been increased too much too fast, unfortunately dictated by the time available within the filming schedule. However the breech was returned to Chris Topp's workshop to be studied and rebuilt.

**Results: 4 August 1998**
*Round 2. Shot 2 (Mary Rose shot)*
It appeared as though the shot cracked in two as (or before) it hit the oak target and splintered to the left. Two holes were created in the third and fourth planks from the top (Fig. 3.147a). The impact hole on the right appears to be the major point of contact, with a near round hole of 160 x 140mm height and a clear hole through the target. The second hole is 100mm left of the first and on the plank below. It is 130mm in height and damaged up to 250mm of the width of the plank. On the inside (Fig. 3.147b), the second stringer from the bottom was glanced by both portions of the shot and splintered over a distance of two frames and over one-third of its width. The shot actually missed the frames and continued behind the target where it imbedded (in bits) in a bank.

*Round 3 Shot 5 Limestone ball*
This also went through the oak target, creating a hole in the top plank and partial damage to the second (Fig. 3.147c). The hole was 165mm high with splintering up to 220mm. It was 420mm above the aiming mark placed on the target. The ball actually hit in front of a

frame, the top of which splintered completely for the height of the top plank, displacing each frame on either side of it (both outwards to each side of the target), and blew off the internal stringer (Fig. 3.147d). This resulted in substantial damage to all three layers of the structure, the outer planking, frames and inner stringer. It actually pushed the stringer off at the treenails. The stringer came to rest just behind the target, intact. Iron bolts may have kept it in position and it would have splintered rather than 'popped' out of position. The shot split into pieces.

*Round 4 shot 13. Lantern shot 160mm diameter, 500mm long*

White paper was spread over the oak target in order to gauge the spread of the shot. When the gun was fired the carriage moved 200mm back and 110mm to the right. The paper was effectively torn to shreds and the spread of shot was 4m with a concentration over the centre of the base of the target in an area of 1.30m, where the gun was aimed (Fig. 3.147e, f). A fragment of the base of the lantern was found here.

*Round 5 Shot 9 Distance shot*

The gun was elevated to 3° with the peg positioned in the eighth hole from the bottom of the elevating post. It was aimed at the centre of the concrete bunker 131m away, just above the sand which half filled it (Fig. 3.147g, h). This is not the maximum range achievable, as the shot hit the roof, just missing the aiming mark (18in left by 18in below) and went into the sand. At 124m from the muzzle the shot was still travelling at 284mps.

The bed suffered damage with the last firing. The front of the carriage had a slight split appearing and two wheels came apart at the fellies by 30mm. The recoil moved the ¾tonne anchoring concrete blocks 230mm to the side and 330mm back from their original positions. The fact that the bed was already compromised with the first firing sequence with a powder charge which was probably too great for the gun was probably a contributory factor.

# 4. Cast Iron Guns

## Alexzandra Hildred

## Introduction

The availability and relative cheapness of iron made it an obvious choice for casting items required in large numbers, such as shot and guns. The commercial exploitation of this alloy was only really possible after the development of the blast furnace, that produced liquid cast iron as the normal product. Early studies include Schubert (1957).

The finding of four cast iron guns from the *Mary Rose* is a most significant discovery and demonstrates how the arms industry was attempting to harness advances in technology to provide more efficient weapons and production methods. These are the earliest securely dated examples of cast iron used to form guns rather than projectiles. These guns have been identified as '*hailshot pieces*' because their projectiles, dice of iron fired through a rectangular bore, caused a devastating scatter at short range. The *Anthony Roll* lists 20 for the *Mary Rose* and 459 on the 58 vessels, providing the earliest recorded evidence for the use of cast ordnance in iron on a large scale. The use of iron in the manufacturing of guns was to soon become of great commercial and military importance to the nation.

Blast furnaces were used in Sussex towards the end of the fifteenth century and utilised some cast iron to produce castings directly from the furnace. Some was used to make cast iron shot. Documentary evidence for early gun casting in the Weald has been expanded upon by Teesdale (1991) and summarised by Hodgkinson (2000) but there is little evidence for the casting of ordnance until the mid-sixteenth century. The mass production of hailshot pieces pre-dates the casting of quantities of iron guns and can be seen as an attempt to experiment with this accessible and cheap alternative to the continental copper required for the bronze guns. These are, therefore, significant not only for those interested in the development of ordnance, but also in the history of the metallurgical industry.

The introduction of iron, a metal which does not lend itself readily either to incised decoration or delicate relief work, eventually resulted in the mass production of plainer guns with cheapness and bulk replacing the individuality and artistic expression of their bronze predecessors. By 1546 iron demi culverins were being cast in Buxted, Sussex and, by 1553, a full culverin, probably at Worth. By the end of the century guns of over three tonnes were being cast in iron. These were not merely experiments and, even by the time of the *Anthony Roll*, cast iron demi culverins, sakers and falconets were already on vessels, albeit in small numbers (Knighton and Loades 2000). The ships had five cast iron guns and 145 bronze, '*galleasses*' had 25 cast iron and 123 bronze, and pinnaces 2 cast iron and 42 bronze, making 32 cast iron guns in total. The types are illuminating: 19 are demi culverins, 14 sakers and one a falconet. The assignment of these is also interesting, only one ship carries an iron demi culverin, the *Trinity Harry*, with one saker each carried by the *Christopher of Bremen*, the *Lartique*, the *Mary Thomas* and the *Mary James*. On the other hand the '*galleasses*' have 16 demi culverins and nine sakers. The *Hart* carried three demi culverins and two sakers, the *Antelope* three demi culverins and two sakers, the *Tiger* and the *Bull* four demi culverins each; the *Swallow* and *Jennet* one demi culverin each, the *Unicorn* one saker and the *Dragon* and the *Jennet* two sakers each. The *Less Pinnace* carried the single saker and a falconet. By this period the size of the casting could already cope with demi culverins.

As we have seen (Chapter 2), during this period bronze ordnance outnumbers cast iron. The situation is changing and, by 1646, when Eldred wrote *The Gunner's Glasse*, many tables are clearly for iron guns. In a discourse regarding the difference between bronze and iron, discussing the relative merits of brass and iron (pp. 10–11), he responds to the view that brass ordnance was to be preferred because unlike iron it would not explode under stress, but merely split:

> '*I have heard many men say so also, but I have seen it otherwise, for being at Rye in the yeere 1613 among other Ordnance, a Brasse Saker being fired, did break and some pieces of great weight did fly a great way, whereof one piece of above a hundred weight struck Mr. Gunner in two parts stark dead ... and divers other Brasse peeces hath bin known to fly in many pieces when they break*'.

He goes on to give instances of the greater weights of all cast iron ordnance over bronze (Chapter 3).

Inventories listing guns within the Tower of London (Blackmore 1976) provide a means of assessing what was around at any particular time. However the Tower, together with other land installations (for example Boulogne, Berwick and Calais) functioned also as repositories which means that they retained obsolete as well as 'state of the art' weapons. Care must be taken when using these as an indication of what was in active service, especially at sea. It is however a useful starting point for gathering information. Iron guns are mentioned as early as 1388 (*ibid.*, 254): 87 '*canones de cupro et ferro*', but with no qualifications. The first mention of guns by name is the inventory of 1495 (*ibid.*, 256–9) which classifies for the first time guns under the headings of '*brasse*' and '*yron*' (*ibid.*, 260). The named bronze guns are the bombards, curtows, serpentines and falcons. Only one named type of iron gun is listed, a serpentine (with chambers). More intriguing are the listings below the heading '*Broken gonnes of brasse callid with chambres of yron*' (*ibid.*, 257), which include '*Hole chambers of yron for gonnes*' possibly suggesting cast iron breeches ('*whole*') for curtows, demi curtows and harquebuses. The next inventory is 1514 (*ibid.*, 259). For the first time we find familiar names such as culverin and minion, although there is no differentiation between iron and bronze. There are however 20 tons of '*Shotte of Iron of sundry sortes being in the Towre*'.

Iron guns are mentioned in 1523 (*ibid.*, 260). Interestingly we see the sling group, but the '*Hoole welslanges*' of iron are listed separately from the '*Slanges with chambres*', suggesting that the former may not have chambers. Alternatively this may relate to size, '*whole*', '*demi*', '*quarter*'. Other named iron guns include stone guns, serpentines and falcons. All those listed as iron in the 1540 inventory are probably wrought: port pieces and fowlers, slings, bases and harquebuses. Corn powder is listed, possibly for the first time. The 1547 inventory (Starkey 1998) divides guns into guns made of bronze, then French and Scottish. The iron guns include bombards, port pieces, slings, fowlers, bases, top pieces and hailshot pieces. Both fine and gross corn powder are listed. There is no specific listing for cast guns. The rest of this extensive inventory suggests that cast iron is being used for larger guns, but most are still of cast bronze and wrought iron. The makers of these cast iron guns are often listed (reinforcing novelty value) as in the case of the demi culverin of Peter Baude's making in Southsea Castle, Portsmouth (Starkey 1998, 4428). Also in Portsmouth (*ibid.*, 4252), is an iron bastard culverin by William Levett as well as three iron sakers, a demi culverin (4294), four iron falcons (4310, 4312) two sakers (4315) and a demi culverin and saker (4396).

The 1559 inventory (Blackmore 1976, 262) has, for the first time, a separate heading for cast iron. Guns include demi cannons, culverins, demi culverins, sakers, and mortar pieces, totalling 28 iron guns against 167 of bronze. The greatest number of any named piece is the saker with 43, 38 of which are bronze. The 1568 inventory (*ibid.*, 264) lists cast iron guns (including hailshot pieces listed as cast for the first time) but still in small amounts. In the 1589 inventory (*ibid.*, 264–6) there are still more bronze and forged iron guns (at least in the Tower) than cast iron. In the detailed 1595 inventory (*ibid.*, 268–80) the weights of guns are given and they are divided into their individual names. Here we begin to see the rise in number of cast iron pieces against bronze, for example there are 13 bronze culverins against 12 cast iron; 41 demi culverins in bronze and 34 in iron; 24 bronze sakers and 17 iron; 8 bronze falcons and 3 iron. Both these inventories list 149 '*ffloukemouth*' or '*ffloukmouthed*' pieces among the cast iron guns (*ibid.*, 266, 275). These may be hailshot pieces (R. Smith, pers. comm.), although the 1595 inventory also lists one wrought iron hailshot piece (*ibid.*, 275). It is only in the inventory of 1617 (*ibid.*, 282) that we see some of the more common iron guns outnumbering bronze. There are still no cast iron cannon and the largest iron guns are the 13 culverins, the same number in bronze are listed. The smaller guns see a dramatic rise in cast iron with 37 iron demi culverins against 17 bronze and 35 iron sakers against 11 bronze.

## Hailshot Pieces

The four hailshot pieces from the *Mary Rose* (80A0544, 79A1088, 81A1080, 90A0028) are all cast iron muzzle-loading small guns presumed to be used by one individual in an anti-personnel role (Fig. 4.1). They have a bore length of 12in (300mm) and a rectangular bore of about 2¼–2½ x ⅞–1⅜in (57–63 x 22–35mm) which tapers towards the breech. The gun extends beyond the breech to form a cylinder into which a short wooden stock is located (Fig. 4.2). The gun and stock have a total length of just under 3ft (715mm) and an approximate total weight of 31lb (14kg) (Table 4.1). A fin-like projection ('*hook*', '*croc*', '*croque*', termed '*fin*' hereafter) from the underside of the piece enables the barrel to be supported over a rail, while the stock could be held firmly beneath the operator's arm. These are, therefore, designed exclusively to be supported by some form of support (Fig. 4.3). The shape of the projecting

*Figure 4.1 Hailshot piece 90A0028*

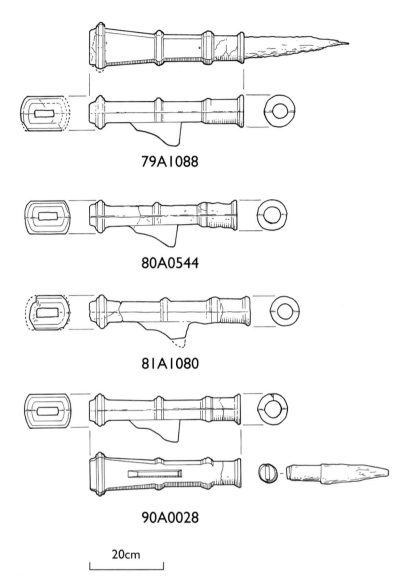

Figure 4.2 All four hailshot pieces

fin suggests a rail support rather than the (wider) sill of a gunport. The morphology of the piece therefore bears a relationship to its position of use within the ship and suggests development designed predominantly for shipboard use. Three had their bores sealed with rectangular tampions followed by wads. One was loaded with 20 iron cubes shaped like 'dice', (79A1088, Fig. 4.4) and the radiograph of 90A0028 shows that this contains about 30 iron dice (Fig. 4.10a). It is therefore likely that all four were loaded and that the dice had corroded within the bores of 80A0544 and 81A1080. The most exciting feature of these with respect to the *Mary Rose* assemblage is that they are exclusively of cast iron and they are the only cast iron guns as yet recovered from the ship. Initial recognition of these as four of the 20 hailshot pieces listed onboard *Mary Rose* was based on the assumption that the iron dice excavated from within their rectangular bores would

Figure 4.3 Reconstruction of the use of a hailshot piece over rail

### Table 4.1 Dimensional data for hailshot pieces (metric)

| Object | 80A0544 | 79A1088 | 81A1080 | 90A0028 |
|---|---|---|---|---|
| Sector | C7 | C8/SS8 | SS9 | SS10 |
| Length gun along cast-line | 417 | 423 | 444 | 415 |
| Total length stock *in situ* | 670 | 715 | 534* | 606* |
| Length exposed stock | 255 | 294 | 90* | 190* |
| Length to back of reinforce behind touch hole* | 330 | 336 | 351 | 335 |
| Internal bore width | 57 | 58 | 57 | 63 |
| Internal bore height | 26 | 22 | 22 | 35 |
| External bore width | 91 | 95 | 90 | 89 |
| External bore height | 56 | 57 | 54 | 54 |
| Wall thickness | 15 | 18–21 | 15 | 15 |
| Back of base ring to end gun | 86 | 87 | 85 | 80 |
| Internal diameter stock socket | 50 | n/a | 45 | 50 |
| External diameter stock socket | 86 | n/a | 66 | 85 |
| Diameter touch hole | 10 | 5 | obscured by fibre | 10 |
| Fin length | 143 | 134 | 141 | 144 |
| Fin height maximum | 66 | 66 | 52* | 65 |
| Fin width rear | 27 | 27 | 24 | 24 |
| Fin width front | 21 | 20 | 21 | 17 |
| Length muzzle lip | 20 | 20 | 20 | 18 |
| Length muzzle moulting (asymmetrical) | 24 / 5+14+5 | 24 / 6+13+5 | 24 / 4+17+3 | 25 / 4+18+3 |
| Height at apex of muzzle moulding* | 94 | 94 | 93 | 90 |
| Length end muzzle moulding to start medial moulding* | 135 | 135 | 135 | 132 |
| Length medial moulding (asymmetrical) | 22 / 5+12+5 | 22 / 5+12+5 | 22 / 5+12+5 | 23 / 5+13+5 |
| Length end medial to start breech moulding* | 101 | 105 | 103 | 102 |
| Length breech moulding (asymmetrical) | 29 / 4+4+19+1 | 32 / 5+7+15+5 | 33 / 6+5+17+5 | 32 / 6+8+15+4 |
| End breech moulding to start stock moulding | 69 | 67 | 65 | 63 |
| Stock pin diameter | 8 | 6 | n/a | 5 |
| Length stock moulding (asymmetrical) | 17 / 6+11 | 18 / 6+12 | 19 / 6+13 | 16 / 6+10 |
| Centre touch hole from bore | 297 | 295 | 311 | 295 |
| Centre touch hole to back breech moulding | 35 | 38 | 37 | 40 |
| Wood for stock | beech | beech | no ident. | beech |
| Stock length | 347 | 393 | 173* | 190* |
| Length stock shoulder | 92 | 93 | 83 | 85 |
| Diameter stock at shoulder | 62 | 64 | 57 | 62 |
| Internal diameter gun for stock | 46 | 44 | n/a | n/a |
| Wedge length | 31 | 25 | 36 | 35 |
| Wedge width | 9 | 10 | 11 | 12 |

* incomplete

scatter as hail upon firing. Further investigations of other contemporary sources adds strength to the identification, with clues as to manufacture, form, size of gun and shot type.

To date these are the only extant examples of these guns. The large number listed in the 1546 inventory suggests either the acceptance of a tried and tested piece of ordnance, or an extremely good marketing technique. Seemingly appearing from nowhere, they form 26% of the iron guns listed for the *Mary Rose* in 1546. On some of the smaller vessels this is higher than 50% (Table 4.3, below).

### Distribution

Three hailshot pieces were recovered from just inboard or just outboard of the Sterncastle (Fig. 4.5). The fourth (90A0028) was recovered just outboard of the bow on the starboard side. This was excavated and lifted during

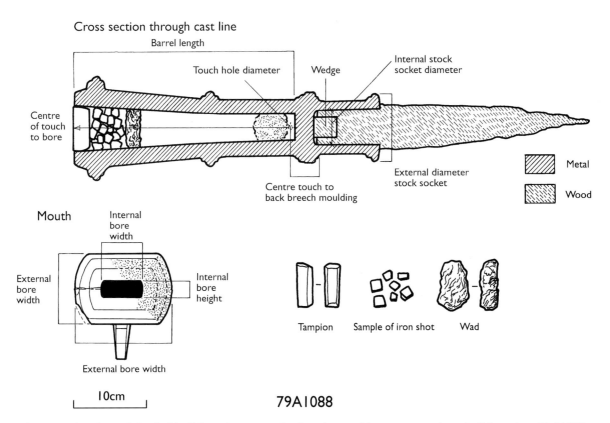

*Figure 4.4 Cross-section through loaded hailshot piece, muzzle elevation and bore contents from hailshot piece 79A1088*

post-excavation monitoring of the hull depression. All locations can be considered to reflect guns in use rather than in store, with original locations presumably over a rail on the castle decks in the bow and stern. One was within primary Tudor sediments (80A0544), one was concreted to the external hull (81A1080) and the others were either directly associated with or concreted to other Tudor artefacts (details in archive).

The only hailshot piece found inboard was recovered 320mm below the top of one of the starboard frames (S22) within the Sterncastle area (80A0544). This was at the junction of the Upper deck and the starboard side just north of a bronze culverin *in situ* on its carriage on the Upper deck at the front of the Sterncastle (80A0976; Fig. 4.5b). 79A1088 was found just behind the cascabel of the bronze bastard demi culverin (79A1232) at the front of the Sterncastle (Chapter 2). As the ship had come to rest on its starboard side, some items fell towards the outside of the ship. This location places it just outside the starboard frames. 81A1080 was resting on the outside of the ship just north of starboard frame S15, outboard

of the Carpenters' cabin (which was on the Main deck) but at Castle deck level. It was aligned vertically with its muzzle against a collapsed Castle deck plank.

90A0028 was excavated from a concretion containing a conglomeration of sixteen ordnance-related items, found during post-excavation monitoring of the site. It was first noted in 1989 and, following unsuccessful attempts to stabilise the area, was excavated and lifted in 1990. It was first separated into three underwater, the portion containing the hailshot piece measuring <860 x 640 x 540mm. The location (SS1–2) suggests ordnance derived from the Forecastle.

## Morphology

The guns have pronounced mouldings at the muzzle, around the mid-point and at the end of the barrel (Fig. 4.6). The touch hole is centrally located just forward of the moulding marking the end of the barrel. The fin projects downwards from beneath the central moulding.

**Table 4.2 Incidence of hailshot pieces recorded in the Tower inventories**

| Inventory | 1547 | 1559 | 1568 | 1589 | 1595 |
|---|---|---|---|---|---|
| No. | 41 | 80 | 173 | 1 | 1 |
| Description | of iron | forged iron | 168 cast, 5 forged 4 'chambers' | forged iron | forged iron |

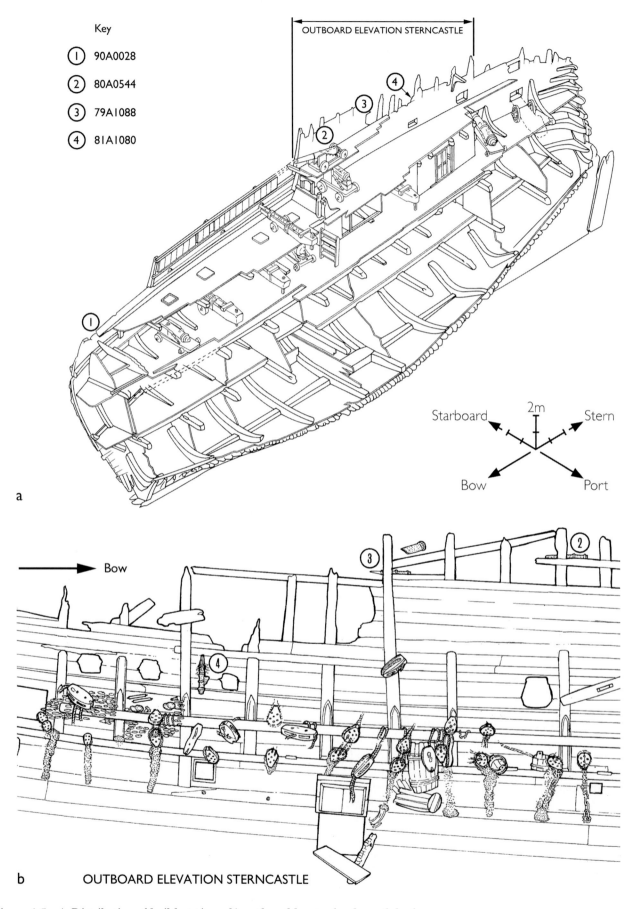

*Figure 4.5 a) Distribution of hailshot pieces; b) outboard locator for three of the four guns*

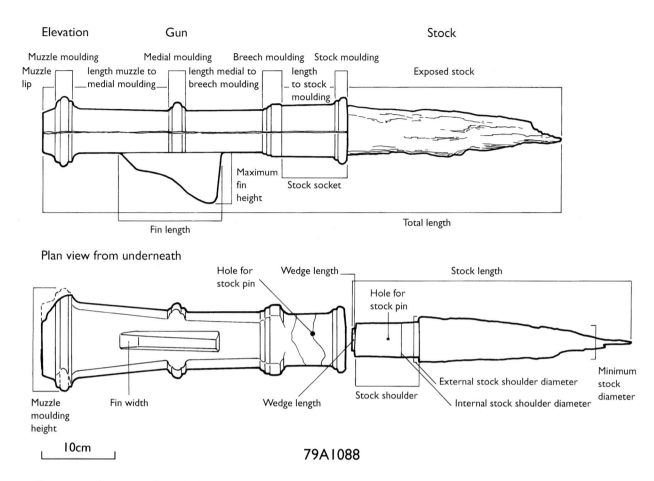

*Figure 4.6 Anatomy of a hailshot piece*

The muzzle and medial mouldings are in the form of a central astragal with fillets while the breech moulding is a ring with a fillet towards the back and ogee towards the bore. Behind the blocked end of the barrel, the gun extends to form a cylinder defined by an astragal with fillet towards the bore. This accepts a wooden stock designed to fit beneath the arm of the user. The stock is shouldered to fit into the gun and tapers away from it. The stock is split and wedged as it locates into the gun and reinforced by an iron pin which goes through the cylinder and stock. Three of the four stocks are made of beech.

The cast line is clearly visible, showing that the top of the gun was cast in a separate mould to the underside. A slight scar on the muzzle may represent the remains of an ingate, suggesting that the gun was cast vertically, with the mouth upwards. The fin appears to be cast in one piece with the lower section. In three instances the iron has corroded severely and partial graphitisation has taken place. 90A0028 is less corroded, so the weight is slightly more reliable. The iron portion weighed 17lb 7oz (8kg) and with the stock the total weight was 19lb (8.7kg). The original weight is far more likely to be similar to the weight of the cast replica (31lb; 14.1kg). Dimensions are given in Table 4.1.

*Radiographic study*

All four guns were coated in concretion. Radiographic examination was undertaken before removal of concretion both for the purposes of artefact identification, for information regarding condition and to formulate the least invasive method of extracting it from the concretion. Details on the internal structure of the guns was a bonus, but was not always forthcoming because of the thickness of the concretion. As the concretions were amorphous lumps, the direction of the source relative to the object was uncontrolled and so the radiographs offer different views, each affording some unique constructional detail and all serving to identify the objects within.

80A0544 is represented by a poor quality radiograph taken directly from below (Fig. 4.7a). It shows the cast-line and confirms that the wooden wedge entered into the stock is fixed with an iron pin. 79A1088, a side view, enables measurement of the internal bore length and confirms the taper of the height to be less than the width (Fig. 4.7b). Both the tampion and iron dice are clearly visible. The dice vary in size from 16mm to 10mm square. The 5mm touch hole is also evident. The fin appears to be integral to the barrel. The thickness of metal between the back of the gun and the stock (the

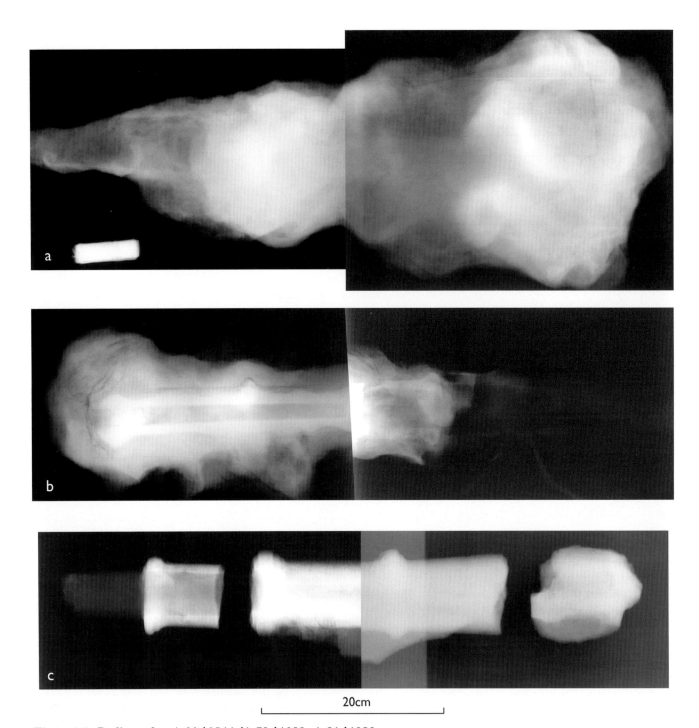

*Figure 4.7 Radiographs: a) 80A0544; b) 79A1088; c) 81A1080*

breech moulding) is 29mm. The stock is not clearly defined.

81A1080 represents a view at an oblique angle to the stock (Fig. 4.7c). Several breaks are visible. The most reliable accessible measurement of the wall thickness is the breech moulding (23mm). A wooden wedge can be seen protruding from the stock into the back of the gun at the same position as one of the breaks. No detail regarding the integrity of the fin with the lower portion of the gun is visible.

The radiograph of 90A0028 revealed a damaged hailshot piece with stock, a large shot and a breech chamber for a base (Fig. 4.8). Excavation began by removal of the shot, which appeared as the most solid (densest) object. The muzzle of the hailshot pointed away from the mouth of the base chamber (see Chapter 3).

The external concretion was very hard and of a buff/cream colour with adhering oyster shells and flints. The oyster shells, some in excess of 50mm length, indicated that the top surface of the concretion had been exposed for some time whilst the outer layer was forming. This was 50mm thick and when removed using a lump hammer directed at 90° (Fig. 4.9), revealed a

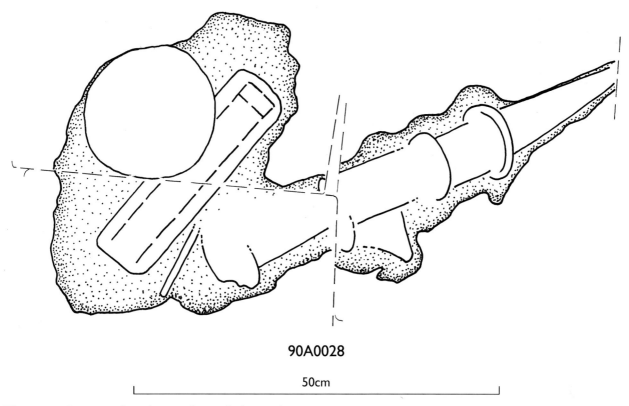

90A0028

|—————————— 50cm ——————————|

*Figure 4.8 Interpretation of concretion enclosing 90A0028*

*Figure 4.9 Hailshot piece 90A0028 emerging from concretion*

**Table 4.3 Incidence of hailshot pieces against total iron guns on vessels in 1546**

| Vessel name | Vessel type | Tonnage | Hailshot/ total iron | % of total iron |
|---|---|---|---|---|
| Henry Grace à Dieu | ship | 1000 | 40/130 | 30.8 |
| Mary Rose | ship | 700 | 20/76 | 26.3 |
| Peter | ship | 600 | 20/112 | 17.9 |
| Great Bark | ship | 500 | 20/71 | 28.2 |
| Matthew | ship | 600 | 2/72 | 2.8 |
| Jesus of Lübeck | ship | 700 | 20/52 | 38.5 |
| Pauncy | ship | 450 | 20/70 | 28.6 |
| Morian (of Danzig) | ship | 500 | 6/21 | 29.0 |
| Struse (of Danzig) | ship | 450 | 12/37 | 32.4 |
| Mary (of) Hamburg | ship | 400 | 12/35 | 34.3 |
| Christopher of Bremen | ship | not given | 12/29 | 41.4 |
| Trinity Harry | ship | 250 | 12/45 | 26.7 |
| Small Bark | ship | 400 | 12/76 | 15.8 |
| Sweepstake | ship | 300 | 12/50 | 24.9 |
| Minion | ship | 300 | 12/65 | 18.5 |
| Grand Mistress | galley | 450 | 12/31 | 38.7 |
| Anne Gallant | galley | 450 | 12/26 | 46.2 |
| Hart | galley | 300 | 12/23 | 51.0 |
| Salamander | galley | 300 | 12/43 | 27.0 |
| Galley Subtle | galley | 200 | 12/28 | 42.9 |
| Swallow | galley | 260 | 12/43 | 27.0 |
| Unicorn | galley | 260 | 8/29 | 27.6 |
| Antelope | galley | 300 | 8/32 | 25.0 |
| Hoy Bark | ship | 80 | 6/18 | 33.3 |
| New Bark | galley | 200 | 6/43 | 14.0 |
| Greyhound | galley | 200 | 6/22 | 27.3 |
| Jennet | galley | 180 | 6/20 | 30.0 |
| Dragon | galley | 140 | 6/34 | 17.6 |
| Roo | pinnace | 80 | 6/18 | 33.3 |
| Lartique | ship | 100 | 4/20 | 20.0 |
| Mary Thomas | ship | 90 | 4/23 | 17.0 |
| George | ship | 60 | 4/15 | 26.7 |
| Mary James | ship | 60 | 4/22 | 18.2 |
| Tiger | galley | 200 | 4/26 | 15.4 |
| Bull | galley | 200 | 4/26 | 15.4 |
| Lion | galley | 140 | 4/36 | 11.0 |
| Falcon | pinnace | 80 | 4/26 | 15.4 |
| Hind | pinnace | 80 | 4/21 | 19.0 |
| Phoenix | pinnace | 40 | 4/18 | 22.2 |
| Merlin | pinnace | 40 | 4/12 | 33.3 |
| Trego Renneger | rowbarge | 20 | 4/16 | 25.0 |
| Hare | rowbarge | 15 | 4/16 | 25.0 |
| Double Rose | rowbarge | 20 | 4/10 | 40.0 |
| Flowre deluce | ?rowbarge | 20 | 4/7 | 57.1 |
| Portcullis | rowbarge | 20 | 4/10 | 40.0 |
| Rose in the Sun | rowbarge | 20 | 4/10 | 40.0 |
| Falcon in the Fetterlock | rowbarge | 20 | 4/12 | 33.3 |
| Less Pinnace | pinnace | 40 | 3/13 | 23.1 |
| Harp | rowbarge | 20 | 3/6 | 50.0 |
| Cloud in the Sun | rowbarge | 20 | 3/10 | 30.0 |
| Hawthorn | rowbarge | 20 | 3/9 | 33.3 |
| Three Ostrich Feathers | rowbarge | 20 | 3/9 | 33.3 |
| Maiden Head | rowbarge | 20 | 3/9 | 33.3 |
| Rose Slype | rowbarge | 20 | 3/9 | 33.3 |
| Gilly Flower | rowbarge | 20 | 3/8 | 37.5 |
| Sun | rowbarge | 20 | 3/10 | 30.0 |
| Saker | pinnace | 80 | 2/16 | 12.5 |
| Brigandine | pinnace | 40 | 2/12 | 16.7 |

grey compact silt encasing the gun. The area to the underside of the hailshot was selected as the best place to open the concretion. This was removed to reveal 10–12mm of black corrosion products, forming a cast over the gun. As the surface of the gun became visible, a sulphurous smell was obvious, and a very thin green layer was noted with small barnacles which began to oxidise and turn orange after about 1½ hours. An old fracture at the rear of the stock came away with a 6mm portion of the rear of the barrel. An extant crack on the fin also separated.

## The importance of hailshot pieces

Inventory records provide valuable information as to the relative importance of these guns merely by their number expressed against the percentage of other guns (Table 4.3). This is particularly valuable in the context of inventories listing guns carried on board ships. Although guns listed in castles, bulwarks or forts might be for immediate defence, items were clearly stored ashore. One would expect that those onboard a warship were all for use.

In some instances descriptive terms, such as material (bronze, iron), method of manufacture (forged, cast), method of loading (chambered), method of deployment (stocked, stocked upon miches), or clues as to age (old and worn, new from the mint) can help with identification. In addition to the Tower inventories (Blackmore 1976) including that of 1540 (PRO E 101/60/3, naval elements summarised in Anderson 1920), information is provided by the 1514 inventory and the *Anthony Roll* of 1546 (Knighton and Loades 2000), the 1547 inventory of the King's possessions (Starkey 1998) and the unpublished 1555 inventory at Longleat.

Blackmore deduced that hailshot pieces were exclusively of iron, but classified them as breech-loading, possibly because of the single entry in the 1568 Tower inventory for four hailshot piece chambers (Blackmore 1976, 265; see also below Table 4.2). The 1547 inventory gives no information regarding form, but under the classification '*Gonnes of yron*' merely lists 41 '*Haile shotte peices*' (*ibid.*, 262; Starkey 1998, 3723). The 1559 inventory clearly shows them under the category '*fforged ordinance of yron*' (Blackmore 1976, 263) where '*Baces*' and '*Harquebusses a crocke*' are followed by 80 '*Haileshotte peces*' and a further five '*Hailleshot peces upon mytches*'. No special shot is listed. By 1568 the number of cast iron hailshot pieces has increased to 168. Five more appear under the forged iron category, and followed by the four '*Haileshot piece chambers*' (*ibid.*, 265). Also listed are '*Dice shott ... of iron, leade and stone and of diverse quantities.*' (*ibid.*, 266).

In the 1589 Tower inventory printed by Blackmore just one hailshot piece is listed (*ibid.*, 267), under the

**Table 4.4 Incidence of hailshot pieces listed on vessels 1546–1555**

| Vessel | 1546 | 1547 | 1555 |
|---|---|---|---|
| Henry Grace à Dieu | 40 | 29 | – |
| Mary Rose | 20 | – | – |
| Peter | 20 | 0 | – |
| Matthew | 2 | 0 | – |
| Great Bark | 20 | 8 | – |
| Jesus of Lübeck | 20 | 0 | – |
| Pauncy | 20 | 0 | – |
| Morian (of Danzig) | 6 | 0 | – |
| Struse (of Danzig) | 12 | 0 | – |
| Mary (of) Hamburg | 12 | 0 | – |
| Christopher of Bremen | 12 | 0 | – |
| Trinity Harry | 12 | 12 | 12 |
| Small/Lesser Bark | 12 | 4 | – |
| Sweepstake | 12 | 10 | 0 |
| Minion | 12 | 0 | – |
| Hart | 12 | 0 | – |
| Lartique | 4 | – | – |
| Mary Thomas | 4 | – | – |
| Hoy Bark | 6 | 0 | – |
| George | 4 | 0 | 0 |
| Mary James | 4 | – | – |
| Grand Mistress | 12 | 8 | – |
| Anne Gallant | 12 | 0 | 0 |
| Antelope | 8 | 4 | 1 |
| Tiger | 4 | 2 | 0 |
| Bull | 4 | 3 | 0 |
| Salamander | 12 | 0 | 0 |
| Unicorn | 8 | 0 | – |
| Swallow | 12 | 0 | 0 |
| Galley Subtle | 12 | 0 | – |
| New Bark | 6 | 3 | 0 |
| Greyhound | 6 | 0 | 0 |
| Jennet | 6 | 0 | 0 |
| Lion/Rose Lion | 4 | 3 | – |
| Dragon | 6 | 0 | – |
| Falcon | 4 | 0 | – |
| Saker | 2 | 0 | 0 |
| Hind | 4 | 0 | – |
| Roo | 6 | 0 | – |
| Phoenix | 4 | 0 | 0 |
| Merlin | 4 | 0 | – |
| Less Pinnace | 3 | 0 | – |
| Brigandine | 2 | 0 | – |
| Hare | 4 | 0 | – |
| Trego Renneger | 4 | 0 | – |
| Double Rose | 4 | 0 | 0 |
| Portcullis | 4 | 0 | – |
| Harp | 3 | 0 | – |
| Cloud in the Sun | 3 | 0 | – |
| Rose in the Sun | 4 | 0 | – |
| Hawthorn | 3 | – | – |
| Three Ostrich Feathers | 3 | 0 | – |
| Falcon in the Fetterlock | 4 | 0 | – |
| Maiden Head | 3 | 0 | – |
| Rose Slip | 3 | 0 | – |
| Gillyflower | 3 | 0 | – |
| Sun | 3 | 0 | – |
| Flower de Luce | 4 | 0 | – |

– vessel not listed, 0 vessel listed, no hailshot

class '*Forge'd iron Ordnance*' and following the bases and slings. After '*Dice shott*' comes 9 hundred pounds of '*Burres alias Haileshott*'. The last of Blackmore's sixteenth century Tower inventories (1595) lists one hailshot piece following the quarter slings, under '*fforged Iron Ordinance*' (ibid., 275). The entry for '*Burre Shott ales haileshott*' gives a total weight of 9107 pounds (expressed as '*ixmlCvijll*') estimated at 49,300 shot (ibid., 276). Again dice shot is listed separately.

Compared with other iron guns (port pieces, fowlers, slings and bases), hailshot pieces were comparatively few in number and short-lived. The inventories printed by Blackmore suggest a huge drop from 168 pieces in 1565 to a single one in 1589. This might seem to reflect a general clear-out following the Armada campaign of the previous year; however the real explanation has recently been noticed elsewhere (Smith, pers. comm., 2002). A further Tower inventory for 1589–91 preserved among Pepys's manuscripts in the Bodleian Library lists 149 '*flukemouthed*' pieces '*alias*' hailshot under the cast iron ordnance (MS Rawlinson A.207, f. 2v). '*Fluke*' is the common name for the flounder, suggesting an oval or purse-shaped mouth, a fair description of the bore of a hailshot piece. These 149 hailshot pieces must be the same as the 149 pieces simply called '*flukemouthed*' in the 1589 and 1595 inventories printed by Blackmore (1976, 266, 275). Even though this shows that hailshot pieces remained in store longer than previously realised, they still disappeared as suddenly as they had arrived. The last mention is in 1606 when they were disposed of (BL Harleian MS 309, f. 70). Although this possible re-identification of '*floukemmouth*' pieces as hailshot suggests that they were in store for longer than originally anticipated, it does not alter the fact that they appeared quite suddenly, were placed on many ships in large numbers seemingly for an extremely short period.

The *Anthony Roll* lists the ordnance within 58 vessels in 1546 (Table 4.3), and includes 459 hailshot pieces. Every vessel has a minimum of two hailshot pieces, with a maximum of 40 listed for the *Henry Grace à Dieu*. The larger ships, the *Mary Rose*, *Peter*, *Great Bark*, *Jesus of Lübeck* and the *Pauncy* all have 20. The *Matthew* lists two, possibly an error as the profiles of the other guns are otherwise in line with the rest of the large ships.

All hailshot pieces are listed under the broad category of 'guns of iron', and their importance must be seen within the context of the number of iron guns. The *Anthony Roll* lists a total of 1814 iron guns, excluding handguns. The hailshot pieces are actually the second most prevalent type, forming 25.3% of the iron gun assemblage (n=1814) within the fleet, never less than 11% per ship and usually around 25%. Bases are most numerous with (49.2%), next most numerous are the larger port pieces (10.9%). The entire sling category comprised only 6.7% and the fowlers represented only 4.3%. The ships carry a total of only 257 guns of bronze, iron guns being over seven times more prevalent. As Table 4.3 shows, hailshot pieces represent a substantial portion of all iron guns carried in 1546.

In comparison, the inventory undertaken following the death of the King in the next year reveals startling differences in the numerical occurrence of hailshot pieces (Table 4.4). Of the 52 vessels with inventories (Starkey 1998, 147–57) only eleven carry a total of 86 hailshot pieces between them, all classified under the heading '*iron pecis*'. The largest number is 29 on the *Henry Grace à Dieu*, followed by 12 on the *Trinity Harry* (Table 4.4). The entries offer little description, they are iron, and their size is reflected by their position with respect to the other iron guns. In all cases they are within the groupings for '*hagbushes*' and handguns, sometimes before and sometimes after top pieces, but always after the base category. If nothing else, this suggests that top pieces were smaller than single bases. West Tilbury (3984) does not even include them with the other guns, but they occur as if by afterthought following entries for shot and powder. East Tilbury is more fulsome, listing ten '*Small cast peces of yron for haileshotte whereof one broken*' (3929). It is likely that this means the guns themselves, as shot is usually listed by weight, and the entry occurs within the list of guns rather than under the separate heading of shot.

Analysis of the gun profile for each vessel does not seem to indicate why some carry hailshot pieces and others do not. The vessel listed as the *Rose Lion* or the *Rose in the Sun* is a rowbarge of 20 tons carrying three hailshot pieces; this seems enigmatic. '*Galliasses*' seem to have between two and four, and the ships (with the exception of the *Less Bark*) more than four. By 1555 (Longleat Miscellaneous MS V, ff. 1, 53–73v) there are only thirteen listed, all '*forged iron*', all on only two out of thirteen vessels, the *Trinity Harry* and the *Antelope*.

Hailshot pieces are far less common in castles and shore fortifications and may have been particularly manufactured for use at sea. In the entire 1547 inventory the only establishments to have them are the Tower (41), Milton Bulwark adjoining Gravesend (11), East Tilbury Bulwark (10, listed as cast), West Tilbury Bulwark (15) and Calshot Castle on the Isle of Wight (10).

Thus in 1547, the total number of hailshot pieces on land establishments listed in the inventory was only 86.

## Analysis of a hailshot piece
by D. Starley and Alison Draper

Hailshot pieces provide the first tangible evidence for the mass production of cast iron guns and their analytical study is important in demonstrating the maturity of the cast iron industry at a specific point in time. One of the *Mary Rose* hailshot pieces (90A0028), and two other cast iron objects, a shot and a piece

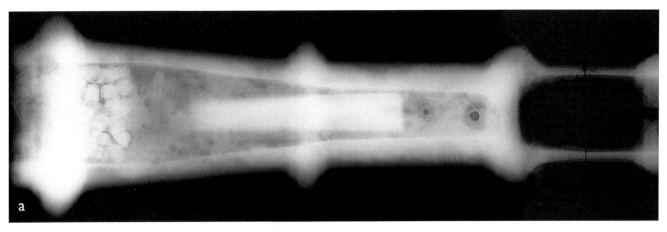

*Figure 4.10 Hailshot piece 90A0028: a) X-radiographic image of hailshot piece after removal of concretion. This shows most internal features more clearly and some, such as the position of a casting chaplet, for the first time; b) existing fracture surface on bottom of underside of projecting fin; c) existing fracture surface through stock socket; d) metallographic section through the stock socket. Etched in 1% nitric acid. Note that the columnar crystals converge from the external surfaces on the lower and right-hand sides but no equivalent growth is visible from the internal socket at the top of the image, where the cooling rate was slower*

resembling a mortar bomb that was thought to represent the nineteenth century salvaging activities of the Deanes were studied at the Royal Armouries, Leeds (Starley and Hildred 2002).

The hailshot piece was radiographed to provide details of the casting. This revealed the weapon to be loaded with 30 cuboidal 'dice' projectiles, presumably of wrought iron (Fig. 4.10a). Interpretation of the radiograph confirms that the gun is a single casting of cast iron. Testing with a magnet showed that some metallic iron remained, although the object's relatively low density (8kg compared with 14kg of the replica, see below) suggests that, despite the well preserved appearance of the surface, the surviving structure is extensively corroded.

Examination of areas on the gun gave further indications of the method of manufacture. The fractured surface of the fin on its underside (Fig. 4.10b) reveals columnar grains which can be seen to meet in the centre of the section, the result of a high thermal gradient during cooling. This suggests both a high pouring temperature, significantly above the liquidus of the alloy, and rapid heat extraction from the mould. Examination of the fractured socket (Fig. 4.10c, d) provided further evidence of the casting history: columnar grains extend inwards, from the outside front and side of the socket, but equivalent grains, radiating from the inside of the socket, are absent. This may indicate that a core of a less thermally conductive medium, such as clay or sand, was used to form the socket for the handle of the gun. Presumably, the bore of the gun was also formed using a similar core material. In contrast, the mould material from the outer surface of the guns may have been of metal.

A casting flash was observed around the outside of the gun (Fig. 4.6 and 4.11) showing that this metal mould was of two parts, unlike the method used for casting large guns (see Chapter 3). Aligning the core within the centre of the metal mould and holding it in place during filling would normally have involved the use of chaplets, small metal pins or wire, which then became incorporated in the casting. The presence of one of these was noted in the radiograph, where preferential corrosion had led to a small square pit in front of the touch hole. The use of chaplets to hold the core vertical during casting was common practice in casting bronze guns (Chapter 2). No other evidence for casting guns in metal moulds is known at this date. However Biringuccio (1540, 319), describes the production of cast iron cannon balls in moulds of brass or cast iron.

Further examination enabled identification of the microstructure confirming white cast iron suggestive of rapid cooling in a metal mould. Whilst white cast iron is known for its great hardness, it is also inherently brittle. Analysis of the cast iron shot from the *Mary Rose* show this to be of similar form and suggest the adoption of the technique of casting shot applied to a gun. Although the study shows that a considerable degree of technological innovation in both the use of materials and methods, the properties required of iron shot are considerably different from those required for a gun. There is no evidence of any refinement in the microstructure suggesting the implementation of any process to toughen the casting. The combination of chill cast iron and thin sections suggests that these guns were in danger of shattering on discharge. Although demonstrating the desire to mass-produce a gun in cast iron, these cannot be seen as a successful evolutionary step forward in the casting of ordnance, but more of an experimental step borne out of necessity. What they do suggest is a desperate attempt to secure a cheaper and more readily available resource than bronze, which was less labour intensive than manufacturing in wrought iron.

*Figure 4.11 Replica hailshot piece*

## Manufacture of a replica hailshot piece

A copy of a cast iron hailshot piece was made by Ron Spencer of Rayne Foundry, Rayne, Essex. The resultant piece well represents the guns found (Fig. 4.11, top). Before casting the reproduction, a small sample from the broken fin of unconserved hailshot piece 90A0028 was sent to the foundry to be analysed for any information on the metal used and in order to ascertain whether it would be possible to replicate the iron. The casting structure was clearly evident in section on visual examination. The iron was identified as white iron (cementite) which would be hard but extremely brittle. The crystalline nature of the material was evident and it was shown to be a high phosphorus iron, which would also make it very weak (see above). The poor quality of the metal combined with the shape of the bore would result in an unstable gun. It was therefore decided to cast the replica in standard grey iron. The poor quality is likely to have resulted from too rapid cooling on casting into a very cold mould, probably of stone or brass. '*Moulders of brass*' for shot for named classes of guns are listed in the inventories (eg, Starkey 1998, 4970, 4971). It was felt that a project to test both brass and sandstone moulds for the chill factor and for longevity of mould following multiple castings would be beneficial.

Modern technology was employed in casting the gun. A bipartite wooden pattern comprising the external form of the gun was made based on the archaeological drawings and the physical study of one of the originals (80A0544). This was placed in two halves of a box which was coated internally with graphite. The two halves were then filled with sand containing an organic resin which hardened to withstand the casting temperature of 1450°C. Once hardened the sand was tapped out of the box, producing a bipartite mould aligned around the cast-line. The bottom mould contained the fin and the top the feeding head, channels and top of the gun. The mould was sprayed with graphite and flamed.

A similar process was undertaken to form two further sand moulds in the form of positive rather than negative items. The rectangular bore and the cylindrical slot for the stock were formed by manufacturing sand cores which were placed within the mould. The molten iron was poured into the feeding head and led through a ceramic filter taking up the space between the mould and the cores, which formed the bore and the stock hole. The cores were held in position by tinned steel chaplets nailed into the base of the bottom mould which eventually became incorporated into the molten iron. The mould was held together during the casting by means of a step on one half of the mould locating with a plinth on the other, reinforced with glue. Sash weights were also placed on top of the mould to hold it down.

During the pouring, the feeder head was overfilled so that, when the metal cooled and shrank, it could draw further molten metal downwards, ensuring a complete casting. The mould was left for 24 hours before being broken open to reveal the gun within. In time the mould breaks down to sand which is cleaned and recycled. Similarly the sand cores also break down so that all elements must be regenerated from the pattern each time a casting is made. This would not be the case if the elements were of brass or stone. The number of times a stone or brass mould could withstand the casting of a hailshot piece is not known, but it is possible that small amounts of damage could be repaired by the use of a clay sleek. This might change subtly the form of the mould, and therefore subsequent guns would be similar, but not identical. The *Mary Rose* hailshot are all very similar, but not identical, suggesting that the moulds used were different.

The gun was cleaned and the touch-hole drilled. Originally the touch hole may have been cast with the gun by means of a core. Other than the time saved by the use of resin-hardened sand to produce a mould, the major difference would be that the original piece was probably cast vertically.

### Firing a replica hailshot piece

One of a number of research projects undertaken with the Royal Armouries was the firing of a reproduction hailshot piece. We wanted to assess the physical scatter of the dice shot from the rectangular bore against a target designed to be a representation of the hull of the Mary Rose. Firing trials took place at British Aerospace Defence Division Ltd, Royal Ordnance Division, at Bellingham, Cumbria (Fig. 4.12a).

Twenty cast iron dice were fabricated by breaking a cast metal plate into pieces with a hammer. The shot weight was established by weighing a representative number of shot from 79A1088, which weighed 10g each. The total shot weight was estimated at 200g or 7oz. A powder charge of 2oz (57g) and proof charge of 4oz (113g) was proposed.

*Figure 4.12 Firing trial of the replica at Ridsdale: a) gun on bed; b) scatter of shot against target*

The powder was placed in the muzzle and tamped down. This was followed by sawdust to fill the void, then the shot, just inside the muzzle. A rectangular willow tampion was placed in the bore to seal it. For proofing it was loaded with 4oz (113g) of commercially available black powder (ICI NPCG) of the coarsest grain available and 4oz (113g) of dice. The powder was chosen because it is relatively slow burning, to replicate as far as possible the original powder without the inconsistency of result. It was primed with NPCG priming powder and electronically ignited. Because of the immediate scatter of the dice it was impossible to track with radar.

The gun was supported over the end of a metal stand and fired (for proofing) at a distance of 5.5m at a target which copied the structure of the *Mary Rose* at Main deck level. The target was covered with a sheet of thick white paper to that it would be easy to see the scatter of the dice. At this short distance the dice bounced off the target (penetrating and tearing the paper) and scattered backwards to a distance of 12m at an angle of 25° (Fig. 4.12b).

The gun is heavy and quite difficult to handle. The recoil, as observed, was strong and undoubtedly would not be comfortable. Further trials with reduced powder charges are certainly required before any attempts are made to fire it manually. Finding a volunteer is also a pre-requisite. It is probably as cumbersome as a large crossbow, and may not be a weapon designed for multiple rapid firing.

The gun has since been returned to the foundry for non-destructive testing before further trials. Should further trials be undertaken, we would increase the weight of the shot to 7oz (200g) and move the target further away in order to establish the greatest effective range for anti-personnel use. Once this is established, the powder charge would be reduced gradually until it becomes ineffective against the target. This should establish the optimum fighting range and powder charge. It is most unfortunate that the spread of shot prevented the muzzle velocity from being established.

# 5. Munitions and Accessories

## Shot

by Alexzandra Hildred with Robert Smith

Over 2000 individual shot were recovered. The diversity in form and the large numbers present mean that this is an extremely important assemblage, inviting detailed study of manufacturing techniques and consistency in manufacture, as well as ascertaining what was acceptable practice regarding the size of shot relative to recovered guns. The types represented include: solid cast lead round shot; cast iron round shot; cast iron round shot with a wrought iron spike (possibly wrapped at the tips); wrought iron dice within a cast lead shot; wrought iron dice for hailshot; cast iron round shot within a cast lead shot; stone shot; and pebbles, iron and broken flint inside canisters as well as small numbers of unusual composites (Fig. 5.1). By far the largest group is round, cast iron shot (Fig. 5.2a) followed by round stone shot (Fig. 5.2b). What is surprising is the enormous range of different types – a fact not reflected in many of the contemporary inventories and documentary sources, which often list just iron, stone and lead. There are few very sharp, clear peaks denoting shot diameters of a particular size, but a considerable spread and there are shot of almost every size from smallest to largest. This suggests that the tolerance in manufacture was much larger and much less precise than we might expect today. This has implications for the range and accuracy of the gun – the larger windage (the difference between bore size and shot size) will not only reduce the range but greatly influence the accuracy of the shot. This suggests that the ballistic potential of the gun could not be maximised, and the likely result was a reduction in range. This has direct implications for the fighting capabilities and the range at which ships engaged (see Chapter 11).

Matching shot to guns is particularly problematic as most of the shot was stored centrally rather than beside individual guns. The bores of wrought iron guns are difficult to measure precisely owing to differential corrosion; consequently windage is hard to calculate and seemingly quite variable.

*Figure 5.1 Types of shot and moulds*

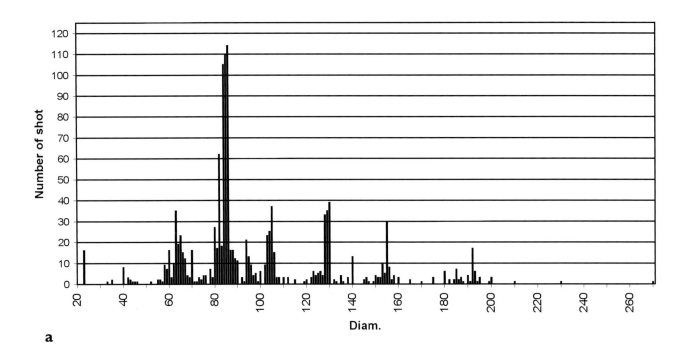

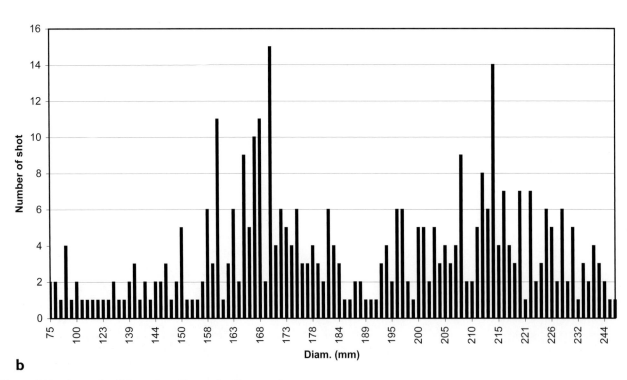

*Figure 5.2 a) graph all iron shot, b) graph all stone shot*

## Supply of shot

There is little published information on the manufacture and supply of shot and powder in this period. Inventories are a source of information regarding the numbers held, but offer little regarding provenance. Some indication of the sources of supply can be found in the accounts for invasion of France in 1513–14, where providers of shot included Lord Curson, Sir Stephen Bull, Robert Scorer and John Bowyer (both of Hartfield), Richard Fermer and Humfrey Lightfoot. Scorer had been appointed the King's gunstone maker in 1510, with a fee of 6*d* a day (*LP* 1, 587(17); he was also a gunner in the Tower, dying by January 1514 when his successor there was named (*ibid.*, 2617(32)). Richard Scorer, probably

Robert's son, was appointed to follow his father as the Royal gunstone maker on 15 May that year (*ibid.*, 2964(41)). Bowyer is also described as a gunstone maker. Fermer was a London grocer and merchant of the Staple, who also supplied corned powder (see below). There is nothing remarkable about the variety of his business; the same account which mentions Fermer and Lightfoot (Fermer's agent) features a draper who supplied bowstaves and a mercer who traded in saltpetre (*ibid.*, 3613, p. 1509). Curson and Bull were not of course tradesmen let alone manufacturers; the former was Master of the Ordnance in the rearward of the expeditionary force, and Bull was captain of the *Regent*. From this is evident a variety of ways in which shot was procured: in-house manufacture, purchase from other makers or tradesmen, or by reimbursing military commanders who had found their own supplies. A development of special interest was the appointment on 27 July 1543 of Thomas Goore as '*le gonneston maker of our shotte of stone of our iren gonnes of our shippes*' (*LP* 18/1, 981(102)). Some at least of the stone shot recovered from the *Mary Rose* must surely have been made by this man.

There is one specific reference, from 1513, to the supply of shot direct to the *Mary Rose*:

'Mary Rose. *Gunstones. Also the said John Hopton hath paid unto Nicholas Sesse for gunstones of iron with crossbars of iron in them at £19 the ton tight, one ton tight - £19. For one ton tight half of round gunstones of iron - £10. And in like wise to a man of Maidstone for gunstones of hewed stone at 13*s *4*d *the 100, for 2,700 of them - £18. So by him paid and delivered the foresaid gunstones to the* Mary Rose *in the time of this account, in all amounting to - £37.*' (*LMR*, 11; spelling there modernised).

The 1514 inventory for the *Mary Rose* lists shot sent from the vessel to the Ordnance Office. These are described as shot of iron, shot of iron with cross bars, pellets of lead great and small, pellets for harquebuses, and dice of iron (Knighton and Loades 2000, 139–43).

## *The* Anthony Roll

Shot is listed in the *Anthony Roll* under the two headings '*shotte of yron*' and '*shotte of stoen and leade*' (Knighton and Loades 2000, 43; Fig. 5.3). Under each heading the number of shot is listed for each named type of gun in descending order of gun size; this usually equates to bore and therefore shot diameter. This evidence has been extremely useful; in some cases the type of shot has helped identify certain classes of gun, such as the wrought iron slings (the only wrought iron guns listed with iron shot) and the hailshot pieces (see Chapters 3 and 4).

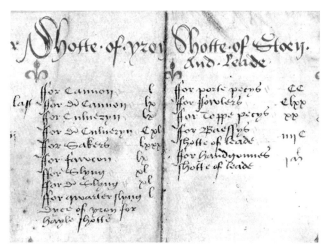

*Figure 5.3* Anthony Roll *heading listing shot (© Pepys Library, Magdalene College, Cambridge)*

Having a closely dated inventory of weapons designated for ships and an archaeological assemblage which includes both guns and the shot is a great benefit. However, as expected, there are difficulties. The weapons listed represent those designated and are not necessarily those carried at any particular time. The number of iron shot excavated is over twice that listed, though the recovered stone shot is very close to the listing; the bulk of the shot listed as '*lead for bases*' is actually made of a wrought iron dice within a cast lead shot, which we have termed '*inset shot*', though in some contemporary inventories this is termed '*shot of dice and lead*'. None of the varied shapes or contents of canister shot found on the ship is listed, nor are any of the odd composites, which include jointed and cross bar shot.

Similarly, the list of guns in the inventory does not exactly match that recovered from the ship (as discussed elsewhere in this volume). Four guns of the correct dimensions for culverins were recovered instead of the two listed, and only two demi culverins instead of the six listed. To date no sakers or falcons have been recovered although these are listed. However what the inventory does offer is the ability to analyse the data and derive information about which guns were expected to fire most. This has important implications for the amount of powder required; the optimal fighting distance; and the expected damage. This helps in our understanding of the ship as a fighting machine (Chapter 11), and enables comparison with other vessels within the fleet (Fig. 5.4). Furthermore the shipboard location of shot of specific types and sizes can be used to predict what guns could have been present and, potentially, where they may have been stationed.

The *Anthony Roll* lists a total of 580 iron shot divided up for each type of gun (Table 5.1). Interestingly this is less than half the recovered assemblage of iron shot. Iron dice for hailshot is listed, though without a figure. By reference to other ships in the inventory it is suggested that the 20 hailshot pieces could have had 1000 dice, about 2.5 rounds per gun. A

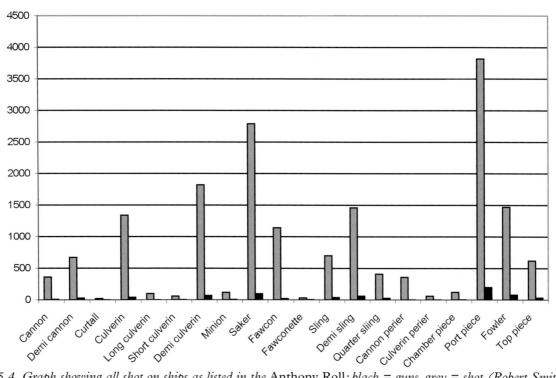

*Figure 5.4 Graph showing all shot on ships as listed in the* Anthony Roll*; black = guns, grey = shot (Robert Smith)*

total of 390 stone shot are listed, to serve the wrought iron port pieces, fowlers and top pieces together with 400 shot of lead for the bases and 1000 for handguns (Table 5.1).

**Table 5.1 Shot as listed in the *Anthony Roll***

| Gun type | Gun material | No. guns | No. shot | Shot per gun |
|---|---|---|---|---|
| *Iron shot* | | | | |
| Cannon | bronze | 2 | 50 | 25 |
| Demi cannon | bronze | 2 | 60 | 30 |
| Culverin | bronze | 2 | 60 | 30 |
| Demi culverin | bronze | 6 | 140 | 23 |
| Saker | bronze | 2 | 80 | 40 |
| Falcon | bronze | 1 | 60 | 60 |
| Sling | iron | 2 | 40 | 20 |
| Demi sling | iron | 3 | 40 | 13 |
| Quarter sling | iron | 1 | 50 | 50 |
| Total | | | 580 | |
| *Stone shot* | | | | |
| Port piece | iron | 12 | 200 | 16 |
| Fowler | iron | 6 | 170 | 28 |
| Top piece | iron | 2 | 20 | 10 |
| Total | | | 390 | |
| *Lead shot* | | | | |
| Base | iron | 30 | 400 | 13 |
| Handgun | iron | 50 | 1000 | 20 |
| Total | | | 1400 | |

## Shot type, sizes and materials: evidence from other sources

Although the *Anthony Roll* lists the quantities of shot and the material from which they were made, it does not specify the actual size of the shot itself nor does it reflect the diversity of type, material and form found within the excavated assemblage. Other historical sources can be used to help us identify the assemblage and link the shot to the guns.

The 1514 inventory (Knighton and Loades 2000, 141) lists what remained on the mothballed *Mary Rose*. The large guns were left on the ship in the hands of the Master (John Browne) and the Purser (John Bryarley), as was a quantity of lead ('*ij sowes j quarter and certeyn cast*'), together with 20 '*Parchement skynnes*', 9 '*Bagges of leder* [leather]' and 22 '*Pykes for stone*', presumably representing relatively immobile or low value items. Items delivered to John Millet and Thomas Elderton, representing the Ordnance Office, included 91 '*Hacbusshes*', 457 '*Shott of yron of dyverse sortes*', 120 '*Shott of yron with crosse barres*', 1000 '*Pellettes of lede grete & smale*', 900 '*Pellettes for hacbusshes*' (10 per gun), 1100 '*Dyce of yron*' and after 74 '*Arrowes of wyld fyre*', 2 '*Balles of wyldfyre*' (ibid., 142). Bows, arrows, gunpowder (both loose and in cartridges) and body armour were also returned. Interestingly, this suggests that the shot for the large guns would be circulated elsewhere. Although '*pykes for stone*' are left, there is no mention of any stone shot which would have served some of the guns.

The inventory of 1547 has little extra detail within the vessel section (Starkey 1998, 144–57), with the exception of the inclusion of '*crossbarre shot*' in quite impressive numbers (134) on the *Henry Grace à Dieu* (*ibid.*, 7242) and in lesser numbers, usually under ten, on smaller vessels (7708, 7905, 7935, 7948). Some of these seem unusual composites: '*Crosebarre shott of Irone and stone*' on the *Great Bark* (7344), '*Crosse barre shot of all sortes*' on the *Antelope* (7576) and for the *Hart* (7601). Cross bar shot is specified for sakers on the *Mary (of) Hamburg* (7624), although the vessel is not listed with such pieces, and for falcons on the *Greyhound* (7803). The *Henry Grace à Dieu* had '*Shott of wildfier*' (7250) but unusual '*Hollowe shott*' (6141) and '*Hollowe shottes of wildfier*' (4530) are not found on ships.

This inventory unusually provides the sizes of shot for named guns in a few instances. These dimensions are as close as we can get to the historical list of shot sizes for *Mary Rose* guns. Weights are not given.

Shot diameter together with gun names include the following:

Demi culverin shot of 3¾in and 4in (*c.* 95mm and 102mm)
Saker shot of 2¾, 3, 3¼, 3½in (*c.* 70–89mm)
Falcon shot of 2½in (*c.* 64mm)

Shot of iron of these diameters is listed for Guisnes: 8, 7½, 7, 6¾, 4, 3¾, 3½, 3¼, 3, 2½, 2¾, 2⅝in (*c.* 203–50.8mm) (5537–47)

## Matching shot with guns

The fundamental problem with linking the shot to the gun for which it was intended rests on the amount of windage. Trying to define accurately the windage allowed at this period is impossible. It is often stated to have been, for the larger guns, ¼in (6mm) while later it is generally accepted to have been one-twentieth the diameter of the shot. However, from the recovered shot these rules do not seem to have been hard and fast and varied from 5mm to 25mm across the whole assemblage. This makes the whole situation very much more complex and less easy to define accurately.

One of the earliest comprehensive lists of shot sizes for specified muzzle-loading guns is that of James Burton, preserved in the Compton Place archives and dated about 1550 (Awty 1989, 143). This includes bore, size of shot, weight of shot, weight of powder, ladle length and breadth and number of shots per barrel of powder. Windage is listed consistently as ¼in (Table 5.2). Unfortunately, there is nothing so comprehensive for the wrought iron guns.

This seemingly accurate and definitive list of specifications does not, however, reflect the actual contemporary situation. For example, though Burton states that a saker had a bore of 3¾in (95mm), other lists show a wide range of bore size. For example the 1547 inventory of the King's possessions (Starkey 1998) lists saker shot of 2¾–3½in (70–89mm) and demi culverin shot of 3¾–4in (95–102mm). Other sources dating from between 1540 and 1587 list saker shot at 3¼–3¾in (83–95mm) and demi culverins 4–4½in (102–114mm; Blackmore 1976, 392–3). It must be remembered too that modern manufacturing tolerances were unachievable in the sixteenth century. These variations, although they seem small, make accurate assignation of shot to gun very difficult – one shot could have been used for a number of different guns. This is not to say that some deductions cannot be made.

## Shot material

The *Anthony Roll* only lists shot of iron, stone and lead and states which guns used which type of shot. Other inventories, particularly that of 1547, confirm the general use of cast iron shot for the cast bronze guns, that is the cannon, demi cannon, culverins, demi culverins, bastard culverins, sakers and falcons, together with some information about sizes. Wrought iron port pieces and fowlers usually fired stone shot, slings (of all sizes) are usually listed with iron shot and bases seem to have used a variety of shot, on balance most are lead or '*of dice and lead*', but can also be of lead, iron or composites of these and stone. The inventory, however, also shows that though these guns fired this shot most of the time, they did, on occasion, use other shot – especially the composite shot of iron and lead and stone

Table 5.2 Burton list of about 1550 (weight in lb, size/length in inches)

| Gun | Chamber cannon | Cannon | Demi cannon | Culverin | Demi culverin | Saker | Falcon | Falconet |
|---|---|---|---|---|---|---|---|---|
| Bore | 8 | 7¾ | 6½ | 5¾ | 4½ | 3¾ | 2¾ | 2¼ |
| Size iron shot | 7¾ | 7½ | 6¼ | 5½ | 3¼★ | 3½ | 2½ | 2 |
| Weight shot | 64 | 59 | 31 | 18 | 8½ | 5½ | 2½ | 1¾ |
| Weight powder | 21 | 38 | 21 | 18 | 8½ | 5½ | 2½ | 1¾ |
| Ladle length | 22¾ | 24 | 22½ | 24 | 19 | 16½ | 14 | 12 |
| Ladle breadth | 12 | 14 | 12 | 9½ | 8 | 5¾ | 4½ | 4 |

★possibly a mistake for 4¼

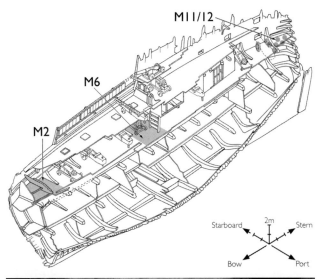

### Table 5.3 Shot material from 1547 inventory (Starkey 1998)

| Gun | Material/description |
|---|---|
| Demi cannon | **iron**, dice & lead |
| Culverin | **iron**, stone |
| Demi culverin | **iron**, lead; dice & lead |
| Saker | **iron**, dice of iron, lead, iron & lead, stone covered with lead |
| Falcon | **iron**, dice of iron, dice & lead, iron & lead, lead, stone covered with lead |
| Port piece | **stone**, iron, cases of hailshot |
| Fowler | **stone**, iron, lead & stone |
| Sling | **iron**, lead, stone, dice & lead, lead diced with stone & iron |
| Demi sling | **iron**, lead, dice & lead, lead diced with stone & iron |
| Quarter sling | **iron**, stone cast about with lead, iron & lead |
| Base | **lead**, dice & lead, stone cast about with lead, shot diced & undiced, lead diced with stone, iron & lead |

Bold = predominant material listed

*Figure 5.5 above) Location of M2/M6 shot stores; below) inside of U7 gunport showing shot behind wood*

covered with lead. Shot of lead with iron dice is listed in the following sizes: 3½in, 3in, 2¾, 2½in, 1¾in, 1⅛in, 1½in (Starkey 1998, 5549–55). In the absence of specific descriptions of size and type of shot for particular guns, this inventory evidence is vital in suggesting the variance in material and size of shot for named guns (Table 5.3).

### Storage of shot

There is little written about the storage of shot on board vessels at this time. On the *Mary Rose* the majority of shot was stored on the Main deck in the area just before the Forecastle (M2/3) and M6, with access to the Upper deck and stern (Fig. 5.5). There was a smaller selection in M10–12. In only one instance was there clear evidence for the storage of substantial numbers of iron shot to suit a particular gun. Shot was found tucked between starboard frames beside a gun port on the Upper deck held in position by a simple plank nailed to the outside faces of the two frames (Fig. 5.5). Of the eighteen cast iron shot found here, fourteen fit the culverin situated at that port (80A0976). There are two concretions (78A0601, 78A0602) which contained a number of shot either fortuitously enclosed on two sides with wood, or within some form of small box. There are no further details but these observations are important. Some of the iron guns had a small supply of stone shot close to them. The majority of stone shot was unfinished, and in store in M6. Cross bar shot seems to

*Figure 5.6 All the iron shot*

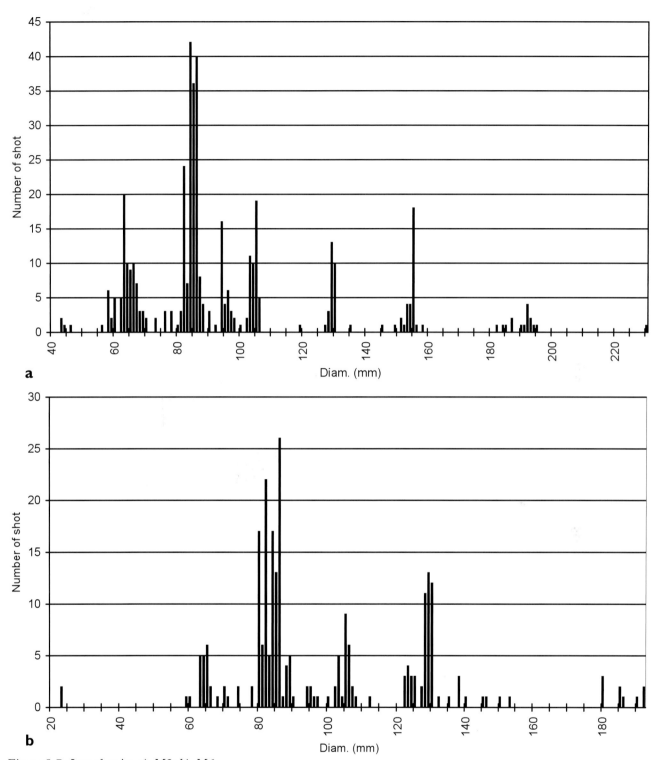

*Figure 5.7 Iron shot in: a) M2; b) M6*

have been stored in M11/12 or M2/3. Small lead round shot and dice for hailshot or canisters may have been supplied in casks. This was certainly the case later on. There is one listing of dice of iron by barrel ('*oone br*') in the 1547 inventory (Starkey 1998, 5786). A basket found in M11 (79S0059) contained three small solid lead shot (79A1207) and a ram (79A1208). Under it was a ladle (79A1211) and another shot (79A1214).

This is the only possible example of a dedicated container for shot.

*Shot of iron*

**Cast iron round shot**

The majority of shot recovered from the *Mary Rose* is cast iron (Fig. 5.6). The *Anthony Roll* lists 580 cast iron shot: 450 for the bronze guns and 130 for the wrought

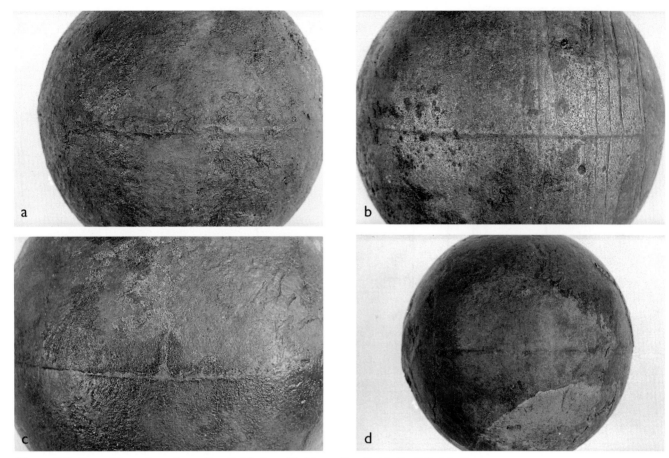

*Figure 5.8 Round iron shot: a) castline finish (rough); b) castline finish (well finished)*

iron slings (Table 5.1). The assemblage recovered includes 1248 shot. The largest groups were recovered from the shot lockers in M2 and M6 (Fig. 5.7), 1200 of these were studied in detail. A recording sheet was prepared with 50 attributes covering: measurements, weight, condition (13 states), the visible presence of a castline and its prominence and finish, the presence and description of the sprue and whether it was single or double, with measurements. The stamp, its location and description were included. Seabed location, ultimate tank destination during conservation and identity number were also noted. All shot were tried against the nine 'shot gauges' recovered (see below). It did not appear that these were obviously for use with the iron shot.

The castline was visible in most cases, displaying differing degrees of surface finishing (Fig. 5.8). Similarly the casting sprue, which was always on the castline, displayed different degrees of care in surface finish, size and shape (Fig. 5.9). This position suggests

Table 5.4 Expected variability in bore and shot diameter

| Gun | Bore max. (mm) | (in) | Bore min. (mm) | (in) | Shot max. (mm) | (in) | Shot min. (mm) | (in) | No. shot in range |
|---|---|---|---|---|---|---|---|---|---|
| Cannon | 215 | 8.46 | 200 | 7.87 | 210 | 8.26 | 190 | 7.48 | 37 |
| Demi cannon | 165 | 6.49 | 160 | 6.29 | 160 | 6.29 | 150 | 5.90 | 72 |
| Culverin | 140 | 5.51 | 132 | 5.19 | 135 | 5.31 | 122 | 4.80 | 159 |
| Demi culverin | 115 | 4.53 | 112 | 4.41 | 110 | 4.33 | 102 | 4.02 | 118 |
| Saker | 101 | 3.98 | 89 | 3.50 | 96 | 3.78 | 79 | 3.11 | 558 |
| Falcon | 70 | 2.76 | 57 | 2.24 | 65 | 2.56 | 47 | 1.85 | 128 |
| Sling | 107 | 4.21 | ? | ? | 102 | 4.10 | 97 | 3.82 | 25 |
| Demi sling | 88 | 3.46 | ? | ? | 83 | 3.26 | 78 | 3.07 | 134 |
| Quarter sling | 69 | 2.72 | ? | ? | 64 | 2.52 | 59 | 2.32 | 90 |

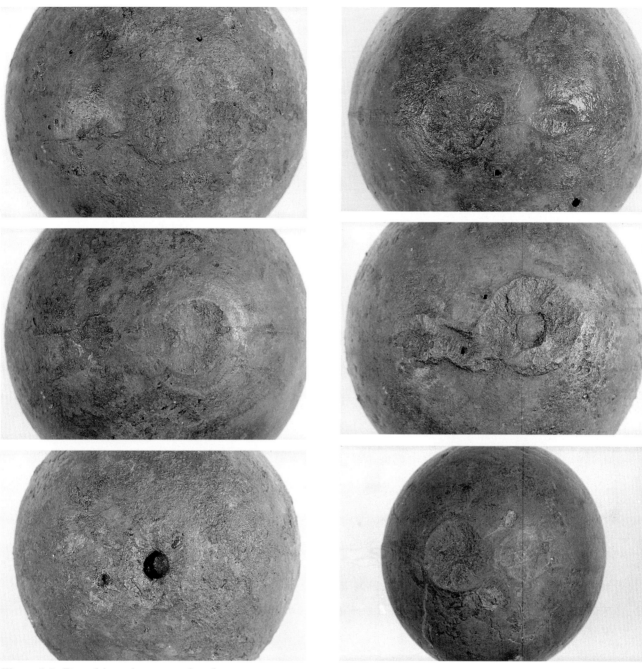

*Figure 5.9 Round iron shot: examples of sprue*

a bipartite mould, similar to those recovered for the casting of composite shot, with the joint aligned vertically (below). In 431 instances an 'H' is present, raised above the level of a circular depression with a diameter of 15–18mm. The location of the 'H' is always polar, if the castline is described as the equator. This had obviously been cut into the original mould and cast with the shot (Figs 5.10–12). Tracings taken of 92 examples show that here are stylistic variations of the 'H': straight-sided (81A2731) concave-sided (82A4277); straight (81A2731) or inverted 'V' shaped central bar (81A4237); and flared (81A3631) or straight-ended (81A4237) side bars (Figs 5.11–12). Rubber castings were made of some better examples,

*Figure 5.10 Iron shot 81A3666 showing 'H' mark*

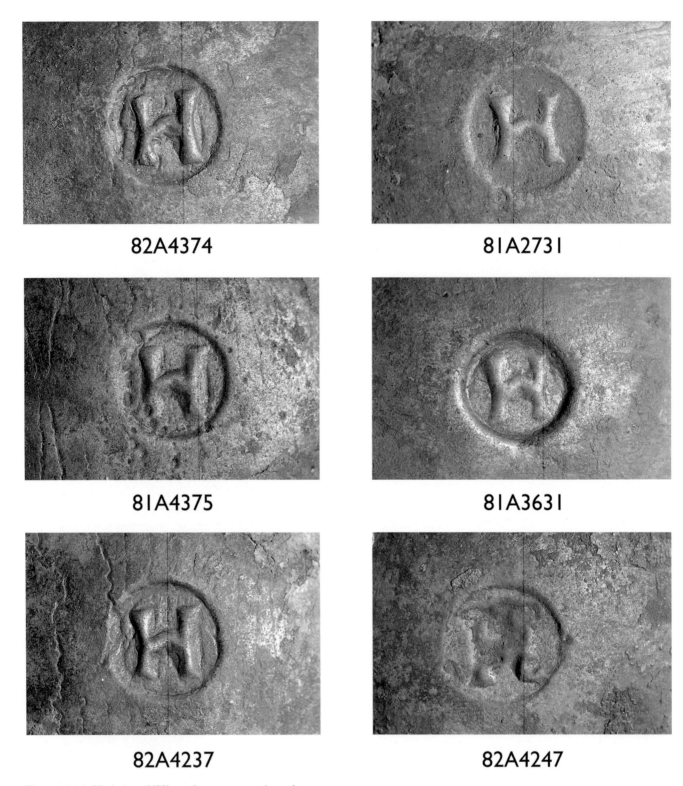

*Figure 5.11 Varieties of 'H' marks present on iron shot*

but no detailed study has been undertaken to ascertain how many moulds are represented. No other markings of any kind were clearly identified.

*Metallography*

One cast iron shot was sectioned and examined (Starley, Royal Armouries AM 1617, 13/6/2000; Fig. 5.13). Polishing and etching revealed large columnar grains, visible by eye, which extended from the surface towards the centre of the casting, indicating a high pouring temperature. The alloy is hyper-eutectic, having a carbon content in excess of 4.3%. Occasional inclusions were identified as manganese sulphide. The presence of sulphur suggests that coal may have been used in the smelting or remelting of the metal. The

83A0444

81A4277

81A3641

81A3411

*Figure 5.12 Varieties of 'H' marks present on iron shot*

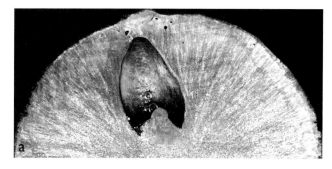

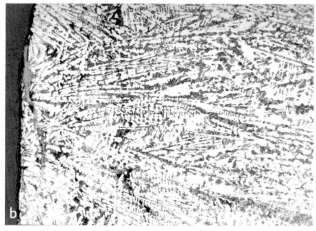

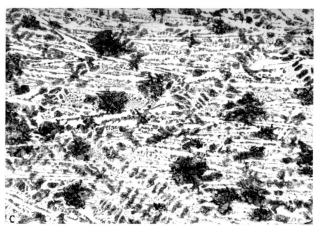

*Figure 5.13 a) Sectioned Royal Armouries shot; b) outer surface (view width 0.75mm); c) inner surface (view width 1.5mm)*

micrographs of the outer surface show white primary crystals of iron carbide in pearlite with an iron carbide matrix (Fig. 5.13b). The centre comprised kish graphite surrounded by pearlite in a matrix of iron carbide and pearlite (Fig. 5.13c). This is consistent with historical accounts of shot production in metal moulds (Biringuccio 1540, 319). Brass moulds are listed for a number of different types of bronze guns within the 1547 inventory, for example within the shot house in Calais (Starkey 1998, 4970–2).

*Shot size*
The shot recovered vary in diameter between 23mm and 268mm and in weight between 32g (1.1oz) and 27,000g (59lb 7oz). Weights vary tremendously

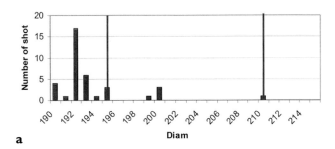

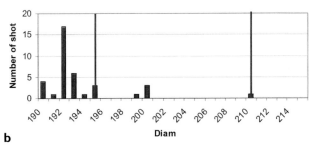

*Figure 5.14 Round iron shot: a) cannon shot; b) demi cannon shot. Vertical lines indicate bores of guns recovered minus 5mm for minimum windage*

depending on the condition of the shot. Within any group of a specified diameter there are examples which are equal to the sixteenth century listed weight for diameter.

*Assigning shot to the guns and identifying the missing guns*
The quantity of shot recovered has enabled an analysis which attempts to identify the shot for specific guns. For the reasons given above this has not proved easy but has yielded considerable information and allowed us to make some suggestions as to the sizes and types of missing guns.

The method used was first to take the bore dimensions of the recovered guns. The bore sizes of those guns which are not present in the assemblage but were listed in the *Anthony Roll* (sakers and falcons) were then added to this list, the information taken from contemporary sources and manuals (1547 inventory, in Starkey 1998; Bourne 1587; Burton in Awty 1989; Norton 1628). Shot sizes for each type were then assumed to be from 5mm less than the largest bore size to 10mm less than the smallest bore size. This assumption reflects both the contemporary statements that windage was around ¼in (*c.* 6mm) and the evidence from the assemblage which suggests that it could be considerably greater (Table 5.4).

In the absence of specific details regarding bore size and shot size for the sling group these were assigned based on the peaks present in the assemblage after the allocation of the cast iron shot to the bronze guns. The information was then plotted onto graphs of the frequency of each size of shot for that size of gun. Nine wooden objects tentatively identified as shot gauges for rapid selection of shot were recovered. Copies of these were made and tested against the shot by passing the shot through the holes within the gauge closest in diameter to the shot.

*Cannon shot*
Two cannon are listed in the *Anthony Roll* with 50 shot. Two cannon were recovered with 43 shot of appropriate diameter (Fig. 5.14a). Individual shot ranged in weight between 8kg (17lb 9oz) and 27kg (59lb 7oz). Shot for the largest gun recovered (bore 8½in/215mm) should have originally weighed in the region of 84lb (38kg). Only one shot was recovered which closely fits this port side gun (79A1276) although a further four shot of 199mm and 200mm may have been appropriate. The remaining 38 shot would all have fitted the second cannon (81A3003, bore 200mm). The most frequent diameter was 192mm, suggesting a windage of 8mm. Twenty-three shot of 180–189mm were recovered. These can only have fitted the two cannon and highlight the problem that windage was probably more than the figures given in contemporary documentary sources. Eight shot of 180–182mm cleared the largest wooden gauge, 81A1287, of 184mm.

In the range 190–210mm, three shot were in sector M6 where both cannon were found. Extending the minimum size of shot down to 160mm (the maximum shot for demi cannon) includes a further 32 cast iron shot, six of which were found in M6 and the largest number of which was found in M2/3 (21 examples; Fig. 5.15). The distribution of all shot above 160mm is relatively limited, but there are interesting incidences in SS10 and M11.

*Demi cannon shot*
The *Anthony Roll* lists two demi cannon with 60 shot. Two demi cannon were recovered, both probably from M10. These have bores of 165mm (81A3000) and 160mm (79A1277). 81A3000 was loaded with an iron shot of 160mm (82A4124). Some 72 shot were recovered within the predicted range of 150–160mm (Fig. shot14b). Weights ranged from 2.5kg (5lb 7oz) and 14.5kg (32lb); the expected weight is 32lb (14.5kg).

A large peak at 155mm of 30 shot would serve both the large gun, with 10mm windage, and the smaller piece with just 5mm windage. A further 20 shot were recovered between 140mm (the bore size of the largest culverin recovered) and 149mm. The closest gauge is 82A4510 at 163mm.

The distribution is relatively limited, with the majority in M2/3 or M6. Five shot of the predicted size were found in the stern, but only one in M10 (Fig. 5.16).

*Culverin shot*
The *Anthony Roll* lists two culverins with 60 shot. Four guns of culverin size were recovered, two with bores of 140mm (81A1423, 81A3002) one of 138mm (80A0976) and one of 132mm (79A1278). 80A0976 was loaded with an iron shot of 128mm (80A1941).

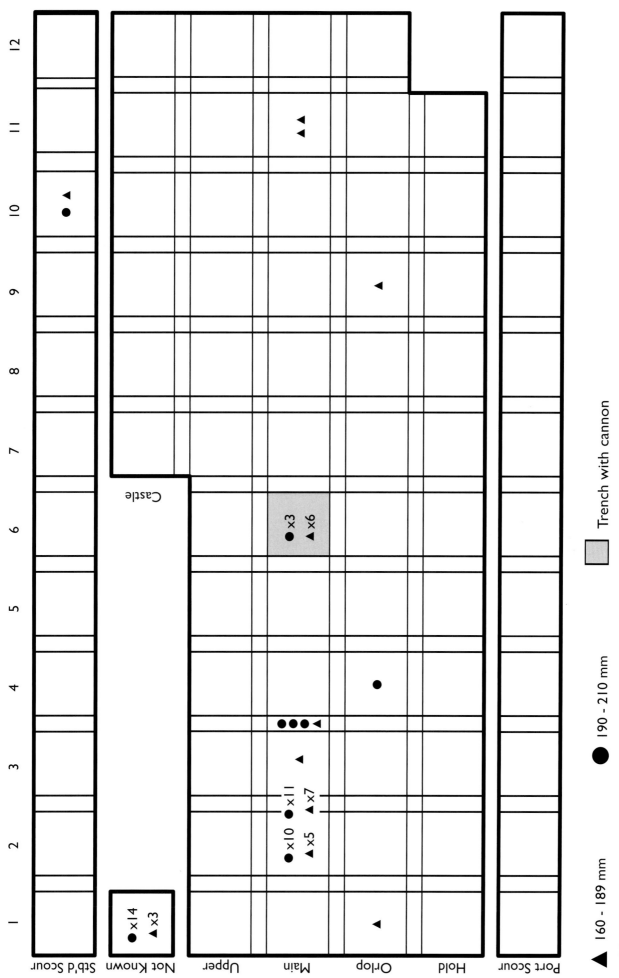

*Figure 5.15 Distribution of cannon shot*

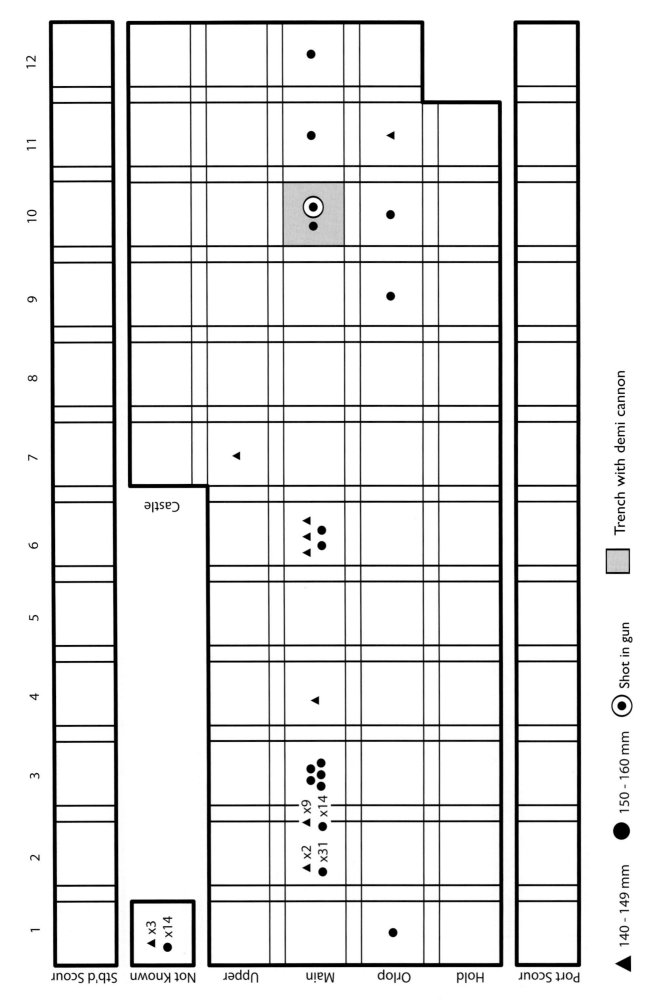

Figure 5.16 Distribution of demi cannon shot

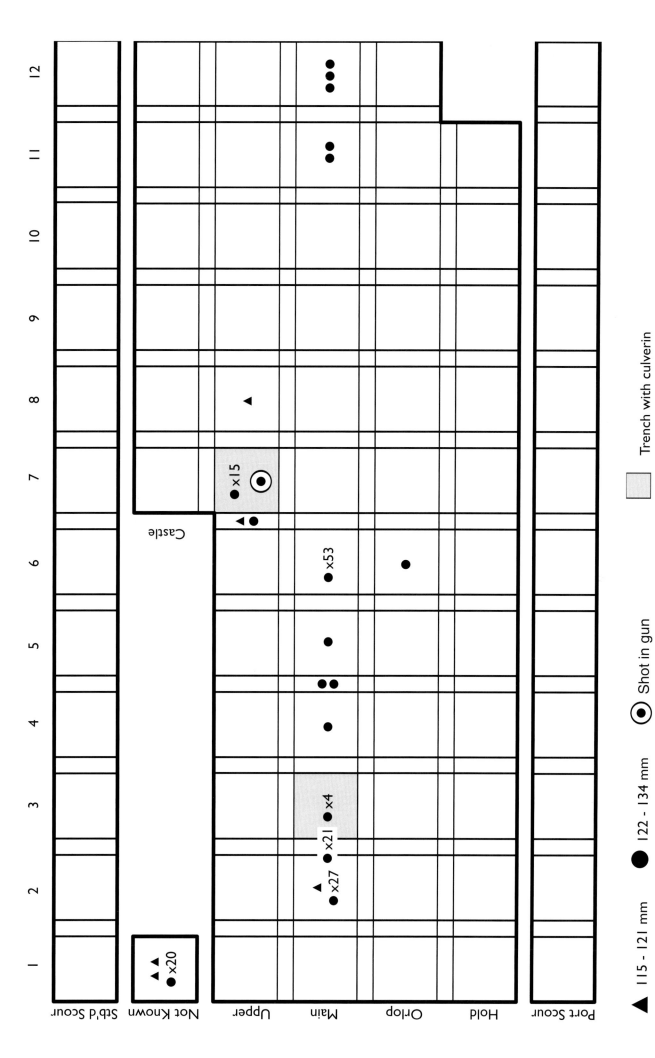

*Figure 5.17 Distribution of culverin shot*

322

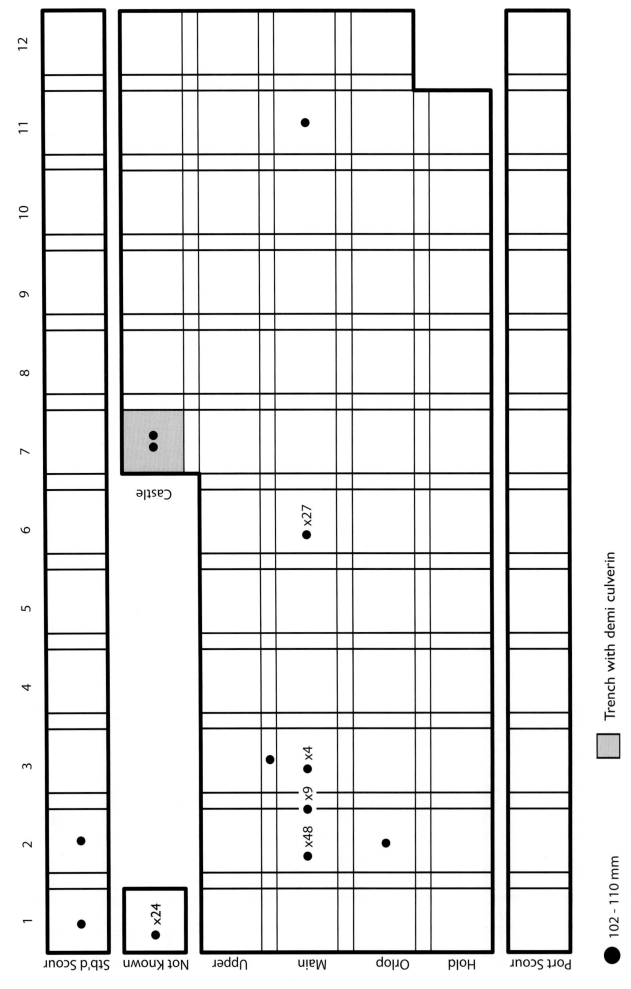

Figure 5.18 Distribution of demi culverin shot

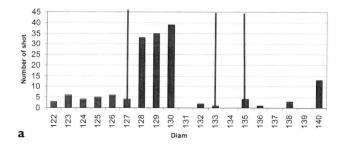

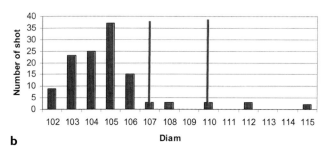

*Figure 5.19 Round iron shot: a) culverin shot; b) demi culverin shot. Vertical lines indicate bores of guns recovered minus 5mm for minimum windage*

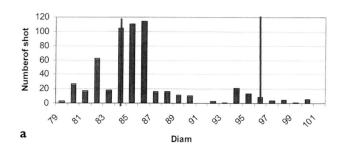

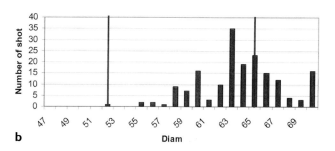

*Figure 5.20 Round iron shot: a) saker shot; b) falcon shot*

The expected number of shot to cover the increased number of guns recovered within the predicted shot range of 122–134mm is 120. In fact, 157 were recovered, the majority between 128mm and 130mm (Fig. 5.19a). Between 121mm and the largest demi culverin there are a further five shot. The range in weight is 2500–8750g (5lb 8oz–19lb 3oz). The expected weight is 18lb (8100g).

Only one gauge was recovered within this range (81A5691) of 138mm diameter. This gauge could be used for the majority of the shot recovered, although only really suitable for the largest gun.

Distribution includes 48 in M2/3 and 53 in M6. Four were found in M3 where two of the guns were positioned. Fifteen were found beside the U7 culverin (80A0976), all 123–140mm, the bore of this gun is 138mm. The other interesting outlier is the group of five found in M11–12 (Fig. 5.17).

*Demi culverin shot*

The *Anthony Roll* lists six demi culverins with 140 shot, just over 23 per gun. Two were recovered, one with a bore of 115mm (79A1279) and the other 112mm (79A1232). The predicted shot range is 102–110mm and there are 118 within this band (Fig. 5.19b). These range in weight from 1150g (2lb 7½oz) to 4500g (9lb 13oz) with an expected weight of 9lb (4050g). Analysis of the shot overall suggests that if any further demi culverins were on board, they are likely to have been smaller rather than larger than those recovered.

One gauge was recovered which suits the shot recovered (81A1285) of 110mm diameter. Four gauges of 100mm diameter were found but fall outside the predicted shot range.

The distribution shows concentrations in both M2/3 (61) and M6 (27), the two principal shot lockers (Fig. 5.18). There are two in C7, where the *in situ* gun was found (79A1232). Single finds in M11, SS1 and SS2 may suggest further guns in either or both areas, but the evidence is slight. A single example in O2 may originally have been in M2 (part of the carriage for the M3 gun was found in O3).

*Saker shot*

Two sakers with 80 shot are listed in the *Anthony Roll* but no sakers were found, so contemporary documents were used to provide a predicted range in shot size of 79–96mm. There are 558 shot in this range (Fig. 5.20a). This 'extra' 478 shot can be explained in a number of ways. First there may have been more sakers on the ship than listed. Other large ships, the *Henry Grace à Dieu* and the *Peter* were each armed with four sakers (Knighton and Loades 2000, 41–2). The second explanation is that these shot were used by another type of gun of similar size to the sakers. The best candidates for this are the wrought iron slings (see Chapter 3 for a fuller discussion). The third possibility is a mixture of both. Unfortunately we have little dimensional information regarding slings and it has always been a problem to identify them in archaeological assemblages. Modern scholars have tended to regard this class of gun as being small and relatively insignificant. However, the one fact that we can see from the contemporary documents is that they fired iron or, occasionally, composite shot – not stone. A single wrought iron gun, found by Alex McKee on the Mary Rose site in 1970

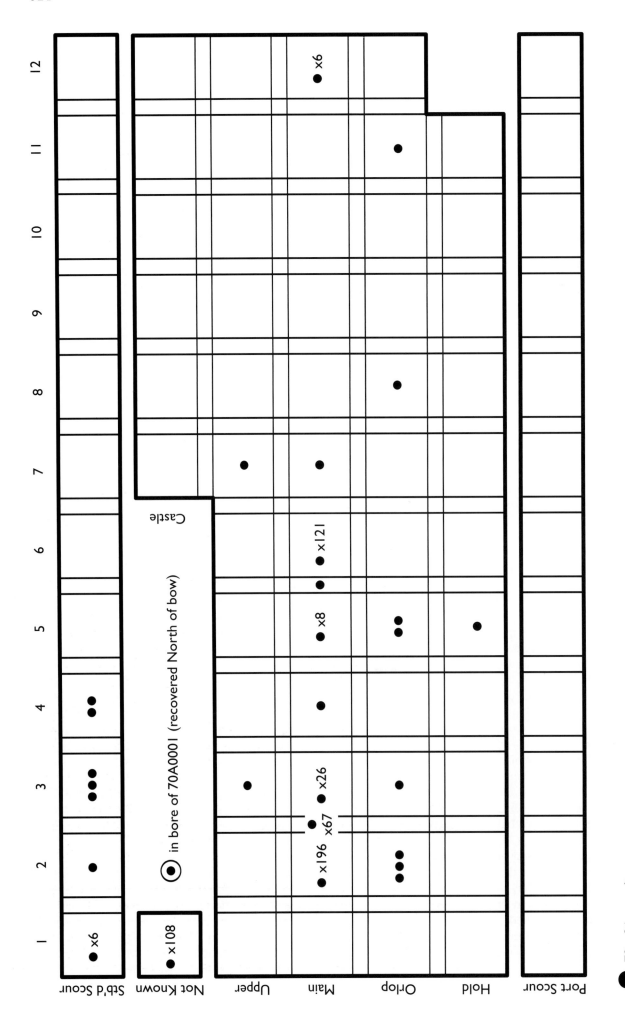

Figure 5.21 *Distribution of saker shot*

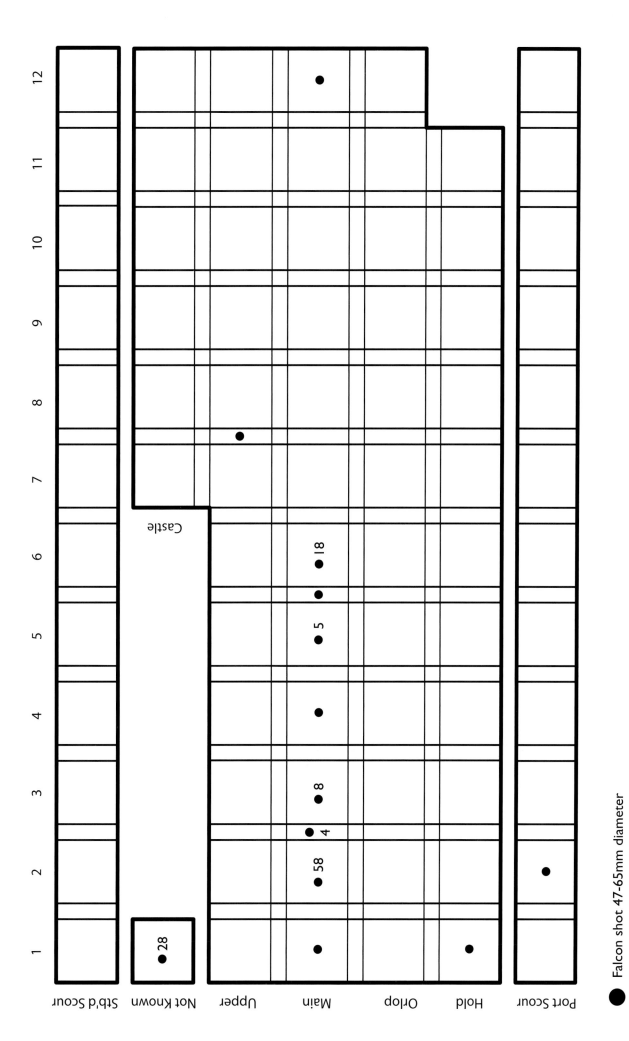

*Figure 5.22 Distribution of falcon shot*

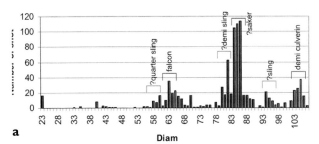

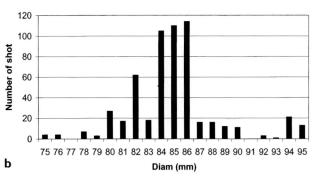

*Figure 5.23 Round iron shot: a) shot less than 110mm; b) possible sling shot*

(70A0001), was loaded with an iron shot and led to this gun being designated as a sling. It has a bore of 95mm (3¾in), which gives a predicted shot maximum of 90mm, within the predicted shot size for sakers. The gun is also very long in relation to its bore and, although incomplete, has a length in calibres of over 24 – putting it in the culverin class of gun. The *Anthony Roll* lists two slings, three demi slings and one quarter sling. The shot listed for these three types totalled 130. This still does not account for all of the surplus, but it is likely that a proportion of this excess was used for the slings (see below).

The weight of shot within this range varied from 401g (14.4oz) to 3628g (7lb 1oz) with an expected weight of 2475g (5lb 7oz). The evidence suggests that any sakers carried would have been of the smaller calibre: 89mm (3½in). Any shot smaller than the limit of 96mm fit through the three gauges of 100mm (81A1228, 81A5662, 83A0019). The cluster of 43 shot between 95mm and 97mm fit this size of gauge well. The bulk of the shot is 80–90mm (508 examples) and these are an extremely loose fit.

The largest concentration is 289 shot in M2/3, followed by 121 in M6. Yet again there is a small outlier in M12 and an interesting scatter between SS1 and SS4. This may reflect the positions of slings or sakers (Fig. 5.21).

*Falcon shot*
Only one falcon with 60 shot is listed in the *Anthony Roll* and none was recovered, so a predicted shot size range of 47–65mm was allocated, based on historical records. There are 128 shot that may have fitted this type of gun (Fig. 5.20b). Weights ranged from 500g (1lb 1oz) to 1133g (2lb 7oz) with an expected shot weight of 900g (2lb 4oz). The graph also suggests that the falcon may have been of the larger size – with a bore of 70mm (2¾in). There are no gauges which can sensibly size these shot. There is a scatter of shot on the Main deck from M1–6, with 70 in M2/3 and eighteen in M6 and only isolated examples elsewhere (Fig. 5.22).

*Slings*
The *Anthony Roll* lists two slings, three demi slings and one quarter sling with 130 iron shot. As discussed in Chapter 3, very little is known about slings, save that, at this period they were wrought iron breech-loading guns, had long chambers and were long in relation to their bore. They fired metal shot, usually iron. In terms of powder, the listed charge is greater in weight than the projectile, which is extremely rare (Norton 1628, 58; see Chapter 3).

Four long, slim guns were recovered from the *Mary Rose* with bores of 60–95mm. Two of these were loaded with iron shot (81A0645, 70A0001). They are all similar and have been identified as slings. Comparison with the culverin class of guns (also long in relation to bore) suggests that the demi sling fired a shot of half the weight and the quarter sling a quarter the weight of the sling. Bore size is never provided, although the diameter of a sling chamber is given as 2½in (63mm; Caruana 1992, 21). In the *Anthony Roll*, in which the guns are apparently arranged in size order from largest to smallest, they are usually listed between the port piece and the fowler. The fowler has a 110–120mm bore while the port pieces are considerably larger. Looking at the complete range recovered there are surprisingly few clusters of larger shot that cannot be ascribed to particular known or recovered guns. However below the demi culverin size of shot, especially around that of the saker, there are a number of peaks and quantities of shot that cannot be easily ascribed nor accounted for by

**Table 5.5 Proposed bore sizes for slings**

| Gun | Bore (mm) | (in) | Shot size (mm) |
|---|---|---|---|
| Sling | 95–103 | 3¾–4 | 92–98 |
| Demi sling | 83–88 | 3¼–3½ | 78–83 |
| Quarter sling | 60–65 | 2¼–2½ | 55–60 |

**Table 5.6 Identification of slings in the *Mary Rose* assemblage (metric)**

| Object | Sector | Length | Bore | Poss. ident. |
|---|---|---|---|---|
| 82A2700 | O3 | 1270 | 60–70 | quarter sling |
| 79A1295 | ?/Deane | 3000 | 85 | demi sling |
| 81A0645* | U6 | 2240 | 90 | demi sling |
| 70A0001 | N of bow | 2295 | 95 | sling |

* iron shot recovered in bore

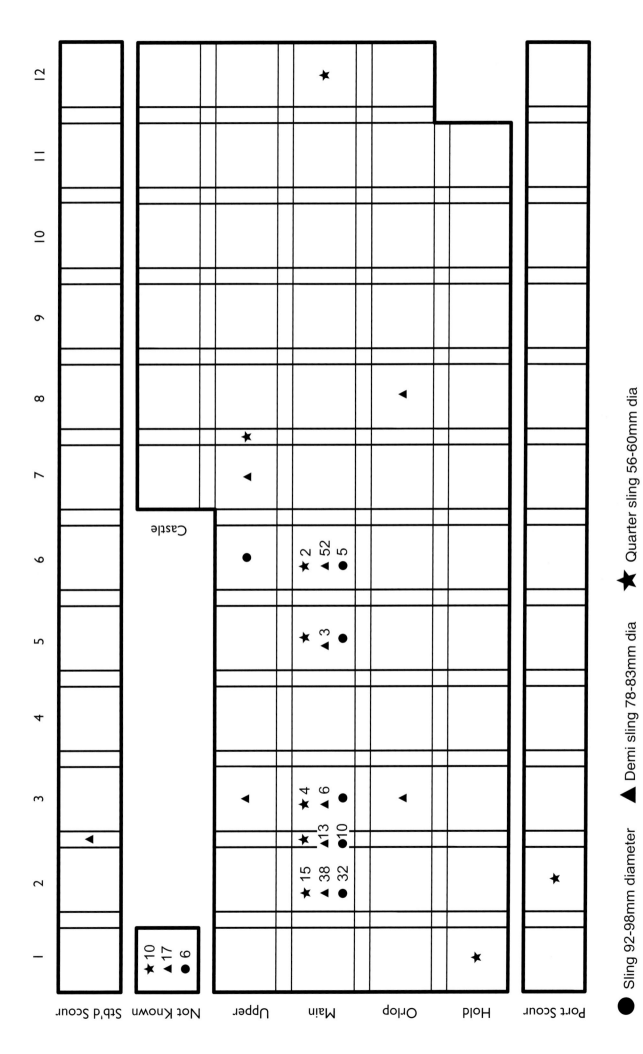

Figure 5.24 Distribution of shot less than 110mm

328

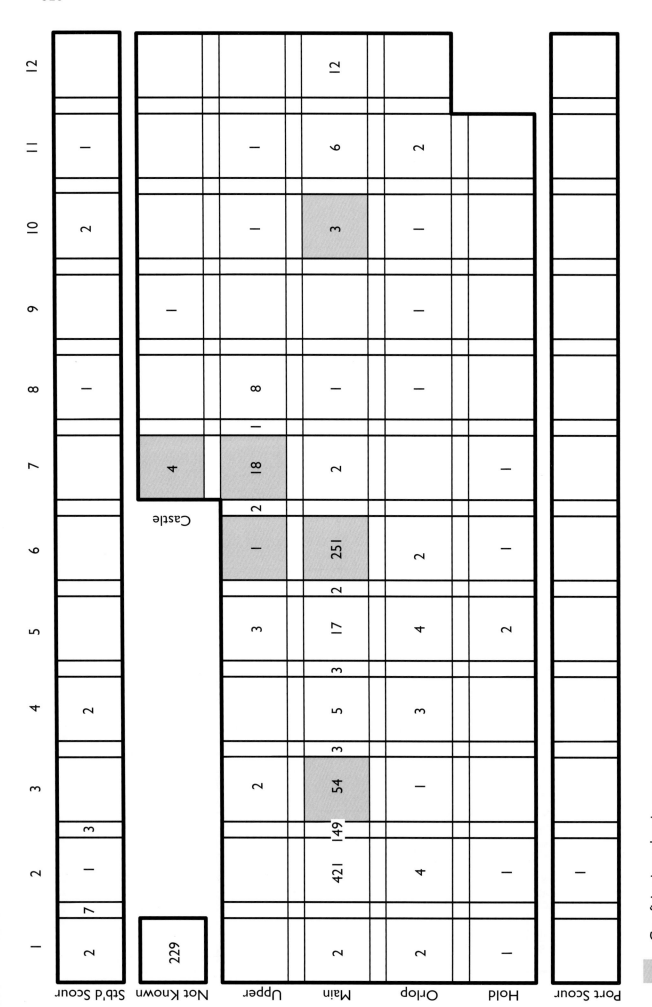

Figure 5.25 Distribution of all iron shot

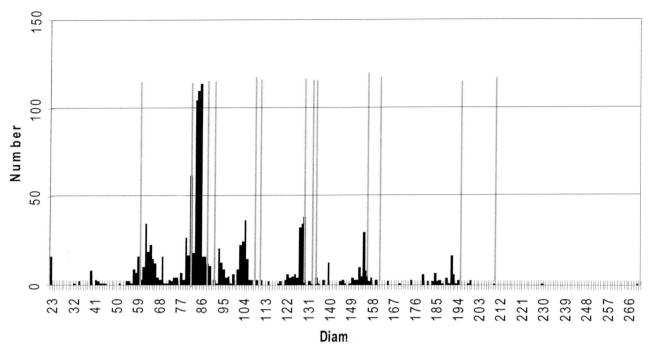

*Figure 5.26 All iron shot. Vertical lines indicate bores of guns recovered minus 5mm for minimum windage*

the known gun assemblage (see saker above). Following a suggestion that the sling may be similar to the demi culverin in size with the demi and quarter sling half of that size, the distribution of shot below the demi culverin was analysed (Fig. 5.23a).

If the peaks for demi culverin (around 105mm), saker (around 84mm) and falcon (around 63mm), are removed then there are a number of unaccounted for concentrations, in particular at 95mm, 82mm and 60mm. These correspond to approximate shot weights of 7lb 7oz, 4lb 10½oz and 2lb. While not in the exact proportions they are in the approximate expected proportions of 1:½:¼ and though not a complete proof, this analysis strongly suggests that these shot were for the slings. If this is the case, then expected bores of the three sizes can be proposed (Tables 5.5–6).

The shot recovered which would fit the sling include 56 found mostly in M2/3 and M6. The 134 shot for the three demi slings have a slightly wider distribution, with several on the Upper deck but most in the shot lockers. The 37 for the quarter sling show a similar distribution, with one in M12 (Fig. 5.24). Slings may also have used composite shot. The distribution of all iron shot of appropriate size is shown in Figure 5.24 and of all iron shot in Figure 5.25. Figure 5.23a proposes guns for shot of 50–105mm and Figure 5.23b shows clusters of shot which may relate to the slings and demi slings. All iron shot together with gun bores minus 5mm for minimum expected windage is shown in Figure 5.26.

*Unattributable iron shot*
The analysis above accounts for the greater proportion of the shot recovered but there are some shot which do not fit. First are those which are too small to fit any of the listed guns, in particular a cluster of sixteen shot of 23mm diameter and eight of 40mm. These may well have been the iron cores of the composite shot for which the outer covering of lead has disappeared or not been cast. The smaller size may also have been used in the handguns or as part of other, scatter type, composite shot. There are also a few shot of larger sizes that do not seem to be suitable for the guns listed but, as we have seen, the windage was not seen as critical and these were probably used as and when necessary. It is also worth noting that there is a strong possibility that shot of odd sizes were left on the ship and accumulated over its long service life.

*General distribution of iron shot*
Most of the 1248 iron shot were found on the Main deck (931) and only 37 on the Upper deck and nineteen in the Starboard scour. The M2 shot locker contained most shot, with 421 examples and most sizes represented (see Fig. 5.7a) There were also 149 in M2/3 and 54 in M3, possibly also from the locker. M6 contained 251 shot, again of mixed sizes (Fig. 5.7b). Much smaller quantities were found on the other decks, as can be seen from Figure 5.25 and there are a large number (229) from unknown locations. There appeared to be no rational manner of storage or evidence of containerisation: the shot had rolled towards the starboard side and become concreted between the Main deck and the side, contained only by large knees to the north and south. In M6, the shot was between and around the cannon, underneath the bed and around the front wheels and axle.

There are a few areas where either single spot finds or groups of shot were found. Some of these can be

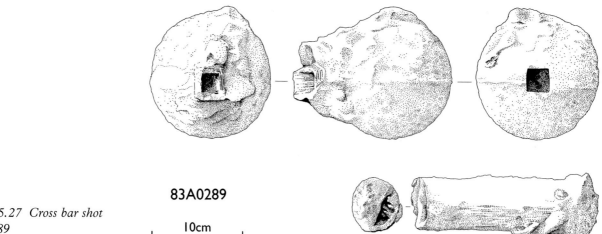

*Figure 5.27 Cross bar shot 83A0289*

explained as deriving from the Main deck storage areas or collapsed Castle/Upper deck/portside structure, or movement as a result of the sinking. A few concentrations may be more significant, including 21 shot from the Main deck in the stern (M10–12). The three in M10 were all large (135mm, 152mm, 165mm) and two may have served the two demi cannons from this area (bores 165mm, 160mm.) A further six were found in M11, three were very large, with two at 175mm. M12 contained twelve shot, of a variety of sizes, including some of the smallest recovered (56mm) with six which would have served a saker or one of the slings. In this position, possibly on an upper deck, a saker would be a useful chase gun. Three shot were found on the castle deck (C1), of 100mm, 104mm and 105mm; these could have provisioned the demi culverin (79A1232) with a bore of 112mm. These and a fourth (size unrecorded) were the only shot in this sector. These groups, together with that from U7 (with shot specific to culverin 80A0976) provide the only clue to gun identification through shot distribution inside the ship.

There are nineteen shot recorded as being in the Starboard scour. A group in the bow (sectors 1–4 with a concentration in 1–2) included twelve shot of 79–104mm which could provision either a saker, demi culverin or sling. This shot is unlikely to have derived from the shot lockers and should be considered as evidence of guns from the Forecastle. Sector SS4 also has two isolated shot of saker size. There is one small shot of 67mm in SS8. Shot of this size was found in U8 and 7/8. Two large shot were found in SS10 (175mm and 182mm). These cannot easily be explained, but it is interesting that shot of a similar size was found in M10. These are too large for the demi cannons and well away from the cannons in M6.

In all but one instance M2/3 had the majority of shot for any given shot/gun band. The shot found with the U7 culverin and the rest of this size show that the majority of culverin shot was stored in M6, or by the one gun. This U7 group suggests that this is one area which may have been provisioned during the battle. A concentration of stone shot was found in M8 and the *in situ* port piece (81A3001) was possibly being reloaded at the time the vessel sank – there was shot in the barrel of the gun but the powder chamber was beside it. The shot found in trenches with guns firing iron shot suggests that many have the same type of concentration as U7 and M8. In M3 there are 54 iron shot, but there are only five which may relate to the *in situ* culverin (81A1423, bore 138mm), these range between 128mm and 135mm. This is not because of a general lack of shot of these sizes – there are 107 of 128–130mm – but they were not in this area. Even within M6 there are few shot appropriate to the *in situ* cannon (81A3003) with a bore of 200mm and expected shot size of 195mm. There are two shot at 192mm and one at 190mm. Even allowing for increased windage, there are still only five shot between 180mm and 189mm. The majority of shot for the two cannons was in M2 or M2/3, with 20 shot over 190mm.

One group of shot was excavated ashore from the remains of a concretion found in storage on the seabed after the main lift (raised in 1986). This included fragments of wood, concretion and at least 11 shot (81A6897–6920), all in the 80–90mm range. Its original location is unknown, but it is similar to the concretions found in both M2 and M6.

The huge variation in weights for any size of shot demonstrates the effects of corrosion (graphitisation). The upper limits come close to the listed weights for particular sizes of shot. It is reasonable to assume that originally there was consistency in casting weights.

**Cross bar shot**

There are no cross bar shot listed for the *Mary Rose* in the *Anthony Roll*, however 26 have been identified (Fig. 5.27). Although the specific name '*cross bar shot*' does not appear in inventories for the Tower until 1559, others suggest that it was in use before this. As a

132 cross bar shot and lists the *Henry Grace à Dieu* with 100, the *Matthew* with 20 and the *Sweepstake* with 12. Of the fifteen 'galleasses', only the *Hart*, the *Antelope* and the *Swallow* carry cross bar shot, a total of 44. There are none listed for the pinnaces and rowbarges.

The extensive inventory of 1547 (Starkey 1998) shows the bias towards the use of cross bar shot on ships. There are none listed for the Tower and Gravesend is the only fortification for which they are listed (four shot). The ships which carry cross bar shot include the *Henry Grace à Dieu* (ibid., 7242), *Peter* (7273), *Swallow* (7509), *Antelope* (7576), *Hart* (7601), *Mary (of) Hamburg* (7624) *Struse of Danzig* (7667) *Jennet* (7708), *Greyhound* (7803), *George* (7905), *Falcon* (7935) and the *Saker* (7948). The guns carried by the *Saker* include only sakers, fowlers and bases. It must be assumed that the sakers used the cross bar shot in this instance. This shot is always at the top of the list or at the bottom. This probably reflects its 'form' rather than size. Other shot are listed in descending order of size.

Twelve of the seventeen vessels in the 1555 inventory are listed with 'crosse barred shot'. They include the *Seven Stars*, *Phoenix*, *George*, *Jerfalcon*, *Jennet*, *Greyhound*, *Bull*, *Tiger*, *Swallow*, *Trinity Henry*, *Salamander* and *Antelope*.

By 1559 cross bar shot appears in the Tower inventories, listed for demi cannon, culverins, demi culverins, sakers, minions and falcons (Blackmore 1976, 263). The amounts have increased by 1568 (ibid., 262) with 170 for demi cannon. The contrast with 4716 solid iron shot is still marked. Yet again these are exclusively for the bronze guns of all sizes except cannon. By 1589 cannon are included (ibid., 266), with 120 cross bar shot and 13,679 solid cast iron shot. Cross bar shot of cannon size was supplied to the *Mary Rose* 44 years earlier.

All have a square hole penetrating the entire diameter of the shot, in most cases bisecting the castline. The holes vary in size from 20mm to 47mm. The majority show clean holes (Fig. 5.28a), however, those excavated from within larger iron concretions had remains of what appeared to be wrought iron bars or spikes set within a cast iron shot. Although most did not

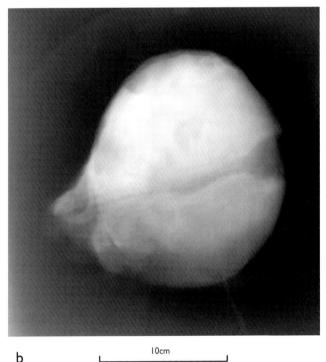

*Figure 5.28 a) Cross bar shot with clean hole; b) radiograph of 83A0289 showing structure*

projectile primarily for use at sea, it is not surprising that it is rarely found within fortifications. Of the thirteen ships listed in the 1514 inventory (Knighton and Loades 2000, 113–58) four have '*shot of iron with pykes*': the *Trinity Sovereign*, the *Kateryn Forteleza*, the *Great Nicholas*, and the *John Baptist*. Entries usually follow (although not necessarily immediately) '*shott of yrone of diverse sortes*'. The *Mary Rose* has 457 '*shot of yron of diverse sortes*' followed immediately by 120 '*shott of yron with crosse barres*' but no listings for '*shot of iron with pykes*'. The Smith description (1627, 86) and use of the specific term binding the '*pikes*' at each end of the cross bar shot suggest that these two names are probably interchangeable.

The *Anthony Roll* lists 176 cross bar shot (Knighton and Loades 2000). The inventory of the 20 ships totals

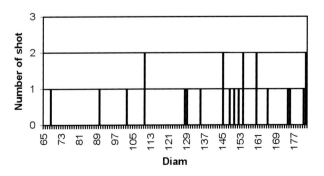

*Figure 5.29 Cross bar shot*

## Table 5.7 Cross bar shot

| Object | Sector | Diam. (mm) | Length bar (mm) | Diam. bar (mm) | Weight (kg) | Description |
|---|---|---|---|---|---|---|
| 82A4156 | ? | 68 | 20 | 25–20 | ? | radiographic study; sphere of corrosion 20mm around shot; central bar 20mm in centre, flaring to 25mm as exits each side |
| 82A4176 | ? | 90 | n/a | n/a | 2.27 | square hole bisects cast line centrally |
| 81A3106 | M11 | 102 | n/a | 20 | 2.27 | square hole bisects cast line centrally |
| 83A0703 | M12 | 110 | 100 | 18 | ? | square hole bisects cast line centrally; radiographic study only, v. badly corroded; ordinary shot (83A0702) also on radiograph |
| 83A0342 | M12 | 110 | 125 | 20 | 1.5 | square hole bisects cast line below centre; centre shot to end spike 260mm |
| 83A0289 | M12 | 128 | 32 | 20 | 2.5 | study from good radiograph; bar appears to be on well-finished cast line |
| 80A2017 | ? | 129 | n/a | 28 | 1.5 | square hole bisects cast line centrally |
| 81A0530 | O1 | 135 | – | – | – | no data |
| 83A0704 | M12 | 145 | 125 | 25 | 3.6 | excavated; spike bound with twine; square hole bisects cast line centrally |
| 79A0508 | ? | 145 | n/a | 25 | 1.8 | square hole bisects cast line centrally |
| 82A2684 | M2/3 | 148 | n/a | 26 | 3.4 | square hole bisects cast line centrally |
| 83A0559 | M12 | 150 | n/a | ? | 2.4 | square hole bisects cast line centrally |
| 82A5175 | ? | 152 | n/a | 20–25 | 3.0 | square hole bisects cast line centrally; good radiograph; poss. second shot, diam. 135mm & bar in concretion, 105mm |
| 89A0040* | ?PS1 | 154 | 10 | 25 | ? | square hole bisects cast line centrally |
| 89A0041* | ?PS1 | 154 | n/a | 24 | 3.5 | square hole bisects cast line centrally |
| ???? | M2/3 | 160 | 30 | ? | 0.5 | |
| 81A3349 | M10 | 160 | – | – | – | no data |
| 82A4406 | ? | 165 | – | – | – | no data |
| 89A0042* | ?PS1 | 172 | 300 | 24 | 8.0 | square hole bisects cast line centrally |
| 81A3820 | M11 | 175 | n/a | 47 | – | square hole bisects cast line centrally |
| 81A2998 | M10 | 181 | n/a | 25.5 | 5.5 | square hole bisects cast line centrally |
| 81A6716 | M2/3 | 182 | n/a | 29 | 5.4 | square hole bisects cast line centrally |
| 82A2686 | M2/3 | 182 | n/a | 30 | 5.0 | square hole bisects cast line centrally, bar associated |

*found together in concretion, 89A0008 containing a single spike 330 mm in length & 35mm diameter, tapering to 5–7mm

survive, radiographic evidence reveals some idea of the structure (Fig. 5.28b).

Diameters range between 68mm and 182mm (Fig. 5.29, Table 5.7). All but six (unlocated) were found either in O1 (2), M2/3 (4), M10 (3), M11 (2) or M12 (6) (Fig. 5.30). Three (89A0040–2), with diameters of 154mm and 172mm, were found just off the port side (PS1) in 1989 in a large concretion that may have originated in the storage area in O1. Of the two examples found in O1 the diameter of one is unrecorded and the second in 135mm. The group found in M2 were of 148mm, 160mm, and two of 182mm. Those from M10–12 have the greatest range in size (102–181mm). All examples from M12 were recovered from concretion on the deck that was removed following recovery of the ship and these provided the most complete examples. The concretion (83A0342) clearly contained two 'iron' shot (Fig. 5.31a). Radiography revealed two cross bar shot (83A0342, 83A0704), and one 90mm solid lead jointed shot (Fig. 5.31b). Both ends of the spikes on 83A0342 (diameter 110mm) are discernible, with a length from shot centre to tip of 260mm. 83A0704 has a diameter of 145mm and contains a 25mm bar of maximum length 125mm. The base of the bar where it joined the shot was clearly wrapped with twine (Fig. 5.32a). This was visible for 95mm along the spike, with traces at the other end. Wrapping or 'arming' of shot was to prevent it jamming in the bore. An associated concretion (83A0343) included the remains of a 200mm length of 25–28mm bar, pointed at one end and clearly armed with a ball of organic material with a diameter of 40mm (Fig. 5.32b). A portion of 30mm bar (82A2687) from M2/3 was also recovered; this may be part of associated shot 82A2686. The spike did not survive in most cases. In all examples evidence came from concretions which

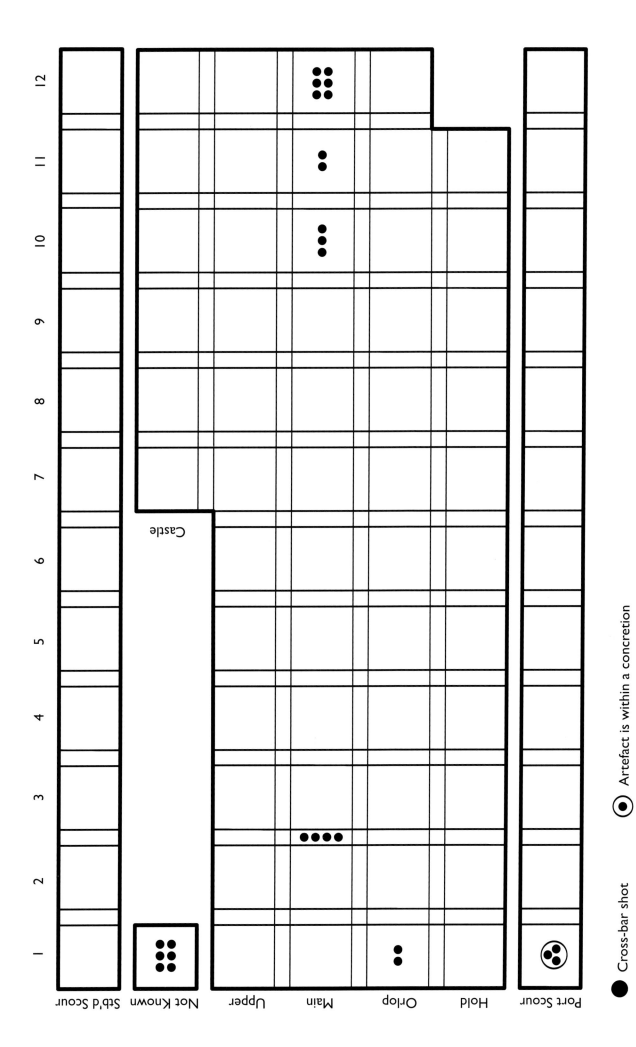

*Figure 5.30 Distribution of cross bar shot*

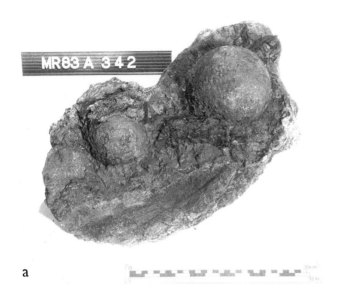

a

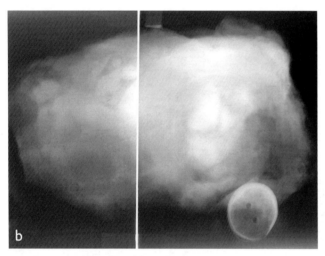

*Figure 5.31 Cross bar shot: a) 83A0342; b) radiograph*

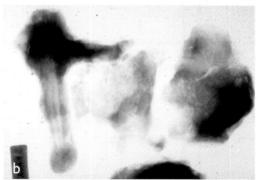

*Figure 5.32 a) Spike shot 83A0704; b) radiograph of armed shot 83A0343; c) woody nature of spike*

were lifted, radiographed and carefully worked to remove the concretion. The radiographs confirmed by excavation (Fig. 5.32c) clearly show that the spike is wrought iron, the visibly 'woody' grain contrasting with the denser, whiter cast iron of the shot. In all but one instance (83A0342) the castline is bisected and is central to the square spike. If the spike is placed in the mould before the iron is poured, the mould must be aligned horizontally, as suggested by McElvogue (1999, 11) for the Alderney wreck (*c*. 1590). Smith's description (1627, 86) specifically states that cross bar shot:

> *'hath a long spike of Iron cast with it as if it did goe thorow the middest of it, the ends whereof are commonly armed for feare of bursting the Peece; which is to binde a little Okum in a little Canvasse at the end of each Pike'.*

.The holes vary between 20mm (82A5175) and 47mm (81A3820). The longest portion of spike is 330mm with a maximum width of 35mm narrowing to a point of 5–7mm (89A0008, Figs 5.33–4). As there are no examples with a complete spike it is not certain whether these were pointed at each end, as in the Alderney examples (McElvogue 1999, 11) or blunt at one end as in the sixteenth century Texel examples (Puype 2001, 121–2). It would be interesting to see whether a metallographic study of the cross bar shot could suggest whether they were cast in a metal or sand bipartite mould. As the hole does not change in size, and as the spikes seem to share the same profile as they exit each side of the shot, it is suggested that *Mary Rose* examples are pointed at each end (Fig. 5.35). Currently many of the shot are inaccessible for more detailed study.

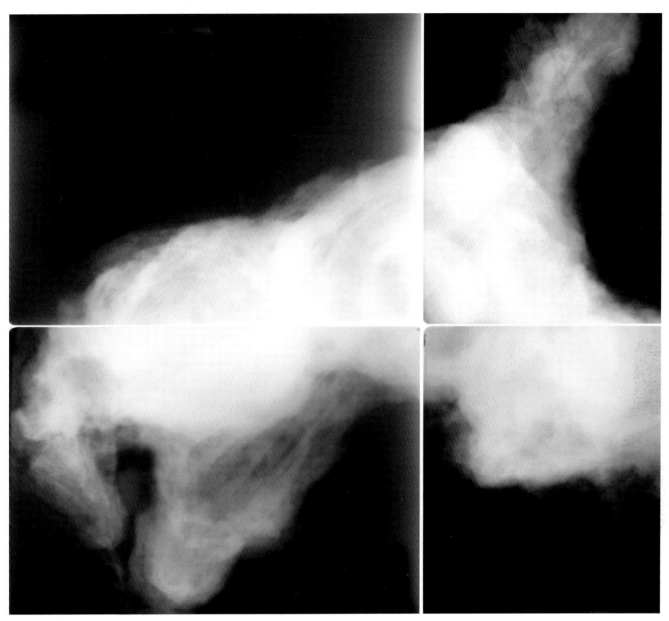

*Figure 5.33 Cross bar shot: composite radiograph of 89A0008 concretion*

Shot which outwardly looks like the shot from the *Mary Rose* is illustrated in Blackmore (1976, 192, 193, 197, 201) where it is listed as incendiary shot. It is described as:

> '*a solid iron ball from which issues two long quadrangular spikes. These shot were wrapped with incendiary material, the spikes being intended to stick fast in a ship's woodwork*'.

The examples illustrated (*ibid*., 329) are listed as being of 533mm and 305mm, 3500g and 3000g. Blackmore goes on to say that these are described as spike or star shot in Skinner's *History of London* of 1795. This does not seem to have been identified as cross bar shot, and the glossary entry gives Smith's description (1627, see above). Peterson (1969, 28) provides an illustration similar to the *Mary Rose* example, although the spike is drawn as though cast with the shot. It is merely described as '*incendiary spike shot*'.

Shot found on the *Scheurrak SO1* wreck near Texel, Netherlands, dated 1593, include ordinary round cast iron shot and what has been termed '*spike*' shot (Puype 2001, 121). These have one blunt and one pointed end: a version of cross bar shot. Puype suggests these were made in a bipartite mould but with a channel half the width of the bar left out of each hemisphere. The square section bar would be hammered into place once the shot had cooled. Puype suggests a consistently tapering spike, slightly larger than the square section to accommodate it. This process is possible as the wrought iron is far less brittle that the cast iron and so could be driven without damage. Those examples from the *Scheurrak* are of 56mm (1lb), 61–2mm (2lb), 71mm

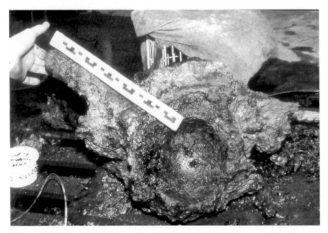

*Figure 5.34 Cross bar shot 89A0008 during excavation*

and 73mm (3lb), and 75mm and 77mm (4lb) and there is a fragment corresponding to a shot diameter of 90mm (6lb). The longest recovered spike is 283mm. In one case the shot as well as the spike is wrought.

*Relating the shot to the guns*
Although comparatively rare, inventories that do list these shot suggest they were used with cast bronze guns. Some include the weights for shot without the bar, which may suggest shot cast separately from the bar which was inserted later. The weight for a demi cannon is 28lb (12.6kg), for a culverin 16lb (7.2kg), a demi culverin 7lb (3.15kg) and a saker 4lb (1.8kg) (Brown, pers. comm. 2002). This is at variance with the *Mary Rose* examples, fifteen of which are for larger guns.

The smallest shot, of 68mm (82A4156) would fit a falcon and the 90mm shot (82A4176) would fit a saker. The locations of these are unknown. The shot of 102mm found in M11 and the two of 110mm found in M12 would fit a demi culverin, the one gun of this type was found in C1. However, it is possible that there was another pair of demi culverins on the Upper deck in the stern (see Chapter 2). The shot of 128mm in M12 is too large for a demi culverin but would fit a culverin. These were situated in U7 and M3. The shot of 145mm found in M12 would fit a demi cannon, although with substantial windage. The closest is 79A1277, of 160mm, in M10 port. The shot of 150mm and 152mm would also fit this gun while that of 160mm in M10 could possibly have been used with the *in situ* demi cannon, 81A3000 (bore 165mm).

Shot found in the Port scourpit at the bow includes two of 154mm and one of 174 mm. The smaller would fit the demi cannon recovered from M10, but the larger would only fit a cannon. The windage is very great: the cannon bores are 200mm and 215mm and both are situated in M6. This is also the case for the largest shot of 200mm found in the Port scourpit. It seems likely that the smaller shot are derived from the store in M2. The two of 182 mm were also found in M2/3.

Two further examples found on the Main deck stern are of 175mm and 181 mm. These would also only fit the cannons in M6.

The distribution suggests storage in O1 (with other gun furniture), M2/3 and also another area in M10–12. The area on the Main deck in the stern is very interesting, supporting an unusual mixture of sizes and types of shot, guns, fireworks and loading implements. This area also contained a range in size of solid iron shot.

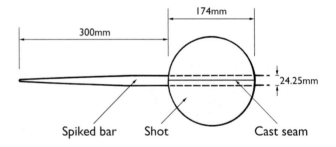

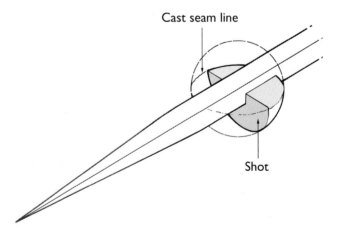

*Figure 5.35 Reconstruction of spiked shot 89A0042 from 89A0009 and diagram to show method of construction*

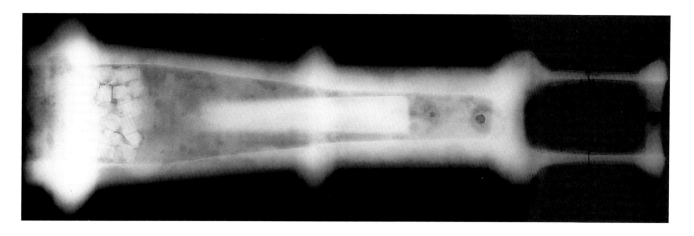

*Figure 5.36 Hailshot: above) radiograph of hailshot piece 79A1088 showing shot in position; below) shot from 79A1088*

## Hailshot

Hailshot was identified as the dice-shaped pieces of iron found within the bores of four small muzzle-loading guns with rectangular bores (for locations see Chapter 4). This led to the identification of these unusual guns as the '*hailshot pieces*' listed as iron guns carried on board most vessels including the *Mary Rose* (Fig. 5.36a). The Anthony Roll does not give the number of dice for the 20 hailshot pieces (Knighton and Loades 2000, 43), but 1000 are listed for the 20 carried on the *Peter*, and 4000 for the 40 hailshot pieces carried on the *Henry Grace à Dieu*, listed as '*dyce of yron for hayleshotte*'. This suggests about 50 dice per gun. The hailshot pieces recovered were loaded with 20–30 iron dice, suggesting two and a half rounds for the *Peter* and 66 dice per gun for those on the *Henry*, giving about three rounds. The 1547 inventory (Starkey 1998, 147–57) shows that hailshot pieces were carried, but no shot is specified. Hailshot pieces are rarely listed within land fortifications, and the only dice listed for them seems to be associated with the casting of dice with lead for other types of shot (*ibid.*, 3942, 3943): '*Dice of Irone*', '*Lead to make shott with alle*'. The small number of dice (100) and the listing of lead in weight (4cwt) strengthens this view.

Blackmore (1976, 243) cites an early usage of the term '*hailshot*' in a 1474 list of ordnance accessories which includes '*Pellettes de ferro pro serpent.*', '*Hailshott de ferro*', '*Gonystonys de ferro*', and '*Lapides gunnorum pro gross bombardes ... pro bumbardelles*'. The Anthony Roll lists '*dice of iron for hailshot*' under the category of iron shot, but after the smallest round shot. In the 1589 Tower inventory dice shot is distinguished from '*Burres alias Haileshott*' (Blackmore 1976, 267), and a solitary hailshot piece is listed, though there are also 149 '*ffloukemouth pieces*' (*ibid.*, 266, 267) which are elsewhere said to be hailshot pieces (see above).

It is suggested that the dice shot may refer to shot inset with an iron dice, and the hailshot to small dice or flakes of iron which are to be used in vast quantities rather than as the centre of another shot. Hailshot continues to be listed, and the 1603 inventory (*ibid.*, 280) includes cases filled with square shot for the muzzle-loaders as well as:

'*Hayleshot viz.
Burres and Diceshott* (weight in tons)
*Base shott* (weight in tons)'

The entry for dice shot is right beneath that for '*shot covered with lead for falcons*'.

By 1620 (*ibid.*, 286–7) the Tower has '*Square shott of Iron In 148 barrells*' and '*Loose without barrells*' in addition to '*Base & Burr*'. We then see entries for cases of wood for loose shot of different kinds, but no specific mention of hailshot.

The small iron dice found inside the hailshot pieces (Fig. 5.36) vary from 10mm to 16mm. A representative sample from 79A1088 gave an average size of 10mm. Modern dice of this size were manufactured and gave a weight of 200g for 20, important for working out the powder charge for test firing. One would assume that, like lead round shot, these would be supplied in casks, or in cases as suggested by Brown (pers. comm. 2002) for hailshot in 1622. It is possibly that, by this stage, the term was less specific than in the sixteenth century.

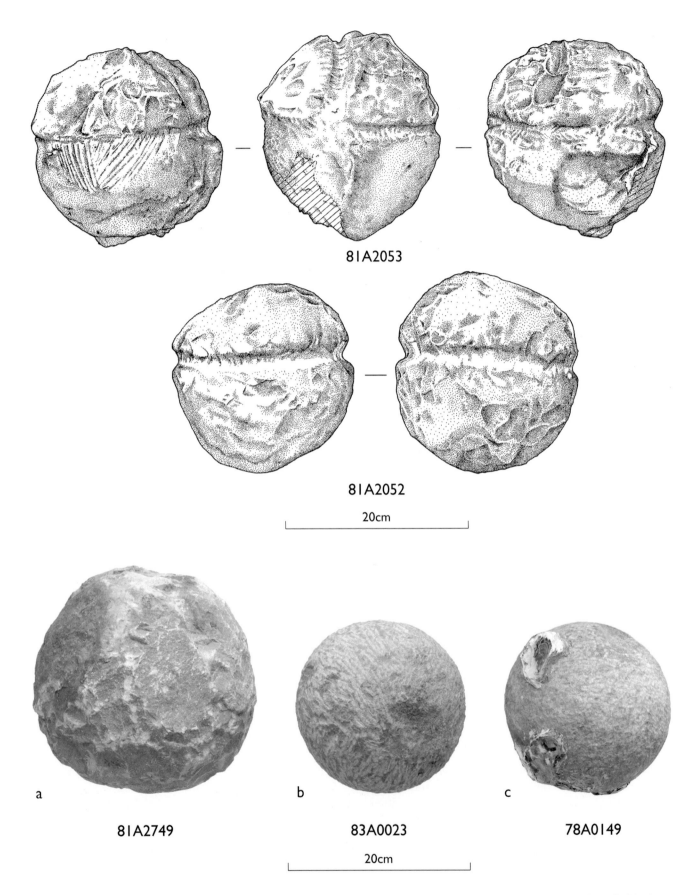

*Figure 5.37 Stone shot: above) marking out lines for unfinished shot; below: a) unfinished; b) semi-finished; c) finished*

## Shot of stone

In total, 387 stone shot were recovered, close to the 390 listed in the *Anthony Roll*, exclusively listed for use with the wrought iron guns. All are handmade, the majority hewn from Kentish Ragstone, a hard limestone (see below) (Fig. 5.37). The stone shot ranges in size from 75mm (3in) to 267mm (10½in) with weights varying between 650g (1lb 7oz) and 24.5kg (54lb) (Fig. 5.38). The total weight of the assemblage is 3385kg (3.4 tons).

It was obvious from the outset that the shot reflected different stages in the manufacturing process, with roughly hewn, semi-finished and finished shot all represented (Fig. 5.37). A high proportion of these (178), are unfinished (Fig. 5.38) and 138 of these are over 198mm. The largest wrought iron gun recovered in 1979–82 has a bore of 195mm. The port piece in the RAI Woolwich has an estimated bore of 200mm. Windage for the recently recovered examples is a minimum of 8mm. The fact that so many are unfinished makes matching shot to named guns difficult. It also means that predicting guns listed but absent from the assemblage is not as straightforward as with the cast iron shot.

The *Anthony Roll* lists sixteen shot for each of the twelve port pieces, 28 for each of the six fowlers, and ten for each of the two top pieces. Where there is shot directly associated with guns, there seem to be 4–13 finished shot for each. This is most clearly demonstrated in M5 (21 finished shot) and M8/O8 (24). In both cases the size range suggests port pieces of slightly differing bore on either side of the deck.

The largest single number of any size is fifteen shot at 170mm (Fig. 5.38). As these are handmade, millimetric accuracy would probably be difficult to achieve and where a group of shot clearly represent projectiles for one particular gun there is a variation of up to 20mm (the shot for the two M8 port pieces ranges between 150mm and 188mm). The weight of shot for a given diameter varies, for example shot of 170mm has a weight variation of 6226–74000g.

Of the total, 155 are too large for the largest gun recovered, estimated at 200mm bore diameter (Woolwich) using an estimated shot diameter of no more than 194mm. Most of these are in M6 (102), M5 (23), M4/5 (14), with small numbers occurring elsewhere (Figs 5.39–40).

The eighteen finished shot larger than 194mm include ten from M6, two from M5/6, one from M5 and two from M4/5 (two unknown). This suggests that, if there was a gun larger than about 200mm, it could have been positioned as the port side pair to the M5 port piece. However, the distribution reflects the general areas for storage of stone shot. The range in size of these oversized shot is great, from 197mm to 267mm, with the only cluster being four at 220mm (Fig. 5.38).

**Petrological analysis**

Visual identification (Owen, pers. comm., 1984) indicates Kentish Ragstone, a localised, rubbly, sandy

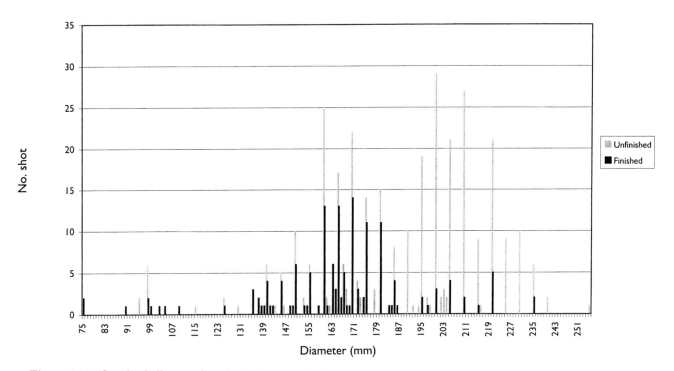

*Figure 5.38 Graph of all stone shot, finished and unfinished*

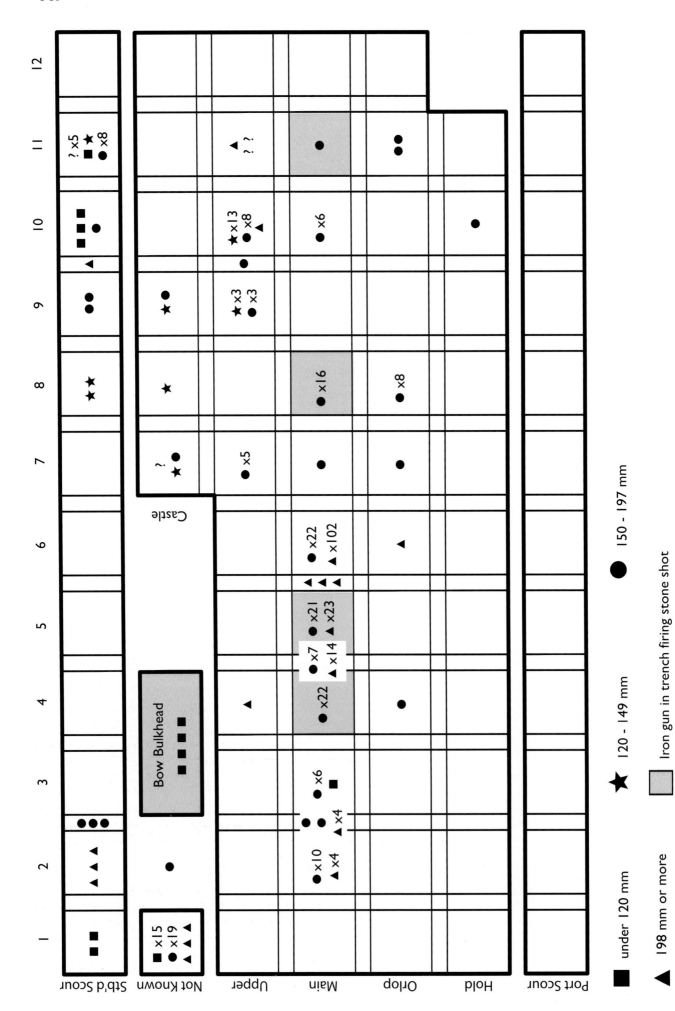

Figure 5.39 Distribution of all stone shot

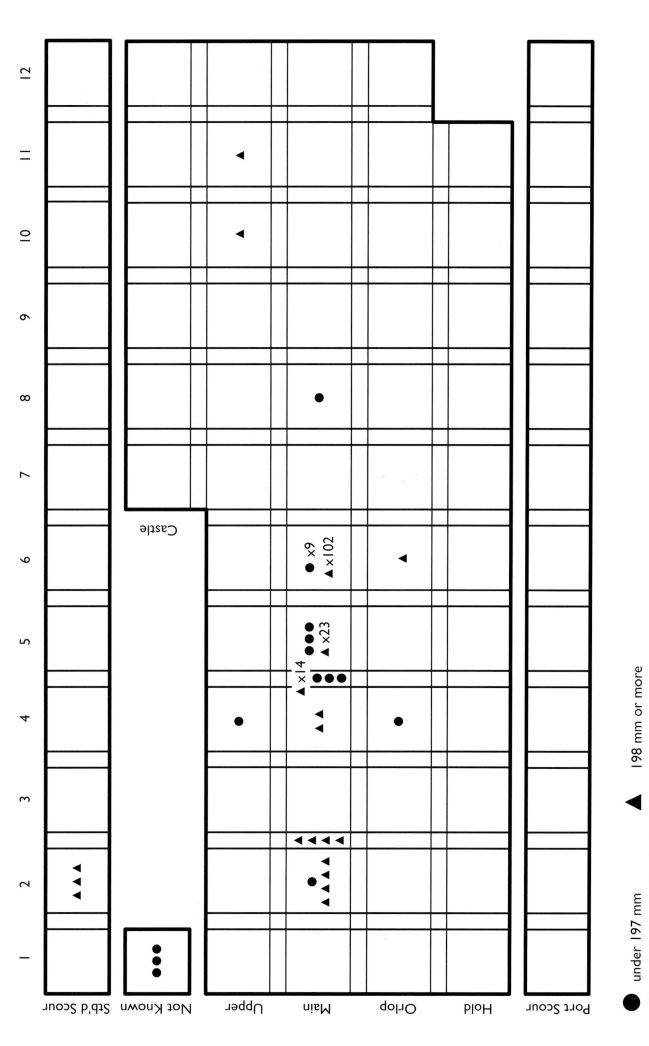

Figure 5.40 Distribution of all unfinished shot

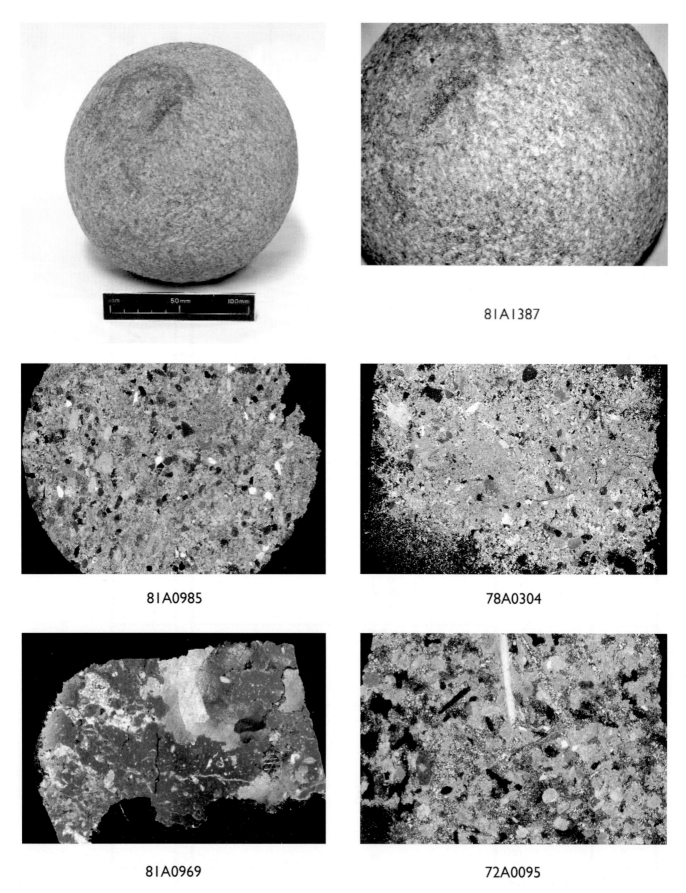

*Figure 5.41 Muscovite granite shot 81A1387 and thin sections of stone shot: see text for details (© Southampton University)*

Table 5.8 Gun bores and stone shot sizes

| Gun | Bore max. | min. | Shot size max. | min. | No. shot in range |
|---|---|---|---|---|---|
| Port piece | 200 | 150 | 185 | 125 | 135 |
| Fowler | 150 | 110 | 115 | 85 | 72 |
| Top piece | ? | ? | ? | ? | ? |

limestone which may be ferruginous or glauconitic and which is very hard and difficult to work. Its specific gravity as calculated from *Mary Rose* shot is as high as 2.87. It runs in a bed from the coast at Folkestone, north through Maidstone, around Guildford and then back down to Eastbourne. It is part of the Hythe beds in the Lower Greensand division of the Lower Cretaceous formations. The stone had been quarried sporadically since Roman times and has been used, for example, in the Tower of London, Canterbury Cathedral and portions of Westminster Abbey. Rock from the western end of the seam is fossil rich, but the area around Maidstone contains many small shells. Maidstone Heath was used for quarrying of stone shot for Henry V in 1414 (Archibald 1840, 385), possibly at Borough Green and Postling. Both are basically 75% calcium carbonate with silica varying between 16% (Postling) and 19% (Borough Green) and alumina iron oxide between 3% (Borough Green) and 6% (Postling).

Petrological analysis, undertaken by David Peacock (Southampton University; Fig 5.41), indicated that several types and sources of stone are represented (in addition to the visually identified Ragstone). Four (78A0304, 81A0895, 82A4250, E261) are of similar Glauconitic limestone containing fossil remains and shells (eg, 81A0985). The amount of glauconite varies, 78A0304 having less than the substantial amounts present in the others. The source could be rounded beach cobbles from the Folkestone area which would require minimal finishing. 81A0969 is a silicified glauconitic limestone containing calcite and probably originates from the Lower Greensand Hythe Beds. 72A0095, a crystalline carboniferous limestone, is silica rich but with no glauconite. Areas of origin could be Plymouth, Bristol or Lancashire. In contrast 81A1387 is a muscovitic granite either from Devon, Cornwall, Brittany, the Channel Islands or Scotland. These shot were compared for size and weight with the rest of the assemblage and show no unusual variance (Martin 1997, figs 44–5; Fig, 5.41).

### Matching shot to guns

Although there are small clusters, the majority of the shot is over 160mm in diameter with the major peaks between 160mm and 170mm. These clusters reflect finished shot for the twelve large wrought iron port pieces and the 200 shot listed for them. The number of finished shot does not match, but the raw materials very nearly equal, the listed figure.

Analysis of size and distribution gives some indication of the possible position of fowlers and port pieces, including guns lifted in the nineteenth century (1836–40). However, as we only know that port pieces were generally larger than fowlers, distinguishing the shot for each is difficult. There are, however, clear gaps where there are no finished shot, at 75–95mm and 115–123mm. The single examples at 112mm, 115mm, 123m, and 130mm confuse a picture which otherwise suggests that top pieces may take shot of *c.* 75mm, fowlers of 95–102mm, and port pieces generally over 135mm. Large windage might cover the few 'strays' listed above. Using the shot data and the bores of all

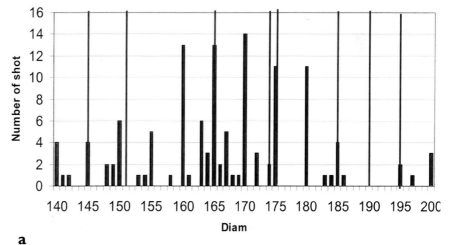

a

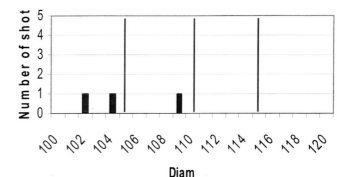

b

*Figure 5.42 Stone shot: a) port piece shot; b) fowler shot*

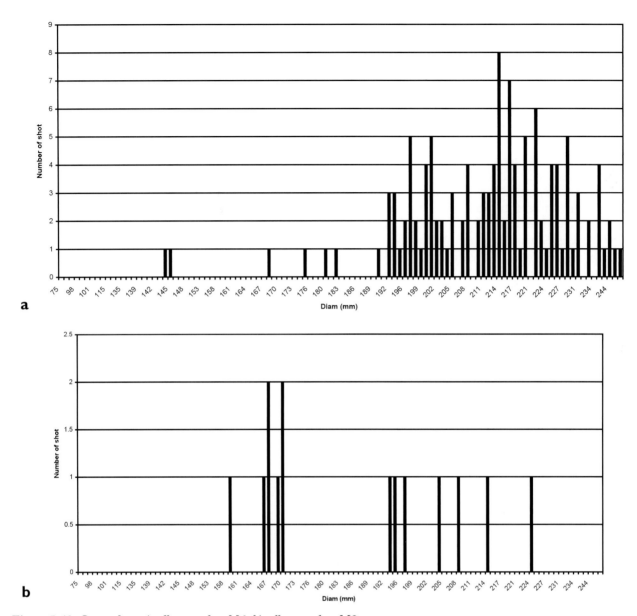

*Figure 5.43 Stone shot: a) all stone shot M6; b) all stone shot M2*

recovered guns, a table of probable shot sizes can be suggested and the number of shot within that band ascertained (Table 5.8).

**Port piece shot**

The *Anthony Roll* lists twelve port pieces with 200 shot. Remains of up to eight port pieces have been recovered with 135 shot between 125 and 200mm (Fig. 5.42a; Table 5.8). The distribution suggests that stone shot was stored beside the port pieces at gun stations as well as in the shot lockers of M2 and M6.

**Fowler shot**

The *Anthony Roll* lists six fowlers with 170 shot, 28 per gun. Four guns with bores of 110mm and 112mm have been tentatively identified as fowlers. Only seven finished stone shot of the correct size have been found (Fig. 5.42b) (only fourteen stone shot less than 110mm were recovered in total). Although it would seem wasteful, it seems that much larger unfinished stone shot was worked to size when required, suggesting a bulk order of rough-cuts. The distribution is limited to SS10 and SS1 and the bow partition excavation with none within the shot lockers.

Three of the four fowlers were found in the bow and would take the 95–100mm shot. The presence of three small shot (75mm, 75mm and 102mm diameter) in SS10/11 suggests the possibility of another pair there.

**Top pieces shot**

The *Anthony Roll* lists 20 stone shot for the two top pieces listed. The only possible evidence for top pieces are the two shot of 75mm from SS10, possibly for a top piece in the bonaventure mast. However, without detailed knowledge of the size of these guns we cannot be sure that these are not for a small fowler. If the shot had been stored with the guns in the tops, they would have fallen clear of the site.

**Distribution**

The distribution of stone shot, both in terms of number and size range, is particularly informative. The bulk of the shot (266) were recovered from the Main deck, with just under half (124) stored amidships in M6 (Figs 5.39 and 5.43a). The majority of these (102, 82%) are too large for the recovered guns and only ten of the 124 are finished (Martin 1997). Other sectors on the Main deck with multiples of stone shot include M2–M5, M8 and M10 (see Fig. 5.39). As one moves away from the M6 storage area, the ratio of finished to unfinished shot increases (eg, 9:14 (64%) in M2; see Fig. 5.40). Wrought iron guns were found *in situ* in M4, M5, M8 and part of a carriage in M11. For the most part, the shot correspond with the position of guns for which they were probably intended.

Most of the 38 shot from the Upper deck were finished. Only two were too large for guns, both unfinished. The shot may, therefore, represent the sizes of guns from this deck or the deck above. Clusters were found in U7, U9 and U10. The structure between U7 and U9 does not support large (vacant) gun ports, however the size of the shot here would fit small port pieces. Similar concentrations (of slightly different sizes) exist outside the ship in the same areas. The size range for each group is an indicator for the bore of gun. The shot recovered with the port piece in M8 was 8mm smaller than the bore, but shot diameters cover a 20mm range. The shot found suggest that both U10 and U11 ports may have had small port pieces.

The fourteen shot in M2 include four larger than the bore of any recovered gun, all of these are unfinished. The range of the finished shot is 159–197mm with two distinct sizes represented: 159–170mm and >193mm (Fig. 5.39). Stratigraphically, the smaller cluster was found first and the larger closer to the starboard side. Although M2 was a store area for iron shot, the finished nature of most of the stone shot poses the question as to whether there were port pieces on the Main deck in the bow. It is also, however, possible that the shot (and therefore a port piece) was originally on the Upper deck as there is access from this area. A port piece chamber was found further forward during the recovery of the stem in 2005.

Of six shot from M2–3, four are unfinished and oversize. The two finished examples, 81A6229 and 81A6232, fit in with the smaller M2 cluster, providing eight shot for a gun of bore *c*. 176mm bore. The seven shot in M3 are all finished and are of limited sizes (115mm, 3 x 160mm, 168mm, 170mm and 173mm). Three shot found in the Starboard scourpit in SS3 (90A0057, 62, 129) are of similar sizes, also finished.

The 22 shot in M4 are all under 180mm, Table 5.9. The *in situ* port piece (81A2650) has a bore of 179mm and one would expect shot of about 170mm. All but two are finished. Most are 160–172mm with one at 150mm (81A1002) and one at 180mm (81A1131). The single shot found in O4, 81A3236 is 166mm. This also may have derived from the Main deck on the port side.

M4/5 had 21 shot, seven below 198mm and fourteen above (200–233mm), only two of these larger shot are finished. The smaller include five at 183–196mm (three unfinished). The remainder include one at 142mm and one at 150mm, both finished. The 44 shot in M5 clearly included a number in store, with 24 shot oversize for the gun, all but one of these unfinished. Of the 20 which fit the gun, only three were unfinished. The shot which might fit the gun range 175–191mm. Thirteen are 181–189mm, the projected shot diameter for the M5 (81A2640) port pieces, bore 195mm. The smaller shot (175–178mm) may be for a port piece of smaller bore, just larger than the M4 gun. The three shot in M5/6 are all large, but two of these (82A3682, 212mm and 82A3683, 214mm) are finished. This may suggest a large port piece as the pair to M5 starboard (Table 5.9).

**Table 5.9 Stone shot groups associated with Main deck guns (only shot smaller than bore listed)**

| M4 bore 179mm | M5 bore 195mm | M8 bore 180mm | O8 ? | M10/11 ? |
|---|---|---|---|---|
| 150 | – | 150 | – | – |
| 160 | – | 160 x 2 | – | – |
| – | – | 163 x 2 | – | – |
| 164 | – | – | – | – |
| 165 | – | 165 x 3 | – | – |
| 166 x 2 | – | – | – | – |
| 167 x 2 | – | 167 x 2 | – | 167 |
| 168 x 4 | – | 168 x 2 | – | – |
| 170 x 5 | – | 170 | 170 | 170 |
| 171 x 2 | – | – | 171 | 171 |
| 172 x 2 | – | 172 | 172 | – |
| – | – | – | 173 x 2 | – |
| – | – | 174 | 174 | – |
| – | 175 | – | – | – |
| – | – | – | 176 | 176 |
| – | 177 x 2 | – | – | – |
| – | 178 x 2 | – | 178 | – |
| 180 | – | – | – | 180 |
| – | 181 | – | – | 181 |
| – | 182 x 4 | – | – | 182 |
| – | 183 x 3 | – | – | – |
| – | 184 | – | – | – |
| – | 186 | – | – | – |
| – | 187 x 2 | – | – | – |
| – | 189 | – | – | – |
| – | 191 | – | – | – |

M6 must be considered the main storage area for unfinished stone shot (Fig. 5.40). It was also one of two areas for storing iron shot, and its position in the centre of the ship is sensible for both access and weight distribution. Although the 124 stone shot range from 144mm to 267mm, most are over 198mm and only ten are finished. Of the 22 below 194mm only two are finished; 81A1813 at 145mm; 81A2046 at 168mm. None of these appears to be suitable for the M5 port piece, bore 195mm. One unfinished 203mm shot was found in O6. M7 contained one finished stone shot, 80A0415 (170mm). It was at a high level, well above that of the port piece in M8 (81A3001, bore 180mm). The shot size is comparable with a group from above this gun and from O8. A similar sized shot was found in O7, 80A1684, 175mm.

The shot from M8 provides us with the best information regarding that for a specific gun (see Table 5.9). The gun itself was loaded: 82A4213 of 172mm diameter, 8mm less than the bore. There were another fifteen shot in a very small, enclosed compartment, many close to the starboard side and between the wheels of the gun carriage and deck. The shot ranged from 150mm to 188mm, predominantly 160–174mm. All studied examples (nine of fifteen) are finished with the exception of the largest 81A5683 at 188mm, which is too large for the gun. Eight finished shot in O8 are 170–178mm and, together with the three found at a high level in M8 those in M/O7 were probably for a port piece originally on the port side in M8.

There were six shot in M10, all finished and 167–182mm in diameter. The closest iron gun was in M11, but had been recovered in the nineteenth century leaving just the spoked wheels and axle *in situ*. One finished shot (176mm) was found in H10 with one in M11 (176mm) and two in O11 (176–177mm). This suggests that at least one gun had a bore of about 189mm and the second of about 179mm. As only smaller shot were found on the Orlop deck and in the Hold, this suggests that the smaller gun could have been on the port side. All of the shot were excavated at around the same time, this area was disturbed and, in this position, the ship is quite narrow.

The only evidence for guns in the very bow of the ship comes from underneath, in the bow excavation or from the Starboard scourpit. There are four stone shot from the bow area, three of 98mm and one of 99mm. All are finished. The two from the SS1 are 95mm and 100mm. These would fit the two fowler barrels found here (90A0126, 90A0131). Shot of this size would also fit 77A0143 from above O1/O2 or the fowler fragment (78A0614) recovered in the nineteenth century. The internal diameters of these are all around 100–110mm. No shot was recovered from U2, but three unfinished, oversized shot came from SS2. These are of comparable size to shot found in M2, some of which was also unfinished.

Shot has been found underneath gun ports in the Starboard scourpit, having fallen through them. Unless there was a port in M1, the closest is M3. The three shot in SS2/3, 165mm, 157mm and 168mm, fit in with those found in M3. All of this shot suggests that there were two port pieces, either in M1 or on the Upper deck in U2–3. Weight distribution and description of these as '*port pieces*' would favour a Main deck location at a lidded port, but it is not impossible that one was situated on the Upper deck in the bow. One shot of 196mm was found in U4, this was unfinished. The finding of a port piece chamber associated with the end of the stem in 2005 reinforces the idea of port pieces being positioned forward of sector 3.

The Upper deck, Castle and stern scourpits on the starboard side have a great deal of shot which form sized groups, but these are difficult to interpret. Four out of five in U7 were all finished (155–159mm). Of the two in C7, directly above, one is 160mm, the other has an unknown diameter. This suggests a large gun, of about 165mm, in the Sterncastle either on the Upper or Castle deck. At the top of the companionway from the Main deck, may have been a convenient place to store shot for use further astern, but the small range of size appears gun specific. C7–9 produced three shot of

**Table 5.10 Stone shot groups on the Upper deck**

| U10/11 | SS10/SS11 |
|---|---|
| unknown diam. x 2 | unknown diam. x 5 |
| – | 75 x 2 |
| – | 100 |
| – | 102 |
| 123 | – |
| 130 | – |
| 135 x 2 | – |
| 139 x 2 | – |
| – | 141 |
| 143 | – |
| 144 | – |
| 146 x 3 | – |
| 148 | – |
| 150 x 2 | – |
| – | 158 x 2 |
| 159 | – |
| – | 160 |
| – | 162 x 3 |
| 163 unfinished | 163 |
| 165 | 165 x 2 |
| 166 | – |
| 168 | – |
| 169 | – |
| 184 | – |
| 214 unfinished x 2 (1xU10, 1 x U11) | – |

140mm, with two in SS8 of 137mm and 138mm. These are of culverin size, the bores of the recovered guns were 138mm and 140mm. In U9 there is evidence for both a small and a large gun, with six shot of 142–173mm, although one at 167mm is unfinished. The presence of two shot of 151mm and 163mm in SS9, both finished, adds credence to this. Between these areas the clusters suggest the possibility of three guns, one of culverin size bore and two port pieces of slightly different sizes. However, it must be borne in mind that some of the shot may be for the port side guns and that stone shot could have been used with bronze guns to increase structural damage.

U10 contains 22 shot ranging 123–214mm. Two are unfinished; 81A1790 (214mm) and 81A0683 (163mm). The only observable clusters are five at 130–139mm, eight at 143–150mm and five at 169mm. U11 contains three shot, two of unknown diameter and one of 214mm. Taking U10–11 together with SS10–11 (Table 5.10) a pattern begins to emerge. Although the shot range in SS11 is great (fifteen ranging from 100mm to 165mm, the bias is towards a larger size of *c.* 160–165mm. There are large shot in U10, but the bias here is towards those <150mm. This may suggest larger guns on the Upper deck U11, with smaller ones in U8–10, or C8–10, with slightly different bore sizes across the deck for each. The recovery of an incomplete port piece from the stern, probably from the port side (71A0169) around U10–11 may indicate that the largest shot in M10 (182mm), derived from M11 port side. The finding of a portion of gun carriage bed for a bronze gun in U10 along with loading instruments for muzzle-loading guns, indicates that there may have been a bronze gun at one of these two ports. Alternatively, port pieces or culverins/demi culverins could have acted as stern chasers positioned either side of the rudder as illustrated on the *Anthony Roll*. The port piece RA XIX.2 (80A2003) was found with stone shot. This has a bore of *c.* 152mm and a projected shot range of 132–44mm. RA XIX.1(80A2004) has a bore of 150mm and expected shot range of 130–142mm. The concentration of shot in U10 (123–150mm) makes this a likely position for at least one of these guns. The finished shot from U9 and U10 together is 123–173mm but only three are >150mm, all in U9. It seems more sensible generally to have the smaller port pieces on the Upper deck and the larger (heavier) examples on the Main deck.

The port piece from M8 was found together with suitable shot. The bore is 180mm and it contained a shot (82A4213) of 172mm. There are 75 shot which would fit this gun in M4/5, M2, M8, SS10–1 and U10. Those found in M8 range from 150mm to 188mm. The two large port pieces in M4 and M5 (81A2650 and 81A2604) have bores of 179mm and 195mm and a shot band of *c.* 160–190mm There are over 40 appropriate finished stone shot in M4–5.

The Royal Artillery Institution port piece, raised in August 1836, has a bore of *c.* 200mm, shot band of 175–194mm and expected shot size of about 190mm. Shot within this range was found in M2, M4, M5, M8, and M10. The excellent condition of the bed, however, favours a starboard gun. The obvious place would be M11 starboard, but the largest shot here is 182mm (M10) and the width across the ship at this point questions the positioning of such a long gun here.

71A0169 is a large, incomplete port piece with bore 190mm. It was recovered from the stern port quarter and may well represent a Main deck port side gun. Its position suggests that it is the port side pair to the M11 gun. The shot band covers 170–84mm and the expected shot size is *c.* 180mm. Shot of this size was found in O8, M8 and M10, M11 and O11. The salvors worked this small area at least twice in 1836 and it is known that they recovered stone shot.

Very simply, the distribution of finished shot suggests another pair of port pieces in the bow, either in M1/2 or U1/2, although the distribution of shot favour the positioning of fowlers in the bow on the Upper deck. The distribution amidships suggests four port pieces, M4 and M5, port and starboard. The total number of finished shot in the stern, if divided by sixteen, comes to 4.9 guns. As the number of finished shot attributed to a single port piece does not equal sixteen, but favours thirteen, six guns can be postulated for the stern. This equates to two in each of M8, M11 and U10/11, probably U10. Alternatively, having two at main deck level firing out of the stern is also a possibility. This accounts for the twelve port pieces listed. Caution is necessary, however, because so many shot are unfinished.

There is little evidence for stone shot for top pieces. The two of 75mm in SS10 may have served them. Alternatively their shot may have fallen very far off site. The height of the masts suggests that the tops, and their guns, could be at least 30m away from the hull, beyond the limit of our excavation. The initial salvaging of items included portions of rig, these may have included the fighting tops, their guns and shot.

Over 300 stone shot were tested against the nine gauges recovered (see Chapter 6). Most could be selected using the gauges carried and the fit is more convincing than for iron shot. It is still possible that some of these functioned as ring gauges or holdfasts to help with the final finishing of the shot.

Twenty-seven shot are from unknown locations, three of these are too large for recovered guns (227mm, 234mm, 237mm), but two are finished.

*Composite shot*

**Composite round shot**

A total of 235 shot are classified as lead/iron composite round shot. When recovered, all were grey and their relative weight suggested that they were lead. It was not until these started to display unusual rust deposits that it became obvious that the bulk of the shot between 20mm and 75mm (1–3in) in diameter were lead/iron composites. The symptoms varied and obviously related to different states of corrosion of an iron core (Walker *et al.* 1989). The features included the cracking or 'peeling' of what was thought to be a wrapped lead layer of less than 2mm in thickness to reveal a round iron shot (Fig. 5.44); 1–4 small triangular corrosion spots (Fig. 5.45a); corrosion spots joined by cracks (Fig. 5.45b); the peeling of these cracks to expose an iron square within (Fig. 5.45c); the extrusion of iron from within the centre of the shot (Fig. 5.45d); and clean empty holes within a cast lead shot (Fig. 5.45e). Occasionally a 'flap' was observed within the lead from which emerged a short fibre. In several instances shot which exhibited these features also had a separate, ambiguous, hole, opposite the dice hole.

Shot were initially sorted visually and then subjected to more detailed description (including corrosion features), response to magnetic stimuli and a roll test (three times, marking with chalk the uppermost area), diameter, weight, seabed location and radiographic study.

The following groups emerged through the visual sorting: 26–30mm (1–c. 1⅛in), 30.2–37mm (1¼in) 37–42 mm (c. 1½in), 42–49 mm (1¾–2in), 50–68 mm (2–2¾in), 70–75mm and over (2¾–3in). It seemed apparent that the majority of the largest shot (70mm or more) were formed by what appeared to be the wrapping of a layer of lead around a misshapen wrought iron core. About one-third of the 50–68mm shot also suggested a round iron shot within a lead coating. This was the most difficult group to distinguish, as the dice is often quite large and exhibits corrosion in a similar fashion to the round inset shot (Fig. 5.46). These dice may reach 45mm (81A0649), with several at 30mm (81A6141, 82A3046). The remainder (206), consist of what is thought to be a wrought iron dice within a cast lead round shot. The term inset shot is used to describe these, with subdivisions of inset dice and inset ball. Both types were studied at the University of Surrey (Dunham 1988; Laughton 1989).

A number of theories exist as to why shot should be made in this way: relative costs of raw materials, ease and speed of manufacture, cushioning the shot down the bore (Wignall in Olds 1976). It could equally be to increase or decrease the weight of a projectile of a specific size. It is the fact that, in all instances, the shot is eccentric which suggests that they may have been particularly manufactured with the desire to induce spin. The effort taken to produce this shot suggests that it was of inherent value, possibly the iron portion would penetrate structure, whereas lead alone would deform against the hull.

The lead coating would be more tolerant of inaccuracies in the bore of the gun, especially a wrought iron gun such as a sling or base, than a similar shot of solid iron. It would also be less subject to corrosion. As the shot is basically used with anti-personnel guns, this is interesting. The real question is whether the offsetting of the dice is a prerequisite of the shot, in order to impart a spin in flight, or whether it is simply a by-product of a manufacturing process dictated by cost and availability of raw materials, or a combination of the two.

The use of wrought iron and lead for the inset dice shot would make these a cheap and easily produced projectile. The temperature required to cast lead (327°C) is far below that required to melt iron (1539°C). The fact that the lead is 44% heavier than iron for the same dimensions would possibly make up in range that lost by the imparted spin. Similarly the re-use of 'reject' or 'old' iron round shot with lead would produce a usable commodity cheaply.

A requirement which might be fulfilled by this type of shot is the need to produce and replenish shot to fit the variety of sizes of bore found within the assemblage, or within assemblages from this period where standardisation was not fully achieved. Although, traditionally, guns were provisioned with shot, depending on the particular battle, some sizes of shot may be exhausted before others. On a long voyage replenishing exhausted shot would have been a constant problem. Whereas the casting of solid iron shot on a vessel is impractical, the casting of lead shot with an inset dice or inset round shot is possible, as the moulds for the smaller shot bear witness. The finding of a 58mm round iron shot inside bronze shot mould

*Figure 5.44 Shot 82A4199*

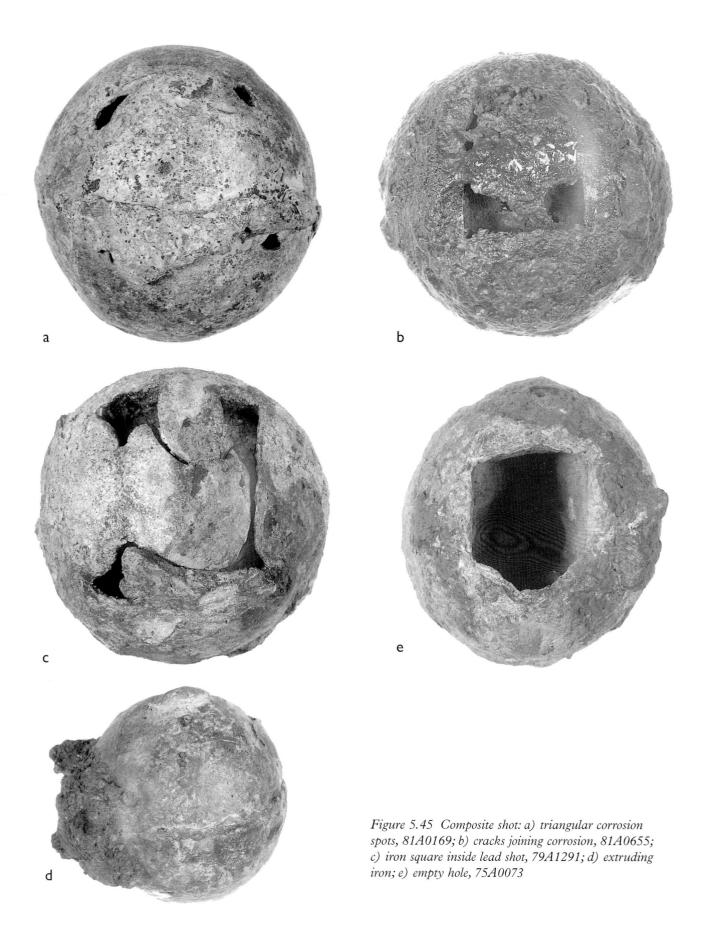

*Figure 5.45 Composite shot: a) triangular corrosion spots, 81A0169; b) cracks joining corrosion, 81A0655; c) iron square inside lead shot, 79A1291; d) extruding iron; e) empty hole, 75A0073*

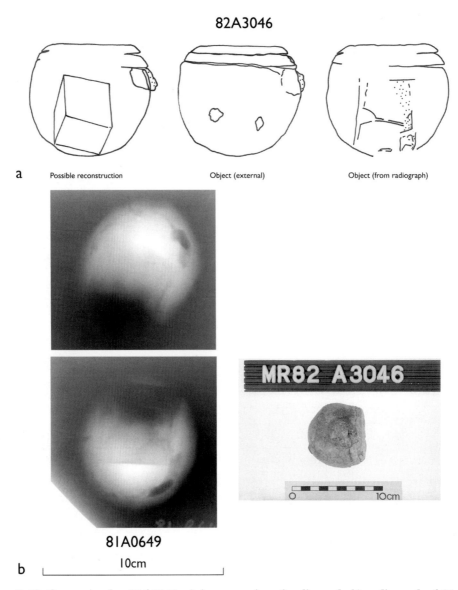

*Figure 5.46 Composite shot 82A3046: a) interpretation of radiograph; b) radiograph of 81A0649*

with an internal diameter of 71mm on the sixteenth century Mollasses Reef wreck (Simmons 1991, 5) may provide the archaeological proof of this practice, just as the presence of the shot on the *Mary Rose* provides tangible results of this procedure.

> '*Why a too small cast-iron shot was found within a bronze mould, the exterior of which apparently exhibited traces of lead casting has been the subject of much discussion, but this is possibly another story, best left for another time*'. (Simmons 1991, 5)

If this was being practiced, one should possibly see clusters of iron round shot which cannot be explained within the assemblage of guns. Conversely, the capability to alter diameter should be considered when using a shot assemblage to predict the sizes of guns.

### Inset dice shot

Most of the composite shot (206) consist of an internal iron dice offset to one hemisphere within a lead shot. These vary in size from 26mm (1in) to 67.5 mm,

*Figure 5.47 Dice straddling castline, 81A0483*

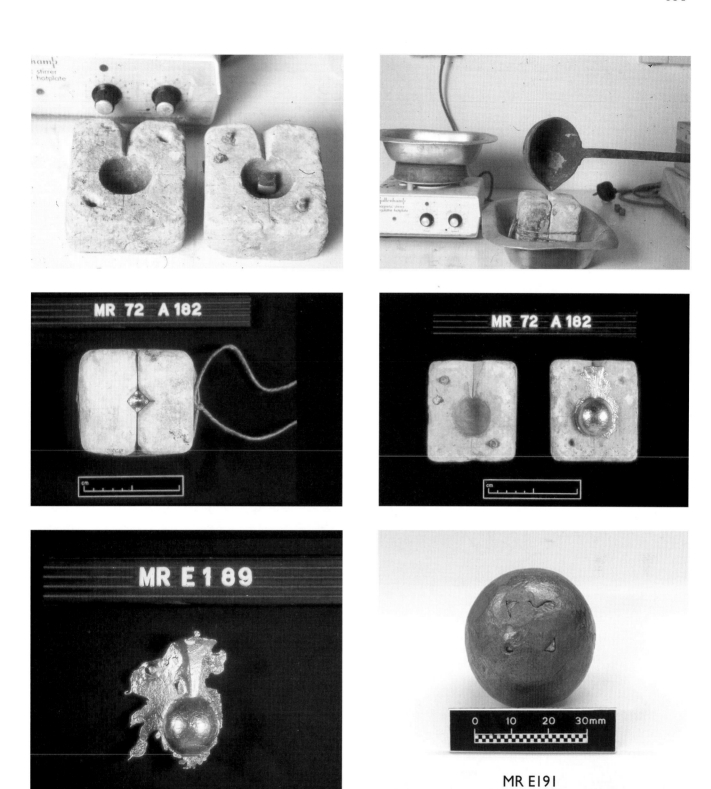

*Figure 5.48 Dice shot experiment: sequence to show finished shot with triangular holes*

(2.65in) with the majority under 50mm (2in). Certain examples retain a clear castline within the lead outer and show the corrosion holes revealing the corners of the dice straddling the castline (81A0483, Fig. 5.47). This suggests that the lead was poured into a mould already containing the iron dice. Three sandstone moulds (72A0067, 42mm diameter, 72A0182, 38mm, and 80A1847) would all fit composite shot with inset dice. Experimentation with an iron dice made from 10mm square steel and modern lead was undertaken within an original *Mary Rose* mould (see below). Contrary to our expectations, the dice became pinned against the bottom of the mould and the shot manufactured (MRE0190, MRE191) revealed tiny

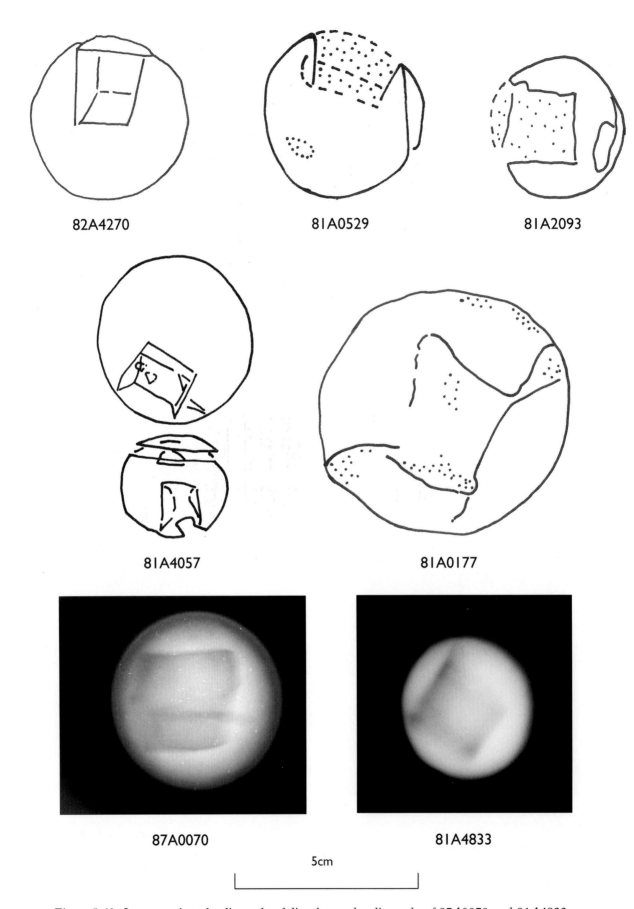

Figure 5.49 Interpretation of radiographs of dice shot and radiographs of 87A0070 and 81A4833

spots showing the eccentric dice (Fig. 5.48) where its corners obviously touched the mould. These tiny spots eventually allowed seawater into the iron centre and promoted corrosion of the internal dice. Even when there is no obvious castline the dice is offset. This was confirmed by using a magnet over the surface of the shot and marking the area of strongest magnetic response. The shot was then rolled to check eccentricity, the marked area invariably faced upwards. Fifteen shot were radiographed, revealing dice of different sizes increasing with the diameter of the shot (Fig. 5.49). The shot was hand-positioned so that the area which displayed the greatest magnetic response (previously marked with chalk) was directly in line with the cobalt 60 source (eg, 87A0070, 81A4833, 82A4270, 81A4057, 81A0177, see Fig. 5.49). The dice are not always square; 72A0143 is 25 x 30mm with a diameter of *c*. 46mm. As so few were radiographed, it is not possible to draw any conclusions regarding the pre-selection of sizes of iron bar for specific shot diameters. Generally, the larger the shot, the larger the dice. Dice measured from this technique varied with examples of 13mm (81A0508), 25mm (82A4270), 30mm (82A3046) and 45 mm (81A0649).

Weight varied depending on size and condition of the dice. Between 26mm and 30mm weight varies between 116g (4.05oz) at 29mm (81A0215) and 152g (5.4oz) at 27.7mm (81A0511). The smallest, at a fraction over 26 mm, weighs 120g (4.2oz; 80A0118), and the largest, at 30mm, 123g (4.3oz; 82A5293). The range between 30mm and 50mm shows a gradual increase overall, with the heaviest (81A4057), only 39mm but weighing 589g (1lb 4oz) and the largest, just under 50mm (82A0838) weighing 569g (1lb 4oz, obviously corroded). There are 21 or 23 (two are uncertain) shot of dice and lead between 50mm and

*Figure 5.50 Inset shot 82A4339*

### Table 5.11 Least weight for composite dice shot

| Size band diam. (in) | Diam. (mm) | Greatest weight (g) | Shot diam. (mm) | Object | Weight (oz) |
|---|---|---|---|---|---|
| 1 | 25.40 | 122.30 | 27.72 | 79A0976 | 4.31 |
| 1¼ | 31.75 | 168.80 | 30.21 | 79A1093 | 5.95 |
| 1½ | 38.10 | 298.40 | 37.61 | 81A1175 | 10.52 |
| 1¾ | 44.45 | 383.20 | 44.00 | 72A0080 | 13.31 |
| 1¾ | 45.00 | 423.00 | 45.00 | 79A0911 | 14.91 |
| 2 | 50.80 | 699.00 | 50.00 | 81A3273 | 24.65 |
| 2¼ | 57.15 | 973.90 | 56.65 | 82A4157 | 34.35 |

Based on greatest current weight or recovered shot of closest size and greatest weight in band

70mm. Weight generally increases with size, with the smallest at 50mm (81A0111) weighing 644g (2lb 4oz) and the largest (81A0649) of 68mm weighing 1670g (3lb 9oz). The heaviest is 81A6141 of 60mm and weighing 2077g (4lb 8oz).

*Composite lead and wrought iron*
One of the most unusual items is 82A3046 (Fig. 5.46). It is 86mm in diameter, containing a corroding dice of 30.5 x 22.3mm. One face has been flattened, puckering up as though it has impacted against something. This is very large for an inset dice shot and weighs 1225g.

Without knowing the precise size of the internal dice it is impossible to suggest an original weight for any one shot. A 'least possible original weight' can be suggested by looking at the greatest weight for each size of shot. There are two methods, comparison within the shot bands as sorted visually, covering a spread in millimetres; or by looking at the greatest weight for shot smaller than exact diameter in quarter-inch increments between 1in and 2¼in (Table 5.11). The latter offers some capability of comparison with historical data. Whatever the original weight, it is greater than for iron shot of the same diameter. Bourne (1587, 14) suggests 1lb 1¼ oz (432g) for a 2in (51mm) diameter iron shot. The heaviest composite shot of this size recovered exceeds this (81A3273, 1lb 8½ oz; 640g). An attempt to work out the sizes of cores has been undertaken using information obtained through a study of the gunner's rule (see below).

Contemporary listings of this type are familiar but the historical weights for composite shot are difficult to

### Table 5.12 Bases recovered with shot

| Gun | Bore (mm) | Shot (mm) | No. shot of recovered |
|---|---|---|---|
| 79A0543 | 45 | 38 | 20 |
| 79A1075 | 37–40 | 30 | 8 |
| 82A4076 | 60 | 48 | 22@45mm |

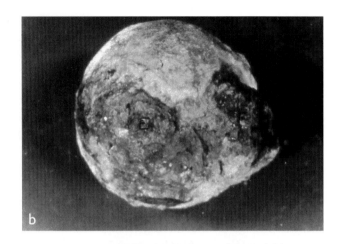
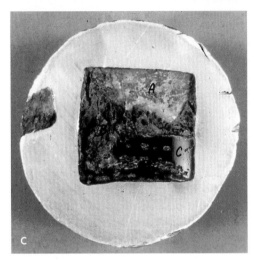
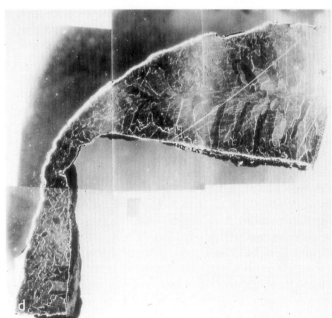

*Figure 5.51 Analysis of composite shot: a) shot 81A4958; b) 81A4436; c) 81A4958 section showing core; d) cross-section of lead; e) lead cube and whisker crystals*

find. In the Richard Wright manuscript of 1563 (Society of Antiquaries MS 94) one of the first entries (f. 7v) is for falcon shot of 2in: '*Facon shote of ij enches hie* [high] *is vj enches in compaes and wayeth in dyce and lede ij pounde v ounces and of yornne a pounde iiij gr*'.

The Wright weight of solid iron shot (1lb 12oz) is more than that listed by Bourne (1587, 14) of 1lb 1¼ oz. The composite shot weight given by Wright for 2in is 9oz (255g) more than iron shot of the same diameter and 13oz (369g) more than the greatest recorded *Mary Rose* weight for that size (81A3273 at 1lb 8oz; 624g). Interestingly, the 2¼in (63.5mm) dice shot recovered from the *Mary Rose* is 9oz (255g) heavier than the historical weight of solid iron given at just over 1lb 9oz

(Bourne 1587, 14). Although this can only be for comparison, it demonstrates the difficulty when dealing with a corroding dice of unknown dimensions. Had the shot been solid lead, it may have been possible to suggest which size groups could be attributed to the demi base, single base and double base by using the half weight principal. Table 5.12 highlights the difficulties with this. It is tempting to suggest that the double bases are over 2in (51mm), the single bases 1½in (38mm) and the demi bases about 1in (25mm).

An additional feature of a 'flap' within the lead was noted on 31 examples. In 11 cases what appeared to be a short fibre was present within the flap. One of these (82A4339) was excavated from within the bore of a base (82A1603); the fibre is no longer visible (Fig. 5.50). In seven examples a fibre was noted adhering to the surface of the shot but without an obvious flap. Where noted, the flap was in the opposite hemisphere to the corroding dice. Whether this is a fortuitous remains of packaging, or has some particular function is not known. It does not appear that these penetrate through to the dice hole. This feature does not seem to coincide with particular size or location of shot.

*Analysis of shot*

Analysis of two imbedded dice shot was undertaken to establish materials and method of manufacture. Both were c. 1½in (38mm) in diameter and dull grey in surface colour. 81A4958 (385.3g; Fig 5.51a), recovered from the Castle deck in the stern, displayed four localised areas of rust to one hemisphere. 81A4436, (381g) from the Upper deck in the stern, displayed two large zones of rust. The castline was visible for about one-third of the circumference in both cases. The density of the shot was 8.8 for 81A4436 and 9.1 for 81A4958, below the density of lead (11.36g/cm$^3$) and above that of iron (7.88g/cm$^3$). The more corroded example (81A4436; Fig. 5.51b) had the lower density. Both were cross-sectioned using a hacksaw and cleaned with acetone. This immediately confirmed the presence of an eccentric iron cube with three visible zones of iron exhibiting differential corroding (Fig. 5.51c). The core was still hard, reflective and in good condition.

The lead displayed a typically cast structure. Cubic and whisker-shaped grains of some of the lead are typical of single crystals produced by the deposition of metal vapour on to a cooler substrate (Fig. 5.51d). The thicker regions showed more gradual cooling, with columnar grains, with the smallest grains in the area which came into immediate contact with the cold iron dice (Fig. 5.51e). This suggests that the molten lead was poured into an unheated mould containing the iron. Physical study of the iron suggested that it was wrought. Electron micro-analysis revealed some silica inclusions containing iron and potassium, This could either be wrought iron containing slag (Dunham 1988) or crude cast iron (Barker, pers. comm., 2009).

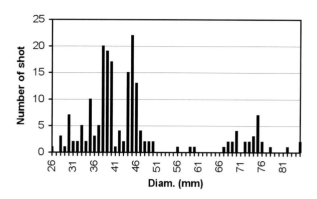

*Figure 5.52 Graph of composite inset shot*

*Matching shot to guns*

The majority of inset dice shot range between 26mm and 50mm in diameter (Fig. 5.52). Although there is usually 1–3 shot for each millimetre between 26mm and 50mm, there are obvious peaks at 29mm (1¼in), 35mm (1⅜in), 37–39 mm (1½in), and 44–46mm (1¾in). The largest number of shot for any of the peaks are 20 at

**Table 5.13 Diameter and weight of inset ball shot**

| Diam. (mm) | Weight (g) | Sector | Object |
|---|---|---|---|
| 52 | 768.6 | M2 | 81A5532 |
| 52 | 828.7 | M2 | 81A3366 |
| 56 | 933.7 | M2 | 81A3390 |
| 60 | 2076.9 | H5 | 81A6141 |
| 66 | 1567.3 | M2 | 81A3404 |
| 66 | 1636.3 | M2 | 81A3357 |
| 66 | 2167.0 | M2 | 81A3405 |
| 66 | 1814.0 | M2 | 81A3367 |
| 67.23 | 1613.0 | M3 | 83A0179 |
| 67.62 | 1618.8 | ? | 82A4399 |
| 69.05 | 1633.8 | M3 | 83A0183 |
| 69.09 | 2018.5 | M2/3 | 81A6091 |
| 72 | 2168.9 | ? | 82A4230 |
| 72 | 2196.1 | ? | 82A4229 |
| 73 | 1732.2 | M2 | 81A3502 |
| 73.16 | 2193.3 | ? | 82A4190 |
| 74 | 1551.4 | M6 | 81A2814 |
| 74.16 | 2031.0 | U6 | 81A0956 |
| 74 | 2174.6 | M2 | 81A3506 |
| 74.53 | 2216.3 | M6 | 83A0238 |
| 74.68 | 2047.0 | U6 | 81A0960 |
| 74.70 | 1615.1 | O2 | 81A0041/2 |
| 74.73 | 1676.7 | M2/3 | 81A6153 |
| 75.01 | 2093.6 | M6 | 81A2362 |
| 75.12 | 1621.2 | M2/3 | 81A6094 |
| 75.43 | 2009.3 | M6 | 83A0239 |
| 75.71 | 1566.5 | M3/3 | 81A6157 |
| 76.31 | 1915.6 | U6 | 81A0957 |
| 78.14 | 1654.9 | ? | 82A4189 |

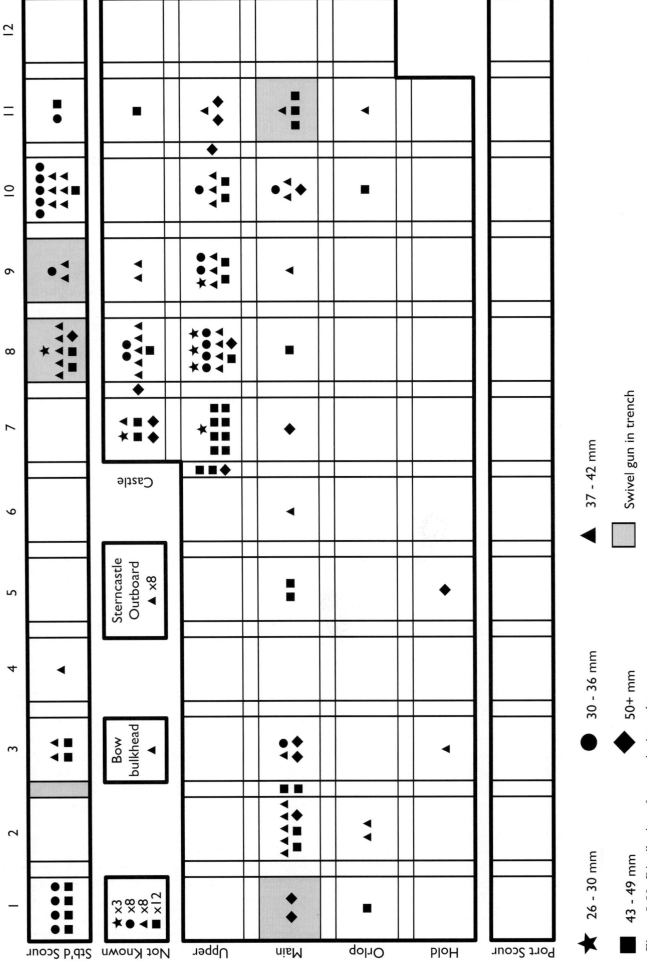

Figure 5.53 *Distribution of composite inset shot*

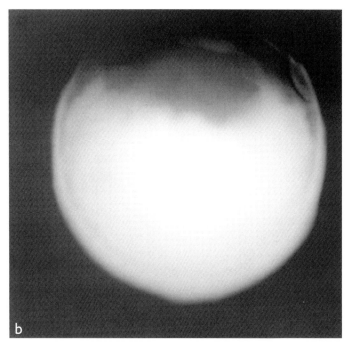

|— 10cm —|

*Figure 5.54 a) Interpretation of b) radiograph of 81A6157; c) inset ball shot 81A6094*

38mm and 22 at 54mm. Over 25% of the assemblage (52) measure 43–46mm. A further five, probably all inset dice, range in diameter from 50mm to 70mm, with two at 50mm and two at 70mm.

Inset dice shot were recovered from inside the bores of three of the four most complete bases (swivel guns) Table 5.12. Several base barrel fragments were loaded with inset shot (85A0024, 90A0027, 82A1603). The difference between bore and shot diameter varies between 7mm and 12mm although the original windage is very difficult to predict with certainty because of internal corrosion of the bores and expansion of the shot due to the corroding internal core. A possible figure would be about 6mm. Where groups of this type of shot have been found together (for example 81A1036–8 in SS10) they can be quite close in size (38–40 mm). Six further shot recovered in SS10 (81A4133–38) are all 36–40mm. The largest range of shot found together is from the bow area, at 28–47mm and it is assumed that these were for a number of bases.

The shot peaks suggest up to four different bore sizes between 26mm and 50mm, with another four from 50mm to 67mm, although these larger examples are represented by a very small number of shot (Table 5.13). Although the *Anthony Roll* does not differentiate within the base group, listing the *Mary Rose* with 30 bases and 400 lead shot for them, the 1547 inventory (Starkey 1998) includes double bases, bases and demi bases. This suggests only three size groups, but we must expect some variation within these.

Most of the guns identified as bases have a bore of 50mm or less, however one complete gun (82A4067) has a bore of 60mm, and one incomplete barrel, possibly representing a base, (81A1082), 60–70mm. Four others have bores of 50mm or slightly more (82A4077, 81A5152, 82A5096, 82A4080); one of *c.* 45mm (79A0543), and four of *c.* 40mm (82A1603, 79A1075, 90A027 and 85A0024). The number of expected shot per gun is thirteen. The number of shot which would serve the identified bases is less than half the listed amount, however, barely half the listed number of bases has been identified. It is likely that additional shot and guns survive in the Starboard scourpit in the areas of the Fore- and Sterncastles. By 1628 Norton lists bases as having a bore of 1¼in (33–34mm) and weighing 20lb (91.2kg with a shot weight of ½lb (227g). We may be observing composite shot for the smaller guns of the sling group as well as the bases.

*Distribution*

The distribution of inset shot and bases can be seen in Figure 5.53. There are 40 shot which are unlocated, eight of which are listed as '*outboard stern castle*', with over 100 in sectors 7–11, mostly on the Upper deck. Two guns were found on the Main deck, one in M11 (79A0543) and one in M1. Both were at a high level in the trench and may have derived from collapsed Upper deck structure. All sectors containing bases had some shot which they could have been used. The other remains of bases were from the starboard side under the Sterncastle and from a large concretion found outboard of sector 3 (starboard side) in 1989/1990.

All the guns recognised as bases were loaded with inset iron dice rather than the lead shot listed in the *Anthony Roll*. Shot of dice and lead are listed in the 1547 inventory (Starkey 1998) for demi culverins (*ibid.*, 4529), sakers (4532, 4964), falcons (4534, 4798, 4967, 4969, 5275), falconets (5318) and slings (4538, 4601, 4665). The entry for Guisnes Castle lists iron shot in descending diameters from 8in to 2⅝in (Starkey 1998, 5537–5547). '*Shot of lead with dice of yrone*' is given in various dimensions between 3½in and 1⅛in (*ibid.*, 5549–5553). '*Hagbusshe shot with dice of yrone*' is specified (5556), and '*Dice of yrone for sacres Fawcons Fawconetes Robynettes Rede bases and hagbusshes*' (5557). '*Sowes of leade for shott*' (5558), '*Mouldes of brasse*' (5562) and '*Mouldes of stone*' (5563) indicate the facility to cast more of these types of shot.

Guisnes was not alone. Pendennis is provisioned with: '*Sling shott of dice and leade*' (4665), '*Demi slinges shot of dice and leade*' (4666), '*Shott of dice and leade for doble basys*' (4667) and '*Hagbutt shott of dice and leade*' (4668). Replenishment is suggested by '*A Sledge of yrone ij° Crowes of yrone and iiij$^{or}$ pickehammers to make stone shott*' (4469) while raw materials include '*Dice of yrone yet vncaste*' (4678) and '*Stone shott vnhewed*' (4679). Elsewhere there are more particular entries for '*Slinge shot of leade diced with stone and yron*' (4691), '*Shotte for doble basses of dice and leade*' (4602) with the like for single bases (4603), and '*Shotte for basis diced and vndiced*' (6337).

Similar examples have been found on other wreck sites including the sixteenth century Cattewater wreck (Redknap 1984), *St Anthony* (K. Cammidge, pers. comm.), the Mollasses Reef wrecks (Olds 1976; Guilmartin 1988; Simmons 1988), *San Pedro* (Peterson 1969), possibly *El Gran Griffon* (classified as '*socketed lead hemispheres*'; Martin 1972), the Padre Island wrecks (Arnold and Weedle 1978) the wreck in Villefranche, France (Guérout *et al.* 1989) and an unnamed wreck in Oman (D. Mearns, pers. comm.).

**Inset ball shot**

A total of 29 shot has been identified as being composites of iron and lead with a round iron centre and lead outside. By 1986 most had become red/brown with a yellowish rust over much of the surface. Most displayed 'tearing' to one hemisphere to reveal a corroding iron sphere within (Fig. 5.54c). Radiographic study (Fig. 5.54a–b) suggested the presence of a roughly spherical iron core rather than a dice. Eight are 50–c. 67mm, and the rest 67–8mm (Fig. 5.55) with peaks at 2¼in (57mm), 2½in (63mm), 2¾in (69mm) and about 2$^{14}$/₁₆in (71–2mm). Weights vary considerably but not obviously in direct relation to size, although generally weight increases with size (Table 5.13). The effects of corrosion make precise measuring and matching shot with guns difficult.

*Analysis*

A single inset ball shot of 70mm (2¾in) diameter was studied at the University of Surrey (83A0239). The shot weighed 2009g (4lb 6oz) and had a measured volume of 203cm³. The lead had split exposing a corroding iron core (Fig. 5.56a). Roll tests suggested that the iron ball was eccentric and sections were taken first through the core, revealing it to be spherical and off-centre; the more corroded section was then cut through the cavity in the lead surround. The iron core exhibited what appeared to be 'blow holes' and a 'pipe' which appear to have been filled with molten lead during manufacture (Fig. 5.56b). This indicated that the core was likely to

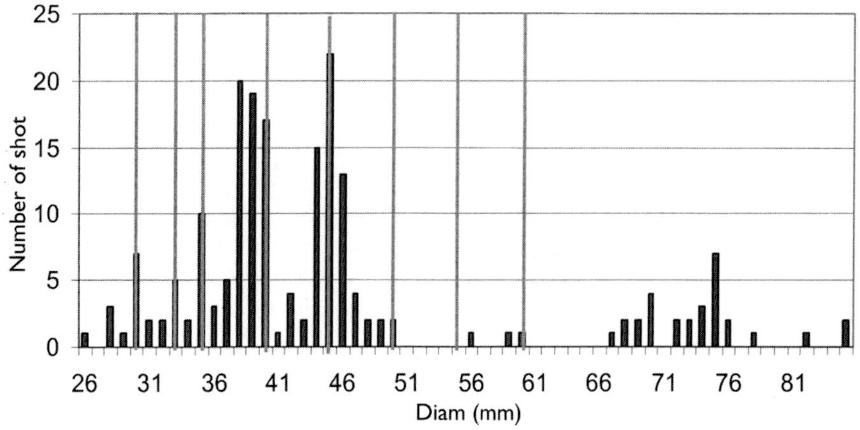

*Figure 5.55 Graph of inset and dice and ball shot*

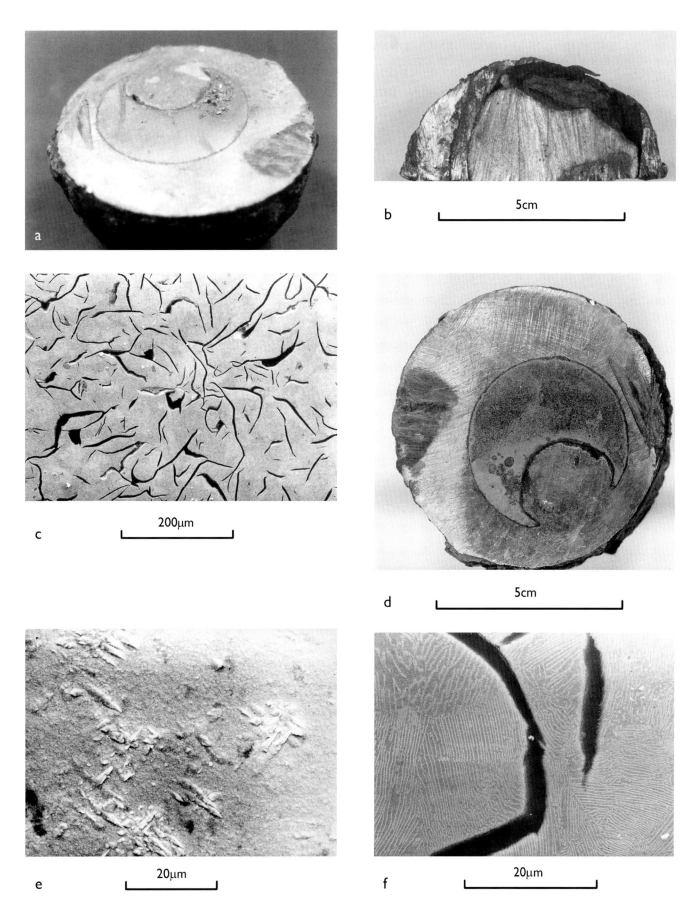

*Figure 5.56 Analysis of 83A0239: a) cross-section across core; b) blow holes, c) cast iron core; d) cross-section; e) SEM of the lead; f) perlite matrix of lead core*

be of cast iron surrounded by cast lead. Metallographic analysis confirmed that both were typical cast structures (Fig. 5.56c, d). Small (chill) grains were present at the iron/lead interface followed by larger columnar grains which grew from the interface. The same features were observed going from the surface of the shot which had been in contact with the mould. Scanning electron microscope photographs of the lead revealed the presence of flow marks at high magnification (Fig. 5.56e). Bulk chemical analysis of the iron using both electron probe micro-analysis (qualitative) and then inductively coupled plasma mass spectroscopy (semi-quantitative) revealed the elements iron, magnesium, phosphorus, silicon, and sulphur, all consistent with cast iron. A general scan of the lead showed it to be pure lead. Metallographic examination of the iron showed what appeared to be graphite flakes within a pearlite matrix with additional non-metallic inclusions (Fig. 5.56f) containing manganese sulphide and small amounts of titanium. Inclusions near the surface tended to be corrosion products containing chlorine, or lead and iron. Further investigation identified additional trace elements and products compatible with saltwater corrosion.

The shot was manufactured using a cast iron core, possibly a reject cast iron shot. The large quantities of magnesium, phosphorus and titanium suggest the use of ores from Berkshire, Norfolk, Wiltshire or Sussex but the evidence is not conclusive (R. Smith, pers. comm., 2008). This was placed into a mould and molten lead poured in. This would initially chill and solidify on the cold iron core. Either the downward pressure of the molten lead would push the core towards the bottom of the mould and keep it pinned towards the base (as with the dice shot) or the fluid would cause the core to float. Both would result in an eccentric centre of gravity, leading to corrosion where the iron was close to the mould, as previously described.

These are the shot described in the 1547 inventory (Starkey 1998) as '*shot of iron and lead*', and listed for quarter slings (*ibid.*, 5321), falconets (5276), falcons (5275) and sakers (5274, 5317, 5348). The inventory of items in the Tower for 1595 (Blackmore 1976, 276) includes 34 '*shott covered with lead for ffalcon*'. A note appended to this states that there are examples in the Beauchamp Tower Museum of stone and iron cannon balls excavated from the Tower including examples of shot consisting of rough chunks of iron covered with lead, some of falcon size. The largest shot mentioned is for the saker, 75–89mm (3–3¾in; weight of iron shot, 4lb 10½ oz–7lb 3½oz). The other named guns include falcons, 65mm (2½in, 2lb 1½oz (64mm/836g) in iron shot), falconets, 50mm (2in, *c*. 1lb 7oz (51mm/595g) in iron), quarter slings (uncertain, but probably 1½–3½in; 38–89mm) and bases (generally less than 60mm). The sectioned shot is 2¾in (70mm), too large for a falcon. It may be for one of the wrought iron slings. The weight is also extremely heavy for its size, even with the corroding core it is 4lb 6oz (1758g). A shot of this size in iron should weigh about 3lb (*c*. 1.4kg). The powder type or charge is unrecorded for a shot of this size and weight.

*Matching shot to guns*
The peaks shown at 65mm and 74 mm could relate to shot for falcons, a small saker or one of the sling groups respectively. The few smaller shot could be for the largest of the bases recovered, 82A4076 and 81A1082 with bores of about 60mm.

*Distribution*
Distribution is extremely limited (Fig. 5.57), with a similar bias as seen with the inset dice shot (Fig. 5.53). Seven shot were found in M6 and U6. M6 is a main storage area for iron and stone shot and U6 had an *in situ* sling (81A0645, shot 82A4215, of 80mm). All of these shot were between 74mm and 76.31mm and could have been used with this gun. In these areas the shot could have served either a sling on the Upper deck or a saker in the Sterncastle.

The second concentration is limited to M2–3, another general storage area for shot with access to the Upper and castle decks in the bow. There are fifteen shot here (and one below in O2) with bore sizes of 52mm, 54mm, 56mm, 66mm (4 shot), 67mm, 69mm, and 73–76 mm. Inset dice shot (occasionally of similar diameter) was also found in this area (Fig. 5.53). The clusters suggests shot for at least three guns, possibly a quarter sling or large base, a falcon (or possibly a demi sling) and a small saker (or sling). The *Anthony Roll* lists iron shot for sakers, falcons and slings and lead shot for bases.

**Parallels with other sites**
The (Dutch) *Scheurrak SO1* wreck site of 1593 has a number of composite shot (Puype 2001) including cast iron shot with a lead 'skin' (32763) of 64mm, corresponding to a Dutch 2-pounder cannon. The surface of the skin is 5mm thick, and there is no indication of whether it is cast or sheet. There is no indication of whether the iron core is cast or wrought and Puype questions the reasons behind increasing the weight of a projectile so much as, if this required a larger powder charge, it might endanger a gun (*ibid.*, 121).

**Odd shot**
There are fourteen unusual shot (Table 5.14) as described below. Most have been radiographed in order to try and understand their method of manufacture. Most are from MU11/12 (Fig. 5.58).

**Hemispherical shot**
There are six examples of lead shot which consist of single hemispheres, or two hemispheres held together by concretion (Fig. 5.59). All show the remains of a small amount of iron within the lead. These vary in

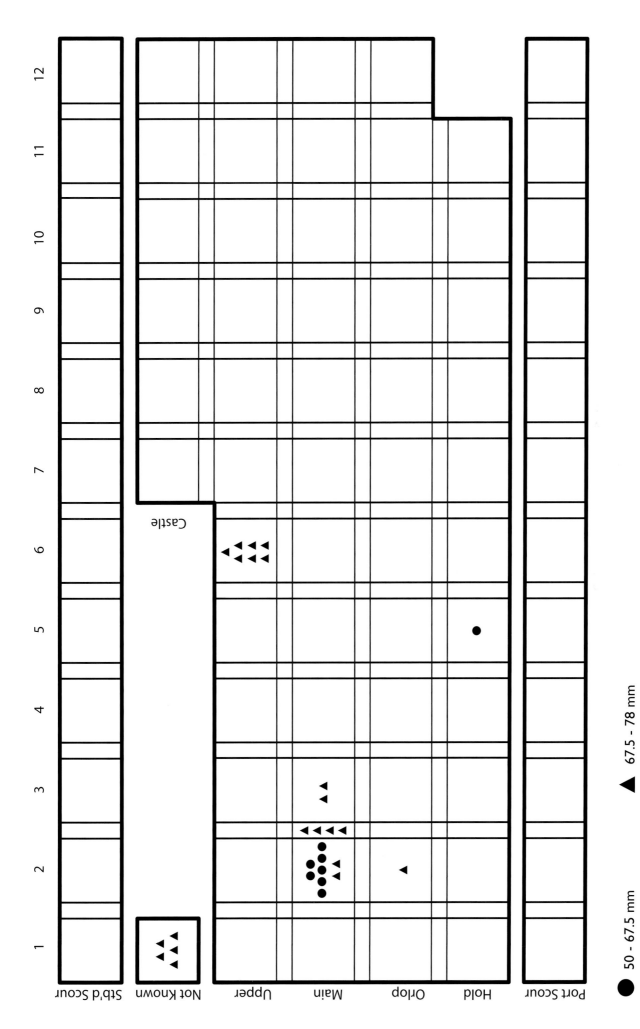

*Figure 5.57 Distribution of inset ball shot*

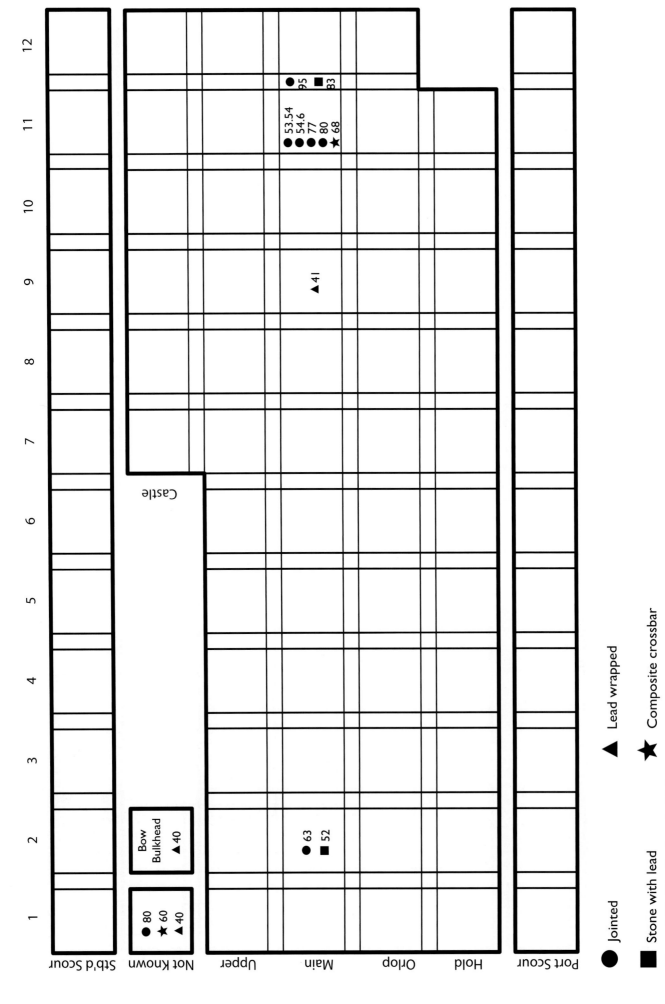

Figure 5.58 Distribution of odd shot

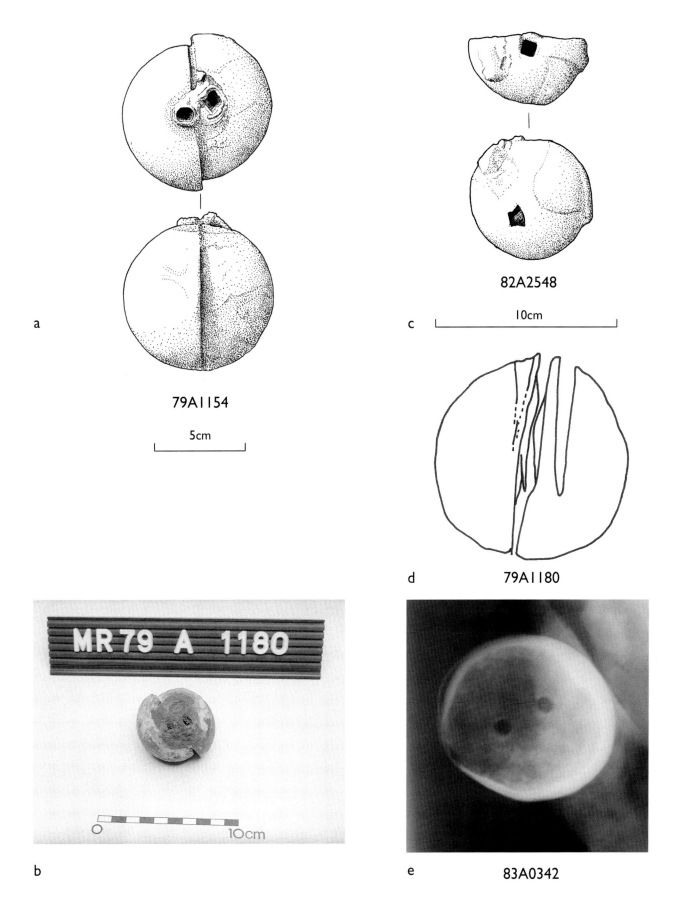

*Figure 5.59  Hemispherical shot: a) 79A1154; b, d) 79A1180; c) 82A2548; e) radiograph of 83A0342*

## Table 5.14 Odd shot

| Object | Sector | Diam. | Weight | Description |
|---|---|---|---|---|
| 79A1154 | M11 | 80.00 | 2583.4 | jointed; 2 hemispheres concreted along flat faces; each has 8mm square hole parallel with flat face; 1 hemisphere slipped relative to other; radiograph suggests either bent spike or 1 spike held by another smaller 1 crossing over it, or an anchor shape |
| 79A1161 | M2 | 52.00 | 876.0 | stone with lead outer |
| 79A1180 | M11 | 53.54 | 902.5 | jointed; as 79A1154; radiograph shows tapering spike hole 32mm long |
| 79A1185 | M11 | 54.60 | 895.7 | jointed; 2 separate hemispheres each with 5–7mm square hole parallel to flat face; radiograph suggests 'Y'-shaped anchor mechanism |
| 79A1186 | M11 | 77.00 | 2516.6 | jointed; as 79A1154, 7.4mm holes; corrosion circle encompasses both holes; radiograph suggests spike in each hemisphere flaring from flat face outwards to circumference |
| 81A0141 | M11 | 68.00 | 1505.0 | composite cross bar; single square hole through shot |
| 81A4060 | SS8 | 34.00 | 195.3 | stone with lead outer |
| 81A4851 | M9 (cabin) | 41.00 | 500.0 | void of small shot partly folded in lead |
| 82A1571 | bow | 40.00 | 850.0 | small shot partly folded in lead, possibly dice; non-magnetic but appears to be lead so probably inset; weight includes lead |
| 82A2548 | M2 | 63.00 | 716.2 | jointed; single hemisphere with 8mm square hole parallel to flat face; radiograph down onto flat face shows 'S' void; side view shows straight void parallel to face with single void going from each end of this towards circumference |
| 82A2588 | ? | 82.00 | ? | jointed; noted in museum |
| 82A4156 | ? | 60.00 | ? | composite cross bar; tapering single hole through lead shot |
| 83A0342 | M11/12 | 95.00 | ? | jointed; in concretion; radiograph shows 2, 9–9.5mm, holes for spikes with gap between of 20mm |
| 83A0411 | M11/12 | 83.00 | 1324.7 | stone, lead outer; spaces on equator for 4 tapering wooden plugs, 2 *in situ* |

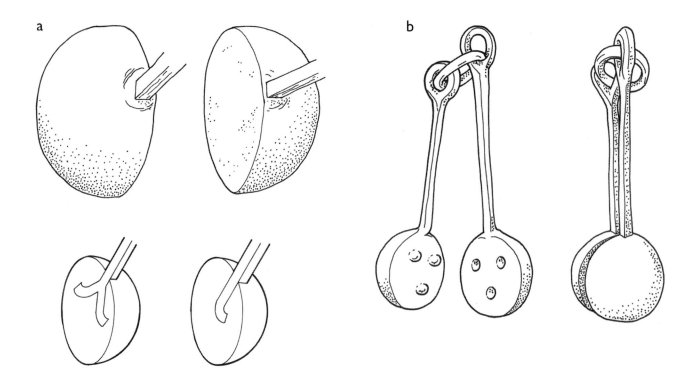

*Figure 5.60 a) Suggested reconstruction of hemispherical shot; b) after Blackmore 1976*

diameter from 50mm to 95mm. All retain a small (5–9mm) hole in each hemisphere close to the flat face, which does not penetrate the entire shot and was often found empty. It is suggested that each half contained an anchoring spike close to and parallel with the flat face that extended either to a bar or chain which connected the two hemispheres (Figs 5.59 and 5.60a) and was similar to the expanding bar shot shown in Blackmore (1976, 192, 200; Fig. 5.60b). As the bar is not present, it is impossible to confirm whether this was expanding bar shot, chain, or merely a jointed bar.

The manner of securing the spike/bar into the shot is difficult to interpret. The shot were radiographed (details in archive) and all show white voids for securing an iron spike, some tapering, and up to 32mm in length, but the shape is difficult to interpret. A tapered bent spike is suggested by 79A1186, and a deep 'S' by 79A1180 (Fig. 5.59). Corrosion products suggest that the wrought iron bar was part of the casting rather than a later addition. Casting wrought iron with lead in an open mould has been suggested and illustrated for the bar shot recovered from the sixteenth century wreck off Alderney (McElvogue 1999). These have diameters of 46.3mm, 76mm, 77mm, 80mm, 83mm and 94mm (Monaghan and Bound 2001, 59). The form is similar to the illustration of *Scheurrak SO1*(27025) drawn by Puype (2001, 121–3). *Scheurrak* examples are of 75mm and incomplete. Puype describes these as having their bars in the plane of the base and suggests that the bar is within one hemisphere only, seriously offsetting the balance of the shot. In the case of the *Mary Rose* examples, (which seem identical) there is evidence for bars in the planes of the bases of each hemisphere which make up the complete shot. We are fortunate that some of these are still concreted together.

Unlike those listed as '*langrel*' shot on the *Batavia* (Green 1989, 59) the *Mary Rose* examples are not cast iron, nor do they have lugs and depressions on their flat faces to ensure a tight fit. They may be anchored and made in a similar fashion though, attached by a wrought iron bar, hinged at the joint. Possible reconstructions are suggested in (Fig. 5.60).

*Distribution*
Five were from MU11 (79A1154, 79A1180, 79A1185, 79A1186, 83A0342) and one from M2 (82A2548, 60mm, 2½in) (Fig. 5.58). Those found in M11 are of three sizes, 50–4mm (79A1185, 79A1180), 77–80mm (79A1186, 79A1154) and 95mm (83A0342). All are too large for the base recovered here (79A0543). Weights vary between 895g and 2583g (Table 5.14).

Chain ('*cheyned*') shot is not mentioned in inventories of items in the Tower until 1589. In 1568, however, there are '*Joincted shot*' (jointed shot; Blackmore 1976, 265), and these may be what are present on the *Mary Rose* assemblage. These may be similar to some of the types of expanding or link shot shown by Blackmore (1976, 192, 200) that are of similar dimensions and weight to the larger *Mary Rose* examples. Alternatively this could be the '*langrel shot*' referred to by Smith (1627, 86). Chain is usually easy to recognise on radiographs, but it is not visible.

**Composite cross bar shot (81A141, M11)**
One shot (81A0141; Fig. 5.61) has been identified as lead with two 9 × 9mm square holes on the castline, diametrically opposed, for a corroded iron spike. It has a diameter of 68mm and weight of 1505g (3lb 4oz).

This is similar in construction to the lead bar/spike shot illustrated by Puype (2001, 123) for the *Scheurrak* wreck (33605) of 75mm diameter. As the *Mary Rose* example has holes of equal size, this suggests that the bar was cast with the shot as opposed to hammered in later. In the latter instance one hole is usually found to be smaller and the spike tapers. A variation may be seen in 82A4156. Radiography of the concretion suggests lead, but it may be good cast iron. This has a diameter of 68mm with a tapering hole of 20mm in the centre and 25mm at the circumference.

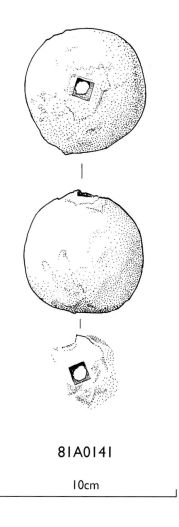

*Figure 5.61 Lead shot 81A0141*

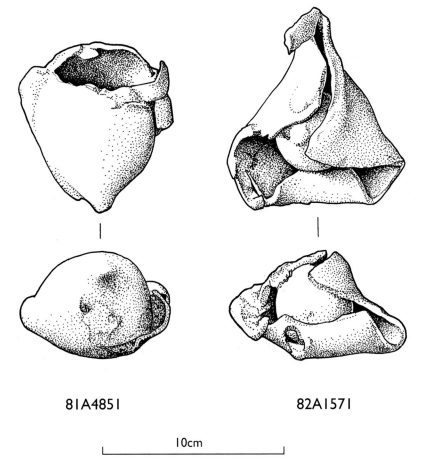

*Figure 5.62 Lead composite shot 82A1571 and 81A4851*

### Other composites

Two items were found which may represent round shot loosely folded in lead. One contains a 40mm shot which appears to be of lead, but may be composite. The lead is folded over it and has a weight of 850g. It was found in the bow (82A1571). The second (81A4851) looks similar, but represents a spherical void of 43mm and weighs 500g and retains the outward appearance of wrapped shot (Fig. 5.62). This was found in the Carpenters' cabin in M9. A similar 'half sphere' of lead of 3mm in thickness with a diameter of 70mm was found in the cabin (81A2536). This looks like a ladle, with a lead handle apparently attached to the outer surface (Fig. 5.63). It was hoped that this might be some remnant of a partial casting from within a corroded casting ladle, but it may be some form of as yet unidentified shot. Rolls and small ingots of lead were found in the cabin and in chest 81A5783 together with a single cast lead shot of 10mm (81A4431). Analysis by XRF indicated a slightly impure lead for both the handle and outer surface of the sphere (96.1/98.5 lead respectively).

There are several shot which appear to be some form of stone covered with

*Figure 5.63 Unidentified hemisphere 81A2536*

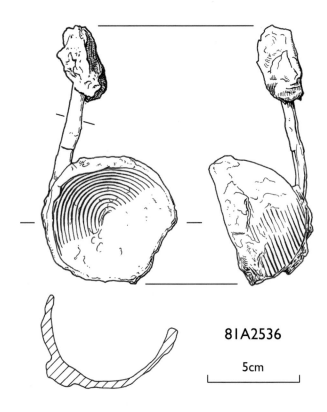

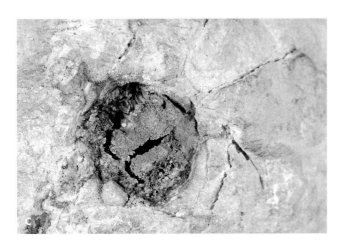

Figure 5.64 Stone and lead shot: a) 81A4060; b) 79A1161; c–d) 83A0411 and close-up of wooden plug

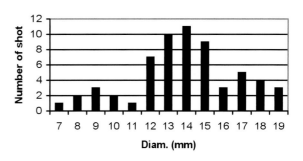

Figure 5.65 Graph of lead shot

lead. 81A4060 from SS8 has a diameter of 34mm and weight of 195g (6.9oz) with several gouges into the lead (Fig. 5.64a). The lead (not analysed) appears to be rough and hammered and the stone looks like a beach pebble. It was found together with five other inset shot with diameters of 46mm, 29mm, 39mm, 40mm and 38mm. A 52mm covered stone shot was recovered from M2 (79A1161) weighing 876g (1lb 13oz; Fig. 5.64b).

83A0411 is extremely unusual (Fig. 5.64c–d). Found in M11/12, it appears to be stone roughly hammered over with lead. It has a diameter of 83mm and weight of 1325g. On a virtual equator are four equally spaced tapering holes of 7–8.5mm diameter; one has a depth of 13.5mm. Two are filled with tapering wooden plugs, almost fuse-like in appearance. A test with nitric acid showed the presence of carbonate, which may indicate that the core is limestone.

## Shot of lead

Four hundred lead shot are listed in the *Anthony Roll* for the 30 bases, thirteen per gun, but the *Roll* does not differentiate between double, single or demi bases. Norton (1628, 23) lists bases as having a bore of 1¼in (32mm) and firing a ½lb (225g) shot. As we have seen, most recovered shot of this size is composite. Fifty handguns with 1000 lead shot (20 each) are also listed.

Figure 5.66 Sprue and casting on 81A6823

*Figure 5.67 Lead shot: a) holes in 79A1181; b) concentric circles on 81A5655*

These rounds would also have served the gun-shields (see Chapter 7). The 1547 inventory includes '*hagbuttes*' '*hagbushes*', '*half hacks*' and '*handgonnes*' (Starkey 1998), but not all vessels are listed with handguns and there is no specified shot. Many entries read simply '*Hagbuttes with their apparell*' (*ibid.*, 7521) or '*Handgonnes complete*' (7388). Sometimes details from the fortifications indicate what this could mean; for example at the Tower '*Demy hakes or handgonnes*' (3768) is followed by '*Flaxes* [flasks] *and towche boxes of iche*' (3769). '*Flaskettes for powder*' are also mentioned (3982). When specified, shot for handguns is invariably of lead (3914, 3939). Hagbusshes are usually listed with lead shot, but there are a few exceptions: '*Shot for hagbusshes of leade diced with stone & yrone*' (4694); '*Hagbusshe shot with dice of yrone*' (5556).

Sixty-two solid round lead shot have been identified, only 6% of the 1000 listed. These range in size from 7mm to 19mm, with at least one example per millimetre (Fig. 5.65). Weights vary between 9g (7mm, 79A1207)

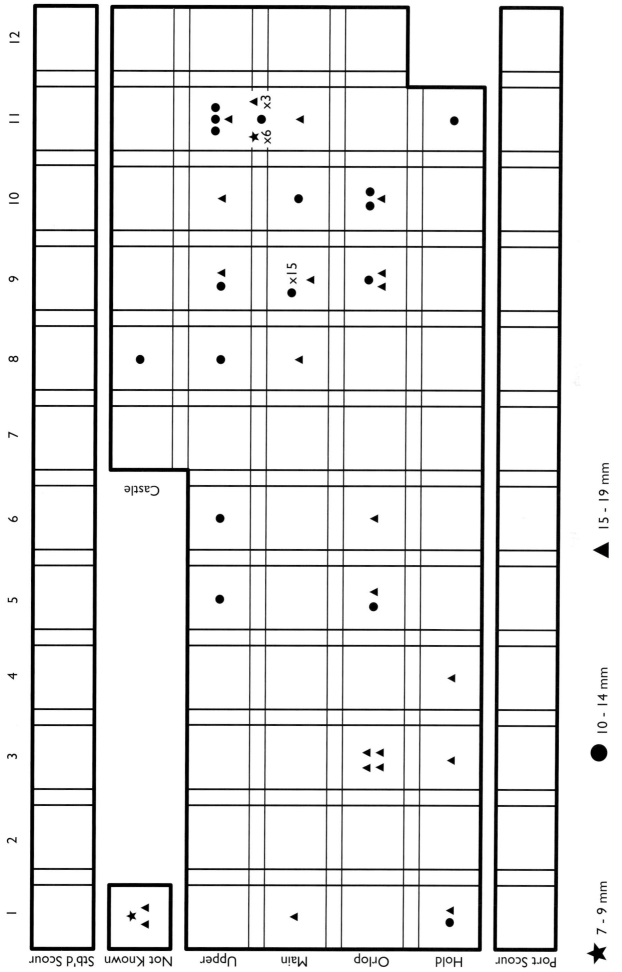

Figure 5.68 Distribution of lead shot

and 37g (18mm, 81A0676). Shot of 15mm weighs about 25g (79A1172). These may have been used with pistols, gun-shields, muskets or heavy harquebuses. Small shot may have been containerised as grapeshot. Seven of the 50 handguns listed were found, one in M9 (81A3884), three in U9 (81A2405, 81A4554, 81A1842) and two in U11 (81A2675, 78A0333) and one in SS8/9 (82A4582). Remains of up to eight gun-shields were also found, mostly in O10 (see Chapter 7, Figs 7.17–18).

The shot was made in a bipartite mould. All show a visible castline and an attempt to neaten the sprue hole by clipping or cutting (Fig. 5.66). In some instances (79A1181) there are two opposing holes on the castline (Fig. 5.67a) of which one is small and square and the other larger and round (the sprue). In some cases concentric circles are visible perpendicular to the castline (81A5655). This suggests lead was poured into a metal mould and chilled rapidly (Fig. 5.67b). Although the post-casting finishing appears to be rudimentary, these are smooth enough to be used with handguns rather than contained within a larger charge as in grapeshot.

As the shot may have been wrapped, estimates of the number of guns from shot size is dubious. Clusters found within the same area of the ship may be more meaningful (Fig. 5.68) and the distribution includes one from the Castle deck, ten from the Upper deck, ten between the Main and Upper decks (all sector 11), 20 from the Main deck (mostly M9), thirteen on the Orlop deck and five in the Hold, and three unknown (Fig. 5.68). Three sizes seem to be represented, 7–9mm, 10–14mm and 15–19mm. The 7–9mm shot is too small for the six handguns recovered, all with bore estimates of over 12.5mm. These may have either been wrapped or for pistols. All occurred in M/U11, three of them (79A1207/1–3) in a small basket (79S0059). Loose cask staves found in this area could have been a container.

The Castle deck example (83A0652) is 11mm. The closest other shot were found in U8–U9 (13.6/14mm and 18mm) suggesting at least two handguns. Upper deck shot can be separated by size and location. Two were found amidships (79A0444 in U5, and 79A0493 in U6); both are 13mm. The others were found in U8–11 and are 12–18mm. Remains of three muskets were found in U9 and U11; shot of 12–14mm would have been used for these. Those in M/U11 include three of 7mm and three of 8–9mm, four between 10mm and 14mm and five between 16mm and 19mm.

There is one shot on the Main deck in the bow (79A1139), of 17.5mm and another of 15mm in M8 (83A0478). The largest concentration on this deck is in the Carpenters' cabin in M9, with fifteen of 10–14mm including two (81A5735, 81A5736) of 12mm diameter, directly associated with the handgun found there, 81A3884. Six (81A6822-7) were excavated from a sample (81S0560) taken from the top of the eastern end of the northern bench within the cabin.

The shot on the Orlop deck are the most difficult to explain. The main areas of use would be the Upper and Castle decks. The four shot in O3 were found high up in the Tudor levels and may relate to port side derived material from the upper decks. The similar sizes of groups in O/H3 and 4 suggest a similar source (14–18mm). There is also a companionway in this area giving access between decks. The other cluster is in O9–10, with six shot of 12–19mm. These may have been used for the gun-shields stored in this area. There is a single shot of 13mm in H11. No specific metallurgical analysis was carried out on the lead shot. Interestingly no container filled with lead shot survived.

## Canister shot

There are 52 artefact numbers allocated to elements which make up three different forms of wooden canister. The species of wood varies, even within a single canister, and includes ash, elm, oak, pine, and Scots pine. They contain flint, some of it smashed to form extremely sharp 'shrapnel'. These probably functioned as anti-rigging and anti-personnel ammunition.

These are described as '*case shot*' by Smith (1627, 86):

> '*A* Case *is made of two peeces of hollow wood joined together like two halfe cartrages, fit to put into the bore of a Peece; & a* Case Shot *is any kinde of small Bullets, Nailes, old iron or the like, to put into the case, to shoot out of the Ordnances or Murderers. These will doe much mischiefe when wee lie boord and boord*' concluding '*All these are used when you are neere a ship, to shoot downe Masts, Yards, Shrouds, teare the sailes, spoile the men, or any thing that is above the decks.*'

Norton (1628, 58) suggests that '*Base and Burre, Musket or any other kind of Murthering Shot, being put vp in bagges or Lanthornes fitted to their Calibres*' are to be used with slings and portingale bases.

By this time both of these guns are yoke-mounted swivel guns but still of wrought iron. It is clear from the tables he includes that port pieces, bases and fowlers were still firing stone shot and slings cast iron shot. The large size of some *Mary Rose* canisters suggest that they were used with large guns like port pieces.

References to payments in 1596 (PRO WO 49/20, f. 58v) include 40 '*cases of wood for base shot*' and 22 '*cases of wood for burr shot*'. This suggests either different contents, or possibly particular uses (Brown, pers. comm.). By the time Smith is writing in the 1620s they are listed together as though they may be completely

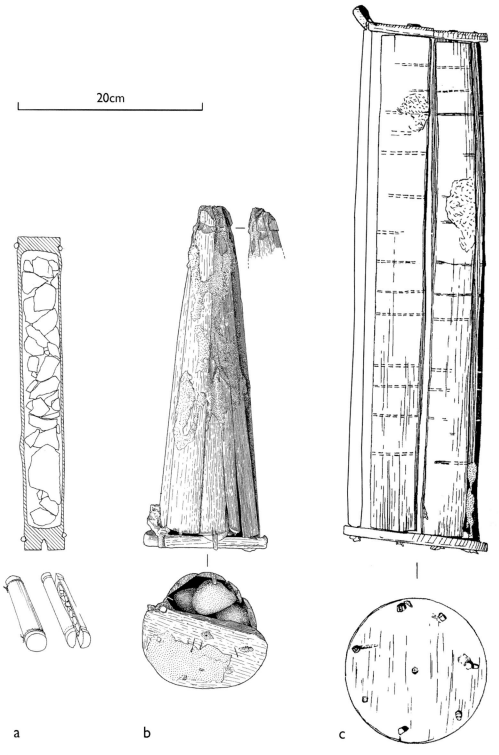

*Figure 5.69 Three forms of canister shot: a) cylindrical canister; b) conical lantern; c) cylindrical lantern*

interchangable. In 1622 (PRO WO 49/52, f. 146v) there is a delivery to ships including cases of wood for stone and burr. Within the next few years the list includes cases of wood for hailshot (Brown, pers. comm.).

Blackmore (1976, 243) quotes an inventory of around 1600 (PRO E 101/67/7) which describes different types of shot: '*Hollow shott*' is described as '*full of small shott*', '*Dise shott*', is '*made of Iron square in forme*' and '*Burr shot alias Haileshott*' is '*made of fragments of Iron roughe and ragged*'.

The 1586 inventory for the Tower includes dice shot '*of iron, leade and stone and of diverse quantities*'

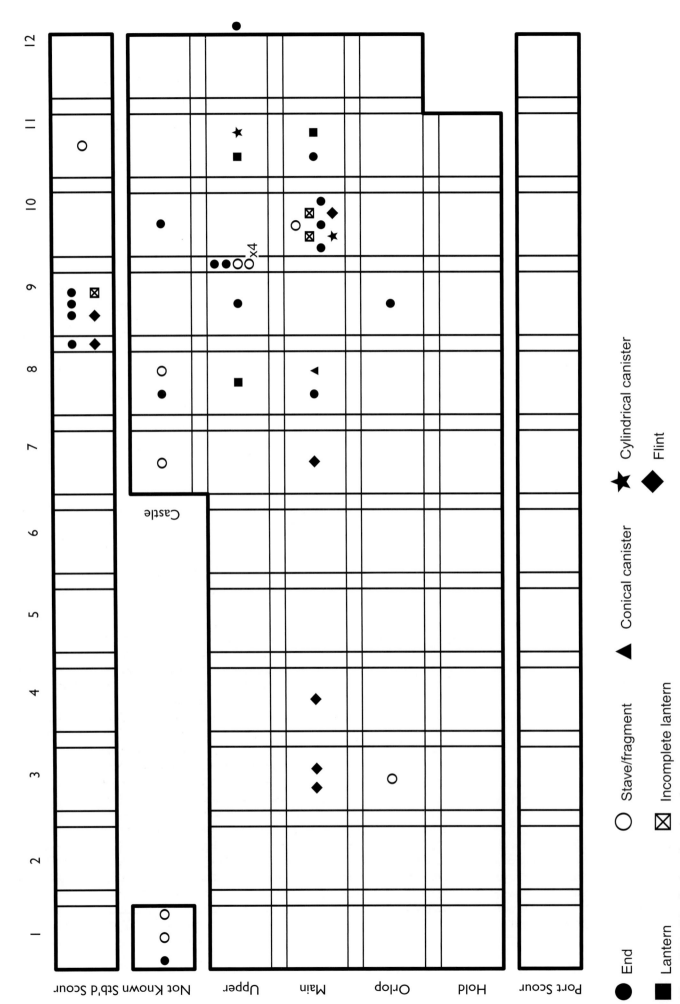

*Figure 5.70 Distribution of canister shot*

(Blackmore 1976, 266). In 1589 (*ibid.*, 267) '*Dice shott*' is listed separately from '*Burres alias Haileshott*'. This distinction is repeated in the 1595 inventory (*ibid.*, 276). A further division is made in 1603 (*ibid.*, 281): '*Cases filled with square shott*' (for demi cannon, culverins, demi culverins and sakers) and '*Hayleshott viz. Burres and Dice shott*' and '*Bace shott*').

The 1547 inventory (Starkey 1998) demonstrates that lead shot for bases (swivel guns) often contained iron dice, so the dice shot can be explained as possibly being square dice to be made into base shot, whereas the rougher dice fragments may have been known as burr shot and used as hailshot, hence the term hailshot pieces, and hailshot viz burr. In 1620 (Blackmore 1976, 286–7) '*hollow shott*' is listed for demi cannons, culverins, demi culverins and sakers; '*Cases of wood with Iron Shott*' for the same guns and '*Square shott of iron In 148 barrells*'. There is a separate entry listing '*Base and Burr*'. Cases of wood for burr shot are listed in the extensive inventory of 1635 (*ibid.*, 289), but that of 1665 (*ibid.*, 308–9) lists '*Tyn Cases fill'd with Musquett Shott*' for culverins, demi culverins, sakers, minions and 3-pounders. Beneath these are simple listings of '*Bace & Burr*', '*Barrs of Iron*' and '*Handgranadoes*'. A separate entry lists '*Tynn Cases filled with iron pestles for Demy Cannon*'. Wooden cases filled with '*Nailes & Stones*' are listed in the 1675 inventory (*ibid.*, 313).

An entry in the 1547 inventory for '*Shotte of stone for port peices*' (Starkey 1998, 4535) is directly followed by '*Cases of haileshotte for the same*' (4536), directly under '*stone shot for port pieces*', so it is clear that port pieces used cases filled with hailshot. The *Anthony Roll* lists '*Dyce of iron for hayleshot*' for every vessel listed with hailshot pieces. These were found on the *Mary Rose* loaded with small dice.

Case or canister shot were also recovered from the *Scheurrak SO1* wreck of 1593 (Puype 2001, 116–28). The assemblage includes three complete examples and fragments of six others. All are of wood, in two longitudinal halves 300mm and 314mm in length with a diameter of 65/62mm, the flat faces of which have been hollowed out. The two halves were probably held together by rope slotted into a groove at each end, similar to *Mary Rose* examples. These would fit a Dutch 3-pounder gun. The contents are assumed to have been stone pellets or musket shot, lead round (but unpolished) shot of 17mm, hailshot and buckshot, of which there are examples around the site. It is possible that some may have been used within cartridges in larger guns to produce '*evil shot*' or '*quaed scherp*' (Puype 2001, 124). None has been found with contents *in situ*.

The *Mary Rose* assemblage includes simple cylindrical canisters (two) conical '*lantern*' shot (one) and cylindrical '*lantern*' shot (three complete and seventeen loose bases) (Fig. 5.69).

*Cylindrical canister*: These are made out of a single piece of wood cut in half lengthways and hollowed out to provide a central container. The ends are grooved to take a securing string. Both found are small, possibly for top pieces, bases or small slings. One is pine.

*Conical lantern shot*: There is one example consisting of tapering staves bound together to form a cone. The narrow end is grooved to take a securing string. A separate top is pegged to the ends of the staves.

*Cylindrical lantern shot*: Manufactured in a similar fashion to lanterns, with disks forming the ends pegged to staves. The staves touch each other. There is a central hole in each end of the canister, possibly for an additional securing string. The staves show discolouring suggestive of binding. Seventeen separate ends were found. Those identified include five in oak, and three each of pine, elm and ash. Wood identification of staves to date are all pine or Scots pine.

**Distribution**

The distribution shows a predominance for the stern in sectors 7–11 of the Main, Upper and Castle decks and the Starboard scourpit (Fig. 5.70), with only an isolated

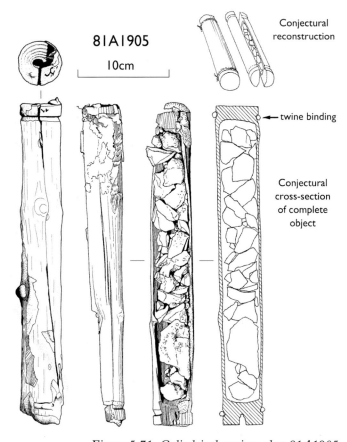

*Figure 5.71 Cylindrical canister shot 81A1905*

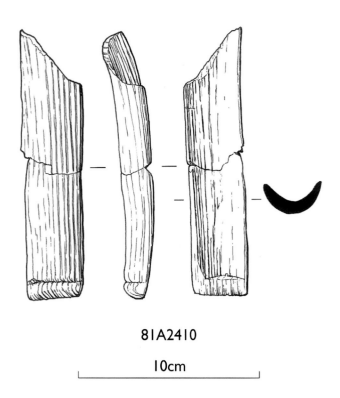

*Figure 5.72 above) Interior of cylindrical canister shot 81A1905; below) fragment 81A241*

long and 45mm across, retaining the original timber at each end (weight 457g; Figs 5.71 and 5.72, above). The contents are shattered, possibly burnt, flint fragments. The longitudinal split is not absolutely central, but twists to avoid a knot in the wood. An 8mm wide, 3mm deep, groove for a binding rope is positioned 12mm from each end. The outside of the wood is stained black and there is gritty material within the flints which may include an incendiary mixture (not analysed, 82S1230).

An incomplete half of a canister, hollowed out from a single branch of pine was found in M10 (81A2410) (Fig. 5.72, below). It is 143mm long, 30mm wide and weighs 12.7g. The base is 3mm thick. Towards the finished end the depth of the hollow decreases from 10mm to 8mm. There is no indication of contents. The only guns recovered small enough for these are the bases. 81A2410 would fit base 79A0543 found in M11 with a bore of 45–50mm and *in situ* shot of 38mm.

### Conical lantern shot

A single, conical container, 82A2298 from M8, is made of seven staves of differing widths which taper to a point (Fig. 5.73). It has an overall length of 365mm and diameter of 132mm with a dry weight of 2500g. The wide end of each stave has an integral tang, up to 20mm long and 8mm wide, which locates into the end piece. The staves are slightly convex towards the outside of the canister, with flat faces facing inwards. The edges overlap one another, but there is no discernible pattern. The staves taper from 45mm at the top to 18mm at the base. A very rough groove ('*croze*') has been cut into the outside of the staves 22mm from the end to take a binding. The end sits over the top of the staves and is not fitted into this croze. The outer edge is flat and 15mm thick. There is a central hole and four circumferential ones positioned about 20mm in from the edge. The largest has a diameter of 10mm. There is black concretion staining to the top and sides. A portion of the top is missing to reveal the tightly packed flint beach pebbles within (flint size typically 5 × 3mm, contents 82S1131).

### Cylindrical lantern shot

These are of similar construction to the lanterns recovered from the *Mary Rose* (*AMR*, Vol. 4, chap. 8) and references to '*lanthorn shot*' (Smith 1628, 58) describes them well. They consist of two end disks (including five in oak, and three each of pine, elm and ash) pegged to lath-like pine/Scots pine staves (Fig. 5.74). The number of staves per container varies as well as their width and length (422–540mm) and thickness

base fragment from a cylindrical lantern example in O9 (79A0684). There is only one stave fragment in the bow (79A0288, O3) with three loose flints from M3 and M4 (80A0821, 81A0999) and one from M7 (81A3225). Perhaps significantly, there are none from the shot stores in M2 and M6.

### Cylindrical canister

Two were found. A complete example, 81A1905, from U11, is made from a longitudinally split branch hollowed out to form two semicircular troughs 336mm

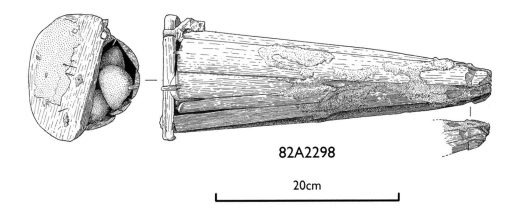

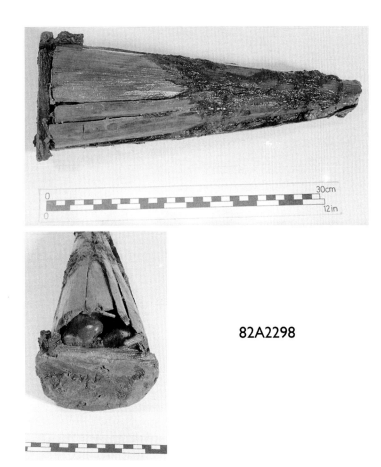

*Figure 5.73 Conical lantern shot 82A2298*

and diameter of the ends (120–170mm). There is a central hole in each end, possibly for an additional securing string. The staves show discolouring suggestive of binding.

Seventeen separate ends and three complete canisters were recovered which, together with loose staves that are not easily assigned to individual canisters, suggest a minimum of fourteen examples. The bases indicate six diameters 120mm, 130mm, 150mm, 165mm, 170mm and 175mm (4in, 5¼in, 6in, 6½in, 6¾in and 7in).

*81A0101; U11*
This was the first complete canister found (length 540mm). Elements comprise seven slightly concave pine staves of varying widths and two centrally pierced, 10–12mm thick, ash ends with a diameter of 152mm (Fig. 5.74) that overlap them. The central holes may be for a suspension cord or possibly a fuse. The ends are pierced with circumferential holes to accept the tangs from the staves which are split and wedged and of which one is missing. Small fragments of woven fabric were noted around the central end-holes and there is staining

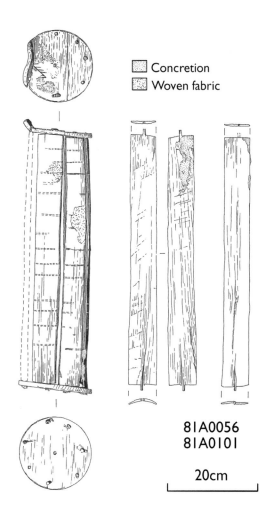

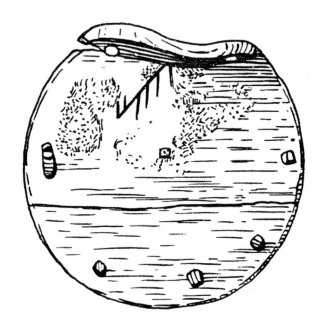

*Figure 5.74 Cylindrical lantern shot 81A0101*

*Figure 5.75 Cylindrical lantern shot 81A0101: a) binding fragment; b) mark on end; c) object in situ in seabed*

suggestive of rope binding over the entire length of the canister, with 15 turns. A small fragment of binding twine was recovered (Fig. 5.75a). There is an incised mark of one long diagonal line with two parallel lines to one side and four to the other on the outside of one end (Fig. 5.75b) with a similar mark on one stave. A separate stave (81A0056) from U11, with a similar marking, may belong to this canister. The canister was full of flint nodules and flakes (Fig. 5.75c) (see below) weighing 7760g. It was found by a vacant gunport in

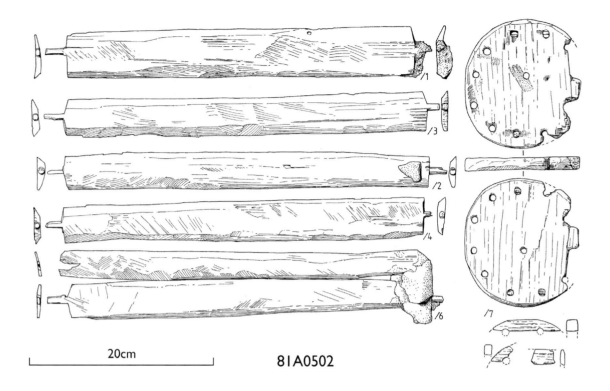

*Figure 5.76 Cylindrical lantern shot 81A0502 and flint contents*

U11 associated with gun carriage elements and shot ranging from 123mm to 214mm.

*81A0502; U8*
This nearly complete canister (length 422mm) comprises one end and a fragment of the other (diameter 131mm, thickness 13mm) with six slightly convex pine staves (Fig. 5.76). There are eleven circumferential and one nearly central hole in the complete end. The staves are 370mm in length with tangs of 25–30mm and are 30–60mm wide. The tang holes are 8–10mm in diameter. The contents included three concreted flints and 36 cobbles weighing 2001g (Fig. 5.76). This is of a similar size to an oak half base (81A5702), which may be the missing end to this canister, and conical canister (82A2298) found in M8.

*81A1183; M11*
Both ends and eight pine staves were recovered of this 540mm long canister (Fig. 5.77). The 170mm diameter, 15mm thick ends have eight circumferential holes of 8mm diameter 10mm in from the edge and two central holes in each end. One end has a similar, but not identical, incised mark to that on 81A0101 (above) also found on 81A2019 and 81A3061 (M10). The staves vary in width from 42mm to 55mm. The canister contained flint cobbles, pebbles and several flakes weighing 8883g.

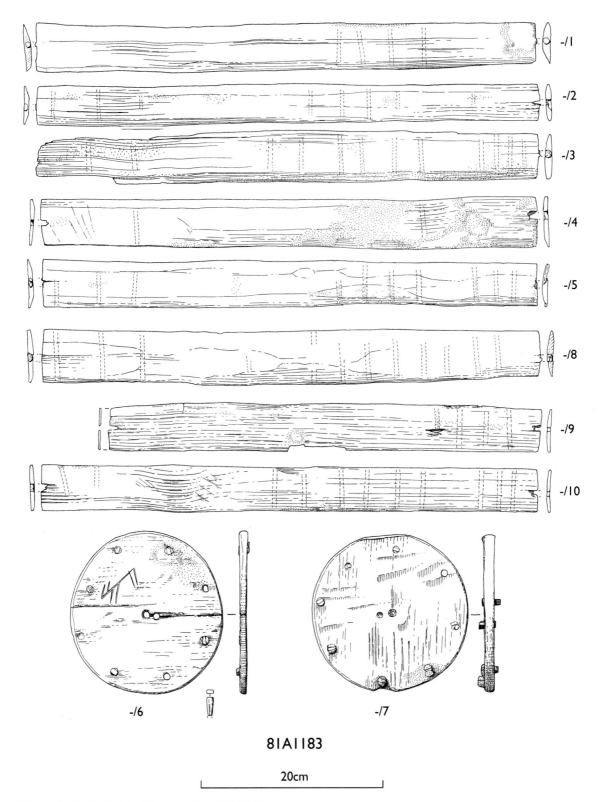

*Figure 5.77 Cylindrical lantern shot 81A1183*

*Partial cylindrical canisters*
Three 'half canisters' were recovered. 82A2696, from the Starboard scourpit (SS9) comprises the central portion of a pine canister, 480mm long and *c.* 120mm diameter, with six *c.* 55mm wide, 5mm thick staves, containing flint pebbles. There are no ends, although tangs are present on one end of four staves. Two pieces of string binding were attached (Fig. 5.78). Four ends were found in SS9, but, at 165–7mm, the diameters of these are too large.

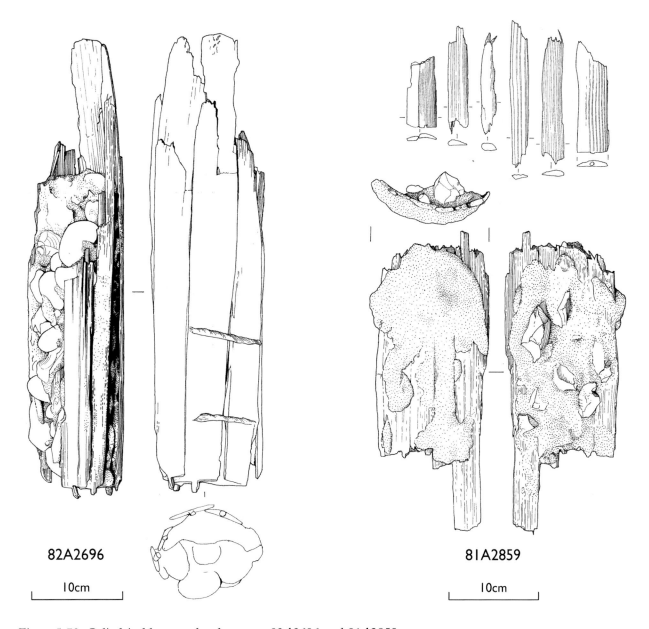

*Figure 5.78 Cylindrical lantern shot fragments 82A2696 and 81A2859*

81A2859 (M10) comprises nine pine staves, four of which are 21–33mm wide and are held together by concreted flints (Fig. 5.78). The rest are separate fragments (length 75–154mm, width 40mm, thickness 5mm). The overall maximum length is 320mm and the diameter c. 130mm. This object was recovered from concretion between the front right wheel of bronze demi cannon 81A3000 and the Main deck. Removal of the front axle and wheels revealed another portion of concreted flints (81A2936). A larger canister end, 81A2431 (149mm diameter), was also recovered from this area.

81A3061 (Fig. 5.79) was also found in M10 and comprises three staves of 290 x 50mm and flint held in concretion with a fragment of base showing incised lines, with a loose stave (81A3043) nearby (length 190mm, width 49mm, thickness 7mm).

*Loose ends and other fragments*

Seventeen loose ends were recovered as well as various staves and collections of flints. With the exception of a few of the latter, all were from the stern and details are given in Table 5.15 for those not already mentioned above. Examples are shown in Figure 5.80.

**Contents**

The contents of the canisters have not been studied in detail. Certain containers appeared to have a very dark, compacted, gritty sediment associated with the pebbles or flints, which may indicate that they had been

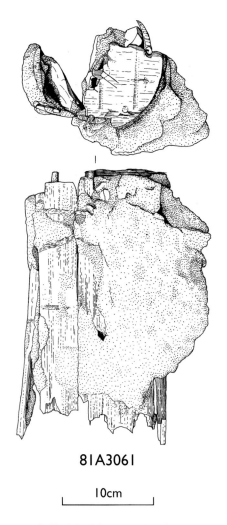

*Figure 5.79 Cylindrical lantern shot fragment 81A3061*

enriched with gunpowder as an incendiary. A small sample (details in archive) from several locations was briefly examined by Julie Gardiner who suggested that the raw material seems mostly to consist of cobbles typical of the Sussex Raised Beaches (nodule diameter up to c. 160mm; eg, 81A1905, Fig. 5.71; 81A0101, Fig. 5.74), with some smaller river cobbles (eg, 81A5028 from 81A0502, Fig. 5.76) and gravel flint, all of which would have been available within a few kilometres of Portsmouth. Some nodules are whole, others are naturally broken pieces, but most are deliberately smashed cobbles that have clearly been struck with a metal hammer. As flint bears the sharpest natural edge of any rock other than obsidian, these angular fragments would have been lethal.

**Matching shot to guns**

Although we cannot be certain which guns used lantern shot, the limited documentation favours wrought iron guns. As many of these were designed for medium or close range, this is an obvious choice of projectile. Experiments with a reproduction port piece loaded with cylindrical lantern shot at close range (see Chapter 3) showed an impressive scatter. The smaller cylindrical canister shot would have fitted some of the bases recovered and, if wrapped, some of the slings. The finding of material associated with some examples (eg, 81A0101) suggests that these may have been contained in cloth bags, or wrapped. This would enable an adjustable and quite accurate fit and might account for the variation in sizes and the lack of consistency between diameter of lantern ends and stone shot peaks. The largest end recovered was 175mm (78A0153) in diameter and the largest port piece bore was 203mm (71A0169). This may be indicative of the non-survival (or non-presence) of large canisters, or of greater windage, or external cloth wrapping. All port pieces found *in situ* were on the Main deck (M4, M5, M8, M11). The sectors where port pieces and shot occur together include M8 and M11. M8 contained a conical lantern shot (82A2298, diameter 132mm) which matches that of a single base for a cylindrical lantern canister also found in M8 (81A5702). Just above, in U8, a similar sized cylindrical lantern shot was found, 81A0502. The port piece in M8 had a bore of 180mm and an in situ stone shot of 172mm. All the stone shot found in this sector had diameters of 150–188mm, most were 160–167mm.

Six elements of canister shot were found between the front wheels of the *in situ* starboard side demi cannon (81A3000) in M10. The gun had come off the deck and a small wicker basket lay between the wheels and the deck. This area also yielded rammers and powder ladles of varying sizes and a large concretion containing wood which may be part of the larger cylindrical lantern shot 81A2936 or 81A2431. The ends range from 130mm to 170mm in diameter which together with a large end (78A0153, diameter 175mm) found outboard of the rudder, were probably used with port pieces in M11. Also in M11 was an incomplete cylindrical canister (81A2410), small enough to be used with the base found in M11 (79A0543) with a bore of 45–50mm and *in situ* shot of 38mm.

The complete cylindrical canister (81A1905) from U11 is the correct size for some of the bases recovered, with bores between 35mm and 80mm. The complete cylindrical lantern canister 81A0101 (diameter 152mm) was leaning against the inside of the starboard side of the ship amongst gun carriage elements in U11 and suggests that a port piece may have been stationed at one of the gun ports in U10/11. Four loose ends for cylindrical lantern shot were recovered with staves and flints in the Starboard scourpit (SS9), directly beneath an Upper deck gunport. The ends were found in pairs, two to the south side of the port and two to the north, with the staves aligned north–south between the ends, but at a higher level. It appears that these had fallen out of the gunport. The artefacts include 82A0944 (two ends, diameter 165mm) 82A0960 (flints, though not obviously associated) 82A0961 (one ash end, diameter 165mm) and, nearby, another ash end, 82A5030 (diameter 167mm) that may be its pair. Obviously there

Table 5.15 Cylindrical lantern loose ends and staves

| Object | Sector | Base diam./thickness | Stave length/width | Wood | Comment |
|---|---|---|---|---|---|
| 79A0684 | O9 | 100/10 | – | oak | tang *in situ*; Th. 10mm, L. 40mm |
| 81A5702 | M8 | 132 | | oak | half end, 5 tang holes |
| 81A2431 | M10 | 149 | | oak or ash | end frag. |
| 81A2109 | M10 | 170/14 | | oak | complete end, 8 tang holes & 1 central |
| 81A2019 | M10 | 150/8 | | ash | end, 5 tang holes, 1 central; mark like 81A3061 |
| 81A3043 | M10 | – | 190/49 | – | stave |
| 81A0032 | M11 | 150 | – | oak | half end, found near 81A1183 (complete) |
| 81A1117 | U9 | 150/18 | 470/43 | pine | frag., 5 holes, 1 stave frag. |
| 78A0212 | U9/10 | 150/20 | | pine | end, 14 associated stave frags |
| 78A0232 | U9/10 | 150/18 | | pine | end, 8 tang holes & central hole, prob. part of 78A0212 |
| 79A1364 | U9/10 | – | longest 200/26 | – | 3 stave frags, longest with tang; poss. part of 78A0212 |
| 78A0274, 0288, 0298, 0300, 0302 | U9/10 | | 277/60, 440/25, 450/43, 140/25, 121/20 | – | stave frags, poss. associated with 78A0212 |
| 79A1105 | C7 | | 405/30 | pine | stave frag. with tang |
| 79A1179 | C8 | 150/22 | – | elm | end, 3 tang holes, 1 tang *in situ* |
| 80A0165 | C8 | | longest 390/10 | pine | 2 stave frags |
| 73A0055 | C10 | | – | elm | end, 3 tang holes, 2 tangs *in situ* |
| 82A0944 | SS9 | 165/15 | 340–425 | – | pair of ends, each has 9 tang holes of which 7 align; 4 stave frags; 15 flints found beneath = ?part of |
| 82A0961 | SS9 | 165/11 | | ash | end, 9 tang holes & one central, poss. pair with 82A5030, with which holes match |
| 82A5030 | SS8/9 | 167 | | ash | end, in 2 halves, 11 tang holes & 1 central (8mm diam.), with 2 flints; poss. pair with 82A0961 |
| 82A2661 | SS11 | – | | – | 1 complete stave & 6 frags, L. between shoulders 34mm |
| 78A0153 | outboard rudder | 175/28 | | elm | end, 10 tang holes & 1 central;, 2 12mm diam. tangs present |
| 79A1293 | ? | 130/17 | | pine | half end, Th. 17mm |
| 82A4481 | ? | | 130/17 | | single stave in concretion |
| 79A0919 | ? | | no data | | |

were originally two lantern shot in this area possibly on the Upper deck. Two stone shot of a similar size were also recovered here.

The distribution of stone shot suggests the possibility of additional port pieces and possibly fowlers in the stern (in addition to port piece in M11). This is borne out by the presence of canister shot of 130mm and 150mm diameter, similar in size to some of the finished shot from this area.

## Manufacture and Selection of Shot
by Alexzandra Hildred

*Shot production in the sixteenth century*

**Inventory evidence**
Inventory evidence suggests that metal shot was cast in moulds (the smaller sizes using tongs) and that much of the stone shot was delivered unfinished and then finished to the correct size using picks. The 1559 inventory for ordnance within the Tower of London (Blackmore 1976, 264) lists '*Casting ladells of yron*' and '*Mouldes of Brasse*'. Brass moulds for shot for named guns are listed in the 1547 inventory (Starkey 1998, 5894–99) where they seem to be listed against the larger guns, whilst the sakers, falcons and falconets are sometimes listed with both brass and iron moulds (*ibid.*, 6530–36). One interpretation is that solid iron shot were cast in brass moulds, but lead and composite shot were cast in iron moulds. Stone moulds are listed at Guisnes (5563), along with '*Sowes of leade for shott*' (5558). Other items occurring in various places include '*Tonges*' (4978), '*Pincers of yrone*' (6537) and '*Sheares to cutt shotte*' (4979). Ladles '*gret and smalle*' are used to melt lead (5001).

Out-stations needed to be more self-reliant and therefore provide a basic list of what was required to cast shot. The fullest range of equipment is detailed at Boulogne, including '*Ladells of iron to cast shott*' (6069),

382

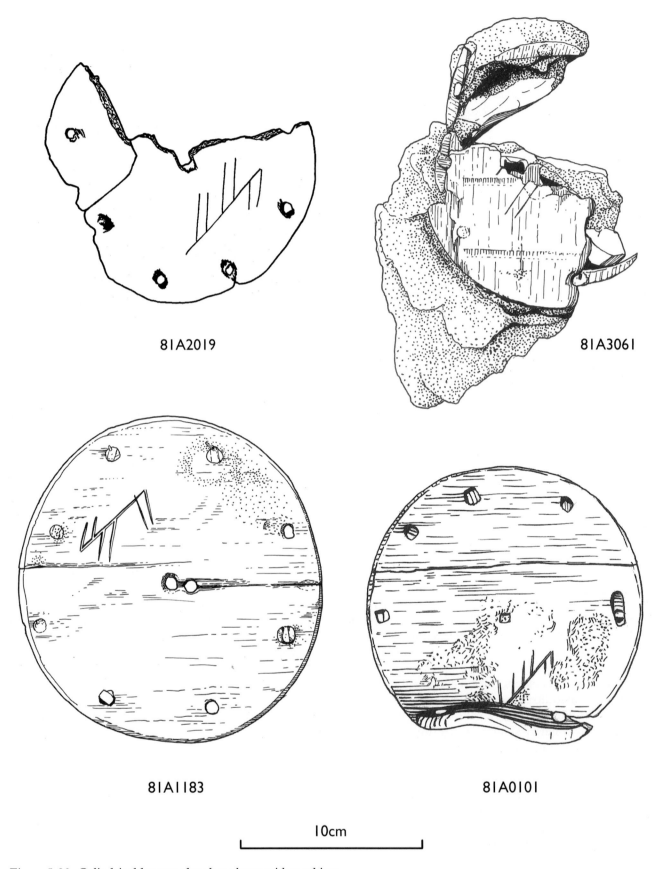

81A2019

81A3061

81A1183

81A0101

10cm

*Figure 5.80 Cylindrical lantern shot: loose bases with markings*

while Carlisle is supplied with 10 '*endes*' of '*Iron for making of dice for leade shott*' (6281). Calais had a shot house with moulds of brass specifically for sakers (4970), falcons (4971) and falconets (4973).

The inventory generally lists iron shot against specific guns but merely '*Stone shotte*' (eg, 4286) or '*Shotte of Stone*' (4306). The most useful references are those for shore establishments since those for ships merely state '*Stone shott of alle sortes*' (eg, 7251). Within fortifications there are numerous references to '*pykes*' (picks), '*pickhammers*' and '*sledgis*' (sledgehammers), and, for larger forts such as Carlisle '*Pickes mattockes and mason axes*' (6298) and '*Great hammers*' with '*watler hamers*' (6299). These are all suggestive of the manufacture, or at least the finishing, of stone shot. This is corroborated by entries such as '*Shott of stone redye hewed and wrought*' (4274) and '*Shotte of stone hewen and wrought*' (4317), '*Stone shott ready hewed*' (4415) and '*Shott of stone for port peices redy made*' (4632). Calshot is equipped with '*Sledgis of Irone*' (4585) and '*Hammers to hewe shotte*' (4586). Portland has '*Crowes of yrone*' (4734), '*Pickhammers to make stone shott*' (4735) and a '*Sleidge of yrone*' (4736). Inventory evidence referring to specific ships is discussed below.

Metallographic study of the iron shot from the *Mary Rose* suggests that metal moulds were used. Two complete and one incomplete stone moulds were recovered which were used to cast lead or composite shot (see below). There is no evidence for casting lead shot for handguns on board but any iron moulds, ladles or tongs would not necessarily survive.

**Casting shot**
by Robert Smith
There is little in the way of contemporary reference to the casting of shot in the sixteenth century. Biringuccio, in his *Pirotechnia* of 1540 (a treatise on mining and the working of metals), is perhaps the only source that we have. He describes a relatively simple method to produce a shot mould in the section entitled '*How iron balls are to be made by casting, for shooting with large and small guns*' (Biringuccio 1540, 319–20):

> '*For these balls, first make a ball of wood or of clay, or have one made of lead or iron, round and of exactly the dimensions that you wish. This is if you wish to make only one in your mould; otherwise take as many as you wish it to contain. Bury half of this, or these, in a board or in clay and, having greased them with oil or lard, make a mould over them of plaster of Paris, or of clay if you have no plaster, exactly as you wish it to be in iron or bronze. Then make the other half opposite this. After having taken out the balls make little openings for the gates and vents and likewise four holes for pegging the parts together. At the back make a stud or a recess to provide a hold for the tongs. When they have been made exactly, cover them with ashes or grease them with oil, and mould them individually with moulding clay, each half by itself. Then make their carriages [sic] and when these have been made and baked, fill them with bronze or molten iron, as you prefer. Thus you will have moulds for casting balls which serve very well and in which one ball or three, five, seven, or more if you wish can be arranged to be cast at one time. Always remember to apply some wash ashes on the inside of the moulds when you cast. Furthermore, arrange a large pair of tongs which have their jaws pierced with a square hole, into which the little stem that you left at the back of the mould enters. Alternatively the tongs are made to enter into the grasp of the recess. Handle the moulds with these tongs as you have occasion.*'

Put simply, using a ball of the required size, a two-piece pattern in plaster or clay is formed around it. The ball is removed and a mould is then made of the pattern produced. The patterns are then removed and the shot is cast by pouring metal into the resultant mould – it can be made of '*bronze or iron, as you prefer*'. The two parts of the mould are fitted with locating pegs for registration. Interestingly, Biringuccio also includes lugs at the back of the moulds and advocates the use of a large pair of tongs into which the lugs fit in order to hold the two halves of the mould together while the molten metal is poured in (Fig. 5.81). Moulds of this form have been identified including a single example from the Molasses Reef wreck (Simmons 1991) and a group of six moulds in Churburg Castle, northern Italy (Smith 2000) that range up to 3½in (90mm) in diameter.

This type of mould is illustrated in contemporary sources, for example in the *Zeughaus* books produced for the Emperor Maximilian I, from the early sixteenth century and in a manuscript of 1582 in the Bodleian

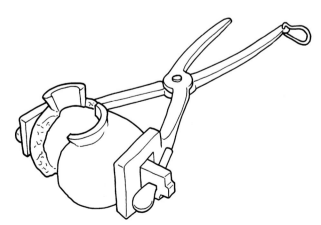

*Figure 5.81 Drawing of tongs with moulds, after Simmons (1991)*

Library, Oxford '*Machines et ustenciles de guerre*' (MS Douce b.2; Barker 2002). Moulds were also made of stone in the sixteenth century. Hemispheres were cut into two rectangular blocks of stone into which, when fitted together, molten metal was poured. These can be multiple types where several shot are cast in one pour.

Biringuccio goes on the describe how to make the furnace to melt iron to cast into the moulds (Biringuccio 1540, 321). When the iron is ready he says:

> '*When you have the necessary quantity of iron well melted and liquefied, in order to proceed more easily arrange a little iron channel long enough so that when the outlet of the hearth is opened, it leads the molten iron to the moulds for the balls. Bring up each pair of moulds that you have with those large tongs that I told you of before and, holding them in the stream, fill them all until the iron overflows. In this way the iron balls that are used in guns are made*'.

**Finishing and making shot on board**
There is evidence for casting metal shot. Two complete and one partially complete sandstone shot moulds for the making of composite shot for the bases were found. There was also the raw material in the form of lead rolls. The making of handgun shot in the field (or on ship) was not uncommon, but as yet no tools have been identified for this purpose. If casting tongs for lead shot were made of iron, they may not have survived.

Evidence for the second phase in the manufacture of composite shot, the pouring of molten lead into a mould containing either an iron dice or smaller iron round shot, has been recovered from a number of wrecks. The spare iron dice for the cubes are not in evidence on the *Mary Rose*, these may have corroded away. Moulds large enough to make the larger, lead-coated, round iron shot found were not found.

The section entitled '*munitions*' in the *Anthony Roll* includes some items which suggest manufacture or at least the finishing of stone shot on board. Included for the *Mary Rose* are twelve '*pyckehamers*' and eight '*sledgys of yron*'.

*Metal shot*
The casting of lead shot on board ships, or the covering of cast iron shot with molten lead, is indicated by the recovery of both lead and moulds. The casting of iron on board would not be expected because of the high temperature required. However, it is clear that relatively large composite shot was cast. The 1514 inventory for the *Gabriel Royal* lists one item under '*Mouldes of brasse for culveryn shott*' and two '*Ladelles for meltynge of lede*' (Knighton and Loades 2000, 126). The *Great Barbara* has '*Mowldes of brasse*' and the *Peter Pomegranate* '*Mouldes of brasse*', both listed just above either picks for stone shot or gun hammers.

As culverins are listed with iron shot cast about with lead, the casting of lead shot around an existing iron ball could be achieved on board though this would, of course, necessitate carrying a mould of the correct size in which to place the iron shot. A mould containing lead residue with a smaller iron round shot inside it was found on the sixteenth century Mollasses Reef wreck and could indicate just this type of activity (Simmons 1991, 5). Composite shot of this form was recovered from the *Mary Rose* (Fig. 5.54).

*Stone shot*
A detailed study was undertaken of 335 of the 387 stone shot recorded in 1997 (Martin 1997). This included study of the surface of each shot for tool marks (see Fig. 5.37a–c). Adhering concretion and shells were noted as was staining, as these provide information on exposure and/or burial conditions. Although it was noted previously that some shot appeared to be more finished than others, this study quantified the situation and proved the majority to be unfinished. Attributes chosen to describe finish included how spherical the shot appeared and how smooth was the surface. Unfinished shot were generally uneven and pitted and described as '*scalloped*'. Tool marks were noted where present, but not all the finished shot had visible marks, for example 80A1767 was spherical and smooth and obviously finished, but displayed no obvious tool marks. Two types of marks were isolated during this study, shot with a '*pecked*' appearance, and shot with linear '*gouges*'. Over half the stone shot recovered are larger than any of the largest port pieces. This high percentage suggests that the shot was being finished as required and that they represent the '*unhewen*' shot listed in inventories. The large flint nodules recovered may represent the raw materials for canister/lantern shot, although those studied suggest collection of beach deposits of uniform size.

It is likely that the pickhammers listed in the Tudor inventories were for the finishing of shot (perhaps a combination of the gun hammers and stone picks featured in earlier documents). The 1514 inventory for the *Mary Rose* specifies 12 '*Pykes for stone*' (Knighton and Loades 2000, 141). The *Anthony Roll* lists pickhammers only on vessels which have guns firing stone shot.

A Tower account of 1455 (cited Blackmore 1976, 243) mentions stone and lead shot as well as eight '*mouldes de tabulis quercus pro facture rotunditate petrarum canonum*' ('*moulds of oaken boards for making round the stones of cannon*'). All but the very smallest shot referred to are stone. Could these '*boards*' be what we have identified as shot gauges (see below)? The majority of the stone shot were restricted to the Main deck, mostly in M5–6 and four of these 'gauges' were found together in this area (see below). There are also references to '*finished*' and '*rough*' stone shot in inventories listing items within the Tower. In 1568 (Blackmore 1976, 266) shot for named guns is listed as '*rough*' or '*polished*'. In 1635 stone shot is stored in sizes of 14in, 10in and 8in (356mm, 254mm, 203mm) (*ibid.*, 292). The same inventory lists 9344 '*Rough stone shott*' (*ibid.*, 301).

The 1514 inventory specifically lists 14 '*Stonepykes* [miswritten *stokepykes*] *of yron*' for the *Henry Grace à Dieu* following '*Hamers for gonnes*' and '*Crowes of yron*' (Knighton and Loades 2000, 119). The *Gabriel Royal* has six '*Stone pykes for makyng of shott*' (*ibid.*, 126), and the *Kateryn Forteleza* has 16 '*Pykeaxes to hew stone*' (*ibid.*, 129). The *Great Barbara* lists '*Pyckes for stone shott*' (*ibid.*, 132); interestingly, this follows the listing for guns in the high deck of the ship, as though the picks, gun hammers and shot were stored there along with the '*Mowldes*' of brass. The *Great Nicholas* has '*Shott of stone rugh and hewen*' (*ibid.*, 134). The *Mary Rose* is listed with 500 '*Shott of stone grete and smale*', 13 '*Hammers for gonnes*' and 22 '*Pykes for stone*' (*ibid.*, 141). Her sister the *Peter Pomegranate* groups 20 '*Pykes of yron*', 2 '*Mouldes of brasse*' and 20 '*Gonnehamers*' (*ibid.*, 146). The *Great Elizabeth* carries '*Stone pykes*' and '*Stone hamers*' (*ibid.*, 150), the *Christ of Greenwich* has '*Pykes for stone*', '*Gonnehamers*' and '*Crowes of yron*' together (*ibid.*, 152); there is nothing equivalent for the *Kateryn Galley* but final vessel inventoried. the *Rose Galley*, has '*Gonnehamers*' and '*Stone pykes*' (*ibid.*, 156).

The *Anthony Roll* lists '*pyckhammers*' for every vessel which carries stone shot; all the ships and '*galleasses*' have these in numbers varying between 20 (*Henry Grace à Dieu*) and two (*Mary James, Jennet*). Of the pinnaces only the *Saker*, the *Roo* and the *Phoenix* have pickhammers, four each. All of these carry stone throwing guns. All the vessels carry '*sledgys*' (sledgehammers) ranging from twelve on the *Henry Grace à Dieu* to one on some of the pinnaces and rowbarges. This is the same with '*crowes of yron*' ,or crowbars. The totals (as listed at the end of the sections in the *Roll*) are 205 sledgehammers and 176 crowbars. There are usually more pickhammers than sledgehammers, with only six instances where the numbers of each are equal. However, sledgehammers are carried on vessels where there is no stone shot (nor picks), suggesting that they were not solely for working stone shot. Unless sledgehammers have changed dramatically since 1545, this is not unexpected and these must also be considered as general vessel tools. None of these tools has been identified, although the sizes of pick blade can be suggested by the markings on the stone shot. One artefact was found in M9, resting on a chest lid (81A4864); this resembles a pick. It was assessed by a mason and thought not to be for working this stone. None of the chisel handles look like they belonged to masonry chisels (see *AMR* Vol. 4, chap. 8) nor as if they had been hit with a metal hammer (Venables, pers. comm.). Masonry chisels today do not have wooden handles as they are usually used with an iron hammer for all but the finest work. These usually have a simple iron shaft tempered to a cutting edge at one end with an abrupt end (iron hammer) or bulb end (rounded so as not to damage a softer mallet). If these were of iron, they may have degraded completely. It is unlikely that the crowbars or sledgehammers would be used for the secondary working of shot.

**Finishing stone shot**
by Clare Venables

The stone shot from the *Mary Rose* has been studied with a view to reconstructing the techniques of masonry used and the tools required to produce the assemblage. Most are of Kentish Rag (see above), chosen for its hardness and because it is a good 'freestone'– a stone which can be worked easily in all directions, with few shelly inclusions. Those studied were in varying stages of manufacture, described as 'primary working' and 'secondary working' to produce the finished shot.

The collection studied also included one granite shot (81A1387, see Fig. 5.41) and one of soft limestone (81A2061). The granite shot is well finished and extremely hard and difficult to work. The limestone is extremely soft.

*Primary working*

A rough sphere would be achieved by striking a stone block (already split or sawn into rough cubes) with an iron hammer causing scallop-shaped flakes to fly off by percussion fracture. None of the shot displays mechanical cut marks. The tool used could be a pickhammer's blunt side or a simple hammer of at least 2lb (around 1kg). The same tool would be used for both large and small shot. Some of the assemblage showed this typical 'scalloping' indicative of primary working only. This shot had discoloured to dark grey in the divets (scallops) with edges of lighter stone possibly resulting from 'polishing' through rolling. This process would have been undertaken on shore, probably by skilled masons and the time taken is expected to be about 30 minutes per shot (based on a 200mm shot),

Although the finish appears rough, it is important to be relatively accurate as off-spheres would take a lot more time to smooth with secondary working. Some shot have been left with sharper edges and deep divets (81A1516). Smaller shot would take longer to work than large shot.

*Secondary working*

The shot displaying secondary working is a variable assemblage with evidence for different hands and varying degrees of skill and level of finish. The skill of the mason determines the time it takes to achieve a good finish, but about 45 minutes per shot is expected. The shot needs to be secured for working, a pyramid of unfinished shot could be used, a box of sand, or possibly with the piece held between the feet.

*Figure 5.82  Stone shot 72A0051 showing effects of blunt point used on shot*

Current masonry practice prescribes two methods for cutting a sphere from a cube. The first is by cutting narrow chamfers which are eventually smoothed together. The second is by cutting two channels with circular profiles, at right-angles. Shot 81A2052, 81A1839, 81A2384, 81A2387 (M6); 80A1898 (M4); and 82A1627 (SS9/10) display a single channel, and 81A2053 (M4/5) (see Fig. 5.37a).and 81A1283 (M5) show two bisecting channels. Additionally on 81A2053 one quarter of the surface of the shot is smooth and finished, whilst the rest is rough. Guides for size and sphericity would not be unexpected. Five of the nine objects classified as 'shot gauges' could be used for this purpose, four were found together (81A1285–1288) in M5.

Two methods of secondary working are present; shot worked using a pointed tool such as a pick and shot worked using a chisel edge or sharpened point tool hit by a hammer. Within each group there is a range in finish because the stone wears the tools quickly. Sharper tools do not necessarily lead to the most spherical shot.

The effect of using a pointed tool suggests that the sharp side of a pick was used to strike tangentially. This bruises the stone off in a pattern of small circular dished cups. There are occasional accidental drags across the surface where the working area has been over-reached. The result can be spherical, but not smooth. Some have been finished using sharper tools than others, for example 72A0051 has been cut with a blunt point (Fig. 5.82). Today the equivalent is a pointed chisel, hammer struck (possibly the '*pickhammers*' listed).

Chisel-working leaves fine scratches. As the cuts are long, they are likely to have been made by a chisel edge or a sharpened point tool hit by a hammer, leaving grooves 15–25mm in length. It may be possible to achieve this effect with a sharpened pick hammer, but it would be difficult to obtain the sort of glancing blows needed to create this effect on a convex surface with a pick. Alongside chisels and wooden mallets used in finer masonry, smaller picks with various shapes are illustrated from the medieval period onwards. The *Book of Trades* (Sachs and Amman 1568, 93) shows a stonemason using a pick with a finely shaped blade on the end cutting a moulding (Fig. 5.83).

Some of the cut marks are made by very sharp blades, such as those on 81A1002, and others with

*Figure 5.83  Stonemason from Sachs and Amman (1568)*

*Figure 5.84 Finishing stone shot: a) sharp blade cuts on 81A1002; b) blunt blade cuts on 81A0968*

blunt blades, for example 81A0968 (Fig. 5.84). Some shot display larger cuts than others, up to 34mm. The resulting surface is finer and smoother than that made by a pick. Although not random, the tool marks show that the masonry methods described above have not been applied. 81A0968 has been made to a size where some of the deep dents from the primary workings have been retained.

The finished shot is spherical, smooth and a dark grey with lighter cut marks. It is possible that this indicates that a long enough period of time elapsed between primary and secondary working for the stone to discolour.

## Shot moulds

Two complete and one partly finished moulds were recovered that appear to have been used for the casting of small lead or composite lead/iron shot (72A0067, 72A0182, 80A1847; Figs 5.85–6), possibly for the 30 bases or swivel guns listed in the *Anthony Roll*. The two finished examples (72A0067, 72A0182) are made from Malmstone, a calcareous glauconitic sandstone from the Upper Greensand deposits which form a low, 20km long, 2km wide scarp below the Chalk escarpment on the edge of the Weald in Hampshire. The hardest form of Malmstone (Reigate Stone) has an extensive history

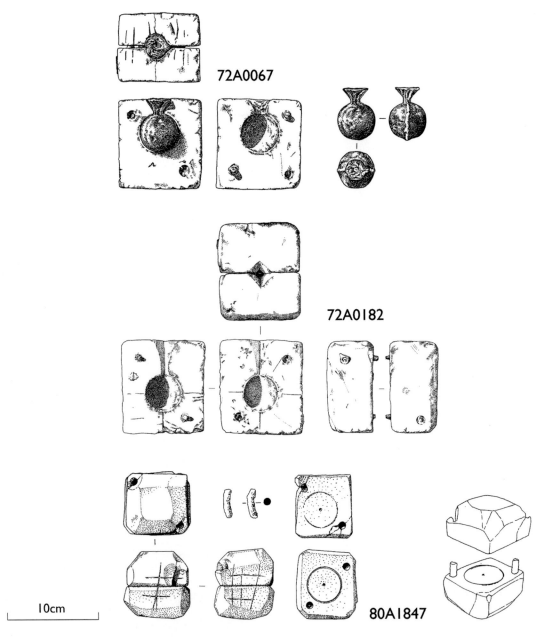

*Figure 5.85 All shot moulds*

both as a building stone and as refractory stone, especially for lining furnaces, leading to a common name of '*firestone*' (Gallois and Edmunds 1965, 38).

The finished moulds (72A0067, 72A0182) are cubes made from two pieces of Malmstone, each with a hemisphere removed from one face. A channel leads from the hemisphere to the top face of each half, forming a pouring channel when the two are joined. Fixing is through lead lugs set in one face locating into holes in the opposite face. The diameters of the hemispheres range from 36mm to 42mm ($1^{7}/_{16}$–$1^{11}/_{16}$in). Both were recovered outboard of the starboard/stern frames at a depth of about 0.90m below the top of the frames, corresponding with the stern panel (where diagonal planking forms the flat surface of the stern above the sternpost). Just inboard on the Main deck (M11) was found a swivel gun (79A0543), two powder ladles (79A1211, 79A1152), a rammer (79A1208) and several composite lead/iron shot. 80A1847. The unfinished mould (Figs 5.85 and 5.87) was found inside a chest (80A1413) recovered from the Orlop deck (O4). Contents of the chest included a knife and sword (80A1851 and fragments) together with personal items (*AMR* Vol. 4, appx 1).

72A0067 consists of a well-worn block 95mm high by 90 x 71mm, cut into two halves of 35mm and 36mm. It weighs 1070g. Each portion has a hemisphere with a diameter of 42mm ($1^{11}/_{16}$in) hollowed out of the centre, offset towards the top of the mould. The pouring channel is 25mm wide at the top tapering to 13mm.

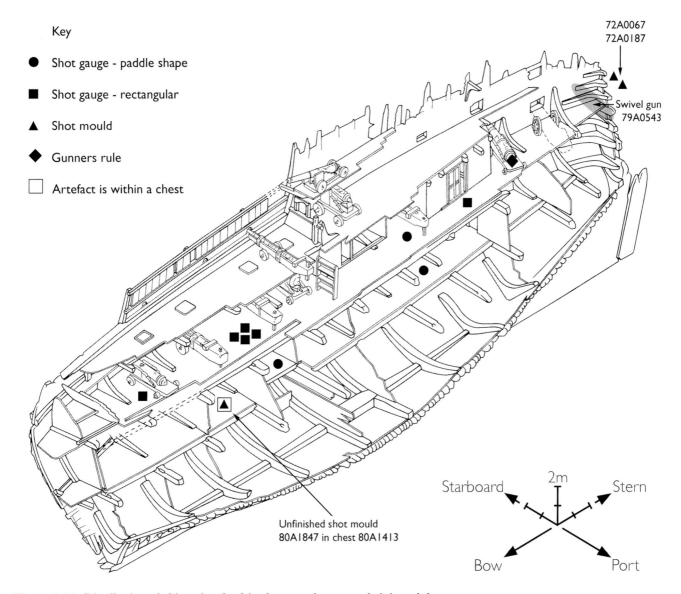

Figure 5.86 Distribution of objects involved in the manufacture and sizing of shot

Diametrically opposed, 10mm diameter, lead pegs protrude 8mm from the inside of one face to locate into 12mm deep holes in the other. The pegs are bent through from the side of the mould to the inside face, but do not protrude from the side. There is a third vacant hole in the bottom of the inside face of one half. Scored lines, some crossed, on each face of the two halves may have been guide lines used in cutting the block.

A solid lead shot was cast using this mould (Fig.5.85) in 1972. Shot of this size was recovered from the ship, including one at 41.5mm diameter found in M1 (79A1007), one of 41.8mm from the bow bulkhead (82A4265) and one of 42mm from SS10 (82A2403) with two more from unknown locations. 72A0182 is also cuboid with champhered edges, measuring 100 x 90 x 99mm and weighing 1557g. The hemispherical hollows are 38mm in diameter, offset towards the bottom of the mould. The pouring channel is 28mm wide at the top and there are two 6mm long pegs set in 8mm deep holes in the inner face of one half, locating with vacant holes in the other, to keep the two halves of the mould together. The locating lugs are entered from the side (as above), but on this example they protrude out of the side of each half, almost at right-angles. Scratches on the outside faces are remnants of aids to cut the hemispheres. An incised line crosses the centre of the hemisphere on the inside of each half marks the centre of the shot, not of the mould (Fig. 5.88). Nine shot of appropriate (35–35.5mm) diameter were recovered (see above).

80A1847 is a rougher, comprising two rather different pieces of stone, akin to quarry rubble. Its measures 70 x 65 x 73mm, weighing 487g. Each half is scored on the sides with what appear to be alignment marks (Fig. 5.85). The inner faces are scored with a

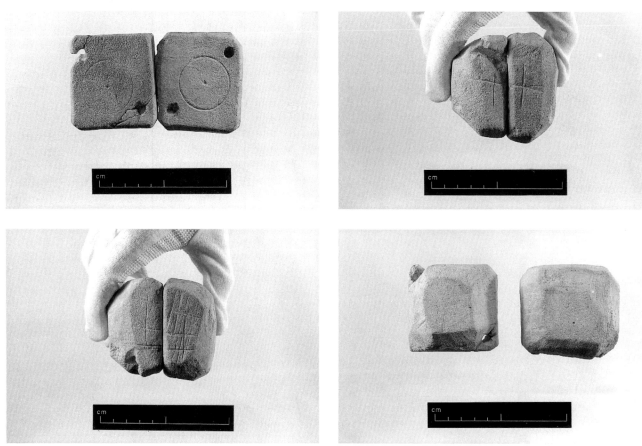

*Figure 5.87 Details of shot mould 80A1847, marking out lines*

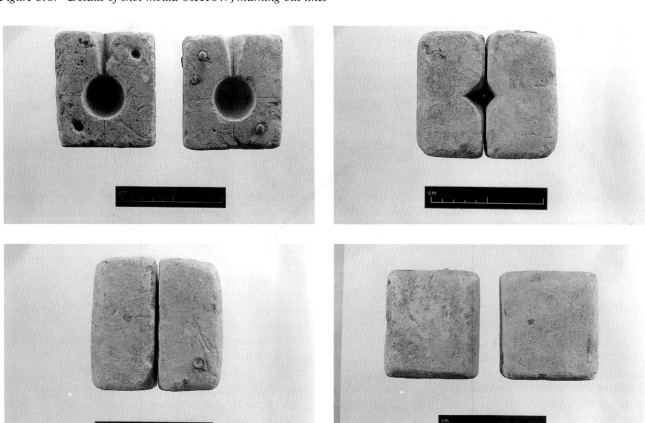

*Figure 5.88 Details of shot mould 72A0812*

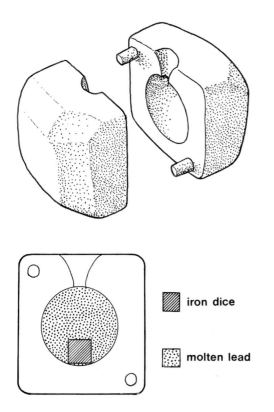
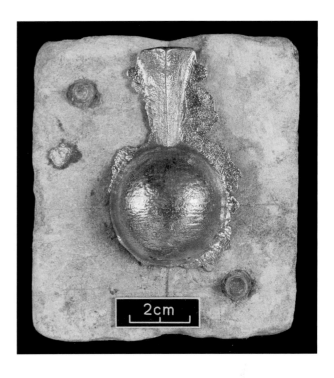
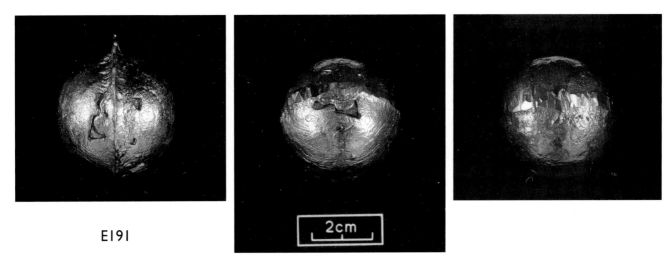

*Figure 5.89 Experimental shot cast in mould 72A0812*

central point and inscribed circle of 36mm (1⅞in). Two lead pegs, 8mm in diameter and 32mm long, are set in diagonally opposed depressions in one inside face but do not penetrate the sides of the block and they are longer than those on finished examples, perhaps they represent part of the production process. The other face is incomplete and the pegs locate into broken holes which penetrate the entire face (Fig. 5.85). The breaking of the mould at the peg insertion points may have occurred while the side holes were being drilled and rendered the object useless, hence its unfinished form. This mould would have accommodated shot of similar size to 72A0182.

Because of the high temperature required to cast iron, it was obvious that the moulds were for casting lead shot. The assemblage was studied with a view to determining whether shot had been cast in similar moulds. All the shot of appropriate size proved to be of 'dice and lead', an iron dice within a cast lead shot. The results of this work are described above under inset shot. Both shot examined (81A4958 and 81A4436) exhibited what appeared to be casting ridges encircling

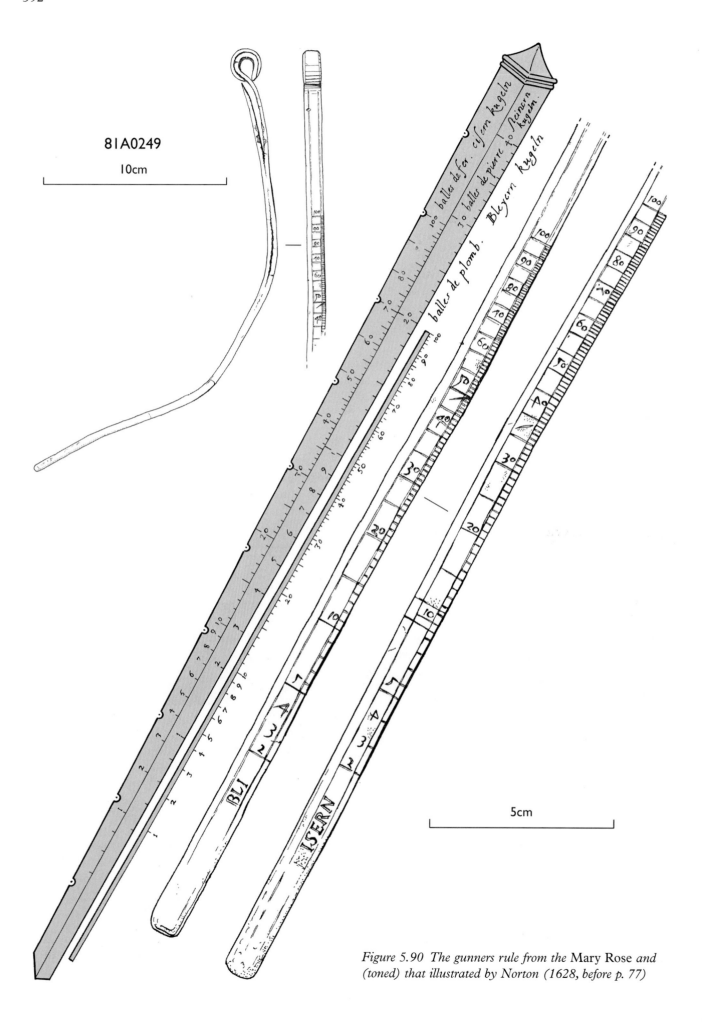

*Figure 5.90 The gunners rule from the* Mary Rose *and (toned) that illustrated by Norton (1628, before p. 77)*

the shot approximately along its centre line, indicative of manufacture in a bipartite mould.

In order to test the possible use of the moulds, a 10mm steel cube (cut from bar) weighing 14.1g was placed into mould 72A0182. The mould was pegged together and tied and gradually heated in an oven to 240°C so as not to expose the sandstone to sudden heat. Molten lead was poured into the warmed mould through the channel. When poured, the lead pushed the iron core to the base of the mould and it flash-set, forcing the dice to penetrate the lead, leaving four triangular holes similar to those observed on composite shot recovered from the ship (Fig. 5.89). The experiment was repeated three times, producing E189 (314.5g), E190 (310.3g) and E191 (313g). The diameter of the hole in the mould was 38mm and the shot were all 37.5mm. Assuming this shrinkage for the other moulds, shot of 41.5mm diameter could have been cast from 72A0067.

Neither moulds nor spare lead are listed in the *Anthony Roll*. Iron dice is, but specifically for use with hailshot pieces, not as a spare commodity. Small pieces of lead were recovered, but not in any regularly sized ingots. Larger rolls of lead were also carried (presumably for ship repairs) so it is feasible that extra base shot could have been manufactured on board. The shot to gun ratio only allows thirteen rounds, suggesting that the bases were used no more frequently than the large guns, even though there was the facility to cast more.

## Gunner's rule

One of the many objects associated with the ordnance carried onboard the *Mary Rose* has been identified as a 'gunner's rule' or scale (Blackmore, pers. comm. 1981; Fig. 5.90). Gunner's rules are designed to provide the weight of round shot for a particular gun either by direct conversion of the bore of the gun into shot weight, or by conversion of the diameter of shot to weight. Simple forms are like wooden rules or scale rules, but by the seventeenth century many were refined to form a knife or stiletto, with calibrations on the blade. Most specimens stop at the 100lb or 120lb mark (45.36–54.32kg) and they are in the order of 12in (304mm) in length. Complex examples are triangular or rectangular in section (providing individual faces for different scales) and later examples have been found combined with quadrants or levels. These refined versions are complex mathematical instruments (eg, Wallace Collection A 1246). All have irregularly spaced lines and (or) numerals that are usually incised providing the weight of one or more material (iron, stone, lead). The weight is then read off the scale from the point of origin of the rule: therefore these provide a converted measurement of distance.

## Historical evidence

Towards the end of the sixteenth century Master Gunners carried instruments. A manuscript treatise on the use of measuring instruments by Wenzel Jamnitzer of Nuremberg (1585), in the Victoria and Albert Museum, shows an illustration of a gunner measuring the bore of his cannon across the muzzle with a rule. Most early descriptions (such as Wright (1563), Lad (1586) Bourne (1587) and Norton (1628)), describe rules as being used with callipers against the diameter of shot. This is important in decoding an individual scale, for if used to convert the bore of the gun into shot weight the calculations used in calibrating the instrument will have to take into account windage, the difference between shot size and bore size. The accepted figure for this is one-twentieth of the bore, so this in itself is a changing unit. If it is to be used against shot, this should represent a mathematical formula based on diameter (or radius) and relative density (see below) of a particular material. Scales on the rules can differ, accounted for by the different materials and the wide variation in the weight of the European pound. Slight differences may be as a result of how the rule was produced; if the rule is actually made by merely noting diameter and weight of real shot rather than calculated mathematically, differences in weight relative to diameter reflect the relative density of the original molten metal and the processes of casting and cooling. The differences in weights and measures as found on rules is discussed by Green (1990, 25).

A nearly contemporary description of the conversion of shot diameter into weight is given by Wright (Society of Antiquaries MS 94, f.11v):

*'Allso another waye ther ys to take the eight [height] of your shotes wher by you shall evydently knowe the eight and the waight of your shot in this forsaid reull [rule] you shall take a payre of callapares compasces and take the eight of your shote [three words repeated] and then take the eight by a reull and so mannye enches as you do fynde by your reull that your shot is of waight and then like [three words repeated] so many inches as the forsaid reull is marked and so many pounde justly wayeth your shote'.*

An illustration (*ibid.*, f.15v) shows a gunner holding a rule bearing two sets of calibrations and pierced at on end for a cord (Fig. 5.91).

John Lad (1586) writes:

*'A Rule to know the height and weight of your shot.*

*This rule declareth the height and weight of your shot being of two inches and from thence to the height of 9 inches, of which height commonly all pieces are made by. Also who will know the height*

*Figure 5.91 Gunner holding rule, after Wright (1563)*

and weight of your shot, first take the shot and measure it round about, and when you have done so divide the said compass into three parts, the which third part I take for the height of the shot, and so many inches as the said third part is in height measured by an inch rule, and look so many inches and a quarter as the foresaid rule is marked in number, so many pounds doth your shot weigh.

Also another manner to take the height of your shot as thus; take a pair of calliper compasses, and take the thickness of your shot which is called the height, measure your height by an inch rule, and so many inches as you find by an inch rule that your shot is of height and so many pound justly weigheth your shot as by this proof. I take a shot and I measure the height and I find it to be six inches, then I go to the foresaid rule and look upon six inches and I find marked that my shot weigheth 26 pound, which is just of all your shot of height if it be full cast and not blown in the ramming'. (Caruana 1992, 28–9; spelling there modernised)

Both are describing a straightforward method of obtaining the weight of shot by direct measurement of the diameter and reading this distance off on the rule. It is at least a two-stage process, using callipers against the shot and placing these against the rule. There is no suggestion of converting bore diameter to shot weight.

One of the most comprehensive early English sources was written by Robert Norton in 1628. Regarding the necessity of requiring the weight of a particular shot to work out the powder charge, Norton states that one needs to be able to confirm this by a swift examination of the shot without having to weigh it. It does have to be assumed, however, that the exact parameters of the gun are known together with the type of powder. The ratio of the amount of powder to weight of shot varies depending on the strength of the powder and the type and wall thickness of the gun. The Norton text is comprehensive, and a 'scale' is described alongside a table for weights. The figure is entitled:

> 'A Table, shewing the Height and Weight of Iron, Lead, and Stone-shot, accurately and newly calculated by the Author, and applyed to our assize of English Measure of Inches and Parts, and to the Haberdepoize [avourdupois] Weight of 16 Ounces to the Pound'.

The scale is '*to be made in Brasse by M. Allen: And in Wood, by M. Nathaniell Gors of Ratcliffe*' and is very similar to the *Mary Rose* example:

> 'It is a square Rule of one foote in length, made eyther of Brasse, Boxe, or other fine grayned Wodde that will not warpe: Vpon one side or square whereof I haue set the height of all sorts of iron shot, from 1 pound to 100 pound weight: And of Stone shot to 37 pound: And of Lead shot to 150 pound weight: Each distinguished from an other by the letter I. for Iron, S. for S[t]one, and L. for Lead-shot, and their Weights and Measures accomodated vnto our English Haberdepoiz weight of 16 ounces to the pound, and to our Foote of assize of 12 Inches to the Foote'. (Norton 1628, 94)

Norton describes the scale as being used to:

> 'finde the waight of any Shot by the Diametre thereof ... because a Gunner, cannot at all times when the weight of a Shot is required, haue Ballance and weights about him, to waigh the same ... it shall not bee a misse ... to set downe, how hee may know the same, to a sufficient neerenesse, by the height of the Dimitre [diameter], or circuit of the circomference thereof; And that allso for varietie sake, as well Arithmetically, and Geometrically, as Tabularly, and by Scale, and Compas.' (Norton 1628, 87)

Table 5.16 *Mary Rose* rule specific gravity calculations (iron)

| Wt (lb)* | Radius (cm)** | Wt (g) | Vol. (⁴/₃πr³) | Unit of RD |
|---|---|---|---|---|
| 2 | 3.10 | 907.2 | 121.80 | 7.27 |
| 3 | 3.50 | 1360.8 | 179.61 | 7.58 |
| 4 | 3.90 | 1814.4 | 248.51 | 7.30 |
| 5 | 4.25 | 2268.0 | 321.60 | 7.05 |
| 6 | 4.43 | 2721.6 | 354.21 | 7.47 |
| 7 | 4.68 | 3175.2 | 429.42 | 7.40 |
| 8 | 4.90 | 3628.8 | 492.88 | 7.36 |
| 9 | 5.13 | 4082.4 | 565.84 | 7.22 |
| 10 | 5.35 | 4536.0 | 641.51 | 7.07 |
| 11 | 5.50 | 4989.6 | 697.00 | 7.16 |
| 12 | 5.60 | 5443.2 | 735.71 | 7.40 |
| 13 | 5.70 | 5896.8 | 775.84 | 7.60 |
| 14 | 5.80 | 6350.4 | 817.39 | 7.77 |
| 15 | 6.10 | 6804.0 | 950.90 | 7.16 |
| 20 | 6.60 | 9072.0 | 1204.42 | 7.53 |
| 25 | 7.15 | 11,340.0 | 1531.31 | 7.41 |
| 30 | 7.70 | 13,608.0 | 1912.57 | 7.12 |
| 35 | 8.05 | 15,876.0 | 2185.40 | 7.27 |
| 40 | 8.30 | 18,144.0 | 2395.41 | 7.58 |
| 45 | 8.70 | 20,412.0 | 2758.69 | 7.40 |
| 50 | 9.00 | 22,680.0 | 3054.02 | 7.43 |
| 55 | 9.30 | 24,948.0 | 3369.71 | 7.40 |
| 60 | 9.70 | 27,216.0 | 3823.49 | 7.12 |
| 65 | 9.90 | 29,484.0 | 4064.91 | 7.25 |
| 70 | 10.20 | 31,752.0 | 4380.69 | 7.14 |
| 75 | 10.40 | 34,020.0 | 4712.43 | 7.22 |
| 80 | 10.70 | 36,288.0 | 5132.11 | 7.07 |
| 85 | 10.90 | 38,556.0 | 5425.31 | 7.11 |
| 90 | 11.10 | 40,824.0 | 5729.46 | 7.13 |
| 95 | 11.30 | 43,092.0 | 6044.78 | 7.13 |
| 100 | 11.60 | 45,360.0 | 6539.11 | 6.94 |
| A | 5.20 | 4536.0 | 589.05 | 7.70 |
| B | 5.50 | 4536.0 | 697.00 | 6.51 |

* Listed on rule; ** measured from origin to Lb

This further reinforces the idea that the rule is used directly against the shot to tell its weight and not across the mouth of the gun. Norton's figure (Fig. 5.90, toned) illustrates a '*scale*' described in French as a '*mesre qui montre le pois des balles selon leur calibre*' ('*a measure which shows the weight of the balls by their diameter/calibre*'). The scale provides weights for lead, iron and stone. This illustrated rule and the accompanying tables have been used to provide a comparison with the rule recovered from the *Mary Rose*.

A number of authors have come to different conclusions regarding the use of recently excavated rules. Colin Martin, in decoding the rule found on the *Trinidad Valencera* (Rodríguez-Salgado 1988, item 9.16) states that '*This device allowed the gunner to measure the bore of his piece and determine the weight of the shot appropriate to it*'. Konstam (1989, 23) takes the actual measured distance from the origin of the rule recovered from the Tal y Bont wreck (*c.* 1703) to be the diameter of the ball and using the formula volume ⁴/₃πr³, multiplied by the relative density (grams per cubic centimetre) of iron and lead, gives the weight of the shot in grams. Green (1990, 25) states that the rule, or tally stick (Bat 4497) found on the *Batavia* (1629) '*gives the relationship between the diameter in duim (average of 25.0mm) and weights of iron, stone and lead shot*'.

Table 5.17 *Mary Rose* rule specific gravity calculations (lead)

| Wt (lb)* | Radius (cm)** | Wt (g) | Vol. ⁴/₃Pir³ | Unit of RD |
|---|---|---|---|---|
| 2 | 2.70 | 907.2 | 82,45 | 11.00 |
| 5 | 3.75 | 2268.0 | 220.92 | 10.27 |
| 10 | 4.70 | 4536.0 | 434.94 | 10.43 |
| 15 | 5.40 | 6804.0 | 659.66 | 10.32 |
| 20 | 6.05 | 9072.0 | 927.71 | 9.78 |
| 25 | 6.50 | 11,340.0 | 1150.49 | 9.86 |
| 30 | 6.90 | 13,608.0 | 1376.23 | 9.89 |
| 35 | 7.35 | 15,876.0 | 1663.43 | 9.55 |
| 40 | 7.55 | 18,144.0 | 1802.96 | 10.06 |
| 45 | 7.90 | 20,412.0 | 2065.50 | 9.88 |
| 50 | 8.30 | 22,680.0 | 2395.40 | 9.47 |
| 55 | 8.55 | 24,948.0 | 2618.44 | 9.53 |
| 60 | 8.85 | 27,216.0 | 2903.85 | 9.37 |
| 65 | 9.10 | 29,484.0 | 3156.96 | 9.33 |
| 70 | 9.35 | 31,752.0 | 3424.36 | 9.27 |
| 75 | 9.55 | 34,020.0 | 3648.84 | 9.32 |
| 80 | 9.75 | 36,288.0 | 3882.92 | 9.35 |
| 85 | 9.95 | 38,556.0 | 4126.80 | 9.34 |
| 90 | 10.15 | 40,824.0 | 4380.70 | 9.32 |
| 95 | 10.35 | 43,092.0 | 4644.79 | 9.28 |
| 100 | 10.50 | 45,360.0 | 4849.68 | 9.35 |

* Listed on rule; ** measured from origin to Lb

**Description and location**

The *Mary Rose* rule was found on the Main deck in M10 in the same area as area as parts of three incendiary missiles, bronze demi cannon (81A3000) recovered on its carriage, six linstocks and shot of differing types and sizes. A former for making cartridges was found close by (see Fig. 5.106, below).

The rule is made of brass, comprising 77.6% copper, 18.4% zinc, 2.8% tin, 0.68% lead and less than 0.25% each of antimony, silver, nickel and iron. It measures 11¾in (298mm) and is ³/₈in (9mm) wide. Lines and numbers are incised on both faces, one face bearing the letters '*ISERN*' and the other '*BLI*'. Comparison with other gunners' rules and with Norton suggests these relate to gradations for iron (*isern*) and lead (*bleyern*) as can be seen from Figure 5.90. The numbers on both faces are 2–5 and then in tens up to

100 and appear to correspond with lines incised on both faces across the centre of the rule. The face marked ISERN bears two extra horizontal lines that extend above and below the line and number 10, across the entire rule to the left edge. The lower of these corresponds with the incised line for 9, and the upper is partway between 11 and 12, these are identified as A and B in Tables 5.16–18. The 4 at both 4 and 40 is a 'mirror' image on either side of the rule, the 5s and 7s are angled. The space between 35 and 40 bears a diagonal line (top right to bottom left) incised into the rule. The face marked BLI bears all the arabic characters in modern form, but there is an incised diagonal line between 45 and 50, this cuts into the gradations on the right side of the face, and does not fit neatly into the space. A small quadrilateral indentation bisects the left vertical line close to the loop.

To each side of these horizontal lines is a single vertical line, almost forming a border down each side of the rule. The border to the left of each face is blank, and to the right there are horizontal lines marking single pound units for the entire length of the rule. The spaces between these are irregular and completely different between the two faces. It is suggested that these relate to weight, and in both cases (iron and lead) the rule has been tested as though it relates to English pounds *avoirdupois*, as although the rule may be of German manufacture, it was found within an English context. In point of fact the divisions are quite crude. The principal marks have been scratched in and the intervening ones merely seem to be equally divided between these and the densities obtained vary.

At the end closest to the 100 the gauge is bent backwards to form a suspension loop. For purposes of measuring the point of origin is taken as the extreme end of the rule opposite the 100.

**Establishing the purpose and accuracy of the rule**
It is assumed that purpose of the rule was to identify the correct weight of shot required for an individual gun. The gunner needs to know the weight of the shot appropriate to the bore of his gun in order to prepare the powder charge. The diameter of the shot at the time of loading is actually a secondary consideration (although this is required for the calculation of weight in the first instance), provided it is small enough not to get jammed in the bore, although velocity, distance and impact are all compromised with increased windage. The accuracy required is in providing a fit powder charge for the weight of the ball. As illustrated above, a gunner's rule could provide weight based upon either the diameter of the shot, or possibly provide the weight of shot mathematically deduced to take into account windage from direct measurement of the bore of the gun. Callipers would be required to provide the diameter of the shot, but the bore diameter could be obtained either by callipers or by putting the rule directly against the bore and reading the weight off the rule, the latter being the most useful. There are a number of ways one can test the numbers on the rule.

To test whether the numbers are weights one can measure the distance to an inscribed unit and suggest that this relates to a shot diameter. One can then calculate weight mathematically using an accepted figure for relative density of the metal and compare this with the mark on the rule and also with historical weights given for shot of that diameter. If they are similar, then it may be assumed that they represent weights.

If the figures on the rule are consistently lower than the historical or mathematical figure for weight of shot of that diameter, then one might assume that the rule is for measuring the bore, and converting it to the weight of a shot of smaller diameter than the bore. In that case, the weight should be equal to the distance – one-twentieth of the diameter (although on larger guns ¼in was accepted). This should apply for both lead and iron units on the *Mary Rose* rule.

Before commenting either on the accuracy of the rule or its probable use, it was decided to look at it in isolation to try and ascertain the unit used within the rule without external comparison. The only assumption made was that the numerals and horizontal lines represented weight in pounds, and that the rule was intended to convert a distance to a weight. The distance was taken from the very end of the rule (hereafter the Origin) as there was no other indication for a place of origin.

This relied on the following:

- converting (assumed) weights as marked on the rule into grams (one pound was taken to equal 453.6g)
- measuring the distance from the origin to the 'weight' mark to obtain an (assumed) diameter (eg, the 11lb mark is at 11cm on the iron scale)
- halving that value to obtain the radius (in the example above giving a radius of 5.5cm)
- applying the formula $\frac{4}{3}\pi r^3$ (in this example 697cm³) to get volume*
- dividing the weight by the volume to give relative density

* Note this is only one formula, $\pi d^3/6$ to get volume direct was not used in this instance.

The relative density obtained by applying the above formula to the figures on the rule provides fluctuating figures for relative density. For iron this varies between 7.05 and 7.77, averaging 7.29, which is less than the accepted modern standard of 7.43. The variation for the specific gravity of lead ranges between 9.27 and 11, with the modern accepted figure as 11.35. Averaging all the figures listed, gives a value of 9.74, still substantially less than the accepted modern standard (Tables 5.16 and 5.17).

Comparing the *Mary Rose* rule with that displayed by Norton (Fig. 5.90) using the iron scale, they are

Table 5.18 Comparison of *Mary Rose* and Norton rules for iron and lead

| | | IRON | | | | | LEAD | | |
|---|---|---|---|---|---|---|---|---|---|
| Norton | | Norton | | *Mary Rose* | Modern | Norton | | *Mary Rose* | Modern |
| in | quarter | lb | oz | lb | (RD 7.43) | lb | oz | lb | (RD 11.35) |
| 1 | 3 | 1 | 0 | | | 0 | 13 | | |
| 2 | 0 | 1 | 1 | | | 1 | 11 | | |
| 2 | 1 | 1 | 9 | | | 2 | 0 | | |
| 2 | 2 | 2 | 2 | 2+(½) | 2.1960 | 3 | 6 | 3+(½) | 3.3700 |
| 2 | 3 | 2 | 14 | 3 | 2.9170 | 4 | 3 | 4+(¼) | 4.4560 |
| 3 | 0 | 3 | 12 | 3+(¾) | 3.7950 | 5 | 0 | 5 exactly | 5.7975 |
| 3 | 1 | 4 | 12 | 4+(¾) | 4.7990 | 6 | 9 | 7 exactly | 7.3300 |
| 3 | 2 | 6 | 1 | 6 exactly | 6.0402 | 8 | 1 | 8+(½) | 9.2260 |
| 3 | 3 | 7 | 5 | 7+(¼) | 7.4008 | 9 | 14 | 10 exactly | 9.7100 |
| 4 | 0- | 8 | 15 | 9 exactly | 8.9960 | 11 | 5 | 13 | 13.7420 |
| 4 | 1 | 10 | 10 | 10+(¼) | 10.7600 | 15 | 15 | just under 16 | 16.4375 |
| 4 | 2 | 12 | 10 | just under 13 | 12.8310 | 17 | 15 | just under 18 | 19.6008 |
| 4 | 3 | 14 | 14 | 15 exactly | 15.0450 | 21 | 5 | just over 20 | 22.9836 |
| 5 | 0 | 17 | 15 | just under 18 | 17.5700 | 24 | 12 | 23 exactly | 26.8403 |
| 5 | 1 | 20 | 1 | just under 20 | 20.2940 | 30 | 0 | 27 exactly | 31.0012 |
| 5 | 2 | 23 | 2 | just over 23 | 23.4360 | 35 | 10 | 31 exactly | 35.8000 |
| 5 | 3 | 26 | 6 | just over 26 | 6.7040 | 39 | 9 | 35 exactly | 40.7929 |
| 6 | 0 | 30 | 0 | 30 exactly | 30.3600 | 45 | 0 | 40 exactly | 46.3800 |
| 6 | 1 | 34 | 0 | 33 exactly | 34.3300 | 51 | 0 | 45 | 52.4550 |
| 6 | 2 | 38 | 0 | 38 exactly | 38.6400 | 57 | 0 | 50 | 59.0390 |
| 6 | 3 | 42 | 0 | 42+(½) | 43.1900 | 63 | 0 | 55 exactly | 65.9790 |
| 7 | 0 | 48 | 0 | 48 exactly | 48.2100 | 72 | 0 | 62 | 73.6500 |
| 7 | 1 | 53 | 0 | 53 exactly | 53.4300 | 79 | 8 | 70 | 81.6260 |
| 7 | 2 | 58 | 0 | 58 exactly | 59.3900 | 87 | 0 | 75 | 90.7200 |
| 7 | 3 | 64 | 0 | 64 exactly | 65.3800 | 96 | 0 | 83 | 99.8741 |
| 8 | 0 | 72 | 10 | 70 | 71.9680 | 106 | 8 | 90 | 109.93 |
| 8 | 1 | 78 | 0 | 77 | 78.9660 | 117 | 0 | 95 | |
| 8 | 2 | 87 | 3 | 88 | 89.6330 | 130 | 8 | over 100* | |
| 8 | 3 | 95 | 0 | 90 | 94.1200 | 142 | 8 | | |
| 9 | 0 | 101 | 0 | 98 | 102.4700 | 150 | 0 | | |
| 9 | 1 | 109 | 6 | over 100 | | | | | |

* Scale ends at 100. Adapted from Norton (fig. 18). Formula used: (radius cm$^3$ x 3.142 x 4) divided by (3 x SG) divided by 453.6. Specific gravity for iron taken as 7.43, for lead as 11.35

remarkably similar. If the '0' on both rules is taken as the very end, the calibrations are very close. This suggests that the Norton rule was reproduced in print to the correct scale. The calibrations are identical in the pound divisions to 10, thereafter all the lines at the 5 and 10 marks are identical until 75 where the *Mary Rose* rule starts to lag behind slightly. This culminates in the 100 mark on the Norton rule being only 97 on the *Mary Rose* one. The description against the Norton rule suggests that it is to read off shot weight from diameter, and not to be placed against the bore of a gun. The figures are reproduced for iron and lead in Table 5.18.

The scale is very accurate indeed for iron shot up to 8in (203.2mm), far too accurate to be able to suggest that the function of the rule was anything except for measuring the diameter of the shot and converting this to weight, in spite of the variation in the figure for the specific gravity. However, this variation is not unexpected. The relative density for grey cast iron fluctuates from 7.03 to 7.13 and for white cast iron from 7.58 to 7.73 (Green 1990, 26). When a *Mary Rose* cast iron shot (81A4844) was sectioned at the Royal Armouries, it was found to contain a both types of iron, the outer regions were found to be pure white cast iron, but the inner regions (where cooling had been slower) had the microstructure of grey cast iron (Starley and Hildred 2002, 143). As cooling rates would differ with mass, one would expect different relative amounts of grey and white iron depending on size. The suggestion can therefore be proposed that, rather than using a

**Table 5.19 Listed historical weights for shot of a given diameter compared with *Mary Rose* and Norton rules**

| In | Quarter | *Mary Rose* rule (oz) | Norton rule | Norton table 1628 | Sloane MS* | Lad 1587 | Bourne 1563 | Burton 1587 | Harrison |
|---|---|---|---|---|---|---|---|---|---|
| 1 | 3 | | 1lb 0oz | | | | | | 2lb 0oz |
| 2 | 0 | | 1lb 1oz | | 1lb 4oz | | 1lb 1.25oz | 1lb 12oz | |
| 2 | 1 | | 1lb 9oz | | | | 1lb 9.33oz | | 2lb 0oz |
| 2 | 2 | 2+(½) | 2lb 2oz | 2b 2oz | 2lb 0oz | | 2lb 2oz | 2lb 8oz | |
| 2 | 3 | 3 | 2lb 14oz | | | | 2lb 14oz | | |
| 3 | 0 | 3+(¾) | 3lb 12oz | 3lb 12oz | 4lb 0oz | | 2lb 12oz | | 4lb 8oz |
| 3 | 1 | 4+(¾) | 4lb 12oz | 5lb 0oz | | | 4lb 12oz | | 5lb 0oz |
| 3 | 2 | 6 exactly | 6lb 1oz | 5lb 4oz | 5lb 0oz | | 6lb 1oz | 5lb 8oz | |
| 3 | 3 | 7+(¼) | 7lb 5oz | 5lb 8oz | | | 7lb 5oz | | |
| 4 | 0 | 9 exactly | 8lb 15oz | 9lb 0oz | | 9lb 8 oz | 8lb 15oz | | 9lb 0oz |
| 4 | 1 | 10+(¼) | 10lb 10oz | 10lb 8oz | | | 10lb 10oz | 8lb 8oz | |
| 4 | 2 | just under 13 | 12lb 10oz | 12lb 8oz | 9lb 0oz | | 12lb 10oz | | |
| 4 | 3 | 15 exactly | 14lb 14oz | 15lb 0oz | | 18lb 0oz | 14lb 14oz | | |
| 5 | 0 | just under 18 | 17lb 15oz | 17lb 8oz | | | 17lb 5oz | | |
| 5 | 1 | just under 20 | 20lb 1oz | 20lb 0oz | | | 20lb 1oz | | 18lb 0oz |
| 5 | 2 | just over 23 | 23lb 2oz | | | 28lb 0oz | 23lb 2oz | 18lb 0oz | |
| 5 | 3 | just over 26 | 26lb 6oz | | 18lb 0oz | | 26lb 6oz | | |
| 6 | 0 | 30 exactly | 30lb 0oz | | | | 30lb 0oz | | |
| 6 | 1 | 33 exactly | 34lb 0oz | 30lb 0oz | 44lb 0oz | | 34lb 0oz | 31lb 0oz | 30lb 0oz |
| 6 | 2 | 38 exactly | 38lb 0oz | | | | 38lb 0oz | | |
| 6 | 3 | 42+(½) | 42lb 0oz | 39lb 0oz | | | 42lb 0oz | | 42lb 0oz |
| 7 | 0 | 48 exactly | 48lb 0oz | | | | 48lb 0oz | | |
| 7 | 1 | 53 exactly | 53lb 0oz | 63lb 0oz | | | 53lb 0oz | | |
| 7 | 2 | 58 exactly | 58lb 0oz | | | | 58lb 0oz | 59lb 0oz | |
| 7 | 3 | 64 exactly | 64lb 0oz | | | | 64lb 0oz | 64lb 0oz | 60lb 0oz |
| 8 | 0 | 70 | 72lb 0oz | | 63lb 0oz | 68lb 0oz | 71lb 0oz | | |
| 8 | 1 | 77 | 78lb 0oz | | | | 78lb 0oz | | 60lb 0oz |
| 8 | 2 | 88 | 87lb 3oz | | | | | | |
| 8 | 3 | 90 | 95lb 0oz | | | | | | |
| 9 | 0 | 98 | 101lb 0oz | | | | 101lb 0oz | | |
| 9 | 1 | over 100 | 109lb 6oz | | | | | | |

* Sloane MS 2497/1580

single unit of relative density and diameter to mathematically formulate the tables used to make the rule, the table was actually made by weighing shot of all diameters which would naturally reflect the fluctuating figures of relative density inherent in the casting process. Another method (known by 1545) would be to take an example of a large ball, deemed typical weight for its nominal size, measure it, and adjust by the cube. Alternatively, it may just be a slightly inaccurate copy of an existing rule, or an accurate copy of an inaccurate rule.

A comparison of the lead faces shows a marked difference. If one takes '0' as the end of the rule, every single line and number is out of sequence, with the 2 on the Norton rule less than 2 on the *Mary Rose* example and the 5 on the *Mary Rose* rule registering nearly 6, with the 10 registering 11. Even if the 2lb marks are aligned, the deviation starts at 16, so that the 100 mark on the Norton rule reads only 85 on the *Mary Rose* rule. As the '0' should have the same origin for both faces, this exaggerates the difference, with the 100 on the Norton rule registering only 80 on the *Mary Rose* rule. Although the scale for the lead shot is variable from the beginning, from 4¾in the weights are so far below the Norton predicted weights as to beg the question as to whether this is not for lead shot, but for some form of composite shot. The *Mary Rose* carried very few solid lead shot larger than ¾in (18.06mm). This is perhaps reinforced by the complete lack of any gradations on the rule below about 50mm diameter. The one exception is a single shot of 2.67in (68mm). Composite lead shot containing an internal iron dice ranging in diameter from 1in to 2¾in (25–68mm) and larger composite shot of 2–3in (52–78.14mm made by casting lead over an existing smaller round shot (inset ball shot) were found (Dunham 1988, 3).

These types of shot, described as '*shot of dice and lead, shot of iron cast about with lead*' are listed in inventories for guns as large as demi culverins with bores varying between 4in and 4¾in (101.6–120.7mm; Starkey 1998, entry 4529) and common for guns up to the size of sakers at 3–4in (76.2–101.6mm; Starkey 1998, 4532) and for the smaller guns (under 3in). Unfortunately contemporary tables of weights for these composite shot have not been found and weighing the *Mary Rose* examples would be futile due to differential corrosion of the iron.

Solid lead shot of the larger sizes shown on the rule are rarely listed in contemporary inventories, but this does not explain why the rule was inaccurate. If the rule had been made for solid lead, it would be easier to merely stop the calculations at four inches.

Lastly, the figures on the rule were compared with the mathematical method of predicting the weight of shot (Norton 1628, 87). The method relies on 'finding the solid square' contained within the ball. It gives cast iron as 4oz per cubic inch. The formula is diameter x diameter x diameter x 11 divided by 21 to give the number of squares. This is to be multiplied against the weight of 4oz per cubic inch. This was tested at a number of positions against the *Mary Rose* rule and the rule illustrated by Norton. The figure of 6in came to 30 on both *Mary Rose* and Norton rules, and by arithmetical calculations, to 28.28. For 2½in the result was 2.04lb; for 5¼in, 18.94lb (20 on rules); 6½in, 35.96lb (38 on rules); 7in, 44.91lb (48 on rules); 8in, 67.04lb (70/72 on rules); 9in, 95.46lb (101/98 on rules).

Both the rules and the arithmetical calculations are fairly similar, and perhaps the observed differences are an indication of the range of what was considered acceptable. The ratio of iron to lead is listed at 30/45, stone to lead 18/72 and stone to iron 18/48.

The conclusion must be that the rule excavated from the *Mary Rose* is designed to calculate the weight of a round shot from the direct measurement of its diameter. The rule is calibrated in pounds of sixteen ounces and compares favourably with the iron shot weights given in extant sixteenth century tables and with the illustration of the rule in Norton (Table 5.19). Compared with modern projections for the volume of spheres of known size using the figure for the relative of modern iron, the scale is close. Compared with the projections for lead there is an observable difference. The suggestion is that the scale indicated as lead was made for providing the weight of composite shot rather than solid. This is explored below.

### The density of iron and lead shot based on the gunner's rule
by Keith Watson

Hildred (2003) has demonstrated that density values derived from the graduated scales marked on the gunner's rule show considerable variation for both cast iron and lead shot (Fig. 5.92). The plot of density against shot radius shows a marked downward trend for lead but remains more or less horizontal for iron. In both cases, though particularly for iron, the overall

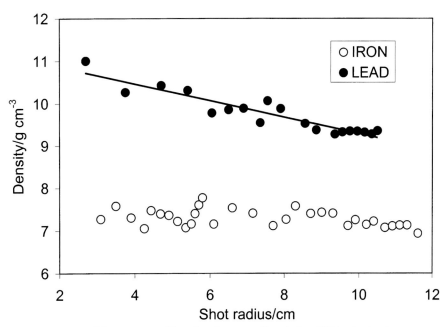

*Figure 5.92 Shot density vs radius (after Hildred 2003)*

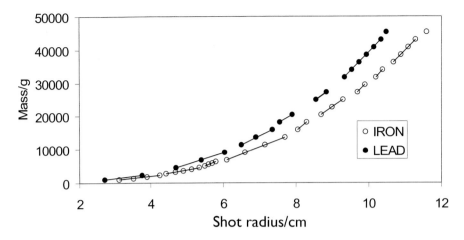

*Figure 5.93 Shot mass vs radius (after Hildred 2003)*

pattern appears to be broken down into short, randomly orientated sections. This is reflected rather more systematically in plots of shot mass against radius (Fig. 5.93); the curvature in both cases seems to be built up in segments. This patterning is consistent with the interpolation of evenly spaced scale divisions between more precisely determined graduations – a feature of a non-linear scale from the late sixteenth century noted by the author (in prep.). Alternatively, it suggests the possibility that the calibration of the gunner's rule may have involved discrete batches of castings that differed slightly from one another. However, any such hypotheses must remain conjectural without further evidence.

Much of the lead shot excavated from the *Mary Rose* contained an iron core, either in the form of a cast iron sphere or a wrought iron dice (Walker and Hildred 2000a). Calculations based on the maximum sized cube that can fit inside a sphere show that it is not possible to achieve a composite density lower than 10.04g/cm$^3$ using a wrought iron dice as the core; this assumes density values for wrought iron and lead of 7.8 and 11.34g/cm$^3$ respectively (Kaye and Laby 1995, 43 and 213).

On the other hand, the complete range of composite densities shown in Figure 5.92 can be produced using spherical cast iron cores of density 7.29g/cm$^3$ (the average value of those in the illustration), again assuming 11.34g/cm$^3$ for lead. The volume fraction of cast iron needed to produce each density value can be used to find the actual volume of iron in each shot.

The radius of each iron core calculated from its volume lies between 44–80% of the total radius of the shot itself. Figure 5.94 shows the relationship between the core radius and the shot radius, tabulated in Table 5.20.

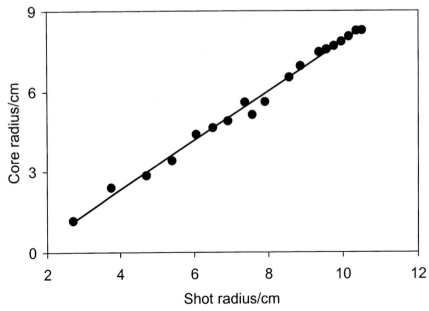

*Figure 5.94 Radius of iron core vs shot radius*

**Table 5.20 Iron core radius for composite inset ball shot (mm)**

| Ext. shot radius | Inner iron core radius |
|---|---|
| 2.70 | 1.1784 |
| 3.75 | 2.4082 |
| 4.70 | 2.8572 |
| 5.40 | 3.4150 |
| 6.05 | 4.4017 |
| 6.50 | 4.6493 |
| 6.90 | 4.9004 |
| 7.35 | 5.6035 |
| 7.55 | 5.1363 |
| 7.90 | 5.6178 |
| 8.55 | 6.5380 |
| 8.85 | 6.9557 |
| 9.35 | 7.4712 |
| 9.55 | 7.5676 |
| 9.75 | 7.6978 |
| 9.95 | 7.8593 |
| 10.15 | 8.0489 |
| 10.35 | 8.2635 |
| 10.50 | 8.2793 |

## Shot gauges

Nine objects were originally identified as gauges through which shot were passed to ensure that they were of the appropriate size for specific guns. The diameters of the holes range from 3$^{15}$/$_{16}$in (100mm) to 7$^{5}$/$_{16}$in (184mm) (Table 5.21). Distribution is shown in Figure 5.86.

To gauge shot, the hole must represent either the bore of the gun, or be smaller than the bore. As discussed previously in this volume, windage for bronze or cast iron guns using cast iron shot at this time has been taken to have been ¼in (c. 6mm) for large guns, although one-twentieth of the bore diameter was preferred (see above). There are no defined rules regarding windage for stone shot for wrought iron guns, but up to 1in has been tested (8mm difference found 81A3001, see Table 3.3). As each stone shot was being recorded it was tried through the gauges nearest in size and the results noted (see archive). Most required 5–10mm leeway between shot and gauge, although some did go through gauges that were only 1mm larger (eg, 73A0006 through gauge 82A4510).

**Historical references**

The earliest illustration the author has found is inscribed in French and German '*Patron duquel en haste on peult Calibrer diuerses sortes de balles*'/'*Model durch welchen in cill underschiedlichen kugeln können Calibrert werden*' ('*pattern with which in haste it is possible to measure different sorts of shot*') (Norton 1628, before p.

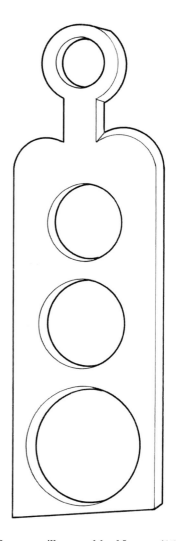

*Figure 5.95 Gauge as illustrated by Norton (1628)*

77). This is a long flat cricket bat shaped object with four holes of different sizes (Fig. 5.95). Eldred (1646, 46) lists these as instruments belonging to a Gunner; '*a height board to try and passe your shot thorrow*'.

Three ring gauges are illustrated in the *Armada* Catalogue (Rodríguez-Salgado 1988, items 9.20–2). These are wooden rings with butterfly shaped handles incorporated within the external circumference. The caption merely states that they were used in conjunction with the gunner's rule (*ibid.*, item 9.16) to size shot of different calibres and that they were all recovered '*in close proximity*'. One gauge has a hole in the handle suggesting that it may have been suspended. These are similar to the paddle-form gauges from the *Mary Rose*.

A copper shot gauge, also a ring but with a simple triangular extension for handling and suspension (Bat 3336) was recovered during the excavation of the Dutch East Indiaman *Batavia* (1629). This is stamped with '*+11*', the diameter (113mm) corresponds with the '*11*' mark on the gunner's tally stick. This corresponds exactly with the *Mary Rose* inscribed weights and measured distance for the iron scale on the gunner's rule (see above).

**Table 5.21 Shot gauges: summary data as measured (imperial measurements to nearest obvious unit)**

| Diameter | | Object | Thickness | | Form | Wood | Sector | Attributes |
| In | mm | | In | mm | | | | |
| --- | --- | --- | --- | --- | --- | --- | --- | --- |
| 3¹⁵⁄₁₆ | 100 | 81A1288 | ³⁄₈ | 10 | rectangular | oak | M5 | scribing circle 1 face; chamfered hole; split |
| 3¹⁵⁄₁₆ | 100 | 81A5662 | ³⁄₈ | 10 | paddle | oak | M8 | inscribed letters |
| 3¹⁵⁄₁₆ | 100 | 83A0019 | ¼ | 11 | paddle | oak | O8 | hole 10mm at base of handle; split to top; integral button at top in line with hole |
| 4¼ | 110 | 81A1285 | ³⁄₈ | 10 | rectangular | oak | M5 | poss. 2 'X' marks; chamfered hole |
| 5⁵⁄₁₆ | 138 | 81A5691 | ³⁄₈ | 10 | paddle | oak | O5 | in halves; looks cut; deep 'V' indentation across join on 1 face, notch on handle outer edges so could be tied; 6 nail holes 1 face, 1 side only; chamfered central hole |
| 6¼ | 163 | 82A4510 | ¹⁄₁₆–¼ | 2–6 | rectangular | oak | M3 | edges cut to form 8 sides; substantial gap (40mm) at top of circle to outer edge, looks cut rather than split |
| 6⁷⁄₁₆ | 170 | 81A1286 | ⁷⁄₈–1 | 22–24 | rectangular; like 81A1408 | elm | M5 | thick chamfer to circle both faces 11mm; cut from circle to outer edge |
| 6⁷⁄₈ | 175 | 81A1408 | ⁷⁄₈–1 | 22–24 | rectangular, like 81A1286 | elm | M9 | thick; chamfered circle; nail hole one corner, 5mm |
| 7⁵⁄₁₆ | 184 | 81A1287 | ³⁄₈ | 10 | rectangular | oak | M5 | inscribed circle at 11mm around central hole on 1 face only; cut from circle to one edge |

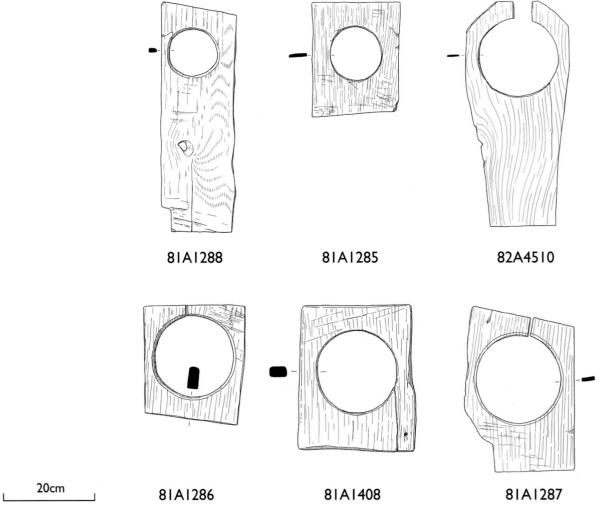

*Figure 5.96 Rectangular gauges*

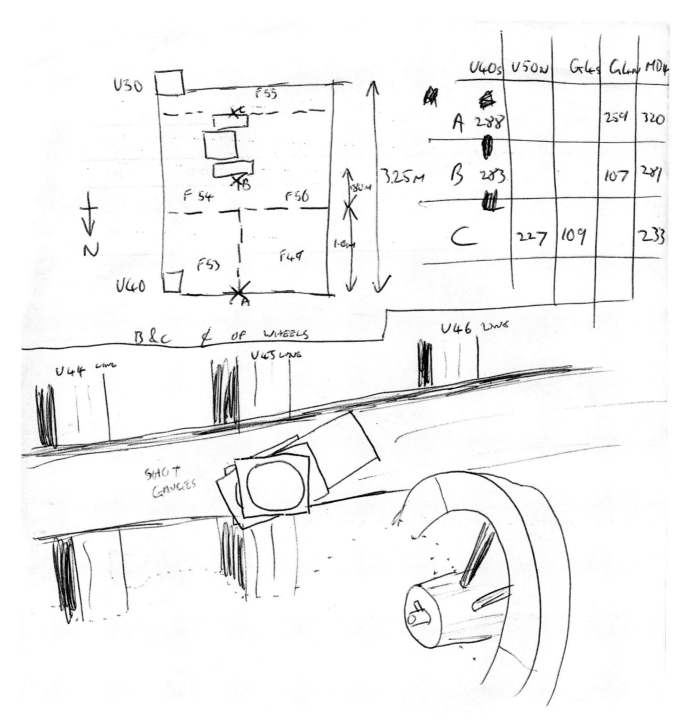

*Figure 5.97 Divelog 81/5/306 showing stack of gauges in M5*

Although contemporary documents indicate the use of boards to measure '*height*' of shot, it has been suggested here that these were used as formers during the secondary working of stone shot. One of the earliest references to stone shot stored in the Tower of London ('*stones for cannon called gunstones*') also mentions 8 moulds of oak for making the cannon balls round (Blackmore 1976, 243, 392, citing PRO E 364/90/D, the account of William Hickling dated 1455; see above, p. 384).

**Square and rectangular gauges**

There are six examples of this form, which are little more than sawn planks of wood with circular holes of varying sizes (Fig. 5.96). Care has been taken to ensure a circular hole, finished with a knife to form a chamfered edge (81A1288, 81A1285, 81A1286, 81A5691). Two display 'scribing circles' on one face (81A1287–8), possibly guides for bevelling. One has edges cut to form eight sides of unequal length (82A4510). Four are oak and two are elm (81A1286,

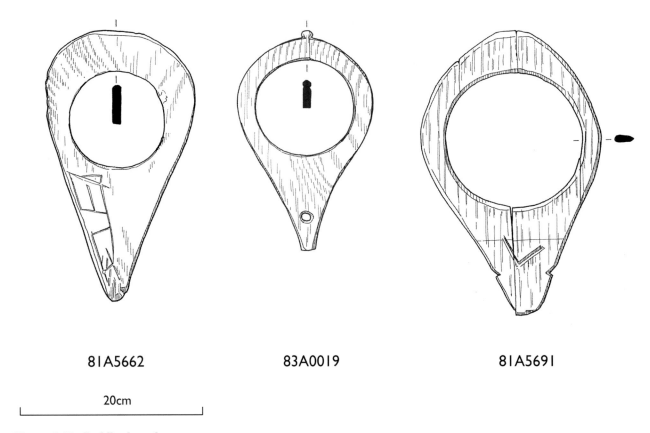

|81A5662|83A0019|81A5691|

20cm

*Figure 5.98 Paddle-shaped gauges*

81A1408). Three appear to have been deliberately cut between the circle and the closest outer edge (81A1286, 81A1287, 82A4150) and two are split (81A1408, 81A1288), though it is not known if this happened post-deposition.

Gauge 82A4510 was found in M3 and 81A1408 in M9. The other four (81A1285–8), were found stacked just above an *in situ* wrought iron port piece (81A2604) on the Main deck in M5 (Figs 5.97 and 5.86). Each had a different diameter hole ranging from 100mm to 184mm. This area served as a store for shot (17 iron, 44 stone). Stone shot ranged from 175mm to 232mm in diameter, of which 26 were unfinished (see above). The M5 port piece has a bore of 195mm, 11mm larger than the largest gauge. Twenty-four of the shot were larger than the gun but all except one of these was unfinished.

It has already been argued that stone shot was finished on board so it is particularly interesting to note that stone shot in this group could only have been used with the largest gauge, 81A1287, at 184mm. There are fourteen stone shot of 175–184mm, but only three are unfinished. There are not enough of the appropriate size to suggest that this was a sorting area (unless most of the sorted finished shot had been removed) and it may be that this was a stone working area using the unfinished stone shot from M6 and delivered to appropriate guns. It may be no coincidence that a grinding wheel (81A1823) was found on the Orlop deck below in O3.

The 17 iron shot range from 58mm to 129mm but only eight of these are larger than 80mm. Seven are between 81mm and 96mm and could be used with the 100mm gauge, 81A1288; one is 110mm and does not fit any gauge. Clearly there is no match between the shot recovered from this area and the gauges found, so why they came to be stacked here is still a matter for debate.

Gauge 82A4510 from M3 has a diameter of 163mm and is too large for the *in situ* culverin (81A1423) found here (bore 140mm). It could possibly have been used to gauge shot for the adjacent port piece in M4 (81A2650) with a bore of 179mm. The seven stone shot in this area ranged from 145mm to 173mm, with three finished shot 168mm, 170mm and 173mm (81A1190, 82A4517, 81A1127); all too large for this gauge. Gauge 81A1408 from M9 has a hole of 175mm diameter and could have been used to select shot for the port piece in M8 (81A3001). This has a bore of 180mm but the shot recovered from within it is 172mm. A total of 24 stone shot came from M/O8, ranging in size from 150mm to 188mm; half of these were unfinished.

**Paddle-shaped gauges**

Three paddle-shaped examples appear more finished than those just described (81A5662, 81A5691, 83A0019; Figs 5.98 and 5.86). One has a circular hole in the 'handle' (83A0019), possibly for suspension. The others each bear incised marks on one surface.

Table 5.22 Recovered guns (metric) and shot gauge sizes

| Object | Gun type | Bore | Shot type | Expected shot size | Gauge | Type |
|---|---|---|---|---|---|---|
| 79A1295 | WI sling | 85 | iron | 75 | | |
| 81A0645* | WI sling | 90 | iron | 80 | | |
| 70A0001* | WI sling | 95 | iron | 85 | | |
| 77A0143 | WI fowler | 110 | stone | 85–102 | ?100 | 2 x P, 1 x R |
| 90A0126 | WI fowler | 110 | stone | 85–102 | ?100 | 2 x P, 1 x R |
| 78A0614* | WI fowler | 110 | stone/90mm | 85–102 | ?100 | 2 x P, 1 x R |
| 90A0131 | WI fowler | 112 | stone | 87–104 | ?100 | 2 x P, 1 x R |
| 79A1232 | demi culverin | 112 | iron | 102–107 | ?100 | 2 x P, 1 x R |
| 79A1279 | demi culverin | 115 | iron | 105–110 | 110 | R |
| 78A1278 | culverin | 132 | iron | 122–127 | | P |
| 80A0976* | culverin | 138 | iron/126–130mm | 128–133 | ?138 | P |
| 81A1423* | culverin | 140 | iron | 130–135 | ?138 | P |
| 81A3002 | culverin | 140 | iron | 130–135 | ?138 | P |
| 80A2004 | WI port piece | 150 | stone | 125–142 | ?138 | P |
| 80A2003* | WI port piece | 152 | stone | 127–144 | ?138 | P |
| NMM* | WI port piece | 160–180** | stone/155mm | 135–152 | 138 | P |
| 81A1277 | demi cannon | 160 | iron | 150–155 | | |
| 81A3000 | demi cannon | 165 | iron/160mm | 155–160 | ?163 | R |
| 81A2650 | WI port piece | 179 | stone | 154–172 | 163, 170 | R |
| 81A3001* | WI port piece | 180 | stone/172mm | 155–172 | 163, 170, ?175 | R |
| 71A0169* | WI port piece | 190 | stone/165/mm | 165–182 | 163, 170, ?175, 184 | R |
| 81A2604 | WI port piece | 195 | stone | 170–187 | 175, 184 | R |
| Woolwich* | WI port piece | c. 200 | stone | 175–192 | 184 | R |
| 81A3003 | cannon | 200 | iron | 190–195 | | |
| 79A1276 | cannon | 215 | iron | 205–210 | | |

\* Records suggest object found with shot *in situ*; dimensions given where known
\*\* This was markedly different at each end, the small dimension is probably the result of corrosion. A shot of 155mm was within the bore of the gun
WI = wrought iron; P = paddle; R = rectangular

81A5662 and 83A0019 both have 100mm diameter holes and were found in M8 and O8 respectively. The only gun in this area was a port piece of 180mm bore so these gauges were not for use with this gun. There was only one iron shot in M8, 79A0980, with a diameter of 140mm and one from O8, diameter 80mm (80A1777). Although this would go through the gauge, the gap is considerable.

These may have been used to select shot for smaller guns on the Upper or Castle decks: the demi culverins (bores up to 115mm) or fowlers (110–128mm). Fowlers used stone shot, so would require larger windage. All fowler remains were found in the bow and there were only four small stone found in the stern. Of the ten stone shot of 95–100mm, all of which are finished, the known distribution is in trench 1, the bow bulkhead, or the Starboard scourpit in the stern (SS10–11).

The smallest shot that would serve the *in situ* demi culverin (79A1232, C7) was 100mm, too large for the gauges. The two sakers listed in the inventory (Chap. 4), needed shot of 65–95mm and the closest iron shot peak is 94–5mm, with an extremely large peak at 84–6mm. So although these gauges could have been used to select shot of this size, the shot distribution does not clearly indicate the position of a suitable gun.

The third paddle-shaped gauge was found in O5 (81A5691). This has a diameter of 138mm, clearly too small to be used with either the M4/M5 port pieces or the cannon in M6. It is 2mm smaller than the bore of the M3 culverin and could possibly be used to size shot for this gun. There were four iron shot from O5, all less than 86mm, and no stone shot

### Matching gauges to guns

There does not seem to be any accepted formula for gauge size relative to bore size. Common sense suggests that the gauge size was between actual bore size and bore size less one-twentieth, this being the accepted allowance for windage. As any loss of gas compromises efficiency, one would expect the gauge to be just large enough to take the shot with minimum windage. The

**Table 5.23 Gauge sizes against iron and stone shot diameters**

| Gauge diam. (mm) | Gauge | Shot size (mm) | No. shot |
|---|---|---|---|
| Iron shot | | | |
| 100 | 81A1288, 81A5662, 83A0019 | 94–96 | 43 |
| 110 | 81A1285 | 100–106 | 114 |
| 138 | 81A5691 | 128–130 | 111 |
| 163 | 82A4510 | 155–156 | 38 |
| 170 | 81A1286 | 160–170 | 5 |
| 175 | 81A1408 | 170–175 | 3 |
| 184 | 81A1287 | 180–184 | 10 |
| Total/% | | | 324/3.85% |
| Total iron shot recovered | | | 1248 |
| *Stone shot* | | | |
| 100 | 81A1288, 81A5662, 83A0019 | 75–98 | 8 |
| 110 | 81A1285 | 99–108 | 6 |
| 138 | 81A5691 | 110–138 | 12 |
| 163 | 82A4510 | 150–161 | 31 |
| 170 | 81A1286 | 161–168 | 38 |
| 175 | 81A1408 | 169–173 | 32 |
| 184 | 81A1287 | 174–182 | 30 |
| Total/% | | | 157/75% |
| Total finished | | | 209 |
| Total stone shot recovered | | | 387 |

The maximum size of shot to fit any gauge was taken to be 2mm less than the gauge diameter

tightest fit of shot to gun found was 5mm for iron shot to bronze gun and 6mm for stone shot to iron gun. The loosest fit recovered was 10mm smaller than bore size for iron shot found within bronze guns. The problems with internal corrosion means that bore measurements of iron guns are often smaller than they were originally, providing a closer fit than would be expected given the unevenness of both bore and shot, compared with bronze guns and cast iron shot. Therefore, a maximum of 25mm was tested for the stone shot. The expected shot range size for guns recovered based on the above is shown in Table 5.22, together with the closest 'match' of gauge. The results suggest the best match between gauges and guns is most convincing for the wrought iron guns. There are complete groups of bronze guns (culverins, cannon, sakers, falcons) which do not appear to have complementary gauges. There is none small enough to be used with the wrought iron slings or the bases. The three 100mm gauges (77A0143, 78A0614, 90A0126, 90A0131) could have served the fowlers, with bores of about 110mm, or the demi culverins (79A1232, 79A1279), but with more than ¼in windage. The 110mm gauge may just have served 79A1232 (bore 112mm) and the 138mm gauge could have been used with three of the culverins, but without much windage. It could, however, have served port piece (KPT0017), but with a large gap. The gauges of 163–184mm could have serviced the range of port pieces. 82A4510 (163mm) might just clear the shot for 81A3000 (165mm), but the windage would be very small. The conclusion must be that if used to select shot for bronze guns, there are a number of gauges missing.

**Matching gauges to shot**

Shot gauge diameters were studied against the peaks for shot (see also above). The largest diameter listed is 2mm below gauge diameter, as shown in Table 5.23. This suggests that 75% of the finished stone shot should pass through the gauges recovered. However, when 335 of the 387 stone shot (87%) were physically tried through the closest sized gauges (Martin 1997), there were a number of discrepancies. It was clear the shot were not perfectly round because they had been hand finished and so fitting them through the gauge closest to their recorded diameter was not always possible. It is suggested that surface irregularities in the shot resulted in some unusual results. However, the following were obtained (in the statistics given below the diameter cited is the largest of three taken on each shot).

- The 100mm gauges were suitable for the 75mm shot but would not fit shot of 96mm (82A1693), although one of the 99mm passed through.
- The 110mm gauge (81A1285) was suited to five shot between 99mm and 102mm and one at 96mm.
- The 138mm gauge (81A5691) fitted three of the five shot between 110mm and 138mm. The smallest one which that not fit through was 81A1895 at 135mm.
- The gauge at 163mm (82A4510) fitted 26 of the 40 shot between 138mm and 163mm and three smaller shot which got stuck in the 138mm gauge. The smallest shot not to clear this gauge was 90A0129 at 157mm.
- The 170mm gauge (81A1286) fitted eleven shot of less than 163mm and 16 of the 43 shot in the range between 163mm and 170mm.
- The 175mm gauge (81A1408) fitted nine shot varying between 163mm and 169mm and 10 of the thirteen shot between 163mm and 168mm.
- The 184mm gauge (81A1287) accepted six shot below 170mm (the smallest at 166mm), and seven between 170 and 175mm. The gauge was tried on the 25 shot between 176mm and 184mm and fitted 23 shot, one (81A2049) was unfinished. Six shot over 185mm fitted through this gauge.

Gauge size was also compared with stone shot with diameters of nearest size to see whether these could have been used as guides to finish shot. The presence of two forms of gauge may suggest two possible uses and the paddle form fits well in the hand, and is therefore better suited to selecting shot than the rectangular

**Table 5.24 Shot size in relation to gauges and/or formers**

| Gauge diam. (mm) | Form | Stone shot peaks (mm) if former | Stone shot peaks (mm) if gauge | Iron shot peaks |
|---|---|---|---|---|
| 100 | R | 98 | 75, 90 | 85–87, 95, 96 |
| 100 | R | 98 | 75, 90 | 85–87, 95, 96 |
| 100 | P | 98 | 75, 90 | 85–87, 95, 96 |
| 110 | R | 108 | 98 | 102, 105 |
| 138 | P | 135, 137, 138 | 125 | ?128–130 |
| 163 | R | 160, 163 | 140, 145, 150 | 155, 160 |
| 170 | R | 165, 170 | 145, 150, 155, 160 | none |
| 175 | R | 170, 175 | 155, 160, 165 | none |
| 184 | R | 180, 184 | 165, 170, 175 | none |

P = paddle; R = rectangular

form. Table 5.24 demonstrates the relationship between the gauges and closest shot peaks. Although it is impossible to be certain, there seems a strong link between the objects and the stone shot, both as guides for manufacture and as gauges. There is no iron shot which would fit the four larger gauges, although there are bronze guns larger than the largest gauge. It is possible that the paddle shaped gauges were used to select iron shot for the bronze guns (and slings) and that the rectangular paddles were used in the finishing of stone shot. In both scenarios the assemblage of objects is incomplete.

## Loading and Firing
by Alexzandra Hildred

The firing of '*great ordnance*' required co-ordination of a gun crew, attention to detail and the performance of well-rehearsed steps, later formalised as 'drill', or 'gun drill'. In addition to trained personnel, specific items of equipment were necessary. These remained similar throughout the smooth bore era (1520s–1860s) and many can be seen in the *Mary Rose* assemblage. Others are absent. There is some difference in what is required for the breech-loaders and the muzzle-loaders.

### Muzzle-loading guns

Typically, a drill for smooth bore ordnance would require the following to be undertaken after a round had been fired. The first would be the cleaning of the bore. This is undertaken with a *worm* or corkscrew-like iron head on a long wooden handle, designed to remove any loose wadding or the remains of the cartridge or powder-bag (if used) from the previous round. A wet sheepskin *sponge* on a long wooden handle was then pushed down the bore to touch the breech to further clean out the bore and extinguish any sparks. The gun would then be ready to load. Loose powder was loaded from a cask into an appropriately sized *ladle* on a long wooden pole, pushing this into the bore and turning it to release the powder. This was sometimes repeated several times. If the powder was already in a *cartridge* (a pre-packed canvas or paper bag), a ladle was not always required (and some authors suggest specific ladles of a slightly different form). The bag was pushed gently into the bore using a *rammer*, a simple, round wooden head on a long handle. A *wad* of hay, cloth or rope was further rammed over the charge (or over a round wooden *tampion*). This wad functioned to prevent an excess of gas escaping from around the projectile which would reduce the force of the explosion and decrease the range.

The projectile was then rammed onto the powder charge. If round shot was used, another wad often termed the top wad (or a wooden tampion) was rammed onto it to prevent the ball from rolling out of the tube. If the powder was cartridge loaded, a wire termed a *priming wire* or *pricker* was inserted into the vent to puncture the bag and insure ignition of the main powder charge. This is not necessary, but is useful. Biringuccio (1540) merely states that the cartridge is rammed to break it. If loose powder was used, this was unnecessary. The priming wire also functioned to clean the touch hole, or vent. *Priming powder* was then poured into the touch hole to prime the gun, which was then ready for firing. On the command to fire, the priming powder was lit by holding the *linstock*, a wooden stick with a bifurcating end which held a saltpeter-impregnated slow burning match. Other items necessary to manoeuvre the gun include iron *crowbars* and wooden *handspikes* with wooden *quoins* for elevating the gun. Extra tampions were often used to plug the gun muzzles when not in use to keep them dry. (See below '*How to loade a Peece of Ordnance Gunner-like*'.)

### Breech-loading guns

As the charge was separately contained in a chamber, the bore of the gun would not need worming or sponging between shots. The pre-loaded chambers would require cleaning after firing. Typically two (sometimes three) chambers are listed with each of the large wrought iron guns, affording two rounds before cleaning the chamber. Practical experience (see Chapter 3) suggests that these would be carefully cleaned, possibly by filling them with water and sponging, and would be loaded with loose powder through a funnel, or possibly with a pre-loaded cartridge. Decanting powder from a ladle into a

### *How to loade a Peece of Ordnance Gunner-like*

*The Peece being mounted, and duly planted on his Platforme, and well prouided with all things in readinesse for seruice, as of powder, Shot, Ladle, Spunge, Ramer, Wadhooke, Wadds' and Tampions. The Gunner must place his Linstock to Lee-wards, or vnder the winde, and hauing cleared the peece and Touch-holes, he must spunge his peece very well standing on the right side of the peece, and drawing out his Spunge, let him giue it two or three blowes vpon the Mouth on the out-side of the peece, to beate off the foulenesse and dust it hath gathered within: then his assistant declining the powder or boudge barrell aside, he shal thrust in his Ladle to fill it, striking off the heaped powder, giuing a shogge to shake downe the surplusse: and it being so filled and striked, put in the Ladle downe to the bottome of her Concaue Cillinder vnto the Touch-hole, but at the first putting of the Ladle (so filled) into the Mouth of the peece, slide in the Ladle staffe, so that the vpper side may keepe vppermost all the way, and when it is ariued to the bottome of the Bore, then hee laying his right thumb vpon the said vpper side of Ladle staffe, neere the Mouth of the peece, and turning the said staffe so much, vntill the said thumb vnremoued vpon the staffe be found directly vnder the same; then giuing two or three shakes and bearing vp the Ladle, that the powder may bee turned out, or goe out cleane, and that the Ladle bring no powder back therewith, for that were a foule fault for a professed Gunner to commit. Then shall he put the powder home softly, with the Rammer that is at the end of the Ladle staffe; putting in a good Wadd, and thrust it home to the powder, giuing three or foure hard strokes, which wil gather the scattered pouder together, and driue close the same, and the rest to the bottome of the Chamber the Assistant, hauing a finger vpon the Touch-hole all the whilst: And then put in the Shot which with a Rammer he must put softly home, and afterwards another Wadd of Hay, Grasse, Weedes, Okham , or such like: And againe, giue two or three good strokes with the Spunges Rammer head: But if the Peece doe require two or three Ladles of powder, it must bee all put in before any Wad, in that maner as the first was mentioned, and so in all other things. And then place the Boudge or powder barrell to wind-wards, and couer it safe with some Hide, garment, or cloath: alwayes auoyding to stand before the mouth of the Peece, but on the right side thereof in loading her for feare of further danger. And lastly, the Peece is to bee layd to the marke and prymed and fired, and so will his Peece said to be loaded Gunner-like.*

(Norton 1628, 101)

chamber on a rolling deck would not be an easy task. Excavation of breech chambers suggests that, following the powder, a wad of hay or oakum was placed in the chamber which was then sealed with a tampion. This is similar to that suggested for chamber pieces by Norton (1628). The shot would be placed in the rear of the gun hall, and the loaded chamber located in after this. The mouth of the gun would also be sealed with a tampion. Cleaning the touch hole and priming the gun would be similar to the muzzle-loaders.

## Positioning of equipment

Norton (1628, 101) states the importance of standing and placing loading implements in the correct place for firing the gun. The following items should be ready: powder, shot, ladle, sponge, rammer, wadhook, wads, tampions. All equipment is to be stored on the right side of the carriage with ladles and sponges facing the mouth and rams and '*wadhooks*' (worms) facing the breech. The linstock should always be stored safely, away from the wind.

The earliest word used to describe the worm in English sources is the wadhook, mentioned in the Tower of London inventories for 1595, 1637 and 1635, and defined by the *Gentleman's Dictionary* of 1705 as 'Wadhook' or 'Worm'... *a small Iron, turned Serpent-wise like a Screw, and put upon the end of a long Staff, to draw out the Wad of a Gun*' (quoted Blackmore 1976, 248).

These would be basically iron screws on wooden handles but no possible examples have been found on the *Mary Rose* (although four staves tentatively identified as '*darts*' may be the wooden ends of wadhooks). Their locations should be at least similar to the rammers or ladles (Figs 5.11–12). It is difficult to gauge at what time their use was routine. Eldred (1646, 42) in discussing '*the manner of lading your Ordnance*', states that the ladle is used to search and clean the piece:

> '*Then comely and gracefully take your Ladle and put it into the Peece, and turn it and move it too and fro, to search if there be any Stones or gravell in the same, that you may bring it out with your Ladle, and so cleere the Peece, and when you have well cleered the same from Stones and gravell with your Ladle, set by your ladle in his place, take your Spunge in comely manner like a souldier, and Spunge well your Peece*'.

No complete sponges have been found, though sheep skin wrapped round the head of a rammer may have served that purpose (the *Anthony Roll* listed twelve '*Shepe skynnys for spongys*'; see also Blackmore 1976, 245). Specific requirements for sponges are mentioned by Norton:

> '*their buttons or heads are to be made of soft fast wood, as Aspe, Birch, Willow, or such like, and to be one dyametre, and ⅔ in length, & not aboue ⅔ of a dyametre of the shots height: The rest being couered with rough Sheepes skinne wooll, and all, be nayled thereon with Copper nayles, so that together they may fill the Soule or Cauity of the Peece*'. (Norton 1628, 111)

This was written a century after the *Mary Rose* was afloat, and so it is unreasonable to expect items of precisely that kind. To date we have not found anything which meets the description. The heads of the rams are in all cases not much longer than their diameters and there is no evidence of copper nails, however a combination rammer/sponge has been considered (see below).

Ladles, rammers, shot, wads (through excavation of breech chambers or of gun barrels), tampions, priming wires and linstocks have been recovered. Handspikes have not been identified, and neither have crowbars. The iron bars found outside the hull by gunport lids but may have been used as lid supports, or may be incomplete crowbars. Gunpowder has only been found through the excavation of gun barrels and breech chambers, although analysis of the contents of oak storage casks is not complete (see also *AMR* Vol. 4, chap. 10). If a cartridge was used, the other necessary item would be the pricker, or priming wire; five of these were recovered (below). These, together with any aiming devices or instruments, can be considered as personal possessions of the gunner.

Eldred (1646, 46) lists the following as the instruments which should belong to a gunner:

> '*first a Quadrant, secondly a Sight-rule, 3ly. A payre of Compasses with crooked points called Calipers, 4ly. A payre of streight pointed Compasses, 5ly. A height bord* [shot gauge] *to try and passe your shot thorrow, 6ly. A case of priming Irons of all sorts, 7ly. A Horn or two for priming powder, 8ly. A Steele to strike fire, 9ly. Two or three Linstocks to give fire well armed with Match, these things no Gunner can well be without*'.

Tthe items listed for the *Mary Rose* in the *Anthony Roll* under munitions include 8 '*Sledgys of yron*' [sledge hammers for driving and removing wedges of all sorts], 12 '*Crowes of yron*' [crowbars for manoeuvring ordnance], 12 '*Comaunders*' [large mallets for knocking in/removing breech chamber wedges], 4000 '*tampions*' [for sealing the bores of guns and the mouths of breech chambers], and quantities of canvas, '*Paper ryall*' [royal]' and '*Fourmes*' [formers] for '*cartowches*' [cartridges].

Under '*Habillimentes for warre*' are 10 coils of '*Ropis of hempe for woling and brechyng*' [the securing of guns to beds and of guns to the side of the ship], 1000 '*Naylis of sundere sortes*', 8 '*Bagges of ledder*', 6 '*Fyrkyns with pursys*' [barrels with leather dispensers, possibly forerunners of budge barrels for gunpowder] ... 12 '*Shepe skynnys for spongys*', and 100ft of '*Tymber for forlockes and koynnys*' [for breech chamber wedges and quoins].

The 1514 inventory is slightly more specific, the *Mary Rose* entry listing 2 '*Chargynge ladelles of coper*' left in the ship and 7 '*Charging ladelles for fawcons and curtowes*' delivered to John Millet and Thomas Elderton [5 great curtows and 2 falcons being listed], with 6 '*Sponges to the same*' and 3 '*Stampes*' [probably rammers]. Other references to ladles are few: the *Henry Grace à Dieu* had 6 '*Chargyng ladylles for gonnes with staves*' but 8 '*Staves withowt ladelles*', and the *Trinity Sovereign* had 11 '*Gonne laddelles of brasse*'.

It may be that, as so many of these items are gun specific, they are considered as accompanying the guns and, therefore, are not listed separately. Certainly the 1547 inventory (Starkey 1998) lists guns '*furnyshed*' (eg, 4109), which may indicate with appropriate loading equipment. Other phrases suggesting this include '*Bastard culveryn of yrone of parson Levettes making with thabulyamentes*' and '*Sacre of yrone with thabillementes*' (4252–4253).

There are entries for guns specifically listing certain pieces of gun furniture, (eg, 4573) '*Cannon of brasse mounted vppon vnshodd wheles with ladells and Rammers*', and some separate entries, (eg, 4589–90) '*Ladells of dyuers sortes*', '*Staves for sponges and Rammers*'. There are in all 81 entries for ladles and eleven for rammers with six separate entries for ram heads and one for staves. Whereas entries for specifically sized ladles and sponges are relatively common, (eg, 3915) '*Demy Culveryns of yron with ladle and sponge to the same*', entries for specifically sized rammers are rare (eg, 5653) '*Rammers for Fawcons i^o*'. The size relative to the bore is not as critical as for a sponge where a near vacuum is achieved during the sponging. Similarly the amount of powder is critical and gun specific. Copper plate for making ladles is sometimes separately recorded (5111, 5900, 5988) and ladles of '*latteyne*' [latten] are also mentioned (4343). There are some separate listings for heads (5041, 5992).

Interestingly there are a few entries for ladles for chamber pieces, (eg, 4722): '*Sling chambers of yrone stocked and mounted with new Axeltries and troulces for euery chambre a ladell and a Rammer*'. One would expect these to be specifically small and short, as they are primarily for cleaning and loading the chambers. There are two entries for sheep skins (6074, 6165), and 47 for sponges, (eg, 6193): '*Ladells and sponges for the Fawcone and Fawconet*'. There are seventeen entries for tampions (none for any wadding) and one entry for a '*Scrapers*' (5651), which may be another term for worm. The entries for Boulogne, which needed to be self-reliant, are the most useful; here (6076–8) we find 12 '*Plates for ladells of all sortes*', 24 '*Rammers and Rammers heades*' and 100 tampions.

Linstocks do not seem to be listed and may be personal, hence the individuality within the *Mary Rose* assemblage. Linstocks are not itemised in the vessel listings.

## *Gun drill on the* Mary Rose
by Nicholas Hall

Reconstructing gun drill from the period when the *Mary Rose* set sail for the last time cannot be based wholly on documentary sources. There is no written contemporary surviving gun drill. Niccolò Tartaglia's seminal work of 1546, dedicated to King Henry VIII, contains nothing so mundane as a specimen gun drill (Tartaglia 1546 but, unless stated otherwise, references here are to the English translation – Lucar 1588). However, Tartaglia does discuss some of the technicalities of loading an artillery piece. An example, incidentally showing that gunners were not always aware of why they were performing certain actions, is found in the 20th colloquy (1588, 38–9).

A bombardiero asks:

> '*for what cause ... are two waddes of hay or of toe put into a Peece at euery time when it is charged ... one wadde after the powder is put in betweene the powder and the pellet, and an other wadde after that the pellet is put in?*'.

Tartaglia, who was not a gunner but an interested mathematician, returns the question to the practical gunner, who replies '*I will confesse ... that I am vnlearned, and that I haue vsed to do so because I haue seene all other Gunners vse to doe the same.*' Tartaglia answers, with his usual good sense, that the first wad sweeps the powder together (when using a ladle for loading) while the second is inserted '*that the pellet may not fall out*'. Luis Collado also contains much practical advice. He stresses that artillery should be loaded '*con garbo e ragione*' (1641, 11), that is, with grace and '*reason*'. But the English '*reason*' is hardly a sufficient translation of ragione – it implies experience and a thoughtful, logical approach.

William Bourne's preface to his work on gunnery (published 1587 but written 1572–3) observes that:

> '*we English men haue not beene counted but of late daies to become good Gunners, and the principall point that hath caused English men to be counted good Gunners, hath been, for that they are hardie or without fear about their ordnance: but for the knowledg in it, other nations and countries haue tasted better therof, as the Italians, French and Spaniardes, for that English men haue had but*

*little instructions but that they haue learned of the Doutchmen or Flemings in the time of King Henry the eight. And the chiefest cause that English men are thought to be good Gunners, is this: for that they are handsome about their Ordnance in ships, on the Sea &c.'* (sig. A iii)

So, grace was not lacking, but according to Bourne, '*reason*' was. Eldred (1646) was also to stress the importance of well-executed drill that would please any observers. But what did the gun drill actually entail?

The guns and their associated equipment as recovered from the *Mary Rose* are vital evidence, but by themselves they cannot tell us much about how they were used. In order to reconstruct possible ways in which the heavy guns were served, two approaches are considered.

*1. The adoption of specific manoeuvres which form an essential part of firing artillery using black powder.* These are in response to the nature of the powder, and the type and size of gun. Without correct implementation of these, the gun will not fire reliably and serving the gun could become unacceptably dangerous to the crew, the gun or even to the ship. This approach draws on experimental archaeology with replica guns, and leads us to a gun drill that works, but cannot be proved to be exactly the Tudor form. The size of the gun and the requirements of the manoeuvres dictate the size of crew required; which in turn enables an estimation of the number of ship's big guns which could be fought at the same time. It hardly needs to be said that this gives a basis for estimating the effectiveness of the *Mary Rose* as an artillery platform engaging enemy ships before closing for boarding.

*2. The use of later sources.* This might provoke the immediate response that this is unacceptable in a serious historical work. However, we should consider that the works of those who wrote within living memory of the loss of the *Mary Rose* or who were trained in the later sixteenth century: Cyprian Lucar, translating and commentating on Tartaglia; William Bourne; William Eldred; Girolamo Ruscelli and others. These authors will be examined for evidence that might illuminate, even by reflection, the state of affairs on board *Mary Rose* in the 1540s. Cyprian Lucar's 1588 English edition of Tartaglia's *Colloquies* with his own useful Appendix, put together from various sources, touches on the proper handling of guns without giving too much away to the modern reader. William Eldred recorded his enormous experience of artillery in *The Gunners Glasse* published in London in 1646. He was then 83 years old, indicating that he might well have begun to learn the gunner's art in the early 1580s – from men that surely would have remembered the days of the great Henry VIII. Where these two approaches tend to lead in the same direction, it may be that we are some way to an understanding of gun drill of the time.

Gun drill on two very different kinds of gun will be discussed. The guns are those that were replicated for firing trials in a joint Royal Armouries/Mary Rose Trust research project (see also Chapters 2 and 3). The guns were:

- The wrought iron port piece; a breech-loading gun without trunnions, fitted into a single piece elm bed, mounted on an axle and two spoked wheels. Elevation of the gun is by means of lifting/lowering the entire carriage. A rectangular slot through the back of the carriage accepted a post with a number of holes at varying heights. A movable peg inserted through one of the holes supported the carriage at the desired height. The port piece fires only stone projectiles.
- The cast bronze culverin; a muzzle-loading gun with trunnions, mounted on a complicated carriage constructed of several major elements and having four solid wood trucks. Elevation is by quoins. The gun fired a cast-iron projectile.

The inherent differences between the two guns require the drills to follow very different patterns. It is hoped that the two selected will serve for others of related type.

**Port piece gun drill**

As far as the present author is aware, the only attempt at a reconstruction of the gun drill of a wrought iron breech-loader on board ship is by Guilmartin (1988) in '*Guns from the Sea*'. In his paper on early modern naval ordnance, he discusses the problems involved in loading a wrought iron breech-loader, in this case a '*bombardeta*' based on the Molasses Reef wreck finds, thought to be a few decades earlier than the sinking of the *Mary Rose*. This bombardeta is smaller than the port piece and, unlike it, fired a cast-iron ball of about 4lb (1.8kg), not dissimilar to the *Mary Rose* slings. Two of these guns were found and are believed to be the complete heavy armament of the ship – the other piece was a smaller '*cerbatana*'. Having only these 'heavy' guns, Guilmartin assumed that the ship was 'overmanned' to the extent that there would have been no shortage of gun crew. Therefore, in his 'walk through' drill experiments with members of the Institute of Nautical Archaeology, he was prepared to accept a large gun crew in his first hypothesis (Fig. 5.99). This proposed a crew of nineteen members (or '*numbers*') including five to reload chambers, but ignoring two to routinely collect powder. Realising then that manpower could, in fact, have been in short supply, he suggested '*the absolute minimum crew needed to work the bombardeta under battle conditions*' (1988, 45–6). This amounted to a total of thirteen, including three to reload chambers, but not counting one to get powder.

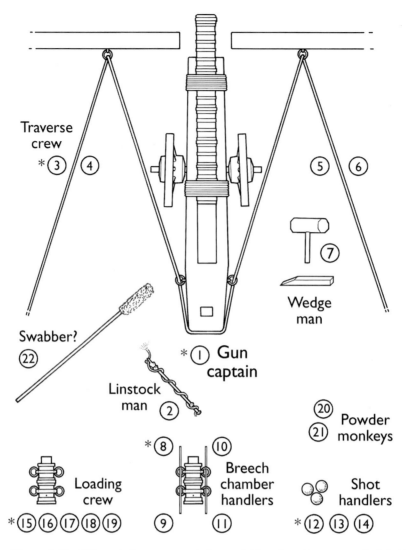

*Figure 5.99 Gilmartin's recommended gun drill and equipment needed (1988)*

In the case of the *Mary Rose*, the number of men available for the number of guns does not allow the luxury of such a large gun crew. It is pertinent to exclude the people counted by Guilmartin as reloading chambers, on the basis that, with two or three chambers, no reloading might be needed during a short action, or that reloading during a longer engagement would take place during those times when no target was available, or the pace of action slowed. To allow for reloading as quickly and as safely as possible, cartridges may have been used. Using pre-loaded chambers, the ship as a fighting machine could fire quickly three rounds per port piece without calling on supernumeraries. There is evidence for this, both in inventory entries (see Chapter 11) and within the written sources. As Lucar states (1588, appendix, 47), '*Euery chamber peece ought to haue three chambers*'. This volume of fire could have a real tactical effect, even if there was some delay in reloading for another firing sequence.

An alternative is that, after the loaded chamber was inserted, two of the chamber team could reload the previously fired chamber. This reduces Guilmartin's total to ten. Guilmartin is unwilling to see members of his crew performing different tasks during the course of the drill, however the four persons involved in chamber insertion and removal are not needed for most of the time. Guilmartin has two extra individuals for training left or right and another for sponging. It is unlikely that the barrel was sponged unless the action was very prolonged and probably not at all until firing ceased. Utilising the chamber-men reduces the crew to seven.

The dedicated wedge man can also be one of the four chamber men, reducing the total required to six. This is the number that was found to be manageable in the reconstruction. Seven could make a useful size crew, especially in view of the heavy work involved and if repeated firing was required.

During the reconstruction the drill was devised as a reaction to the tasks as they were encountered. Like

Guilmartin, the work was carried out ashore; frequently though, some of the 'crew' being sailors, it was possible to imagine the added difficulty of serving the gun in a seaway. Unlike Guilmartin, for safety reasons and to try to save time during the trials, a drill diagram was prepared before going onto the range. It must be said that at that stage the author was chiefly concerned to get the rounds away as safely and expeditiously as possible. However, as it happened, it is considered that the drill evolved to achieve this could approximate to sixteenth century practice, given that the objective and the equipment were similar.

Lucar in his fifty-third chapter (1588, appendix, 47) mentions one hazard to the gunner when firing a breech-loading gun. In fact, he is probably referring to the kind with a laterally-inserted wedge, while that on the port piece is knocked in from above but, nevertheless, it served as a warning to us:

'How you may duely charge any Chamber peece of Artillery, and ... any Cannon Periero.
Pvt into euery chamber so much powder as his peece requireth for a due charge, and with a rammer beate a tampion of softe wood downe vppon the gunpowder. Moreover, put a bigge wadde into the peece at that ende where the mouth of the chamber must goe in, and after the wadde thruste into the peece at the sayde ende a fitte pellet, & when you haue done all this, put the chamber into the lowest ende of the peece, lock them fast togeather, and cause the sayde tampion to lie harde vppon the powder in the sayde chamber, and the pellet to touch the tampion, and the wadde to lie close by the pellet ... and when a Gunner will geue fire to a chamber peece, he ought not to stande vppō that side of the peece where a wedge of yron is put to locke the chamber in the peece, because the sayde wedge may through the discharge of that peece flie out, and kill the Gunner.'

The port piece chamber was so heavy and the proximity of the chamber team members to each other made it impracticable to heave it safely over the tail of the bed without first elevating the gun – dropping the trail onto the ground (Fig. 5.100). The height of the tail of the bed was dictated by the height of the lower cill of the gunport above the deck in the original gunbay, and the height of the top cill of the port. As with the rest of the drill, there is no evidence for this manoeuvre, but it seemed essential, even on shore. So this appears, in the gun drill description, at the commencement of the loading sequence. Any attempt to reconstruct the orders used on board the *Mary Rose* seemed doomed to failure and the unwelcome sound of cod Tudor language. The first recorded orders known to the author are contained in Eldred (1646, 44); they might represent Elizabethan practice, but are better discussed below under muzzle-loading drill.

*Description of gun drill as performed with replica port piece (Figs 5.99–100)*

*Elevate gun by lowering back of sledge on to ground, remove elevating post, check barrel retaining ropes*
*Remove forelock, wedge and chamber*
*Place chamber to rear of gun (left)*
*Clean chamber*
*Load with cartridge, wadding*
*Ram*
*Insert tampion and hammer flush with chamber mouth*
*Lay down chamber, insert handspikes laterally through lifting rings*
*Load projectile into breech end of barrel and hold*
*Lay chamber in sledge, remove handspikes*
*Use handspikes to drive chamber forward; insert forelock and wedge*
*Drive home chamber by striking wedge*
*Check alignment and fit of chamber to barrel*
*Place elevating post in socket*
*Depress the gun by raising the trail*
*Lay for line (aim)*
*Elevate or depress as required and position retaining peg at position requested by gun captain Aiming*
*Check recoil ropes and barrel retaining ropes*

*GIVE FIRE*

*Knock out forelock and remove wedge*
*Insert handspikes and remove chamber*
*Elevate and remove post*
*Upend chamber and sponge*
*Examine for any residue*
*Clear vent*

*Gun crew*

A gun crew of six was adequate to fire the gun (Fig. 5.101). This is considerably fewer than the number suggested by Guilmartin (1988) who suggested 15–22. Although our working crew of six did not allow for the fetching of shot or powder from elsewhere in the ship, it is assumed that some of these tasks were undertaken by individuals servicing a number of guns, and so they cannot be directly attributed to a specific firing crew. Spatial constraints around the individual guns would severely limit the number of workers. Spare shot, breech chambers and breech chamber wedges were located beside individual port pieces, so at least two rounds could be fired without supplementing the crew. During initial trials we tried to load loose powder which proved difficult on land, let alone on a rolling vessel. As the inventory lists paper for cartridges, it is assuming that the breech-loading guns were cartridge loaded, although no archaeological evidence for this has been found.

Further questions to address with future firings will include the maximum rate of fire and how long the crew

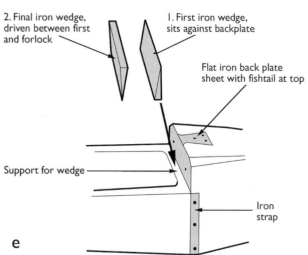
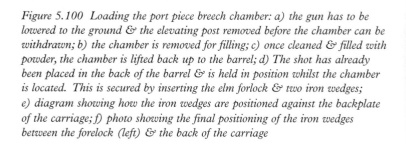

*Figure 5.100 Loading the port piece breech chamber: a) the gun has to be lowered to the ground & the elevating post removed before the chamber can be withdrawn; b) the chamber is removed for filling; c) once cleaned & filled with powder, the chamber is lifted back up to the barrel; d) The shot has already been placed in the back of the barrel & is held in position whilst the chamber is located. This is secured by inserting the elm forlock & two iron wedges; e) diagram showing how the iron wedges are positioned against the backplate of the carriage; f) photo showing the final positioning of the iron wedges between the forelock (left) & the back of the carriage*

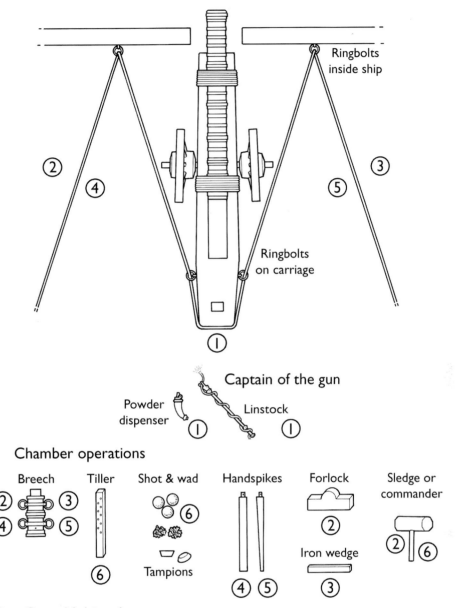

*Figure 5.101 Port piece gun crew positions*

of six can keep up a maximum rate before tiring. The effects of environmental constraints (light, space) as well as the effect of a moving platform and target need consideration and trials, all of these will have an affect on rate and sustainability of fire. Inventory evidence suggests 13–28 rounds per gun, the *Mary Rose* 1546 inventory lists only 16–17 rounds per port piece. The gun crew required was dictated by a mixture of tasks and available space. 39 carriage mounted guns are listed in the near contemporary *Anthony Roll*, with 30 listed '*Gunners*'. By 1582 payments record a hierarchy of Master Gunner, Gunner's Mate, Quarter Gunner and Quarter Gunner's Mate (Rodger 1997, 500). The distribution of gunners to guns would be an interesting study as there are a total of 91 guns (excluding handguns). In addition to the 39 carriage mounted guns, each requiring up to six individuals (although not simultaneously), there are 30 swivel guns, two top pieces and 20 hailshot pieces. It is not certain whether these smaller guns would require only one or more than one individual to fire them.

The six person crew can be broken down as follows:

Breech chamber insertion and removal: Four individuals.
Placing of shot/removal of elevating peg/removal and replacement of elevating post: One
Prime/lay/fire: One (Captain).

Numbered crew
    1: Gun Captain
    2, 3, 4, 5: Breech crew. 2 and 4 to left of gun, 3 and 5 to right
    6: Ball/peg person
    2 is second in command to Gun Captain
    The chamber is loaded by 6 assisted by 2 who rams after the straw and wad is placed
    6 then goes to gun and holds shot in position in barrel. Gun carriage is resting on the ground and the post is out. 2–5 present breech into position (Fig. 5.100).
    2 and 6 start to reload chamber whilst 4 and 5 handspike. 5 places elm forelock and hammers assisted by 3 who offers up iron wedge
    Gun primed by 1
    Rear of gun lifted up by 4 and 5
    Post put in by 6, followed by peg at position determined by 1 who sights gun
    Gun fired by 1
    6 and 2 start on another chamber

Task specific

| | |
|---|---|
| Remove wedges (1 iron and 1 elm) possibly using mallet: | 5 & 3 |
| Put two handspikes through the rings: | 4 & 5 |
| Slide back breech chamber: | 2, 3, 4, 5 |
| Lift out breech chamber: | 2, 3, 4, 5 |
| Lift up back of carriage to take weight off peg: | 4 and 5 |
| Remove elevating peg: | 6 |
| Place trail on ground: | 4 and 5 |
| Pull out elevating post: | 6 |
| Clean breech chamber: | 2 |
| Load breech chamber with powder in cartridge: | 6 |
| Wad of hay/oakum to fill chamber: | 6 |
| Ram down straw: | 2 |
| Tampion to seal chamber: | 6 |
| Handspikes through chamber: | 4 and 5 |
| Place ball in back of gun: | 6 |
| Lift breech into position and push forwards: | 2, 3, 4, 5 |
| Start to reload spare chamber: | 6 and 2 |
| Swabbing the barrel | |

It is necessary to place the back of the carriage on the ground in order to lift the breech chamber into position, it is too high otherwise. It is also necessary to remove the post as it is impossible to lift the chamber over it. This allows for the breech crew to rest the breech chamber on the top of the back of the bed briefly if necessary. The wad, tampion, ball, shot, and peg are kept directly behind gun. When found, the original gun (M8) may have just been reloaded. There was shot in the barrel, but neither elevating post, chamber nor forelock were in position (see above, 81A3001), all were found beside the gun.

**Culverin drill**

Although this section is based on trials undertaken using a reproduction of bronze culverin 81A1423 (Chapter 2), it is applicable to all the large smooth bore muzzle-loading guns on board; the cannon, demi cannon, culverins, sakers and falcons.

Culverins were long guns intended to fire cast iron solid shot weighing about 18–20lb (8.2–9.1kg), calibre of around 5½in (140mm). Unlike the port piece bed, the carriage for the culverin is made up of several distinct elements; a bed with separate vertical cheeks or side pieces which are relieved to accept the pivots, or trunnions on the gun (see Chapter 2 for details). The height of the cheeks can be made to suit the appropriate gunport, while the breech end of the gun is supported on wooden wedges (quoins) making alteration of the elevation of the barrel fairly straightforward. Elm, the timber of the original carriage, was also used for the replica.

The culverin is much heavier than the port piece (the reproduction culverin weighed approximately 2.1 tonnes without bed whilst the port piece weighed approximately 1 tonne). It also presented a contrast in drill requirements. It is more difficult to load, so its rate of fire is likely to have been much slower. On land a culverin could fire up to 70 times a day; the implication is that this would be during siege operations so the whole day, however long daylight happened to be, was available, while a sea fight was likely to take place over a shorter period. On land this might give a rate of fire of a little faster than one round every ten minutes; this would certainly reduce the risk of premature ignition when reloading. Over a shorter action at sea, maybe this rate could be increased.

The drill on the *Mary Rose* is hampered by lack of space in which to retract the guns inboard fully, making sponging and ramming extremely awkward. Withdrawing and turning the guns and outboard loading must be considered (Chapter 11). Either operation would effectively slow down the reload rate. It would appear that one round would take several minutes to fire, especially if Bourne's description of waiting for your ship to be in the best position relative to the target ship for firing held true (1587, chap. 15).

As with any muzzle-loading gun, there are certain commonsense precautions to be taken with the culverin; these can form the basis of a workable drill. Additionally, the kind of manoeuvres and orders used can be drawn from somewhat later printed sources.

Eldred (1646) published a list of '*postures*' somewhat in the nature of those devised for small arms drill, for the better remembrance of his student. However, one surprise in Eldred's postures that might be taken as a form of loading orders, is the lack of a fire order. Possibly this reflects the authority that was delegated to the skilled master gunner or captain of each gun once action had been ordered and suggests, therefore, that the gunner himself fired the gun. This appears to fit with Bourne, for example, in his section on ship to ship actions (1587, chap. 15). Good timing of the moment of discharge was vital to success and required great experience, often, probably, possessed only by the gunner.

Eldred (1646, 44) gives the instructions:

1. *Put back your Peece,*
2. *Order your Peece to Load,*
3. *Search your Peece,*
4. *Spunge your Peece;*
5. *Fill your Ladle,*
6. *Put in your Powder,*
7. *Empty your Ladle:*
8. *Put up your Powder,*
9. *Thrust home your wad,*
10. *Regard your shot.*
11. *Put home your shot gently,*
12. *Thrust home your last wad with three strokes.*
13. *Gage your Peece.*

When firing nowadays it is accepted practice to add some kind of warning order followed by:

## GIVE FIRE

While the exact actions intended under 1 and 2 might be debated, the instructions seem practical. While 3, the use of the worm or wadhook to check for any foreign matter and to remove this, seems essential, there is no definite evidence for worms on board the ship. Several unidentified staffs are currently under consideration. Collado is interesting here (1641, 119). He states that the gun must be clean inside and out before loading '…*se il pezzo sarà netto, & sano, si di dentro, come di fuora …*'

Nowadays, for safety reasons, worming is done at the commencement of proceedings and then after every round. But in the heyday of smooth bore naval ordnance it appears that worming was only done every few rounds (Lavery 1989). The worm was certainly an accepted part of the gunner's equipment by this date (see above). Its Italian name was '*cava fieno*'; the English '*wadhook*' was possibly a rough translation and, interestingly, is the term used at the time. From the insistence, in the Italian sources, on cleanliness and from the absence of any indication of worming after every round, it might be safely deduced that sponging was carried out for each round. This would have been a vital safety precaution.

Eldred's orders give the use of the ladle and all the writers of artillery manuals cover this item in detail. Gunners apparently were obliged to be familiar with this most awkward piece of equipment and know how to use it well. Collado (1641, 123) suggests that dry drill using fine sand instead of gunpowder in the ladle should be practised '*servisi dell'arena minuta*'. In the mid-sixteenth century it could be argued that it was only used on land if really necessary; at sea, its use seems exceptionally demanding if not actually impossible. Bourne (1587, 30) condemns the ladle roundly: '*I do see that there is no worse lading or charging of Ordnaunce, than with a Ladell.*' In brief, Bourne's main and very practical-sounding objections to charging with a ladle are that:

The ladle is '*unhandy*'; it is almost impossible to get an accurately measured charge of powder loaded into the barrel for successive shots, making consistent shooting impossible. It is dangerous.

Ships might have carried them for emergencies and perhaps for use ashore if required. Collado discusses cartridges of linen or paper ('*sacchetti*' or '*scartocci*' '*di tela*' or '*carta reale*') and recommends them for faster firing, although he is not so keen on the idea suggested by others of making up the projectile with the cartridge. This proved to be an idea that was adopted during the Napoleonic wars.

When firing muzzle-loading guns today, the vent is always served during the loading process. This is to avoid the rush of air when ramming causing any remains of the previous round from flaring up and igniting the new powder charge, whether loaded loose or in a cartridge. This practice is described by Norton (1628, 101) '*the Assistant, hauing a finger vpon the Touch-hole all the whilst*' (see above; see Fig. 5.102).

When loading was complete, the pricker was inserted into the vent to pierce the cartridge to allow the priming powder to ignite the main charge. Priming would follow the use of the pricker allowing the gunner to fire the piece. If the gun was not fired quickly for some reason, the primer would be removed and an '*apron of lead*' (or some covering) would be placed over the vent, unless the gun had its own built in, hinged, vent cover, as is present on cannon 81A3003 and indicated on several others. Ignition was provided by means of a burning match held in a linstock. The match cord might not have been carried lit the whole time, although large amounts of match was supplied. It is possible that a candle lantern was used to light it. In any case, it was vital to keep the source of ignition separate from any powder until the gun was fired. This cannot have been easy in the confined spaces on board ship and, at least to this writer, the danger of having to have a light of some kind burning constantly is a powerful argument in favour of cartridges instead of loose loading.

*Figure 5.102 Loading and firing the culverin*

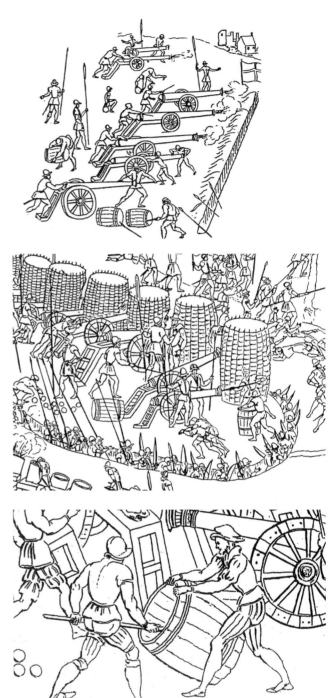

*Figure 5.103 Loading and firing muzzle-loaders from the Seige of Boulogne*

The sequence adopted for firing the replica began with worming, followed by sponging, loading using a cloth cartridge, wadding, projectile and final wad, serving the vent throughout (Fig. 5.102). The cartridge was pricked and priming was completed using fine powder from a horn flask. Although the minimum number required to fire the culverin was three, it appears that even for a smaller gun, a team of three in addition to the captain was recommended. For heavier guns, a larger number must usually have been allowed as William Eldred (1646, 41) explained:

*'And first you must understand that to handle your Ordnance, you ought to have 3 men at least, but to greater Peeces more men, as the weightinesse of the Peeces shall require; but to any small Peece you are to be allowed three men, one to handle the Ladle and Spunge, and the other two to provide wad and shot and to hold the budge-barill, and to serve with their hand spikes at all occasions when they are commanded'.*

The illustrations showing the loading and firing of muzzle-loaders from the *Siege of Boulogne* (Fig. 5.103) show very few people serving each gun, the maximum that appear to be dedicated here are three persons.

**Conclusion**
Although there is no clearly written gun drill with precise and detailed instructions surviving from this period, there is a strong emphasis on careful drill and safety procedures in the near contemporary sources. It would be a failing in the gunner's duty if he risked the gun, crew and indeed the ship, through poor drill; he would have failed in his duty to his sovereign, just as if the mariner's standard of seamanship was not at the highest level possible. To what extent the ideal was approached in practice must remain a matter for speculation, at least as far as the gun drill is concerned.

*Gunpowder*
by Bert Hall and Alexzandra Hildred

**Composition and type**
by Bert Hall

*Composition*
As every schoolboy knows, gunpowder is made of three things: saltpetre, sulphur, and powdered charcoal. The optimal proportions are approximately 75% saltpetre to 12.5% each of sulphur and charcoal although, in fact, the mixture is not very critical and many low-powered gunpowders are closer to 60% saltpetre. Robert Norton (1628, 144) noted:

> 'There are infinite receipts [recipes] *for making of Powder, but most States haue enioyned a certaine proportion amongst themselues, although much different from one another: wherefore no certainty can bee herein generally concluded, but euery man must practise for his experience.*'

Gunpowder is not a chemical compound, which means that its preparation more often resembles cooking rather than an exact science. As in cooking, the quality of raw materials, the skill of the maker, and the methods of preparation often have more to do with the outcome than do precise recipes.

Gunpowder was invented in China although it is not, as often claimed, of extreme antiquity there or initially developed for ordnance of any size. The earliest written formula dates from AD 1044. By the mid-thirteenth century, at least, a few Europeans were aware of gunpowder from Oriental sources. The development of gunpowder in the late Middle Ages and early post-medieval period was a complex process. Europeans had to adapt their local raw materials to its reproduction and had to learn how to make gunpowder that would burn rapidly enough to explode. Poorly compounded gunpowders, or those using inferior raw materials, would burn slowly or prove hard to ignite. Several centuries of trial-and-error experience lay behind the powder stores of the *Mary Rose*.

The most important component of gunpowder is saltpetre (ideally potassium nitrate). This is the oxidising agent, providing the oxygen essential for combustion from the oxygen atoms within the potassium nitrate. Atmospheric oxygen is not involved and, indeed, gunpowder burns in the absence of air just as it does in the open. This was one feature that made it attractive in the sixteenth century. Incendiary devices made with gunpowder as a component were extremely difficult to extinguish once they were ignited. If they hit their intended target, it was very likely to be set ablaze.

Making and purifying saltpetre was one of the first tasks Europeans had to master before they could successfully 'domesticate' gunpowder. In China, gunpowder emerged against a background of centuries of alchemical interest in nitrates; the Chinese had mastered how to extract and purify nitrates from rotting organic matter. Europeans had no experience of this and it took more than a century after gunpowder first appeared in Europe before saltpetre could be grown on special beds. Like bread, wine and cheese, gunpowder ultimately owes its existence to the action of microbes. When living tissue decays (or is digested by some other organism), the amino acids are broken down into simpler and simpler compounds. One important by-product of this decomposition is ammonia, often as the end result of several stages of this breakdown. Ammonium ions ($NH_4+$), oxidise to form nitrites and nitrates, chiefly through the combined action of two genera of bacteria: *nitrosomonas* and *nitrobacter*. Serge Winogradsky, who first discovered the role of these two micro-organisms, also discovered that calcium or magnesium carbonate were necessary for them to grow properly.

Saltpetre '*plantations*' (as they were euphemistically called) consisted of large stone-lined beds in which animal manure could be left to rot under controlled conditions to generate '*nitrous earths*' (another euphemism), whose nitrate content was high enough to produce adequate yields. In Europe, these first appear in the Rhineland in the late 1370s and quickly spread over the rest of western Europe. Saltpetre was extracted by washing these '*earths*' with water, concentrating the soluble salts in a hot, supersaturated solution called '*mother liquor*', and then allowing it to cool. Potassium saltpetre ($KNO_3$) has a solubility of some 247g/100cc in water approaching 100°C, but only 13.3g/100cc at just above 0°C. This means that a hot solution almost saturated with saltpetre will precipitate most of its $KNO_3$ merely by allowing the vessel to cool to the ambient temperature of a workshop (c. 10–15°C). Iterated several times, this is a crude but powerful technique for producing fairly pure nitrate crystals.

Unfortunately it does not work at all to discriminate between various nitrate salts. As noted above, *nitrosomonas* and *nitrobacter* require calcium or magnesium carbonate to be present in order to grow properly, which means calcium nitrate, $Ca(NO_3)_2$, and magnesium nitrate, $Mg(NO_3)_2$, will make up the predominate form of nitrates that these bacteria excrete. These salts will precipitate nearly as readily as potassium nitrate when the concentrated mixed nitrate solution is cooled. Neither $Mg(NO_3)_2$ nor $Ca(NO_3)_2$ is a bad oxidising agent under what we might call 'laboratory' conditions but having either present in large amounts in gunpowder creates a difficult problem. They attract atmospheric moisture to an extreme degree, making gunpowder containing them highly prone to spoilage through dampness. Therefore it became the practice (described by Biringuccio in 1540) to mix a solution of partly refined saltpetre with potash, $K_2CO_3$, to produce potassium nitrate saltpetre, by the equation: $Ca(NO_3)_2 + K_2CO_3 \rightarrow CaCO_3 + 2\ KNO_3$. Calcium carbonate ($CaCO_3$), a highly insoluble salt that is the principal constituent of limestone; its counterpart, or magnesium carbonate or magnesite ($MgCO_3$) is almost equally insoluble. Naturally occurring nitrates '*earths*' (including potassium) would be variable, therefore performing differently.

However, analysis of gunpowder samples from the *Mary Rose* suggests that there may have been widespread differences in how saltpetre was refined in the 1540s. A sample analysed by the Royal Armouries (MR84GPS0001) showed a significant concentration of potassium nitrate (62.8%), with charcoal at 12.4% and sulphur at 17.5%. Correcting for losses (the percentages do not add to 100), this would yield ratios roughly consistent with a recipe calling for 6 parts saltpetre, 2 parts charcoal and 3 parts sulphur. Ratios like this can be found in several gunner's manuals of the mid-sixteenth century. Samples 80S0011–S0128, processed by Portsmouth Polytechnic (Matron 1982), show quite different results on emission spectrophotometric analysis. This yields strong to very strong readings for magnesium and calcium but only weak readings for potassium and variable readings for sodium. This is consistent with saltpetre that had not been processed with potash to substitute potassium for calcium and magnesium. All the samples were from bronze guns. However, as saltpetre is soluble, the likelihood of finding any sample with original proportions is unlikely.

The conclusion that saltpetre refining practices varied so widely would be surprising to some historians and must be treated as highly tentative pending further work. The analysis of submerged gunpowder samples is a difficult art at best, and the results could be skewed dramatically by the action of seawater. Still, the results are tantalising and demand further effort to resolve the questions they raise. There are at least nine other gunpowder samples from *Mary Rose* that have not been analysed chemically, and these might yield data of significance. They should be the focus of further research.

## Production

As gunpowder is not a chemical compound but a mechanical mixture of three solids, the gunpowder maker must somehow arrange for the particles of the three ingredients to occupy the closest possible physical position. Ideally, any cubic millimetre of gunpowder should contain the same proportions of saltpetre, charcoal, and sulphur, in the finest possible form. All early gunpowder recipes stipulate that the raw materials must be as finely pulverised as possible before being mixed together. Yet one cannot make satisfactory gunpowder merely by stirring these fine powders together in a bowl. In practice, the powder maker mixed the ingredients together in a mortar under the repeated blows of a pestle, a process known as 'incorporation.' The shearing action of the pestle head on the walls of the mortar is important in consolidating the ingredients, and much trial-and-error went into determining the optimal shape for both mortar and pestle.

The stamping-and-grinding process had to continue without cessation for many hours. Norton (1628, 144) put it thus: '*worke those three Materialls well together: for therein consisteth a greater difference of force, by the difference in working of them, then is credible without experience*'. Norton recommended six, seven, or eight hours under the stamp. Surirey de Saint-Rémy's treatise, published in 1702, recommended a full 24 hours under the stamp for the best quality gunpowder, and he noted that the resultant mix should be as fine as women's face powder (ie, ground talc). This is not merely some sort of alchemical superstition, but is a fundamental aspect of 'incorporation'. Tiny particles of the main ingredients are pushed together by the shearing action of the pestle. The amount of time over which this shearing action takes place, and the consistency with which it is applied, both contribute to the quality of the finished product.

One critical aspect concerns the origin of the carbon. It has to come from powdered charcoal and the species of wood used for the charcoal affects the quality of the gunpowder. Incorporation forces tiny particles of pulverised sulphur and saltpetre into the pores of the particles of charcoal. Wood species whose cell structure preserves millions of tiny pores even after being turned into charcoal and then ground to fine powder make the best gunpowder; and hazel and willow were the favourites.

Incorporation is such a fundamental determinant of gunpowder's ultimate ballistic performance (as Norton's remark implies) that its technical history deserves more attention than it has received hitherto. Virtually all scholars have assumed that the earliest method of making gunpowder was to grind the ingredients together in a dry state. There is really no evidence to support this assumption – though evidence for early gunpowder preparation is very fragmentary and may well have omitted every detail, and there may have been a number of methods. Later practice was quite different. Enough water was added to the mortar '*that the coales dust not, and so little that the composition in working become not paste*' (Norton 1628, 144-5). In this wetted mixture, still being ground and pressed by the pestle, the charcoal and the sulphur are insoluble but saltpetre will start to dissolve in the modest amounts of liquid required. As a slurry, the saltpetre and the sulphur start to coat the grains of the charcoal, binding the entire mass together as a semi-solid 'cake'. Gunpowder that has been wet-incorporated is mechanically different from what it would be if the raw materials were left dry. The change is irreversible, for once the saltpetre and sulphur have dried on the charcoal granules, the mixture has achieved the necessary degree of homogeneity and intimately blended particles that makes for good gunpowder.

If dry incorporation was the early standard practice, then the question arises of just when did wet incorporation begin. There are shadowy hints from the late fourteenth century, but the definitive answer lies in a nameless codex in the Austrian National Library, dated 1411 (*Codex Vindobona* 3069) which contains a recipe for '*plain gunpowder*' that advises the maker: 'Crush these [ingredients] *together* [in a mortar] *along with good wine in which you have dissolved some camphor, and* [afterwards] *dry it out in the sun.*' The anonymous author believed that camphorated wine would prevent the powder from spoilage (another bit of evidence suggesting that calcium saltpetre was problematic at this date) but, as wine is about 90% water, the liquid will serve just as well to incorporate the ingredients as described above.

From this point onward, wet incorporation becomes more and more the norm. In fact, Biringuccio could admonish his reader in the following terms (1540, 413):

> '*since these powders are things that ignite very easily when they are being made, they would not be without danger to the one who makes them if he did not prevent this by moistening. Therefore take care not to crush them while dry, both to avoid this danger and also because they crush better. Moisten it with ordinary water to a certain degree of moistness so that it sticks together when squeezed in the hand*'.

Biringuccio goes on to recommend that, although some powder makers use vinegar as the moistening agent and others even resort to camphorated *aqua vitae*, he recommended just '*common water*'. Biringuccio's words were available to English readers from 1562, in which year Peter Whitehorne added to his translation of Machiavelli's *The Art of War* an appendix that Whitehorne claimed as his own but which was, in large part, plagiarised from Biringuccio and other Italian sources. However there is evidence for the use of vinegar in the preparation of fine powder for handguns for the *Mary Rose* in 1513 (Knighton and Loades 2002, 60).

The ground (wet) powder must be dried and crushed before it can be loaded into a gun. Most of the surviving texts are surprisingly reticent about this step, unless some special process or intermediate step is being recommended. Because the incorporating mortar (made of wood, brass or stone known not to spark) becomes warm with the mechanical action of the pestle, one can simply continue working the mixture until it is nearly completely dry, leaving the remaining moisture to dry out by spreading the powder in the sun or on trays in special heated sheds. Shaping the semi-pasty mass of gunpowder into lumps or spheroids that look like loaves of bread and then drying them was known by the German term for dumpling (*Knollen*). This made powder that would store better by being moisture resistant (still more evidence suggesting calcium saltpetre). Obviously the loaves were crushed into powder before loading a gun. '*Knollenpulver*,' as this type of gunpowder was known, is mentioned in about 1420 and continues to be recommended by artillery writers for more than two centuries, at least for circumstances where storage might prove to be problematic.

*Knollenpulver* never seems to have become the mainstream product of the powder maker. Instead it was shaped into proper grains or corns of gunpowder, each one like the other in size, shape and composition. Such '*corned*' gunpowder is considerably more powerful than uncorned because of the way it burns. Each grain of gunpowder burns from the outside to its core, so the size of the grain acts as a powerful determinant on the burning characteristics of a gunpowder charge. From solid geometry we know that for an infinite volume of space, larger-diameter solids (like more coarsely grained powders) will fill the space with the same total mass of solids as will an equal volume of smaller-diameter solids (small-grained gunpowder with all grains of equal size, for instance). Remarkably, the grains will only occupy slightly less than 75% of the available space, whatever their size. However, the larger-diameter solids will have a smaller total surface area (the sum of all the grains' surface areas) than will the smaller-diameter solids. Thus grain size means more or less surface area for the same mass and volume of gunpowder.

This packing order, called a face-centred cubic lattice (or close-packed hexagonal) approximates how many ordinary everyday objects actually work – stacks of apples at the greengrocer's for example. But at a certain threshold of small particle size, the lattice collapses into a simple heap, like a heap of baker's flour. If a gunpowder charge is granulated so that it approximates a face-centred cubic lattice (or close-packed hexagonal) there is adequate space between the grains for combustion to propagate and the whole charge burns quite quickly. (Relative terms like 'quick' and 'slow' refer to the millisecond time scale on which gunpowder does in fact burn, but the difference of a few milliseconds may make a large difference in macroscopic performance.) If the grains are large, the total inter-grain space remains the same, but the total surface area of all grains is less and combustion, which depends on surface area, is slower. Large grained charges burn more slowly. If the grain size is extremely fine, the packing order collapses, leaving no space for combustion to propagate and burning slows down as well. For maximum burn speed and thus maximum gas pressure and maximum ballistic effect, one needs just the right size of gunpowder grain or '*corn*', the 'optimum' size.

Wet incorporation and corning go hand-in-hand for two reasons. First, wet incorporation produces a gunpowder mix that burns faster owing to its more intimate blend of ingredients. Each particle will contain the appropriate ratio of ingredients mingled together as closely as mechanical mixing can achieve. This is why recipes for *Knollenpulver* (which is necessarily wet-mixed) warn that two units of such gunpowder it is equivalent to three units of mealed (possibly dry incorporated?) gunpowder. Note that the grains of smashed lumps of gunpowder could hardly have been very regular, yet this warning still stands. The second reason is that corning offers the powder maker the opportunity to regulate the ballistic effect of his mixture by controlling grain size.

How this was achieved is described by Peter Whitehorne in 1562 (f.28):

> '*The maner of cornyng all sortes of pouder is with a Seeue made, with a thicke skinne of Parchement, full of little rounde holes, into the whiche seue the powder must bee put, while it is danke, and also a little bowle, that when you sifte, maie rolle up and doune, upon the clottes of pouder, to breake theim, that it maie corne, and runne through the holes of the Seeue*'.

As noted above, Whitehorne plagiarised much of his description of powder making directly from Biringuccio, but he interpolated this possibly original passage. There are earlier references to corning gunpowder using a colander or sieve, specifically from Franz Helm's *Buch von der probierten Künsten* in 1535, but no earlier description of the process has come to light. Indirect evidence suggests that controlled grain corning was probably practiced from about 1500 onward.

Forcing the wet mixture through a perforated surface allowed the powdermaker to control the size of the grains by choosing the size of perforations. Powdermakers used smaller or larger diameters to produce controlled-diameter gunpowder grains, which could then be dried and stored until needed. This appears to be what lies behind the numerous sixteenth century references in inventories to '*grosse corne powder*,' and '*ffyne corne powder*.' What remains unexplained is the mysterious term '*serpentine*.'

To quote William Bourne's description (1587, 6): '*if it be Sarpentine pouder, then the pouder will be as fine as sande, and as soft as floure*'. This and similar characterisations of '*serpentine*' have led English scholars since the publication of Col. Henry Hime's *The Origin of Artillery* in 1915 to conclude that England continued to produce dry incorporated gunpowder throughout the sixteenth century – and in copious quantities, to judge from the amounts of '*serpentine*' that appear in English inventories. If this opinion is true it would set English practice apart from Continental powdermakers, for whom dry incorporated powder seems to have been a rare speciality, produced only for making rockets and fireworks. In the current state of research, this commonplace opinion seems less and less likely.

Although some sixteenth century writers, like Whitehorne use the term '*sarpentine*' as a label for the weakest and most primitive gunpowders, Whitehorne treats this type as a historical curiosity, not something that modern, mid-sixteenth century gunners are likely to encounter. Whitehorne's more modern recipes are for '*grosse*' or '*fine*' corned powders '*as used now-adayes*,' powders that are necessarily wet-incorporated. Whitehorne also interpolates the word '*serpentine*' into a passage he plagiarises from Biringuccio (1562, 27v): '*For that if Serpentine pouder, should be occupied in hand gunnes, or Harkebuses, it would scant be able to drive their pelletes a quaites caste*.' Biringuccio has the same result, but does not include the word '*serpentine*.' '*For, if you should use that for heavy guns in arquebuses and pistols, it would throw the ball scarcely ten braccia out of the barrel*' (1540, 412). He simply refers to gunpowder like that used for heavy guns '*artigliaria grossa*'. What this means is explained in the same text quoted by Hildred (below) (Biringuccio 1540, 413):

> '*Now to make common powder for heavy guns* [artigliaria grossa], *take three parts of refined saltpetre, two of willow charcoal, and one of sulphur. Everything is incorporated well together by grinding finely, and then dried of all moisture*'.

The last phrase implies quite strongly that powder for heavy guns, what Whitehorne interpolates as '*serpentine*' powder, is wet incorporated and then dried. To reach the state described by Bourne as '*fine as sande, and as soft as floure*,' this would necessarily have to be reground.

But should this fine powder have been present in such staggering quantities as the available inventories indicate for Elizabethan ships? The answer must lie in the burning qualities of powders whose grain size is too small to support a face-centred cubic lattice structure. Twentieth century research has shown that in the open air, well-incorporated (that is, wet-incorporated) gunpowder will ignite quickly but the same powder, packed tightly into an enclosed chamber, will sustain combustion only along a burning front that works its way through the enclosed mass of powder. The effect is rather like the glowing coal on a cigarette, but speeded up dramatically. Elizabethan gunners recognised this; serpentine powder, '*if it be rammed in hard ... it is not so quicke in þe fiering*' (Bourne 1587, 12). By contrast, combustion propagates through an enclosed mass of corned gunpowder very quickly, within about three milliseconds, and the charge begins to generate its maximum volume of propellant gases almost immediately. This is, however, exceedingly dangerous for the gun itself, in that the shot is unable to move so quickly down the bore as to relieve the pressure at the breech. Using a finely granulated corn powder in a large artillery piece could seriously damage the gun as Biringuccio's advises: '*if you should use that made for arquebuses in heavy guns without great discretion, you might easily break them or spoil them for other uses*' (1540, 412).

The author suggests that the English retention of '*serpentine*' gunpowder throughout the sixteenth century is not the retention of obsolete dry-incorporation powder, but the use of finely ground wet-incorporated gunpowder in an attempt to create powder which could be used safely in large artillery pieces. Customarily ,gunpowder designated for larger pieces of artillery contained less saltpetre than that for smaller pieces and was therefore weaker. This reflects the fact that big guns were actually operating much closer to their limits than smaller pieces. For all their imposing bulk, big guns were more fragile – relative to the task demanded of them – than their smaller brethren. It was an effort to control the power inherent in wet-incorporated gunpowder and the danger that this type of powder presented to 'great ordnance' that led to all such efforts to slow down the burn rate of gunpowder in bigger guns.

Another view is that the retention of old guns, and the continued practice of forging wrought iron guns, both evidenced by the inventories of the period, meant that serpentine powder continued to be made.

In effect, English gunners made cannon gunpowder too small in the grain – so it would burn slowly. Although Continental gunners did much the same in the early part of the century (recall Biringuccio's recipe for uncorned '*common powder*' for '*artigliaria grossa*', in time they began to make cannon gunpowder too big in the grain – so it would burn slowly. By the time of Bourne's treatise in 1587 *The Arte of Shooting in Great Ordnaunce* (pp. 12–13), the disadvantages of the older '*base powder*' was becoming apparent:

> '[Serpentine] *lyeth and bloweth in the breech of the peece, before it can take fire* [ie, before the charge is entirely aflame], *so by that meanes it heateth and streineth the peece, and halfe of the force of the pouder is gone, before the shotte be deliuered* [clears the muzzle] ... *Nowe whereas they shoote good pouder, or cornepouder, they take much lesse pouder, and it sendeth the shotte quicker awaye, and it dothe not heate the peece so fast: for this we doe see by common experience, that a little heat by long continuance, doth heat more than a great heat by little continuance.*'

Corn powder, made in appropriately sized grains and loaded in lesser amounts than serpentine, rectified these problems, shooting at higher muzzle velocities at less danger to the gun and its crew.

## Gunpowder from the *Mary Rose*
by Alexzandra Hildred

*Historical evidence: forms of gunpowder*
The *Anthony Roll* demonstrates that, by 1545, three forms of gunpowder were in use: serpentine, '*gross*' corn and '*fine*' corn (Table 5.25). Many vessels carry all three forms and it is clear that most was serpentine, followed by gross and then fine corn powder, others carry only two forms. The *Mary Rose* is listed with two lasts of serpentine powder in barrels and three barrels of unspecified corn powder to supply her 15 bronze and 24 large wrought iron guns, 20 cast iron hailshot pieces, two top pieces, 30 bases and 50 handguns.

What remains uncertain is which powder was used with which type of gun. Hall (above) suggests that wet-incorporated and finely ground serpentine powder was used because it was safer than larger grained (corned) wet-incorporated powder when used in large artillery. '*Serpentine*' does suggest an affiliation with the gun of that name. Blackmore (1976, 242) describes '*serpentine*' guns as being made of bronze or iron, often breech-loading and in a variety of sizes, all of which are long in relation to calibre. Twenty-eight serpentines are listed for the *Mary Rose* in 1514 with 21 half barrels of powder and one '*chestfull*' of '*cartouches*' (cartridges). Inventories of items in the Tower between 1495 and 1523 include entries for powder; the latter includes the first mention of serpentine powder in the Tower stores, but serpentine powder continues long after the guns. Bourne (1587, 10–13) specifies which guns are loaded with serpentine powder: cannon, demi cannon, culverins, falcons, falconets, sakers and minions. The amount is expressed as ratio of weight of shot, determined by the thickness of metal at the breech for each type of gun.

The three types of powder listed in the *Anthony Roll* probably coincide with those described by Biringuccio (1540, 413):

> '*Now in order to make the common powder for heavy guns, take three parts of refined saltpeter, two of willow charcoal, and one of sulphur. Everything is incorporated well together by grinding finely and then dried of all moisture as I*

Table 5.25 *Anthony Roll*: gunpowder summary

| Ship | Total bronze guns | Total iron guns | Serpentine powder | Gross corn powder | Fine corn powder (lb) |
|---|---|---|---|---|---|
| *Harry Grace à Dieu* | 21 | 130 | 2 last | 6 barrels | – |
| *Mary Rose* | 15 | 76 | 2 last | 3 barrels | – |
| *Peter* | 14 | 112 | 2 last | 2 barrels | – |
| *Matthew* | 10 | 72 | 1½ last | 2 barrels | 40 |
| *Great Bark* | 12 | 71 | 30 demi barrels | 2 barrels | 2 demi barrels |
| *Jesus of Lübeck* | 6 | 52 | 1 last | 1 barrel | 40 |
| *Pauncy* | 13 | 70 | 30 demi barrels | 2 demi barrels | 40 |
| *Morion* | 4 | 27 | 8 barrels | 1 demi barrel | 24 |
| *Struse* | 4 | 37 | 12 demi barrels | 1 demi barrel | 24 |
| *Mary (of) Hamburg* | 5 | 35 | 27 demi barrels | 1 demi barrel | 24 |
| *Christopher of Bremen* | 4 | 29 | 14 demi barrels | – | 24 s |
| *Trinity Henry* | 5 | 45 | 1 last | 1 barrel | 24 s |
| *Small Bark* | 11 | 76 | 1 last | 1 demi barrel | 24 s |
| *Sweepstake* | 8 | 50 | 1 last | – | 24 s |
| *Minion* | 4 | 65 | 1 last | – | 24 |
| *Lartyque* | 1 | 21 | 10 demi barrels | – | 16 |
| *Mary Thomas* | 1 | 23 | 10 demi barrels | – | 20 |
| *Hoy Bark* | 1 | 18 | 8 demi barrels | – | 16 |
| *George* | 2 | 15 | 5 demi barrels | – | 8 s |
| *Mary James* | 1 | 22 | 9 demi barrels | – | 20 |
| *Grand Mistress* | 7 | 31 | 10 barrels | – | 14 |
| *Anne Gallant* | 7 | 26 | 2 last | – | 24 |
| *Hart* | 4 | 36 | 12 barrels | 2 barrels | 24 |
| *Antelope* | 4 | 32 | 10 barrels | 2 barrels | 24 |
| *Tiger* | 3 | 26 | 12 demi barrels | 1 demi barrel | 24 |
| *Bull* | 5 | 26 | 12 demi barrels | 1 demi barrel | 16 |
| *Salamander* | 8 | 43 | 1 last | – | 24 |
| *Unicorn* | 5 | 29 | 6 barrels | 1 demi barrel | 16 |
| *Swallow* | 6 | 43 | 23 barrels | – | 30 |
| *Galley Subtle* | 31 | 28 | 18 demi barrels | – | 1 demi barrel |
| *New Bark* | 5 | 43 | 12 | 1 demi barrel | 24 |
| *Greyhound* | 8 | 22 | 20 demi barrel | 1 demi barrel | 24 |
| *Jennet* | 3 | 20 | 10 demi barrels | – | 24 |
| *Lion* | 3 | 36 | 20 demi barrels | – | 16 |
| *Dragon* | 2 | 34 | 16 demi barrels | – | 20 |
| *Falcon* | 4 | 26 | 8 demi barrels | – | 12 s |
| *Saker* | 2 | 16 | 8 demi barrels | – | 8 |
| *Hind* | 1 | 21 | 6 demi barrels | – | 12 |
| *Roo* | 5 | 18 | 8 demi barrels | – | 12 |
| *Phoenix* | 2 | 18 | 6 demi barrels | – | 12 |
| *Merlin* | 4 | 12 | 6 demi barrels | – | 12 |
| *Less Pinnace* | 1 | 13 | 8 demi barrels | – | 8 |
| *Brigandine* | 3 | 12 | 4 demi barrels | – | 12 |
| *Hare* | 1 | 16 | 6 demi barrels | – | 12 |
| *Trego Renneger* | 0 | 16 | 4 demi barrels | – | 12 |
| *Double Rose* | 1 | 10 | 2 demi barrels | 3 demi barrels | 12 |
| *Flower de Luce* | 2 | 11 | 2 demi barrels | 3 demi barrels | 12 |
| *Portcullis* | 1 | 10 | 1 demi barrels | 2 demi barrels | 8 |
| *Harp* | 1 | 9 | 1 demi barrels | 2 demi barrels | 6 |
| *Cloud in the Sun* | 2 | 10 | 1 demi barrels | 3 demi barrels | 6 |
| *Rose in the Sun* | 11 | 10 | 2 demi barrels | 2 demi barrels | 8 |
| *Hawthorn* | 1 | 9 | 1 demi barrels | 2 demi barrels | 6 |
| *Three Ostrich Feathers* | 1 | 9 | 1 demi barrels | 2 demi barrels | 6 |
| *Falcon in the Fetterlock* | 3 | 12 | 2 demi barrels | 3 demi barrels | 8 |
| *Maiden Head* | 1 | 9 | 1 demi barrels | 2 demi barrels | 6 |
| *Rose Slip* | 1 | 9 | 1 demi barrels | 2 demi barrels | 6 |
| *Gillyflower* | 1 | 8 | 1 demi barrels | 2 demi barrels | 6 |
| *Sun* | 2 | 10 | 3 demi barrels | – | 8 |

*have told you. If you wish to make powder for medium guns, take five parts of refined saltpeter, one and a half of charcoal, and one of sulphur; these are mixed by grinding finely, and then granulated and dried. To make that for arquebuses and pistols, take ten parts of saltpeter, one of clean hazlenut-twig charcoal, and one of sulphur; these are mixed by crushing or milling them very fine and then they are granulated and dried'.*

This suggests that powder for large guns was wet-incorporated and ground finely and merely dried, which in other parts of the text is described as *'common'*, *'serpentine'*, or *'base powder'* whilst that for medium guns had more saltpetre, was ground, granulated and dried. Handgun powder has a larger proportion of saltpetre, is also wet-incorporated, but very finely ground and then granulated and dried. This gives us the three types of powder listed on the roll, but still does not help us differentiate between medium and large guns! When Captain John Smith wrote *A Sea Grammar* in 1627 the situation seems to have changed, with all guns using corned powder (Smith 1627, 90):

*'Note that seldome in Ships [do] they use any Ordnance greater than Demy Canons, nor have they any certainty either at point blanke or at any random. Note your Serpentine powder in old time was in meale, but now corned and made stronger, and called* Canon corne powder. *But that for small Ordnance is called* corne Powder fine, *and ought to have in strength a quarter more, because those small Peeces are better fortified than the greater'*

This is hinted at by Bourne (1587, 12):

*'Nowe whereas they shoote good pouder, or cornepouder, they take much lesse pouder, and it sendeth the shotte quicker awaye, and it dothe not heate the peece so fast: for this we doe see by common experience, that a little heat by long continuance, doth heat more than a great heat by little continuance.'*

The question still remains as to which guns used gross corn at the time of the *Mary Rose*. This is further complicated by the fact that many guns were still individual, and old guns were replaced slowly. It may be that the stock of older guns used serpentine powder or corned powder with a reduced powder charge, thereby safely obtaining the higher peak pressure.

Gunpowder is mentioned in the Tower inventory of 1353, the earliest in Blackmore's selection (1976, 252): *'xvj libros dj.* [16½lb] *pulveris pro gunnis'*. By the sixteenth century powder was usually reckoned in lasts, barrels and half barrels. In 1523 (*ibid.*, 260) 13 lasts of *'serpentyne gonpoudre'* is listed, with guns including bombards, curtows, serpentines and stone guns, but also slings, culverins and falcons. The first mention of corn powder is in 1540, and distinguished from serpentine powder (*ibid.*, 261): *'Sarpentyne powder lxxx lastes. Cornepowder iiij$^{ml}$v$^c$l$^{li}$'* [4550lb, assuming c=100 (see below)]. With the exception of the 1547 inventory for the Tower (see below), serpentine powder continues to be more abundant than corn powder. In 1589 (PRO WO 55/1659; Blackmore 1976, 266), and again in 1595, the constituents to make powder are included, but only corn powder listed. By 1635 powder is listed as *'cannon powder'* stored in 1219 barrels to equal 50 last 19 hundredweight (121,900lb; 55.28 metric tonnes). This is important in that it suggests that each barrel contained 100lb (*c.* 45.4kg) of powder. Fine powder is also listed (22 lasts 9 hundredweight in 537 barrels) or also around 100lb per barrel. It seems this was the norm by the seventeenth century. By looking at the amounts of different types of powder carried on vessels, it is clear that most guns still used serpentine powder, with two lasts listed for *Mary Rose*. In order to work out which powder was used for particular guns an attempt was made to calculate the amount of powder required to fire specified listed guns until their listed shot was exhausted. The amount of powder required per round, the total amount of gunpowder carried, and the number of rounds, must all be known; this relies in turn on knowing the weights represented by *'barrels'* and *'lasts'*. Most early sources agree that a last of gunpowder weighed a hundredweight, but that was itself a variable measure, and interpretation is further complicated by the way such quantities were expressed. Though the modern reckoning of a hundredweight (cwt) as 112lb was already widely used, the older and simpler meaning of 100lb is still found. Because in any permutation the Roman numeral *'c'* followed by *'li'* for *'libri'* could stand for 100lb or 112lb, *'cc$^{li}$'* might mean either 250lb or 274lb [(112 x 2) + 50]. The usage in any particular case can only be certainly known if it is explicit, or if it can be deduced from a given calculation. The naval accounts of Henry VII's reign show gunpowder purchased in multiple hundreds and odd pounds at 6*d* the lb (Oppenheim 1896a, 13, 20), and the sums indicate that the hundreds are just that. The Burton list of 1550 (Awty 1989, 143) likewise uses the cwt of 100lb, giving a barrel of 200lb and a last of 2400lb:

*'Every barrel of powder is ever understood 2 cwt. of powder at 5 score the hundred and every pound is 16 ounces and every last of powder is 24 cwt. and neither more nor less'.*

Bourne (1587, 72) uses the cwt of 112lb, but explains that this includes an allowance of 12lb for the container:

*'A last of Pouder is 24 hundred weight, caske and all, and euerye hundred weight to contayne. 112.*

*pound, so ... that you haue 24. hundred pounde of Pouder in euery last, and so is allowed 12 pound in euery 100. weight, for the caske, which is in al allowed for the caske of a last of Powder, 288. pound.'*

That, of course, gives a net cwt of 100lb, and this is assumed in the figures which follow.

Oppenheim (1896a, 20, 52, 117)) records payments for gunpowder during the reign of Henry VII. He lists the number of barrels and total weight in pounds and the cost. All the figures suggest 2 cwt as being the weight of a powder barrel containing 200lb of gunpowder and also suggest a practice of coating the barrels with tar as a waterproofing agent. An early Portuguese account lists lead-lined barrels (Barker, pers. comm. 2009).

One assumption might be that the 4800lb (2 last) of serpentine powder listed for the *Mary Rose* was intended for the wrought iron guns with chambers (of similar manufacture to serpentine guns). This would allow for the 300lb or 600lb of corned powder being for the fifteen cast bronze muzzle-loaders, and possibly the cast iron hailshot pieces. This theory has been tested against the number of shot and guns listed for the *Mary Rose* in the *Anthony Roll*. This is extremely tentative, as the amount of powder required for the different types of guns is not always stated, and the literature suggests that although not routine in the mid-sixteenth century, corned powder could be used with even the largest guns, but with a reduced powder charge. As suggested by the three barrels of corn powder is not enough for the bronze guns, unless they each only fired four times. The only guns which can be accommodated within the total corn powder carried (leaving enough for handguns and priming) are either the sakers or falcons, but not both. These are potential candidates as both might qualify as the '*medium guns*' mentioned by Bourne, which were described as taking a stronger saltpetre mix of gunpowder which was finely ground, granulated and dried (see above). When one looks at the possible total amount of gunpowder needed to exhaust all the shot carried for particular guns, it is clear that the gunpowder listed is insufficient (Table 5.27). An estimated 448lb 14oz (203.6kg) is needed to fire both the bronze and iron guns once. To fire all shot needs 9055lb 8oz (4107.6kg). The total carried is only 5976lb (2716kg), 66% of the gunpowder required.

Comparisons were made with two less well armed vessels, *Sun* and *Gillyflower* (20 ton rowbarges) to see if a clearer comparison between amount of gunpowder and types of guns could be determined (Table 5.28). As the table shows, the *Sun* (carrying a saker and falcon) is under-resourced for powder, requiring 560lb (254kg) to fire all her shot. She is carrying just over half of her requirement, with only enough corn powder for her handguns. The *Gillyflower* has only one mounted gun, a saker. Although also under-resourced it appears that the gross corn powder is for her saker and the fine corn for her handguns, with the serpentine possibly for bases and hailshot pieces. The smallest vessels are usually only furnished with bases, hailshot, handguns and one or two bronze guns. In four cases out of 24, the presence of demi culverins coincides with gross corn, and in eleven the presence of sakers or falcons is accompanied by gross corn powder. There are eight instances of sakers but no gross corn powder and instances where there are bronze and wrought iron guns but only serpentine and fine corn, with only enough fine corn for the handguns. No vessel appears to be provisioned with enough powder for all the shot listed and often there is no differentiation between fine and gross corn powder. Where the only guns listed are bases and hailshot and handguns, the powder is listed as serpentine and fine corn only (see the *Trego Renneger*; Knighton and Loades, 2000, 91).

The same disparity between shot and powder, the predominance of serpentine over corn and the general under-resourcing of powder, can be seen in the 1547 inventory (Starkey 1998). It was an expensive commodity, mostly obtained from abroad and whereas powder deteriorates (especially in wet conditions) shot do not. Despite the difficulties and the inevitable ambiguities, it seems reasonable to suggest that

**Table 5.26 Estimated gunpowder expenditure for bronze guns (weight in lb)**

| Gun | No. listed | Total shot | Shot per gun | Shot Weight | Powder per round (est.) | Total powder to fire guns once | Total powder to fire all shot |
|---|---|---|---|---|---|---|---|
| Cannon | 2 | 50 | 25 | 44 | 30 | 60 | 1500 |
| Demi cannon | 2 | 40 | 20 | 30 | 20 | 40 | 400 |
| Culverin | 2 | 40 | 20 | 18 | 14 | 28 | 560 |
| Demi culverin | 6 | 140 | 23 | 9 | 9 | 54 | 1260 |
| Saker | 2 | 80 | 40 | 5 | 5 | 10 | 400 |
| Falcon | 1 | 60 | 60 | 2 | 2½ | 2½ | 150 |
| Total | 15 | 410 | | | | 194½ | 4270 |

Figures taken from the *Anthony Roll*, not from excavated assemblage. The weights of gunpowder & shot are taken from British Sloane MS. 2497 (*c.* 1580) reprinted in Blackmore (1976, 392)

Table 5.27 Estimated gunpowder expenditure for all *Mary Rose* guns and shot (weight in lb)

| Gun | No. listed | Total shot | Shot per gun | Powder per round | Total powder to fire guns once | Total powder to fire all shot |
|---|---|---|---|---|---|---|
| Cannon | 2 | 50 | 25 | 40 | 80 | 2000 |
| Demi cannon | 2 | 40 | 20 | 20 | 40 | 800 |
| Culverin | 2 | 40 | 20 | 15 | 30 | 600 |
| Demi culverin | 6 | 140 | 23 | 9 | 54 | 1260 |
| Saker | 2 | 80 | 40 | 5¼ | 10½ | 420 |
| Falcon | 1 | 60 | 60 | 2½ | 2½ | 150 |
| Port piece (24in chamber ext.) | 12 | 200 | 17 | 9 | 108 | 1800 |
| Sling | 2 | 40 | 20 | 8★ | 16 | 320 |
| Demi sling | 3 | 40 | 14 | 4★ | 12 | 160 |
| Quarter sling | 1 | 50 | 50 | 2★ | 2 | 100 |
| Fowler | 6 | 170 | 29 | 5★ | 30 | 850 |
| Base | 30 | 400 | 14 | 1½★ | 45 | 600 |
| Top piece | 2 | 20 | 10 | 7oz★ | 14oz | ? |
| Hailshot piece | 20 | ?1000★★ | 20 per round | 4oz★ | 5 | 13 |
| Handgun | 50 | 1000 | 200 | 1oz★ | 13 | 62½ |
| Total | 139 | 3330 | | 121¾ | 449lb 14oz | 9135½ |

★These are estimates using the size of chambers where known & Norton's figures (1628, 58).
★★ The *Mary Rose* entry for this is blank. The *Peter* is listed with 1000 dice for her 20 hailshot

'serpentine' was used for most of the wrought iron guns and the cannons and larger culverins, and that the demi culverins, sakers and falcons could have used '*gross corn*', preserving the '*fine corn*' for handguns and priming.

Certainly, by the time Eldred wrote *The Gunners Glasse* in 1646 he was able to dismiss serpentine powder in a single sentence (pp. 25–6):

*'There was in ancient time a kind of Powder called Serpentine-Powder; ... this powder being the first, was made in a small kind of dust like meale, and was but of a weak receipt in comparison of that we now use, and neither was it corned as our Powder that we use in these dayes, for which cause, though it were then but weake and now the strength in a manner doubled ... a pound of the powder which is now in use ... is as strong as two pound of the old Serpentine Powder'.*

In attempting to understand the reasons for the different forms of gunpowder and what they were used

Table 5.28 Estimated gunpowder expenditure of all *Sun* and *Gillyflower* guns and shot (weight in lb)

| Gun | No. listed | Total shot | Shot per gun | Powder per round | Total powder to fire guns once | Total powder to fire all shot |
|---|---|---|---|---|---|---|
| *Sun* | | | | | | |
| Saker | 1 | 40 | 40 | 5¼ | 5¼ | 210 |
| Falcon | 1 | 40 | 40 | 2½ | 2½ | 100 |
| Base | 7 | 160 | 23 | 1½ | 10½ | 240 |
| Hailshot piece | 3 | 300/20 | 7 rounds | 4oz | 12oz | 5 |
| Handgun | 4 | 80 | 20 | 1oz | 4oz | 5 |
| *Gillyflower* | | | | | | |
| Saker | 1 | 70 | 70 | 5¼ | 5¼ | 367½ |
| Base | 5 | 180 | 36 | 1½ | 7½ | 270 |
| Hailshot piece | 3 | 400/20 | 7 rounds | 4oz | 12oz | 5¼ |
| Handgun | 3 | 60 | 20 | 1oz | 3oz | 3¾ |

The powder estimates for the larger guns are based on using serpentine powder

**Table 5.29 Degree of fortification and gunpowder charge**

| Ratio of metal to shot | Types of gun listed | Weight of powder |
|---|---|---|
| 1–½cwt metal per 1lb shot | double cannon | ⅔ weight of shot |
| 1½–2cwt metal to 1lb shot | demi cannon | wt in powder as shot weighs –⅙th |
| >2cwt metal to 1lb shot | saker, minion | wt in powder as iron shot weighs |
| 3cwt metal to 1lb shot | double culverin, demi culverin, falcon, falconet | wt in powder as shot weight+⅙th |

with, it is important to understand the criteria upon which the amount of powder needed to propel a particular type and weight of shot from a specific gun was judged. Bourne (1587, 10, 13) is extremely clear: the thickness of the walls of the gun relative to the bore is the crucial factor. This 'degree of fortification' can be deduced from the weight of the gun expressed as the number of pounds per pound of shot weight (Table 5.29). This indicates that the strongest guns, those with the most fortification, are the culverins, sakers, minions and falcons.

In his treatise *The Gunner*, written in 1628, Robert Norton discusses all aspects of gunnery including 'ancient' and 'foreign'. He states that cast ordnance in the past was weaker and less fortified, partially due to the weakness of the powder (Norton 1628, 42):

> 'Saltpeter eyther being ill or not refyned, their sulphur vnclarified, their coales not of good wood, or else ill burnt; making therewith also their powder euilly receipted, slenderly wrought, and altogether vncorned, made it prooue to be but weake (in respect of the corned powder made now a dayes) wherefore they also made their Ordnance then accordingly, (that is much weaker then now:) for the powder now being double or treble more then it was in force of rarifaction and quicknes; requireth likewise to encrease the Mettall twice or thrice more then before for each Peece. For whereas then they allowed for the Canon 80 pound of Mettall for each pound that the Shot wayed, now they allow 200 pound & more for each pound of the Shot; and for Culuerings then they allowed but 100, and for Saker, Falcon and lesser Peeces they were wont onely to allow 150 for one. But now for the Culverings they allow 300, and for the small Ordnance 400 pounds for each pound their seuerall shots of cast yron is to weigh.'

This suggests that the overall weight of a particular gun should increase with the introduction of the new powder; this is needed to control recoil resulting from higher muzzle velocity. A study of gun weights against bore may give an indication as to when the change occurred, provided that the length of the gun remains within the original parameters.

The table Norton produces for '*Ordinary proportions allowed for English Ordnance*' (ibid., 53) shows the type of bronze guns, sizes, sizes of ladles, and weight of shot and both serpentine and corn powder (Table 5.30). The table was actually produced in order to indicate correct ladle sizes for serpentine powder (see section on ladles, below), the corn powder seems to be an after-thought, and is out of sequence. He covers this saying:

> 'the powder here mentioned is for Serpentine powder, which being now out of vse, the Corne powder being ½ stronger, therefore ½ of these weights is to be abated, as in the last Colume appeareth.'

The table clearly demonstrates that the guns capable of taking the most powder relative to shot weight are the

**Table 5.30 Weight of serpentine and corn powder for named guns (weight in lb)**

| Gun | Bore (in) | Shot weight | Weight serpentine | Weight corned | Length gun (ft) | Weight gun | Fortification* |
|---|---|---|---|---|---|---|---|
| Cannon Royal | 8 | 63 | 40 | 27 | 12 | 8000 | 127 |
| Cannon 'of 7' | 7 | 39 | 25 | 18 | 11 | 7000 | 180 |
| Demi cannon | 6½ | 30 | 20 | 14 | 10 | 6000 | 200 |
| Culverin | 5–5½ | 15–20 | 14, 15, 16 | 10, 12, 15 | 12 | 4300–4600 | 286–230 |
| Demi culverin | 4¼–4¾ | 9–12v | 8, 9, 10 | 6–8 | 12 | 2200–2500 | 244–200 |
| Saker | 3½–4 | 5–5½ | 5–5½ | 5–5½ | 9 | 1400–1600 | 280–290 |
| Minion | 3¼ | 3¾ | 3¾ | 3 | 7½ | 1200 | 320 |
| Falcon | 2¾ | 2½ | 2½ | 2 | 7 | 700 | 280 |
| Falconet | 2¼ | 1½ | 1½ | 1 | 6 | 500 | 285 |
| Rabinet | 1½ | ¾ | ¾ | ½ | 4 | 300 | 400 |
| Base | 1¾ | ½ | ½ | ?½ | 3½ | 201 | 402 |

* weight of gun divided by weight of shot

culverins, sakers, minions and falcons. Sakers are the most numerically significant bronze guns listed in the *Anthony Roll* (one-third of the total). The effect of a greater powder charge should include increased muzzle velocity, range and impact.

One of the earliest English sources which includes range consists of a series of notes compiled by John Lad (1586) as shown in Caruana (1992). This includes sketches of different types of gun with length, weight, bore, shot size and weight, powder charge and point blank range in feet and paces. The ranges, together with gun, projectile and powder data from Lad (Carauna 1992), Wright (1563), Bourne (1587), Norton (1628) have been amalgamated in Table 5.31. The weights quoted by Lad seem consistently high. The cannon class is significantly less fortified than the culverin to adequately fortify a cannon would result in an unworkable weight. The demi culverins, sakers and falcons all seem to be able to take their full weight of shot in powder, or more if serpentine, and over three-quarters of their shot weight in corned powder. This does not seem to occur with the cannon class. If range is considered, although the greatest range observed (at point blank) is for the culverin, the saker achieved

Table 5.31 Gun and projectile sizes, gunpowder charge and range

| Gun | Source | Bore (in) | Shot diam. | Shot wt (lb) | Powder serp. | Powder corned | Gun wt (lb) | Gun L. (ft) | Lb metal: 1lb shot | Range (ft/m) |
|---|---|---|---|---|---|---|---|---|---|---|
| Cannon royal | Lad | 8 | 7¾ | 63 | 40 | – | 8000 | 12 | 126 | 1500/457.2 |
| | Wright | 8 | 7¾ | 63 | 40 | – | – | – | – | – |
| | Bourne | 8 | 7¾ | 64 | 42 | – | 7500 | 11–12 | 117 | – |
| | Burton | 8 | 7¾ | 59 | 38 | – | – | – | – | – |
| | Norton | 8 | 7¾ | 63 | 40 | ?27 | 8000 | 12 | 126 | – |
| Demi cannon | Lad | 6¼ | 6 | 30 | – | ?25 | 5800 | 12 | 193 | 1525/465 |
| | Wright | 6 | – | 33 | 20 | – | – | – | – | – |
| | Bourne | 6 | 5 | 25 | 23 | – | 5000 | – | 200 | – |
| | Burton | 6½ | 6 | 31 | 21 | – | – | – | – | – |
| | Norton | 6½ | 6 | 30 | 20 | 14 | 6000 | 10 | 200 | – |
| Whole culverin | Lad | 5¼ | 5 | 18 | 15 | – | 4770 | – | 261 | 2000/610 |
| | Wright | 5 | 5 | 18 | 18 | – | – | – | – | – |
| | Bourne | 5¼ | 5 | 17 | 18 | – | 4400 | 12–13 | 259 | – |
| | Burton | 5¾ | 5½ | 18 | 18 | – | – | – | – | – |
| | Norton | 5¼ | 5 | 17 | 15 | 12 | 4400 | 12 | 259 | – |
| Demi culverin | Lad | 4 | 4 | 9s | 9s | – | 4760 ?iron | 13 | 528 ?iron | 1900/579 |
| | Wright | 4½ | 4 | | 8–9 | – | – | – | – | – |
| | Bourne | 4¾ | 4½ | 12½ | 12 | – | 3200 | 10–12 | 256 | – |
| | Burton | 4½ | 3¼* | 8½ | 8½ | – | – | – | – | – |
| | Norton | 4½ | 4¼ | 10 | 9 | – | 2500 | 11 | 200 | – |
| Saker | Lad | 3½ | 3¼ | 5¼ | 5 | – | 2000 | 9 | 380 ? | 1450/442 |
| | Wright | 3½ | 3¼ | 5 | 5 | – | – | – | – | – |
| | Bourne | 3½ | 3¼ | 4¾ | 5½ | – | 1400 | 10 | 248 | – |
| | Burton | 3¾ | 3½ | 5½ | 5½ | – | – | – | – | – |
| | Norton | 3½ | 3¼ | 5 | 5 | 4 | 1400 | 9 | 290 | – |
| Falcon | Lad | – | – | – | – | – | – | – | – | – |
| | Wright | 2¾ | 2½ | 2¼ | 2 | – | – | – | – | – |
| | Bourne | 2¾ | 2½ | 2⅛ | 2½ | – | 750 | 7 | 340 | – |
| | Burton | 2¾ | 2½ | 2½ | 2½ | – | – | – | – | – |
| | Norton | 2¾ | 2½ | ½ | 2 | – | 700 | 7 | 280 | – |

* Possibly an error for 4¼, noted by Awty (1989). Eldred (1646) has been omitted as it is difficult to ascertain which of his figures refers to iron guns and which to bronze guns

1450ft (*c.* 442m) for a gun weighing only 1600lb (726kg) at the most, whereas a cannon only achieved 50ft more (457m) but weighed up to 8000lb (over 3.5 tonnes). Firing trials with a reproduction of a *Mary Rose* culverin (Chapter 2) have shown that the damage to a wooden target consists of a very neat hole, hardly larger than the shot. A large number of smaller guns can therefore inflict a greater number of hits.

Perhaps the trade off between muzzle velocity, impact, weight and production unit cost was epitomised in guns such as culverins, sakers and falcons. Where gross corned powder appears, these are the guns most likely to have used it. The 'mixed arsenal' of Henry VIII included both old and new guns, the older perhaps unable to take corned powder. It would appear that during this period most of the large guns still used the powder '*as fine as flour*' – ie, serpentine.

*Storage of gunpowder*
The best source for information regarding the varying units of measurement regarding gunpowder is the 1547 inventory (Starkey 1998). Units of measurement include barrels (the term '*barrel*' is used for containers of specific sizes so the term '*cask*' is used generically here) lasts and pounds. The types of containers include barrels, demi barrels, firkins (*ibid.*, 7389), small firkins (6777) firkins with purses (6472), grand or great barrels (7390), half grand barrels (7263) and hogsheads (5984). Barrels from Antwerp, Nuremberg and Spain are also mentioned. Historical sources suggest that containers for cartridges included chests, tubs, pewter canisters, and barrels. All of these types of object were found on the *Mary Rose*. A brief look at distribution and contents of casks and containers that might have contained gunpowder is given below.

There seems to be little contemporary documentation regarding where or how to store gunpowder on vessels. Commonsense would dictate that storage should not be close to the galley, nor require an access route which passed an open fire. Later developments show that a great deal of care was taken in the choice of powder store, considering both security and fire. As spoilage due to damp is frequently discussed one would also assume that inherently damp areas would be avoided if possible.

Reference in a late fifteenth century account (Oppenheim 1896a, 20) to tar in a barrel of gunpowder suggests an attempt at water-proofing the container. Lucar (1588, appendix, 2) advises on the safe storage of gunpowder and cartridges:

> '*Also if a Gunner will charge his peece with Cartredges, he ought to set them vpright in a tubbe or some other woodden vessel, which (though it shall seeme to stande in a place out of danger for fire) should neuer be vncouered for any longer time than while the same cartredges are taken out one by one to charge the peece.*'

Captain Smith (1627, 15) suggests a powder room in the stern by inference whilst discussing the positioning of the galley. He suggests that putting a galley in the stern of a warship is '*very troublesome to the use of his Ordnance, and very dangerous lying over the powder-roome*'. This suggests that, by 1627, the powder store was beneath the gun deck in the stern. However, early records suggest an area deep beneath the bow (eg, Cleves (1456–1528); see Oudendijk 1941).

*Casks:* There is a major problem in attempting to establish which of the casks recovered from the *Mary Rose* might have been gunpowder barrels. The Mary Rose archive contains 3759 objects identified as staved containers or components of them. Although intact or partly intact containers and closely related groups of staves were recovered from the ship, most have disintegrated so that any attempt to reconstruct the original number of individual vessels would necessitate the laying out and ordering of all the pieces in the assemblage. Since the vast majority of this material remains individually packed and in wet storage, not yet having been permanently conserved, such a massive undertaking is simply not yet feasible (if ever). The original recording of the objects does, however, allow us to suggest that the number of staved containers resembling any form of a cask on board the *Mary Rose* is in the region of 120–150. These figures are based on database calculations of staves of similar lengths from their designated areas and the number of possible staves for each container. Of these, only a small fraction was found with any substantial original contents that could be identified with certainty, or with minimal residues that could be described with circumspection. A detailed study by Jen Rodrigues of a sample of 21 more or less complete casks – with and without contents – is presented in *AMR* Vol. 4 (chap. 10). Among those studied by Rodrigues is one that was thought to have contained gunpowder (81A1169 – see below for a further consideration).

Although it has not yet been possible to reconstruct many casks from the ship we can at least make an approximation of the size of barrels we should be looking for. If, as the historical sources intimate, a barrel of gunpowder holds 200lb (90.7kg), the *Mary Rose* should have carried 24 barrels of serpentine powder and three of corned, 27 in all. By 1635 (see above) these sizes were standardised at 100lb of powder. (Various scenes from the *Siege of Boulogne* suggest the use of barrels of fairly regular size; see Fig. 5.103). A gunpowder barrel containing 100lb is about 540mm in height with a diameter of 400mm. Barrels of twice this capacity would need to be substantially larger. Rodrigues uses the standard formula for calculating the capacity of a column ($\pi r^2 h$) with an error margin of 3% to approximate the capacity of casks. This suggests that the *Mary Rose* casks fall into three general size groups: small (16.84–150 litres, 3.7–33 gallons), medium (151–250 litres, *c.* 33–55 gallons), and large (251–448.57 litres, *c.* 55–98 gallons). The capacity of a 100lb gunpowder barrel is 68±2 litres (14.5–15.4 gallons). A 200lb barrel would also fall within Rodrigues' small size.

The physical remains of such a significant number of similarly sized barrels or staves should be immediately apparent, even if broken up, but there is no one sector which can be identified as a central store for 27 gunpowder barrels. Neither the small or medium casks show any obvious patterning in distribution across the main storage deck (the Orlop) and the Hold, though the large casks were predominantly concentrated in the Hold (76.5%), mostly in H8 and H9.

Residues within or adhering to various cask elements were sampled but few of these have yet been analysed. Only two casks have been potentially associated with gunpowder (81A1169 and 81A4586). The first of these was in the forward Hold. Rodrigues describes this cask thus (*AMR* Vol. 4, 415–16):

'Cask 81A1169, *the only one with black powdery residue that may have been gunpowder, was verybadly damaged but it was recovered from the forward Hold of the ship in H1, which itself had suffered severe disintegration.* [There are no scorched marks on the interior of the staves which sometimes result from the 'firing' process during cask assembly.] *While bevelling the circumference of the headpieces, the cooper had made some in-cuts into the wood grain, creating the only interruptions along the otherwise smooth bevelled surface.* [The purpose of these is unknown.] *The only mark present is a branded symbol on the exterior of one headpiece: a circle at the centre of what vaguely resembles an 'X' with three of its points protruding from the circle. The chivs of the staves have been hollowed by an adze and, in many instances, the chimes have broken off completely. The twelve staves recovered for this cask have sapwood along one side. The full length of any of the staves could not be determined but the croze distance is 650mm, giving the cask an approximate total height of 680mm. Fragments of wooden hoops were recovered with this cask, all with bark still attached. Vague hoop impressions on stave exteriors suggest there were hoops around the chime and bilge sections. Metal stains at the ends on both the interior and exterior faces of one stave would have resulted from contact with other metallic objects and not stains from iron hoops. Iron hoops were not yet introduced* [in England] *during this period and they were never used on gunpowder casks, as sparks caused from knockings could cause an explosion.*

*This was thought to be a gunpowder cask because it contained a fine black powdery residue. However, the substance possessed an unusually small amount of sulphur for gunpowder and could not be positively identified as such (samples 81S1089, 81S1263). The low sulphur content could, of course, have been the result of chemical changes in the post-depositional environment.*'

Two other casks of potential interest here were found in H2 (81A0762, 81A1001) as they are the only two from the ship to have contained traces of pitch/tar. They are both small (*c.* 83 litres and 58 litres respectively), but of distinctly different form to other casks on the ship, being amongst the heaviest and bulkiest, even without the contents. This suggests that they were specifically chosen to contain pitch/tar rather than being simply coated with it to hold gunpowder, as described above.

The Orlop deck had a large number of casks. The greatest numbers were in O7–O10, with approximately eight, thirteen, five and eleven respectively. Areas further aft (sectors 10/11) are candidates for the gunpowder store according to the historical sources and a sample (81S0120) from cask 81A4586 (H9), taken during a pilot study to determine the preservation of pollen on the ship, proved to be probably gunpowder (Grieg, *AMR* Vol. 4, chap. 16). Another sample (81S0441), of sediment surrounding the bilge pump that was also found in O8, contained fine black material, possibly charcoal, which may also have been gunpowder (Scaife, *AMR* Vol. 4, chap. 15). Other casks in this area contained salt beef (five casks), salt fish (one), and fruit/wine (one). Only three casks were found aft of this (O11–12), one of them (an open-ended container) possibly containing tallow. In view of the lanterns stored in this area, the former for making cartridges found in M11 (81A2278) and the reference to open tubs (see above) for the storage of cartridges, this is unexpected as it suggests the concentrated storage of flammable items

Only twelve casks have been identified from the whole of the Main deck, at least one of which (in M7) contained salt beef. Three 'gun deck' storage areas for shot and gun furniture have been identified in M2–3, M5–6, M10–11. There were elements of up to three casks in M5 and possibly three in M6 but their contents are unknown. The Upper deck had even fewer casks, five in U9 and at least one in U8, as well as one containing salt pork that was lashed to the rigging. Again, there is no indication as to content though this was both a living area for the crew and contained many pieces of archery equipment and hand weapons such as pikes and bills.

One further possible clue to the presence of gunpowder barrels was explored, namely the marks that

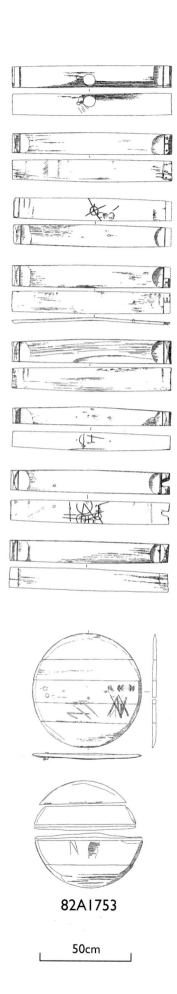

*Figure 5.104 Possible gunpowder casks: above) marking on the head piece of cask 81A2610; left) various marks, including race marks on cask 82A1753*

appear on over 100 cask elements. Many cask heads and some staves carry incised or branded marks. These are of several types. Some are thought to be indications of content or batch numbers and/or are indicative of officially issued supplies to the ship. These include various combinations of Roman numerals and other letters, an 'HR', 'H' or 'B' brand, a 'W' incised mark and Admiralty arrows (Fig. 5.104). The first recorded use of the broad arrow is in the Tower of London inventory for 1495 (Blackmore 1976, 257) which included '*marking yrons with the brode arow hed*'. Other object groups on board ship possess similar types of markings (see *AMR* Vol. 4, chap. 11 for examples of non-ordnance related objects and elsewhere in this volume). The 'H' brand is especially familiar as an official mark of the King though Rodrigues states (*AMR* Vol. 4, chap. 10) that is possible that it is an assize or search mark of the Coopers' Company, as ordered by the Court of Aldermen of the City of London in 1514–1518 for all vessels holding beer or ale. Where similar brand marks occur on more than one vessel (fourteen of the 33 vessels examined by Rodrigues (42%) have brand marks) they suggest use of the same branding iron. The 'W' incised mark, however, while appearing on various artefacts, was incised freehand. The brand 'C' and a pair of tongs appear on individual cask elements from M9 (81A4563), O4 (81A3214), H4 (81A2199) and H6 (81A5731). The tongs may be a pair of shot making tongs (see Chapter 5) and the 'C' could indicate 100lb.

A few cask elements carry more complex designs. At least six have combinations of circular marks and/or lines created by a tool called a scribe. The latter have been identified as shippers' marks in a study of casks from the Red Bay wreck (Bradley nd) which, together with the branded ones, are thought to indicate ownership. Assembly marks (race marks) on pieces which have corresponding lines to match on adjacent components are also common (Fig. 5.104), especially on headpieces. Brands include the 'H', crown, crown and sword, double crown, arrow, c with tongs and a bill head (80A0891, O5).

Of the two probable gunpowder barrels identified from their residual contents, 81A1169 bears only what appears to be a branded shipper's mark of an 'X' set off-centre within a circle, and /x/ on the centre headpiece, while 81A4586 is unmarked. The distribution of marked cask elements was examined. The 'H', and 'HR' marks all occur on casks in H9, apart from one in M6, though where size can be determined, the casks in this area of the Hold are of Rodrigues' large size. Marks with the crown are restricted to H1 and O1, a storage area for gun furniture. The arrowmark is more widespread, it is on other classes of object, but interestingly occurs on four casks in O11. The 'W' is also widespread. At present there is insufficient information to point to a single place with enough casks of similar size or marking to suggest the position of a powder store. If an area deep within the bow is the case, this may lie within the unexcavated areas of the site.

*Chests*: It is possible, though perhaps unlikely, that some of the simple crate-like chests recovered from the ship (see *AMR* Vol. 4, chap. 10) could have held powder-filled cartridges. These crates seem to have been part of the ship's issue. The two empty examples of this type that could be identified (81A0608 and 81A1681) were both close to major storage areas for gun furniture, but no more can be said.

*Tubs*: Rodrigues also examined the remains of at least five tubs (*AMR* Vol. 4, chap. 8). All were made by cutting down full-sized oak casks and are, therefore, wider at the open top than at the base and their shape reflects the curvature of the original casks. They were recovered from the Hold (H5), Main deck (M9) and Upper deck (U7, U9). They do not have handles which suggests either that they contained objects that were not very heavy (though the tubs themselves certainly are) or that they generally remained in one place, in which case, their locations are unhelpful in the present context. No contents survived.

*Canisters*: Two pewter canisters were recovered from O10 (81A5981, 82A0976). They are closely similar to examples found in the Barber-surgeon's cabin and it is unlikely that these were intended for gunpowder unless they were the personal possession of an individual officer. They are probably English made, in high tin pewter and carry identical stamps of a lion stamp on the base tentatively identified as the Lion Rampant of the Duchy of Cornwall (identified by Rosemary Weinstein, *AMR* Vol. 4, 460). The two seem to be a pair and, given that they were found in one of the main storage areas of the ship (O10) they probably originated from one of the personal chests found here. Later examples for cartridges had loops to hold a suspension string, these did not.

*Composition of gunpowder samples*

A small number of samples, or series of samples, of probable gunpowder were analysed using a variety of methods (details in archive). Apart from 82S1263, which was taken from incendiary projectile 81A2088 and 84GPS2001, whose provenance is unclear, all the samples are from the chambers or barrels of guns (gunpowder analysis tables are in Appendix 2).

*84GPS001*: This sample was analysed at the Royal Armouries Armament Research and Development Establishment EM2 Branch. Microscopic examination revealed a granular form with a random distribution of yellow and white particles on a black background indicative of gunpowder. Electron microprobe analysis confirmed the presence of the constituents of gunpowder, potassium (62.8%), sulphur (17.5%) and carbon (12.4%) and 4.4% volatile matter. The presence of hydrogen in a form other than as water (0.6%) was not considered unusual in the presence of charcoal. Hall (above) suggests that this might reflect a recipe of 6 parts saltpetre, 2 parts charcoal and 3 parts sulphur. Smith (pers. comm. 2006) advises caution with this interpretation due to the solubility of the components.

*80S0117–S0128* (gunpowder analytical samples MR1–5): A series of samples was taken from bronze culverin 80A0976 and analysed at Portsmouth Polytechnic by P.S. Maton as a BSc project. From breech to muzzle these are 80S0117/119/120/124 (= analytical sample MR4), 123 (MR1), 125 (MR2), followed by the shot, then samples S0126 (MR3) and S0122 (MR5) towards the mouth. Samples were dried and passed through a 20 mm mesh sieve to separate the powdery material from any fibres. Samples from behind the shot broke down to powders of varying colours from grey-green to black; most contained a few fibres. Sample S0125 (behind the shot) and S0126 (in front of the shot) both consisted of fibrous masses, possibly remnants of wadding placed before and after the shot.

The presence of free elemental sulphur was confirmed in samples directly behind the ball, (where one would expect the best chance of finding evidence for gunpowder) using both the Ingram and Toms method (Table Appendix 2.1) and bomb calorimetric determinations (Table Appendix 2.2) with values of 3.15–11.7% and 1.22–16.14% sulphur indicated respectively. This showed higher levels of sulphur in the breech in the two samples immediately before the shot, with the amount decreasing after the shot and the lowest amount closest to the mouth.

Quantative analysis was made for the elements of iron, copper, calcium and manganese. The samples showed a range of values for iron consistent with the position of the samples within a bronze barrel and close to an iron shot (Tables Appendix 2.3–4). This analysis has been used by Hall (above) to suggest that the combination of strong readings of magnesium and

Magnification ×1500

*Figure 5.105 Micrographs of: a) gunpowder from demi cannon 81A3000; b) gunpowder from port piece chamber 82A0972; c) modern gunpowder*

calcium, weak readings for potassium and varying readings for sodium, may reflect saltpetre which has not been processed with potash to substitute potassium for calcium and magnesium. Smith (pers. comm. 2006) advises caution here too, as current experiments suggest that potash was necessary to make viable gunpowder, and that this is extremely soluble. The percentage of carbon (Table Appendix 2.5) is higher behind the shot, probably a reflection of the powder.

*81S1409/15/1 and 82S1207/9/1*: Two further samples were submitted to the Royal Armaments Research and Development Establishment EM2 branch for analysis in 1985. 81S1409 was excavated from bronze demi cannon (81A3000) and consisted of several hard black lumps contaminated with sandy coloured particles. 82S1207/9/1 was excavated from the breech chamber of a wrought iron port piece (81A3001) and consisted of a wet brown/black substance representing the wad. The samples were prepared by drying a weighed amount of each to a constant weight and the percentage of water content, and total solids calculated. The dried samples were extracted with distilled water and filtered. The dried residue was weighed to give the percentage of insoluble solids and, by difference, the percentage of soluble solids in the original samples (Table Appendix 2.6). The soluble solids were analysed by ion chromatography. The soluble solids present in the samples originally received proved to be mostly sodium chloride. Some sulphate was found to be present in the sample from the bronze gun (81S1409), and potassium ions were found in both. Some ammonium ions were present in the iron gun sample (82S1207; Fig. 5.105b)). An elemental analysis of the insoluble solids was made by ignition of the sample at 500°C (Table Appendix 2.7).

The visual appearance of the residual ash suggested that it was mainly iron. The ash from the port piece was much darker than from the bronze gun and was the colour usually associated with $Fe_3O_4$ (magnetite) suggesting that the sample could have been part of an intense rusting process. The ratio of sulphur to carbon is very low for each sample and one would expect more equal amounts in gunpowder; however the sulphur may have reacted over the years. The dried residue was analysed by emission spectroscopy. Both samples contained iron as a major constituent together with silicon. Appreciable amounts of copper and calcium were found in the bronze gun sample (over 1%). Both contained trace quantities of twelve other elements often associated with non specifically defined steels (ie, magnesium, sodium, potassium, aluminium, arsenic, titanium, bismuth, silver, lead, antimony and boron).

A portion of the dried residue was further examined using a scanning electron microscope fitted with an X ray analysis attachment. The powder from the bronze gun (Fig. 5.105a, 81S1409) bears most resemblance to modern powder (Fig. 5.105c). All show a particle size in the order to 10mm. The x-ray analysis indicated the presence of the elements presented in (Table Appendix 2.8) in order of the amounts present. Elements of atomic weight less than oxygen (eg carbon) would not be detected by this method. A portion of the samples was dried and a weighed amount extracted with ether. The percentage of ether soluble material was determined by weighing the dried extracted material and the ether solution was submitted to infra-red analysis. For comparative purposes a modern military gunpowder was included in the exercise. The charcoal in this modern gunpowder was made from beech and alder (Table Appendix 2.9). The infra-red spectra of the extracts were broadly similar but some absorption peaks were not common to all. All extracts gave strong absorption peaks at wave numbers 2850 cm$^{-1}$, attributed to C-H stretching frequencies. Other peaks in common were observed at 1730, 1460, and 1270 cm$^{-1}$ and were attributed to esters, alcohols, aldehydes or ketones. Other peaks attributable to these structures were not common to all extracts. These were compared with infra-red spectra from available essential oils derived from pine and cedar wood. These were found to be similar but to contain fewer absorption peaks. The spectra obtained from the ether soluble components of the samples would be consistent with the hydrocarbons (terepenes, sesquiterpenes) and the alcohol, esters, aldehydes and ketones one would expect to find in products such as charcoal derived from wood.

The conclusions were that the samples contained gunpowder degraded by time and sea water. The low sulphur content (relative to carbon) may be attributable to this degradation rather than to the original composition. Sample 81A1207/9/1, from the iron gun, was heavily contaminated with iron. All ether soluble extracts gave rise to infra-red spectra consistent with the presence of oils which could have been derived from wood. Hall (*ibid.*) has suggested that the results indicate differential processes in the refining of the saltpetre and has indicated the importance of these samples as demonstrating techniques as used in the sixteenth century.

## *Formers*

One object has been identified as a former, a wooden pattern around which paper or canvas is folded to form a gunpowder cartridge (81A2278). Formers are listed for the *Mary Rose* in the *Anthony Roll* (Knighton and Loades 2000, 43): '*Canvas for cartowches xx$^{ti}$ ellys. Paper ryall for cartowches j qwayer. Fourmes for cartowches vj*'.

### Inventory and historical evidence
The word '*former*' or '*form*' appears to be used to denote a pattern or model. Lucar (1588, appendix, 26) gives detailed instructions for the making of copper powder ladles from wooden patterns:

> '*make a modell or forme of a long and rounde peece of wood ... and let the model be in his rounde compasse so much more lesse than the circumference of a pellet that will fitte the peece for which you doe intende to make a Ladle, as the plate of the Ladle is thick: then bende rounde that parte of the plate which shall holde the charge of gunpowder vppon the sayde modell*'.

The text is accompanied by an illustration (p. 27) (Fig. 5.106).

Formers were also used to make cartridges, as described by Norton (1628, 127–8):

> '*Cartredges are either to bee made with Canuas Fustian, or other linnen cloath, or with thicke strong Paper, especially of Paper Royall: which prepared, take the height of the bore of the Peece, without the vent of the Shott, and cut the cloath or paper of the breadth of three such heights; and in length, for the* Cannon *3, for the* Culuering *4, and for the Saker Falcon, &c. 4½ of the heights of their proper Bores, and leauing in the midst at the top and bottom one other such height, at each place to make a couer and botto me for the* Cartredge, *cutting each side and end, somewhat larger, then the strict measures appointed for the sowing or glewing of the seames thereof, so much as will counteruaile the same, hauing also a respect for augmenting and deminishing those measures, as the powder shall bee better or worse then ordinary, and also abating with discretion, when as your Peeces shall be already heated in fight, least else you endanger the breaking or splitting of your Peece. Hauing resolued then for what sort of Ordnance your Cartredges are to serue, you are accordingly to haue a Modell or Former of wood turned of the height of the Shot, and of a conuenient length, longer then the Cartredge is to be. Then if you make them of Canuas, halfe a dyametre is to be allowed more in breadth for the seames: but if they be made of Royall paper, then hauing lapped it once about the Former, leaue about ½ inch surplussage more then will compasse it, which with Starch, Paste, or mouth Glew, close about the said Former, hauing some part of the same substance, fitted vpon the end of the Former; first for a bottome, which must also be pasted or glewed close, and fast to the side of the* Cartredge, *so that being dry, it may hold the Powder fast, and sure from spilling. And you must remember to tallow the said Former, so that the* Cartredge *being so moulded thereon, it may be easily and without tearing, slipped off againe ... if the Peece*

*Figure 5.106 Illustration of a former by Tartaglia (Lucar 1588)*

*be equall bored, and the* Cartredge *made of Paper, then there is no more to doe, but to put the* Cartredge *into the mouth of the Peece, and with the Rammer-head, to put it home, to the bottome of the bore of the Peece, with two or three easie stroakes: and then with a sharpe three squared Pryming Iron, to cut and pryme the* Cartredge, *that the Powder prymed at the* touch-hole, *may giue fire to the quick powder thereby. In all other things for wadding before and after the Shot, and ramming home the Shot, you are to performe the vsuall manner taught in his proper place.'*

This reveals a number of important points. First, the former is actually the size of the shot, not the bore of the gun. Secondly, it suggests that ladles were not routinely used to deliver paper cartridges, however this cannot discount their possible use to fill cartridges. Thirdly, the types of gun listed are all muzzle-loaders (cannon, culverins, sakers and falcons). Fourthly, the use of tallow is described. Tallow was found in the same area as the *Mary Rose* former (see below). This adds credence to the idea that some of the casks stored in O10 and O11 may have held gunpowder (discussed above). The *Anthony Roll* suggests that cartridges were used with some of the guns. Formers are listed on all vessels with the exception of the Christopher of Bremen (Knighton and Loades 2000, 51). These are listed together with other items required to manufacture cartridges: paper royal and canvas (Table 5.32).

The presence of both ladles and cartridges poses the question as to which guns were loaded with loose gunpowder and which with cartridges or, indeed, whether the ladles present were for loading cartridges rather than loose powder. Although ladles are not listed in 1546, the 1514 inventory for the *Mary Rose* suggests that both were carried (Knighton and Loades 2000, 141–2):

*'Cart tuches of gonnepowder j chestfull'* along with 20 *'Parchement skynnes'*, 2 *'Chargynge ladelles of coper'*, 7 *'Charging ladelles for fawcons and*

**Table 5.32 Gunpowder and elements for cartridge making in the *Anthony Roll* (ships)**

| Vessel | Serpentine | Grosse corn | Fine corn (lb) | Former | Paper royal (quire) | Canvas (ells) |
|---|---|---|---|---|---|---|
| Henry Grace à Dieu | 2 last in barrels | corned only 6 B | | 6 | 1 | 20 |
| Mary Rose | 2 last in barrels | corned only 3B | | 6 | 1 | 20 |
| Peter | 2 last in barrels | corned only 2B | | 5 | 2 | 20 |
| Matthew | 1½ last | 2 B | 40 | 6 | 1 | 8 |
| Great Bark | 30 DB | 2 B | 2 DB | 4 | 1 | 12 |
| Jesus of Lübeck | 1 last | 1 B | 40 | 3 | 1 | 10 |
| Pauncy | 30 DB | 2 DB | 40 | 4 | 1 | 12 |
| Morian | 8 DB | 1 DB | 24 | 2 | 1 | 6 |
| Struse | 12 DB | 1 DB | 24 | 2 | – | 4 |
| Mary (of) Hamburg | 27 DB | 1 DB | 24 | 3 | 1 | 4 |
| Christopher of Bremen | 14 DB | – | 24 | – | 1 | 4 |
| Trinity Harry | 1 last | 1 B | 24 | 2 | 1 | 7 |
| Small Bark | 1 last | 1 DB | 24 | 4 | 1 | 12 |
| Sweepstake | 1 last | – | 24 | 4 | 1 | 12 |
| Minion | 1 last | corned only 24lb | | 3 | 1 | 8 |
| Lartique | 10 DB | corned only 26lb | | 2 | – | 4 |
| Mary Thomas | 10 DB | – | 20 | 2 | – | 3 |
| Hoy Bark | 8 DB | – | 26 | 1 | – | 2 |
| Mary James | 9 DB | – | 20 | 2 | – | 2 |
| George | 5 DB | – | 8 | 2 | – | 3 |

Paper royal and canvas have not been identified within the *Mary Rose* assemblages
Key: DB = demi or half barrel. B = barrel. An ell is 45in (1143mm). A quire relates to a number of pieces of parchment (normally four but up to 24)

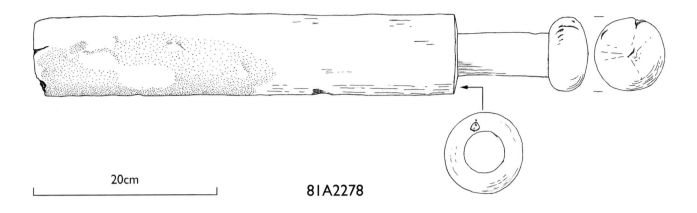

*Figure 5.107 Former 81A2278*

curtowes', and 'Gonnepowder xxj di barelles' [assumed to be 21 half barrels rather than 21½ barrels]. *'the marke the better, and not so apte to be shedde or spylled for when the cartredged bee'*

By the time of William Bourne published his work in 1587, the advantages of cartridges over loose loading were well known (p. 31):

*'And also it is unprofitable and daungerous to lade or charge a peece wyth a Ladell, for that the Pouder is apte to bee shedde or spylled beeyng hastyly done, and then it is apte to bee fiered ... Wherefore if youre Pouder bee in Cartredges and also weyed, the peece is more sooner and easilyer laden or charged, and hee shall keepe the length of the marke the better ... and not so apte to be shedde or [sp]ylled, for when that the Cartredges bee fylled, then they may bee set uprighte in some Tubbe or Barrell, and then they maye take out one by one as neede shall require, and so couer the Barrell close againe, that it maye bee wythout daunger.'*

Although cartridges could be used for most of the larger guns listed for the *Mary Rose*, the six formers listed suggests that these were used primarily with the muzzle-loaders. Although the loose loading of breech chambers is extremely difficult, funnels would make this relatively straightforward. The chamber neck is restricted, so any stiff cartridge would have to have a much smaller diameter than the inside of the chamber.

Amongst the ordnance and munitions at Berwick recorded in the 1547 inventory (Starkey 1998) are '*Formes of woode for Cartouches for Cannons*' (6419). Formers are also specified for demi cannon, culverin, demi culverin, saker, falcon and falconet (6420–6425). '*Chardging ladells*' are also listed at Berwick for all those gun types (6386–6391, 6393). In the 1595 inventory of the Tower of London, items among '*Turners woorke*' at the Tower of London incude '*formers for Cartowches*' for cannon perriers, demi cannon, culverin, demi culverin, saker, minion and falcon (Blackmore 1976, 279–80).

This confirms that cartridge loading was used for the bronze guns, possibly with the aid of a ladle. There is a rough correlation between number of formers and the different sizes of muzzle-loaders. If one looks at the lightly armed pinnaces, for instance (Table 5.33), many vessels listed with one former have two differently sized breech-loading iron guns, and only one muzzle-loading gun (*Falcon, Saker, Hind, Phoenix*). The number of types of muzzle-loaders often equals the number of formers.

The number of formers listed for the *Mary Rose* (6) agrees with the listed types of cast bronze gun and it is tempting to suggest cartridge loading for these guns with delivery possibly being through the ladles found. No other evidence for cartridges has been found. These are known to have been stored in casks, chests and tubs (see above).

Table 5.33 Gun types and numbers of formers in the *Anthony Roll* (Pinnaces)

| Vessel | Muzzle-loaders | Former | Paper royal (quire) | Canvas (ells) |
|---|---|---|---|---|
| *Falcon* | 4 sakers | 1 | – | 4 |
| *Saker* | 2 sakers | 1 | – | 2 |
| *Hind* | 1 saker | 1 | – | 3 |
| *Roo* | culverin perrier, 2 demi culverins | 2 | – | 4 |
| *Phoenix* | 2 sakers | 1 | – | 2 |
| *Merlin* | 3 minions, 1 falcon | 2 | – | 4 |
| *Less Pinnace* | 1 saker, 1 falcon, 1 falconet | 2 | 1½ | – |
| *Brigandine* | 1 falcon, 1 falconet | 2 | – | 4 |
| *Hare* | 1 saker | 1 | ½ | – |
| *Trego Renneger* | blank | 1 | 1½ | – |
| *Double Rose* | 1 demi culverin | 2 | 1½ | – |
| *Flower de Luce* | 1 demi culverin, 1 saker | 2 | 1½ | 6 |
| *Portcullis* | 1 saker | 1 | 1½ | 2 |
| *Harp* | 1 saker | 1 | 1½ | – |
| *Cloud in the Sun* | 1 demi culverin, 1 saker | 2 | 1½ | 4 |
| *Rose in the Sun* | 1 demi culverin | 1 | – | 4 |
| *Hawthorn* | 1 saker | 1 | 1½ | – |
| *Three Ostrich Feathers* | 1 saker | 1 | – | 3 |
| *Falcon in the Fetterlock* | 1 demi culverin, 2 falcons | 2 | 1½ | 6 |
| *Maiden Head* | 1 saker | 1 | 1½ | – |
| *Rose Slip* | 1 saker | 1 | 1½ | – |
| *Gillyflower* | 1 saker | 1 | 1½ | – |
| *Sun* | 1 saker, 1 falcon | 1 | 1½ | – |

**Description and location**
The only former identified, 81A2278 (Fig. 5.107) is turned out of a single piece of alder. It has a length of 595mm, with a diameter at the base of 75mm, increasing to 82mm at the junction with the handle. The handle is 100mm long with a diameter of 45mm and has a pronounced pommel 34mm in length.

A former could be used to make cartridges for all guns larger than its own diameter (75–82mm) maximum (3⅛in) but, in practice, they were used to make cartridges of similar size to their own. The recovered bronze ordnance ranges from 120–210mm (4½–8¼in). Applying Norton's rule (above), the former length should be six-and-a-half times its diameter. This is between 6.0 and 6.6 (depending on which diameter is used). This former could have been used for the two sakers listed. Although none has been recovered, several hundred appropriately sized cast iron shot were found (Chapter 2).

The former was found concreted to the Main deck in the stern M11 (Fig. 5.108). The area included a mass of concretion surrounding two spoked wheels representing the remains of a port piece carriage. Artefacts in close proximity included a shovel (81A2194) and a powder ladle (81A2212) found concreted next to each other. The ladle had a head diameter of 4½–4¾in (115–120mm) so it would have been capable of delivering a cartridge made around the former. The condition of the ladle is good and not corroded. Clearly a ladle of this size is too great to decant powder into the breech chamber of even the largest port piece. The proportion of ladles (six out of thirteen) and rammers (ten of 33) in this area is very high. What is most interesting is the lack of other formers in such an otherwise rich assemblage of ladles and rammers. Whereas the rammers might be expected to have a wider distribution, serving guns around the ship, perhaps the ladles were restricted to a specific 'safe' location. Good candidates would be on the port side in the stern (the single former being the sole remnant of a larger concentration) or further forward in the bow.

**Wood turning**
by Robin Wood
The piece is spindle turned out of a piece of roundwood, possibly from a coppiced woodland. It seems unusual that roundwood has been used rather than the cleft section of a larger tree. Cleft timber is less susceptible to splits, but is more likely to distort to the oval during drying. Roundwood was still used to make large cotton bobbins for the Lancashire mills until the twentieth century. Perhaps this relates to function: a requirement for a round rather than oval section so that the cartridge fitted the bore safely.

The process would probably have included the cutting of the piece and the partial removal of bark. The section would then have been stood upright for the sap

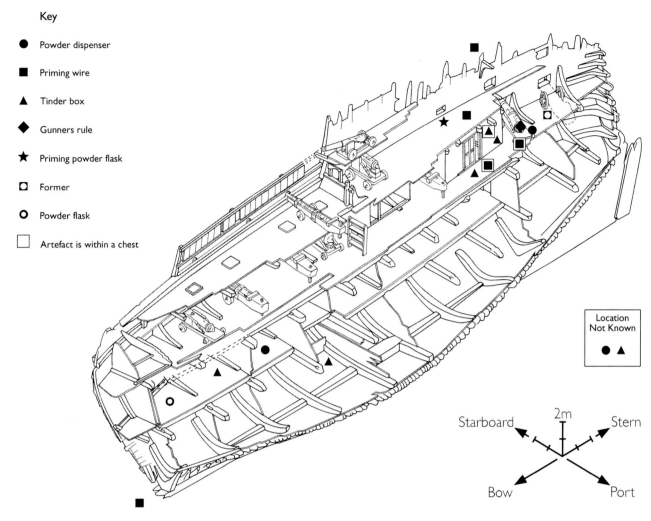

*Figure 5.108 Distribution of small items associated with loading and firing*

to drain and for controlled drying to take place. The removal of some of the bark would inhibit rot, and inhibit excessively swift drying. The piece retains several knots near the handle. These would have hindered the turning process. The piece would have been turned between the centres of a pole lathe without a mandrel as the drive strap would simply wrap around the piece. The evidence for this is not present as both ends of the piece are degraded. There are fine tool marks over most surfaces.

## Powder ladles

### Identification and historical significance

Elements of up to 21 powder ladles have been identified (Fig. 5.109). These are not listed in the *Anthony Roll*. One of the best contemporary descriptions of the powder ladle is found in the Pirotechnia of Vannoccio Biringuccio (1540, 417–18, Fig. 5.110c and d).

'In order to load the guns make an instrument which gunners call a loading ladle. This is like a gutter pipe of sheet iron or copper, three times as long as the diameter of the ball. It is bent to correspond exactly to half the circumference of the diameter of the hollow of the gun mouth. This is fastened at one end to a round piece made like the knob of a flail handle, with an empty hole at the back where the end of a pole is put as a handle. At the foot of this, on the other end of the same pole, another similar piece is put. The pole with the said sheet-iron ladle full of powder is put in the gun and carries it to the end. Then, turning the hand over, it is emptied inside and the powder is struck with the foot of the pole, which presses and sends it into place. Thus the guns are loaded, with this manner and method. This ladle is thrust into the powder keg and filled very well, then placed into the gun and pressed as I have told you above. The first load is pressed lightly and the same amount of powder

*Figure 5.109 Selection of ladles*

*is again taken with the same ladle, and the process is repeated as before. In short, in two or three times (according to the fineness and quality of the powder and the capacity of the ladle) you proceed to put in as much as you see by experience is equal to the weight of the ball, or at least two-thirds of this'.*

This is an important description, detailing how to make appropriately sized ladles and suggesting that both ladle and rammer are on either end of a single stave. It also alludes to the dual function of the ladle; first to provide the correct amount of a particular powder for a specific gun and, secondly, to deliver it to the breech portion of the gun without loss of contents. Sometimes both functions were required, sometimes only one. Identifying their function within the *Mary Rose* assemblage is a factor in understanding the particulars of the practice of warfare at sea at this time. Their function may also help to determine whether they were used with all guns, or just the bronze muzzle-loaders. This is useful when trying to match ladles with particular guns.

As early as 1377 the Tower of London inventory shows six ladles, with the same number given in 1381 and 1388 (Blackmore 1976, 253–4). In 1495 (*ibid.*, 257) there are in all 49 '*Charging ladilles of sundry sorts*' for specified guns, both small wrought iron breech-loaders and larger pieces ('*Bumbardelles*', curtows, demi curtows, serpentynes and falcons). Additionally, under '*sundry stuffes of Ordynaunce*', there are six '*Charging ladilles without stafes for serpentynes*' and four '*Boxes for charging ladilles*'. The same inventory (*ibid.*, 258) includes in substantial quantities '*Double plates of yron for charging ladilles for the great ordinance*' and '*Sengle plates of yron for charging ladilles for the small ordinance*'.

In the extensive 1595 Tower inventory under '*Smythes woorke*' (*ibid.*, 277–8) appear '*Ladles of copper with Staffs & Ramers*' and separately '*Ladles of copper without staves*' (both groups for named guns). Among '*turners worke*' (*ibid.*, 279) are tabulated '*Ladle hedds*', '*Spunge hedds*' and '*Ramer hedds*' for named guns, together with '*Staves of sundry sortes*' distinguished as '*With Sponge heds only*', '*With Ramer hedds only*' and '*with Ramer & sponge hedds*'. One entry is specific as to wood: '*Ladle staves spare, viz. of Ashe viij$^c$lx* [860] *and of firre CCCx* [310]'. By the seventeenth century these were being purchased separately (Brown, pers. comm., 2001).

There is little question of the need for a ladle when loading loose powder through the muzzle. There is question of their use with breech-loading ordnance, where a funnel directly into the chamber would be easier. The use of pre-prepared cartridges, especially on an unstable platform, seems more sensible. Bourne (1587, 30–1) reinforces this view:

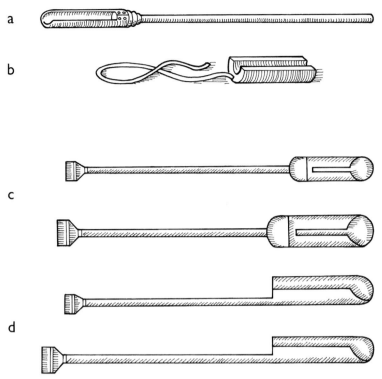

*Figure 5.110 Representations of ladles: a–b) by Tartaglia (Lucar 1588); c–d) Wright (1563)*

*'I do see, that there is no worse lading or charging of Ordnaunce, than with a Ladell, whether that it be by Sea or lande, for by the lading with a Ladell, it muste bee twice filled, and then at euery tyme that the Pouder is putte into the peece, it muste bee put uppe with the Rammer heade, so that they muste eyther turne the other ende of the Ladell, or else if that the Rammer heade bee uppon the spondge staffe, then he muste change the staues, whiche is a greate cumber to doe in a narrowe roome. And also in the chargyng of a peece wyth a Ladell, hee cannot fill it so equally, but that the Ladell shall haue sometyme more Pouder, and sometyme lesse Pouder, by a good qu?titye, and especially if that hee dothe it hastely as in the tyme of seruice it alwayes requireth haste, and that may cause hym that gyueth leuell, to shoote under or ouer the marke ... And also it is unprofitable and daungerous to lade or charge a peece wyth a Ladell, for that the Pouder is apte to bee shedde or spylled beeyng hastyly done ... Wherefore if youre Pouder bee in Cartredges ... the peece is more sooner and easilyer laden or charged.'*

The use of cartridges at sea as early as 1514 negated the need for a measured ladle, unless it was used to fill the cartridge or as a back-up. Detailed historical descriptions appear to be lacking in vital detail, encouraging supposition. As formers are listed in the inventory and are represented within the assemblage, it is likely that cartridges were used. Merely ramming a cartridge can tear the container (especially after swabbing). Biringuccio (1540, 418) following the description of loading loose powder with a ladle, suggests the use of a ladle to deliver a cartridge:

*'Guns are also loaded in another way, which experts call the cartridge method. A tube is made of paper folded two or three times and wrapped on a round piece of wood as long and as thick as you think is required for your gun, or as you wish. This is closed at the foot and filled with all the powder that it can contain. Then it is put in the gun with the said ladle and pushed so hard with the rammer that is bursts and scatters the powder in the gun'.*

Tartaglia (Lucar 1588, appendix, 27) illustrates two different forms of ladle, one for loose loading and one for cartridge loading of chambered pieces (Fig. 5.110a and b). The latter is termed a *'scaffetta'* and resembles a trough with a rope attached to enable it to be withdrawn from the gun following delivery of the cartridge. Although not similar in form, Wright (1563) also shows two forms, open and closed ladles with specific dimensions for named guns. (Fig. 5.110c and d). Both authors show a combined ladle and rammer, although Lucar (1588, 31) states that this is only acceptable for smaller guns, up to demi culverin size.

Only one type of ladle is represented in the *Mary Rose* assemblage: open for nearly half of the circumference of the face of the ladle, similar to examples illustrated by Tartaglia (Lucar 1588; Fig.

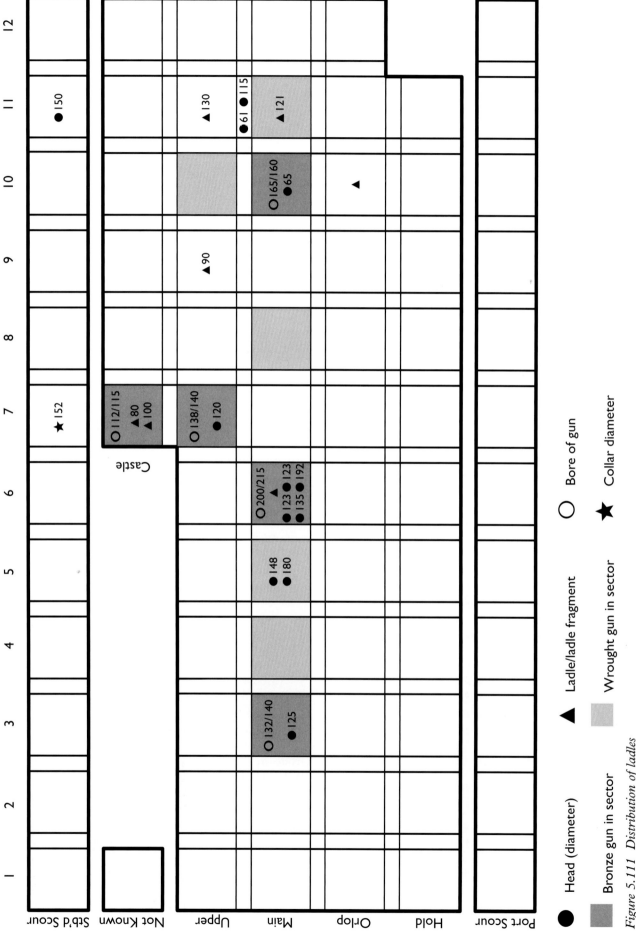

*Figure 5.111* Distribution of ladles

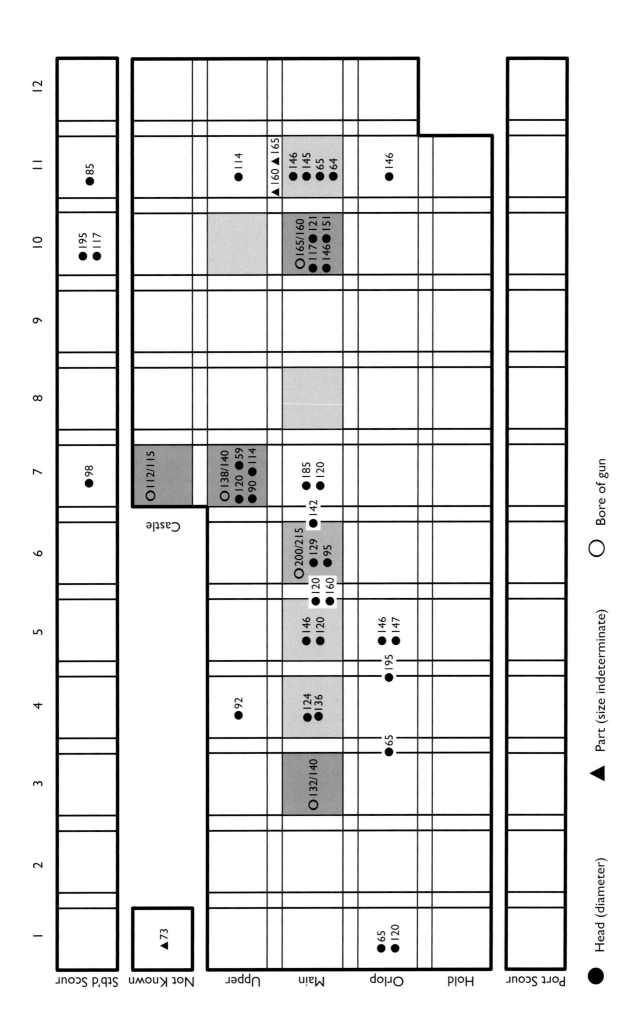

*Figure 5.112 Distribution of rammers*

Table 5.34 Components and materials of *Mary Rose* ladles

| Object | Sector | Handle | Head | Collar | Ladle Length | Diameter head | Length head | Handle hole diam. | Nearest gun/bore size (mm) | Comments |
|---|---|---|---|---|---|---|---|---|---|---|
| 79A1021 | C7 | silver fir | np | np | Cua 470 | 100 | | | demi culverin 79A1232 (121) | beneath gun; handle detached, octagonal, 1420 x 40mm, poss not part of ladle; 2 holes each side for collar attachment |
| 81A0832 | C7 | np | np | np | Cua 408 | 80 | | | close to demi culverin 79A1232 but would fit saker | similar to 81A0868; close to gun; 2 longitudinally aligned holes each side for ladle attachment to collar & 3 to attach ladle to head |
| 79A0931 | U7 | unid. | ash | Cua | Cua* | 120 | 121 | 28 | culverin 80A0976 (138) | found during excavation of guns 79A1232/ 80A0976; discolouration suggests collar overlapped ladle 25mm; hole for handle reinforcement; 50mm wide, collar held to ladle with 2 fixings each side; 3 nails to hold collar to head & 3 to attach ladle; collar bent over face 5mm |
| 81A0868 | U9 | np | np | np | Cua 420 | 90 | | | would fit saker | 1 hole each side for attachment to ladle & 3 to attach to head, 6mm diam.; similar to 81A0832 |
| 81A0241 | U11 | np | np | np | Cua 410 | Max. 130 | | | ?culverin | found when excavating collapsed portside carriage for gun 79A1277; 2 fixing points to collar each side, 6mm; 2 x 3mm holes for head attachment towards middle of ladle |
| 79A0895 | M3 | ash | poplar | Cua | Cu inc. corrosion | 125 | 127 | 30 | 81A1423 (M3) (140) | ladle overlaps collar by 5–18mm; collar width 53mm; ladle held with 2 x 5mm fixings each side; 3 nail-holes for fixings front of collar to face & 2 at back; nail to reinforce handle; 6 holes for nails to retain ladle to head |
| 80A0635 | M5 | np | ash | np | np | 180 | 190 | 44 | 81A3003/79A1276 (M6/M5) (203, 220) | Between collapsed gun carriage & planking; 7mm shoulder for collar; 14 holes for collar & ladle attachment |
| 80A1480 | M5 | unid. | poplar | Cua | Cu 570 | 148 | 135 | 40 | as above | Collar width 75mm, 2 fixings each side; held to head with 5 fixings; 3 fixings for handle to head |
| 80A0695 | M6 | ash | poplar | Cu | Cu incomplete | 123 | 135 | 29 | 80A0976/ 81A3002 (U7/8) (146, 140) | found between deck beams, poss. from U7; collar 72mm, 2 fixings each side for ladle, 3 fixings for collar to head; similiar to 79A0893, 79A0895 |
| 80A0938 | M6 | Ash | np | np | np | | | 35 (handle) | | found in area of rammers & ladles, 817mm long with finished end |

445

| Object | Sector | Handle | Head | Collar | Ladle Length | Diameter head | Length head | Handle hole diam. | Nearest gun/bore size (mm) | Comments |
|---|---|---|---|---|---|---|---|---|---|---|
| 80A0955 | M6 | ash | pyrus or maple | unid. | not analysed 539 | 135 | 135 | 29 | ?culverin | incised 'XI' on head; 210mm of handle survives; 11 holes; width collar 53mm, 4 holes; side holes for ladle staggered |
| 81A1551 | M6 | ash | poplar | lost | Cu 590 | 192 | 200 | 45 | cannon 81A3003 (200) | above gun 81A3003, with rammer 81A1550; collar rebate 75mm, 5mm deep; 6 holes for collar attachment & 2 each side for ladle to collar; 7 holes for ladle to head fixings |
| 81A1711 | M6 | ash | ash | Cua | Cu | 123 | 129 | 28 | culverin? | behind right rear wheel of gun carriage for 81A3003; handle frags 115mm & 140mm length; collar 52mm wide, 4 fixings, bent over face; 1 reinforcing nail for handle |
| 81A2855 | M10 | unid. | poplar | Cua | Cua 320 | 65 | 90 | 22 | ?falcon | same level as *in situ* gun 81A3000; collar width 45mm; 2 x 10mm holes for collar attachment to ladle, 1 for collar to head |
| 81A2212 | M11 | np | np | Cua | Cua 555 | 121 | | | ?culverin | collar width 70mm; ladle attachment of 3 x 5mm diam.; on rojecting tabs; discolouration suggests collar overlapped ladle 20–25mm |
| 80A1328 | M/U7 | np | ash | np | np | 120 | 135 | 29 | 80A0976/81A3002 (U7/8) (146, 140) | collar rebate 60mm |
| 79A1152 | M/U11 | np | ash | not analysed | Cu Incomplete | 115 | 103 | 34 | ?demi culverin | collar rebate 42mm; 2 x 10mm holes to attach to ladle; similar to 79A1211 |
| 79A1211 | M/U11 | pine | poplar | Cua | Cua 315 | 61 | 7??? | 25 | ?falcon | collar width 40mm 2 x 10mm fixings for collar to ladle; collar held with 1 x 5mm fixing; similar to 81A2855 |
| 81A2705 | O10 | np | np | not analysed | | | | | | possible collar, copper strip 160 x 90 x 4mm, no fixing holes |
| 82A1624 | SS07 | np | np | Cua | np | | | | ?demi culverin 79A1277 (M10) (160) | collar only, 152mm diam., 66mm width overlapped by ladle by 25mm; 2 fixing holes (on side present) for ladle; 1 central hole to fix collar to head |
| 82A2674 | SC11 | np | alder | np | np | 150 | 145 | 39 | Demi cannon 81A3000 (178) | outside M11 gunport; brass band around neck; collar rebate 65mm |

np = not present; Cu(a) = copper (alloy)

5.110a) and Wright (1563; Fig. 5.110d) Commonsense suggests that, if for loading chambers, these would have short handles to save space, and be small enough to fit into the neck of the chamber without spillage. Similarly, if used just for filling and delivering cartridges, size would not be quite as critical as if used as a measure required in loose loading. Unfortunately nearly all ladles seem to be incomplete and are often broken just behind the head. The longest single fragment is 80A0955, which retains 210mm of handle (Fig. 5.116). Although the end is not absolutely finished, is does look as though it was smoothed in antiquity.

Most early illustrations depict ladles with rammers on the opposite end of the handle (Biringuccio, Wright, Norton). As late as 1646 Eldred (p. 44) describes the use of the combined tool – '*Ladle and Spunge*'. Lucar, however (1588, 31), suggests that only smaller guns such as falcons, sakers, and small demi culverins can use conjoined rammer/ladles and that larger guns require separate staves. The possible incidence of separate and combined examples, and their relationships with specific guns, is discussed below.

**The scope of the collection**
It is estimated that 21 ladles are represented based on the components recovered (Fig. 5.111). These include thirteen turned wooden heads, sixteen sheets forming the ladles and eleven head collars. (Table 5.34). There are also portions of eleven handles, some held into the head with a small wedge. One example has a 14mm wide metal reinforcing band around the neck (82A2674). Metallurgical analysis shows that the ladles are either virtually pure copper or leaded brass (average 72.6% copper, 24.7% zinc, 1.1% lead; Table 5.35).

**Distribution**
As most ladles and rammers recovered have incomplete handles, the distribution of both ladles with handles (eleven examples) and rammers without handles (fifteen) has been studied (Figs 5.111-12). As can be seen the distributions coincide in only five cases (in most cases of items brought up in different years, making the possibility of their being parts of the same object less likely). Furthermore there are instances of ladles and rammers from the same area each retaining handle fragments. Loose handle fragments cannot be matched with heads on the basis of size, wood species, or the diameter of the handle socket hole in the head.

Several divelogs for the area around the cannon amidships (81A3003) clearly depict a rammer (81A1500) and a ladle (81A1551) lying beside each other above the gun (Fig. 5.113). The ladle has a diameter of 192mm with a poplar head and an ash handle. The rammer has a diameter of 95mm and has an ash head and a poplar handle. This is the only clear incidence of a pair found at the same time which are clearly two separate entities. The likelihood that these items relate to the gun is discussed below.

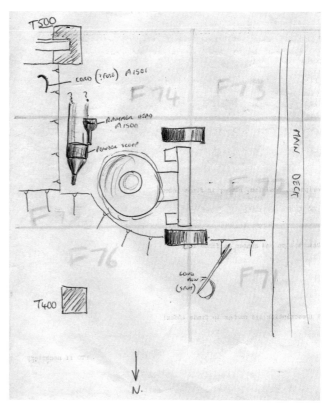

*Figure 5.113 Divelog sketch of ladle and rammer for M6 cannon: ladle 81A1551, rammer 81A1550*

Overall, the distribution of ladles demonstrates clearly an affinity with the bronze muzzle-loaders (Fig. 5.111), with the largest number (ten) on the Main deck, all in areas with *in situ* guns. Seven were found amidships, as were many of the rammers, 'shot gauges' and much shot. The three found between the Main and Upper decks, and could have derived from either deck as there were guns in appropriate positions on both. There was access between the Main, Upper and Castle decks in sectors 6/7 and many of the smaller ladles found in M5/6 could have been for guns situated elsewhere. Three were found on the Upper deck, two close to ports. One was found on the Castle deck close to the *in situ* demi culverin and two outboard in the Starboard scour. There were no ladles found in the Hold, and only one fragment on the Orlop deck suggesting that these objects were not in storage at the time of the sinking.

**Relating ladles to specific guns**
*Historical evidence*
If a ladle is for loose loading or for the preparation of a charge for a cartridge, the amount of powder required is specific to the dimensions of the gun and to the weight of the shot. The size of the ladle and the number of times it is filled, is therefore predetermined by the gun and shot. Wall thickness at the breech is obviously the critical factor when it comes to loading the gun with powder. Guns are therefore described as '*ordinary*' (wall

## Table 5.35 Metal analysis of ladle components

*Analysis of collars (XRFS)*

| Object | Sector | Sb | Sn | Ag | Pb | As | Zn | Cu | Ni | Fe |
|---|---|---|---|---|---|---|---|---|---|---|
| 79A0895 | M3 | 0.09 | 0.02 | 0.04 | 1.03 | Nd | 24.8 | 73.3 | <0.30 | 0.42 |
| 79A1480 | M5 | <0.2 | <0.1 | 0.2 | 1.0 | Nd | 26.0 | 71.0 | 0.4 | 1.4 |
| 80A0695 | M6 | 0.05 | Nd | 0.05 | 0.86 | 0.15 | 23.6 | 74.8 | 0.24 | 0.30 |
| 81A2855 | M10 | <0.1 | 2.0 | <0.1 | 1.4 | <0.1 | 24.0 | 71.5 | 0.2 | 0.7 |
| 81A2212 | M11 | 0.06 | 0.01 | 00.1 | 1.1 | 0.02 | 26.9 | 71.4 | 0.27 | 0.27 |
| 79A1211 | U/M11 | 0.02 | 0.10 | 0.02 | 0.22 | 0.33 | 25.0 | 72.5 | 1.43 | 0.33 |
| 79A0931 | U7 | 0.09 | 0.06 | Nd | 1.56 | Nd | 22.5 | 74.9 | 0.26 | 0.66 |
| 82A1624 | SS07 | 0.29 | 0.05 | Nd | 1.07 | 0.28 | 26.3 | 71.0 | 0.24 | 0.87 |
| 82A2674 band | SC11 | 0.28 | 0.60 | Nd | 1.18 | Nd | 25.1 | 71.5 | 0.23 | 1.18 |
| 81A2855 | O10 | <0.1 | 2.0 | <0.1 | 1.4 | <0.1 | 24.0 | 71.5 | 0.2 | 0.7 |

*Ladle portion*

| Object | Sector | Sb | Sn | Ag | Pb | As | Zn | Cu | Ni | Fe |
|---|---|---|---|---|---|---|---|---|---|---|
| 79A0895 | M3 | 1 | <<1 | <1 | 3 | <1 | <1 | 100 | <1 | 4 |
| 79A1480 | M5 | <<1 | Nd | <<1 | <1 | <<1 | <<1 | 100 | <,1 | 2 |
| 80A0695 | M6 | 1 | <<1 | <<1 | <1 | <1 | <<1 | 100 | <1 | <1 |
| 81A1551 | M6 | .1 | Nd | <<1 | 3<1 | <1 | <<1 | 100 | <1 | 1 |
| 81A1711 | M6 | <1 | Nd | <<1 | <1 | <<1 | <<1 | 100 | <1 | <1 |
| 81A2855 | M10 | <0.01 | Nd | 0.03 | 1.12 | Nd | 24.7 | 73.0 | 0.32 | 0.72 |
| 81A2212 | M11 | 0.04 | 0.33 | 0.02 | 0.95 | 0.06 | 24.9 | 79.9 | 0.30 | 0.50 |
| 79A1152 | U/M11 | <1 | <<1 | <<1 | 1 | <1 | <1 | 100 | <<1 | 1 |
| 79A1211 | U/M11 | 0.1 | Nd | <0.1 | 1.0 | Nd | 25.0 | 72.0 | 0.8 | 0.9 |
| 81A0868 | U9 | 0.2 | <0.1 | 0.1 | 1.4 | 0.1 | 22.5 | 71.0 | 0.1 | 4.0 |
| 81A0241 | U11 | 0.36 | 0.23 | 0.11 | O.76 | 0.51 | 26.1 | 70.8 | 0.11 | 0.95 |
| 79A1021 | C1 | 0.3 | 0.3 | <0.1 | 1.6 | Nd | 25.0 | 69.5 | 0.2 | 3.1 |
| 80A0832 | C1 | 0.13 | 0.07 | 0.03 | 0.98 | Nd | 23.4 | 74.8 | 0.44 | 0.13 |
| 81A2855 | M10 | <0.01 | Nd | 0.03 | 1.12 | Nd | 24.7 | 73.0 | 0.32 | 0.72 |

thickness at breech equal to the diameter of the bore, '*reinforced*' (wall thickness over diameter of bore) or '*lessened*' (wall thickness less than diameter of bore). The size of the ladle has to cater for these differences and so is very much 'gun specific'. This relationship is expressed by historical authors in slightly different ways. At this period, and extending from as early as 1563 (Wright) to 1628 (Norton), the matter is further complicated by the use of either '*serpentine*' or '*corned*' powder (see above for possible definitions); less was required for any gun when corned powder was used. Norton clearly states amounts, but it is obvious that he is working from tables originally produced for serpentine powder with corned powder added as an

Table 5.36 Estimates of ladle sizes in relation to estimates of powder weights required for different gun sizes (after Wright 1563, Bourne 1587 (listing serpentine powder only), and Norton 1628)

| Source | Gun | Bore (in) | Shot diam./ no in lable length (in) | Shot wt (lb) | Wt serp. (lb) | Wt corned (lb) | Ladle length (in/mm)* | Ladle breadth (in/diam.** (mm) | Gun length (ft) | Gun wt (lb) |
|---|---|---|---|---|---|---|---|---|---|---|
| Bourne | cannon | 8¼ | 8/ | 70 | – | – | 24/610 | 15½/203.2 | 12 | 8000 |
| Norton | cannon royal | 8 | 7¾/3 | 63 | 40 | 27 | 24/610 | 14/197 | 12 | 8000 |
| Bourne | cannon | 8 | 7¾ | 64 | 42 | – | 23/584 | 15¼/197 | 11–12 | 7500 |
| Norton | cannon | 7 | 6¾/3.3 | 39 | 25 | 18 | 22/559 | 12¾/171 | 11 | 7000 |
| Norton | demi cannon | 6½ | 6¼/3.4 | 30 | 20 | 14 | 21/533 | 11½/159 | 10 | 6000 |
| Bourne | demi cannon | 5¾ | 5½ | 38 | 25 | – | 23/584 | 12⅕/140 | 11–12 | 5000 |
| Norton | whole culverin | 5 | 4¾ | 15 | 14 | 10 | 19/483 | 9/121 | – | 4300 |
| Bourne | whole culverin | 5 | 4¾ | 15 | 16 | – | 24/610 | 9/121 | – | 4300 |
| Wright | culverin | | 5/4.4 | 18 | 16 | – | 22/559 | 10/127 | – | – |
| Norton | culverin | 5¼ | 5/4 | ? | 15 | 12 | 20/508 | 10/127 | 12 | 4400 |
| Bourne | ordinary culverin | 5¼ | 5/ | 17 | 18 | – | 25/635 | 9/127 | 12 | 4500 |
| Norton | culverin | 5½ | 5¼/4 | 20 | 16 | 15 | 21/533 | 11/133 | – | 4600 |
| Bourne | elder culverin | 5½ | 5¼ | 20 | 20 | – | 23/584 | 10/133 | 12–13 | 4800 |
| Norton | demi culverin | 4¼ | 4/4.2 | 9 | 8 | 6 | 17/431 | 7/102 | – | 2200 |
| Wright | demi culverin | – | 4/6.7 | 12* | 12s | – | 27/685 | 13/102 | – | – |
| Bourne | demi culverin | 4¼ | 4/ | 9 | 10 | – | 20/508 | 7¾/102 | 9–10 | 2200 |
| Norton | demi culverin | 4½ | 4¼ | 10½ | 9 | 7¾ | 18/457 | 8/108 | 11 | 2400 |
| Bourne | demi culverin | 4½ | 4¼ | 10¾ | 11–12 | – | 21¼/540 | 8/108 | 10 | 2700 |
| Norton | demi culverin | 4¾ | 4½ | 12½ | 10 | 8 | 19/423 | 9/114 | – | 2500 |
| Bourne | elder demi culverin | 4¾ | 4½ | 12 | 12 | – | 22/559 | 8/114 | 12 | 3200 |
| Norton | saker | 3½ | 3¼ | 5 | 5 | 4 | 30/762 | 9/82.5 | – | 1400 |
| Bourne | lower saker | 3½ | 3¼ | 4¾ | 5 | – | 15/381 | 5½/82.5 | 8 | 1300–1400 |
| Wright | saker | – | – | 5 | 5 | – | 16/406 | 7/ | – | – |
| Norton | saker | 3¾ | 3½/8.85 | 5¼ | 5¼ | 4¾ | 31/787 | 9½/90 | 9 | 1500 |
| Bourne | saker | 3¾ | 3½/ | 6 | 6 | – | 15½/394 | 6¾/90 | 8 | 1500 |
| Norton | saker | 4 | 3¾/8.53 | 5½ | 5½ | 5½ | 32/813 | 10/95.3 | – | 1600 |
| Bourne | saker | 4 | 3¾/ | 7½ | 7¼ | – | 17/432 | 7¼/95.3 | 10 | 1800 |
| Norton | falcon | 2¾ | 2½/8.8 | 2½ | 2½ | 2 | 22/559 | 5/63.5 | – | 700 |
| Bourne | falcon | 2¾ | 2½ | 2⅛ | 2½ | – | 12/305 | 4¼/63.5 | 7 | 700–750 |
| Wright | falcon | – | – | 1 3oz | 3 | – | 12/305 | 3/ | – | – |
| Norton | base | 1¼ | 1/9.01 | ½ | ½ | ? | 9/229 | 2/25.4 | 3½ | 201 |

*Length of ladle is based on numbers of diameters as listed in sources. ** diameter based on twice the histroical shot diameter converted to metric

extra column (Norton 1628, 53). He does not give new measurements for ladle sizes needed to cater for the smaller amounts. Bourne (1587, 12–13) suggests that the transition to corned powder was relatively recent:

*'And furthermore, now of late yeares, they haue deuised a more stronger sorte of pouder... Nowe whereas they shoote good pouder, or cornepouder, they take much lesse pouder ... to charge a peece w$^t$ cornepowder , or any other good pouder ... for the most parte, thereof two pounde will goe as farre as three pound of Serpētine pouder'.*

Wright (1563) provides lengths and breadths for ladles and formers needed for cartridges for specific guns. The technique he cites for manufacturing the ladles relies on its length and breadth being a set number of times the size of the shot. The ladle has to be filled a specific number of times to equal the weight of the shot.

Bourne explains the rationale behind loose loading (1587, chap. 3). Ladles are designed to be filled twice with serpentine powder to fire a shot of correct size for a particular gun. If the compass (circumference) of the shot is divided into five equal parts, the breadth of the ladle should be equivalent to three parts, the space at the top being left to turn the ladle and spill the powder. This actually suggests a ladle which (if the head is the same size as the shot) is slightly enclosed at the top rather than being equal to only half of the circum-

Table 5.37 Comparison of *Mary Rose* ladle dimensions and historical closest fit with gun class (Bourne 1587)

| Object | Sector | Ladle diam. (mm) | Ladle length (mm) | No. shot diam in length | Bourne gun class |
|---|---|---|---|---|---|
| 79A1211 | M11 | 61 | 315 | 5.16 | falcon |
| 81A2855 | M10 | 65 | 320 | 4.92 | falcon |
| 81A0832 | C1 | 80 | 408 | 5.10 | saker |
| 81A0868 | U9 | 90 | 420 | 4.70 | saker |
| 79A1021 | C1 | 100 | 470 | 4.70 | demi culv. |
| 81A2212 | M11 | 121 | 555 | 4.60 | culverin |
| 81A1711 | M6 | 123 | c.600 | 5.00 | culverin |
| 80A0955 | M6 | 135 | 539 | 3.99 | culverin* |
| 80A1480 | M5 | 148 | 570 | 3.85 | culverin |
| 81A1551 | M6 | 192 | 590 | 3.07 | cannon |
| 81A0241 | U11 | 130 | 410 | 3.20 | culverin (v. short) |

*This from Norton (1628)

ference of the shot, thereby providing a safer option for loose loading, as illustrated slightly earlier by Wright (Fig. 5.111). To get weight in powder equal to the shot, the ladle would have to be nine times the breadth (diameter) of the shot. As this could be quite long, calculations can be made using different multiples, such as four-and-a-half times the diameter of the shot in length, but requiring two fills to equal the weight of the shot. The ratio of serpentine powder required expressed relative to shot weight depends initially on the ratio of the weight of gun to the weight of the shot, both of which must be known when making the ladle. (Table 5.36) gives an indication of the weight of powder and consequent length of ladle needed for guns of different weight/shot weight ratio.

In 1628 Norton includes tables which demonstrate that the amount of powder still depended very much on the thickness of the gun at the breech and therefore the ability to withstand the explosion (1628, 53). The tables list ladle sizes and the number of times they are filled based on whether the gun is '*reinforced*', '*lessened*' or '*ordinary*' (see above). Using information from Wright (1563), Bourne (1587), and Norton (1628), it is possible to suggest the size of ladle needed for different types and weights of guns (Table 5.37).

According to Norton, the '*button*' (head) of the ladle must be of a diameter and height or thickness so that, together with the brass plate (collar) it is equal to the height of the shot, which is generally ¼in less than the bore of the gun. Basically, it has to fit in the bore. Norton states that one of the attributes of a gunner is that he should be able to '*trace, and cut out, and also make up, and finish, all manner of ladles*'. The importance of obtaining the correct powder charge is reinforced:

'*It is not only shamefull, but dangerous also for a Gunner, to imploy an improper Ladle, unless upon necessity, and then also with great care and judgement, being either too long, too shorte, too lowe, or too high for the Peece, for being too high, it will not enter, too low, it will not fill, but spoill, and too long over, and too short under-charge, each of which should bee accounted for an absurd fault*'.

Norton is also specific regarding the positioning of ladles, rammers, sponges and wadhooks (*ibid.*, 74):

'*hee is to place them all on the right side of the Cariages of the Peeces they belong vnto, so that the* Ladles *and* Spunges *be turned towards the mouthes, and the* Rammers *and* Waddhookes *towards the Breech of their proper Peeces*'.

**The *Mary Rose* ladles**

A comparison between the size of the *Mary Rose* ladles and figures given in Table 5.37 (allowing for ¼in windage) shows that, while there is no perfect fit, there are clear groupings which seem to relate to the demi culverin, culverin, demi cannon, and cannon carried (Fig. 5.114). The smaller ladles fit well with Bourne parameters in terms of actual length but there is almost twice the difference in given lengths between the parameters given by the three authors for the smaller ladles Table 5.36). This may well reflect the number of times the ladle is filled rather than the use of either serpentine or corned powder, although it cannot be discounted that different guns used different types of powder. This would, of course, have an effect on the volume required. By the time Eldred was writing, in the 1640s, the larger ladles were filled twice and those for the smaller guns only once. The *Anthony Roll* lists the *Mary Rose* with two sakers and one falcon. The ladles suggest the possibility of two falcons, both in MU10/11 and possibly two sakers on the Upper or Castle deck in the stern (U9/C7), or on a second Castle deck (see Chapter 12).

It is clear from the sizes of some of the ladles recovered from the *Mary Rose* that they were used with the in situ bronze guns in the areas in which they were found. This is clear in M3, M6, U7 and C7 and it seems likely that they were used primarily to load these guns, either loose powder or cartridges. A more restricted distribution might be expected if they were used for the pre-filling of cartridges or breeches. The area in M10/M/U11 included a single former for making cartridges and the smaller ladles found here (81A1211, 81A2855) may be associated with cartridge preparation, although they fall within the historical size parameters for falcons (bore *c.* 70mm; see Table 5.37 and Fig. 5.114) and one would have expected a contained safe area for this activity.

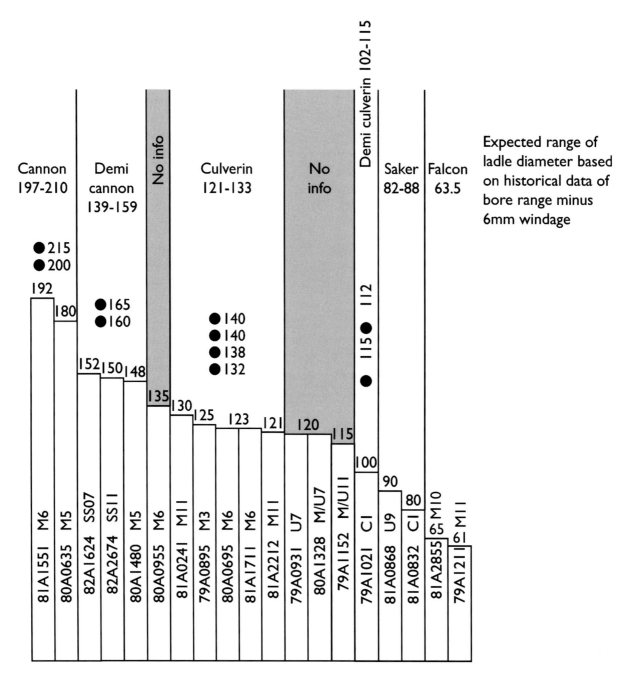

*Figure 5.114 Matching ladle to guns*

The cannon in M6 (81A3003) has a bore of 200mm (*c.* 7in), suggesting a difference of 8mm (³⁄₈in) between the ladle (81A1551) found near it and the bore. This ladle is the 'best fit' for this gun of those found. Clearly the tolerance is very important and this is a good indicator. The associated rammer (81A1500) is smaller than expected and does not even fit with the historical parameters for a sponge head for this type of gun (128mm/5in), so only circumstantial evidence points to this being the rammer for this particular gun. One cannot discount the wrapping of this with sheepskin (also listed on the *Anthony Roll*) for sponging, although there is no sign of any fixing point.

**Method of construction**

The head of the ladle is conical, tapering to form a neck into which the handle is fitted (Fig. 5.115). This is pierced with a hole for the handle insertion. The head is of turned wood, with a rebate for a collar 40–75mm wide (average 61mm) and 4–6mm deep, completely encircling the outer circumference of the head. All but one ladle is made of a sheet which is rounded at the far

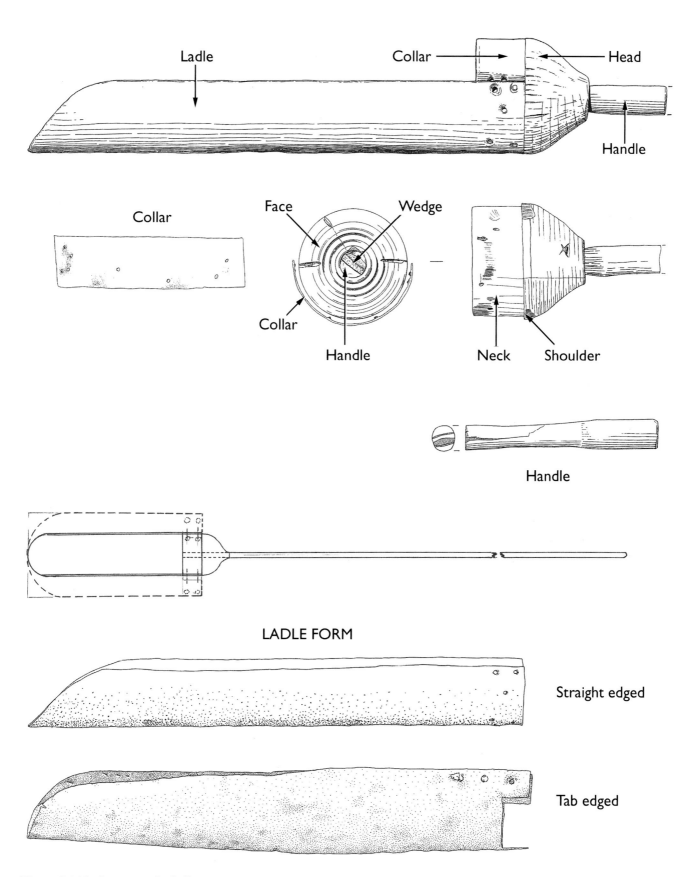

*Figure 5.115 Anatomy of a ladle*

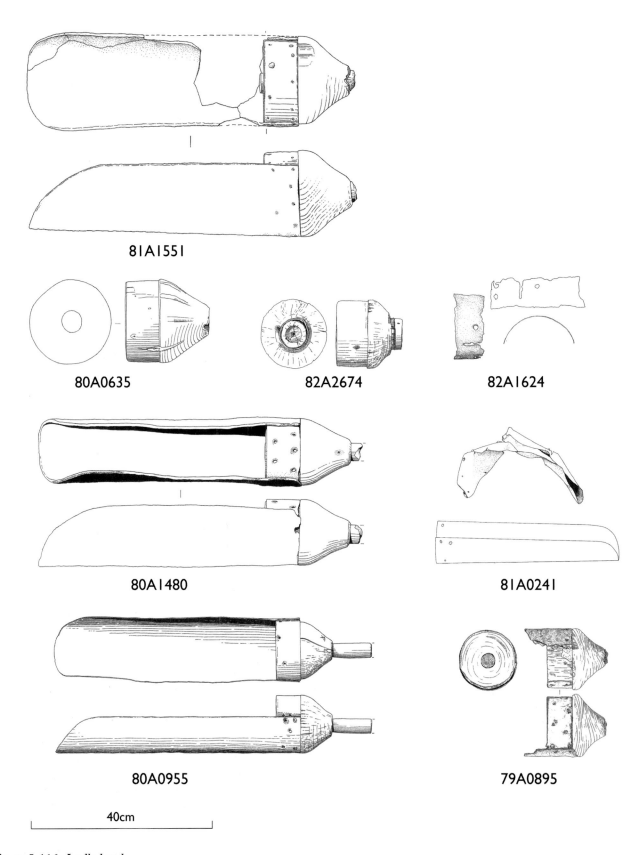

Figure 5.116 Ladle heads

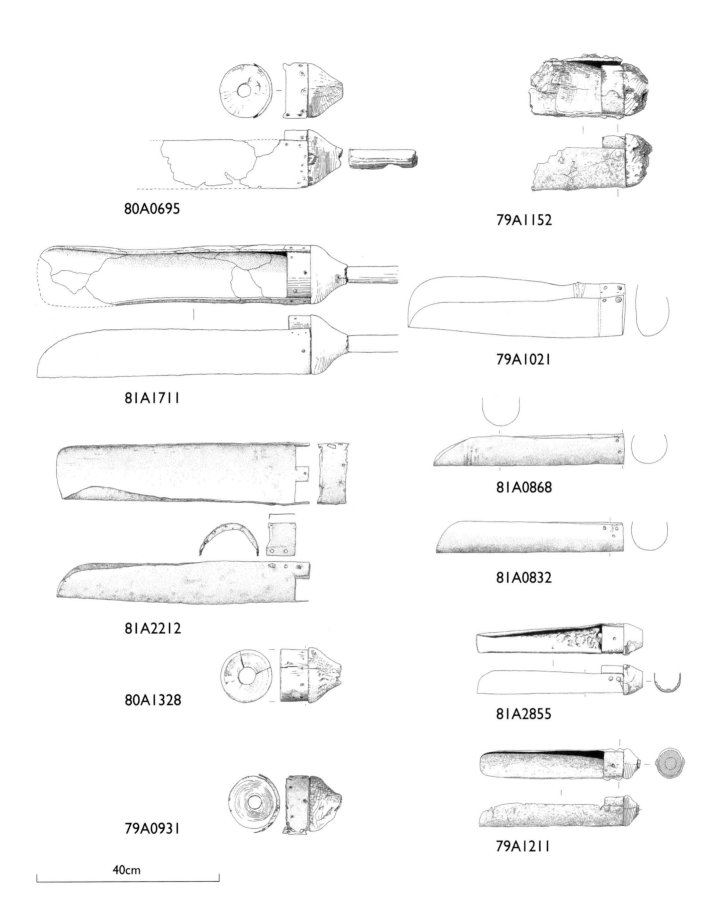

*Figure 5.117 Ladle heads*

end and cut straight at the head end and which overlaps a metal collar. Ladle 81A2212 (Fig. 5.117) has the head end cut to form three tabs, each of which is penetrated with a nail. The collar is rectangular and either encircles the head with five fixing points, or is bent to cover a portion of the face of the head, generally with two fixing points on each side. The ladle and collar are riveted together, in all instances with the holes punched towards the inside. Each side retains two holes, placed longitudinally along the ladle, which are matched on each side of the collar. The head is entered into the circle formed by the ladle and collar. In some cases a small portion of the collar is bent over the face (possibly because the collar is wider than the rebate for it) and the lip hammered.

The collar and ladle are then secured to the head with iron nails. In certain examples, notably 80A0635 (Fig. 5.116), the nails appear to be rectangular in section. There are substantial differences in the number, size and distribution of these fixing points. Most nail-holes appear to be about 5mm in diameter and made from the outside. Those which attach the collar look more like rivets, and are often larger (up to 10mm; 79A1021, 79A1211, 81A2855; Fig. 5.117). The handle is chamfered where it fits into the head and is often held by splitting the end at the face and inserting a wedge. In some instances a few additional nails have been added, spaced equally around the rear of the head. All the handles appear to have broken so a complete length for individual ladles is not possible; each should have been at least 300mm longer than the gun for which it was intended.

Seven of the thirteen ladles are thicker than the others and are degraded to a dark blue, displaying a brittle and crystalline form with distinct laminations. It is possible that this represents a double thickness of original metal. Various studies have been made in an attempt to identify the cause of this degradation (eg, Kendall 1981), the presence of tar or gunpowder being two of the theories examined, but all have proved inconclusive (details in archive).

**Turned ladle heads**
by Robin Wood

Nine powder ladles were studied. All the heads are turned out of roundwood and include a complete cross-section of a small stem of a tree rather than a split section. The diameters range from 61mm to 192mm. Wood of this size could have come from coppice woodland grown on a 10–20 year rotation. The centres are bored through with an auger to take the handle, this hole was centred in the middle of the mass of wood rather than through the pith, in most heads the pith is well off to one side, which again could indicate coppice growth. It most likely that this hole was bored first and then the profile of the head turned. It is possible to bore a hole perfectly through the centre of a turned head but it would require a specialised jig to hold auger and head in alignment. A much simpler method is to bore the hole through the raw log, jam fit it onto a cylindrical mandrel which is turned to the same diameter as the augur, and then fit the whole assemblage to the lathe and turn the profile of the head. A blow from a mallet removes the finished head from the mandrel, which can then be reused indefinitely. This method ensures that the finished head is in perfect alignment with its handle.

The outer surface of the heads is very smoothly finished as would be expected if using a wide flat chisel cutting down the grain. The faces of the heads show the toolmarks more clearly as the turner is cutting across the end grain. Here there is conclusive proof that the heads were turned on reciprocating lathes as it is sometimes possible to see where the tool stopped cutting at the end of a stroke of the lathe. A step is cut into the profile to allow the fitting by numerous nails of the metal part of the ladle. Eight of the nine ladles share a similar profile and similarities with the largest group of parallel-sided rams and could have come from the same workshop (see below). One ladle is very different having a raised lip at the top, which was originally fitted with a metal band.

## Rammers

### Identification and historical significance

Thirty-three turned wooden heads identified as rammer heads were found during the excavation (Fig. 5.118). These are not listed in the *Anthony Roll*. Their function is to accurately position all items put into the gun barrel; specifically ensuring that the tampion, wad and shot are pushed against the charge. They are also used as a tool to investigate whether a bore is straight, tapered, chambered or loaded, merely by pushing it into the bore. This requires a relatively good fit (Norton 1628, 74).

Lucar describes the construction and use of a rammer (1588, appendix, 31):

> '*Make at the ende of a strong staffe (which ought always to be two foote or thereabout longer than the peece for which the rammer shall be made) a round bobbe of wood, and let that bobbe be equal in thicknesse to the heigth of a fit pellet for that peece, and in length so much as once and a halfe the heigth of the same pellet. This bobbe of wood among Gunners is named a rammer because it serueth to thrust home the powder which shal lie loose and dispearsed within a peece, to driue a Tampion close vnto a Cartredge, or to a charge of loose gunpowder within a peece, & to put a pellet close vnto the Tampion*'.

They must be close enough to the bore diameter to ensure that the powder and wad are completely pushed forwards without leaving a residue on the sides and the

*Figure 5.118 Selection of rammers*

handle must be long enough to reach the length of the bore. The text above suggests a size similar to shot, approximately one-twentieth less than the bore, as suggested by the entries in the 1595 Tower inventory (Blackmore 1976, 277–80). As discussed above, it is clear that the combined rammer/ladle was used, and it is in these instances that rammers are listed for specified gun (eg, p. 277: '*Ladles of copper with Staffs & Ramers viz. for Canon viij ladles*'). These are shown in contemporary illustrations (see above, Fig. 5.110), though Lucar states that the larger guns (culverin and larger) require separate items (Lucar 1588, 31).

### The scope of the collection

Only 22 of the 33 identified rammers retain associated handle fragments, most are broken where the handle emerges from the head. Attempts to find handles either solely for rammers or for a conjoined rammer/ladle have been inconclusive (see above). One complete rammer was found, 80A1416. This has a length of 3440mm and appears to be finished. The handle is consistently 50mm in diameter throughout its length and circular in cross-section.

A similar range of woods to those used for ladles is represented, and sizes are similar (compare Table 5.34 with Table 5.38) although twice as many rammers were found. The most obvious reason for this is their potential use for loading chambers for wrought iron breech-loading guns, and possibly the wrapping of the rammer heads with sheepskin to make sponges for use with all guns. Sheepskin has been found, and is listed within the *Anthony Roll*.

### Form

Three head forms have been identified (Fig. 5.119). The classic rammer shape, present on fourteen rammers, has a defined straight edge, or shoulder, leading back from the face for 30–80mm, which then slopes gradually back to meet the handle (eg, 80A1416). The second form (thirteen examples) has a less defined shoulder sloping gradually back from the face to the back (eg, 80A0895), and the third (six examples) is more or less conical without any defined shoulder area (80A1332).

Approximately 50% show nail-holes close to the insertion point of the handle, possibly as additional handle supports. These are not associated with any particular form, nor can they be considered as conclusive evidence to suggest the wrapping and nailing of sheepskin to make sponges, in many cases they are single holes which go through the head and handle (see Table 5.38). In some instances the handles are split and wedged at the face with an oak wedge.

### Rammers or sponges?

There are more rammers than muzzle-loaders. Only one-third are small enough to fit the breech chambers of iron guns (50–120mm, majority below 100mm, Tables 5.39 and 5.40). In some instances it is possible that items classified as rammers may have functioned as sponges, simply wrapped and nailed. A number have nail holes around the neck, but in several instances these have been noted as merely reinforcing nails to add strength to the handle/head junction and corrosion suggests that these are of iron, not the copper advised. Rammer 80A1318 from U7 (Fig. 5.120) fits well with the projected sponge size for a saker (bore 88–95mm). There is nothing small enough for a falcon sponge (estimate 42mm). Table 5.39 lists the historical parameters for accessories together with bore size of recovered guns and diameters of all recovered rammers (and possible sponges) and ladles. Of the ten best candidates, all but one (81A2680) are of conical or semi-conical form. Only two are of poplar, whereas half

456

Table 5.38 Components of *Mary Rose* rammers compared with historical and actual gun sizes

| Object | Sector | Handle | Diam. | Head | Diam. | Length | Handle hole diam. | Gun size (Bourne 1587) | Closest *Mary Rose* gun (bore) | Comment |
|---|---|---|---|---|---|---|---|---|---|---|
| 80A1318 | U7 | np | – | poplar | 59 | 63 | 30 | ?falcon | demi culverin 79A1232, C1, (112); 79A1279 (115) | behind *in situ* culverin 80A0976; conical |
| 81A0663 | M11 | ash | 40 | ash | 64 | 80 | 30 | falcon | none, ?Sterncastle saker | above M11 gunport; handle 388mm surviving |
| 81A1755 | O3/4 | poplar | 35–38 | ash | 65 | 64 | 25 | falcon | none, ?Forecastle saker? | 2 frags handle 215mm & 185mm, wedged above M11 gunport; similar to 81A0531 |
| 81A0662 | M11 | np | – | ash | 65 | 71 | 24 | falcon | none, ?Sterncastle saker | above M11 gunport; similar to 81A0531 |
| 81A0531 | O1 | np | – | ash | 65 | 75 | 22 | Falcon | none, ?storage area for saker | in gun furniture store; similar to 81A0532 |
| 82A4086 | ?SS/SC | ash | 30 | ash | 73 | 80 | 28 | minion | none | handle frag. 224mm; with 82A2673, ?would fit chamber outside M11 gunport but must have fallen from above; |
| 82A2673 | SC11 | poplar | 35 | ash | 85 | 89 | 25 | saker | none, ?Sterncastle saker | handle frag. 257mm; with 82A4086, ?would fit chamber |
| 80A1053 | U7 | poplar * | – | ash | 90 | 100 | 27 | saker | demi culverin 79A1232, C1 (112), Culverin 80A0976, U7 (138), 79A1279 (115) | loose just behind *in situ* culverin 80A0976; conical |
| 81A4066 | U4 | ash* | 30 | poplar | 92 | 111 | 24 | saker | culverin 81A1423, M3 (140) | 8mm nail-hole 23mm below handle insertion for added support |
| 81A1500 | M6 | poplar* | 40 | ash | 95 | 100 | 30 | saker | cannon 81A3003, M6 (200), culverin 80A0976, U7 (138) | lying over cannon 81A3003 beside ladle 81A1551; poss. 1 nail-hole 4mm for handle support, 24mm below neck |
| 81A0889 | SS7 | np | – | ash | 98 | 110 | 22 | demi culverin | demi culverin 79A1279 (115) | outside Sterncastle, poss assoc. with demi culverin 79A1232; 6mm groove extends half way around head, possibly for reinforcing band |
| 81A0944 | U11 | poplar | 40 | ash | 114 | 125 | 38 | demi culverin | none, ?demi culverin Sterncastle | N of spoked wheels on Upper deck; handle in frags, length 1910mm; nail-hole through head |
| 82A2636 | SS10 | unid. * | – | maple | 117 | 122 | 29 | demi culverin | none, ?demi culverin Sterncastle | beneath M10 gunport lid, outboard, with rammer 81A2640 & tampion 82A2639, same size; conical |
| 81A0710 | M10 | np | – | elm | 117 | 105 | 43 | demi culverin | demi cannon 81A3000, M10 (165 demi cannon 79A1277, M10 (160) | beside right rear wheel of demi cannon 81A3000 |
| 80A1355 | M7 | unid. * | – | poplar | 120 | 132 | 29 | culverin | culverin 80A0976, 81A3002 (both 140) | found during excav. of culverins 80A0976 & 81A3002; handle wedged. 3 equally spaced holes round head |
| 81A0532 | O1 | np | – | poplar | 120 | 116 | 36 | culverin | none, ?store for culverin | in storage area for gun carriage elements with rammer 81A0531 |
| 80A0405 | M5/6 | unid.* | 38 | ash | 120 | 114 | 40 | culverin | culverin 80A0976, U7 81A3002, U7 port | close to port side gun carriage; 2 handle frags, 1 wedged; two score lines from neck to face, 2 nail- holes |
| 81A3030 | M10 | np | – | poplar | 121 | 125 | 34 | culverin | demi cannon 81A3000,M10 79A1277 (165, 160) | beside gunport after recovery of demi cannon 81A3000 |
| 81A1019 | M4 | ash* | 32 | poplar | 124 | 110 | 33 | culverin | culverin 81A1423, M3 (140) 79A1278 (132) | close to gun carriage wheel of M4 iron gun; handle frag. 155mm in head, reinforced with nail entered at angle through collar |
| 81A1849 | M6 | poplar | 31 | poplar | 129 | 120 | 44 | culverin | culverin 80A0976, U7 (138), 81A3002, U7 port (140) | in shot behind cannon, found same time as ladle 81A1711 |
| 80A0525 | U7 | poplar | 35 | ash | 130 | 149 | 35 | culverin | as above | just behind culverin 80A0976; handle frags 100mm, 250mm; nail-hole in neck |
| 80A1332 | M4 | unid. | 30 | elm | 136 | 120 | 38 | culverin | culverin 800976, ?81A3002 | just above iron port piece, aligned across gun; 3 handle frags, 95mm, 30mm, 25mm; conical; 1 nail-hole in neck |

| Object | Sector | Handle | Diam. | Head | Diam. | Length | Handle hole diam. | Gun size (Bourne 1587) | Closest Mary Rose gun (bore) | Comment |
|---|---|---|---|---|---|---|---|---|---|---|
| 80A0895 | M6/7 | poplar | 32 | ash | 142 | 150 | 39 | none; demi cannon (158) | cannon 81A3003, 79A1276, M6 (200, 215) | handle frags. 3 equidistant nail-holes 18mm below neck |
| 81A0829 | M11 | np | – | poplar | 145 | 125 | 35 | demi cannon ? | demi cannon 81A3000, 79A1277, M10 (165, 160) | near spoked wheel by gunport; conical; similar to 82A2636 |
| 81A0823 | O11 | unid.* | – | elm | 146 | 135 | 33 | demi cannon | as above | found on border M/011, poss. originally port side |
| 81A0778 | M10 | unid. | 35 | poplar | 146 | 134 | 43 | demi cannon | as above | by right wheel of demi cannon 81A3000; 2 circular score marks round head |
| 80A0141 | M5 | np | – | ash | 146 | 150 | 45 | demi cannon | cannon 81A3003, 79A1276, M6 (200, 215) | near collapsed port side gun carriage for cannon 79A1276 |
| 81A2680 | O5 | np | – | poplar | 147 | 155 | 39 | demi cannon | as above | square 5mm hole for iron fixing near neck |
| 81A2409 | M10 | ash frag. | – | poplar | 151 | 140 | 32 | demi cannon | demi cannon 81A3000, 79A1277, M10 (165, 160) | between collapsed portside carriage and in situ demi cannon; 1 nail, 4mm |
| 80A1416 | M5/6 | unid. | 34–50 | unid. | 160 | 165 | 34 | demi cannon | cannon 81A3003, 79A1276, M6 (200, 215) | head towards bow, handle towards stern; behind M5 iron gun (81A2604) handle complete, 3440mm |
| 80A1324 | M7 | fir * | 35–50 | ash | 185 | 180 | 35 | ?cannon | as above | handle frag. 578mm; 8 staggered holes around head |
| 81A3234 | O4/5 | poplar | 30 | ash | 195 | 195 | 40 | cannon | as above | poss. derived from Main deck. 3 handle frags, 1 wedged; 2 nail-holes, 4mm square |
| 82A2640 | SS10 | np | – | poplar | 195 | 205 | 45 | cannon | none; demi cannons too small | beneath M10 gunport lid, outboard. with rammer 82A2636; 2 nail-holes in neck |

of the shouldered style are made of this wood. It might be expected that a less defined or finished form might suffice for a sponge rather than a rammer.

Norton advocates that a sponge should be either aspen (poplar), birch or willow and be one and two-thirds the size of the shot in length and not above two-thirds the diameter in height, the rest being made up by the sheep skin cover, nailed on with copper nails. As the diameter of the shot for large pieces is often taken as ¼in less than the bore of the gun (windage), it is possible to use this formula to look for potential sponge heads. These should be slightly longer than rammer heads, and approximately two-thirds less than the bores of the extant guns allowing for windage. For example, a cannon such as 81A3003, with a bore of 194mm ($7^5/_8$in allowing for ¼in windage) would require a sponge of 129mm ($5^1/_8$in) diameter (194 x $^2/_3$). Matching archaeological and historical evidence is however, as we have seen elsewhere, not an exact science, large shot, for instance, can be up to an inch small in diameter than the bore, however, calculations have been made on the basis of size location and redundancy and suggest that there are 11–16 sponge candidates amongst the *Mary Rose* assemblage. (Table 5.39, Fig. 5.120). The size match between historical sources and *Mary Rose* candidates is not consistent, so the data is presented for consideration only.

**Distribution**

Rammer distribution is similar to that of ladles, although there are two on the Orlop deck with stored gun carriage extras (81A0531, 81A0532, 01) (Fig. 5.112). Just over half (17 = 51%) were found on the Main deck. The four in M10 (81A0710, 81A0778, 81A2409, 81A3030) and three in M11 (81A0662, 81A0829, 81A0663) suggest the presence of more guns than merely the two muzzle-loading demi cannons recovered. Five were found on the Upper deck, one in U4, three in U7 and one in U11. Three came from the starboard scourpit in the stern, suggesting that at least one of the empty Upper deck gun ports on the starboard side in U10–11 may have had a muzzle-loader (Table 5.38). One is of uncertain location, but most likely from beneath the Sterncastle (82A4086). The fact that so many more heads have survived for rammers than ladles is interesting as there are no reasons for selective preservation of one above the other. This, together with the presence of one former (described above) suggests the use of cartridges in some cases. Rammers would need to be used for both loose powder and cartridge loading but ladles are not a pre-requisite for cartridge loading. Rammers can be used for loading breech chambers, although if using serpentine powder hard ramming is not advised and the distribution of small rammers does not coincide with breech-loading guns, except for the two in M11. The manufacture of cartridges and loading of chambers may have taken place in a specified location away from the guns.

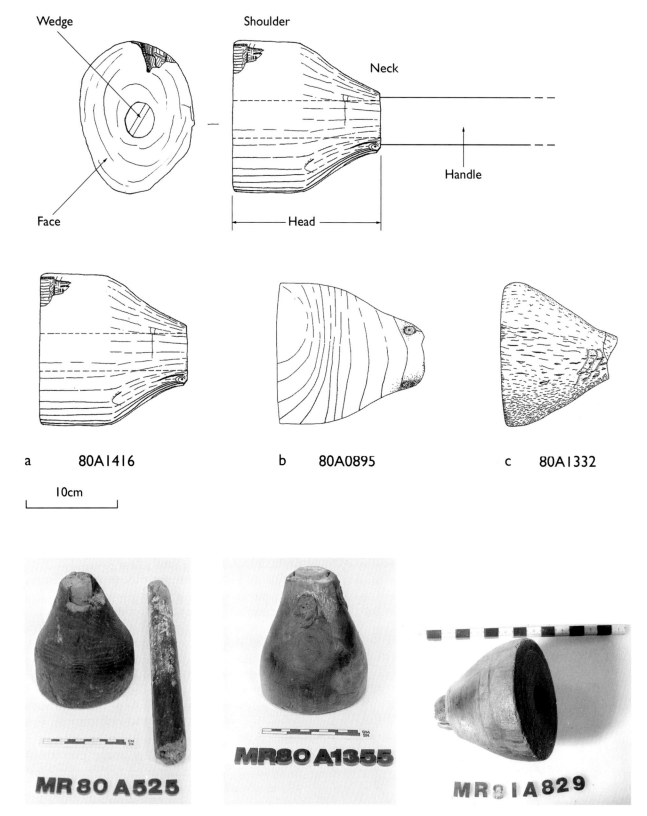

*Figure 5.119 Rammer head forms*

### Relating rammers to specific guns

The assumption must be made that the rammers served (at least) the muzzle-loading guns and that their distribution, as recovered, will at least partly reflect the area of the ship in which they were used. Clearly a rammer or ladle no matter where is it on the ship, will not fit a gun with a smaller bore. As with the ladles, discussed above, the dimensions of the rammers (allowing for ¼in 'windage') have been compared with the evidence from historical authors and the bore sizes

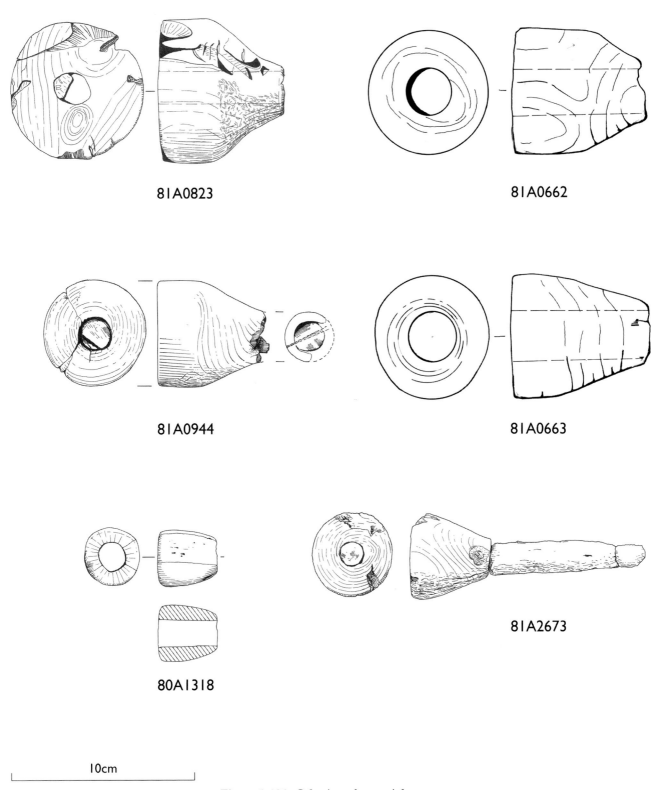

*Figure 5.120 Selection of potential sponges*

of the guns recovered or inferred on the *Mary Rose* (Table 5.39, Fig. 5.122). Any rammers which cannot easily be reconciled with either class of evidence may be considered candidates for sponges, or rammers/spongers for chambers for the breech-loaders. It is indeed unfortunate that the fit of the rammer need not be bore specific, although clearly this was preferred. Ideally each gun should have one rammer, one ladle and one sponger. As many of these guns are from opposite sides of the ship, sharing certain items cannot be completely discounted, although not preferred.

Table 5.39 Comparison between historical referents (Bourne 1587) for head diameters of rammers, ladles and sponges and *Mary Rose* examples (metric)

| Gun | Bore | Ladle head est. diam. | *Mary Rose* ladle | Rammer size | *Mary Rose* Rammers | Sponge size | *Mary Rose* poss. sponges * |
|---|---|---|---|---|---|---|---|
| Cannon | 215 | 203 | 192 (81A1551) | 209 | 195 (82A2640) | 139 | 147 (81A2680) |
| | 200 | 197 | 180 (80A0635) | 194 | 195 (81A3234) | 129 | 146 (81A0778) |
| | | | | | 185 (80A1324) | | 146 (80A0141) |
| | | | | | | | 146 (81A0823) |
| | | | | | | | 142 (80A0895) |
| | | | | | | | 136 (80A1332) |
| Demi cannon | 165 | 146 | 150 (82A2674) | 159 | 160 (80A1416) | 106 | 117 (81A0710) |
| | 160 | | 148 (80A1480) | 155 | 151 (81A2409) | 103 | 117 (82A2636) |
| | | | | | 147 (81A2680) | | 115 (81A0944) |
| | | | | | 146 (80A0141) | | |
| | | | | | 146 (81A0778) | | |
| | | | | | 146 (81A0823) | | |
| | | | | | 145 (81A0829) | | |
| Culverin | 140 | 136–121 | 135 (80A0955) | | 136 (80A1332) | 89 | 98 (81A0899) |
| | 140 | | 125 (79A0895) | | 130 (80A0525) | 89 | 95 (81A1500) |
| | 138 | | 123 (81A1711) | | 129 (81A1849) | 88 | 92 (81A4066) |
| | 132 | | 123 (80A0695) | | 124 81A1019) | 84 | 90 (81A1053) |
| | | | 121 (79A1211) | | 121 (81A3030) | | 85 (82A2673) |
| | | | 120 (79A0931) | | 120 (80A0405) | | |
| | | | 120 (80A1328) | | 120 (81A0532) | | |
| | | | | | 120 (80A1355) | | |
| Demi culverin | 115 | 114–102 | 100 (79A1021) | 109 | 98 (81A0889) | 72 | 73 (82A4076) |
| | 112 | | | 106 | 95 (81A1550) | 70 | |
| | | | | | 92 (81A4066) | | |
| | | | | | 90 (80A1053) | | |
| Saker | 95–88 | | 80 (81A0832) | 89–82 | 85 (82A2673) | 55–54 | 59 (80A1318) |
| Falcon | c. 69 | | 65 (81A2855) | 64 | 65 (81A1755) | 42 | - |
| | | | 61 (79A1211) | | 65 (81A0662) | | |
| | | | | | 65 (81A0531) | | |
| | | | | | 64 (81A0663) | | |

Obvious associations by both size and location are the rammer 80A1332 (Fig. 5.122) of 136mm diameter together with ladle 79A0895 (125mm) with either the in situ culverin (81A1423) or its potential pair on the port side (79A1278), with bores of 140mm and 132mm respectively. Rammer 81A1019 (M4) would also fit a culverin. The larger rammers in O5 and M5–M7, where the largest ladles were also found, can only have been used with the cannons situated amidships. The area contained both ladles and rammers for other guns, in particular for guns of the culverin size 120–136mm. The large number of both rammers and ladles of this size (eight of each) is unexpected as there is not enough shot on board for eight culverins (see Chapter 2 and above) and these are too large for any of the chambers recovered. The distribution of both ladles and rammers in the stern of the Main deck and Starboard scourpit reflects the locations of the *in situ* demi cannons (81A3000, 79A1277).

As Table 5.39 shows, the assemblage more or less caters for the guns recovered. It suggests the possibility of more culverins rather than the demi culverins listed in the *Anthony Roll* and that there may have been be at least one saker and one or two falcons – unless the rammers and ladles found in M10/M11 were for loading cartridges or breech chambers. Table 5.39 and Figure 5.121 show all the rammers and all of the bronze guns and chambers (less windage of bore: one-twentieth), together with the peaks of cast iron shot.

### Method of manufacture
by Robin Wood

Twenty-six ram heads were studied. All are turned out of roundwood and contain a complete cross-section of a small stem of a coppiced woodland tree grown on a 10–20 year rotation. The method and sequence of manufacture seems to have been identical to that used for the ladles (see above). Some of the heads show the

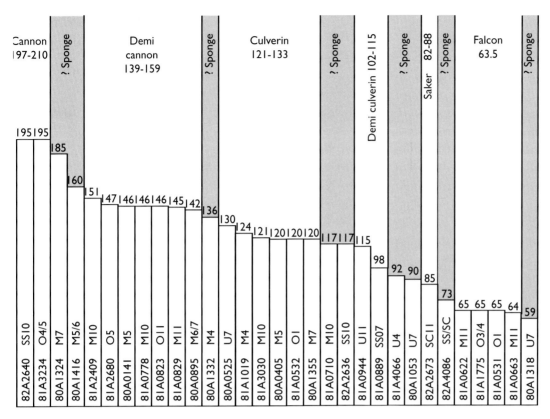

Figure 5.121 Matching rammers to guns

bark surface on part of their circumference indicating that the head was made as large as could be from a given piece of wood. The fact that this creates a slight flat on the circumference confirms that the ram heads did not have to be a tight fit within the bore.

All the handles were made from cleft sections of straight-grained timber. These were carefully shaped and smoothed. Several taper towards both ends with the thickest part near the centre, although this is not obvious in the complete example. They also have a shoulder at the point where the head fits. The tapering profile and the choice of light wood species (Table 5.38) suggests that reducing weight was important. For most tool handles ash would have been chosen because of its elasticity and strength but, here, lightness seems to have been more important. The one ash handle is very slow grown whereas those selected primarily for strength should be fast grown.

A broken section of handle was still in place in five heads and this showed two methods of attachment. In the two larger examples the handles were held securely in place by a large wedge driven into the end. On the smaller rams, handles were still perfectly round, tight fitting and had no wedges. These handles must have been very carefully seasoned before being precisely rounded and fitted to the head, otherwise they would have shrunk across the grain. If the head was only partly dried when fitted, it would shrink on to the handle producing an extremely tight fitting joint. This type of joint is still extensively used by turners in chairmaking today.

Some of the heads have holes where nails had been driven through into the handle, sometimes three small nails, sometimes one large nail which went right through the handle and into the other side of the head. There was no obvious pattern and many heads had no nails, so it is assumed that this represents repair rather than design.

**Stylistic analysis**

The largest stylistic group has a pronounced flat, nearly parallel side for nearly half the height before tapering in a smooth ogee curve (Fig. 5.122). The others are all rather more conical in form, the taper starting almost immediately in a convex curve. This group, however, are less homogeneous than the first and could have been subdivided; some could be described as intermediate forms which start to taper more slowly near the base though are never parallel sided while another two (eg, 81A0829) have the conical form but also a distinctive raised lip at the top as they meet the handle. It is assumed that this is a purely aesthetic difference rather than functional and may reflect the work of two or more different turners or workshops.

Table 5.40 Rammer diameter, gun and chamber bores and cast iron shot peaks

| Gun S (starboard) P (port) side | Bore (less $^1/_{20}$) | *Mary Rose* rammers (all) | Sector | Chambers (all) bore (less $^1/_{20}$) | Sector | Cast iron shot size range | No. shot |
|---|---|---|---|---|---|---|---|
| Cannon          |     | 195 (82A2640) | SS10 |               |       |         |     |
| M6 (P)          | 204 | 195 (81A3234) | O4/5 |               |       | 190–210 | 37  |
| M6 (S)          | 190 | 185 (80A1324) | M7   |               |       |         |     |
|                 |     | 160 (80A1416) | M5/6 |               |       |         |     |
| Demi cannon     |     | 151 (81A2409) | M10  |               |       |         |     |
| M10 (S)         |     | 147 (81A2680) | O5   |               |       |         |     |
| M10 (P)         | 157 | 146 (80A0141) | M5   |               |       | 150–160 | 72  |
|                 | 151 | 146 (81A0778) | M10  |               |       |         |     |
|                 |     | 146 (81A0823) | O11  |               |       |         |     |
|                 |     | 145 (81A0829) | M11  |               |       |         |     |
|                 |     | 142 (80A0895) | M6/7 |               |       |         |     |
|                 |     | 136 (80A1332) | M2   |               |       |         |     |
| Culverin        |     | 130 (80A0525) | U7   | 124 (71A0169) | Stern |         |     |
| M3 (S)          | 133 | 129 (81A1849) | M6   | 114 (82A4081) | U11   |         |     |
| U7 (S)          | 133 | 124 (81A1019) | M4   |               |       | 122–135 | 159 |
| U7 (P)          | 131 | 121 (81A3030) | M10  |               |       |         |     |
| 79A1278 (?)     | 125 | 120 (80A0405) | M5/6 |               |       |         |     |
|                 |     | 120 (81A0532) | O1   |               |       |         |     |
|                 |     | 120 (80A1355) | M7   |               |       |         |     |
|                 |     | 117 (81A0710) | M10  |               |       |         |     |
|                 |     | 117 (82A2636) | SS10 |               |       |         |     |
|                 |     | 114 (81A0944) | U11  |               |       |         |     |
| Demi culverin   |     | 98 (81A0889)  | SS7  | 109 (81A2604) | M5    |         |     |
| 79A1279         |     | 95 (81A1550)  | M6   | 104 (80A2000) | ?     |         |     |
| (?C1)           | 109 | 92 (81A4066)  | U4   | 104 (KPT0017) | ?     | 102–110 | 118 |
|                 | 106 | 90 (80A1053)  | U7   | 95 (90A0001)  | ?     |         |     |
|                 |     |               |      | 95 (82A5176)  | ?     |         |     |
|                 |     |               |      | 95 (82A5172)  | ?     |         |     |
|                 |     |               |      | 95 (82A5097)  | ?     |         |     |
|                 |     |               |      | 95 (81A0772)  | U6    |         |     |
|                 |     |               |      | 91 (85A0028)  | M2    |         |     |
| Saker           | c. 86 | 85 (82A2673) | SC11 | 84 (83A0513) | M4    |         |     |
| Minion (?)      | c. 73 | 73 (82A4086) | SS/SC| 84 (82A0792) | M8    | 79–96   | 558 |
|                 |     |               |      | 84 (81A1084)  | M1    |         |     |
|                 |     |               |      | 81 (81A4890)  | ?     | (peak 83) |   |
|                 |     |               |      | 76 (80A1952)  | ?     |         |     |
|                 |     |               |      | 69 (81A0027)  | ?     |         |     |
|                 |     |               |      | 69 (77A0143)  | 01/02 |         |     |
| Falcon (?)      | c. 65 | 65 (81A1755) | O3/4 | 57 (81A0645) | U6    |         | –   |
|                 |     | 65 (81A0662)  | M11  | 48 (90A0099)  | SS10  | 47–65   | 128 |
|                 |     | 65 (81A0531)  | O1   | 48 (80A2001)  | ?     |         |     |
|                 |     | 64 (81A0663)  | M11  |               |       |         |     |
|                 |     | 59 (80A1318)  | U7   |               |       |         |     |

One ram head (81A2680, Fig. 5.123) has clearly been whittled down by hand with a knife, perhaps to fit a smaller gun. This head also has a poorly fitting whittled section of handle with a strip of organic material, probably leather, wrapped between head and handle, presumably in an attempt to secure a tighter fit.

## Tampions
by Alexzandra Hildred, Robin Wood and Rowena Gale

### Identification and historical significance
Tampions are turned disks of wood, several centimetres thick. At least one was used each time a gun was fired.

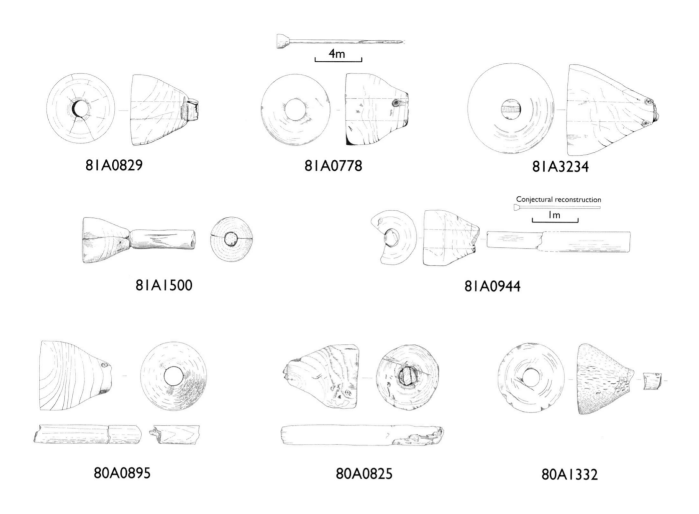

*Figure 5.122 Selection of rammers*

*Figure 5.123 Rammer 81A2680 showing whittling by hand in manufacture*

It was rammed down after the charge to maintain a certain level of compression. The edge of the tampion is bevelled to allow it easier entry to the bore (Fig. 5.124). Tampions were clearly mass-produced items. Examples were found loose on the decks, within the muzzles of the muzzle-loading guns, sealing the chambers of the wrought iron guns and sealing the muzzles of the hailshot pieces. They were also recovered as sets on 'reels', as manufactured (Fig. 5.124b). There is evidence to suggest that some were stored in casks. Four thousand are listed for the *Mary Rose* in the *Anthony Roll*.

Tampions are listed in inventories for ordnance in the Tower of London from as early as 1388. They are described as '*tamppones de ligno*' in great numbers (Blackmore 1976, 254). Blackmore (*ibid.*, 246), following E.F. Tout, suggests that they may serve to separate the powder from the shot within a gun. Lucar (1588, appendix, 31) described a particular rammer '*which serueth to beat downe a Tampion of wood, close vnto a cartredge lying within the chāber of a cānō periero*', both separating the powder from the shot and 'sealing' the chamber. In later years their sole function was to seal

*Figure 5.124 Tampions: a) 'end-on' view of tampion 81A1110 showing bevelled edge; b) reel of tampions 81A0322*

the muzzle, as a form of protection. The *Mary Rose* assemblage demonstrates both the separator of charge and projectile, and the muzzle/chamber sealer.

The chambers for wrought iron guns had the mouth of the chamber sealed with a tampion which separated the charge (in the chamber) from the shot in the back of the gun. Behind the tampion was a wad and then the powder (Fig. 5.125a). In the case of the muzzle-loading hailshot pieces, the rectangular muzzle was sealed with a specifically made poplar tampion. Behind this was the iron dice shot, followed by a wad and, lastly, by the powder (Fig. 5.125b). A similar observation made by those cleaning the guns (under direction of Portsmouth City Museum Conservation Dept) on some of the bronze guns indicates their use in sealing the muzzles. From muzzle to breech the observed sequence was tampion, followed by shot, a second tampion and the powder, but this was not recorded and is by word of mouth. Without physical evidence, we cannot be sure that the second tampion was not a wad. When the practice of sealing muzzles became routine for larger guns is uncertain. The vessel inventory of 1514 (Knighton and Loades 2000) does not list any tampions. That undertaken at the end of Henry's reign contains only seventeen entries for tampions, all ashore (Starkey 1998). Occasionally there is some description (eg, 6005): '*Tampions of woode*'.

The *Anthony Roll* lists over 35,000 tampions carried by all ships though numbers vary considerably from vessel to vessel, as do the guns. The smallest consistent unit appears to be 200, on the rowbarges. A typical rowbarge, the *Rose Slip*, has one muzzle-loader (a saker), five swivel guns (bases) and three hailshot pieces, in addition to handguns. Dividing the number of shot by guns gives 28 rounds for the bases, 70 for the saker, and (approximately, if we take 20 dice per round) 15 for the hailshot piece, making 113 rounds in total.

The 200 tampions covers this and allows for a second tampion in each saker round, should this be the practice. As far as the *Mary Rose* is concerned, the 4000 tampions well cover the 1290 shot listed (not including hailshot or handgun shot). This number can provide two tampions per round for the 370 shot listed for muzzle-loaders, if necessary.

Norton, in advising '*How to loade a Peece of Ordnance Gunner-Like*' (1628, 101), mentions both wad and tampion along with shot, powder, ladle, rammer, wadhook and linstock, but does not explain where or when the tampion is used during loading. However in an earlier passage on '*Gunner-like*' loading of mortars (*ibid.*, 60), he says:

> '*the Chamber is to bee well sponged and cleansed before the putting in of the powder, whether you loade it with loose powder or Cartouch turning the Mouth neere vpright, the powder being so put into the Chaber, ther must be a wad put in either of hay or Okam, & after a Tampkin of Willow or other soft Wood; such as may, together with the powder that was first put in, fully fill vp the whole Chamber thereof, that there may bee no vacuity betweene the powder and wadd, or wadd and shot; after which the shot shall be also put in at the Mouth with a wadd after it; especially if the Peece be not much mounted, least the shot goe out too soone, and the wadd between the Tampkin and the Shot, is not onely to saue the shot from the Tampkins breaking of it, but also to auoide vacuities, which are very dangerous for the Peece by second expansions.*'

This is actually using hay to fill the chamber and the tampion to seal the chamber from the shot within a chambered piece. When excavating breech chambers

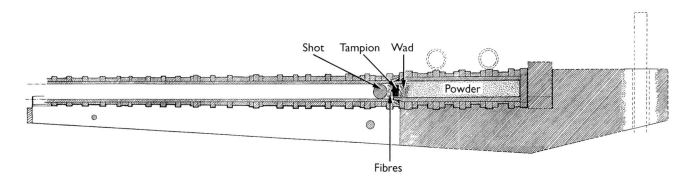

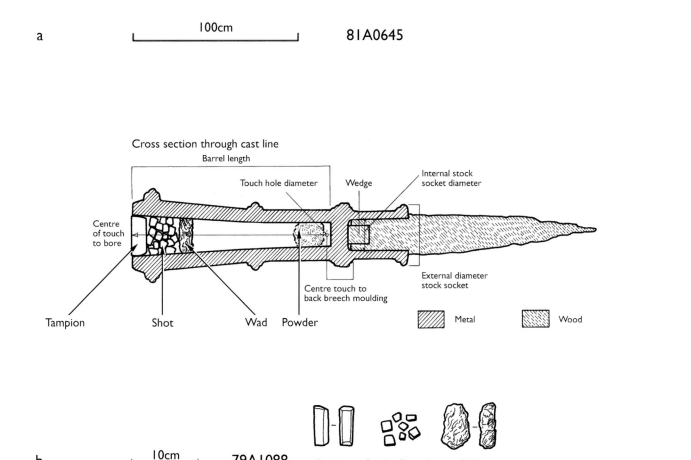

*Figure 5.125 Tampions: a) cross-section of wrought iron sling 81A0645 to show position of the tampion; b) cross-section of hailshot piece 79A1088 showing position of tampion*

the sequence observed was a tampion, followed by a wad, followed by the powder. Both tampions and wads have been found in other archaeological contexts. A number of guns recovered from the East Indiaman *Batavia* (1629) were loaded. Excavation of bores in most instances revealed a bore plug (tampion), a wad, shot and powder (Green 1989, 54).

### Scope of the collection

Fewer than 380 of the 4000 listed tampions were recovered. Other than those found *in situ* within guns or breech chambers (32) tampions were either found loose (157), often individually, or as reels on a 'stalk' (Fig. 5.124b). A total of 57 reels was found, representing 187 tampions, the majority (39) in O1 (Fig. 5.127), which

**Table 5.41 Tampion reels from O1–2/H1 (metric)**

| Object | No. tamps | Wood | Diam. wide end | Diam. narrow end | Thickness |
|---|---|---|---|---|---|
| *Reels in O1 possibly associated with cask 81A0361* | | | | | |
| 81A0321 | 6 | unid. | 62–82 | 50–70 | 22–37 |
| 81A0322 | 1+4 | poplar | 127–158 | 100–113 | 38–47 |
| 81A0352 | 5 | poplar | 68–73 | 48–62 | 20–40 |
| 81A0364 | 3 | unid. | 111–115 | 83–101 | 29–46 |
| 81A0366 | 2 | poplar | 80, 84 | 65 | 27, 37 |
| 81A0367 | 2 | poplar | 83, 85 | 68 | 33 |
| 81A0383 | 2 | poplar | 78, 80 | 63, 67 | 28, 24 |
| 81A0384 | 2 | poplar | 74, 75 | 57, 52 | 21 |
| 81A0385 | 2 | unid. | 76, 79 | 65, 67 | 23 |
| 81A0386 | 2 | unid. | 82, 83 | 65, 66 | 27, 21 |
| 81A0387 | 6 | poplar | 101–108 | 87–99 | 31–39 |
| 81A0388 | 4 | poplar | 74 | 59 | 36 |
| 81A0389 | 4 | unid. | 129 | 110 | 42–60 |
| 81A0390 | 5 | willow | 120–126 | 101–112 | 32–41 |
| 81A0413 | 4 | unid. | 52 | 44 | 23 |
| 81A0414 | 2 | unid. | 67, 75 | 58, 65 | 29, 24 |
| 81A0415 | 2 | unid. | 57 | 48 | 26 |
| 81A0416 | 4 | unid. | 63–68 | 51–58 | 21–25 |
| 81A0417 | 4 | poplar | 75–80 | 67–69 | 23–25 |
| 81A0418 | 6 | unid. | 75–80 | 59–68 | 22–31 |
| 81A0454 | 2 | poplar | 148, 155 | 130 | 31, 34 |
| *Other reels in O1 and H1\** | | | | | |
| 81A0122\* | 5 | unid. | 119–121 | 100–106 | 33–51 |
| 81A0293 | 2 | poplar | 65, 78 | 65, 74 | 27, 28 |
| 81A0308 | 8 | poplar | 77–87 | 70–75 | 21–38 |
| 81A0528 | 4 | unid. | 133 | 129 | 33 |
| 81A0686 | 4 | unid. | 83–87 | 71–72 | 25 |
| 81A0687 | 4 | unid. | 80 | 70 | 24 |
| *Tampion reels in concretion in O1 under cask 81A0750* | | | | | |
| 81A0534 | 2 | unid. | 91, 92 | 73, 79 | 25–32 |
| 81A0535 | 4 | unid. | 116 | 92 | 39–43 |
| 81A0575 | 12 | poplar | 68 | 58 | 24 |
| 81A0577 | 2 | Unid. | 82, 84 | 67, 70 | 22, 27 |
| 81A0579 | 7 | unid. | 81–84 | 66–71 | 25–32 |
| 81A0585 | 2 | poplar | 158 | 80 | 24 |
| 81A0590 | 2 | poplar | 78 | 65 | 27 |
| 81A0621 | 1 | unid. | 77 | 63 | 24 |
| 81A0622 | 4 | unid | 83–93 | 61–78 | 23–25 |
| *Tampion reels in O1 with 'straw' and rope* | | | | | |
| 81A0562 | 7 | poplar | 63 | 56 | 20 |
| 81A0563 | 7 | poplar | 64 | 54 | 23 |
| 81A0564 | 2 | unid. | 125 | 118 | 28, 37 |
| *Individual tampions in O1 possibly associated with cask 81A0361* | | | | | |
| 81A0363, 0365, 0368–82, 0395, 0403–12 | 28 | 13 x poplar; elm; 14 x unid. | 66–135 | 55–130 | 21–50 |
| *Individual tampions in O1 under cask 81A0750* | | | | | |
| 81A0533, 0580, 0586, 0668 | 6 | poplar; 5 x unid. | 68–134 | 55–106 | 27–41 |
| *Other tampions in O1/H1* | | | | | |
| 77A0083, 81A0121, 0183–5, 0523, 0501, 0541, 0544, 0735, 1030, 82A0020 | 10 | 6 x poplar; 4 x unid. | 57–112 | 47–106 | 22–45 |

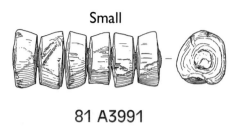

81 A3991

81A0387

81A0122

20cm

*Figure 5.126 Examples of tampion reels*

functioned as a storage area for gun carriage spares and mostly amongst broken casks which may originally have held them. Diameters range from 47mm (1⅞in) to 158mm (6in) and thickness from 11mm (⅜in) to 55mm (2⅛in) (although there is one unusual example of 88mm (3½in). Eleven sizes seem to be represented (cited in inches): 5¾ x 4¾; 5¼ x 4½; 5 x 4¼; 5 x 4; 4¾ x 4¼; 4 x 3½; 3½ x 3; 3¼ x 2¼; 3 x 2½; 2¾ x 2¼; 2½ x 2. Thickness generally increases with diameter, but there are several examples of very thin tampions with large diameters. Most tampions have a tapering profile, with one face wider than the other (10–50mm, with a few smaller). These were designed to be placed in the

Table 5.42 Tampion reels not in storage, by sector (wood unidentified) (metric)

| Sector | Object | No. tamps | Diam. wide end | Diam. narrow end | Thickness |
|---|---|---|---|---|---|
| C8 | 80A0204 | 3 | 57 | 48 | 19 |
| U/C8 | 81A0517 | 3 | 64 | 57 | 20 |
| U8 | 81A0519 | 10 | 66 | 52 | 23 |
| U7 | 80A1137 | 4 | 142 | 114 | 31 |
| U9 | 81A0738 | 4 | 87–110 | 82 | 28 |
| M9 | 81A1215 | 3 | 133 | 110 | 29–32 |
|  | 81A3905 | 5 | 128–140 | no record | 52–56 |
|  | 81A5701 | 5 | 130 | 115 | 25 |
| O2 | 81A0834 | 10* | 70–85 | 60–75 | 25–30 |
|  | 81A4445 | 2 | 75 | 50 | 29, 43 |
| O3 | 80A0829 | 3 | 69, 98 | 62, 69 | 23 |
| O6 | 81A3991 | 6 | 60 | no record | 22 |
|  | 83A0404 | 4 | 75 | 60 | 25 |
| O7 | 80A1781 | 4 | 69 | 54 | 24 |
| H3 | 81A1120 | 4 | 85–110 | 90–95 | 30–35 |
| H4 | 80A0393 | 2 | 120 | 90 | 24 |
| ? | 77A0162 | 5 | 139 | 120 | 50 |

\* poplar

gun/breech with the largest face outwards. Examples of different sizes are shown in Figure 5.126).

**Tampions found on reels**

Tampions were produced in reels, most complete examples suggest six on each reel. They were stored on board in reels and then individual tampions were separated before use. Those still on a reel are usually of very similar size and are arranged with similar sized faces together. The central 'stalk' is 14–33mm in diameter, usually with gaps between the tampions of 12–15mm for easy access in order to break off individual items (Fig. 5.124b; Table 5.41).

**Distribution**

The distribution of tampions is shown in Figure 5.128 and in Tables 5.41–4.

*Tampion reels in store*

Sector O1 and parts of H1 served as store areas for ordnance related items, including truck wheels, spoked wheels, axles and parts of a trailing carriage in addition to rammers and iron shot. Two metal sheets (79A1218,

Table 5.43 Numbers and sizes of loose tampions by sector (all poplar) (metric)

| Sector | No. tamps | Diam. wide end | Diam. narrow end | Thickness |
|---|---|---|---|---|
| Castle | 7 | 62–144 | 50–105 | 33–38 |
| U1 | 1 | 47 | 42 | 11 |
| U6 | 1 | 100 | 62 | 30 |
| U7 | 1 | 65 | 55 | 19 |
| U8 | 5 | 47–144 | 42–125 | 16–43 |
| U9 | 3 | 122–135 | 100–110 | 25–33 |
| U10 | 8 | 75–140 | 65–124 | 27–41 |
| U11 | 2 | no record | no record | no record |
| M1 | 2 | no record | no record | no record |
| M3 | 3 | 120, 160 | ?, 145 | ?, 88 |
| M4 | 7 | 89–145 | 80–130 | 25–45 |
| M5 | 5 | 60–130 | 45–110 | 22–54 |
| M6 | 1 | 90 | 65 | 32 |
| M7 | 1 | 130 | 113 | 35 |
| M8 | 5 | 60–128 | 54–111 | 15–37 |
| M9 | 9 | 53–153 | 50–132 | 20–37 |
| M10 | 13 | 51–145 | 35–125 | 24–51 |
| M10/11 | 1 | 116 | 102 | 32 |
| M/U11 | 1 | 93 | 78 | 29 |
| O2 | 5 | 75–120 | 70–107 | 26–31 |
| O4 | 4 | 70–140 | 55–115 | 25–35 |
| O5 | 6 | 82–130 | 69–110 | 26–45 |
| O7 | 4 | 82–122 | 69–107 | 22–45 |
| O8 | 3 | 57–150 | 51–110 | 22–50 |
| O9 | 1 | 101 | no record | 31 |
| H2 | 1 | 75 | 70 | 27 |
| H3 | 1 | 119 | 97 | 41 |
| H4 | 1 | 75 | 40 | 28 |
| SS1 | 1 | 95 | 70 | 25 |
| SS2 | 1 | no record | no record | no record |
| SS9 | 1 | no record | no record | no record |
| SS10 | 1 | 125 | 100 | 43 |
| ?/offsite | 15 | 51–130 | 48–124 | 22–40 |

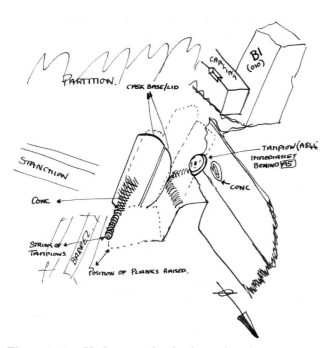

*Figure 5.127 Underwater sketch of tampions in storage in O1*

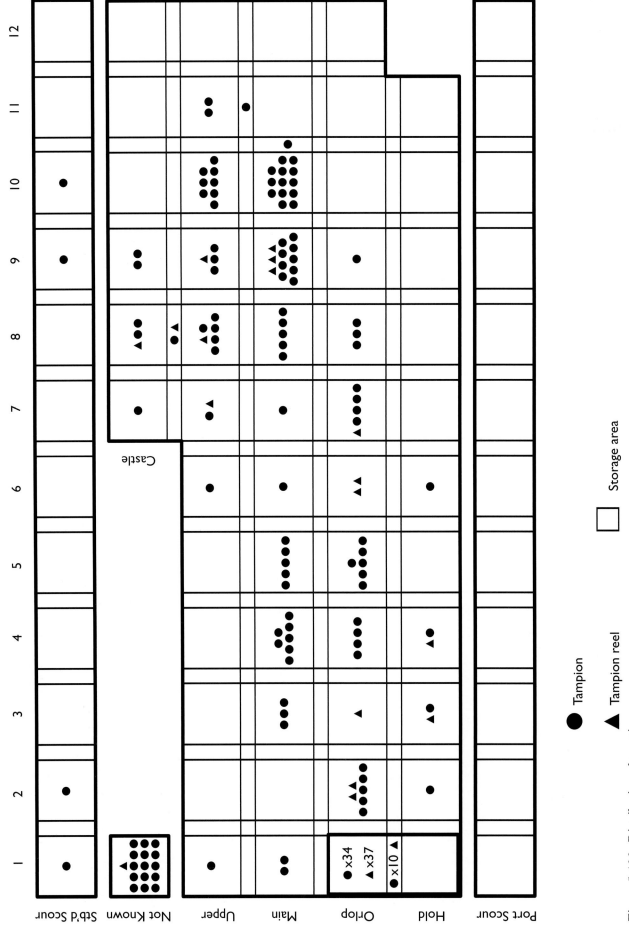

*Figure 5.128 Distribution of tampions*

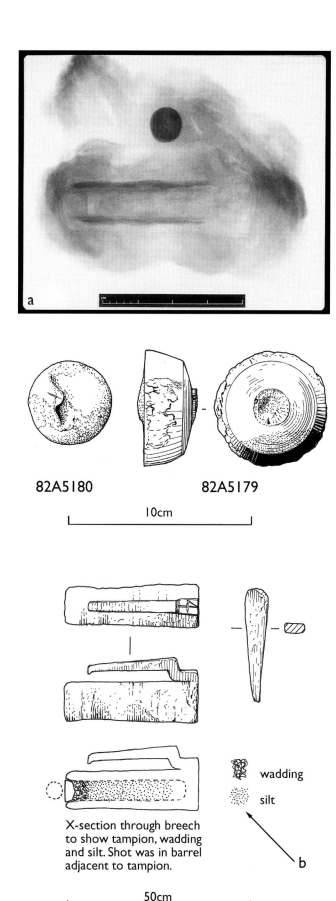

*Figure 5.129 Tampions: a) radiograph of breech swivel (89A0005) with tampion in situ; b) tampion and wad from 82A4076*

*Figure 5.130 Examples of tampions from port pieces: 71A0169 and 81A2604*

81A0478) and two lead patches (77A0152, 81A0295) may have functioned as aprons or touch hole covers for some of the guns. Forty-one tampions were also present, mostly in the southern part of O2, all loose, some of which may have been stored in casks (several dive logs record tampions as being inside casks but this has not been verified). The best evidence for tampions deriving from a cask comes from cask 81A0361 in O1 (Table 5.41). This small cask (capacity *c*. 134 litres; Rodrigues, *AMR* Vol. 4, chap. 10) was incomplete but the archive states that 28 loose tampions and 21 reels were within it. There were fragments of at least three other casks in this area, any of which might also have originally contained tampions. Some of the tampions not directly associated with the casks were found in a mass of straw and rope close to the partition to O2; this may have provided the raw material for wads, both oakum and straw or hay were used (Fig. 5.127).

*Tampion reels and loose tampions found elsewhere*
Seventeen reels were found outside the storage area (Fig. 5.128; Table 5.42). Although a wide range of sizes is represented, there is some patterning. For instance, all the reels on the Orlop deck are less than 98mm in diameter and would be suitable for swivel guns or chambers for wrought iron guns. All three on the Main deck were found outside or within the Carpenters' cabin in M9 and all are 130–140mm in diameter and thus of culverin size. Mixed sizes occur on the Upper deck, as do a variety of guns.

### Table 5.44 Tampions in wrought iron breech-loading guns and cast iron muzzle-loading hailshot pieces

| Gun | Type/element | Tampion | Dimensions | Wad | Shot | Rope | Description |
|---|---|---|---|---|---|---|---|
| 71A0169 | port piece chamber & barrel | yes | D. 100 T. 50 | | yes, stone | | hardly any chamfer; heavy iron staining & splitting; hole where stalk to reel was |
| 78A0246 | gun frag. | yes | no record | | | | gun frag. missing |
| 81A2604 | port piece chamber & barrel | 81A2604/4 | 114 x 106 x 32 | yes | | yes | tampion in excellent condition, poplar |
| 81A2650 | port piece chamber & barrel | 82S1208/1 | no record | 82S1208/1–6 | | yes | sequence on excavation: tampion, wad, rope, sediment, wad |
| 82A5097 | port piece chamber | yes | no record | | | | in wet storage, no details |
| 85A0028 | port piece chamber | 85A0031 | 85 x 85 x 30 | 85A0032 | | yes 'cushion' | |
| 81A0645 | sling chamber & barrel | in chamber 82A4222 | 65 x 55 x 27 | 81A4222 | yes | | against ball, muzzle side |
| 82A2700 | sling chamber | 82A5998 | 80 x 70 x 30 | 82A5999 | | | |
| 81A0722 | sling, chamber neck | yes | no record | | | | in wet storage, no details |
| 90A0125 | fowler chamber | yes | D. 75 | | | | in wet storage, no details |
| 90A0127 | fowler chamber | yes | D. 70-75 | yes | | | neck broken showing tampion & wad; in wet storage |
| 90A0126 | fowler chamber | poss. in concretion | no record | | | | chamber still attached to gun & in wet storage |
| 78A0611 | ?fowler chamber | ?yes | no record | | | | recorded as having tampion |
| 79A1075 | base chamber | 79A1075/5 | 40 x 40 | | | | frags.; rough edges, iron stained |
| 82A5179 | base chamber for 82A4076 | yes | 56 x 47 x 21 | | | | v. rough edges; iron impregnated. Stalk from reel visible on smaller face |
| 79A0543 | base chamber | 79A0543/11 | 43 x 38 x 20 | | | | bulged outwards on smaller face; more iron staining on larger face; poss pine |
| 85A0026 | base chamber | yes | D. c. 47 | | | | shows in radiograph |
| 89A0109 | base chamber | 89A0140 | D. 40 | | | | heavily iron stained |
| 89A0108 | base chamber | 89A0142 | D. 60 | | | | heavily iron stained |
| 89A0007 | base chamber | yes | D. 45 | | | | heavily iron stained |
| 89A0091 | base chamber | yes | D. 45 | | | | |
| 90A0122 | base chamber | yes | no record | | | | in wet storage, no details |
| 81A2640 | loose | yes | no record | yes | | | in concretion close to U11 gun no record port D 100 max. T. 53 |
| 79A1088 | hailshot | 79A1088/5 | 60 x 19 x 21 | 79A1080/4 | 20 iron dice | | rectangular. chamfer of 10mm |
| 81A1080 | hailshot | 81A1080/3 | no record | | | | rectangular |
| 80A0544 | hailshot | 80A0544/3 | no record | 80A0544/4 | yes | | possibly pine |

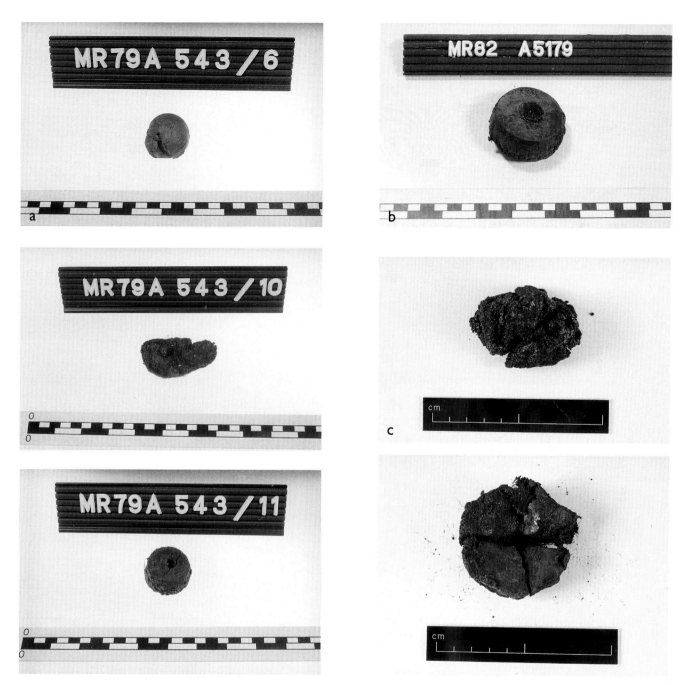

*Figure 5.131 Examples of tampions from swivel guns: a) 79A0543; b) 82A5179 from breech of 82A4076; c) wad and tampion (89A0142) from swivel gun breech 89A0108*

There are 116 records for individual tampions (Fig. 5.128; Table 5.43). Other than stating that the majority were found in trenches with guns or gunports, little can be said. There appear to be more in the trenches with iron guns (M4, M5, M8, U10; the exceptions are M9 and M10) and these seem to have two sizes, large and small. This may relate to a practice whereby as well as sealing the chamber with a tampion, and extra tampion might be placed either in the back of the barrel before the shot, or in the mouth of the bore to seal it against the damp. There is however no archaeological evidence to support this. In only one instance (M9) is a loose tampion of a similar size to a reel. The Carpenters' cabin is an interesting place to find both tampions and reels, as these were brought on board as manufactured goods.

*Tampions found* in situ

Although most of the guns found with shot should have contained tampions, not all are documented, and not all have survived the process of extraction. Available information is summarised in Table 5.44. Iron guns would have had tampions sealing the mouth of the chamber, the bronze guns would have had these in the

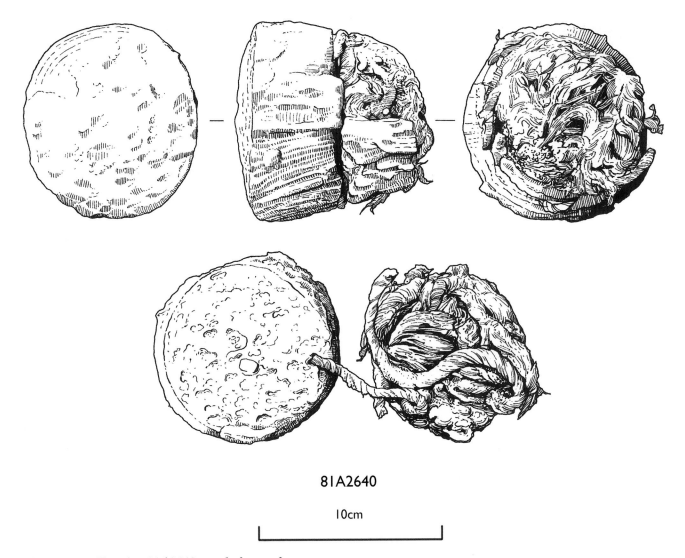

*Figure 5.132 Tampion 81A2640 attached to wad*

bore. If these are situated between the powder and the projectile, they are some way down the gun and are extremely difficult to extract. The processes involved in cleaning the bores often included pressure washing, the fragments resulting would be mixed with the general detritus from the barrel and lost. All of the four cast bronze muzzle-loading and the complete wrought iron guns recovered in the nineteenth century were apparently loaded (see Chapters 3 and 4) but precise data are not available.

There are no extant tampions recorded from the bronze guns, indeed the recording of the excavation of the bores of these is extremely poor. Excavation of the barrels of wrought iron guns is physically easier as there is access from both sides. As many of the iron guns were radiographed for identification and to aid in the removal of concretion, this sometimes revealed the presence of items such as tampions and their precise position within the gun (Figs 5.125 and 5.129). The excavation of port piece chamber 85A0028 revealed a rope cushion, a tampion attached to a wad and then the powder. This also revealed a rope cushion adhering to the outer face of the tampion at the junction between chamber and barrel. Examples from the carriage-mounted guns are shown in Figure 5.130, and from the swivel guns (bases) in Figure 5.131. Examples from swivel guns tend to be extremely thick for their size and often quite angled, depending on the mouth of the chamber. The outward face always appears more finished, with any stalk remains showing predominantly on the unseen, inner face. These are always an extremely close fit, and must have been put in place by being struck with a commander, or mallet.

The three cast iron hailshot pieces from the stern were each loaded with a rectangular tampion followed by small iron dice shot, a wad and then powder. There are several excellent examples of wads attached to tampions, for instance 81A2640, and provides a diameter of 100mm for that chamber (Fig. 5.132). It is extremely iron stained.

*Figure 5.133 Possible tampions 81A1426 and 81A1075*

Two items originally identified as tampions (81A1426 from M9 and the other (81A1075) chipped out of concretion around gun 81A1423 in M3) are turned, tapering flat-faced discs, respectively 150–132mm and 160–145mm in diameter. Each have a lip (6mm and 32mm) in width and 25–30mm in height and then slope inwards for about 37mm to the smaller face. Both appear turned, but did not have any 'stem' residue identifiable in many tampions. Their function is unknown but they may be muzzle tampions (Fig. 5.133).

**Associating tampions with guns**

Associating either reels or loose tampions with guns is difficult. Although those on the Main and Upper decks were generally found in sectors containing guns (Fig. 5.128), few are the correct size for the gun or chamber found in that area. What is clear, however, is that there are very few for the larger guns.

**Manufacture of tampions**
by Robin Wood

Reels were made from short sections of roundwood of various species. If manufacturing thousands, it would be sensible to use coppiced timber, which fits with the hardwood identification of willow and poplar.

The wood was not especially straight grained and many reels have large knots in them. The log was fitted to the lathe with the use of a mandrel which had metal spikes. This was driven into the end of the log and the whole assembly mounted between the centres of the pole lathe ready for turning with the drive strap wound around the mandrel. The outside of the log would be turned to a rough cylinder of the required outer diameter then a series of grooves cut to separate the individual tampions. About 10mm of wood is left in the centre attaching the tampions together, this was normally snapped by hand leaving characteristic torn fibres, although a few separated tampions show that they were cut from the reel with a knife.

The way in which these grooves are cut varies from one reel to the next showing that several different turners were making them. Some complete reels show that the cuts were made in a clean 'V' shape leaving tampions with sloping sides, fatter in the centre. These cuts would have been made with a skew chisel cutting down each side of the 'V' and meeting at a point. Other reels have parallel sides. Here, a parallel-sided groove, about 10mm wide, has been cut out using a gouge or a hook tool; the curved profile of the cutting tool can be clearly seen at the bottom of this cut. Some tampions have a straight cut on one side of the separating groove and a sloping cut on the other. The difference in the way these cuts are made is interesting since the variation has no effect on function, it is just the way each turner chose to work. It is also interesting that such variety was allowed by those commissioning the work, it is difficult

*Figure 5.134 Indications of manufacture on tampion reel 83A0404*

to imagine this degree of variation in a naval order today or even 200 years ago.

Further proof that many turners were working producing reels of tampions is evident in the marks left by the spikes of the mandrels on the tampion at the ends of the reel. They are least distinct on end grain as a result of cracking during drying, however, five mandrel marks are very clear and all have four 'spikes' 7–10mm long by 2–3mm wide. The interesting point is that these five marks were made by five different mandrels and the slots cut by the spikes are all in slightly different orientation to each other. This means that at least five turners were making tampions and probably many more. On the other end of the reels from the mandrel spikes and on some individual tampions we see the characteristic conical hole created by the point centre of the lathe (Fig. 5.134).

As the tampions are spindle turned, and the grain runs through, rather than across, the flat surface, the wood at the edge is very weak, compressing easily when entered into the bore or chamber. A reproduction was made using poplar. A pole lathe was used because the lathe attachment marks on the originals suggest a pole or bow lathe rather than a water powered one. The poplar was rough cut to a prepared section with an axe and then was turned, first to a cylinder, then making the cuts between the pairs (larger faces meeting) and the individual tampions and finally cutting each tampion to a tapered profile. A large flat or 'skew' chisel was used to form the cylinder and then to taper the individual tampions. A small 'hook' tool was used to cut the grooves between the tampions but any small parting tool would have sufficed. The process was easy and quick and it was considered possible to train an apprentice with no previous experience to make them within half a day. The speed of production is estimated one to two minutes for a reel of six or seven tampions of 2in (50mm) diameter.

### Wads

Smith (1627, 86) explains '*Waddings*' as '*Okum, old clouts* [rags], *or straw, put* [in] *after the powder and the Bullet*'; presumably he is describing their use in muzzle-loaders. The *Gentleman's Dictionary* of 1705 is more specific:

> '*Wadd, is a Stopper of Hay or Straw forced into a* Gun *upon the* Powder, *to keep it close in the* Chamber; *when it is home at the* Powder, *the* Gunner *gives it generally three thumps with the* Rammer *Head*'.

These entries are quoted by Blackmore (1976, 248), whose own definition of 'wad' is '*A plug of tow, hay, wood or any other material used to keep the powder and shot in position when charging a gun*'.

The wadhook is a tool for withdrawing the wad when unloading the gun. Wads are generally not listed in inventories as they are 'made up' just before loading by simply grabbing suitable material and scrunching it up. Biringuccio (1540, 418) simply states that:

> '*When the powder has been put in the gun as I have taught you above, put in a wad made with some hay. Press this in firmly with the rammer at the end of the ramrod, and then put in the ball without force … In order that the ball may stay close to the hay and press the powder, strike it with the ramrod with all your force and strength. Thus you will have your gun loaded*'.

Wads are, however, listed by Norton (1628) as being one of the items kept '*in readiness for service*', along with the shot, powder, ladle, rammer, wadhook, linstock and tampions. The loading procedure he describes used the

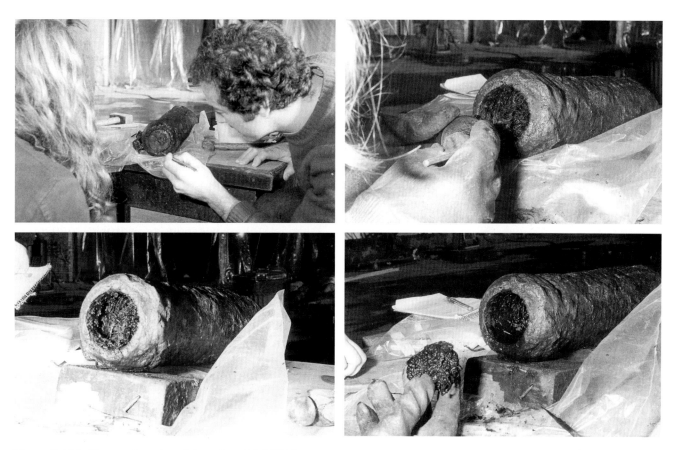

*Figure 5.135 Excavation of wad from base 82A4076*

## Tables 5.45 Wads

| Gun | Wad | Description |
|---|---|---|
| 79A0543 Base chamber | 79A0543/10 | just behind tampion; fibrous |
| 82A4076 Base chamber | 82S1228/1–3 breech 82S1229/1–4 barrel | tampion with fibres from wad adhering to inner face; wad removed, described as 'plug of wadding' 50g; tangle of thin fibres like ball of hair; excavation of barrel yielded composite shot (82A5180) of 43mm; fibres adhering to iron of gun around shot (82S1229/1); wad recovered towards bore end of shot, described as 'tangled cord' (82S1229/3); only recorded instance of wad in front of shot |
| 80A0544 Hailshot | 80A0544/4 | cigar shaped, dark & horse-hair like; 55 x 25 x 15mm; in bore after tampion & shot |
| 80A1080 Hailshot | 80A1080/4 | cigar shaped, light grey, like fine wool; 65 x 25 x 35mm |
| 79A1088 Hailshot | 79A1088/4 | in bore; excavation revealed tampion, 20 iron dice, wad then powder; felted appearance; brown/gold–grey, 70 x 30 x 20mm |
| 82A5172 | 82S1209/1–8 | no tampion visible; neck incomplete; wadding visible at 80mm in bore; removed as complete unit (82S1209/1) following wad; poss. threads noted (S3) |
| 81A2604 Port piece | 82S1106 | 1 litre black wadding recovered from barrel, wad & silt from breech |
| 81A2650 Port piece chamber | 82S1208/1–6 | chamber blocked with tampion 81S1208/1, followed by dark wad (S2) then dark gritty powder (S3/5), progressively wetter towards breech; clots of 'wadding' in (S4) |
| 82A0792 Port piece chamber | 82S1207/1–9 | broken & degraded chamber; mouth incomplete, no tampion; piece of 'unspun yarn' attached to front of chamber (S1); further excavation revealed powder (S3–4); 'wool' remains noted in powder |
| 85A0028 Port piece chamber | 85A0032 | bunch of small twisted rope, 165 x 3–5mm; wt (dry) 262.3g; directly behind tampion |
| 82A2700 Sling chamber | 85A5999 | bunched group of twisted small rope, 150 x 3–5mm = 10 thin twisted fibres |
| 81A0645 Sling chamber | 81A4222 | tobacco like wadding found after tampion; not recorded |

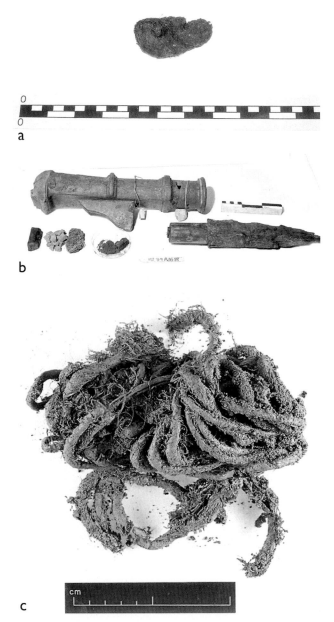

*Figure 5.136 Examples of wads: a) from 79A0543; b) from hailshot piece 79A1088; c) from port piece chamber 85A0028*

'The pouder too loose, and not well put up with the rammer head, and also the wadde too slacke in like manner, will make the shotte to come short of the marke by the meanes of the loosenesse: you must therefore put up the pouder with the rammer head somewhat close, and the wadde to go close in, and driue it home unto the pouder, but beate it not in too hard'.

Only a few wads were recognised during the excavation of chambers, and none was recorded with the excavation of muzzle-loading guns (with the exception of the hailshot pieces, see Chapter 4 and Table 5.45). In only two instances (base 82A4076 and port piece 81A2604) was a wad found in both the chamber and the barrel, possibly functioning as a top wad on the muzzle side of the shot. The excavation of the base chamber (see Fig. 3.115) is shown in Figure 5.135; there is no visual record of the wad from the barrel of 81A2604. As yet no container storing the twisted ropes utilised has been recorded, although loose rope and straw in O1 with other gun furniture is a possible source. Analysis of the fibres is ongoing and many breeches are still in wet storage. Some have radiographs which show tampions, but the definition is not good enough to reveal wads. In two instances wads were found attached to tampions by a single strand, 81A2640 and during the excavation of chamber 85A0028 (Table 5.45). The few recorded examples are listed in Table 5.45 and examples can be seen in Figure 5.136.

### Priming wires

Five copper allow wires have been identified as priming wires (Figs 5.137–8, below). These had two functions: to clean the vent or touch hole of a gun between firings and to prick the paper or linen cartridge containing gunpowder prior to priming the gun. Names such as '*reamers*' or '*vent prickers*' or merely '*prickers*' are common in the literature. A loop or other '*handle*' is required at one end with a sharp point at the other to penetrate the cartridge. A screw thread is often added to help pick up any debris left in the touch hole following firing.

**Inventory and historical evidence**

Priming wires do not occur within the inventories for the Tower of London until 1675 (Blackmore 1976, 316) where they are listed as '*priming irons*', together with rope sponges, powder horns and linstocks Blackmore (*ibid.*, 240) describes priming iron or primer as '*An iron spike or wire to clear the vent and to pierce or open the powder bag or cartridge when it was in the breech ready for firing*'. Their absence from the contemporary inventories suggests that they may have been person-

wad after the powder and before the shot. He states that the wad is rammed onto the powder (softly), then the shot is put in and then another wad of '*hay, grass, weedes, okum or such like*'. Eldred (1646, 43), also describes the use of the wad immediately after the powder '*a good stiffe wad*' which is put on to the powder with several good strokes of the rammer. The shot is then put in, and then the last wad and then given three good strokes with the rammer head.

Bourne (1587, 2) under '*Considerations to be had in shooting of Ordinance*' has a special section entitled '*Of the wadde or the pouder rammed in too hard or too loose*'. The first section deals with bad powder, but the second explains the importance of the wad:

*Figure 5.137 Use of priming wires from the* Siege of Boulogne

al possessions and if, as described above, they were generally made of iron, more may have been lost from the ship through corrosion.

Lucar (1588, appendix, 46) explains the procedure after a gun is loaded with a cartridge:

> 'put a long pricker into the touchhole of the Peece so charged, and with the same pricker pearce diuers holes thorow the cartredge lying within the Peece ... Finally, fill the touchhole of this Peece with good and dry corne gunpowder, and make about the touchhole a little trayne of powder'.

Priming wires and horn containers for priming powder are described by Captain John Smith (1627, 88) as being among the possessions of a naval Gunner: '*A* Horne *is his touch box; his* Primer *is a small long peece of iron, sharpe at the small end, to pierce the Cartrage thorow the toutch hole*'. This application is depicted in the engraving *The Siege of Boulogne* (Fig. 5.137). A figure on the right is placing a wire into the touch hole of a gun whilst the others are demonstrating different parts of the loading process.

**Table 5.46 Priming wires (metric)**

| Object | Sector | Length | Diam. | Head length | Head width | Twist length | Description/analysis (XRFS) |
|---|---|---|---|---|---|---|---|
| 81A1291 | M9, inside chest 81A1429 | 265 | 1 | 10 | 19 | – | single wire; figure-of-eight head, no screw |
| 81A0083 | U9 | 362 | 1–3 | 40 | 35 | 25 | single wire with 2 side elements; heart-shaped head; stem alloy = tin (2.53%), lead (1.47%), zinc (13.9%), copper (81.8%) |
| 81A0835 | M10, inside chest 81A0923 | 433 | 3 | 40 | 26 | 34 | two wires; 1 bent to form shepherd's staff, second woven within to form reef knot; stem alloy: tin (0.10%) lead (1.43%), zinc (24.5%), copper (73%) |
| 79A1011 | SS10 | 236 | 1 | 21 | 28 | 32 | single wire; figure-of-eight head; leaded brass: copper (70.4%), zinc (26.8%), lead (2.01%) |
| 04A0018 | bow, W. of stem | 350 | 4 | 40 | 30 | 16 | single wire; shepherd's staff; not sampled |

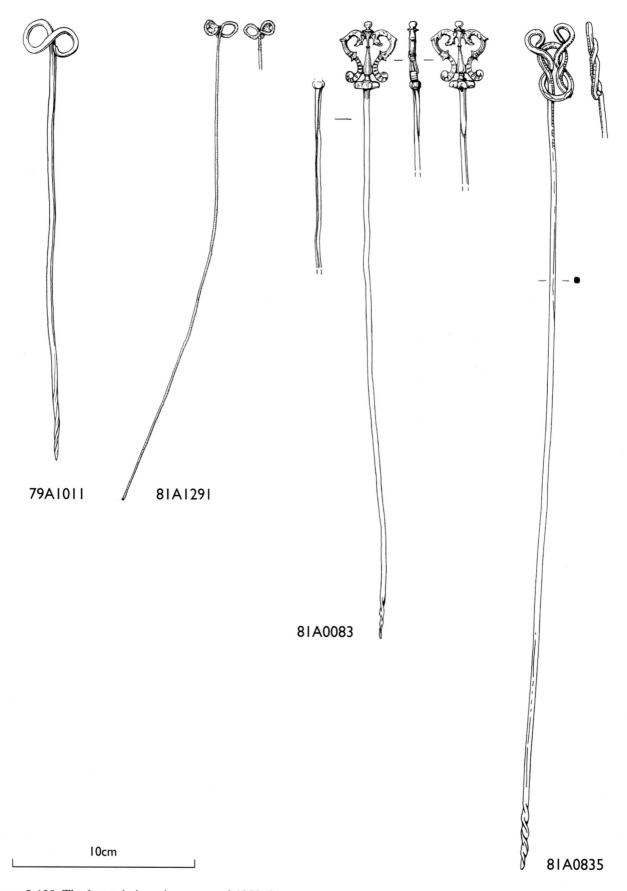

Figure 5.138 The four priming wires recovered 1979–81

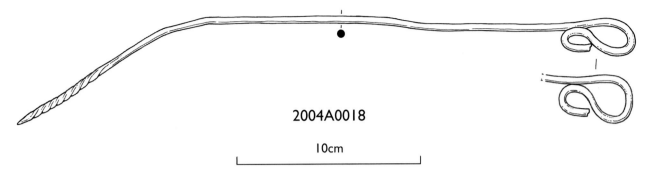

*Figure 5.139 Priming wire found in 2004*

**Description and location**
The five identified wires are 1–3mm in diameter with lengths of 236–433mm (Table 5.46). Three are of simple form, made of one thick wire with the upper end bent to form a figure of eight aligned horizontally (Fig. 5.138, 81A1291, 79A1011) or a simple shepherd's staff (04A0018; Fig. 5.139). Two are more ornate (81A0835, 81A0083; Fig. 5.138); the former with a complex loop resembling a shepherd's staff and the latter with a finely embellished heart design. The heads vary in length from 19mm to 35mm. Four have a simple twist at their lower end and the fifth (81A1291) may be broken. Of the four sampled, three are quaternary alloys of tin, lead, zinc and copper; one is leaded brass (79A1011). Four examples of copper alloy priming wires have been recovered from the sixteenth century *Scheurrak SO1* wreck near Texel (Puype 2001, 119 items 14245 and 90001). These are almost identical to the simple figure of eight forms from the *Mary Rose*.

All except one of the priming wires were found in the stern (Fig. 5.108), two in chests on the Main deck. 81A1291 was found in chest 81A1429, outside the Carpenters' cabin. This sophisticated, lockable oak chest is the most highly decorated chest recovered (see *AMR* Vol 4., chap. 10 and appx 1). It has a carved panel with a shield and a lock-plate. Other contents included a kidney dagger, jewellery, silver coins, a leather book cover, two bosun's calls (silver whistles), woodworking tools and a broken linstock (81A1753). These contents suggest that it was personal, belonging to someone of high status, possibly a Master gunner.

Wire 81A0835 was found in an oak bench chest (81A0923) in M10, together with a pewter plate, two book covers, a knife, linstock (81A0837), saw handle and small wooden box. The sector contained a bronze demi cannon (81A3000) and also the only gunner's rule (81A0249, see above). This priming wire is one of the more ornate examples and the contents of the chest suggest a gun captain, or Master gunner.

The most ornate wire (81A0083) was found on the Upper deck in U9, apparently associated with human remains. This area contained a high number of individuals, many personal possessions including chests and navigation equipment. A powder dispenser used to hold priming powder was also recovered from this area (81A0143). Another (79A1011) was recovered from the Starboard scourpit (SS10); there was a gun inboard of this area on the Upper deck. The final priming wire (04A0018) was found just to the west of the stem timber, close to the reburied port side lower hull structure found in 2004/2005. There were no artefacts associated closely with this object.

The different lengths of the wires may correspond to specific guns, relating to the practice of levelling the gun by means of inserting the wire into the touch hole and marking off the height of the barrel then moving to the front of the gun and levelling using the mark. No other aiming devices have yet been identified, and with the constraints imposed by the height of the gunports, both sighting and access to the muzzle is difficult. This usage is mentioned within contemporary treatises (Bourne 1587, 15):

'To dispart any peece of Ordnaunce truely ... Some also wyll take a priming yron, and put it into the tutchole, and then lay it unto the mouth of the peece, and looke what it commeth unto more than the measure, they will take that for their dispart: but that maye deceive them, as it is generally false.'

*Powder dispensers*

Four ovoid copper alloy lids fitted with funnels have been identified as the top portions of gunpowder dispensers. These could be used for three purposes: to provide either the main propellant or priming charge for a handgun; or to provide the priming charge for a large gun. This action is termed '*priming the gun*' and requires the pouring of a small amount of powder into the vent or touch hole. The containers which delivered the main charge for a handgun often incorporated a measure, usually a spring which operated a cover to allow only a funnel-full of powder through, enough for a single charge. These typically had end caps to seal the funnel and are known by a variety of names, most commonly a powder flask. Dispensers holding priming powder are commonly referred to as priming flasks, touch-boxes or

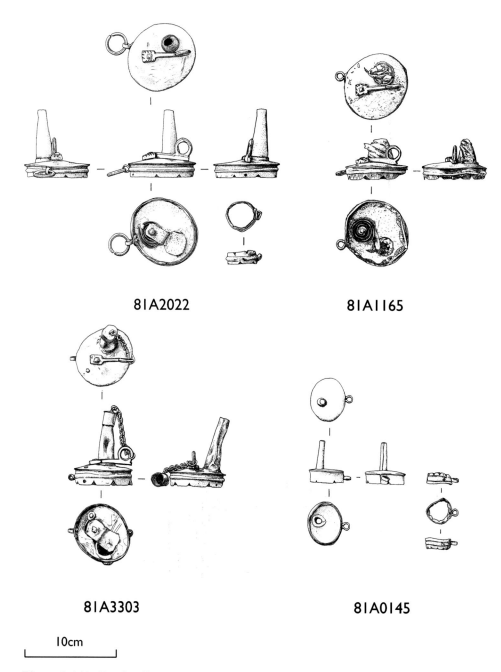

81A2022

81A1165

81A3303

81A0145

10cm

*Figure 5.140 Powder dispensers*

simply touches. The use of animal horn as a conveniently sized container has resulted in the common name '*powder horn*' or simply '*horn*'. Horn was used for both powder and priming flasks but this material has not survived well on the *Mary Rose*. Later examples tended to use the same material at each end – usually copper alloy or wood.

Generally the tops were of copper alloy or iron but the containers could be made from a variety of materials including horn, ivory, leather, (often reinforced with steel strips), wood, or leather covered with wood and bronze. Many were highly decorated, often gilded. Shapes also varied: gourd, triangular, circular, oval, rectangular, fiddle, shield and horn shaped forms can be found. Sixteenth century examples in England exist in the Royal Armouries, the Victoria and Albert Museum, and the Wallace Collection. Both powder and priming flasks have been recovered from the Alderney wreck (Monaghan and Bound 2001, 91–9). These flasks there are trapezoidal in form and are made of wood, in some cases covered with leather or textile with, apparently, square lids.

Most examples incorporate lifting lugs or rings for suspension and were suspended from a belt often called a bandolier.

## Table 5.47 Powder dispensers

| Object/functon | Sector | Description/analysis (XRFS) |
|---|---|---|
| 81A2022<br>Handgun main powder charge or primer large gun | O4 | lid, funnel, suspension loop with ring; lever activated spring, funnel plate; lower band with eye suspension loop<br>*lid*: 72 x 63mm, side with medial ridge 12mm overlapped & welded to lid; 3 nail-holes 2mm diam.; eye suspension loop; lower edge slashed<br>*funnel*: Ht 50mm, D base 15mm, top 10mm, overlapped & welded<br>*lever*: L 53mm, ring Ht 22mm, retaining nail D 3mm<br>*lower band*: D 35mm, Ht 10mm, overlapped & welded with medial ridge; slashed top & bottom with inverted 'V's; suspension loop with larger ring; body, brass; copper (80.8%), zinc (16.8%) |
| 81A3303<br>Handgun main powder charge or primer large gun | M10 | lid, funnel, suspension loop; lever activated spring, funnel plate; 10 link chain from lever to funnel cap<br>*lid*: 63 x 60mm, side with medial ridge 15mm overlapped & welded to lid; nail holes; lower end slashed, inverted 'V's; eye suspension loop<br>*funnel*: Ht 50mm, D base 12mm, top 10mm, overlapped & welded<br>*lever*: L 52mm, ring Ht 22mm, retaining nail D 4mm; chain, 's' shape from lever ring to top of cap<br>*cap*: 12mm width strip overlapped & welded; cap sides welded beneath top, D 10mm<br>*fulcrum*: copper alloy; tin (7.65%), lead (4.63%), zinc (3.27%), copper (83.6%)<br>*top*: copper alloy; tin (1.30%), lead (2.58%), zinc (17.4%), copper (77.7%) |
| 81A0145<br>Handgun primer | U9/10 | lid, funnel, suspension ring<br>*lid*: 40 x 38 mm; straight side 12mm of overlapping strip welded beneath top; no nail-holes or medial ridge; lower end cut with 4 inverted 'V's<br>*funnel*: Ht 32mm, D base 6mm, top 6mm; overlapped & welded; funnel pushed from top of lid through<br>*lower band*: D 30mm, Ht 10mm with eye loop; 2 incised lines around centre; 1 end slashed, 10 inverted 'V's<br>*body*: brass; copper (76.4%), zinc (20.4%) |
| 81A1165<br>Handgun main powder charge or primer large gun | ?Starboard scour | lid, funnel (inc.); suspension loop; lever activated spring<br>*lid*: 72 x 63 mm, side Ht 12mm made from overlapping strip; medial ridge, slashed at lower end, inverted 'V', with decorative marks & the letter 8 retaining holes 5, D 2mm; suspension loop<br>*funnel (inc.)*: Ht 20mm D base 20mm<br>*lever*: L 50mm, ring Ht 18mm, retaining nail D 3mm. Brass; copper (73%), zinc (25%) |

### Inventory and historical evidence

Two types were found on the *Mary Rose* (see below). There are three larger examples (81A2022, 81A1165; 81A3303) that have tapered funnels and a lever-activated spring to access it, and one simpler, smaller example (81A0145), incorporating a straight pouring funnel (Fig. 5.141). One of the larger examples (81A3303) retains a cap to the funnel which is attached by a copper alloy chain to a loop formed within the lever. The other two larger examples also retain the loop suggesting that they may have had funnel caps. This difference in form may relate to function; the larger examples, which had the ability to produce a measured volume, would be used to provide the precise charge for a particular gun (probably a handgun) with the smaller, simpler device being used as a pourer to prime guns or handguns, probably the latter. The use of primers is described in sixteenth century texts such as Lucar (1588, appendix, 1), in Chapter 1, entitled '*The properties, office, and duetie of a Gunner*':

> '*And when a Gunner shall be appoynted to doe an exployte, he ought to want neither a fire stone, nor a tyndar box with a good steele, nor flintstones, nor tindar, nor gunmatches, nor a flaske full of good touchpowder to kindle his gunmatch and fire, when neede shall require.*'

Handguns of this period would normally have a dedicated horn or flask for the main charge, a priming flask or touch-box and a purse for shot, as indicated by the following entry from the 1547 inventory; '*Twoo Italion Peces parcell guilte and varnysshed couered with vellet with flaskes touche boxes and purses*' (Starkey 1998, 8210). Clearly horns were also used with handguns, in this instance for the main propellant (*ibid.*, 8314):

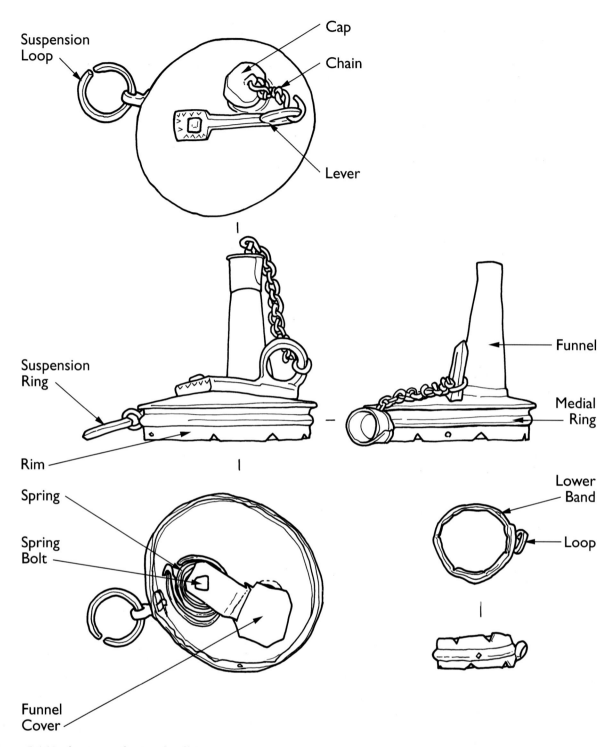

*Figure 5.141 Anatomy of a powder dispenser*

'one Chamber grauen and guilte with A fier Locke and A blacke stocke couered with blacke Lether A purse of blacke vellet A white horne garnisshed with Copper and guilte and A Touche boxe of copper and guilte'.

The *Anthony Roll* does not list dispensers in any form. The section of the 1547 inventory (Starkey 1998) dealing with ships suggests their inclusion with handguns: eg, '*Handgonnes complete*' (7388) and '*Hagbusshes with their apparell*' (7410).

Inventories of items in the Tower of London for 1675 suggest their use with large guns (Blackmore 1976, 316). Serviceable examples include 4431 powder horns, 51 priming irons and 1008 linstocks.

**Description and distribution**
Details of the four powder dispensers are provided in Table 5.47. The four examples (Fig. 5.140) exhibit differing degrees of completeness. The lids are made up of a number of different elements (Fig. 5.141). All include a simple copper alloy ovoid cut from a sheet. This is pierced from the underside and cut to allow the funnel to be entered from below. The upward turn of the sheet has been slashed to allow it to conform to the base of the funnel. The funnel is made from an overlapped sheet of copper alloy. In one instance (81A1165), an extra retaining ring has been placed over the base of the funnel either for extra strength or neatness. This is loosely butted.

The rim is formed from a thin strip of copper alloy soldered to the underside of the flat top. In the three larger dispensers this has a medial ridge. All display triangular cuts at the lower edge with the points uppermost. The larger examples have small concretion-filled holes interspersed with the slashes. These would have held iron retaining pins. They also retain the mechanism for the spring operation. This consists of a bolt going through the lever, lid and spring cover which is bent and soldered to the funnel cover. All appear to be cut from sheet. The rim side (in all four cases) supports a suspension loop made by piercing the rim with a loop made from a clenched strip. In one instance (81A2022) a larger ring pierces the loop.

All of the tops have lipped rims which are cut to enable them to be compressed into a tubular container. Two have additional smaller bands, suggesting attachment to a tapering container, such as a horn. The absence of the containers supports the idea of horn, as little horn survives on the wreck. A brass coloured powder dispenser top to a horn container is shown in the painting *The Conversion of St Paul*, by Peter Breughel (1567). The possibility of wood or of leather reinforced with steel has been considered and discarded, there is no archaeological evidence to support this.

The lower grip band for 81A2022 is slashed at both top and bottom, and bears a medial ridge. This suggests that it formed a joining portion between two other elements, both pushed over the band, thus leaving only the medial ridge visible. It is possible that the tips were of another, more perishable, material such as silver. This band retains a loop, presumably for an attachment cord. In contrast, the lower band of 81A0145 (the simple funnel) is only slashed at one edge of the strip and instead of a medial ridge it displays two incised lines. This may have just been pushed over the end of the 'horn' rather than providing a joining portion.

The probable method of operation is to invert the flask (first capping the end with the thumb), then pushing the ring on the lever. This activates the spring beneath and moves the funnel cover over the hole, sealing a small portion of powder within the funnel. This would then be turned upright and the cap replaced keeping the powder in the funnel ready for use. On

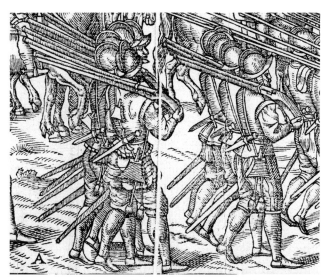

*Figure 5.142 Powder dispensers being carried by troops in engraving* English Troops on the March *in* The Image of Ireland *by John Derricke (1581) (Edinburgh University Library De.3.76, reproduced by permission)*

81A3303 the funnel cap consists of a strip of copper alloy overlapped and soldered. This is soldered to the underneath of a copper alloy circle and the seam reinforced with a small split ring, which accepts the end of a copper alloy chain originating on the ring incorporated within the lever.

Examples showing both types represented on the *Mary Rose* can be seen in the engraving dated 1581 entitled *English Troops on the March* in *The Image of Ireland* by John Derricke. This depicts harquebusiers carrying large powder dispensers slung from a diagonal band over the right hip and in several instances a smaller priming flask slung across the back (Fig. 5.142).

81A2022 was found in O4 (Fig. 5.108), resting on the right front segment of a leather jerkin (81A1963) surrounding a ribcage (Fairly Complete Skeleton (FCS) 16), in the south-western portion of the trench. An impression of a belt was found around the waisted portion of the jerkin. A sundial, thimble ring, comb, shoe, rope and copper alloy aiglets and parts of a second skeleton (FCS14) were also found in close proximity (*AMR* Vol. 4, chap. 2).

Another large dispenser (81A3033) was within the mass of objects surrounding bronze demi cannon (81A3000) and carriage on the Main deck in the stern. Two linstocks and the gunner's rule (81A0249) were also found close to this gun (81A0914, 81A2864) with a third linstock within the chest nearby (81A0837) which also held a priming wire (see above). 81A1165 is from an unknown location.

The smaller dispenser (81A0145) was found among a jumble of personal chests, objects (including a priming wire) and human remains in U9. There is no large gun in this area, however the partial remains of two handguns were recovered (81A2405, 81A4554). It is probable that this was a primer for handguns.

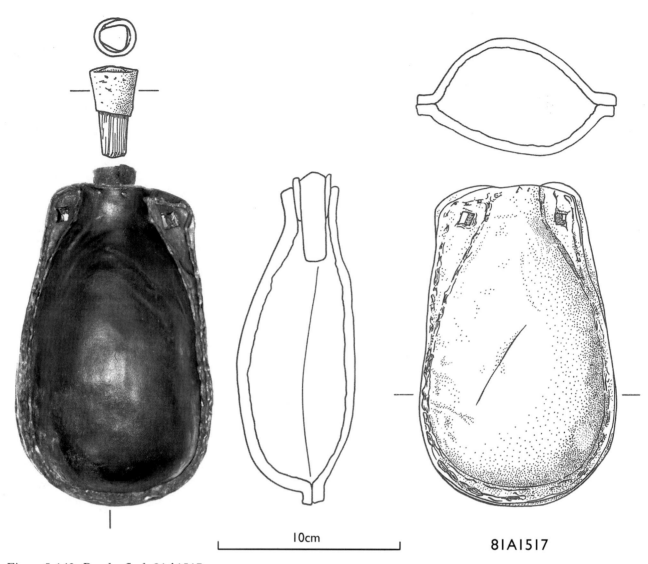

*Figure 5.143   Powder flask 81A1517*

The form of the three larger ones suggest that they were for dispensing a measured charge, such as would be needed for a handgun. However, the location of those in O4 (via ladder to M4) and M10 with respect to large guns means that we cannot rule out the possibility that they were used to prime these guns, especially as later patterns of cannon priming horn are found with springs for measured charges (eg, RA XIII.107; Blackmore 1976, 182). A similar question is posed regarding several of the Alderney examples (Monaghan and Bound 2001, 91).

*Powder flask*
by Alexzandra Hildred

One leather flask has been tentatively identified as a powder flask (81A1517; Fig. 5.143). It has a length of 170mm a width of 100mm. It retains a stopper and has two holes cut to accept a cord. These are 7mm square and have been made with a knife. It is made from two pieces of leather, 4.4mm thick at the base, with the flesh side inwards and has no waterproofing pitch lining. It was probably made of boiled leather either placed whilst damp over a wooden or sand former and allowed to dry slowly to retain its form and then saddle-stitched along each side, or stiched and then filled with sand and allowed to dry. The stitch-holes are 6mm apart. The stopper (46 x 16.9mm at top, narrowing to 12.5mm) is formed of a rough wood dowel axed so that it is faceted. This has been wrapped with a piece of leather which is overlapped and chamfered. It is undecorated. It was found on the Orlop deck in the bow (O2). It shares some similarities with other flasks recovered, notably 81A2218 and 79A1213 (see *AMR* Vol. 4, 452–6). A companionway in this sector led to the gun deck above. Residue on the surface of the flask has been identified by D. Adamson (University of Portsmouth 1985) as gunpowder (83S423 1–2), although there appears to be no archive record of this work.

**Top of a powder horn**

One object has been recently identified as the top of a powder horn. It was on display in the Mary Rose Museum as a mystery item and identified by a visitor as the top of a powder horn, although incorrectly assembled. 81A2896 (Fig. 5.144), consists of a roughly circular elm disc, 93mm in diameter but wedge-shaped in profile with a thickness ranging from c.16 mm to 36mm. It has a chamfered edge and an approximately square hole 18mm across in the centre. The hole is plugged by a wooden stopper, made of ash, one end of which is carved into four ridges and tapers to form a flat-topped pyramid, as does the other. There is a drilled rectangular hole through the stopper between the second and third ridges. The other edge of the disc is punctuated by twelve more-or-less regularly spaced copper alloy nails or pins, with two additional pin-holes containing concretion. These would secure the top to the horn container. The stopper would have a string through the top, provided it was turned around. This item, together with three similar objects, is listed as a unidentified form of bung or stopper (*AMR* Vol. 4, 371–2, fig. 8.67). The description of a crudely fitting bung or lid to a container from which the contents could be poured through the hole in the centre on removal of the stopper is perfectly suited to this new identity.

All four of the objects are circular or ovoid, all imperfect, suited to individual horns. 81A2852 is poplar, wedge-shaped in profile (16–36mm thick) with chamfered edges (three holes evident) and is oval in outline (c. 80 x 72 on its upper face) with a square central hole 18mm across. It has a plain flat-topped pyramid stopper with a drilled rectangular hole. 81A1448 is also oval (72 x 64mm) with chamfered edges with three small holes in the circumference and a rectangular hole in the centre. 81A3257 has a diameter of 70mm and also retains a pyramid shaped stopper through a central hole. In addition there is 81A2301, a stopper similar to that of 81A2852. 81A2852, 81A2896 and 81A1448 were found in the Carpenters' cabin (M9) and 81A3257 and 81A2301 were found in O4. All of these objects should be considered as possible tops for powder horns.

## Linstocks

> '... *with linstock now the devilish cannon touches, and down goes all before them*'
> (Shakespeare, *Henry V*, III, Prologue)

A linstock is a wooden pole to which a smouldering match is fixed. The pole enables the gunner to stand at twice his arm's length in order to bring the match to the touch hole of a gun. This ignites the priming powder which, in turn, ignites the main powder charge and fires the gun. Illustrations demonstrating linstocks in use

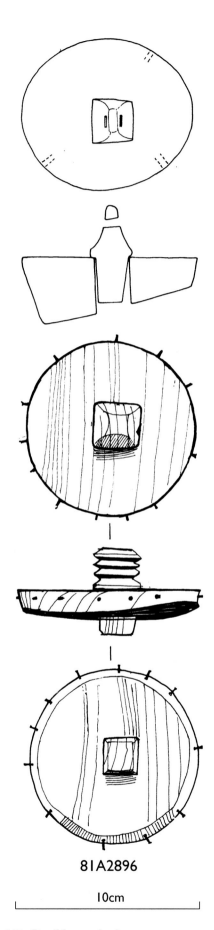

*Figure 5.144 Possible powder horn top*

include the Cowdray Engraving, an eighteenth century copy of a contemporary painting which depicted the sinking of the *Mary Rose* (Fig. 5.145, top) and *The Embarkation of Henry VIII at Dover*, dated c. 1545 (Fig. 5.145, bottom).

**Inventory and historical evidence**
The first reference to linstocks within the gun furniture held at the Tower of London is from the 1675 inventory (Blackmore 1976, 233). Linstocks are listed under the heading '*Habilliaments of War*' along with rope sponges, powder horns and priming irons (*ibid.*, 316). The earliest inventories, of 1388, 1396, 1399 and 1405, list '*toucha*' or '*touches*' '*de ferr*o' (*ibid.*, 246, 254–6). These may be heated iron rods for lighting the guns.

Sixteenth century references include the *Tartaglia* of Cyprian Lucar (1588, appendix, 1) where the '*properties, office and duetie of a Gunner*' include the recommendation:

> '*Also a Gunner ought alwayes to haue a gunners staffe, or a partisant, or a halbert sticking by him for a part of his defence, and he ought to put into the cocke of his Gunners staffe a gunmatch, or wrappe about the lower end of his staffe, partisant and halbert a good gunmatch, which may geue fire vnto his peeces of artillerie when neede shall require.*'

Descriptions of linstocks used at sea include Smith (1627, 88):

> '[The Gunner's] Lint stock *is a handsome carved stick, more than halfe a yard long, with a Cocke at the one end to hold fast his Match, and a sharp pike* [spike] *in the other to sticke it fast upon the Deck or platforme upright*'.

There are a number of recipes for making the match, but basically they are saltpetre or gunpowder impregnated cords. Lucar (1588, appendix, 20) gives the following method:

> '*Make small ropes or cordes of bumbas, or of cotton wooll, put the same into an earthen pot or pan which must haue in it so much strong vineger or rather* aqua vitae, *brimstone, and saltpeeter, or in steade of saltpeeter, grosse gunpowder, mingled together, as will couer the same ropes, and seeth all those things together in the same pot ouer a fire vntil the* aqua vitae, *brimstone, and saltpeeter, or grosse gunpowder shall waxe thicke and incorporate, and then pull the same ropes well soked in that composition one after another out of the pot, and hang one of them from another vppon a pole to dry in the sunne, so as when they are thorow dry you may winde or role them vp for gunmatches to geue fire vnto great & smal peeces of artillerie, mynes, trunkes, pikes, dartes, arrowes, pottes, hollow pellettes, and all other firewoorkes.*'

*Figure 5.145 Use of linstocks from (top) the Cowdray Engraving and (bottom)* The Embarkation of Henry VIII at Dover

An illustrative contemporary reference (Biringuccio 1540, 420) to loading and shooting guns describes the use of the linstock thus: '*Then they fill the touch-hole with fine powder and ignite the fire with a pole that has a piece of lighted rope on the end.*'

Inventory entries use both '*lint*' and '*match*' suggesting that these terms are one and the same. Specific entries in the 1547 inventory (Starkey 1998) list them together, eg '*Lyntes or matchies*' (3761), '*Lyntes and Matches*' (5561), '*Matches alias Lyntes*' (5970, 6053, 6144). Sometimes a quantity is expressed, eg '*Matches DCCC*' (5771), sometimes a measure. This is can be the

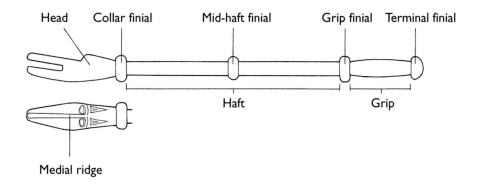

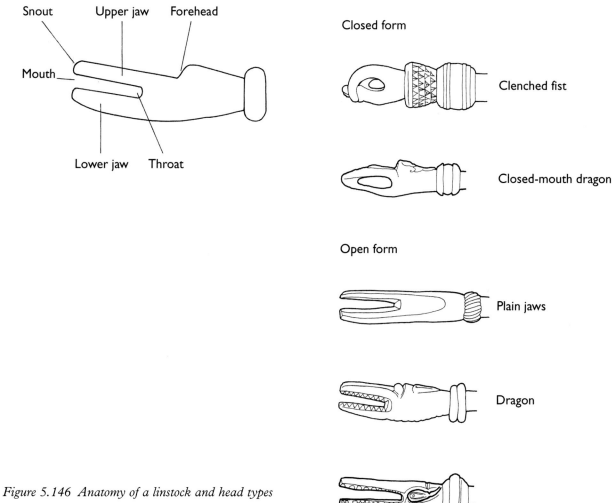

*Figure 5.146 Anatomy of a linstock and head types*

thousandweight (3761) or hundredweight (7158) (assuming '$M^l$ weight' and '$C$ weight' are 1120lb and 112lb respectively), or a simple multiple of hundreds and thousands of pounds, as '*Matches ijM$^l$iij$^c$lib*' [2300lb] (6471). Where '*lint*' is referred to listings include coils (4513), rolls (5701, 5751), rolls and bundles (5095), and weight (3761, 5095, 6293, 6841, 7090). Although we might suggest a distinction between '*raw lint*' and '*prepared match*', this cannot be verified.

There is also an indication of the storage of lint in casks, eg '*Lynte barrells oone*' (5399).

No casks with prepared match, cord or cord lengths have been identified, although loose cordage was found amongst casks in the gun furniture store area in O1.

**Description and distribution**

The *Mary Rose* assemblage of ancillary gun furniture contains 44 such '*poles*', eighteen of which are complete.

Table 5.48 Linstocks (metric)

| Object | Sector | No. on Fig. 5.154 | Type | Description | Wood | Length | Head Length | Haft Length | Haft diam | Grip length & shape | Heel spike |
|---|---|---|---|---|---|---|---|---|---|---|---|
| 79A0953 | H3 | 1 | – | haft frag. & finial; conical finials | beech | 180 | – | 165 | 21 | – | n/a |
| 81A1922 | H4, in conc. F099 | 2 | dragon | flat nosed, geometric decoration, bulbous eyes forehead, ears to side | ash | 690 | 180 | 330 | 23 | 102 P | Yes |
| 81A3858 | H5 | 3 | jaws | plain, similar to 81A3772; no mid-haft finial, finials decorated, scalloped | poplar | 860 | 135 | 509 | 30 | 110 R | no |
| 80A1544 | H5 | 4 | dragon | plain; head roughly carved | ash | 256 | 133 | 95 | 26 | – | n/a |
| 82A0974 | H8 | 5 | dragon | broken at throat; intricately carved; burning each side, bulbous eyes, ears on top | ash | 753 | 102 | 445 | 30 | 118 R-P | 6 x 4 |
| 82A1701 | H8 | 6 | ?dragon | carved medial ridge with slashed lines; bulbous eyes & brows, ears | poplar | 670 | 110 | 353 | 23 | 140 | no |
| 81A3031 | O4 in conc. F144 | 7 | – | part grip and terminal finial only | ash | – | – | – | – | 60 P | 5x5 |
| 80A1352 | O5 | 8 | clenched fist | plain cuff; mid-haft finial; carved grip | ash | 770 | 64 inc | 519 | 26 | 115 P | 4x4 |
| 81A5601 | O5 | 9 | dragon | flat nosed, carved zig zag; no eyes, ears top of head above mouth; all over decoration, ornate carved grip, zig-zags | ash | 737 | 140 | 413 | 25 | 96 P | 4x4 |
| 81A3896 | O5 | 10 | dragon | closed; carved; eyes, forehead, ears above eyes; mid-haft finial | ash | 631 | 120 | 481 | 26 | – | ? |
| 80A0717 | O6 | 11 | – | 2 frags haft & grip with carved finials | beech | n/a | 132 | 12 | 25 | 68 | no |
| 81A3060 | O8 | 12 | dragon | flat nose, no eyes or teeth; carved top mouth, incised waves; ears top of head | pine | 653 | 110 | 397 | 26 | 80 P | yes |
| 82A0995 | O8 | 13 | dragon | teeth; arch eyes, forehead, ears above eyes. | ash | 808 | 113 | 468 | 26 | 90 | no |
| 81A2892 | O9 in rope F127 | 14 | dragon | pursed lips; no teeth or eyes; forehead; rhomboid finials ?unfinished; octagonal haft | ash | 592 | 114 | 417 | 21 | – | ? |
| 81A2545 | O9 F127 | 15 | jaws | plain; no decoration, jaws nearly closed; mid-haft finial & grip carved, geometric | ash | 757 | 60 | 515 | 17 | 100 R/P | no |
| 81A3266 | O9 F127 | 16 | dragon | plain; no eyes, ears, teeth; carved medial ridge & finials | poplar | 741 | 142 | 385 | 25 | 120 R | no |
| 81A3693 | O9 F127 | 17 | dragon | dragon, flat nose; carved slashed lines, eyes to side of head; teeth; carved finials | ash | 800 | 130 | 470 | 25 | 110 P | yes |
| 81A3772 | O9 F127 | 18 | jaws | plain jaws, no detail; carved finials | ash | 695 | 153 | 375 | 25 | 100 R/P | no |
| 81A3082 | O9 F127 | 19 | jaws | jaws, teeth suggested; finials semi-carved blocks; lint/match 81S1164 | beech | 790 | 475 | 475 | 28 | 88 P | no |
| 80A0969 | M7 | 20 | – | terminal finial only | ash | 100 inc | – | – | – | Finial & grip P | yes |
| 81A3029 | M8 | 21 | – | grip, terminal & grip finials, part haft | ash | 370 | – | 160 | 24 | 115 P | no |
| 81A3831 | M8 | 22 | – | plain haft, grip finial, grip, terminal finial | maple | 447 | – | 305 | 26 | 91 R-P | 8mm |
| 81A1447 | M9 | 23 | – | haft, incised 'W'; grip finial, grip, terminal finial; carved finials | ash | 615 | – | 445 | 27 | 85 P | 6mm |
| 81A1332 | M9 by chest 81A1337 | 24 | clenched fist | carved lace cuff & finials | poplar | 710 | 95 | 385 | 27 | 110 P | 4mm |
| 81A1753 | M9 in chest 81A1429 | 25 | dragon | teeth, eyes, brow ridges, ears on head. finials decorated with 4 rhombuses, 8 triangles | poplar | 523 | 140 | 320 | 21 | n/a | n/a |
| 81A3903 | M9 | 26 | dragon | teeth, eyes, brow ridge, ears on top of head, carved finials, grip with central finial, scroll carving; copy made | poplar | 819 | 135 | 496 | 25 | 125 P | yes |

| Object | Sector | No. on Fig. 5.154 | Type | Description | Wood | Length | Head Length | Haft Length | Haft diam | Grip length & shape | Heel spike |
|---|---|---|---|---|---|---|---|---|---|---|---|
| 81A1333 | M9 by chest 81A1429 | 27 | dragon | incomplete, plain; possibly unfinished | larch | 786 | 70 inc. | 530 | 26 | 75 R | no |
| 81A4479 | M9 in chest 81A5967 in cabin | 28 | – | grip frag. & terminal finial, scored | ash | 100 | n/a | n/a | – | 55 R | no |
| 81A3872 | M9 in chest 81A5967 in cabin | 29 | dragon | intricately carved head & collar; teeth, eyes, ears | ash | 183 | 122 | 36 | 25 | n/a | n/a |
| 81A1523 | M9 | 30 | dragon | no teeth, or eyes; b row ridge; multi-faceted finials | ash | 750 | 140 | 420 | – | 110 R | 7x4 |
| 81A4356 | M9 | 31 | – | finial, probably for a grip, possibly for 81A3872 D 40 | ash | 30 | – | – | 25 | n/a | n/a |
| 81A1929 | M10 | 32 | – | 2 parts; collar finial, haft, grip, terminal finial; scored lines; probably two linstocks | poplar | 620, 593 | n/a | 591, 406 | 25–28 | 94 P | 8x5 |
| 81A2864 | M10 by gun 81A3000 | 33 | jaws | plain, no detail | poplar | 680 | 150 | 360 | 25 | 95 P | no |
| 81A3700 | M10 | 34 | – | marking out lines, defined head, cut jaws & brow | birch | 648 | 148 | 273 x 273 | 50–52 | n/a, inc. | n/a |
| 81A0837 | M10 in chest 81A0923 | 35 | dragon | closed; carved, eyes, ears, no teeth; mid-haft finial | ash | 745 | 102 | 224+ 230 | 25 | 88 P | no |
| 81A0914 | M10 by gun 81A3000 | 36 | dragon | ornately carved scales, no teeth, eyes, ears to side of head | ash | 698 | 140 | 345 | – | 100 P | yes 4 |
| 79A0182 | U3 | 37 | – | 2 frags; possibly mid-haft finial, grip & terminal finial; zig-zag decoration | ash | 393 | n/a | 54, 26 | 26 | 105 R | yes 12 |
| 80A0503 | U8 | 38 | – | 2 frags; collar finial & part haft, haft & grip finial; carved finials | ash | 114+ 210 | n/a | 68+ 168 | 21 | n/a | |
| 80A1193 | U8 | 39 | dragon | 2 frags; carved scales, teeth, eyes, ears; multi-faceted finials, mid-haft finial | pear | 405 | 136 | 191 | 22 | n/a | |
| 81A0193 | U11 | 40 | – | portion of grip, terminal finial with scored lines | ash | n/a | n/a | n/a | – | 99 R–P | |
| 81A2821 | SS10 | 41 | – | incomplete haft with small terminal finial with spike in situ; unusual; no defined grip; terminal finial | beech | 120 | n/a | 100 | 25 | 10 R | |
| 82A2165 | ? | | – | single length, long unusual collar, haft & grip finial | ash | 375 | n/a | 260 | 24 | – | |
| 72A0156 | ? | | – | collar finial, haft, haft & grip finial; zig zag incised decoration | birch | 298+ 171 | n/a | 245+ 140 | – | – | |
| 79A0899 | Port scour | | – | grip frag. with terminal finial; scale decoration | beech | 140 | n/a | n/a | – | 105 P | |

R = round; P = polygonal

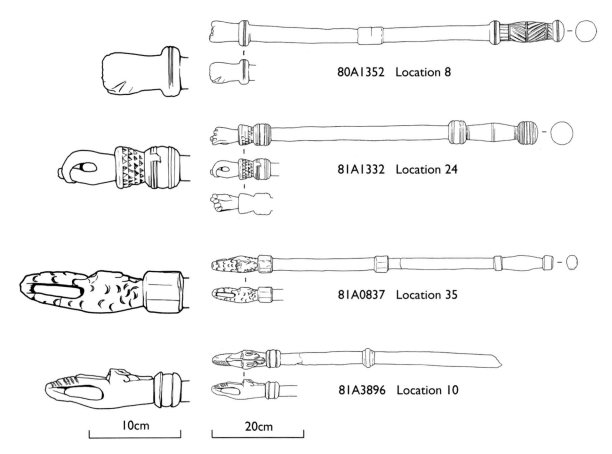

*Figure 5.147 Linstocks, closed forms (location number refers to Figure 5.154)*

All but three are in secure Tudor contexts with known locations. There is enough variation within the assemblage to suggest that these were personal items, perhaps belonging to the 30 'gunners' listed on the *Anthony Roll*. These may have belonged to the master gunners, as suggested by Stow 1598, 116): '*Their master gunner confronts me with his linstock, readie to give fire*'.

Linstocks are carved from a single piece of wood and all have a defined area to take the match separated from the main haft by a 'collar' which is carved or faceted (Fig. 5.146). The haft provides the main length and is carved to include a hand-grip, both ends of which have carved or faceted finials. In certain examples the haft carries a carved finial around the mid-point (eg, 80A1352, 81A0837, 81A3896, Fig. 5.147). Just over half of those with complete grips have a square hole with remains of iron concretion in the end of the grip, possibly the remains of an iron spike used to secure the linstock to the deck.

Of the relatively complete linstocks, chip carving to form a stylised dragon's head is the most common design (nineteen examples). Five others have plain jaws and two others are represented by clenched fists. Eighteen are of unknown design (usually because the head is not present). A number bear scorch marks from the match. One portion of short cord was found wrapped around a linstock (81S1164 on 81A3082 from inside a coil of rope in O9). This has not yet been analysed.

The linstocks have been recorded by overall length, dimensions of the head, finials, haft and grip, wood type, head morphology and distribution (Hildred 1997, 56–7). These details are summarised in Table 5.48. A variety of woods has been used all from trees that grow naturally, or can be coppiced, to produce strong, straight, flexible poles. Ash is the most common (25, 57%) with eight poplar (18%) and five beech (11%), five other woods being used for one or two examples each (Table 5.48). Ash would have been easily and widely available and its inherent strength and flexibility made it ideal for handles of all sorts as well as for staff weapons such as pikes and bills (see also *AMR* vol. 4).

*Size and proportion*
Where it is possible to measure complete, or almost complete, lengths most are 650–800mm in overall length (mean 698mm) with one very long example (81A3858) at 860mm. Head lengths vary from 60mm to 180mm, hafts between 330mm and 530mm, and grips between 80mm and 141mm. Grips are either round in section (13 examples) or polygonal (18). In some cases they are intricately carved (81A5601, Fig. 5.148; 81A3903, Fig. 5.150).

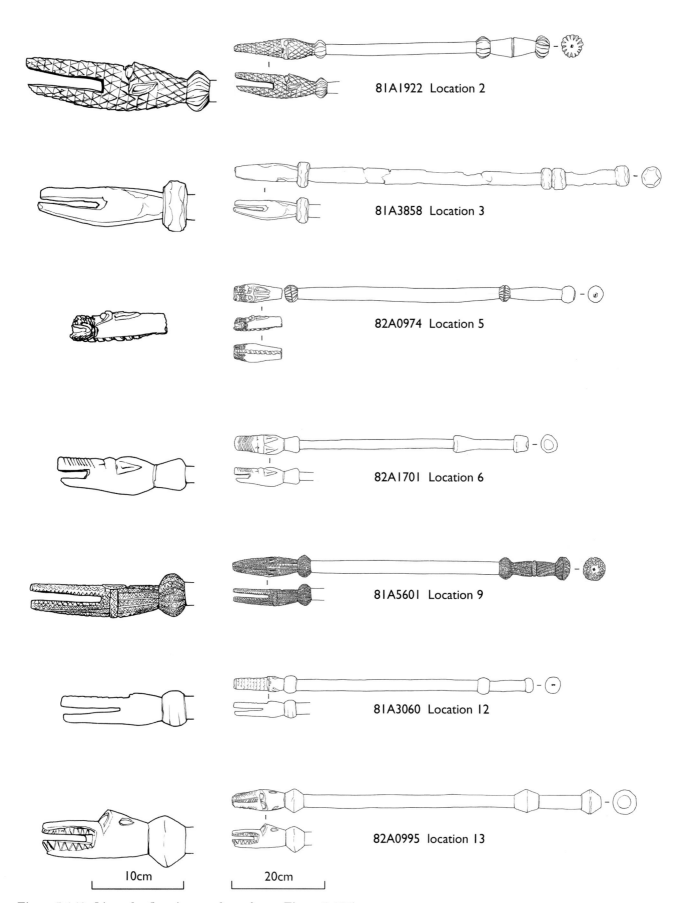

*Figure 5.148 Linstocks (location number refers to Figure 5.154)*

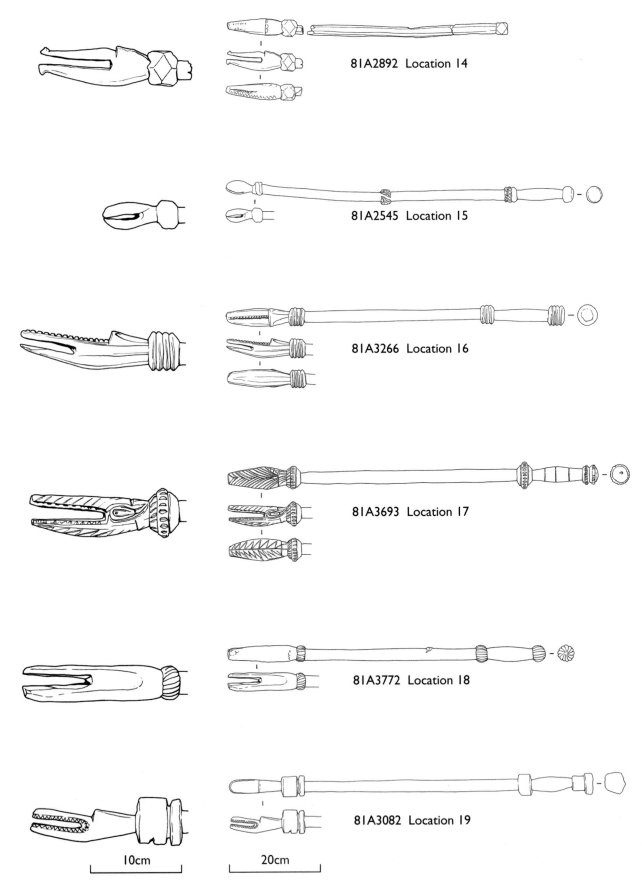

Figure 5.149 Linstocks (location number refers to Figure 5.154)

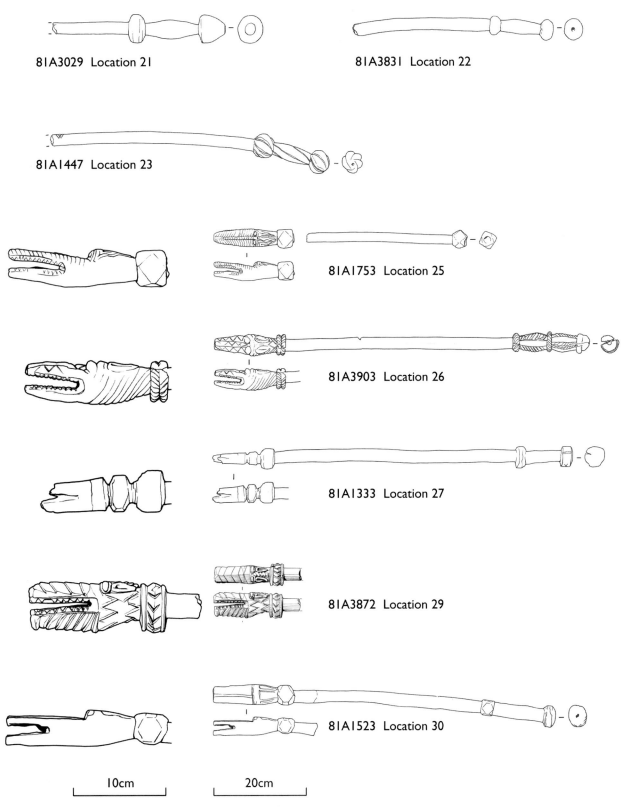

Figure 5.150 Linstocks (location number refers to Figure 5.154)

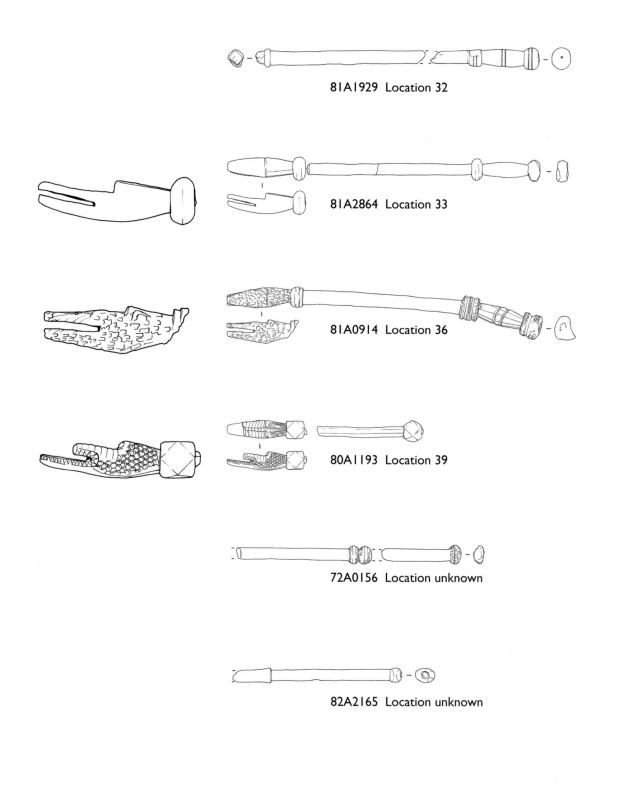

*Figure 5.151 Linstocks (location number refers to Figure 5.154)*

An attempt to classify linstocks based on length proved inconclusive; essentially these are about the length of a man's arm. Haft diameters are more consistent. Out of the 25 linstocks which retain measurable portions of the haft, none is less than 20mm in diameter, 23 (92%) are 25–30mm (1–1¼in), and only one is over 30mm (31mm). One example, 81A3700 (Fig. 5.153) appears to be an unfinished, roughed-out example and gives an indication of the size and shape of wood used. It has a square section

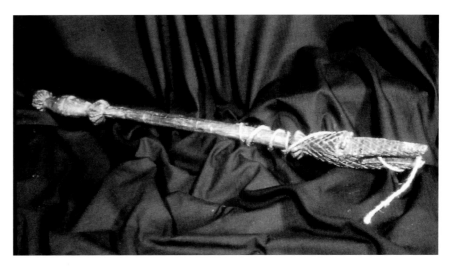

*Figure 5.152 Linstock 81A1922*

50–52mm across with plain jaws. It was found just outside the Carpenters' cabin, though it was also close to a gun so we cannot say whether or not it was in use or still being worked on when the ship sank.

The widest section of the linstock (giving an indication of the minimum diameter of wood necessary) is either the diameter of the collar or the grip finial. Of 28 measureable linstocks, eleven (39%) are largest at the collar finial, seven (25%) at the grip, and five (18%) at the terminal. Four show the widest portions being equal at the collar and grip and one has equal diameters at collar, grip and terminal finials. In three instances the jaw spread proved to be the widest measurable point.

A comparison of other dimensions proves to be as inconclusive, indicating that exact measurements were not adhered to, and perhaps not necessary (details in archive). The distance between the start of the grip finial and the end of the terminal finial in the majority of instances is more or less balanced by the distance between the end of the mouth and the back of the collar finial. The grip was probably made for the comfort of the user, and the head length probably depended on what was being carved. Together, the lack of consistency in the measurements and proportions suggests that linstocks were not supplied as bundles of equal-length staves for finishing on board.

*Forms*

Although most linstocks are individually carved, there are several that seem to be duplicates and may have been manufactured by the same individual or to a set design. In design terms a few basic styles emerge, based on the method of grasping the lint and degree and nature of decoration. Two methods for clasping the '*lint*' or slow match have been identified (Fig. 5.146): 'closed', where the lint has to be passed through a hole in the linstock and 'open', where the match is gripped within a 'V'.

There are four closed forms, two clenched fists (80A1352, 81A1332; Fig. 5.147) and two closed dragon jaws (81A0837, 81A3896; Fig. 5.147). 81A1332 (M9) consists of a right hand with index finger crossed over the thumb (Fig. 5.147). The collar finial resembles a lace-cuff, the fingers enclose the lint, which is wound round the cuff. 80A1352 (O5) is also a right hand but is broken at the knuckles. It has a mid-haft finial. Both are considerably longer (710mm, 770mm) than a similar linstock recovered from the *Trinidad Valencera* (Rodríguez-Salgado 1988, item 9.19). That example is 410mm long with a hole for the insertion of a heel spike.

Both the dragon forms have carved heads with a crest on top, carved collar finials and mid-haft finials. On 81A0837 (M10) the eyes are not defined, but the head is decorated with incised semi-circles. The finials are roughly carved hexagons with a polygonal section grip. 81A3896 (O5) is broken before the grip, and has defined eyes, forehead and ears on top of the head. Decoration to the top of the snout includes incised geometric lines. Each has a tapered snout enclosed at the front, so the lint must pass sideways through the elliptical hole formed by the mouth.

There are 23 linstocks which are based upon using open jaws to hold the match. Surface decoration ranges from plain to ornately carved representations identified as fire-breathing dragons. Dragons occur frequently in manuscripts, paintings or engravings, and their choice for the burning match is not surprising.

A study of the attributes and stylistic features of the dragon heads did not indicate there was a distinct formula which could be applied across the assemblage; the styles appear to be purely a matter of choice and the use of design is not based on specific wood types, location within the ship, length or handle form. Figures 5.147–51 indicate the range of forms present, grouped according to the following criteria:

- Eyes, prominent forehead and medial ridge, ears above eyes
- Prominent forehead, plain
- Lack of forehead, eyes, ears moving towards side of head
- Lack of forehead, ears behind eyes

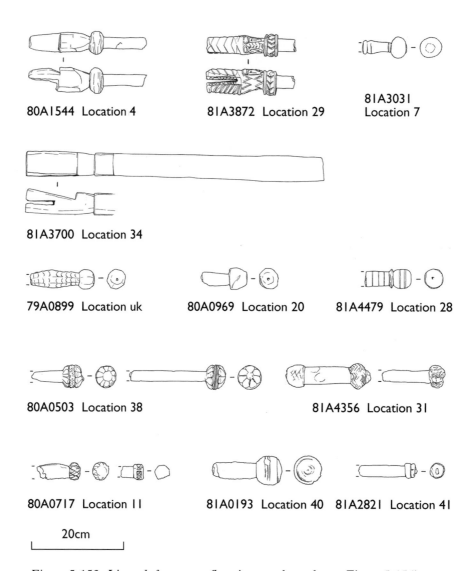

*Figure 5.153 Linstock fragments (location number refers to Figure 5.154)*

81A1922 is a particularly fine example of an open-mouthed dragon head (Figs 5.148 and 5.152). It is 690mm long with the head carved with geometric decoration to represent scales. It has bulbous eyes below a prominent brow ridge. The ears are ranged to the side of the head and the collar, grip and terminal finials are scalloped.

Sixteen fragments of hafts or grips exist which cannot be associated with any heads. A selection is shown in Figure 5.153. There are two 'pairs' of linstocks (81A0837/81A3896 and 81A1753/ 80A1193). These share such similar features that they may have been made either by one individual or as copies, though in neither case was the pair found together (Table 5.48 and Fig. 5.154).

Linstocks carved to represent dragons are found in other contexts: a portion of a head of a dragon style linstock was recovered from the *Trinidad Valencera* (Rodríguez-Salgado 1988, 9.18) with one from another sixteenth century wreck off Villefranche sur Mer, together with armaments and other gun furniture (E. Reith, pers. comm. item 0167). An example dated 1590 was recovered together with a fine collection of gun furniture from the *Scheurrak SO1* wreck near Texel, Holland (Puype 2001, 117–8, no. 23222). Many plain 'jaw' style linstocks have been recovered from the seventeenth century Swedish warships *Vasa* (1628) and *Kronan* (1676).

*Distribution*

The distribution of the 44 linstocks is shown in Figure 5.154. As with all objects on the ship, linstocks may have been subject to some movement during the sinking and subsequent collapse of the ship's structure. Being wooden it is likely that any linstocks on the open and Upper decks floated away as the ship sank. Only four were found – one in the bow and three in the stern with one more in the Starboard scourpit.

There are six linstocks from the Hold, thirteen from the Orlop deck and seventeen from the Main deck (three locations are unrecorded). As Figure 5.154 shows, the majority were found in the stern of the Main deck (M9/10) and on the Orlop deck, in a principal stowage area of O8–10. A smaller concentration occurs in H/O 4–5, where there was also a ladder (companionway) that ran between the Hold and Main deck. Of the six linstocks in the Hold, four were found in and forward of the galley where they were mixed with fuel logs, rope and the remains of several skeletons. The two found in the galley (81A3858, 80A1544) are both damaged, the former broken into five pieces and just beneath the Orlop deck carling (the longitudinal support for the half beams supporting the deck). It is possible that these originated on the decks above as there is evidence for the displacement of many items through the companionway and hatches here (see *AMR* Vol. 4, chap. 10). Indeed, the only linstock from the bow was from a concretion lens in O4 (81A3031). This lense was given a feature number (F144) allocated to cover all items within the concretion at the south-east corner of the deck that was thought to be the remains of a small gun. The rear stepped cheek to the *in situ* culverin found in M3 (81A3090), three tampions, the wheel of a bronze gun carriage, and several concretions were found within or close to this concretion. The area was part of the main store for the galley and contained large quantities of pewter and wooden serving and eating/drinking vessels, cooking pots, fuel logs, casks, a basket and at least one personal storage chest. Human remains including several fairly complete skeletons were found here, either washed through the companionway from decks above or, perhaps, trapped as they tried to escape. The gun/gun furniture is likely to have fallen from the deck above.

Of the two linstocks from H8, 82A1701 was recovered just below the deck beam (O80) and was excavated with associated leather and 82A0974 was found halfway between rider 8 and deck beam O80 associated with skeletal material. The presence of an incomplete, empty and badly damaged chest found in O9 (81A3088) and the relatively large number of linstocks found there (six of the thirteen on this deck) could suggest that these items were originally all within this chest. O9 was a partitioned area used for the storage of rope, beer/wine casks and chests, both personal and containing supplies such as longbows, arrows, spiles and shives.

Two linstocks were excavated from the Orlop deck in the stern (O8), an area mostly used for the storage of food supplies and chests and which produced the remains of at least thirteen of the crew. 82A0995 was in the south of the trench, close to the underside of the main deck beam. It was found broken just behind the head over a displaced plank, in the same position as the fracture, suggesting that this occurred during the sinking.

Four of the six linstocks from O9 were just above a large rope coil (81A0254, 81A3082) or in sediment within the coil. The archaeological interpretation is that this represented a layer between what had originally been two coils of rope, one on each side of a central access route. Within this layer were two linstocks (81A3772, 81A3266), personal items, clothing and a large number of tools. Above the rope there were two arrow chests (81A2582, 2398) and the remains of two other chests. The northern chest (81A3088) was standing on end with its sides askew with a linstock (81A3082) resting on one end. Its measures 936 x 275 x 295mm, long enough to accommodate the linstocks. The position and orientation of the chest relative to the linstocks just above it (M9, M10) and below it suggests a potential association. It is possible that this had originally been stored on the main deck, and that it had been upturned and fallen into the Orlop deck spilling its contents. Alternatively we may be looking at the storage of linstocks on the Orlop deck together with the arrows.

All seventeen linstocks (39% of the assemblage) from the Main gundeck were found in the stern (M7–10), with only two of the seven certain gun stations having linstocks in the area immediately surrounding them. Two linstocks were found in M8 (81A3029, 81A3831), beside the port piece (81A3001). Other gun furniture recovered here included shot, gauges for rapid matching of shot to gun bore, and spare breech chambers and wedges. It is suggested (see Chapters 3 and 11) that this gun was in the process of being reloaded as the ship sank. Both linstocks were broken. 81A3831 was broken just behind the collar finial, the haft, grip and terminal finial were present and it displayed a spike hole. It is the only maple linstock within the collection. The second (81A3029) is broken mid-way along the haft. This had not been fitted with a spike, and was found close to concretion against the northern partitioning. The lack of heads to these linstocks suggest that they broke at or shortly after the sinking, before being incorporated into the sediments.

The Carpenters' cabin was adjacent the M8 gun, separated only by a small partitioned bay. At least eight chests containing woodworking tools and personal possessions were found within the cabin or had fallen and broken open outside, against the cabin wall. Eight linstocks (and one finial, 81A4356) were recovered from this area, five outside the cabin and three, plus a finial, inside. Linstocks were associated with two chests outside the cabin. Linstock 81A1753 was in several pieces inside chest 81A1429 with the dragon head separated from the haft. The chest, decorated with a shield design, also contained jewellery, coins, shoes and clothes fastenings, knives, woodworking tools and a priming wire (see above). It clearly belonged to an individual of high status, presumably an officer and possibly a gun captain or master gunner. 81A1333 was found just beside the chest. It is damaged at the mouth, with a small fragment missing and had not been fitted with a spike. 81A1332 was also found close to this chest and is possibly derived from it, or from adjacent chest,

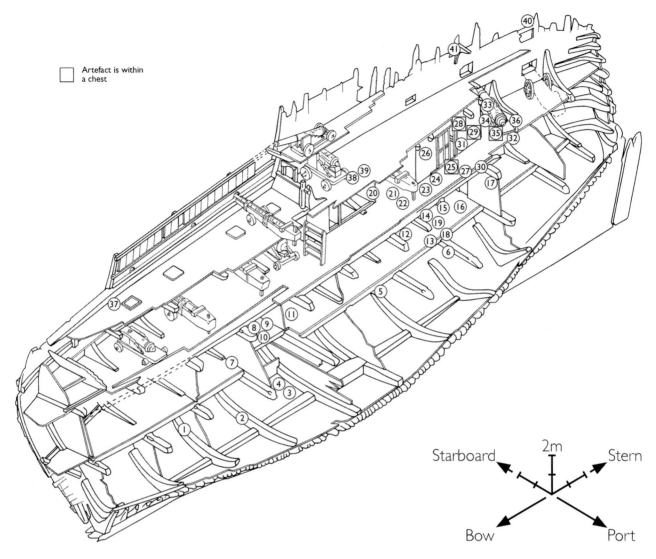

*Figure 5.154 Distribution of linstocks (see Table 5.48)*

81A1337, also a personal chest that had been smashed open as the ship listed. Linstocks 81A1447 and 81A1523 were found loose in this area.

Linstocks 81A4356 and 81A3903 were both found loose within the Carpenters' cabin. 81A3903 was close to chests 81A5783 and 81A5967 and was lying north–south over the northern bench or sleeping platform. It lay beside a handgun (81A3884). Chest 81A5967, stowed under the southern bench, contained two linstocks. This is a damaged, simple crate (of chest subtype 1.1/1: see *AMR* Vol. 4, chap. 10) that was largely filled with concreted woodworking tools, including planes, augers, axe/adzes, rules and a mallet that were probably officially issued supplies. It is suggested that linstock 81A3872, represented only by its dragon head and collar, had broken and was in the cabin for repairs or reuse.

Three linstocks in M10 (81A2864, 81A0837, 81A0914) were associated with demi cannon 81A3000. Linstock 81A3700 was also in close proximity but is an unfinished example so may have been undergoing finishing by one of the Carpenters rather then being in use. All the parts of the gun carriage were recovered, together with a breech chamber wedge from an iron gun, a powder dispenser (81A3303), a powder ladle for a small gun (81A2855), two rammers (81A2409 and 81A3030), tampions, a broken canister with cracked flints, shot, a barrel stave, a wedge, and three incendiary projectiles. The diversity in sizes of rammers, shot and tampions indicate that they served different guns. Linstock 81A0837 was found in chest 81A0923 in M10. The chest lay on its side resting on the Main deck to the west of deck beam M90. As it also contained, among other things, a pewter plate stamped with the initials 'HB', two book covers and a priming wire (81A0835) this chest probably also belonged to an officer, probably a gunner.

Four linstocks were recovered from the Upper deck. 79A0182 was found in two, badly degraded, portions in U3. Two were from U8 (80A503, ash, unknown style and 80A1193, open-mouth dragon). 80A0503, in two pieces, was found close to *in situ* culverin 80A0976. The terminal finial and grip portion of linstock 81A0193 was found beside the last gun port in the stern. There was no gun found for this port and this area was worked during salvage episodes between 1836–40, but the presence of gun furniture and shot suggests that both the M8 and U8 gun stations were active at the time of sinking.

*Finished and unfinished linstocks*
The probability that the linstocks were individually owned naturally begs the question as whether they are carved by their owners or to order. The distribution of linstocks with similar features has demonstrated that these are from different locations, and therefore may not belong to the same person. The observed similarities may indicate production by the same workshop or craftsman, or that specific designs were commonly used and ordered. It is also possible that the ship's carpenters decorated linstocks to order.

The presence or absence of the spike on the terminal finial was compared with the completeness of overall design in order to see whether there was the possibility that other linstocks were still incomplete. Thirty of the 44 linstocks (68%) retained terminal finials but only sixteen (53%, 36% of total assemblage) of these had spike holes. Those without included five of the 'plain jaws', and a number of quite decorative linstocks but without teeth or eyes. Possibly these were incomplete. All but one of the plain linstocks are centred around M9/O9, close to the Carpenters' cabin (Fig. 5.154), the other was in H5. One of these linstocks has burn marks around the mouth (81A3266), suggesting use. It may be that the linstocks were here for the fitting of the heel spike or to be further carved. Linstocks with spike holes have a wider distribution. Seven cluster around the Carpenters' cabin but the other nine are from all over the ship; most are decorated.

There is, however, evidence for finishing items onboard. The finding of the roughout (81A3700) on the Main deck (M10) suggests that some may have been made or at least finished on board. An experiment was undertaken to copy an intricately carved linstock (81A3903) though not by an experienced wood carver and using a readily available piece of teak and a Stanley knife. The section was square cut, similar to the roughout. Outlines of the head, grip and finials were transferred to the wood from the archaeological drawing by pencil and a saw used to cut the angle of the head and to remove material from the mouth. The haft was easy to make round and smooth as was the twisted rope shape of the grip using just the knife with a rasp and needle file to smooth the inside of the mouth and to enlarge the eyes, nose, ears and twisted rope. Decorative incisions were obtained using a fine-headed gouge and surface decoration to the head with the knife.

The experiment took ten hours and was undertaken by Clare Venables. It is a fine copy.

The unique designs of the linstocks seem to reflect personalisation and ownership in the same way as do the carved wooden knife sheaths from the ship (*AMR* Vol. 4, chap. 3). There are no broad arrow or 'H' marks denoting property of the King, nor any reference to the supply of linstocks in any of the ships' inventories of the period. There is a 'W' mark on 81A1447, which could denote ownership or be more symbolic (*AMR* Vol. 4, chap. 11). On the other hand, very few were found in personal chests. Of those that were, two were in chests that also contained priming wires and may have belonged to gunners and two were in chests inside the Carpenters' cabin may have been there for repair. What is most surprising is the lack of linstocks associated either with skeletal remains or directly with other guns. Unlike longbows, arrows, bracers or ballock daggers, most do not appear to have been in the hands of individuals but rather stored together. The concentration around the Carpenters' cabin may indicate that they were stored here, possibly in one or more open boxes or chests, or if they were being carried, it is possible that their location corresponds to an accessway off the Main deck to the Upper deck, though no evidence for a companionway survives in this area. Potentially the distribution suggests that the arena of operations was restricted to the stern at this point in the battle. The Cowdray Engraving shows the French galleys firing towards the stern of the *Mary Rose*, she may have been engaging from her stern just before she sank (see Chapter 11).

*Tinder boxes*

Four unusual boxes have been tentatively identified as tinder boxes, with the possibility of a fifth (see Chapter 8) of a somewhat different design and about half the size. Traditionally these would contain items required to make fire: dry, flammable substances together with a flint and a steel. Although the identification is tentative, these objects are of competely different construction to any other boxes found on the *Mary Rose* (see *AMR* Vol. 4, chaps 8 and 10). Each is carved out of a single block of oak rather than being made of boards or laths. All are rectangular, 310–385mm in length, 175–212mm in width, and 90–105mm in height with a wall thickness of *c.* 25mm (Table 5.49). All have sliding lids, or slots to accept them (Figs 155–6). All are divided into a large compartment and one or two small compartments. Most bear cut marks inside the large compartment; one contained a knife. All have circular impressions (one or more) on the inside of the large compartment and two have holes through the base. These are darkened as though slightly scorched. Three of the boxes contained what was thought to be ' felt' or soft contents which

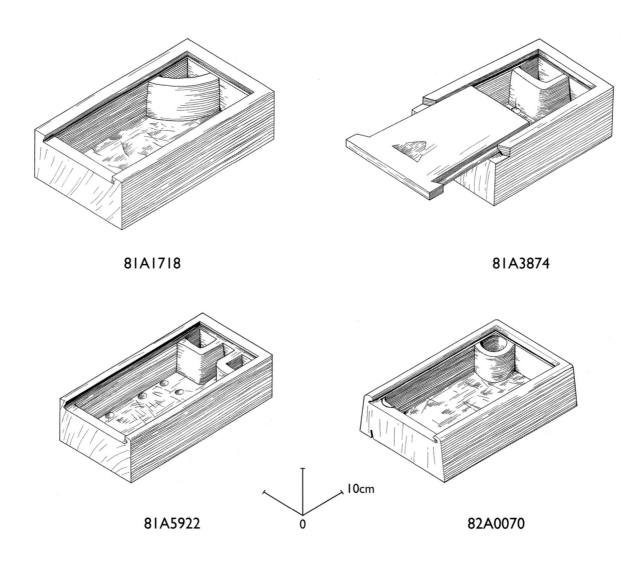
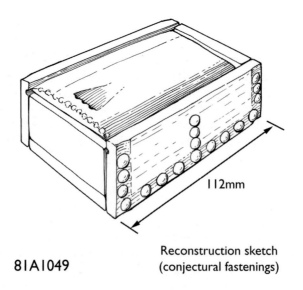

Figure 5.155 Tinder boxes

Table 5.49 Tinder boxes

| Object | Sector | Length ext. | Length int. | Width ext. | Width int. | Height ext. | Height int. | Description | Samples |
|---|---|---|---|---|---|---|---|---|---|
| 82A0070 | H4 | 315 | 277 | 210 | 165 | 85 | 62 | circular corner compartment; hole in base, 25mm diam. bung carved for handle; lid frag. | 'sootlike material from outside & fibrous 'matting' inside inc. prob. human hair (82S0038–9; 1003; 1337) |
| 81A1718 | O3 | 375 | 385 | 210 | 175 | 100 | 75 | | |
| 81A3874 | M9 | 295 | 240 | 190 | 98 | 98 | 70 | square compartment; hole in base; knife cuts on base & circular depressions; complete lid; contained knife | |
| 81A5922 | M9 | 360 | 325 | 185 | 90 | 90 | 75 | 2 rectangular compartments; rebate for lid; 9 circular depressions in base 22mm diam; knife marks | 81S1217 sediment |

81A3874    81A5922    82A0070    81A1718

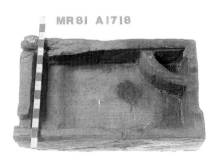
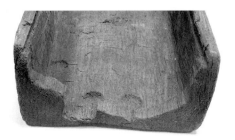

81A5922

*Figure 5.156 Tinder boxes*

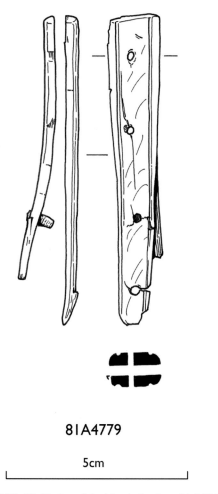

**81A4779**

|———— 5cm ————|

*Figure 5.157 Knife found inside tinder box 81A5967*

*meanes aboue declared made touchwood) rubbe well that parte betweene your handes for to make it softe and apte to take fire. But when you will make tinder for a Gunners tinder boxe, take peeces of fustian, or of olde and fine linnen clothe, make them to burne and flame in a fire, & suddenly before the flame which is in thē doth die, choke their fire, & keepe their tinder so made in a boxe lined within with clothe, to the ende it may not be moyste at any time'.*

### Description and distribution

Two of the boxes were found inside the Carpenters' cabin (81A3874, 81A5922), as was the smaller example, 81A1049, another was in the Hold just forward of the galley (81A0070) in H4 and the last on the Orlop deck in an area (O3) that seems to have been largely a rigging and meat store (81A1718) (Fig. 5.108). The locations of the last two suggest that these may have been used to transport smouldering substances from the fire in the galley to designated areas around the ship, a task also undertaken by using lanterns (Barker, pers. comm., 2009).

Summary information on the four boxes is presented in Table 5.49. 81A3874 was within chest 81A5967, the crate that contained a variety of woodworking tools, linstocks and personal items such as a wooden bowl. This box is the most complete of the four. The lid is almost 'T' shaped with a small groove to act as a finger hold for opening. The handle of a small boxwood scale tang knife (81A4779), originally held

were sampled for analysis. The boxes appear similar to boxes used in the eighteenth century to contain linseed soaked cloths used in caulking. The method of manufacture and distribution would accommodate both uses, so identification is still uncertain.

### Historical evidence

By the sixteenth century gunners were expected to have a touch box with tinder. Lucar (1588, appendix, 20–1) recommends the following '*To make touchwood and tinder for a Gunners Tinder boxe*':

*'Take those great things which are called olde Todestooles growing at the bottomes of nuttrees, beechtrees, okes, and such other like trees, drye them with the smoke of fire, & then cut them into so many peeces as you will, and hauing well beaten them, boyle thē in strong lie with waule floure, or saltpeeter, till all the lie shal be consumed. After this laying them in a heape vppon a boorde, drie them in an ouen which must not be made verie hotte, and after you haue so done, beate them well with a woodden mallet, and when you shall haue cause to vse any parte of those Todestooles (now by the*

*Figure 5.158 Tinder boxes depicted in the Cowdray Engraving*

together by four rivets, was found inside (Fig. 5.157). The handle measures 82mm by15mm tapering to 8mm and carries two incised lines. Box 81A5992, found in the north-east part of M9, has two rectangular compartments and several circular depressions in the base (Fig. 5.155).

Although the identification must remain tentative, these boxes are extremely unusual in form. Although larger than later tinder boxes, these are tantalizingly similar to boxes shown beside gunners in contemporary engravings such as the Cowdray Engraving (Fig. 5.148).

The hair and 'matting' recovered from samples within 82A0070 could represent tinder. Another possible source of tinder, in the light of Lucar's comment (above), could be dried fungus. Three fragments of bracket fungus (81S1275), of a variety that typically grows on beech trees, were found in chest 81A1328, which contained mostly woodworking tools, outside the Carpenters' cabin (*AMR* Vol. 4, chap. 8).

## Gun flints

Three gun flints (81A0179, 81A0115, 81A1710) have been identified, two on the Upper deck in the stern (U10) and one close to a cannon in M6 (81A1710). The first two are typologically eighteenth century, one of platform type and the other rather more wedged in form (S. de Lotbiniere, pers. comm.) and were found at the upper limits of the Tudor deposits. The only flint found apparently associated with a gun has been lost.

The photograph shows a longer rectangular flint of 75mm length of eighteenth or nineteenth century type. This area was subject to disturbance during salvage operations between 1836–40.

## Fuses

Four fragments of cord have been identified as 'fuses' (81A1501, 81S1164, 83S0381, 83S0384), of which only one survives. Identification has been made on the basis of association with a gun, chamber or linstock. Recording is rudimentary and analysis has not taken place

Three fuses were described as either being close to the touch holes of guns, or actually protruding from them. One of these (81A1501) was described as being in situ on the barrel of bronze cannon 81A3003 (Deane 81/6/106). This poses the question as to whether the priming of guns included the placement of a short piece of saltpetre impregnated fuse cord within the priming powder in the touch hole, or whether these arecord has a length of 300mm with a double 'Z'twist (81A1501).

83S0381 was removed from the touch hole of hailshot piece 81A1080, recovered from the Starboard scourpit SS6/7. The excavation of swivel chamber 82A4367 from concretion also revealed fibre around the touch hole (83S0384, 50g).

One linstock (81A3082), found in a coil of rope in O9, retained a cord or slow match wound between its jaws (81S1164). This cord is described as having an exterior coating not unlike pitch or tar (81/09/959).

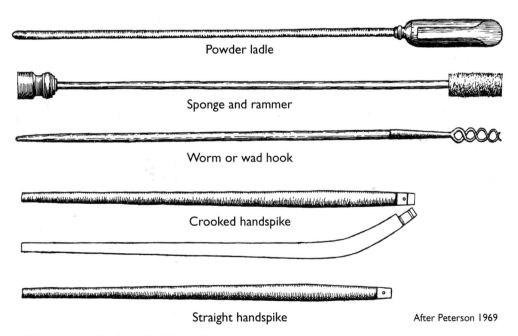

*Figure 5.159 Tools needed for working guns*

**Table 5.50 Objects possibly associated with working the guns (metric)**

| Object | Sector | Head | | | Handle | | Material/head shape | Associated with/comment |
|---|---|---|---|---|---|---|---|---|
| | | length | width | thickness | diam. | length | | |
| *Commander* | | | | | | | | |
| 81A2617 | O1 | 180 | 120 | 110 | | | elm/rectangular | gun furniture |
| 81A0452 | M2 | 440 | 180 | 180 | | 440 | oak/octagonal | close to shot store & companionway |
| 81A0569 | M2 | 175 | 70 | 70 | 25 | 200 | oak/square | as above |
| 80A1743 | M7 | 200 | 90 | 90 | | 310 | elm head, oak handle/rectangular | |
| 81A0887 | M8 | 225 | 100 | 100 | | 320 | oak/rectangular | port piece (81A3001) |
| 79A0782 | U4 | 290 | 140 | 140 | | 930 | oak/rectangular | no gun but may have been a wrought iron sling, gunport (unlidded) |
| 81A1876 | U10 | 260 | 155 | | | 845 | oak head, poplar handle/cylindrical | broken carriage for wrought iron gun, gun furniture & gunport (unlidded) |
| *Crowbars* | | | | | | | | |
| 79A0901 | M3 | no data | | | | | | |
| 81A1514 | M3 | | | | 34–39 | 210–215 | iron | inboard |
| 83A0214 | M4 | | | | 180 | 410 | iron | frag. outboard in concretion, gunport in this sector |
| 83A0592 | M11 | | | | 140 | 200 | iron | as above, gunport (lidded) |
| 83A0561 | M12 | | | | 55 | 155 | iron | as above, associated with poss. gun frags 83A0555, gun furniture inboard in this sector |
| *?Handspikes* | | | | | | | | |
| 82A0119 | H4 | | | | 78 | 2370 | | galley fuel store |
| 79A1229 | O1 | | | | 55 | 860 | | |
| 81A2879 | O3 | | | | 50 | 1600 | | |
| 81A3089 | O4 | | | | 56 | 1810 | | |
| 81A0716 | M10 | | | | | 770 | ash | demi cannon (81A3000), shot, gunport (lidded) |
| 80A1305 | U6 | | | | 50 | 1910 | | worked end; wrought iron sling (81A0645) & other gun frags in sector, gunport (unlidded) |
| 80A0870 | U7 | | | | 26-30 | 420 | | culverin (80A0976) & gunport (unlidded) |
| 80A0509 | C8 | | | | 30-55 | 190 | | frags of wrought iron gun & canister shot in this sector |
| 82A2601 | ? | | | | 45 | 1730 | birch | |
| 87A0109 | ? | | | | 50 | 350 | | |

## Working the Guns
by Alexzandra Hildred and Nicholas Hall

In addition to the specialised items required for loading and firing the guns, appropriate general 'tools' would include mauls, mallets/commanders, sledge hammers, and handspikes and crowbars. The latter two are used for moving guns and to help disengaging chambers and forelocks from wrought iron guns. Decks would need to be kept clean and clear of debris, and therefore brushes, buckets and shovels should be expected to be on the gun decks. In addition, Lucar (1588) specifically mentions mattocks, handaxes, iron hoops and *'little handbaskets'*. Axes, adzes and hand axes on the *Mary Rose* are restricted almost exclusively to the Carpenters' cabin and tool chests and are assumed to be woodworking tools. Many of the objects discussed here could have been used for a variety of purposes and most are described in detail in *AMR* Vol. 4, chapters 8 and 10, where examples are more fully illustrated. Illustration and distribution of tools which may have been associated with the practice of gunnery is shown in Figures 5.159–62 and they are summarised in Table 5.50.

Twelve *'crowes of iron'* are amongst the items listed for the *Mary Rose* in the *Anthony Roll*. All vessels listed with any carriage mounted guns are provisioned with at least one crowbar and sledge hammers (Knighton and Loades 2000). The number per vessel does not seem to have any obvious proportional relationship with the number or type of guns carried, although in general the more guns, the more tools there are. The 39 carriage

505

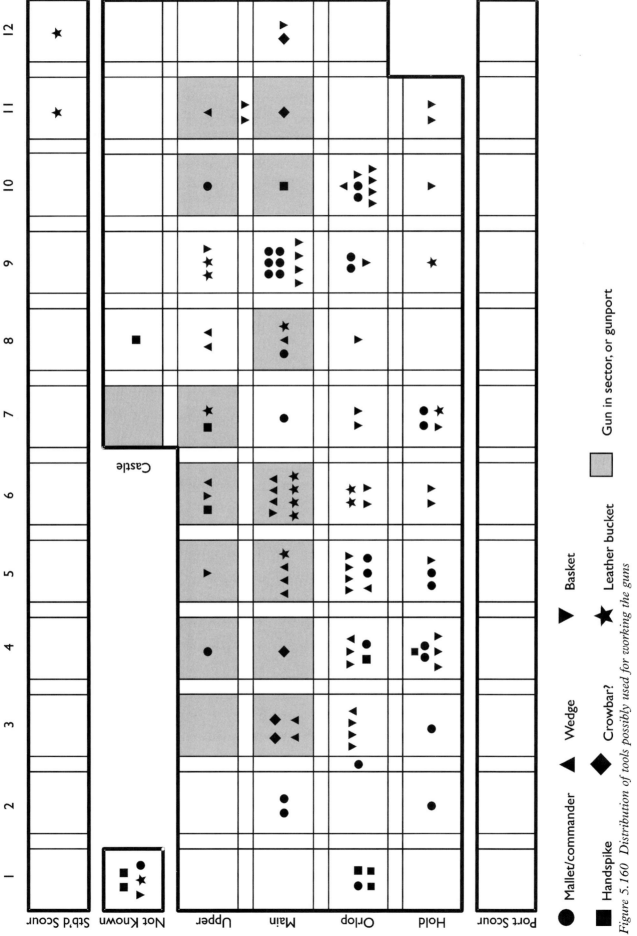

Figure 5.160 Distribution of tools possibly used for working the guns

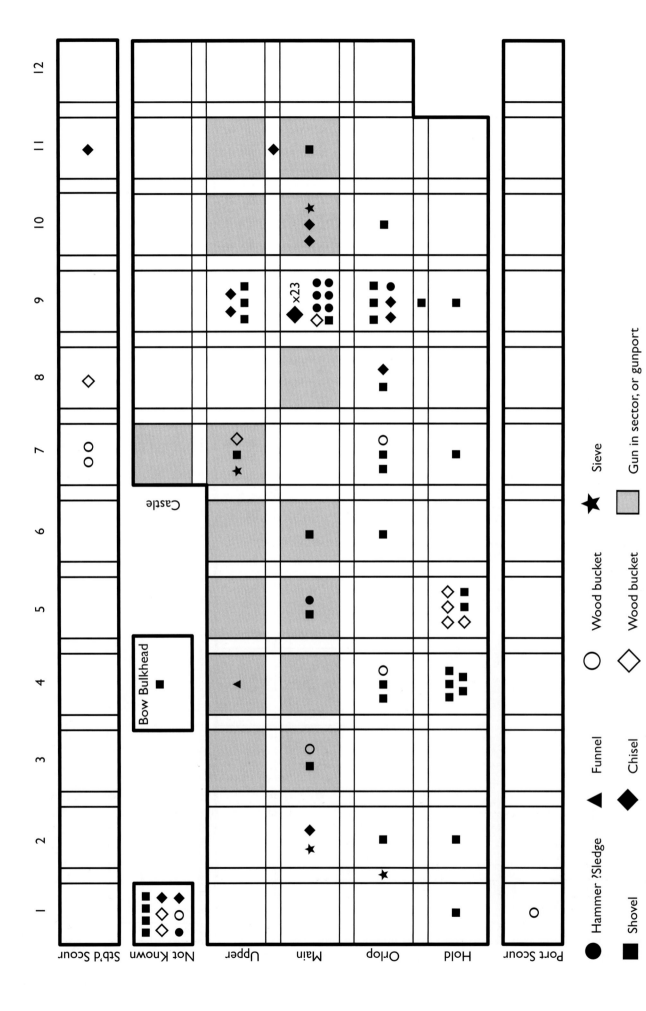

*Figure 5.161 Distribution of tools possibly used for working the guns*

mounted guns on the *Mary Rose* had to share twelve crows, and the 24 iron guns, twelve commanders. Other '*habillements for the warre*' included eight bags of leather (possibly to carry priming powder), six firkins with purses (probably '*budge*' barrels), twelve sheepskins for sponges and timber for forelocks. Finally wedges or quoins would be used to elevate the muzzle-loading guns and wedges possibly for fine-tuning the forelock system on the larger breech-loaders as well as for chocking carriage wheels and '*filling in*'. Lead patches, termed aprons, were often put over the touch holes to keep these dry. None has been positively identified, but a bronze vent cover was found with cannon 81A3003 (see Chapter 2).

## Commanders or mallets

Wooden mallets were a common tool in the sixteenth century as they are today and would have had a variety of uses (*AMR* Vol. 4, 309–10, fig. 8.11). Commanders or '*beetles*' were large-headed mallets used to knock in wedges, to knock out the forelocks on wrought iron guns and for a multitude of other purposes. Twelve are listed for the *Mary Rose* in the *Anthony Roll* and, indeed, they are listed for most of the vessels, even those without wrought iron guns. The 30 mallets from the *Mary Rose* vary in size and shape and only seven were found in areas which had or may have had guns or gun furniture and could be considered as possible commanders. One is from O1 (81A2617) where much of the spare gun furniture was stored. There are two from M2 (81A0452 and 81A0569), close to the shot store and accessible to the Upper and Main deck guns in sectors 3–6, and one from M7 (80A1743). 81A0887, from M8 was in an area that had one of the port pieces with two forelocks, both beside the gun (Fig. 5.160). The position and associations of the mallet suggest that it may have served this gun which may have been in the process of being reloaded, as the barrel contained a shot but the chamber was beside it. A mallet in U4 (79A0782) may have been used as a commander. Although there was no gun, it is expected that a wrought iron sling, similar to the U6 gun, would have been situated here. This would require a mallet for insertion and removal of the

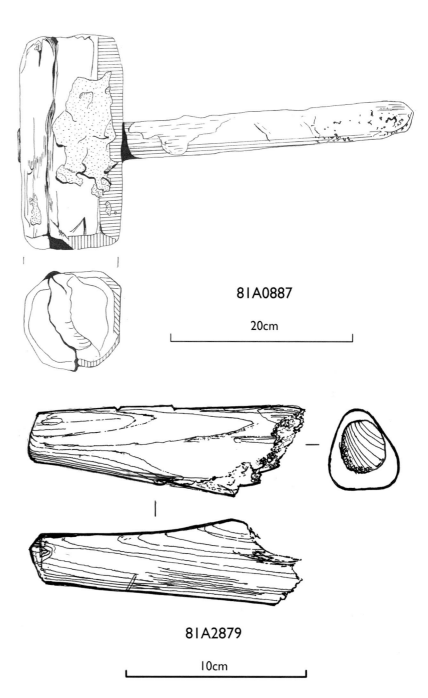

*Figure 5.162 Top) mallet 81A0887; bottom) possible handspike 81A2879*

forelock. The mallet recovered from U10 (81A1876) was amongst the broken carriage and broken elements of a wrought iron gun (Feature 2). This and the example from M8 are the only mallets directly associated with guns.

## Sledgehammers

Eight '*sledgis*' of iron are listed under the munitions for the war but, again, could have had a variety of uses. Although increasing in number with guns taking stone shot, they also occur on lists for vessels which do not carry guns firing stone shot. Nine wooden hammer handles were recovered from the ship but these all have slender handles suggesting that they were used with lightweight hammers and most were from the Carpenters' cabin (*AMR* Vol. 4, 306–7, fig. 8.8). One example (82A4467) is from M5, a gun station, but has a handle only 25mm in diameter.

## Chisels

The ship's carpenters brought an array of woodworking chisels (35) with them, but some chisels could have been used for finishing shot. Nine wooden handles have been identified as flat ended chisels (*AMR* Vol. 4, 307–8, fig. 8.10, chisel type 2), designed to be hit with a mallet and possibly for such use. However, only three of these were found elsewhere than in the Carpenters' cabin or a chest just outside it, and one of the remaining two is of a different form to all the others and from an insecure context in the Starboard scourpit (78A0072). Chisel handle 79A1153, found in M/U11, is only 77mm in length. This probably makes it too short to hold and hit with a mallet. The distribution of all chisels and mallets is shown in Figures 5.160–1.

## 'Crows' *(crowbars) of iron*

On shore '*crowes*' are listed with equipment for the manufacture of shot where they would also have been used for the removal and handling of stone blocks. Onboard, they are more likely to have been used for moving heavy loads, or pushing or propping open gunport lids. Five were found (Table 5.50): two from M3 (79A0901, 81A01514) and the rest from within concretion against the side of the ship during washing of the hull following recovery (83A0214, 83A0592 and 83A0561. The latter, in the area of M12, was associated with possible gun fragments (83A0555). The proximity of these items to the Main deck gunports suggests that they may have been used in the operation of the guns or ports.

## Wooden handspikes

Ten wooden 'unidentified' poles are possible candidates for handspikes for the moving and elevating of gun. One of these (82A0119) was in H4, which was partly used as a fuel store for the galley, and is described in the archive as a '*worked log*' (note that the fuel logs were cut to lengths of 800–1240mm and this 'log' measures 2370mm). Three were on the Orlop deck in O1 (79A1229), O3 (81A2879s, Fig. 5.162, bottom) and O4 (81A3098) with only one on the Main deck, an ash 'handle' in M10 (81A0716). Two were found on the Upper deck, in U6 (80A1305) and U7 (80A0870); the former has an end carved into a wedge shape and is the most likely contender. There was one in C8 (80A0509) and two others from unknown locations (Table 5.50; Figs 5.159 and 5.160).

## Baskets

Wicker baskets of different sizes were used for a variety of purposes from storage of fish to tools in the Carpenters' cabin. Most recovered from the ship were very fragmentary and some disintegrated on recovery. Contemporary battle scenes suggest these were used for a multitude of purposes including for holding gun related items (Fig. 5.163, top). One, from M6 (81A2299), was close to the *in situ* cannon (no associated objects) and one fragment of possibly four baskets in the Carpenters' cabin in M9 (82A2300, now disintegrated) apparently contained brass nails, possibly from gunshields (82A1586, 82A5048–51). Only one basket contained gunnery related items, a small wicker example in M/U11 (79A1266/79A1270; Fig. 5.163, bottom) which came up in pieces. This contained small lead shot (79A1269) but also woodworking tools and personal items including handles, pins and several unidentified objects (79A1273–4). A small powder ladle (79A1211) and lead shot (79A1214) found just below it are thought to have originally derived from this basket, with a second (79A1152) found close by. Baskets from the Upper deck included fragments of fine and unusual examples that were probably for specific uses unrelated to the guns. As the nature and associations of most of the basket fragments recovered are unclear, only those mentioned here are shown on Figure 5.160 (for a full distribution list and discussion, see *AMR* Vol. 4, 400–8, figs 10.23–31 and table 10.1).

## Leather buckets

Seven near-complete handled leather buckets were recovered from the ship along with fragments of perhaps eight more (*AMR*, Vol. 4, 359–67, figs 8.56–61). These are robust and of fairly uniform size (Fig. 5.164, Table 5.51). The capacity of one reconstructed example (80A1946) is estimated to be 6500 ml (just under 12 pints). The buckets are made of cattle hide and have two basic components, a circular base and rectangular panel for the body. The base is stitched to the body with a closed sewn seam which incorporates a thin leather band for support and strengthened by a wider band stitched to the outside of the body at the base, incorporated within the seam. Base support is provided by a pair of wide crossing straps bracing the base from beneath and also incorporated within the seam. The straps are further reinforced with a smaller set of crossed straps. The body

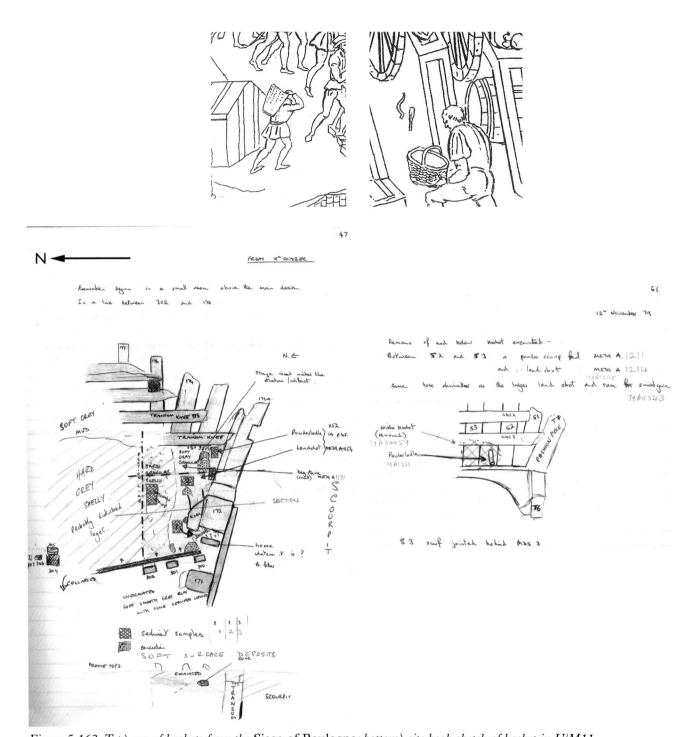

*Figure 5.163 Top) use of baskets from the* Siege of Boulogne; *bottom) site book sketch of basket in U/M11*

panel is seamed (usually double lapped) and covered by one of two vertical strips tapering inwards towards the top. The mouth of the bucket has a thin supporting hoop of wood covered by a leather band. The handle is formed from two straps, one wrapped inside the other and stitched. This is attached by simple bent over leather strips stitched to the top and reinforced with a metal rivet.

Although they could have been used for many purposes, it is noticeable that the distribution of the leather buckets does correspond quite well with that of the gun stations on the Main deck and they may have been used for holding water or vinegar required to clean the gun bores. Examples were found in M6 (four, 81A4149, 81A1582 and two stacked, 80A1316 and 80A1946). Fragments of possibly two more were found in M5 (81A1196, 81A1248), and one in M8 (81A3252). Three buckets were found on the Upper deck, 79A0948 from U7 where culverin 80A0976 was positioned, and two fragments in U9: 81A1635 on top of shot and 81A1907. U9 sported three very small gun ports, probably for hand weapons and there was a

Table 5.51 Buckets and possible tubs (original measurements in parenthesis) (metric)

| Object | Sector | Material | Base diam. | Mouth diam. | Height | No. Staves | Comments and associations |
|---|---|---|---|---|---|---|---|
| 78A0165 | SS2 | leather | 126 (153) | 236 (250) | 260 (265) | | pitch/tar and sheepskin patch; typologically post-Tudor, found in upper levels |
| 80A1316 | M6 | leather | 155 | 218 | c. 258 (265) | | stacked with 80A1946; cannon in sector |
| 80A1946 | M6 | leather | 135–170 | 218 | 265 (260) | | stacked with 80A1316; cannon in sector |
| 81A1582 | M6 | leather | 160 | – | 239 (250) | | cannon in sector |
| 81A1907 | U9 | leather | 158 (150) | – | 240 | | staff weapons, archery equipment, human remains |
| 81A3252 | M8 | leather | 158 (160) | – | 237 (240) | | pitch inside; gunbay with port piece being reloaded |
| 81A4149 | M6 | leather | 170 (185) | – (240) | (280) | | cannon in sector |
| 82A1758 | H9 | leather | 158 (180) | – | 260 (285) | | storage, barrels |
| 72A0086 | ? | wood: unid. | – | – | – | ? | lost |
| 79A1283 | M3 | wood: oak | – | – | 200 | 8 | ?wider top, smaller base |
| 80A0220 | O7 | wood: unid. | 328 | – | 242 | c.10 | wider base, smaller top |
| 81A0922 | PS1 | wood: oak | 253 | – | 350 | 15 | wider top, smaller base |
| 82A0095 | SS7 | wood: ash | 290 | – | 256 | 10 | wider base, smaller top |
| 82A2667 | SC7 | wood: willow | 300 | – | 170 | 8 | wider base, smaller top |
| 81A3239 | O4 | wood: oak | 298 | – | 250 | 12 | wider base, smaller top |
| 81A0874 | U7 | wood: unid. | 485 | | 640 | 21 | wider top, smaller base (with 5 cobbles/flints) |
| 81A550/ | M9 | wood: oak | – | | 150 | ? | wider top, smaller base |
| 81A658 | SC8? | wood: oak | | | | | |
| 82A1763/ | H5 | wood: unid. | | | | | wider top, smaller base |
| 82A1869/ | | wood: oak | 615 | | 480 | 19 | under conservation (part of 82A1869) |
| 82A1883/ | | wood: oak | | | | | |
| 82A1763 | | wood: unid. | | | | | |
| 91A0002/ | ? | wood: oak | – | | 372 | 2 found | wider top, smaller base |
| 91A0003/ | ? | wood: oak | | | | | |
| 81A0557/ | U9 | wood: unid. | | | | | wider top, smaller base |
| 81A1203/ | | wood: unid. | 600 | | 580 | 11 | |
| 81A1726/ | | wood: unid. | | | | | |
| 81A2574 | | wood: oak? | | | | | |
| 81A0786 | U9 | wood: unid | – | | – | ? | 200mm long, lost |

concentration of gun furniture in the area (see Chapter 11 for a summary).

Bucket fragments from the Orlop deck and Hold were not associated with guns but the former (two fragments from O6, 80A1609, 80A1891) were associated with a linstock and may have fallen through a central hatch from the Main deck. The two in H8 and H9 (79A1085, 82A1758) may also have been associated with the weaponry stored there or with the linstocks found in H8.

An example from SS12 (78A0165) has a broad arrow mark associated with Crown ownership and a sheepskin patch. It was a particularly exciting find. Unfortunately this differs typologically from the rest of the assemblage and has been dated as later – a probability reinforced by its having been found in the upper levels of the excavation and, therefore, in an unsecure context (see *AMR* Vol. 4, 365).

## Wooden tubs and buckets

The remains of parts of five tubs were recorded, two staves from an unknown location and the others from H5, M9, U7 and U8, the latter being a muster station for archers. Although these are similar to mess tubs (*AMR* Vol. 4, 356–9, fig. 8.55) tubs were often used to hold cartridges containing gunpowder (Bourne 1587, 31 quoted above). One tub contained a number of flint cobbles, though these were possibly only used to weigh down the container. Six staved wooden buckets were also found and others possibly remain to be identified

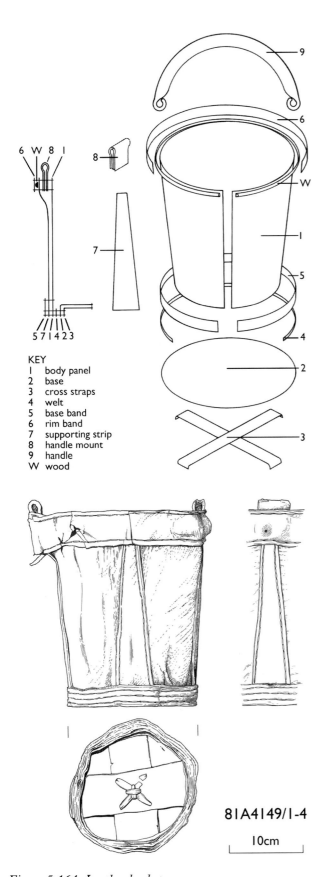

Figure 5.164 Leather bucket

among the many staves still in wet storage. These taper inwards from the base which would have made them more stable and less liable to spillage on a moving deck

(Figs 5.164–5; Table 5.51). Although not associated with guns, the recovery of four of them from the scourpits suggests that they were on the open weather deck or castle deck when the ship sank and they could have been used in a similar manner to the leather examples.

## Funnel and sieves

A very fragmentary, staved and hooped item from U5 (81A2469) has been identified as part of a wooden funnel (*AMR* Vol. 4, 358–9). Funnels were used for decanting loose powder into cartridges, or chambers. Funnels are listed in inventories alongside powder ladles and termed '*ffunnells of plate*' (Blackmore 1976, 283), suggesting pewter or tin.

Six objects have tentatively been identified as parts of sieves, of which two copper alloy strips and a handle (81A0699) were found in M10, close to the *in situ* demi cannon, with handles possibly from another sieve in M2 (79A1254) and a third in U7 (82A0883; *AMR* Vol. 4, fig 8.50), close to a bronze culverin. Two pewter fragments from M5 have recently been reidentified as potential sieve fragments (80A1140, 80A1118). Reconstruction of the remains suggests two D-shaped handles placed opposite to each other, attached to a central circular or cone-shaped wire sieve with diameters of between 320mm and 360mm (Fig. 5.166). Sieves are comparatively common in inventories for shore-based facilities where they are always associated with the preparation and storage of gunpowder along with mortars and pestles (pewter would not be a suitable metal for this purpose). None of the potential sieve fragments was from the galley area so they are not obviously associated with food preparation.

## Shovels and brushes

The gun deck would need to be clean, and we might expect a coincidence in the distribution of shovels, besom brooms and water containers. Shovels are commonly listed for shore-based facilities within the overall classification of ordnance. Thirty-six beech shovels were recovered, all but two having an iron sheath to the edge of the blade (*AMR* Vol. 4, 348–9, fig. 8.46; Fig. 5.166). They were widely distributed on the ship, mostly in the Hold and on the Orlop deck in close association with casks, but four of five found on the Main deck, were in sectors containing guns: M3, M5, M6, and M11. Four or five fragmented besom ('witches') brooms (*AMR* Vol. 4, 355–6), made from probably elm twigs tied to a hazel handle, were also found, but the number is too small to make meaningful comparison with the distribution of possibly related items and it is likely that other examples have simply disintegrated and become dispersed. One fragment

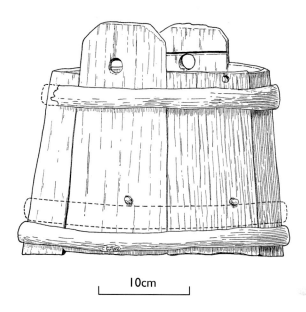

82A2667/1-13

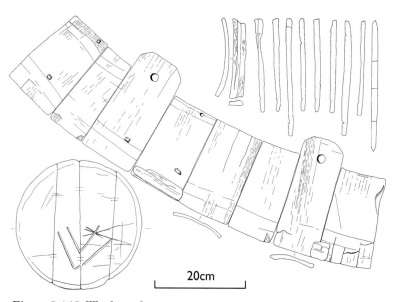

*Figure 5.165 Wooden tub*

(80A0569), with short, straight twigs still bound in a round bundle, was found in M6.

## Wedges

Triangular, tapering wedges were a vital part of ship's equipment with many uses, including several associated with guns. In total, 84 have been recorded and these seem to fall into three distinct size groups (*AMR* Vol. 4, 349–4, figs 8.47–8):

small: length 55–170mm, width 18–60mm, thickness 9–35mm

medium: length 217–330mm, width 50–105mm, thickness 37–75

large: length 335–893mm, width 25–199mm, thickness 40–260mm

Wedges or quoins were used to elevate guns, chock wheels, and 'fill in' various elements of both carriage and gun, but few have been found in direct association. Some of these have handles, a morphological feature that provides some clue to purpose, particularly in the case of quoins used to raise the gun cascabel where it would be dangerous to attempt to position a wedge without a handle or some form of handgrip. Indistinguishable wedges without handles could be

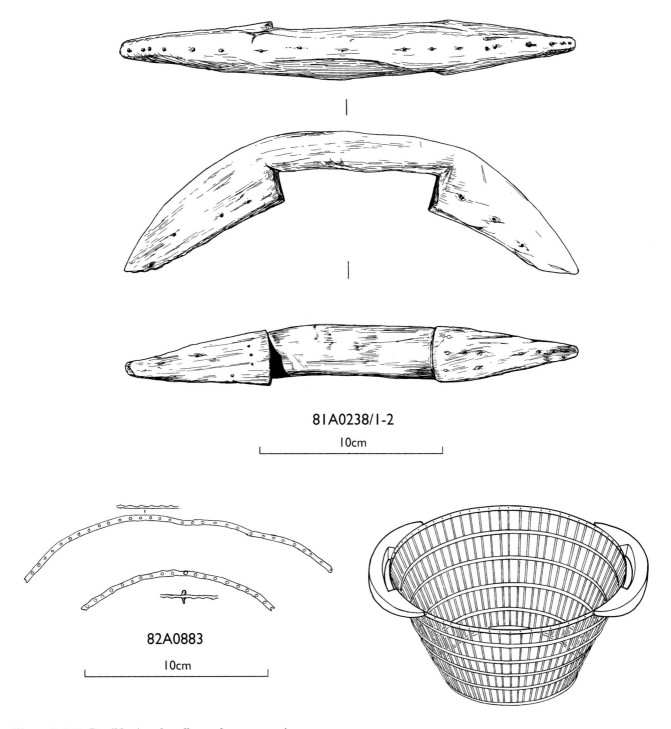

*Figure 5.166 Possible sieve handles and reconstruction*

used for other purposes and matching possible examples with ordnance is very difficult unless they are found *in situ*.

The description of quoins is given by Smith (1627, 84): '*great wedges of wood with a little handle at the end, to put them forward or backward for levelling the Peece as you please*'. All of the bronze guns require a quoin to raise the cascabel so that the muzzle is lowered and the gun lies horizontally within the carriage (the morphology of the gunports dictates that little elevation above horizontal was possible; see Chapter 11).

There are 17 wedges which have been considered as possible quoins (Figs 5.160 and 5.168). All fall into the large category, or just below, but only four have any form of handle. Nine were on the Main deck (six in M5/6), four on the Orlop deck and four on the Upper deck. Their distribution and sizes, together with suggested associations with guns, are indicated in Table

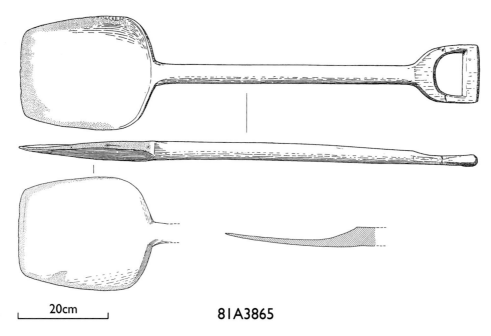

*Figure 5.167 Wooden shovel*

5.51 and on Figure 5.160. One handled wedge (81A2238) was found in situ beneath the cascabel of the M6 cannon (81A3003; Fig. 5.168). This has a short, inserted wooden handle of 350mm length and 50mm diameter and has a 'Z' incised into the rear face. At this point the wedge has elevated the cascabel 100mm off the carriage bed. The only other example found in direct association with a gun is 80A1158, also a large example, recovered with the collapsed carriage for demi cannon 81A3002 found in U8. This has an integral handle. Examples are shown in Figure 5.168.

The distance between the rear cheeks of the gun carriages ranges from 300mm (demi culverin) to 500mm (cannon). All but one of the possible quoins (80A1254) would fit. In several instances the cascabel of the gun cannot fit lower than a certain position on the carriage as it is too wide to clear the rear cheeks (79A1232), therefore dictating the need for a quoin. The *in situ* quoin has a very finished form. It has a flat base, is steeply sloping (210–10mm) with the back face angled towards the front and has a rectangular rear face. Many of the others are less well finished, are not very

**Table 5.52 Wedges, possible gun-related quoins**

| Object | Sector | Length | Width | Height | Comments |
|--------|--------|--------|-------|--------|----------|
| 81A2238 | M6* | 590 | 260 | 210–20 | handle, stick form. *In situ* under cannon 81A3003; elm |
| 80A1473 | O5 | 695 | 465 | 140 | handle, stick form; chamfered at end as though 'cut off' |
| 80A1158 | U8* | 730 | 310 | 100–80 | integral handle; found in direct assoc. with carriage elements for 81A3002 (demi cannon); v. little elevation offered |
| 82A4468 | M5 | 450 | 170 | 65 | integral handle; v narrow, little elevation |
| 81A0970 | M3* | 580 | 190 | 152–20 | no handle. similar to 81A2238; elm |
| 81A2206 | O3 | 893 | 200 | 300–20 | no handle. Similar to 81A2238 |
| 80A1365 | U6 | 465 | 185 | 140 | no handle. Similar to 81A2238 |
| 81A1536 | M6* | 330 | 90 | 50 | no handle; narrow & thin, rough finish |
| 81A1624 | M6* | 325 | 106 | 45 | no handle; narrow & thin, rough finish |
| 81A1397 | M5 | 418 | 110 | 70 | no handle; narrow & thin, rough finish |
| 81A1334 | M5 | 480 | 135 | 75 | no handle; narrow & thin, rough finish |
| 80A0985 | M8 | 500 | 140 | 190 | no handle |
| 80A1254 | O4 | 340 | 260 | 670 | no handle; v. high |
| 81A5879 | O10 | 300 | 100 | 100 | no handle; elm |
| 81A1932 | U11** | 605 | 330 | 120 | no handle; elm |
| 80A0495 | U8* | 565 | 268 | 215 | no handle |
| 82A4515 | M3* | 595 | 150 | 75 | no handle; little elevation |

* Presence of bronze gun in sector. ** Possible bronze gun in sector

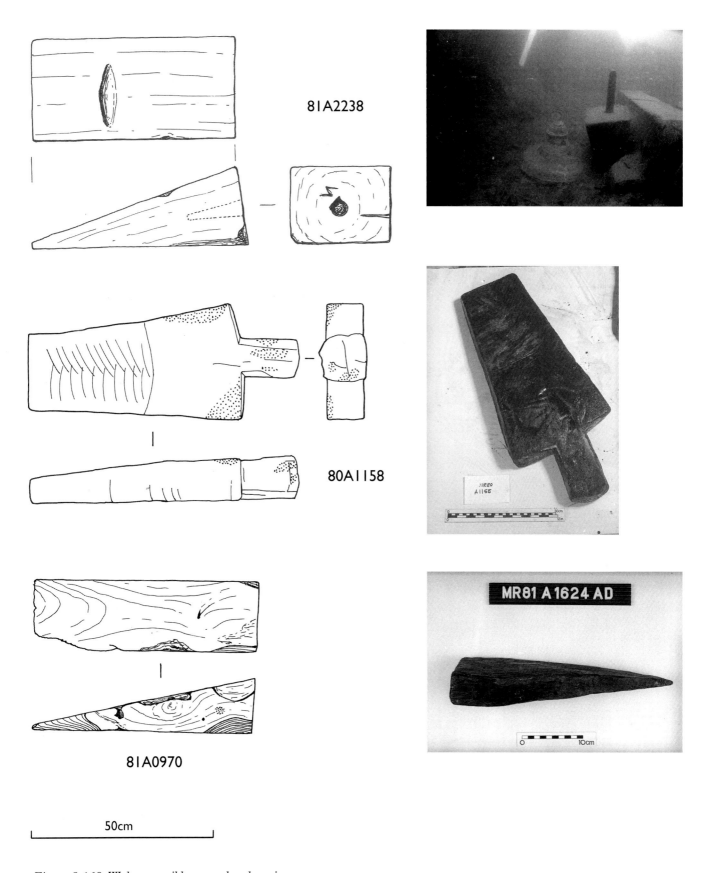

*Figure 5.168 Wedges, possible gun-related quoins*

thick, and do not slope enough to be of much use in this position. The area where the greatest concentration occurs (M5/6) is also where the main mast would have penetrated the deck and some could, therefore, be mast wedges.

# WEAPONS OF WARRE:
## The Armaments of the *Mary Rose*

edited by Alexzandra Hildred

# WEAPONS OF WARRE:

The Armaments of the *Mary Rose*

edited by Alexzandra Hildred

with major contributions from

Dominic Fontana, Nicholas Hall, Robert Hardy, C.S. Knighton, David Loades,
Robert Smith, Hugh Soar, Keith Watson and Guy Wilson

and contributions from

D. Adams C. Bartlett, C. Boyton, Ruth Rynas Brown, Hector Cole, A. Crowley,
Laura Davidson, Alison Draper, Andrew Elkerton, Nick Evans, Rowena Gale, Bert Hall,
Peter Holt, Mark Hopkins, Adam Jackson, S. Jackson, the late R. King, Kirran Louise Klein,
B.W. Kooi, Douglas McElvogue, J.P. Northover, Maggie Richards, Thom Richardson,
N.A.M. Rodger, David Starley, Ann Stirland, Clare Venables, John Waller,
Alan Williams and Robin Wood

The Archaeology of the *Mary Rose*
Volume 3
Parts 3 and 4

2011

Published 2011 by The Mary Rose Trust Ltd
College Road, HM Naval Base, Portsmouth, England PO1 3LX

Copyright © 2011 The Mary Rose Trust Ltd
*All rights reserved*

**British Library Cataloguing in Publication Data**
A catalogue record for this book is available from the British Library

ISBN 978-0-9544029-3-8

Series Editor: Julie Gardiner
Series Editor (graphics): Peter Crossman

Designed by Julie Gardiner and Peter Crossman
Produced by Wessex Archaeology

Printed by Short Run Press Ltd, Exeter, England

The publishers acknowledge with gratitude a grant from the
Heritage Lottery Fund towards the cost of publishing this volume

*Cover: Device and surface decoration from bronze guns 81A3000 and 79A1232 with cast iron shot.*
*Photographs: Peter Crossman, Dominic Fontana and Stephen Foote. © Mary Rose Trust Ltd.*
*Designed by Stephen Foote with Alexzandra Hildred*

The Mary Rose Trust is a Registered Charity No. 277503

# Contents of Parts 3 and 4

List of Figures . . . . . . . . . . . . . . . . . . . . . . . . . . . .vii
List of Tables . . . . . . . . . . . . . . . . . . . . . . . . . . . . . .xi
List of Colour Plates . . . . . . . . . . . . . . . . . . . . . .xiii
Contents of DVD . . . . . . . . . . . . . . . . . . . . . . . .xiv

**PART 3: HAND-HELD WEAPONS**

6. **Incendiary devices**, by Alexzandra Hildred
    Introduction . . . . . . . . . . . . . . . . . . . . . . . .519
Historical importance . . . . . . . . . . . . . . . . . . . .521
    *Mary Rose* incendiary projectiles . . . . . . . . . .523
        Distribution of other incendiary items . . .529
        Analysis of incendiary mixture . . . . . . . . .531
        Mortar bomb, by Alexzandra Hildred and
            Andrew Elkerton, with a contribution
            by David Starley . . . . . . . . . . . . . . . .531

7. **Hand-held firearms**
    Handguns . . . . . . . . . . . . . . . . . . . . . . . . . .537
    Scope and importance of the assemblage,
        by Guy Wilson . . . . . . . . . . . . . . . . . . .537
    Inventory evidence, by Alexzandra Hildred . .543
    Distribution of handguns, by
        Alexzandra Hildred . . . . . . . . . . . . . . . .545
    The assemblage, by Douglas McElvogue
        and Guy Wilson . . . . . . . . . . . . . . . . . .545
    Matching the shot to the guns, by
        Alexzandra Hildred . . . . . . . . . . . . . . . .551
    Associated objects, by Alexzandra Hildred . . .551
    Replication and firing, by Alexzandra Hildred 553
    Gun-shields: '*Targettes steilde with gonnes*',
        by Alexzandra Hildred . . . . . . . . . . . . . .553
    Scope and significance of the assemblage . . .553
    Morphology . . . . . . . . . . . . . . . . . . . . . . . .555
    Distribution of the gun-shields . . . . . . . . . .560
    Leather elements, with a contribution
        by Nick Evans . . . . . . . . . . . . . . . . . . .561
    The gun-shields . . . . . . . . . . . . . . . . . . . . . .564
    Comparative study of two gun-shields in
        the Royal Armouries Collections . . . . . . .573
    Suggested method of manufacture . . . . . . . .577

8. **Archery**
    Scope and importance of the assemblage,
        by Alexzandra Hildred . . . . . . . . . . . . . .528
    Archery of Henry VIII's vessels: inventory
        evidence, by Alexzandra Hildred . . . . . . .579
        The inventory of 1547 . . . . . . . . . . . . .581
    Distribution of the assemblage,
        by Alexzandra Hildred . . . . . . . . . . . . . .583
    Distribution of archers based on associated
        artefacts, by Alexzandra Hildred . . . . . . . .585

Historical importance and assessment of the
    longbow assemblage, by Robert Hardy . . .586
Longbows, by Robert Hardy . . . . . . . . . . . . .589
    Archery at sea . . . . . . . . . . . . . . . . . . . . .593
    The longbow assemblage, by Clive Bartlett,
        Chris Boyton, Steve Jackson, Adam Jackson,
        Douglas McElvogue, Alexzandra Hildred
        and Keith Watson . . . . . . . . . . . . . . . .594
    Statistical analysis of the longbow data,
        by Keith Watson . . . . . . . . . . . . . . . . .618
    Assessment of the working capabilities of the
        longbows Robert Hardy, the late R. King,
        B.W. Kooi, D. Adams and A. Crowley .622
    Horn nock, by Alexzandra Hildred
        with the late Roy King . . . . . . . . . . . . .632
    Bowstrings, by Alexzandra Hildred and
        John Waller . . . . . . . . . . . . . . . . . . . .632
Longbow chests, by Alexzandra Hildred . . . .641
Wristguards, by Hugh Soar . . . . . . . . . . . .644
    Distribution . . . . . . . . . . . . . . . . . . . . . .647
    Form . . . . . . . . . . . . . . . . . . . . . . . . . .647
    Fastening . . . . . . . . . . . . . . . . . . . . . . .650
    Decoration, by Matthew Champion
        and Hugh Soar . . . . . . . . . . . . . . . . . .651
    Non-leather wristguards . . . . . . . . . . . . . .665
Arrows by Alexzandra Hildred with
    Keith Watson, Mark Hopkins, Adam Jackson
    and John Waller . . . . . . . . . . . . . . . . . . .665
    Introduction . . . . . . . . . . . . . . . . . . . . .665
    Distribution . . . . . . . . . . . . . . . . . . . . . .666
    Form and function . . . . . . . . . . . . . . . . .669
    Scope and importance of the assemblage .672
    Analysis, by Keith Watson . . . . . . . . . . . . .677
    Other features of the shafts,
        by Adam Jackson . . . . . . . . . . . . . . . .686
    Manufacture, by Keith Watson . . . . . . . . .688
    Conclusions, by Alexzandra Hildred . . . . .688
Arrow chests, by Alexzandra Hildred . . . . . . .689
Arrow spacers, by Alexzandra Hildred . . . . . .693
Arrow bag, by Alexzandra Hildred
    and John Waller . . . . . . . . . . . . . . . . . . .698
Handgun bolts, incendiary arrows or top
    darts?, by Alexzandra Hildred . . . . . . . . .699
Identification of archers from skeletal
    remains, by Alexzandra Hildred and
    Ann Stirland . . . . . . . . . . . . . . . . . . . . .702

9. **Hand-to-Hand Fighting**
    Staff weapons, by Guy Wilson with
        Alexzandra Hildred . . . . . . . . . . . . . . .713
    Scope and importance of the assemblage . . . .713
        The *Anthony Roll* . . . . . . . . . . . . . . . . .718
        Bills, with a contribution by Hector Cole .719

|   |   |
|---|---|
| Pikes | 735 |
| Halberd | 736 |
| Hafts, possibly for throwing darts | 740 |
| Edged weapons | 740 |
|     Swords and daggers, by Guy Wilson with Alexzandra Hildred | 740 |
|     Sword and dagger grips, by Guy Wilson with Laura Davidson, Alexzandra Hildred and Douglas McElvogue | 765 |
|     Daggers | 771 |
|     Ballock daggers, by Maggie Richards with Alexzandra Hildred and Guy Wilson | 776 |
|     Scabbards, sheaths, belts and hangers, by Alexzandra Hildred, Laura Davidson, Douglas McElvogue and Guy Wilson | 793 |

## 10. Armour and Personal Protection

Armour for warfare, by Alexzandra Hildred and Alan Williams ..................820
    Armour worn ......................820
Protection at sea, by Alexzandra Hildred ....825
    Armour worn ......................825
    Armour carried ....................826
    The metallurgy of sixteenth century armour, by Alan Williams ..........827
The *Mary Rose* armour, by Alexzandra Hildred ...............828
    Possible armour and armour-related fragments, with contributions by Kirran Lousie Klein, Thom Richardson, Alan Williams and Guy Wilson ......829
Mail, by Thom Richardson .............842
    Types and distribution of *Mary Rose* mail, by Alexzandra Hildred and Nick Evans ...................844
Skeletal remains suggesting the distribution of armour clad men, by Alexzandra Hildred and Ann Stirland ..................850
    The Upper deck ...................850
    The Main deck ....................851
    The Orlop deck ...................852
    The Hold .........................852

## PART 4: FIGHTING THE SHIP

## 11. Fighting the ship

Historical evidence .....................855
    Fighting tactics, by N.A.M. Rodger .....855
    Documentary evidence, by Alexzandra Hildred with a contribution by Dominic Fontana ............860
    The theatre of war: geographical evidence from the Cowdray Engraving and GIS, by Dominic Fontana and Alexzandra Hildred ...............871
Archaeological evidence, by Alexzandra Hildred ...............886
    Gunports .........................891
    Layout and operation of the gundecks ...899
Overloaded with ordnance?, by Alexzandra Hildred ...............917
Playing with arcs of fire: how well defended was the *Mary Rose*?, by Alexzandra Hildred ...919
Fighting with the ship .................924
    The grapnel, by Douglas McElvogue ....924
    Shear hooks, by Douglas McElvogue ....925
Rules of engagement: what does the ammunition tell us?, by Alexzandra Hildred .........925

## 12. The Beginning of the end: some reflections on the ordnance of the *Mary Rose* and avenues for future research, by Alexzandra Hildred

Introduction .........................932
Back to the future: opening the door .......934
The 2003–5 excavations: an interim statement, by Alexzandra Hildred and Peter Holt ...938
Postscript: document CP 201/127, by C.S. Knighton and Alexzandra Hildred ...............946

**Technical appendices**

*1. Gun Metallurgy* ......................950
    1.1 Average metal composition for the bronze guns ......................950
    1.2 Analysis of guns from the Mary Rose, by J.P. Northover .................951
*2. Gunpowder (GP) analysis: analytical tables* ...965

Bibliography ..........................967
Index, by Barbara Hird ..................976

# List of Figures

Fig. 6.1 Types of wildfire
Fig. 6.2 Use of shot of wildfire at sea
Fig. 6.3 Fire trunk/tubes serving as a light gun and guns firing trunks of wildfire
Fig. 6.4 Musket firing an incendiary arrow
Fig. 6.5 Arrows, trunks and garlands of wildfire
Fig. 6.6 Incendiary projectile 81A2866
Fig. 6.7 Location of incendiary darts
Fig. 6.8 Incendiary projectile 81A2866 as found
Fig. 6.9 Flight, shaft and remains of bag containing incendiary mixture
Fig. 6.10 Proposed reconstruction of the stages of manufacture and final composition of an incendiary dart
Fig. 6.11 Firepots with slings and incendiary arrows
Fig. 6.12 Fragments of incendiary darts
Fig. 6.13 Incendiary dart at sea in fighting tops
Fig. 6.14 Incendiary dart in Alnwick Castle
Fig. 6.15 Reconstruction of 13in mortar shell
Fig. 6.16 Reconstruction of mortar bomb
Fig. 6.17 Distribution of probable mortar bomb fragments
Fig. 7.1. Anatomy of a handgun
Fig. 7.2 Royal Armouries matchlock RA.XII
Fig. 7.3. Gun with fishtail butt of ?1588
Fig. 7.4 Matchlocks in the Royal Armouries and GARDO crest
Fig. 7.5 Distribution of handguns
Fig. 7.6 Harquebus stock 82A4502
Fig. 7.7 Harquebus stock 81A2679
Fig. 7.8 Gardo crest on 81A2679
Fig. 7.9 Harquebus stock fragments
Fig. 7.10 Butt fragments
Fig. 7.11 Stock fragment 81A3884 and partial stock reconstruction
Fig. 7.12 Graeme Rimer firing a replica of the *Mary Rose* harquebus
Fig. 7.13 Gun-shield 81A5771 as lifted
Fig. 7.14 Gun-shield forms and anatomy of the outside
Fig. 7.15 Anatomy of the inside of a gun-shield
Fig. 7.16 Gun-shields in the Walters Art Gallery, external view and Art Institute of Chicago, internal view
Fig. 7.17 Distribution of gun-shield elements
Fig. 7.18 Distribution of numbered gun-shields in O9/10
Fig. 7.19 Radiograph of 82A1530 gun and Royal Armouries gun
Fig. 7.20 Pouch-pocket 82A0577
Fig. 7.21 Grip 82A0940 and strap 82A0948
Fig. 7.22 Gun-shield 81A5771
Fig. 7.23 Pouch-pocket and spring elements associated with 81A5771
Fig. 7.24 Gun-shield 82A0992
Fig. 7.25 Details of boss on 82A0992 and 81A5771
Fig. 7.26 Gun-shield 82A1530
Fig. 7.27 Gun-shield 82A1586
Fig. 7.28 Gun-shield 82A2060 with radiograph
Fig. 7.29 Gun-shield 82A2954
Fig. 7.30 Gun-shield 88A0009
Fig. 7.31 Probable gun-shield fragment 82A4465
Fig. 7.32 Royal Armouries gun-shield V34
Fig. 7.33 Royal Armouries gun-shield V38
Fig. 8.1 Distribution all archery related major groups in relation to that of Fairly Complete Skeletons
Fig. 8.2 Distribution of longbows
Fig. 8.3 Finger ring 81A4394 decorated with sheaf of arrows
Fig. 8.4 Typical longbow and drawing showing belly and back, upper and lower limb
Fig. 8.5 Distribution of heartwood/sapwood in bows
Fig. 8.6 Fluting on bow
Fig. 8.7 Pale bow tip with ridge at end
Fig. 8.8 Sections of longbows
Fig. 8.9 Graph showing range of complete lengths of bows studied
Fig. 8.10 Graph showing range (of complete bows from study) found in chests 81A1862 and 81A3927
Fig. 8.11 Bows set back in the handle
Fig. 8.12 Graph showing lengths of bows set back in the handle
Fig. 8.13 Graph showing lengths of bows set back in the handle from chests
Fig. 8.14 Reflex and deflex bows
Fig. 8.15 Deflex bows
Fig. 8.16 Trapeziodal bow 81A1607
Fig. 8.17 Nock slots upper tip, ridge
Fig. 8.18 Upper limb with double nock and right side lower limb, single slot
Fig. 8.19 Double nock on upper limb
Fig. 8.20 Comparison between nock slots in the upper and lower limbs
Fig. 8.21 Length range of double nocked bows
Fig. 8.22 Length range of double nocked bows from chests
Fig. 8.23 Bowmarks
Fig. 8.24 Bowmarks
Fig. 8.25 Catalogue of bowmarks
Fig. 8.26 Size ranges of 102 bows
Fig. 8.27 Draw-weights
Fig. 8.28 Constructs for the average bow
Fig. 8.29 Graph of mass versus maximum width
Fig. 8.30 Comparisons of constructs
Fig. 8.31–4 Selection of longbows
Fig. 8.35 Growth rings on sectioned bow

| | | | |
|---|---|---|---|
| Fig. 8.36 | Bow 81A1648 on tiller | Fig. 8.84 | Location of divots in relation to annual growth |
| Fig. 8.37 | Roy King bending fresh stave | | |
| Fig. 8.38 | Reproduction A1 straight and bent | Fig. 8.85 | Saw marks on face in slot |
| Fig. 8.39 | Reproducion A4 on tiller | Fig. 8.86 | Exploded view of arrow chest construction |
| Fig. 8.40 | Simon Stanley shooting the reproduction | Fig. 8.87 | Underwater photographs of arrow chests |
| Fig. 8.41 | Nock; original and replica and bow tips | Fig. 8.88 | Spacers 79A1204 and 82A2026 |
| Fig. 8.42 | Nock 97A0003 and possible bowstring | Fig. 8.89 | Reconstruction drawing of spacer 79A0653 with arrows |
| Fig. 8.43 | Bowstrings; fastenings and strung bow | | |
| Fig. 8.44 | Bowstrings; details of attachments | Fig. 8.90 | Extract from the Cowdray Engraving showing archers with spacer/bag |
| Fig. 8.45 | Distribution of longbow and arrow chests | | |
| Fig. 8.46 | Exploded view of longbow chest construction | Fig. 8.91 | Bag 816844 and associated arrows 81A0116 |
| Fig. 8.47 | Underwater photograph of bows in chest and on deck as recovered | Fig. 8.92 | Bag ancient quiver |
| | | Fig. 8.93 | Tip and end of of possible handgun bolt 81A1093 |
| Fig. 8.48 | Newport ship wristguard and a medieval sea battle from the *Beauchamp Pageant* with armguard on an archer's arm | Fig. 8.94 | Group of possible handgun bolts 81A1093 as recovered |
| | | Fig. 8.95 | Handgun bolts 81A1093 and unidentified possible ramrods 81A1022 |
| Fig. 8.49 | Wristguards found with the burial of the Amesbury Archer (2400–2200 cal. BC) and ivory wristguard from Southend | | |
| | | Fig. 8.96 | Sitebook showing contents of chest 81A1328 |
| Fig. 8.50 | Distributon of wristguards and related items | | |
| | | Fig. 8.97 | Unusual small handplanes |
| Fig. 8.51 | Wristguard shapes | Fig. 8.98 | Lidded box 81A1049 |
| Fig. 8.52 | Fastenings on wristguards | Fig. 8.99 | Scapula and example with *os acromiale* |
| Fig. 8.53 | Reconstuction of 81A4341 showing the position of the wristguard on the archer's forearm | Fig. 8.100 | Divelog recording discovery of human remains FCS82, a possible archer |
| | | Fig. 8.101 | Jerkin 81A0090 |
| Fig. 8.54 | Decorated leather armguard in the British Museum bearing Plantagenet insignia | Fig. 8.102 | Underwater photograph of 'Little John' and detail of arrows and arrow bag |
| | | Fig. 8.103 | Personal possessions associated with the remains of FCS 74 and 75 |
| Fig. 8.55 | Wristguard 82A0943 | | |
| Fig. 8.56 | Book covers 81A0895, 81A4131 | Fig. 8.104 | Jerkin 825026 and comb 82A5064 associated with human remains |
| Fig. 8.57–62 | Decorated wristguards | | |
| Fig. 8.63–4 | Undecorated wristguards | Fig. 8.105 | Divelog 82/8/58 showing 82H1000 and associated objects |
| Fig. 8.65 | Laced armguards 81A1863 and 81A4675 | | |
| Fig. 8.66 | Ivory wristguard 81A0815 | Fig. 9.1 | Examples of staff weapons from the Cowdray Engraving |
| Fig. 8.67 | Signet ring decorated with letter 'K' | | |
| Fig. 8.68 | Reconstruction of a spacer of arrows and illustrations of arrows | Fig. 9.2 | Anatomy of a bill, pike and halberd |
| | | Fig. 9.3 | Bill from Coventry and use of bills from the *Beauchamp Pageant* |
| Fig. 8.69 | Distribution of arrows and spacers | | |
| Fig. 8.70 | Anatomy of an arrow | Fig. 9.4 | Ceremonial halberd in the Royal Armouries |
| Fig. 8.71 | Types of leaf-shaped and bodkin arrowheads | Fig. 9.5 | Bill from Jamestown, USA |
| | | Fig. 9.6 | Anatomy of a billhead and 'T' shaped cross-nail |
| Fig. 8.72 | Overlap of different arrow studies | | |
| Fig. 8.73 | XRF analysis for copper and zinc of arrow shaft and arrow analysed | Fig. 9.7 | Tops of *Mary Rose* bill hafts |
| | | Fig. 9.8 | Bill hafts |
| Fig. 8.74 | Principal length measurements of the arrow shafts | Fig. 9.9 | Marked tops of bill hafts |
| | | Fig. 9.10 | Bill haft 79A1170 showing brass nails in cross formation |
| Fig. 8.75 | Histogram of total length of the arrow shafts | | |
| | | Fig. 9.11 | Bill in Higgins Armory Museum |
| Fig. 8.76 | Graphs of arrow data | Fig. 9.12 | Bill 82A3590 |
| Fig. 8.77 | Arrow shaft profiles by wood type | Fig. 9.13 | Bill 82A2637 |
| Fig. 8.78 | Graphs of arrow nock data | Fig. 9.14 | Bill haft 81A1646 with concretion and possible textile |
| Fig. 8.79 | Nock orientation relative to annual growth | | |
| Fig. 8.80 | Tip cone types | Fig. 9.15 | Distribution of staff weapons |
| Fig. 8.81 | Arrow shaft profiles | Fig. 9.16 | The 'pike garden' |
| Fig. 8.82 | Arrow shaft data | Fig. 9.17 | Stages in forging a bill |
| Fig. 8.83 | Arrow mass vs total length by wood type | | |

|  |  |
|---|---|
| Fig. 9.18 | Stages in forging a bill |
| Fig. 9.19 | Anatomy of a pikehead and pike from Royal Armouries |
| Fig. 9.20 | Remains of pike staves |
| Fig. 9.21 | Pike stave 81A4965 showing cheek rebates |
| Fig. 9.22 | Haft and collar, probably of a halberd |
| Fig. 9.23 | Halberds in Royal Armouries |
| Fig. 9.24 | Haft fragments possibly from throwing darts |
| Fig. 9.25 | Anatomy of a sword and scabbard |
| Fig. 9.26 | Anatomy of a dagger and sheath |
| Fig. 9.27 | Basket hilt sword 82A3589 and details of hilt |
| Fig. 9.28 | Radiographs of 83A0048, 82A1915 and 82A1932 |
| Fig. 9.29 | Stain of a third basket hilt sword on seabed |
| Fig. 9.30 | Pommel and grip fragments |
| Fig. 9.31 | Anatomy of a by-knife |
| Fig. 9.32 | Drawing taken from effigy of Lord Parr |
| Fig. 9.33 | Dagger from Hook Court, sword hangers and girdle end |
| Fig. 9.34 | Distribution all sword finds |
| Fig. 9.35 | Distribution of Fairly Complete Skeletons and with catalogue entries numbered |
| Fig. 9.36 | Basket-hilted sword 82A3589 after removal of concretion |
| Fig. 9.37 | Guard of basket-hilted sword 82A3589 after removal of concretion |
| Fig. 9.38 | Metallography of the blade on basket-hilted sword 82A3589 |
| Fig. 9.39 | Similar basket-hilted swords |
| Fig. 9.40 | Reconstructions of hilts |
| Fig. 9.41 | Grip in concretion showing wire |
| Fig. 9.42 | Grips 82A2054 and 82A4412 |
| Fig. 9.43 | Selection of wooden grips |
| Fig. 9.44 | Selection of wooden grips |
| Fig. 9.45 | Pommels and pommel fragments |
| Fig. 9.46 | Dagger from the Thames foreshore at the Tower of London |
| Fig. 9.47 | Pommel fragments and sheathing |
| Fig. 9.48 | Drawing of radiograph of 82A0027 and of pommel from sword 82A3589 |
| Fig. 9.49 | Rondel dagger 82A1914 in concretion |
| Fig. 9.50 | Rondel dagger excavated from the foreshore of the Thames at Queenhithe |
| Fig. 9.51 | Radiograph of 82A0975/2 |
| Fig. 9.52 | Dagger 83A0022 |
| Fig. 9.53 | Construction of ballock dagger grip, lobes with pins |
| Fig. 9.54 | Ballock dagger tang hole shapes |
| Fig. 9.55 | Distribution of ballock daggers and associated by-knives |
| Fig. 9.56 | Chest 81A1429 and objects |
| Fig. 9.57 | The variety of pommel shapes found on *Mary Rose* ballock daggers |
| Fig. 9.58 | The variey of grip sections found on *Mary Rose* ballock daggers |
| Fig. 9.59 | Diamond-sectioned ballock dagger grips |
| Fig. 9.60 | Octagonal-sectioned ballock dagger grips |
| Fig. 9.61 | Hexagonal, round, ovoid and lobe-shaped ballock dagger grips |
| Fig. 9.62 | Honed ballock dagger grips and examples with one lobe larger than the other |
| Fig. 9.63 | Faceted ballock dagger grip, plain pommel 81A0811 |
| Fig. 9.64 | Staining under lobes for ballock dagger guard of 81A1412 |
| Fig. 9.65 | Radiographs showing pinning on 81A1845 and 82A0890 |
| Fig. 9.66 | The variety of tang buttons found on *Mary Rose* ballock daggers |
| Fig. 9.67 | Dagger 81A0862 |
| Fig. 9.68 | Hilt 80A1350 showing the guard cut-out to take a double-edged blade and radiographs |
| Fig. 9.69 | Diamond grip ballock daggers together with by-knives |
| Fig. 9.70 | Octagonal, hexagonal, round and ovoid shaped ballock dagger grips with by-knives |
| Fig. 9.71 | Scabbards, complete examples |
| Fig. 9.72 | Decorated scabbards |
| Fig. 9.73 | Scabbards 80A0549 and 79A1146 |
| Fig. 9.74 | Details of seam location and stitches |
| Fig. 9.75 | Stitch types found on *Mary Rose* scabbards and sheaths |
| Fig. 9.76 | Scabbard 83A0151 |
| Fig. 9.77 | The range of chapes found on *Mary Rose* scabbards and sheaths |
| Fig. 9.78 | Radiographs of chapes |
| Fig. 9.79 | Sheaths 79A1146 and 81A1567, concretion stain on tip from corroded iron chape |
| Fig. 9.80 | Sheaths |
| Fig. 9.81 | Radiographs of 82A1870, 82A1916, 82A1697 |
| Fig. 9.82 | Albrecht Durer's engraving *Knight, Death and Devil* and drawing from engraving by Giovanni Battista Palimba |
| Fig. 9.83 | Sword hanger and sheath 80A1709 |
| Fig. 9.84 | Reconstruction of bifurcating belt |
| Fig. 9.85 | Sword belt, hanger and scabbard 81A1567 |
| Fig. 9.86 | Sword belts and hangers |
| Fig. 9.87 | Sword hangers |
| Fig. 10.1 | Mail garments, Brigandine (Jack) and plate armour |
| Fig. 10.2 | Comb morion; Spanish morion with pear stalk finial; sketch of helmets from Boulogne, Marquisson/Portsmouth |
| Fig. 10.3 | Almain rivets |
| Fig. 10.4 | Drawing of armour, from the *Siege of Boulogne* |
| Fig. 10.5 | 82A4096 breastplate |
| Fig. 10.6 | Distribution of armour and possible fittings |
| Fig. 10.7 | Breastplate RA 111.4572 |
| Fig. 10.8 | A sample from the breastplate |
| Fig. 10.9 | Sample 81S0167, stack of breastplates |
| Fig. 10.10 | Buckler 82A4212 |
| Figs 10.11–13 | Possible armour fragments |

Fig. 10.14 Range of sixteenth century buckle types and those actually found
Fig. 10.15 Selection of buckles
Fig. 10.16 Types of mail and methods of manufacture
Fig. 10.17 Examples of mail fragments
Fig. 10.18 Examples of mail fragments
Fig. 10.19 Distribution of mail
Fig. 10.20 Distribution of Fairly Complete Skeletons
Fig. 10.21 Distribution of human bones
Fig. 11.1 Tactics: gaining the weather gauge
Fig. 11.2 Artist's reconstruction of demi culverin facing forwards in the Sterncastle; front of Sterncastle as it survives in museum
Fig. 11.3 Canted guns on the *Henry Grace à Dieu*, as showing in the Cowdray Engraving
Fig. 11.4 Sailing formation: line abreast
Fig. 11.5 Dardanelles gun
Fig. 11.6 Various schemes for naming of the decks
Fig. 11.7 Outboard elevation of the *Mary Rose* in 1514
Fig. 11.8 Outboard elevation of the *Mary Rose* in 1545
Fig. 11.9 Internal elevation of the *Mary Rose* showing final arrangement of decks
Fig. 11.10 The Cowdray Engraving showing the sinking of the *Mary Rose*
Fig.11.11 Details from the Cowdray Engraving
Fig.11.12 Digital terrain models of the Solent; aspect of the battle as in the Cowdray Engraving and with the English and French fleets
Fig.11.13 The position of the fleets plotted on an old chart
Fig.11.14 Positions of the opposing fleets with 250m range rings
Fig. 11.15 Fleet with ships' tracks
Fig. 11.16 Charts showing the tidal changes throughout the day
Fig. 11.17 External elevation of the ship
Fig. 11.18 Gunports
Fig. 11.19 Internal structure of the ship
Fig. 11.20 Diagonal braces
Fig. 11.21 Line of the Main deck and interpretation showing positions of rising knees and photograph of large knees on Main deck showing how they constrain gun movement
Fig. 11.22 Tiers of guns and tiller/rudder attachments from the *Anthony Roll*
Fig. 11.23 Gunport lids, external
Fig. 11.24 Gunport lids, internal
Fig. 11.25 Fittings found beside gunports
Fig. 11.26 Views of gunports and swivel holes in inner wale at Upper deck level
Fig. 11.27 Operation of gunport lids
Fig. 11.28 Location of armament stores
Fig. 11.29 Isometric with all guns found in position and spot locations for recovered handguns and hailshot pieces
Fig. 11.30 Main deck showing all guns in position
Fig. 11.31 Securing iron guns to the side of the ship
Fig. 11.32 Main deck plan with possible methods of bringing in bronze guns to reload
Fig. 11.33 Positioning of alternative types of gun out of the stern at Main deck level
Fig. 11.34 Stern elevation showing possible location of port piece or demi culverin
Fig. 11.35 Outboard elevation of Sterncastle showing gun below port and details of structure around M8 port
Fig. 11.36 Plan of the Upper deck with all postulated guns in position
Fig. 11.37 The Upper deck in the waist
Fig. 11.38 Reconstruction outboard view of upper part of ship's side in waist and outboard view showing blinds
Fig. 11.39 Possible distribution of ordnance on Castle deck in stern, bases on either side of rudder
Fig. 11.40 Possible distribution of ordnance on Castle deck in stern, demi culverins on either side of rudder
Fig. 11.41 Approximate intersection of arcs of fire of Main deck ordnance at various ranges
Fig. 11.42 Approximate intersection of arcs of fire of Upper deck ordnance at various ranges
Fig. 11.43 Protective zones afforded by proposed ordnance on the Castle deck
Fig. 11.44 Protective zones afforded by the intersections of arcs of fire on all decks
Fig. 11.45 Zones of coverage afforded by the longest range guns on all decks
Fig. 11.46 Zones of coverage afforded by weapons on a modern warship
Fig. 11.47 Detail of grapnel on *Anthony Roll* and reconstruction of grapnel
Fig. 12.1 Solent chart with Historic Wreck circle and proposed line of MOD dredge
Fig. 12.2 Data provided by remote sensing
Fig. 12.3 The ROV vehicle and sieving spoil
Fig. 12.4 The positions of the trenches into spoil and area worked north of bow
Fig. 12.5 The Site Recorder displays
Fig. 12.6 Location of ordnance recovered 2003–5
Fig. 12.7 Scabbard strap and top and ballock dagger
Fig. 12.8. Distribution of ordnance in the Forecastle area
Fig. 12.9 Tritech parametric profiler on frame
Fig. 12.10 Priming wire and dagger quillion block or part of gunner's stiletto
Fig. 12.11 Sand barge and hopper used in reburial of wreck site
Fig. 12.12 Sketch of bow showing two tiers of forward facing guns in Sterncastle and *Anthony Roll* image of *Mary Rose* for comparison
Fig. 12.13 Simple plan of *Mary Rose* with alternative gun arrangements on Castle decks
Fig. 12.14 Bowstring found in 2003 and previously recorded example

Fig. Appendix1.1 Segregation; 81A1423 and 79A1232
Fig. Appendix 1.2 81A1423 correlation of tin and antimony
Fig. Appendix 1.3 *Mary Rose* bronze ordnance segregation
Fig. Appendix 1.4 Analysis of European bronze ordnance 1500–1560
Fig. Appendix 1.5 lead isotope analysis of *Mary Rose* ordnance

# List of Tables

Table 6.1 Analysis of the incendiary mix from projectile 81A2866
Table 6.2 Fragments of 'charge' associated with Tudor artefacts
Table 6.3 Mortar bomb fragments by sector
Table 7.1 Common surviving measurements of handgun stocks
Table 7.2. Size, number, and distribution of small lead round shot suitable for handguns, and estimated bore sizes
Table 7.3 Components of gun-shields recovered
Table 7.4 Principal measurements of gun-shields
Table 7.5 XRF results of metal components of gun-shields
Table 7.6 Comparative data for the Royal Armouries gun-shields
Table 8.1 Archery equipment listed in the *Anthony Roll*
Table 8.2 Distribution of longbows
Table 8.3 Contents of chest 81A5783
Table 8.4 Location of 104 complete longbows
Table 8.5 Longbow data
Table 8.6 Distribution of longbows of reflexed and deflexed shape
Table 8.7 'Slab-sided' longbows; location and length
Table 8.8 'Slab-sided' longbows; width and depth along length
Table 8.9 Round D-sectioned longbows; width and depth along length
Table 8.10 Flat D-sectioned longbows; width and depth along length
Table 8.11 D-sectioned longbows; width and depth along length
Table 8.12 Handled bow 81A3949; width and depth along length
Table 8.13 Longbows with a trapezoid section and relieved grip area
Table 8.14 Distribution of longbows with double nocks
Table 8.15 Distribution of marked longbows
Table 8.16 Timber grade and uniformity
Table 8.17 Distribution of used longbows
Table 8.18 Draw-weights of *Mary Rose* bows by location
Table 8.19 Grouped frequency table for bow weight by wood type
Table 8.20 Variables used in the preliminary analysis
Table 8.21 Mean length, mass, maximum width and maximum depth by category
Table 8.22 Testing of *Mary Rose* longbows
Table 8.23 Length and elastic modulii for selected longbows
Table 8.24 Weight and density of selected longbows
Table 8.25 Weight and density of different woods
Table 8.26 Estimated draw-weights of *Mary Rose* longbows
Table 8.27 Experimental shooting data
Table 8.28 Range and velocity tests of MRA4 Harlington
Table 8.29 Measurement of horn nocks from arrows
Table 8.30 String fit to arrow nock
Table 8.31 Contents of longbow chests
Table 8.32 Wristguards
Table 8.33 Distribution of arrows by deck
Table 8.34 Arrows in chests
Table 8.35 Arrows; wood analysis
Table 8.36 Arrows: wood analysis arrows in chests
Table 8.37 Arrows; wood identification in arrow 'bundles'
Table 8.38 Density tables of woods used in *Mary Rose* arrow assemblage
Table 8.39 Density classification
Table 8.40 Mechanical properties
Table 8.41 Specific gravity and modulus of elasticity
Table 8.42 Distribution of wood types
Table 8.43 Two-way distribution of location and length group

Table 8.44 Mean total length and 95% confidence limits by arrow chest
Table 8.45 Percentiles of binding length plus nock end length
Table 8.46 Mean and 95% confidence limits for total length and for binding length plus nock length by wood type
Table 8.47 Two-way distribution of wood type and arrow chest
Table 8.48 Distribution of binding frequency
Table 8.49 Two-way distribution of nock end diameter and tip end diameter
Table 8.50 Distribution of arrow profiles
Table 8.51 Two-way distribution of profile and arrow chest
Table 8.52 Two-way distribution of wood type and profile
Table 8.53 Arrow mass (g) by wood type
Table 8.54 Arrow mass (g) by wood type, length group and profile
Table 8.55 Contents of arrow chests
Table 8.56 Arrow spacers
Table 8.57 Contents of chest 81A1328
Table 8.58 Details of short arrows 81A1093
Table 8.59 Contents of chests possibly belonging to archers
Table 9.1 Total numbers of Morris pikes, bills and longbows listed by vessel type in the *Anthony Roll*
Table 9.2 Summary of basic measurements for bill hafts
Table 9.3 Grips and attached pommels from swords and daggers
Table 9.4 Pommel sheaths with metal analysis
Table 9.5 Ballock daggers
Table 9.6 By-Knives associated with ballock daggers
Table 9.7 Unassociated by-knives
Table 9.8 Scabbards
Table 9.9 Chapes
Table 10.1 Composition (inclusions) of breastplate 82A4096
Table 10.2 Leather straps and other items possibly for armour
Table 10.3 Tudor buckles
Table 10.4 Analysis of mail samples
Table 10.5 Type 1 (butt link) mail
Table 10.6 Type 2 and other mixed rivet/butt link mail
Table 10.7 Type 3 (rivet link) mail
Table 10.8 Mail of uncertain link form
Table 11.1 Comparison of inventory records for *Mary Rose* guns
Table 11.2 Extroplated location of guns on named ships in 1514
Table 11.3 Gun ranges from historical sources
Table 11.4 Historical gun range (average point blank in metres)
Table 11.5 Range of culverin calculated in degrees of the quadrant
Table 11.6 Gunport dimensions
Table 11.7 Gunport lid dimensions
Table 11.8 Ordnance on the Main deck
Table 11.9 Possible ordnance on the Upper deck
Table 11.10 Ordnance on the Castle deck (postulated)
Table 11.11 Arcs of fire: Main deck
Table 11.12 Arcs of fire: Upper deck
Table 11.13 Weapons and ammunition
Table 11.14 Weapon '*preference*' based on number of rounds per gun
Table 11.15 Daily expected firing capability on land
Table 11.16 Sequence based on total number shot supplied
Table 11.17 Preference list based on number guns carried
Table 11.18 Weapons and ammunition according to range
Table 11.19 Ship supported/carriage-mounted guns and shot
Table 11.20 Bronze guns/shot listed and recovered
Table 11.21 Endurance based on shot and reload rates
Table 11.22 Endurance based on volley firing (all guns loaded)
Table 12.1 Artefacts recovered from spoil mounds
Table 12.2 Artefacts recovered from the Timber Park
Table 12.3 Artefacts found on stem or western timbers
Table 12.4 Artefacts found east of stem

Table Appendix 1.1 Guns sampled for metal analysis
Table Appendix 1.2 Segregation in the bronze guns
Table Appendix 1.3 Sample analysis of the bronze guns from the *Mary Rose*
Table Appendix 1.4 Compositions of English guns of the reign of Henry VIII
Table Appendix 1.5 Analysis of European bronze ordnance 1500–1560
Table Appendix 1.6 Lead isotope analysis of bronze guns
Table Appendix 1.7 Detailed analysis of the bronze guns
Table Appendix 2.1 Gravimetric and Titrimetric determination of sulphur
Table Appendix 2.2 Determination of sulphur content (bomb calorific determinations)
Table Appendix 2.3 Emission spectophotometric analysis of powder samples
Table Appendix 2.4 Determination of metals by atomic absorbtion spectrophotometry
Table Appendix 2.5 Determination of carbon content
Table Appendix 2.6 Percentage of soluable solids in samples 81S1409 and 82S1207
Table Appendix 2.7 Elemental analysis of insoluble solids, samples 81S1409 and 82S1207
Table Appendix 2.8 Elements present in samples 81S1409 and 82S1207
Table Appendix 2.9 Ether soluble material present in samples 81S1409 and 82S1207

# List of Colour Plates

Plate 1    The *Mary Rose* as she appears in the *Anthony Roll* of 1546; painting by Geoff Hunt *The Mary Rose. Henry VIII's flagship 1545*

Plate 2    Cannon 81A3003 on carriage underwater; culverin 80A0976 on its carriage on the Upper deck in the stern; detail of Arms and Garter, bronze demi culverin 79A1232; detail of device, bronze demi cannon 81A3002

Plate 3    The bellfoundry during the casting of the reproduction culverin; firing the reproduction port piece

Plate 4    Copper alloy priming wire; gunner's rule, shot gauge and muzzle of demi culverin; scorched head of linstock; selection of objects used in firing the guns

Plate 5    Remains of wooden gun-shield; decorated copper alloy gun-shield boss; underwater shot of chest 81A1862 full of longbows; decorated leather wristguard 80A0901

Plate 6    Arrow spacer containing arrows; stain of a basket-hilted sword on the seabed; ballock dagger grip; copper alloy fittings for sword belts and scabbards

Plate 7    The only complete sword recovered; remains of the only identifiable armour breast plate; view along the surviving portion of the Main deck showing knees and gunports

Plate 8    Internal elevation of the starboard side showing decks and gunports; external view of Main deck gunport

# Contents of DVD

**Image gallery (PDF)**

*Bronze guns*
    Line drawings
    Colour photographs
*Iron guns*
    Line drawings
    Colour photographs
*Gun furniture*
    Line drawings
    Colour photographs
        Colour photographs (shot)
*Handweapons*
    Edged weapons
        Line drawings
        Colour photographs
    Handguns and gun-shields
        Line drawings
        Colour photographs
    Staff weapons
        Line drawings
        Colour photographs
*Archery*
    Line drawings
    Colour photographs
*Protective*
    Line drawings
    Colour photographs
*Ship structure*
    Line drawings
    Colour and black and white photographs
*Paintings*

**Spreadsheets, databases and recording forms**
    *Pro-forma recording sheets (PDF)*
        Arrows
        Linstocks
        Powder dispensers
        Shot
        Staff weapons
    *Longbow data sheets (PDF)*
    *Ordnance database (PDF)*
    *Site Recorder interactive database*

**Videos (three short films)**

Mary Rose *2003*
A short film about the 2003 site investigations and excavations. It outlines the Ministry of Defence's potential plans to dredge a new deeper and straighter channel into Portsmouth for the new aircraft carriers. This would have cut into the 500m zone of protection around the wreck site, and provided funding for specific work onsite. Written by Alex Hildred, narrated and produced by Stephen Foote.

*Reunion and Reburial 2003–2005*
A short film put together from 2004 site film footage and extracts from the BBC *Timewatch* programme *Secrets of the Mary Rose*. It introduces the plans for the 2005 season to recover the stem and anchor and rebury the site following the excavations of 2003–2005. Written and partly narrated by Alex Hildred, produced by Kester Keighley. Reproduced by permission.

*Making and firing* Mary Rose *guns*
This film is an excerpt of a documentary produced by Yorkshire Television and the Royal Armouries. The excerpt features the building and firing of reproductions of a wrought iron and a cast bronze gun between 1998 and 2000. Included by permission of the Board of Trustees of the Royal Armouries. Produced by Kester Keighley.

# PART 3: HAND-HELD WEAPONS

# 6. Incendiary Devices
## Alexzandra Hildred

*'Their saile to burne, we shoot our arrowes of wilde fire'*
(Hakluyt 1588–1600, vol. 5)

## Introduction

Incendiary weapons were an important component in medieval and post-medieval warfare. Both sixteenth and seventeenth century military manuals list numerous types of missiles combined or coated with combustibles. These can be held, thrown, fired from a gun or shot from a longbow or crossbow. Recipes are many and varied but all rely on gunpowder, which, together with a range of flammable materials, make up the incendiary. They are highly inflammable, readily ignited and very difficult to extinguish.

True *'wildfire'* is the incendiary *'Greek Fire'* of the Byzantines, and is liquid or semi-liquid. Projectiles carrying wildfire are described in detail in the military manual *De Re Militari* by Flavius Vegetius Renatus dated AD 384–9. The *Dromons* or oared galleys of the Byzantine Imperial Fleet used Greek Fire as a main attack weapon. By the time of the Crusades, Greek Fire was part of an army's standard equipment, propelled at the enemy by means of loaded arrows, large open firepots, sealed clay grenades and a siphon from which a flaming jet was emitted. In sixteenth century inventories, in particular that of the King's possessions taken after the death of Henry VIII (Starkey 1998), the term *'wildfire'* appears to be applied to any incendiary composition.

Wright's illustrated treatise of 1563 (Society of Antiquaries MS 94) describes in detail the manufacture of incendiary weapons (Figs 6.1–2). One method was:

*'A ryceate* [receipt, ie recipe] *for fyere workes and the temperaunces for speres, harroes, hopes, balles, quarrelles and polles, as here after followithe. Take corn powder xxxj pounde and viij pounde of sullfer and myngle them together, then take iiij pounde of swet oyll and iij pounde of turpentyn and iij pounde of camfher and pute them in a pote over the fyere mylke warme or more, then strayne it over the receates and myngell them together and rube them betwen your handes halfe a nower, then take mercury subley* [subtilely, ie thinly] *and iiij pound of fynde powder and ij pounde of verdegrace and ij pounde of arsennyke and bet all thes together in powder and then take salt peter iiij pounde so that it be above them, take canvas and cut it after this sorte the lenthe of xiiij enches and vij enches in breadthe then take marlyn corde or whipe corde to bynde them with all and iiij pounde of pyche to border this receates. But remember the toche holl and to make strong worke in your workyng. This is for speres, trunkes, arrowes'.*

Other contemporary descriptions are found in the Tenth Book of the *Pirotechnica* of Vannoccio Biringuccio (1540), Norton (1628) and Lucar (1588), though the most famous work is the beautifully illustrated *Firework Book* of c. 1400, an anonymous book written in Germany. Fifty-five copies of this are extant, one of which was translated into English and printed in the *Journal of the Arms and Armour Society* (Kramer 2001). A thorough modern study is found in Partington (1960).

Although many ingredients are listed, the most frequently cited appear to be: ammonia, animal fat, arsenic, *assa fetida* (a resinous gum), *aqua vitae*, bay salt (originally from the Bay of Biscay, but generally used to describe sea salt), brandy (Kramer 2001), camphor, egg, juniper oil, linseed oil, mercury, naphtha, petroleum, pitch, quicklime, rosin (a particular type of resin), sal ammoniac (ammonium chloride), saltpetre, oil of sulphur (sulphuric acid), sulphur (often termed brimstone) turpentine, undistilled wine, verdegris (copper acetate), vinegar, wax and gunpowder (Partington 1960, 42–90; Biringuccio 1540, bk 10). Flax, or linen thread/cloth, acted as a wick or a wrapping to bind in the flammable fluid mixture. The distinctive materials for Greek Fire include liquid petroleum, pitch, rosin, oils, sulphur, tar or bitumen and quicklime; all these ingredients had to be imported from the Middle East.

A fair selection of 'recipes' can be found in Norton (1628, 149–57) from which it is clear that manufacture was a two-stage process: mixing the incendiary composition (basically gunpowder but incorporating

sulphur, rosin, linseed oil and turpentine) followed by 'arming' or 'capping' – applying an outer coating incorporating the fuse. Items listed for this included pitch, rosin, sulphur, tallow, marline, canvas and thread.

Some incendiaries are, in themselves, containers (tubes, balls, sacks, pots, shot, garlands) filled with an incendiary mixture and lit with a fuse (Fig. 6.1). These can either be projected as individual items, or be combined with another missile or weapon such as an arrow, pike or lance (Fig. 6.1). 'Garlands' are linen or hemp bags filled with an incendiary mix and stitched to an iron hoop which was thrown over the enemy's head. 'Shot of wildfire' are hollow cast iron or copper alloy shot filled with an incendiary mix (Fig. 6.2) Derivations include spike shot, where the ends of the spikes are bound with an incendiary coated material. One of the spike shot recovered from the *Mary Rose* looks as though it has been armed (Fig 5.32b).

The most common incendiaries are '*fire trunks*' or '*tubes*' (Fig. 6.3, top). These are reinforced cylinders of wood, paper, iron or sheet copper filled with an incendiary mix. All require a fuse and can be either attached to the ends of pikes or lances, to thrust a tongue of fire at the enemy at close quarters, or, if loaded with shot in addition to the incendiary mix, they can be attached to the end of a pole to act as a small portable gun. Without the pole they can be fired from a gun (Fig. 6.3, bottom). The 1547 inventory lists 269 trunks, making these the most numerically significant

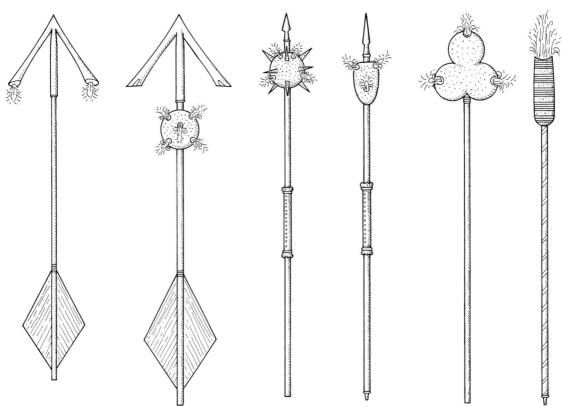

*Figure 6.1 Above) ball, sack, pots garlands of wildfire; below) incendiary darts, pikes, trunk (after Wright 1563)*

*Figure 6.2 Use of shot of wildfire at sea (after Wright 1563)*

incendiaries recorded (Blair 1999). Wright (1563) describes these as being 20in (508mm) in length and 10in (254mm) in circumference and 1½in (38mm) in the mouth on the end of a long staff. Biringuccio's description of fire tubes details the method of manufacture and use (1540, 425–6):

> 'Fire tubes are commonly made in order to frighten horses or to harm enemy soldiers ... they are beautiful things to see and when the name tubes of fire is heard it terrifies those who do not have a defense ready ... Above all, they are good in naval battles.
>
> Some are also made to not only to vomit fire but to shoot certain balls which ignite when they issue and burst in the air. I once made some of these like a gun, which I caused to shoot stone balls able to break any good thick wooden gate, and they served me admirably for the purpose that I made them for.'

Sir Henry Mainwaring, in the earliest *English Seaman's Dictionary*, written between 1620 and 1623, describes fireworks as:

> '*Fireworks are any kind of artificial receipts applied to any kind of engine, weapon or instrument, whereby we use to set on fire the hull, sails, or masts of a ship in a fight; whereof there are many sorts, but the most commonly used at sea are these:* fire-pots, fire-balls, fire-pikes, trunks, brass-balls, *arrows with firework, and the like*'. (in Manwaring and Perrin 1922, 147).

Archaeological evidence for the use of incendiaries at sea include matchlocks which could have used fire arrows (Fig. 6.4) recovered from the wrecks *Association* of 1707 (Morris 1984) and *Batavia* (1629) (Green and North 1984). Grenades/fire pots were recovered from the sixteenth century Alderney wreck (Davenport and Burns 1995; Monaghan and Bound 2001); firepots or '*alacancias*' from the 'Church Rocks' site, Teignmouth, Devon (Preece and Burton 1993); *alacancias* and a '*bomba*' or trunk from *La Trinidad Valencera* (Martin 1994), both sixteenth century sites. A large collection of sixteenth century incendiary devices, including garlands, grenades, fire pots and exploding shot is in the collection in Coburg Castle, Germany (Geibig 2001). A survey of military fireworks concentrating on English manuscript sources is summarised by Ruth Brown (2005, 25). Figure 6.5 (Norton 1628, tract 3, chap. 23) shows a selection including trunks, wreaths and arrows designed to be shot from crossbows and longbows. Although probably not to scale, note should be taken of the longbow (probably length *c.* 2000mm) and the incendiary arrow shown below.

## Historical Importance

The 1514 inventory (Knighton and Loades 2000, 113–58) details thirteen vessels thoroughly. The armaments carried suggests an emphasis on close quarters engagements (as reflected by the number of hand weapons and size of guns). Crossbows are present in addition to longbows, arrows, morris pikes and bills. Javelins are listed, for example for the *Henry Grace à Dieu* (nine dozen), *Kateryn Forteleza* (100) and the *Great Barbara* (150). Darts are listed in '*bundles*' or in dozens, with the largest number on the *Henry Grace à Dieu* (57 dozen).

The *Gabriel Royal* carries five balls of wildfire with hooks of iron and 56 '*Boltes of wyldefyre to cast at shippes*' (Knighton and Loades 2000, 126). The *Great Nicholas* is supplied with five arrows of wildfire and eight darts (*ibid.*, 135). The *Mary Rose* carries eight heads for arrows of wildfire and 29 hooks for arrows of wildfire together with one lime pot, 74 arrows of wildfire, two

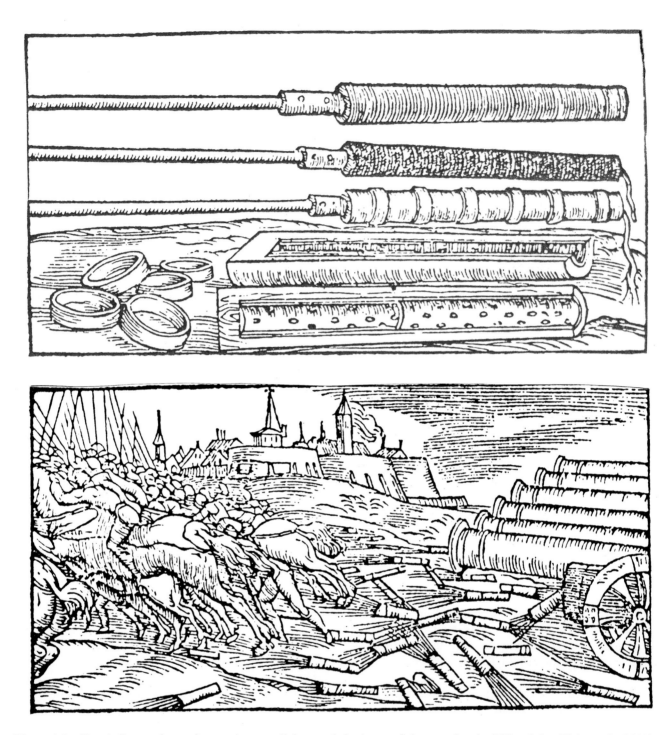

*Figure 6.3 Above) fire trunk or tubes serving as a light gun; below) guns firing trunks of wildfire (after Biringuccio 1540)*

balls of wildfire, '*xxj di barelles*' of powder (assumed to 21 half barrels) and a chest of cartridges (*ibid.*, 141–2). The *Peter Pomegranate* carried eight darts of wildfire, twelve great and 20 small balls of wildfire and two chests with quarrels, arrows and trunks ('*trompes*') of wildfire. The *Great Elizabeth* is listed with 40 '*Cast of wyldfyre*' (*ibid.*, 150).

The 1546 *Anthony Roll* (Knighton and Loades 2000) contains reference to two items which could be considered incendiary and caustic, darts for tops and lime pots. The darts for tops are routinely listed in dozens; the *Mary Rose* carries 40 dozen. This is an incredible number compared with the 150 bills, 150 pikes, 250 longbows and 400 sheaves of arrows. The sheer number suggests that these must be small, they also must be easily transported to and handled within the fighting tops, already furnished with the wrought iron top pieces (see Brown, Chapter 3). There is no mention of darts or arrows of wildfire, nor any specific reference to incendiaries.

*Figure 6.4 Musket firing an incendiary arrow (after Smith 1627)*

All nineteen ships have darts for tops, with the *Henry Grace à Dieu* having 100 dozen, the *Peter* 30 dozen. The total number is 3624. The fifteen galleys have 70 dozen with individual vessels averaging 2–6 dozen. The ten pinnaces all have top darts, with 1–4 dozen, totalling 24 dozen. Of the thirteen rowbarges only the *Double Rose* is listed with one dozen. The vessel total is 23,124. The 1547 inventory (Starkey 1998, 102–44) includes detailed armament inventories for over 70 establishments. Items listed include trunks, fire balls, pikes, arrows, lances, hoops, and great and small pots. Also listed are quarrels and sheaf arrows with wildfire. Many of these are entered beside lanterns, candles and tallow. Establishments with substantial amounts of incendiaries include the Tower of London, Bullingham and Blackness forts, in Calais, Roxburgh, Tynemouth and Carlisle. Of particular interest are the following: God's House in Portsmouth, with linseed, turpentine, rosen, salt peter pitch, tar, canvas, marlin, packthread, twine and pack needles; Southsea Castle holding gunpowder and '*Certeyn stuf for fireworke*' (ibid., 4457); Yarmouth (Isle of Wight) supplied with shot of wildfire; and Calshot with trunks for wildfire and pitch and tar. At Calais there is a '*Wildfyre house*' (4991–8). Calais and Boulogne with their outliers have over five times the number of incendiary items than in all of England. Calais would have to be able to manufacture incendiaries on site rather than rely totally on pre-made imported devices.

Caustic substances which are designed to be thrown as anti-personnel missiles are often covered by the term '*lime pots*' or '*casting pots*'. If linseed oil is mixed with the quicklime this becomes a caustic and flammable oil which would be difficult to extinguish or remove from the skin and would cause severe burns, scalds and shock.

Lime pots (but not specifically '*firepots*'), under the heading of '*habillements for warre*' are mentioned in 1546 (Knighton and Loades 2000) for the *Mary Rose* and nineteen of the 20 ships have lime pots, the exception being the *Matthew*. The highest number is 20 dozen (*Henry Grace à Dieu*). Most have three dozen, with two dozen as the least (*Hoye Bark*). The total for ships is 91 dozen. For the fifteen galleasses, with most vessels having 2–4 dozen, the total is 1176 dozen [14,112]. The *Mary Rose* is listed with ten dozen. The ten pinnaces have 2–3 dozen per vessel, total 23 dozen. All thirteen rowbarges have lime pots, totalling 23 dozen. The total carried in the fleet is 2580. Both lime pots and incendiary darts can be considered as important additions to the weapons based on numerical incidence alone. This suggests that close combat warfare was still a vital part of any engagement.

The high numbers of both these types of object suggest that they must be relatively small, simple unglazed pots as described by Giebig (2001, 88–9). The incendiaries described below are therefore probably not the 40 dozen darts for tops listed for the *Mary Rose*. The durability of ceramics suggests that had there been lime pots on board, this should be reflected in the archaeological record unless they had all been on the Upper or Castle decks and dispersed. The one pottery vessel found in the same area as the incendiaries described below (81A2103) has now been identified as a glazed cooking pot. All the ceramics recovered from the ship are of recognised types from known pottery production centres and all appear to be either medicinal or domestic in function (Brown and Thomson, AMR Vol. 4, chap. 11).

## *Mary Rose* Incendiary Projectiles

Three items identified as incendiary darts were recovered from the *Mary Rose* (81A2866 (Fig. 6.6), 81A6703, 81A6931). Described at the time as '*paddles*', these were uncovered during excavation around bronze demi cannon 81A3000 on the Main deck in the stern (M10) (Fig 6.7). As the vessel sank, the back of the gun carriage had lifted up off the deck towards the underneath of the Upper deck and the space created beneath the axle and wheels became a collection area for stone shot (81A3602–4, 2270) and iron shot which

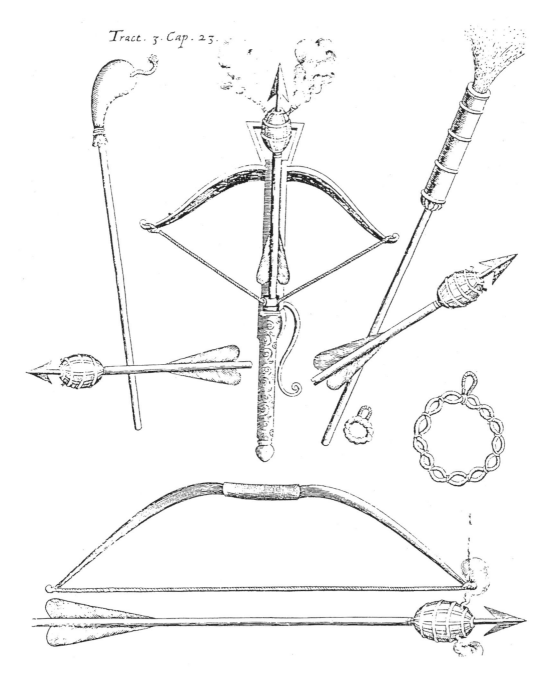

*Figure 6.5 Arrows, trunks and garlands of wildfire (after Norton 1628)*

*Figure 6.6 Incendiary projectile 81A2866*

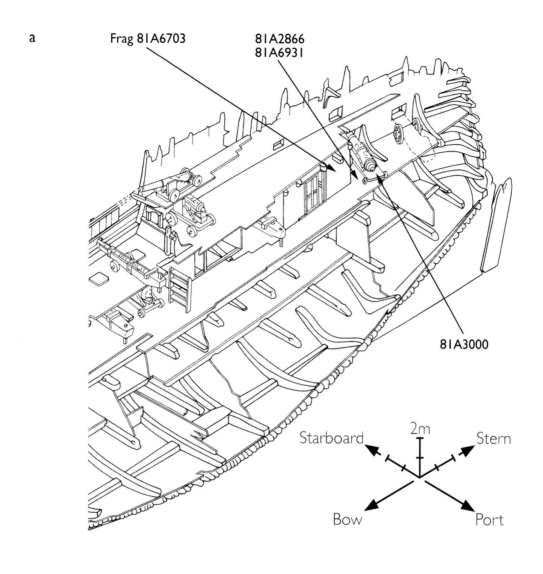

*Figure 6.7 a) Isometric showing location of incendiary darts; b) divelog showing the seabed location of incendiary darts*

*Figure 6.8 Photograph of 81A2866 as found*

subsequently corroded into a mass of concretion. Two of the flights were excavated from between the side cheeks of the carriage and the partition forming the southern end of the Carpenters' cabin, with flight ends closer to the Main deck and heads towards the Upper deck, from where they may have fallen. Other artefacts associated with the carriage included linstocks (81A2002; 81A2864), lantern shot (81A2019, 81A2431, 81A3061, 81A3043), brush (81A2208), curved brick (81A2275), tampions (81A2400–3, 81A2865), wooden bowl (81A2396), powder ladle (81A2855), rammers (81A2410, 81A3030) and the broken projectile shaft (81A2866); not identified at the time. A chest (81A2035) found near the gun, contained, among other things, the pottery vessel mentioned above.

Two of the flights, one of beech and the other oak, were brought up together (numbered 81A2521) with six fragments of haft, five semicircular in cross-section and one circular, 25–40mm in diameter. Several weeks later a '*pole*' was raised from the same area, circular in section, 1265mm in length and 40mm in diameter. The location, type of wood and matching of the break enabled this to be identified as the missing haft to flight 81A2521. The head end to this was also identified, having been raised as an unidentified pole with a collared end (81A2866) of 580mm, diameter 40mm.

The photograph taken before conservation (Fig. 6.8) shows a dark mass compressed around the collar extending for nearly half the surviving length of the haft. This material was separated before treatment (81S1166). On first appearance it appeared to be a soft black mass containing large solid lumps that were curved in section and made of hardened pitch on the outer side. The inside appeared to consist of a congealed semi-solid black waxy substance resembling gunpowder. Cleaning of the lumps revealed a fabric lining adhering to the inner surface of the pitch with lengths of cord protruding at right-angles to the shaft from the broken edges of the pitch and detectable behind the fabric layer. Two of the lumps had carved oak pegs of inserted through the pitch and fabric into the '*gunpowder*' (Fig. 6.9). It appeared to be the remains of a bag which had been filled, bound and then coated with rosin.

The reconstructed object (81A2866; Fig. 6.10) therefore resembles a long, large wooden dart. It has an oak flight 375mm long by 175mm wide, supported on haft made from silver fir (*Abies* sp.) 1265mm in length and 40mm in diameter, with a head end 580mm in length. The reassembled length is 2220mm. The flight is formed of a single, 6-sided piece of wood that slotted into the split end of the haft. Both ends of the flight are cut to form 45mm ends to accept the split haft. The sides are angled to form a dart, with edges of 125mm (base) and 320mm (towards the head). The maximum width is 175mm. The end of the haft supporting the flight is split for 200mm to form a rib on each side of the flight. There are remains of a tar coating and a nail hole at the top and bottom of the split to retain the flight. At the upper end there is a score mark 8mm in width, around the shaft. The flight is 5mm thick and has nail holes which match those on the haft. The 40mm diameter haft tapers to a collar 12mm wide, 30mm in diameter. The face of the collared end has an iron stained 10mm square hole recessed to a depth of 60mm, presumably to accept some form of iron spike. Four 2mm diameter holes are equally spaced around the circumference just below the collar with four more a further 300mm along the pole. This set is in pairs and similar in appearance to small staple holes. All carry evidence of iron residue that retained impressions of fabric. The haft is heavily iron stained and a small amount of fabric can be seen protruding from beneath a layer of pitch attached to the haft below the pairs of holes.

Identification of this object as a large incendiary projectile was based on its similarity to several illustrations in the Wright manuscript (1563) and in Biringuccio (1540) and the probable remains of gunpowder.

The proposed method of manufacture is as follows (Fig. 6.10):

1. Preparation of the shaft with flight and iron tip.
2. The nailing of a linen bag around the haft.
3. Filling the bag with an incendiary mix, nailing the top below the collar and binding with twine.

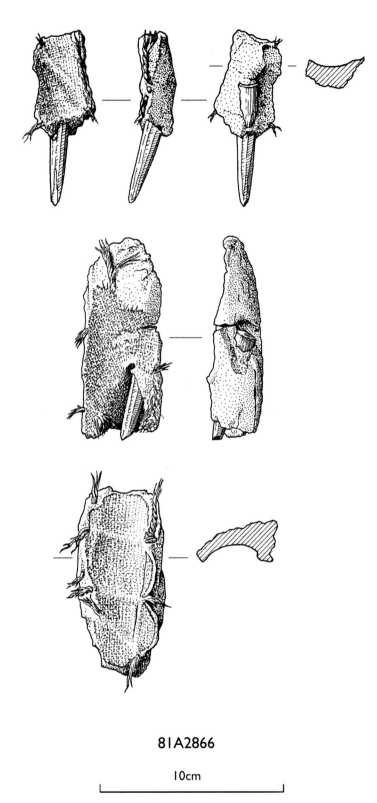

Figure 6.9 *Flight, shaft and remains of bag containing incendiary mixture, 81A2866*

4. Covering the bound bag with pitch and nserting the oak fuse pegs.

Investigation of the cords suggests the possibility of two applications of pitch, one bound closer to the vertical and another to the horizontal. It would be easier, safer and cleaner to pierce the bound bag of incendiary mixture with the wooden fuse pegs after an initial application of pitch had been allowed to harden. The extra cords, if attached at one end only to the tip of the projectile, may have served as conduits for the incendiary mixture and burning pitch. They would quickly fall away from around the head of the projectile as the pitch softened with the heat and remain attached or hang close to the side of the target, thus ensuring that the incendiary mixture did not merely spread backwards along the length of the shaft. The fuse pegs are placed perfectly to direct the thrust of the incendiary mixture towards the target.

Biringuccio describes the manufacture and use of similar items (1540, 434–5; Fig. 6.11):

> '*A liquid composition is also made in a kettle in which are put pork fat, petroleum oil, oil of sulphur, live sulphur, doubly refined saltpeter, aqua vitae, Grecian pitch, turpentine, and some coarse gunpowder. Having liquified everything well over a fire, stir it thoroughly with a stick so that it may be well incorporated, and proceed to fill all the pots or vessels that you have with this composition in the desired amounts, Then put a layer of good gunpowder over them so that they will ignite easily when desired, and let them stand. Use them when you wish, throwing them with slings, or tied with rope or iron wire like a whip handle, or in other ways of throwing by hand.*
> 
> *With this same composition you can also fill certain linen cloth sacks. These are wrapped with ropes and are shaped like balls. They are shot from iron* cerbottana *like the balls from fire tubes. You could also smear this composition on everything that you wish to burn easily…..when inflamed, it penetrates and maintains a strong fire. You could also bind this at the end of the iron on the pikes of foot soldiers by filling with it a purselike bag, made for carrying the fire among the enemy or any other place at which you wish to throw so that it may ignite and burn.*'

Two other, similar objects, have been identified as possible incendiary projectile flights (Fig. 6.12). 81A6931 (raised as 81A2521/2, see above) is a beech flight with eight sides and of similar size to 81A2866, with maximum length of 350mm and central (widest) width of 170mm. It retains staining on the flight for a split haft of 40mm and four iron nail holes for fixing the haft to the flight. 81A6703 is a

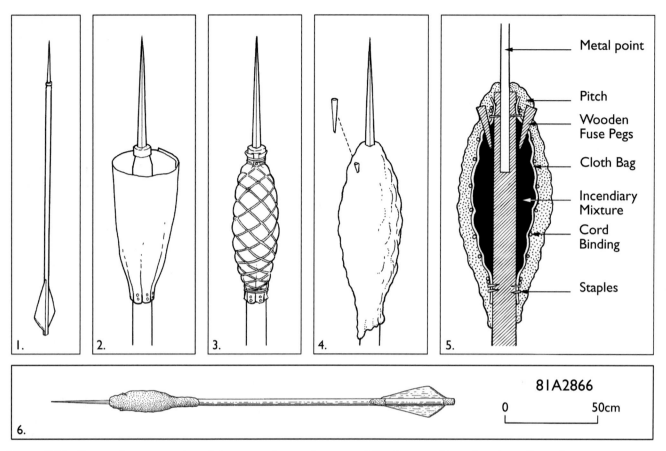

*Figure 6.10 Proposed reconstruction of the stages of manufacture and final composition of an incendiary dart*

*Figure 6.11 Fire pots with slings and incendiary arrows (after Biringuccio 1540)*

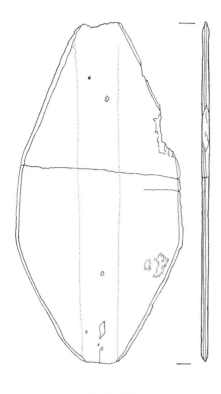

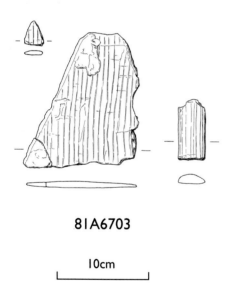

*Figure 6.12 Fragments of incendiary darts 81A6703 and 81A6931*

quadrilateral beech fragment 160mm in length, 8mm thick and 40mm wide at the top and 110mm at the broken end (projected width of 180mm). It was found with a poplar haft fragment, 30mm wide and 52mm long inside the Carpenters' cabin, by the south wall which divides this area from the M10 gunbay, where the other darts were found. A number of elements which might suggest a toolkit for making incendiaries was found in chest 81A1328 in the Carpenters' cabin (see Chapter 8).

The size of all the flights would enable the darts to be inserted into the bore of the adjacent demi cannon (81A3000). There is no evidence as to whether the items were thrown from the fighting tops, as is suggested by the arrow in the foremast seen in the WA Kraek engraving (Fig. 6.13), or fired from a gun. Lucar (1588, appendix, 64) suggests the latter, but notes the use of iron flights rather than wood: '*An iron dart shot out of any great piece of artillery or thrown out of your hands against a wooden object will burn and consume the same object.*' The object described here is over 2ft (610mm) in length with iron wings placed a little below the upper end like the feathers of an arrow.

> '*put vppon the dart neare vnto his point a bagge wide in the middest and narrow towards both his endes ... & with a mixture made of 12 parts of Saltpeeter, eight partes of Brimstone brusied grosselie like pepper cornes, and foure partes of grosse gunpowder mingled togeather, fill that bagge ... binding well togeather both the endes of the said bagge so filled, and nailing the full bagge vnto the dart ... cote them with canuas and winde packthreede very hard vppon the same canuas, and then couer the said canuas all ouer with paste made of meale sod in water ... charge not the peece out of which this dart shall be shot with so much gunpowder as is his ordinarie charge, nor with any tampion or wadde.*' (ibid., 64–5)

## Distribution of other incendiary items

The distribution of objects relating to gunpowder or requiring ignition indicates that all gunpowder samples (with the exception of those excavated from casks or gun chambers) were recovered from the Hold and Orlop deck in the bow (eg, 81S1169). However the distribution of candles (81A4192, 81A4840, 81A2106) and a barrel of tallow (81A1017) show that these are all in the stern (O/H11) within 3m of the incendiary projectiles. This also coincides exactly with the distribution of lanterns. The greatest concentration of linstocks (Chapter 5) is also within 3m. It seems that the limited distribution relates to the storage of items requiring fire within one area, rather than these necessarily being the raw materials to manufacture incendiary devices. The small arrows found in chest 81A1328 in the Carpenters' cabin, together with flint, tinder and small planes have been identified as being used to manufacture firesticks. The only recovered former for making cartridges was found in M11 (Chapter 5). Storing powder well away from either the galley fire or the gundecks was common practice later on, so we may be seeing a similar precaution. A pot containing pitch and several casks of pitch/tar were found in the Hold, forward of the galley.

*Figure 6.13 Incendiary dart at sea in fighting tops (by 'WA' (© Ashmolean Museum, Oxford PA1310)*

The distribution of all forms of lantern shot showed that the highest concentration was within M10 (six examples), with the remainder relatively evenly spread across the stern. It is not impossible that these projectiles carried an incendiary mixture within the flints/stones contained in the canisters, though no samples were processed.

Three unusual hafts of similar dimensions to the fire arrow, also from the stern of the ship, (79A1102, 79A0681, 81A4435) retain spike holes 10mm in diameter, and the ends are fashioned to take an iron collar. There is no evidence for bags at the head ends but all are broken (180mm, 200mm 540mm) so it is not impossible that they retained an incendiary bag further down the haft. One has a groove of 5mm width 150mm from the top, but this is clean. There is no evidence of any form of pitch coating. Although atypical, these are considered with the other staff weapons (Chapter 9).

The collection at Alnwick Castle in Northumberland contains an object from High Bridge, Lincoln, similar to the *Mary Rose* fire arrow, dated 1420–50 or earlier (Fig. 6.14). The arrow consists of a barbed,

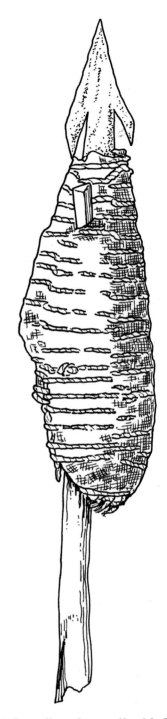

*Figure 6.14 Incendiary dart at Alnwick Castle*

265mm long, metal foreshaft made from a flattened strip of 12mm wrought iron with a thickness of 3mm. The head portion is 54mm long and 24mm wide. At the distal end the metal is flattened to form a socket 45mm long for arrow insertion. Behind the barbed head is a mass of combustible material wrapped in canvas 132mm long with a diameter of 64mm and bound by single-ply cord. A wooden peg is inserted at the front of the bundle, presumably to take a fuse. A canvas sleeve seems to have been wrapped around the metal head into which gunpowder, saltpetre, sulphur and charcoal was placed. This was then reinforced by the cord wrappings. The cover was dressed with pitch and a wooden peg for the fuse inserted through the canvas bag. The fuse was probably of tow soaked in saltpetre and dried. Other comparative arrows exist in Museum collections in Geneva, Zurich and Berne, but all are less than 2ft (610mm) in length. Coburg Castle (Giebig 2001, 88–98) holds an extensive collection of pots, grenades and garlands, but no incendiary arrows or darts.

Although mentioned in artillery treatises, the actual existence of incendiaries is proven by descriptive entries in inventories. The amounts of each type suggest their relative importance. The location is also valuable, suggesting which battle zones were provisioned with incendiaries, or which had to be self-sufficient producers, as reflected by the items listed. In many cases a study of inventories will enable an artefact to be identified and named.

## Analysis of incendiary mixture

Three samples thought to contain gunpowder were analysed from different sources for comparison. These were from the incendiary projectile (81A2866, 87S0210), from the barrel of the bronze demi cannon (81A3000; 87S0211) close to the incendiary, and from a breech chamber (82A0792; 87S0212) for a large wrought iron port piece (81A3001). The samples were taken to the Royal Armaments Research and Development Establishment at Fort Halstead in Kent.

Traditional gunpowder for this period for large ordnance consists of (although there are a number of variations) approximately 4–5 parts saltpetre to 1 part coal to 1 part sulphur (Norton 1628; and see Chapter 5). Parts of the residues from the dried samples were analysed using a combination of ion chromatography, elemental anaysis, emission spectro-scopy, scanning electron microscopy, x-ray analysis and infra-red spectroscopy (details of methods and results in archive). All the samples were found to contain gunpowder degraded by time and seawater. The low sulphur content (relative to carbon) observed has been attributed to degradation and is not considered to reflect the original composition. Emission spectroscopy showed all samples to contain iron as a major constituent together with silicon. In the case of samples 87S0210 and 87S0211, appreciable amounts (>1%) of copper and calcium were found. All samples contained trace quantities of twelve other elements often associated with non-specifically defined steels (magnesium, sodium, potassium, aluminium, arsenic, titanium, bismuth, silver, lead, antimony and boron). The samples from the incendiary and the bronze gun had the lowest sulphur to carbon ratios and were found to contain sulphate iron. That from the iron breech was heavily contaminated with iron. In addition, the sample from the incendiary had a high ether soluble content.

**Table 6.1 Analysis of the incendiary mix from projectile 81A2866 (sample 87S0210)**

*a) Soluble solids (%)*

| Water | Soluble solids | % Insoluble solids |
|---|---|---|
| 68.2 | 0.2 | 31.6 |

*b) Elemental analysis of insoluable solids (%)*

| Ash | C | S | N | H |
|---|---|---|---|---|
| 20.1 | 68.3 | 5.8 | 0.4 | 3.2 |

*c) Ether extracts of using infra-red spectroscopy compared with modern gunpowder*

| 87S0210 | Modern |
|---|---|
| Sulphur | Sulphur |
| Silicon | |
| Copper | |
| Iron | |
| Calcium | Calcium |
| Potassium | Potassium |

*d) Ether extracts (% wt total solids) compared with modern gunpowder*

| Sample | Ether soluble material | Ether insoluble material |
|---|---|---|
| 87S0210 | 12.5 | 87.5 |
| Modern gunpowder | 2.0 | 98.0 |

This could well represent some of the organic matter quoted in the contemporary recipes such as rosin, petroleum oil, Grecian pitch or turpentine. All the other soluble extracts gave rise to infra-red spectra consistent with the presence of oils which could have been derived from wood (see Chapter 5). Results from the incendiary projectile are summarised in Table 6.1.

## Mortar bomb
by Alexzandra Hildred and Andrew Elkerton

Fifty-two curved fragments of cast iron with walls of 43–78mm thick were recovered during the excavation. The concave appearance and angle of slope suggested a hollow cast iron sphere, larger than any of the cast iron shot recovered from the vessel (Chapter 5). None of these displayed the amount of concretion that the Tudor shot generally displayed, and all were heavy. They were registered as '*charge fragments*'.

On 13 October 1840 John Deane and William Edwards applied to the Board of Ordnance for condemned bombshells for the purpose of exploding gunpowder to make craters within which to dig for objects on the site (Bevan 1996, 171). The entry is annotated on the 16 October with the authorisation that

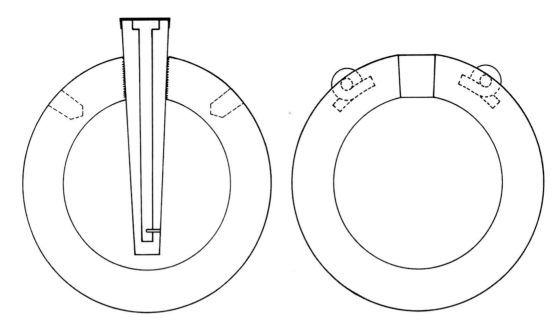

*Figure 6.15 Reconstruction of 13in mortar shell (based on illustrations provided by the Royal Armouries)*

six unserviceable 13in (330mm) shells should be issued, without charge.

These are hollow spheres of cast iron fitted with a tapering hole to take a fuse plug and filled with gunpowder (Fig. 6.15). The '13in' is a nominal diameter, the actual size is slightly smaller, c. 12¾in (324mm). They carry 6lb 8oz (2.9kg) of bursting powder, containing 11lb 4oz (5.1kg) of powder in total, with a full weight of 194lb (88kg). Most have a provision for lifting loops or sockets set into the surface of the shell. They were in service for many years and were stored in large quantities, becoming obsolete by 1880.

Not all the recorded fragments are currently available for study. Some have been treated by hydrogen reduction during conservation and so are inappropriate for chemical or metallurgical analysis. Two portions have been fitted together convincingly, 80A0696 from M6 and 80A0348 from M5 and look like a mortar shell (Fig. 6.16, above). A broad arrow mark is clearly visible on one of them. Twenty-one fragments were studied (Bevan and Towse, internal report, April 2008) and offered up against a mould of the correct size. Although the edges did not fit perfectly, the components formed the majority of one a sphere of the correct size (Fig 6.16, below). The weight of these pieces was 156lb (78.9kg), although some of the weight will have been reduced due to corrosion. Interestingly, nothing similar was found in the recent excavations around the stem (see Chapter 12).

**Distribution**

The distribution of the possible mortar bomb fragments is shown in Figure 6.17. There is a definite cluster around midships. Certainly Deane and Edwards concentrated in this area, recovering the cannon 79A1276 and its collapsed carriage (Chapter 2). There was damage and erosion to the trunnion support cheeks of the carriage; they had split vertically down from the middle of the trunnion recess. This is exactly where force would be needed to pull the gun off the bed, breaking the already formed concretion of the wrought iron cap squares (see Chapter 2 for details). A number of post-Tudor finds were recovered from in and around this area including blue and white floral pattern china, a modern bottle and coal (80A0772). Divelogs record other post-Tudor objects including several pipe stems and a number of pieces of concretion, which were discarded. The timber, T55, had also been wrenched east by about 40° and T209 displaced both north and east. The carriage was partly in the hard grey shelly layer (context 6), partly in its own scour (context 9) and partly within the Tudor context 10.

Just south of the broken carriage there was a substantial disturbance of the sediments (Feature 2; see sketch, 1980 site book, 29, M5/6). This was a '*slimy black deposit*' between O6 and M6. Old breaks were recorded in several Orlop deck planks (80T36 and T37). The feature was interpreted as being twentieth century anchor drag but could equally represent Deane activity. The presence of eighteen of the 'shell' fragments in Sector 6 suggests that this area may have been selected for the trailing of mortar shells. What is clear is that the cannon was lifted in 1836, so if a charge was placed here this would require two invasive operations by Deane. This might account for the two noted disturbances. There are three fragments in M8, one piece (81A3265) found under the axle of an *in situ* port piece (81A3001) close to a base of a bottle of questionable date. The axle had a number of turns of

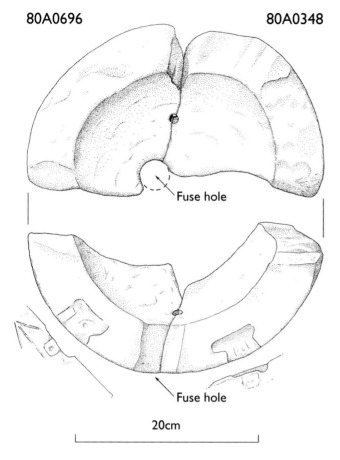

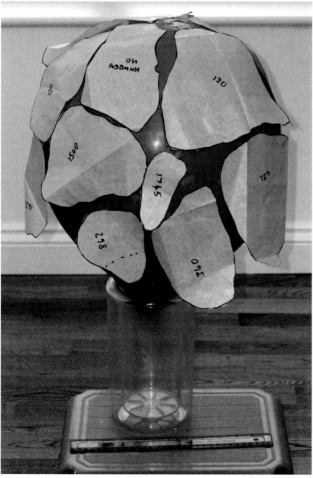

rope wound around it and it is just possible that the Deanes were trying to recover the gun. The excavation reports show that there was a vast amount of hard, thin concretion around the gun and the three pieces identified may therefore be only a small reflection of the true picture (see Chapter 11).

**Context**

At first inspection the contexts of the charge fragments appear to be mostly Tudor. Two are recorded as loose; three are assigned to context 6; ten to context 10; three in concretion (context 28) and seven in other specific contexts listed as Tudor (see the introduction to this volume and *AMR* vol. 1, chap 9 for a summary of contexts). Upon closer inspection of the diving reports it appears that, where a context can be assigned with any assurance, it is actually post-Tudor. It is unfortunate, but entirely understandable, that the level of seabed recording favours the more intrinsically interesting artefacts. Examination of the list of artefacts associated with these fragments indicates that some have been associated together because they look similar, and not because they were excavated beside one another. For example, 81A0024 is associated with 80A0130; they were excavated in different sectors in different years although they are both '*charge*' fragments. Similarly 80A0432 is associated with 80A1500, but these are, again, from different sectors.

Some charge fragments were associated with Tudor objects, although usually within context 6, the hard grey shelly layer that sealed most of the secure Tudor layers (see *AMR* Vol. 1, chap 9 for a summary of site stratigraphy). In most instances the excavation of the 'charge' is not represented, it is just appended as an artefact to that particular dive. In these instances it is not possible to distinguish between an excavated find and a loose find which could have fallen in the trench from a higher level. Table 6.2 lists all those fragments that were associated with Tudor objects or ship's timbers. Table 6.3 lists all fragments of mortar bomb by sector.

**Analysis of a mortar bomb**
by David Starley

In an attempt to see how similar or different these fragments were from cast iron objects undoubtedly of the Tudor period, a single fragment (81A5709/6) of a possible mortar bomb, together with a cast iron shot and a hailshot piece, were submitted for metallography at the Royal Armouries, Leeds. An external diameter of 300mm was calculated from the single fragment, 220mm in length, 43–54mm in thickness and with a mass of 4.52kg.

*Figure 6.16 Above) reconstruction of mortar bomb from fragments 80A0696 and 80A0348; below) other fragments*

**Table 6.2 Fragments of '*charge*' associated with Tudor artefacts**

| Object | Sector | Associated objects | Comment/dive report assessment |
|---|---|---|---|
| 80A1326 | M9 | 80A1323–5, staved container, ram | associated with port side Deane worked carriage and F2 ('anchor drag') |
| 80A1432 | H7 | 80A1428 sword handle; 80A1433 staved container; 80H0171, H0174 | log does not show excavation of charge; objects undamaged |
| 80A1745 | ? | 80A1746–7 unident. concretion | no information, not shown |
| 80A2062 | H6 | 80A1590–1 iron bar | no information, not shown; shelly deposit |
| 81A1730 | C8/9 | 81A4958–9 inset shot | no informaion, not shown |
| 81A2863 | M8 | by rear of M8 gun sledge | appears to be only identified frag. in a Tudor deposit; rear of sledge quite close to layer 6 (interface Tudor:modern layers); rope around carriage axle, possible assoc. with 19th century salvage |
| 81A3778 | O7 | adhering to plank | no information regarding excavation |
| 82A0756 | O10 | by S end of chest 82A0946 | concerted to deck beam; within localised scour layer assoc. with deck beams |
| 97A0009 | H6 | by 81A1973 fiddle | charge not positively identified in divelog |

The results revealed a coarse grained structure visible without magnification. It comprises a narrow chill margin on the exterior (convex surface) of the casting. Columnar grains extend from this surface almost to the inner surface (45mm of the 54mm thickness). The remaining 9mm contains grains of varying orientation. Metallography revealed that the structure is white cast iron, comprising primary iron carbide laths and pearlite, although almost totally obscured by corrosion which penetrates the entire section. Areas of phosphide eutectic (ferrite iron carbide and iron phosphide) were also identified. Occasional angular, dove-grey inclusions probably indicate manganese sulphide. Sulphur and phosphorus were detected in the alloy. These two elements would have helped increase liquid metal fluidity in the production of a relatively complex casting. The orientation of the metallic crystals indicates that solidification proceeded much more rapidly from the outer mould face than from the inner. This suggests that the outer mould was constructed of a much more thermally conductive medium, such as stone or metal, whereas the inner core would have been of clay or sand. This structure has considerable similarities to the hailshot piece and shot which could be due to the choice of similar alloy and production technique for the casting of similar objects, rather than from contemporaneity or for reasons of similar geographical source.

**Discussion**
by Alexzandra Hildred

Portions of the casting were studied by Nicholas Hall, Royal Armouries Museum of Artillery, Fort Nelson, where a number of extant examples of mortar bombs can be found. The dimensions and overall form are entirely consistent with an eighteenth or nineteenth century mortar bomb.

The similarity in metallography to the Tudor objects has caused some revision of the simplistic mortar bomb

**Table 6.3 Mortar bomb fragments by sector**

| Object | Sector | No. elements |
|---|---|---|
| 81A6771 | ? | 1–5 |
| 82A1860 | H5 | |
| 82A4349 | H5 | |
| 80A0423 | H6 | |
| 80A1590 | H6 | 1–2 |
| 80A1591 | H6 | 1–3 |
| 80A2062 | H6 | |
| 97A0009 | H6 | |
| 80A1432 | H7 | |
| 80A0633 | H8 | |
| 81A2085 | O6 | |
| 81A2261 | O6 | |
| 81A3071 | O6 | |
| 81A5709 | O6 | |
| 83A0116 | O6 | |
| 80A0130 | O7 | |
| 80A1500 | O7 | 1–2 |
| 81A3778 | O7 | |
| 80A1745 | O8 | 1–2 |
| 81A4912 | O9 | |
| 82A0756 | O10 | |
| 81A0024 | bow, Main deck | |
| 80A0348 | M5 | |
| 80A0696 | M6 | 1–2 |
| 81A0986 | M6 | |
| 81A2808 | M6 | |
| 81A2809 | M6 | |
| 80A1326 | M7 | |
| 80A0340 | M8 | |
| 81A2863 | M8 | |
| 81A3256 | M8 | |
| 81A1395 | M9 | |
| 81A3799 | M10 | |
| 81A1562 | U10 | |
| 81A1730 | C8 | |
| 72A0076 | SS10 | |

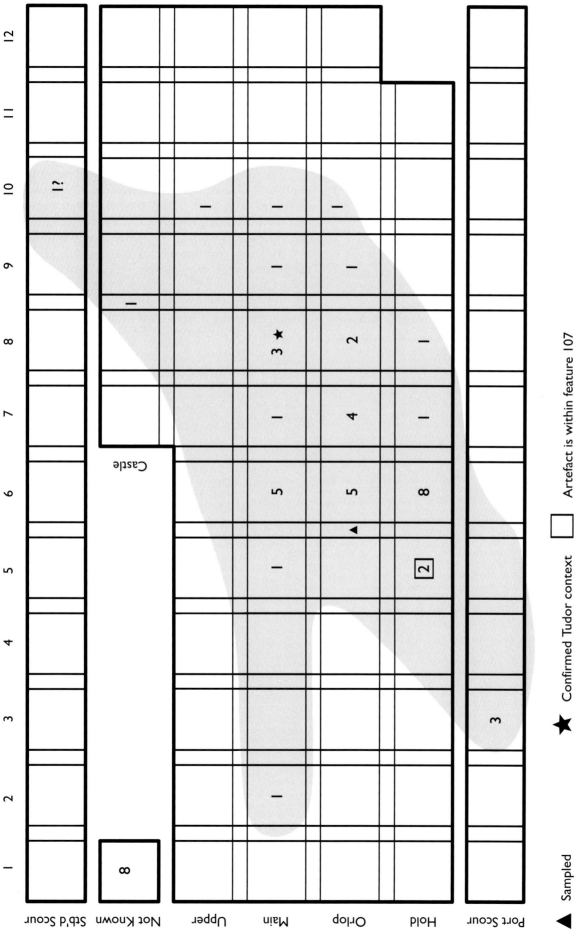

*Figure 6.17 Distribution of probable mortar bomb fragments*

theory. The metallographic study of an extant mortar bomb is suggested (although corrosion may mask certain features). The matter is further confused by the fact that Tudor incendiary devices were recovered, and incendiary shells were known at this time. Although some are described as being of cast iron, bronze was used more. The wall thickness, in general, appears to be thinner.

Chapter 6 of Book 10 in Biringuccio's *Pirotechnica* (1540, 428–32) is entitled '*The Method of Making Metal Balls Which Burst in Many Parts, for Shooting on Armies Lined Up in Battle*'. He describes these as being '*of metal*', probably bronze (*ibid.*, 429):

> '*Not yet satisfied even with these* [cast iron shot] *in order to injure men more than the aforesaid ones do, the good men of intelligence have thought of making those hollow balls of metal, and have given them a way of bursting into many pieces so that each piece may strike a blow. And whereas, with the shooting of an ordinary ball with a gun only one arrived among the people, this ball that breaks in pieces is equivalent to many. Thus they proceeded to make these hollow balls of metal and filled them with strong powder through a little hole. The fire is introduced with a fuse when they are shot with a gun or launched in some other way by men. Thus the powder that is inside ignites on arriving and, since it has no escape, bursts the ball into many pieces among the enemy*'.

The manufacturing process he describes necessitates both a fuse hole and core supports, which would all leave traces in the form of channels within the walls of the shot. He does, however, discuss casting both inside and outside in sand, as with small iron round shot.

All things considered, the distribution, little direct association with other objects, general lack of decay and degradation, and comparison with extant mortar shells fit well with the identification of these fragments as being the remains of mortar bombs placed by Deane and Edwards in 1840.

# 7. Hand-held Firearms

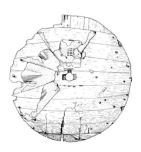

## Handguns

*Scope and importance of the assemblage*
by Guy Wilson

Hand-held firearms had been gradually becoming more used in Europe for almost 200 years before the reign of King Henry VIII, basically as a relatively crude weapon of the common soldier (Blair 1983, 33). It was only from the early sixteenth century that firearms had become practical and effective enough to make a major impact on warfare. It was during the Italian Wars of the first half of that century that firearms rose to dominate the battlefield. The battle of Cerignola, fought in 1503, is widely regarded as the first occasion when victory was won by hand-held firearms, but it was the battles of Bicocca (1522) and Pavia (1525) that confirmed that firearms had become a major and enduring force in warfare. At the same time the rapidly developing efficiency of firearms ensured that they began to be used by the nobles and monarchs of Europe. For most of his reign King Henry VIII was a major patron of gunmakers, encouraging the development of the craft in England, employing craftsmen to make him guns for his own sporting use and importing military guns from the continent in large numbers to equip his forces (Held 1957, 61; Blackmore 1999, 7–9). But though firearms might have been increasingly important in royal armies and good for royal sport, they were a danger in the wrong hands. As early as 1508 there had been an attempt to control the problem in England by forbidding their use without a royal patent or licence. Henry VIII reinforced this attempt at control in 1515 by forbidding the shooting of firearms by anyone with an income of less than £200 (the vast majority of the population). But the growing popularity of firearms could not be so easily controlled, and in 1542, a new law was passed allowing those with £100 a year income or more to shoot and keep firearms (Held 1957, 65).

The earliest evidence for any use in quantity of both handguns and pikes in England is the great muster of the citizens of London at Mile End on the 8 May 1539. Some 36,000 men appeared '*in harnys, wythe morys pykes, and handgonys and bowys*' (Blair 1999, 21). The number of handguns recorded in forts, garrisons and ships increased dramatically towards the end of Henry's reign. In part this must have been as a direct result of the 1540 Ordinances whereby:

> '*Every man* [of the garrisons] *must furnish himself with convenient weapons, as a dagger and sword, a halberd or bill, and harnesss, and every gunner a handgun or hagbush at his own charge, between this and midsummer, and for every day without them after that time to forfeit 3 days' wages*'. (*LP* 14/2,785 art. 10)

Early handguns, known as '*Schotbussen*', were in use on Hanseatic ships in the late fourteenth century (Ellmers 1994, 45). From that time, the use of handguns aboard ships steadily increased until they outnumbered, then totally superseded, the longbow and crossbow. Handguns had begun to find their way onto English ships in the fifteenth century. In 1485 the *Martyn Garsia* had four handguns as part of its official armament and the *Gouernor* seven '*hakebusses*' (Oppenheim 1896b, 69, 73). But it is clear that the main hand-held projectile armament remained the traditional English longbow. However, the days of the longbow were coming to an end while those of the gun were only just beginning.

The evidence of ships' inventories alone does not show the increasing importance assumed by handguns during the reign of Henry VIII. The 1514 inventory of the King's fleet, taken near the end of two years of war with France, lists the *Mary Rose* as having 91 '*hacbusshes*' and 900 pellets for them (Knighton and Loades 2000, 142). The *Anthony Roll* (1546) lists only 50 handguns and 1000 lead shot for them, as opposed to 250 longbows and 400 sheaves of arrows (see below). However, when the handguns and longbows listed in the *Anthony Roll* are compared with the evidence of where the actual weapons were found and how they could be used on board the *Mary Rose*, the growing importance of handguns in battle becomes very clear. Whereas many of the longbows and arrows were in store beneath decks, only one handgun was not found in a position suggesting imminent use. Alongside evidence demonstrating no great change in arming Henrician fighting men must be considered the well-documented and very considerable attempts by Henry (later in his reign) to acquire substantial numbers of handguns,

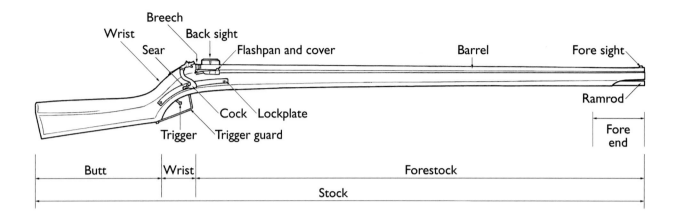
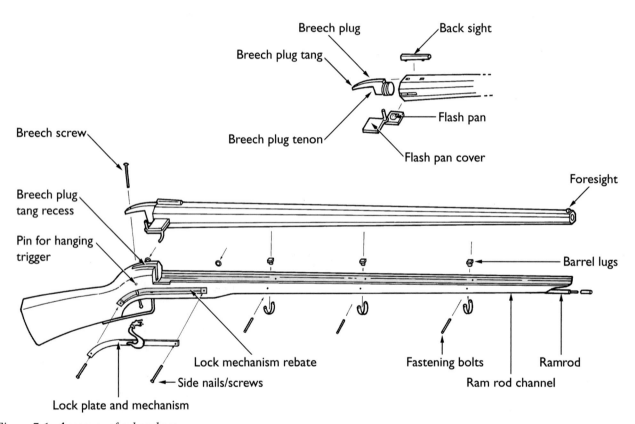

*Figure 7.1 Anatomy of a handgun*

many from Venice and Brescia, for his troops (Wilson 1988, 8–9).

With very few exceptions, and excluding the gun-shields (see below), all that survive of hand-held firearms from the *Mary Rose* are fragments of stocks (Fig. 7.1). These fall into three distinct types. First, there are a number of fragments of the stock of one gun (81A3884) that indicate a butt form that is very similar to the cheek stock of the modified matchlock (originally a wheel-lock) breech-loading gun of Henry VIII, dated 1537, in the Royal Armouries (XII.1; Fig. 7.2). Henry's gun, however, is equipped with a patch box, has a differently-positioned and more pronounced finger-rest cut on the underside of the stock and has a concave profile to the end of the butt whereas the example from the ship is straight. Both have a pronounced downward angle to the fore-end in front of the wrist that serves to strengthen the stock around the lock. The breech-loading gun of King Henry is stamped on the breech block with the initials '*WH*' above a *fleur-de-lys* and it

*Figure 7.2 Royal Armouries matchlock RA.XII*

has been suggested that this is the mark of William Hunt who was appointed Keeper of the King's Handguns and Demi Hawks in 1538 (*LP* 13/2, 491(22); Norman and Wilson 1982, 73–4). The similarities between the King's gun and the fragments of the stock found in the Carpenters' cabin raises the intriguing possibility that there may have been some English-made firearms on board the *Mary Rose* (but see below). There is also an extraordinary similarity between the decoration on the barrel of the King's gun, which takes the form of a chiselled column and the barrel of demi culverin 79A1232 (also dated 1537).

The second type is exemplified by a butt (81A1842; Fig. 7.10, top) shaped in the form of a fishtail. This form of stock became the most popular for military matchlock muskets by the end of the sixteenth century and survived into the second half of the seventeenth century (Fig. 7.3) but, if this is indeed the butt of a musket or other military long gun, it is by far the earliest datable evidence for it. Its development is usually dated to the last decade of the sixteenth century (Blair 1983, 53). This form of butt is clearly illustrated in Jacob de Gheyn's famous drill manual *Wapenhandelinghe van Roers, Musquetten ende Spiessen*. This was first published in The Hague in 1607 but had been commissioned earlier by Count Jan II of Nassau, nephew of Prince Maurits. It has been suggested that the fishtail-butted matchlocks illustrated were drawn from five model weapons issued to regiments by the Dutch States General in 1599 in an attempt to standardise military equipment (Hoff 1978, 8–9). An alternate suggestion is that the original drawings were commissioned in 1597 by Count Jan, pre-dating these models (Sotheby's London sale catalogue 12 Nov. 1990, lot 68).

However, there remains the fascinating possibility that this is not the stock of a long gun at all but of a pistol. A close parallel to it in both form and size can be found on the decoratively carved 'fishtail' butt of a three-barrelled revolving pistol in the Civico Museo delle Armi Luigi Marzoli, Brescia. This relates closely to another three-barrelled revolver with a hooked butt in the Palazzo Ducale, Venice (no. N.30). This can be identified with some degree of certainty as the gun described in the 1548 inventory of the ducal armoury as '*uno scioppo da serpa con tre cane*' (a gun like a serpent with three barrels). If this attribution is correct then this revolving gun is the earliest datable revolver known

(Blackmore 1965, 80). Both it and the example in Brescia appear likely to be of Italian, and probably Brescian, origin. It is known that later in the century Brescia was producing guns with fishtail butts (Carpegna 1997, 427) and these may be early examples

*Figure 7.3 Gun with fishtail butt from from* The True Portraiture of the Valiant English Soldiers *of ?1588 (© Pepys Museum, Magdalene College, Cambridge)*

of a stock form that was later to become more popular with Brescian makers and their clients. The barrels of another revolving gun were in the collections of Elias Ashmole by the time of the 1685 catalogue and remain in the Ashmolean Museum, Oxford (MacGregor 1983, no. 89, 197–9). If the fishtail butt found on the *Mary Rose* is related to this group it is both additional evidence of the importance of Venice and Brescia in supplying the weapons needed by King Henry for his war with France and the only evidence discovered of the carrying and use of pistols on board the ship. This is potentially very important, as in 1545 the pistol was not the common weapon that it was to become in the second half of the century.

The first evidence for the possible use of pistols is contained in a letter sent to the Emperor Maximilian by the Ausschuss Landtag of his Austrian lands in 1518 making recommendations for new laws. This includes mention of highwaymen and other criminals who '*carry guns secretly under their clothing.*' Similar evidence comes from an edict issued in Brescia in 1532 prohibiting the use of guns that are small enough to be carried under clothing. The earliest datable pistol known is one dated 1534 that belonged to the Emperor Charles V. It is now in the Real Armeria, Madrid (no. K.44) accompanied by a number of other undated pistols of comparable age. While most of these seem to be of south German manufacture, it has long been believed, both from the evidence of an obviously early group of Italian-made wheel-lock pistols and the etymology of the word '*pistol*' itself, that the origins of the pistol must be sought in Italy (Blair 1965, 4).

Pistols with fishtail butts of less pronounced form were also made and used in Germany in the mid-sixteenth century. The earliest evidence for them comes from a painting dated 1549 by Lucas Cranach the elder in the Germanisches National Museum, Nuremberg (no. GM 226). This shows three armoured cavalrymen each with a single fishtail-stocked pistol in a holster. The earliest surviving pistols of this form than are the Brescian group discussed above are in the Waffensammlung of the Kunsthistorisches Museum, Vienna (nos A438, A439, A 439a) and are dated 1555. Such fishtail-butted pistols remained popular in German lands for several decades but were gradually supplanted in the 1570s by the 'puffer' type of pistol with a large, spherical butt. Seventeen of them survive in the Landeszeughaus in Graz and are believed to have been made in Nuremberg in the 1570s and 1580s (Brooker 2007, 40, 61–2, 112–24).

The remainder of the stocks found on the *Mary Rose* belong to a third, and very distinctive, type of military firearm, the snap matchlock harquebus (81A2679, 81A2405, 82A4502). One (81A2679) was found complete with the iron oxide residue of its distinctive octagonal barrel which still had an inlaid copper alloy mark in the form of a shield partly visible on its top flat near the breech (Figs 7.7, 7.8, inset) as described below.

Fortunately three snap matchlocks acquired by the Royal Armouries in 1986 (XII 5313–5) proved to be of exactly the same form as the remains from the *Mary Rose*. One of these supported an identical, but complete and fully legible, brass shield-mark that clearly spelt out the word '*GARDO*' (Fig. 7.4). There is no doubt that this otherwise unrecorded mark relates to the town of Gardone in the Trompia valley some 50 miles (*c.* 80km) east of Milan and 11 miles (*c.* 18km) from the city of Brescia (Carpegna 1997, 65).

From 1426 to 1797 this area of Lombardy formed part of the Venetian republic. It was an important centre for the mining and smelting of iron ore throughout the Middle Ages and became especially famous for the production of guns. Brescia itself specialised in making locks, stocks and mounts, and assembling parts into finished guns. Craftsmen of Gardone specialised in making barrels and the town was given a legal monopoly of barrel making by the Venetian Republic after 1542. Vast quantities of barrels were made in Gardone, up to 300 a day it was reported in 1572. Even more interesting in the present context is the report of one Captain Marino Cavalli in 1544 that '*a mountain of guns are produced in Gardone, and they do it so easily that in two or three workshops they make 400 barrels a day*' (Carpegna 1997, 427). Most of the barrels went to Brescia for stocking but some to Milan and other gun-making centres and Gardone soon came to make complete guns itself. Given this it is perhaps surprising that this mark is so rare, but confirmation of its association with the town comes from the appearance of the letters GARDO on the mark of the mid-sixteenth century Gardonese gunmaker Venturi or Venturini. The name may be correct as recorded on the mark, an alteration of the Christian name Ventura, or an abbreviation of the name Venturini. A stockmaker Felice Venturini is recorded in Brescia in 1627 (Carpegna 1997, 97–8). Its rarity, though, may suggest that it is the mark of a maker or factory not that of the town. In this context it is also interesting to note that it has been tentatively suggested that the other marks found on the Royal Armouries harquebuses (the initials '*LO*' and '*BA*') must have been the marks of one of the Gardonese forges or '*fucine*' where iron plates were wrought into barrels (Wilson 1988, 6–9).

Certainly the scale of production in Brescia and Gardone was quite extraordinary. In 1562 the historian Paolo Paruta sent a description of the Trompia valley to the Venetian Senate in which he stated that '*every year the said Valley produces XXV thousand guns that are fetched off by merchants into foreign lands.*' In 1537 the Venetian Senate prohibited the export of arms and armour without licence from the Republic and the surviving records of the grants of these licences, although incomplete, show that Brescian firearms were exported to all corners of Italy and Europe, including, most interestingly in this context, to King Henry VIII of England (Wilson 1988, 8–9).

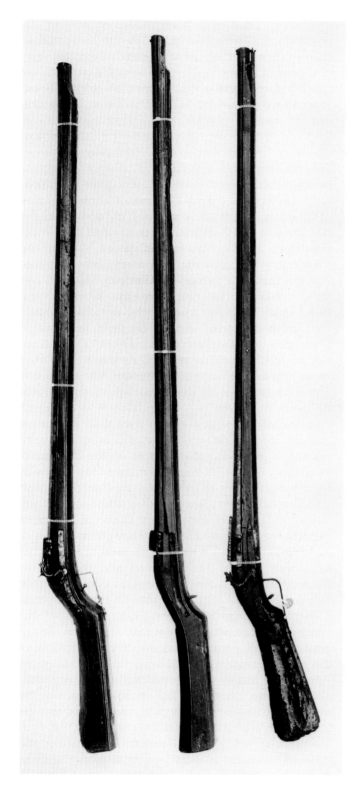

*Figure 7.4 Matchlocks in the Royal Armouries (RA X.11 5313–15 D) and GARDO crest*

In 1544–5, in preparation for war with France, Henry acquired very large numbers of harquebuses from Italy. Although not always specified in the records, it is likely that most if not all of these came from Brescia. On 14 June 1544 the Doge and Senate of Venice informed the Governor of Brescia that they had agreed, at the request of the King of England's ambassador, to allow the export from the city and territory of Brescia to England of 1500 harquebuses of various sorts. It seems very possible that the *Mary Rose* guns may have formed part of this shipment.

Certainly the capacity of the Trompia valley to produce large quantities of guns must have been of interest to the English King in his rush to re-arm, especially as it was difficult for him to find adequate numbers of guns elsewhere. On 29 August 1544, for instance, the King's agent in Antwerp reported that whereas the King wanted to buy 2000 harquebuses, there were no more than 200 wheel-locks and matchlocks available in the town. Early in 1545 King Henry was also trying to buy even larger numbers (apparently up to 9000) of Italian guns from the Milanese merchant Christopher Carcano. It seems that he had received about half of them by August 1545, but numbers of others were seized during their journey through Germany and the Netherlands. A year later, in April 1546, King Henry's agents were still trying to obtain the release of some of the guns seized *en route* to England in 1545 and it is possible that the '*discharge for the custom of 7,516 hackbuts*' bought from Christopher Carcano, dating from August 1546, may relate to the guns ordered in 1545 and thus may accurately reflect the number eventually received in England (Wilson 1988, 6–9).

There can be little doubt that the fragments of harquebuses found on the *Mary Rose* are the remains of guns made in the Trompia valley and exported from Italy to England in 1544 or early 1545. It is possible, however, bearing in mind that Carcano was a Milanese merchant, that only the barrels were made in the valley and were sent to Milan for stocking.

Matchlocks of this type so far identified are restricted to the three in the Royal Armouries, the surviving fragments from the *Mary Rose*, a decorated matchlock (D 156) in the Leibrustkammaren of the Kunsthistorisches Museum, Vienna (see below) and a gun in the Birmingham Museum and Art Gallery (no. 1885.5.24; see Blackmore 1999, 39). This latter differs from the others in being shorter and heavier (with a barrel just under 33in (838mm) long, flared at the muzzle, and with a bore of just over 0.56in (14mm)) and in having a very unusual external safety catch. Springing from the front of the serpentine of the Birmingham gun is a forward extension, chiselled as a naturalistic arm and hand that engages with a pivoting rectangular plate set through the lock plate ahead of it when the plate is turned edge-on to the hand. At the breech are two unidentified marks. On the top flat is an inlaid copper alloy shield mark bearing the letters 'CR/EM/A', and on the two adjoining side flats inlaid copper alloy shields bearing the letters '*Z*' (or 3) / *B* and then an upside down comma following and *passim* round inscribed letters. Although so far unidentified these marks are clearly related to the brass marks on the Royal Armouries gun no XII.5315. The similarity of treatment and placement of the five letters CREMA to the five letters GARDO, on the mark on the Royal Armouries' snap matchlock suggests that this, too, may be a town mark. If this is the case it must mean that the Birmingham barrel was made in Crema, 25 miles (40km) south-west of Brescia on the River Serio that had been part of the Republic of Venice since 1449. If this attribution is correct, it is evidence that this type of snap matchlock was not so much a speciality of Gardone as of the whole Brescian region. Research into the early arms industry of Crema remains to be undertaken, but there is at least some evidence for the production of crossbows there in the mid-fifteenth century. Two brothers from Rheims worked there as crossbow and clock makers and took the name of the town before moving to Ferrara in 1466 (Royal Armouries Museum Leeds MS De Cosson Dictionary, entry '*Crema*').

These finds confirm that the snap matchlock was of considerable significance in the development of military firearms. The snap matchlock mechanism works in basically the same way as the later flintlock. Constructed around a plate that was secured to the side of the stock, an arm (the cock, see Fig. 7.1) holding the burning match is held away from the pan of priming powder by a spring under compression. Pulling the trigger activates a sear (Fig 7.1) and releases the spring. The spring propels the cock into the pan. The later, and far more common, form of military matchlock operated in the opposite way and was sometimes called a sear-lock, or, in seventeenth century England, as a '*tricker*' (or trigger) lock (Blackmore 1965, 10–11). On a tricker lock the cock is held away from the pan by a relaxed spring. Pulling the trigger compresses the spring and moves the cock, against the force of the spring, towards the pan.

Some of the earliest known matchlock guns have snap matchlocks. Early military examples survive in the arsenal at Graz in Austria and in the Historical Museum, Basel. In the Hermitage Museum, St Petersburg, there is a German gun of about 1500 with a snap matchlock mounted on the outside of the stock (Blair 1983, 31). The butt bears the arms of Behaim of Nuremberg and is of the same general form as that of 81A3884 (Fig. 7.11). A similar gun, missing its external lock, with a rifled bronze barrel and the arms of Maximilian I on the stock, that indicate it was made between 1493 and 1508, was in the Renwick Collection (Tuscon, Arizona; Blackmore 1965, pl. 43). It has generally been considered, however, that the use of the snap matchlock declined quite rapidly in the first half of the sixteenth century, though it continued in limited use well beyond this date in a sophisticated form for target shooting and hunting (Hoff 1963). However, the evidence of the finds on the *Mary Rose* shows that the snap matchlock was still one of the most, perhaps even the most, common type of military small arm in the 1540s. This fact, now clearly established, also helps to explain why the Japanese, who were introduced to European firearms by Portuguese traders in 1543, adopted the snap matchlock as their standard weapon. It is also interesting to note that the Japanese may have copied very specific features from the European guns that they first saw. The pans of Japanese matchlocks tend to be of a double recurved form like one half of a violin body. This feature is clearly seen, too, on the harquebuses in the Royal Armouries of the type that were on board *Mary Rose*. This form of pan and the other major distinctive features of the *Mary Rose* harquebuses (with the exception of the pronounced down-curve at the wrist of the stock) are also found on one known luxury arm – a snap matchlock of about 1560 in the Leibrustkammaren of the Kunsthistorisches Museum, Vienna (D 156). This has its lock and barrel, richly damascened in gold, and its stock covered with black velvet and secured by silver-gilt studs. It came from the collections of Archduke Ferdinand of Tyrol in Schloss Ambras and for many years has been thought to have been of Spanish origin (Thomas and Gember 1990, 43, pl. 16). However, there seems no doubt that it belongs to the same group and has the same origin as the *Mary Rose* harquebuses.

The finding of both an early and a late form of matchlock on board the *Mary Rose* is of great importance in understanding the development of the use of the handgun at sea. As only one example of the former has been identified, it could be considered a 'hand me down', carried as a personal weapon, or was possibly left on board from a previous war (*c.* 1514 or 1526; see *AMR* Vol. 1, chap. 2). It might even represent the personal firearm of someone on board who was of German descent. It is even possible that the piece reflects Henry's difficulties in obtaining enough handguns for the war and that he was, therefore, reliant on issuing older weapons from the stores for use at sea, if not also on land. The use of both early and 'state of the art' handguns could, therefore, be taken to imply simply that this was a period of transition or that the basic inherent concept (the use of black powder to fire a projectile) meant that any form of handgun was considered to be a suitably effective weapon. It was not until the end of the sixteenth century that the form and size of handguns and their general use in the English army was standardised.

*Inventory evidence*
by Alexzandra Hildred

**The *Anthony Roll***
All vessels are listed with handguns and shot of lead. The number of handguns varies between three and 100. The largest ship, the *Henry Grace à Dieu*, carries 100. The *Peter* carries 60 and the *Mary Rose*, 50. The *Matthew*, *Great Bark*, *Jesus* and *Pauncy* carry 20 each. Others vary between 12 and four. The '*galliasses*' carry between six and twelve, the exception is the *Galley Subtle*, with 50. Pinnaces and rowbarges are listed as carrying between three and six per vessel. The total for all vessels is 729, 422 on the ships, 204 on the '*galleasses*' and 103 between the rowbarges and pinnaces. Beneath the heading of '*Shotte of stoen and leade*' is usually the listing '*shotte of leade for handgonnes*'. The entry for the *Mary Rose* is 1000, providing 20 shot per gun. The small amount of cornpowder carried on the *Mary Rose* suggests that this was used primarily for handguns (see Chapter 5).

The number of rounds on the ships varies between fifteen (the *Great Bark*) and 40 (the *George*). Most vessels carry 20–25 rounds per gun, with the exception of several of the '*galleasses*' with over 33 rounds per gun (the *Grand Mistress*, the *Greyhound*). Two vessels have 28 and the rest 16–20. The exception again is the *Galley Subtle*, with only eight rounds for each of her 50 handguns. The pinnaces and rowbarges carry 16–33 rounds, with the majority having 20 each. The average number of arrows per bow is approximately 33–4 (Chapter 8). It is clear that both handguns and longbows were an important part of any ships armament, with the longbow still considered elemental in any engagement. This must be partly due to the higher rate acheived by the bow, averaging 12–20 arrows per minute, though achieving this with the heaviest *Mary Rose* bows must have required years of training. Although the effectiveness of the handgun against armour was realised (Richardson 1998, 50), clearly the problems and costs of acquiring handguns (and indeed the nature of the engagement) were factors contributing to the continued use of the longbow at this period.

**The 1547 inventory**
The detailed 1547 inventory (Starkey 1998) provides information about ancillary items accompanying hand-held firearms. The section dealing with vessels lists '*handgonnes*', '*hagbusshes*', and '*half hackes*' (or '*demi hacks*'), the heaviest of which required some form of support. Whilst shot is not listed specifically for handguns, a number of entries list either handguns or hagbusshes as '*complete*' (eg, 7388, 7564) or '*with their apparell*' (7521, 7612). Nearly half the vessels listed have no portable firearms. This is startlingly different from the 1546 inventory.

The section dealing with fortifications is more detailed. The heavier handguns are variously termed '*harquebusses*', '*hagbuts*', '*hackebuts*', '*archabus*', double, half, or demi '*hacks*'. More specifically there are '*Beamyshe*' '*Bemes*' or '*Boymysshe*' [Bohemian] examples, '*Hagbusshes of yrone with tailes*' (4867, 4880), '*Hagbusshes a croke*' and '*Hagbuttes a croke of brasse*' (6126). Entries such as '*Boymyshe hagbusshes vppon stockes*' (5345), '*Tristells to levell hagbusshes oon*' (6573), and the many references to '*trestells*', '*tristels*' or '*trindells*' indicate the heaviest examples, which required support and possibly a second operator to light the match. Where mentioned the material is usually iron, but sometimes brass. Many of these will have had wooden stocks like a rifle, but with an oblong butt held against the cheek when firing. Some supported pieces were not fully stocked, but had a recess behind the breech into which a wooden butt was fitted, this form of mounting can be seen in the hailshot pieces (Chapter 4). Many of the hackbutts may have originated from Germany, hence the name, '*hackenbusche*', corrupted to '*handbuchse*'. Nuremburg was a centre of production.

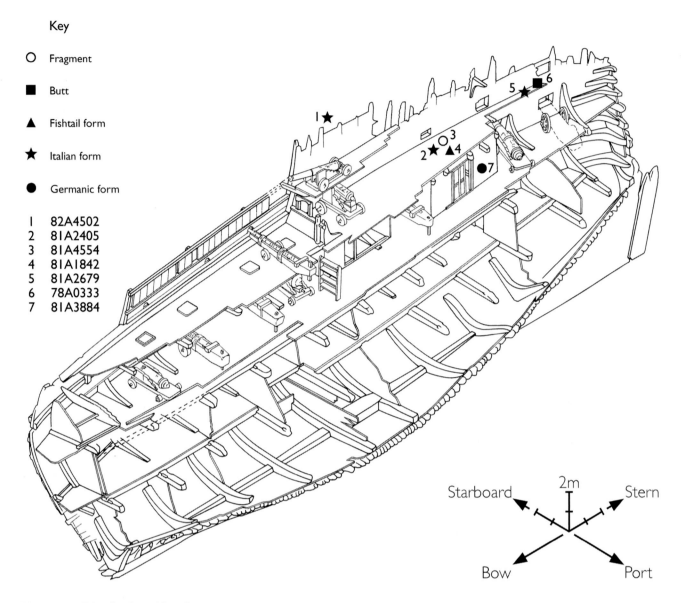

*Figure 7.5 Distribution of handguns*

English terms include '*hagbus*', '*hagbut*', '*hackbut*' and the French '*arquebus*' or '*harquebus*'. The weight of these varied between 9lb and 49.5lb (*c.* 4.1–22.5kg), so most required a parapet or trestle. The total number listed is 2369, including 140 for sea service. The entries of '*hagbusshes of iron*' may well be describing something similar to a hailshot piece, similarly the '*haquebushe a' croc*', furnished with a hook.

Handguns are also sometimes identified as '*Italyan peicis*' (4043), '*of thitallyon fasshion*' (5087) or '*of Italyon making*' (5843). Henry started importing these as early as 1513, the majority shipped through Antwerp. The total number of handguns in the inventory is 8399, 185 of them for use at sea. Most will have been fitted with a matchlock (called a lint). These are the lightest of portable hand firearms listed, weighing about 5lb (2.3kg) (Blair 1999; pers. comm.).

In contrast to the large numbers of long guns listed in the inventory, there are only nine '*tacks*' or pistols (8315–21) and a further 277 short guns (3822, 8174). Today these might be categorised as carbines or pistols. This emphasises the rarity of pistols at this time and reinforces the importance of the possible pistol butt (see above) on the ship.

Ancillary items include '*Flaxes* [flasks] *and towche boxes*' (3769), '*Flaskettes for powder*' (3982), '*Hornes for powder*' (3983), and '*Purses of blacke lether to put pellettes in*' (4345). One entry includes '*Handgonnes with boxes belonging vnto them*' and '*mouldes for them*' (4012). The numbers suggest that each gun is furnished at least with a flask and a touch box. It is less clear what is meant by the 150 '*Chardge of plate*' (5090) for a large quantity of demi hakes. Also featured are '*Measures for hagbusshes*' (5373), along with several entries for lint and one for

Table 7.1 Common surviving measurements of handgun stocks (metric)

| | 82A4502 | 81A2679 | 81A3884 | 81A2405 | 81A4554 |
|---|---|---|---|---|---|
| Sector | SS8/9 | U11 | M9 | U9 | U9 |
| Wood | unid. | walnut | poplar | poplar | – |
| Surviving length | 710 | 1115 | 495 | 242 | 195 |
| Butt width | max. 48 | 48 | 45–37 | – | – |
| Butt height | 48–60 | 60 | 102–44 | 76 | – |
| Stock length | 500 | 460 | – | 242 | 195 |
| Stock width | 48–29 | 52–36 | – | 50 | ?24 |
| Stock height | 72 | 60 | – | 81 | ?29 |
| Barrel groove length | 504 | 1050 | – | 128 | – |
| Barrel groove width | 44–32 | 40–28 | 26 | 42 | – |
| Barrel length | – | >445 | – | – | – |
| Barrel diameter | – | 16 | – | – | – |
| Bore | 13 | 12 | – | – | – |
| Trigger recess | – | – | 125 x 15 | 48 | – |
| Lockplate recess | 170 x 13–12 | 151 x 16–12 x 9 | max. 150 x 26 x 4 | 196 x ? x 12 | – |
| Flashpan recess | 38 x 4 | 51 x 7 | | | |
| Breech plug diameter | 26 | 4 | – | – | – |
| Breech plug tenon | 36 | – | – | 29 x 21–18 x 8 | – |
| Breech plug tang | 81 x 5–2.5 | 90 x 14–12 | – | 77 x 20–18 x 8 | – |
| Ramrod diameter | – | 9 | 12 | 13 | – |

matches (6471) which appears to relate to '*Handgonnes complete*' just before (6468).

Where powder is mentioned, it is invariably corned, as '*Hagbuttes furnyshed*' followed by '*Corne powder for the same*' (4489–90). Also occurring after '*Hagbusshe shotte*' are '*Bowgies* [purses] *for powder*' and '*Bowgies full of powder*' (4850–2). Most of the shot is of lead, although '*Shotte of yron for hagbusshes*' is also found (3981). References are also made to the casting of lead shot (4737) and the storage of lead (4923, 4951; Chapter 5).

The number of handguns increased in numbers dramatically between 1540 and 1547, from 268 in 1540 to 8399 in 1547. In part this must have been as a direct result of the 1540 Ordinances dictating that every gunner must equip himself with a handgun (see above).

## Distribution of handguns
by Alexzandra Hildred

All the identified remains of handguns were found in the stern (Fig. 7.5). Three were found in U9 (81A2405, 81A4554, 81A1842) and the best-preserved example, 82A4502, was found in concretion containing a swivel gun (82A4080) in the Starboard scourpit in the same general area (U8/9). The fishtail butt/possible early pistol (81A1842) was trapped within the 'v' created by the junction of the Upper deck and the starboard side within a large concretion (81A1865) which also included shot, two sword fragments (81A1747, 81A1759), and a human arm bone (81H0215). 81A2679 was found in U11. A small fragment of butt had been recovered from this area in 1978 at a higher level (78A0333). 81A3884 was found inside the Carpenters' Cabin on the Main deck (M9), suggesting perhaps that it was a personal possession (see below). With the exception of the last, the distribution is identical to that of the staff weapons and a portion of the archery assemblage. The finding of handguns inboard of the three small ports in the upper deck (U9–10) suggests that these ports may have been used by handgunners (see Chapter 11).

## The assemblage
by Douglas McElvogue and Guy Wilson

Most surviving examples that illustrate the early development of handguns are hunting weapons, or the highly decorative personal arms of royalty and aristocracy. Few military firearms survive from the early sixteenth century, a period of transition in the development of the construction, use, and standardisation of firearms. Therefore, the remains of any standard munitions weapons are of great importance. Although the general condition of the surviving handguns is relatively poor, the collection as a whole offers some intriguing and important details on the construction, form, and type of firearms being used on board a mid-sixteenth century naval vessel.

The assemblage includes small fragmentary portions of ramrods, separate, broken and degraded parts of stocks, and several complete butts and fore-ends, some of which include barrel portions. The number of

546

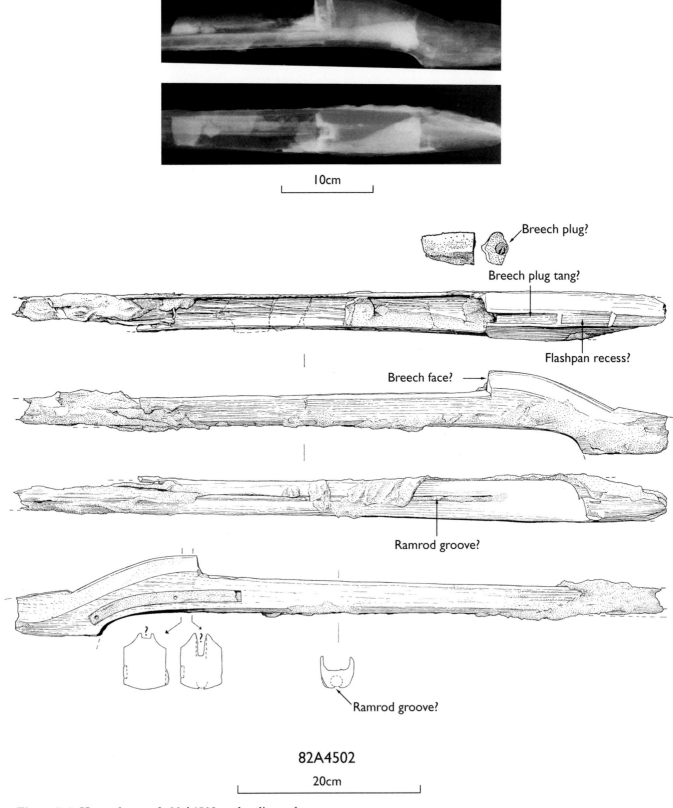

*Figure 7.6 Harquebus stock 82A4502 and radiograph*

securely identified handguns is seven. Each surviving stock was manufactured from a single piece of wood, radially converted – probably but not necessarily cleft – with the heartwood oriented to the top. Poplar, beech and walnut have been identified.

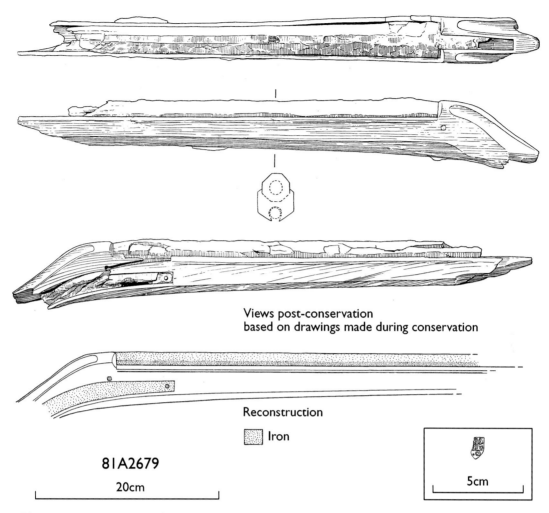

*Figure 7.7 Harquebus 81A2679*

## Stocks

The basic components of the handgun are illustrated in Figure 7.1. Principal measurements of the stocks are given in Table 7.1.

### 82A4502

This nearly complete harquebus stock (Fig. 7.6) was found in concretion (82A4080) in the Starboard scourpit (SS8/9). It is missing the butt from the end of the wrist and the muzzle of the fore-end (see Fig. 7.1 for an explanation of terms). The condition of the piece is good with little or no degradation. The wrist (or small) of the stock curves down smoothly towards where it steps to the butt. The stock is slab-sided with a gently rounded underside with chamfered edges to the fore-end. The underside is drilled and slit for a ramrod groove and contains fragments of the wooden ramrod. The top edges of the barrel groove are chamfered. The partial remains of the barrel were *in situ* when first excavated, suggesting an octagonal barrel. A 38mm long and 4mm deep rebate has been cut for the pan into the right-hand side chamfer, just ahead of the breech.

The breech face is 38mm high and 38mm wide and the square tenon for the 12mm diameter, 26mm long breech plug is evident just behind it. A single pin (5mm diameter) held the breech in place. The tang of the breech was fastened by a single 5mm nail. Two incised lines run from the tang rebate to the step at the butt. The lockplate recess curves smoothly down the neck. A rebate, 2.5mm deep and 2.5mm wide, ensured that the lockplate would sit flush against the side of the stock. Three 3mm diameter screws fastened the lockplate to the stock from the lockplate side. A single 8 x 11mm trigger hole is evident in the underside of the stock. Radiographs show details of the sear spring and breech plug (Fig. 7.6). There is a 12mm solid lead shot within the barrel.

### 81A2679

Fragments of the fore-end of the stock and the barrel survive of this harquebus (Fig. 7.7). Like the other harquebuses from the ship, it is slab-sided, with a large chamfer on the lower edge and a slight one to the top of the barrel groove. The estimated reconstructed length is 1350mm. There is an open ramrod groove along the bottom that still houses the remains of the ramrod, the end of which has a copper alloy reinforcing sleeve attached.

*Figure 7.8 GARDO crest on 81A2679*

The barrel is in very poor condition. It has an octagonal cross-section, the faces alternating between 14mm and 16mm in width. A partly missing and deformed broken copper alloy maker's stamp is positioned 180mm from the end of the breech. This consists of a shield, the upper part of which is divided into 4 squares, each containing a Gothic letter. Only the right half survives intact, with a letter '*A*' in the top square above a mis-shapen '*D*'. The letter '*O*' is placed centrally below, where the shield tapers (Fig. 7.8). Although broken, this is clearly the GARDO mark seen on at least one example of a harquebus in the Royal Armouries mentioned above (Fig. 7.4). The barrel was attached to the stock by two 4mm diameter wooden pins (one surviving) and a single breech plug pin of 4mm diameter. Radiographs of the concretion on this piece show that the pan cover is 50mm long and 16mm wide. It is shaped like a small '*B*' or one half of a waisted stringed instrument as are the pans of the similar harquebuses in the Royal Armouries. There is also evidence for a back sight.

The lockplate recess curves downwards, following the shape of the stock. The recess has a maximum depth where the trigger sear sits of 22mm. The trigger recess has two small holes in the bottom, probably to secure a trigger guard. In front of this is a hand rest. The plate is fastened to the stock by 3mm diameter pins from the lockplate side. The wrist is decorated with 15mm wide grooves running along the length of the piece.

*81A2405*

This fragment, from U9, represents the rear part of a stock, consisting mainly of the wrist and lock mechanism recess, in a very poor state of preservation (Fig. 7.9, top). It follows the general form of the harquebuses, being slab-sided and has a 60° chamfer at the top of the wrist. The barrel groove bears evidence for a breech tenon and tang. A single, 3mm breech screw is evident. The lockplate has a slight curve downward throughout its length. The actual plate recess of the lock is 220mm long though the lock recess is smaller. Its depth increases to 36mm at the recess for the trigger spring. The remains of the ramrod can be seen in its open channel. The barrel groove shows evidence for a pan rebate.

All these remains are clearly directly related to three snap matchlocks in the collection of the Royal Armouries (XII.5313–5) acquired in 1986. The main diagnostic features are: the slab-sided stocks with pronounced down-curved wrist and open rammer channel; the octagonal barrels (or cut-outs for them) with alternately wider and narrower flats; and the long thin, usually down-curved lockplates (or the cut-outs for them). A general description of the Royal Armouries guns will give a clearer picture of the fragmentary remains from the ship (see Fig. 7.4).

The two lockplates that survive are thin and flat, curving down at the rear to follow the line of the stock, to which they are attached by two wood screws entering from the face of the lockplate: 5315 tapers to front and back, 5313 does not. The ends of the serpentines are formed as dogs' heads, bifurcated as match-holders operated by a ring-headed screw at the neck of each. The serpentines are held away from the pan, against the action of an internal main spring of 'U' form which presses down on the tumbler, by a horizontally-acting sear which protrudes through a hole in the lockplate and bears on the heel. They are attached to tumblers by a short pin riveted at both ends. Directly beneath the sear hole a rectangular-headed pin has been riveted through the plate to act as a stop for the serpentine.

All the stocks are slab-sided, cut with incised lines at top and bottom borders. They have a pronounced down-curve at the wrist from which springs a straight butt, angled down from the axis of the barrel. The butts have eight irregular sides. A rammer channel has been drilled in the fore-end of each stock which has then had a groove cut from beneath into the channel to allow for shrinkage and expansion and to facilitate the removal of a jammed rammer. Both sides of this groove have plain, raised mouldings. 'L' form trigger guards of iron spring from the down-curved small. The rammer of 5315 survives complete with its steel tip formed as a bifurcated button with a forward pointing central tongue. It may be to hold a cleaning cloth or may be

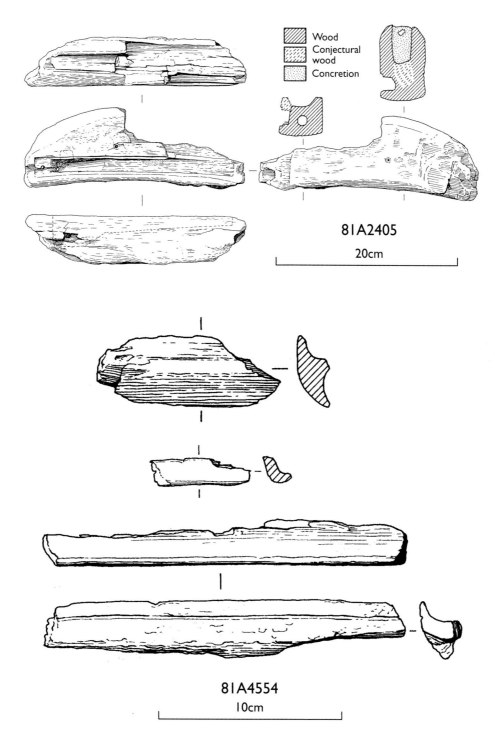

*Figure 7.9 Harquebus stock fragments: top) 81A2405; bottom) 81A4554*

intended as a sprung barrel scraper. These heads may relate to the '*sguradori*' (borescrapers) mentioned in *Il Catastico Bresciano*, a manuscript volume dated 1608–1610 that gives details of the geography and industry of the Brescia area, including a full list of all gun parts produced (Morin and Held 1980, 31).

The barrels are nearly 40in (1020mm) long with calibres ranging from 0.44 to 0.47 in. They are of irregular octagonal section, with alternately wide and narrow flats; all have belled muzzles; and all are attached to the stock by pins passing through two lugs, a screw rising vertically from the rear of the rammer channel and a screw at the end of the tang. All have peep rear sights and block or blade foresights and two are pierced midway along with vertically tapering slots, perhaps intended to take an adjustable plate drilled with a pierced hole or cut with a sighting notch. All the barrels are marked: XII.5313 underneath with '*LO*' in a pearled circle; XII.5314 at the breech on the flat right of the top flat with the intials '*BA*'; and XII.5315 with on

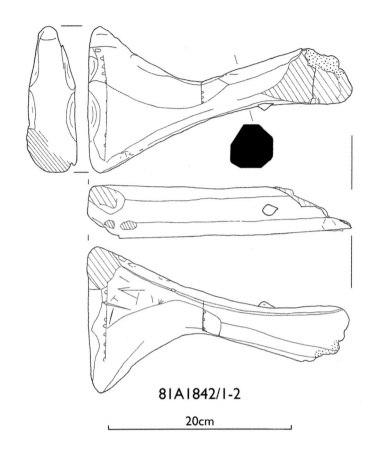

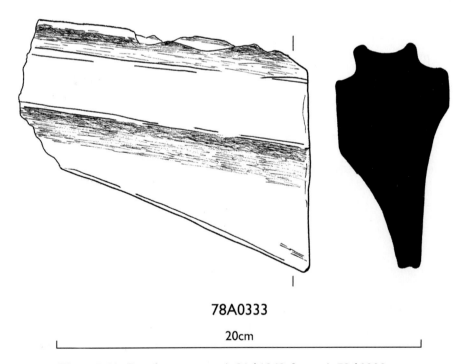

Figure 7.10 Butt fragments: top) 81A1842; bottom) 78A0333

the top flat an inset copper alloy shield bearing the initials GARDO with incised trefoils to front and rear, and on the side flats St Barbara in incised line borders with trefoils to front and rear. Attached to the right side-flat beneath the peep sight is a thick plate recessed towards the rear as a pan, and pierced vertically at the front by a pin attaching the pivoting pan-cover. The plate is cut decoratively in plan to resemble the body of a waisted stringed musical instrument. The pan cover has an extended operating lever at the end opposite the pivot.

Apart from the structural similarities what confirms the link between these guns and the remains on the *Mary Rose* is the GARDO mark that is quite clearly the same as the now deformed mark on 81A2679.

**Other firearms**

*81A4554*

The highly fragmented remains of the fore-end of another stock were found in U9 (Fig. 7.9, bottom). It is slab-sided in a shallow 'V'-form and the lower part of the stock is rounded. There is evidence for a ramrod groove. It has a total length of 195mm and could be part of 81A2405.

*81A1842*

This beech fishtail butt from U9 is broken forward of the swell into the wrist (Fig. 7.10, top). It is in fair condition though somewhat dried-out. It has a length of 254mm and width of 53mm, reducing to 46mm at the break. The butt is 150mm high curving down to make a wrist of 40mm, 52mm from the butt, before swelling back up into the wrist. The cross-section of the wrist is roughly octagonal. On the underside of the stock is a single 3mm diameter screw, probably for fastening a trigger guard. It shows evidence of surface decoration, most probably made by the owner rather than the manufacturer. It is fashioned from radially split beechwood, with heartwood to the top. As suggested above, its small size raises the possibility that this may be the only evidence that pistols were carried on the ship. A close parallel to this form can be found on a small group of Italian, probably Brescian, three-barrelled revolving pistols dated to the 1540s (see above).

*78A0333*

The remains of a butt broken behind the neck were found in U11 (Fig. 7.10, bottom). The piece is cracked and worn with, rounded edges. It has a maximum length of 162mm. The end of the butt has a height of 162mm and the broken end (towards the wrist) is only 55mm high. There are no fastenings or constructional details evident although there is possible evidence of a channel. As it is fragmentary, we cannot be certain that this does not represent another fishtail butt. It is fashioned from a radially converted piece of wood with the heartwood uppermost.

*81A3884*

Fragmentary remains of a firearm in very poor condition with flaking surfaces were found in the Carpenters' Cabin in M9 (Fig. 7.11). It has a conjectural reconstructed length of 495mm (Fig. 7.11, lower). Separately excavated elements include the butt, parts of the fore-end (81A5811) and fragments of ramrod (81A5734). The overall form of the piece is slab-sided with large chamfers to the bottom and small ones along the top of the stock. It is bulkier than the rest of the assemblage. It has been fashioned (although not necessarily cleft) from a radially orientated piece of poplar with heartwood uppermost. It retains evidence for side branches in the lower forward part and is drilled with an enclosed rammer channel. This has two small holes in the bottom, probably to secure a trigger guard. In front of this is a hand rest.

There are no exact parallels for this form of stock. It appears to be closest to the stock of the breech-loading gun dated 1537 in the Royal Armouries (XII.1) that belonged to King Henry VIII. (see above and Fig. 7.2).

*Matching the shot to the guns*
by Douglas M<sup>c</sup>Elvogue

Sixty-two of the 1000 lead shot listed for the *Mary Rose* in the *Anthony Roll* have been recovered (Chapter 5; Fig. 5.69). Sizes range from 7mm to 19mm in diameter, with weights of 8.5–32g (Table 7.2). Due to deterioration of the barrels, only two bore sizes could be measured, at 13mm and 12mm. The surviving ramrods are 9mm, 12mm and 13mm in diameter, with grooves of approximately 14mm. This suggests a standard bore size of ½in (12.7mm), with a slightly narrower ramrod contained in a wider groove. If this was indeed the standard bore size then analysis of the shot suggests that few fit the guns. Instead, it points towards three probable bore sizes with a range of shot provided to supply each. The postulated original bores are ⅜in (*c.* 9.5mm), ½in (12.7mm) and ¾in (19mm) in diameter (Table 7.2). Shot made for ½in bores are, therefore, 10–12mm in diameter. There was no 'standard' for casting shot and while the observed variation might reflect a percentage of below-standard shot that might be expected in large orders, it could equally indicate that the tolerance for casting shot was 1/16in (1.6mm) in diameter. However the fifteen lead shot found together inside the Carpenters' cabin containing gun stock 81A3884 ranged from 10mm to 14mm, although thirteen were of 12mm or 13mm.

If three sizes of handgun were being used aboard the *Mary Rose* these are likely to have comprised small pistol sized weapons and two larger sizes of harquebus. The smaller size of shot might require wrapping before loading, ensuring a snug fit, however variation in weight would have an effect on the impact of the ball.

*Associated objects*
by Douglas M<sup>c</sup>Elvogue

An individual hand-gunner might have been expected to carry a powder flask, a priming flask, and a pouch containing lead shot of the required size. It is also

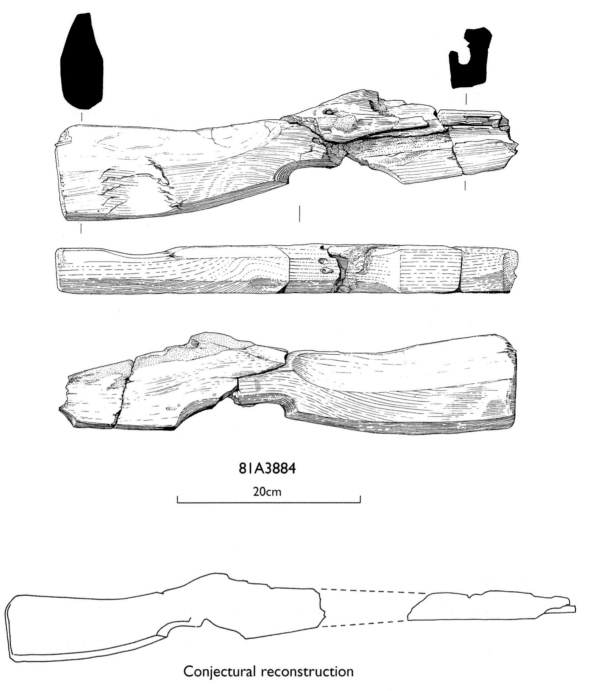

*Figure 7.11 Stock fragment 81A3884 and partial reconstruction*

**Table 7.2 Size, number and distribution of small lead round shot suitable for handguns, with estimated bore sizes**

| Bore | ⅜in (9mm) | | | ½in (13mm) | | | | ¾in (19mm) |
|---|---|---|---|---|---|---|---|---|
| Shot diameter | 7mm | 8mm | 9mm | 10mm | 11mm | 12mm | 13–14mm | 15–19mm |
| No. | 3 | 2 | 2 | 1 | 1 | 8 | 21 | 24 |
| Main concentration | U11 | U11 | U11 | M9 | C8 | M9/11 | M9/11 | stern area |

possible that they would have carried their own shot tongs, or iron moulds to cast shot. Leather drawstring pouches have been found in association with gun-shields (see below), and a variety of often highly decorated pouches and purses have also been found (*AMR* Vol. 4, 108–11), though not physically associated with handguns. One of the four powder dispensers recovered (81A0145) has been tentatively identified as a priming flask for a handgun (it is smaller than the others and has no spring cut-off for dispensing a measured charge); this was also found in U9 and is described in Chapter 5. No casks have been identified which contained either lead shot or cornpowder (*AMR* Vol. 4, chap. 10).

*Replication and firing*
by Alexzandra Hildred

The Royal Armouries undertook to replicate and trial a number of historical weapons to assess their ballistic performance, including a copy of one of the snap matchlock harquebuses owned by the Royal Armouries similar to 81A2679, 81A2405 and 82A4502. It was proofed with a charge of 120 grains but, for the tests, was loaded with incremental charges from 50–90 grains of modern black powder. The projectile was a ½in (12.7mm) solid lead ball. The results were compared with those of a replica medieval handgun, a *Mary Rose* style longbow, a crossbow, a sling and a spear (Richardson 1998, 50–2).

The medieval handgun achieved a velocity of 181.8m/s (50g charge) and the harquebus velocities varied between 378.7m/s (50g charge) and 520m/s (90g charge). The longbow (90lb draw-weight) reached 45.6m/s. Penetration tests were made against a 500mm square shield made of varying numbers of sheets of 2mm mild steel. Whereas similar tests carried out using replica longbow and crossbow failed to penetrate a single 2mm sheet, the harquebus, loaded with 90g, penetrated a 4mm sheet and nearly pierced a 6mm shield (Fig. 7.12).

## Gun-shields: '*Targettes steilde with gonnes*'
by Alexzandra Hildred

*Scope and significance of the assemblage*

On 2 December 1981, during the final week of a nine month diving season, a convex circular object of 560mm diameter, formed of two cross-layers of wooden laths (*c.* 3–5mm thick) was excavated from the Orlop deck in the stern (O10) (Fig. 7.13). Strips of what appeared to be copper in the form of oak leaves radiated from a central hole surrounded by an ornate copper-alloy boss. These appeared to be seam-covering pieces for iron plates which had disintegrated (Fig. 7.14). A rectangle (35 x 37mm) had been cut out of the wood above the central hole and had partial remains of concretion around it. All the copper alloy decoration was held onto the wood by dome-headed copper alloy nails, clenched on their inner surface. The object (81A5771) was lifted together with its encasing sediments which included leather and corroded iron

*Figure 7.12 Graeme Rimer firing a replica of the* Mary Rose *harquebus*

*Master Lord Ambassador*

*The immense liberality and sum of gratitude which everyone universally reports of the most high and famous King your lord, that his Majesty has continually in remembrance men of expertise* (vertuosi), *especially of the Italian nation, has moved we the three under-written, namely myself Master Giovanbattista, painter of Ravenna, with two other companions, to come to serve His Majesty in such manner that I offer to make artificial fires of divers sorts to annoy* (offender) *the enemy in pots* (vasi) *of terracotta of several sorts to throw* (tirare) *with the hands; likewise pastils* (pastelli) *of fire which are thrown with the hands to burn bridges* (ponti) *and other wooden structures of the sea, and pikes and darts, arrows with fire and guns* (schioppi) *within, which annoy the enemy with great force and damage* (danno); *item several roundels* (rotelle) *and bracers* (imbracciadore) *with guns within are thrown* (tirano) *on the enemy and pierce any armour. Powder, again in several forms; one that makes no gun* (non far schippo) *which serves well for ambushes; and is thrown at the enemy without being heard from afar* (da' lontano), *'passa' [sense unclear] like the other fine powder. Item, certain balls with guns within, which are thrown with the hand and pierce the enemy, and in each ball are four guns within. With other secrets and expertise* (virtu), *the which I reserve to myself to be able to accomplish* (reuscire) *better when I am in His Majesty's presence.*

PRO SP1/184, f. 7 (*LP* 19/1, 219)

*Figure 7.13 Photograph of gun-shield 81A5771 as lifted*

concretion. Controlled excavation ashore revealed a coiled copper alloy spring and a small leather pouch-pocket (Figs 7.15 and 7.23). Confirmation that the object was a shield (also called a '*target*') which would have held a centrally mounted breech-loading matchlock gun, was made by Claude Blair. During 1982, remains of a further six examples of varying completeness were recovered from the same area and one from an adjacent trench. Their location on the Orlop deck indicated that they were in storage at the time of the sinking.

The description of a combination shield/target with a gun is first known from a petition of 19 March 1544 from an Italian painter, Giovanbattista of Ravenna. In this letter he offers to make for Henry '*round shields with arm pieces with guns inside them that fire upon the enemy and pierce any armour*' (see extract above). Henry was already procuring handguns from Italy at this time, so it seems plausible that this offer was accepted. Recent study and detailed analysis of paint and fabric suggests that some examples may be foreign imports, but that others (particularly those similar in size and form to the *Mary Rose* examples) may be English. This is based on the type of cloth lining and also the particular colours, the red and yellow. These have been identified as livery colours issued to the Royal Household during the 1544 Boulogne campaign (Metcalf *et al.* 2006, 80). This is another strong indicator of a *Mary Rose* link with Henry VIII's guard.

Identification of these unusual objects was made possible by the presence in the collection of the Victoria & Albert Museum, London and the Royal Armouries of a number of similar objects dating to the reign of Henry VIII. Moreover, the inventory of Henry's possessions of 1547 (Starkey 1998) lists targets with (3787–91, 3828, 3850) and without guns, mostly within the Tower of London. Two forms of shielding device are mentioned: targets and the smaller bucklers. Both were generally circular or oval. Targets were secured to the forearm by means of straps on the inside ('*brases*'), and bucklers had simple handles. Bucklers were used more in non-military contexts and for ceremonial use (see also Chapter 10). Infantry armed with swords and targets were used as a back-up for pikemen and harquebuisiers.

The inventory lists 874 '*targets*', 88 combined with guns. Many had been at Westminster during the reign of the King, but 353 without guns and 35 with were taken to the Tower Armouries in July 1547 (Richardson 2002, 28). The 46 extant gun-shields are the only known survivors from this large collection. These are apparently unique to Henry's armoury (Blair 1999) and finding eight of these on board in a military context is unexpected and intriguing. Many of the previously recorded examples are ornately decorated, possibly even furnishing Henry's personal bodyguard (Blair 1958, 182). This attribution partially arises from their location at Westminster, an armoury which was used mainly for Court arms and armour, the King's own possessions and those of royal guard and other Court officials. The inventory suggests great diversity, both in size of gun, and decorative elements such as fringing, painting, gilding and different materials such as silk and velvet used within the manufacture. This diversity may be an indication of different places of manufacture or different purposes.

In addition to the eighteen examples belonging to the Royal Armouries and a further 20 in other institutions, we can now add eight from the *Mary Rose*, which include decorated examples. More may exist within private collections. Of these 46, two may never have been fitted with guns, and two others are made entirely of metal (see below). Forty-four are likely to date from the mid-1540s. The *Mary Rose* assemblage, although comprising only the organic and some copper-alloy elements, is unique in that the items have not been subject to repair or replacement of worn out portions.

*Morphology*

The component parts of the gun-shield are illustrated in Figs 7.14 and 7.15 (see also Figs 7.32–3, below). Most of the surviving sixteenth century examples are made of two or more layers of wooden laths, each approximately 3mm in thickness, laid perpendicular to each other and covered with steel plates on their convex, external face. Diameters with guns range from 440mm to 600mm and weigh 4.5–5kg (10–11lb). The inside of the shield is sometimes covered with a woven cloth (wool or hemp) possibly glued to the laths and held in position by a circle of bias-binding held with iron tacks. In a number of instances there is evidence for a semicircular cushioned pad ('*enarme*'), also fastened with bias, which may have supported the left forearm. The shields are pierced to take a gun and also for sighting (Fig. 7.15). Detailed scientific analysis has recently been undertaken on the construction of some of these within other exhibitions (Metcalf *et al.* 2006, 76–85) and suggests that some may be of English manufacture.

In provenanced examples the gun is a breech-loader with a matchlock ignition system and separate chamber. The guns weigh *c.* 1.6–2.3kg (3½–5lb), are 220–320mm long, and have internal bores of 12–16mm, accepting a solid lead round shot. On the inside of the shield (Fig. 7.15), the round barrel is squared off and has a hinged cover with a backsight which opens to allow insertion of the chamber. The cover is opened and closed by levering a curved steel spring which disengages or engages a lug into a depression on the left side of the breech. In some cases the chamber cover is secured by inserting a steel pin through two eyed-lugs on the base-plate which project through two slots in the chamber cover. The gun is secured to the base-plate which is secured to the shield. The barrel protrudes through an aperture cut through the shield either centrally, or above the centre. Sighting is through one or more small holes cut above or around the barrel of the gun. In some instances there is an additional rectangular viewing grille immediately above the gun. The chamber varies in size but is typically around 81mm in length with an external diameter of 22mm and internal diameter of 18mm; it is blocked at the end facing the user.

The matchlock ignition system uses a slow match clamped in a serpentine and secured by a wing screw. This is a single bar, bent to form the serpentine at one side (left) and a lever on the right. The centre is secured on to a plate (100–115mm long) which is nailed to the inside of the shield. The powder is ignited by depressing the lever, drawing the lighted match to a touch hole which penetrates the cover and cartridge. A coiled

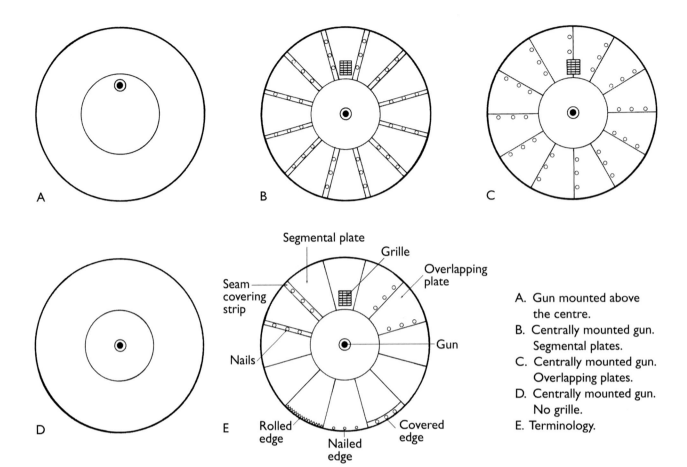

*Figure 7.14 Gun-shield forms and anatomy of the outside*

spring prevented premature firing by holding the serpentine away from the touch hole (Fig. 7.15).

There are two principal methods of holding the object, sometimes only interpretable from the distribution and arrangements of nail-holes: a) a simple 'dustbin lid' handle which originates from beneath the base-plate and extends downwards, bifurcating within the lower third of the target and nailed to the inner laths (Fig. 7.16, bottom), and b) a strap and buckle which was attached around the left forearm with the fingers of the left hand holding a leather grip. In some cases both options are present. All the *Mary Rose* gun-shields were probably held by straps (Fig. 7.15). Metcalf *et al.* (2006, 79, 83) suggest that examples similar to those found on the *Mary Rose* may have required some form of support, such as a bulwark, in order to steady the gun for firing, on the basis of damage evident on an example from the Victoria and Albert Museum (V&A M.507-1927), which they attributed to the absorbing of recoil stress resulting from repeated firing from a fixed position. Gun-shields were normally supported by means of a braced strap ('*guige*') around the neck and this seems a likely form of additional support.

A detailed study of the construction, dimensions, and physical arrangement of parts on the gun-shields from both the ship and in the Royal Armouries revealed that two forms existed: those with a centrally mounted gun and those with a gun mounted above the centre. This in turn affected the sighting arrangement and the position, form, and attachment of the external plates (Fig. 7.14). All the *Mary Rose* examples are of the former type.

### Gun-shield with gun mounted above centre of shield

This group are quite finely made and often carry surface decoration. The diameter is about 490mm. The gun is set vertically above the centre of the shield. Sighting is through a series of small holes around the gun. They do not carry a boss. There is an outer covering of eight segmental iron plates covered by a central disc (Fig. 7.14). The joints between the plates and the outer edges are covered by seam-covering strips and the base plate penetrates the internal slats of the shield. Most surviving objects in this group have elaborate patterns etched into the iron plates, for example Royal Armouries (RA V.39, V.40, V.41, V.42, V.48, V.80, V.99), Royal Collection Windsor (acc. no. 72764) and Higgins Armoury Museum (acc. no. 2533) but none of the *Mary Rose* examples is of this type.

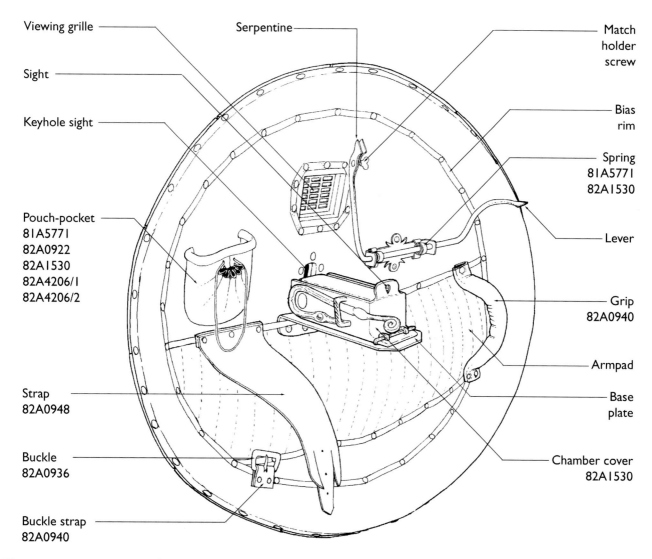

*Figure 7.15 Anatomy of the inside of a gun-shield*

**Gun-shield with centrally mounted gun**

All examples of this type have a square or rectangular hole cut vertically above the gun fitted with an iron grille for sighting, in addition to a small round hole just above the barrel (Fig. 7.14). Apart from a smaller '*buckler*' in the Palace Armoury in Valletta, Malta (435A; see below), external diameters are in the range 445–600mm. These are generally heavier and of larger diameter than those with the gun set above the centre.

In addition to the eight from the *Mary Rose*, 24 examples have been identified. The plates are segmental and have either embossed decoration, usually almond-shaped, along the centre of each segment, or surface etching onto a plain or embossed plate. There are usually 10–13 plates. Three methods of attachment have been noted: overlapping plates nailed together at the overlaps; direct nailing along the sides and each end of the plate; or tensioning the plates by positioning a seam-covering strip over the junction and nailing the strip to the wood beneath. Examples with no seam-coverings may have had decorative fringes at the plate junctions (eg, Art Institute of Chicago (AIC) 1278; Walters Art Gallery (WAG) 54 1414; Fig. 7.16). '*Targettes paynted and gilte fringed with silke*' are mentioned in the ordnance and munitions held in the Tower of London (though not with a gun) (Starkey 1998, 3818).

None of the *Mary Rose* gun-shields has overlapping plates, nor do RA V.34, V.35, V.38, V.115; WAG 54 1414, or AIC 1278. Those with overlapping surface plates include RA V.37, V.43, and V.81; Metropolitan Museum of Art (MM) 14.25 746 and possibly 745.

The outer edge of the plates are either rolled backwards onto the top of the plate to form a neat lip (eg, RA V.34, V.35, V.38, WAG 54 5414), covered with a strip at the outer edge (RA V.37), or are left bare (RA V.115; AIC 1278).

The viewing grilles are divided into eighteen perforations, either aligned six across and three down, or three across and six down. This does not appear to be related to any other feature or with the method of orienting the plates to accommodate the grille. The corners of the grille can either form the edges of the

plate junction (RA V.43). The only instance where the positioning appears to be vital is that of the Metropolitan Museum example, where the plate junctions are covered by small bosses. Thus the grille has to be central to the plate to allow room for them.

In this type, the base-plate is turned downwards against the inner face of the shield and is nailed in position with three iron nails entering from the outside just beneath the base-plate. This has the effect of bracketing the base-plate to the inner surface (Figs 7.15, 7.16 bottom). On all the *Mary Rose* gun-shields the gun aperture is rounded on the area beneath the gun but relieved to form lobes on each side and directly above the gun. This may be for some bayonet securing system (eg, 81A5771, Fig. 7.22).

The gun-shields with centrally mounted guns that retain evidence for handles such as nail-holes, in addition to a buckle strap and hand-grip (see above) include WAG 54 1414 and AIC 1278. RA V.34 and V.38 display holes and staining suggestive of having supported a handle towards the lower segment of the shield. On V.34, the extension of the base plate behind the breech cover has a circular hole of 7mm diameter which may have been a location point for the origin of the handle, as seen on the Art Institute of Chicago example (Fig. 7.16 bottom). One *Mary Rose* example (82A1586), displays nail-holes at the lower part of the shield which may have retained the bifurcating handle (Fig. 7.27).

Most of the *Mary Rose* gun-shields are too degraded to be able to unequivocally state exactly what forms of enarmes they supported, although all display the nail-holes suggestive of the bias ring just inside the outer circumference, and a semicircular set of nail-holes suggestive of the arm pad. To date two straps (82A0948, 82A0967) have been identified and one hand grip (82A0940, Figs 7.15 and 7.21).

### Shields without guns

Royal Armouries V83 and Palace Armouries, Valletta 435A (illustration available at http://www.edu.um.edu.mt/malta/html/shields.html) are similar to the shields with centrally mounted guns, but lack both gun and grille, although they have central apertures. Both have twelve segmental truncated surface plates with seam-covering strips over the plate junctions, fastened with brass nails. Their external embossing is similar as is their elongated pyramidical form (see Metcalf *et al.* 2006, table 1 for a detailed description of known extant gun-shields). Additional fastenings on the inner and outer edges of the plates secure each to the wood beneath. There is engraving on the surface of the Royal Armouries piece. The small Valletta example is described as a '*buckler*' in the object caption. It is likely that this originally came from the Tower of London in the sixteenth century. The foundation is of oak laths with the remnants of internal backing cloth, cushioned arm pad, and matchlock ignition system in evidence.

*Figure 7.16 Top) Walters Art gallery gun-shield (Acc. no. 54 1414), external view; bottom) Art Institute of Chicago gun-shield (AIC 1278), internal view. Reproduced by permission*

plates (as on RA V.34; Fig. 7.32), be central and not touching the edges (RA V.115), or utilise the side of the grille as part of the plate seam (AIC 1278), or a combination. Grille sizes vary between 35mm and 55mm in either direction.

A similar pattern is seen on gun-shields with overlapping plates: in two cases the grille forms the seam of two plates (RA V.37, V.81), in one it is central (MM 14 745), and in another grille centre forms a

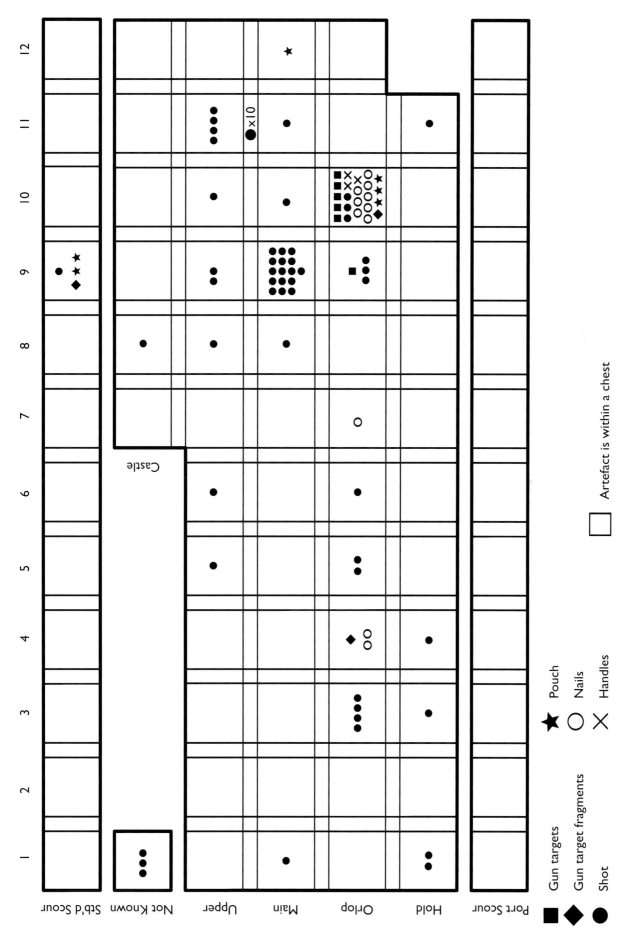

*Figure 7.17 Distribution of gun-shield elements*

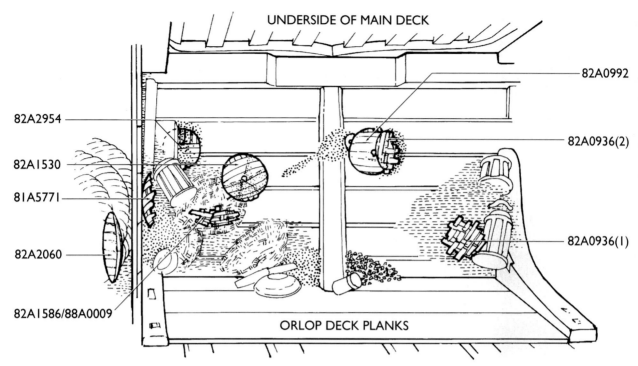

*Figure 7.18 Distribution of numbered gun-shields in O9/10*

**Other related forms**

Item 30.3430 in the Hermitage Museum, St Petersburg has a wheel-lock mechanism while Item 238 in the Bargello Museum, Florence is a 6-barrel wheel-lock. Both of these appear to be made entirely of iron and are internally different from those discussed above.

*Distribution of the gun-shields*

Seven relatively complete gun-shields were recovered (Fig. 7.17) with a number of laths which may be parts of an eighth (82A0936). Most were found on top of one another against the starboard side of the ship (Fig. 7.18) in O10. Four were relatively complete, (81A5771, 82A1530, 82A0992, 82A0936) three required substantial reconstruction after conservation based on wood type and lath width. 82A2060 (relatively complete) was found nearby in O9 within a large coil rope. O10 was part of one of the principal storage areas of the ship and also contained baskets, lanterns, bowls, a drum, pewter canisters, and several large chests containing personal possessions (see *AMR* Vol. 4, chap. 10). Many casks were also found here.

Although the gun-shields were associated with iron corrosion products, gamma radiography of some of the concretions from O10 revealed only one substantial barrel fragment of a breech-loading gun (82A1530, Fig. 7.19a) and a possible serpentine ignition system with screw (82A0754, see 7.15) as described above. The question arises, therefore, as to whether the guns may have been removed for ease of stacking and whether a substantial amount of concretion removed from the junction of the starboard side and the Orlop deck in this area (not yet radiographed) may have contained the guns.

Only three lead shot were found in this sector (70A0661, 81A0576, 83A0482) of 12mm, 19mm, and 13mm respectively. One lead shot of 15mm (81A0309) was found in the same sector as gun-shield 82A2060 but not in direct association.

Three other isolated fragments may be from gun-shields. A fragment of lath with a nail (82A5154–5) came from within a concretion beneath the Sterncastle. Interestingly this contained the remains of an incomplete handgun, (82A4502) and swivel gun (82A4080). This could suggest that one or two gun-shields were already on the Upper or Castle deck when the ship sank. A fragment of lath with nails (82A4465) was recovered from the Orlop deck in the bow during the washing of the ship following recovery, and a small lath fragment came from an unknown location (82A5248).

Analysis was undertaken of the distribution of other elements that could suggest the presence of disintegrated or concreted gun-shields, namely the distinctive leather pouch-pockets, dome headed copper alloy nails and copper alloy springs. The springs (eg 81A5771, Fig. 7.23) are made from a cut strip of brass sheet 1.5mm thick and 10mm wide. The finished coil is 26mm in height and is coiled three times. The central hole to accommodate the serpentine arm is 5mm. A flat face, 20mm in length, rests against the bar, as shown in Figure 7.15.

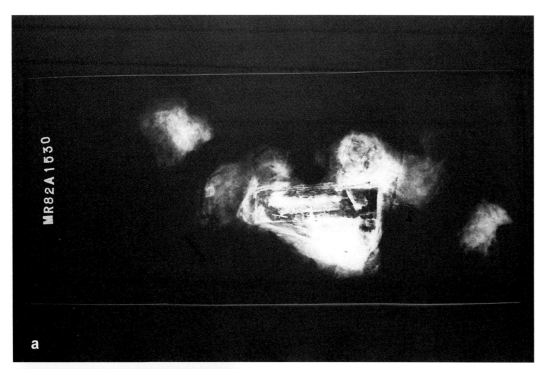

*Figure 7.19
Radiograph of: a) on
82A1530 gun; b) on
Royal Armouries gun*

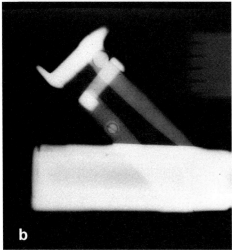

Nineteen nails were studied in detail to look at integrity within the group (archive report). This included nine which separated from shields during conservation, and ten that could not be associated directly with any object. All have domed heads with a diameter of 7–8.5mm and with a height of 4.5mm. The shaft is rounded, tapering to a point away from the head. These served to attach the iron plates to the outer surface of the shield. If complete (and used on a shield) the nail is bent at a right-angles 13–17mm below the top of the head (eg, 82A1586, Fig. 7.27). A second bend is observed a further 6mm along the shaft to bury the pointed tip within the wooden laths. Where complete, the tip is only 1mm. In some instances the shaft is central to the head, in others it appears to be offset away from the curve (towards the back), or towards the curve (to the front). These appear to have been hammered.

Examples of nails, bosses, seam-covering strips and springs were analysed using XRFS (full details, E. Pitt, in archive). The nails and springs appear to be quaternary alloys of tin, lead, zinc and copper. The boss and strips were made up of a thinner, softer copper foil which would be easier to emboss over a wooden former (Table 7.6, below).

Of twelve loose nails, only one (83A0383) found in O7, during the cleaning of the ship after recovery, was not found in direct association or very close to identified gun-shield parts. This could have washed in from elsewhere. Only two springs were identified, both directly associated with gun-shields. In contrast, three unassociated pouch-pockets were recorded: two from under the Sterncastle (82A42006, 82A4206) and one from M12 (83A0577), recovered after raising of the hull. Overall, the evidence suggests a possible maximum of fourteen gun-shields, most of which seem to have been in storage at the time of the sinking.

## Leather elements

### Pouch-pockets
by Alexzandra Hildred and Nick Evans

The first gun-shield found (81A5771) had leather fragments adhering to its inside. When cleaned, this was found to consist of a semicircle together with a fragmentary pouch-pocket. Similar leather fragments were found with 82A0992 and 82A1530, with a further three examples represented by the unassociated finds, 82A4206/1 and 2, and 83A0577. A replica was made.

Each is made up of three elements. A back-patch, 90–120mm square (the final size of the pouch) a

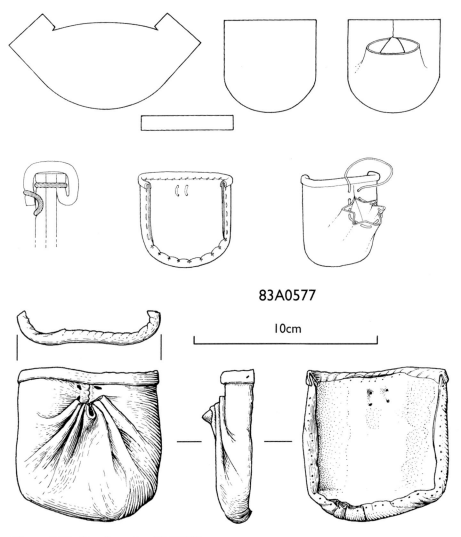

*Figure 7.20 Pouch-pocket 82A0577*

semicircular front-piece (radius *c.* 50mm), cut to shape, and a seam-covering strip of 20 x 90mm (Fig. 7.20). The front-piece is slashed to accommodate a central seam, and the larger portion of the semicircle is gathered, with the sides seamed to the back-patch. The gathered portion is covered with the strip which is turned over the top of both the front and back-patch and is over-stitched to the latter to bind the two pieces at the top. Both are then turned towards the back-patch and invisibly stitched to the lining of the gun-shield.

The front-piece thus forms a drawstring pouch held tightly with a cord attached through four holes (which provides the running stitch to gather the pouch). A pocket is formed between the back of the back-patch and the lining of the shield. The shape of 83A0577, the largest example at 120 x120mm (Fig. 7.20), suggests that it could have accommodated two chambers side by side. The drawstring pocket is a potential candidate for carrying shot. None of these pockets has been identified from examples within other collections.

*Straps and grips*

Two leather fragments from O10 have been identified as parts of gun-shield enarmes (Fig. 7.21): a grip portion (82A0940) and a strap (82A0948). An iron buckle with strap (82A0936) was found with gun-shield 82A0992.

Grip 82A0940 is a single leather strip 224mm in length, 20mm wide, and 3mm thick. Each side is folded and over-stitched to the other forming a central seam at the back. At each end there are two 4mm holes surrounded by concretion, suggesting iron. At the ends, the leather edges overlap with a single nail apparently penetrating both sides.

Strap 82A0948 comprises three pieces of leather. The first is an elongated triangle, 320mm long and 111mm wide at the base. This end has three evenly spaced nail-holes (2mm diameter). The other end is slashed and bevelled to form a central tab and two side strips. The end of the tab is cut at an angle to the base as on many belts. This is 32mm wide at the origin of the cut and 20mm at the end. It is 4mm thick. Three 2mm

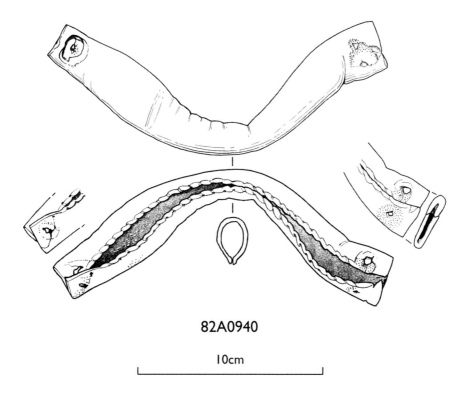

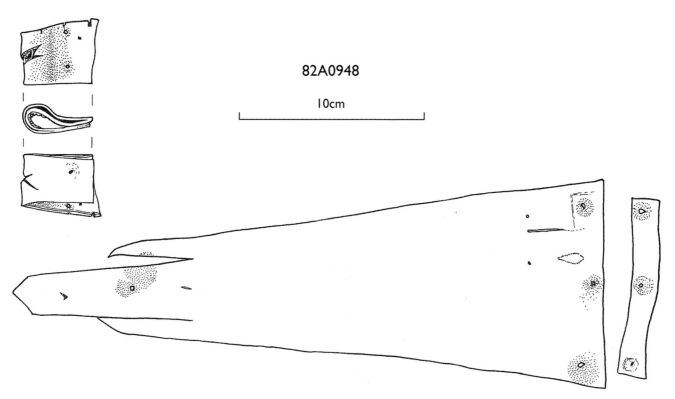

*Figure 7.21 Grip 82A0940 and strap 82A0948*

diameter nail-holes are set at 30mm, 70mm, and 100mm from the end. The second portion is a strip 96 x 16mm designed to strengthen the nailed end. This retains nail-holes matching those on the strap. The third element is a buckle retainer formed of a single piece of folded leather which is pierced to take the loop and pin on a buckle. This measures 30 x 40 x 3mm. It is iron stained.

The buckle (82A0936) is the only iron buckle surviving from the ship. It is very corroded and

Table 7.3 Components of gun-shields recovered

| Component | 81A5771 | 82A0992 | 82A1530 | 82A1586 | 82A2060 | 82A2954 | 88A0009 | 82A0936 | Isolated element |
|---|---|---|---|---|---|---|---|---|---|
| Sector | (O10) | (O10) | (O10) | (O10) | (O9) | (O10) | (O10) | (O10) | |
| Shield/frags | ✓ | ✓ | ✓ | ✓ | ✓ | ✓ | ✓ | ✓ | 82A4465 82A5155 82A5248 |
| Boss | ✓ | ✓ | | | | | | | |
| Seam strip | ✓ | | | | | | | | |
| Spring | ✓ | | ✓ | | | | | | |
| Handgun | | | ✓ | | | | | | |
| Serpentine | ✓ | | | | | | | | |
| Dome-headed nails | ✓ | ✓ | ✓ | ✓ | ✓ | ✓ | ✓ | ? | |
| Buckle | | | | | | | | ✓ | |
| Strap | | | | | | | | ✓ | 82A0948 |
| Grip | | | | | | | | | 82A0940 |
| Pouch-pocket | ✓ | ✓ | ✓ | | | | | | 82A4206 83A0577 |

Table 7.4 Principal measurements for gun-shields (metric)

| | 81A5771 | 82A0992 | 82A1530 | 82A1586 | 82A2060 | 82A2954 | 88A0009 |
|---|---|---|---|---|---|---|---|
| % Complete | 75% | 60% | 90% | 60% | 90%+ | 70% | 60% |
| Wood | beech | oak | oak | oak | oak | oak | oak |
| Diameter | 562 | 492 | 470 | 480 | 460 | 475 | 480 |
| No. plates | 8–10 | 10 | 10 | 10 | 10 | 10 | – |
| Boss diameter | 140–165 | 162 | ?100 | – | – | – | – |
| No. ext. laths | 9 | 6 of 7 | 10 | – | 7 of 8 | 6+ | 5+ |
| Direction | horizontal | horizontal | horizontal | horizontal | horizontal | horizontal | horizontal |
| Ext. lath width | 40–65 | 40–80 | 35–58 | 40–60 | 35–78 | 38–52 | 43–65 |
| Ext. lath length | 275–470 | 230–380 | 240–465 | 220–365 | 275–460 | <220 | <310 |
| Grille width/height | 37 x 35 | 42 x 45 | 40 x 41 | 40 x ? | 43 x 45 | – | – |
| No. int. laths | 7 | 7–9 | 9 | – | 12 | 4+ | 5+ |
| Direction | vertical | vertical | vertical | vertical | vertical | vertical | vertical |
| Int. lath width | 48–60 | 35–75 | 30–75 | – | 30–50 | 35–54 | 40–65 |
| Int. lath length | <220 | – | – | – | 300–450 | – | <420 |
| Pouch-pocket height/width | 90 x 98 | 80 x 80 | 83 x 90 | – | – | – | – |

incomplete and has a length of 20mm and width of 28mm. It appears to be a single loop rectangular buckle (see Chapter 10). There is no evidence of a pin, only a hole of 2.5mm. The strap is 50mm wide, 90mm long and 3mm thick.

## The gun-shields

Table 7.3 lists the various elements represented for each gun-shield. Principal measurements for the seven certain examples are given in Table 7.4. Unless otherwise stated, the widest and longest laths are in the central parts of the gun-shield, internally and externally, the largest overall is usually that just below the gun. Details of XRF analysis of metal components is provided in Table 7.5.

### 81A5771

This object, the first of the gun-shields to be positively identified, was *c.* 75% complete as found, with copper seam-covering strips over the remains of corroded iron plates. The boss was found nearby (Figs 7.13, 7.22). Of the nine surviving external beech laths, the uppermost one is cut with the rebate for the viewing grille and two are cut to accept the gun; parts of three external laths survive below the gun aperture.

The internal vertical section comprises seven laths, in 70 fragments, varying in width from 48mm to 60mm and in length up to 220mm. The central lath is cut to

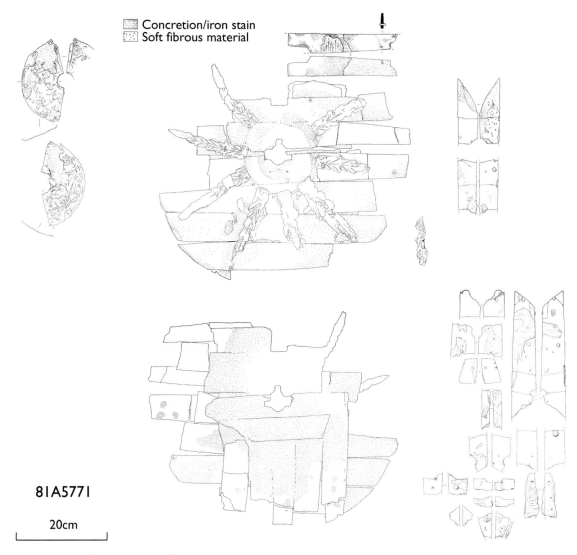

*Figure 7.22 Gun-shield 81A5771*

accept the grille and the major portion of the gun. The three holes to accept the base plate are visible beneath the gun aperture.

The inner layer of laths consists of fragments which reinforce the area of the gun aperture. They do not seem to exist above the centre of the gun-shield. The spring (Fig. 7.23) was found underneath. It consists of a thin, narrow (10–17mm wide) strip of copper alloy bent back upon itself for 20mm to hook over the serpentine retaining plate. The remainder was coiled around the lever and does not seem to be fixed. It has a coiled length of 26mm.

The incomplete leather purse-pocket (Fig. 7.23) has a maximum width of 100mm and depth of 90mm. It is covered with concretion on its inner surface. The binding at the top of the pocket is visible, as are the gathers forming the drawstring pouch to the front. Stitch-marks around the pocket indicate where it was sewn on to backing material.

The boss is identical to that on 82A0992 (Figs 7.24–25). Both seem to be formed of thin (1–2mm thick) copper sheet. Although in poor condition, there is clear evidence for *repoussé* decoration incorporating human and fantastic heads with foliate decoration, decorative elements that can be seen in woodcuts, leatherwork, and metalwork of this period, including on the ship (for instance, on book covers; see *AMR* Vol. 4, chap. 3). Its details are discussed below

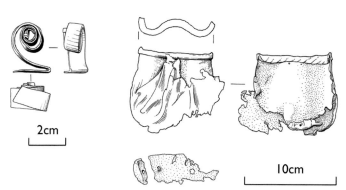

*Figure 7.23 Pouch-pocket and spring elements associated with 81A5771*

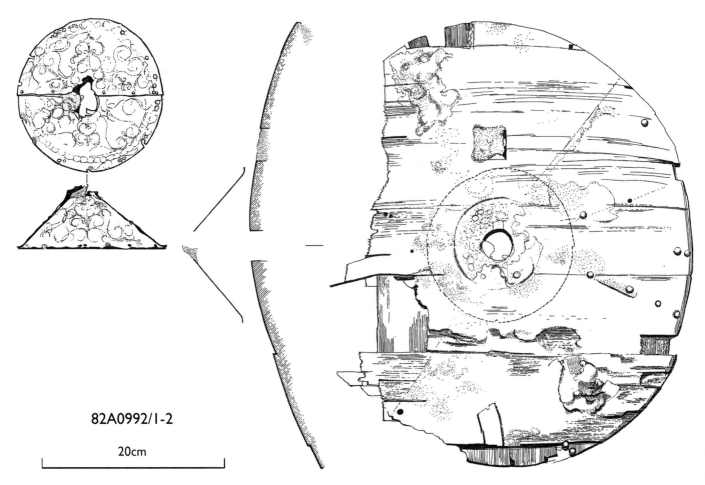

*Figure 7.24 Gun-shield 82A0992*

under object 82A0992, on which they are much clearer. The two halves forming the boss on each gun-shield are identical, all four segments probably having been made using the same former. The seam-covering pieces were attached over the iron plates by two nails under the outer circumference of the boss, and by nails staggered alternately across the centre line of the tapering (30–10mm wide) strip, which forms the stem of the decorative leaf, and of which 75mm survives. The outer edge resembles an elongated oak leaf.

The laths are too degraded to ascertain how the object was held. The surface iron forming the outer plates is only fragmentary, with no decorative features visible. An incomplete fragment of leather with a length of 220mm and width of 120mm (81A5771/8) remains unidentified.

**82A0992**

This object (Fig. 7.24) is nearly complete on the right side (viewed from outside) but has a number of lath ends absent on the left. Six of the seven external laths are present. The side lobes of the gun aperture are *c*. 15mm measured horizontally and 24mm deep, with a top lobe of 18 x 15mm. They carry vestigial remains of iron.

Corroded remains of a circular iron plate (diameter 160mm), possibly a backing for the boss, cover the area immediately to the left of the gun aperture. Three staggered nail-holes below the gun aperture are visible for the attachment of the base plate (one centrally above the other two). The viewing grille is cut out of two horizontal laths, possibly to avoid weakening only one.

The external surface supports the remains of corroded iron plates and holes for the attachment of a seam-covering piece. Two nails are *in situ*, clenched on the inside. Although not immediately visible, there appear to be three or four attachment nail-holes along each segment. In addition there is one hole just inside the outer circumference (30mm) placed centrally on each segment.

The boss is formed of two nearly identical hemispherical elements the edges of which overlap each other. These are held with three fastenings on each side of the aperture to form a cone with central hole for the gun barrel. The cone is attached via seven fastenings on each portion of the boss.

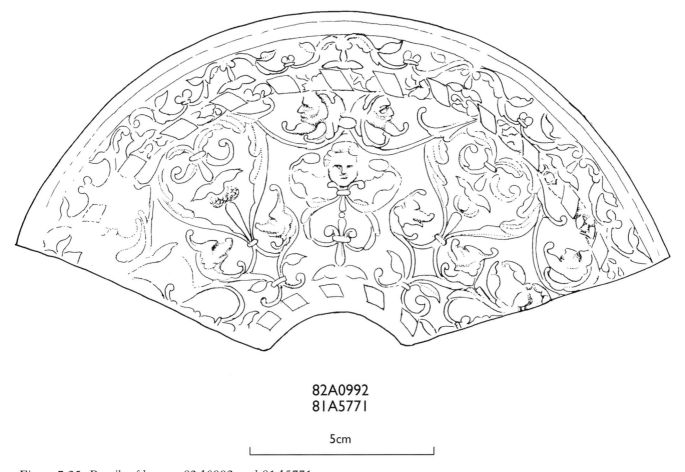

*Figure 7.25 Details of boss on 82A0992 and 81A5771*

The design, which is identical to that on 81A5771, includes a central portrait incorporated into a foliate pattern above which two bearded profiles are situated, facing away from the centre and towards a heart shaped foliate motif (Fig. 7.25). Identical profiles face inwards across the foliate design. Below the portrait and to each side, also facing away from it, are the profiles of two fantastic beasts, possibly Griffins or Wyverns. Each faces its identical twin, and together these form the base of the foliate design. The pattern is bounded by a stylised twisted rope design. There is no evidence for any copper-alloy seam-covering pieces.

Twenty-one unattached lath fragments were found with the main part of this object but cannot be pieced together. Approximately 30% of the purse-pouch survives in fragments, with the twelve largest including the front-piece with two drawstring holes, and the front binding strip.

**82A1530**
Approximately 90% of this gun-shield survives including fragments of the external iron plates and internal mechanism (Fig. 7.26). The gun aperture has a diameter of 30mm, extending to 50mm at the outer edge of the side lobes which contain the remains of concretion. Below the gun aperture are the three nail-holes for the breech retaining plate.

The viewing grille appears to be formed of three rows of five vertical slots, but it is possible that a sixth row is obscured within the concretion. Around the viewing grille are small holes which may have held it in place, or be the fixing points for the usual bias strip. There is one hole at each corner of the grille, and one centrally along each side.

Four of the segmental iron plates are distinguishable, three with a well-defined raised almond-shaped keel (similar to RA V.34, V.35, and V.38, AIC 1278(2256), and WAG 54 1414, discussed above). The plates are secured by three dome-headed copper alloy nails equally spaced along the edges of the segments, 25mm in from the circumference and one nail in the centre.

The gun-shield was recovered with associated corrosion products. These suggested some form of central boss or backing plate on the outside of the object, with concretion, fabric and leather concreted to the inside which included a well-preserved pouch-pocket, situated to the left of the gun (Fig. 7.26). Stitch-holes in the leather suggest that it was sewn on to an internal backing to the gun-shield which has not survived.

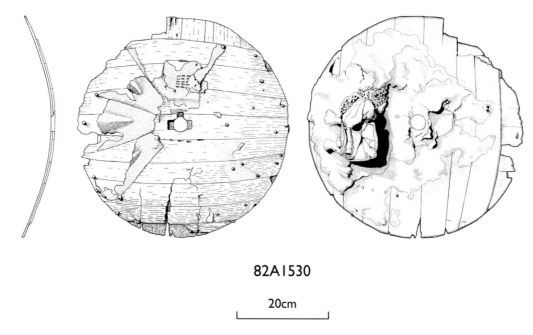

*Figure 7.26  Gun-shield 82A1530*

An inner ring of small tack-holes (with iron corrosion products), 50mm in from the outer edge of the gun-shield, is similar in position, size, and distribution to tacks holding the bias which secures the backing material in place on many of the Royal Armouries examples (see below). The semicircular distribution of holes along the lower segment of the gun-shield is also suggestive of the bias seen holding cushioned pads on extant gun-shields.

Radiographic examination revealed what are probably the breech cover and base-plate. The cover appears to be rectangular and still supported on its base-plate, with the lugs for securing the cover to the base-plate present (see Fig. 7.19a). It has an estimated length of 100mm and width of 40mm, comparable with the dimensions of RA V.34 (Fig. 7.19b). It appears to have a hollow centre (for the chamber). This identification confirms that at least one of the gun-shields originally held a gun and could be considered operational rather than purely decorative. The thin, rolled copper alloy spring is 8mm wide, coiled, with one end bent double to allow it to be held behind the serpentine bracket. Associated iron corrosion products could be the remains of the serpentine.

### 82A1586

This object comprises three segments including most of the bottom four laths and parts of the upper right and upper left quarters (Fig. 7.27). The layers of laths are held by four dome-headed nails (*in situ*), with nail-holes for a further nine. The central internal (vertical) lath is visible as the external lath is absent. It reveals three staggered nail-holes for the attachment of the base-plate supporting the gun, and has been relieved to accept the barrel. Surface concretion and staining indicate the former presence of iron plates and there is a substantial amount of corroded iron and iron staining in the central area, suggestive of either a boss or a central plate.

Each plate was held by four nails along the edges. It is not possible to determine whether the plates overlapped, or were held by seam-covering pieces. In addition each plate was held at the circumference by a centrally placed nail. Four nail-holes lie close to the centre of the bottom of the gun-shield internally and probably secured a handle. A nail-hole beneath those for the attachment of the base-plate may have accommodated a fixing for the top of the handle.

The top right quarter is represented by parts of two laths, the horizontal lath immediately above the gun aperture and the next lath upwards including the remaining portion of the grille opening. The top left quarter of the gun-shield is represented by six horizontal and three vertical laths. The gun-shield has been broken along the left side of the grille, but the segment shows the corner of the grille opening.

### 82A2060

Found within a coil of rope in O9, this is the most complete gun-shield recovered, and the only substantial example not in O10 (Fig. 7.28). The plates are held by three (possibly four) dome-headed nails along the segments. In addition there are two nails equally spaced along the outer edge of the circumference. There were originally ten plates, one being substantially smaller than the others accommodating only the viewing grille (as on RA V.38, Fig. 7.33). The radiograph of 82A2060 clearly shows nail-holes for the grille and/or bias around the grille opening (Fig. 7.28). This consisted of a nail at each corner, and one placed centrally along each side. It also shows an inner circle of small tack holes, suggestive

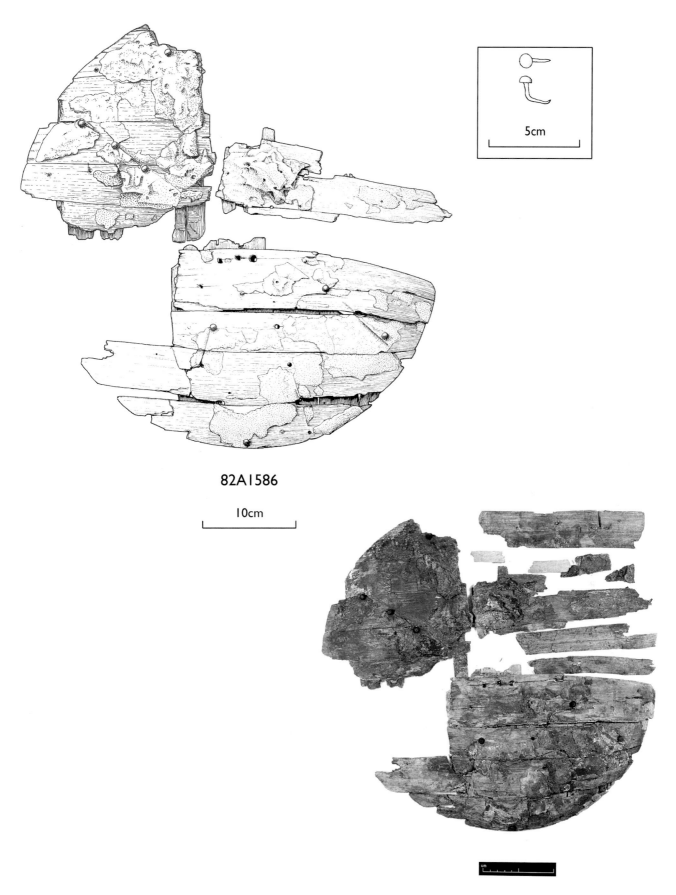

*Figure 7.27  Gun-shield 82A1586*

570

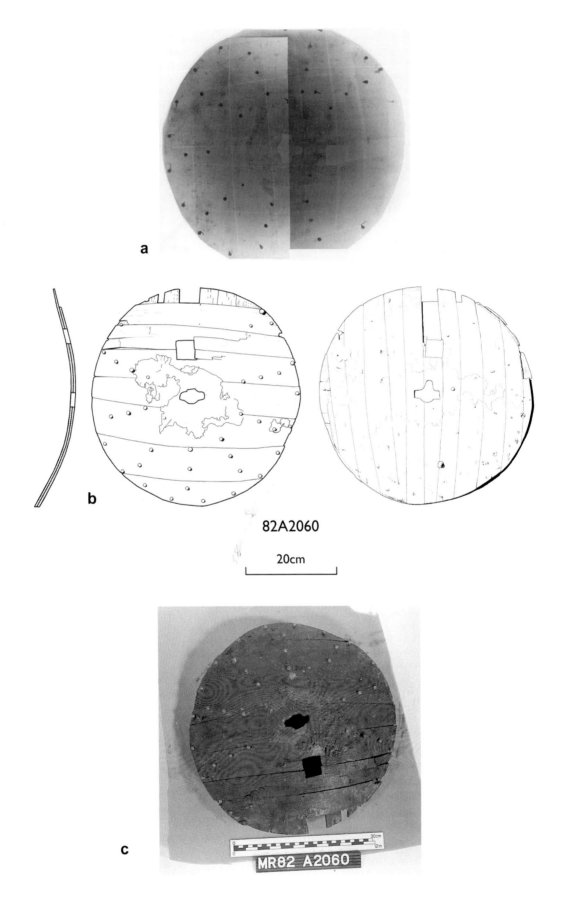

*Figure 7.28 Gun-shield 82A2060: a) radiograph; b) line drawing, left (outside), right (inside); c) photograph outside*

of the bias strip seen on other examples, with a semicircular area showing clearly as having been protected underneath the gun. The area around the gun aperture shows a thin layer of concretion, with more than thirteen holes in a circular pattern at a distance of 75mm from the centre of the aperture, presumably representing the corroded remains of an iron boss or circular plate.

A number of holes to the left of the gun aperture may be for attachment of the serpentine, while three staggered holes beneath it suggest that the base-plate was bent inwards towards the gun-shield and secured. An inner circle of nail-holes about 40mm from the centre of the aperture may provide attachment points for an iron disc (as seen on RA V.38).

The internal laths are more consistent in width over their lengths than on most of the gun-shields. The central lath accommodating most of the gun aperture is, unusually, not the widest, being 35mm at the top and bottom and 40mm in the centre. It is not possible to be certain which of a number of nail-holes are for attachment of the buckle and strap and handgrip, but there are none at the bottom towards the centre which could have supported an iron handle.

### 82A2954

Two non-conjoining parts of this object were recovered and associated during this study. The major portion of the right hand side of the gun-shield is present, the outer laths having become separated from the inner ones. A smaller section of the left hand side is present. Only two external laths have surviving outer edges, all other lath ends are ragged (Fig. 7.29).

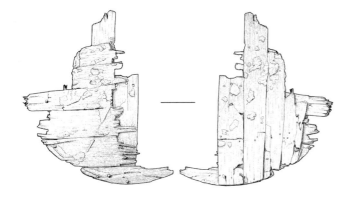

82A2954

20cm

*Figure 7.29  Gun-shield 82A2954*

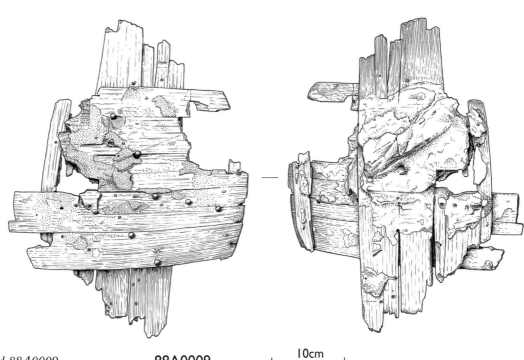

*Figure 7.30  Gun-shield 88A0009*

88A0009

10cm

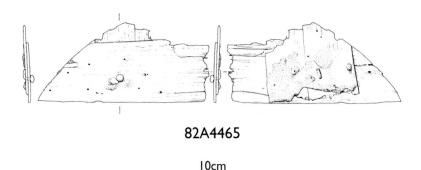

*Figure 7.31 Probable gun-shield fragment 82A4465*

On the right hand side, the object is broken just to the right of the centre line. Features such as a portion of the gun aperture and the right side of the grille opening are distinguishable on the surviving segment. There are two in situ dome headed nails. The presence of very light bands at what appear to be segmental joins suggest the former presence of seam-covering pieces. A circle of nail-holes 55mm from the centre of the gun aperture suggests the presence of a central iron boss or plate. The left hand portion is also broken just right of the centre, revealing the internal, central vertical lath with the lower part of the gun aperture and lobes. This has the three distinguishing nail-holes for the attachment of the base-plate. There are three dome headed nails *in situ*.

This segment also shows an impression of concreted fabric on the inner surface.

### 88A0009

This consists of the central part of a gun-shield with its gun aperture and concreted remains of the viewing grille (Fig. 7.30). Surface staining indicates the corrosion of iron plates laid over the oak laths. The seven surviving nails are clenched on their inner surface and show the original plate junctions. The plates were held by three nails along the junctions and possibly by another one at the circumference, centrally placed. The nail centres are 48mm apart along the plate junctions. The gun aperture is round with projecting side lugs of 20mm. There are the three holes for the base-plate fixings beneath the gun aperture. Internally, the remains of a woven material were visible within concretion on the inside of the gun-shield before conservation.

### Possible fragments
*82A4465*

This fragment, found in the bow area of the Orlop deck after the ship was raised, comprises two layers including a portion of an outside curved edge which must represent either the top right or bottom left of a gun-shield (Fig. 7.31). The single, external (horizontal) lath has a length of 185mm, a maximum width of 65mm and thickness of 3mm. It retains one 8mm diameter brass nail. There are fragments of two internal vertical laths, one 70 x 44mm and the other 80 x 50mm, both are 3mm thick.

*82A0936*

Two groups of laths (not illustrated) were found amongst the gun-shields and other objects in O10 that could be part of a gun-shield: 1) 38 beech lath fragments, maximum width 44mm, longest fragment 185mm, found among the remains of a lantern in the southern end of the sector against the partition planking, and 2) 35 beech lath fragments associated with oak-lathed gun-shield 82A0992. One fragment retains concreted remains of iron buckle. Radiography of iron corrosion products show what may be part of a serpentine mechanism. A loose brass dome-headed nail was also recovered.

There are no features among the laths found in group 1 that enable definitive identification. Although there are numerous nail-holes, these are not arranged in any characteristic pattern. Many of the laths are parallel-sided, 44mm wide and 4mm thick, and the holes seem to be surrounded with rings of concretion, suggesting iron washers. The Higgins Armoury example (see above) retains iron washers, but it is not certain whether these are original. The maximum lath length obtainable by joining the broken elements together is 240mm.

The group 2 laths vary in width along their lengths (35–72mm), and one fragment shows cross layering of laths held by concretion. The radiograph of this element before conservation shows what appears to be a gun-shield portion, with associated metal and a pierced leather strap with the remains of an iron buckle (30 x 24mm) concreted to one of the fragments.

The possible serpentine seems to be represented by an 88mm long arm, bifurcating at one end with a gradual arch over its length. A metal plate at one end may be the remains of the securing bracket, as on RA V.34. There may also be a winged screw also within the associated concretion.

*82A5154–5*

A number of small laths found within a large concretion (82A4080; not illustrated) were apparently recovered from beneath the Sterncastle following lifting of the ship. Two copper alloy dome headed nails are also present.

**Table 7.5 XRF results of metal components of the gun-shields**

| Object | Sample | Description | Sb | Sn | Ag | Pb | As | Zn | Cu | Ni | Fe | Type |
|---|---|---|---|---|---|---|---|---|---|---|---|---|
| 81A5771 | S0217 | spring | 0.08 | 2.01 | 0.03 | nd | 2.05 | 16.3 | 71 | 0.64 | 8.08 | 1 |
| | S0218 | seam cover strip | <<1 | <1 | <<1 | 2 | nd | 24 | 100 | <<1 | 22 | 3 |
| | S0219 | boss | <<1 | <1 | <<1 | <1 | nd | 19 | 100 | <<1 | <1 | 3 |
| | S0220 | seam cover strip | <<1 | <1 | <1 | 2 | nd | 24 | 100 | <1 | 25 | 3 |
| 82A2060 | S0221 | nail head* | <1 | <1 | <1 | 4 | nd | 100 | 46 | 1 | 2 | 3 |
| | S0222 | nail head* | <1 | <1 | <1 | 7 | <1 | 100 | 93 | 2 | 7 | 2 |
| 82A1586 | S0223 | nail* | 0.13 | 0.18 | 0.03 | 1.04 | nd | 18.1 | 79.8 | 0.22 | 0.69 | 2 |
| 82A1530 | S0238 | shield spring[1] | 0.04 | 2.17 | 0.01 | 0.89 | nd | 19.2 | 77.1 | 0.04 | 0.49 | 1 |
| 82A0992 | S0239 | boss[1] | 0.10 | 0.20 | <0.10 | 0.20 | nd | 21 | 78 | <1 | 0.60 | 2 |
| 82A1586 | S0306 | nail[2] | <0.10 | 0.10 | <0.10 | 1.60 | 0.20 | 18 | 77.5 | 0.30 | 2 | 2 |
| 82A0992 | S0307 | nail[3] | <0.10 | 0.30 | <0.10 | 0.80 | 0.10 | 20 | 76 | 0.30 | 2.4 | 2 |

* = corroded
[1] Trace Cd, Co
[2] Trace Ba, Bi, K, Ca
[3] Trace Ba, Cd, K, Ca, Co

*82A5248*
Identified in archive as a fragment from an unknown location. Details unknown.

## Comparative study of two gun-shields in the Royal Armouries Collections

During this study, two gun-shields were loaned to the Mary Rose Trust for radiography and comparative study (RA V.38 and V.34). Both are the heavier types of gun-shield with centrally mounted guns, similar to *Mary Rose* examples. V.34 preserves the serpentine mechanism with spring but V.38 does not. This study has aided interpretation of the remains and helped to clarify the structure and composition of the *Mary Rose* gun-shields and the radiographs have been particularly helpful in elucidating hidden detail. Comparative measurements are provided in Table 7.6.

**V.34**
This example is composed of internal and external laths disposed diagonally and positioned at right angles to one another (Fig. 7.32). The 13 iron outer plates do not overlap and are covered with seam-binding strips. The plates are painted in different colours (white, green, blue, red, yellow) embossed centrally with a tear-drop shape. As on V.38, the central boss does not hold the plates leaving an irregular gap. Two small screws with washers protruding from the outside of the gun-shield beneath the base-plate form fixing points for the boss. The viewing grille is similar in size to V.38 and is offset 15° from horizontal. The rough edges of this are covered with a bias tape and held with seven tacks. A slot has been cut above the aperture for sighting and location of the gun, and there is an eyed lug (10mm square) which helps to retain the gun.

The gun changes section from circular to rectangular towards the breech (Fig. 7.32). It is breech-loading and the muzzle protrudes 129mm beyond the shield through the central boss. The gun is made with a cut-out on its underside to house a loose lug (8 x 10mm) which enables the gun to be hooked over the outside of the gun-shield. This lug will only pass through the shield (from the inside to the outside) if it is aligned so that the lug faces upwards and is pushed through the key shaped opening at the top of the central aperture for the gun and then rotated 90°. Thus this opening has two functions, to serve as a line of sight, and to enable to gun to locate through the shield.

The base-plate is 130mm long and 56mm wide. It is bent downwards at the junction with the shield and is held by two iron screws with small washers on the inside with an empty central hole. It has two projecting eyed lugs on its upper surface which locate into slots cut out of the chamber cover. These are 180mm (left) and 150mm (right) in length, 5mm and 7mm wide respectively. A simple pin would have held the cover in position.

The gun-shield has an inner lining of wool. It carries bias binding to the inside edges of the grille as well as the cushioned pad. Portions of three large nails or staples (13–21 x 5–7mm) can be seen. These may have been a refurbishment to keep the laths together and were added after the lining. The woollen lining, which appears to have been stiffened with glue, is old and worn. Occasional staining and textile fragments on *Mary Rose* examples may be indicative of similar arrangements.

It is turned over to the front and gripped under the plates. A bias strip, 10mm wide, is placed 16mm inside the rim and held with 5mm diameter iron tacks. Below the gun a semicircular cushioned pad has been attached over the bias circle and held with iron tacks. The ends of

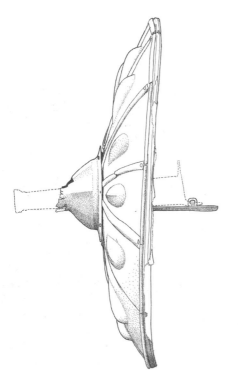

20cm

*Figure 7.32 Royal Armouries gun-shield V.34: radiograph, drawing, side view and photograph outside*

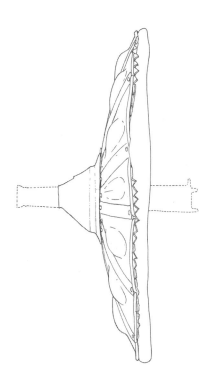

20cm

*Fig. 7.33  Royal Armouries gun-shield V.38: radiograph, drawing, side view and photograph outside*

leather enarmes are visible over this. There are three tacks with vestigial leather remains below them spaced 23mm apart above the pad and three closely spaced tacks over the leather beneath it. This is similar spacing to that seen on 82A0936.

The matchlock ignition system is visible on the inside of the shield in the form of a pivoting axle held on a retaining bracket set 67mm to the right of the gun aperture, in line with the sighting slot. This is angled at 5° above the horizontal and is 100mm long, held with two 7mm diameter tacks at the centre of the bracket. At each extremity the bracket is turned inwards. This supports an axle on its left arm and a pull cord on its right, tied through a 3.5mm hole. The axle retains a length of cord which would have been soaked in saltpeter so that it would smoulder slowly when ignited. The left arm is curved, 100mm long, and bifurcates to hold the match. This is held by a winged screw. An iron spring (7mm in diameter) is held in position by being bent over the top of the bracket and tensioned against the gun-shield. This is similar to the *Mary Rose* examples. The chamber cover pivots over the breech by depressing a domed lug located on its upper left side. The top of the breech cover has been cut away to form a slot (14mm square) to allow access to the (missing) cartridge. The base of the chamber cover extends behind the breech for 24mm to accommodate the lugs protruding through from the base-plate. The underside of the breech is inscribed with '+*IIII*'. A rectangular depression on the breech top again carries incised six-point stars.

## V.38

This is a circular gun-shield, convex to the outside, with a centrally mounted gun (Fig. 7.33). The wooden frame has internal and external laths positioned at right-angles to one another but, unlike on all the *Mary Rose* examples, this (and V.34, above) has them aligned on the diagonal. Together the layers have a thickness of 15mm.

The thirteen segmental iron plates butt against each other. These are rolled at their outer edges, and are held with iron seam-covering strips (155–160mm long, 15–20mm wide). The outward edges carry a fringed leather band beneath and a glued lining, possibly canvas, edged with leather strips. This lining covers the external surface of the gun-shield and is placed directly on the laths. The plates have been embossed with a central tear-drop shape. The central iron boss is now fastened with modern screws. The boss does not function to hold the plates, and displays an unsightly gap. It is cut to provide a line of sight for the gun, as is the iron retaining plate beneath it. This, turn, overlies a leather collar. A similar notched plate is fastened to the inside, again with modern screws. The radiograph (Fig. 7.33 above) shows a darker area behind it indicating the metal washers on the inside and outside.

The rectangular grille is directly above the boss. It is attached by three (one missing) modern 4mm diameter screws in three of the four corners. A red woollen lining, overlies two layers of linen sandwiching a layer of animal hair and is held on by tacked leather strips and a cushioned pad has been tacked over this below the gun. The vestiges of leather which formed the original enarmes are present in the form of leather fragments in the correct position for a strap on the left and a grip on the right. Three holes on the right side suggest where the serpentine-holding bracket would have been located as well as a pair below the gun where the end of a handle would have been attached. Its origin would have been at the base-plate.

The gun is breech-loading and protrudes 100mm from the front of the shield (Fig. 7.33 below). It has a hinged chamber cover and removable chamber. This is 82mm in length, with an external diameter of 22mm and a bore of 14mm. The gun is held in position by means of a rebate cut in its underside which may have had a projecting lug (as observed on V.34), though this is not now present. Towards the breech, the gun retains a lug with a hole on its underside. This would have located into a base-plate and have been pinned, but the base-plate was removed at some stage. The radiograph reveals three empty holes for base-plate nails in the external central iron plate. These penetrate the wood. The base-plate would have folded back and downwards following the internal lines of the shield. Although none of the *Mary Rose* examples retains a base-plate, dark staining in this area on some of the gun-shields recovered may be indicative of their former presence.

The gun has a smooth bore, tapering externally outwards to form a rectangular breech. Hinged over this is a cover, which lifts upwards by the outward movement of a lever disengaging a domed lug which locates into a depression on the left side of the gun breech. The breech cover has a small 'V'-shaped sight on its top surface, and a 10mm lip extending inwards to enable the operator to lift the chamber cover. The upper surface is cut away to reveal a 2mm touch hole in the chamber. A rectangular (11 x 5mm) depression towards the muzzle carries two four-pointed stars, possibly a maker's mark.

Removal of the central internal backing plate revealed a number of internal layers (linen, hair, linen) and then the internal wooden laths. A key-shaped sight, similar to those found on *Mary Rose* examples, is cut through the wooden laths and is covered by the new lining which must represent one of a number of phases of restoration. The fact that this sighting device has been covered suggests that this restoration took place after the functional life of the object. The cutting of a groove under the gun to rest it on the laths may be contemporaneous with the lining. Numerous cropped nails are visible, in particular a line of three (possibly five) 110mm beneath the centre of the gun-shield. These

**Table 7.6 Comparative data for the Royal Armouries gun-shields (metric)**

|  | V.38 | V.34 |
| --- | --- | --- |
| Wood | oak | unid. |
| Diameter | 450 | 475 |
| Weight (g) | 3405 | 4086 |
| Total thickness | 15 | – |
| No. plates | 13 | 13 |
| Plate length | 160 | – |
| Plate width | 35–110 | – |
| Boss diam. | 125 | – |
| No. external laths | 12 | 12 |
| Direction | \ | \ |
| Grille width/height | 65 x 60 | – |
| No. internal laths | 12 | 12 |
| Direction | / | / |
| Gun aperture diam. | 34 | 30 |
| Gun length | 248 | 225 |
| Gun bore | 16 | 14.5 |

may form the remains of an earlier pad. The corresponding upper line is not visible. All the *Mary Rose* nails are copper, none of those on V38 appears to be.

The removal of two of the seam covering pieces and one of the segmental plates revealed a coarse canvas like fabric, heavily impregnated with a glue-like substance covering the external surface and applied directly over the laths.

*Suggested method of manufacture*

Comparative study of the two Royal Armouries pieces and the more fragmentary remains from the ship suggests the following method of manufacture. A former, the size of the internal finished object would be manufactured (probably from clay) with a slightly concave upper surface. This would be hardened and oiled or waxed and then covered with a linen cloth and covered with glue. Heated, pre-bent laths would then be glued onto this forming the inner layer. These, in turn, would be coated with glue and the external laths heated and pressed over them (*Mary Rose* examples show colour differences between the internal and external laths as well as a precipitant between them). The wooden shield could then be removed from the former and inverted. The inside would be lined. On the Royal Armouries examples the lining consisted of two layers of glued linen with one of animal hair between them. Possibly an extra lining would be positioned (perhaps held by a leather strip tacked around the circumference) and the gun aperture and grille cut.

Finally the internal fittings would be added and the outer plates positioned so as to grip the edge of the final layer of internal lining. An iron backing plate could be used beneath the boss, presumably as a reinforcement. *Mary Rose* gun-shields show a circle of darker, iron stained residue which could be remains of a flat retaining plate.

# 8. Archery

## Scope and Importance of the Assemblage
by Alexzandra Hildred

The archery assemblage from the *Mary Rose* constitutes the only precisely dated archery assemblage surviving from any historical period. It is also a contained group: all the items present were designed to function together under the stressful conditions of war. The richness of this collection is unparalleled, and includes the remains of 172 longbows and 2303 complete arrows (7834 fragments), together with four chests for longbows and seven for arrows. Wristguards are also present, 22 in leather (many bearing decorative stamps), together with one of horn and one of ivory. Remains of up to eighteen leather arrow spacers were found, each pierced with holes to take 24 arrows and one with a leather fragment which may be the partial remains of an arrow bag. One horn nock (see below for definition and description) was excavated from corroding armour, confirming our assumption that the bows did have horn nocks, most of which have not survived within the particular environmental conditions on the wreck site. In

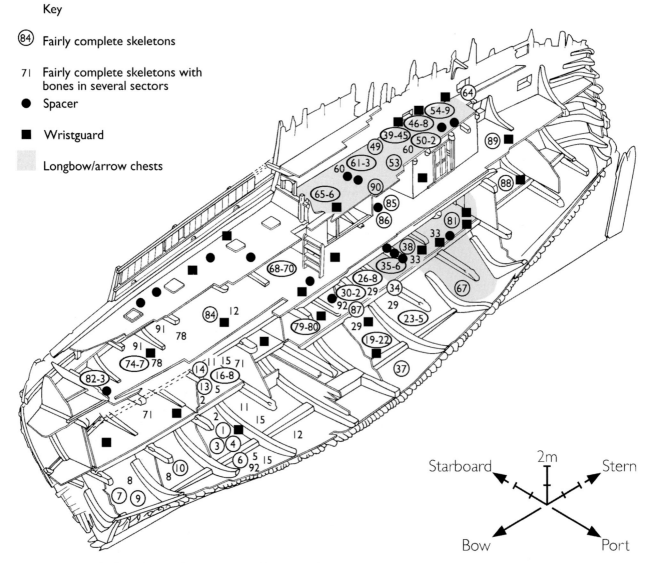

*Figure 8.1 Distribution of all archery related major groups in relation to that of Fairly Complete Skeletons (FCS)*

February 2009 the remains of a bowstring was tentatively identified from the sieving of a sample of the spoil mounds in 2003 (Chapter 12). The distribution of archery equipment (Fig. 8.1) suggests two distinct areas: a storage area in the stern on the Orlop deck and a mustering area on the Upper deck beneath the Sterncastle. The most likely areas for use would be on the open portion (waist or '*weather deck*') of the Upper deck between the bow and the stern and on the Castle decks in the bow and stern. Loose finds including longbows, arrows and wristguards, often in association with human remains, may reflect the positions of archers at the time of the sinking (Fig. 8.1).

The generally excellent condition of the wood of the surviving *Mary Rose* longbows has enabled a number of diverse studies. Measurements of the longbows, including length, width and depth along the limbs, have provided information on form and enabled the prediction of bracing height (the curvature of the bow before it is drawn and when the arrow leaves the bow during release). Microscopic study of broken bows has enabled a study of the growth rings, offering some indications of the environmental conditions needed for growth. Analysis of the relative density of the bows has provided further information on the nature and quality of the wood. These studies have enabled interpretations of inherent strength/flexibility and also of the amount of physical force required to draw back the bowstring and the weight held by the archer momentarily before releasing the arrow (the draw-weight). These reflect the skill, stature and strength of the archer.

Measurement of the arrow tips has enabled suggestions on head form, and the length and diameter of the arrow nock slots and of the few horn inserts has also fuelled discussions as to possible string diameters and their potential strength (another way of suggesting the draw-weight of the bow). Analysis of the arrows has allowed identification of some of the wood and feathers used, together with the glue and binding and even some of the preparatory methods employed during manufacture.

The sheer number of items has huge implications for understanding the techniques of manufacture and the ability to produce, in large numbers, a hand-crafted weapon and projectiles. Similarities and differences within the assemblage must be quantified to try and ascertain whether these were functional requirements, or merely reflect natural differences which occur when so many items are ordered. The items missing from the assemblage include the metal heads of the arrows and the majority of the six gross of bowstrings listed in the 1546 inventory. The size, shape and materials used for these are of fundamental importance in the performance of the bow and continue to promote much discussion.

Naturally, early in the study of the *Mary Rose* assemblage, there was an awareness of the need to replicate and test the weapons and their projectiles to assess their functional envelope and the strength and skills required of the operator. Whilst many tests have been undertaken, as is common practice within experimental archaeology, these should be ongoing. It is hoped that the questions posed here will encourage further investigations, both practical and theoretical.

The richness of the archive and the willingness of individuals to contribute to our understanding of the assemblage have resulted in a multi-authored section, with some conflicting interpretations. The fact that the collection goes so far beyond our previous understanding and evidence from archives has provided us with more questions than answers. It has fuelled a healthy but, as yet, unresolved academic debate about particular issues, the most notable being the draw-weights of the bows – are the bows really as powerful as some of the scientific test results suggest? – and the variety in the types of wood, shape and length of the arrow assemblage – are we seeing acceptable variations within a single weapon system (as the archives suggest) or are the bows and arrows designed for very different functions (as the physical evidence may suggest)? It is hoped that some of the suggestions proposed and techniques used will encourage further interpretation.

The Mary Rose Trust has benefited from the knowledge and experience of Robert Hardy, both as a student of the history of the longbow and as an active archer and bowyer. Since the first longbow was found in 1979, he has guided the identification, recording and conservation of the longbows and arrows and has himself undertaken research and experimentation in an attempt to uncover their secrets. We are delighted to include his authoritative account of this most important assemblage and its place in the wider history of the development of the longbow.

## Archery on Henry VIII's Vessels: Inventory Evidence
By Alexzandra Hildred

As we have seen in earlier chapters, there are a number of inventories which cover the life of the *Mary Rose* and these can be used to gather information about the importance of archery as is reflected in the numbers of items listed and their fluctuation over time. Some inventories are more descriptive than others and can be used to gather information about the artefacts and how they were transported and stored.

The earliest inventory which lists the *Mary Rose* was compiled at the end of the war with France in the late summer of 1514 (Knighton and Loades 2000, 113–58). It includes thirteen vessels which were being decommissioned. In addition to pure numerical incidence, there is information on the nature of the materials used and how the items were stored. All thirteen vessels carry longbows of yew, but some carried bows of other materials as well. The *Trinity Sovereign* carried 78 yew bows in two chests and two chests of witch hazel bows; the *John Baptist* carried 151 of yew and one chest and 84 of witch hazel in two chests; the *Katherine Forteleza*, 150 yew and 119 witch hazel; the *Rose Galley*, 55 yew and 40 of elm. Not only does this show some diversity in the woods used which does not appear in later inventories, but also provides information regarding transportation and storage. Eight of the thirteen vessels state the number of bows and follow this with the number of '*chestes to the same*'. Arrows are listed as being in sheaves in eight out of thirteen entries, and with chests in eleven. This suggests that the arrows in sheaves may actually also be inside chests. Two entries for arrows of wildfire occur: the *Great Nicholas* (5) and the *Mary Rose* (74). There is no mention of the term '*livery arrows*' which is so common in the later inventories.

Strings are only listed in seven instances, either merely as numbers (presumably of individual strings) or in '*dozens*' – for example the *Great Barbara* fifteen dozen and the *Katherine Galley* three dozen. One must remember that these are items were being recorded following the war with France and so loss must be taken into account. In this context the lack of bowstrings cannot be considered as unusual. Later inventories list bowstrings described as '*decayed*' as a routine; perhaps during a campaign these were discarded rather than returned.

There are four instances where some of the archery equipment is listed as still being on the vessel. The *Henry Grace à Dieu* is listed as having 124 bows with two chests and three whole chests of arrows '*in the storehouse of the shipp*'. The *Great Barbara* carried twelve bows and arrows (12 sheaves), bills, stone shot and picks for shot. The *Great Nicholas* carried a similar assemblage. The *Christ of Greenwich* had ten bows and six sheaves of arrows. Because locations were known by different names from ship to ship, it is difficult to relate them to particular areas of the *Mary Rose* (see also discussion on deck names in Chapter 11). What may be suggested from the numbers on the *Henry Grace à Dieu* and delivered to the Master and Purser for safekeeping, is that, in this instance, this is a storage location on the vessel as distinct from what appears to be an 'in use' location in the other cases. This distinction is observable in the distribution of 'stored' and 'ready for action' locations on the *Mary Rose*.

The 1514 entry for the *Mary Rose* includes eight items under '*Hede*[s] *for arrowes of wyldefyre*', 29 '*Hockes for arrowes of wyldefyre*' and 600 '*strynges*' left in the ship in addition to a number of items of ordnance. Twenty bows and 20 arrows were handed over to John Hopton (Clerk controller) and the 123 bows of yew, with two chests, and 504 sheaves of arrows with eleven chests '*to the same*', to Millet and Elderton of the Ordnance Office, together with 74 arrows of '*wyldefyre*'.

The inventory of vessels undertaken in 1540 (PRO E 101/60/3) only lists the guns; even shot is unrecorded. No significance should be attached to the absence of archery equipment from this particular list. The *Anthony Roll* of 1546 (Knighton and Loades 2000) and the inventory of 1547 (Starkey 1998) demonstrate the continued importance of longbows on vessels. Blair (1999) points out that the number of bows in the 1547 inventory (especially in shore establish-ments) is relatively small, with a number of references being to bows in need of repair or described as '*nothing worth*' and suggests that these stocks had been allowed to dwindle, having been replaced with portable firearms: '*a process that had clearly started before the reforms of 1538*' (Blair 1999, 24). The number of bows relative to handguns on representative vessels from the *Anthony*

Table 8.1 Sample of archery equipment listed in the *Anthony Roll*

| Vessel | Tonnage | Vessel type | No. handguns/ pikes/bills | No. bows of '*Eugh*'/ SPB | No. '*groce*' bowstrings/ indiv. strings | No. '*lyvere arrows in sheyvs*'/indiv. arrows | No. arrows per bow |
|---|---|---|---|---|---|---|---|
| *Henry Grace à Dieu* | 1050 | ship | 100/200/200 | 200/2.9 | 10/1440 | 750/18,000 | 36.0 |
| *Mary Rose* | 700 | ship | 50/150/150 | 250/3.4 | 6/864 | 400/9600 | 38.4 |
| *Peter* | 600 | ship | 40/200/200 | 200/4.3 | 6/864 | 300/7200 | 36.0 |
| *Matthew* | 600 | ship | 20/150/150 | 200/2.9 | 4/576 | 300/7200 | 36.0 |
| *Great Bark* | 500 | ship | 20/100/100 | 150/2.9 | 3/432 | 225/5400 | 36.0 |
| *Grand Mistress* | 450 | 'galleas' | 12/100/100 | 150/1.9 | 2/288 | 225/5400 | 36.0 |
| *Anne Gallant* | 450 | 'galleas' | 12/100/100 | 140/2.1 | 2/288 | 210/5040 | 36.0 |
| *Hart* | 300 | 'galleas' | 12/100/100 | 100/2.8 | 2/288 | 150/3600 | 36.0 |
| *Antelope* | 200 | 'galleas' | 12/100/100 | 100/2.8 | 2/288 | 150/3600 | 36.0 |
| *Tiger* | 200* | 'galleas' | 12/50/50 | 60/2.4 | 1/144 | 100/2400 | 40.0 |
| *Falcon* | 80* | pinnace | 6/30/30 | 40/3.6 | 1/144 | 60/1440 | 36.0 |
| *Hare* | 15* | rowbarge | 6/20/20 | 20/3.6 | ½/72 | 40/960 | 48.0 |

SPB = strings per bow; * vessels without soldiers listed. Source: Knighton and Loades 2000

*Roll* inventory can be seen in Table 8.1. This has a separate column detailing the number of bows, bowstrings, arrows, morris pikes, bills and darts for tops carried on each of the 58 vessels. In 1555 (Longleat Miscellaneous MS V, ff.1, 53–73v) these are grouped under the heading of '*Artillery*', with the pikes, bills and darts under a separate heading of '*Munitions*'. All the vessels listed in 1546 carry '*bowes of yew*', and the word longbow is not in evidence. All carry '*bowstrynges*', by the '*groce*' (144) or '*demi groce*' (72), with the exception of the *Portcullis* (20 tons), the *Hawthorn* (20 tons), the *Maiden Head* (20 tons) and the *Rose Slip* (20 tons) which all carry 5 dozen. The *Gillyflower* (20 tons) lists 6 dozen (72). All the vessels also carry arrows. All are prefixed with the term '*lyvere*' [livery]. By this period this term referred to any equipment supplied by the King.

The term '*livery*' is not as widely used in relation to fortifications as it is for vessels. In 1547 most vessels are provisioned with '*livery arrows*'. The five exceptions merely list arrows in sheaves. On land, only thirteen establishments have '*livery arrows*' out of over 70 separate entries for arrows (Starkey 1998). All except one vessel (The *Cloud in the Sun*) have arrows listed in '*sheaves*'. The sheaves of arrows recovered from the *Mary Rose* contained 24 arrows each. One entry is very illuminating, the *Swallow* with '*100 sheafs in two chests*'. This suggests 50 sheaves per chest: 1200 arrows. The fullest arrow chest from the *Mary Rose* carried 48 sheaves and the inventory lists a total of 400 sheaves, which would be eight chests if they were all so shipped; remains of seven were recovered.

All the vessels listed in the *Anthony Roll* carry handguns, however the number of handguns does not seem to be proportional to the number of bows carried. All carried staff weapons. All hand-held weapons seem to decline in numbers as a reflection of size of vessel and total crew rather than in any discernable proportions. The largest number of bows is on the biggest vessel, the *Henry Grace à Dieu*, with 500 bows and 750 sheaves of arrows (18,000), providing 36 arrows per bow. This ratio provides 1.8–3 minutes of continual shooting at a rate of 12–20 arrows per minute, should all the bows be used at once. Although this is an unlikely scenario, this ratio does not change with the size or type of vessel. The *Portcullis, Harp, Hawthorn, Maiden Head* and *Rose Slip* all carry the smallest number of bows (10 each) with an arrow:bow ratio of 36:1. This ratio provides the same number of minutes shooting (assuming all the bows are in use) as the *Henry Grace à Dieu*. The ratio of arrows to bows does not vary a huge amount, with 36:1 being the norm. The variance is 32:1 (ship, the *Mary* [of] *Hamburg*) to 48:1 on two of the rowbarges (the *Hare* and the *Falcon in the Fetterlock*).

The number of strings is variable, the ten gross of bowstrings listed for the *Henry Grace à Dieu* (1440) providing 2.88 strings per bow. The string to bow ratio is nearly consistent at 6:1 for the smallest vessels, the rowbarges (the *Harp* has seven). This may reflect the capabilities of the crew, which may decline with the size of vessel.

The total number of longbows on 58 vessels in 1546 was 4835, with 7335 sheaves of arrows (176,040 arrows). This works out at 36.4 arrows per bow. The total number of strings is 15,144, just over three per bow. The situation a year later is similar, with 3473 bows over 52 vessels and 126,424 arrows (36.4) arrows per bow.

Details of the crew are given for all vessels in 1546, divided into soldiers, mariners and gunners for the ships and for the '*galleasses*' the *Grand Mistress, Anne Gallant, Hart* and *Antelope*. All the rest have mariners and gunners, but no soldiers. If the role of archers was specialised, it is interesting that there is no dedicated, named portion of the crew. One might assume that the soldiers would take on (at least some of) this responsibility but, of the 34 vessels without soldiers, many carry considerable numbers of longbows. Therefore we have to assume either that there was an elite corps not listed or that (at least in some instances) the mariners and/or gunners were expected to use the bows. The latter is a difficult consideration as historical accounts suggest that archers were a highly trained elite and the primary role of the mariner was to work the vessel, and the gunners to man the guns. One wonders whether the increase in number of strings to bows observed on smaller vessels without soldiers suggests that mariners/gunners were using longbows and breaking more strings. However the personalised nature of many of the wristguards and particular skeletal abnormalities suggests the presence of at least some professional archers on board the *Mary Rose*. It may be that professional archers were expected to be among the personal retinues of the officers. Archers may also have been supplied through the livery companies and livery affiliation is suggested by some of the decorated wristguards (see below). A recently noticed victualling account for the 1545 campaign (Cambridge University Library MS Dd.13.25; *LP* 21/1, 1256) allows daily rations to 500 men aboard the *Mary Rose*; that is 85 more than the complement given in the *Anthony Roll*, and might indicate an additional body of archers.

The longbow and arrow assemblage recovered is close to the *Roll*, with 172 of the 250 listed yew longbows accounted for. Fewer than half the listed arrows were recovered but, if all of the seven arrow chests had been full, this would account for 8400 of the 9600 arrows listed (assuming 50 sheaves to a chest).

## *The inventory of 1547*

Although it would be difficult to estimate the total number of bows produced for war, the comprehensive inventory of Henry VIII's possessions (Starkey 1998) mentions thousands; among the main stores the Tower of London had 3060, Newcastle 2000, Calais 1500 and Berwick 1300. From these places bows, arrows and strings were despatched as necessary to smaller fortresses and to ships.

Longbows and arrows are found on all the vessels in varying numbers, except the *Mary Willoughby* (Starkey 1998, f. 424r). Bows are listed separately, except for the *Greyhound* (7791) and the *Hoy Bark* (7994) and several establishments on shore (eg, *ibid.*, 4544, 5791, 7151). In a number of instances these are in, or with, chests (eg, 3986–7, 4358, 4516, 6450, 7096). There is one specific reference to direct supply from the Tower of London. Ten chests of bows, 20 '*Chestes with sheif arrowes*' and one barrel of bowstrings were among munitions at God's House, Portsmouth, which had been sent '*from the Towre*' on 27 September 1547 by the Lord Great Master of the Household (4358–9, 4365). During the campaign in Scotland at the start of Edward VI's reign, the Admiral commanding at Broughty Craig (on the mouth of the Tay) was supplied with 100 '*Bowes of eugh*', 400 chests of livery arrows and 10 gross of bowstrings (6720–2). Arrows are most frequently listed as being in sheaves (of 24); the '*galleass*' *Swallow* has 100 sheaves in two chests (7500).

There is no suggestion that bows, arrows or strings were supplied in specific sizes or weights, unlike all manner of other weapons or projectiles. One possibility is that longbows and arrows – even strings – were tailored to specific tasks, or for men of particular physique or skill. Or it may be that we may are too precise with our measurements, and that greater tolerance was accepted.

In the inventory arrows are categorised in a number of ways. They are quantified in simple totals, in chests or in sheaves, and are themselves described as '*livery arrows*' or '*sheaf arrows*' (individually, in chests or in sheaves). '*Sheaf*' is sometimes used both descriptively and numerically as in an entry for Calshot Castle (4605): '*Sheif arrowes Clxiij sheif*'. What the actual difference was between arrows, livery arrows and sheaf arrows remains uncertain.

The contents of the arrow loft at Calais are noted in particular detail (5047–5051): '*Arrowes reddy Fethered*', '*Arrowes reddy fethered and cased*', '*Arrowes vnfethered*', '*Arrowe cases of red lether with girdells*' and '*Heades for lyvery arrowes*'.

At outstations such as Calais, which needed to be self reliant, we find the most detail regarding spares. In the '*Crosbowe Chambre*' (5098–5133) were '*Packe threde*' and '*Glewe for bowes and arrowes*' (5108–9). A further entry (5132) reads '*Bowstringes moost of them decaied*'. One entry for Guisnes (5577) reads '*Sheif arrowes with wild fire liij$^{th}$*'. Several entries (eg, 6199) mention '*Arrowes which be old and nought*'. Pontefract has '*Lyvery arrowes good and badde*' (6221), '*Lyvery arrowes without heddes*' (6222) and '*Tymbre for arrowes*' (6223). Although there is no description of the chests, one entry (5202) from the Morris Pike House at Calais describes '*Chestes with lockes for bowes and arrowes*'. None of the bow or arrow chests recovered from the *Mary Rose* had a lock.

Some information is given for bowstrings. Many are merely reckoned in dozens (eg, 3946, 4056, 4129, 4163, 4175, 4640) or by the gross (eg, 4250, 4606, 5973, 6056); elsewhere they are listed by the barrel (eg, 4517), half barrel (eg. 4148, 5363), firkin, or half firkin (eg, 4075). An entry for Yarmouth (4494) reads '*Bowstringes oone firkyne conteynyng ij$^o$ grosse*'. At the Tower (3743) there were eight barrels of bowstrings containing 80 gross, and at Newhaven (Ambleteuse) (5877) two barrels containing 20 gross. It is fair to assume from this evidence that bowstrings were despatched in casks, and that 10 gross to the barrel was a standard measure.

The publication of this inventory (Starkey 1998) has encouraged study and analysis. Davies (2005) has examined the crew profiles listed, looking at the numbers of soldiers and mariners and the total number of hand weapons (including longbows) compared with the number of longbows alone in order to work out a ratio of weapons to crew. This varies between 1.1:1 and 3.1:1. He has used this to question whether the longbow was a principal or secondary weapon at sea and, in looking at the number of arrows provided (36–48 per bow), has questioned just what the role of the longbow might have been. In so doing, this in turn questions the possibility of 'blanket fire' with so few arrows listed. The suggestion that the longbow might be a secondary weapon leads to the obvious conclusion that the operator might not be a specialist archer which, in turn, has been used to question who would be capable of drawing longbows with the substantial draw-weights proposed for some of the *Mary Rose* longbows (Hardy 1992, 213; 2005, 15–18). To achieve the predicted average draw-weights of 145lb (*c.* 65.8kg) we may have to consider that, in general, the male population was stronger than we might expect. Skeletal abnormalities observed in the shoulders of a number of crew members are consistent with the use of bows of the predicted weights (Stirland below and *AMR* Vol. 4, chap. 13). However, ongoing analysis of skeletal pathologies should consider other occupations subject to repetitive stress – such as routine climbing or hauling of ropes. The inventory evidence, interesting and valuable though it is, cannot itself answer all questions. We remain uncertain of the extent to which soldiers and mariners supplemented the listed gunners, or how many of them would have used longbows. The problem is most acute in the case of the smaller vessels, all of which carried bows but not many them having an established quota of soldiers. Nevertheless the 1555 Longleat inventory, albeit for a smaller selection of vessels, shows that bows were still then carried aboard. Thus we can say that during the active life of the *Mary Rose* (1512–1545) longbows remained an important part the armament of the King's Ships.

## Distribution of the Assemblage
by Alexzandra Hildred

The total number of longbows recovered from the *Mary Rose* in 1979–82 is 172, of which 138 are complete. Four bow chests were recovered, two from the Upper deck in the stern and two from the Orlop deck in the stern. Loose bows were scattered around the empty bow chests on the Upper deck. The chests on the Orlop deck held 40 bows (81A1862, F28) and 50 bows (81A3927, F159) respectively. The chests are nearly identical in size and manufacture and it is likely that, when full, they each contained 50 longbows. The *Anthony Roll* lists 250 longbows for the *Mary Rose*; of which at least 50 could well have been accommodated in a fifth chest. The chests are discussed in more detail below.

As the ship sank loose bows could have floated away. If an elite corps of archers were on board, some may have been carrying their own bows. Therefore, while the majority of the recovered bows represent military issue from the Tower, a few may be personal (for example 79A0807, not found within a chest). Eight longbows are recorded as having been recovered by John Deane in 1840 (one illustrated in Bevan 1996, pl. 23) and sold at auction (Norman and Wilson 1982, 106). Two of these are now owned by the Royal Armouries and another is in the National Army Museum, London.

The greatest concentration of longbows is in the stern (Fig. 8.2, Table 8.2), but it is relatively limited, with the bulk of the assemblage being in or around the chests on the Orlop deck, where they presumably reflect a storage location, and around the two open chests on the Upper deck. Although little is written about archery at sea in the sixteenth century, space and height would presumably dictate that the uppermost Castle deck or the '*weather deck*' would be the most suitable fighting positions. The chests on the Upper deck would be within reach of both areas.

There is only one longbow from the Hold, 81A2257 from H8, directly below the cluster from the stored chests. One hundred and four longbows were recovered from the Orlop deck, two in the bow in O3 and O3/4 and the rest in the area around the stored chests in O7–O9, of which 90 were inside the chests (Table 8.2).

The distribution of the sixteen longbows on the Main deck is widespread, with 1–3 bows in most sectors and a slightly higher concentration in M7–9 (Fig. 8.2; Table 8.2). On the Upper deck, apart from the concentration around the open chests in U7–9, there is, again, a wide spread of examples with one or two in each sector. There are no longbows from the Castle decks, but two from the Starboard scourpit (SS3–4) may have been displaced from the Forecastle.

This distribution suggests a storage area on the Orlop deck and a secondary area on the Upper deck where bows were kept (or brought) for immediate use in the Sterncastle and on the weather deck – in areas where archers would be expected to be at action stations. The isolated examples in other areas may be attributed, in most cases, to longbows being carried by individual archers. The distribution of wristguards associated with human remains broadly follows the same trends (see Figs 8.1 and 8.50).

Table 8.2 Distribution of longbows

| Sector | Object | Context | No. |
|---|---|---|---|
| H8 | 81A2257 | loose | 1 |
| O3 | 81A1697 | loose | 1 |
| O3/4 | 80A1298 | loose | 1 |
| O7 | 80A1353, 81A3770, 81A3790, 81A562, 81A5763; 82A1583 | loose | 6 |
| O8 | 81A3928–3977 | chest 81A3927 | 50 |
| O8 | 82A0922, 82A1590, 82A1607, 82A2268 | loose | 4 |
| O9 | 81A1597–1648, 81A1654–57, 81A1767 | chest 81A1862 | 40 |
| O9 | 80A1072, 81A2553 | loose | 2 |
| M3/4 | 80A0298 | loose | 1 |
| M5 | 80A1528 | loose | 1 |
| M6 | 80A1291, 81A0050, 81A1490 | loose | 3 |
| M7 | 80A0907, 80A01468, 80A01469 | loose | 3 |
| M8 | 80A0754, 80A0763, 80A01940, 81A2949 | loose | 4 |
| M9 | 81A2303, 81A5953 | loose | 2 |
| M10 | 81A0874 | loose | 1 |
| M11 | 79A1192 | loose | 1 |
| U3 | 79A0614 | loose | 1 |
| U4 | 79A0807, 79A0855 | loose | 2 |
| U5 | 79A0812 | loose | 1 |
| U5/6 | 81A5877 | loose | 1 |
| U7 | 79A0939, 79A0997, 80A0513, 80A0652, 80A0672, 80A0834, 80A0911, 80A0981 | loose | 8 |
| U7/8 | 80A0451, 80A0473, 80A0912 | loose, chest 80A0602 nearby | 3 |
| U8 | 80A0265, 80A0282, 80A0444, 80A0445, 80A0449, 80A0450, 80A0452, 80A0548, 80A0608, 80A0642, 80A01066, 80A01195, 81A2005, 81A2018, 81A2269, 81A2284 | | 16 |
| U8/9 | 80A1142 | | 1 |
| U9 | 80A0191, 81A0609, 81A0643, 81A0667, 81A0712, 81A0747, 81A0850, 81A01256, 81A01312, 81A01558, 81A01559, 81A01699 | loose (empty chest 81A1325 nearby) | 12 |
| U10 | 79A0741, 81A0213, 81A1140 | | 3 |
| SS3 | 81A1338 | | 1 |
| SS4 | 81A1310 | | 1 |
| Unknown | 82A4812 | | 1 |

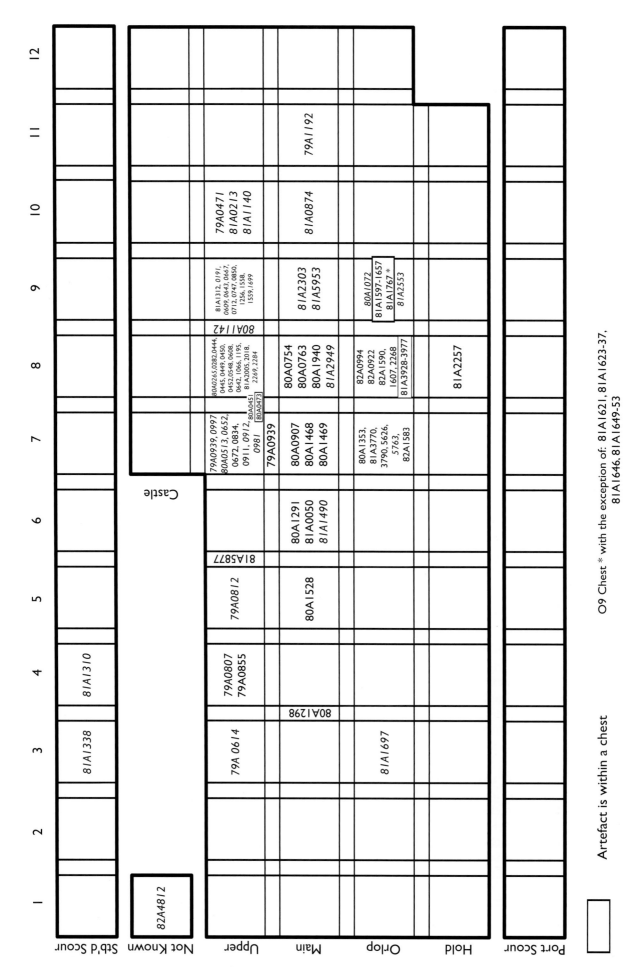

Figure 8.2 Distribution of longbows (italic text denotes incomplete object)

*Figure 8.3 Finger ring 81A4394 decorated with a sheaf of arrows*

## Distribution of archers based on associated artefacts
by Alexzandra Hildred

Several groups of artefacts found together are suggestive of items belonging to archers. Chest 81A5783 from the Carpenters' cabin (Table 8.3) seems to have belonged to an individual of some wealth and status. It contained domestic items including three pewter saucers (bearing a maker's mark of a crowned pewterer's hammer), a pewter plate and a highly decorated tankard that may be an example of a '*dessert*' or banqueting pewter (Weinstein, *AMR* Vol. 4, 450). Personal items included a knife, book, embroidered leather pouch, and a sundial in a case made from a reused book cover. The pouch contained a bone dice and concreted silver items including a finger ring and coins. Post-excavation cleaning of the ring revealed a heraldic shield containing a sheaf of arrows tied at the centre (Fig. 8.3); this was the emblem of the London fletchers (see below, p. 663). Items relating to ordnance included an arrow fragment and a decorated leather wristguard with a stamped floral decoration (see Fig 8.62).

Two longbows and arrow sheaves 82A1587 and 82A1589 were recovered from the southern end of O8 together with the remains of a number of individuals (F133) and a highly decorated wristguard (82A1524; Fig. 8.61) as well as woodworking objects.

Objects and clothing were found together in the bow of the Main deck in M3 (Feature 018). At the time of excavation this was described by the archaeological supervisor as 'two skeletons trapped against the starboard side of the main deck'. Amongst the mass of objects associated with this skeletal material several items appear to be associated most closely together, including a leather jerkin (81A0743), a pair of shoes (81A4063/81A4539), a belt buckle (81A0812), a signet ring bearing the letter 'K' (81A0810; see Fig. 8.67,

**Table 8.3 Contents of chest 81A5783**

| *CHEST 81A5783. (F128) TYPE 2.4. M9. ELM/WALNUT* | |
|---|---|
| Standing on end against Upper deck planks. Intact, contents excavated on shore. 1140x455x385mm | |
| UNIDENTIFIED | 81A5819 |
| UNIDENTIFIED | 81A5821 |
| UNIDENTIFIED | 81A5825 |
| COIN+RING(S) | 81A4394 |
| PLATE, PEWTER | 81A5823 |
| PLATE, PEWTER | 81A5827 |
| PLATE, PEWTER | 81A6849 |
| PLATE, PEWTER | 81A6850 |
| ARROW | 81A5829 |
| WRISTGUARD | 81A5826 |
| ?CLOTHES FASTENING | 81A4392/1-6 |
| ?CLOTHES FASTENING | 81A4393/1-5 |
| SHOT, LEAD | 81A4432 |
| POUCH | 81A5818/1-11 |
| BOOK | 81A5817 |
| DICE | 81A4391/1-2 |
| SUNDIAL+CASE | 81A5681/1-2 |
| KNIFE | 81A5828 |
| CABLE | 81A5815 |
| CHALK-LINE-REEL(LONG-) | 81A5686 |
| GIMLET? | 81A5820 |
| GIMLET? | 81A5824 |
| WEIGHT | 81A4431 |
| WEIGHTS | 81A4433-4 |
| TANKARD, PEWTER | 81A5654 |

below), a silver coin (81A0814), a carved knife handle (81A0766), a comb (81S184), a pewter spoon (81A0666), and a wooden bowl (81A4069). The wooden bowl has two double-headed arrows in the centre of the inside and a double-headed arrow bisecting what appears to be an arrow with fletchings, or a dart, on the centre of the outside. Various forms of arrow were also used to indicate official issue but these markings could denote ownership (see *AMR* Vol. 4, chap. 11 for a discussion), perhaps by an archer. The only ivory wristguard recovered (81A0815) was also associated with this group of objects (Figs 8.66 and 8.103). One longbow was found on each of the Main and Orlop decks in M3/O3 and a spacer with arrows was recovered from M2 (81A0116). These skeletons have not been isolated as exhibiting features indicative of long-term usage of the longbow, although two skeletons (FCS82/83) in the adjacent sector (M2) have.

## Historical Importance and Assessment of the Longbow Assemblage
By Robert Hardy

Very much has been written, especially in the last 20 years, on English and Welsh military archery between the conquest of England in 1066 and the last occasion on which archer recruits were accepted into the Tudor armies, during the reign of Elizabeth 1. The detail of that history goes far beyond the scope of this introduction, but the aim is to bring to the report on the *Mary Rose* archery equipment a reliable background which will reveal how vitally important to the understanding of medieval military history is the fact that such a quantity of archery equipment in such remarkable condition was safely recovered from the *Mary Rose* wreck.

Anyone who sees the bows from the *Mary Rose*, and compares them with war bows in the museums of Denmark or Schleswig-Holstein will realise that the Scandinavian weapons, through often from 4–500 years earlier, and some much earlier than that, are the direct ancestors of the weapons from the *Mary Rose*. Most of these early bows are made from yew timber, the finest material for a simple, self-bow; a bow that uses no laminations, and requires minimal adjustment from the stave first cut from the parent tree.

More than 6000 years ago men were hunting with bows of no great strength; a rare example, known as the Rotten Bottom Bow, radiocarbon dated to the first half of the 4th millennium BC is to be seen in the National Museum of Scotland, Edinburgh (info. on radiocarbon date supplied by A. Sheridan). But as bows began to be used in warfare, to resist the effects of arrow shot, protective clothing began to be worn; the bows increased in strength so that arrows would penetrate protection; the protection became more resistant; the bow weights increased. So it continued through the development of mail armour, padded jackets, heavier bows, heavier arrows, jackets with bits of plate armour sewn or riveted to the padding, mail and part plate armour until, by the end of the fourteenth century, those who could afford it, or whose magnates could afford to equip their men-at-arms, were armed '*cap-a-pie*' (head to foot) in plate armour. To cope with that the bow had to be extremely heavy, up to the furthest level of a man's ability to draw it, and the arrow heavy enough to strike with force enough to penetrate such armour. Armour was more and more designed to deflect attack, whether by sword, lance, bill or arrow, so even the best trained archers with the strongest bows and the sharpest arrowheads could not be guaranteed success; ricochet and bounce could greatly help the armour wearer. At the battle of Poitiers in 1356, according to the chronicler Geoffrey le Baker, the English were increasingly able to drive their arrows through French armour the closer they drew to the enemy (Strickland and Hardy 2005, 301). In any major conflict during the 100 or more years of war between the French and the English in the late middle ages this was the case; the shorter the range the more deadly the arrows.

Accounts recalling the strategic use and often battle-winning capability of the longbow in military engagements on land during the late medieval period are many and variable. We know that at Hastings in 1066, when the Normans were well provided with archers while Saxon archers were mostly still in the north, nothing could shift the English from their lines until the Normans started shooting with a high trajectory. This not only began to cause casualties behind the Saxon shield-wall, but felled their King Harold, with an arrow in his eye, as the Bayeux Tapestry appears to show.

Historical accounts exist which intimate the power of the fourteenth and fifteenth century war bows. An earlier account by Gerald of Wales in his *Iterarium Kambriae* of 1188 describes the shooting and bow making skills of the men of Gwent:

> 'The bows they use are not made of horn, nor of sapwood, nor yet of yew. The Welsh carve their bows out of the dwarf elm-trees in the forest. They are nothing much to look at, not even rubbed smooth, but left in a rough and unpolished state. Still, they are firm and strong. You could not only shoot far with them, but also they are powerful enough to inflict serious wounds in a close fight'. (Strickland and Hardy 2005, 43)

He also tells of an incident at the siege of Abergavenny Castle in 1182 where two Anglo-Norman soldiers fled to the keep:

*'The Welsh shot at them from behind, and with the arrows which sped from their bows they actually penetrated the oak doorway of the tower, which was almost as thick as a man's palm'.* (*ibid.*, 44)

To penetrate seasoned oak to nearly the depth of a hand (100mm) from no more than 23m requires not only a yew bow of some 100lb (46kg) draw-weight, but an arrowhead of a shape and temper that seems not to have been in general use among English military archers until a good while later. Furthermore, these bows were not of yew, but of witch elm, which is a good but inferior bow timber. These Welsh archers must have been pulling bows of well over 100lb. In England there were bows of good yew, home grown or imported, and of a '*longbow*' Scandinavian pattern that came with the Normans (Norse Men or Vikings originally) in quantities 120 years earlier. In effect we can envisage the possible beginnings of a new breed of bow: the rough powerful Welsh weapon and the fine northern longbow, stallion and mare so to speak, whose mating would produce a really powerful, finely made weapon in choice timber: the great warbow of the English in whose many armies using such weapons were, through the years, very many Welsh archers.

Arguments are plentiful regarding the validity of the historical accounts describing the damage inflicted by medieval war bows (Hardy 1976; Bradbury 1999; Strickland and Hardy 2005). Until now we have had no physical evidence of these weapons. Battles of particular significance where English archers are reported to have played a vital role in the outcome of the engagement include Dupplin Muir in 1332, Halidon Hill (1333), Crécy (1346), Poitiers (1356), Agincourt (1415), Towton (1461) and Bosworth in 1485. The tactics employed at Dupplin Muir were fundamental to the success of English archers in later battles (Bradbury 1999, 88; Strickland and Hardy 2005, 182–6). The English were arranged into a single '*battle*' of men at arms with large flanking masses of archers forward from the central battle. The archer's volleying was so deadly that great numbers of their enemy fell from the arrow storm, until quarters were so close that the archers had to discard bows and take to side arms. The Scots' losses were enormous, as the Lanercost Chronicle claims, the crush of the dead and dying in the centre grew '*to the height of a spear*'. The English lost '*not one archer*'. At Agincourt an eyewitness speaks of the piles of dead being higher than the height of a man. At Dupplin the proportion of archers to men-at-arms was 3:1, at Agincourt, 5:1. At Agincourt the French losses were possibly 10,000, the English probably no more than a few hundred (Strickland and Hardy 2005, 302).

When the young Edward III faced the Scots on Halidon Hill by Berwick, he had with him the commanders who had been at Dupplin the year before. He drew up his men in three battles with flanking archers between each division. Thus, if three Dupplin formations are put together side by side, the forward horns of archers on both flanks of the central and each inward flank of the outer divisions, or battles, join together in a sort of rough triangle or semi-circular wedge or block. So arranged, each battle forms a re-entrant into which the attacking forces must ride or run, exposed all the time to an increasingly close and murderous arrow storm, first to their front, then to their flanks. The English achieved in this a formation which, with variations, lasted the long years of the conflict, until the French learned how to avoid the frontal arrack and how to invalidate the English strength of archery. It is a formation which, if the enemy may be persuaded to attack, can cover the whole front and flanks of the attackers with arrow-shot. Froissart speaks again and again, as do others, of the English arrow storm darkening the sky '*like snow in winter*', or '*as it were winter rain*' or '*blotting out the sun*'. Equally there are many explicit records of men pierced by arrows, attesting to the strength of the bow. Three of the most famous being Philip of France at Crécy who left the bloody fight with an arrowhead in his jaw; David of Scotland at Neville's Cross took one arrow in his face which required two surgeons from York to draw it out, and another in his skull which could not be removed, remaining there until his death; and Prince Hal at Shrewsbury whose arrowhead in the face would probably have proved lethal had not a surgeon invented a new screw machine which, not without appalling pain, finally gripped the head and withdrew it (Strickland and Hardy 2005, 285).

By the time of Agincourt armour was greatly improved, but even so, Enguerrand de Monstrelet allows that even at long range many French were killed and wounded by English volleys, and Waurin (who was present) speaks of many French being disabled and wounded before they came to close quarters (*ibid.*, 333–6).

Towton, fought in a snowstorm in March 1461, started out as an archery duel which was initially responsible for the huge casualties (*ibid.*, 374–7). The Yorkist archers, facing roughly north, with wind and snow behind them, and positioned probably right across their battles' front, shot a volley into the Lancastrians opposite, evidently gauging the range well enough through the flurrying whiteness, then stepped back several paces, so that when the Lancastrians, blinded by the snow, shot back, their arrows fell short. The Yorkists advanced, picked up the arrows that stuck up from the ground and shot them back. How long this grim line-dance lasted no one knows, but the final outcome and the casualty conjectures are fearsome to read.

Space does not allow for an examination of the men of the bow, but there are many studies which cover recruitment in all its aspects, social provenance, pay and conditions (see Prestwich 1996). We can see that, under repeated enforcement of archery practice and the banning of most rival sports, with the enticement of good pay and possible riches to be acquired in foreign wars, with pardons granted to archers under sentence for crimes, and '*protection*' offered for the families and goods of those who would join the gathering armies, there was a growing reservoir of eager young men, who were taught to rely on each other, to obey their commanders and to develop their skills. Their disciplines of shooting in ranks could be practiced at the butts, by churches or village greens throughout the country; their battlefield organisation was in groups, platoons of 20, including a '*vintainer*' in charge, then in hundreds, companies with a '*centenar*' in command, and then in thousands under a marshal of archers, derived from Roman infantry practice. Each company of a 100 might have, in some cases, a surgeon, a priest and a '*crier*'. So orders of range, changes of direction, and movement could be quickly transmitted from marshal's trumpeter to centenar to vintainer and thus readily understood by each small division of the larger formations of archers. These archers, some in uniforms like the men of Cheshire in green and white (Strickland and Hardy 2005, 201), some in half armour and some in habergeons, some in jacks and iron-bounded '*cuirbouilli*' helmets (which archers called '*Querbole*'), seem to have been well supplied. In 1359 a million arrows, 25,000 bows and 75,000 bowstrings were added to the stocks in the Tower of London, the principal arsenal; in May 1360, 10,000 bows, 390,000 arrows; in June 20,000 bows, 850,000 arrows – such indications of 'military back up' speak for themselves.

The importance of the longbow in medieval warfare is indisputable. Within the *Mary Rose* assemblage we are looking at a refined weapon which was manufactured in large numbers and must have necessitated substantial support industries. In the past, when bow timber was in military demand and large orders went from the Royal Wardrobe and the Armouries at the Tower of London to Europe for thousands and thousands of staves annually, yew trees were specifically grown for the purpose. They were closely planted in large plantations. Tended by foresters, they grew straight and without pins and shoots to spoil the grain of the timber.

Wood for bows in sufficient quantities to furnish the industry had been imported from as early as the thirteenth century, probably from Spain, the Baltic and the Adriatic or even Ireland (Strickland and Hardy 2005, 24, Wadge 2007, 216–35). As merchants increased their prices, import duties were progressively imposed. In 1472 Parliament enacted that:

> '*every Merchant Stranger* [foreigner] ... *which ... shall bring, send or convey into this Land any Merchandidse in Carrack, Galley, or Ship* [of] *the City or Country of Venice, or other City, Town, or Country, from whence any such Bow-Staves have been before this time brought ... at every Time of their bringing ... any such Merchandises into this Realm, shall bring ... for every* [Tun Weight] *of such Merchandises ... Four Bow-staves*'. (12 Edw. IV *c.* 2; *Statutes of the Realm*, 2, 432; Megson 1993, 97)

During the reign of Richard III the price of Lombardy staves rose from £2 to £8 per 100. To combat this a duty of ten bowstaves for every cask of Malmsey or Tyre wine was imposed (1 Ric. III c. 11; *Statutes of the Realm*, 2, 494; Megson 1993, 98).

Yew from Spain was considered to be of such high quality that the term '*Spanish yew*' (and less often '*Italian yew*') was used to describe superior grades of stave. Yew (*Taxus baccata*) does not have any subspecies, therefore physical study of the wood cannot determine origin with any certainty, but certainly the yew is likely to have been of European origin. By the sixteenth century many bowstaves were being imported through Venice but their origins could have been much wider (Strickland and Hardy 2005, 25). In 1510 alone Henry VIII was granted leave from the Doge of Venice to import 40,000 staves for his first French war. Throughout his reign Henry sought to preserve the longbow against the increased demand for handguns; indeed, it was to Henry that Roger Ascham dedicated his book '*Toxophilus: The School of Shooting*', written in 1544 (Ascham 1545), the first English book to be written on archery. Although all of the longbows recovered from the *Mary Rose* are of yew, although Ascham lists a number of woods which can be used to make bows:

> '*As for Brazil, elm, wych and ash, experience doth prove them to be but mean for bows; and so to conclude, yew, of all other things, is what whereof perfect shooting would have a bow made*'. (1545, 106)

A larger variety of wood is listed for arrows and it is possible that particular woods were chosen for their inherent features, taking into account what was desired of the arrow and the capabilities and preferences of the archer. The book also includes instructions on practical shooting techniques and within its pages *Toxophilus* ('*bow-lover*') justifies his love of archery and argues for retaining the bow as a weapon of war above the handgun.

To implement this, and also to preserve the best quality staves for adults and trained archers, Henry maintained a strict pricing policy. As the *Mary Rose* was being built during 1511 Henry re-affirmed the law of 1369 whereby all subjects under the age of 60, who were not '*lame, decrepit or maimed*' (excepting the Clergy and Judges) were required to practise in the use of the longbow. Parents were to provide every boy from 7–17 years with a bow and two arrows. After 17, every man was to find himself a bow and four arrows; every bowyer for every yew bow he made was to make '*at least two bows of wych elm or other wood of mean price*' under penalty of imprisonment for eight days. Butts were to be provided in every town. The maximum price of a yew bow was 3s 4d (one sixth of a £) (33 Hen. VIII c. 9; *Statutes of the Realm*, 3, 837–41; Megson 1993, 61).

At the age of 24, every man was expected to shoot to a mark at 220yd (*c.* 200m) distance. Despite these incentives and statutes, the decline of the longbow and the rise of the handgun continued and, by 1595, longbows were removed from the official lists of weapons issued to ships.

The longbow in some form or other has lasted into the present as a weapon for sport. For a long time it was assumed that the medieval bow was much of the same strength as the eighteenth and nineteenth century longbow used by gentlemen of leisure at the Butts, at Rovers (or field shoots) or the prescribed distances still shot by the Woodmen of Arden, from 100yd to 240yd (*c.* 90–220m), or '*twelve score*'. To reach twelve score with light clout arrows requires a bow of 60–70lb (*c.* 27–32kg) draw-weight, which today requires much practice and hard work.

The first of 172 longbows recovered from the *Mary Rose* excavations came up in 1979. The excavation had already yielded staffs from both pikes and bills from the same area but this rather knobbly object, blotched with oyster-spat, was carefully tapered from the centre to both tips and was identified by the author as the first complete Tudor longbow to see the light of day since the Deane brothers had recovered a small number from the site in 1836–40 (see above). Accompanied by John Levy (Professor of Wood Science) and Peter Pratt (Professor of Crystal Physics) both of the Imperial College of Science and Technology, London, this was to be the first of many exciting discoveries. By the end of 1979 eight longbows had been recovered, mostly from the Upper deck, displaying varying degrees of completeness and condition. These included an example of formidable size, 79A0807, one of the first complete bows. Rough as it was, this must have had a draw-weight of well over 100lb (45.3kg). The knobs on this bow displayed the skill of the bowmaker in leaving proud the knots in the timber but our ideas of late medieval or Tudor warbows were conditioned by fine fifteenth century illustrations, such as that depicting the battle of Poitiers in Froissart, at the Bibliothèque Nationale (MS. fr. 2643, f. 207), the St Sebastian paintings of Hans Memling (*The Resurrection with the Martyrdom of St Sebastian and the Ascension*, Musée du Louvre, Paris), or the marvellous drawings of bowmen in battle from the *Beauchamp Pageant* in the British Library (BL MS Cotton Julius E. IV, art, 6, f. 18v; see Fig. 8.48b).

## Longbows
by Robert Hardy

A longbow consists of a stave which is tapered from the middle towards each end to form two tips. Grooves, termed '*nocks*', are cut into the tips of the longbow to take the bowstring (often with the addition of a strengthening cone of horn on each tip, see below). When the bow is drawn, by pulling the string backwards towards the archer and pushing the bow forwards, the limbs are bent. At full extent these describe a near perfect semicircle. Bows have two limbs, described as '*upper*' and '*lower*'. In some longbows a defined area near the centre of the bow is fashioned to form a grip or handle. Longbows also have a '*front and back*', described as a '*belly*' and a '*back*' (Fig. 8.4). The belly portion faces the archer's belly. When the bowstring is released the energy stored in the limbs is transferred to the arrow through the bowstring. The weapon therefore consists of the bow, the string and the arrow and all of their component parts. More than any other hand-held weapon, there is a direct relationship between the longbow and the user. Today bows are tailor-made to suit the stature, strength and skill of their owners.

> '*The length of the arrow is determined by the height and proportion of the archer, the distance between the hand that holds the bow and the one that draws the string ... to his best length. Between the arrow length and the sensible length of the bow there is a direct relation. The bow should be long enough to bend safely, allowing the arrow to be drawn to its head, and, at a given draw-weight, or strength to drive that arrow with sufficient force and speed. There is no upper limit to the length of a bow. The longer the bow, the safer it is to bend it but, within certain limits, the longer a bow the weaker it is at a given length of draw*'. (Hardy 1992, 7)

In producing a '*self bow*' (made of a single unjointed, unpieced stave) the bowyer will use a stave that combines both the strength of the old wood contained within the centre of the tree or branch – the heartwood – with the natural elasticity contained within the new wood – the sapwood. The orientation of the sapwood and heartwood within the bow limbs is crucial. The majority of the stave consists of the heartwood, including the belly of the bow, whilst the sapwood forms the thinner section placed at the back of the bow. The limbs are shaped by the bowyer to increase the bend in the sapwood whilst maintaining the strength of the heartwood, creating a perfect natural spring, Traditionally this is achieved through the creation of a D-section longbow, with the sapwood back comprising the upright whilst the heartwood belly includes the curve of the 'D'. All of the longbows recovered are made from a single stave of yew and nearly all have a D-shaped cross-section, with no routinely defined grip area.

All of the longbows recovered in 1979–80 were found loose, mostly from the Upper deck in the stern around collapsed chests (Figs 8.1–2). Many were

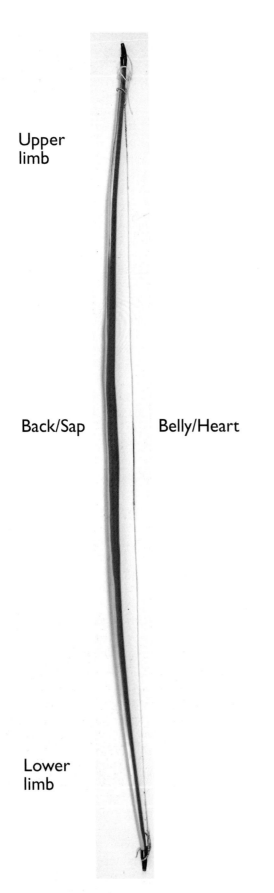
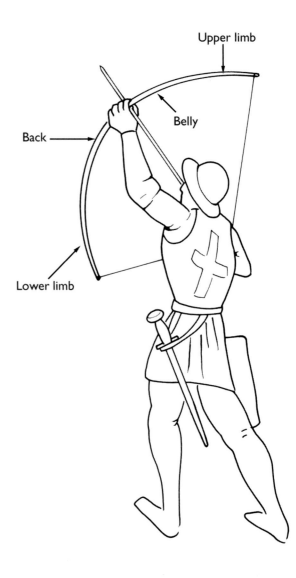

*Figure 8.4 Photograph of a* Mary Rose *longbow and drawing showing belly and back, upper and lower limb*

incomplete, oyster-spatted and, to varying degrees, encrusted and dilapidated but their discovery was only a prelude to the recovery of two complete chests from the Orlop deck that contained almost perfectly preserved longbows showing clearly the distinction between the sap and heartwood (Fig. 8.5) (the distinction and relative uses of heartwood and sapwood are discussed above). As bows grow old, the sapwood is liable to fracture and, once cracks are noticed, repairs can be made to a shooting bow. The sapwood of the *Mary Rose* bows is still, in some cases, astonishingly firm and elastic, though in others it is perished and untrustworthy.

The bows are superbly, even daringly, crafted; in some cases knobs are excised with a brave disregard to the dangers of weakness – no doubt the result of long practice and absolute confidence; in other cases wood is left proud over pins, knots and swirls in the grain. The finish has a fine dash to it and it is not too perfect. The wood is not reduced to the immaculate smoothness we would ask for nowadays – again indicating sureness in

a

b

*Figure 8.5 Longbows: a) heartwood and sapwood; b) cross-section showing distribution of heartwood/sapwood in bows*

the makers' craft – and often it is easy to see the clear fluting of the surface left by the long, swift strokes of the '*float*', the bowyers' finishing tool (Fig. 8.6). The quality of the timber, its density, in many cases the extreme fineness of the grain, suggests that the timber was of a straighter and finer quality than could have been found in the British Isles. It is likely that most of it was imported from the Continent where such timber certainly could be found.

The bows in the finest condition were those found in chests. Other bows and part-bows exhibit varying amounts of breakage (probably as a result of the sinking) and degradation as a result of their more exposed locations. Since the same bow can show good preservation at one end and bad deterioration at the other, it is likely the poorly preserved ends were exposed for a longer period. The surface condition of the bows, with a very thin layer of partial decay and slight discoloration, is strikingly different from the condition of the surface of the conical tips to each. These display an average of 50mm of paler wood suggesting that these surfaces had not been exposed for the same length of time as the rest of the bow. Horn can be destroyed through microbial action and the evidence of the condition of these conical ends has been interpreted to suggest that they had been fitted with horn nocks at the time and that the nocks have been subsequently destroyed by such action (horn is very poorly preserved on the *Mary Rose*).

Sometimes there is a slight ridging at the lower part of this coned tip (Fig. 8.7). This feature is probably the result of whittling the bow wood for a short space below the position of the horn nock, possibly carried out by individual archers who wanted their weapons to be '*whip ended*'. This can increase the '*cast*' of the bow by speeding the return of the limbs from full draw to the braced position on release of the arrow. Also at the bow tips there are visible remains of slots or notches that were cut into the timber to retain the string. These are found on the left hand side of the upper tip and on the right hand side of the lower. In a number of cases double slots have been observed, always on the upper limb (Figs 8.18 and 8.20). These may be the remains of '*tillering nocks*'. Tillering is the process of tapering the bow from the centre to the ends, during which the bow has frequently to be held on the tiller by some grip at its centre, then drawn to a gradually increasing depth with the aid of a pulley, to ensure that the weapon is properly balanced. In the earlier stages of this process the stave is much stronger than it is to be when fully tillered; a 150lb (67.95kg) bow might well be 200lb (90.65kg) or more at its early bendings. This requires very strong bracing strings to fit into the tillering nocks at the bow ends which are left thick enough to withstand such stress until the last stage of bowmaking. When the bow is finally tillered the tips are '*coned*' to fit into the drilled-out horn nocks, made from the tips of cow or stag horn, and in this process the tillering nocks are more or less worked out, depending on their depth and placement in the first instance.

*Figure 8.6 Fluting on bow*

*Figure 8.7 Pale bow tip showing ridge (79A0812)*

In some cases the very tip of the wood has been cleanly cut off at a forward angle, proof that the old practice was (as is still sometimes the case) to have neat little horns, more like a deep ring, to protect the wood under it from the working of the string, but not to weight the limb-ends. A glance at the bows in Memling's *Martyrdom of St Sebastian*, painted probably some 75 years before the *Mary Rose* sank, shows just such neat little horns. The historical depiction of horns together with this archaeological suggestion of tip protection does much to counteract the affirmation among some that these were unfinished staves awaiting the bowyers' hands.

The definitive proof of the existence of horn nocks came in 1997 with the identification of a complete example (97A0003, see Fig. 8.42a), slotted for the string on one side only. Diagonally grooved horns can prove difficult to use fitted with a modern string, which is traditionally knotted at the lower nock and looped for easy bracing and unbracing at the upper. There is a tendency for the loop (on the upper limb) to slip too far towards the single groove, resulting in an uneven brace, and a more uneven draw, diagonally across the length of the bow's belly. The archer Simon Stanley has found one answer; no loop, but a bowyer's knot, or timber hitch at each end. This keeps the string straight and central down the belly, but is a devil to unbrace because, in shooting, the knots become extremely tight.

Three basic forms based on cross-section have been identified (see below). Most can be said to have the basic 'D' exhibiting a rounded belly and flatter back. These can be sub-divided based on the roundness of the back, into those with a rounded back (almost circular in cross-section), the traditional flat back 'D', and a deeper 'D' where the cross-section is squarer as the depth of the belly is exaggerated. There are a number which are much longer than the rest and a few large bows which exhibit marked 'edges' which can be said to be '*slab-sided*' (Fig. 8.8).

The bows exhibit a variety of shapes. Some lie straight, or almost straight, but a number show a slight bend or deflex towards the belly, or a '*decurve*' (see Figs 8.14 (right side) and 8.15). This is also known as '*string follow*' and is found in bows which have been shot over a long period. Other bows show a reflex curve, or a '*recurve*,' where the limbs bend towards the back (see Figs 8.11 and 8.14 (left side)). This is termed '*set back in the handle*'. This has an advantage in that it helps to reduce string follow in used bows, which results in a longer distance to be travelled by the bow's outer limbs from the drawn to the release position, and possibly a faster return and better '*cast*'. This also increases the draw-weight or power of the bows. As with deflexion, this feature will eventually diminish the more the bow is shot and may, therefore, be an indicator of whether, and how often, a bow has been used. The cause of recurve in the *Mary Rose* bows has led to much controversy. One theory suggests that, in water, the sapwood shrinks faster than the heart and pulls the heart forward. Another suggestion is that it is the result of heat treatment during the bowery process and a third is that distortion may be induced from pressure of either sediment or associated objects during burial, or merely the effects of submersion. Alternatively the staves could have been cut from slightly bent trees or taken from the bole which, when cut, would be likely to bend outwards towards the sap resulting in a slight recurve in staves cut from the whole circumference of the log (Hardy in archive). None of the bows retains any evidence of binding at the centre of the bow and this area is not

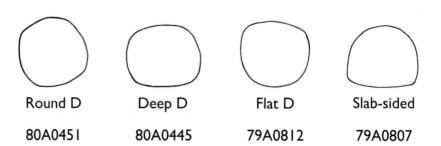

*Figure 8.8 Sections of longbows*

shaped to make the grip easier nor the middle of the bow stiffer.

The majority of the bows have a mark on the bow-hand side of the upper limb and several have marks on the lower limb. Most consist of small 'prick' holes, with a long-tailed triangular cross-section; a smaller number are stamped or incised (details below).

As to the quality of the bow-making: it is wonderful. Plainly these craft masters knew the tolerances of their materials so well that, for them, a bow had no need to be made a museum specimen. In the skill of their making, again, we are in personal touch with the men themselves who were under constant pressure to produce fine weapons in enormous numbers. If we allow that 30,000 bows a year might be made in times of war or national danger, which a broad inspection of contemporary orders conservatively suggests, we are speaking of the manufacture of some 7,000,000 bows in the 250 years of the military longbow's greatest usage (AD 1300–1500).

The bows from the *Mary Rose* form a unique collection and offer many answers to questions that historians and archers (and even archer-historians) have been asking, or guessing at, ever since Bishop Latimer complained about the decline of archery in a sermon before the boy King Edward VI, four years after the *Mary Rose* sank, and William Harrison in 1577 cried that '*Our strong shooting is decayed and laid in bed*'. On the other hand, Giovanni Michele, reporting to the Pope some years before the Armada, states that the English '*draw the bow with such force and dexterity that some are said to pierce corselets and armour*' (Hardy 1992, 139). Slightly earlier, in 1482–3, we have the description of a large force summoned by summoned by Richard III and the Duke of Buckingham, written by the Italian observer Dominic Mancini (1483, 99):

*'There is hardly any without a helmet, and none without bows and arrows: their bows and arrows are thicker and longer than those used by other nations, just as their bodies are stronger than other peoples, for they seem to have hands and arms of iron. The range of their bows is no less than that of our arbalests'.*

Until now we have not known what the medieval longbow was like: the bow of Bosworth, Agincourt, Poitiers and Crécy. We now have, from 60 years after Bosworth, well over 100 examples of longbows intended for warfare at sea. There is no cogent reason which should incline us to believe that these Tudor bows are different in kind, make or purpose from the bows of 100 or 200 years before. Nor is there any reason to suppose that they are unfinished bow staves. They are not sophisticated weapons, in the sense that the development away from the longbow marked a major step forward in sophistication, so we can assume that their predecessors were not markedly less sophisticated than the *Mary Rose* bows. In this assemblage we have a collection that is truly representative of the medieval war bow and that, once we have properly analysed all the lessons they can teach us, we shall be able to throw a strong beam of light on those dark battlefields of the fourteenth and fifteenth centuries that have lain in the dusk of uncertainty and guesswork.

## Archery at sea

We have already discussed the importance of the longbow in the context of medieval warfare on land but it is clear that bows were also used on English ships throughout the period. They became of increasing importance from the end of the thirteenth century as projectile weapons became more dominant all over Europe, both on land and at sea (Lane 1969, 161–71). During Edward I's Welsh campaigns in the last quarter of that century, for instance, both crossbowmen and archers were drafted from his armies to serve as marines in his ships (Morris 1901, 173). More than a century later a list of the armament of the warship *Holyghost*, made in 1416, shows the same mix of weapons – 14 bows, 91 sheaves of arrows and six crossbows. But by now these ancient projectile weapons had been joined by cannon – seven breech-loaders with 12 chambers (Friel 1995, 150). This is not surprising as naval warfare was conceived as being a series of sieges, with a ship attacking and taking an opposing ship. Just as bows were used in England to attack and defend castles and walled towns, so they were used to attack and defend castellated ships. The English attack on the island of Cadsand in the estuary of the river Scheldt (Flanders) in 1337 is a reminder that those who fought on board ship were also those who fought in field armies. The fleet contained about 500 men-at-arms and some 2000 archers. The island's defenders were lined up on the shore. Archers stationed in the forecastles of the advancing fleet devastated the defending army with volleys of arrows. This enabled the English to leap from their ships to the land and continue the fight with men-at-arms in the centre and archers on the wings (Hardy 1976, 57).

At the massive sea battle off Sluys (Flanders) in 1340 it is claimed that the English fleet had about 16,000 soldiers on board, some 4000 men-at-arms and 12,000 archers (Hardy 1976, 58) but was heavily outnumbered in ships and men by the French forces sent against it. Nevertheless Edward III won what Froissart called a '*murderous and horrible battle*' in which '*archers and crossbowmen shot with all their might at each other*' (Heath 1980, 130).

By the reign of Henry VII cannon and handguns had joined bows on English ships and the crossbow was hardly used at all. When the *Grace Dieu* delivered up its stores and armaments on the Hamble in October 1485 there were 140 bows on board but only twelve crossbows, described as being in '*feble*' condition (Oppenheim 1896b, 38–9). It appears that bows were standard issue on warships of the time, along with arrows in sheaves and bowstrings, often, as described above, issued in chests with bowstrings supplied by the barrel. In 1497 two new barks, *Sweepstake* and *Mary Fortune*, were commissioned and equipped for the navy. The Master of the King's Ordnance supplied 30 bows and 60 sheaves of arrows, together with four dozen strings for each ship. The ships were immediately involved in the Scottish war and, at the conclusion of this, accounts were made of what had been used, broken or lost in defending the ships during the war. The *Sweepstake* had used up 24 bows, 36 strings and 25 sheaves of arrows; the *Mary Fortune* had used up 14 bows, 36 strings and 30 sheaves. It appears that, unless both ships had been very careless with their stores of weapons, they had seen action and used their bows (Oppenheim 1896b, 303, 309, 329, 334). For this campaign the Master of the King's Ordnance had supplied a total of 1456 bows for equipping ten ships in the fleet, along with thirteen gross of strings and 1412 sheaves of arrows. The largest ships received 200 bows (the *Regent*), the smallest 20 (the *Bark of Penzance*), with the strings and arrows distributed roughly pro rata Oppenheim (1896b, 339–43); although the individual numbers assigned to each vessel fall well short of the totals listed). The same account lists each vessel as having '*Gonne Poudre*', '*Dyce of yron*' and '*Tampiones*', but none of the vessel listings includes guns, and those received at the Blockhouse in Portsmouth include only one great serpentine, two murderers and two stone guns (*ibid.*, 338). This is surely good evidence for the enduring importance of bows as ship-board weapons, although their day was coming to an end in the face of the rapid improvements being made in portable firearms.

The continuity in ship's armament during the reign of Henry VIII seems clear. The chronicler Holinshed wrote of the sea battle off Brest in 1511 that it was a cruel fight '*for the archers of the English part and the crossbows of the French part, did their uttermost*'. But Cardinal Wolsey, writing of the same battle, attributes victory to the shot of '*both gonnys and arrow*' (Heath 1980, 158). The instructions once attributed to Thomas Audley and presumed to date from the middle years of Henry VIII's reign suggest that crossbows at least were still an important part of any boarding action:

> '*In case you board* [close in to another ship] *your enemy enter* [board] *not till you see the smoke gone and then shoot off all your pieces, your port-pieces, the pieces of hail-shot,* [and] *crossbow shot to beat his cage deck, and if you see his deck well ridden* [cleared] *then enter with your best men, but first win his tops in any wise if it be possible*'. (Corbett 1905, 15–16)

The future was to lie with the gun not the bow.

## The longbow assemblage
by Clive Bartlett, Chris Boyton, Steve Jackson, Adam Jackson, Douglas M<sup>c</sup>Elvogue, Alexzandra Hildred and Keith Watson

A detailed study was undertaken during 2001 and 2002 of 140 of the total assemblage of 172 longbows. Of these, 104 were relatively complete. Their distribution on the ship is given in Table 8.5 (see Fig. 8.2). The primary objectives were to provide a thorough and consistent descriptive and dimensional record and to make a detailed record of the position and nature of the marks. The process of recording took two distinctive forms: specific linear measurements together with the intensive recording of the width and depths of the upper and lower limbs at 100mm intervals from the perceived centre of the bow through to the tips, and recording of particular features and their inter-relationships in terms of linear measurement, orientation and contextual deposition. These features included overall shape, appearance, marks, indications of horn fitting to the bow tips, presence or absence of notches at the tips, the quality of the grain and an assessment of the probable draw-weight of the bow. The study included 56% of the total bows listed for the ship. Linear measurements and recorded features can be found in Table 8.5. As with all wooden objects recovered from the ship and conserved, there is the potential for alteration in size and shape, often by swelling and then shrinking. As the longbow wood is incredibly fine-grained and was conserved immediately without long term soaking in water or PEG (polyethylene glycol, a chemical wax) and as these were air-dried rather than freeze dried, less distortion is expected within the longbow assemblage than amongst the arrows. Observations (in particular of condition) made by visual analysis alone are often misleading, especially as regards the strength of the wood.

**Table 8.4 Location of the 104 complete longbows studied in 2001–2**

| Deck | Context | No. studied | Total bows | % of assemblage |
|---|---|---|---|---|
| Hold | loose | 1 | 1 | 100 |
| Main | loose | 5 | 16 | 33 |
| Orlop | chest 81A1862 | 30 | 40 | 75 |
| Orlop | chest 81A3927 | 41 | 50 | 82 |
| Orlop | loose | 3 | 14 | 19 |
| Upper | loose | 20 | 48 | 42 |
| SS/unknown | | 3 | 3 | 100 |
| Un-numbered | | 1 | 0 | |
| Total | | 104 | 172 | |

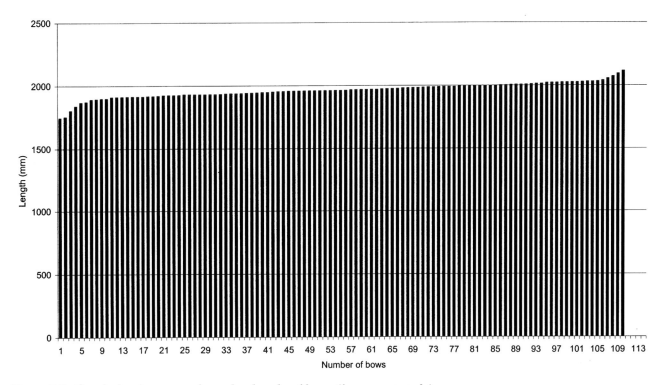

*Figure 8.9 Graph showing range of complete lengths of bows (from recent study)*

## Length

A wide range of complete longbow lengths was recorded, from 1839mm to 2113mm without obvious geographical clustering of particular lengths (Table 8.5 and Fig. 8.9). The graph for 73% of the complete bows recovered (101 of 139) shows a gradual increase in length from 1895mm to 2038mm, with peaks at 2038–113mm and 1839–95mm (Fig. 8.9). If the data

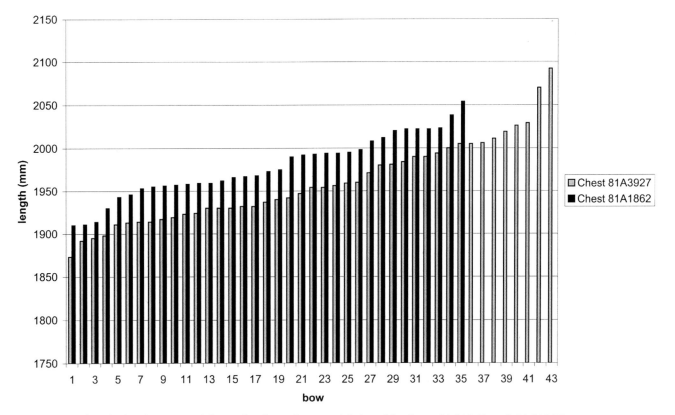

*Figure 8.10 Graph showing range (of complete bows from study) found in chests 81A1862 and 81A3927*

Table 8.5 Longbow data

| Bow No. | Sector | Length inches | Length mm | Centre W/D | LL 500mm W/D | LL 800mm W/D | UL 500mm W/D | UL 800mm W/D |
|---|---|---|---|---|---|---|---|---|
| 79A0807 | U4 | 83.1875 | 2113 | 38.8/35.6 | 34.4/29.5 | 26.8/23.8 | 34.3/29.2 | 26.8/22.8 |
| 79A0812 | U5 | 78.1875 | 1986 | 36/35 | 32.6/28.4 | 24.1/20 | 31.4/27.4 | 21.9/20.4 |
| 79A0855 | U4? | 77.5 | 1968 | 37.1/31.8 | 32/27.4 | 23.6/19.7 | 29/26.3 | 23.6/20 |
| 80A0282 | U8 | 79.24(e) | 2013(e) | /33.7 | 32.5/26.3 | 21.9/20.6 | 29.2/28.2 | 22/22 |
| 80A0444 | U8 | 78.3125 | 1989 | 35.1/32.8 | 30.2/27 | 21.8/20.3 | 29.2/26.2 | 20.7/22.2 |
| 80A0445 | U8 | 77.875 | 1978 | 36.3/31.2 | 31/27.5 | 21/21 | 31/28.6 | 21.7/20.2 |
| 80A0449 | U8 | 78.625 | 2003 | 37.5/32.3 | 32/26.7 | 23.2/20.4 | 31/26.7 | 22.3/19.2 |
| 80A0450 | U8 | 79.6875 | 2024 | 34.6/35.9 | 29.5/26.4 | 22.3/20.9 | 29.2/25.9 | 21.7/19.7 |
| 80A0451 | U7/8 ? | 79.875 | 2029 | 34.7/33 | 29.5/26.3 | 21.9/20.1 | 29.9/25.4 | 21.4/19.7 |
| 80A0473 | U8 | 77.375 | 1965 | 36/33.6 | 30.5/28 | 20.6/20 | 29.5/25.4 | 20.4/18 |
| 80A0608 | U8 | 76 | 1931 | 34.8/34.3 | 28.9/27.4 | 20.3/20.4 | 27.8/25.5 | 20.0/18.3 |
| 80A0642 | U8 | 77.5 | 1967 | 35.4/32.1 | 30.7/26.3 | 23.1/21.3 | 28.5/25.5 | 19.8/18.7 |
| 80A0652 | U7 | 68.75 | 1746 | 36.5/33.3 | 30.3/25.3 | Broken | 30.6/26 | 21.5/20.2 |
| 80A0672 | U7 | 77.75 | 1974 | 35.2/31.3 | 31.3/26.4 | 22.8/20.2 | 30.2/26.3 | 21.4/19.6 |
| 80A0763 | M8 | 72.4375 | 1839 | 33.5/30.2 | 27.5/23.7 | 17.2/16.2 | 27.8/23 | 18/15.5 |
| 80A0834 | U7 | 79.375(e) | 2016(e) | 39.8/34.9 | 33.9/30.3 | 25.3/23.5 | 33.5/30.5 | 24.6/21.7 |
| 80A0907 | M7 | 77.125 | 1960 | 34.7/29.2 | 29.4/25.1 | 19.9/19 | 28.1/24.4 | 18.3/17.7 |
| 80A0911 | U7 | 76.75 | 1950 | 35.3/30.5 | 30/26.3 | 20.4/18.9 | 29.6/24.6 | 221/18.3 |
| 80A1066 | U8 | 80.0625 | 2031 | 36.7/34.7 | 30.7/27.3 | 23/20.8 | 32.2/28.8 | 23.1/21.7 |
| 80A1142 | U8/9 | 80.5625(e) | 2038 | 35.9/32.8 | 30/26.6 | 21.9/22 | 29.8/27.5 | 22/20.2 |
| 80A1195 | U8 | 76.25 | 1937 | 32.7/31.9 | 26.4/26.5 | 20.1/21 | 26.5/25.4 | 19.1/18.4 |
| 80A1291 | M6 | 75.625(e) | 1921(e) | 35.4/34.5 | 30.5/28.3 | 22/22.3 | 28.4/27.2 | 19.6/18.1 |
| 80A1298 | M3/4 ? | 73.5 | 1867 | 35.4/30.6 | 30.2/24.3 | 18.8/17.1 | 29.4/23.3 | 17.6/16.1 |
| 80A1353 | O7 | 75.8125 | 1925 | 34.9/31.5 | 27.4/25.5 | 18/19.1 | 30.2/26 | 20.4/18.7 |
| 80A1468 | M7 | 76.625(e) | 1921(e) | 33.4/31.9 | 28.2/25.4 | 19.4/19.3 | 26.6/25.7 | 18.1/18.7 |
| 80A1469 | M7 | 78(e) | 1981(e) | 36.5/33.6 | 30.5/26.3 | 22.2/20.5 | 31.2/27.9 | 21.7/18.6 |
| 80A1528 | M5 | 78.75 | 2000 | 36.1/34.5 | 31.7/28.3 | 21.7/21 | 30.2/28.1 | 22/20.6 |
| 81A0050 | M6 | | 2006 | | | | | |
| 81A0712 | U9 | | 2000 | | | | | |
| 81A0850 | U9 | 69.125 | 1754 | 35/27.2 | 29.6/23.9 | 16.6/13.6 | 24.4/16,6 | 16.5/16.3 |
| 81A1312 | U9 | | 1920 | | | | | |
| 81A1597 | O9 Ch | 77.4375 | 1967 | 36/33 | 30.1/26.6 | 20/19.4 | 29/25.4 | 20.2/19.7 |
| 81A1598 | O9 Ch | 79.625 | 2022 | 38.8/33.4 | 32.4/27.3 | 23/21.4 | 31.5/27 | 22/20.6 |
| 81A1599 | O9 Ch | 78.425 | 1992 | | | | | |
| 81A1600 | O9 Ch | 77.0625 | 1957 | 35.1/32.2 | 29/26.1 | 21.2/19 | 29.4/27 | 21.8/20.9 |
| 81A1601 | 09 Ch | 77.6875 | 1973 | 38.4/35.2 | 32.5/29.1 | 23.2/22.8 | 33.7/30.1 | 23.6/22.8 |
| 81A1602 | O9 Ch | 80.25 | 2038 | 35.6/34.8 | 31/28.8 | 23.4/24.1 | 30.4/28 | 22.5/21.7 |
| 81A1603 | O9 Ch | 75.25 | 1911 | 34.6/32.7 | 29.4/27.2 | 19.4/18.9 | 28.2/26 | 20/18.9 |
| 81A1604 | O9 Ch | 76.5 | 1943 | 36.8/33.5 | 30.2/28.4 | 20.8/19 | 31.5/27.6 | 21.4/20.9 |
| 81A1605 | O9 Ch | 77.25(e) | 1962(e) | 35.8/33.3 | 31.1/28.9 | 21.3/20.5 | 29.8/26.8 | 21/20.9 |
| 81A1606 | O9 Ch | 75.2 | 1910 | | | | | |
| 81A1607 | O9 Ch | 78.5 | 1994 | 33.4/32 | 28.4/25.1 | 21.1/20.4 | 27.9/26 | 20.8/18.6 |
| 81A1608 | O9 Ch | 78.9375 | 2008 | 36.3/33.8 | 31.3/28.4 | 22.9/21.4 | 30.6/26.5 | 22.4/20.7 |
| 81A1609 | O9 Ch | 77.4375(e) | 1967(e) | 38.3/24.2 | 32.3/29.3 | 22.8/21.4 | 32.7/28.1 | 22.8/21.9 |
| 81A1610 | O9 Ch | 76.125(e) | 1932e | 35.9/33.9 | 31.4/26.9 | 21.5/22.2 | 30.2/26.9 | 20.7/19.8 |
| 81A1611 | O9 Ch | 77.4375 | 1966 | 35.1/34.4 | 30.6/26.9 | 22/20.9 | 30.7/27.7 | 21.1/20.6 |
| 81A1612 | O9 Ch | 79.5625 | 2020 | 35.6/31.6 | 30.4/26.5 | 22.3/20.4 | 30.3/25.3 | 21.8/19.3 |
| 81A1613 | O9 Ch | 76 | 1930 | 36.8/33 | 30.7/28.4 | 21.4/22.3 | 30.2/26.9 | 20.8/21.9 |
| 81A1614 | O9 Ch | 76.875 | 1953 | 34.1/33.6 | 29.5/29.2 | 21.3/20.1 | 28.8/26.3 | 20.7/20.9 |
| 81A1615 | O9 Ch | 77.25 | 1962 | 34.9/32.8 | 29.8/26.6 | 20.3/19.8 | 29.3/26.1 | 20.6/20.8 |
| 81A1616 | O9 Ch | 75.375 | 1914 | 34.1/31.8 | 29/25.7 | 19.4/19.3 | 28.4/25.4 | 19.2/19.2 |
| 81A1617 | O9 Ch | 78.46 | 1993 | | | | | |
| 81A1618 | O9 Ch | 77.125 | 1959 | 34.9/34 | 29.7/28.4 | 20.9/20 | 30.4/28.2 | 22.3/21.5 |
| 81A1619 | O9 Ch | 81(e) | 2058E | 35.9/33.9 | 32.3/27.9 | 23.7/21.2 | 31.7/27.3 | 23.3/21.1 |
| 81A1620 | O9 Ch | 77 | 1956 | | | | | |
| 81A1622 | O9 Ch | 78.6875 | 1998 | 37.2/36 | 33.6/29.7 | 22.6/22.9 | 32.5/29.8 | 22.2/23.6 |
| 81A1638 | O9 Ch | 77.125 | 1958 | 36.1/32.6 | 31.1/25.2 | 21.1/19.8 | 29.5/26 | 21.3/19.7 |
| 81A1639 | O9 Ch | 76.625 | 1946 | 36/33.1 | 30.6/25.4 | 20.5/20.2 | 29.2/26.8 | 20.7/19.8 |
| 81A1640 | O9 Ch | 78.5625 | 1995 | 36.1/35.3 | 31.9/29.2 | 22.3/24.3 | 31/27.6 | 21.9/21.5 |
| 81A1641 | O9 Ch | 79.625 | 2023 | 35/33.3 | 29.8/27.3 | 21.4/22.1 | 29/26.1 | 21.3/21.8 |
| 81A1642 | O9 Ch | 79.25 | 2012 | 35/332.5 | 31.8/21.8 | 23.7/22.6 | 32.2/27.8 | 23.4/21.2 |
| 81A1643 | O9 Ch | 80.875 | 2054 | 39.7/34.3 | 33.5/28.5 | 24.4/20.9 | 32.6/26.5 | 23.2/21.4 |
| 81A1644 | O9 Ch | 79.5625 | 2022 | 33.7/32.5 | 29.9/27.5 | 22.3/20.8 | 28.4/26.5 | 21.4/20.1 |
| 81A1645 | O9 Ch | 77.12 | 1959 | | | | | |
| 81A1647 | O9 Ch | 77.75 | 1975 | 36.2/32.4 | 30.9/26.8 | 21.7/20.3 | 27.4/25.4 | 21.4/20 |
| 81A1648 | O9 Ch | 77.48 | 1968 | | | | | |
| 81A1654 | O9 Ch | 79.52 | 2020 | | | | | |

| Bow No. | Mark Desc. | Set back handle | UL double notch | Grain | Comment |
|---|---|---|---|---|---|
| 79A0807 | 3 Chevron | Y | NO | F | Slab sided, almost square x-sect. Dark, knotty wood |
| 79A0812 | 4 Chevron | ? | ? | M | |
| 79A0855 | | ? | NO | C | *Very knobbly clear nocks, appears old* |
| 80A0282 | 2 Chevron | ? | O | | |
| 80A0444 | 2 Chevron | ? | YES (LL) | M | *Double nock upper end, single lower. Some recurve* |
| 80A0445 | | Y | YES | M | *?mark. Limb dimensions similar* |
| 80A0449 | 3 Chevron | ? | YES | M | *Tool marks visible. Sapwood deterioration* |
| 80A0450 | 3 Chevron | ? | YES (e) | | *Nock one end barely visible, clear other, 1 inch (c. 25mm) from tip* |
| 80A0451 | 4 Chevron | Y | ? | M | |
| 80A0473 | 4 Chevron | N | NO | M | *Clear mark. Lower nock barely visible. Dry and light* |
| 80A0608 | Circle | Y | O | | |
| 80A0642 | Cross | ? | O | F | |
| 80A0652 | | Y | Damaged | C | Uncertain. Broken tip |
| 80A0672 | | Y | NO | M | Clear nocks either end. Knobbly, medium strength |
| 80A0763 | | Y | NO | F | Uncertain |
| 80A0834 | | ? | O | C | |
| 80A0907 | 8 Chevron | ? | NO | ? | Mark both upper & lower limbs !! Flat sided. Chrysals |
| 80A0911 | 2 Chevron | Y | NO | M | Slightly damaged lower limb |
| 80A1066 | | ? | NO | M | Uncertain limb orientation |
| 80A1142 | | ? | O | M | *Nock barely visible 1 end, other 1 1/4 inches (c.32mm) from tip* |
| 80A1195 | | | | M | Many knots & pins |
| 80A1291 | Circle | ? | YES | C | *Nock barely visible 1 end, other 1 inch (c. 25mm) from tip* |
| 80A1298 | 3 Chevron | ? | YES | M | *Deflex greater in upper limb. Braced bow* |
| 80A1353 | | ? | ? | C | Damaged lower limb |
| 80A1468 | 4 Chevron | ? | Damaged | C | Damaged lower limb |
| 80A1469 | | Y | O | M | *Nock 1 end barely visible, other 3/4 inch (c. 21mm) from tip* |
| 80A1528 | 5 Chevron | ? | NO | F | *Used bow. Dry, worn, gribble, sapwood less than one-quarter* |
| 81A0050 | | Y | YES | | |
| 81A0712 | | Y | NO | | Good example of waisting on lower limb |
| 81A0850 | | ? | NO | F | *Gribbled. One end good, nock clear. Length conjecture* |
| 81A1312 | 2 Chevron | Y | Damaged | | Bent |
| 81A1597 | 5 Chevron | ? | O | F | *Cross mark! Lively bow, gribbled end* |
| 81A1598 | 2 Squares | Y | YES(e) | F | *Clear nocks, gribble damage* |
| 81A1599 | | Y | NO | | Appearance of having been strung |
| 81A1600 | Circle | ? | d | M | *Straight* |
| 81A1601 | | N | NO | M | Used big bow. Between slabsided and round section |
| 81A1602 | 2 Chevron | ? | YES | M | Mark below centre. Used, 2 nocks upper tip 2mm apart |
| 81A1603 | Complex | Y | NO | F | Also complex mark. Used, central mark, arrow over top |
| 81A1604 | 4 Chevron | N | NO | F | Big bow. Mark a handspan above centre for arrowpass |
| 81A1605 | 8 Chevron | ? | NO | M | Clear mark, used bow .First rate timber. Slight upper nock |
| 81A1606 | | N | NO | | Appears chot in, chatter marks. Used, followed the string |
| 81A1607 | 3 Chevron | Y | YES | F | Slabsided, handled, heavy. Flat back, big belly |
| 81A1608 | | ? | O | F | *Nock upper tip barely visible* |
| 81A1609 | | ? | ? | F | |
| 81A1610 | 2 Chevron | Y | YES(e) | M | Used bow.Straight, upper nock barely visible |
| 81A1611 | 2 Chevron | Y | YES | F | Fine strong bow,beautifully tillered over bumps |
| 81A1612 | | N | NO | F | Used bow, clear nocks,fine dark timber |
| 81A1613 | 4 Chevron | N | NO | F | Used bow, both nocks clear, bumpy timber |
| 81A1614 | | Y | YES | C | Knobbly, open grain, 20 lines to inch, English yew? |
| 81A1615 | Complex | Y | YES | F | Lower limb wider but depth almost same as upper |
| 81A1616 | Complex | Y | O | M | Double mark, odd bowyer's 'thumb print' lower limb |
| 81A1617 | 5 Chevron | Y | NO | | Used bow |
| 81A1618 | 5 Chevron | Y | Damaged | M | Tip damage. Pins used to define grip area. Used bow |
| 81A1619 | 4 Chevron | Y(e) | Damaged | M | *Very clear mark from deterioration of horn* |
| 81A1620 | 3 Chevron | Y | YES | | *Upper nock barely visible* |
| 81A1622 | | N | NO | F | Big handled bow, handle 98mm above centre, 59mm below |
| 81A1638 | Cross | N | NO | F | Used. Very handsome, both nocks clear |
| 81A1639 | 7 Chevron | Y | O | M | Lower limb wider but shallower |
| 81A1640 | 7 Chevron | Y | YES | M | Knobbly, used . Slight nocks either end |
| 81A1641 | | ? | O | F | *Recurved, upper tip cut* |
| 81A1642 | | N | NO | F | *Very clear tool marks. Small handle* |
| 81A1643 | | Y | YES | C | *Big bow, double nocked, handled section, flat belly* |
| 81A1644 | Square | Y | NO | C | *Slender tip lower limb. Used* |
| 81A1645 | | | | | Large bow, used. Upper nock barely visible |
| 81A1647 | 3 Chevron | Y | NO | M | UL string notch absent |
| 81A1648 | | | | | *Replica Concord ii. Drawn repeatedly to 30 inches (762mm)* |
| 81A1654 | | | | | *Replica Concord I. Drawn to 28 inches (711mm), broken and mended* |

| Bow No. | Sector | Length inches | Length mm | Centre W/D | LL 500mm W/D | LL 800mm W/D | UL 500mm W/D | UL 800mm W/D |
|---|---|---|---|---|---|---|---|---|
| 81A1655 | O9 Ch | 77 | 1955 | 37.6/35.7 | 32.4/29.4 | 23.6/22 | 31.6/29.2 | 22.1/21.7 |
| 81A1656 | O9 Ch | 78.375 | 1990 | 37.7/36.4 | 31.1/26.6 | 21.5/20.2 | 32.1/28.8 | 23.6/21.4 |
| 81A1657 | O9 Ch | 73.625(e) | 1870(e) | 37.5/33.4 | 32.2/27.3 | Broken | 34.6/28.8 | 23/22.3 |
| 81A1767 | O9 Ch | 78.5 | 1994 | 38.7/36.7 | 33.5/30.8 | 23.2/22.6 | 32,4/28.8 | 23.2/21,4 |
| 81A2005 | U8 | | | | | | | |
| 81A2018 | U8 | 76.1875 | 1935 | 34.2/32.6 | 28.7/26.8 | 20.6/19.7 | 30.4/26 | 21/19.3 |
| 81A2257 | H8 | 76 | 1931 | 34.5/32.4 | 29.6/27.5 | 20.5/20 | 28.6/25.5 | 20/19.6 |
| 81A2949 | M8 | 71 | 1803 | 31.2/29.8 | 25.9/23.9 | 17.6/14 | 28.2/25.5 | 18.1/16.9 |
| 81A3944 | O8 Ch | 76.9375 | 1954 | 37.5/34 | 33.3/27.6 | 22.2/22,2 | 31.4/28.7 | 21.4/21.2 |
| 81A3770 | O7 | 75.75 | 1924 | 34.4/35.6 | 27.7/25.6 | 19.2/20 | 35.8/25.4 | 18.9/20 |
| 81A3790 | O7 | 75.5 | 1917 | 34.6/34.6 | 28.7/27.7 | 19.4/21.5 | 28.6/28 | 20.3/19.2 |
| 81A3928 | O8 Ch | 79.1875 | 2011 | 34.1/33.5 | 28.5/26.1 | 21.5/20 | 28.4/25.6 | 21.5/20.5 |
| 81A3929 | O8 Ch | 74.75 | 1898 | 34.6/34.1 | 28.1/25.1 | 18.4/18.6 | 26.8/24.8 | 17.6/17.8 |
| 81A3930 | O8 Ch | 76 | 1930 | 35.3/32.7 | 30.4/27.0 | 21.5/19.9 | 29.6/26.2 | 20.2/19.3 |
| 81A3931 | O8 Ch | 78.375 | 1990 | 37.4/35.6 | 32.3/28.4 | 22.8/21.9 | 31.2/28.2 | 22.3/21 |
| 81A3932 | O8 Ch | 76.5 | 1942 | 35.1/30.7 | 29.5/25.2 | 19.6/18.7 | 28.7/25.9 | 19.9/19.6 |
| 81A3934 | O8 Ch | 73.75 | 1873 | 36.1/33.1 | 30/27.4 | 19.6/19.8 | 29.2/26.9 | 19.1/19.7 |
| 81A3935 | O8 Ch | 74.625 | 1895 | 33.6.30.4 | 27.8/24.9 | 17.8/17.8 | 27/24.2 | 17.8/17.7 |
| 81A3936 | O8 Ch | 79.875 | 2029 | 37.4/35.3 | 32/28.6 | 23.4/21.3 | 30.8/27.7 | 22.8/20.9 |
| 81A3937 | O8 Ch | 75.375 | 1914 | 35.1/31.6 | 29.8/25.7 | 20.8/19.6 | 28.4/25.2 | 19.7/18.8 |
| 81A3938 | O8 Ch | 75.25 | 1911 | 34.5/30.9 | 28/24.8 | 19.2/18.2 | 27.9/23.9 | 18.4/18.8 |
| 81A3939 | O8 Ch | 76 | 1930 | 38.2/34.8 | 31.7/29 | 22.1/20.2 | 31.6/27.2 | 21.8/19.8 |
| 81A3940 | O8 Ch | 79 | 2006 | 37.6/34.5 | 33/28.6 | 23.5/21.5 | 32.5/27.8 | 23.4/20.9 |
| 81A3945 | O8 Ch | 76.6875 | 1947 | 33.8/32.5 | 28.8/26.2 | 20.4/20.2 | 27.9/25.3 | 20/18.2 |
| 81A3946 | O8 Ch | 77.375 | 1960 | 35.1/31.4 | 29.5/25.4 | 19.9/19.9 | 28.6/26 | 19.7/19.3 |
| 81A3947 | O8 Ch | 75.3125 | 1913 | 32.9/31.9 | 28.7/24.7 | 19.3/19 | 29/24.2 | 19.9/18.4 |
| 81A3948 | O8 Ch | 78.5 | 1994 | 38.9/33.4 | 33.1/27.1 | 24.4/22.4 | 31.1/27.4 | 22.3/21.5 |
| 81A3949 | O8 Ch | 78.375 | 1990 | 31.8/31.7 | 29.1/24.5 | 20.1/20.1 | 28.3/24 | 20.2/18.9 |
| 81A3950 | O8 Ch | 75.5 | 1917 | 34/32.9 | 28.9/25.4 | 19.4/19.7 | 27.3/25.3 | 18.9/17.9 |
| 81A3951 | O8 Ch | 76 | 1930 | 35.2/32.4 | 29.8/26.2 | 21/20.5 | 28.5/25.7 | 20.3/19.4 |
| 81A3952 | O8 Ch | 74.5 | 1892 | 34.2/33.1 | 28.9/26.2 | 19.4/19.4 | 27.8/25.4 | 19.5/19.3 |
| 81A3953 | O8 Ch | 77.75(e) | 1975(e) | 36.6/33.8 | 29.8/27 | 20.8/19.5 | 31.4/29.7 | 22.6/22 |
| 81A3954 | O8 Ch | 78.75 | 2000 | 35.2/33.2 | 30.2/27.4 | 20.9/20.2 | 28.4/25.1 | 20.3/19.4 |
| 81A3955 | O8 Ch | 78 | 1981 | 34.4/34 | 30.5/27.1 | 21.5/19.6 | 28.7/26.3 | 20.7/20 |
| 81A3956 | O8 Ch | 77.625 | 1971 | 37.1/32.5 | 31.8/28.2 | 22.5/20.3 | 30.7/25.9 | 21.6/21 |
| 81A3958 | O8 Ch | 77.125 | 1959 | 37.6/32.8 | 31.8/27.3 | 21.9/20.6 | 31.3/27.4 | 22/19.9 |
| 81A3959 | O8 Ch | 78.125 | 1984 | 36.7/32.4 | 31.7/26.8 | 22.7/21.4 | 30.8/27.4 | 21.8/20.3 |
| 81A3960 | O8 Ch | 78.9375 | 2005 | 36.9/33.4 | 32.5/28 | 23.4/21.8 | 32.5/26.1 | 22.4/20.7 |
| 81A3961 | O8 Ch | 76.25 | 1937 | 37.4/33 | 31.2/27.4 | 21.7/21.1 | 31.4/26.7 | 21.7/20 |
| 81A3962 | O8 Ch | 78 | 1980 | 37.4/33.9 | 33.1/29 | 23.4/22.7 | 31.7/27.2 | 22.6/21.7 |
| 81A3963 | O8 Ch | 78.9375 | 2005 | 36.3/33 | 31.8/29 | 22.8/22.3 | 31.1/28.8 | 23.2/21.4 |
| 81A3964 | O8 Ch | 79.8125 | 2026 | 38.4/35 | 34.1/30.1 | 23.9/21.6 | 32.1/26.7 | 23.2/21.1 |
| 81A3965 | O8 Ch | 75.75 | 1924 | 35.4/32.7 | 31/29.8 | 21/22 | 28.9/27.3 | 20.2/18.5 |
| 81A3966 | O8 Ch | 79.5 | 2019 | 37.3/34 | 30.9/27.4 | 22.1/22.6 | 30/25.7 | 21.8/20.8 |
| 81A3967 | O8 Ch | 75.375 | 1914 | 35.5/31.6 | 30.5/28 | 20.8/20.3 | 28.9/26.5 | 19.6/18.2 |
| 81A3968 | O8 Ch | 77.125 | 1958 | 34.7/33 | 30.2/28.3 | 20.5/21.2 | 28.7/25.3 | 19.8/20 |
| 81A3969 | O8 Ch | 81.5 | 2070 | 38.1/34.1 | 30.1/27.7 | 20.1/21.7 | 31.8/27.7 | 24.3/21.3 |
| 81A3970 | O8 Ch | 75.5625 | 1919 | 36.1/32.6 | 30.7/26.1 | 20.6/20.7 | 30/25.9 | 20.1/18.8 |
| 81A3971 | O8 Ch | 75.75 | 1923 | 35.9/33.6 | 29.3/27.9 | 20.4/20.2 | 29.2/26.1 | 20.3/20 |
| 81A3972 | O8 Ch | 76.0625 | 1932 | 36.5/32.7 | 32/27 | 21.8/21.3 | 30.1/26.9 | 21.3/21.7 |
| 81A3973 | O8 Ch | 76.375 | 1940 | 32.7/34.2 | 27/24.6 | 19.2/18.8 | 27.9/25.4 | 19.5/18.7 |
| 81A3974 | O8 Ch | 82.36 | 2092 | | | | | |
| 81A3975 | O8 Ch | | | | | | | |
| 81A3976 | O8 Ch | 76.0625 | 1932 | 36.1/34.8 | 29.5/26.9 | 19.9/20.1 | 28.1/26.8 | 18.9/19.6 |
| 81A3977 | O8 Ch | 77 | 1956 | 33.5/30.5 | 28,5/27.2 | 19.7/19.9 | 27.9/25.1 | 20.1/18.7 |
| 81A5763 | O7 | 78.1875 | 1986 | 35.7/32.5 | 30.9/27.7 | 22.6/20.4 | 30.1/27.6 | 21/19.4 |
| 81A5877 | U5/6 | 74.625 | 1898 | 34.1/31.9 | 27.2/25 | 17.5/17.5 | 29.2/25.7 | 19.1/19.4 |
| 82A0994 | O8 | 78 | 1981.2 | 33.6/33.7 | 28.8/25.9 | 20.4/20.2 | 27.7/26 | 19.7/18.7 |
| 82A1607 | O8 | 76.3125 | 1938 | 34.6/33.5 | 30/27.1 | 19.7/20.5 | 28.5/26.5 | 19.3/19.3 |
| 82A2268 | O8 | 72.75(d) | 1848(d) | | | | | |
| 01A0708 | ? | 78.4375 | 1992 | 35.2/31.1 | 30.2/26.9 | 21.2/20.5 | 29/26.2 | 21.2/20.6 |

| Bow No. | Mark Desc. | Set back handle | UL double notch | Grain | Comment |
|---|---|---|---|---|---|
| 81A1655 | | Y | NO | F | *Strong nocks, supurb bow* |
| 81A1656 | 2 Chevron | ? | YES ? | F | |
| 81A1657 | | Y | NO | F | Broken lower limb |
| 81A1767 | 4 Chevron | Y | NO | C | *Very large. Used? Mark 6 7/8 inches (c. 175mm) above centre* |
| 81A2005 | Flower | Y | Damaged | | Broken tip/damaged |
| 81A2018 | 7 Chevron | N | NO | M | Mark on wrong side". *Used* |
| 81A2257 | 2 Chevron | Y | NO | F | |
| 81A2949 | | N | | M | Damaged - scarf joint? Poss. arrow pass marks |
| 81A3944 | 7 Chevron | ? | ? | C | Number misread as 3644, reassigned based on RH data |
| 81A3770 | Flower | N | NO | F | |
| 81A3790 | Complex | Y | Damaged | F | Chevron 6 and star of dots near centre |
| 81A3928 | Complex | Y | YES | M | Complex of 14 chevrons, geometric design |
| 81A3929 | 2 Chevrons | N | YES | F | *Clear mark, both nocks slight, dry wood, differentiation* |
| 81A3930 | 5 Chevron | Y | O | M | *Recurve* |
| 81A3931 | | Y | YES | F | Shake at grip. *Mark uncertain, big bow lateral stress grip* |
| 81A3932 | 7 Chevron | Y | NO | C | *Lower limb shallower in places* |
| 81A3934 | 2 Chevron | ? | NO | F | *Used* |
| 81A3935 | | ? | NO | M | *Used* |
| 81A3936 | 5 Chevron | Y | NO | F | |
| 81A3937 | | Y | NO | F | *Used. Very fine, highly coloured bow* |
| 81A3938 | 3 Chevron | Y | YES | F | *Fine bow, clear nocks, slight recurve, signs of cambium* |
| 81A3939 | 5 Chevron | Y | NO | M | *Used bow, straight. Some recurve* |
| 81A3940 | | ? | ? | F | *Very heavy bow, stunning piece of bowmaking* |
| 81A3945 | | Y | NO | F | |
| 81A3946 | 7 Chevron | ? | ? | M | |
| 81A3947 | | Y | NO | C | |
| 81A3948 | 5 Chevron | Y | NO | F | |
| 81A3949 | | ? | YES | F | *Double nock upper limb. Long streaks cambium on back* |
| 81A3950 | 5 Chevron | Y | YES | C | *Large bowyers indent above mark, fine example expertise* |
| 81A3951 | | ? | YES | F | |
| 81A3952 | | ? | O | M | *Used. Roy King studied this for approximations 1, 2* |
| 81A3953 | | ? | O | F | *Used* |
| 81A3954 | 3 Chevron | Y | YES | F | *Straight* |
| 81A3955 | 3 Chevron | ? | NO | M | *Used, signs of arrow wear. Bottom tip cut for horn* |
| 81A3956 | Cross | N | NO | | *Possibly used* |
| 81A3958 | 5 Chevron | Y | NO | C | |
| 81A3959 | | ? | NO | M | Compare workmanship to 81A0994. *Used* |
| 81A3960 | 7 Chevron | Y | NO | C | *Big bow, handled, broken tip, squared* |
| 81A3961 | 7 Chevron | Y | NO | M | |
| 81A3962 | | Y | | F | |
| 81A3963 | | Y | NO | F | Transverse fractures on back - shrinkage ? |
| 81A3964 | | ? | YES | M | *Fine bow* |
| 81A3965 | | Y | YES | M | |
| 81A3966 | Cross | ? | YES | M | |
| 81A3967 | | ? | YES | F | |
| 81A3968 | 4 Chevron | Y | NO | F | |
| 81A3969 | Cross | Y | NO | C | Interesting bowyers mark |
| 81A3970 | 5 Chevron | ? | YES | M | Double nock upper limb |
| 81A3971 | 2 Chevron | N | NO | F | One string notch only (lower limb ) |
| 81A3972 | 5 Chevron | Y | YES | C | Second nock lower limb |
| 81A3973 | 6 Chevron | ? | YES | M | Second mark barely visible, lower mark is hand position |
| 81A3974 | | | | | |
| 81A3975 | | N | | | Tested by Pratt. Modern nock fitted |
| 81A3976 | | N | YES | M | Used, obvious hand grip |
| 81A3977 | 6 Chevron | Y | NO | F | Very light bow, like Oregon yew |
| 81A5763 | 3 Chevron | Y | NO | M | |
| 81A5877 | 5 Chevron | Y | NO | C | Broken tip. Bad chrysal |
| 82A0994 | Circle/5ch | ? | YES | M | Interesting bowyer's mark. Circle with 5 chevron mark |
| 82A1607 | 2 Chevron | ? | YES | F | *Used. Damage to back/side.Both nocks barely visible* |
| 82A2268 | | N | Damaged | M | *Used bow Lower tip cut for horn. Possible arrow wear* |
| 01A0708 | 5 Chevron | Y | NO | M | No original number. Number assigned 7/8/2001 |

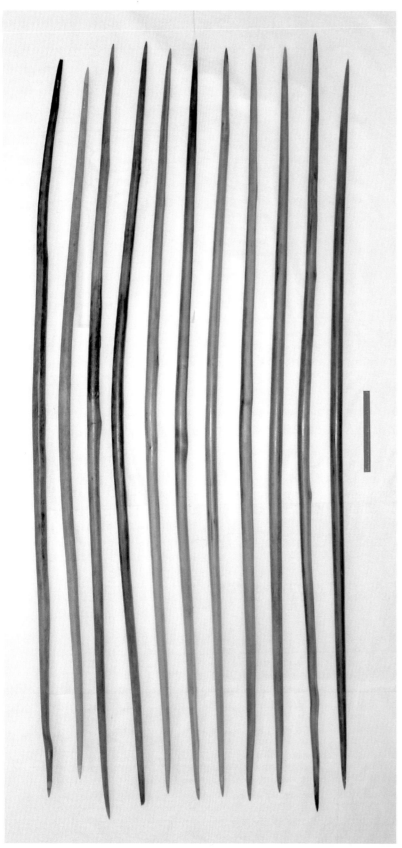

*Figure 8.11 Bows set back in the handle (81A1647, 81A1641, 81A1610, 81A3968, 81A3932, 81A3947, 81A3930, 80A0282, 80A0450, 81A0763, 81A3790)*

are grouped and sorted according to location on the ship, the results clearly indicate that a range of lengths is present even within the two relatively full longbow chests (81A3927 and 81A1862) stored on the Orlop deck (Fig. 8.10).

## Shape

Very few of the bows can be described as '*straight*'. Most exhibit some form of curvature. The most striking feature is a reflexed shape centred at the handle, where the limbs bend towards the back, away from the natural shape. This feature is observed in 97 (64%) of the 140 longbows recently studied, of which 76 are complete (Fig. 8.11). As discussed above, the shape can be achieved either through choosing timber which has a curvature within the grain or by the introduction of heat during bow manufacture. It is possible that, if the timber was chosen to give maximum straightness after use, these bows could represent 'unused' examples. However, this feature is found throughout the assemblage in terms of bow length (1839–2113mm) and in both loose and stored bows (Figs 8.12–13). Although there is a high incidence of these stored in and found loose around chests on the Upper deck (Table 8.6), distribution alone cannot be used to suggest that they are unused. The incidence of this feature is 55% of the total bows in chest 81A1862 and 70% in chest 81A3927.

The artificial setting of recurves into bow limbs has been a common practice

**Table 8.6 Distribution of longbows of reflexed and deflexed shape**

| Deck | Context | Reflexed | Deflexed |
|---|---|---|---|
| Hold | loose | 1 | 0 |
| Orlop | chest 81A1862 | 29 | 5 |
| Orlop | chest 81A3927 | 35 | 4 |
| Orlop | loose | 4 | 2 |
| Main | loose | 6 | 3 |
| Upper | loose | 16 | 3 |
| Upper | chest 81A3927 | 1 | 0 |
| Unknown | | 5 | 1 |

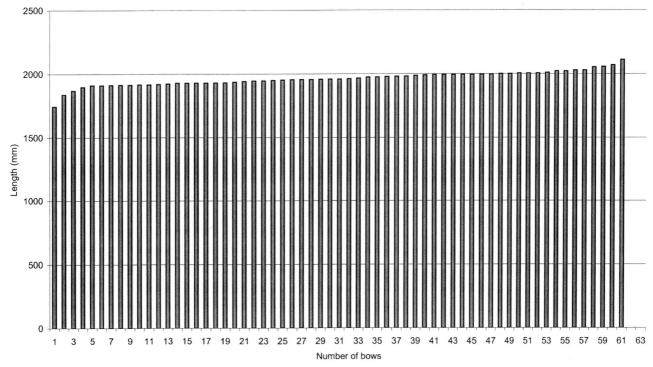

*Figure 8.12 Graph showing lengths of bows set back in handle*

throughout the recorded history of bow making but has never been observed in the English medieval tradition, and there are very few objective sources available which can be studied empirically. It is currently impossible to ascertain whether this form was induced accidentally (through pressures endured during burial) whether it was a design feature, or whether it was inherent within the wood used.

There is also evidence for a distinct backward curve towards the belly, or deflex shape, in some longbows (Figs 8.15). This '*string follow*' can, as mentioned above, form naturally when a bow has been shot over a long period of time and may, therefore, provide an indication of used bows. Again, the distribution of these is relatively widespread and is also found on examples in the chests (Table 8.6). A number of bows appear to have been immersed fully strung and must have remained so until the string gave way; some appear to represent their full bracing height (see Table 8.17).

**Cross-section**
As stated above, differing cross-sections to the bows were noted early in the studies (Hardy 1992, 201), with forms ranging from nearly circular to rectangular or trapezoidal. The majority display the traditional 'D' or oval section with a slightly rounded back and well-rounded belly. The cross-sections may be described as '*slab-sided*' (Tables 8.7 and 8.8), '*rounded D*' (Table 8.9), '*flat D*', (Table 8.10), '*D*' and '*deep D*' (Table 8.11; Fig. 8.8). The 'rounded' form is where the edges of the back are scarcely discernable, having been shaved off to achieve an almost circular form. Statistical evaluation of the measurements of the cross-sections highlighted the eight slab-sided longbows as being the most substantial recovered. Where observed, the grain is variable; two were classified as '*fine*' and one '*coarse*' (Table 8.5). Of these, 79A0807 contains many imperfections and is darker than most. Its location, loose in U4, together with its size and distinctive features, make it a candidate for a personally owned bow. It was one of the heaviest recovered with a dead-weight of 1kg (Hardy, pers. comm, 1984). The mark, however (three chevrons), is found throughout the assemblage, especially on boxed bows.

Table 8.7 '**Slab-sided**' **longbows, location and length**

| Object | Sector | Context | Length (mm) | Mark |
|---|---|---|---|---|
| 81A1607H | O9 | Chest 81A1862 | 1993 | chevron 3 |
| 81A1622H | O9 | Chest 81A1862 | 1998 | chevron 7 |
| 81A1642H | O9 | Chest 81A1862 | 2012 | none |
| 81A3944 | O8 | Chest 81A3927 | 1957 | none |
| 81A3960 | O8 | Chest 81A3927 | 2005 | chevron 7 |
| 81A0050 | M6 | loose | 2006 | none |
| 79A0807 | U4 | loose | 2113 | chevron 3 |
| 79A0855 | ?U4 | loose | 1968 | indecipherable |

H denotes bow that is also handled

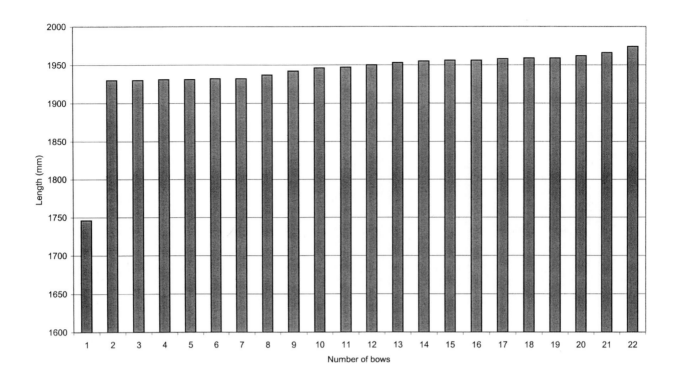

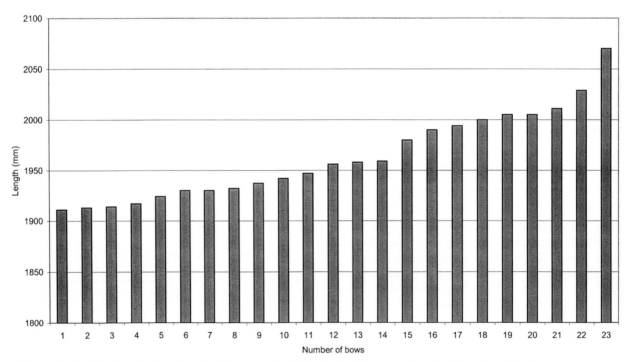

*Figure 8.13 Graphs showing lengths of bows set back in the handle from chests 81A1862 and 81A3927*

**Grip area**

There are seven longbows which possess limbs of a distinctive shape and cross-section, four noted during the 2001 study and another three suggested by earlier studies. The limbs are flat-sided to give a trapezoid section at the grip where the section abruptly changes to a rounded D-section, leaving a small triangular-shaped protuberance at each corner (Fig. 8.16). These bows have been described as being '*handled*', although not in the modern form, where there is a rise/covering or stiffening at the centre. Three of these (81A1622, 81A1642, 81A3960) were classified as slab-sided (Table 8.8) and one not (81A3949) (Table 8.12). When these were first identified by Robert Hardy (archive report 1984/5) it was realised that these also seemed to be some of the largest and sturdiest bows with high projected draw-weights. One suggestion was that the handled area might be used as a foot-rest to aid with

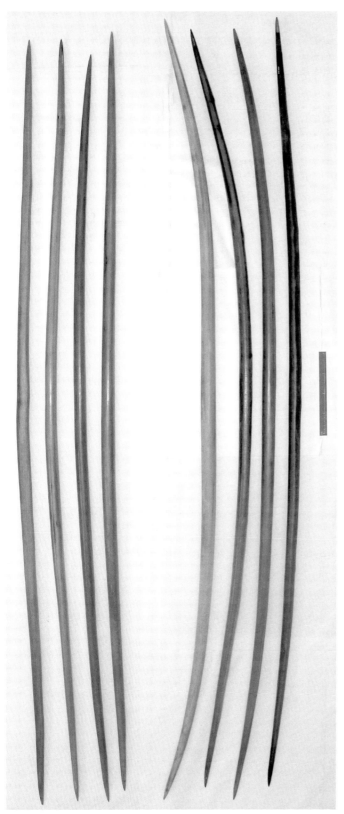

*Figure 8.14 Reflex bows (left) 81A3968, 81A3932, 81A1647, 81A1610; deflex bows (right) 81A1599, 81A1604, 81A3953, 80A0907*

bracing but as it was always in the order of 150mm, too small for a pair of feet, this was rejected. Another suggestion was that it was a cut-out to accept a binding which might be a requirement for the shooting of 'fire-

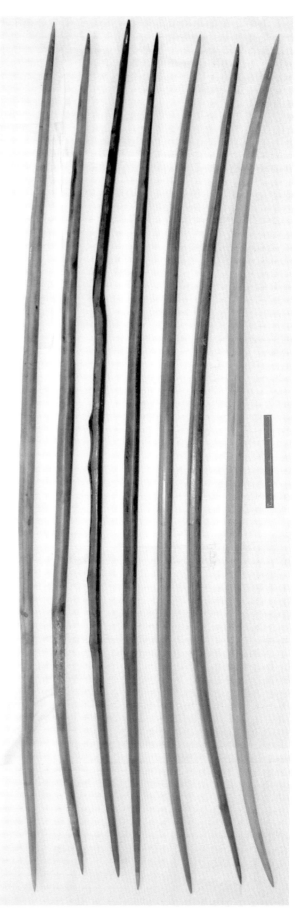

*Figure 8.15 Deflex bows (82A994, 81A3770, 79A0855, 80A0907, 81A3953, 81A1604, 81A1599)*

Table 8.8 'Slab-sided longbows, width and depth along length (metric)

| Object | 79A0807 | 79A0855 | 81A1607H | 81A1622H | 81A1642H | 81A3960H |
|---|---|---|---|---|---|---|
| *Length** | *2113/2089* | *1968/1953* | *1994/1994* | *1998/1996* | *2012/2003* | *2005/2002* |
| Interval | width/depth | width/depth | width/depth | width/depth | width/depth | width/depth |
| Centre | 38.9/35.6 | 37.1/31.8 | 33.4/32.0 | 37.2/36.0 | 35.0/32.5 | 36.9/33.4 |
| | | | *Lower limb* | | | |
| 100 | 38.8/34.4 | 36.0/29.5 | 32.9/30.9 | 38.8/37.0 | 35.7/31.5 | 36.9/33.4 |
| 200 | 38.4/33.1 | 35.5/29.1 | 32.2/28.9 | 37.8/33.7 | 36.5/29.3 | 37.0/32.7 |
| 400 | 36.0/31.1 | 33.5/29.6 | 29.8/26.3 | 35.5/30.6 | 33.7/28.7 | 34.3/29.0 |
| 600 | 32.9/27.3 | 29.3/24.9 | 26.8/24.1 | 31.2/27.9 | 29.9/26.3 | 30.1/25.9 |
| 800 | 26.8/23.8 | 23.6/19.7 | 21.1/20.4 | 22.6/22.9 | 23.7/22.6 | 23.4/21.8 |
| 900 | 21.8/20.6 | 18.1/16.9 | 16.1/16.4 | 16.8/18.0 | 18.3/18.2 | 18.4/17.8 |
| L. nock stain | 44.8 | 39.4 | 44.0 | 48.5 | 43.8 | damaged |
| L. limb tillering nock | 13.2/14.5 | 14.0/13/6** | 11.5/12.4** | 12.6/13.7** | 12.2/13.4 | 13.3/13.6 |
| | | | *Upper limb* | | | |
| 100 | 39.0/35.0 | 36.1/30.7 | 33.5/30.3 | 37.3/33.3 | 36.0/30.4 | 38.0/32.5 |
| 200 | 38.1/32.7 | 35.0/30.0 | 32.2/29.5 | 37.3/33.2 | 35.6/29.8 | 37.2/30.9 |
| 400 | 35.8/30.2 | 32.1/28.3 | 29.6/28.1 | 35.5/31.7 | 32.7/28.0 | 34.5/28.1 |
| 600 | 32.0/27.2 | 27.8/23.4 | 25.9/23.5 | 30.7/28.6 | 30.1/24.6 | 30.2/25.8 |
| 800 | 26.8/22.8 | 23.6/20.0 | 20.8/18.6 | 22.2/23.6 | 23.4/21.2 | 22.4/20.7 |
| 900 | 21.6/18.8 | 17.9/16.0 | 15.9/15.5 | 16.2/17.9 | 18.1/16.3 | 16.8/16.1 |
| L. horn stain | 55.5 | 42.4 | 50.0 | 52.5 | 45.0 | 48.4 |
| U. limb tillering nock | 13.6/16.6 | 13.4/12.7** | 12.0/13.1** | 12.4/14.0** | 12.4/13.5 | 13.1/13.4 |

\* Length: along back/tip to tip. H denotes longbow that is also handled. ** at stain (see also Tables 8.9–11

Table 8.9 Round D-section longbows, width and depth along length

| Object | 80A04501 | 80A0451 | 81A1601 | 82A0994 |
|---|---|---|---|---|
| *Length** | *2024/2016* | *2029/2027* | *1973/1964* | *1981/1975* |
| Interval | width/depth | width/depth | width/depth | width/depth |
| Centre | 34.6/35.9 | 34.7/33.0 | 38.4/35.2 | 33.6/33.7 |
| | | *Lower limb* | | |
| 100 | 35.4/33.0 | 34.8/31.4 | 38.4/37.4 | 33.6/31.9 |
| 200 | 33.8/32.4 | 34.0/30.2 | 37.6/32.8 | 32.9/30.2 |
| 400 | 30.9/28.3 | 31.3/27.1 | 34.3/30.0 | 29.4/27.4 |
| 600 | 27.5/25.4 | 27.3/24.6 | 30.2/29.0 | 27.2/24.5 |
| 800 | 22.3/20.9 | 21.9/20.1 | 23.2/22.8 | 20.4/20.2 |
| 900 | 17.5/16.7 | 17.3/16.8 | 17.6/17.7 | 15.6/12.3 |
| L. horn stain | 43.2 | 46.4 | damaged | 43.9 |
| L. limb tillering nock | 13.2/12.0** | 12.0/12.4** | damaged | 11.3/12.3** |
| | | *Upper limb* | | |
| 100 | 34.2/31.2 | 34.5/31.4 | 38.1/34.7 | 32.8/31.7 |
| 200 | 33.5/31.1 | 34.5/30.7 | 37.8/33.0 | 32.2/31.5 |
| 400 | 30.6/28.5 | 32.0/27.4 | 35.7/32.3 | 29.6/27.2 |
| 600 | 27.3/24.3 | 28.5/24.3 | 31.3/28.6 | 25.8/23.5 |
| 800 | 21.7/24.3 | 21.4/19.7 | 23.6/22.8 | 19.7/18.7 |
| 900 | 17.1/16.1 | 16.2/16.3 | 17.8/17.8 | 14.4/14.3 |
| L. horn stain | 41.9 | 48.6 | 52.0 | 45.7 |
| U. limb tillering nock | 11.5/11.6** | 12.0/12.9** | 14.5/14.8 | 10.9/12.6** |

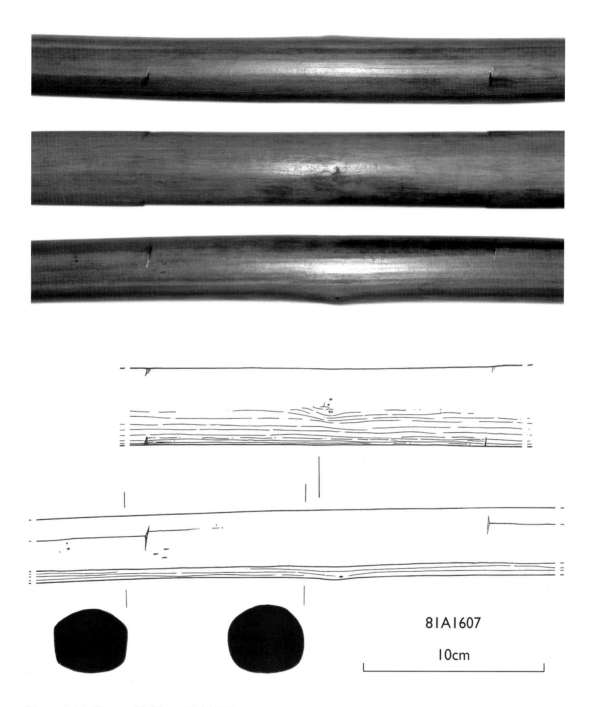

*Figure 8.16 Trapezoidal bow 81A1607*

*arrows*', ie, to protect the bow-wood from the burning arrow-tip. Six were in chests on the Orlop, and one loose on the Main deck (81A0050) (Table 8.13). Widths and depths are shown in Tables 8.8 and 8.11.

**Nocks**

Horn nocks were attached at the limb tips to accommodate the bowstring. Although only one was found there is evidence on every longbow of their existence, as described above. This evidence includes an observed colour change at the tip, where the nock was fitted, often accentuated by the presence of a ridge that was created through the removal of material after the nocks were fitted. The second feature may be the presence of slots (also termed nocks) cut into the upper and lower limb tips which could be to accommodate the bowstring, cut through the horn and into the wood. Both features are shown in Figure 8.17. These occur on the left-hand side of the upper tip and the right-hand side of the lower tip (Fig. 8.18). On 30 bows studied in detail, widths are 11.6–13.6mm and depths 11.9–14.6mm on the upper limb. On the lower limb widths

Table 8.10 Flat D-section longbows, width and depth along length

| Object | 79A0812 | 80A0907 | 80A1142 | 80A1298 |
|---|---|---|---|---|
| *Length** | *1986/1981* | *1960/1954* | *2038/2023* | *1867/1813* |
| **Interval** | **width/depth** | **width/depth** | **width/depth** | **width/depth** |
| Centre | 36.0/35.0 | 34.7/29.2 | 35.9/32.8 | 35.4/30.6 |
| | | *Lower limb* | | |
| 100 | 35.7/32.9 | 34.7/28.7 | 35.4/32.4 | 35.0/29.9 |
| 200 | 35.2/31.5 | 33.5/27.0 | 34.5/31.0 | 34.2/29.2 |
| 400 | 32.9/29.7 | 30.7/26.3 | 31.4/27.9 | 31.8/26.4 |
| 600 | 30.7/30.0 | 27.1/23.4 | 28.3/25.0 | 28.0/22.2 |
| 800 | 24.1/20.0 | 19.9/19.0 | 21.9/22.0 | 18.8/17.1 |
| 900 | 18.1/16.9 | 14.6/14.0 | 17.8/17.8 | n/a |
| L. horn stain | 43.1 | 44.0 | 40.0 | 50.1 |
| L. limb tillering nock | 13.4/16.1** | 12.2/11.8** | 12.8/13.2** | 12.6/11.6** |
| | | *Upper limb* | | |
| 100 | 35.8/33.0 | 33.7/26.7 | 35.8/34.5 | 34.2/30.9 |
| 200 | 34.9/31.0 | 31.7/27.4 | 34.8/33.0 | 33.8/31.4 |
| 400 | 32.2/28.6 | 29.8/25.6 | 31.5/27.7 | 30.7/25.3 |
| 600 | 29.0/25.3 | 26.0/22.9 | 27.8/25.0 | 26.7/22.5 |
| 800 | 21.9/20.4 | 18.3/17.7 | 22.0/20.2 | 17.6/16.1 |
| 900 | 16.5/16.1 | 13.1/14.2 | 18.0/19.4 | n/a |
| L. horn stain | 47.1 | 47.5 | 49.3 | damaged |
| U. limb tillering nock | 13.6/13.4** | 11.6/13.0** | 13.7/15.7** | damaged |

are 12–13.9mm and depth 12.6–14.9mm. These are always slanted rather than at right-angles to the length of the bow. Some are scarcely visible. The interpretation of these as slots for the bowstrings would fit in with the type of nock recovered, a side nock, where the string groove is positioned on the side of the bow tip, not the back. However this does not refute the suggestion that they may represent remains of tillering slot (see above).

There is a distinct difference in both shape and form between the upper and lower limb nock slots. The nocks of the upper limbs have a characteristic hook shape (Fig. 8.20a and b) whilst the lower slot is more open (Fig. 8.19c). A possible reason for this could be that the

*Figure 8.17 Nock slots upper tip, ridge 79A0855 (above) and 81A3928 (below)*

*Figure 8.18 above) Upper limb with double nock (bow 81A3928); below) right side lower limb, single slot*

## Table 8.11 D-sectioned longbows, width and depth along length

**A.**

| Object | 80A0445 | 80A0449 | 80A0608 | 80A0672 | 81A1602 |
|---|---|---|---|---|---|
| Length* | *1978/1950* | *2003/2003* | *1931/1902* | *1974/1975* | *2038/2038* |
| Interval | Width/depth | Width/depth | Width/depth | Width/depth | Width/depth |
| Centre | 33.6/31.2 | 37.5/32.2 | 34.8/34.3 | 35.2/31.3 | 35.6/34.8 |
| | | | *Lower limb* | | |
| 100 | 36.0/31.9 | 37.5/32.6 | 35.1/31.1 | 35.6/32.0 | 35.6/33.3 |
| 200 | 35.4/31.4 | 26.9/29.6 | 34.1/31.5 | 34.8/30.6 | 35.0/33.7 |
| 400 | 33.0/28.0 | 33.9/27.9 | 31.0/29.4 | 32.8/30.1 | 32.7/30.5 |
| 600 | 28.6/23.9 | 30.0/25.5 | 26.9/26.1 | 28.5/24.5 | 28.9/28.6 |
| 800 | 21.0/21.0 | 23.2/20.4 | 20.3/20.4 | 22.8/20.2 | 23.4/24.1 |
| 900 | 15.8/15.0 | 17.5/19.1 | 13.8/14.9 | 17.7/16.8 | 18.8/19.5 |
| L. horn stain | 44.0 | 43.0 | 35.0 | 42.4 | 48.7 |
| L. limb tillering nock | 13.9/13.3 | 12.7/13.4** | 11.0/12.5** | 13.7/13.5** | 13.0/14.3** |
| | | | *Upper limb* | | |
| 100 | 35.3/31.2 | 36.6/31.1 | 33.8/30.7 | 35.2/30.0 | 34.7/32.5 |
| 200 | 34.6/29.8 | 35.9/29.6 | 32.8/30.7 | 34.4/29.2 | 33.9/32.5 |
| 400 | 32.1/27.9 | 32.9/26.9 | 29.7/27.7 | 31.8/26.3 | 31.8/29.5 |
| 600 | 27.9/24.3 | 28.6/24.5 | 25.7/23.7 | 28.3/23.8 | 28.5/26.2 |
| 800 | 21.7/20.2 | 22.3/19.2 | 20.0/18.3 | 21.4/19.6 | 22.5/21.7 |
| 900 | 16.7/14.4 | 17.9/15.8 | 14.0/13.6 | 16.4/15.3 | 17.7/17.7 |
| L. horn stain | 50.0 | 37.0 | 32.7 | 45.5 | 46.7 |
| U. limb tillering nock | 13.2/13.4 | 13.3/13.4** | 12.1/12.2** | 13.4/12.8** | 12.6/14.0** |

**B.**

| Object | 81A1603 | 81A1614 | 81A3934 | 81A3935 | 81A3948 |
|---|---|---|---|---|---|
| Length* | *1911/1900* | *1953/1953* | *1873/1868* | *1895/1892* | *1994/1989* |
| Interval | Width/depth | Width/depth | Width/depth | Width/depth | Width/depth |
| Centre | 34.5/32.7 | 34.1/33.6 | 36.1/33.1 | 33.4/30.4 | 38.9/33.4 |
| | | | *Lower limb* | | |
| 100 | 34.6/32.8 | 33.9/31.5 | 35.7/32.8 | 32.9/30.3 | 38.7/32.5 |
| 200 | 34.1/31.9 | 33.4/30.9 | 35.2/31.7 | 32.2/28.4 | 37.8/31.7 |
| 400 | 32.4/28.4 | 31.2/28.4 | 31.7/29.1 | 29.4/26.0 | 34.0/28.2 |
| 600 | 27.2/24.3 | 27.8/25.0 | 27.4/26.4 | 25.4/23.2 | 31.0/25.2 |
| 800 | 19.4/18.9 | 21.3/20.1 | 19.6/19.8 | 17.8/17.8 | 24.4/22.4 |
| 900 | 12.9/14.5 | 15.4/16.8 | n/a | 12.5/13.2 | 18.5/17.5 |
| L. horn stain | 46.2 | 47.9 | 38.0 | 42.2 | 47.2 |
| L. limb tillering nock | 11.9/14.1** | 12.6/13.6** | 12.2/12.8** | 11.4/12.3** | 14.1/15.1** |
| | | | *Upper limb* | | |
| 100 | 33.6/31.0 | 33.7/30.2 | 35.9/32.2 | 32.6/30.3 | 38.3/32.8 |
| 200 | 32.8/29.7 | 33.1/30.4 | 34.5/30.6 | 31.3/28.6 | 36.8/31.1 |
| 400 | 30.1/26.9 | 30.5/28.1 | 30.9/28.0 | 28.6/25.4 | 33.2/27.4 |
| 600 | 26.3/24.4 | 27.4/24.5 | 26.9/25.1 | 24.5/22.3 | 28.8/25.5 |
| 800 | 20.0/18.9 | 20.7/20.9 | 19.1/19.7 | 17.7/17.7 | 22.3/21.5 |
| 900 | 13.0/13.8 | 14.8/15.5 | n/a | 12.0/12.7 | 17.6/16.6 |
| L. horn stain | 52.0 | 50.8 | 44.3 | 38.5 | 48.5 |
| U. limb tillering nock | 12.5/13.4** | 12.3/13.4** | 12.5/13.2** | 11.3/12.3** | 13.1/13.7** |

**Table 8.11 continued**

C.

| Object | 81A3952 | 81A3973 | 81A5763 | 81A5877 | 82A1607 |
|---|---|---|---|---|---|
| *Length\** | | | | | |
| **Interval** | **Width/depth** | **Width/depth** | **Width/depth** | **Width/depth** | **Width/depth** |
| Centre | 34.2/33.1 | 32.7/34.2 | 35.9/32.9 | 34.1/31.9 | 34.6/33.5 |
| *Lower limb* | | | | | |
| 100 | 34.3/31.7 | 31.8/29.8 | 35.7/32.5 | 33.1/29.3 | 35.0/32.4 |
| 200 | 33.6/30.2 | 31.0/29.0 | 35.1/31.6 | 32.2/28.3 | 34.7/30.5 |
| 400 | 30.7/27.4 | 28.7/26.2 | 32.2/29.6 | 30.0/25.3 | 31.7/28.5 |
| 600 | 27.1/23.6 | 25.7/22.9 | 29.0/25.6 | 24.6/22.6 | 27.2/25.3 |
| 800 | 19.4/19.4 | 19.2/18.8 | 22.6/20.4 | 17.5/17.5 | 19.7/20.5 |
| 900 | 12.3/13.1 | 13.7/14.7 | 17.7/16.8 | 11.0/12.1 | 14.5/15.6 |
| L. horn stain | 36.6 | 42.2 | 46.3 | 41.0 | 41.8 |
| L. limb tillering nock | 11.0/11.2** | 11.3/11.7** | 13.5/13.1 | damaged | 12.0/12.7** |
| *Upper limb* | | | | | |
| 100 | 33.3/30.8 | 32.7/31.4 | 35.8/33.5 | 33.7/31.6 | 34.1/30.9 |
| 200 | 32.4/29.2 | 32.4/29.9 | 34.8/32.4 | 33.3/30.9 | 32.9/29.8 |
| 400 | 29.8/26.4 | 29.5/27.2 | 32.8/29.1 | 30.7/26.5 | 30.4/27.4 |
| 600 | 26.5/23.9 | 25.9/23.8 | 27.9/25.3 | 26.6/23.7 | 26.1/24.5 |
| 800 | 19.5/19.3 | 19.5/18.7 | 21.0/19.4 | 19.1/19.4 | 19.3/19.3 |
| 900 | 12.2/12.8 | 14.5/14.4 | 15.6/15.8 | 13.3/13.5 | 13.7/14.4 |
| L. horn stain | 41.0 | 45.2 | 39.7 | 28.3 | 53.2 |
| U. limb tillering nock | 11.2/12.2** | 11.5/12.5 | 11.5/12.5 | damaged | 12.5/13.2** |

**Table 8.12 Handled bow 81A3949, width and depth along length**

| Object | 81A3949H |
|---|---|
| *Length\** | *1990/1987* |
| **Interval** | **Width/depth** |
| Centre | 31.8/30.7 |
| *Lower limb* | |
| 100 | 32.0/30.7 |
| 200 | 32.2/29.8 |
| 400 | 30.3/26.3 |
| 600 | 26.9/23.4 |
| 800 | 20.3/20.1 |
| 900 | 15.2/16.7 |
| L horn stain | 45.4 |
| L. limb tillering nock | 12.3/12.8** |
| *Upper limb* | |
| 100 | 32.2/29.5 |
| 200 | 32.0/28.4 |
| 400 | 30.1/26.1 |
| 600 | 26.1/22.8 |
| 899 | 20.2/18.9 |
| 900 | 15.0/15.5 |
| L. horn stain | 48.3 |
| U. limb tillering nock | 12.2/12.6** |

upper slot accepted a pre-made loop on the bowstring, whilst an archer's knot would be tied off on the lower one.

On 30 of the 104 complete bows studied there is a second slot on the upper limb (Fig. 8.19). Those complete longbows that have double nocks cover a range of lengths from 1898mm to 2038mm (Fig. 8.21) but also are distributed throughout the ship (Table 8.14). This feature is apparent on longbows found both loose and in chests (Fig. 8.22). One of the possible reasons for this would be to accommodate a second, longer string, known as a *'bastard string'* sometimes used as an aid to stringing the bows.

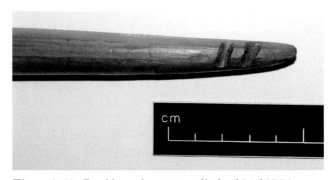

*Figure 8.19 Double nock on upper limb of 81A3954*

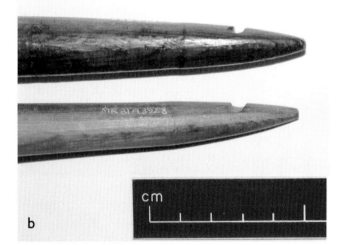

*Figure 8.20 Comparison between nock slots in the upper and lower limbs demonstrating that the lower slot is more open: a) 79A0855; b) 81A3928*

Table 8.13 Longbows with a trapezoidal section and relieved grip area, location and length

| Object | Sector | In chest | Length (mm) | Mark |
|---|---|---|---|---|
| 81A3949* | O8 | 81A3927 | 1993 | chevron 2 |
| 81A3960 | O8 | 81A3927 | 2005 | chevron 7 |
| 81A3974* | O8 | 81A3927 | 2092 | chevron 7 |
| 81A1607* | O9 | 81A1862 | 1993 | chevron 3 |
| 81A1622 | O9 | 81A1862 | 1998 | chevron 7 |
| 81A1642 | O9 | 81A1862 | 2012 | none |
| 81A0050 | M6 | loose | 2006 | none |

Table 8.14 Distribution of longbows with double nocks

| Deck | Context | No. | Total studied | % of total |
|---|---|---|---|---|
| Orlop | chest 81A1862 | 9 | 30 | 30 |
| Orlop | chest 81A3927 | 15 | 41 | 38 |
| Orlop | loose | 1 | 3 | 33 |
| Main | loose | 1 | 5 | 20 |
| Upper | loose | 3 | 19 | 16 |
| Unknown | unknown | 1 | 4 | 25 |

## Longbow marks

The recent study of 104 longbows revealed 91 with easily visible marks. These are positioned around the geographic centre of the longbow (most are on the upper limb, just above the centre of the bow) and are

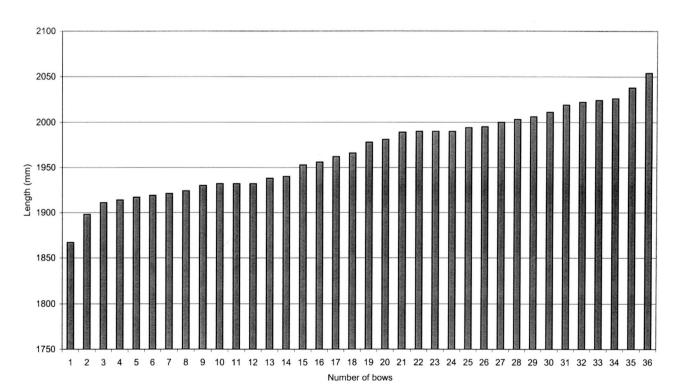

*Figure 8.21 Length range of double nocked bows*

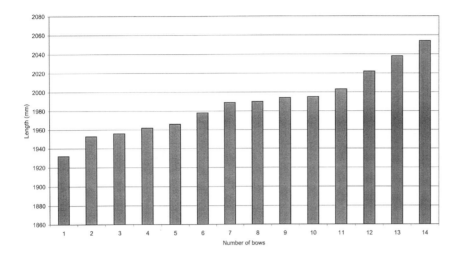

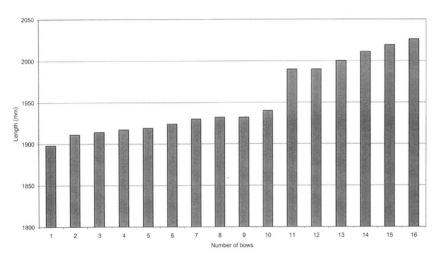

*Figure 8.22 Length range of double nocked bows from chests 81A1862 (top) and 81A3927 (bottom)*

incised or pricked rather than stamped or burnt (Figs 8.23–5). Most consist of a series of chevrons or small triangular indentations, possibly made with the edge of a '*float*' blade (Hardy 1992, 204). They occur in multiples of 2–5, 7 and 8. The seven chevron marks are shaped like a 'fir-tree', with tips pointing either inwards or outwards. The individual chevrons can be aligned in a number of ways, as can the orientation of the mark itself. This results in what can be interpreted as a huge number of individual marks, or a lesser number of similar but not identical, ones. In four instances multiples of chevrons have been grouped to form complex marks with up to 47 impressions (eg, 81A1603, Fig. 8.25c). These complex multiples are only found on bows in the chests. Circles also occur, some possibly made with dividers. These include plain incised circles; circles filled with dots, lines or tiny chevron-like marks; circles containing crosses; and chevrons. There are a number incorporating a complex series of chevrons, pin pricks and tiny incised dashes.

The most common mark consists of five chevrons, usually four aligned cardinally around a

*Figure 8.23 Bowmarks*

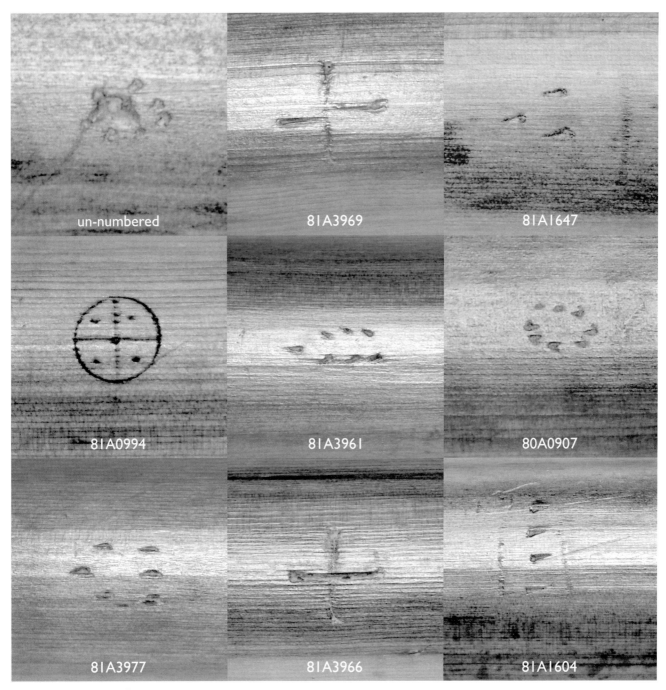

*Figure 8.24 Bowmarks*

central one. Eleven are recorded from chest 81A3927 (O9), four from chest 81A1862 (O8), one loose on the Main deck (80A1528, M5), one loose on the Upper

Table 8.15 Distribution of marked longbows

| Deck | Context | Frequency | % |
|---|---|---|---|
| Hold | loose | 1/1 | 100 |
| Orlop | chest 81A1862 | 25/40 | 62.5 |
| Orlop | chest 81A3927 | 34/50 | 68 |
| Orlop | loose | 6/14 | 42.9 |
| Main | loose | 8/16 | 50 |
| Upper | loose | 17/48 | 35.4 |

deck (81A5877, U5/6) and one from an unknown location (81A0708). Three chevrons occur on 21 bows both in chests and loose on all decks Figure 8.25 includes the location of all marked bows. Deck distribution is shown in Table 8.15.

Of the 82 bows found loose, 32 bear marks (39%), mostly similar to those found in the chests. However a few are unique. Three bows with circular incisions (80A1291, 80A0608, 82A0994), and one with a single mark consisting of a series of dots within a rectangle (81A1140) were found loose. There does not seem to be a 'standard mark' for loose bows, but nor does there appear to be more individuality within the marks on the

| Type | Description | Upper deck | Main deck | Orlop deck | Orlop chest 81A1862 | Orlop chest 81A3927 | Hold |
|---|---|---|---|---|---|---|---|
| 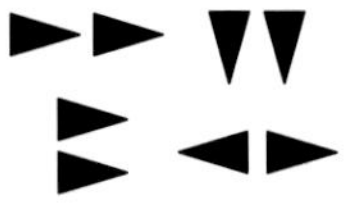 | 2 chevrons any configuration | 80A0282 81A1312 80A0444 80A0911 | | 82A1607 | 81A1656 81A1598 81A1610 81A1602 | 81A3937 81A3971 81A3934 81A3949 | |
| 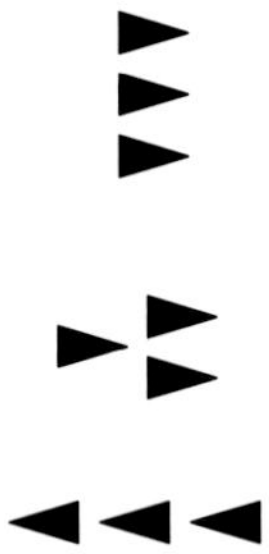 | 3 chevrons any configuration | 79A0807 81A1559 80A0449 80A0450 | 80A1298 80A0907 | 81A5763 82A0994 | 81A1607 81A1620 81A1647 81A1598 81A1604 81A1611 81A1619 81A1767 | 81A3938 81A3968 81A3954 81A3955 81A3929 | 81A2257 |
|  | 4 chevrons any configuration | 79A0812 80A0451 80A0473 | 80A1468 80A0763 | | 81A1613 81A1619 81A1767 81A1604 | | |
| 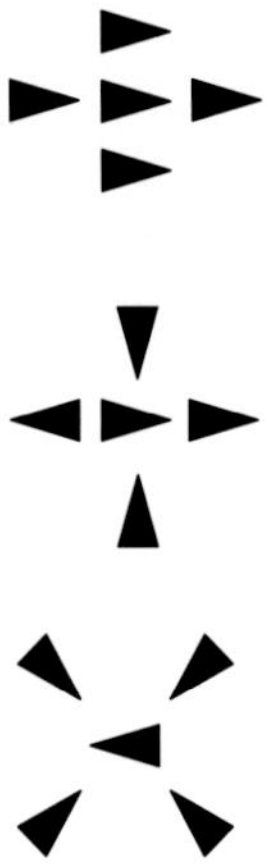 | 5 chevrons any configuration | | 80A1528 81A2303 | | 81A1617 81A1597 81A1616 81A1618 | 81A3930 81A3936 81A3941 81A3943 81A3948 81A3950 81A3959 81A3972 81A3933 81A3958 81A3939 | |

*Figure 8.25 a–c) Catalogue of bowmarks*

| Type | Description | Upper deck | Main deck | Orlop deck | Orlop chest 81A1862 | Orlop chest 81A3927 | Hold |
|---|---|---|---|---|---|---|---|
| ▲▲ ▶▶<br>▲▲ ▶▶<br>▲▲ ▶▶<br>▼▼▼▼▼▼ | 6 chevrons<br><br>any configuration | | 81A2303 | 81A3970 (additional mark at centre) | | 81A3973<br>81A3977 | |
| ▼<br>▶▶ ◀◀<br>▲<br>▲▲<br>▲▲▲ | Fir tree or 7 chevrons<br><br>any configuration | | | | 81A1639<br>81A1640<br>81A1622 | 81A3960<br>81A3944<br>81A3946<br>81A3962<br>81A3974<br>81A3961<br>81A3964<br>81A3932 | |
| ∪∪ ∪∪<br>∪∪ ∪<br>• • •<br>• • •<br>• •<br>✷ | 7 marks | 81A2018 | | 81A3770 | 81A1655 | | |
| ○ | circles | | 80A1291 | | | | |
| ⊙ | | 80A0608 | | | | | |
| ⊛ | | | | 82A0994 | | | |
| ⌒ | | | | | 81A1644 | | |
| ✝<br>┼<br>┼ | crosses, varying | 80A0642<br>81A2284<br>80A0473 | | | 81A1638 | 81A3942<br>81A3956<br>81A3966<br>81A3969 | |

613

| Type | Description | Upper deck | Main deck | Orlop deck | Orlop chest 81A1862 | Orlop chest 81A3927 | Hold |
|---|---|---|---|---|---|---|---|
| (complex marks) | Complex marks, multiple chevrons or pricks | 81A1140 | | | | | |
| (dots pattern) | | | | | | 81A1600 | |
| (triangles and dots) | | | | | | 81A1611 | |
| (arrow of dots and triangles) | | | | | | 81A1603 | |
| (scattered triangles) | | | | | | | 81A3928 |
| (fir-tree of triangles) | | | | | | 81A1615 | |

loose bows than those in boxes. The only mark that occurs exclusively in chests is the '*fir-tree*' of seven chevrons (eleven examples).

In each case, over half of the bows within the chests carry marks (eg, 25 out of 40 (62.5%) in chest 81A1862). The most common mark in this chest was three chevrons (eight examples) but they are not all identically aligned. This chest also contained four bows with extremely complex multiple chevron markings. Thirty-four of the 50 bows in chest 81A3927 (68%) had marks. There was one with multiple markings (81A3928) but the most numerous were the five point chevron and seven chevron 'fir-tree'. There were also three '*crosses*' (81A3956, 3966, 3969) and one circle with dashed vertical lines (81A3945).

Eight bows were identified as being of greater size than the rest, some visually and others by a statistical analysis of their profiles. All eight are of the 'slab-sided' profile, and five are 'handled', five were found in chests and three loose. There are no unique marks on these, although all that occur are chevrons and several are unmarked. One appears to have two marks (81A3960), but examples of both can be found elsewhere (see Table 8.7).

The location of the marks, predominantly on the upper bow hand just above the centre, has been interpreted in several possible ways: as being marks to suggest the arrow pass (Hardy 1992, 204), bowyers' makers' marks (as found on other objects recovered), or merely to identify the upper limb. Other suggestions include '*guild marks*'. The similarity of so many of the marks is not what one would expect if these were marks of personal ownership (see linstocks for instance, Chapter 5).

## Grain and timber quality

The number of annual rings per inch was recorded on 138 bows to provide an indication of the uniformity and quality of the timber used. This was recorded as: '*coarse grain*' (up to 40 rings to the inch), '*medium grain*' (41–60 rings) and '*fine grain*' (61–100+ rings) (Tables 8.5 and 8.16).

These results show that bows with differing grain size were distributed all over the ship with equal numbers of medium and fine grained examples, irrespective of length (Fig. 8.26). The finer-grained bows were twice as common as coarse grained examples. This suggests that grain was not a major consideration when choosing staves suitable to make longbows of specific lengths.

## Bow making: general observations

The bows do not exhibit the level of finish that we find on modern examples. The backs display a better finish than the bellies. As the raising of the grain on the back of a bow can lead to failure, this attention to the back demonstrates good practice. All the bows still display small patches of the '*bast*' or '*cambrium*' layer, the layer

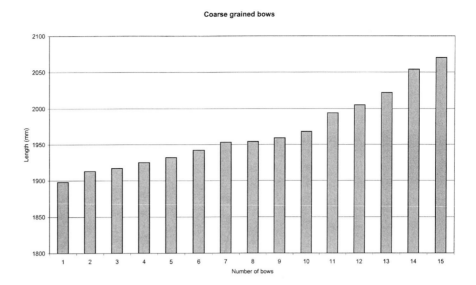
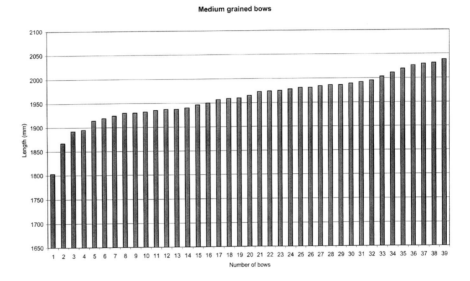
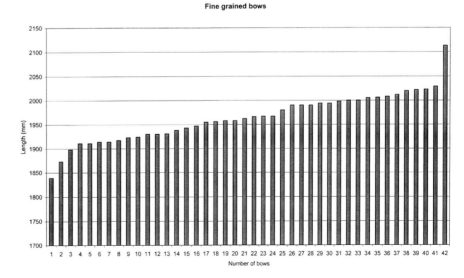

*Figure 8.26 Size ranges of 102 bows: coarse, medium and fine grained*

that lies between the sapwood and the bark. The bows are not perfectly smoothed. They all exhibit a series of

**Table 8.16 Timber grade and uniformity**

| Deck | Coarse[1] | Medium[2] | Fine[3] | Total |
|---|---|---|---|---|
| Orlop | 1 | 3 | 4 | 8 |
| Orlop: chest 81A3927 | 8 | 15 | 18 | 41 |
| Orlop: chest 81A1862 | 4 | 13 | 20 | 37 |
| Hold | 0 | 0 | 1 | 1 |
| Upper | 9 | 17 | 7 | 33 |
| Main | 2 | 5 | 4 | 11 |
| Starboard scourpit | 0 | 0 | 1 | 1 |
| unknown/illegible no. | 2 | 3 | 1 | 6 |
| Total | 26 | 56 | 56 | 138 |

[1]<40 annual rings per inch; [2]41–60 annual rings per inch; [3]61+ annual rings per inch

**Table 8.17 Distribution of used longbows (after Hardy, in archive 1984)**

| Sector | Context | Objects |
|---|---|---|
| O7 | loose | 81A1583 |
| O8 | loose | 82A0992, 82A0994, 82A1607, 82A2268 |
| O8 | chest 81A3927 | 81A3934–5, 81A3937–8, 81A3955, 81A3956, 81A3976 |
| O9 | chest 81A1862 | 81A1597–9, 81A1601–3, 81A1606, 81A1610, 81A1612–13, 81A1617–18, 81A1638, 81A1640, 81A1644 |
| M4 | loose | 80A1528 |
| U4 | loose | 79A0855 |
| U8 | loose | 81A2018 |

facets, or flat surfaces, around the circumference of the belly which suggests the use of an edged tool, possibly a plane or a '*float*'. Particularly noticeable is that fact that these facets appear to run unbroken over the length of the bow. Small pins have been left in place and 'worked around' (as in modern practice) with a resultant slight 'swelling' caused by the extra material being left in place. In some instances larger pins (knots) have been removed leaving significant (and to modern eyes, dangerous) depressions in the limbs.

One striking feature of the bows is the similarity of taper over the last 16in (406mm) or so of each limb (see Tables 8.9–12). Tips on both limbs on most bows have been reduced to the same width and depth (0.5in/12.7mm) where the nock discoloration commences. This suggests that the horn nocks were bored out with a standard diameter auger or bit. Some bows have tips that remain completely undamaged. On a few of these the extreme tip is chamfered, possibly as a result of cutting the ends off the tips to accommodate particularly curved horns.

Some bows have a noticeable depression around the circumference immediately below the area of the horn nocks. This was termed '*waisting*' (report in archive) and is taken as evidence that the final tillering took place after the horn nocks were fitted. The work appears to have been by an edge, perhaps a '*scrape*' or similar tool, as opposed to a rasp or file.

It is highly likely the bow staves were cleft from the parent tree or branch using an axe and wedges as opposed to being sawn. The re-appearance of natural deviations in the wood (twisting, etc) supports this suggestion. It cannot, of course, be determined in what condition the bowmaker originally received the stave.

**Used bows**

Twenty-nine of the longbows have been noted as 'used' (Hardy in archive). Thirteen are from chest 81A1862 in O9 and seven from chest 81A3927 in O8. A further five were loose in O8, possibly having derived from this chest. The others are widely distributed (Table 8.17).

**The draw-weight of the bows**

Estimates were made of draw-weight based on the limb dimensions of 48 bows (Table 8.18). Although somewhat arbitrary, these reinforce previous studies (Hardy 1992, 213; 2005, 13–19 and Fig. 8.27) predicting greater draw-weights than we would expect

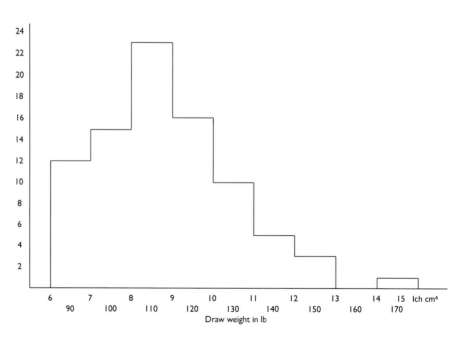

*Figure 8.27 Draw-weights (source: Kooi and Pratt)*

Table 8.18 Draw-weights of *Mary Rose* longbows by location

| 160lb | 140lb | 130lb | 120lb | 110lb | 100lb | 90lb | 80lb | 70lb | 65lb |
|---|---|---|---|---|---|---|---|---|---|
| *U4 SS* | *M6 SS H* |  | U8 |  | U8 | O7 | U5/6 | *M8* | U8 |
|  | O9 | O9 | O9 | O9 | U9 | U7 | O7 | O8 | *M9* |
|  | O9 | O9 | O8 | O9 | O8 | U7 | O8 | O8 | O9 |
|  |  | O9 SS H | O8 | O8 |  | H8 | U9 | O9 |  |
|  |  | O8 SS H | O8 | O8 |  | U9 | U10 |  |  |
|  |  |  | O8 | O8 |  | U9 |  |  |  |
|  |  |  | O8 | O8 |  | O8 |  |  |  |
|  |  |  | O8 |  |  | O8 |  |  |  |
|  |  |  |  |  |  | O8 |  |  |  |
|  |  |  |  |  |  | O9 |  |  |  |
|  |  |  |  |  |  | O9 |  |  |  |
| 1 | 3 | 7 | 7 | 4 | 3 | 11 | 5 | 4 | 3 |

Loose bows in italics; SS = slab-sided; H = handled

to see today (see below). Just over half (25 = 52%) are estimated at 100lb (46.36kg) or more, nineteen of which are from chests (mostly in chest 81A3927 in O8). Only 23 are estimated at 90lb (40.82kg) or less, of which only nine are in chests. Distribution suggests that within this sub-section of the assemblage there is no obvious difference in draw-weight based on location, so predicted draw-weight alone cannot be used to suggest the presence of a group of professional archers with personal bows of a greater draw-weight than the rest.

The spread of draw-weights suggested by this visual study is between 65lb (29.5kg) and 160 lb (72.6kg). Although all of the draw-weights suggested are lower than those predicted by the statistical analysis undertaken by Hardy and Kooi (see below), the range is still as large. The finding of some of the (predicted) heaviest bows within chests suggests that either the variation within the weapons system was acceptable (all bows could be used by all those who needed to use them) or that professional archers had to equip (or re-equip) themselves from the general stores; this infers that a sorting exercise had to be undertaken when choosing bows and in turn matching them with arrows.

**Estimating longbow weights from arrow shaft data**
by Keith Watson

The finding of five different arrow forms together with nine different wood species and two different lengths (see below) has prompted questions as to whether there would be significant differences in the properties of the finished arrows. Weight will have an effect on both range and penetration but weight and resistance to bending of the different woods used for arrow shafts might dictate the strength of the bow required to optimise the qualities inherent in any particular arrow. Whilst the size of the assemblage available for scrutiny for any specific form/length/species is too small to be reliable, there are overall predictions which can be obtained from studying specific qualities of particular species of wood used for arrows.

The forces acting on an arrow released from a longbow should cause the shaft to buckle so that it 'snakes' around the bow stave and flies on towards the target. In general the arrow is matched to the bow and has the appropriate '*spine*' rating (resistance to bending) for the weight of bow. Generally the arrow needs to be stiffer for more powerful longbows and for longer draw lengths.

The AMO (Archery Manufacturers and Merchants Organisation USA) spine rating for wooden shafts is determined by measuring the deflection in inches produced by a 2lb (0.91kg) weight at the centre of the span of an arrow shaft supported horizontally at two points 26in (660mm) apart (AMO Standards Committee 2000, 6–7). The AMO standards provide tables of spine ratings (as ranges of deflection values) for two types of arrow (target arrows and field/hunting arrows), different ranges of bow weight and different draw lengths.

In the present exercise, the bow weight corresponding to each arrow shaft was estimated by first calculating the deflection under the AMO test conditions, then using the AMO tables for field and hunting arrows, in reverse, to find the corresponding bow weight.

Shafts of 30in (the 755–854mm group) were chosen since these are the most abundant within the *Mary Rose* assemblage (details below). The mean radius of each shaft was calculated from the values taken at 100mm intervals along its length. The chosen arrows were measured soon after recovery, and before conservation. Therefore these are the least likely to have suffered excessive swelling, distortion or shrinkage. The mean was then used in the following equation (Weast 1986, F-78) with the AMO test dimensions converted to SI units:

$$s = \frac{mgl^3}{12\pi r^4 E}$$

s (m) represents the deflection, mg (kg m s$^{-2}$ = N) the load, l (m) the span, r (m) the radius and E (N m$^{-2}$) the Young's modulus of the wood used to make the shaft.

The Young's modulus values (GN m$^{-2}$), taken from Farmer (1972), are as follows: poplar 8.6; birch 13.3; alder 8.8; willow 6.5. (Microscopic examination of 646 samples (8.25% of the assemblage) suggests that poplar, birch and alder are the most prevalent woods in the *Mary Rose* assemblage). Where different species are possible expert advice was taken (G. McConnachie, pers. comm.) and, in the case of willow, an average value was used.

The deflection values mostly fell below the range provided by the AMO standards so it was necessary to extrapolate the AMO data. A number of mathematical models were used to describe the relationship between AMO bow weight in pounds and deflection in inches (using range mid-points in both cases). The power function below provided the closest fit of those tested:

Bow weight = 18.047 x Deflection$^{-1.1086}$ ($R^2$ = 0.9895)

The deflection value for each shaft was used in this equation to estimate the corresponding bow weight. The bow weights were tabulated as grouped frequencies for each type of wood (Table 8.19).

It is clear that birch stands out from the rest but it is important to remember that the bow weights are estimated and subject to a number of potential errors. Some could not be quantified, for example in assuming the published Young's modulus values, and in extrapolating the AMO data. Although the arrows used for this equation had not been soaking for a long time in water and were well-preserved due to burial, some swelling may have occurred which might increase the shaft radii by 5–10% (M. Jones, pers. comm.). If this is the case, it would make the true bow weights about 20–35% lower than the values given in Table 8.19. The results suggest that optimum longbow weights for the poplar arrows (77% of those identified) lie between 50lb and 90lb (22.7–40.8kg); for birch (14%) 100–130lb (45.4–59.0kg); for alder (6%) 60–70lb (27.2–31.8kg); and for willow (1%) 50–70lb (22.7–31.8kg). The species of wood used in the production of the AMO tables is not known, although cedar is a preferred choice in the American market. What effect this may have on the predictions is currently unknown and further research is suggested.

*Statistical analysis of the longbow data*
by Keith Watson

The data obtained from the recent study of 116 of the 172 longbows (see above) were presented for statistical analysis. The sample was not randomly drawn, but included all complete and accessible bows. The study was undertaken in three stages:

- The variables used in the first stage of analysis are listed in Table 8.20. These can be described as either quantitative (measured on a continuous scale) or qualitative (divided into separate categories). The length measured along the bowstave was used, rather than the shortest distance from tip to tip, since the bows vary in curvature. Width and depth are the maximum values, from bowhand to shafthand side

**Table 8.19 Grouped frequency table for bow weight by wood type**

| Bow weight interval (lb) | Frequency | | | |
|---|---|---|---|---|
| | Poplar | Birch | Alder | Willow |
| 40–50 | 2 | | | |
| 50–60 | 19 | | 2 | 2 |
| 60–70 | 21 | | 6 | 2 |
| 70–80 | 17 | | 2 | |
| 80–90 | 12 | | 2 | |
| 90–100 | 1 | 4 | 1 | |
| 100–110 | | 12 | 1 | |
| 110–120 | | 15 | | |
| 120–130 | | 10 | | |
| 130–140 | | 4 | | |
| 140–150 | | 6 | | |
| 150–160 | | 2 | | |
| 160–170 | | 2 | | |
| 170–180 | | 1 | | |
| 180–190 | | 1 | | |

**Table 8.20 Variables used in the preliminary analysis**

| Quantitative (measurements) | |
|---|---|
| Length (mm) | |
| Mass (g) | |
| Maximum width (mm) | |
| Maximum depth (mm) | |
| **Qualitative (categories)** | |
| Appearance | robust (74)★ |
| | fine (42) |
| Cross-section | rounded (9) |
| | D (9) |
| | deep D (76) |
| | flat D (14) |
| | slab-sided (8) |
| Location | chest 81A1862 in O9 (36) |
| | chest 81A3927 in O8 (42) |
| | loose (38) |

( )★ = No. of longbows in category

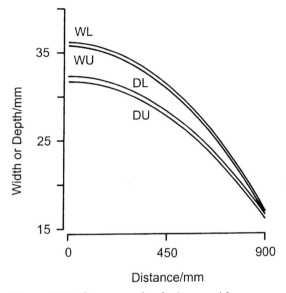

*Figure 8.28 Constructs for the 'average' bow*

### Table 8.21 Mean length, mass, maximum width and maximum depth by category

**Appearance**

|  | Robust | Fine |
|---|---|---|
| Length (mm) | 1975 (1.01)[a] | 1956 |
| Mass (g) | 774 (1.11) | 700 |
| Max. width (mm) | 36.5 (1.04) | 35.0 |
| Max. depth (mm) | 33.9 (1.03) | 32.8 |

**Cross-section**

|  | Slab-sided | Other |
|---|---|---|
| Length (mm) | 2010 (1.02) | 1965 |
| Mass (g) | 893 (1.21) | 737 |
| Max. width (mm) | 38.4 (1.07) | 35.8 |
| Max. depth (mm) | 35.0 (1.05) | 33.4 |

**Location**

|  | Chest 81A1862 | Chest 81A3927 |
|---|---|---|
| Length (mm) | 1979 (1.01) | 1957 |
| Mass (g) | 790 (1.12) | 703 |
| Max. depth (mm) | 34.0 (1.02) | 33.3 |

[a] Number in parentheses = ratio of larger to smaller value in that row

and from belly to back respectively. A subjective distinction was made by the group regarding cross-section and whether the longbow was robust or fine. '*Location*' distinguishes between the two longbow chests from the Orlop deck (81A1862 and 81A3927) and loose bows found in various places around the ship.

- The second stage of the analysis employed width and depth measurements taken at 100mm intervals from the centre to the tip of the bow along the lower and upper limbs, using a linear scale placed alongside the bowstave. These data were used to produce constructs which provide a simple graphical comparison between the 'average' shape of longbows within each category under consideration.
- The final stage was aimed at finding whether the bow marks could be related to any dimensional differences.

Table 8.21 compares mean values of length, mass, maximum width and maximum depth within categories. Only statistically significant differences are shown (p<0.05). The ratio in parentheses indicates the magnitude of the difference between the pair of means in that row, for example between the mean lengths of the robust and fine bows. Mass is clearly the strongest indicator of differences in appearance, cross-section and location. The upper part of the table makes an objective distinction between robust and fine longbows; clearly, on average, the robust longbows are longer, heavier, wider and deeper than the fine. Preliminary analysis had shown that the slab-sided bows stand out from the rest; the middle part of the table shows that they are markedly longer, heavier, wider and deeper than the other cross-section categories combined. Finally, the lower part shows that the longbows from chest 81A1862 are longer, heavier and deeper than those from chest 81A3927; however, there is no significant difference in width. The loose longbows are not included since they were dispersed around the ship.

The next stage of the study produced constructs providing graphical representations of longbow geo-

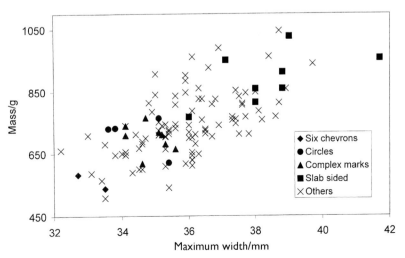

*Figure 8.29 Mass of bow versus maximum width*

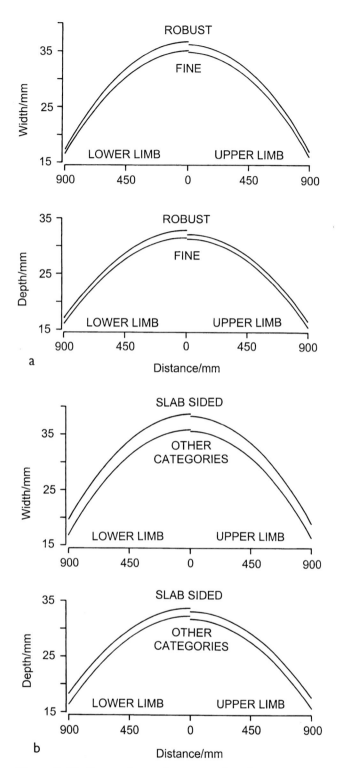
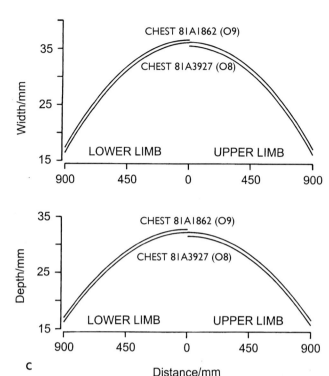

*Figure 8.30 Comparisons of constructs by: a) appearance; b) cross-section; c) location*

metry for comparison. Figure 8.28 shows the constructs for width and depth up to 900mm along the upper and lower limbs from the centre of the bow towards the tip; these are averaged across all 116 cases[1]. Clearly, for the average longbow, width is greater than the corresponding depth along each limb; furthermore, the lower limb is slightly wider and deeper than the upper, which is usual in good bowyers' practice. Figure 8.30 shows comparisons based on appearance, cross-section and location respectively. Width is shown in the top half of each figure, and the lower limb is on the left hand side.

The difference between robust and fine longbows (Fig. 8.30a) is immediately obvious. The slab-sided longbows are considerably more substantial than the other cross-section categories combined (Fig. 8.30b). The difference between chests is relatively subtle (Fig. 8.30c) although the longbows from chest 81A1862 are clearly the more substantial.

The longbows were divided into groups depending on the type of bow mark. Every group was then compared with each of the others in terms of length, mass, maximum width and maximum depth. With three exceptions, the different types of bow mark do not appear to reflect dimensional differences. This suggests that they may have non-functional significance, for example identification marks of different bowyers. However, '*circles*', '*complex marks*' and particularly '*six chevrons*' lie at the lower end of the plot of mass vs maximum width; this is evidently in contrast to the more substantial slab-sided bows, which tend to lie at the upper end (Fig. 8.29). A selection of longbows of varying cross-sections is shown in Figures 8.31–4.

**Endnote**
[1] Linear regression equations were found for the variation in width and in depth with distance along both the upper and lower limb of each bow. These were obtained from plots of the width and depth values at 100mm intervals against the square of the distance from the centre of the bow. The lowest coefficient of determination was 0.926, with only fifteen of the

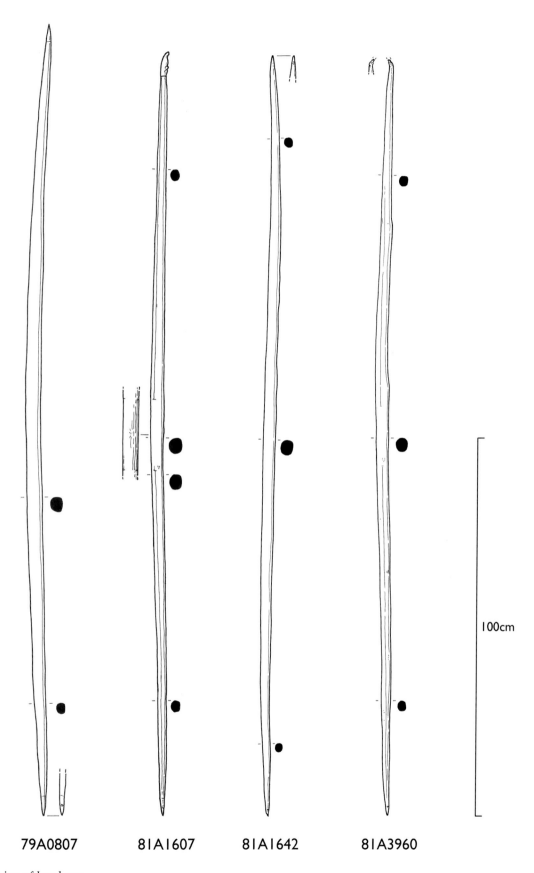

*Figure 8.31 Selection of longbows*

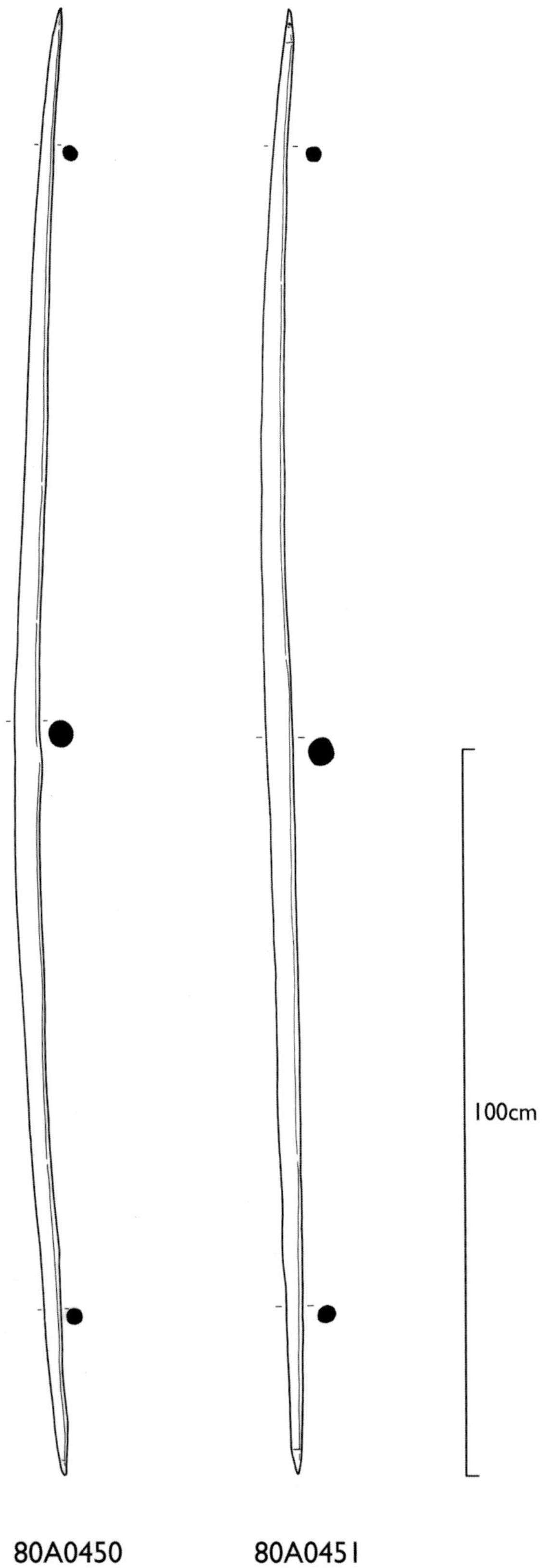

80A0450    80A0451

*Figure 8.32 Selection of longbows*

464 equations (ie 116 x 4) having values below 0.950. The mean intercept and gradient were found for each category under comparison, and the resulting regression equation used to produce the construct.

## Assessment of the working capabilities of the longbows
by Robert Hardy, J. Levy, P. Pratt, the late Roy King, S. Stanley, B.W. Kooi, D. Adams and A. Crowley

A number of projects have been instigated aimed at developing an understanding of the functional capabilities of the longbows. Initially this was a study of the bows themselves and comprised detailed post-conservation study of the wood and the dimensions of each stave. The staves of yew (*Taxus baccata*) from which the longbows were made were cut or cleft radially from a log across the sapwood/heartwood boundary. When finished, each bow had sapwood on the back (the convex side) whilst the belly consisted of heartwood.

The anatomical characteristics of broken fragments of bows were examined. The cross-sections show the growth rings and from the narrow radial increment between each, it is clear that the trees had been very slow growing (Fig. 8.35). This is likely to have been the deliberate choice of the bowmakers since, in general, slow growth softwoods have greater strength properties than fast grown. From the curvature of the growth rings, the diameter of the log from which the staves were cut is likely to have been at least 8in (200mm). The fact that, in every case, the more perishable sapwood is present on the back of the bow is evidence of the craft skill of the bowyer. The outer sapwood of a tree has a higher tensile strength along the grain than the inner heartwood, whilst the latter has a greater strength in compression than the former. Drawing a bow made in this way uses both properties to their best advantage. The high natural durability of the yew heartwood will, undoubtedly, have been a major factor in their present

**Table 8.22 Testing of *Mary Rose* longbows**

| Object | Tip deflection (mm/in) | Load (kg/lb) | Description |
|---|---|---|---|
| 81A1654 | 203/8 | 26.3/58 | longitudinal crack; compared well with 24.5kg (54lb) at 203mm deflection in earlier tests |
| 81A1607 | 203/8 | low | large; investigation revealed local degradation inside bow at fracture; repaired & on display |
| 81A3975 | – | 19.1/42 | braced to 178mm/7in & drawn to 559mm (22in) at load of 19.1kg before sapwood cracked |
| 81A1648 | – | 27.2/60 | force 24.9kg (55lb) Jan. 1982 & by April 1983 increased through exercising to 27.2kg; throughout following period actual weight of bow remained constant |

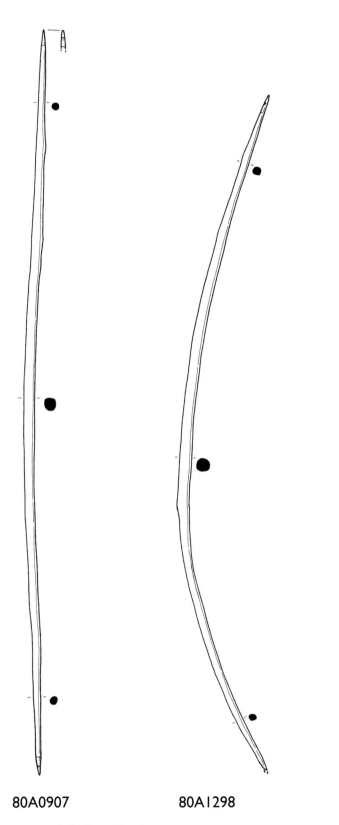

*Figure 8.33 Selection of longbows*

bow – the tension face – than on the belly which is under compression.

In order to try and ascertain the draw-weight of these large bows (an attribute critical to the performance of the bow) it was decided to test a sample using a mechanical tiller to bend the bows. Before attempting this it was necessary to test whether the yew had similar properties to modern yew with respect to bending. Several broken longbow parts were tested (eg, 80A0513, 79A1192) to measure the '*modulus of elasticity*', ie, the resistance of the yew to deformation (details in archive). Initial results suggested that the wood was comparable with modern yew and therefore it was decided to test physically several complete and near complete bows (81A1648, 81A1654, 81A1607, 81A3975). Using a mechanical tiller to draw the tips of the bows downwards (to simulate the action of drawing the bow) the wood, which appeared to be in near-pristine condition, proved to have nowhere near its expected original strength (Table 8.22). The full draw-length of 762mm (30in) was achieved on one example (81A1648) but with a draw-weight of only 27kg (60lb) (Fig. 8.36). Another broke and exposed internal decay and the third (81A1654) developed a small crack across the grain in the sapwood. The fourth showed failure by cracks developing along the grain at one end. This suggested that, although the sapwood had retained sufficient tensile strength not to fail across the grain, it was failing in bending by a '*sheer failure*' along the grain in the earlywood of the growth ring. The low draw-weights achieved and the damage sustained suggested that the bows were damaged in the sapwood. Any further, purely mechanical, tests on the originals to predict draw-weight would be both dangerous and erroneous. Microscopic examination of the broken bow confirmed a wide range of cellular deterioration. To estimate the amount of degradation the elastic modulus of three of the complete bows was measured. The results demonstrated that the modulus of the *Mary Rose* bows is some 50% lower than that of a modern bow, confirming the serious nature of the degradation.

condition. The sapwood, which has a much lower natural durability, was seen to have deteriorated to some degree in a number of instances. Small cracks across the grain are evidence of this deterioration and are more likely to cause serious loss of strength on the back of the

In order to try and ascertain original draw-weight, two further avenues were explored; computer modelling of the critical dimensions of three bows (79A0812,

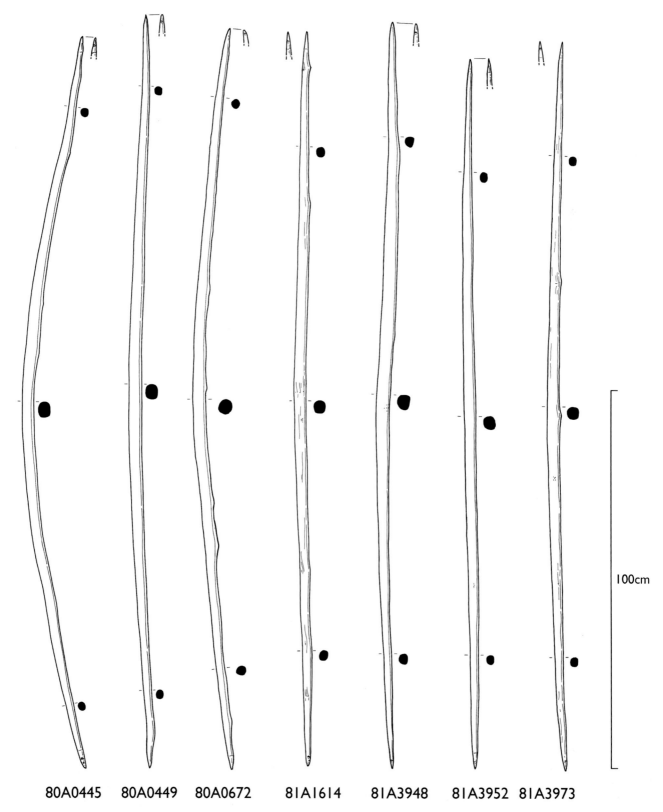

80A0445    80A0449    80A0672    81A1614    81A3948    81A3952  81A3973

*Figure 8.34 Selection of longbows*

81A1648, 81A1607) to predict draw-weight; and the manufacturing and drawing of reproduction bows (MRA1–MRA3) to ascertain true draw-weight. The dimensions of MRA1 were then subjected to the same mathematical computer modelling. As every timber stave dictates its own best form so that no two bows can be exactly alike, the term '*approximation*' was used to describe the reproductions.

One of the authors (BWK) had already developed a method for computing the mechanics of bows and arrows (Kooi 1983; Kooi and Bergmann 1997). The draw-weight of a bow is proportional to the cube of the

*Figure 8.35 Growth rings on sectioned bow*

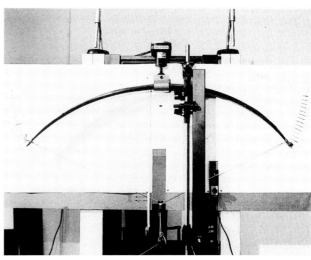

*Figure 8.36 Bow 81A1648 on tiller*

height of its cross-section, while it is only proportional to the width of the limb. This means that the weight is about three times more sensitive with respect to errors in height than in width. Critically, the original modulus of elasticity of the wood of the *Mary Rose* bows had to be estimated (see Table 8.23 for explanation and data).

**Table 8.23 Length and elastic modulii for selected longbows (based on 1984 measurements)**

| Object | Length (mm)* 1984 | 1998** | B=L(3) 200(3) | 1 Ch cm (4/) | 1 Ch/B | E GN/m(3) | P |
|---|---|---|---|---|---|---|---|
| 81A1607 | 2010 | 1994 | 1.010 | 14.60 | 14.50 | 4.52 | 0.62 |
| 80A0834 | 2020 | 2016 | 1.030 | 12.70 | 12.30 | | |
| 81A1767 | 2000 | 1994 | 1.000 | 12.50 | 12.50 | 4.80 | |
| 79A0807 | 2120 | 2113 | 1.190 | 11.40 | 9.50 | | |
| 80A0451 | 2050 | 2029 | 1.080 | 11.40 | 10.50 | | |
| 80A1469 | 1960 | 1981E | 0.940 | 11.25 | 12.00 | | |
| 80A1528 | 1990 | 2000 | 0.986 | 10.80 | 10.90 | | 0.56 |
| 80A1066 | 2020 | 2031 | 1.030 | 10.50 | 10.50 | 10.2 | |
| 81A1648 | 1990 | 1968 | 0.986 | 10.50 | 10.60 | 4.15 | 0.63 |
| 80A0473 | 1980 | 1965 | 0.970 | 10.20 | 10.50 | | |
| 80A0608 | 2020 | 1931 | 1.030 | 10.10 | 9.80 | | |
| 81A3975 | 1960 | – | 0.940 | 10.04 | 10.70 | | 0.53 |
| 80A1291 | 1910 | 1921E | 0.870 | 10.00 | 11.50 | | |
| 81A1654 | 2010 | 2020 | 1.010 | 9.80 | 9.70 | | |
| 80A0449 | 2000 | 2003 | 1.000 | 9.65 | 9.65 | | |
| 80A0652 | 1980 | 1746 | 0.970 | 9.40 | 9.70 | | |
| 80A1353 | 1910 | 1925 | 0.870 | 9.35 | 10.75 | | |
| 80A0642 | 1970 | 1967 | 0.950 | 9.20 | 9.70 | | |
| 80A0445 | 1970 | 1978 | 0.950 | 9.10 | 9.60 | | |
| 80A0450 | 2020 | 2024 | 1.030 | 8.70 | 8.40 | | |
| 80A0672 | 1970 | 1974 | 0.950 | 8.60 | 9.00 | | |
| 80A0444 | 2000 | 1989 | 1.000 | 8.30 | 8.30 | | |
| 79A0812 | 1980 | 1956 | 0.970 | 8.30 | 8.60 | | |
| 80A0282 | 1980 | 2013E | 0.970 | 8.30 | 8.60 | | |
| 80A1195 | 1930 | 1937 | 0.900 | 7.60 | 8.40 | | |

* see text for details of how measurements were taken; ** used in Table 8.5

Table 8.24 Weight and density of selected *Mary Rose* longbows

| Object | Sector/context | Weight (g) | Volume (cc) | Density |
|---|---|---|---|---|
| 81A3975 | O8 chest 81A3927 | 701.7 | 1331 | 0.53 |
| 81A3952 | O8 chest 81A3927 | 630.8 | 1155 | 0.55 |
| 80A1528 | M5 loose | 737.1 | 1325 | 0.56 |
| 81A3973 | O8 chest 81A3927 | 637.9 | 1119 | 0.57 |
| 81A3976 | O8 chest 81A3927 | 722.9 | 1255 | 0.58 |
| 81A1655 | O9 chest 81A1862 | 843.4 | 1450 | 0.58 |
| 81A3935 | O8 chest 81A3927 | 616.6 | 1046 | 0.59 |
| 81A1638 | O9 chest 81A1862 | 829.2 | 1395 | 0.59 |
| 81A3956 | O8 chest 81A3927 | 836.3 | 1365 | 0.61 |
| 81A3940 | O8 chest 81A3927 | 900.1 | 1460 | 0.62 |
| 81A1622 | O9 chest 81A1862 | 999.4 | 1606 | 0.62 |
| 81A1607 | O9 chest 81A1862 | 1050.0 | 1685 | 0.62 |
| 81A1603 | O9 chest 81A1862 | 737.1 | 1178 | 0.63 |
| 81A1648 | O9 chest 81A1862 | 878.9 | 1400 | 0.63 |
| 81A3974 | O8 chest 81A3927 | 1141.1 | 1696 | 0.67 |
| 81A1645 | O9 chest 81A1862 | 1063.1 | 1532 | 0.69 |
| 81A3953 | O8 chest 81A3927 | 1013.5 | 1378 | 0.73 |
| 81A1599 | O9 chest 81A1862 | 1091.5 | 1423 | 0.76 |

- *The length of the working limbs and the shape of the unbraced bow* was obtained from the archive drawings. The length of the working limbs in the case of the *Mary Rose* bows is the length from nock to nock since no increase in section is found at the midpoint of the bow to serve as a handle. Since the horn nocks have vanished, the position was estimated from the slots on the bow tips.
- *The shape and size of the cross-section of the limbs along the bow length* was obtained by measuring the archaeological drawings with a quantitative image analyser (details in archive) and reworking the data to produce a section stiffness every 100mm along the length of the bow. This was undertaken for 30 bows.
- *The draw length* was obtained from the measurement of 262 complete arrows, constituting 11% of the assemblage (undertaken by A. Elkerton, 1980). Modelling was based on a 762mm (30in) draw length. Further studies on the arrows suggest that the arrows of 762mm draw length are the most common by a figure of 4:1 (see below).
- *The bracing height*, or the distance between the string and the hand-position measured to the back of the bow, was based on measurements taken from bows which appeared to retain their shape. The figure 160–78mm (6¼–7in) was used.
- *The stiffness of the string* was obtained from the mechanical testing of a hemp bowstring at the Instron testing facility in High Wycombe (Instron is a company specialising in precision testing of range of properties inherent in any given material, such as tensile or mechanical strength).
- *The effective elastic modulus of the whole bow* was obtained by using data from the section stiffness (details in archive).

To compute the dynamic properties the following were required:

- The weight and density of the bow (see Table 8.24)
- The weight and stiffness of the arrow (see Table 8.25)

The weights and densities of a number of the bows were obtained. The density was obtained by weighing each bow in air and measuring the volume of water displaced by each from a narrow cylindrical container (Table 8.24). The range of densities from 0.53 to 0.76 is greater than that found in modern yew (0.53–0.69). Since the elastic modulus of wood increases linearly with density, the existence of very dense yew (>0.70) requires further study.

The weight and stiffness of the arrows was harder to estimate in view of the degradation of the wooden shafts and lack of steel heads. The measured diameters on a group of unconserved arrows (Elkerton 1983, 80A0762, as above) suggested an upper limit of 12.5mm and a lower limit of 10mm. Associating 12.5mm diameter with 800mm length and 10mm with 750mm length, the maximum volumes of wood are 98.2cc and 58.9cc respectively. The range of weights for the shaft including fletching is 42–64g for the larger

Table 8.25 Weight and density of different woods

| Wood | Density | E, GN/m$^2$ |
|---|---|---|
| Poplar, grey | 0.43 | 9.5 |
| Alder | 0.45 | 8.8 |
| Ash | 0.60 | 11.9 |
| Birch | 0.60 | 13.3 |
| Oak | 0.61 | 10.1 |
| Hornbeam | 0.65 | 11.9 |

Table 8.26 Estimated draw-weights of *Mary Rose* longbows

| Object | Location | Estimate (lb) | Estimate (kg) |
|---|---|---|---|
| XI-2 | Tower of London | 98 | 44.4 |
| XI-1 | Tower of London | 101 | 45.8 |
| 79A0812* | Mary Rose Trust | 110 | 49.9 |
| 81A3952 | Mary Rose Trust | 115 | 52.1 |
| 81A1654 | Mary Rose Trust | 124 | 56.2 |
| 81A1648* | Mary Rose Trust | 136 | 61.7 |
| 81A3975 | Mary Rose Trust | 137 | 62.1 |
| 81A1607* | Mary Rose Trust | 185 | 83.9 |
| MRA1* |  | 102–104 | 46.4–47.2 |
| MRA4 |  | 150 | 69 |

* Bows with measurements analysed by B. Kooi

arrows and 25–38g for the smaller (for either ash or birch); the weights are 58.8g and 35.3g respectively. From a previous study of medieval arrowheads, the average weight of the small English warhead, Type 16 in the *London Medieval Catalogue* of Ward Perkins (1942; now reclassified as M4 by Jessop (1996)) is 7g. The total weight of 35.3+7 = 42.3g is remarkably close to the 42.6g of the Type 16 head found on the Westminster Abbey arrow, the only other extant arrow available for study. For the longer shaft a Type 7 armour-piercing long bodkin (for examples see arrows below) was suggested as appropriate (Hardy 1992, 229). The average weight of these heads is 13g, leading to a total weight of 71.8g. The densities of the different woods reflected within the arrow assemblage are shown in Table 8.26.

**Estimates of the strength of the *Mary Rose* bows**

Kooi's computer modelling gave bow-strengths ranging from 110lb (49.9kg) draw-weight to an incredible 185lb (83.9kg) at 30in (76.2mm; Table 8.26). Extrapolation of this provides the graph shown in Figure 8.27. The largest group comprised those in the 150–160lb range (68kg–72.5kg). Our later work (below) suggests that this is the weight of bow needed to shoot the types of late medieval arrows required to inflict the sort of damage evidenced in contemporary descriptions.

Kooi's computer analysis of 79A0812 produced a figure of 144lb/65kg. However, this was greater than that suggested for the approximation MRA1 (modelled after 79A0812) for the modulus of elasticity. Using the MRA1 modulus (7.5 GN/m²) the weight of 79A0812 would be reduced to 109.5lb/49kg. Either figure is surprisingly high, especially since 79A0812 appears to be one of the weaker bows. Accepting a realistic draw-weight of 110lb/49kg for this example, the figures for the bows have been calculated as ranging from 98lb (44.4kg) to 185lb (83.9kg; Table 8.26).

The estimates for the bows recovered in 1836–1840 came out as the lowest, but these were made without the full data that was gathered from the Portsmouth bows. Pratt felt that the huge *Mary Rose* bow 81A1607, rated by Kooi at 185lb (83.9kg), would be likely to break at full draw and suggested that it might survive a lesser draw of perhaps 28in (711 mm), which would reduce the draw-weight to 172lb (77.9kg). Archer Simon Stanley can command this weight completely but finds that that anything greater than 190lb (86.07kg) is uncomfortable for sustained use.

The experimental draw-weight (obtained by the drawing of the bow using a tiller) of 81A1654 was 65% of Kooi's predicted value, while that of 81A1648 is only 44%. The extent of the reduction in modulus for 81A1648 appears to be almost 50% and it is significant that both bows broke across the sapwood). If the sapwood makes no contribution at all to the bending stiffness of the bow, and if we take the typical figure of one quarter of the thickness of the bow as sapwood, the bending stiffness and thus the strength, would be reduced to 42% of the theoretical figure. This is close to the reduction actually found on 81A1648, although in this case the sapwood has not cracked. Probably both sapwood and heartwood have become degraded, the sapwood more than the heart.

It had already been realised that to come up with real answers, copies of *Mary Rose* bows should be made in new and comparable wood. This was undertaken by the late Roy King, Bowyer to the *Mary Rose* Trust. Using some staves of English yew but mostly staves from Oregon, he began to make Mary Rose-type bows (MRA1–3; Fig. 8.37). He was joined later in making approximations by John Cave of Ludlow, also using Oregon timber.

Oregon timber was felt most closely to resemble the type and quality of the timber imported from the Continent in such enormous quantities throughout the Middle Ages, and to have grown in roughly similar climatic conditions to trees grown in Italy, Spain, Austria, Switzerland, Poland and Scandinavia (ie, timber growing through colder winters and hotter summers than are usual in the softer climate of the British Isles). Pragmatically, yew from this source was also used because of the paucity of available Continental yew.

The first bow was made from an Oregon yew stave donated by Don Adams, to the overall design features of the *Mary Rose* bows. Under test at the Instron testing facility the bow weighed 102.8lb/46.6kg at 30in (762mm) draw-length. The computed draw-weight is in remarkably close agreement at 102.4lb/46.5kg. This lends considerable weight to the high figures predicted for the *Mary Rose* bows in Table 8.25. In addition to the static force-draw curve, Kooi has calculated the dynamic properties of MRA1 for arrows of various mass. His results for the efficiency of the bow and for the initial velocities of these arrows are given in Table

*Figure 8.37 Roy King bending fresh stave*

8.26 and velocity figures are reinforced by further trials (Table 8.27). King made three bows of different weights, from 102.4lb (46.43kg) to 135lb (61kg) which Hardy presented to the Simon Stanley to break in (Figs 8.37–8). These were new bows, their vital, verifiable statistics known to the last decimal point.

A straight stave of Oregon yew 75¼in (1911mm) in length was obtained for the manufacture of the '*approximation*'. The yew was at its very best and strongest; it was hard, heavy and rang well when struck. This wood appeared to be stiffer and less flexible than Italian yew, which, if used, might have resulted in a lighter weight bow of the same size, certainly not any more. The bow produced is the ultimate that could be expected of a yew bow of its design, size and weight. The exercise suggested that it would be possible to create a formula which could be used to create a large number of very similar bows.

The sapwood was worked down slightly to remove any small bumps. As the stave was of good quality, there was no requirement to follow the bumps. The sides were worked down, square to the back, to the required finished width, leaving just a little extra at the extreme

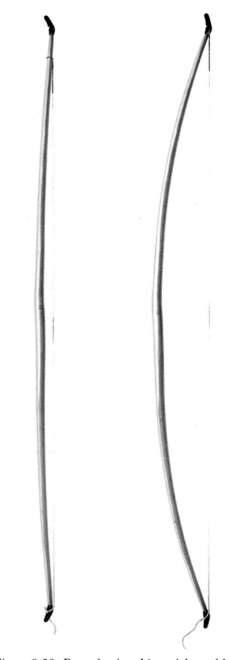

*Figure 8.38 Reproduction A1 straight and bent*

tips. The back sapwood was further worked square and flat to the sides and the back was generously rounded. This resulted in a 'sharp edge'. The outer edges of the back sloped inwards from almost mid-limb, to produce a narrow back approximately 1.4in (6mm) narrower than the limb width. A little rounding was carried out on the back edges. The bow was tapered in thickness and required nothing further to be removed except a few shavings at the centre and tips. In this state it was almost totally inflexible. Shaping the bow (see above) resulted in the profile being too thin, and a section of yew from the over-large handle was glued (9in; 228mm in length) onto the belly.

After 24 hours the belly was worked 'round' and all symmetrical operations of bowmaking forgotten, the

*Figure 8.39 Reproducion A4 on tiller*

wood dictated where scratching and scraping took place. The bow was held in a vice and each limb then pulled back and held behind a stop 7in (178mm) rear of the bow tip. This is equal to a 7in (178mm) brace. At this stage the bow was still very inflexible. A 10st (63.5kg) man just managed to brace the bow in a swivel vice, but only once. The string (made of the terylene fibre Dacron) stretched, then settled down to 5in (127mm) bracing.

The bow was then mounted on a tiller and repeatedly drawn, unbraced, braced and redrawn over time to bend it gradually until the 28in/711mm desired draw-length was achieved (details in archive; Fig. 8.39). A straight edge across the tips illustrates what an 11in (279mm) bracing height would be. The Dacron string was exchanged for Kevlar which proved easier to brace with less stretch. When first drawn to 28in (711mm) it gave 94lb (42.6kg) at 28in (711mm). It was left braced and drawn to 20in (508mm) for 2½ hours. It weighed 93lb (42.2kg at 28in (711mm). It was then left braced overnight for 10 hours, then unbraced for 5 hours, rebraced and drawn to 28in (711mm) and registered 94lb (42.6kg). The weight appeared to be static.

The bottom limb was weakened slightly, for better balance. It was drawn to 28in (711mm) twelve times, then weighed at 91lb (41.4kg) at 28in (711mm), which appeared to be the maximum draw of the bow. It had no chrysals (these are visible lines indicating areas of compression where collapse and fracture are imminent) in it and it could perhaps have been coaxed to 29in (736mm). A 30in (762mm) draw was considered unsafe.

To pull the limbs into shape the bow was left at 26in (660mm) draw on the tiller for 30 minutes and then drawn to 28in (711mm) giving 87lb (39.4kg). The 28in (711mm) draw-length could just be achieved and the bow shape was perfect. The bracing height was still 7½in (190mm). It was left braced for 2½ hours, then it weighed 88lb (40kg) at 28in draw-length (711mm). The bow was sanded and the first coat of sealer applied.

It was then left unbraced for 20 hours, showing a string follow (backward curve towards the belly) of only 2in (50mm). It was braced to 7¼in (184mm) and given ten draws to 80lb (36kg) on the spring balance. When drawn to 28in (711mm) six times it gave 91lb (41kg). When it was quickly drawn to 30in (762mm), or slightly less, for three draws 102lb (46kg) was registered and the belly still appeared to have no chrysals in it (a feature which plagues American yew).

The bow was left unbraced for a further 18 hours and then braced and weighed to give 92lb (42kg) at 28in draw-length (711mm). Still braced, it was left for a further 3 hours and gave 92lb at 28in (42kg x 711mm). At this point, horns were fitted with 6ft 2½in (1841mm) between the underhorn grooves.

The bow was then left unbraced for 5 days and then braced to 7¼in (184mm). It was then drawn to 28in (711mm) ten times and then ten times to 30in (762mm). Weights were 92/93lb at 28in, 104/105lb at 30in (42–42.2kg and 47.2–47.67kg respectively). The belly showed no chrysals. It was then polished and finished. The 105lb (47.67kg) seemed to be the stable draw-weight at 30in (762mm). However the bow would be safest at 28in (711mm) arrow length. Although it could be drawn to 30in (762mm) and appeared still to retain flexibility, it would probably not stand repeated use. Beyond this draw it would probably break (Fig. 8.40).

## Velocity, range and impact

Much remains to be done on the *Mary Rose* bows themselves, although their general features are now clear. With an average length of 6ft 5½in (1968mm) these are suitable for archers 5ft 11in (1.8m) tall. This puts them at the very top of the height range for the 104 *Mary Rose* men whose stature could be gauged (see *AMR* Vol. 4, chap. 13). A draw-length of 30in (762mm) is appropriate for such a height and this implies an average draw-weight of 130lb (59kg) with a spread of 95–165lb (43–74.9kg). In addition, two very large bows are estimated with draw-weights of 185lb (84kg) at 30in (762mm), if they could be drawn so far, although 172lb (78kg) at 28in (711mm) seems a more likely maximum for them in view of the acceptable strains in the outer tensile fibres (see Fig. 8.27).

*Figure 8.40 Simon Stanley shooting the reproduction (photographs supplied by R. Hardy)*

Two types of arrow are appropriate for these bows. A light arrow with a short barbed head (Ward-Perkins Type 16/Jessop M4), with a draw-length of 28–29in (711–736mm) and a total weight of 40–45g, and a heavy long bodkin (arrowhead specifically designed to penetrate plate and mail; see Fig. 8.71 below) with a draw-length of 30–31in (762–787mm) and a total weight of 70–75g.

Trials have been undertaken using *Mary Rose* '*approximation*' bows with arrows made of different types of wood and using several types of head. These suggest a release velocity of 55.8–64.6m per second at ranges of 226–244m (heavy/light arrows; see Table 8.27).

A level of uncertainty in the predicted draw-weights arises from lack of knowledge of the elastic modulus of individual bows; further measurements of density could help lessen this uncertainty. No allowance has yet been made for the shape for the bow limb before bracing, although many of the bows have a noticeable reflex shape. This reflex may be a natural consequence of the internal stress distribution in the growing tree which is released when the stave is cleft from the trunk. The increase in draw-weight for those reflexed bows found on the *Mary Rose* might be as much as 10% over that of a similar straight bow.

The extent of the agreements of the dynamic properties such as initial velocity, efficiency and range, need further consideration. If these are capable of prediction to the same sort of accuracy then the true behaviour of the medieval longbow at the end of its period of greatness will be better understood.

Recent tests have concentrated on tracking arrows shot from *Mary Rose* approximations (and other bows) using Doppler radar and deducing impact force and thence penetration possibilities. Using an approximation with a draw-weight of 150lb (68kg) at 32in (812mm) for the majority of the tests, these obtained release velocities of 52.28–70.07m per second at ranges of up to 328m (Strickland and Hardy 2005, 409, quoting A. Crowley) (Table 8.28).

The tests are designed to include various distances for ranges of arrow flight at various bow draw-weights versus arrow weights and dimensions; speed of flight

**Table 8.27 Experimental shooting data: arrow mass, efficiency and velocity**

| Mass (g) | Efficiency (%) | Initial velocity (m/sec) |
|---|---|---|
| 38.88 | 49.4 | 59.2 |
| 42.27 | 52.2 | 58.4 |
| 58.32 | 64.9 | 55.4 |
| 87.48 | 78.0 | 49.6 |
| 116.64 | 84.4 | 44.7 |

Table 8.28 Range and initial velocity tests of MRA4 Harlington (22/07/1993)

| Arrow | Round 1 distance (m)/ time (sec) | Round 2 distance (m)/ time (sec) | Round 3 distance (m)/ time (sec) | Average distance (m) | Round 4 distance (m) | Initial velocity (m/sec) |
|---|---|---|---|---|---|---|
| Poplar, ½in diam., light bodkin 42.5g | 245 / 9 | 246 / 8 | 239 / 8 | 244 | 194 | 64.6 |
| Ash, ⁷⁄₁₆in medium bodkin 56.7g | 245 / 7 | 245 / 7 | 235 / 7 | 242 | – | 62.2 |
| Ash, ½in head, ⅞in centre, ⅜in before feathers 63.8g | 234 / 8 | 222 / 8 | 222 / 8 | 226 | – | 55.8 |
| Ash, light head | 255 / 8 | – | – | – | – | – |

Round 4 was undertaken in the afternoon after the bows had been sitting in the sun. An increase in temperature may account for the 4–5% reduction in range. The wind at ground level was 8–10 mph but the arrows were wobbling/flirting in the wind at the top of flight at 61–64m above ground. Wind speed increases with height above ground

figures; figures for the success and failure of target penetration versus bow weights, arrow weights and dimensions, arrow fletchings, and arrowhead design against targets of various compositions, thicknesses, shapes and angles of strike. In all tests it is difficult to simulate medieval attack and defence properties; furthermore many other tests are funded by television which, sadly, can result in the visual requirements overriding the scientific. For example, targets manufactured to show whether medieval plate armour could be pierced by longbow arrows need to be manufactured of the same materials, in the same manner and to the same specifications as the original, otherwise the results are meaningless. The target also needs to be properly supported, as it would have been by a human body, but allowed to give and so absorb a good deal of the striking force before penetration.

Simon Stanley trained himself to these huge bows and began to discover extraordinary things: arrows of the weight and type that we know were used throughout the late medieval period could not be shot with good effect from bows weighing less than about 46kg and the heavier types of arrows, weighing 100g or more, demanded bows in the 60–75kg range (132–165lb). With this kind of partnership, arrows of 100–115g shot from such bows would fly at least as far as 220m. Test shooting has enabled us to assess that such arrows from such bows would, over this range, only lose some 16% of their initial velocity, regaining at extreme range some of the speed lost during the middle and height, of their flight, and thus regaining some of their power of impact. This is because of the element of gravity added to their downward trajectory (Strickland and Hardy 2005, 408–14).

Arrows that are too light for a bow will either break on release, the shaft not being able to withstand the force imparted by the bowstring, or, if they survive, will be deflected from true flight and proper speed; their important initial velocity will be reduced. It seems probable that, by the time plate armour was in partial use such as at Crécy and Poitiers, the mid-fourteenth century arrow weights took a leap, and bow-weights with them. By the time of Agincourt in 1415, when plate armour was more complete, the necessary weight of arrows to achieve dangerous penetration must have been up to 4oz (113g). Charles II, 250 years later, is reputed to have complained: '*No one is left to shoot a quarter-pound arrow!*' (Strickland and Hardy 2005, 26).

An arrow tends to strike from above and, to reach a vital organ, would need to have great penetrative force. Experience showed that the greatest effect was had at such ranges as 100yd, 80yd, 60yd, 40yd (91.4m, 73.2m, 54.9m, 36.6m) down to point blank (20yd/18m). At 18m heavy arrows from heavy bows would be able to penetrate most targets, as long as deflection did not cause them to fly off (Strickland and Hardy 2005, 412). Deflection became more and more likely as part-plate and whole-plate armours developed. The arrowhead was shortened, thickened and made heavier, and shaped so that it had cutting edges, usually three or four in a trihedral or tetrahedral head. As the head became heavier, a thicker and heavier shaft was needed to carry it and, of course, a heavier bow to discharge the arrow. The heavier, thicker shaft requires a bigger socket to the arrow. In response to the need to penetrate mail came the Type 16/M4, a much narrower two-sided, bladed and barbed head, ballistically much more effective and able to penetrate mail (see Fig. 8.71). At short range it could be effective against plate, deadly at almost any range against lightly armed men and fearsome against horses.

If weight estimates for the *Mary Rose* bows are accurate then we must question what sort of men were able to master them. The greater part of the men who provided the archer corps from the reign of Edward I to that of Richard III (1272–1485) were drawn from those who worked on the land. They were men who, in their daily lives, were accustomed to unremitting hard physical labour. They were used to hard living conditions, to surviving cold, flood, disease, times of dearth, but they also enjoyed times of plenty, a ruthless mix of Nature's riches and privations. It is possible that their metabolism was more efficient than ours; if so it would mean that they converted their diet more immediately into strength and energy. Together with hard and frequent practice with the longbow they would become strong and efficient archers.

Gradually the use of the bow declined in favour of the handgun. The population was falling. There had been bubonic plague in the 1540s and '50s and a '*sweating sickness*' seems to have killed about a third of the able-bodied population. There was a lowering of living standards, a general loss of sheer physical strength among men '*pinched and weaned from meat*' (Strickland and Hardy 2005, 30–1). Physical strength, maintained by good diet and the robust work of old agriculture, was a necessity for the men who used the great bow. Even the wages of the archers were whittled away by inflation. The social changes that were taking place, the inflation, the extravagance that was much complained of, the reduction in the size of households and the loss of belief in the ideas of duty and service, all worked together against the continuation of great numbers of strong and willing archers. Evidence of childhood malnutrition has been observed by Stirland in the skeletal remains of a number of the *Mary Rose* crew:

> '*Given their average ages this is hardly surprising. Most of them will have been born in the 1520s, a time when famine was periodically widespread … in particular the severe winter famine of 1527–28*'. (Strickland and Hardy 2005, 30; see also *AMR* Vol. 4, chap. 13)

She points out that most of the bones affected by osteomalacia were well healed and that in general the young men seem to have been relatively healthy.

The exceptional state of so many of the bones afforded evidence of high muscle and tendon activity. Stirland's detailed analysis of the human remains from the ship indicted the presence of a skeletal anomaly in 26 shoulder blades, including ten pairs, that seem most likely to have resulted from a repeated activity that put great stress on the muscle attachments of the shoulder (see below for further discussion). These findings led Stirland to suggest the presence of a group of men '*composed of specialist or professional archers*' (*AMR* Vol. 4, 537). Like the majority of men on board for whom age

*Figure 8.41 Nock: a) the original nock and a replica; b) bow tips showing lighter area with angled tip (81A3952, 81A1647)*

could be determined, these were young adults (aged 18–30) of an average height comparable with that of modern Europeans (*AMR* Vol. 4, 520).

A study of the growth rings on some of the *Mary Rose* bows (see Fig. 8.35) suggests a denser and therefore stronger wood than the Oregon timber used in the approximations, probably due to a longer growth period, It is likely, therefore, to have produced even higher draw-weights than predicted. This conclusion, which means the best European yew is stronger than that from America, and future tests should be undertaken using this timber.

## Horn nock
by Alexzandra Hildred with the late Roy King

The 'nock' is the place where the string is fitted both on the bow and on the arrow. Almost always a horn tip is placed on the bow tip with a groove cut into it to accept the string (Fig. 8.41a). It will also protect the bows delicate tips, particularly on the lower limb, from the strangulation damage caused by a self-tightening timber hitch. Ascham states (1545, 104) that '*you must look that your bow be well nocked, for fear the sharpness of the horn sheer asunder the string*'. It seems likely that all the *Mary Rose* bows were nocked and that these were made of horn (see above). On an arrow, an insert of horn may be placed into the nock slot to prevent the arrow splitting. A number of these remain *in situ*, protected within

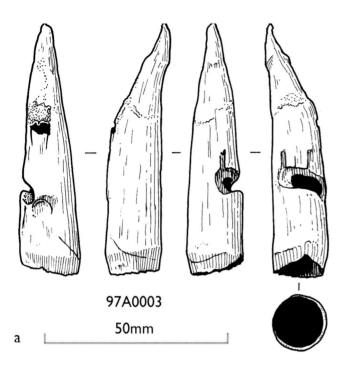

the wooden nock slot. The majority of these had been further protected from seawater because they were contained in elm chests. Horn nocks on the tips of the bows would not be as well protected as those on arrow shafts and their absence is, therefore, understandable.

In 1997 a single horn nock (97A0003) was recovered from concretion in U5 (Figs 8.41–2a), finally proving that nocks existed. This area has other archery related material and the concretion was thought to contain corroding armour around a human pelvis (81H0097) and included a leather strap (97A0004) as well as indeterminable corroded iron (97A0005). One bow was recovered from the area (79A0812) with another from U5/6 (81A5877).

The nock is carved from the natural tip of a cow horn and is badly degraded. It is 70mm long with an internal diameter of 14mm and wall thickness of 1–2mm. The tip is eroded but intact and tapers at an angle to 2mm. There is a single angled groove on one side to take the bowstring, corresponding to about a quarter of the circumference of the nock. This groove is 4–8mm wide and 16mm long. There is also a small hole at the tip. It is a '*side-nock*', reinforcing the suggestion, made as early as 1983, that the presence of a groove in the wood on one side only on most of the bow tips indicated the use of side-nocks, with the string scarring the wood beneath the horn. The authors consider that the nock also assists in the stringing of the bow by enabling the use of a larger string loop, which in turn allows the string to be slipped further down the bow (see working capabilities above) and is especially suited to heavy bows. The top groove on bows with side-nocks is generally cut to the archer's left hand side when shooting the bow. The string then takes a path diagonally across the bow to the opposing bottom groove on the bow's right-hand side. This feature was observed to be of Scottish tradition as late as 1801 (Roberts 1801). Presumably, as bows were made smaller, the requirement for the side-horn was reduced and generally became obsolete.

A number of *Mary Rose* bows show a small 45% angled flat to the extreme tip of the wood that would have been covered by the nock ('*underhorn*') (Fig. 8.41b). One interpretation is that this shows that the limb-tip had protruded through a tapered hole within a 'forward sloping' horn nock. On more modern examples of bows, in some instances the hole in the horn has punctured the tip. This could be a '*purge hole*' to enable the escape of air when the horn is glued to the bow. A simpler explanation is that is merely a by-product of the raw material, some horns being more naturally curved than others. The extra depth of the drilled horn in the heavy *Mary Rose* bows means that bursting through the tip may be a possibility.

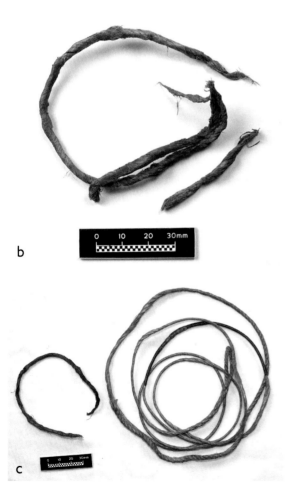

*Figure 8.42 a) Nock 97A0003; b) possible bowstring (03A0166); c) the longest fragment of 03A0166 beside a modern bowstring (single-looped hemp string, Artemis Archery Collection, courtesy of Hugh Soar)*

A copy of the *Mary Rose* horn was made by Roy King and tested for fit on eighteen bows. It fit very well on one bow (81A1614), from a closed chest in O9, and relatively well on two others (81A1647 from the same chest and 81A3969 from a chest in O8). Four more showed a good fit to one end, but were either too loose (81A3947, 81A3943, 81A3946) or too tight (81A3944) on the other.

**Method of manufacture**

The bowyer would have a collection of finished horn tips and would probably take out two that seemed visually to be a pair and fit them to the bow. This would entail altering the wood of the bow tip to fit the horn, as can be seen in the paring down of bow tips. When the hole in the horn was not true to centre, the bow tip would be cut to suit and to bring it back in line. A number of glues could have been used, most likely hide glue which is water soluble.

## Bowstrings
by Alexzandra Hildred and John Waller

> 'The bow and arrow contribute but one complete instrument; of which each is a component part. The same observation, indeed, applies to the string'.
> (Roberts 1801, 129)

**Function**

The function of the bowstring is to draw back the arrow and bend the wooden bow so that, when released, the bow flexes back to its original shape and the arrow will fly a specific for a specific distance based on the draw weight of the bow, the weight of the arrow, the wind conditions, and the weight of the bowstring.

The *Anthony Roll* lists six gross of bowstrings (864) for the 250 bows listed, providing an average of 3.4 strings per bow. A 200mm portion of a possible string has recently been identified (03A0166) and is currently being analysed. It is made of a number of hair-like fibres twisted together (Fig. 8.42b). It has a maximum diameter of 3mm which fits the horn inserts on the arrow nocks, although if served (see below) it probably would not. It was found during the remote excavation of a series of trenches into the spoil moulds to assess the potential for remains therein in 2003 (all spoil was diverted to the surface and sieved). The location was port spoil midships (50.45.7978N 001 06.2675W). Other objects recovered during the same exercise included a ballock knife (03A0104), a comb (03A0094), two aiglets (lace-ends) (03A0096, 03A0099) and an arrow fragment (03A00125). Two fragments of barrel were also recovered, 03A0152 (hooping) and 03A0116, a stave. Figure 8.42c shows this beside a nineteenth century hemp bowstring to demonstrate the similarity in form. Inventory evidence suggests that these would probably have been shipped in barrels, a full barrel containing ten gross, or 1440 strings (Starkey 1998, 5132, 5877, 4494). This means that the strings allocated to the *Mary Rose* would fit comfortably in a single barrel, possibly even a demi barrel. Casks were distributed throughout the Hold of the ship, with a concentration in H7. This area also contained nine fairly complete skeletons with clothing and personal possessions, including two wristguards (see Fig. 8.1 and below). It is not unexpected that archers would carry one, if not two, spare strings about their person (Smythe 1590, 70). The closest arrow chest was found in H9, about 6m aft.

The strength of the string is a vital element in the understanding of the operational mechanics of the *Mary Rose* bows, as used. The inter-relationship of the length and profile of both bow and arrow, the density of the woods used, the type, length and height of the fletchings, and the form of the arrowhead (therefore the weight), together with the strength and weight of the string and the strength and skill of the archer determine the functional capability of the weapon. Recent experiments in the making and using of approximations based on the *Mary Rose* bows, and a variety of statistical studies, have been undertaken as discussed elsewhere in this chapter, but in very few of the published results of these projects is there any real discussion of the strings. The exception to this is Soar (2006, 153–9). Many of the tests are made using man-made materials of different specifications to natural fibres. Yet the string is potentially the weakest link within the weapon – it is a well known fact that a poor or inadequate string can either break itself or cause the bow to break, both render the weapon useless.

There is ample literature to support this statement; however we need look no further than Ascham (1545, 102–3):

> 'The next indeed; a thing though it be little, yet not a little to be regarded. But herein you must be content to put your trust in honest stringers. And surely stringers ought more diligently to be looked upon by the officers, than either bowyer or fletcher, because they may deceive a simple man the more easilier. An ill string breaketh many a good bow, nor no other thing half so many. In war, if a string break, the man is lost, and is no man, for his weapon is gone; and although he have two strings put on at once, yet he shall have small leisure and less room to bend his bow; therefore God send us good stringers both for war and peace. Now what a string ought to be made on, whether of good hemp, as they do now-a-days, or of flax, or of silk, I leave that to the judgement of stringers of whom we must buy them ... Great strings and little strings be for divers purposes: the great string is more surer for the bow, more stable to prick withall, but slower for the cast. The little string is clean contrary, not so sure, therefore to be taken heed of,

*lest with long tarrying on it break your bow, more fit to shoot far, than apt to prick near; therefore, when you know the nature of both big and little, you must fit your bow according to the occasion of your shooting.'*

The bows on the *Mary Rose* were government issue, and there is no archival evidence to suggest that different types or even sizes were intentionally produced. This is in conflict with the assemblage, which includes a number of slab-sided bows that are significantly larger than the rest, and a range in draw-weight which might require different sized strings (though this seems unlikely as it is never mentioned in any historical reference). Whatever their range and draw-weight, one therefore is led to the assumption that either strings of differing sizes were supplied, which would be difficult to manage and is not suggested by any of the inventories, or one size of string would have the strength to stand in all the bows – but this seems equally unlikely. Perhaps the observed differences within the bows were just an accepted variation within the weapons system, or perhaps it points to the fact that the bows are not as heavy as calculated, although the big bows are 'big' bows.

*'In stringing of your bow (though this place belong rather to the handling than to the thing itself, yet because the thing, and the handling of the thing, be so joined together, I must need sometimes couple the one with the other) you must mark the fit length of your bow. For, if the string be too short, the bending will give, and at the last slip, and so put the bow in jeopardy. If it be long, the bending must needs be in the small of the string, which being sore twined, must needs snap in sunder, to the destruction of many good bows. Moreover, you must look that your bow be well nocked, for fear the sharpness of the horn sheer asunder the string. And that chanceth oft when in bending, the string hath but one wap to strengthen it withal. You must mark also to set your string straight on, or else the one end shall writhe contrary to the other, and so break your bow. When the string beginneth never so little to wear, trust it not, but away with it; for it is an ill saved halfpenny, that costs a man a crown. Thus you see how many jeapardies hangeth over the silly poor bow, by reason only of the string. As when the string is short, when it is long, when either of the nocks be naught, when it hath but one wap, and when it tarrieth over long on. … But again, in stringing your bow, you must look for much bend or little bend, for they be clean contrary. The little bend hath but one commodity, which is in shooting faster, and farther shoot, and the cause thereof is, because the string hath so far a passage or [before] it part with the shaft. The great bend hath many commodities; for it maketh easier shooting, the bow being half drawn before. It needeth no bracer, for the string stoppeth before it come at the arm. It will not so soon hit a man's sleeve or other gear, by the same reason. It hurteth not the shaft feather, as the low bend doth. It suffereth a man better to espy his mark. Therefore let your bow have good big bend, a shaftment and two fingers at the least, for these which I have spoken of.'* (ibid., 103–4).

**Material**

The material and size of the string is of great importance in understanding the performance of the bow. The string is a controlling factor in the strength that can be asserted by the bow which, in turn, affects performance. Ashcam and Smythe, both writing in the sixteenth century, advocate hemp made of the longest strands, twisted very tightly and rubbed afterwards with a water glue (Roberts 1801, 126, quoting Smythe 1590). More detail is provided by *Noire in L'Art d'Archerie* written in 1515 (Elmer 1946, 266):

*'Strings should be made of silk or fine female hemp, which is finer than male hemp. They should be gummed and not glued.'*

The native country of the hemp plant (*Cannabis sativa*) is thought to be temperate Asia, probably near the Caspian Sea, although it is found throughout Europe, including in England (Hartley 1986). Some of the finest hemp still comes from the province of Piedmont, Italy, and this was recognised by 1792 (Elmer 1952, 267, quoting Mosley 1792):

*'The most general material of which Strings are now made in England, is hemp; of which the Italian answers the best; this substance possesses many advantages over all other sorts.'*

Although there are references which suggest that English hemp should be used (Oxley 1968; Soar 2006, 157) and, during the mid-fifteenth century the Tower was supplied with bowstrings made of English hemp (Richardson pers. comm., 2006). However trade with the Baltic countries, through the Hanse, and with Italy was routine and a substantial portion of yew longbows were sourced abroad; therefore the importation of strings is a possibility, although there is little documentary evidence to date (Wadge 2007, 232). In the later Middle Ages Flanders was a major centre of production of bow and crossbow strings. Mons and Anvers were famed for their hempen crossbow strings, many of which were exported, however Italian hemp was still considered one of the best to make strings, having long and fine yet strong threads. European hemp can reach great heights, up to nearly 5m, which makes it ideal for bowstrings as they can be made of fibres which run from one end to the other without a break.

This may be the major distinction between hemp and linen, which has shorter strands which require twisting together to form a thread to produce the required length for a longbow string. It seems likely, therefore, that long-stranded hemp was used for the *Mary Rose* bowstrings.

Other materials than hemp or linen have been suggested, namely nettle fibre and silk. Nettle fibre is strong and thread has long been made from retted nettles (steeped in water to season). Fifteenth century sources allude to a specific '*bowstring hemp*' called lilachae, a cross between nettle and hemp grown in England especially for bowstrings (Hardy, pers. comm. 2006), and indeed hemp nettle is still found in England today (*Galeopsis tetrahit*). Silk is also strong enough although it stretches more than hemp or linen and, in the sixteenth century, appears to have been sometimes used for strings on recreational bows. The preservation of even small portions of silk in the *Mary Rose* assemblage (the flight bindings on the arrows for example) suggests that some vestigial remains of silk bowstrings might have survived had they been present. There is a plentiful assemblage of surviving rigging and cordage on the ship which suggests that the lack of bowstrings is the result of either their location – perhaps in a barrel on the open Castle deck – or because any applied glue or gum may have altered the preservation qualities of the hemp – just as the appliance of tar aided with the preservation of the rope. The small string tentatively identified as a bowstring, has, in particular, the compactness of the twist and appears to be treated with some form of 'wax' or other substance to bind it together, which is needed for a bowstring but not so vital for a rope made up of a number of more loosely twisted fibres (Fig. 8.42 b, c).

A relatively thorough study undertaken by Maurice Taylor in 1940 calculated the strength coefficient for 40 materials including gut, silk, fourteen different mechanically spun linen textile threads, India hemp, Manilla hemp and jute (Hickman *et al.* 1992, 257–8). Silk had the best strength coefficient followed by linen; however silk is subject to excessive stretch whereas linen is not. A strength co-efficient ('*strength of thread*' or '*strength at rupture*') was calculated (see Hickman *et al.* 1992 for details) and showed that linen was the best choice for bowstrings as it retained a high rupturing strength (over five times greater than silk) without permanent stretch. However, Elmer (1952, 285) noted that the hemp studied by Taylor was of an inferior nature and that a comparison should be made with true hemp of the best quality which, he felt, would be comparable with, if not superior to, linen.

### Form and fit
Self evidently, any string would have to be of sufficient length to reach between the limbs of the longbow. Documentary and pictorial evidence suggests a fixed

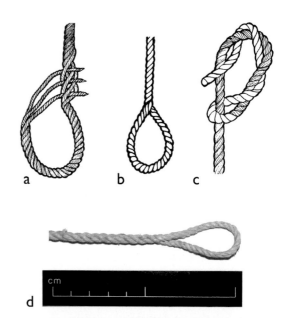

*Figure 8.43 Bowstrings: a, b, c) fixed loop; d) bowyers knot; e) strung bow with knot, side view*

loop at one end with the other end free. This would be tied with a timber hitch (the traditional bowyer's knot) which would tighten as the bow was drawn (Fig. 8.43). Recent shooting experiments using single-slotted side horns similar to the one found on the *Mary Rose* have suggested that using timber hitches on both ends prevents slippage and undue pressure from the string on one area of the bow tip; the knot spreads the pressure

Table 8.29 Measurement of horn nocks from arrows (metric)

| Object | Width (1) | Width (2) | Length slot | Length insert | Width insert | Thickness insert |
|---|---|---|---|---|---|---|
| 81A0991* | 2 | 2.8 | 8 | 48 | 8 | 1 |
| 82A2229 | 2.8 | 3 | 6 | 47 | 7.8 | 1.5 |
| 82A4430 | 2.8 | 3 | 5 | 30 | 8 | 1.2 |
| 82A4431 | 3 | 4 | 5 | 32 | 8 | <1 |
| 82A2232 | 2.8 | 3 | 6 | 41 | 8 | 1 |
| 82A4461 | 2.2 | 3.8 | 7 | 35 | 8 | 1.2 |
| 82A4497 | 2.8 | 3 | 6 | 30 | 9 | 1.3 |

Width (1) = between prongs inner end; width (2) between prongs outer end)

All examples from H9 chest 83A1761 except 81A0991 from M4

around the entire tip (Hardy 2005, 13). However, as a self-tighening knot, replacing a string in combat held with this knot at both ends might be difficult.

The string is made up of a number of fibres or threads, twisted and possibly glued or waxed. Depending on fibre length, these could be single fibres (hemp) or spun threads (linen). An area near the centre (the nocking point which makes contact with the arrow) might be wrapped ('*served*') with a thinner thread to strengthen it, ensure a good fit in the arrow nock, protect the fingers and act as a 'location guide' for the correct positioning of the arrow. Indeed, Ascham suggests serving the string but as a bowstring on a warbow would not be used as many times as a string made for a recreational bow, serving may not have been necessary, although Smythe (1590) advocates the practice. An archer-seaman with hardened hands may not have needed finger protection and, with shooting experience, would be able to position his arrow correctly without a guide.

Experience in making strings from spun linen thread suggests that an allowance of about 15% is needed for 'inefficiency of manufacture' and up to 25% for longevity of the string. Most practical archers agree that the tensile strength of the string should be equal to at least three times the draw-weight of the bow. Therefore, for a 50lb (22.7kg) longbow the string must have a tensile strength of a minimum of 150lb (68kg, excluding the 15%) and a 180lb (81.7kg) bow would require a string with a tensile strength of 558lb (253.1kg). Some archers prefer a margin of four to five times the draw-weight to ensure the safety of the bow, and seven to ensure string longevity.

The diameter of the finished string is critical. Contributory factors include: the raw material and initial processing; tightness of the twist; application of glue or wax; and whether or not the string is served. On the basis of diameter and material the maximum number of strands can be postulated, increased slightly by waxing and decreased by wrapping the nocking point to prevent wear. The maximum tension that any material of a specified diameter will take can therefore be calculated, and provide a maximum draw-weight for the bows. However, there are no strings available to test and we cannot be certain that the materials available today have identical properties to those used in the past. It is hoped that the portion of string recently identified (03A0166) may be able to be copied and tested in the future.

### *Mary Rose* bowstring size

We have several ways of duducing string diameter. The first is the size (depth and width) of the horn inserts in the arrow nocks, where the arrow sits on the string (Fig. 8.44a). The second is the size of the slots cut into the wood of the arrows (Fig. 8.44b). A small number of dry horn inserts removed from arrows measure 1.8–2mm ($^1/_{14}$–$^1/_{13}$in) in width and provide the maximum upper limit of string diameter for these arrows (Table 8.29). A study of nock widths and depths on 184 arrows (Jackson 2003) showed a range in slot width of 2.5–5mm ($^1/_{10}$–$^1/_{5}$in) with the majority being 3–4mm (c. $^1/_{8}$in). Of course, an archer can, if required, shoot an arrow with a nock bigger than his string, but not one with a smaller nock. This variation may, therefore, reflect variation in string diameter. However, problems arise in the possible alteration of dimensions as a result of shrinkage or expansion of the wood (and to a lesser extent, horn), the extent of which is unpredictable, so that the observed variation may actually reflect post-depositional changes. It is assumed that the majority of arrows contained horn inserts that have deteriorated. The measurements of the horn inserts (Table 8.29) are, therefore, possibly a more reliable indicator of original size as horn is not as vulnerable to shrinkage or expansion as wood. Those studied suggest a maximum non-compressible string diameter of 2mm, although the 3mm fragment found recently does compress into the smallest horn insert (Fig 8.44e). This is only slightly smaller than the $^1/_{8}$in which many archers would advocate using when using hempen strings and heavy warbows today. The third potential indicator of size is from the slot on the one recovered longbow horn nock (97A0003, Fig. 8.41). Although in poor condition, this varied from 5mm within the groove, to a maximum of 10mm across the outside, allowing for angled edges (Fig. 8.44c, d and 8.42). This is less reliable, as this part of the string is usually thicker towards the ends than at the arrow-nocking point.

Although the available sample size is not statistically viable and there are a number of imponderables, it seems reasonable, on the basis of the apparently conflicting dimensions of the nock slots and the horn inserts, to consider two maximum diameters for bowstrings to fit 2mm and 5mm nocks. It should be noted, however, that there appears to be no inventory documentation to suggest varying sizes in either

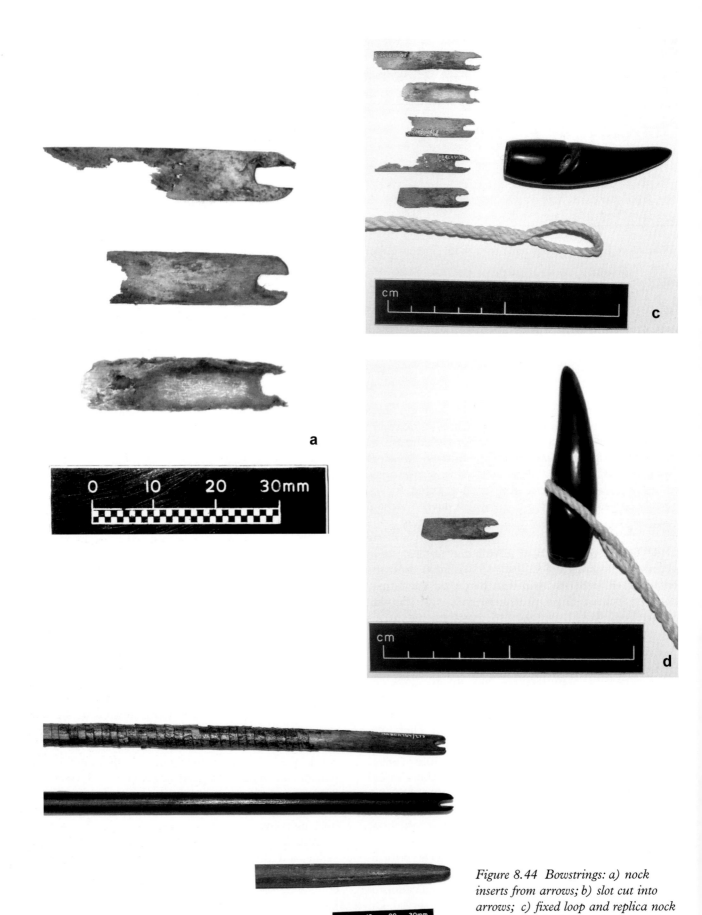

Figure 8.44 Bowstrings: a) nock inserts from arrows; b) slot cut into arrows; c) fixed loop and replica nock and inserts; d) detail string around replica nock

longbows, strings or arrows in fortifications or on board ships but much of the literature concerning the practice of recreational archery clearly indicates that a string should be manufactured to suit the bow (length and density) and the purpose for which it was made: 'You must fit your bow [string] *according to the occasion of your shooting; and to the strength of your bow*' (Roberts 1801, 126).

> 'The thickness of the string must of course depend upon the power of the bow for which it was designed; and the only rule that can be laid down on this head is, that the string should be of sufficient consistency to insure the safety of the bow, which will frequently fly on the failure of the string'. (from *The Archer's Guide by an Old Toxophilite*, 1833, 126).

Even as early as 1544, Ascham advocates two strings: '*little*' and '*great*' (Ascham 1545, 102), so perhaps this is what is reflected within the assemblage. It must be stressed however, that many historical sources are concerned with non-military archery.

Basically a '*great*' string would provide more certainty of shot, better safety of the bow and better penetration, whereas a '*little*' string would provide a longer cast, but be less safe for the string or bow by being long used it breaks the bow' (roberts 1801, 127), or reduces efficiency.

**Matching the strings to the longbows**
Available literature suggests the use for recreational archery of different sized bows with different sized strings and lengths of arrow to suit particular purposes. This equation has also to consider the capabilities of the archer – his strength and technique. This is apparent throughout Ascham (1545), where the qualities of the archer and the components of his weapon are personalised. But how possible could this have been on the battlefield where most of the bows were not owned by the individual but were provided by the state? The 250 listed bows, the arrows and the strings for the *Mary Rose* were not personal items.

Watson's statistical analysis of a sample of arrows confirmed the presence of two groups at lengths of 740mm and 790mm, providing draw-lengths of 28.1in (*c*. 710mm) and 30in (*c*. 760mm), as previously published (Pratt 1992). The 30in arrows formed 80% of the assemblage scrutinised. The significant differences in length are at the tip, but there is also a significant increase in diameter, depth and width of the nock slot on the longer arrows. It has been suggested that this may reflect the use of larger strings to shoot the longer arrows. As described above, recent longbow studies reinforced the conclusions of earlier work (Hardy 1992, 197) suggesting two distinct groups: termed '*fine*' and '*robust*'.

Overall, then, assessment of the collection suggested just two different 'types' of longbows (fine and robust), provided with two differing arrow lengths and perhaps potentially different diameters of string based on the variation in nock slot size. However a small string will sit in a larger nock slot and there is no archival evidence to suggest any variation in bowstring size. The one potential string (3mm) does compress into the smallest horn insert. Whether this observed difference relates to function or skill/size/strength or more directly to bow weight, is extremely difficult to ascertain but the association between arrows of differing lengths in the same arrow bag suggests that it was primarily functional (see below).

We cannot be absolutely certain that there was a large enough dedicated group of archers on board to use the 250 longbows, or whether soldiers or mariners were expected to undertake this role – all men would have undertaken routine archery practice since childhood (see above). '*Archers*' are absent from all crew profiles on vessels of this period and the addition of a large dedicated corps of archers might have a substantial affect on both space and stability, however, it is possible that some archers may have been onboard as part of personal retinues, indentured individuals or those provided by the livery companies (see Wadge 2007, 35). This is an interesting suggestion with respect to some of the decorated wristguards recovered (see below).

The archery equipment listed for the *Mary Rose* in the *Anthony Roll* provides 38 arrows and three strings per longbow which raises questions about the apparent high usage of strings in relation to the number of arrows (>1:12). Even allowing for redundancy and the possibility of breakage (three strings to one bow would seem not unreasonable in combat), this is quite high. Could this reflect the weakness (ie, small diameter) of the string or is this related to function? Were strings routinely changed before they weakened with use? Can the diameters observed be used to suggest the use of lighter strings, for a further cast, used by a non-specialist force at a particular point during a battle, or were these tailor-made for the finer bows with the heavier strings being used with heavier bows at shorter range using shorter arrows for greater penetration? The complication with this is that, on balance, the longer arrows had the wider and deeper nock slots (see Watson below).

There is information on the performance of $1/8$in (3mm) strings, but most often experiments have used linen as export of traditional hemp strings to America ceased after World War I. Early American experiments included those of Saxton Pope (1925, 35). These included making a linen bowstring of $1/8$in from 60 strands of Irish linen (flax) which weighed $3/4$oz (21g). The diameter, when well waxed, was $1/8$in (3mm) and the length 5ft 8in (1727mm). Each strand of thread had a tensile strength of 6lb (*c*. 2.7kg), and the entire string

Table 8.30 String fit to arrow nock

| String | A | B | C | D | E | F | G |
|---|---|---|---|---|---|---|---|
| Description | 3 pieces, fragile | 1 piece with eye 1 end, loose at other | 1 piece with eye 1 end, knot at other | double looped | single loop, poss. side nock | single loop, frayed other end | single loop, incomplete |
| Material/no. strands | hemp | hemp 'blued' ?Belgian | hemp | hemp, ?Belgian | hemp, Scottish or Flemish | linen, 30 strands | hemp, 33 strands |
| Notes | 19th C gents; glued | 19th C ladies; glued | early/mid-19th C | mid-19th C ladies; glued | early 19th C | 180–200lb '*not shot it*'; glued | 70lb; waxed |
| Length | 210, 860, 450 | 1685 | 1690 | 1520 | 1980 | 1875 | 480 |
| Twist | Z, 3 sets, v. loose | Z, 3 sets, tight | Z, 3 sets, loose | Z, 3 sets, tight | Z, 3 sets, loose | S, loose | Z, 3 sets, loose |
| Loop length | 45 | 30 | 40 | 40/4, 30/3 | 45 | 40 | 45 |
| String diam. | 4 | 2 | 2.2 | 1.9 | 3 | 5 | 2.2 |
| Served | for 190 | 690–842 | 720–740, 780–832 | 125, 3 sets | 79–81, 81–83, 83–90 | not served | not served |
| Served diam. | 2 | 2–3 | 1.8 | 1.9–3 | 2.8 | n/a | n/a |
| Fit to nock, string | will squash to fit | will fit | will compress | will fit | will compress | not at all | will fit |
| Fit to nock, served area | fits up to halfway | fits halfway, can push down | both areas fit easily | first set only | not at all | n/a | n/a |

A = nineteenth century Scottish, Artemis Archery Collection; B, D = Artemis Archery Collection; C = Artemis Archery Collection Lacey Loan; E = Col. G. Simpson R.C.A.; F = Gareth Ross; G = John Waller, Royal Armouries

had a breaking point of 360lb (163.3kg); this suggests that it could have been safe with a bow up to 120lb (54.4kg). However, it was made for a 65lb (29.5kg) hunting bow, using the ratio of approximately 6lb tensile strength of the string for each pound of the bow. If industrial spun linen is inferior to raw hemp, which was used to make medieval and Tudor bowstrings, in terms of tensile strength then larger draw-weights must be considered. With strings of up to 5mm diameter, draw-weights of at least 100lb (*c.* 45.4kg) should be feasible and up to 180lb (*c.* 81.7kg) if the string had a strength three times the weight of the bow (see above). The average for the studied sample of nock slots on the arrows was 3–4mm (Jackson, below).

Data on 2mm strings is less easy to find in the available literature, however hempen strings of this size exist in archery collections, many of them for lightweight 'ladies" bows. Seven bowstrings of different materials and form were tested for fit against a copy of a *Mary Rose* arrow nock made in boxwood (Table 8.30), though the tensile strength was not tested nor the strands counted in case of damaging the strings. The maximum width at the top of the nock was just over 2mm and, at the base, 1.8mm. A looser twist to the string will compress and fit into the slot easily, allowing even a larger diameter string to fit. The tighter the twist, the more exact the sizing has to be between the size of string and the arrow nock. Similarly if the area is served, this will not compress, whereas an unserved string will compress into the nock slot. All the strings fitted the reproduction bow nock. There seemed to be huge variability in the size of the individual fibres of different strings. The finest (thinner than fine hair), was observed on an early nineteenth century Scottish gentleman's bow and the coarsest (resembling horse hair) was a linen string (Table 8.30, F). The conclusion was that more leeway for string diameter would be possible if the string was left unserved.

**Conclusions**

Studies of the horn inserts for the arrows and the nocks in the arrow shafts have provided diameters of 2–5mm for the bowstrings. We are uncertain about dimensional changes which may have affected either the arrow shafts or the horn nocks. Currently it is impossible to be certain whether the observed difference relates to two different string diameters, or to one. While standardisation might be expected for large shipments of weapons, the information so far available suggests that either the standardisation was 'looser' than we might accept today, and that the different arrow lengths recovered were related to function; or that there were strings of different sizes with distinct relationships between sizes of longbows, arrows and strings which related either to function, skill/size/strength. Leaving the string unserved would confer greater flexibility in matching string to arrow.

As the diameter of a given material will have a specific tensile strength, it was hoped that this might be used to reinforce the data on the draw-weight of the

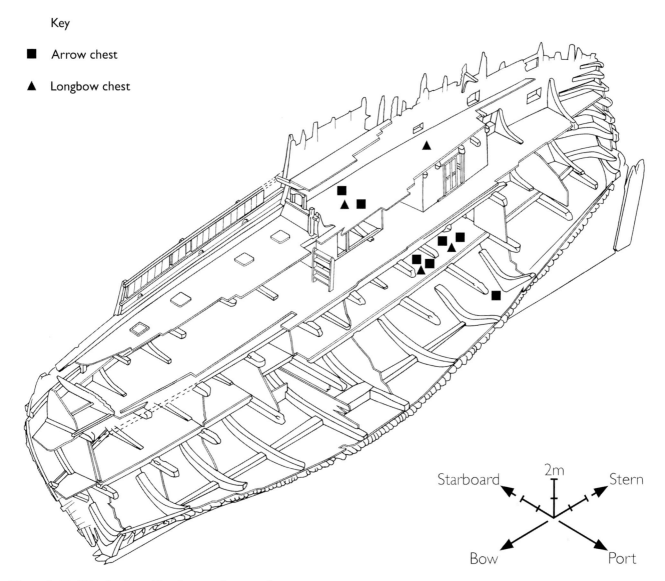

*Figure 8.45 Distribution of longbow and arrow chests*

longbows. This is not possible at the present time. Continuing work analysing the length, mass and spine of the arrows is proving more illuminating (see below).

To date, not enough experimentation has been done using unspun hemp or in testing of tensile strength of European hemp from different geographical areas. As a result we cannot be sure how strong an approximation of a Tudor hemp string of given diameter will be. It is not known where the bowstrings may have been kept on the vessel, but they were probably stored in a barrel. Many casks of various sizes were recovered in areas of the ship where the longbow and arrow chests were found, as well as smaller, stave-built tubs (81A1449 in U8 and 81A1203 in U9) found beneath the empty longbow chests.

The possible portion of a bowstring (see above) derived from one of the storage areas. As yet the number of strands forming this string have not been counted, nor has specific material identification taken place. Residue of glue (if not water soluble) or wax (if originally present) might confirm this tentative identification. At 3mm diameter this fits the arrow nock slots. It may be possible to replicate and test a string of this size.

## Longbow Chests
by Alexzandra Hildred

Four chests for longbows have been identified, two from the Upper deck and two from the Orlop deck (Fig. 8.45). These are simple elm crates manufactured of 19mm elm board nailed together without any hinges or locks (*Mary Rose* chest type 1.1/2.2, see *AMR* Vol. 4, chap. 10). The average dimensions are 2235 x 381 x 357mm, although there is variation. The lid is larger

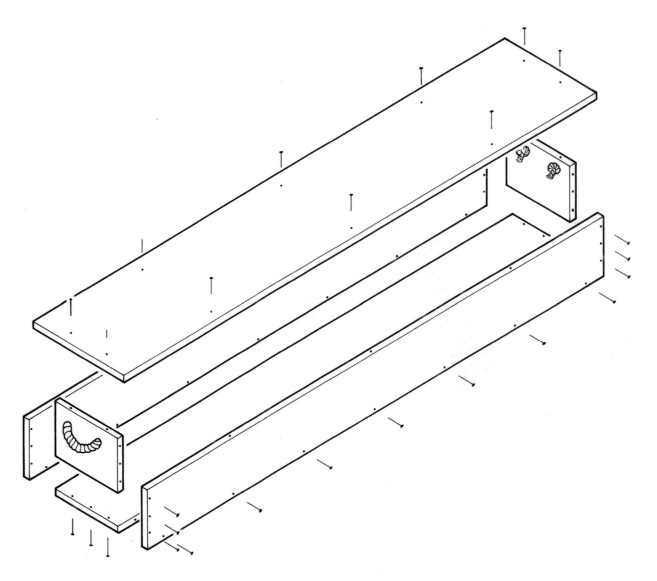

*Figure 8.46 Exploded view of longbow chest construction*

than the base by 25mm, possibly to facilitate levering off. The ends were nailed to the base from beneath, then the sides were nailed onto the ends and finally the lids were nailed to both sides and the ends from above. The ends were drilled with two holes to support simple rope handles held with knots inside the chest (Fig. 8.46). Although longbow chests are not mentioned in the *Anthony Roll*, two chests are listed for bows in 1514 (Knighton and Loades 2000, 142). The inventory of 1547 (Starkey 1998) records many of these and it is clear that this was a common method of storage and transport.

The two chests from the Upper deck in the stern (80A0602 in U7/8, 81A1325 in U9) did not contain any longbows and were found without their lids, although longbows, arrows and arrow chests were found in these areas. Two separate lids were found nearby (80A1044 in U8 and 80A1172 in U7) but it cannot be proven that these were the appropriate lids. This area contained many hand weapons including longbows, staff weapons, handguns and hailshot pieces. This may not relate to an original location, as many items were found above a bronze culverin which had fallen across the Upper deck from the port side (81A3002) and came to rest aligned north–south along the deck. It is possible that some of these were originally on the Castle deck. In either position it seems likely that the chests of longbows had been brought out of storage ready for combat and their contents distributed. Forty-three longbows were recovered from the Upper deck in U7–10 (Fig 8.2).

The first chest found (80A0602, U7/8 Box 5) was not intact, with the sides displaced from the ends. It measures 2190 x 350 x 278mm. Leather (80A0596, 80A0606), a bill (80A603) and human remains (80H25-7) were found close by. Chest 81A1325 (U9, F59) had all its parts, but had suffered damage from a large coil of rope falling on to it. It was lying north–south along the deck. Two longbows were found just north of it but items found inside are likely to have

*Figure 8.47 a) Underwater photograph of bows in chest; b) on deck as recovered*

fallen there (Table 8.31). It measures 1910 x 285 x 304mm, but is damaged. An unidentified object (81A6858) and an ankle boot (81A4750) were inside it.

Two other chests were found in the area identified as the archery store on the Orlop deck: 81A3927 (F159) in O8 and 81A1862 (F28) in (O9). The lids of both were nailed into position. 81A3927 (2240 x 330 x 310mm) contained 50 bows and 81A1862 (2250 x 360 x 280mm) contained 40 (Table 8.31). 81A1862 was aligned north–south along the deck. It was excavated underwater, with the longbows brought up over six dives (Fig. 8.47a). The bows appeared to have been packed in tar (samples 81S297, 81S300, 81S302).

The most exciting chest was 81A3927, aligned nearly east–west and resting on one end. It was lifted with its contents intact. It had been partially damaged by falling against a stanchion which had broken about a third of one side, causing it to collapse, exposing the bow tips. It was lifted by sliding polythene under the chest and bunching the ends together so that it could be placed into a specifically constructed wooden crate, which was then hauled by derrick to the deck of the recovery vessel *Sleipner*.

The chest was excavated on the recovery vessel by removing the bows layer by layer, and observing their orientation (heartwood/sapwood) with respect to the top, base or sides of the chest (Fig. 8.47b). This appeared to be random (the chest had fallen over at some stage and so the sides were originally mistaken for the lid; Hildred, archive report, 9 November 1981). Samples were taken of each layer (81S537–542), in particular looking for the tar packing noted during the underwater excavation of 81A1862. None was seen. The 50 bows were in seven layers, the first four with eight bows each and the remainder with seven, six and five respectively (Table 8.31). One bow retained an unidentified '*fatty substance*' over the ends. All the tips were lighter than the rest of the bows. This was more

### Table 8.31 Contents of longbow chests

| *LONGBOW CHEST 81A1862. (F028) TYPE 1.1/2.2 09. ELM* | |
|---|---|
| Recovered intact, full of longbows. 2250x360x280mm | |
| LONGBOWS | 81A1597-1620 |
| LONGBOW | 81A1622 |
| LONGBOWS | 81A1638-48 |
| LONGBOWS | 81A1654-7 |
| LONGBOW | 81A1767 |

| *LONGBOW CHEST 81A3927. (F159) TYPE 1.1/2.2 08. ELM* | |
|---|---|
| Complete, lidded chest. Had broken open and partially collapsed but contained full complement of 50 bows. 2240x330x310mm | |
| UNIDENTIFIED | 81A3978 |
| LONGBOWS (layer 1) | 81A3928–35 |
| LONGBOWS (layer 2) | 81A3936–43 |
| LONGBOWS (layer 3) | 81A3944–51 |
| LONGBOWS (layer 4) | 81A3952–59 |
| LONGBOWS (layer 5) | 81A3960–66 |
| LONGBOWS (layer 6) | 81A3967–72 |
| LONGBOWS (layer 7) | 81A3973–77 |

| *CHEST 80A0602. (BX05) TYPE 1.1/2.2 U7/8. ELM* | |
|---|---|
| Longbow type. Chest sprung open and incomplete, unlidded though 'spare' lid nearby. Items may have fallen in. 2192x350x278mm | |
| UNIDENTIFIED | 80A0606/1-4 |
| SHOE | 80A0604/1-2 |
| ARROW | 80A0607 |
| BILL | 80A0603/1-2 |

| *CHEST 81A1325. (F059) TYPE 1.1/2.2 U9. ELM* | |
|---|---|
| Longbow type chest with no lid. Some longbows nearby. Objects in chest probably fell in. 1910x285x304mm | |
| UNIDENTIFIED | 81A6858 |
| ANKLE-BOOT | 81A4750/1-3 |

obvious at the eastern end of the chest where the tips had been exposed and the ends had taken up some of the colour from the surrounding silts. Residue was also noted on 81A3948–50.

It is likely that the 50 bows represented a full chest, and that the 250 longbows listed would, therefore, have been supplied in five chests, four of which have been recovered. Elm chest 81A3824, found lidless and broken in the archery store area of O8 and containing a shoe, a knife sheath and some line which many have fallen in, is the only other chest on board of identical construction to the longbow/arrow chests. It measures 2880 x 330 x 350mm and so is considerably larger than the others. It is possible that it accommodated bills, also centrally stored in the Tower of London and possibly therefore shipped in similar crates. A few broken bill hafts were found in this area.

## Wristguards
by Hugh Soar

The purpose of the wristguard, armguard or bracer is two-fold. As an arm-/wrist-guard it serves to take and absorb the impact of the bowstring when released, where shooting technique is imperfect, or when the bow is low or inadequately '*braced*', that is, when the string is too close to the bow limbs (the modern use of the term '*brace*'). The secondary function is to gather and hold loose clothing from the path of the bowstring.

Ascham (1545, 100) describes its use thus:

> '*Little is to be said of the bracer. A bracer serveth for two causes, one to save his arm from the stripe of the string, and his doublet from wearing; and the other is, that the string gliding sharply and quickly off the bracer, may make a sharper shot*'.

However, he does not personally advocate the need for a bracer, so we may assume that it is a desirable, but not essential, archery accessory:

> '*But it is best, by my judgement to give the bow so much bent, that the string need never touch a man's arm, and so should a man need no bracer, as I know many good archers which occupy none*'. (ibid.).

The noun '*bracer*' (brace) derives from the Latin '*bracchia*' (arm), from whence also comes the French '*bras*' and, by extension, '*gardebras*' – alternatively '*wardebras*' (or '*warbrasse*'). It is also the same root that gives us '*embrace*'. Anglicised, the term '*bracer*' has become '*armguard*'. Both '*armguard*' and '*bracer*' are used today. Within the *Mary Rose* assemblage we use the name '*wristguard*', for consistency as this has been used throughout the archive (although we acknowledge that these various terms are not entirely synonymous). The 24 examples recovered form the largest known single, closely-dated collection of post-medieval English wristguards and the only collection recovered from a shipwreck. One decorated leather wristguard has been recovered from close to the keelson/maststep of the late fifteenth/sixteenth century Newport Ship, found in the Usk at Newport, Wales (Fig. 8.48a). Examples are visible on one of the few known illustrations of archers at sea, depicting a battle dated around 1480–90, illustrated in the *Beauchamp Pageant* (BL MS Cotton Julius E. IV, art. 6, f. 18v; Fig. 8.48b).

Archers' wristguards are known from as early as the latest Neolithic–Early Bronze Age (c. 2200–1700 BC), occurring particularly in graves associated with a specific type of pottery ('*Beakers*') and barbed-and-tanged flint arrowheads (eg, Clarke 1970; Fig. 8.49a). Prehistoric examples (that survive) are invariable made

of stone, often of exotic origin or appearance, and may be oval or rectangular in shape, usually with a drilled hole at either end. Their use as functional wristguards, however, has recently been challenged and they may have been, effectively, ceremonial representations of organic examples employed in burials to indicate the social status of the deceased (Fokkens *et al.* 2008). In medieval, Tudor and renaissance England, leather is the most common material, though examples are known in bone, ivory and horn (Fig. 8.49b). Twenty-two of the *Mary Rose* wristguards are leather; one is ivory (81A0815) and one horn (81A2219).

Many wristguards bear some form of ornamentation. Such markings or deliberate decoration of any artefact, whether for civil or military use, may be for a variety of reasons; to denote function, ownership, hierarchical rank, badge of office, affiliation (a 'trademark'); or they may be symbolic – purely for

*Figure 8.48 a) Replica of the Newport ship wristguard ©Hugh Soar; b) a medieval sea battle from the* Beauchamp Pageant *(BL MS Cotton Julius EIV, art. VI). Notice the armguard on the arm of the archer on the extreme left (highlighted)*

*Figure 8.49 a) Two stone wristguards found with the burial of the so-called 'Amesbury Archer' (2400–2200 cal. BC; © Wessex Archaeology, photograph by Elaine A. Wakefield); b) ivory wristguard from Southend (© Southend Museum)*

aesthetic or talismanic properties. The symbols used can be secular or religious, simple or complex (see *AMR* vol. 4, chap. 11 for a discussion of marks found on *Mary Rose* objects). Sixteen of the 22 leather wristguards are decorated (see below).

Various aspects of the assemblage have been considered: material; form (including shape and method of attachment), decoration, distribution and associations. Summary information on each is presented in Table 8.32.

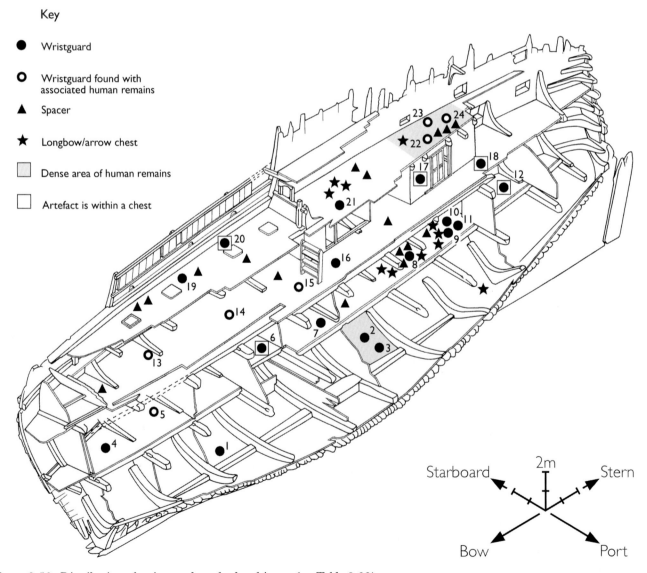

*Figure 8.50 Distribution of wristguards and related items (see Table 8.32)*

## Distribution

Three wristguards were found in the Hold, nine on the Orlop deck, six on the Main deck and six on the Upper deck (Fig. 8.50). Their distribution broadly mirrors that of other archery related objects; the Orlop deck in the stern, where the majority of stored chests of longbows and arrows were found, and the open chests on the Upper deck beneath the Sterncastle. Some were directly associated with skeletal remains (81A1185, 81A1158, 81A4639, 81A0815, 81A146) and others were found in areas where a large number of individuals were found (U9, H7), or at access ways. In eight instances they were found in the same area as leather spacers containing arrows.

Four were found in what are thought to be personal chests on the Main and Orlop decks (Fig. 8.50, nos 6, 12, 17, 18; Table 8.32). Three of these were decorated (Table 8.32). Contents within the chests suggest that those from the Main deck belonged to individuals of high status (see *AMR* Vol. 4, chap. 10 for details). Two wristguards from the Main deck (not in chests) have Royal emblems, one supports a crowned Tudor rose (80A1973) and one bears the Arms of England (81A1460); the latter was found amongst skeletal remains beside one of the two cannon carried, possibly suggesting dual roles for these individuals. The single ivory example (81A0815) is of very high quality and is also from the Main deck (M3).

## Form

The wristguard is both a functional and a personal object. The overall profile is dictated by function – to exactly fit the contour of an arm. Historically English examples have been of three shapes: elliptical, rectangular and a tapered or mitred rectangle. Sizes have varied both in length and width; with width identified as being '*half*' or '*three-quarter sleeve*'. The *Mary Rose* examples range in length from 111mm (81A2670) to 183mm (81A4675); most are 120–150mm (Table 8.32).

Most of the leather ones are rectangular in shape and worn with the flesh side against the arm. In several instances they are made of two identical pieces of leather, one of which may bear decoration. Shape varies: straight-edged rectangles, which can be either ornately decorated (81A1460, 81A4241, 82A0943, 82A2282) or plain (81A1863, 81A2192, 81A2670); with mitred edges (80A0901, 80A1308, 80A1239, 80A1973, 81A1173, 81A2845, 81A4675, 81A5826,

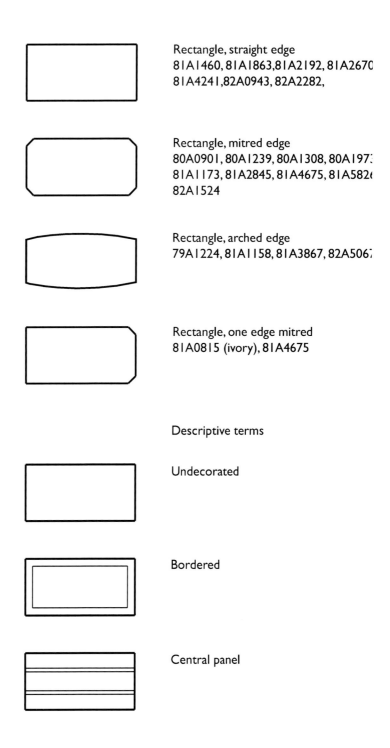

Figure 8.51 Wristguard shapes

82A1524); or with slightly convex long edges (79A1224, 81A1158, 81A3867, 82A5067; Fig. 8.51). Several fragments (81A1863, 81A4639) are of indeterminable form.

Eight have borders (including an undecorated example, 80A1308), defined by a single incised line close to the edge, and thirteen do not. The bordered examples are well finished and do not seem to suggest

## Table 8.32 Wristguards

| Object | No. on Fig. 8.50/ Sector | Material | Shape | Length | Width | Th. | Fastening | Strap 1 dimensions | Strap 2 dimensions | Decoration | Notes |
|---|---|---|---|---|---|---|---|---|---|---|---|
| 80A1308 | 1: H4 | leather, bovine | rectangle mitred? bordered | 144 | 118 | 3 | bifurcating straps, 1 pair | 120x12, knotted end; split 40mm | 61x20; split 45mm, staining from buckle plate | Plain with border, dyed red; no central panel | 6 FCS in area and companionway |
| 80A1239 | 2: H7 | leather, calf | rectangle, bordered | 142 | 118 | 5 | single straps, 1 pair | | | 8 Tudor Roses, 4 each side of central panel | lying on ballast; dense human remains; 6 FCS nearby |
| 82A5067 | 3: H7 | leather | rectangle, mitred | 122 | 87 | 3 | single straps, 1 pair | | | plain but cut marks on surface & stitch-holes | |
| 81A4241 | 4: O2 | leather, calf | rectangle, mitred, bordered | 145 | 117 | 2 | Bifurcating straps, 1 pair | 200x18, split 45mm, rivets 10mm from end | 60x18, split 20–30mm; before 20mm buckle plate | Royal Arms, Marian Ave; central panel | ?high status |
| 81A2219 | 5: O3 | horn | frag. only | 70 | | | | | | | high status |
| 81A2845 | 6: O5 in chest 81A2706 | leather, bovine | rectangle, convex, bordered | 147 | 134 | 1 | single straps, 1 pair | | | 10 *fleurs de lys* & 17 irregular, unident. stamp/seal; central panel | 2 lead weights in chest, loose arrows in area |
| 81A1863 | 7: O6 | leather, bovine | frag., rectangle? | 166 | 166 | 3 | laced | | | plain | 2 FCS nearby |
| 82A1524 | 8: O8 | leather | rectangle, mitered | 140 | 120 | 3 | 1 single strap 1 bifurcating | | | gridiron; wiredrawers gauge; crowned eagle or griffin in shield; central panel | archery store |
| 81A2670 | 9: O9 | leather | rectangle convex, bordered | 111 | 115 | 3 | single straps, 2 pair | 60x9 | | ?traces of decoration | archery store |
| 81A3867 | 10: O9 | leather, calf | rectangle, convex | 140 | 120 | 1.5 | single straps, 1 pair | | | *fleurs de lys*; central panel | archery store |
| 82A2282 | 11: O9 | leather, bovine | rectangle | 147 | 115 | 7 | single straps, 1 pair | | | 33 stamps, instruments of the passion; central panel | archery store |
| 82A0943 | 12: O10 in chest 82A0492 | leather, bovine | rectangle | 145 | 120 | 1 | 1 single strap 1 bifurcating | | 205x20–8, split 30mm, rivets 12mm from end | book stamp of bearded faces in profile in foliate border; central panel | |
| 81A0815 | 13: M3 | ivory | | 114 | 67 | 3 | bifurcating straps, 1 pair | | | 2 parallel incised lines | high status |

649

| Object | No. on Fig. 8.50/ Sector | Material | Shape | Length | Width | Th. | Fastening | Strap 1 dimensions | Strap 2 dimensions | Decoration | Notes |
|---|---|---|---|---|---|---|---|---|---|---|---|
| 81A1173 | 14: M5 | leather, bovine | rectangle, convex | 120 | 120 | 3 | single straps, 1 pair | 110x20, 3 eyelet holes | | secular, 15 flowerheads; central panel | skeletal remains, behind iron gun |
| 81A1460 | 15: M6 | leather | rectangle bordered | 146 | 127 | 5 | bifurcating straps, 1 pair | 240x20, 4 holes, split 60mm | 80x20, split 40mm; evid. for 20mm buckle plate, rivet-holes 20mm from end | Royal Arms; Marian Ave | skeletal remains, behind cannon |
| 80A1973 | 16: M7 | leather, bovine | rectangle, mitered | 140 | 100 | 2 | single straps, 1 pair | | | 9 crowned Tudor Roses; central panel | |
| 81A5826 | 17: M9 in chest 81A5783 | leather, bovine | rectangle, mitered | 130 | 115 | 1 | single straps, 1 pair | 80x15 | | secular, 15 flowerheads; central panel | high status objects |
| 81A2192 | 18: M10 in chest 81A2035 | leather, bovine | rectangle | 145 | 135 | 1 | single straps, 1 pair | | | | chest includes low countries glazed pot, razor & whetstone |
| 79A1224 | 19: U4 | leather, bovine | rectangle, convex, bordered | 150 | 132 | 2 | single straps, 1 pair | 95x20 | | 24 Tudor Roses; central panel | spacer nearby |
| 81A4675 | 20: U5 | leather, calf | rectangle, mitered 1 end | 183 | 108–95 | 2 | laced | | | | skeletal remains |
| 80A0901 | 21: U7 | leather, bovine | rectangle, mitered | 142 | 125 | 3 | single straps, 1 pair | | | 20 nimbuses; central panel | spacer in sector |
| 81A1158 | 22: U9 | leather, calf | rectangle, mitered | 123 | 109 | 1 | single straps, 1 pair | 100x15 | | 15 St Peter's keys & sword | |
| 81A1185 | 23: U9 | leather, calf | uncertain bordered | 84 | 54 | 1.5 | uncertain | | | secular, 12 flowerheads; central panel with X & circle decoration | dense human remains |
| 81A4639 | 24: U9 | leather | frag. | 58 | 45 | 1.4 | uncertain | | | 3 sunbursts | |

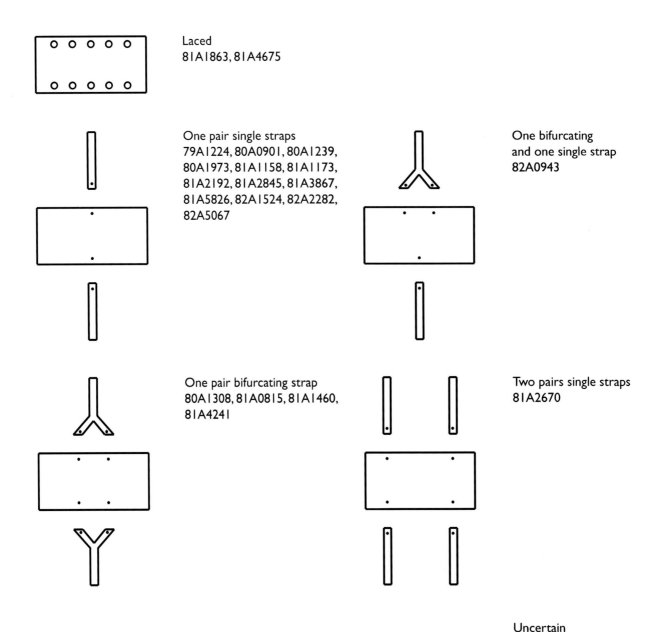

*Figure 8.52 Fastenings on wristguards*

the reworking of other objects into wristguards. Decoration is contained within the border with only the occasional overlap. One, 81A4241, has been reworked by cutting into the borders on the long edges. The nature of the decoration, religious or secular, is not influenced by the presence or absence of bordered edges. One very ornately decorated example, 82A0943, has no border and apparently uses a book cover stamp (see below). Ornamented wristguards often display an undecorated central panel (width 18–42mm) defined on each side by single or double incised lines. This panel corresponds to the area where the bowstring would pass (Fig. 8.51). The exception is 81A1460, which bears decoration on only one side.

Shape does not seem to bear any relation to the degree of ornamentation either; the most elaborately decorated examples (82A0943, 81A1460, 81A4241) are plain rectangles with an undefined central space. One of the most 'finished' and clean-edged examples is plain (80A1308). Despite a varying degree of deterioration to the leather, some suggestion of higher status may be suggested on the basis of form and quality of decoration (79A1224, 80A1239, 80A1308, 81A0815, 81A1460, 81A2845, 81A3867 and 81A4241) or by location (81A5826, within a chest in the Carpenters' Cabin).

## Fastening

*'In a bracer a man must take heed of three things; that it have no nails in it, that it have no buckles, and that it be fast on with laces without agglets'.*
(Ascham 1545, 100)

For effective use, a wristguard must remain in contact with the inner forearm and be held securely in position. Examples from the *Mary Rose* include: single strap; bifurcated single strap; multiple straps; laced. Laced examples use either single or multiple laces or thongs (Fig. 8.52). Many are fragmentary and the fastening type can often only be determined from the relative positions of attachment points. Experimentation has shown that curved or mitred rectangular wristguards with individual straps remain closer to the arm in use than they do if the strap bifurcates. There is no direct evidence for the method by which the strap was fastened on the arm but an elliptical example found on a spoil mound during excavations at Billingsgate, London, retains a socketed buckle, a strap with decorative eyelets and a strap loop to retain the tang (Egan and Pritchard 1991, 229, 231, fig. 143). A painting depicting the Martyrdom of St Sebastian (*Sebastian's Altar*, by the Meister der Hellige Sippe, c. 1483, Richartz Museum, Cologne) illustrates a wristguard which clearly shows a buckle attached to a bifurcated single strap. None of the copper alloy buckles from the *Mary Rose* has been identified as either a wristguard or armour buckle.

Two examples (81A1863, 81A4675) are laced. 81A4675 has five pairs of holes suggesting either a continuous lace or five pairs of individual laces (Fig. 8.52). In practice it is notoriously difficult to tighten a single continuous lace with one hand. 81A1863 is fragmentary, but shows three small holes along one side. Examples fastened by a single strap have one hole in the centre of each long edge (Fig. 8.52). Seven of the twelve recorded retain evidence of a rivet (Table 8.32) (79A1224, 80A0901, 80A1239, 81A1158, 81A1173, 81A2192, 81A5826, 82A2282) (80A1973, 81A1185, 81A3867, 82A1524). Only one example probably had two single leather straps (81A2670). It retains a fragment centrally on one side, with a rivet hole opposite, but there are also holes asymmetrically on each side of the wristguard at one end that could be for an additional strap.

Three wristguards were held to the arm by means of a single strap fastened to each long side which splits into two where it is attached to the body of the wristguard and is held by single or multiple rivets. Three are leather (80A1308, 81A1460, 81A4241) and one is ivory (81A0815). 82A0943 has one bifurcating and one single strap. Of these, 80A1308 is plain, the others carry some of the most elaborate decoration recorded. Fastening variants include 81A2192, where four additional laces may have supplemented a single strap.

On decorated wristguard 81A1460, the decorated portion faced uppermost when worn, the undecorated portion taking the wear of the bowstring. Evidence for the former presence of buckles can be seen on examples with bifurcating straps (Klein, *AMR* Vol. 4, 99–105). One strap is longer than the other (the portion going under the arm) and carries pin-holes. The shorter has

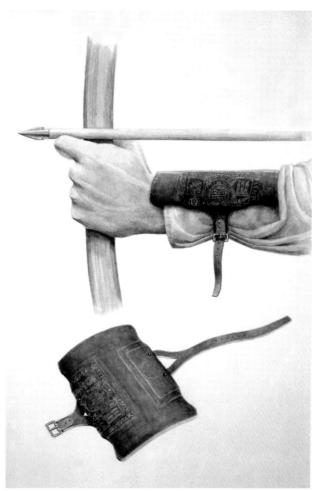

*Figure 8.53 Reconstuction of 81A4341 showing the position of the wristguard on the archer's forearm (painting by Nick Evans)*

two holes for the attachment of a small, flat buckle with plate and all examples show ferrous staining, suggesting an iron buckle (as depicted in Fig. 8.53).

## Decoration
by Matthew Champion

Most wristguards are decorated. All known decorated British sixteenth century examples, with the exception of one in the British Museum decorated with a rose and crown (BM PY1922; Fig. 8.54), show the use of stamped rather than tooled decoration. In many cases the stamped designs have been edged with a single or double line, which has then been further tooled onto the leather. This can be achieved with speed by running a fine blunt object across the surface of the damp leather. The decoration often leaves a central band clear so as not to obstruct the path of the bowstring. Most of the stamps used on *Mary Rose* wristguards appear to have been applied haphazardly. In some cases the stamps overlap or are obscured by the straps (eg,

*Figure 8.54 Decorated leather armguard in the British Museum bearing Plantagenet insignia. This half-sleeve armguard has reputed connections with Henry VI's retinue (BM PY1922: reproduced by permission of the British Museum)*

81A1460; Fig 8.57). Three notable exceptions are 81A4241, 81A4160 and 82A0943.

The use of a similar stamp design on several examples (eg, 81A4160, 81A4241) and of a probable book cover stamp (82A0943) suggests that some may have been produced in the same workshop. It follows that a single workshop was capable of producing both fine quality and 'cheap' decorated leather; the wristguard is not a reused book cover but utilises a book stamp. The leather used to decorate bookbindings is by necessity thinner than that used for wristguards and may be calf or sheepskin (see *AMR* Vol. 4, chap. 3), wheras all identified wristguards are cow leather.

No identical stamp appears on more than one wristguard, although the two with *fleurs de lys* are very similar (81A2845, 81A3867). None can yet be identified with a particular workshop or craftsman. A book cover (81A0895) supports a rondel decoration and also carries the letters 'MD' (Martin Doture) the initials of the bookmaker or binder. The initials are not present on the wristguard bearing part of the same design. It may be that some of the stamps are commonly used on a variety of objects and the stamps may be stock items readily available which cannot therefore be tied down to a particular workshop. An example found in Finsbury Fields, London, and now in the British Museum (BM PY1961, 2–2, 48) is so similar to the *Mary Rose* ones that it suggests a similar date, if not production within the same workshop. The *fleur de lys* stamp, while not exactly the same as either used on the *Mary Rose* examples, is extremely similar. It is possible that all are the work of the same craftsman.

**Leather decoration techniques**
The three main methods (discounting the application of metal fittings) used to decorate leatherwork in the sixteenth century were cutting (carving), stamping (blind stamping) and tooling (embossing). For practical purposes the decoration was only applied on the grain side of the leather.

Cutting involves cutting into the surface of the leather. One side of the cut is then pushed down into the body of the leather leaving the other edge standing proud. This method is normally only found around the outline of lettered inscriptions, for instance on belts and girdles, and, although becoming common in later centuries, it is the least commonly found technique used in this period. As the surface of the leather is cut it tends to allow the damp to seep into the fibres and these gradually expand to fill in the design. The technique is not found on *Mary Rose* wristguards.

Blind stamping is cheap and effective and involves the application of pre-carved stamps to the surface of the leather to leave an impression. This is done when the surface of the leather is damp. The stamp is applied, usually with a blow from a mallet or press (if the stamp is large) and, when the leather dries, the design is set into the surface. There is no obvious advantage to heating the tools (darkening of the leather is natural and not necessarily due to the application of heat).

The surface of the leather is dampened and must remain damp when tooling is used. The craftsman begins to work the leather using similar tools to the goldsmith. Some areas are pressed down, whilst others are lifted forward, fine contours and graduated depths can be achieved with relative ease. The background can be pricked or stippled in, to push it further backwards and leave the design standing proud of the surface. In more complex designs tooled motifs can be either raised up by pressure from the back of the leather or the design can be moulded over some sort of former (embossing). This is more advanced, and more complex. This method is rarely found on mundane items. An example of a tooled wristguard is the British Museum Rose and Crown example (BM PY 1922; Fig. 8.54). This would also have been coloured and possibly gilded. It would not be surprising if some of the *Mary Rose* wristguards were originally coloured with vegetable dye, which has since decayed. The colour most often used is red. The British Museum wristguard shows evidence of gold leaf. This is usually restricted to highly decorated pieces of leatherwork. Although expensive, it is easy to apply. The leather must first be treated with a 'size' to glue the leaf, usually egg albumen. A warm iron can be used to apply the leaf, which is finished by burnishing. Experimentation suggests that the British Museum example would take a day to manufacture, whereas the *Mary Rose* examples would take about 10 minutes to produce.

**Stamps used on the *Mary Rose* wristguards**
All of the decoration was achieved by blind stamping on damp leather. The stamps show a variety of designs and, when used singly or in conjunction with others, have produced some highly decorative effects. Stamps, which

are rare survivals from this period, are usually described as 'book binders' stamps and are fabricated in copper alloy and are generally of a higher standard of workmanship than most of the stamped wristguards.

The stamps used to decorate the wristguards, although distinctive and utilising symbols which may have particular meanings (such as the *fleur de lys*), are often widespread and are merely preferred decorative features. The stamps have been produced to a high quality, but in all likelihood were probably not created purely for use on wristguards. Some are not particularly suited to them and (for example) 82A0943 is very similar to an impression on two books recovered (81A0895, 81A4131). The gridiron is more commonly found on other garments, and even the decorative stamps displaying the Royal Arms may not have been designed exclusively for wristguard decoration. In the case of 81A4241 the stamp extends over the edge of the leather in an attempt to leave the central section free from obstructions, which might impede the passage of the bowstring. On 81A1460 decoration extends into the central portion. In both instances the design is damaged and partially obscured by the application of the straps (see Fig. 8.57). This suggests that the stamps used were merely at hand rather than created specifically.

## The *Mary Rose* decorated wristguards
by Hugh Soar

The subject matter of ornamentation and the degree to which the surface of the object has been decorated varies. The stamped decoration may consist of a single stamp (or several single stamps) applied a number of times, or a more ornate single imprint. The care with which the stamp has been applied in relation to the edges of the object, the central panel or indeed other stamps, varies. One example may have an additional imprint from a seal (81A2845).

The significance of the individual decorations is as yet unclear, although they do distinguish each object, so may merely reflect ownership. Some clearly have Royal emblems (81A1460, 81A4241, 81A2845, 81A3867, 80A1973, 79A1224, 80A1239) and others are clearly religious (81A1158, 80A0901, 82A2282). It is possible that the symbols used may reflect authority, symbols of status, badges of office, talismans, links with mustering authorities, with livery companies or even a combination of these. Whilst the mustering of men for armed service was substantially a matter for parishes, City Guilds also contributed to the numbers. Wristguards bearing stamps which may be guild insignia include 82A1524 and possibly 92A0943. Religious symbols, which may suggest affiliation with parish churches, include 80A0901, 81A1158, 82A2282 and 80a0901. Purely 'floral' designs are found on 81A4639, 81A5826, 81A1173 and 81A1185.

A hierarchy existed within companies of archers engaged in land battles, and may have existed at sea, thus some decorated wristguards might indicate the authority or status of the owner. Reasons for adornment may be as simple as custom for those who practised archery on Sundays and holy days and it is possible that religious symbols actually had no deeper significance than to be a constant reminder of the faith of the shooter. The recovery of undecorated examples from the *Mary Rose* and elsewhere shows that ornamentation

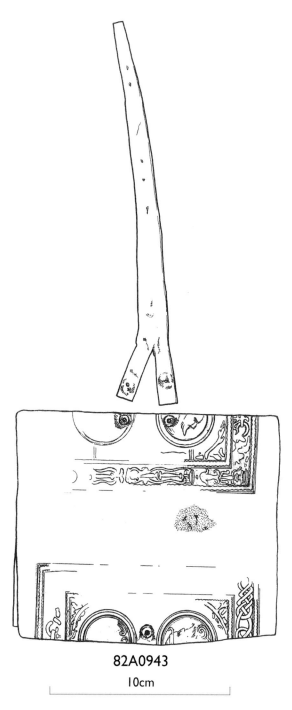

*Figure 8.55 Wristguard 82A0943 (compare with Fig. 8.56, book cover 81A4131)*

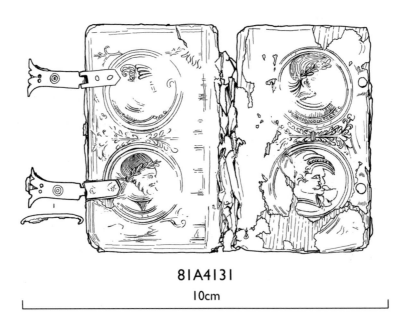

*Figure 8.56 Book covers 81A0895, 81A4131 (see AMR Vol. 4, chap 3 for full description and discussion)*

was not universal, and by implication, not a requirement.

Decoration is summarised in Table 8.32. Seven examples bear ornamentation suggestive of Royal affiliation and three stand out as having extremely refined stamps which incorporate a number of images. These are described in detail below.

*Book stamp decoration (82A0943)*
82A0943 is a plain rectangle without defined borders, made of bovine leather and found in chest 82A0942 in O10, where a number of personal chests were stored (Fig. 8.55). It was secured with one tapered bifurcating strap. There are five holes in the bifurcating strap with rivets positioned 12mm from the end.

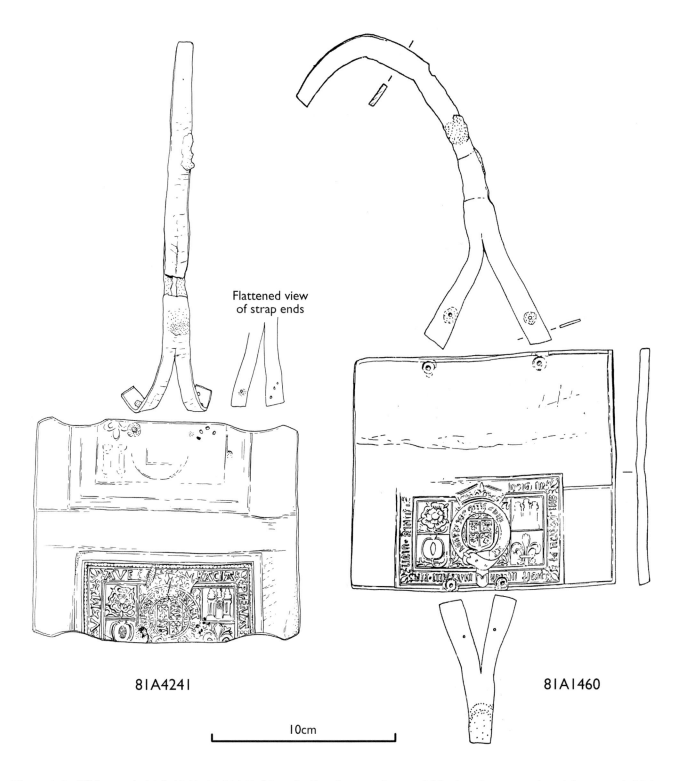

*Figure 8.57 Wristguards 81A 4241, 81A1460. Note the Royal quarterings and Marian Aves suggestive of the wearer's high status, and the bifurcating fastenings*

The wristguard is decorated using what appears to be a book cover stamp used on either side of the wristguard. However, unlike the case covers of a sundial (81A5681) and a coin balance (81A4701) from the *Mary Rose*, the wristguard is not a reused book cover.

The foliate and roundel design containing faces in profile is a common book cover design of the sixteenth century and this example is similar (though not the same) as designs found on at least two book covers from the *Mary Rose*, one of which was made in London (see

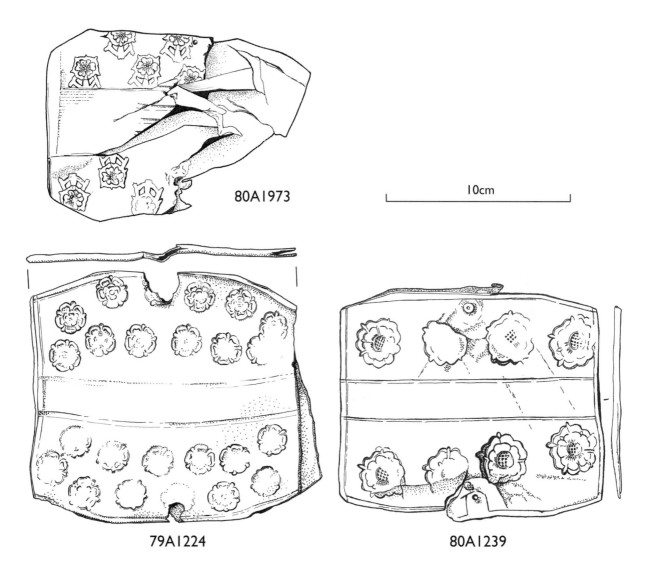

*Figure 8.58 Wristguards 80A1973, 79A1224, 80A1239 with decorative representations of the Tudor Rose*

*AMR* Vol. 4, chap. 3). It may suggest an association with the Stationers' Company, within whose orbit the Bookbinders fell (Fig. 8.56). In order to keep a central plain panel the stamp has been placed so that the lower portion is missing on both sides, again suggesting the use of a stamp not specifically designed for this type of object.

The object was found in an elm bench chest that also contained personal items including clothing, a dagger sheath, comb and a knife. There was nothing else to indicate the status or position of the owner.

*Royal ornamentation*
Seven wristguards suggest royal affiliation based on ornamentation. Two stamps are almost identical (81A1460 and 81A4241; Fig. 8.57). Both bear the Royal Arms comprising the *fleur de lys* for France (1st and 4th quarters) and the three lions of England (2nd and 3rd quarters), positioned centrally within a shield encircled by the Garter bearing the words *Honi Soet Qui Mala Pens* and surmounted by the Royal Crown. In both cases these Royal Arms are incorporated into a bordered design which includes the quartered Royal Arms and other dynastic emblems: the Tudor Rose, the pomegranate of Granada (badge of Henry's first wife, Catherine of Aragon), the triple turret of Castile (also borne by Catherine), the portcullis of the Beaufort family (through whom the Tudors descended from Edward III) and the *fleur de lys*. Both examples are elegantly made and held with two bifurcating straps. Both carry inscriptions within the border identified as invocations to the Virgin Mary.

These are the most likely candidates for special status and may indicate the presence on board of members of the Yeoman of the Guard. The Musee de L'Armée in Paris has a buckler or small shield, (1/6) which bears similar ornamentation. This is a Wrexham buckler of the type shown being carried by Henry's

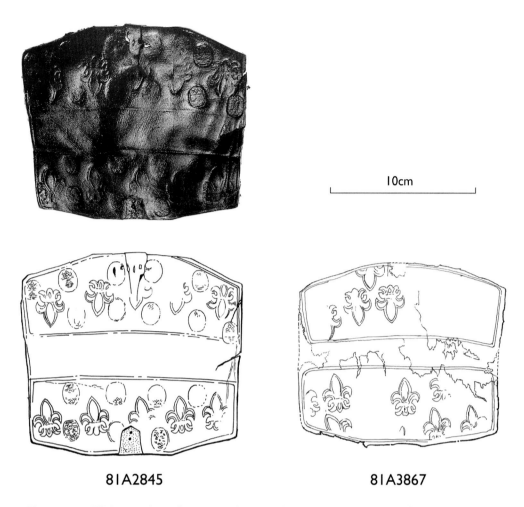

*Figure 8.59 Wristguards 81A2845, 81A3867. The representations of the* fleur-de-lys *suggest a high status for the wearer*

guard in the *Embarkation at Dover* and the *Field of Cloth of Gold* paintings. It may have been presented by Henry to Anne de Montmorency in 1532 (Edwards and Blair 1982, 74–115).

81A1460 (Fig. 8.57), found with human arm bones (81H0171) in M6, is a plain rectangle with a bordered edge but no central panel. It is fastened by two bifurcating leather straps held with rivets. The longer strap is pierced with four holes; the shorter strap, which attaches to the decorated panel, has rivet holes 20mm from the end, and shows evidence for an iron buckle plate.

Ornamentation consists of a single stamp, used once. It is relatively complete but only certain words within the border can be read and may represent an *Ave Maria* or prayer. Certain words can be deciphered, '*post*', '*et*' and '*amen*'. It is possible that this contains one line from the antiphon *Salve Regina* (pers. comm. Dom Daniel Rees, Librarian, Downside Abbey): '*Et Jesum benedictum fructum ventris tui nobis post hoc exilium ostende*' ('*And Jesus, the blessed fruit of your womb, show us after this exile*').

The lettering is indistinct and some words may be abbreviations. The other great prayer to the Blessed Virgin '*Mary Alma Redemptoris Mater*' contains the line '*Virgo prius ac posterius*' ('*Virgin previously and afterwards*'). Another possibility is '*Virgo ante partum, et post partum permanet*'. '*Post*' may be an abbreviation of '*posterius*' and '*et*' an alternative to '*ac*'.

81A4241 (Fig. 8.57), from O2, is a plain rectangle of calf leather with mitred corners, a bordered edge and a central panel. It was fastened with two bifurcating straps, one of which bears evidence for possible stitching. The longer strap is split for 45mm and has rivets 10mm from the end. The shorter strap is split for 30mm and has evidence for a 20mm long buckle plate. The ornamentation is only partial. It is a single stamp, used twice and it is similar to 81A1460, with variations in the rose and the fleur de lys. The border carries the commonest of all prayers to the Virgin: '*AVE [MARIA G]RACIA PLENA [DOMINVS TEC]VM*' ('*Hail Mary, full of Grace, the Lord is with thee*'). The centre bears the crowned garter within which parts of the inscription '*Hony Soet Qvy Mala Pens*' are visible. This object was recovered in a sample (81S0275) consisting of '*straw*',

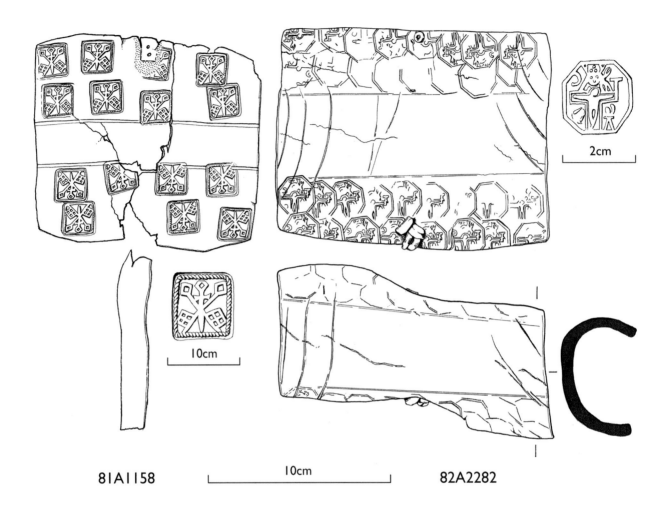

*Figure 8.60 Wristguard 81A1158 with religious symbolism of St Peter's keys and 82A2282, perhaps tools of the Passion*

leather, textile, a comb, aiglet, shoe fragments and 39 beads making up a paternoster.

Five other wristguards incorporate Royal insignia in a lesser form. Three (79A1224, 80A1239, 80A1973) bear Tudor roses (Fig. 8.58) and two *fleur de lys* (81A2845, 81A3867; Fig. 8.59). 81A3867 carries an additional unidentified seal or initials which are hardly visible and cannot be described. Both are similar in shape to a wristguard in the British Museum (BM 1961, 0202, 48) recovered from Finsbury Fields.

All five carry central panels and are fastened by a single strap. Three are convex-sided rectangles and the fourth (80A1973) a mitred rectangle. The distribution of these is widespread (Fig. 8.50), but corresponds in general to places with stored archery equipment, potential action stations (Upper deck in the waist) and close to access routes to the Upper deck via a companionway in M6/7. The distribution and associations suggest that they were being worn by men engaged in the fighting of the ship and the royal affiliation may reflect a specialised role, if not authority.

Of the three with Tudor roses, 80A1973 is creased and damaged with evidence of alteration by cutting and no evidence of fastenings, while 79A1224 was originally fastened by two centrally placed straps, both torn away and 80A1239 has centrally positioned rivet holes suggest a pair of single straps. The first was found outside the Barber-Surgeon's cabin in M7 close to a companionway leading to the Upper deck; 79A1224 was in U4 and 80A1239 was lying on the ballast in H7. Although not associated with any specific objects, the remains of up to six individuals were recovered from this area (FCS 19–22, 34 and 37).

The two wristguards decorated with multiple *fleurs de lys* are 81A2845 (Fig. 8.59), found in a broken chest in O5 which also contained two lead weights and 81A3867, from O9. Perforations and evidence for stitching suggest fastening by single leather straps.

*Religious symbols*

In addition to the Marian invocations on the two wristguards bearing royal insignia, and the possible association of the *fleur de lys* with the cult of the Virgin (see Weinstein, *AMR* Vol. 4, chap. 11 for a discussion), four others may have religious affiliations.

81A1158 (Fig. 8.60), from U9, shows evidence of recutting and has centrally positioned rivet holes to one side suggesting a single strap fastening. It is ornamented with fifteen depictions of St Peter's keys crossed and bisected with a sword, each contained within a corded rectangle, four of these overlap the central panel. The crossed keys are specific to St Peter with the sword representing the inverted cross of his martyrdom. It undoubtedly represents the Bishopric of Exeter. The captain of the *Mary Rose* was Sir George Carew, a West Country man.

82A2282 (Fig. 8.60), from O9, bears the end of a 10mm wide strap/thong, knotted through a hole in the centre of one long edge, with a single rivet hole on the opposite edge suggesting one strap. Thirty-three overlain stamps may represent the instruments of the Passion – hammer, nails and pincers are depicted. Other interpretations include the crucifixion, incorporating the figures of Mary and John at the foot of the Cross surmounted by God the Father accepting the sacrifice. Alternatively it may be that this represents a depiction of the 'Sunday Christ'.

80A0901 (Fig. 8.61) is an unbordered rectangle, mitred at the corners, that has been cut to a lozenge-shape. An asymmetrically positioned rivet on one edge, and a tear on the other suggest centrally placed straps. It is decorated with 20 nimbuses. A nimbus represents a cloud-like splendour investing a God. It was found in U7 close to arrow and longbow chests and a sheaf of arrows together with a portion of a spacer (80A0877).

*Possible association with guilds*
City Guilds were required to provide men and equipment. An extract from the records of the Goldsmiths' Company includes the following for 1544:

> 'On Sondaye the xxvj$^{ti}$ daye of October anno xxxvij° regni regis Henrici octavi *a commyssyon was sent from my lord the Mayre ffor xxiiij$^{ti}$ harnesse men with russett cotes which commyssyon hereafter ensuyth.*
> Goldsmithes xxiiij$^{ti}$ *wherof bowemen iiij billmen xx well harnessed and appoyntted in every degree as the[y] were at the last tyme and to bryng theym tomorowe at x of the clock to Leden Hall'.*
> (Wardens' Accounts and Court Minutes, Book H (p. 12)

In addition to the possible association between 82A0943 and the Stationers' Company, 82A1524 may have a connection with one of the city guilds (Fig. 8.61). It is an unbordered rectangle with mitred corners. A centrally positioned rivet hole and tear opposite suggest two single straps, which do not survive. Ornamentation consists of three parallel lines of three stamps on each side of a central panel. The stamps

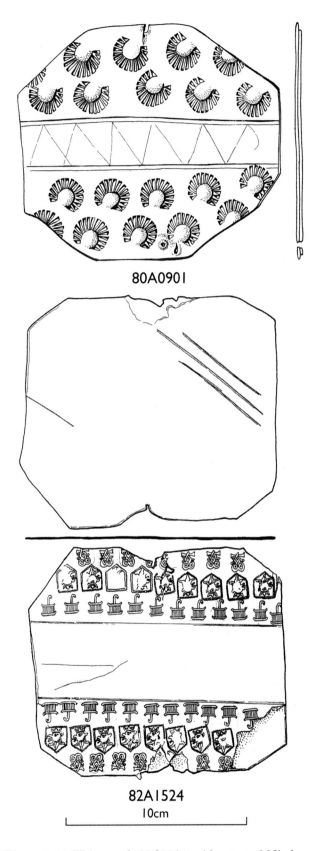

*Figure 8.61 Wristguards 80A0901, with repeated Nimbus religious symbol, and 82A1524 bearing symbols suggesting association with the Worshipful Company of Girdlers*

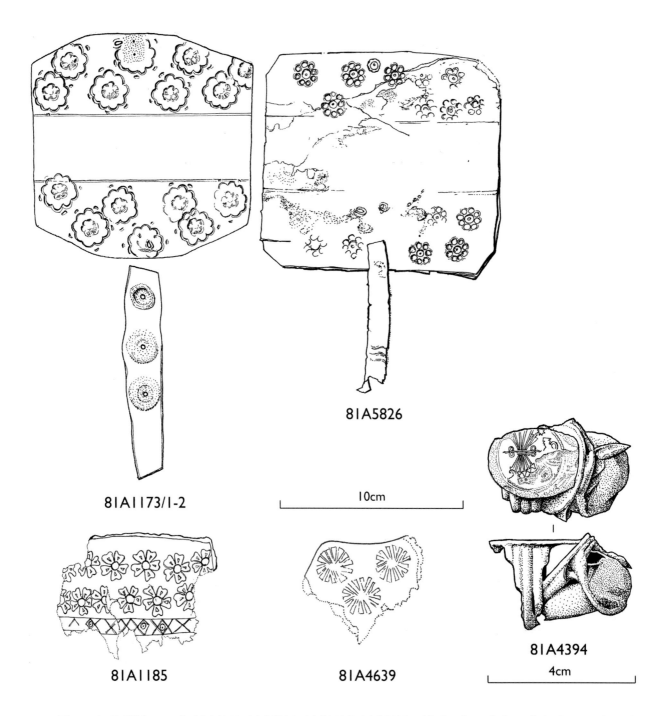

*Figure 8.62 Wristguards 81A1173, 81A5826, 81A1185, 81A4639 with floral symbols and single strap fastenings, and sheaf ring 81A4394 decorated with a sheaf of arrows*

include a gridiron, an eagle or a griffin regally crowned within a shield and a wiredrawer's gauge which appears to be coupled to a pin. The gridiron is associated with the Girdlers' Company and with St Lawrence, their patron. Together these stamps may relate to the informal association of the Girdlers' Company with that of the Wiredrawers' Company and the Pinners' Company formalised in 1568. This symbol also appears on the boundary markers of the Parish of St Lawrence Jewry, so the owner may have had a connection with the Girdlers'/Wiredrawers' Companies, with that Parish, or perhaps with the Ward (Cheap) in which the Church was situated. It is possible that the lowest symbol is a monogram of 'I [J] B', a John Bartholomew who, in 1560, is listed as a benefactor to the Girdlers' Company. Although the gridiron has affiliation with the martyrdom of St Lawrence and thus the Worshipful Company of Girdlers, the presence of the two other motifs suggest a strong affiliation to at least one other Guild.

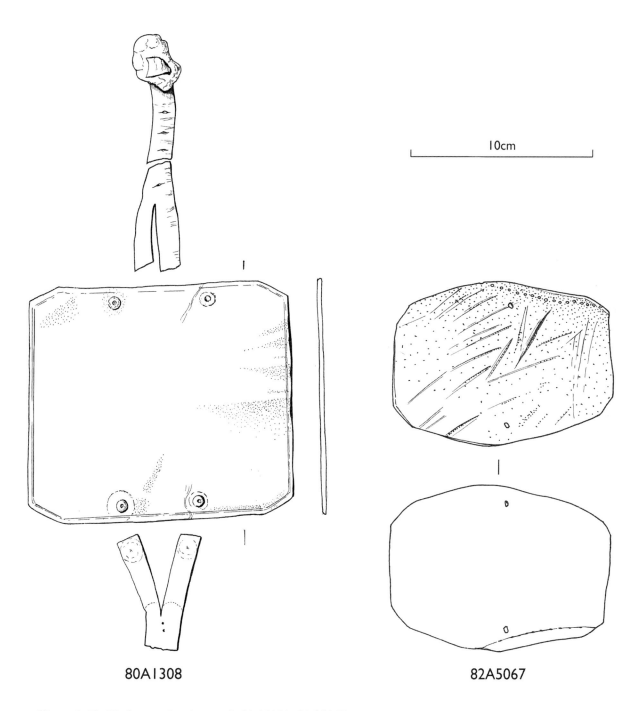

*Figure 8.63 Undecorated wristguards 80A1308, 82A5067*

This wristguard was found in O8, directly associated with human remains (81H0046). The area contained a large number of archery related objects including longbow and arrow chests, sheaves of arrows, loose arrows and a spacer (81A1522). Other items included casks and further human remains.

*Other decoration*
Purely secular decoration is found on four wristguards (81A1173; 81A5826; 81A1185; 81A4639; Fig. 8.62). All bear floral stamps – almost childlike flower impressions. On 81A1173; 81A5826 these are arranged either side of a central panel bordered by pairs of incised lines while the surviving fragment of 81A1185 appears to have a central panel bordered by a frieze of 'X's and circles. 81A4639 is too fragmentary for its form, overall decoration or fastenings to be determined.

81A1173 and 81A5826 appear to have had just a single pair of centrally positioned straps – that on 81A1173 bears three concretion stained holes suggesting the former presence of reinforced eyelets set 25mm apart between centres.

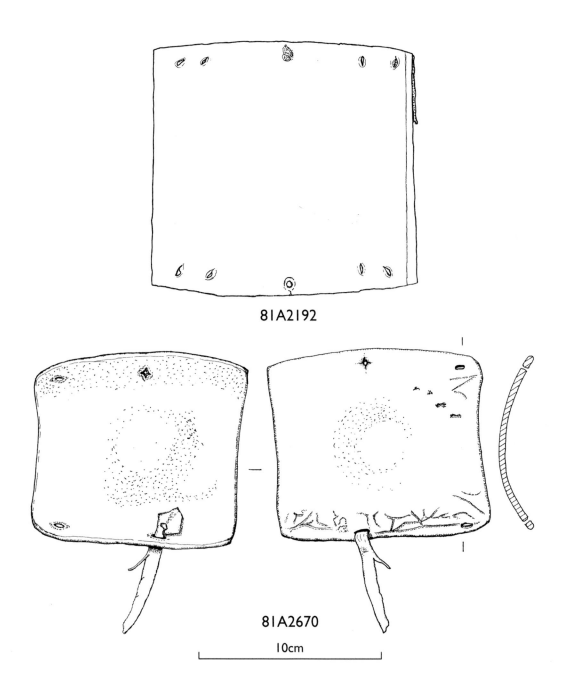

*Figure 8.64 Undecorated wristguard 81A2192, with possible supplementary lacing holes, and 81A2670*

Similar stylised floral symbols are associated with each of the archers represented on the '*Flodden Window*' of Middleton Church, Lancashire. This was made in 1515 to commemorate the part played by Middleton bowmen at Flodden and displays seventeen archers with their priest. *Mary Rose* men, possibly even archers, also had a role in this battle (above, p.14). A symbol of the stylised corolla is associated with each. The scene depicted is purported to show the archers praying prior to their departure and it is possible that the decorations formed part of their insignia.

81A1173 was found associated with human remains (81H0145) beside the left wheel of a carriage for an iron port piece in M8, with loose arrows nearby. 81A5826 was found in chest 81A5783, standing upright in M9. This is perhaps the closest we can get to recreating the possessions of an archer. First the chest was intact and excavated ashore. It contained a number of high status personal objects including a pewter tankard, unusual in both shape and in the '*wriggled*' decoration covering most of the surface in bands (81A5654). It is the only known sixteenth century tankard bearing this decoration and may be part of

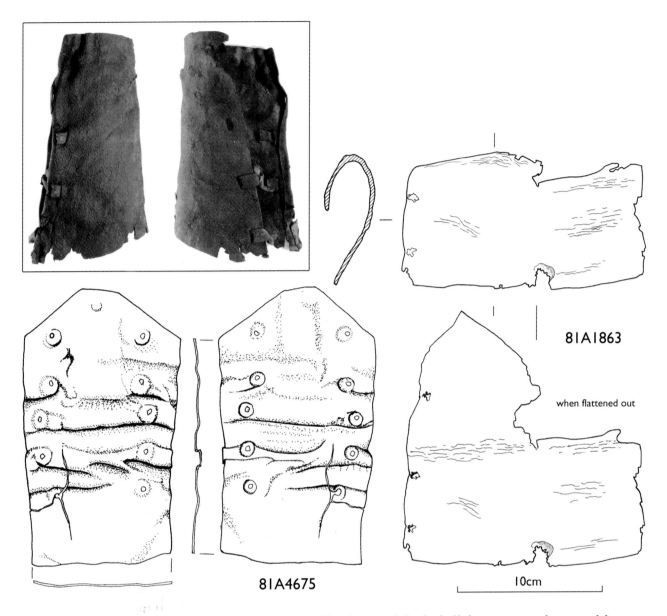

*Figure 8.65 Laced armguards 81A1863 and 81A4675. The photograph is of a half-sleeve armguard recovered from Stogursey Castle, Somerset (reproduced by permission of Taunton Museum)*

'*dessert*' or banqueting pewter set. It also contained three similar, high quality pewter saucers (81A5827, 81A6849, 81A6850), all bearing the mark of the crowned pewterer's hammer, and a pewter dish (all discussed and illustrated in *AMR* Vol. 4, chap. 11). Other objects (*AMR* Vol. 4, chap. 3) included a sundial in a case (81A5681) and fragments of a book (81A5817). A leather pouch (81A5818) contained clothing fastenings, tiny lead weights, bone die, a single lead shot and a coin concreted to a silver ring (81A4394) which, perhaps significantly, bore relief decoration showing a sheaf of six arrows (Fig. 8.62; unfortunately no longer visible). It may be that this represents the armorials of the Fletchers' Guild; a bundle of arrows is included on the London Fletchers' shield. A single arrow (81A5829) was also found in the chest. On the other hand there are also items indicative of the chest having belonged to a carpenter (two gimlets and a chalk-line reel). Although not in the same chest, two small thumb planes were found in a chest just outside the cabin. It is suggested that these may have been used to make incendiary firesticks (an incendiary projectile similar to an arrow; see Chapter 10) or '*sprites*' – wooden projectiles fired from handguns (see below and Figs 8.93–5) for a description of some possible examples).

The other two wristguard fragments bearing floral decoration were found in the U9 associated with human remains.

**Undecorated wristguards**

Seven leather wristguards are undecorated (81A2192, 81A1863, 81A2670, 81A4675, 80A1308, 82A5067, 80A1308). They were widely distributed around the

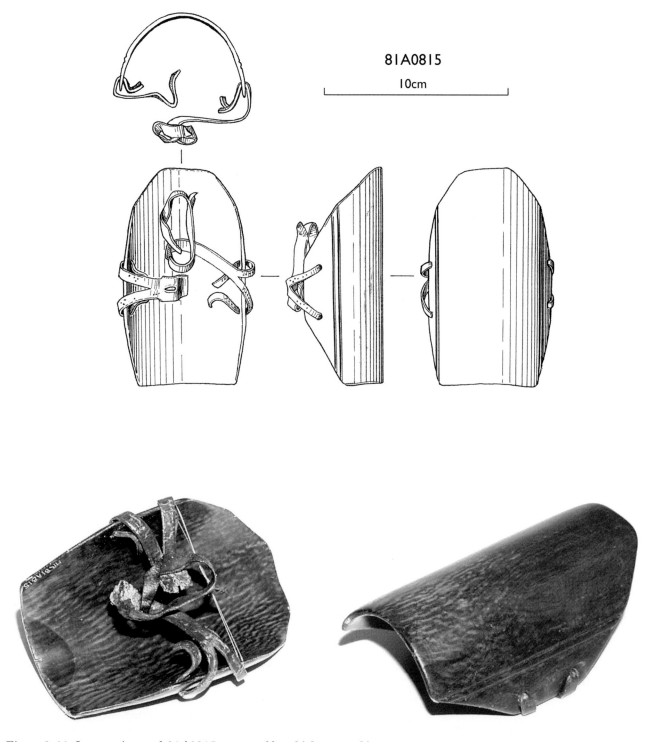

*Figure 8.66 Ivory wristguard 81A0815, presumably a high status object*

ship (Fig. 8.50; Table 8.32). Only one was found in a chest (81A2192 in chest 81A2035 in M10 along with an unidentified, unstitched piece of circular leather (81A2191), diameter 130mm that could be a strengthener for an arrow bag base. Most of the remaining wristguards were associated with human remains.

80A1308 (Fig. 8.63) appears to have been dyed red and is finely finished, albeit undecorated. 82A5067 (Fig. 8.63) bears sixteen stitch-holes and various small slashes or cuts. It may have been made from a re-used shoe. 81A2670 (Fig. 8.64) bears possible traces of indecipherable decoration while 81A2192 (Fig. 8.64) has both centrally positioned rivet holes and two perforations at each corner, suggesting a pair of straps with additional lacing. The fragmentary wristguard, 81A1863 (Fig. 8.65), is of indeterminate shape but displays three holes to one edge, which indicates lacing

*Figure 8.67 Signet ring decorated with letter 'K' (see AMR Vol. 4, chap. 3 for full description)*

as a means of fastening and one larger, torn hole to another, suggesting the presence of a fastening strap (depending on the original orientation of the object). This may be similar to a 'rustic' half sleeve wristguard recovered from an excavation at Stogursey Castle (Somerset) which retains portions of leather thonging. 81A4675 (Fig. 8.65) has five symmetrically placed rivet holes on each side possibly for a single or, more likely, five pairs of laces/thongs (simply laced, two crossing, or one long one).

## Non-leather wristguards

Wristguard 81A0815 (Fig. 8.66; Table 8.32) is one of only two objects recovered from the *Mary Rose* to be positively identified as being made from ivory (the other is a comb; 82A0945). It is demonstrably superior in quality to all the other wristguards. It appears have been made by using the natural curvature of the tusk to form a half-sleeve guard that has been cut to fit a small arm. It was fastened with bifurcating leather straps fed through symmetrically positioned holes, two on each side, secured by a buckle or 'D' ring, indicated by concretion. It bears two parallel longitudinal striations 20mm and 16mm from the edge, possibly to provide flexibility.

It was found on a human radius (81H0109) in close proximity to a finger bone (81H0104) bearing a signet ring in copper alloy with the letter 'K' in relief (81A0810; *AMR* Vol. 4, fig. 3.9; pl. 12), and next to a culverin in M3 (Fig. 8.67). There are eighteen ivory wristguards in the Longman Collection of the Pitt Rivers Museum, Oxford (1927 24.1–24.18) from Amiens, France (dated 1500–1700). An English example is in Southend Museum (Fig. 8.49b).

Wristguard 81A2219, from O3, is made of horn. Very little horn has survived the burial conditions of the *Mary Rose* and this object was fragmentary and is now missing.

Clearly the variation in style and decoration found among the wristguards suggests that the archers on board may not have been a single corps, although the seven bearing Royal ornamentation, particularly the two bearing the Royal Arms, are noteworthy. It is clear that men functioning as archers were moving throughout the ship and the association of several of the wristguards with guns or individuals on the Main gun deck provokes questions as to their role on board. The fact that four were found in chests when the ship was actively prepared for engagement which anticipated an element of archery is interesting, possibly lending strength to the argument that at least some of those expected to use the bows might have had other primary tasks.

## Arrows

by Alexzandra Hildred with Keith Watson, Mark Hopkins, Adam Jackson and John Waller

*'Draw archers draw you your arrows to the head'.*
Shakespeare, *Richard III*, v.3

### Introduction

The *Anthony Roll* lists 400 '*Lyvere arrowes in shevis*' for the *Mary Rose*. This equates to 9600 shafts, just over 38 arrows per bow. Although has been suggested that '*livery*' indicates manufacture by the Worshipful Company of Fletchers, a livery company formed in 1371 (Soar 2004, 74, 217) it is understood here to mean a type of arrow supplied for military service. Other inventories, in particular that of 1547 (Starkey 1998), show that arrows were issued either in dozens or in sheaves (of 24) and contained in bags, chests and cases. However, as with the longbows, variation in size is more limited than it is with other weapons, ammunition, and types of gunpowder. This implies that there was a known and upheld standard (possibly incorporating a range of sizes within it) for longbows, strings and complete arrows. It is rare to find components of arrows listed separately but arrows in varying stages of assembly are identified in Arrow Loft at Calais: '*Arrowes reddy Fetherede*', '*Arrowes reddy fethered and cased*', '*Arrowes vnfethered*', '*Arrowe cases of red lether with girdells*' and '*Heades for lyvery arrowes*' (*ibid.*, 5047–51). Pontefract Castle has '*Lyvery arrowes good and badde*' and '*Lyvery arrows without heddes*' as well as a cartload of timber for them (*ibid.*, 6221–3). Whereas the longbows are often prefixed by the preferred wood, yew, this is not the case with arrows. This raises the question as to whether this was not a consideration: a war arrow was a war arrow, regardless of wood and whatever arrows were supplied could be used with all strings and all bows. However, this goes against all of the extant (recreational) treatises (and common practice for sporting archery) which advocates

products of a range of different craftsmen and artisans into a highly effective and efficient weapon of war. The arrows from the *Mary Rose* form the only substantial sample available for study and interpretation.

The recovered assemblage includes 7834 allocated artefact numbers, representing a minimum number of 3738 arrows. Of these, 2303 (29.4%) are listed as complete shafts (Fig. 8.68a, b), constituting 24% of the total as listed in the *Anthony Roll*. None of the iron heads was recovered, however, there is one half of an arrowhead socket, and iron staining and concretion indicate that they were *in situ* when the ship sank (Fig. 8.68c).

The excellent condition of some of the arrow shafts has enabled detailed studies of specific groups. The total number studied to date is 2203 (28.12%) and includes arrows from closed contexts, such as chests, and from a cross-section of other contexts and locations (see Fig. 8.69 and Tables 8.33 and 8.34). A small portion of the arrows have been conserved: subjected to desalination, soaked in a solution incorporating a chemical wax followed by freeze drying to remove the liquid and treated by further applications of wax once dried. The condition of these varies and many are twisted, warped, fractured or collapsed, though some are straight and appear to retain good form. When this study was undertaken the majority of the arrows were soaking in reinforced plastic corrugated trays in 50 gallon (*c.* 227 litres) tanks containing a 10% solution of PEG 600, awaiting freeze-drying. Although their form appeared to be less distorted than the fully conserved shafts, post-excavation damage was evident, especially at the nocks and tips. They were extremely fragile and handling was minimal. Most are no longer suitable for wood analysis. Some dimensional changes are inevitable and the amount of alteration may relate to the original material and burial environment as well as to post-excavation storage, handling and treatment. Although dimensions along the grain will have changed very little, swelling is likely to have increased shaft diameters by 5–10% depending on the degree of degradation of the wood (Waller and Jones, pers. comms). On conserved examples, some collapse of cell structure has actually caused shrinkage in diameter which is particularly evident over the area of the flights where the glue and bindings are now larger than the arrow and form a loose

*Figure 8.68 a) Reconstruction of a spacer of arrows; b) arrows 80A0764/160 and /273; c) iron stained tips of same arrows; d) underwater photograph of arrows arranged tip to tip in chest*

that the weight and strength/spine of the arrow relate directly to the power of the bow.

The processes involved in sourcing the raw materials and manufacture of arrows are diverse and complex. The numbers of military arrows required suggest that this was a major industry. A study of this assemblage provides an insight not only into military archery in the sixteenth century but also the levels of administration and organisation required to bring together the

Table 8.33 Distribution of arrows by deck

| Deck | Complete arrows | Complete arrows in chests | Tip frags | Nock frags | Shaft frags | Min. no. arrows | Total |
|---|---|---|---|---|---|---|---|
| Castle | 0 | 0 | 1 | 0 | 1 | 1 | 2 |
| Upper | 347 | 262 | 344 | 375 | 1159 | 792 | 2225 |
| Main | 2 | 0 | 37 | 30 | 126 | 47 | 195 |
| Orlop | 1611 | 1574 | 894 | 374 | 1429 | 2512 | 4308 |
| Hold | 343 | 338 | 159 | 148 | 454 | 547 | 1104 |
| Site total | 2303 | 2174 | 1435 | 927 | 3169 | 3899 | 7834 |
| % of listed total* | 24 | 23 | 40 | 34 |  | 41 |  |

* 9600 listed in *Anthony Roll*

outer skin. Therefore, any cross-section of the assemblage studied now must not only include arrows from a range of burial environments and locations around the ship, but also examples that have undergone the different post-excavation methods of storage and treatment.

*Distribution*

Unless otherwise stated, numbers quoted below are for arrow fragments rather than complete arrow shafts. Arrows were found throughout the ship, however the majority (4837, 61.7%) were found in or around the seven arrow chests found on the Upper deck, Orlop deck and Hold (Figs 8.45, 8.68d, 8.69; Tables 8.33 and 8.34) (see below). Arrows were found either individually, as multiples tied with leather straps, or inside chests (either loose or tied). Arrows were also found within pierced circular leather disks, interpreted as spacers used to contain arrows for immediate use, some of these were also tied with leather straps. None of these was in a chest (see below). The spacers and chests are described later in this chapter. Those arrows found in chests were stored tip to tip (Fig. 8.68c). The chests on the Orlop deck and in the Hold were clearly in storage areas, while those on the Upper deck suggest a muster station for archers with easy access to the weather deck and to the castles, both areas where one would expect archers to be stationed during battle (Fig. 8.45). The deck totals provide two for the Castle deck, 2225 for the Upper deck (688 in chests), 195 from the Main deck, 4308 from the Orlop deck (3585 in chests) and 1104 from the Hold (564 in chests). Because of the large number of fragments arrows found in groups were registered under one number and further subdivided into part numbers; therefore the numbers shown on the schematic trench plan (Fig. 8.69) are less than the actual totals (apart from chest 80A0726 on the Upper deck: contents 1–288). Many of the arrows in the stored chests were still visibly separated into sheaves, however the ties had largely disintegrated. Although attempts were made to lift these bundles as discrete units, the difficulties involved in doing so both underwater and throughout the post-excavation handling ashore means that we cannot be absolutely certain that items listed as sheaves underwater are still in their original groups.

**The Upper deck**

In addition to the arrows associated with chest 80A0726, arrows are found in most sectors on the Upper deck (Fig. 8.69). Although it is clear that U7–8 represents the remains of a muster station, with the loose arrows described by excavators as being '*in bundles*' possibly deriving from a second empty chest (80A0476), the finding of arrows and arrow sheaves as well as loose longbows and wristguards between U3 and U6 suggest actual positions for archers. Nine of the 17 arrow spacers were found on the Upper deck, four in the waist (see below). Three others were found in U9, associated with the features containing large numbers of human remains, possibly suggesting a bottleneck in an access route. So, although there is ongoing debate regarding the use of the blinds in the waist (see *AMR* Vol. 2), the distribution of archery-related material suggests that archers were using this area.

**The Main deck**

There were isolated spot finds of arrows and bows at the forward end of the Main deck, associated with human remains identified as potential archers (FCS 81 and 82, see Stirland below) (Fig. 8.69). These finds included arrows, spacers, longbows and a wristguard in M3. The individuals were found at the base of a companionway to

Table 8.34 Arrows in chests

| Sector | Chest | Complete arrows | Tip frags | Nock frags | Shaft frags | Min. no. arrows | Total |
|---|---|---|---|---|---|---|---|
| U7/8 | 80A0726 | 262 | 148 | 112 | 166 | 410 | 688 |
| O8 | 81A5638 | 4 | 11 | 3 | 85 | 15 | 103 |
| O9 | 81A2398 | 943 | 174 | 49 | 179 | 1117 | 1345 |
| O9 | 81A2582 | 627 | 536 | 215 | 759 | 1163 | 2137 |
| H9 | 82A1761 | 338 | 31 | 77 | 118 | 415 | 564 |
| Site total |  | 2174 | 900 | 456 | 1307 | 3120 | 4837 |
| % of listed total |  | 23 | 32 | 27 |  | 32.5 | 50 |

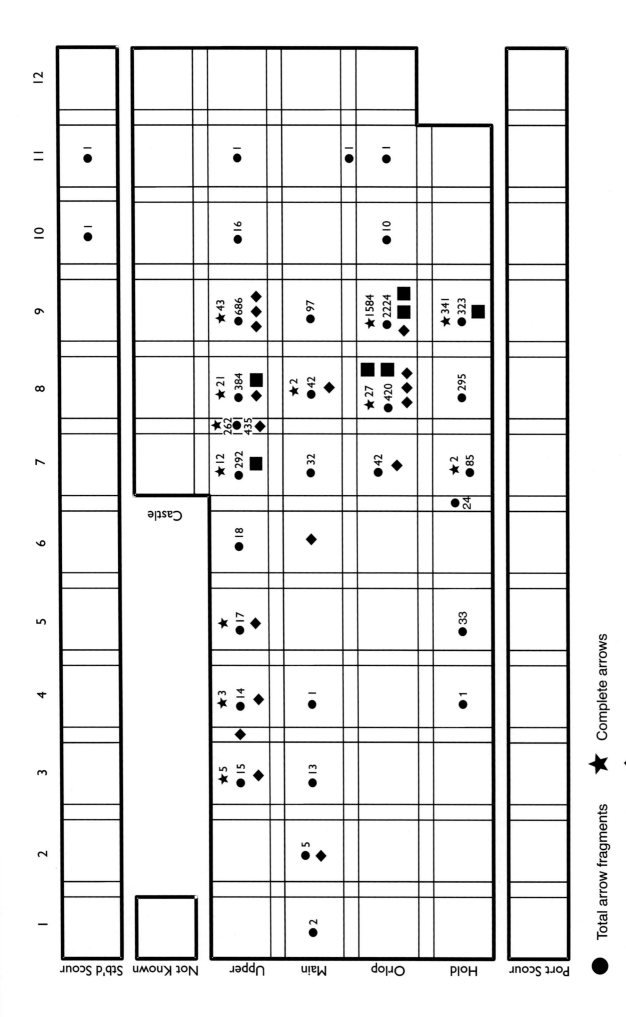

Figure 8.69 *Distribution of arrows and spacers*

the Upper deck. A sheaf and spacer found in M8 (81A2828) were also found with human remains. A single arrow was found in a personal chest in the Carpenters' cabin in M9 (81A5829), together with a wristguard (81A5826) and a concretion containing coins and a pouch (81A5818) which, itself, contained a finger ring displaying a sheaf of arrows (81A4394; Fig. 8.3 above).

**The Orlop deck**
The distribution of arrows is more restricted on the Orlop deck than on any other, being basically confined to O8–9 and the stored chests themselves (Figs 8.45 and 8.69). This area also had a high density of human remains, some of which were closely associated with loose arrows and arrows within spacers (81A1523, 82A1587). We can, therefore, perhaps suggest that archers were actively moving between the stores and other areas of the ship. One of the chests was empty (81A3824) and a second only partly full (81A5638).

**The Hold**
Spot finds of archery equipment were found in H4 and 5, possibly having drifted down the companionway in this area at or shortly after the sinking. Those found in H7–9 may have originally been in the chest found in H9, though more probably in O9. Many of these finds were associated with human remains, possibly suggesting active use of the stores here too. The relatively high density of unboxed arrows in H9 also presumably derived from the upturned, unlidded chest found in that area (82A1761).

## Form and function

In its most basic form an arrow consists of a shaft upon which flights (fletchings) are attached (Fig. 8.70). These are designed to stabilise the arrow during flight. A fine slot is cut into the base of the shaft, in line with the grain, to accommodate a tapered horn insert. Then, across the grain, a larger slot is cut to take the bowstring. Both these actons serve to strengthen the nock against the force of the bowstring when the arrow is shot. Feathers for the fletchings are glued to the shaft and bound with thread. The tip cone is cut to a point to accommodate a socketed arrowhead which is probably glued into position. Shaping the cone and ensuring that the arrowhead fits securely leaves cut marks which form well-defined facets creating a distinctive shoulder.

The *Mary Rose* assemblage includes complete shafts as well as portions of shaft, tips and nock ends. The materials, form and dimensions of each of these elements and how they are incorporated within a single arrow affect the weight and therefore the ballistic potential of the projectile. The material, profile, length, weight and spine (flexibility) of the arrow; the draw-weight of the bow and the diameter, material and strength of the string; combined with the skill and strength of the archer, govern the performance of the longbow (described above).

## Historical information

*'A shaft hath three principal parts, the stele, the feathers and the head.'* (Ascham 1545, 117)

*The shaft or 'stele'*
Ascham states that ash is the best wood overall, especially for war arrows, however, he adds the rider that '*Asp*' (poplar) is more commonly used. As arrows for military issue were required in huge numbers, availability and cost were probably the overriding factors:

*'Steles be made of divers woods: as, Brazil, Turkey wood, Fustic, Sugar-chest, Hardbeam [Hornbeam], Birch, Ash, Oak, Service-tree, Hulder [Alder], Blackthorn, Beech, Elder, Asp [Poplar], Sallow'.* (ibid.)

*'... alder, blackthorn, service tree, beech, elder, asp, and sallow, either for their weakness or lightness, make hollow, starting, studding, gadding shafts ... birch, hardbeam, some oak, and some ash, being both strong enough to stand in a bow, and also light enough to fly far, are best for a mean, which is to be sought out in everything'.* (ibid., 119)

*'... let every man, when he knoweth his own strength, and the nature of every wood, provide and fit himself thereafter. Yet as concerning sheaf arrows for war, (as I suppose) it were better to make them of good ash, and not of asp, as they be now-a-days. For of all other woods that ever I proved, ash being big is swiftest , and again heavy to give a great stripe withal, which asp shall not do'.* (ibid., 120)

*'A stele ... must be made as the grain lieth, and as it groweth, or else it will never fly clean ... A knotty stele may be suffered in a big shaft, but for a little shaft it is nothing fit, both because it will never fly far; and besides that it is ever in danger of breaking.'* (ibid., 118)

*'... let every man, when he knoweth his own strength, and the nature of every wood, provide and fit himself thereafter. Yet as concerning sheaf arrows for war, (as I suppose) it were better to make them of good ash, and not of asp, as they be now-a-days. For of all other woods that ever I proved, ash being big is swiftest , and again heavy to give a great stripe withal, which asp shall not do.'* (ibid., 120)

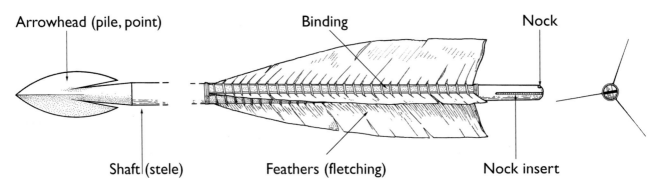

*Figure 8.70 Anatomy of an arrow*

> 'A stele ... must be made as the grain lieth, and as it groweth, or else it will never fly clean ... A knotty stele may be suffered in a big shaft, but for a little shaft it is nothing fit, both because it will never fly far; and besides that it is ever in danger of breaking.' (ibid., 118)

> 'Now how big, how small, how heavy, how light, how long, how short, a shaft should be particularly for every man ... it is better to have a shaft a little too short than over-long, somewhat too light than over-lumpish, a little too small than a great deal too big.' (ibid., 118–19)

### Length, diameter and weight

There is little in contemporary historical sources specifying length, diameter or weight of the arrow; however there are references to arrows being '*three quarters of a standard*'. If this means three-quarters of an English ell, then this would be 33¾in (851mm). If this refers to the English yard, this is 27in (686mm). There is also documentary evidence for arrows with 7in (178mm), 8in (203mm) and 9in (229mm) fletchings in 1475 specifically for warfare (Wadge 2007, 178). Accounts also suggest some variation in cost, whether this is related to length, length of fletching, type of wood or quality of wood or manufacture is not known and there are references to quality with standard and 'best' arrows (ibid., 163–75).

It is implicit that length (and therefore, in part at least, weight) of the arrow was governed by the size and strength of the bow which, in turn, fitted the size, strength and skill of the shooter. The length of the arrows drawn is also governed by the stature and length of the archer's arms. It is however possible for an archer to draw a longer arrow (if required, the extra length of the bow would increase its power, which could increase the range). It was taken for granted, however, that maximum range was achieved with a lighter arrow, whereas better penetration could be achieved with a heavier arrow:

### Shape

Ascham discussed shape with respect to the strength of the bow and nature of the shooting. Several forms are apparent, termed '*taper fashion*' or '*bobtails*' and those which are '*big-breasted*' (ibid., 120) (see Fig. 8.81). For a discussion on shape and how it affects performance, see Soar (2006, 89).

### The nock

The slot whereby the arrow sits on the string also appears to have been variable:

> 'The nock of the shaft is diversely made; for some be great and full, some handsome and little; some wide, some narrow, some deep, some shallow, some round, some long, some with one nock, some with a double nock, whereof everyone hath his property. The great and full nock may be well felt, and many ways they save a shaft from breaking, The handsome and little nock will go clean away from the hand; the wide nock is naught, both for breaking of the shaft and also for sudden slipping out of the string, when the narrow nock doth avoid both those harms. The deep and long nock is good in war for sure keeping in of the string.' (ibid., 121)

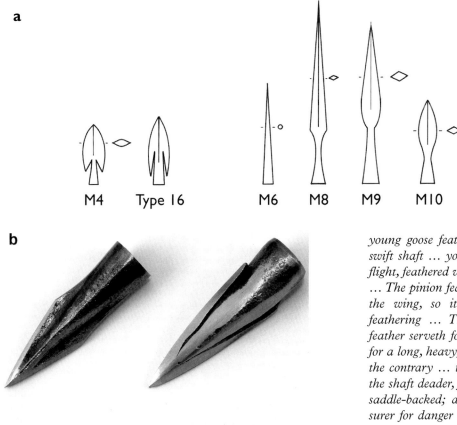

*Figure 8.71 a) types of leaf-shaped and bodkin arrowheads most probably used on the* Mary Rose *(typologies after Ward Perkins and Jessop, see text); b) reproduction bodkin and leaf-shaped arrowheads*

*The flights*
Much discussion is given over to the feathers, with goose feathers being paramount. *'there is no feather but only of a goose that hath all commodities in it'* (ibid., 124). However the age, sex and type of goose is discussed; whether the wing be left or right; which feather is used and whether the rib is thick or thin. This may relate to preparation; if the feather is cut this results in a thicker rib or spine (giving more stability to the feather but taking more time to prepare) but if it is paired or peeled this produces a thinner spine (and is a much quicker process), but requires steaming and pressing to regain shape before fixing to the shaft. The length of the flights are discussed as well as orientation to the nock. Finally, the form of the flight is discussed, whether this is '*swine-backed*' or '*saddle-backed*', whether '*round or square shorn*' (ibid., 123).

*'... you may fit your shafts according to your shooting ... The old goose feather is stiff and strong, good for a wind, and fittest for a dead shaft: the young goose feather is weak and fine, best for a swift shaft ... you must have divers shafts of one flight, feathered with divers wings, for divers winds ... The pinion feathers, as it hath the first place in the wing, so it hath the first place in good feathering ... The length and shortness of the feather serveth for divers shafts, as a long feather for a long, heavy, or big shaft, the short feather for the contrary ... the swine-backed fashion maketh the shaft deader, for it gathereth more air than the saddle-backed; and therefore the saddle-back is surer for danger of weather, and fitter for smooth flying. Again, to sheer a shaft round ... or after the triangle fashion, which is much used now-a-days, both be good. For roundness is apt for flying of his own nature, and all manner of triangle fashion, (the sharp point going before) is also naturally apt for quick entering.'* (ibid., 125–8)

*The heads*
By the sixteenth century wrought iron was the accepted raw material for arrowheads, however, the form and weight varied with function *'But, of all other, iron and steel must needs be the fittest for heads'* (ibid., 130). Three basic forms are described; the broad arrowhead or swallow tail, with two barbs going backwards towards the stele; the second termed the '*forked head*' (or crescent form) with two points stretching forwards, which were for hunting, and the third, the boring head, termed the '*bodkin*':

*'Thus heads which make a little hole and deep, be better in war, than those which make a great hole and stick fast in ... But, to make an end of all heads for war, I would wish that the head-makers of England should make their sheaf-arrow heads more harder pointed than they be.'* (ibid., 131–2)

*'Short heads be better than long: for first, the long head is worse for the maker to file straight compass every way; again it is worse for the fletcher to set straight on; thirdly, it is always in more jeopardy of breaking when it is on.'* (ibid., 134) (Fig. 8.71)

There are two current classification systems for arrowheads. Ward Perkins (1940) lists 21 different types while Jessop (1996) describes 28 types, the latter subdivided into four major groups: Tanged, Military, Hunting and Practice. The two which are the thought to be the most likely within the *Mary Rose* assemblage are the small barbed, lightweight, broad-head designed to penetrate and tear flesh (Ward Perkins Type 16, Jessop M4) and the piercing/penetrating small bodkin head, designed primarily to penetrate body armour as it increased in thickness and strength (Ward Perkins Type 7–10; Jessop type M6, 8, 9, 10; Fig. 8.71, top). Figure 8.71b shows recreations of a bodkin (left) and small barbed broad-head (right). Whatever the type, it is suggested that both types would need to be able to fit into the holes in the spacers and that the barbed heads were of a form which allowed withdrawal without catching on the leather. A small bodkin, or a small Type 16, would fit through the arrow spacers recovered and is also faster to produce these types of head than a larger type of barbed head; they would lend themselves to mass production (Jones in Hardy 1992, 121). It is possible to make a small bodkin or a small Type 16 in six minutes, whereas a Type 16 takes 20 minutes (Waller, pers. comm. 2009). An example of a small light-weight barbed head designed to pierce and cut flesh is found in the arrow (of uncertain date) in Westminster Abbey. (For a discussion on the manufacture and trialing of different forms of medieval arrowhead, see Stretton in Soar 2006, 102–52).

The weight of the head also varies, whereas M4/Type 16 examples weigh 6.45–7.51g, the Jessop M6/Type 7–8 heads weigh 10–20g (Pratt in Hardy 1992, 229). Tests on the M4/Type 16s suggest a composite structure comprising a hardened steel tip with a softer steel shank produced through forging, a heat treatment producing a hard and tough head (Jones in Hardy 1992, 232). A recent study of M4/Type 16 arrowheads (Starley 2005, 214) confirmed an iron socket with steel point quenched and hardened. It is suggested (Wadge 2007, 183) that hardening the arrowheads to improve penetration of armour may have been ordered and specified by the Crown as early as 1368. However the specifications for arrowheads '*according to the method and form which was delivered to me on the King's behalf*' is mentioned in a letter of 1336 from the sheriff of Norfolk and Suffolk to John Fletcher and Robert Seward, fletcher of Norfolk. This asks them to arrange for the seizure of all the flooks of anchors for making the heads '*as the custom has been*' (Wilson, pers. comm. 2009).

In 1406 Parliament enacted that:

> '*Because the Arrowsmiths do make many faulty Heads for Arrows and Quarels, defective, not well, nor* [lawful] *nor* [defensible,] *to the great Jeopardy and Deceit of the People, and of the whole Realm; It is ordained and established, That all the Heads and for Arrows and Quarels after this Time to be made, shall be* [well boiled] *or brased, and hardened at the Points with Steel..* (7 Hen. IV *c*. 7; *Statutes of the Realm*, 2, 153)

## Scope and importance of the assemblage

The large size of the assemblage (7834 numbered fragments including 2303 complete shafts) has meant that it has been impossible to record every fragment comprehensively, however 2203 (28%), including most of the complete examples, have been subject to detailed recording and the wood used for 647 (8.25% of recovered arrows) has been identified. The continued wet storage of the major portion of the assemblage has had a huge affect on accessibility as far as recording and analysis are concerned.

A number of studies have been undertaken by Elkerton (1983), Woodgate (1983a; 1983b), Randall (1985), Hopkins (1998; 1999) and Jackson (2003; 2005) (Fig. 8.72). All the studies are dominated by measurements, both linear between topographically defined features and diameter measurements at topographically defined points and there is considerable overlap. Various recording methods have been adopted. Unlike ceramic and lithic artefacts, for example, a clearly defined methodology for recording arrow shafts has not yet been developed, since prior to the excavation of the *Mary Rose* there were very few archaeologically recorded examples. An evolving methodology is reflected throughout the studies from the recording forms through the actual process of recording to the analysis and interpretation of the acquired data. The data collected in each case stimulated particular questions which were then addressed in later studies; therefore it is difficult to compare results from each study directly, especially when looking for the presence or absence of certain features. A feature prevalent within the sample of arrow shafts examined for one study may have been equally

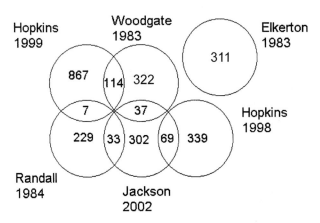

*Figure 8.72 Overlap of different arrow studies*

prevalent in another assemblage, but may have been either overlooked, or simply not identified. For example analysis for evidence of coppiced timber was not specifically addressed until the study of 2003, while evidence of tip cone working was restricted to the studies of 1998 and 1999.

All of the available data were reworked and subject to analysis by Keith Watson for this publication. Where conclusions are based only on specific studies rather than combined data, this is clearly stated. Amalgamation of the data has shown us where potential gaps within our knowledge lie and it is hoped that some of these may be addressed by future investigations. In studying the assemblage we must never lose sight of the reasons for recording in the first place; otherwise it is easy to get embroiled in minutiae.

The primary reasons for studying the arrow assemblages are to try and understand the function of archery on the *Mary Rose* by interrogating the characteristics of both the weapon and projectile. Was the function to harass, through volley-shooting (if that was possible on ship), or to snipe at targeted individuals? What strength/skill level was required and at what distances could these weapons be effective? Where were these found and does this add to our knowledge about the specific final engagement? Can the specifications of the arrows be used to help work out the maximum strength of the bows from which the arrows available must have been shot?

There are many features of an individual arrow. All are important and most are interrelated. Therefore, each should be recorded and studied with respect to every other feature. One of the most critical factors is complete weight, as this (probably more than any other single element) affects range and impact. Weight must be estimated based on the wood species (using density tables) for a given length and profile, hence the need to measure the assemblage fully. Complete metal heads are, however, absent from the assemblage. The arrow is a both vehicle to transport the head to the target and a stabilising feature during flight and penetration. While form and weight determine the aerodynamic qualities of the projectile, the shape of the arrowhead and the strength of the material determine the potential damage that can be inflicted upon the target. The shape, size (and therefore weight) of the heads must be extrapolated based on tip length, form and finishing, as must the length and weight of the fletchings, horn nock and binding. An excellent technical explanation is given by Pratt (in Hardy 1992, 226–32).

## Features studied

*Wood species*: Qualities inherent in different woods include variation in flexibility (spine), weight and ability to take impact. These may affect the function. Interrelated with both species and function may be length, diameter, shape and type of head. We can ascertain species but it is only with interrogation of the assemblage that we may be able to ascertain any patterns of other features. Other features within the wood include grain anomalies and choice of wood with respect to knots or divots. Whether the wood is coppiced or not obviously relates to supply, this distinction has not yet been achieved.

*Length*: In addition to complete length, the length of individual parts of the arrow can provide important information:
- The maximum draw-length (taken from the shoulder to the tip end of the nock slot). This relates to the length of the bow (and in turn the stature, strength/skill of the archer). It may also relate to function.
- The length of the tip upon which the iron head sits; variation in length within the collection as a whole can provide information on the length of head socket and possibly the head type.
- The width plus depth of the nock slot, providing information on string diameter.
- The length of the fletchings based on any imprints of binding or discoloration of this area. The length of the fletchings gives some indication as to the requirement to counterbalance the weight of the arrow, possibly suggesting function. Analysis of accounts suggest that 7in (178mm) was more prevalent for warfare (Wadge 2007, 178). Experiments with domestic goose feathers (Waller, pers. comm. 2008) suggests that 7¼in (183mm) is the longest vane readily produced.
- The height of the fletching is also very important in ensuring the stable flight of the arrow. As no complete fletchings survive from the *Mary Rose* an indication of the height of the fletchings on arrows of this period must be estimated from illustrations and paintings.
- The length of the portion of the stele, or shaft, between the shoulder and end of the fletchings.

*Diameter*: Diameters taken along the length of the stele provide a profile of the arrow. Profile obviously affects the arrow during flight, and particular profiles may relate to function. Diameters can either be taken at pre-determined places relative to other features; such as the shoulder, midway between shoulder and the start of the fletching, the start of the fletching, the end of nock and the end of the arrow; or they can be taken at pre-determined units, such as every 10mm along the shaft. Different approaches were taken by the varying studies.

*The arrow tip*: The length, diameter at shoulder and shape of the tip can provide information about the diameter and length of the cone supporting the arrowhead. By reference to extant collections of arrowheads, specifically the Museum of London and the British Museum, and resultant typologies (Ward Perkins 1940; Jessop 1996) types can be suggested

Table 8.35 Arrows: wood analysis

| No. | Species | % of sample |
|---|---|---|
| 2303 (complete) | 646 samples taken | 28% (of complete), 6.73% of listed |
| 501 | poplar | 77 |
| 90 | birch | 14 |
| 38 | alder | 6 |
| 7 | willow | 1 |
| 3 | elder | 0.50 |
| 2 | hornbeam | 0.30 |
| 2 | birch/poplar | 0.30 |
| 1 | hawthorn | 0.15 |
| 1 | ash | 0.15 |
| 1 | walnut | 0.15 |

based on 'best fit' (ie, date and form). These can be weighed and the materials and finishes analysed. This can provide information on potential damage achievable on a variety of targets. If any residue is noted beneath the concretion stained tips, this can be analysed for the potential presence of glue.

*The fletchings*: The material remains of any flights can be analysed together with any residue of glue used to affix the flights or binding. The binding thread can be analysed and the consistency in length, direction of twists and number of twists can be recorded. This can be used to provide an insight into consistency in manufacture, or perhaps suggest any formula that may have been applied.

*The nock*: The width and depth of the nock is a critical measurement. The least of these must be taken as an indicator of the maximum diameter of string, taking into account the possibility of any dimensional change. Any observed differences may relate to string sizes and therefore bow size and draw-weight. Any differences need to be noted against other features such as raw material, profile and length. The cutting of a slot in the tapered circular end of the shaft for the insertion of a tapering horn strip requires skill and precision. Obviously, where nock inserts were found separate from arrows (nine examples) these were also studied.

## Material analysis
### Wood species
Different types of wood were found in the *Mary Rose* assemblage and it is clear that selection was based on a number of factors. Ease of working into the required form, strength, spine and the weight of the wood are all inherent factors to be taken into consideration. External factors include availability and cost. Thin-sections taken from 646 out of 2303 complete arrows (28%; 6.7% of all arrows listed) and examined at up to x 400 magnification indicate the presence of nine species of wood (Table 8.35). The samples taken include examples from chests and from other contexts, including single spot finds and from sheaves. A large sample derives from chest 80A0726 on the Upper deck. This contained 688 arrow fragments of which 269 were analysed (39%); 267 were identified as poplar (*Populus* sp.), one alder (*Alnus glutinosa*) and one birch (*Betula* sp.), although the majority appear to have been identified visually rather than microscopically. As can be seen from Table 8.35, over 75% of all sampled arrows are poplar, with birch and alder between them accounting for a further 20%. Other woods identified include willow, hornbeam, elder, hawthorn and walnut. Ash (the most common wood suggested by Ascham) is only recorded from a single sheaf, 81A5643. The few sheaves examined seemed to contain mixed species, for example, 81A5643, 1–42 and 81A5849, 1–61 contained alder, ash, birch and poplar; but as all the contents of each sheaf have not been analysed, the proportions are, as yet, unknown.

Five chests containing arrows have had varying percentages of their contents analysed (Table 8.36). Most of these were recovered in what the excavators thought were '*bundles*', some possibly tied together (site books H/O 9–11, 1981–1982). In almost every case where '*bundles*' have been sampled these have contained

Table 8.36 Arrows: wood analysis of arrows in chests

| Chest | Sector | No. arrows | No. sampled | % of whole | Wood (no.) | % of sample |
|---|---|---|---|---|---|---|
| 80A0726 | U7/8 | 688 | 271 | 39 | poplar (267) | 99.0 |
| | | | | | birch (3) | 1.0 |
| | | | | | alder (1) | 0.4 |
| 81A5638 | O8 | 103 | 3 | 3 | poplar (3) | 100.0 |
| 81A2398 | O9 | 1345 | 179 | 13 | poplar (96) | 55.0 |
| | | | | | birch (48) | 27.0 |
| | | | | | alder (33) | 17.0 |
| | | | | | hornbeam (2) | 1.0 |
| | | | | | walnut (1) | 0.5 |
| | | | | | willow (1) | 0.6 |
| | | | | | elder (1) | 0.6 |
| 81A2582 | O9 | 2137 | 81 | 4 | poplar (56) | 69.0 |
| | | | | | birch (17) | 21.0 |
| | | | | | willow (5) | 6.2 |
| | | | | | elder (2) | 3.0 |
| | | | | | alder (1) | 1.2 |
| 82A1761 | H9 | 564 | 56 | 10 | poplar (25) | 44.0 |
| | | | | | birch (18) | 32.0 |
| | | | | | hawthorn (1) | 2.0 |
| | | | | | alder (1) | 2.0 |
| | | | | | willow (1) | 2.0 |

### Table 8.37 Arrows: wood identifications for arrows in 'bundles' in chests

*1. Chest 81A2398 (O9). Total arrows/frags in chest = 1345*

| Object | Total arrows | No. samples | Poplar | Birch | Alder | Elder | Willow |
|---|---|---|---|---|---|---|---|
| 81A2472 | 55 | 51 | 14 | 25 | 12 | 0 | 0 |
| 81A2449 | 63 | 66 | 38 | 11 | 15 | 1 | 1 |
| 81A2453 | 30 | 20 | 10 | 7 | 3 | 0 | 0 |
| 81A2462 | 29 | 21 | 19 | 2 | 0 | 0 | 0 |

*2. Chest 81AS2582 (O9). Total arrows/frags in chest = 2137*

| Object | Total arrows | No. samples | Poplar | Birch | Alder | Elder | Willow | Hornbeam |
|---|---|---|---|---|---|---|---|---|
| 81A2559 | 36 | 7 | 3 | 3 | 0 | 1 | 0 | 0 |
| 81A2556 | 23 | 10 | 4 | 5 | 1 | 0 | 0 | 0 |
| 81A2564 | 31 | 7 | 3 | 1 | 0 | 1 | 2 | 0 |
| 81A2520 | 39 | 10 | 9 | 0 | 0 | 0 | 1 | 0 |
| 81A2561 | 34 | 10 | 8 | 1 | 0 | 0 | 1 | 0 |
| 81A2494 | 22 | 5 | 5 | 0 | 0 | 0 | 0 | 0 |
| 81A2498 | 23 | 12 | 9 | 2 | 0 | 0 | 1 | 0 |
| 81A2491 | 26 | 3 | 2 | 0 | 0 | 0 | 0 | 1 |
| 81A2493 | 23 | 2 | 0 | 2 | 0 | 0 | 0 | 0 |

*3. Chest 82A1761 (H9). Total arrows/frags in chest = 564*

| Object | Total arrows | No. samples | Poplar | Birch | Alder | Willow |
|---|---|---|---|---|---|---|
| 82A1755 | 41 | 41 | 23 | 16 | 1 | 1 |

arrows mixed wood species (Table 8.37). Density tables for the woods commonly found are shown in Tables 8.38–41.

Tie-mark impressions covering a portion of the surface of arrow shafts were noted by Woodgate (1983b) confirming the tied bundles noted above, together with the remains of red thread on two shafts, which were assumed to be some form of colour coding. Ties were also noted by Elkerton (1983) from post-excavation analysis of arrows from an open chest on the Upper deck (80A0726, U7). Archaeologists excavating this chest underwater noted that all of the arrows were also in bundles but that there was no longer any evidence of ties. The bundles were lifted as units and decanted into containers (site book 1980, 7/8, 1, 157).

### Chemical and microscopic analysis

A number of other studies have been commissioned specifically to identify the materials present in the excavated arrow shafts through the application of chemical and microscopic analysis. The material associated with the binding for the fletchings has been subjected to a range of different analyses. A sample of feather was examined at the Royal Scottish Museum and tentatively identified as coming from a large species of waterfowl, either goose or swan, but more probably the latter (O'Berg, pers. comm., 1983).

Samples from the bindings and tips of six shafts were sent to North East London Polytechnic for analysis. Spectroscopic and chromatographic analysis of material obtained by solvent extraction from the bindings indicated the presence of beeswax and, probably, animal fat. Similar analysis of samples from the tip cones suggested a fish-based glue, but there was no wax (*AMR* Vol. 4, 632–3). Nothing was found on the surface of the shafts between the tip cone and the binding, suggesting that the materials found at either end were part of the original construction rather than subsequent contamination. The binding thread was found to be silk and the wax appeared to have been applied in a series of layers (Evans and Card 1984, *AMR* Vol. 4, 633 and table 16.2). Experimentation using goose feathers and water soluble hide glue suggests the following as a possible method (Waller, pers. comm. 2008): sticking the (peeled as opposed to cut and ground) feather by eye using hide glue, coating with a beeswax and tallow mixture to protect the glue against water, binding with silk to secure the feathers, painting with a second coat of the beeswax mixture to protect the silk thread (tallow acts as a moth deterrant). This attaches the feathers securely and provides the layering observed. A single arrow shaft (80A0476/3) was examined using X-ray fluorescence

### Table 8.38 Density tables of woods used in the *Mary Rose* arrow assemblage

| Wood | Density | BS | St | CS | SR |
|---|---|---|---|---|---|
| Alder | | | | | |
| Ash | He | M | L | M | M |
| Birch | He | H | M | H | M |
| Elder | | | | | |
| Elm (wych) | He | M | L | M | M |
| Hawthorn | | | | | |
| Hornbeam | He | M | M | H | M |
| Poplar | Li | L | v. L | M | v. L |
| Willow | Li | L | v. L | L | L |
| Walnut | | | | | |
| Ramin | He | H | M | H | L |

SG = Average specific gravity @ 12% moisture content; BS = bending strength; St = stiffness; CS = crushing strength; SR = shock resistance (see also Tables 8.40–1) He = heavy; Li = light; H = high; M = medium; L = low

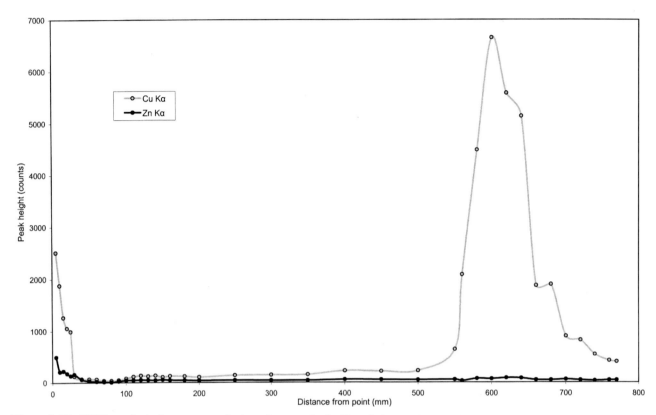

*Figure 8.73 XRF analysis for copper and zinc of arrow shaft (Royal Armouries) and arrow analysed*

**Table 8.39 Density classification**

| | |
|---|---|
| Light | 300–450 weight kg/m³ @ 12% moisture content |
| Medium | 450–650 weight kg/m³ @ 12% moisture content |
| Heavy | 650–800 weight kg/m³ @ 12% moisture content |
| V. heavy | 800–1000 weight kg/m³ @ 12% moisture content |

(XRF) analysis by the Royal Armouries in 2000 (Starley, AM1613). This included repeated analysis at 5mm increments along the length of the arrow and revealed a pattern in the presence of non-ferrous metal compounds. The cone, where the arrowhead had been attached, showed enhanced levels of copper and zinc (Fig. 8.73). At the nock, even higher levels of copper were found, associated with an applied layer where the fletchings had been attached. Previous analysis of this material by gas chromatography and gas chromatography/mass spectrometry had revealed the presence of beeswax and a series of triacylglycerides, thought likely to derive from animal fat (*AMR* Vol. 4, 632–3, table 16.2). An interpretation could be the treatment of fletchings with verdegris to protect them or the glue from fungal or bacterial attack. Excavations at Holm Hill, near Tewkesbury, unearthed 72 arrowheads of which 57 had been coated with variations of copper alloys to produce a plated arrowhead. Copper is a well-known poison when introduced directly into the

**Table 8.40 Mechanical properties**

| Property | BS | St | CS | SR |
|---|---|---|---|---|
| Unit | N/mm² | kN/mm² | N/mm² | m |
| V. low | <50 | <10 | <20 | <0.6 |
| Low | 50–85 | 10–12 | 20–35 | 0.6–0.9 |
| Medium | 85–120 | 12–15 | 35–55 | 0.9–1.2 |
| High | 120–175 | 15–20 | 55–85 | 1.2–1.6 |

**Table 8.41 Specific gravity and modulus of elasticity (after Farmer 1972)**

| Wood | Av. SG | BS N/mm² | MoE N/mm² | Notes |
|---|---|---|---|---|
| Alder | 0.53 | 80 | 8800 | |
| Ash | 0.69 (av) | 116 | 11,900 | V. tough, used for sports equipment |
| Birch | 0.66 | 123 | 13,300 | Tough as ash, similar to hornbeam |
| Elm (wych) | 0.55 (0.60) | 68 | 7000 | Distorts easily |
| Hornbeam | 0.75 | 119 | 11,000 | Hard, tough, similar to birch |
| Poplar | 0.45 (av) | 41 | 8600 | Tough for weight, similar to willow |
| Willow | 0.45 | 63 | 5800–7000 | Lightweight, easily worked, poor bending, similar to poplar |

MoE = Modulus of elasticity

Table 8.42 Distribution of wood types (Randall 1985 (table 5 of Cooke's data)

| Wood | No. arrows/frags | % |
|---|---|---|
| Poplar | 148 | 55.8 |
| Birch | 75 | 28.3 |
| Alder | 31 | 11.7 |
| Willow | 7 | 2.6 |
| Elder | 3 | 1.1 |
| Hornbeam | 1 | 0.4 |

bloodstream and this may be one reason why it was used to attach the heads, or else as a plating to the iron head. There is some evidence to suggest that the French claimed the English arrows to be poisonous, however there is very little evidence to support the use of verdegris as protection against fungal or bacterial attack. Analysis of further arrows for confirmation of the presence of copper associated with the fletchings is recommended, although there are comments of a green tinge to the vestigial fletchings before conservation.

The fragmentary remains of a small, possibly barbed, arrowhead and the iron oxide surface film of another (identified as being of Ward Perkins (1940) Type 16/Jessop (1996) Type M4) were examined at Imperial College, London (Mary Rose Trust archive). In the second case the iron itself appeared to have dissolved away leaving – more or less intact – the oxide film formed when the metal was forged (Figs 8.68 and 8.71a).

*Analysis*
by Keith Watson

The data from all studies were combined to give a sample of 1051 arrow shafts (46% of the complete shafts recovered, 11% of the total listed for the vessel). The criteria for selection included completeness (the

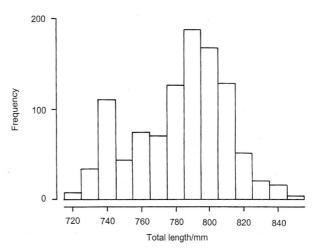

*Figure 8.75 Histogram of total length of the arrow shafts*

ability to provide all the required measurements) and that the arrows had been measured before conservation. Where the arrows had been studied more than once, the earliest set of measurements was used. The majority of arrows used in this analysis were distributed between chests 80A0726 (U7), 81A2582 (O9), 81A2398 (O9) and 82A1761 (H9). The principal length measurements used are shown in Figure 8.74. The sub-set used for any information involving wood species consisted of 265 arrows taken from Randall's study of 1985 (Table 8.42).

**Complete length**

A wide spread of complete arrow shaft lengths was recorded ranging from 667mm to 880mm. Most lie in the range 715–854mm in a bimodal (double peaked) distribution with modes at 740mm and 790mm (Fig. 8.75). Subtracting the median nock depth (6mm) and the median tip length (22mm) from the two modal values gives estimated draw lengths of 712mm (28.03in) and 762mm (30.00in) respectively. This is in agreement with previously published values (Pratt 1992, 208) although the present assemblage contains a higher proportion of 30in arrows, providing a ratio of 4½:1 relative to the 28in arrows, compared with 3:1 suggested by Pratt.

Table 8.43 Two-way distribution of location and length group

| | Length group (mm) | | | | |
|---|---|---|---|---|---|
| Sector | <700 | 700–714 | 715–754 (740mm) | 755–854 (790mm) | >854 |
| U7 chest 80A0726 | | 8 | 45 | 258 | |
| O9 chest 81A2582 | | | 9 | 153 | |
| O9 chest 81A2398 | | | 18 | 290 | 9 |
| H9 chest 82A1761 | 6 | | 113 | 119 | |
| Loose | | | 2 | 21 | |

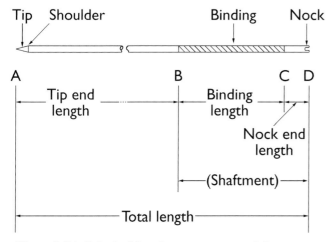

*Figure 8.74 Principal length measurements of the arrow shafts*

**Table 8.44 Mean total length and 95% confidence limits by arrow chest (from Randall's data)**

| Chest | Total length (mm) | |
|---|---|---|
| | Mean | 95% |
| 81A2582 (O9) | 788.8 | 784.5–793.1 |
| 81A2398 (O9) | 800.8 | 795.6–806.0 |
| 82A1761 (H9) | 759.6 | 751.8–767.4 |

There is a strong linear relationship between total length (Fig. 8.74, AD) and tip end length (Fig. 8.74, AB) for the combined data (Fig. 8.76a). There are three small outlying clusters which occupy specific locations (Table 8.43): <700mm in chest 82A1761 (H9), 700–714mm from the 'in use' chest on the Upper deck (80A0726) and those >854mm in chest 81A2398 (O9). The 790mm modal group (755–854mm) predominates in all locations except that from H9 where the 740mm modal group (715–754mm) is also present in almost equal numbers. The 95% confidence limits of total length by chest are displayed in Table 8.44 (calculated from Randall's 1985 study of shafts from chests 81A2582, 81A2398 and 82A1761).

The significance of these findings is unclear but it may indicate that two main draw-lengths were requested or possibly that two different head forms were used for different purposes.

**Binding length and nock length**

The regression equation in Figure 8.76a indicates that total length increases by close to 1mm for every 1mm increase in tip end length. Also, the coefficient of determination ($R^2$) indicates that 97.7% of the variation in total length is related to the variation in tip end length, leaving only 2.3% related to other factors. Elkerton's data from 311 shafts from Upper deck chest 80A0726 (Fig. 8.76b) suggests that the binding length (Fig. 8.74, BC) is inversely related to the nock end length (Fig. 8.74, CD). The equation suggests that the binding length decreases by about 1mm for every 1mm increase in nock end length, keeping their sum (Fig. 8.74, BD) more or less fixed. The constants in both regression equations (Fig. 8.76a and b) suggest that the sum is in the region of 220–30mm. The percentiles of this combined length were therefore computed (Table 8.45). They show that 50% of the values lie between 220mm and 225mm and 90% between 213mm and 228mm, with the median (mid-point) value at 223mm. Clearly considerable effort was made to keep this length constant (see below).

**Table 8.45 Percentiles of binding length + nock end length (from Elkerton's data)**

| % | 5 | 10 | 25 | 50 | 75 | 90 | 95 |
|---|---|---|---|---|---|---|---|
| mm | 213 | 216 | 220 | 223 | 225 | 228 | 228 |

**Table 8.46 Mean and 95% confidence limits for total length and for binding length+nock end length by wood type (from Randall's data)**

| Wood | Total length (mm) | | Binding+nock end length (mm) | |
|---|---|---|---|---|
| | Mean | 95% | Mean | 95% |
| Poplar | 785.8 | 780.3–791.3 | 223.2 | 222.5–223.9 |
| Birch | 788.7 | 783.1–794.2 | 221.9 | 220.9–222.8 |
| Alder | 812.6 | 794.6–830.6 | 218.0 | 216.3–219.7 |

Figure 8.76c, plotted from Randall's data, shows the relationship between total length and tip end length by wood type. Alder appears in the upper part of the main group and forms a separate cluster at around 875mm total length. Table 8.46, calculated from Randall's data, shows the 95% confidence limits for total length and for binding plus nock end length for poplar, birch and alder (there is insufficient data for the other wood types to make significant comparisons with them). The alder shafts have the largest mean total length and the smallest mean binding plus nock end length. Table 8.47 shows that the alder shafts were mostly concentrated in chest 81A2398 (O9), presumably contributing towards the greater mean total length of the shafts in this chest (see Table 8.44). Once again, the significance of these observations, if any, is not clear and further research is required.

Figure 8.77 was plotted from Randall's diameter data and suggests that, towards the tip end, the poplar and alder shafts tend to be thin and, noting the very small sample size, the willow shafts tend to be thick. The shape of these profiles is discussed in more detail below.

A visual study of the bindings (Woodgate 1983a; 1983b) suggests that there was a staged approach with the feathers being glued into position and then secured with the thread, followed by application of another coat of glue. In most of the 322 shafts used in her study, the binding just covered the front end of the slot for the nock insert. Randall (1985) noted in sixteen examples (out of 269 shafts and fragments studied: 5.9%) impressions of thread around the nock end consistent with thread binding which could have been used to hold the nock insert in place as the glue was setting, or

**Table 8.47 Two-way distribution of wood type and arrow chest (from Randall's data)**

| Wood | Chest 81A2582 (O9) | Chest 81A2398 (O9) | Chest 82A1761 (H9) | Total |
|---|---|---|---|---|
| Poplar | 33 | 34 | 17 | 84 |
| Birch | 12 | 35 | 16 | 63 |
| Alder | 1 | 17 | 0 | 18 |
| Willow | 3 | 1 | 1 | 5 |
| Elder | 1 | 0 | 0 | 1 |
| Hornbeam | 1 | 0 | 0 | 1 |
| Total | 51 | 87 | 34 | 172 |

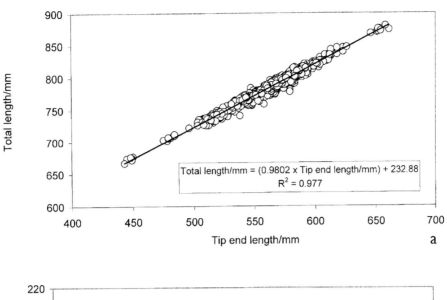

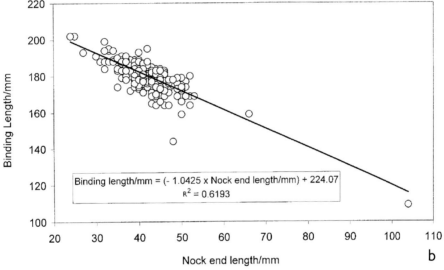

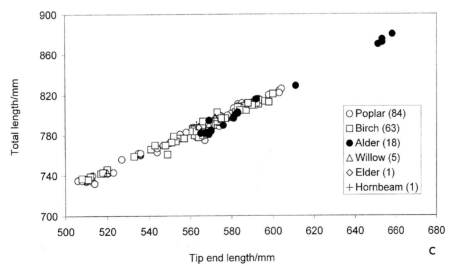

*Figure 8.76 a–b) Arrow total length vs tip end length, different variables; c) binding length vs nock end length (data after Elkerton 1983)*

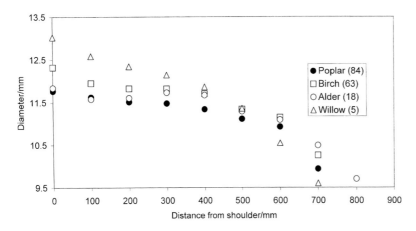

*Figure 8.77 Arrow shaft profiles by wood type (data after Randall 1985). N.B. The number in parentheses indicates sample size*

possibly added as a reinforcement. A survey of late fifteenth century and earlier paintings including arrows reveals definite and regular binding along the feathers (Hardy, pers. comm., 2008). Jackson (2003) recorded bindings which overlapped the nock insert in 52% of the sample from 441 arrows, suggesting that the nock insert had been fitted before the binding was applied. Randall also found two instances of double impressions within the bindings which suggested that the position of the fletching was altered before being finally bound in place. One example showed additional impressions as though the thread had been removed and replaced. He noted eleven examples where the flights had been set at a slight angle to the long axis of the shaft, possibly to increase the spin of the arrow during flight. Its presence among the assemblage is in less than 5% of the recorded shafts.

Hopkins (1998) studied 408 shafts from chest 81A2582 (O9) and recorded that, in every case, the binding thread had been wound in a clockwise direction from the tip end of the shaftment (ie, the portion of the arrow where the fletching is placed) to the nock end (ie, 'Z' wound). This indicates that the arrows have been bound by a right-handed person (Waller, pers. comm. 2009). Jackson recorded the binding frequency for 434 out of the 441 shafts he studied, 84.1% having values of five or six turns per inch (Table 8.48).

**Table 8.48 Distribution of binding fequency (from Jackson's data)**

| Turns per inch | No. arrows | % |
|---|---|---|
| 4 | 56 | 12.9 |
| 5 | 164 | 37.8 |
| 6 | 201 | 46.3 |
| 7 | 12 | 2.8 |
| 8 | 1 | 0.2 |

### Nock length, width and orientation

Nock width and depth were measured in two studies (Hopkins 1998; Jackson 2003; 2005). Widths varied between 1.8mm and 5.3mm and depths between 2.0mm and 10.0mm (combined data), giving overall 95% confidence intervals of 3.09±0.08mm and 6.19±0.14mm for mean width and depth respectively (see Fig. 8.78a). This variation must be seen in context; the nock is cut into the base of the arrow to enable it to fit on the bowstring, therefore a range of nock widths may be indicative of a range of bowstring diameters. Analysis aimed at identifying any correlation between the width of the nock, its depth, the diameter at the nock and draw length proved inconclusive (Fig. 8.78a–c). Analysis of the orientation of the nock relative to the annual growth was undertaken for 441 arrows (Jackson, 2003; 2005) and produced a range of orientations (Fig. 8.79). Apart from a large number of arrows that could not be measured, the majority of cases either followed the annual growth (44.8%) or ran perpendicular to it (33.7%), with a small proportion (13.9%) aligned at 45°.

### Nock inserts

The evidence for nock inserts consists of the presence of a slot cut across the width of the nock to enable the insertion of a tapered horn strip to strengthen and reinforce the nock. There are nine extant horn nocks recovered separately from arrows and one *in situ*. These vary in width between 2.8mm and 4.0mm, which would have to accommodate any string (see below). As the horn is less likely to have altered in size post-depositionally than wood, the confidence in these measurements is high.

Two inserts have been identified as horn (Armitage 1983). None was found among the 311 shafts studied from chest 80A0726 in U7, although Elkerton recorded a dark glutinous substance within the nock slots (Elkerton 1983).

### Arrow tips

The most detailed work on the tip cones was undertaken by Hopkins (1998) and concentrated on 408 shafts from chest 81A2582 in O9. He identified four 'types' of tip cone (Fig. 8.80). Type 1, the least common, was smooth. However, Hopkins noted that six of the eight Type 1 shafts were the only ones in his study to have been conserved, possibly accounting for their smoothness. Type 2 cones appeared to have been whittled to a point, leaving 4–8 facets. Since no 'nicks' were evident at the shoulder, Hopkins suggested that the cones had been whittled away from the shoulder towards the tip – like sharpening a pencil. In contrast, Soar (2006, 96) describes a method which involves forming the shoulder by cutting a groove around the

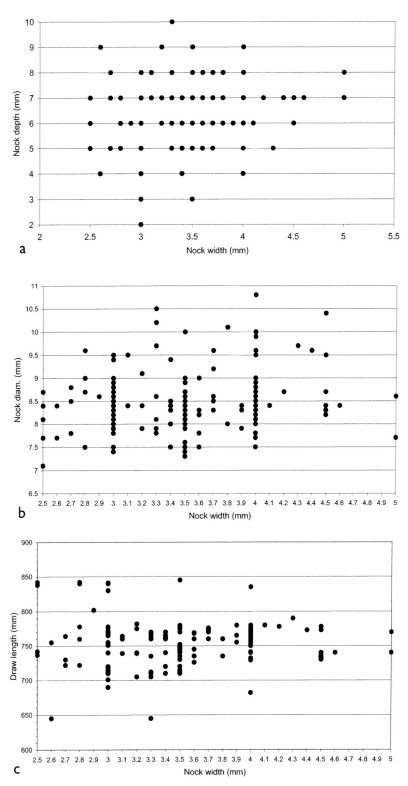

*Figure 8.78 a) Nock depth vs width; b) nock diameter vs nock width; c) draw length vs nock width (data after Jackson 2003)*

circumference and, initially, cutting back towards it from the tip direction, then forming the cone by cutting forwards towards the tip, as before, and finishing it with a file. Types 3 and 4 are both derivatives of Type 2 that have been shortened either by cutting a secondary, steeper cone or by slicing the tip off diagonally, presumably because it was originally too long to fit the arrowhead socket. The various types of tip cone were evenly distributed among the shafts, irrespective of length and thickness.

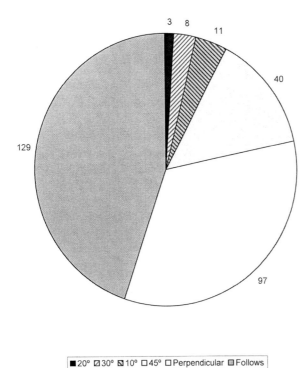

*Figure 8.79 Nock orientation relative to annual growth (data after Jackson 2003)*

The length of arrow tips found within the assemblage varied between 8mm (81A2509/28, 81A2458/18) and 29mm (82A1888/13, 82A1889/1) with the majority in the 15–26mm range.

Table 8.49 is a two-way frequency table of nock end diameter and tip end diameter values. Most of the arrows fall into the 11–13mm range of tip end diameter values and the 9–10mm range of nock end diameter values. Furthermore, the tip end diameter is invariably larger than the nock end diameter.

Double shoulders were noted by Woodgate (1983a; 1983b) in one instance (out of 322). Randall found seven instances (out of 269). The presence of a double shoulder suggests that the base of the cone may have been too wide for the open end of the socket to reach the shoulder.

Jackson (2003) noted that the tip cones were very crudely fashioned compared with other parts of the shafts. He referred to a fourteenth century document which mentioned headless arrows and arrowheads purchased separately. This suggests the possibility that the *Mary Rose* arrowheads might not have been fitted by the fletcher but separately by somebody else. During the manufacture of a medieval arrowhead the ferrule would be beaten around a mandrel to ensure conformity of its taper, but the overlap of the metal inside each arrowhead means it needs individual fitting to the arrow shaft. The tip of the shaft would be pared with a knife (as in sharpening a pencil) until a reasonable fit was obtained. When the glue was applied it filled any gaps and ensured a good fit (Waller, pers.

**Table 8.49 Two-way distribution of nock end diameter and tip end diameter (from Elkerton's data)**

| Nock end diam. (mm) | Tip end diameter (mm) | | | | | Total |
|---|---|---|---|---|---|---|
| | 10 | 11 | 12 | 13 | 14 | |
| 8 | 1 | – | 2 | 4 | – | 7 |
| 9 | 12 | 16 | 29 | 34 | 1 | 92 |
| 10 | – | 24 | 101 | 68 | 4 | 197 |
| 11 | – | – | 2 | 8 | 2 | 12 |
| Total | 13 | 40 | 134 | 114 | 7 | 308 |

comm. 2009). Despite the apparently crude workmanship of the tip cones, the mean angle at the tip lies close to 30° regardless of the shaft profile.

**Combined data profiles**

Four types of arrow shaft profile are generally recognised today (Soar 2003, 5; 2006, 71–2). Two of these were described by Ascham in 1544–5, at about the time the *Mary Rose* sank:

1. *'Barrelled'* shafts, which taper to each end from a maximum diameter roughly halfway along the shaft. This profile reduces vibration in flight giving greater stability, which is particularly helpful for shooting at a distance.

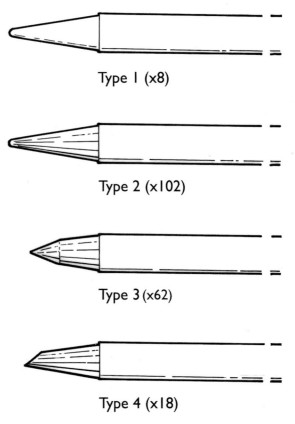

*Figure 8.80 Tip cone types (after Hopkins 1988)*
*N.B. number in parentheses indicates sample size*

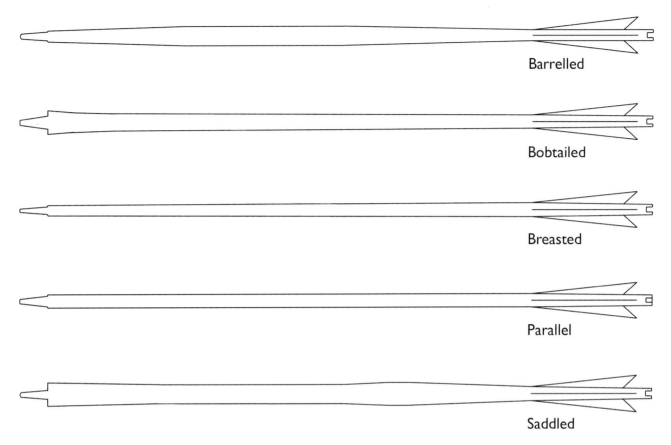

*Figure 8.81 Arrow shaft profiles*

2. *'Bobtailed'* shafts which taper from shoulder to nock. These were regarded as appropriate for distance shooting (Ascham 1544–5, 120) – and they could carry a heavier arrowhead (Soar 2006, 71).
3. *'Big-breasted'* shafts, which taper from nock to shoulder; Ascham viewed these as *'fit for him which shooteth right afore him'* (ibid.), in other words at point blank range.
4. *'Parallel'* shafts, of more or less constant diameter, which are the ordinary shafts used for general purposes.

Walrond (1894, 309) classified arrows in much the same way: *'bobtail'*, *'barrelled'* and *'parallel'* – but instead of *'breasted'* he described *'chested'* arrows as *'largest at from 12 to 18 inches [c. 300–460 mm] from the nock'*. Randall (1985) and Woodgate (1983a; 1983b), between them, recognised barrelled, bobtailed and parallel types of shaft – or *'tapered'*, *'torpedo'* and *'ordinary'* respectively, as they described them. However, they used rather stringent numerical criteria, largely based on the first 200mm from the shoulder, which restricted classification. The present analysis has used their original data and that of Hopkins (1998; 1999) to classify the profiles of individual shafts according to the four types described above. In addition to these, a number of *'saddled'* shafts were identified that had a slightly reduced diameter towards the middle (Figs 8.81–2) – as observed by Woodgate (1983b). As far as current practice is concerned, there appears to be no technical justification for this type of profile (Boyton, pers. comm. 2009); it is possible that it may have resulted from shrinkage due to differential moisture content when the shafts were made or, as Soar suggests (2006, 72), inept profiling – or it may even have been some kind of novel design that has not stood the test of time.

In the present analysis a longitudinal profile was produced for each of the shafts based on their diameters at 100mm intervals from the shoulder. These were then categorised (Table 8.50) and *'average'* profiles plotted (Fig. 8.82a). The *'breasted'* profile tallies very closely with Walrond's definition of *'chested'* shafts.

Clearly the greatest differences occur near the shoulder but the profiles converge so that, by 500mm, they are all very similar. Table 8.51 and Figure 8.82b show how the shafts with different profiles were distributed between the arrow chests (chest 80A0726 is not included since profile data is not available). If we consider the vertical columns in Table 8.51, the total sample (right hand column) contains, in decreasing order of frequency, mostly bobtailed shafts, then parallel, then barrelled, then saddled and, finally,

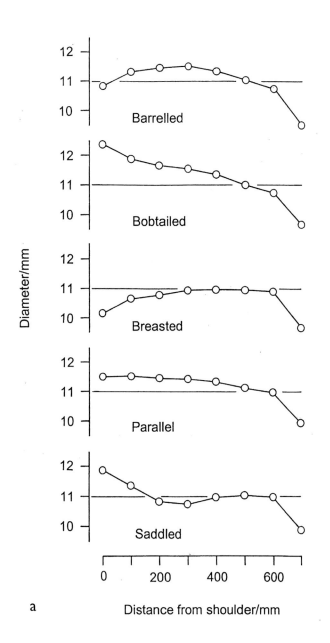

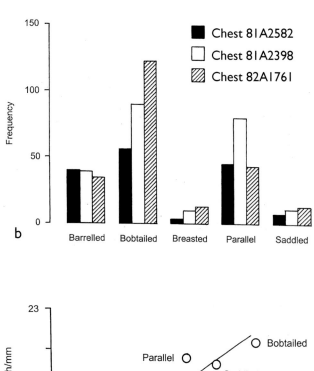

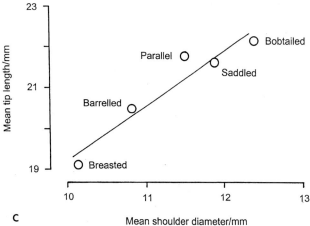

*Figure 8.82 a) arrow shaft profiles; b) distribution of arrow shaft profiles by location; c) mean tip length vs mean shoulder diameter*

breasted. This order is maintained within each chest, with the exception of chest 82A1761, which contains equal numbers of saddled and breasted shafts. The distribution of profile types within each chest therefore reflects the overall distribution.

The same is not true if we consider the horizontal rows shown in Table 8.51. The bottom (total) row shows that there are roughly equal number of shafts from chests 81A2398 and 82A1761 but considerably fewer from chest 81A2582. However, Table 8.51 and Figure 8.82b clearly show that these proportions are not maintained if we consider each type of profile separately. For example, there are more or less equal numbers of barrelled shafts in all three chests – rather more than would be expected in chest 81A2582. But more significantly, in chest 82A1761 there is a markedly higher number of bobtailed shafts than expected – also a markedly lower number of parallel shafts (to the benefit of chest 81A2398). This, and the unexpectedly high proportion of 740mm shafts in chest 82A1761 (see Table 8.43), suggests the possibility of a relationship between profile type and length group. However, further investigation failed to find a significant association between them.

Table 8.52 shows the two-way distribution of wood type and profile. Poplar and birch are mostly associated

**Table 8.50 Distribution of arrow profiles**

| Profile | No. | % |
|---|---|---|
| Barrelled | 119 | 18.9 |
| Bobtailed | 272 | 43.1 |
| Breasted | 28 | 4.4 |
| Parallel | 180 | 28.5 |
| Saddled | 32 | 5.1 |
| Total | 631 | 100.0 |

**Table 8.51 Two-way distribution of profile and arrow chest (expected frequencies in parentheses)**

| Profile | Chest 81A2582 (O9) | Chest 81A2398 (O9) | Chest 82A1761 (H9) | Total |
|---|---|---|---|---|
| Barrelled | 40 (28.6) | 39 (43.1) | 35 (42.3) | 114 |
| Bobtailed | 56 (67.3) | 90 (101.2) | 122 (99.5) | 268 |
| Breasted | 4 (6.8) | 10 (10.2) | 13 (10.0) | 27 |
| Parallel | 45 (42.2) | 80 (63.4) | 43 (62.3) | 168 |
| Saddled | 8 (8.0) | 11 (12.1) | 13 (11.9) | 32 |
| Total | 153 | 230 | 226 | 609 |

with bobtailed and parallel shafts, then barrelled, saddled and breasted – in that order, in keeping with the vertical trend in Table 8.51. This is reflected in Figure 8.77, which shows that the profiles for poplar and birch are more or less parallel to each other and taper from the shoulder to a relatively straight section around 200–300mm before tapering to the nock. Although the sample size for alder is small, the profile is initially very similar to poplar, then barrelling becomes evident between 200mm and 500mm, reflecting the preponderance of bobtailed and barrelled shafts. The willow sample is extremely small but there is some statistical evidence that the willow shafts tend to be the thickest towards the shoulder.

Clearly two main draw-lengths are evident within the assemblage with the 762mm (30in) being over four times more prevalent then the 712mm (28in) examples. With the exception of one chest in H9 (82A1761), where there are similar numbers of these draw-lengths, the 762mm group predominates in all locations.

Whether the different draw-lengths were to suit individual archers or for particular functions or for differing types of heads is not known. The first interpretation implies that there was a method of assigning and matching both longbows and arrows to individuals, and that there were individuals on board who possessed the knowledge to do so. The second suggests sorting based on function, possibly with different arrow lengths for differing ranges. What is interesting is the mixing of these within the chests, suggesting that whatever sorting took place before an engagement only required a limited number of chests to be taken out of store. Otherwise, if all the arrows of one size were being stored in a single chest, then either pre-sorting in the storage area or the bringing out of more chests would be required. This mixing suggests the latter: an individual may choose a bow out of an open chest which contains a range of longbows that, at first glance, suits his build, but takes (or is given) a sheaf of arrows of mixed proportions. How, why and when he uses them may be a matter of personal choice. Pre-sorting into sheaves containing a mixture of arrows would obviate any last-minute attempts to try and source specific arrows from chests under battle conditions.

There are significant differences in mean values of tip length between profiles and it is tempting to suppose that this reflects the use of different types of arrowhead. However, mean tip length forms a strong linear relationship with mean shoulder diameter (Fig. 8.82c). Simple trigonometry shows that the mean angle subtended at the tip varies only between 29.6° and 31.2° for the five types of profile. It therefore seems likely that the relationship merely reflects a constant tip cone angle for the different profiles. Although the sample of alder shafts is small, these seem to differ from the others and it is tempting to suggest that they may have had a different purpose. They are, on average, longer than the poplar and birch shafts and have shorter binding plus nock end lengths – and some of them occupy a discrete cluster of shafts that stands out from all the rest (Fig. 8.76c). Furthermore, they tend to be concentrated in one chest (81A2398; O9).

The majority of arrows studied are of poplar and birch. Of the five profiles observed, bobtailed predominate followed by parallel and then barrelled. In general there is no evidence to suggest that there was anything special about the shafts made from poplar or birch which, between them, comprised 84% of Randall's sample of 265 arrows and fragments. The extremely small sample of willow shaft hints that they were thicker towards the shoulder than the rest.

**The effect of species and profile on arrow weight**

Examination of the arrow assemblage revealed that there are two main length groups of arrow shaft, five different arrow profiles and nine different woods. Combined with the potential for at least two different arrowheads (Ward Perkins (1940) type 16/Jessop (1996) type M4 and types 8–10) there appears to be a great potential for a huge variance in projectile weight based on the different relative density for the most predominant woods within the assemblage.

**Table 8.52 Two-way distribution of wood type and profile**

| Wood | Barrelled | Bobtailed | Breasted | Parallel | Saddled | Total |
|---|---|---|---|---|---|---|
| Poplar | 19 | 34 | 1 | 28 | 2 | 84 |
| Birch | 7 | 33 | 1 | 20 | 2 | 63 |
| Alder | 7 | 6 | 1 | 3 | 1 | 18 |
| Willow | – | 3 | – | 2 | – | 5 |
| Elder | – | 1 | – | – | – | 1 |
| Hornbeam | – | 1 | – | – | – | 1 |
| Total | 33 | 78 | 3 | 53 | 5 | 172 |

**Table 8.53 Arrow mass (g) by wood type**

| Parameter | Alder | Birch | Poplar |
|---|---|---|---|
| No. shafts | 18 | 63 | 84 |
| Arithmetic mean | 42.3 | 52.4 | 33.5 |
| Standard deviation | 3.6 | 5.1 | 3.7 |
| 95% conf. limit | 40.5–44.1 | 51.1–53.6 | 32.7–34.4 |
| Range of values | 37.9–49.1 | 39.7–64.6 | 25.3–41.6 |
| Mean+ Type 16 (7g av.) | 49.3 | 59.4 | 40.5 |
| Mean + Type 7 (15g av.) | 57.3 | 67.4 | 48.5 |

Ignoring profile in the first instance, shaft volume was calculated using the mean of the radii measured at 100mm intervals from the shoulder assuming the nock end radius to be the same as the last recorded diameter and adding the volume of the tip cone.

For each arrow the total volume ($m^3$) was multiplied by the density ($kg/m^3$) to give its mass ($m^3 \times kg/m^3 = kg$). This was multiplied by 1000 to give the mass in grams. The density values used were 450kg/$m^3$ for poplar, 660 kg/$m^3$ for birch and 530 kg/$m^3$ for alder (Farmer 1972).

Wood type seems to be the predominant factor, and Table 8.53 shows that its overall effect is quite large with 33.5g as the mean for poplar and 52.4g for birch. Including the weight of the lightest head on the poplar shaft gives 40.5g, and a heavy head on the birch shaft gives 67.4g.

The five profiles and different lengths for each of the major wood types were also examined with respect to mass (Table 8.54). Because of the small sample size it is impossible to compare both lengths of shaft over all wood types and profiles and a formal overall analysis is virtually impossible. However, a more detailed examination of the predominant length group (755–854mm) shows statistically significant differences ($p<0.05$) between the mean arrow shaft masses for:

- Alder, between barrelled and bobtailed (+11%);
- Birch, between barrelled and bobtailed (+9%);
- Poplar, between barrelled and parallel (+8%).

The values in brackets are the percentage differences relative to the barrelled shafts and are possibly no more than a reflection of the fact that barrelled shafts taper from the middle towards both ends and therefore tend to be on the light side.

With this in mind shaft volumes were tested (pooled across wood type and length group) producing the following means:

Barrelled: $7.31 \times 10^{-5}$ $m^3$
Bobtailed: $7.73 \times 10^{-5}$ $m^3$
Parallel: $7.93 \times 10^{-5}$ $m^3$

Comparing the differences between these means demonstrates the following:

Barrelled vs bobtailed, significant ($p<0.05$) (6%)
Barrelled vs parallel, very highly significant ($p<0.001$) (8%)
Bobtailed vs parallel, not significant ($p=0.17$) (3%)

In other words, the mean volume of the barrelled shafts, taken as a whole, is less than the mean volume of bobtailed and parallel shafts.

Arrow mass and total length were plotted by wood type (Fig. 8.83). This demonstrates a clear discrimination between wood types.

## Other features of the shafts
by Adam Jackson

A range of different grain anomalies within the timber was identified in a study of 441 shafts (Jackson 2003; 2005). Features such as knots and examples of grain breakout appear to have had a significant impact on the arrow shafts, particularly the resulting orientation of the feature in relation to the nock and, consequently, its orientation when the arrow was shot. They have the potential to provide some further insights into the quality of the timber used.

There were many different types of knots observed; complete knots, dead knots, clusters of knots and some that were so close to the surface that they had been removed altogether. While knots are the result of the natural growth of the timber, grain breakout (or a divot) is associated with the working of the timber with hand

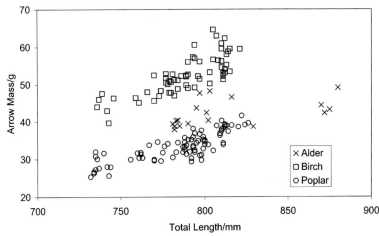

*Figure 8.83 Arrow mass vs total length by wood type (data after Randall 1985)*

**Table 8.54 Arrow mass (g) by wood type, length group and profile**

|  |  | Barrelled | Bobtailed | Breasted | Parallel | Saddled |
|---|---|---|---|---|---|---|
| *Alder (N=18)* | | | | | | |
| Overall length 755–853mm | No. | 5 | 6 | 0 | 2 | 1 |
| | Mean | 39.0 | 43.4 | | 39.6 | 48.2 |
| | Std Dev. | 0.9 | 3.3 | | 1.1 | |
| | Range | 37.9–40.2 | 39.4–47.7 | | 38.9–40.4 | |
| >854mm | No. | 2 | 0 | 1 | 1 | 0 |
| | Mean | 43.8 | | 42.4 | 49.1 | |
| | Std Dev. | 0.9 | | | | |
| | Range | 43.2–44.4 | | | | |
| *Birch (N=63)* | | | | | | |
| Overall length 715–754mm | No. | 1 | 5 | 0 | 0 | 0 |
| | Mean | 39.7 | 45.4 | | | |
| | Std Dev. | | 1.9 | | | |
| | Range | | 43.0–47.6 | | | |
| 755–854mm | No. | 6 | 28 | 1 | 20 | 2 |
| | Mean | 49.6 | 54.0 | 50.1 | 53.0 | 56.5 |
| | Std Dev. | 3.1 | 4.7 | | 4.4 | 4.3 |
| | Range | 45.1–53.0 | 47.8–64.6 | | 45.6–60.5 | 53.4–59.5 |
| *Poplar (N=84)* | | | | | | |
| Overall length 715–754mm | No. | 3 | 6 | 1 | 1 | 1 |
| | Mean | 29.3 | 27.3 | 28.0 | 27.2 | 32.0 |
| | Std Dev. | 2.7 | 2.0 | | | |
| | Range | 26.3–31.6 | 25.3–30.8 | | | |
| 755–854mm | No. | 16 | 28 | 0 | 27 | 1 |
| | Mean | 32.9 | 34.4 | | 35.4 | 32.8 |
| | Std Dev. | 2.4 | 2.7 | | 3.7 | |
| | Range | 29.9–38.1 | 29.3–39.2 | | 29.5–41.6 | |

tools. In this particular context breakout was the direct result of shaping the arrow shaft from the processed raw material. During this process of shaping the shaft, part of it becomes detached from the main body of the shaft along the grain, leaving a distinct depression on the surface.

Analysis of the orientation of the breakout in relation to the grain illustrates that the majority either followed the grain or ran perpendicular to it, thereby enabling a distinction to be made as to the original cause (Fig. 8.84). Examples of breakout that follow the grain can be attributed to the process described above, whereas, those example that are perpendicular to the annual growth could have resulted from the removal of knots.

Whilst the presence of these features in the assemblage may come as a surprise, it seems that it was acceptable for an arrow shaft to contain knots and divots, providing they did not compromise its structural integrity. Several breakouts were overlapped by the binding, suggesting that they had occurred during manufacture Therefore, instead of seeing these features as a negative aspect, they should be seen as an indicator of the skill of the artisan in being able to utilise material

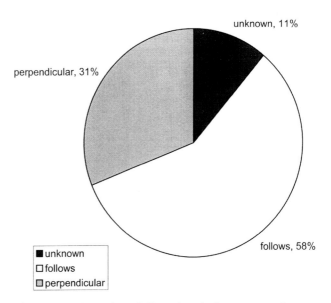

*Figure 8.84 Location of divots in relation to annual growth*

*Figure 8.85 Saw marks on face in slot*

with particular defects inherent within its structure, and thereby turning it into a respectable war shaft that was probably intended only to be used once.

**Tool marks**

A range of tool marks reflecting varying levels of care or skill was observed. These ranged from relatively crude marks left from working of the tips (Hopkins 1998) to the more precise saw marks located on the inner face of the slot cut for the nock insert (Jackson 2003; see Fig. 8.85). These observed differences may reflect processes undertaken by different craftsmen. In particular, the relative lack of tool marks on the shaft suggests that the shaft and the horn insert were made probably by the same individual, whereas the tips were shaped to fit the arrowheads at a later stage.

*Manufacture*
by Keith Watson

It is not unreasonable to suppose that the first stages in the manufacture of the arrows involved shaping the shaft into one or other of the five profiles. Whether saddled shafts were produced deliberately or accidentally must remain conjectural pending further evidence. Most of the other arrows would have been suitable for shooting at a distance, from ship to ship for example. However, a small percentage were breasted which, according to Ascham (1544–5, 120), makes them suitable for shooting at point blank range – possibly to repel boarders. However, we should note that breasted arrows are used for distance shooting nowadays (Soar, pers. comm.).

The apparent constancy in length from start of fletching to the end of the shaft provides possible insight into the fletching process. First, the fletcher gauged a point some 223mm (8¾in) from the nock end to mark the tip end of the binding, then applied the binding by working backwards towards the nock. In this way, any variation in binding length would be compensated by the residual variation in the remaining nock end length, as indicated by the inverse relationship between the two. The average binding covered about 7⅛in (181mm).

The fletchers seem to have been allowed some latitude in their work. For example the presence of grain breakouts and knots seems to have been acceptable, at least to a certain extent. Furthermore, nock inserts were not always accurately located through the centre of the shaft cross-section, and very few nocks were cut perpendicularly to the insert. The orientation of the nock relative to the end grain does not seem to have been important, presumably because of the reinforcing effect of the horn insert.

*Conclusions*
by Alexzandra Hildred

It is tempting to suggest that much more recording is needed in order to understand the complex mixture of wood type, profile and length. However the absence of any conclusive information on head form and the condition of the arrows might mean that, even if all the arrows were recorded in detail, we might be no further forward. The form and weight of the arrowhead will affect both distance and penetration, when one combines this with shaft form and length and wood species there appears to be a huge variety within the assemblage. There are studies which look at profile, species and arrowhead type with respect to initial velocity, range and impact (for example Hardy, in Strickland and Hardy 2005, 409–14, and Stretton, in Soar 2006, 127–53) but none which actually tests the multitude of variables observed (wood species, shaft profile and length, binding length) with a constant head type and weight, and then varying head type and weight. Many experiments are also undertaken using composite bows and man-made string. Although these would be viable merely as vehicles to propel the varieties of arrow, to understand performance of our assemblage experimentation should include self yew bows with hemp string as close to the originals in size and material as possible.

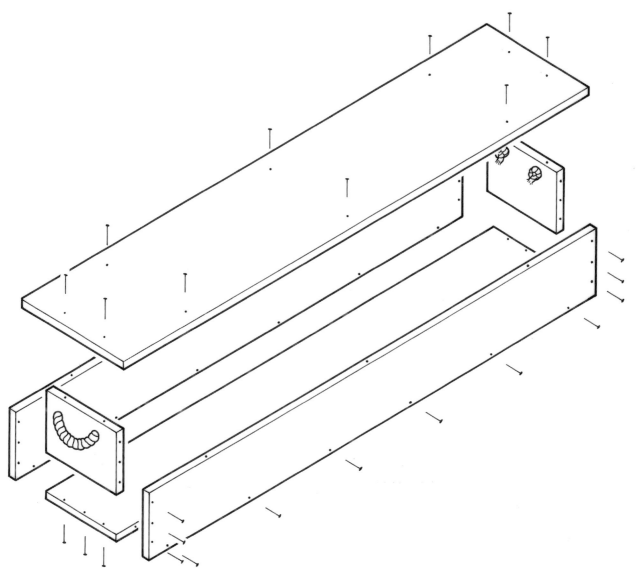

*Figure 8.86 Exploded view of arrow chest construction*

This must be left as a future challenge and until then we can only wonder whether we are measuring something which is, or was, an important factor, or which was insignificant and merely reflected what was easily sourced. We may find that, with the different shaft length, head type (and weight) and shaft weights that we have no real distinctions, just a linear progression in weight from a lightest to a heaviest, and that profile combined with weight mean that real distinctions in performance are almost impossible to ascertain. One would also expect to find some evidence for fire arrows, which would be exceedingly useful at sea and are listed in the 1547 inventory as '*Sheif arrowes with wild fire*' (Starkey 1998, 5577). Such usage may well be indicated by stele length and tip form; but not all our shafts have been comprehensively studied, so this observation is one of many which are still speculative.

## Arrow Chests
by Alexzandra Hildred

Seven closely similar chests were found which are of a type that seems to have been specifically made for carrying and storing arrows (chest type 1.1/2.1). These are all of very similar construction to the longbow chests (see above and Fig. 8.46), made simply of 19mm thick elm planks nailed together, but are rather shorter (1930 x 381 x 305mm average). Having neither hinges nor locks, the lids are made 25mm larger all round than the rest of the chest, and are merely nailed in position (see *AMR* Vol. 4, chap. 10; Fig. 8.86). Three may be better described as crates since they were found lidless with no sign of any nail-holes in the carcass to show where a lid had been fitted. Only five of the chests actually contained arrows (Table 8.55), but the type and

**Table 8.55 Contents of arrow chests**

| ARROW CHEST 80A0726. (BX02) TYPE 1.1/2.1 U7/8. ELM | |
|---|---|
| Almost complete, lidded and full. Broken open, one end crushed. 1920x400x340mm | |
| SHERD, CERAMIC | 80A1867 |
| ARROWS | 80A0764/001-268 |

| ARROW CHEST 81A2398. (F036) TYPE 1.1/2.1 09. ELM | |
|---|---|
| Lidded and full of arrows though crushed at one end. Rope handles found inside. 1935x385x340mm | |
| ARROW | 81A4285 |
| ARROW | 81A4289 |
| ARROW | 81A4294 |
| ARROW(S) | 81A2415/1-29 |
| ARROW(S) | 81A2416-7 |
| ARROW(S) | 81A2418/1-24 |
| ARROW(S) | 81A2419/1-28 |
| ARROW(S) | 81A2420-1 |
| ARROW(S) | 81A2422/1-25 |
| ARROW(S) | 81A2423/1-30 |
| ARROW(S) | 81A2424/1-24 |
| ARROW(S) | 81A2425/1-36 |
| ARROW(S) | 81A2426/1-19 |
| ARROW(S) | 81A2427/1-27 |
| ARROW(S) | 81A2428/1-25 |
| ARROW(S) | 81A2429/1-24 |
| ARROW(S) | 81A2430 |
| ARROW(S) | 81A2443/1-19 |
| ARROW(S) | 81A2444/1-14 |
| ARROW(S) | 81A2445/1-14 |
| ARROW(S) | 81A2446/1-20 |
| ARROW(S) | 81A2447/1-18 |
| ARROW(S) | 81A2448/1-19 |
| ARROW(S) | 81A2449/1-63 |
| ARROW(S) | 81A2450/1-22 |
| ARROW(S) | 81A2451/1-36 |
| ARROW(S) | 81A2452/1-28 |
| ARROW(S) | 81A2453/1-30 |
| ARROW(S) | 81A2454/1-27 |
| ARROW(S) | 81A2455/1-23 |
| ARROW(S) | 81A2456/1-18 |
| ARROW(S) | 81A2457/1-26 |
| ARROW(S) | 81A2458/1-27 |
| ARROW(S) | 81A2459/1-24 |
| ARROW(S) | 81A2460 |
| ARROW(S) | 81A2461/1-29 |
| ARROW(S) | 81A2462/1-29 |
| ARROW(S) | 81A2463/1-25 |
| ARROW(S) | 81A2464 |
| ARROW(S) | 81A2465/1-24 |
| ARROW(S) | 81A2466/1-24 |
| ARROW(S) | 81A2467/1-17 |
| ARROW(S) | 81A2468/1-17 |
| ARROW(S) | 81A2471/1-21 |
| ARROW(S) | 81A2472/1-55 |
| ARROW(S) | 81A2473/1-24 |
| ARROW(S) | 81A2474/1-44 |
| ARROW(S) | 81A2475/1-24 |
| ARROW(S) | 81A2476/1-24 |
| ARROW(S) | 81A2477/1-14 |
| ARROW(S) | 81A4298 |

| ARROW CHEST 81A2582. (F029) TYPE 1.1/2.1. 09. ELM | |
|---|---|
| Recovered intact but damaged, standing on end. Incised III on one side. 1935x380x365mm | |
| ARROW | 81A4299 |
| ARROW-SHEAF | 81A2489/1-19 |
| ARROW-SHEAF | 81A2490/1-31 |
| ARROW-SHEAF | 81A2491/1-26 |
| ARROW-SHEAF | 81A2492/1-21 |
| ARROW-SHEAF | 81A2493/1-23 |
| ARROW-SHEAF | 81A2494/1-22 |
| ARROW-SHEAF | 81A2495/1-30 |
| ARROW-SHEAF | 81A2496/1-29 |
| ARROW-SHEAF | 81A2497/1-42 |
| ARROW-SHEAF | 81A2498/1-23 |
| ARROW-SHEAF | 81A2499/1-22 |
| ARROW-SHEAF | 81A2500/1-20 |
| ARROW-SHEAF | 81A2501/1-28 |
| ARROW-SHEAF | 81A2502/1-21 |
| ARROW-SHEAF | 81A2504/1-28 |
| ARROW-SHEAF | 81A2506/1-31 |
| ARROW-SHEAF | 81A2507/1-26 |
| ARROW-SHEAF | 81A2508/1-28 |
| ARROW-SHEAF | 81A2509/1-23 |
| ARROW-SHEAF | 81A2510/1-79 |
| ARROW-SHEAF | 81A2511/1-23 |
| ARROW-SHEAF | 81A2512/1-29 |
| ARROW-SHEAF | 81A2513/1-27 |
| ARROW-SHEAF | 81A2514 |
| ARROW-SHEAF | 81A2515/1-27 |
| ARROW-SHEAF | 81A2518 |
| ARROW-SHEAF | 81A2519/1-38 |
| ARROW-SHEAF | 81A2520/1-39 |
| ARROW-SHEAF | 81A2556/1-23 |
| ARROW-SHEAF | 81A2559/1-36 |
| ARROW-SHEAF | 81A2560 |
| ARROW-SHEAF | 81A2561/1-34 |
| ARROW-SHEAF | 81A2562/1-28 |
| ARROW-SHEAF | 81A2563/1-38 |
| ARROW-SHEAF | 81A2564/1-31 |
| ARROW(S) | 81A2440 |
| ARROW(S) | 81A2546/1-113 |
| ARROW(S) | 81A2557/1-108 |
| ARROW(S) | 81A2558/1-99 |
| ARROW(S) | 81A2565 |

| | |
|---|---|
| ARROW(S) | 81A2566/1-75 |
| ARROW(S) | 81A2567/1-136 |
| ARROW(S) | 81A2568/1-18 |
| ARROW(S) | 81A2588 |
| ARROW(S) | 81A2589/1-88 |
| ARROW(S) | 81A4293 |

*ARROW CHEST 81A5638. (F169) TYPE 1.1/2.1 08. ELM*
Lidded. Complete but broken open and partially empty but arrows lying around. 1885x340x364mm

| | |
|---|---|
| ARROW | 81A5616 |
| ARROW(S) | 81A3910/1-49 |
| STAVED-CONTAINER | 81A6700 |

*ARROW CHEST 82A1761. (F201) TYPE 1.1/2.1. H9. ELM*
Lidless. Found upside down over barrels. One end empty of arrows, other contained arrows. Bucket and staved container probably fell into chest. 1890x345x365mm

| | |
|---|---|
| ARROW | 82A4428/1-2 |
| ARROW | 82A4429/1-2 |
| ARROW | 82A4430/1-2 |
| ARROW | 82A4431 |
| ARROW | 82A4432/1-2 |
| ARROW | 82A4461 |
| ARROW | 82A4497/1-2 |
| ARROW-SHEAF | 82A1887/1-49 |
| ARROW-SHEAF | 82A1888/1-55 |
| ARROW-SHEAF | 82A1889 |
| ARROW-SHEAF | 82A1890 |
| ARROW-SHEAF | 82A1891 |
| ARROW-SHEAF | 82A1892 |
| ARROW(S) | 82A1703 |
| ARROW(S) | 82A1755 |
| ARROW(S) | 82A1756 |
| ARROW(S) | 82A1757 |
| ARROW(S) | 82A2023 |
| STAVED-CONTAINER | 82A1728 |
| BUCKET | 82A1758 |

location of the other two indicates that these were arrow chests also. One spare lid with holes was found (82A1611, sector O8), suggesting an eighth. The *Anthony Roll* lists 400 sheaves (of 24 arrows), or 9600 arrows (of which more than 7000 fragments were recovered, see above). If these were accommodated in chests, 40 sheaves to a chest would equate to ten chests.

The distribution of these chests corresponds with that of much of the rest of the archery equipment (see Figs 8.1, 8.45, 8.50). However, one found lying upside down on the ballast in H9 (82A1761) had probably been displaced to this position during or after the sinking.

Two arrow chests were excavated from the Upper deck in the stern (U7–8). One (80A0476) lay directly behind the rear wheels of a bronze culverin (80A0976). The second (80A0726), lay adjacent to longbow chest 80A0602 (Sitebook 1980 M/U7/8, pp. 133 and 147). All of these objects were aligned nearly North–South along the deck. The area contained what the excavators described as '*bundles*' of arrows. Remains of the stepped cheeks for the carriage for a bronze gun (80A0948, 80A0949) lay between the boxes, as did human remains. Individual longbows and bills were recovered and, the following year, as the excavation trenches deepened, a concentration of staff weapons was revealed.

80A0476 (Box 4; 1900 x 350 x 335mm) was unlidded and had no nail-holes. It was broken and found with a single longbow (80A0473) inside (presumably post-depositional contamination). '*Bundles*' of arrows lay around it (divelog 80/8/227; 80A0690). Human remains directly associated with the chest included two fairly complete skeletons (FCS63 and 65). 80A0726 (Box 2; 1920 x 400 x 340mm) was

*Figure 8.87 Arrow chests: a–b) underwater photographs of arrow chest 81A0267; c) underwater photographs of chest 81A2582 showing arrows stored tip to tip; d) Detail of broken end of chest*

lidded and apparently full. It appeared to have smashed against a heavy object damaging the chest and breaking some of the arrows, which were arranged in layers, tip to tip. The space between the box and the planks of the Upper deck (100mm) was filled with arrows and disarticulated human remains. The arrows were excavated using a water jet to loosen the sediments, aided by gentle brushing with a paintbrush, and an airlift to remove the debris. They were then decanted into gutter pipes, covered in a netlon sleeve and placed in a tray for eventual lifting. All the arrows were given the same number (80A764), grouped 1–268, according to the batches in which they were lifted, with a total of 936 individual objects or 39 sheaves.

Although many of the dive logs and site book references refer to '*bundles*', the observation was made that:

> '*Although apparently in bundles, there seems to be nothing holding them together. Perhaps this has gone. Maybe they were simply stored in quantity, all were fairly tightly packed in the box.*' (Barak, 80/MU/7/8, p. 151)

The first chest found in storage in O9 (81A2398, F36, 1935 x 385 x 340mm) was lidded and apparently full with its rope handles lying inside the chest. There are three vertical lines and a crude star incised on the chest. It was aligned horizontally along the deck (Figs 8.45 and 8.87 a–b). The northern end was fractured and overlain by longbow chest 81A1862. A fairly complete skeleton (FCS42), wearing a leather jerkin (81A4531–2) and apparently a scabbard (81A1521) lay on the exposed arrow chest at the North end, possibly accounting for the damage. A small amount of tar was noted in the chest (81S1211). Inside, the arrows were aligned tip to tip, two arrows along the length of the box, with concretion formed by the corroding tips in the centre (a few of which were recovered, 81A2404, 81A4298, 83S0371 concreted tips, 81A4289, single tip). Archaeological supervisor Faye Kert notes in the archaeological diary for that area (Kert: 1981, H/O 9–11, p. 181) that the tips appeared to have been interleaved, and of bodkin shape (armour piercing). The arrows at the northern end were badly fractured. The arrows were excavated in layers underwater, decanting each layer into gutter pipes (eg, 81A2415–81A2430;

81A2433; 81A2444–2468; 81A2471–2477) of 36 arrows each and placing these into a large tray. The tray was then taken using a lifting bag to the collection area in preparation for a direct lift from above.

Although the excavators were looking for evidence of bundle ties (eg, samples 83S0280, 83S0370), there is confusion: Kert (Kert: 1981 H/0 9/10/11, p. 183) notes: '*No trace of string tieing* [sic] *up bundles of arrows observed*'. Dobbs, however, adds '*The binding is very thin, 1.2mm*'. Other items within the chest included some metal fragments (81S1189), possible decomposing horn (81S1311,1312) and some fragments of leather (83S0436, 0437). Some of the arrows have been sampled for analysis of feathers (83S0430, S0433, S0434, S0435) and the horn nock inserts (83S0007) (see above and *AMR* Vol. 4, chap. 15).

81A2582 (F29; 1935 x 380 x 365mm; Figs 8.45 and 8.87c–d) was also lidded and full (arrows, eg, 81A2440, 81A2513–2515; 81A2518–2520; 81A2556; 81A2559–2564) and standing on end in O9. The arrows from the uppermost portion were badly shattered. They were removed underwater. Yet again, there is no absolute evidence for these being tied in sheaves, but one possible piece of binding was sampled (82S1280) and 0.25 litres of tar/pitch was recovered (81S1219). Preservation was such that samples could be taken of shaft fragments with flakes of adhesive together with possible horn nocks (82S1280), and what was thought to be silk binding on some of the flight ends (83S0003).

Two arrow chests were recovered from O8. 81A3824 (2880 x 330 x 350mm) an empty chest of longbow type, but too long for bows and therefore assumed to be an arrow chest, was found towards the North of the trench resting on top of one end of a chest containing arrows, 81A5638 (F169). There was also a chest containing longbows (81A3927; F159) standing vertically nearby. All were exposed at the same time during the excavation and the area contained a somewhat confused mixture of material, including remains of many casks and a spare ship's pump which hindered excavation.

81A3824 was a lidless crate and was empty. 81A5638 (1885 x 340 x 364) had a lid, but was damaged and the lid displaced. It contained arrows (81A3910, 81A5616, 81S109) but was not full, though arrows were also found under it. Tip concretions were recovered (81A5616). A fairly complete skeleton (FCS36) was found at the north end of the chest along with a kidney dagger handle (81A3894) associated with leather (81A3895).

A further lidless arrow chest (82A1761, F201) was lying upside down on top of two casks which, in turn, rested on the ballast in H9. One end had been displaced and cask stave fragments fallen in (82A1728). The North end of the chest was completely empty but the Southern part appeared to be full (82A1703, 82A1755–7, 82A1887–92, 82A2023, 82A4428–33, 82A4461, 82A4467). Some arrows had slipped out and were lying on the ballast. Tied sheaves were not observed, although the excavators describe lifting '*sheaves*'. Tar was found inside the chest (82S0082) and *samples* of arrow shafts were taken for analysis of the bindings and glue (83S0263–76; 83S0292–9; 83S0300, 83S0354–6, 83S0006), feathers (82S1260–1), thread and nock fragments (82S4429, 82S4431, 83S0014). A chest of a similar construction to the arrow chests was found in H7, 82A0842 (700 x 300, but this is too short for conventional arrows.

## Arrow Spacers
by Alexzandra Hildred

Seventeen leather discs were recovered, varying in diameter from 105mm to 160mm and pierced with holes. Eleven of these enclosed arrows (Figs 8.88–9; Table 8.56). The diameters and thicknesses of the nine complete disks vary (see Table 8.56) but each has 24 holes of 12–15mm diameter. These appear to have been cut from the grain side towards the flesh. All identified leather is bovine, with some more specifically identified as cattle. Their distribution (Fig. 8.50) is predominantly on the Upper deck, with nine recovered from areas where other archery objects were found; three were found on the Main deck and five on the Orlop deck. In the nine instances with arrows in position the disks were close to the upper end of the flight binding.

These discs were a neat device for keeping arrows secure without damaging the flights and provide an upper limit (15mm) for the diameter of the corroded iron heads. In use the arrows would have to be withdrawn backwards to avoid pulling the flights through the holes and damaging them. This means that any small barbed head would have to be small enough to avoid being caught in the leather when being withdrawn. These have been termed '*spacers*'. It is impossible to state with any certainty whether the grain or the flesh side faced the flights, although one would assume that the arrows would be most easily pushed through from the grain to the flesh.

No spacers were found inside chests, although some were associated with chests on both the Orlop and Upper deck (see Fig. 8.1). As there are so few (and they are quite distinctive even when fragmentary) it is difficult to see them as being pre-prepared 'rounds' of ammunition. Small numbers of arrows are often shown tucked under archer's belts (eg, in a border decoration of an archer in the *Luttrell Psalter* (BL Add MS 42130, f. 56) and various depictions of both English and French longbowmen and crossbowmen and in the Bibliothèque Nationale and the British Library). In most instances these are passed through the belt on the right hip with their heads pointing forwards and downwards, in a position for the archer to see the heads, as depicted in the Cowdray Engraving (Fig. 8.90). In this position the flights are uppermost with the tips closest to the ground. These are probably the '*arrow*

Table 8.56 Arrow spacers

| Object | Sector | Complete/incomplete | Diam. (mm) | Thickness (mm) | No. holes/diam (mm) | Arrows *in situ*? | Arrows nearby? | Strap | Stitch-holes | Comment/associations |
|---|---|---|---|---|---|---|---|---|---|---|
| 79A0653 | U4 | C | 110 | 4–5 | 24/12 | yes, 23 | | no | not visible | Large coil of rope. Bill 79A0630; longbows 79A0870/0855 nearby; wristguard 79A1224 found later; binding 79S0032 |
| 79A0673 Cattle | U3 | C | 130 | 6–8 | 24/12–15 | no | 79A0672 | yes 79A0676 | edge | Found beside bulwark rail with arrows but not clear if these inside spacer. Bills 79A0674–5; longbow 79A0614 found previously; some human bone |
| 79A0784 Mark X | U3/4 | C | 115 | 6–8 | 24/12–15 | yes | 79A0784 | no | edge | Large coil of rope. Mallet 79A0782; bowl 79A0783; bills 79A0785/0790; longbow 79A0807 or 79A0790 nearby; human bone 79H0063 |
| 79A1204 Bovine | U5 | C | 120 | 10 | 24/12 | yes, 7 | | no | edge | Wristguard 81A4675; longbow 81A5877 found here later. Jerkin 79A1203; bowl 79A1205; human bone 79H0106 |
| 80A0877 Bovine | U7/8 | I (4 frags) | n/a | 4 | 11/10–12 | yes | | 10 frags 10mm wide 1.5 thick | flesh to edge | Dense area of archery related material |
| 80A1146 Cattle | U8 | I (6 frags) | n/a | 5 | 4/13 | no | 80A1147/1149/1152 | no | in top out side | Bowl 80A1156; bill, 80A1151; human bone 81H0115 |
| 81A4800 Cattle | U9 | I (8 frags) | n/a | 3 | 3/12–13 | no | yes, frag. | 10 x 2mm 22mm long | flesh to grain | Area of longbows; FCS39 |
| 81A1341 Bovine, inscribed circles | U9 | C | 135–160 | 4–6 | 24/12–15 | ? | 81A1340 | 5 frags, 13mm wide, 1mm thick | not visible | Area of bows & human remains |
| 81A1255 Cattle | U9 | I (3 frags) | n/a | 1 | 5/10 | yes | others around | 2 frags 10–12mm wide | edge to flesh | Aligned N–S along deck beside daggers 81A1251–2; comb 81A4655; wristguards 81A1158/81A4639) from same area & 14 longbows |
| 81A0116 Cattle | M2 | C | 120 | 6 | 24/12 | yes | no | yes well strapped 12mm wide | flesh to edge | Tips to flesh side; remains poss. arrow bag 81A6844; FCS82 |

| Object | Sector | Complete/ incomplete | Diam. (mm) | Thickness (mm) | No. holes/ diam (mm) | Arrows in situ? | Arrows nearby? | Strap | Stitch-holes | Comment/associations |
|---|---|---|---|---|---|---|---|---|---|---|
| 80A1376 Cattle | M6 | I (2 frags) | n/a | 5 | 5/12 | no | no | no | flesh to edge | None but wristguard 81A1460 later found nearby & 3 longbows. Much of area taken up by gun; several FCS nearby |
| 81A2828 Cattle | M8 | I (7 frags) | n.a | 4 | 13/11–15 | no | 81A2827, 17 arrows | no | flesh to edge 3mm | Between *in situ* iron gun & partition of Carpenters' cabin. FCS85; 4 longbows in area; much 'straw' (prob. hay) 81S0393; arrow tips pointing towards Main deck, flights towards Upper deck |
| 82A2026 Bovine | O7 | C | 130 | 4 | 24/12 | yes | | well tied. 5 x 40mm | edge 4mm apart | 82H0084/0087; comb and scabbard; thong around flights |
| 82A1587 | O8 | C | 105 | 8 | 24 | yes | 82A1589 | well tied | edge 6mm apart | Longbow 82A1590 |
| 82A1523 | O8 | C | c. 140 | 5 | 24 | yes | 82A1522 | well tied | edge | Lying N–S, tips to S, with 81A1522. Wristguard 82A1524; dagger 82A1525; pouch 82A1526; cask (82A1592–4); FCS33, FCS36; longbows |
| 82A1522 Cattle | O8 | I (5 frags+) holes torn | n/a | 3 | 7 | yes 40 frags | 82A1523 | yes 9mm wide, 200mm long | not visible | As above |
| 82A2281 Bovine | O9 | | | | | yes 18 | | | | |

695

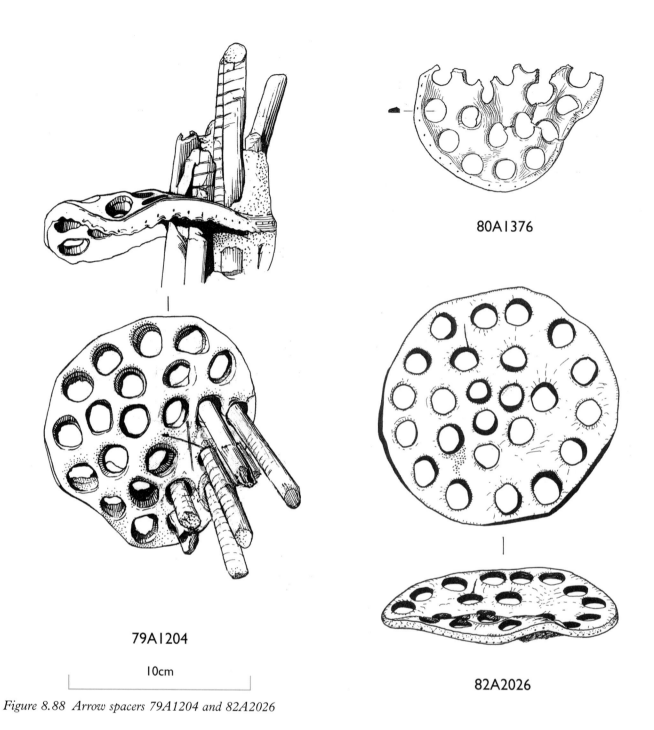

*Figure 8.88 Arrow spacers 79A1204 and 82A2026*

*Figure 8.89 Reconstruction drawing of spacer 79A0653 with arrows*

*Figure 8.90 Extract from the Cowdray Engraving showing archers with some form of spacer/bag*

*girdles*' listed in inventories. The small number recovered may indicate personal property, as with the wristguards.

Leather '*straps*', 10–15mm wide, were found with eleven spacers, three of them wound around the bundles approximately half way between the spacer and the arrow tips (81A0116, 82A1523, 82A1587). Arrows held in bundles by leather straps can be seen in medieval illustrations, for example, *Les Passages faits Outremer* from the series *The Siege of Jerusalem* of c. 1490 (Bibliothèque Nationale, Paris), and wrapped bundles held at the waist (possibly even with a spacer) can be seen in the *Beauchamp Pageant* of c. 1480–92 (BL Cotton MS Julius E. IV, f. 20v).

A series of small stitch-holes is visible 2–3mm apart on the edge of thirteen of the spacers (Fig. 8.88). The thread seems to have gone from the flesh side of the leather to the edge, suggesting that something was stitched to the edge. As leather survives extremely well, it is likely that this would be a fibre and that the spacers could, therefore, have functioned as the top of a quiver, or more descriptively, an arrow bag. The presence of fibres was noted under the leather strap/straps in one instance (81A0116). This would suggest that the spacer was stitched to a cloth bag and tied on the outside with the strap.

There are illustrations that indicate a number of forms of arrow container carried by archers. A simple waist quiver can be seen in a Flemish illustration of 1460–1480 (BL Royal MS E.I, f.12). This appears to be slung over the right hip, possibly from a leather strap over the right shoulder, with the flights up and slanted backwards. It appears to be conical, flaring towards the arrow flights. Illustrations in the *Chronicle of Diebold Schilling* (Bibliothèque de la Bourgeoisie de Berne) depict a stiff conical bag which has the heads protruding through the smaller, tied end, again held at the right hip. The bag appears to have some form of internal framework, although the texture of the outer surface suggests textile.

Simple loose bags containing arrows can be seen covering the ground in Froissart's *Chronicles*, and the depiction of English longbowmen at Poitiers in the Bibliothèque Nationale, shows a soft bag similar to a stocking, with the top turned down.

Many of the loose arrows were found in bundles of 24 (although little evidence of the ties were found). These may correspond to '*sheaves*' described in inventories of the period. The extensive and descriptive inventory prepared after the death of Henry VIII does not list any arrow bags, and has only one reference to a quiver in the Tower of London (Starkey 1998, 3870): '*Quyver for pricke arrowes for crosbowes oone*'.

Objects associated with the spacers are listed in Table 8.56. Five of the nine spacers from the Upper deck were in the stern, beneath the Castle deck, in and around open chests of both longbows and arrows (Figs 8.1 and 8.50). The other four are from the open weather deck towards the bow, isolated from collections of archery equipment, but where archers may have been stationed during an engagement. Although not directly associated with the spacers, the presence of five isolated longbows and two wristguards also on the weather deck certainly suggests that archers were at arms here. It may be significant that, in three out of four cases, the spacers were also associated with two staff weapons.

Three spacers were found on the Main deck. 81A0116 (M2) contained arrows (16 nock ends, 7 tips, 27 fragments, diameter 8mm) and was tied with a leather strap. Associated leather may be the remains of an arrow bag (see below). It was found with a fairly complete skeleton (FCS82, Figs 8.91 and 8.100). No longbows or wristguards were found in M2, but an ivory wristguard (81A0815) was found close by, just in M3, and a longbow had previously been recovered from the southern end of M3, close to the companionway to the deck above (80A0298).

Of the five spacers on the Orlop deck, 81A1522 and 81A1523 were found at the same time in O8. Two days later another spacer, 82A1587, was found in the south of the same trench. The arrows and spacer were on top of a cask, with other arrows visible beneath it (82A1589). This also had a longbow lying across it, 82A1590 with another bow lying parallel to the Main deck. It is convenient to see these as personal items brought onboard, similar to the wristguards. The small number recovered and the distribution (reflecting in use rather than in store) favours this suggestion.

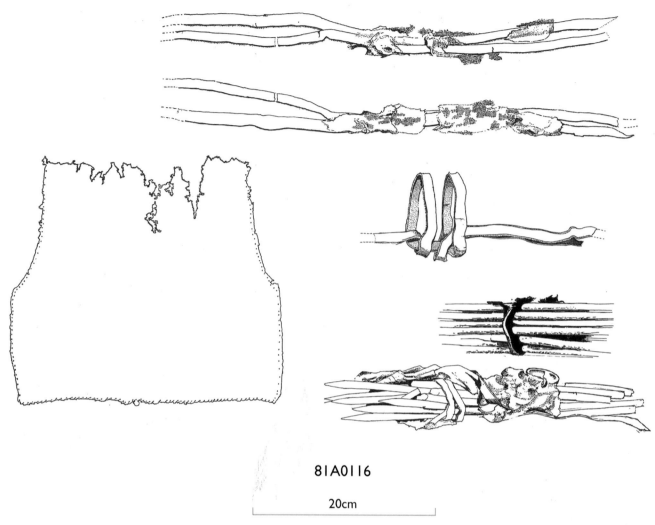

*Figure 8.91 Arrow bag 816844 and associated arrows 81A0116*

## Arrow Bag
by Alexzandra Hildred and John Waller

The fragmentary remains of one possible leather arrow bag were found (81A6844) in M2, together with a bundle of arrows (81A0116) held in a spacer and tied with four turns of a leather strap (see above). The piece of leather is rectangular with two side edges relieved and looking like sleeves. (Fig. 8.91). It measures 285mm by 250mm with the relieved edges having a length of 220mm. These edges and the straight edges below all carry stitch-holes. The lower edge also retains stitch holes. The upper edge is somewhat torn.

The bag is incomplete but matching up the stitch-holes from the lower sections and presenting one seam central to one hemisphere produces a bag, one end of which fits the circumference of the spacer to which it may have been stitched. Overhand stitches entered through the leather top, the spacer and a [cloth] bag, would provide a 'flush' seamless inside with the rough edges outside. A shoulder strap/waist strap would be a convenient addition and several spacers were found with fragments of leather strap (see above).

This reconstruction drawing is based on a drawing from the notebook of the antiquarian Francis Grose (Bodleian Library MS. Top. Gen. e. 70, f. 24) (Fig. 8.92) captioned '*Ancient quiver about the size of a stocking kept at Canterbury*'. It clearly shows a tied bag with a leather disk at about the centre through which an arrow is inserted. The flight is uppermost, with what is described as '*loose leather*' at the flight end. The disk is captioned '*stiff leather full of holes*'. The tip of the arrow is resting in hay, just above the tied end of the bag, but the head depicted is a broad head of some size and could not have been pushed through the hole illustrated and was possibly drawn as an '*arrow head*'. In use, the tie would presumably be released and the arrow withdrawn head first from the bottom of the bag, pulling the flights through the spacer and down the entire length of the bag. However, if the arrows were held tip upwards, there would be the danger of damaging the flights. An anonymous Flemish painting

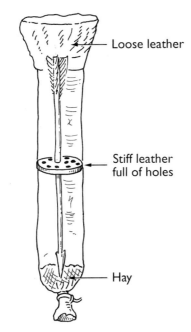

Figure 8.92 *Bag ancient quiver (after drawing in Bodleian Library)*

of about 1560, depicting the martyrdom of St Sebastian, shows a number of well represented items of archery equipment, including two arrow bags which are open at both ends and wider in the centre, possibly to accommodate a spacer. One is slung across the back of an archer suspended from some sort of shoulder strap and is discussed in detail by Credland (1982). Although lost, a photograph of the painting is reproduced and discussed by Rimer (1998, 56).

llustrations depicting archers with arrows tucked through their belts show the flights uppermost (see above). Other illustrations of arrow bags include the battle of Poitiers from Froissart's *Chronicles* (Bibliothèque Nationale MS fr. 2643, f. 207). This appears to be a soft stocking holding arrows with their flights just outside the bag. This is shown within the arrow bags carried by archers in the Cowdray Engraving (Fig. 8.90). Close examination of the Cowdray would favour placing the spacer closer to the top of the bag, closer to the flights than the example described by Grose. It is also the position in which many were found (Fig. 8.89). This keeps the flights separate and close to the top of the bag so that the archer could see (or feel) the flights. As this would work most effectively by drawing the arrowheads back through the spacer, it also gives a maximum diameter for the arrowheads of between just under 12mm (the minimum diameter of spacer hole) to just under 15mm (the maximum recorded hole diameter from a spacer). The iron heads must be compact and relatively small, and suggest a narrow head, or one with small barbs, rather than a large barbed head. A small barbed arrowhead of diamond or almond section with close-fitting barbs would pass through the holes of the spacers.

Current experimentation is ongoing (Waller and Waller, in. prep. 2009) using white canvas and leather spacers based on *Mary Rose* dimensions. The bag produced incorporated a top portion based on the dimensions and form of the leather from the *Mary Rose*. Copied in canvas, these were stitched to the top end of a canvas tube. The spacer was stitched just above this seam, in this way the top (in the case of the *Mary Rose*, the leather portion) can be rolled back to completely reveal the flights. The shape of the leather is such that it expands towards the top, enabling a drawstring placed through the very top of the upper portion to be tightened to close the top over the flights for protection. The bottom of the bag can be tied to keep the tips together, as is shown in the Grose example. Simple loops on the side of the tube towards the top and bottom enable the bag to be slung across the back when not in immediate use (for marching) or enable it to be moved down to the waist so that the arrows lie over one hip for ease of use, both these attitudes are shown in a number of illustrations. Experimentation with constrictor knots such as the '*marlin hitch*' in leather straps tied on the outside of the bag helped to keep the arrows from moving within the bag and looked very similar to the example found with 81A0116. It is possible that these are for the '*girdles for sheave arrows*' mentioned in contemporary inventories.

## Handgun bolts, Incendiary Arrows or Top Darts?
by Alexzandra Hildred

Nine arrow-like wooden objects (81A1093/1–11, visually identified as ash) were found together in a chest (81A1328; Table 8.57) just outside the Carpenters' cabin on the Main deck (Figs 8.93–5). Identified underwater as '*crossbow bolts*', and later re-identified as handgun bolts, they are round in section, have a whittled tip of 18–25mm length (Fig. 8.93a), a maximum diameter of no more than 12.6mm and length of no more than 525mm. In profile they increase from the shoulder to the mid-point (taken as 250mm from the tip) and decrease in diameter towards the end. There is a defined area at the end where a lozenge-shaped section has been relieved from opposing hemispheres, creating two flattened surfaces (Fig. 8.93b). This varies in length between 22.6mm and 25.4mm and is rounded at the top. The width of this chamfered area varies between 6mm and 10mm. The ends vary slightly in diameter with an average of 10mm (Table 8.58). Binding is noted as having been present

### Table 8.57 Contents of chest 81A1328

| CHEST 81A1328. (F043) TYPE 2.2. M9. ELM/OAK Found in jumble of chests outside Carpenters' cabin. Complete but one side collapsed inwards. Lid decorated with circles containing stylised flowers. 965x355x340mm ||
|---|---|
| UNIDENTIFIED("THINGS") | 81A1022-3 |
| UNIDENTIFIED | 81A1047 |
| UNIDENTIFIED("THING") | 81A1048/1-2 |
| UNIDENTIFIED | 81A1067 |
| UNIDENTIFIED | 81A1052 |
| HANDLE? | 81A1042 |
| HANDLE? | 81A1045 |
| HANDGUN-BOLT(S) | 81A1093/1-11 |
| FLOAT | 81A1046 |
| COMB | 81A4234 |
| BOOK | 81A1062 |
| INK-POT? | 81A1054 |
| KNIFE | 81A1065 |
| BASKET | 81A1055 |
| BOX | 81A1049 |
| BRUSH | 81A1061/1-2 |
| HANDLE+A395 | 81A1043 |
| HANDLE | 81A1044 |
| HAMMER? | 81A1041 |
| PLANE, MOULDING/THUMB | 81A1039 |
| PLANE, MOULDING/THUMB | 81A1040 |
| BALANCE? | 81A1053 |
| WHETSTONE | 81A1066 |
| ANTLER (WORKED) | 81S1272 |
| BRACKET FUNGUS | 81S1275 |

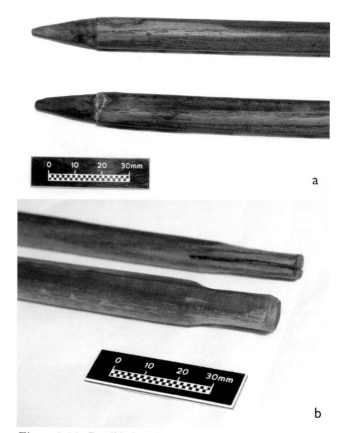

*Figure 8.93 Possible handgun bolt 81A1093 a) tip and end; b) nock end*

but there is no longer any visible evidence of this and there are no details. Although only five were available for detailed study, only one of these had a slot (81A1093/2; Fig. 8.95, b), 32.5mm long and 1–2mm wide.

The excavation of the chest contents (Fig. 8.96) suggested a mix of personal objects and tools (Table 8.59). Personal possessions included a very fine comb (81A4234) and a book (81A1062, identified underwater as a pouch). Associated with the book (but not thought to be necessarily part of it) was a piece of paper with the writing '*W Wosell*'; this no longer survives. A simple inkpot (81A1054, identified underwater as a hammer blank; *AMR* Vol. 4, fig. 3.23) was stored close to the book. Originally the inkpot had a wooden stopper but this is no longer present. Tools included a brush with an alder handle (81A1061) of 'paintbrush' type (*AMR* Vol. 4, 355). What was thought to be a balance arm (81A1053) was found together with a number of unidentifiable handles of willow and alder, and three turned ash rods (81A1022) with rammer-like heads and burnished concentric lines around the heads (Fig. 8.95). Although identified underwater as '*ramrods*' for guns, the heads vary in diameter between 23.6mm and 25.9mm with the longest recorded length at 580mm. These are too large for anything currently within the assemblage, although they do look like ramrods. Other as yet unidentified objects include four small willow twigs with whittled ends, a possible fishing float, a few rectangular blocks of wood and a piece of worked red deer antler, together with leather fragments and further (as yet) unidentified wood. Two small planes were also

### Table 8.58 Details of short arrows 81A1093 (metric)

| Part no. | Length | Tip length | Diam. at shoulder | Diam. at 250mm | Diam. at chamfer | Length at chamfer | Length slot | Diam. end |
|---|---|---|---|---|---|---|---|---|
| 1 | 508 | 23 | 11.6 | 12.6 | 12.4 | 254 | n/a | 9.1 |
| 2 | 493 | 24 | 11.3 | 11.5 | 9.4 | 248 | 325 | 8.7 |
| 3 | 512 | 20 | 10.9 | 11.4 | 11.1 | 251 | n/a | 9.2 |
| 4 | 510 | 25 | 12.7 | 13.7 | 11.9 | 226 | n/a | 9.2 |
| 5 | 515 | 18 | 9.9 | 12.4 | 12.3 | 242 | n/a | 9.1 |

*Figure 8.94 Group of possible handgun bolts 81A1093 as recovered*

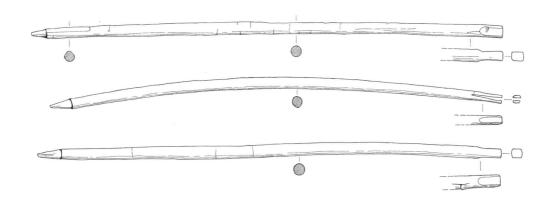

*Figure 8.95 Handgun bolts 81A1093 and unidentified possible ramrods 81A1022*

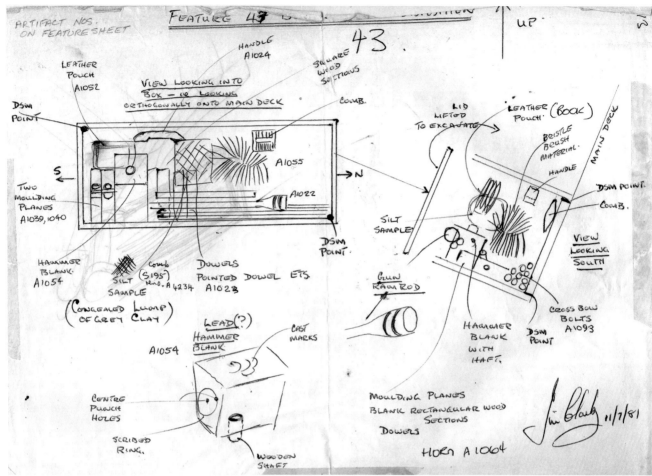

*Figure 8.96 Sitebook drawing (MU 9-11, 11/7/81) showing contents of chest 81A1328*

found in the chest, lying side by side (Figs 8.96–7). These have been tentatively identified as thumb planes used to manufacture fire sticks or incendiary arrows for use with matchlock muskets (*AMR* Vol. 4, 316).

Perhaps even more enlightening is the presence of a small box (81A1049) with two compartments both with internal handled lids. This may be a small tinder box (Fig. 8.98) although it is of slightly different construction to the larger oak tinder boxes recovered (the identification of which is now questioned, they resemble later boxes holding linseed impregnated cloths used in caulking). The presence of a flint (81A1051), a knife handle (81A1065) and bracket fungus (81S1275), which could be used as tinder (see *AMR* Vol. 4, 318), within the chest add further credence to the idea of a the presence of a '*kit box*' for incendiary arrows.

It is tentatively suggested that the arrow-like objects could be incendiary projectiles (arrows of wildfire) used with either a handgun or a longbow. Recent experimental work (Stretton in Soar 2006, 152 and fig. 24) has suggested that a long (6in/152mm) bodkin arrowhead on an ordinary but 'short' arrow will take an incendiary mix, in the form of a linen bag placed over the spike tied with a leather strap and coated with rosin (hence the need for a brush). Shot from a conventional longbow, this was effective (ie, achieving ignition) against a bale of wet hay and a wooden stable door. Draw length was not compromised, as the length of the arrow was made up with the 6in (152mm) bodkin head. If 152mm is added to the range of lengths of the unidentified arrows, the lengths of those examined range from 645mm to 667mm, just within the range of the shortest 'normal' arrows in the assemblage.

However there are significant differences in the ends of these when compared with the rest of the arrow assemblage, the notched taper at the end is typical of crossbow bolts and the single one which has a cut is probably the only nearly finished example cut to take a bone or horn reinforce. These are, however, longer than the average crossbow bolts which are 370mm, and never more than 400mm, so as yet these must remain unidentified.

## The Identification of Archers from Skeletal Remains

by Alexzandra Hildred and Ann Stirland

The largest concentrations of Fairly Complete Skeletons (FCS) included 29 individuals in the area of

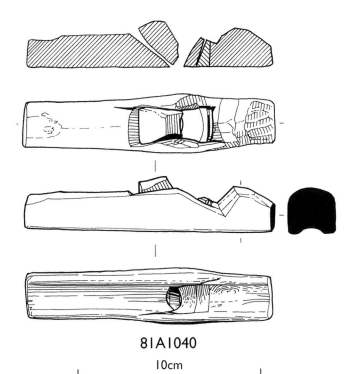

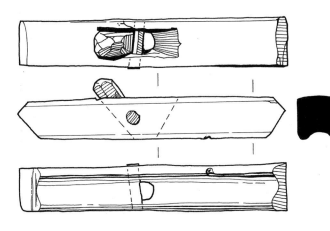

### Table 8.59 Contents of chests possibly belonging to archers

*BENCH CHEST 82A0942. (F186) TYPE 3.2. 010. ELM*
Recovered complete but in fragile condition. 1125x425x390mm

| | |
|---|---|
| UNIDENTIFIED | 82A0947 |
| UNIDENTIFIED | 82A0967/1-2 |
| UNIDENTIFIED ("THING") | 82A0969 |
| AIGLETS | 82A5002–3 |
| BOOT (THIGH) | 82A5014/1-2 |
| SHOE | 82A0845 |
| BRAID | 82A5000 |
| WRISTGUARD | 82A0943 |
| DAGGER-SHEATH | 82A0970 |
| INSET-SHOT | 82A0799 |
| COMB | 82A0945 |
| KNIFE | 82A5001 |
| BESOM | 82A0946 |

*CHEST 81A2035. (F041) TYPE 2.4. M10. ELM*
Standing upright. Virtually complete but broken open. Internal compartment. Cooking pot in one end of chest, Rest at other end. 1090x0380x325mm

| | |
|---|---|
| UNIDENTIFIED | 81A1994 |
| UNIDENTIFIED ("THING") | 81A2191 |
| SHOE | 81A1682–3 |
| JETTON | 81A1903 |
| JAR, CERAMIC | 81A2103/1-25 |
| WRISTGUARD | 81A2192 |
| COMB | 81A2190 |
| RAZOR | 81A2101/1-2 |
| LINE-ADJUSTER (RIGGING) | 81A2104 |
| WHETSTONE | 81A2102 |

the Upper deck (U7–9), where, as we have seen, there was a concentration of archery equipment. Ten individuals were found in O7–8, where archery equipment was stored (Fig. 8.1).

The best indicators for the identification of archers are the association between wristguards and/or arrow spacers and human remains, and specific features (morphological changes) of the skeletal remains (see Stirland, *AMR* Vol. 4, chap. 13 for a full discussion of the latter). Chests containing wristguards may also indicate ownership by archers.

Wristguards are physically attached to the bow arm in use while spacers indicate the presence of individual quivers of arrows that would be carried by the archer for his personal use. The distributions are shown together with the relevant skeletal remains in Figure 8.1. As can be seen, there are FCS in every coincidental area except on the open weather deck (U2–6) and it seems

*Figure 8.97 Unusual small handplanes 81A1040 and 81A1039*

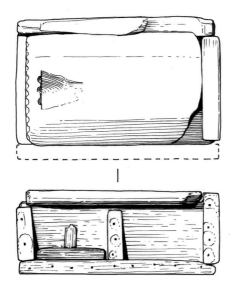

Box without lid

81A1049

10cm

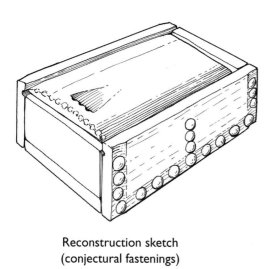

Reconstruction sketch
(conjectural fastenings)

*Figure 8.98 Lidded box 81A1049*

reasonable to suggest that some of these individuals may have been specialist archers or crew members using bows. In some instances FCS, or other substantial collections of human bone (indicated below as, eg, 81H0145), were found in areas where the only other archery finds are wristguards.

Wristguards associated with human remains include the ivory example in M3 (81A0815/FCS75), a decorated wristguard in M5 (81A1173/81H0145), one decorated with the Arms of England in M6 (81A1460/81H0171, now thought to be part of FCS70), and one in U5 (81A4675/81H1075), together with a spacer. Only one wristguard from U9 was in direct contact with skeletal remains (81A4639/81H1071). Upon examination these represented a number of individuals and no specific assignation can be made. Seven arrow spacers were directly associated with human remains; that in O7 (82A2026) was found with a comb, a scabbard and human bone (82H0084). Two of the three found in O8 were found with human remains – 82A1523 (associated with 82H0141–4), and 82A1587 associated with longbow 82A1584 and arrows 82A1590. This area yielded six FCS (33–6, 38, 90). A single spacer in M2 (81A0116) was found with FCS83, with FCS82 in the same area while 81A2828 was found in M8 amongst '*straw*' and human remains (81H0322) with FCS86 recorded from M7/8. Although all five spacers found between U7 and U9 were found within a dense area of skeletal material, only 81A4800 was directly associated with skeletal remains (81H0192).

Ann Stirland's analysis of the human remains from the *Mary Rose* identified in 26 out of 207 scapulae (shoulder blades) recovered (12.5%) an anomaly known as *os acromiale*. In this condition, the final

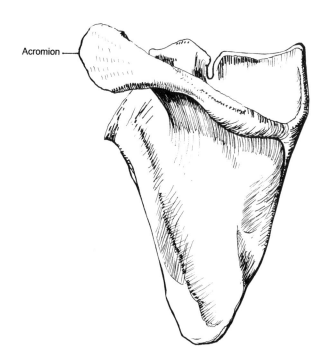

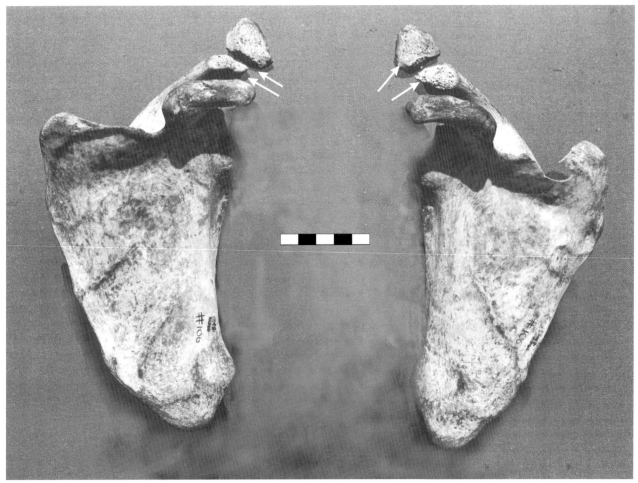

*Figure 8.99 above) Normal scapula; below) one with os acromiale*

epiphyseal element (the 'growing end' of the bone) of the acromial process fails to unite to the bone (Fig. 8.99, above), leaving a characteristic incomplete bone or pair of bones (Fig. 8.99, below). In the example shown here, both the ununited fragments were also recovered. The epiphysis normally unites to the bones in males by 18–19 years of age. This union does not occur, however, in about 3–6% of individuals in a modern population (Stirland 1984). The frequency on the *Mary Rose* is, therefore, significant.

Many of the affected scapulae are single bones, lacking their matching partner, so it is unclear whether

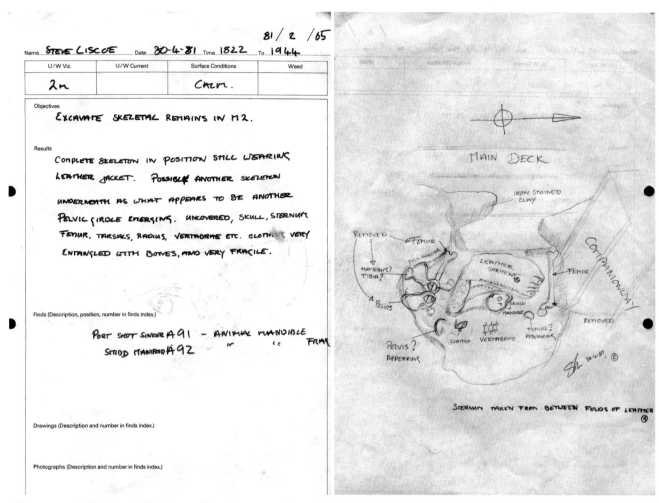

*Figure 8.100 Divelog 81/2/05 recording discovery of human remains FCS82, a possible archer*

the anomaly affected both bones of a pair or only one. However, there are 52 complete pairs of scapulae in this sample, of which, ten pairs have *os acromiale* (19%). Six pairs have the anomaly bilaterally, three on the left side only and one on the right side only. Looking at the ten pairs as a group, nine scapulae have it on the left and seven on the right while, of the entire sample of 26 scapulae, 15 (14.7%) have the anomaly on the left and 11 (10.5%) have it on the right. Thus, there is not only a considerably increased prevalence of *os acromiale* in the group when compared with the modern incidence but there is also an increased prevalence of the condition on the left side. Although *os acromiale* affects an epiphysis, all the affected bones belong to fully adult men, so the high prevalence is not simply caused by immature bones, and a further explanation is needed. *Os acromiale* has often been considered as a developmental anomaly, the epiphysis failing to unite to the main bone. There have been suggestions, however, that it can be caused by a trauma to the shoulder (Miles 1994), and that it is more common bilaterally (Liberson 1937; Sterling et al. 1995). It is certainly more common bilaterally among the *Mary Rose* men, although its cause was unclear. Accidental shoulder trauma is unlikely to affect so many men in the same way but a common pattern of activity might do so. A full discussion is presented in *AMR* Vol. 4, chap. 13, where the principal conclusion drawn is that the unusually high frequency of *os acromiale* in the men of the *Mary Rose* may indicate the presence of specialist archers.

In terms of specific skeletons, FCS82 and FCS83 in M2 were both young adults – FCS82 was a little older than FCS83. He was also considerably shorter at about 1.68m (5ft 6in), compared with FCS83 who was about 1.82m (5ft 11in). The best evidence for an archer is on the bones of FCS83. All muscle attachments are very well developed and the spine is noticeably stressed in the mid-section. Both clavicles (collar bones) have deep impressions for one of the principal ligaments, the costoclavicular ligament, which is particularly severe on the right side, while the left clavicle has very strong attachment areas for other ligaments. Simon Stanley (pers. comm. 2000) reports sore clavicles after shooting the bow repeatedly and it may be that FCS83 was an archer. He could have been situated in the Forecastle; both men could have fallen down the companion way as the ship heeled over.

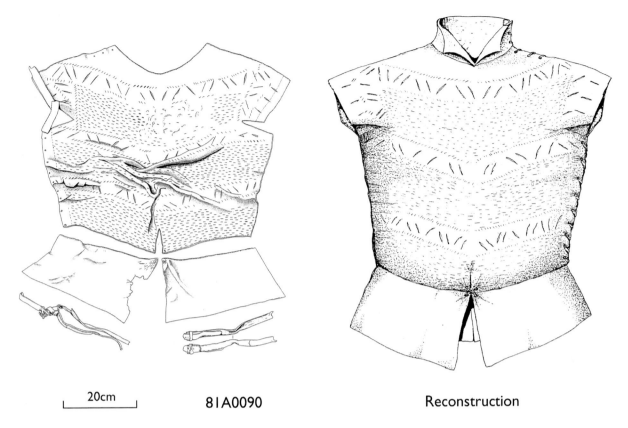

Figure 8.101 Leather jerkin 81A0090

FCS7, who was found in H2, may also have been an archer. This skeleton is one of the most complete from the ship. He was probably in his mid-twenties and about 1.76m (5ft 9in) tall. There is *os acromiale* of both scapulae and strong development of humeral entheses (the points of muscle attachments on the upper arm). The left humerus has an avulsion of the medial epicondyle (where the part of the upper arm bone to which one of the muscles attaches at the elbow has been pulled off) and a similar avulsion fracture of the left elbow occurred in a skeleton from the battle of Towton with *os acromiale*, also assumed to be an archer (Boylston *et al.* 1997, 38). FCS7 also has some spinal changes, particularly affecting the thoracic-lumbar junction including twisting of the vertebral articulations (a feature of other spines in this group of FCS) that may be associated with the use of the longbow. Very few items were found with this FCS, however the location (just beneath deck beam O20) may suggest an original location of higher up. One of the most highly decorated wristguards, 81A4241 (Fig. 8.57) bearing the Royal Arms, was found on the orlop deck just above O20. The geographical association is very close and the association of a Royal emblem with the skeleton of what might be a professional archer is tantalising.

FCS82 was found lying against the Main deck orientated north–south to the west (port side) of the companionway (MU 81, 1–2 site book, 43, Fig. 8.100). The skeleton was described as being as complete (analysis proved it not to be) and appeared to be wearing a jerkin (81A0090/93; divelog 81/2/64–66). The cattle and goatskin jerkin is one of the best preserved and most elaborate recovered (Forster, *AMR* Vol. 4, chap. 2). It is side fastening and would have been laced up the left side with silk thread, some of which survived and is of chest size 44in (1120mm) with a waist of 42½in (1080mm). The entire bodice is elaborately pinked and snipped (Fig. 8.101) and it also had a leather belt around the waist with a buckle (81A0093). The back is more deteriorated than the front, suggesting that the body lay face downwards. Other items directly associated include a comb (81A0094) and shoe fragments (81A0095, shoe type 1, see Mould, *AMR* Vol. 4, chap. 2).

FCS83 was identified as an archer by the excavators and is christened '*Little John*' in the records (1981, MU 1–2, 60, divelog 81/2/79; Fig. 8.102, top). He was also found south of the companionway associated with unidentified concretion and arrows in a spacer (81A0116) tied with a strap (Fig 8.102, bottom). A sediment sample (81S0021) contained rope (81A4688), thread (81A4689 and 81S1384) and further arrow fragments (81A4690).

Two or more skeletons in M3 were identified during the excavation as being '*trapped against the main deck and the starboard side*'. FCS 74 was excavated over a period of 18 days in June 1981. The ribcage was associated with a jerkin (81A0743), a belt buckle (81A0812) and

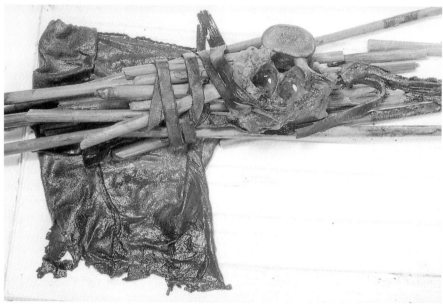

*Figure 8.102 top) underwater photograph of 'Little John' (FCS83); bottom) detail of arrows 81A0116 with arrow bag 81A6844*

shoes (81A4063/4559, type 2.1). FCS75, close by, wore a finger ring bearing the letter 'K' (81A0810, see Fig. 8.67, above) fully documented in (*AMR* Vol. 4, plate 12) and an ivory wristguard (81A0815; see Fig. 8.66, above). Further analysis of this individual suggests that he had extremely limited use of his right arm, with severe damage to the right elbow as a result of a trauma to the radius before the age of fourteen. It is unlikely that he would be able to draw a bow and even many of the tasks required of a gunner would be difficult if not impossible (R. Drew, pers. comm. 2008). Other associated small objects with these FCS included a silver coin (81A0814), a comb covered in tar (81A4231), a decorated knife sheath (81A0766), a pewter spoon (81A0666) and a wooden bowl (81A4069) with two intersecting arrows carved in the bottom on the inside and two much larger ones across the whole of the outside, nearly obscuring carved geometric marks or letters on the centre of the base (Fig. 8.103). This area contained six FCS (74–8, 91), five of which have been identified as potential gunners due to particular skeletal pathology (Stirland, *AMR* Vol. 4, chap. 13). The sixth is less robust, and relatively short.

FCS 70 was found close to a bronze cannon in M6. He was 1.78m tall and between 18 and 25 years old. He had a twisted spine and grooves on one of his hands. He was wearing a jerkin (81A1573, see *AMR* Vol. 4, chap. 2) and carried a comb in pouch. A longbow (81A1490) was beside the body as was the scabbard and sword

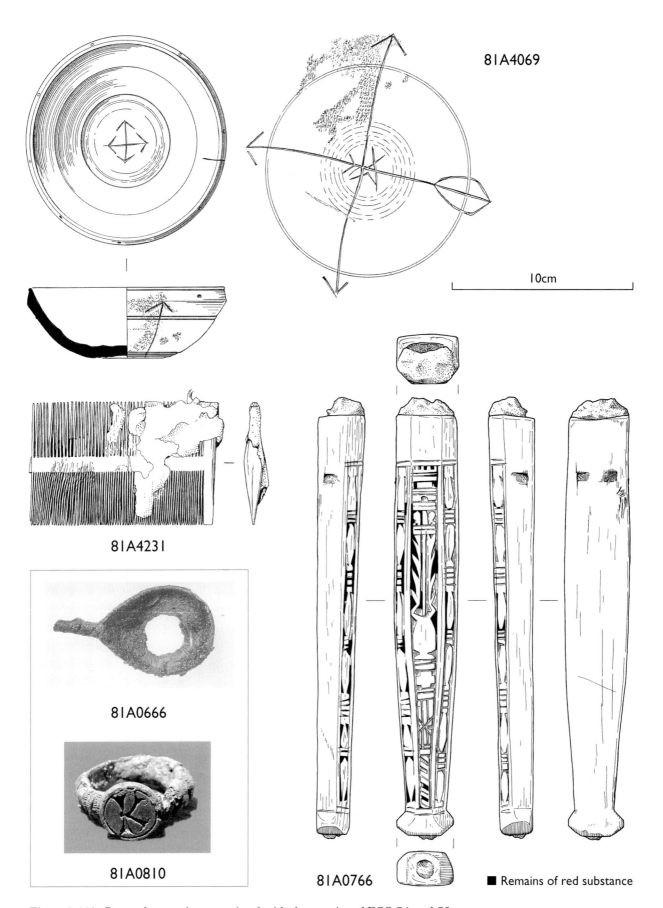

*Figure 8.103 Personal possessions associated with the remains of FCS 74 and 75*

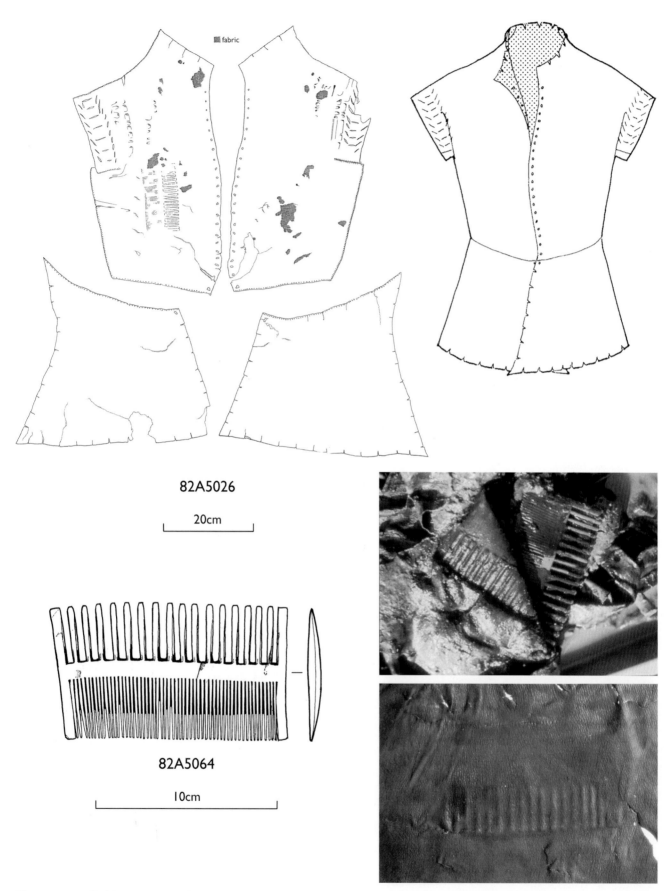

*Figure 8.104 Jerkin 825026 and comb 82A5064 associated with human remains (82H1000) with detail of jerkin showing comb impression*

| DIVE NUMBER 58 82/8/58 | NAME: KEIGHLEY | DATE 19/5/82 | TIME: DOWN 17.19 | TIME: UP 18.31 |
|---|---|---|---|---|

| U/W VISIBILITY | U/W CURRENT | SURFACE CURRENT | WEED |
|---|---|---|---|
| 2-3m | W | SLIGHT | LOTS |

OBJECTIVES

Excavate further in 08 – S end, trying to get S end to same level.

SECTOR WORKED: 08

RESULTS

Removed more bones as shown below, two lots of arrows removed, skull taken out of situ & placed temporarily where barrel was; area of leather also removed.

SKETCH: (FOUR CORNER SQUARES = POINT OF REFERENCE e.g. MAIN BEAMS

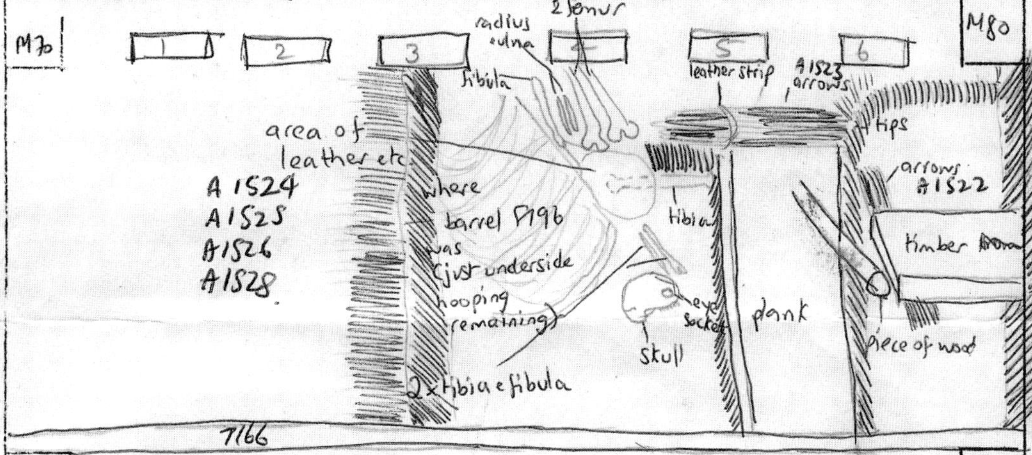


070 ← N Skull near SW end of F196. 080

FIND: DESCRIPTION: IDENTITY NUMBER:

A1522 – arrow frags. A1523 – arrow frags with spacer bones
A1524 – wristguard/strap L135 x W120mm. H46.
A1525 – kidney dagger L.120mm
A1526 – pouch & comb A1528 – shoe sole – frags x2.

| CONTEXT | DSM | NOT DSM | LOCATION DESCRIPTION: |
|---|---|---|---|
| | | | |

| DATUM | DISTANCE | DATUM | DISTANCE | DATUM | DISTANCE | DATUM | DISTANCE |
|---|---|---|---|---|---|---|---|
| | | | | | | | |
| | | | | | | | |

PTO IF NECESSARY

*Figure 8.105 Divelog 82/8/58 showing 82H1000 and associated objects*

hanger with attached pomander (81A1567, see Fig. 9.85). Although not articulated with the humerus, a radius and ulna (81H0171), together with wristguard 81A1460 stamped with the Royal Arms (see Fig. 8.57, above), is assumed to be part of this skeleton (Dean, divelogs 81/6/20, 81/6/33).

A group of bones (82H1000) found in the south end of O8 may also represent the remains of an archer. Four FCS were found in this area, which functioned at least partly as an archery store and which probably had an access way to the Main deck via a ladder. Artefacts associated with 82H1000 include a calfskin jerkin (82A5026) of chest size of 38in (965mm) and waist of 35in (889mm) with a comb (82A5064) possibly from inside a pocket in an inner garment. This left an imprint on the inside of the left bodice. The jerkin fastens at the centre front and would have been laced. It is pinked and snipped around the arms and pinked around the skirt flaps and neck (Fig. 8.104). Arrow fragments (82A5028), shoes (82A1515, 1528), a ballock knife (82A1525), a wristguard bearing stamped decoration suggesting affiliation with several liveries (82A1524) and a sheepskin pouch (82A1526) were found in the same area. Two longbows (82A1590 and 82A1607) were also recovered as well as two sheaves of arrows (82A1587, 82A1589), possibly derived from the chest in O9 (81A1862) (Fig. 8.105). There is no bulkhead between the two sectors.

Three chests contained wristguards and may, therefore, have been the property of archers. The contents of chest 81A5783 from M9, that contained wristguard 81A5826, have already been discussed (see above and Table 8.3). Elm bench chest 82A0942 contained one of the most highly decorated wristguards (82A0943; Tables 8.59, 8.32, Fig. 8.55, above). The presence also of the only ivory comb from the ship and a pair of thigh boots suggests an individual of some rank or status. Elm chest 81A2035 was found standing upright near a bronze gun in M10. A mixed group of objects was found in it, in addition to (undecorated) wristguard 81A2192 (Tables 8.59, 8.32). A pottery vessel in this chest is a small, loop-handled redware cooking pot made in the Low Countries.

# 9. Hand to Hand Fighting

## Staff Weapons
by Guy Wilson with Alexzandra Hildred

*Scope and importance of the assemblage*

Staff weapons, weapons which were mounted on long hafts of wood or metal, were known as '*staves*' from the late Middle Ages until the seventeenth century. Many staff weapons seem to have been adapted from agricultural implements and there are a huge number of variants. In England the '*call to arms*' was '*Bills and Bows*' for spearmen and archers, reflecting the importance of these weapons. Most were utilitarian and were carried by ordinary foot soldiers. However, decorative examples from the sixteenth century are also quite common. Some of these would have been ceremonial weapons, others made for wealthy commanders and some for the guards of the great magnates and status-conscious monarchs like Henry VIII. Both richly decorated and more humble plain examples were imported from Italy during Henry's reign. Major suppliers were the Florentine merchants Leonardo Frescobaldi, Francis de Barde and Franciscus Taunell (Norman and Wilson 1982, 65–71). It is the decorated examples that tend to have been preserved.

The *Statute of Winchester* in 1285 decreed that the able-bodied had to give military service to the Crown when required and to equip themselves with arms and armour related to their wealth. This statute was reaffirmed by Henry VIII in 1511. The Crown only provided arms and equipment for the Royal bodyguard, the Yeomen of the Guard (founded by Henry VII), the crews of ships and the garrisons of the royal castles, forts and bulwarks (Blair 1999, 6). When Henry VIII came to the throne there was no standing army. Although men were required to provide arms, the government made provision for additional stocks before the French war of 1512–14, making large purchases of arms and armour. This was also the case before the invasions of France in 1522 and 1544.

Despite the major tactical changes that had been taking place on the Continent since the fifteenth century as a result of the advent of trained pikemen used en masse and the increased use of improved hand firearms, the English continued to employ battle tactics utilising infantry armed with longbows and bills during most of Henry's reign. In 1538 the perceived threat of an invasion by the alliance of the Emperor Charles V, Francis I of France and Pope Paul III caused Henry to review the nation's defences. This resulted in a massive campaign of building and arming forts and increasing the fleet and the armaments carried. The shire musters before the invasion of France in 1544 showed an increase in both pikes and '*hagbuts*' ('*harquebus*', '*arquebus*', a handgun; see Chapter 7), but longbows and bills still outnumbered the other arms. Vessel inventories (see below) suggest that both pikes and bills were of equal importance at sea and the most numerous of the staff weapons.

The Royal Armouries retains in its collections a considerable number of both decorative and plain staff weapons that can be identified as coming from the royal stores amassed during Henry VIII's reign. This extensive resource can be compared with the fragments found on the *Mary Rose*. Beyond these, and examples of Henrician staff weapons that have been dispersed over the years, there are many surviving examples of pikes, bills and halberds in British, European and North American museums. However, in general '*their often exceptional technical and artistic quality makes them less representative of the arms that were actually used in battle by ordinary combatants*' (Contamine 1984, 122). The *Mary Rose* is, therefore, an important resource for studying the ordinary government issue of weapons for the army by sea. The only identifiable staff weapons from the *Mary Rose* are bills (*c.* 100), pikes (*c.* 20), a single halberd and five haft fragments from unidentified weapons. Examples of these can be seen in the *Cowdray Engraving* (Fig. 9.1) which depicts the sinking of the *Mary Rose*, and in the painting *the Embarkation of Henry VIII from Dover for the Field of Cloth of Gold* (Royal Collection RCIN 405793). Very few original hafts survive and the recovery of so many presents a unique opportunity to study the form and variety of datable, genuine hafts.

### The bill
The munition bill was a derivation of the agricultural hedging-bill or pruning hook used to lop off unwanted branches. It was mounted on a long haft. Measurements of a bill of similar head form and haft diameter to the *Mary Rose* examples include an overall length of 2045mm with head length of 488mm and weight of 2.3kg (RA.VII 1493; Norman and Wilson 1982, 68).

*Figure 9.1 Examples of staff weapons from the Cowdray Engraving: a) bill; b) pike; c) halberd; d) dart*

The bill is illustrated as a weapon as early as the thirteenth century (Blair 1962, 23). The iron head is forged with a single cutting edge. This consists of a convex or flattened 'S'-shape which separates at the top to form a vertical spike and a forward curved hook. Towards the centre at the back is a single back spike or fluke (Figs 9.1a; 9.2a).

A distinctive form of English bill seems first to emerge in the mid–late fifteenth century. It is illustrated in the manuscript 'Pageant of the birth, life and death of Richard Beauchamp Earl of Warwick' (the *Beauchamp Pageant*) in the British Library (BL Cotton MS Julius E. IV, ff. 5, 5v, 7, 8v, 9v, 21, 25; Fig. 9.3b). This is dated to between 1483 and 1492 (Sinclair 2003). This is recognisably related to later English bills but has a longer and thinner blade than became common a little later. By the reign of Henry VIII, although of basically similar design, Italian bills usually had a longer head and were longer and thinner than the English ones (Norman and Wilson 1982, 68, cat. no. 52), as well as often being decorated. Relatively few English billheads survive.

All English bills have plain top and back spikes and a blade, ending at the top in some sort of forward hook. This blade may be vertical or angled forward, and is sometimes straight and sometimes curved. The head may spring at right-angles from the socket but is often at an acute angle. Occasionally, as on those from St

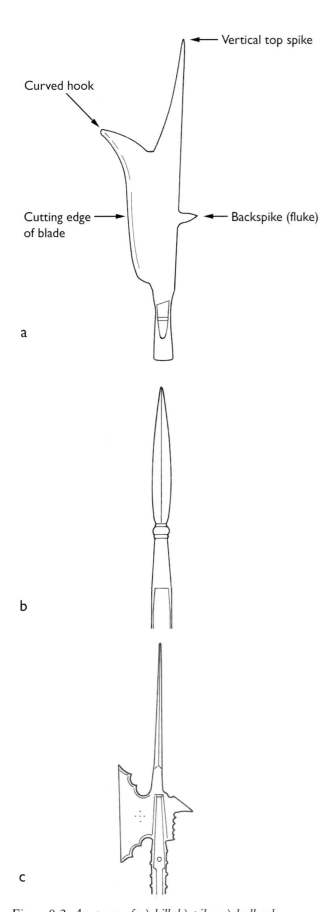

*Figure 9.2 Anatomy of a) bill; b) pike; c) halberd*

Mary's Hall, Coventry, the shoulder of the blade gently curves into the main edge (Fig. 9.3a). There is every reason to suppose that these are some of the 22 bills recorded in the Coventry City armoury in 1589 (Coventry City Record Office, MS W1696) and/or of the total of 48 recorded there in 1640 (MS W989). A variant of this form is seen in 82A2637 (Fig 9.13) where the front of the head forming the blade springs from the socket in an obtuse-angled curve and continues generally to expand, forming a slightly down-turned blade, until the hook, or flukes, curves away from it in a greater angle.

The English (black or fighting) bill was very commonly used in Tudor England. Many of Henry's troops who beat the Scottish army at Flodden in 1513 were armed with it (Oman 1937, 302) and it continued to be important in English armies until the end of the century. The 1633 inventory of the armoury of the Earl of Pembroke at Wilton still included 60 black bills among the weapons. Part of the bill's popularity may have been its cheapness. In 1542 Henry VIII adopted a policy of controlling the price of weapons in an attempt to ensure that properly equipped soldiers presented themselves at musters. While the price of arming swords was set at 2s 8d, halberds at 16–20d, and javelins at 10–14d, bills '*ready helmed*' were to cost only 12d (Wilkinson 2002, 37–8).

Inventories of arsenals give only an insight into what they contained at moments in time: moments which, for various reasons, may be atypical, so that this evidence is often difficult to interpret. However, the various Tudor inventories of the arsenal in the Tower of London seem to tell a consistent and perhaps reliable story. They suggest that bills became more widely used in the sixteenth century than before and remained of military importance until well after the reign of Henry VIII. In the twelfth year of Henry VII (1496–7) some 7349 new bills were bought from various makers, though only 42 were in the Tower when a new master of the King's Ordnance took office (PRO E 36/8, pp 166, 183). By 1514 the number in the Tower had increased to 4000 (PRO SP 1/7, f. 319v [*LP* 1, 2812]), and by 1523 to 7000 '*billys redy helved*' and another 7000 '*bylheddes not helvyd*' (PRO SP 1/28, f. 223 [*LP* 3, 3351]). The closest inventory prior to the date of the sinking of the *Mary Rose* (1540), shows 8000 bills, nearly double the 4500 '*morris pikes*' (see below), though the numbers of each in the various castles and forts mentioned are generally roughly equal (PRO E 101/60, no. 3).

The 1547 inventory of the possessions of Henry VIII lists over 28,000 bills, black bills or fighting bills on the King's land defences and ships (Starkey 1998), well over twice the number of pikes or morris pikes recorded. On only about 17% of the land fortifications were there more pikes than bills. At sea the difference is even starker – only about 11% of Henry's warships had more pikes than bills. Some caution about the exact

*Figure 9.3 a) Bill from Coventry; b) use of bills from the* Beauchamp Pageant *(BL Cotton MS Julius E. IV, f.25)*

numbers of pikes listed in this inventory is necessary. They are usually listed as either '*pickes*' or '*Moris pickes*' with only one entry (*ibid.*, 6998) linking them definitely together as '*pickes called Morispickes*'. But the work '*pick*' could also mean '*pickaxe*', as in entry 6298, where it occurs in association with mattocks and mason's axes. However, there is no doubt that bills appear in far larger numbers than pikes. Nor did the bill go out of use rapidly after the death of Henry. Even the inventory taken towards the end of the reign of Queen Elizabeth I, in 1599, shows nearly 6000 bills and billheads still stored in the Tower (PRO WO 55/1673).

However, bills were not just munition weapons handed out to troops at need. Many Englishmen acquired and used their own or, if they were wealthy enough to have an obligation to provide troops, issued them for use by others. In 1566 Sir John Wentworth of Gosfield in Essex bequeathed to his nephew a dozen pikes, six bills, six halberds and six javelins; weapons that may have related to his obligations to provide troops (Emmison 1978, 47). It would be wrong to think, as scholars have perhaps been inclined to when speculating on the derivation of the weapon from the peasant's agricultural tool, that all those who used bills for fighting came from the lower classes of society. The bill could be owned and used by a gentleman, as well. In 1570 Katherine Forde, widow of John Ford Esq of Great Horkesley in Essex left to her daughter, the wife of '*Richard Smyth, gentleman*', one black bill. This was one of several staff weapons that she had owned, others included a pike and '*my*' halberd, that went to her son and a pollaxe that she left to a friend, the wife of another gentleman (*ibid.*, 83–4).

The Venetian ambassador Daniel Barbaro gave a lively account of English billmen in 1551:

*'their weapon being a short thick staff (asta), with an iron like a peasant's hedging bill, but much thicker and heavier than what is used in the Venetian territories. With this they strike so violently as to unhorse the cavalry; and it is made short because they like close quarters.'* (Brown 1864–5, 703 [p. 350])

When King Henry VIII led an English army to France in 1513 the vast majority of the infantry were equipped with bills or bows (Oman 1937, 291). At the siege of Boulogne a year before the *Mary Rose* sank there were more bills in use in the English army than pikes (*ibid.*, 332–3) and the English army that fought the Scots at Pinkie in 1547 was also still dominated by bow and bill (*ibid.*, 359). It seems clear, however, that gradually, as firearms came to dominate the battlefield, the pike became more important in English armies and the use of the bill somewhat declined. Lord Audley's *ABC for the Wars*, a manuscript written for the education of the young King Edward VI (reigned 1547–1553) stressed the importance of ensuring that each company in an army 'be like appointed to so many shot, so many pikes, so many bills' (Cruickshank 1990, 94). When the obligation of landed gentry to hold arms and armour for the defence of the Kingdom, first codified in 1181, was revised in 1557/8, among the equipment that those worth £1000 or more had to maintain were 30 longbows, 20 hackbuts, 40 pikes and 20 bills or halberds (Wilkinson 2002, 44).

The relative decline in importance of the bill is confirmed by the authors of late Elizabethan military manuals. Sir Roger Williams, writing in 1590, suggested that for every 1000 pikes in an army there should be 200 shorter staff weapons, either bills or halberds. Similarly, Humphrey Barwick (*c.* 1594), suggested a ratio of fifteen bills or halberds to every 35 pikes. However, Robert Barret (1598), makes no mention whatsoever of bills, only of pikes and halberds. Roger Williams himself prefers the pike to the bill, considering it more destructive and versatile, and complains that bills are now made '*for the most part all of yron with a little steel or none at all*' (Evans 1972).

Bills had been in use in England from at least the thirteenth century (Laking 1920–2, 3, 111) and seem to have been carried by a wide variety of people. It seems clear from such evidence as the Paston Letters that, in the mid-fifteenth century, bills were used both by bands of criminals and by ordinary citizens protecting their rights (Gairdner 1904, 2, 267; 4, 169). Billmen '*shoke theire bylles*' at the battle of Bosworth in 1485 (Young and Adair 1964, 87) and at other conflicts during the Wars of the Roses. Certainly before and at the beginning of the Tudor period the bill was carried and used frequently enough to come under legal restriction. In 1483 the carrying of bills (as well as '*glaives*' [a staff weapon mounted with a long blade], swords and bucklers) in London was forbidden (St Aubyn 1983, 166–7). In 1487, in an effort to make life in English urban areas less violent, Henry VII forbade all civilians, except civic officials, from carrying '*bills, bows, arrows, swords, long or short*' in any town or city (Wilkinson 2002, 23, 26). However, it should not be assumed that at this time the bill was as popular in England as it clearly became in the sixteenth century. A detailed account of the arms and equipment of 201 troops mustered at Bridport, Dorset, in September 1457 in response to a French raid on the port of Sandwich records that only three were armed with bills, while listing eleven glaives, ten pollaxes, ten axes and five spears (Richardson 1997, 46), though this was intentionally a muster principally of archers.

## The pike
The pike is a spear-like staff weapon with a long haft and small, single tipped, leaf- or lozenge-shaped blade, usually called the '*morris pike*' (Moorish) in England, traditionally used en masse by squares of foot soldiers (Figs 9.1b; 9.2b). It was commonly in use from the fifteenth to seventeenth centuries. The role of pikemen was to maintain a barrier of points to protect halberdiers and musketeers from cavalry attack. At sea the '*half-pike*' was used to repel boarders right up to the twentieth century. There is no information about the length of the pikes at this period issued for use on ships, and the term '*half pike*' does not appear to be this old.

The pikes found on the ship came mostly from the Upper deck in the Sterncastle and the open weather deck in the waist. Is this, perhaps, evidence that they were ready to repel boarders? Twenty head ends have been recovered that can definitely be identified as pike hafts. They are all of ash except for one that has, surprisingly, been identified as fruitwood (Pomoideae). From the fragments of pike haft surviving from the *Mary Rose* it has so far proved impossible to determine their original length. The assumption, probably correct, must be that they were normal, long pikes with an overall length of some 18ft (*c.* 5.5m). However, until there is definite proof there remains an alternate possibility. Henrician inventories give few details of the weapons listed and almost never dimensions. The 1547 inventory of the King's possessions, that lists the weapons on all his land forts and ships as well as in his palaces, is no exception. However, when recording what was in store at Southsea Castle, the clerk, uniquely, differentiated between long and short pikes (Starkey 1998, 4451–2). We should not, therefore, be too certain that a '*pike*' is always a long one. Indeed, common sense and later practice suggests that short pikes would have been easier to handle and therefore more useful on the crowded and confined decks of a warship.

The small leaf- or lozenge-shaped head was attached by steel side pieces (termed '*cheeks*' or '*langets*') rebated into the wooden haft. Nearly 100 survive in the Royal Armouries, though most have new hafts and many may be Italian imports. 'Morris' pikes are listed within the Tower armouries from the late fifteenth century until 1559, after which the term '*pike*' alone is used (Norman and Wilson 1982, 71).

The earliest evidence for the use of pikes in quantity by native English troops is at the great muster of citizens of London held at Mile End on 8 May 1539. This lists 36,000 men '*in harnys, wythe morys pykes, and handguns and bowys*', (Blair 1999, 21, citing Stow 1603, 1, 103; 2, 284). Henry realised the importance of this weapon. In 1513 he hired foreign mercenaries armed with pikes and hand firearms for his invasion of France. This was repeated for his next two French campaigns (Cruickshank 1990, 68). The importance of the pike during sea engagements as an anti-boarding weapon is reflected by the numbers listed in contemporary inventories (see below).

**The halberd**

Like the bill, the halberd is a form of axe. Conventionally it had a length of 2.0–2.3m (6ft 6in–7ft 6in) so that the fighting head was above the head of the carrying soldier. The head has a flat, usually triangular blade balanced at the rear by a short fluke and surmounted by a top spike which is sometimes shaped like a spearhead (Figs 9.1c and 9.2c). It possibly developed out of the long-handled axe and may be Swiss in origin. It was widely used all over Europe from the fifteenth century and was the principal dismounted weapon of the Yeomen of the Guard until the late seventeenth century, when it was replaced by the partisan, which is still carried on State occasions (Blair 1999). Henry's troops certainly carried halberds and there are records of considerable numbers being ordered from Italy (Wilson 1985, 17). Some of the halberds illustrated in the painting *The Embarkation of Henry from Dover for the Field of Cloth of Gold* are similar to an etched halberd, probably of Italian manufacture, that survives in the Royal Armouries (VII. 962; Fig. 9.4). This is of very fine quality and may have been made for the King's own soldiers, or guard (Norman and Wilson 1982, 67–8).

**The dart**

The dart was a light throwing spear carrying a barbed head and was often fletched. It was an ancient weapon of both the Scottish and Irish Gaels (Fig, 9.1d). Top

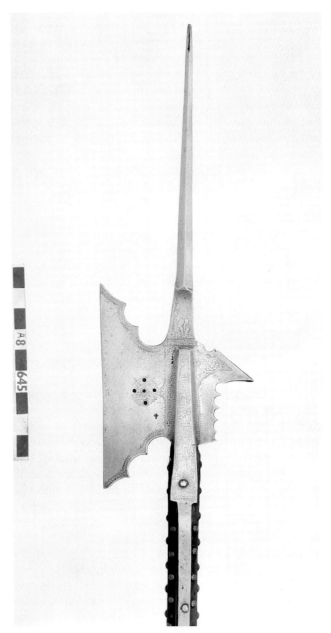

*Figure 9.4 Ceremonial halberd in the Royal Armouries (RA VII.962; ©Royal Armouries)*

darts are listed in the *Anthony Roll* and were possibly thrown from the fighting tops of vessels. The *Beauchamp Pageant* (BL Cotton MS Julius E. IV, f. 18v) shows a man about to cast a dart from a fighting top (Sinclair 2003, 122; see Fig. 1.3). The weapon appears to taper to a smooth point without a barb or visible head and without feathers and also seems to have a small spike in the centre of the bottom end.

## *The* Anthony Roll

The *Anthony Roll* gives us a view of the importance of staff weapons within the armament of the 58 vessels forming the fleet. All ships, galleys, rowbarges and pinnaces carry morris pikes, bills and longbows. The

Table 9.1 Numbers of Morris pikes, bills and longbows listed by vessel type in the *Anthony Roll*

| Vessel type | Pikes | Bills | Longbows | No. pikes/bills per vessel |
|---|---|---|---|---|
| Ship | 2590 | 2560 | 3090 | 30–200 |
| Galley | 1560 | 1570 | 1320 | 40–100 |
| Pinnacle & rowbarge | 394 | 332 | 465 | 8–30 |
| Fleet total (1546) | 4544 | 4462 | 4875 | |
| 1547 fleet total | 2188 | 3408 | 3173 | 20–500 |

totals per vessel type are given in Table 9.1. The number of morris pikes and bills listed for each vessel is frequently the same. The *Mary Rose* is listed with 150 morris pikes and 150 bills (as against 250 longbows). The largest number on any ship is 200 of each for the 1000 ton *Henry Grace à Dieu* (500 longbows), and the smallest is 30 (of each) on the *Mary Thomas* (40 longbows). The total for the fleet indicates the scale of production (over 9000). By 1547 the number of bills for sea service had dropped by about 23% but morris pikes by some 51% and longbows by 34% (Table 9.1). The inventory clearly demonstrates that, in spite of being armed with substantial guns of varying ranges, close range defence using projectiles shot from bows (and handguns) was an important element of an engagement and hand-to-hand combat was widely practiced.

Halberds are not mentioned in the *Anthony Roll* and as the one possible surviving haft (80A1288, see below) is ornate and supports an incised copper alloy collar it may be a badge of office. The 1547 inventory lists only 940 halberds within the castles and fortifications including the English footholds in France. Entries for the Tower of London include very ornate examples: '*White halberdes garnyshed with crymsen velvet*' (Starkey 1998, 3784), '*Halberdes gilte with staves couered with purple velvet and fringed with gold and silke*' (3840; see also 3746, 3783, 3841). Calais is listed with 200 '*New halbardes with redde fustyan of Naples*' (5094). Clearly these are not mundane munitions.

Forty dozen '*Daertes for toppys*' are listed for the *Mary Rose* in the *Anthony Roll*. By the sheer quantity listed, it is unlikely that these are represented by the three large incendiary projectiles recovered (see Chapter 6), although some of the unidentified hafts may be darts. It is possible that they were stored in areas of the ship that have not survived.

## Bills

### Identification and form

Although distinctive (Fig. 9.5), no firmly datable typology has yet been established for the English bill. It

*Figure 9.5 Bill from Jamestown, USA © Colonial National Park, Jamestown Museum*

is generally accepted as being shorter and thicker than its continental counterpart, typically with an open socketed head at least some of which were attached by means of a 'T'-shaped nail which seems to have been known as a '*cross nail*' in Tudor England. In 1599 the Tower of London contained 6600 '*Crossenailes for blackbilles as in former Remaines*' (PRO WO 55/1673) (Fig. 9.6b).

As the survival of iron components from staff weapons from the *Mary Rose* is frustratingly poor, identification of bill hafts among the many hundreds of possible fragments has been made on the basis of a comparison with the proportions recorded for surviving English bills (which, unfortunately, are not numerous) and is confined to those examples that retain the end of the haft which has been worked to accept the iron head. This analysis indicates a minimum number of 105 though only twelve retain evidence of the iron head. There are 29 wood fragments which have been identified as the possible butt ends of bills, a further 836 haft fragments which may be from pikes or bills. Identification of these fragments is based on wood type, location, and association as well as form and is uncertain.

Examination of the head ends shows that they are all made to accept the open side-sockets that are a feature of English bills and 20 show definitive staining suggesting a 'T'-shaped nail (Fig. 9.6; details in archive). Two bill hafts show evidence of having heads

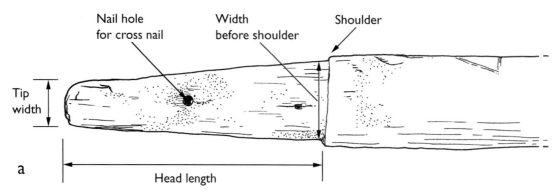

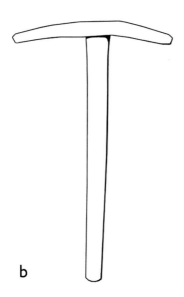

*Figure 9.6 a) Anatomy of a billhead; b) 'T' shaped cross-nail*

attached by two 'T'-shaped cross-nails (79A1222, Fig. 9.9 and 79A0677).

**Hafts**

The assemblage of bill hafts found on the *Mary Rose* is in itself of very great importance. Very few original hafts survive and the recovery of so many presents a unique opportunity to study the form and variety of datable, genuine hafts. Features of the head are shown in Figure 9.6 and a selection of hafts, chosen to illustrate the range of sizes, variations in form and intrinsically interesting features, is provided in Figures 9.7–9. The wood used for 83 out of the 88 hafts which have been analysed (95%) is ash. The few exceptions include spruce (79A0510), willow (82A1556), pine (80A1283), hazel (81A1796) and oak (81A0303). As these are grip fragments (and definitive identification is therefore uncertain) they have been excluded from further discussion.

The end of the round–ovoid section haft which supports the iron head is generally conical with a chamfer on each face and often a more pronounced or secondary chamfer on the front face, although collapse of the wood structure tends to make them appear flatter. The latter can vary between being hardly discernable (eg, 81A2241, Fig. 9.7) to very steep (82A2050, Fig. 9.7) and this does not appear to coincide with size or number of fixing nails. The pointed tips vary between squared off (eg, 79A1109, Fig. 9.9) and rounded, almost like a ducks bill (eg, 79A0748, Fig. 9.9). At least three examples appear to be unfinished as they are without a shoulder (81A0258, 79A0655, and 81A1688, Figs 9.7 and 9.9). A 2–3mm shoulder allows the iron head to lie flush with the haft. The diameter of

**Table 9.2 Summary of basic measurements for bill hafts (all identified wood = ash)**

| Head socket (head length) | Total in group/no. wood idents | Shoulder width (average) | Tip width (average) | Tip to nail 1 (average) | Tip to nail 2 (average) | Shoulder to last nail (average) |
|---|---|---|---|---|---|---|
| 100–114mm (4–4½in) | 9/3 | 34–50 (42.4) | 18–26 (22.6) | 30–80 (39.1) | 78–96 (86.0) | 17–24 (19.0) |
| 115–127mm (4½–5in) | 14/5 | 30–50 (43.4) | 15–28 (22.9) | 20–70 (43.2) | 85–105 (93.8) | 20–100 (40.3) |
| 128–134mm (5–5¼in) | 20/9 | 1@26 then 40–50 (46.4) | 19–22 (23.9) | 28–80 (51.9) | 75–120 (99.6) | 15–128 (52.5) |
| 135–140mm (5¼–5½in) | 15/6 | 30–50 (44.2) | 10–35 (23.4) | 46–112 (63.0) | 105–127 (113.0) | 17–86 (41.8) |
| 141–150mm (5½–6in) | 10/6 | 40–50 (45.2) | 20–30 (24.7) | 40–80 (58.5) | 105–108 (112.0) | 24–90 (47.0) |
| 151–156mm (6in+) | 6/1 | 38–50 (42.6) | 22–30 (26.2) | 50–80 (60.4) | 112–128 (120.8) | 30–95 (45.0) |

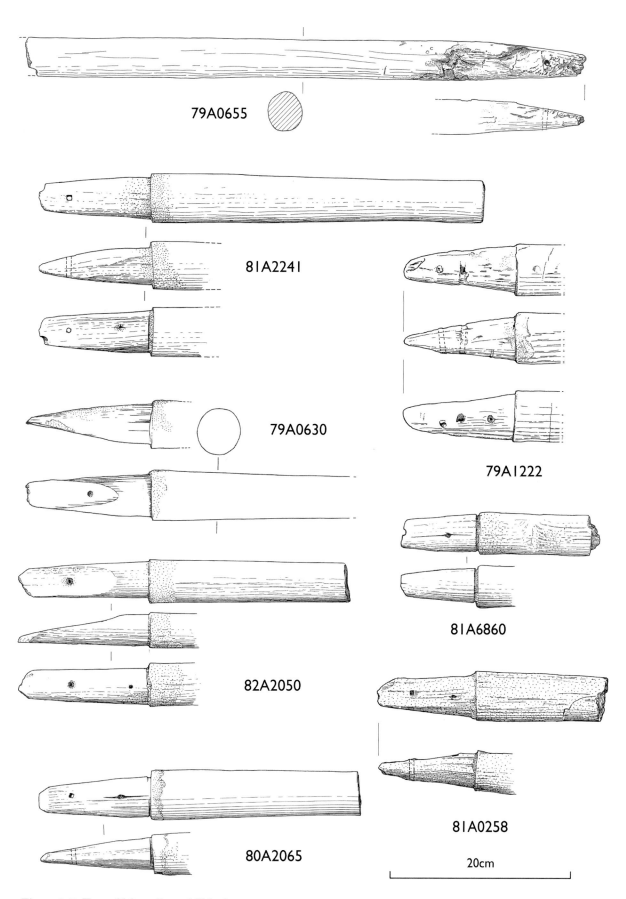

*Figure 9.7 Tops of* Mary Rose *bill hafts*

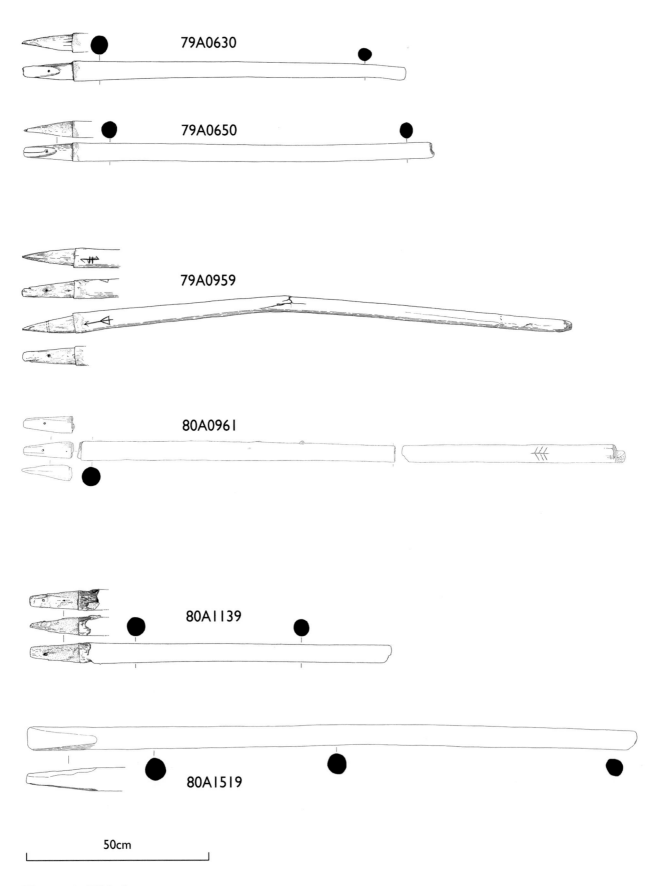

Figure 9.8 Bill hafts

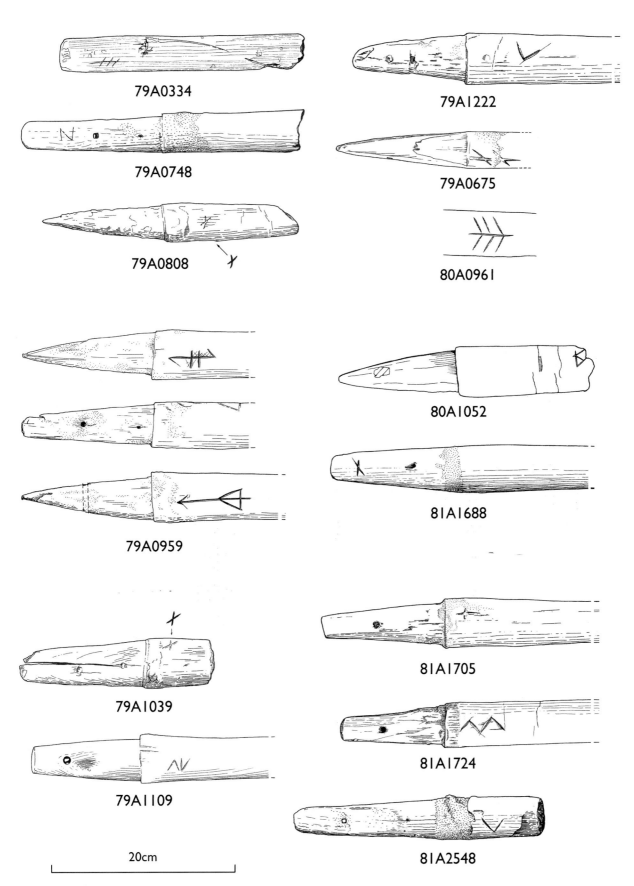

*Figure 9.9 Marked tops of bill hafts*

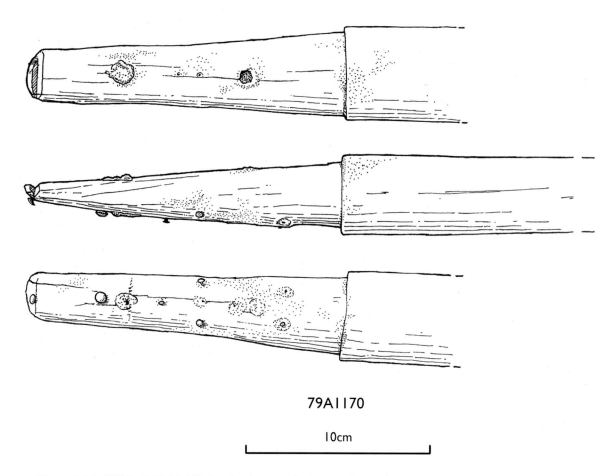

*Figure 9.10 Bill haft 79A1170 showing brass nails in cross formation*

the haft is greatest at the shoulder and tapers gradually (Fig. 9.6), though some have a very marked enlargement just behind the head which reduces almost immediately (eg, 81A1705, Fig. 9.9).

Most of the hafts are broken within 50mm of the tip, only fifteen have a surviving length of over one metre, the longest (80A1519) being 1670mm (Fig. 9.8). All of the hafts fashioned to accept the head were studied and measurements taken of the length of the haft-head (as indicated by the fashioning of the tip), the widths of the tip and shoulder of the socket, and the position and shape of the nailholes (Fig. 9.6). However, many of the tips are eroded and accurate measurements could not, therefore, be obtained (eg, 79A0655, Fig. 9.7).

Seventy-four measured examples could be divided into six size categories based on the length of the haft-head (Table 9.2), with 12% being less than 114mm (5in) in head length, 66% between 115mm and 140mm (5–6in) and just 21% longer than 140mm. There is no direct correlation between the three measurements taken (details in archive) and Table 9.2 shows that the width of the shoulder does not vary much according to the length of the head. The width of the tip, however, shows considerable variation with inconsistency both within and across the size ranges. The position of the nail holes marking the attachment point of the cross-nail is also very variable though, on average, distance from the tip increases with the length of the head, as might be expected.

Taken together, these observations must reflect in part on the size and proportions of the billhead itself. There is no consistency in the position of the nail hole for the securing the cross-nail in relation to the shoulder, for instance, in the case of bills 79A0650 (Fig. 9.8) and 79A0808 (Fig. 9.9) which are of very similar proportions, the distances are 85mm and 59mm respectively. The former has a round nail hole and the latter a square one. Of 68 examples where the shape of the nail hole could be determined, 36 (53%) are square and 32 (47%) round. The exit hole of the cross-nail (where the shape can be determined) is round, suggesting that this hole was pre-drilled and in many cases fitted with a square nail. On a small number of examples (eg, 81A1646) the hole is surrounded by a small round iron stain suggesting merely a round head rather than a cross-nail. A number are marked either on the tip or on the haft just behind the shoulder (Fig. 9.9).

The cross-nail staining (where observed) shows that the horizontal bar of the 'T' is consistently positioned within the open part of the socket and overlaps and secures the metal edges of the socket. Its position is often marked by a dark stain on the wood and, in few cases, a linear impression ¼–⅜in (6–9mm) wide, may also mark the position of the 'T'.

Many of the bills retain two holes for holding the iron heads (eg, 81A0258, 82A2050, Fig. 9.7 and Table 9.2). The lower hole penetrates the surface for only a few millimetres and, in many instances, is too small to be noticed, or is described as a '*tack*'. It is entered from the back of the head in the area completely encased in the socket. This, and the fact that it is not always present, suggests that it may be a secondary fixing put in through the head if it became loose through usage. In many cases the wood around it is split and it does not look like a pre-drilled hole. In the cases of obviously broken bill heads, breaking at the first nail hole is not uncommon (eg, 81A6860, Fig. 9.7). In examples with only one hole it is usually positioned closer to the centre of the head than where there are two (eg, 79A0630, Fig. 9.7), and it is generally over 5mm in size. Perhaps a more centrally placed fixing point is less likely to allow the head to work loose. One example (79A1222: Fig. 9.7) has three nail holes, two of which completely penetrate the head.

The diameter of the haft was taken at intervals along the length of twelve examples whose surviving length was greater than 380mm (examples illustrated in Fig. 9.8). These indicate a gentle taper from the shoulder: the five examples with surviving lengths of 380–500mm show a taper of 73–88% and those surviving beyond one metre taper by 70–83% (average diameters at shoulder and end are 53mm and 42.8mm respectively).

Two objects tentatively identified as bills are of unusual proportions. 79A1170, from U4, is 1510mm long with a head length of 179mm and 80A1519, found in M5, has the longest surviving length of 1670mm with a head length of 190mm (Fig. 9.8). While the shoulder and tip widths of the former fall within the range observed on the measured examples summarised in Table 9.1 (45mm and 25mm), those for the latter fall at the very top of the range for shoulder width, at 50mm, and well outside that for tip width (50mm). For a variety of reasons (for instance, on 80A1519 there is no discernible shoulder or nail holes) both identifications are open to question and while U4 produced a number of bills (see below) 80A1519 would be the only example from M5. 79A1170 has nine brass nails making a cross formation around the 'T' bar (Fig. 9.10). These appear to be decorative, do not penetrate the entire head and are in addition to the expected cross-nail and additional lower nail holes.

**Heads**

Only twelve examples retain evidence of the iron head (81A0911, 81A0916, 81A1594, 81A5790, 81A6911, 81A6917, 82A2637, 82A3590, 82A3899, 82A4504, 90A0064, 90A0107), of which one (82A2637) is virtually complete. This was discovered in concretion and could only be studied in any detail by radiography (see below). Two variations of blade form are evident; both display an obtuse angle of exit from the socket, but one variation retains a straight neck canting up to the bottom edge of the blade (82A3590 (Fig. 9.12), and 81A6917); and the second variant continues in a curve to join the main edge of the blade, as exemplified by 82A2637 (Fig. 9.13).

Bills of the first type include one in the collection of the Cambridge University Museum of Archaeology and Anthropology (No Z11431), another, from the collections of Guy Laking and Kretzschmar von Kienbusch, is in the Philadelphia Museum of Art and another, with an '*RH*' mark on the right face of the blade, was sold at auction by Christie's in London in 1979 (23 May, lot 39). A bill with both this feature and two side nails (evidence for which appears on a number of the *Mary Rose* bills) is in the collection of the Ingleby family at Ripley Castle, Yorkshire. Two more, but with much shorter necks, come from Farleigh Hungerford Castle, Somerset (now in the care of the Royal Armouries). There is evidence of an active armoury in the castle in the mid-fifteenth century (Jackson nda, 43) which continued into the Tudor period. An inventory taken of it in 1524, after Lady Hungerford was attainted for murder, lists 80 bills (Nichols and Jackson 1862). Unfortunately, however, after falling into disrepair when the castle ceased to be occupied in the late seventeenth century (Jackson nda, 165) the armoury was revived in the early nineteenth century with major additions coming from Exeter Castle (Jackson ndb, 66). This makes it impossible to be certain from whence any individual item came. This feature is also found on one bill from Jamestown Virginia (J12422; Fig. 9.5), founded in 1607 (see below).

Bills of the second type include: one in the Higgins Armory Museum, Worcester, Massachusetts (Fig. 9.11; Inv. No. 857); one in the collection of Lord Armstrong at Bamburgh Castle, Northumberland, that has a slightly downward canted blade like this head from the *Mary Rose*; three in St Mary's Hall, Coventry; and one excavated in the ruins of the third–fourth Statehouses at Jamestown (COLOJ-10952). However, all these, along with the generality of English bills have their spikes set considerably lower on the back than appears to be the case with this example.

That a similar bills of both sorts should have been discovered in a seventeenth century context in North America is especially intriguing, and that it is not chance is suggested by the fact that the slightly different form of head also found on the *Mary Rose* (82A3590

*Figure 9.11 Bill in Higgins Armory Museum (Worcester, Massachusetts)*

and 81A6917 q.v.) also finds parallels in finds at Jamestown. It is recorded that, in 1623, a consignment of 1000 brown bills was sent over to the colony from the Tower of London. Of these only 950 arrived as 50 were left in Bermuda at the request of the authorities there. In all six bills have been excavated at Jamestown and it is reasonable to assume that they formed part of this supply (Peterson 1956, 96, 101, 322). At present it is not possible to determine whether the patterns of bill found on the *Mary Rose* were made for 80 years or more or whether among the weapons supplied to the colonists at Jamestown were some very old weapons from the Tower.

The best example with a head for study is 82A3590 (Fig. 9.12), which survives to 1420mm in length (described in detail below). There can be no certainty that its form is entirely typical of the assemblage as a whole, although the form of the hafts indicate that they were mostly, if not all, made to take the open-socketed heads that were a feature of English bills. As stated in the introduction to this chapter, 82A3590 has a blade that springs straight from the socket but at an obtuse angle. It is impossible to know how common this form might have been but certainly only a few examples survive. In addition to those examples mentioned above, a straight neck canting up to the bottom of the edge of the blade is a feature of a tall, thin bill from the Thames at Chiswick in the collection of the Museum of London (A 17294). This is close in form to one illustrated in the *Beauchamp Pageant* (Fig. 9.3b) and is therefore thought to be of late fifteenth century date. Curiously, three other similar bills have been excavated from the site of the English colony at Jamestown (see above: COLOJ-10951, 1129111384; Fig. 9.5, Peterson 1956, 96, 101, 332).

*82A3590*

The socket and lower part of the blade are the only parts of this head to survive (Fig. 9.12). The length of the socket is 115mm (4½in), making it one of the smaller examples found (Table 9.2). The socket is open, formed by simply wrapping the metal round on itself to join at the socket's mouth, while the rest remains open on one side. The remaining thickness of the metal at this point is 3mm (⅛in). The 'T'-shaped cross-nail is set through the haft and riveted to the socket on which it rests 52mm (2in) from the socket end. The diameter of the nail hole is 6mm (¼in) and it can be seen clearly on the radiograph (Fig. 9.12, top). A second nail is positioned 25 mm (1in) closer to the socket. In common with other examples on *Mary Rose* bills, this does not penetrate through the entire haft. The blade thickness varies from 1–3mm midway on the cutting edge to 9–10mm at the junction with the socket neck. Where the head is inserted into the socket the blade thickness is 13mm. The back of the head appears to spring straight up from the socket, but little of it remains. The front of the head springs from the socket at an obtuse angle and, after about 32mm (1¼in) angles up more steeply in a sweeping curve.

The remains of the ash haft show and end tapering from approximately 50mm in diameter to a broken point to fit the socket. The haft appears to have been roughly faceted into four faces and tapers from 55mm at the socket to 35mm at the break. The total haft length is 1180mm.

*82A2637*

The concreted remains of this nearly complete billhead were found outboard of the Sterncastle (SS7) associated with haft 82A2608 (Fig. 9.13). The maximum width of the socket is 51mm with a length of 103mm. The concreted blade has a width of 205–265mm with a thickness of 50mm and length of 365–390mm. The socket is an open socket formed by wrapping the sheet of metal of which the head is made round on itself to join at the socket's mouth, while the rest remains open one side. To secure the head firmly, a 'T'-shaped rivet, known as a cross-nail, was set through the haft and rivetted to the socket. The ends of the

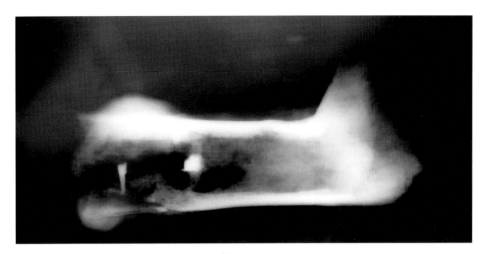

10cm

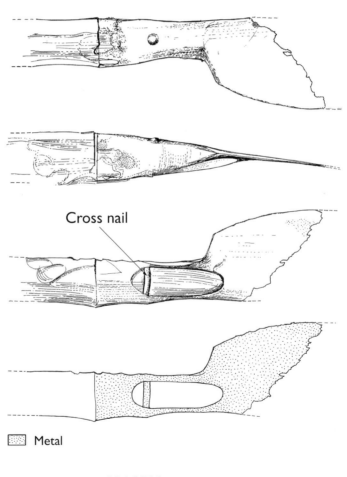

Cross nail

Metal

**82A3590**

20cm

*Figure 9.12 Bill 82A3590 with radiograph*

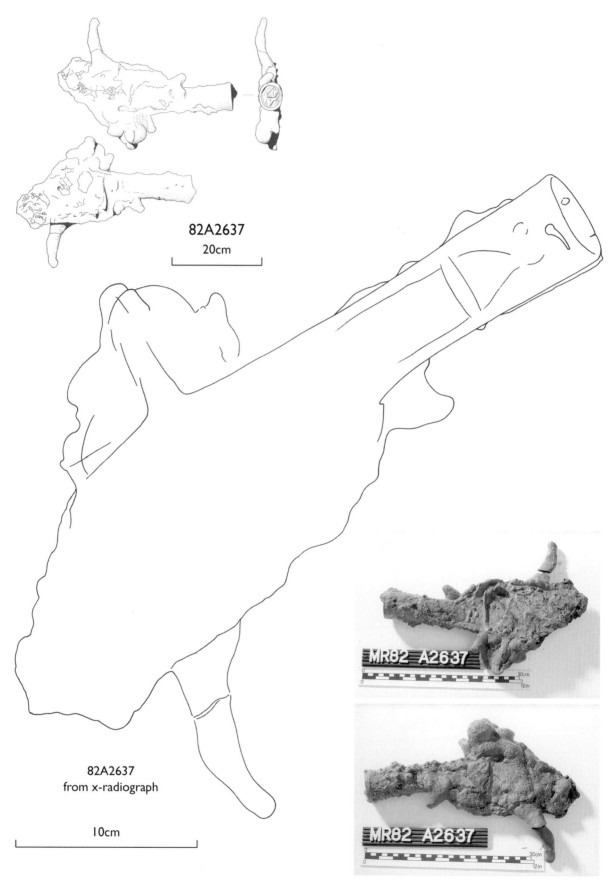

*Figure 9.13 Bill 82A2637 in concretion with interpretation from radiograph*

cross-piece at the other end rested on the edges of the socket 76mm (3in) from its end. The front of the head forming the blade springs from the socket in an obtuse-angled curve and continues to expand gradually, forming a slightly down-turned blade, until the hook, or fluke, curves away from it at a greater angle. The top of the head is damaged and the upper, curved edge of the hook cannot be seen in the radiograph. The back of the blade rises vertically from the socket. A straight-sided spike, tapering to a point, emerges from the back almost opposite where the fluke springs from the blade on the front of the head. The top spike appears to have been bent forward towards the fluke and its exact form remains uncertain. The two associated haft fragments are of 125mm and 188mm in length.

*Other billheads in concretion*
A study of the other portions of iron blade was undertaken in order to ascertain if any other forms were present. This was achieved through looking at the radiographs taken of the concretion surrounding some of the billheads, or by examining the unconserved wet heads. Radiography primarily used an Iridium 192 source with a source to film distance of 1000mm at 12.5ci for 30–60 minutes. The results were largely inconclusive, but showed that 81A6911 (comprising a portion of haft and head 265mm in length) supported a blade 70mm wide and 5mm thick which had an angle similar to 82A3590. The head-end is broken at the first nail hole but the estimated length of the head is 134mm (diameter 39mm at the shoulder and 47mm for the haft with a slight taper). 81A6917 has length of 240mm and maximum width of 110mm (blade width 75mm) with a socket width of 60mm. In form it exits the socket at an angle similar to 82A3590. After about 32mm it angles up more steeply in what now appears to be a straight line, though this may well have been broken back from the original edge. 90A0064 also shows a similar blade form and angle to 82A3590, with the exception that this appears to start much closer to the shoulder, within 30mm. The edge of the blade is present for 50mm away from the head. A 5mm nail hole is visible 30mm from the tip.

81A1594, found in M6, is also a haft fragment, 210mm in length and 90mm (maximum) in width, with its head-end partially encased in concretion which varies in thickness from 3.5–16.8mm. The only notable feature is a possible weld line for the back spike. This appears as a concentric feature and looks separate from the rest of the iron. The head length is 117mm with a tip width of 25mm.

*Possible lining to a billhead*
Bill 81A1646 retains an area of concretion on the head (Fig. 9.14). The upper nail hole penetrates this and has what appears to be remains of textile in this area. This suggests that the socket had an internal fabric lining, presumably to ensure a snug fit.

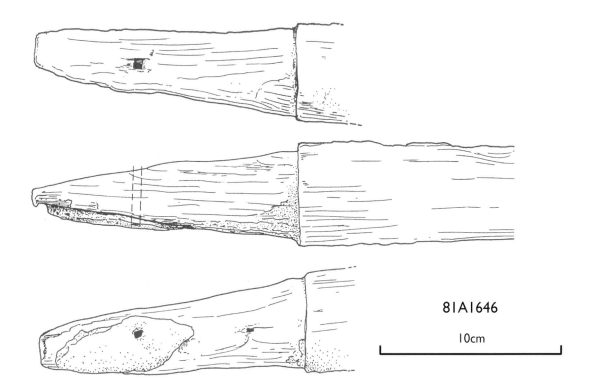

*Figure 9.14 Bill haft 81A1646 with concretion and possible textile*

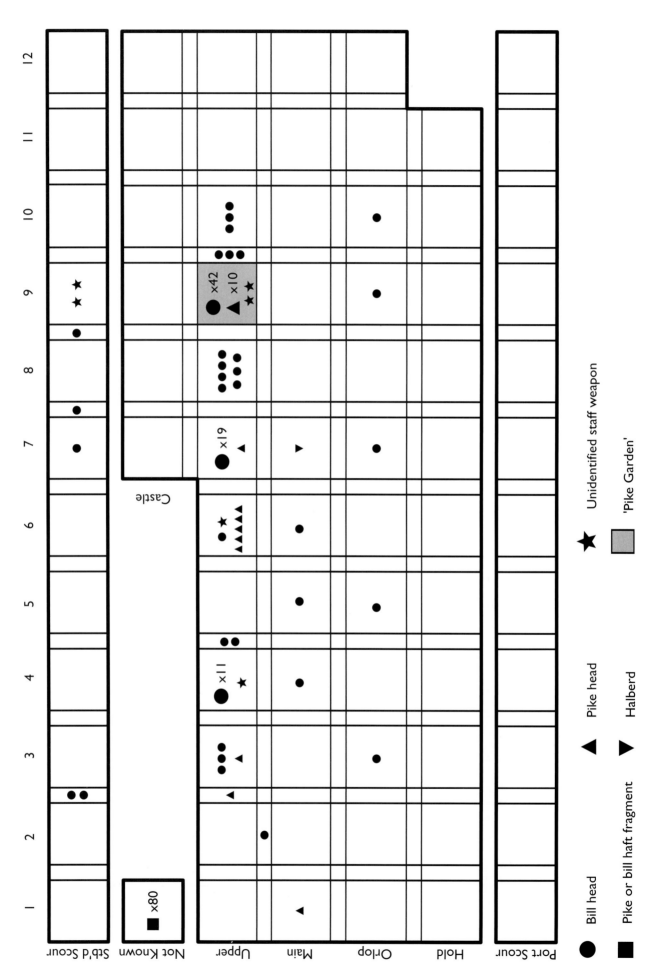

*Figure 9.15 Distribution of staff weapons*

*Figure 9.16 The 'pike garden' in U9*

## Markings

Twenty bills bear incised marks on either the haft or the head portion. A selection is illustrated in Figure 9.9. Of those on the heads, a number were on the backs (eg, the Z on 79A0748) so these would not be visible once the iron head was in place. All appear to be either crude initials (single examples of A, O, W, Z, an A or K, two Vs and several Xs) or geometric markings incised into the wood.

## Distribution

The largest single concentration of billheads is in U9 (42), with a cluster of nineteen close to the start of the open waist (U7) and in U4 (11). The total on the Upper deck was 92 (Fig. 9.15). Because of the numbers of heads and haft fragments found in U9 (over 140) the area was referred to as the *'pike garden'* by the regular excavators (Fig. 9.16). Five examples in the Starboard scourpit reflect similar locations outboard. Very few were found on the lower decks indicating that they were not in storage below decks at the time of the sinking (unlike about half of the longbows and arrows, see Chapter 8) and most could have been displaced from original locations at or after the sinking. All the staff weapons would need to be used either on the open weather deck, or castle decks as they are too long to be used anywhere else. The Upper deck in the Sterncastle would have been a convenient 'at hand' storage area. The butt ends and haft fragments mirror the locations of the heads.

## Method of manufacture and fixing

The hafts were made from a log cut to the required length and then further cut lengthways into planks roughly 60mm thick. These were again cut lengthways into billets of approximately 60 x 60mm and rounded off. The wrought iron metal heads would have the spikes welded on. An iron billet was beaten to form a sheet, wrapped and hammer-welded at one end to form a socket. The edges were then ground to form the cutting edge.

It is probable that the metal billhead and the wooden haft were made separately, possibly by different individuals, and later assembled. The haft may have been pre-drilled to accept the cross-nail, which probably had a $^5/_{16}$in (4–6mm) square shank at the head, tapering to a rounded, smaller end. The head would be offered into the socket and the 'T'-shaped nail driven through the haft and into and through the closed portion of the socket and then flattened with the 'T' bar overlapping the metal edges of the socket. Occasionally an additional nail was used, either at the time of manufacture or added later. It has been suggested that the lower hole, which (if present) is always visible on the enclosed side of the head and does not always penetrate the entire head, may be an initial locator point for the head. Therefore it may represent a particular technique of manufacture not employed universally. There is no indication of holes for possible spikes on the ends of the T-bar.

*Figure 9.17 Forging a bill: a) front and; b) back of finished bill; c) starting the socket; d) socket blank; e) forming the socket; f) completed socket*

## The forging of a *Mary Rose* billhead
by Hector Cole

Radiographs and examination of the surviving remains of billheads and staffs were used in conjunction with surviving bills from the Royal Armouries in Leeds to determine the most likely and best-fit example for replication (Fig. 9.17a, b). Materials and forging techniques available to the Tudor blacksmith were used where possible. The major exception to this was the use of a forging coal/charcoal mix on the fire instead of charcoal.

The head was made up of three separate pieces of metal: 1) the main body of wrought iron; 2) the front spike of medium carbon steel (0.45% carbon); 3) the top spike of wrought iron. The billet for the main body was a piece of wrought iron measuring 380 x 13mm and 0.2kg in weight. The front spike was 130 x 13 x 13mm and weighed 0.2kg. The top spike was 75 x 20 x 20mm and, again, weighed 0.2kg. The total weight of material before forging was 3.75kg.

The first part of the bill to be forged was the socket (Fig. 9.17 c–f), which required sixteen heats for the

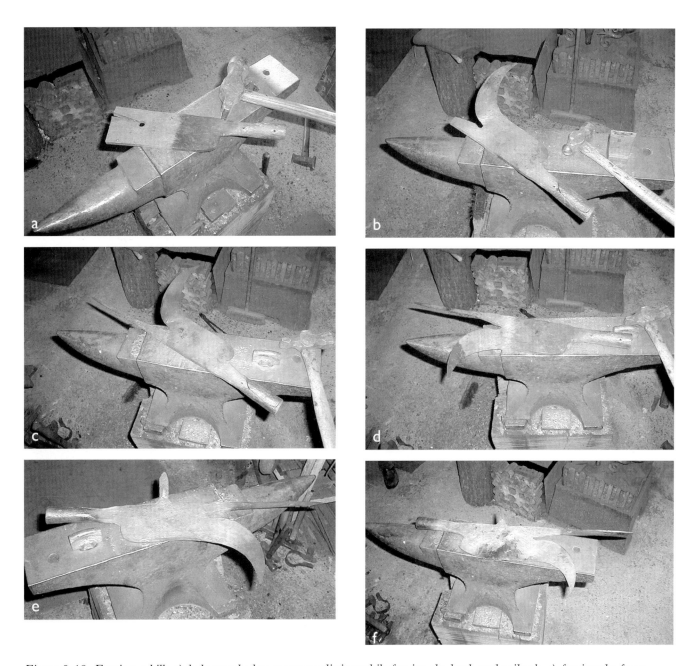

*Figure 9.18 Forging a bill: a) hole punched to prevent splitting while forging the hook and spike; b–c) forging the front spike; d) completed hook and spike; e) top spike blank welded on to blade; f) forged top spike*

socket blank. Forging the socket itself, including the punching of holes to secure the shaft, took thirteen heats and a total time of 2½ hours.

The next stage was to form the hook and the base of the front spike. A hole was first punched through the plate at the base of the junction of the hook and the spike to prevent the iron splitting along its length when forging these parts to shape (Fig. 9.18b). The front spike base was then bent out of the way so that the hook could be forged to shape. The forging of the hook took nineteen heats and 1½ hours (Fig. 9.18, c). The base of the front spike was then forged to shape to allow the steel to be fire welded on to form the point. (thirteen heats; 1 hour (Fig. 9.18d). The main body of the blade was then forged to shape by thinning the front edge down to form the cutting edge and forming the scarf to take the top spike (sixteen heats; 1½ hours). The top spike was welded onto the main body using a split scarf weld (Fig. 9.18e). On some of the originals the weld was often just a flat scarf weld, which is quicker but not as strong as a split weld. Preparing the end of the bar for welding and then welding it to the main body took four heats and half an hour. The forging of the spike to its correct shape took a further three heats and half an hour (Fig. 9.18f). Another six heats and half an hour was needed for the final truing up of the head ready for grinding and finishing, which, itself, took 1½ hours (see Fig. 9.6).

The weight of the forged head was 3.2kg and, after grinding, it weighed 2.9kg. This gives a total material

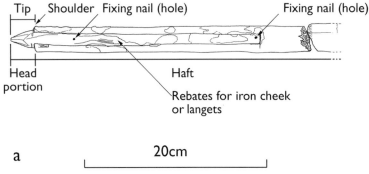

*Figure 9.19 a) Anatomy of a pikehead; b) pike from Royal Armouries (RA VII.1715; ©Royal Armouries)*

loss of 0.85kg, of which 0.55kg was lost in the forging and 0.3kg in grinding and finishing.

The 'T'-nail for securing the shaft was forged from 20 x 6mm wrought iron and took 20 minutes to make. It has a 5mm diameter shank and a 40 x 12mm head.

Twenty-five kilogrammes of fuel was used to forge the head. The equivalent amount of charcoal needed would be in the region of 75kg. The total working time was 9 hours 50 minutes. This does not include time spent selecting materials, lighting fires, mixing coal and other forge tasks. The medieval smith would have had a striker for some of the work which would speed up production a little so an estimate of production rate would be at the most two heads per day. In a large workshop specialising in arms production the grinding and finishing of the head would have been carried out by another member of the team. This time does not include the making of the haft and the fitting of the head to the haft as this would have been carried out by other member of the work force.

The haft was a 51 x 51mm (2 x 2in) ash pole which was shaped to accept the head. The fitting of the head including the socket, nail and 'T' nail took 1 hour.

*Tools*

The forge used double acting bellows and a 16mm diameter iron nozzle hole. This nozzle hole was only just sufficient to give a fire with the necessary heat to carry out all the forging operations. A 25mm diameter hole would have been better as this would have given a broader fire to heat up the main body of the bill to a more even forging temperature. The tools used in the forging were:

- Anvil, 2½cwt (126kg) London pattern
- Hammers, assorted, ranging from 1lb to 7lb (450–3150g)
- Tongs, open mouthed, box and socket
- Leg vice for holding metal while splitting, grinding and filing
- Round punches, 16mm diameter and 6mm diameter
- Hot set for splitting
- Socket stake for forming socket
- Belt finisher for finishing (equivalent of medieval grinding wheel)
- Assorted files
- Nail header for making 'T' nail

## Pikes

**Identification and form**

An example of a pike head which may be representative of those on board the *Mary Rose* is item VII.1715 in the Royal Armouries collection (Fig. 9.19b). Although the haft is incomplete, it has a steel head with a narrow, leaf shaped blade of flattened diamond section, 203mm (8in) in length. This supports a round-sectioned, tapering socket to which the '*langets*' (steel cheekpieces) are attached. Between the head and the socket is a shoulder (Fig. 9.19a).

The archaeological evidence for the 150 listed pikes listed is sparse. This could be due to a number of factors: there may have been fewer on board than are listed; they may have been stored in areas that were badly eroded; the pikes themselves may have been preferentially eroded; or they may simply have been light enough to have floated away on the currents. Misidentification of broken elements is also a possibility, of course; only those items recorded as possible pikes and bills were studied in addition to a search made for sticklike objects within the areas where the majority were found. The size of the hafts (about 25mm at the haft behind the head) means that they are relatively small and fragile compared to bills.

The assemblage consists of tips (usually with the shoulder and a portion of the haft), broken portions of haft which include the rebates for the cheeks, portions of haft that include the end of the cheek rebates, and a variety of broken bits of round haft, some of which could be smaller portions of bills (Fig. 9.20, 9.21). None of the pikes was recovered with enough concretion to allow any interpretation of the iron portion of the head, so examples must be sought from extant collections. Fewer than 100 fragments are

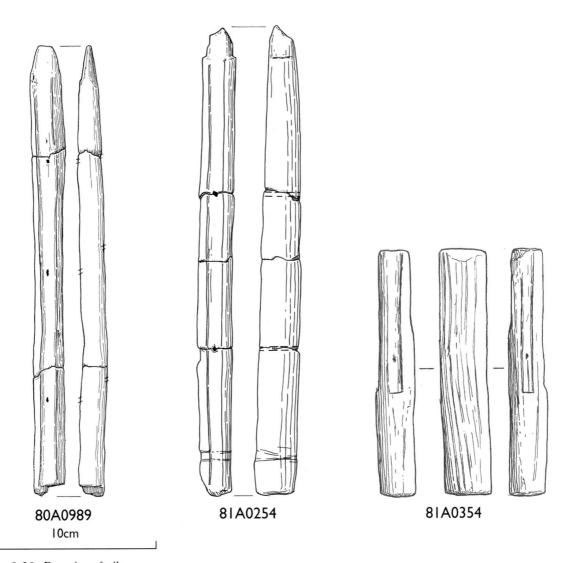

*Figure 9.20 Remains of pike staves*

*Figure 9.21 Pike stave 81A4965 showing cheek rebates*

recorded. The tip and head and the very end of the rebate to hold the iron langets are the only elements which can categorically constitute one pike. As there are more heads than cheek rebate ends, heads were used and only twenty can be positively identified. In addition, there are 31 items listed in the archive as pike fragments. These are mostly degraded fragments but have the remains cheek rebates, or the end of the rebate (eg, 81A0989, 81A0354, Fig. 9.20). A further eight items may also be haft portions of pikes as well as the thirteen portions that have been classified as either pikes or bills. These are extremely fragmentary and collapsed. Only six hafts have been identified as to wood species and, not surprisingly, all are ash. No marks were observed.

The head portion of the pike comprises a pencil-like tip, 12–37mm in length (average 23.2mm), 14–35mm wide at the shoulder junction (average 19mm) with an increase of 2–3mm for the haft, the difference between the last two measurements constituting the thickness of the missing iron at this position (Fig. 9.19a). Continuing from the tip, there are two 6–13mm wide, 0.5–3mm deep, opposing rebates to accept the langets which hold the iron head in position. These extend down the haft for 255–400mm, and one may be up to 50mm longer then the other. Few pikes retain both tip and end of cheek rebates (eg, 81A0354 (Fig. 9.20, end only) and 81A4965 (Fig. 9.21, tip and end)). The langets were held with very small nails, only 1–2mm in diameter, and the holes (sometimes with concretion, but no actual nails survive) vary from round (eg, 81A0354, Fig. 9.20) through oval to square (81A025). In most cases they do not go through the haft and are staggered between the sides (80A0989), though some do penetrate (81A0254; see Fig. 9.20, full details in archive).

The tip shapes vary. Some are almost perfectly pencil shaped but without a point (81A4965, Fig. 9.21), others have a blunt end (80A0989, Fig. 9.20) and some examples have been obviously shaved in flakes with a small knife (79A1013). There does not seem to be enough variation in form or size to suggest different types. The slightest example overall is 81A0254, and the largest is 80A0989 (Fig. 9.20).

There are only five pikes which have heads and hafts over 300mm in length (79A1013, 80A0989, 81A4965, 81A0625, 79A1354). Measurements of diameter taken along the lengths of five hafts indicated that they were could be either of consistent diameter or taper slightly away from the head. The longest, 79A1354, which measures 1495mm, increases in diameter from 22mm at the shoulder to 37mm at the end – an increase of some 68%, but this is an extreme case (details in archive).

**Distribution**

Virtually all the identified pikes come from the Upper deck, especially in U9 (twelve heads) and in the open waist in U6 (five heads). Only one head was found on the Main deck (Fig. 9.15). As with the bills, the Upper deck is the most obvious place for both the use and the storage of these weapons which would have been very unwieldy below.

## Halberd

No halberds are listed as being on board the *Mary Rose* in the *Anthony Roll*, but one object (80A1288) has been identified as the haft for a halberd because of the similarity of its decorated brass collar to a halberd in the Royal Armouries (VII.1516), a German example of about 1520.

**80A1288**

The slender ash haft has a square section and is 30mm in diameter under the collar expanding to 38mm at the end (Fig. 9.22). It is in two non-conjoining parts, the upper portion, comprising the tip that went into the socket, being 289mm and the lower, which formed the upper part of the stave, 299m long (detailed measurements are in the archive). The length of the haft head above the collar is 56mm. Two opposing sides are recessed to take the langets attached to the metal head and the other two, which are shorter and narrower, are similarly recessed to take metal reinforces. One of these latter clearly does not start until some 2mm ($^1/_{16}$in) below the collar. All the rebates are 1.5–2mm deep. The langet rebates each have six nail holes and a single hole at the top and bottom of the rebate for rivets. The rebates for the metal reinforces have either two or three nail holes in each rebate, and each have a hole at the top and bottom for a rivet.

At the top of the haft is a brass collar (76.0% Cu, 9.28% Pb, 12.2% Zn), also square-sectioned. It has concave, chamfered sides and is decorated with a lightly roped incised pattern. It appears to fit tightly on the sides with pinholes and more loosely on those drilled through for langets, and there seem to be the rusted remains of the langets between the haft and the collar.

Above the collar the haft tapers to a broken point. A rivet hole, 19mm above the collar, goes through the haft in the same plane as the langet holes to secure the head socket.

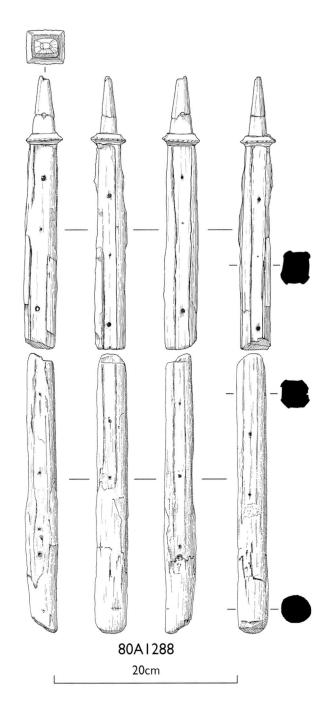

*Figure 9.22 Haft and collar, probably of a halberd (80A1288)*

The fragments of this haft were found on the Main deck in M7, outside the Barber-surgeon's cabin. It is ornate and is presumed to be personal rather than a standard issue weapon. The decorated halberd in the Royal Armouries (VII.1516, Fig. 9.23a) with which it has been compared has a straight-edged, slightly down-curved blade with a down-curved beak-like fluke, the upper shoulder of which is cut out with two semicircular cusps with a blunt tooth-like projection left between them. Below the fluke, running down the back of the socket, is a cusped edge that forms sharp teeth. The top spike is unusually long and has a stout, shouldered base. Just below the head there is the brass collar that is almost identical to that on *Mary Rose* example.

Halberds were certainly carried by Henry's soldiers. They are shown in the paintings within the collection of HM The Queen of the embarkation of English troops at Dover in 1520 (RCIN 405793), and of the *Field of Cloth of Gold* (RCIN 405794) and on the eighteenth century Cowdray Engraving (Society of Antiquaries of London; see Figs 1.2 and 9.1). Some of those illustrated are similar to a fine etched halberd, probably of Italian manufacture, that survives in the Royal Armouries (VII.962; Fig. 9.4, above; Norman and Wilson 1982, 67–8). There is documentary evidence, too, to suggest that decorated halberds were produced for use by Henry's guard. The Italian merchant Leonardo Frescobaldi was paid for supplying weapons in 1513 and these included a number of halberds for the Royal guard. In 1520 Sir William Skevington, Master of the Ordnance, was paid for supplying '*gilt halberds and javelins for the Guard*' (Wilson 1985, 17). A blued and gilt head of a halberd, decorated with the punched pointillé decoration that is found on so many of the Henrician staff weapons in the Royal Armouries, was excavated from the Thames foreshore at Queenhithe in 1977 and is now in the Royal Armouries (VII.1717; Fig. 9.23b). It is of exactly the same form as another, apparently plain halberd (VII.970) from the old Tower of London collections that close examination showed had originally been similarly decorated, to one in the collection of the Rotunda Museum, Woolwich (XII.28; Wilson 1985, 18–19) and to one sold at auction in 2008 (Wallis and Wallis, Lewes, 1 April, lot 869). These heads are not of the same form as Royal Armouries VII.1516 but there are some similarities. They all have flat-sided sockets suggesting that their hafts would have been of squared section like that from the *Mary Rose*, and they all have cusping forming teeth on the shoulders of, and beneath, the fluke. There is little, if any evidence that halberds were imported into England from Germany at this time, although the inventory of all royal munitions stored in the kingdom of 1540 (PRO E 101/60/3) lists 80 '*fflemyshe*' halberds in the Tower. However, there is equally no proof that Royal Armouries VII.1516 is of German manufacture. It would fit easily with examples that can be said with some confidence to be of Italian origin.

However, before asserting positively that the haft fragments from the *Mary Rose* are those of an imported Italian halberd, a number of possibilities need to be explored. This haft shows clear signs of having been made to take metal reinforces on all four of its faces: two langets attached to the head and two separate reinforcing strips. Halberds with this arrangement on a square haft do exist and a good many of these have

*Figure 9.23 Halberd in Royal Armouries: a) VII.1516 with a collar like that on 80A1288; b) VII.1717 from the Thames at Queenhithe, decorated to match other wearpon's in Henry VIII's guard (©Royal Armouries)*

collars but there are many more that exhibit neither of these features. Of course, this may be due simply to the fact that the majority have had their hafts replaced, at which stage the two reinforces separate from the head could easily have been discarded. On the other hand, there is a type of staff weapon that does usually have metal reinforces to all four sides of its haft, even despite re-hafting; namely the pollaxe. The possibility must be considered, therefore, that this is the haft of a pollaxe not a halberd. Pollaxes are known to have been carried by Henrician troops and specifically by units of Henry VIII's bodyguard. A portrait of William Palmer, dated to about 1540, that shows him with a basket-hilted sword similar to that found on the *Mary Rose* (82A3589) has him dressed as a member of the Royal bodyguard known as the Gentlemen Pensioners and holding a pollaxe (Blair 1981, 173–5, fig. 77). Together with the presence of gun shields on the ship, a weapon traditionally associated with Henry VIII's guard, this haft, whether it be for a halberd or a pollaxe, might raise interesting speculation as to the types of troops on board when the *Mary Rose* sank. However, no firm identification is yet possible.

The haft from the *Mary Rose* may appear rather slender for that of a halberd, but it is by no means unusual and would fit the group with Henrician

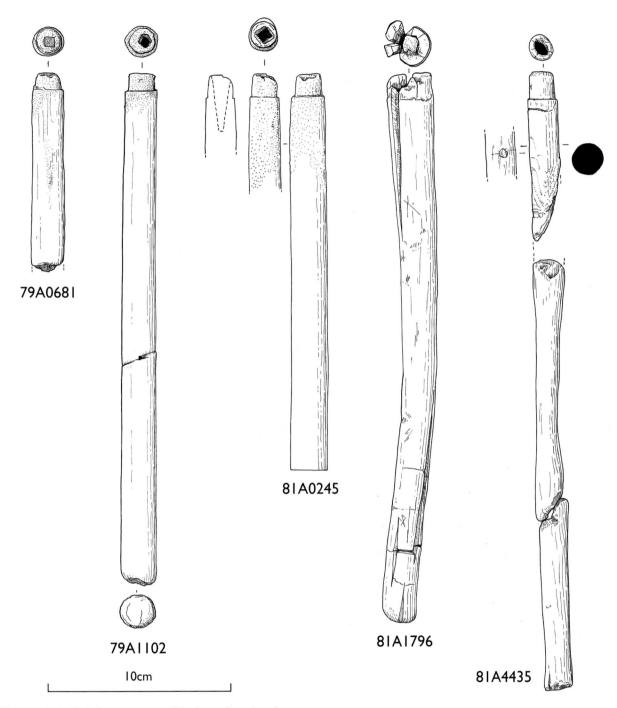

*Figure 9.24 Haft fragments possibly from throwing darts*

associations discussed above. But slender hafts also seem to be common on pollaxes. Three examined in the Royal Armouries (VII.1352, 1661, 1827) had hafts (replacements but fitting the sockets well) measuring in section at the socket about 32mm (1¼in) square. They were, therefore, of about the same dimensions as the haft from the *Mary Rose*. Most pollaxes that take square-sectioned hafts seem to have had reinforces to all four sides, some of which run the full length of the haft while some do not. The lower portion of the *Mary Rose* haft indicates that two of the reinforces stop while the others continue. This may be a factor in favour of it being a halberd. The rivet through the socket to strengthen the join that is clearly present on 80A1288 is also a feature found on a considerable number of pollaxes, as well as on some halberds and also on the form of mace found in early Tudor Britain that is known as a holy-water sprinkler. An example of the latter in the Royal Armouries (VII.1642) has a square-sectioned haft reinforced on each side by rebated metal strips, except that all of these, as on many pollaxes, protect the entire length of the haft. The haft of this example and others, however, are stouter than the haft from the *Mary Rose* as they have to take the considerable weight of the head. The possibility that this haft is that of a holy-water sprinkler may, therefore, be discounted.

There is one final type of weapon that this slender haft might, in theory, have belonged to, namely a short-hafted hammer or mace. However, no examples of the correct conformation with wooden haft extending to a short socket seem to be known. Most that have wooden handles have them attached at the end of long metal sockets that form a good part of the haft. In conclusion, therefore, although it is impossible to be certain, the weight of evidence suggests that this is the haft of a halberd, probably one imported from Italy, and perhaps a decorated one intended for one of the King's guard.

### Hafts possibly for throwing darts

Five haft fragments may belong to unidentified staff weapons (79A0681, 79A1102, 81A0245, 81A1796, 81A4435) which may be for throwing darts. They are illustrated in Figure 9.24. The hafts are all cylindrical (35–40mm diameter) with no obvious taper, and are shouldered with a space for a collar which may have been of iron. The 'tip', which is only 17–20mm long, is 27–30mm in diameter and appears to be rounded off. All retain a 10mm square hole in the end to retain some form of iron fixing. Where clean, this has a depth of 60mm (eg, 81A0245). This is similar to the hole in an incendiary device from the ship, 81A2866 (see Chapter 6), but this had a lip under which the incendiary bag was attached rather than a collar. There is only one possible nail hole on 81A4435. This is 90mm down the haft and is 5mm in diameter. There are no traces of tar as might be found on an incendiary dart. Two are ash (81A4435, 79A0681), one hazel (81A1796) and the others not identified. Two were found in the Starboard scourpit (SS9; 79A0681, 79A1102), two in U9 (81A1796, 81A4435) and one in U4 (81A0245).

## Edged Weapons

### Swords and daggers
by Guy Wilson with Alexzandra Hildred

#### Scope and historical importance of the assemblage

Swords and daggers were the most important weapons of personal defence in Tudor times, both for civilians and those serving in the military force (Fig. 9.25). Prohibitions on the carrying of swords and daggers seem to have had little effect (Wilkinson 2002, 52). Illustrative evidence might suggest that either a sword, dagger or knife were normal parts of male attire, more often worn than not. However, it is possible that this evidence overstates the case. Nevertheless the evidence of inventories, wills and court records and the developing popularity of manuals of sword fighting and of fencing schools throughout Europe emphasises the importance of bladed weapons for personal defence. In London, fencing and fencing schools were traditionally prohibited unless properly licensed, but Tudor law could not stop fencing masters from plying their trade. By the reign of Henry VIII these masters seem to have been organised into something resembling a craft guild (Anglo 2000, 8). In 1540 Henry had to establish a commission to look into their widespread malpractices. William Harrison, in *Description of England*, first published in 1577 as an introduction to Holinshed's *Chronicles*, says that in England it is very rare to see any man '*above eighteen or twenty years old*' without at least a dagger and that both noblemen and their servants usually wore a sword or rapier as well. As a final comment he writes that '*no man travelleth by the way without his sword or some such weapon*' (*ibid.*, 36, 324; Scott 1976, 244; Wilkinson 2002, 34). Corroboration for the widespread ownership of swords and daggers by all classes in society comes from surviving wills. For instance, in 1568 Paul Bartlette, a labourer from East Ham, left his sword to a friend and '*my dagger with my velvet girdle*' to one of his brothers (Emmison 1983, 70).

In London, the most important manufacturing centre in England, the production of swords and daggers was split between cutlers, who made the hilts and assembled the weapons, bladesmiths, who made the blades (and controlled the grinders who finished and sharpened them) and sheathers who made the scabbards or sheaths. From the beginning of the fifteenth century at the latest the cutlers had achieved a formal dominance over the whole trade and process (Cowgill *et al.* 1987, 32–3). In theory, all London-made blades had to be marked with the mark of the master of the workshop from which they came. Swords and daggers also came into London from other towns and cities in Britain but there is little evidence to suggest that large numbers were imported from abroad for sale to the general public, as opposed to the mass purchases for equipping royal forces that are well recorded (*ibid.*, 34).

The 1547 inventory of the possessions of King Henry VIII (Starkey 1998) contains significant detailed information about swords and daggers made for the King and his personal troops. Given that there is a grey line between what we might define as a short sword and a long dagger or knife, and the lack of certainty about some Tudor terminology, the inventory lists over 100 daggers or knives for fighting or hunting and almost 600 swords. Of these latter over 300 are described simply as

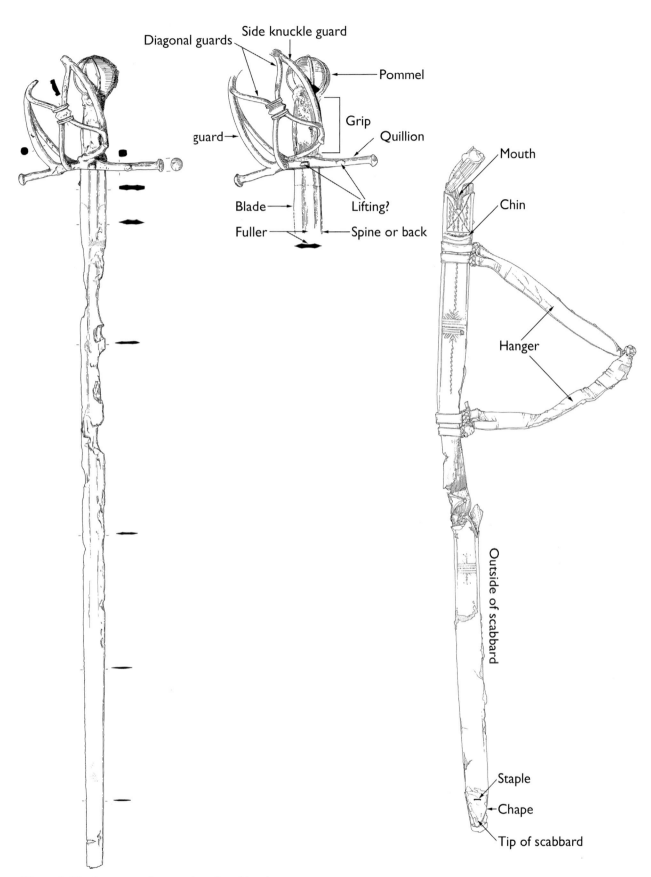

*Figure 9.25 Anatomy of a sword and scabbard*

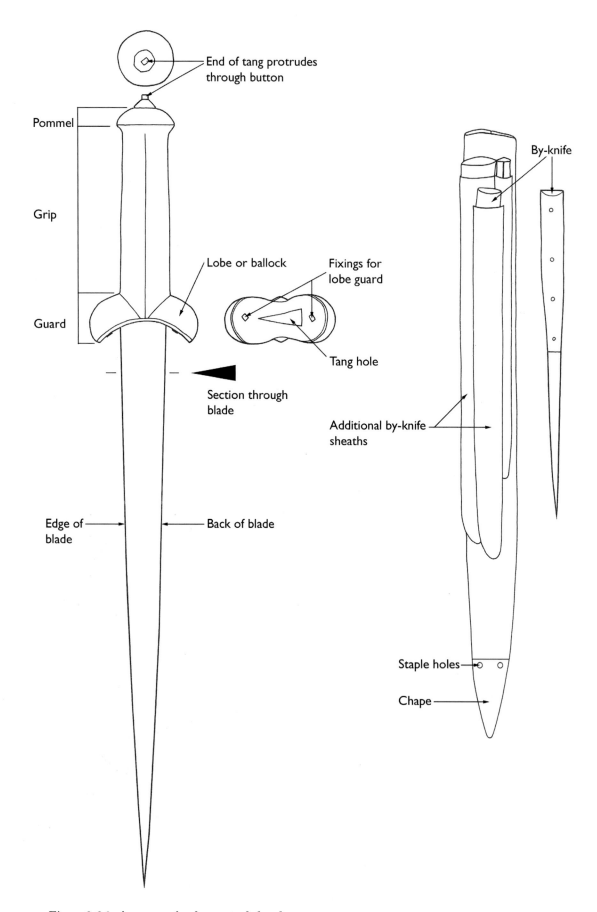

*Figure 9.26 Anatomy of a dagger and sheath*

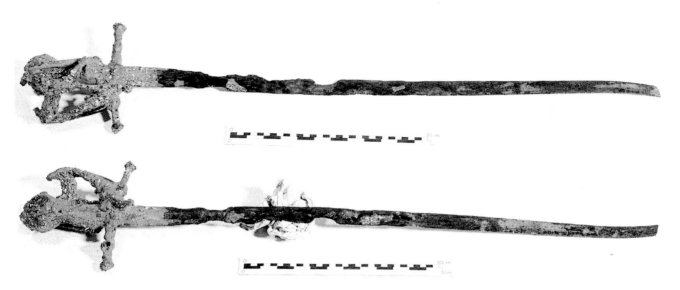

*Figure 9.27 Basket hilt sword 82A3589 and details of hilt*

'*swords*' or '*arming swords*' and no further details are given. These were, presumably, the munition weapons of Henry's guard or others whom he equipped directly. A further 120 are described as being blunted for tournament use (Starkey 1998, 5133). The rest are described in greater detail and, for the most part,

appear to be ornate and high quality weapons produced for the King himself. They include two-handed, stabbing swords (tucks and rapiers) and ceremonial bearing swords, swords from many areas of Europe and the Middle East, those with double-edged blades, swords (back swords) with single-edged blades and specialised hunting swords of various sorts. The daggers vary from one '*poinado*' (presumably a stabbing dagger) to short knives, and hunting knives, but most are listed simply as '*daggers*', '*skeins*' or '*woodknives*'.

There is no doubt that the sixteenth century saw an explosion in the variety of swords and daggers being made and used, and not only by monarchs and the nobility. In 1559 Thomas Collte, a gentleman of Waltham Cross, left his arming sword to one nephew, his two-handed sword to another, another sword (with accompanying buckler, dagger and cloak) to a friend and his rapier and dagger to his brother (Emmison 1978, 184–5). Clearly the swords and daggers used throughout society were becoming more specialised than ever before. This variety, supporting a bewildering number of different hilt forms, had to be described and throughout Europe there was a sudden explosion in terminology that seems to have confused people at the time and has certainly continued to confuse scholars since (Anglo 2000, 95–102).

Even today we cannot always distinguish between swords and daggers carried and used by civilians or soldiers. For example, over this period the rapier gradually developed into the most fashionable of civilian swords. It is shown in many portraits of members of the nobility dressed in their finest and its use was taught in fencing schools throughout Europe. However, it was not only a civilian sword. Sir John Smythe, writing at the end of the sixteenth century, makes it clear that both pikemen and cavalrymen were carrying long rapiers into battle, a practice he disagreed with for the very practical reason that, in the press of battle on foot, such long swords were difficult to draw and broke when used against armour. On horseback they could not be drawn without letting go of the bridle.

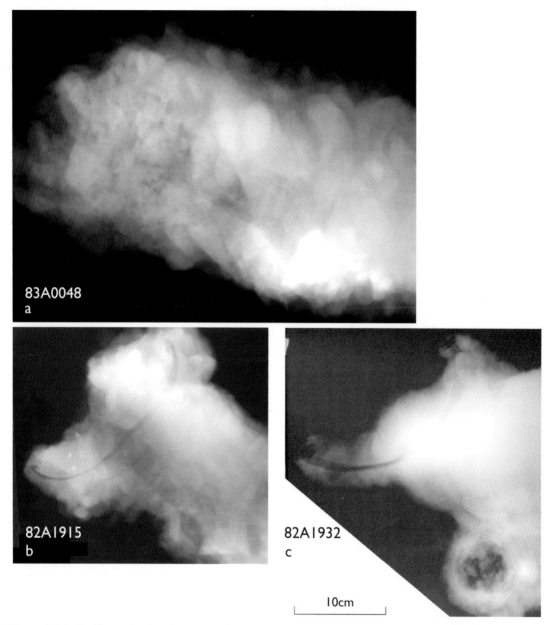

*Figure 9.28 Radiograph of 83A0048, 82A1915 and 82A1932*

Long swords had been causing other problems in civilian society since at the latest the middle of the century. In 1557 it was made unlawful to sell, wear or use swords or rapiers with blades longer than '*a yard and a half-quarter*' (45in or 1143mm).

Neither the 1514 inventories of Henry's warships nor the *Anthony Roll*, lists any swords or daggers among the official stores and equipment. It seems that the supply of these was always regarded as the responsibility of individual members of the crew. This is broadly in line with the 1540 Ordinances for equipping garrisons which make it clear, *inter alia*, that soldiers should supply their own swords and daggers at their own cost (*LP* 14/2, 785). However, this seems curiously at odds with, for instance, the similarity of the vast number of dagger finds from the *Mary Rose*. Most are of a very distinctive form of ballock dagger (Fig. 9.26). This type is found elsewhere in England at this time and it may just have been so fashionable that most of the crew felt they must have one. On the other hand, it does raise interesting questions about how such items were supplied and whether there was ever any centralised or bulk ordering.

Very few of the remains identified as swords give any indication as to hilt form. Those which do include the basket-hilted sword 82A3589, excavated from beneath the Sterncastle in 1982 during work to dig the tunnels to secure the lifting wires needed to raise the hull (Fig. 9.27). It seems to have survived more intact than almost any other small iron object because it was crushed and buried within anaerobic sediments beneath the ship. This is a unique example of a stage in the development of a distinctive form of British basket-hilted sword, an account of which had been published by Blair in 1981, the year before this sword was found. From it there is a clear line of development through more complex basket-hilts of the Elizabethan period to the type of basket-hilted sword commonly found in the early seventeenth century. It is not yet clear where this constantly developing but distinctive type of sword was made, whether in one or more towns or workshops, though there are such similarities between many of them that production in one place seems a distinct possibility, at least for the earlier part of their development.

Despite the usual association of basket-hilts with Scotland and documentary evidence from the sixteenth and early seventeenth centuries that this type of sword may have been known at the time as '*Irish*' (which could have meant Scottish as much as Irish), the majority of excavated examples come from England and the few portraits and illustrations in which they appear are of Englishmen or have an English subject; the balance of probability seems to favour them being made in England. The sword from the *Mary Rose* is critical to an understanding of the chronology of the evolution of this type as it is the only surviving example that can be proved to have been made by a certain date. It is also important because it suggests that what we are looking at may be as much, or more, a military than a civilian type. Intriguingly, a number of Tudor depictions show swords of this general type being worn by members of various Royal bodyguards. It appears that this was not the only sword of this distinctive type aboard the *Mary Rose*. A similar pommel has been discovered in concretion (82A0027) further suggesting that some of the Royal bodyguard may have been on board.

The form of the hilts of several others are known only from radiography of elements of concretion (83A0048, 82A0027, 82A1932, 82A0191). These provide exceptionally important evidence of types of hilt that were in use in England in the 1540s and help to date and confirm the English connections of extant examples from this country and of others with no known provenance. One of these (83A0048, found in M7) has a guard formed of recurved quillons which, at the front of the hilt, form a knuckle guard (Fig. 9.28a). A good number of this type come from English contexts and it has been suggested that it is from this type that developed the British basket-hilted swords like that found on the *Mary Rose* (Wilson 1986, 1–2; 1990, 23–4). The group includes a sword found on the probable site of the 1460 battle of Wakefield (Wilson 1990, 24) and parts of another in Strangers Hall, Norwich (no. 76–64; Wilson 1986, 3–4) that is said to have come from Mousehold Heath, outside the City, and may have been left there during Ket's rebellion in 1549. The presence of this sword hilt on the *Mary Rose* makes it all the more likely that the Norwich example indeed dates from the 1540s.

Two other sword hilts found by radiography are characterised by their arched quillons. One, 82A1915 (H7; Fig. 9.28b), has a triangular-shaped escutcheon that would have fitted a scabbard with a cut-out mouth, like 80A1709 (M3), and a disc or spherical pommel apparently with a large and pronounced tang button. The other, 82A1932 (found in O7 with a scabbard) is generally similar but has a semicircular escutcheon and a spherical pommel (Fig. 9.28c). These two belong to a group of swords with English associations that include a number from churches, most notably one in Westminster Abbey, which, since its discovery in 1869, has been believed by many to have been Henry V's own sword (Marks and Williamson 2003, 195).

The discovery of these two swords on the *Mary Rose* also shows is that this type continued to be used for warfare as well as chivalric ceremonies. What we cannot know is whether these were newly-made or treasured heirlooms. However, although we should beware of assuming that everything on the ship was new and fashionable, the presence of these swords should prompt some redating of the remainder of this general type, including the late fifteenth century date traditionally given to a group of funerary swords of similar form found in English churches (Cripps-Day 1922, 151–273).

Another similar example was recognised as a stain on the seabed, associated with an extant grip (81A0719, Fig. 9.29). Although vestigial, it is clearly a sword with a front quillon that curves back to form a knuckle-guard and a straight rear quillon. The presence on the *Mary*

*Figure 9.29 Stain of a third basket hilt sword on seabed*

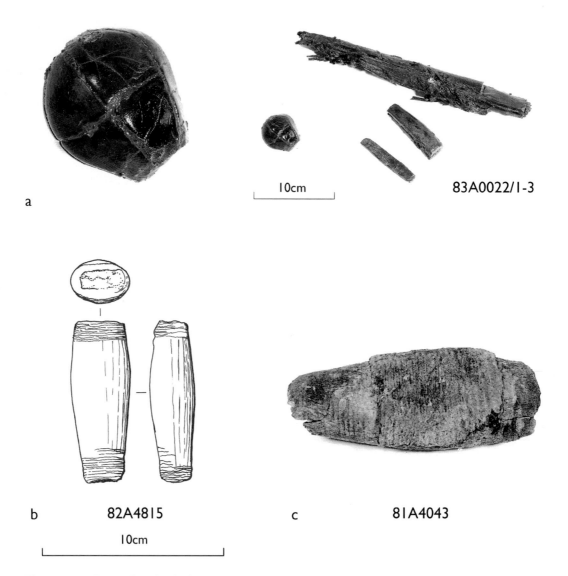

*Figure 9.30 Pommel and grip fragments: a) 83A0022 (from dagger); b) 82A4815; c) 81A4043*

*Rose* of two of the type of swords from which the form of basket-hilt (also found on the ship) is believed to have developed may suggest that the development was far more rapid than has previously been considered.

It is not just the evidence of sword hilts that is significant about the assemblage found on the *Mary Rose*. There is also a very important group of detached sword pommels, or rather the copper alloy sheathing for them, that represent a previously unrecorded type of spherical pommel with a recessed central band parallel to the blade (see Figs 9.45–7). Also of interest is the radiating copper alloy banding to go round a sphere of *c*. 50mm (2in) diameter that has previously been thought to be the remains of another pommel (81A0133; Fig. 9.45). There seems some reason to question this and it may, perhaps, be the remains of pomander, the only recovered example of which (81A4395) was found in the same concretion as a sword scabbard (though it is not known if they were physically attached to one another).

The assemblage of 65 ballock daggers is by far the largest and most important so far discovered. The wooden grips are uniformly priapic. Most are made of boxwood ('*dudgeon*'), the preferred wood for the manufacture of dagger and knife grips throughout the medieval period. The hilts of the ballock daggers are all clearly of one general and very distinctive pattern but they show considerable variety of detail in the faceting and reworking of the both pommel and grip. Although the blades do not survive, the evidence of their form is unequivocal and is confirmed by the radiography of a single blade within a concretion (see Fig. 9.68). The cut-out at the blade end of the hilt to take the forte of the blade and the tang shows that they were, with one exception, single-edged. This conflicts with Peterson (1986, 37) who says that on:

*'almost all late* [ie, 'sixteenth century'] *ballock knives the blades were slender, double-edged, and so thickly diamond shaped in cross-section that they can almost be considered four sided'.*

Ballock daggers were not the only daggers to be used on the *Mary Rose*. Still surviving in concretion is the grip, the outline of the guards and a portion of the blade of a rondel dagger (82A1914, H7, see below). This had been one of the most common types of dagger for some 200 years but was nearing the end of its popularity when the ship sank. Its presence on the *Mary Rose* is another reminder that not everything in use would necessarily have been new or of the latest type and fashion.

One of the most fascinating and unusual finds is that of a burrwood hilt and matching wooden pommel and sheath (83A0022) (Fig. 9.30a). It has been identified as a '*Landsknecht*' dagger, a nineteenth century term used rather loosely to describe a variety of usually later sixteenth century quillon and rondel daggers. But even given this variety, there is some reason to question this identification and, without the evidence of the guard, it is not possible to be certain.

The wooden grips are also of considerable interest (Fig. 9.30b and c). No grips of any other material were found. This fact can be set alongside 1547 inventory that shows Henry's luxury arms to have had grips made of many materials, including precious and other metals (some enamelled), stone, crystal, mother-of-pearl and bone. On the *Mary Rose* is a unique assemblage of dismounted grips from swords and daggers without the binding that would have covered many of them (of steel or precious metal wire or thread). Some can be recognised as coming almost certainly from specific types of dagger but there remain a large number of plain and ordinary grips that cannot be categorised. What is certain from other evidence is that length alone cannot be used to deduce whether a grip was made for a sword or a dagger. What is also clear, and only becomes apparent when a large number of dismounted grips are examined, is that the normal way to make a grip in Tudor times was to take a single piece of wood, shape it, then drill, file and/or chisel a slot through its long axis to take the tang of the blade (Table 9.3, below).

At least 43 by-knife hafts were also recovered, most made of boxwood and nearly all of scale-tang form (Fig. 9.31; Tables 9.6, 9.7). No blades survive. Most were directly associated with ballock daggers, carried in the dagger's sheaths which may seem surprising. However this confirms other evidence to suggest that, in England at least, it was not an uncommon practice. A ballock dagger of *Mary Rose* type is clearly shown with its by-knives and a simple cord suspension from the belt in the effigy of William, Lord Parr (d. 1546) and his wife, Mary (d. 1555) in Horton church, Northamptonshire (Blair 1962, 109f; Hartsthorne 1891, 340) (Fig. 9.32). At least one ballock dagger of this type has been excavated from an English site with its sheath and two by-knives, from the moat of Hooke Court near Beaminster, Dorset (Dorset County Museum no. 1944-23-1). The sheath has an iron top locket with a decoratively fretted plate running down it that protects the mouths of the slits to take the by-knives. It also has an iron staple set across the back of the sheath just below the mouth, for attaching it to the belt. The by-

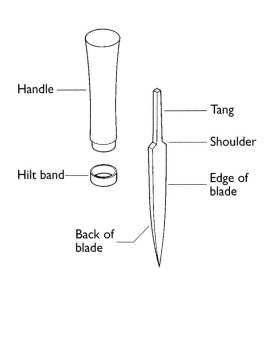
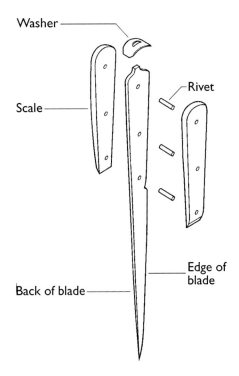

*Figure 9.31 Anatomy of a by-knife*

*Figure 9.32 Drawing taken from effigy of Lord Parr, St Mary Magdalene church, Horton, Northamptonshire*

knives have wooden grips and iron flat cap pommels surmounted by iron shaped finials. This may give a very good idea of the original complete form of the ballock daggers from the *Mary Rose* (Fig. 9.33a).

Some of the recovered objects have three small grips remaining in the dagger sheaths, one much smaller than the others. This may be evidence for the carrying of a bodkin (a stout needle with a grip) together with two by-knives. Both by-knives and bodkins are often listed together with daggers in the 1547 inventory (Starkey 1998) and the practice of carrying them in dagger sheaths was not restricted to the luxury arms of the very wealthy. For instance, in 1565 Edward Trappes, a yeoman from Chigwell, left his son '*my dagger with the black silk handle and iron hilt and pommel, being black varnished, having 2 knives and a bodkin*' (Emmison 1983, 16).

In total there are about 200 finds that were tentatively identified as being parts of, or accessories to, swords or daggers. These include the 65 easily recognisable hilts of ballock daggers, together with over 50 detached grips, and remains of scabbards, sheaths and hangers. Detached grips are difficult to classify as being either from a sword or a dagger, and so most are considered together. The same can be said of many of the nondescript and fragile remains of scabbards, which may include dagger sheaths. The leather and wooden fragments from as many as 60 scabbards were found, and eight additional loose straps. In many cases the stitching has gone and the objects have disintegrated. Some items only exist as vestigial remains within concretion and these can be difficult to identify with any certainty.

Accoutrements might include a belt, often with a buckle and ancillary leather straps to suspended the scabbard. Other evidence is provided by the copper alloy chapes placed around the end of the scabbard to protect the blade tip, and copper alloy devices to aid with suspending the sword from the girdle. These include one ornate leaded-brass tripartite hanger (81A0787; Fig 9.33b), a strap with a copper alloy hook end (81A3089; Fig. 9.33c) and a strap with a decorated copper alloy end (81A3117) which may be a tassel (Fig. 9.33d). All the buckles recovered from the *Mary Rose* have been studied (*AMR* Vol. 4, chap. 2) but only those found in direct association with scabbards or swords will be considered here. Many more leather and wooden items relating to edged weapons are probably lost among the many disintegrated fragments found in excavation. Only two scabbards retain attached straps for suspension, 80A1709 and 81A1567, the latter recovered with the pomander mentioned above. Although slightly different, both use straps which have been split at one end and further split, to form four equal straps. Decorated bifurcating straps have, therefore, also been considered here.

**Distribution**

The largest concentrations of swords, daggers and related finds were on the Upper deck in the stern, in areas with human remains and fairly complete skeletons (Figs 9.34–5) and on the Orlop deck, in particular associated with human remains in O7. Concentrations in the Hold include ballock daggers in H5 (the area directly supporting the galley), and in storage areas of H7 and H8. Swords and daggers were found on the Main (gun) deck in M1–7, directly associated with human remains in M2 and M6, in particular with those of a possible gun crew found close to the cannon in M6 (Fig. 9.35a).

Four chests contained sword or daggers (see *AMR* Vol. 4 for descriptions of most items and appendix 1 for full contents). The remains of a blade grip and pommel (80A1405) together with a second grip (80A1838) and scabbard remains (80A1551) were found in chest 80A1413 in O4 together with four combs, a comb-case, leather flask, beads, a pendant, aiglets, a pair of shoes, leather straps, a shell, a knife, a pin, a mould for casting inset dice shot (80A1847) and a concretion containing silver coins and a clasp. A scabbard and grip (81A2977) with degraded pommel fragments was found in a chest (81A2941) in O10. This also contained shoes, aiglets, a ribbon, beads, a paternoster, die and what is possibly a key. A second chest in O10 (82A0942) contained a dagger and sheath (82A0970), aiglets, a thigh boot, a shoe, an archer's wristguard (82A0943), inset dice shot (82A0799), a comb, knife and besom broom. The only ballock dagger definitely excavated from within a chest was 81A1304, which was found in a locked chest with a carved panel found leaning against the Carpenters' Cabin in M9 (81A1429). As well as a selection of woodworking tools and clothing, items included silver coinage, two boatswain's calls, silver jewellery, a book

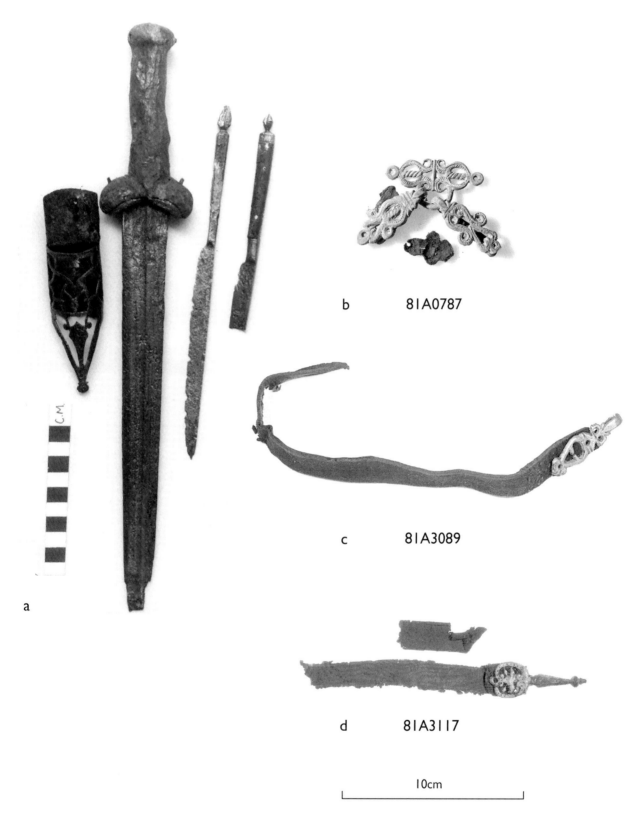

*Figure 9.33 a) Dagger from Hook Court, near Beaminster, Dorset (© Dorset County Museum; 1944-23-1; b–c) sword hangers 81A0787 and 81A3089; d) girdle end 81A3117*

and one of only two engraved copper alloy topped knives (which may be Flemish), a priming wire and a linstock, possibly suggesting that this was owned by a gunner. The only other element associated with a chest was a strap with a copper alloy hook identified as a sword hanger (81A3089). This was found resting on the

### Table 9.3 Grips and attached pommels from swords and daggers

| Sector/location on Fig. 9.35b | Object | Wood | Tang | Coated | Length (mm) | Width (mm) | Notes |
|---|---|---|---|---|---|---|---|
| H1 | 81A0120 | beech | corroded, ovoid | poss. linen with silver wire over | 90 | 30, 35, 20 | loose find; poor condition, broken; wire binding noted in '84 top & base; suggest this was all over |
| H7 | 80A1428 | – | yes | wire impression | 85 | 22 | complete; poss, wire covered, stained; associated 81H0171, 0174 |
| H7/4 | 82A1915 | – | conc | string & cloth? | 85 | 25, 30, 23 | complete; from radiograph, may be wire bound; found with pommel |
| H7/28 | 82A1914 | – | ovoid | no, in concretion | 90 | 20, 22, 22 | complete; rondel dagger in concretion; 2 vertical slit each side for insertion of metal strip |
| O2/30 | 83A0022 | burr | triangle | no | 84.5 | 43 | complete; dagger; with sheath & wooden pommel; hexagonal section, incised decoration on flats |
| O2/3 /18 | 81A2487 | beech | ? | looks coated | 86-89 | 30, 36, 25 | complete; surface impregnated with metal; poss. wire covered |
| O4/12 in chest 80A1413 | 80A1405 | beech | ? | none visible | 78-80 | 26, 30, 21 | complete; blade, grip & pommel; with scabbard 80A1851, leather strap/strips 80A1846, 1945 & grip 80A1838 |
| O4/9 in chest 80A1413 | 80A1838 | willow | oval | iron binding visible top & base | 80 | 28, 31, 20 | incomplete; prob. fully wrapped; with 80A1405 |
| O4 | 81A3739 | – | rect. | no | 76 | 25, 30, 20 | loose; prob. originally wire bound |
| O7/5 | 82A1932 | – | ? | ? | | | complete; in concretion; radiograph suggests pommel, grip, hilt, blade, scabbard; downward turning hilt |
| O7/6 | 82A2054 | maple or elm burr-wood | tapering | copper strips each side | 85.5 | 35 mid | complete dagger handle; rectangular section; one strip *in situ* |
| O8 | 82A1613 | beech | ? | yes, all over. | 95 | 27, 17 inc | incomplete; coating like fish scale |
| O10/29 | 82A0975 | – | ? | ? | 63 | 47* | complete; * in concretion, with pommel |
| O10/21 in chest 81A2941 | 81A2979 | – | ? | not visible now | 85 | 25, 30, 20 | bad condition, fragmented; with scabbard 81A2977; v. degraded pommel frags |
| O10/22 | 82A0758 | willow/poplar | ? | top, centre & base | 80 | 30, 35, 30 | complete; half pommel present with part of hilt; 2 two centrally placed grooves running around circumference & 1 at each end; heavily concreted |
| O11 | 81A1980 | beech | rect. | iron stained, inc. | 51 | 30, 22 | broken in half horizontally |
| O11/7 | 82A4412 | pomoideae (pear) – burr-wood | tapering | decorated, incised lines | 77.5 | 31 | rectangular; different from other handles; 4 longitudinal lines, 2 on edges 1 in centre 'back' & 'front' |
| M1/10 | 81A0133 | – | rect. | no evidence in 1984 | 76 | 24, 32 | incomplete; found with pommel, blade & scabbard frags |
| M1/14 | 79A1112 | beech | rect. | yes, now gone | 82.5 | 21, 30, 16 | binding noted in past |
| M2 | 83A0436 | – | ? | evidence of leather | 86 | 30 | at base of companionway in concretion with pommels 83A0432/0433 & scabbard 83A0435; possibly 2 swords on top of each other crushed by iron shot; suggest pommel 0433, handle 0436 & scabbard 0435 may represent parts of 1 sword |
| M6/15 | 81A4180 | – | corroded | silver wire, central strip? | 82.5 | 30 | wood laminating; iron staining both ends |
| M6 | 81A4181 | beech | ? | & | 78 | 25, 30, 22 | complete; red/brown textile cover; plant bast fibre |
| M7/2 | 83A0048 | – | ? | ? | 85 | 30 | from radiograph; pommel, hilt, blade, S-shaped guard; scabbard frags |
| U3 | 79A0483 | – | ? | too fragmentary | 80 | 30, 20 | only half grip present |
| U4 | 81A4086 | beech | ? | linear iron stain | 80 | 30, 24 | poss. iron wire |
| U8/9 /13 | 81A4098 | – | rect. | iron staining all over | 76 | 25, 30, 22 | complete; more iron on top; heavily concreted |
| U9/47 | 81A0688 | hazel/willow | rect. | remains all over | 79 | 30, 30, 30 | poss. silver plating around base |

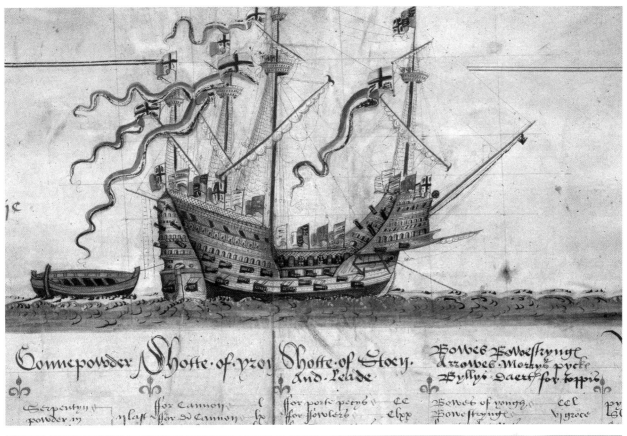

*Plate 1 Top) The* Mary Rose *as she appears in the* Anthony Roll *of 1546; bottom) painting by Geoff Hunt PPRSMA: The Mary Rose. Henry VIII's flagship 1545*

*Plate 2 Top left) Cannon 81A3003 on carriage underwater with wedge* in situ*; top right) culverin 80A0976 on its carriage on the Upper deck in the stern; bottom left: detail of Arms and Garter, bronze demi culverin 79A1232; bottom right) detail of the device of bronze demi cannon 81A3002  (Courtesy S. Foote)*

*Plate 3 Above: 19 July 1999, John Taylor Bellfoundry Loughborough: the 3 ton tilter furnace and ladle used during the pour of the reproduction culverin, a copy of 81A1423; right: burning out the wooden core used to support the staves during the manufacture of the barrel of the reproduction port piece (photo courtesy C Topp and Co); reproduction port piece firing trials at Ridsdale firing range: the gun before insertion of the breech chamber and forelock; using handspikes through the rings of the breech chamber to lift it out for reloading; positioning the breech chamber*

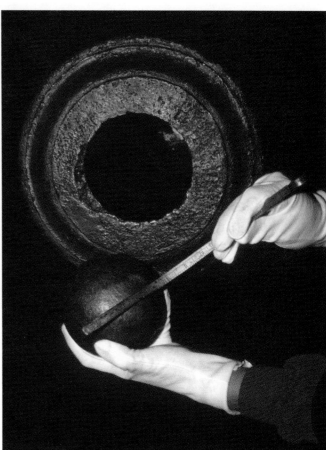

*Plate 4 Top left): copper alloy priming wire with intricate heart-shaped top; top right) gunner's rule, shot gauge and muzzle of demi culverin; centre left) scorched head of a linstock in the form of a fire breathing dragon (courtesy S. Foote); below) selection of objects used in firing the guns*

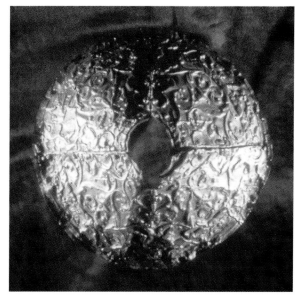

*Plate 5 Top left) Remains of wooden gun-shield copper alloy boss in place; top right) detail of decorated copper alloy gun-shield boss; bottom left) underwater shot of chest 81A1862 full of longbows; bottom right) decorated leather wristguard 80A0901*

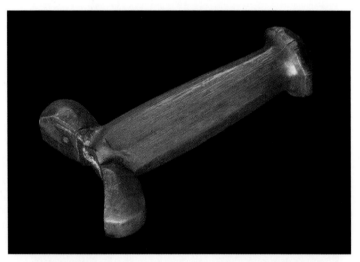

*Plate 6 Top) arrow speacer containing arrows, and associated leather straps; centre left) just a stain in the silt – all that remains of the iron blade and hilt from a sword; the pommel and wooden grip survived; centre right) boxwood ballock dagger grip (courtesy S. Foote); bottom) copper alloy fittings for sword belts and scabbards*

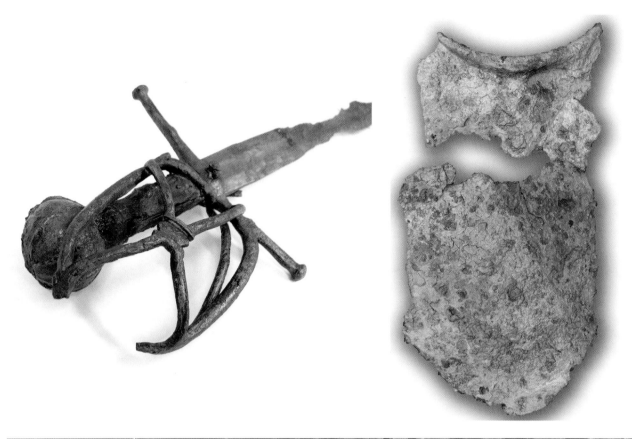

*Plate 7 Top left): The only complete sword recovered from the ship. This single example of a basket-hilted sword has had a huge impact on the understanding of the origin of these types of hilt; top right) remains of the only identifiable armour breast plate; bottom) view along the surviving portion of the Main deck stern in the museum showing knees and gunports*

*Plate 8 Above) view of the internal elevation of the starboard side of the* Mary Rose, *showing the Hold, Orlop deck, Main deck and Upper deck. The gunports are clearly visible on the Main deck and an archer assumes position on the open Weather deck. The front of the Sterncastle is visible to the right (courtesy S Foote); below: External view of one of the Main deck gunports. Iron nails reinforce the ends of the carvel planks where the gunports have been cut, and the rebate to take the gunport lid can be seen. Part of the iron strap hinge for the lid is visible to the left*

**Table 9.3 (continued)**

| Sector/ location on Fig 9.35b | Object | Wood | Tang | Coated | Length (mm) | Width (mm) | Notes |
|---|---|---|---|---|---|---|---|
| U9/3 | 81A0718 | – | ? | leather cover, fragmented | 82 | 27 | iron pins noted in '84; silver noted, but no leather; pommel associated hilt form from sediment stain in seabed (81A0719) |
| U9/11 | 81A2041 | – | ? | plain, coated unknown layer | 76 | 28, 30, 22 | complete; no marks of binding |
| U9/16 | 81A1222 | – | corroded | ? | 89 | 25, 30, 20 | badly degraded surface |
| U9/17 | 81A1763 | beech | ? | none visible now | 89 | 30, 30, 25 | |
| U9 | 81A4310 | beech | ? | textile covering | 90 | 26 | brown wool, plain woven tabby weave |
| U9 | 81A4326 | – | corroded | vestigial binding | 80 | 31 | complete; more binding over top portion |
| U10 | 78A0299 | – | ? | eroded | 72 | 30, 30, 20 | complete, cracked |
| SS1 | 78A0199 | – | ? | ? | 80 | 28, 28, 20 | |
| SS1 | 78A0101 | beech | rect. | binding frags | 81 | 30, 35, 20 | iron stained |
| C2/8 | 80A1041 | willow | ? | poss. central | 76 | 31 | complete; well cleaned in '84 |
| O2 | 83A0022 | box | ovoid | 2 recesses | 85 | 30, 32, 20 | dagger handle |
| SS11 | 82A3589 | poplar or alder | blade *in situ* | textile at top & base, leather | 80 | 30, 35, 30 | complete sword; poss. calf leather also |
| ? | 78A0272 | – | ? | ? | 81 | 30, 30, 20 | pommel like 823589, suggesting at least 2 basket-hilted swords on board |
| ? | 81A4043 | – | ? | ?wire binding | 80 | 25, 35 | incomplete at base |
| ?/27 | 82A0027 | – | ? | ? | | | radiograph only; pommel & grip & part of guard in concretion |
| ? | 82A1866 | beech | ? | yes & wire binding | 75? | 30, 20 | trace of coating covered by binding |
| ? | 82A4815 | beech | corroded | binding evident | 85 | 32 | binding at top and base |
| ? | 82A4816 | willow or poplar | ? | ? | 85 | 31 | |
| ? | 89A0082 | | | no data | | | |

Where three dimensions exist for width, this is width at hilt, centre & at top. A single measurement is usually around the centre. Two measurements represent hilt and centre

collapsed and broken portions of a chest in O9 (81A3088). There were no other objects recovered with this chest.

Although the association of many of these objects with human remains suggests that most were being worn, there is little conclusive proof as to whether it was normal to carry both sword and dagger or whether they were considered alternate weapons – examples found together are rarities; for example in H8 there were four ballock daggers but no swords. There is, however, other evidence that suggests that at least the nobility of Henrician England could be armed with both sword (sometimes a cross-hilt) and dagger (sometimes with a ballock hilt). A cross-hilted sword and ballock dagger are clearly shown on the effigy of Lord Parr of Horton, mentioned above, and there are several instances of close associations, for example ballock dagger (with sheath and by-knives) 81A0239 was found close to the bifurcated sword belt 81A0240 in U4, and dagger sheath 83A0151 was found together with fragments of a larger scabbard in O8. Locations of all swords, daggers, scabbards and sheaths are shown on Figure 9.34. Figure 9.35b shows the distribution of objects discussed within the text. Summary information on the grips and attached pommels of both swords and daggers is provided in Table 9.3.

**Basket-hilted sword**
*82A3589, SS11* (Fig. 9.36)
As stated above, the sword was found outboard beneath the Sterncastle. It has been crushed, is mis-shapen and broken in places. The description is therefore based on a physical examination and a comparison with other similar swords (Fig. 9.35b, nos 1, 12, 13 and see Fig. 9.27). The overall length is 1005mm the blade being 865mm and the grip 83mm (3¼in).

The pommel is spherical and divided into eight facets by longitudinal raised mouldings. Radiography has shown that it was made in two halves, joined horizontally, probably by brazing. Scientific analysis supposedly showed that, while the ribs were iron, the body of the pommel were made of a copper alloy with a core of a substance less dense than iron (Wilson 1986, 7). Re-examination suggests that the core may have been a hollow iron sphere.

The guard is made of round-section iron bars. The quillons are straight with knob terminals. The knuckle bow is joined by a saltire of bars to a corresponding bow

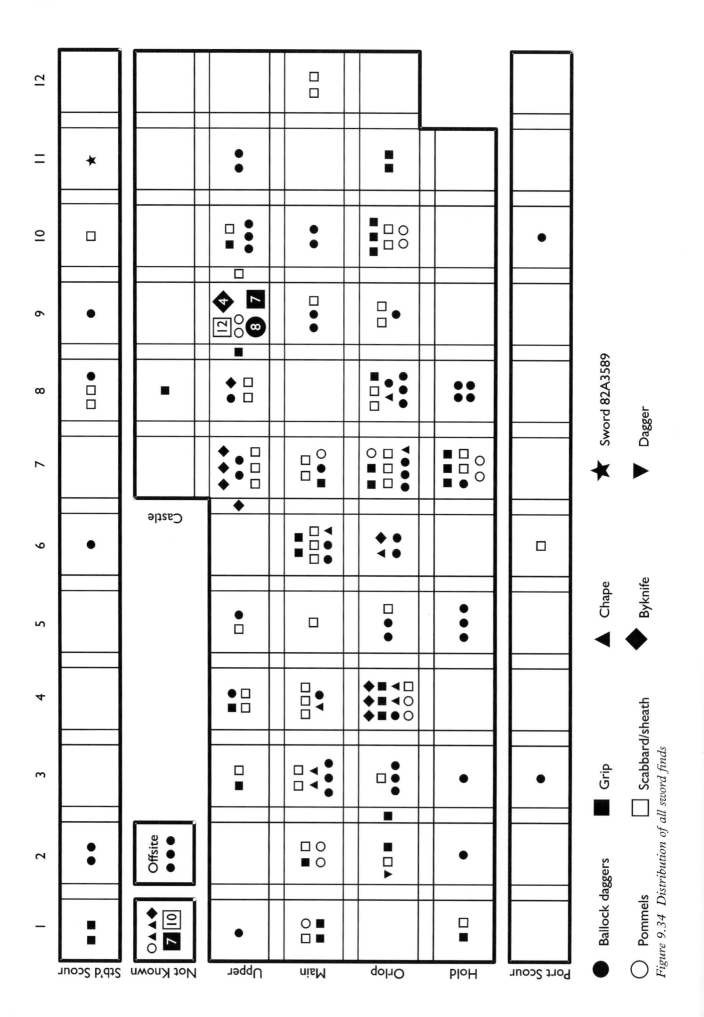

*Figure 9.34 Distribution of all sword finds*

on the outside of the guard that originally would have been in line with the centre of the blade. In the centre of the saltire, the bars expand into a small plate which is chiselled in relief with a flattened baluster moulding. The ends of both bows appear to have stopped at some distance from the pommel and there is no evidence that they ever touched, were attached to, or were set into the pommel. A diagonal bar, now missing, would have joined the tip of the side bow to the centre of the rear quillon, where the quillon is clearly marked by a broken stub though there is no visible evidence of the weld joint at the tip of the side bow. There is some evidence (in the expansion of the quillon and the existence of a stub at the blade end of the side bow) to suggest that, springing from the root of the forward quillon, there would have been a serpentine fore guard which curved under the quillon towards the inside before curving back under itself to join the side bow at its root. The inner guard consists of one bar, now fragmentary, which springs from the end of the knuckle bow, curves back on itself and which must originally have joined the fore guard either where it joined the front quillon, or before this (Figs 9.27 and 9.37). The wooden grip (beech or alder) is heavily encrusted but shows traces of fibres at each end. These could either be the remains of fibre bindings or evidence of a fabric covering and there are also traces of calf leather.

The blade is straight and double-edged with a wide central fuller starting at the hilt and turning into a flat just under halfway along its length (the precise position cannot now be determined). It is broken off just before the point. The blade has a flat-edged ricasso (an unsharpened portion of the blade closest to the hilt) c. 32mm (1¼in) long that tapers slightly to the hilt. Metallographic analysis by Alan Williams showed that there was an outer layer (2mm thick at the edge) of a different material to the central core. The microstructure of the outer layer showed dark-etching martensite without visible ferrite, but with numerous small slag inclusions. These were concentrated in two converging lines, interpreted as weld boundaries. The core showed mostly ferrite, the proportion increasing towards the centre of the blade. A micro-hardness survey of one side of the cross-section at the tip end of the blade revealed that the cutting edge had been hardened (sample 87S0657; details in archive). The results suggest that the blade was made by forge welding together a piece of iron and a much smaller piece of the more expensive medium-carbon steel. This composite bar was then forged out into the shape of a sword blade, leaving a thin steel surface layer around the edges. The blade was then hardened by a process of quenching and tempering in which the steel (perhaps 0.5% carbon) formed martensite of hardness over 500 VPH while the iron (0% carbon) was left as unhardened ferrite. This steel edge was, however, only 2mm thick and would not have survived many sharpenings. (Fig. 9.38). No records could be found to substantiate early suggestions that the pommel was made of copper alloy and experience suggests that it is more likely to have been made of iron parts brazed together. Analyses of similar pommels have proved them all to be hollow.

The decoration on the plate in the centre of the saltire is similar to, though larger than, that on a more developed basket-hilt in the Royal Armouries (IX.2574) dating to c. 1560–70 (Blair 1981, 212–4, figs 113–5). Recent x-ray analysis of similar, but later, swords in the same collection suggests that the saltire and central plate were generally cut from a single piece of iron.

The origin of this type may, perhaps, be found in a group of late fifteenth century swords with knuckle bows and a lug on the outside of the hilt. It is possible that the lug may have been extended into another bow and that then the space between the bows filled with additional bars for greater protection (Wilson 1986, 1–4; 1990, 23–4). While this remains speculation, it is certain that, by about 1520, swords with two bows and additional guards between were being produced. There is a complete example of this type in the Royal Armouries (IX.3615) and another, broken example in the collection of the Duke of Buccleuch at Boughton House, Northamptonshire (Wilson 1990, 24–6). Both have baskets on the outside of the hilt formed by joining the two bows with two transverse bars that, nearer the quillons, join to the front quillon by a central longitudinal bar. Both already have a serpentine fore guard of the general type that we can be certain (see below) formed part of the guard of the sword from the *Mary Rose*. All that would be needed to change them into a clear prototype of the *Mary Rose* form of guard would be the substitution of a saltire of bars for the transverse and longitudinal ones between the two bows. From these swords, then, there appears a clear line of development to the *Mary Rose* sword (though actual examples of the intermediate forms are lacking) and beyond it (where numbers of actual examples survive) to the type of basket-hilted sword commonly found in the early seventeenth century which was, for example, sent in numbers to the first successful English colony in North America established on Jamestown Island in 1607.

A group of four swords and a detached guard, none of which can be dated accurately from their form and provenance alone, seem to represent the next stage of the evolution of this type of basket-hilt (Wilson 1986, 4–8). These are a sword found near Bradford (Bolling Hall Museum no. 175/25); a sword found in the River Can at Chelmsford (Chelmsford & Essex Museum no. N–22263) (Fig. 9.39a); an almost identical example to it, excavated but of unknown provenance in a French private collection; and a sword excavated from the Thames foreshore under Southwark Bridge, together with two ballock daggers of the type found on the *Mary Rose* (Royal Armouries (IX.4427) (Fig. 9.39b); and a detached hilt from the moat of Pembridge Castle, also in the Royal Armouries (IX.1404).

Instead of an inner guard consisting of one bar as on the *Mary Rose* hilt these all have inner guards in the

Key

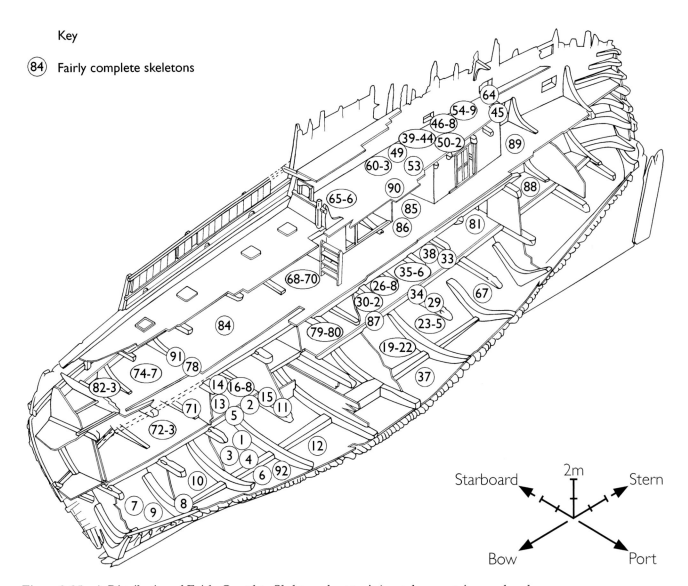

*Figure 9.35 a) Distribution of Fairly Complete Skeletons; b, opposite) catalogue entries numbered*

form of an inverted 'T' made up of a diagonal bar joining the end of the knuckle bow to the centre of the rear quillon exactly opposite the diagonal bar on the outside, from the centre of which springs a curved bar which joins the fore guard where it begins to bend back on itself. These guards join the fore guard before they join the front quillon and the inner guard of the *Mary Rose* sword was probably similarly attached. Unlike the sword from the *Mary Rose* those from Chelmsford and Pembridge and the 'French' example all have recurved rather than straight quillons. The hilts of these three are also different from the rest in being made of ribbon-like bars decorated with longitudinal ribbing. The Chelmsford and French examples, together with the Bradford sword, all have a large, decoratively writhen knop on the plate in the centre of the saltire of bars on the outside of the hilt. The first two, in particular, are elegant and very well made.

Basket-hilted swords were clearly made for a wide variety of owners and purposes and were not exclusively utilitarian and military weapons (Blair 1981, 154). By about the 1560s yet more complex and developed hilts of this form of sword had evolved. These include two in the Royal Armouries (IX.2574, IX.5517), one in a private collection and one in the collection of the Naseby (1645) Battlefield Project (NBP/2008/0001; Fig. 9.39c; Blair 1981, 212–6; Wilson 2002, 149–51). This latter sword is especially interesting because of the possible longevity of its use. It was found during the First World War in a field on the site of the battle of Naseby (14 June 1645) and may have been carried during that decisive battle.

The finding of this type of basket-hilted sword within a precisely dated military context on the *Mary Rose* is important because it suggests that it may be a military rather than a civilian weapon. What may be an earlier form of this type of basket-hilt is worn by one of the Yeoman of the Guard in the painting of the *Field of Cloth of Gold* of 1520 (RCIN 405794). What is certainly a very closely related hilt to that of the *Mary Rose* sword

| Deck | 1 | 2 | 3 | 4 | 5 | 6 | 7 | 8 | 9 | 10 | 11 | 12 |
|---|---|---|---|---|---|---|---|---|---|---|---|---|
| Stb'd Scour | | | | | | | | | | | — | |
| Not Known | 27, 40 | | | | | | | 8 | | | | |
| Castle | | | | | | | | | | | | |
| Upper | 10, 14, 36 | 23, 24 | 31 | 36 | | | 35 | 33, 39 | 3, 11, 16, 17, 19, 20, 38, 47 | 45 | | |
| | | | | | | | | 13 | 34 | | | |
| Main | | 30 | | | 37 | 15, 32 | 2 | | | | | |
| Orlop | | 18 | | 9, 44, 25, 12, 42, 46 | | | 5, 6 | | 41, 43 | 21, 22, 29 | | |
| Hold | | | | | | | 4, 28 | | | | 7 | |
| Port Scour | | | | | | | | | | | | |

☐ Artefact is within a chest

756

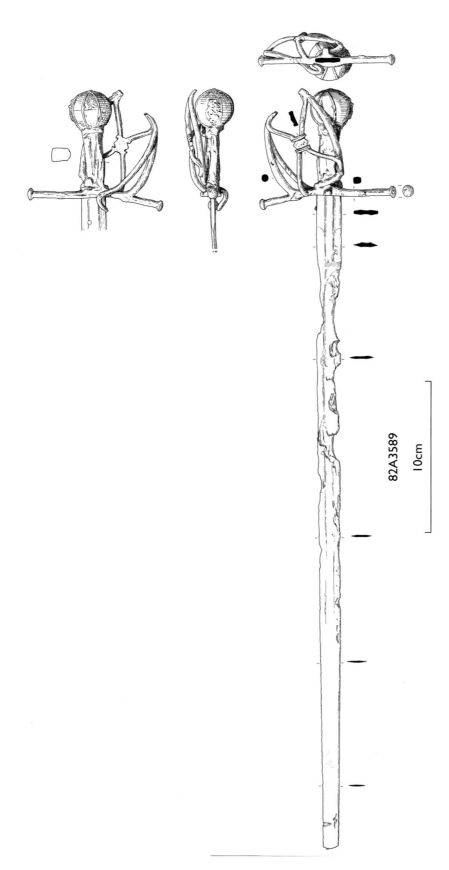

Figure 9.36 Basket-hilted sword 82A3589 after removal of concretion

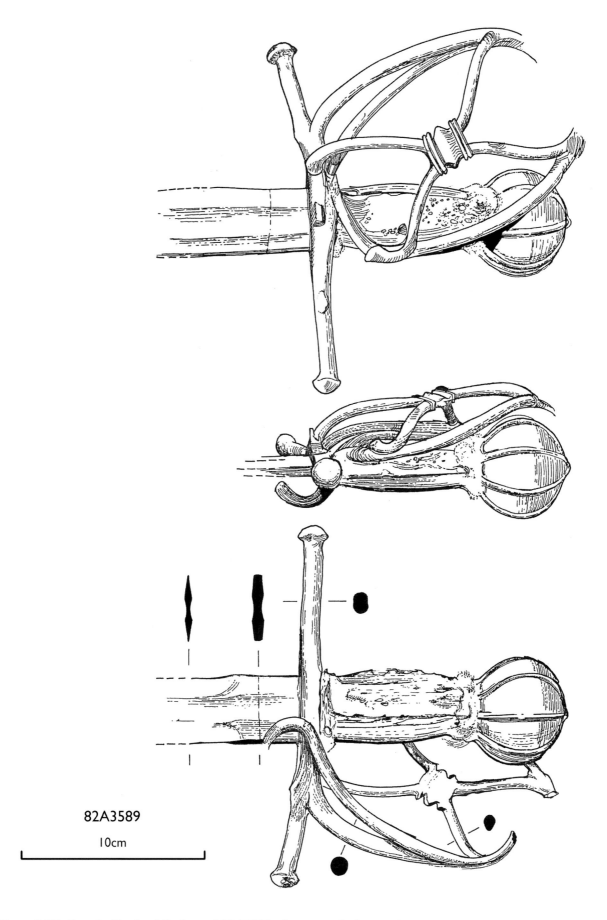

*Figure 9.37 Guard of basket-hilted sword 82A3589 after removal of concretion*

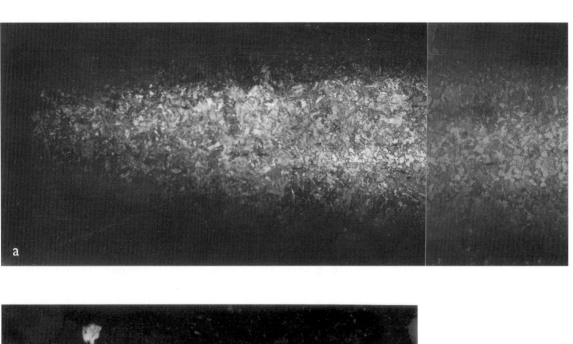

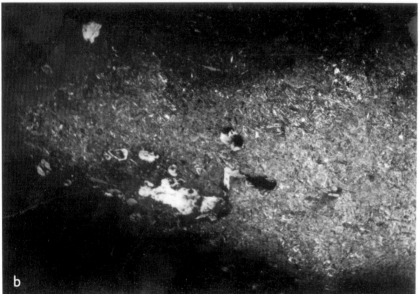

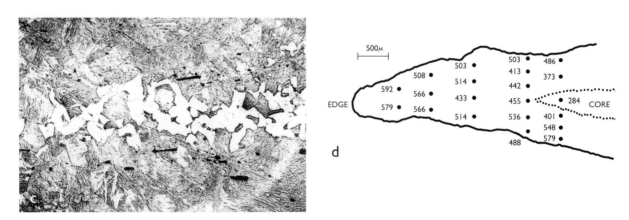

*Figure 9.38 Blade basket-hilted sword 82A3589: a–c) photomicrographs of blade, section at broken end: (a) section across blade, moving from the edge to the centre (to the right), x 60; (b) section across blade, at edge, showing martensite and corrosion products, x60; (c) microstructure of the edge showing where the steel outer layer was joined to the iron core. The iron appears as white grains (ferrite) and the hardened steel as dark areas (martensite); there are also some elongated slag inclusions. Magnification x 160; d) a map of microhardness measurements showing that the blade possessed a hardened outer edge, albeit one only 2mm deep (500 microns = 0.5mm) (photomicrographs by Alan Williams)*

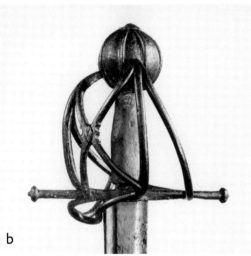

*Figure 9.39 Similar basket-hilted swords: a) from the River Cam at Chelmsford (© Chelmsford & Essex Museum, N-22263); b) Royal Armouries sword RA IX 4427 (© Royal Armouries); c) sword found at the site of the Battle of Naseby (courtesy Battlefields Trust)*

appears on the portrait of about 1540 by an unknown artist of a member of the Palmer family dressed as one of Henry VIII's Gentlemen Pensioners (Blair 1981, fig. 77). Both John Palmer (d. 1553) and his son William were members of this bodyguard. A similar type of basket-hilted sword is worn by a Yeoman of the Guard in an illustration by Lucas de Heere dating 1567–1574 (BL Add MS 28300). These illustrations add weight to the suggestion that such swords were favoured by Tudor Royal guards.

A pommel of the same distinctive type as that on this sword has been found by radiography of an element of concretion found on the wreck site in 1982 (82A0027; see Fig. 9.48). This provides evidence that there was more than one English basket-hilted sword on board the ship when it sank.

### Sword with curved forward quillon
*81A0719, U9* (Figs 9.29; 9.35b, 3)

This object was observed as a dark stain in the Tudor silts and was photographed *in situ* (Figs 9.29). Attempts were made at taking a plaster of Paris cast of the blade from the impression in the silts left on removal of the corroded iron but this failed. The grip was recovered separately, length 82mm.

The surviving photograph clearly shows a sword with the front quillon bent to form a knuckle-guard and a straight rear quillon. That this is the intended and original shape of the guard rather than the result of post-sinking deformation is evident from the fact that the front quillon is considerably longer than the rear one (which it would not have been had the sword been made as a cross hilt). The stain made by the quillons is considerably fatter than the quillons themselves would have been (the result, presumably of the stain running or migrating away from the metal itself as this dissolved). At the centre of the guard the staining expands yet further. This may be pure chance or evidence of a lug or additional side guard that appears on some swords of this sort. The pommel appears as a roughly ovoid impression, flatter on the top than on the underside and inclines towards the tip of the front quillon. It is impossible to be certain now whether it was originally ovoid or, like other surviving similar swords, had a beaked pommel. The stain made by the blade is also very wide, which may be caused either by '*running*' or because the sword was in its scabbard when the ship sank. From the photograph the blade appears to curve slightly and to expand, like a falchion, towards its tip. It is impossible to be certain whether or not this is a result of deformation after deposition.

The sword clearly belongs to a group that seems to have been in use for over 150 years from about 1430 to the end of the sixteenth century (Norman 1980, 68–9). A good number of this type, including some believed to be English, also have lugs in the centre of the cross and it has been suggested that from these the British basket-hilted swords like that found on the *Mary Rose* may have

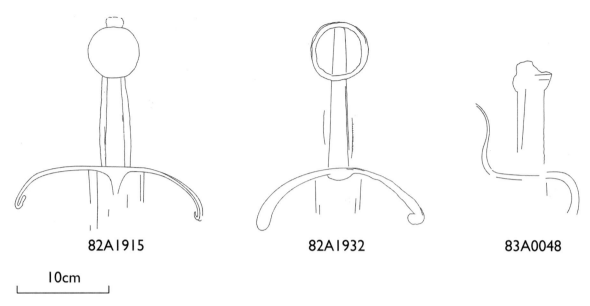

*Figure 9.40 Reconstructions of hilts for 82A1915, 81A1932, 83A0048*

developed – the lug gradually extending to form a side bow and then the space between the side bow and knuckle bow being filled in by other bars to form a proto-basket (for this argument and details of swords of this type with English provenances see Wilson (1986, 1–2; 1990, 23–4 and Mazansky 2005, 44–6). The majority of this group have cap pommels of truncated wedge form with a thinner extension towards the tip of the front quillon that gives them a beaked appearance. They also have rear quillons not straight, as on this example, but curving towards the blade.

However, the most firmly dated surviving sword of this general type does have a straight rear quillon. It was preserved at Clarke Hall, Wakefield until it was sold at auction by Christie's, London (14 April 1966, lot 176). It was found early in the twentieth century during the cutting of a new drain on what is believed to be the site of the battle fought at Wakefield in 1460 (Wilson 1990, 24). It has a flat disc-like pommel with, at the front, a centrally pierced lobe in the plane of the blade that matches a similar pierced expansion of the end of the front quillon. Another hilt of the same form, with a straight rear quillon, survives in Warwick Castle (no. E008), decorated with motifs associated with the Earls of Warwick (Wilson 1986, 4). This also has a side-guard that extends from the pronounced triangular quillon block two-thirds of the way along the tang. The guards are all formed as ragged staffs. The pommel is in the form of a muzzled bear's head. It has been suggested that this may be a later associated one (Mazansky 2005, 46) but its general similarity to the beaked pommels of this group make it possible that it is the original. The sword has been tentatively dated to the late sixteenth century. While this is very possible, given the appearance of a sword of this type with a straight rear quillon in Pieter Brueghel's *The Wedding Dance* dating from about 1567 (Kunsthistorisches Museum, Vienna, inv. no. 1059), there is no reason why it should not be considerably earlier in date.

The swords of this type generally have straight blades and the sword in the photograph probably was also straight-bladed. However, the photograph does show a slight curvature to the blade and the possibility that this might have been an original feature not a deformation should not be entirely discounted. Intriguingly, in 1979, a falchion, with recurved quillons broadly comparable to those on the wider group of these swords and a gently curving tang, was recovered by a young boy from a stream in the city of Durham (current whereabouts unknown). The general consensus at the time was that it was a later, eastern weapon, probably of Chinese origin, but its form did not fit it neatly into any category and a possibility remains that it may be a European or even an English weapon of the Tudor period.

**Sword hilt and sword with recurved quillons**
*83A0048, M7* (Figs 9.35b, 2; 9.40; 9.41)
Sword hilt 83A0048 was identified by radiography within concretion found inside the Barber-surgeon's Cabin. The radiograph shows a guard formed of recurved quillons. The serpentine front quillon bends down at right-angles towards the pommel to form a knuckle guard. The rear quillon gently curves away from the grip towards the blade. The tip cannot be seen and where it ends it appears either to be bending more tightly in towards the blade or beginning to form a droplet finial. Only a fragment of the base of the pommel remains and it is impossible to be certain of its form. It seems to curve away from the grip at too steep an angle to be a spherical or disc pommel, though it is of curved not angular form. There appears to be some

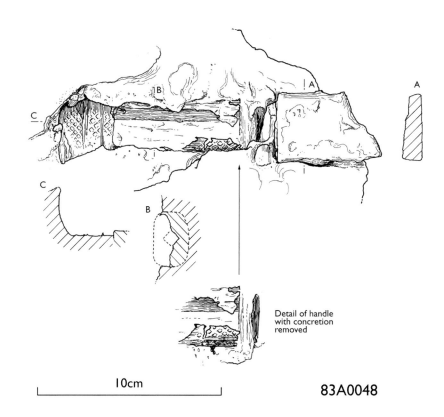

*Figure 9.41 80A0048 grip in concretion showing wire*

evidence for a band of some sort running across the pommel. The grip appears to be straight sided. A drawing of it (Fig. 9.41) seems to show a lattice of wire covering both the grip and the remains of the pommel (Fig. 9.40).

This is a hilt of Norman's type 5 (Norman 1980, 68–9) for which he found evidence from around 1430 to the end of the sixteenth century. Most illustrative evidence, however, comes from the fifteenth century with later sixteenth century examples largely confined to German lands where they often appear as peasant weapons, for example in the paintings of Pieter Brueghel. A good number of this type, including some believed to be English, also have lugs in the centre of the cross and it has been suggested the British basket-hilted swords like that found on the *Mary Rose* may have developed from it, as described above (Wilson 1986, 1–2; 1990, 23–4). The most firmly dated sword of this general type (but without a lug and with a straight rear quillon) is that from the Battle of Wakefield site mentioned above (Wilson 1990, 24). The guard of another sword of the same general group, with recurved quillons survives in Strangers' Hall, Norwich (no. 76–64), together with part of the blade (Wilson 1986, 3–4). It is said to have come from Mousehold Heath, outside of the City. While the heath area seems to have served for a long time as a mustering point the most likely time for the sword to have been lost or thrown away was in 1549 when the rebel leader Ket had some 16,000 men under his command there. The presence of this sword hilt on the *Mary Rose* proves that this type continued to be used in England until Ket's time and makes it more likely that the Norwich example may date from the middle of the sixteenth century. Swords of this type are generally plain and utilitarian. The apparent decorative wire lattice over the pommel is, therefore, both unusual and interesting, though it may simply be part of the grip binding that has migrated post-deposition.

### Sword hilts with curved quillons
82A1915, H7; 82A1932, O7 (Fig. 9.35b, nos 4 and 5)

Hilt 82A1915 consists of an imprint of a bound grip, the side of a pommel and the top of a blade within concretion which mave been associated with silk, possibly part of a cover. The shoulder and a short length of one flat side of the ricasso of the blade can be seen. The disc pommel is *c*. 50mm (2in) in diameter with large concave chamfers to the rims. The centre of each face may have been relieved by a circular indent (Fig. 9.40).

A radiograph, taken before the concretion was broken, shows a guard formed of ribbon quillons, set at right-angles to the plane of the blade, curving towards the blade at an ever-increasing angle (Fig. 9.28b). The ends are formed by turning the ribbon of metal tightly under itself. In the centre of the visible side, the quillons expand into a narrow, concave-sided triangular escutcheon that extends up over the centre of the blade. The grip appears to be of normal truncated elliptical plan, the maximum width being nearer the blade than the pommel and the width at the blade being greater than at the pommel. The pommel is shown to have a large and pronounced button.

With its ribbon-like quillons and triangular escutcheon, 82A1915 bears general comparison with the Sword of Mercy in the English Crown Jewels, made for the coronation of Charles I in 1626. An example from the Royal Armouries of a generally similar sword with a large, somewhat flattened, spheroidal pommel (IX.5427) was excavated from the Thames foreshore at Queenhithe and has, up to now, been thought to date to the last quarter of the fifteenth century. According to Norman (1980, 239), who cites both illustrative evidence and one actual example (Real Armeria, Madrid, no. G 46) this type of pommel (Norman type 4) is found until about 1550. A large wheel pommel is clearly shown in Holbein's portrait of Sir Nicholas

Carew in the collection of the Duke of Buccleuch at Drumlanrig Castle, dated around 1532–6. With its triangular escutcheon this sword is the type for which the scabbards 80A1709, 81A1734 and 81A1885 were made.

The pommel, grip, guard, scabbard and blade of another sword (82A1932) were identified in concretion. The radiograph shows a guard formed of a rectangular bar, thinner in the plane of the blade. It curves towards the blade in a wide arc, gradually expanding in thickness in that plane from a central quillon block with a semicircular edge towards the blade to button finials, only one of which is visible. The grip again appears to be of normal truncated elliptical plan, the maximum width being nearer the blade than the pommel and the width at the blade being greater than at the pommel, which is spherical. The maximum length of this sword is 240mm and the blade has a maximum width of 50mm (Fig. 9.40).

As stated in the introduction, these two swords belong to a group that include a number associated with tombs in English churches most notably the sword in Westminster Abbey thought to have belonged to Henry V (Marks and Williamson 2003, 195). Traditionally, none of these has been dated later than the last quarter of the fifteenth century. A more recent find is a sword (now in the Royal Armouries, IX.5427) with a guard of similar arched form to those of these two swords from the *Mary Rose* and a large, somewhat flattened, spheroidal pommel. It was excavated from the Thames foreshore at Queenhithe along with some coins from the reign of Edward IV so it, too, has been dated to the last quarter of the fifteenth century. Its quillons are formed in one with a broad triangular escutcheon and it has a straight but short blade that has led to the suggestion that it may have been made for a child.

Blair (1998, 342–4) has gathered evidence to show that the cruciform 'arming' sword never went completely out of use and that cross-hilted swords with ribbon-like quillons, either straight, or arched as on the swords from the *Mary Rose*, were widely used until the early sixteenth century and continued in use even later for ceremonial and chivalric purposes. However, he also drew attention to the curious fact that no sixteenth century swords of this type with English associations had yet been recognised. He showed that, while they are frequently shown on English armoured monumental effigies of the sixteenth and early seventeenth centuries, they do not appear in portraits, with the exception of Holbein's portrait of Sir Nicholas Carew. This shows a sword with arched quillons and a disc pommel that is clearly linked to the group under discussion. Sir Nicholas served as Henry VIII's Master of the Horse from 1522, until his execution for treason in 1539. He was no relation of Vice-Admiral Sir George Carew, the last captain of the *Mary Rose*.

These swords from the *Mary Rose* now provide the evidence of English association that should allow others of the group to be identified. They allow us to be more certain that the Royal Armouries sword found in the Thames was of a type not uncommonly used in England at least until the middle of the sixteenth century, in warfare as well as for wearing at chivalric ceremonies and should lead to consideration of some redating of other examples, such as those mentioned above (Cripps-Day 1922, 117–273).

Funerary swords of this group have arched quillons and most have small shield-shaped escutcheons in the centre of the cross. One, at Dunsford church in Devon, the only one of the group which has been given a sixteenth century date, has a globular pommel similar to that on 82A1932 (*ibid.*, 173, fig. 1607), while the rest have disc pommels with chamfered edges and a small circular recess in the centre of each face, similar to that illustrated on the portrait of Sir Nicholas Carew and to 82A1915 (Blair 1998, 344; Mann 1940, 376–7, pl. lxii). Examples at Whitelackington church, Somerset (Cripps-Day 1922, 236, fig. 1730) and Standon church, Hertfordshire (Victoria and Albert Museum M48–1980) have been associated with burials from the late sixteenth and early seventeenth centuries; Cripps-Day suggests that the Whitelackington sword, though late fifteenth century, had a late sixteenth century blade associated with it, but Blair cautions against accepting this as evidence for dating as, by then, heralds were known to be using old arms and armour as funerary achievements.

This group is similar to the sword in Westminster Abbey long believed to be that of Henry V which has a wheel pommel and a very small triangular quillon block (Mann 1962, 405–9). With their ribbon-like quillons and triangular escutcheons, 82A1915 and others of the group of funerary swords, also bear general comparison with both the Swords of Spiritual Justice and the Sword of Mercy in the English Crown Jewels, though both these have straight quillons with the tips scrolling towards the blade. A number of cross-hilted swords with very pronounced buttons to their wheel pommels (like that on 82A1915) also provide parallels. These include an unprovenanced example in the Royal Armouries (IX. 1426); a sword in the Real Armeria, Madrid (G 23) with a gilt hilt and a blade etched with religious inscriptions in Latin (possibly made in the Flanders in the late fifteenth century); and a sword thought to be either French or Italian in the Musée de l'Armée, Paris (no. J 27) with gilded hilt and blade, etched and gilt on one side of the forte with the Judgement of Paris and on the other with Venus at the Forge of Vulcan. These last two are reminders that this general type of sword does not appear to have been restricted to England. However, there is plenty of evidence, including excavated examples, to show that spherical pommels were particularly popular in England.

*Sword and dagger grips*
by Guy Wilson with Laura Davidson,
Alexzandra Hildred and Douglas McElvogue

As grips found in isolation are difficult to identify, all have been included here. Forty-six wooden grips for either swords or daggers have been recovered, many as separate elements. The grips are made exclusively in wood and are of single piece construction. Of the eighteen studied for wood species, most (thirteen) are of beech, but willow or willow/poplar, burrwood, hazel/willow, alder and fruitwood grips have also been identified (see Table 9.3).

This represents a unique assemblage of dismounted grips from swords and daggers without the binding that would have covered many of them. Examination of 25 of them in 1984 revealed traces or imprints of wire binding (iron, copper plated, copper or possibly silver) in at least fourteen, but most of this evidence has since been lost. Some evidence has been found of textile and leather coverings. Where this is observed, the binding is wound tightly around the grip. In addition to sword 82A3589 and grip 82A1915, discussed above, textile remains, apparently a plain woven tabby weave wool, were noted on grip 81A4310 and 81A4181, recovered with bast fibres attached. Some of the grips were obviously not covered. The burrwood grip, 82A2054, has longitudinal recesses to take decorative copper alloy bands and 82A758 has a moulding running around its circumference at its mid-point, formed by incising a line on either side. Similar, single, incised lines run around the grip at either end. This may have been a practical attempt to improve grip as well as a decorative treatment. A ballock dagger in the Museum of London (no. 25124) from Paul's Stairs, Blackfriars, London, has a raised moulding around the grip two-thirds of the way from the pommel to the guard. This is similar, but not identical in form, to the type of ballock daggers found on the *Mary Rose*, having a grip that tapers from pommel to guard rather than the straight ones typical of finds on the ship.

Most of the grips are similar in form with a sub-rectangular cross-section with rounded corners and slightly convex sides (Figs 9.25, 9.42–4). They have a convex vertical profile with maximum width at the centre (*c*. 35mm) and towards the quillons and narrowest at the top, close to the pommel (*c*. 20mm). The tang recesses are rectangular, 20mm wide at the top reducing to 13–15mm at the bottom, with a half circle in the middle. This could have accommodated the tang and wedge helping to keep it in place. The ends of the grips are flat cut. The variable condition of the conserved (dry) grips makes detailed measurements and study of shape somewhat unreliable.

There are a large number of plain and ordinary grips that cannot be categorised and, as stated in the introduction, length does not indicate whether a grip was made for a sword or a dagger. Length varies, with a single long grip of 95mm (82A1613), and clusters at 86–89mm (3⅜–3½in), 82.5mm (3¼in) and 76mm (3in).

**Distribution and associations**
The distribution of the grips is given in Figure 9.34 with selected examples shown on Figure 9.35b. The largest concentrations are found in areas with human remains including the assemblage of Fairly Complete Skeletons, particularly on the Upper deck (U9) and on the Main deck (M1–2), strongly indicating that they were being worn at the time of sinking (Figs 9.34–5). The twelve on the Orlop deck could have been either in use or in storage though only three were found inside chests, two in the same one (80A1413) in O4 (80A1838, 80A1405) and 82A2979 from 81A2941 in O10. The most complete daggers (excluding ballock daggers) include 82A1914 (H7), 83A0022 (O2) and 82A0975 (O10). Associated objects are discussed above and in many instances there are fragments of scabbards, sheaths, pommels or chapes found within the same sector (see Fig. 9.34). Grips with specific features of interest are described separately, otherwise details are in Table 9.3 which also indicates where examples were found with other parts of the same sword or dagger.

*Decorated grips: 82A2054, O7; 82A4412, O11*
(Figs 9.35b nos 6 and 7; 9.42)
Grip 82A2054 is a one-piece, burrwood grip of ovoid section, tapering slightly (Fig. 9.42). The centres of the flattened sides are recessed longitudinally. In one of the recesses are remains of a thin copper alloy strip attached by four copper alloy pins. The grip is cut out to take a substantial tapering tang, the width of the hole measuring 19mm (¾in) at one end and 15.5mm (⅝in) at the other.

At 85.5mm (3⅓in), this is one of the longest detached grips found on the *Mary Rose* and this might suggest that it was for a sword not a dagger. However, uncovered burrwood grips with longitudinal decorative metal strips are a feature of dagger hilts, as, for instance, on one excavated from the Thames foreshore at Queenhithe (now in the Royal Armouries: X.1716). The fruitwood grip of this ends in two protruding horn and is decorated on each side with a single longitudinal band of copper alloy. This band runs over the end, which is also decoratively incised with a pattern of gently curved lines forming sunburst segments. Some scale-tang daggers have the exposed sides of the tang covered with decorative copper alloy plates, or are otherwise decorated, and on some matching bands run down the centre of the scales themselves (for instance, a ring-pommelled dagger in the British Museum, no. 83, 4–1, 636). This practice may be the origin of the adoption of the same decorative technique on whittle tang daggers. Occasionally the decorative metal strip(s) is/are set in a spiral. Examples of this include two long-bladed rondel daggers excavated from the Thames foreshore, at Queenhithe (RA X.599) and Bull Wharf

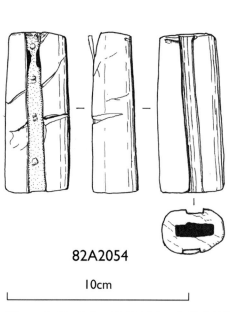
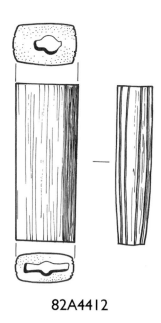
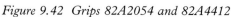

*Figure 9.42 Grips 82A2054 and 82A4412*

(private collection). The former has two blued iron strips edged with twisted copper alloy wire and the latter two copper alloy strips decorated with religious inscriptions.

A one-piece, burrwood slab-sided grip of rounded rectangular section from O11 (82A4412) also tapers slightly (Fig. 9.42). One of the narrower sides is decorated with four longitudinal incised lines. The other was probably similarly decorated but only two of the lines are now visible. The cut-out for the tapering tang has clearly been started by drilling, with presumably chisels and/or files then being used to expand it laterally into a slot. It measures 21mm (1³⁄₁₆in) across at one end and 11mm (⁷⁄₁₆in) at the other.

As the wood has a decorative grain and the grip is decorated, we can suggest that it was not intended to be bound. A similar grip on a quillon dagger in the Museum of London (no. 7672) is from a dagger, missing its pommel, excavated at Brook's Wharf on the Thames in the City of London in 1866. Another quillon dagger in that museum (no. 25129), complete with a cap pommel, has a similar tapering grip but of ovoid section with a concave gutter running the length of each of the wider sides. It was found on the Thames foreshore at Paul's Stairs during the Blackfriars improvement scheme of 1969.

*Group of wooden grips: 79A1112\*, 80A1041, 80A1838\*, 81A0133/1\*, 81A0688\*, 81A1222\*, 81A1763\*, 81A2041, 81A2487, 81A3739\*, 81A4098, 81A4180\**
(Figs 9.34; 9.35; 9.43)
Twelve one-piece wooden grips were available for detailed examination. A selection is shown in Figure 9.43. All are of ovoid section and truncated elliptical plan. On each, the maximum width is nearer the blade than the pommel and the width at the blade is greater than at the pommel. They are relatively crude and all seem intended to take wire or cord bindings, though only one (80A1838) now exhibits any direct evidence (traces of narrow lateral banding in concretion that look to be the imprint of binding). Nine of the twelve (indicated by \*) show clear evidence that the longitudinal chiselled cut-out for the tang has been started by drilling centrally. Distribution is shown on Figures 9.34 and 9.35b. The grips range from 76mm (3in) to 89mm (3½in) in length fall into three distinct groups according to length.

Group 1: *c.* 76mm (3in) long: 80A1041, 80A1838, 81A0133/1, 81A2041, 81A3739, 81A4098, 81A0688 (Fig. 9.35b, nos 8–13 and 47).
Group 2: *c.* 82mm (3¼in) long: 79A1112, 81A4180 (Fig. 9.35b, nos 14–15)
Group 3: 86–89mm (3⅜in–3½in) long: 81A1222, 81A1763, 81A2487 (Fig. 9.35b, nos 16–18).

It has already been stated that length alone is no guide as to the function of the grips – indeed the length of the grip on sword 82A3589 puts it into the shortest category (and is by no means unusual for its type), whereas that on the so-called 'Landsknecht' dagger fits into Group 2 and that of the rondel dagger (82A1914) is at the upper end of the length of Group 3. Details of all the grips are shown in Table 9.3. Other wooden grips are shown in Figure 9.44.

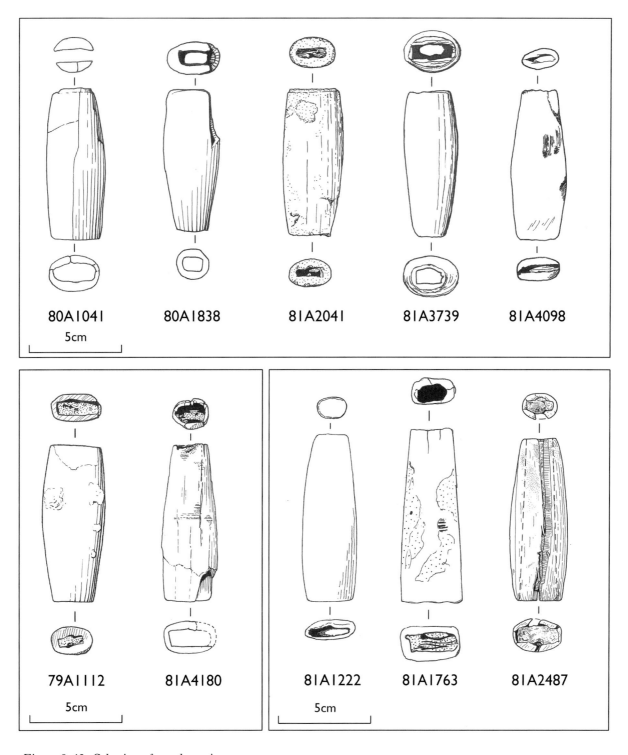

*Figure 9.43 Selection of wooden grips*

### Detached pommels
Remains of the copper alloy outer sheathing for at least twelve, possibly thirteen, detached sword pommels were recovered with radiographic evidence for three more still within concretion (Table 9.4). With two exceptions they seem to be of an identical and hitherto unknown type. All must have contained iron which has deteriorated. Seven were analysed by X-ray fluorescent spectrometry and all are quaternary alloys of copper, mostly containing an average of *c.* 60–80% copper, *c.* 1–23% iron, *c.* 12–33% zinc and 1–5% tin (full details in archive). At least two styles have been found: one with segmental, copper alloy covering strips (82A3589, 82A0027 and 81A133 if it is a pommel), and six with two hemispheres of copper joined by a wide band (83A432, 83A0433, 82A0758, 81A0919, 81A2979, 81A0718). Three are visible only on radiographs (82A1932, 82A0975, 82A0027) and one is a negative

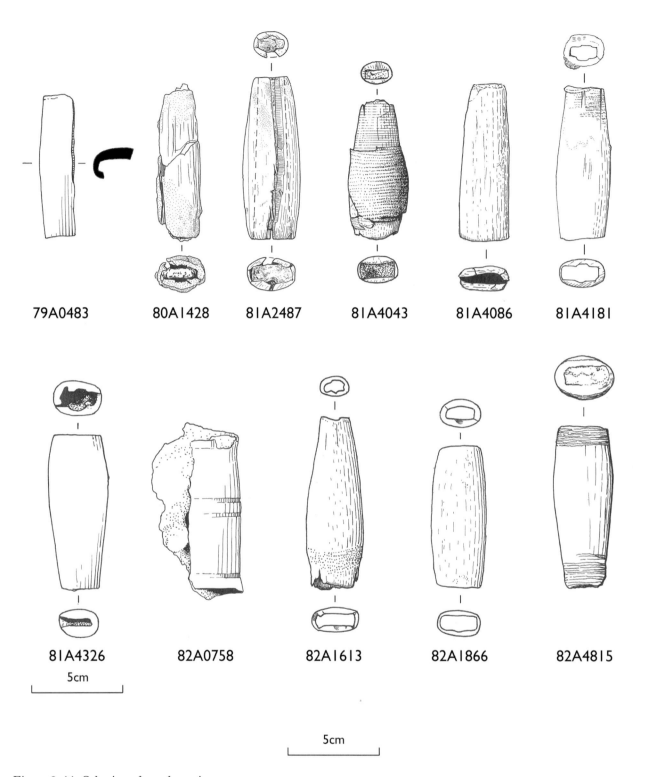

Figure 9.44 Selection of wooden grips

cast (83A0494). Nine (75%) were found with other elements of the same sword (Table 9.3).

*Distribution*
All pommels were found in sectors with other sword fragments Figure 9.35b, nos 19–27) and exhibit similar associations with skeletal remains (Fig. 9.35a). Two were found with other sword remains in chests (see above and Fig. 9.35b).

*Spherical sword pommels with a central recessed band: 81A0718, 81A0919, 81A2979, 82A0758, 83A0432, 83A0433 (Figs 9.35b, nos 19–24, 9.45, 9.47)*
Thin copper alloy sheathing is all that survives of six pommels of this type. Of these, 83A0433 is the most

## Table 9.4 Pommel sheaths with metal analysis

| Object | Sector/location on Fig. 9.35b | Diam. (mm) | Association | Description |
|---|---|---|---|---|
| 82A1915 | H7/4 | ? | hilt with curved quillions | disc pommel, large pronounced button; in concretion, radiograph only |
| 83A0494 | H7 | ? | loose | negative cast only |
| 80A1405 | O4 in chest 80A1413 | ? | grip & blade frags | missing; listed as frags; quaternary alloy Cu 60%, Zn 19%, Sn 2%, Fe 17.5% |
| 81A0133 | M1/26 | 60 | grip | segmental copper strips, width 4.5mm; poss. 8 originally each side of horizontal band; quaternary alloy Cu 83%, Zn 8%, Sn 3%, Fe 4%, Pb 1%. Similar to sword 82A3589 & 82A0027, but thinner; poss. pomander |
| 81A2414 | O4/25 | 35 | loose | single ring, poss. pomander; corroded; quaternary alloy Cu 100%, Zn 33%, Sn 5%, Fe 23% |
| 82A1932 | O7/5 | ? | grip, guard, blade frag. | Spherical; in concretion, radiograph only |
| 81A2979 | O10/21 in chest 81A2941 | 57–8 | grip & scabbard 81A2977 | 2 recurving dish-shaped spherical caps joined by central band |
| 82A0758 | O10/22 | 57–8 | grip, found next to skull & comb | 1 hemisphere almost complete, but several holes, 1 ring from other portion; 2 recurving dish-shaped spherical caps joined by central band; quaternary alloy Cu 79%, Zn 15%, Sn 2.4%, Pb 2%, Fe 1.8% |
| 82A0975 | O10/29 | ? | grip | missing |
| 83A0432 | M2/23 | 57–8 | in concretion at base of companionway | 2 recurving dish-shaped spherical caps joined by central band; contaminated quaternary alloy Cu 80%, Zn 2.6%, Sn 2%, Fe 13% |
| 83A0433 | M2/34 | 57–8 | in concretion at base of companionway | 2 recurving dish-shaped spherical caps joined by central band; best example of type; leaded brass Cu 74%, Zn 21%, Pb 2% |
| 83A0048 | M7 | ? | in concretion in Barber-surgeon's Cabin | in concretion, radiograph only |
| 81A0718 | U9/19 | 50 | skeletal remains, grip; against side of box & human remains; blade & guard present as stains in sediment | 2 recurving dish-shaped spherical caps joined by central band |
| 81A0919 | U9/20 | 57–8 | loose | 2 recurving dish-shaped spherical caps joined by central band |
| 82A3589 | SS11/1 | 57–8 | on complete basket-hilted sword | complete; made in 2 halves & faceted, divided by segmental strips; identical to 82A0027 |
| 82A0027 | UK/27 | 50 | in concretion | 6 or 8 facets separated by raised longitudinal mouldings; made in two halves; identical to 82A3589 & 81A0133; from radiograph only |

Not all are detached. Analysis of 82A3589 excluded

complete Fig. 9.45). It has approximately half of the two dishes and attached central band surviving. The rest of the group consists of fragments of dishes and/or bands.

Both visual examination (some of the group are heavily stained with what looks like iron rust) and the conservation records suggest that the sheathing originally covered an iron core. The analysis suggests contamination with iron, with varying percentages. The pommels appear to have had a diameter of *c*. 58mm (2¼in) confirming that these are sword, not dagger pommels. The iron core would presumably have been hollow as a solid core of this size would be likely to have unbalanced the sword.

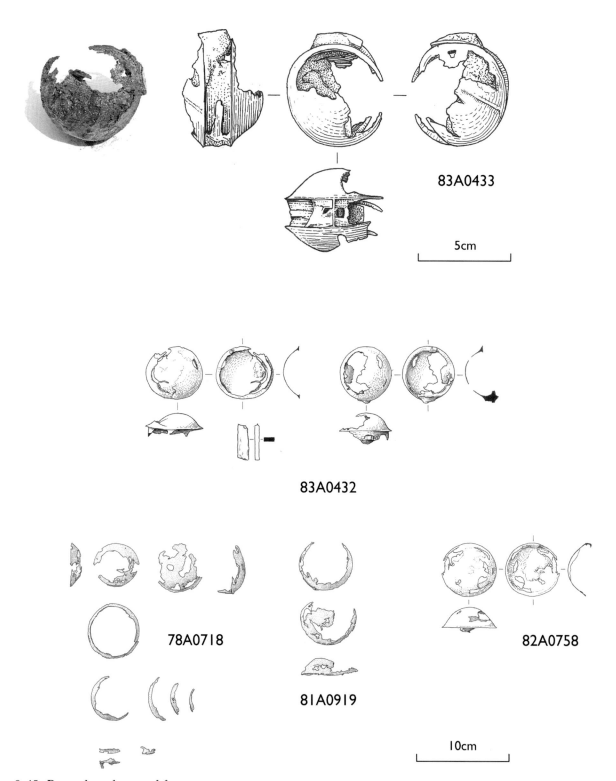

*Figure 9.45 Pommels and pommel fragments*

The pommels consist of two hemispherical caps, recurving like dishes just before the rim. These are joined to a central recessed band just over 15mm (½in) wide that runs parallel to the blade. Examination of 83A0433 suggests strengthening of the band by layering. It also contained holes for locating the pommel on the tang – one squared hole survives with a pin crossing the band just beside it.

This group is especially interesting for two reasons. First, it represents an otherwise unrecorded pommel type. Norman (1980, 237–83) discusses all types of sword pommels seen by or known to him and this type is not mentioned. Close in form, though not in construction, is the copper alloy pommel of a dagger found with a copper alloy grip on the Thames foreshore at the Tower of London (Museum of London no.

A10738) (Fig. 9.46). This consists of two spherical caps split into eight radiating segments decorated with quatrefoils and joined by pronounced central mouldings.

Secondly, this assemblage provides the first evidence to suggest that some iron pommels may have been normally, rather than occasionally, sheathed in another metal. The copper alloy may have covered the iron for decorative reasons – it would facilitate decoration by mercuric gilding, and is itself a good medium for engraved decoration. This latter use of copper alloy is not unknown on dagger pommels. A ballock dagger in the Royal Armouries (X.8) that comes from the old Tower collection has an open-bottomed, rounded cap-shaped copper alloy pommel. This is split longitudinally into three segments by double engraved lines, with lateral decorative lines highlighting the waist before the open bottom. Visual inspection reveals a residue of iron rust attached to the inside and a hole on one side, possibly a piercing for a securing pin. This may be a decorative sheath to a plain iron pommel, now lost. Copper alloy sheathing found on *Mary Rose* examples may, therefore, have a purely decorative function. Equally, it might be an attempt to protect the exposed iron from the salt-laden air at sea. If this were the case, logic suggests sword guards would be similarly treated, but currently this is unsubstantiated.

Additionally there is a smaller copper alloy 'ring' (c. 35mm (1³⁄₈in) diameter), 81A2414. This may represent a smaller example of this group, or an accompanying dagger (Fig. 9.35b, no. 25; 9.47).

*Sheathing bands for a spherical pommel or pomander: 81A0133/2; M1 (Figs 9.35b, no. 10, 9.47)*
This object consists of eight narrow bands, 4.5mm wide, radiating from a small circlet to a waistband, representing just over half of the copper alloy sheathing of a pommel or a pomander. Four bands are nearly complete and four are fragmentary. The roots of three similar bands (providing a mirror image) can be seen emerging from the other side of the waistband. Currently the waistband has jagged edges and a maximum width of 7mm. Its original width and form cannot be confirmed. It has an upright central ridge, perhaps formed by brazing together two bands, each bent longitudinally at right-angles (Fig. 9.47). Found with a grip (81A0133/1) and scabbard (81A0133/3), it has been identified as a pommel. Although possible, the thin, fragile central ridge to the waistband is unusual on sword pommels. It is similar in size to the wooden pomander recovered (81A4395), interestingly found in concretion with a sword scabbard (81A1567).

*Pommel in concretion: 82A0027, Sector unknown*
(Fig. 9.48)
This pommel is spherical (diameter 50mm (2½in) and is divided into either six or eight facets by longitudinal raised mouldings. It is made in two halves, joined horizontally, probably by brazing. There appears to be a bun-shaped button on the top to serve as a washer for the tang. Part of the hilt may also survive in the concretion but the radiograph image is too indistinct for description. However, the pommel is of the distinctive type found on basket-hilted sword of the type found on the *Mary Rose* (82A3589) and indicates that there was more than one basket-hilted sword on board the ship (Fig. 9.48). The implications for ownership have been discussed above.

*Figure 9.46 Dagger from the Thames foreshore at the Tower of London (© Museum of London; no. A10738)*

## Daggers

Most of the artefacts that have been clearly identified as daggers are ballock daggers with just a few notable exceptions.

**Rondel dagger**
*82A1914, H7 (Figs 9.34, 9.35b, no. 28, 9.49)*
All that remains of this object are impressions in concretion of the guard, wooden grip and c. 50mm of the blade still embedded in concretion. The wooden grip is of round section with slightly convex sides. It is broader at the guard than at the pommel. The rondels forming the guard and pommel would have been of iron and were made as truncated, concave-sided cones with their bases facing away from the grip. The rondel that forms the pommel is the smaller of the two, with a diameter of c. 43mm, and is round. Still surviving in the concretion is a small copper alloy washer around the end of the blade tang. The rondel forming the guard is slightly larger and oval in plan (46 x 32mm), its longer axis set in the plane of the sides of the blade. As far as can be ascertained the blade is straight and single-edged with a flat back and flat sides c. 26mm across at the hilt.

Rondel daggers were one of the most common forms of dagger used in Europe from the fourteenth century on but, by the mid-sixteenth century, were coming to the end of their popularity. They appear in a wide variety of types, but this one is rather unusual. Despite the apparent need to maximise protection to the hand by making the guard as big as possible,

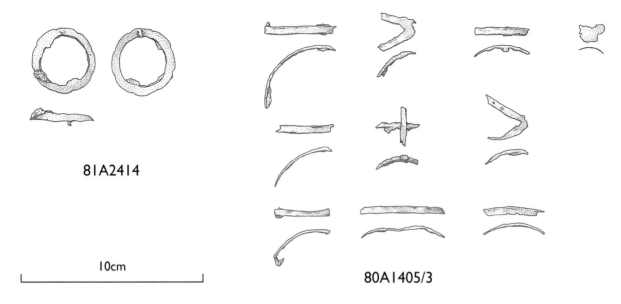

*Figure 9.47 Pommel fragments and sheathing from 81A0133*

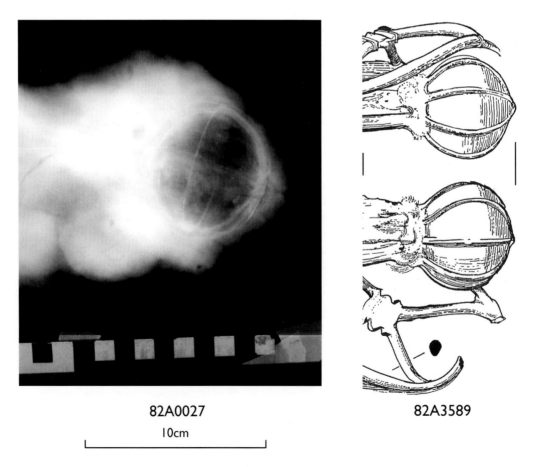

*Figure 9.48 Radiograph of 82A0027 and drawing of pommel from sword 82A3589 for comparison*

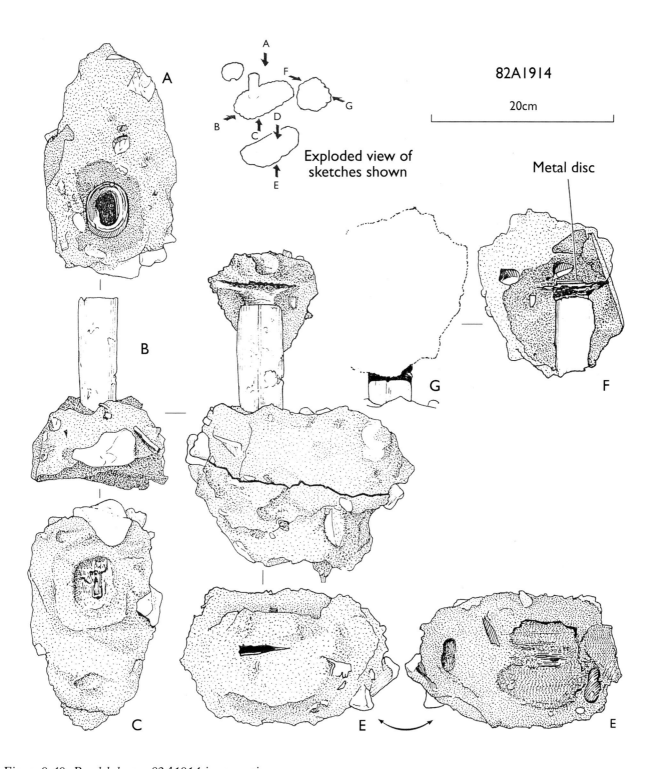

*Figure 9.49 Rondel dagger 82A1914 in concretion*

daggers with rondels of different sizes usually have the larger one mounted as the pommel and the smaller as the guard. The opposite is true of the *Mary Rose* dagger. Also, most that have rondels in the form of truncated cones mount them with the bases of the cones towards the grip. Again the opposite is true of this dagger. Both these features can be seen, for example, on a rondel dagger in the Royal Armouries which was excavated from the foreshore of the Thames at Queenhithe (X.1708, Fig. 9.50). This has wooden rondels reinforced on the faces away from the grip by copper alloy plates. The guard is a truncated convex-sided cone of the same form as those on the *Mary Rose* dagger, but with the wider base facing the grip. The pommel is a

*Figure 9.50 Rondel dagger excavated from the foreshore of the Thames at Queenhithe Royal Armouries X.1708; © Royal Armouries)*

larger rondel formed as a very shallow cone. It has a very stout, diamond-sectioned blade and is clearly a stabbing dagger.

### Dagger in concretion
*82A0975/2, O10* (Figs 9.34, 9.35b, no. 29, 9.51)
The radiograph of this object shows a wire-bound grip with the base of a pommel, probably ovoid, springing from it, and concretion around the centre of the guard (the extremities do not survive) that suggests that it consisted of straight quillons angled slightly towards the blade. The grip is of typical form, *c*. 63mm long, tapering from a maximum girth at about the middle of its length to the pommel and the guard

The exact form is unclear. If the description just given is correct, it would belong to a type common throughout Europe in the sixteenth century including in England. Examples include a number of daggers excavated from the Thames foreshore, currently in museum and private collections, together with a detached copper alloy quillons of a dagger found in the River Ouse, Cambridgeshire (Cambridge Archaeological Museum no. 32.242). The brass of Philip Chatwyn (*c*. 1524), in Alvechurch church, Worcestershire, shows a generally similar dagger, though with arched quillons, providing evidence of their use in England during the reign of Henry VIII (London Museum 1940, fig. 9).

### Dagger with wooden pommel, grip, sheath and by-knife hilt
*83A0022, O2* (Figs 9.34, 9.35b, no. 30, 9.52)
The burrwood pommel and the grip were found together. They have an irregular hexagonal section with the two main flats larger than the four side ones. The flats are edged and separated by incised longitudinal lines. In plan the pommel, which is 47mm long, is in the form of an inverted droplet or a hen's egg with its point to the grip. The 85mm long grip is convex in outline, slightly wider at the blade end than at the pommel. The two smaller flats on one side have broken off and are missing. A chamfered piece of wood found with the hilt does not fit the break and probably does not belong. No trace of a guard remains.

Associated with the hilt elements are a sheath, 305mm in length and a by-knife hilt. The sheath consists of a wooden core of two laths covered by leather which butts longitudinally down the back. There is no evidence that it was ever sewn together. About 64mm down from the mouth of the scabbard the leather cover is slit to take a by-knife. The tip of the sheath is missing. Rusted remnants of the blade remain within. The by-knife hilt (94mm long) consists of two triangular-sectioned scales of wood tapering from the chamfered butt to the blade end and joined by four rivets (presumably copper alloy) that would have passed through the now missing tang.

The pommel is closest in shape to Norman's type 16 (Norman 1980, 245), which he says was in use from 1470–1585, though he does not mention any examples with a hexagonal section. It is, therefore, unusual both in its form and in the material of which it is made. Indeed this appears to be one of only a very few separate, wooden pommels known to exist. There is a group of ring-pommeled daggers with scale tangs, some but not all with rondel hilts, of which a few have pommels formed around an expanded ringed end to the tang by two plates of wood separate from the grip plates. These include one excavated from the Englischer Garten in Munich (Münchener Stadtmuseum A

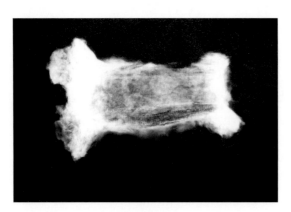

*Figure 9.51 Radiograph of 82A0975/2*

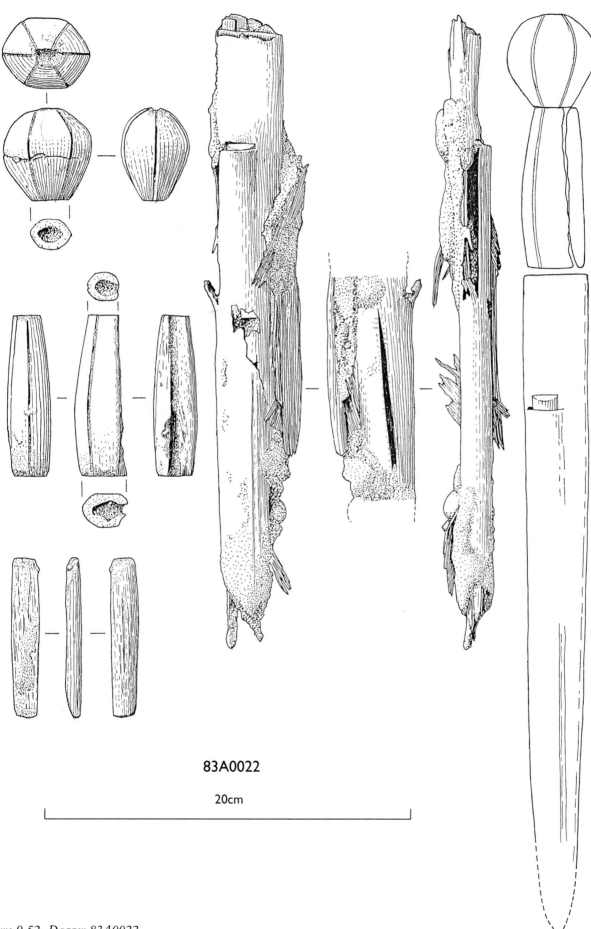

*Figure 9.52 Dagger 83A0022*

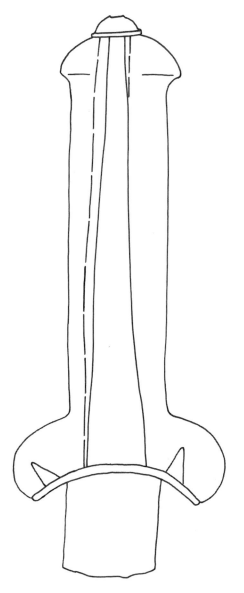

*Figure 9.53 Construction of ballock dagger grip, lobes with pins*

having a separate wooden pommel. The general form of this pommel seems to have been used for a very long time. One of similar octagonal section appears on a rondel dagger carved on an effigy in Much Marcle church, Herefordshire that has been dated to about 1400 (London Museum 1940, fig. 8, no. 6).

It seems clear from the incised line borders to the flats on both elements of the *Mary Rose* dagger that they were not intended to be covered. In 1983 A.V.B. Norman (pers. comm.) identified the pommel and grip as belonging to a '*Landsknecht*' dagger, a modern term often very loosely applied to a considerable variety of daggers which have been split into three main types (Peterson 1970, 44–6) and which were used especially in Germanic lands during the sixteenth century. Of the three types, the one that these parts could resemble is a form with the guard consisting of short quillons and a side ring and a pommel, often rather large, shaped like an urn, an egg, a pear, or an inverted cone.

However, even though the pommel under discussion may seem to fit this description, as this type of dagger seems to have been used predominantly in eastern Germanic lands in the last quarter of the century it should not be too readily assumed that the wooden pommel and grip from the *Mary Rose* belong to this type. Indeed, the pommel is large enough to be that of a sword and the grip fits within the largest group of detached grips found on the ship so the possibility that they may come from a sword should not be entirely dismissed. However, although the wooden pommel is quite heavy it might not be weighty enough to provide a good balance to the sword so the likelihood remains that the pommel and grip belong to a dagger, perhaps one with short quillons. Nevertheless, without the evidence of the guard it seems that a definitive identification is not currently possible.

## Ballock daggers
by Maggie Richards with Alexzandra Hildred and Guy Wilson

Ballock daggers were used in England from the second half of the fourteenth century. They can be seen in the *Luttrell Psalter* (BL Add. MS 42130) dating to the 1330s. One can be seen on the Wardieue brass of about 1360 in Bodiam church, Sussex (Mann 1962, 2, 373). It appears clear that, from the outset, ballock daggers were used in civilian as well as military contexts. Towards the end of the fifteenth century an accentuated phallic-form appears with a bulbous pommel and the ballock daggers from *Mary Rose* are of this general type (Fig. 9.53).

This is a form found frequently in England, usually with single-edged blades. Cut-outs in the guards on the *Mary Rose* examples shows them to have had single-edged blades. Similar examples been excavated from the Thames foreshore, including two in the Royal

72/335) and another in the National Museum of Wales that was found on the mound between Dolwyddelan Castle and Tomen-y-Mur, Merionethshire. On another of the group in the Royal Armouries (X.284) the pommel plates are integral with the grip. Wooden pommels made in this way, integral with the grip and guard, are not uncommon and are found, for instance, on a group of baselard daggers dating to about 1400 with pommels matching the short thick quillons of the guard (eg, Philadelphia Museum of Art no. 683 and another in the Metropolitan Museum of Art, New York; acc. no. 29.158.660; Peterson 1970, pl. 19). They also appear on many ballock daggers, including those of the distinctive English type found on the *Mary Rose*, on one from the Thames at Queenhithe that has a very distinctive spherical pommel (Museum of London) and on one in the Musée D'Armes, Liège that has a trifurcated crown-like pommel. What is unusual is

775

**Table 9.5 Ballock daggers (all boxwood unless otherwise indicated) (metric)**

| Object (by-knife) | Sector/ no. on Fig. 9.55 | Length | Pommel shape, width at lobes | Button, diam. | Tang hole top diam. | Tang hole lobe shape, size | Grip length, shape | Blade type, lobe fixings |
|---|---|---|---|---|---|---|---|---|
| 81A4568 | H2/1 | 122 | plain, 58 | staining | | triangular, 12 | 84, octagonal | single, 1 pin each |
| 81A2927* (81A2940) | H5/2 | 127 | plain, 56 | staining | 12 | triangular, 13 x 8 | 87, diamond | single, 1 pin each |
| 81A3914 | H5/3 | 121 | plain, 57 | round, 15 | concreted | key, 16 x 7 | 86, diamond | single, 1 pin each |
| 81A5642 | H5/4 | 119 | plain, 8 | round | 4 | 19 x 7 | 77.89, diamond | single,1 through each |
| 82A1510 | H7/5 | 125 | plain, 58 | staining | 8 | key, 13 | 85, diamond | single, 2 small pin holes under each |
| 81A2286 | H8/6 | 114 | plain, serrated decoration to edge, 65 | square | button | 17 x 10 | 77, octagonal | single, 1 diagonally drilled through |
| 82A0890 | H8/7 | 129 | plain, twisted segmentation at edge, 62 | round ?*in situ* | | 16 x 7 | octagonal | single, 1 drilled through |
| 82A0912 | H8/8 | 132 | hexagonal, 73 | – | – | – | hexagonal | single, 2 small pins |
| 82A1720 | H8/9 | 116 | plain, 59 | – | 10 | key, 16 x 9 | 79, octagonal | single, 1 diagonally drilled |
| 81A1914 (81A1915–6) | O3/10 | 120 | plain, 59 | staining | 7 | 29 x 11 | 79, diamond | single, 2 small pins underneath |
| 81A2577 | O3/11 | 126 | octagonal, fluted, 60 | staining | not visible | 21.5 x 10 | 81, octagonal | single,1 pin underside only |
| 81A4297 | O3/12 | 123 | plain, 61 | staining | 9 | 28 x 12 | 80, diamond | single, 2 pin holes outer edge |
| 80A1350 | O4/13 | 117 | plain, incised lines, 59 | staining | 8 | diamond, 20 x 11 | 85, octagonal | double,1 drilled each lobe through |
| 80A1337 | O5/14 | 105 | plain, 57 | staining | 10 | key, 19 x 9 | 74, diamond | single, 1 nail underneath |
| 80A0766 | O5/15 | 125 | plain, 71 | staining | 7 | key, 17 x 9 | 85, diamond | single, 2 pins underneath |
| 81A3174 (81A5666) | O6/16 | 120 | plain, 57 | staining | 10 | key, 16 x 10 | 84, diamond | single, 1 each lobe, drilled through |
| 81A3983 | O6/17 | 122 | plain, 55 | 14, height 8 | nail *in situ* | triangular, 17 x 8 | 87, diamond | single, 1 pin, underside only |
| 82A0760** | O7/18 | 113 | octagonal/fluted, 51 | square, 13 | 5 | triangular 18 x 10 | 80, diamond | single, 1 drilled through |
| 82A1916 (82A1916) | O7/19 | 124 | plain, 69 | staining | 6 | triangular, 25 x 11 | 86, octagonal | single, 1 drilled through |
| 82A2053 | O7/20 | 118 | plain, 60 | staining | 8 | triangular, 17 x 9 | 79, diamond | single, 1 drilled through |
| 82A0111, (82A0112–3) | O8/21 | 114 | plain, 54 | staining | 9 | rectangular, 17 x 11 | 85, ovoid | single, 1 drilled through |
| 82A0114 | O8/22 | 117 | plain, 48 | staining | 9 | key, 24 x 10 | 88, diamond | single, 2 pins underneath |
| 82A1525 | O8/23 | 119 | plain, 66 | staining | | 18 x 9 | 82, octagonal | single, 1 drilled through |
| 81A3894 | O8/24 | 108 | fluted edges, 63 | staining | 10 | 18 x 10 | 77, octagonal | single, 1 drilled through |
| 81A3882 | O9/25 | 122 | fluted/petals, 64 | staining | 9 | 29 x 12 | 84, octagonal | single, 3 pins 1 lobe, 2 other, under only |
| 82A4509 | M3/26 | 120 | plain, 69 | – | 14 | kite, 28 x 10 | 86.32, diamond | single, 2 pins each, under only |
| 83A0242 | M3/27 | 115 | fluted, octagonal | staining | 7 | 22 x 9 | 80.81, octagonal | single,1 drilled through |
| 80A1705 | M3/28 | 126 | petals, notches carved on 1 lobe, incised lines, 57 | *in situ*, octagonal, domed | – | 19 x 8 | 83.79, octagonal | single, 1 large each, under only |
| 81A0862 | M4/29 | 121 | plain, 57 | *in situ*, circular 13, height 6 | – | – | 86.03, diamond | single, blade frag *in situ* L. 32.72, W 25.89, Th 11.12 |
| 80A1378 | M6/30 | 122 | plain | *in situ*, square, 12 | – | – | 89.11, octagonal | single, 2 pins each, underneath |

776

**Table 9.5 (continued)**

| Object (by-knife) | Sector/ no. on Fig. 9.55 | Length | Pommel shape, width at lobes | Button, diam. | Tang hole top diam. | Tang hole lobe shape, size | Grip length, shape | Blade type, lobe fixings |
|---|---|---|---|---|---|---|---|---|
| 81A1570 | M6/31 | 125 | plain, chamfer underside, 59 | – | 8 | key, 16 x 7 | 90.04, diamond | single, small pin each lobe |
| 80A0423, (80A1221)** | M7/32 | 123 | plain | staining | | | 82.13, diamond | single, 2 pins underneath |
| 81A1304, @ (81A4324, 81A6955) | M9/33 in chest 81A1429 | 120 | octagonal, 64 | staining | 9 | 21 x 10 | 83.21, octagonal | single, 2 pins underneath |
| 81A3768, (81A3828)** | M9/34 | 116 | hexagonal, 55 | staining | 8 | 15 x 6 | 81.78, hexagonal | single,1 drilled through |
| 81A0811 | M10/35 | 117 | octagonal | *in situ*, 10, height 5 | | 16 x 8 | 82.57, octagonal | single, 1 drilled through; rivet *in situ* diam. 4.56, L. 16.74 |
| 81A0818 | M10/36 | 127 | octagonal | – | 11 | | 86.36, octagonal | n/a, half haft only |
| 78A0215, (78A0216)** | ?U1/37 | 119 | plain 60, 35 | | 7 | 29 x 10 | 84.38, diamond | single, 3 pins each, under only |
| 81A0239 | U4/38 | 125 | plain, 61 | 12 | 8 | | 89.63, diamond | n/a |
| 79A1000 sheath also | U5/39 | 126 | plain 44 | *in situ*, square, domed, 13, height 8 | | 29 x 10 | 90.03, diamond | single, 2 pins each, under only |
| 80A0583 | U7/40 | 123 | octagonal, 67 | – | 15 | 14 x 6 | 82, octagonal | ?double, 1 diagonal drilled through |
| 80A0585 | U7/41 | 124.62 | plain, chamfer under, 69 | staining | 8 | triangular, 31 x 12 | 86.47, diamond | single, 2 pins each, under only |
| 80A1109 | U8/42 | 115.53 | plain, chamfer under, 51 | staining | 10 | 29 x 12 | 79.51, diamond | single, 2 pins under only |
| 81A0734, 81A4104 | U9/43 | 125.10 | octagonal, 61 | staining | 7 | 17 x 9 | 76.21, hexagonal | single, 1 drilled through |
| 81A1251 | U9/44 | 120 | plain, chamfer under | staining | 8 | 19 x 10 | 88.40, diamond | single, NV iron staining |
| 81A1252 | U9/45 | 116 | plain, 64 | square, 12 | 7 | triangular, 29 x 11 | 79.78, circular | single, 6 pins around lobe edges |
| 81A1290 | U9/46 | 126 | octagonal, 64 | staining | 7 | rectangular, 18 x 8 | 85.42, octagonal | ?double1, 2 drilled through |
| 81A1422 | U9/47 | 110 inc. | plain, 60 | staining | 6 | 20 x 8 | 70, circular | single, 1 pin hole 1 lobe, 2 on other |
| 81A1665 | U9/48 | 110 | plain, 45 | square stain | 6 | 10 x 7 | 78.14, circular | single, 1 drilled through |
| 81A2197 | U9/49 | 109 | fluted, petals, 56 | staining | 8 | triangular, 16 x 9 | 79.19, diamond | single, 1 pin hole underside |
| 81A4517 | U9/50 | 117 | plain, 58 | staining | 7 | 16 x 8 | 89.02, diamond | single, 1 pin underside |
| 78A0334 | ?U10/51 | 134 | plain, 67 | – | 8 | 32 x 11 | 105.53, diamond | single, 2 pins, under only |
| 81A1412 | U10/52 | 124 | plain, 71 | round, domed | | 19 x 7 | 89.03, diamond | single, 1 pin centre each lobe |
| 81A0073, (81A0233) | U10/53 | 123. | plain, 60 | staining | 13 | 18 x 7 | 86.30, diamond | single,1 pin underside |
| 78A0124** | U11/54 | 133 | octagonal, 67 | *in situ*, square | | 18 x 9 | 77.02, octagonal | single, hole diagonally through |
| 81A0051 | U11/55 | 128 | octagonal, 67 | staining | 9 | 23 x 8 | 82.09, octagonal | ?double, 1 drilled through |

Table 9.5 (continued)

| Object (by-knife) | Sector/ no. on Fig. 9.55 | Length | Pommel shape, width at lobes | Button, diam. | Tang hole top diam. | Tang hole lobe shape, size | Grip length, shape | Blade type, lobe fixings |
|---|---|---|---|---|---|---|---|---|
| 82A1572 | Bow bulkhead/ 56 | 129 | octagonal, 74 | staining | 8 | 18 x 10 | 84.30, octagonal | ?double, 1 drilled through |
| 82A3257 | ?PS3/57 | 115 | plain, 59 | staining | 13 | 19 x 8 | 84.93, diamond | double, 1 pin underside |
| 82A1704 | PS10/58 | 121 | plain, 47 | staining | 6 | 17 x 9 | 85.24, diamond | single, 1 pin, under only |
| 81A0426 | SS2/59 | 125 | plain, 58 | in situ 10, height 5 | 6 | 21 x 8 | 84.56, octagonal | single, 1 pin, under only |
| 81A1845, (81A1845) | SS2/60 | 132 | plain, 63 | – | 9 | 31 x 13 | 87.3, diamond | single, frag in situ, 2 pins underneath |
| 82A0006 | SS6/61 | 134 | plain, 62 | staining | 7 | 29 x 11 | 91.94, octagonal | single, 2 pins under only |
| 81A5873 | ?SS8/62 | 124 | plain, 56 | staining | 10 | 16 x 6 | 89.56, diamond | single, 1 pin under only |
| 82A0958 | ?SS9/63 | 108 | plain, 49 | staining | 6 | 15 x 6 | 80.08, diamond | single, not recorded |
| 82A0015 | offsite/64 | 114 | plain | | 7 | 27 | 76.70, octagonal | single, 1 pin under only |
| 82A1604 | offsite/65 | 110 | plain, 54 | staining | 10 | 18 | 82, octagonal | single, drilled through |

\* = ash; \*\* = maple; @ = Pomoideae

Armouries discovered beneath Southwark Bridge (X.1704–5) together with a basket-hilted sword (RA IX.4427; Fig. 9.39b) similar to the *Mary Rose* example (82A3589), two in the Museum of London (nos 57.82, 7644 and A2445) and others in private collections. One, of octagonal section, in the Cambridge University Museum of Archaeology and Anthropology (no. 23.1156) is believed to have been found either in Hitchin or in London; another found on the site of Liverpool Street Station in 1865 is unusual in having a

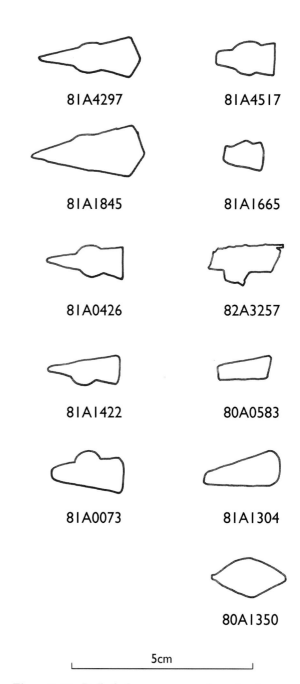

*Figure 9.54 Ballock dagger cross-sections showing tang hole shapes and wedge positions*

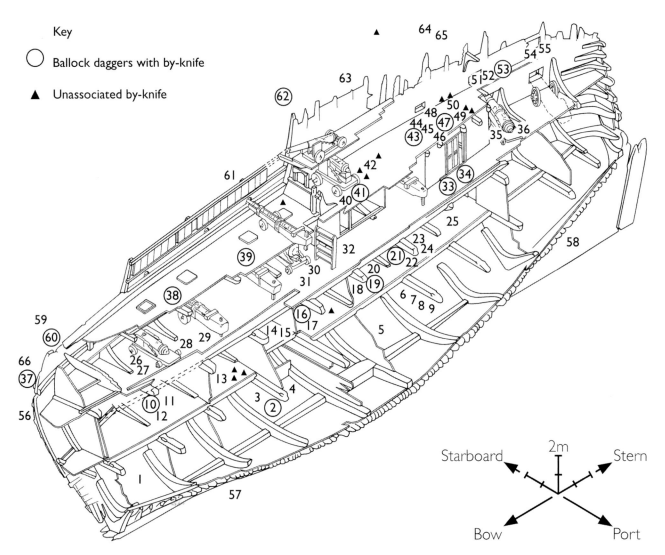

*Figure 9.55 Distribution of ballock daggers and associated by-knives*

bone hilt (Museum of London no. 7649) and the fragment of another, found at Westminster, is made of jet (London Museum [catalogue] 1940, 48, fig.10, 8). Where blades survive on these finds they are single-edged as are most on the *Mary Rose*, but one octagonal section dagger of this type in the Castle Museum, York has a double-edged blade with a ricasso. Only one ballock dagger recovered from the *Mary Rose* has a surviving double-edged blade (80A1350).

The hilts of ballock daggers are usually turned and carved from a solid piece of wood to include a bulbous pommel at one end and a pair of lobes at the blade end joined by the grip. A hole was drilled centrally through the entire length of the haft to take the blade tang. This hole is usually covered by a button at the top and an iron or copper alloy lobe guard under the base of the lobes that functions to secure the blade and prevent the haft from splitting. This is held with small pins or nails or two large rivets passing though each lobe (Fig. 9.53). The tang hole at the base is shaped to that of the blade and, in some most cases, provides the only information we have on the size and shape of the blades. Most are wedge shaped, with only one cutting edge, however double-edged blades do exist and leave diamond shaped holes. In some instances small iron wedges were used on either side of the blade to insure a tight fit to the haft (Fig 9.54).

The ballock dagger was worn within a leather sheath, sometimes strengthened with thin laths of wood, usually beech. A good quality sheath would have a decorated cap (chape), fitted on the end to prevent the tip of the blade piercing the leather. The sheath was usually suspended from a belt or girdle and occasionally had extra pockets to store ancillary knives or by-knives as discussed above (see Figs 9.26 and 9.32). Many of the *Mary Rose* examples were accompanied by by-knives carried in special pockets on the outside of their sheaths. In all, 29 by-knife hilts were found with or close to ballock daggers, enabling these to be paired with sixteen daggers (see Tables 9.5 and 9.6). By-knives not directly associated with ballock daggers are described in Table 9.7.

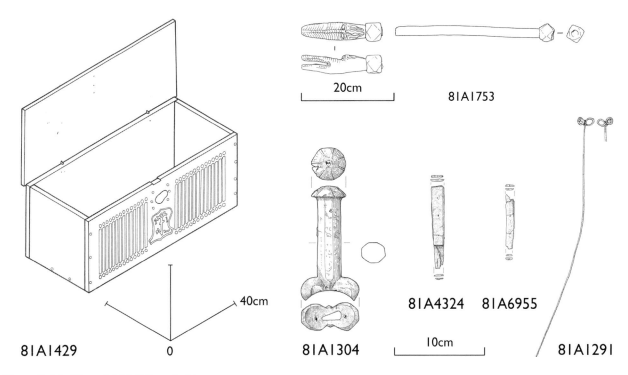

*Figure 9.56  Chest 81A1429 and objects*

**Table 9.6  By-knives associated with ballock daggers (metric; all boxwood unless otherwise stated)**

| Object | Sector/ location on Fig. 9.56 | Length | Width | No. rivets in scales | Notes | Associated ballock dagger (see Table 9.5) |
|---|---|---|---|---|---|---|
| 81A2940/1 | H5/2 | 82 | 14–7 | 4 | | 81A2927* |
| 81A2940/2 | H5/2 | 48 | 14–7 | 2 | ?pricker | 81A2927* |
| 81A1915 | O3/10 | 87 | 18–7 | 3 | | 81A1914 |
| 81A1914 | O3/10 | 54 | 12–7 | 3 | ?pricker | 81A1914 |
| 81A5666 | O6/16 | 78 | 14–8 | 3 | | 81A3174 |
| 82A1916/2 | O7/19 | 88 | 19–12 | 3 | | 82A1916/1 |
| 82A1916/3 | O7/19 | 88 | 13–7 | 3 | | 82A1916/1 |
| 82A1916/4 | O7/19 | 42 | 10–5 | 3 | ?pricker | 82A1916/1 |
| 82A0112 | O8/21** | 68 | 13–11 diam. | none | whittle tang knife | 82A0111 |
| 82A0113 | O8/21** | 68 | 14–10 diam. | none | whittle tang knife | 82A0111 |
| 80A1221 | M7/32 | 90 | 17–8 | 3 | | 80A423** |
| 81A4324 | M9/33@@ | 87 | 12–5 | 4 | | 81A1304@ |
| 81A6955 | M9/33@@ | 52+ | 10–5 | 3? | incomplete haft | 81A1304@ |
| 81A3828 | M9/34# | 87 | 12–7 | 4 | | 81A3768** |
| 78A0216 | U1?/37@@ | n/a | n/a | n/a | haft missing | 78A0215** |
| 81A0239/2 | U4/38 | 95 | 14–8 | 4 | | 81A0239/1 |
| 81A0239/3 | U4/38 | 41 | 13–5 | 3 | ?pricker | 81A0239/1 |
| 81A0239/4 | U4/38 | 83 | 20–8 | 4 | | 81A0239/1 |
| 81A0239/5 | U4/38 | 76 | 16–9 | 4 | | 81A0239/1 |
| 79A1000/2 | U5/39 | 215 | 15–12 | none | knife carved from solid piece of wood handle & blade = ?fake knife | 79A1000/1 |
| 79A1000/3 | U5/39 | 49 | 10–3 | 2 | ?pricker | 79A1000/1 |
| 79A1000/4 | U5/39 | 83 | 10–8 | 4 | | 79A1000/1 |
| 80A0588 | U7/41## | 87 | 15–8 | 3 | | 80A0585 |
| 81A4104 | U9/43 | 83 | 14–9 | 3 | | 81A0734 |
| 81A1437 | U9/47** | 51 | 10–5 | 3 | ?pricker | 81A1422# |
| 81A0233 | U10/53** | 52 | 13–10 | 4? | ?pricker | 81A0073 |
| 81A1845/2 | SS2/60 | 83 | 17–10 | 4 | | 81A1845/1 |
| 81A1845/3 | SS2/60 | 83 | 17–10 | 3 | ?pricker | 81A1845/1 |
| 81A1845/4 | SS2/60 | 75 | 13–10 | 4 | | 81A1845/1 |

* = ash; ** = maple; @ = Pomoidaea; @@ = beech; # = unidentified; ## = willow

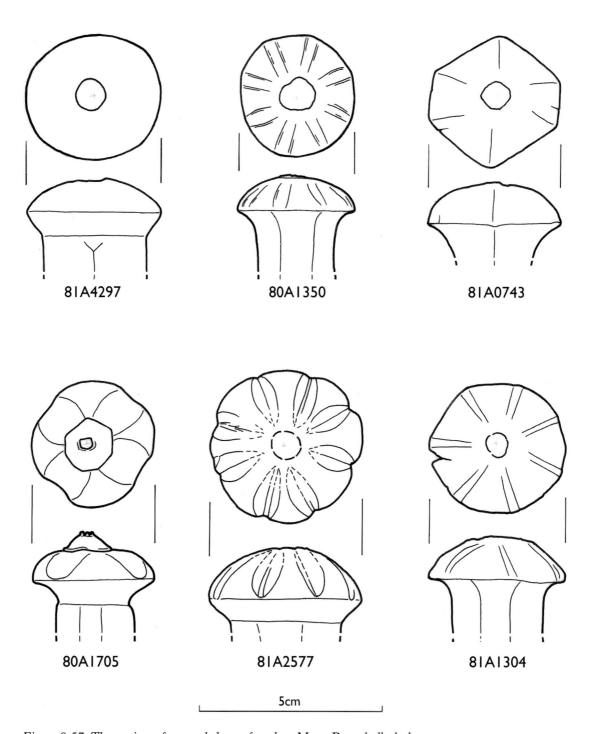

*Figure 9.57 The variety of pommel shapes found on* Mary Rose *ballock daggers*

The majority of sources depict the ballock dagger positioned behind the right hip of the wearer. For a right-handed man the left hip was the position in which to carry the primary weapon, a sword, with the right hip used for the secondary weapon, the dagger. Although found in a military context, the evidence suggests that these were personal possessions and not military issue. As stated above, both swords and daggers were considered to be items of dress. Weapons used in hand-to-hand combat were of necessity tailored to individual requirements. Size of haft, length of blade and weight of dagger were important matters of personal preference and the variation in size and form within the assemblage reflects this. However, there are some similarities suggesting the hand of the same cutler.

Sixty-four of the 65 ballock daggers were excavated in association with human remains indicating that they were being worn at the time of sinking (Figs 9.34,

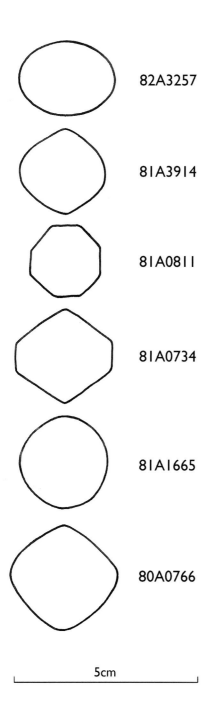

*Figure 9.58 The variety of grip sections found on* Mary Rose *ballock daggers*

Table 9.7 Unassociated by-knives (metric; all boxwood)

| Object | Sector | Length | Width | No. rivets on scales | Notes |
|---|---|---|---|---|---|
| 81A3243 | O4 | 90 | 10–8 | 4 | |
| 81A4254 | O4 | 85 | 25–19 | ?3 | incomplete haft |
| 81A4255 | O4 | 87 | 16 | 4 | |
| 81A5666 | O6 | 78 | 14 | 3 | |
| 82A4424 | U6/7 | 58 | 15–10 | 3 | incomplete haft |
| 80A0689 | U7 | 86 | 20–5 | 4 | |
| 80A1970 | U7 | 95 | 15–8 | 3 | found together |
| 80A1971 | U7 | 81 | 14–10 | 4 | |
| 80A1084 | U8 | 55 | 11–7 | 3 | ?pricker |
| 81A0849 | U9 | 84 | 13–8 | 4 | |
| 83A0571 | U9 | 50 | 12–5 | 3 | ?pricker |
| 83A0686 | U9 | 86 | 15–5 | 4 | |
| 81A2300 | spoil | 83 | 17–8 | 3 | |

A single dagger (81A1304) was found inside a locked, carved oak chest that may have belonged to a master gunner, found just outside the Carpenters' Cabin on the Main deck (81A1429), (Fig. 9.56). This was the only dagger haft to be made of fruitwood. The faceted shape of the haft is similar to an example found on the site of Salisbury House, London (Museum of London no. 48.9) and another in the Metropolitan Museum, New York (acc. no. 25.188:5). This example has a blade decorated at the forte and an ornate '*cuir bouilli*' sheath decorated with punch-work. It also has an English provenance and is recorded as having been found at a cottage at Wheldrake in Yorkshire (Dean 1929, no. 58, pl. xx).

Of the 65 ballock knives, 57 are made of boxwood, five maple (81A3983, 80A0423, 81A3768, 78A0215, 78A0124), one ash (81A2927), one of wood from the Pomoideae group which includes hawthorn, pear, apple and rowan (81A1304) and one remains unidentified (81A1422). There is variation in shape and decoration of pommel and grip and this is reflected in the dimensions. Pommel diameters range from 26mm to 41mm, grip lengths 76–105mm and mid-point diameters 19–33mm. The haft length varies between 105mm and 127mm (Table 9.7). Pommels are either round or are faceted to be hexagonal or octagonal. Round pommels are either plain (80A1350) or decorated with raised ridges (81A1304) or incised lines (80A1350) dividing the pommel into sections. The underside of these may be chamfered. Faceted pommels include complex floral patterns achieved by cutting and re-cutting the original hemisphere to form leaves or petals (81A3882, Fig. 9.60; 81A2577, Fig. 9.57).

9,35a, 9.55). This suggests that 36% of the individuals whose remains have been recovered were in possession of a ballock dagger. They were found on all decks in a similar distribution to other sword and dagger-related items, with the largest concentration, again on the Upper deck in U7–11 (sixteen ballock daggers and thirteen by-knives). Those on the Main deck correspond almost exactly to the Fairly Complete Skeletons identified by Stirland (Fig. 9.35a), with three in M3 associated with the large culverin stationed there.

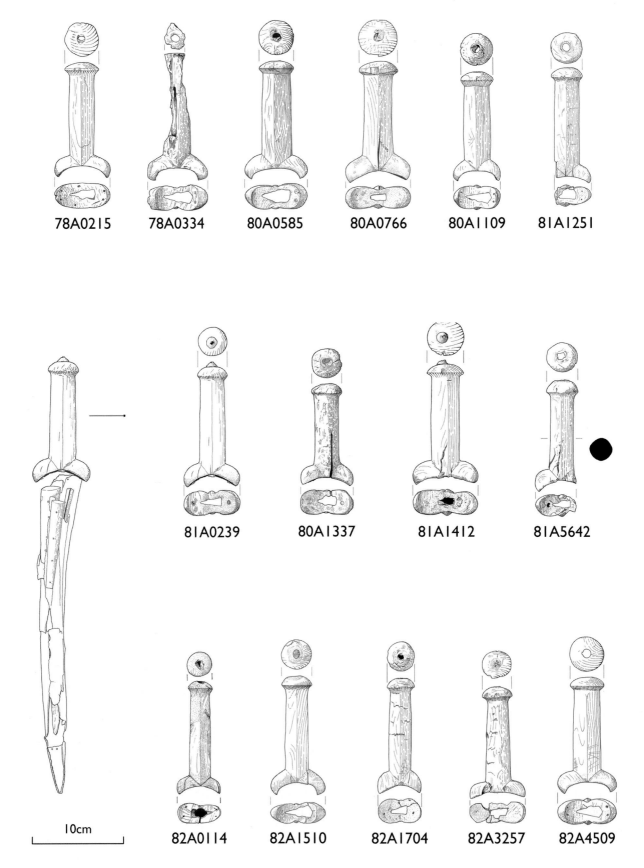

Figure 9.59 Diamond-sectioned ballock dagger grips

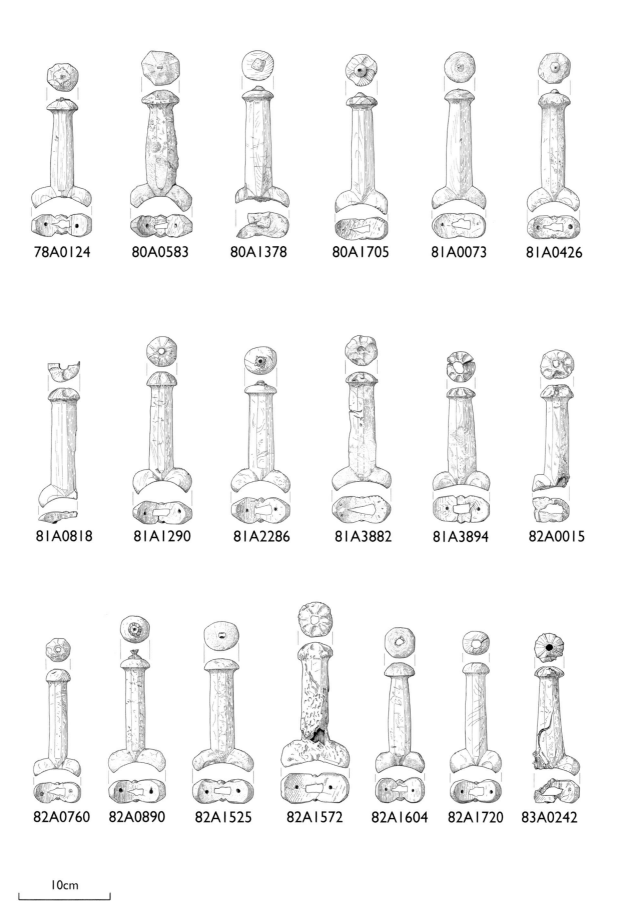

*Figure 9.60 Octagonal-sectioned ballock dagger grips*

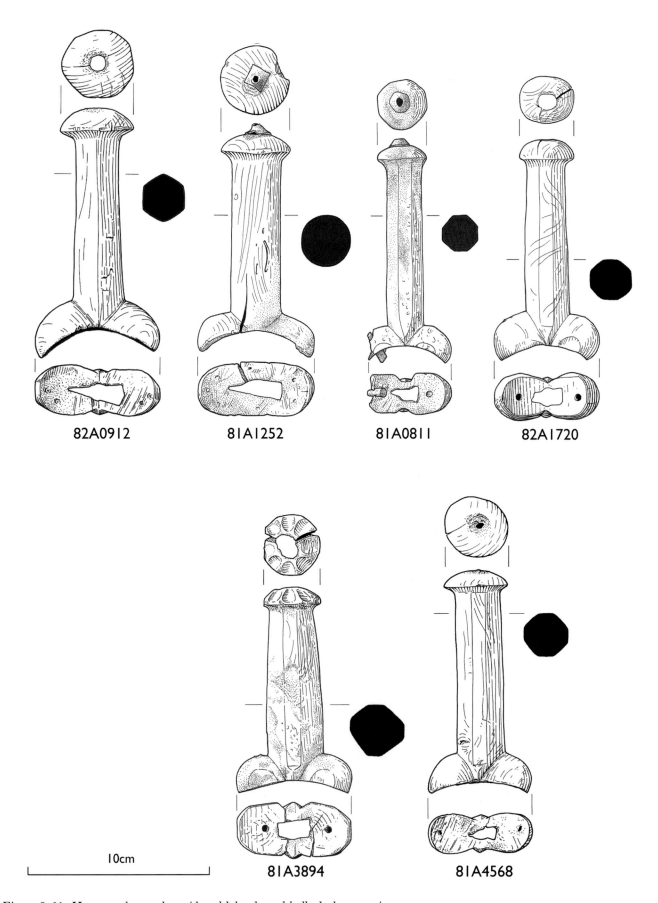

*Figure 9.61 Hexagonal, round, ovoid and lobe-shaped ballock dagger grips*

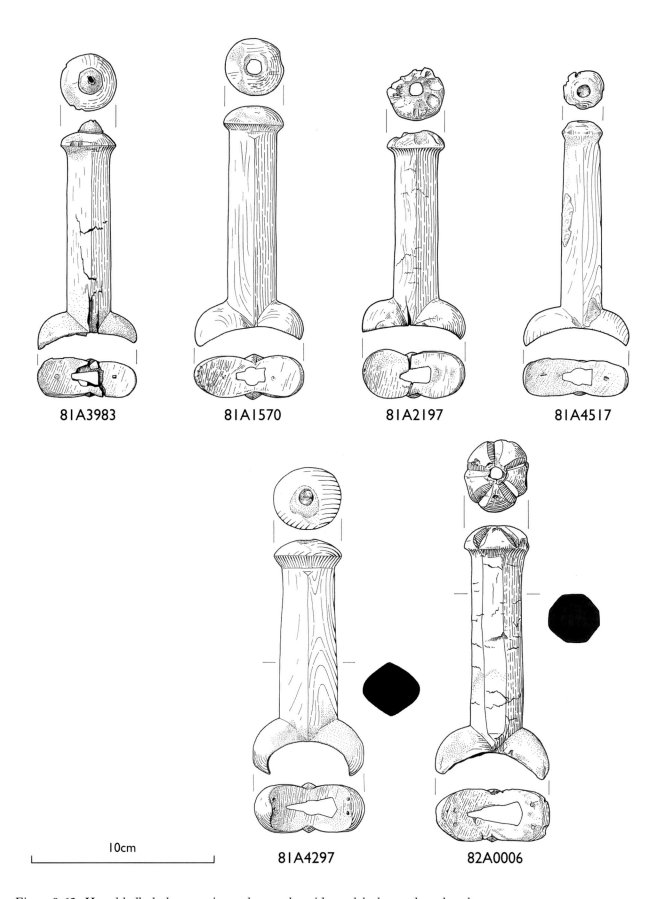

*Figure 9.62 Honed ballock dagger grips and examples with one lobe larger than the other*

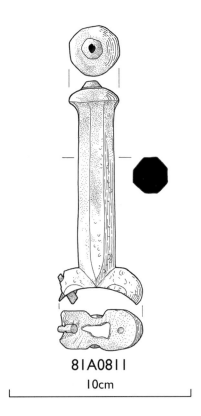

*Figure 9.64 Staining under lobes for ballock dagger guard of 81A1412*

*Figure 9.63 Faceted ballock dagger grip, plain pommel 81A0811*

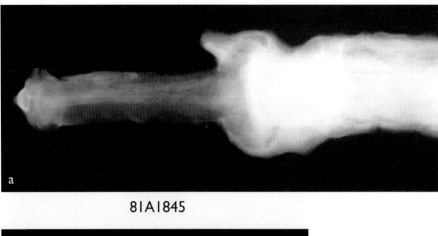

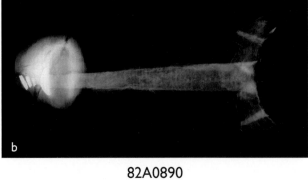

*Figure 9.65 (below) Radiographs showing pinning on a) 81A1845; b) 82A089*

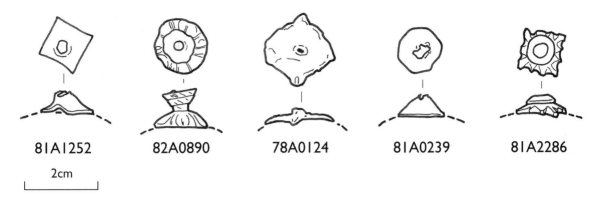

*Figure 9.66 The variety of tang buttons found on* Mary Rose *ballock daggers*

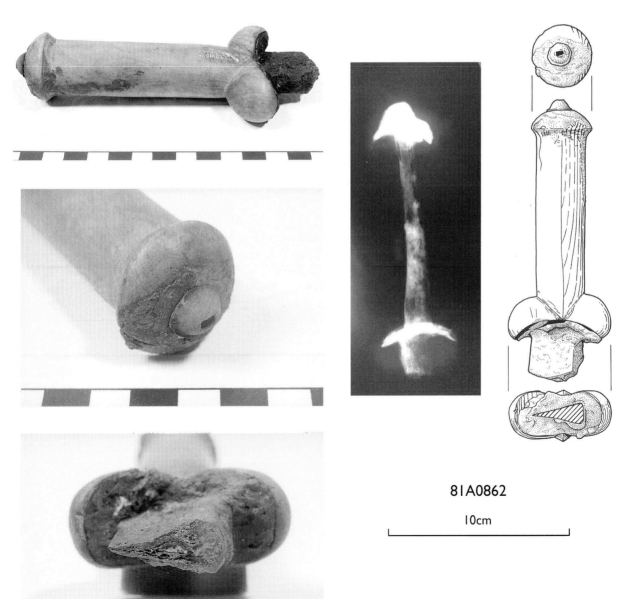

81A0862

*Figure 9.67 Dagger 81A0862*

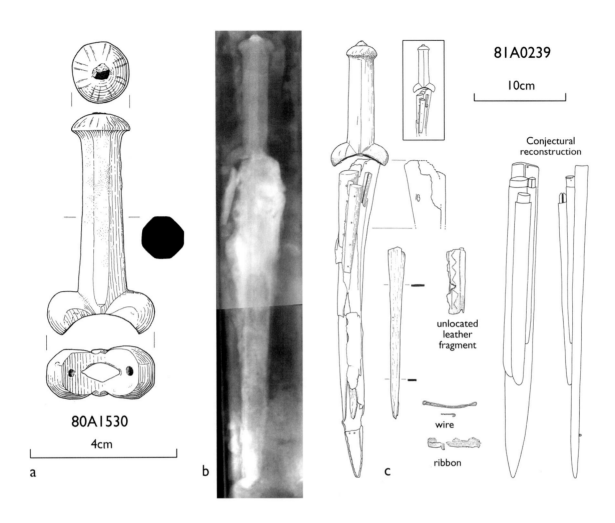

*Figure 9.68 a) drawing of hilt 80A1350 showing the guard cut-out to take a double-edged blade; b) radiograph of 79A1000; c) 81A0239*

Five distinct grip shapes are present, reflected in the cross sections (Fig. 9.58). Thirty-six examples (55%) have a diamond-shaped cross-section. The top and base points of the diamond form a central raised ridge. This runs down the length of the grip both front and back. The 'side points' of the diamond are rounded off for comfort. Twenty-four are faceted, resulting in an octagonal cross-section. Two are faceted but hexagonal and two have simple featureless round grips and one is ovoid. Diamond and faceted grips would provide a non-slip grip and these dominate the assemblage. The most comfortable is the diamond grip, any extra faceting causes the grip to sit less easily within the palm. The shape and size are ultimately related to the size of the owners hand and personal preference (Figs 9.59–61).

Two lobe shapes are present, round and faceted (Fig. 9.61). The round lobes have two variations depending on whether the tip of the lobe is blunt, resulting in a globular lobe (82A0006) or honed to give a sharpened appearance (81A4297) (Fig. 9.62). In the honed examples the points curve slightly inward toward the blade, producing the effect of an inverted pair of bull/buffalo horns. Usually the lobes are symmetrical, but there are four examples where one lobe is larger than the other (81A3983, 81A1570, 81A2197 and 81A4517 (Fig. 9.62). Faceting of the grip is not always reflected in the faceting of the pommel and lobes. There is one example where a faceted grip is topped by a plain pommel (81A0811; Fig. 9.63). The more prevalent grip form, the diamond shape, is usually topped with a plain pommel and horned lobes.

Residual iron-staining to the underside of the lobes suggests that most would have had guards of iron (Fig. 9.64). The presence of holes for either rivets or pins on each example confirms this. Radiography of a concreted example demonstrates the presence of a guard which covered the underneath of the lobes, but did not extend beyond the wood. Variations include: a single small nail hammered through the guard into the lobe for up to 8mm; two nails/pins into each globe; and six nails/pins at regular intervals around the underside of the edges of the lobes (Fig. 9.65a).

In multi-pinned examples it is not clear whether nails were entered through the lobe guard or if cuts were

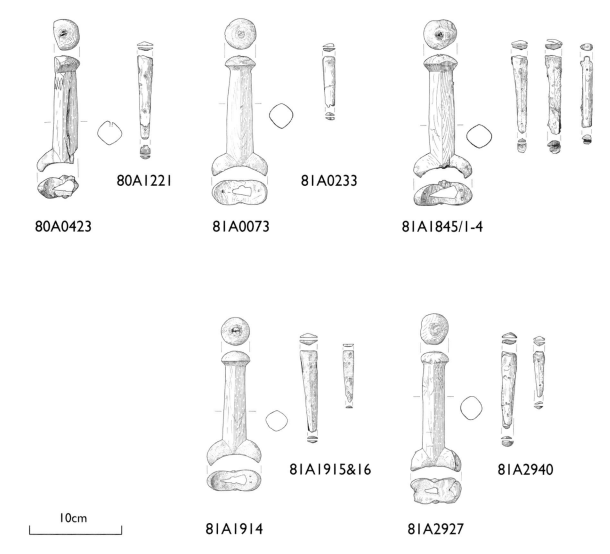

*Figure 9.69 Diamond grip ballock daggers together with by-knives*

made in the guard to accommodate them (Fig. 9.65b). These cuts could then be bent at right-angles to form the attachment points.

Rivetted lobe guards used one rivet through each lobe, in most cases through the entire lobe. One 15mm long rivet of 4mm diameter was recovered *in situ* from 81A0811. It is possible that riveted examples may have had decorative buttons (Dean 1929, pl. xx). In a few examples a drilled hole, comparable in size and appearance to the riveted examples, is found only on the underside of the lobe. Radiography indicates that the hole stops just short of the uppermost side of the lobe.

Buttons were either iron or copper alloy, a few of the latter survive. Shape can be ascertained from the pattern left in the wood. Examples suggest domed, square, round or faceted. The diameter is consistent at 12mm. Only two buttons retained any surface decoration (Fig. 9.66).

In most instances the dagger blade had totally disintegrated. One example retains a concreted blade portion (81A0862; Fig. 9.67). Examination of the shape and size of the tang holes reveals the maximum width of the blade and also whether it had a single (wedge-shaped), or a double (diamond) edge, or whether a single edge required additional supporting wedges (keyhole- shaped). Blade widths varied from 15–32mm. All but one are single edged. 80A1350 has a double-edged blade 20mm in width (Fig. 9.68a).

Radiography of concretions has revealed a few incomplete sections of ballock dagger blades. Together with the dimensions for sheaths this suggests a maximum blade length of 335mm, (Fig. 9.68b).

Most of the sheaths found with ballock daggers are fragmentary. Many were equipped with two by-knives, like the dagger from Hooke Court (see Fig. 9.33a, above) and some had more than two. The radiograph of 79A1000 (Fig. 9.68b) shows a sheath with one scale-tang by-knife, one possible pricker and a third, all-wood, by-knife. The best surviving sheath is 81A0239, although there remains some doubt about the association of the hilt with the blade and sheath. This bovine leather sheath is equipped with four by-knives,

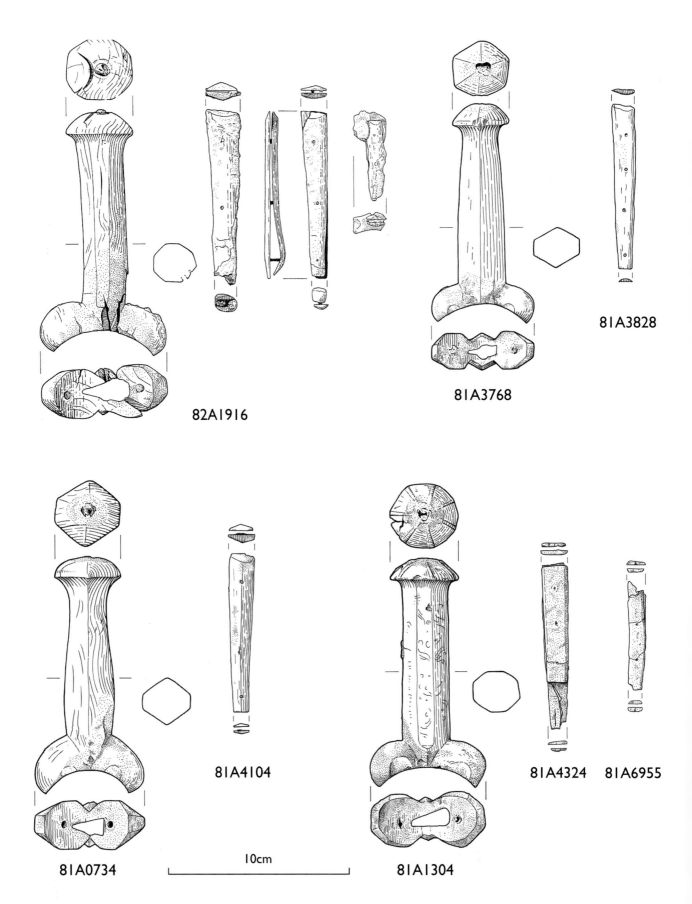

Figure 9.70 Octagonal, hexagonal, round and ovoid shaped ballock dagger grips with by-knives

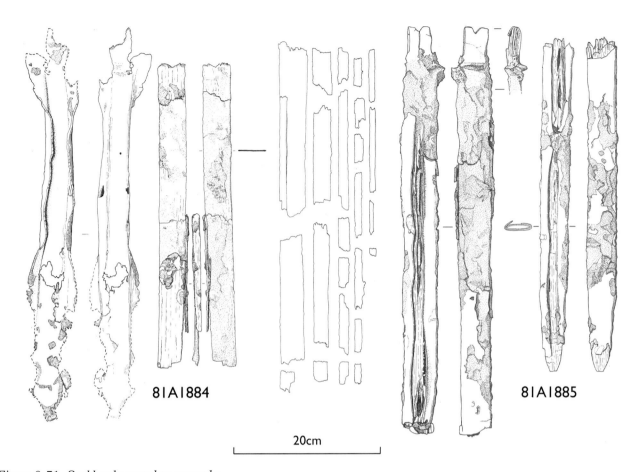

*Figure 9.71 Scabbards: complete examples*

although one is much smaller than the others and may be a bodkin (Fig. 9.68c). It is 335mm long and 30mm wide and bears a central panel 10mm in width carrying an incised geometric design comprising two parallel lines with a zig-zag design between them. It has two holes for a fixing staple on the inside near the tip. The leather is slit just below the mouth to accept by-knives/bodkins, the handles of which have survived (Tables 9.6 and 9.7).

Diamond shaped grip ballock knives are shown with their by-knives in Figure 9.69, and octagonal and hexagonal in Figure 9.70.

## Scabbards, sheaths, belts and hangers
by Alexzandra Hildred, Laura Davison,
Douglas M<sup>c</sup>Elvogue and Guy Wilson

### Scope and distribution
There are over 70 artefact numbers attributed to scabbards or sheaths. This total excludes any fragments thought to be ballock dagger sheaths (see above). Most are incomplete and consist of fragments of leather, wood and concretion. Some can be directly associated with hilts or grips (Table 9.8). Their distribution is similar to that of scabbards and grips (Figs 9.34, 9.35b).

The most complete examples include 80A1709 (M3), 81A1580 (M6), 80A0549 (U8), 81A1883 (U9), 81A1884 (U9) and 81A1885 (U9) (Figs 9.34, 9.71–4, 9.83, 9.85). These range in length from 635mm (80A0549, tip missing) to 955mm (81A1885) whilst the sword (82A3589) has a nearly complete blade 865mm in length. Most scabbards are between 710mm and 820mm in length (Table 9.8). The width at the top is 40–60mm. Two retain straps by which the scabbard is suspended; 80A1709 and 81A1567 (see Fig. 9.83, 9.85). Most of the identified leather is cowhide. The best quality came from calf leather, followed by bovine and finally cattle. Some of these were mixed within individual scabbards, for example 80A1709 contains elements of all three. Only one scabbard is not cowhide, 81A2932 (O4) is listed in archive as being of sheep skin but no details are available.

Eight are decorated: 80A0549 (U8), 81A1709 (M3), 81A1580 (M6), 81A0795 (U9), 81A1734 (U9/10), 81A1883 (U9), 81A4846 (unknown), 83A0151 (O8). Engraved decorations include lines of

## Table 9.8 Scabbards

| Object | Sector/Location on Fig. 9.35b | Leather | Wood | Decoration | Second sheath? | Hump | Length (mm) | Width (max) (mm) | Associations/description |
|---|---|---|---|---|---|---|---|---|---|
| 78A0338 | H1 | calf | * | no | no | – | 258 | – | 3 frags; poss side-stitched, small stitches to both sides of 1 frag; soft & smaller than most, poss. liner or dagger sheath |
| 80A1427 | H7 | – | elm | – | – | – | 800 | 40 | slat only |
| 82A1697 | H7 | bovine | * | no | no | no | 190 | 110 | 7 frags, leather & slats; inc tip, poss. with iron chape; back centre stitch visible |
| 83A0022 | O2/30 | * | * | incised | yes | no | 305 | 30 | dagger sheath; with grip (hexagonal section), wooden pommel; n o sign of seam, leather butts together at centre back; 64mm down sheath is slit for by-knife; tip missing |
| 81A1917 | O3 | none | beech | – | – | – | 80 | 28–33 | wood slats only, iron stained |
| 81A2932 | O4 | sheep | – | no | no | no | 440 | 60 | leather only; 7 frags; resting on top of skull 81H0346; incomplete |
| 82A1870 | O7 | bovine | * | – | – | yes | 275 | – | tip portion, showing shoulder for metal chape; slats present on 1 side |
| 82A1916 | O7 | * | not visible | no | yes | – | 108 | 25–12 | from radiograph; upper portion in concretion; 3 by-knives |
| 82A1932 | O7 | * | * | no | ? | ? | 240 | 50 | grip, guard, pommel, blade in concretion |
| 83A0151 | O8 | bovine | * | engraved | no | yes | 465/310 | 42/30 | 2 frags; lines in geometric design; concreted |
| 83A0699 | O8 | – | – | – | – | – | – | – | concreted; frags with chape |
| 81A1521 | O9 | bovine | maple | no | poss. | no | – | – | leather & wood slats; by 81H0173 above chest full of arrows; iron stained; back centre seam; poss. length of strap also associated, strap 81A4536 found in same area, could be part of same hanger |
| 81A3208 | O9 | * | * | – | – | – | 445 | 40 | 4 pieces leather & slats, 16 frags; in front of chest 81A3285; records suggest handle present & thin long green material strip poss. for suspension of sword; ?centre stitched (part of sample 81S1234) |
| 81A2977 | O10 in chest 81A2941 | bovine | * | – | – | – | 370 | 30 | frags, mostly wood slats; with grip 81A2979; braid 81A475; 1leather originally, now lost |
| 82A0970 | O10 in chest 82A0942 | cattle | wood | no | – | – | 150 | 32 | 14 wooden slats, 12 frags; leather; leather over 3 slats; with wristguard, inset dice shot, knife |
| 81A0133 | M1/10 | bovine | * | no | no | no | 370 | 45 | leather, grip, pommel, scabbard, slats; centre seam; 2 holes near top on opposite sides with 2nd hole 2, further down on 1 side |
| 83A0435 | M2 | bovine? | * | no | – | – | 410 | 60 | in concretion with two pommels 83A0432–3 & grip 83A0436; at base of companionway, remains of 2 swords; prob. part of 83A0433 & 0436 |
| 80A1865 | M3 | – | – | – | – | – | 24 | – | 2 frags metal & leather frags., corroded scabbard |
| 80A1709 | M3/31 | bovine | beech | engraved | no | yes | 820 | 40 | hanger, chape, helt buckle, hanging straps; associated with 80H0223; linear decoration on front; back centre stitched; no staining of wood, poss. sword not in sheath; complete |
| 80A1817 | M4 | none | elm | – | – | – | 800 | 40 | slats, leather & chape only |
| 81A0864 | M4 | * | beech | no | – | – | 70 | 30 | 8 frags; associated with ballock dagger 81A0862 |
| 81A1006 | M4 | calf | * | no | – | – | 230 | 30 | flesh side insid; incomplete; stitch-holes on edges of some frags, wide apart |
| 83A0572 | M5 | * | * | – | – | – | 330 | – | 4 frags |
| 81A1567 | M6/32 | calf | * | no | no | no | 510 | 40 | iron staining to tip suggesting iron chape; pomander attached (81A4396) & bifurcating slings & belt; stitching back centre; poor condition but relatively complete in length |

| Object | Sector/Location on Fig. 9.35b | Leather | Wood | Decoration | Second sheath? | Hump | Length (mm) | Width (max) (mm) | Associations/description |
|---|---|---|---|---|---|---|---|---|---|
| 81A1580 | M6 | bovine | * | engraved | no | no | 820 | 50 | many crossing lines; Star of David top centre; centre back seam |
| 81A1621 | M6 | – | – | – | – | – | 220 | 65 | iron frags with chape |
| 81A5605 | M7 | calf | beech | – | – | – | 75 | 26 | fragmentary |
| 83A0050 | M7 | – | – | – | – | – | 188 | 80 | concretion with wood, associated with grip 83A0048 |
| 83A0678 | M9 | * | * | – | – | – | 140 | 30 | fragmentary |
| 83A0544 | M12 | * | * | – | – | – | 60 | 25 | frags in concretion |
| 83A0547 | M12 | * | * | – | – | – | 450 | 40 | leather, concretion wood; stitching centre back |
| 79A0483 | U3 | * | * | – | – | – | 270 | 40 | scabbard frags, grip |
| 79A1058 | U4 | bovine | * | – | – | – | 47 | 35 | ?throat frag |
| 79A1111 | U4 | bovine | * | – | – | – | 470 | 36 | slats only present now, originally leather & concretion |
| 79A1146 | U5 | bovine | * | – | – | yes | 765 | 45 | nearly complete scabbard (slats & leather, some leather missing) hump just below slit at top for by-knife; centre back stitching & secondary hump towards tip, below is concretion outside leather suggesting corroded chape |
| 80A0558 | U7 | cattle | * | no | – | – | – | – | 3 frags, iron corrosion |
| 80A0592 | U7 | cattle | – | – | – | – | 95 | – | incomplete, 'tubular' |
| 83A0510 | U7 | goat | – | – | – | – | 215 | 22 | 2 frags |
| 80A0330 | M/U8 | cattle | – | incised | – | – | 770 | 20 | lines cut into edges; v. poor |
| 80A0549 | U8/33 | * | beech | embossed | ?yes | yes | 635 | 44 | nearly complete, leather, slats, tip missing; hump 60mm below top, 2 by-knife pockets; similar design to hanging straps 80A800 (U7); circles in sets of 4 or forming flower & wavy pattern on outside edge |
| 80A1968 | U8 | calf | wood | – | – | – | 770 | 35 | slat only remaining |
| 81A0795 | U9 | bovine | * | engraved | – | – | 350 | 40 | 5 frags leather & slats; geometric double lines forming series of 'X's underscored; v. similar to 81A1734 |
| 81A0808 | U9 | calf | * | – | – | – | 160 | 48 | 2 tip frags, 4 strap frags, prob. hanging straps; centre back stitching; strap 220x30mm (max.) |
| 81A0846 | U9 | * | beech | – | – | – | 60 | 33 | fragmentary, no leather; near 81H0113/4, knife handle 81A847 & leather 81A0848 |
| 81A1421 | U9 | bovine | * | – | – | – | 300 | 35 | 7 frags, slats only remaining |
| 81A1759 | U9 | bovine | * | – | – | – | 330 | 40 | wood, leather, slats; lower portion with broken tip frag. showing angular cut back from tip & hole on each side of seam 40mm up from tip |
| 81A1780 | U9 | bovine | – | – | – | – | 230 | 35 | 3 frags, poss. tip & including strap frags; found near arrows |
| 81A1882 | U9 | * | * | – | – | – | 393 | 44 | wooden fragments |
| 81A1883 | U9 | cattle | * | stamped | – | – | 700 | 40 | nearly complete scabbard, stitched centre back, spur 25mm below top; many motifs, possibly heraldic |
| 81A1884 | U9 | cattle | * | – | – | – | 710 | 44 | nearly complete scabbard; poor condition; slats separate (24 frags); centre back seam; with textile, near arrows |
| 81A1885 | U9 | bovine | * | – | – | – | 955 | 42 | heavily encrusted, nearly complete scabbard; centre back seam/laths 39mm wide 3mm thick |
| 81A2255 | U9 | bovine | * | – | – | – | 300 | 43 | leather degraded & orange stained, tip end present, lower section only |
| 81A6782 | U9 | none | * | – | – | – | 35–210 | 35 | 10 slat frags; associated with grip/pommel 81A0718–9 |

Table 9.8 (continued)

| Object | Sector/Location on Fig. 9.35b | Leather | Wood | Decor-ation | Second sheath? | Hump | Length (mm) | Width (max) (mm) | Associations/description |
|---|---|---|---|---|---|---|---|---|---|
| 81A1734 | U9/10/34 | calf | – | engraved | – | – | 675 | 60 | 'X's similar to 81A0795 with underscoring |
| 78A0267 | U10 | * | – | – | – | – | – | – | frags only |
| 79A1365 | ? | cattle | – | – | – | – | 185 | 90 | frags only |
| 81A0242 | ? | – | – | – | – | – | 298 | 43 | slat |
| 81A4846 | ? | bovine | * | stamped | – | – | 175 | 40 | decoration no longer visible |
| 82A4352 | ? | bovine | – | – | – | – | 340 | 50 | frags |
| 81A5743 | ? | * | – | – | – | – | 285 | 35 | metal, leather & rope |
| 82A4090 | ? | – | none | – | – | – | 220 | – | 10 frags |
| 82A4347 | ? | * | – | – | – | – | 220 | 80 | leather & metal |
| 83A0334 | ? | – | – | – | – | – | 222 | – | concretion |
| 83A0500 | ? | * | – | – | – | – | 395 | 25–60 | concretion & leather |
| 81A6337 | ? | – | * | – | – | – | 79 | – | leather with wood, 5 frags |
| 82A5162 | SS8 | * | * | – | – | – | – | – | 9 frags, not recorded |
| 82A5163 | SS8 | * | – | – | – | – | 260 | 110 | concretion, with 82A5152 from 82A4080 |
| 90A0112 | SS10 | * | * | – | – | – | 280 | 36–20 | 13 sheath & blade frags |
| 82A2671 | PS6 | – | – | – | – | – | – | – | scabbard with blade in concretion |

* Present, but not identified to species

dots, starbursts, dots arranged to form a diamond, a 'Star of David', wavy lines and simple lines arranged to make geometric patterns. Two examples, 81A1883 and 81A4846, appear to be decorated with a series of stamped motifs. No two are identical (Fig. 9.72–4). Only nine have had the wooden slats identified: six are beech, one maple, two are elm.

The degraded nature of many of the pieces makes even partial reconstruction of individual objects difficult, nevertheless an attempt has been made to show the distribution of relevant fragments (Fig. 9.34). Once again, the overall distribution mirrors that of other sword and dagger related items though few were associated with grips, pommels or chapes. The largest group was from U9 (twelve). Remains of up to fifteen scabbards were found widely scattered on the Main deck but mostly in M3–6 where four large guns were stationed. Eight were found on the Orlop deck, and only two in the Hold. Dimensions and associations can be seen in Table 9.8, and particular examples are discussed more fully below. Scabbard 80A1851 was found in chest 80A1413 in O4 together with two grips (80A1405 and 80A1838) and loose straps (80A1846, 80A1945). Chest 81A2941 in O10 contained scabbard 81A2977 and sword remains 81A2979.

### Historical importance of the assemblage
by Guy Wilson

The assemblage of scabbards and sheaths found on the *Mary Rose* and the related swords belts and hangers, are of considerable interest as so few, what might be called ordinary ones, survive from this period or are described in documents. Many of the swords and daggers recorded in the 1547 inventory (Starkey 1998) are listed with their accompanying scabbards and sheaths, the former term being used mainly for swords and the latter for daggers (see Figs 9.25–6). However, some swords have sheaths and some daggers have scabbards, as the terms seem to have been at least partly interchangeable. Most are listed as being of velvet (presumably a velvet cover over wooden laths) and only very few of leather. There is also one recorded as being made of fish-skin.

What survives on the *Mary Rose* are leather scabbards and sheaths, presumably an indication of a difference in normal practice between luxury weapons and those carried less as fashion accessories and more for use. As already described, a sword scabbard (80A0549, U8) is formed to accept two by-knives and another (79A1146, U5) to take a single by-knife while the dagger scabbards frequently housed two or more

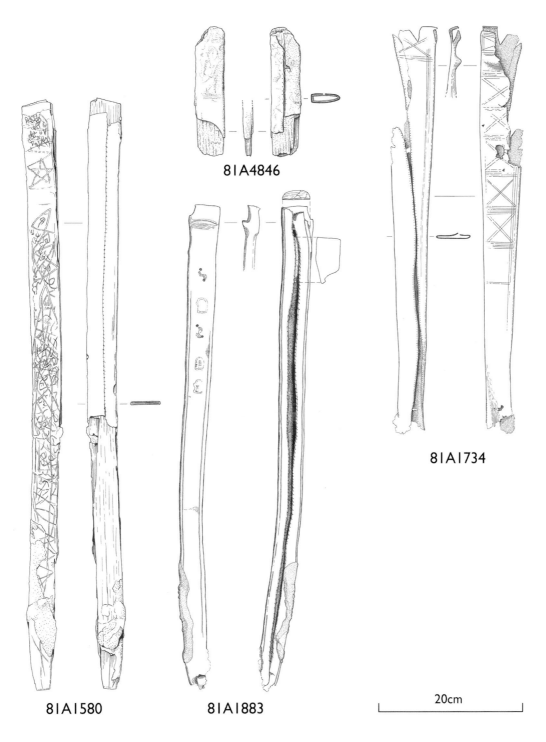

*Figure 9.72 Decorated scabbards*

(Fig. 9.73; 43 in total). The accompanying by-knives, etc, are accommodated by transversely slitting and longitudinally moulding the leather on the front of the scabbard to form a sheath between the leather and the wooden lath. There is evidence on at least five scabbards that the moulding was reinforced and prevented from deforming by supporting metal staples.

There is evidence from other surviving examples, from documentary evidence and from art, that confirms that this is not untypical. In the 1547 inventory about 20% of the scabbards and sheaths described had by-knives and bodkins and a very few were also equipped with a variety of other instruments including scissors, compasses, tweezers, a ruler, and a whetstone (Starkey

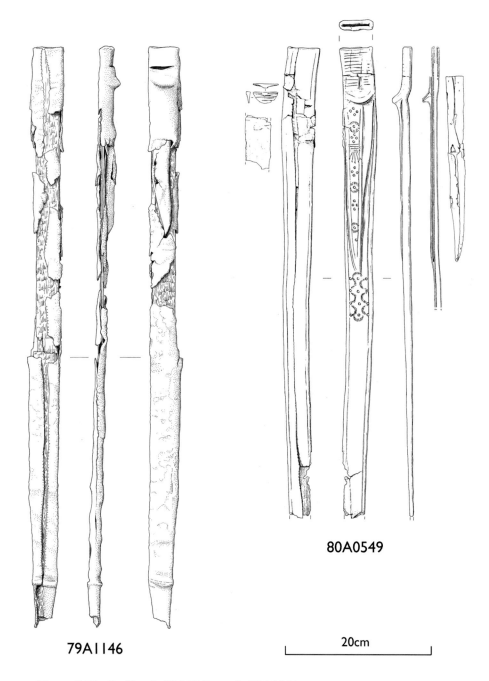

*Figure 9.73 Scabbards 80A0549 and 79A1146*

1998; see also discussion of leather and wood toolholders in *AMR* Vol. 4, chap. 8). It was not just the very wealthy who had daggers with by-knives and bodkins in their sheaths (see the bequest of the Essex yeoman Edward Trappes, quoted above, p. 560). Norman (1980, 310) suggested that where sword and dagger were worn together and one had by-knives, the other normally did not. There is no current evidence from the record of the assemblage on the *Mary Rose* to contradict this.

The main component of a scabbard is its outer leather sheath. Most, but not all of those from the *Mary Rose*, are lined with laths of wood. Wooden laths are present on some of the dagger sheaths, for example 81A0239 and 81A1780. Most, but not all, are stitched together on the inner face. The evidence suggests that both double and single lengths of thread were used to sew the butt seams with longitudinal stitches, the double giving a stronger and tidier seam that was traditionally more popular (Cowgill *et al.* 1987, 37). On some of the *Mary Rose* examples (81A1580, 81A0808), however, the butt seam is secured by transverse stitches using a single thread. These seams run down the middle of the back, except on one sword scabbard (81A1734,

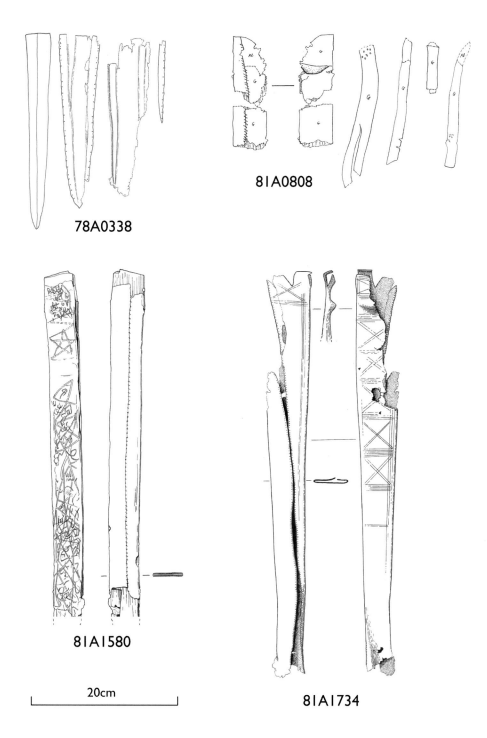

*Figure 9.74 Details of seam location and stitches 81A1580, 81A0808 and 78A0338, 81A1734*

U9/10) where it runs diagonally from one edge to another in its length. A second calf-skin example (78A0338, H/O1) appears to show stitch-holes along both sides, but this is fragmentary and thinner than the rest of the assemblage (Figs 9.74–5).

During the medieval period the most common form of decoration on English daggers and sheaths was engraving (de Neergaard 1987, 40). This was evidently still the case at the time that the *Mary Rose* was lost.

About 20% of the sword scabbards are decorated with engraved lines and punchings, the rest are either plain or too fragmentary to tell. Twenty-two per cent have a large transverse bump, or chin, just below the mouth to prevent the loops of the hanging straps from riding too high up or coming off altogether (Fig. 9.73, 80A0549). This chin is formed by inserting a shaped piece of wood between the leather and the wooden lath. Of the remainder, another 22% do not appear to have this

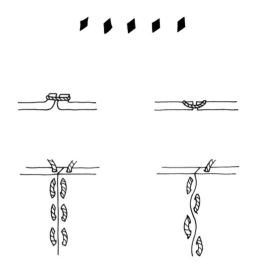

*Figure 9.75 Stitch types found on Mary Rose scabbards and sheaths (after Cowgill et al. 1987, fig. 9/10)*

expanded neck and it is impossible to be certain about the others. This transverse neck is also visible on at least one of the dagger sheaths (83A0151) and may be either for securing hangers like those used for swords or, more likely, a single cord or chain by which they were suspended vertically from the waist belt (Fig. 9.76).

The use of chapes to protect the ends of scabbards and sheaths seems to have been normal practice at the time – most of those described in any detail in the 1547 inventory have chapes (Starkey 1998) – so we should expect to find them on the scabbards and sheaths from the ship. Although the assemblage is fragmentary and interpretation difficult, some generalisations can be made. There is considerable evidence that all sword scabbards and some dagger sheaths had metal chapes to protect their ends, Table 9.9. Three chapes were found loose (81A2534 and 81A1851 (O4) and 83A0699 (O8)). Three survive on scabbard or scabbard fragments (80A1709 and 80A1866 (M3) and 80A1817 (M4)) and four were revealed by radiography (82A1697 (H7), 81A1621 (M6), 82A1870 (O7) and 82A1916 (O6). Of the three analysed, 80A1851 is leaded brass (75.5% Cu, 3.2% Fe, 19.5% Zn, 1.6% Pb), and two (80A1866, 80A1709) are quaternary alloys of copper (84.5/79.0% Cu, 3.2/3.0% Fe, 9.6/9.8% Zn; details in archive). The chape of 80A1817 is fixed by a copper alloy staple and similar staples, or the holes in the leather made by them, can be found on radiographs of scabbards and sheaths in concretion and of actual examples. Elsewhere there is some possible evidence to suggest that chapes could also be fixed by compression alone, or held with a simple staple which has deteriorated. There are several forms, depending on whether the join is in the centre of a flat face (83A0699), or at the edges (80A1817, 81A2534) (Fig. 9.77). These may be rounded over the blade tip with a straight top edge (80A1817), or shaped in the form of the tip and elongate with a scalloped upper edge (81A2534). Radiographs suggest a third variant, with a simple half semicircle cut across the top and blade tip cover showing an extended knob more exaggerated in form than 83A0699, as illustrated by 82A1697 (Figs 9.77, 9.78). Some are fragmentary (80A1866, 80A1851). Many of the scabbards show differential colouring or retain concretion on the tip suggesting that

**Table 9.9 Chapes**

| Object | Sector | Length | Width | Thickness | Metal analysis | Associations, description |
|---|---|---|---|---|---|---|
| 80A1709 | M3 | 50 | 25 | 10 (base) | quaternary alloy: Cu 79%, Zn 10%, Sn 4%, Pb 4%, Fe 3% | on scabbard; fragmentary |
| 80A1817 | M4 | 35 | 25 | 5 | | on scabbard; rounded base, straight top |
| 80A1851 | O4 | 54 | 22 | 10 | leaded brass; Cu 76%, Zn 20%, Pb 2%, Fe 3% | loose, on chest |
| 80A1866 | M3 | 11 | 3 | – | quaternary alloy Cu 85%, Zn 10%, Sn 3%, Fe 3% | fragmentary with sword frags; associated with 80A1709 & 80H0023 |
| 81A2534 | O4 | 50 | 25 | 3 | | loose; tapered with scalloped top |
| 83A0699 | O8 | 52 | 32 | 2 | | loose; tapered with central join |
| 81A1621 | M6 | – | – | – | | from radiograph; semi-circular, elongate end; looks similar to 80A1709 |
| 82A1697 | ? | 70 | 35 | – | | from radiograph, semi-circular, elongate end |
| 82A1870 | O7 | – | – | – | | from radiograph, sword scabbard with iron chape |
| 82A1916 | O6 | – | – | – | | from radiograph, on ballock dagger sheath; like 80A1709 |
| 04A0038 | Bow excav. | 48 | 26 | 11 | | found east of stern |

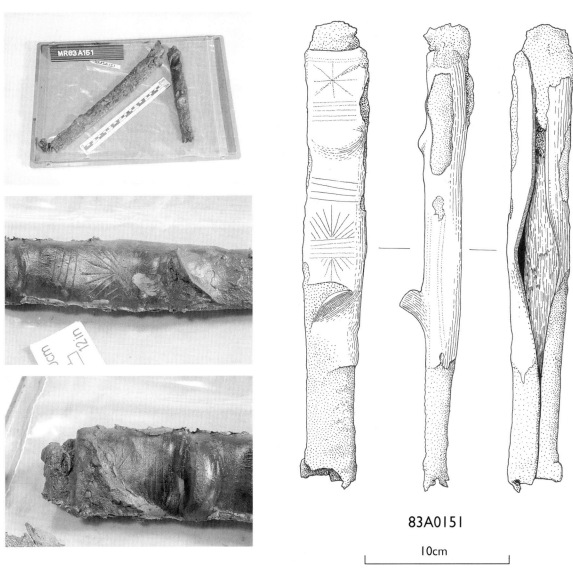

*Figure 9.76 Scabbard 83A0151*

chapes were originally present (for example 79A1146, 81A1567; Fig. 9.79) and, indeed, many are broken at this junction (80A0549 (Fig. 9.73), 81A1521, 81A1734). Concretion staining on the tip of scabbard 81A1759 (Fig. 9.80) could be interpreted as coming from an iron chape enclosing it rather than corrosion from the blade itself. Further confirmation are the fixing points, the inside of the scabbard is pierced near the tip with two holes to take a staple.

On at least two scabbards/sheaths (81A1759, 81A1780) the sides have been cut-away at the tip (Fig. 9.80). This would make no sense unless it was done to allow the fixing of a chape, as it would have exposed the edges of the blade to dirt and wet. What survives of 81A1780 is now in two parts and, as the upper part reaches a maximum width of only 32mm, it is almost certainly the sheath of a dagger not the scabbard of a sword. A radiograph of scabbard 82A1870 clearly shows the straight upper edge of a chape, seemingly of iron, running across the leather and a copper alloy fixing staple (Fig. 9.81a). A little below this there is a curious transverse raised moulding in the leather that may have been formed by pressure of a moulded chape or made intentionally to help secure one.

At least one dagger sheath (81A0239) seems to have been equipped with an iron chape. A radiograph of it in concretion shows a copper alloy staple passing through the outer side of the sheath near the tip, the only likely purpose of which must have been to secure a chape. The tip is missing and the sheath below the staple more damaged than above it, both consistent with the effect of additional rust damage from an iron chape.

However, certainly not all the dagger sheaths from the *Mary Rose* had chapes. The wood-reinforced leather tip of one (81A2255) shows no evidence and has been clearly carefully wrapped round to form a protective

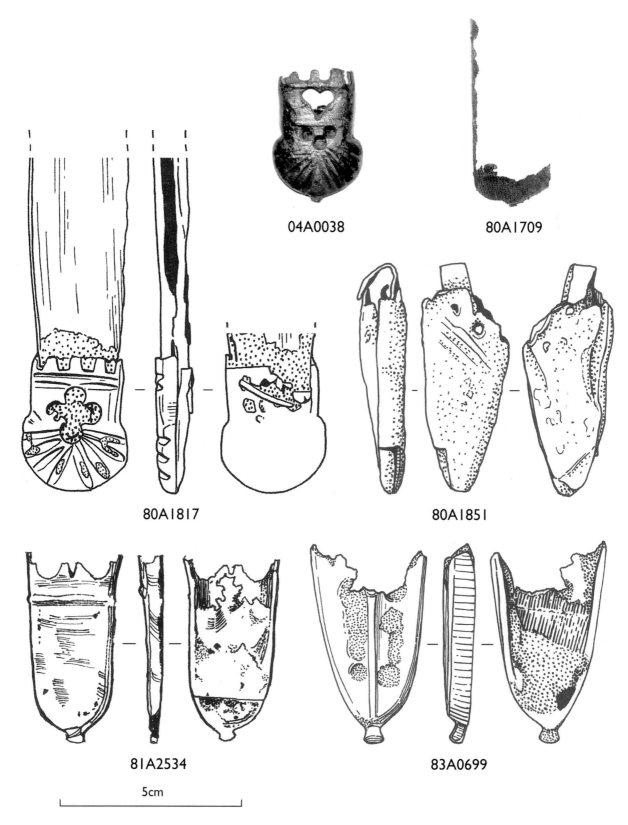

*Figure 9.77 The range of chapes found on* Mary Rose *scabbards and sheaths*

end in contrast to that described above that had been cut away (Fig. 9.80).

The mouths of the scabbards and sheaths, do not, generally, seem to have had similar metal reinforcement to that of their tips. There is little evidence for metal lockets to protect them although these had been in use from the fourteenth century (Blair 1962, 17). Where scabbard and sheath mouths survive in good condition

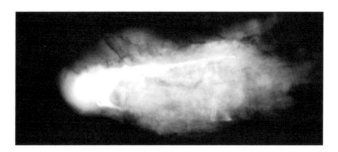

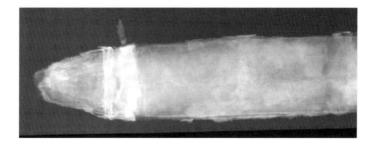

*Figure 9.78 Radiographs of chapes*

the leather is seen to be formed over the end of the laths, often cut into triangular tongues, and tucked down the inside (and presumably glued) to form a neat and secure entrance for the weapon. This can be seen, for instance, on the dagger sheath 83A0151 (Fig. 9.76).

There is, however, evidence that lockets were reasonably common in England, at least for the scabbards and sheaths of luxury arms. A royal cutler's bill dating from 1547 includes an item for gilding the new locket of a wood knife as well as its chape and pommel (Blair 1954, 109). In the 1547 inventory of Henry VIII's possessions about half of the scabbards and sheaths that are listed with chapes also have lockets or '*lockers*'. Holbein's cartoon for the lost wall painting in the Privy Chamber of Whitehall (painted 1536–7, destroyed 1698) in the National Portrait Gallery,

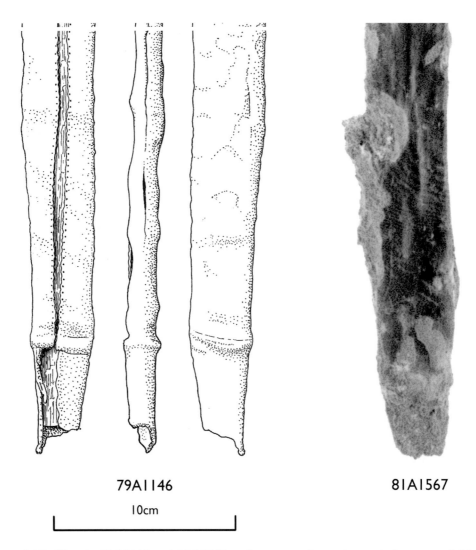

*Figure 9.79 Sheaths 79A1146 and 81A1567 to show concretion stain on tip from corroded iron chape*

London (4027), shows Henry VIII wearing a dagger with a sheath which has a metal chape and locket decorated to match the hilt (Roberts 1979, 82). This documentary and illustrative evidence is confirmed by at least one surviving example. A ballock dagger of the same type as those found on the *Mary Rose* was excavated with the iron top locket of its sheath from the moat of Hooke Court near Beaminster, Dorset (Dorset County Museum 1944–23–1; see Fig. 9.33a). The locket is of iron and has a decoratively fretted plate running down the sheath that protects the mouths of the slits for the by-knives. It also has an iron staple set across the back of the sheath just below the mouth for attaching the sheath to the belt by means of a cord or sash. Iron lockets for daggers equipped with by-knives have also been found on the Thames foreshore (eg, Royal Armouries X.588 and X.594(d)) but these cannot be dated securely and may well be Elizabethan.

Close examination of the finds from the *Mary Rose* reveals evidence to suggest that some scabbards and sheaths may have had lockets. 81A1734 has a mouth reinforced by the leather being bent over on itself but is weakened by triangular cut-outs in the centre (only that on the back survives) to accept a cross hilt with a triangular escutcheon, and by slits down each side (Fig. 9.74). Especially as no evidence of strengthening laths survives this would seem to have left the mouth impossibly weak and suggests that it must have originally been fitted with a strengthening metal locket. This interpretation is supported by the evidence of another sword scabbard (81A1885; Fig. 9.71). This also has triangular cut-outs in the centre of the mouth but, instead of the sides being slit, they have been cut away, the wooden laths and rusted remains of the blade being clearly visible from side-on at this point. Both scabbards have expanded chins to secure hanging straps and the slits or cut-outs stop above this as would be expected if they were associated with a metal locket. The width of 81A1885 just below the chin is some 50mm, above the chin where the sides have been cut away it reduces to

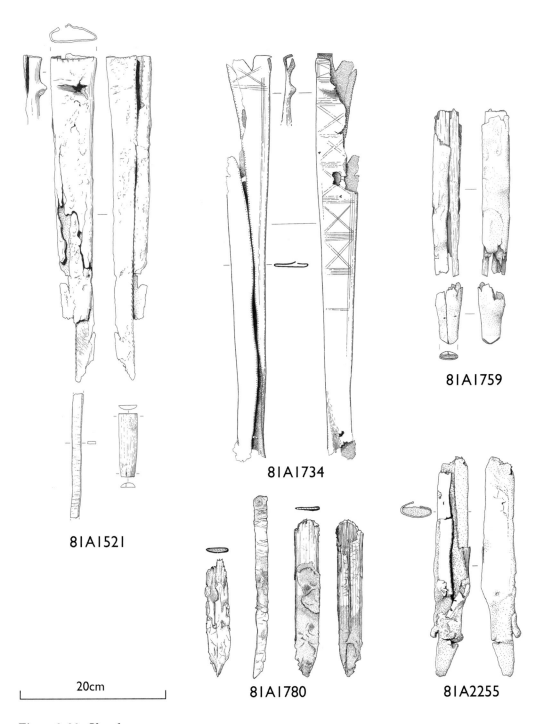

*Figure 9.80 Sheaths*

32mm showing that about 10mm has been removed from each side.

There is also evidence of one surviving locket on the radiograph of the blade and sheath of ballock dagger 82A1916. This shows the upper portion of the sheath with the hafts of two or more by-knives (Fig. 9.81b). Around this area, strengthening the sheath as on the Hooke Court dagger, can be seen the moulded rim of a locket and a large copper alloy staple either for fixing the locket or for attaching a suspension cord (see Fig. 9.33a). There is no evidence to suggest that it is fretted like the Hooke Court example and it may be more like the iron locket, with a mouth for an accompanying by-knife, found on the sheath of the woodknife made late in his reign for Henry VIII by Diego da Çaias (Royal Collection, RCW 61316).

Despite a considerable amount of evidence from early sixteenth century art, little is known in detail about how swords and daggers were secured to the clothing in such a way as to allow the user maximum

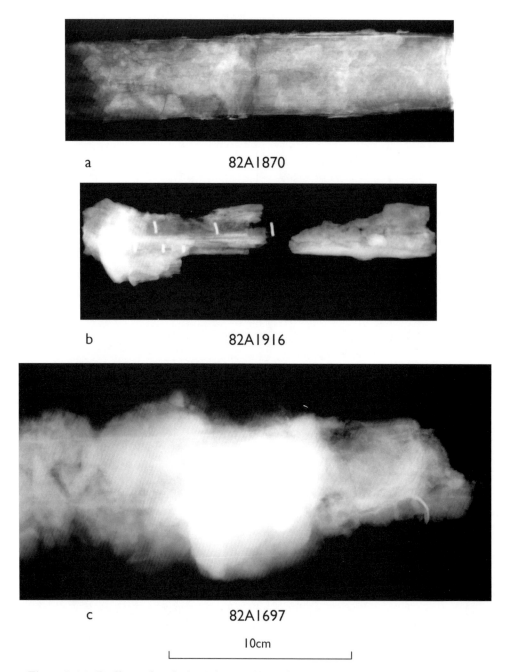

*Figure 9.81 Radiographs of: a) 82A1870; b) 82A1916; c) 82A1697*

mobility and comfort, whilst allowing for rapid deployment of the weapon. This is partly because, on a portrait or an effigy a hand, arm or piece of clothing can often obscure a vital part or detail of the suspension system; and partly because of the organic nature of many elements that has resulted in only a few examples surviving for study and comparison. The survival of a considerable amount of leatherwork on the *Mary Rose* offers an opportunity to learn a good deal more about this subject and the importance of the assemblage cannot be overstated. It includes, for instance, one hanger and belt (80A1709) that clearly represents an earlier stage in the development of what may be called the standard hanger of late Elizabethan and early Stuart times (Fig. 9.83).

In general terms it seems clear that there was a wide variety of suspension methods in use for swords and daggers. Swords were usually suspended from a waist belt, but the form of suspension varied considerably. The very old method of suspending the scabbard directly from a, usually rather loose, belt seems to have continued. It can be seen, for example, on the brass of John Ansty, dating from about 1460, in Stow-cum-Quy church in Cambridgeshire (Hartsthorne 1891, 338, fig.

68), on a Dürer engraving, *Young Couple Threatened by Death*, of 1498 (Strauss 1972, 40–1) and on an etching by Daniel Hopfer of Kunz van der Rosen that has been dated to about 1515 (National Gallery of Art, Washington, B 87) and that is also known from copies by other artists (Levenson *et al.* 1973, 522–5). That this simple method of suspension continued into the second half of the sixteenth century, at least for the lower classes, is shown by the paintings of peasant scenes by Pieter Brueghel the Elder (for instance *The Peasant's Wedding Dance* of 1566, Detroit Institute of Arts). What is often unclear is how the scabbard was attached to the belt as this is often hidden behind the scabbard itself.

Sometimes the belt was wound round the scabbard and secured in some way. This is clearly shown on the discarded scabbards in Pollaiuolo's engraving *Battle of the Nudes* that dates to the early 1470s (Levenson *et al.* 1973, 66) and on a Holbein pen and watercolour dating to the 1520s of *Sapor and Valerian* in the Basel Offentliche Kunstsammlung, Kupferstichkabinett (Roberts 1979, 27). Erhard Altdorfer's pen and ink sketch of two '*Landsknechte*' from the Robert Lehman collection (exhibited Metropolitan Museum of Art, New York, 1987), dated 1513, seems to show both a waist belt and a separate, looser, belt, which must surely have been attached to the waist belt in some way to prevent it falling off. This loose belt is wound round the scabbard and secured by a form of Turk's head lashing. A similar arrangement, but with a plain binding around the scabbard, can be seen in an early sixteenth century Franco–Netherlandish tapestry in the Burrell Collection, Glasgow (Reg. no. 46/60), depicting heron hunting with falcons.

Sometimes it was not the belt itself that was wound round the scabbard but a strap attached to it at both ends. It could be a single cord or leather attached at one end to the belt and then wound round the upper scabbard and secured. This is seen especially in illustrations of German *Landsknechts*, for instance Dürer's engraving '*Five Landsknechts and an Oriental on Horseback*' of 1495 (Strauss 1972, 12–13). This method of attachment had a long life and is clearly still shown in a number of woodcuts by Jost Amman that mostly date to the 1570s (Werner 1968, ix).

However, according to Norman (1980, 293) the '*commonest late medieval method*' of attaching the scabbard to the wearer was not by a single strap but by suspending it from the belt by two or more '*hangers*', usually leather straps tied, buckled or hooked to the left side of the belt and similarly tied round or hooked to the scabbard. Usually these hangers spring from the belt at one point. The *Beauchamp Pageant* (BL Cotton MS, Julius E. IV, f. 4r and v) of the late 1480s clearly shows them attached to the belt of both an archer and a fully armoured figure by a ring on the left side (see Sinclair 2003, 65–6). The kneeling, armoured figure in Piero della Francesca's painting *Pala dei Montefeltro* (Pinacoteca di Brera, Milan) dated 1466–74 shows two hangers wrapping round the scabbard that attach to the belt at the same point on the left side, and a strap running diagonally down from the right side of the waist to the top of the scabbard. Sometimes, however, two hangers appear in illustrations hanging down parallel to each other, showing that each was attached to the belt at a different point. This is shown, for example in the engraving by Jacopo Francia of the *Patron Saints of Bologna* (1515–20; Levenson *et al.* 1973, 500–1). They also appear on Albrecht Dürer's famous engraving *Knight, Death and Devil*, dated 1513, where the rear, or lower, one can be clearly seen to be a thong or cord, one end of which loops over the belt and the other circles the scabbard before passing though its own looped end (Fig. 9.82). Occasionally, instead of leather hangers metal chains were used that attached to upper and lower lockets on the scabbard. These can be seen on some of a group of Antwerp wood carvings for altars that have been dated to 1510–1520 (Musée Royaux d'Art de d'Histoire, Brussels, inv. nos 8653, 2515).

Where hangers are used to suspend the scabbard it was usually balanced and secured by a further, retaining strap attached near the top of the scabbard at one end and near the centre of the belt at the other. This method of suspension is shown clearly on a devotional portrait of Margaret of Denmark, Queen of Scotland (1456–86) by Hugo van der Goes in the Royal Collection (on loan to the National Galleries of Scotland, Edinburgh, no. G 13). It forms one of the three '*Trinity*' panels thought to have been painted for the Collegiate Church of the Holy Trinity, Edinburgh, between 1473 and 1478. Behind the kneeling figure of Margaret is a fully armoured, though bare-headed, St George. His sword is suspended from the left side of his waist by two hangers formed either from one or two cords, that are looped round the top and middle of the scabbard. How this cord or cords is secured to the belt is obscured. In addition there is a retaining strap that seems to be secured at one end to the inside of the scabbard near its mouth. At the other it ends with a buckle that attaches it to a shorter strap that somehow attaches to the waist belt near the centre of the wearer's waist.

Nearer in date to the *Mary Rose* Holbein's portrait of Sir Nicholas Carew, mentioned above, probably completed 1532–6, shows a single leather hanger on the left side (a second may be obscured by the hand) but, instead of a securing strap running up to the centre of the belt, the waist belt itself passes round the outside of the scabbard just below its mouth. According to Norman (1980, 293) the exception to this configuration of hangers and securing straps is when hangers are suspended from the left side of a loose hip belt, a fashion particularly associated with the '*Landsknechts*'. In this case he has found illustrative evidence that the belt itself, rising diagonally across the wearer's groin from left to right, took the place of the additional retaining strap. Norman calls the retaining strap the 'front sling' and provides evidence to suggest that in the

*Figure 9.82 above) Albrecht Dürer's engraving* Knight, Death and Devil, *dated 1513 (New York, Metropolitan Museum of Art); below) drawing from engraving of about 1500 by Giovanni Battista Palimba of St George fighting the dragon in the National Gallery of Art, Washington (© No B-8301)*

late fifteenth century this normally attached to the waist belt to the left of centre, but that gradually its point of attachment moved to the right. The fashion seems to have been set by Italy, where Norman cites illustrative evidence dating from 1525, and not to have become common in northern Europe before the middle of the century (*ibid.*, 293–4).

By Elizabethan times hooking and buckling had largely taken over from tying. Hooks were certainly in use in Europe to suspend hangers from the belt during Henry VIII's reign. The front retaining strap shown on the painting of a halberdier by Jacopo Carucci in the J. Paul Getty Museum, California (no. 89.PA.49), from the late 1520s, ends just below the right side of the waist belt with a triangular metal mount with a button finial. This clearly implies a hook and eye attachment and Norman records that this method of attachment is shown in Agnolo Bronzino's portrait of Guidobaldi II della Rovere of the early 1530s (Pitti Palace, Florence, no. 149). That this was also not uncommon in Tudor England seems to be proved by analysis of the 1547 inventory (Starkey 1998). Some of the sword belts (or girdles as they are called) are recorded as having buckles and pendants and a smaller number with studs as well. While it is uncertain what exactly the pendants referred to it is possible to hazard a guess. They seem to be decorated to match the buckles (eg, Starkey 1998, 14462, which had a '*pendaunte and buckle silver guilte*') and they seem to have been made of metal, either iron (*ibid.*, 14467 which had '*pendauntes locker and chape of Iron gilte*'), copper (*ibid.*, 14382: '*buckelles and pendauntes of copper guilte*') or silver (*ibid.*, 14380: '*Buckelles & pendauntes of Silver and guilte*'). This suggests that they may be the suspension hooks not the fabric or leather hanging strap.

Among the items found on the *Mary Rose* is a fixed ring that forms part of a decoratively pierced copper alloy plate with six holes by which it could have been attached to a belt. When found, hooked to the fixed ring were the two copper alloy finials with long, spatulate hooks that appear to have been intended to attach to straps. This could be the single point belt fixing for two sword hangers discussed above. A similar finial and hook were found attached to a tapering leather strap (81A3089) that could be a sword hanger (Fig. 9.87). In addition, a radiograph of the sword scabbard 82A1697 shows the lower part of the scabbard clearly with a chape at the end and, at the top of the surviving portion, what appears to be a wide loop of leather with a ring attached, (Fig. 9.81c).

However, the evidence of what was found on the *Mary Rose* suggests that tying leather straps together was still one of the most popular methods of strap attachment in the 1540s. Some of the belts and hangers also show evidence of attachment by buckles, rivets or eyelets and either Turk's heads, monkey's tails or manrope-type knots. Fragments that may or may not be parts of sword belts (for instance 79A1223 and 79A1174) also show that removable fixings through eyelets were combined in some way with permanent fixing by rivets. The girdle with buckle, pendant and three studs of gold that belonged to Henry VIII (Starkey 1998, 14362) may be evidence that attachment by studs not hooks was also still found on luxury arms to a limited extent.

The strapwork surviving from the *Mary Rose* also confirms the evidence of near contemporary portraiture, effigial sculpture and other illustrations that hangers were sometimes made integral with the sword belt by bifurcating the leather strap of which the belt was made. The alabaster effigy of Sir Edward Redman (c. 1510), in Harewood church, Yorkshire, shows his sword attached by a loose diagonal belt, that bifurcates into two hangers that wrap around the scabbard before (in some way hidden from sight) joining another strap that rises diagonally to the left of the wearer's waist (Routh and Knowles 1983, 46–51). A similar arrangement appears on one of two armoured figures on the brass of Alice Lawrence (d. 1531) and her two husbands in Middleton church, Lancashire; on the brass of William Ashby and his wife Jane in Harefield church, Middlesex, that dates from after her death in 1537; and in illustrations of the alabaster effigy of William, Lord Parr, in Horton church, Northamptonshire (Fig. 9.32). Lord Parr died in 1546 but the effigy shows both him and his wife and (probably, but not necessarily) from her death in 1555 (Blair 1962, 109f; Hartsthorne 1891, fig. 69). Nor was this arrangement a new idea. There had been a vogue for split belts for about 50 years beginning around 1270. One of the earliest is a Purbeck marble effigy in Braunston church, Northamptonshire that shows a trifurcated belt wrapping round the scabbard (Hartsthorne 1891, 328–31, figs 16–19). Some of the fixing arrangements were very complex and effigies and brasses show a wide range of methods of lacing, tying or just wrapping these belt ends to and around the sword scabbard. The style seems eventually to have given way to belts with metal ends hooking onto lockets on the scabbards. But it may be that the old methods of tying and lacing did not entirely die out.

These bifurcated belts work in reverse to those found on the *Mary Rose*, in which the belt bifurcates on the left side of the wearer's waist. A simple version of this arrangement on a loose hip belt can be seen in the engraving *Judgement Hall of Pilate* in the Museum of Fine Arts, Boston (H. AH9, illustrated by Levenson *et al.* 1973, 19) by Baccio Baldini (d. 1487) and also in an engraving of about 1500 by Giovanni Battista Palumba, of St George fighting the dragon, (National Gallery of Art, Washington B-8301; illustrated *ibid.*, 447; Fig. 9.82). The brass of Sir John Ansty, mentioned above, also shows a belt split in this way at the left of the waist, although the bifurcation is hidden by a large stud. As with the other type of split belts this type seems to have a long history.

A simple form appears on at least thirteen of the mid-thirteenth century illustrations in the '*Maciejowski*' Bible (Pierpoint Morgan Library, New York). Cornelius Cort's engraving *The Conversion of St Paul* (1576) shows a bifurcated belt very similar to those found on the *Mary Rose*, worn by a Roman soldier (British Museum V.8-146; Bury 2001, 111). The print is reversed so that the soldier wears his sword on the right side but clearly shows the waist belt bifurcating at the side of the waist, the lower bifurcation splitting again just before it attaches, by some means, to rings bound to the scabbard. They were also used for the suspension of daggers. The stone effigy at Whitchurch, Shropshire, of John Talbot, Ist Earl of Shrewsbury (d. 1453) shows a loose hip belt bifurcating from the left and the two ends joined by metal links to either side of a wide plate or leaf suspending the quillon dagger from the right side of the wearer.

What is curious about the *Mary Rose* belts is that many of them show signs that, after bifurcating to form a waist band and a hanger, the hanger bifurcates again and then each of these two hangers themselves bifurcate. Experiments to construct a practical hanging system from this physical arrangement that actually works has suggested that what wraps round the scabbard are the bifurcated hangers before they split for the final time, and that their split ends are used to bind them tight onto the scabbard. Binding the hangers in this way also serves to keep them rigid and apart and prevents them slipping together. This simple method of attachment proves to be very secure and effective. However, the most complete surviving scabbard from the *Mary Rose* (80A1709) has a slightly different arrangement in which, instead of splitting for the final time, the bifurcated hangers that pass round the scabbard are bound together by a separate leather thong. It is therefore important to recognise that the suspension arrangements for scabbards on the *Mary Rose* varied in detail and may have been more varied than the surviving fragmentary evidence suggests.

It seems reasonable to suggest that these bifurcated belts were a step on the way to developing the sophisticated hangers that came into common use in England later in the sixteenth century. As such they would not have been perfect. One problem would have been the inevitable puckering of the leather as it sagged away from its natural direction under the weight of the sword. This could have been partly overcome, in theory, by cutting the straps from wide piece of leather with the hangers angled appropriately, but it would have been expensive and wasteful. The other possibility is that, after bifurcating a strap of leather, water and heat were applied to stretch it so that it bent, without puckering, in the required direction. Some possible corroborative evidence comes from 80A0800 (Fig. 9.86), on which the broad fretted strap that bifurcates from the waist belt clearly curves away from it. However this may be a result of subsequent deformation. It is also important to recognise that, in parts of Europe, other precursors of the later sophisticated hangers were in use during Henry VIII's reign, and that England may have been backward in accepting the new type. An Antwerp wooden altar carving in the Musée Royaux d'Art et d'Histoire, Brussels (inv. no. 8653), dated to around 1515, clearly shows a cross-hilted sword (now missing its cross) suspended from a belt to which is attached a single triangular piece, presumably of leather, with three loops through which the scabbard passes.

Few tips survive on the bifurcated ends of the *Mary Rose* belts and there is no evidence of buckles towards the end, perhaps suggesting that they were tied or bound round the scabbard. However, there are numbers of fragmentary strap ends found associated with scabbards that clearly show that they have been riveted into a loop (for instance 81A2255) and one (part of the same find) that has an eyelet hole at the tip that could suggest that it was formed into a loop secured by a stud or monkey's tail.

It is clear that waist belts for swords were known in Tudor England as '*girdles*'. All the sword belts mentioned in the 1547 inventory of Henry VIII's possessions are called girdles, and the use of this term for a waist belt for a sword or dagger in conjunction with the term '*hanger*' for the straps suspending the weapon itself continued into Elizabethan and Stuart times (Beard 1940, 249–54). In 1568, for instance, Paul Bartlette, a labourer from East Ham, left to one of his brothers '*my dagger with my velvet girdle*' (Emmison 1983, 70). At the other end of the social spectrum the terms '*girdles*' and '*hangers*' were still being used to describe sword belts made for Prince Charles in 1618–19 (Southwick 2006, 223–4)

No evidence of shoulder belts for swords has so far been found among the material recovered from the *Mary Rose*. However, there is at least some evidence for the use at this time of shoulder belts with an additional hanger strap or some other means of securing the sword and scabbard (Waterer 1981, 88). One is clearly shown on an engraving *Hercules and the Giants* dating to the last quarter of the fifteenth century in the Boston Museum of Fine Arts (H2; Levenson *et al.* 1973, 70) and another in Mantegna's *The Bearers of Trophies and Bullion*, one of the nine panels of *The Triumph of Caesar* in the Royal Collection (IN 403960) that date from the late 1480s. Interestingly, an engraving of this, perhaps executed by Zoan Andrea, shows the shoulder belt secured under a waist belt (Levenson *et al.* 1973, 214–17). On one of Albrecht Dürer's engravings, *Christ Before Pilate* (1512), parallel hangers are bound round the scabbard and attach to a shoulder belt through two rings (Strauss 1972, 126–7).

Only one scabbard survives from the *Mary Rose* with its hangers and sword-belt still attached (80A1709; Fig. 9.83, reconstructed on Fig. 9.84). It does not appear typical of the majority of belt and hanger finds from the *Mary Rose* as the waist belt and hanging straps are separate, rather than integral, bifurcating belts. It does, however, offer the only real evidence that the sword hangers to the left side were assisted by a retaining strap like those that appear in contemporary or near contemporary art, running from the scabbard to the front centre, or as was often later the case, the front right of the waist belt (described below). The arrangement is identical to that of later hangers which also have the securing strap but which replace the open triangle created by the two (or more) separate hanger straps joined at the belt by a plate with an attachment hook, expanding towards the scabbard where they end in multiple buckled hanging loops through which the scabbard passes. The advantage of this developed version is that the balance of the sword in its scabbard cannot be altered by the two, thin hanger straps of the earlier version riding closer together during use, a problem highlighted by reconstruction and experiment. However, the older arrangement seems to have had a long life and is illustrated in Amman's *Künstbüchlin* (1599, fig. 5) but containing illustrations mostly from 20 years before or more (Werner 1968, fig. 53).

The dagger sheaths found on the *Mary Rose* show little evidence to confirm how they were attached to the wearer. However, Tudor illustrations show clearly that they were normally suspended from the waist belt by means of a cord, sash or chain. The 1547 inventory shows that the finest quality daggers were most frequently suspended from the belt by gold or silver chains. Sometimes these seem to have been suspended from a ring on the belt (see Starkey 1998, 2754). Illustrative evidence shows the dagger either hanging centrally, to the left (if no sword is worn) or to the right. Holbein's portrait of Charles de Solier (1534–5), in the Dresden Staatliche Gemäldegalerie, shows a chain looped over a fabric waist belt and, apparently looping round the metal sheath of a quillon dagger beneath a wide '*puffed and slashed*' moulding (Roberts 1979, 77). The cartoon for the lost wall painting in the Privy Chamber of Whitehall (painted 1536–7, destroyed 1698) in the National Portrait Gallery (NPG 40277) shows Henry VIII with a dagger suspended just to the left of the centre of his waist on a chain or cord that passes through a loose ring on a fabric belt that loops across his front under the fabric waist belt (Roberts 1979, 81–2). The chain or cord appears to attach in some way to the back of the sheath which has a metal chape and locket decorated to match the hilt. A similar arrangement is also shown on the painting by Remigius van Leemput (Royal Collection, RCIN 405750) showing Henry VIII and Jane Seymour with Henry VII and Elizabeth of York. A ballock dagger with two by-knives is clearly shown on the effigy of Lord Parr of Horton, suspended from the left side of waist belt of his armoured body by a double cord that is attached in some way to the back of the sheath near the mouth.

Such evidence is confirmed by a ballock dagger of the same type as those found on the *Mary Rose*, excavated with the iron top locket of its sheath from the moat of Hooke Court (Dorset County Museum, Dorchester 1944–23–1). The locket is iron and has a decoratively fretted plate running down the sheath that protects the mouths of the slits to take the by-knives. It also has an iron staple set across the back of the sheath just below the mouth, for attaching the sheath to the belt by means of a cord or sash (see Fig. 9.33a). That very occasionally swords could be similarly attached may be suggested by one entry in the 1547 inventory which describes a Flemish sword with '*a cheine with a hooke of silver & gilte*' (Starkey 1998, 14500).

The *Mary Rose* dagger sheaths, however, do not seem to conform to this clear pattern of suspension. No evidence has been found near the mouth of staples or other attachments to secure a chain or cord. However, at least one sheath (83A0151, Fig. 9.76) has an expanded transverse neck below the mouth, like those found on numbers of the sword scabbards for securing hanger loops. While this could suggest that daggers were suspended in the same way as swords, the weight of evidence suggests that it is more likely that the vertical suspension cord or chain was secured around this expanded neck.

It was also common in the late fifteenth and sixteenth centuries to attach a dagger sheath directly to the belt of the right side or back of the wearer so that it sat almost horizontally, sometimes with the hilt slightly raised to prevent the dagger falling out. This arrangement is commonly seen on illustrations of '*Landsknechte*' and other northern European soldiers, both mounted and on foot. A rondel dagger is worn in this way by one of the figures in Erhard Altdorfer's 1513 pen and ink sketch of two '*Landsknechte*' (Robert Lehman collection, exhibited Metropolitan Museum of Art, New York, 1987) and by a cavalryman with lance from Amman (1599; Werner 1968, fig. 225). This method does not seem to have been popular in England at this time and there is no evidence for it on the *Mary Rose*. Details of all scabbards can be found in Table 9.9 and the most complete examples are illustrated in Figures 9.71–2. Selected examples are described in full below.

**Scabbard 80A1709**
*M3* (Figs 9.35b, no. 31; 9.83)
This is a nearly complete scabbard with hanging straps, belt and buckle fragment, and chape, 820mm in length, 40mm wide at the top. The scabbard is formed of two laths of wood covered by a single piece of cut and shaped leather sewn together down the back. Most of the leather is missing over the central portion but it is complete to the tip, where the remains of a copper alloy chape (25mm wide, 50mm long) attached to the leather by a transverse staple, can be traced. The front and back of the mouth are not straight but are cut in downward arcs from each edge with a small gap between them in the centre. Some distance below the mouth is the transverse expanded neck, reinforced with wood that prevents the scabbard from slipping through the suspension hangers. The uppermost hanger is situated just behind this feature, 53mm from the top of the scabbard. The second is 430mm from the tip (the distance between the two being 200mm). There seems never to have been a metal top locket, as between the mouth and the expanded neck is a panel of incised decoration matching that found lower down. The scabbard is decorated with rectangular panels delineated by incised lines containing a variety of incised transverse and radiating lines and rows of interlocking 'S' scroll stampings.

The scabbard is suspended on two straps (maximum length 250mm, width 21–5mm), The one that survives to its upper end tapers and is then trifurcated and bound in a Turk's head, monkey's fist or manrope knot by which it can be secured to the waist belt through an eyelet hole (a form of fixing still used in falconry). Emerging from the base of this knot is a flap of torn leather. It seems possible that this may be the end of the other strap which was trifurcated in the same way and bound in one standing knot that joined the two straps together at one suspension point. The lower ends of these straps are bifurcated, the bifurcations looping round the scabbard and then being bound onto themselves by separate thongs to form two fixed loops. The result is a 'V' shaped hanger giving two clear lower suspension points to balance the sword and scabbard. It would have been slid on from the tip of the scabbard and the upper loop located just beneath the expanded neck. Through this upper loop a further strap is looped round and secured on itself by a rivet. This strap runs longitudinally up the scabbard, linked firmly to it by the hanger loop. The end is missing but the strap clearly extends beyond the mouth of the scabbard and is correctly orientated to help secure and balance the sword by attaching to the front of the waist belt in some way. Experiment has shown that the tension provided by this additional strap also prevents the sword from riding up out of the scabbard (Fig. 9.84).

The waist belt appears to be almost complete at 825mm approximate length but to have lost both ends. It has a width of 25mm and retains diamond holes at one end that may have served as fixing holes for the buckle tongue. The other end is pierced with small stitch holes for attaching a narrower strap of leather that springs downwards. Apparently surrounding the fixed end of this strap, but perhaps on the waist belt beyond it, is a much corroded metal loop that may be to retain the end of the waist belt once it has passed through the buckle. Some 30mm below the belt this strap passes into an incomplete gunmetal buckle (13.6 x 6.4mm) and continues beyond it.

However, this buckle must, in fact, have been originally attached to the now missing upper end of another strap, presumably that which runs longi-

810

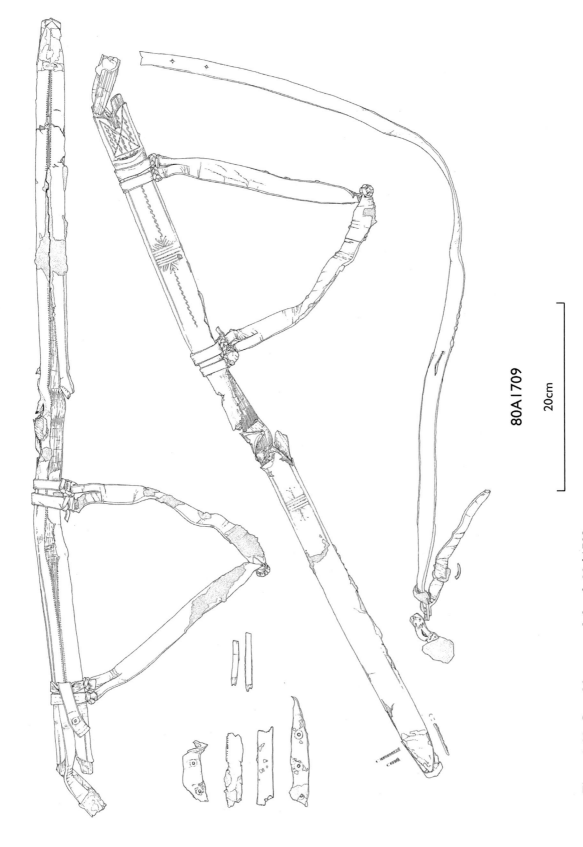

Figure 9.83 Sword hanger and sheath 80A1709

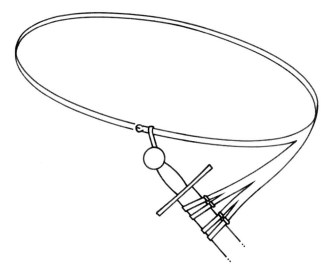

*Figure 9.84 Reconstruction of bifurcating belt (from an original sketch by Guy Wilson)*

tudinally up the back of the scabbard. What seems to be the very end of the waist belt and the remains of the central buckle were found as a separate piece with the rest of the sword belt and scabbard. The piece now consists of a lump of metallic concretion (the buckle) from which emerges a short leather strap bent over on itself and pierced with sewing holes for attachment. Where it bends back on itself it is pierced with an eyelet that can only be explained as the hole for the tongue of the buckle. Some way along from the buckle end of the belt is the eyelet through which the hanger straps can be attached by their knot.

This is the only scabbard from the *Mary Rose* to survive with its hangers and sword-belt accompanying it and offers the only evidence for an extra strap from the scabbard to the front centre of the waist belt to help secure and balance the sword by direct attachment to the front of the waist. As discussed above, it is not typical of the majority of belt and hanger finds as the waist belt and hanging straps are separate rather than integral, bifurcating belts.

The form of the mouth of the scabbard clearly indicates that this was made for a sword with a langet-shaped quillon block at the centre of the guard, as on 82A1915.

**Scabbard 81A1567**
*M6 (Figs 9.35b, no. 32; 9.85)*
This incomplete scabbard was accompanied by portions of a belt and parts of four straps, as well as a boxwood pomander (81A4395; see *AMR* Vol. 4, fig. 3.53) found attached to the scabbard by a silk tassel within attached concretion. The scabbard is missing the top end and tip but survives to a length of 510mm and is 50mm wide at the top, tapering to 40mm to where concretion suggested the presence of an iron chape.

The bifurcated end of one of the straps, a thick piece of leather, was turned over itself and then wrapped round and buttoned to another, thinner, strap, broken off at each end so that its original length cannot be determined. This bifurcated strap is *c.* 300mm long, 31mm wide and ends in blunt tapered points, pierced with two buckle holes. A similar, thick, bifurcated strap 280mm long was found with no evidence of any leather from another strap through its turned and buttoned end. The fourth strap is a longer portion (130–200mm), with possibly two other fragments, and may be the waist belt. It has three buckle holes in the centre. What may be two other parts of the same strap also survive, together with another strap fragment, measuring 80mm, 95mm and 100mm in length. There is also a small oblong of leather with one canted end like a plough, but with a rounded forward corner pierced with three holes. The bifurcated straps with tapered ends appear to be suspended from a belt and to be intended to buckle to another strap to secure whatever was attached to that. While in theory they could have buckled to straps going round a scabbard, the sword would appear to have hung too low to be practical. The purpose of these straps must, therefore, remain a mystery. Their similarity to straps that do seem to be for suspending swords suggest that caution is required in attributing a definite purpose to fragmentary elements of leatherwork from the *Mary Rose*.

**Scabbard 80A0549**
*U8 (Figs 9.35b, no. 33; 9.73)*
This sword scabbard is formed of two laths of wood covered by a single piece of cut and shaped leather sewn together down the back. Just below the straight mouth the leather is cut transversely to accept two by-knives between the outer skin and the wooden lathes, the leather being stretched and formed around the shape of their blades. Below this is the transverse expanded neck, reinforced with wood that prevents the scabbard from slipping through the hangers by which it would have been suspended to the belt of the wearer. Above the neck the scabbard is incised with a number of lateral lines, whether as decoration or to provide purchase for a missing top locket, is uncertain. Below the neck an incised line runs down the borders of the front of the scabbard. In addition the leading by-knife sheath is decorated for its length with small punched circles, parallel and radiating lines and sunbursts. Immediately below the sheath the scabbard is similarly decorated for a space with small punched circles and semi-circles of the radiating lines of the sunbursts. This decoration then stops and the scabbard is plain for the rest of its length except for the border lines. The tip of the scabbard, the chape and a top locket (if there ever was one) are missing.

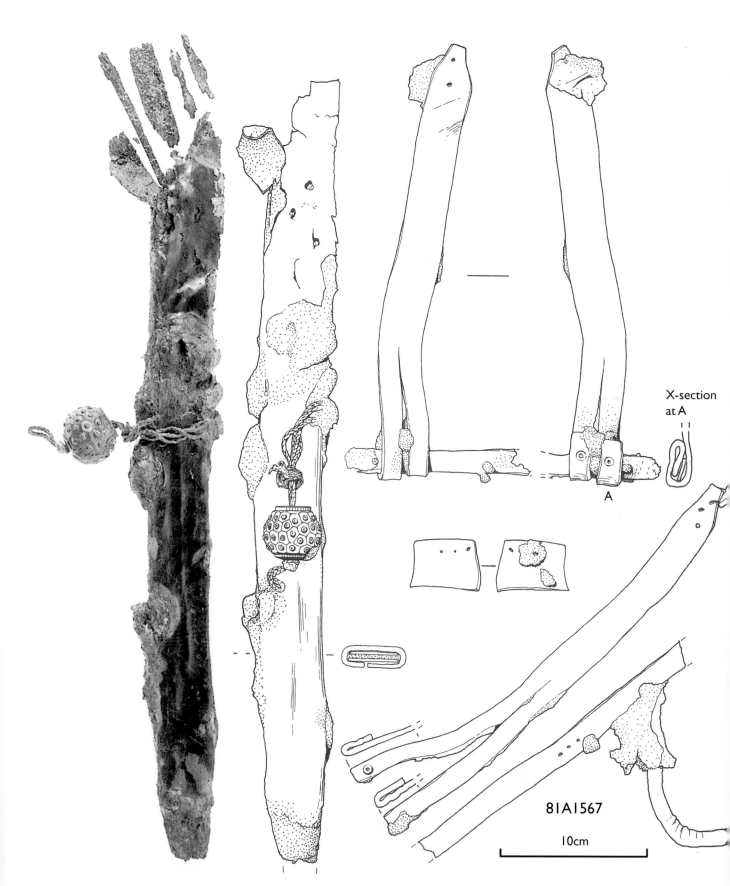

*Figure 9.85  Sword belt, hanger and scabbard 81A1567*

## Scabbard 81A1734

*U9/10* (Figs 9.35b, no. 34; 9.80)

No evidence of wooden lath reinforcement was found with the substantial remains of this sword scabbard. Below the mouth is an expanded moulded chin to secure the hangers and the scabbard tapers from 60mm at the mouth to 40mm at the surviving base. On the back or inside the seam runs diagonally from one side to the other along the scabbard's original length. Regular and frequent holes along either edge show that the seam was sewn together. The evidence of the incised-line decoration is conclusive that the scabbard has not been post-depositionally deformed as it works perfectly in the present configuration of the scabbard. The front is decorated longitudinally down each edge with multiple straight lines. Transverse multiple lines divide the length of the scabbard into panels. With the exception of that containing the expanded neck the upper panels from the mouth for over half the surviving length have crossed diagonal lines (pairs on all but that nearest the mouth that has only single lines). The back is decorated by longitudinal multiple lines down one edge and by one panel just below the mouth formed as on the front by transverse lines and containing pairs of crossed diagonal lines. These would have been covered if a strengthening locket were fitted.

## Decorated belt 80A0800

*U7* (Figs 9.35b, no. 35; 9.86)

This object is a tapering bovine leather sword waist belt with integral hangers, 752mm in length and 40–15mm wide. The waist belt is decorated for its entire surviving length with repeated lines of three small circles punched diagonally across its width. It expands constantly in width from its right hand end where three holes show that it was attached to the, now missing, left end by a buckle, also now missing. As it comes round the left side of the wearer it bifurcates into straps 30mm and 10mm wide, the thinner strap, plain but for punched lines of circles, continuing round the waist and the broader angling down and bifurcating again after 110mm to form the two main hanger straps, 15mm wide (on the fragment) and then again to form four hangers to go round the sword scabbard. The decoration of the broader strap begins just before the waist belt bifurcates and continues until just after the broader arm splits into the two hanger straps. It consists of a central line of cut out lozenges (diaper frets) aligned tip-to-tip with a line of cut-out ellipses on either side positioned in the interstices between the diapers. Around the ellipses is punched a chain of little circles. At the second bifurcation the diapers cease and each narrower strap is decorated only with ellipses and punched chains of circles.

Two further fragments of the waist belt survive, neither displays evidence for the attachment of a buckle. At 178mm total length they give a credible minimum length for the waist belt itself of something over 762mm.

This type of bifurcating belt would only work if the sword was further supported (as on 80A1709, though this has different hangers) by another strap going in the opposite direction to the hanger strap from near the mouth of the scabbard to somewhere near the front centre of the waist belt. This is found on 80A1709.

## Decorated belt 81A0240

*U4* (Figs 9.35b, no. 36; 9.86)

This is the most complete strap recovered. It is made of calf leather and has a total length to the finished end of 650mm and tapers out to a 50mm width.

It is incised for its entire surviving length, except where it expands just before the main hanger strap bifurcates, with a central serpentine line created by repeated punched dots and a dot in each semicircle this creates. Where the belt expands as it comes round the left side of the wearer, the greater width is used by expanding the serpentines and replacing the single dots with scrolling dotted lines springing from the central serpentine. The strap bifurcates after 270mm into a thin (15mm) strap that continues round the waist and a broader (35mm) strap, decorated with fretting and punching, that angles down. This is then split centrally after a further 270mm to form the two main hanger straps both of which bifurcate again after 50mm to form four hangers, each less than 10mm in width and 110mm in length.

The decoration of the broader strap begins just after the waist belt bifurcates and continues until just before the broader arm splits into the two hanger straps decorated to match the waist belt. It consists of quadrilateral and triangular frets with concave side and elliptical 'eye' frets, all bordered by lines of punched dots. Of the lower pair of hanger straps the lower has lost its end while the upper appears to be complete and is slit near its end for a fixing and shows signs of having been bent back on itself (whether in use or after the wreck is impossible to tell). The upper pair are looped through a thicker strap of leather and tied back on themselves by leather thongs. The thicker leather strap is itself formed into a loop by two iron staples and a leather thong and is slit longitudinally for some distance in its mid-section. The leather seems to be broken here and what happened beyond this point remains uncertain.

A single (140mm) loop, of unknown function, has been threaded through one of the hanger straps secured by a knot. An additional small fragment of the belt is decorated with two pierced eyes between opposed pierced triangles. It is only 35mm long. As in the case of 80A0800, an additional strap would have been needed to balance the sword.

This belt was found with the ballock dagger and accompanying by-knives 81A0239 (Fig. 9.68c). This does not mean, however, that this is a dagger belt. The

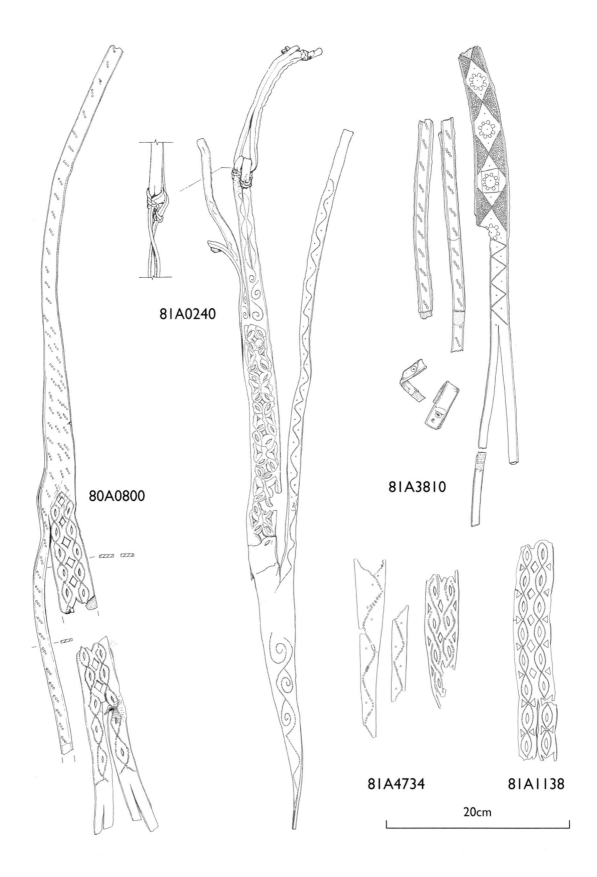

*Figure 9.86 Sword belts and hangers*

effigy of Lord Parr, discussed above indicates that daggers were worn together with swords and were usually suspended directly from the right side of the front of the belt by cord or sash. This belt would also require another hanger going in the opposite direction from near the mouth of the scabbard to somewhere near the front centre of the waist belt as discussed above, and found with 80A1709.

**Decorated belt 81A3810**
*O5* (Figs 9.35b, no. 37; 9.86)
Five fragments of bovine leather form part of the decorative front of a sword belt and hanger straps, with a combined length of 535mm. The largest part consists of the decorated and expanding part of the belt before the bifurcation and the lower bifurcation (the upper is lost) until well after it bifurcates again. The decoration consists of large lozenges aligned tip- to-tip with a large, circular flower pattern stamped in the middle consisting of a central dot surrounded by a large circle around the edge of which are eight, uniformly spaced, much smaller circles. To either side of the flower pattern is a stamped dot. The opposed, triangular interstices between the diapers are filled with very small circles. Where the belt bifurcates the decoration changes to a zigzag of small circles with a single dot in the centre of each triangular interstice created. This decoration ends at the second bifurcation.

The belt tapers from 30mm to 35mm over a length of 180mm, then splits to form one 20mm strap and one of 15mm. The narrow strap has broken off, and is represented by fragments only. These have widths of 15mm and lengths of 244mm and 210mm. These are decorated with incised line borders and diagonal lines of four small stamped circles. The strap is again split centrally at 110mm from the first to form two equal, undecorated ends 12mm and 8mm wide.
Additionally there are two 20mm wide fragments of the end. Each has been bent to form a loop and each shows evidence of being held with a single rivet.

**Decorated belt 81A1138**
*U9* (Figs 9.35b, no. 38; 9.86)
Two fragments of another sword belt made of bovine leather and of very similar form to 80A0800 were recovered. One 204mm fragment has been split centrally at 190mm from the single end. At the split the width is 40mm, each strip being 20mm in width. The other fragment, 107mm in length, tapers from 40mm to 28mm and probably represents the end, which would taper to a tip.

The decoration is identical and they appear to be parts of the same belt. They form the expanding part of the belt on the left side just before and including the bifurcation. The decoration consists of a central line of lozenges cut through the leather with a row of fretted eyelets on either side, surrounded by criss-crossing serpentine lines of small punched circles. On either side of these near the edges of the belt is a line of fretted triangles, their apexes pointing inward to the centre of each diaper. At the bifurcation the same scheme continues, half on each strap but with what was the central diaper being replaced by another triangle to oppose that on the outer edge.

At the other end, where the belt is narrower, decoration consists of central pierced eyelets surrounded by a line of small circles alternating with two opposed triangles. One of these central eyelets appears on the central section and two more on the final fragment.

**Decorated belt 81A4734**
*U8* (Figs 9.35b, no. 39; 9.86)
Three fragments (lengths 85mm, 93mm 105mm, widths *c.* 15–40mm) of an expanding sword waist belt decorated with a serpentine line formed of repeated small, punched circles. One fragment (140mm in length) of the full width (30mm) carrying fretted decoration consisting of a central line of lozenges cut through the leather, alternating with two lines of 'eye' shaped frets, all surrounded by criss-crossing serpentine lines of small, punched circles. All are made of cattle or goat leather.

**Decorated belt 81A4189**
*Location unknown* (Fig. 9.35b, no. 40)
This fragment is of a trifurcating sword belt. It has a length of 330mm and a width of 32mm. It is unusual because it does not expand gradually from a narrow to broad belt but rapidly at one end with pronounced shoulders from which it runs parallel-sided to the point where it splits. Here, instead of bifurcating and then bifurcating again, it splits into three straps, the central one being bound around with a Turk's head, monkey's fist or manrope knot just after the split. The longest is 160mm.

**Decorated belt 81A4536**
*O9* (Fig. 9.35b, no. 41)
Three sword belt fragments (80mm, 50mm 45mm) comprise parts of a 20mm wide strap, split centrally and additional fragments from secondary splits (10mm wide). There is a triangular pattern in circles incised on the unsplit portion.

**Strap fittings and buckle plate 81A3089**
*O9* (Figs 9.33c; 9.35b, no. 43; 9.87)
A calf leather strap, 260mm long, 20mm wide and 2mm thick, found beside a broken and empty chest (81A3088). It is torn at one end, perhaps from a central buckle hole, but seems to taper to a damaged tip. At the other it is riveted to a decoratively cast and pierced

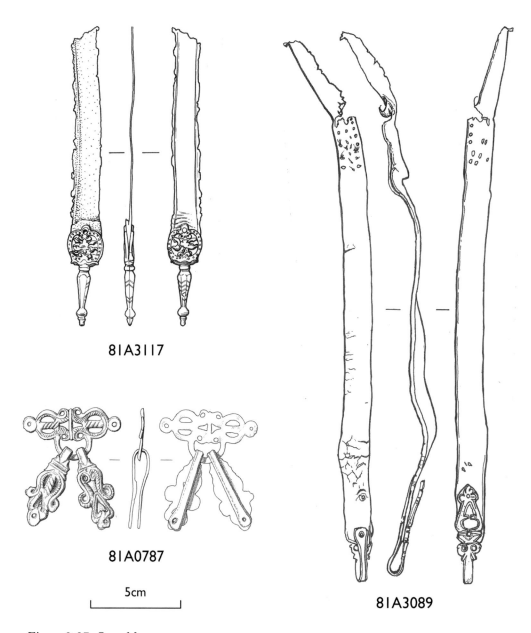

*Figure 9.87 Sword hangers*

copper alloy plate with a long spatulate hook held with a rivet. This is similar in general form to the two hooks found on belt plate 81A0787 (below). There seems reason to believe that this is a sword hanger. Leather strap 81A3080 may have been associated with it.

**Strap fragment 81A3117**
O4 (Figs 9.33d; Figs 9.35b, no. 44; 9.87)
This strap fragment (length 120mm, width 18mm) has an unusual decorative copper alloy end piece in the form of a rondel (length 55mm, diameter 25mm), which, like a cloth seal, grips the strap and is riveted by a single rivet centrally above it. The rondel terminates in a single finial. This looks as though it was designed to hang vertically. It is possible that this could slot horizontally into a partner piece. It is copper alloy, Cu 84%, Sn 2%, Zn 13%, Pb 1% (details in archive).

**Belt plate 81A0787**
U10 (Figs 9.33b; 9.35b, no. 45; 9.87)
This is a belt plate comprising a tripartite element of leaded brass formed of one flat plate (520 x 280mm) with decorative cut-out pattern . It was designed to lie flat against a leather garment and is held with stitches through four central and two end holes. Also incorporated within the design is a semicircular loop which suspends two decorative loops, each 450mm in length and 20mm wide, designed to grip a leather strap and hold this with a single rivet at the far end of each. These also support a foliate decoration with cut out

portions and are similar in form to the termination on strap 81A3089: Cu 69%, Zn 19%, Pb 8%, Fe 4% (details in archive).

**Buckle frame 80A1709**
*O4* (Fig. 9.35b, no. 46)
An incomplete, cast, leaded gunmetal buckle frame was found together with scabbard and straps (80A1709) human remains (80H0223) and an aiglet (80A1684). The surviving portion measures 13.6 x 6.4mm; the pin and plate are missing.

**Buckle frame 82A5170**
*Location unknown*
A leaded brass, single loop, trapezoid buckle frame was found with scabbard 82A4352, and metal concretion. It is flat, 14.2 x 21.4mm and decorated with geometric lines on the outside bar and has a possible tinned surface coating. The strap width is 13.5mm. The pin and plate are missing.

# 10. Armour and Personal Protection

When the *Mary Rose* was laid up in 1514, towards the end of two years of war with France, she was carrying 146 gorgets, 173 pairs of '*splints*' (a type of arm guard), 180 '*Salettes*', 206 breastplates and so many backs (Knighton and Loades 2000, 142). Yet only a single piece of armour, a breastplate, has been positively identified (though other items imply the presence of more). This may indicate an increase in stand-off fighting rather than hand-to-hand combat, or merely the difference between the requirements of a sea battle and for the transport of an invading army. For once the *Anthony Roll* is no help, since its concern is only with ordnance and associated equipment.

In addition to the material described and discussed below there are also several gun-shields – circular shields fitted with an integral handgun. These are described with other handguns in Chapter 7.

## Armour for Warfare
by Alexzandra Hildred and Alan Williams

### Armour worn

Armour offered protection for an individual. The type of armour used reflected the type of weapons that it was intended to protect against and the nature of the engagement – how much mobility was required and whether the wearer had to support the armour himself. There are three types of armour based on methods of construction: mail, large plate and small plate (Brown 1997, 101). Mail was made from linked metal rings and could be transformed into numerous garments (Fig. 10.1a–b). It was flexible and easily repaired, but offered less good protection against missiles such as arrows. This was the main type of armour used until the beginning of the fourteenth century, when the suit of articulated plate armour began to be manufactured, which provided better protection for those who could afford them.

Armour made of small plates linked to one another or to a fabric base was also used. This offered some of the strength of large plate together with the flexibility afforded by the small size of the plates. Garments include '*brigandines*' or '*jacks*', basically reinforced tunics which could be sleeved or sleeveless (Fig. 10.1c–d). Both afforded protection to the breast and back. Brigandines consisted of small plates of iron or steel overlapping in vertical strips, riveted onto a canvas garment and fastened either at the front by laces or buckles and straps, or across the shoulders and down the sides. The jack is a simpler version of the brigandine, consisting of small overlapping iron plates fixed between layers of fabric using trellis pattern stitches (Blair 1979). Hardy (1992, 119) notes that, at Agincourt in 1415, archers wore loose fitting jacks belted at the waist. A mail shirt could be worn underneath this. Both were garments commonly worn by infantry.

By the beginning of the sixteenth century plate armour was widely employed, with special forms for both tournament and war. Cavalry might wear more than the infantry. Armour made of either iron or (the much more expensive) steel plates was usually held together either by leather straps, riveted to the armour and held fast with a buckle or by hinges with pins. Most of the body, except armpits and buttocks, could be protected in this way (Fig. 10.1e).

Helmets were made in a wide variety of styles including the '*sallet*', '*cabasset*', and '*morion*', although some of these names seem to be interchangeable (Blair 1999, 223). Some had leather chin straps with buckles. All would have had textile linings padded with wool or cotton. The comb morion is an open-faced helmet, with a brim forming peaks at the front and back, with a raised central comb (Fig. 10.2a). This was common headwear for pikemen throughout Europe. The Spanish morion, called in France a '*cabasset*', had a similar brim, or a flat one, and a skull rising to a 'pear stalk' finial (Fig. 10.2b). It is so-called from its close resemblance to characteristic form of the later sixteenth century Spanish kettle hat, called a '*capacete*'. Many of the pikemen, billmen and archers shown in the contemporary depictions of the *Siege of Boulogne* and the camps at Marquison and Southsea (Fig. 10.2c) wear morions of different types.

### The mass-production of armour
Most infantrymen endeavoured to wear some armour and although they could not readily bear the same weight as a horseman, a market for mass-produced armour developed. Infantry half-armour, which com-

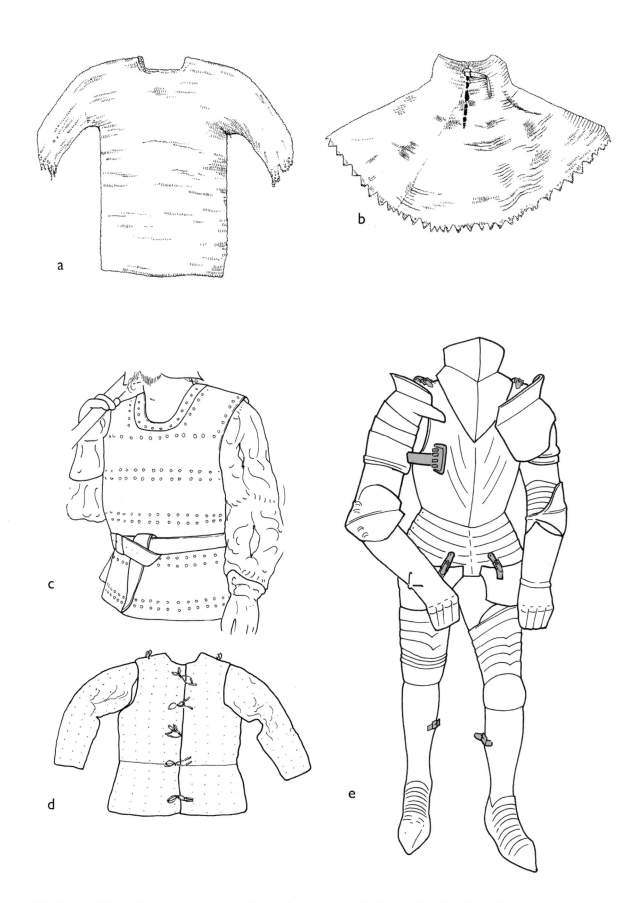

*Figure 10.1* Types of sixteenth century armour: *a–b) mail garments; c–d) Brigandine (Jack); e) plate armour*

*Figure 10.2 Helmets: a) comb morion; b) Spanish morion with pear stalk finial; c) sketch of helmets from Boulogne, Marquisson/Portsmouth*

a breastplate, backplate, tassets to protect the thighs, and splints for the arms, was extremely common. Many of these were termed '*Almain Rivets*' whether or not they were imported from Germany. The plates were articulated on rivets and internal leathers (Fig. 10.3). They could be fitted with '*tassets*' (plates suspended over the thighs) from the lower edge of the skirt of the breastplate. One set in 1512 included '*a salet, a gorjet, a breastplate, a backplate and a pair of splints*'. The characteristic arm defences of the almain rivets protected the outer sides of the arm only. These comprised two gutter-shaped plates linked to a shell-shaped elbow piece by internal riveted leathers, with shoulder-defences of overlapping '*lames*' (narrow plates allowing articulation) that capped the points of the shoulders. The hand-defences comprised overlapping lames protecting the back of the hands only, held by internal straps across the palms and fingers. The helmet of the almain rivet is usually described as a '*sallet*'. This was a close-fitting open helmet covering the cheeks but keeping the front of the face exposed, with a pointed peak, sometimes pivoted at the sides and often worn with a separate face-defence (Blair 1999, 44). These were the lightest and cheapest form of infantry armour.

From the first years of Henry VIII's reign, arms and armour were regularly imported from Germany, Italy, and Flanders. In 1511 Henry despatched Richard Jerningham and two other gentlemen of his court to Germany and Italy to buy arms and armour. Some of this was for his own personal use but, in addition, Jerningham bought (in Milan in 1513) 5000 foot soldiers' suits, or almain rivets. At about the same time, Henry had bought from a Florentine merchant, Guido da Portenari, 2000 almain rivets for 16 shillings the set. In 1539, by contrast, Henry looked to northern Europe and bought 1200 '*complete harness*' (horsemen's armours, with protection for the arms and legs) at Cologne for £454 (carriage paid) and 2700 at Antwerp for £630, so the cost had fallen to about 7s 6d or less. Since the material must have cost at least 3s and

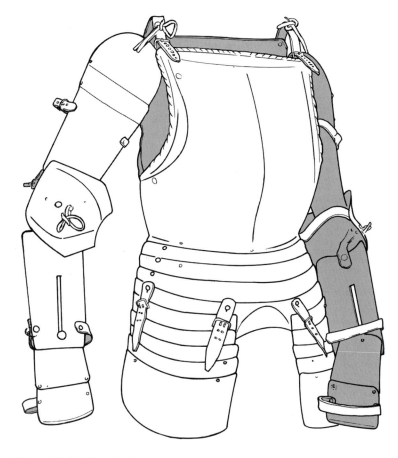

*Figure 10.3 Almain rivets*

workmen were paid around 9*d* a day, then the cost of making the armour must have been two and six days' wages, not counting fuel or overheads.

At this time, a suit of fine armour might have cost £8–12 (perhaps 200 days' wages), so evidently not only were techniques of mass-production being employed for manufacturing munition armours, but they would have been made of the cheapest metal available – wrought iron made from refined cast iron. No attempt was made to harden them; there would have been no point, since the hardness of iron or steel of low carbon content is unaffected by quenching (as described below).

At the same time, the increasing threat from firearms meant that armour had to protect its wearer against ever more powerful weapons. Armour of quality might be of a harder steel but munition armour could only be made bullet-proof by being thicker, and so less easy to wear. Perhaps unsurprisingly, its use started to decline in the last quarter of the sixteenth century, when the infantry were the first to abandon it (Williams 2003, chap. 9.5).

Armour was normally worn over a padded garment, a doublet or arming-doublet. Mail would be used where there were gaps in the plate armour, usually in the form of a skirt below the waist, sleeves or gussets covering the armpits and insides of the elbows, and sometimes a 'standard' of mail at the neck. All of these could be secured to the doublet by points, knotted laces with pointed metal tips (aiglets) termed '*arming points*'. In addition to being used for attaching the mail reinforcements, these could be threaded through the doublet to secure some parts of the plate armour. Arming hose were worn with the doublet combined with shoes which were also often equipped with points to hold down any foot defences. Five sets of arming points are listed in the 1547 inventory (Starkey 1998, entries 14569–72).

Buff jerkins were in themselves protective garments. Although sleeveless, these were made from thick leather (originally from the buffalo). Often worn over doublets, these would offer some body protection with lower weight.

## Protection at Sea
by Alexzandra Hildred

### Armour worn

There is relatively little recorded about the use of armour at sea in the sixteenth century. Illustrations such as the archers depicted on vessels from the *Beauchamp Pageant* in the British Library (Cotton MS Julius E. IV) (see Fig. 8.48b) suggest that archers wore mail shirts and possibly jacks and close-fitting skull-cap helmets. The cross-bowmen from the opposing ship appear to be similarly armed, but with broader brimmed helmets. There is a fair amount of pictorial evidence of land engagements where archers are wearing plate armour over mail. However, many of the most famous depictions of the sixteenth century lack detail regarding armour. *The Embarkation of Henry VIII at Dover* (dated *c*. 1545; see Fig. 1.6) does not show any armour-clad men at all, either on land or at sea. On board ship, none of the few mariners or gentlemen officers illustrated is wearing armour. *The Departure of Henry from Calais* in 1544 (Royal Library, cat. no. 24, Windsor Castle) shows archers and billmen with morion-style helmets and what appear to be buff jerkins, many with the Cross of St George, but only one jerkin-clad individual is shown on a vessel. Several show what could be fringed edges to the skirt, which could represent slashed hose or mail. There is consistency within the style of the helmets and jerkins worn by all infantry. In *The Siege of Boulogne* (Royal Library cat. no. 26, Windsor Castle), similar attire is shown for the infantry, but those on horse appear to be wearing breastplates (Fig. 10.4). The men shown in small boats appear to be wearing a different style of helmet to the

*Figure 10.4 Drawings of armour, from* The Siege of Boulogne

reasonably consistent. The *Henry Grace à Dieu* had 200 each '*Backes and brestes of almyne ryvettes*', '*Salettes*' and '*Standardes of mayle*', together with 198 pairs of splints (*ibid.*, 119). The *Mary Rose* had a slightly less tidy collection comprising 180 sallets, 206 breastplates, the same number of backs, 146 gorgets, and 173 pairs of splints (*ibid.*, 142).

This equipment was what remained in the ships at the end of a campaign in France, so it cannot be conclusively stated that the armour was for use on board, or even to furnish the ships' company during skirmishes ashore. The inventoried weapons include items used against cavalry horses – stakes for the field and '*casting caltrops*' (iron balls with spikes, which also had uses in sea combat: see below) Nevertheless the figures for body armour are consistent with the hand weapons (bills, pikes, bows, javelins and darts) also listed.

Neck protectors or '*gorgets*' have been recovered from the *Batavia* (1629) and the *Kronan* (1676). Brass mail links have been found within concretion on the sixteenth century wrecks in the Texas tidelands (Olds 1976, 99–101). A selection of plate armour (10 breastplates, 1 back and 10 helmets) was recovered from a late sixteenth or early seventeenth century wreck off Alderney (Richardson in Monaghan and Bound 2001, 105–17). Both the helmets and breastplates were close together and thought to have been stored in stacks rather than being worn (Monaghan and Bound 2001, 63).

## Armour carried

Protection for an individual was augmented by the use of portable defences such as shields. By the mid-sixteenth century the cavalry shield had been replaced by the target and buckler. These were usually circular or oval and secured to the forearm by means of internal leather straps ('*target*') or handles ('*buckler*'). The buckler was usually smaller than the target, and was essentially a non-military shield used by civilians and bodyguards. It was used as a parrying device with a sword held in the other hand. Targets of a slightly later period are described by Sutcliffe in 1593 (Blair 1999, 68) as being made of wood, either hooped or barred with iron, oval and 3½ft by 2½ft (1067 x 762mm).

Bucklers are likely to have been circular, concave at the front, made of leather and faced with numerous riveted contiguous iron rings centring on a pear shaped iron boss with a projecting point, and sometimes crossed by strips to produce a trellis pattern. These can be seen being carried by the Yeomen of the Guard in the *Field of Cloth of Gold* paintings at Hampton Court (Royal Collection, cat. no. 23). The area around Wrexham in Wales was a major centre of production (Edwards and Blair 1982, 74–85).

infantry, but none appears to be wearing armour. The Cowdray Engraving, depicting the Battle of the Solent, shows very few individuals on ships, none of them with armour (Fig. 1.2).

Inventories for vessels in the sixteenth century rarely list armour. However, every vessel inventoried in 1514 (Knighton and Loades 2000, 113–58) has some. Items listed included back and breastplates, splints, sallets, brigandines and standards of mail. The quantities were not great – mostly less than 300 items even for the larger ships, though within each vessel the assemblages are

*The metallurgy of sixteenth century armour*
by Alan Williams

**The raw material: making the iron**
In a simple iron making furnace, iron ore (iron oxide) would be heated with charcoal (carbon). The iron ore was reduced to iron, but not melted. The non-metallic oxides present from the ore and from the furnace lining, as well as any unreduced iron oxide, would react together to form a slag, a glass-like material with a free-running temperature of around 1200°C. The products at that temperature would be a '*bloom*' or porous lump of solid iron and a liquid slag. All medieval iron (often called '*wrought iron*') contains a greater or lesser amount of slag entrapped in the iron, which affects its mechanical properties. The slag inclusions might be elongated during forging, but could not be eliminated.

By the sixteenth century a good deal of iron was being made in blast furnaces. In these the iron was melted but, in so doing, it absorbed 2–3% of carbon, which lowered its melting point but made it brittle and useless for forging. It might then be converted to forgeable iron in the '*finery*' where it was melted and most of the carbon burned out again. Fined cast iron would also contain a quantity of slag and so would be generally similar to wrought iron from the bloomery, except in its cost (Buchwald and Wivel 1998).

*Steel*
The advantage of employing iron for weapons and tools is that iron ores are very widespread (unlike those of copper or tin, for instance) and so iron is comparatively cheap to make. It can be hardened by the absorption of some carbon (usually *c*. 0.2–0.6%) to form a harder alloy; steel. This is not, however, any harder than bronze, unless another technique is employed, namely that of '*quenching*'. By plunging it red-hot into water, a different crystalline material forms within the steel, which becomes extremely hard but brittle.

It is hardened steel, not iron, which is markedly superior to bronze in making cutting and piercing tools and weapons. The reliable production of this, however, is problematic. It is difficult to control the carbon content of a bloom which never melts, and the successful heat treatment of steel artefacts requires accurate control of both time and temperature. This was extremely difficult in the sixteenth century when neither quantity could be measured accurately.

Steel might have been made by selecting higher-carbon fragments from a heterogeneous bloom after breaking it up, or by heating pieces of bloomery iron in carbon for many hours and then forging them together. Variations in the carbon content are due to the fact that medieval '*steel*' was such a very heterogeneous material, even if some craftsmen attempted to treat it in a consistent way. Some attempt might be made to homogenise it by folding and forging it out, perhaps more than once. Such banded steels are frequently found in medieval artefacts, as well as heat-treatments which do not correspond to modern procedures.

Steel remained a luxury product throughout the Middle Ages, and the price at which it was sold in England was approximately four times that of iron. It is not, therefore, surprising that only armour of the best quality was made of steel, and all-steel weapons were less common than iron weapons with a steel edge (Williams 2003).

**Metallographic analysis of armour**
Metallography is the microscopic examination of a prepared metal surface, usually by sampling. On armour, the rim of a plate is particularly suitable; on swords, a broken end is preferable. The sample is embedded, polished until it is optically flat, and etched to reveal the crystalline structure of the metal. If the metal is simply iron, then crystals of iron alone (called '*ferrite*') will be seen microscopically (in fact, it is the boundaries between the crystal grains that are visible, since the individual atoms are too small). If the metal is steel, or steel in parts, other constituents will be visible, depending on how the steel has been heat-treated.

If a steel is allowed to cool in air after forging, equilibrium conditions will prevail and the carbon dissolved in the iron above 900°C comes out of solution as a mixture of iron carbide and ferrite ('*pearlite*') displaying a distinctive lamellar appearance. If the steel is cooled more rapidly, or quenched, and equilibrium is not attained, other crystalline products may form:

1. pearlite in a nodular form, almost irresolvable.
2. a material of acicular (needle-like) appearance, harder than pearlite, called bainite.
3. very rapid cooling may form martensite, a material of lath-like appearance and great hardness.

Much depends on the carbon content and the dimensions of the object being treated, but quenching in water without delay generally results in such rapid cooling that an all-martensite structure is obtained (this is called full-quenching). Quenching in oil, molten lead, or some other less drastic coolant than water, may produce a mixture of microconstituents – pearlite and bainite as well as martensite – as may an interrupted or delayed quench. These procedures are collectively called '*slack-quenching*' and are avoided nowadays. The customary modern procedure for hardening a plain carbon steel is to fully quench it and then to reheat it carefully to '*temper*' it.

Tempering reduces the hardness of martensite somewhat but, by removing most of the internal stresses, it reduces its brittleness and hence increases its impact strength. Over-tempering (for too long or at too high a temperature) can reduce the hardness so far that it undoes all the benefits of quenching. Successful tempering requires a steel of consistent composition as well as accurate control of temperature and time. These requirements would have been difficult to meet in the sixteenth century and so it is not surprising that slack-quenching seems to have been the preferred heat-treatment of many armourers, presumably on the grounds that only one operation was involved. Although the product would have been of inferior strength to a fully-quenched and tempered one, there was less danger of cracking or warping it.

The hardness of a metal can be measured by making a microscopic indentation upon the surface with a diamond under a known load, and then measuring the size of the indentation. This measurement can be converted to hardness on the Vickers Pyramid Hardness (VPH) scale, whose units are kg/mm$^2$. The hardness of wrought irons are usually between 90 and 120 VPH. The hardness of a medium-carbon (0.5%) steel would be around 250 VPH if air-cooled. Slack-quenching might raise the hardness to 300–400 VPH, but full-quenching up to 700 VPH. Tempering might then reduce this to 400–500 VPH.

*Italian armour*

The large plates needed to make 'suits' of armour first appear in the fourteenth century, so Italian bloomeries must have been large enough by then to make very substantial blooms. This may well have been the result of the application of water power to bellows and (perhaps more importantly for making sheet metal) trip hammers. Northern Italy became an important centre for the manufacture of arms and armour. It would take another century for the Germans to catch up with the Italian industry. The best Italian armour made for knightly customers of the fourteenth and fifteenth centuries was made of steel, frequently hardened by heat-treatment, as indicated by examples from the armoury of Churburg, where 34 specimens from 21 armours have been examined and most found to be steel. Furthermore, 18 of them were found to be of steel quenched to a typical hardness of 300–350 VPH (Williams 1986).

Armour of lower cost for infantrymen could be made of steel, but less commonly, and generally would not have been hardened. Surprisingly, later Italian armour does not maintain the high metallurgical standards of the fifteenth century. The best armour was still made of steel (of perhaps 250 VPH), but never quench-hardened after about 1510 (Williams 2003, section 4). It might, however, have been decorated with patterns of etching and gilding; the heat-treatment for one process was evidently not compatible with the heat-treatment for the other.

*German armour*

After the middle of the fifteenth century, the armourers of southern Germany began to rival those of northern Italy. The most important centres of production were Nuremberg, Augsburg and Landshut. Their products were marked with a city mark to show that they satisfied guild requirements and also, especially in the case of fine armour, with a craftsman's mark. Armours made by Lorenz Helmschmied in Augsburg after 1480 were generally made of steel hardened by quenching and tempering, to a typical hardness of 350–400 VPH (Williams 2003, chap. 5.4). Other craftsmen in Nuremberg and Landshut were achieving similar results before 1500.

The Imperial workshop in Innsbruck had an excellent raw material in the form of the medium-carbon steel that had been made at Leoben since Roman times, but techniques of heat-treatment were introduced by craftsmen from Augsburg during the late fifteenth century. Innsbruck then produced the best armour in sixteenth century Europe, quenched and tempered to a typical hardness of 500 VPH, as well as this process being successfully combined with decorative gilding.

The English Royal Workshop was set up at Greenwich by Henry VIII in the early years of the century, partly in response to a diplomatic present of Innsbruck armour (Williams 1979). Its products were made of (imported) medium-carbon steel but this was not successfully hardened until the middle of the century. In the second half of the century Greenwich armours hardened their steel by quenching and tempering, combining gilded decoration with a regular hardness of 360 VPH (Williams and de Reuck 1995).

## The *Mary Rose* Armour
by Alexzandra Hildred

The only piece of armour positively identified from the *Mary Rose* is a single breastplate (82A4096; Fig. 10.5). Unfortunately, it is from an unknown location and was recovered together with an incomplete circular corroded iron object, which has tentatively been identified as a buckler (82A4212; Fig. 10.10). Two items previously identified as the rims of helmets (80A1140 and 80A1118) are in fact of pewter and so are likely to be from domestic objects.

Part of the forward area of the Orlop deck (O2–3) seems to have been a major rigging store. Amongst the rigging elements and sailcloth were some 26 items of ordnance, including tampions, daggers, shot, a wristguard, and shot. A sample was taken from an elongate, convex, layered concretion (81S0167; Fig. 10.9) in this area and could be a stack of corroded breastplates, while a large number of leather straps with evidence for corroded rivets may represent the remains of buckles and straps from armour.

In total 32 items have been listed as possible armour. Most (18) are leather straps, six of which were attached to, or were directly associated with, buckles (80A0612, 80A0714, 81A0230, 81A4873, 82A5085, 82A5131).

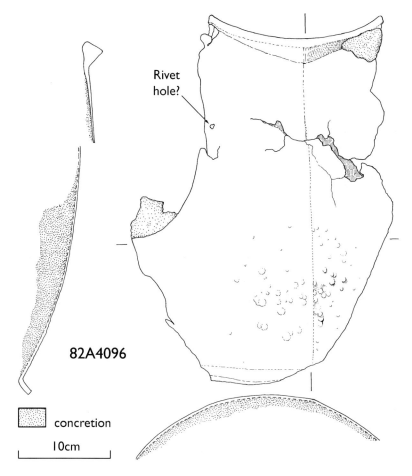

*Figure 10.5 Breastplate 82A4096*

directly on to the plates or on to a small rectangular plate which was then riveted on to the armour. The distribution of possible armour related items is shown in Figure 10.6. Among the 18 straps is 81A4676, from an archer's leather wristguard (81A4675) found in U5, which retains a series of rivets down each long edge (Fig. 8.65). This is completely different from the other wristguards recovered.

## Possible armour and armour-related fragments

### Breastplate 82A4096

*Historical analysis*
by Thom Richardson

The breastplate is of globose form with a slight medial ridge. It has a boxed inward turn at the neck and a plain edge at the surviving arm opening (on the right) with an adjacent rivet for a missing gusset (Fig. 10.5). There are no holes for the attachment of a bracket for a lance rest, indicating that the breastplate formed part of a munition armour for infantry, most probably an Almain Rivet (see Fig. 10.3).

The general form of the breastplate is characteristic of a series of north-west European breastplates of the early sixteenth century. However, all but one of the other known examples are from lance armours, bearing attachments for lance rests, and have integral turns at the arm edges rather than gussets. These include an example in the Royal Armouries (III.4572), attributed to Flemish production of about 1500, but without provenance (Fig. 10.7), and one formerly in the Barry and Gwynn collections (Sotheby's London 5 July 1965, *Important early armour formerly in the Collection of Sir Edward Barry of Ockwell's Manor, Bray*; Christie's London *Antique arms and armour* 16 July 2002, lot 285, Sotheby's 1965, lot 6). The exception is a breastplate with single articulating gussets at each shoulder that was offered for sale by Peter Finer in 2005 (Finer, no. 11). Examples of gussets at the arms of these defences appears around 1510 on Italianate munition breastplates from the armoury of the Knights of St John (such as Museum of the Order of St John, London, no. 2612) and on early Italian breastplates from the same provenance (Karcheski and Richardson 2000, 73–4, no. 4.30; 77–83, nos 4.35, 36, 39, 42, and 44). The most distinctive feature of the breastplate is the boxed turn at the neck (Fig. 10.5) with the central point. This, too, is found on a group of north-west European breastplates of the early sixteenth century. It first appears as a

Another six of these consist of metal with leather (81A1166, 81A0664, 81A0875, 81A4085, 81A4842, 82A1762). Five are metal only (82A1566, 81A5646, 82A1685, 81A0873, 81A0977), one leather only (81A4675), and two are wooden fragments thought to have been elements of helmet lining because of their proximity to human skulls and the nature of the surrounding concretion (81A4515, 81A4522). The radiograph of 82A1566 looks like a helmet and has a width of 280mm and height of 270mm. It was found with buckle 82A5131. Most are extremely fragmentary and cannot be positively identified. In many cases, the identifications were based on the observations of the divers and archaeologists who reported that they were working amongst the remains of corroded armour. Many of the concretions recovered are fragile and fragmentary and have not been worked.

The paucity of physical remains of armour has led us to look for other indications, such as the straps, buckles and aiglets, all of which can be used as fastenings for armour. Straps which had obvious remains of concretion and, in particular, those with holes which may have held rivets, have been identified tentatively as armour straps. Typically, straps were riveted either

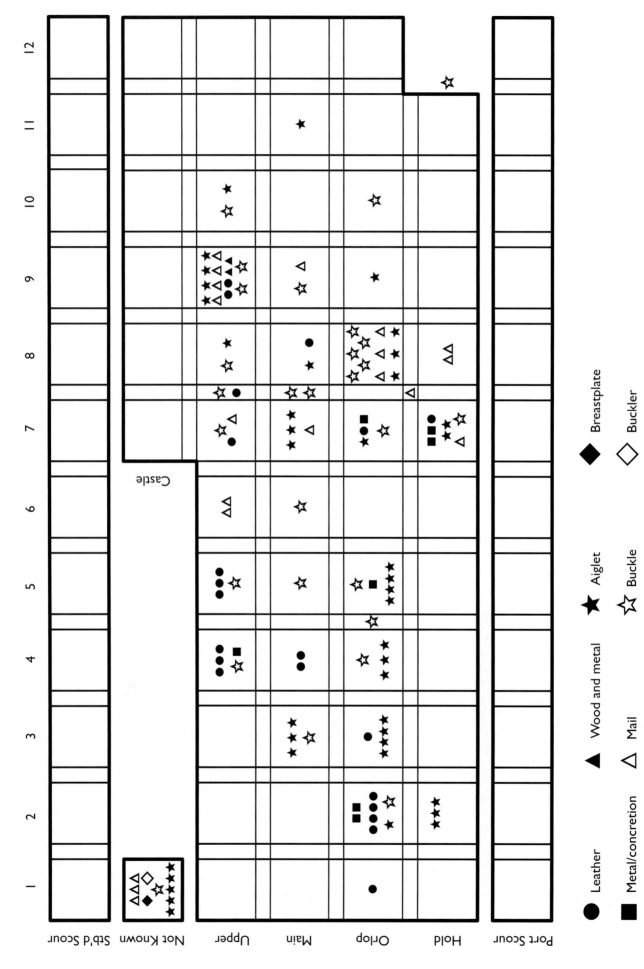

*Figure 10.6 Distribution of armour and possible fittings*

feature on several Italian backplates of the later fifteenth century (eg, Real Armeria, Turin, no. B.19; Boccia 1982, tav. 183). By about 1490 it appears on Italian breastplates (eg, Mantua, Sanctuary of Madonna della Grazie, no. B5; Boccia 1982, tav. 360–75). It can also been seen in a less pronounced form on some, probably Italian, munition armour of the period, such as the breastplate from Rhodes in the Metropolitan Museum of Art, New York (Karcheski and Richardson 2000, 70, no. 4.26). The feature disappears after about 1500 from most Italian production, but persists in the armour made for export in Milan by Niccolo (called Nicholas in the documents) Silva (Pyhrr 1984, figs 11–12), and in contemporary armour made in Flanders and England. It is found on two pieces of English provenance and possibly of English manufacture: the Silvered and Engraved armour of Henry VIII (RA II.5), and a breastplate in the Royal Armouries (III.71), probably from the armoury of Henry VIII, and also on an armour in the Hungarian National Museum, Budapest. In addition to these is an unmarked, unprovenanced piece sold at auction (Christie's London, Antique arms and armour 16 December 2002, lot 66).

The first of these has been attributed to various armourers in the employ of Henry VIII at Greenwich. Blair (1965, 35-6) favoured the Italians Filippo de Grampis and Giovanni Angelo de Littis, identified with the armourers of Milan whose salaries and expenses were paid in 1511 by John Blewbery, yeoman of the armoury (*LP* 1, 3608 [p. 1496]). An alternative attribution is to Peter Fevers of Brussels, who was employed at Greenwich from 1511 (or before), probably to 1518 when he was recorded as lately dead (*ibid.*, 1, 3608 [p. 1497]; 2/2, p. 1479; Blair 1965, 34-5; Richardson 2002, 11).

The stamped mark on the armet of this armour, a helm, corresponds to the mark of Peter Fevers on a document (Blair 1965, 34–5). The armour was complete by May 1516 when the King commissioned his harness gilder, Paul van Vrelant, to decorate its matching bard '*like sample according to a compete harness which of late he made for our body*' for £200 paid, in three instalments (*LP* 2, 1950 [undated; calendared under May 1516 by association with annuity to Vrelant granted on 28th of that month]).

The second breastplate (RA III.71) is stamped with a crowned '*W*'. This mark has been ascribed to the second of the Brussels armourers employed by Henry VIII, Jacob, or Copyn, de Watte, who entered Henry's employ along with Peter Fevers in about 1511, and was still being paid £20 per annum in 1520. The ascription of the mark to him was first made by Cripps-Day (1944, 11–12) and followed by Blair (1965, 34). De Watte also appears to have made armour for the imperial court before moving to England, for there are examples of work marked with the crowned '*W*' in the Real Armeria, Madrid (a pair of vambraces of a common Low Countries style: no. A277, Valencia 1898,

*Figure 10.7 Breastplate RA III.4572 (© Royal Armouries)*

90–1, fig. 57) and a gauntlet, probably from the same armour, in the Instituto de Valencia de Don Juan, Madrid (Arizcun and Sanchez 1927, 26, no. 25). These, in turn, are closely comparable with other pieces with both English and Low Countries connections, such as the fragments of an armour of about 1500 from the Church of St Lawrence, Hatfield, the armour of the '*Man in the Cupboard*', associated with the tomb of Karel von Egmond, Duke of Gelre, in the Church of St Eusebius, Arnhem (Richardson 2000; van der Sloot 1985), and, again, the Silvered and Engraved armour of Henry VIII.

The third breastplate forms part of an incomplete armour transferred from the Kunsthistorisches Museum, Vienna (no. A77) to the Hungarian National Museum in Budapest in 1934 (Blair 1965, pl. xiv, c–d). It bears the Roman '*M*' and crescent mark formerly ascribed tentatively to Martin van Royne/de Prone, Master armourer of Henry VIII's royal workshop at Greenwich. This has recently been reappraised by Pierre Terjanian and attributed to Guille Margot of Brussels, who produced armour for imperial commissions, including the bard of the Silvered and Engraved armour (mentioned above) and the embossed '*Burgundian bard*' presented by Maximilian I to Henry VIII in 1514, both of which bear the same mark (respectively RA VI.1–5, 6–12; Blair *passim*; Terjanian 2006).

In contrast to these breastplates of royal and noble quality, one quite closely comparable munition breastplate of similar form is known, in the collection of the Laing Art Gallery, Newcastle upon Tyne (no. M23220). It is globose, with a slight hint of a medial ridge, a flange at the waist with a rivet hole at the left and a damaged one to the right for attachment of a fauld (horizontal plates below the breastplate to protect the abdomen). It has no holes for a lance rest, is somewhat crude, with boxed, outward turns at the neck and arms (without the central point found on the examples mentioned above), a slightly depressed neck, and a rivet hole at the left shoulder for a strap/buckle (damaged at the right). It is corroded overall and areas at the right side and shoulder are lost. It is comparable with several examples from Rhodes (such as RA III.1084; Museum of the Order of St John (nos 2630, 2632) and Higgins Armory Museum, Worcester, Mass. (no. 808; Karcheski and Richardson 2000, 71–6)). It should be attributed to Italian or north-west European manufacture, about 1510. It was presented to the Laing by F.H. Cripps-Day in 1934 as '*from Settle, possibly from Flodden*', though the association with the battle of Flodden is not otherwise supported.

The comparison with other pieces of similar form and English provenance suggests that the *Mary Rose* breastplate is most probably of English or Flemish, or conceivably Italian, manufacture. However, it is of munition quality, unlike the most closely comparable pieces which form parts of armours made for royalty and nobility. It is best dated to the period 1515–25.

*Metallurgical analysis of the breastplate*
by Alan Williams
A sample (87S0656) was taken from the left side of the turn at the top of the breastplate. The microstructure consists of ferrite and slag only (Fig. 10.8, section at x35 magnification). The average hardness (determined on a GKN microhardness tester with a 100g load) was 175 VPH. This is typical of an item of munition armour of low cost. X-ray microanalysis shows that the armour fragment consists of virtually pure iron. It appears to contain two sorts of slag inclusions, evidently relics of two events (presumably smelting and forging). The elongated inclusions present consist mainly of iron silicates with considerable phosphorus and traces of manganese, potassium and calcium. Some round inclusions are present, these consist mostly of iron phosphide (Table 10.1). This appears to be a typical

**Table 10.1 Composition (inclusions) of breastplate 82A4096 (sample 87S0656)**

|  | Fe | Si | P | Mn | Ca | K | Total |
|---|---|---|---|---|---|---|---|
| Elongated | 65.6 | 19.6 | 8.6 | 4.2 | 1.3 | 0.5 | 99.8 |
| Round | 63.7 | 0.8 | 29.0 | 6.4 | 0.0 | 0.0 | 99.9 |

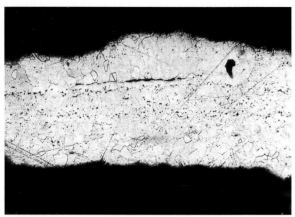

*Figure 10.8 A sample from the breastplate (viewed in section x35): ferrite and slag*

bloomery or finery iron, smelted from a high phosphorus ore.

The analysis indicates that the armour fragment from the *Mary Rose* contains little carbon, but considerable slag, so its fracture toughness would have been poor. The inclusions present suggest it originated in a high phosphorus ore body, which makes its production from a blast furnace and finery likely as the expected procedure for mass-produced armour of minimal cost.

### Possible pile of breastplates (81S0167)
As described above, a large, laminated concretion thought to be a pile of breastplates was found in O2 amongst piles of rigging, sailcloth and gun furniture.

*Metallurgical analysis: corroded sample, thought to be a pile of breastplates 81S0167*
The sample (examined by University of Portsmouth, 1986, 1987), measuring 100 x 150 x 200mm, had a rusty appearance and appeared to be formed of six thin layers of curved iron interspersed with compacted layers of sand containing pebbles of varying sizes. Leather straps could be seen protruding from some of the layers (Fig. 10.9). One radiograph (Archive file 64/1) showed the laminar appearance visible in the photograph but another taken from above showed what appeared to be broken chunks rather than anything identifiable. Analysis of surface products was compared with a sample of seabed sediment from the same sector (Majewski 1986). The marine sediments from the trench showed quartz, illite, sodium chloride, calcium carbonate and calcium hydride. The surface of the object did not contain sodium chloride nor calcium hydride but was otherwise similar. In addition it showed iron oxide, iron carbonates, iron sulphide, and calcium sulphate. This suggested that iron had diffused into the sediments encasing the object. On the basis of this result the item was sectioned vertically through the layers to try and ascertain the internal structure (Vernon 1987). Scanning electron microscopy revealed

*Figure 10.9 Sample 81S0167, possibly a stack of breastplates (84A0172–5)*

iron and sulphur with traces of silicon and calcium. The iron content was roughly similar throughout; the sulphur was constant in the outer layers but reduced towards the centre. A distinction was found between the metallic surface and the compact sandy layers. No base metal was found within the sample but the compounds were felt to be consistent with corroded iron. While the results do not contradict the possibility of this being a stack of breastplates, they do not confirm the identification.

### Possible Buckler 82A4212

This object, like the breastplate with which it was found, is from an unknown location. It was originally circular with a diameter of 280mm and convex but is incomplete and corroded (Fig. 10.10). There appears to be a (100mm diameter) central round boss with a flattened face, but there are no distinguishing features, nor evidence of rivets bordering the edge.

### Straps, wood and metal fastenings

All of the leather straps not directly associated with identifiable objects were studied and compared with those on armour in the collection of the Royal Armouries, where most examples retaining original buckles are 12–15mm in width. The range of *Mary Rose* straps is wider (Table 10.2). Examples in bold in Table 10.2 seem most likely to be for armour and examples are illustrated in Figures 10.11–13.

The distribution of straps shows concentrations on the Upper deck, particularly in U4–5, the open waist of the ship where men at arms would have been stationed ready for battle. Several items were found in U7–9, beneath the Castle deck (Fig. 10.6). These include wooden lath fragments and metal found with human skulls which could indicate the presence of helmets. This area of the ship contained many of the personal items of the crew, including those stored in chests, but also much archery equipment and staff weapons.

Only three items have been tentatively identified from the main gundeck. Of these, 81A0875, from M4, is a lump of concretion which includes three strap

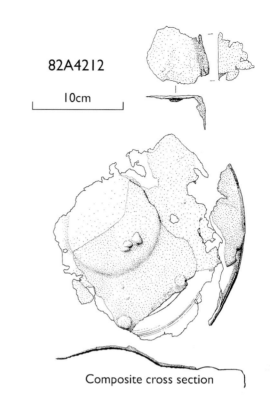

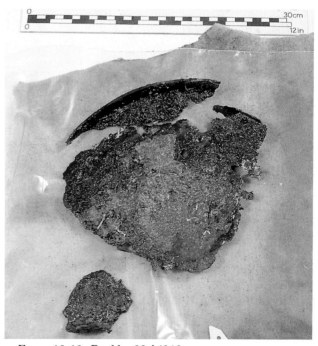

*Figure 10.10 Buckler 82A4212*

fragments possibly riveted to iron plates. A radiograph shows the straps lying parallel, 30mm apart and nine rivet holes are visible, as well as a slit in one end.

The main concentration of possible armour fragments from the Orlop deck was in O2–3, where both rigging and armour may have been in storage, including the possible stack of breastplates (above). Many of the pieces listed in Table 10.2 were found

### Table 10.2 Leather straps and other items possibly for armour (metric)

| Object | Type | Location | Length | Width | Thickness | Comment/description |
|---|---|---|---|---|---|---|
| 79A0686 | strap | U4 | 145 | 25 | – | remains of 4 x 5mm diam. central rivet holes with staining around to 20mm; centres 40mm apart; poss. for belts riveted to backs or breasts |
| 79A1223 | strap | U4 | 130 | 20 | – | central rivet hole 4mm diam., staining to 8mm; 35mm buttonhole-type slits cut lengthways in each end with 5mm holes at far end |
| 81A4085 | strap | U4 | – | – | – | leather & concretion, object missing |
| 79A1174 | strap | U5 | 360 | 20–25 | – | in 3 pieces; sets of stitch-holes each side, 12mm spacing; no rivets |
| 81A4675 | strap/wristguard | U5 | 164 | 88 | 1 | hexagonal frag, 5 x 3mm diam. rivet holes, 1 close to each long edge; staining to 18mm |
| 81A4676 | strap/wristguard | U5 | 114 | 22 | 2 | triangular at wider end, 3 x 3mm diam. rivet holes; staining to 18mm |
| 80A0612 | strap with buckle | U7/8 | – | – | – | for buckle see Table 10.3 |
| 80A0714 | strap with buckle | U7 | 295 | 10–15 | – | with buckle 80A0712 (see Table 10.3); single 3.4mm diam. rivet hole |
| 81A0230 | strap with buckle | U9 | 270 | 11–19 | 2 | 3 frags; no features; for buckle see Table 10.3 |
| 81A2247 | strip | U9 | 222 | 13–23 | 2 | tapering strip |
| 81A4515 | wood | U9 | 60 | 27 | 3 | 2 x wooden lath frags associated with human skulls |
| 81A4522 | wood & metal | U9 | 130 | 37 | – | 3 x wooden lath frags attached to metal associated with human skulls |
| 81A0863 | strap | M4 | – | – | – | concretion stained, associated with human bone (81H0086) by N wheel of port piece 81A2650 |
| 81A0875 | concretion/strap | M4 | 240 | <20 | 2 | tapering strip, single 8–10mm rivet in 3–5mm hole at butt end; see text for further details |
| 81A3832 | strap | M8 | 460 | 18-30 | – | single rivet hole, 5mm diam., staining to 15mm at wide end; end cut to 3 equal faces, narrow end cross-cut |
| 79A1216 | strap | O1 | 120 | 25 | 2 | 3 centrally placed rivets, 5mm diameter, staining to 25mm |
| 81A0664 | straps & metal | O2 | 130<br>200<br>150 | 100<br>30<br>25 | 80 | concretion containing 3 parallel straps, all with rivets; 2 loose straps, longest has rivet hole at finished end, shorter has 3 placed centrally 25mm apart |
| 81A0871 | strap | O2 | 155 | 15–?? | 1.5 | prob. part of 81A0872, tapering, single rivet hole at end |
| 81A0872 | strap | O2 | 308 | 15–32 | – | prob. part of 81A0871, iron stained at wide end |
| 81A0977 | concretion | O2 | – | – | – | 5 frags largest 130 x 120mm, rounded edge to pie-shaped frag.; see text |
| 81A1166 | strap | O2 | – | 21 | – | 31 frags of varying length, many with central 3mm diam. rivet holes 30mm apart, iron staining to 15mm |
| 81A4842 | straps & metal | O2 | – | 25 | – | 13 concretion stained frags of varying length, longest 135mm; 3 have 2/3 central 3–4mm diam. rivet holes 25–30mm apart, staining to 18mm, others have 1 at end |
| 81A2298 | strap | O3 | 250<br>300 | 20<br>20 | – | shorter strap has ends folded back & double-stitched, longer has 1 folded end with 3mm diam. rivet hole, staining to 8mm, other end concretion stained |
| 82A1762 | concretion<br>strap(s) | O7 | 250<br>120 | 200<br>12–15.5 | 35 | with buckle 82A5085 and aiglet 82A5086; no holes or staining on strap; concretion associated with 80 x 80mm stitched cloth; radiograph shows straps, 1 with hole, & cloth |
| 82A1685 | metal | H7 | 254 | 162 | – | curved element, poss. leg or arm protection or cast of same |
| 82A1566 | cast | H7 | – | – | – | convex, almost circular, poss. helmet?; 270mm diam., v. thin; Cua buckle (82A5131) visible in attached by thin iron |
| 82A5131 | straps & buckle | H7 | 110 | 13–15 | – | inside 82A1566. 2 x 2.7mm rivet holes, 1 pin-hole |

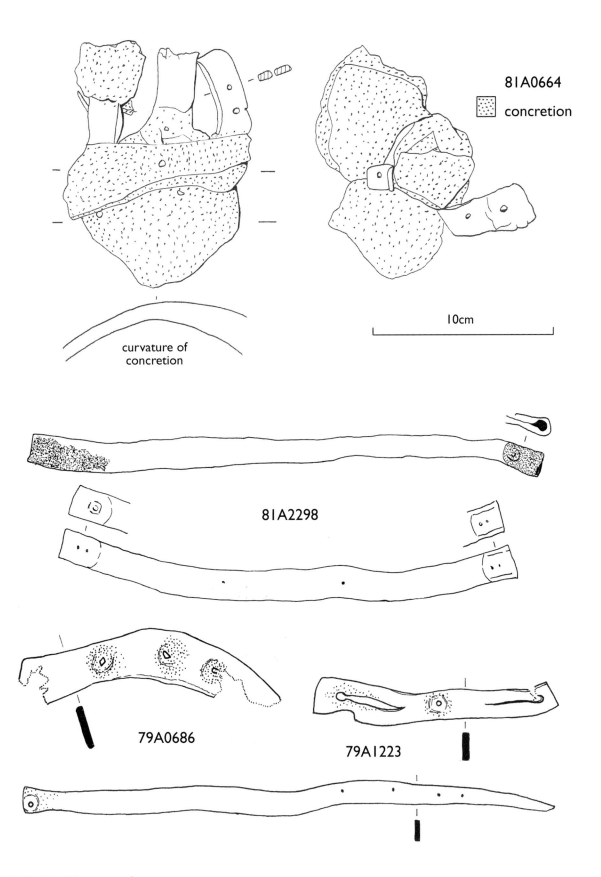

*Figure 10.11 Possible armour fragments*

*Figure 10.12 Possible armour fragments*

together there, some in concretion. Of these, radiography of 81A0977 shows what might be five fragments possibly from a single broken piece of plate armour. One fragment is pie-shaped with a rounded edge and another appears to be keeled. Loose iron wire is also visible.

Three possible armour fragments were found in the Hold, all in H7, one of the major storage areas probably accessed via a hatch either in O7 or O8, as suggested by the presence of a ladder in O8.

### Aiglets

Aiglets, lace-ends or '*points*' are tapered tubes 16–28mm long made from copper alloy sheeting that was either bent or rolled into a tapered tube over the end of a lace (see *AMR* Vol. 4, chap. 2). They had a variety of uses. In a domestic context they were attached to each end of a lace to prevent fraying, and to make it easier to thread through the holes of a garment for fastening. Aiglet-tipped laces were also used to fasten shoes and possibly to tie pouches, knife-sheaths and other suspended objects to a belt or girdle. In a military context they were used to fasten padded garments worn under armour and to attach those garments to armour.

More than half of the 165 aiglets recovered from the ship were inside chests where they were associated with textile remains and shoes but not with any concretion suggesting the presence of armour. Elsewhere, they were generally also found, singly or in groups, with shoes and leather or cloth jerkins, and human remains,

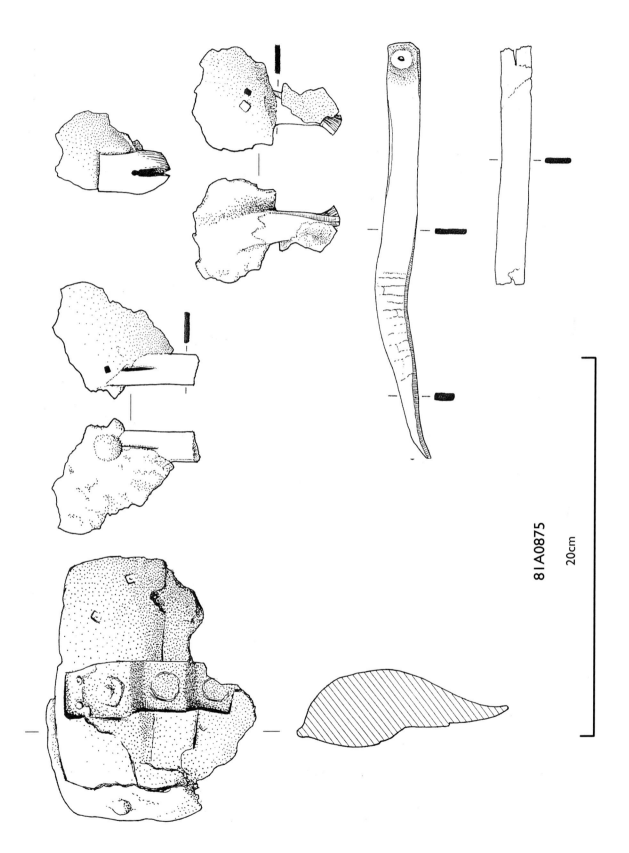

*Figure 10.13 Possible armour fragments*

Table 10.3 Tudor buckles (excluding one in a chest and one shoe buckle) (metric)

| Object | Sector | Material (all cast) & type | Dimensions | Strap width | Description/comment/association |
|---|---|---|---|---|---|
| 79A0257 | H12 | leaded brass, single-loop oval | 34.0 x 36.0 | 28.0 | frame distorted; undecorated; broken lip; pin missing, no plate |
| 79A0442 | U5 | brass, double-loop rectangular | 38.9 x 27.1 | 19.0 | flat profile, transverse ridge decoration on inside & outside bar; central bar cast with frame; pin missing, no plate |
| 79A1233 | U4 | gunmetal, single-loop trapezoid | 19.7 x 31.6 | 20.0 | flat profile; geometric lines on outside bar & sides; pin notch; pin missing, no plate |
| 80A0612 | U7/8 | leaded brass double-loop asymmetrical | 27.2 x 22.8 | 16.7 | leather remains *in situ*; similar to 79A0712; resting on arrow box, associated arrow & leather clothing (80S0611) & 80H0028 |
| 80A0712 | U7 | leaded brass, double-loop asymmetrical | 25.1 x 23.4 | 15.9 | angled profile; geometric lines on inside & outside bars & knops either end of central bar, which cast with frame; pin missing, no plate; leather *in situ*, similar unpublished example from Southampton; associated arrow 80A0713, strap 80S0714, 80H0040 |
| 80A0992 | M7/8 | gunmetal, double-loop oval | 28.8 x 25.5 | 13.0 | flat profile; undecorated, but hand tooled; pin missing; central bar cast with frame; leaded brass folded sheet plate, unrecessed with slot for pin; rectangular with angled end & 2 rivet holes, 20.1x13.1mm; either *in situ*, common late medieval & early post-medieval type (Margeson 1993, 30; Egan & Pritchard 1991, 83) |
| 80A0996 | M7/8 | leaded gunmetal, double-loop oval | 24.9 x 20.3 | 11.9 | similar to 80A0992 |
| 80A1709 | O4 | incomplete, leaded gunmetal | 13.6 x 6.4 | – | only small piece of frame remains, no decoration; pn missing, no plate, poss. not buckle; attached to leather scabbard & straps 80A1865–6 and 80H0223 |
| 81A0109 | U10 | incomplete, concretion covered copper single-loop trapezoid | 25.5 x 27.5 | – | pin missing, no plate |
| 81A0230 | U9 | | 11.5 x 21.2 | 13.0 | flat profile & undecorated; leather *in situ*; similar to 81A4905; associated with wood 81A0229, originally thought to be helmet lining |
| 81A0812 | M3 | leaded gunmetal, double-loop oval | 29.1 x 25.1 | 13.6 | flat profile & undecorated; pin missing; central bar cast with frame, both have black lacquer surface coating; similar to 79A0992; leather *in situ* |
| 81A1307 | M5 | gunmetal double-loop oval | 32.5 x 30.3 | 18.9 | angled profile; geometric lines on inside & outside bars; pin missing; central bar cast with frame, no plate |
| 81A2528 | U8 | gunmetal, double-loop oval | 38.2 x 23.5 | 11.2 | angled profile with knops in each corner & either side of central bar; hand-tooled; pin missing; central bar cast with frame, no plate; associated arrow frags 81A2525 & concretion 81A2523 |
| 81A2955 | M9 | brass, double-loop oval | 36.6 x 31.2 | 19.1 | angled profile; geometric lines all around frame; cast brass pin tapers continually; central bar cast with frame, no plate |
| 81A3235 | O4/5 | frame missing | – | – | cast brass pin tapers continually; gunmetal folded sheet plate, unrecessed with slot for pin; rectangular with convex end & one rivet hole, 28.2x14.7mm; leather strap *in situ* |
| 81A3864 | O8 | brass, double-loop oval | 25.1 x 23.5 | 18.4 | angled profile with geometric lines on inside & outside bar & sides; pin missing; central bar missing, holes in frame for it, no plate |
| 81A4186 | O2 | leaded brass, single-loop trapezoid | 13.9 x 21.8 | 13.5 | similar to 81A4905 |
| 81A4874 | U9 | brass, single-loop D-shaped | 15.3 x 23.2 | 15.0 | flat profile & undecorated, inside bar split; pin missing, no plate; associated concretion 82A1126, arrows 81A4871, staff weapon 81A4872, leather strap 82A4873 |
| 81A4898 | O8 | leaded gunmetal, single-loop trapezoid | 13.6 x 21.5 | 13.5 | similar to 81A4905; associated with buckles 81A4903–5, 6854 & rivets 81A4904 |
| 81A4903 | O8 | leaded brass double-loop oval | 25.8 x 20.3 | 11.1 | flat profile & undecorated; hand tooled; pin missing; central bar cast with frame, no plate; see 81A4898 |
| 81A4905 | O8 | leaded brass single-loop trapezoid | 13.7 x 21.8 | 13.5 | flat profile & decorated with geometric lines on outside bar & poss. tinned surface coating; pin missing, no plate; see 81A4898 |
| 81A6854 | O8 | cast leaded brass single-loop trapezoid | 13.5 x 21.1 | 13.5 | similar to 81A4905; see 81A4898 |
| 82A0936 | O10 | heavily corroded iron | – | – | found attached to inside of gun-shield; no data |

| Object | Sector | Material (all cast) & type | Dimensions | Strap width | Description/comment/association |
|---|---|---|---|---|---|
| 82A5069 | ? | gunmetal double-loop oval | 44.0 x 30.8 | 14.5 | flat profile; rose on inside & outside bar & knops either end of central bar; central bar cast with frame; pin missing, no plate; similar examples from Southampton & Winchester, unpublished (also Margeson 1993, cat. no. 174; Crummy 1988, cat. no. 1758) |
| 82A5085 | O7 | brass double-loop asymmetrical | 26.1 x 22.1 | 15.0 | similar to 80A0712; attached to leather strap & concretion 82A1762, 82H1011, jerkin 82A5084, aiglet 82A5086 & other personal objects |
| 82A5131 | H7 | leaded brass double-loop oval | 29.3 x 23.6 | 14.0 | similar to 80A992; attached to strap, found with concretion 82A1566, concretion & textile 82S0119 |
| 82A5170 | ? | leaded brass single-loop trapezoid | 14.2 x 21.4 | 13.5 | similar to 81A4905; leather between plate; associated with scabbard 82A4352, concretion & twisted link 81A4811 |
| 83A0197 | M6 | leaded brass double-loop oval | 27.4 x 19.5 | 10.6 | flat profile & undecorated; cast Cua pin tapers continually; folded-sheet Cua plate, unrecessed with slot for pin; rectangular with straight end & 2 rivet holes, with irregular shaped rivets *in situ*, 25.5x10.8mm; leather strap *in situ* |

indicating that they were being worn at the time of the sinking. The jerkins could, of course have been worn under armour but there is no direct evidence for this (see *AMR* Vol. 4, chap. 2). Only one (82A5086) was directly associated with a concretion (82A1762) that might represent armour. This was found in O7 together with human remains, leather fragments and leather straps (Table 10.2).

The distribution of all aiglets not within chests or directly associated with shoes is shown in Figure 10.6. There is nothing to distinguish these physically from those in storage. A few elements of the distribution are worthy of comment. The main concentration on the Upper deck is in O9 (81A0801, 81A4039, 81A4188, 81A4286), where there was a jumble of objects and human bones, including several fairly complete skeletons, as well as the remains of at least five leather jerkins. Several possible armour related items were also found here (Table 10.2) as well as fragments of mail (see below and Fig. 10.6). The longest aiglet from the ship (82A5086; 35mm) is perhaps the most likely of all to be armour related. It was found within concretion in O7 (82A1762) which also contained several strap fragments, cloth and a buckle (82A5085).

Among those recovered from the Main deck, three loose examples (80A1864) were found close to a bronze culverin (81A1423) in M3. Six fairly complete skeletons were found close to this gun and analysis of their remains suggested that they were the gun crew (Stirland, below; see also *AMR* Vol. 4, chap. 13). Parts of two leather jerkins and several shoes associated with them but no possible armour. One loose aiglet (83A0480) was found in M8 which also produced a strap with a rivet hole.

A single aiglet (81A4199) was found in the same area as the possible armour related fragments in O2, where it was associated with a piece of plain woven woollen cloth, possibly the remains of a jerkin. Several groups were found in O3 where all five of the positively identified cloth jerkins from the ship were recovered as well as several fragments of strap (Table 10.2). Three aiglets associated with human remains and parts of three leather jerkins were also found here. If nothing else, this indicates that there were a number of men in this part of the Orlop deck when the ship went down – possibly even on their way to retrieve body armour from the stores.

Few aiglets were found in the Hold and these, again, are from areas (H2 and H7) that contained skeletal remains, the latter also producing fragments of possible armour, a strap and buckle and fragments of mail suggesting that at least one armour clad individual may have been in this sector.

## Buckles
by Kirran Louise Klein

Like aiglets, buckles were used in both domestic and military contexts. In the former case this was mostly on

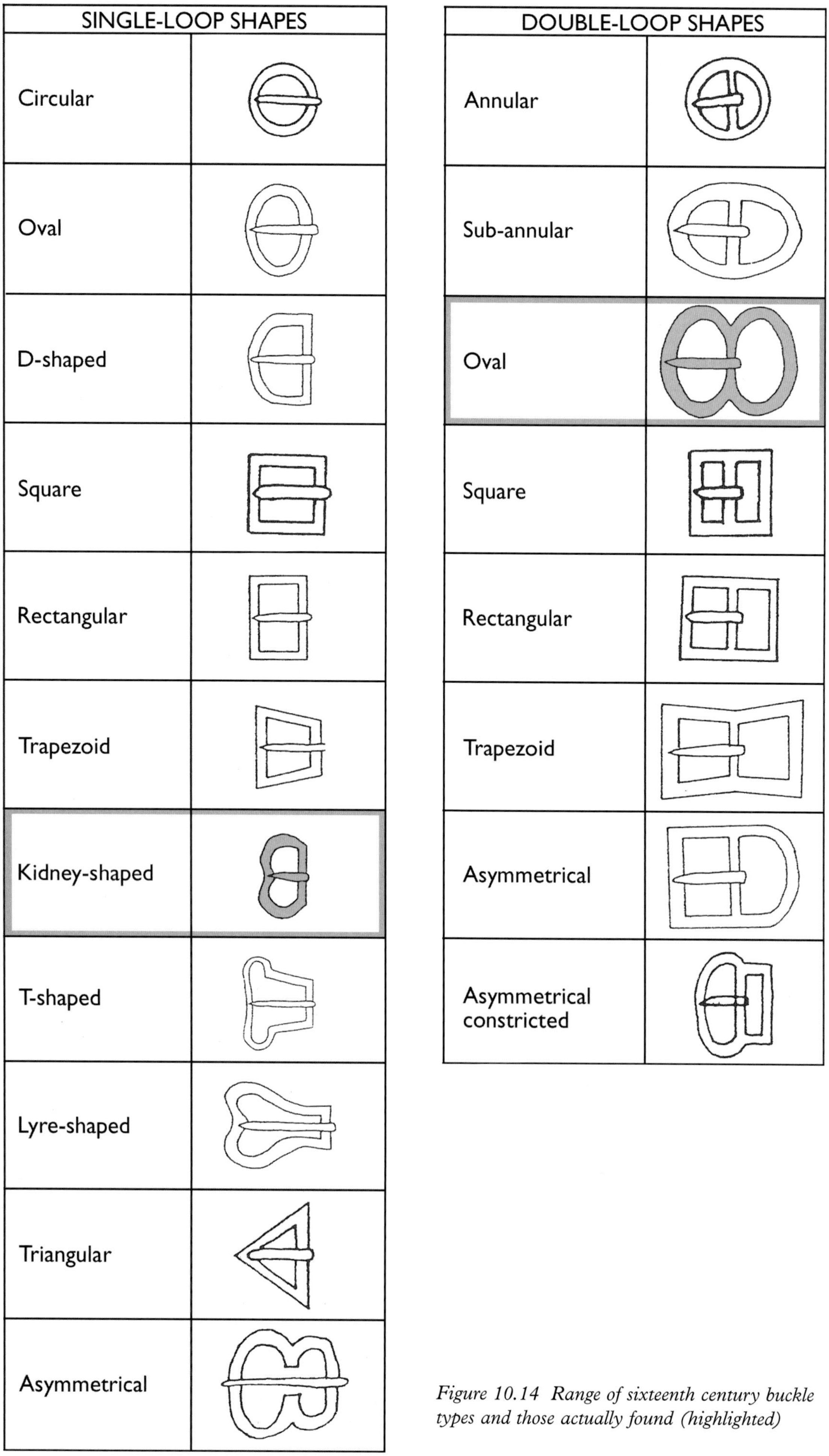

| SINGLE-LOOP SHAPES | |
|---|---|
| Circular | |
| Oval | |
| D-shaped | |
| Square | |
| Rectangular | |
| Trapezoid | |
| Kidney-shaped | |
| T-shaped | |
| Lyre-shaped | |
| Triangular | |
| Asymmetrical | |

| DOUBLE-LOOP SHAPES | |
|---|---|
| Annular | |
| Sub-annular | |
| Oval | |
| Square | |
| Rectangular | |
| Trapezoid | |
| Asymmetrical | |
| Asymmetrical constricted | |

*Figure 10.14 Range of sixteenth century buckle types and those actually found (highlighted)*

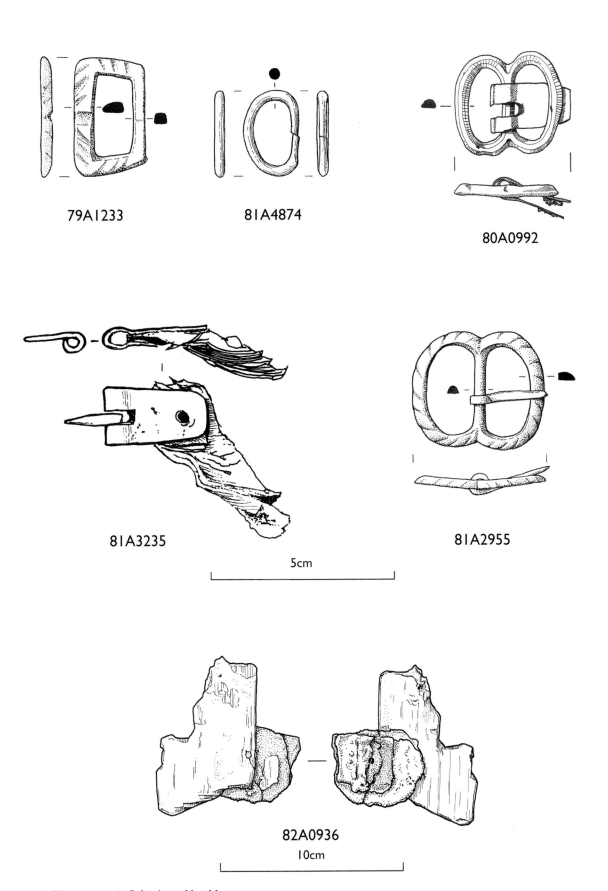

*Figure 10.15 Selection of buckles*

shoes as lower body clothing was generally held up with aiglets and laces (see *AMR* Vol. 4, chap. 2), though personal knives and daggers could also be hung from buckled waist belts (see Chapter 9). In military terms they were used on belts which suspended swords or daggers; on baldricks or belts worn diagonally across the shoulder down to the waist to suspend swords; on arm guards or wrist bracers worn by archers; and on straps to secure gun-shields. They were used in combination with a leather strap and sometimes a small rectangular buckle plate to secure elements of armour to each other. The breastplate and backplates often had shoulder straps with buckles and a waist belt riveted to the backplate (Blair 1979, 124). In this instance, either the breastplate or backplate would have two buckles and the other would hold the riveted strap. The average half armour of this period would, therefore, carry up to ten buckles. Buckles were also used to fasten jacks or brigandines (see Fig. 10.1).

A full description and discussion of the buckles appears in *AMR* Vol. 4 (chap. 2). Here we assess the likelihood of specifically armour related buckles being present in the assemblage. Twenty-eight of 33 surviving buckles are considered to be certainly Tudor in date. Of these, all but one are made of a copper alloy, ranging from leaded brass to gunmetal (Table 10.3), with a single, surviving iron example. The iron buckle (82A0936) is unequivocally military as it was attached to a gun-shield.

All leather straps found with buckles were studied and a random selection of undesignated 'straps' were also looked at to see whether there was evidence of buckles. Five wristguards (Chapter 8) show evidence of having had iron buckles.

Apart from 82A0936 there are no clearly military examples in the collection. Comparison was made with buckles on plate armour at the Royal Armouries in Leeds and civilian buckles found in London and housed in the Museum of London. Although there are some problems in determining which buckles in these collections are original to the armour, it is clear that the *Mary Rose* assemblage contrasts markedly with the buckles in the Royal Armouries but has more affinity with those in the Museum of London. All those buckles in the Royal Armouries that are believed to be original are iron, consisting of a frame and folded sheet plate with one rivet hole. Both single and double-loop buckles are represented, predominately single-loop kidney shaped and double-loop oval frames. In all cases the buckle-plate is riveted directly onto the armour, with no leather between the two halves of the plate. All the *Mary Rose* buckles with surviving plates have leather between them.

The Museum of London, in contrast, houses a collection of copper alloy and lead/tin buckles from the Thames foreshore, identified as Tudor and assumed to be civilian. Twenty-three of these were studied and double-loop examples found to be most common. Single-loop examples are most commonly lead/tin and annular in shape and double-looped ones copper alloy, with oval most dominant, followed by annular. Both single and double-loop examples usually have a flat profile and few have any features or decoration.

Of the 28 copper alloy buckles from the *Mary Rose* 16 (55%) are double-looped with oval the most common shape. The single-looped examples are predominately trapezoid in shape (8) with just one D-shaped example (81A4874) and all are flat in profile, whereas the double-loop buckles more often have an angled profile (Figs 10.14–15).

The obvious conclusion must be that the surviving buckles from the ship are civilian dress rather than military. However, the assemblage is biased because the many iron buckles that must have been present have simply been lost to corrosion. It is possible, also, that copper alloy buckles are indicative of higher status individuals so that their distribution might relate to the location of officers and gentlemen on board the ship, some of whom at least would have been armed and possibly armoured.

The distribution of buckles in relation to possible armour related fragments is shown on Figure 10.6 (see Chapter 9 for the distribution of edged weapons). This excludes the only buckle (80A1612) found in a chest – that of the Barber-surgeon in the M7 cabin – and shoe buckle 81A1713. As with the aiglets, the distribution essentially follows that of the armour related pieces but, again, that also partly mirrors the main concentrations of human remains and is thus similarly inconclusive as to the function of the buckles that have survived. Only three (81A0230, 82A5085, 82A5131) were found directly associated with leather or concretion that were thought to be the remains of armour during excavation (Table 10.2).

## Mail
by Thom Richardson

Mail is a form of armour constructed of interwoven circular metal links designed to combine maximum flexibility with protection. During the Middle Ages mail formed the standard defence for men-at-arms in Europe but, with the development of plate armour, became less important. By the end of the fifteenth century mail was restricted to shirts (called '*hauberks*' or '*habergeons*'), sleeves, collars (called '*gorgets*' or, more commonly '*standards*'), skirts and capes (called '*bishop's mantles*'). Small sections of mail, such as gussets or full sleeves, were sewn to arming doublets to provide protection for vulnerable areas which normal plate defences could not protect, such as the armpit and inside of the elbow joint. Complete mail shirts were still worn by light troops during the sixteenth century.

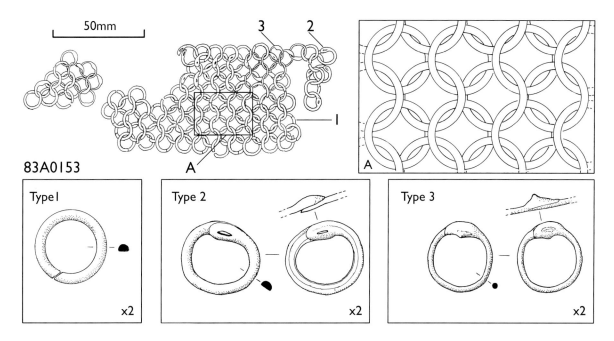

*Figure 10.16 Types of mail and methods of manufacture*

Standards, sleeves, and shirts could have been worn in conjunction with other types of armour such as breastplates and backplates, brigandines, quilted jacks, or jacks of plate. Standards of mail, breastplates, backplates, and plate arm defences ('*splentes*'), and '*salettes*' (helmets) are all mentioned in the 1514 Inventory (Knighton and Loades 2000, 113–58).

The construction of mail is almost universal. Commonly, each link passes through four others, two in the row above and two in the row below (Fig. 10.16). The shape of the garment is formed by the addition or subtraction of links. Links of differing diameters and thickness can be used within one garment to give added protection or flexibility. They are either solid (punched from solid plate); welded (where the ends of the links are welded together to form a circle without a joint) or riveted together. Solid links are incorporated as alternate rows with riveted links. The rivets are usually of iron and were generally made in the form of small wedges passed through rectangular holes formed from the rear of the link, so the flat end of the wedge-shaped rivet was at the rear of the link and the point of the wedge came out through the narrower hole at the front where it could be riveted either flush or in a rounded dome, depending on the riveting pliers used (Metcalf, pers. comm. 1999). Butted links are usually found in Asian mail of the eighteenth century and later, and are almost unknown in European mail, except in fakes made in the nineteenth century. It is extremely interesting to find examples of European butted mail, therefore, in a secure archaeological context within the assemblage from the *Mary Rose* (Fig 10.16).

Riveted mail links were constructed by winding soft iron wire, drawn through a series of successively smaller holes, around an iron mandrel (of appropriate size to the desired finished ring) and cutting it into links with a chisel or wire cutters. The link was often pressed, using a punch and former, to give a semicircular section. The flat portion produced was designed to lie facing outwards in the finished garment, as did the points of any rivets. Alternatively, wire could be produced with a semicircular section from the outset, by drawing the wire between a flat and a notched plate which were successively brought closer together (Burgess 1953). Draw marks on both sides of the faces of some of the better examples of *Mary Rose* mail suggest that this technique was employed (Eaves, pers. comm. 1991). The ends of the link were then overlapped and placed in a pair of pliers cut with a groove to produce the desired section of overlap, and hammered. The overlapped section was then pierced for the rivet. Once the link was threaded through two others in the row above and two more in the row below, the overlapped ends were riveted together. Where solid links are found arranged in rows alternating with riveted links this is thought to be a sign of early mail, made before about 1400, though in Asia, particularly in India, this style of construction continued until the seventeenth century.

Although the bulk of any mail garment would be of iron links, borders of brass or latten (copper alloy) links, either riveted with brass or iron rivets or simply butted together, were not uncommon. Strips of brass butted mail edging does exist (Green and Thurley 1987).

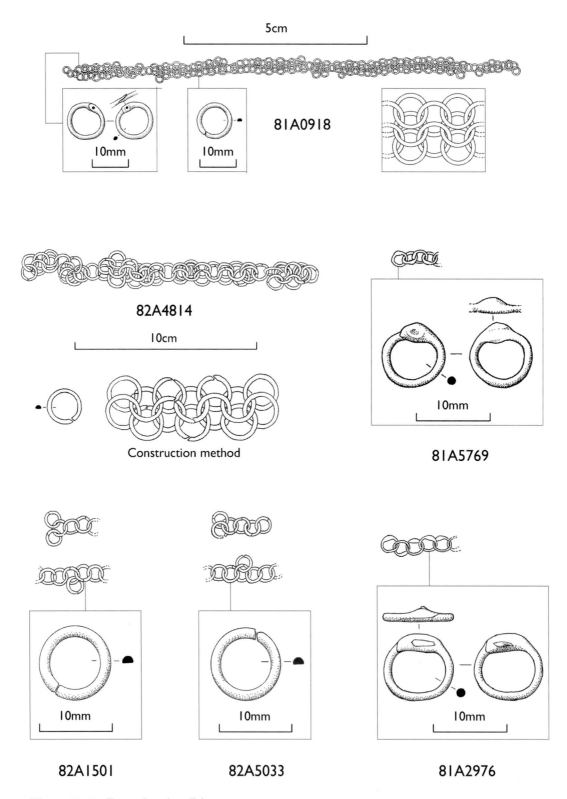

*Figure 10.17 Examples of mail fragments*

## Types and distribution of Mary Rose mail
by Alexzandra Hildred and Nick Evans

Nineteen fragments of mail were excavated from the *Mary Rose*. Additional mail links were recovered during the sieving of spoil heaps in 2003. Some of the recoveries comprise several pieces. Nine samples were analysed by E.E.H. Pitt for composition using XRFS (Table 10.4; Pitt 1989). All proved to be leaded brass.

No complete garments are present. Many fragments are only one link deep, linked laterally and resembling a chain. It is likely that these fragments are decorative

## Table 10.4 Analysis of mail samples

| Object | Link type | Sb | Sn | Ag | Pb | As | Zn | Cu | Ni | Fe | Trace |
|---|---|---|---|---|---|---|---|---|---|---|---|
| 81A0918 | butt | – | – | 0.06 | 1.08 | – | 27.30 | 71.40 | 0.07 | 0.05 | – |
| 81A2254 | butt | 0.02 | 0.36 | 0.05 | 2.08 | 0.14 | 21.40 | 75.40 | 0.38 | 0.14 | Cd Bi |
| 81A4999 | butt | 0.03 | <0.01 | 0.07 | 0.80 | 0.27 | 28.00 | 70.50 | 0.16 | 0.13 | Cd ?Bi |
| 82A0752 | butt | 0.06 | <0.01 | 0.04 | 2.52 | – | 28.40 | 68.70 | 0.34 | 0.38 | – |
| 82A4507 | butt | 0.06 | 0.04 | 0.04 | 2.51 | 0.05 | 15.50 | 81.80 | 0.36 | 0.34 | Bi Cd |
| 82A5174 | butt | 0.02 | 0.03 | 0.17 | 1.85 | 0.08 | 26.00 | 68.80 | 0.02 | 3.06 | – |
| 83A0153 | butt | <0.01 | – | 0.05 | 1.28 | 0.05 | 24.70 | 73.70 | 0.11 | 0.10 | – |
| 81A1664 | rivet | 0.02 | 0.03 | 0.22 | 1.37 | – | 22.00 | 75.30 | 0.47 | 0.60 | Cd Bi |
| 81A5769 | rivet | 0.03 | 0.03 | 0.30 | 1.63 | 0.67 | 20.40 | 76.00 | 0.25 | 0.69 | Cd Bi |
| 82A5174 | rivet | 0.03 | – | 0.14 | 2.06 | 1.88 | 14.00 | 79.40 | 1.00 | 1.46 | Bi |

Composition (wt%) % metal present; all samples were massive in form & wre composed mostly of metal rather than corrosion products

edging of larger items of mail made of iron which have corroded away. Several fragile examples are still within concretion and radiography suggests the presence of iron links (81A2254). Ongoing radiography of concretions may yet provide more tangible remains of iron mail.

Decorative borders are commonly found at the neck, cuffs (or shoulder edges of sleeveless shirts), front openings, and bottom edge of garments. Since the mail edges were worked in rows in the same direction as the garments, a distinction can be drawn between edgings of linear rows, which normally appear at the neck and lower edges, and those with transverse rows which appear at the cuffs and front openings. Most of the borders recovered seem to be lower edges.

### Link type and size

Three different methods for closing the link-ends have been noted. In certain examples there is evidence for the three types within one portion of mail (83A0153), possibly resulting from piecemeal repair (Fig. 10.16). The external diameter of the links range from 7mm to 11mm with a wire diameter of just under 1mm to 2mm. There is some variance with condition.

*Type 1: Butted together with angled ends*
The links are semicircular in cross-section with the flat surfaces all on the same side to enable the mail to present a flat face inwards (Fig. 10.16; Table 10.5). The ends are pressed together along angled edges to form butt joints.

*Type 2: Overlapped, semicircular cross-section, riveted*
The rings are semicircular in cross-section with all the flat surfaces on the same side (Fig. 10.16; Table 10.6). This suggests drawing through a flat and a notched plate, or by being pressed. The ends are overlapped and riveted together. In many instances the rivets have corroded away.

*Type 3: Overlapped, circular in cross-section, riveted*
The rings still retain their circular cross-section and are generally thinner than Types 1 and 2 (Fig. 10.16; Table 10.7). They are overlapped and riveted. The rivets are mostly present, exhibiting a localised blue-black staining and residue over the riveted area. Outwardly this looks like a lead solder, but it may be discolouration due to the presence of the rivet. In one instance (81A1664, Fig. 10.18) there is a portion of an upstanding rivet. This is a flat, triangular shape and appears to have been passed through a slot and pinched over each face.

### Object form

By far the most common form of the recovered fragments is the single length strip, one link deep, of butted (Type 1) semicircular section links (Fig. 10.17). In two instances (82A5033, 82A1501) a second link has been added at intervals (non-regular) along the length to give two-link depth in one or two places. These are best interpreted as two-row borders; when arranged in this way, the length of each is approximately half its present expanded length. The lengths of a number of these pieces suggest the bottom edges of mail shirts.

The fragments of mail which are riveted with a semicircular cross-section (Type 2) are only found in mixed examples; they are never the only form of link present. Those which are round in cross-section (Type 3) do occur as strips of one link deep (81A2976) (Fig. 10.17).

There are three examples with mixed link types, all more than one link deep. One is a mixed butted and riveted piece with circular cross-section (81A0918; Fig. 10.17) and the other two are mixes of the three types (83A0153, 82A5174, Table 10.6). The largest fragment recovered is (83A0153; Fig. 10.18). It was found in concretion during the cleaning of the ship. It was recovered from the area of the Carpenters' cabin (M9)

### Table 10.5 Type 1 (butt link) mail

| No. on Fig.10.19 | Object | Sector | Form | Length | Weight | Link ext diam. | Link int. diam | Wire thickness | Comment |
|---|---|---|---|---|---|---|---|---|---|
| 1 | 82A0752* | H7 | 11 links | 80 | 3.2 | 10 | 7.5 | 1.25 | associated with arrows & human remains, similar to 82A1501 |
| 2 | 82A1501 | O8 | 68 links, 1–2 deep | 450 | 18.2 | 10 | 7 | 1.5 | 4 strips found on ribcage of FCS33 over arrows; similar to 82A0752 |
|   |   |   | 42 links, 1–2 deep | 280 | 10.8 | 10 | 7 | 1.5 |   |
|   |   |   | 37 links, 1–2 deep | 230 | 10.1 | 10 | 7 | 1.5 |   |
|   |   |   | 17 links | 115 | 5.3 | 10 | 7 | 1.5 |   |
| 3 | 82A5033 | O8 | 65 links, 2 of 2 links deep | 420 | 15 | 10 | 7 | 1.5 | associated with human remains & arrows, close to 82A1501 |
|   |   |   | 8 links | 50 | 2.1 | 10 | 7 | 1.5 |   |
| 4 | 82A4814 | H/O7/8 | length 53 links, 3 deep | 130 | 13.8 | 10 | 7 | 1.5 |   |
| 5 | 82A4496 | M7 | 3 links | 24 | 1 (link wt 0.2g) | 10 | 6 | 2 | found in dry dock during washing; similar to 82A4507 |
|   |   |   | 2 links | 15 |   | 10 | 6 | 2 |   |
| 6 | 82A4507* | U7 | 4 links, 1 loose butt link | 28 | 1 | 10 | 6 | 2 | found in dry dock during washing; similar to 82A4496 |
| 7 | 81A4964 | U6 | 10 links, 1:4 formation, 1–2 deep | 35 | 2.1 | 9.5 | 7 | 1.25 |   |
|   |   |   | 8 links, as above | 30 | 1.7 | 9.5 | 7 | 1.25 | concreted to unidentified wood 81A0804; smaller than 81A0988 |
|   |   |   | 6 links, as above | 25 | 1.6 | 9.5 | 7 | 1.25 |   |
|   |   |   | 10 loose links |   | 1.6 | 9.5 | 7 | 1.25 |   |
|   |   |   | 5 frags |   | 0.7 | 9.5 | 7 | 1.25 |   |
| 8 | 81A0988* | U6 | 34 links, 1–2 deep at end | 334 | 7.8 | 10 | 7.5–8.5 | 1–1.25 | between deck planks & starboard side, concreted, larger than 81A4964 |
| 9 | 81A2254 | U9 | 4 links | 30 | 0.6 | 10 | 7 | 1–2 within each link | found with human remains, bills, in concretion with belt & scabbard 81A2255; twisted wire loops of different diams nearby (9–12mm) & different alloy |
|   |   |   | 3 links | 25 | 0.5 | 10 | 7 | as above |   |
|   |   |   | 3 links | 25 | 0.5 | 10 | 7 | as above |   |
|   |   |   | 21 loose links |   | 7.7 | 10 | 7 | as above |   |
|   |   |   | 61 frags |   |   | 10 | 7 | as above |   |
|   | 81A4999* | ? | 3 links | 15 | 0.9 | 10 | 7.5 | 1.5 | similar to 82A5033 & 82A1501 in O8 |

* = metal analysed

and was associated with iron concretion. Unclassifiable forms and types are listed in Table 10.8.

**Distribution**

The distribution of mail by Type and object number is given in Figure 10.19, and in relation to other possible armour in Figure 10.6. Tables 10.5–10.8 give the associations of each object. Two distinct clusters were found: on the Upper deck in the stern and in the Hold and Orlop deck. Whilst the Upper deck had the majority of staff weapons, both areas contained quantities of staff weapons and archery equipment and had significant concentrations of human remains. It is certain that any hand to hand combat would have to be either from

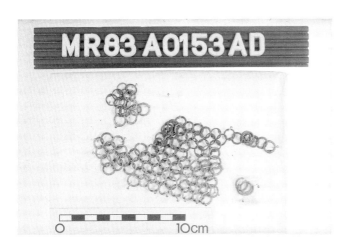

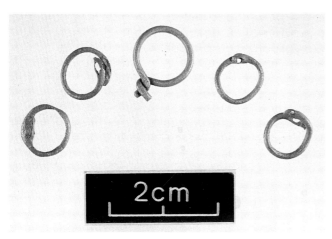

*Figure 10.18 Examples of mail fragments: above) 83A0153 before conservation; centre) after conservation; below) twisted rings and riveted nail links 81A1664*

positions within the Castle decks or on the open waist of the Upper deck (U3–6). Therefore the enclosed area of U7–9 cannot be considered a fighting post for soldiers, but undoubtedly functioned as a muster point for collection of hand weapons.

Six portions of mail were found on the Upper deck. The two of Type 1 links from U6 (81A4964, 81A0988) were concreted to timber. The Type 1 example from U7 (82A4507) is very similar in form and size to one found directly below in M7 (82A4496). All four examples

from U9 were found in an area containing many skeletal remains. 81A2254 was found within a concretion containing a belt with a scabbard (81A2255) and 81A4613 within a concretion carrying a stud and leather. Although three of these were all found close to starboard frame 15, each is a different type.

All four fragments of mail on the Orlop deck were found close to the archery store in O8 (see Chapter 8 and *AMR* Vol. 4, chap. 10) The mail was most probably associated with the skeletal remains found here. Two

Table 10.6 Type 2 and other mixed rivet/butt link mail (semi-circular cross-section unless otherwise stated)

| No. on Fig. 10.19 | Object | Sector | Form | Length | Weight | Link ext diam. | Link int. diam | Wire thickness | Comment |
|---|---|---|---|---|---|---|---|---|---|
| 14 | 81A0918* | U9 | 177 butted links with 1 riveted link attached, 3:2 formation, max. 3 deep | 510 | 36.6 | 9 | 6.5 | 1–1.5 | associated with staff weapons, human remains, wooden bowl 81A0921 |
| | | | 10 butted links, 1 smaller at each end | 65 | 3.2 | 8 | 6 | 1 | |
| | | | 7 butted links | 48 | 1.8 | 10 | 7 | 1.5 | |
| | | | 8 loose links, fragile & worn | | 3.9 | 10 | 7 | 1.5 | |
| 15 | 83A0153* | M9 | frag. of 92 links, 1:4 formation, max. 14 x 13 links, rivet slot 3 x 1mm. Comprising: | 115 x 100 | 24.8 | 11 | 7 | 2 | garment frag., 2 sections of loose mail; excavated from concretion recovered during washing of ship, could have come from Upper deck |
| | | | 70 butted | | | 11 | 7 | 2 | |
| | | | 12 riveted | | | 11 | 8 | 1.5 | |
| | | | 10 riveted, round cross-section | | | 11 | 8 | 1.5 | |
| | | | frag. 15 links, 7 butted, 7 riveted, round cross-section & rivet present; max. 4 deep | 45 x 30 | 3.8 | 8.5-9 | 8 | >1-11 | |
| | | | 3 loose links: 1 riveted, round section, 2 butt | – | 0.9** | – | 7 | 1.5 | |
| | 82A5174* | ? | 20 links: 10 butted, 4 riveted, 6 riveted with round cross-section; max, 4 deep | 155 | – | – | – | – | in concretion |
| | | | 34 links: 3 butted, 6 riveted, 14 riveted with round cross-section; max. 3 deep | 105 | – | – | – | – | in concretion |
| | | | 16 links: 3 butted, 13 riveted with round cross-section 1 loose, open | 45 | – | – | – | – | in concretion |

* = metal analysed; ** link weight 0.3g

Table 10.7 Type 3 (rivet link) mail

| No. on Fig. 10.19 | Object | Sector | Form@ | Length | Weight | Link ext. diam. | Link int. diam. | Wire thickness | Comment |
|---|---|---|---|---|---|---|---|---|---|
| 10 | 81A2976 | H8 | 73 links; flattened over rivets which in situ, localised solder-like staining | 465 | 17.5 | 10 | 7–8 | 1–1.5 | associated with human remains; v. similar to others but larger links |
| 11 | 81A5769* | O8 | 94 links; flattened over rivets; rivet holes?empty (3 x 1mm) | 575 | 23.1 | 8 | 6.5 | >1 | associated with human remains 81H0470, concretion 81A5770, gold half sovereign 81A2976 |
| 12 | 81A5958 | H8 | 68 links; rivets in situ; localised solder-like staining | 520 | 17.1 | 10 | 7.5 | 1.5 | associated with human remains 81H0493,0497 inc. skull with chain 'around neck', whetstone holder 81A5959 |
| 13 | 81A1664* | U9 | 4 loose links, evid. for triangular flat rivet through slot & pinched over each edge; poss. soldered | – | 0.2g each | 10 | 6–7 | 1.5–2 | associated with human remains & staff weapons; 1 loose twisted wire |

* metal analysed; @ single strip unless stated otherwise

### Table 10.8 Mail of uncertain link form

| No. on Fig. 10.19 | Object | Sector | Form | Length | Weight | Link ext. diam. | Link int. diam. | Wire thickness | Comment |
|---|---|---|---|---|---|---|---|---|---|
| 16 | 81A4613 | U9 | 2 broken links | 14 | 1.1 | 10 | – | – | in concretion with leather stud 81A4608 |
|  | 82A6000 | ? | concretion inc. 4 lengths, all apparently butt linked |  |  |  |  |  |  |
|  |  |  | 6 links deep | 115 |  | 10 | 8 | 1 |  |
|  |  |  |  | 35 |  | 10 | 8 | 1 |  |
|  |  |  | 6 links deep | 135 |  | 10 | 8 | 1 |  |
|  |  |  | frag. | 40 |  | 10 | 8 | 1 |  |
|  |  |  | 6 loose links |  |  | 10 | 8 | 1 |  |

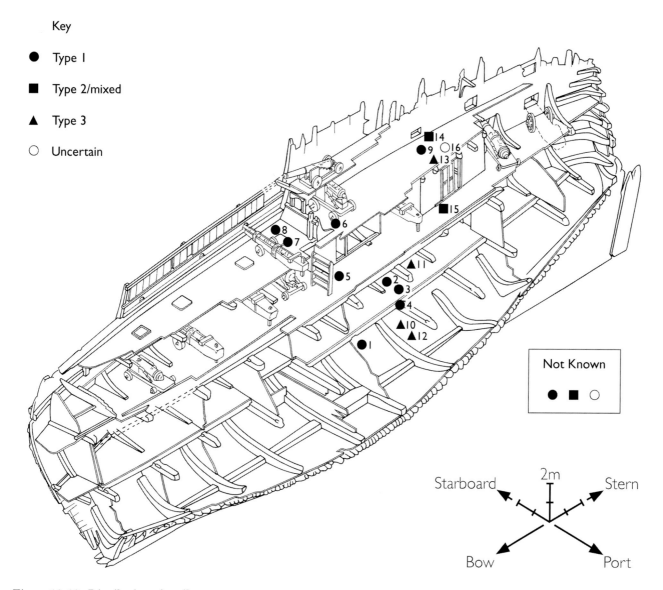

Figure 10.19 Distribution of mail

846

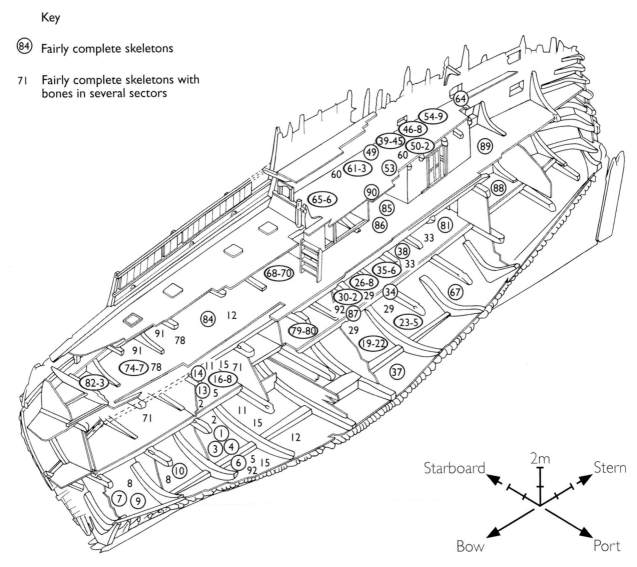

Figure 10.20 Distribution of Fairly Complete Skeletons

Type 1 fragments (82A5033, 82A1501) of similar size and form, were found directly associated with both arrows and human remains. The butted fragment from H8 (82A0752) was again associated with both arrows and human remains.

## Skeletal Remains Suggesting the Distribution of Armour-clad Men
by Alexzandra Hildred and Ann Stirland

The lack of evidence of plate armour means that the possible distribution of men who may have been wearing armour has to be based on the remains of the various elements discussed earlier in this chapter. The one identifiable item, a breastplate (82A4096), unfortunately came from an unknown location. The distribution of mail is shown in Figure 10.19, with the distribution of Fairly Complete Skeletons (FCS) in Figure 10.20 and the overall distribution of human remains in Figure 10.21. When compared with Figure 10.6, it is clear that the majority of objects that are most likely to be armour related come from where there are concentrations of bodies, though direct evidence for armour-clad men is lacking. Nevertheless, some aspects of the distribution are worth highlighting.

### The Upper deck

There is a concentration of human remains representing 28 individuals on the Upper deck beneath the Sterncastle. Most of the Castle deck(s) and a good portion of the Upper deck in this area were destroyed at or after the sinking, and bodies not trapped by wreckage may have floated. Therefore the concentrations of human remains found may not reflect where individuals originally were at the time of sinking. This area was an accommodation area for many of the crew but it also functioned as the storage area (or more likely, a muster

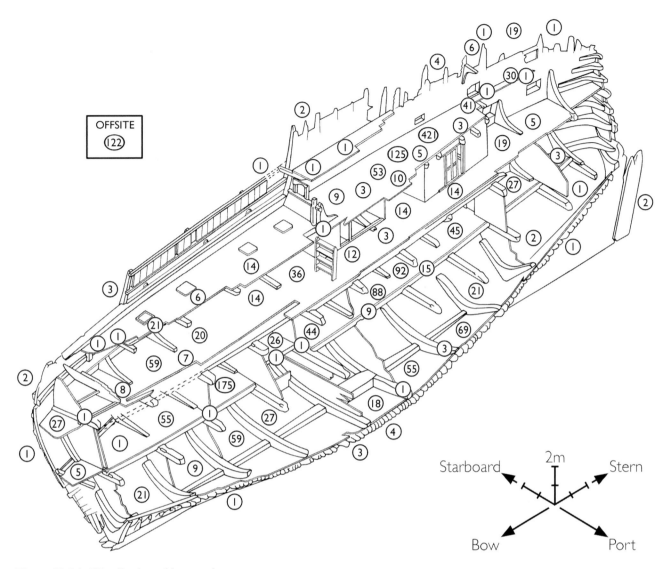

*Figure 10.21 Distribution of human bones*

station) for staff weapons and archery equipment. Potential armour remains include leather straps, buckles, aiglets and mail fragments as well as two wood fragments associated with skulls. When found, these were thought to be potential helmet liners (81A4515, 81A4522), though this identification has not been confirmed. A single piece of mail was recovered from U6 (81A4964) but no human remains are recorded from this area.

## The Main deck

The Main deck produced 15 FCS, mostly from the forward part of the deck. This small number of individuals is not surprising: at battle stations this deck must have retained only those required to work and service the guns and ship. If all the guns were to be fired at once on one side (see Chapter 11), the maximum expected would be fewer than 60. Access for escape off this deck would certainly have been be easier than from the Orlop deck beneath. Guns were positioned in most sectors (see Chapters 2–4). A potential gun crew has been identified from M3 consisted of three bodies found directly behind the gun and two just south in M3/4. Vertebral pathologies suggestive of excessive manual labour and, in particular load-bearing on the lower spine, were evident in five out of six of these skeletons. The sixth was shorter and of a lighter build than the others and showed little sign of excessive strain on the lower vertebrae. Whether these men died at their station, or while trying to escape up the companionway in M4, is not clear.

Gunners would not be expected to wear armour, which may be why so few items were found on this deck. The M3 'gun crew' seem to have been wearing leather jerkins. Possibly one of these originally supported the single aiglet found beside the gun; otherwise there was a single buckle. Apart from two fragments of mail (M7, M9, neither of which are gun stations) both of which could be intrusive on this deck, the only other possible candidates for armour are buckles and aiglets.

## *The Orlop deck*

A minimum of 25 men died on the Orlop deck. At least twelve, possibly fourteen, were in O8–9, where we might assume they were collecting archery equipment and other items from the stores. Two mail fragments (probably the remains of decorative lower borders to shirts), three aiglets, and five buckles were recovered from O8. Four of the buckles were found together, also associated with rivets suggestive of armour. O10 contained one individual and a possible armour fragment, but these were not associated with each other.

There are single elements which may relate to armour in most other sectors of the Orlop deck. Notable amongst them are the nine straps found in the same area (O2) as the possible stack of breastplates. However, only a few human bones were found in this area. Just aft of this, in O3, a larger number of aiglets were found, some attached to laces and jerkins and five within a concretion.

## *The Hold*

At least 19 individuals were recovered from the Hold, with bodies concentrated in the bow of the ship and just aft of the galley – both areas containing numerous casks. The group of bodies in H4 may represent men trying to escape up the companionway as this was the only access in the bow. Perhaps these men had been working in the galley – there were no possible armour fragments found here. H2 contained aiglets associated with cloth, but not directly with the skeletal remains, although a possible armour strap was recovered from this area. There were three individuals in H7 and several metal fragments thought to be armour, in addition to two aiglets and one piece of mail.

# PART 4: FIGHTING THE SHIP

# 11. Fighting the Ship

In order to attempt to understand how the *Mary Rose* may have functioned in battle, it is important to have a knowledge not only of the ship and her armaments, drawn from both historical and archaeological sources, but also of the practice of warfare at sea in the sixteenth century. In addition it is vital to include knowledge amassed of engagements in which the *Mary Rose* participated and any descriptive narratives thereof.

## Historical Evidence

### Fighting tactics
by N.A.M. Rodger

The wartime functions of northern European sailing fleets in the later medieval period were predominantly logistical; naval battles were unusual and relatively few of those which did occur were described by informed eye-witnesses. Contemporary writers took little interest in naval warfare (Mediterranean galleys excepted), and virtually no naval professional literature survives. Fortunately the situation was changing by the early sixteenth century, and we have a number of manuals dealing with the handling of fleets, which give us a fairly clear idea of contemporary tactics in battle. Of these the most important is the *Instruction de Toutes Manieres de guerroyer, tant par terre que par mer, & des choses y servants* by Philip of Ravenstein, Duke of Cleves (1456–1528), a Burgundian nobleman whose chequered military career was spent partly in Burgundian and partly in French service. Although first printed in 1558, the only modern edition is Oudendijk (Oudendijk 1941). Besides much soldiering on land, he served as Admiral of Flanders from 1485 to 1488, and commanded a French fleet in the Levant in 1499–1502. He possibly wrote his book soon afterwards and between 1516 and 1519 (by which time he had returned to Burgundian service) he dedicated a copy to the future Emperor Charles V. He may well have revised it at this date, or even at some later date before his death (Oudendijk 1941, 3–63; Sicking 1988, 43). More-or-less contemporary with this book are a number of manuals, including the works of Jehan Bytharne (1543), Antoine de Conflans (Mollat de Jourdin, M., 1984), Alonso de Chaves (1530) and an anonymous English manuscript which deals with the same subject, and may be at least in part derived from it[1]. There are some independent passages on tactics in Fernando Oliveira's *A arte da guerra do mar* of 1555, which was based on the author's knowledge and experience in sea service between the 1520s–1540s. We have also the orders issued by the Burgundian admiral Henry of Borssele in 1474 (De Jonge 1858–62, I, 736–40; Weber 1982, 23–7) and the English Lord Lisle in 1545 (Corbett 1905, 20–4).

All of these give a clear picture of late medieval naval tactics. The first essential was to gain the weather gage (ie, get upwind of other vessels), which was the position of moral as well as practical superiority, always conceded to the flagship of a fleet (Fig. 11.1). To take the weather gage was a gross breach of respect in a friend, and a clear sign of hostile intent in a stranger. From the windward position, the attacking ship ran down on her opponent, rounded up alongside, fired all her guns and instantly entered her boarders. It was assumed that the guns were aimed at the enemy's men on deck, and were effective only at short range, from their smoke and concussion as much as their shot. They were part of a brief but intense bombardment by all the missile weapons, which also included longbows, crossbows, lances, darts, javelins, iron bars, boiling oil and water. Particular attention was paid to missiles hurled or dropped from the tops, to protect against which heavy netting was stretched from Forecastle to Sterncastle above the waist (Oudendijk 1941, 123, 131). This also protected against falling spars and other rigging elements as well as making boarding and entering the ship difficult. For a fuller discussion of tactics, see Rodger (1996, 302).

All these were standard late medieval tactics for sailing ships in action, and do not appear to have been fundamentally changed by the general adoption of guns during the fourteenth century. The 'artillery revolution', if we may call it that, occurred in the very different naval world of the galley. As early as 1445 Burgundian galleys were built at Antwerp armed with five '*veuglaires*' each, which were forged iron breech-loaders firing stone shot of about a 4in (*c*. 100mm) bore (Paviot 1995, 294–9; DeVries 1990, 822–3; Lehmann 1984, 19–33). These are thought to represent guns similar to the '*fowlers*'

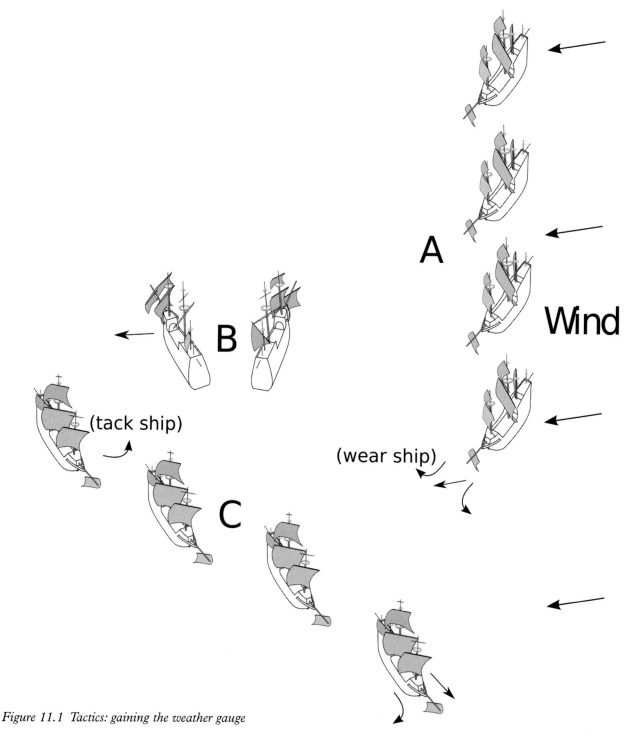

*Figure 11.1 Tactics: gaining the weather gauge*

Sailing seamanship was, in many ways, as important as gunnery. There are parallels with modern sailing races but without the guns. Pre-battle manoeuvring to gain the weather gauge on an enemy offered numerous advantages: 1) The fleet holding the weather gauge (A) was more able to dictate the engagement because they could choose their speed and point of attack, with a greater range of attack courses than the enemy. Even if the enemy changed tack (B) it was the fleet with the weather gauge who had most options. 2) The fleet to leeward (C) cannot head into wind to engage their opponents, and sailing close hauled (or close to the wind) to present their broadside will slow them down. 3) Tacking (taking the bow of the ship across the wind) is not easy with a square rig vessel and will lose much of the vessel's momentum in the process. Wearing ship (taking the stern of the ship across the wind – known to dinghy sailors as 'gybbing') is much easier but would initially head the fleet downwind and away from the enemy. 4) The weather gauge fleet can move onto the same course as the enemy or wear ship onto the opposite broadside (in spite of the tricky lateen sails). 5) They can aim at the lower hull and waterline of the enemy, exposed by the list caused by the wind. 6) In most cases the smoke of battle will drift away from them. 7) At close quarters their sails will take the wind out of those of the enemy. 8) There were really only two advantages to holding the lee gauge: a) due to the list caused by the wind the guns could be more easily trained on the enemy's rigging, if desirable and b) it was easier to flee the engagement (of particular interest for a smaller and/or faster force).

*Figure 11.2 Left) artist's reconstruction of demi culverin 81A1232 facing forwards in the Sterncastle line drawing; right) the front of the Sterncastle as it survives in the museum*

listed in the *Anthony Roll* (see Blackmore 1976, 230). It is usually said that Mediterranean galleys began to ship one or two heavy guns in the '*eyes*' (front) of the ship in the latter part of the fifteenth century. Initially these too were wrought iron breech-loaders, but early in the sixteenth century galleys began to adopt bronze muzzle-loaders which afforded increased range and penetration (Olesa Muñido 1971; Guilmartin 1971; 1974; Boyer 1989; Mollat du Jourdin 1992; Fernández Duro 1895–1903, 1, 323–4). These weapons allowed galleys to stand off and sink ships out of range of small arms. The thick sides and high freeboard of a carrack had hitherto been relatively invulnerable to attack by galleys, whose usefulness was limited to short-range, inshore and amphibious operations. Now galleys presented a serious threat to ships, to which there was no obvious counter. The English had an unpleasant lesson in the new realities of naval warfare in the 1513 campaign, when the French Mediterranean galley squadron drove the English fleet away from Brest in confusion, sinking one ship outright and badly damaging another. The Lord Admiral of England was killed in a counter attack (La Roncière 1899–1932, 3, 104–6; Spont 1897, 121–33, 159).

By this time ships were already mounting large guns in their gunrooms, firing astern from ports cut in the stern panels, and they may already have been mounting broadside guns below decks, firing out of ports. Neither of these, however, solved the pressing tactical problems presented by galleys with heavy guns. The only known tactical formation at sea remained the line abreast, as it always had been. Though galleys with their heavy guns, and carracks with their '*forestages*' full of men, fought in different ways, both attacked with the bow. To counter the galley, the naval men of Henry VIII's reign were looking for a vessel which combined the strength, speed, seaworthiness and capacity of the ship with the powerful ahead-firing armament of the galley. Clearly this is a problem which was under review, as is suggested by the order to place culverins on either side of the foremast in a number of vessels, including the *Mary Rose* (see Chapter 12), regardless of the potential consequences for stability and handling. Moreover they wanted their heavy guns to fire from low down (like the galleys) in

*Figure 11.3 Canted guns (highlighted) on the* Henry Grace à Dieu, *as showing in the Cowdray Engraving (©Society of Antiquaries of London)*

order to hit the enemy on the waterline. It has been argued elsewhere that the solution to this tactical problem lay in the development of what we have come to call the '*galleon*', probably beginning in the 1540s. This new type combined the after-part of a ship with the low forecastle and heavy bow chasers of a galley (Rodger 1996, 304–6; 1997, 205–13).

The *Mary Rose*, in her final state after rebuilding, is a ship of the transitional era. She was armed with heavy guns, but she remained essentially a carrack or great ship, and her manning and fitting were clearly designed in the traditional manner to achieve a decision by boarding (for a full description of the ship's anatomy see *AMR*, Vol. 2). She represents an era when heavy guns were treated with respect, but it was not yet clear exactly how they might be used decisively. Cleves, for example, still assumes that battles are normally settled by boarding, but he already regards heavy guns as serious weapons threatening both men and ship. He requires his officers to keep the boarders below decks until the last moment to protect them from gunfire; he provides splinter protection using mattresses and bales of wool; and he recommends a damage control party to tackle hull damage from gunfire (Oudendijk 1941, 131). He explicitly warns that galleys are now so powerfully armed that they may sink even big ships (*ibid.*, 137). This warning may derive from his experience in the Mediterranean campaign, or from the war of 1512–13, though it is also possible that his stress on heavy guns represents a late revision of his text incorporating lessons of the 1520–22 campaigns in which English and Burgundian squadrons operated together (Sicking 1999). Moreover, he discusses the possibility of attacking with a ship, faster and better gunned than the enemy, but less well manned and lower. In this case he recommends avoiding hand-to-hand fighting: '*en tousjours tourpiant authour, en thirant de mon artillerye pour les tousjours grever*' (Oudendijk 1941, 134–5). '*Tourpiant authour*', '*spinning around*' to bring other guns to bear, is perhaps what the *Mary Rose* was doing when she foundered. Certainly we have here both an explicit discussion in theory, and an unambiguous example in practice, of a recent phenomenon in history: a naval action fought entirely by heavy gunfire, without the two sides making any physical contact.

The tactical problems of handling a ship like the *Mary Rose* in such an action sprang largely from the want of guns bearing right ahead. The *Mary Rose* had two bronze demi culverins on the first level of the Sterncastle, firing forward on arcs clearing the Forecastle on either side (Fig. 11.2), and possibly two sakers on a deck above (see Chapter 12; Fig. 12.13). This is exactly what Cleves recommends:

'*Sur le premier estaige du chasteau derrière, là où on thyre le cabestan, devez avoir deux grosse*

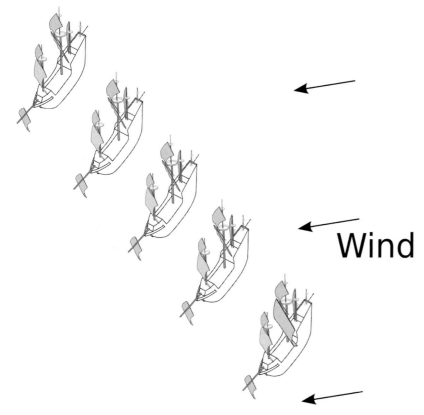

*Figure 11.4 Sailing formation: line abreast*

*colevryne, à chacun costé du matz une, lesquelz thyrent en devant, car pour leur grande longheur ne scaroyent tourner pour thyrer de costé.'* (Oudendijk 1941, 122)

She could also have canted several of her broadside guns to fire as far forward as possible, an established English technique later in the century, and seemingly what the *Henry Grace à Dieu* is doing in the Cowdray Engraving (Fig. 11.3; Rodger 1996, 312). The latter method would have obliged the ship to advance on the enemy in a series of awkward diagonal movements punctuated by sharp turns. It is fairly certain, moreover, that heavy guns (or at least, muzzle-loading guns) were not reloaded in close action (*ibid.*, 312–13) and, to achieve the maximum fire, a ship mounting broadside guns had to 'spin around' to fire each side in turn before withdrawing to reload. The dangers of sharp turns in a big ship heavily loaded with ordnance and men with possibly an unpractised crew and open gunports near the waterline, were perhaps not so obvious at a time when a series of broadside gunports cut low in the hull may still have been a novelty.

The fleet commander of this period had the unenviable task of integrating into his formation ships which could not advance straight at the enemy, and had to make sudden sharp turns. Most authorities agreed that the only proper fighting formation was a tight line abreast (Fig. 11.4). This protected both ships and galleys from attack on the beam, which was their most vulnerable point, and presenting their bows towards the enemy (Oliveira 1937, 127; Cerezo Martínez 1983; Olesa Muñido 1971, 2, 83–211). It was impossible for ships like the *Mary Rose* to take part in such a formation at the same time as sheering to one side or the other to fire guns ahead. An alternative was the wedge formation suggested by Chaves and actually adopted by Lisle (Corbett 1898, 1, 54–8; 1905, 6–24) or the parallel columns used by the Danish admiral Herluf Trolle in 1564 (Probst 1993, 177–9)2 but neither could have been more than a palliative to the lack of ahead fire, and Lisle's orders clearly do not envisage a gunnery action of any sort, but rather a brief melée followed by boarding in the traditional fashion. Galleys were still indispensable for ahead fire, but it was extremely difficult to integrate into a sailing fleet vessels so limited in speed, endurance and seaworthiness. Oliveira, who had experience of galley operations in the Channel, realised their limitations within this arena, although he accepted the necessity of some sort of oared vessels (Oliveira 1937, 66–7, 119). These were realised in the successful rowbarges. In practice the French solution, in the calm waters of Spithead, was to send the galleys ahead as a skirmishing line, while Lord Lisle, off Shoreham, disposed his oared vessels in wings covering the flanks of the main body of ships (La Roncière 1899–1932, 3, 419–8; Corbett 1898, 1, 49–58).

All these unsatisfactory expedients were soon abandoned when the development of the galleon allowed English squadrons to be made up of sailing ships alone, mounting a powerful ahead-firing armament and able to attack in a line abreast. By then the *Mary Rose* had gone, preserving for a distant generation the form and armament of a big ship of the transitional era.

**Endnotes**
1. Extracts of Chaves are printed in Fernández Duro (1895–1903, 1, 379–91) and translated by Corbett (1905, 6–13). Davies prints the anonymous orders (1962, 223–4). Corbett (1905, 14–17) wrongly attributes them to Lord Audley.
2. The notion of Trolle's formation as a wedge is discussed and now abandoned (Probst 1993, 177–9).

## Documentary evidence

### Changes in armaments
by Alexzandra Hildred

The archaeological evidence for armament obviously relates to the ship in her final state but that was the product of more than three decades of service and considerable changes in the armaments supplied and the structure of the ship in order to accommodate them. Moreover, significant changes in the ordnance carried reflect not only developments in the number and variety of the weapons being manufactured, and in the manner they were deployed within the fleet, but also to the fighting tactics, as just described. In order to understand the vessel as a fighting ship we need to consider her armaments and capabilities throughout her career, not just at the last battle. Throughout this volume we have drawn on the evidence from a variety of manuscript sources, including inventory evidence for both ships and land installations of various dates. The ships' inventories of 1514 (Knighton and Loades 2000, 113–58), 1540 (PRO E 101/60/3), and 1546 (Knighton and Loades 2000, 41–106) are used here to examine changes in the armament of the *Mary Rose* herself and of other ships in the fleet in order to build up a picture of the development of naval firepower and tactics during Henry's reign (Table 11.1). To these three principal sources can be added information taken from the 1547 inventory (Starkey 1998).

Vessels appear to be using some of the most advanced weapons available and, therefore, a study of their armament pin-points changes within a defined period of time. By contrasting the 1540 and 1546, or 1546 and 1547 inventories we can suggest which weapons may have been durable and operationally sound, and those which might have only been short-lived. It is also possible to postulate when new weapons came into use. The relative importance of specific gun types can also be surmised merely by their number, expressed as a percentage of all gun types listed. The differing capabilities of these types of gun and the number of rounds carried enables some interpretation of the nature of engagement with respect to distance, damage and duration. Comparing vessel inventories with those for fortifications may also show which artillery or munitions may be specifically used in the context of warfare at sea.

### The Mary Rose in 1514

The 1514 inventory was undertaken after the end of the first war with France. It lists thirteen ships and concerns mostly to the rigging and the armaments. On decommissioning, armaments were either handed over to the Ordnance Office representatives, John Millet and Thomas Elderton (five vessels), kept on board and handed over to the named masters/pursers of individual ships (the *Henry Grace à Dieu* and the *Great Elizabeth*), or divided between the two (six vessels). It is the

**Table 11.1 Comparison of inventory records for *Mary Rose* guns**

| Gun | 1514 | 1540 | 1546 |
|---|---|---|---|
| *Guns of 'brass'* | | | |
| Great Curtow | 5*AS | | |
| Murderer | 2?*AS?AP | | |
| Cannon | | | 2*AS |
| Demi cannon | | 4*AS | 2*AS |
| Culverin | | 2*AS | 2*AS |
| Saker | | 2*AS | 6*AS |
| Falcon | 2*AP | 5*AS | 2*AS |
| Falconet | 3?*AP | 2*AP | 1*AP |
| Little brass gun without chamber | 1AP | | |
| Total bronze | 13 | 15 | 15 |
| Anti-personnel (AP) | 6–8 | 2 | 1 |
| Anti-ship (AS) | 5–7 | 13 | 14 |
| *Guns of iron* | | | |
| Great Murderer | 1*AS | | |
| Murderer | 3?*AS?AP | | |
| Cast piece | 2AP | | |
| Port piece | | 9*AS | 12*AS |
| Sling (various sizes) | 2*AS | 6*AS | 6*AS |
| Serpentine** | 28*AP | | |
| Fowler | | 6*AS?AP | 6*AS?AP |
| Base | | 60AP | 30AP |
| Stone gun** | 26*?AP | | |
| Top gun/piece | 3AP | | 2AP |
| Hailshot piece | | | 20AP |
| Total iron | 65 | 81 | 76 |
| Total guns | 78 | 96 | 91 |
| Carriage-mounted | 10–18 | 36 | 39 |
| Anti-personnel | 65–68 | 62–66 | 53–59 |
| Anti-ship | 8–13 | 28–38 | 32–38 |

\* = carriage-mounted guns; \*\* = serpentines and stone guns may be on wood stocks but ship-mounted/supported, possibly on swivel mountings; AP = anti-personel; AS = anti-ship

powder, smaller shot, staff weapons, handguns, armour, and archery equipment in this last group which is handed over to the Ordnance Office.

Unusually this inventory not only names gun types but gives their positions on five vessels (the *John Baptist*, *Great Barbara*, *Great Nicholas*, *Great Elizabeth* and *Peter Pomegranate* – but sadly not the *Mary Rose*). Unfortunately the terminology used to describe gun locations between the five vessels differs greatly, making interpretation difficult (see below and Table 11.2).

Items are listed (Knighton and Loades 2000, 142) which do not appear in other inventories for the *Mary Rose*. These include cross-bar shot, balls and arrows of wildfire, '*cartouches*' (cartridges) of gunpowder in barrels, charging ladles, sponges, leather buckets,

Table 11.2 Extrapolated location of guns on named ships in 1514; decks in most likely descending order within ship

| Ship | Bow | Uncertain | Waist | Stern |
|---|---|---|---|---|
| Great Nicholas | Forecastle | | | Somercastle Somerdeck |
| Peter Pomegranate | Upper Lop Upper Forecastle Forecastle Brest of Forecastle | | Waist | Deck Brest of Somercastle |
| Great Elizabeth | Forecastle Nederdeck Middle Deck Upper Lop | | | Stern (?transom) Middle deck Neder deck Upper deck |
| Great Barbara | Forecastle | Upper lop Hoile | Waist | Barbican Middle deck High deck |
| John Baptist | Forecastle | Upperlop | | Somercastle Stern (?transom) Middle deck Upper deck |

Source: PRO, E36/13 (Knighton and Loades 2000)

scoops and shovels. There is also armour – '*Salettes*' (helmets) '*Brestes*' (breastplates), '*Backes*' (backplates), '*Gorgettes*' (neck armour) and '*Splentes*' (arm armour) – for around 200 men. These items were not necessarily for use at sea. The '*Stakes for the feld*', were certainly not; and the many '*shod*' wheels for gun carriages (ie, with iron tyres) are a further reminder that the ship had been transporting land forces.

The gun names and their sizes in 1514 differ markedly from both the 1540 and 1546 inventories (Table 11.1). The emphasis is on the lighter anti-personnel guns, traditionally positioned high up in the castles. The consistent lack of significant numbers of large guns suggests that there is no requirement for a lower gundeck (corresponding to the extant Main deck). The various types of ambiguous guns are described briefly below in order of their numerical incidence/descending order of size as listed for the *Mary Rose*. The guns are either wrought iron, cast bronze (listed as brass) or occasionally cast iron.

*Serpentines* (28, iron): The most numerous gun, present on every vessel. The term is used to describe a gun long in relation to calibre, characteristics found in culverins. Usually listed as iron, but occasionally of brass. Given the numbers present they must be small: The *Henry Grace à Dieu* had 130 iron and five bronze serpentines, including eight for the ship's boat (Table 11.1). They are usually denoted single or double, and with two chambers, '*miches*' and '*forelockes*' (possibly swivel pegs and wedges). Less common is evidence of carriages: the *Henry* (Knighton and Loades 2000, 119) had examples '*apon wheles shod with yron*' and '*apon wheles unshodd*'.

Blackmore (1976, 242) noted that serpentines appear in the Tower inventories of 1495, 1514 and 1523. He also quoted details from Exchequer records in the Public Record Office: an account of 1455 describes four serpentines of 7½–8½ft (*c.* 2.3–2.6m) in length with chambers and throwing a 2½–3½lb (*c.* 1.1–1.6kg) lead ball. In 1474 one example with chambers, bolts, pouches and locks weighed 770lb (349kg); there were also three small serpentines and a long red one. A record from 1492 showed them shooting 6lb (2.7kg) iron shot with 7lb (3.2kg) powder. A reference in 1508 to '*one hole pece of iron calld a serpentyne alyas aslange*', suggests may be long thin guns, akin to slings. In 1513, 50 are listed for ships, weighing 261¼lb (300kg) each, with chambers of 41lb (18.6kg) (BL Stowe MS 146, f. 41; also cited Blackmore, *ibid*.). Ammunition for serpentines is mentioned only in the inventory of 1547, where lead, or dice and lead shot, is specified. Their position in the list of iron guns is always around the sling/base/falconet grouping. By 1547 only two ships carry serpentines: the *Struse of Danzig* (3) and the *Mary Willoughby* (1) (Starkey 1998, 7649, 8140). Entries for land fortifications mention serpentines stocked, as at Pontefract: '*Serpentynes of yrone stocked and bound with yron with xij chambers to the same*' (6206) and Berwick: '*Serpentynes of yrone with ij Chambers a pece with forlockes stocked with Miches of yrone*' (6569), and similar entries (6599, 6607, 6614, 6620, 6629, 6636, 6643, 6653, 6660, 6669, 6680, 6687).

These descriptions suggest similarities with the sixteenth century swivel guns recovered from a wreck in the Cattewater, Plymouth, one of which is mounted on an oak bed (Redknap 1984; 1997, 76–9). There are parallels from Europe, notably a collection in the Danish Royal Arsenal Museum recovered from the Kattegat, with examples from Sweden and the Goodwins (Redknap 1997, 76). Positions given for serpentines include the forecastle, upper forecastle, somercastle, waist, '*brest*' (of the castle), stern, upper deck (in the castles) and in the '*holle*' (see below).

*Stone guns* (26, iron): On every vessel except the *Trinity*, *John Baptist* and *Katherine Galley*, always iron chamber pieces. Listed '*upon trotill wheels* [possibly small solid trucks] *with miches, bolts and forelocks*', and are sometimes listed as '*great*'. Often they are listed together with top guns and serpentines and it is convenient to see these as being similar but firing stone shot. On the *Great Elizabeth* they are positioned in the '*foretop*' and '*meson*' top, so they cannot be very large. Blackmore (1976, 245) notes an entry as early as 1474 to eleven stone guns of copper bought for Edward IV's invasion of France. These may well be have been the eleven '*Stone gonnes ... ych of them with his miche & foreloke*' carried by the *Sovereign* in 1495 (*ibid.*; Oppenheim 1896a, 194). The Tower inventory of 1523 includes ten stone guns, listed after slings and before serpentines (*ibid.*, 260). Their location where stated in 1514 was variously given as the '*holle*' (*Great Barbara*), '*somer deck*' (*Great Nicholas*), forecastle (*John Baptist*, *Peter*), '*forecastell in the neder deck*' (*Great Elizabeth*) and '*barbican*' (*Christ of Greenwich*).

*Murderers* (six: two of bronze and four of iron): Nine vessels carry murderers. These are either brass or iron, are breech-loading and may be prefixed with '*Great*'. The *Mary Rose* has six, two of which are bronze (Table 11.1). The greatest number on any ship is fourteen. The 1547 inventory includes '*Mortherers of yrone to shotte haile shotte*' (Starkey 1998, 4013). Aboard the *Peter* in 1514 six great murderers of iron are listed on the '*upper lop*' while the *Great Elizabeth* houses two on of brass on the '*upper lopp*'. They can be described as '*square*', possibly akin to a hailshot piece (Norton 1628, 59). Murderers are listed in land fortification entries between 1495 (a single '*Gonne named mumdrer withoute chamber and forelock*') and 1677 (Blackmore 1976, 236, 256). Norton describes them in 1628 as the fourth kind of ordnance, that '*either shoote stone shot, Fireballes, Murthering shot, or els no shot at all*'.

Murdering shot is described by Norton as '*base, burr of any kind of murthering shot put up in bagges or lanterns fitted to the bore*' (containing sharp small fragments which can maim or kill):

> '*The sorts of this kind are the morter peeces, murtherers, petttards etc, being in length under six times the height of the bore at their mouthes, of each sort of this kind some are bigger, and some leser according as the assigned service requireth*'.

They are listed as being shorter than the port pieces, fowlers, slings, perriers or bombards, but all are made of iron and can be used with murthering shot. By 1670 they are either bronze or iron and are described as swivel guns firing anti-personnel (scattering) shot (Blackmore 1976, 236). It is convenient to see these as being short guns, shorter than port pieces and fowlers. There is little evidence as to how they are mounted and no real data on their size is available.

*Curtows* (five, bronze): Curtows, described as '*great*', '*demi*', '*half*', or merely '*curtows*'. These are usually large guns. Five vessels carry them, and the fleet total is fourteen. The *Mary Rose* is provisioned with five of the great curtows, the *Henry Grace à Dieu* with only one, the *Trinity Sovereign* with three curtows '*stocked apon trotill wheles*', one demi curtow, similarly stocked and two half curtows '*apon wheles shodd with yron*'. The *Gabriel Royal* has one upon bare wheels, and the *Kateryn Forteleza* also has one. It is most unfortunate that none of the vessels for which gun locations are given carries either of these two classes.

Curtows are generally accepted as being the forefathers of the classic bronze muzzle-loading cannon. Most references are to bronze curtows, and rarely are there references to chambers. However guns listed or the Tower in 1495 include one '*Curtow*' and one '*Dyme curto*' among '*Broken gonnes of brasse callid* [called] *with chambres of yron*' (Blackmore 1976, 257). In 1523 there are '*Courtowes*' and '*Doble Courtowes*' (*ibid.*, 260). Shot is either stone or iron. In *c.* 1492 curtows were listed as shooting an iron shot of 100lb (45.4kg) and demi curtows one of 40lb (18.1kg) (*ibid.*, 225). In 1513 curtows are listed as shooting 60lb (27.2kg) shot with 40lb (18.1kg) of powder 40 times per day (*ibid.*). Charging ladles are listed for these in 1495. Curtows are listed in 1523 but not in 1540, when cannon are listed. Teesdale (1991, 21) equates the weight of these with bronze demi cannon.

*Falconets* (three, bronze): Eight are listed for the fleet in 1514. They fired shot of about 2in (51mm) diameter and were sometimes on '*treseles*'. These are long in relation to the bore and are considered to be of the '*culverin*' type. The *Great Barbara* has six falconets (five on trestles and one on wheels) in the '*hygh deck*' of the ship. In 1587 (Blackmore 1976, 229) they are listed as weighing 360–400lb (164–81kg) and one is shown on a carriage in a manuscript dated 1563 (Wright 1563, 8).

*Cast guns* (two, iron): Cast iron is relatively rare at this time, but there are some '*cast pieces of yron*'. In addition to the two on the *Mary Rose* (with four chambers) there are two with four chambers on the *Great Nicholas* in the forecastle and one on the *John Baptist* on the middle deck (possibly in the stern).

*Top guns* (three, iron): Top guns are rarely listed, with only two on the *John Baptist*, three on the *Mary Rose* and one each on the *Peter* and *Christ of Greenwich*. Used in the fighting tops, these are small guns, possibly akin to, but smaller than, the bases carried in 1545.

*Slings* (two, iron): Slings (slings, and demi slings or half slings) are usually made of iron and listed with chambers. They appear on all vessels (except the *Great Barbara*) and are sometimes described as being '*on trotill wheels*' or '*stocked*'. Thirty-four are listed in total, firing iron shot. The *Great Elizabeth* carries two bronze half slings in her stern while the *Mary Rose* has two demi slings. They are listed for the Tower between 1523 and 1676 and may represent the wrought iron equivalent of the culverin class, long in relation to bore. On the *Great Nicholas* these are on the '*somer*' deck (two), on the *John Baptist* on the middle deck (two) and on the *Peter* one in the breast of the forecastle and two (great) in the breast

*Figure 11.5 The Dardanelles gun: a 'bombard' (©Royal Armouries)*

of the sterncastle. The *Christ of Greenwich* has one in the barbican and one in the '*highest*' deck in the stern. These guns are probably not dissimilar from the slings carried on the *Mary Rose* when she sank. Their positions, primarily fore and aft, suggest that they may represent the longer range of guns at this time.

*Falcons* (two, bronze): Twenty-five '*brass*' falcons are listed in 1514. The *Christ of Greenwich* carries one iron half falcon and the *Gabriel Royal* two iron falcons. These are accepted as being small muzzle-loading guns on carriages, firing a 2½–3in (63–76mm) lead or composite shot. They occur in Tower inventories between1495 and1726. Usually they occur in pairs; only the *Henry Grace à Dieu* (eight) and the *Gabriel Royal* (six bronze and two iron) have more than a single pair of falcons. Her two falcons are on shod wheels in the middle deck. On the *John Baptist* the single falcon is on the upper deck (in the stern), and the *Peter* has two in the '*brest of the somer castell*'.

*Little bronzs gun without a chamber* (one): There is no information regarding the size of this, but it is unlikely to be large and appears not to have been in service as it occurs within the list of accoutrements rather than with the guns.

The descriptions attributed to many of the bronze guns in this inventory suggest that they were still relatively uncommon and that there was little formal classification especially of the 'long thin' bronze guns. This is not unexpected as the mould cannot be reused, so mass production was not easy to achieve. There are, however, some names which continued to be used for several hundred years and some which are nearly obsolete by this time. Bronze guns listed on other vessels are listed below.

*Large bronze guns ('bombards')*: Other large pieces include '*Great bumberds of brass*'. One of these is listed on the *Henry Grace à Dieu* as '*Grete bumberds of brasse apon iiij trotill wheles of Herberd makyng j*' (probably referring to the gunfounder Herbert de Pole, who worked with Simon Giles of Mechlin). The term '*bombard*' was originally used to describe the stone throwing engines used in medieval sieges but, following the invention of gunpowder the term was used to describe the largest battering siege pieces (Smith and Brown 1989, v). These are accepted as being short in relation to their bore and fired either iron or stone shot. Blackmore (1976, 220) quotes a document of 1513: '*Every bombard shotith of Iron ... ij$^c$lx lb.* [260 lb/117.9kg] *And of powder at every shot ... lxxx lb.* [80lb/36.3kg] *which may be shot v tymes a day every pece*'. This would require a bore of 12–13in (305–330mm). Extant examples are muzzle-loaders, made with separate powder chambers which locate into the barrel but are not designed to be removed to reload. Illustrations suggest a large carriage with two cheek pieces on two pairs of spoked wheels, however '*truck*' wheels are specifically mentioned. These occur in listings for the Tower between 1514 and 1675. By 1675 (*ibid.*, 310) descriptions include '*bombard or port peece 10 inches* [254mm] *Diam 01 on an Uns. Shipp Carr.*' This suggests the possibility that the roles of the two types of gun may have been the same (battering at close range). The major differences between the port piece and bombard are the calibre and the fact that the port pieces are breech-loading (and wrought iron). The weight of the largest extant bombards include the bronze Dardanelles gun at over 16 tons (Fig. 11.5), the iron *Mons Meg* at just under 6 tons and *Dulle Griet* at over 16 tons. Smaller examples exist (the smaller at Mont Saint Michel is about 2¾ tons) but this wide range makes the suggestion of a weight or size unreliable though one would not expect the largest of these to be put aboard a ship.

*Culverins* (bronze): *The Henry Grace à Dieu* is listed with two great culverins unshod, the *Trinity Sovereign* with three, the *Gabriel Royal* with one great culverin and the *Katherine Galley* with one with shod wheels. Culverins are listed in inventories for the Tower between 1514 and 1726. In 1513 it was said that '*Every Culveryn Shotith of Iron and of powder at every shot xxij lb., which may be shot xxxvj tymes a day every pece*' (Blackmore 1976, 224). A 22lb (10kg) iron shot would require a bore of at least 5½in (140mm). The Tower inventories often mention the gunfounder, again suggesting that these may be 'novel' or unusual enough to be recognised by the name of the maker.

Other unusual items include one entry for '*Smale vice peces of brasse apon shodd wheles of Symondes* [Simon Giles] *makyng*' and three '*Longe vice pece*[s] *of brasse of the same makyng*', all on the *Henry Grace à Dieu*. ('Vice' pieces may be guns which retain a vice or screw, a large extant example of which is the Dardanelles gun [1464] currently at the Royal Armouries Museum of Artillery in Fort Nelson, Fareham. In 1562 an ironmonger supplied the Navy with a small quantity of '*vizes for the beastes to stand on*' [Bodleian MS Rawlinson A. 200, f. 37], perhaps some way of securing ordnance.) In 1514 the *Henry* also has a '*fayre pece of brasse of Arragows makyi*ng' (presumably Aragonese). The *Peter Pomegranate* has a single piece of brass '*with a chamber*' and the *Mary Rose* a '*lytell gonne of brasse without a chamber*'.

*Great iron guns*: There are 35 listings for '*great iron guns*'. There are twelve on the *Henry Grace à Dieu* of '*oon makyng and bygnes*' and four more '*Grete yron gonnes of oone sort that come owt of Flaunders with myches boltes and forelockes*'. This second description suggests that these may not be as '*great*' as at first thought. Two others are listed as '*Grete Spanysh peces*'. Great iron guns are listed on the *Henry Grace à Dieu* (eighteen), the *Trinity Soverign* (seven) and the *John Baptist* (ten); the latter are located on the upper lop (four), stern (five) and the stern middle deck (one).

*Capstan guns, top guns, organ and pot guns*: One vessel, (the *Peter Pomegranate*) lists two capstan guns of iron with chambers. Only the *Gabriel Royal* lists '*organs*' on three stocks. These are multi-barrelled small guns (similar in shape to a church organ). Four '*Pott gonnes*' are also listed in the *Peter*, seemingly at the stern and in '*the brest of the smale deck*'. These were probably small mortars.

Another useful source of information is the list of ordnance shipped to Scotland in 1497, among the naval accounts of Henry VII's reign printed for the Navy Records Society (Oppenheim 1896a, 82–132). Falcons, serpentines and demi curtows (in small numbers) are the most prevalent mounted gun with the occasional bumbardell and curtow. All those on vessels appear to be in bronze, with shot of iron, stone and sometimes pellets of lead. Artillery in store in Portsmouth includes serpentines, stone guns and murderers of iron. Many of the larger guns have names, such as the '*Curtowe of brasse ... cald the Dragon*' and the '*Demy Curtowe of brasse cald Berwick*' (*ibid.*, 85, 86). The list includes most things necessary for warfare at this period and what they were carried in, for example chests of arrows and longbows, dice of iron in baskets, nails, candles and halberds in chests, gunpowder, tallow and bowstrings in barrels. It is also clear that most items are for use ashore.

The inventory of weapons in the Tower in 1523 can be used to estimate the relative sizes of these early guns. They appear to conform to a hierarchy in size under the headings of bronze or iron, althouth there are still guns of unknown size, and specific diameters are still unknown. The '*bronze*' heirarchy includes bombards, bombardels, double curtows, curtows, culverins, serpentines, chamber pieces, falcons and falconets. The iron guns include slings, stone guns with chambers, serpentines and falcons. By inference this suggests that serpentines are smaller than culverins but larger than falcons, and stone guns are larger than serpentines but smaller than slings. This suggests that the majority of the guns carried were therefore relatively small, with possibly no more than a 3in (76mm) diameter bore.

*Gun locations, identifying the gundecks*: The words used to describe the locations of guns in 1514 vary considerably but include the '*forecastle*', '*somercastle*', '*barbican*', '*holle*' and '*stern*'. A barbican can be a gatehouse or towered gateway, or a temporary wooden tower. An early *OED* description (de Pisan 1489) specifically states that '*In the greatest vessels of were* [war] *men make towris and barbacanes*'. A somercastle is similarly described as '*a movable tower used in a siege*'. It seems that, for the most part the bowcastle is described as the forecastle and the sterncastle as the somercastle. On the *Great Barbara*, which lists the holle and barbican (and the waist), neither the stern nor the somercastle are mentioned. Separate locations are given within the bow and stern for different ships but it is not clear from the various terms used where each deck, or part deck, sits in relation to the others. '*Neder*', for instance, means lower or lowest but, if the heaviest guns are stationed lowermost for each area, then the '*upperlop*' should be the lowest deck as this location is often listed with the heaviest guns. Perhaps this term is interchangeable with '*neder deck*'. It could relate to the Upper deck on the *Mary Rose*. The forecastle would then relate to the first castle deck, and the upper forecastle to a second; neither survive, but both are depicted on the *Anthony Roll* illustration. Alternatively the '*upperlop*' could relate to the Main deck of the *Mary Rose* in 1514, if it were high enough above the waterline to accommodate guns without lidded ports (possibly the two stern most ports on the extant main deck and perhaps one or two in the bow, as yet not excavated). The '*breast*' may relate to the guns firing above the breast hooks or the transom knees (or even each 'side' of the ship). Other locations include the forward facing sides of the castles, however, the specific deck cannot be determined, although one higher than the waist is suggested. Table 11.2 presents a possible arrangement of decks.

The 1495 inventory for the *Sovereign* (Oppenheim 1896a, 194–5) also gives locations for the guns. These include the waist, which carries 20 '*Stone gonnes*' and the somercastle, the stern (possibly on either side of the rudder), the deck over the somercastle and the poop. The forecastle appears to have two decks above the '*waist*'. The guns attributed to the waist are on miches as well as all the others and there appear to be no carriage mounted guns. This may be unusual, as she was being outfitted primarily for a trading voyage. The 'waist' may mean a deck going around the whole ship to accommodate the stone guns, or it could represent the waist as we understand it, between the castles on the upper deck. The Upper deck of the *Mary Rose* shows evidence for blocked swivel holes all along the stern. This would give two decks and the poop above the Upper deck, as shown in the *Anthony Roll*.

The armaments of the *Mary Rose* in 1514 show a very different vessel from the one which sank 31 years later (Table 11.1). Although there are thirteen bronze guns, only three fewer than listed in 1546, three of these (the falconets) are smaller than any bronze guns in 1546. What appears to be absent in 1514 are the ten longer range guns, the sakers and culverins. Although there are five '*great curtows*' and two '*murderers*' – neither can be specifically identified. The curtows were probably the equivalent of later cannon providing some medium range firepower. Most of the iron guns are also imprecisely known but the serpentines may be functionally the forerunners of the bases which occur in the later inventories (Table 11.1).

The guns which were probably mounted on carriages, rather than on stock-like beds with swivel yokes, are the curtows, falcons, falconets, and slings, a total of twelve guns. There is little information on the mounting of murderers, but as these evolved into standard swivel guns they may not be mounted at this stage. Even if these were mounted swivel guns, the total is still only eighteen mounted guns (including the two listed as '*brass*'; Table 11.1), compared with 36 in 1540 and 39 in 1546. Only the five curtows can certainly be accepted as being large guns, comparable with the larger bronze guns carried in 1540 and 1546. Two of the five vessels in 1514 suggest guns on three decks, the *Great Elizabeth* and the *Peter Pomegranate* with the majority, being the lighter, iron guns, predominantly on the upper decks (see above and Table 11.2). We may assume a similar overall deployment on the *Mary Rose*. Various authors have attempted to rationalise the possible deck names and locations, including Oppenheim (1896b) and Carr-Laughton (1960) as shown in Figure 11.6 (and see *AMR* Vol. 2, chap. 20). The starboard elevation of the ship as she may have looked in 1514 is shown in Figure 11.7.

*The* Mary Rose *in 1540*
The 1540 inventory is interesting in that it has so few ships listed (ten) and lists guns only. Shot is not mentioned, neither are any munitions or hand weapons. The inventory is dated 13 February and the vessels listed are virtually the same as those mentioned a year earlier (26 January 1539) as being in the Thames (*LP* 14/1, 143; *AMR* Vol. I, appx, 139). These are described as being '*new made, standing in their docks there, masts ready but not set up, who cannot be made ready to sail under three months time after commandment given*'. The only vessel missing from the earlier letter is the *Sweepstake*. This '*new making*' must be no later than January 1539 and must have accommodated the guns listed in 1540, and no earlier than the first mention of port pieces in 1535. There is little mention of the *Mary Rose* over this period. The rebuild referred to by Cromwell in 1536 as among his achievements since coming into the King's service (*LP* 10, 1231; *AMR* Vol. I, appx, 138) may well have occurred some years earlier. A further refit for the *Mary Rose* might be indicated by the £1000 given on 8 March 1538 to Henry Huttoft '*towards the new making of a certain ship for the King*' (*LP* 13/2, 457 [p. 179]). (See *LMR*, p. 96, *AMR* Vol. 1, chap. 1 and *AMR* Vol. 2, chap. 20 for discussion). Equally it may relate to the refits which were still incomplete in Febuary 1539.

By 1540 we find names which are comfortably recognisable; including cannon, demi cannon, culverin and sakers. Iron guns, sometimes for the first time, include port pieces, fowlers and bases, together with slings. The 1540 inventory for the *Mary Rose* demonstrates that ships were also being provisioned with these new guns, and she is listed with fifteen bronze guns: in place of the five great curtows are four demi cannon. Other bronze guns include two culverins, two demi culverins, five sakers and two falcons, all requiring carriages. The three falconets and the little bronze gun are absent, perhaps their function was covered by the additional slings and bases. For the first time we are witnessing the introduction of a number of the longer range guns firing iron shot, the culverins, and the introduction of '*corn powder*', a stronger type of gunpowder.

The greatest change however is among the wrought iron guns, where there is an addition of nine large wrought iron port pieces, all over a ton in weight (including carriage) and it is estimated that in 1540 she carried fifteen guns of over a ton and four over two tons. There is also an increase in six smaller wrought iron guns, the fowlers, and the addition of four wrought iron slings. The number of swivel-type guns increases from 54 (serpentines and stone guns) in 1514 to 60 (bases) and while the murderers, cast pieces and top pieces are

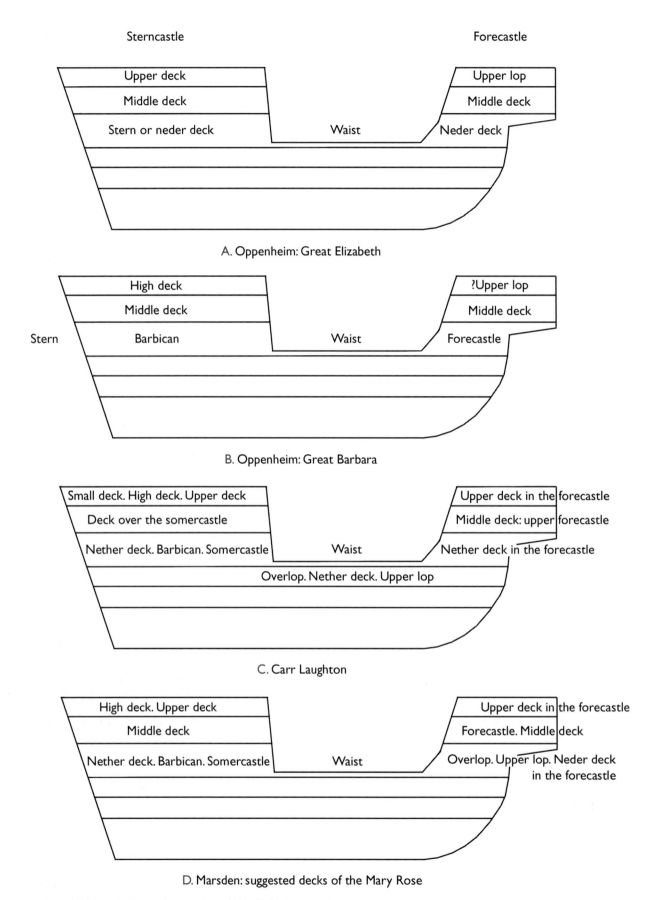

*Figure 11.6 Various schemes for naming of the decks (see text)*

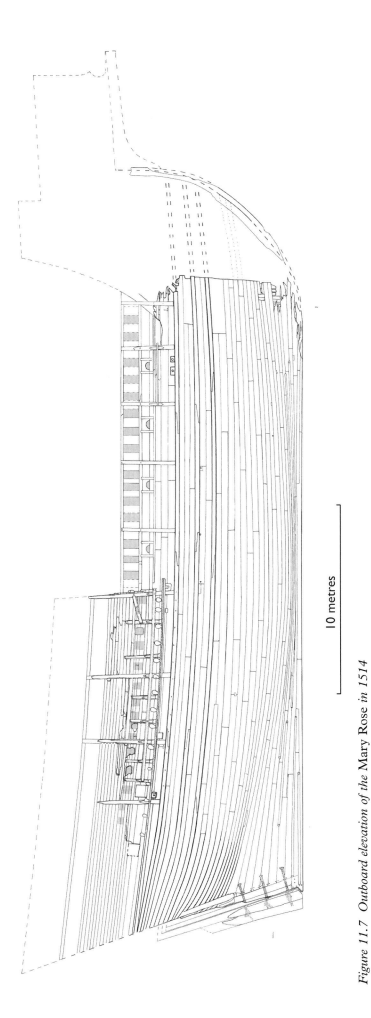

Figure 11.7 Outboard elevation of the Mary Rose in 1514

gone, six fowlers and four more slings are added. Where found *in situ*, these were Upper deck guns. The number of wrought iron guns increases from 65 to 81. This increase is in the largest iron guns.

All the *in situ* port pieces found on the *Mary Rose* were stationed at lidded ports along the main gun deck. These are new guns, first mentioned in the 1530s (see Chapter 3). Although wrought iron is still predominant, with the number of bronze guns remaining relatively static, the general trend is towards larger, carriage-mounted guns. The addition of nine port pieces in 1540 could not be achieved before cutting a row of gunports in the lowest deck above the waterline, the Main deck. This, in turn, could not be undertaken before the acceptance of the gunport lid as an integral part of shipbuilding. Although attributed to the Breton shipbuilder Descharges, sometime around 1501, this modification does not seem to have been adopted before the refit (probably in 1536) that resulted in the 1540 inventory.

The types of bronze guns suggest a rationalisation and a size sequence which stay basically the same for several hundred years. In place of the five great curtows are four demi cannon. The ship is also provisioned with two culverins and two demi culverins. It is likely that the demi cannon and culverins would be accommodated on the Main deck giving a very similar distribution to that of 1546. The smaller bronze guns include the addition of five sakers. Even a conservative estimate for the additional weight of the new guns would be in the region of 23 tons without carriages, shot or powder, 19 tons of which is expected to be on the Main deck. The only other vessel which has a list of ordnance in all three inventories is the *Peter Pomegranate*. The types of gun and how they change in time is very similar to the *Mary Rose*. In 1540 she has a radical upgrade in the number of large guns, with ten port pieces, four slings, five fowlers and five quarter slings. Similarly her bronze artillery changes to the cannon and culverin classes.

All ten vessels in the 1540 inventory carry the cannon and culverin class of guns, smaller vessels (the *Lion*) having only the smaller types such as sakers and falcons. All carry port pieces with a minimum of five on the *Small Galley* and twelve on the *Great Galley*.

*The* Mary Rose *in 1545: a summary*
The 1546 inventory shows little change from 1540 in any of the ships described in both inventories. The emphasis is an upgrade in the size of the weapons. The number of guns on the

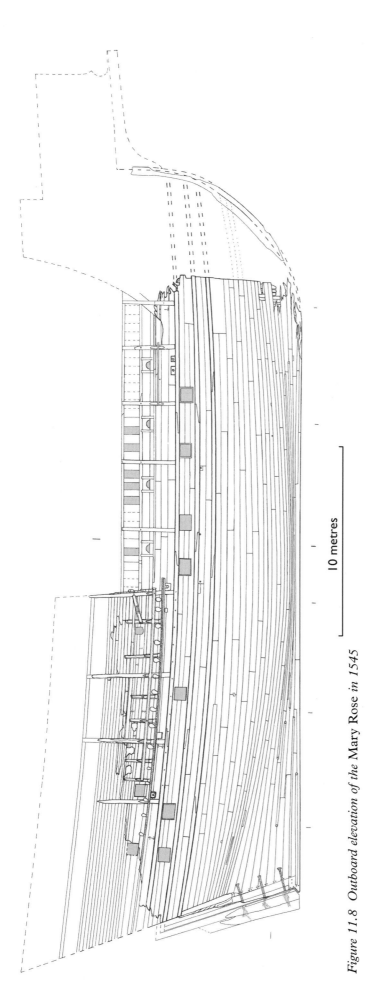

Figure 11.8 Outboard elevation of the Mary Rose in 1545

*Mary Rose* is actually reduced from 96 to 91. Two cannon are added but two demi cannon are taken away (Table 11.1). The addition of bronze cannon in the centre of the ship flanked by two wrought iron port pieces of similar bore affords the midships section of the vessel an armoury of heavy, medium range guns with large bores, with the longer range culverins and demi cannon at the bow and stern (see below). Three sakers and one falcon are lost but four larger bronze guns, demi culverins, are gained. The number of wrought iron port pieces is increased and two top pieces are added. The number of bases is halved but the shortfall is presumably made up by the 20 newly cast hailshot pieces, firing small dice as particularly nasty anti-personnel weapons (Chapter 4). A balance appears to be achieved offering long, medium and short range with all round coverage (see below).

At the start of Henry VIII's reign bronze ordnance was imported from abroad but during his reign he established bronze foundries in and around London and bought in the best practitioners. One founder in particular, Peter Baude, is thought to have been a key person in the development of the cast iron industry. Copper still had to be imported and the cost for producing a bronze gun varied between seven and twelve times the cost of producing an iron gun. It is interesting that, with the perceived threat of invasion, Henry sought to arm his ships and forts with the wrought iron guns reflecting an older technology but yielding a cheaper and more rapidly produced weapon.

The plethora of new guns indicated by the changing names and quantities in the various inventories, regardless of the technology used to produce them, must reflect the political arena and the perceived threat of invasion by the European alliance. It is clear that, certainly by the 1540s if not earlier, experimentation was being carried out in casting iron guns (Teesdale 1991, 11). In the 1555 inventory (Longleat Miscellaneous MS V, ff.1, 53–73v) there are 55 cast iron carriage-mounted guns in a fleet of 24 vessels, none larger than demi culverins. This figure gradually increases, with both the bronze and wrought iron guns eventually being largely replaced by castings in iron. The reasons for this are undoubtedly financial. The inventories indicate not so much a decrease in the number of guns overall, although this is the case in some instances, but an increase in the number of large guns. This may coincide with the requirement for a gun deck closer to the waterline that was fitted with lidded gunports, possibly constructed between 1523 and 1539–40. However, the lengthy delay in the cutting of gunports along

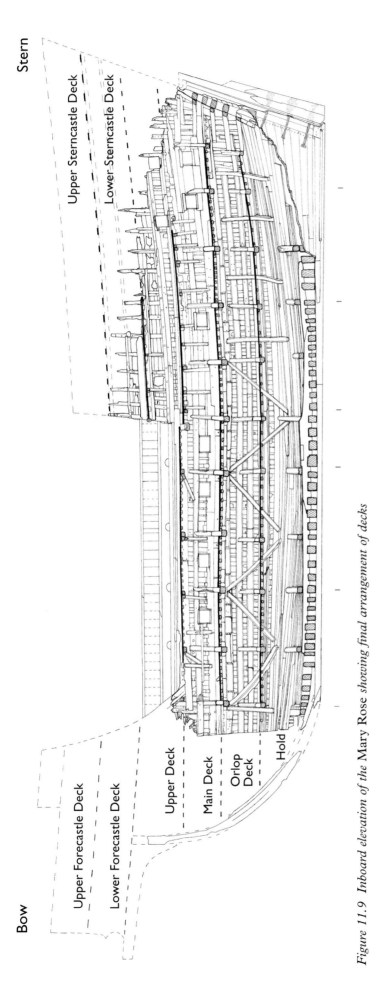

*Figure 11.9 Inboard elevation of the Mary Rose showing final arrangement of decks*

the lowermost decks of vessels indicates some conservatism. If gunport lids can be attributed to 1501, then why does it take until the middle of the century to fit them in large numbers into capital vessels? Figure 11.8 shows the starboard elevation of the ship in 1545 and Figure 11.9 the probable final layout of the decks.

**The ship in battle**
by Dominic Fontana and Alexzandra Hildred
There is little historical documentation regarding the battles in which the *Mary Rose* fought (see Chapter. 1; *AMR* Vol. 1; *LMR*). There is one early reference to her qualities as a sailing vessel in a general description of the fleet's movement from the Thames to the Downs, in a letter dated 22 March 1513 from Sir Edward Howard, Lord Admiral to the King. In this account the *Mary Rose* is faster than a number of other vessels and is highly praised: *'the flower, I trow, of all ships that ever sailed'* ... '*Sir, she is the noblest ship of sail* [and] *great ship at this hour that I trow be in Christendom. A ship of 100 ton will not be sooner at her ... about than she*' (*LMR*, 13; *AMR* Vol. 1, appx, 55). This suggests that she was, at least in 1513, fast and able to turn swiftly and safely. This seems still to have been the case in 1522, following a visit by the King to Dover where he boarded both the *Mary Rose* and the *Henry Grace à Dieu*. Sir William Fitzwilliam, Vice-Admiral, wrote to the King on 4 June:

> '*and to advertise your grace of the sailing of the* Henry Grace à Dieu, *the same day we made sail, she sailed as well as any ship that was in the fleet, and rather better, and weathered them all save the* Mary Rose. *And if she go by a wind, I assure your grace there will be hard choice between the* Mary Rose *and her, and next them the Galley.*' (*LMR*, 47; *AMR* Vol. 1, appx, 115)

There is little mention of an active role for the *Mary Rose* between 1525 and 1542, during much of which period she was laid up '*in ordinary*' in Deptford, although there is evidence for extensive refurbishment and rebuilding during that period (see *AMR* Vol. 1, chap. 1; Knighton and Loades 2002, 95–106). Although it is probable that *Mary Rose* was with the fleet during the campaign which resulted in the reduction of Boulogne in 1544, there is little evidence to suggest that she actively engaged in warfare during this period.

One of the last documents which refers to the *Mary Rose* by name is entitled '*The order of battle upon the sea*' dated sometime in June 1545 (*AMR* Vol. 1, appx, 145). This may be part of Lord Lisle's attack plan against the French invasion fleet mustering in the mouth of the Seine. It lists 24 vessels by name prefixed by the letters '*b*' for battle and '*w*' for wing. The *Mary Rose* is one of fourteen warships, together with '*all the ships of Bristol*'. Of these, six are '*great ships*', including the *Henry Grace à Dieu*, the *Peter*, the *Jesus*, the *Matthew* and the *Pauncy*. Three are '*galliasses*'; including the only true Mediterranean style galley, the *Galley Subtle*, together with the *Swallow* and the *Unicorn*. Vessels of uncertain type include the *Venetian*, the *Argosy*, the *Saviour*, the *Painted Bull* and the *Ship of Montrego*. The wing includes ten vessels comprising four ships and five '*galliasses*' together with '*all the king's small ships*'. Named ships include the *Struse*, the *Mary* [of] *Hamburg*, the *Sweepstake* and the *Minion*. '*Galliasses*' include the *Great Galley*, the *Salamander*, the *Less Galley* and the *Less New Galley*. There are two of uncertain specification (the *Ship of Dover* and the *New Galleon of Kent*). Certainly the battle held the most heavily armed vessels and the wing the smaller, lesser armed but more manoeuvrable, largely oared vessels. It may be assumed that a similar formation was adopted less than a month later in the Solent.

There are a number of accounts of the sinking of the *Mary Rose*, a few contemporary and several published a number of years later (see *AMR* Vol. 1, chap. 13 and *AMR* Vol. 2, chap. 2). The most authoritative is given in a despatch, partly ciphered, from the Imperial Ambassador François van der Delft to Charles V (*LMR*, 62; *AMR* Vol. 1, appx, 149) which clearly states that the events leading up to the sinking occurred on a single day.

The news that the French were approaching came while the King was at dinner (then meaning the midday meal) on the flagship *Henry Grace à Dieu*; and within two hours the entire enemy fleet was sighted outside the harbour. The Ambassador's report says that five French galleys had '*entered well into the harbour*' and that as the English fleet approached they '*kept up a cannonade against the galleys*'. But the English '*could not get out for want of wind, and in consequence of the opposition of the enemy*'. The term '*harbour*' can be taken to include the Spithead anchorage where many of the ships could lie, protecting the narrow entrance to the inner harbour. The sinking of the *Mary Rose* is said to have occurred '*towards evening*' that day. Van der Delft's explanation ('*misfortune and carelessness*') was based on an interview with one of the few survivors, who was his fellow-countryman. What the Ambassador relayed to the Emperor is the closest we have to an immediate eye-witness account:

> '*The disaster was caused by their not having closed the lowest row of gunports on one side of the ship. Having fired the guns on that side, the ship was turning in order to fire from her other, when the wind caught her sails so strongly as to heel her over, and plunged her open gunports beneath the water, which flooded and sank her*'.

Lord Russell, in a letter dated 23 July 1545 to Sir William Paget, Principal Secretary, also considered that it was through '*rashness and great negligence*' that the ship '*in such wise cast away, with those that were within her*' (*LMR*, 61; *AMR* Vol. 1, appx, 148). The French side is represented by the account of soldier Martin Du Bellay, first printed in 1569 (Stone 1907; *LMR*, 63; *AMR* Vol. 1, appx). This describes the French fleet as arriving on 18 July suggesting that the English fleet came out of Portsmouth to oppose them and that there was a lengthy but inconclusive artillery engagement that evening. The next day the English fleet were becalmed and would have therefore remained at anchor in Spithead. The English were harassed by the artillery fire from a party of French galleys for, as Du Bellay reports, '*over an hour*' although it was probably somewhat longer. This situation, conditioned by the lack of wind and the prevailing tidal currents, continued until later in the afternoon when both wind and tide changed altering the nature of the battlefield and the tactics which could be employed. Du Bellay claims that the *Mary Rose* was sunk by cannon shot and the *Henry Grace à Dieu* badly damaged. Towards the end of the afternoon the wind came up (described as being from the land) which together with the tide enabled the English fleet to move and bear down on the French.

Sir Peter Carew's narrative, written up 40 years or so later by John Hooker, gives a different account of events again (*LMR*, 82; *AMR* Vol. 1, appx, 168). Carew recalls that his brother Sir George (captain of the *Mary Rose* and newly appointed Vice-Admiral) first saw the French fleet approaching. As Sir George commanded his men to hoist sail, the *Mary Rose* began to heel. His uncle Gawain Carew, commanding another ship, called across to Sir George, who shouted back that '*he had the sort of knaves, whom he could not rule*'. Sir Peter interprets this as too many skilled men ('*a hundred mariners, the worst of them being able to be a master in the best ship within the realm*') each disdainful of the others and refusing to obey orders. An alternative explanation might be that mariners panicked when they realised that too much was being asked of the ship, and that she had become dangerously unstable.

The briefer account transmitted through the *Chronicles* of Hall (1548: *AMR* Vol. I, appx, 151), Wriothesley (written by 1559: *LMR*, 64), Holinshed (1577: *AMR* Vol. I, appx, 169) and Stow (1603) places the main action within a single day, though where specified this is 20 July.

Taken together there seem to be a number of potential causes for the sinking, all of which should be examined and tested against the evidence we have, historical, archaeological and geographic. The suggested causes include:

- Overloaded with too much ordnance.
- Gunports 'left' open.
- Gunports too low to the waterline.
- Guns unbreeched.
- Negligence or insubordination.
- Enemy fire.

A combination of these is likely to have contributed to the sinking and the reader is referred to chapters elsewhere in this volume, and in *AMR* Vol. 2 for consideration of the structure and loading of the ship, and for the archaeological evidence for her condition and battle state.

## *The Theatre of War: geographical evidence from the Cowdray Engraving and GIS*
by Dominic Fontana and Alexzandra Hildred

In addition to the archaeological evidence gained from excavation and study of the physical remains of the *Mary Rose* herself, the geographical setting within which she fought her last action is still extant and potentially can tell us much about what happened to the *Mary Rose*. Spithead, the Solent, Portsmouth and the Isle of Wight provide 'The Theatre of War' staging the actions and events of July 1545. There are various written accounts of the battle provided by observers viewing the encounter from both the English and French sides and further documentary sources, such as the 1547 inventory of Henry VIII's possessions (Starkey 1998), include significant information about the weaponry available within the forts at Portsmouth. There are several large-scale Tudor maps of Portsmouth which show both the fortifications and the town in some detail. There is even an early sea chart depicting the entrance channel to Portsmouth Harbour and this clearly shows the navigational hazards of the surrounding sandbanks (UK Hydrographic Office Chart Ref D623). The Cowdray Engraving too provides a detailed visual representation of the opposing fleets, land fortifications, the town of Portsmouth, the Isle of Wight and the battle itself. Consequently, there is considerable data from diverse sources, in various formats, which are both individually and collectively useful in assisting our understanding of the events of the battle and the circumstances under which the *Mary Rose* was lost.

There is also an engraving entitled '*The Encampment of the English Forces near Portsmouth Together with of View of the English and French fleets at Commencement of the Action Between Them on the 19th July 1545*' (Fig. 11.10). It is derived from a contemporary mural that decorated the wall of the dining parlour at Cowdray House, Midhurst, Sussex. The original painting was one of a set of five which included three pictures recalling Henry VIII's campaign in France of summer 1544, the panoramic painting showing the events in Portsmouth and the Solent in 1545 and the Coronation procession of Edward VI in 1547. Sir Anthony Browne was the owner of Cowdray House between 1542, when he inherited it from his half brother William Fitzwilliam (Earl of Southampton and Lord Admiral of England), and his own death in 1548. Browne probably commissioned the works: certainly he plays a prominent role in all the scenes, where his closeness to the King and hence his importance to the events should be noted.

The Society of Antiquaries of London commissioned Samuel Hieronymus Grimm to draw the Portsmouth painting in the 1770s and the ensuing engraving by James Basire (published 1778) proved to be a fortuitous commission as the original paintings were lost in a fire in 1793. A written description of the original paintings was made by Sir Joseph Ayloffe (Ayloffe 1775). Ayloffe considered the painting to be an accurate representation of the scene and was fulsome in his praise. '...*is evidently handled with the greatest attention to truth; all is regular, circumstantial, and intelligible, nothing misrepresented, disguised, or confused.*' . However, although he provides some useful information regarding details, Ayloffe does not go beyond a fairly basic interpretation of the events and there remains much in the picture to be investigated. Similar engravings, also by Basire, published around 1774, of the meeting between Henry VIII and Francis I in 1520 – *The Field of Cloth of Gold* – and *The Embarkation of King Henry VIII at Dover 1520*, are accurate reproductions (given the technological limitations of the engraving process) of their respective painted originals, still extant at Hampton Court. Therefore there can be some confidence in the veracity of the translation from painted original to copper engraving of the Cowdray pictures.

Three of the engravings cover the events of the French campaign during the summer of 1544 and provide a visual record of the King's departure from Calais on his way to join the siege of Boulogne, the encampment of the King at Marquison (*en route*) and the siege of Boulogne itself. The fourth engraved image concerns the events of 1545 in the Solent. Each of the images present their landscapes from a high viewpoint, giving the impression of a panorama. These are not actual viewing points such as hilltops from which one could in reality study the vista but have been constructed by the artist to present the landscape in such a way that the events they contain should be clear within their geographical contextual setting. Indeed, for the 1544 pictures in France in particular, the artist has employed the technique of vertical exaggeration of the land surface to present the hills more clearly than they

*Figure 11.10 The Cowdray Engraving showing the sinking of the* Mary Rose *and disposition of the fleets (© Society of MR = Mary Rose; HGD = Henry Grace à Dieu; EF = English fleet; TP = the Platform; ST = Square Tower; B =*

would be if true scale had been used. Such an approach is used frequently today when modelling three-dimensional computer based landscapes as this makes the features and slopes much easier to understand. Each of the engravings contains considerable detail regarding costume, weaponry, tactics, storage and transport of ammunition, gunnery practice, defensive fortifications and provisioning of an army. However, there is no accompanying annotation so that the images need to be interpreted in the light of the documentary and archaeological evidence.

It is suggested (Friel, pers. comm. 1992) that the Portsmouth scene represents a number of episodes around the staged battle, including the fleet engaging the French on the 19 July and the sinking of the *Mary Rose*. The sinking is in the centre of the image. In the foreground below is Southsea Castle, before which the King passes on horseback, facing forward and apparently oblivious to the disaster happening behind him. Sir Anthony Browne is in immediate attendance alongside the Duke of Suffolk, commander of the land forces which were massed at Portsmouth to repel the invasion (Fig. 11.11a). The picture also the French raids on the Isle of Wight, depicting Bembridge alight and French raiding parties retreating from the eastern shores of the island. In the distance the French galley attack on Sandown Bay is shown. The English are on the march near Sandown and are shown defending the bridge at Yarbridge (Fig. 11.11b).

The town of Portsmouth is shown in some considerable detail in the Cowdray Engraving, including the defences along the walls and those at Point and the Camber. Much of this detail concurs with three sixteenth century manuscript maps of Portsmouth (BL Cotton MSS Augustus I.i.81 [c. 1545–6], I.ii.15 [c. 1552], and I.ii.117 [c. 1584]). Insignificant features such as a small wall and gate between the Square Tower and the Round Tower have been clearly shown in the Cowdray Engraving and on the maps, and there are a number of other buildings which may be similarly identified in these sources including the enclosed site of the mediaeval hospice known as the *Domus Dei*. Its precinct wall is visible with a gateway in *Penny Street*. Also clearly shown is the nearby small chapel which was built to atone for the murder of Adam Moleyns, Bishop of Chichester on the 9th of January 1411 (Wright 1873). The shape of the Camber Quayside too is recorded in the three maps and the Cowdray Engraving in a consistent way. Such agreement between disparate sources gives confidence in the Cowdray portrayal of the scene at Portsmouth and the battle itself.

In the engraving the English army is encamped on Southsea Common, stretching from the town walls to Eastney, probably along the slightly higher, and hence less marshy, ground of what is now *King's Road*, *Elm Grove* and on to *Highland Road* in the East. At the entrance to the inner harbour there are two ships passing the Round Tower with little sail – just a square sail on the foremast and a lateen on the mizzen mast (see Fig. 11.10). Those which have already exited the narrows are shown with more sail and towing their boats. They appear to be closer to the Gosport side,

*Antiquaries of London). Key (left to right): FF = French fleet; FG = French galleys; SC = Southsea Castle;* baskets; RT = Round Tower

possibly using the Swashway channel which is itself clearly marked on the early chart of the entrance to Portsmouth Harbour (UK Hydrographic Office Chart Ref D623). Most of the rest of the English fleet is stationary and without sail. A number of vessels face the shore.

The depictions of individual ships vary in their level of structural detail but although in some cases useful, it has not been possible to firmly identify many of them. However, in the row of vessels behind the *Henry Grace à Dieu* are three large galleys; the central vessel showing a lateen sail on a single mast is possibly the *Galley Subtle* (see above), the smaller oared vessel in the foreground (which might correspond to the wing) may be the *Less Galley*. A command list of late June or early July (*LP*, appx, 1697), which distinguishes merely between '*battle*' and '*wing*' places this vessel in the latter. No attempt can yet be made to identify other ships. On 10 August, in the aftermath of the battle, Lord Admiral Lisle issued more detailed instructions for command structure and its signalling (*LP*, 20/2, 88; Knighton and Loades 2000, 159).

In the Cowdray Engraving the *Mary Rose* is flying a pennant with the cross of St George from her foremast. The Orders specifically state that no vessel should act without the word of the admiral of each ward. One wonders whether Sir George Carew acted out of turn during the battle (possibly to show his skills as newly made Vice-Admiral) and consequently it was to address this problem that Lisle felt the immediate need to reinforce the command structure. This could account for some of the remarks of '*too much haste*' and '*folly*' and even '*negligence*'. There are also some alterations to the positions of key vessels between the two sets of orders, with five strong ships moving from the wing into the battle, suggesting a concentration of forces (Knighton, pers. comm. 2007, citing *LP*, 20/2, 88).

Most of the French fleet appear to be using St Helen's Road anchorage, with four galleys having moved forward into the Solent itself to harass the English fleet, probably with the intent of drawing them out into the open waters of the English Channel where they could be engaged by the entire French fleet. The English ships are predominantly shown with little sail, the *Henry Grace à Dieu* is engaging with a French galley in a nearly head-on attitude, firing one gun only from her port bow (see Figs 11.3 and 11.10). Interestingly, with that exception, all of the gunports on the port side of the *Henry* are shown as being closed. Another French galley is displayed firing towards the sunken *Mary Rose*, with no other vessels in the line of fire. When considering the action undertaken by the advance party of French galleys the Cowdray engraving includes a rather tantalising vignette where of one of the oared galleys in the French fleet off St Helen's, just to the rear of the galley flying a Papal flag, appears to be partly sunken (Fig. 11.11c). There is a statement from the record of the Privy Council meeting held in Portsmouth on 22 July 1545 (*LP* 20/1, 1244): '*A Frenchman saved in a ship sunk by Blakye of Rye sent to my lord Admiral to be examined*'. This implies that a French vessel was sunk at some stage during the action. William Blakye was

*Figure 11.11 Details from the Cowdray Engraving: a) Southsea Castle with Henry VIII, Sir Charles Brandon and Anthony Browne; b) the English defending the bridge at Yarbridge, Isle of Wight; c) sunken vessel within the French fleet at St Helen's (© Society of Antiquaries of London)*

captain of the *Magdalen*, one of the boats of Rye. It was probably quite a small vessel having a crew of just 37 men. It is recorded as having taken part in an action on 10th August 1545 off Camber Castle near Rye (Cooper 1864)). The loss of a French vessel is also reported in a letter from Lord Russell to Sir William Paget, Principal Secretary to King Henry VIII, written from Bodmin on 23 July 1545 where he states:

> '... at the writing of your letters, 17 of the galleys came in the order of battle to the fight, of the which one was sunk, and the ships began to retire, which I believe will not come again...' (LMR, 61)

Van der Delft recorded that five French galleys entered the Solent (*LMR*, 62; *AMR* Vol. 1, appx 149), however the Cowdray Engraving shows only four attacking the English. Perhaps the apparently partly sunken galley is the fifth one, which, having attacked the English ships, sustained damage and returned to the safety of the French fleet, only to sink at St Helen's.

With little or no wind and with a west-east tide flowing on the morning of the battle, the Solent was a confined arena and movement for the English ships would have been considerably restricted and the majority of vessels would need to be anchored if they were not to drift towards the French on the current.

Although the accuracy of many of the events shown in the engraving cannot be directly tested, the geographical positions of the depicted features and the positions and potential movements within the opposing fleets may be usefully investigated. The viewpoint displayed by the image was examined and reconstructed using Digital Terrain Modelling (DTM) and Geographic Information System (GIS) technology to create a computer model of the landscape surface (Fig. 11.12a) which can then be manipulated by rotating and scaling the model to recreate the view of the landscape as it appears in the Cowdray Engraving (Fig. 11.12b). In turn, this enables the known geographical positions of the surviving and identifiable landmarks, such as Southsea Castle and Lumps Fort, to provide a framework by which the geographical coordinates of the ephemeral events shown in the image, including ship positions, might be determined using GIS (Fig. 11.12c). Geographical coordinate cover was therefore extended from known positions to other non-permanent features recorded in the Engraving including the ships at the anchorages of Spithead and St Helen's.

In common with the other pictures from Cowdray House the artist has developed a reconstructed landscape displaying it as though it were a panoramic vista. This would be easily achievable using Tudor technology by laying a map of Portsmouth, the Isle of Wight and the Solent on the floor and then viewing the map from the desired angle and altitude, thereby enabling the artist to sketch the natural features and coastline into their panoramic viewpoint position. It would then be a simple matter to add the prominent buildings, events and people into their appropriate locations within this panoramic landscape. Considerable care has been taken to present the geography accurately and this is exemplified by the artist's

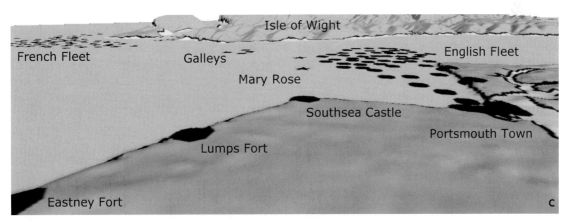

*Figure 11.12 Digital terrain models: a) of the Solent; b) aspect of the battle as in the Cowdray Engraving; c) as above with the English and French fleets*

inclusion of a small inlet on the north coast of the Isle of Wight which plays no part whatsoever in the events of the battle but has the correct location, orientation and size for Wootton Creek. Similarly, the angle created by the Gosport eastern shoreline and the position of Haslar Creek is also as it should be depicted to create an accurate geographical setting.

The positions of each vessel, including the *Henry Grace à Dieu* and the *Mary Rose*, were estimated from the Engraving into a GIS digital map data layer and then superimposed over a computer based DTM landscape view. By a process of iterative adjustment of individual ship locations the relative positions of the ships may be determined in concert with the known

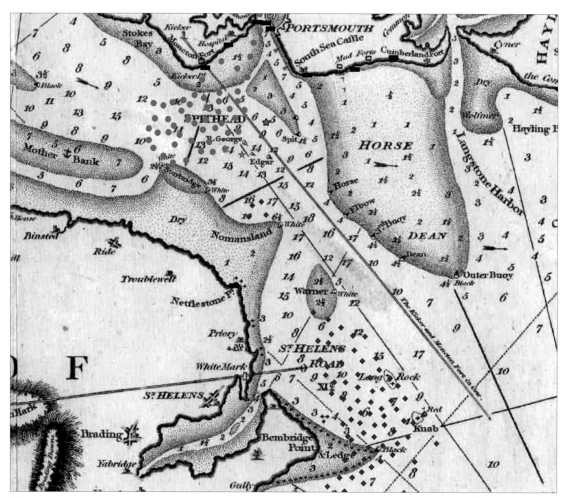

*Figure 11.13 The positions of the fleets on historical chart of the Solent, by William Heather, 1797*

positions of the extant fortifications, the seabed position of the wreck and the coastline. This develops a digital map of the individual ship positions derived from the Cowdray view and when this is employed within the GIS it becomes possible to measure the distances between any of the vessels allowing an assessment of the geography of the battlefield and the constraints on tactics or manoeuvres that it imposed. It is also possible to calculate the time that a ship's passage might take between any two points within the battlefield space and to study the potential gunnery ranges that may have been required for both attack and defence.

The GIS also makes it practicable to use various charts or other maps as background images which further illustrate the positions of the fleets against modern or historical map data, such as that describing the past topography of the seabed (Fig. 11.13). There appears to be little difference in the general pattern of the bathymetry in the area of the battle between 1545 and today apart from, perhaps, some increased depth of the maintained channels. However, despite this apparent accuracy it is not meaningful to provide a simple quantitative measure of plannimetric accuracy for all objects within this representational space as this will vary considerably depending upon what the object or event is and where it is located although, when combined with data from other sources, reasonable estimates of the geographical positions of many of the features may be made. In the case of the positions of the English ships when allowance is made for the bathymetry, the number of ships in the fleet, their individual need for sufficient room to swing at anchor with the tide, their general positional descriptions provided by Lisle's June orders and the size of the ships themselves their positions may be considered reasonably accurate to within about ±200m. It is acknowledged that this is by no means perfect but it absence of better data it provides something of a geographical framework from which to consider the events of the battle.

The GIS can be thought of as a means of exploring the geographical nature of the 'Theatre of War' and may be used as a virtual model of the battlefield to help visualise conjectural scenarios thereby, assisting the development of an improved understanding of the battle with its component engagements and geographical and technological constraints. By understanding the interaction of these constraints and influences deductions may then be made about the likely course of events. For example, there are

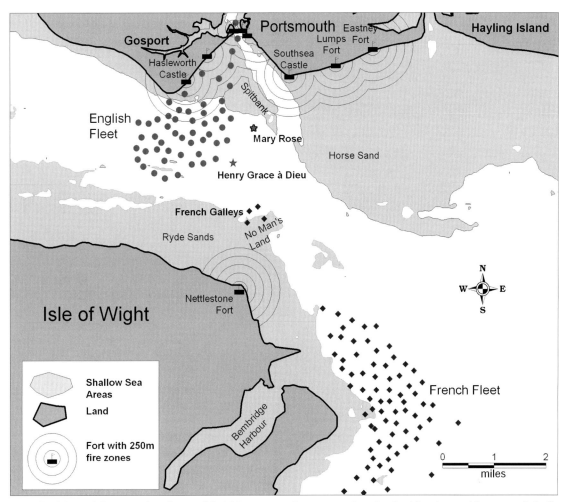

*Figure 11.14 Position of the opposing fleets with 250m range rings around major fortifications with areas of shallow water*

significant areas of Spithead where the large English ships could not go as they required too much depth of water, whereas the French galleys could traverse some of these shallow areas with ease. The GIS makes it a simple matter to map these conditions and interestingly, the advance party of the French Galleys are shown on the Cowdray Engraving as occupying the shallows of 'No Man's Land' at the north-eastern extent of Ryde Sands. This point lies directly between the two fleets and could certainly be considered an area held by neither protagonist. This geographical factor was probably important to the conduct of the battle as the French galleys could occupy this location without too much risk of direct close-range attack from the large English ships and they were probably also out of range of the English guns based at Nettlestone fort on the north coast of the Isle of Wight.

Within the GIS it is possible to study the potential influence of gunnery on the battlefield. Range indication rings can be placed around all the fortifications to show the defence of the harbour approaches (Fig. 11.14). These are shown in 250m increments. They suggest that to defend the navigable channel into the harbour closest to Southsea, the guns at Southsea Castle must have had ranges not less than 750–1000m (to reach the far side of the channel) and those on Haselworth Castle (on the Gosport shore) of up to 1250m, to defend the shallower, smaller Swashway Channel. The presence of Spitsand in the centre of the approach to Portsmouth means that the gun ranges from the Portsmouth and Gosport sides of the approach do not have to intersect; this distance is about 2 miles (3.2km). The distance between Southsea Castle and the French galleys is about the same. Some of the bronze guns positioned at Southsea Castle are shown in an elevated condition and are firing at the French, who appear to be just out of range (see Tables 11.3–11.5 and Figs 11.3 and 11.10).

This data, showing the coastline, fortifications and potential gunnery ranges, may then be viewed in conjunction with the bathymetry of the seafloor. This demonstrates the extremely restricted area within which the English fleet is operating. The navigable corridor between the two anchorages of Spithead and St Helen's are constrained by three sandbars, Spitsand/Spitbank in the north-west, Horse and Dean Sand to the north-east and Ryde Sand to the south, as well as other smaller shoals. The main channel that leads between Horse and Dean Sand and Spitsand/Spitbank is a submerged riverbed that once flowed into the Solent. It is the only

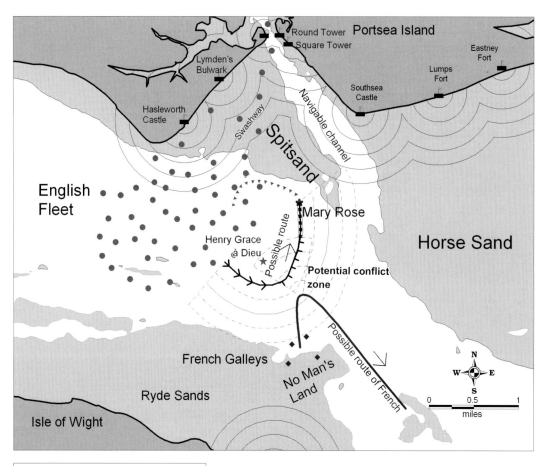

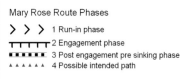

Figure 11.15 Fleet with ships tracks (the end position for the Mary Rose line is defined by her archaeological position on the seabed, and her start position is assumed to be from within the fleet)

deep water access route from Spithead and the Solent into Portsmouth Harbour and, as it approaches the harbour entrance passing close to the Southsea shore where it is less than 500m wide at a navigable depth. Twice a day these rivers still 'flow', as the tide is funnelled into the narrow channels causing a very strong tidal current capable of preventing sailing ships from travelling against the flow. If a ship's passage is correctly timed this current can be used either to take vessels out of the harbour under ebb flow or into the harbour on a flood tide. The narrowness of the navigable channel and these strong currents would have been very important considerations for both potential attackers and defenders of Portsmouth.

Having determined aspects of the geography of the battlefield, including the positions of the fixed onshore fortifications, the relative positions of individual ships derived from the Cowdray Engraving, the bathymetry gained from modern hydrographic survey and historical charting of the seabed; it then becomes possible to develop conjectural models of the progress of the battle whilst taking these factors into account. Possibly the most important scenario to consider is the final passage of the *Mary Rose* herself which ended when she sank. We know from archaeological excavation her final position within the Solent and we know too that her wreck was oriented facing northwards in the direction of Spitsand, which was 600m directly in front of her bow. If we assume that the ship did not relocate significantly from the position when she sank until her excavation we can trace back the path she must have taken across the Solent to arrive at this point (Fig. 11.15).

Clearly, the degree of uncertainty about this route will increase the further we project back along her potential path however, it is unlikely to be more than perhaps two or three hundred metres from her true position as her movements would have been very much constrained within the waters of Spithead and consequently there are only a very limited range of possibilities available. Also from the Cowdray Engraving we can determine the positions of the party

Table 11.3 Gun ranges from historical sources

| Gun type | Lucar 1588 | Lad 1587 | Reynolds Rule | Bianco in Norton 1627 | Eldred 1646 |
|---|---|---|---|---|---|
| Cannon | | PB: 1500ft/460m<br>R: 6750ft/2060m | PB 400yd/365m<br>(20 score) | PB: 300yd/274m | |
| Demi cannon | | PB: 1525ft/465m<br>R: 6840ft/2084m | PB: 560yd/512m<br>(28 score) | PB: 260yd/237m | |
| Culverin | PB: 200 paces<br>R: 800 paces | PB: 2000ft/609m<br>R: 9000ft/2743m | PB:500yd/457m<br>(25 score) | PB:300yd/274m | PB: 460yd/420m<br>R: 2650yd/2422m |
| Demi culverin | | PB: 1800ft/549m<br>R: 8900ft/2733m | PB:400yd/366m<br>(20 score) | PB: 290yd/265m | PB: 400yd/365m |
| Saker | | PB: 1450ft/441m<br>R: 6840ft/2084m | PB: 360yd/329m<br>(18 score) | PB: 250yd/228M | PB: 360yd/329m<br>R: 2170yd/1984m |
| Falcon | PB: 320 yd/292m<br>R: 1204yd/1170m | PB: 1350ft/411m<br>R: 6594ft/2029m | PB: 320 yd/292m<br>(16 score) | PB 216yd/197m | |

PB = point blank; R = random (elevated to 45°).
The source for Lad is Caruana (1992). The transcription gives units in both paces and feet. The pace used appears to be 5ft. There are several mistakes and the figures for Lad are high and may represent double paces

of French galleys which had advanced into Spithead to attack the vanguard of the English fleet. Obviously, as these galleys did not sink the certainty of their positions is not as good as for the *Mary Rose* herself but they are probably quite reasonable estimates and provide a suitable starting point for consideration of the actions described by Du Bellay.

The French account of the battle by Martin Du Bellay (1495/8–1559; account written after 1546) has the French fleet of 150 ships setting sail from France on 16 July and arriving off the east coast of the Isle of Wight on the 18th.

Du Bellay's description of the English fleet (Stone 1907, 2) is:

*'The enemy's fleet consisted of 60 picked ships, well manned and equipped, forty of which, favoured by the land breeze, came out of Portsmouth with great promptitude and in such good order that one might say they awaited with assurance the advance of our force to give it battle. But the Admiral, advancing against them with the rest of the galleys, the remainder of their fleet came out of the harbour to oppose him. After a lengthy artillery engagement the enemy began to edge away to the left, under shelter of the land, to a place where they were protected by some forts which were on the cliff, and on the other side by sandbanks and rocks covered with water lying across the fairway, leaving only a narrow and slanting entry for but few ships abreast* [perhaps this is describing the fortifications of Gosport and the Hamilton Bank or more probably the fort at Nettlestone on the Isle of Wight and the channel between the shallows of No Man's Land and Horse Sand]. *This retreat and the approach of darkness put an end to the fighting for this day, and in spite of the amount of cannon and other artillery which were fired we received little appreciable hurt'.*

Plans were made for the next day, whereby the Admiral D'Annebault would lead 30 ships. Galleys were dispatched to engage the enemy in the morning with the intent of enticing them outside the narrows of Spithead, towards the waiting French ships off St Helens:

*'The changeable weather so varied ... in the morning, favoured by the sea, which was calm, without wind or strong current, our galleys were able to manoeuvre at their pleasure and to the disadvantage of the enemy who, not being able to move for want of wind, remained exposed to our artillery, which had greater power on the ships than they had on it [the galleys], the ships being higher out of the water and more bulky, adding to which, our galleys, by using their oars, could retire out of danger and gain the advantage. In this manner fortune favoured our arms for more than an hour, during which time, in addition to other damage received by the enemy, the* Mary Rose, *one of their principal ships was sunk by cannon shot, and of the five or six hundred men who were in her, only thirty-five were saved.* [Du Bellay was probably part of the delegation led by Admiral D'Annebault to conduct the ratification of the Treaty of Camp, which declared peace between England and France in August 1546 at Hampton Court. He would have met many of the English delegates with whom he no doubt discussed the events of the previous summers at great length.]. *The* Great Harry, *which carried their Admiral, was so damaged that*

Table 11.4 Historical gun ranges (average point blank in metres)

| Gun type | Lad | Reynolds Rule | Lucar | Bianco | Eldred | Average | Ave. used in text |
|---|---|---|---|---|---|---|---|
| Cannon | 460 | 365 | – | 274 | – | 366 *(289) | 300 |
| Demi cannon | 465 | 512 | – | 237 | – | 404 *(327) | 350 |
| Culverin | 609 | 457 | 200 paces/ ?(400yd)/ ?365m | 274 | 420 | 425 *(364) | 420 |
| Demi culverin | 549 | 366 | – | 265 | 365 | 386 *(317) | 380 |
| Saker | 441 | 329 | – | 228 | 329 | 325 *(273) | 300 |
| Falcon | 411 | 292 | 292 | 197 | – | 298 *(231) | 250 |

*Where these occur the lowest figure has been used. The figures quoted are high, and it may be that the paces represent double paces and that the final figures should be halved

*had it not been aided and supported by the neighbouring ships, it would have come to a like end. They would have sustained still more notable losses had not the weather turned in their favour, which not only freed them from this peril, but also enabled them to attack us, by the rise of a land wind, which bore them with the tide full sail against our galleys; and so sudden was this change that our men scarcely had time or opportunity to turn their prows; for during the calm I have mentioned, and in the heat of the engagement, the galleys were so hotly attacked, the ships coming down on them so suddenly and at such speed, that unavoidably they would have struck them below the water line and sent them to the bottom ...'*

Du Bellay's report suggests that there had been considerable exchange of fire resulting in damage to both the *Mary Rose* and *Henry Grace à Dieu* and that the French Galleys had been able to shoot at the much larger and at that point stationary, English ships with some impunity and having done this '*by using their oars, could retire out of danger*'. This implies that the galleys had advanced into danger in the first place, possibly by approaching the English vanguard as closely as they dared before firing their bow mounted guns then turning quickly to make a speedy exit away from any pursuing English shot.

This tactic makes sense given that the French were using oar-powered galleys, which were able to move independently of the wind, whilst the English ships could not at that stage easily manoeuvre to bring their main armament to bear on the attackers. Du Bellay's description of the weather '*the sea, which was calm, without wind or strong current*' describes typical high-pressure summer conditions in the Solent, where through much of the day there is little or no wind until mid afternoon when a sea breeze often occurs. This breeze could have provided the *Mary Rose* with sufficient power to get underway. Indeed, we might conjecture that if the *Mary Rose* had been taking incoming fire from a series of harrying attacks by the galleys the command on board would have had good reason to attempt a manoeuvre designed to enable the *Mary Rose* to return fire with her most powerful armament mounted along her main gun deck, firing to either port or starboard.

When considering any potential scenario for the battle it is very important to understand the tidal movements for the day as these would have had a significant impact on the way that the ships as weapons systems could be deployed. The tides can be calculated from the Moon phase for 19 July 1545 as the tide in the Solent is directly related to each stage of the lunar cycle. There is a complication imposed by the change of calendar which took place in Britain in 1752, changing from the Julian Calendar to the modern Gregorian Calendar. 10 days must be added to the Tudor date of 19 July 1545 to compensate for the change. Therefore, the date used for the lunar phase and tidal calculation is 29 July 1545. The Moon was full on 23 July 1545 (Gregorian) and consequently the day of the battle was six days after the full Moon. This suggests (K. Carter, former Queen's Harbour Master, pers. comm. 2007) a high tide at around 16.00 GMT in the afternoon. (Local solar noon is about 12.11 GMT so local time is approximately 11 minutes later than GMT, which would affect the exact timing, but this adjustment is probably rather too precise for the calculation here). Such a lunar condition would provide a tidal range between the maximum of a Spring tide and minimum of a Neap tide so that the tidal currents would be reasonably strong but not quite as fast as they would be for a Spring tide. Low tide would have been at around 08.45 in the morning and that consequently the tidal current would have flowed from west to east from around 10.00 to around 13.00, with slack water around 14.00 at approximately two hours before high tide, the current would then turn to flow from east to west, beginning to run quite strongly (about 1 knot in the

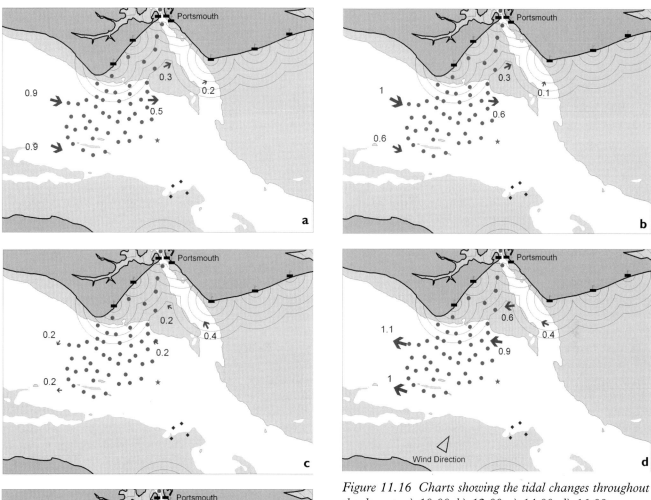

*Figure 11.16 Charts showing the tidal changes throughout the day at: a) 10.00; b) 12.00; c) 14.00; d) 16.00; e) 17.00 hours*

vicinity of the larger English ships) by about 16.00, the current flow increasing to around 1.4 knots at about 17.00 hours (Fig. 11.16a–e).

Under these conditions throughout the morning and early afternoon the French galleys would have had a considerable advantage as they would have been able to bear down on the anchored English ships with reasonable speed provided by their own oar power, attack the English ships and then speed away to the east assisted by the tidal current. At the same time the English vessels would have been held in a '*stern-on*' attitude to the attacking French galleys by the flow of the tidal current. The English ships, swinging on their anchor cables, would have had their bows held into the current leaving their sterns facing towards the enemy. Consequently, the English would not have been able to aim easily their broadside firing guns at the French as the whole ship would need to be brought '*beam-on*' to the tidal current. Conceivably, this could have been achieved by springing the anchor cable or using the ship's boats to pull them round against the tidal current but there is no mention of this being done in any of the historical records nor is there any indication on the Cowdray engraving of such practice. This situation would have been exacerbated by the lack of wind required by the English vessels for their directional control and propulsion. In other words, the English ships were unable to move or control their direction of fire thereby, rendering them relatively defenceless against the swift head-on attacks from the French galleys.

Throughout the morning the English ships could only reply with their rearward firing guns mounted in the stern. The Cowdray engraving shows that some, but by no means all of the English ships had stern firing guns. Both the *Mary Rose* and the *Henry Grace à Dieu* were so equipped. If Du Bellay's account is accepted, that there was considerable exchange of fire, it would be not unreasonable to assume that successful hits on vessels were achieved by both sides and although at present, there does not appear to be any direct physical evidence within the archaeological remains of the *Mary Rose* herself, damage received from the French galley fire may have been a factor involved in the chain of events which ultimately led to her sinking and cannot be conclusively discounted.

For the English to employ their greatest firepower against the French, the English warships needed to raise their anchors and get underway. Given that the *Mary Rose*, the *Henry Grace à Dieu* and most of the English fleet were entirely driven by sail the ships' masters would need to be certain that the passage of their ship would be sufficiently fast to ensure directional control could be maintained, whilst pursuing a path that would enable the main armament to be brought into full effect. In other words they would need to make a speed with their own vessels in excess of the tidal current flow so that they were travelling fast enough to control their direction of travel. This could not be achieved during the morning or early afternoon as the tidal current was unfavourable and sufficient wind was unavailable. However, in the mid-afternoon the situation changed as the tidal current turned to flow from east to west and the afternoon breeze, began to blow. At this stage it would be both possible and desirable for the English to turn the tables on the French and hit back.

With the tidal current running from east to west at around 15.00 hours in the afternoon the English ships, still swinging on their anchor cables, would have changed direction such that their sterns no longer faced towards the enemy. This meant that incoming French galley attacks were now directed at the English bows where they were not protected by any large forward-firing ordnance and in these circumstances the French galleys could make fast, current-assisted runs in towards the English ships with a reasonable degree of certainty that the English could not strike back with powerful guns. Consequently, there would be extremely good reason for the English ships to get underway as quickly as they possibly could (after around 15.00 hours) with the intention of bringing their main armament into action. This change in the tidal flow occurred at around the same time that a sea breeze would be expected to begin to blow on the hot summer's afternoon and the English would probably have considered this to be a most sensible course of action as with 2 or 3 knots of forward speed they would be able to make a passage across the Solent from which they could engage the French galleys using their main armament thereby significantly outgunning their opponents.

Whatever happened at this stage in the battle we know for certain that the *Mary Rose* sank, that her archaeological remains were found some 600m to the south of the shallows of Spitsand and that her wreck position suggests that she was making a northwards passage at the point where she sank. This implies that her route would bring her across the front of the main English fleet, presenting her starboard side armament towards the incoming attacks of the advance party of French galleys. In all probability this manoeuvre would have begun from the point where she had been at anchor in a central position within the vanguard of the English fleet as indicated by Lisle's June Orders. This could only have been in an area of water sufficiently deep to accommodate the *Mary Rose* consequently, there are a very limited range of possibilities for the location from which she began her last fateful passage.

After the *Mary Rose* had sunk Du Bellay reports that the weather changed enabling the English to bear down on the French galleys and force them out of the Solent to rejoin their fleet at St Helen's. Clearly, the exact timing of this change of weather is important in the sequence of events. It would seem likely that the change in the weather and tidal conditions actually began before the *Mary Rose* sank rather than after as suggested by Du Bellay although the magnitude of such change may have increased throughout the period. To make any manoeuvre whatsoever the *Mary Rose* required sufficient wind to move from her anchored position in the vanguard of the English fleet. Her anchored location is very unlikely to have been the place at which she foundered, therefore, she must have made passage to arrive at the point where she sank and to do so she must have made use of the available wind.

The GIS may be used to examine the circumstances potentially surrounding her loss further. If we conjecture that the *Mary Rose* was making her passage, from within the fleet to the point of her sinking with the intention of enabling her broadside to engage the French galleys, then the *Mary Rose*'s potential route may be divided into four phrases, a run-in phase, a conflict phase, a post conflict phase and finally an extension of the *Mary Rose*'s possible route had she not sunk enabling her to reach the safety of the English fleet aided by the current. There is also the possibility of an alternative route, beyond her point of sinking, which could represent the path her sailing master may have wished to follow in order to ground the *Mary Rose* on Spitsand. This route might be chosen if the sailing master felt that she was not handling as he would normally expect possibly resulting from overloading of men and armaments or because her hull had been holed and was shipping water from damage received much earlier in the day.

Assuming that the *Mary Rose* could make about 3 knots, which would be quite a practical speed given the weather conditions and the space available in which to manoeuvre, the GIS calculates that the run-in phase would have taken approximately 12 minutes, the engagement phase would have taken around nine minutes, the post engagement prior to sinking phase would have taken just 6¾ minutes. Had the *Mary Rose* been able to follow its possible intended path taking a turn to port back into the main body of the English fleet this section would have taken around 15 minutes. The potential path from the point at which she sank to that at which she could have run aground on Spitsand would have taken around 6 minutes at 3 knots (Fig. 11.15).

To organise and navigate the ship in these very difficult conditions would not have been easy. To add to the complications there would have been the problem of Vice-Admiral Sir George Carew as he had been appointed to his post only the day before, and would not have had time to develop his managerial relationship with the ship's master nor would he have had the opportunity to rehearse any manoeuvre or commands with the crew. Henry VIII was watching the engagement from Southsea Castle and had a clear view of the entire battlefield and Sir George Carew knew that the King could see any action that he took and would, most likely, have been anxious to impress with suitably aggressive posturing. On the other hand, the master of the *Mary Rose* would have known the waters in which they were sailing well and would undoubtedly have known that the ship's northerly passage would shortly ground the vessel unless a turn to starboard or to port was made very rapidly. He would also have been aware that to turn to port would have been easier for the ship than to turn to starboard given the direction of the prevailing current, wind and the possible set-up of the sails at that time. He would also have been aware that turning to starboard, whilst presenting an aggressive stance, could have left the *Mary Rose* isolated, separated from the main English fleet and in a position where the attacking galleys would have probably had the upper hand at least for a short while.

Once she was underway there were 12 minutes in which to plan these manoeuvres followed by around 9 minutes of intense action and then about 6 minutes in which to secure the safety of the ship. This would be extremely difficult to achieve under optimal conditions with a well-organised, practised crew and command. Clearly, something went wrong at this point because the ship sank. It is probable that the ship heeled to starboard bringing the open gunports below the sea surface such that water came onto the main gun deck. We can be reasonably certain that once seawater had begun to enter the gunports it would take only a very short while for sufficient weight of water to flow into the hull creating a '*free surface effect*' similar to that which capsized the roll-on-roll-off ferry, *Herald of Free Enterprise*, off the Belgian port of Zeebrugge in 1987. In this condition the rapidly moving mass of flowing water would dramatically alter the balance point of the vessel and it would take only a few seconds to severely degrade the stability of the ship sinking her very rapidly.

Why, after 34 years of successful sailing, was the *Mary Rose* overcome at this particular moment? Her position within the battlefield suggests that she may have been attempting to make a sharp turn to port which would have pushed her gunports close to the sea although this must have happened many times before in her service career without mishap. However, the danger of this manoeuvre would have been magnified if the hull had been holed earlier in the action and water had accumulated in her bilges or if there was significant overloading of men and armaments on her upper deck. Either of these factors would have caused the *Mary Rose* to sail in a '*tender*' and unstable condition. It should be recognised that water occupies far less space than an equivalent mass of people and their arms and when allowed to flow freely can move its considerable weight very rapidly. It is possible that ingress of water into the hull had compromised the ship's stability and that the potential overloading, the probable confusion and lack of clear lines of command during the action were also significant elements in the combination of unfortunate circumstances that resulted in the final sinking event. Undoubtedly, this is an area for further research.

Although the *Mary Rose* had sunk it was not the end of the battle. There was a significant French invasion fleet moored off the Isle of Wight with its army more than twice the size of Henry VIII's force still threatening England.

Du Bellay's description continues: the French turned and headed for the shallows and St Helen's, still intent on drawing the English closer. This failed and, after a chase by the English galleys, the English sensibly retreated back to their anchorage rather than being drawn out into the more open water of the English Channel where the French advantage of greater ship numbers would pay dividends. This is when the French Admiral switched tactics and started, instead, to burn villages on the Isle of Wight.

*'Having news of the King of England's arrival at Portsmouth, he was of the opinion that by landing, wasting and burning his country in his sight, and slaying his men almost within his reach, indignation at such an injury, compassion for the*

**Table 11.5 Range of culverin calculated in degrees of the quadrant (Eldred 1646)**

| Elevation (°) | Yards | Metres |
|---|---|---|
| Level | 460 | 420 |
| 1 | 630 | 576 |
| 2 | 900 | 823 |
| 3 | 1120 | 1024 |
| 4 | 1330 | 1216 |
| 5 | 1460 | 1335 |
| 6 | 1770 | 1618 (over 1 statute mile) |
| 7 | 1990 | 1820 (just under 1 nautical mile) |
| 8 | 2210 | 2021 |
| 9 | 2430 | 2222 |
| 10 | 2650 | 2422 (1.3 nautical, 1.5 statute miles) |

*blood and death of his subjects ... would move him to such an extent that he would dispatch his fleet to their assistance, especially as they were but two cannon-shot distant'.*

Clearly, achievable gunnery ranges are also important to our understanding of the battlefield and the interaction between the opposing forces. From Du Bellay's description it is not known whether '*two cannon-shot*' means the distance between the encampment at Southsea and the Isle of Wight (over 4 statute miles; *c.* 6.5km), or from the encampment to the Spithead anchorage to (1½–2 miles; 2.2–3.2km). Either way, this suggests a shot range of 1–2 miles (*c.* 1.6–3.2km). Although the use of the phrase here is most probably a poetical description rather than a quantitative measurement it is interesting to note that the distance to the anchorage of two cannon shot is achievable at maximum elevation of the gun or by taking advantage of ricochet (as practised by the Portuguese by 1495 (Barker 1998) and certainly understood by Bourne (1587). He states that, if aimed point blank (by eye, with the muzzle just higher than the breech so that the muzzle ring and breech rings are at the same height, about 1° of the quadrant) from a low position, a shot can glance and bounce a number of times, provided that the surface is flat (or calm), thereby extending its distance, up to '*three quarters of the best of the randar*', (random = 45°), '*the oftener it it doth glaunce or graze, and the furthyr it flieth etc*' (Bourne 1587, 88). Distances of a mile (1.6km) were not considered unusual and firing trials on the reproduction *Mary Rose* culverin at 7° of elevation recorded 1300–1400m to the first bounce (Hall 2001, 115; see Chapter 2). Bourne refers to '*almost a mile*' (Bourne 1587, 37).

Range is also discussed in a number of sixteenth century treatises, but there are potential pitfalls regarding the actual units used. Distances are frequently cited in '*paces*' (sometimes described as single, sometimes double), and these vary in number of feet depending on whether they are English, Venetian or Portuguese paces. Other terms used include feet and braccio (an arm's length) and occasionally miles. There is an early unpublished Genoese source of 1469 for a bombard test-firing over 2 '*miglia*' (3km) and one '*miglia*' on successive shots out to sea, firing a 130kg shot with one-sixth weight of powder (Barker 1998). Tartaglia (Lucar 1588, 9) suggests a four-fold increase in distance between point blank and 45° elevation (6 points of the quadrant), giving point blank ranges for a culverin of 200 paces, a saker 360yd (329m) and a falcon 320yd (292m) (Table 11.3). The '*utmost randon*' for each is exactly four times point-blank. Some authors suggest a ten-fold (Cataneo 1584) or even eleven-fold increase. A slightly later volume of Brazilian State Papers, covering the period 1607–1633, contains a detailed section on Portuguese ordnance of the period (Barker and Castro 1995, 18–26). Putting exact distances to all of these is inherently difficult. Details are given of the length and weight of guns, weight of shot and powder, and three ranges; line of metal (point blank, where the breech and muzzle mouldings line up by sight), level bore (bore horizontal) and greatest elevation (muzzle raised 45° from horizontal). The ranges quoted are impressive, with the double culverin firing an 18.6kg shot achieving 1053m at point blank and 6248m at greatest range, around six times point blank. However achieving 45° at sea, even on a roll, seems extreme. These figures cannot be directly equated with English guns nearly a century earlier but the details certainly suggest that the few earlier texts (which may seem excessive) should not be discounted.

The specific theatre of the war must also be taken into account with these potential ranges in mind when considering whether the *Mary Rose* could have engaged the French or, indeed, could have been hit by French fire. Table 11.3 includes figures taken from a number of early dated sources which show considerable variation. Figures have been averaged for point blank (Table 11.4) and those are used hereafter. Table 11.5 shows the increased ranges possible with elevation of a culverin.

The French account by Du Bellay (Stone 1907) continues describing the events of the following day, 20 July. Having failed to entice the fleet out, the Admiral became intent on attacking Portsmouth. The French pilots and captains responded in the negative:

*'The entrance was by a channel where only four ships could go abreast and could be easily defended by the enemy bringing forward a like number to oppose them: besides, one could only enter on a favourable wind and tide, and when the said four ships should be obstructed the current would carry those following upon them and scatter them. Added to this, the enemy would be fighting on their own ground, whence they would be aided by their artillery, to our detriment'.*

This is an accurate assessment of the passage that attacking ships would have to make when attempting to assault Portsmouth Harbour. On entering the Solent the ships would need to avoid the shallows of Ryde Sand and No Man's Land to the south and Horse Sand to the north; they would then need to navigate to the deep-water channel between the eastern end of Spitsand and Horse Sand opposite. This would require that they sailed directly towards the guns of Southsea Castle until they were well within the artillery range and then make a turn to port running parallel with the Southsea shoreline which was again heavily defended with English artillery, passing the guns of the Green Bulwark at the eastern extent of Portsmouth town (now known as the King's Bastion), then the saluting platform battery followed by the Square Tower, then the Hot Walls and finally the Round Tower (see Fig 11.10). If they had succeeded in passing the English guns they would then be faced with the possibility of becoming entangled with the harbour boom chain, which could be used to close the harbour entrance. The chain is referred to thus in Leland's itinerary of 1535–43:

> *'There is at this point of the haven Portesmuth toun, and a great round tourre almost doble in quantite and strenkith to that that is on the west side of the haven right agayn it: and heere is a might*[y] *chaine of yren to draw from tourre to towrre'*.

The Cowdray Engraving shows a capstan for raising it from the sea bed just to the right of the Round Tower complete with a figure gesturing towards it somewhat forlornly. Interestingly, Williams (1979) suggests that the chain was not available for the 1545 battle which may have given rise to considerable anxiety within the English command and could explain why the capstan is shown in the Cowdray image adjacent to the Round Tower but that the chain itself is not evident in the picture. The boom chain is prominently displayed in the 1584 map (BL Cotton MS Augustus I.ii.117) of Portsmouth and was clearly seen as an important defensive asset at that time.

The geography of the battlefield also provides the possibility of considering the practical useful ranges of the English guns (rather than their extreme or point-blank ranges) by examining the defences of Portsmouth that guard the harbour entrance and the main approach channel. The Round Tower defends the narrow entrance into the inner harbour, the Camber dock and dockyard.

The 1547 inventory lists the armaments as two sakers, one falcon and three fowlers, all of which have been dismissed at times as not being 'anti-ship guns'. The average point-blank range for the saker is about 300m and the falcon 250m. These ranges cover the entrance which is about 250m. A 'platform' runs between the Round Tower and the eastern edge of the town (see Fig. 11.10). On the Cowdray Engraving two guns can be seen between the baskets beside the Round Tower these are possibly the listed single brass culverin and iron demi culverin (Starkey 1998, 4251, 4252), with average point blank ranges of 420m and 380m. Guns situated in this position have to defend the main navigation channel into the harbour, hence the requirement for long range guns. A maximum of 750m is needed to completely cover the width of the channel in this area, this would require only 2° of elevation for the culverin (Table 11.5). The rest of the platform is provisioned with an artillery array not unlike the Main deck of the *Mary Rose*, with nine guns of differing qualities and ranges; including three bombards (medium range guns designed primarily for smashing), port pieces, demi cannons, culverins and two sakers, with average point blank ranges of 300–350m respectively. Nine guns are visible on the Engraving, although there may have been more. Presumably these were designed with the intention of inflicting serious damage, hopefully preventing an attacking vessel from entering the harbour and not just to frighten or slightly damage an enemy ship.

Southsea Castle is clearly the pre-eminent shore defence and is listed with seventeen guns, again very mixed types including the longest range culverins and demi culverins and the shorter range smashing guns, such as port pieces. The list includes a cannon, demi cannon, culverin, two demi culverins, four sakers, a falcon, a falconet, a sling, two port pieces and a mortar and two listed *'of nothing worth'*. On the Gosport side, Lymdens bulwark is listed with two demi culverins and two port pieces with four broken serpentines. Hasleworth (*'Hasilworth'*, no longer extant) has only one culverin and one saker. Clearly the defences of Southsea Castle are for guarding the main navigable channel directly in front of it thereby controlling the entrance to Portsmouth Harbour.

The Engraving shows that preparations were in place to repel an invasion anywhere along the southern Portsea Island and Gosport shore. Individual guns are located all along the coast eastwards from the town to prevent landings on the beaches. A large group of soldiers carrying staff weapons stands in formation behind Lumps Fort and smaller groups of soldiers stand with pikes, bills and longbows, several are armed with handguns. The distribution of guns and troops among the shore defences suggests a good degree of organisation and understanding of the capabilities of each element within the combined defensive weapon system and the GIS visualisation demonstrates a reasonably comprehensive coverage of the important locations. This visualisation also enables us to suggest, for any predicted route, the distances between vessels. The distance between the *Henry Grace à Dieu* (shown firing a starboard bow gun, see Figs. 11.3 and 11.10) and the French galleys can be deduced as being between 1000–1350m and possibly less than 700m as the galleys approached. This is within range of some of the guns: 1350m requires 5.25° elevation for a culverin (Eldred 1646, 89; Table 11.5; Fig. 11.15).

Du Bellay's account of the battle concludes by recording that Admiral D'Annebault dispatched three pilots to sound the channels into the harbours that night. They reported that the channel was not straight *'but winding and bending towards the enemy, so that an enemy's ship could scarce enter without a pilot or without the certainty of being discovered and having to fight'*. This surveying activity was also discussed at the Privy Council meeting held in Portsmouth on 21 July 1545: *'Letters to the earl of Arundel that the constable of Bosome* [Bosham] *hundred reported that the enemy with two boats sounded the haven of Chichester'* (*LP* 20/1, 1235). Such intelligence suggests that the English were well aware of what the French were attempting.

During the 21st the French considered the options of either trying to obtain the Isle of Wight as a French stronghold on English soil, or to return to France to prevent access of English supplies to Boulogne. It seems that the latter was decided, and the fleet was to set sail for Dover and then crossed to Boulogne, abandoning the invasion but leaving a legacy in the form of the *Mary Rose*. Ultimately she sank due to an ingress of water through her open gunports. What caused these to dip below the water may be as a result of a culmination of a number of seemingly insignificant factors. The distances involved and the weapons carried suggest the *Mary Rose* could have sustained damage from, French gunfire. Although not visible in the hull as recovered, it is possible that portions of the structure still buried within the seabed may yield further information. Damage to the bow might have affected handling, causing confusion in the least. A sudden gust of wind may have exacerbated the situation. Alterations which may have been made to the ship just months before (see Chapter 12) may have weakened her. A bold manoeuvre by a newly appointed Vice-Admiral combined with a relatively recent and little-tested array of weapons combined with additional soldiers may have endangered the ship within the confines of this particular 'theatre of war'. The operational capabilities of the weapons can only be fully understood with analysis of the ship as a platform for the guns, the evidence for which is discussed below.

## Archaeological evidence
by Alexzandra Hildred

The archaeological resource available includes the ship and the weapons carried, as presented on the seabed, as well as others known to have been recovered during earlier salvage (Chapters 2–4 and see *AMR* Vol. 1, chap. 2). The inventory of guns suggested by the *Anthony Roll* provides additional information about the number of particular types of guns and shot which the ship may have been carrying, but nothing about their shipboard positions nor functions.

In order to attempt to understand how the ship may have been fought, we need to assess not only the array of weapons she supported, their requirements and specific functional capabilities, but also the physical constraints dictated by the structure of the vessel. On a large scale this includes hull form and handling but, in order to understand how the weapons were used, the physical constraints of the structure around each gun station must be considered. Storage areas for items such as shot and gunpowder must be identified and access routes from these to the guns suggested. Operational prerequisites for the use of longbows, handguns and rail-mounted guns must be known and best-fit locations suggested, taking into account the archaeological remains.

An important factor in any engagement must be an efficient command structure, communication and allocation of the tasks needed to fight the ship within the limitations of the size and structure of the crew. According to the *Anthony Roll*, the *Mary Rose* carried 30 gunners, 185 soldiers and 200 mariners. Any scenario for the operation of the ship in her final battle, and for her sinking, must include the scant historical data we have, the new appointment of Sir George Carew as Vice-Admiral, and his experience of both the vessel and of the theatre of the war in Portsmouth Harbour. Unfortunately, these are areas where archaeology alone cannot easily provide tangible answers.

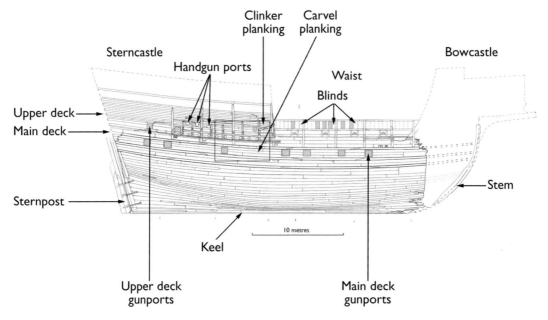

*Figure 11.17 External elevation of ship*

Questions already raised above concerning the circumstances leading to the sinking include whether or not the *Mary Rose* was inherently stable, whether she was hit by enemy fire and whether the crew were experienced in the use of the vessel as she had been rebuilt. This, in turn, raises questions as to the nature and extent of differences, both in tonnage and in handling, between the ship as built, and as she was at the end of her life. Necessarily this must include changes in armaments and the accommodation of these on board the vessel, together with the ease of both sailing and fighting the ship.

The structure of the hull and layout of the ship (Figs 11.17–20) are described and illustrated in detail in *AMR* Vol. 2, where matters of stability, structural strength, rigging, storage and the physical effects of refitting are discussed. All of these aspects impact on the handling and operation of the ship and her armaments and, in particular, it is important to try and correlate changes to the ship structure with those in the ordnance carried – though this is not as simple as it may appear given the paucity of relevant documentary evidence and ambiguity in the names and sizes of many of the guns (see Chapters 4 and 5).

According to the *Anthony Roll*, the *Mary Rose* in 1545 carried 91 guns (including 50 handguns; see Chapter 7 for handguns). Even without the handguns this is a lot of fire-power to find room for, especially given that each (or each pair at least, see below) of the large carriage mounted guns required a team of gunners (see Chaps 2, 3 and 5) and space to bring some of the guns inboard to reload, and that all the associated gun furniture, powder and ammunition also had to be readily available. Some areas of the main storage deck (the Orlop deck) seem to have been set aside for the storage of gun-related equipment (eg, gun furniture in O1, gun-shields in O10) but much of the shot would have needed to be closer to hand. Some probable shot storage areas have been identified on the Main deck (Chapter 5). The location of the gunpowder store is still a matter for conjecture (Chapter 5 and below) and, as we have seen in earlier chapters, there was a considerable array of other weaponry on board. Chests containing archery equipment were located on both the Orlop (O7–9) and the Upper decks (O7–9), the latter alongside staff weapons, presumably positioned ready for use.

The structure of the ship imposed considerable constraint on the already limited space available. The differences in supporting structure affects the weight-carrying capacities of these decks and therefore the number, size and weight of guns which could be carried at any time. Some of the substantial rising knees were part of the original build and some were later additions. They posed considerable constraint to the movement of guns longitudinally along the ship (Fig. 11.21). The Main deck is further provisioned with horizontal lodging knees which stiffen the ship while, at Upper

a

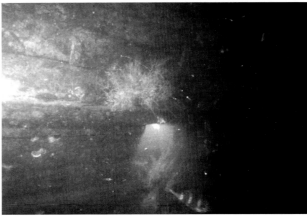

b

c

*Figure 11.18 Gunports: a) internal view of Main deck showing gunports; b) Upper deck gunport (U7), external underwater view; c) gunport lid (inside)*

deck level, the beams are attached to the hull by hanging knees, so here there is no obstruction at deck level imposed by rising knees. Many of these important structural features also imposed constraints on the cutting of gunports. Central hatches along the Orlop, Main and probably the Upper deck provided further restrictions, as did the positioning of masts (and

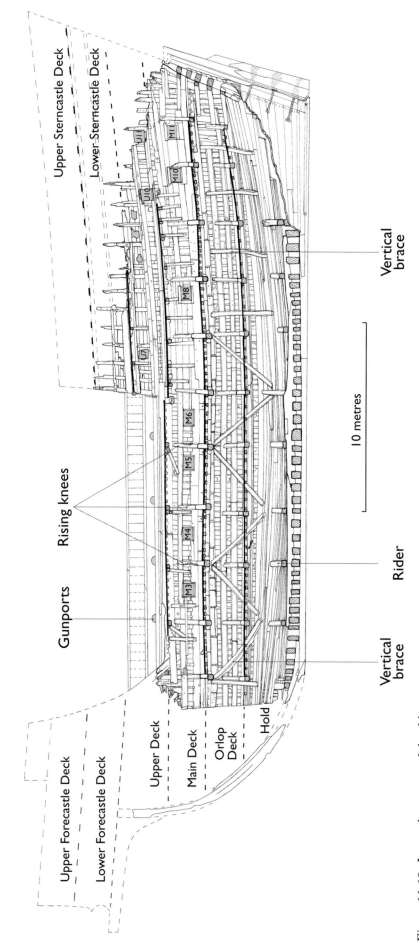

*Figure 11.19 Internal structure of the ship*

Figure 11.20 Diagonal braces

associated rigging), capstan, anchor handling and any cabins.

One of the most important restrictive devices was the operation of the rudder and the imposition on space that the tiller (or possibly even a prototype whipstaff) would have had on the placement and access to sternchase guns (Fig. 11.22a and b). These were crucial to the defence of the ship: 16 out of 20 ships depicted in the *Anthony Roll* carry a pair of guns situated below the top of the rudder, very close to the waterline, apparently lower than the guns on the lowermost gundeck. Eight of these vessels have clearly depicted gunport lids and all appear to keep the area where the rudder enters the ship clear of guns, showing only hawse holes at this level. Guns above the rudder are depicted in tiers usually of pairs of guns (Fig. 11.22c), the *Henry Grace à Dieu* with four tiers above the rudder, the *Mary Rose* with three, the *Matthew* with two and the *Mary* [of] *Hamburg* with one. The *Jesus* is shown with low guns but no gunport lids and with three tiers of guns, two as pairs and a single on the highest level.

While no direct archaeological evidence exists to support the positioning of large carriage-mounted guns in the stern, these are proposed. It is even possible that the level of erosion in the stern was exacerbated by the existence of ports on the Main deck. Although no evidence for lids has been found in the excavation area

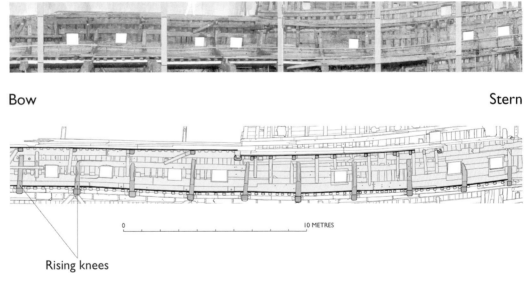

Figure 11.21 above) Line of Main deck and interpretation showing the positions of rising knees (toned); left) large knees on Main deck showing how they constrain the gun movement

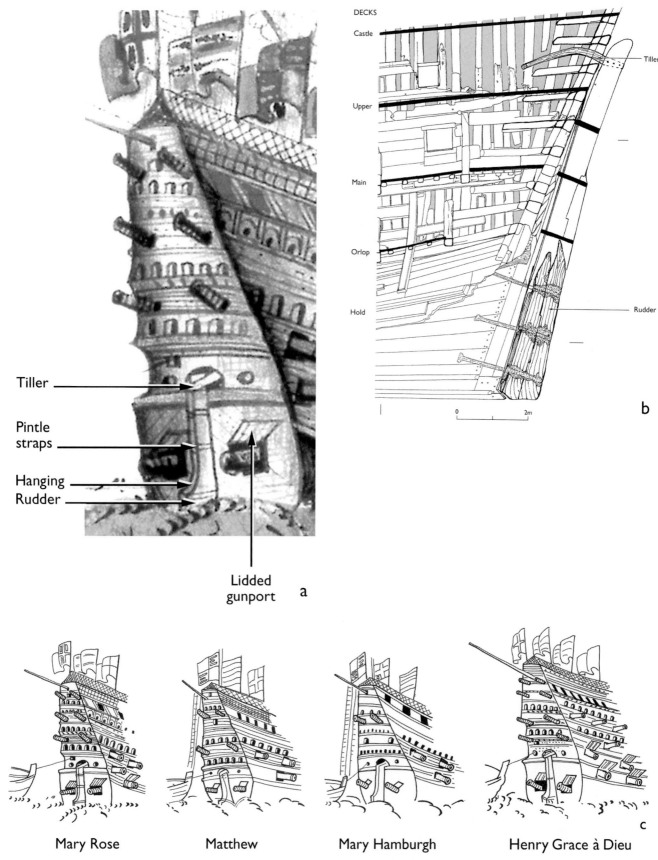

*Figure 11.22 a) Tiers of guns and tiller/rudder attachments from the Anthony Roll; b) elevation of the port quarter showing possible tiller position; c)* Anthony Roll *stern views of the* Mary Rose, Matthew, Mary *[of]* Hamburg *and* Henry Grace à Dieu, *showing guns*

outside the ship around the sternpost and rudder, it is assumed that they existed. How the operation of the lids was achieved is still controversial and uncertain. The lack of refined tackle to operate the gunport lids seems to indicate (albeit inconceivably) that the lids could not be fully opened from the inside and required the use of overhead ropes on the outside of the ship (Fig. 11.27 below), presumably running from the upper rail above the Upper deck (see *AMR* Vol. 2, chap 11). This required good communication between decks and may be partly the reason for the small hatches in the Upper deck that lie above the principal gun stations on the deck below, in addition to providing ventilation and light. Though seemingly cumbersome, this could be the method described by Philippe Duc de Cleves (*c*. 1506–1516, published in part by 1558; Barker 1998). This apparent lack of refinement of an operation fundamental to the fighting, and indeed to the safety, of the ship reinforces the notion that the provision of lidded gunports, at least on the sides of the ship, was a modification to the original hull.

## Gunports

Fourteen gunport lids were recovered in total, seven from outside the starboard side, two in the ship (M5, M6), and the rest in the port scourpit (a selection is shown in Figs 11.23–4 and a selection of associated fittings in Fig. 11.25). There are 12 gunports for carriage-mounted guns in the extant hull and three for smaller guns. The loss of portions of hull structure, in particular the bow and Forecastle, the Sterncastle including the stern panel, mean that we cannot be sure of the total number (see *AMR*, Vol. 1, figs 11.26–7 for illustrations of all the surviving lids). Dimensions are given in Table 11.6.

The main extant gundeck supports seven ports (Fig. 11.26a–b) all of which had lids. There are three gunports for carriage-mounted guns on the Upper deck, all unlidded, and possibly two on the Castle deck, one firing forward and the other possibly in the starboard side aft of the front of the Sterncastle. In addition, there are three small irregularly-shaped ports

### Table 11.6 Gunport dimensions (mm)

| Port | Max. width | Min. width | Min. height | Eye bolt centre–centre | LSDP fore | LSDP aft | Swivel pintle hole | Bolt notes | Notes |
|---|---|---|---|---|---|---|---|---|---|
| M3 | 810 | 700 | 650 | 1100 | 410 | 400 | – | | abrasion to top sill |
| M4 | 790 | 690 | 600 | 950 | 450 | 440 | – | | as above |
| M5 | 790 | 760 | 610 | 1100 | 500 | 500 | – | | as above |
| M6 | 900 | 850 | 650 | 1190 | 460 | 450 | – | | |
| M8 | 700 | 690 | 600 | bolts 65 below port, 650 apart | 465 | 450 | – | below port | deck height above at this point 2m |
| M10 | 840 | 735 | 590 | 1000 | 540 | – | – | | aft ringbolt virtually on knee, poss. in centre of port |
| M11 | 760 | 730 | 590/580 (aft) | 1150 | 640 | 640 | – | | |
| U7 | ? | 500 | 540/480 (aft) | 940 to sill centre | 1030 to port | – | in sill 55 | | |
| S1/U9* | 300 | ? | 250 | – | central, 1100** | – | 60 | | |
| S2/U9* | 310 | ? | 430 | – | 1100** | – | 50 | | |
| S1/U10* | 330 | ? | 330 | – | central, 1100* | – | eroded 50 | | |
| U10 | 690 | ? | 540/560 (aft) | larger 1100, smaller 830 | 170 to large sill, 110–120 for inner sill lip | – | – | | handgun above 190mm to edge sill; central sill lip 110 L, 120 R |
| U11 | 640 | ? | top not present | – | min. 350, max. 450–500 | – | – | | chain plate visible outside bottom sill, 170mm beneath. This directly sbove M11 port |

LSDP = Lower sill to top of deck plank;
* Handgun; ** measurements taken to outer edge of sill, not to extreme outside

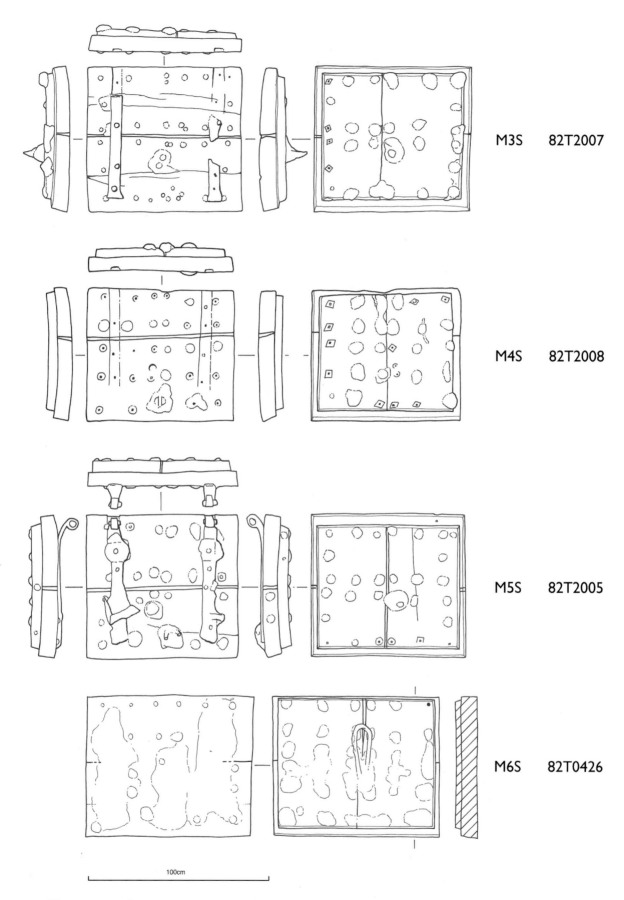

*Figure 11.23 Gunports: external elevations*

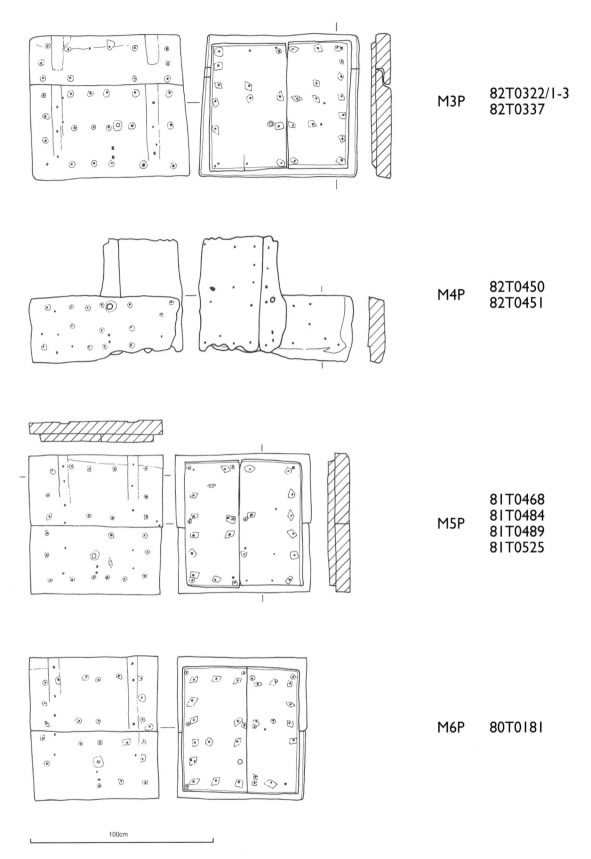

*Figure 11.24 Gunports: internal elevations*

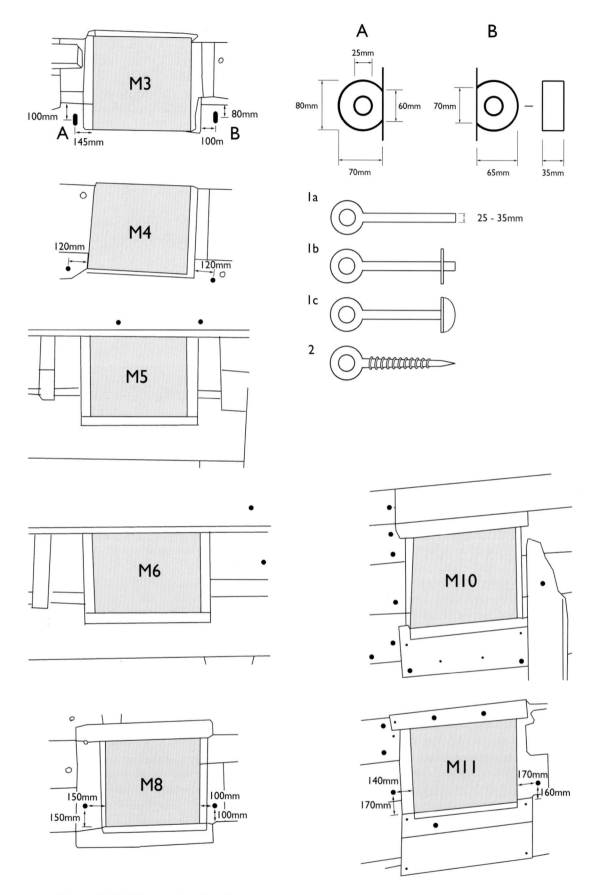

*Figure 11.25 Fittings found beside gunports*

Table 11.7 Gunport lid dimensions (mm)

| Gunport sector | Timber no. | Width top | Width bottom | Sector recovered | Diagonal L–R x R–L | Thickness | Condition |
|---|---|---|---|---|---|---|---|
| *Starboard* | | | | | | | |
| M3 | 82T0007 | 830 | 740 | SS3 | 1100 x 1100 | 130 | complete 1–4 |
| M4 | 82T0008 | 790 | 670 | SS3 | 1030 x 1030 | 130 | complete 1–4 |
| M5 | 82T2005 | 830 | 740 | *in situ* | 1100 x 1100 | 140 | complete 1–4 |
| M6 | 82T0426 | 920 | 730 | *in situ* | 1170 x 1170 | ? | complete 1–4 |
| M8 | 81T1453, 82T0002–3 | 750 | 695 | *in situ* | 1000 x 1040 | 155 | complete, in frags |
| M10 | 82T0428 | 810 | 680 | SS10 | 1020 x 1110 | 105 | complete, in frags |
| M11 | 82T0429 | 810 | 675 | SS11 | 1000 x 1085 | 110 | complete, in frags |
| *Port* | | | | | | | |
| M3 | 82T0322, 82T0337 | 820 | 745 | bow scour | 980 x 1000 | ? | complete, in frags |
| M4 | 82T0450–1 | – | 690 | bow scour | 1000 x ? | 165 | 2 parts missing |
| M5 | 81T0468, 82T0484–9, 82T0525 | 820 | 760 | M5 | 1095 x 1145 | 110 | complete, in frags |
| M6 | 80T0181, 80T0243 | 685 | 730 | M6 | 990 x 990 | 100 | complete, in frags |
| M8 | 83T0001, 83T0009, 83T0012, 83T0098 | 790 | 710 | PS7–8 | 1020 x 1050 | 110 | complete, in frags |
| M10 | 76T0204 | 780 | 655 | PS10 | 980 x 1000 | ? | fragment |
| M11 | 76T0180 | 732 | 656 | PS11 | 920 x 850 | 100 | fragment |

cut into the clinker planking at Upper deck level (Fig. 11.26c). Thirteen holes for swivel guns are visible in the Upper deck inner wale, or gunrail (Fig. 11.26d). Three are just beneath the small ports and one within the lower sill of a port housing a culverin (U7, 80A0976). The others although present in the wale, are essentially 'blocked in' as there are no ports above them. All were inaccessible to swivel guns in 1545 and relate to an earlier system of arming the ship or possibly represent reused timbers. There is evidence for four small unlidded ports in the waist situated within the horizontal boards below the blinds. Interpretation is made from one *in situ* gun (U6, 81A0645) that these supported the six wrought iron slings listed in the *Anthony Roll*.

The seven Main deck gunports are all cut into carvel hull planking through the frames and inner planking. Some are lined with timbers and retain a lintel above and sill below. Ports in the stern of the ship are markedly different to those in the stern (Figs 11.21, 11.25). On the Upper deck, the U7 port and the gunports for handguns, together with those in the waist, are merely cut into the hull structure, without sills or lintels. The U10 and U11 larger ports share many similarities with those on the Main deck below.

The importance of gunport lids is dealt with briefly in the introduction to this volume and their structure is discussed in detail in *AMR* Vol. 2 (Chap. 11). Briefly, each has four elements: two vertical planks on the inside and two horizontal ones on the outside, held with spikes driven from the outside and clenched on their inner faces (Fig. 11.18c). The outer planks are generally thicker than the inner. The lids are shouldered so that the outside is flush with the outer hull planking creating a watertight seal (Figs 11.24, 11.26b). Dimensions are given in Table 11.7. The sizes of the gunports relative to the guns which were placed in them means that the guns cannot be elevated much more than a few degrees above the horizontal, affording only the shortest range. Sighting the gun through the port was also difficult, suggesting that aiming was undertaken on the roll of the vessel rather than by the use of any levelling devices, as indicated by Bourne (1587) reinforced by the lack of levelling devices.

As far as the mounting of ordnance is concerned there are a number of factors which must be considered. These include the height above the deck, which puts constraints on the gun and mounting, and the height of the port; the upper sill being the constraint to the maximum elevation from horizontal that can be achieved by any particular gun. As discussed above, this has an effect on expected range, although firing on the upward roll or bouncing the shot can be used to increase range. The physical constraints of the vessel regarding the position of ports relative to stanchions, knees and hatches and height between decks all have an effect on the working of the guns. Physical evidence around the ports for fittings for recoil and breeching provide the most conclusive evidence we have for the restraining of guns.

All the gunports and lids are individual in size and they vary in height above the deck (Tables 11.6 and 11.7). Although we are uncertain as to the height of the lowermost sills of the gunports above the water, the

*Figure 11.26 Gunports: a) internal view of Main deck gunport in M5; b) external view of Main deck gunport; c) internal view of handgun ports on the Upper deck; d) swivel holes in inner wale at Upper deck level*

freeboard (the vertical distance between the water and the gunwale) was calculated as being between 990mm and 1060mm based on preliminary survey of the hull (Coates 1985). Our current estimation puts the waterline at about the position of the lower wale, about a metre below the lowermost gunport sill. Undoubtedly the open gunports allowed the ingress of water into the ship once she had heeled, but we cannot state that the sills were actually too low. Although the amount of ordnance carried as a percentage of tonnage was high, the recovered assemblage fits relatively well with the inventory so it does not appear that the *Mary Rose* significantly overloaded with ordnance on the day she sank. This does not however discount the fact that the number of guns she was expected to carry (or their disposition) was not too much for her 35 year old hull.

The height above the deck of Main deck gunports varies from 410mm to 640mm, depending on the type of gun and size of wheels. Bourne (1587) advocates specific positioning of guns in vessels. His chapter 16 deals exclusively with '*In what order to place Ordnaunce in Shippes*':

*'And furthermore, I do think it conueniente to shew you how to fit or place Ordnaunce in any Shippe: & this is to be considered, first that þe cariag be made in such sort, that þe peece may lie right in the middle of the port, & that the trockes or wheeles be not too hygh, for if þe trockes be too high, then it will keepe the cariage that it will not goe close vnto the Shippes side, and by that meanes the Peece will not scant go out of the porte, excepte that the Peece be of some reasonable length: and also, if that the Shyppe doe holde that waye, the Trockes will alwayes rūne close to the Shyppes side, so that if you haue any occasion to make a shotte, you shall not bring the Trockes off from the Shyppes side, but that it will rūne too again. And the wheele or Trocke beyng very hygh, it is not a small thing vnder a Trocke wyll stay it but that it may runne ouer it, &c.*

*And also, if that the Trocke be hygh, it wyll cause the peece to haue the greater reuerse or recoyle, therefore, the lower that the wheeles or Trockes be, it is the better and so forth.*

*Alwayes prouided, that the peece bee placed in the verye middle of the porte, that is to saye, that the peece lying leuell at poynte blancke, and the Shyppe, to bee vprighte, wythout any helding [heeling], that it be as many inches from the lower syde of the porte beneath, as it is vnto the vpper part aboue iustely. And the deeper or hygher that þe portes bee vp and downe, it is the better to make a shot, for the heldyng of the Shyppe, whether that it bee the lee syde, or the weather syde of the Shyppe, for if you haue anye occasyon to shoote eyther forwardes or backwards, the steeradge of the Shyppe wyll serue the turne, but if that the Shyppe dothe heelde much, then if that the peece bee lette by the lower parte of the porte, then you muste needes shoote ouer the marke, and if it be lette by the upper syde of the porte, then you shall shoote shorte of the marke. &c. Wherefore, when that the Carpenters dothe cutte out anye portes in a Shippe, then lette them cutte them out deepe ynough uppe and downe. &c.*

*And also, it is verye euyll, for to haue the Orloppe or Decke too lowe under the porte, for then the carriage muste bee made verye hygh, and that is verye euill in dyuers respectes, for then in the shootyng off the peece, it is apte to ouerthrowe, and also by the labouring and seelyng [lurching] of the Shyppe, and so foorth.* [This appears to have been what happened with the U8 port side bronze gun 81A3002]

*And furthermore, you muste haue consyderation for the fytting of youre Ordnaunce in the Shippes, as thys, the shorter Ordnaunce is beste to bee placed out at the Shippes syde, for two or three causes, as this.*

*Fyrste, for the ease of the Shyppe, for theyr shortenesse they are the lyghter: and also, if that the Shyppes shoulde heelde wyth the bearyng of a Sayle, that you muste shutte the portes, especially if that the Ordnaunce bee vppon the lower Orloppe, and then the shorter peece is the easyer to bee taken in, both for the shortenesse and the weyght also.* [This appears to be the case with the exception of the first bronze gun in the bow, a culverin, and the last in M10, a foreshortened demi cannon. All guns on the Main deck (the lowest 'orlop' with ports) are short and four of the seven are breech-loading. Because of the morphology of the sledge only a certain portion can protrude through the port. The U7 culverin is extremely short, and is on very high trunnion support cheeks. It may have been designed to fire a shot at an angle.]

*In lyke manner, the shorter that the peece lyeth oute of the shyppes syde, the lesse it shall annoy them in the tacklyng of the Shyppes Sayles, for if that the peece doe lye verye farre oute of the Shyppes syde, then the Sheetes and Tackes, or the Bolynes wyll alwayes bee foule of the Ordnaunce, whereby it maye muche annoy them in foule weather, and so foorth.*

*And it is verye good for you to haue long Ordnaūce to bee placed righte oute of the Sterne of the Shyppe for two causes: the one is this.*

*The peece muste lye verye farre oute of the porte, or else in the shooting, it may blowe up the Counter of the Shyppes sterne.*

*And also, the peece had neede be very large, for else it will not go very farre out, for the worke of a ships sterne hangeth very farre outwards from the decke or Orloppe vp to the port, so that the carriage may be close belowe, but not aloft, &c. And also if you haue any chasing peeces to shoote right forwardes, then they must bee long Ordnaunce in like manner, so that you must fitte your Ordnaunce, according vnto the place that it must lye in, and also (as is before rehearsed) that it is not good for to haue the mountance or carredge to high. Therefore, if that the Orloppe or decke be too lowe vnder the porte, then it is good for you to make a platforme vnder the port, that the trockes of the carredge may stand vpon. And also, when you doe take the measure of the porte, from the decke or Orloppe, to the end to fitte the mountance or carredge in height, that the peece may lye right in the middle of the porte, then you viewing the decke or Orlop, and considering what height you will haue the wheele or Trocke, and also marke whether or how that the Ships side doth hang inwards, or outwards, and also the Cambring of the decke or Orloppe, and then you perceiuing where the formost trockes doth or must stande, when that the carredge doth go close to the porte. Then where as the very middle of the foremost trockes dothe stande, there take the true measure in heygth from the Decke or Orloppe, upwards, and so shall you knowe iustly howe many ynches will laye the peece righte in the very middle of the porte: for if you doe take the measure of the heygth of the porte from the porte downe vnto the Decke or Orloppe, then by the meanes of the Cambering, the Decke or Orloppe, and also the wheeles or Trockes doth not come to stand right vnder the porte, so by that meanes the Decke or Orlop is higher inwards, and that shal cause you to make the mountance or carriage too high, for that the wheeles or Trockes that the carriage lyeth vpon, shall be a foote more or lesse into the Shipwards, and then look into the Cambering of the Decke or Orloppe, that it riseth inwardes more, than it is righte vnder the Porte, you shall take the measure so much too high for the peece to lay her right in the middle of the Porte &c.'*

By 1627, when Captain Smith wrote *A Sea Grammar*, the emphasis was different:

'*Lay the Orlop with good planke according to her proportion. So levell as may be is the best in a man of Warre, because all the* Ports *may be of such equall height, so that every peece may serve any Port without making any* Beds *or platformes to raise them*'. (Smith 1627, 6)

What we cannot be sure of is how much the *Mary Rose* and her crew had seen action with the weapons system apportioned to her in 1545, or how experienced they were with the operation of the gunport lids.

### The operation of gunport lids

One of the unanswered questions is how the lids were opened and closed and how they were maintained in the open position. Evidence is limited as most of the lids are still in conservation and their metal parts remain in concretion, obscuring detail. One lid (82T0426, M6 starboard) retains evidence of a centrally placed oval handle on the inner face while a radiograph of 82T2007 (M3 starboard) suggests a ring and a central bolt, though this is not certain. Concretions which may be

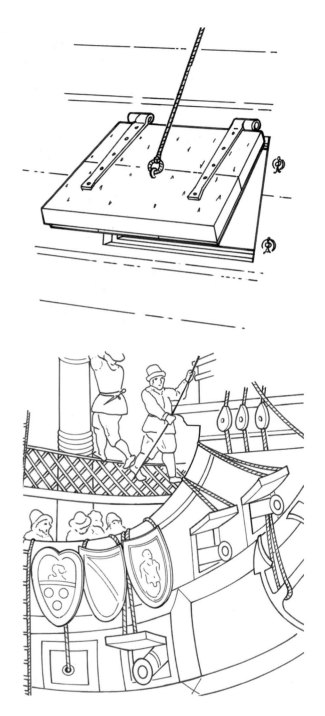

*Figure 11.27 a) The method of opening gunport lids from the seventeenth century onwards, was by rope through a hole in the hull just above each gunport. The absence of such a rope hole in the* Mary Rose *requires an alternative method; b) from a sixteenth century print of gunport lids being opened by ropes from the deck above*

rings have been found outboard in SS03 (81A1526, 2216, 2645), possibly associated with lid 81A4897 (see Figs 11.23 and 11.27).

The central handle on the inside has been interpreted as a simple means of securing the lid shut using a wooden pole of approximately 140mm in diameter. This passed through the handle and was secured, possibly with a wedge (von der Porten 1979). Smith (1627, 79) describes such an arrangement: Smith (1627, 88) describes such an arrangement:

> 'Shackels *are a kinde of Rings, but not round … fixed to the middest of the ports within boord; through which wee put a billet to keepe fast the port for flying open in foule weather, which may easily indanger, if not sinke the Ship*'

The means of opening the ports suggested by von der Porten was by a central ring on the outside being attached to a rope going through the hull directly above the centre of the port, as in the case of the *Vasa* (almost a century later). Despite intensive searches, no such holes have been found for the *Mary Rose* lids. There is also no conclusive evidence for the external ring, although 82T0426 (M6 starboard; Fig 11.23) has concretion in this area of the correct size. In some cases iron bars were found close to the lids and it is possible that these were used to keep the ports open by bracing them: they may also have been used to prize them open or to keep them closed. *The Embarkation of King Henry VIII at Dover* of 1520 shows what may be similar items situated on each side of lidded ports on several vessels. The objects recovered however may be the remains of the crowbars supplied for moving guns.

There are few documentary sources, but contracts for a group of Great Ships drafted on 16 May 1589 (PRO SP 12/224, nos 45, 83, 84) offer some evidence; in the version signed by Matthew Baker on 3 June (no. 84):

> 'All the portes to be furnishede with ringe boultes and all such as be uppon the lower overlopp or elsewher that hath a dimie colveringe to have iiij$^{or}$ ringe boultes to a porte, the reste 2 ringe boultes to a porte, with shackels and ringes accordinglie'

## *Layout and operation of the gundecks*

The three gundecks enable an overview of the positioning of most of the larger guns with respect to space around them and access about the ship. Some questions regarding storage and use of the portable weapons can also be addressed and this information is summarised in Figure 11.28.

We know where the archery equipment (O8–9, H9), gun-shields (O10) tampions and spare gun carriage parts (H/O1/2) were stored. We know that shot was stored in M2/3 and M5/6, the latter containing the largest concentration of stone shot and about a fifth of the iron shot as well as shot gauges, rammers and powder ladles. Both these areas had access to the Upper deck and castles via companionways in M2 and M6. We know that the muster point for the collection of staff weapons and archery equipment was between U7 and U9, although it is possible that the latter was originally stored on the Castle deck as it was found above a collapsed port side gun which had overturned and fallen against the starboard side (81A3002). The distribution of handguns suggests that they may have been used on the Upper deck in the stern (U9–11). The hailshot pieces were also probably used from the Castle decks along with the ship-mounted swivel guns. Archers and soldiers with pikes and bills can only have operated on the Castle decks or in the waist on the Upper deck facilitated by the removal of some of the blinds (Fig. 11.29) as the Main and Upper deck under the Castles are too low.

No casks have been confirmed as containing gunpowder or ready-made cartridges, nor any chests (the 1514 inventory for the *Mary Rose* suggests chests; Knighton and Loades 2000, 142). Common sense suggest that these would not be on the gundecks because of the risk of fire, nor in the Hold unless on some form of platform as spoilage due to damp was a major problem. Early sources do, however, suggest that gunpowder in casks was stored deep below the Forecastle (Cleves *c.* 1516, see Oudendijk 1941). It seems most sensible for the powder store to be away from any open fires with ready access but also partitioned off: Sectors O1, O4, O6 and O10 would be suitable candidates. Distance from the galley (H5) would favour partly partitioned locations either in O1 or O10, or H1 beneath the gun furniture stores. Possible access routes would be via the Main deck hatches, via companionways in O4 and O8 or through a passage from the missing port side. Over half of the linstocks, used to ignite the powder charges, were in M/O 8–10, suggesting that they were stored there (a smaller group of seven in O/H 5–6 may have fallen through Main deck hatches), but the lack of linstocks beside most of the guns is slightly puzzling, though they may have floated away.

### The Main gundeck

The Main deck has a length of *c.* 37.5m of maximum width 10.5m amidships. At the last gun bay in the stern (M11) this is reduced to 7m. The planks are made of 70mm thick oak. The assumption of a series of 1.5m wide hatches from M4–M11 means that space, particularly in the bow and stern, is extremely limited and M11 is very constrained, with some portion of the gun carriage having to rest on the hatch covers. Even amidships, the position of the main mast dictates that the cannon (the shortest guns on this deck) can only just be brought inboard. The hatches in M2–3 are over 2m wide. A fore-and-aft ridge, 40mm high and 60–160mm wide, running the length of the deck from M3–8 (possibly to prevent water entering through the hatches) would have been a further slight obstruction. There is also a rebate for a stanchion on one of the deck

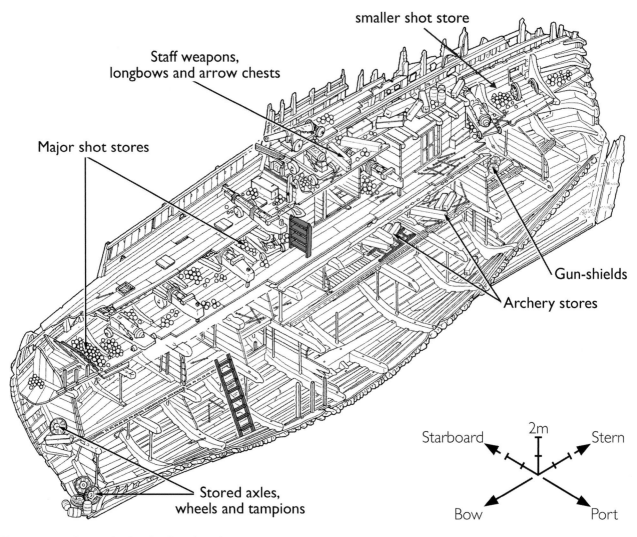

*Figure 11.28 Isometric showing location of armament stores*

beams which would have impeded the guns, especially if it was one of a row. The waterline is estimated to have been about 1m below the lower gunport sills. The height from the deck to the half beams supporting the Upper deck varies from 1.75m in the bow and midships to only 1.65 under the Sterncastle. There is evidence for four partitioned cabins on the starboard side in M2, M7–8 and M9. The guns were positioned between these cabins and it is likely that the cabins were duplicated on the port side. Access between decks was afforded by companionways to the Upper deck in M2 and M6 and to the Orlop deck and Hold in M4. Portions of a collapsed companionway found in O7–9 suggest access in the stern, although the exact orientation cannot be confirmed. Additional access would be possible through the hatches for the hauling of equipment and supplies (Figs 11.29–30).

The seven gunports in the hull on the starboard side are in sectors M3–6 below the waist, and in sectors M8, M10 and M11. Their heights above the deck and dimensions (Tables 11.6, 11.7) indicate a maximum 3° elevation (see above for discussion). The gunport lids were either found open or just outside the ship in the scourpit. Ring bolts are positioned inboard on the hull just beside or beneath the lower corners of each port (see Fig. 11.25). In the absence of any further bolts, it is assumed that these were used to lash the guns inboard when the ports were shut, to restrain movement during firing (to retard or possibly prevent recoil, especially given the potential hazard of the central hatches) and as tackle for running the gun out. All the gun carriages exhibited scars and marks indicative of rope attachments and rope may also have attached the back of the carriages to a fitting in the deck, providing a purchase point to haul the guns inboard and restrain them (Fig. 11.31). Retarding recoil was particularly important with the carriages for the wrought iron guns which, when in use, had an elevating post going through the back of the carriage. This was not designed to move. Two nearly complete port side carriages for bronze guns were found in the upper sediments in M5 and M10 suggesting that the carriages may have been held to the side of the ship until the sediment built up beneath them as they did not fall to the starboard side; though a

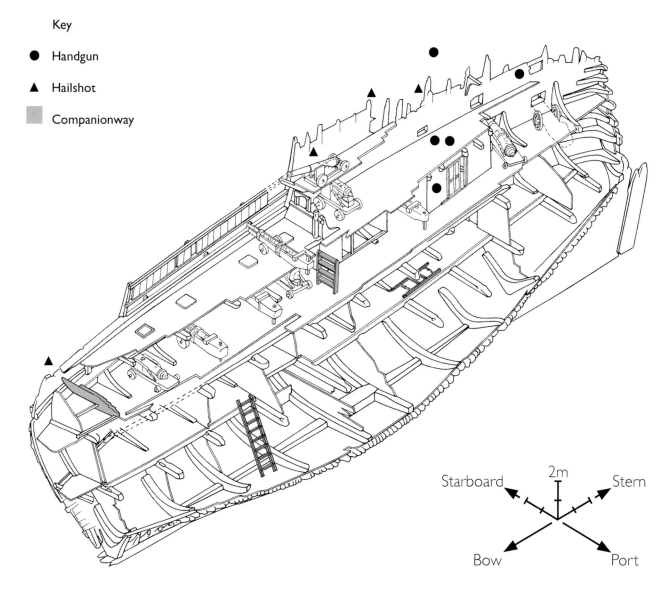

*Figure 11.29 Isometric with all guns found in position and spot locations for recovered handguns and hailshot pieces*

single gun on the Upper deck did (81A3002). There were no guns in M7 or M9 where cabins were situated. The gun in M4 was a port piece, a breech-loader that did not need to be brought inboard to reload. Whilst we cannot be sure if the bronze guns were loaded inboard or out, it is suggested here that the deck hatches may not have been present in every gunbay. Hatch covers were found loose in M3, and in each of M7–11).

Although we cannot accept the *Anthony Roll* as being entirely accurate, the largest ships are all depicted with gunports on the Main deck forward of the rise to the Forecastle. This is where the *Mary Rose*'s structure is incomplete, but the discovery of a powder chamber for a large port piece on the starboard side of the bow in 2005 might suggest that an iron gun was situated further forward on the Main or Upper deck. Similarly, all the large vessels are depicted with gunports at the stern on either side of the rudder at a low level commensurate with the Main deck. This would give a Main deck armament of eighteen guns, eight along each side and one either side of the sternpost, although movement in this area could be restricted if the steering tiller (shown on the *Anthony Roll* at a higher level) was on the Main deck (see *AMR* Vol. 2) and Fig. 11.34.

Six of the seven starboard side Main deck guns were found *in situ* on their carriages, providing vital information about their mounting and operation. Of the fourteen guns known to have been on the deck, six were bronze, with three on each side. These included two culverins situated at the forward end of the waist area in M3 (81A1423 and probably 79A1278), a pair of cannon amidships in M6 (81A3003, 79A1276) and a pair of demi cannon towards the stern in M10 (81A3000, 79A1277). Culverins were long range guns, long in relation to their bore and long range guns were traditionally placed in the bow and the stern. The *Henry Grace à Dieu* is shown firing a gun in a similar position in the Cowdray Engraving (Fig. 11.3). The cannon are

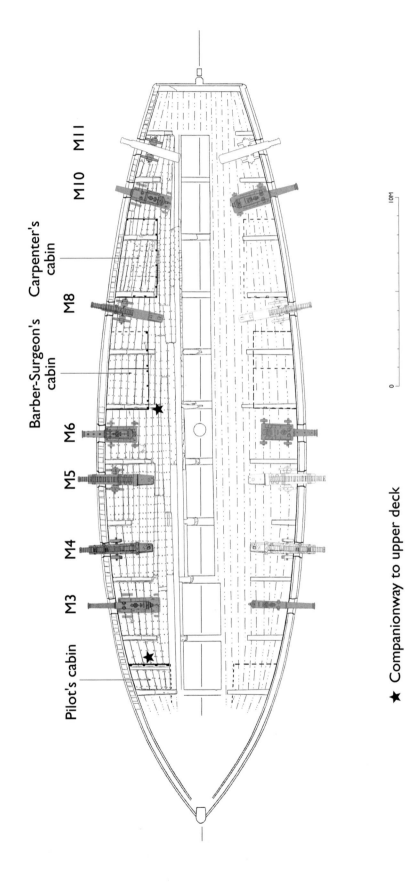

*Figure 11.30 Main deck showing all guns in position with restrictions to movement and access afforded by hatches, cabins and access routes via companionways*

★ Companionway to upper deck

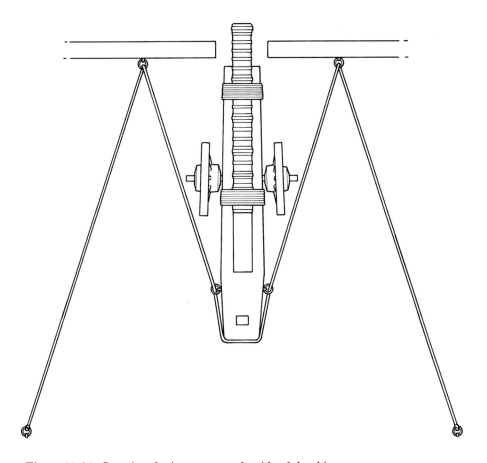

*Figure 11.31 Securing the iron guns to the side of the ship*

shorter bronze guns with large bores, traditionally smashing or battering pieces. Their position in M6, suggests that they may have been for battering an enemy ship to starboard or port. It was necessary to have short guns in this position anyway because of presence of the main mast. The demi cannon in the stern carried a heavier shot than the culverin, and can be considered to be medium to long range guns.

The bronze guns were interspersed with eight large wrought iron breech-loading port pieces (four on each side). Little is known about these guns and authors still dismiss them as small iron pieces incapable of inflicting injury to structure (for a recent example see Parker 1996, 270). Port pieces are rarely (if ever) mentioned before 1535, but are listed for all the ships in the 1540 inventory, with nine in the *Mary Rose*. If the association of the term '*port piece*' with guns used at large gunports is valid, this suggests that at least some alterations to the Main deck to accommodate them occurred before 1540. Perhaps these were the result of the immediate requirement to increase arms both ashore and afloat due to the political climate, the technology to make them being readily available and relatively cheap. As breech-loaders they were easily and rapidly reloaded without the requirement to bring them in from the gunport. They could be tied to the side of the ship and

**Table 11.8 Ordnance on the Main deck**

| Gun station | Gun | Object | Weight (kg) | Location |
|---|---|---|---|---|
| M3S | culverin | 81A1423 | 2071 | *in situ* |
| M3P | culverin | 79A1278 | 2195 | Deane gun |
| M4S | port piece | 81A2560 | c. 620 | *in situ* |
| M4P | port piece | ? | c. 620 | |
| M5S | port piece | ? | c. 620 | |
| M5P | port piece | ? | c. 620 | |
| M6S | cannon | 81A3003 | 2463 | *in situ* |
| M6P | cannon | 79A1276 | 2180 | Deane gun |
| M8S | port piece | 81A3001 | c. 620 | *in situ* |
| M8P | port piece | ? | c. 620 | |
| M10S | demi cannon | 81A3000 | 2513 | *in situ* |
| M10P | demi cannon | 79A1277 | 2740 | Deane gun |
| M11S | port piece | ? | c. 620 | wheels only |
| M11P | port piece | ? | c. 620 | |
| M1S | port piece | ? | c. 620 | no structure |
| M1P | port piece | ? | c. 620 | no structure |
| Stern S | port piece | ? | c. 620 | no structure |
| Stern P | port piece | ? | c. 620 | no structure |
| Extra chambers | | | 2160 | |
| Total | | | 23,762 | |

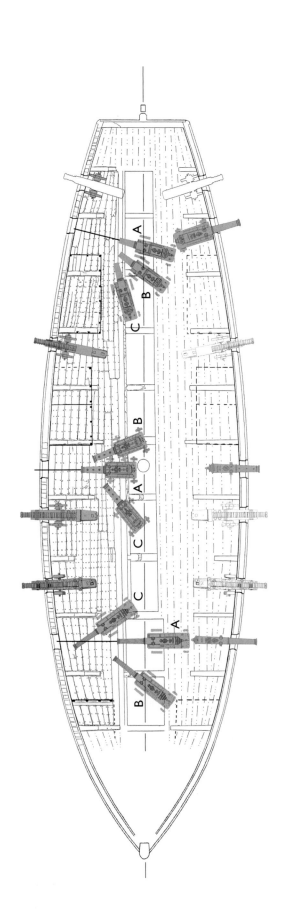

*Figure 11.32 Main deck plan with bronze guns brought in to reload. Position A assumes that the gun is brought straight back, B is a gentle turn (either to bow or stern depending on other structural constraints) and C is a hard turn*

the use of an elevating post suggests that they were not made to utilise any recoil. The weight of the gun in relation to the size and weight of the stone projectile they could throw is light. Furthermore these were dual function guns. Their stone shot could smash a ship's side, shattering on impact and creating a large and jagged hole and also disturbing seams well below the waterline and tearing butt ends open. But they were also devastatingly effective anti-personnel weapons when loaded with sharp cracked flint or beach pebbles. If the guns were fired in turn while approaching an enemy, this would enable the early use of the culverins to be followed up by either a battering broadside of all of the port pieces and cannon at a closer range or even alongside an enemy vessel, or by firing the port pieces and cannon in turn along one side (or even while standing off beside an enemy vessel in order to reload the breech-loaders without the need to disengage) followed up by the longer range demi cannon (M10) as the ship continued to pass the enemy vessel.

All of the guns found *in situ* on the Main deck are anti-ship guns but the height restriction severely limited their possible elevation (see above) and range, while the view from the gunports with the guns in position was also extremely limited. The estimated range is 225–413m while trials of the port piece suggest that 200m would be possible (see Chapter 3), although the full battering effect would be much more pronounced the shorter the range. An effective fighting range to include all the broadside guns together would be about 100m or less.

The positioning of the four port pieces on either side of the Main deck in M4 and M5 may have been tactical, as suggested above, but may also have been influenced by the positioning of the hatches (which are necessarily less robust than the deck planking) making it difficult to reload guns inboard. Indeed, to reload any of the bronze guns they would need to rest in part (if a portion of the loading implements were put outside the port), or entirely on the central hatches, or pivoted to avoid them (Fig. 11.32). This would require moving the gun on an arc once the muzzle was clear of the port by handspiking the rear trucks. Although not inherently difficult, it does have an impact on space. This is one reason why ships

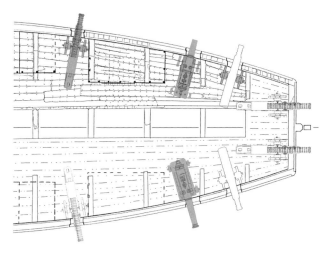

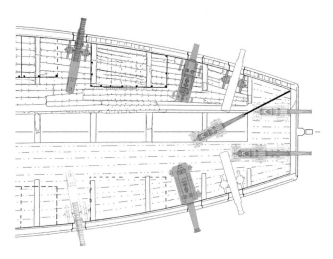

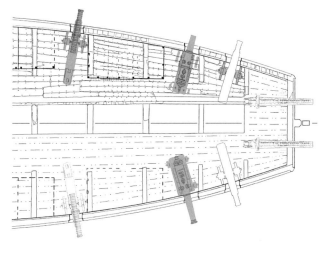

*Figure 11.33 Positioning of alternative types of gun out of the stern at Main deck level: a) ship with port pieces in stern; b) with demi culverins; c) with slings*

often opted to break and turn the vessel to fire the other broadside. An axle and pair of spoked wheels at the vacant gunport in M11 confirms the former presence of a port piece here. This would have needed to be a short gun as the space is confined and it would constitute a permanent obstruction to the stern and conflict with use of a hatch in this area.

The stern panel of the *Mary Rose* is eroded, with only one metre of diagonal planking surviving above the fashion piece, so we cannot be certain that there were gunports cut on either side of the rudder as depicted in the *Anthony Roll* (see above). A swivel gun found in M11, between the wing transoms and the starboard side at a high level (79A0543), has a relatively short tiller and so operating this in a confined space would be possible. It is possible that this gun was positioned on the Main deck in the stern (although it may have fallen from above). Sense and historical sources would favour a larger gun in this area. Long-range capability would suggest culverins, but bringing these inboard would be impossible unless the port pieces in M11 were extremely short, or angled outwards so as to give more coverage at the stern starboard and port quarters and more room inboard. The ship is adequately provisioned with culverins (Figs 11.32–3). Of the four port pieces listed but not recovered, two might be accommodated here but, again, they would have to have been extremely short and these guns are medium range. There are 2–4 demi culverins missing from the assemblage and these long range guns would fit the space, however, the only demi culverins recovered were on the Castle deck and the limitations on elevation posed by a Main deck position would not make best use of their long range capability (Fig. 11.34).

A conservative estimate of the weight of guns on the Main deck in 1545 (without carriages) is in the order of 21 tons, excluding possible extra guns either in the bow forward of sector M3 or out of the stern. If port pieces were positioned in these areas they would bring the total to just under 24 tons (Table 11.8). Positioning demi culverins in the stern would not alter this figure greatly.

*Gun stations on the Main deck*
Each gun station is considered in turn from fore to aft. For details of gunport sizes and heights above the deck see Tables 11.6 and 11.7. Details of associated objects and their locations are given in preceding chapters and Figure 11.30.

*M3*: The first gunport of the Main deck is midway between the M2 and M3 rising knees. The carriage can be angled towards either bow or stern without obstruction offering a wide arc of fire, but little elevation. A loaded bronze culverin (81A1423) of 2071kg (length 3350mm and bore 140mm) was found on its carriage (missing the left rear cheek) at this port (see Fig. 11.30) The back of the carriage had been forced off the deck as the ship heeled, with shot falling

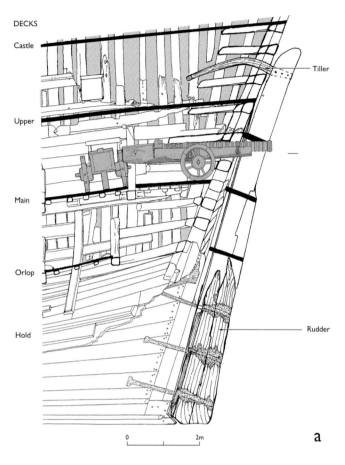

a

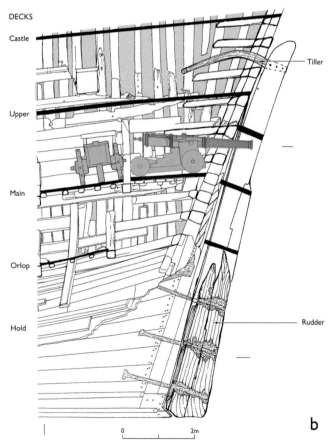

b

*Figure 11.34 Stern elevation showing possible location of: a) port piece and; b) demi culverin. Note the lack of space with either type of gun*

beneath the rear wheels. A ladle (head diam. 125mm) and a rammer (head diam. 136mm) were found here but no linstock. Two possible crowbars recovered nearby could have been used to move heavy loads or to secure the gunport open or shut (see above). A nearby shot gauge was too large for the gun. Six FCS (see *AMR* Vol. 4, 540) were found associated with this gun. All were in their twenties, and three had pronounced spinal changes consistent with regular lifting and hauling, perhaps from loading the gun. Although 53 cast iron shot were found in this area, only five were suitable for use with this gun (diam. 128–135mm). This is a muzzle-loading gun that would either have had to be loaded outboard, or brought inboard partially (if implements were pushed out of the port) or fully. To achieve the latter, the gun would have to be turned or run across the central hatch (if present; Fig. 11.32).

Culverins had the longest range of the guns carried, with a point-blank range of about 420m. As the aft rising knee over deck beam M30 was about a metre away, it was possible to angle this gun so that the arc of fire was forward.

*M4 and M5*: The guns stationed at these two ports were both large port pieces (81A2650, 81A2604) (see Fig. 11.30). The height restrictions imposed by the gunports mean that the guns could not be elevated more than a few degrees above horizontal. Firing trials successfully reached a target 130m away at 3° of elevation, at 124m from the muzzle the shot was still travelling at 284m per second so a greater range is assured. The M4 port is between the M3 and M4 rising knees but the M4 knee has been champhered to give less of an obstruction, almost suggesting the creation of a combined work area rather than individual gunbays. The M5 gunport is very close to the rising knee and the gun could only be used directly out of the port – its manoeuvrability further restricted by the difficulties of 'training' the type of carriage used for these guns (see Chapter 4). The combined weight of these port pieces is *c.* 1.24 tons without carriages or spare breech chambers. It is proposed that there were two similar guns on the port side in M4 and M5.

The barrel lengths are around 2400mm and bores 179mm and 195mm respectively. Both guns were ready to be fired: they had their breech chambers loaded with powder and sealed with a tampion, and the wedges (forelocks) between the chambers and the back of the carriages were in position. The elevating post for the gun in M4 was not *in situ*, but there were two spare ones nearby. Spare chambers were found beside the right wheel of each gun and there were two additional forelocks within the area. As both guns are breech-loading there was no problem with any central hatches until they have to be secured inboard. There were no linstocks in this area (see Chapter 5. Fig. 5.152 for distribution), but several on the deck below could have drifted there after the sinking, perhaps down the nearby ladder. No shot or commanders (mallets used to drive home the forelocks) were found nearby. A crowbar, a powder flask, seven loose tampions, two leather buckets, two powder ladles and three rammers were recovered. These, and others found in M6, may have been stored here. Four of the nine recorded 'shot gauges' were found in M5, with a fifth in O5 beneath, but none of them fitted the port pieces (see Chapter 5).

Both guns would have used either solid stone or lantern shot. Of 22 stone shot recovered from M4, 19 would have fitted the gun here, while 22 of 42 found in M5 would have fitted that gun. However, half of them were unfinished (see Chapter 5). M5 also contained seventeen iron shot, all very small for these guns. One nearly complete rammer (80A01416) lay along the deck across both areas.

Other weapons in the vicinity included a single longbow in M4, with two arrows, a ballock dagger and sheath and two sword scabbards. M5 contained a longbow, a bracer and scabbard. Anti-boarding implements included up to 14 bills and one pike, all from the upper levels suggesting an original location either on the port side or on the Upper deck. The collapsed remains of a port side gun carriage, probably for the cannon (79A1276) was recovered in 1836 (see Chapter 2). Gunport lids were found inboard in M4 and M5, and it is assumed that these were from the port side gunports. These are extremely important when considering the wrecking process and interference.

*M6*: The gunport is positioned almost centrally between the rising knees so the gun could be angled slightly towards the stern or bow. A bronze cannon (81A3003) was found *in situ* at this port. It weighs 2463kg and it has a bore of 200mm. As the ship sank the gun fell further out of the port, so that the lifting dolphins were outside the ship. The back of the carriage had been forced off the deck and shot had become trapped between the underside of the carriage and the deck. The gun appears not to have been loaded, so perhaps it had been fired. Three FCS (68–70) were found in this area, at least one of which was pinned under the gun carriage. The cascabel of the gun was raised to the horizontal by a wedge (quoin). The area is constrained by the forward wall of the Barber-surgeon's cabin and a companionway (thought to have been nailed into position) leading to the Upper deck. This left little room on the aft side of the gun, and made bringing it inboard difficult. Two leather buckets, one stacked inside the other, lay close to the gun. Five powder ladles and two rammers, 199 iron shot (only eight of which fitted the cannon) and 123 stone shot, mostly unfinished, indicate that this was a store for gun furniture. Two ballock daggers, two sword handles and two bill fragments were found between M6 and M7. Another rammer, the remains of five bills, a highly decorated halberd, a ballock dagger, by-knife, sword handle and the end finial of a linstock were found in M7. Very little shot was recovered. Three longbows and a wristguard from a high level possibly derived from the Upper deck.

*M8*: The gun bay in M8 was very small, just 1.75m wide, sandwiched by a partition towards the bow and the Carpenters' cabin towards the stern. The gun situated here (81A3001) was a wrought iron port piece 2870mm long with a bore of 180mm, weighing *c.* 1 ton, including the carriage (see Fig. 11.30). The breech chamber was not in position, though two breeches were nearby, along with two breech chamber wedges. The rear of the gun contained a stone shot (diameter 172mm) and so may have been in the process of being loaded. The elevating post was close by, but not *in situ*. The gun had actually been forced off the carriage, breaking the ropes binding the barrel to the carriage, and was found beneath the ship following recovery of the hull along with a portion of the front of the carriage (Fig. 11.35, a). The gunport lid was in its correct position, above the port with the inside facing outwards as if it was open. The presence of two linstocks and a shot gauge of 100mm are further possible indicators of loading in progress. A gauge of a better size for this gun, of 175mm, was found in the adjacent sector (M9). Most of the sixteen stone shot (150–185mm diam.) were finished and there were two broken canisters with shot, both of 132mm diameter. One leather bucket and an oak mallet with a rectangular head were also found, the latter beside the gun. This may have functioned as a commander. Hand weapons included four longbows, a loose arrow and a sheaf of arrows. One longbow and a sheaf were found at a similar level to the gun. As these objects were found at a high level, they may have derived from the Upper deck, where archers and stored equipment were located.

Excavation outboard of this gunport revealed a good deal of cordage and several barrels above the gunport wedged beneath the chain wale and jammed against the side of the ship. A cable was noted amongst the cordage just north of the gunport (Fig. 11.35, b). As the excavation progressed a change in sediment type was noted (crushed shelly sand with fragments of rope) and the observation was made that the layer containing the rope was deposited after the ship sank (Adams, archive report 81/7/8, 59). Two cables were identified going under the ship beneath the barrels under frame S14, and by this stage they had been tentatively identified as Tudor salvage cables and their importance noted (*ibid.*, 77). As excavation revealed the gunport lid, hinged open above the port, a mass of smaller cordage was noted around the lid, with the cable still visible at the

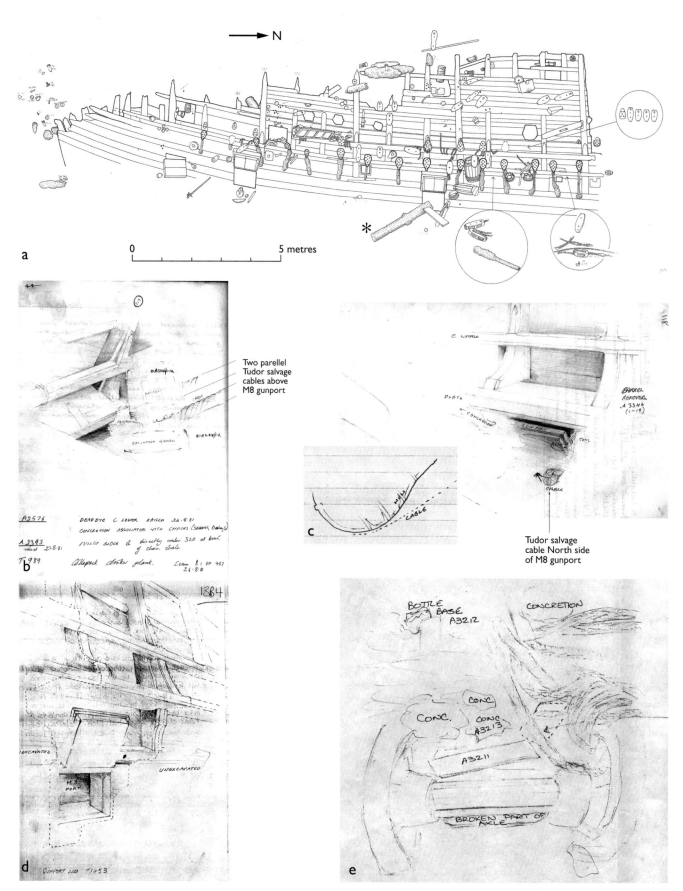

*Figure 11.35 a) Outboard elevation Sterncastle showing gun (81A3001) below port; b) cable above M8 gunport; c) cable beside and below M8 port; d) structure around port following excavation; e) rope around the wheels of the gun carriage in M8*

northern edge (Fig. 11.35, c). The cables continued under the hull below the gunport (Fig. 11.35, c, inset). Many of the dive logs record that there was a great deal of rope behind and to the north of the carriage (and possibly also the remains of a decomposed sack).

Amongst the rope and concretion inboard were human remains, arrows and personal possessions with a great number of fragments of concretion. When the wheels and axle were exposed, it was clear that rope was wound around the rim of the northern wheel with a number of turns (Fig. 11.35, e). The cordage binding the wheel was not noted on other guns, and it is suggested that this was also related to the attempts to salvage the ship, or the gun. Although the majority of the objects within the trench at this level were Tudor, one glass bottle base of questionable date (81A3212) was found adjacent to the axle as well as one fragment tentatively identified as part of a 13in mortar bomb purported to have been used by John Deane. It is possible that some of the as yet unidentified concretions within this area represent more. The rope binding the wheels may represent a Tudor salvage attempt to recover the gun or to remove it to put salvage cables through the gunport. Alternatively it may represent a later attempt to recover the gun by Deane and Edwards. Activity in this area, either Tudor or later, may have contributed to the gun falling (or being pushed) through the gunport to rest beneath the ship and the breakage to the front of the carriage.

*M9*: Two longbows, a wristguard, a sheaf of arrows, three loose arrows and nine linstocks were found in sector M9, outside the Carpenters' cabin where a number of chests had smashed against the cabin partitioning as the ship sank. A priming wire from one of these chests is one of very few recovered. The concentration of linstocks suggests that they may have been kept here. A shot gauge (175mm diam.), the remains of an incendiary arrow, nine loose tampions and three sets of semi-finished tampions reels were also in M9. Two of the latter were inside the cabin which also contained a handgun together with nineteen lead shot (diam. 10–14mm) and the remains of probably two bills. Personal weapons included two ballock daggers and remains of several scabbards. A chest containing items which may have been used to manufacture small incendiary arrows was also found in this area (F43; see Chapter 8).

The remains of ladder were found in O7–O9 and a store of lanterns was identified in O10 and O11 which suggest that there was access between the Main and Orlop decks.

*M10*: This gun station contained a bronze demi cannon 2690mm in length with a bore of 165mm and weight of 2513kg (81A3000; see Fig. 11.30). It was loaded with a cast iron shot of 160mm diameter. The gun would have had to be angled slightly sternwards rather than facing directly out of the port because of a combination of the shape of the hull at this point, the close proximity of the gunport to the aft rising knee, and the stern end of the Carpenters' cabin. The gunport lid was found outside the ship, above the gunport in an open position.

The area was confused by the presence of the carriage for a port side bronze gun which had ended up with its rear wheels nearly touching the rear wheels of the demi cannon. Among the axles and wheels were six finished stone shot, too large for the demi cannon, a spare breech chamber for a wrought iron gun, two bar shot (diam. 160mm, 180mm) and fragments of canister shot (diam. 149mm, 150mm). Only three cast iron shot were found (of 135mm, 165mm and 152mm). The only calibre gauge (81A0249, see Chapter 5) was found beside the demi cannon and a powder flask was also associated with the gun, together with one FCS (89). A priming wire was recovered from a chest nearby. A powder ladle, parts of three rammers, thirteen tampions (55–125mm) and five linstocks were found around the gun.

Other weapons include a longbow, two ballock daggers, and the remains of two large incendiary darts, possibly originally positioned in a fighting top, lying beside the demi cannon (see Chapter 6 for a description and use).

*M11*: The area around the gunport in this sector was significantly damaged, undoubtedly by the nineteenth century salvaging of the gun and carriage, possibly exaggerated by later anchor drag. The port piece that would have been stationed here had been pulled outwards, leaving a pair of spoked wheels and axle. The port is close to the M11 rising knee so could only fire directly out of the port or angled to the stern. A swivel gun found here (79A0543) had probably collapsed from a higher level.

Five small inset dice shot for the swivel gun (of 38–45mm diam.) were recovered as well as a small cross-bar and four jointed shot (55–95mm diam.) and an unusual stone shot with lead plugs (83mm diam.) all possibly for guns on the Upper deck. Larger shot, more suited to a port piece, included one stone and two bar shot. These were found at the junction of the Main and Upper decks and are more likely to relate to guns that were on the Upper deck. A pattern to make gunpowder cartridges (a former) was found, along with two rammers and a ladle. All ranged from 80mm to 90mm in diameter and are, again, likely to have belonged with smaller guns. The ladle may originally have been in a small wicker basket that contained lead shot. A third rammer – of culverin size – had a diameter of 138mm. Parts of two canister shot were also recovered and a single longbow at a high level may have come from the deck above.

*M12*: Five bar shot were found in M12, with diameters of 110–150mm. In addition eleven cast iron and one

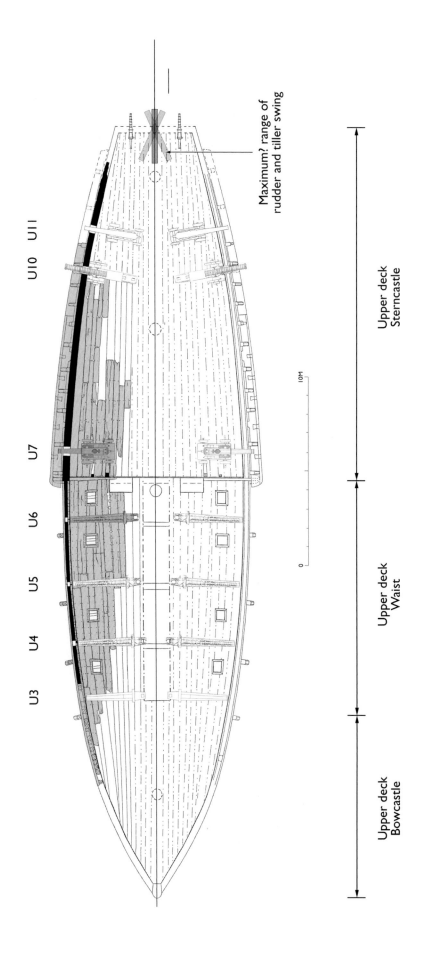

*Figure 11.36 Plan of the Upper deck with all postulated guns in position. Note the restriction on space at the stern afforded by the tiller swing*

inset shot were recovered, with diameters of 82–157mm. The ladle and small rammers from M11 and five of the cast iron shot of 86mm diameter recovered from M12 would serve a saker, so perhaps a gun position out of the stern should be considered. Fragments of a scabbard were also recovered.

## The Upper deck

The Upper deck is incomplete and there is little structure relating to the bow. The predicted length is about 37m with a maximum width of 9.50m. This deck is of a lighter construction than the Main deck, having 45mm thick planks (see *AMR* Vol. 2, Chap. 12 for details) with the area under the Sterncastle different to that in the waist and also stepped down by 270mm, resulting in a maximum headroom of 1.72m (Fig. 11.22a). The open waist is approximately 14m long. The side of the ship in the waist stood about 2.2m above the Upper deck. It is a complex structure that includes gunports, '*blinds*' (some of which could be removed so as to afford protection and access for archers) and an overhead protective covering that supported horizontal, anti-boarding netting. The tops of the frames in the waist rise to about 400mm above the deck and are covered by a longitudinal '*capping*' timber 400mm wide and 120mm thick.

Four small, covered hatches, each 500mm square, are cut through the Upper deck planking in the waist in U3–6. (Figs 11.36–7). These are positioned directly above the guns on the Main deck and could have provided light, facilitated the escape of smoke and possibly aided with communication. Access was by means of companionways from M2 and M6. Fragments of the cheeks of a companionway were found in U10. It is assumed that it provided access to the Castle deck but there are structural indicators (C. Dobbs pers. comm. 2006) suggesting access to the Main deck.

### Semicircular gunports

Between the '*capping*' timber and a second, longitudinal timber termed the '*waist rail*' are two horizontal softwood planks on edge, one above the other, the lower being the '*sill plank*' (200mm wide and 40mm thick). Above this is the row of '*gunport planks*' 3m long x 400mm wide x 40mm thick These originally contained four semicircular gunport openings, each 420mm wide. They are about 500mm above the deck. Each gunport in the waist is positioned forward of one on the Main deck. These gunports exist at the forward ends of sectors U3–6, with a centre to centre distance of 3–3.6m (Figs 11.38).

*Figure 11.37 The Upper deck in the waist looking forward*

### Gunports beneath the stern

Three very small gunports, in U9, are cut into the overlapping weatherboarding of the Sterncastle. They are less than 40mm in width and about a metre off the deck. Although it is tempting to see these as ports for bases on the strength of the stirrup holes in the rail beneath each port, there is a lack of evidence inboard to confirm this and the holes are, in any case, too small. The presence of handgun stocks on the Upper deck close to these ports favours their use for handguns in 1545 (see Figs 11.20, 26c, 29).

Two other trapezoidal gunports exist on the starboard side at Upper deck level towards the stern, in U10 and U11 (500mm high by 620mm wide; Fig. 11.20). That in U10 is directly above a drainage dale. In order to access these there should be a drop in the Upper deck, but even so, the gunports are only about 300–350mm above the deck. The concreted and fragmentary remains of one iron gun (1981, F2 including 81A2097/2636) together with very eroded carriage components, suggests that at least one of these ports serviced an iron gun, and it is possible that this was a fowler stationed in U10. However, a number of combinations of guns can be suggested for these gunports, ranging from small port pieces to fowlers, or a mixture of port pieces, fowlers, demi culverins or sakers (see below).

*The 'blinds' and the anti-boarding netting*
The *'waist rail'*, roughly 210 x 170mm in cross-section, lay above the gunport planks. Along the inboard side of its upper face are slots, 40mm wide but of varying length, to accommodate the lower edges of the blinds, which were positioned between the waist rail and the *'upper rail'* (Fig. 11.38). Fifteen rectangular blinds, made of poplar were found *in situ* in two groups forward of the Sterncastle on the starboard side, each blind being about 420mm wide, 800mm high and 400mm thick. Other blinds were found loose in and around the ship. Based on the number of rebates which enabled removal of the blinds, four types can be suggested.

Some of the blinds appear to be removable and others fixed. Marsden (*AMR* Vol. 2, Chap. 12) considers that, as built, the requirement was for removable blinds for the archers, but that this requirement was later reduced with the increased reliance on gunpowder weapons, leaving only a small number that could be easily removed so that the position could be used by an archer. The blinds are shown as highly coloured panels in the *Anthony Roll*, and decoration was clearly part of their function. None of the coloured decoration of any part of *Mary Rose* has survived.

Fragments of heavy-duty, rope netting were found over the Upper deck. The rope is 16mm in diameter and was knotted into a diamond pattern (diamonds 450–530mm long). This is thought to be part of a horizontal net used to deter boarding.

*The Upper deck beneath the Forecastle*
There is not enough excavated bow structure to suggest the positions of guns in the forecastle, but the remains of a carriage from the bow excavations suggest that a saker or a falcon was positioned there. Guns firing forward would be expected in U1 but must remain conjectural. Isolated gun carriage components such as trucks, wheels and axles were found here (for a discussion of this, see Chapter 12).

*Guns and weapons in the open waist*
Because it was unenclosed, a great many objects will have either been lost from the waist at and after the sinking, or have moved considerable distances from their original positions. It was also the most vulnerable area of the ship to natural decay on the seabed. Consequently comparatively few finds were recovered overall.

A wrought iron sling (81A0645) was the only nearly complete gun found in the waist, in U6 (Fig. 11.36). It

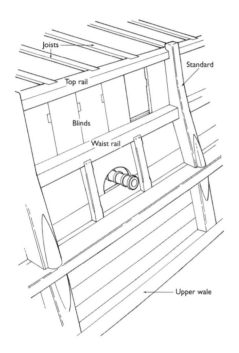

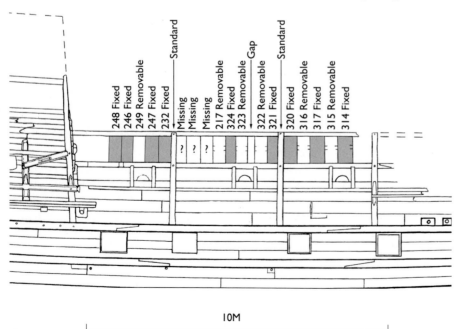

*Figure 11.38 Gunports in the waist: above) reconstruction outboard view of the upper part of the ship's side in the waist. Except for the 'joists' which had collapsed, the remaining timbers were found in relation to each other; below) reconstruction outboard view of the upper part of the ship's side in the waist showing blind*

was *in situ*, with its muzzle going out through a semicircular hole in the blinds and was loaded with iron shot. This has a bore of 90mm and incomplete length of 2240mm. It weighs 1200kg. A breech chamber, probably for a sling (81A0772) was found in U5/6 with further iron gun fragments in U6. There was an axle and one wheel from a small gun carriage in U1, two further wheels and a breech chamber wedge in U2 and a second wedge in U2/3, suggesting that at least one of the gunports here had a wrought iron gun, possibly similar to the sling. A possible commander for a gun was found in U4 as was a small diameter rammer and several small iron and lead shot suitable for a sling (there was also one large stone shot). A single solid lead shot for a handgun was also raised. The presence of an anchor cable stretching between U3 and U6 and an anchor found in U6 may compromise the use of some of these ports. This cable is currently a mater of study.

Loose arrows, two spacers and two sheaves of arrows, a wristguard and three longbows (one retaining the only surviving horn nock from the longbow assemblage) attest to the presence of archers, while pike fragments and numerous broken bill hafts probably indicate that men were stood at arms, or at least mustering ready for action. The number of bill fragments decreases towards the stern. Personal arms such as ballock knives, sword grips and scabbard fragments were found here (Chapter 9). The single iron shot (diameter 88mm) would have suited a sling or a saker. Rather little human bone (and no FCS) was found in the waist though this is not surprising as many bodies will have floated away or become carrion.

### The Upper deck beneath the Sterncastle

A culverin (80A0976) was positioned at a starboard gunport in U7 towards the front of the Sterncastle (Fig. 11.36). It was on a carriage that had high trunnion support cheeks so as to enable increased elevation. The gunport retains a stirrup hole in the lower sill suggesting that a swivel gun may have been positioned there at some time previously. A second bronze gun (81A3002) lay fore-and-aft along the deck in U8–10, having fallen, presumably as the ship heeled. Carriage components found directly above this gun, and the location of the gun itself, suggest that it may have fallen backwards from the port side. The height of both gun carriages dictate that the guns could not have been used in any other Upper deck gun station as all other ports are too low.

The ports in U10 and U11 contained evidence for both bronze and iron guns. Fragments of an iron gun, two breech chamber wedges, a serving mallet, a single axle and wheel fragments suggest that an iron gun was situated at U10. A very small portion of a bronze gun carriage bed was found in U11 as well as a stepped cheek for a bronze gun found in U10. This cannot be associated with any known stern gun carriage and suggests that one of the stern ports on the Upper or

**Table 11.9 Possible ordnance on the Upper deck**

| Gun station | Gun | Object | Weight (kg) | Location |
|---|---|---|---|---|
| U1S | saker* | ? | 680 | no structure |
| U1P | saker* | ? | 680 | no structure |
| U3S | fowler | ? | ?300 | |
| U3P | fowler | ? | ?300 | |
| U4S | demi sling | ? | ?400 | |
| U4P | demi sling | ? | ?400 | |
| U5S | quarter sling | ? | ?300 | |
| U5P | demi sling | ? | ?400 | |
| U6S | sling | 81A0645 | ?500 | *in situ* |
| U6P | sling | ? | ?500 | |
| U7S | culverin | 80A0976 | 1353 | *in situ* |
| U7P | demi cannon | 81A3002 (culverin size) | 1393 | fallen from P to S |
| U10S | fowler | concretion & frags | ?300 | port v. low to deck |
| U10P | fowler | ? | ?300 | |
| U11S | demi culverin | ? | ?1193 | |
| U11P | demi culverin | ? | ?1193 | |
| Stern | fowler | ? | ?300 | |
| Stern | fowler | ? | ?300 | |
| Extra chambers | sling (6) fowler (6) | | 990 | |
| Total | | | 10,792 | |

* may be on Castle deck (see Chapter 12)

Castle deck may have carried a bronze gun. Similarly there are finds of shot which could be used with fowlers, falcons or sakers, notably a collection of twelve small shot in M12. Powder ladles (traditionally used with muzzle-loaders) of the size for sakers and falcons were found in M11, M10 (falcons), C7 (saker) and U9 (saker). It is uncertain as to whether these were used for filling cartridges, for loading loose powder or for the positioning of pre-prepared cartridges. A former for making cartridges, also found in sector M11, is of the correct size for a saker. Small numbers of tampions of assorted sizes were found in the stern and two linstocks.

A partly empty box of arrows with loose arrows beside it was found in U7/8 (Fig. 11.28). Three loose longbows were found alongside another box that had a single longbow inside and loose longbows and many arrows (and two spacers) were found spread along the deck. Two bills and 26 fragments of staff weapons were present along with several small cast lead shot for handguns, while edged weapons included a ballock dagger with several sheaths and a by-knife. The absence of gunports aft of U7 may indicate some other usage. The capstan still needs to be accommodated and it is suggested that this may have been between U7 and U9 in the centre of the ship.

Whatever combination of guns is suggested for the Upper deck, it is likely that, to accommodate the eleven carriage-mounted guns for which positions have not yet been identified (two demi culverins, two sakers, one falcon and six fowlers), the vessel required two ports at Upper deck level on either side of the bow (as shown in the *Anthony Roll*). If it is impossible to accommodate the large port pieces in the stern on the Main deck, firing aft, we also need two ports for carriage mounted guns out of the stern. If we cannot accommodate two more port pieces on the Main deck in the bow, then two of the bow Upper deck ports need to support these. In this case we would need to accommodate two more carriage mounted guns on the Castle deck, probably in the stern. This would account for ten of the eleven carriage mounted guns. It is suggested that at least some of the fowlers were on the Upper deck, leaving one carriage mounted gun without a pair, the single falcon. This would have to be accommodated on the Castle deck in the bow or stern. Single guns are shown out of the stern of the *Jesus of Lübeck* in the *Anthony Roll* (Knighton and Loades 2000, 47).

Given this uncertainty, estimated weights for all guns on the Upper deck can only be calculated to be about 11 tons (Table 11.9).

*Gun stations in the Upper deck stern*
U7: This port is of a different size and shape to those on the Main deck and is cut through the overlapping weatherboarding. The lower sill of this port still retains a hole to support a swivel gun possibly suggesting an earlier phase of armaments. The culverin (80A0976) from this gun station is 2955mm long and weighs 1353kg. It has a bore of 138mm and was loaded (128mm shot). The carriage has trunnion support cheeks higher than the rear stepped cheeks to accommodate the distance between the deck and the bottom of the gunport one metre above it. To access the port the gun would require several quoins (wedges) under the cascabel. A wedge was found loose, and is similar in size to the quoin found *in situ* beneath the cannon amidships (81A3003), but it does not have a handle. A ladle and three rammers were found beside the gun. Two of the rammers are too small (although they could have been wrapped for sponges) but the third is appropriate for this gun. Nineteen iron shot are all 123–140mm in diameter and some had been placed between the frames, held by planks. Four small inset dice shot were also recovered and five large stone shot (150–160mm). Hand weapons included seven longbows, arrows, a wristguard, two ballock daggers and scabbards and 42 fragments of bill.

U9: Three handguns were in U9 and U10 (See Figs 11.26c and 11.29) and it seems most likely that the tiny ports (see Fig. 11.26c) were used for these weapons. A small numbers of cast lead shot (14–18mm diam.) were also found here and a powder flask. Items relating to larger guns include a small ladle and priming wire, inset dice shot (32–46mm diam.), six stone shot (137–74mm) and a single bar shot (145mm).

It seems certain that men were mustered here as the remains of nineteen FCS (39–45, 50–2, 46–4, 54–9) were found, some associated with mail armour, and much disarticulated bone. Twelve loose longbows, many arrows, both in sheaves and loose, and three wristguards indicate the presence of archer. Edged weapons included ballock daggers, by-knives, nine sword grips and scabbard and sheath fragments. As 128 fragments of bills and 20 pike fragments were in this area it is likely that this was a store for staff weapons affording immediate access by the crew.

U10: It is difficult to see how a carriage mounted gun of the types recovered could have functioned at this port. Perhaps a wrought iron gun on a very low carriage was mounted here, since a single axle, wheel fragments, a serving mallet, and portions of wrought iron barrel were found with two breech chamber wedges. Twenty-four stone shot were also recovered (135–210mm diam.), although the larger shot was unfinished. A single stepped rear cheek for a bronze gun carriage was also found in U10 together with a large rammers (195mm) in the Starboard scourpit. A small number of inset dice shot was also recovered, with three ballock daggers, by-knives, a sword handle and scabbard fragments, and a sword hanger.

U11: A single fragment of the bed of a gun carriage for a bronze gun was found in U11 together with a large ladle and rammer (165mm, 114mm diam.). Two fragments from a wrought iron gun were found, but these are too incomplete to enable any interpretation. A linstock, portions of handguns, two ballock daggers, elements of canister or lantern shot, and several inset shot were also present.

**The Castles**
There is little detail of Forecastle structure. The collapsed timbers recovered in 2004 in the bow may be all that survives, though it is possible that a significant portion may still lie more deeply buried to the north-east of the wreck site. In 2004 an incomplete swivel gun was recovered in this area along with inset and iron shot which could have served sakers or slings.

*The Sterncastle deck*
Very little of the sterncastle deck has survived (Fig. 11.39–40); see *AMR* Vol. 2, Chap. 13) but it would have been over 15m long with a maximum width of over 8m. Access was by a companionway from U6 and possibly from U10.

The frames at this height in the ship are 400–820mm apart. In the lower portion of the Sterncastle these are covered with overlapping weatherboarding on the outside to a height of about 900mm above the deck.

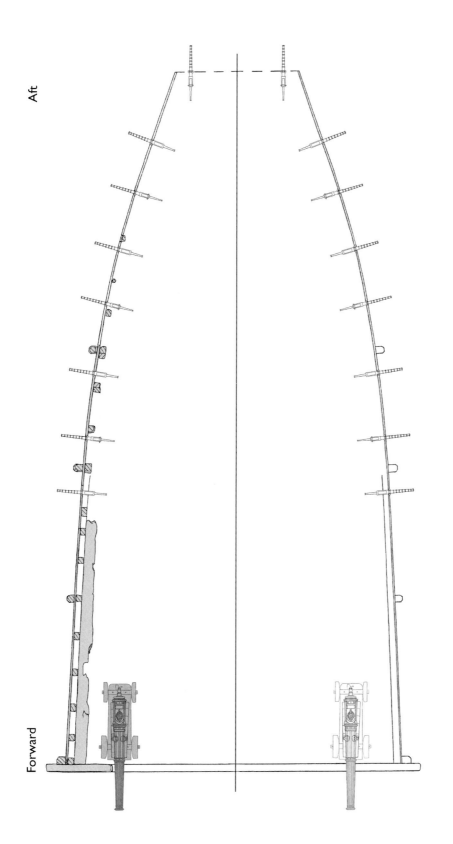

*Figure 11.39 Possible distribution of ordnance on the Castle deck in the stern, bases on either side of the rudder*

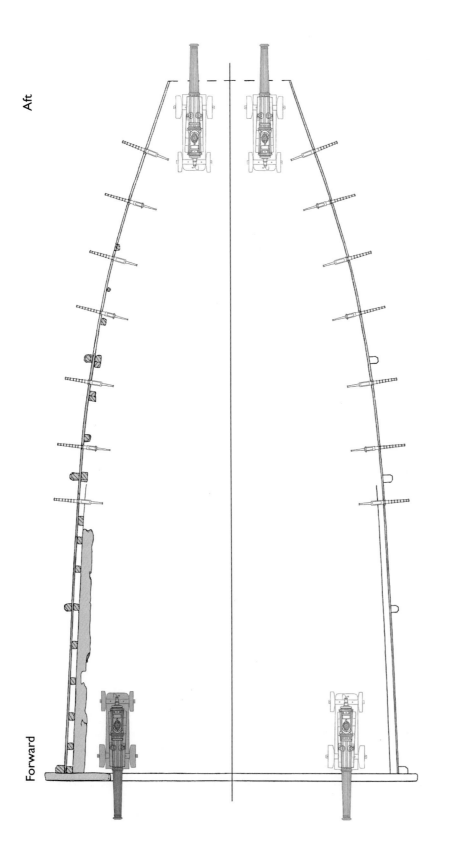

*Figure 11.40 Possible distribution of ordnance on the Castle deck in the stern, demi culverins on either side of the rudder*

Vertical standards have been rebated to fit over the boards and above this planking a horizontal sill supported swivel guns, with elm panels with semicircular cut-outs for guns positioned between the frames. There is evidence on some of the frames for a second rail.

*Possible guns on the Castle decks*
The Castle decks in the bow and stern must have supported most of the 30 bases and 20 hailshot pieces listed in the *Anthony Roll*, together with either the single falcon or a pair of demi culverins, a pair of sakers and the falcon depending on whether or not more ports existed on the Main deck that could have taken the larger guns. The distance between the series of blocked stirrup holes for swivel guns is generally less than a metre (there is a gap of nearly 2m between the first and second at the forward end of the Sterncastle). A demi culverin found *in situ* (79A1232) was a major constriction that obscured two of the castle deck stirrup holes, and they must represent an earlier phase of armament. There is still ample room for the 15 swivel guns required on each side, especially if the hailshot pieces (portable ship supported firearms) could have utilised the space between the bases.

Figure 11.39 shows a possible distribution of guns with bases along each side and in the stern. Figure 11.40 shows a similar arrangement but with demi culverins in the stern, replicating the long range capability afforded by the two found *in situ* in the front of the Sterncastle. The estimated weight of ordnance on the Castle decks to meet with the inventory is around 10 tons (Table 11.10). One should not dismiss the possibility of a second Castle deck, in which case this weight could be distributed over the two decks.

*Weapon finds on the Castle decks*
A single demi culverin (79A1232) was found on the Sterncastle deck facing forwards and lying fore-and-aft against the starboard side in C1. The gun, which is of 3210mm long, with a bore of 112mm and weight 1293kg, was found on its carriage. It is assumed that it fired through a port cut in the front of the Sterncastle (see Figs 11.2, 11.28, 11.39–40), at the widest point of the ship. This gun was intended to be used for high angle, long range fire. The carriage had been adapted to suit its position by having very large foretrucks, both axles positioned above the bed and rear trucks some 100mm smaller than the foretrucks, increasing the possible elevation of the muzzle by inclining the bed. The breech must have been slightly elevated, with the cascabel button possibly resting on the rear axle, as the carriage is too narrow to enable the complete lowering of the base ring to the bed. Although large foretrucks would lower resistance to recoil, they would enable the gun to be moved around the ship more easily. As the structure is incomplete in this area, it is not known whether there were any gunports in the starboard side

**Table 11.10 Ordnance on the Castle deck (postulated)**

| Gun station | Gun | No. | Weight (kg) each | Location |
|---|---|---|---|---|
| Forecastle | base | 14 (7S+7P) | 225 | no structure |
| Forecastle | hailshot piece | 8 (4S + 4P) | 16 | no structure |
| Sterncastle | base | 16 (7S + 7P + 2 stern) | 225 | some structure |
| Sterncastle | hailshot piece | 12 (6S + 6P) | 16 | some structure |
| C7S | demi culverin | 79A1232 | 1043 | *in situ* |
| C7P | demi culverin | 79A1229 | 1293 | |
| Tops | top piece | 2 | 180 | |
| Extra chambers for bases | | 30 | 10 | |
| Total | | | 10,066 | |

alongside this gun. Although these would be obscured when the gun was facing forward, they could be accessed by pulling in the gun (or even utilising any slight recoil) and turning it to face out of the starboard side, providing another broadside gun. This manoeuvre would enable access to the swivel holes in the upper rail not usable when the gun is facing forwards. The positioning of guns in this location was advocated by Philippe of Cleves in his manual *Toutes manieres de guerroyer* written around 1500 (Rodger above) and illustrated in a drawing of a caravela at Lisbon of *c.* 1550–5 (R. Barker, pers. comm. 2006).

Two ladles were found with the gun together with four iron shot of the correct size. A hailshot piece was found close by (79A1088) and two fragments of wrought iron gun with a second hailshot piece towards C8 (80A0544). Fragments of canister shot were recovered along with sword remains and inset dice shot (35–50mm diam.). In addition, two swivel guns were recovered from the scourpit beneath the Sterncastle in SS9 and SS10.

## Overloaded with Ordnance?
by Alexzandra Hildred

The total predicted weight of the ordnance (including spare breech chambers, but excluding carriages and shot) is about 45 tons (Tables 11.8–10). This is distributed over the three gun decks with *c.* 24 tons on the Main deck, 11 tons on the Upper and 11 tons in the Castles fore and aft (on either one or two decks). Following the fomat adopted by Parker (1996, 269–300) for the Elizabethan period (noting that the

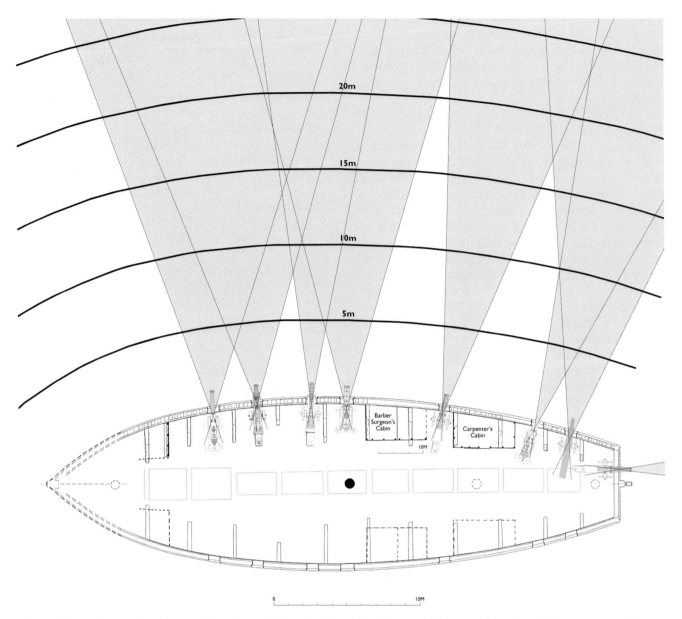

*Figure 11.41 Approximate intersection of arcs of fire of Main deck ordnance with ranges from the ship's side at 5m, 10m, 15m and 20m*

factors used there for displacement are contentious) the *Mary Rose* ordnance would represent 4.3% of total displacement estimated at 1050 tons (listed 700 tons burthen x 1.5). This is close to the figure of 4.5% quoted as being '*unprecedented*' in the *Dreadnought* in 1573. However, slightly different displacement figures (based on provisional data) were calculated by Coates for the *Mary Rose* (Coates 1985). His initial calculations were based on a multiplication factor of 1.64, implied from 700 tons (1150 tons displacement with 150 tons ballast at 42in (1067mm) freeboard), with [30 tons of] ordnance therefore representing 2.6% of 1150 tons in 1514. For the modified ship in 1545 he used 1.91 implied from 800 tons (1530 tons including 230 tons ballast with a final freeboard of 39in (991mm) and allowing for other changes) with the 45 tons of ordnance representing 3%.

The *Mary Rose* tonnage was also calculated using the rule conventionally ascribed to Matthew Baker. Assuming a keel length of 106ft (32.3m), moulded breadth of 39ft 4in (12m) and depth in hold of 16ft (4.9m) (± a foot (0.30m) for uncertainty in the measuring points). This gives 667 tons (±40) and provides a ratio of guns of 0.0674 tons/ton burthen. This approximates nearly the data for Parker's estimations for tonnage and percentage displacement for ships in 1595 (Parker 1996, 287). Conversely, if the declared burthen of the *Mary Rose* in 1545 was 800 tons, then the same data would allow at least 0.1046 tons, 1.7 tons more than the actual figure carried. Equally, Coates' early estimate of 1530 tons displacement in 1545 gives a much lower ratio of guns: 3% of displacement. Thus various assumptions about displacement of the ship as rebuilt, or on the day of the sinking, lead to ratios significantly lower than those presented by Parker for the next generation: an average of 4.8%, though the range includes 2.2–7.3 %. Some of the lowest ratios are from the older ships, and ships larger than the *Mary Rose*.

The proportion of 'ship-smashing guns' (minimum 9lb (4.1kg) shot as suggested by Parker (1996)) is less than the Elizabethan examples provided, with only 32 of the 91 *Mary Rose* guns qualifying. The total number of guns is still very high, and indeed the number of 'heavy' guns on the *Mary Rose* is more than on any ship listed in the 1595 inventory, with the exception of the *Merhonour* (1400 tons displacement, 38 heavy guns) and the *Vanguard* (1500 tons, 33 heavy guns; Parker 1996, 287).

Although the total weight of the guns carried on the *Mary Rose* appears high, it is also well within the limits suggested by a Dutch rule dated 1650–1660 (Barker 1999, 21–9). Using the formula cited there the lowermost gundeck of the *Mary Rose* would be able to safely accommodate 101.66 tons, with more guns allowed on other decks (compared with the actual total weight of ordnance of 45 tons). While the rule was for differently-shaped ships and conditions, it again suggests that the weight of ordnance on the *Mary Rose* was not exceptional.

Indeed, the largest ships of the Henrican period carried a substantial number of ship-smashing guns and were capable of stand-off gunnery. The number of smashing guns within the Henrician fleet estimated by Parker as 200 guns and 15,000 tons for the entire fleet does not take into account the 213 port pieces and 32 slings. The port pieces, as demonstrated by those on the *Mary Rose* often have an 8in (203mm) bore and fire a 15lb (6.8kg) solid stone or 25lb (11.3kg) lantern shot. The slings are the wrought iron equivalent of the demi culverin and must also be considered as ship-smashers. A more realistic number, therefore, is 445 ship-smashing guns in 1547 for 15,000 tons of vessel. This is proportional to the figure of 600 ship-smashing guns listed in 1595 for the fleet of 20,000 tons.

In short, there is little or no evidence for overloading of the ship in terms of weight of guns alone, though issues of instability resulting from where these were located with respect to the stability and strength and design of the ship is even more intractable and beyond discussion here.

The major difference in ballistic capability between the fleets of 1545 and 1595 and later was probably not in the number of guns capable of firing a shot of a certain size, but in the muzzle energy achieved by the guns. Firing trials carried out on the *Mary Rose* port piece demonstrated a best muzzle energy per ton of gun of 130ft tons against the *Mary Rose* culverin which achieved a muzzle energy per ton of gun of 437ft tons. The port piece affords excellent results for a gun which can be relatively cheaply and easily produced, rapidly reloaded, is light for its bore. It can be tied to the side of a ship to retard recoil and as a perrier its projectile has a devastating effect on wood. However, it is suggested that the muzzle energy afforded by the cast muzzle-loaders ultimately led to these guns being chosen for preference, and the most obvious reason for this must be increasing ranges at which ships engaged. What is interesting is that, for some time later, breech-loading port pieces made in bronze continued to be manufactured and used on ships (Blackmore 1976, 271, 273). However one must not disregard the fact that the guns and ships of the Henrician navy formed an important step in this process and it is to this time period which we should be looking in order to assess the origins of the broadside engagement, the desire for increasing the distances at which ships engaged and the realisation that consistency in gun and projectile manufacture was vital to this success.

## Playing with Arcs of Fire: How Well Defended was the *Mary Rose*?
by Alexzandra Hildred

Despite having a mixture of guns of differing capabilities, it appears that the *Mary Rose* was relatively well defended, especially on the Main deck. Most of the side of the ship is covered by the arcs of fire of her Main deck guns at a point about 25m away, where the arcs of the shortest range Main deck guns cross (Fig. 11.41). The angles of fire afforded by the Main deck ports, the particular guns and the structural features around each port, give arcs of fire which vary between 39.1° (M3 culverin) and 19.6° (M5 port piece; Table 11.11). The arcs of the culverin in M3, the cannon in M6, and the demi cannon in M10 (longer range guns) cross at about 65m. The port pieces in M4, 5, 8 and 11 all have comfortable ranges of over 100m, but it is suggested that, for maximum shattering effect, this would be extreme, therefore the 65m mark of the three longer range guns is probably important. The two longest range guns, the culverin in M3 and the demi cannon in

**Table 11.11 Arcs of fire: Main deck**

| Gunport | Arc of fire (°) | Distance(1) | Distance(2) |
|---------|-----------------|-------------|-------------|
| M3 | 39.1 | M3–4: 3.59 | M3–6: 14.41 |
| M4 | 38.3 | M4–5: 8.34 | |
| M5 | 19.6 | M5–6: 4.47 | |
| M6 | 31.4 | M6–8: 23.35* | M6–10: 65.54 |
| M8 | 22.8 | M8–10: 23.34** | |
| M10 | 20.8 | M10–11: 3.39 | M3–10: 99.44 |
| M11 | 30.3 | M8–11: 17.14 | |

Distance: distance of intersections of arcs from side of ship (m). (1) all guns; (2) bronze guns. Points of origin of arcs are taken from the back of the cascabel/breech chamber
* Barber-surgeon's cabin; ** Carpenters' cabin

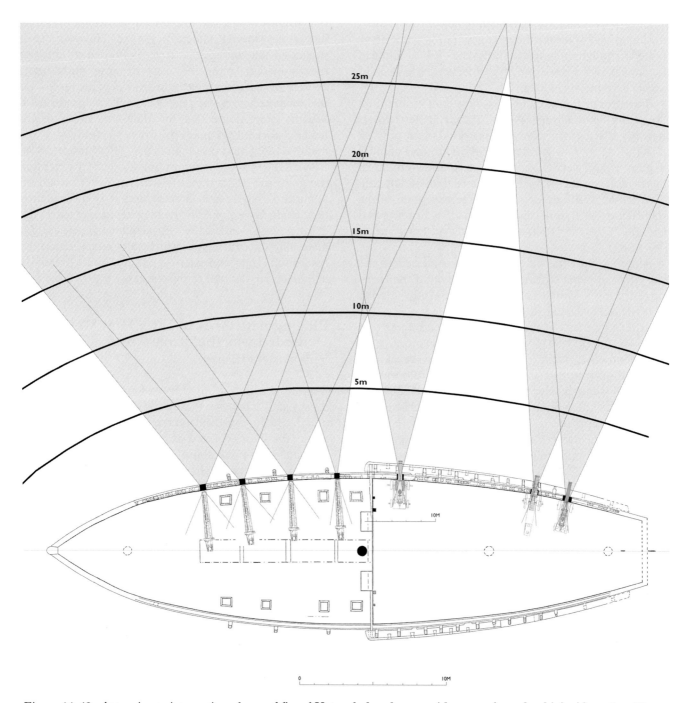

*Figure 11.42 Approximate intersection of arcs of fire of Upper deck ordnance with ranges from the ship's side at 5m, 10m, 15m, 20m and 25m. This assumes slings in the waist, an iron gun in U10 and a bronze gun in U11*

M10, afford some longer range protection in quite wide arcs around the bow and stern as a result of the curvature of the hull at these positions and the ability to angle the guns against the port. The only forward firing gun is the demi culverin firing from the Castle deck for a distance of up to 320m forward of the bow. This provides some protection to the bow even without adding further guns to this large and seemingly undefended area. The stern, however, is impossible to defend without placing guns on either side of the rudder, probably on the Main deck and two further levels, as suggested by the *Anthony Roll*. Either port pieces or demi culverins would suffice in this position and it is difficult to suggest which would be used. Traditionally these were long guns in order to clear the structure. The ship-smashing capability of the port pieces and the creation of a neat working area for four of these, two out of the stern and M11 port and starboard, is appealing. Furthermore, the positioning of the demi culverins in U11 in the stern would be a significant contribution to the Upper deck armament (Fig. 11.42).

The majority of ship-smashing firepower is contained in the middle of the ship and might suggest a

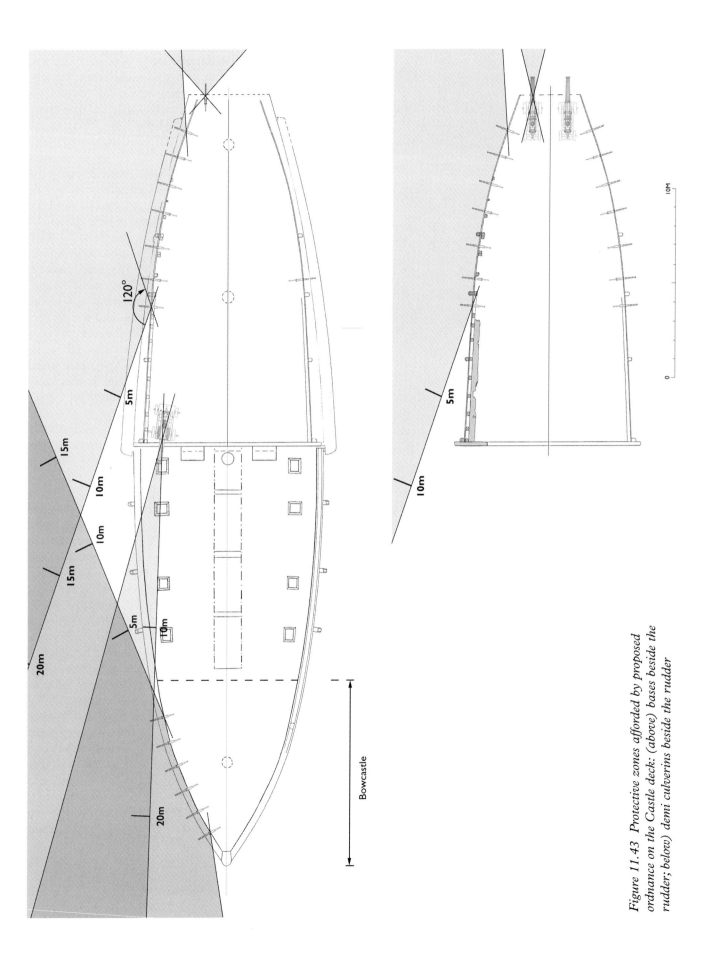

Figure 11.43 Protective zones afforded by proposed ordnance on the Castle deck: (above) bases beside the rudder; (below) demi culverins beside the rudder

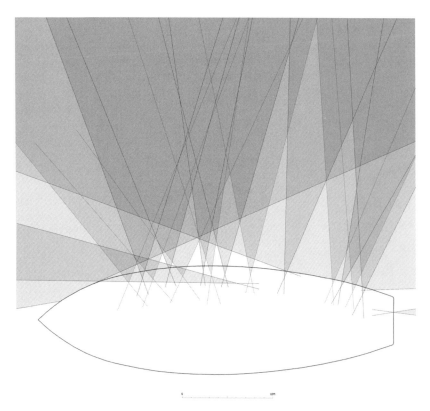

*Figure 11.44 Protective zones afforded by the intersections of arcs of fire on all decks*

the hull (Table 11.12 and Fig. 11.42). Here slings are classed as 'ship-smashing' guns. Although range is unrecorded and trials have not been undertaken, the size of the chamber, length of barrel and weight of projectile suggest that these would be medium to long range weapons, so 100m is suggested. The gunport in U7, just aft of amidships, supported a culverin with a predicted range of about 413m. Another bronze gun may have been situated at the U11 port, offering longer range potential towards the stern. The position of the U7 culverin, amidships, reinforces the notion that engaging with broadside guns at range was an important consideration. The handguns found inboard on the Upper deck in the stern, and the archers stationed in the waist, suggest that this was an important location for anti-boarding activities. The deck in the waist is far less obstructed than is the Main deck. At least one saker on the Upper deck on either side of the bow would enable longer distance protection (est. 323m) and offensive capability in this important area. Bases were probably situated on the Upper deck at the stern, depending on the space available around the steering gear, though it is possible that there were no guns at this level. In that case, to match the *Anthony Roll*, there would need to be another small deck above the Sterncastle deck. The range of bases is uncertain, but is expected to have been short and the guns used principally as anti-personnel weapons.

The Castle deck is even more difficult to provision with guns as so little survives. It is anticipated that most bases were situated in the castles, with perhaps a single falcon provisioning both sides or the stern. The range of the bases is, again, uncertain, but these are anti-personnel weapons. The estimated point blank range of a falcon is up to 287m. The spaces between the stirrup holes for bases on the Upper deck suggest 0.5–1.0m spacings between guns. If four were housed in the stern, fourteen in the Sterncastle distributed port and starboard and twelve in the Forecastle on the sides, the full 30 can be accommodated. Portable handguns and hailshot pieces and longbows could be used between them. The bases provides almost blanket cover along the length of the Castle decks with arcs of up to 150° for each gun. Unfortunately range is not known. The only areas not well covered are the fore and aft quarters of the castles. The demi culverin at the front of the Sterncastle has an arc of 14–21° and range of about 380m (Fig. 11.43). However, an additional deck in the stern would provide more space for handgunners and archers and additional fire-ahead capability. The arcs of

stand-off engagement alongside at a range of probably less than 100m using the battering pieces. Longest-range capability from Main deck guns is afforded by the two guns situated close to the each end of the ship, the arcs of which intersect well within the range of the port pieces and cannon amidships.

The arcs of fire and defence of the ship at Upper and Castle deck level is much harder to predict, although some suggestions can be made as to the best potential positions for absent guns. The side of the ship is again well defended with wrought iron slings, the largest of which carried cast iron shot of 4in (102mm) diameter (nearly 8lb 15oz; 4kg). The arcs of fire range from 27.5° to 63° and the centre of the ship is protected by the intersections of the arcs at a distance of about 27m from

**Table 11.12 Arcs of fire: Upper deck**

| Gunport | Arc of fire (°) | Distance of intersections of arcs (all guns) from side of ship (m) at 27° |
|---|---|---|
| U3 | 63 | U3–4: 1.76 |
| U4 | 63 | U4–5: 2.64 |
| U5 | 63 | U5–6: 3.54 |
| U6 | 27 | U6–7: 10.69 |
| U7 | 27 | U7–10: 26.71 |
| U10 | 27 | U10–11: 3.7 |
| U11 | 30 | |

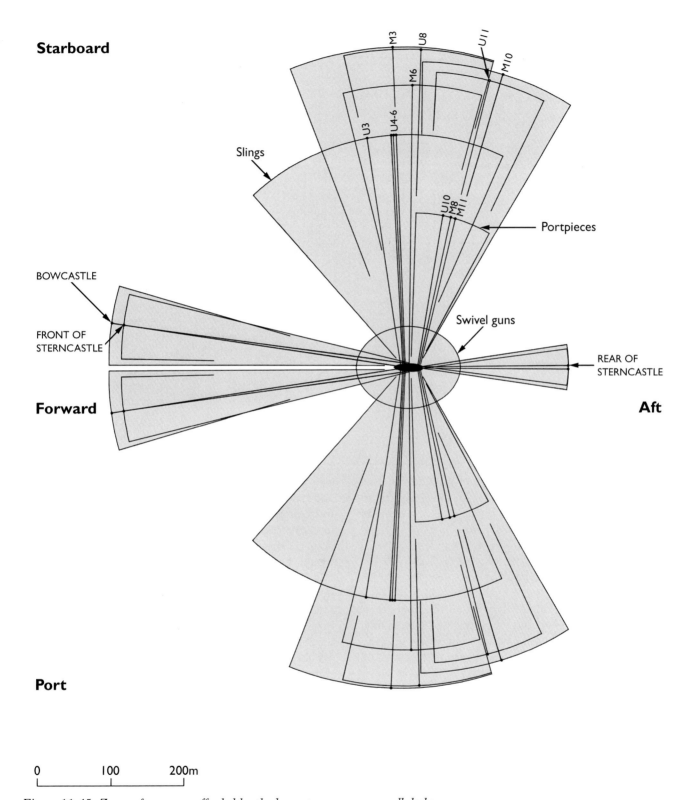

*Figure 11.45 Zones of coverage afforded by the longest range guns on all decks*

fire for the ship afforded by all guns over the three known gun decks is shown in Figure 11.44. The hand-held weapons are not factored into this illustration.

Figure 11.45 shows the zones covered by the longest range guns on every deck, assuming a long range gun (demi culverin or sling) in the Forecastle. This clearly demonstrates the importance of the demi culverins firing forward from the front of the Sterncastle. The ship's defences would be enhanced by the presence of another port piece on the Main deck in the bow, where there is zone of 35° which is only covered by the bases. In the stern this is 50°. The *Anthony Roll* displays the

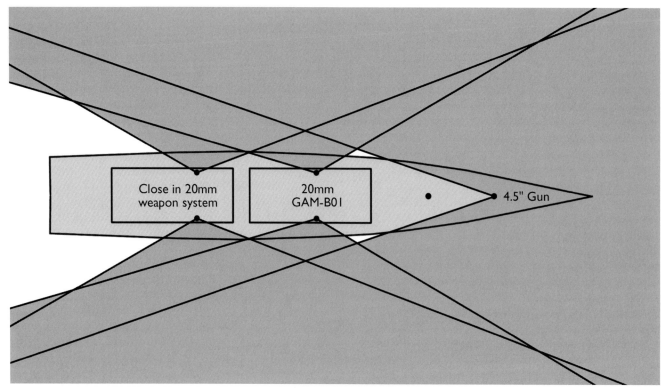

*Figure 11.46 Zones of coverage afforded by weapons on a modern warship*

ship with guns in the back of the Sterncastle, accommodating these would greatly enhance the protection of the stern. Figure 11.46 shows a similar arrangement for a modern warship.

## Fighting with the Ship

Although it has been demonstrated that the *Mary Rose* was capable of engaging weapons at increasing distances, the continued inclusion of numbers of non-gunpowder weapons and staff weapons, and listings of items such as '*lime pots*', imply that fighting at close quarters was still important. One of the best methods to achieve this was to '*grapel*' or make fast to another vessel using ropes or purposely designed '*grappling irons*' often termed '*grapnels*'.

### *The grapnel*
by Douglas M<sup>c</sup>Elvogue

In the *Anthony Roll* a bright red grapnel hangs at the end of the *Mary Rose*'s bowsprit (Fig. 11.47a). No grapnel was found during the archaeological excavations but John Deane did recover a large grapnel in the 1830s. This is illustrated in what was meant to become '*John Deane's Cabinet of Submarine Recoveries, Relics and Antiquities, etc*', instigated in 1836 but never completed (Bevan 1996, 114).

In Deane's illustration the grapnel is shown next to an onion bottle and conglomerate concretion, lying above a longbow. As published (*ibid.*, pl. 23) it appears to be drawn to the same (but unidentified) scale as the first two objects, with the longbow at a different scale (Fig. 11.47a, inset). If this was the case, the grapnel would be of a size that could be thrown by hand. However the original artwork has very faint pencil markings, representing one-eighth and one-half: all the objects were drawn at 1:8 scale apart from the onion bottle which was 1:2. It is known that a number of longbows were recovered by the Deanes; two of these are in the Royal Armouries. One looks remarkably similar to the illustrated bow and, at 1:8, would be 1.87m – the same length as the Royal Armouries longbow. At similar scale, the grapnel measures 1.4m long with a 2.6m chain made from 25 links, each 100mm wide internally. This makes it quite substantial (Fig. 11.47b) and more akin to the object depicted in the *Anthony Roll*.

During Tudor times the '*grapnel*' (variously spelt: '*Crapnolles*', '*grapeyrons*', '*grappers*', '*grapilles*', '*grapulles*', '*grapyrons*' and '*gavulles*') was generally listed in association with a '*cheyne, of yron*'. Six vessels are listed with grapnels in 1514; *Henry Grace à Dieu*, *Trinity Sovereign*, *Gabriel Royal*, *Kateryn Forteleza*, *John Baptist* and the *Christ of Greenwich*. The *Henry Grace à Dieu* has one '*Grapilles with the cheyne hanging apon the bowsprit with apole having acolk of brasse*' (Knighton and Loades 2000, 113). From this description we can surmise that the *Mary Rose* would have had a grapnel with an iron chain hanging off the bowsprit from a block, or possibly a pole, with a bronze sheave. The chain would have been sufficiently long to be out of the reach of the axes used by the opposing force, and would be fastened to a rope which led back into the forecastle where it would be made fast and probably fed around a small windlass, so that the grapple and the ship it had caught could be winched in.

The 1514 inventory also describes '*Crapnolles for the bote to ryde by*'. These may have been similar but were employed as anchors for the ship's boat. These are listed separately from any '*grapyrons*' or '*grapilles*' associated with the bowsprit.

**Description of use in battle**
Grappling was the established tactic (see below) but its perils were dramatically demonstrated at the start of Henry VIII's first French war in the attack on the Breton flagship *Cordelière* during the action of 10 August 1512. After the *Mary James* had fired a prolonged cannonade, standing off to avoid entanglement with a ship twice her size ('*senza pertanto incatenarsi*': Spont 1897, 62). The battle was taken up by the much larger *Regent*, which grappled and boarded the *Cordelière*. But on the point of victory she was destroyed alongside her prey:

> '*sodenly as they war yelding themsylf, the caryke was one a flamyng fyre, and lyke wyse the Regent ... so ankyrryd and fastyd to the caryke that by no meanys possybyll she mygth ... depart.*' (ibid., 50)

We can surmise that the grapnel was used when boarding was considered to be the main tactic. The two ships would steer for each other and attempt to drop their grapnel into one another, thus binding them together.

## Shear hooks

A second feature seen on the *Anthony Roll* is the presence of shear hooks, '*Sherehokes*' (Knighton and Loades 2000, 120). These were scythe-like hooks attached to the yards that could slash through another ship's rigging at close quarters. They are illustrated in pairs on the end of the main and fore yards of the *Henry Grace à Dieu, Mary Rose, Peter Pomegranate, Matthew, Jesus of Lübeck, Pauncy*, and on the end of the bowsprit on the *Jesus of Lübeck* and *Pauncy*. No shear hooks was found during the excavation. A single possible shear hook was recovered from the site of the Armada ship *El Gran Griffon* (Martin 1972, 68), being a curved iron blade 0.38m long. Pairs of '*Sherrhokes*' are featured in the 1514 inventory. These are listed for the main yard on the *Trinity Sovereign* (4) and the bowsprit for the *Kateryn Forteleza* (2), and on the main yard for the *John Baptist* (4).

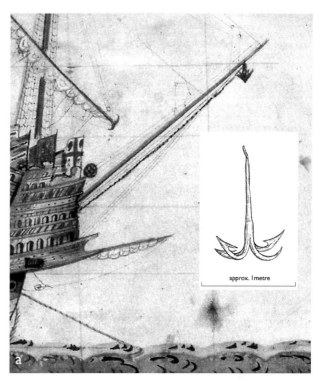

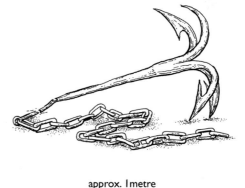

*Figure 11.47 Grapnel: a) detail shown on* Anthony Roll*; b) reconstruction*

# Rules of Engagement: What Does the Ammunition Tell Us?
by Alexzandra Hildred

We know relatively little about how the *Mary Rose* fought. We are uncertain of the distance at which she engaged or whether guns were fired singly or in volley, so there is still debate as to whether the objective was to sink enemy ships or merely to capture them. We know little about the duration of battles. Can the assemblage as listed in the *Anthony Roll* suggest battle range, duration or even profile?

The *Anthony Roll* lists the number of shot carried for each type of gun. By dividing this by the number of named guns of the relevant type, the number of rounds per gun can be calculated. This was analysed with a view to questioning whether '*preferred*' gun types would

Table 11.13 Weapons and ammunition

| Weapon (bronze then iron) | Breech-/muzzle-loading | Range point blank | No. weapons listed | Total ammunition listed | Rounds | Preference |
|---|---|---|---|---|---|---|
| Cannon | ML | medium | 2 | 50 | 25 iron | 7 |
| Demi cannon | ML | long | 2 | 60 | 30 iron | 5 |
| Culverin | ML | long | 2 | 60 | 30 iron | 5 |
| Demi culverin | ML | long | 6 | 140 | 23 iron | 8 |
| Saker | ML | medium–long | 2 | 80 | 40 iron | 3 |
| Falcon | ML | medium | 1 | 60 | 60 iron | 1 |
| Port piece | ML | medium | 12 | 200 | 16.66 stone | 10 |
| Sling | | medium–?long | 2 | 40 | 20 iron | 9 |
| Demi sling | BL | medium | 3 | 40 | 13.3 iron | 11 |
| Quarter sling | BL | ?short | 1 | 50 | 50 iron | 2 |
| Fowler | BL | ?short–medium | 6 | 170 | 28.3 stone | 6 |
| Lead for bases | BL | short | 30 | 400 | 13 | 12 |
| Top piece | BL | short | 2 | 20 | 10 stone | 13 |
| Dice for hailshot | ML | short | 20 | not listed | – | – |
| Lead for handguns | | short | 50 | 1000 | 20 | 9 |
| Longbows | | ?short | 250 | 9600 | 38 | 4 |

emerge, ie, guns which, by the number of rounds supplied, might be expected to be used more frequently than other guns, and whether this might have a bearing on the nature of any engagement. An example might be that if only the anti-personnel guns had large amounts of shot and the larger guns very few, than the most prevalent portion of an engagement might be anti-personnel, and therefore, at this period, at close range. One of the inherent difficulties is deciding which guns are anti-personnel and which are anti-ship. For example, a port piece of 8in (200mm) bore, loaded with a lantern shot of 25lb (11.4kg) filled with cracked flint could be anti-rigging (if elevated), but also anti-personnel. Loaded with a Kentish ragstone round shot of only 15lb (6.8kg) it definitely becomes a ship-smasher (Hall 1998, 57–66). A saker firing a cast shot of 5–7lb (2.3–3.2kg) is traditionally used as anti-rigging and to break masts, spars and disable rudders. Although this is a long range gun (which was to become an extremely important part of a vessel's armament) it would not match Parker's quoted figure of 9lb (4.1kg) that would qualify it as an anti-ship gun (Parker 1996, 270). The *Anthony Roll* figures for projectiles per weapon are given in Table 11.13.

The '*preference*' list thus generated (Table 11.14) demonstrates the importance of portable, easy to use, light-weight weapons. Although many would have ranges of up to 100m, 50m would be a more sensible range to achieve maximum success. We must bear in mind that, with all our trials (real or hypothetical), although a weapon may be capable of achieving a particular range, that does not mean this is the range at which it was routinely used. Most of the tables included here have been created using the results of trials on land not at sea. Trollope (1994, 27) alludes to this and comes up with some interesting suggestions regarding reducing '*quoted*' range at sea and considering instead '*effective*' range. His conclusions were startling: an effective range for any gun capable of penetrating a wooden hull actually hitting the target with a slow roll would be only 75yd (68.6m).

It is possible to look at the listed historical ranges for the guns and see whether the '*preference*' sequence based on number of rounds can suggest a preference for close, medium or long range engagements. This is extremely tenuous as the ranges given vary considerably and there are a number of guns for which it has been impossible to find historical information. Nevertheless, the following is put forward as our best attempt with the information available. This should be tested and challenged by future work.

Table 11.14 Weapon '*preference*' based on no. of rounds per gun

| Short range (<100m) more likely 0–50m | Medium range (expected pb) more likely 50–150m | Long range (expected pb) more likely 150m+ |
|---|---|---|
| Quarter sling | falcon (*c.* 250) | ?saker (220–440) |
| Longbow | ?saker (220–400) | culverin (400–600) |
| ?Fowler | ?fowler | demi culverin (250–550) |
| Handgun | cannon (280–350) | demi cannon (200–500) |
| Base | ?sling (300) | |
| Top piece | port piece (200) | |
| | ?demi sling | |

pb = point blank, no elevation

Table 11.15 Daily expected firing capability on land (after Lucar 1588, chap. 55)

| Gun | Fired daily | Minute per round | No. gunners | No. assistants/labourers |
|---|---|---|---|---|
| Ordinary double cannon to batter | 30 | 10.00 | | |
| Demi cannon (eldest & biggest) to batter | 80 | 3.75 | | |
| Ordinary demi cannon to batter | 108 (in 5 hrs) | 2.77 | 3 | 15 |
| Ordinary quarter cannon | 110 | 2.72 | | |
| Eldest & biggest old culverin | 60 | 5.00 | | |
| Ordinary culverin | 60 | 5.00 | | |
| Eldest demi culverin | 70 | 4.28 | 2 | 10 |
| Eldest & biggest saker | 80 | 3.75 | 1 | 5 |
| Falcon | 120 | 2.50 | | |

Table 11.16 Sequence (within ranges) based on total no. shot supplied

| Short range | Medium range | Longe range |
|---|---|---|
| Longbow | port piece | demi culverin |
| Handgun | ?fowler | saker |
| Base | ?saker | culverin/demi cannon |
| ?Fowler | falcon | |
| Quarter sling | cannon | |
| Top piece | sling/demi sling | |

The straightforward '*preference*' list generated from the actual assemblage (Table 11.14), differs significantly from the order (by gun size) provided by the *Anthony Roll* (Table 11.13). However when considering the number of rounds supplied for each gun, we must also take into account the ease (and speed) of loading and firing. The shot per gun is more for the muzzle-loading guns for both medium and long range firing. It might be expected that a breech-loader would be much easier and quicker to reload and that this would be reflected in the number of rounds supplied; this is not obvious. The size and weight of any gun will have a bearing on the number of men who will be needed to reload it – especially a large muzzle-loader which may need to be brought inboard to reload. We need to consider which of the other pieces might be engaged to maintain a bombardment without a substantial drain on man-power while muzzle-loaders were being reloaded. Some weapons are clearly more labour intensive than others.

While there is the view that guns were not reloaded immediately during battle, there are historical sources which suggest that guns (especially at sea) were routinely kept loaded (Bourne 1587, 56). There are several near contemporary treatises which give some indication as to how many times a gun could be fired per day (eg, Lucar 1588, chap. 55. His figures are assimilated in Table 11.15).

The crew listed for the *Mary Rose* included 200 mariners, 185 soldiers and 30 gunners. The total number of carriage mounted guns is 39, fifteen are similar to the smooth bore cast bronze guns listed in Table 11.15. Although the man-power listed needed to serve the guns as shown in Table 11.15 is not comprehensive, it does demonstrate that three times more personnel are required to maintain a demi cannon than a saker. These appear to be extremely high rates of fire and it must be remembered that these are rates for use on land where many factors will be different.

If one just looks at total number of projectiles for each weapon yet another sequence is produced: longbow, handgun, base, port piece, fowler, demi culverin, saker, demi cannon, culverin, falcon, cannon, quarter sling, sling, demi sling, top piece. These can be subdivided into ranges to give a slightly different sequence (Table 11.16). Hailshot pieces are not listed here as amount of shot is not listed for the *Mary Rose*.

Another way of attempting to recreate a blueprint of any battle is to consider not only the number of times a single gun is fired, but the total number of particular types of gun carried and total amount of ammunition. This gives a different '*preference of use*'. The importance of any single type of gun, based on its numerical superiority in the *Anthony Roll* (and therefore slightly different from the assemblage recovered), provides the following sequence: longbow; handgun; base; hailshot piece; port piece; demi culverin/fowler; demi sling;

Table 11.17 Preference list based on no. guns carried (sequence based on total no. type)

| Weapon | Sequence | Range (m) pb |
|---|---|---|
| Longbow | 1 (250) | 220–350 on land; use on ship ?short |
| Handgun | 2 (50) | 60; short |
| Base | 3 (30) | short |
| Hailshot piece | 4 (20) | short |
| Port piece | 5 (12) | medium |
| Fowler | 6 (6) | short–medium |
| Demi culverin | 6 (60) | 250–550; long |
| Demi sling | 7 (3) | medium |
| Top piece | 8 (2) | short |
| Sling | 8 (2) | ?medium/long |
| Saker | 8 (2) | 220–400 ?medium/long |
| Cannon | 8 (2) | 280–350; medium |
| Demi cannon | 8 (2) | 200–500; long |
| Culverin | 8 (2) | 400–600; long |
| Falcon | 9 (1) | *c.* 250 |
| Quarter sling | 9 (1) | ?short |

### Table 11.18 Weapons and ammunition according to range

| Range | Short | Medium | Long |
|---|---|---|---|
| No. weapons | 353 | 28 (26 exc. sakers) | 10 (12 inc. sakers) |
| Total ammunition | 11,069 (exc. hailshot) | 640 (560 exc. sakers) | 260 (340 inc. sakers) |

Fowlers & slings considered medium range, sakers shown for both medium & long range

culverin/cannon/demi cannon/sling/saker/top piece; falcon. These can then be further subdivided based on range (long, medium, short) and function (anti-ship, anti-personnel). There are few early English references to range and a series of range tables is given above (Table 11.3–5). The English examples have been used to compile Table 11.17, which demonstrates the numerical predominance of short range weapons. When the amount of ammunition is taken into account this becomes less overpowering (Table 11.18). If all portable arms are discounted the number of short to medium range guns is broadly similar (Table 11.18). Table 11.19 demonstrates a similarity in the number of medium and short range weapons but suggests a slightly higher usage of the former based on number of shot provisioned. However, this is based on the inclusion of fowlers as medium range guns. As their range is uncertain, if they are placed into the short range guns, the balance is altered slightly (Table 11.19). This shows a predominance for shorter range guns, but the number of shot per gun is still highest for the medium range weapons. Although the number of long range guns is small, the shot for them suggests 22–28 per gun. A similar amount is carried for the medium range guns (21–23 per gun). The amount carried for short range is dramatically lower, 14–16 per gun, but there are many more portable weapons.

Scrutinising the recovered assemblage in a similar manner encounters many potential problems. The assemblage is incomplete, at least as far as the guns carried and some of the smaller shot are concerned. The guns recovered differ slightly from the *Anthony Roll* inventory, and some of those listed have not been found at all. To compensate, hypothetical bands have been created to cover the smallest and largest bore size for any listed guns based on historical sources (Table 5.4). The fact that some of the wrought iron guns used cast iron shot makes the analysis more challenging, especially as the parameters for some of these guns is uncertain (for example the slings, Table 5.5). This is further compromised by the fact that most stone shot is unfinished and, therefore, cannot be closely matched to guns. Taken as a whole, the cast iron shot fit within the historical parameters for the listed guns and the increased number of small iron shot recovered enable larger shot to be made by placing these in a mould and pouring in lead (see Chapter 5).

Table 11.20 displays bronze guns based on shot per gun ratio and shows the number of shot expected and recovered within the wide historical bore bands for that type of gun. The shot expected is based on the number listed in the *Anthony Roll* per type of gun but has been increased to meet with the excavated assemblage where there are discrepancies, for example the culverins (2 listed, 4 recovered) and demi culverins (6 listed, 2 recovered, 4 expected in total).

The totals for shot recovered from the *Mary Rose* are: iron 1248, stone 387, composite lead 235 and lead 61. This gives 668 more cast iron shot than listed, though actual stone shot is similar, though largely unfinished. There are 155 fewer composite shot than listed and small lead shot is hardly represented, though this is probably a reflection of the wrecking/salvage rather than what was actually on board.

Using the rates of fire suggested by Lucar for smooth bore muzzle-loaders, it seems that the vessel was only provisioned for between one half and one day of action using her longer range guns assuming that all guns were being used to their maximum capability. The exception is the sakers; these appear to be provisioned for 3½ days (Table 11.20). This large amount of saker shot may relate to another use, however, such as for signalling, though sakers were to become some of the most frequently carried guns in later years. Alternatively this small shot may have been placed inside larger moulds into which lead was poured to create shot of greater (and variable) sizes.

Overlaps in diameter of shot to bore exist between sakers and demi culverins and demi culverins and culverins, when adopting the widest historical margins. Where more shot is recovered than listed, it has been difficult to work out the day capability as the shot is shared with so many other types of guns. So, the totals expected for the iron guns firing iron shot have been deducted and the number of days expected for the remainder calculated. However, by this method there is still considerably more shot than listed for most guns.

### Table 11.19 Ship supported/carriage-mounted guns and shot

| Range | Short | Medium | Long |
|---|---|---|---|
| No. guns | 33 | 28 (26 exc. sakers) | 10 (12 inc. sakers) |
| Shot | 469 | 640 (560 exc sakers) | 260 (240 inc. sakers) |
| *Fowlers considered short range* | | | |
| No. guns | 39 | 21 (19 exc. sakers) | 10 (12 inc. sakers) |
| Shot | 639 | 477 (397 exc. sakers) | 237 (317 inc. sakers) |

Excludes hailshot pieces as shot figures not given for the *Mary Rose*

Table 11.20 Bronze guns/shot listed and recovered

| Gun | Bore (in) | Shot size (in) | Fired daily | Shot required* | Shot listed per gun/total expected | Day capability listed | Shot recovered | Day capability recovered* | Range |
|---|---|---|---|---|---|---|---|---|---|
| Falcon | 2¾ | 2½ | 120 | 120 | 60/80 | ½ | 128 | 1 | medium |
| Saker | 3½ | 3 | 80 | 160 | 40/80 | ½ | 558 | 3½ | medium/long |
| Culverin | 5¼ | 5 | 60 | 240 | 30/120 | ½ | 159 | +½ | long |
| Demi cannon | 6¼ | 6 | 80–108 | 188 | 30/60 | ⅓ | 72 | −½ | long |
| Cannon | 8 | 7½ | 30 | 60 | 25/50 | ⅘ | 37 | ½ | medium |
| Demi culverin | 4¼ | 4 | 70 | 280 | 23/92 | ⅓ | 118 | −½ | long |

* using all guns

The shortfall for the smaller shot may relate to where these guns were located, probably on Upper or Castle decks from where it would have been more easily lost. Iron shot was found beside the U8 culverin (80A0976), tucked into the structure of the ship, suggesting that this may have been a practice for some guns positioned away from the main shot stores in M2 and M6. Losses of this size of shot would therefore be expected.

Despite the limitations, how can this help us to understand the way the *Mary Rose* might have fought? A model for endurance has been developed, based on the number of guns to which we have apportioned certain attributes, including range and estimated time of reloading. The most striking observation is the huge importance of engaging at middle distance. Using the calculated gun to shot ratios (Table 11.19), and reload rates (Table 11.15) the potential endurance (if fired sequentially) in minutes of engagement can be formulated for short, medium and long ranges (Table 11.21).

The twenty hailshot pieces have been excluded as there is no information about shot, although c. 1000 might be expected (as listed for the 20 hailshot pieces on the *Henry Grace à Dieu*). The excavated bores have yielded c. 20 dice, suggesting 50 rounds, 2.5 per gun. At 1.5 minutes per round this would add another 75 minutes per gun to the short range anti-personnel capability, increasing this to a total of 23,605 minutes.

This would increase the short range total to 3785 minutes, just over five days. This reinforces the continuing importance of fighting at close range.

However, although the number of projectiles per type of gun give an idea of the importance of that particular gun within any engagement, the numbers of guns and how they might be used together need to be considered. Sequential firing of all the guns is unlikely, especially at close range when as a prelude to boarding firing as many small arms as possible would be more likely. What is suggested is that particular types of guns would be used at specific times within a battle, dictated by range and the tactics employed by both vessels. The potential variables are enormous but another simple method of looking at endurance is by estimating how long it would take to exhaust shot if all the guns of a particular range on one side of a ship were used at once. Using the figures provided and assuming that the first round is already loaded (excluding handguns, hailshot and longbows) the endurance is markedly different; short range are exhausted in 19.5 minutes, medium range in 61.6 minutes and long range in 94.5 minutes (Table 11.22). Even if one includes the hand weapons the maximum estimated time would be for the hailshot, 75 minutes. As poorly cast guns one wonders whether they would be able to withstand repeated firing. The handguns would be depleted in 20 minutes and the longbows in under three. This again reinforces the

Table 11.21 Endurance based on shot and reload rates

| Range | Handgun(AP) | Longbow(AP) | Short | Medium | Long |
|---|---|---|---|---|---|
| No. weapons | 50 | 250 | 33 guns | 28 guns | 10 guns |
| No. rounds | 20 | 38 | 14 | 23 | 22 |
| Reload rate (minutes) | 1 | 16 arrows each | 1.5 | 2.8 | 4.5 |
| Minutes per gun | 50 | 2.37 | 21 | 64.4 | 99 |
| Total minutes | 1000 | 592.5 | 693 | 1803.2 | 990 |
| Total minutes | | | 2285.5 | 1932 | 792 |

AP = anti-personnel. Table includes sakers as medium range guns

Table 11.22 Endurance (minutes) based on volley firing (all guns loaded)

| Range | No. rounds | Reload time | Endurance |
|---|---|---|---|
| Short | 22 | 1.5 | 19.5 |
| Medium | 21 | 2.8 | 61.6 |
| Long | 13 | 4.5 | 94.5 |

importance of the medium range capability and indeed the growing importance of long range weapons.

It appears, then, that *Mary Rose* was provisioned adequately for engaging at a range of distances. The number of guns carried and the number of rounds suggests that a significant portion of any engagement was carried out at medium range. The largest single number of medium range guns are the port pieces, and indeed the greatest amount of medium range shot is the Kentish ragstone provided for them. The positions of these guns, with at least four each side of the Main deck, with three port pieces and a cannon spanning the midships area, suggests that the ship in her final stages included a design of hull and dispersal of weapons that enabled a substantial semi-broadside using battering pieces, with the primary task of sinking ships. We should consider the possibility that, in order to facilitate this, reloading without turning the ship would have been practiced, especially as the port pieces are breech-loading. If sinking the enemy vessel was not achieved, the fowlers, also stone throwing guns, could be used as the range lessened, followed by the ship supported anti-personnel weapons (for a much shorter period), augmented by hand held weapons. Finally the pikes, bills and swords were used in an anti-boarding manoeuvre. The port pieces of the *Mary Rose* illustrate the beginnings of the attempt in England to standardise the use of a particular gun in large numbers against the hull of an enemy vessel, at predicted extreme ranges up to 200m though more probably at less than 100m where the damage inflicted would be greatest. We are also witnessing the appreciation of the perrier possibly for the large amount of damage which a stone shot could inflict.

Tactics which employ guns at different ranges for specific purposes using particular types of shot are alluded to in the few sixteenth century treatises available. Bourne (1587, 61 and 55–6), referring to an approaching enemy, states the following:

> 'with their great Ordnaunce, as Cannons, or Culuerings, at a great distance, to shoote the whole yron shot as you doe at battery, & as they doe aproch neere, then to shoote Faucon shotte, and as they doe come neerer, Faconetshotte, or smal base shotte, and at hand all manner of spoylying shot, as chayne shotte, or cliue [cleft] shot and dise shot, and such other like.

> And furthermore, for the Sea fight, if the one doe meane to lay the other aboorde, then they doe call vp their company, eyther for to enter or to defend: and first, if that they doe meane for to enter (as you may knowe) that hee will p[e]rcase [perchance] to laye you aboorde, then marke where that you doe see anye Scottles for to come vppe at, as they will stande neere there aboutes, to the intente for to bee readie, for to come vppe vnder the Scottles: there giue leuell with your Fowlers, or Slinges, or Bases, for there you shall bee sure to doe most good, then furthermore, if you doe meane for to enter him, then giue leuel with your Fowlers and Portpeeces, where you doe see his chiefest fight of his shippe is, and especially be sure to haue them charged, and to shoote, them off at the first boording of the Shippes, for then you shall be sure to speede. And furthermore, marke where his men haue most recourse, there discharge your Fowlers and Bases. And furthermore, for the annoyance of your enemie, if that at the boording that the S[h]ippes lye, therefore you may take away their steeradge with one of your great peeces that is to shoote at his Rother [rudder], and furthermore at his mayne mast, and so foorth.'

The term '*board*' is taken to mean '*closing with a ship*'; what we now consider boarding was then termed '*to enter*' (Corbett 1905, 15).

There are some earlier references to the sequence of firing guns. De Chaves clearly advocated a strategy comprising a wedge formation, strongest in the centre (three quarters of the vessels) and most manouverable and smallest to either side. Vessels in the centre are arranged in rows, with the strongest first and the most heavily armed behind. De Chaves lists a number of tactical alterations to the formation as a response to that of the opposing fleet. Gaining the weather gage is listed as being paramount, but the reason given is to keep smoke from the enemies guns away from the home fleet, whilst engulfing the enemy fleet. Regarding '*Battle*', he wrote:

> 'Then the flagship shall bid a trumpet sound, and at that signal all shall move in their aforesaid order; and as they come into range they shall commence to play their most powerful artillery, taking care that the first shots do not miss ... when the first shots hit, inasmuch as they are the largest, they strike great dread and terror into the enemy; for seeing how great hurt they suffer, they think how much greater it will be at close range ... Having so begun firing, they shall always first play the largest guns, which are on the side or board towards the enemy, and likewise they shall move over from the other side those guns which have wheeled carriages to run on the upper part of the

*deck and poop. And then when nearer they should use the smaller ones, and by no means should they fire them at first, for afar off they will do no hurt, and besides the enemy would know there is dearth of good artillery and will take better heart to make or abide an attack. And after having come to closer quarters then they ought to play the lighter artillery. And so soon as they come to board or grapple all the other kinds of arms shall be used ... first, missiles, such as harpoons ['dardos'] and stones, hand-guns ['escopétas'] and cross-bows, and then the fire-balls aforesaid, as well from the tops as from the castles, and at the same time the calthrops, linstocks, stinkballs ['pildoras'], grenades, and the scorpions for the sails and rigging. At this moment they should sound all the trumpets, with a lusty cheer from every ship at once they should grapple and fight with every kind of weapon, those with staffed scythes or shear-hooks cutting the enemy's rigging, and the others with the fire instruments ['trompas y bocas de fuego'] raining fire down on the enemy's rigging and crew ... The ships of support in like manner should have care to keep somewhat apart and not to grapple till they see where they should first bring succour. The more they keep clear the more will they have opportunity of either standing off and using their guns, or of coming to close range with their other firearms.'* (Corbett 1905, 10–12)

The orders formerly attributed to Thomas Audley refer to port pieces and hailshot pieces, not generally listed before the mid-1530s:

> *'In case you board your enemy enter not till you see the smoke gone and then shoot off all your pieces, your port-pieces, the pieces of hail-shot* [and] *cross-bow shot to beat his cage deck, and if you see his deck well ridden* [cleared] *then enter with your best men, but first win his tops in any wise if it be possible.'* (ibid., 15–16)

The weapons listed and indeed recovered suggest that the *Mary Rose* carried a layered system of guns which could be used at differing ranges or, by altering the projectile, for differing purposes. It also suggests that a significant portion of any battle was expected to be fought at middle range and that the guns chosen for this were port pieces, for their ship-smashing qualities. The carrying of unfinished stone shot allowed for last minute adjustments to suit specific bore and militate against running out of shot of a specific size. The same can be said of the lead covered dice or lead covered round shot, all of which could be completed to the required size from the semi-worked materials found on board.

# 12. The End of the Beginning: Some Reflections on the Ordnance of the *Mary Rose* and Avenues for Future Research

## Alexzandra Hildred

## Introduction

*'Having through the grace of God arrived at the appointed end of my voyage and neither knowing nor perceiving a way of going farther, I had decided to furl the sails and cast the heavy anchors into the water in order to enjoy in tranquillity respite from my sailing, when I was warned by my pilot that before I could disembark I should turn my glance backward and reexamine my work in detail to see if there were any part that had by chance remained undescribed in the obscurity of silence.'* (Biringuccio 1540, 444)

When I first joined the Mary Rose Project as a volunteer diver in 1979, I never for a second realised that it would dominate my life for even the following three years, let alone the next 30. As fate would have it, my first experience of the *Mary Rose* was not one of sight, for the visibility that October was notoriously bad, but of touch: the cold bronze of the cascabel of the (then *in situ*) enigmatic bronze bastard culverin on the Castle deck, the first of six bronze guns that would later be recovered. When in 1982, following the lift, Margaret Rule suggested that I begin to catalogue the ordnance rather than write up a more general report for the stern of the ship (an area where I had shared the archaeological supervision) I was at first sceptical; I knew far more about the excavation of the stern than I did about guns, shot and the art of war.

I now realise that I had been presented with the most wonderful gift: the chance to record, almost without boundaries (with the exception of finance) the *raison d'être* of the ship, a fighting vessel, which through her lifetime, bridged the gap between the medieval and modern worlds. In addition, she was the favourite vessel of a famously belligerent King, who took a personal interest in the design and manufacture of arms, armour, artillery and ships.

When Henry VIII took over the throne in 1509 England was not a sea power. His meagre inheritance consisted of only five ships and one officer, the Clerk of the Ships. Within the year two new ships were under construction, one of them destined to become the only Tudor warship most people can now name, the *Mary Rose*.

During the 35-year life of the *Mary Rose* we can trace a complete revision of the use of gunpowder weapons at sea. At the beginning of Henry's reign the majority of these were small anti-personnel guns located high in the castles, and medieval tactics of grappling and boarding were still predominant. By 1540 the balance of guns had changed and, by 1545 the *Mary Rose* carried a minimum of 14 heavy guns mounted on carriages, and stationed on a gun deck running the entire length of the ship just above the waterline. Accommodation of these weapons required a complete revision of shipbuilding tradition so that numerous reasonably watertight lids could be fitted at ports cut within the hull.

The significance of the *Mary Rose* is not only that she is the only Tudor warship ever recovered, but that in the sixteenth century she had been one of the first integrated fighting machines capable of powerful stand-off warfare. Guns similar to the cast bronze muzzle-loaders on four-wheeled carriages were to remain the dominant weapons of naval warfare for the next two centuries. Within the ordnance carried on the *Mary Rose* we see the beginning of the capability for a broadsides engagement. Although it was some decades before this became settled practice, the armament of the *Mary Rose* was to define the nature of sea combat in the age of sail.

By looking at historical sources, such as weapons inventories, alongside the archaeology we have begun to chart these changes. Alterations to the structure of the hull, including the blocking of some gunports and the cutting of others, are among the more visible evidence; in the inventories, development can be tracked in the changing names of the guns. By the time Henry died, in 1547, this transformation was, in essence, complete; he left a fleet of over 50 vessels, including twenty 'Great Ships', which were large fighting platforms. But they were not merely that; they were both symbols and physical manifestations of the power and organisational abilities of the emerging Tudor nation state. This is reflected in the embellishments on the ten recovered bronze guns, many of which bear Henry's monogram and the names of the founders he set up to cast guns for him in London. At the beginning of his reign bronze guns had to be imported from the Low Countries. By the end of the 1520s they were being cast at Houndsditch, helped by the French founder Peter Baude, who may have been head-hunted by the King for that purpose. Before his last French war Henry promoted the application of Kentish iron manufacture to the casting of ordnance; from this developed the thriving gunfounding industry of the Weald.

To furnish these home-made bronze guns, the casting of iron shot had to increase, and this was related to the first attempts to cast iron guns. This is most importantly evidenced in the finding of four of the 20 listed cast iron hailshot pieces, muzzle-loading anti-personnel weapons with a rectangular bore, firing a hailstorm of wrought-iron dice at close range. These are the only examples we have of the second most numerous of the iron guns carried in the fleet. The commonest of all was the wrought iron swivel gun, also an anti-personnel weapon.

After these in quantity come the large wrought iron breech-loading port pieces, capable of firing a 200mm stone shot over 100m and with a muzzle velocity of 338m/s. Port pieces were found on the main gun deck, interspersed with muzzle-loading bronze guns.

The obvious importance of the wrought iron breech-loader, confirmed by its presence within the main armaments of the *Mary Rose*, has revolutionised our dating of wrought-iron ordnance. Many sites thought to be fifteenth century are now being reviewed in the light of the *Mary Rose* assemblage.

Just as important as the guns are the carriages on which they stood. Excavation of these increases our understanding of the operation of weapons themselves within the confines of the gun bays. Each type of gun can now be seen as one layer of an integrated weapons system.

Long-range capability was afforded by the cast-bronze muzzle-loaders, whose cast iron shot could punch a clean hole through the side of a wooden ship. From short range the twelve port pieces fired Kentish ragstone shot, which shattered timbers and splintered into sharp flakes, acting as shrapnel on impact. These guns were multi-purpose; at shorter range still, and loaded with wooden canisters packed with lethal iron or flint fragments, they were effective as anti-personnel weapons. Yet smaller are the 30 wrought-iron swivel guns using lead shot inset with offset iron dice which imparts a spin. These, listed in inventories as '*shot of dice and lead*', have been found on a number of sites worldwide; but they often remain unrecognised because on many of these sites the iron dice has eroded away, leaving a void. Positive identification as shot has only been achieved through their remarkable preservation on the *Mary Rose*.

Other gunpowder weapons include handguns, imported in bulk from the famous gun-producing towns in Italy, together with the first known instance of shields carrying small breech-loading guns found in a fighting context.

The assemblage includes over 2000 shot, together with the loading implements and even the individually carved wooden linstocks used by the gun captain to fire the guns. Every object type provides extremes, from the highly decorated and expensive preserve of the officers to the mundane tools of the crew. Although many museums have ornate weapons of the period, it is rare to find dated examples in such profusion. At the other end of the scale the *Mary Rose* has yielded the few dated examples of the wooded hafts of pikes and bills, fundamental staff weapons used by infantry on land and important anti-boarding devices aboard a ship. Yet again, the scope of the collection enables a study of differences in style and manufacture. A single, highly decorated halberd might have been carried by a member of the King's bodyguard; the several ornate gun-shields might indicate the presence of a whole detachment of the Yeomen of the Guard or Gentlemen Pensioners.

The single, simple basket-hilted sword recovered from beneath a gun port on the starboard side provides the earliest precisely dated example of the type and encapsulates the importance of every object on the *Mary Rose*. The entire typology for British basket-hilted swords now revolves around this miracle of survival. The *Mary Rose* is a vessel that truly crosses the watershed from medieval to early modern, both structurally and within the assemblage and technology of her weapons.

Over 170 of the 250 yew longbows listed as being on board have been recovered. They are the only examples of precisely dated English longbows, and are probably similar to those used at Crécy and Agincourt. Remarkably, nothing tangible was previously known about these, arguably the most 'English' of any historical weapons. The study of this group alone exemplifies the potential of every object group – studies in size and wood density enabling the prediction of draw weight and therefore suggesting the strength, skill and stature of the men who used them. Studies continue into the manufacturing techniques, the raw material and the marks made by the bowyers; so does testing of the originals and shooting trials using approximations.

Some of the bows were found in chests stored deep within the ship, with others on the Upper deck in the positions where they were used. Over 2000 complete arrows were recovered of the 9600 listed as being on board. The assemblage includes arrows of two distinct draw lengths, at least four different profiles, and nine woods. They bear the remains of glue and flights and vestigial information regarding the tips; enough to replicate and test.

The weapons are the reason for the building of the ship and are inexorably linked with its structure, the artefact assemblage, and the men who lived and died working them. They are also the means by which the *Mary Rose* was identified. John Deane was the first to put a name to the wreck, on the recovery of a bronze gun in 1836. Latterly, it was the finding of a wrought iron gun by Alexander McKee's team in 1970 that pinpointed the centre of the search which was to result, a year later, in the finding of the first timbers. These were later identified as the port frames of the *Mary Rose*.

## Back to the Future: Opening the Door

The magnitude of the collection means that this publication can represent only the very beginning of research. We have just begun to realise the potential information that future analysis and interpretation may, indeed must, realise. To facilitate this we must ensure that as much of the collection as possible is accessible. This ethos is enshrined in the philosophy for the planned new Museum, due to open in 2012. Alongside the display of objects, however, must be the availability of the most detailed information we have: the archive. A priority must therefore be the digitisation and presentation of those currently restricted resources. This is perhaps one of the most pressing future requirements.

In spite of the importance of the *Mary Rose* and the gathering together of much historical evidence (for example in *Letters from the Mary Rose*: Knighton and Loades 2002), there are still many years when we have little or no information about where she was or what she was doing. There are unanswered questions from the start about precisely where, when and by whom she was built (see Chapter 1). We have nothing like the detailed accounts which document the building of other ships, for example the *Sweepstake* and the *Mary Fortune* in 1497 (Oppenheim 1896b). Between 1524 and 1531 there is little recorded, understandably so since she was probably laid up at Portsmouth throughout that time. But there is always the possibility that new evidence of her activity will come to light. We are not certain about Cromwell's reference to her being '*new made*' by 1536. Was this a major refit to accommodate the ordnance listed in 1540? Or could the guns specified there have been placed on the ship at any time after 1514? At the end of the story there are still many gaps in the sources, most notably in reports of the sinking.

Although the fleet was active during 1543 and 1544, there is only one mention of the *Mary Rose*. This referred to her being equipped with six double cannon (and the '*Great Harry*' (*Henry Grace à Dieu*) with twelve) during mobilisation in August 1543 when, according to the Imperial ambassador, and '*incredible number of guns*' was mounted on each ship. This is striking because no double cannon are listed for any ship in the *Anthony Roll*, and only three in any location feature in the 1547 inventory (Starkey 1998, 4267, 6733, 6761). Some caution is necessary, because this is not direct or expert information, though it would seem that the *Mary Rose* carried at least one such weapon to the end (79A1276). Perhaps, therefore, it was in 1543–4 that further alterations were made, as suggested by the dendrochronological dating of some of the major internal frames (see *AMR* Vol. 2, chap. 19). As will be described below, documentary evidence has only just come to light which confirms that Henry VIII ordered the *Mary Rose* and other ships to be adapted to carry the heavier ordnance. There may be other historical sources as yet unrealised, as likely to be found by luck as by dedicated searching.

Although we have charted structural changes within the ship throughout her life, and attempted to date them from dendrochronology, this is an avenue of work which should continue once the ship is dry and more accessible. We can date the diagonal braces to later than the build and can date alterations to structure beside Main deck lidded gunports to after 1540, but we have not yet had any success in dating the port lids or sills themselves. In the absence of archaeological dating we should look more closely at any documentary sources for the adoption of lidded gunports on the broad sides. The lack of structural evidence for gunports within the transom is extremely frustrating. The level of erosion here is just below where ports would be expected but there appeared to be no timbers in any of the stern trenches that could have been eroded lids. It is possible that this portion of the ship was damaged before or during the sinking, or, indeed during later salvage operations. A detailed assessment of all timbers from the 1975 and 1976 excavations outside the stern might help here.

Another particular question relevant to our understanding of fighting capability is the actual number of decks the ship supported at various times in her life. Was there a Main and Upper gun deck at the build, and how many decks were there in the castles when she sank? If we are considering the addition of a second Castle deck (see below) then the distribution of ordnance has to be changed accordingly. The dendrochronology dates for the Main deck beams show that some were felled as late as 1535, but it is not impossible that some of the earlier dates are simply from older wood. Similarly, the only date for an Upper deck beam is 1520, with differing dates for the hanging knees, some of which were obviously reused (see *AMR* Vol. 2, chap. 19). The visible design difference in the Upper deck in the waist and within the Sterncastle begs interpretation. The ship is a reservoir of information; we still need to find ways of fully tapping into it.

Trials have been undertaken in the manufacturing and firing of two types of gun: the port piece and the culverin. Using a combination of historical information and scientific data, some basic parameters have been established regarding firing distance, rates of fire and damage capabilities. However, these experiments were not undertaken at sea, nor under battlefield conditions, nor using black powder, therefore the results must be interpreted with this in mind. Achieving particular results with our reproduction weapons does not necessarily mean that the same performance was required in the Tudor period. We are still using modern eyes and modern standards.

We know nothing about the functional qualities of the six fowlers or, perhaps more importantly, those of the slings. The slings may well be the wrought iron counterparts of the culverins, with their large reinforced powder chambers and iron shot. Therefore their potential should be tested (see below). The fact that many of the wrought-iron breech chambers and incomplete fragments (including all of the carriages) are still undergoing conservation means that detailed examination is not yet possible. As stated in Chapter 3; the wrought iron assemblage is the most challenging and, arguably, the most important. More detailed radiography, metal analysis can be tried but since sophisticated techniques do not always supply the answer we must sometimes resort to a simple physical comparison of disparate elements.

There are missing guns. Some of these, such as '*top guns*' that the inventories list for the *Mary Rose* have not been identified here or elsewhere, so finding them would be of immense significance. Those listed are likely to have fallen clear of the tops and may still remain within the seabed surrounding the site.

It may even be possible to find some of the bronze guns known to have been recovered by earlier salvors. According to the *Anthony Roll* these should include demi culverins, two sakers and a falcon. Some of the guns immediately recovered from the wreck may have been redeployed in other ships. Alternatively they may have gone to castles and other fortifications where such weapons are plentifully recorded in 1547. It may yet be possible to identify guns from the *Mary Rose* in establishments still extant.

The dated assemblage of cast iron shot and the small hailshot pieces (Chapters 4 and 5) includes only a handful of items that have yet been examined metallographically. Further work, especially on the hailshot pieces, must continue. As early examples of the attempted mass production in iron of guns rather than projectiles, these are of special importance to all students of historic gunfounding and metallurgy.

The location of many of the guns in the ship is based on a best fit with the structure and distribution of known guns, shot (Chapter 5) and gun carriage elements (Chapters 3 and 4). Assigning positions to the guns as listed in the *Anthony Roll* is important, since the distribution of the guns can suggest how the ship was fought. Although different arrangements are suggested (Chapters 2, 3 and 11) it may be that the true disposition has not yet been elucidated. In the light of historical evidence which only came to light when this report was being completed, alternatives may need to be considered (see Postscript).

It seems inconceivable that it has not yet been possible to identify positively a single one of the many barrels which should have contained gunpowder (Chapter 5). The two casks which appear to retain traces of gunpowder residue were from different sectors of the Hold: H1 and H9. H1 was directly below the store of tampions and other gun furniture and this, or an area forward of it, certainly fits with the historical sources. Cask components were recovered from the excavations in 2003–5 (see below), all east of the stem, but not in the amounts required. Gunpowder barrels were of specified sizes and capacities and, when conservation is complete, study of all cask components might highlight these similarities.

The lack of sponges and wadhooks or worms to clean the gun barrels is unexpected. It is possible that items identified as potential '*darts*' (Chapter 9) could be the hafts for wadhooks, but there are very few. Similarly, some of the rams have one or more nail-holes which might have been used to secure a sheepskin, but this is by no means certain.

The fact that so many important sword handles have been identified through radiography of tiny amorphous portions of concretion (see below and Chapter 9) is a tribute to our work of recovery and examination. But it is also a cautionary lesson, for it is so easy for these fragments to be forgotten, to dry out and to fall apart. We have not yet radiographed the many concretions raised in 2003–5, a number of which appeared to be edged weapons. This an obvious avenue for further research. The finding of a sword pommel and handle (04A0110) in concretion and a chape (04A0038) which are undoubtedly Tudor in origin are helpful indicators in the dating of at least some of the sediments in the eastern area of the trench across the bow. An, as yet undated, dagger quillion or possibly part of gunner's stiletto (05A0023), found near the northern fluke of the newly discovered anchor, may provide another secure point of reference. Study of the 600 artefacts recovered during the recent excavations has barely started (see below).

The archery assemblage is a prime candidate for further research (Chapter 8). One might presume that there should be an intimate correlation between bow and arrow, yet this has not been demonstrated. There are numerous questions (with some answers suggested but by no means proven) regarding the differing length and profile of the arrows and the longbows. The apparent mixture of arrows of different lengths within a single arrow spacer was unexpected, and no convincing explanation has been yet been offered. The vast majority of the arrows are currently inaccessible within conservation tanks. Many are bloated beyond recognition and were certainly too fragile to study

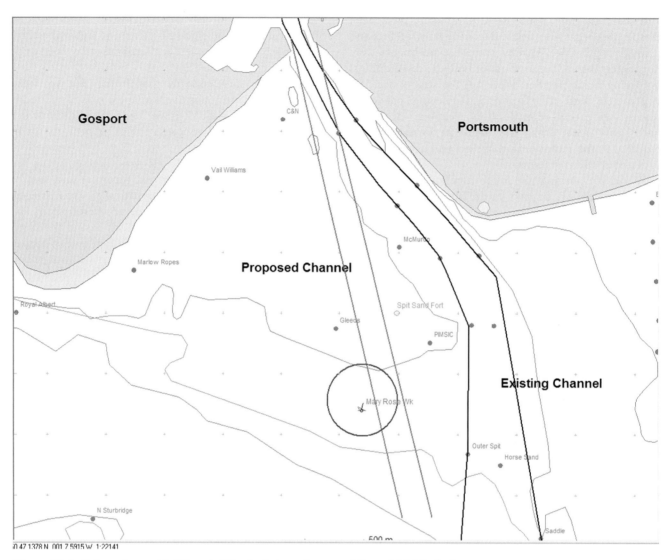

*Figure 12.1 Solent chart with Historic Wreck circle and proposed line of MOD dredge*

whilst wet, although ingenious devices have been manufactured and used to mitigate the risk of damage (Jackson, archive). Study of the arrows has had to be based on accessibility, with as little damage as possible. Statistically, therefore, the data is likely to be flawed. Some further work is still viable. When arrows are taken out of wet tanks for freeze-drying they should be examined, and again when dry. The relative dimensions should then be compared. This process has a bearing on the validity of comparative measurements already taken of immediately recovered, saturated and dry arrows. There has been huge interest in the arrow assemblage, particularly from researchers active in replication and shooting.

An obvious area for further work would be the nature of the bowstrings (ideally with some institutional partner looking at the varying qualities of hemp), culminating in strictly conducted and recorded trials using the best recreation of bow, arrow and string. The potential identification of a bowstring from 2003 (03A0166) means that this may, at last, be possible.

There are many objects within the assemblage still awaiting interpretation. What once appeared to be nothing more than sticks are now thought to be either crossbow bolts or incendiary arrows (Chapter 8). Further insights might well be aided by a comprehensive study of the three-dimensional spatial associations of the entire site. As yet this is not possible but should be relatively straightforward were the thousands of measurements collected using the Direct Survey Measurement method to be processed. The proposed new museum displays will be based on a spatial rather than a typological approach, and it is hoped that a better identification of objects and their function may be achieved as a result. Large numbers of wooden artefacts are still awaiting microscopic identification of species. As material is selected for a more comprehensive display collection, further identification of objects and fragments should result. Particular items which may remain unidentified include portions of crossbows; and numerous hafts may well be parts of top darts, wadhooks, sponges or handspikes.

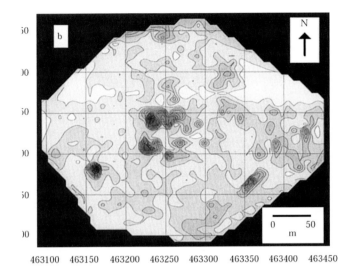

*Figure 12.2 Data provided by remote sensing (Chirp sub-bottom profile): above) anomalies (dark tone) around the hull depression (courtesy of National Oceanographic Centre, Southampton); below) repositioning the hull into the depression using Site Recorder, showing the anomalies to the west at bow and stern (courtesy of 3H Consulting)*

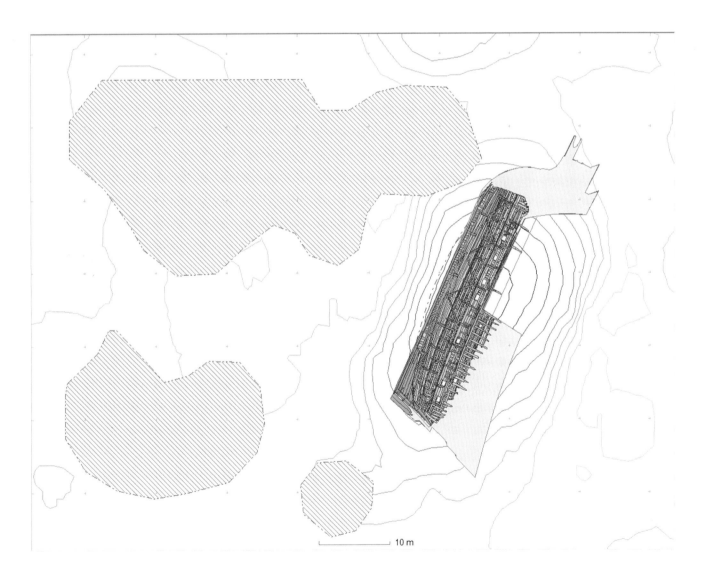

Although the armour recovered is extremely limited (Chapter 10), it includes items rare in a dated British context. Such are the mere fragments of mail. These include butt links presumed to be of European manufacture.

Small items such as these could accidentally have been redeposited on the spoil mounds during the excavation through the use of airlifts without sieves. Recent re-excavation of these deposits mounds yielded a surprising wealth of objects, emphasising the continuing archaeological importance of the wreck site.

There are also still uncertainties about the actual sinking of the ship, explored in Chapter 11. The contemporary descriptions are conflicting, although both the immediate report of the diplomat François van der Delft and the later French account make it clear that the *Mary Rose* was engaged to the last. The importation of the battlefleets into a GIS package and the modelling of the tracks of the *Mary Rose* and French galleys, has enabled a multitude of ideas to be explored. Since it appears that the ship passed within range of the galleys, it remains possible that she sustained some damage from them. This may have compromised her stability, the '*gust of wind*' combined with the open gunports, providing the fatal blow. Further explorations of the battle, the stability of the ship, alterations to her structure and ordnance will undoubtedly lead to further speculation.

The orientation of the guns, powder chambers, and the presence of shot and loading equipment strongly suggest that the M8 starboard port piece at least was in the process of being reloaded, reinforcing the view that the *Mary Rose* had engaged her guns. In addition, both M10 and U10 gun bays contained objects suggesting gun crews in action (linstocks, powder dispensers, the only gunner's rule). The ship could, therefore, have been fighting with her stern when she turned and sank. This fresh look at the theatre of war must be considered as a starting point for further investigation.

A shipwreck is a living community immediately and catastrophically lost at a precise moment in time. The story of the *Mary Rose*, woven together from the tangible evidence of the ship herself, the artefacts and documents, is that of the most powerful fighting machine of the time, a predecessor of the aircraft carriers of the modern Navy. It is therefore ironic that the prospect of a new generation of carriers should have threatened the integrity of the wreck site, and that in response to that threat we were allowed and enabled, through MoD funding to undertake a further series of excavations in 2003–5. The results have left us with even more tantalising questions; including the state of the ship before she sank, the nature of the sinking, and the effects of both Tudor and nineteenth century salvage operations.

*Figure 12.3 Remote excavation: above) the crawling ROV; below) sieving the spoil*

## The 2003–5 Excavations: an Interim Statement
by Alexzandra Hildred and Peter Holt

The area within a 300m radius around the depression which originally contained the hull of the *Mary Rose* is still designated as a site of archaeological and historical importance under the *Protection of Wrecks Act* 1973 (Fig. 12.1). The Mary Rose Trust is responsible for the curation of everything deriving from the vessel, including what is still left within the seabed.

When the intact remains of the hull of the *Mary Rose* were raised in 1982 it was known that timbers lay to the north of the site and that these should represent the stem and portions of the bow, potentially including the Forecastle. In 1981–2 a trench was excavated beneath the Sterncastle towards the east for about 6m. This revealed some of the most intact rigging elements found, including the chains for the main and mizzen masts, swivel guns and hailshot pieces which had fallen

Table 12.1 Artefacts recovered from spoil mounds

| Object | Type | Location | Description | Dimensions (mm) |
|---|---|---|---|---|
| 03A0011 | arrow | S stern | frag | L.42 D.14 |
| 03A0125 | arrow | P midships | 3 frags | L.57–80 D.10–11 |
| 03A0183 | arrow | P midships | 2 frags | L.47, 55 D. 10, 11 |
| 03A0184 | arrow | P midships | frags | L19 D.9.2 |
| 03A0185 | arrow | P midships | head fragment | L.55 D.12 |
| 03A0188 | arrow | P midships | 6 frags including fletching | L.52–173 D.10.5–12.1 |
| 03A0189 | arrow | S stern | 40 frags | L.< 244 D.12 |
| 03A0190 | arrow | P midships | nock end | L.61 D.11 |
| 03A0203 | arrow | P midships | 43 frags | |
| 03A0204 | arrow | P midships | nock with fletchings | L.80 D.14 |
| 03A0205 | arrow | P midships | 4 head fragments | L.32072 D.11–14 |
| 03A0233 | arrow | P midships | 6 frags | |
| 03A0236 | arrow | P midships | nock end, fletchings | L.13 D.11 |
| 03A0274 | arrow | P stern | 22 frags, 3 fletched ends | L.13 D.10 |
| 03A0285 | arrow | P stern | 35 frags, 7 heads, 5 nocks | L.45–68 D.9–12 |
| 03A0347 | arrow | S stern | 40 frags | |
| 03A0166 | ?bowstring | P midships | 2 frags | L.200 D.3 |
| 03A0215 | bill | P midships | head end | L. 123 D.31 |
| 03A0313 | bill | S stern | head end | L. 130 D.40 |
| 03A0314 | bill | S stern | haft fragment | L.130 D.45 |
| 03A0315 | bill | S stern | head end | L.73 D.25 |
| 03A0317 | bill | S stern | head end | L.105 D.28 |
| 03A0322 | bill | S stern | 2 frags, fit | L.200 D.38 |
| 03A0349 | bill | S stern | 3 frags haft | L.105–173. D. 39 |
| 03A0299 | pike | P midships | haft, grooves | L 148 W.35 |
| 03A0213 | pike | P midships | haft.possibly bill | L.45 D.35 |
| 03A0277 | pike | P stern | haft | L.145 W.30 |
| 03A0305 | pike | P stern | haft | L.94 W.24 |
| 03A0341 | pike | S stern | 2 frags | L. 83,97 D. 27,31 |
| 03A0104 | ballock knife | P midships | complete | L.120 W. 53 H.25 |
| 03A0198 | sheath | P midships | frags | L.21 W.23 |
| 03A0270 | knife handle | P midships | frags | L.55 W.14 |
| 03A0279 | bi-knives | P midships | 2, back to back 2-3 rivet holes | L.70 W. 23 Th. 5 |
| 03A0319 | scabbard top | S stern | top with fixing point | L.110 W.23–30 |
| 03A0357 | strap, sword hanger or scabbard frag. | S stern | decorated tiny circles, spiral pattern | |
| 03A0137 | mail | P midships | 1 link, hole for rivet | D.6.2 |
| 03A0223 | mail | P midships | 2 links | D.9 W.10 Th. 1–2 |
| 03A0231 | mail | P midships | links | L.27 W 14. Ht 3 |
| 03A0323 | mail | S stern | link | D.12.4 |
| 03A0333 | mail | S stern | links x 10, attached | L. 71 D. 9.3 |
| 03A0316 | tampion | S stern | incomplete | L.54 W.4 Th.16 |
| 03A0339 | chamber plug? | S stern | incomplete. Base chamber frag? | L 10.5 Th.6.8 |

P = port; S = starboard

from the Upper and Castle decks. The wealth of material in this single trench reflects the potential for artefacts and structure which may have fallen outside the ship eastwards as she heeled. Post-excavation seabed searches in 1983–5 along the starboard side found Tudor artefacts up to 18m away from the depression.

Site monitoring since 1983 has revealed objects and structure eroding out of the sides of the depression, with concentrations along the starboard side and the bow. One concretion alone yielded the remains of twelve different guns together with shot and a portion of a chest (Chapter 3). Remote sensing using a sub-bottom profiler indicated the presence of layers of greater density than the seabed to the west of the site with defined lenses at the bow and stern (Fig. 12.2, above; Quinn et al. 1997). Although interpreted as being a particularly shell-laden sedimentary layer, it is possible that this may contain fragments pertaining to the port side of the vessel.

The MoD funding was for a defined programme of site clearance and investigation in 2003–5. These works included the sampling of a percentage of the silts excavated out of the hull in 1979–82 that had been redeposited by tidal action as linear mounds on either side of the hull depression (Fig. 12.2, below).

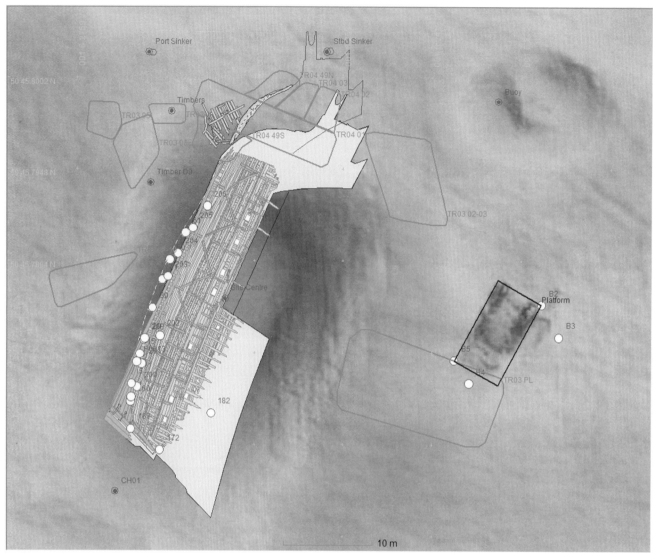

*Figure 12.4 The positions of the trenches into the spoil and the area worked north of the bow with position of the ship superimposed and fixed points indicated. The area to the right is the diving platform recovered in 2003*

Sampling the spoil was achieved using a remotely operated crawler vehicle (ROV) fitted with an airlift suction head and a water jet, which recovered all sediments to a sieve on the surface (Fig. 12.3). The ROV was fitted with two cameras, providing a complete video archive of the excavation. It was also fitted with an acoustic positioning system (with a fix point also on the airlift head) so that its location could be tracked and recorded. The system used on the ROV and carried by divers was a Fusion Long Baseline (LBL) underwater acoustic positioning system (APS) provided by Sonardyne International Ltd. As there were no survey control points left on the site (all had been on the structure of the ship), these were recreated from the corner positions of a sunken diving platform which had been surveyed during the original Partridge Rangemeter survey to position key points on the ship in 1975 (Kelland 1976). This enabled the 'ship' to be placed back into the depression so that our excavations

**Table 12.2 Artefacts recovered from the Timber Park**

| Object | Type | Description | Dimensions (mm) |
|---|---|---|---|
| 03A0070 | breech chamber | port piece chamber in concretion | L.620 ED.200 |
| 03A0072 | axle or bed frag. | single bolt hole; eroded/incomplete | L.210 W.110 Ht.70 |
| 03A0073 | axle frag. | bolt hole 35–90mm | L.1100 W.130 |
| 03A0074 | ?axle frag. | eroded | L.66 W.105 Ht.80 |
| 03A0075 | ?axle frag. | eroded | L.440 W.120 Ht.85 |
| 03A0066–8, 79–80 | concretion | uncertain | – |

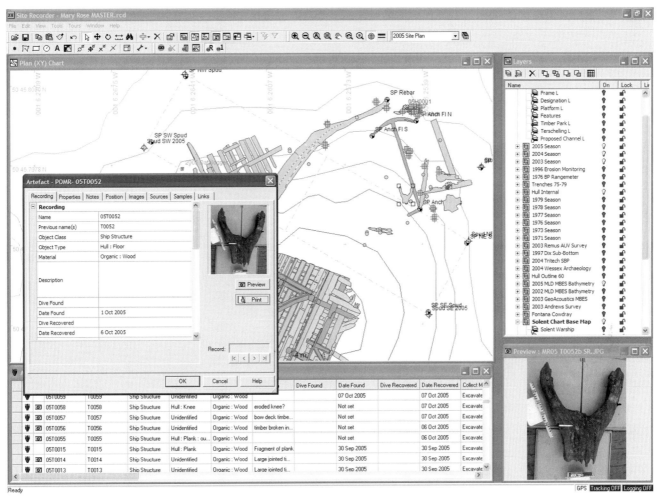

*Figure 12.5 The Site Recorder displays*

and new trenches could be related to previous work (Figs 12.2, below and 12.4).

The information management system Site Recorder (Holt 2004; 2007) was chosen as the data management system for all phases of the work. During the planning phase, in the absence of survey control points from the 1982 excavation, the program was used to establish the most likely location for the recovered hull timbers on the seabed. Site Recorder was also used to collate data from previous geophysical surveys using sidescan sonar, multibeam echo sounders, magnetometers and sub-bottom profilers. By correlating the datasets with the existing wide-area site plan it was possible correlate some of the targets and to plan investigative dives.

During the excavation Site Recorder was used for real time data collection and as a decision support tool. The divers doing the excavation work were using surface-supplied diving equipment fitted with helmet cameras, lights and voice communications. They could also use a Sonardyne Fusion APS to position finds and structure on the seabed (Holt 2004). The APS fed position information directly to Site Recorder so that the position of the diver or ROV could be seen on the site plan. Finds and structure could then be added to Site Recorder in real time, positioned by the APS and including details and descriptions given by the diver. The integration of all the sources of information helped

**Table 12.3 Artefacts found on stem or western timbers**

| Object | Type | Descripton | Dimensions (mm) |
|---|---|---|---|
| 04A0018 | reamer | W. of W. timbers | L.350 D.4 |
| 04A0014 | shot lead | ?has been fired | D.14 |
| 04A0033 | sword blade | concreted, 205 x 65mm, broken to reveal void | blade W.33–35 |
| 04A0080 | ?powder ladle | corroded frag. | L.235 W.155 |
| 03A0087 | ?gun carriage frag. | S. end of stem | L.57 W.75 Th.9 |
| 04A0121 | lead shot | | D.14 |
| 03A0083 | cast iron shot | W. edge, W. timbers, upper levels | D.140 |
| 05A0052 | bill | head, nail-hole 100mm from end | L.280 D.34 |
| 05A0011 | sword blade | in concretion | L.90 W.45 |

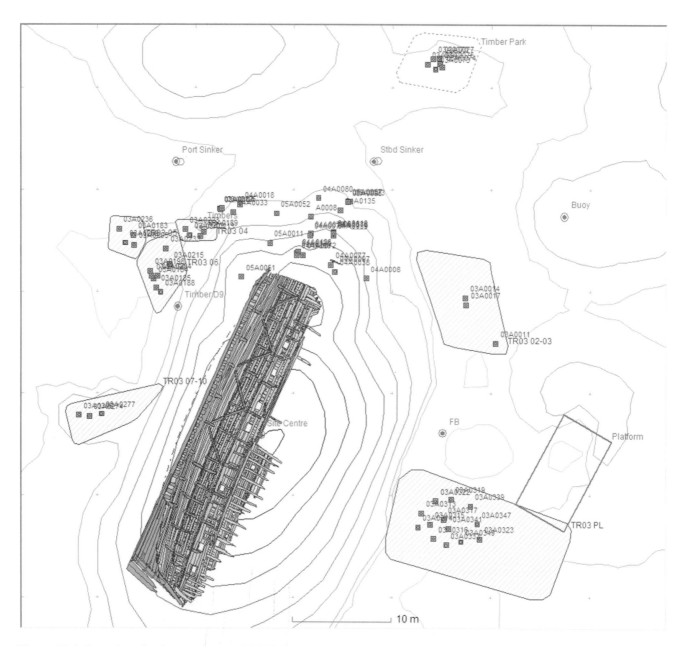

*Figure 12.6 Location of ordnance recovered 2003–5*

ensure that a complete and accurate record was captured in real time in an efficient manner. The archaeologists on the vessel could see the site plan as it developed and could use the information from the GIS to aid decision making and plan further work (Fig. 12.5). The site plans and records for the 2003–5 period can be accessed online via the 3H Consulting website (www.3hconsulting.com/SitesMaryRoseMain.htm).

Approximately 10% of the total site spoil was excavated by the ROV from five trenches, three on the port side and two larger ones on the starboard side (Fig. 12.3). The spoil was diverted to the surface for screening (Fig. 12.3). Over 300 artefacts were recovered, for the first time enabling us to quantify what may be contained in the spoil. Finds included complete knife handles, a pocket sundial, a gold coin and a wealth of organic material. The distribution of all ordnance-related artefacts is shown in Figure 12.6 and these are listed in Table 12.1. As expected, many were small items, such as mail links (03A0223, 03A0323, 03A0333), broken arrow fragments, portions of pikes or bills, and scabbard and bi-knife fragments. The distribution of these with respect to known concentrations of similar objects recovered from inside the wreck was as predicted, for example, the decorated leather strap (03A0357; Fig 12.7a) and scabbard top (03A0319; Fig. 12.7b) were both in the stern trench, outboard of U9, where many similar items were found inboard during earlier excavations (Chapter 9). What was surprising was the recovery of a complete ballock

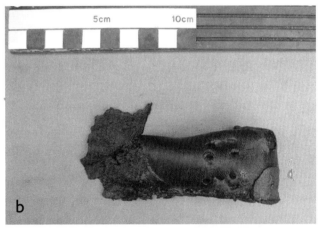

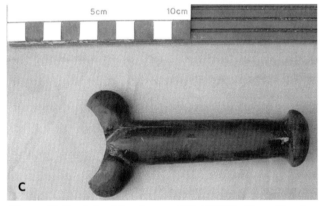

*Figure 12.7 a) Scabbard strap (03A0357); b) scabbard top (03A0319); c) ballock dagger (03A0104)*

dagger (03A0104; Fig. 12.7c) and the possible bowstring fragment (03A0166, see Fig. 8.42b).

Clearance of a seabed storage area to the north-east (labelled 'Timber Park' on Fig. 12.6), also produced ordnance, including a large breech chamber for a port piece (03A0070), four objects which look like gun carriage axles (03A0072–5) and a possible carriage bed fragment (03A0077) (Table 12.2).

One cast iron shot was recovered (diameter 140mm; 03A00083), associated with a possible gun carriage element (03A0087). Both were found in the port bow area, close to the inboard end of a large timber, later identified as the stempost (03T0049; described and illustrated in *AMR* Vol. 2, chap. 6). Cleaning of this area revealed the ends of several frame-sized timbers to the west, suggesting associated structure. This important

### Table 12.4 Artefacts found east of stem

| Object | Type | Description | Dimensions (mm) |
|---|---|---|---|
| 05A0057 | breech chamber, port piece | W of anchor centre, 2 sets lifting rings | L.650 ext. D.265 |
| 04A0135 | swivel gun breech plug? | N of anchor, T3 | D.45 |
| 04A0097 | swivel gun incomplete | T1 E, centre | L. 1100 W. 360, IB 45 |
| 04A0072 | spoked wheel hub frag. | T1 E 2 spokes 40 x 30mm, 2 more holes | L.250 W.270 Th. 98 |
| 04A0077 | gun carriage frag. | T1 E. iron hole D.28 | L. 801 W.155 Ht 140 |
| 04A0076 | axle frag | T1 E | L. 430 W.138 |
| 04A0125 | ?bill /halberd head | T1 E, centre in concretion | L. 295 W.220 Th.80 |
| 04A0008 | pike/Bill frag. | T1 E, SE quadrant | L.96 W.41 |
| 05A0095 | scabbard frag. | anchor N, in concretion | L.135 W.140 Th.50 |
| 05A0023 | quillion block | near anchor, fluke; square hole centre with dome surrounding; acorn finials, leaf decoration | L.70 W.13 |
| 04A0086 | blade | T1 E, NE, concretion | L. 70 blade W.32 |
| 04A0038 | chape | T1 E, centre, decorated | L.48 W.26 Th.11 |
| 05A0037 | concreted blade | near anchor | L190 W.40 |
| 04A0126 | concreted blade | T1 E centre. 5 frags | |
| 04A0110 | sword handle & pommel | T1E, centre, in concretion | L.115 W.95 pommel D.54 handle L.85 W.35 |
| 05A0043 | ?arrow head | anchor, in concretion | L.230 W.20 |
| 05A0033 | cast iron shot | NE sinker | D.85 |
| 04A0132 | composite dice | T3, near anchor | D.45 |
| 04A0102 | composite dice | T1 E, centre | D.35 |
| 04A0127 | composite dice | T1 E, centre | D.45 |
| 04A0030 | composite dice | T1 E/SE | D.48 |
| 04A0118 | composite dice?/iron | in concretion | D.?60 |
| 04A0012 | composite dice | T1 E | D.45 |
| 04A0017 | composite dice | T1 E | D.45 |
| 04A0025 | 2 composite dice | T1 E, SE | D.45 |
| 04A0026 | concretion with iron shot? | T1 E/SE | D.66 |
| 04A0127 | stone shot | T1 E, finished | D.100 Wt 1.5kg |
| 04A0031 | composite dice | T1 E | |
| 04A0088 | composite dice | near swivel 04A0097 | D.48 |
| 04A0096 | cast iron | T1 E centre/N | D.85 |
| 04A0070 | composite dice | T1 E | D.41 |

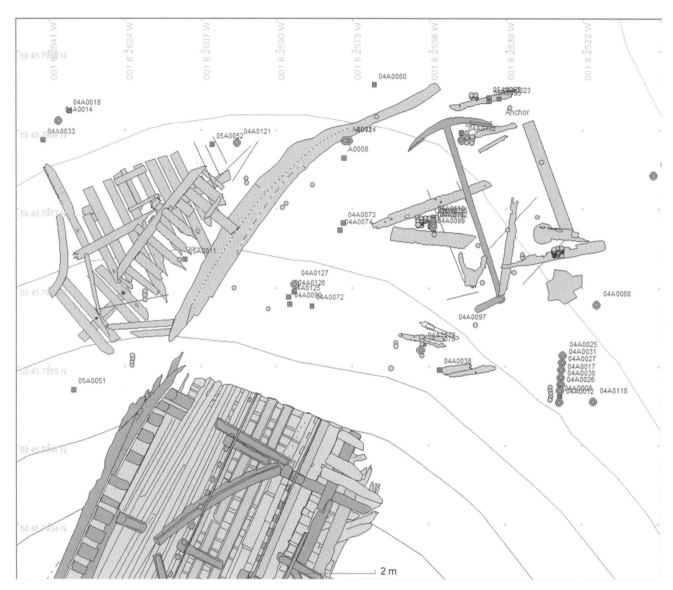

*Figure 12.8 Distribution of ordnance in the Forecastle area*

discovery led to funding for another season of excavation in 2004. This uncovered a substantial portion of hull framing and inner and outer planking from the port side of the vessel to the west of the stem, and collapsed remains of timbers and Tudor objects associated with the starboard side of the ship to its east. The eastern excavation also yielded a number of timbers (including a Y-shaped floor timber, 05T0053, Fig. 12.5), a large wrought iron bow anchor, a powder chamber from a wrought iron gun, sections of a swivel gun and eleven shot of appropriate size (Table 12.3; Fig. 12.8). During 2004 a 16 x 4m aluminium grid frame was placed over the area within which remains of the Forecastle were expected to lie (the east–west rectangle immediately north of the ship shown in Fig. 12.4). The grid helped to support divers and airlifts but, more importantly, it was used to suspend a parametric sub-bottom profiler (loaned by Tritech) at a fixed level above the site. The profiler was used to try and ascertain the full extent of the buried remains (Fig. 12.9). The stem was clearly visible on the sonar trace, but the remains to the east were far harder to interpret, and excavation yielded several layers of collapsed timbers. Whilst the stem and collapsed timbers to the east had been anticipated, the port side structure had not. The comparatively good condition of this structure demonstrates the potential for survival of elements from the port side. As the end of the stem was beyond the grid boundary, three parallel trenches were worked northwards from the northern boundary (Fig.12.4).

All of the identifiable objects found west of the stem were typologically Tudor (see Table 12.4), for example the priming wire (04A0018, Fig. 12.10a). However, many objects associated with the collapsed timbers, swivel gun, anchor and the area immediately to its north were post-Tudor. A quillion block, either from a dagger or possibly a gunner's stiletto (05A0023, Fig. 12.10b) was found here; further study of this item might show it

*Figure 12.9 Tritech parametric profiler on frame*

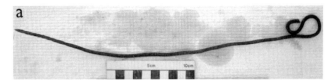

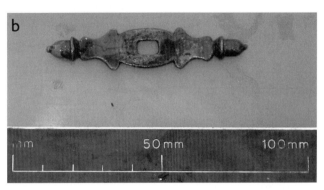

*Figure 12.10 a) Priming wire (04A0018); b) dagger quillion block or part of a gunner's stiletto (05A0023)*

to be later than the sinking. This area of water has been an anchorage for 500 years or more, so objects of widely ranging date can be found on the seabed. The stratigraphy east of the stem was more complex and included evidence from excavations of 1971–1982. Although the seabed immediately beneath the stem and the anchor was remarkably similar to that found underneath the hull in 1982, we cannot be sure that there is no more structure in this area.

The decision was made to record, photograph and preserve the western timbers *in situ*, and to raise the stem, anchor and any associated timbers east of the stem. This was successfully achieved in 2005, with funding from the MoD and support from English Heritage. The remit did not include any further excavation of the eastern area, so it was impossible to confirm how much of the starboard bow or Forecastle may remain buried.

As the MoD had, by this time, decided against dredging a new channel into Portsmouth Harbour, the site cannot any longer be considered to be under direct physical threat from that source. But there are other forces at work. Analysis of several of the timbers raised in 2004 showed colonisation by shipworm larvae including the very virulent *Lyrodus pedicellatus*. Although these only attack exposed timbers, their presence here gives cause for concern. A programme was immediately instigated to monitor the infestation. This included the placing of sacrificial timber blocks on the seabed with a data-logger to record a range of environmental conditions, including the preservation qualities of the sediment. Recovered at pre-determined intervals, analysis of these blocks can be used to monitor the rate of infestation on exposed wood by marine borers. This work is continuing.

Integral to the project design was a reburial policy. This included covering the entire area excavated during 2003–5 with a fabric membrane. This was held down by a layer of sandbags, then topped with 110 tons of loose sand deployed through a hopper and a pipe, with further sand delivered to the seabed in ton bags which were slashed over specific areas (Fig. 12.11).

Inevitably, the excavation has left us with a number of unanswered questions. The most significant concerns the separation of the stem from the keel. Detailed structural analysis undertaken by Doug M<sup>c</sup>Elvogue has helped to address some of these (see *AMR* Vol. 2, chap. 6), but the key questions must be why and when did this occur. Both the stem and western timbers seem to be on or within Tudor layers, so was it during the sinking, possibly exacerbated by battle damage? Or was it the

*Figure 12.11 Sand barge and hopper used in reburial of the 2003–2005 trenches*

Tudor salvage attempt which broke the stem and pulled some of the port side to the west, causing it to lie at opposing angles to the rest of the hull? New questions emerged with the anchor raised in 2005. Although seemingly identical with the two bower anchors recovered from beneath the bow in 1982, this is not necessarily from the *Mary Rose*. It might merely be an anchor which was dragged into the site at a later date, causing some damage. Quite possibly, though, it came from one of the Tudor salvage barges. Perhaps it holed the ship, or broke the stem and tore away the port side. This may help to explain why the 1545 salvage attempt was abandoned, though the main reason reported was '*dislodging of the foremast*' (Knighton and Loades 2002, 129–30).

Another possible cause is the nineteenth century salvaging of objects, which might have been more invasive than previously assumed. Although there is some wear on the stem and the inside of the port side structure, this does not seem to be as exaggerated as would be expected if the damage was caused in the nineteenth century. The (few) objects directly associated with the port side timbers were all Tudor.

The same applies to the eastern portions of the site. Only further excavation will establish if there are indeed any coherent remains of bow castle structure, or whether this was largely destroyed, so that the pieces we recovered constitute the bulk of the remains. Hopefully time, and future opportunities, will tell.

## Postscript: Document CP 201/127
by C.S Knighton and Alexzandra Hildred

Other questions may yet be solved (or new ones raised) from documentary sources, though discoveries are more likely to come by chance than through any dedicated programme of research. So it was in December 2007 when the document transcribed below was first placed in its proper context. It was noticed by one of us (CSK) while looking in the Hatfield House archives for material on the navy of Edward VI and Mary. When the Hatfield papers were arranged in the nineteenth century this item was provisionally dated 1557, and the printed calendar did not mention the ships it concerned; so nobody had thought to consult it in connection with the *Mary Rose*. There are many reasons for disregarding the calendar date, which has been subsequently pencilled on to the manuscript (Knighton and Hildred forthcoming). Most obviously, among those reporting to the sovereign is James Baker, who is known only as Henry VIII's master shipwright. Moreover he is documented at Portsmouth, working on the *Jennet* and the *New Bark* in August 1545 immediately after the loss of the *Mary Rose*. At the same time Benjamin Gonson, named after Baker, was acting Surveyor of the Navy. So this report on modifications to those three ships was probably submitted just before the Battle of the Solent, and is therefore the last report from the *Mary Rose*.

Although only half a sheet of paper, part of which is torn away, it is full of new and challenging information. Some of the terminology remains obscure, but the meaning is clear enough, and quite stunning. The King had ordered these three ships (and doubtless others) to carry extra guns forward, and these were to be canted as finely as possible to match the firing capacity of the galleys while presenting the narrowest target. This was, as Rodger has argued, a matter of preponderating concern to naval tacticians (Chapter 11); now we have clear evidence of the resulting conflict between operational requirements and the dictates of naval architecture. It appears that anything, even major structural alteration, was being considered to achieve the desired effect. In order to place the guns slant-wise the shipwrights would have to remove knees or even masts; they told the King sensibly but bravely that if they followed his commands they would be weakening the very structure of his ships.

The document (Cecil Papers 201/127) is here given in modern spelling, extending the abbreviations and supplying as much of the lost text as can plausibly be done:

King's]s Majesty of ships following by John
..]ke, James Baker, Benjamin Gonson and
...that i]s to say:

| | |
|---|---|
| *The Mary Rose.* | First the[*re can be*] no more [*or*]dnance laid at the luff without the taking away of 2 kn[*ees*] and the spoiling of the clamps that beareth the bits, which will be a great weakening to the same part of the ship. |
| | Item, she hath right over the luff two whole slings lying forwards over quarter-wise, and at the barbican head likewise forward over 2 culverins, and the decks over the same shooting likewise forward over 2 sakers. |
| *The Jennet* | Item, her foremast cannot well be otherwise translated for lack of breadth on the bow, and now shooteth forward over of either side the foremast one saker. |
| *The New Bark.* | Item, she hath forwards over of one side the foremast a demi culverin and over the other side a saker lying, as we suppose, ... er to pass than to translate the said foremast where is [*room for*] for 2 demi culverins, if it may please Your Grace. |

The first point is clear enough; there was only just enough room between the knees to position her existing breech-loading guns. Here it is suggested that to accommodate muzzle-loading cast bronze demi culverins at the '*luff*' (the bulkhead of the Forecastle), two knees would have to be removed. This would also affect the deck clamps ('*thick timbers which lie for and aft, close under the beams of the first orlop, and do bear them up at either end*': (Mainwaring 1644 in Manwaring and Perrin 1922, vol. 2, 126) and the '*bitts*' (two square pieces of timber which stand like pillars in the bow). They are used to belay the cables when riding an anchor, but were known to be points of weakness:

'*In extraordinary storms, we are fain to make fast the cable to the main mast for the better relieving of the bitts and safety of the bows, which have in great roadsteads been violently torn from the other part of the ship.*' (ibid., 99)

The 1545 report is stating that the strength of the bow would be severely compromised if these features were moved.

It is clearly stated that the *Mary Rose* already had two '*whole slings*' (rather than demi slings) at the luff, angled as far forward as possible. The description of the weapons may be imprecise, but it could simply indicate similarities in size and power between whole slings and demi culverins. This needs to be tested by the replication and trials of slings. It also calls for a reconsideration of the demi sling found north of the site in 1970 (70A0001). Although it has been suggested that this gun was lifted and dropped by Deane and Edwards, we cannot be certain. It may have come from the Forecastle.

Until the recovery of the stem there was no data on which to base the length or breadth of the bow. Archival and archaeological evidence suggests that slings are long guns and positioning them in the Forecastle had not been considered. Based on this document an illustration has been undertaken showing what the *Mary Rose* might have looked like with a second castle deck in the bow and stern (Fig. 12.12a). This compares well with the ship as portrayed in the *Anthony Roll* (Fig. 12.12b). Maritime artist Geoff Hunt has included all of the new information in his recent painting of the ship, which also shows her with a second castle deck in the bow and stern.

The majority of cast iron shot was stored in M2, with easy access to the Forecastle. Only three cast iron shot were recovered in 2003–5, two of 85mm in diameter (04A0096, 05A0033) and one of 140mm (03A0083). None of these is a perfect match for the predicted sizes of sling shot (92–8mm) or demi culverin shot (102–10mm). Elements of carriages that might relate to Forecastle guns include a small stepped check found in sector O1 (81A0107), similar in size to the carriage supporting the *in situ* demi culverin, and several wheels and axles.

The '*barbican*' (forward part of the Sterncastle) is described with two culverins firing as far forward as possible, and above these two sakers, also firing forward

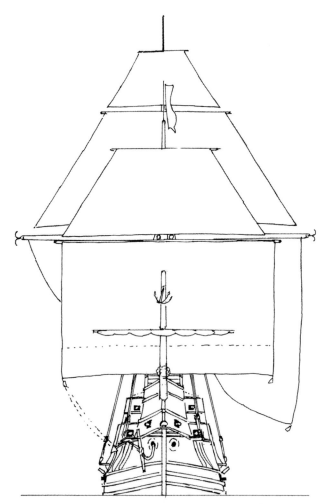

(Fig. 12.13). However, the only fragment of a trailing carriage of suitable size to support a saker was found in the bow (81A0451). But since two tiers of guns are described, and the only forward-firing capacity is from the Castle deck, we must consider the possibility of two Sterncastle decks. The gun actually found in a forward-firing position on the Castle deck was a demi culverin rather than a full (or '*whole*') culverin, but it would be plausible to locate a saker above. The 121 shot of saker size found in M6 (Chapter 5, table shot 21) would be within easy reach of this position.

The *Jennet* is described as having one saker firing forward, probably moved to either side of the foremast as required. It is pointed out that it would be impossible to alter the position of the mast (to accommodate larger guns) because the ship is not wide enough at this point. The *New Bark* is described with a demi culverin on one side of the foremast and a saker on the other. If the mast were to be moved it would be possible to accommodate two demi culverins.

It is possible that the *Mary Rose* had already been altered to take the equivalent of two long-range guns, one each side of the foremast; but the language of the report and everything we can establish about its context suggests that it is responding to very recent orders. If they were trying to modify all ships to take demi culverins firing forward from the Forecastle, the slings mentioned in the report may have been replaced by the demi culverins. Not only would the structure have to be altered to accommodate the length required to bring these inboard to reload, but there would be an additional increase in weight of *c.* 700kg per gun. Figure 12.13a shows the ship with a single Castle deck with slings in the bow and Figures 12.13b–c with a second Castle deck supporting sakers in the stern.

Geoff Hunt's preliminary sketch (Fig. 12.12a) shows what the *Mary Rose* would have looked like from the bow with a second tier of guns in the front of the Sterncastle facing forwards. Her slender Forecastle illustrates the lack of breadth at the bow to accommodate guns but also demonstrates that the guns facing forward from the Sterncastle can be accommodated within the structure, reinforcing the importance of these guns and of the second Castle deck.

It could be that Henry VIII heeded the warning, and held back from the intended new armament. Perhaps he had so much admiration for the technical expertise of James Baker and his colleagues that he accepted their advice on how to fight his battle. The probability is that he just told them to get on with it. If, then, the proposed modifications had been made to the *Mary Rose*, could this have so weakened her structure as to explain the break in the stem or the detachment of the bow? It does not explain why the ship sank, but it may well help us to understand what we have recently uncovered.

The chance discovery at this late stage of a document which challenges our original views on the

*Figure 12.12 (above) Preliminary sketch by Geoff Hunt in preparation for his recent painting. This shows the slender lines of the Forecastle and demonstrates the capability to accommodate two tiers of guns facing forward in the front of the Sterncastle; (below) the* Anthony Roll *image of the* Mary Rose *for comparison, also showing at least two decks in the castles*

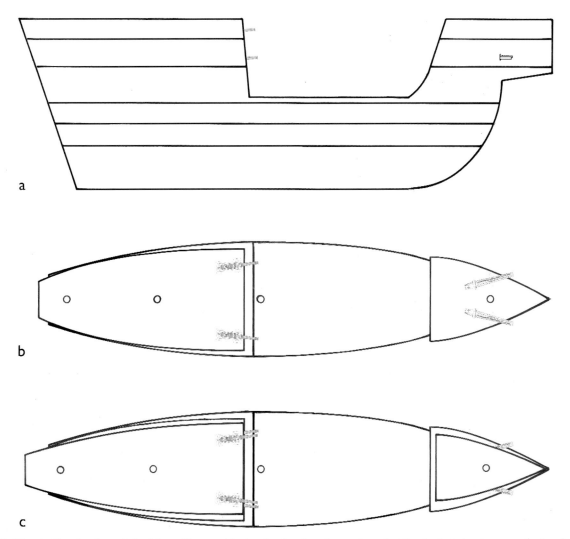

*Figure 12.13 a) Simple plan of the* Mary Rose *with demi culverins situated on the Castle deck in the stern facing forward, and slings situated in the bow; b) with slings and a second Castle deck and two layers of guns in the stern, demi culverins on the lower Castle deck and sakers on the upper; c) elevation with a second Castle deck in the bow and stern*

number of decks and the distribution of the guns is exciting but also chastening. The recent excavations have raised many further questions about the formation of the site.

New work (Sanders 2010) on the cables questions our identification of the cable outboard M8 as Tudor salvage cable, indeed the *Mary Rose* may have sunk on her own anchor cables. Re-examination of the cordage following the potential identification of a bowstring from 2003 has revealed another (Fig. 12.14). This was associated with arrows within a secure context inboard. We can be assured that there is much more to do: historical research, re-examination and further identification of the assemblage, and at some stage, another excavation of the site. The message is clear: '*Work unfinished*'.

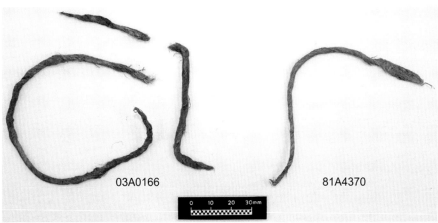

*Figure 12.14 Bowstring found in 2003 compared with previously recorded example*

# Appendix 1: Gun Metallurgy

*1.1 Average metal composition for the bronze guns*

| Gun type | Object | Fe | Co | Ni | Cu | Zn | As | Sb | Sn | Ag | Bi | Pb | Au | S |
|---|---|---|---|---|---|---|---|---|---|---|---|---|---|---|
| Cannon Royal | 79A1276 | 0.03 | 0.01 | 0.21 | 95.03 | 0.02 | 0.26 | 0.59 | 3.63 | 0.05 | 0.01 | 0.10 | 0.05 | 0.00 |
| Cannon | 81A3003 | 0.04 | 0.01 | 0.20 | 94.46 | 0.01 | 0.18 | 0.34 | 4.56 | 0.05 | 0.02 | 0.12 | 0.01 | 0.01 |
| Demi cannon | 81A3000 | 0.03 | 0.01 | 0.18 | 92.06 | 0.01 | 0.30 | 0.62 | 6.58 | 0.04 | 0.02 | 0.12 | 0.03 | 0.01 |
| Demi cannon | 79A1277 | 0.02 | 0.01 | 0.24 | 93.98 | 0.00 | 0.30 | 0.61 | 4.62 | 0.03 | 0.06 | 0.10 | 0.02 | 0.01 |
| Demi cannon | 81A3002 | 0.09 | 0.01 | 0.19 | 88.71 | 3.15 | 0.23 | 0.51 | 6.81 | 0.05 | 0.02 | 0.20 | 0.02 | 0.00 |
| Culverin | 81A1423 | 0.05 | 0.01 | 0.09 | 84.71 | 0.04 | 0.28 | 0.63 | 5.60 | 0.07 | 0.03 | 0.10 | 0.04 | 0.01 |
| Culverin | 80A0976 | 0.03 | 0.01 | 0.12 | 93.80 | 0.02 | 0.40 | 1.33 | 4.04 | 0.08 | 0.01 | 0.14 | 0.02 | 0.00 |
| Culverin | 79A1278 | 0.00 | 0.01 | 0.33 | 92.71 | 0.04 | 0.98 | 0.60 | 4.98 | 0.05 | 0.00 | 0.27 | 0.03 | 0.01 |
| Demi culverin | 79A1232 | 0.04 | 0.01 | 0.13 | 92.94 | 0.13 | 0.43 | 1.52 | 4.47 | 0.10 | 0.02 | 0.17 | 0.03 | 0.02 |

*1.2 Analysis of guns from the* Mary Rose
by Peter Northover

A programme of analysis was instituted to determine the alloys used for the bronze guns from the *Mary Rose* and how they compared with contemporary English and European guns. The assemblage includes three guns cast by the Owen brothers, three cast by the Arcano/Arcana family and one by Peter Baude, cast over a period of only nine years. One of the aims therefore was to assess the consistency of a foundry's output and working practices. Of further interest was the relationship between the alloys actually used and those referred to in sixteenth century treatises, together with postulating the origins of the metals used. In line with the methods of sampling carried out on 22 bronze guns covering the period 1464–1870 from the collections of the Royal Armouries (Blackmore 1976, 407–9), and 41 samples from western, central and northern Europe (see Northover and Rychner 1998) the *Mary Rose* samples were taken from the outer surface of the gun. It is understood that to produce a detailed analysis of any gun not only multiple samples need to be taken, but these would have to penetrate the entire wall of the gun. This was rejected as being too destructive.

Eight guns were sampled (Table Appendix 1.1), all at least three times, and two guns (81A1423, 79A1232) were multiply sampled to study segregation within these large castings (Table Appendix 1.2a). Most samples were bored using a handheld modelmaker's electric drill with a 1mm diameter bit. One sample was cut with a jeweller's piercing saw from a trunnion of gun 81A1423 to obtain evidence of microstructure. Analysis was normally by electron probe microanalysis with wavelength dispersive spectrometry[1] (details in archive). A ninth bronze gun, recovered in 1840 and now in the collections of the Royal Armouries, (XIX 167) had already been analysed as part of an earlier programme (Blackmore 1976, 407).

**Homogeneity and sampling strategy**
Where castings as large as 2000kg are being sampled for analysis, choosing sample locations to give a representative analysis is bound to be challenging.

Table Appendix 1.1 Guns sampled for metal analysis (metric)

| Object | Gun type | Founder | Date | Length | Bore | Weight |
|---|---|---|---|---|---|---|
| 81A1423 | culverin | Peter Baude | 1543 | 3655 | 140 | 2071 |
| 81A3003 | cannon | Peter Baude | ? | 2960 | 200 | 2463 |
| 81A3000 | demi cannon | John & Robert Owen | 1542 | 2960 | 165 | 2513 |
| 80A0976 | culverin | unknown | 1535 | 2955 | 138 | 1353 |
| 81A3002 | demi cannon | Francesco Arcanus | 1535 | 2635 | 140 | 1393 |
| 79A1232 | demi culverin | John & Robert Owen | 1537 | 3210 | 112 | 1293 |
| 79A1276 | cannon royal | John & Robert Owen | 1535 | 2977 | 215 | 2180 |
| 79A1277 | demi cannon | Arcangelo Arcanus | 1542 | 3730 | 160 | 2740 |

## Table Appendix 1.2 Segregation in the bronze guns

### a) 81A1423

| Sample | Location | | Fe | Co | Ni | Cu | Zn | As | Sb | Sn | Ag | Bi | Pb | Au | S |
|---|---|---|---|---|---|---|---|---|---|---|---|---|---|---|---|
| MRT1 | Trunnion | Mean | 0.01 | 0.01 | 0.10 | 92.08 | 0.05 | 0.31 | 0.66 | 6.52 | 0.09 | 0.03 | 0.08 | 0.06 | 0.01 |
| | | Std. Dev. | 0.01 | 0.01 | 0.01 | 0.54 | 0.02 | 0.05 | 0.07 | 0.32 | 0.03 | 0.03 | 0.05 | 0.04 | 0.01 |
| MRT2 | Cascable/10mm | Mean | 0.10 | 0.01 | 0.10 | 90.68 | 0.05 | 0.38 | 0.93 | 7.52 | 0.11 | 0.02 | 0.06 | 0.90 | 0.01 |
| | | Std. Dev. | 0.04 | 0.01 | 0.01 | 0.79 | 0.02 | 0.05 | 0.05 | 0.14 | 0.04 | 0.03 | 0.05 | 0.04 | 0.01 |
| MRT3 | Ring/320mm | Mean | 0.03 | 0.01 | 0.10 | 91.80 | 0.05 | 0.34 | 0.81 | 6.59 | 0.07 | 0.03 | 0.09 | 0.07 | 0.01 |
| | | Std. Dev. | 0.01 | 0.01 | 0.02 | 0.33 | 0.02 | 0.05 | 0.09 | 0.23 | 0.03 | 0.03 | 0.06 | 0.06 | 0.01 |
| MRT4 | 820mm | Mean | 0.05 | 0.01 | 0.10 | 92.85 | 0.05 | 0.28 | 0.55 | 5.88 | 0.06 | 0.03 | 0.06 | 0.05 | 0.01 |
| | | Std. Dev. | 0.04 | 0.00 | 0.02 | 1.35 | 0.02 | 0.06 | 0.15 | 1.15 | 0.03 | 0.04 | 0.06 | 0.06 | 0.01 |
| MRT5 | 1300mm | Mean | 0.08 | 0.01 | 0.10 | 92.82 | 0.06 | 0.27 | 0.62 | 5.77 | 0.07 | 0.03 | 0.11 | 0.06 | 0.01 |
| | | Std. Dev. | 0.02 | 0.01 | 0.02 | 0.39 | 0.02 | 0.03 | 0.03 | 0.14 | 0.02 | 0.04 | 0.04 | 0.04 | 0.01 |
| MRT6 | Dolphin/1600mm | Mean | 0.04 | 0.02 | 0.09 | 92.46 | 0.05 | 0.33 | 0.72 | 6.05 | 0.08 | 0.03 | 0.08 | 0.05 | 0.01 |
| | | Std. Dev. | 0.02 | 0.01 | 0.02 | 0.39 | 0.02 | 0.03 | 0.03 | 0.14 | 0.02 | 0.04 | 0.04 | 0.04 | 0.01 |
| MRT7 | 1780mm | Mean | 0.10 | 0.01 | 0.09 | 92.75 | 0.04 | 0.28 | 0.66 | 5.77 | 0.07 | 0.02 | 0.33 | 0.08 | 0.01 |
| | | Std. Dev. | 0.04 | 0.01 | 0.02 | 1.02 | 0.02 | 0.04 | 0.07 | 0.37 | 0.02 | 0.03 | 0.71 | 0.07 | 0.01 |
| MRT8 | Ring/2170mm | Mean | 0.05 | 0.00 | 0.10 | 92.47 | 0.04 | 0.32 | 0.72 | 6.06 | 0.08 | 0.05 | 0.08 | 0.03 | 0.01 |
| | | Std. Dev. | 0.03 | 0.01 | 0.02 | 0.98 | 0.02 | 0.04 | 0.11 | 0.53 | 0.02 | 0.05 | 0.06 | 0.04 | 0.01 |
| MRT9 | 2750mm | Mean | 0.06 | 0.01 | 0.10 | 91.97 | 0.04 | 0.35 | 0.74 | 6.47 | 0.09 | 0.03 | 0.08 | 0.06 | 0.01 |
| | | Std. Dev. | 0.04 | 0.01 | 0.01 | 0.97 | 0.02 | 0.08 | 0.20 | 0.94 | 0.02 | 0.03 | 0.05 | 0.05 | 0.01 |
| MRT10 | 3170mm | Mean | 0.03 | 0.01 | 0.09 | 93.28 | 0.05 | 0.27 | 0.60 | 5.44 | 0.06 | 0.03 | 0.09 | 0.05 | 0.01 |
| | | Std. Dev. | 0.01 | 0.01 | 0.02 | 1.28 | 0.01 | 0.05 | 0.15 | 0.82 | 0.03 | 0.03 | 0.05 | 0.05 | 0.01 |
| MRT11 | Muzzle/3540mm | Mean | 0.03 | 0.01 | 0.10 | 93.59 | 0.05 | 0.29 | 0.56 | 5.16 | 0.07 | 0.03 | 0.10 | 0.01 | 0.01 |
| | | Std. Dev. | 0.02 | 0.01 | 0.02 | 0.91 | 0.02 | 0.07 | 0.13 | 0.70 | 0.04 | 0.03 | 0.08 | 0.01 | 0.01 |

### b) 79 A 1232

| Sample | Location | Distance (mm) | Cu | Sb | Sn | Pb |
|---|---|---|---|---|---|---|
| S5/S0202 | Cascable button | 0 | 94.93 | 1.28 | 3.79 | |
| S15/S0209 | Cascable button | 0 | 92.18 | 1.79 | 5.37 | 0.67 |
| S4/S0201 | Cascable ring | 35 | 90.50 | 2.06 | 6.36 | 1.29 |
| S9A/S0205 | Base ring | 350 | 94.01 | 1.27 | 4.73 | |
| | Base ring | 350 | 92.29 | 1.82 | 4.94 | 0.96 |
| S6/S0203 | Under 1st ring | 910 | 92.40 | 1.72 | 5.07 | 0.81 |
| S7/S0204 | Left lion | 1344 | 91.89 | 1.92 | 5.51 | 0.64 |
| S11A/S0206 | Left trunnion | 1638 | 92.61 | 1.75 | 5.01 | 0.64 |
| S13/S0207 | Chase | 1988 | 93.44 | 1.48 | 4.35 | 0.74 |
| S2/S0199 | Muzzle ring | 3136 | 94.73 | 0.99 | 3.53 | 0.75 |
| | Muzzle | 3198 | 95.75 | 1.05 | 3.20 | |

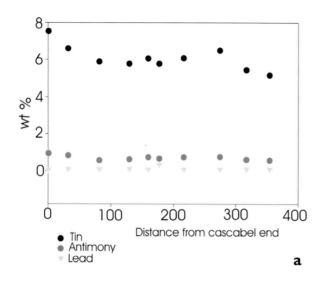
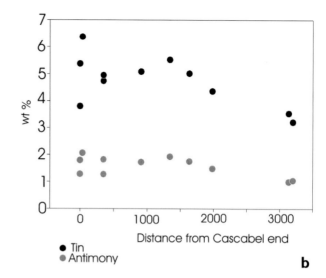

*Figure Appendix 1.1  Segregation a) 81A1423; b) 79A1232*

There are several competing mechanisms which decide the overall distribution of alloying elements in the finished casting. Which is predominant depends on the rate at which the casting is freezing, and on the direction of freezing. It would require detailed knowledge of the thermal properties of the metal and the mould together with computer simulation to attempt to determine that.

In the casting of a bronze cannon the principal factors are thermal convection and gravity segregation in the liquid, the direction in which the solidification front is moving, and more locally, dendritic segregation (for a review of these questions see Barker 1983, Campbell 1991). The first two are the most difficult to predict because it is necessary to know at any stage how the remaining liquid is distributed within the mould/casting assembly. In the case of gravity segregation it might be expected that, in the liquid as in the solid, the first material to freeze will be the low-tin crystals and the last will be those most enriched in tin. This will be reinforced by the tendency for the solidification front to move from the breech towards the muzzle rather than from the surface to the centre of the tube wall. If, as many writers describe (eg, Biringuccio 1540, 255–60), the guns were cast muzzle upwards it might be expected that the muzzle region would be richer in tin. Finally, natural processes of segregation within a dendritic structure will tend to drive solute rich liquid towards the bases of the dendrites. A segregation of tin can build up in the surface zone (in its more extreme forms known as '*tin sweat*'). In other words, concentration gradients can exist through the thickness of the tube wall as well as along it. Most sampling drills penetrate a relatively short distance from the surface and it may be the effects of this segregation that is being sampled. The pattern of segregation will also be modified by surface excrescences such as the cascable, the dolphins and the trunnions. These develop their own local pattern of segregation but have little impact on the strength of the barrel.

Two guns were sampled at several points along their length to measure linear segregation – the first (81A1423) as part of this project, and the second (79A1232) as part of an undergraduate dissertation study at the University of Surrey (Muskett 1987). It was not possible to measure concentration through the thickness of the tube. The results of the two studies are summarised in Table Appendix 1.2 and plotted in Figure Appendix 1.1, with all the individual analyses listed in tables Appendix 1.7–1.8, below. The standard deviations given in Table Appendix 1.2a offer a measure of the homogeneity of individual samples; the same was not possible in Table Appendix 1.2b because there were

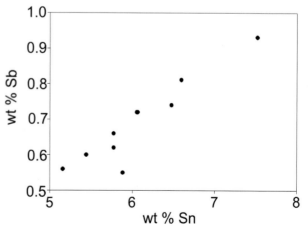

*Figure Appendix 1.2  81A1423 correlation of tin and antimony*

## Table Appendix 1.3 Sample analysis of the bronze guns

| Sample | Gun | Location | Fe | Co | Ni | Cu | Zn | As | Sb | Sn | Ag | Bi | Pb | Au | S |
|---|---|---|---|---|---|---|---|---|---|---|---|---|---|---|---|
| MRT 1 | 81A1423 | Trunnion | 0.01 | 0.01 | 0.10 | 92.08 | 0.05 | 0.31 | 0.66 | 6.52 | 0.09 | 0.03 | 0.08 | 0.06 | 0.01 |
| MRT 2 | | Cascable/10mm | 0.10 | 0.01 | 0.10 | 90.68 | 0.05 | 0.38 | 0.93 | 7.52 | 0.11 | 0.02 | 0.06 | 0.02 | 0.01 |
| MRT 3 | | Ring/320mm | 0.03 | 0.01 | 0.10 | 91.80 | 0.05 | 0.34 | 0.81 | 6.59 | 0.07 | 0.03 | 0.09 | 0.07 | 0.01 |
| MRT 4 | | 820mm | 0.05 | 0.01 | 0.10 | 92.85 | 0.05 | 0.28 | 0.55 | 5.88 | 0.06 | 0.03 | 0.09 | 0.05 | 0.01 |
| MRT 5 | | 1300mm | 0.08 | 0.01 | 0.10 | 92.82 | 0.06 | 0.27 | 0.62 | 5.77 | 0.07 | 0.03 | 0.11 | 0.06 | 0.01 |
| MRT 6 | | Dolphin/1600mm | 0.04 | 0.02 | 0.09 | 92.46 | 0.05 | 0.33 | 0.72 | 6.05 | 0.08 | 0.03 | 0.08 | 0.05 | 0.01 |
| MRT 7 | | 1780mm | 0.10 | 0.01 | 0.09 | 92.52 | 0.04 | 0.28 | 0.66 | 5.77 | 0.10 | 0.02 | 0.33 | 0.08 | 0.01 |
| MRT 8 | | Ring/2170mm | 0.05 | 0.00 | 0.10 | 92.47 | 0.04 | 0.32 | 0.72 | 6.06 | 0.08 | 0.05 | 0.08 | 0.03 | 0.01 |
| MRT 9 | | 2750mm | 0.06 | 0.01 | 0.10 | 91.97 | 0.04 | 0.35 | 0.74 | 6.47 | 0.09 | 0.03 | 0.08 | 0.06 | 0.01 |
| MRT 10 | | 3170mm | 0.03 | 0.01 | 0.09 | 93.28 | 0.05 | 0.27 | 0.60 | 5.44 | 0.06 | 0.03 | 0.09 | 0.05 | 0.01 |
| MRT 11 | | Muzzle/3540mm | 0.03 | 0.01 | 0.10 | 93.59 | 0.05 | 0.29 | 0.56 | 5.16 | 0.07 | 0.03 | 0.10 | 0.01 | 0.01 |
| MRT 12 | 80A0976 | Cascable | 0.01 | 0.00 | 0.12 | 93.21 | 0.02 | 0.45 | 1.54 | 4.32 | 0.10 | 0.00 | 0.18 | 0.04 | 0.00 |
| MRT 13 | | Trunnion | 0.04 | 0.01 | 0.13 | 93.38 | 0.03 | 0.41 | 1.45 | 4.30 | 0.09 | 0.02 | 0.13 | 0.02 | 0.00 |
| MRT 14 | | Tube | 0.03 | 0.01 | 0.12 | 94.80 | 0.00 | 0.34 | 1.00 | 3.51 | 0.05 | 0.01 | 0.11 | 0.00 | 0.00 |
| MRT 15 | 81A3000 | Cascable | 0.01 | 0.00 | 0.19 | 92.34 | 0.04 | 0.28 | 0.47 | 6.48 | 0.04 | 0.01 | 0.11 | 0.02 | 0.01 |
| MRT 16 | | Trunnion | 0.03 | 0.01 | 0.19 | 91.12 | 0.00 | 0.27 | 0.67 | 7.50 | 0.02 | 0.02 | 0.12 | 0.04 | 0.01 |
| MRT 17 | | Tube | 0.04 | 0.00 | 0.16 | 92.72 | 0.00 | 0.35 | 0.70 | 5.78 | 0.04 | 0.03 | 0.14 | 0.04 | 0.01 |
| MRT 18 | 81A3002 | Cascable | 0.10 | 0.00 | 0.19 | 87.65 | 3.55 | 0.22 | 0.60 | 7.12 | 0.08 | 0.01 | 0.43 | 0.04 | 0.01 |
| MRT 19 | | Trunnion | 0.08 | 0.01 | 0.19 | 88.82 | 2.91 | 0.26 | 0.52 | 7.03 | 0.04 | 0.04 | 0.07 | 0.02 | 0.00 |
| MRT 20 | | Tube | 0.10 | 0.01 | 0.20 | 89.65 | 2.98 | 0.21 | 0.42 | 6.28 | 0.04 | 0.00 | 0.10 | 0.00 | 0.00 |
| MRT 21 | 79A1232 | Cascable | 0.02 | 0.01 | 0.13 | 91.87 | 0.06 | 0.47 | 1.83 | 5.09 | 0.13 | 0.01 | 0.28 | 0.04 | 0.04 |
| MRT 22 | | Trunnion | 0.05 | 0.01 | 0.13 | 92.90 | 0.15 | 0.46 | 1.57 | 4.47 | 0.08 | 0.03 | 0.12 | 0.02 | 0.00 |
| MRT 23 | | Tube | 0.05 | 0.01 | 0.14 | 94.05 | 0.17 | 0.34 | 1.15 | 3.87 | 0.08 | 0.01 | 0.12 | 0.02 | 0.01 |
| MRT 24 | 81A3003 | Cascable | 0.07 | 0.01 | 0.21 | 95.17 | 0.01 | 0.13 | 0.26 | 3.92 | 0.04 | 0.03 | 0.14 | 0.01 | 0.01 |
| MRT 25 | | Trunnion | 0.01 | 0.00 | 0.19 | 92.96 | 0.01 | 0.24 | 0.51 | 5.81 | 0.09 | 0.02 | 0.13 | 0.03 | 0.00 |
| MRT 26 | | Tube | 0.03 | 0.01 | 0.20 | 95.26 | 0.00 | 0.19 | 0.23 | 3.95 | 0.04 | 0.01 | 0.08 | 0.00 | 0.01 |
| MRT 27 | 79A1277 | Cascable | 0.05 | 0.01 | 0.23 | 92.76 | 0.00 | 0.43 | 0.78 | 5.51 | 0.02 | 0.09 | 0.09 | 0.01 | 0.01 |
| MRT 28 | | Trunnion | 0.01 | 0.01 | 0.24 | 94.65 | 0.00 | 0.24 | 0.53 | 4.03 | 0.06 | 0.05 | 0.16 | 0.02 | 0.01 |
| MRT 29 | | Tube | 0.00 | 0.00 | 0.25 | 94.52 | 0.00 | 0.23 | 0.53 | 4.32 | 0.03 | 0.03 | 0.07 | 0.01 | 0.00 |
| MRT 30 | 79A1276 | Cascable | 0.03 | 0.01 | 0.19 | 95.85 | 0.01 | 0.24 | 0.51 | 3.02 | 0.03 | 0.02 | 0.07 | 0.04 | 0.00 |
| MRT 31 | | Trunnion | 0.03 | 0.01 | 0.22 | 94.43 | 0.03 | 0.32 | 0.71 | 4.06 | 0.05 | 0.02 | 0.06 | 0.05 | 0.00 |
| MRT 32 | | Tube | 0.01 | 0.01 | 0.23 | 94.81 | 0.01 | 0.23 | 0.55 | 3.81 | 0.09 | 0.01 | 0.18 | 0.05 | 0.01 |

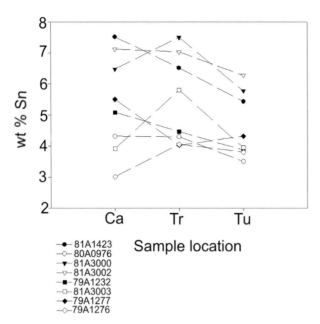

*Figure Appendix 1.3* Mary Rose *bronze ordnance segregation*

only four analyses per sample, rather than ten. The graphs for tin and antimony in both cases follow the same profile. This is typical for cast bronzes where tin and antimony segregate together; this correlation is demonstrated for 81A1423 in Figure Appendix 1.2.

Both guns show a variation in composition of the outermost 2–5mm along the length of the gun. In neither case is the trend linear: very clearly in the case of 81A1423 there is an increase in tin content at the breech end and a decrease at the muzzle with a more uniform concentration along the mid-length. 79A1232 seems similar although there is a substantial scatter of results towards the cascable. The rise at mid-length is because the samples were cut from one of the dolphins and one of the trunnions, both of these probably solidified early, as with the cascable, and have been subject to a degree of tin sweat. The drop off in tin content at the muzzle is also interesting in the light of Biringuccio's statement that it is desirable to add extra tin to the melt in the latter stages of the pour when the metal level in the mould is approaching the level of the muzzle (*ibid.*, 136–79, 250–60). This was intended to combat shrinkage by reducing the liquidus temperature and viscosity of the remaining metal and so increase its ability to feed cavities within the dendrite mesh. Exothermic vents were used to achieve the same results in the casting of the culverin (Chapter 2).

The rest of the guns were sampled at the muzzle, by the trunnion and at the cascable (Table Appendix 1.3). These compositions are plotted in Figure Appendix 1.3 for all eight guns to show the proportion that follow the same pattern of segregation as 81A1423 and 79A1232. The results reflect a division into two groups based on tin content (see below). Five share a similar pattern of segregation, one the reverse and one shows the highest concentration of tin in the trunnion area, possibly the result of local tin sweat. This pattern was also observed in an extensive analysis of guns within the Swedish Army Museum (Forshell 1992). Similarly, all three patterns of segregation were found but with a strong tendency for the lowest tin concentration to be at the muzzle (61% of guns analysed). The highest tin concentrations were equally distributed between the chase near the trunnions and the vent area (43% of guns analysed in each case). Although the guns analysed were mainly seventeenth century rather than sixteenth, the similarities must be the result of comparable foundry practice.

These data prompted us to sample the *Mary Rose* guns at three locations and select the mean of the three analyses for each as providing a linear representation of that gun. More accurately, the mean is representative of the surface regions of the gun since we lack any detailed analysis of a cross-section through the tube, but there is no reason to suppose that segregation through the thickness is more severe than along the length. While the cooling time was too short for segregation along the length of the gun to be homogenised, it is possible that the cross-section concentration gradients were significantly reduced while the gun cooled from, say, 900°C down to 600°C. Having reached this conclusion we can now compare the compositions of the eight guns analysed (Table Appendix 1.3).

**The alloys**

All the guns were cast in what can be described as unleaded low tin bronzes, although one (81A3002) contains a low alloy level of zinc. The graphic in Figure Appendix 1.3 suggests that they can be divided into two groups, one averaging around 4–5% tin and the other around 6–7% tin. Their overall mean compositions are listed in Table Appendix 1.4 together with date and maker; also included are the few available analyses of contemporary guns by the same makers (it should be noted that re-analysis has shown that the data published in Blackmore (1976) are erroneous). These were analysed from large samples drilled close to the mid-length of the guns. Simple inspection of the table suggests that 4–5% tin was indeed the norm for bronze guns cast in England during the reign of Henry VIII. A few show features indicating the use of scrap, while the compositions of the majority are compatible with being cast from newly made bronze, most probably from imported copper and English tin.

Allowing for the different sampling strategies, the three guns cast in 1542 by Arcangelo Arcanus (one from the *Mary Rose* and two from the Armouries) are very similar in composition and can be used as a standard with which to compare the others. The three guns gave mean tin contents of 4.5–5%. One part tin to 20 parts copper equates to 4.8% tin by weight, which broadly relates to one part in 20, so this could represent

Table Appendix 1.4 Composition of English guns of the reign of Henry VIII

| Sample | Gun | Date | Country | Foundry | Gun type | Fe | Co | Ni | Cu | Zn | As | Sb | Sn | Ag | Bi | Pb | Au | S |
|---|---|---|---|---|---|---|---|---|---|---|---|---|---|---|---|---|---|---|
| MRT12 | 80A0976 | 1535 | English | ?? | culverin | 0.03 | 0.01 | 0.12 | 93.80 | 0.02 | 0.40 | 1.33 | 4.04 | 0.08 | 0.01 | 0.14 | 0.02 | 0.00 |
| MRT18 | 81A3002 | 1535 | English | Francisco Arcanus | demi cannon | 0.09 | 0.01 | 0.19 | 88.71 | 3.15 | 0.23 | 0.51 | 6.81 | 0.05 | 0.02 | 0.20 | 0.02 | 0.00 |
| gun3 | RA XIX.18 (Cat. 29) | 1542 | English | Arcangelo Arcanus | cannon | 0.00 | 0.00 | 0.18 | 93.94 | 0.03 | 0.36 | 0.47 | 4.81 | 0.04 | 0.03 | 0.10 | 0.03 | 0.01 |
| gun26 | RA XIX.167 (Cat 30) | 1542 | English | Arcangelo Arcanus | culverin | 0.00 | 0.01 | 0.33 | 92.71 | 0.04 | 0.98 | 0.60 | 4.98 | 0.05 | 0.00 | 0.27 | 0.03 | 0.01 |
| MRT27 | 79A1277 | 1542 | English | Arcang. Arcanus | demi cannon | 0.02 | 0.01 | 0.24 | 93.98 | 0.00 | 0.30 | 0.61 | 4.62 | 0.03 | 0.06 | 0.10 | 0.02 | 0.01 |
| MRT30 | 79A1276 | 1535 | English | J. & R. Owen | cannon | 0.03 | 0.01 | 0.21 | 95.03 | 0.02 | 0.26 | 0.59 | 3.63 | 0.05 | 0.01 | 0.10 | 0.05 | 0.00 |
| MRT21 | 79A1232 | 1537 | English | J. & R. Owen | demi culverin | 0.04 | 0.01 | 0.13 | 92.94 | 0.13 | 0.43 | 1.52 | 4.47 | 0.10 | 0.02 | 0.17 | 0.03 | 0.02 |
| MRT15 | 81A3000 | 1542 | English | J. & R. Owen | demi cannon | 0.03 | 0.01 | 0.18 | 92.06 | 0.01 | 0.30 | 0.62 | 6.58 | 0.04 | 0.02 | 0.12 | 0.03 | 0.01 |
| gun4 | RA XIX.19 (Cat. 31) | 1546 | English | J. & R. Owen | demi cannon | 0.04 | 0.00 | 0.48 | 92.06 | 0.49 | 0.80 | 0.73 | 5.20 | 0.06 | 0.03 | 0.08 | 0.02 | 0.00 |
| gun1 | RA XIX.17 (Cat. 28) | 1535 | English | Peter Baude | 3 barrels | 0.23 | 0.00 | 0.21 | 89.39 | 0.15 | 0.23 | 0.88 | 6.77 | 0.13 | 0.00 | 1.92 | 0.00 | 0.08 |
| MRT1 | 81A1423 | 1543 | English | Peter Baude | dulverin | 0.05 | 0.01 | 0.09 | 84.71 | 0.04 | 0.28 | 0.63 | 5.60 | 0.07 | 0.03 | 0.10 | 0.04 | 0.01 |
| MRT24 | 81A3003 | ? | English | Peter Baude | dannon | 0.04 | 0.01 | 0.20 | 94.46 | 0.01 | 0.18 | 0.34 | 4.56 | 0.05 | 0.02 | 0.12 | 0.01 | 0.01 |

the intended recipe. The presence of 3.2% zinc in 81A3002 might be answered by the incorporation of brass or gunmetal (latten) scrap in place of some of the copper. The use of scrap could also account for the rather higher tin content. The higher tin content of gun XIX.17 from the Royal Armouries may also be associated with the use of some scrap, accounting for the zinc (0.15%) and iron (0.23%) impurities and the low alloy level of lead, a feature absent in the other guns. The 0.49% zinc in RA XIX.19 may also be the result the use of scrap, although the tin content is not far out of line. The only remaining exception to the 4.8–5.0% tin rule is therefore 81A3000, with 6.6% tin but normal levels of other elements; this mixture includes tin as one part in fifteen of the alloy. This is one of the larger guns with muzzle damage (see Figs 5.39 and 5.41, above). It has been suggested (P. Jones, pers. comm. 2000) that this is a result of a casting flaw.

The impurity pattern observed is very consistent, the principal impurities being arsenic, antimony, nickel and silver with smaller traces of iron, cobalt, zinc, bismuth, lead and sulphur. The microprobe tends to exaggerate the presence of gold but there is very probably a minor trace in most guns. The low levels of iron, zinc and lead encourage the hypothesis that most of the copper was new, ingot metal. Scrap alloys would (unless well sorted and correctly identified) tend to add zinc and lead to the alloy; iron could come from poorly refined metal or even from unnoticed fragments of iron scrap which would dissolve readily in molten bronze, or even suggest the use of old or failed guns which would include iron from the original core supports (crown). The similarity in impurity patterns should not be used to imply a common origin for the copper. These similarities are more likely to reflect trends in mining and smelting or the operations of the metal market resulting in a relatively homogeneous product.

*Comparison with continental ordnance*
There have been three large programmes of analysis of copper alloy ordnance in recent years; in Berlin (Riederer 1972; 1977; *Berliner Datenbank*); in Stockholm (Forshell 1984; 1992) and in Oxford (Northover and Gilmour, unpublished data). Although the different databases have not been explicitly standardised against each other, recent comparative analyses of bronze demonstrate that, with due care, results from different laboratories can be combined (Northover and Rychner 1998). The point of concern here is to ensure that similar modes of sampling were used; this can be confirmed. The analyses include examples from western, central

**Table Appendix 1.5 Analysis of European bronze ordnance 1500–1560**

| Date | Country | Foundry | Founder | Type | Fe | Co | Ni | Cu | Zn | As | Sb | Sn | Ag | Bi | Pb | Au | S |
|---|---|---|---|---|---|---|---|---|---|---|---|---|---|---|---|---|---|
| 1535 | English | London (Houndsditch) | ?? | culverin | 0.03 | 0.01 | 0.12 | 93.80 | 0.02 | 0.40 | 1.33 | 4.04 | 0.08 | 0.01 | 0.14 | 0.02 | 0.00 |
| 1535 | English | London (Salisbury Place?) | Franc. Arcanus | demi cannon | 0.09 | 0.01 | 0.19 | 88.71 | 3.15 | 0.23 | 0.51 | 6.81 | 0.05 | 0.02 | 0.20 | 0.02 | 0.00 |
| 1542 | English | London (Salisbury Place?) | Arcang. Arcanus | cannon | 0.00 | 0.00 | 0.18 | 93.94 | 0.03 | 0.36 | 0.47 | 4.81 | 0.04 | 0.03 | 0.10 | 0.03 | 0.01 |
| 1542 | English | London (Salisbury Place?) | Arcang. Arcanus | culverin | 0.00 | 0.01 | 0.33 | 92.71 | 0.04 | 0.98 | 0.60 | 4.98 | 0.05 | 0.00 | 0.27 | 0.03 | 0.01 |
| 1542 | English | London (Salisbury Place?) | Arcang. Arcanus | demi cannon | 0.02 | 0.01 | 0.24 | 93.98 | 0.00 | 0.30 | 0.61 | 4.62 | 0.03 | 0.06 | 0.10 | 0.02 | 0.01 |
| 1535 | English | London (Houndsditch) | J.&R. Owen | cannon | 0.03 | 0.01 | 0.21 | 95.03 | 0.02 | 0.26 | 0.59 | 3.63 | 0.05 | 0.01 | 0.10 | 0.05 | 0.00 |
| 1537 | English | London (Houndsditch) | J.&R. Owen | demi culverin | 0.04 | 0.01 | 0.13 | 92.94 | 0.13 | 0.43 | 1.52 | 4.47 | 0.10 | 0.02 | 0.17 | 0.03 | 0.02 |
| 1542 | English | London (Houndsditch) | J.&R. Owen | demi cannon | 0.03 | 0.01 | 0.18 | 92.06 | 0.01 | 0.30 | 0.62 | 6.58 | 0.04 | 0.02 | 0.12 | 0.03 | 0.01 |
| 1546 | English | London (Houndsditch) | J.&R. Owen | demi cannon | 0.04 | 0.00 | 0.48 | 92.06 | 0.49 | 0.80 | 0.73 | 5.20 | 0.06 | 0.03 | 0.08 | 0.02 | 0.00 |
| 1535 | English | London (Houndsditch) | Peter Baude | 3 barrels | 0.23 | 0.00 | 0.21 | 89.39 | 0.15 | 0.23 | 0.88 | 6.77 | 0.13 | 0.00 | 1.92 | 0.00 | 0.08 |
| 1543 | English | London (Houndsditch) | Peter Baude | culverin | 0.05 | 0.01 | 0.09 | 84.71 | 0.04 | 0.28 | 0.63 | 5.60 | 0.07 | 0.03 | 0.10 | 0.04 | 0.01 |
| <1546 | English | London (Houndsditch) | Peter Baude | cannon | 0.04 | 0.01 | 0.20 | 94.46 | 0.01 | 0.18 | 0.34 | 4.56 | 0.05 | 0.02 | 0.12 | 0.01 | 0.01 |
| 1507 | French | | | barrel | 0.01 | 0.00 | 0.01 | 93.00 | 0.01 | | 0.46 | 3.10 | 0.14 | | 2.40 | | |
| 1515 | French | | | siege gun | 0.01 | 0.00 | 0.13 | 87.00 | 0.28 | | 0.25 | 9.30 | 0.09 | | 1.20 | | |
| 1520 | Portuguese | | | 9 pdr breech loader | 0.01 | 0.00 | 0.09 | 94.57 | 0.01 | <0.20 | 0.25 | 4.78 | 0.02 | 0.00 | 0.13 | 0.02 | 0.00 |
| 1537 | Austrian | Vienna | L. Mairhofer | falconet | 0.07 | 0.00 | 0.48 | 91.50 | 0.00 | | 2.10 | 4.50 | 0.35 | | 0.40 | | |
| 1543 | Austrian | Vienna | J. Perger | falconet | 0.05 | 0.00 | 0.30 | 93.00 | 0.05 | | 0.95 | 4.70 | 0.47 | | 0.40 | | |
| 1543 | Austrian | Vienna | J. Perger | falconet | 0.06 | 0.00 | 0.31 | 91.00 | 0.05 | | 1.30 | 5.70 | 0.43 | | 0.60 | | |
| 1550 | Austrian | Vienna | | falconet | 0.03 | 0.00 | 0.19 | 93.00 | 0.00 | | 0.82 | 5.00 | 0.39 | | 0.10 | | |
| 1554 | Austrian | Venice | D. Dinckheim | howitzer | 0.08 | 0.00 | 0.97 | 93.00 | 0.00 | | 1.13 | 3.40 | 0.23 | | 0.60 | | |
| 1550 | Austrian | Innsbruck | H. Ch Löffler | serpentinel | 0.02 | 0.00 | 0.15 | 89.50 | 0.00 | | 0.21 | 8.70 | 0.10 | | 0.60 | | |
| 1558 | Austrian | Innsbruck | G. Löffler | demi cannon | 0.05 | 0.00 | 0.10 | 91.00 | 0.00 | | 0.33 | 6.70 | 0.15 | | 0.60 | | |
| 1500 | German | | | demi culverin | 0.09 | 0.00 | 0.15 | 93.00 | 0.12 | | 0.61 | 5.80 | 0.07 | | 0.25 | | |
| 1500 | German | | | demi culverin | 0.92 | 0.00 | 0.20 | 93.00 | 0.01 | | 0.68 | 5.50 | 0.07 | | 0.25 | | |
| 1524 | German | | | double cannon | 0.02 | 0.00 | 0.62 | 89.00 | 0.57 | | 0.52 | 7.50 | 0.17 | | 1.40 | | |
| 1525 | German | | | double cannon | 0.01 | 0.00 | 0.43 | 89.50 | 0.03 | | 1.02 | 8.00 | 0.25 | | 1.00 | | |
| 1526 | German | | | demi culverin | 0.06 | 0.00 | 0.24 | 92.00 | 0.02 | | 0.44 | 6.40 | 0.10 | | 1.00 | | |
| 1543 | German | | | cannon | 0.01 | 0.00 | 0.34 | 91.00 | 0.01 | | 0.33 | 7.90 | 0.20 | | 0.50 | | |
| 1544 | German | | | cannon | 0.01 | 0.00 | 0.35 | 88.00 | 0.05 | | 0.62 | 9.10 | 0.31 | | 1.20 | | |
| 1553 | German | | | ¾ cannon | 0.01 | | 0.28 | 91.00 | 0.05 | | 0.44 | 6.80 | 0.14 | | 1.30 | | |
| 1554 | German | | | ¾ cannon | 0.02 | | 0.16 | 91.00 | 0.01 | | 0.44 | 6.70 | 0.09 | | 1.30 | | |

| Date | Country | Foundry | Founder | Type | Fe | Co | Ni | Cu | Zn | As | Sb | Sn | Ag | Bi | Pb | Au | S |
|---|---|---|---|---|---|---|---|---|---|---|---|---|---|---|---|---|---|
| 1531 | Bavaria | Bavaria | | field gun | 0.03 | | 0.96 | 87.63 | 0.42 | 0.66 | 0.01 | 9.25 | 0.23 | | 0.71 | | |
| 1521 | Italian | | | barrel | 0.01 | 0.00 | 0.15 | 90.00 | 0.15 | | 1.00 | 6.50 | 2.40 | | 2.40 | | |
| 1521 | Italian | | | barrel | 0.01 | 0.00 | 0.05 | 87.00 | 0.07 | | 0.63 | 10.80 | 0.70 | | 0.70 | | |
| 1524 | Italian | | | demi-culverin | 0.25 | 0.00 | 0.20 | 88.00 | 0.19 | | 0.56 | 6.40 | 0.08 | | 4.50 | | |
| 1550 | Venetian | Venice | | culverin | 0.02 | 0.00 | 0.06 | 94.50 | 0.61 | | 0.15 | 3.80 | 0.06 | | 0.50 | | |
| 1550 | Venetian | Venice | | culverin | 0.09 | 0.00 | 0.20 | 93.50 | 0.59 | | 0.25 | 4.60 | 0.11 | | 0.40 | | |
| 1535 | Sweden | | | 12 pdr | 0.10 | | 0.03 | 87.80 | 0.40 | 0.06 | 0.10 | 7.60 | 0.09 | 0.01 | 2.60 | | |
| 1546 | Sweden | Oxelösund | | | 0.02 | 0.02 | 0.02 | 92.75 | 0.01 | 0.42 | 0.03 | 5.73 | 0.09 | 0.12 | 0.81 | 0.00 | |
| 16th C | Norway | | | | 0.07 | 0.04 | 0.23 | 90.18 | 0.14 | 0.44 | 0.83 | 5.86 | 0.09 | 0.04 | 2.06 | 0.00 | |
| 16th C | Norway | | | | 0.27 | 0.05 | 0.44 | 88.61 | 0.22 | 0.69 | 1.44 | 4.26 | 0.10 | 0.02 | 3.93 | 0.00 | |

and northern Europe, Italy and Russia and the Far East. Unfortunately for considering the technology of ordnance in the sixteenth century for Spain and Portugal. A total of 41 guns cast between 1500 and 1560 have been analysed (Table Appendix 1.5).

It was suggested above that English bronze guns appear to have been cast to a standard of one part tin to 219 or 20 parts of copper, with lead as an impurity only. The exceptions observed are accounted for by the inclusion of scrap in the alloy. Allowing for the uncertainties resulting from differing patterns of segregation in different guns, and possible systematic differences between the results from different laboratories, the plot of lead against tin for these 41 guns demonstrates clear regional differences. England is not alone in using a low tin bronze. The Austrian foundry in Vienna and the arsenal in Venice also fall into this category; interestingly the Austrian foundry in Innsbruck does not. Higher tin and higher lead content is common in Germany and Italy; from France and Sweden the data is too limited and too scattered to enable sensible conclusions.

In analysing the Berlin database Riederer (1972; 1977) found that the plot of silver against lead was a useful tool for discriminating between outputs from different foundries. There are three factors contributing to this: the possible addition of lead to the alloy as lead metal or leaded scrap; the original composition of the copper; and the possibility that a large quantity of European copper (especially from Germany and central Europe) was de-silverised by liquation with lead. A plot of silver against lead for the 41 guns is given in Figure Appendix 1.4. This shows a remarkable separation between the different groups. With one exception, the English guns cluster close to the origin and imply the use of low silver or de-silverised copper. By contrast, the products of the Vienna arsenal may not have been de-silverised. One reason for the difference might be state control of the copper in Austria contrasting with the need for England to buy copper on the international market. At this time Venice may also have needed to buy copper on the open market, although in earlier times it would have had control of Cypriot copper. The two guns analysed from Sweden have low levels of antimony and nickel and less than 0.1% silver; it is possible that Swedish copper was being exploited here.

**Lead isotope analysis**

Although it was not expected that lead isotope analysis of the bronze in the *Mary Rose* guns would yield direct information about the provenance of the metal, it was thought that this analysis would give a measure of the uniformity of the metal supply for Henry VIII's armament programme. Samples from five guns were therefore analysed at TU-Bergakademie Freiberg, Saxony. The data are given in Table Appendix 1.7 and are plotted in Figure Appendix 1.5. The grouping of the points on the graph is very close and could suggest a

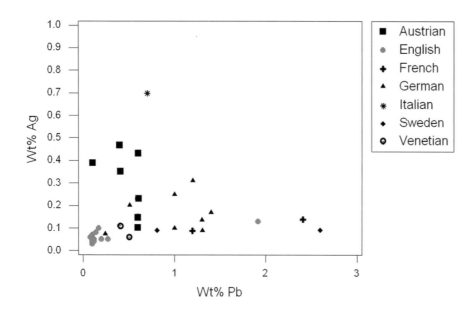

*Figure Appendix 1.4 Analysis of European bronze ordnance 1500–1560: plot of silver against lead*

very similar history. The point that is slightly detached from the others is the 81A3002 cast in 1535 by Francesco Arcanus and described above as having some scrap incorporated into the melt as evidenced by the zinc content. This alone would be sufficient to modify the lead isotope ratios. The other four points are then very close together indeed; however, they may well be measuring the isotope ratios of the lead used in the liquation for removing silver rather than the original lead in the copper. If this was so, it is then tempting to see a single channel for the purchase of copper by the English government.

Even this may be misleading. When the isotope ratios of European ore sources are mapped (eg, Rohl and Needham 1998, plots 19–20), it is apparent that the signatures of many European ore deposits overlap, or indeed, overlay each other. There are, in fact, data entirely consistent with a British origin for the lead, but copper both smelted and de-silverised on the continent is a more likely origin during the 1530s and 1540s. If this is accepted, and in view of the available published data, it is possible to suggest, that Germany and central Europe are a more likely source than France. It would be interesting to pursue this research further in the future to find out more about the economics of Henry VIII's re-armament programme.

**Metallography**

Understanding of the macro- and microstructures of large cast bronze guns is hampered by the same problem of sample size and location as is determining overall alloy composition. This is enhanced by the physical problem of cutting samples from large cylindrical shapes without causing excessive damage. A coring drill could be used but then there is the challenge of detaching the core from such a ductile metal as a 5% tin bronze. However, it was felt that some microstructural data were desirable; consequently a single sample was cut with a small razor saw from one trunnion on 81A1423. The sample was examined in both polished and etched state, but the structure revealed after etching is the most significant. This displayed large homogenised grains uniformly crossed by many deformation markings, most probably

**Table Appendix 1.6 Lead isotope analysis of bronze guns**

| Sample | Gun | Date | Country | Foundry | Type | Cu | Zn | Sb | Sn | Pb | 208Pb/ 206Pb | 207Pb/ 206Pb | 206Pb/ 204Pb |
|---|---|---|---|---|---|---|---|---|---|---|---|---|---|
| MRT12 | 80A0976 | 1535 | English | ?? | culverin | 93.80 | 0.02 | 1.33 | 4.04 | 0.14 | 2.0811 | 0.84669 | 18.4713 |
| MRT18 | 81A3002 | 1535 | English | Franc. Arcanus | demi cannon | 88.71 | 3.15 | 0.51 | 6.81 | 0.20 | 2.0902 | 0.85249 | 18.3457 |
| MRT27 | 79A1277 | 1542 | English | Arcang. Arcanus | demi cannon | 93.98 | 0.00 | 0.61 | 4.62 | 0.10 | 2.0860 | 0.84888 | 18.4205 |
| MRT21 | 79A1232 | 1537 | English | J.&R. Owen | demi culverin | 92.94 | 0.13 | 1.52 | 4.47 | 0.17 | 2.0832 | 0.84763 | 18.4435 |
| MRT24 | 81A3003 | ? | English | Peter Baude | cannon | 94.46 | 0.01 | 0.34 | 4.56 | 0.12 | 2.0868 | 0.84997 | 18.3970 |

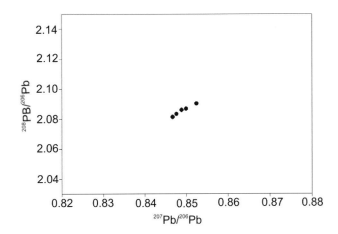

*Figure Appendix 1.5 Lead isotope analysis of* Mary Rose *ordnance*

deformation twins, but perhaps including some slip traces. The homogeneity of the grains can be explained by the slow cooling of the bulk of the casting providing opportunity for the dendritic segregation in the metal already solidified at the surface of the casting to be homogenised. The deformation of the surface comes from the all-over peening of the surface (ie, striking with the wedge-shaped end of a hammerhead) after the casting has been removed from the mould and cleaned, as described by Biringuccio (1540, 307):

> 'Then with a scouring brush and water uncover the emblems, coats of arms, borders and cornices that you made as ornaments. Striking it with a hammer, push inwards every superfluous bit of bronze. Finally, having cleaned the bronze by washing, scraping, and cutting away every bit of clay, strike its surfaces gently with a large hammer and thus make it very smooth. Likewise finish the mouth and all its cornices as exactly and completely as possible with files and any other instrument that serves best'.

That this was already old practice with guns is demonstrated by the microstructure of the so-called 'Dardanelles Gun' in the collections of the Royal Armouries and on display at Fort Nelson (Williams and Patterson 1986). Exactly the same features of homogenised grains and deformation twins are exhibited by the majority of the samples (in the publication the deformation twins are erroneously described as slip bands). The technique of peening the surface of castings was used by gunfounders in other areas as well. A statue of the head of Henri II of France, one of many memorials cast to the order of Catherine de Medici after the death of Henri II in 1569, utilised recycled ordnance bronze with 6.7% tin and 0.37% lead. The actual casting of the head was done in a gun foundry and this also demonstrates deformed homogenised grains resulting from slow cooling followed by surface peening.

## Recipes

It is to be regretted that the handbook of technology closest in date to the casting of the *Mary Rose* guns, that of Vannoccio Biringuccio published in 1540, is not specific about the alloy used for guns. In the third chapter of the fifth book of the Pirotechnia he distinguishes between bronzes for statues, guns, mortars, etc, on one hand and those for bells on the other. For the former he gives a range of 8–12lb (*c.* 3.63–5.45kg) tin per 100lb (45.4kg) copper (7.4–10.7%), and for the latter 23–26lb (10.43–11.79kg) tin (18.7–20.6%). Beyond that he leaves it to the judgement of the founder as to what alloy should be adopted. In the sixth chapter of the seventh book he also discusses ternary and quaternary alloys, the gunmetals and leaded gunmetals of today. He does not give any specific alloy compositions, again leaving the matter to the founder, but does say that the additions could be made to decrease the viscosity and improve the castability of the melt. Certainly he never specifically mentions a bronze with the 4–5% tin alloys used by Henry VIII's founders.

Later recipes were summarised by writers in the first half of the seventeenth century (eg, Lucar (1588); Capo Bianco (1598); Ufano (1613); Norton (1628), Hexham (1643)):

| Date | Copper | Tin | Lead | Other |
|------|--------|-----|------|-------|
| 1588 | 100 | 8 | | 10 (latten) |
| 1598 | 100 | 20 | | 5 (latten) |
| 1613 | 100 | 8 | 5 | 5 (latten) |
| 1643 | 75 | 11 | | 25 (bell metal) |

Even allowing for some loss of tin by oxidation during melting, these recipes are going to give tin contents of 7–12% or more, once again comfortably exceeding the recipes used in London, Vienna and Venice. The quantities for latten and bell metal may represent the upper limits to avoid the addition of too much scrap of unknown quality. A more usual source of scrap would be failed and worn guns and the headers from the foundry's own castings. Depending on the resources available, the incorporation of scrap could lead to slight variations in tin content. On the other hand, the consistency suggested by the analysed output of Arcangelo Arcanus in 1542 is best accounted for by the use of new metal, the rate of production at that time having outstripped the supply of scrap guns.

## Conclusions

Augmenting the guns analysed from the *Mary Rose* with data from contemporary English guns, we can achieve a coherent picture of policies for naval guns in the later part of the reign of Henry VIII.

## Table Appendix 1.7 Detailed analysis of the bronze guns

| Sample | Gun | Location | Fe | Co | Ni | Cu | Zn | As | Sb | Sn | Ag | Bi | Pb | Au | S |
|---|---|---|---|---|---|---|---|---|---|---|---|---|---|---|---|
| MRT 1/1 | 81A1423 | Trunnion | 0.01 | 0.00 | 0.09 | 91.45 | 0.08 | 0.39 | 0.76 | 6.85 | 0.13 | 0.05 | 0.10 | 0.07 | 0.02 |
| MRT 1/2 | | | 0.01 | 0.01 | 0.12 | 92.08 | 0.03 | 0.29 | 0.70 | 6.47 | 0.11 | 0.02 | 0.04 | 0.13 | 0.00 |
| MRT 1/3 | | | 0.02 | 0.04 | 0.09 | 92.79 | 0.01 | 0.24 | 0.57 | 6.02 | 0.04 | 0.07 | 0.01 | 0.07 | 0.03 |
| MRT 1/4 | | | 0.02 | 0.00 | 0.13 | 92.65 | 0.04 | 0.26 | 0.58 | 6.03 | 0.08 | 0.09 | 0.12 | 0.00 | 0.00 |
| MRT 1/5 | | | 0.01 | 0.00 | 0.09 | 92.13 | 0.08 | 0.32 | 0.63 | 6.46 | 0.12 | 0.00 | 0.11 | 0.03 | 0.03 |
| MRT 1/6 | | | 0.02 | 0.02 | 0.11 | 92.01 | 0.07 | 0.28 | 0.64 | 6.55 | 0.08 | 0.00 | 0.17 | 0.06 | 0.00 |
| MRT 1/7 | | | 0.01 | 0.01 | 0.10 | 91.76 | 0.04 | 0.38 | 0.69 | 6.76 | 0.09 | 0.00 | 0.09 | 0.07 | 0.00 |
| MRT 1/8 | | | 0.01 | 0.00 | 0.09 | 92.06 | 0.07 | 0.31 | 0.64 | 6.65 | 0.10 | 0.00 | 0.07 | 0.00 | 0.00 |
| MRT 1/9 | | | 0.00 | 0.00 | 0.12 | 92.21 | 0.03 | 0.33 | 0.62 | 6.43 | 0.08 | 0.02 | 0.03 | 0.12 | 0.00 |
| MRT 1/10 | | | 0.01 | 0.01 | 0.10 | 91.59 | 0.05 | 0.27 | 0.78 | 6.92 | 0.07 | 0.04 | 0.07 | 0.07 | 0.00 |
| MRT 1/11 | | | 0.03 | 0.00 | 0.04 | 81.42 | 0.06 | 0.00 | 0.20 | 1.78 | 0.04 | 0.00 | 0.00 | 0.00 | 16.43 |
| MRT 1/12 | | | 0.05 | 0.00 | 0.01 | 79.89 | 0.04 | 0.00 | 0.01 | 0.13 | 0.04 | 0.06 | 0.00 | 0.02 | 19.75 |
| MRT 2/1 | 81A1423 | Cascable/ 10mm | 0.08 | 0.00 | 0.08 | 91.02 | 0.09 | 0.29 | 0.90 | 7.41 | 0.09 | 0.03 | 0.00 | 0.00 | 0.02 |
| MRT 2/2 | | | 0.13 | 0.03 | 0.12 | 90.77 | 0.04 | 0.38 | 0.94 | 7.46 | 0.10 | 0.00 | 0.00 | 0.02 | 0.00 |
| MRT 2/3 | | | 0.17 | 0.00 | 0.10 | 90.51 | 0.06 | 0.32 | 0.93 | 7.57 | 0.18 | 0.09 | 0.08 | 0.00 | 0.00 |
| MRT 2/4 | | | 0.13 | 0.00 | 0.10 | 91.14 | 0.04 | 0.36 | 0.86 | 7.27 | 0.07 | 0.00 | 0.05 | 0.00 | 0.00 |
| MRT 2/5 | | | 0.08 | 0.01 | 0.09 | 90.88 | 0.07 | 0.34 | 0.88 | 7.41 | 0.08 | 0.00 | 0.11 | 0.04 | 0.01 |
| MRT 2/6 | | | 0.09 | 0.00 | 0.10 | 90.38 | 0.07 | 0.41 | 0.91 | 7.63 | 0.13 | 0.06 | 0.10 | 0.11 | 0.00 |
| MRT 2/7 | | | 0.05 | 0.01 | 0.11 | 90.90 | 0.04 | 0.39 | 0.95 | 7.46 | 0.07 | 0.02 | 0.00 | 0.00 | 0.00 |
| MRT 2/8 | | | 0.06 | 0.00 | 0.10 | 90.46 | 0.00 | 0.43 | 0.97 | 7.74 | 0.11 | 0.00 | 0.11 | 0.00 | 0.02 |
| MRT 2/9 | | | 0.06 | 0.01 | 0.07 | 90.53 | 0.06 | 0.45 | 0.96 | 7.57 | 0.14 | 0.02 | 0.05 | 0.06 | 0.01 |
| MRT 2/10 | | | 0.17 | 0.02 | 0.10 | 90.25 | 0.06 | 0.40 | 1.03 | 7.69 | 0.14 | 0.00 | 0.11 | 0.00 | 0.02 |
| MRT 3/1 | 81A1423 | Ring/ 320mm | 0.03 | 0.00 | 0.09 | 92.07 | 0.09 | 0.27 | 0.74 | 6.32 | 0.04 | 0.03 | 0.17 | 0.10 | 0.03 |
| MRT 3/2 | | | 0.03 | 0.00 | 0.10 | 91.64 | 0.06 | 0.34 | 0.77 | 6.66 | 0.09 | 0.00 | 0.18 | 0.11 | 0.00 |
| MRT 3/3 | | | 0.03 | 0.00 | 0.10 | 91.85 | 0.05 | 0.43 | 0.80 | 6.62 | 0.03 | 0.00 | 0.06 | 0.02 | 0.02 |
| MRT 3/4 | | | 0.01 | 0.02 | 0.11 | 91.48 | 0.03 | 0.30 | 0.95 | 6.73 | 0.07 | 0.08 | 0.09 | 0.13 | 0.00 |
| MRT 3/5 | | | 0.02 | 0.00 | 0.14 | 91.55 | 0.06 | 0.35 | 0.87 | 6.68 | 0.06 | 0.05 | 0.15 | 0.06 | 0.02 |
| MRT 3/6 | | | 0.01 | 0.03 | 0.10 | 91.27 | 0.04 | 0.33 | 0.93 | 7.00 | 0.07 | 0.02 | 0.02 | 0.19 | 0.00 |
| MRT 3/7 | | | 0.01 | 0.00 | 0.10 | 91.23 | 0.06 | 0.35 | 0.96 | 7.05 | 0.12 | 0.01 | 0.02 | 0.06 | 0.02 |
| MRT 3/8 | | | 0.02 | 0.00 | 0.08 | 91.32 | 0.07 | 0.38 | 0.95 | 6.91 | 0.10 | 0.06 | 0.06 | 0.04 | 0.01 |
| MRT 3/9 | | | 0.02 | 0.01 | 0.08 | 91.79 | 0.02 | 0.38 | 0.79 | 6.69 | 0.09 | 0.00 | 0.04 | 0.08 | 0.01 |
| MRT 3/10 | | | 0.01 | 0.00 | 0.11 | 91.93 | 0.04 | 0.31 | 0.77 | 6.53 | 0.05 | 0.09 | 0.16 | 0.00 | 0.01 |
| MRT 4/1 | 81A1423 | 820mm | 0.15 | 0.01 | 0.11 | 93.60 | 0.08 | 0.32 | 0.43 | 5.23 | 0.03 | 0.00 | 0.04 | 0.00 | 0.00 |
| MRT 4/2 | | | 0.03 | 0.02 | 0.07 | 91.76 | 0.04 | 0.32 | 0.66 | 6.88 | 0.07 | 0.02 | 0.13 | 0.00 | 0.00 |
| MRT 4/3 | | | 0.03 | 0.02 | 0.12 | 91.78 | 0.05 | 0.36 | 0.65 | 6.87 | 0.09 | 0.00 | 0.01 | 0.00 | 0.02 |
| MRT 4/4 | | | 0.02 | 0.00 | 0.09 | 92.46 | 0.08 | 0.23 | 0.62 | 6.27 | 0.04 | 0.00 | 0.17 | 0.00 | 0.02 |
| MRT 4/5 | | | 0.03 | 0.01 | 0.10 | 92.26 | 0.03 | 0.32 | 0.55 | 6.49 | 0.04 | 0.01 | 0.10 | 0.06 | 0.00 |
| MRT 4/6 | | | 0.03 | 0.01 | 0.10 | 94.41 | 0.05 | 0.29 | 0.35 | 4.48 | 0.10 | 0.05 | 0.12 | 0.00 | 0.01 |
| MRT 4/7 | | | 0.05 | 0.01 | 0.08 | 94.58 | 0.04 | 0.16 | 0.38 | 4.40 | 0.03 | 0.11 | 0.09 | 0.07 | 0.01 |
| MRT 4/8 | | | 0.04 | 0.01 | 0.12 | 94.39 | 0.09 | 0.24 | 0.40 | 4.39 | 0.08 | 0.10 | 0.00 | 0.13 | 0.02 |
| MRT 4/9 | | | 0.05 | 0.01 | 0.09 | 91.45 | 0.05 | 0.28 | 0.76 | 7.04 | 0.07 | 0.00 | 0.12 | 0.07 | 0.01 |
| MRT 4/10 | | | 0.02 | 0.02 | 0.10 | 91.80 | 0.04 | 0.28 | 0.68 | 6.75 | 0.03 | 0.00 | 0.13 | 0.14 | 0.01 |
| MRT 5/1 | 81A1423 | 1300mm | 0.03 | 0.00 | 0.09 | 93.41 | 0.06 | 0.25 | 0.57 | 5.48 | 0.01 | 0.00 | 0.06 | 0.02 | 0.01 |
| MRT 5/2 | | | 0.02 | 0.02 | 0.10 | 93.20 | 0.07 | 0.22 | 0.58 | 5.36 | 0.03 | 0.01 | 0.17 | 0.20 | 0.02 |
| MRT 5/3 | | | 0.03 | 0.01 | 0.11 | 92.90 | 0.00 | 0.22 | 0.66 | 5.82 | 0.08 | 0.02 | 0.05 | 0.13 | 0.00 |
| MRT 5/4 | | | 0.03 | 0.00 | 0.09 | 92.21 | 0.06 | 0.28 | 0.68 | 6.22 | 0.10 | 0.09 | 0.18 | 0.06 | 0.01 |
| MRT 5/5 | | | 0.03 | 0.02 | 0.10 | 93.38 | 0.04 | 0.30 | 0.51 | 5.35 | 0.04 | 0.01 | 0.17 | 0.00 | 0.04 |
| MRT 5/6 | | | 0.03 | 0.01 | 0.12 | 93.64 | 0.05 | 0.25 | 0.51 | 5.18 | 0.09 | 0.00 | 0.10 | 0.00 | 0.01 |
| MRT 5/7 | | | 0.19 | 0.02 | 0.08 | 92.72 | 0.03 | 0.26 | 0.63 | 5.81 | 0.06 | 0.00 | 0.19 | 0.00 | 0.00 |
| MRT 5/8 | | | 0.13 | 0.00 | 0.09 | 92.52 | 0.09 | 0.33 | 0.64 | 5.97 | 0.07 | 0.05 | 0.03 | 0.07 | 0.02 |
| MRT 5/9 | | | 0.15 | 0.02 | 0.09 | 92.16 | 0.07 | 0.34 | 0.66 | 6.14 | 0.10 | 0.05 | 0.15 | 0.06 | 0.00 |
| MRT 5/10 | | | 0.18 | 0.02 | 0.09 | 92.12 | 0.08 | 0.28 | 0.72 | 6.30 | 0.11 | 0.05 | 0.03 | 0.00 | 0.02 |

| Sample | Gun | Location | Fe | Co | Ni | Cu | Zn | As | Sb | Sn | Ag | Bi | Pb | Au | S |
|---|---|---|---|---|---|---|---|---|---|---|---|---|---|---|---|
| MRT 6/1 | 81A1423 | Dolphin/1600mm | 0.01 | 0.00 | 0.06 | 92.71 | 0.08 | 0.34 | 0.68 | 5.86 | 0.06 | 0.08 | 0.04 | 0.07 | 0.01 |
| MRT 6/2 | | | 0.03 | 0.03 | 0.09 | 92.60 | 0.07 | 0.31 | 0.71 | 5.99 | 0.06 | 0.00 | 0.05 | 0.06 | 0.00 |
| MRT 6/3 | | | 0.04 | 0.00 | 0.09 | 92.61 | 0.02 | 0.31 | 0.71 | 5.99 | 0.10 | 0.07 | 0.03 | 0.02 | 0.01 |
| MRT 6/4 | | | 0.07 | 0.02 | 0.10 | 92.55 | 0.06 | 0.36 | 0.69 | 5.95 | 0.08 | 0.00 | 0.10 | 0.00 | 0.02 |
| MRT 6/5 | | | 0.07 | 0.02 | 0.09 | 92.66 | 0.07 | 0.35 | 0.69 | 5.93 | 0.08 | 0.00 | 0.03 | 0.02 | 0.00 |
| MRT 6/6 | | | 0.04 | 0.03 | 0.12 | 92.43 | 0.04 | 0.32 | 0.69 | 5.97 | 0.06 | 0.02 | 0.13 | 0.14 | 0.01 |
| MRT 6/7 | | | 0.05 | 0.01 | 0.10 | 92.39 | 0.03 | 0.33 | 0.78 | 6.09 | 0.06 | 0.00 | 0.08 | 0.06 | 0.03 |
| MRT 6/8 | | | 0.07 | 0.02 | 0.10 | 92.12 | 0.06 | 0.36 | 0.76 | 6.28 | 0.11 | 0.00 | 0.11 | 0.00 | 0.00 |
| MRT 6/9 | | | 0.03 | 0.00 | 0.09 | 92.28 | 0.04 | 0.27 | 0.75 | 6.22 | 0.09 | 0.06 | 0.07 | 0.09 | 0.00 |
| MRT 6/10 | | | 0.01 | 0.04 | 0.05 | 92.20 | 0.06 | 0.34 | 0.75 | 6.22 | 0.08 | 0.08 | 0.10 | 0.07 | 0.00 |
| MRT 7/1 | 81A1423 | 1780mm | 0.10 | 0.00 | 0.09 | 92.87 | 0.08 | 0.27 | 0.62 | 5.87 | 0.05 | 0.00 | 0.04 | 0.00 | 0.01 |
| MRT 7/2 | | | 0.08 | 0.02 | 0.10 | 92.87 | 0.04 | 0.35 | 0.65 | 5.74 | 0.04 | 0.00 | 0.11 | 0.00 | 0.00 |
| MRT 7/3 | | | 0.10 | 0.02 | 0.06 | 92.76 | 0.01 | 0.27 | 0.65 | 5.84 | 0.06 | 0.00 | 0.14 | 0.09 | 0.01 |
| MRT 7/4 | | | 0.06 | 0.02 | 0.10 | 92.99 | 0.03 | 0.26 | 0.65 | 5.61 | 0.10 | 0.00 | 0.15 | 0.04 | 0.00 |
| MRT 7/5 | | | 0.10 | 0.03 | 0.09 | 92.50 | 0.02 | 0.22 | 0.70 | 5.91 | 0.09 | 0.06 | 0.08 | 0.18 | 0.02 |
| MRT 7/6 | | | 0.20 | 0.00 | 0.12 | 92.19 | 0.03 | 0.29 | 0.72 | 6.10 | 0.09 | 0.00 | 0.15 | 0.09 | 0.02 |
| MRT 7/7 | | | 0.14 | 0.00 | 0.10 | 92.19 | 0.05 | 0.29 | 0.75 | 6.15 | 0.05 | 0.05 | 0.07 | 0.18 | 0.00 |
| MRT 7/8 | | | 0.12 | 0.01 | 0.09 | 91.19 | 0.02 | 0.31 | 0.53 | 5.17 | 0.05 | 0.00 | 2.34 | 0.14 | 0.03 |
| MRT 7/9 | | | 0.09 | 0.02 | 0.11 | 93.57 | 0.06 | 0.25 | 0.55 | 5.17 | 0.06 | 0.03 | 0.07 | 0.00 | 0.02 |
| MRT 7/10 | | | 0.07 | 0.00 | 0.09 | 92.20 | 0.05 | 0.30 | 0.74 | 6.16 | 0.10 | 0.09 | 0.12 | 0.09 | 0.00 |
| MRT 8/1 | 81 A1423 | Ring/2170mm | 0.05 | 0.00 | 0.14 | 92.28 | 0.05 | 0.24 | 0.73 | 6.33 | 0.11 | 0.05 | 0.01 | 0.01 | 0.01 |
| MRT 8/2 | | | 0.04 | 0.00 | 0.09 | 92.23 | 0.03 | 0.38 | 0.73 | 6.31 | 0.07 | 0.01 | 0.03 | 0.08 | 0.00 |
| MRT 8/3 | | | 0.00 | 0.00 | 0.11 | 92.54 | 0.02 | 0.33 | 0.68 | 6.08 | 0.11 | 0.00 | 0.09 | 0.02 | 0.00 |
| MRT 8/4 | | | 0.03 | 0.01 | 0.10 | 92.96 | 0.06 | 0.29 | 0.67 | 5.78 | 0.08 | 0.03 | 0.00 | 0.00 | 0.00 |
| MRT 8/5 | | | 0.06 | 0.02 | 0.07 | 92.52 | 0.01 | 0.29 | 0.71 | 6.15 | 0.09 | 0.00 | 0.08 | 0.00 | 0.00 |
| MRT 8/6 | | | 0.04 | 0.01 | 0.09 | 92.70 | 0.04 | 0.30 | 0.64 | 5.79 | 0.04 | 0.07 | 0.20 | 0.09 | 0.01 |
| MRT 8/7 | | | 0.04 | 0.00 | 0.09 | 94.11 | 0.03 | 0.33 | 0.51 | 4.74 | 0.05 | 0.05 | 0.03 | 0.00 | 0.01 |
| MRT 8/8 | | | 0.09 | 0.00 | 0.10 | 91.82 | 0.06 | 0.32 | 0.84 | 6.44 | 0.07 | 0.12 | 0.13 | 0.02 | 0.00 |
| MRT 8/9 | | | 0.10 | 0.00 | 0.08 | 91.35 | 0.05 | 0.37 | 0.93 | 6.70 | 0.10 | 0.13 | 0.12 | 0.07 | 0.00 |
| MRT 8/10 | | | 0.05 | 0.00 | 0.12 | 92.31 | 0.01 | 0.32 | 0.76 | 6.25 | 0.08 | 0.00 | 0.07 | 0.00 | 0.03 |
| MRT 9/1 | 81 A1423 | 2750mm | 0.01 | 0.01 | 0.08 | 91.01 | 0.04 | 0.45 | 1.00 | 7.02 | 0.10 | 0.08 | 0.12 | 0.06 | 0.03 |
| MRT 9/2 | | | 0.03 | 0.01 | 0.09 | 92.15 | 0.04 | 0.36 | 0.69 | 6.38 | 0.06 | 0.00 | 0.05 | 0.12 | 0.01 |
| MRT 9/3 | | | 0.05 | 0.03 | 0.12 | 92.77 | 0.05 | 0.30 | 0.56 | 6.03 | 0.10 | 0.01 | 0.00 | 0.00 | 0.01 |
| MRT 9/4 | | | 0.10 | 0.02 | 0.12 | 91.14 | 0.04 | 0.43 | 0.82 | 7.00 | 0.10 | 0.06 | 0.12 | 0.05 | 0.00 |
| MRT 9/5 | | | 0.10 | 0.00 | 0.10 | 91.42 | 0.02 | 0.37 | 0.81 | 7.00 | 0.08 | 0.00 | 0.02 | 0.08 | 0.00 |
| MRT 9/6 | | | 0.07 | 0.01 | 0.10 | 93.27 | 0.01 | 0.28 | 0.53 | 5.57 | 0.08 | 0.00 | 0.06 | 0.00 | 0.03 |
| MRT 9/7 | | | 0.05 | 0.01 | 0.09 | 91.44 | 0.06 | 0.34 | 0.77 | 6.87 | 0.09 | 0.00 | 0.14 | 0.15 | 0.00 |
| MRT 9/8 | | | 0.04 | 0.02 | 0.09 | 91.79 | 0.05 | 0.26 | 0.76 | 6.74 | 0.12 | 0.05 | 0.09 | 0.00 | 0.01 |
| MRT 9/9 | | | 0.03 | 0.01 | 0.09 | 90.71 | 0.02 | 0.44 | 0.97 | 7.46 | 0.08 | 0.00 | 0.08 | 0.10 | 0.00 |
| MRT 9/10 | | | 0.12 | 0.00 | 0.08 | 94.23 | 0.09 | 0.24 | 0.41 | 4.55 | 0.06 | 0.07 | 0.12 | 0.02 | 0.00 |
| MRT 10/1 | 81 A1423 | 3170mm | 0.05 | 0.02 | 0.08 | 92.13 | 0.04 | 0.28 | 0.78 | 6.45 | 0.10 | 0.05 | 0.03 | 0.00 | 0.00 |
| MRT 10/2 | | | 0.03 | 0.02 | 0.07 | 92.10 | 0.06 | 0.33 | 0.79 | 6.23 | 0.08 | 0.07 | 0.15 | 0.06 | 0.01 |
| MRT 10/3 | | | 0.02 | 0.00 | 0.12 | 92.93 | 0.05 | 0.28 | 0.62 | 5.62 | 0.08 | 0.08 | 0.08 | 0.11 | 0.01 |
| MRT 10/4 | | | 0.01 | 0.01 | 0.10 | 94.26 | 0.02 | 0.22 | 0.49 | 4.69 | 0.03 | 0.00 | 0.11 | 0.05 | 0.00 |
| MRT 10/5 | | | 0.02 | 0.02 | 0.10 | 93.86 | 0.06 | 0.20 | 0.52 | 4.97 | 0.02 | 0.00 | 0.14 | 0.07 | 0.01 |
| MRT 10/6 | | | 0.04 | 0.01 | 0.08 | 93.30 | 0.05 | 0.27 | 0.59 | 5.36 | 0.07 | 0.05 | 0.14 | 0.02 | 0.01 |
| MRT 10/7 | | | 0.03 | 0.02 | 0.08 | 92.03 | 0.05 | 0.33 | 0.75 | 6.53 | 0.05 | 0.00 | 0.13 | 0.00 | 0.00 |
| MRT 10/8 | | | 0.04 | 0.00 | 0.08 | 94.78 | 0.04 | 0.25 | 0.39 | 4.22 | 0.02 | 0.00 | 0.05 | 0.13 | 0.00 |
| MRT 10/9 | | | 0.03 | 0.00 | 0.11 | 94.25 | 0.03 | 0.21 | 0.43 | 4.87 | 0.04 | 0.00 | 0.01 | 0.00 | 0.01 |

| Sample | Gun | Location | Fe | Co | Ni | Cu | Zn | As | Sb | Sn | Ag | Bi | Pb | Au | S |
|---|---|---|---|---|---|---|---|---|---|---|---|---|---|---|---|
| MRT 11/1 | 81 A1423 | Muzzle/ 3540mm | 0.01 | 0.00 | 0.09 | 94.37 | 0.05 | 0.28 | 0.42 | 4.43 | 0.11 | 0.07 | 0.14 | 0.02 | 0.03 |
| MRT 11/2 | | | 0.03 | 0.00 | 0.13 | 95.09 | 0.04 | 0.25 | 0.34 | 3.91 | 0.04 | 0.00 | 0.16 | 0.01 | 0.00 |
| MRT 11/3 | | | 0.02 | 0.01 | 0.08 | 94.32 | 0.07 | 0.21 | 0.55 | 4.70 | 0.02 | 0.00 | 0.00 | 0.01 | 0.00 |
| MRT 11/4 | | | 0.03 | 0.02 | 0.11 | 93.76 | 0.04 | 0.23 | 0.57 | 5.13 | 0.06 | 0.00 | 0.00 | 0.03 | 0.02 |
| MRT 11/5 | | | 0.03 | 0.02 | 0.11 | 92.60 | 0.06 | 0.37 | 0.70 | 5.82 | 0.11 | 0.04 | 0.11 | 0.04 | 0.00 |
| MRT 11/6 | | | 0.04 | 0.02 | 0.13 | 92.47 | 0.03 | 0.35 | 0.71 | 5.97 | 0.12 | 0.02 | 0.13 | 0.00 | 0.01 |
| MRT 11/7 | | | 0.05 | 0.02 | 0.07 | 92.56 | 0.03 | 0.35 | 0.71 | 5.95 | 0.10 | 0.05 | 0.05 | 0.03 | 0.02 |
| MRT 11/8 | | | 0.04 | 0.00 | 0.08 | 92.95 | 0.06 | 0.36 | 0.60 | 5.57 | 0.05 | 0.03 | 0.25 | 0.00 | 0.01 |
| MRT 11/9 | | | 0.04 | 0.03 | 0.09 | 93.59 | 0.09 | 0.28 | 0.54 | 5.19 | 0.03 | 0.00 | 0.11 | 0.00 | 0.00 |
| MRT 11/10 | | | 0.00 | 0.01 | 0.11 | 94.03 | 0.06 | 0.20 | 0.51 | 4.93 | 0.07 | 0.05 | 0.01 | 0.00 | 0.01 |
| MRT 12/1 | 80 A976 | Cascable | 0.00 | 0.00 | 0.16 | 93.34 | 0.01 | 0.48 | 1.37 | 4.08 | 0.15 | 0.00 | 0.32 | 0.08 | 0.00 |
| MRT 12/2 | | | 0.04 | 0.00 | 0.11 | 93.40 | 0.00 | 0.43 | 1.45 | 4.29 | 0.03 | 0.00 | 0.24 | 0.00 | 0.01 |
| MRT 12/3 | | | 0.03 | 0.00 | 0.12 | 93.35 | 0.08 | 0.44 | 1.49 | 4.34 | 0.09 | 0.00 | 0.02 | 0.03 | 0.00 |
| MRT 12/4 | | | 0.00 | 0.02 | 0.12 | 92.96 | 0.00 | 0.47 | 1.69 | 4.53 | 0.10 | 0.01 | 0.02 | 0.08 | 0.01 |
| MRT 12/5 | | | 0.00 | 0.00 | 0.09 | 93.01 | 0.00 | 0.43 | 1.67 | 4.37 | 0.12 | 0.00 | 0.30 | 0.00 | 0.00 |
| MRT 13/1 | 80 A976 | Trunnion | 0.03 | 0.01 | 0.13 | 92.31 | 0.00 | 0.47 | 1.90 | 4.85 | 0.10 | 0.04 | 0.15 | 0.00 | 0.01 |
| MRT 13/2 | | | 0.04 | 0.03 | 0.11 | 93.09 | 0.07 | 0.41 | 1.53 | 4.53 | 0.07 | 0.00 | 0.11 | 0.00 | 0.00 |
| MRT 13/3 | | | 0.00 | 0.01 | 0.13 | 93.69 | 0.00 | 0.34 | 1.37 | 4.18 | 0.13 | 0.01 | 0.09 | 0.06 | 0.00 |
| MRT 13/4 | | | 0.05 | 0.01 | 0.14 | 93.79 | 0.03 | 0.46 | 1.26 | 3.99 | 0.09 | 0.00 | 0.18 | 0.00 | 0.00 |
| MRT 13/5 | | | 0.06 | 0.00 | 0.12 | 94.00 | 0.07 | 0.36 | 1.21 | 3.94 | 0.07 | 0.03 | 0.11 | 0.03 | 0.01 |
| MRT 14/1 | 80 A976 | Tube | 0.03 | 0.00 | 0.13 | 94.89 | 0.02 | 0.29 | 0.97 | 3.45 | 0.05 | 0.00 | 0.15 | 0.02 | 0.00 |
| MRT 14/2 | | | 0.04 | 0.01 | 0.11 | 93.16 | 0.00 | 0.44 | 1.47 | 4.44 | 0.05 | 0.00 | 0.27 | 0.00 | 0.01 |
| MRT 14/3 | | | 0.03 | 0.00 | 0.13 | 95.00 | 0.00 | 0.36 | 0.91 | 3.53 | 0.03 | 0.00 | 0.00 | 0.00 | 0.00 |
| MRT 14/4 | | | 0.02 | 0.03 | 0.12 | 96.42 | 0.00 | 0.23 | 0.63 | 2.42 | 0.05 | 0.00 | 0.06 | 0.00 | 0.00 |
| MRT 14/5 | | | 0.05 | 0.00 | 0.09 | 94.55 | 0.00 | 0.38 | 1.01 | 3.70 | 0.08 | 0.04 | 0.08 | 0.00 | 0.01 |
| MRT 15/1 | 81 A3000 | Cascable | 0.00 | 0.00 | 0.21 | 92.38 | 0.10 | 0.24 | 0.48 | 6.42 | 0.00 | 0.02 | 0.07 | 0.07 | 0.01 |
| MRT 15/2 | | | 0.02 | 0.00 | 0.19 | 92.41 | 0.00 | 0.27 | 0.48 | 6.47 | 0.03 | 0.02 | 0.10 | 0.01 | 0.00 |
| MRT 15/3 | | | 0.00 | 0.01 | 0.18 | 92.18 | 0.06 | 0.29 | 0.44 | 6.53 | 0.07 | 0.00 | 0.22 | 0.00 | 0.02 |
| MRT 15/4 | | | 0.00 | 0.00 | 0.16 | 92.40 | 0.00 | 0.31 | 0.50 | 6.49 | 0.06 | 0.02 | 0.06 | 0.00 | 0.00 |
| MRT 16/1 | 81 A3000 | Trunnion | 0.02 | 0.01 | 0.20 | 89.97 | 0.00 | 0.34 | 0.92 | 8.23 | 0.00 | 0.02 | 0.19 | 0.08 | 0.02 |
| MRT 16/2 | | | 0.04 | 0.02 | 0.19 | 90.60 | 0.00 | 0.31 | 0.72 | 7.98 | 0.04 | 0.00 | 0.08 | 0.02 | 0.00 |
| MRT 16/3 | | | 0.03 | 0.02 | 0.20 | 90.91 | 0.00 | 0.22 | 0.65 | 7.66 | 0.05 | 0.04 | 0.19 | 0.02 | 0.00 |
| MRT 16/4 | | | 0.05 | 0.00 | 0.18 | 92.99 | 0.00 | 0.19 | 0.39 | 6.13 | 0.00 | 0.00 | 0.01 | 0.04 | 0.02 |
| MRT 17/1 | 81 A3000 | Tube | 0.08 | 0.00 | 0.14 | 94.02 | 0.00 | 0.45 | 1.10 | 4.04 | 0.05 | 0.00 | 0.08 | 0.00 | 0.03 |
| MRT 17/2 | | | 0.00 | 0.00 | 0.20 | 91.29 | 0.00 | 0.36 | 0.56 | 7.24 | 0.05 | 0.06 | 0.23 | 0.00 | 0.00 |
| MRT 17/3 | | | 0.03 | 0.00 | 0.15 | 92.84 | 0.00 | 0.25 | 0.44 | 6.05 | 0.02 | 0.02 | 0.09 | 0.11 | 0.00 |
| MRT 18/1 | 81 A3002 | Cascable | 0.09 | 0.00 | 0.19 | 88.14 | 3.55 | 0.22 | 0.57 | 6.73 | 0.05 | 0.00 | 0.42 | 0.04 | 0.01 |
| MRT 18/2 | | | 0.12 | 0.01 | 0.18 | 87.34 | 3.52 | 0.23 | 0.62 | 7.56 | 0.03 | 0.00 | 0.31 | 0.07 | 0.01 |
| MRT 18/3 | | | 0.10 | 0.00 | 0.22 | 87.86 | 3.64 | 0.17 | 0.56 | 6.75 | 0.02 | 0.06 | 0.62 | 0.01 | 0.01 |
| MRT 18/4 | | | 0.09 | 0.00 | 0.16 | 87.64 | 3.54 | 0.22 | 0.61 | 6.76 | 0.23 | 0.01 | 0.68 | 0.03 | 0.02 |
| MRT 18/5 | | | 0.10 | 0.01 | 0.21 | 87.27 | 3.52 | 0.24 | 0.62 | 7.78 | 0.08 | 0.00 | 0.13 | 0.05 | 0.00 |
| MRT 19/1 | 81 A3002 | Trunnion | 0.11 | 0.04 | 0.18 | 88.81 | 3.40 | 0.22 | 0.47 | 6.66 | 0.04 | 0.03 | 0.05 | 0.00 | 0.00 |
| MRT 19/2 | | | 0.08 | 0.00 | 0.16 | 88.18 | 3.62 | 0.29 | 0.57 | 6.89 | 0.00 | 0.03 | 0.16 | 0.00 | 0.00 |
| MRT 19/3 | | | 0.06 | 0.01 | 0.18 | 88.96 | 2.19 | 0.26 | 0.56 | 7.59 | 0.09 | 0.06 | 0.03 | 0.00 | 0.01 |
| MRT 19/4 | | | 0.08 | 0.01 | 0.23 | 89.43 | 2.01 | 0.30 | 0.52 | 7.31 | 0.06 | 0.00 | 0.05 | 0.01 | 0.00 |
| MRT 19/5 | | | 0.09 | 0.00 | 0.20 | 88.71 | 3.35 | 0.24 | 0.48 | 6.69 | 0.00 | 0.08 | 0.06 | 0.11 | 0.00 |
| MRT 19/6 | | | 0.09 | 0.00 | 0.17 | 90.66 | 2.54 | 0.18 | 0.38 | 5.88 | 0.03 | 0.00 | 0.07 | 0.00 | 0.00 |
| MRT 20/1 | 81 A3002 | Tube | 0.10 | 0.04 | 0.25 | 89.36 | 3.09 | 0.17 | 0.41 | 6.43 | 0.01 | 0.00 | 0.14 | 0.00 | 0.01 |
| MRT 20/2 | | | 0.10 | 0.00 | 0.19 | 90.06 | 3.20 | 0.22 | 0.33 | 5.70 | 0.00 | 0.03 | 0.16 | 0.00 | 0.00 |
| MRT 20/3 | | | 0.11 | 0.00 | 0.23 | 89.46 | 3.01 | 0.22 | 0.42 | 6.39 | 0.11 | 0.00 | 0.05 | 0.00 | 0.00 |
| MRT 20/4 | | | 0.13 | 0.00 | 0.17 | 89.45 | 2.96 | 0.24 | 0.46 | 6.43 | 0.06 | 0.00 | 0.10 | 0.00 | 0.00 |

| Sample | Gun | Location | Fe | Co | Ni | Cu | Zn | As | Sb | Sn | Ag | Bi | Pb | Au | S |
|---|---|---|---|---|---|---|---|---|---|---|---|---|---|---|---|
| MRT 21/1 | 79 A1232 | Cascable | 0.08 | 0.00 | 0.18 | 88.92 | 3.10 | 0.23 | 0.52 | 6.85 | 0.03 | 0.00 | 0.10 | 0.00 | 0.00 |
| MRT 21/2 | | | 0.01 | 0.01 | 0.12 | 91.72 | 0.10 | 0.53 | 1.73 | 5.31 | 0.09 | 0.00 | 0.35 | 0.00 | 0.04 |
| MRT 21/3 | | | 0.03 | 0.00 | 0.12 | 90.43 | 0.04 | 0.58 | 2.47 | 5.88 | 0.19 | 0.01 | 0.23 | 0.03 | 0.00 |
| MRT 21/4 | | | 0.02 | 0.01 | 0.19 | 91.88 | 0.08 | 0.49 | 1.80 | 4.88 | 0.13 | 0.00 | 0.31 | 0.13 | 0.09 |
| MRT 21/5 | | | 0.04 | 0.01 | 0.10 | 93.47 | 0.03 | 0.30 | 1.32 | 4.27 | 0.13 | 0.04 | 0.25 | 0.00 | 0.04 |
| MRT 22/1 | 79 A1232 | Trunnion | 0.05 | 0.01 | 0.11 | 92.95 | 0.10 | 0.47 | 1.39 | 4.69 | 0.09 | 0.00 | 0.14 | 0.02 | 0.00 |
| MRT 22/2 | | | 0.03 | 0.02 | 0.11 | 91.86 | 0.17 | 0.49 | 1.84 | 5.12 | 0.08 | 0.10 | 0.18 | 0.00 | 0.00 |
| MRT 22/3 | | | 0.02 | 0.01 | 0.18 | 91.27 | 0.14 | 0.61 | 2.07 | 5.37 | 0.12 | 0.02 | 0.16 | 0.04 | 0.01 |
| MRT 22/4 | | | 0.06 | 0.00 | 0.14 | 91.10 | 0.16 | 0.59 | 2.18 | 5.56 | 0.13 | 0.01 | 0.04 | 0.03 | 0.00 |
| MRT 22/5 | | | 0.10 | 0.02 | 0.13 | 97.33 | 0.20 | 0.17 | 0.39 | 1.58 | 0.01 | 0.00 | 0.06 | 0.00 | 0.01 |
| MRT 22/6 | | | 0.03 | 0.02 | 0.14 | 93.44 | 0.19 | 0.46 | 1.50 | 4.04 | 0.11 | 0.00 | 0.06 | 0.00 | 0.01 |
| MRT 22/7 | | | 0.05 | 0.01 | 0.13 | 92.35 | 0.22 | 0.43 | 1.84 | 4.48 | 0.08 | 0.00 | 0.34 | 0.05 | 0.02 |
| MRT 23/1 | 79 A1232 | Tube | 0.03 | 0.01 | 0.16 | 93.08 | 0.00 | 0.16 | 0.54 | 5.95 | 0.01 | 0.00 | 0.04 | 0.00 | 0.01 |
| MRT 23/2 | | | 0.05 | 0.01 | 0.13 | 94.48 | 0.14 | 0.32 | 1.18 | 3.56 | 0.11 | 0.00 | 0.02 | 0.00 | 0.00 |
| MRT 23/3 | | | 0.01 | 0.01 | 0.12 | 93.32 | 0.22 | 0.44 | 1.45 | 4.02 | 0.17 | 0.07 | 0.18 | 0.00 | 0.00 |
| MRT 23/4 | | | 0.11 | 0.01 | 0.16 | 96.41 | 0.18 | 0.26 | 0.56 | 2.09 | 0.06 | 0.00 | 0.08 | 0.08 | 0.00 |
| MRT 24/1 | 81 A3003 | Cascable | 0.10 | 0.02 | 0.12 | 95.25 | 0.21 | 0.31 | 0.96 | 2.91 | 0.02 | 0.00 | 0.08 | 0.01 | 0.01 |
| MRT 24/2 | | | 0.07 | 0.00 | 0.19 | 95.50 | 0.00 | 0.11 | 0.23 | 3.73 | 0.05 | 0.02 | 0.09 | 0.00 | 0.01 |
| MRT 24/3 | | | 0.09 | 0.00 | 0.22 | 95.95 | 0.00 | 0.10 | 0.21 | 3.11 | 0.03 | 0.02 | 0.25 | 0.00 | 0.02 |
| MRT 24/4 | | | 0.07 | 0.03 | 0.22 | 96.09 | 0.02 | 0.14 | 0.22 | 3.07 | 0.03 | 0.00 | 0.11 | 0.00 | 0.01 |
| MRT 24/5 | | | 0.04 | 0.00 | 0.19 | 93.15 | 0.00 | 0.16 | 0.40 | 5.77 | 0.05 | 0.08 | 0.09 | 0.05 | 0.00 |
| MRT 25/1 | 81 A3003 | Trunnion | 0.01 | 0.02 | 0.21 | 93.72 | 0.00 | 0.23 | 0.39 | 5.36 | 0.01 | 0.00 | 0.05 | 0.00 | 0.00 |
| MRT 25/2 | | | 0.02 | 0.00 | 0.22 | 92.19 | 0.00 | 0.20 | 0.64 | 6.16 | 0.25 | 0.05 | 0.28 | 0.00 | 0.01 |
| MRT 25/3 | | | 0.00 | 0.00 | 0.19 | 93.09 | 0.07 | 0.25 | 0.49 | 5.82 | 0.08 | 0.00 | 0.02 | 0.00 | 0.00 |
| MRT 26/1 | 81 A3003 | Tube | 0.00 | 0.00 | 0.14 | 93.13 | 0.00 | 0.26 | 0.47 | 5.71 | 0.04 | 0.03 | 0.08 | 0.12 | 0.01 |
| MRT 27/1 | 79 A1277 | Cascable | 0.02 | 0.00 | 0.18 | 92.66 | 0.00 | 0.26 | 0.56 | 6.02 | 0.07 | 0.00 | 0.21 | 0.01 | 0.00 |
| MRT 27/2 | | | 0.01 | 0.00 | 0.19 | 95.88 | 0.00 | 0.12 | 0.18 | 3.54 | 0.05 | 0.00 | 0.01 | 0.00 | 0.01 |
| MRT 27/3 | | | 0.03 | 0.02 | 0.22 | 96.46 | 0.00 | 0.18 | 0.13 | 2.82 | 0.02 | 0.02 | 0.10 | 0.00 | 0.00 |
| MRT 27/4 | | | 0.06 | 0.01 | 0.19 | 93.45 | 0.00 | 0.26 | 0.38 | 5.47 | 0.04 | 0.00 | 0.14 | 0.00 | 0.01 |
| MRT 27/5 | | | 0.05 | 0.01 | 0.23 | 92.76 | 0.00 | 0.43 | 0.78 | 5.51 | 0.02 | 0.09 | 0.09 | 0.01 | 0.01 |
| MRT 28/1 | 79 A1277 | Trunnion | 0.02 | 0.00 | 0.22 | 94.75 | 0.00 | 0.22 | 0.49 | 4.11 | 0.03 | 0.00 | 0.10 | 0.06 | 0.00 |
| MRT 28/2 | | | 0.00 | 0.00 | 0.26 | 94.80 | 0.00 | 0.24 | 0.43 | 3.91 | 0.05 | 0.13 | 0.14 | 0.02 | 0.02 |
| MRT 28/3 | | | 0.00 | 0.03 | 0.25 | 95.08 | 0.00 | 0.22 | 0.42 | 3.66 | 0.00 | 0.09 | 0.26 | 0.01 | 0.00 |
| MRT 28/4 | | | 0.01 | 0.01 | 0.19 | 94.61 | 0.00 | 0.22 | 0.62 | 4.13 | 0.00 | 0.00 | 0.18 | 0.01 | 0.00 |
| MRT 28/5 | | | 0.01 | 0.01 | 0.26 | 94.01 | 0.00 | 0.30 | 0.69 | 4.37 | 0.19 | 0.06 | 0.10 | 0.00 | 0.01 |
| MRT 29/1 | 79 A1277 | Tube | 0.01 | 0.00 | 0.18 | 92.91 | 0.00 | 0.23 | 0.57 | 5.94 | 0.01 | 0.00 | 0.12 | 0.05 | 0.00 |
| MRT 29/2 | | | 0.00 | 0.00 | 0.27 | 94.52 | 0.00 | 0.31 | 0.58 | 4.25 | 0.04 | 0.00 | 0.02 | 0.01 | 0.00 |
| MRT 29/3 | | | 0.00 | 0.01 | 0.26 | 94.43 | 0.00 | 0.32 | 0.62 | 4.24 | 0.02 | 0.00 | 0.09 | 0.00 | 0.02 |
| MRT 29/4 | | | 0.00 | 0.00 | 0.27 | 94.51 | 0.00 | 0.15 | 0.60 | 4.17 | 0.07 | 0.10 | 0.10 | 0.01 | 0.00 |
| MRT 29/5 | | | 0.00 | 0.00 | 0.27 | 96.25 | 0.00 | 0.17 | 0.29 | 2.99 | 0.00 | 0.02 | 0.00 | 0.00 | 0.00 |
| MRT 30/1 | 79 A1276 | Cascable | 0.01 | 0.01 | 0.22 | 96.65 | 0.00 | 0.18 | 0.46 | 2.42 | 0.04 | 0.02 | 0.00 | 0.00 | 0.00 |
| MRT 30/2 | | | 0.03 | 0.03 | 0.22 | 96.30 | 0.00 | 0.21 | 0.48 | 2.50 | 0.02 | 0.00 | 0.10 | 0.10 | 0.00 |
| MRT 30/3 | | | 0.02 | 0.00 | 0.18 | 95.94 | 0.00 | 0.24 | 0.60 | 2.87 | 0.00 | 0.06 | 0.04 | 0.04 | 0.01 |
| MRT 30/4 | | | 0.05 | 0.00 | 0.15 | 93.81 | 0.00 | 0.35 | 0.62 | 4.84 | 0.06 | 0.00 | 0.09 | 0.04 | 0.00 |
| MRT 30/5 | | | 0.04 | 0.00 | 0.16 | 96.55 | 0.03 | 0.21 | 0.41 | 2.48 | 0.01 | 0.00 | 0.10 | 0.00 | 0.00 |

| Sample | Gun | Location | Fe | Co | Ni | Cu | Zn | As | Sb | Sn | Ag | Bi | Pb | Au | S |
|---|---|---|---|---|---|---|---|---|---|---|---|---|---|---|---|
| MRT 31/1 | 79 A1276 | Trunnion | 0.04 | 0.01 | 0.11 | 92.25 | 0.15 | 0.51 | 1.73 | 5.06 | 0.03 | 0.00 | 0.05 | 0.06 | 0.00 |
| MRT 31/2 | | | 0.00 | 0.00 | 0.29 | 94.21 | 0.00 | 0.28 | 0.58 | 4.33 | 0.09 | 0.00 | 0.11 | 0.11 | 0.00 |
| MRT 31/3 | | | 0.04 | 0.01 | 0.22 | 95.56 | 0.00 | 0.26 | 0.36 | 3.33 | 0.08 | 0.05 | 0.02 | 0.07 | 0.00 |
| MRT 31/4 | | | 0.06 | 0.01 | 0.25 | 95.26 | 0.00 | 0.26 | 0.47 | 3.67 | 0.00 | 0.03 | 0.00 | 0.00 | 0.00 |
| MRT 31/5 | | | 0.02 | 0.00 | 0.23 | 94.90 | 0.00 | 0.28 | 0.44 | 3.92 | 0.04 | 0.02 | 0.12 | 0.03 | 0.00 |
| MRT 32/1 | 79 A1276 | Tube | 0.01 | 0.01 | 0.26 | 94.79 | 0.02 | 0.22 | 0.56 | 3.98 | 0.10 | 0.00 | 0.00 | 0.04 | 0.00 |
| MRT 32/2 | | | 0.00 | 0.02 | 0.24 | 95.06 | 0.02 | 0.22 | 0.43 | 3.74 | 0.09 | 0.01 | 0.12 | 0.04 | 0.00 |
| MRT 32/3 | | | 0.00 | 0.00 | 0.22 | 95.12 | 0.00 | 0.13 | 0.49 | 3.73 | 0.11 | 0.01 | 0.15 | 0.03 | 0.01 |
| MRT 32/4 | | | 0.06 | 0.02 | 0.20 | 94.37 | 0.00 | 0.36 | 0.81 | 3.82 | 0.13 | 0.00 | 0.17 | 0.06 | 0.01 |
| MRT 32/5 | | | 0.00 | 0.00 | 0.23 | 94.71 | 0.02 | 0.23 | 0.46 | 3.79 | 0.02 | 0.00 | 0.44 | 0.09 | 0.01 |

- Guns were cast in an unleaded low tin bronze with one part tin to either 24 or 25 parts copper.
- Most of the guns analysed were cast from new metal or from recycled guns of the same composition. A smaller proportion incorporated other scrap metal such as latten.
- The alloy choice is at variance with contemporary treatises. It finds parallels in the foundries of Vienna and Venice.
- The source of the copper might have been Germany/central Europe rather than France. It is probable the copper had been de-silverised.
- In common with other research programmes, no consistent pattern of segregation was found in the analyses.
- The microstructures showed slow cooling followed by cleaning and then peening of the surface by hammering. This process has been recorded in contemporary treatises.

The results of this project are of sufficient interest to justify further research on the guns of the reign of Henry VIII, both in England and elsewhere in Europe. The primary aim would be to confirm or reject the conclusions concerning the alloy composition. If confirmed, then it is to be hoped that documentary research might reveal the origins of the policy.

**Endnote**

1. All the drilled samples from the guns in the care of the Mary Rose Trust were bored with a 1mm bit in a hand-held modelmaker's electric drill. The cut sample was removed with a small razor saw. The samples from guns in the Royal Armouries collection were taken in the same way but with a larger, 2.5mm, bit. All samples were hot-mounted in a carbon-filled thermosetting resin, ground and polished to a 1μm diamond finish.

All analyses were by electron probe microanalysis using wavelength dispersive spectrometry; owing to a temporary unavailability of the electron microprobe in Oxford assistance was sought from the laboratories of the Natural History Museum, London. The method used there was essentially the same as that adopted in Oxford with thirteen elements analysed using pure element and mineral standards with count times of 20s per element. The operating parameters were generally an accelerating voltage of 20kV, a beam current of 30nA, and an X-ray take-off angle of 40EC. Detection limits were of the order of 100ppm for most elements, but 300–400ppm for gold.

In the analyses made at the Natural History Museum a search was made for phosphorus in the bronze. However, an interference between the phosphorus and copper X-ray spectra means a detection limit of 0.15%; no phosphorus was found above that limit. The comparative data quoted from laboratories in Stockholm and Berlin were obtained by atomic absorption.

# Appendix 2  Gunpowder (GP) analysis: analytical tables

In Tables Appendix 2.1–5 samples are arranged in order breech to mouth, with samples MR4, MR1 and MR2 coming from behind the shot and MR3 and MR5 from the barrel of culverin 79A0976

### Table Appendix 2.1 Gravimetric and Titrimetric determinations of suphur

| GP sample no. | Archive sample no. | Weight (g) | Gravimetric weight (g) | Det. % sulphur by weight |
|---|---|---|---|---|
| MR4 | 80S0117, 119, 120, 124 | 0.2891 | 0.1893 | 8.98 |
| MR1 | 80S0123 | 0.1522 | 0.0889 | 8.00 |
| MR2 | 80S0125 | 0.1226 | 0.0947 | 11.70 |
| MR3 | 80S0126 | 0.1966 | 0.0562 | 3.90 |
| MR5 | 80S0122 | 0.3099 | 0.0714 | 3.15 |

### Table Appendix 2.2 Determination of sulphur content (bomb calorific determinations)

| GP sample | Weight sulphur (g) x $10^{-3}$ | % weight sulphur |
|---|---|---|
| MR4 | 25.84 | 8.930 |
| MR1 | 6.35 | 4.200 |
| MR2 | 19.79 | 16.140 |
| MR3 | 3.87 | 1.968 |
| MR5 | 3.79 | 1.220 |

### Table Appendix 2.3 Emission spectrophotometric analysis of powder samples

| Element | MR4 | MR1 | MR2 | MR3 | MR5 |
|---|---|---|---|---|---|
| Mg | strong | strong | strong | strong | strong |
| Mn | medium | medium | medium | medium | medium |
| Pb | weak | medium | medium | weak | weak |
| Sn | – | weak | medium | weak | – |
| Ga | – | very weak | – | – | – |
| Si | medium/strong | strong | strong | strong | strong |
| Fe | medium/strong | medium/strong | strong | medium | medium/strong |
| N | weak | weak | very weak | very weak | weak |
| Al | weak | strong | strong | weak/medium | weak |
| Ca | strong | very strong | strong | strong | strong |
| Cu | strong | strong | very strong | strong | strong |
| Na | medium | weak | medium | strong | weak/medium |
| K | weak | weak | weak | weak | weak |
| Cr | – | – | weak | – | – |
| Ti | – | – | – | weak | – |
| Ag | very weak | – | – | very weak | very weak |

**Table Appendix 2.4 Determination of metals by atomic absorbtion spectrophotometry**

| GP sample | Fe mg/l | Mn mg/l | Ca mg/l | Cu mg/l |
|---|---|---|---|---|
| MR4 | 50.0 | 2.5 | 135.0 | 60.0 |
| MR1 | 50.0 | 5.5 | 120.0 | 30.0 |
| MR2 | 30.0 | 4.0 | 85.0 | 72.5 |
| MR3 | 60.0 | 0.0 | 75.0 | 45.5 |
| MR5 | 750.0 | 2.5 | 130.0 | 32.0 |

**Table Appendix 2.5 Determination of carbon content**

| GP sample | Weight mg | Corrected time | NcHCo$_3$ content | Equivalent weight carbon mg | %c |
|---|---|---|---|---|---|
| MR4 | 418.00 | 25.70 | 25.60 | 307.30 | 73.50 |
| MR1 | 337.00 | 18.25 | 18.18 | 218.20 | 64.70 |
| MR2 | 632.00 | 27.90 | 27.90 | 333.60 | 52.80 |
| MR3 | 197.00 | 11.10 | 11.12 | 133.50 | 67.70 |
| MR5 | 516.00 | 6.80 | 6.77 | 81.29 | 15.75 |

**Table Appendix 2.6 Percentage of soluble solids in samples 81S1409 and 82S1207**

| Sample | From | % Water | % Soluble solids | % Insoluble solids |
|---|---|---|---|---|
| 81S1409/15/1 | demi cannon 81A3000 | 17.2 | 3.9 | 78.9 |
| 82S1207/9/1 | breech chamber, port piece 82A3001 | 32.8 | 7.3 | 59.9 |

**Table Appendix 2.7 Elemental analysis of insoluble solids, samples 81S1409 and 82S1207**

| Sample | Ash | C | S | N | H |
|---|---|---|---|---|---|
| 81S1409 | 36.6 | 54.3 | 1.1 | 0.5 | 2.0 |
| 82S1207 | 60.1 | 20.3 | 7.9 | 1.5 | 0.2 |

**Table Appendix 2.8 Elements present in samples 81S1409 and 82S1207 compared with modern sample (see Fig. 5.103, above)**

| 81S1409 | 82S1207 | Modern |
|---|---|---|
| Silicon | iron | sulphur |
| Sulphur | sulphur | potassium |
| Calcium | chlorine | calcium |
| Iron | silicon | |
| Potassium | aluminium | |
| Aluminium | | |
| Magnesium | | |

**Table Appendix 2.9 Ether soluble material present in samples 81S1409 and 82S1207 compared with modern sample**

| Sample | % ether soluble material* | % ether insoluble material |
|---|---|---|
| 81S1409 | 0.1 | 99.9 |
| 82S1207 | 5.2 | 94.8 |
| Modern | 2.0 | 98.0 |

# Bibliography

**Manuscripts**

Public Record Office (PRO)

E 36 Exchequer, Treasury of Receipt, Miscellaneous Books
E 101 Exchequer, King's Remembrancer, Accounts Various
E 315 Exchequer, Augmentations Office, Miscellaneous Books
E 364 Exchequer, Pipe Office, Rolls of Foreign Accounts
E 404 Exchequer, Treasury of Receipt, Writs and Warrants for Issues
SP1 State Papers, Domestic Series, Henry VIII
SP 12 State Papers, Domestic Series, Elizabeth I
SP 16 State Papers, Domestic Series, Charles I
WO 55 War Office, Ordnance Office, Miscellanea

**British Library (BL)**
*Additional MSS*
22047 (*Anthony Roll* pt 2)
28300 (Notes by David Rhunkenius)
42130 (Luttrell Psalter)
48092 (Muster and Ordnance Records 1569–1574)

*Cotton MSS*
Augustus I.i.18, I.i.29, I.i.81, I.ii.15, I ii.64, I, ii.117 (Maps)
Julius E. IV [item 6] (*Beauchamp Pageant*)
Otho E. IX (Naval papers)

*Lansdowne MSS*
118 (Memoranda of W. Cecil, Lord Burghley)

*Sloane MSS*
3651 (Draft of W. Bourne's *The Arte of Shooting at Great Ordnaunce*)

*Stowe MSS*
146 (Orders and Warrants for Military Supplies 1512–1515)

**Bodleian Library, Oxford**
*Ashmole MSS*
343.VII / ff.128–139v (*Secrets of Gunmen*)
824.XXII / ff. 217v–240v

*Douce MSS*
b.2 (*Machines et ustenciles de guerre*)

*Rawlinson MSS*
A. 192, art. 4/ff. 22–36
A. 207 art. 1

*Topography MSS*
Top. Gen. e.70

**East Sussex Record Office, Lewes**
*Compton Place Archives*
SAS-CP/5/184 (formerly SAS CP/184 (Memoranda of James Burton)

**Hatfield House**
CP 201/51 (Command list July 1545)
CP 201/127 (Shipwrights' report 1545)

**Longleat House**
Miscellaneous MS V, ff.1, 53–73v (Inventory of 1555)

**Pepys Library, Magdalene College, Cambridge (PL)**
2876, 2878 (Miscellany of Matters Historical and Naval vols VIII and X)
2991 (*Anthony Roll* pts 1 and 3)

*Other manuscripts are cited in the text*

**Works in print, typescript or electronic fomat**
Amman, J., 1599. *Künstbuchlin*. Frankfort
AMO Standards Committee, 2000. *AMO Standards, Field Publication FP-3*. Gainesville, Florida: Archery Manufacturers & Merchants Organisation
Anderson, R.C., 1920. Armaments in 1540. *Mariner's Mirror* 6, 281
Anglo, S., 2000. *The Martial Arts of Renaissance Europe*. New Haven & London: Yale University Press
Anscombe, A. 1914. Prégent de Bidoux's raid in Sussex in 1514 and the Cotton MS *Augustus I (i)*, 18. *Transcriptions of the Royal Historical Society* 3rd series 8, 103–11
Archibald, C.D., 1840. Observations on some ancient pieces of ordnance, and other relics, discovered in the Island of Walney, in Lancashire. *Archaeologia* 28, 373–92
Arizcun, Flority J.M. & Sanchez Cantón F.J., 1927. *Catálogo de las armas Instituto de Valencia de Don Juan*. Madrid
Armitage, P., 1983. Analysis of Horn and Bone Objects from the Mary Rose. Portsmouth: Mary Rose Trust, archive report
Armstrong, D.R., 1997. A wrought iron gun for early 16th century sea service. *Journal of the Ordnance Society* 9, 27–49

Arnold, J.B. & Weddle, R., 1978. *The Nautical Archeology of Padre Island. The Spanish Shipwrecks of 1554.* New York & London: Academic Press/Texas Antiquities Committee Publication 7

Ascham, R., 1545. *Toxophilus; The Schole of Shootinge*. Frome: Butler & Tanner (facsimile, ed. Hodgkinson 1985)

Awty, B.G., 1988. The Arcana Family of Cesena as gunfounders and military engineers. *Newcomen Society Transactions* 59, 61–80

Awty, B.G., 1989. Parson Levett and English cannon founding. *Sussex Archaeological Collections* 127, 133–45

Ayloffe, Sir Joseph, 1775. An account of some ancient English historical paintings at Cowdry, in Sussex. *Archaeologia* 3, 239–72

Baker, R., 1563. The relation of one William Rutter to N. Anthony Hickman his master touching a voyage set out to Guinea in the yeare 1562 ... also written in verse by Robert Baker. In Hakluyt 1598–1600, vol. 6, 258–61

Barker, R.A., 1983. Bronze cannon founders: comments upon Guilmartin 1974, 1982. *International Journal of Nautical Archaeology and Underwater Exploration* 12(1), 67–74

Barker, R.A., 1998. A glance at ricochet. *Journal of the Ordnance Society* 10, 1–16

Barker, R.A., 1999. A proportional ordinance for ordnance? *Journal of the Ordnance Society* 11, 21–9

Barker, R.A., 2002. Gun carriages in a notebook of Portuguese field artillery. *Journal of the Ordnance Society* 14, 19–30

Barker, R.A. & Castro A.H.F., 1995. Livro Primeiro do Governo do Brazil, 1607–1633. *Journal of the Ordnance Society* 7, 18–26

Barnard, F.P., 1914. Henry the Eighth's navy. In Elton, O. (ed.), *A Miscellany. Presented to John Macdonald Mackay, LL.D.* 132–41. Liverpool

Barret, R., 1598. *The Theorike and Praktise of Moderne Warres*. London

Beard, C.R., 1940. Girdles, shoulder-belts and scarves. *The Connoisseur* 106(472), 249–54

Bevan, J., 1996. *The Infernal Diver*. London: Submex

Biringuccio, V., 1540. *The Pirotechnia of Vannoccio Biringuccio* (eds Smith & Gnudi 1990). Mineola NY: Dover

Blackmore, H.L., 1965. *Guns and Rifles of the World*. London: Batsford

Blackmore, H.L., 1976. *The Armouries of the Tower of London. 1. Ordnance*. London: HMSO

Blackmore, H.L., 1988. Master Jacobo's culverin. *Arms and Armour Society Journal* 12(5), 312–44

Blackmore, H.L., 1999. *Gunmakers of London Supplement 1350–1850*. Bloomfield, Ontario & Alexandra Bay, NY: Museum Restoration Service

Blair, C., 1954. A royal cutler's bill of 1547. *Journal of the Arms and Armour Society* 1(6), 104–11

Blair, C., 1958. *European Armour circa 1066 to circa 1700*. London: Batsford

Blair, C., 1962. *European & American Arms c 1100–1850*. London: Batsford

Blair, C., 1968. *Pistols of the World*. London: Batsford

Blair, C., 1979. *European Armour circa 1066 to circa 1700* (2nd edn). London: Batsford

Blair, C., 1981. The early basket-hilt in Britain. In Caldwell, D. (ed.), *Scottish Weapons and Fortifications*, 153–252. Edinburgh: John Donald

Blair, C. (ed.), 1983. *Pollard's History of Firearms*. Feltham: Country Life Books

Blair, C., 1998. The sword catalogue. In Blair, C. (ed.), *The Crown Jewels: the history of the Coronation Regalia in the Jewel House of the Tower of London* vol. 2, 317–83. London: HMSO

Blair, C., 1999. *Henry VIII Inventory: Arms and Armour*. Unpublished manuscript

Boccia, L.G., 1982. *Le Armature di. S. Maria delle Grazie di Curtatone di Mantova, e l'Armatura Lombarde del '400*. Milan

Bourne, W., 1587. *The Arte of Shooting at Great Ordnaunce*. Amsterdam: Da Capo (1969 facsimile)

Boyer, P., 1989. Artillerie et tactique navale en Méditerranée au XVIe siècle. *Revue Historique des Armées* 174, 110–21

Boylston, A., Novak, S., Sutherland, T., Holst, M. & Coughlan, J., 1997. Burials from the battle of Towton. *Royal Armouries Yearbook* 2, 36–9

Bradbury. J., 1999. *The Medieval Archer*. New York: Barnes & Noble

Bradley, C., n.d.. *Preliminary Analysis of the Staved Container Remains Recovered from the 1981 Underwater Excavations at Red Bay*. Ottowa: Parks Canada Microfiche report 260

Brewer, J., Gairdner, J. and Brodie, R.H. (eds), 1862–1932. *Letters and Papers, Foreign and Domestic of the Reign of Henry VIII* [cited as LP with entry number]

Brooker, R., 2007. *Landeszeughaus Graz*. Austria: Radschloss Sammlung, Graz

Brown, K.L., 1999. *Buckles from the Mary Rose*. Unpublished undergraduate dissertation, University of Bradford

Brown, R.L., Brown, H.F. & Hinds, A.B (eds), 1864. *Calendar of State Papers and Manuscripts, relating to English affairs existing in the archives and collections of Venice, and in other Libraries of Northern Italy*. London: Longman, Roberts & Green, later HMSO

Brown, R.R., 1997. Arms and armour from wrecks: an introduction. In Redknap 1997, 101–9

Brown, R.R., 2005. Troncks, rockets and fiery balls: Military fireworks of the early modern period. *Journal of the Ordnance Society* 17, 25–37

Buchwald, V.F. & Wivel, H., 1998. Slag analysis as a method for the characterisation and provenancing of ancient iron objects. *Materials Characterisation* 40, 73–96

Burgess, M., 1953. The mail maker's technique. *Antiquaries Journal* 33, 48–55

Bury, M., 2001. *The Print in Italy 1550–1620*. London: British Museum Press

Bytharne, J., 1543. Book of War by Sea and Land. In Laughton, J.K. (ed.), *The Naval Miscellany* 1, 1–21. London: Navy Records Society 20 (1902)

Campbell, J., 1991. *Castings*. Oxford: Butterworth-Heinemann.

Capo Bianco, A., 1598. *Corona e palma militare di Artiglieria. Nella quale si tratta dell'inventione di essa, e dell'operare nelle fattioni da Terra, e Mare, fuochi artificiati da Giuoco, e Guerra; & d'un nuovo Instrumento per misurare distanze...* Venice

Carpegna, N. di., 1997. *Brescian Firearms from Matchlock to Flintlock: a compendium of names. marks and works together with an attempt at classification.* Rome: Edizioni de Luca

Carr Laughton, L.G., 1957. Clove-board. *Mariner's Mirror* 43, 247–9

Carr Laughton, L.G., 1960. Early Tudor ship-guns. *Mariner's Mirror* 46, 242–85

Caruana, A.B., 1992. *Tudor Artillery 1485–1603.* Bloomfield, (Ontario): Historical Arms Series 30

Cataneo, G., 1584. *Dell'arte militare libri cinque, ne' quali si tratta il modo di fortificare offendere, et diffendere una fortezza: et l'ordine come si debbano fare gli alloggiamenti campali; & formare le battaglie, con l'essamine de' bombardieri, & di far fuochi arteficiati.* Brescia

Cerezo Martínez, R., 1983. La táctica naval en el siglo XVI. *Revista de Historia Naval* 1(2), 29–61

Chaves, A. de., 1530. *The Seaman's Glass*

Cipolla, C.M., 1965, *Guns and Sails in the Early Phase of European Expansion.* London: Collins

Claerbergen, E.V. van, 1999. *The Portrait of Sir John Luttrell: a Tudor mystery.* London: Courtauld Institute

Clowes, W.L., 1897–1903. *The Royal Navy: a history from the earliest times to the present.* London: Sampson Low, Marston & Co

Coates, J.F., 1985. Flower of the fleet. A naval architect's view of the *Mary Rose. Naval Architect*, June 1985, E280–1, 285

Collado, L., 1641. *Prattica Manuale dell'Artiglieria, opera historica, politica, e militare. Composta da Luigi Colliado Ingegnero del Real Essercito di S. Maesta Catolica in Italia.* Milan

Colvin, H.M., 1963–82. *The History of the King's Works.* Oxford: HMSO

Contamine, P., 1984. *War in the Middle Ages* (trans. Jones). Oxford: Blackwell

Cooper, W.D., 1864. Notices of Winchlesea in and after the fifteenth century. *Sussex Archaeological Collections* 8, 201–34

Corbett, J.S., 1898. *Drake and the Tudor Navy.* London

Corbett, J.S. (ed.), 1905. *Fighting Instructions 1530–1816.* London: Navy Records Society 29 (reprinted 1971)

Cowgill, J., de Neergaard, M. & Griffiths, N., 1987. *Knives and Scabbards.* Medieval finds from Excavations in London 1. London: HMSO

Credland, A.G., 1982. The medieval war arrow. *Journal of the Society of Archer-Antiquaries* 25, 28-35

Cripps-Day, F.H., 1922. On armour preserved in English Churches. In Laking 1920–2, vol. 5, 149–273

Cripps-Day, F.H., 1944. *Fragmenta Armamentaria, Vol. 1. An Introduction to the Study of Greenwich Armour. Part III. Documentary Evidence.* Privately Printed,

Cruickshank, C., 1990. *Henry VIII and the Invasion of France.* Stroud: Alan Sutton

Dasent, J.R. (ed.), 1890–1907. *Acts of the Privy Council of England*, new series. London

Davies, C.S.L., 1962. Naval discipline in the early sixteenth century. *Mariner's Mirror* 48, 223–4

Davies, C.S.L., 1965. The administration of the Royal Navy under Henry VIII: the origins of the Navy Board. *English Historical Review* 80, 268–88

Davies, J., 2001. Military archery and the Inventory of King Henry VIII. *Journal of the Society of Archer-Antiquaries*, 44, 31–8

Davies, J., 2005. 'We do fynde in our countre great lack of bowes and arrows': Tudor military archery and the Inventory of King Henry VIII. *Journal of the Society for Army Historical Research* 83, 11–29

Davenport, T.G. & Burns, R., 1995. A sixteenth century wreck off the island of Alderney. In Bound, M. (ed.), *The Archaeology of Ships of War*, 30–40. Oswestry: International Maritime Archaeology 2

Dean, B., 1929. *Catalogue of European Daggers.* New York: Metropolitan Museum of Art

De Jonge, J.C., 1858–62. *Geschiedenis van het Nederlandsche Zeewesen* (2nd edn). Haarlem

Derricke, J., 1851. *The Image of Irelande, with a Discouerie of Woodkarne* (with notes of Sir Walter Scott) (ed. Small, Edinburgh 1883)

DeVries, K., 1990. A 1445 reference to shipboard artillery. *Technology & Culture* 31, 818–29

DeVries, K., 1998. The effectiveness of fifteenth-century shipboard artillery. *Mariner's Mirror* 84, 398–99

Du Bellay, M., 1569. *Memoirs ... de plusieurs choses advenues au Royaume de France depuis l'an MDIII jusques au trespass du roy François premier* (reprinted in *Memoirs de Martin et Guillaume du Bellay*, ed. Bourrilly & Vindre, Paris 1908–19)

Dunham, R.J., 1988. *Lead Shot from the* Mary Rose. Unpublished thesis, University of Surrey

Edwards, I. & Blair, C., 1982. Welsh bucklers. *Antiquaries Journal* 62, 74–115

Egan, G. & Pritchard, F., 1991. *Dress Accessories c. 1150–c. 1450.* London: Medieval Finds from Excavations in London

Eldred, W., 1646. *The Gunner's Glasse.* London: Robert Boydel

Elkerton, A., 1983. *Notes on Arrows (80A764) From Chest in U7.* Portsmouth: Mary Rose Trust, archive report

Elmer, R., 1946. *Target Archery.* New York: Alfred A. Knopf

Elmer, R., 1952. *Target Archery* (2nd edn). London: Hutchinson

Ellmers, D., 1994. The Cog as cargo carrier. In Gardiner, R. & Unger, R.W. (eds), *Cogs, Caravels and Galleons*, 29–46. Chatham: Conway Maritime Press

Emmison, F.G., 1978. *Elizabethan Life: Wills of Essex Gentry Merchants.* Chelmsford: Essex County Council

Emmison, F.G., 1983. *Essex Wills (England), Volume 2 1565–1571.* Boston, Mass.: New England Historic Genealogical Society

Evans, J. & Card, M., 1984. *Analysis of Glue from the Mary Rose Livery Arrows.* North-East London Polytechnic: unpublished report

Evans, J.X. (ed.), 1972. *The Works of Sir Roger Williams.* Oxford: University Press

Farmer, R.H., 1972. *Handbook of Hardwoods* (2nd revised edn). Building Research Establishment, Princes Risborough Laboratory. London: HMSO

Fernández Duro, C., 1895–1903. *Armada Española desde la unión de los reinos de Castilla y de León*. Madrid

Fokkens, H., Achterkamp, Y. & Kuipers, M., 2008. Bracers or bracelets? About the functionality and meaning of Bell Beaker wrist-guards. *Proceedings of the Prehistoric Society* 74, 109–40

Forshell, H., 1984. *Bronze Cannon Analysis: alloy composition related to corrosion picture*. Stockholm: Armémuseum, Rapport 2

Forshell, H., 1992. *The Inception of Copper Mining in Falun*. Stockholm: Archaeological Research Laboratory, Stockholm, University Theses and Papers in Archaeology B:2

Friel, I., 1995. *The Good Ship: ships, shipbuilding and technology in England 1200–1520*. London: British Museum Press

Gairdner, J., 1904. *The Paston Letters*. Edinburgh: John Grant

Gallois, R.W. & Edmunds F.H., 1965. *British Regional Geology The Wealden District* (4th edn). London: HMSO

Geibig, A., 2001. Pyrotechnic devices from Coburg Castle. *Royal Armouries Yearbook* 6, 88–97

Grancsay, S.V., 1966. *The Illustrated Inventory of the Arms and Armor of Emperor Charles V*. Madrid

Green, J.N., 1989. *The Loss of the AVOC Retourschip Batavia, Western Australia 1629. An Excavation Report and Catalogue of Artefacts*. Oxford: British Archaeological Report S489

Green, J.N., 1990. Further information on gunners rules or tally sticks. Journal of the Ordnance Society 2, 25–32

Green, J. & North, N., 1984. Further comment on the handguns ex Association. *International Journal of Nautical Archaeology* 13(4), 334–7

Green, H.J.M. & Thurley, S.J., 1987. Excavations on the west side of Whitehall, 1960–2 Part 1: From the building of the Tudor palace to the construction of the modern offices of state. *Transactions of the London & Middlesex Archaeological Society* 38, 59–130

Guérout, M., Rieth, E. & Gassend, J., 1989. Le Navire Génois De Villefranche un naufrage de 1516? *Archaeonautica* 9. Paris: CNRS

Guilmartin, J.F. jnr, 1971. The early provision of artillery armament on Mediterranean war galleys. *Mariner's Mirror* 59, 257–80

Guilmartin, J.F. jnr, 1974. *Gunpowder and Galleys: changing technology and Mediterranean warfare at sea in the sixteenth century*. Cambridge: University Press

Guilmartin, J.F. jnr, 1988. Early modern naval ordnance and European penetration of the Caribbean: the operational dimension. *International Journal of Nautical Archaeology* 17(1), 31–53

Hakluyt, E., 1598–1600. *The Principal Navigations, Voyages, Traffiques, and Discoveries of The English Nation* (ed. Raleigh, Glasgow 1903–5)

Hall, E., 1548. *The Union of the Two Noble and Illustre Families of Lancastre and Yorke* (reprinted ed. Ellis as *Hall's Chronicle*, London 1909)

Hall, N., 1998. Building and firing a *Mary Rose* port piece. *Royal Armouries Yearbook* 3, 57–67

Hall, N., 2001. Casting and firing a *Mary Rose* culverin. *Royal Armouries Yearbook* 6, 106–1

Hardy, R., 1976. *Longbow: a social and military history*. Cambridge: Patrick Stevens

Hardy, R., 1992. *Longbow: a social and military History* (3rd edn). Sparkford: Patrick Stevens

Hartley, P., 1986. Further speculation on the nature of bowstrings. *Journal of the Society of Archer-Antiquaries* 29, 14–15

Hartsthorne, A., 1891. The sword belts of the Middle Ages. *Archaeological Journal* 48, 320–40

Heath, E.G., 1980. *Archery: a military history*. London: Osprey

Held, R., 1957. *The Age of Firearms*. New York

Hexham, H., 1643. *The Third Parto of Principles of the Art Military* III. Rotterdam

Hickman, C.N., Nagler, F. & Klopsteg, P.E., 1992. *Archery: the technical side*. Milwaukee: Derrydale Press

Hildred, A., 1997. The material culture of the *Mary Rose* (1545) as a fighting vessel: the uses of wood. In Redknap (ed.) 1997, 51–72

Hildred, A. & Rule, M., 1984. Armaments from the *Mary Rose*. *Antique Arms & Militaria* May 1984, 17–24

Hildred, A., 2003. A gunner's rule from the *Mary Rose*. *Journal of the Ordnance Society* 15, 29–40

Hodgkinson, J.S., 2000. Gunfounding in the Weald. *Journal of the Ordnance Society* 12, 31–47

Holinshed, R., 1577. *Chronicle* (reprinted ed. Ellis as *Holinshed's Chronicles*, London 1807–8)

Hoff, A., 1978. *Dutch Firearms*. London

Hoff, A., 1963. *Late Firearms with Snap Matchlocks. Four Studies on History of Arms*. Copenhagen: Tøjhusmuseets Skrifter 7.

Holt, P., 2004. The application of the Fusion Positioning System to marine archaeology. In Akal, T. (ed.), *Congress on The Application of Recent Advances in Underwater Detection and Survey Techniques to Underwater Archaeology*, 231–8. Bodrum. Uluburun

Holt, P., 2007. *The Site Recorder Database Schema* http://www.3hConsulting.com/Research/research_schema.htm

Hope W.H. St J., 1919. *Cowdray and Easebourne Priory in the County of Sussex*. London: Country Life Books

Hopkins, M., 1998. *Tudor Longbow Arrows; a comparative study of* Toxophilus *by Roger Ascham and the arrow collection from the site of the* Mary Rose. Winchester, King Alfred's University College, unpublished undergraduate dissertation

Hopkins, M, 1999. *Arrows analysis*. Portsmouth: Mary Rose Trust archive dataset

Horsey, S., 1842. *Narrative of the Loss of the* Mary Rose. Portsea

Hutchinson, G., 1994. *Medieval Ships and Shipping*. London: Leicester University Press

Jackson, A,. 2003. *Arrow Analysis and Report*. Portsmouth: Mary Rose Trust archive dataset and report

Jackson, A., 2005. *Arrow Report (revised)*. Portsmouth: Mary Rose Trust archive report

Jackson, J.E., nda. *MS Notes for a History of the Hungerford Family*. Unpublished Manuscript Devizes Museum

Jackson, J.E., ndb. *Farleigh Hungerford MS Collections*. Unpublished Manuscript, Devizes Museum

Jessop, O., 1996. A new artefact typology for the study of medieval arrowheads. *Medieval Archaeology* 40, 192–205

Kaestlin, J.P., 1963. *Catalogue of the Museum of Artillery in the Rotunda at Woolwich. Part 1. Ordnance*. London: HMSO

Karcheski, W.J. & Richardson, T., 2000. *The Medieval Armour from Rhodes*. Leeds: Royal Armouries

Kaye, G.W.C. & Laby, T.H., 1995. *Tables of Physical and Chemical Constants* (16th edn). Harlow: Longman

Kelland, N., 1976. A method for carrying our accurate planimetric surveys underwater. *Hydrographic Journal* 2, 17–31

Kendall, K. 1981. *Analysis of a Powder Scoop from the Mary Rose*. Portsmouth Polytechnic: unpublished BSc dissertation

Kennard, A.N., 1986. *Gunfounding and Gunfounders: a directory of cannon founders from earliest times to 1850*. London: Arms & Armour Press

Kenyon, J.R., 1982. Ordnance and the King's fortifications in 1547–48. Society of Antiquaries MS. 129, Folios 250–374. *Archaeologia* 107, 165–213

Knighton, C.S., 1981. *Catalogue of the Pepys Library at Magdalene College, Cambridge 5.2, Modern Manuscripts*, Woodbridge

Knighton, C.S., 2004. A century on: Pepys and the Elizabethan Navy. *Transactions of the Royal Historical Society* 6th series 14, 141–51

Knighton, C.S. & Hildred, A.M.V., forthcoming. Overgunning the *Mary Rose*: the King was warned. *Journal of the Ordnance Society*

Knighton, C.S. & Loades, D.M., 2000. *The Anthony Roll of Henry VIII's Navy*. Aldershot: Navy Records Society Occasional Publication 2

Knighton, C.S. & Loades, D.M., 2002. *Letters from the Mary Rose*. Stroud: Sutton [cited as *LMR with entry number*]

Knighton, C.S. & Loades D., forthcoming. *The Navy of Edward VI and Mary I*. Aldershot: Navy Records Society

Knighton, C.S. & Wilson, T., 2001. Serjeant Knight's discourse on the cross and flags of St George (1678). *Antiquaries Journal* 81, 351–90

Kooi, B.W., 1983. *The Mechanics of the Bow and Arrow*. Groningen Ryksuniversitet: unpublished PhD thesis

Kooi, B.W. & Bergmann, C.A., 1997. An approach to the study of ancient archery using mathematical modelling. *Antiquity* 71, 124–34

Konstam, A., 1989. A gunner's rule from the 'Bronze Bell' wreck, Tal-y-bont, Gwynedd. *Journal of the Ordnance Society* 1, 23–6

Kramer, G.W., 2001. *The Firework Book: Gunpowder in Medieval Germany* (trans. Leibnitz). London: Arms & Armour Society

Lad, J., 1586. *A Rule of Gunners Art*, notebook

Laking, G.F., 1920–2. *A Record of European Armour and Arms Through Seven Centuries*. London: Bell

Lane, F.C. 1969. The crossbow in the nautical revolution of the Middle Ages. *Explorations in Economic History* 7, 161–71

La Roncière, C.M. 1899–1932. *Histoire de la Marine Française*. Paris

Laughton A.H.T., 1989. *Lead Shot from the Mary Rose*. University of Surrey: unpublished BSc thesis

Lavery, B., 1987. *The Arming and Fitting of English Ships of War 1600–1815*. London: Conway Maritime Press

Lavery, B., 1989. *Nelson's Navy*. London: Conway Maritime Press

Lehmann, L.Th., 1984. *Galleys in the Netherlands*. Amsterdam

Levenson, J.A., Oberhuber, K. & Sheenan, L., 1973. *Early Italian Engravings from the National Gallery of Art*. Washington: Koedler

Liberson, F., 1937. Os acromiale – a contested anomaly. *Journal of Bone & Joint Surgery* 19, 683–9

*LMR see* Knighton and Loades 2002

Loades, D., 1992. *The Tudor Navy*. Aldershot

Lloyd, C. & Thurley, S., 1990. *Henry VIII: images of a Tudor king*. Oxford: Phaidon

London Museum, 1940. *Medieval Catalogue*. London: London Museum

*LP = see* Brewer *et al.* (eds), 1862–1932

Lucar, C., 1588. *Translation of Tartaglia, Three Bookes of Colloquies Concerning the Arte of Shooting in Great and Small Peeces of Artillerie with appendix*. London

MacGregor, A., 1983. *Tradescant's Rarities*. Oxford: Essays of the Foundation of the Ashmolean Museum

McElvogue, D.M., 1999. Ordnance, shot and artefacts from a sixteenth century wreck off Alderney. *Journal of the Ordnance Society* 11, 1–17

McKee, A. 1973. *King Henry VIII's Mary Rose*. London: Souvenir

McKee, A., 1982. *How we Found the Mary Rose*. London: Souvenir

McKee, A., 1986. The Mary Rose's complement. *Mariner's Mirror* 72, 74

Machiavelli, N., trans P. Whithorn, 1560–2. The Arte of Warre (Certain Waies for the Orderyng of Souldiers in Battleray). Amsterdam: Da Capo (1969 facsimile)

Mackie, J.D., 1951. The English army at Flodden. *Miscellany of the Scottish History Society* 8, 33–85

Mainwaring, Sir H., 1644, *The Seaman's Dictionary*. In G.E. Manwaring & W.G. Perrin, 1922. *The Life and Works of Sir Henry Mainwaring* Vol 2, 83–260. London: Navy Records Society

Majewski, A. 1986. *Analysis of a Metal Artefact from the Mary Rose*. Portsmouth Polytechnic: unpublished BSc dissertation

Mancini, D., 1483. *The Usurpation of Richard the Third*. (1969 edn, ed. Armstrong) Oxford: Clarendon Press

Mann, J.G., 1940. A tournament helm in Melbury Sampford Church. *Antiquaries Journal* 20, 368–79

Mann, J.G., 1962. *Wallace Collection Catalogues: European Arms and Armour*. London: Trustees of the Wallace Collection

Marks, R. & Williamson, P. 2003. *Gothic: Art for England 1400–1547*. London: Victoria & Albert Museum

Martin, C.J.M., 1972. *El Gran Grifon*. An Armada wreck on Fair Isle. *International Journal of Nautical Archaeology* 1, 59–71

Martin, C.J.M., 1994. Incendiary weapons from the Spanish Armada wreck *La Trinidad Valencia*, 1588. *International Journal of Nautical Archaeology* 23, 207–17

Martin, P., 1997. *Balls! Stone shot from the* Mary Rose. University of Southampton: unpublished MSc thesis

Matron, S., 1982. *Gunpowder Samples from the* Mary Rose. Portsmouth Polytechnic: unpublished BSc dissertation

Mazansky, C., 2005. *British Basket-Hilted Swords.* Woodbridge: Boydell/Royal Armouries Museum

Megson, B., 1993. *Such Goodly Company*. Privately printed for the Worshipful Company of Bowyers

Metcalf, S., North, R.E. & Balfour, D., 2006. The conservation of a gun-shield from the arsenal of Henry VIII. In Smith, R.D (ed.), *Make All Sure: the conservation and restoration of arms and armour*, 76–91. Leeds: Basiliscoe

Miles, A.E.W., 1994. Non-union of the epiphysis of the acromion in the skeletal remains of a Scottish population of ca. 1700. *International Journal of Osteoarchaeology* 4, 149–63

Mollat du Jourdin, M., 1984. Le livre des 'faiz de la Marine et navigages' d'Antoines de Conflans. In Mollat du Jourdin, M. & Chillaud-Toutée, F. (eds), *107e Congrès Nationale des Sociétés Savantes, Collection d'Histoire Maritime (Brest)*, 9–44. Paris: CTHS

Mollat du Jourdin, M., 1992. 'tre roi sur la mer': naissance d'une ambition. In Contamine, P. (ed.), *Histoire Militaire de la France* 1, 279–301. Paris

Monaghan, J. & Bound, M., 2001. *A Ship Cast Away About Alderney. Investigations of an Elizabethan Shipwreck*. Alderney: Alderney Maritime Trust

Morin, M. & Held, R., 1980. *Beretta: la dinastia industriale più antico al mondo*. Chiasso: Acquafresca Editrice L.

Morris, J.E., 1901. *The Welsh Wars of Edward I*. Oxford: Clarendon

Morris, R., 1984. The 15th century hand-gun ex *Association*. *International Journal of Nautical Archaeology* 13(1), 76–7

Mosley, W.M., 1792. *An Essay on Archery*. Worcester: J. & J. Hall

Murphy, S., 2001. The technology of casting cannon. *Journal of the Ordnance Society* 13, 73–95

Muscat, J., 2000. *The Carrack of the Order*. Malta: A.C. Aquilina

Muskett, R.J., 1987. *Determination of the Origin and Manufacturing Techniques of the Guns of the* Mary Rose. University of Surrey: unpublished BSc Dissertation

Neergaard, M. de, 1987. The decoration of medieval scabbards, in Cowgill *et al.* 1987, 41–3

Nichols. J.G. & Jackson, J.E., 1862. Inventory of the Goods of Dame Agnes Hungerford Attainted of Murder, 14 Henry VIII. *Archaeologia* 38, 369–79

Northover J.P. & Rychner, V., 1998. Bronze analysis, experience of a comparative programme. In Mordant, C., Pernot, M. & Rychner, V. (eds), *L'Atelier du bronzier en Europe du XXᵉ au VIIIᵉ siècle avant Notre Ère 1 (Session de Neuchâtel)*, 19–40. Paris: Éditions du CTHS

Norman, A.V.B., 1980. *The Rapier and Small Sword 1460–1820*. London: Ayer

Norman, A.V.B. & Wilson, G.M., 1982. *Treasures from the Tower of London*. London: Arms and Armour Press

Norton, R., 1628. *The Gunner. Shewing the Whole Practise of Artillerie*. London

OED = Oxford English Dictionary

Olds D.L., 1976. *Texas Legacy from the Gulf. A Report on C 16th Shipwreck Materials Recovered from the Texas Tidelands*. Austin: Texas Memorial Museum. Miscellaneous Paper 5

Olesa Muñido, F-F., 1968. *La organización naval de los estados Mediterráneos y en especial de España durante los siglos XVI y XVII*. Madrid

Olesa Muñido, F-F., 1971. *La galera en la navegación y el combate*. Madrid

Oliveira, F., 1937. *A arte da guerra do mar*, ed. Quirino da Fonseca and Botelho de Sousa (Lisbon 1937, originally Coimbra 1555)

Oman, C., 1963. *Medieval Silver Nefs*. London

Oman, C.W.C, 1937. *A History of the Art of War in the Sixteenth Century*. London: Methuen

Oppenheim M., 1896a. *Naval Accounts and Inventories of the Reign of Henry VII, 1485–8 and 1495–7*. London: Navy Records Society 8

Oppenheim, M., 1896b. *A History of the Administration of the Royal Navy and of Merchant Shipping in Relation to the Navy from 1509 to 1660*. Aldershot: Temple Smith (1988 reprint)

Oudendijk, J.K., 1941. *Een Bourgondisch Ridder over ten Oorlog ter Zee: Philips van Kleef als Leermeester van Karel V*. Amsterdam: Nederlandsche Akademie van Wetenschappen, Commissie voor Zeegeschiedenis, Werken 7

Oxley, J.E., 1968. *The Fletchers and Longbowstringmakers of London*. London: Unwin

Parker, G., 1996. The Dreadnought Revolution of Tudor England. *Mariner's Mirror* 82, 269–300

Partington, J.R., 1960. *A History of Greek Fire and Gunpowder*. Cambridge: Heffer

Payne, A., 2000. An artistic survey. In Knighton & Loades 2000, 20–7

Paviot, J., 1995. *La politique navale des ducs de Bourgogne, 1384–1482*. Lille

Peterson, H.L., 1956. *Arms and Armor in Colonial America 1526–1783*. Harrisburg PA: Stackpole

Peterson, H.L., 1969. *Round Shot and Rammers. An Introduction to Muzzle-loading Land Artillery in the United States*. New York: Bonanza

Peterson, H.L., 1970. *Daggers and Fighting Knives of the Western World from the Stone Age till 1900*. New York: Bonanza

Pisan C. de., 1489. *The Boke of the Fayt of Armes and of Chyvalrye*. Amsterdam: Theatrum Orbis Terrarum (translated Caxton, 1489, facsimile edition 1976)

Pitt, E., 1989. Short *Report on the Analysis of Non Ferrous Based Alloy Artefacts Salvaged from the* Mary Rose. Coventry Polytechnic: unpublished report

Pope, S., 1925. *Hunting with the Bow and Arrow*. New York: G.P. Putnams

Porten, E. von der, 1979. *The Invention of the Gunport.* Portsmouth: Mary Rose Trust archive report

Pratt, P.L., 1992. Testing the bows. In Hardy 1992, 205–17

Preece, C. & Burton, S., 1993. Church Rocks, 1975–83: a reassessment. *International Journal of Nautical Archaeology* 22, 257–65

Prestwich, M., 1996. *Armies and Warfare in the Middle Ages: the English experience.* Newhaven & London: Yale University Press

Probst, N., 1993. Nordeuropaeisk spanteopslagning 1500-og 1600-tallet. Belyst ud fra danske kilder. *Maritim Kontakt* 16, 7–42

Puype, J.P., 2001. Arms, munitions and artillery equipment from a shipwreck of the early 1590s. *Royal Armouries Yearbook* 6, 117–28

Pyhrr, S.W., 1984. Some elements of armour attributed to Niccolo Silva. *Metropolitan Museum Journal* 18, 111–21

QinetiQ Ltd., 2001. *Replica Gun Trial.* Shoeburyness: unpublished report

Quinn, R., Bull, J.M., Dix, J.K. & Adams, J.R., 1997. The Mary Rose site – geophysical evidence for palaeo-scour marks. *International Journal of Nautical Archaeology* 26, 3–16

Randall, J., 1985. *Preliminary Report on Mary Rose Arrows studied from 3.1.85–8.3.85.* Portsmouth: Mary Rose Trust archive report

Redknap, M., 1984. *The Cattewater Wreck: the investigation of an armed vessel of the early sixteenth century.* Oxford: British Archaeological Report 131/National Maritime Museum Archaeological Series 8

Redknap, M. (ed), 1997. *Artefacts from Wrecks. Dated Assemblages from the Late Middle Ages to the Industrial Revolution.* Oxford: Oxbow Monograph 84

Richardson, T., 1997. The Bridport muster roll of 1457. *Royal Armouries Yearbook* 2, 46–52

Richardson, T., 1998. Ballastic testing of historical weapons. *Royal Armouries Yearbook* 3, 50–3

Richardson, T., 2000. Armour in the church of St. Lawrence, Hatfield, South Yorkshire. *Royal Armouries Yearbook* 5, 11–18. Leeds: Royal Armouries

Richardson, T., 2002. *The Armour and Arms of Henry VIII.* Leeds: Royal Armouries

Riederer, J., 1972. Metallanalysen von Geschützbronzen. *Waffen und Kostümkunde* 14, 49–56

Riederer, J., 1977. Die Zusammensetzung der Bronzegeschütze des Heeresgeschicghtlichen Museums in Wiener Arsenal. *Berliner Beiträge zur Archäometrie* 2, 27–40

Rimer, G., 1998. An ancient quiver. *Royal Armouries Yearbook* 3, 53–7

Roberts J., 1979. *Holbein.* London: Oresko

Roberts, T., 1801. *The English Bowman.* London

Rodger, N.A.M., 1996. The development of broadside gunnery, 1450–1650. *Mariner's Mirror* 82, 301–24

Rodger N.A.M., 1997. *The Safeguard of the Sea. A Naval History of Great Britain. Volume 1 660–1649.* London: HarperCollins

Rodríguez-Salgado, M.J. (ed.), 1988. *Armada 1588–1988. An International Exhibition to Commemorate the Spanish Armada. The Official Catalogue.* Harmondsworth: Penguin

Rogers, T.D., 1980, *Sir Frederic Madden at Cambridge.* Cambridge: Cambridge Bibliographical Society Monograph 9

Rohl, B. & Needham, S., 1998. *The Circulation of Metal in the British Bronze: the application of lead isotope analysis.* London: British Museum Occasional Paper 102

Ross, L., 1980. *16th-century Spanish Basque Coopering Technology: A report of the staved containers found in 1978-79 on the wreck of the whaling galleon San Juan, sunk in Red Bay, Labrador, AD 1565.* Ottawa: Parks Canada Manuscript Report 408

Routh P. & Knowles, R., 1983. *The Medieval Monuments of Harewood.* Wakefield

Rule, M., 1982. *The Mary Rose: the excavation and raising of Henry VIII's Flagship.* London: Conway Maritime Press

Rule, N., 1989. The Direct Survey Method (DSM) of underwater survey, and its application underwater. *International Journal of Nautical Archaeology* 18, 157–62

St Aubyn, G., 1983. *The Year of Three Kings: 1483.* London: Collins

St John Hennesey, N., 1991. The Dublin breach-loading swivel gun. *Journal of the Ordnance Society* 3, 1–3

Sachs, H. & Amman, J., 1568. *The Book of Trades: Ständebuch.* New York: Dover Press (1973 facsimile)

Scarisbrick, J.J., 1997. *Henry VIII* (2nd edn). New Haven & London: Yale University Press

Scott, A.F., 1976. *Everyone a Witness: The Tudor Age.* New York: Thomas Crowell

Schubert, H.R., 1957. *History of the British Iron and Steel Industry from c. 450 B.C. to A.D. 1975.* London: Routledge & Keegan Paul

Sicking, L., 1998. *Zeemacht en onmacht: Maritieme politiek in de Nederlanden 1488–1558.* Amsterdam: De Bataafsche Leeuw

Sicking, L., 1999. Meesters van de zee: Maritieme samenwerking tussen Karel V en Hendrik VIII. *Tijdschrift voor Zeegeschiedenis* 18, 3–12

Simmons, J.J., 1988. Wrought iron ordnance: revealing discoveries from the New World. *International Journal of Nautical Archaeology* 17(1), 25–34

Simmons, J.J., 1989. Lidded-breech wrought-iron swivel gun at Southsea Castle, Portsmouth, England. *Journal of the Ordnance Society* 1, 63–7

Simmons, J.J., 1991. Replication of early 16th century shot-mould tongs. *Journal of the Ordnance Society* 3, 5–10

Sinclair, A., (ed.), 2003. *The Beauchamp Pageant.* Donington: P. Watkins/Richard III & Yorkist History Trust

Sloot, R.B.F., van der, 1985. Het belang van de wapenfabricage in de Nederlanden gedurende de 16e en 17e eew, met bijzondere aandacht voor harnassen en helmen. *CL Themadag* 8, 5–17

Smith, J., 1627. *A Sea Grammar.* New York: De Capo (1968 facsimile)

Smith, R.D., 1988. Towards a new typology for wrought iron ordnance. *International Journal of Nautical Archaeology* 17(1) 5–16

Smith, R.D., 1997. A 16th-century bronze cannon from London. *Royal Armouries Yearbook* 2, 107–12

Smith, R.D., 2000. Casting shot in the 16th century. *Journal of the Ordnance Society* 12, 88–93

Smith, R.D., 2001. A carriage and cast-iron cannon at Windsor Castle. *Journal of the Ordnance Society* 13, 25–38

Smith, R.D. & Brown, R.R., 1989. *Bombards Mons Meg and her Sisters*. Leeds: Royal Armouries Monograph 1

Smythe, Sir J., 1590. *Certain Discoveries Concerning the Formes and Effects of Divers Sorts of Weapons*. London

Soar, H.D.H., 2003. *Of Bowmen and Battles*. Tolworth: Glade

Soar, H.D.H., 2004. *The Crooked Stick: a history of the longbow*. Yardley: Westholme

Soar, H.D.H., with Gibbs, J., Jury, C. & Stretton, M., 2006. *Secrets of the English Warbow*. Yardley: Westholme

Southwick, L., 2006. Robert South 'His Majesties Cutler' and aspects of the sword trade in London in the Seventeenth Century. *Journal of the Arms & Armour Society* 18(5), 196–231

Spont, A., 1897. *Letters and Papers Relating to the War with France of 1512–1513*. Navy Records Society 10

Starley, D., 1989. appendix 2: metallography. In Smith & Brown 1989, 90–7

Starley, D., 2005. What's the point? A metallurgical insight into medieval arrowheads. In Bork, R. (ed.), *De Re Metallica: the uses of metal in the Middle Ages*, 207–18. Aldershot: Ashgate

Starley, D. & Hildred, A., 2002. Technological examination of a *Mary Rose* hailshot piece; new evidence for early iron gun casting. *Royal Armouries Yearbook* 7, 139–46

Starkey, D. (ed.), 1998. *The Inventory of King Henry VIII*. London: Harvey Miller

Sterling, J.C., Meyers, M.C., Chesshir, W. & Calvo, R.D., 1995. Os acromiale in a baseball catcher. *Medicine & Science in Sports & Exercise* 27, 795–9

Stirland, A., 1984. A possible correlation between os acromiale and occupation in the burials from the *Mary Rose*. *Proceedings of the 5th European Meeting of the Palaeopathology Association*, 327–34

Stirland, A., 2002. *Raising the Dead. The Skeleton Crew of King Henry VIII's Great Ship, the Mary Rose* (2nd edn). Chichester: Wiley

Stone, P., 1907. *Account of the French Descent on the Isle of Wight, July 1545, under Claude D'Annebault*. Newport: Isle of Wight County Press

Stow, J., 1598. *A Survey of London* (1908 reprint ed. Kingsford, Oxford: Clarendon)

Stow, J., 1603. *The Annales of England, faithfully collected out of the most autenticall authors etc. by John Stow, citizen of London*

Strauss, W.L. (ed.), 1972. *The Complete Engravings, Etchings and Drypoints of Albrecht Dürer*. New York: Dover

Strickland, M. & Hardy, R., 2005. *The Great Warbow*. Stroud: Sutton

Tartaglia, N. 1546. *Quesiti e inventioni diverse*. Venice: Venturino Ruffinelli/Niccolo Tartaglia

Tartaglia, N., 1588. *Nuova Scientia*. Venice (trans Lucar and incorporated into *Three Books of Colloquies*). London: John Harrison (*see also* Lucar 1588)

Teesdale, E.B., 1991. *Gunfounding in the Weald in the Sixteenth Century*. Leeds: Royal Armouries Monograph 2

Terjanian, P., 2006. *The Master of the Roman M under a crescent identified. New light on the career and achievements of the author of two horse bards of Henry VIII in the Royal Armouries*. Unpublished lecture to the Arms & Armour Society, London 6 July 2006

Thomas, B. & Gamber O., 1990. *Kunsthistorisches Museum, Wien, Waffensammlung: Katalog der Leibrüstkammer 2*. Busto Arsizio

Tomalin, D., Cross, J. & Motkin, D., 1988. An Alberghetti bronze minion and carriage from Yarmouth Roads, Isle of Wight. *International Journal of Nautical Archaeology* 17(1), 75–86

Topham, J., 1782. A description of an ancient picture in Windsor Castle, representing the embarkation of King Henry VIII at Dover, May 31, 1520, preparatory to his Interview with the French King Francis I. *Archaeologia* 6, 179–220

Toxophilite, Old, 1833. *The Archer's Guide. By an Old Toxophilite*. London: T. Hurst

Trollope, C., 1994. The guns of the Queen's ships during the Armada Campaign, 1588. *Journal of the Ordnance Society* 6, 22–38

Ufano, D., 1613. *Tratado de la Artilleria y uso della Platicado por el Capitan Diego Ufano el las Guerras de Flandes*. Brussels

Valencia, Conde de Don Juan., 1898. *Catálogo historico-descriptivo de la Real Armeria*. Madrid

Vernon, K., 1987. *Continuing work on a Metal Artefact from the Mary Rose*. Portsmouth Polytechnic: undergraduate dissertation

Wadge, R., 2007. *Arrowstorm. The World of the Archer in the Hundred Years War*. Stroud: Spellmount

Walker, R. & Hildred, A., 2000a. Manufacture and corrosion of lead shot from the flagship *Mary Rose*. *Studies in Conservation* 45(4), 217–25

Walker, R. & Hildred, A., 2000b. Quantitative non-destructive analysis of metals in artefacts. *International Journal of Nautical Archaeology* 29(1), 163–6

Walker, R., Dunham, R., Hildred, A. & Rule, M., 1989. Analytical study of composite shot from the *Mary Rose*. *Journal of the Historical Metallurgy Society* 23, 84–90

Walrond, H., 1894. The arrow. In Longman, C.J. & Walrond, H. (eds), *The Badminton Library Archery*, 304–15. London: Longman

Walton, S.A., 1999. *The Art of Gunnery in Renaissance England*. Univeristy of Toronto: unpublished PhD thesis

Ward Perkins, J., 1940. Arrowheads. *London Museum Medieval Catalogue*, 65–73. London: HMSO

Warnicke, R., 2008. *The Marrying of Anne of Cleves*. Cambridge: Cambridge University Press

Waterer, J.W., 1981. *Leather and the Warrior: an account of the importance of leather to the fighting man from the time of the ancient Greeks to World War II*. Northampton: Museum of Leathercraft

Weast, R.C. (ed.), 1986. *CRC Handbook of Chemistry and Physics* (67th edn). Boca Raton, Florida: CRC Press

Weber, R.E.J., 1982. *De Seinboken voor Nederlandse Oorlogsvloten en Konvoien tot 1690*. Amsterdam: Nederlandsche Akademie van Wetenschappen, Commissie voor Zeegeschiedenis, Werken 15

Werner, A., 1968. *293 Renaissance Woodcuts for Artists and Illustrators: Jost Amman's Kunstbüchlin*. New York: Dover

Wilkinson, F., 2002. *Those Entrusted with Arms*. London: Greenhill

Williams, A.R., 1979. A technical note on some of the armour of King Henry VIII and his contemporaries. *Archaeologia* 106, 157–66

Williams, A.R., 1986. Fifteenth century armour from Cherbourg; a metallurgical study. *Armi Antiche* 32, 3–82

Williams, A.R., 1999. The steel of the Negroli. *Metropolitan Museum Journal* 34, 101–24

Williams, A.R., 2003. *The Knight and the Blast Furnace*. Leiden: Brill

Williams, A.R. & Patterson, A.J.R., 1986. A Turkish bronze cannon in the Tower of London. *Gladius* 23, 185–205

Williams, A.R. & Reuck, A. de, 1995. *The Royal Armoury at Greenwich, 1515–1649*. Leeds: Royal Armouries Monograph 4

Williams, G.H., 1979. *The Western Defences of Portsmouth Harbour 1400–1800*. Portsmouth Papers 30

Williams, Sir R., 1590. *A Briefe Discourse of Warre*. London

Wilson, G.M., 1985. A halberd head from the River Thames. *Park Lane Arms Fair* 2, 15–20

Wilson, G.M., 1986. Notes on some early basket-hilted swords. *Journal of the Arms & Armour Society* 12(1), 1–19

Wilson, G.M., 1988. Some important snap matchlock guns. *Canadian Journal of Arms Collecting* 26(1), 3–10

Wilson, G.M., 1990. Further notes on early basket-hilted swords. *Journal of the Arms & Armour Society* 13, supplement, 23–6

Wilson, G.M., 2002. *The Ambler Sword*. Royal Armouries Yearbook 7, 149–50

Woodgate, J., 1983a. *Arrow Study Report 25.7.83–19.8.83*. Portsmouth: Mary Rose Trust archive report

Woodgate, J., 1983b. *Arrow Study Report 6.10.83–1.12.83*. Portsmouth: Mary Rose Trust archive report

Wright H.P., 1873. *The Story of the* Domus Dei *of Portsmouth*. London: James Parker

Wright, R., 1563. *Gunnery*. London: Society of Antiquaries of London SAL/MS 94

Young, P. & Adair J., 1964. *Hastings to Culloden*. London: Wrens Park

# Index

## by Barbara Hird

NOTE: Locators in *italic* type denote illustrations; those in **bold** type, main references.

2003–5 excavations 931, **934–42**
    ordnance-related finds 935, 937, *938*, 938–9, 940
    sampling of spoil mounds 933, *934*, **935–9**, 936–7
        methodology *934*, 935–8, *937*
    structural timbers found 939–40, *941*

access between decks 895, 896, *898*, 907
accuracy of guns 307
acoustic positioning system (APS) 936, 937
Admiralty organisation 1–2
adze 94, *95*
Agincourt, battle of 587, 631, 818
aiglets 233, 821, 832, 835
    distribution 243, *826*, 835, 847, 848
*alacancias* (firepots) 521
alder
    arrows 618, 669, 674–5, 676, 677, 678, 686
        profile types *680*, 685
    sword/dagger grips 763
Alderney, wreck off 39, 129, 334, 480, 484, 822
*All Hallows Cog* 12–13
alloys *see under* bronze
almain rivets 820, *821*
Altdorfer, Erhard; illustrations of *Landsknechte* 806, 811
alterations *see* rebuilding of *Mary Rose*
ambassadors
    Imperial *see* van der Delft, François
    Venetian *see* Barbaro, Daniel
America, North
    basket-hilted swords 753
    bills 725–6, *726*
Amman, Jost 386, *386*, 806, 809, 811
anchors 165, 940, *940*, 941, 942
Annebault, Claude d', Admiral of France 11–12, 875, 879–80, 880, 881
*Anne Gallant* 580, 581
    guns 74, 199, 225, 300, 301
    munitions 199, 200, 225, 301, 424
Ansty, Sir John, of Stow-cum-Quy, Cambridgeshire; memorial brass 805, *808*
*Antelope* 580, 581
    guns 58, 199, 225, 291, 300, 301, 302
    munitions 199, 311, 331, 424
Anthony, Anthony 2, 3, 7, 9, 48
    *see also Anthony Roll*
Anthony, William 3

*Anthony Roll* **2–7**
    archaeological evidence compared 4–7, 309
    description of manuscript 2–4, *3*
    illustration of *Mary Rose* 2, 3, *3*, 4, 6–7
        castle decks 918, 943, *944*
        fighting tops 272
        grapnel 920, *921*
        gunports
            lidded 6, 14, 138, 885, 886, 887
            Upper Deck 6, 140, *141*
        guns forward of rise to Forecastle 138
        shear hooks 921
        sternchaser guns 138, 235, 885, *886*, 910, 919–20
    importance and accuracy **4–7**
    inventory 2, 3, *3*, 23
        and changes in armament 7, 8
        gunpowder expenditure indicated by 426, 427
        guns 856, 863–4, 882
            classification of bronze 35–6
            discrepancy with assemblage 5–6, 36, 104–5, 309
            firing rate 309
            items for working 385, 409–10, 464, 504, 507
            Main Deck positions as indicated in *141*
            numbers 22, 25, 214, 883
            order of list 202, 214
            preferred types 921–4
    *see also under* archery equipment; arrows; bases; bills; bowstrings; cannon; cartridges; crew; culverins; falcons; forelocks; fowlers; gunpowder; hailshot pieces; halberds; handguns; *Henry Grace à Dieu*; incendiary devices; longbows; port pieces; shot; slings; staff weapons; top pieces
Antwerp 541, 544, 851
APS (Acoustic Positioning System) 936, 937
Arcana family
    cascabel shapes *34*
    gun carriages with jointed beds 40, 42, 43
Arcano, Arcangelo 22, 27, 29
    alloys 946, 950–1, 952, 954, 955
    *see also* cannon (79A1277; 79A1278)
Arcano, Francesco 19, 21, 22
    alloys 21, 946, 951, 952, 954
    guns in *Mary Rose* by *see* cannon (81A3002)
        possible *see* culverins (79A1279; 80A0976)
Arcano, Rafaelo 22

archers
  *Anthony Roll* 581
  armour 818, 821
  chests with possessions *585*, 585, 662–3, 699–702, *700–2*
    wristguards *646*, 647, 650, 703, 712
  livery company affiliations 581, 585, *585*, 639, 712
  locations *578*, 579, **585–6**, 647, 667, 669, 691, 697, 702–3, 704, 895
  physical and mental qualities 629, 631–2, 929
  skeletal remains 629, 632, **702–12**, *705–6*, *707*, *708–11*
    abnormalities 581, 582, 586, 704–6, *705*, 707, 708
    distribution *578*, 579, 585–6, 647, 667, 697, 702–3, 895
  soldiers and mariners possibly act as 582, 639, 647
  training and organisation 587–8
archery and archery equipment **578–712**
  development of archery at sea **593–4**
  distribution *see under* Hold; Main Deck; Orlop Deck; Upper Deck
  future research 931–2
  government control 588–9, 639
  import of yew and hemp 588, 591, 635–6
  inventory evidence **579–82**
    1546, *Anthony Roll* 5, 6, 537, 543, 580, 581, 665, 691, 718–19
  scope and importance of assemblage 578–9
  storage 579, 895, *896*
  Welsh production 586–7
  *see also* archers; arrow bag; arrow chests; arrow girdle; arrow spacers; arrows; bowstrings; longbow chests; longbows; wristguards; *and under* chests (personal); replication
Archibald, C. D. 131–2
archive, *Mary Rose*; digitisation 930
arcs of fire 914–20
  canting of broadside guns 854, 855, 942
armguards *see* wristguards
armour and personal protection **818–48**
  assemblage 824–46
  distribution 826, 835, 846–8, *846–7*
    *see also under* Hold; Main Deck; Orlop Deck; Upper Deck
  and firearms 821
  future research 933
  human remains associated with 842, 843, *845*, 846–8, *846–7*
  imports from Continent 820–1, 824, 825, 827, 828
  inventory (1514) 818, 822, 857
  metallurgy 818, 821, **823–4**, 828–9, *828–9*
  plate 818, *819*, 824
    *see also* breastplates
  production 822, 824
    mass- 818, 820–1
  worn at sea 821–2
  worn for warfare 818–21
  *see also* aiglets; almain rivets; breastplates; brigandines; bucklers; buckles, armour; helmets; jerkins; mail; shields; straps, wood and metal fastenings for armour

arrow bag 578, 694, 697, 698–9, *698–9*
arrow chests 578, 689–93, *689*, *692*
  arrangement of arrows 666, 667, 692
  contents list 690–1
  distribution *641*, 667, 668, 669, 691, 883
    Hold 667, 668, 669, 691, 693
    and human remains 578, 579
    Orlop Deck 667, *668*, 669, 690–1, *692*, 692–3
    Upper Deck 667, *668*, 669, 690, 691–2, *692*, 909
    and wristguards *646*, 647
  inventory entries 580, 582
  mixture of types within 683–5
arrow girdles 693, 697, 699
arrow marks, broad
  on casks 432, *432*, 433
  on guns 218, 249, *249*
arrow spacers 578, *666*, **693–7**, *696–7*
  associated objects 697
    arrow bag 578, 694, 697, 698–9, *698–9*
    arrows 693, 694–5, *696*, 707, *708*, 931
    leather straps 694–5, 697, 707, *708*
    wristguards *646*, 647
  and arrowhead diameter 672, 693, 699
  distribution *578*, 693
    on Main Deck *668*, 669, 693, 694–5, 697, 704
    on Orlop Deck 693, 695, 697, 704
    on Upper Deck 693, 694, 697, 704
  and human remains *578*, *668*, 669, 703–4, 707, *708*
  mixture of arrows within 931
  personal ownership 697
  reconstruction *666*
arrows **665–89**
  bindings 579, 669, 675, 678, *679*, 680, 688
  and bowstring diameters 579, 673, 674
  bundles *see* sheaves *below*
  combined data profiles 682–5
  conservation 666–7
  distribution 537, 667, *668*, 669
    *see also under* arrow chests
  draw-lengths, predominant 685
  experimental shooting data 630–1
  flights 579, 669, *670*, *671*, 673, 674, 693
  form and function **669–72**
  future research 688–9, 931–2
  glues 579, 669, 675, 693
  heads 579, 666, 669, *670*, *671*, 671–2, 677
    maximum diameter 672, 693, 699
    vestigial remains 579, *666*, 666, 667, 677, 692, 693
    *see also* tips *below*
  historical information **669–72**
  human remains associated with *578*, 579, 691
  incendiary *see* darts and arrows, incendiary
  inventory entries 580, 581, 582, *582*
    1546, *Anthony Roll* 5, 6, 537, 580, 581, 665, 691
  knots and grain breakout 673, 686–8
  length 579, 669, *670*, 673, 677–8, 688
  livery 581, 582, 665

and longbows
- matching to bow 579, 617–18, 630, 665–6, 929, 931–2
- number per bow 580, 581, 582
- manufacture 579, **688**
- mass, efficiency and velocity 630–1
- material analysis 674–7
- nocks 669, *670*, 670, **674**, **680**, *681–2*
  - and bow string diameter 579, 673, 674
  - horn inserts 632–3, 637, *638*, 639, 640, 669, *670*, **680**, 688, 693
  - length, width and orientation 678, *679*, 680, *681–2*, 688
- range 631, 670, 688
- scope and importance of assemblage 578, 579, **672–88**
- shafts 669, 670, *670*, 672
  - deflection values 617–18
  - diameter 669, 670, 673
  - profiles *680*, 682–5, *683–4*, 685–6, 687, 688
- sheaves 580, 581, 582, 685, 667, 675, 691, 692, 693, 697
- small, used to make firesticks 529
- from spoil mounds 935, 938
- tar/pitch packing 692, 693
- tie-mark impressions 675
- tips 673–4, **680–2**, *682*, 692, 693
  - cones 669, *670*, 680–2, *682*
  - iron staining *666*, 666, 667
  - measurements 579, 685
- tool marks 688, *688*
- variety of types 579, 683–5, 929, 931
- weight 626–7, 669, 670, 685–6, 687
- wood species 579, 669, 673, 674–7
  - and arrow weight 685–6, 687
  - and bow weight 617–18
  - densities 675–6
  - and profile *680*, 684–5, 685–6, 687
  - specific gravity and modulus of elasticity 618, 676
- XRF analysis 675–7, *676*
- *see also* arrow bag; arrow chests; arrow girdles; arrow spacers; darts and arrows, incendiary

artillery revolution 851–2

Ascham, Roger; *Toxophilus* 588, 632
- on arrows 669, 671, 682–3, 688
- on bowstrings 634–5, 637, 639
- on bracers/wristguards 644, 650

ash
- items made of 669, 720, 779, 781
  - gun carriage parts 39, 40, 94
- weight, density and elasticity 626, 676

Audley, Sir Thomas, Lord Audley 594, 717, 927

auger, bull-nosed 94, *95*

Augsburg; armourers 824

Austria 540, 824, 951–3, *954*

axle trees 39, 40, *41*, 42

axles 40, 94, 102, 105, *106*, 107, 109
- *Anthony Roll* compared with assemblage 6
- bolts 41, *41*
- distribution 102–3, 105, *105*, 109
- fragments from Timber Park 936, 939
- loose, not in store 104, *106*, 109, 112, *113*, 114
- stored 109

bags, leather 6, 507
- *see also* arrow bag

Bahamas 239, 254

Bahia Mujeres, Mexico; wreck 223

Baker, James 943, *943*

Baker, Matthew 895, 914

ballistics 127–9, 289, 914

barbican 860

barrels (containers)
- budge 507
- *see also* casks

Barret, Robert 717

Bartlette, Paul 740, 809

bases *132*, **135–6**, *136*, **221–68**
- anatomy 224, 227
- barrels 133, 224, 227, 230, 240
- chamber holders 224, 227, 231, 244
  - construction 236–7, *237*, 242
  - summary table 232–3
- chambers 131, 133, 222, 224, 227, 228–9, 230, 243
  - construction 237–8, *239*, 240
  - dimensions 223, 232–3
  - distribution *231*, 235
  - number in assemblage 133
  - radiography 243, *469*
  - summary table 232–3
- construction **236–43**
- demi bases 224, 357
- dimensions 222, 223, 227, 232–3
- distribution *231*, 261, *261*, *356*, 358
  - *see also under* Starboard scour
- double and single bases 223, 224, 240, 357
- external appearance *132*
- in fighting tops 235
- gunpowder 133, 232–3, 243, 247, 266, 426, 427, 428
- identification and importance 132, **222–7**
- inventory entries 214, 223–4
  - 1540 225, 856, 861, 864
  - 1546, *Anthony Roll* 4, 5, 6, 131, 223–4, 225, 357, 856, 864
    - shot 309, 310, 367, 922
  - 1547 222–3, 224–7, 357
    - shot 222–3, 312
- ladle size needed 448
- loaded 230, 232–3, 243, *469*, 470, *471*, 472
  - *see also* tampions; wads *below*
- method of manufacture **236–43**
- morphology *132*, 135–6, *136*
- mountings *see* swivel mountings *below*
- Norton on 24, 74, 133, 222, 358
- Portingale bases 223, 225
- radiography 236, 243, 245, 252, *252–3*, 264, *264*, 265–6, *265*, 268, *469*
  - *see also under* individual guns *below*
- range 918

scope and distribution of assemblage **227–36**
shot 133, 222–3, 227, 230, 243, 311
    canister 370, 380–1
    inset ball 236, 360
    inset dice 223, 236, *256*, 256, 258, 260, 309, 353, *356*, 357–8
    in inventories 222–3, 227, 309, 310, 312, 367, 922
    matching to guns 357–8, 360
shrimp bases 224
in Sterncastle *235*, 235–6, 918
superseded by bronze guns 270
swivel mountings 132, **135–6**, *136*, 221, *222*, 224, 227, 247, 255, *256*, *257*, 261, *267*, 268, 272
    construction **238–9**, *241–2*
    holes for *234*, 235, *235*, 861, 891, *892*, 918
tampions *in situ* in 232–3, 243, 247, *256*, 258, *469*, 470, *471*, 472
tillers *224*, 227, 236, *236*, 240–1, 244
trunnions *224*, 227, 232–3, 237, *238*, 242–3, 244
typology
    contemporary names 224–5
    Form 1 *243*, 244–54
    Form 2 *243*, 244, 254–66
    unknown 266–8
wads 232–3, 243, 247, *256*, 258, *471*, 475, *475*, 476, *476*
in weapon preference sequences 921–4
wedges 232–3, 236–7, *237–8*, 244
wedge slots *224*, 227, 236–7, *237–8*
weights 223
windage 230
Wright's illustration 132, *132*
yokes *see* swivel mountings *above*
79A0543 *226*, 228, 229, **232**, 244, **247–9**, *248–9*
    construction 237, 247, 249
    contents 230, 232, 247
    dimensions 227, 232
    location 227, 229, 232, 235, 247
    radiography 244, *244*, 247, 249, *249*
    swivel mounting *241*, 247
    tiller *241*, 247
79A1075 *226*, 229, *229*, **233**, 244, **250–1**, *250–1*
    construction 237, 250–1, *251*
    contents 230, 233, 250
    dimensions 227, 238
    location 227, 233, 235, 250, *250*
    radiography 250–1, *251*
81A0051 235
81A1082 230, 236, 254, **264**, *264*
81A1852 230, 235
81A2408 230, 235
81A5152 230, *266*, **268**, *268*
82A0024 **254**, *254*
82A1603 222, 229–30, 254, **259–60**, *259–60*
82A2589 233
82A4076 *226*, 228, 229, **232**, 254, **255–8**, *256–8*
    construction 237, 255–6
    dimensions 222, 227, 232
    fibres near touch hole 243

    handle 238, *240*, *258*, 258
    location 227, 229, 232, *255*, 255
    radiography 255, *257–8*
    shot 230, *256*, 256, 258
    swivel mounting *241*, 255, *256*, *257*
    tampion and wad 232, *256*, 256, 258, *469*, *471*, 475, *475*
    tiller shape 241
    typology, Form 2 *243*
82A4077 *226*, 228, 229, **232**, **245–6**, *246*
    construction 237, 245–6
    contents 232, 243, 245
    dimensions 227, 232, 238
    location 227, 232, 245
    radiography 244, *244*, 245
    tiller shape 241
    typology, Form 1 *243*, 244
82A4080 227, 228, 230, **232**, 236, 254, **265–6**, *265*
82A4357 *229*, 232, 243
82A4395 *229*, 233
82A5095 230
82A5096 227, *229*, **232**, 244, 252, *252–3*
82A5152 240, 265
85A0024 227, 229, *229*, 230, **233**, 244
85A0026 *229*, 230, **232**, 235, 237–8, 243
89A0005 230, *469*
89A0007 *229*, 230, **232**, 235, 243
89A0051 230
89A0090 230, *242*
89A0091 *228*, **232**, 235, 243
89A0108 *229*, **233**, 235, 243, *471*
89A0109 *229*, **232**, 235, 243
89A0113 *229*, **233**, 235, 243
90A0027 227, *229*, 230, 243, **261**, *263*
90A0050 227, *229*, 230, 232, 243, 254, **261**, *261*, 263
90A0052 227, *229*, **232**, *241*, 243, *266–7*, **268**
90A0063 230
90A0106 *229*, 230, **232**, 243
90A0122 *229*, **232**, 243, 261, *261*
90A0123 *229*, **232**, 243, 261, *261*
90A0124 *228*, **232**, 261, *261*
90A0130 *228*, **232**, 243
90A0132 227, **261**, *263*
04A0097 227, *266*, **266**, *268*
baskets associated with guns 504, *505*, 508, *509*
    in M/U11 247, 313, 370, *505*, 508, *509*
Bassett-Lowke model of *Mary Rose* 21, 204
Batavia wreck 365, 395, 401, 465, 521, 822
Batcock, Thomas 116
bathymetry of Solent 872, *872*, 873, *873*, 874, 880, 881
battle, *Mary Rose* in 865–82
    *see also* manoeuvring in battle; Solent, battle of the
Baude, Peter 22, 27, *34*, 51, 929
    alloys 21, 946, 951, 952, 954
    iron casting 292, 864
    *see also* cannon (81A3003); culverins (81A1423)
*Beauchamp Pageant* 7
    archers 589, 644, *645*, 697, 821

bill 726, 714, *716*
darts 269, *269*, 718
sword hangers 806
Bedford, 1st Earl of (John Russell) 866, 870
beech, items of 763, 779
belt plate *816*, 817
belts, sword 805, 808–9, *814*, 815–17
    buckles 838
    with scabbards 809, *810*, 811, *812*, 813
Berwick upon Tweed Castle *20*, 216, 437, 582, 857
bills 719–34
    American examples 725–6, *726*
    anatomy *715*, 714–15
    in concretion 726, *728–9*, 729
    fragments from spoil mounds 935, 938
    hafts 719, 720, *721–4*, 724–5, *729*, 731, 734, 929
        distribution *730*, 909
    heads 719, 725–6, *728*, 729, *730*, 731
        anatomy 714–15, *715*, *720*
    identification and form 719–20
    inventory entries
        1546, *Anthony Roll* 580, 581, 718–19
        1547 715–16
    markings 731
    measurements 713
    method of manufacture 731
    nails, T-shaped *715*, 719–20, *720*, 724–5, 726, *727*, 731, 734
    numbers compared with pikes 715–16, 717
    personal ownership 716–17
    radiography 732, 726, *727–8*, 729
    replication 732–4, *732–3*
    scope and importance of assemblage 713–17, *714–16*
    wood species 720
birch, arrows of 669, 674–5, 677, 678, *680*, 684–5
    technical characteristics 618, 626, 676, 686
Biringuccio, Vannoccio; *Pirotechnia*
    on alloys 955
    on cannon 45
    on cartridges 441
    on casting of guns 114, *116*, 116, 948
    on casting of shot 304, 383, 384
    on gun carriages 40, 54
    on gunpowder 423
    on gunpowder production 421, 422, 423, 425
    on incendiary devices 519, 526, 527, *528*, 536
    on ladles 439–40, 441, 446
    on linstocks 486
    on peening of surface of gun 955
    on reviewing work 928
    on types of guns 23
    on wads 474
Blackness Fortress, France *20*, 201, 523
Blakye, William (captain of the *Mawdalen* of Rye) 869–70
blinds, Upper Deck *198*, 199, 895, 907, 908, *908*
boarding *see under* tactics, naval
bodkins 748, 789, 791
bodyguard, royal 555

possible presence in *Mary Rose* 656, 738, 929
    weapons 713, 738
        bucklers 822
        halberds 718, 737, 929
        swords and daggers 740, 743, 745, 754, 759
bomb *see* mortar bomb
bombards 19, 24, 45, 859, 860, 881
bombardels 860
book cover stamp *see under* wristguards (decoration)
boom chain, Portsmouth harbour 880–1
Bosworth, battle of 587, 717
Boulogne *20*
    inventory (1547) 216, 224, 381, 410, 523
    sieges 10, 12, 717, 865, 882
    *see also* Siege of Boulogne
Bourne, William 23
    on cartridges 437, 440–1
    on gun drill 410–11, *411*, 416, 417, 479, 891, 923
    on gunner's rule 393
    on gunpowder 422, 423, 425
        weight of charges 428, 448, 449
    hundredweight as unit 425–6
    on ranges of guns 880, 926
    on short guns amidst rigging 74
    statistics on guns and shot 21, 46, 58, 89, 101, 102, 214, 318, 428, 429, 460
        weights of shot 353, 354, 355
    on wads 476
bow 934, 942
    possible extra guns in 944, *945*
    stone shot 346
    *see also* Forecastle
bow bulkhead excavation
    axle 109, 112
    base elements 232, 233, 261, *261*
    shot *340*, *356*, *362*
    trucks 108
bowl, wooden 708, *709*
bowstrings *633*, **634–41**, 949, *949*
    excavated fragment 578–9, *633*, 634, 641, 935, 939, *949*
    form and fit 636–7, *636*
    function 634–5
    future research 932, 949, *949*
    inventory evidence 580, 582
        *Anthony Roll* 5, 6, 580, 581, 634, 639
    matching to bows 639–40
    material 579, 635–6
    served 637, 640
    size 579, 637–9, *638*, 673, 674
    storage 634, 641
Bowyer, John, of Hartfield (gunstone maker, *fl.* 1513–14) 308, 309
box, wooden; possible shot storage 312
boxwood by-knives 779, 781
brace and bit 94, *95*
bracers *see* wristguards
braces, structural *884*, 885, 930
Brandon, Charles (Duke of Suffolk) 868, *868*, 870

Bray, Edward (captain of *Mary Rose*) 8
breastplates
    excavated example 818, 824, 825–8, *825–6*, *828*, 846
    possible pile of corroded 824, 828–9, *829*
breech chambers *see* chambers
Brescia; handgun production 537–8, 539–40, 540–1
Brest 8, 853
    sea battle off (1511) 594
Brigandine, Robert 1, 8
*Brigandine*
    gunpowder 424
    guns 101, 102, 225, 300, 301, 438
brigandines (jacks, protective jackets) 818, *819*, 838
Brighton, Sussex 3, 7, 223
bronze
    alloys 19, 117, 950–3, 954–5
        Continental 951–3, *954*
            imports to England 953–4, 960
    casting 27, **112–16**, *115*, *116*
        flaws 60, *60*, 122
        replica culverin 121–3, *122*, *123*
        techniques 112–16, *115*, *116*, *118*, 119
        wire used to wrap core 48, 52
    London foundries 22, 864
    properties of cast 22
    *see also* gunfounders; guns, bronze; *and under* metallurgy
Browne, Sir Anthony 867, 868, *868*, 870
Browne, John (master of *Mary Rose*) 9, 310
Brueghel, Pieter the Elder 760, 761, 805
brushes 504, 511–12
Bryarley, John (purser of *Mary Rose*) 310
buckets 504
    leather *161*, 162, *505*, 508–10, *511*
    wooden *506*
bucklers 555, 822, 824, *826*, 829, *829*
buckles, armour 748, 818, 824, 834–8, *836–7*
    distribution *826*
    frames 817
    plate *816*, 817
    in sword suspension systems 748, 807, *810*, 811, 813, 838
budge barrels 507
building of *Mary Rose* 8, 928, 930
Bull, Sir Stephen (captain of the *Regent*) 308, 309
*Bull*
    guns 225, 269, 291, 300, 301
    munitions 331, 424
Bullingham Bulwark, Devon *20*, 523
burrwood sword/dagger grips 763, 772, *773*, 774
Burton List 24, 318, 429
by-knives
    anatomy *742*, *747*
    associated with daggers *742*, 746–7, *747–8*, *747–9*, 778, *778*, 779, *788*, 789, *790*, 791
        accommodated in sheaths 794, 796, 798
    distribution *752*, *778*, 779
    radiography *788*, 789
    significance of assemblage 747–8, *747–9*

    from spoil mound 935, 938
    in sword scabbards 794, 796, *798*, 798, 813
    unassociated with daggers 781
    wood species 779, 781
Bytharne, Jehan 851

cabins
    and location of guns 896, 898
    *see also* Carpenters' cabin, items from
cables, Tudor salvage 50, *50*, 903, *904*, 905, 949
Calais, France; items in inventory (1547) 216, 317, 383, 523, 582, 665, 719
Calshot Castle, Hants *20*, 302, 523, 582
Camber Castle, Sussex *20*, 201
candles 529
canisters, pewter 433
cannon **45–73**
    arcs of fire *915*, 915–16
    cannon royal/double cannon/cannon of 8 24, 45, **47–50**, 930
        dimensions and weight 21, 429
        gunpowder charge 428, 429
        *see also* 79A1276
    carriages
        distribution 26, *26*
        radiography 57, *57*, 61
        replication 71–3, *71–3*
        *see also under individual guns below*
    chamber cannon 311
    demi cannon 24, **29**, 45, **58–73**, 925
        dimensions and weight 21, 29, 58, 314
        fortification 428
        gunpowder expenditure 426, 427, 428, 429
        introduction and development 292, 863
        inventory entries 5, 7–8, 25, 36, 58, 310, 312, 856, 861, 922
        ladles for 448, 449, 450, *450*, 460
        on Portsmouth fortifications 881
        rammers for 457, 460, 462
        range 429, 875, 876, 881
        shot 310, 312, 314, 318, *318*, *320*, 336, 429, 462, 922, 925
        sizes 311, 462
        uses 897, 899, 921–4
        *see also* 79A1277; 81A3000; 81A3002 *below*
    dimensions 21, **29**, 45, 46, 47, 58, 429
        bore sizes 314, 462
    double cannon *see* cannon royal *above*
    elevation 43, 54–5
    fortification 45, 428
    gunpowder expenditure 426, 427, 428, 429
    introduction and development 292, 856, 863
    inventory entries 925
        1540 7–8, 45, 46, 58, 861
        1546, *Anthony Roll* 5, 7–8, 25, 36, 46, 58, 310, 856, 861, 864, 922
        1547 930
    ladles for 448, 449, 450, 450, 460

Norton on 21, 24, 45, 58, 429, 875
perriers 24
rammers for 450, 457, 460, 462
range 314, 875, 876
shot 45, 310, 311, 318, *318*, *319*, 330, 462, 922
shot gauges not found 405–6
types 24, 45, 46
uses 138, 897, 899, 921–4
vent covers 51, 417, 507
in weapon preference sequences 921–4
whole cannon/cannon of 7 21, 24, 46, 428
    see also 81A3003 *below*
79A1276 *24*, **29**, 45, **47–50**, *47–50*, 930
    carriage 26, *26*, 42, 43, 48–9, *49*
    cascabel *34*
    contents 29, 48, 318
    date 29, 46, 48
    device and decoration 29, 31, 30, 33, 47–8, *47*, *48*
    dimensions and weight 21, 29, 46, 47
    dolphins *30*, 48, *48*
    fortification 428
    founders 21, 29, 47, 48
    location *26*, 26, 29, *47*, 47, 49
    metallurgy 21, 946, 949, *950*, 951, 959–60
    salvage 47, 48, *49*, 49–50, *50*
    trunnion mark 33, *35*
79A1277 *24*, **29**, **64–9**, *64–9*
    carriage 26, *26*, 36, 40, 42, 43, 65–9, *65–9*
    cascabel *34*
    device and decoration 29, 31, *32*, 64, *64*, 65
    dimensions and weight 21, 29, 58, 64
    dolphins *30*, 64, *64*, 65
    founder 29, 64
    lead isotope analysis 954
    location, probable *26*, 26, 29, 58, 64, *64*, 897, 899
    metallurgy 946, 949, *950*, 951, 959
    salvage 64–5, 69
    shot 29, 64, 318
    trunnions 64, *64*, 65
81A3000 *24*, **29**, **59–63**, *59–63*
    arc of fire *915*, 915–16, *919*
    associated items 460, 498, 504, 523, *525*, 526
    incendiary devices *525*, 527, 529, *529*, 531
    carriage 37, *38*, 42, 43, 69, 60–1, *61–4*
    cascabel *34*
    casting flaw 60, *60*
    device and decoration 31, *32*, 33, *33*, 59, *59*, *60*
    dimensions and weight 21, 29, 58, 59
    dolphins *30*, 59, *60*, 60
    founder 29, 59, 60
    gunpowder 60, 434–5, *434*
    gun station described 905
    lifting 37, *38*, 60, *61*
    location *26*, 26, 29, 58, 59, *59*, 60, *61*, 63, 897, 899, 905
    metallurgy 946, 949, *950*, 951, 958
    radiography 61
    shot 60, 318
    trunnions *59*, *60*, 60
    use 897, 899
81A3002 *24*, **29**, 36, **70–3**, *70–3*, 895
    carriage 26, 37, *39*, 40, 42–3, 71–3, *71–3*, 86, 123, 893
    cascabel shape *34*
    classification 27, 35–6, 58
    device and decoration 29, 30, 31, 33, 35, *36*, 70, *70*, 71
    dimensions and weight 21, 27, 29, 35–6, 58, 70, 74, 87
    excavation *39*
    faceting 35, *36*, 70, *70*, 71
    founder 21, 29, 33, *33*, 36, 70
    inscription 33, *33*, 36, 70, *70*, 71
    lead isotope analysis 954
    location
        likely original 26, *26*, 27, 29, *39*, 70, *70*, 73, 74, 86
        when found 26, *71*, 895, 897, 909
    metallurgy 21, 946, 949, 950, *950*, 951, 954, 958
    probable pair 37, 58, 71, 86
    scar 71, *71*
    shot 71, 318, 323
    trunnions 33, 35, 70, *70*, 71
81A3003 *24*, **51–7**, *51–7*
    arc of fire 915, *919*
    associated items 55, 55, 56, 450, 503, 903
    carriage 37, *38*, 42, 43, 52–5, *52–4*, 55–7, *56*
    cartridge, possible 56
    cascabel *34*, 35, 51, 52, *52*, 54
    date 29, 46
    device and decoration 29, 31, 33, 51, *51*, *52*
    dimensions and weight 21, 29, 46, 51
    dolphins *31*, 51, *51*, 52
    elevation 54–5
    founder 22, 29, 51
    gun position described 903
    inscription 51, *51*, *52*
    lifting 37, *38*, 55–7, *56*
    location 26, *26*, 29, 55, 51, *51*, 138, 897, 899
    metallurgy 946, 949, *950*, 951, 954, 959
    quoin beneath cascabel 52, *52*, 127, 514, *515*
    radiography 57, *57*
    shot 52, *55*, 55, 56, 318
    trunnions 51, *51*, *52*
    use 138, *141*, 897, 899
    vent cover 51, 417, 507
canting of guns *854*, 855, 942
canvas for cartridges 6, 56
cap squares 37, *39*, 40, *41*
Capo Bianco, A. 875, 876, 955
capstan guns 860
capstans
    on *Mary Rose* 909
    for Portsmouth boom chain 881
captains of *Mary Rose see* Bray, Edward; Carew, Sir George
Carew, Gawain 866
Carew, Sir George 9, 10, 659
    and sinking of *Mary Rose* 866, 869, 879, 882
Carew, Sir Nicholas 762, 807
Carew, Sir Peter 866

Carlisle, Cumbria *20*, 225, 227, 383, 523
Carpenters' Cabin, items from
    archery items in personal chest *585*, 585, 650, 669
    arrows used to make firesticks 529
    handgun 370, *544*, 545
    linstocks 489, 497–8, *498*
    powder horn tops 485
    shot 366, 370
    tampions and tampion reels 471
    tinder boxes 502
carpentry tools 94, *95*, 97
carracks, depictions of *269*, 270
carriages *see* gun carriages
cartouches 31, 33
cartridges
    formers 6, 409, **435–9**, *436–7*, *439*, 529
    inventory entries for materials 409
        1546, *Anthony Roll* 6, 126, 435, 436
    ladles used to load 441, *441*
    storage 430, 433
    use 126–7, 407, 409, 412, 413, 417, 436–7, 440–1
    *see also* canvas; paper; priming wires
carvel building *13*, 14
cascabels *23*, *24*, *28*, *34*
casks
    for bowstrings 641
    distribution 431, 433, 931
    future study 931
    for gunpowder 425–6, 430–3, *432*, 931
    marks 431–3, *432*
    for shot 313
    for tampions 463, 466
    as unit of measurement 430
cast guns (cast pieces) 856, 858
casting of metal *see under* bronze; iron
castles *see* Forecastle; Sterncastle
Cataneo, Girolamo 880
*Catastico Bresciano, Il* 549
Cattewater wreck 227, 270, 358, 857
caulking 9, 502
Cerignola, battle of 537
chains for main and mizzen masts 934–5
chambers, breech
    for bronze guns 292
    for wrought iron guns *130*, 131, 133, *134*, 135
        *see also under* bases; port pieces; slings
changes in armament **851–65**
    general developments 851–5, 928
    in *Mary Rose* 7–8, 10, 12–15, 35–6, 854–5, 856–65
    *see also* Hatfield House document
chapes 748, *752*, 778, 800–804, *801–4*, 931
    metal analysis 798, 799, 800
Charles V, Holy Roman Emperor and King of Spain 1–2, 3, 9, 10, 540, 713
Charles II, King of Great Britain 631
charts of Solent, early 867, 869, *872*
Chaves, Alonso de 851, 855, 926–7
Chelmsford sword 753, 754, 759

chests
    gunpowder 433
    linstocks in 488, 489, 497–8, *498*
    personal
        archery equipment *585*, 585, 662–3, 699–702, *700–2*
            wristguards 646, 647, 703, 712
        linstocks 497–8, *498*
        swords, daggers and related finds 748, 751, 763, *779*, 781, 794
    priming wires in 479
    *see also* arrow chests; longbow chests
Chicago, Art Institute of; gun-shield 557, *558*
China; gunpowder 419
chisels 94, *95*, 385, 386, *506*, 508
*Christ of Greenwich* 269, 385, 580, 858, 859, 920
*Christopher* (of Bremen)
    fighting tops 269, 272
    guns 137, 225, 269, 291, 300, 301
    munitions 424, 436
*Christopher of Danzig* 137
*Chronicle of Diebold Schilling* 697
Churburg Castle, Italy; shot moulds 383
Church Rocks site, Teignmouth, Devon 101, 223, 521
Clerk of the Ships 1, 2, 9, 928
Cleves, Philip of Ravenstein, Duke of 10, 93, 851, 854, 887, 913
clinker building *13*, 14
*Cloud in the Sun* 200, 225, 300, 301, 424, 438, 581
Coburg Castle, Germany; incendiary devices 521, 531
*Codex Vindobana* 3069 421
coin, gold 938
Collado, Luis 410, 417
combs 708, *709*, *710*, 712
command structure 879, 882
commanders (large mallets) 6, 409, 504, *505*, **507**, *507*
common piece (type of gun) 25
communications on board 882, 887, 895, 907
    *see also* companionways
companionways 896, *898*, 907
concretion removal 298, *299*, 301
Conflans, Antoine de 851
conservation methods
    arrows 666–7
    longbows 594
    wrought iron guns 134
*Cordelière* (Breton flagship) 921
Coventry; bills from St Mary's Hall 714–15, *716*, 725
Cowdray Engraving *10–11*, 14, 867–70, *868–70*
    accuracy 868, 870–1
    armour not in use 822
    arrow spacer/bag 693, *697*, 699
    fighting tops 272, *272*
    gunports 14
    *Henry Grace à Dieu* 854, 855, 869, *868–9*, 897, 899
    linstocks 486, *486*
    Portsmouth fortifications 868, *869*, 870, 880, *881*
    Southsea Castle 868, *868*, 870
    staff weapons 713, *714*, 737

tinder boxes *502*, 503
vessel positions 499, 869, 876, *868–9*
Crécy, battle of 587, 631
Crema, Italy 542
crew 581, 883
    alleged insubordination 866, 867
    *Anthony Roll* evidence 4–5, 882
    *see also* archers; gun crew; human remains; soldiers
Cromwell, Thomas (Earl of Essex) 2, 9, 861, 930
crossbows 553, 594, 932
crowbars 6, 385, 407, 409, 504, *505*, 507, 508
crown (casting support) 114, 116
cruzeta (casting support) 114
culverins 29, **74–100**, 860
    arcs of fire *915*, 915–16, *917*, 918, 919, *919*
    carriages 89, 108
        replication **76–82**, *77–82*, **94–100**, *94–100*, 123, 129
        *see also under individual guns below*
    casting 60, **112–23**
    demi (bastard) culverins 29, **89–100**
        carriage elements 89, 108
        dimensions and weight 21, 314, 429
        fortification 428
        gunpowder expenditure 426, 427, 428, 429
        inventory entries
            1514 89
            1540 89, 861, 863
            1546, *Anthony Roll* 5, 8, 25, 89, 864, 922
            compared with assemblage 5, 89, 931
        iron, introduction of 89, 292
        ladles for 448, 449, *450*
        Norton on 24, 74
        on Portsmouth fortifications 881
        rammers for 456, 460, 462
        range 314, 429, 875, 876, 881
        shot 312, 429, 925
            composite 312, 336
            iron 310, 311, 330, *322*, *323*, 323, 922
        shot gauges 405–6
        sling class equivalents 201
        in stern on Main Deck, possible *901*, 901, *902*
        in Sterncastle 89
            forward-firing 89, 90–100, 853, *853*, 854, *911–12*, 913, *945*; arcs of fire 916, *917*, 918, 919, *919*; Hatfield House document on 943–4, *944*, *945*
        terms used of 89
        in weapon preference sequences 921–4
        *see also* 79A1232; 79A1279 *below*
    dimensions and weight 21, 314, 860, 893
    double culverins 428, 880
    elder culverins 448
    faceted 35, 36, 75, *76*, 83, *83*, *84*, 90, *90*, *91*
    firing, gun drill **123–9**, 411, **416–19**, 449
    gunpowder 426, 427, 428, 430
    introduction and development 291, 292, 863
    inventory entries
        1514 74, 89
        1523 860

        1540 7–8, 74, 89, 856, 861, 863
        1546, *Anthony Roll* 5, 8, 25, 36, 74, 89, 310, 856, 864, 922
    ladles for 448, 449, *450*
    lead isotope analysis 954
    on Main Deck 74, 897, 899
        *see also* 79A1278; 81A1423 *below*
    metallurgy *see under individual guns below*
    muzzle energy 128, 914
    Norton on 21, 24, 74, 89, 429, 875
    on Portsmouth fortifications 881
    rammers for 456, 460, 462
    range 314, 429, 875, 876, 879, 880, 881
    rate of fire 416, 925
    replica *see under* 81A1423 *below*
    shot 925
        composite 312, 336, 384
        iron 310, 311, 312, 318, *321*, 323, *323*, 330
        lead 312
        stone 312, 347
        *see also under individual guns below*
    shot gauges for 405–6
    slings compared 200, 201–2
    in Sterncastle *see* 79A1232; 79A1279 *below and under* demi (bastard) culverins *above*
    types 24, 74, 89
    on Upper Deck 74, 893, 909, 918
    *see also* 80A0976
    uses 93, 138, 897, 899, *915*, 915–16, 921–4, 925
    whole culverins 24, 29, **74–88**
        dimensions and weight 21, 29, 448
        shot, gunpowder charge and range 429
        *see also* 79A1278; 80A0976; 81A1423 *below*
    79A1232 15, *24*, 29, 92–100, *92–100*
        carriage 37, 40, 42, 43, *44*, 45, **94–100**, 105, 108, 107, *110*, 112
        cascabel *34*
        casting 92, 93
        device and decoration 31, *32*, 35, 92–3, *92*, *93*
        dimensions and weight 21, 29, 92
        dolphins *30*, 92, *93*, 93
        faceting 92, *92*, 93, *93*
        founders 29, 92–3
        gun station *911–12*, 913
        inscription 31, *32*, 92–3, *92*, *93*
        lead isotope analysis 954
        lifting 37
        location *26*, *27*, 27, 29, 89, 92, *92*, 93
        metallurgy 946, 947, 950, 951, 948, 949, 950, 954, 959
        possible pair 36–7, 89, 91, 103
        shot 93, 323, 330
        trunnions *92*, 93, *93*
        use 93
    79A1278 *24*, **29**, **87–8**, *87–8*
        dimensions and weight 21, 29, 70, 87
        founder 29, 87–8
        carriage not found 42

cascabel *34*
Deane and Edwards' salvage 87–8, 165
device and decoration 29, 31, *32*, 87, *87*, *88*
dolphins 87, *88*
gunpowder 87–8
inscription *32*, 87, *87*, *88*
location, possible *26*, 26, 87, *87*, 897, 899
metallurgy 946
possible pair 74, 88
rammer and ladle for 460
shot 29, 87–8, 318, 323
trunnions 87, *88*
use 897, 899
wadding 87–8
79A1279 24, 29, **90–1**, *90–1*
carriage elements possibly from 91, 103
cascabel *34*, 35
casting scars 90, 91, *91*
device and decoration 31, *32*, 35, 90, *90*
dimensions and weight 21, 29, 90
Edwards' salvage 90, 91
faceting 35, *36*, 90, *90*, *91*
founder, possible 22, 29, 36
inscription *32*, 89, 90, *90*
location, possible *26*, 26, 27, 29, 36–7, 89, 90, *90*, 91
possible pair 36–7, 89, 91, 103
shot 29, 323
trunnions 33, *35*, 90, *90*, *91*
80A0976 24, 29, **83–6**, *83–5*
carriage 37, 42–3, *44*, 71, 84, *85*, 86, 105, *110*
cascabel *34*
casting scar 83
dimensions and weight 21, 27, 29, 74, 893
device and decoration 31, *32*, 83, *83*, *84*
faceting 35, *36*, 83, *83*, *84*
founder, possible 22, 29, 36, 86
gun station 909, 910
gunpowder 433–4
handspike associated with 504
inscription *32*, 83, *83*, *84*
lead isotope analysis 954
lifting 37, 84
location *26*–7, *26*, *27*, 74, 83, *83*, 893
metallurgy 946, 949, *950*, 951, 954, 958
possible pair 37, 58, 71, 74, 86
shot 84, 318, 323
shot storage between frames near *312*, 312
trunnions 33, *35*, 83–4, *83*, *84*
81A1423 24, 29, **75–82**, *75–82*, *141*, 901–2
arc of fire 915–16, *919*
associated finds 460, 902
carriage 37, *38*, 42, 43, *44*, 75–6, 76–82, *77–82*
cascabel *34*, 35
casting 75
device and decoration 29, 31, *32*, *34*, 35, 75, *75*, *76*
dimensions and weight 21, 29, 75, 93
dolphins *31*, 75, *76*, *77*
faceting 35, 75, *76*

founder 21, 29, 75
gun station 901–2
gunpowder 75
inscription *32*, 75, *75*, *76*
lifting on carriage 37, *38*, 75–6
location 25, 26, *26*, 29, 75, *75*, 138, 897, 899, *915*, 915–16
metallurgy 21, 946, 947, *950*, 951, 948, 949, 950, 954–5, 956–8
possible pair 74, 88
replica **112–29**, **416–19**
carriage **76–82**, *77–82*, **94–100**, *94–100*, 123, 129
casting 60, 112–23
choice of gun for replication 116–17
firing **123–9**, 880, 930–1; gun drill 41, 127, 411, **416–19**, *418*
gunpowder 126–7
shot 29, 75, 318, 323
trunnions 75, *76*, *77*
underwater illustration *in situ* 25
use 138, 897, 899, *915*, 915–16
wadding 75
*see also* gun crew
*see also* falconets; falcons; sakers
Curson, Lord 308, 309
curtall (type of cannon) 45
curtows 4, 7, 19, 858, 860
inventory entries (1514) 7, 856, 858, 861
cutlers 740, 780

daggers 740–8, 750–1, **769–91**
anatomy *742*
ballock 774–91
blades 747, *787–8*, 789
by-knives 746–7, 747–8, *747–9*, 778, *778*, 779
complete, from spoil mound 935, 938–9, *939*
cross-sections *777*
distribution *752*, *778*, 780–1
grips 763, 774, 778, *781–3*, 786, 788, *790*, 791
list with details 775–7
lobes, lobe guards and pinning 774, 778, *784*, *786*, 788–9
pommels *780*, *786*, 781
similarity of finds on *Mary Rose* 744
tang buttons *787*, 789
blades *742*, 747, *787–8*, 789
chapes *742*
distribution 748, 751, *752*, *754–5*, 763, *778*, 780–1
grips *742*, 763–4, *764–6*, 778, *781–3*, 786, 788, *790*, 791
human remains associated with 751, *754–5*, 780–1
*Landsknecht* type 747, 763, 764, 772, *773*, 774
lockets; Hook Court dagger 809
personal ownership 780
production 740
provision 744
radiography 747, *786*, *788*, 788–9
rondel 747, 764, 769, *770*, 771–2
for royal guard 740, 743

scope and importance of assemblage 740–8
straight quillon type 763, 772, *772*
swords worn as well as 751
undated possible quillon (05A0023) 931
*see also* sheaths *and under* by-knives; pommels
Dardanelles Gun *859*, 859, 860, 955
darts and arrows, incendiary 520, *520*, **523–9**, *524–30*, 530, **699–702**, *700–2*, *714*, 718
analysis of incendiary mixture 531
distribution 272, *525*
fired by matchlocks 521, *523*
future research 689, 932
inventory entries 310, 521–2, 523, 580, 719
longbows suitable for firing 603, 605
wadhook hafts possibly mistaken for 409, 931
data management system; Site Recorder 937, *937*
Deane, John, and associates (Charles Deane and William Edwards)
bomb used by 157, **531–6**, *532–3*, *535*, 905
damage to site 82, 103, 942
items recovered
debate over ownership 64–5
guns
bronze *25*, *26*, 26, *28*, 47, 64–5, 69, 87–8, 930; *see also* cannon (79A1276; 79A1277)
wrought iron *28*, 131–2, 134, 165, 168, 171, 180–1, 202, 217; *see also under* port pieces
iron wedges 195, 285
other items 165, 583, 920
decks, number and layout of *862*, 865, *865*, 930
*see also individual decks and under* Forecastle; Sterncastle
dendrochronology 579, 930
Deptford dockyard 1, 2, 8, 9, 865
Descharges (Breton shipbuilder) 8, 14, 863
development of naval ordnance 15, 19–22, 851–5, 928
cast iron guns 22, 89, 101, 132, 291, 292, 864
importance of *Mary Rose* to 12–15, 854–5, **856–65**
*see also* tactics, naval
devices 29–35, 924
Garter 29–31, *32*, *33*
rose and crown 29, 31, *32*
shield 29–31, *33*
*see also* royal emblems *and under* wristguards (decoration)
digitisation of archive 930
Direct Survey Measurement data 932
disease 12, 632
dockyards, naval 1, 2, 8, 9, 865
dolphins (lifting lugs) 23, 27, *30*, 31, *31*
casting 112, 113, *116*
*Double Rose* 225, 300, 301, 424, 438
Dover 20
Henry VIII and Charles V at (1522) 3, 9, 865
swivel gun from sea in 223
*see also Embarkation of Henry VIII from Dover...*
Downs, the 9, 865
*Dragon* 225, 291, 300, 301, 424
drake (type of cannon) 24
draw-knife 94, *95*

dredging in Solent, proposed new *932*, 934
Du Bellay, Martin 866, 875–6, 877–8, 879–82
Dudley, Sir John *see* Lisle, Viscount
Dürer, Albrecht 805, 806, *806*, 809

East Tilbury Bulwark, Kent *20*, 302
edged weapons 740–817
*see also* belts, sword; by-knives; daggers; hangers; scabbards; sheaths; swords
Edward I, King of England 593
Edward III, King of England 30, 32, 593
Edwards, William
guns raised by 47, 64–5, 69, 87–8, 90, 91, 131–2, 134, 137–8
mortar bomb 157, **531–6**, *532–3*, *535*, 905
*see also* Deane, John, and associates
elder, arrows of 669, 674–5, 677, 678, 685
Elderton, Thomas, of Ordnance Office 310, 856
Eldred, William; *The Gunner's Glasse*
on bases 223
on bronze vs. iron guns 291
on gun drill 127, 411, 413, 417
on gunner's equipment 401, 409, 446, 449, 476
on gunpowder 427
on ranges of guns 875, 876, 879, 881
elevating posts 135, *135*, **195**, *196–7*, **197**, *199*
in experimental gun drill 413, *414*, *415*, 416
elevation of guns
gun carriage design affects 43–4, 54–5
physical constraints 287, 891, 893–4, 896, 900
and range 875, 876, 879, 881
elevations of *Mary Rose* 861, *863*, *864*, 865, 882
elm
gun carriage parts 39, 40, 94, *96*
technical characteristics 675, *676*
*Embarkation of Henry VIII from Dover for the Field of Cloth of Gold, The* (painting) 7, 867
buckler carried by royal guard 656
lidded gunports 14, *14*, 895
linstocks 486, *486*
staff weapons 713, 718
endurance of guns 924, 925–6
Erith storehouse 2, 200
experimentation *see* replication
eyelets 807, 808

faceting of gun barrels 35, *36*, 75, *76*, 83, *83*, *84*, 90, *90*, 91
*Falcon*
archery equipment 580
guns 101, 199, 225, 300, 301, 437, 438
munitions 199, 331, 424, 437, 438
*Falcon in the Fetterlock* 424, 581
guns 101, 102, 225, 300, 301, 438
falconets **858**
cast iron 291
fortification 428
inventory entries 7, 856, 861
Norton classifies as culverins 24, 74

relative size 74, 201, 860
shot size and weight 214, 311
weight of gunpowder 428
falcons **102**, *102*, **859**
associated items 247
ladles 448, 449, *450*, 460
rammers 456, 460, 462
shot 102, 310, 311, 312, *325*, *326*, 326, 336, 360, 429, 462, 925
sponges 460
carriages 102
dimensions and weight 21, 314, 428, 429, 462
gunpowder 426, 427, 428, 429, 430
introduction and development 292
inventory entries 7, 102, 856, 861
*Anthony Roll* 25, 310, 922
archaeological evidence compared 5, 27, 36, 925, 931
shot 310, 312
Norton on 21, 24, 74, 102, 429, 875
on Portsmouth fortifications 881
possible locations 27, 908, 909, 910
range 314, 429, 875, 876, 880, 881, 918
relative size 201, 860
shipped to Scotland (1497) 860
in weapon preference sequences 921–4
feathers for arrow flights 579, 671, 693
Feature 13 *140*, *141*
Fermer, Richard (merchant, *fl.* 1513–14) 308, 309
Fevers, Peter (armourer at Greenwich) 827
Field of Cloth of Gold 9
painting, *The Field of Cloth of Gold* 656, 737, 754, 759, 822, 867
*see also* Embarkation of Henry VIII from Dover...
fighting the ship **851–927**
archaeological evidence 882–913
arcs of fire 914–20
gundeck layout and operation 895–913
gunports and 887–95, *888–90*, *892*
historical evidence 851–82
changes in armament 856–65
performance of *Mary Rose* in battle 865–82
tactics 851–5
rules of engagement 921–7
shot as evidence 921–7
stability issue 14, 867, 882, 883, 913–14, 934
storage of gun-related equipment 895, *896*
*see also* sinking of *Mary Rose*; Solent, battle of the; tactics, naval
fighting tops 235, *271*, *272*, 851
*see also* top darts
fire trunks or tubes *520*, 520–1, *522*
firearms, hand-held *see* gun-shields; handguns
fireballs 24, 520, *520*
firepots *520*, 521, *528*
firesticks 529
fireworks 521
firing of guns *see* loading and firing of guns
firkins with purses (barrels for gunpowder) 6, 507

fitting out of *Mary Rose* 8
flagship, *Mary Rose* as 8, 9
Flanders 19, 820–1, 825, 827
*see also* Antwerp
flask *see* powder flask
Fletchers, Worshipful Company of *585*, 585, 663, 665
flint
fragments in canister shot 370, *372*, 379–80, 384
gun flints 503
Flodden campaign 8–9, 715
Flodden Window, Middleton, Greater Manchester 662
floor timber, Y-shaped *937*, 940
*Flower de Luce* 225, 300, 301, 424, 438
Forecastle
*Anthony Roll* illustration *3*, 6, 918, 943, *944*
gun carriage elements 27, 103–5, *105*, *106*, *107*
guns possibly located in 6, 89, 227, 229, 908, 909, 910, 913
Hatfield House document on 199, 204, 943, 944, *944*, *945*
potential location on seabed 934
second deck, postulated *3*, 943–4, *944–5*, 945
terms used for 860
forelocks 135, *135*, **192–5**, *192–4*
*Anthony Roll* lists timber for 6, 507
distribution 192, 195, *197*
for replica port piece *284*, 285
in gun drill 413, *414*, *415*, 416
for 81A2604 151, *154*
for 81A2650 143, *145*, *146*, 147, 192, *193*, *194*, 195
foresights *130*, 131
formers
for cartridges 6, 409, **435–9**, *436–7*, *439*, 529
for finishing stone shot 384, 403
Fort Nelson, Hampshire 123
fortification of guns (wall thickness) 24, 25, 428
fortifications, land
and battle of the Solent 868, *868*, 870, *870*, *871*, 874, 879, 881
Henry VIII's development of coastal 2, 13, 19, *20*
inventory (1547) 200–1, 216, 224–7, 302
*see also individual establishments and under individual types of armament*
foundries 864
Houndsditch royal 19, 22, 929
Italian, Salisbury Place, London 22
fowlers *213–21*
barrel and chamber fragments 217–21, *217–21*
carriages 216
chambers *216*, 216, 219
in concretion 219, *219–21*, 221, *261*
dimensions 216–17
Deane's find 217
development 213
identification and historical importance 132, 213–17
inventory entries 131, 213–16, 312, 856, 861, 863
1546, *Anthony Roll* 5, 131, 213–14, 922
shot 310, 339, 344

in land fortifications 216, 224, 881
location on Upper Deck 347, 909, 910
morphology and construction 134–5, *134*, *135*, 217
radiography *215*, *216*, 217, *218*, 218–19
range 880, 881
shot *215*, 218, 311, 312, *343*, 343–4, 346, 347
    in inventories 214–15, 310, 311, 312, 339, 344
shot gauges for use with 405–6
stock-fowlers 24
superseded by bronze guns 270
tampions *in situ* in chambers 219, 470
uses 921–4, 931
*veuglaires* similar to 851–2
wad 219, *220*
in weapon preference sequences 921–4
France
    alloys 19, 951–3, *954*
    Edward III's claim to throne 30, 32
    English fear of invasion (1538) 713
    Henry VIII's wars against 1, 2, 8–12, 13, 308, 594, 717, 853
        *see also* Solent, battle of the
    naval capabilities 10–11
    *see also* Field of Cloth of Gold
Francis I, King of France 2, 10–11, 713
free surface effect 878–9
Frescobaldi, Leonardo 713
Froissart, Jean; illustrated manuscript 589, 697, 699
fruitwood sword/dagger grips 763, 781
Fuenterrabia, gun cast in 114, 116
fungus, dried; use as tinder 502, 503
funnel *506*, 511
fuse 488, 490, 503
future research 930–4

*Gabriel Royal* 384, 385, 521, 920
    guns 74, 859, 860
galleon, development of 853–4, 855
galley of *Mary Rose* 502
*Galley Subtle*
    on Cowdray Engraving *869*, 869
    gunpowder 424
    guns 46, 225, 300, 301
    handguns and ammunition 543
    in Lisle's orders of June 1545 866
galleys 851–2
    *see also* Solent, battle of the (French galleys' movements)
'galliasses'
    bronze guns 74, 89, 101, 102
    handguns 543
    iron guns 137, 199, 214, 302
    tactical use 866
GARDO crest 548, *548*, 550, 551
garlands (incendiary devices) 520, *520*, 521, *521*, *524*
gauges *see* shot gauges
Gentlemen Pensioners 738, 929
geographical information on battle of the Solent 867–82
    *see also* GIS

*George*
    guns 137, 225, 269, 300, 301
    munitions 331, 424, 436, 543
Gerard of Wales 586–7, *586*
Germany
    alloys used in gunfounding 951–3, *954*
    armour 820–1, 824
    handguns 540
    *see also* Landsknechte
*Gillyflower*
    archery equipment 581
    guns 225, 300, 301, 438
    munitions 424, 426, 427, 438
Giovanbattista of Ravenna 554–5
Girdlers' Company 659–60, *659*
girdles
    for sheave arrows 693, 697, 699
    sword belts 808–9
GIS (Geographical Information System) **870–4**, 934
    creation 870–2
    and geography of battlefield 872–4
    positions of vessels 871–2, *873*, 875, 881
    routes for final passage of *Mary Rose* 874, 878–9
glues for arrow-making 579, 669, 675, 693
Goldsmiths' Company 659
Gonson, Benjamin (acting Surveyor of the Navy) 942, 943
Gonson, William (Keeper of Erith storehouse) 2, 9
Goodwin Sands gun 227, 270, *270*
Goore, Thomas (king's gunstone maker) 309
*Grace Dieu* 13, 593–4
Grampis, Filippo de (armourer at Greenwich) 827
*Gran Griffon, El* 358, 921
*Grand Mistress* 580, 581
    guns 58, 74, 225, 300, 301
    munitions 214, 424, 543
grapnel 920–1, *921*
grappling *see under* tactics, naval
*Great Barbara* 521, 580
    deck names 860, *862*
    guns 856, 857, 858
    items for finishing shot 384, 385
*Great Bark* 543, 580
    guns 58, 89, 137, 225, 300, 301, 302
    munitions 311, 424, 436, 543
*Great Elizabeth* 385, 522
    deck names 861, *862*
    guns 200, 269, 856, 857, 858, 861
*Great Galley* 866
    guns 46, 89, 101, 863
*Great Nicholas* 521, 580
    guns 856, 857, 858
    shot 331, 385
Great Ships 13, 928–9
Greek fire 519–20
Greenwich; royal armoury workshop 824, 827
grenades (firepots) 520, 521, *528*
*Greyhound* 582
    guns 102, 137, 225, 269, 270, 300, 301

munitions 269, 311, 331, 424
grips, wooden *741*, **750–1**, **763–4**, *764–6*, 931
   distribution *752*
   observed as stain *745*, 746, 751, 759
   significance of assemblage *746*, 746–7, 748
   scabbards associated with 791–4
   wood species 763
Grose, Francis 698, *699*
guard, royal *see* bodyguard, royal
Guido da Portenari 820
guilds and livery companies 581, 639, 712
   Fletchers' *585*, 585, 663, 665
   insignia on wristguards 653, *659*, 659–61
   Stationers' 656, 659
Guilmartin, J. F. 411–12, *412*, 413
Guisnes Castle, France 216, 358, 381, 582
gun carriages and elements **37–45**, **102–12**
   anatomy 39, *41*, *138*
   beds 39, 40, *41*, 41–2, 45, 103, *104*
   for bronze guns 26, *26*, 37–45
      loose parts 89, 102–12
      reconstruction and replication 71–3, *71–3*, 76–82, *77–82*, 94–100, *94–100*, 123, 129
      *see also under* cannon; culverins
   cap squares *37*, 39, 40, *41*
   cheeks 39, *41*, 104, *105*
      rear stepped 39, 40, *41*, 42, 43, 44–5, 103, 104, *106*
      single piece 129
      trunnion support 39, 40, *41*, 42, 43, 44–5, 103, *104*
   dimensions 42, 142
      affected by location 44–5, 55
   distribution 27, 43–4, 89, 91, 102–3, *105*, 943
      axles 102–3, *105*, 105, *106*, 107, 109, 936, 939
   elevation of gun affected by 43–4, 54–5
   experimental assembling 41
   fitting to gun 43, 54, 55, 61, 134–5, *135*, 284, *284*
   fragments from Timber Park 936, 939
   guns *in situ* on 25, *26*, 134, 137–8, 897
   importance of *Mary Rose* carriages 15, 929
   inventory entries 36, 37, 39
   for iron guns 134, 216
      fitting of gun to carriage 134–5, *135*
      morphology 135, *135*
      *see also under* fowlers; port pieces; slings
   isolated elements **102–12**
   missing guns' locations indicated by 26, *26*, 91, 103–5
   Norton on 132–3
   parts 39, *41*
   radiography *57*, 57, 61, 97, 82
      replication 41
      for cannon 71–3, *71–3*
      for culverins 76–82, *77–82*, 94–100, *94–100*, 123, 129
      for port piece *284*, 284–5
   re-use 36
   ropes 49, 133
      marks 896, *899*
   storage 107–8, 109
   *see also under* wheels, spoked

transom bolts *41*
wedges 39, 40, *41*, *284*, 285
wood species used 39, 40
*see also* axle trees; axles; elevating posts; forelocks; trucks; wheels, spoked; *and under* cannon; culverins; fowlers; guns, bronze; guns, wrought iron; port pieces; slings
gun crew
   size and composition 411–12, 413, 415–16, 418, 647, 923
   skeletal remains 708, 835, 847, 902
gundecks
   identification and names 860–1, 862
   layout and operation 895–913
   Castles 910–13
   Main Deck 895–907
   Upper Deck 907–10
gun drill **410–19**
   culverins **123–9**, 411, **416–19**, 449
   port pieces 285–7, **411–16**
gun flints 503
gunfounders
   alloys 117, 950–4, 955
   continental, working in England 19, 22
      *see also* Arcano family *and members*; Baude, Peter
   Henry VIII's support to 928–9
   inscriptions with names 29, 31, 33
   of Kent/Sussex Weald 19, 22, 291, 929
   *see also* Owen, John and Robert
gunmetal *see* bronze; iron
gunner's rule *392*, *393*, 393–400
   location *389*, *439*, 934
gunports **887–95**
   *Anthony Roll* illustrations 6, 14, 138, 140, *144*, 885, *886*, 887
   Bassett Lowke model *204*
   broadside 13, 14
   dating impossible 930
   dimensions 887
   elevation of guns constrained by 287, 891, 893–4, 896
   elevations *888–9*
   and fighting the ship 887–95, *888–90*, *892*
   fittings *890*, 896, *899*
      for operating lids 147, 894–5, *894*
   for handguns *882*, 891, *892*, 907, 910
   inventory evidence 6, 7
   lids
      *Anthony Roll* illustrations 138, 885, *886*, 887
      construction *883*, 891
      dimensions 891
      and fighting the ship 887–95, *888–90*, *892*
      introduction and development 8, 13, 14, *13*, *14*, 863, 887, 928
      open at sinking 12, *25*, *26*, 855, 866, 867, 878, 879, 882, 892
      operation 147, 887, 894–5, *894*
      and placement of guns on lower decks 19, *21*, 864–5, 885, *886*, 887
   location 6, 883, 887, 891, 930

        Main Deck 6, 8, *13*, 14, 853, *882*, *883*, 887–95, *888–90*, *892*, 903
        Sterncastle 6, 887
        Upper Deck 6, *13*, 14, *883*, 887, 891, *892*, 907, *908*
    for port pieces 137, 138, 863
    for sternchaser guns 8, 885, *886*, 887
    semi-circular *198*, 199, *204*, 892, 907, *908*
    sighting of guns hindered by 891
    waterline in relation to 892, 896
gunpowder **419–35**
    composition and type 419–20
        analyses 433–5, *434*, 961–2
    corned 5, 8, 292, 421–2, 423–30, 447–8, 543, 545, 861
        corning process 285
    distribution 529
    estimated expenditure 426, 427
    for handguns 426, 427, 543, 545
    historical evidence 423–30
    inventory entries
        1540 8, 861
        1546, *Anthony Roll* 5, 423, 424, 426, 436, 543
        1547 426–7, 430
    in *Mary Rose* 423–35
    method of use 407, 408, 409
    production 420–3
    for replica guns 126–7, 285, 287
    serpentine 5, 422–3, 423–30
    storage 425–6, 430–3, *432*, 436, 883, 895, 931
    units of measurement 425–6, 430
    weight ratio of powder to shot 124, 126, 127, 173, 200, 202, 449
    *see also* ladles, powder; powder dispensers; powder flask; powder horn; *and under individual types of gun*
guns 19–306
    classification; alternative names or types 24–5
    *see also* guns, bronze; guns, cast iron; guns, wrought iron; individual types of gun, *and individual topics throughout index*
    guns, bronze **19–129**
    anatomy *23*
    cascabels *23*, *24*, *28*, *34*
    casting techniques **112–16**, *115*, *116*
    classifications 22–5, 35–6
    Continental production 19, 951–3, *954*
    costs 21–2, 130, 864
    dating 28, 29, 33
    decoration and devices 27–8, 29–35, *32–3*, 112, 113, *116*, 116, 929
    development 19–22
        supersede wrought iron guns 136, 200, 213, 216, 270
    dimensions and weight 21, 29, 36
    distribution 25–9
        Main Deck *26*, 26, *27*, 28, *28*, 897, 899, *900*
        possible pairings 35–7
        Sterncastle *26*, 27–8, *27*, *28*, 36–7, 103
        Upper Deck *26*, 26–7, *27*, *28*, 36–7
    fortification 24, 25, 428
    gunpowder expenditure 426, 427
    powder to shot ratio 428
    inscriptions 31–3, *32*, *33*, 929
    inventories 35–6
        1514 7, 35–6
        1540 7–8, 36, 856
        1546, *Anthony Roll* 8, 22, 25, 35–6, 856
            comparison with assemblage 5, 6, 36, 104–5
    iron gun types later cast in 136, 200, 213, 216, 270
    little brass gun without chamber 856, 859
    mountings *see* gun carriages
    parts *23*
    proofing 114, 116
    replication *see under* culverins
    salvage
        Tudor 36–7, 931
        19th century *25*, *26*, 26, *28*, 47, 64–5, 69, 87–8, 930
    symbolic importance 929
    uses 15, 897, 899, 921–4, 929
    *see also* bombards; cannon; culverins; curtows; serpentines *and under* gun carriages; metallurgy
guns, cast iron **291–306**
    development and production 19, 21–2, 130, 200, 291–2, 864, 929
    inventory entries 22, 131, 292, 300, 864
    *see also* hailshot pieces
guns, wrought iron 22, **130–290**
    accessibility problems 134
    breech-loaders 131
    bronze guns supersede 136, 200, 213, 216, 270
    component parts *130*, 131
    conservation methods 134
    costs 130, 864
    dating 929
    distribution 134, 192
    elements raised since 1982 134
    gunpowder type and expenditure 426, 427
    identification and historical importance 22, 131–3, 929
    increase in production 19
    inventory entries 214
        1514 7, 856
        1540 7, 8, 137, 300, 856, 861, 863
        1546, *Anthony Roll* 5, 6, 22, 137, 214, 856
        1547 131, 137, 292
    order of types 214
    manufacturing method 130–1, 274–84
    material 274, 283
    metallurgical composition 130, 131, 170
    morphology 134–6
    predominance in Henry VIII's navy 22, 300
    radiography 133
        *see also under* bases; fowlers; port pieces; slings
    and rediscovery of wreck 930
    reload speed 131
    salvage, 19th century *28*, 131–2, 134
    scope of assemblage 133–4
    size limited by method of construction 283–4
    uses 15, 921–4
    welding process 130–1

*see also* bases; fowlers; gun carriages (for iron guns); port pieces; slings; top pieces
gun-shields **553–77**
    assemblage 564–73, *565–72*
    buckles 563–4, 838
    comparison of RA V.34 and RA V.38 573–7, *574–5*
    distribution of elements 554, *559–60*, 560–1
    handgun elements 564, 567
    measurements 564
    method of manufacture 576–7
    morphology 555–8, *556–7, 558*
    nails 564, 568, *559, 560–1, 569*
    pouch-pockets 557, *559*, 560, 561–2, *562*, 564, *565*, *565*, 567
    radiography 560, *561*, 568, *570*
    and royal guards' possible presence 929
    scope and significance of assemblage 553–5, 929
    shot 368, *559*
    springs *555, 557*, 560, 561, 564, *565*, *565*, 568, 573
    in storage 554, 560, 895, *896*
    straps and grips *556, 557*, 562–4, *563*
    XRFS analysis 561, 573, 577
gun stations 895–913
    Castles 910–13
    Main Deck 895–907
    Upper Deck 907–10

H mark 314–15, *315–17*, 432, *432*, 433
hafts 737
    possible confusion with darts 409, 530, 931
    unusual, possibly incendiary darts 530
    *see also under* bills; pikes
hagbusshes 368
hailshot 337, *337*
    *see also* shot (canister)
hailshot pieces 132, **292–306**, 929
    analysis 302–4, *303*
    concretion removal 298, *299*, 301
    dimensional data 294
    distribution 219, 294–5, *296*, 895, 913, 934–5
    endurance 925
    fuse in touch hole 503
    future research 931
    gunpowder type and expenditure 305–6, 426, 427
    identification 132, 291, 301
    importance 132, 301–2
    introduction and development 292, 856, 864
    inventory entries 300–2, 856, 864
        1546, *Anthony Roll* 5, 131, 214, 291, 294, 300, 301, 302, 922
    manufacture 303–4
    metallurgy 302–4, *303*
    method of use 292–3, *293*, 465
    morphology 295, 297, *297*
    numbers 214, 300, 302
    radiography 293, *297–301, 298–9, 303*, 303, *337*, 337
    replication 304–5, *304*, 305–6, *305*
    shot 132, 291, 293, *295*, 309, *337*, 864
    powder to shot weight ratio 305, 306
    radiography 303, *303*
    tampions 293, *295*, 465, 470, 472, *472*
    in Tower 292, 295
    wads 293, *295*, 472, 475, *476*
    79A1088 292, *293*, 294, 295, *295*, 296, 297–8, *298*, *337*, *465*
    80A0544 292, *293*, 294, 295, *296*, 297, *298*, 304
    81A1080 292, *293*, 294, 295, *296*, 298, *298*, 503
    90A0028 219, 292, *293*, 298, *299*, 303, *303*
halberds 715, 736–40, *737–8*
    anatomy *715*, 718
        example in *Mary Rose* 714–15, 718, *718*, 719, **736–40**, *737*, 929
        absence from *Anthony Roll* 5, 6, 719
        location *730*, 737
        possible ceremonial nature 719, 737
hammers 6, 740
    *see also* picks and pick-hammers; sledgehammers
hand to hand fighting
    weapons 713–817
        edged 740–817
        staff 713–40
    *see also individual types of weapon and under* tactics, naval
handguns **537–53**
    anatomy *538*
    assemblage 545–53
    associated objects 544–5, 551, 553
    bolts, possible 699–702, *700–2*
    control of ownership 537, 545
    distribution 235, *544*, 545, 895
    fishtail butt type 539–40, *539*, 544, 545, *550*, 551
    German forms 542, *544*, 545
    gunports for *882*, 891, *892*, 907, 910
    gunpowder 426, 427, 543, 545
    Henry VIII's purchases 537, 540–2, 544
    incendiary projectiles 521, *523*
    inventory entries 368, 543–5
        *Anthony Roll* 5, 537, 543, 580, 581, 922
    Italian forms 540–2, *544*, 545, 549, 929
    longbow superseded by 537–8, 543, 588–9, 632
    powder dispensers 479–84, *480, 482, 483*, 544, 551, 553
    radiography *546*, 547
    replication and firing 553, *553*
    scope and importance of assemblage 537–43
    shot 310, 367–8, 551, *552*, 922
    stocks **538–43**, 545, 546, 547–51, *546–50*
        *see also* harquebuses; matchlock; pistols
handsaw, frame 94, *95*, 97
handspikes 409, *503*, 504, *505*, 508, 932
    function 407, 413, *415*, 416, 504
    hangers *741*, 806, 809, 815, 935
    with human remains 708, *812*
    significance of assemblage 748
    with scabbard 808, 809, *812*, 811, 813, 814
Hanseatic League 537
*Hare* 424
    archery equipment 580, 581

guns 225, 300, 301, 438
*Harp* 581
    guns 225, 300, 301, 438
    munitions 424, 438
harquebuses 540–3, *541*, 543–4, *546–7*, 547–8, 548–51
    GARDO crest 540, *541*, 550, 551
    location *544*, 545
    replication and firing 553, *553*
*Hart* 580, 581
    guns 58, 199, 225, 269, 291, 300, 301
    munitions 199, 311, 331, 424
Hasleworth Castle *873*, *874*, 881
Hastings, battle of 586
hatches
    central 883, 895, 896, *898*
    small, in Upper Deck above guns 887, 907
    Hatfield House document (Cecil Papers 201/127) 10, 12, 199, 200, 204, 930, 942–5
Hawthorn 225, 300, 301, 424, 438, 581
hawthorn arrows 674, 675
hazel 720, 763
helmets 819, 820, 821, 825, 847
hemp for bowstrings 635–6
Henry V, King of England 343, 587, 745, 762
Henry VII, King of England 29, 31, 593–4
Henry VIII, King of England
    at battle of the Solent 12, 879
    devices 29, 31–3, *32*, 59–60, *59*, *60*
    divorce 1–2
    foreign policy 10
        *see also* Field of Cloth of Gold; France (Henry VIII's wars against)
    fortification of coast 2, 13, 19, *20*
    handgun purchases 537, 540–2, 544
    interest in armaments 19, 48, 588–9, 713, 928–9
    inventory (1547) 7, 19, 23
        *see also under individual types of object*
    and Ireland 32–3
    in *Mary Rose* at Dover (1522) 3, 9, 865
    naval development programme 2, 13, 19, 22, 928–9
        orders adaptation of ships for heavier ordnance 199, 200, 204, 930, 942–5
    royal style 31–3, 59–60, *59–60*
    Silvered and Engraved armour 827
    *see also* bodyguard, royal
*Henry Eights Carbine* 92
*Henry Grace à Dieu*
    Anthony Roll illustration 4, 14, 138, 272, 921
    archery equipment 580, 581, 719
    armaments from decommissioned ships 856
    armour 822
    at battle of Solent *874*, 875–6, 881
        Cowdray Engraving *854*, 855, *868–9*, 869, 897, 899
    fighting tops 272
    grapnel 920
    gunports with lids 14
    gunpowder 424, 436
    guns 860

    bases 225
    bombards 859
    cannon 46, 58, 930
    culverins 74, 89, 860
    curtows 858
    falcons 102, 859
    forward of rise up to Forecastle 138
    hailshot pieces 300, 301, 302
    handguns 543
    port pieces 137
    sakers 101, 323
    serpentines 857
    sling 200
    sternchaser 885, *886*
    handguns 543
    hand weapons 521, 543, 719
    Henry VIII boards at Dover (1522) 865
    incendiary devices 521, 523, 311
    inventories 7
    ladles 410
    in Lisle's orders of June 1545 866
    shear hooks 921
    shot 214, 310–11, 331, 337, 385, 436
    staff weapons 521, 719
*Herald of Free Enterprise* sinking 878–9
hilts *see* grips, wooden
*Hind* 225, 300, 301, 424, 437, 438
Historic Wreck site, protection of *932*, 934
Holbein, Hans 762, 805, 806, 807, 809
Hold
    archery equipment 583, *584*
        arrows and chests 667, *668*, 669, 691, 693
        longbow 611, 612
    armour *826*, 832, 835, 842, *845*, 846, 848
    base elements 230, 231, 235
    casks 431, 433, 931
    gun carriage elements 102–3, 108, 109, 112, 190, *191*, 192
    gun-shield elements 559
    gunpowder and incendiary items 529, 931
    human remains 706–7, 848
    linstocks 488, *498*, 496, 497
    scabbards 792, 796
    shot
        composite inset *356*, *361*
        iron 102, *324*, *325*, *327*, *328*
        lead *369*, 370
        stone *340*, 346
    swords, daggers and related finds 748, *752*, 748, 750
    tampions 466, 467, *468*, 469
    tinder box 502
    tools for working guns *505–6*
    wristguards *646*, 647
holy water sprinkler (form of mace) 739–40
Honourable Artillery Company 3
Hooke Court, Dorset; ballock dagger 748, *749*, 789, 805, 809
Hopton, John 8, 9, 309
horn, items of
    wristguard 578, 645, 665

*see also* arrows (nocks, horn inserts)
hornbeam
    arrows 669, 674–5, 677, 678 685
    technical characteristics 626, 676
Horton, Northants; effigy of William, Lord Parr 747–8, *748*, 751, 808, 809, 815
Houndsditch; royal gun foundry 19, 22, 929
Howard, Sir Edward 2, 8, 865
Howard, Thomas (Earl of Surrey) 8, 9
*Hoy Bark*
    archery equipment 582
    guns 101, 137, 214, 213, 225, 269, 300, 301
    incendiary devices 523
    munitions 424, 436
hull *882–5*, 883, 928
human remains
    items associated with
        aiglets 832, 835
        buckles 838
        grips 763
        jerkins 707, *707*, 708, *710*, 712, 847
        powder dispenser 479, 483
        priming wire 479
        signet ring 707–8, *709*
        *see also under* armour; arrow chests (distribution); arrow spacers; arrows; daggers; hangers; knives (sheath); linstocks; longbow chests; longbows (distribution); scabbards; swords; wristguards
    *see also* archers (skeletal remains); gun crew; pathology; *and under* Hold; Main Deck; Orlop Deck; Upper Deck
Hundred Years' War 30
Hunsdyche *see* Houndsditch

illness in fleets 12, 632
imports from Continent
    armour 820–1, 824, 825, 827, 828
    daggers 747, 763, 764, 772, *773*, 774
    handguns 540–2, *544*, 545, 549, 929
    metal for guns 953–4, 960
    staff weapons 713, 718
    yew and hemp for longbows 588, 591, 635–6
incendiary devices **519–36**
    historical importance 521–3
    ingredients of combustibles 519–20, 531
    inventory entries 519, 520, 521–3, 531
        1546, *Anthony Roll* 522–3, 719
    in *Mary Rose* **523–36**
    shot 311, *334*, 520, *521*, 521
    staff weapons combined with 520, *520*
    types of weapon *520*, 520–1
    *see also* darts and arrows, incendiary; mortar bomb
information management system; Site Recorder 937, *937*
Innsbruck; armour production 824
insignia *see* devices *and under* guilds and livery companies; wristguards (decoration)
intelligence in battle of Solent 881
inventories 23
    absence of items suggests personal ownership 499
    changes in armament 7–8, 35–6, 856–65
    size hierarchy of types of gun 202, 214, 860
    1495, of *Sovereign* 861
    1514 7, 23, 856–61
        items unique to 856–7
        location of guns on named ships 856, 857, 858
        and mothballing of *Mary Rose* 309, 310
        *see also under individual types of object*
    1540 9, 23, 580
        *see also under individual types of object*
    1546 see *Anthony Roll*
    1547 (of Henry VIII) 7, 19, 23
        *see also under individual types of object*
    1547 (Roxburgh) 523
    1555 (Longleat Miscellaneous MS V) 23, 301, 302, 331, 582, 864
    1558 270
    1574 and 1576 lists of ordnance 270
    1633, of Earl of Pembroke 715
    *see also under* archery and archery equipment; armour and personal protection; arrow chests; arrows; bases; bills; Boulogne; bowstrings; cannon; culverins; curtows; darts and arrows, incendiary; falconets; falcons; fortifications, land; fowlers; gun carriages and elements; gunports; gunpowder; guns, bronze; guns, cast iron; guns, wrought iron; hailshot pieces; handguns; incendiary devices; linstocks; longbow chests; longbows; pikes; port pieces; rammers; sakers; scabbards; serpentines; shot; shot moulds; slings; soldiers on *Mary Rose*; Solent, battle of the; Southsea Castle; sponges; swords; tampions; top pieces/top guns; Tower of London
Ireland 32–3
iron
    cast
        development of industry 291, 864
        introduction of guns of 89, 101, 132, 200, 292
        process and properties 22
    steel, making and hardening of 823–4
    wrought; properties 130–1, 170
    *see also* guns, cast iron; guns, wrought iron
Ironbridge Gorge Trust; port piece chamber 181
Italy
    alloys used in gunfounding 951–3, *954*
    armour 820, **824**, 825, 827, 828
    handguns 929
    staff weapons 713, 718
ivory objects
    comb 712
    *see also under* wristguards

jacks (brigandines) 818, *819*, 838
Jamnitzer, Wenzel 393
javelins 521, 715
*Jennet* 9
    guns 101, 225, 291, 300, 301
        proposed increase in weight 942, 943, 944

munitions 331, 385, 424
jerkins, leather 821
    aiglets associated with 832, 835
    comb impression *710*, 712
    on bodies of
        archers 707, *707*, 708, *710*, 712
        gun crew 847
Jermyn, Thomas (Clerk of the Ships) 9
*Jesus (of Lübeck)* 543, 866, 921
    guns 46, 137, 200, 215, 225
        hailshot pieces 300, 301, 302
        sternchaser 885, 910
    munitions 424, 436
*John Baptist* 920. 921
    guns 200, 269, 580, 858, 859, 860
        locations 856, 857
    shot 331, 385
Johnson, Cornelis 43, 269

*Kateryn Forteleza* 521, 580, 920, 921
    guns 74, 858
    munitions 331, 385
*Katherine Galley* 580, 860
keel bolt 50
Kentish Ragstone 339, *342*, 343, 385
Ket, Robert; rebellion 745, 761
knees 883, *884–5*, 930
knives 935, 938
    sheath, associated with human remains 708, *709*
    in tinder box 499, 501, *502*, 502–3
    *see also* by-knives
knots in sword suspension systems 807
Kraek, W.A. 529, *530*
*Kronan* 496, 822

Lad, John and Christopher 23, 223, 393–4, 429, 875, 876
ladles, metal-working 381, 383
ladles, powder **439–54**, *440–1*, *446*, *450–3*, 503
    and cartridges 435, 436, 441, *441*
    components and materials 444–5
    construction method 450, *451–3*, 454
    distribution *442*, 446
        in M11 247, 313, 438, *439*
    identification and historical significance 439–41, 446
    inventory entries 410, 436–7
    matching to guns 446–50, 460
    metal analysis 446, 447
    method of use 407, 408, 409, 417, 418
    rammers combined with 441, *441*, 446
    scope of *Mary Rose* collection 444–5, 446
    turned heads 454
*Landsknechte* 806, 811
    *Landsknecht* style daggers 747, 763, 764, 772, *773*, 774
lanterns 502, 529
*Lartique*
    fighting tops 272
    guns 137, 225, 291, 300, 301
    munitions 424, 436

last (unit of weight for gunpowder) 425–6
lead rolls (raw material for shot) 384, 393
legitimate pieces 25
Leland, John 881
*Less Bark* 269, 301, 302
*Less Galley* 866, 869, *869*
*Less New Galley* 866
*Less Pinnace* 214, 424, 438
    guns 102, 214, 225, 291, 300, 301, 438
lessened pieces 25
levelling guns using priming wire 479
Levett, William 292
lifting of guns 37
Lightfoot, Humphrey (*fl.* 1513–14) 308, 309
lime pots 6, 521, 522, 523
Lincoln, incendiary dart from 530–1, *530*
linstocks **485–99**, *486–7*, *490–6*, 929
    cord or slow match attached to 488, 490, 503
    descriptions 487–96, *490–5*
    distribution 161, 162, 488–9, 496–9, *498*, 529, 895
    finished and unfinished 499
    human remains associated with 497, 499
    inventory and historical evidence 410, 486–7, 499
    method of use 407, 408, 409, *415*, 417
    personalisation and ownership 410, 497–8, 499
    wood species 488–9, 490
*Lion/Roselion* 424
    guns 101, 102, 199, 214, 225, 300, 301, 302, 863
Lisle, Viscount (John Dudley) 2, 9, 11, 12
    The order of battle upon the sea (June 1545) 851, 855, **865–6**, 869, 872, 878
Littis, Giovanni Angelo de (armourer at Greenwich) 827
livery arrows 581, 582, 665
livery companies *see* guilds and livery companies
loading and firing of guns **407–503**, 923, 929
    distribution of items associated with *439*
    loading sequence for chambers 178, 464–5, *465*
    powder to shot weight ratio 124, 126, 127, 173, 200, 202, 449
    reloading 131, 412, 413, 415, 899–900, *900*, 900–1, 926
    rate 2, 131, 309, 923, 925
    replica guns
        culverin **123–9**, 880, 914, 930–1
        gun drill 127, 411, **416–19**, *418*
        port piece 195, 283, **285–90**, *286–8*, 411, 412, 413, 415, 899–900, 930–1
        two compared 123–4, 127, 129
    windage and 124
    *see also* fuses; gun drill; gun flints; gunpowder; ladles, powder; linstocks; powder dispensers; priming wires; rammers; ranges of guns; shot; sponges; tampions; tinder boxes; wads; worms
London
    Billingsgate, wristguard from 651
    bronze foundries 19, 22, 864, 929
    Mile End muster (1539) 537, 718
    Mousehold Heath mustering ground 745, 761
    sword and dagger production 740

*see also* Deptford dockyard; Greenwich
longbow chests **578, 641–4**, *641–3*
    construction 641–2, *642*
    distribution *641*, 583, *584*, 883
        and human remains *578*, 579
        and wristguards *646*, 647
    inventory entries 580, 582, 642
longbows **586–641**
    anatomy 589, *590*
    and arrows
        matching to bow 579, 617–18, 630, 665–6, 929, 931–2
        number per bow 580, 581, 582
    assemblage **594–618**
    and bowstrings 639–40
    bracing height 579, 601, 617
    catalogue of bows examined 596–9
    conservation methods 594
    cross-sections 589, 592, *592*, **601, 604, 606–8**, 619, 620
    data, table of 596–9
    Deane's recovery 583
    dendrochronology 579, *625*
    density 579, 626–7, 632
    description 589–93
    distribution **583–6**, *584*, 594, 910
        and human remains *578*, 579
        marked 610–11
        used 616
    draw-weights 579, 582, 589, **616–17**, *616*, 622, 623–6, 629, 637, 929
    fine and robust 619–20, 639
    form 579
    grip area 602–3, 605, *605*
    growth rings 579, *625*
    handguns supersede 537–8, 543, 588–9, 632
    handled 589, 601, 604, 602–3, *605*, 605, 608, 609
    historical significance of assemblage **586–9**
    importance of assemblage **578–9, 586–9**, 929
    and incendiary arrows 603, 605
    inventory entries 580, 582
        *Anthony Roll* 5, 6, 537, 543, 580, 581, 718–19, 922
    large 589, 629
    manufacture 579, 589, 590–1, 593, **615–16**, 634, 929
    marks 593, 597, 599, 609–15, 610–14, 620, 929
    measurements 579, 595, 596–9, 600
    nocks 578–9, 589, 591–2, **605–6**, *606*, **608–9**, *609*, **632–4**, *632*, *633*
        double 597, 599, *606*, *608*, 608–9, *609–10*
        replication *636*, *638*
    personalisation 583
    pricing policy 588–9
    rate of fire 543
    recording of assemblage 594–618
    reflexed and deflexed 592–3, 600–1, *603*
    replication and firing 553, 627–9, *628–9*, *636*, *638*, 929
    scope of assemblage 578–9
    set back in the handle 592–3, 597, 599, 600, *600–2*
    size ranges by fineness of grain *615*
    slab-sided 592, *592*
    statistical analysis **618–22**
    in store 537, 583
    tillering process 591, 616
    used 616
    weight and density 626–7
    wood
        grain and quality 591, 594, 615, *615*, 616
        imported 588, 591, 635–6
        sapwood and heartwood 589, *590*, 590, *591*, 622–3
        species used 580, 588–9
    working capabilities **622–32**
        computer modelling 623–7
        strength 627–9
        velocity, range and impact 629–32
    *see also* longbow chests
lost wax process 112, 116, *118*, 119
Lucar, Cyprian
    on alloy recipes 955
    on daily firing capability 923
    on formers 435, *436*
    on gun drill 411, 412, 413
    on gunnery ranges 875, 876, 880
    on gunpowder and cartridge storage 430
    on incendiary weapons 519, 529
    on ladles 441, *441*
    on linstocks 486
    on priming 477, 481
    on rammers 454, 463
    on tampions 463
    on tinder 502
Lumps Fort, Hants 870, *871*, *874*, 881
Luttrell, Sir John; portrait 269
*Luttrell Psalter* 693, 774
Lymdens Bulwark, Hants *874*, 881

mace (holy-water sprinkler) 739–40
McKee, Alexander 199, 207, 323, 326, 930
Madden, Sir Frederic 4
magnafluxing technique 239
*Maiden Head* 225, 300, 301, 424, 438, 581
Maidstone Heath, Kent 343
mail (armour) **838–46**, *839–40*, *843*
    construction 839, *840*
    distribution *826*, 846, 842–3, *845*, 846
    fragments from spoil mounds 935, 938
    and human remains 842, 843, *845*, 846, 847, 848
    object form *840*, 841–2, *843*
    types *839*, 841, 842, 844–5
    XRFS analysis 840, 841
Main Deck **895–907**
    *Anthony Roll* illustration 6
    alterations, dating of 930
    archery equipment
        arrows 667, *668*, 669
        associated with human remains 585–6, 647, 667, *668*, 669, 704, 706, 707–8, 712
        longbows 583, *584*, 611, 612–13
        spacers *668*, 669, 693, 694–5, 697, 704

wristguards *646*, 647, 704, 712
armour *826*, 829, 835, 847–8
companionways *898*
dendrochronology 930
dimensions 895
gun carriages and elements 102, 108, 109, 112, *186–8*, 189, 896–7
gundeck layout and operation **895–907**
gunports 6, 8, *13*, 14, 49, *882*, *883*, 887–95, *888–90*, *892*
guns
    in action at time of sinking 934
    *Anthony Roll* indication of positions 6, *141*
    arcs of fire 914–19, *915*, *918*, *919*
    bronze
        positions 26, *26*, *27*, *28*; missing falcon 102, 247
        *see also* cannon (79A1276; 79A1277; 81A3000; 81A3003); culverins (on Main Deck; 79A1278; 81A1423)
    complement in 1540 863
    forelocks indicate positions 195, *197*
    gun stations 901–7
    introduction 7, 863
    physical constraints 883, 893, 895–6, *898*, 899, 900, 901
    reloading 900, 900–1
    in stern, possible 138, *141*, 897, *901*, 901, 916
    weight 863, 901
    wrought iron
        bases 231, 232, 338, 356, 358; *see also* bases (79A0543, 79A1075)
        *see also* port pieces (distribution, Main Deck; 81A2604; 81A2650; 81A3001)
gun-shield elements *559*
halberd haft 737
hatches, central 883, 895, 896, *898*
height 891, 896
human remains
    archers, possible 585–6, 667, 669, 706, 707–8, 712, 812
    armour-clad 847–8
    *see also* gun crew
knees 883, *884–5*, 930
munitions and gun-related items 901–7
    basket 247
    gun flint 503
    gunner's rule 395
    gunpowder and incendiary items 529, 530
    incendiary darts 272
    ladles 247, *442*, 446, *446*, 449
    linstocks 488–9, 496–7, 497–8, *498*
    powder dispenser 481, 483–4
    priming wires 477, 479
    rammers 247, *443*, 446, 457
    tampions 467, *468*, 469, 471
    tools for working guns 247, *505–6*, 509
    *see also* shot *below*
reconstruction of midships area, in museum *43*

shot
    canister 380, *372*, 373–4
    cross bar 332, *333*, 336, *362*
    composite inset 247, *356*, 360, *361*
    for falcons 102
    hemispherical *362*, 365
    iron 318, *319*, *320*, *321*, *322*, 323, *324*, 325, 326, *327*, *328*
        discussed 329, 330
    items for manufacturing and sizing of *389*, 402
    lead 247, *369*, 370
    lead wrapped *362*
    for sakers/demi slings 101
    stone 339, *340–1*, *344*, 345–6, 347
        pick for finishing 385
    stone with lead *362*, 367
    storage *312*, 312, 313, *313*, 329, 883, 895, *896*
shot gauges 384, *389*, 402, *403*, 404
staff weapons 725, *730*
swords, daggers and related finds 748, 750, *752*, 763, 781, 792–3, 794
*see also* cabins
Mainwaring, Sir Henry 521
mallets *161*, 162
    *see also* commanders
Malmstone shot moulds 387–8
manoeuvring in battle 854, 855, 866, *900*, 900–1
maple by-knives 779, 781
maps, Tudor
    pictorial, *plats* 3, 7
    of Portsmouth 867, 868, 869, 881
Margot, Guille, of Brussels (armourer) 827
*Marie of Weymouth* 13
*Marlion* 101, 102
*Martyn Garsia* 537
*Mary Fortune* 594
*Mary (of) Hamburg* 272, 581, 866
    guns 137, 225, 300, 301, 885, *886*
    munitions 311, 331, 424, 436
*Mary James* 921
    guns 137, 225, 269, 291, 300, 301
    munitions 385, 424, 436
*Mary Thomas* 272, 719
    guns 137, 225, 291, 300, 301
    munitions 424, 436
*Mary of the Tower* 13
*Mary Willoughby* 857
mass-production of armour 818, 820–1
masters of *Mary Rose* 8, 9, 310, 879
masts 934–5
    constrain gun positions 883, 895, 899
    main 50, 165, 515, 895, 899, 934–5
    mizzen 934–5
matches 486–7
matchlocks
    possible early (81A3884) 370, 538–9, *539*, 543, 551, 552, *544*, 545
    RA XII.1 538–9, *539*, 551

*Matthew* 543, 580, 866, 921
    guns 58, 89, 137, 225, 885, *886*
        hailshot pieces 300, 301, 302
    munitions 331, 424, 436
mattocks 504
mauls 94, *95*
Mechelen; gunfounders 19
Meister der Hellige Sippe 651
Memling, Hans 589, 592
mercenaries 718
merchantmen, hired 13
*Merlin* 300, 301, 225, 424, 438
metallography
    armour 823–4
    bronze guns 954–5
    mortar bomb fragments 533–4
    shot 316–17, *317*
    swords 753, *758*
metallurgy
    armour 818, 821, **823–4**, 828–9, *828–9*
    bronze guns 19, 946–60
        alloys 117, 950–3, 954–5
        comparison with European ordnance 951–3, *954*
        homogeneity and sampling strategy 946–50
        lead isotope analysis 953–4
        metallography 953–4
        provenance of raw materials 953–4, 960
    cast iron hailshot pieces 302–4, *303*
    chapes 798, 799, 800
    ladle components 446, 447
    pommels 767
    wrought iron guns 130–1, 170
    *see also* metallography
Mexico, Gulf of 203, 219, 223
Michele, Giovanni 593
Middleton, Greater Manchester; Flodden Window 662
Milan 540, 542
military service, law on 713
Millet, John, of Ordnance Office 310, 856
Milton Bulwark, Kent 20, 302
*Minion* 9, 866
    gunpowder 424, 436
    guns 58, 89, 101, 137, 225, 269, 300, 301
minion (type of cannon) 24, 104, 428, 456, 462
Ministry of Defence *932*, 934
mobilisation 2, 8, 9
    *see also* musters, shire
model of *Mary Rose*, Bassett-Lowke *21*, *204*
Molasses Reef wreck finds
    inset dice shot 358
    shot mould 348, 350, 383, 384
    wrought iron guns 223, 239, 245, 411
*Morian (of Danzig)* 424, 436
    guns 137, 225, 300, 301
morions (helmets) 818, *820*
Morlaix 9
mortar bomb 157, **531–6**, *532–3*, *535*, 905
mortar pieces (cast iron guns) 24, 292

mortise chisel 94, *95*
mothballing of ships 2
    of *Mary Rose* 9, 309, 310, 865
Mousehold Heath mustering ground 745, 761
murderer (type of gun) 4, 24, 856, **858**, 860, 861
Museum, new Mary Rose 930, 932
muskets 521, *523*
muster station, Upper Deck 579, 846–7, 895, 910
musters, shire 713, 717, 718
    Bridport, Dorset 717
    Mile End, London 537, 718
    Mousehold Heath, London 745, 761

nails, T-shaped, in bills *715*, 719–20, *720*, 724–5, 726, *727*, 731, 734
Naseby Battlefield, sword from 754, *759*
navy
    Admiralty organisation 1–2
    hiring of merchantmen 13
    in peacetime 2
        *see also* mothballing of ships
    *see also* development of naval ordnance; Henry VIII (naval development programme); Ordnance Office
nef, Nuremberg silver 270
netting, anti-boarding 851, 907, 908
nettle fibre for bowstrings 636
*New Bark* 214, 225, 300, 301, 424
    proposed modifications 942, 943, 944
*New Galleon of Kent* 866
New York, Metropolitan Museum of Art; gun-shield 557, 558
Newcastle upon Tyne 8–9, *20*, 582
Newhaven Fortress 20, 582
Newport Ship; wristguard 644, 645
nocks *see under* arrows; longbows
Norton, Robert **23**, 318
    on alloy recipes 955
    on canister shot 370
    on cartridge-making 435–6
    on casting of guns 60, 113
    on formers 435–6
    on fortification of guns 24, 45, 428–9
    on fowlers 24, 213, 216
    on gun drill 408, 409, 417, 464, 474, 476
    on gunner's rule *392*, *393*, 394–5, 396–7, 398, 399
    on gunpowder production 420
    on incendiary weapons 519–20, 521, *524*
    on iron guns 24, 132–3
    on ladles combined with rammers 446
    on powder to shot weight ratio 133, 173, 326, 428–9
    on powder weights for particular gun sizes 133, 447–8, 449
    on shot gauges 401, *401*
    on sponges 409, 457
    on tampions 464
    on types of bronze guns 24, 74
    *see also under* bases; cannon; culverins; falconets; falcons; gun carriages; port pieces; sakers; slings

Norwich; sword in Stranger's Hall 745, 761
Nottingham Castle 20, 227

oak
   items of 39, 40, 669, 720
   weight and density 626
old woman's tooth (hand router) 94, *95*
Oliviera, Fernando 851, 855
*order of battle upon the sea, The* (Lisle, June 1545) 851, 855, **865–6**, 869, 872, 878
Ordinances (1540) 537, 545, 744
ordinary, *Mary Rose* in (1525–42) 9, 309, 310, 865
Ordnance Office 3, 309, 310, 856
organ (type of gun) 860
Orlop Deck
   archery equipment 579
      arrow chests 667, *668*, 669, 690–1, *692*, 692–3
      longbows 583, *584*
         in chests 583, *584*, 611, 612–14, *641*, *643*, 643–4
         marked 611, 612–14
         skeletal remains associated with 702–3, 704, *710–11*, 712
      spacers 693, 695, 697, 704
      wristguards *646*, 647
   armour 824, *826*, 829, 832, 835, 842, 843, *845*, 846, 848
   casks 431, 433, 436
   gun carriage elements 102–3, 104, *106*, 108, 109, 190, *190–1*
   gun-shield elements 559, 560, *560*
   guns 202, 219, 230, 235
   hatches 883
   human remains 702–3, 704, *710–11*, 712, 848
   munitions and gun-related equipment 883
      gunpowder and incendiary items 436, 502, 529, 895
      ladle *442*, 446
      linstocks 488, 496–7, *498*
      powder dispenser 481, 483–4
      powder flasks 484, *484*, 485
      rammers *443*, 457
   shot
      canister *372*, 373–4
      composite inset *356*, 360, *361*
      cross bar 336, 332, *333*
      iron 101, *319*, *320*, *321*, *322*, 323, *324*, 327, *328*
      lead *369*, 370
      stone 339, *340*, *341*, 345, 346, 347
   shot gauges *389*, 402
   shot mould 388, *389*
   tampions 466, 467, *468*, 469
   tinder box 502
   tools for working guns *505–6*
   staff weapons *730*
   swords, daggers and related finds 748, 750, *752*, 763, 792, 794
*os acromiale* 704–6, *705*, 707
osteomalacia 632
Ostrich Feathers see (*Three*) *Ostrich Feathers*

overloading of *Mary Rose*, possible 867, 879, 882, 892–3, 913–14
Owen, John and Robert 22, 27, 31, 33, 35
   alloys 21, 946, 951, 952, 954
   see also cannon (79A1276; 81A3000); culverins (79A1232)

Padre Island, Texas; wrecks 223, 244, 358
Palumba, Giovanni Battista *806*, 808
panelling, decorative, on Sterncastle *235*, 236
paper for cartridges 6, 409, 435, 436, 438
Parr, William, Lord Parr; tomb effigy 747–8, *748*, 751, 808, 809, 815
pathology
   childhood malnutrition 632
   of shoulder, in archers 581, 582, 586, 704–6, *705*, 707
   spinal, of gun crew 847
*Pauncy* 543, 866, 921
   guns 58, 101, 102, 137, 225
      fowlers 213, 214, 215
      hailshot pieces 300, 301, 302
   munitions 424, 436
peacetime, navy in 2
   see also mothballing of ships
peening 955, 960
Pembroke, Earl of (Philip Herbert); inventory (1633) 715
Pendennis Castle, Cornwall 20, 227, 358
Pepys, Samuel 3–4
performance of *Mary Rose* in battle 865–82
perriers (guns for stone shot) 24
personal ownership of arms 581, 697, 716–17, 744, 780
   linstocks 410, 497–8, 499
   see also provision of arms
personnel see crew
*Peter* (*Pomegranate*)
   Anthony Roll illustration 4, 272, 921
   archery equipment 580
   building 8
   fighting tops 269, 272
   guns 7, 858, 860, 863
      bases 223, 225
      cannon 46, 58
      culverins 74, 89
      falcons 859
      hailshot pieces 300, 301, 302
      locations 856, 857, 861
      port pieces 137
      sakers 101, 323
      top guns, slings 269, 858–9
   handguns 543
   incendiary devices 522
   in Lisle's orders of June 1545 866
   munitions 384, 385, 424, 436
      shot 214, 331, 337, 385
   shear hooks 921
petrological analysis of shot 339, *342*, 343
pettard (type of gun) 24

*Phoenix*
    guns 101, 225, 300, 301, 437, 438
    munitions 331, 424, 437, 438
photogrammetric study 170
picks and pick-hammers 6, 381, 383, 384, 385, 386
    inventory entries 384, 385
pikes *734–6*, **735–6**
    anatomy *715, 734*, 735, 736
    distribution *730*, 736
    fragments from spoil mounds 935, 938
    hafts 717, 929
    heads *730, 734*, 735–6
    identification and form *734–6*, 735–6
    incendiary 520, *520*
    inventory entries 580, 581, 715–16, 718–19
    numbers in comparison with bills 715–16, 717
    'pike garden' *730, 731*, 731
    scope and importance of assemblage *714–15*, 717–18, 929
    staves *735*, 735–6
pine bill haft 720
pistols 539–40, *539*, 544, *550*
pitch/tar 529, 692, 693
'plague' 12, 632
planes
    jack 94, *95*
    thumb 701–2, *702–3*
*plats* (pictorial maps) 3, 7
Poitiers, battle of 586, 589, 631, 697, 699
pollaxes 738–40
pomanders 746, 769, *770, 812*, 813
pommels
    on daggers 742, 750–1
        ballock *780, 786*, 781
        *Landsknecht* type *772, 773*, 774
    detached 746, *746*, **764–9**, *768–70*
    distribution *752*, 766
    radiography 764–5, 769, *770*
    X-ray fluorescent spectrometry 765
    *see also under* swords
Pontefract Castle, Yorks *20*, 582, 665, 857
poplar
    arrows 618, 669, 674–5, 677, 678, *680*, 684–5, 686
    sword/dagger grips 763
    technical characteristics 618, 626, 676
*Portcullis* 581
    guns 225, 300, 301, 438
    munitions 424, 438
port pieces **136–99**
    anatomy *138*
    arcs of fire 915, *919*
    barrels *138, 139*, 199
        fragments *182*, 183, 186
    bore size 142, *199*, 343
    bronze, later casting in 136
    carriages 132–3, 137–8, 142
        anatomy *138*
        fragments 107, 171, *171*, 187–99
        replication *284*, 284–5

        statistics 142
        81A2604 148, *148–9*, 151, *154–5*, 155
        81A2650 143, *143, 145, 146*, 147
        81A3001 156, *156*, 157, *158–60, 284*, 284–5
        for RA XIX.1 171, *171*
        *see also* elevating posts; forelocks *and under* wheels, spoked
    chambers **173–81**
        anatomy *138*
        construction 174–7, *175*
        distribution 140, *173, 197*, 897
        in experimental gun drill 413, *414, 415*, 416
        fragments 181, *181, 182*, 183, 185–6
        identification 173–4, 175
        *in situ 139*
            81A2604 148, *151–2*, 153
            81A2650 153
            82A0792 *139*
        lifting rings 174, *175–6, 181*, 183, 185–6
        loading sequence *175*, 178, 464–5, *465*
        number per gun 143, 412
        radiography 177, 178
        Royal Armouries 171, 172, 180–1
        statistics 153
        2003/2005 finds 134, 138, 140, 174, 180, 897, 936, 939, 940
        71A0169 *163*, 163, 164, 174, *176*
    Deane and Edwards' salvage 137–8, 142
        *see also* NMM KPT1836; RA.1; RA XIX.1 *below*
    dimensions 142
    distribution 26, *26*, 132, 138, 140, *140, 141*
        fragments *197*, 199
        Main Deck 132, 138, 140, *141*, 142, 169, 347, 863, 899, *901, 902*
            at ports with limited space 897, 900, 901
        in stern 138, *141, 901*, 901, *902*
        Upper Deck *141*, 345, 347, 897, 910, *919*
    firing rate 412, 413, 415, 899–900
    fragments 181–6, *181, 182*
        distribution *197*, 199
    functions 137, 138, 900, 914, 921–4, 929
    gun crews 413, 415–16
    gun drill 285–7, **411–16**
    gunports 137, 138, 863
    gunpowder 133, 143, *144*, 163–4, 285, 287, 427
    historical significance 136–7, 929
    identification 132, 136–7
    introduction 863, 899–900
    inventory entries 136
        1540 7, 8, 137, 856, 861
        1546, Anthony Roll 5, 131, 136, 137, *141*, 214, 856, 864
        shot 310, 344, 922
        1547 131, 137, 312
    metallurgical examination 170
    morphology 134–5, *134, 135*, 140, 142
    muzzle energy 914
    Norton on 24, 132–3, 136–7, 142

photogrammetric study 170
on Portsmouth fortifications 881
radiography 148, *152*, *157*, 160–1, 163, 177, 178
range 881, *919*, 926
reloading 412, 413, 415, 899–900
replication 127, 129, 157, **272–90**, 411, 930–1
    see also under 81A3001 below
scope of assemblage 137–8, *139*
shot
    *Anthony Roll* list 310, 339, 344, 922
    canister 136, 312, 370, 373, 380–1, 900, 929
    stone 133, 142, 311, 312, *343*, 343–4, 345–6, 347
shot gauges 405–6
statistics 136–7, 142, 153
tampions 143, *144*, 148, 163, 177–8, 464–5, *465*, *469*, 470, 472
    replica *286*, 287, 413, *415*, 416
wads 148, 177, 178, 179, 475, *476*, 476
in weapon preference sequences 921–4
wedges *138*, 195, 413, *414*, *415*, 416
71A0169 139, 142, 153, **163–4**, *163–4*
    chamber 153, *163*, 163, 164
81A2604 *139*, **148–55**, *148–52*, *154–5*, 902–3
    arc of fire 915
    axle 107, *110*
    bore size relative to barrel length *199*
    carriage 107, *110*, 148, *148–9*, 151, *154–5*, 155
        elevating post 195, *196*, *197*, 197
        forelock 151, *154*, *192*, *194*
    chamber *139*, 148, *151–2*, 153, 280
    gun station 902–3
    location *140*, *141*, 142, 148, *148*
    radiography 148, *152*
    statistics 142, 153
81A2650 *139*, **143–7**, *143–6*, 902–3
    arc of fire 915
    carriage 143, *143*, *145*, *146*, 147
        elevating post 195, *196*, 197
        forelock *192*, *193*, *194*, 195
    chambers *139*, 143, *143*, *144*, *145*, *146*, 147, 153, 280
    gun station 902–3
    location *140*, *141*, 142, *143*, 143, 897
    statistics 142, 153
81A3001 *139*, **156–62**, *156–61*
    arc of fire 915, *919*
    associated items *161*, 162, 497, 504
    barrel *139*, *157*, 160–1
    carriage 156, *156*, 157, *158–60*, **161–2**, *274*
    chamber *139*, 156, *157*, 161, 273, 280
    dimensions and construction 273
    elevating post 195, *196*, 199
    forelock *192*, *193*, *194*
    gun position 903, *904*, 905
    gunpowder 434–5, *434*, 531
    location *140*, *141*, 142, 156, *156*, 273, *273*, 899
    photographs and drawings *274*
    radiography *157*, 160–1
    range *919*
    reloading when sunk 497, 934
    shot in barrel 156, *157*, *161*, 162, 273
    replica **272–90**, 930–1
        barrel 274–9, *277–8*
        carriage *284*, 284–5
        chamber 279–83, *280–1*, 289
        choice for replication 157
        discussion of techniques 283–4
        firing 123–4, 127, 129, 195, 283, **285–90**, *286–8*, **411–16**, *414–15*, 914, 930–1
        gunpowder 285, 287
        materials 274, 283
        method of manufacture 274–84
        shot 285–7, *286*
        size limitation 283–4
        wedges *284*, 285, 413, *414*, *415*, 416
    statistics 142, 156, 160–2
    treatment 273
NMM KPT0017 *139*, 153, **165–7**, *165–6*
    chamber 142, 174, *176*
RA XIX.1 (80A2004) 142, 153, 171, *171*, 180, 347
RA XIX.2 (80A2003) *139*, 142, 153, 172, *172*, 180, 347
Woolwich 1/10 *139*, **168–70**, *168–9*
    carriage 168, *168*, *169*, 170
    chamber *139*, *168*, *169*, 170, 174, *176*
    location 168, 169, 347
    recovery 131–2, 168
Port scour 489, *506*, 752, 796
    shot *325*, *327*, *328*, 332, *333*, 336
port side
    guns see cannon (79A1277; 79A1276; 81A3002); culverins (79A1278)
    structural timbers 940, *940*
Portsmouth *20*
    Cowdray Engraving 868, *869*, *870*
    dockyard, Tudor 1, 2, 8, 9
    fortifications and guns *20*, 292, 860, 868, *869*, *870*, *874*, **880–1**
    French consider attacking 880, 881–2
    God's House 201, 523, 582
    Tudor maps and charts 867, 868, *869*, 881
*Portsmouth Independent* 165, 171
post-Tudor artefacts 940–1
    see also mortar bomb
pot guns 860
pots of wildfire 520, *520*, 528
pouches, leather drawstring 551, 553
powder bag, possible 210
powder dispensers (for priming powder) **479–84**, *480*, *482*, *483*
    distribution *439*, 481, 483
    in gun drill *415*, 418
    for handguns **479–84**, 544, 551, 553
powder flask *439*, **484**, *484*
powder horn, top of 485, *485*
powder ladles see ladles, powder
Powell Cotton chamber E31/104 *176*, 181
pricing of weapons 588–9, 715

prickers *see* priming wires
priming powder 407, 409, 418
    *see also* powder dispensers
priming wires **476–9**, *477–9*, 940, *941*
    distribution *439*, 477, 479
    use 407, 409, 417, 418
*Primrose* 89, 101
profiler, parametric *933*, 935, 940, *940*, **941**
Prone, Martin de (Martin van Royne, Master armourer at Greenwich) 827
proofing of guns 114, 116, 124
*Protection of Wrecks Act* (1973) *932*, 934
provision of arms 713, 716, 717, 740, 744
    Ordinances of 1540 on 537, 545, 744
    *see also* personal ownership of arms

quillon block, possible post-Tudor 940–1, *941*
quivers 697, 698–9, *698–9*
quoins 6, 52, *52*, 127, 407

rabinet (unit of weight of gunpowder) 428
radiography *see under* bases; bills; by-knives; cannon; daggers; fowlers; gun carriages; guns, wrought iron; hailshot pieces; handguns; pommels; port pieces; scabbards; sheaths; shot (composite, iron and lead); slings; swords; tampions
rammers 409, **454–62**, *455*, *458–9*, *461*, *463*, *503*
    associated with cannon 55, *55*, 450
    distribution 247, *438*, *443*, 446, 457
    in gun drill *125*, 126, 407, 408, 409, 413, 416, 418, *418*
    inventory entry (1547) 410
    items originally identified as 699–702, *700–2*
    ladles combined with 441, *441*, 446
    manufacture 460–1
    matching to guns 455–7, 458–60, *461*, 462
        missing falcon 247, 313
    replication *125*, 126, 418, *418*
    scope of assemblage 455, 456–7
    for slings 201
    stylistic analysis 461–2, *463*
ranges of guns 875, 876
    in battle of Solent *874*, 875, 876, 879, 880, 881
    elevation affects 875, 876, 879, 881
    historical information 875, 876
    quoted and effective 922
    shot and 307, 314, 926–7
    shore-based guns *874*, 881
    and tactics 851, 921–7, 929
rate of fire 2, 131, 309, 923, 925
rebuilding of *Mary Rose* 19, 854–5, 865
    and change in armament 10, 15, 861, 863, 942–5
    Cromwell refers to 861, 930
    dendrochronological dating 930
    gunports 8, 13, **13**, 14, *14*, 863, 887, 928
    Hatfield House document on plans for (1545) 10, 12, 199, 200, 204, 930, 942–5
    and sinking 882, 883
    reburial of site of *Mary Rose* 941, *942*
    recoil ropes and fittings 49, *890*, *891*, *896*, *899*

*Regent* 1, 13, 594, 921
reinforcement
    chambers 131
    guns *23*, 25, 446–7, 449
reloading *see under* loading and firing of guns
remote sensing; sub-bottom profiler 933, 935, 940, *940*, *941*
remotely operated crawler vehicle (ROV) *934*, 936
replication
    archery equipment 579, *666*
        firing of arrows 630–1
        longbows 627–9, *628–9*, 929
        nocks *636*, *638*
    billhead 732–4, *732–3*
    gunners' tools 125–6, *125*, 418, *418*
    gunpowder 126–7, 285, 287
    *see also under* culverins (81A1423); gun carriages; hailshot pieces; port pieces (81A3001); shot; slings
revolver, possible earliest 539
Reynolds Rule 875, 876
rigging 74, 934–5
ring, fixed, in sword suspension system 807
rings, finger
    decorated with sheaf of arrows *585*, 585
    signet, with letter K *665*, 665, 707–8, *709*
rivets 807, 808, 825
robinet (type of gun) 24, 74
*Rode Slip* 581
*Roo* 101
    guns 225, 300, 301
    munitions 424, 438
rope 6
    guns secured by *56*, 56–7, 410
    recoil *56*, 56–7
    in experimental gun drill 413, *415*, 415
    wads in port piece chambers 177, 178, 179
    *see also* wooling
*Rose Galley* 385, 580
*Rose Lion see Lion/Roselion*
*Rose Slip/Slype*
    archery equipment 581
    guns 225, 300, 301, 438
    munitions 424, 438, 464
*Rose in the Sun* 424
    guns 101, 225, 300, 301, 302, 438
Rotten Bottom Bow 586
router, hand 94, *95*
ROV (remotely operated crawler vehicle) *934*, 936
rowbarges
    archery equipment 580, 581
    guns 74, 89, 101, 102, 137, 200, 214, 300, 302
    handguns 543
    munitions 331, 385, 426, 464, 523
    staff weapons 718, 719
    tactical use 855
Roxburgh inventory (1547) 523
Royal emblems 29–33, *32*, *33*, 59–60, *59–60*, 929
    *see also under* wristguards (decoration)
Royal style 31–3, 59–60, *59*, *60*

Royne, Martin van (Martin de Prone, master armourer at Greenwich) 827
rudder 885, *886*
rules of engagement 921–7
    preferred weapon types 921–4
    *see also* ranges of guns
Russell, John (1st Earl of Bedford) 866, 870

Sachs, Hans 386, *386*
sacks of wildfire 520, *520*
sailing qualities of *Mary Rose* 9, 12, 865
St Anthony wreck 223, 358
St John, Order of 270
*Saker*
    guns 101, 214, 225, 300, 301, 437, 438
    munitions 331, 424, 437, 438
sakers 22, 101, *101*
    associated items 27
        former 438
        ladles 448, 449, *450*, 460
        rammers and sponges 456, 460, 462
    carriages 102
    dimensions and weight 21
    fortification 428
    gunpowder 426, 427, 428, 429, 430
    inventory entries 101
        1540 7–8, 101, 856, 861, 863
        1546, *Anthony Roll* 5, 25, 36, 101, 310, 856, 922
        1547 312
    iron supersede bronze 101, 291, 292
    locations, possible 101, 908, 909
        Sterncastle 27, 854, 943–4, *945*
    missing, possibly removed by salvors 931
    Norton on 21, 24, 74, 101, 429, 875
    on Portsmouth fortifications 881
    range 314, 429, 875, 876, 880, 881
    shot suitable for 101, 310, 312, *323*, 323, *324*, 326, 330, 462, 922, 924, 925
        cross bar 311, 331, 336
        inset ball 360
        sizes 311, 429
    sling class equivalents 201
    in weapon preference sequences 921–4
*Salamander*
    guns 101, 225, 300, 301
    in Lisle's orders of June 1545 866
    munitions 331, 424
Salisbury Place, London; Italian foundry 22
sallets 819, 820
saltpetre 419–20
salvage
    Tudor 3, 49–50, *50*, 941–2
        cables 50, *50*, 903, *904*, 905, 949
        guns removed by 36–7, 89, 217, 272, 931
    19th century *see* Deane, John, and associates
*San Estaban*, wreck off Padre Island 223
*San Pedro* 358
Sandown Castle, Isle of Wight *20*, 201

*Santa Anna* 270
*Santa Maria de Ycar*, wreck off Padre Island 223
*Saviour* 866
scabbards *741*, 793–819
    anatomy *741*, 798
    by-knives in 794, 796, *798*, 798, 813
    chapes *741*, 798, 700–804, *801–4*
    chins *741*, *798*, 799–800, *801*, 805
    complete examples *791*, 791
    decorated 798, 791, 794, *795–8*, 813
    descriptions of selected 811–15
    distribution *752*, 794
    with hanger and sword-belt attached 808, 809, *812*, 811, 813
    with human remains 708, *812*
    importance of assemblage 748, 794–805
    inventory entries (1547) 794, 798, 799, 805
    listing with details 792–4
    lockets *804*, 804–5
    mouths *741*, 804–5
    radiography 798, 800, *801*, 802, 803, *806*
    scope of assemblage 791–4
    seams and stitching 799, *799–800*
    from spoil mounds 935, 938, *939*
    with sword elements 791–4
    suspension systems 748, 805–17
        bifurcating 807–8, *808*, 811, 812, 813, *813*, 814, 815–17
        trifurcating 817
    transverse neck type *797*, 799–800, 801
    use of term 794
    *see also* sheaths (for daggers) *and under* buckles
*scaffetta* (type of ladle) 441, *441*
*Scheurrak S01* wreck 39, 479, 496
    shot 335, 360, 365, 373
Scorer, Richard (king's gunstone maker) 308
Scorer, Robert (king's gunstone maker) 308
Scotland 1, 10
    Flodden campaign 8–9, 715
    guns shipped to (1497) 860
scribing gauge 94, *95*
seabed; sub bottom profiler *933*, 935, 940, *940*, *941*
serpentines 13, 15, 19, 200, **857**, 860
    inventory entries 423, 857
service career of *Mary Rose* 8–10
Sesse, Nicholas (supplier of shot) 309
*Seven Stars* 331
shear hooks 921
sheaths (for daggers) 778, *788*, *789*, 791
    anatomy 742
    by-knives in 778, 788, 789, 794, 796, 798
    chapes 778, 798, 800–804, *801–4*
    distribution *752*
    importance of assemblage 748, 794–805
    lockets *804*, 804–5
    mouths 804–5
    production 740
    radiography 804

from spoil mound 935
suspension systems 809, 811
use of term 794
sheepskins 6, 455, 507
shields 822
*see also* gun-shields
*Ship of Dover* 866
*Ship of Montrego* 866
shipbuilding
techniques *13*, 14
transitional era in design 854–5, 928
shipkeepers 9
shipworm larvae 941
shot **307–407**
and accuracy of guns 307
and battle range, duration and profile 921–7
canister **370–81**
conical lantern *371*, *372*, *373*, *374*, *375*, 380
contents *372*, 373, 379–80, 384
cylindrical canister *371*, *372*, *373*, *373–4*, *374*, 380
partial 378–9, *379*
cylindrical lantern *371*, *372*, *373*, 374–7, *376–8*, 380–1, 530
distribution *372*, 373–4, 380–1, 530
guns used with 136, 312, 370, 373, 380–1, 900, 929
inventory entries
1546, *Anthony Roll* 5, 6, 309, 373
1547 337, 373
loose ends and other fragments 379, *380*, 381, *382*
storage 312–13
composite, iron and lead 348–67
casting on board 348, 350, 384, 927
core and shot radius, and density 400–1
corrosion 348, *349*
cross bar 364, **365**, *365*
distribution 247, *356*, 358, 360, *361*, *362*
eccentricity 348, 360
gunner's rule and 398–9, 401
guns used with 102, 247, 384, 927
radiography *331*, 331, 332, *334*, 334, 335
hemispherical (jointed) 309, 364, **360**, *362–4*, **365**, 366, *366*
inset ball 348, *357*, **358–60**, *361*
guns used with 236, 360, 384
inventory (1547) 360
making on board 350, 384
inset dice **350–8**
analysis *354*, 355, 357–8, *359*, 360
casting; on board/on site 348, 350, 358, 927; experimental *351*, 351, 353, *391*, *393*
diameters 133, 312, 355, *355*, *358*
eccentricity 348, 929
guns used with 201, 927, 929; bases 222–3, 227, 236, 247, 250, *256*, 256, 258, 260, 309, 311, 353, *356*, 357–8, 929
inventory entries 227, 311; 1546, *Anthony Roll* 5, 6, 309; 1547 311–12, 358
moulds for *351*, *351*, 353, *391*, *393*

radiography 303, *303*, *350*, *352*, 353, *357*, 358
weights 133, 353–5, *353*, 358
windage 357
odd shot 360, 362–7
parallels with other sites 360
radiography 331, *331*, 332, 334, *334*, 335, 357, 358, *363*
composite, other 5, 6, 311–12
documentary evidence 310–11
*see also* inventory entries *below*
evidence on fighting 921–7
finishing and making on board **384–5**, 924, 927
*see also under* composite *above* and stone *below*
inventory entries 4, 308, **309–11**
evidence on production 381, 383
on preferred gun types 921–4
1514 309, *310*
1546, *Anthony Roll* 214, **309–10**, *309*, 311, 921–4
assemblage compared 5, 6, 202, 309, 313
1547 214, 222–3, **310–11**, 311–12
*see also under individual types of shot*
identification of guns by 4, 222–3, 309, 318–19
incendiary 311, *334*, 520, 521, *521*
iron **313–37**
cast, round 307, **313–30**
attributes recorded 314
density vs. radius 399–400
distribution 101, *102*, 313, *313*, 318–30 *passim*, 937, 939, 943
future research 931
gauges tested against 314, 318, 323, 326, 329, 406–7
guns used with 102, 311, 318–20; cannon 45, 48, 52, 60, 64, 71, 318, *318*, *319*, *320*; culverins 75, 84, 87–8; slings 101, 132, 199, 201, 205, *205*, 206, 311
H stamp 314–15, *315–17*
identification of guns by 132
inventory entries: 1514 309, 310; 1546, *Anthony Roll* 5, 6, 199, 309, 310, 313
and location of missing guns 102, 318–29, 943
manufacturing process 304, 314, *314*, 317, 383–4, 929
metallography 316–17, **317**
and range 314
sizes *308*, 317–18, *329*
storage 101, 313, *313*, 329
unattributable to named guns 329
windage 318, 323, 326, 329
cross bar *330*, **330–6**
distribution 312, 332, *333*
inventory entries 309, 310–11, 330–1
radiography 331, *331*, 332, 334, *334*, 335
relating shot to guns 336
dice (hailshot) 132, 291, 309, 310, *337*, *337*, 864
guns loaded with 293, 295, 303
jointed *see* composite (hemispherical) *above*
lantern *see under* shot, canister

lead **367–70**, *367–9*
   density vs. radius 399–400
   distribution 313, *369*, 370
   experimental casting 389
   finishing and making on board 384, 545
   guns used with 201, 370
      bases 311, 358, 367
      falcons 102, 247
      handguns 5, 6, 537, 545, 551
      inventory entries 545
         1514 309, 310
         1546, *Anthony Roll* 5, 6, 102, 357, 309, 310, 357, 367, 537, 551
         for bases 311, 358, 367
   and location of missing falcon 102, 247
   storage 312–13
lead wrapped *362*, 364, *366*, 366
location of guns by 102, 247, 309, 318–29
manufacture **381–9**
   methods 381–7
   on board 384–5, 924, 927
      *see also* replication *below and under* composite *above and* stone *below*
   tolerances 307, 311
matching to guns 307, 309, 310, **311**, 924
materials 311–12
murthering (scatter) 24, 136, 858
   *see also* iron dice *above*
Ordnance Office receives shot from mothballed ships 309, 310
powder to shot weight ratios 124, 126, 127, 173, 200, 202, 449
preferred gun types indicated by 921–4
   and range 307, 314, 926–7
replenishment 348, 350, 358, 384–5, 924, 927
replication
   cast iron, for culverin 124, 125, 127
   composite inset dice *351*, 351, 353, *391*, 393
   lead 389
   stone, for port piece 285–7, *286*
selection 392–407
sizes for named guns 311
spike (term for cross bar) *334*, 335–6, *336*, 520
stone 307, **339–47**
   associated with cannon 55, *55*
   diameters 142, **308**, 339, *339*, 343, *343*
   distribution *149*, 272, 339, *340–1*, 345–7
   finishing *338*, 339, 381
      on board 384–7, *386–7*, 404, 508, 924, 927
   gauges for 347, 386, 406–7
   guns used with 339, **343–5**
      fowlers 214–15, *215*, 218, 311, *343*, 343–4
      port pieces 131, 142, 311, 163, *343*, 343–4, 900; 81A3001 156, *157*, *161*, 162, 273, 285–7, *286*
      top pieces 269, 272, 343–4, 344–5, 347
   inventory entries 214–15, 311
      1546, *Anthony Roll* 5, 6, 269, 309–10, 339, 344
   materials 339, *342*, 343, 385
   on mothballed ship 309, 310
   oversized 339
   petrological analysis 339, *342*, 343
   replication 285–7, *286*
   unfinished 217, 312, *338*, 339, *341*, 344, 345–7
   weights 339
   windage 339
stone and lead 201, 214, 362, 364, 366–7, *367*
storage *312*, **312–13**
   access 882, 883
   central 101, 307, *312*, 312–13, 329, 883
   other locations 75, 312, *312*, 329, 883, 895, *896*
supply 308–9
   *see also* replenishment *above*
variety within assemblage 307, *307*
weight to diameter ratio 393–400
of wildfire 520, *521*, 521
windage; variability 307, 311
   *see also under individual types of gun*
shot gauges **401–7**
   alternative identification as formers 384, 403
   distribution 384, *389*
   and finishing of stone shot 386
   historical sources on 401, 409
   matching to guns 55, *55*, *161*, 162, 405–6
   matching to shot 314, 318, 323, 326, 329, 347, 406–7
   paddle-shaped 401, *404*, 404–5, 406, 407
   rectangular 401, *401–2*, 403–4, 406, 407
   testing shot against 314, 318, 323, 326, 329, 347, 406–7
shot moulds 350, 383, **387–93**, *388–91*
   brass and iron 381
   distribution *389*
   experimental use *351*, 351, 353, 389, *391*, 393
   inventory (1547) 381, 383
   stone 381, 384, 387–8
   tongs for use with 383, *383*
shovels 504, *506*, 511–12, *514*
Shrewsbury, battle of 587
shrimp bases 224
*Siege of Boulogne, The* (engraving)
   armour 821–2, *822*
   guns and gun equipment *223*, 224–5, *418*, 419, 430, *477*, 477, *509*
sieves *506*, 511, *513*
sighting of guns 55, 891
silk for bowstrings 635, 636
sinking of *Mary Rose* **866–82**, *868–9*
   contemporary descriptions **866–7**, 869, 875–6, 877–8, 879–82, 934
   final passage of ship *874*, 874, 878–9
   geographical evidence 867–82, 934
   ingress of water as cause 12, *25*, 26, 855, 866, 867, 878, 879, 882, 892
   possible or alleged factors 12, 13–14, 882
      alterations to ship 882, 883, 942–5
      command structure 879
      damage by enemy fire 866, 867, 876, 878, 879, 882, 883, 934

free surface effect 878–9
    human haste, folly, or negligence 866, 867, 869, 882
    manoeuvre 866, 876, 879, 882, 934
    overloading 867, 879, 882, 930, 942–5
site of *Mary Rose*
  protection *932*, 934
  reburial *941*, *942*
Site Recorder data management system 937, *937*
sledgehammers 6, 385, <u>506</u>, 507–8
  gunners' use 409, 415, 504
slings **199–213, 858–9**
  arcs of fire 918, *919*
  barrels/barrel and chamber fragments 202, *204*, 205–10, 205–10
  bore size 199, *199*, 314
  carriages 108, 132–3, 199, *199*, 200, 201, 205, *206*, 207–8, *208*
    parts for 107, 108
  cast, bronze or iron 200
  chambers 173, 175, *176*, 199, 200, 201, 202–3, *203*
    distribution *204*
    fragments 203, 210–13, *211–12*
  construction 134–5, *134–5*, *201*, 202
  culverin equivalents 200, 201–2
  demi-slings 200
    bore size and expected range 314
    gunpowder expenditure 427
    inventory entries 214, 310, 922
    possible location 101, 943
    shot 101, 310, 312
    *see also* 70A0001; 79A1295 *below*
  description and discussion **858–9**
  dimensions 173, 202, 203, 205, 860, 909
  distribution 202, 203–4, *204*
  gunports for *198*, 891
  gunpowder 133, 200, 202, 205, *205*, 206, 326, 427
  historical significance **199–202**
  identification 132, **199–202**, 309, 323, 326
  inventory entries
    1514 200, 856, 858, 861
    1523 860
    1540 856, 861, 863
    1546, *Anthony Roll* 4, 5, 6, 131, 199, 214, 310, 856, 922
    1547 131, 200–1, 224, 312
  locations 203–4, *204*
    Hatfield House document on 101, 943–4, *945*
    Main Deck in stern *901*, 901
    Upper Deck *198*, 199, 202, 203–4, *204*, 205, 210, *210*, 212, *212*, 909, 918, *919*
  swivel holes for *234–5*, 235, 861, 891, *892*
  morphology 134–5, *134–5*, *201*, 202
  Norton on 24, 132, 133, 200, 326
  powder to shot ratio 200, 326
  quarter slings 200, 310, 312, 314, 427, 922
    *see also* 82A2700
  radiography *198*, 199, 202, 205, 207, *208–9*, 208–9, 210, *210*

relative size 200, 201–2, 860
replication *201*, *202*, 202
scope and distribution of assemblage **202–4**
shot
  canister 370
  dice and lead 201
  inventory entries 101, 199, 310, 311, 312
  iron 311, **326**, *326*, *327*, **329**, 330
    in individual guns 199, *205*, 205, 206, *206*, 207, 326, *465*, 909
      identification of guns by 132, 201, 309, 323
      ratio per gun 199
  lead, and stone covered with lead 201
  Norton on size and weight 133
shot gauges not found 405–6
swivel mountings 200
tampions in 202, 205, *205*, 206, 210, *465*, 470
types 200
uses 314, 914, 921–4, 931
wads 202, 205, *205*, 206, 475
in weapon preference sequences 921–4
70A0001 *198*, 199–200, *201*, 202, 207, 943
  identification 323, 326
79A1295 *201*, 202, *202*, 207–8, *208*, 326
80A1952 202–3, *203*, 210–11, *211*
80A2000 172
80A2001 171, 202–3, *203*, 211
81A0645 *198*, 199, 199–200, **205–7**, *205–6*
  carriage 108, 192, 195, 197, 199, *199*, 205, *206*, 207
  chamber 173, 199, 202, 203, *203*
  construction *201*, 202
  dimensions 173, 199, 202, 203, 205, 909
  gunport *198*, 891
  handspike associated with 504
  identification 326
  loading sequence *465*
  location *198*, *204*, 891, 909
  radiography 205
  range and arc of fire *919*
  shot 205, *205*, *206*, 207, 326, *465*, 909
  underwater photograph *in situ* 198
81A0708 203, *204*, 212, *212*
81A0756 203, *204*, *212*, 213
81A0772 203, *204*, 208–10, *208–9*, *210*
81A0783 203, *204*, 210, *228*, 232, 235
81A0803 203, *204*, *212*, 213
82A2700 202, *204*, **208–10**, *208–9*, 326
90A0099 203, *203*, *204*, 211, *211*
Sluys, battle of 13, 593
*Small/Lesser Bark*
  guns 58, 101, 137, 225, 300, 301
  munitions 424, 436
*Small/Less Galley* 46, 89, 101, 863
Smith, Captain John; *A Sea Grammar*
  on gunport lids 895
  on gunpowder 425, 430
  on height of gundecks 894
  on linstocks 486

on musket with incendiary arrow *523*
on priming wires 477
on quoins 513
on shot 334, 335, 370
Smythe, Sir John 635, 743
soldiers on *Mary Rose* 882, 895
    inventory evidence 856–7, 882
    possible royal guards 656, 738, 929
    *see also* archers
Solent
    bathymetry 872, *872*, 873, *873*, 874, 880, 881
    battle of *see following entry*
    early charts 867, 869, *872*
    recent proposed dredging *932*, 934
Solent, battle of the (19 July 1545) 10–12, *10–11*, 867–82
    contemporary accounts **866–7**, 869, 870, 875–6, 877–8, 879–82, 934
    events 874–82
    final passage of *Mary Rose* 874, 878–9, 934
    French galleys' movements 869–70, *870*, 873, *873*, *874*, 874–5, 876, 877, 878
    geographical evidence 867–82, 934
    gunnery ranges *874*, 875, 876, 879, 880, 881
    inventory evidence 867, 881
    King watching 12, 879
    land fortifications and 868, *868*, 870, *870*, *871*, 874, 879, 881
    Lisle's command lists 851, 855, **865–6**, 869, 872, 878
    wind and currents 870, 876–7, *877*, 878
    *see also* sinking of *Mary Rose* and under *Henry Grace à Dieu*
somercastle 860
Southend Museum; ivory wristguard *646*, 665
Southsea Castle *20*
    and battle of the Solent 868, *868*, 870, *871*, 874, 879
    inventory (1547) 523, 717–18, 881
*Sovereign* 1, 13, 14, 861
spacers *see* arrow spacers
spatial associations of site, three-dimensional 932
Spert, Thomas (master of *Mary Rose*) 8
spider (casting support) 114
spoil mounds *see* 2003–5 excavations
sponges 125, *126*, 201, 407, 408, 409, 412, 503
    comparison with gun bores 460
    in culverin gun drill 127, 416, 417, 418, *418*
    future research 931, 932
    identification 455, 457
    inventory entries 410
        sheepskins for 409, 410, 507
spoon, pewter 708, *709*
stability of *Mary Rose* 14, 867, 882, 883, 913–14, 934
staff weapons **713–40**
    *Anthony Roll* 5, 6, 580, 581, **718–19**
    distribution *730*, 883, 895, *896*, 909, 910
    incendiary devices combined with 520, *520*
    scope and importance of assemblage 713–18
    storage 895, *896*
    *see also* bills; halberds; pikes; pollaxes
standardisation of guns 926

Starboard scour, finds from
    bases 230, 231, 232, 233, 235, 254, *254*, 259, *259*, *261*, 261, *264*, 264, 268, *356*, 358
        relatively complete 227, 229, 245, 255, *255*
    darts 740
    fowlers 217, 219
    gun-shield elements *559*
    ladles *442*, 446
    linstocks 489, 496
    longbows 583, *584*
    post-excavation searches (1983–5) 935
    priming wire 477, 479
    rammers *443*
    shot 367
        canister 380–1, *372*, 373
        composite inset *356*, 358
        iron 318, *319*, 322, 323, *324*, 326, *327*, *328*, 330
        stone 272, *340*, *341*, 344, 346, 347
    sling 203, 204, *204*
    staff weapons *730*, 731
    swords, daggers and related finds 751, *752*, 794
    tampions 467, *468*
    tools for working guns *505–6*
    trucks 108
Stationers' Company 656, 659
*Statute of Winchester* 713
steel; making and hardening 823–4
stempost 934, 937, 939, 941–2, 944
Sterncastle 910–13
    *Anthony Roll* illustration 6, 918, 934, *944*
    archers 895
    decks
        Castle 910–13
        postulated upper 101, 918, 930, 943–4, *944–5*, 945
    excavations beneath and to east 934–5
    gun carriage elements 27, 103
    gunports 6, 887
    guns 910, *911–12*, 913
        arcs of fire 916, *917*, 918, 919, *919*
        bronze 26, 27–8, *27*, *28*, 36–7, 103
            demi culverins 89; possible forward-firing 853, *853*, 854, *911–12*, 913, 918, 919, *919*, 943–4, *945*; *see also* culverins, demi (79A1232; 79A1279)
            falcons 27, 910
            sakers 27, 943–4
        iron
            bases 234, *235*, 235–6, 895, 918
            fowlers 217
            hailshot pieces 295, *296*, 895, 913
            slings 203, 204, *204*, 210, 945
        postulated forward-firing 853, *853*, 854, *911–12*, 913, 918, 919, *919*, 943–4, *945*
        postulated sternchasers *911–12*, 913, 916, 919–20
        on Upper Deck under Castle deck 26–7, *26*, *39*; *see also* cannon (81A3002); culverins (80A0976)
    gun-shield elements *559*, 560
    handspike *505*

ladles *442*, 446, 449
muster station under 579, 846–7, 895, 910
panelling, decorative *235*, 236
rail *234–5*, 235
shot 102, *322*, 323, *328*, 328, 330, *340*, 346–7, 346, *356*, *369*, 370, *372*, 373
soldiers 895
sword grip *752*
tampions 467, *468*
terms used for 860
sternchaser guns, possible 347, 853, 885, *886*, 916
*Anthony Roll* illustration 138, 235, 885, *886*, 910, 919–20
gunports 8, 885, *886*, 887
Main Deck *141*, 897, *901*, 901, 916
Sterncastle *911–12*, 913, 916, 919–20
stiletto, gunner's 931, 940–1, *941*
stock-fowlers 24
stocks (gun carriages for wrought iron guns) 135
stone guns 7, 13, 19, 856, **857–8**, 860
storage 895, *896*
access 882, 883, 895
archery equipment 537, 579, 583, 667, 895, *896*
bowstrings 634, 641
see also arrow chests; longbow chests
*see also* Orlop Deck *and under* cartridges; gun carriages; gun-shields; gunpowder; shot; staff weapons; trucks; wheels, spoked
Stow-cum-Quy, Cambridgeshire; Ansty memorial brass 805, 808
straps, wood and metal fastenings for armour 818, 824–5, 829–32, *831–3*
with arrow spacers 694–5, 697, 707, *708*
arrows tied in bundles with 667, 675
decorated, from spoil mound 935, 938, *939*
distribution *826*, 829, 832, 848
metal fittings *816*, 817, 824, 825, 830,
*see also* aiglets; buckles
from wristguard 825
*see also* scabbards (suspension systems)
structure and layout of ship *882–5*, 883
braces *884*, 885
and positioning of guns 883, 885, *884–6*, 891
*see also* gundecks; timbers, structural
*Struse (of Danzig)* 866
guns 137, 225, 300, 301, 857
munitions 331, 424, 436
sub-bottom profiler, parametric 933, 935, 940, *940*, *941*
Suffolk, Duke of (Charles Brandon) 868, *868*, 870
Sun 225, 300, 301, 424, 426, 427, 438
sundial, pocket 938
Surrey, Earl of (Thomas Howard) 8, 9
Surveyor of the Ordnance, Anthony as 3
Sussex Weald; gun casting 19, 22, 291
*Swallow*
archery equipment 581, 582
guns 58, 101, 137, 225, 270, 291, 300, 301
in Lisle's orders of June 1545 866
munitions 270, 331, 424

swan neck (tool) *95*, 97
Sweden; gunmetal alloys 951–3, *954*
*Sweepstake* 9, 594, 866
guns 58, 101, 137, 215, 225, 300, 301
munitions 331, 424, 436
swivel guns 882, 929
*see also* bases; slings; swivel mountings
swivel mountings 200, 272
*see also under* bases
swords **744–6**, **751–62**
anatomy *741*
arched/curved quillons 761–2
82A1915 *744*, 745, 760, **761–2**
82A1932 *744*, 745, 762
basket-hilted
historical development 745, 746, 753–4, 759–60, 761, 929
81A0719, stain on seabed *745*, 746, 751, 759–60
82A0027, pommel 745, 751, 769, *770*
82A3589, early example *743*, 744–5, **751**, *752*, **753–4**, *756–8*, **759**, 929
ceremonial 761, 762
civilian and military 743–4, 745, 754
in concretions 931, 937
daggers worn with 751
descriptions 751–62
distribution 748, 751, *752*, 754–5
of Henry V 745, 762
human remains associated with 751, *754–5*
inventory entries, 1547 740, 743, 747, 748, 808–9
metallographic analysis 753, *758*, 767
pommels *741*, 745, 750–1, *752*, 931
basket hilt type 745, 751, 769, *770*
detached *746*, 746, **764–9**, *768–70*
pricing 715
production 740
provision as individual responsibility 744
radiography *744*, 745, 751, *758*, 760, 761, 931
recurved quillons (83A0048) *744*, 745, 750, 760–1, *760–1*
for royal guard 740, 743, 745, 754, 759
scope and importance of assemblage 740–8
from spoil heap 931
suspension systems *see under* scabbards
*see also* grips, wooden; scabbards
Sydney, Sir William 9

*tabellae* (cartouches) on bronze guns 31
tactics, naval 15, **851–5**
broadsides engagement 13, 14, 853, 926, 928
formations 853, 855, *855*, 926
grappling, boarding and hand-to-hand combat *12*, 13, 594, 717, 718, 851, 920–1, 928
manoeuvring to fire from each side in turn 854, 855, 866, *900*, 900–1
weather gauge, gaining of 851, 852, 926
*see also* development of naval ordnance *and under* ranges of guns
tallow 435, 436, 529, 675

tampions 6, 410, **462–74**, *464–9*, *471–4*
   distribution 467–73, *468*
   identification and historical significance 462–5
   inventory entries 410, 464
   manufacture 473–4, *474*
   method of use 407, 409, 462–5, *465*, 470
      experimental *286*, 287, 413, *415*, 416
   radiography *469*
   reels 465–6, *466*, 467, *468*, 469, 473–4, *474*
   replication *286*, 287, 413, *415*, 416
   from spoil mound 935
   in store 463, 466, 467, *468*, 469
   wads attached to *472*, 476
   wood species 473
   *see also under* bases; fowlers; hailshot pieces; port pieces; slings
tar/pitch as packing material 643, 692, 693
targets 822
Tartaglia, Niccolo 19, 410, 411, *441*, 441, 486, 880
Taylors (Bellfounders) of Northampton *117*, 119
Teignmouth, Devon; Church Rocks wreck site 101, 223, 521
Texel *see* Scheurrak SO1 wreck
(*Three*) *Ostrich Feathers* 225 300, 301, 424, 438
*Tiger* 580
   guns 225, 269, 291, 300, 301
   munitions 331, 424
Tilbury, Kent *20*, 302
tiller 885, *886*, 897, *906*
timber
   sources 94
   types *see* wood species
Timber Park 936, 938
timbers, structural 937, *937*, 939–40, *940*, 941
   *see also* stempost
tinder boxes 499–503, *500–1*
   distribution *439*, 501, 502
   knife inside 499, 501, *502*, 502–3
   possible small 702, *704*
   possibly holders for caulking cloths 502, 702
tongs, casting 381, 383, *383*, 384
tools
   carpenters', used to make replica carriage 94, *95*, 97
   gunners' 125–6, *125*, *503*, **504–15**
      distribution *505–6*
   *see also* baskets; buckets; chisels; commanders; crowbars; funnels; handspikes; picks and pick-hammers; shovels; sledgehammers; wedges
top darts 5, 6, 269, 270, 718, 719, 932
top pieces/top guns **269–72**, 856, *858*, 931
   carracks, depictions of *269*, 270
   gunpowder type and expenditure 426, 427
   inventory entries 269–70, 856, 858, 922
      ammunition 269, 310, 339, 344
         archaeological evidence compared 5, 132
         1546, *Anthony Roll* 131, 214, 269, 344
   morphology 136
   shot 269, 272, 310, 339, 343–4, 344–5, 347
   in weapon preference sequences 921–4

tops, fighting 235, *271*, 272, 851
touch boxes for handguns 544
Tower of London; items in inventories
   archery equipment 581–2, 588
   gunners' equipment 409, 437, 440, 463, 476, 482, 486
   gunpowder 425
   guns 857, 858, 859, 860
      cast iron 291–2, 295, 301–2
      size hierarchy 860
      wrought iron 200, 213, 223, 269, 858
   incendiary devices 523
   shot 331, 337, 360, 371, 373, 384–5
      items for making or finishing 381, 384, 403, 437
   staff weapons 715, 716, 718, 719
Towton, battle of 587
transitional era, *Mary Rose* as ship of 854–5, 928
transom, guns in 138, *141*, 235, 930
   *see also* sternchaser guns
Trappes, Edward, of Chigwell 748, 798
*Trego Renneger* 101, 225, 300, 301, 424, 438
*Trinidad Valencera* 395, 495, 496, 521
*Trinity Harry* 272, 579–80, 920, 921
   guns
      bronze 58, 74, 101, 102, 858, 860
      cast iron demi culverin 291
      wrought iron 137, 200, 215, 225, 269, 300, 301, 302
   munitions 269, 331, 410, 424, 436
Tritech parametric sub-bottom profiler 933, 935, 940, *940, 941*
trucks 102, 105, **108**, 109
   attached to loose axles 109, 112, *113*
   for carriages for bronze guns 39, 40–1, *41*, 42, 44, 45
   for carriages for wrought iron guns 135, *135*, 142, *146*, 147, 201
   distribution in ship 102–3, *105*, 107–8
   loose 102–3, *105*, 105, 107–8
   storage 107–8
trunnions *23*, *37*, 224
   casting 112, 113, *116*
   marks 31, 33, 35
   support cheek 40, 45
tubs, wooden 433, *506*, 510–11, *512*
Tynemouth Castle, Northumbria 20, 523

Ubsdell, R.A.C. 168, 169, 170
Ufano, Diego 955
*Unicorn* 866
   guns 200, 225, 291, 300, 301
   munitions 200, 424
Upper Deck 907–10
   access to armaments storage 895
   *Anthony Roll* illustration 6, 140, *144*
   archers/soldiers on 579, 658, 667, 691, 697, 702, 704, 895
   archery equipment 579, 583, *584*, 883, 895, 909, 910
      arrow chests 667, *668*, 669, 690, 691–2, *692*
      longbows and chests 611, 612–14, 641–3, *641*, 644
      spacers 693, 694, 697, 704
      wristguards *646*, 647, 658, 704

armour *826, 829, 835, 842–3, 845*, 846–7
beam U55 50
blinds *198,* 199, 895, 907, 908, *908*
casks 431
darts 740
dimensions 907
gun carriages and parts 43–4, 104, *106*, 108, 109, 112
gun-shield elements *559*, 560
gundeck layout and operation 907–10, 934
gunports *13,* 14, *198,* 199, *882, 883,* 887, 891, *892,* 907, *908*
    *Anthony Roll* illustration 6, 140, *144*
guns 863
    arcs of fire and coverage *916,* 918, *919*
    bronze
        actual and inferred positions *26,* 26–7, *27, 28, 906,* 908, 909
        demi culverins/demi cannon 74, 89, *312,* 312, *893,* 918
        falcon 102, 908, 909
        loading instruments for 347
        possible Tudor salvage 36–7
    iron
        forelocks 195, *197*
        fowlers 217, 347
        port pieces 140, *141,* 345, 347, 910, *919*
        slings *198,* 199, 202, 203–4, *204,* 205, *210,* 210, *212,* 212, 891, 909, 918, *919*; swivel holes for *234–5,* 235, 861, 891, *892*
        in stern 26–7, *26,* 910
        weight 910
    handguns and ports 235, *882,* 891–2, 895, 907, 910
hatches
    central 883
    small, above Main Deck guns 887, 907
human remains 667, 691, 702, 704, 846–7
ladles *442,* 446, 449
leather buckets *505,* 509–10
linstocks 489, *498,* 498–9
muster station in stern 579, 846–7, 895, 910
netting 851, 907, 908
overhead protective covering 907, 908, *908*
possible overloading 879
priming wire 477, 479
rammers *443,* 457
reconstruction in museum *43*
scabbards 793–4, *794*
shot
    canister *372,* 373
    composite inset *356,* 360, *361*
    iron 102, *320, 321,* 323, *324, 325, 327, 328,* 329, **330**
    lead *369,* 370
    stone *312,* 312, *340,* 341, 345, 346, 347
staff weapons 717, 725, *730,* 731, 736, 883, 895, 909
swivel holes *234–5,* 235, 861, 891, *892*
swords, daggers and related finds 748, 750–1, *752,* 763, 781
tampions 467, *468,* 469, 471
tools for working guns *505–6*
waist *43,* 908–9
upperlop; meaning of term 860–1, *862*

Valletta, Malta; Palace Armoury 557, 558
van der Delft, François (Imperial ambassador) 866, 870, 930, 934
*Vanguard* 914
*Vasa* 496, 895
*Venetian* 866
Venice
    ambassador in London 717
    bowstaves 588
    gunmetal alloys 951–3, *954*
    handgun production and export 537–8, 539, 540, 541
    vent covers, cannon 51, 417, 507
verdegris, treatment of arrows with 676–7
verso (type of gun) 223
veuglaires (iron guns) 213, 851–2
victualling account (1545) 581
Vienna; handguns in Kunsthistorisches Museum 540, 542, 543
Villefranche sur Mer, France; wreck 358, 496
Vrelant, Paul van (royal harness gilder) 827

W mark on casks *432, 432,* 433
W.A. (artist; illustration of carrack) *269,* 270
wadhooks *see* worms
wads **474–6**, *475–6*
    *in situ*
        in bronze guns 48, 75, 87–8
        *see also under* bases; fowlers; hailshot pieces; port pieces; slings
    method of use
        in breech-loaders *286,* 287, 408, 409, 413, *415,* 416, 464–5, *465*
        in muzzle-loaders 407, 418
        with replica guns 126–7, *286,* 287, 413, *415,* 416, 418
        use of two 410
    tampions attached to *472,* 476
Wakefield, battle of; sword from site 745, 760, 761
Wales 395, 586–7, 593, 822
Walmer Castle, Kent *20,* 201
walnut arrows 674, 675
Walters Art Museum, Baltimore; gun-shield 557, *558*
Wark Castle, Northumbria *20,* 227
Watte, Jacob, or Copyn, de (royal armourer) 827
Weald, Kent/Sussex; gunfounding 19, 22, 291, 929
weather gauge 851, *852,* 926
wedge formation 855, 926
wedges, iron 40, *41, 505,* 507, **512–15**, *515*
    *see also under* port pieces
weight of ordnance 910
    Hatfield House document and 12, 930, 942–5
    inscriptions with 33
    as proportion of tonnage 892–3, 913–14
    and sinking of *Mary Rose* 867, 879, 882, 892–3, 913–14

West Tilbury Bulwark *20*, 302
wheels, solid *see* trucks
wheels, spoked 6, 184, *184–91*, **187–92**
    arrow marks *191*, 184, *185*, 189
    construction *184–5*, 187–8
    distribution 102–3, *105*, 107–8, *197*
    loose 102–3, *103*, *105*, 107–8
    measurements 184
    on port piece carriages 102, 135, 142, 184, *184–91*, 187–92
        for 81A2604 148, *148–9*, 151, *155*, 155
        for 81A3001 156, 157, *159*
    in store **103**, 107–8, 190, *190–1*, 192
Whitehorne, Peter 421, 422
Wight, Isle of 11, 12, 868, *868*, *870*, 879–80, 881–2
wildfire (combustible material) 519–20
    *see also* darts and arrows, incendiary; incendiary devices
willow
    items of 720, 763, 779
    arrows 618, 669, 674–5, 677, 678, *680*, 685
    technical characteristics 618, 676
*Winchester, Statute of* 713
windage 124, 230, 318, 401
    inset dice shot 357
    iron shot 318, 323, 326, 329
    stone shot 339
    variability 307, 311
Winter Proportions 24
Wolsey, Thomas, Cardinal 2, 9, 594
wood species
    technical characteristics 626, 675–6, 685–6
    *see also individual species and under* arrows; bills; by-knives; grips; gun carriages; linstocks; tampions
wood turning 438–9, 454, 473–4, *474*
wooling (ropes binding guns to carriages) 133, *135*, 135, *145*, 147, 151, *154*
    *Anthony Roll* entry 410
    in experimental gun drill 413, *415*
Woolwich dockyard 1, 2
working life of *Mary Rose* 8–10
worms (wadhooks) **125–6**, *125*, *503*
    lack of finds 417, 931, 932
    method of use 127, 407, 408, 409, 418, *418*
wreaths (incendiary devices) *see* garlands
Wrexham area; production of bucklers 822
Wright, Richard; Society of Antiquaries MS 94 23

    on bases 132, 222, *222*
    on cannon 45, *45*
    on gun, projectile and charge sizes 353–4, 429, 447, 448, 449
    on gun ranges 429
    on gunner's rule 393, *394*
    on incendiary devices 519, 520–1, 526
    on ladles 449
        combined with rammers *441*, 446
wristguards (bracers) 578, **644–65**
    bordered 647, *647*, 650
    catalogue 648–9
    in chests *646*, 647, 650, 703, 712
    decoration 578, 645–6, 650, **651–63**
        book cover stamp 650, *652*, 653, *653–4*, 654–6
        floral 653, *660*, 661–3
        guild insignia 581, 653, *659*, 659–61, 712
        religious symbols 653, *655*, 657, *658*, 658–9
        royal emblems 647, 653, 654, *655–7*, 656–8, 704, 707
        stamps used 652–3
        techniques 652
    distribution 583, 665, *646*, 647, 658, 707, 910
    in relation to human remains *578*, 579, 704, 712
    fastening *650*, 650–1, *651*, 825, 838
    form 647, *647*, 650
    high status examples 647, 650, 653, 656
    historical development 644–5, *645–6*
    horn 578, 645, 665
    human remains associated with *578*, 579, 586, *646*, 647, 703–4, 708, 712
    ivory 578, 586, 645, *646*, 647, *664*, 665, 704, 708
    materials 645
    personalised nature 581
    prehistoric 644–5, *646*
    related items; distribution *646*, 647
    scope and importance of assemblage 578
    undecorated 653–4, *661–3*, 663–5

X-ray fluorescent spectrometry (XRFS) 675–7, 765, 840, 841

Yarbridge, Isle of Wight 868, *870*
Yarmouth, Isle of Wight 20, 523, 582
Yarmouth Roads, Isle of Wight; wreck site 101, 104, *107*
Yeomen of the Guard *see* bodyguard, royal
yokes and pegs *see* swivel mountings